CALCULUS

with Early Transcendentals

Paul Sisson · Tibor Szarvas

Executive Editor: Claudia Vance

Executive Project Manager: Kimberly Cumbie

Vice President, Research and Development: Marcel Prevuznak

Editorial Assistants: Danielle C. Bess, Doug Chappell, Susan Fuller, Margaret Gibbs, Robin Hendrix, Barbara Miller, Nina Waldron, Barry Wright, III

Copy Editors: Phillip Bushkar, Taylor Hamrick, Mary Katherine Huffman, Rebecca Johnson, Justin Lamothe, Sojwal Pohekar, Eric Powers, Kara Roché, Joseph Tracy

Answer Key Editors: Taylor Jones, Jason Ling, Jake Stauch

Review Coordinator: Lisa Young

Senior Designer: Tee Jay Zajac

Layout & Original Graphics: Tee Jay Zajac

Graphics: Robert Alexander, Margaret Gibbs, Jennifer Moran, Tee Jay Zajac

Quant Systems India: E. Jeevan Kumar, D. Kanthi, U. Nagesh, B. Syam Prasad

Cover Design: Tee Jay Zajac

Cover Sculpture:
Arabesque XXIX
12″ H × 10½″ W × 9½″ D
Bubinga Wood
by Robert Longhurst
www.robertlonghurst.com

Chapter Opening Artwork:
Calculus Series
4.625″ H × 8.315″ W
Digital Painting
by Jameson Deichman
www.jamesondeichman.com

A division of Quant Systems, Inc.
546 Long Point Road, Mount Pleasant, SC 29464

Printed in the United States of America

Mathematica is a registered trademark of Wolfram Research, Inc.
Maple is a registered trademark of Waterloo Maple Inc.

Library of Congress Control Number: 2014957010

ISBN: 978-1-935782-21-6

In Memoriam, Carol Ann Sisson and Dr. István Tibor Szarvas

TABLE OF CONTENTS

Preface

From the Authors...xi

Acknowledgements..xii

About the Cover xiii

Features xv

Chapter 1

A Function Primer

1.1 Functions and How We Represent Them...3

1.2 A Function Repertory...21

1.3 Transforming and Combining Functions...41

1.4 Inverse Functions..58

1.5 Calculus, Calculators, and Computer Algebra Systems 80

  Review Exercises..96

  Project...99

Chapter 2

Limits and the Derivative

2.1 Rates of Change and Tangents...103

2.2 Limits All Around the Plane 117

2.3 The Mathematical Definition of Limit...134

2.4 Determining Limits of Functions..146

2.5 Continuity 159

2.6 Rate of Change Revisited: The Derivative...179

  Review Exercises..192

  Project...196

Chapter 3

Differentiation

3.1 Differentiation Notation and Consequences 199

3.2 Derivatives of Polynomials, Exponentials, Products, and Quotients 216

3.3 Derivatives of Trigonometric Functions 239

3.4 The Chain Rule 253

3.5 Implicit Differentiation 266

3.6 Derivatives of Inverse Functions 277

3.7 Rates of Change in Use 293

3.8 Related Rates 304

3.9 Linearization and Differentials 315

Review Exercises 324

Project 328

Chapter 4

Applications of Differentiation

4.1 Extreme Values of Functions 331

4.2 The Mean Value Theorem 344

4.3 The First and Second Derivative Tests 354

4.4 L'Hôpital's Rule 370

4.5 Calculus and Curve Sketching 381

4.6 Optimization Problems 394

4.7 Antiderivatives 407

Review Exercises 417

Project 421

Chapter 5

Integration

5.1 Area, Distance, and Riemann Sums 425

5.2 The Definite Integral 437

5.3 The Fundamental Theorem of Calculus 451

5.4 Indefinite Integrals and the Substitution Rule 465

5.5 The Substitution Rule and Definite Integration 475

Review Exercises 485

Project 488

Chapter 6

Applications of the Definite Integral

6.1 Finding Volumes Using Slices 491

6.2 Finding Volumes Using Cylindrical Shells.. 505

6.3 Arc Length and Surface Area... 515

6.4 Moments and Centers of Mass.. 525

6.5 Force, Work, and Pressure 539

6.6 Hyperbolic Functions 552

Review Exercises.. 564

Project.. 569

Chapter 7

Techniques of Integration

7.1 Integration by Parts 573

7.2 The Partial Fractions Method 581

7.3 Trigonometric Integrals.. 592

7.4 Trigonometric Substitutions 599

7.5 Integration Summary and Integration Using Computer Algebra Systems.............. 608

7.6 Numerical Integration 616

7.7 Improper Integrals... 627

Review Exercises.. 637

Project.. 641

Chapter 8

Differential Equations

8.1 Separable Differential Equations... 645

8.2 First-Order Linear Differential Equations 654

8.3 Autonomous Differential Equations and Slope Fields........................... 663

8.4 Second-Order Linear Differential Equations... 672

Review Exercises.. 682

Project.. 686

Chapter 9

Parametric Equations and Polar Coordinates

9.1 Parametric Equations 689

9.2 Calculus and Parametric Equations 702

9.3 Polar Coordinates 713

9.4 Calculus in Polar Coordinates 724

9.5 Conic Sections in Cartesian Coordinates 733

9.6 Conic Sections in Polar Coordinates 752

Review Exercises 760

Project 764

Chapter 10

Sequences and Series

10.1 Sequences 767

10.2 Infinite Series 781

10.3 The Integral Test 795

10.4 Comparison Tests 802

10.5 The Ratio and Root Tests 809

10.6 Absolute and Conditional Convergence 816

10.7 Power Series 824

10.8 Taylor and Maclaurin Series 834

10.9 Further Applications of Series 847

Review Exercises 859

Project 864

Chapter 11

Vectors and the Geometry of Space

11.1 Three-Dimensional Cartesian Space 867

11.2 Vectors and Vector Algebra 873

11.3 The Dot Product 884

11.4 The Cross Product 895

11.5 Describing Lines and Planes 905

11.6 Cylinders and Quadric Surfaces 916

Review Exercises 925

Project 930

Chapter 12

Vector Functions

12.1 Vector-Valued Functions 933

12.2 Arc Length and the Unit Tangent Vector 945

12.3 The Unit Normal and Binormal Vectors, Curvature, and Torsion 954

12.4 Planetary Motion and Kepler's Laws 967

    Review Exercises 975

    Project 980

Chapter 13

Partial Derivatives

13.1 Functions of Several Variables 983

13.2 Limits and Continuity of Multivariable Functions 994

13.3 Partial Derivatives 1002

13.4 The Chain Rule 1017

13.5 Directional Derivatives and Gradient Vectors 1026

13.6 Tangent Planes and Differentials 1036

13.7 Extreme Values of Functions of Two Variables 1045

13.8 Lagrange Multipliers 1058

    Review Exercises 1066

    Project 1072

Chapter 14

Multiple Integrals

14.1 Double Integrals 1075

14.2 Applications of Double Integrals 1086

14.3 Double Integrals in Polar Coordinates 1097

14.4 Triple Integrals 1105

14.5 Triple Integrals in Cylindrical and Spherical Coordinates 1117

14.6 Substitutions and Multiple Integrals 1129

    Review Exercises 1138

    Project 1143

Chapter 15
Vector Calculus

15.1 Vector Fields 1147

15.2 Line Integrals 1156

15.3 The Fundamental Theorem for Line Integrals 1166

15.4 Green's Theorem 1177

15.5 Parametric Surfaces and Surface Area 1192

15.6 Surface Integrals 1203

15.7 Stokes' Theorem 1212

15.8 The Divergence Theorem 1219

Review Exercises 1227

Project 1234

Appendices

A   Fundamentals of *Mathematica* A-2

B   Properties of Exponents and Logarithms, Graphs of Exponential and Logarithmic Functions ... A-8

C   Trigonometric and Hyperbolic Functions A-9

D   Complex Numbers A-14

E   Proofs of Selected Theorems A-21

Answer Key AK-1

Index I-1

Table of Integrals TI-1

FROM THE AUTHORS

This book arises from our deeply held belief that teaching and learning calculus should be a fascinating and rewarding experience for student and professor alike, playing a major role in the student's overall academic growth.

This is true not only because calculus ranks among the monumental achievements of the human intellect, thus providing an excellent introduction to higher-order thinking, but because of its wide-ranging applications in mathematics, the sciences, the business world, and the social sciences.

Our goal was to produce a text that builds on the natural intuition and curiosity of the reader, blending a student-friendly style of exposition with precision and depth. We believe that if done well, the calculus sequence should be a highly enjoyable journey of discovery and growth for the student, reflecting the journeys of discovery experienced by those who originally developed calculus centuries ago.

In other words, we strived to produce a book that is not only instructive, but also enjoyable to read—one that takes its readers from intuitive problem introductions to the rigor of concepts, definitions, and proofs in a natural manner.

Some of the distinctive features of our text include the following:

- A large number of examples and exercises in each section that demonstrate problem-solving techniques, reinforce conceptual understanding, and stimulate interest in the subject

- Carefully selected exercises that gradually increase in level of difficulty, ranging from skill-building "drill-and-practice" problems to less routine, more challenging, and occasionally deep theoretical questions

- Multistep, guided exploratory exercises that allow students to discover certain principles and connections on their own

- A rich variety of application problems from within and outside of mathematics and the sciences

- A constant emphasis on modern technology and its potential to enhance teaching and problem solving, as well as being a tool for investigation, reinforcement, and illustration

In summary, we have aimed for a comprehensive, mathematically rigorous exposition that not only uncovers the inherent beauty and depth of calculus, but also provides insight into the many applications of the subject.

We hope you enjoy the journey to the fullest. Let us know how we did and where we can improve!

Paul Sisson and Tibor Szarvas

ACKNOWLEDGEMENTS

We are grateful to all who have guided and assisted us throughout the long, satisfying process of writing this text. Thank you to Dr. James Hawkes, Marcel Prevuznak, Emily Cook, Kim Cumbie, and all the people at Hawkes Learning for their dedication to this project. In particular, we would like to thank our editor, Claudia Vance, and her editorial team members for their hard work and commitment: Robert Alexander, Danielle Bess, Doug Chappell, Robin Hendrix, Barbara Miller, Nina Waldron, and Tee Jay Zajac.

We are deeply appreciative of the efforts of our reviewers for their many insightful comments and reviews:

Carryn Bellomo-Warren   *University of Nevada, Las Vegas*

John F. Beyers   *University of Maryland–University College*

Mariah Birgen   *Wartburg College*

Douglas K. Brown   *Catawba College*

Julian M. Buck   *Francis Marion University*

Teena Carroll   *Emory & Henry College*

Brian Dalpiaz   *Spoon River College–Canton*

David Dixon   *Mountainair High School*

Sarah Duffin   *Southern Utah University*

Vincent Ferlini   *Keene State College*

Thomas L. Fitzkee   *Francis Marion University*

Roy Harris   *Stephen F. Austin State University*

Pramod Kanwar   *Ohio University–Zanesville*

Noureen Khan   *University of North Texas at Dallas*

Ravinder Kumar   *Tougaloo College*

Zsolt Lengvárszky   *Louisiana State University–Shreveport*

Richard Mabry   *Louisiana State University–Shreveport*

Rita Marie O'Brien   *Navarro College–Corsicana Campus*

Jennie Pegg   *Holmes Community College–Grenada*

Stanley Perrine   *Georgia Gwinnett College*

Paul Rokicky   *Cuyahoga Community College–Western Campus*

Melinda Rudibaugh   *Chandler-Gilbert Community College–Pecos*

Jacob Siehler   *Washington and Lee University*

John Taylor   *University of North Carolina–Charlotte*

Ruth Trubnik   *Delaware Valley College*

Robert P. Vilardi   *Troy University–Montgomery Campus*

We are indebted to uncounted family members, teachers, colleagues, and students for helping us along the path that led to the writing of *Calculus with Early Transcendentals*; the list is far too long for us to acknowledge them individually. But we are particularly grateful to our parents, William Reid Sisson and Carol Ann Sisson and Dr. István Tibor Szarvas and Dr. Katalin I. Tiborné Szarvas, for setting us on this path initially.

Finally, Paul thanks his wife Cindy and Tibor thanks his wife Anita and sons David, Daniel, and Gergely for their unstinting support and understanding at all times, especially those many early-morning and weekend hours spent writing!

ABOUT THE COVER

Arabesque XXIX

The sculpture on the cover of this text is a piece entitled *Arabesque XXIX* by the American artist Robert Longhurst (b. 1949). Although not a mathematician, Longhurst explains his work in this way: "It just so happens that what I produce sometimes turns out with an orientation toward math. My interest is in creating pieces that have appealing form. Proportion and scale are very important as is the craft aspect of my work." His pieces, which can be found at www.robertlonghurst.com, have captured the attention of many mathematicians over the years, and his work has appeared or been discussed in several math texts.

The object depicted by *Arabesque XXIX* is an example of an **Enneper surface**, a simpler version of which was introduced by the German mathematician Alfred Enneper (1830–1885) in 1864. Enneper, with his contemporary Karl Weierstrass (1815–1897), made great strides in understanding and characterizing *minimal surfaces*. A minimal surface is one that spans a given boundary curve with, locally, the least surface area, meaning that the surface area cannot be decreased by deforming any part of the surface slightly (a soap film spanning a boundary formed by a simple closed loop of wire is an example of a naturally occurring minimal surface). Except for the simplest cases, the area of a surface is a concept that is difficult to even define, let alone calculate, without the benefit of calculus; you will learn how calculus is used to define and determine surface area in Chapter 6.

An alternate, but equivalent, characterization is that a minimal surface is one for which each point has zero *mean curvature*, meaning that the maximum curvature of the surface at the point is equal in magnitude (but opposite in sign) to its minimum curvature. The *curvature* of a given curve at a particular point is a characteristic with precise meaning, which you will learn about in Chapter 12. The surface in Figure 1, a portion of the original Enneper surface, has zero mean curvature at every point; this is illustrated for the central point shown in red, as the curvatures of the two dashed curves add to zero. This particular configuration is also an example of a *saddle point*, which you will study further in Chapter 13.

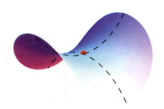

Figure 1

The characterization of minimal surfaces that Enneper and Weierstrass discovered is expressed in terms of integrals of functions of the complex plane, and their work falls into the categories of *differential geometry* and *calculus of variations*, two branches of mathematics that build upon the calculus you'll learn in this text. But the formulas resulting from their integrals are examples of *parametric surfaces*, which you'll study in detail in Chapter 15.

Figure 2

Specifically, the surface of *Arabesque XXIX* is similar to that defined parametrically by

$$x = r\cos\theta - \frac{r^5}{5}\cos 5\theta$$

$$y = -r\sin\theta - \frac{r^5}{5}\sin 5\theta$$

$$z = \frac{2r^3}{3}\cos 3\theta$$

where $0 \le r \le 1.38$ and $0 \le \theta \le 2\pi$. The graph of the surface defined by this parametrization appears in Figure 2.

FEATURES

Chapter Introductions

Each chapter begins with a brief introduction to the historical context of the calculus concepts that follow. Calculus is a human endeavor, and knowledge of how and why a particular idea developed is of great help in understanding it. Too often, calculus is presented in cold, abstract chunks completely divorced from the rest of reality. While a few students may be able to master the material this way, most benefit from an explanation of how calculus originated and how it relates to what people were doing at the time it was created. The introductions connect material learned in previous chapters to upcoming concepts, and illustrate to students the practical and historical importance of the calculus theories they are about to study.

Introduction

Chapter 2 introduced the notion of differentiation by means of several elementary applications. In this chapter we learn additional notation, develop techniques that allow us to find derivatives of many classes of functions, and explore many more applications. As we will see, the applications arise from a broad array of disciplines and collectively illustrate the profound power of the derivative.

Each bit of notation found in this chapter reflects the thinking of the mathematician responsible for introducing it, and it is frequently the case that a given mathematical concept can be expressed in a number of equivalent ways. While this may seem needlessly redundant at first, we'll see that the alternatives all offer their own unique advantages, and hence their use persists. Such giants as the Swiss mathematician Leonhard Euler (1707–1783) and the Italian French mathematician Joseph-Louis Lagrange (1736–1813), as well as Sir Isaac Newton (1643–1727) and Gottfried Wilhelm Leibniz (1646–1716), are among those who devised ways of succinctly writing the new ideas of calculus. Each had certain objectives in mind, usually

At the same time, mathematicians were discovering general rules of differentiation that allow derivatives of entire classes of functions to be determined relatively easily. In this chapter, we'll learn how to differentiate polynomial, rational, trigonometric, exponential, and logarithmic functions, among others.

We'll also see how arithmetic combinations and/or compositions of functions can be differentiated if we know the derivatives of the component pieces. These techniques are all ultimately based on the difference quotient definition of derivative seen in Chapter 2, but they enable us to differentiate

> In this chapter we learn additional notation, develop techniques that allow us to find derivatives of many classes of functions, and explore many more applications.

Joseph-Louis Lagrange
(1736–1813)

functions without having to laboriously calculate limits of difference quotients.

Finally, we'll learn how derivatives allow us to solve problems from disciplines such as biology, physics, chemistry, and economics.

The ability to efficiently and accurately calculate instantaneous rates of change, in all these different settings, allows us to solve problems that would otherwise be intractable.

TOPICS
1. The meaning of indefinite integration
2. The Substitution Rule

Substi

The Fundam
between integ
tool for calcu
that the notat
integral sign.
techniques fo

TOPICS

1. Each section begins with a list of topics. These concise objectives are a helpful guide for both reference and class preparation. The topics are clearly labeled throughout the section as the relevant material is covered.

Definitions, Theorems, Proofs, Properties, and Procedures

Definitions, theorems, proofs, properties, and procedures are clearly identified and stand out from the surrounding text for easy reference. Step-by-step procedures aid students in learning important problem-solving techniques. The procedures are then clearly illustrated in the examples that follow to help students apply the same steps in exercises at the end of the section. Selected proofs are provided in the text when necessary to establish meaningful connections between mathematical concepts, and then students are asked to prove additional theoretical statements in the exercises.

Theorem

The Substitution Rule for Definite Integrals

If g' is a continuous function on the interval $[a,b]$, and if f is continuous on the range of $u = g(x)$, then

$$\int_a^b f(g(x))g'(x)\,dx = \int_{g(a)}^{g(b)} f(u)\,du.$$

Proof

First, we know that both integrals exist since both integrands are continuous on their respective intervals: $(f \circ g)g'$ is a product of two continuous functions on $[a,b]$, and f is continuous on the interval $[g(a), g(b)]$. Our task is simply to prove that the integrals are equal.

As in the first version of the Substitution Rule, let F be an antiderivative of f.

$$\int_a^b f(g(x))g'(x)\,dx = F(g(x))\Big]_{x=a}^{x=b} \qquad \frac{d}{dx}F(g(x)) = f(g(x))g'(x)$$
$$= F(g(b)) - F(g(a))$$
$$= F(u)\Big]_{u=g(a)}^{u=g(b)}$$
$$= \int_{g(a)}^{g(b)} f(u)\,du$$

Finding Absolute Extrema

Assume that f is continuous on the interval $[a,b]$. To find the absolute extrema of f on $[a,b]$, perform the following steps.

Step 1: Find the critical points of f in (a,b).

Step 2: Evaluate f at each of the critical points in (a,b) and at the two endpoints a and b.

Step 3: Compare the values of the function found in Step 2. The largest is the absolute maximum of f on $[a,b]$ and the smallest is the absolute minimum of f on $[a,b]$.

Definition

Critical Point

Assume that f is defined on an open interval containing c. We say c is a **critical point** of the function f if $f'(c) = 0$ or $f'(c)$ does not exist.

Caution Notes ⚠

Many common errors or difficulties that students encounter when learning calculus are highlighted throughout the text, along with explanations of how to correct them. These warnings help students avoid common mistakes, and aid students in relieving frustration when working through problems.

similar pattern, but that is not the case—the truth is a bit more interesting.

Caution ⚠

The derivative of a product is *not* the product of the derivatives.

$$\frac{d}{dx}\big[f(x)g(x)\big] \neq f'(x)g'(x)$$

Similarly, the derivative of a quotient is *not* the quotient of the derivatives.

$$\frac{d}{dx}\left[\frac{f(x)}{g(x)}\right] \neq \frac{f'(x)}{g'(x)}$$

To see why this is so, it may be helpful to use a geometric interpretation; such an approach is very much in keeping with early Greek mathematics, which almost always

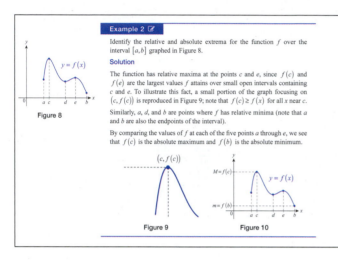

Example 2 ✏

Identify the relative and absolute extrema for the function f over the interval $[a,b]$ graphed in Figure 8.

Solution

The function has relative maxima at the points c and e, since $f(c)$ and $f(e)$ are the largest values f attains over small open intervals containing c and e. To illustrate this fact, a small portion of the graph focusing on $(c, f(c))$ is reproduced in Figure 9; note that $f(c) \geq f(x)$ for all x near c.

Similarly, a, d, and b are points where f has relative minima (note that a and b are also the endpoints of the interval).

By comparing the values of f at each of the five points a through e, we see that $f(c)$ is the absolute maximum and $f(b)$ is the absolute minimum.

Figure 8

Figure 9   Figure 10

Examples ✏

Each section contains numerous examples that illustrate the concepts presented and the skills to be mastered. The examples, which are clearly set off from the surrounding text, demonstrate problem-solving techniques and reinforce conceptual understanding. Many examples contain real-world data to stimulate student interest and illustrate practical applications of calculus.

Applications

The examples and exercises throughout the text contain a rich variety of application problems from within and outside the fields of mathematics and the sciences. These practical applications of the fundamentals of calculus engage students and illustrate the importance of these skills in the real world.

19. A cistern in the form of an inverted circular cone is being filled with water at the rate of 75 liters per minute. If the cistern is 5 meters deep, and the radius of its opening is 2 meters, find the rate at which the water level is rising in the cistern half an hour after the filling process began. (**Hint:** 1 m³ = 1000 L.)

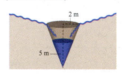

20. Repeat Exercise 19, this time assuming that the cistern is in the form of a pyramid with a 4-by-4-meter square opening.

21. A ship passes a lighthouse at 3:15 p.m., sailing to the east at 10 mph, while another ship sailing due south at 12 mph passes the same point half an hour later. How fast will they be separating at 5:45 p.m.?

22. A tourist at scenic Point Loma, California uses a telescope to track a boat approaching the shore. If the boat moves at a rate of 10 meters per second, and the lens of the telescope is 35 meters above water level, how fast is the angle of depression of the telescope (θ) changing when the boat is 200 meters from shore?

23. When preparing cereal for her child, a mother is pouring milk into a bowl, the shape of which can be approximated by a hemisphere with a radius of 6 in. If milk is being poured at a rate of 4 in.³/s, how fast is the level of milk rising in the bowl when it is 1.5 inches deep? (**Hint:** The volume of fluid of height h in a hemispherical bowl of radius r is $V = \pi h^2\left(r - \frac{1}{3}h\right)$.)

24. Suppose that in Exercise 29 of Section 3.7, the sand is being poured at a rate of 8 cubic inches per second. Find the rate of change of the height of the cone when it is 4 inches tall.

25. When finished playing in the sand, the child of Exercise 24 takes advantage of a nice wind and starts flying his kite on the beach. When the kite reaches an altitude of 60 feet the wind starts blowing it horizontally away from the child at a rate of 15 feet per second while maintaining the altitude of the kite. How fast does the child have to be letting out the string when 100 feet are already out?

26. A passenger airplane, flying at an altitude of 6.5 miles at a ground speed of 585 miles per hour, passes directly over an observer who is on the ground. How fast is the distance between the observer and the plane increasing 3 minutes later?

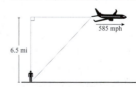

27. A military plane is flying directly toward an air traffic control tower, maintaining an altitude of 9 miles above the tower. The radar detects that the distance between the plane and the tower is 15 miles and that it is decreasing at a rate of 950 miles per hour. What is the ground speed of the plane?

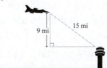

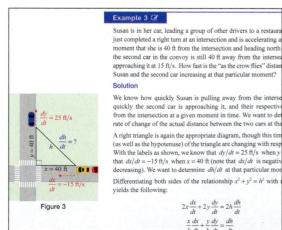

Example 3 ✏

Susan is in her car, leading a group of other drivers to a restaura... just completed a right turn at an intersection and is accelerating a... moment that she is 40 ft from the intersection and heading north... the second car in the convoy is still 40 ft away from the intersec... approaching it at 15 ft/s. How fast is the "as the crow flies" distan... Susan and the second car increasing at that particular moment?

Solution

We know how quickly Susan is pulling away from the interse... quickly the second car is approaching it, and their respectiv... from the intersection at a given moment in time. We want to de... rate of change of the actual distance between the two cars at tha...

A right triangle is again the appropriate diagram, though this tim... (as well as the hypotenuse) of the triangle are changing with resp... With the labels as shown, we know that $dy/dt = 25$ ft/s when y... that $dx/dt = -15$ ft/s when $x = 40$ ft (note that dx/dt is negativ... decreasing). We want to determine dh/dt at that particular mom...

Differentiating both sides of the relationship $x^2 + y^2 = h^2$ with ... yields the following:

$$2x\frac{dx}{dt} + 2y\frac{dy}{dt} = 2h\frac{dh}{dt}$$

$$\frac{x}{h}\frac{dx}{dt} + \frac{y}{h}\frac{dy}{dt} = \frac{dh}{dt}$$

Figure 3

Technology

The text emphasizes the capabilities of modern technology as a tool for investigation, reinforcement, and illustration without distracting from the importance of knowing the fundamental theories of calculus. Technology notes are included throughout to highlight ways in which technology can help to solve problems or explain the concepts taught. Instructors are given the flexibility to include graphing calculator or computer algebra system (CAS) instruction as appropriate. The exercise sets contain separate sections of problems that focus on using technology to perform more advanced computations.

Technology Note 🖵

Computer algebra systems such as *Mathematica* provide additional tools for determining limits, but it should always be remembered that software has limitations and can be fooled. *Mathematica* contains the built-in command **Limit** that uses the same mathematical facts we will learn in the next two sections to correctly evaluate many types of limits. Its use is illustrated below. (For additional information on *Mathematica* and the use of the **Limit** command, see Appendix A.)

```
In[1]:=  Limit[(Sqrt[x^2 + 4] - 2) / x^2, x → 0]

            1
Out[1]=  ---
            4
```

Figure 17

3.3 Technology Exercises

65–70 Use a computer algebra system to find the derivative of $f(x)$. Then graph f along with its derivative on the same screen. By zooming in, if necessary, find at least two x-values where the graph of f has a horizontal tangent line. What can you say about f' at such points? (Answers will vary.)

65. $f(x) = \dfrac{x}{1 + \cos x}$ **66.** $f(x) = \dfrac{1 - \sec x}{1 + \sec x}$

67. $f(x) = \dfrac{\csc x}{x}$ **68.** $f(x) = \dfrac{\sin x}{\cos x + \tan x}$

69. $f(x) = \cos x(\cot x + \tan x)$

70. $f(x) = \dfrac{\cot x}{\sec x + x \cos x}$

71. Find the maximum velocity and acceleration values in Exercise 62 by using a graphing calculator or computer algebra system to graph the velocity and acceleration functions of the oscillating object.

Exercises

Each section concludes with a large selection of exercises designed to allow the student to practice skills and master concepts. The exercises gradually increase in level of difficulty, ranging from skill-building "drill-and-practice" problems to less routine, more challenging, and occasionally deep theoretical questions. The sets are designed to allow instructors to assign exercises based on a wide variety of course structures and to encourage students to master elementary skills or challenge themselves, as appropriate. Throughout the text, multistep, guided exploratory exercises give students the opportunity to discover certain principles and connections on their own. Many of the more difficult problems include helpful hints to point the student in the right direction, without giving away the solution. The most challenging exercises are marked with an asterisk (*). There are also a wide variety of true/false exercises that encourage students to think critically about the concepts learned in the section, requiring them to articulate why a particular statement is invalid.

4.2 Exercises

1–4 Use the graph of the function to visually estimate the value of c in the given interval that satisfies the conclusion of the Mean Value (or Rolle's) Theorem; then check your guess by calculation. If such a c doesn't exist, explain why.

1. $f(x) = -x^2 + 6x - 4$ on $[1,4]$

2. $g(x) = -2\sqrt{|x-2|} + 2$ on $[-2,6]$

3. $h(x) = \dfrac{x^3}{3} - 2x^2 + \dfrac{11x}{3}$ on $[1,4]$

4. $k(x) = -\dfrac{1}{(x-3)^2} + 3$ on $[0,6]$

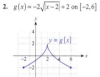

5–8 Prove that the equation has exactly one real solution on the given interval.

5. $x^3 - 3x^2 = 25$ on $[2,3]$ **6.** $5x^3 + 7x = 9$ on $\mathbb{R}$

7. $\arctan x = 3 - x$ on $[0,3]$ **8.** $\tan x = \cos x$ on $\left(0, \dfrac{\pi}{2}\right)$

9–20 Determine whether Rolle's Theorem applies to the function on the given interval. If so, find all possible values of c as in the conclusion of the theorem. If the theorem does not apply, state the reason.

9. $f(x) = -x^2 + 4x - 3$ on $[1,3]$ **10.** $g(x) = x^3 - 5x^2 + 2x + 10$ on $[-1,4]$

11. $h(x) = 2x^4 - x^3 + 6x^2 + x - 8$ on $[1,3]$ **12.** $F(x) = \dfrac{3}{(x-1)^2}$ on $[0,2]$

116.* Consider the following function:

$$f(x) = \begin{cases} 2x^2 & \text{if } x \text{ is irrational} \\ 4x^2 & \text{if } x \text{ is rational} \end{cases}$$

Use f to explain why the changing of signs of the derivative is not necessary for a function to have a local extremum.

117. By considering $f(x) = \begin{cases} x & \text{if } x \le 1 \\ 3 - x & \text{if } x > 1 \end{cases}$, explain

why the First Derivative Test can't be used for a discontinuous function.

118. Suppose that $f(x)$ and $g(x)$ are at least twice differentiable and that both their first and second derivatives are positive everywhere on an interval I. Which of the following can you prove from these conditions? Prove those statements that are true, and provide counterexamples for the rest.

a. $f(x) + g(x)$ is increasing on I.

b. $f(x) + g(x)$ is concave up on I.

c. $f(x) \cdot g(x)$ is increasing on I.

d. $f(x) \cdot g(x)$ is concave up on I.

e. $f(g(x))$ is increasing on I.

f. $f(g(x))$ is concave up on I.

119.* Use mathematical induction to prove the following generalization of the Second Derivative Test: Suppose that the derivatives of all orders of the function f exist at c, up to $f^{(2k)}(c)$, and that $f'(c) = f''(c) = \cdots = f^{(2k-1)}(c) = 0$, but $f^{(2k)}(c) \ne 0$. Then if $f^{(2k)}(c) < 0$, f has a relative maximum at c; if $f^{(2k)}(c) > 0$, f has a relative minimum at c.

120.* Use mathematical induction to prove that if f is $(2k+1)$-times differentiable at c, $f'(c) = f''(c) = \cdots = f^{(2k)}(c) = 0$, but $f^{(2k+1)}(c) \ne 0$, then f has a point of inflection at c.

121–128 *True or False?* Determine whether the given statement is true or false. In case of a false statement, explain or provide a counterexample.

121. Not all fourth-degree polynomials have inflection points.

122. A function with no inflection points cannot change concavity.

123. If $f'(x)$ is negative on $(-\infty, c)$ and positive on (c, ∞), then f has a minimum at c.

124. If $f(x)$ and $g(x)$ are decreasing, then so is $(f + g)(x)$.

125. If $f(x)$ and $g(x)$ are decreasing, then so is $(f \cdot g)(x)$.

126. A polynomial of degree n cannot have more than $n - 1$ extrema on $\mathbb{R}$.

127. If $c \in \mathbb{R}$ is a critical point, then the function has a local minimum or a local maximum at c.

128. If $f''(c) = 0$, then c is an inflection point.

Chapter 1
Review Exercises

1–4 Find the domain and range of the given relation and determine whether the relation is a function.

1. $R = \{(-2,9),(-3,-3),(-2,2),(-2,-9)\}$

2. $3x - 4y = 17$

3. $x = y^2 - 6$

4. $x = \sqrt{y - 4}$

5–8 Identify the domain, codomain, and range of the given function.

5. $f : \mathbb{N} \to \mathbb{R}, \quad f(x) = \dfrac{3x}{4}$

6. $g : \mathbb{R} \to \mathbb{R}, \quad g(x) = 5x + 1$

7. $h : \mathbb{R} \to \mathbb{R}, \quad h(x) = \dfrac{1}{x^2 + 1}$

8. $k : \mathbb{N} \to \mathbb{R}, \quad k(x) = 2 + \sqrt{x - 1}$

9–12 Find the value of the given function for **a.** $f(x-1)$, **b.** $f(x^2)$, and **c.** $\dfrac{f(x+h) - f(x)}{h}$.

9. $f(x) = (x+5)(2x)$

10. $f(x) = \sqrt[3]{x} + 6(x+4)$

11. $f(x) = \dfrac{3}{x+2}$

12. $f(x) = \sin 2x$

13–14 Find all intervals of monotonicity (intervals where the function is increasing or decreasing) for the given function.

13. $f(x) = (x-2)^4 - 6$

14. $R(x) = \begin{cases} (x+2)^2 & \text{if } x < -1 \\ -x & \text{if } x \geq -1 \end{cases}$

19–20 Discuss the symmetry of the given equation and then graph it.

19. $y = |5x|$

20. $x^2 + y^2 = 25$

21–34 Graph the given function. Locate the x- and y-intercepts, if any.

21. $f(x) = 7x - 2$

22. $f(x) = \dfrac{2x-6}{3}$

23. $f(x) = (x-1)^2 - 1$

24. $f(x) = -x^2 - 6x - 11$

25. $f(x) = 4x^3$

26. $f(x) = -\dfrac{2}{x^2}$

27. $f(x) = \dfrac{-x^3 + 7x + 6}{2}$

28. $f(x) = \dfrac{x+1}{x^2 - 4}$

29. $f(x) = \dfrac{\sqrt[3]{x}}{2}$

30. $f(x) = 5|-x|$

31. $f(x) = \left\lfloor \dfrac{2x}{3} \right\rfloor$

32. $f(x) = \begin{cases} x^2 & \text{if } x < 1 \\ \dfrac{1}{x} & \text{if } x \geq 1 \end{cases}$

33. $f(x) = -\cot x$

34. $f(x) = \log_{1/2} x$

35–36 Sketch the graph of the given function by first identifying the more basic function that has been shifted, reflected, stretched, or compressed. Then determine the domain and range of the function.

35. $f(x) = (x-1)^3 + 2$

36. $f(x) = 4|x+3|$

37–38 Write an equation for the function described.

37. Use the function $f(x) = 1/x$. Move the function 2 units to the right and 3 units up.

38. Use the function $f(x) = \sqrt[3]{x}$. Move the function 1 unit to the left and reflect across the y-axis.

Chapter
Review Exercises

Each chapter features a review section that contains a comprehensive overview of exercises from the chapter. These review exercises challenge students to utilize a combination of the skills and concepts learned throughout the preceding sections. Every review section contains true/false exercises to help students master the concepts of that chapter. The medium difficulty level of the review exercises is intended to help students effectively prepare for tests.

Chapter
Projects ✎

Each chapter contains a project that extends beyond the fundamental concepts covered to a more challenging application of calculus. These projects are suitable as individual or group assignments and allow students to extend their knowledge of calculus beyond the typical theories studied. The projects also provide opportunities to explore the aids of technology in understanding and performing more rigorous computations.

Project ✎

Chapter 5

The topic of this project is the so-called *sine integral function*, which is important for its applications, most notably in electrical engineering and signal processing.

1. Consider the following piecewise defined function:

$$f(t) = \begin{cases} \dfrac{\sin t}{t} & \text{if } t > 0 \\ 1 & \text{if } t = 0 \end{cases}$$

Prove that for any $x \geq 0$, $f(t)$ is integrable on $[0, x]$.

2. The **sine integral function** is defined as follows:

$$\text{Si}(x) = \int_0^x f(t)\, dt, \text{ for } x \geq 0$$

Prove that $\text{Si}(x)$ is continuous.

3. Find the derivative $\dfrac{d}{dx}\text{Si}(x)$.

4. Without graphing first, write a short paragraph on why you would expect the graph of $\text{Si}(x)$ to be oscillating. Explain why its amplitude is expected to decrease as $x \to \infty$.

5. Find the x-values where the relative maxima and minima of $\text{Si}(x)$ occur.

6. Extend the definition of $\text{Si}(x)$ to negative x-values and prove that for any $a > 0$,
$$\int_{-a}^{a} \text{Si}(x)\, dx = 0.$$

7. Use a computer algebra system to plot the graph of $\text{Si}(x)$ on the interval $[-8\pi, 8\pi]$.

8. Use a computer algebra system to approximate the range of $y = \text{Si}(x)$ to four decimal places.

A Function Primer

1.1 Functions and How We Represent Them   3

1. Relations and Functions
2. The Language of Functions
3. Ways of Describing Functions
4. The Power of Graphs

1.2 A Function Repertory   21

1. Algebraic Functions
2. Transcendental Functions
3. Piecewise Defined Functions

1.3 Transforming and Combining Functions   41

1. Shifting, Reflecting, and Stretching Functions
2. Combining Functions Arithmetically
3. Composing and Decomposing Functions
4. Interlude: Recursive Graphics

1.4 Inverse Functions   58

1. Inverse Relations and Inverse Functions
2. Finding Inverse Functions
3. Logarithms as Inverse Functions
4. Inverse Trigonometric Functions

1.5 Calculus, Calculators, and Computer Algebra Systems   80

1. Graphs via Calculators and Computers
2. Animations and Models
3. Least-Squares Curve Fitting

Introduction

This chapter begins with the study of *relations* and then moves on to the important class of relations called *functions*. As concepts, relations and functions are powerful and useful generalizations of the equations that one typically studies in introductory math courses. Functions, in particular, lie at the heart of a great deal of the mathematics that you will encounter throughout calculus.

The history of the function concept serves as a good illustration of how mathematics develops. One of the first people to use the idea in a mathematical context was the German mathematician and philosopher Gottfried Wilhelm Leibniz (1646–1716), one of two people usually credited with the development of calculus. Initially, Leibniz and other mathematicians tended to use the term to indicate that one quantity could be defined in terms of another by some sort of algebraic expression, and this (incomplete) definition of function is often encountered even today in elementary mathematics. As the problems that mathematicians were trying to solve increased in complexity, however, it became apparent that functional relations between quantities existed in situations where no such algebraic expression was possible. One example came from the study of heat flow in materials, in which a description of the temperature at a given point at a given time was often described in terms of an infinite sum, not an algebraic expression.

> ## The proof of the power of functions lies in the multitude and diversity of their applications.

The result of numerous refinements and revisions of the function concept is the definition that you will encounter in this chapter, and is essentially due to the German mathematician Peter Gustav Lejeune Dirichlet (1805–1859). Dirichlet also refined our notion of what is meant by a *variable* and gave us our modern understanding of *dependent* and *independent* variables, all of which you will read about soon.

The proof of the power of functions lies in the multitude and diversity of their applications. The subtle and easily overlooked advantage of function notation deserves special mention; innovations in notation often go a long way toward solving difficult problems. In fact, if the notation is sufficiently advanced, the mere act of using it to state a problem does much of the work of actually solving it. As you work through Chapter 1, pay special attention to how function notation works—a solid understanding of the notation is a necessary prerequisite to success in calculus.

Peter Gustav Lejeune Dirichlet
(1805–1859)

---

1. Adapted from illustration published in Peter Gustav Lejeune Dirichlet, *Werke, Herausgegeben auf Veranlassung der Königlich Preussischen Akademie der Wissenschaften* (Berlin, 1889).

1.1 **Functions and How We Represent Them**

TOPIC 1 **Relations and Functions**

TOPICS

1. Relations and functions
2. The language of functions
3. Ways of describing functions
4. The power of graphs

Functions lie at the heart of nearly every aspect of calculus and a solid understanding of them is therefore a necessary step toward success in calculus. This chapter reviews the language and notation of functions, the different ways that functions may be represented, frequently seen functions, and techniques for using and modifying functions.

To begin, it is useful to first define what we mean, mathematically, by the word *relation*; as we will soon see, functions are specialized relations.

Definition 💡

Relation, Domain, and Range

Given two sets A and B, a **relation** R from A to B is a set of ordered pairs of the form (a,b), where $a \in A$ and $b \in B$. (An *ordered pair* is just what it sounds like—a pair of elements written in such a way that it is obvious which is the first element and which is the second.)

Given any particular relation R from A to B, the set of all the first elements (the *first coordinates*) of the ordered pairs is called the **domain** of R, and the set of all second elements (the *second coordinates*) is called the **range** of R. Using set-builder notation, we can write

$$\text{Domain of } R = \left\{ a \middle| (a,b) \in R \text{ for some } b \in B \right\}$$

and

$$\text{Range of } R = \left\{ b \middle| (a,b) \in R \text{ for some } a \in A \right\}.$$

Although it may not have been stressed at the time, you have undoubtedly had much experience with relations in previous math classes. Indeed, the graph of any equation in x and y, where x and y represent real numbers, is a relation from $\mathbb{R}$ to $\mathbb{R}$, where $\mathbb{R}$ stands for the set of real numbers. More generally, any subset of $\mathbb{R} \times \mathbb{R}$ (the *Cartesian plane*) represents a relation—the subset does not have to arise from an equation or have any obvious mathematical basis.

We will provide four examples of relations, some of which can be defined by a formula or at least a picture, while others will, at first, seem to be "nonmathematical."

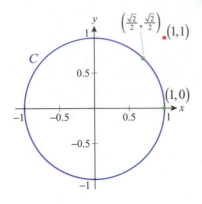

Figure 1

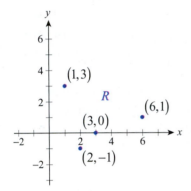

Figure 2

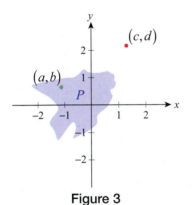

Figure 3

Example 1 ✎

a. For real numbers x and y, suppose they are in relation C (i.e., $(x, y) \in C$) if they satisfy the equation $x^2 + y^2 = 1$. For example, $(1, 0) \in C$, $(\sqrt{2}/2, \sqrt{2}/2) \in C$, but $(1, 1) \notin C$. Figure 1 shows the graph of the equation defining C. (Can you see why we called the relation C?) Note that in this example, the domain of C (the set of x-coordinates) is the set

$$\text{Domain of } C = \{x \in \mathbb{R} \mid -1 \le x \le 1\},$$

and similarly,

$$\text{Range of } C = \{y \in \mathbb{R} \mid -1 \le y \le 1\}.$$

b. The following set of ordered pairs is a relation from $\mathbb{R}$ to $\mathbb{R}$.

$$R = \{(1, 3), (2, -1), (3, 0), (6, 1)\}$$

Examining the first and second coordinates of the ordered pairs, respectively, we see that

$$\text{Domain of } R = \{1, 2, 3, 6\},$$

while

$$\text{Range of } R = \{-1, 0, 1, 3\}.$$

Notice that just as we did in part a., we can represent this relation in the Cartesian coordinate system (see Figure 2). Also note that relations do not necessarily have any immediate mathematical meaning such as the one in part a. For example, part c. below shows a relation "defined by a picture."

c. Figure 3 can be used to define relation P. Note that two real numbers x and y are in relation P if the point (x, y) falls in the shaded area or its boundary. For example, as shown by the illustration, $(a, b) \in P$, but $(c, d) \notin P$.

This example is a bit more general than that in part b., but it still satisfies the definition. See part d. for an even more general and nonnumeric example.

d. Suppose that the student population on a particular college campus speaks a total of five languages: English, Chinese, German, Japanese and Spanish. If we represent the above set of languages by $\{E, C, G, J, S\}$, the following relation L from the set of students to the set of languages arises quite naturally.

$$(x, y) \in L \text{ if student } x \text{ speaks language } y$$

Notice that the domain of L is the set of students at the college, while the range is the set of the five given languages. Supposing that Chihiro speaks English and Japanese, but not Spanish, we can express this in symbols by writing

$$(\text{Chihiro}, E) \in L, \quad (\text{Chihiro}, J) \in L, \quad \text{and} \quad (\text{Chihiro}, S) \notin L.$$

As Example 1 indicates, relations can arise in a multitude of ways. We may encounter relations defined by an equation or a formula, by a picture, by data collected from an experiment, or in words. This characteristic will carry over to our study of functions, which we are now ready to define.

Definition 💡

Functions and Function Notation

A **function** f from a set A to a set B is a relation from A to B with the additional property that <u>each</u> element $a \in A$ is related to <u>exactly one</u> element $b \in B$. Instead of writing $(a,b) \in f$, we typically use **function notation** and write $b = f(a)$. In other words, given $a \in A$, $f(a)$ is the unique element of B associated with a.

The notation $f(a)$ is read "f of a," and when A and B are sets of numbers, $f(a)$ is often referred to as the *value of f at a*. The notation $f : A \rightarrow B$ is used to indicate that f is a function (not just a relation) from the set A to the set B.

The underlined words in the above definition are critical. First, note that since *each* element of A is related to an element of B by f, the domain of f is all of A. Second, the requirement that *exactly one* element of B is related to each $a \in A$ means that there is a completely unambiguous dependence in the relationship. This dependence (which has come to be known as *functional dependence*) is the key to the great utility of functions in mathematics.

Caution ⚠

Don't be misled by the parentheses in the function notation to think that $f(a)$ might represent the product of f and a! While parentheses do often indicate multiplication, they are also used in defining functions.

TOPIC 2   The Language of Functions

A few additional observations and definitions are in order. As noted above, the notation $f : A \rightarrow B$ automatically indicates that the domain of f is A. In this context, the set B is called the **codomain** of f. As with relations, the range of f is the set of elements of B that are "used" by f. Using function notation, we can quickly note that the range of f is thus the set $\left\{ f(a) \,\middle|\, a \in A \right\}$.

The nature of functional dependence is also made clear by two other phrases. If, as above, a is a symbol representing an arbitrary element of the domain of f, then a is called an **independent variable**. And, as above, if b is used to represent an arbitrary element of the range of f, then b is called a **dependent variable**. These phrases often arise when we want to make use of the functional dependence of one variable upon another in a given equation. For example, in the equation $y = 3x^2 - 5$, the value of the variable y is completely determined by whatever value is assigned to the variable x, so y would be labeled the *dependent* variable and x the *independent* variable.

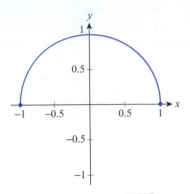

Figure 4 $y = \sqrt{1-x^2}$

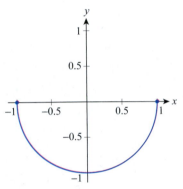

Figure 5 $y = -\sqrt{1-x^2}$

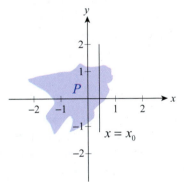

Figure 6

Example 2 ✐

Indicate which of the relations from Example 1 are functions, and for those functions make the distinction between the codomain and the range.

Solution

a. The way in which y depends on x in part a. of Example 1 does not demonstrate functional dependence. Consider, for example, the fact that both of the pairs $(0,1)$ and $(0,-1)$ are in the relation C, and thus 0 is related to both 1 and -1. However, if we express y from the equation by considering only the positive square root as follows:

$$y = \sqrt{1-x^2}$$

then the above difficulty is eliminated. Indeed, we obtained a function whose graph is the upper semicircle of radius 1, centered at the origin (see Figure 4). Notice that we can still consider $\{x \in \mathbb{R} | -1 \le x \le 1\}$ to be the domain of our function; however, the set of elements "used" by f, the range of f, is the following smaller set.

$$\{y \in \mathbb{R} | 0 \le y \le 1\}$$

Finally, let us also note that in the case of $y = -\sqrt{1-x^2}$ we obtain a different function whose graph is the lower semicircle (see Figure 5).

b. The relation defined in part b. of the previous example is a function. It is easy to check by inspection that each domain element is related to exactly one value.

The codomain of R is $\mathbb{R}$ while the range of R is $\{-1, 0, 1, 3\}$.

c. Examining the graph of the relation given in part c. of Example 1, it is easily concluded that this relation is not a function. Indeed, picking an appropriate x-value $x = x_0$ and drawing a vertical line through it, we see that the y-coordinate of any point falling on the line segment in the interior of the shading is related to x_0 (see Figure 6), thus leaving us with a multitude of choices for y_0.

d. With regards to relation L given in part d. of Example 1, since there are students on campus (such as Chihiro) who speak more than one language, the relation L is not a function either.

Example 3 ✐

As we have seen in the preceding examples, we can define a relation or even a function by writing an equation in two variables and considering the set of ordered pairs satisfying our equation. Find the function defined by the given equation, and note the dependent and independent variables.

a. $xy = x - 2y$ b. $y^2 - x - 1 = 0$ c. $dt - 2t \sin \pi t = 0$

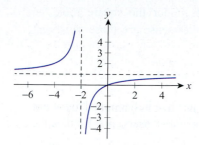

Figure 7 $y = \dfrac{x}{x+2}$

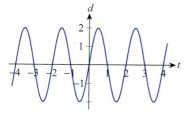

Figure 8 $x = y^2 - 1$

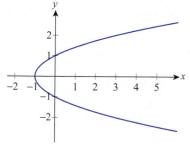

Figure 9 $d = 2\sin \pi t$

Solution

a. Moving the terms containing y to the left side, we obtain

$$(x+2)y = x, \text{ that is, } y = \frac{x}{x+2}.$$

We see that y is completely and unambiguously determined by x, so this is clearly a functional relationship with y being the dependent and x the independent variable. We often put it this way: y is a function of x.

b. Trying to express y as in part a. leads to the ambiguity

$$y = \pm\sqrt{x+1}.$$

However, it is rather straightforward to see that we can express x as a function of y in this case: $x = y^2 - 1$ defines x as a function of y; that is, y is the independent variable, and x is the dependent variable.

c. We divide both sides of the equation by t for $t \neq 0$ and obtain

$$d = 2\sin \pi t,$$

with d being the dependent, and t the independent variable. (Note that if you ever studied simple harmonic motion, you undoubtedly worked with functions similar to this one.)

In some situations, a function is given to us by means of a formula with no information regarding the domain. In such cases, by convention, the **implied domain** is the largest collection of points for which the formula returns a real number.

Example 4 ✐

Find the implied domain for each of the following functions, and also determine the range in each case.

a. $f(x) = \dfrac{x}{x+2}$ **b.** $g(x) = \sqrt{4 - x^2}$

Solution

a. The formula for $f(x)$ certainly returns a well-defined real number for all x except for $x = -2$, which makes the denominator 0. Hence the domain of f is the following set.

$$\text{Domain of } f = \{x \in \mathbb{R} \mid x \neq -2\}$$

As for the range, it is easy to see that f never assumes $y = 1$, for this would imply the equality of the numerator and the denominator in the formula, which never happens. In fact, being a rational function with the horizontal asymptote of $y = 1$, and having "opposing behavior" on the two sides of its vertical asymptote $x = -2$, we see that f assumes every real value except $y = 1$ (see Figure 10). Thus, the range of f is as follows.

$$\text{Range of } f = \{y \in \mathbb{R} \mid y \neq 1\}$$

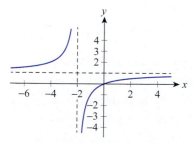

Figure 10 $f(x) = \dfrac{x}{x+2}$

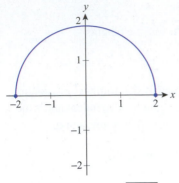

Figure 11 $g(x) = \sqrt{4-x^2}$

b. Since we cannot have negative values under the square root, the only real numbers for which $g(x)$ makes sense are those satisfying $4 - x^2 \geq 0$.

$$\text{Domain of } g = \left\{ x \in \mathbb{R} \mid x^2 \leq 4 \right\} = \left\{ x \in \mathbb{R} \mid -2 \leq x \leq 2 \right\}$$

Notice that the graph of g is a semicircle just like in Example 2a, but this time the radius is 2 (see Figure 11). This latter observation implies the following:

$$\text{Range of } g = \left\{ y \in \mathbb{R} \mid 0 \leq y \leq 2 \right\}$$

TOPIC 3   Ways of Describing Functions

As we will see, functional relationships arise in many situations that can be described or analyzed with calculus, but the exact nature of the relationships can be portrayed in a variety of ways. The four most common ways are mathematical formulas, tables of data, graphs, and verbal descriptions.

You have undoubtedly seen many examples of mathematical formulas already: such familiar equations as $A = \pi r^2$ and $C = \frac{5}{9}(F - 32)$ serve as illustrations. The first equation relates the area A of a circle to its radius r, and the second equation exactly describes the conversion of Fahrenheit temperature F to Celsius temperature C. Note that the form in which these equations are written leads to a natural interpretation of r and F as independent variables and of A and C as dependent variables.

The functional relationships in these two equations can be made more explicit by writing them in the form $A(r) = \pi r^2$ and $C(F) = \frac{5}{9}(F - 32)$, but it is important to realize that these functions express exactly the same relationships as the equations in the preceding paragraph. Again, context is everything: here, $A(r)$ and $C(F)$ are notational shorthand indicating that A is a function of r and that C is a function of F, not an indication that A is to be multiplied by r or C by F. In a formula such as $A(r) = \pi r^2$, the variable r is sometimes referred to as the **argument** of the function, and it is instructive to realize that the exact symbol used for the argument is irrelevant. That is, $A(s) = \pi s^2$, $A(radius) = \pi (radius)^2$, and $A(\square) = \pi \square^2$ all describe exactly the same function A.

Example 5 ✐

Find the following formulas and rewrite them so as to explicitly express the functional relationship between the variables. Identify the argument in each case.

a. The area of an equilateral triangle

b. The stopping distance required by a car traveling at speed v from the moment the brakes are applied

c. The volume of a circular cylinder

Solution

a. The area of an equilateral triangle of side length s is $A = \dfrac{1}{2}s\left(\dfrac{\sqrt{3}}{2}s\right)$

(using the well-known fact that the height h of an equilateral triangle is $h = \dfrac{\sqrt{3}}{2}s$). This formula can be rewritten as

$$A(s) = \frac{\sqrt{3}}{4}s^2,$$

a functional relationship with s being the argument, explicitly showing how the area of an equilateral triangle depends on its side length.

b. The stopping distance d required by a car traveling at speed v when the brakes are applied is estimated by the formula $d = v^2/(2\mu g)$, where μ is the coefficient of friction between the tires and the pavement, and g is the gravity constant. Since μ can also be considered a constant (at least in the short term, when conditions are nearly unchanging) we see that stopping distance depends on initial speed only. This can be expressed by the function

$$d(v) = \frac{v^2}{2\mu g},$$

with v being the argument of function d. (Note the quadratic dependence of d on v! It certainly pays to observe all speed limit signs and warnings.)

c. The volume of a circular cylinder is calculated by $V = \pi r^2 h$, where r is the radius of the base, and h stands for the height of the cylinder. Note that the volume here is a function of two independent quantities. We can express this functional relationship as follows.

$$V(r,h) = \pi r^2 h$$

Both r and h are considered arguments here, and thus $V(r,h)$ is our first example of a *function of two variables*.

Technology Note 🖵

Functions and function notation are immensely important not just as mathematical concepts but also in the way we use technology as a mathematical aid. Graphing calculators and computer algebra systems (CAS) all have methods for defining functions and then making use of them. Figure 12 shows how the function $f(x) = x^3 + 5x^2 - 7x$ is defined using a graphing calculator and how the function is then evaluated at several points and graphed for $x \in [-10, 5]$.

Figure 13 shows how the same function $f(x) = x^3 + 5x^2 - 7x$ is defined in *Mathematica*, evaluated, and then graphed over the specified region. (See Appendix A for a brief introduction to using *Mathematica*.)

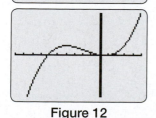

Figure 12

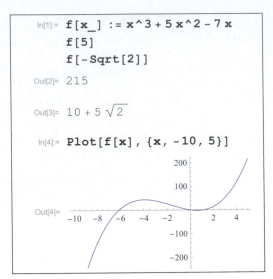

Figure 13

Tables of data, regardless of the setting, almost always implicitly describe a function. For example, any table of population data over two or more points in time (possibly for a city, for a country, or for the number of bacteria in a petri dish) defines population P as a function of time t. Table 1 is a table of the US population in every census year from 1950 to 2010.

Using function notation (and using the label P as in the preceding paragraph), we would write, for instance, $P(1950) = 151,325,798$ and $P(2010) = 308,745,538$. The first column in Table 1 represents the possible values for the independent variable t, and the second column represents the corresponding values of $P(t)$.

Table 1 only defines seven values for the function $P(t)$, and it is very often the case that we need to be able to extend the knowledge presented in a table to other values of the independent variable. For example, we might want estimates of the US population for every year between 1950 and 2010, or we might desire a projected population for the year 2050. The acts of estimating values of a function based on given data are called **interpolation** and **extrapolation**, and are perhaps best introduced with a picture. Figure 14 contains a plot of the seven data points from Table 1, along with a line that seems to approximate the general behavior of the points well.

US Population As Measured by the Census Bureau

Year	Population
1950	151,325,798
1960	179,323,175
1970	203,211,926
1980	226,545,805
1990	248,709,873
2000	281,421,906
2010	308,745,538

Source: www.census.gov

Table 1

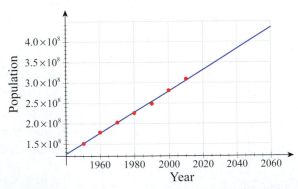

Figure 14 Graph of US Population Data, 1950–2010

To **interpolate** a value from a table or graph means to arrive at an estimate for the value of the implied function *between* two known data points, while to **extrapolate** means to guess at a value *beyond* the given points. In our example, we might interpolate a value for $P(1985)$ to be close to 240 million (2.4×10^8 in scientific notation) and extrapolate a value for $P(2050)$ to be approximately 410 million. However, this census data example is also a good illustration of the limitations of tables and graphs. Figure 15 contains the plot of an expanded version of the US Census table, along with a curve that approximates the behavior.

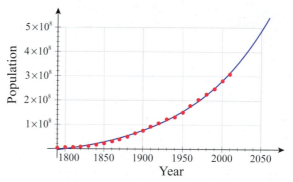

Figure 15 Graph of US Population Data, 1790–2010

If we were asked to extrapolate a value for $P(2050)$ on the basis of this expanded graph, we would probably guess a value of 480 million, significantly different from our first extrapolation of 410 million.

Technology Note 💻

You may be wondering how the blue line and blue curve in Figures 14 and 15 were drawn and whether they can be improved upon. There is no definitive answer to the second question, as *curve fitting* (the act of constructing a curve that closely approximates given data) depends upon making some assumptions about the mathematical nature of the data, and it is often the case that different but equally defensible assumptions can be made. In this case, it was assumed that the seven data points in Figure 14 lay approximately along a straight line (that is, that $P(t) = a + bt$ for some choice of a and b) and that the expanded data set in Figure 15 could be approximated by an exponential curve (specifically, that $P(t) = a + b^t$ for some a and b).

The answer to the first question above (the *how* question) is that the parameters a and b were chosen in each case so that the sum of the squares of the differences between the given data and the fitted curve was as small as possible. This method, called the *least-squares* method of curve fitting, will be fully explored later in this text, but for the moment we will focus on the results rather than the method. Many software packages (such as *Mathematica*, Maple, and spreadsheet programs such as Microsoft Excel) and graphing calculators are capable of calculating a least-squares fit. Using the command **Fit** from *Mathematica*, we arrive at the linear function

$$P(t) = -4.87678 \times 10^9 + 2.57841 \times 10^6 t$$

(whose graph is the blue line in Figure 14)

and the exponential function

$$P(t) = -4.6544 \times 10^7 + 1.00985^t$$

(whose graph is the blue curve in Figure 15).

(See Appendix A for a brief introduction to using *Mathematica*.)

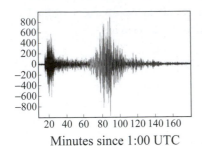

Figure 16

Seismogram of December 25, 2004
Sumatra Earthquake

Source: Pennsylvania Geological Survey's
website: http://www.dcnr.state.pa.us/topogeo/
hazards/earthquakes/sumatra/index.htm
(2 Apr. 2014)

The US Census information might have been presented to us visually as nothing more than the exponential graph seen in Figure 15 (without the red data points), and we could then use the graph to estimate values of the implied function for various years.

As another example, the output of a seismograph (such as Figure 16) is a function of time representing the vertical motion of the ground at a given location on Earth.

Recall that, mathematically, the **graph** of a function f that maps real numbers to real numbers is the collection of all points of the form $(x, f(x))$, where x takes on all the values in the domain of f. In some cases, such as the seismogram of Figure 16, information is presented to us first in the form of a picture, and we then find it useful to infer a functional relationship from the picture. In other cases, a function is presented in the form of a table or formula, and we find it useful to construct the graph of the given data and possibly even (as in the US population example) to "fill in" the parts of the implied function that might be missing.

The preceding paragraph gives rise to several questions. The first might be: Is it possible to graph functions other than just real-valued functions of real numbers? And the second might be: Given a picture (such as Figure 16), can we always assume that it does indeed represent the graph of a function?

We will return to the second question when we discuss further the power of graphs. As to the first question, we will see numerous examples throughout this text of ways to graph functions that turn real numbers into ordered pairs of real numbers (that is, functions from $\mathbb{R}$ to $\mathbb{R}^2$) and functions that do the opposite (functions from $\mathbb{R}^2$ to $\mathbb{R}$), along with a few more exotic variations of these ideas. But there is one more visualization tool that is commonly used to give a sense of what a specific function does, sometimes referred to as an *arrow diagram*. We will introduce its use here along with one more example of a function, this time described verbally.

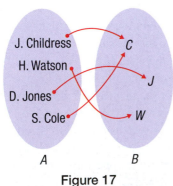

Figure 17

Arrow Diagram of Filing Function F

Consider the act of filing correspondence in a filing cabinet according to the last name of the sender. This act represents a function, which for convenience we might name F for *File*. Thus, for example, a letter from James Childress would be filed in the $F(\text{Childress}) = C$ folder, and one from Deanna Jones would be filed in the $F(\text{Jones}) = J$ folder. Figure 17 is an **arrow diagram** illustrating how F relates the domain set A of correspondents to the codomain set B of file folders.

Note that this act F does indeed satisfy all the requirements to be a function, and that for any specific set of correspondents it would be easy to identify the domain and range of the function. It is also worthwhile to note, however, that the inverse act of relating folders to the correspondents it contains is not likely to be a function, as even the small example depicted in Figure 17 demonstrates. For instance, folder C contains both S. Cole and J. Childress, so it is not the case that

<u>each</u> element of set B is related to <u>exactly one</u> element of set A. We will return to this important point in Section 1.4 when we discuss inverse functions.

Example 6 ✎

Below, we will give examples of functions that are easily described verbally, the likes of which you meet on a daily basis, perhaps without even thinking of them as functions.

a. In California, license plates of passenger vehicles follow the pattern of a number followed by three letters and then by three digits, such as, "4ZNR915." Since every registered car has a unique license plate with no duplications allowed, we can think of this as a functional relation between the set of registered California passenger cars as the domain and the set of strings following the above pattern as the codomain.

b. When a class of calculus students walks into a classroom and sits down, the students (perhaps unknowingly) define a function with its domain being the students themselves and the codomain being the chairs in the room. Since (at least in the ideal case) each student is able to sit and do so on no more than one chair at a time, we see that this is indeed a functional relationship between the two sets. Note that coming back to class two days later and sitting down in a slightly different order amounts to no more than redefining the function. Also, since every chair may not always be used, there is a clear distinction here between codomain and range.

c. The temperature on a given day in Shreveport, Louisiana, is a function of time, with its values being numbers from the Fahrenheit temperature scale. This is a function we can graph on a particular day, as shown below.

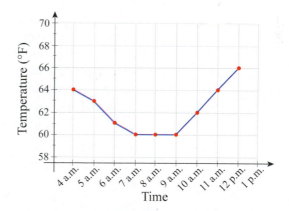

Figure 18

TOPIC 4 **The Power of Graphs**

We still have an unanswered question from above: given a graph of a relation, how do we determine whether it represents a functional relationship? For graphs in $\mathbb{R}^2$, there is a simple test to apply.

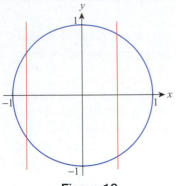

Figure 19

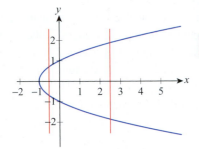

Figure 20

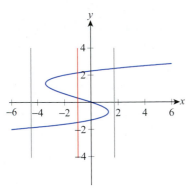

Figure 21

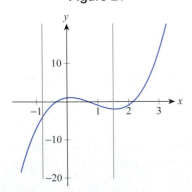

Figure 22

The Vertical Line Test

Given a graph of a relation in the Cartesian plane, the graph represents a function if no vertical line passes through the graph more than once. If even *one* vertical line intersects the graph in two or more points, the relation fails to be a function.

The vertical line test is nothing more than a restatement of the definition of function as it applies to relations from $\mathbb{R}$ to $\mathbb{R}$, but it is worth mentioning because it gives us a quick visual criterion to apply to a large number of graphs. It also immediately answers, in the affirmative, the question of whether graphs generated by mechanical means (such as the seismogram in Figure 16) represent functions. Because the instruments used to generate such graphs will only return one reading for any given point in time, a functional relationship is guaranteed.

Example 7 ✎

Below, we provide examples of the use of the vertical line test.

a. The unit circle centered at the origin does not define y as a function of x, since there are plenty of vertical lines intersecting the graph more than once (see Figure 19). In fact, as we have seen in Example 2a, the equation of the unit circle $x^2 + y^2 = 1$ defines both functions $y = \pm\sqrt{1 - x^2}$, each having a semicircle for its graph, and thus both satisfying the vertical line test.

b. An easy application of the vertical line test on the graph of $y^2 = x + 1$ shows that y is not a function of x (see Figure 20). We note, however, that again both functions $y = \pm\sqrt{x+1}$ are defined by the original equation. Furthermore, as we noted in Example 3b, our equation does actually define x as a function of y. (Note that because of the change in roles between the variables, from the perspective of the dependent variable x, "vertical lines" are the ones that appear horizontal in the traditional xy-coordinate system. With this interpretation, the "vertical line test" still applies to x as a function of y!)

c. The graph in Figure 21 is not a function either. Note that while a lot of vertical lines intersect the graph only once, it is enough to find one that doesn't.

d. Finally, we mention that the graphs of all functions you worked with in your precalculus course (such as polynomials, rational and trigonometric functions, etc.) do pass the vertical line test, as you can convince yourself easily. As an example, the function $f(x) = 2x^3 - 5x^2 + x + 1$, shown in Figure 22, intersects every vertical line exactly once.

The power of a visual depiction of a relation cannot be overstated (though pictures can be deceptive as well). We will close this section with a brief catalog of function properties that all have strong visual connotations.

Increasing/Decreasing

A real-valued function f is said to be **increasing** on an interval I if whenever x_1 and x_2 are in I,

$$x_1 < x_2 \implies f(x_1) \le f(x_2).$$

Similarly, f is said to be **decreasing** on I if whenever x_1 and x_2 are in I,

$$x_1 < x_2 \implies f(x_1) \ge f(x_2).$$

In either case, f is said to be **monotone** on I, and if the inequality that compares the function values is strict, the function is said to be **strictly increasing** or **decreasing** on I.

Example 8 ✍

Identify intervals of *monotonicity* (intervals where the function is increasing or decreasing) for the following functions.

a. $f(x) = x^2$ **b.** $g(x) = x^3 - 3x$ **c.** $h(x) = \cos x$

Solution

a. The graph of $f(x) = x^2$ is a prototypical parabola, opening upward, centered at the origin (see Figure 23). It is clear that f is decreasing on the interval $(-\infty, 0)$ while it is increasing on $(0, \infty)$. Visually, this means that the graph is "falling" as we scan it from left to right along the negative x-axis, while it is "rising" for positive x-values.

b. Examining the graph of $g(x) = x^3 - 3x$ in Figure 24, we see g is increasing on $(-\infty, -1)$ and $(1, \infty)$, while it is decreasing "in between" these half lines, that is, on the interval $(-1, 1)$.

c. From the graph of $h(x) = \cos x$, we conclude that h is decreasing on the interval $(0, \pi)$ and increasing on $(\pi, 2\pi)$ (see Figure 25). More than that, because of 2π periodicity, $\cos x$ is decreasing on all intervals of the form $(2k\pi, (2k+1)\pi)$ for an arbitrary integer k, while it is increasing on all intervals $((2k-1)\pi, 2k\pi)$.

Graphs of functions often display some sort of symmetry, and recognizing the symmetric nature of a given function can provide a great deal of insight into its behavior. The two most elementary kinds of symmetry are termed *even* and *odd*.

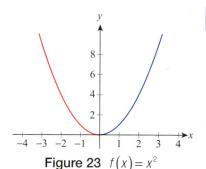

Figure 23 $f(x) = x^2$

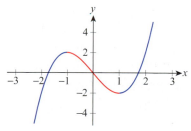

Figure 24 $g(x) = x^3 - 3x$

Figure 25 $h(x) = \cos x$

Definition 💡

Even and Odd Functions

If a real-valued function f has the property that $f(-x) = f(x)$ for all x in its domain, f is said to be an **even function**. Its graph will display symmetry with respect to the y-axis, as shown in Figure 26.

If a real-valued function f has the property that $f(-x) = -f(x)$ for all x in its domain, f is said to be an **odd function**. Its graph will display symmetry with respect to the origin, as shown in Figure 27.

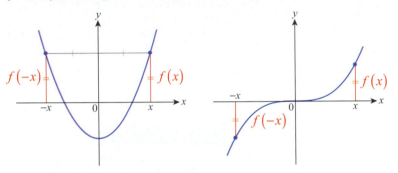

Figure 26 An Even Function   **Figure 27** An Odd Function

More generally, relations between two variables (say, x and y) can also display symmetry. When such relations are graphed in the Cartesian plane, the three most elementary kinds of symmetry are symmetry with respect to the axes and the origin.

Definition 💡

Symmetry of Equations

An equation in x and y is **symmetric with respect to**

1. the **y-axis** if replacing x with $-x$ results in an equivalent equation;

2. the **x-axis** if replacing y with $-y$ results in an equivalent equation;

3. the **origin** if replacing x with $-x$ and y with $-y$ results in an equivalent equation.

Example 9 ✏️

Discuss the symmetry of the following functions and equations.

a. $y - x^2 = 0$ **b.** $y^2 + 3x^2 = x + 2$ **c.** $y + 2x = x^3$

d. $h(x) = \dfrac{x^3 + 2x}{2x^4 - x^2 + 7}$ **e.** $k(x) = \sin x$

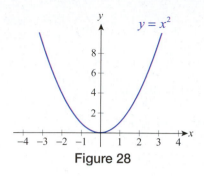

Figure 28

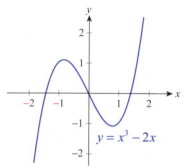

Figure 29

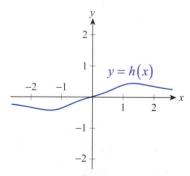

Figure 30

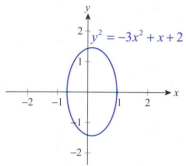

Figure 31

Solution

a. Since $(-x)^2 = x^2$, we see that replacing x by $-x$ yields an equivalent equation, so the graph must be symmetric with respect to the y-axis. Note that we can rewrite this equation as $y = x^2$, and so its graph is the graph of the function $f(x) = x^2$. This graph, the prototypical parabola, indeed displays symmetry with respect to the y-axis (see Figure 28).

b. If we replace y with $-y$, we obtain an equivalent equation, since $(-y)^2 = y^2$. The same is not true of x, since replacing it by $-x$ turns the right-hand side of the equation into $-x + 2$. Hence we conclude that the graph of this equation is symmetric with respect to the x-axis, but not the y-axis (see Figure 29).

c. If we simultaneously replace x with $-x$ and y with $-y$ in this equation, we obtain

$$-y + 2(-x) = (-x)^3$$
$$-y - 2x = -x^3,$$

and this latter equation is seen to be equivalent to the original, so we conclude that its graph is symmetric with respect to the origin. Finally, note that this graph is the same as that of the function $g(x) = x^3 - 2x$ (see Figure 30).

d. This example is more easily done algebraically instead of graphically. By substituting $-x$ for x we obtain

$$h(-x) = \frac{(-x)^3 + 2(-x)}{2(-x)^4 - (-x)^2 + 7} = \frac{-x^3 - 2x}{2x^4 - x^2 + 7} = -\frac{x^3 + 2x}{2x^4 - x^2 + 7} = -h(x),$$

and conclude that h is an odd function; thus its graph is symmetric with respect to the origin (see Figure 31).

e. The trigonometric function $k(x) = \sin x$ is odd, since $\sin(-x) = -\sin x$, and so its graph displays symmetry with respect to the origin (see Figure 32), a fact you have undoubtedly seen before. Finally, we also note the well-known fact that $g(x) = \cos x$ is even, for we know $\cos(-x) = \cos x$ to be true for all x. These facts are also clearly seen by visually examining these basic trigonometric graphs.

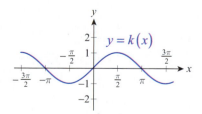

Figure 32

1.1 **Exercises**

1–22 Describe the domain and range of the given relation.

1. $R = \{(-3,1),(-3,5),(-3,-1),(0,0),(1,2)\}$

2. $S = \{(3,-1),(2.6,6),(\pi,0.5),(e,100)\}$

3. $T = \{(4,5.98),(-2,-8),(-2,0),(3,\cos 3)\}$

4. $U = \{(4,4),(4,\pi),(\pi,4),(4,0)\}$

5. $F = \{(\text{Tanisha},\text{swimming}),(\text{Don},\text{biking}),(\text{Peter},\text{skating}),(\text{David},\text{skateboarding})\}$

6. $L = \{(\text{Lin},\text{Chinese}),(\text{Chuck},\text{English}),(\text{Sarah},\text{German}),(\text{Daniel},\text{Hungarian})\}$

7. $A = \{(x,y)\,|\,x \in \mathbb{Z},\ y = 2x+3\}$

8. $B = \left\{(x,y)\,\middle|\,x \in \mathbb{R},\ y = \dfrac{x}{2}\right\}$

9. $C = \{(x,-2x+7)\,|\,x \in \mathbb{Z}\}$

10. $D = \{(2x,5y)\,|\,x \in \mathbb{N},\ y = x+1\}$

11. $3x = y+5$

12. $\sqrt{2}x - 1.2y = 3$

13. $x = 5$

14. $y = \pi$

15. $x = 3y^2 - 1$

16. $y = |x| - 2$

17.
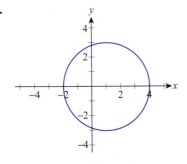

18.
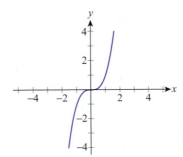

19.
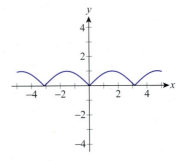

20.
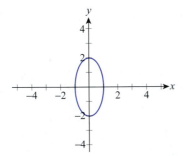

21.
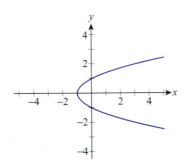

22.
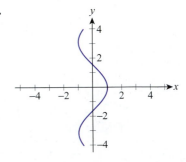

23–26 Describe the domain and range of the given relation. Choose an appropriate domain on which the given relation makes sense. (Answers will vary.)

23. $\{(x,y)|\text{student } x \text{ is registered for course } y\}$

24. $\{(x,n)|x \text{ wears size } n \text{ shoes}\}$

25. $\{(x,y)|y \text{ is the father of } x\}$

26. $\{(P,n)|\text{person } P \text{ weighs } n \text{ pounds}\}$

27–30 List the ordered pairs in the given relation R on the set $A = \{1,2,3,4,5\}$.

27. $(a,b) \in R$ if and only if $a = b$

28. $(a,b) \in R$ if and only if $a < b$

29. $(a,b) \in R$ if and only if $a \mid b$

30. $(a,b) \in R$ if and only if $a + b = 5$

31–34 Determine whether the given relation is a function. If the relation is not a function, explain why.

31. $A = \{(1,3),(-2,4),(0,4)\}$

32. $B = \{(0,0),(0,1),(2,3),(4,5),(6,7)\}$

33. $C = \{(-1,2),(\pi,3),(-1,0),(1,2)\}$

34. $D = \{(1,2),(2,1),(3,4),(4,3)\}$

35–40. Determine whether the relations given in Exercises 17–22 are functions. For those that are not, explain why.

41–52 Determine whether the given equation is a function. If the equation is not a function, explain why.

41. $y = 3x - 4$

42. $x = 3y - 4$

43. $x^2 + y^2 = 9$

44. $x + y^2 = 9$

45. $x^2 + y = 9$

46. $y = \sqrt[3]{x}$

47. $x = x^3 - y$

48. $xy = 4$

49. $x = \pi$

50. $y = \dfrac{3x}{x^2 + 1}$

51. $F = 5r^2\pi$

52. $V = \dfrac{4}{3}r^3\pi$

53–58 Express y explicitly as a function of x from the given relation.

53. $\dfrac{x + 3y}{2} = 5$

54. $\dfrac{x - 3y}{5} = \dfrac{2y + 7x}{3}$

55. $3x^2 - y = 5 - x + 2y$

56. $x + 7 - 3y = (x - 2)^2 + y$

57. $yx^2 - y = 3x + 1$

58. $x + 1 = yx^2$

59–66 Find the value of the given function for
a. $f(-2)$, **b.** $f(x+1)$, **c.** $f(x+h)$, and **d.** $\dfrac{f(x+h) - f(x)}{h}$.

59. $f(x) = \dfrac{1}{3}x + 2$

60. $f(x) = \dfrac{5x - 3}{2}$

61. $f(x) = x^2 - 3$

62. $f(x) = 3x^2 - 5x + \dfrac{1}{2}$

63. $f(x) = \sqrt{x}$

64. $f(x) = \dfrac{1}{\sqrt{x + 2}}$

65. $f(x) = \dfrac{1}{x + 1}$

66. $f(x) = (x - 1)^3 + 5$

67–72 Identify the domain, codomain, and range of the given function.

67. $f: \mathbb{N} \to \mathbb{N}, \quad f(x) = x + 1$

68. $g: \mathbb{N} \to \mathbb{Z}, \quad g(x) = 3x - 2$

69. $h: \mathbb{Z} \to \mathbb{Z}, \quad h(x) = x^2$

70. $F: \mathbb{R} \to \mathbb{R}, \quad F(x) = 2x^4 + 1$

71. $G: [0,\infty) \to \mathbb{R}, \quad G(x) = \sqrt{x}$

72. $H: \mathbb{Q}^+ \to \mathbb{Q}, \quad H(x) = \dfrac{1}{x}$ (Note that $\mathbb{Q}^+$ stands for the set of positive rational numbers.)

73–82 Find the implied domain of the given function.

73. $f(x) = \dfrac{x + 1}{x^2 - x - 6}$

74. $g(x) = \sqrt{3x + 2}$

75. $h(x) = \dfrac{2}{\sqrt{x^2 - 4x + 3}}$

76. $F(t) = \dfrac{1}{\sqrt{4 - t^2}}$

77. $G(s) = \sqrt{2 - s} + \sqrt{s}$

78. $D(h) = \dfrac{\dfrac{1}{\sqrt{1 + h}} - 1}{h}$

79. $R(x) = \dfrac{1}{|2x+3|}$ **80.** $H(z) = z^{3/2} - 2$

81. $F(\theta) = \dfrac{2}{1-\cos\theta}$ **82.** $\varphi(x) = \dfrac{5}{\sin x - \dfrac{\sqrt{2}}{2}}$

83–88 Turn the formula into a function by finding the argument(s) of the function. Identify any functions of two variables.

83. $C = 2\pi r$ **84.** $V = \dfrac{4}{3}r^3\pi$

85. $C = \dfrac{5}{9}(F - 32)$ **86.** $A = 6a^2$

87. $V = \dfrac{1}{3}b^2 h$ **88.** $E = \dfrac{1}{2}mv^2$

89–94 Use the vertical line test to decide whether y is a function of x.

89. $y^3 + 1 = x$ **90.** $2x^2 + 2y^2 = 18$

91. $y^2 + 1 = x$ **92.** $x = (y-2)^2$

93. $x = y^3 - 2y$ **94.** $yx^2 = 1$

95–101 Find all intervals of monotonicity (intervals where the function is increasing or decreasing) for the given function.

95. $f(x) = (x-1)^2$ **96.** $g(x) = 4x - x^2$

97. $h(x) = x^3 - 12x$ **98.** $k(x) = \dfrac{x^2}{x^2 + 1}$

99. $F(x) = |x-1|$ **100.** $G(x) = 2x + |3x - 1|$

101. $H(x) = |x+1| + |x-2|$

102–110 Discuss the symmetry of the given equation. Give reasons. (**Hint:** See Example 9.)

102. $y = x^2 - 1$ **103.** $x = y^2 - 1$

104. $x^4 + y^4 = 5$ **105.** $|x| + |y| = 2$

106. $x - |y| = 2$ **107.** $xy = 2$

108. $y = \dfrac{2x^3 - x}{x^4 + x^2}$ **109.** $y^2 + 6x = x^3$

110. $y = (x-1)^2$

111. Express the perimeter of a square as a function of its area.

112. Express the area of an equilateral triangle as a function of its perimeter.

113. An open-top box is constructed from a 20 in. by 30 in. piece of cardboard by cutting out a square of side length x from each of the four corners and folding up the sides, as shown in the figure below. Express the volume of the box as a function of x.

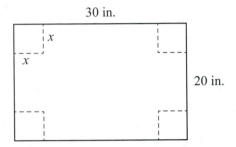

30 in.

20 in.

114. Express the surface area A of a cube as a function of its volume V.

115. The height of a circular cone is equal to the diameter of its base. Express its volume V as a function of the radius r of the base.

116. Express the volume of a sphere as a function of its surface area.

117. Knowing that water boils at 212 °F, which corresponds to 100 °C, and the fact that freezing occurs at 32 °F, which is 0 °C, obtain the linear function $C(F)$ that expresses the Celsius temperature C as a function of the Fahrenheit reading F.

118.* The organizers of an educational leadership seminar series have found that the seminar attracts 100 participants when the registration fee is set to $150. They estimate that for each increase of $10 in the registration fee, they will end up with 5 fewer registered participants. Express the revenue R as a function of the registration fee F.

1.2 **A Function Repertory**

TOPICS

1. Algebraic functions
2. Transcendental functions
3. Piecewise defined functions

The development of calculus can be viewed historically as the creation of an ever-expanding collection of mathematical tools and methods used to model and understand natural phenomena. This use of calculus remains hugely important, and indeed most of this book will focus on just that—the development and application of the tools of calculus. One important subset of these tools consists of the basic functions that, through centuries of experience, have proven to be good models of natural behavior. This section will present a quick overview of the functions we will be using repeatedly. As you will see, familiarity with them and a reasonable level of comfort in their use is very beneficial.

TOPIC 1 **Algebraic Functions**

The basic functions we use can be split up into two large classes: algebraic functions and transcendental functions. We will begin by providing examples of algebraic functions and work toward a definition of this class.

Definition 💡

Polynomial Functions

A **polynomial function** f of one variable, say the variable x, is any function that can be written in the form $f(x) = a_n x^n + a_{n-1} x^{n-1} + \cdots + a_1 x + a_0$, where each a_i is a constant and n is a nonnegative integer. Each a_i is called the **coefficient** of its associated **term** $a_i x^i$. The highest power of the variable is called the **degree** of the polynomial, and the coefficient of the term containing that highest power is called the **leading coefficient** of the polynomial.

Although polynomial functions with complex number coefficients are useful in some branches of mathematics, every polynomial in this book will be assumed to have real coefficients unless otherwise stated, and hence polynomial functions turn real numbers into real numbers. And since any real number can be raised to any positive integer power, a moment's thought will convince you that the domain (and codomain) of every polynomial function is all of $\mathbb{R}$. The range, however, may not be all of $\mathbb{R}$, and indeed the determination of the precise range for a given polynomial function often lies at the heart of a problem.

Polynomial functions of degree 1 are called **first-degree** polynomials, and their graphs are straight lines. This follows from the fact that such a polynomial has the form $f(x) = a_1 x + a_0$, so the graph is a line with slope a_1 and y-intercept a_0. A polynomial of the form $f(x) = a_0$ is a **constant** polynomial, and its graph is also a line (a horizontal line running through a_0 on the y-axis). The degree of a constant polynomial is 0, unless a_0 happens to be 0 (in which case the polynomial has no nonzero terms and is said to have no degree). Since their graphs are lines, first-degree and constant polynomials are together termed **linear** functions.

Second-degree polynomials have the form $f(x) = a_2x^2 + a_1x + a_0$, and their graphs are parabolas. Such functions also go by the name of **quadratic** functions. Similarly, **third-degree** polynomials are also known as **cubic** functions, and **fourth-degree** polynomials as **quartic** functions. Example 1 illustrates some typical shapes of these frequently occurring functions.

Example 1 ✎

Graph the following polynomial functions, and briefly discuss their most notable characteristics.

a. $f(x) = -\dfrac{3}{2}x + 3$ **b.** $g(x) = 2x^2 - 4x - 6$

c. $h(x) = \dfrac{1}{2}(x+2)(x+1)(x-3)$ **d.** $k(x) = -\dfrac{1}{3}x^4 + 2x^3 - \dfrac{1}{3}x^2 - 8x + \dfrac{20}{3}$

Solution

a. The function $f(x)$ is easily recognized as a first-degree function. The slope, which is also the leading coefficient, is negative, meaning that the graph "falls" as we scan it from left to right (see Figure 1). The graph confirms the y-intercept being 3, while the x-intercept can quickly be found to be 2 by solving the equation $0 = -\frac{3}{2}x + 3$. Finally, notice that the range of this function is all of $\mathbb{R}$, as is the case for all linear functions, unless they are constant.

b. The function $g(x)$ is a quadratic polynomial with a positive leading coefficient of 2. This means that the graph is an upward-opening parabola (see Figure 2). The y-intercept of -6 is easily found by substituting $x = 0$ in the formula. Note that the constant term turns out to be the y-intercept, which is no accident and happens with all polynomials, since substituting $x = 0$ will eliminate all terms but the constant. The x-intercepts, on the other hand, are the two solutions of $2x^2 - 4x - 6 = 0$, namely, -1 and 3. Finally, we note that since the vertex of the parabola is at $(1, -8)$, the range of g is the set $\{y \in \mathbb{R} \mid y \geq -8\}$.

c. The function $h(x)$, being a product of three linear polynomials, is certainly cubic, which is readily seen upon expanding and obtaining the standard form of $h(x) = \frac{1}{2}x^3 - \frac{7}{2}x - 3$. As before, the y-intercept is the constant term, -3, and we see from the factored form of $h(x)$ that the x-intercepts are -2, -1, and 3 (these values being the *zeros* of the polynomial). Finally, notice that, unlike the parabola, this graph "falls" to the left and "rises" to the right, as is the case with all polynomials of odd degree and positive leading coefficient, so the range of $h(x)$ is the entire set of $\mathbb{R}$ (see Figure 3).

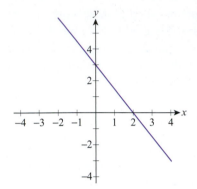

Figure 1 $f(x) = -\dfrac{3}{2}x + 3$

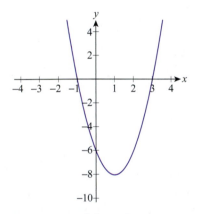

Figure 2 $g(x) = 2x^2 - 4x - 6$

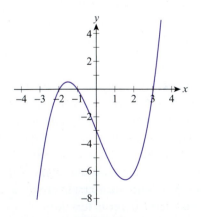

Figure 3

$h(x) = \dfrac{1}{2}(x+2)(x+1)(x-3)$

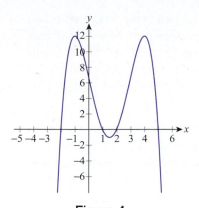

Figure 4

$$k(x) = -\tfrac{1}{3}x^4 + 2x^3 - \tfrac{1}{3}x^2 - 8x + \tfrac{20}{3}$$

d. The function $k(x)$ is an example of a quartic polynomial with a negative leading coefficient, so we expect the graph to be "falling" to both the right and the left, implying that the function's range will be something less than the entire set of $\mathbb{R}$. As in parts b. and c., the y-intercept is the constant term, $\tfrac{20}{3}$. To find x-intercepts, we can use, say, the Rational Zero Theorem along with long or synthetic division to obtain the fully factored form $k(x) = -\tfrac{1}{3}(x-2)(x+2)(x-1)(x-5)$ and conclude that the x-intercepts are ± 2, 1, and 5. From the graph shown in Figure 4, we also see the range to be the set $\{y \in \mathbb{R} \mid y \le 12\}$.

Many familiar natural phenomena can be modeled with relatively simple polynomial functions, and we encounter them frequently. Consider, for example, the following two models from the realm of kinematics (the study of the motion of objects).

Model 1: If an object is moving in a given direction with constant acceleration a, its velocity v (in that direction) at time t is given by the formula $at + v_0$, where v_0 is its initial velocity at time $t = 0$. Using function notation, we would write $v(t) = at + v_0$. Note that v is a linear polynomial.

Model 2: If an object is moving in a given direction with constant acceleration a, its position p (in that direction) at time t is given by the function $p(t) = \tfrac{1}{2}at^2 + v_0 t + p_0$, where p_0 is its position at time $t = 0$ and v_0 is its initial velocity. Note that p is a quadratic polynomial.

As we will see, calculus provides exactly the tools we need to derive these models and fully understand the relationship between position, velocity, and acceleration.

Example 2 ✏

A diver jumps from a 24-foot-high diving board so that her position above the water at any given time during the jump is represented by the function $p(t) = -16t^2 + 8t + 24$, where p is measured in feet (ft), and t in seconds (s). Her velocity as a function of time is given by $v(t) = -32t + 8$ ft/s. How much later and with what velocity does the diver hit the water?

Solution

First of all, since the position function $p(t)$ returns the altitude of the diver above the water surface for any specified t-value, we note that even though the diver may not move along a linear path, the position function only applies to one dimension. The same is true of velocity, whereas $v(t)$ returns the instantaneous velocity of the diver at any given time t. Also notice how the constant acceleration caused by gravity, -32 ft/s^2, the initial upward jumping speed of 8 ft/s, and the height of the platform, 24 ft, are "hidden" in the formula for $p(t)$! Indeed, we see from the formula for $v(t)$ that the velocity decreases by 32 ft/s with every passing second; thus there is a negative acceleration of -32 ft/s^2. Substituting $t = 0$ into the formula for $v(t)$ yields the initial velocity $v(0) = 8$ ft/s, while the height of the platform is the initial position $p(0) = 24$ ft.

To find out when the diver hits the water, we turn to the position function and note that at the moment of impact, the position over the water becomes 0, so we solve the quadratic equation $-16t^2 + 8t + 24 = 0$. The solutions are $t = 1.5$ and $t = -1$. We can safely discard the latter, since time cannot be negative, so we conclude that the diver hits the water 1.5 seconds after leaving the diving board.

As for the velocity of impact, we turn to the velocity function $v(t)$. When $t = 1.5$, the velocity of impact is obtained by evaluating $v(1.5) = -40$ ft/s. Note that the negative sign indicates that the direction of motion at the time of impact is downward.

Definition 💡

Rational Functions

A **rational function** f is a function that can be written as a ratio of two polynomials, say $f(x) = p(x)/q(x)$, where p and q are polynomial functions and q is not the zero polynomial. Even though q is not allowed to be identically zero, there may certainly be values of x for which $q(x) = 0$, and at those values f is undefined. Consequently, the domain of f is the set $\{x \in \mathbb{R} \mid q(x) \neq 0\}$.

Chances are, you have spent some time in a previous math class learning techniques for graphing rational functions and identifying characteristics of particular interest. For instance, it is often useful to identify the y-intercept, the x-intercept(s), and all *vertical*, *horizontal*, or *oblique asymptotes* of a rational function. The following example contains illustrations of these objects.

Example 3 ✏️

Sketch the graphs of the following rational functions, noting all intercepts and asymptotes.

a. $f(x) = \dfrac{x^2 - 1}{x - 1}$

b. $g(x) = \dfrac{x^2 + 1}{x^2 + 2x - 15}$

c. $h(x) = \dfrac{x^2 + 1}{x - 1}$

d. $k(x) = \dfrac{x^3 + x^2 + 2x + 2}{x^2 + 9}$

Solution

a. First of all, note that the domain of f is all real numbers except for 1, and that after canceling the common factor of $x - 1$, the function reduces to $f(x) = x + 1$. This holds whenever $x \neq 1$ and thus the graph of f is simply that of $y = x + 1$ with one point missing from it, since f is not defined at $x = 1$. We denote this with an open circle at the point on the graph corresponding to $x = 1$. Note the x- and y-intercepts of -1 and 1, respectively; there are no asymptotes.

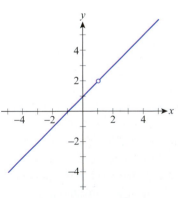

Figure 5 $f(x) = \dfrac{x^2 - 1}{x - 1}$

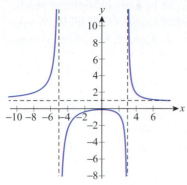

Figure 6 $g(x) = \dfrac{x^2+1}{x^2+2x-15}$

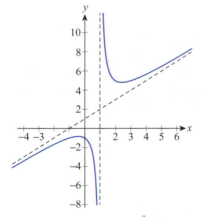

Figure 7 $h(x) = \dfrac{x^2+1}{x-1}$

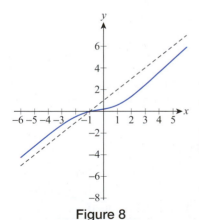

Figure 8

$k(x) = \dfrac{x^3+x^2+2x+2}{x^2+9}$

b. $g(x) = \dfrac{x^2+1}{x^2+2x-15} = \dfrac{x^2+1}{(x+5)(x-3)}$

Since the numerator cannot be factored over the real numbers, we find two vertical asymptotes with equations $x = -5$ and $x = 3$. Also since the numerator and denominator have the same degree and the ratio of the leading coefficients is 1, the line $y = 1$ is the horizontal asymptote. Since the numerator is never 0, there are no x-intercepts, while the y-intercept is the value of the function at 0: $g(0) = -\frac{1}{15}$.

c. Since $x = 1$ makes the denominator 0, and the numerator is not factorable over the real numbers, $h(x)$ has the vertical asymptote $x = 1$ (in stark contrast with $f(x)$ discussed in part a.). Furthermore, since the degree of the denominator is less than that of the numerator, there is no horizontal asymptote. However, since the difference in degrees is just one, we anticipate an oblique (or slant) asymptote, which is revealed by dividing $x - 1$ into $x^2 + 1$ by long or synthetic division.

$$\frac{x^2+1}{x-1} = (x+1) + \frac{2}{x-1}$$

Thus we have obtained that the line $y = x + 1$ is a slant asymptote for h. As far as intercepts, there are no x-intercepts since the numerator is never 0, but the function value $h(0) = -1$ yields the y-intercept of -1.

d. Like in part c., the degree of the numerator of $k(x)$ is one more than the degree of the denominator and therefore, k has an oblique asymptote. Polynomial division yields

$$k(x) = (x+1) - \frac{7x+7}{x^2+9},$$

so we conclude that $y = x + 1$ is the oblique asymptote. Finding $k(0)$ reveals that the function crosses the y-axis at $\frac{2}{9}$. To find x-intercepts, we need the zeros of the numerator polynomial. One of the potential rational zeros is -1, which actually is a zero, and thus the numerator factors as $(x+1)(x^2+2)$, so -1 is the only x-intercept.

Definition 💡

Power Functions

A **power function** in the variable x is a function of the form $f(x) = x^a$, where $a \neq 0$. Note that if a is a positive integer, such a function is an example of a single-term polynomial. If a has the form $1/n$, where n is a positive integer, such a function is a **root function** (and $x^{1/n}$, also written $\sqrt[n]{x}$, is called the nth **root** of x). If $a = -1$, the function is the **reciprocal function** and is an example of a rational function.

While the exponent a in the previous definition can be any nonzero real number, the cases where a is an integer or the reciprocal of an integer are of special interest. Figure 9 contains the graphs of $f(x) = x^a$ for $a = 1$, 3, and 5. Note that the graphs exhibit symmetry with respect to the origin.

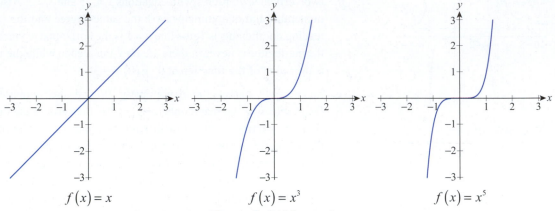

$f(x) = x$ $f(x) = x^3$ $f(x) = x^5$

Figure 9 Odd Exponents

Figure 10 contains the graphs of $f(x) = x^a$ for $a = 2$, 4, and 6. The functions are even and therefore each graph displays symmetry with respect to the y-axis.

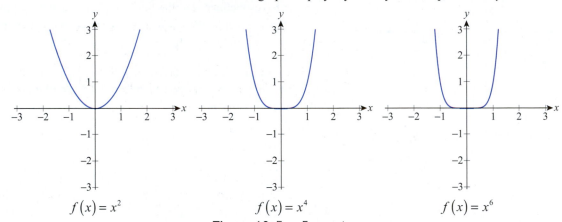

$f(x) = x^2$ $f(x) = x^4$ $f(x) = x^6$

Figure 10 Even Exponents

The graphs of the square root, cube root, fourth root, and fifth root functions are shown in Figure 11. Note that the domain of even root functions is $[0, \infty)$, while the domain of odd root functions is all of $\mathbb{R}$.

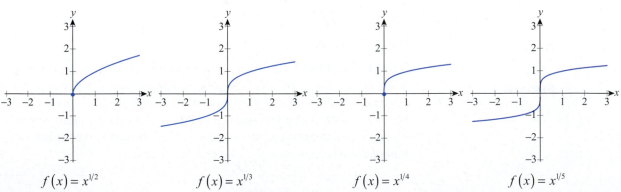

$f(x) = x^{1/2}$ $f(x) = x^{1/3}$ $f(x) = x^{1/4}$ $f(x) = x^{1/5}$

Figure 11 Root Functions

The graph of the reciprocal function $f(x) = x^{-1}$ appears in Figure 12; note that it is another example of an odd function. The function $f(x) = x^{-1}$ also frequently appears in equation form as $y = 1/x$ or $xy = 1$, and the graph of this function (or of the equation) is an example of a hyperbola (whose asymptotes are the coordinate axes in this case). For the sake of comparison, the graphs of $f(x) = x^{-2}$ and $f(x) = x^{-3}$ are also shown.

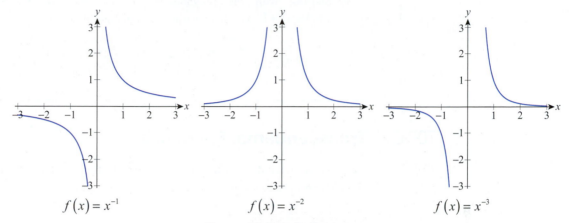

$$f(x) = x^{-1} \qquad\qquad f(x) = x^{-2} \qquad\qquad f(x) = x^{-3}$$

Figure 12 Negative Exponents

With these examples in mind, we can now define the class of algebraic functions. Polynomial functions, rational functions, and power functions of the form x^r where r is rational are all constructed using only algebraic operations (as defined below), and this characteristic is the basis of our definition.

Definition 💡

Algebraic Functions

A function f is in the class of **algebraic functions** if it can be constructed, starting with polynomials, in a finite number of steps using only the algebraic operations of addition, subtraction, multiplication, division, and the taking of roots.

Example 4 ✎

We provide a few examples of algebraic functions and one way each can be constructed.

a. Consider the algebraic function

$$F(x) = -4x^6 + 7x^2 - \sqrt[3]{2x^2 - 3}.$$

It can be constructed in a finite number of steps as shown below. For example, using the notation $f(x) = x^2$, we recognize $F(x)$ as

$$F(x) = -4\left[f(x)\right]^3 + 7f(x) - \sqrt[3]{2f(x) - 3}.$$

Note that this is not the only way to think about constructing F. Can you think of an alternate one?

b. Starting with the "easy" polynomials $g(x) = x$, and $h(x) = 5$, we can construct

$$G(x) = \frac{\sqrt{h(x)g(x) - 2}}{\left[g(x)\right]^3 + 4} + h(x) = \frac{\sqrt{5x - 2}}{x^3 + 4} + 5.$$

c. Starting with the polynomials $p(x) = x + 1$, $q(x) = x - 1$, and $r(x) = x^2 + 1$, we can construct the algebraic function

$$H(x) = \frac{\sqrt[3]{p(x)q(x) + \sqrt{3r(x)}}}{2r(x)q(x)} = \frac{\sqrt[3]{x^2 - 1 + \sqrt{3x^2 + 3}}}{2x^3 - 2x^2 + 2x - 2}.$$

TOPIC 2 Transcendental Functions

Now that we have defined the class of algebraic functions, the definition of the class of *transcendental* functions is deceptively simple.

Definition 💡

Transcendental Functions

A function f is in the class of **transcendental functions** if it is not algebraic. Such a function cannot be constructed with a finite number of applications of algebraic operations and in this sense *transcends* elementary algebra.

Without previous experience, a student new to mathematics would likely be hard-pressed to come up with an example of a transcendental function. Based only on the definition above, one approach might be to construct a function using an *infinite* number of algebraic operations, and (as we will see) this approach can actually be made to work. But you have doubtless already encountered three important examples of transcendental functions in the past: trigonometric, exponential, and logarithmic functions.

Definition 💡

Trigonometric Functions

The three fundamental trigonometric functions are **sine**, **cosine**, and **tangent** (abbreviated as sin, cos, and tan). Their reciprocals are named, respectively, **cosecant**, **secant**, and **cotangent** (abbreviated as csc, sec, and cot).

Appendix C contains a refresher of the important properties and historical derivation of the trigonometric functions, but a few key facts will be cited here. Recall that, unlike any of the functions seen thus far, trigonometric functions are *periodic*; that is, the complete graph of each one consists of an infinite repetition of one basic shape. Also recall that the argument of a trigonometric function represents an angle measure and that, unless specifically indicated otherwise, angles are assumed to be measured in *radians* (not degrees). Figure 13 illustrates the graphs of the sine, cosine, and tangent functions.

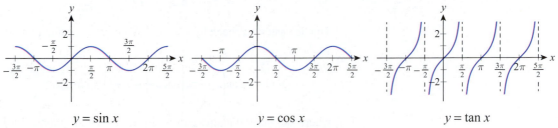

$$y = \sin x \qquad\qquad y = \cos x \qquad\qquad y = \tan x$$

Figure 13 Graphs of Sine, Cosine, and Tangent

Note that the domain for both sine and cosine is $(-\infty,\infty)$, and that they both have a range of $[-1,1]$. Tangent is defined for all x except $\pm\{\pi/2, 3\pi/2, 5\pi/2, \ldots\}$ (tangent has vertical asymptotes at these values of x), while its range is $(-\infty,\infty)$. The graphs of cosecant, secant, and cotangent can be found in Appendix C.

Example 5 ✐

When working with trigonometric functions in calculus, it often surprises students that equivalent answers to particular problems may appear to be completely different. This phenomenon is related to the inherent features of trigonometric functions and to the ways they are related to each other. To illustrate, let us simplify the following trigonometric expressions.

a. $\sin(x-4\pi)\sec(-x)$ **b.** $\dfrac{1}{\tan^2(t-\pi)+1}$ **c.** $\dfrac{\cos\alpha\cot\alpha}{1-\cos\left(\dfrac{\pi}{2}-\alpha\right)}-1$

Solution

a. We start by noting that $\sin x$ and $\cos x$ are 2π-periodic, so $\sin(x-4\pi)$ simply equals $\sin x$ for all x. Furthermore $\sec x$, being the reciprocal of $\cos x$, is an even function, so $\sec(-x)=\sec x$. Thus we obtain the following.

$$\sin(x-4\pi)\sec(-x) = \sin x\sec x$$
$$= \sin x\frac{1}{\cos x}$$
$$= \tan x$$

b. As above, we begin by making use of the observation that since $\tan t$ is π-periodic, $\tan(t-\pi)=\tan t$ and so we can write the following.

$$\frac{1}{\tan^2 t+1} = \frac{1}{\sec^2 t}$$
$$= \cos^2 t$$

Note that here we took advantage of the identity $\tan^2 t+1=\sec^2 t$, which can easily be checked.

c. We begin by recalling the following identity.

$$\cos\left(\frac{\pi}{2}-\alpha\right)=\sin\alpha$$

This is easiest to check by noting that since the cosine function is even, we have

$$\cos\left(\frac{\pi}{2}-\alpha\right)=\cos\left(\alpha-\frac{\pi}{2}\right)=\sin\alpha,$$

where the latter equality can easily be checked graphically by noting that if we shift the graph of the cosine function to the right by $\pi/2$ units, we obtain the graph of the sine function (see Figure 13). Using this, and the well-known trigonometric identity of $\sin^2 x + \cos^2 x = 1$, our derivation is as follows.

$$\frac{\cos\alpha\cot\alpha}{1-\cos\left(\dfrac{\pi}{2}-\alpha\right)}-1=\frac{\cos\alpha\dfrac{\cos\alpha}{\sin\alpha}}{1-\sin\alpha}-1$$

$$=\frac{\cos^2\alpha}{\sin\alpha\left(1-\sin\alpha\right)}-1$$

$$=\frac{1-\sin^2\alpha}{\sin\alpha\left(1-\sin\alpha\right)}-1$$

$$=\frac{\left(1+\sin\alpha\right)\left(1-\sin\alpha\right)}{\sin\alpha\left(1-\sin\alpha\right)}-1$$

$$=\frac{1+\sin\alpha}{\sin\alpha}-1$$

$$=\frac{1}{\sin\alpha}+\frac{\sin\alpha}{\sin\alpha}-1$$

$$=\frac{1}{\sin\alpha}+1-1$$

$$=\csc\alpha$$

If not explicitly stated, it is understood that equivalences are only valid where the two expressions are both defined. In this case, since $\cot\alpha = 1/\tan\alpha$ and $\csc\alpha = 1/\sin\alpha$, neither expression is defined when α is a multiple of π (that is, for $\alpha = k\pi$, where k is any integer). In addition, the original expression is not defined when $1-\cos\left(\dfrac{\pi}{2}-\alpha\right)=0$ (that is, for $\alpha = \dfrac{\pi}{2}+2k\pi$).

Example 6 ✍

The following examples illustrate how trigonometric functions can be used to model well-known real-life phenomena.

a. If an object attached to a spring is pulled down A inches from equilibrium, the resulting oscillating motion is called *simple harmonic motion*. The position function of such a motion can be either $f(t) = A\sin(\omega t)$ or $g(t) = A\cos(\omega t)$, depending on whether the object is in its downward position or at equilibrium when $t = 0$. Here ω is a constant that is related to the characteristics of the spring, and A is the amplitude. Notice how the periodicity of the position functions reflects the oscillating nature of the motion.

b. The change in temperature at a particular location on Earth is referred to by meteorologists as *diurnal temperature variation*. Simply put, the surface receives its *heat influx* from the sun, through what is called *solar radiation*, but the surface also constantly loses heat as a result of its own so-called *terrestrial radiation*. During daytime hours, solar radiation is stronger, causing the Earth to warm up. At night, however, there is a drop in temperature as a result of terrestrial radiation, a process that lasts until about one hour after sunrise, at which time the cycle starts again.

Because of the near-periodic nature of these events, it is reasonable to conjecture that trigonometric functions play a role in the modeling of diurnal temperature variation. British meteorologist Sir David Brunt (1886–1965) gave the following mathematical model for the heat influx F into the ground at the equinoxes.

$$F = \begin{cases} F_0\left(\cos(\omega t) - \dfrac{1}{\pi}\right) & \text{if } -\dfrac{\pi}{2} < \omega t < \dfrac{\pi}{2} \\[2ex] \dfrac{F_0}{\pi} & \text{if } \dfrac{\pi}{2} < \omega t < \dfrac{3\pi}{2} \end{cases}$$

where F_0 is a constant, ω is the Earth's angular velocity, and t stands for time, with $t = 0$ at noon. Notice that the first line in the model represents daytime hours, while the second line corresponds to nighttime, characterized by constant heat loss. Also notice the periodicity of the model.

Finally, observe that the above model consists of two separate "rules," each applying to daytime or nighttime hours, respectively. These so-called "piecewise defined" functions will be discussed in more detail at the end of this section.

Definition 💡

Exponential Functions

Given a fixed positive real number a not equal to 1, $f(x) = a^x$ is the **exponential function** with base a.

Note that the distinguishing feature of an exponential function is that the variable occurs in the exponent. The graphs of two particular exponential functions, $f(x) = \left(\frac{1}{2}\right)^x$ and $g(x) = 2^x$, are shown in Figure 14.

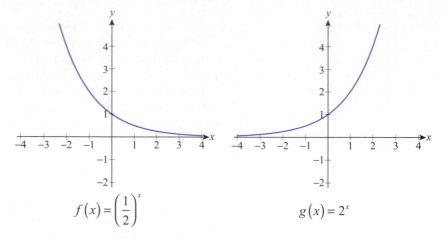

$$f(x) = \left(\frac{1}{2}\right)^x \qquad\qquad g(x) = 2^x$$

Figure 14 Graphs of Exponential Functions

Although exponential functions are not elementary in the sense of, say, polynomial functions, their evaluation is not necessarily difficult. For example, it is easy to compute (and compare) values of the two functions above.

x	$f(x) = \left(\frac{1}{2}\right)^x$	$g(x) = 2^x$
-1	$f(-1) = \left(\frac{1}{2}\right)^{-1} = 2$	$g(-1) = (2)^{-1} = \frac{1}{2}$
0	$f(0) = \left(\frac{1}{2}\right)^0 = 1$	$g(0) = 2^0 = 1$
1	$f(1) = \left(\frac{1}{2}\right)^1 = \frac{1}{2}$	$g(1) = 2^1 = 2$
2	$f(2) = \left(\frac{1}{2}\right)^2 = \frac{1}{4}$	$g(2) = 2^2 = 4$

One simple but important observation is that, for every base, the domain of an exponential function is $(-\infty, \infty)$ and the range is $(0, \infty)$. Another is that if the base a lies between 0 and 1, the exponential function is strictly decreasing, while an exponential function with base a greater than 1 is strictly increasing. For ease of reference, Appendix B contains a summary of the basic properties of exponential functions.

We study exponential functions because exponential behavior is exhibited in so many natural and man-made phenomena: radioactive decay, rates of temperature change, spread of epidemics, population growth, and compound interest are all good examples. The base e (an irrational number whose first few digits are 2.71828...) is particularly useful in modeling such behavior, a fact that is reflected, as we will soon see, in the nice calculus properties it possesses.

Example 7 ✎

Populations in nature, as mentioned previously, are known to grow following an exponential pattern if there are no restrictions such as predators and limits in food supply or space. This means the population size P is an exponential function of time, such as $P(t) = P_0 a^t$, where P_0 is the initial population size corresponding to $t = 0$. This simple model can be used, in the short term and with certain restrictions, to study a wide variety of population growth problems, as illustrated by the following example.

Suppose that 1000 rabbits are released on an uninhabited island that provides practically unlimited food supply, at least for the first few months. If the population doubles every month, how many rabbits are on the island two and a half months after their initial release?

Solution

Knowing the initial size of the population, we start out with the population function

$$P(t) = 1000a^t$$

where a is a yet-unknown base that we will try to determine from the given data.

Note that

$$P(0) = 1000a^0 = 1000,$$

so our function nicely captures the fact that the initial population size is 1000 rabbits. Investigating further, since we know that the population doubles after the first month, we can write

$$2000 = P(1) = 1000a^1 = 1000a,$$

so $a = 2$ and thus

$$P(t) = 1000\left(2^t\right).$$

We can now use a calculator to determine that

$$P(2.5) = 1000\left(2^{2.5}\right) \approx 5657.$$

We conclude that there are approximately 5657 rabbits on the island two and a half months after their initial release.

Definition 💡

Logarithmic Functions

Given a fixed positive real number a not equal to 1, $f(x) = \log_a x$ is the **logarithmic function** with base a. The logarithmic function with base 10 often appears simply as $\log x$, and the logarithmic function with base e usually appears as $\ln x$ (the "ln" stands for "natural logarithm").

Unlike the definitions of polynomial, rational, or exponential functions, the above definition doesn't give us much guidance in how to evaluate logarithms or a sense of their overall nature—it really does nothing more than introduce logarithmic notation and nomenclature. True understanding depends upon knowing that $\log_a x$ is the *inverse function* of a^x, and we will return to the topic of logarithms when we discuss inverses of functions in Section 1.4. Until then, it is enough to note that logarithms are one more important subset of transcendental functions, and that they arise in mathematics for the same reasons exponential functions do. For ease of reference, Appendix B also contains a summary of the basic properties of logarithmic functions.

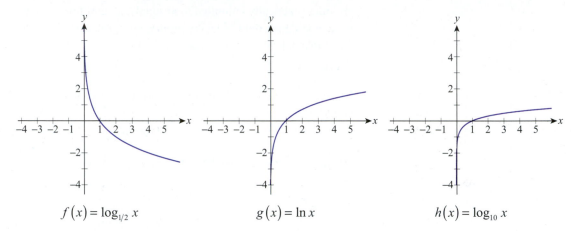

$$f(x) = \log_{1/2} x \qquad\qquad g(x) = \ln x \qquad\qquad h(x) = \log_{10} x$$

Figure 15 Graphs of Logarithmic Functions

TOPIC 3 Piecewise Defined Functions

We finish this section with a short discussion of functions that are, in a particular sense, pieced together from other functions.

Consider the function f defined by $f(x) = \begin{cases} -x & \text{if } x < 0 \\ x & \text{if } x \geq 0 \end{cases}$. To evaluate this function for a given x, the first step is to determine which rule to apply, that is, whether x is less than 0 or not. The graph of f simply consists of two pieces, one for $x < 0$ and the other for $x \geq 0$; both pieces are very basic linear functions, and the graph appears in Figure 16.

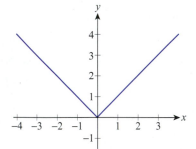

Figure 16
Absolute Value Function

Although we have used f as a simple example of a function defined in pieces, you have probably seen it before under the name of the **absolute value function**. In this guise, the function can be defined as $f(x) = |x|$; note that the piecewise defined rule is nothing more than the definition of absolute value.

With the absolute value function as a familiar guide, we can easily handle a slightly more complicated example.

Example 8 ✐

Sketch the graph of the following piecewise defined function $f(x)$ and find the values $f(-2)$, $f(0)$, $f(1)$, $f(2)$, and $f(3)$.

$$f(x) = \begin{cases} x^2 & \text{if } x < 0 \\ 3x+1 & \text{if } 0 \le x < 2 \\ 7 - \sqrt{x-2} & \text{if } x \ge 2 \end{cases}$$

Solution

This graph is "patched" together from three pieces, as follows. Notice that f is the squaring function for negative x-values, so the first piece of the graph is simply the left-hand branch of the prototypical parabola. For x-values between 0 and 2, the graph is a segment of the straight line $y = 3x + 1$, followed by the graph of $y = 7 - \sqrt{x-2}$ (which is also a half parabola) for all x-values greater than 2.

To find the value of $f(-2)$, we determine that the first of the three rules applies since -2 is negative.

$$f(-2) = (-2)^2 = 4$$

Next, we use the second rule to find the following.

$$f(0) = 3 \cdot 0 + 1 = 1$$

Note that this means that the y-axis as a vertical line meets the graph at $(0,1)$ and misses it at the origin, because $(0,0)$ is not a point on the graph. This is indicated by the empty and full circles in Figure 17.

For $x = 1$ the second rule applies, so finding $f(1)$ is straightforward.

$$f(1) = 3 \cdot 1 + 1 = 4$$

We see that the second rule applies only to x-values strictly less than 2, so we use the third rule to evaluate f at $x = 2$.

$$f(2) = 7 - \sqrt{2-2} = 7$$

Notice, however, that if we had used the second rule with $x = 2$, we would have obtained the same function value! Graphically, this is reflected in the fact that the two "pieces" of the graph meet at the point $(2,7)$, in other words, the graph doesn't "break" like it does at $x = 0$.

To find the last function value, we use the third rule.

$$f(3) = 7 - \sqrt{3-2} = 7 - 1 = 6$$

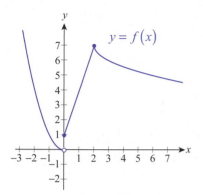

Figure 17

Technology Note 🖵

Computer algebra systems such as *Mathematica* usually have built-in commands or syntax for defining and working with piecewise defined functions. In *Mathematica*, that command is (nicely enough) **Piecewise**. Figure 18 shows its use. Note that, in general, graphing technology will not automatically denote "holes" in graphs with empty circles. (See Appendix A for a brief introduction to using *Mathematica*.)

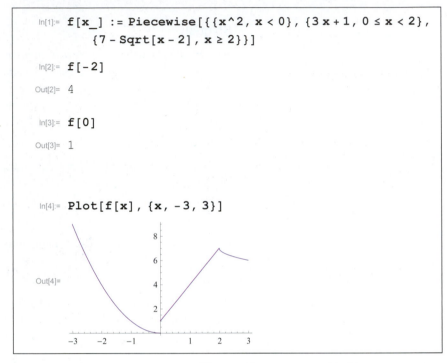

```
In[1]:= f[x_] := Piecewise[{{x^2, x < 0}, {3 x + 1, 0 ≤ x < 2},
            {7 - Sqrt[x - 2], x ≥ 2}}]

In[2]:= f[-2]

Out[2]= 4

In[3]:= f[0]

Out[3]= 1

In[4]:= Plot[f[x], {x, -3, 3}]
```

Figure 18

1.2 **Exercises**

1–8 Identify the degree, leading coefficient, intercepts, and range of the given polynomial function, and then graph the function.

1. $f(x) = \dfrac{1}{2}x - \dfrac{3}{2}$

2. $g(x) = -1.2x + 4.8$

3. $h(x) = 2x^2 - 3x - 2$

4. $u(x) = \dfrac{1}{2}x^2 + x - \dfrac{3}{2}$

5. $v(x) = x^3 - 7x + 6$

6. $F(x) = 10 - 8x + \dfrac{x^2}{2} + \dfrac{x^3}{2}$

7. $G(x) = \dfrac{x^4}{4} - 2x^2$

8. $H(x) = 2x^4 + 12x^3 + 2x^2 - 48x - 40$

9–16 Find all asymptotes and intercepts of the given rational function and then sketch the graph of the function.

9. $f(x) = \dfrac{5}{x-1}$

10. $g(x) = \dfrac{x^2 - 4}{2x - x^2}$

11. $h(x) = \dfrac{x^2 + 3}{x + 3}$

12. $u(x) = \dfrac{x + 2}{x^2 - 9}$

13. $v(x) = \dfrac{x^2 - 2x - 3}{2x^2 - 5x - 3}$

14. $F(x) = \dfrac{3x^2 + 1}{x - 2}$

15. $G(x) = \dfrac{x^2 + 2x}{x + 1}$

16. $H(x) = \dfrac{x^3 - 27}{x^2 + 5}$

17–24 Construct the algebraic function in a finite number of steps. (Answers will vary.)

17. $f(x) = \dfrac{\sqrt{x^2 - 1}}{x + 1}$

18. $g(x) = \sqrt[3]{\dfrac{x - 1}{-2 + x + x^2}}$

19. $h(x) = \sqrt{2x^2 + x + 1} + 3x(2x + 1)$

20. $u(x) = 13x^3(2 - x) + 3\sqrt{x} - 5x^2(2x - x^2)$

21. $v(x) = \sqrt{2} + \dfrac{x + 3}{\sqrt[5]{2x^2 + 2x - 12}}$

22. $F(x) = \dfrac{\left(x^3 - 4x^2 - 7x + 10\right)^{2/3}}{\sqrt[5]{x - 5}}$

23. $G(x) = \left(x + \left(x + \left(x + (x + 1)^3\right)^3\right)^3\right)^3$

24. $H(x) = \sqrt{2x + \sqrt{2x + \sqrt{2x + \sqrt{2x}}}}$

25–34 Simplify the given trigonometric expression.

25. $\dfrac{1 - \cos^2\left(\dfrac{\pi}{2} - x\right)}{\cos x}$

26. $\dfrac{1}{\sec^2 x} + \sin x \cos\left(\dfrac{\pi}{2} - x\right)$

27. $\sin\alpha(\csc\alpha - \sin\alpha)$

28. $\dfrac{1}{1 + \cos\alpha} + \dfrac{1}{1 - \cos\alpha}$

29. $\cot^2\theta - \cos^2\theta\cot^2\theta$

30. $\cos x(1 + \tan^2 x)$

31. $\dfrac{\sin\beta}{1 + \cos\beta} + \cot\beta$

32. $\dfrac{1}{\cos(-t)\csc(-t)}$

33. $\dfrac{1 - \tan^2 x}{\cot^2 x - 1}$

34. $\dfrac{\sin x \tan\left(\dfrac{\pi}{2} - x\right)}{\cos x}$

35–38 Graph the given piecewise defined function. Use empty or full circles as appropriate at the endpoints of the intervals of definition.

35. $F(x) = \begin{cases} -x^2 & \text{if } x \le 0 \\ \dfrac{1}{2}x + 1 & \text{if } x > 0 \end{cases}$

36. $G(x) = \begin{cases} -2x - 4 & \text{if } x \le -2 \\ \dfrac{1}{2}x + \dfrac{3}{2} & \text{if } -2 < x \le 1 \\ \dfrac{1}{x - 1} & \text{if } x > 1 \end{cases}$

37. $H(x) = \begin{cases} -x & \text{if } x \le 0 \\ \sin x & \text{if } 0 < x \le \dfrac{\pi}{2} \\ \sqrt[3]{x - \dfrac{\pi}{2}} & \text{if } x > \dfrac{\pi}{2} \end{cases}$

38. $u(x) = \begin{cases} -\sqrt{-x - 1} & \text{if } x \le -1 \\ \sqrt{1 - x^2} & \text{if } -1 < x \le 0 \\ x^2 & \text{if } x > 0 \end{cases}$

39–42 Create a piecewise defined rule and then graph the function. Use empty or full circles as appropriate at the endpoints of the intervals of definition.

39. $f(x) = |x - 1|$

40. $g(x) = \dfrac{x}{|x|}$

41. $h(x) = |\sin x|$

42. $v(x) = |x + 2| + |x - 3|$

43–45 The greatest integer function is defined as follows: For $x \in \mathbb{R}$, $[\![x]\!]$ is the greatest integer less than or equal to x. For example, $[\![\pi]\!] = 3$, $[\![1]\!] = 1$, $[\![-1.5]\!] = -2$, and so on.

Use the greatest integer function to sketch the graph of the given function.

43. $f(x) = x - [\![x]\!]$

44. $g(x) = [\![x]\!] - x$

45. $h(x) = [\![\sin x]\!]$

46–48 Simple polynomial functions are used to model real-life phenomena. (**Hint:** See Example 2 for guidance as you work through these problems.)

46. Suppose that while vacationing in Europe, one day you feel a bit dizzy and your host hands you a metric thermometer. Upon checking your temperature, the reading is 39.5 °C. Would you call the doctor? (**Hint:** Recall that the conversion formula between the Fahrenheit and Celsius scales is the linear function $C = \frac{5}{9}(F - 32)$. Express F from this formula to answer the question.)

47. Two trains are 630 miles apart, heading directly toward each other. The first train is traveling at 95 mph, and the second train is traveling at 85 mph. How long will it be before the trains pass each other?

48. Jessica started a candle business a few weeks ago, and noticed that the relationship between her total cost in producing the candles and the number of candles produced can be modeled by a linear function. She was able to make 3 candles for a total cost of $29, while 7 candles cost her a total of $41 to produce.

   a. Find a formula for the total investment as a function of the number of candles produced.

   b. Graph the function found in part a. What are the practical meanings of the slope and y-intercept in this particular situation?

   c. How much will be Jessica's total cost in producing 50 candles?

   d. If Jessica plans to invest a total of $320 in the next 3 months, how many candles will she be able to produce?

49–60 Find a formula for the quantity to be optimized, and use the location of the vertex of its graph to solve the problem.

49. A farmer has a total of 200 yards of fencing to enclose a rectangular pen, so that one of the four sides will be the existing wall of a barn. What should the length and width be in order to maximize the enclosed area? (**Hint:** Let x represent the width, and find an expression for the length in terms of x. Then write an expression for the area and analyze the resulting function.)

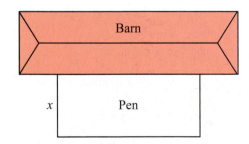

50. A rancher has a rectangular piece of sheet metal that is 20 inches wide by 10 feet long. He plans to fold the metal into a three-sided channel and weld two rectangular pieces of metal to the ends to form a watering trough 10 feet long. How should he fold the metal in order to maximize the volume of the resulting trough?

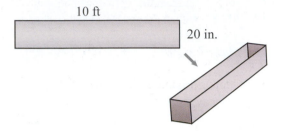

51. Cindy wants to construct three rectangular dog-training arenas side by side using a total of 400 feet of fencing. What should the overall length and width be in order to maximize the area of the three combined arenas? (**Hint:** Let x represent the width, as shown, and find an expression for the overall length in terms of x.)

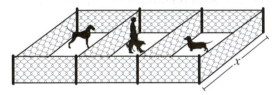

52. Among all the pairs of numbers with a sum of 10, find the pair whose product is maximum.

53. Find the point on the line $2x + y = 5$ that is closest to the origin. (**Hint:** Instead of trying to minimize the distance between the origin and points on the line, minimize the square of the distance.)

54. Among all the pairs of numbers (x, y) such that $2x + y = 20$, find the pair for which the sum of the squares is minimum.

55. Find a pair of numbers whose product is maximum when two times the first number plus the second number is 48.

56. The total revenue for Morris' Studio Apartments is given as the function $R(x) = 100x - 0.1x^2$, where x is the number of apartments rented. What is the number of apartments rented that produces the maximum revenue?

57. The total cost of manufacturing golf clubs is given as the function $C(x) = 800 - 10x + 0.20x^2$, where x is the number of sets of golf clubs produced. How many sets of golf clubs should be manufactured to incur minimum cost?

58. A rock is thrown upward with a velocity of 48 feet per second from the top of a 64-foot-high cliff. What is the maximum height attained by the rock? (**Hint:** Use $h(t) = -16t^2 + 48t + 64$ to describe the height of the rock as a function of time t.)

59. Jason is driving his Mustang GT down a two-lane highway one night, carefully observing the posted speed limit sign of 55 mph. His headlights suddenly illuminate a white-tailed deer, about 120 ft in front of his car, and he immediately hits the brakes. Suppose that the coefficient of friction between his car's tires and the pavement is $\mu = 0.9$. Using the quadratic model from Example 5b in Section 1.1, do you think he will hit the deer? What if he had traveled at 60 mph?

60. A student is throwing a small rubber ball during physical education class at an upward angle so that the horizontal component of the ball's initial velocity is 40 feet per second. If the vertical position function of the ball is given by $h(t) = -16t^2 + 24t + 7$, how far from the student will the ball hit the ground? (**Hint:** First determine how long it will take for the ball to hit the ground. The vertical position h is measured in feet, t in seconds. Ignore air resistance.)

61–72 Trigonometric and exponential functions are used to model real-life situations. (**Hint:** See Examples 6 and 7 for guidance as you work through these problems.)

61. Suppose several potatoes are dumped into the basket of a grocer's scale, which then proceeds to bounce up and down with an amplitude of 4 cm. As discussed in Example 6, a first approximation to this motion can be given by a trigonometric model. Supposing that the constant ω for the above motion is 6π and that $t = 0$ when the potatoes land in the basket, find the position function for this motion. How long does it take for the basket to complete a full period?

62. The size of a local coyote population in a certain California national forest is estimated to cycle annually according to the function $P(t) = 250 + 20\sin(\pi t/6)$, where t is measured in months, starting on March 1st of each year.

   a. What is the approximate size of the population on July 1st?

   b. When is the population expected to be the smallest, and what is its size then?

63. A certain species of fish is to be introduced into a new man-made lake, and wildlife experts estimate that the population will grow according to $P(t) = (1000)2^{t/3}$, where t represents the number of years from the time of introduction.

   a. What is the doubling time for this population of fish?

   b. How long will it take for the population to reach 8000 fish, according to this model?

64. Assuming a current world population of 6 billion people, and exponential growth at an annual rate of 1.9%, what will the world population be in **a.** 10 years and **b.** 50 years?

65. Suppose that a new virus has broken out in isolated parts of Africa, and it is spreading exponentially through tribal villages. The growth of this new virus can be mapped using the following formula where P stands for the number of people in a village, V for the number of infected individuals, and d for the number of days since the virus first appeared:

$$V = P\left(1 - e^{-0.18d}\right)$$

According to this equation, how many people in a tribe of 300 will be infected after 5 days?

66. The radioactive element polonium-210 decays according to the function $A(t) = A_0 e^{-0.004951t}$, where A_0 is the mass at time $t = 0$, and t is measured in days. The fact that $A(140) = A_0/2$ means that the half-life of polonium-210 is 140 days. What percentage of the original mass of a sample of polonium-210 remains after one year?

67. The half-life of a radioactive material is the time required for an initial quantity to decrease to half its original value. In the case of radium, this is approximately 1600 years.

 a. Determine a so that $A(t) = A_0 a^t$ describes the amount of radium after t years, where A_0 is the initial amount at $t = 0$.

 b. How much of a 1-gram sample of radium would remain after 100 years?

 c. How much of a 1-gram sample of radium would remain after 1000 years?

68. When continuous compounding is used in banking, the balance after t years is described by the formula $A(t) = Pe^{rt}$, where P is the initial amount (or principal) at $t = 0$, and r is the annual interest rate. Suppose Mario made a deposit two years ago, which is compounded continuously at an annual rate of 4.5%. If his current balance is $1094.17, how much was his initial deposit? How much longer would he have to wait until his initial deposit doubles?

69. The function $f(t) = C(1+r)^t$ models the rise in the cost of a product that has a cost of C today, subject to an average yearly inflation rate of r for t years. If the average annual rate of inflation over the next decade is assumed to be 3%, what will the inflation-adjusted cost of a $100,000 house be in 10 years?

70. The concentration of a certain drug in the bloodstream after t minutes is given by the formula $C(t) = 0.05(1 - e^{-0.2t})$. What is the concentration after 10 minutes?

71. Carbon-11 has a radioactive half-life of approximately 20 minutes, and is useful as a diagnostic tool in certain medical applications. Because of the relatively short half-life, time is a crucial factor when conducting experiments with this element.

 a. Determine a so that the formula $A(t) = A_0 a^t$ describes the amount of carbon-11 left after t minutes, where A_0 is the amount at time $t = 0$.

 b. How much of a 2-kilogram sample of carbon-11 would be left after 30 minutes?

 c. How much of a 2-kilogram sample of carbon-11 would be left after 6 hours?

72. Charles has recently inherited $8000, which he wants to deposit in a savings account. He has determined that his two best bets are an account that compounds annually at a rate of 3.20% and an account that compounds continuously at an annual rate of 3.15%. Which account would pay Charles more interest?

73–82 *True or False?* Determine whether the given statement is true or false. In case of a false statement, explain or provide a counterexample.

73. The slope of the graph of $y = Ax + B$ is A.

74. The slope of the graph of $y = Ax^2 + Bx + C$ is B.

75. The lines with equations $y = Ax + B$ and $y = -Bx + A$ are perpendicular to each other.

76. A quadratic function can have up to two y-intercepts.

77. If line L_1 has positive slope and L_2 is perpendicular to L_1, then the slope of L_2 is negative.

78. If a polynomial has even degree, then its graph always rises to both the right and the left.

79. All rational functions of the form $p(x)/q(x)$, where $q(x)$ is nonconstant, have at least one asymptote of some kind.

80. Trigonometric functions are transcendental.

81. Logarithmic functions are transcendental.

82. If a population of bacteria grows without restriction from 1000 to 2000 in one hour, then it will grow to 3000 during the next hour.

1.3 Transforming and Combining Functions

TOPICS

1. Shifting, reflecting, and stretching functions
2. Combining functions arithmetically
3. Composing and decomposing functions
4. Interlude: recursive graphics

The repertory of functions in Section 1.2 contains the building blocks for very nearly all the functions we encounter in calculus, but they are just that—building blocks. Typically, we will find it necessary to modify an existing function or combine two or more functions in some way. The most common and useful ways are reviewed in this section.

TOPIC 1   Shifting, Reflecting, and Stretching Functions

Definition 💡

Horizontal Shifting

Let $f(x)$ be a function whose graph is known, and let h be a fixed real number. If we replace x in the definition of f by $x - h$, we obtain a new function $g(x) = f(x - h)$. The graph of g is the same shape as the graph of f, but shifted to the right by h units if $h > 0$ and shifted to the left by h units if $h < 0$.

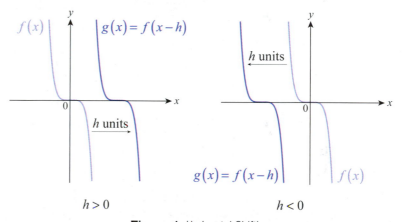

Figure 1 Horizontal Shifting

Caution ⚠

A word of caution is in order here. It is easy to forget that the minus sign in the expression $x - h$ is critical. It may help to remember a few specific examples: replacing x with $x - 5$ shifts the graph 5 units to the *right*, since 5 is positive. Replacing x with $x + 5$ shifts the graph 5 units to the *left*, since we have actually replaced x with $x - (-5)$. With practice, knowing the effect that replacing x with $x - h$ has on the graph of a function will become natural.

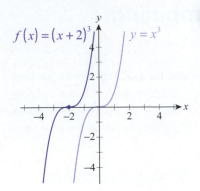

$f(x)=(x+2)^3$ $y=x^3$

Figure 2

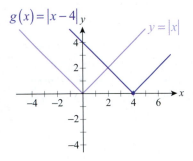

$g(x)=|x-4|$ $y=|x|$

Figure 3

Example 1 ✐

Sketch the graphs of the following functions.

a. $f(x)=(x+2)^3$ **b.** $g(x)=|x-4|$

Solution

a. The shape of $(x+2)^3$ is the same as the shape of x^3, since one expression is obtained from the other by replacing x by $x+2$. We simply draw the basic cubic shape (the shape of $y=x^3$), then shift it to the left by 2 units (see Figure 2). Note, for example, that $(-2,0)$ is one point on the graph of f.

b. The basic function being shifted is $|x|$. The graph of $g(x)=|x-4|$ has the same shape as the graph of the absolute value function, but shifted to the right by 4 units (see Figure 3). Note, for example, that $(4,0)$ lies on the graph of g.

Definition 💡

Vertical Shifting

Let $f(x)$ be a function whose graph is known, and let k be a fixed real number. The graph of the function $g(x)=f(x)+k$ is the same shape as the graph of f, but shifted upward by k units if $k>0$ and shifted downward by k units if $k<0$.

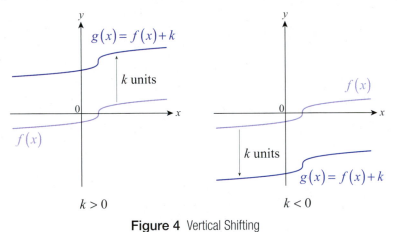

Figure 4 Vertical Shifting

The effect of adding a constant k to a function can be most easily remembered by recalling that every point on the graph of a function f has the form $(x,f(x))$, so adding k shifts every point $(x,f(x))$ to $(x,f(x)+k)$. These new points are above the originals if k is positive and below the originals if k is negative.

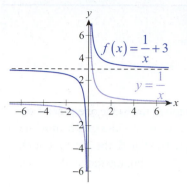

Figure 5

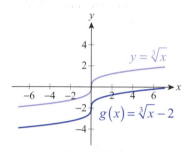

Figure 6

Example 2 ✐

Sketch the graphs of the following functions.

a. $f(x) = \dfrac{1}{x} + 3$

b. $g(x) = \sqrt[3]{x} - 2$

Solution

a. The graph of $f(x) = (1/x) + 3$ is the graph of $y = 1/x$ shifted up by 3 units (see Figure 5). Note that this doesn't affect the domain: the domain of f is $(-\infty, 0) \cup (0, \infty)$, the same as the domain of $y = 1/x$. However, the range is affected. The range of f is $(-\infty, 3) \cup (3, \infty)$.

b. The basic function being shifted is $\sqrt[3]{x}$. To graph $g(x) = \sqrt[3]{x} - 2$, we shift the graph of $y = \sqrt[3]{x}$ down by 2 units (see Figure 6).

Definition ♀

Reflecting with Respect to the Axes

Let $f(x)$ be a function whose graph is known.

1. The graph of the function $g(x) = -f(x)$ is the reflection of the graph of f with respect to the x-axis.

2. The graph of the function $h(x) = f(-x)$ is the reflection of the graph of f with respect to the y-axis.

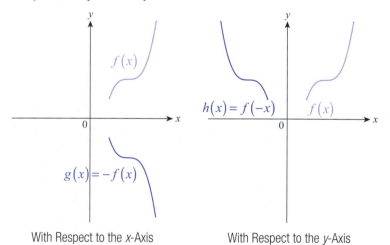

With Respect to the x-Axis   With Respect to the y-Axis

Figure 7 Reflecting with Respect to the Axes

In other words, a function is reflected with respect to the x-axis by multiplying the entire function by -1, and reflected with respect to the y-axis by replacing x with $-x$.

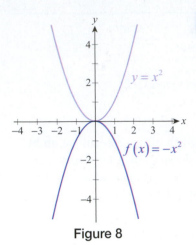

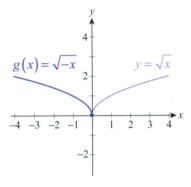

Figure 8

Figure 9

Example 3 ✐

Sketch the graphs of the following functions.

a. $f(x) = -x^2$ **b.** $g(x) = \sqrt{-x}$

Solution

a. To graph $f(x) = -x^2$, begin with the graph of the basic parabola $y = x^2$. The entire function is multiplied by -1, so reflect the graph over the x-axis, resulting in the original shape turned upside down (see Figure 8). Note that the domain is still the entire real line, but the range of f is the interval $(-\infty, 0]$.

b. To graph $g(x) = \sqrt{-x}$, we reflect the graph $y = \sqrt{x}$ with respect to the y-axis because x has been replaced by $-x$ (see Figure 9). Note that this changes the domain, but not the range. The domain of g is the interval $(-\infty, 0]$ and the range is $[0, \infty)$.

Definition 💡

Stretching and Compressing

Let $f(x)$ be a function whose graph is known, and let a be a positive real number.

1. The graph of the function $g(x) = af(x)$ is *stretched* vertically compared to the graph of f if $a > 1$.

2. The graph of the function $g(x) = af(x)$ is *compressed* vertically compared to the graph of f if $0 < a < 1$.

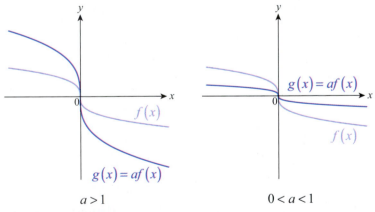

Figure 10 Stretching and Compressing

If the function g is obtained from the function f by multiplying f by a negative real number, think of the number as the product of -1 and a positive real number (namely, its absolute value). The previous definition tells us what multiplication by a positive constant does to a graph, and we already know that multiplying a function by -1 reflects the graph with respect to the x-axis.

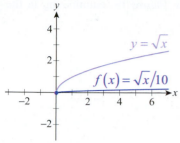

Figure 11

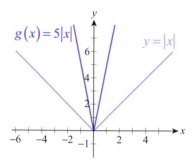

Figure 12

Example 4 ✐

Sketch the graphs of the following functions.

a. $f(x) = \dfrac{\sqrt{x}}{10}$ **b.** $g(x) = 5|x|$

Solution

a. Begin with the graph of $\sqrt{x}$. The shape of $f(x)$ is similar to the shape of $y = \sqrt{x}$, but compressed considerably because the second coordinates have been multiplied by the factor of $\frac{1}{10}$, and are consequently smaller (see Figure 11).

b. Begin with the graph of the absolute value function. In contrast to part a., the graph of $g(x) = 5|x|$ is stretched compared to $y = |x|$ (see Figure 12). Every second coordinate has been multiplied by a factor of 5, and is consequently larger.

We can now put all of the above together and consider functions that have been derived through a sequence of transformations from simpler functions.

Order of Transformations

If a function g has been obtained from a simpler function f through a number of transformations, g can be analyzed by looking for the transformations in this order.

Step 1: Horizontal shifting

Step 2: Stretching and compressing

Step 3: Reflecting with respect to the axes

Step 4: Vertical shifting

Consider, for example, the function $g(x) = -2\sqrt{x+1} + 3$. The function g has been "built up" from the basic square root function through a variety of transformations.

Step 1: First, $\sqrt{x}$ has been transformed into $\sqrt{x+1}$ by replacing x with $x + 1$, and we know that this corresponds graphically to a shift leftward by 1 unit.

Step 2: Next, the function $\sqrt{x+1}$ has been multiplied by 2 to get the function $2\sqrt{x+1}$, and we know that this has the effect of stretching the graph of $\sqrt{x+1}$ vertically.

Step 3: The function $2\sqrt{x+1}$ has been multiplied by -1, giving us $-2\sqrt{x+1}$, and the graph of this is the reflection of $2\sqrt{x+1}$ with respect to the x-axis.

Step 4: Finally, the constant 3 has been added to $-2\sqrt{x+1}$, shifting the entire graph upward by 3 units.

These transformations are illustrated in order in Figure 13, culminating in the graph of $g(x) = -2\sqrt{x+1} + 3$.

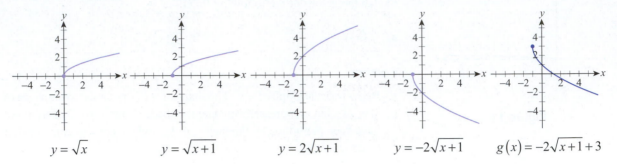

$y = \sqrt{x}$ $y = \sqrt{x+1}$ $y = 2\sqrt{x+1}$ $y = -2\sqrt{x+1}$ $g(x) = -2\sqrt{x+1} + 3$

Figure 13 Building the Graph of $g(x) = -2\sqrt{x+1} + 3$

Example 5 ✎

Sketch the graph of the function $f(x) = 1/(2-x)$.

Solution

The basic function that f is similar to is $1/x$. Now we follow the order of transformations.

Step 1: If we replace x by $x + 2$ (shifting the graph 2 units to the left), we obtain the function $1/(x+2)$, which is closer to what we want.

Step 2: There does not appear to be any stretching or compressing transformation.

Step 3: If we now replace x by $-x$, we have $1/(-x+2)$, which is the same as f. This reflects the graph of $1/(x+2)$ with respect to the y-axis.

Step 4: Since we already found f, we know there is no vertical shift.

The entire sequence of transformations is shown in Figure 14, ending with the graph of f.

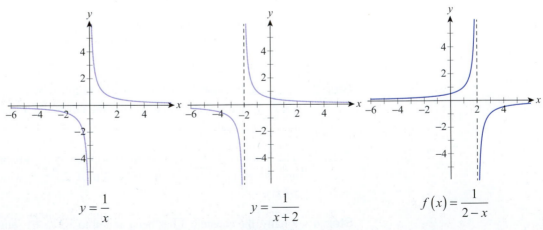

$y = \dfrac{1}{x}$ $y = \dfrac{1}{x+2}$ $f(x) = \dfrac{1}{2-x}$

Figure 14

Note: An alternate approach to graphing $f(x) = 1/(2-x)$ is to rewrite the function in the form $f(x) = -1/(x-2)$. In this form, the graph of f is the graph of $1/x$ shifted 2 units to the right, and then reflected with respect to the x-axis. The result is the same, as you should verify.

TOPIC 2   Combining Functions Arithmetically

Shifting, stretching, and reflecting are all modifications applied to a single function. We also often need to combine two or more functions.

We begin with four arithmetic ways of combining two or more functions to obtain new functions. The four arithmetic operations are very familiar to you: addition, subtraction, multiplication, and division. The only possibly new idea is that we are applying these operations to functions, not numbers. But as we will see, the arithmetic combination of functions is based entirely on the arithmetic combination of numbers.

Definition 💡

Adding, Subtracting, Multiplying, and Dividing Functions

Let f and g be two functions. The **sum** $f + g$, **difference** $f - g$, **product** fg, and **quotient** f/g are four new functions defined as follows.

1. $(f + g)(x) = f(x) + g(x)$

2. $(f - g)(x) = f(x) - g(x)$

3. $(fg)(x) = f(x)g(x)$

4. $\left(\dfrac{f}{g}\right)(x) = \dfrac{f(x)}{g(x)}$, provided that $g(x) \neq 0$

The domain of each of these new functions consists of the common elements of the individual domains of f and g, with the added condition that for the quotient function we have to omit those elements for which $g(x) = 0$.

With the above definition, we can determine the sum, difference, product, or quotient of two functions at one particular value for x or find a formula for these new functions based on the formulas for f and g, if they are available.

Example 6 ✎

Given that $f(-2) = 5$ and $g(-2) = -3$, find $(f - g)(-2)$ and $(f/g)(-2)$.

Solution

By the definition of the difference and quotient of functions, we obtain the following.

$$(f - g)(-2) = f(-2) - g(-2)$$
$$= 5 - (-3) = 8$$

$$\left(\frac{f}{g}\right)(-2) = \frac{f(-2)}{g(-2)}$$

$$= \frac{5}{-3} = -\frac{5}{3}$$

Example 7 ✎

Given the two functions $f(x) = 4x^2 - 1$ and $g(x) = \sqrt{x}$, find $(f + g)(x)$ and $(fg)(x)$.

Solution

By the definition of the sum and product of functions, we obtain the following.

$$(f + g)(x) = f(x) + g(x)$$
$$= 4x^2 - 1 + \sqrt{x}$$

$$(fg)(x) = (4x^2 - 1)(\sqrt{x})$$
$$= 4x^{5/2} - x^{1/2}$$

Note that the domain of f is the entire real line, while the domain of g is $[0, \infty)$.

Since the domain of two functions combined arithmetically is the intersection of the individual domains, $f + g$ and fg both have the domain $[0, \infty)$.

Example 8 ✎

Based on the graphs of f and g in Figure 15, determine the domain of f/g and evaluate $(f/g)(1)$.

Solution

From Figure 15, we can see that the domain of both f and g individually is the set of all real numbers, $(-\infty, \infty)$. To find the domain of f/g, we need to check where $g(x) = 0$. Based on the graph, this occurs when $x = -2$ and $x = 2$, so the domain of f/g is as follows.

$$(-\infty, -2) \cup (-2, 2) \cup (2, \infty)$$

Also based upon the graphs, it appears that $f(1) = 1$ and $g(1) = 3$, so $(f/g)(1) = \frac{1}{3}$.

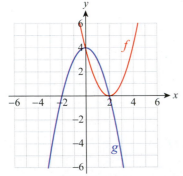

Figure 15

TOPIC 3 Composing and Decomposing Functions

A fifth way of combining functions is to form the *composition* of one function with another. Informally speaking, this means to apply one function, say f, to the output of another function, say g. The symbol for composition is an open circle.

Definition 💡

Composing Functions

Let f and g be two functions. The **composition** of f and g, denoted $f \circ g$, is the function defined by $(f \circ g)(x) = f(g(x))$. (The function $f \circ g$ is read "f composed with g" or "f of g," and $f \circ g$ is also referred to as a **composite function**.)

The domain of $f \circ g$ consists of all x in the domain of g for which $g(x)$ is in the domain of f.

The diagram in Figure 16 is a sort of schematic of the composition of two functions. The circles represent sets, with the leftmost circle being the domain of the function g. The arrows indicate the element that x is associated with by the various functions.

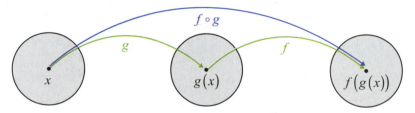

Figure 16 Composition of *f* and *g*

As with the four arithmetic ways of combining functions, we can evaluate the composition of two functions at a single point or find a formula for the composition if we have been given formulas for the individual functions.

Caution ⚠

Note that the order of f and g is important. In general, we can expect the function $f \circ g$ to be different from the function $g \circ f$. That is, the composition of two functions, unlike the sum and product of two functions, is not commutative.

Example 9 ✏

Given $f(x) = x^2$ and $g(x) = x - 3$, find the following.

a. $(f \circ g)(6)$ **b.** $(f \circ g)(x)$

Solution

a. Since $(f \circ g)(6) = f(g(6))$, the first step is to calculate $g(6)$.

$$g(6) = 6 - 3 = 3$$

Then, apply f to the result: $(f \circ g)(6) = f(g(6)) = f(3) = 3^2 = 9$.

b. To find the formula for $f \circ g$, we simply apply the definition of composition and then simplify.

$$(f \circ g)(x) = f(g(x))$$
$$= f(x - 3)$$
$$= (x - 3)^2$$
$$= x^2 - 6x + 9$$

Note that once we have found a formula for $f \circ g$, we have an alternative way of answering the first question: $(f \circ g)(6) = (6)^2 - (6)(6) + 9 = 9$.

· Example 10 ✎

Let $f(x) = x^2 - 4$ and $g(x) = \sqrt{x}$. Find formulas and state the domains for the following functions.

a. $f \circ g$ **b.** $g \circ f$

Solution

a. $(f \circ g)(x) = f(g(x))$
$$= f(\sqrt{x})$$
$$= (\sqrt{x})^2 - 4 = x - 4$$

The answer may seem to indicate that the domain of $f \circ g$ is all real numbers, but this is incorrect. The domain of $f \circ g$ is actually the interval $[0, \infty)$, because only nonnegative numbers can be plugged into g.

b. $(g \circ f)(x) = g(f(x))$
$$= g(x^2 - 4)$$
$$= \sqrt{x^2 - 4}$$

The domain of $g \circ f$ consists of all x for which $x^2 - 4 \geq 0$, or $x^2 \geq 4$. In interval form, the domain is $(-\infty, -2] \cup [2, \infty)$.

Often, functions can be best understood by recognizing them as a composition of two or more simpler functions. We have already seen one instance of this: shifting, reflecting, stretching, and compressing can all be thought of as a composition of two or more functions. For example, the function $h(x) = (x - 2)^3$ can be thought of as the composition of the functions $f(x) = x^3$ and $g(x) = x - 2$.

$$f(g(x)) = f(x - 2)$$
$$= (x - 2)^3$$
$$= h(x)$$

To "decompose" a function into a composition of simpler functions, it is usually best to identify what the function does to its argument from the inside out. That is, identify the first thing that is done to the argument, then the second, and so on. Each action describes a less complex function, and can be identified as such. The composition of these functions, with the innermost function corresponding to the first action, the next innermost corresponding to the second action, and so on, is then equivalent to the original function.

Decomposition can often be done in several different ways. Consider, for example, the function $f(x) = \sqrt[3]{5x^2 - 1}$. The following illustrates some of the ways f can be written as a composition of functions. Be sure you understand how each of the different compositions is equivalent to f.

1. $g(x) = \sqrt[3]{x}$ $g\big(h(x)\big) = g\big(5x^2 - 1\big)$
    $h(x) = 5x^2 - 1$ $= \sqrt[3]{5x^2 - 1}$
                                          $= f(x)$

2. $g(x) = \sqrt[3]{x - 1}$ $g\big(h(x)\big) = g\big(5x^2\big)$
    $h(x) = 5x^2$ $= \sqrt[3]{5x^2 - 1}$
                                          $= f(x)$

3. $g(x) = \sqrt[3]{x}$ $g\big(h(k(x))\big) = g\big(h(x^2)\big)$
    $h(x) = 5x - 1$ $= g\big(5x^2 - 1\big)$
    $k(x) = x^2$ $= \sqrt[3]{5x^2 - 1}$
                                          $= f(x)$

Example 11 ✐

Decompose the function $f(x) = \big|x^2 - 3\big| + 2$ into a composition of **a.** two functions and **b.** three functions.

Solution

a. $g(x) = |x| + 2$ $g\big(h(x)\big) = g\big(x^2 - 3\big)$
    $h(x) = x^2 - 3$ $= \big|x^2 - 3\big| + 2$
                                          $= f(x)$

b. $g(x) = x + 2$ $g\big(h(k(x))\big) = g\big(h(x^2)\big)$
    $h(x) = |x - 3|$ $= g\big(\big|x^2 - 3\big|\big)$
    $k(x) = x^2$ $= \big|x^2 - 3\big| + 2$
                                          $= f(x)$

TOPIC 4 **Interlude: Recursive Graphics**

Recursion, in general, refers to using the output of a function as its input and repeating the process a certain number of times. In other words, recursion refers to the composition of a function with itself, possibly many times. Recursion has many varied uses, one of which is a branch of mathematical art.

Some special nomenclature and notation have evolved to describe recursion. If f is a function, $f^2(x)$ is used in this context to stand for $f(f(x))$, or $(f \circ f)(x)$ (not $\left[f(x)\right]^2$!). Similarly, $f^3(x)$ stands for $f(f(f(x)))$, or $(f \circ f \circ f)(x)$ and so on. The functions f^2, f^3, ... are called **iterates** of f, with f^n being the nth **iterate** of f.

Some of the most famous recursively generated mathematical art is based on functions whose inputs and outputs are complex numbers. Recall that every complex number can be expressed in the form, $a + bi$ where a and b are real numbers and i is the imaginary unit.

> *So, Nat'ralists observe, a Flea*
>
> *Hath smaller Fleas that on him prey,*
>
> *And these have smaller Fleas to bite 'em,*
>
> *And so proceed, ad infinitum.*
>
> *- Jonathan Swift*

A one-dimensional coordinate system, such as the real number line, is insufficient to graph complex numbers, but complex numbers are easily graphed in a two-dimensional coordinate system.

To graph the number $a + bi$, we treat it as the ordered pair (a,b) and plot the point (a,b) in the Cartesian plane, where the horizontal axis represents pure real numbers and the vertical axis represents pure imaginary numbers.

Benoit Mandelbrot used the function $f(z) = z^2 + c$, where both z and c are variables representing complex numbers, to generate the image known as the Mandelbrot set in 1979. The basic idea is to evaluate the sequence of iterates $f(0) = 0^2 + c = c$, $\quad f^2(0) = f(c) = c^2 + c$, $\quad f^3(0) = f(c^2 + c) = (c^2 + c)^2 + c, \ldots$ for various complex numbers c and determine if the sequence of complex numbers stays close to the origin or not. Those complex numbers that result in so-called "bounded" sequences are colored white. We have used similar ideas to generate our own recursive art, as described below.

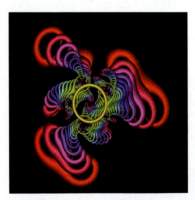

Figure 17 *i* of the Storm

The image "*i* of the storm" reproduced in Figure 17 is based on the function $f(z) = \dfrac{(1-i)z^4 + (7+i)z}{2z^5 + 6}$, where again z is a variable that will be replaced with complex numbers. The image is actually a picture of the complex plane, with the origin in the very center of the golden ring. The golden ring consists of those complex numbers that lie between 0.9 and 1.1 in distance from the origin. The rules for coloring other complex numbers in the plane are as follows: given an initial complex number z not on the gold ring, $f(z)$ is calculated. If the complex number $f(z)$ lies somewhere on the gold ring, the original number z is colored the deepest shade of green. If not, the iterate $f^2(z)$ is calculated. If this result lies in the gold ring, the original z is colored a bluish shade of green. If not, the process continues up to the 12th iterate $f^{12}(z)$, using a different color each time. If $f^{12}(z)$ lies in the gold ring, z is colored red, and if not the process halts and z is colored black.

The idea of recursion can be used to generate any number of similar images, with the end result usually striking and often surprising even to the creator.

1.3 **Exercises**

1–23 Sketch the graph of the given function by first identifying the more basic function that has been shifted, reflected, stretched, or compressed. Then determine the domain and range of the function.

1. $f(x)=(x+2)^3$

2. $G(x)=|x-4|$

3. $p(x)=-(x+1)^2+2$

4. $g(x)=\sqrt{x+3}-1$

5. $q(x)=(1-x)^2$

6. $r(x)=-\sqrt[3]{x}$

7. $s(x)=\sqrt{2-x}$

8. $F(x)=\dfrac{|x+2|}{3}+3$

9. $w(x)=\dfrac{1}{(x-3)^2}$

10. $v(x)=\dfrac{1}{3x}-2$

11. $f(x)=\dfrac{1}{2-x}$

12. $k(x)=\sqrt{-x}+2$

13. $b(x)=[\![x-4]\!]+4$

14. $R(x)=4-|2x|$

15. $S(x)=(3-x)^3$

16. $g(x)=-\dfrac{1}{x+1}$

17. $h(x)=\dfrac{x^2}{2}-3$

18. $W(x)=1-|4-x|$

19. $g(x)=x^2-6x+9$ (**Hint:** Find a better way to write the function.)

20. $h(x)=\dfrac{|x|}{x}$ (**Hint:** Evaluate h at a few points to understand its behavior.)

21. $W(x)=\dfrac{x-1}{|x-1|}$ **22.** $S(x)=[\![x-2]\!]$

23. $V(x)=-3\sqrt{x-1}+2$

24–29 Write an equation for the function described.

24. Use the function $f(x)=x^2$. Move the function 4 units to the right and 2 units up.

25. Use the function $f(x)=x^2$. Move the function 6 units up and reflect across the x-axis.

26. Use the function $f(x)=x^3$. Move the function 1 unit to the left and reflect across the y-axis.

27. Use the function $f(x)=\sqrt{x}$. Move the function 5 units to the left and reflect across the x-axis.

28. Use the function $f(x)=\sqrt{x}$. Move the function 3 units down and reflect across the y-axis.

29. Use the function $f(x)=|x|$. Move the function 7 units to the left, reflect across the x-axis, and reflect across the y-axis.

30–33 Use your knowledge about transformations to find a possible formula for the function $f(x)$ given by its graph.

30.

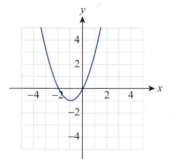

31.

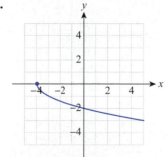

32.

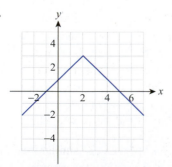

33.

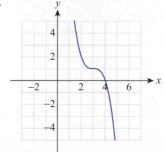

34–45 Use the information given to determine **a.** $(f+g)(-1)$, **b.** $(f-g)(-1)$, **c.** $(fg)(-1)$, and **d.** $(f/g)(-1)$.

34. $f(-1)=-3;\quad g(-1)=5$

35. $f(-1)=0;\quad g(-1)=-1$

36. $f(x)=x^2-3;\quad g(x)=x$

37. $f(x)=\sqrt[3]{x};\quad g(x)=x-1$

38. $f(-1)=15;\quad g(-1)=-3$

39. $f(x)=\dfrac{x+5}{2};\quad g(x)=6x$

40. $f(x)=x^4+1;\quad g(x)=x^{11}+2$

41. $f(x)=\dfrac{6-x}{2};\quad g(x)=\sqrt{\dfrac{x}{-4}}$

42. $f=\{(5,2),(0,-1),(-1,3),(-2,4)\};$
$g=\{(-1,3),(0,5)\}$

43. $f=\{(3,15),(2,-1),(-1,1)\};\quad g(x)=-2$

44.

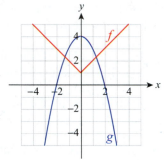

45.

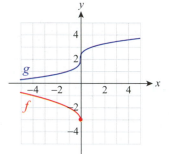

46–53 Find the formula and domain for **a.** $f+g$ and **b.** f/g.

46. $f(x)=|x|;\quad g(x)=\sqrt{x}$

47. $f(x)=x^2-1;\quad g(x)=\sqrt[3]{x}$

48. $f(x)=x-1;\quad g(x)=x^2-1$

49. $f(x)=x^{3/2};\quad g(x)=x-3$

50. $f(x)=3x;\quad g(x)=x^3-8$

51. $f(x)=x^3+4;\quad g(x)=\sqrt{x-2}$

52. $f(x)=-2x^2;\quad g(x)=[\![x+4]\!]$

53. $f(x)=6x-1;\quad g(x)=x^{2/3}$

54–63 Evaluate the expression, if possible, given $f(x)=1/x^2$ and $g(x)=2x+3$.

54. $(f+g)(-7)$

55. $(f+g)(-10)$

56. $(f-g)(-5)$

57. $(f-g)(0)$

58. $(fg)(4)$

59. $(fg)(-3)$

60. $\left(\dfrac{f}{g}\right)(-2)$

61. $\left(\dfrac{f}{g}\right)(0)$

62. $\left(\dfrac{g}{f}\right)(1)$

63. $\left(\dfrac{g}{f}\right)(-6)$

64–73 Use the information given to determine $(f \circ g)(3)$.

64. $f(-5) = 2;\ \ g(3) = -5$

65. $f(\pi) = \pi^2;\ \ g(3) = \pi$

66. $f(x) = x^2 - 3;\ \ g(x) = \sqrt{x}$

67. $f(x) = \sqrt{x^2 - 9};\ \ g(x) = 1 - 2x$

68. $f(x) = 2 + \sqrt{x};\ \ g(x) = x^3 + x^2$

69. $f(x) = x^{3/2} - 3;\ \ g(x) = \left\lVert \dfrac{3x}{2} \right\rVert$

70. $f(x) = \sqrt{x + 6};\ \ g(x) = \sqrt{4x - 3}$

71. $f(x) = \sqrt{\dfrac{3x}{14}};\ \ g(x) = x^4 - x^3 - x^2 - x$

72.

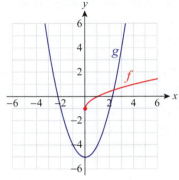

73.

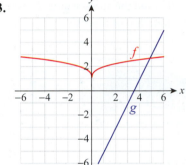

74–87 Find the formula and domain for **a.** $f \circ g$ and **b.** $g \circ f$.

74. $f(x) = \sqrt{x - 1};\ \ g(x) = x^2$

75. $f(x) = \dfrac{1}{x};\ \ g(x) = x - 1$

76. $f(x) = \dfrac{4x - 2}{3};\ \ g(x) = \dfrac{1}{x}$

77. $f(x) = 1 - x;\ \ g(x) = \sqrt{x}$

78. $f(x) = [\![x - 3]\!];\ \ g(x) = x^3 + 1$

79. $f(x) = x^2 + 2x;\ \ g(x) = 3x^2 + 5$

80. $f(x) = x^2 + 1;\ \ g(x) = 3x^2 + 5$

81. $f(x) = \sqrt{x};\ \ g(x) = 2x$

82. $f(x) = \dfrac{1}{x + 7};\ \ g(x) = \dfrac{2}{x}$

83. $f(x) = \dfrac{1}{x};\ \ g(x) = \dfrac{1}{x}$

84. $f(x) = x^2;\ \ g(x) = 3x + 1$

85. $f(x) = \sqrt[3]{x};\ \ g(x) = x^3$

86. $f(x) = \sqrt{x - 4};\ \ g(x) = x^2 + 2$

87. $f(x) = \dfrac{3}{1 - x};\ \ g(x) = 3x^2$

88–93 Write the given function as a composition of two functions. (Answers will vary.)

88. $f(x) = \sqrt[3]{3x^2 - 1}$

89. $f(x) = \dfrac{2}{5x - 1}$

90. $f(x) = |x - 2| + 3$

91. $f(x) = x + \sqrt{x + 2} - 5$

92. $f(x) = |x^3 - 5x| + 7$

93. $f(x) = \dfrac{\sqrt{x - 3}}{x^2 - 6x + 9}$

94. The volume of a right circular cylinder is given by the formula $V = \pi r^2 h$. If the height h is three times the radius r, show the volume V as a function of r.

95. The surface area of a wind sock is defined by the formula $S = \pi r \sqrt{r^2 + h^2}$ where r is the radius of the base of the wind sock and h is the height of the wind sock. As the wind sock is being knitted by an automated knitter, the height h is increasing with time t as defined by the formula $h(t) = \frac{1}{4}t^2, t \geq 0$. Find the surface area S of the wind sock as a function of time t.

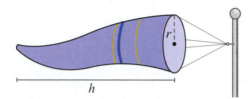

96. The volume of the wind sock described in the previous exercise is given by the formula $V = \frac{1}{3}\pi r^2 h$ where r is the radius of the wind sock and h is the height of the wind sock. If the height h is increasing with time t as defined by the formula $h(t) = \frac{1}{4}t^2$, $t \geq 0$, find the volume V of the wind sock as a function of time t.

97. A widget factory produces n widgets in t hours of a single day. The number of widgets the factory produces is given by the formula $n(t) = 10{,}000t - 25t^2, 0 \leq t \leq 9$. The cost c in dollars of producing n widgets is given by the formula $c(n) = 2040 + 1.74n$. Find the cost c as a function of time t that the factory is producing the widgets.

98. Suppose that $H(x)$ represents the percentage of income spent on a home loan in the year x and $C(x)$ represents the percentage of income spent on a car loan in the year x. If $I(x)$ represents the income in year x, determine the function L that represents the total loan expenses in year x.

99. Given two odd functions f and g, show that $f \circ g$ is also odd. Then verify this fact with the particular functions $f(x) = \sqrt[3]{x}$ and $g(x) = -x^3/(3x^2 - 9)$. (**Hint:** Recall that a function is odd if $f(-x) = -f(x)$ for all x in the domain of f.)

100. Given two even functions f and g, show that the product is also even. Then verify this fact with the particular functions $f(x) = 2x^4 - x^2$ and $g(x) = 1/x^2$. (**Hint:** Recall that a function is even if $f(-x) = f(x)$ for all x in the domain of f.)

101–108 As mentioned in the Interlude, a given complex number c is said to be in the Mandelbrot set if, for the function $f(z) = z^2 + c$, the sequence of iterates $f(0), f^2(0), f^3(0), \ldots$ stays close to the origin (which is the complex number $0 + 0i$). It can be shown that if any single iterate falls more than 2 units in distance (magnitude) from the origin, then the remaining iterates will grow larger and larger in magnitude. In practice, computer programs that generate the Mandelbrot set calculate the iterates up to a predecided point in the sequence, such as $f^{50}(0)$, and if no iterate to this point exceeds 2 in magnitude the number c is admitted to the set. The magnitude of a complex number $a + bi$ is the distance between the point (a, b) and the origin, so the formula for the magnitude of $a + bi$ is $\sqrt{a^2 + b^2}$.

Use the above criterion to determine, without a calculator or computer, if the given complex number is in the Mandelbrot set or not.

101. $c = 0$

102. $c = 1$

103. $c = i$

104. $c = -1$

105. $c = 1 + i$

106. $c = -2$

107. $c = 1 - i$

108. $c = -1 - i$

109–112 *True or False?* Determine whether the given statement is true or false. In case of a false statement, explain or provide a counterexample.

109. The graph of any quadratic polynomial is a transformation of the prototypical parabola.

110. The graphs of $y = f(x)$ and $y = f(-x)$ are reflection images of each other.

111. A cubic function can have up to three x-intercepts.

112. If $f(x)$ is an algebraic function and c is a nonzero constant, then $f(cx) = cf(x)$.

1.3 **Technology Exercises**

113–118 Mentally sketch the graph of the given function by identifying the basic shape that has been shifted, reflected, stretched, or compressed. Then use a graphing calculator or computer algebra system to graph the function and check your reasoning.

113. $f(x) = -2(3-x)^3 + 5$

114. $f(x) = \dfrac{3}{x+5} - 1$

115. $f(x) = \dfrac{-1}{(x-2)^2} - 3$

116. $f(x) = -3|x+2| - 4$

117. $f(x) = -\sqrt{1-x} + 2$

118. $f(x) = \sqrt[3]{2+x} - 1$

119–124 Write a possible equation for the function depicted on the graphing calculator. The function is shown in a $[-10,10]$ by $[-10,10]$ window.

119.

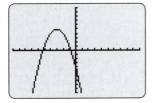

120.

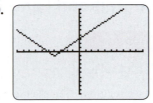

121.

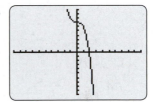

122.

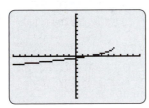

123.

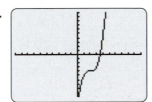

124.

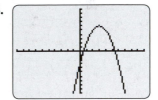

125–127 Use a computer algebra system to determine $(f+g)(x)$, $(fg)(x)$, $(f \circ g)(x)$, and $(g \circ f)(x)$ for the given pair of functions.

125. $f(x) = (3x+2)^2$; $g(x) = \sqrt{x^2 + 5}$

126. $f(x) = \dfrac{1}{3x-5}$; $g(x) = (x+2)^3$

127. $f(x) = \dfrac{x+1}{x-1}$; $g(x) = \dfrac{x-1}{x}$

1.4 Inverse Functions

TOPICS

1. Inverse relations and inverse functions

2. Finding inverse functions

3. Logarithms as inverse functions

4. Inverse trigonometric functions

In much of mathematics, the act of "undoing" one or more mathematical operations plays a critical role. For instance, to solve the equation $3x + 2 = 8$, the first step is to "undo" the addition of 2 on the left-hand side (by subtracting 2 from both sides) and the second step is to "undo" the multiplication by 3 (by dividing both sides by 3). In the context of more complex problems, the "undoing" process is often a matter of finding and applying the inverse of a function.

We begin our discussion with the more general idea of the inverse of a relation. Recall that a relation is just a set of ordered pairs; the inverse of a given relation is the set of these ordered pairs with the first and second coordinates exchanged.

TOPIC 1   Inverse Relations and Inverse Functions

Definition 💡

Inverse of a Relation

Let R be a relation. The **inverse of R**, denoted R^{-1}, is the set

$$R^{-1} = \left\{ (b,a) \big| (a,b) \in R \right\}.$$

Note that this automatically implies the following:

$$\text{Domain of } R^{-1} = \text{Range of } R$$
$$\text{Range of } R^{-1} = \text{Domain of } R$$

Example 1 📝

Determine the inverse of each of the following relations. Then graph each relation and its inverse, and determine the domain and range of both.

a. $R = \left\{ (4,-1), (-3,2), (0,5) \right\}$ **b.** $y = x^2$

Solution

a. $R = \left\{ (4,-1), (-3,2), (0,5) \right\}$ $R^{-1} = \left\{ (-1,4), (2,-3), (5,0) \right\}$

Domain of $R = \left\{ 4, -3, 0 \right\}$ Domain of $R^{-1} = \left\{ -1, 2, 5 \right\}$

Range of $R = \left\{ -1, 2, 5 \right\}$ Range of $R^{-1} = \left\{ 4, -3, 0 \right\}$

In Figure 1, R is in blue and its inverse is in red. The relation R consists of three ordered pairs, and its inverse is simply these three ordered pairs with the coordinates exchanged. Note that the domain of R is the range of R^{-1}, and vice versa.

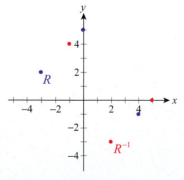

Figure 1

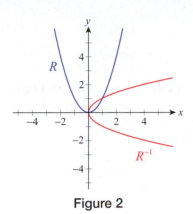

Figure 2

b. $R = \left\{ (x,y) \big| y = x^2 \right\}$ $R^{-1} = \left\{ (x,y) \big| x = y^2 \right\}$

Domain of $R = \mathbb{R}$ Domain of $R^{-1} = [0,\infty)$

Range of $R = [0,\infty)$ Range of $R^{-1} = \mathbb{R}$

In this problem, R is described by the given equation in x and y. The inverse relation is the set of ordered pairs in R with the coordinates exchanged, so we can describe the inverse relation just by exchanging x and y in the equation. The graphs of R and R^{-1} are shown in Figure 2.

Consider the graphs of the two relations and their respective inverses in Example 1. By definition, an ordered pair (b,a) lies on the graph of the relation R^{-1} if and only if (a,b) lies on the graph of R, so it shouldn't be surprising that the graphs of a relation and its inverse bear some resemblance to one another. Specifically, they are mirror images of one another with respect to the line $y = x$. If you were to fold the Cartesian plane in half along the line $y = x$ in the two examples above, you would see that the points in R and R^{-1} coincide with one another.

The two relations in Example 1 illustrate another important point. Note that in both cases, R is a function, as its graph passes the vertical line test. By the same criterion, R^{-1} in Example 1a is also a function, but R^{-1} in Example 1b is not. The conclusion to be drawn is that even if a relation is a function, its inverse may or may not be a function.

With a bit more thought, we can draw a stronger conclusion from Example 1 about when the inverse of a relation is a function.

In practice, we will only be concerned with the question of whether the inverse of a function f, denoted f^{-1}, is itself a function. Note that f^{-1} has already been defined: f^{-1} stands for the inverse of f, where we are making use of the fact that a function is also a relation.

Caution ⚠

We are faced with another example of reuse of notation. f^{-1} does not stand for $1/f$! We use an exponent of -1 to indicate the reciprocal of a number or an algebraic expression, but when applied to a function or a relation it stands for the inverse relation.

Assume that f is a function. The inverse f^{-1} will only be a function in its own right if its graph passes the vertical line test; that is, only if each element of the domain of f^{-1} is paired with exactly one element of the range of f^{-1}. But this criterion is identical to saying that each element of the range of f is paired with exactly one element of the domain of f. In other words, every *horizontal* line in the Cartesian plane must intersect the graph of f no more than once. We say that functions meeting this condition pass the horizontal line test.

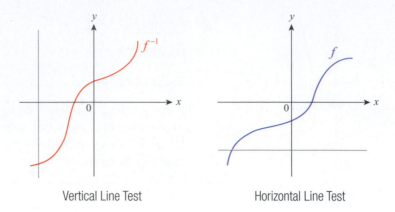

| Vertical Line Test | Horizontal Line Test |

Figure 3

Definition 💡

The Horizontal Line Test

Let f be a function. We say that the graph of f passes the **horizontal line test** if every horizontal line in the plane intersects the graph no more than once. The inverse of f is a function only if f passes the horizontal line test.

The horizontal line test is only useful if the graph of f is available to study. We can also phrase the above condition in a nongraphical manner as follows. The inverse of f will only be a function if for every pair of distinct elements x_1 and x_2 in the domain of f, we have $f(x_1) \neq f(x_2)$. This criterion is important enough to merit a name.

Definition 💡

One-to-One Functions

A function f is **one-to-one** if for every pair of distinct elements x_1 and x_2 in the domain of f, we have $f(x_1) \neq f(x_2)$. This means that every element of the range of f is paired with exactly one element of the domain of f.

If we now examine Example 1 again, we see that the function R in Example 1a is one-to-one, and so we know that its inverse is also a function. On the other hand, the function R in Example 1b is not one-to-one (plenty of horizontal lines pass through the graph twice), so its inverse is not a function.

Example 2 ✍

Determine if the following functions have inverse functions.

a. $f(x) = |x|$ **b.** $g(x) = (x+2)^3$

Solution

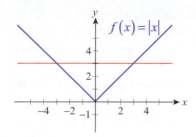

Figure 4

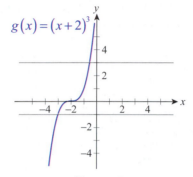

Figure 5

a. The function f does not have an inverse function, a fact easily demonstrated by showing that its graph does not pass the horizontal line test (see Figure 4). An algebraic proof that f does not have an inverse function is the following: even though $-3 \neq 3$, we have $f(-3) = f(3)$. There are an infinite number of pairs of numbers that show f is not one-to-one, but one such pair is all it takes to show that f does not have an inverse function.

b. The graph of g is the standard cubic shape shifted horizontally 2 units to the left. We can see that this graph passes the horizontal line test, so g has an inverse function (see Figure 5). But again, it is good practice to prove this algebraically. Note how each line in the following argument implies the next line.

$$x_1 \neq x_2 \implies x_1 + 2 \neq x_2 + 2$$
$$\implies (x_1 + 2)^3 \neq (x_2 + 2)^3$$
$$\implies g(x_1) \neq g(x_2)$$

This argument shows that any two distinct elements of the domain of g lead to different values when plugged into g, so g is one-to-one and hence has an inverse function.

TOPIC 2   Finding Inverse Functions

In applying the notion of the inverse of a function, we will most often begin with a formula for f and want to find a formula for f^{-1}. This will allow us, for instance, to transform equations of the form

$$f(x) = y$$

into the form

$$x = f^{-1}(y).$$

Before we discuss the general algorithm for finding a formula for f^{-1}, consider the problem with which we began this section. If we define $f(x) = 3x + 2$, the equation $3x + 2 = 8$ can be written as $f(x) = 8$. Note that f is one-to-one, so f^{-1} does exist (as a function). If we can find a formula for f^{-1}, we can transform the equation into $x = f^{-1}(8)$. This is an overly complicated way to solve this equation, but it illustrates the point well.

What should the formula for f^{-1} be? Consider what f does to its argument. The first action is to multiply x by 3, and then the second is to add 2. To "undo" f, we need to negate these two actions in reverse order: subtract 2 and then divide the result by 3.

$$f^{-1}(x) = \frac{x-2}{3}$$

Applying this to the problem at hand, we obtain the following.

$$x = f^{-1}(8) = \frac{8-2}{3} = 2$$

This method of analyzing a function f and then finding a formula for f^{-1} by undoing the actions of f in reverse order is conceptually important, and works for simple functions. For other functions, however, the following algorithm may be necessary.

Finding the Inverse of a Function

Let f be a one-to-one function, and assume that f is defined by a formula. To find a formula for f^{-1}, perform the following steps.

Step 1: Replace $f(x)$ in the definition of f with the variable y. The result is an equation in x and y that is solved for y at this point.

Step 2: Interchange x and y in the equation.

Step 3: Solve the new equation for y.

Step 4: Replace the y in the resulting equation with $f^{-1}(x)$.

Example 3 ✎

Find the inverse of each of the following functions.

a. $f(x) = (x-1)^3 + 2$ **b.** $g(x) = \dfrac{x-3}{2x+1}$

Solution

a. We can find the inverse of this function either by the algorithm or by undoing the actions of f in reverse order. The function f subtracts 1 from x, cubes the result, and adds 2; its inverse will first subtract 2, take the cube root of the result, and then add 1.

$$f(x) = (x-1)^3 + 2$$
$$f^{-1}(x) = \sqrt[3]{(x-2)} + 1$$

b. $g(x) = \dfrac{x-3}{2x+1}$ The inverse of the function g is most easily found by the algorithm.

$y = \dfrac{x-3}{2x+1}$ Step 1: Replace $g(x)$ with y.

$x = \dfrac{y-3}{2y+1}$ Step 2: Interchange x and y in the equation.

$x(2y+1) = y-3$ Step 3: Solve the equation for y.

$2xy + x = y - 3$

$2xy - y = -x - 3$

$y(2x-1) = -x-3$

$y = \dfrac{-x-3}{2x-1}$

$g^{-1}(x) = \dfrac{-x-3}{2x-1}$ Step 4: Replace y in the resulting equation with $g^{-1}(x)$.

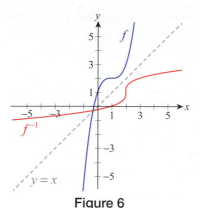

Figure 6

Graph of a Function and Its Inverse

Remember that the graphs of a relation and its inverse are mirror images of one another with respect to the line $y = x$; this is still true if the relations are functions. We can demonstrate this fact by graphing the function and its inverse from Example 3a, as shown in Figure 6.

We can use the functions and their inverses from Example 3 to illustrate one last important point. The key characteristic of the inverse of a function is that it undoes the function. This means that if a function and its inverse are composed together, in either order, the resulting function has no effect on any allowable input.

Properties of Inverse Functions

$f^{-1}\big(f(x)\big) = x$ for all x in the domain of f

$f\big(f^{-1}(x)\big) = x$ for all x in the domain of f^{-1}

For example, given $g(x) = \dfrac{x-3}{2x+1}$ and $g^{-1}(x) = \dfrac{-x-3}{2x-1}$ (from Example 3b), we have the following:

$$g^{-1}\big(g(x)\big) = g^{-1}\left(\frac{x-3}{2x+1}\right) = \frac{-\dfrac{x-3}{2x+1} - 3}{2\left(\dfrac{x-3}{2x+1}\right) - 1}$$

$$= \left(\frac{-\dfrac{x-3}{2x+1} - 3}{2\left(\dfrac{x-3}{2x+1}\right) - 1}\right)\left(\frac{2x+1}{2x+1}\right)$$

$$= \frac{-x+3-6x-3}{2x-6-2x-1} = \frac{-7x}{-7} = x$$

A similar calculation shows that $g\big(g^{-1}(x)\big) = x,$ as you should verify.

TOPIC 3   Logarithms as Inverse Functions

One of the best reasons for the existence of logarithmic functions is simply the usefulness of inverse functions. Consider the two simple equations $2^x = 8$ and $2^x = 9$. The first is quickly solved by inspection, and it's easy to guess and verify that $x = 3$ is a solution. With a little more effort and the knowledge that the graph of 2^x is one-to-one (by the horizontal line test), it's even possible to prove that $x = 3$ is in fact the only solution to the equation $2^x = 8$; any other value for x results in 2^x being either less or more than 8. But how do we solve $2^x = 9$?

We can begin to make progress by rephrasing the equation in function form as follows: define the function f by $f(x) = 2^x$ and let y denote an arbitrary real number. Then what we are actually trying to do is solve the equation $f(x) = y$ for the variable x. But that's exactly the process of finding the inverse formula for the function f! If we can determine f^{-1}, we will use the observation that $f(x) = y \Leftrightarrow x = f^{-1}(y)$ to calculate x for a given y. Note that we've already determined that $3 = f^{-1}(8)$; we just need to calculate $f^{-1}(9)$ in order to solve the second equation.

Unfortunately, we've now progressed as far as we can algebraically—it is not possible to express f^{-1} as an algebraic function. We are forced to introduce the class of transcendental functions called logarithms in order to proceed.

Logarithmic Functions

Let a be a fixed positive real number not equal to 1. The logarithmic function with base a is defined to be the inverse of the exponential function with base a and is denoted $\log_a x$. In symbols,

$$\text{if } f(x) = a^x, \quad \text{then } f^{-1}(x) = \log_a x.$$

In equation form, the definition of logarithm implies that

$$a^x = y \quad \Leftrightarrow \quad x = \log_a y.$$

Note that a is the base in both equations; it serves as either the base of the exponential function or the base of the logarithmic function.

This logarithmic notation does not do much by itself until we build up a degree of familiarity with the behavior of logarithmic functions and discover their properties. For instance, we can begin by noting that since the range of every exponential function is $(0, \infty)$, this set must be the domain of every logarithmic function. Similarly, the range of any logarithmic function is the domain of its corresponding exponential function, which is $(-\infty, \infty)$. But probably the best way to become familiar with logarithmic functions is to gain experience with their graphs.

Recall that the graphs of a function and its inverse are reflections of one another with respect to the line $y = x$. Since exponential functions come in two forms ($0 < a < 1$ and $a > 1$), logarithmic functions fall into two similar categories. In each of the graphs in Figure 7, the red curve is the graph of an exponential function representative of its class, and the blue curve is the corresponding logarithmic function.

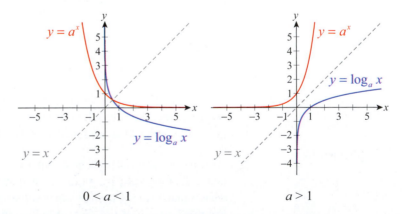

Figure 7 The Two Classes of Logarithmic Functions

Note that the domain of each of the logarithmic functions in Figure 7 is indeed $(0,\infty)$, and that the range in each case is $(-\infty,\infty)$. Also note that the y-axis is a vertical asymptote for both, and that neither has a horizontal asymptote.

We can shift, reflect, stretch, and compress the graphs of logarithms as we can with any other function. The techniques summarized in Section 1.3 are all we need to do the work in Example 4.

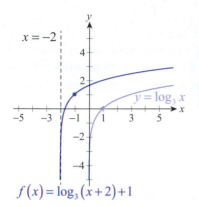

$f(x) = \log_3(x+2)+1$

Figure 8

Example 4 ✎

Sketch the graphs of the following functions.

a. $f(x) = \log_3(x+2)+1$ **b.** $g(x) = \log_2(-x-1)$

c. $h(x) = \log_{1/2} x - 2$

Solution

a. Begin by graphing $y = \log_3 x$, which is the basis for the graph of f. We know several points on the graph of $\log_3 x$ exactly; for instance, $(1,0)$ and $(3,1)$ are on the graph of $\log_3 x$.

Since x has been replaced by $x+2$, we shift this graph 2 units to the left. To find the graph of f, we then shift the result 1 unit up. Note that the vertical asymptote has also shifted to the left (see Figure 8).

b. Again, we start with the basic shape of the graph, and then worry about the transformations. The basic shape of the graph of g is the same as the shape of $y = \log_2 x$. As above, some points on $\log_2 x$ are easily determined, such as $(1,0)$ and $(2,1)$.

To obtain g from $\log_2 x$, the variable x is replaced with $x-1$, which shifts the graph 1 unit to the right, and then x is replaced by $-x$, which reflects the graph with respect to the y-axis (see Figure 9).

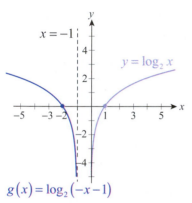

$g(x) = \log_2(-x-1)$

Figure 9

c. We begin with the graph of $y = \log_{1/2} x$. This is a decreasing function, as the base is between 0 and 1. Note that $(1,0)$ and $\left(\frac{1}{2},1\right)$ are two points on the graph of $\log_{1/2} x$.

If we then shift the graph 2 units down, we obtain the graph of the function h (see Figure 10).

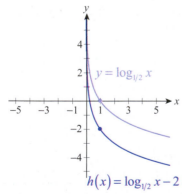

$h(x) = \log_{1/2} x - 2$

Figure 10

Now that we have graphed logarithmic functions, we can augment our understanding of their behavior with a few algebraic observations. First, our work in Example 4 certainly suggests that the point $(1,0)$ is always on the graph of $\log_a x$ for any allowable base a, and this is indeed the case. A similar observation is that $(a,1)$ is always on the graph of $\log_a x$. These two facts are nothing more than restatements of two corresponding facts about exponential functions and are a consequence of the definition of logarithms.

$$\log_a 1 = 0 \quad \text{because} \quad a^0 = 1$$

$$\log_a a = 1 \quad \text{because} \quad a^1 = a$$

More generally, we can use the fact that the functions $\log_a x$ and a^x are inverses of one another to write the following.

$$\log_a\left(a^x\right) = x \quad \text{and} \quad a^{\log_a x} = x$$

We have already mentioned in Section 1.2 the fact that the number e, called the natural base, plays a fundamental role in many important real-world situations, so it is perhaps not surprising that the logarithmic function with base e is worthy of special attention. For historical reasons, the logarithmic function with base 10 is also singled out for further mention.

Definition 💡

Common and Natural Logarithms

The function $\log_{10} x$ is called the **common logarithm** and is usually written as $\log x$.

The function $\log_e x$ is called the **natural logarithm** and is usually written as $\ln x$.

The next task to accomplish regarding logarithms is to review some of their algebraic properties. The following table lists three general properties of logarithms alongside the corresponding properties of exponents.

Theorem 🔍

Properties of Logarithms and Exponents

In each of the following statements a is assumed to be a positive real number not equal to 1. For the properties of logarithms, x and y represent positive real numbers and r is an arbitrary real number. For the properties of exponents, x and y represent arbitrary real numbers.

	Logarithmic Property	Exponential Property
1.	$\log_a\left(xy\right) = \log_a x + \log_a y$	$a^{x+y} = a^x a^y$
2.	$\log_a\left(x/y\right) = \log_a x - \log_a y$	$a^{x-y} = a^x / a^y$
3.	$\log_a\left(x^r\right) = r \log_a x$	$\left(a^x\right)^r = a^{xr}$

We will illustrate the correspondence between the properties of logarithms and the more familiar properties of exponents by proving the first one. The proofs of the second and third will be left as exercises.

Proof ✎

We will start with the right-hand side of the statement and convert the expressions to exponential form. To do this, let $m = \log_a x$ and $n = \log_a y$. The equivalent exponential form of these two equations is

$$x = a^m \quad \text{and} \quad y = a^n.$$

Since we are interested in the product xy, note that

$$xy = a^m a^n = a^{m+n}.$$

The statement $xy = a^{m+n}$ can be converted to logarithmic form, giving us

$$\log_a (xy) = m + n.$$

If we now refer back to the original definition of m and n, we have achieved our goal:

$$\log_a (xy) = \log_a x + \log_a y.$$

The properties of logarithms may appear strange at first, but with time they will come to seem as natural as the properties of exponents. In fact, they are the properties of exponents, simply restated in logarithmic form.

In some situations, we will find it useful to use properties of logarithms to decompose a complicated expression into a sum or difference of simpler expressions, while in other situations we will do just the reverse, combining a sum or a difference of logarithms into one logarithm. Examples 5 and 6 illustrate these processes.

Example 5 ✐

Use the properties of logarithms to expand the following expressions as much as possible (that is, decompose the expressions into sums or differences of the simplest possible terms).

a. $\log_a \sqrt[3]{\dfrac{xy^2}{z^4}}$

b. $\log \dfrac{2.7 \times 10^4}{x^{-2}}$

Solution

a. $\log_a \sqrt[3]{\dfrac{xy^2}{z^4}} = \log_a \left(\left(\dfrac{xy^2}{z^4} \right)^{1/3} \right)$ *Rewrite the radical as an exponent.*

$= \dfrac{1}{3} \log_a \dfrac{xy^2}{z^4}$ *Bring the exponent in front of the logarithm using the third property.*

$= \dfrac{1}{3} \left(\log_a x + \log_a \left(y^2 \right) - \log_a \left(z^4 \right) \right)$ *Expand the expression using the first two properties.*

$= \dfrac{1}{3} \left(\log_a x + 2 \log_a y - 4 \log_a z \right)$ *Apply the third property to the last two terms.*

Note that the base is immaterial: a can represent any legitimate logarithmic base.

b. Recall that if the base is not explicitly written, it is assumed to be 10. This is a convenient base when dealing with scientific notation. Note that we can use a calculator to approximate $\log 2.7$.

$$\log \frac{2.7 \times 10^4}{x^{-2}} = \log 2.7 + \log\left(10^4\right) - \log\left(x^{-2}\right)$$
$$= \log 2.7 + 4 + 2 \log x$$
$$\approx 4.43 + 2 \log x$$

Example 6 ✐

Use the properties of logarithms to condense the following expressions as much as possible (that is, rewrite the expressions as a sum or difference of as few logarithms as possible).

a. $2 \log_3 \dfrac{x}{3} - \log_3 \dfrac{1}{y}$

b. $\ln\left(x^2\right) - \dfrac{1}{2} \ln y + \ln 2$

Solution

a. Note that before a sum or difference of log terms can be combined, they have to have the same coefficient. This is usually straightforward to arrange since the coefficient can be moved up into the exponent.

$$2\log_3 \frac{x}{3} - \log_3 \frac{1}{y} = \log_3\left(\left(\frac{x}{3}\right)^2\right) + \log_3\left(\left(\frac{1}{y}\right)^{-1}\right) \quad \text{Use the third property to make the coefficients appear as exponents.}$$

$$= \log_3 \frac{x^2}{9} + \log_3 y \quad \text{Evaluate the exponents.}$$

$$= \log_3 \frac{x^2 y}{9} \quad \text{Combine terms using the first property.}$$

Note also that the order in which the properties are applied can vary, though all lead to the same final answer.

b. Again, we begin by rewriting each term so that its coefficient is 1 or −1; the sums or differences that result can then be combined.

$$\ln\left(x^2\right) - \frac{1}{2}\ln y + \ln 2 = \ln\left(x^2\right) - \ln\left(y^{1/2}\right) + \ln 2$$

$$= \ln \frac{x^2}{y^{1/2}} + \ln 2$$

$$= \ln \frac{2x^2}{\sqrt{y}}$$

It should be noted that most calculators, if they are capable of calculating logarithms at all, are only equipped to evaluate common and natural logarithms. Such calculators normally have a button labeled "LOG" for the common logarithm and a button labeled "LN" for the natural logarithm.

It is important to note that we can certainly use calculators to find the decimal form of a logarithmic expression with any base a. To illustrate, we will show the thought process for finding $\log_2 9$.

$$x = \log_2 9 \qquad \text{Let } x \text{ stand for the number } \log_2 9.$$

$$2^x = 9 \qquad \text{Convert to exponential form.}$$

$$\ln\left(2^x\right) = \ln 9 \qquad \text{Take the natural logarithm of both sides.}$$

$$x \ln 2 = \ln 9 \qquad \text{Apply the third property of logarithms.}$$

$$x = \frac{\ln 9}{\ln 2} \qquad \text{Solve for } x.$$

$$x \approx 3.17$$

There is nothing special about the natural logarithm, at least as far as this problem is concerned. However, if a calculator is to be used to approximate the number $\log_2 9$, there are (for most calculators) only two good choices: the natural logarithm and the common logarithm. If we had done the work above with the common logarithm, the final answer would have been the same.

$$x = \frac{\log 9}{\log 2} \approx 3.17$$

More generally, a logarithm with base b can be converted to a logarithm with base a through the same reasoning, as summarized below.

Formula ⚲

Change of Base Formula

Let a and b be positive real numbers, neither of them equal to 1, and let x be a positive real number. Then

$$\log_b x = \frac{\log_a x}{\log_a b}.$$

TOPIC 4   Inverse Trigonometric Functions

The rationale for inverse trigonometric functions is the rationale for inverses of functions in general. In many situations, we will want to find an angle having a certain specified property, and our method will be to "undo" the action of a given trigonometric function. As a simple example, suppose we need to find an acute angle θ for which $\sin\theta = \frac{1}{2}$. Chances are you have enough experience with right triangles to recall that $\sin(\pi/6) = \frac{1}{2}$, so it must be the case that $\theta = \pi/6$. But what if we seek an angle φ for which $\sin\varphi = 0.7$? The problem is similar, but we don't yet have a way to determine φ.

Recall, however, that a function will have an inverse only if it is one-to-one. Recall also that this means the graph of the function must pass the horizontal line test; this is something the trigonometric functions fail to do. Fortunately, there is a way out. By restricting the domain of a trigonometric function wisely, we can make it one-to-one and thus invertible. We will go through the process step by step for the sine function and then briefly show how the other trigonometric functions are dealt with similarly.

There are many ways we could restrict the domain of sine in order to make it one-to-one, but we are guided also by the desire to not lose more than we have to in the restriction. For instance, we could specify that we will only define sine over the interval $[0, \pi/2]$, but by doing so we prevent the newly defined function from ever taking on a negative value (note that $0 \leq \sin x \leq 1$ for $x \in [0, \pi/2]$). Figure 11 indicates that $[-\pi/2, \pi/2]$ is the largest interval containing $[0, \pi/2]$ that we could choose for the restricted domain; the red portion of the graph is one-to-one and takes on all values between -1 and 1.

In practice, context will tell us whether we want to think of sine as being defined on the entire real line or only over the interval $[-\pi/2, \pi/2]$, but the biggest hint will be whether we need to apply the inverse of the sine function. If so, the restricted domain for sine is called for.

Two notations are commonly used for the inverse trigonometric functions. In the case of sine, $\sin x = y$ is equivalent to the equations

$$x = \arcsin y \quad \text{and} \quad x = \sin^{-1} y.$$

The arcsine notation derives from the fact that $\arcsin y$ is the length of the arc (on the unit circle) corresponding to the angle x. The $\sin^{-1} y$ notation is in keeping with our use of f^{-1} to stand for the inverse of the function f.

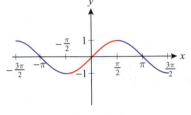

Figure 11
Restricting the Domain of Sine

Definition 💡

Arcsine

Given $x \in [-\pi/2, \pi/2]$, **arcsine** is defined by either of the following.

$$y = \sin x \iff x = \arcsin y \quad \text{or} \quad y = \sin x \iff x = \sin^{-1} y$$

In words, $\arcsin y$ is the angle whose sine is y. Since the (restricted) domain of sine is $[-\pi/2, \pi/2]$ and its range is $[-1,1]$, the domain of arcsine is $[-1,1]$ and its range is $[-\pi/2, \pi/2]$.

The best way to finish up this introduction to arcsine is with a graph of the function. We already saw that the graphs of a function and its inverse are reflections of one another with respect to the line $y = x$, and this is all we need in order to generate the graph of arcsine (see Figure 12).

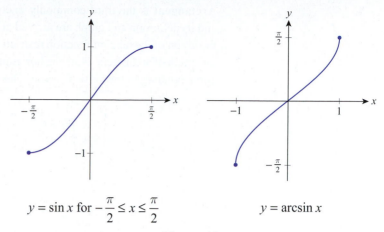

$$y = \sin x \text{ for } -\frac{\pi}{2} \le x \le \frac{\pi}{2}$$ $$y = \arcsin x$$

Figure 12

Example 7

With the derivation of arcsine as a guide, construct a definition of arccosine and plot the resulting function.

Solution

As with sine, we first need to restrict the domain of cosine to an interval over which cosine is one-to-one. Picture the graph of cosine in your mind. Most people would probably say that the natural choice for the restricted domain is the interval $[0, \pi]$, and this is indeed the convention. This is all we need in order to make our definition.

Given $x \in [0, \pi]$, **arccosine** (with its two notations) is defined by

$$y = \cos x \iff x = \arccos y \quad \text{or} \quad y = \cos x \iff x = \cos^{-1} y.$$

To graph the arccosine function, we simply reflect the restricted graph of cosine with respect to the line $y = x$. Since the (restricted) domain of cosine is $[0, \pi]$ and the range is $[-1, 1]$, we know that the domain of arccosine will be $[-1, 1]$ and its range will be $[0, \pi]$. This knowledge serves as a good way to double-check our graph of arccosine.

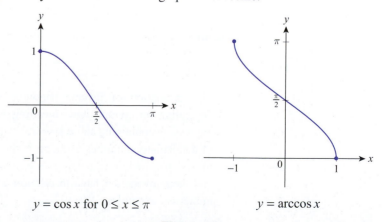

$$y = \cos x \text{ for } 0 \le x \le \pi$$ $$y = \arccos x$$

Figure 13

Arctangent is the third commonly encountered inverse trigonometric function, and its definition and graph are arrived at in a similar manner. Figure 14 illustrates the graph of the tangent function restricted to the interval $(-\pi/2, \pi/2)$ along with the graph of arctangent. Note the horizontal asymptotes in the graph of arctangent, corresponding to the vertical asymptotes in the graph of tangent.

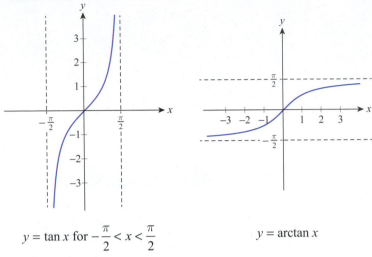

$$y = \tan x \text{ for } -\frac{\pi}{2} < x < \frac{\pi}{2} \qquad\qquad y = \arctan x$$

Figure 14

The following box summarizes facts about the definitions, domains, and ranges of arcsine, arccosine, and arctangent.

Definition 💡

Arcsine, Arccosine, and Arctangent

Function	Notation 1	Notation 2	Domain	Range
Inverse Sine	$\arcsin y = x \Leftrightarrow y = \sin x$	$\sin^{-1} y = x \Leftrightarrow y = \sin x$	$[-1, 1]$	$\left[-\dfrac{\pi}{2}, \dfrac{\pi}{2}\right]$
Inverse Cosine	$\arccos y = x \Leftrightarrow y = \cos x$	$\cos^{-1} y = x \Leftrightarrow y = \cos x$	$[-1, 1]$	$[0, \pi]$
Inverse Tangent	$\arctan y = x \Leftrightarrow y = \tan x$	$\tan^{-1} y = x \Leftrightarrow y = \tan x$	$(-\infty, \infty)$	$\left(-\dfrac{\pi}{2}, \dfrac{\pi}{2}\right)$

The evaluation of inverse trigonometric functions can take several forms, depending on context. One meaning is the actual numerical evaluation of an expression containing an inverse trig function; this may or may not require the use of a calculator. Another meaning is the simplification of expressions containing inverse trig functions, using nothing more than our knowledge of how functions and their inverses behave in relation to one another. We will begin with some numerical examples.

Example 8 ✎

Evaluate the following expressions, if possible.

a. $\arcsin\left(\sin\dfrac{3\pi}{4}\right)$ **b.** $\cos\left(\cos^{-1}(-0.2)\right)$

Solution

a. The potential error in this problem is to assume that $\arcsin\left(\sin(3\pi/4)\right) = 3\pi/4$, since arcsin and sin are inverse functions of one another. But $3\pi/4$ lies outside the range of arcsin, which is $\left[-\pi/2, \pi/2\right]$ so we know this can't be the answer. The key is to evaluate the expressions individually:

$$\sin\left(\frac{3\pi}{4}\right) = \frac{1}{\sqrt{2}}$$

$$\arcsin\left(\frac{1}{\sqrt{2}}\right) = \frac{\pi}{4}$$

b. The number -0.2 lies in the domain of arccosine, and all real numbers lie in the domain of cosine, so all the parts of the expression $\cos\left(\cos^{-1}(-0.2)\right)$ make sense and we are safe in stating $\cos\left(\cos^{-1}(-0.2)\right) = -0.2$. If we wanted to explore the expression a bit further, we could note that, from the graph of arccosine, $\cos^{-1}(-0.2)$ is some positive number, and further that $\cos^{-1}(-0.2)$ must be greater than $\pi/2$ since $\cos\left(\cos^{-1}(-0.2)\right)$ is negative. This is indeed the case: a calculator tells us that $\cos^{-1}(-0.2)$ is approximately 1.8.

The last example demonstrated the evaluation of compositions of trig functions with their inverses, but other compositions are also possible. In many cases, a picture aids greatly in the computation.

Example 9 ✎

Evaluate the following expressions.

a. $\tan\left(\sin^{-1}\left(-\dfrac{4}{5}\right)\right)$ **b.** $\cos\left(\arctan 0.4\right)$

Solution

a. Remember that the range of arcsin is $\left[-\pi/2, \pi/2\right]$, and in particular that $\sin^{-1}\left(-\frac{4}{5}\right)$ will lie between $-\pi/2$ and 0 (the graph tells us that arcsin of a negative number is negative). If we let $\theta = \sin^{-1}\left(-\frac{4}{5}\right)$, then $\sin\theta = -\frac{4}{5}$ and we can sketch the triangle shown in Figure 15 to illustrate the relationship between θ and the given numbers.

The Pythagorean Theorem allows us to calculate x.

$$x = \sqrt{5^2 - (-4)^2} = \sqrt{9} = 3$$

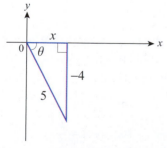

Figure 15

Now we can see that $\tan\theta = -\frac{4}{3}$, so

$$\tan\left(\sin^{-1}\left(-\frac{4}{5}\right)\right) = -\frac{4}{3}.$$

b. We can employ the same method and let $\theta = \arctan 0.4$. This leads to

$$\tan\theta = 0.4 = \frac{4}{10} = \frac{2}{5}$$

and then to the sketch, as seen in Figure 16.

The Pythagorean Theorem gives us $r = \sqrt{5^2 + 2^2} = \sqrt{29}$, and so $\cos(\arctan 0.4) = 5/\sqrt{29}$.

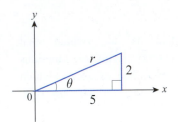

Figure 16

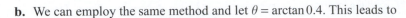

Example 10 ✐

Express $\sin\left(\cos^{-1} 2x\right)$ as an algebraic function of x, assuming $-\frac{1}{2} \le x \le \frac{1}{2}$.

Solution

Let $\theta = \cos^{-1} 2x$. Then $\cos\theta = 2x$ and we are led to consider a sketch like Figure 17.

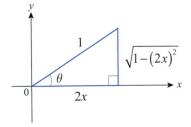

Figure 17

In the sketch, we have chosen the simplest lengths for the adjacent side and the hypotenuse that make $\cos\theta = 2x$, though any positive multiples of these lengths would also work. And as always, once the lengths of two sides of the right triangle have been determined, the Pythagorean Theorem provides the length of the third side. Now we can refer to the sketch to see that

$$\sin\left(\cos^{-1} 2x\right) = \sin\theta = \frac{\sqrt{1-(2x)^2}}{1} = \sqrt{1-4x^2}.$$

1.4 **Exercises**

1–12 Graph the inverse of the given relation, and state its domain and range.

1. $R = \{(-4,2),(3,2),(0,-1),(3,-2)\}$

2. $S = \{(-3,-3),(-1,-1),(0,1),(4,4)\}$

3. $y = x^3$ 4. $y = |x| + 2$

5. $x = |y|$ 6. $x = -\sqrt{y}$

7. $y = \frac{1}{2}x - 3$ 8. $y = -x + 1$

9. $y = [\![x]\!]$

10. $T = \{(4,2),(3,-1),(-2,-1),(2,4)\}$

11. $x = y^2 - 2$ 12. $y = 2\sqrt{x}$

13–22 Determine if the given function has an inverse function. If not, suggest a domain to restrict the function so that it would have an inverse function. (Answers will vary.)

13. $f(x) = x^2 + 1$ 14. $g(x) = (x-2)^3 - 1$

15. $h(x) = \sqrt{x+3}$ 16. $s(x) = \frac{1}{x^2}$

17. $G(x) = 3x - 5$ 18. $F(x) = -x^2 + 5$

19. $r(x) = -\sqrt{x^3}$ **20.** $b(x) = [\![x]\!]$ **37.** $r(x) = \sqrt[5]{2x}$

21. $m(x) = \dfrac{13x - 2}{4}$ **22.** $H(x) = |x - 12|$

38–45 Show that $f^{-1}(f(x)) = x$ and that $f(f^{-1}(x)) = x$.

23–37 Find the inverse of the given function.

38. $f(x) = 2x - 3;$ $f^{-1}(x) = \dfrac{x + 3}{2}$

23. $f(x) = x^{1/3} - 2$ **24.** $g(x) = 4x - 3$

39. $f(x) = x^2, \ x \geq 0;$ $f^{-1}(x) = \sqrt{x}$

25. $r(x) = \dfrac{x - 1}{3x + 2}$ **26.** $s(x) = \dfrac{1 - x}{1 + x}$

40. $f(x) = \dfrac{3x - 1}{5};$ $f^{-1}(x) = \dfrac{5x + 1}{3}$

27. $F(x) = (x - 5)^3 + 2$ **28.** $G(x) = \sqrt[3]{3x - 1}$

41. $f(x) = \dfrac{x - 5}{2x + 3};$ $f^{-1}(x) = \dfrac{3x + 5}{1 - 2x}$

29. $V(x) = \dfrac{x + 5}{2}$ **30.** $W(x) = \dfrac{1}{x}$

42. $f(x) = (x - 2)^2, \ x \geq 2;$ $f^{-1}(x) = \sqrt{x} + 2, \ x \geq 0$

31. $h(x) = x^{3/5} - 2$ **32.** $A(x) = (x^3 + 1)^{1/5}$

43. $f(x) = \sqrt[3]{x + 2} - 1;$ $f^{-1}(x) = (x + 1)^3 - 2$

33. $J(x) = \dfrac{2}{1 - 3x}$ **34.** $k(x) = \dfrac{x + 4}{3 - x}$

44. $f(x) = \dfrac{1}{x};$ $f^{-1}(x) = \dfrac{1}{x}$

35. $h(x) = x^7 + 6$ **36.** $F(x) = \dfrac{3 - x^5}{-9}$

45. $f(x) = \dfrac{1}{1 + x}, \ x \geq 0;$ $f^{-1}(x) = \dfrac{1 - x}{x}, \ 0 < x \leq 1$

46–51 Match the function with the graph of the inverse of the function (labeled A–F).

46. $f(x) = x^3$ **47.** $f(x) = x - 5$ **48.** $f(x) = \sqrt{x - 4}$

49. $f(x) = x^2$ **50.** $f(x) = \dfrac{x}{4}$ **51.** $f(x) = \sqrt[3]{x + 1}$

A.

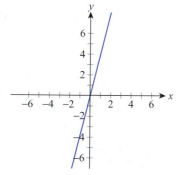

B.

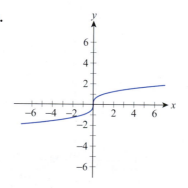

C.

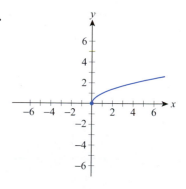

D.

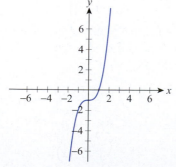

E.

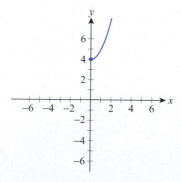

F.

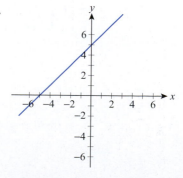

52–60 Match the logarithmic function with its graph (labeled A–I).

52. $f(x) = \log_2 x - 1$

53. $f(x) = \log_2(2-x)$

54. $f(x) = \log_2(-x)$

55. $f(x) = \log_2(x-3)$

56. $f(x) = 1 - \log_2 x$

57. $f(x) = -\log_2 x$

58. $f(x) = -\log_2(-x)$

59. $f(x) = \log_2 x$

60. $f(x) = \log_2 x + 3$

A.

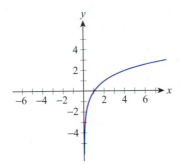

B.

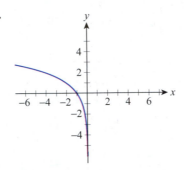

C.

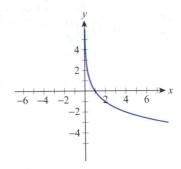

D.

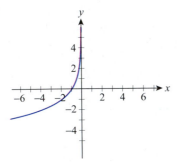

E.

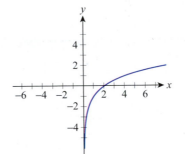

F.

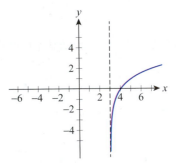

G.

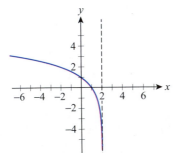

H.

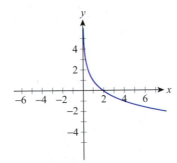

I.

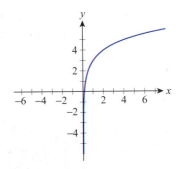

61–72 Sketch the graph of the given function.

61. $f(x) = \log_3 (x-1)$

62. $g(x) = \log_5 (x+2) - 1$

63. $r(x) = \log_{1/2} (x-3)$

64. $p(x) = 3 - \log_2 (x+1)$

65. $q(x) = \log_3 (2-x)$

66. $s(x) = \log_{1/3} (5-x)$

67. $h(x) = \log_7 (x-3) + 3$

68. $m(x) = \log_{1/2} (1-x)$

69. $f(x) = \log_3 (6-x)$

70. $p(x) = 4 - \log_{10} (x+3)$

71. $s(x) = -\log_{1/3} (-x)$

72. $g(x) = \log_5 (2x) - 1$

73–78 Evaluate the given expression without using a calculator.

73. $\log_4 16$

74. $\log_5 (25^3)$

75. $\ln(e^4) + \ln(e^3)$

76. $\log_4 \dfrac{1}{64}$

77. $\ln(e^{1.5}) - \log_4 2$

78. $\log_2 (8^{2\log_2 4 - \log_2 4})$

79–84 Evaluate the given logarithmic expression to two decimal places. (**Hint:** Use the change of base formula.)

79. $\log_6 (3^4)$

80. $\log_7 14.3$

81. $\log_{1/2} (\pi^{-2})$

82. $\log_{1/5} 626$

83. $\ln(\log 123)$

84. $\log_{17} 0.041$

85–90 Use the properties of logarithms to rewrite the given expression as a single term that does not contain a logarithm.

85. $5^{2\log_5 x}$

86. $\log_4 16 \cdot \log_x (x^2)$

87. $e^{2 - \ln x + \ln p}$

88. $e^{5\left(\ln \sqrt[5]{3} + \ln x\right)}$

89. $10^{\log(x^3) - 4\log y}$

90. $a^{\log_a b + 4\log_a \sqrt{a}}$

91–99 Use the properties of logarithms to expand the given expression as much as possible; that is, decompose the expression into sums or differences of the simplest possible terms. Simplify any numerical expressions that can be evaluated without a calculator.

91. $\ln \dfrac{\sqrt{x^3} \, pq^5}{e^7}$

92. $\log_a \sqrt[5]{\dfrac{a^4 b}{c^2}}$

93. $\log\left(\log\left(100x^3\right)\right)$

94. $\log_3 (9x + 27y)$

95. $\log \dfrac{10}{\sqrt{x+y}}$

96. $\ln\left(\ln\left(e^{ex}\right)\right)$

97. $\log_2 \dfrac{y^2 + z}{16x^4}$

98. $\log\left(\log\left(100{,}000^{2x}\right)\right)$

99. $\log_b \sqrt{\dfrac{x^4 y}{z^2}}$

100–105 Use the properties of logarithms to condense the given expression as much as possible, writing the answer as a single term with a coefficient of 1.

100. $\dfrac{1}{5}\left(\log_7 (x^2) - \log_7 (pq)\right)$

101. $\ln 3 + \ln p - 2\ln q$

102. $2\left(\log_5 \sqrt{x} - \log_5 y\right)$

103. $\log(x - 10) - \log x$

104. $2\log(a^2 b) - \log \dfrac{1}{b} + \log \dfrac{1}{a}$

105. $3\left(\ln \sqrt[3]{e^2} - \ln(xy)\right)$

106–111 Evaluate the given expression, if possible.

106. $\cos^{-1}\left(\cos \dfrac{2\pi}{4}\right)$

107. $\sin^{-1}\left(\sin \dfrac{3\pi}{2}\right)$

108. $\tan\left(\tan^{-1}(0.5)\right)$

109. $\sin^{-1}\left(\sin \dfrac{7\pi}{6}\right)$

110. $\cos\left(\cos^{-1}(-0.8)\right)$

111. $\tan^{-1}\left(\tan \dfrac{5\pi}{4}\right)$

112–117 Most calculators are not equipped with arccosecant, arcsecant, and arccotangent buttons, but expressions involving these functions can still be evaluated. For example, to evaluate $\csc^{-1} x$, let $\theta = \csc^{-1} x$.

$$\csc\theta = x$$
$$\frac{1}{\sin\theta} = x$$
$$\sin\theta = \frac{1}{x}$$
$$\theta = \sin^{-1}\frac{1}{x}$$

Use the method described above to evaluate the given expression. (Round your answer to four decimal places.)

112. $\csc^{-1} 5$

113. $\sec^{-1}(-0.5)$

114. $\cot^{-1} 150$

115. $\cot^{-1}(-0.2)$

116. $\csc^{-1}(-8.9)$

117. $\sec^{-1} 2$

118–123 Find the value of the given expression without using a calculator.

118. $\sin\left(\arctan\sqrt{3}\right)$

119. $\cos\left(\sec^{-1}(-2)\right)$

120. $\tan\left(\operatorname{arccot} 1\right)$

121. $\csc\left(\arccos\left(-\frac{\sqrt{3}}{2}\right)\right)$

122. $\tan\left(\sin^{-1}\left(-\frac{\sqrt{2}}{2}\right)\right)$

123. $\sec\left(\csc^{-1}\frac{2\sqrt{3}}{3}\right)$

124–129 Rewrite the given function as a purely algebraic function.

124. $\tan\left(\cos^{-1} x\right)$

125. $\cot\left(\sin^{-1}\frac{2}{x}\right)$

126. $\sec\left(\tan^{-1} 3x\right)$

127. $\tan\left(\sin^{-1}\frac{x}{\sqrt{x^2+3}}\right)$

128. $\sin\left(\sec^{-1} x\right)$

129. $\cos\left(\tan^{-1}\frac{x}{4}\right)$

130–133 Sketch the graph of the given function. Then graph the function using a graphing calculator or computer algebra system to check your answer.

130. $f(x) = \sin^{-1}(x-3)$

131. $f(x) = \sec^{-1} 2x$

132. $f(x) = \arctan\frac{x}{2}$

133. $f(x) = 2\arccos x$

134–137 An inverse function can be used to encode and decode words and sentences by assigning each letter of the alphabet a numerical value (A = 1, B = 2, C = 3, ..., Z = 26). Example: Use the function $f(x) = x^2$ to encode the word CALCULUS. The encoded message would be 9 1 144 9 441 144 441 361. The word can then be decoded by using the inverse function $f^{-1}(x) = \sqrt{x}$. The inverse values are 3 1 12 3 21 12 21 19, which translates back to the word CALCULUS.

Encode or decode the given message using the numerical values A = 1, B = 2, C = 3, ..., Z = 26.

134. Encode the message SANDY SHOES using the function $f(x) = 4x - 3$.

135. Encode the message WILL IT RAIN TODAY using the function $f(x) = x^2 - 8$.

136. The following message was encoded using the function $f(x) = 8x - 7$. Decode the message.

41 137 65 145 9 33 33 169 113 89 89 33

193 9 1 89 89 1 105 25 57 113 137 145

33 145 57 113 33 145

137. The following message was encoded using the function $f(x) = 5x + 1$. Decode the message.

91 26 66 26 66 11 26 91 126 76 106 91

96 106 71 11 61 76 16 56

138–139 The energy released during earthquakes can vary greatly, but logarithms provide a convenient way to analyze and compare the intensity of earthquakes. Earthquake intensity is measured on the Richter scale (named for the American seismologist Charles F. Richter, 1900–1985). In the formula that follows, I_0 is the intensity of a just-discernible earthquake, I is the intensity of an earthquake being analyzed, and R is its ranking on the Richter scale.

$$R = \log\frac{I}{I_0}$$

By this measure, earthquakes range from a classification of small ($R < 4.5$), to moderate ($4.5 \le R < 5.5$), to large ($5.5 \le R < 6.5$), to major ($6.5 \le R < 7.5$), and finally to greatest ($R \ge 7.5$).

Use this information to solve the problem.

138. The 1994 Northridge, California earthquake measured 6.7 on the Richter scale. What was the intensity, relative to a 0-level earthquake, of this event?

139. The April, 2009 Abruzzo earthquake in Italy was 2,000,000 times as intense as a 0-level earthquake. What was the Richter ranking of this tragic event?

140–141 Sound intensity is another quantity that varies greatly, and the measure of how the human ear perceives intensity, in units called decibels, is very similar to the measure of earthquake intensity. If I_0 is the intensity of a just-discernible sound, I is the intensity of the sound being analyzed, and D is its decibel level, we have the formula $D = 10\log(I/I_0)$. Decibel levels range from 0 for a barely discernible sound, to 40 for the level of normal conversation, to 80 for heavy traffic, to 120 for a loud rock concert, and finally (as far as humans are concerned) to around 160, at which point the eardrum is likely to rupture.

Use the decibel formula given above to answer the question.

140. A construction worker operating a jackhammer would experience noise with an intensity of 20 watts/meter^2 if it weren't for ear protection. Given that $I_0 = 10^{-12}$ watts/meter^2, what is the decibel level for such noise?

141. The intensity of a cat's soft purring is measured to be 2.19×10^{-11} watts/meter^2. Given that $I_0 = 10^{-12}$ watts/meter^2, what is the decibel level of this noise?

142–143 Use inverse trigonometric functions to solve the problem. (Round your answer to four decimal places.)

142. Kim is watching a space shuttle launch from an observation spot 2 miles away from the launchpad. Find the angle of elevation to the shuttle for each of the following heights.

   a. 0.5 miles   **b.** 2 miles   **c.** 2.8 miles

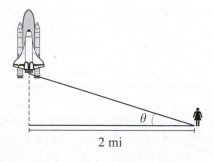

2 mi

143. Jesse is rowing in the men's singles race. The length of the oar from the side of the shell to the water is 7 feet. At what angle is the oar from the side of the boat when the blade is at the following distances from the boat?

   a. 2 feet   **b.** 3 feet   **c.** 5 feet

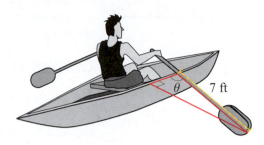

7 ft

144–151 *True or False?* Determine whether the given statement is true or false. In case of a false statement, explain or provide a counterexample.

144. All exponential functions are one-to-one.

145. $\sin(\arcsin x) = x$ for all $x \in [-1,1]$

146. $\arcsin(\sin x) = x$ for all $x \in \mathbb{R}$

147. $\tan(\arctan x) = x$ for all $x \in \mathbb{R}$

148. $\arccos(\cos(3\pi/2)) = 3\pi/2$

149. The domain of $\arcsin x$ is $[-\pi/2, \pi/2]$.

150. The domain of $f(x) = \cot^{-1} x$ is $\mathbb{R}$.

151. The function $f(x) = \sin(\tan^{-1} x)$ can be represented as an algebraic function.

1.5 Calculus, Calculators, and Computer Algebra Systems

TOPICS

1. Graphs via calculators and computers

2. Animations and models

3. Least-squares curve fitting

The technology available at any given time inevitably colors the ways in which mathematics is learned and used. The historical interplay between calculus and technology is extensive, and no discourse on calculus would be complete without a discussion of computational tools. The tools available on graphing calculators and computers are the focus of this section.

TOPIC 1   Graphs via Calculators and Computers

The graphing capabilities of modern calculators and mathematical software are an especially useful technological addition. While pictures can be misleading and must be used with a small amount of skepticism, there is no denying that the ability to quickly sketch curves and surfaces greatly speeds up the process of solving many problems. The details on the use of a particular calculator or software package are best left to the user's manual, but some features are common across all graphing technology.

One of the most basic features is the ability to choose the **display** or **viewing window** when graphing a function. When graphing functions from $\mathbb{R}$ to $\mathbb{R}$, this is simply the choice of the minimum and maximum values for the horizontal and vertical axes. For the purposes of illustration, we will refer to these values as $xMin$, $xMax$, $yMin$, and $yMax$ (on some calculators, these are the exact labels for these quantities). The display window is then a rectangle in the plane bounded by these values, $[xMin, xMax]$ by $[yMin, yMax]$, and their choice determines the portion of the function being graphed. It is important to note that this choice effectively gives the user the ability to zoom in (or out) on a particular part of a graph.

Figure 1

$f(x) = x^3 - 30x + 15$
on $[-10,10]$ by $[-10,10]$

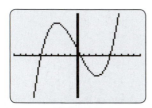

Figure 2

$f(x) = x^3 - 30x + 15$
on $[-10,10]$ by $[-100,100]$

Example 1 ✎

Graph the function $f(x) = x^3 - 30x + 15$ in the following viewing windows using a graphing calculator.

a. $[-10,10]$ by $[-10,10]$ **b.** $[-10,10]$ by $[-100,100]$

Solution

a. We set $xMin = -10$, $xMax = 10$, $yMin = -10$, and $yMax = 10$. The resulting graph is displayed in Figure 1. Note that it appears to be cut off at the top and bottom, indicating that the graph continues vertically beyond the viewing window.

b. For this viewing window we just need to change the range of y-values to be from $yMin = -100$ to $yMax = 100$. Figure 2 shows the graph in this new viewing window. This is a more complete picture of the graph of $f(x)$, revealing the significant parts of the graph.

The next example illustrates the use of a graphing calculator and a computer algebra system to graph an interesting part of the function $f(x) = x^2 \sin(1/x)$.

Example 2 ✎

We will illustrate the graph of the function $f(x) = x^2 \sin(1/x)$, showing the graph at different magnifications or viewing windows.

The first observation we wish to make is that since $\sin(1/x)$ cannot take any values above 1 or below −1, the graph of $f(x)$ cannot go above x^2 or below $-x^2$, as illustrated in Figure 3.

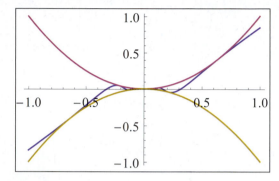

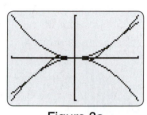

Figure 3a
Graphs of x^2, $f(x)$, and $-x^2$

Figure 3b Graphs of x^2, $f(x)$, and $-x^2$

In order to better understand the graph of $f(x)$, let us take a closer look at $g(x) = \sin(1/x)$. As you might recall from your experience with precalculus, $g(x)$ oscillates between −1 and +1 (like any well-mannered sine function would), but it does so in a surprising way (see Figure 4): it makes infinitely many oscillations near 0 on both sides of the y-axis. This is most easily checked by noting that $g(x) = \sin(1/x) = 0$ for $x = 1/(n\pi)$, $n = \pm 1, \pm 2, \dots$. In other words, $g(x) = 0$ for infinitely many values of x that approach 0 as n grows large. In a similar fashion, one can show that $g(x) = 1$ and $g(x) = -1$ infinitely many times on both sides of 0.

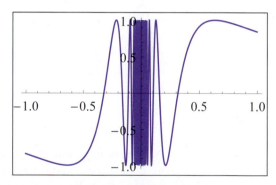

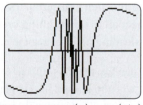

Figure 4a $g(x) = \sin(1/x)$

Figure 4b $g(x) = \sin(1/x)$

The effect of multiplying $g(x)$ by x^2 is that $f(x) = x^2 \sin(1/x)$ will still oscillate in a similar fashion, but now between x^2 and $-x^2$ (again, you might recall "damped trigonometric graphs" from your prior studies). The oscillation of $f(x)$ is not very clear in Figure 3 but is well illustrated if we ask our technology to "zoom in" or, equivalently, choose a "smaller" viewing window, as shown in Figure 5.

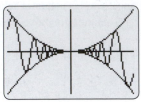

Figure 5a

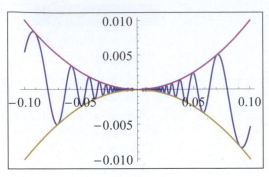

Figure 5b

In order to get a better sense of the infinitude of oscillations, we can zoom in even further, as shown in Figure 6.

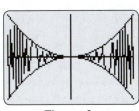

Figure 6a

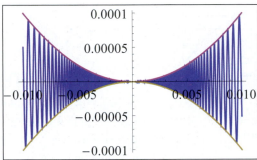

Figure 6b

Finally, we wish to note that $f(x)$ is not defined for $x = 0$; that is, there is no function value for $x = 0$, which is not clear from the graphs we received. Graphing technology, however, is still an extremely useful tool, if we stay aware of its limitations.

All calculators and computers construct graphs by similar means and under similar assumptions, at least until told to do something different. The default behavior, for instance, is to calculate (or *sample*) the value of the function at a certain number of points between *xMin* and *xMax* and then to connect all of the resulting ordered pairs with line segments. This can lead to a very misleading picture in some cases.

As an example, consider the tangent function. Using the identity $\tan x = (\sin x)/(\cos x)$ as a guide, we know that tangent is undefined whenever $\cos x$ is 0, namely $x = \pm\{\pi/2, 3\pi/2, \ldots\}$. But some calculators and software will automatically connect each consecutive pair of sampled points along the graph of tangent and create what appears to be a continuous graph with vertical portions—a result that is not only wrong, but that doesn't even qualify as a function! Figure 7 contains graphs generated by such a calculator and by software without additional guidance.

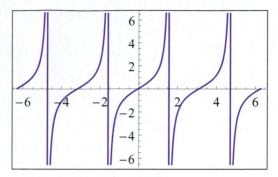

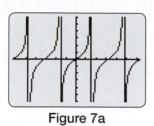

Figure 7a

Figure 7b

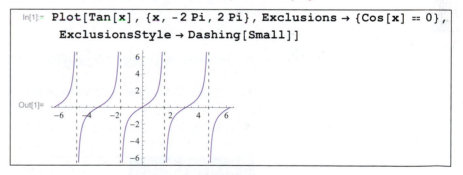

Figure 8

Technology Note 🖵

With guidance by the user, the graphs generated by calculators and software can usually be much improved in the cases where the default algorithm leads to misleading (or even wildly inaccurate) pictures. Most graphing calculators, for example, allow the user to specify that the "sampled" points on a graph not be connected by line segments. This can be done by changing the graphing mode from **CONNECTED** to **DOT**. The result is a picture that may not look very smooth, but at least is free of such erroneous artifacts as vertical segments (see Figure 8).

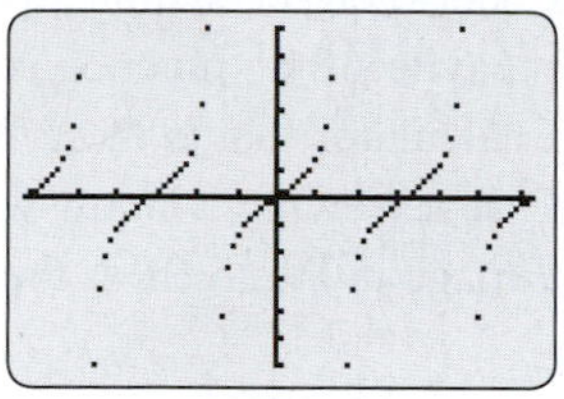

Software such as *Mathematica* allows the user much more control. Figure 9 illustrates the use of options in **Plot** that direct *Mathematica* to skip over values of x for which $\cos x$ equals 0 and to indicate those "exclusions" with dashed lines, a notation we commonly use for asymptotes.

```
In[1]:= Plot[Tan[x], {x, -2 Pi, 2 Pi}, Exclusions → {Cos[x] == 0},
        ExclusionsStyle → Dashing[Small]]
```

Out[1]=

Figure 9

The process of sampling points of a function in order to sketch its graph can lead to other types of errors, and again it is important to temper the output of a calculator or software package with your mathematical intuition and knowledge. It may seem hard to believe, but the different screenshots in Figure 10 are all depictions of the relatively simple function $f(x) = \sin 25x$.

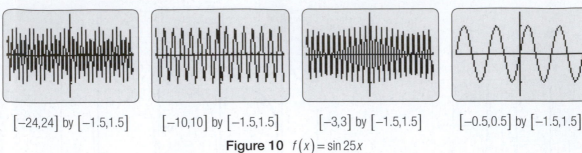

$\left[-24,24\right]$ by $\left[-1.5,1.5\right]$ $\left[-10,10\right]$ by $\left[-1.5,1.5\right]$ $\left[-3,3\right]$ by $\left[-1.5,1.5\right]$ $\left[-0.5,0.5\right]$ by $\left[-1.5,1.5\right]$

Figure 10 $f\left(x\right)=\sin 25x$

What is the explanation for these wildly different graphs? The reason the images in Figure 10 appear to be so different is that the function $f\left(x\right)=\sin 25x$ oscillates many more times than can be captured accurately with sampling in a viewing window that is many times the size of the period of the function. Recall that $\sin x$ has a period of 2π, so the function $\sin 25x$ has a period of $2\pi/25\approx 0.25$. When we use too large a display for this function, the "gaps" between the sampled points hide one or more of the oscillations—most of the behavior of the function is thus missed. If we want to obtain a representation on a calculator that is more likely to be an accurate graph of this function, we should use a viewing window whose horizontal spread is comparable to 0.25, so the last image in Figure 10 is the best choice of the four.

It is easy to conceive of functions that have some attributes best viewed at high magnification, that is, a narrow viewing window and other attributes that show up best from far away (with a wide viewing window). With some technology, the only recourse is to experiment with different magnifications and to examine different regions of the graph until a comprehensive picture has been built up, but computer software today offers us an increasingly large array of tools. We will explore some of these in the next few pages.

TOPIC 2   Animations and Models

As a simple illustration of a tool not readily available until recently, we will examine the graph of $f\left(x\right)=\sin 25x$ once again, this time making use of the **Manipulate** command in *Mathematica*.

Example 3 ✐

As we said previously, the graph of $\sin 25x$ oscillates with a high-enough frequency to warrant caution when using a graphing tool. The following screenshots show the use of the **Manipulate** command in *Mathematica*. If you have access to this software package, type in and execute the command as shown and then use the slider to dynamically zoom in and out of the picture. What we are actually doing is dynamically changing the horizontal extent of the viewing window from a minimum of $\left[-0.1,0.1\right]$ to a maximum of $\left[-10,10\right]$. We have also instructed *Mathematica* to increase the number of sampled points (with the **PlotPoints** option) in order to capture the oscillatory behavior more accurately. (See Appendix A for more guidance on using *Mathematica*.)

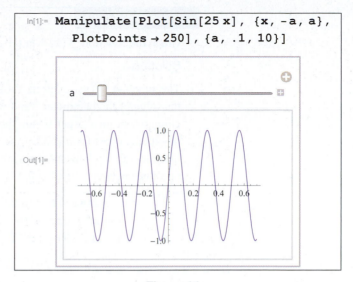

Figure 11a

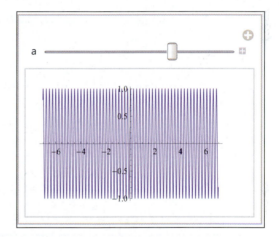

Figure 11b

In Example 3, the variable a is referred to as a *parameter*, a quantity that we want to allow to vary in order to gain insight into some particular behavior. In the usage above, changing a equates to changing the viewing window, but parameters can appear in many different ways. Example 4 illustrates the use of two parameters, one controlling the size of the viewing window and the other appearing in the definition of the function itself.

Example 4 ✎

Graph the function $f(x) = \ln x + \dfrac{\sin cx}{c}$.

Solution

The function $f(x)$ exhibits different characteristics depending on both the value of the parameter c and whether we are interested in the behavior close to 0 or on a larger scale. The next screenshots show just a few of the possible choices for c and for the viewing window. Note that the width of the viewing window can vary from $[0, 0.1]$ to $[0, 20]$, and that c can take on values between 1 and 100.

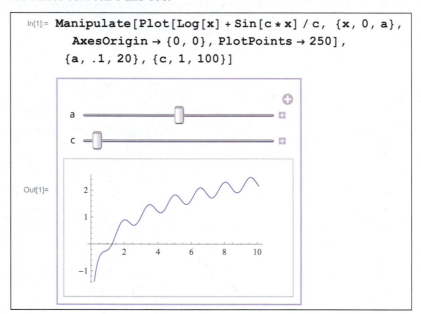

```
In[1]:= Manipulate[Plot[Log[x] + Sin[c * x] / c, {x, 0, a},
          AxesOrigin → {0, 0}, PlotPoints → 250],
          {a, .1, 20}, {c, 1, 100}]
```

Figure 12a

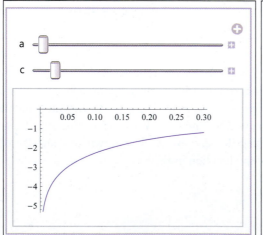

Figure 12b

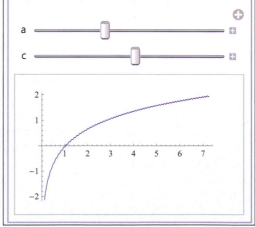

Figure 12c

Throughout this book you will find numerous **mathematical models**, depictions of real-world phenomena using equations, functions, and the tools of algebra and calculus. Many of these models incorporate parameters in a very natural way, and the ability to animate these models, that is, view graphical representations of them for different parameter values, will be especially useful.

Constructing a Mathematical Model

The process of constructing a mathematical model is one you are familiar with—you have worked with many models already, though they may not have been labeled as such. In general, the steps in making a model of a real-world phenomenon are as follows.

Step 1: Identify the measurable or observable quantities and label them as variables.

Step 2: Find mathematical relationships between the variables, making simplifying assumptions about the phenomenon if necessary.

Step 3: Draw mathematical conclusions from the equations, functions, or other mathematical relationships found (this usually amounts to solving equations or performing mathematical operations).

Step 4: Interpret the mathematical conclusions in terms of the real-world phenomenon, and test the descriptive/predictive power of the model against reality. Adjust and improve the model if necessary.

Example 5 ✑

Build a mathematical model to predict an automobile's total stopping distance d (i.e., the total distance traveled from the moment the driver decides to stop until the car comes to a complete stop) as a function of its initial speed v. Then construct a graph with animation that is dependent on road conditions and the driver's response time.

Solution

First, we identify time, speed, and distance as the most important measurable quantities for the purposes of our model. Next, observe that the actual stopping distance will consist of two parts: the distance the car travels after the driver decides to stop but before the brakes start to work, that is, the "response distance," and the actual "braking distance" (the distance traveled while braking). Labeling these by d_1 and d_2, respectively, yields

$$d = d_1 + d_2,$$

where d is the total stopping distance. Next, we will let v stand for the speed of the car before braking, M for the total mass of the car and passengers, and μ for the coefficient of friction between the tires and the pavement.

We will make the simplifying assumptions that the brakes come on instantly and operate with constant force. The road surface is assumed to be uniform and level. Finally, we ignore air resistance and heat loss. Note that d_1 can be calculated as

$$d_1 = vt,$$

where t is the time the driver requires to respond. In order to find d_2 note that the car's kinetic energy (the energy stemming from its forward motion) is dissipated during braking, all the way down to 0. Since the braking force F_b arises as a result of friction between the tires and the pavement, we have

$$F_b = \mu Mg,$$

where g is the gravity constant. By the law of conservation of energy, the total work done by F_b must equal the car's initial kinetic energy of $\frac{1}{2}Mv^2$, so

$$F_b d_2 = \mu Mg d_2 = \frac{1}{2}Mv^2,$$

from which we obtain

$$d_2 = \frac{v^2}{2\mu g}.$$

Thus our mathematical model for the total stopping distance becomes

$$d = d_1 + d_2 = vt + \frac{v^2}{2\mu g}.$$

Since experimental data show the value of t to be approximately 0.75 seconds for the average driver, and $\mu = 0.9$ is realistic on dry pavement, we can complete our model by writing

$$d = 0.75v + \frac{v^2}{1.8g}.$$

In Figure 13, v is assumed to be in meters per second (note that 30 m/s corresponds to 67.1 mph and that $g \approx 9.81\,\mathrm{m/s^2}$ in the metric system). Observe the quadratic functional dependence, which definitely warns us to observe speed limit signs!

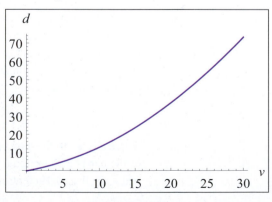

Figure 13

The flexibility of our model is demonstrated by the fact that we can treat the driver's response time t and the friction coefficient μ as parameters, corresponding to the facts that no two drivers are equal and road conditions often change. Exploiting the capabilities of *Mathematica*'s **Manipulate** command, we can easily create animations of the previous graph that are dependent on said parameters.

Notice from the following graphs that while a slower driver response does result in increased stopping distance as predicted by our model, the real drama happens on a slippery road surface. For small μ-values, stopping distance can easily more than double even if the driver has a really short reaction time!

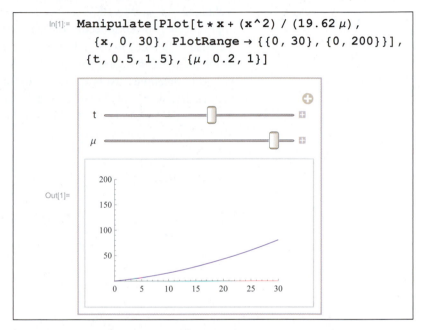

```
In[1]:=  Manipulate[Plot[t * x + (x^2) / (19.62 μ),
           {x, 0, 30}, PlotRange → {{0, 30}, {0, 200}}],
           {t, 0.5, 1.5}, {μ, 0.2, 1}]
```

Out[1]=

Figure 14a

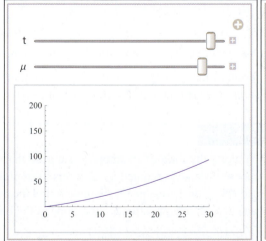

Figure 14b

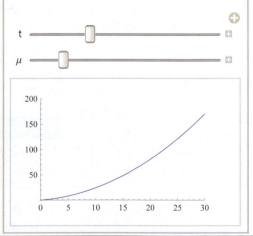

Figure 14c

TOPIC 3   Least-Squares Curve Fitting

We will close this section with one more example of model building. In some cases, especially where the relevant measurable quantities consist only of collected data, the "simplifying assumptions" in the modeling process may be nothing more complicated than something like "assume the data lie along a straight line" or "assume the data exhibit exponential behavior." In such cases, an **empirical model** may be all that is desired; such a model is based only on the data—no underlying physical principles are used.

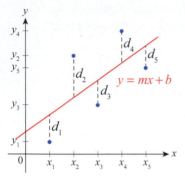

Figure 15

As mentioned briefly in Section 1.1, **least-squares curve fitting** is a useful technique for constructing a function (of a predetermined form) that best approximates a given collection of data points. To illustrate the technique, suppose we wish to fit a line to the n data points $(x_1, y_1), (x_2, y_2), \ldots, (x_n, y_n)$. Every line in the plane has the form $y = mx + b$ for some slope m and y-intercept b, so the task is to determine the best choices for m and b. If we let $d_i = y_i - (mx_i + b)$, then each d_i represents the vertical difference between the ith data point and the line $y = mx + b$ (see Figure 15). The least-squares method gets its name from the objective of choosing m and b so that $d_1^2 + d_2^2 + \cdots + d_n^2$ is minimized.

As we will see later, calculus allows us to fairly easily determine the following formulas for m and b.

$$m = \frac{\left(\sum_{i=1}^{n} x_i\right)\left(\sum_{i=1}^{n} y_i\right) - n\left(\sum_{i=1}^{n} x_i y_i\right)}{\left(\sum_{i=1}^{n} x_i\right)^2 - n\left(\sum_{i=1}^{n} x_i^2\right)}$$

$$b = \frac{\left(\sum_{i=1}^{n} x_i\right)\left(\sum_{i=1}^{n} x_i y_i\right) - \left(\sum_{i=1}^{n} y_i\right)\left(\sum_{i=1}^{n} x_i^2\right)}{\left(\sum_{i=1}^{n} x_i\right)^2 - n\left(\sum_{i=1}^{n} x_i^2\right)}$$

(Recall that the notation $\sum_{i=1}^{n} x_i$ stands for the sum $x_1 + x_2 + \cdots + x_n$.) These are exactly the formulas that graphing calculators and computer software programs use when determining the line of best fit for a given data set.

Example 6 ✎

Continuing along the lines of Example 5, suppose that Table 1 contains driver response distances (denoted by d in this example) as a function of initial speed v. Use the least-squares method to find the best-fitting line for the data, and give an interpretation for the slope of the line.

Solution

v (km/h)	d (m)
40	8.5
48	10.0
56	11.9
64	13.4
72	15.2
82	16.8
88	18.6
96	20.1
104	21.9
112	23.5
120	25.3

Table 1

Note that for the purpose of using our formulas, the numbers in the left column are the values of x_i, while the right column provides the y_i-values. However, using *Mathematica*'s **Fit** command (see Figure 16) or a graphing calculator (see Figure 17), we can obtain the equation of the best-fitting line much quicker:

$$d = 0.21v + 0.008$$

In[1]:= **Fit[{{40, 8.5}, {48, 10}, {56, 11.9}, {64, 13.4},**
    {72, 15.2}, {82, 16.8}, {88, 18.6}, {96, 20.1},
    {104, 21.9}, {112, 23.5}, {120, 25.3}}, {1, x}, x]

Out[1]= 0.00770263 + 0.209881 x

Figure 16

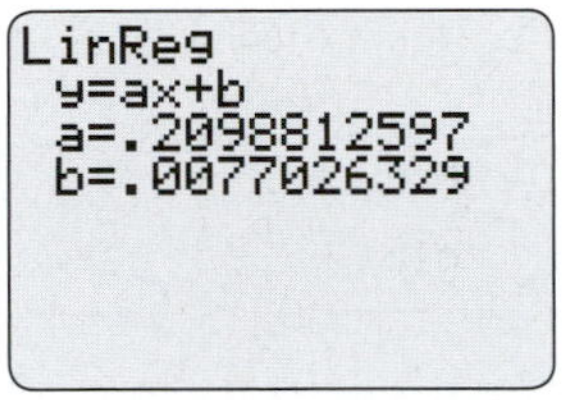

Since the constant term is close to 0, ignoring it we obtain

$$d \approx 0.21v,$$

and thus the slope of the best-fitting line is

$$0.21 \approx \frac{d}{v} = t,$$

```
LinReg
y=ax+b
a=.2098812597
b=.0077026329
```

Figure 17

which is the driver's response time. In order to express t in seconds, we note that since $1\,\text{m/s} = 3.6\,\text{km/h}$, if $\bar{v}$ is the speed expressed in meters per second, we have $\bar{v} = v/3.6$, and thus t (in seconds) can be obtained as

$$t = \frac{d}{\bar{v}} = \frac{3.6d}{v} \approx 3.6 \cdot 0.21 \approx 0.76.$$

In other words, the slope of the best-fitting line indicates that based upon this particular data set, the average driver response time is approximately 0.76 seconds.

1.5 **Exercises**

1–12 Express the given function (using ( ), ^, ×, ÷, etc.) in a format suitable for entering into a graphing utility.

1. $f(x) = 1 + 3x + \sqrt{x}$

2. $g(x) = 3x - 2 + \sqrt[3]{x}$

3. $h(x) = \dfrac{3x}{\sqrt{x-1}}$

4. $k(x) = \dfrac{1 + \sqrt{x}}{2 - 3x}$

5. $u(x) = \dfrac{(-3x + 16)^4}{3x + 6}$

6. $v(x) = \sqrt{2 + 5x + \sqrt{x}}$

7. $F(x) = \left(2x^{2/3} + 3x^{5/3}\right)^5$

8. $G(x) = \left(9x^{1/5} + 2x^{3/5}\right)^{10}$

9. $H(x) = \ln(x^2 + 1) + 2^{\sqrt{x}}$

10. $K(x) = \dfrac{e^{\cos x}}{\sqrt{\log(x^4 + 2)}}$

11. $Q(x) = \dfrac{\arctan x + 1}{\left(\cos(\arcsin x)\right)^3}$

12. $R(x) = \sqrt{(\arccos x)^2 + \log_2(\tan x)}$

13. During the last 5 years, the advertising manager for a corporation has gathered the following data that show the relationship between the advertising budget (in millions of dollars) and the total sales (in thousands of units).

Advertising and Sales

Advertising budget (x) (in millions)	$4.50	$6.50	$3.50	$4.20	$2.60
Sales (y) (in thousands)	37	46	42	32	29

a. Find the least-squares regression line for the data.

b. Estimate the sales if $4 million is budgeted for advertising.

14. Records at a company for the last 5 years show the following relationship between the units sold (in thousands) and the price of a product.

Sales

Price (p)	$8.80	$8.00	$7.50	$6.90	$6.20
Units sold (x) (in thousands)	3.8	5.2	7.3	8.0	9.6

a. Find the least-squares regression line for the price in terms of units sold.

b. Estimate the price that should be charged in order to sell 10,000 units.

15. The following data show the amount spent on office-building construction (in thousands) for a particular county during a 6-month period.

Office Construction

Month	Apr	May	Jun	Jul	Aug	Sep
Amount (in thousands)	$24	$24	$30	$49	$68	$69

a. Find the least-squares regression line for the data. (Let $x = 1$ correspond to January, $x = 2$ to February, etc.)

b. Estimate the amount spent on construction in October.

16. The annual revenue (in millions of dollars) for a corporation is given in the following table.

Annual Revenue

Year	2003	2004	2005	2006	2007	2008
Revenue (in millions)	$66	$82	$127	$201	$310	$315

a. Find the least-squares regression line for the data. (Let $x = 0$ correspond to the year 2003.)

b. Estimate the revenue for 2009.

17. The price of livestock futures is the estimated market price of livestock on the delivery date (end of the indicated month). The cattle futures (in cents per pound) for the months February through July are as follows.

Livestock Futures

Month	Feb	Mar	Apr	May	Jun	Jul
Price (cents per pound)	79.10	76.02	71.80	71.45	71.45	72.50

a. Find the least-squares regression line for the data. (Let $x = 1$ correspond to January, $x = 2$ to February, etc.)

b. Estimate the price for August.

18. The total number of foreign tourists visiting the United States between 2000 and 2004, as reported by the US Travel and Tourism Administration, is shown in the following table.

Foreign Tourists Visiting the United States					
Year	2000	2001	2002	2003	2004
Tourists (in millions)	25.7	26.3	29.7	34.2	38.3

a. Find the least-squares regression line for the data. (Let x represent the number of years passed since 2000.)

b. Estimate the number of foreign tourists that visited the United States during 2006.

1.5 Technology Exercises

19–28 Determine whether there are points where we need to be careful in interpreting the result when using graphing technology to graph the given function. Find all those points and explain. Then use a graphing calculator or computer algebra system to sketch the graph, using various viewing windows.

19. $f(x) = x^4 \cos \dfrac{1}{x}$

20. $G(x) = \cos \dfrac{1}{x-2}$

21. $p(x) = \dfrac{1}{2} \tan(3x - 2)$

22. $g(x) = \sec(2x + 1)$

23. $q(x) = \dfrac{x^2 - 2x - 1}{x + 1}$

24. $r(x) = \dfrac{x^2 + 1}{x^2 - 9}$

25. $h(x) = \dfrac{2x^4 + 1}{2x^4 - 1}$

26. $F(x) = \ln(\cos x)$

27. $s(x) = \cos(\ln x)$

28. $t(x) = \sin(\csc x)$

29–40 Use a graphing calculator or computer algebra system to graph the given function in the window $[-10, 10]$ by $[-10, 10]$. Explain what appears to be wrong with the picture. Then find a more appropriate window, which reveals the significant parts of the graph, and draw the "improved" graph.

29. $f(x) = \dfrac{3x - 25}{\sqrt{x^2 + 5}}$

30. $g(x) = (40 + 3x)\sqrt{16 - x}$

31. $h(x) = (3x + 4)^2 (5x - 25)^2$

32. $F(x) = (6x + 30)^2 (3x - 15)^2$

33. $G(x) = 35 + 17x - x^2 - x^3$

34. $H(x) = 210 - 80x + x^3$

35. $r(x) = \sqrt[3]{x^3 - x^2 - x - 50}$

36. $u(x) = \sqrt[3]{x^4 - 3x^2 - 3x - 30}$

37. $v(x) = \left(12 - 6x - x^2\right)^{4/3}$

38. $f(x) = \left(x^3 - x - 100\right)^{1/3}$

39. $g(x) = x^2 \sin \dfrac{\pi}{x - 12}$

40. $h(x) = \sec^2 \dfrac{x}{10}$

41–46 Use a graphing calculator or computer algebra system to graph the given function in a suitable window and find the smallest y-value possible. (Use only the given interval, if specified. Round your answer to four decimal places.)

41. $f(x) = x^2 - 104x + 2724$

42. $g(x) = \dfrac{-1 - x^2 - 3x^3}{5^x}$

43. $h(x) = x^3 - 17x + 5; \quad -3 \le x \le 5$

44. $F(x) = \dfrac{\sqrt[3]{x} - 150}{5 + x^2}$

45. $G(x) = x^{1.5} - 8x - 15$

46. $H(x) = x^{1.8} - x - 100$

47–52 Use a graphing calculator or computer algebra system to graph the given function in a suitable window and find the greatest y-value possible. (Use only the given interval, if specified. Round your answer to four decimal places.)

47. $f(x) = 50 - 2^x$; $-10 \le x \le 10$ **48.** $g(x) = (x+1)^5 - 1.5^{x+1}$ **49.** $h(x) = x^{17} - 17^x$; $-2 \le x \le 2$

50. $k(x) = x(3^{-x})$ **51.** $F(x) = \dfrac{-2x}{x^2 + 1}$ **52.** $G(x) = \dfrac{3 - 5x}{\sqrt{3x^2 + 2}}$

53–58 Use a graphing calculator or computer algebra system to graph the given function and describe the characteristics of the graph as c varies. Use different viewing windows.

53. $f(x) = x^2 - cx$ **54.** $g(x) = \dfrac{1}{2}x^3 - c(x^2 + x + 1)$ **55.** $h(x) = e^{cx}$

56. $k(x) = \ln(x^2 + cx + 1)$ **57.** $F(x) = \dfrac{x}{c} + \cos\dfrac{c^2 x}{c}$ **58.** $G(x) = \dfrac{cx^2}{x + cx^3}$

59–64 Use a graphing calculator or computer algebra system to approximate the solution(s) of the given equation, rounded to four decimal places. (**Hint:** Zoom in on the x-intercepts or points of intersection as appropriate for each equation.)

59. $x^3 - 20x - 2 = 0$ **60.** $2x^3 = 31x + 2$ **61.** $3\cos x = \sqrt{x}$

62. $\arctan x = \dfrac{1}{100}x^5$ **63.** $\ln x = x - 2$ **64.** $x + 5 = e^x$

65–70 Use appropriately large viewing windows on a graphing calculator or computer algebra system to decide which of the given functions eventually "rises faster" toward infinity.

65. $f(x) = \dfrac{1}{2}x^3$; $g(x) = x^2$ **66.** $f(x) = \sqrt{x}$; $g(x) = x$ **67.** $f(x) = 5\sqrt{x}$; $g(x) = \dfrac{1}{5}x$

68. $f(x) = \dfrac{1}{2}e^x$; $g(x) = x^2$ **69.** $f(x) = 5\log x + 5$; $g(x) = \dfrac{1}{2}x^5$ **70.** $f(x) = 10\arctan x$; $g(x) = 2\sqrt[3]{x}$

71–73 Most graphing calculators and computer algebra systems have regression capabilities to fit curves other than lines to a given data set. Frequently, depending on the tendency of the data, a quadratic, an exponential, or some other type of curve provides for much better approximation. Most often the choice is the modeler's.

Use the regression capabilities of your technology to build a graphical model and then answer the questions.

71. The following table shows daytime temperatures in El Cajon, CA on a particular spring day from 7 a.m. to 12 p.m. Find the best-fitting curve and use it to predict the temperatures at 1 p.m. and 2 p.m.

Daytime Temperatures in El Cajon, CA

Time	6 a.m.	7 a.m.	8 a.m.	9 a.m.	10 a.m.	11 a.m.	12 p.m.
Temperature (°F)	47	50	55	61	68	73	75

72. The following table shows the winning times of the Olympic men's 100 m dash champions. Find the best-fitting curve and use it to predict the winning times at the next three Olympics.

Olympic Men's 100 m Dash Winning Times

Year	Time (s)	Year	Time (s)	Year	Time (s)
1896	12.00	1936	10.30	1980	10.25
1900	11.00	1948	10.30	1984	9.99
1904	11.00	1952	10.40	1988	9.92
1908	10.80	1956	10.50	1992	9.96
1912	10.80	1960	10.20	1996	9.84
1920	10.80	1964	10.00	2000	9.87
1924	10.60	1968	9.95	2004	9.85
1928	10.80	1972	10.14	2008	9.69
1932	10.30	1976	10.06	2012	9.63

73. The following table shows acceleration times for the Ferrari Enzo up to 130 mph. Find the best-fitting curve and use it to predict the acceleration times for the Enzo from **a.** 0 to 150 mph and **b.** 0 to 170 mph.

Acceleration Times for Ferrari Enzo

Speed (mph)	0–30	0–40	0–50	0–60	0–70	0–80	0–90	0–100	0–110	0–120	0–130
Time (s)	1.5	2.0	2.7	3.3	3.8	5.0	5.8	6.6	8.0	9.2	10.3

Source: *Car and Driver*

Chapter 1
Review Exercises

1–4 Find the domain and range of the given relation and determine whether the relation is a function.

1. $R = \{(-2,9),(-3,-3),(-2,2),(-2,-9)\}$

2. $3x - 4y = 17$

3. $x = y^2 - 6$

4. $x = \sqrt{y - 4}$

5–8 Identify the domain, codomain, and range of the given function.

5. $f : \mathbb{N} \to \mathbb{R}, \quad f(x) = \dfrac{3x}{4}$

6. $g : \mathbb{R} \to \mathbb{R}, \quad g(x) = 5x + 1$

7. $h : \mathbb{R} \to \mathbb{R}, \quad h(x) = \dfrac{1}{x^2 + 1}$

8. $k : \mathbb{N} \to \mathbb{R}, \quad k(x) = 2 + \sqrt{x - 1}$

9–12 Find the value of the given function for **a.** $f(x-1)$, **b.** $f(x^2)$, and **c.** $\dfrac{f(x+h) - f(x)}{h}$.

9. $f(x) = (x+5)(2x)$

10. $f(x) = \sqrt[3]{x} + 6(x+4)$

11. $f(x) = \dfrac{3}{x+2}$

12. $f(x) = \sin 2x$

13–14 Find all intervals of monotonicity (intervals where the function is increasing or decreasing) for the given function.

13. $f(x) = (x-2)^4 - 6$

14. $R(x) = \begin{cases} (x+2)^2 & \text{if } x < -1 \\ -x & \text{if } x \geq -1 \end{cases}$

15–18 Determine if the function is even, odd, or neither and then graph it.

15. $f(x) = \dfrac{1}{3}x^3$

16. $f(x) = \sqrt{x}$

17. $f(x) = -2\sin x$

18. $g(x) = -2\sin^2 x$

19–20 Discuss the symmetry of the given equation and then graph it.

19. $y = |5x|$

20. $x^2 + y^2 = 25$

21–34 Graph the given function. Locate the x- and y-intercepts, if any.

21. $f(x) = 7x - 2$

22. $f(x) = \dfrac{2x - 6}{3}$

23. $f(x) = (x-1)^2 - 1$

24. $f(x) = -x^2 - 6x - 11$

25. $f(x) = 4x^3$

26. $f(x) = -\dfrac{2}{x^2}$

27. $f(x) = \dfrac{-x^3 + 7x + 6}{2}$

28. $f(x) = \dfrac{x+1}{x^2 - 4}$

29. $f(x) = \dfrac{\sqrt[3]{x}}{2}$

30. $f(x) = 5|-x|$

31. $f(x) = \left\| \dfrac{2x}{3} \right\|$

32. $f(x) = \begin{cases} x^2 & \text{if } x < 1 \\ \dfrac{1}{x} & \text{if } x \geq 1 \end{cases}$

33. $f(x) = -\cot x$

34. $f(x) = \log_{1/2} x$

35–36 Sketch the graph of the given function by first identifying the more basic function that has been shifted, reflected, stretched, or compressed. Then determine the domain and range of the function.

35. $f(x) = (x-1)^3 + 2$

36. $f(x) = 4|x+3|$

37–38 Write an equation for the function described.

37. Use the function $f(x) = 1/x$. Move the function 2 units to the right and 3 units up.

38. Use the function $f(x) = \sqrt[3]{x}$. Move the function 1 unit to the left and reflect across the y-axis.

39–40 Find the formula and domain for **a.** $f + g$ and **b.** f/g.

39. $f(x) = x^2; \quad g(x) = \sqrt{x}$

40. $f(x) = \dfrac{1}{x-2}; \quad g(x) = \sqrt[3]{x}$

41–42 Use the information given to determine **a.** $(f \circ g)(x)$, **b.** $(g \circ f)(x)$, and **c.** $(f \circ g)(3)$.

41. $f(x) = -x + 1; \quad g(x) = -x - 1$

42. $f(x) = \dfrac{x^{-1}}{18} - 3; \quad g(x) = \dfrac{x - 4}{x^3}$

43–44 Write the given function as a composition of two functions. (Answers will vary.)

43. $f(x) = \dfrac{\sqrt{x + 3} + 2}{x^2 + 6x + 9}$

44. $f(x) = |x + 2| + x^2 + 4x + 4$

45–46 Graph the inverse of the given relation, and state its domain and range.

45. $R = \{(-3, 5), (2, 1), (0, -5), (-1, -2)\}$

46. $y = \dfrac{1}{3} x^2$

47–50 Find the inverse of the given function.

47. $f(x) = \dfrac{2}{7x - 1}$

48. $f(x) = \dfrac{4x - 3}{x}$

49. $f(x) = x^{1/5} - 6$

50. $f(x) = \dfrac{6x - 7}{2 - x}$

51. Show that $f^{-1}(f(x)) = x$ and that $f(f^{-1}(x)) = x$ for the functions given in Exercises 49 and 50.

52–53 Rewrite the given function as a purely algebraic function.

52. $\cos(\sin^{-1} x)$

53. $\tan\left(\sec^{-1} \dfrac{x}{2}\right)$

54–55 Use the properties of logarithms to expand the given expression as much as possible; that is, decompose the expression into sums or differences of the simplest possible terms.

54. $\ln \dfrac{x^2 y - xz}{5y^3}$

55. $\log \sqrt[3]{\dfrac{yz^2}{x^4}}$

56. If a pebble is shot upward with an initial (vertical) velocity of 56 ft/s, how high does it go? (**Hint:** Use the height function $h(t) = -16t^2 + 56t$.)

57. The half-life of actinium-225 is 10 days. Assuming that A_0 is the mass at time $t = 0$, find the function $A(t)$ that gives the mass remaining after t days. What percentage of the original mass of a sample of Ac-225 remains after 25 days?

58. Prove that the product of an odd function and an even function is odd.

59. Among all the pairs of numbers with a sum of 15, find the pair whose product is maximum.

60. The January 2010 Haiti earthquake was initially reported as a 7.2-magnitude quake on the Richter scale, while the March 2011 earthquake off the eastern shores of Japan was magnitude 9.0. According to these numbers, about how many times more intense was the Japanese earthquake? (**Hint:** See the directions preceding Exercise 138 in Section 1.4.)

61–62 Determine if the given complex number is in the Mandelbrot set. (**Hint:** See the directions preceding Exercises 101–108 in Section 1.3.)

61. $c = -i$

62. $c = 2$

63–76 *True or False?* Determine whether the given statement is true or false. In case of a false statement, explain or provide a counterexample.

63. The graph of any function can always be represented by a curve that passes the vertical line test.

64. Any linear function whose graph is a line with negative slope always has an inverse.

65. If f is a function, and $f(a) = f(b)$, then $a = b$.

66. If $(2, 5)$ is a point on the graph of an odd function, then $(-2, -5)$ is also on the graph.

67. The graph of a quadratic function is a parabola, so quadratic functions are even functions.

68. In general, $(f \circ g)(x) = (g \circ f)(x)$ holds for all x if both compositions can be formed.

69. If a function can be decomposed into three functions, it can only be done in one way.

70. The graph of $f(x) + a$ is that of $f(x)$ translated vertically by a units.

71. If $c \in \mathbb{R}$ is a constant, and f is any function, then the graphs of $f(cx)$ and $cf(x)$ are identical.

72. When the graph of any function $f(x)$ is reflected about the line $y = x$, the graph of the inverse is obtained.

73. The graph of any function must pass the horizontal line test.

74. If (a, b) is a point on the graph of an invertible function $f(x)$, then (b, a) is on the graph of its inverse.

75. The common logarithm and $y = 10^x$ are inverses.

76. The function $f(x) = \tan^{-1} x$ has no asymptotes.

Chapter 1
Technology Exercises

77–79 Mentally sketch the graph of the given function by identifying the basic shape that has been shifted, reflected, stretched, or compressed. Then use a graphing calculator or a computer algebra system to graph the function and check your reasoning.

77. $f(x) = \ln(x+1) + 2$

78. $f(x) = -\dfrac{2}{x-3} + 1$

79. $f(x) = \sin \pi x - 1$

80. The annual expenditures (in millions of dollars) for a corporation are given in the table below.

Annual Expenditures

Year	2006	2007	2008	2009	2010	2011
Expenditures (in millions)	$16.2	$17.1	$18.8	$19.6	$21.1	$22.9

a. Find the least-squares regression line for the data. (Let $x = 0$ correspond to the year 2006.)

b. Estimate the expenditures for 2012.

81–82 Use a graphing calculator or computer algebra system to approximate the solution(s) of the given equation, rounded to four decimal places. (**Hint:** Zoom in on the *x*-intercepts or points of intersection as appropriate for each equation.)

81. $x^5 - x^3 - 3 = 0$

82. $x^2 + 6 = 2^{x+1}$

83–84 Use a graphing calculator or computer algebra system to graph the given function and describe the characteristics of the graph as c varies. Use different viewing windows.

83. $u(x) = \dfrac{1 - e^{c/x}}{1 + e^{c/x}}$

84. $v(x) = \dfrac{x}{c^2} \sqrt[4]{c^4 - x^4}$

Project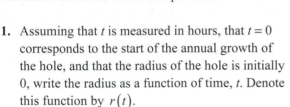

Chapter 1

As time goes on, there is increasing awareness, controversy, and legislation regarding the ozone layer and other environmental issues. The hole in the ozone layer over the South Pole disappears and reappears in a cyclical manner annually. Suppose that over a particular stretch of time the hole is assumed to be circular with a radius growing at a constant rate of 2.6 kilometers per hour.

Photo Courtesy of NASA

1. Assuming that t is measured in hours, that $t = 0$ corresponds to the start of the annual growth of the hole, and that the radius of the hole is initially 0, write the radius as a function of time, t. Denote this function by $r(t)$.

2. Use function composition to write the area of the hole as a function of time, t. Denote this function by $A(t)$. Sketch the graph of $A(t)$ and label the axes appropriately.

3. After finding $A(1)$, the area of the ozone hole at the end of the first hour, determine the time necessary for this area to double. How much additional time does it take to reach three times the initial area?

4. Are the two time intervals you found in Question 3 equal? If not, which one is greater? Explain your finding. (Use a comparison of some basic functions discussed in Section 1.2 in your explanation.)

5. What are the radius and area after 3 hours? After 5.5 hours?

6. What is the average rate of change of the area from 3 hours to 5.5 hours?

7. What is the average rate of change of the area from 5.5 hours to 8 hours?

8. Is the average rate of change of the area increasing or decreasing as time passes?

9. What flaws do you see with this model? Can you think of a better approach to modeling the growth of the ozone hole?

Limits and the Derivative

2.1 Rates of Change and Tangents 103

1. The Velocity Problem
2. The Tangent Problem

2.2 Limits All Around the Plane 117

1. Limits in Verbal, Numerical, and Visual Forms
2. Vertical Asymptotes and One-Sided Limits
3. Horizontal Asymptotes and Limits at Infinity
4. Limits and Technology

2.3 The Mathematical Definition of Limit 134

1. Limits Formally Defined
2. Proving a Limit Exists
3. Proving a Limit Does Not Exist

2.4 Determining Limits of Functions 146

1. Limit Laws
2. Limit Determination Techniques

2.5 Continuity 159

1. Continuity and Discontinuity at a Point
2. Continuous Functions
3. Properties of Continuity

2.6 Rate of Change Revisited: The Derivative 179

1. Velocity and Tangent Recap
2. The Derivative as a Function

Introduction

This chapter opens with a discussion of two broadly defined problems whose solutions turn out to have much in common. Mathematically, the task of determining the instantaneous velocity of an object and the task of finding the line tangent to the graph of a function both depend on the concept of *limit*. Much of this chapter revolves around developing an intuitive sense as well as a rigorous definition of the concept.

The limit idea inherently involves motion and reflects a major difference between the relatively static world of algebra and the dynamic world of calculus. As with so many concepts in mathematics, the evolution of the idea spans cultures and ages. Some early thinkers, such as the philosopher Zeno of Elea (ca. 495–ca. 430 BC) and the mathematician Archimedes of Syracuse (ca. 287–ca. 212 BC), developed an appreciation of the power and depth of the concept centuries before later mathematicians overcame the difficulties of rigorously defining and using limits.

Zeno, in fact, is chiefly remembered today for the many paradoxes he devised, which illustrate the danger of naïve assumptions about limits and infinity—a variation of one of these paradoxes appears as Example 2 in Section 2.2. Archimedes, whose name usually appears on any list of the greatest mathematicians of all time, used methods we would now classify as belonging to calculus to achieve results that were centuries ahead of their time. Formulas for the area under a parabola and for the volumes and surface areas of certain three-dimensional objects are just a few examples.

1700s by a long list of famous mathematicians (many of whom you will read about in the coming chapters), but rigorous definitions of "limit" and related ideas didn't appear until the 1800s. The definition we use today is essentially due to the French mathematician Augustin-Louis Cauchy (1789–1857), who also refined and made rigorous the idea of *continuity* as it applies to functions.

The two problems that open the chapter serve as motivation for the mathematics that follows, but by the end of the chapter it will be apparent

> ## The limit idea inherently involves motion and reflects a major difference between the relatively static world of algebra and the dynamic world of calculus.

Much later, in the seventeenth century AD, the ideas and methods involving limits were brought together and ultimately characterized as "calculus." But even after calculus came to be recognized as an especially rich branch of mathematics, the limit concept retained its somewhat dangerous reputation and continued to trick unwary thinkers. Great discoveries and advances were made using calculus throughout the 1600s and

that they are only representatives of the many problems that can be solved by means of the *derivative* of a function. The definition of "derivative" and the many methods associated with finding and using derivatives constitute a major portion of calculus and of this text.

Archimedes of Syracuse
(ca. 287–ca. 212 BC)

---

1. Adapted from line engraving by Remondini, *Archimedes Siracus*. Wellcome Library no. 569i, http://wellcomeimages.org/indexplus/image/V0000191.html. This file is licensed under the Creative Commons Attribution 4.0 International license.

TOPICS

1. The velocity problem

2. The tangent problem

2.1 Rates of Change and Tangents

The need for calculus can be demonstrated in many ways, ranging from common everyday experiences to problems of a more esoteric nature. We will begin our development of calculus with two easily described problems.

TOPIC 1   The Velocity Problem

The simple relationship *distance = rate × time*, or $d = rt$ is often one of the first applications of basic algebra to be studied. This equation can be taken as the definition of average rate r of travel for an object moving distance d over time t (that is, the **average rate** is the *ratio* of d over t), or it can be used to solve for any one of the variables if the other two quantities are known. It is undeniably useful in situations where an object's average rate is all that is desired (especially so if the object is moving at a constant rate), but it is just as undeniably limited in scope. Most of our everyday experiences involve objects that move at varying speeds, and we don't have to progress very far in mathematics before needing a way to handle more than just *average* rates of travel.

By convention, we often use the word *velocity* or *speed* when discussing a rate of travel that is not necessarily constant (later, we will further refine our usage and differentiate between speed and velocity). To emphasize the distinction with average rate even more, the word *instantaneous* sometimes appears before velocity. One of the best examples of instantaneous velocity is the number you read off your speedometer as you drive—at any given moment, the gauge indicates your velocity *at that instant*.

One form of the "velocity problem" can be put in this context: Suppose you are given the task of calculating a car's instantaneous velocity from perfect knowledge about its distance traveled over various lengths of time but without benefit of the speedometer. How would you do it?

To give us something to visualize and work with, suppose that the car is traveling along a straight road and $d(t)$ represents its location, relative to its starting point, at time t; we can set $t = 0$ as the moment when the car starts moving, so $d(0)$ is its initial location. Units of measurement are immaterial at the moment so they will not be specified—it is important only to remember that $d(t)$ represents distance with respect to time t. Let's suppose further that we want to determine the car's velocity at time $t = 10$. Knowing nothing more than the relationship $d = rt$, we can easily compute some average rates over time intervals that contain $t = 10$. For example, the average rate of travel from $t = 9$ to $t = 11$ is calculated as follows.

$$\frac{d(11) - d(9)}{11 - 9} = \frac{d(11) - d(9)}{2}$$

If we suspect that the instantaneous velocity of the car varies quite a bit over the time interval $[9,11]$ and that the average rate is not a good reflection of the velocity at exactly $t = 10$, we can easily compute the average rate over a shorter interval, say $[9.5, 10.5]$.

$$\frac{d(10.5) - d(9.5)}{10.5 - 9.5} = d(10.5) - d(9.5)$$

And there's no need to stop there. Our intuition tells us that as the time intervals get shorter and shorter, the average rate of travel over those time intervals should get closer and closer to the instantaneous velocity at $t = 10$.

With a bit more thought, our intuition may tell us something else. Namely, it really shouldn't matter much if the small intervals we're considering are centered at $t = 10$ or not, as long as the intervals all contain that point in time. After all, the average rate of travel that we are computing is an approximation to the instantaneous velocity at *every* point of the interval, not just the midpoint. In practice, we may find it easier to use intervals that have our point of interest at an endpoint, rather than the midpoint.

At this time, it will be convenient to introduce some more terms and notation. Given a real-valued function f defined at a point x_0, the **difference quotient of** f **at** x_0 **with increment** h is the following ratio.

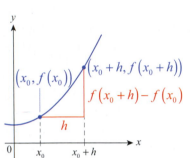

Figure 1

$$\frac{f(x_0 + h) - f(x_0)}{h}$$

Note that if f represents a distance function as described above, this ratio corresponds to the average rate of travel over an interval containing x_0, with x_0 serving as either the left endpoint (if h is positive) or the right endpoint (if h is negative). More generally, if x_0 is contained in the interval $[a, b]$, the ratio

$$\frac{\Delta f}{\Delta x} = \frac{f(b) - f(a)}{b - a}$$

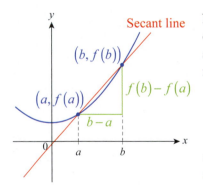

Figure 2

corresponds to the average rate of travel over the interval and also serves as an approximation to the instantaneous velocity at x_0. You have probably seen the symbol Δ before: it is the Greek letter delta and is used to designate a change in the quantity that follows it. The above ratio is read "change in f over change in x" and can be interpreted as the slope of the line (known as the *secant line*) shown in Figure 2.

Example 1 ✐

A piece of rock falls into a 256-foot-deep canyon from its rim. If we ignore air resistance, its distance below the rim at time t is

$$d(t) = 16t^2,$$

where d is measured in feet and t is measured in seconds. Estimate **a.** the instantaneous velocity of the rock after 2 seconds and **b.** the velocity of impact.

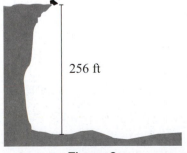

Figure 3

Solution

a. We will start by calculating the average velocity of the rock from $t = 2$ to $t = 3$. Note that the elapsed time is $\Delta t = 3 - 2 = 1$, while the distance traveled is $\Delta d = d(3) - d(2)$, so we can set up the following difference quotient to calculate the desired average velocity over the interval $[2,3]$.

$$\frac{\Delta d}{\Delta t} = \frac{d(3) - d(2)}{3 - 2}$$
$$= \frac{16 \cdot 3^2 - 16 \cdot 2^2}{1}$$
$$= 16 \cdot 9 - 16 \cdot 4 = 80 \text{ ft/s}$$

However, this is not expected to be an accurate reflection of the instantaneous velocity at $t = 2$, since the rock is accelerating significantly between $t = 2$ and $t = 3$. We obtain a more accurate guess if we work over a much shorter time interval, say the one from $t = 2$ to $t = 2.1$.

$$\frac{\Delta d}{\Delta t} = \frac{d(2.1) - d(2)}{2.1 - 2}$$
$$= \frac{16 \cdot (2.1)^2 - 16 \cdot 2^2}{0.1}$$
$$= \frac{16 \cdot (4.41) - 16 \cdot 4}{0.1} = 65.6 \text{ ft/s}$$

This is a much better guess, but there is no need to stop here. The following table summarizes the average velocities over time intervals of decreasing lengths, all the way down to a length of 0.01 seconds.

Time interval	[2,2.09]	[2,2.08]	[2,2.07]	[2,2.06]	[2,2.05]	[2,2.04]	[2,2.03]	[2,2.02]	[2,2.01]
Average velocity $\Delta d/\Delta t$	65.44	65.28	65.12	64.96	64.8	64.64	64.48	64.32	64.16

Since the velocity numbers in the table seem to be approaching 64 ft/s as the intervals are becoming shorter, we might guess that number to be the instantaneous velocity of the rock at $t = 2$. Notice also that in the above table, 2 serves as the left endpoint of all of the intervals. Intuitively, we expect the same instantaneous velocity to be obtained if $t = 2$ were the right endpoint, or if it were in the interior of each interval, as long as the interval's length is decreasing and approaching 0. In our next table, we chose $t = 2$ to be the right endpoint of each time interval.

Time interval	[1.91,2]	[1.92,2]	[1.93,2]	[1.94,2]	[1.95,2]	[1.96,2]	[1.97,2]	[1.98,2]	[1.99,2]
Average velocity $\Delta d/\Delta t$	62.56	62.72	62.88	63.04	63.2	63.36	63.52	63.68	63.84

Notice that the average velocities here are all less than in the first table, but are increasing as the time interval is getting shorter, "shrinking" onto the point $t = 2$. This is precisely what should be expected because of the steady acceleration of the rock. In conclusion, both tables support our guess that the rock's instantaneous velocity at $t = 2$ is 64 ft/s downward. As a final confirmation, let us calculate the difference quotient over the very short interval $[1.999, 2]$.

$$\frac{\Delta d}{\Delta t} = \frac{d(2) - d(1.999)}{2 - 1.999}$$

$$= \frac{16 \cdot 2^2 - 16 \cdot (1.999)^2}{0.001}$$

$$= 63.984 \text{ ft/s}$$

b. To estimate the velocity of impact, we first need to find when exactly impact occurs. Being in possession of the position function and knowing that at impact the position of the rock is exactly the depth of 256 feet, we can find the time of impact by solving

$$d(t) = 256$$
$$16t^2 = 256$$
$$t^2 = 16,$$

which gives us $t = 4$, since we can safely discard the negative root of $t = -4$.

To find the velocity of impact, we use the same strategy as in part a. but this time using $t = 4$ as the right endpoint of our intervals for our difference quotients. The table below summarizes our calculations.

Time interval	$[3.91, 4]$	$[3.92, 4]$	$[3.93, 4]$	$[3.94, 4]$	$[3.95, 4]$	$[3.96, 4]$	$[3.97, 4]$	$[3.98, 4]$	$[3.99, 4]$
Average velocity $\Delta d / \Delta t$	126.56	126.72	126.88	127.04	127.2	127.36	127.52	127.68	127.84

As before, we examine the table and conclude that the velocity of impact is approximately 128 ft/s (you may want to further confirm our finding by calculating difference quotients over even shorter time intervals).

Example 2 ✎

Daniela lives just a few miles down a straight stretch of highway from her college campus. One morning she left for class, and her position relative to her house was given by the function

$$d(t) = 1600t^2 - 16,000t^3 + 40,000t^4,$$

where d is measured in miles, t in hours, with $t = 0$ corresponding to the time when she left her house. Use difference quotients to estimate her instantaneous velocity **a.** 3 minutes after she left, **b.** 6 minutes after she left, and **c.** 7.5 minutes after she left.

From your calculations, can you guess what might have happened?

Solution

a. Just like in Example 1, we will evaluate difference quotients over shorter and shorter time intervals, each using $t = 3$ minutes as the left endpoint. In this example, however, we will denote the length of a time interval by h, and call it an *increment*. With this notation, and by noting that 3 minutes = 0.05 hours, a typical difference quotient becomes the following.

$$\frac{d(0.05+h)-d(0.05)}{h}$$

For example, consider the average velocity on the time interval starting at 3 minutes into the drive and lasting for $h = 0.6$ minutes = 0.01 hours.

$$v_{ave} = \frac{d(0.05+0.01)-d(0.05)}{0.01}$$
$$= 57.24 \text{ mph}$$

We will get a better approximation of the instantaneous velocity at $t = 3$ minutes if we shorten the time intervals over which we calculate the average velocities. The following table summarizes our calculations over several intervals of decreasing length.

h	0.01	0.005	0.001	0.0005	0.0001
v_{ave}	57.24	58.81	59.79	59.90	59.98

So we conclude that at exactly three minutes into the trip, Daniela was driving at approximately 60 mph.

b. Next, we will estimate the instantaneous velocity at $t = 6$ minutes = 0.1 hours. This time, however, we will use negative increments when setting up the difference quotients. For example, with $h = -0.03$, we obtain

$$v_{ave} = \frac{d(0.1-0.03)-d(0.1)}{-0.03}$$
$$= \frac{d(0.1)-d(0.07)}{0.03}$$
$$= 22.92 \text{ mph.}$$

Interestingly, it appears from the above calculation that Daniela is slowing down considerably. In search of the instantaneous velocity, we shorten the time intervals and summarize our calculations in the following table.

h	−0.03	−0.01	−0.001	−0.0005	−0.0001
v_{ave}	22.92	7.96	0.80	0.40	0.08

From the table we must conclude that Daniela slowed down and stopped, for the instantaneous velocity at $t = 6$ minutes appears to be 0. Did she stop due to a traffic jam, or did she turn back? Our next calculation has the answer.

c. If we evaluate several difference quotients with decreasing increments, using $t = 7.5$ minutes $= 0.125$ hours as the left endpoint, using the now-familiar formula

$$v_{ave} = \frac{d(0.125 + h) - d(0.125)}{h},$$

we obtain the following table.

h	0.01	0.005	0.001	0.0005	0.0001
v_{ave}	−43.56	−40.65	−38.15	−37.82	−37.56

We estimate the instantaneous velocity at $t = 7.5$ minutes to be −37.5 mph. The negative sign indicates that the direction changed, so Daniela is actually driving back toward her house. (She might have left something at home.) In fact, the graph of the position function is quite revealing, so we show it in Figure 4.

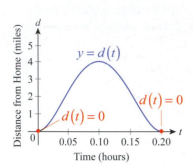

Figure 4

TOPIC 2 **The Tangent Problem**

Our second problem will be presented in a purely mathematical context, but, as we will see, it is closely related to the first.

The word "tangent" comes from the Latin word *tangens*, and translates as "touching" (and while there is a connection with the trigonometric function that goes by the name tangent, that connection is not especially instructive at the moment). In mathematics, a line is tangent to a given curve if it "just touches" the curve. In some cases, it is easy to identify such lines; in Figure 5, each of the two red lines is tangent to its associated curve, while the green lines are not. In these two examples, the tangent line touches the curve at exactly one point and doesn't cross the curve.

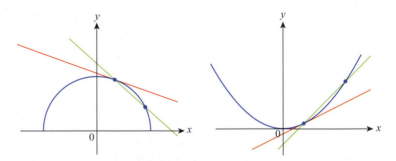

Figure 5 Tangent and Nontangent Lines

In many cases, however, it is not at all clear how to construct the line tangent to a curve at a given point, and, in any event, we will soon need a precise method of construction that results in an equation for a line (as opposed to just a rough sketch). How should we proceed?

Our two simple examples in Figure 5 are perhaps deceptive. If we think a bit harder about what it means for a line to be tangent to a curve, we might be led to the idea that a tangent line should represent the relative rise or fall of the curve at a particular point—in other words, the tangent line should have the same "slope" or trend as the curve at that point. With this interpretation, it's clear that sometimes a tangent line will intersect the curve at more than one point, and that sometimes it is not possible for a unique line to capture the trend at all. Figure 6 illustrates these possibilities.

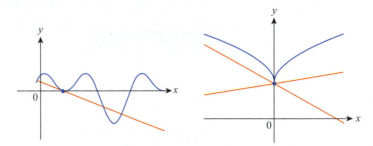

Figure 6 More Tangent and Nontangent Lines

To nail down precisely what we mean by "tangent," and to work toward an equation form for the tangent line, we can follow an approach similar to what we used for the velocity problem. Namely, we will use a process that allows us to construct approximations to the tangent line that are as accurate as we desire.

To set the stage, let's assume that we have a function f for which we want to construct the tangent line at $x = x_0$, that is, at the point $\left(x_0, f\left(x_0\right)\right)$ on the graph of f. Note that all we really need is the slope of the tangent line, since we already know that we want the tangent to pass through the point $\left(x_0, f\left(x_0\right)\right)$. While we don't yet know the slope of the tangent line, we *do* know the slope of any line that passes through $\left(x_0, f\left(x_0\right)\right)$ and a point on the graph of f slightly removed from $\left(x_0, f\left(x_0\right)\right)$; such a line is referred to as a *secant* line (again, not to be confused with the trigonometric function secant). An easy way to denote a "slightly removed" point on the graph is to let $x = x_0 + h$, where h is as close (but not equal) to 0 as we wish it to be. The "slightly removed" point is then $\left(x_0 + h, f\left(x_0 + h\right)\right)$, and the slope of the associated secant line is as follows.

$$\frac{f\left(x_0 + h\right) - f\left(x_0\right)}{\left(x_0 + h\right) - x_0} = \frac{f\left(x_0 + h\right) - f\left(x_0\right)}{h}$$

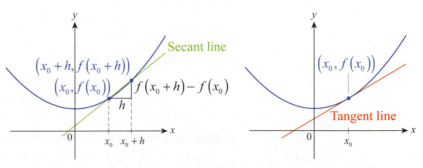

Figure 7

The connection between the tangent problem and the velocity problem is now clear, as we have just constructed a difference quotient again. To make the connection even more apparent, note that in both problems the key concept is "rate of change." In the first, we seek the instantaneous rate of change of the position of an object, and we label the result "velocity." In the second, we seek the rate of change of the graph of a function at a particular point, and we interpret the result as the slope of the line tangent to the graph at that point. The next few examples illustrate how we can actually construct such tangent lines.

Example 3 ✎

Construct difference quotients to approximate the slope of the line tangent to the graph of $f(x) = 0.2x^3 - 1.8x^2 + 3.6x$ at the point $(1, 2)$.

Solution

Note that in this problem, $x_0 = 1$. Just as in Examples 1 and 2, we will construct difference quotients using several h-values that are decreasing in magnitude. To start off, letting $h = 1$ we obtain the following.

$$\frac{f(1+1) - f(1)}{1} = f(2) - f(1)$$
$$= 0.2 \cdot 2^3 - 1.8 \cdot 2^2 + 3.6 \cdot 2 - \left(0.2 \cdot 1^3 - 1.8 \cdot 1^2 + 3.6 \cdot 1\right)$$
$$= -0.4$$

This means that -0.4 is the slope of the secant line corresponding to $h = 1$. However, given that $x_0 = 1$, our choice of $h = 1$ means that the x-coordinate of the "slightly removed" point in this case is $x = 2$, which is a rather big step away from x_0, so we don't yet expect the above slope to reflect the trend of the graph in any meaningful way. Let us check what happens if $h = 0.5$.

$$\frac{f(1+0.5) - f(1)}{0.5} = 2 \cdot \left[f(1.5) - f(1) \right]$$
$$= 2 \cdot \left[0.2 \cdot (1.5)^3 - 1.8 \cdot (1.5)^2 + 3.6 \cdot 1.5 - \left(0.2 \cdot 1^3 - 1.8 \cdot 1^2 + 3.6 \cdot 1\right) \right]$$
$$= 0.05$$

This shows us that the associated secant line is now nearly horizontal, and its slope went from negative to positive. Decreasing h even further, we evaluate several more difference quotients, and the following table displays our results.

h	0.3	0.1	0.05	0.01	0.001
m_{sec}	0.258	0.482	0.5405	0.58802	0.59880

In the second row of the above table, the slope of the secant associated with each corresponding h-value is denoted by m_{sec}, and we conclude that the slope of the actual tangent line approaches $m = 0.6$. (Later we will learn how to confirm that it actually is equal to 0.6.)

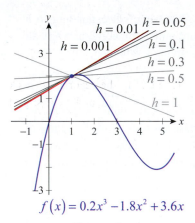

$$f(x) = 0.2x^3 - 1.8x^2 + 3.6x$$

Figure 8

Figure 8 shows the graph of $f(x)$ and the calculated secant lines, as well as the predicted tangent. Note how the secant lines move closer and closer to the tangent as h decreases. Also, the secant line corresponding to $h = 0.001$ is virtually indistinguishable from the tangent line shown in red.

Example 4 ✍

Show that the slope of a tangent line of the absolute value function $A(x) = |x|$ at $(0,0)$ can't be inferred from a table of difference quotients.

Solution

In this example, $x_0 = 0$, so using the formula for the difference quotient with, say, $h = 0.5$ we obtain the following.

$$\frac{A(x_0 + h) - A(x_0)}{h} = \frac{|0 + h| - |0|}{h}$$

$$= \frac{|0.5|}{0.5} = 1$$

On the other hand, if h is negative, say $h = -0.5$, the difference quotient is as follows.

$$\frac{A(x_0 + h) - A(x_0)}{h} = \frac{|0 + h| - |0|}{h}$$

$$= \frac{|-0.5|}{-0.5}$$

$$= \frac{0.5}{-0.5} = -1$$

In fact, if we construct a table like before, but this time alternating between positive and negative h-values, something curious happens.

h	0.1	−0.1	0.05	−0.05	0.01	−0.01
m_{sec}	1	−1	1	−1	1	−1

First of all, it seems that ± 1 are the only possible values for the slope of the secant, and second, it seems to be $+1$ when $h > 0$ and -1 when $h < 0$. Notice that this is an observation we can prove fairly easily.

If $h > 0$, since in this case $|h| = h$, the difference quotient becomes

$$\frac{|0 + h|}{h} = \frac{|h|}{h} = \frac{h}{h} = 1,$$

regardless of the value of h. On the other hand, if $h < 0$, then $|h| = -h$, so

$$\frac{|0 + h|}{h} = \frac{|h|}{h} = \frac{-h}{h} = -1,$$

and again, h can be any negative real number.

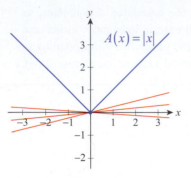

Figure 9

Notice that our results indicate that any secant line corresponding to a positive h-value will coincide with the right branch of the graph, while all secant lines obtained from $h < 0$ coincide with the left branch.

What does all this mean? There is no single real number being approached by the difference quotients or slopes as h is getting smaller. Graphically, this means that there is no tangent, no single line that would "best capture" the trend of the graph, none that would best align itself to the graph of $A(x)$ at $(0,0)$. The graph has a "sharp turn" there, making it impossible for a tangent line to exist, as shown in Figure 9. We also express this fact by saying that the graph is not "smooth" at $x = 0$.

We might summarize the underlying idea in both the velocity and tangent line problems as follows: *The true value of the quantity we seek can be approximated to any desired degree of accuracy through the use of smaller and smaller increments.* Perhaps it won't come as a surprise that the same technique can be used in a wide variety of settings, not just for velocity and the slope of a tangent. For example, we can approximate the length of a curvilinear segment, such as a portion of the graph of a function, or the area of a plane region bounded by the graph of a function by replacing the graph with series of shorter and shorter line segments. Our last example provides an illustration and serves as a preview of topics we will explore in much greater detail later in the text.

Example 5 ✍

Consider the graph of the function $f(x) = x^{3/2}$ on the interval $[0,1]$. By successively dividing $[0,1]$ into smaller and smaller subintervals, estimate

a. the area of the region R between the curve and the x-axis from $x = 0$ to $x = 1$, and

b. the arc length s of the segment of the graph connecting the points $(0,0)$ and $(1,1)$.

Solution

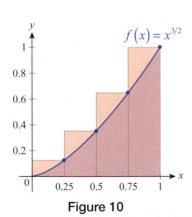

Figure 10

a. First, we divide $[0,1]$ into four subintervals of equal width, by considering $[0,0.25]$, $[0.25,0.5]$, $[0.5,0.75]$, and $[0.75,1]$, respectively. Then we create rectangles of width 0.25 on top of each interval, using the function value at the right endpoint to serve as the height, as shown in Figure 10. Then we simply add up the four areas to serve as our first approximation for the area under the curve.

$$S_4 = A_1 + A_2 + A_3 + A_4$$
$$= 0.25 \cdot f(0.25) + 0.25 \cdot f(0.5) + 0.25 \cdot f(0.75) + 0.25 \cdot f(1)$$
$$= 0.25 \cdot (0.25)^{3/2} + 0.25 \cdot (0.5)^{3/2} + 0.25 \cdot (0.75)^{3/2} + 0.25 \cdot (1)^{3/2}$$
$$\approx 0.5320$$

As is evident from our illustration in Figure 10, this calculation is overestimating the actual area of the region, since the union of the four rectangles completely covers R with lots to spare, but we can produce better estimates if we use more and more rectangles by dividing $[0,1]$

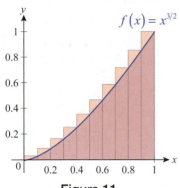

Figure 11

into more and more subintervals. For example, performing a similar calculation with 10 rectangles, each of width 0.1, and again using the function values at the right endpoints for the height of each (see Figure 11), yields the approximation $S_{10} \approx 0.4512$.

We shouldn't be surprised that this number is smaller than our result for S_4, if we compare Figures 10 and 11. S_{10} is still overestimating the true area of R, but as we can see, the error is much less, and we expect it to go down even further if we increase the number of rectangles.

Hand calculations quickly become tedious as n increases, but a computer algebra system or a programmable calculator helps us create a table of values like the one below.

n	50	100	1000	10,000
S_n	0.41005	0.40501	0.40050	0.40005

From the previous table, we conjecture that the true value of the area of R might be $A = 0.4$. Later we will learn how to exactly calculate this type of area.

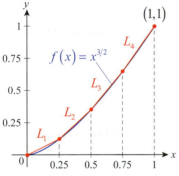

Figure 12

b. As far as the arc length of the curve connecting $(0,0)$ with $(1,1)$, we will illustrate the calculations by first dividing $[0,1]$ into four subintervals, as we did in part a.

If we label the endpoints as $x_1 = 0$, $x_2 = 0.25$, ..., $x_5 = 1$, and connect the points $(x_i, f(x_i))$ on the graph with $(x_{i+1}, f(x_{i+1}))$ for $i = 1, ..., 4$, a "crude" first approximation for the arc length will simply be the sum of the lengths of the four resulting line segments (see Figure 12). This can be calculated using the Pythagorean Theorem as follows.

$$s_4 = L_1 + L_2 + L_3 + L_4$$

$$= \sqrt{(0.25)^2 + \left[(0.25)^{3/2}\right]^2} + \sqrt{(0.25)^2 + \left[(0.5)^{3/2} - (0.25)^{3/2}\right]^2}$$

$$+ \sqrt{(0.25)^2 + \left[(0.75)^{3/2} - (0.5)^{3/2}\right]^2} + \sqrt{(0.25)^2 + \left[1 - (0.75)^{3/2}\right]^2}$$

$$\approx 1.4362$$

Upon dividing $[0,1]$ into 10 equal parts, a similar, but a bit longer, calculation yields the much better approximation of

$$s_{10} \approx 1.4389.$$

As before, with the help of a computer or programmable calculator we can generate a table of values such as the following.

n	50	100	1000	10,000
s_n	1.43966	1.43970	1.43971	1.43971

From the table above, we conclude that the true value of s is approximately $s \approx 1.43971$.

2.1 **Exercises**

1–6 Estimate the slope of the tangent line shown in the given graph.

1.

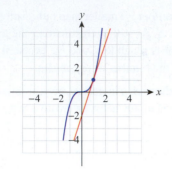

2.

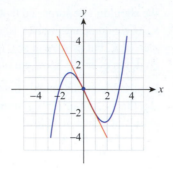

3.

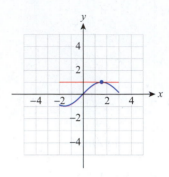

4.

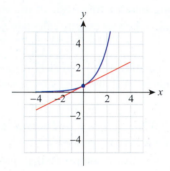

5.

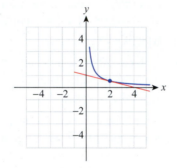

6.

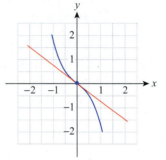

7–18 Use difference quotients to approximate the slope of the tangent to the graph of the function at the given point. Use at least five different h-values that are decreasing in magnitude. (Answers will vary.)

7. $f(x) = 1 - 2x$; $(1, -1)$

8. $g(x) = \dfrac{5}{4}x - 8$; $(8, 2)$

9. $h(x) = \dfrac{1}{3}x^2 - 1$; $(3, 2)$

10. $F(x) = 3 + x - \dfrac{x^2}{2}$; $(4, -1)$

11. $G(x) = \dfrac{1}{4}x^3 - x + 1$; $(-2, 1)$

12. $k(x) = 10 - x^{3/2}$; $(4, 2)$

13. $H(x) = \ln x + 1$; $(e, 2)$

14. $u(x) = \cos x$; $\left(\dfrac{\pi}{2}, 0\right)$

15. $v(x) = \log 2x - 1$; $(5, 0)$

16. $w(x) = \tan x$; $(0, 0)$

17. $p(x) = -x^4 + 1$; $(1, 0)$

18. $q(x) = x^5 - x + 3$; $(0, 3)$

19. An arrow is shot into the air and its height in feet after t seconds is given by the function $f(t) = -16t^2 + 80t$. The graph of the curve $y = f(t)$ is shown.

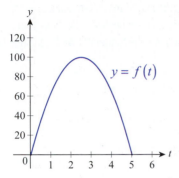

   a. Find the height of the arrow when $t = 2$ seconds.

   b. Find the instantaneous velocity of the arrow when $t = 2$ seconds.

   c. Find the slope of the line tangent to the curve at $t = 2$ seconds.

   d. Find the time it takes the arrow to reach its peak.

20. Suppose that a sailboat is observed, over a period of 5 minutes, to travel a distance from a starting point according to the function $s(t) = t^3 + 60t$, where t is time in minutes and s is the distance traveled in meters.

   a. How far will it travel during the first 6 seconds?

   b. What is the average velocity during the first 6 seconds?

   c. Estimate how fast the boat is moving at the starting point.

   d. Estimate how fast the boat is moving at the end of 3 minutes.

21. A model rocket is fired vertically upward. The height after t seconds is $h(t) = 192t - 16t^2$ feet.

   a. What will be its height at the end of the first second?

   b. What is the average velocity of the rocket during the first second?

   c. Estimate the instantaneous velocity at $t = 0$ seconds.

   d. Estimate the instantaneous velocity at $t = 4$ seconds.

   e. When will the velocity be 0? (**Hint:** You may want to start with the initial velocity you found in part a. and use the fact that under the influence of gravity, when air resistance is ignored, vertical upward velocity decreases by 32 ft/s every second. Once you have a guess, test it by a table of difference quotients.)

22. A particle moving in a straight line is at a distance of $s(t) = 2.5t^2 + 18t$ feet from its starting point after t seconds, where $0 \le t \le 12$. Estimate the instantaneous velocity at **a.** $t = 6$ seconds and **b.** $t = 9$ seconds.

23. The distance, in meters, traveled by a moving particle in t seconds is given by $d(t) = 3t(t+1)$. Estimate the instantaneous velocity at **a.** $t = 0$ seconds, **b.** $t = 2$ seconds, and **c.** at time t_0. (**Hint:** Write the difference quotient corresponding to $t = t_0$, simplify, and try to find the value being approached by the expression as h decreases.)

24. The distance, in meters, traveled by a moving particle in t seconds is given by $d(t) = t^2 - 3t$. Estimate the instantaneous velocity at **a.** $t = 0$ seconds, **b.** $t = 4$ seconds, and **c.*** at time t_0. (See the hint given in part c. of the previous problem.)

25. After start, on a straight stretch of the track, a race car's velocity changes according to the function $v(t) = -1.8t^2 + 18t$, when $0 \le t \le 10$, t is measured in seconds, and $v(t)$ is measured in meters per second.

   a. When does peak velocity occur and what is it? (**Hint:** The graph of $v(t)$ may be helpful.)

   b. When does peak deceleration occur?

   c. Use difference quotients to estimate peak deceleration. Approximately what multiple of $g \approx 9.81$ m/s^2 have you obtained?

26. If we ignore air resistance, a falling body will fall $16t^2$ feet in t seconds.

    a. How far will it fall between $t = 2$ and $t = 2.1$?

    b. What is its average velocity between $t = 2$ and $t = 2.1$?

    c. Estimate its instantaneous velocity at $t = 2$.

27. A student dropped a textbook from the top floor of his dorm and it fell according to the formula $S(t) = -16t^2 + 8\sqrt{t}$, where t is the time in seconds and $S(t)$ is the distance in feet from the top of the building.

    a. If the textbook hit the ground in exactly 2.5 seconds, how high is the building?

    b. What was the average speed for the trip?

    c. What was the instantaneous velocity at $t = 1$ second?

    d. What was the velocity of impact?

28–31 Approximate the area of the region between the graph of the function and the x-axis on the given interval. Use **a.** $n = 4$ and **b.** $n = 5$. (Round your answer to four decimal places.)

28. $f(x) = x^2$ on $[0,1]$

29. $g(x) = 16x - x^3$ on $[0,4]$

30. $h(x) = \sin x$ on $[0,\pi]$

31. $F(x) = e^x + 1$ on $[-10,0]$

32–35 Approximate the arc length of the graph of the function on the given interval. Use **a.** $n = 4$ and **b.** $n = 5$. (Round your answer to four decimal places.)

32. $f(x) = \sqrt{x}$ on $[0,1]$

33. $g(x) = x^3 + x^2$ on $[-1,0]$

34. $F(x) = \cos x$ on $\left[0, \dfrac{\pi}{2}\right]$

35. $G(x) = \ln x + 1$ on $[1,2]$

2.1 Technology Exercises

36–39 Use a graphing calculator or computer algebra system to graph $f(x)$ along with three secant lines at the indicated x-value, corresponding to the difference quotients with h-values of 0.2, 0.1, and 0.01, respectively. Can you come up with a possible equation for the tangent? Use technology to test your conjecture.

36. $f(x) = x^2$; $x = 2$

37. $f(x) = -x^3 + x + 1$; $x = \dfrac{\sqrt{3}}{3}$

38. $f(x) = \sin x + \cos x$; $x = 0$

39. $f(x) = 3\sqrt{x}$; $x = 4$

40–43 Use a graphing calculator or computer algebra system to graph the given function $f(x)$ along with $D(x) = \dfrac{f(x + 0.001) - f(x)}{0.001}$ in the same coordinate system. Explain how the function values of $D(x)$ are reflected on the graph of $f(x)$.

40. $f(x) = x^4$ **41.** $f(x) = x(3 - x)$

42. $f(x) = \sin x$ **43.** $f(x) = \ln x$

44–47 Use a graphing calculator or computer algebra system to find the x-values at which the graph of $f(x)$ does not have a tangent line. Explain.

44. $f(x) = -|x - 1| + 1$ **45.** $f(x) = |x^2 - 4|$

46. $f(x) = |\ln x|$ **47.** $f(x) = (x - 1)^{2/3}$

48–51. Use a computer algebra system to find approximations for the areas in Exercises 28–31 by using **a.** $n = 100$ and **b.** $n = 1000$. (Round your answers to four decimal places.)

52–55. Use a computer algebra system to find approximations for the arc lengths in Exercises 32–35 by using **a.** $n = 100$ and **b.** $n = 1000$. (Round your answers to four decimal places.)

2.2 Limits All Around the Plane

TOPICS

1. Limits in verbal, numerical, and visual forms
2. Vertical asymptotes and one-sided limits
3. Horizontal asymptotes and limits at infinity
4. Limits and technology

In the previous section we studied two broadly defined problems whose solutions called for the ability to determine the trend of some behavior. Namely, we sought to fully understand the nature of a fraction called the difference quotient as the denominator in the fraction approached 0. This need to exactly determine "limiting" behavior is a characteristic of calculus, and we will devote much of the next four sections to refining our understanding of limits.

TOPIC 1 Limits in Verbal, Numerical, and Visual Forms

We encounter the notion of limiting behavior in many different contexts, mathematical and otherwise. In the velocity problem of Section 2.1, we looked at the behavior of the average rate of travel over shorter and shorter time intervals; that is, we were looking for the value of the rate of travel "in the limit" as the length of the interval shrank to 0. Similarly, in the tangent problem we sought the limiting behavior of the secant lines as the horizontal difference between two points on a graph shrank to 0.

Every limit scenario, when described verbally, contains phrases of a dynamic nature. Examples of such phrases are "as h approaches 0," "as x goes to 3," and "as N grows without bound." In a less mathematical setting, the phrases might be something like "mortgage rates are expected to approach 5 percent" or "assuming the national debt keeps getting bigger and bigger." We might also see tables of data that hint at some limiting behavior.

In order to progress mathematically, we need notation that precisely describes whatever limit situation is under discussion; every such mathematical scenario can be expressed in terms of a function and a variable whose value is approaching a given point. We will begin with an informal introduction of limit notation here and follow up with a formal definition in Section 2.3.

For the purpose of discussion, let us suppose we have a function $f(x)$ defined on an open interval containing c, except possibly at c itself (the reason for this exception will be explained shortly). We say that the limit of $f(x)$ is L as x approaches c, and write

$$\lim_{x \to c} f(x) = L,$$

if the value of $f(x)$ is closer and closer to L as x takes on values closer and closer (but not equal) to c.

Note the small but important caveat: the limit of a function $f(x)$ as x approaches c does not depend *at all* on the value of f at c; in fact, the limit may exist when f is not even defined at c. This technical detail highlights one distinction between the mathematical meaning of "limit" and casual usage of the word—in everyday language, the boundary between limiting behavior and behavior at the limit point is often blurred. The distinction is important mathematically because, as we will see repeatedly, the behavior of a function near the point c, but not at c, is the key.

Example 1 ✎

Suppose the graph shown in Figure 1 is that of the function $y = f(x)$. Using your previous experience with graphs of functions, it is reasonable to conclude that the function value at $x = c$ is L, that is, $f(c) = L$. Also, since the graph is a nice, smooth, unbroken curve, we expect $f(x)$ to be close to L if x is close to c. In other words, as x approaches c from either side along the x-axis, $f(x)$ will be approaching L, a fact we can confirm visually. As before, we can express this by writing

$$\lim_{x \to c} f(x) = L.$$

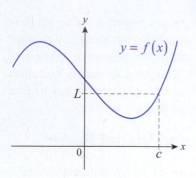

Figure 1

Next, suppose we redefine $f(c)$ to be $L/2$ (or some other number), but do not change anything else about f (let's call the resulting function $\hat{f}$). Graphically, this means that the point $(c, \hat{f}(c))$ "jumps out" of the graph, and moves down to the $L/2$ level, as shown in Figure 2. In other words, by redefining a single function value, we have relocated *one point* $(c, f(c))$ from the graph of f, but we haven't changed a bit the behavior of the function *near* $(c, f(c))$. If x takes on values closer and closer to c, but not equal to c, the function value $\hat{f}(x)$ will still move closer and closer to L, which is the second coordinate of the empty circle (i.e., its altitude over the x-axis). In other words, we can still write

$$\lim_{x \to c} \hat{f}(x) = L.$$

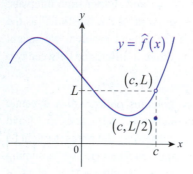

Figure 2

Finally, we will consider one last scenario. Suppose that we delete the function value corresponding to $x = c$ altogether, thus making the function undefined at $x = c$. We will call the resulting function $\tilde{f}$. The empty circle in Figure 3 indicates that one point is "missing" from the graph; that is, the vertical line $x = c$ doesn't meet the graph at all. However, notice that this still doesn't have any influence on the function values *near* $x = c$. As before, when x approaches c, but is not equal to c, the function values $\tilde{f}(x)$ approach L, that is,

$$\lim_{x \to c} \tilde{f}(x) = L.$$

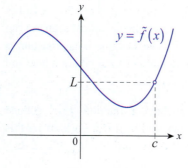

Figure 3

The conclusion to be drawn is that $\lim_{x \to c} f(x)$ may exist without the actual function value $f(c)$ being equal to it or, indeed, even defined.

Example 2 ✎

According to a classic paradox, you can never eat an apple for the following reason. A certain amount of time is needed to consume the first half of the apple, and at that point half the apple remains. Additional time is needed to eat half of the remaining half, which constitutes one-quarter of the original apple. After that, more time is needed to consume half of the remaining quarter, or one-eighth of the apple, and so on. By this reasoning, the process of eating the apple never ends.

Of course, this runs counter to our experience, so where does the problem lie? If you were challenged by an argument such as the one above, could you provide a mathematically correct resolution of the paradox?

Solution

As we shall see, the concept of limits can help resolve the apparent flaw in the argument. Let us assume that a calculus student performs the experiment, and holding a stopwatch in her left hand, eats an apple in exactly 4 minutes. We will make the realistic assumption that the time necessary to eat a certain portion of the apple is directly proportional to its size, and we will monitor the portion of the fruit having been consumed. Proceeding as in the above argument, we first note that half of the apple is gone in the first 2 minutes. At the end of the third minute, another quarter is gone, so we stand at $\frac{1}{2} + \frac{1}{4} = \frac{3}{4}$ of the apple consumed, with 1 minute left on the clock. At 3 and a half minutes, $\frac{3}{4} + \frac{1}{8} = \frac{7}{8}$ is gone, and at 3 minutes and 45 seconds, $\frac{7}{8} + \frac{1}{16} = \frac{15}{16}$ is gone. With half of the remaining time left, that is, at 3 minutes and 52.5 seconds, $\frac{15}{16} + \frac{1}{32} = \frac{31}{32}$ of the apple is gone, and so on. If we examine the fractions we obtain, it is easy to see that they get closer and closer to 1 as time progresses toward 4 minutes. In fact, a little later, we will have the rigorous mathematical tools to *prove* that the portion of the apple consumed, as time approaches 4 minutes, is as close to 1 as we desire. Mathematically, we say that the limit of these values is 1, that is, the student indeed ate the entire apple in 4 minutes, in accordance with what we would expect from our own experience.

As important as it is to understand what it means when we say that a certain limit exists, it is just as important to understand what it means when a limit fails to exist. We will again defer the technical definition of this notion to Section 2.3, but we can develop our intuitive grasp of the idea and introduce some more notation now.

Assume again that we have a function f defined on an open interval containing the point c, except possibly at c itself. The limit of $f(x)$ as x approaches c then fails to exist if the values of $f(x)$ do not approach any particular real number L. This could occur for various reasons: $f(x)$ might be getting larger and larger (in absolute value), the values of $f(x)$ may "jump" as x passes from one side of c to the other, or the values of $f(x)$ may simply not "settle down" to any particular value no matter how narrowly we focus on the region around c. Figure 4 illustrates these three possibilities.

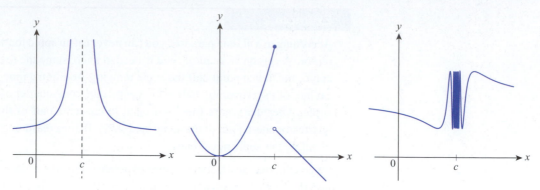

Figure 4 Failure of a Limit to Exist

While potentially confusing, we use the notation

$$\lim_{x \to c} f(x) = \infty \quad \text{or} \quad \lim_{x \to c} f(x) = -\infty$$

to indicate the nonexistence of a limit when it fails to exist because the function f either increases without bound (approaches positive infinity) or decreases without bound (approaches negative infinity) as $x \to c$. It should be remembered, however, that in either scenario the limit does not exist since f is not approaching a fixed real number. The notation simply describes the particular *way* in which the limit fails to exist.

TOPIC 2   Vertical Asymptotes and One-Sided Limits

You have encountered vertical asymptotes before; several examples have already appeared in this text in the review of rational functions and some trigonometric functions such as tangent. The topic under discussion now is not so much the vertical asymptote itself, but the language and notation we use to describe the behavior of a function *near* a vertical asymptote.

A distinguishing feature of a vertical asymptote is that, at least from one side, the function under discussion grows in magnitude without bound. Our first task is to introduce notation that succinctly and accurately describes such behavior. Assume the function f is defined on an open interval with right endpoint c. We write

$$\lim_{x \to c^-} f(x) = \infty$$

if the values of $f(x)$ increase without bound as x approaches the point c from the left. Similarly, we write

$$\lim_{x \to c^-} f(x) = -\infty$$

if the values of $f(x)$ decrease without bound (approach negative infinity) as x approaches the point c from the left. Figure 5 illustrates these two scenarios.

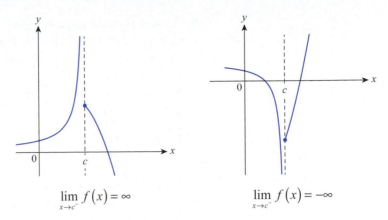

$$\lim_{x \to c^-} f(x) = \infty \qquad\qquad \lim_{x \to c^-} f(x) = -\infty$$

Figure 5 Asymptotic Behavior from the Left

Similarly, we use the notation $\lim\limits_{x \to c^+} f(x)$ in discussing the limit of the function f as x approaches the point c from the right, with the implicit assumption that f is defined on an open interval whose *left* endpoint is c. And as noted before, we need to be careful in interpreting the four statements $\lim\limits_{x \to c^-} f(x) = \infty$, $\lim\limits_{x \to c^-} f(x) = -\infty$, $\lim\limits_{x \to c^+} f(x) = \infty$, and $\lim\limits_{x \to c^+} f(x) = -\infty$; all four indicate that these *one-sided* limits of f at c do not exist, since f is either increasing or decreasing without bound.

You may be wondering why the clause "defined on an open interval" keeps appearing in these limit discussions. The reason is that in order for us to study the limit of a function as x approaches a point c, we need to know that f is defined at all points near c (but not necessarily at c). The exact meaning of *near* will vary from one application to another, but by stipulating that f is defined on an open interval around c, or to one side or the other in the case of one-sided limits, we can be sure that $f(x)$ always represents a real number as x approaches c.

Example 3 ✐

Use one-sided limit notation to describe the behavior of the following functions near their vertical asymptote(s).

a. $f(x) = \dfrac{-x}{x-1}$ **b.** $g(x) = \dfrac{x^2+1}{x^2-x-6}$ **c.** $h(x) = \dfrac{1}{(x-1)^2}$

Solution

a. As we can see from the graph in Figure 6, the sole vertical asymptote occurs at $x = 1$, a fact we can deduce from the formula as well. A further examination of the graph reveals that as x approaches 1 from the left, the function values increase without bound, a fact we can express as follows:

$$\lim_{x \to 1^-} f(x) = \infty$$

On the other hand, as x approaches 1 from the right, the function values decrease:

$$\lim_{x \to 1^+} f(x) = -\infty$$

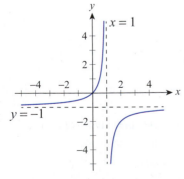

Figure 6 $f(x) = \dfrac{-x}{x-1}$

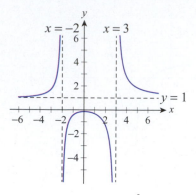

Figure 7 $g(x) = \dfrac{x^2 + 1}{x^2 - x - 6}$

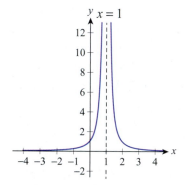

Figure 8 $h(x) = \dfrac{1}{(x-1)^2}$

b. As seen from the graph in Figure 7, $g(x)$ has two vertical asymptotes, at $x = -2$ and $x = 3$, a fact we can confirm algebraically by factoring the denominator of $g(x)$. The function values increase without bound as x approaches -2 from the left and decrease in a similar manner as x approaches -2 from the right. Therefore, we can write $\lim\limits_{x \to -2^-} g(x) = \infty$ and $\lim\limits_{x \to -2^+} g(x) = -\infty$.

Examining the graph on both sides of the vertical asymptote $x = 3$ in a similar manner, we conclude that $\lim\limits_{x \to 3^-} g(x) = -\infty$ and $\lim\limits_{x \to 3^+} g(x) = \infty$.

c. Examining the graph of $h(x)$ in Figure 8, we see that the behavior is the same on both sides of the vertical asymptote $x = 1$; namely, that of the function values increasing without bound as x approaches 1. This makes it unnecessary to distinguish between the two one-sided limits in this case; instead, we can simply write $\lim\limits_{x \to 1} h(x) = \infty$.

Example 4 ✎

Deduce the one-sided infinite limits of $f(x) = 1/(3 - x)$ at its vertical asymptote without graphing the function.

Solution

We see from the formula that the sole vertical asymptote of $f(x)$ occurs at $x = 3$ (which is where the denominator is 0). We will examine the two one-sided limits separately. Let us assume first that x is approaching 3 from the left. This means that the x-values are moving closer and closer to $x = 3$ through values less than 3 (though approaching it in magnitude). For example, when $x = 2.99$,

$$f(2.99) = \frac{1}{3 - 2.99}$$

$$= \frac{1}{0.01} = 100.$$

Throughout this process of x approaching 3, $f(x)$ stays positive, but its denominator decreases; thus the function values themselves increase in magnitude.

A couple of easy calculations underscore this observation:

$$f(2.999) = \frac{1}{3 - 2.999} = \frac{1}{0.001} = 1000,$$

$$f(2.9999) = \frac{1}{3 - 2.9999} = \frac{1}{0.0001} = 10,000,$$

and so on. The closer x is to 3, the greater the function value $f(x)$ is in magnitude (remember, x never actually reaches the value of 3).

We can now summarize our observations by writing

$$\lim\limits_{x \to 3^-} \frac{1}{3 - x} = \infty.$$

If, on the other hand, x approaches 3 from the right, the denominator and thus the function value becomes negative. As an example,

$$f(3.01) = \frac{1}{3 - 3.01}$$

$$= \frac{1}{-0.01} = -100.$$

It is still true that the closer x is to 3, the greater the function values are in magnitude, but this time, they are all negative. For example, as you can easily check,

$$f(3.001) = -1000,$$
$$f(3.0001) = -10,000,$$

and so on. As before, we can express this by writing

$$\lim_{x \to 3^+} f(x) = -\infty.$$

We can briefly summarize the above as follows: the presence of a vertical asymptote $x = 3$ indicates that the values of $f(x)$ increase or decrease without bound near $x = 3$.

In general, these tendencies may or may not be different on the two sides of the asymptote, that is, for $x < 3$ and $x > 3$, respectively. We were, however, able to find out whether the respective one-sided limits were $+\infty$ or $-\infty$ from the sign of the expression as x approached 3 from a fixed side, even though a graph was not readily available.

With two-sided and one-sided limit notation in hand, we can now provide a more formal definition of vertical asymptote.

Definition 💡

Vertical Asymptote

We say that a function f has a **vertical asymptote** at the point c if at least one of the following statements holds:

$$\lim_{x \to c^-} f(x) = \infty \qquad \lim_{x \to c^+} f(x) = \infty \qquad \lim_{x \to c} f(x) = \infty$$

$$\lim_{x \to c^-} f(x) = -\infty \qquad \lim_{x \to c^+} f(x) = -\infty \qquad \lim_{x \to c} f(x) = -\infty$$

And now that one-sided limit notation has been introduced, it is easy to extend its meaning. We write

$$\lim_{x \to c^-} f(x) = L \quad \text{or} \quad \lim_{x \to c^+} f(x) = L$$

if the values of $f(x)$ get closer and closer to L as x approaches c from, respectively, the left or right. The next example illustrates the use of this notation.

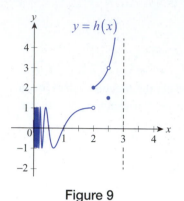

Figure 9

Example 5 ✐

Use the graph of the function h in Figure 9 to decide whether the given one-sided limits exist. For those that do, find their values.

a. $\displaystyle\lim_{x\to 0^+} h(x)$ **b.** $\displaystyle\lim_{x\to 2^-} h(x)$ **c.** $\displaystyle\lim_{x\to 2^+} h(x)$

d. $\displaystyle\lim_{x\to 2.5^-} h(x)$ **e.** $\displaystyle\lim_{x\to 2.5^+} h(x)$ **f.** $\displaystyle\lim_{x\to 3^-} h(x)$

Solution

a. We see from the graph that $h(x)$ oscillates "wildly" between $+1$ and -1 on the right-hand side of $x = 0$. Moreover, the closer x moves to 0, the more oscillations we have; in fact, there are infinitely many oscillations near 0. (Our technology is incapable of graphing infinitely many "waves," which is the reason why we see a "shaded rectangle" immediately to the right of the y-axis.) Therefore, there is no single value being approached by the values of $h(x)$, so we conclude that $\displaystyle\lim_{x\to 0^+} h(x)$ does not exist.

b. As x approaches 2 from the left, the function values are approaching 1, so we conclude

$$\lim_{x\to 2^-} h(x) = 1.$$

Notice, however, that the actual function value is $h(2) = 2$. In general, we cannot expect the limit of a function at c to always equal the function value at c. In other words, as we previously observed in Example 1, what a function does *near c* may not be the same as what it does *at c*.

c. Examining the tendency of the function values as x approaches 2 from the right, we see that

$$\lim_{x\to 2^+} h(x) = h(2) = 2.$$

In other words, the one-sided limit and the function value agree in this case.

d–e. The graph shows that the function values $h(x)$ approach 3 as x approaches 2.5 from either direction, so we can write

$$\lim_{x\to 2.5^-} h(x) = \lim_{x\to 2.5^+} h(x) = 3.$$

Notice that just like in Example 1, since both one-sided limits exist and are equal, we conclude that the *two-sided* limit exists at 2.5, and

$$\lim_{x\to 2.5} h(x) = 3.$$

The last important observation we wish to make here is that the function value at $x = 2.5$ differs from the limit, so once again

$$\lim_{x\to 2.5} h(x) \neq h(2.5).$$

f. Because $x = 3$ is a vertical asymptote, $\displaystyle\lim_{x\to 3^-} h(x)$ does not exist, but we can write

$$\lim_{x\to 3^-} h(x) = \infty.$$

TOPIC 3   **Horizontal Asymptotes and Limits at Infinity**

We have just seen examples of when it is appropriate to say that a given limit (two-sided or one-sided) *is* $\pm\infty$; we will now study limits *at* $\pm\infty$.

Just as an unbounded limit at a point c corresponds to a vertical asymptote at that point, there is a connection between horizontal asymptotes and limits at infinity. Recall that a horizontal line passing through L on the y-axis is a horizontal asymptote for the function f if the graph of f approaches the line as x is allowed to either increase or decrease without bound. A given function can have either 0, 1, or 2 horizontal asymptotes, as illustrated in Figure 10; unlike vertical asymptotes, a function can cross a horizontal asymptote any number of times (or not at all).

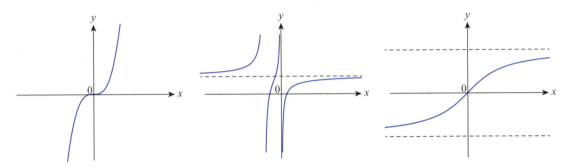

Figure 10 Limits at Infinity

And as with vertical asymptotes, the notion of limit is exactly what is needed to make the definition of horizontal asymptote more precise.

Definition 💡

Horizontal Asymptote

We say that the line $y = L$ is a **horizontal asymptote** for the function f if either of the following statements is true.

$$\lim_{x \to \infty} f(x) = L \quad \text{or} \quad \lim_{x \to -\infty} f(x) = L$$

Example 6 ✍

Identify the horizontal asymptotes of the following functions using their graphs shown in Figures 11 and 12, respectively.

a. $f(x) = \dfrac{4}{\pi} \arctan x$ **b.** $g(x) = e^{-x} \cos 5x$

Solution

a. As shown by the graph in Figure 11, the values of $f(x)$ approach the value of 2 as x increases without bound, and they will be as close to 2 as we wish if we make sure x is large enough. In other words, we can write

$$\lim_{x \to \infty} f(x) = 2$$

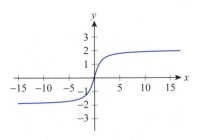

Figure 11 $f(x) = \dfrac{4}{\pi} \arctan x$

and conclude that the line $y = 2$ is a horizontal asymptote for f. Similarly, we conclude that

$$\lim_{x \to -\infty} f(x) = -2,$$

and thus the line $y = -2$ is another horizontal asymptote for f.

b. The graph of g, though oscillating, is approaching the x-axis more and more as x increases (see Figure 12). In other words, as we will soon be able to prove, the function values of $g(x)$ will be as small in absolute value as we wish, if we merely choose x large enough. This is precisely what we need to conclude that the x-axis or $y = 0$ is a horizontal asymptote for g. Using limit notation, this means

$$\lim_{x \to \infty} g(x) = 0.$$

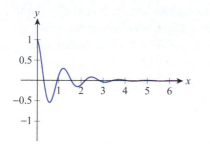

Figure 12 $g(x) = e^{-x} \cos 5x$

As a final remark, since the cosine factor causes $g(x)$ to oscillate, the graph actually crosses its horizontal asymptote infinitely often.

Many functions tend to increase in magnitude without bound as x goes to ∞ or $-\infty$; the first graph in Figure 10 is an example. In such cases, it is natural to use limit notation to provide more information on the behavior of the function. We write

$$\lim_{x \to \infty} f(x) = \infty$$

if the values of f increase without bound as $x \to \infty$, with similar meanings attached to the remaining three possibilities,

$$\lim_{x \to \infty} f(x) = -\infty, \quad \lim_{x \to -\infty} f(x) = \infty, \quad \text{and} \quad \lim_{x \to -\infty} f(x) = -\infty.$$

Example 7 ✏

Decide whether the given functions increase or decrease without bound as x goes to ∞ and $-\infty$. Use limit notation to express your conclusion.

a. $f(x) = -x^2$ **b.** $g(x) = x^3 - 7$ **c.** $h(x) = x^5 - 6x^2 + 3$

Solution

a. First, we will examine the function values $f(x)$ as x goes to ∞. Our first observation is that for any positive real number x, if $x > 1$ then $x^2 > x$, so if x tends to ∞, then x^2 will certainly grow in magnitude, without bound. Therefore, $f(x) = -x^2$ will *decrease* without bound as x tends to ∞. Using limit notation,

$$\lim_{x \to \infty} f(x) = -\infty.$$

Next, let us assume x goes to $-\infty$. Like before, x^2 will grow without bound, and since x^2 is positive for any nonzero x, we conclude that $f(x) = -x^2$ again decreases without bound. Thus,

$$\lim_{x \to -\infty} f(x) = -\infty.$$

b. We again let x go to ∞ first. Then by an argument similar to the one given in part a., x^3 grows without bound. This tendency is not influenced by subtracting 7 from x^3, so

$$\lim_{x \to \infty} g(x) = \infty.$$

If on the other hand, x goes to $-\infty$, $x^3 - 7$ will decrease without bound, since the cube of a negative number is negative. Thus we conclude that

$$\lim_{x \to -\infty} g(x) = -\infty.$$

c. As for the function h, we will first argue that its values grow without bound as x tends toward ∞. The reasoning is as follows. It is clear that the first term, x^5 increases without bound as x goes to ∞. And as x grows, the fifth-degree term increases so much faster than the quadratic term, that their difference $x^5 - 6x^2$ still increases. As an illustration, when $x = 3$, $x^5 - 6x^2 = 3^5 - 6 \cdot 9 = 243 - 54 = 189$, but when, say, $x = 10$, $x^5 - 6x^2 = 10^5 - 6 \cdot 10^2 = 100{,}000 - 600 = 99{,}400$. Adding 3 to the previous expression (or adding or subtracting any constant, for that matter) will not change the increasing tendency of the function values, so we conclude that

$$\lim_{x \to \infty} h(x) = \infty.$$

We can argue in a similar fashion in the case of x going to $-\infty$, but since the fifth power of a negative number is negative, x^5 *decreases* without bound in this case. This time, however, subtracting $6x^2$ only helps (but like before, the quadratic term would not be able to change the process, even if we added it to x^5). Again, the constant term is immaterial from the perspective of the tendency of the function values, so we conclude that

$$\lim_{x \to -\infty} h(x) = -\infty.$$

As a final remark, you may recall having discussed the "end behaviors" of polynomial functions in your precalculus course, though at that time most likely without the limit notation.

TOPIC 4   **Limits and Technology**

x	$f(x)$
0.00001	0.25
0.000001	0.2
0.0000001	0
0.00000001	0

Table 1

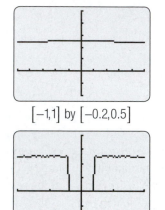

$[-1,1]$ by $[-0.2,0.5]$

$[-0.00001,0.00001]$ by $[-0.2,0.4]$

Figure 13
Exploring $\lim\limits_{x \to 0} f(x)$
Using a Graphing Calculator

To this point, we have relied upon graphs, tables, intuition, and elementary reasoning in order to arrive at guesses for limits that we feel confident about. In the next two sections, we will learn a number of mathematical facts (in the forms of theorems and lemmas) that allow us to accurately determine limits of functions without the need for guesswork. But we will occasionally make use of technological aids such as graphing calculators and computer algebra systems in order to guide our explorations. We will close this section with a few examples of the use and misuse of technology in determining limits.

Consider the function $f(x) = \dfrac{\sqrt{x^2 + 4} - 2}{x^2}$.

If we needed to determine the behavior of this function as x goes to 0, without the benefit of the tools we will soon encounter, we might take several different approaches. One approach would be to note that the denominator of the fraction goes to 0 as x goes to 0, so we might guess that the function "blows up", that is, increases in magnitude without bound. That guess would be wrong—our analysis so far misses an important fact. A closer examination of the fraction reveals that the numerator also approaches 0 as x goes to 0, and this common situation can lead to a variety of possible outcomes. A second approach would be to evaluate this function for a number of values of x that are increasingly small; this may or may not lead to a correct guess for the limit, depending on the specific values chosen for x and the technology used to calculate the values of the function (see Table 1). And a third approach would be to use a graphing calculator (see Figure 13) or computer algebra system (see Figure 14) to graph the function near 0. Again, this may or may not lead to a correct guess, depending on the viewing window specified for the graph.

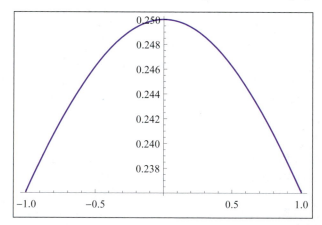
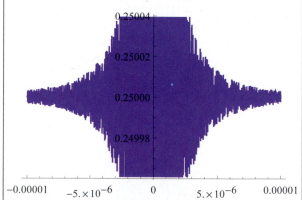

Figure 14 Exploring $\lim\limits_{x \to 0} f(x)$ Using a Computer Algebra System

What is going on, and what results should we trust? The reason we are getting conflicting guidance from our technology is a loss of *significant digits* as the operations of subtraction and division are performed. All numbers used in computation in a calculator or software program are subject to rounding errors, as only a certain number of digits are kept in memory unless specifically directed otherwise. This leads to a loss of precision that is especially problematic in subtracting one number from another that is nearly the same (as in the numerator of $f(x)$) and when calculating the ratio of two small but similar numbers.

A prolonged exploration and a weighing of all the results would probably lead us to eventually guess that the limit of this function as x approaches 0 is $\frac{1}{4}$, but we might be left with a shadow of doubt. Fortunately, the limit theorems of the next two sections will provide us the means to remove all doubt in such situations.

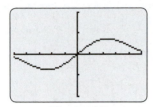

Figure 15

$$g(x) = \frac{1 - \cos x}{x}$$

on $[-5,5]$ by $[-2,2]$

x	$g(x)$
0.1	0.05
0.01	0.005
0.001	0.0005
0.0001	0.00005

Table 2

Example 8 ✎

Use graphing technology to predict the value of $\lim\limits_{x \to 0} g(x)$, where

$$g(x) = \frac{1 - \cos x}{x}.$$

Solution

Since $\cos x$ approaches 1 as x goes to 0, we once again face the ambiguity of both the numerator and denominator approaching 0 as x goes to 0. Again we don't have enough information to guess at any particular number since, as we said before, a "0/0-type" limit can lead to a variety of different answers.

Using a graphing calculator, we can sketch and examine the graph, which appears to be approaching the origin as x goes to 0 (see Figure 15). This is confirmed by a table of values that we can create with the help of a calculator, using decreasing x-values.

From the above observations it is natural to expect that the function values will steadily move toward 0 when x approaches 0, as illustrated by the first graph in Figure 16. However, if we zoom in, again something strange may happen. We used a computer algebra system to illustrate.

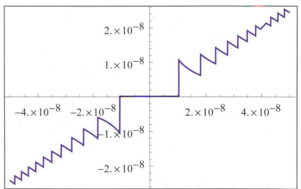

Figure 16 Exploring $\lim\limits_{x \to 0} g(x)$ Using a Computer Algebra System

Again we see that, because of rounding errors and other reasons, our technology can sometimes mislead us by giving seemingly conflicting or inaccurate feedback, and we must be aware of this when using it.

We will stick with our guess that

$$\lim_{x \to 0} g(x) = \lim_{x \to 0} \frac{1 - \cos x}{x} = 0,$$

but realize that we haven't actually proved this at all. We will learn how to do that after discussing limit theorems in upcoming sections.

Technology Note 💻

Computer algebra systems such as *Mathematica* provide additional tools for determining limits, but it should always be remembered that software has limitations and can be fooled. *Mathematica* contains the built-in command **Limit** that uses the same mathematical facts we will learn in the next two sections to correctly evaluate many types of limits. Its use is illustrated below. (For additional information on *Mathematica* and the use of the **Limit** command, see Appendix A.)

```
In[1]:= Limit[(Sqrt[x^2 + 4] - 2) / x^2, x → 0]

Out[1]= 1
         ─
         4
```

Figure 17

2.2 **Exercises**

1–4 Use the graph of the function to find the indicated limit (if it exists).

1. $\lim\limits_{x \to 3} f(x)$

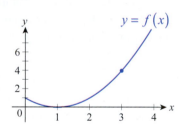

2. $\lim\limits_{x \to 1} g(x)$

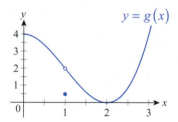

3. $\lim\limits_{x \to 0} h(x)$

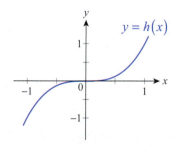

4. $\lim\limits_{x \to 1} k(x)$

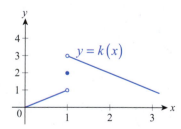

5–12 Create a table of values to estimate the value of the indicated limit without graphing the function. Choose the last *x*-value so that it is no more than 0.001 units from the given *c*-value.

5. $\lim\limits_{x \to \sqrt{2}} \dfrac{x^2 - 2}{x - \sqrt{2}}$

6. $\lim\limits_{x \to 3} \dfrac{x^3 - 9x^2 + 27x - 27}{x - 3}$

7. $\lim\limits_{x \to 1} \dfrac{x^{10} - 1}{x - 1}$

8. $\lim\limits_{x \to 0} \dfrac{4 \sin x}{3x}$

9. $\lim\limits_{x \to \pi} \dfrac{2 \cos x - 1}{1 - \sin x}$

10. $\lim\limits_{x \to 7^-} \dfrac{x^2 - 49}{x - 7}$

11. $\lim\limits_{x \to 7^+} \dfrac{x^2 + 49}{x - 7}$

12. $\lim\limits_{x \to 0^+} \dfrac{\sqrt{4 + x}}{x}$

13–24 Use one-sided limit notation to describe the behavior of the function near its vertical asymptote(s).

13.

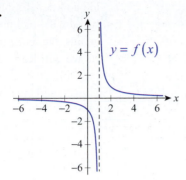

14.

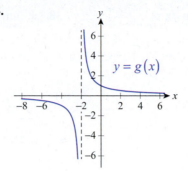

15.

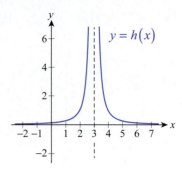

16.

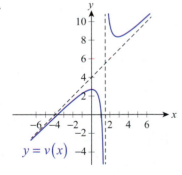

17.

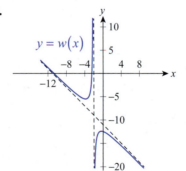

18.

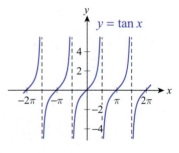

19.

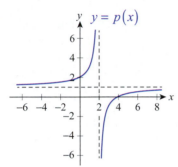

20.

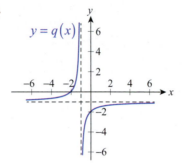

21.

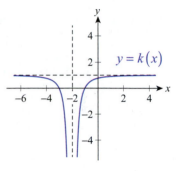

22.

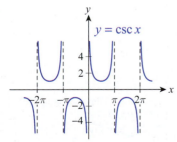

23.

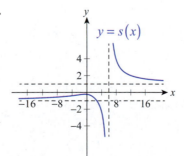

24.

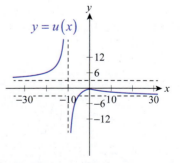

25–36. Consider the functions given in Exercises 13–24. Find their limits at ∞ and $-\infty$ (if they exist). When applicable, use the horizontal asymptote(s) as a guide.

37–46 Use the graph to find the indicated one-sided limits, if they exist.

37. **a.** $\lim\limits_{x \to 2^-} f(x)$ **b.** $\lim\limits_{x \to 2^+} f(x)$ **38.** **a.** $\lim\limits_{x \to 1^-} g(x)$ **b.** $\lim\limits_{x \to 1^+} g(x)$

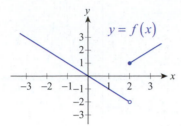

39. **a.** $\lim\limits_{x \to 1^-} h(x)$ **b.** $\lim\limits_{x \to 1^+} h(x)$ **40.** **a.** $\lim\limits_{x \to (-\pi/2)^-} F(x)$ **b.** $\lim\limits_{x \to (-\pi/2)^+} F(x)$

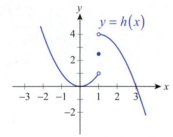

41. **a.** $\lim\limits_{x \to 0^-} G(x)$ **b.** $\lim\limits_{x \to 0^+} G(x)$ **42.** **a.** $\lim\limits_{x \to -1^-} k(x)$ **b.** $\lim\limits_{x \to -1^+} k(x)$

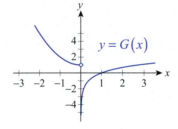

43. **a.** $\lim\limits_{x \to 0^-} H(x)$ **b.** $\lim\limits_{x \to 0^+} H(x)$ **44.** **a.** $\lim\limits_{x \to 2^-} u(x)$ **b.** $\lim\limits_{x \to 2^+} u(x)$

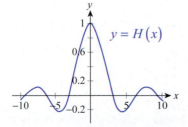

45. a. $\lim\limits_{x\to0^-} v(x)$ **b.** $\lim\limits_{x\to0^+} v(x)$

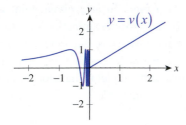

46. a. $\lim\limits_{x\to0^-} w(x)$ **b.** $\lim\limits_{x\to0^+} w(x)$

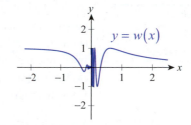

47–58 Use limit notation to describe the unbounded behavior of the given function as x approaches ∞ and/or $-\infty$.

47. $f(x)=x^3$

48. $g(x)=x^2+2.1x-1$

49. $h(x)=-x^4+0.2x^3$

50. $k(x)=-0.35x^5+x+1.35$

51. $F(x)=\sqrt{x+2}$

52. $G(x)=\sqrt[3]{x+1}-2.3$

53. $H(x)=|x+2|$

54. $K(x)=-|x+2|-1$

55. $u(x)=|x-1|+|x+2|$

56. $v(x)=e^{x+2}$

57. $s(x)=-10^{-x}+1$

58. $t(x)=\ln x-1$

59–63 *True or False?* Determine whether the given statement is true or false. In case of a false statement, explain or provide a counterexample.

59. If $\lim\limits_{x\to c} f(x)$ does not exist, then $f(x)$ is undefined at $x=c$.

60. If $f(x)$ is undefined at $x=c$, then $\lim\limits_{x\to c} f(x)$ does not exist.

61. If $f(x)$ is defined on $(0,\infty)$ and $y=0$ is a horizontal asymptote for $f(x)$, then there exists a number M such that if $x>M$ then $f(x)<1/10^6$.

62. If $f(x)$ has a vertical asymptote at $x=c$, then either $\lim\limits_{x\to c} f(x)=\infty$ or $\lim\limits_{x\to c} f(x)=-\infty$.

63. If $\lim\limits_{x\to c} f(x)$ does not exist, then
$\lim\limits_{x\to c^-} f(x)\ne\lim\limits_{x\to c^+} f(x)$ or at least one of $\lim\limits_{x\to c^-} f(x)$ or $\lim\limits_{x\to c^+} f(x)$ does not exist.

2.2 **Technology Exercises**

64–71 Use a graphing calculator or computer algebra system to decide whether the given limit exists by evaluating the function at several x-values approaching the indicated c-value. Then graph the function to confirm your findings. Do you obtain misleading graphs when choosing small viewing windows?

64. $\lim\limits_{x\to2} \dfrac{x^2-5x+6}{x-2}$

65. $\lim\limits_{x\to3} \dfrac{x-2}{x^2-5x+6}$

66. $\lim\limits_{x\to-1.5} \dfrac{x^2-2.25}{x+1.5}$

67. $\lim\limits_{x\to1} \dfrac{x-1}{\sqrt{x}-1}$

68. $\lim\limits_{x\to3} \dfrac{\sqrt{x+1}-2}{x-3}$

69. $\lim\limits_{x\to0} \dfrac{\sin 3x}{2x}$

70. $\lim\limits_{x\to0^+} \cos\dfrac{1}{x}$

71. $\lim\limits_{x\to0^+} x\cos\dfrac{1}{x}$

72. Evaluate the function $f(x)=\left(1+\dfrac{1}{x}\right)^x$ for several consecutive positive integers, and try to observe a tendency. Then use a graphing calculator or computer algebra system to graph $f(x)$ in a large viewing window and try to guess $\lim\limits_{x\to\infty} f(x)$. Have you seen that number before?

73. Write a program in a graphing calculator or computer algebra system to estimate the limit of an input function as x approaches c. (Calculate $f(x)$ successively at x-values increasingly close to c and display the results.) Try your program on Exercises 64–71.

74–81. Use the Limit command specific to your computer algebra system to evaluate the limits in Exercises 64–71. Are your previous results confirmed?

2.3 **The Mathematical Definition of Limit**

TOPICS

1. Limits formally defined

2. Proving a limit exists

3. Proving a limit does not exist

As we have seen, determining the limit of a function at a given point is not necessarily a trivial task. Pictures, tables, or a cursory examination of a formula (if one is available) may not be sufficient to allow us to accurately evaluate a limit, and technological aids can also lead us astray. In this section, we present a rigorous mathematical definition of *limit* that will then let us develop an array of tools that make limit evaluation both precise and tractable.

TOPIC 1 **Limits Formally Defined**

We still have only an informal notion of what the statement $\lim\limits_{x \to c} f(x) = L$ means. Now that our limit intuition has begun to develop, we can refine our informal notion and make the definition precise and mathematically useful. To be exact, we need to capture the idea that all the values of $f(x)$ become as close to L as we care to specify as x gets close enough to c.

The two parts of that previous sentence that need more work are "as close to L as we care to specify" and "as x gets close enough to c." The most common definition of limit refines those two phrases through the use of two Greek letters, epsilon (ε) and delta (δ), and the following is often referred to as the *epsilon-delta* definition of limit.

Definition ⚲

Formal Definition of Limit

Let f be a function defined on an open interval containing c, except possibly at c itself. We say that the **limit of $f(x)$ as x approaches c is L**, and write $\lim\limits_{x \to c} f(x) = L$, if for every number $\varepsilon > 0$ there is a number $\delta > 0$ such that $|f(x) - L| < \varepsilon$ whenever x satisfies $0 < |x - c| < \delta$.

The use of the letters ε and δ originated with the French mathematician Augustin-Louis Cauchy (1789–1857). His choice was deliberate, and served to remind the reader of the words *error* and *difference* (the French spellings are similar); you may find this correspondence useful as well. We may now interpret the statement $\lim\limits_{x \to c} f(x) = L$ in this way: if we wish to guarantee that the *error* between $f(x)$ and L is less than the amount ε, it suffices to make sure that the *difference* between x and c is less than the amount δ (but remember that we do not allow x to equal c).

Figure 1 shows the relationship between L, c, ε, and δ for a given function f. The successive images illustrate the way in which δ depends on ε: as we specify smaller values for the error ε, the difference δ correspondingly shrinks. But in each image, any x chosen between $c - \delta$ and $c + \delta$ (except c itself) results in a value for $f(x)$ that is between $L - \varepsilon$ and $L + \varepsilon$.

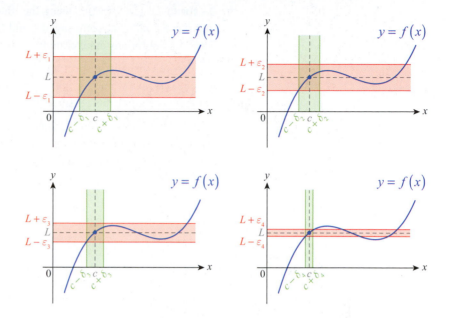

Figure 1   Correspondence between δ and ε

The true value of the formal limit definition is that it allows us to prove limit claims absolutely. In fact, you might consider the correspondence between ε and δ as a series of challenges and responses. If a skeptic wished to challenge the claim that $\lim_{x \to c} f(x) = L$ for the function f shown in Figure 1, each of the four ε's might be presented in turn with the demand "tell me a condition that guarantees $f(x)$ is within ε of L." The response, for each ε, would be the corresponding δ: if $x \neq c$ is chosen between $c - \delta$ and $c + \delta$ then $f(x)$ is guaranteed to fall between $L - \varepsilon$ and $L + \varepsilon$.

Note that the δ in each image is actually the largest possible choice—any positive number smaller than the pictured δ would also serve to guarantee $\left| f(x) - L \right| < \varepsilon$. Also note that, for this function and for the four ε's shown, the largest possible δ is dictated by where the line $y = L - \varepsilon$ intersects the graph of f. In the first and second images, in particular, it is clear that x could be selected from a longer interval to the right of c and still force $\left| f(x) - L \right| < \varepsilon$. Any larger δ than that shown would not suffice, however, for points selected to the left of c.

Example 1 ✎

In our first example, we will use the graph of $f(x) = -0.9x^2 + 2$ with $c = 0.5$ to estimate the value of δ corresponding to **a.** $\varepsilon = 1$, **b.** $\varepsilon = 0.5$, **c.** $\varepsilon = 0.25$, and **d.** $\varepsilon = 0.1$.

Solution

a. Like in our discussion before, we see that for $\varepsilon = 1$, the value of δ is dictated by where the line $y = L - \varepsilon$ intersects the graph of f (see Figure 2). If, for example, we choose $\delta = 0.65$, we can ensure that for any $x \neq c$ between $c - \delta$ and $c + \delta$, $f(x)$ will fall between $L - \varepsilon$ and $L + \varepsilon$.

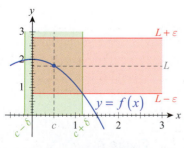

Figure 2

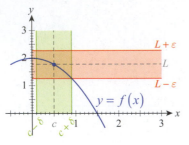

Figure 3

b. As for the case of $\varepsilon = 0.5$, using the illustration in Figure 3, it is clear that if $x \neq c$ falls between, say, 0.1 and 0.9, then $L - \varepsilon < f(x) < L + \varepsilon$. Thus $\delta = 0.4$ (or any smaller number) is a good choice.

c. As before, visually inspecting the graph in Figure 4 leads to the conclusion that when $\varepsilon = 0.25$, the value of $\delta = 0.2$ (or any smaller number) is a good choice, since 0.2 is less than the distance between c and either of the green vertical lines.

d. Again, a visual inspection reveals that $\varepsilon = 0.1$ will require a much smaller δ; $\delta = 0.1$ (or any smaller number) works well (see Figure 5).

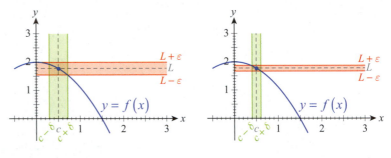

Figure 4   **Figure 5**

Example 2 ✐

Consider the following function:

$$g(x) = \begin{cases} x^3 & \text{if } x \neq 0.7 \\ 1 & \text{if } x = 0.7 \end{cases}$$

The empty and full circles in the graph in Figure 6 reflect the fact that x^3 has been redefined at $x = 0.7$: the coordinates of the "hole" are $\left(0.7, (0.7)^3\right) = (0.7, 0.343)$, while $g(0.7) = 1$.

In addition, we can also learn from Figure 6 that there is a $\delta > 0$ corresponding to $\varepsilon = 0.3$ so that if we choose any $x \neq c$ from the interval $(c - \delta, c + \delta)$, the corresponding g-value will fall within ε of $(0.7)^3 = 0.343$.

Figures 7, 8, and 9 are illustrations of the existence of δ-values corresponding to smaller and smaller ε's:

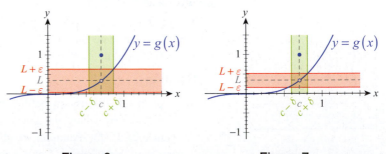

Figure 6   **Figure 7**

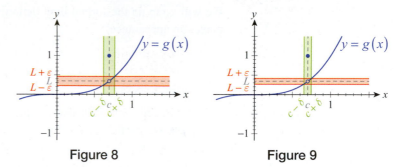

Figure 8   **Figure 9**

From these illustrations, it is at least visually clear that

$$\lim_{x \to 0.7} g(x) = (0.7)^3 = 0.343,$$

even though the function value $g(0.7) = 1$. In other words,

$$\lim_{x \to 0.7} g(x) \neq g(0.7);$$

the limit and the function value at $c = 0.7$ are not equal. More is actually true: we could define $g(0.7)$ to be *any* real number, without affecting the limit. In general, the function value $g(c)$ and the limit $\lim_{x \to c} g(x)$ are entirely independent of each other, that is, the existence and/or value of one does not affect that of the other.

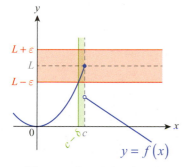

Figure 10 Left-Hand Limit

Our formal definitions for the other varieties of limits are similar. Figures 10 and 11 illustrate the construction of the largest δ corresponding to a given ε when one-sided limits are under consideration.

Definition 💡

Limits from the Left (Left-Hand Limits)

Let f be a function defined on an open interval whose right endpoint is c. We say that the **limit of $f(x)$ as x approaches c from the left is L**, and write $\lim_{x \to c^-} f(x) = L$, if for every number $\varepsilon > 0$ there is a number $\delta > 0$ such that $|f(x) - L| < \varepsilon$ whenever x satisfies $c - \delta < x < c$.

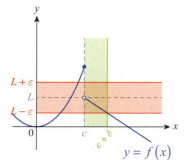

Figure 11 Right-Hand Limit

Definition 💡

Limits from the Right (Right-Hand Limits)

Let f be a function defined on an open interval whose left endpoint is c. We say that the **limit of $f(x)$ as x approaches c from the right is L**, and write $\lim_{x \to c^+} f(x) = L$, if for every number $\varepsilon > 0$ there is a number $\delta > 0$ such that $|f(x) - L| < \varepsilon$ whenever x satisfies $c < x < c + \delta$.

We will wrap up the formal limit definitions with cases in which the symbol ∞ makes an appearance.

Definition 💡

Infinite Limits

Let f be a function defined on an open interval containing c, except possibly at c itself. We say that the **limit of $f(x)$ as x approaches c is positive infinity**, and write $\lim_{x \to c} f(x) = \infty$, if for every positive number M there is a number $\delta > 0$ such that $f(x) > M$ whenever x satisfies $0 < |x - c| < \delta$.

Similarly, we say that the **limit of $f(x)$ as x approaches c is negative infinity**, and write $\lim_{x \to c} f(x) = -\infty$, if for every negative number N there is a number $\delta > 0$ such that $f(x) < N$ whenever x satisfies $0 < |x - c| < \delta$. **Infinite one-sided limits** are defined in a manner analogous to finite one-sided limits.

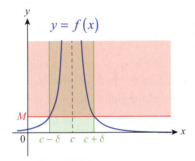

Figure 12

Correspondence between M and δ for an Infinite Limit

If we wish to prove that a given function has an infinite limit at a point c, our task is to prove that given some (presumably large) number M, there is an interval $(c - \delta, c + \delta)$ on which the function is larger than M (except possibly at the point c itself). Figure 12 shows how an appropriate δ can be chosen for the given value of M.

Finally, limits at infinity are defined formally as follows.

Definition 💡

Limits at Infinity

Let f be a function defined on some interval (a, ∞). We say the **limit of f at infinity is L**, and write $\lim_{x \to \infty} f(x) = L$, if for every number $\varepsilon > 0$ there is a number N such that $|f(x) - L| < \varepsilon$ for all $x > N$.

Similarly, for a function defined on some interval $(-\infty, b)$ we say the **limit of f at negative infinity is L**, and write $\lim_{x \to -\infty} f(x) = L$, if for every number $\varepsilon > 0$ there is a number N such that $|f(x) - L| < \varepsilon$ for all $x < N$.

Be sure you understand how these formal definitions of limits at infinity relate to horizontal asymptotes.

TOPIC 2   Proving a Limit Exists

Although the formal definition of limit gives us the ability to prove limit claims beyond a shadow of a doubt, it does not immediately give us the means by which to determine the value of a given limit in the first place. In other words, we need to know what L is before we can prove that $\lim_{x \to c} f(x) = L$. Fortunately, as we will soon see, our limit definition is the basis for a number of theorems that will make determination of limits much easier.

As a stepping-stone toward those theorems, we will develop our ability to use the limit definition to prove some example claims.

Example 3 ✐

Use the ε-δ definition of limit to prove that $\lim\limits_{x\to 3}(2x-1)=5$.

Solution

Suppose an $\varepsilon > 0$ is given. We need to find a $\delta > 0$ such that

$$0<\left|x-3\right|<\delta \;\;\Rightarrow\;\; \left|(2x-1)-5\right|<\varepsilon.$$

Notice that this latter inequality is equivalent to the following:

$$\left|2x-6\right|<\varepsilon \;\;\Leftrightarrow\;\; \left|2(x-3)\right|<\varepsilon$$
$$\Leftrightarrow\;\; \left|2\right|\cdot\left|x-3\right|<\varepsilon$$
$$\Leftrightarrow\;\; 2\left|x-3\right|<\varepsilon$$
$$\Leftrightarrow\;\; \left|x-3\right|<\frac{\varepsilon}{2}$$

Reading the above chain of equivalent inequalities from the bottom up, this means that $\delta = \varepsilon/2$ (or any smaller number) works. In other words, given $\varepsilon > 0$, if we choose $\delta = \varepsilon/2$, then $0<\left|x-3\right|<\delta$ implies

$$\left|(2x-1)-5\right|=\left|2x-6\right|=\left|2(x-3)\right|=2\left|x-3\right|<2\delta=2\cdot\frac{\varepsilon}{2}=\varepsilon,$$

just as we needed to show.

What all this means is that if a skeptic were to challenge you with a small $\varepsilon > 0$ of his or her choosing, you could always respond, for each ε, by picking $\delta = \varepsilon/2$: if $x \neq c$ is chosen between $3 - \delta$ and $3 + \delta$, then $2x - 1$ is guaranteed to fall between $5 - \varepsilon$ and $5 + \varepsilon$. For example, given $\varepsilon = 0.1$, let $\delta = 0.05$. For the reason why x needs to be "twice as close" to 3 as the skeptic's challenge, examine the graph in Figure 13, and note that the slope of the line is 2.

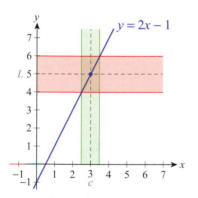

Figure 13

Example 4 ✐

Use the ε-δ definition of limit to prove that $\lim\limits_{x\to 4}x^2=16$.

Solution

Again, supposing that $\varepsilon > 0$ is given, we need to find a $\delta > 0$ such that

$$0<\left|x-4\right|<\delta \;\;\Rightarrow\;\; \left|x^2-16\right|<\varepsilon.$$

Our first observation is that

$$\left|x^2-16\right|=\left|(x+4)(x-4)\right|=\left|x+4\right|\cdot\left|x-4\right|.$$

Next, we will agree to choose a $\delta > 0$ while also making sure that $\delta < 1$. Note that we can do this, since once a successful $\delta > 0$ is chosen, any smaller number works just fine.

Since $|x - 4| < \delta$ translates into $4 - \delta < x < 4 + \delta$, which implies $3 < x < 5$ if $\delta < 1$, we have $|x + 4| = x + 4 < 5 + 4 = 9$ and thus

$$|x^2 - 16| = |x + 4| \cdot |x - 4| < 9|x - 4|.$$

So for the given $\varepsilon > 0$, let $\delta > 0$ be chosen so that it is less than the smaller of the numbers 1 and $\varepsilon/9$.

Then if $|x - 4| < \delta$,

$$|x^2 - 16| = |x + 4| \cdot |x - 4| < 9|x - 4| < 9\delta < 9 \cdot \frac{\varepsilon}{9} = \varepsilon,$$

this latter inequality finishing our proof.

Example 5 ✎

Use the ε-δ definition of limit to prove that $\displaystyle\lim_{x \to \infty} \frac{x^2 + 1}{x^2} = 1$.

Solution

As before, let $\varepsilon > 0$ be given. We need to find a number N such that

$$x > N \quad \Rightarrow \quad \left| \frac{x^2 + 1}{x^2} - 1 \right| < \varepsilon.$$

Notice that since $\dfrac{x^2 + 1}{x^2} = 1 + \dfrac{1}{x^2}$, this latter inequality is equivalent to

$$\left| \left(1 + \frac{1}{x^2}\right) - 1 \right| < \varepsilon \quad \Leftrightarrow \quad \left| \frac{1}{x^2} \right| < \varepsilon$$

$$\Leftrightarrow \quad \frac{1}{x^2} < \varepsilon,$$

since x^2 is nonnegative. Also observe that if we choose N large enough so that $1/N^2 < \varepsilon$, then $x > N$ will imply $1/x^2 < 1/N^2 < \varepsilon$. The above choice is always possible, by simply making sure $N > 1/\sqrt{\varepsilon}$.

We can summarize our observations as follows. For a given $\varepsilon > 0$, choose and fix a positive number N satisfying $N > 1/\sqrt{\varepsilon}$. Then if $x > N$,

$$\left| \frac{x^2 + 1}{x^2} - 1 \right| = \left| 1 + \frac{1}{x^2} - 1 \right| = \left| \frac{1}{x^2} \right| = \frac{1}{x^2} < \frac{1}{N^2} < \frac{1}{\left(\frac{1}{\sqrt{\varepsilon}}\right)^2} = \frac{1}{\frac{1}{\varepsilon}} = \varepsilon,$$

and this latter chain of inequalities finishes our argument.

Example 6 ✐

Prove that $\lim\limits_{x \to 2} \dfrac{1}{(x-2)^2} = \infty$.

Solution

For an arbitrary, but fixed, positive number M, we need to exhibit a number $\delta > 0$ such that $1/(x-2)^2 > M$ whenever x satisfies $0 < |x-2| < \delta$. To that end, let us assume that $M > 0$ is fixed, and choose a $\delta > 0$ small enough so that $1/\delta^2 \geq M$. Note that this is possible; for example, $\delta = 1/\sqrt{M}$ (or any smaller number) works: if $0 < |x-2| < \delta$, then

$$\frac{1}{(x-2)^2} > \frac{1}{\delta^2} = \frac{1}{\left(\dfrac{1}{\sqrt{M}}\right)^2} = \frac{1}{\dfrac{1}{M}} = M,$$

which is precisely the inequality we wanted.

As a final remark, we wish to emphasize that $\lim\limits_{x \to 2} \dfrac{1}{(x-2)^2}$ actually does not exist, since ∞ is not a number. What is happening is that the limit fails to exist because the function $f(x) = 1/(x-2)^2$ grows without bound as x approaches 2. We have established this using techniques as in the previous examples, and expressed the same fact using limit notation and the ∞ symbol.

TOPIC 3   Proving a Limit Does Not Exist

We will close this section with an illustration of how we can use our limit definition to prove that a given function does *not* have a limit at a given point. There are occasions when the ability to prove that something does not happen is just as useful as the ability to prove that it does.

Example 7 ✐

Prove that for the function $f(x) = \cos(1/x)$, $\lim\limits_{x \to 0} f(x)$ does not exist by showing that the function does not satisfy the ε-δ definition for any possible L at $c = 0$.

Solution

We will first need to think carefully about what it means for the definition to *fail* at $c = 0$. This will happen precisely when no number L works as the limit of the function at 0, in other words, when x can move closer to 0 than *any* given $\delta > 0$ without the corresponding function values $f(x)$ approaching any number. You can think about this, too, as a challenge game. Suppose a skeptic challenges you with a very small number δ, and you are able to show that the function values $f(x)$, corresponding to nonzero x-values with $-\delta < x < \delta$, do not approach any number L. If you are always able to respond to the challenge, no matter how small $\delta > 0$ becomes, you will have proved that $\lim\limits_{x \to 0} f(x)$ does not exist. This is exactly what we endeavor to do in this example.

Suppose that $\delta > 0$ is given. Recall that $\cos 2k\pi = 1$ and $\cos((2k+1)\pi) = -1$ for all $k \in \mathbb{Z}$. Choose and fix a big enough positive integer k such that $1/(2k\pi) < \delta$ (since δ is already fixed, you can achieve this by merely choosing k big enough).

Next, we will use the real numbers $x_1 = \dfrac{1}{2k\pi}$ and $x_2 = \dfrac{1}{(2k+1)\pi}$ as follows. Note that both are less than δ in magnitude, but

$$f(x_1) = \cos\left(\frac{1}{x_1}\right)$$

$$= \cos\left(\frac{1}{\dfrac{1}{2k\pi}}\right)$$

$$= \cos 2k\pi = 1,$$

while

$$f(x_2) = \cos\left(\frac{1}{x_2}\right)$$

$$= \cos\left(\frac{1}{\dfrac{1}{(2k+1)\pi}}\right)$$

$$= \cos((2k+1)\pi) = -1.$$

What this means is that we were able to find x-values less than δ in magnitude, namely x_1 and x_2, for which the corresponding function values $f(x_1)$ and $f(x_2)$ are a full 2 units apart. More importantly, we can do the same, no matter how small a $\delta > 0$ is specified. Thus the values of $f(x)$ cannot be approaching any limit L at all.

To put our argument on a more precise footing, we will show that the definition for the existence of $\lim_{x \to 0} f(x)$ fails, by showing that there is an $\varepsilon > 0$ for which no $\delta > 0$ exists to satisfy the definition.

Let us pick, say, $\varepsilon = 1$, and suppose that there is a $\delta > 0$ such that for some L,

$$|f(x) - L| < 1,$$

whenever $0 < |x - 0| < \delta$. Then proceed to find x_1, x_2 as above. Since both $|x_1| < \delta$ and $|x_2| < \delta$, by assumption we have first of all the inequality

$$|f(x_1) - L| = |1 - L| < 1.$$

What this means is that L is less than 1 unit away from 1; in particular, L is positive.

On the other hand, because of the choice of x_2, we also have the inequality

$$|f(x_2) - L| = |-1 - L| < 1;$$

in other words, L is less than 1 unit from -1 and therefore is negative.

Since such a number does not exist, $\lim_{x \to 0} f(x)$ cannot exist either.

2.3 Exercises

1–4 Use the graph to estimate δ corresponding to the given ε satisfying the ε-δ definition of $\lim_{x \to c} f(x) = L$.

1.

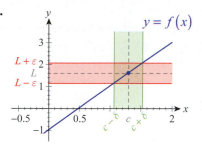

2.

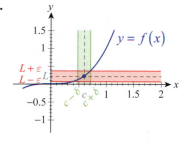

3.

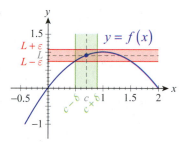

4.

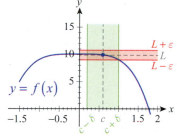

5–10 Calculus students gave the following definitions for the existence of a limit of $f(x)$ at c. Find and correct any errors.

5. "$\lim_{x \to c} f(x)$ exists if for any $\varepsilon > 0$ and real number L there is a $\delta > 0$ such that $0 < |x - L| < \delta$ implies $|f(x) - c| < \varepsilon$."

6. "$\lim_{x \to c} f(x)$ exists and equals L if for any $\varepsilon > 0$ and $\delta > 0$ whenever $0 < |x - c| < \varepsilon$, we have $|f(x) - L| < \delta$."

7. "If there is a real number L such that for an $\varepsilon > 0$ there is a $\delta > 0$ such that whenever $|x - c| < \delta$ and $x \neq c$, we have $|f(x) - L| < \varepsilon$, we say that the limit of the function at c is L."

8. "We say that $\lim_{x \to c} f(x) = L$, if for any $\varepsilon > 0$ there is a $\delta > 0$ such that $|x - c| < \delta \Rightarrow |f(x) - L| < \varepsilon$."

9. "We say that $\lim_{x \to c} f(x) = L$, if for any $\varepsilon > 0$ there is a $\delta > 0$ such that $0 \leq |x - c| \leq \delta \Rightarrow |f(x) - L| \leq \varepsilon$."

10. "If the real number L is such that for any $\varepsilon > 0$ there is a $\delta > 0$ such that $0 < |x - c| < \delta \Rightarrow |f(x) - L| < \varepsilon$, we say that $\lim_{x \to c} f(x) = L$."

11–20 Find a $\delta > 0$ that satisfies the limit claim corresponding to $\varepsilon = 0.1$, that is, such that $0 < |x - c| < \delta$ would imply $|f(x) - L| < 0.1$.

11. $\lim_{x \to 2} (5x - 1) = 9$

12. $\lim_{x \to 1} (3x + 1) = 4$

13. $\lim_{x \to -1} (-x + 2) = 3$

14. $\lim_{x \to 6} \left(4 - \frac{x}{2}\right) = 1$

15. $\lim_{x \to 0} x^2 = 0$

16. $\lim_{x \to 8} \sqrt[3]{x} = 2$

17. $\lim_{x \to 1} \frac{1}{x} = 1$

18. $\lim_{x \to 0} e^x = 1$

19. $\lim_{x \to 1} \ln x = 0$

20. $\lim_{x \to 0} \cos x = 1$

21–26 Find a number N that satisfies the limit claim corresponding to $\varepsilon = 0.1$, that is, such that $x > N$ (or $x < N$, as appropriate) would imply $|f(x) - L| < 0.1$.

21. $\lim_{x \to \infty} \frac{1}{x} = 0$

22. $\lim_{x \to \infty} \frac{x^2 - 1}{x^2 + 2x} = 1$

23. $\lim_{x \to -\infty} \frac{x + 1}{x} = 1$

24. $\lim_{x \to \infty} \frac{2x}{\sqrt{x^2 + x}} = 2$

25. $\lim_{x \to -\infty} e^x = 0$

26. $\lim_{x \to \infty} \arctan x = \frac{\pi}{2}$

27–32 For the given function $f(x)$, find a $\delta > 0$ corresponding to $M = 100$, that is, such that $0 < |x - c| < \delta$ would imply $f(x) > 100$ (let $N = -100$ if the limit is $-\infty$, in which case $0 < |x - c| < \delta$ should imply $f(x) < -100$).

27. $\lim\limits_{x \to 0} \dfrac{1}{x^2} = \infty$

28. $\lim\limits_{x \to 0^+} \dfrac{1}{x} = \infty$

29. $\lim\limits_{x \to -1} \dfrac{-1}{(x+1)^2} = -\infty$

30. $\lim\limits_{x \to 0^+} \ln x = -\infty$

31. $\lim\limits_{x \to (\pi/2)^-} \tan x = \infty$

32. $\lim\limits_{x \to 0^+} \csc x = \infty$

33–46 Use the ε-δ definition to prove the limit claim. (**Hint:** See Examples 3 and 4 for guidance as you work through these exercises.)

33. $\lim\limits_{x \to 1} (2x + 3) = 5$

34. $\lim\limits_{x \to 7} x = 7$

35. $\lim\limits_{x \to c} a = a$

36. $\lim\limits_{x \to 4} \left(\dfrac{1}{4} x + 1 \right) = 2$

37. $\lim\limits_{x \to 0} \left(\dfrac{1}{2} - 4x \right) = \dfrac{1}{2}$

38. $\lim\limits_{x \to 9} \left(5 - \dfrac{x}{3} \right) = 2$

39. $\lim\limits_{x \to 1} x^3 = 1$

40. $\lim\limits_{x \to 0} x^2 = 0$

41. $\lim\limits_{x \to 0} \dfrac{1}{2} |x| = 0$

42. $\lim\limits_{x \to -2} |x + 2| = 0$

43. $\lim\limits_{x \to 1^+} \sqrt{x - 1} = 0$

44. $\lim\limits_{x \to 0^-} \left(\sqrt[3]{x} + 1 \right) = 1$

45. $\lim\limits_{x \to 1} (x^2 + x) = 2$

46. $\lim\limits_{x \to 3} (3x^2 - 9x + 5) = 5$

47–62 Give the formal definition of the limit claim. Then use the definition to prove the claim. (**Hint:** See Examples 5 and 6 for guidance as you work through these exercises.)

47. $\lim\limits_{x \to \infty} \dfrac{1 + x}{x} = 1$

48. $\lim\limits_{x \to -\infty} \dfrac{2}{x^2} = 0$

49. $\lim\limits_{x \to \infty} \dfrac{1}{\sqrt{x}} = 0$

50. $\lim\limits_{x \to \infty} \dfrac{1 + 3x^3}{x^3} = 3$

51. $\lim\limits_{x \to -\infty} 2^x = 0$

52. $\lim\limits_{x \to \infty} \left(e^{-x} - 1 \right) = -1$

53. $\lim\limits_{x \to \infty} \dfrac{\sin x}{x} = 0$

54. $\lim\limits_{x \to \infty} (2 \arctan x) = \pi$

55. $\lim\limits_{x \to 0^+} \dfrac{1}{x} = \infty$

56. $\lim\limits_{x \to 0} \dfrac{1}{x^4} = \infty$

57. $\lim\limits_{x \to -1} \dfrac{-1}{(x+1)^2} = -\infty$

58. $\lim\limits_{x \to 0^+} \log x = -\infty$

59. $\lim\limits_{x \to (\pi/2)^+} \tan x = \infty$

60. $\lim\limits_{x \to 0^+} \csc x = \infty$

61. $\lim\limits_{x \to 2} \dfrac{-3}{(x-2)^2} = -\infty$

62. $\lim\limits_{x \to -2^-} \dfrac{x+3}{x+2} = -\infty$

63–67 Decide whether the given limit exists. Prove your conclusion. (**Hint:** See Example 7 for guidance as you work through these exercises.)

63. $\lim\limits_{x \to 0^+} \sin \dfrac{\pi}{x}$

64. $\lim\limits_{x \to 0^+} x^2 \cos \dfrac{1}{x}$

65. $\lim\limits_{x \to 0} \dfrac{|x|}{x}$

66.* $\lim\limits_{x \to 1} f(x)$, where $f(x) = \begin{cases} 1 & \text{if } x \text{ is rational} \\ 0 & \text{if } x \text{ is irrational} \end{cases}$

67.* $\lim\limits_{x \to 0} g(x)$, where $g(x) = \begin{cases} x & \text{if } x \text{ is rational} \\ 0 & \text{if } x \text{ is irrational} \end{cases}$

68. Use ε and δ to state what $\lim\limits_{x \to c} f(x) \neq L$ means.

69. A piston is manufactured to fit into the cylinder of a certain automobile engine. Suppose that the diameter of the cylinder is 82 mm and that the cross-sectional area of the piston is not allowed to be less than 99.89% of that of the cylinder. If both are perfectly round, what does this mean in terms of maximum tolerance for the clearance between the piston and the cylinder wall? (Be sure to identify which function and data take the roles of $f(x)$, c, ε, and δ in this problem.)

70. The tension in a stretched steel wire (in newtons, N) is calculated by the formula $F = E \dfrac{\Delta L}{L_0} A$, where $E = 2 \times 10^{11} \text{ N/m}^2$ is the elastic modulus (or Young's modulus) of steel, ΔL is the elongation, L_0 the original length, and A the cross-sectional area (in m^2). Suppose a 1-meter-long steel string of radius 1 millimeter is stretched by 2 millimeters when tuning a musical instrument.

   a. Calculate the tension in the string caused by the above tightening.

   b. If we are not allowed to overload the string by more than 100 N, what is the tolerance in the amount of stretching? (Be sure to identify the function and data taking the roles of c, ε, and δ in this problem.)

71–73 *True or False?* Determine whether the given statement is true or false. In case of a false statement, explain or provide a counterexample.

71. If $f(c) = L$, then as x approaches c, $\lim\limits_{x \to c} f(x) = L$.

72. If $\lim\limits_{x \to c} f(x)$ exists and equals L, then $f(c) = L$.

73. If $f(x) < g(x)$ for all $x \ne c$, and both $\lim\limits_{x \to c} f(x)$ and $\lim\limits_{x \to c} g(x)$ exist, then $\lim\limits_{x \to c} f(x) < \lim\limits_{x \to c} g(x)$.

2.3 **Technology Exercises**

74–83 Use a graphing calculator or computer algebra system to estimate the given limit. By zooming in appropriately, find δ-values that correspond to $\varepsilon = 0.1$. (Answers will vary.)

74. $\lim\limits_{x \to 5} \dfrac{x^2 - 5x + 6}{x - 2}$

75. $\lim\limits_{x \to 0} \dfrac{\sqrt{x + 5} - \sqrt{5}}{x}$

76. $\lim\limits_{x \to 3.5} \dfrac{x^2 - 6.25}{x + 2.5}$

77. $\lim\limits_{x \to 0} \dfrac{x - 1}{\sqrt{x} - 1}$

78. $\lim\limits_{x \to 0} \dfrac{\sin 3x}{2x}$

79. $\lim\limits_{x \to -\infty} \dfrac{2x + 3}{\sqrt{x^2 + 1}}$

80. $\lim\limits_{x \to \infty} \dfrac{\sqrt{9x^2 + 1}}{x - 2}$

81. $\lim\limits_{x \to -\infty} \dfrac{2x^2 + 1.5x - 7}{\sqrt{x^4 + 1}}$

82. $\lim\limits_{x \to \infty} \left(\sqrt{x^2 + 3x + 5} - \sqrt{x^2 + 2x + 1} \right)$

83. $\lim\limits_{x \to \infty} \left(1 + \dfrac{1}{x} \right)^x$

84–89 Use a graphing calculator or computer algebra system to locate a vertical asymptote of the given function. Then for such an asymptote $x = c$ find an appropriate value $\delta > 0$ such that $|x - c| < \delta \Rightarrow |f(x)| > 10$. (Answers will vary.)

84. $f(x) = \dfrac{x^2 - 7}{x^3 + x + 1}$

85. $f(x) = \dfrac{3x + 1}{2x^4 + x - 5}$

86. $f(x) = \ln \dfrac{x^2}{x^2 + 1}$

87. $f(x) = \tan\left(\dfrac{1}{2} x + 3 \right)$

88. $f(x) = \csc(2x + 1)$

89. $f(x) = \cot\left(\dfrac{1}{2} \cos x \right)$

2.4 Determining Limits of Functions

TOPICS

1. Limit laws

2. Limit determination techniques

We now have a formal definition of *limit* with which to work, and some experience in proving limit claims with epsilon-delta arguments. What we are lacking is a collection of tools allowing us to determine limits in the first place. The theorems in this section will help us do just that and, at the same time, provide the necessary proof that the resulting limits are correct.

TOPIC 1   Limit Laws

We will begin with the basic limit laws and show how they can be used both in immediate applications and in deriving more powerful laws.

Theorem ⚬

Basic Limit Laws

Let f and g be two functions such that both $\lim\limits_{x \to c} f(x)$ and $\lim\limits_{x \to c} g(x)$ exist, and let k be a fixed real number. Then the following laws hold.

Sum Law
$$\lim_{x \to c} \left[f(x) + g(x) \right] = \lim_{x \to c} f(x) + \lim_{x \to c} g(x)$$

Difference Law
$$\lim_{x \to c} \left[f(x) - g(x) \right] = \lim_{x \to c} f(x) - \lim_{x \to c} g(x)$$

Constant Multiple Law $\lim\limits_{x \to c} \left[kf(x) \right] = k \lim\limits_{x \to c} f(x)$

Product Law
$$\lim_{x \to c} \left[f(x) g(x) \right] = \lim_{x \to c} f(x) \cdot \lim_{x \to c} g(x)$$

Quotient Law
$$\lim_{x \to c} \frac{f(x)}{g(x)} = \frac{\lim\limits_{x \to c} f(x)}{\lim\limits_{x \to c} g(x)}, \text{ provided } \lim_{x \to c} g(x) \neq 0$$

Proof ✎

We will prove the Sum Law here, and present the proofs of the remaining laws in Appendix E.

For ease of exposition, let $L = \lim\limits_{x \to c} f(x)$ and $M = \lim\limits_{x \to c} g(x)$; we need to show that $\lim\limits_{x \to c} \left[f(x) + g(x) \right] = L + M$ using our $\varepsilon\text{-}\delta$ definition of limit. To this end, assume $\varepsilon > 0$ is given. Then by assumption there exist $\delta_1 > 0$ and $\delta_2 > 0$ such that

$$0 < |x - c| < \delta_1 \quad \Rightarrow \quad |f(x) - L| < \frac{\varepsilon}{2}$$

and

$$0 < |x - c| < \delta_2 \quad \Rightarrow \quad |g(x) - M| < \frac{\varepsilon}{2}.$$

Note that we have found δ_1 and δ_2 so that the differences between the functions and their respective limits are smaller than $\varepsilon/2$. We did so in order to obtain the following consequence for all $x \neq c$ chosen within δ of c, where δ is the smaller of δ_1 and δ_2.

$$|f(x) + g(x) - (L+M)| = |f(x) - L + g(x) - M|$$

$$\leq |f(x) - L| + |g(x) - M| \quad \text{Triangle Inequality: } |a+b| \leq |a| + |b|$$

$$< \frac{\varepsilon}{2} + \frac{\varepsilon}{2} = \varepsilon \qquad\qquad \delta \leq \delta_1 \text{ and } \delta \leq \delta_2$$

Since we have demonstrated that the function $f(x) + g(x)$ is within ε of $L + M$ for all x such that $0 < |x - c| < \delta$, our proof is complete.

Intuitively, the limit laws seem reasonable. For instance, if $\lim\limits_{x \to c} f(x) = L$ and $\lim\limits_{x \to c} g(x) = M$, the Product Law points out (in a precise manner) that for values of x close to c, the function $f \cdot g$ assumes values close to $L \cdot M$. As always, though, it is important to remember that statements about the limit at a point c describe behavior *near* c, not *at* c.

Example 1 ✎

For the functions f and g graphed in Figure 1, determine if the following limits exist. If a particular limit exists, evaluate it. If not, give reasons why it fails to exist.

a. $\lim\limits_{x \to 2}\left[f(x)g(x)\right]$ **b.** $\lim\limits_{x \to 2}\left[\dfrac{3}{7}f(x) + g(x)\right]$

c. $\lim\limits_{x \to 0}\left[g(x) - f(x)\right]$ **d.** $\lim\limits_{x \to 0}\left[f(x)g(x)\right]$

Solution

a. As we can see from the figure, the limits of both f and g exist at $c = 2$. In case of g, we note once again that the actual function value at 2 has no bearing on the value of the limit.

$$\lim\limits_{x \to 2} f(x) = 7 \quad \text{and} \quad \lim\limits_{x \to 2} g(x) = 3$$

Now we evaluate the limit using the Product Law.

$$\lim\limits_{x \to 2}\left[f(x)g(x)\right] = \lim\limits_{x \to 2} f(x) \cdot \lim\limits_{x \to 2} g(x) = 7 \cdot 3 = 21$$

b. Here we are going to use the Sum Law and the Constant Multiple Law.

$$\lim\limits_{x \to 2}\left[\frac{3}{7}f(x) + g(x)\right] = \lim\limits_{x \to 2}\left[\frac{3}{7}f(x)\right] + \lim\limits_{x \to 2} g(x) = \frac{3}{7}\lim\limits_{x \to 2} f(x) + \lim\limits_{x \to 2} g(x)$$

$$= \frac{3}{7} \cdot 7 + 3 = 3 + 3 = 6$$

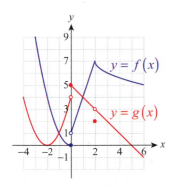

$y = f(x)$

$y = g(x)$

Figure 1

c. We will first examine the behavior of f near 0. It is clear from the graph that both one-sided limits exist, but they don't agree; thus the two-sided limit of f cannot exist at $x = 0$. In fact, $\lim_{x \to 0^-} f(x) = 0$ while $\lim_{x \to 0^+} f(x) = 1$, and so $\lim_{x \to 0} f(x)$ does not exist.

Similarly for g, $\lim_{x \to 0^-} g(x) = 4$ while $\lim_{x \to 0^+} g(x) = 5$, and thus $\lim_{x \to 0} g(x)$ cannot exist.

In words, since the one-sided limits are unequal, the two-sided limit of g cannot exist at 0. However, the limit laws do apply to one-sided limits, so we can determine those for $\big[g(x) - f(x) \big]$.

$$\lim_{x \to 0^-} \big[g(x) - f(x) \big] = \lim_{x \to 0^-} g(x) - \lim_{x \to 0^-} f(x) = 4 - 0 = 4$$

$$\lim_{x \to 0^+} \big[g(x) - f(x) \big] = \lim_{x \to 0^+} g(x) - \lim_{x \to 0^+} f(x) = 5 - 1 = 4$$

Notice from our findings that something interesting is actually going on here. Even though the two-sided limits of f and g do not exist individually, the one-sided limits of $g - f$ exist and agree at $x = 0$. What this means is that the limit of $g - f$ at 0 actually exists.

$$\lim_{x \to 0} \big[g(x) - f(x) \big] = 4$$

d. Using the Product Law for one-sided limits, we obtain

$$\lim_{x \to 0^-} \big[f(x) g(x) \big] = \lim_{x \to 0^-} f(x) \cdot \lim_{x \to 0^-} g(x) = 0 \cdot 4 = 0, \text{ while}$$

$$\lim_{x \to 0^+} \big[f(x) g(x) \big] = \lim_{x \to 0^+} f(x) \cdot \lim_{x \to 0^+} g(x) = 1 \cdot 5 = 5.$$

Since the one-sided limits of the product function $f \cdot g$ are unequal, we conclude that $\lim_{x \to 0} \big[f(x) g(x) \big]$ does not exist.

Example 2 ✎

Let $f(x) = \begin{cases} \dfrac{|x|}{x} & \text{if } x \neq 0 \\ 0 & \text{if } x = 0 \end{cases}$ and $g(x) = \cos x$. Determine whether $\lim_{x \to 0} \big[f(x) g(x) \big]$ exists.

Solution

We will first examine the behavior of $f(x)$ near 0. If x approaches 0 from the left, then since $x < 0$, we have

$$f(x) = \frac{|x|}{x} = \frac{-x}{x} = -1,$$

so f is the constant -1 for all negative x-values.

Therefore,

$$\lim_{x \to 0^-} f(x) = -1.$$

On the other hand, for all positive x-values $|x| = x$, so

$$f(x) = \frac{|x|}{x} = \frac{x}{x} = 1,$$

thus

$$\lim_{x \to 0^+} f(x) = 1.$$

Since the one-sided limits are unequal, $\lim_{x \to 0} f(x)$ does not exist. Consequently, the Product Law is not applicable, but since it does apply to one-sided limits, we can proceed to examine those separately. Recalling the well-known fact that $\lim_{x \to 0} \cos x = 1$, we obtain

$$\lim_{x \to 0^-} \left[f(x) g(x) \right] = \lim_{x \to 0^-} f(x) \cdot \lim_{x \to 0^-} g(x) = (-1) \cdot 1 = -1, \text{ while}$$

$$\lim_{x \to 0^+} \left[f(x) g(x) \right] = \lim_{x \to 0^+} f(x) \cdot \lim_{x \to 0^+} g(x) = 1 \cdot 1 = 1.$$

Thus we conclude that the two-sided limit, $\lim_{x \to 0} \left[f(x) g(x) \right]$, does not exist.

The five limit laws that we have listed so far can be quickly extended to a much larger collection when combined with one another and with a few easily proved statements. For instance, if the Product Law is applied to the product of a function with itself, we obtain the statement

$$\lim_{x \to c} \left[f(x) f(x) \right] = \lim_{x \to c} f(x) \cdot \lim_{x \to c} f(x) = \left[\lim_{x \to c} f(x) \right]^2$$

(assuming $\lim_{x \to c} f(x)$ exists). And the same law can now be applied to the product of $f(x)$ and $\left[f(x) \right]^2$ to reach a similar conclusion about the limit of the function $\left[f(x) \right]^3$. In general, repeated application of the Product Law results in the Power Law, the proof of which can be found in Appendix E.

Theorem ⚲

Positive Integer Power Law

Let f be a function for which $\lim_{x \to c} f(x)$ exists, and let m be a fixed positive integer. Then

$$\lim_{x \to c} \left[f(x) \right]^m = \left[\lim_{x \to c} f(x) \right]^m.$$

The two limit statements

$$\lim_{x \to c} 1 = 1 \quad \text{and} \quad \lim_{x \to c} x = c$$

are certainly reasonable and easily proved, and allow us to extend our list of limit laws to large classes of functions. In particular, we can now state the following two laws, which are examples of what are sometimes referred to as Direct Substitution Laws.

Theorem

Polynomial Substitution Law

Let p be a polynomial function. Then

$$\lim_{x \to c} p(x) = p(c).$$

Theorem

Rational Function Substitution Law

Let p and q be polynomial functions. Then if $q(c) \neq 0$,

$$\lim_{x \to c} \frac{p(x)}{q(x)} = \frac{p(c)}{q(c)}.$$

In Exercises 65–68, you will be guided through the proofs of the above statements.

Example 3

Find the following limits.

a. $\lim_{x \to 2} \left(4x^3 - 5x^2 + 1 \right)$

b. $\lim_{x \to -1} \left(2x^4 + x^3 - 3x^2 + 7.5 \right)$

c. $\lim_{x \to 3} \dfrac{x^2 + 2}{3x - 1}$

d. $\lim_{x \to 1} \dfrac{2x^3 - 5x + 1}{2x - x^2}$

Solution

a. First using the Sum and Difference Laws, we obtain

$$\lim_{x \to 2} \left(4x^3 - 5x^2 + 1 \right) = \lim_{x \to 2} \left(4x^3 \right) - \lim_{x \to 2} \left(5x^2 \right) + \lim_{x \to 2} (1)$$

$$= 4 \lim_{x \to 2} x^3 - 5 \lim_{x \to 2} x^2 + 1,$$

where in the last step we used the Constant Multiple Law. Next, the Positive Integer Power Law comes to bear.

$$4 \lim_{x \to 2} x^3 - 5 \lim_{x \to 2} x^2 + 1 = 4 \left(\lim_{x \to 2} x \right)^3 - 5 \left(\lim_{x \to 2} x \right)^2 + 1 = 4(2)^3 - 5(2)^2 + 1 = 13$$

Notice that our repeated application of the various rules eventually led us to finding the limit by simply substituting $x = 2$ into the polynomial. This is exactly what the Polynomial Substitution Law allows us to do; so henceforth we don't even have to go through the above, somewhat lengthy, process when finding the limits of polynomials.

b. In this case, we will simply use the Polynomial Substitution Law.

$$\lim_{x \to -1}\left(2x^4 + x^3 - 3x^2 + 7.5\right) = 2(-1)^4 + (-1)^3 - 3(-1)^2 + 7.5 = 5.5$$

c. Notice that our third limit is that of a rational function. Since both the numerator and denominator are polynomials, we can apply the Quotient Law combined with a repeated application of the Polynomial Substitution Law as follows:

$$\lim_{x \to 3}\frac{x^2 + 2}{3x - 1} = \frac{\lim_{x \to 3}\left(x^2 + 2\right)}{\lim_{x \to 3}\left(3x - 1\right)} = \frac{3^2 + 2}{3 \cdot 3 - 1} = \frac{9 + 2}{9 - 1} = \frac{11}{8}$$

Notice, however, that what we did was in effect substituting $x = 3$ into the given rational function. This is exactly what the Rational Function Substitution Law says we can always do unless the process results in a 0 denominator.

d. This time, we will simply refer to the Rational Function Substitution Law.

$$\lim_{x \to 1}\frac{2x^3 - 5x + 1}{2x - x^2} = \frac{2(1)^3 - 5 \cdot 1 + 1}{2 \cdot 1 - 1^2} = \frac{-2}{1} = -2$$

In summary, we wish to emphasize that when evaluating limits of a polynomial or rational function, our theorems allow for the calculation to be reduced to a simple matter of evaluating the function at the limit point. The only exception is for those c-values that cause the denominator of a given rational limit to equal 0.

We can gain another significant extension with the addition of the following law, the proof of which is a consequence of the Intermediate Value Theorem of Section 2.5.

Theorem 🔍

Positive Integer Root Law

Let f be a function for which $\lim_{x \to c} f(x)$ exists, and let n be a fixed positive integer. Then

$$\lim_{x \to c}\sqrt[n]{f(x)} = \sqrt[n]{\lim_{x \to c} f(x)},$$

with the assumption that $\lim_{x \to c} f(x)$ is nonnegative if n is even.

Exercise 69 shows how the Positive Integer Power Law and the Positive Integer Root Law we have (and, if necessary, the Quotient Law) can be combined to yield the following.

Theorem 🔍

Rational Power Law

Let f be a function for which $\lim_{x \to c} f(x)$ exists, and let m and n be fixed nonzero integers with no common factor. Then

$$\lim_{x \to c} \left[f(x) \right]^{m/n} = \left[\lim_{x \to c} f(x) \right]^{m/n},$$

with the assumption that $\lim_{x \to c} f(x)$ is nonnegative if n is even.

Example 4 ✎

Use the Rational Power Law to evaluate the following limits.

a. $\lim\limits_{x \to 1} \left(\dfrac{5x+3}{x^2 - 2x + 2} \right)^{7/3}$ **b.** $\lim\limits_{x \to 2} \sqrt{\left(x^4 + 2x^2 + 4 \right)^3}$ **c.** $\lim\limits_{x \to 0} \sqrt[3]{4\cos^2 x}$

Solution

a. First of all, we claim that for $f(x) = (5x+3)/(x^2 - 2x + 2)$, $\lim\limits_{x \to 1} f(x)$ exists. In fact, by direct substitution we have the following.

$$\lim_{x \to 1} f(x) = \lim_{x \to 1} \frac{5x+3}{x^2 - 2x + 2} = \frac{5 \cdot 1 + 3}{1^2 - 2 \cdot 1 + 2} = \frac{8}{1 - 2 + 2} = 8$$

It follows that the Rational Power Law applies.

$$\lim_{x \to 1} \left(\frac{5x+3}{x^2 - 2x + 2} \right)^{7/3} = \left(\lim_{x \to 1} \frac{5x+3}{x^2 - 2x + 2} \right)^{7/3} = 8^{7/3} = 2^7 = 128$$

b. To start us off, notice that $\lim\limits_{x \to 2} \left(x^4 + 2x^2 + 4 \right) = 28$, a fact easily verified by direct substitution. Also since the above limit is positive and

$$\sqrt{\left(x^4 + 2x^2 + 4 \right)^3} = \left(x^4 + 2x^2 + 4 \right)^{3/2},$$

the Rational Power Law applies.

$$\lim_{x \to 2} \sqrt{\left(x^4 + 2x^2 + 4 \right)^3} = \lim_{x \to 2} \left(x^4 + 2x^2 + 4 \right)^{3/2} = \left[\lim_{x \to 2} \left(x^4 + 2x^2 + 4 \right) \right]^{3/2} = 28^{3/2}$$

$$= \left(\sqrt{28} \right)^3 = \left(2\sqrt{7} \right)^3 = 2^3 \left(\sqrt{7} \right)^3 = 8 \cdot 7 \cdot \sqrt{7} = 56\sqrt{7}$$

c. Since $4\cos^2 x = \left(2\cos x \right)^2$, and since $\lim\limits_{x \to 0} (2\cos x) = 2$, the Rational Power Law once again applies.

$$\lim_{x \to 0} \sqrt[3]{4\cos^2 x} = \lim_{x \to 0} \left(2\cos x \right)^{2/3} = \left[\lim_{x \to 0} \left(2\cos x \right) \right]^{2/3} = 2^{2/3} = \sqrt[3]{4}$$

TOPIC 2   Limit Determination Techniques

The limit laws we have stated greatly simplify the determination of many limits and also show that limits of polynomial and rational functions can be found simply by evaluating the function at the limit point (remember that in the case of rational functions the limit point must be in the domain of the function). Functions possessing this property, which we will refer to as the *Direct Substitution Property*, are called *continuous*; much more will be said about these well-behaved functions in the next section. But we will close this section with a discussion of techniques to use when the limit laws and/or direct substitution do not immediately apply.

Many of our techniques make use of the fact that the limit of a function f at a point c is determined entirely by its behavior near, but not at, c. If we can find a function g that is identical to f near c, and if the limit of g at the point c is easy to determine, then we are done. The following examples illustrate the steps typically taken in this process.

Example 5 ✎

Find $\displaystyle\lim_{x \to -3} \frac{x^2 - 9}{x + 3}$.

Solution

First of all we note that we cannot use direct substitution to evaluate this limit, for the denominator equals 0 at $x = -3$. In fact, $f(x) = \left(x^2 - 9\right)/(x + 3)$ is not even defined at $x = -3$. However, as far as the limit is concerned, that is not a problem at all (recall that the existence and/or value of $f(c)$ and $\displaystyle\lim_{x \to c} f(x)$ have no bearing on each other). The limit not only exists, but finding it is surprisingly easy, using a bit of algebra.

$$\lim_{x \to -3} \frac{x^2 - 9}{x + 3} = \lim_{x \to -3} \frac{(x + 3)(x - 3)}{x + 3} = \lim_{x \to -3} (x - 3) = -3 - 3 = -6$$

A few very important remarks are in order. First of all, canceling the factor of $(x + 3)$ is legitimate, since even though x is approaching -3, it never is actually equal to -3 throughout the limit process, so we have not divided by 0. Second, the function $g(x) = x - 3$ is actually different from $f(x)$. In fact, they agree everywhere but at $x = -3$; at which point $f(x)$ is undefined, but $g(-3) = -6$ (see Figure 2). However, the behaviors of f and g are identical as x approaches -3, and therefore, so are their limits. (Examining the chain of equalities above, note that we never actually stated the equality of f and g, but merely the fact that their limits at -3 were equal.)

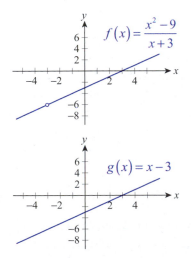

Figure 2

Algebraic techniques similar to the one from Example 5 are common when evaluating limits analytically. We provide further illustrations in the next two examples.

Example 6 ✐

Use algebra to evaluate $\lim\limits_{x \to 1} \dfrac{\sqrt{x+3}-2}{x-1}$.

Solution

The trick with a limit such as this is to use the "conjugate product rule" $(a+b)(a-b)=a^2-b^2$. If we multiply both the numerator and denominator by the "conjugate" of the expression containing the radical (in this case, that is $\sqrt{x+3}+2$), we obtain the following.

$$\lim_{x \to 1} \frac{\sqrt{x+3}-2}{x-1} = \lim_{x \to 1} \left(\frac{\sqrt{x+3}-2}{x-1} \cdot \frac{\sqrt{x+3}+2}{\sqrt{x+3}+2} \right) = \lim_{x \to 1} \frac{\left(\sqrt{x+3}\right)^2 - 2^2}{(x-1)\left(\sqrt{x+3}+2\right)}$$

$$= \lim_{x \to 1} \frac{x+3-4}{(x-1)\left(\sqrt{x+3}+2\right)} = \lim_{x \to 1} \frac{x-1}{(x-1)\left(\sqrt{x+3}+2\right)}$$

$$= \lim_{x \to 1} \frac{1}{\sqrt{x+3}+2} = \frac{1}{\sqrt{1+3}+2} = \frac{1}{2+2} = \frac{1}{4}$$

As in the previous example, the cancellation was legitimate, and even though in the process we did change the function, we did not change the value of the limit.

Example 7 ✐

Find the limit: $\lim\limits_{h \to 0} \dfrac{(h+2)^2 - 4}{h}$

Solution

We will take an algebraic approach very similar to the previous examples, but keeping in mind the fact that in the current problem, the variable is denoted by h, rather than the usual x. This, however, should not cause any difficulties.

$$\lim_{h \to 0} \frac{(h+2)^2 - 4}{h} = \lim_{h \to 0} \frac{\left(h^2 + 4h + 4\right) - 4}{h} = \lim_{h \to 0} \frac{h^2 + 4h}{h}$$

$$= \lim_{h \to 0} \frac{h(h+4)}{h} = \lim_{h \to 0} (h+4) = 4$$

Again, canceling h is legitimate, since h never actually assumes the value 0.

Another technique for determining $\lim\limits_{x \to c} f(x)$ calls for finding two other functions, g and h, such that f is "squeezed between" g and h and for which $\lim\limits_{x \to c} g(x)$ and $\lim\limits_{x \to c} h(x)$ are easier to determine.

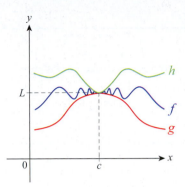

Figure 3 The Squeeze Theorem

Theorem

The Squeeze Theorem

If $g(x) \le f(x) \le h(x)$ for all x in some open interval containing c, except possibly at c itself, and if $\lim_{x \to c} g(x) = \lim_{x \to c} h(x) = L$, then $\lim_{x \to c} f(x) = L$ as well.

The statement also holds for limits at infinity, that is, for $c = -\infty$ or $c = \infty$.

The Squeeze Theorem also goes by names such as the Sandwich Theorem and the Pinching Theorem, and a proof is included in Appendix E.

Example 8

Use the Squeeze Theorem to prove $\lim_{x \to 0} x \sin \dfrac{1}{x} = 0$.

Solution

As a first observation, we recall that the sine value of any angle is never greater than 1 or less than -1; that is, we have the well-known inequalities

$$-1 \le \sin \alpha \le 1$$

for any α. This certainly means that for any nonzero x, choosing $\alpha = 1/x$,

$$-1 \le \sin \frac{1}{x} \le 1,$$

a fact you have seen before. Multiplying all sides of this chain of inequalities by $|x|$, we obtain

$$-|x| \le x \sin \frac{1}{x} \le |x|.$$

In words, we can say that while the sine function oscillates between -1 and 1, $x \sin(1/x)$ will oscillate between $-|x|$ and $|x|$. The previous inequality coupled with the fact that $\lim_{x \to 0}(-|x|) = \lim_{x \to 0}|x| = 0$ means that the functions $f(x) = x \sin(1/x)$, $g(x) = -|x|$, and $h(x) = |x|$ satisfy the hypotheses of the Squeeze Theorem. Thus we can simply invoke the theorem, which ensures that the claim

$$\lim_{x \to 0} x \sin \frac{1}{x} = 0$$

is now proven.

Figure 4 shows the Squeeze Theorem at work. Notice how f is "squeezed between" g and h near the origin.

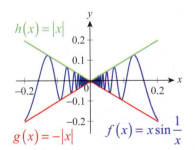

Figure 4

We will end with one last limit theorem that will prove useful in some derivations to follow. (See Appendix E for a proof.)

> **Theorem** 🔍
>
> ### Upper Bound Theorem
>
> If $f(x) \le g(x)$ for all x in some open interval containing c, except possibly at c itself, and if the limits of f and g both exist at c, then
>
> $$\lim_{x \to c} f(x) \le \lim_{x \to c} g(x).$$

2.4 Exercises

1–2 Use the graph to find the given limit.

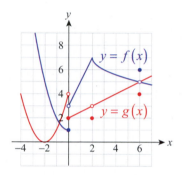

1. a. $\displaystyle \lim_{x \to 0^+} \left[g(x) - 2f(x) \right]$ **b.** $\displaystyle \lim_{x \to 2^+} \left[g(x) f(x) \right]$

2. a. $\displaystyle \lim_{x \to 6} \left[g(x) + f(x) \right]$ **b.** $\displaystyle \lim_{x \to 0^-} \frac{f(x)}{2g(x)}$

3–20 Use appropriate limit laws to evaluate the given limit.

3. $\displaystyle \lim_{x \to 4} 5$

4. $\displaystyle \lim_{x \to 4} 5x$

5. $\displaystyle \lim_{x \to 3} (2x + 1)$

6. $\displaystyle \lim_{x \to 1/2} (3 - 4x)$

7. $\displaystyle \lim_{x \to -3} x^2$

8. $\displaystyle \lim_{x \to -2} (-x^5)$

9. $\displaystyle \lim_{x \to 3} (2x^2 - x + 7)$

10. $\displaystyle \lim_{x \to -1} \left(3 + x - \frac{5}{2} x^2 \right)$

11. $\displaystyle \lim_{x \to 1/2} (2x^3 - 3x^2 + x - 4)$

12. $\displaystyle \lim_{x \to -2} (3x^3 - x^5)$

13. $\displaystyle \lim_{x \to 1} \frac{3x - 7}{x + 1}$

14. $\displaystyle \lim_{x \to -1} \frac{5x + 3}{x^2 - x}$

15. $\displaystyle \lim_{x \to 3} \left(\frac{4x}{11x - x^3} \right)^{1/3}$

16. $\displaystyle \lim_{t \to 1} \left(\frac{2t + t^3}{3t^2 + 1} \right)^{3/2}$

17. $\displaystyle \lim_{x \to -2} \sqrt[3]{5x^4 - x^3 + 3x^2 + 2x + 4}$

18. $\displaystyle \lim_{x \to 4} \sqrt{x^4 + 2x^2 + 1}$

19. $\displaystyle \lim_{x \to -3} \left(\frac{x^4 - 5x}{x^3 + 2x^2 - 4x} \right)^{4/5}$

20. $\displaystyle \lim_{x \to -5} \sqrt[3]{(x^4 + 2x^3 + x^2)^2}$

21–44 Use algebra to evaluate the given limit.

21. $\displaystyle \lim_{x \to 6} \frac{x^2 - 36}{x - 6}$

22. $\displaystyle \lim_{x \to -7} \frac{x + 7}{x^2 - 49}$

23. $\displaystyle \lim_{x \to 3} \frac{3 - 13x + 4x^2}{x - 3}$

24. $\displaystyle \lim_{x \to 4} \frac{x^4 - 256}{x^2 - 16}$

25. $\displaystyle \lim_{x \to 5} \frac{2x^3 - 7x^2 - 14x - 5}{x^2 - 25}$

26. $\displaystyle \lim_{x \to 2} \frac{x^3 - 8}{x^3 - 2x^2 + 2x - 4}$

27. $\displaystyle \lim_{x \to 7} \frac{\sqrt{x + 2} - 3}{x - 7}$

28. $\displaystyle \lim_{x \to 9} \frac{3 - \sqrt{x}}{x - 9}$

29. $\displaystyle \lim_{x \to 0} \frac{\sqrt{x + 5} - \sqrt{5}}{x}$

30. $\displaystyle \lim_{x \to 0} \frac{\frac{1}{4 + x} - \frac{1}{4}}{x}$

31. $\displaystyle \lim_{x \to 2} \frac{\frac{1}{3} - \frac{1}{1 + x}}{x - 2}$

32. If $f(x) = x^2$, find $\displaystyle \lim_{x \to 2} \frac{f(x) - f(2)}{x - 2}$.

33. If $g(x) = x^2 - 2$, find $\displaystyle \lim_{h \to 0} \frac{g(3 + h) - g(3)}{h}$.

34. If $k(x) = 1 - x + x^2$, find $\displaystyle \lim_{h \to 0} \frac{k(2 - h) - k(2)}{h}$.

35. If $p(x) = x^3 + x$, find $\displaystyle \lim_{x \to 1} \frac{p(x) - p(1)}{x - 1}$.

36. If $F(x) = \frac{1}{x}$, find $\displaystyle \lim_{x \to 1/2} \frac{F(x) - F\left(\frac{1}{2}\right)}{x - \frac{1}{2}}$.

37. $\lim\limits_{h \to 0} \dfrac{\sqrt{x+h}-\sqrt{x}}{h}$

38. $\lim\limits_{x \to 3} \dfrac{2x^2-3x-9}{x^2-9}$

39. $\lim\limits_{x \to 2} \dfrac{x^4-16}{3x^2-5x-2}$

40. $\lim\limits_{x \to -3} \dfrac{\frac{1}{3}+\frac{1}{x}}{x^3+27}$

41. $\lim\limits_{x \to 2} \dfrac{3-\sqrt{x+7}}{x-2}$

42. $\lim\limits_{x \to 8} \dfrac{8-x}{\sqrt[3]{x}-2}$

43. $\lim\limits_{y \to 0} \left(\dfrac{1}{y}+\dfrac{1}{y^2-y} \right)$

44. $\lim\limits_{x \to 1} \dfrac{\sqrt{x}-1}{\sqrt[3]{x}-1}$

45–50 Use $\lim\limits_{x \to c} f(x) = 3$ and $\lim\limits_{x \to c} g(x) = -2$ to find the limit.

45. $\lim\limits_{x \to c} \left[2f(x)-g(x) \right]$

46. $\lim\limits_{x \to c} \dfrac{4f(x)+3g(x)}{f(x)-\frac{1}{2}g(x)}$

47. $\lim\limits_{x \to c} \sqrt{\left[f(x) \right]^4 + 10\left[g(x) \right]^2}$

48. $\lim\limits_{x \to c} \left(\left[f(x)-1 \right]^2 \sqrt[3]{g(x)} \right)$

49. $\lim\limits_{x \to c} \left(\left[f(x) \right]^2 + (x-2)g(x) \right)$

50. $\lim\limits_{x \to c} \left(\dfrac{f(x)+g(x)}{\left[g(x) \right]^2} \right)^{3/2}$

51–58 Use the limit laws to find the one-sided limit.

51. $\lim\limits_{x \to 0^+} \dfrac{x}{|x|}$

52. $\lim\limits_{x \to 0^-} \operatorname{sgn} x \cos x$, where $\operatorname{sgn} x = \begin{cases} 1 & \text{if } x > 0 \\ 0 & \text{if } x = 0 \\ -1 & \text{if } x < 0 \end{cases}$

53. $\lim\limits_{x \to 2^+} \dfrac{\sqrt{x-2}}{3x+1}$

54. $\lim\limits_{x \to 1^-} \sqrt{1-x^2}$

55. $\lim\limits_{x \to (1/3)^-} \dfrac{\sqrt{1-3x}}{6x+5}$

56. $\lim\limits_{x \to 1^+} \left([\![x]\!] - x \right)$ (See the definition of $f(x) = [\![x]\!]$ in Section 1.2, before Exercises 43–45.)

57. $\lim\limits_{x \to 2^+} [\![x]\!] e^x$

58. $\lim\limits_{x \to -1^-} \dfrac{[\![x]\!](2x^2+1)}{x+3}$

59–64 Use the Squeeze Theorem to prove the limit claim.

59. $\lim\limits_{x \to 0} x^2 \sin \dfrac{1}{x} = 0$

60. $\lim\limits_{x \to 0} |x| \cos x = 0$

61. $\lim\limits_{x \to \infty} \dfrac{\cos x}{x} = 0$

62. $\lim\limits_{x \to -\infty} e^x \sin x = 0$

63. $\lim\limits_{x \to 0^+} x^{3/2} e^{\cos(1/x)} = 0$

64. $\lim\limits_{x \to \infty} \dfrac{\sin^2 x + 1}{2+x} = 0$

65. Provide a rigorous proof of the limit claim $\lim\limits_{x \to c} 1 = 1$. (**Hint:** Use the fact that for the constant 1 function, $f(x) = 1$ for all x, so in particular, if an $\varepsilon > 0$ is given, $|f(x)-1| = |1-1| = 0$, which makes the choice of δ "easy.")

66. Provide a rigorous proof of the limit claim $\lim\limits_{x \to c} x = c$. (**Hint:** Since $f(x) = x$ in this problem, for a given $\varepsilon > 0$ we need to ensure that $|f(x)-c| = |x-c| < \varepsilon$ as long as $0 < |x-c| < \delta$. This observation makes the choice of δ obvious.)

67. Use Exercise 66 and the basic limit laws to prove the Polynomial Substitution Law. (**Hint:** From Exercise 66 and a repeated application of the Product Law it follows that $\lim\limits_{x \to c} x^k = c^k$. As a next step, from the Constant Multiple Law we can conclude that if $a \in \mathbb{R}$, $\lim\limits_{x \to c} ax^k = ac^k$. From the above claim, a repeated application of the Sum Law will yield the result for a general polynomial.)

68. Use Exercise 67 and the basic limit laws to prove the Rational Function Substitution Law.

69. Combine the Positive Integer Power Law and the Positive Integer Root Law to prove the Rational Power Law. (**Hint:** Assuming first that both m and n are positive, we can write
$$\lim\limits_{x \to c} \left[f(x) \right]^{m/n} = \lim\limits_{x \to c} \left[\left[f(x) \right]^{1/n} \right]^m = \lim\limits_{x \to c} \left[\sqrt[n]{f(x)} \right]^m.$$
Now use the Positive Integer Power Law followed by the Positive Integer Root Law to obtain that the above limit is equal to
$$\left[\lim\limits_{x \to c} \sqrt[n]{f(x)} \right]^m = \left[\sqrt[n]{\lim\limits_{x \to c} f(x)} \right]^m,$$
from which the result follows. If m is negative, note that $\left[f(x) \right]^{m/n} = 1 / \left[f(x) \right]^{-m/n}$, where $-m$ is positive. Thus if we use the Quotient Law along with the previous argument, we obtain

$$\lim_{x \to c} \left[f(x) \right]^{m/n} = \lim_{x \to c} \frac{1}{\left[f(x) \right]^{-m/n}}$$

$$= \frac{1}{\lim_{x \to c} \left[f(x) \right]^{-m/n}}$$

$$= \frac{1}{\left[\lim_{x \to c} f(x) \right]^{-m/n}},$$

from which the result easily follows.)

70. Let $D(x) = \begin{cases} 0 & \text{if } x \text{ is rational} \\ 1 & \text{if } x \text{ is irrational} \end{cases}$.

Does $\lim_{x \to 0} D(x)$ exist? Prove your answer.

71. Let $F(x) = \begin{cases} 0 & \text{if } x \text{ is rational} \\ x^2 & \text{if } x \text{ is irrational} \end{cases}$.

Does $\lim_{x \to 0} F(x)$ exist? Prove your answer.

72.* Prove that if $\lim_{x \to c} f(x) = L$ and $\lim_{x \to c} f(x) = K$, then $L = K$. In words, prove that if the limit of f exists at c, then the limit is unique.

73.* Prove that if n and m are positive integers, then
$$\lim_{x \to 1} \frac{x^n - 1}{x^m - 1} = \frac{n}{m}.$$

74.* Prove that if $\lim_{x \to c} f(x) = 0$, then $\lim_{x \to c} \left| f(x) \right| = 0$.

75.* Prove that if $\lim_{x \to c} f(x) = 0$ and $g(x)$ is such that $\left| g(x) \right| \le M$ for some number M (such functions are called bounded), then $\lim_{x \to c} \left[f(x) g(x) \right] = 0$.

76.* Prove that in Exercise 75, it is sufficient to require the boundedness of g only on an interval around c (except at c itself).

77.* By finding functions f and g such that $\lim_{x \to c} f(x) = 0$ but $\lim_{x \to c} \left[f(x) g(x) \right] \neq 0$, show that it is necessary to impose a boundedness condition on g in Exercise 75.

78.* Give examples of f and g to show that **a.** the existence of $\lim_{x \to c} \left[f(x) + g(x) \right]$ does not imply the existence of $\lim_{x \to c} f(x)$ and **b.** the existence of $\lim_{x \to c} \left[f(x) g(x) \right]$ does not imply the existence of $\lim_{x \to c} f(x)$.

79. A *concave spherical mirror* is a part of the inside of a sphere, silvered to form a reflective surface. The radius r of the sphere is called the mirror's *radius of curvature*. If the size of such a mirror is small relative to its radius of curvature, light rays parallel to its principal axis are reflected through approximately a single point, called *focus*. In the following illustration, C denotes the center, F_d is the focus, while d is the distance between the incoming ray and the principal axis. Note that according to the Law of Reflection, the incoming and reflected rays make the same size angle α with the radius $\overline{CR}$ (this radius is called *normal* to the mirror surface). One way to determine the *focal length* (the distance between the mirror and the focus) is to find the limiting position of F_d as $d \to 0$. Noting that the triangle $\triangle CRF_d$ is isosceles, by similarity we obtain $\dfrac{CF_d}{(r/2)} = \dfrac{r}{\sqrt{r^2 - d^2}}$. Use this observation to express CF_d and determine the focal length of the spherical mirror by taking the limit as $d \to 0$.

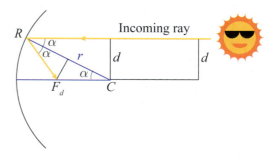

2.5 **Continuity**

TOPICS

1. Continuity and discontinuity at a point
2. Continuous functions
3. Properties of continuity

Continuity of a function f at a point c was defined informally in Section 2.4 as synonymous with the Direct Substitution Property. That is, if $\lim_{x \to c} f(x) = f(c)$, then we say that f is continuous at c. Continuity, as it turns out, is one of those fundamental concepts that serves as the cornerstone for a great deal of other mathematical ideas. We will devote this section to a more thorough and rigorous study of the notion.

TOPIC 1   **Continuity and Discontinuity at a Point**

As is often the case with deep, fundamental concepts, a solid understanding of continuity was not easily achieved, and it long evaded mathematicians. A once-common and imprecise definition, still often seen today, is that a continuous function is one whose graph can be drawn without lifting pen or pencil off the paper. This coincides with the actual definition of continuity in the case of sufficiently simple functions but is inadequate for our purposes in calculus.

Definition 💡

Continuity at a Point

Given a function f defined on an open interval containing c, we say f is **continuous at c** if

$$\lim_{x \to c} f(x) = f(c).$$

If the domain of f is an interval containing c either as a left or right endpoint, we also define f to be **right-continuous** or **left-continuous at c** if, respectively,

$$\lim_{x \to c^+} f(x) = f(c) \quad \text{or} \quad \lim_{x \to c^-} f(x) = f(c).$$

In usage, continuity refers to either the first or the second definition depending on the context.

This definition actually requires f to possess three properties, and it's instructive to break the definition down into these three components. Specifically, in order for f to be continuous at the point c,

1. f must be defined at c;

2. the limit of f at c must exist (one-sided limit when c is an endpoint);

3. the value of the limit must equal $f(c)$.

If any one of the three properties fails to hold, then the function f is **discontinuous** at the point c and we call c a **point of discontinuity** of f. Our first two examples illustrate the identification of points of continuity and discontinuity.

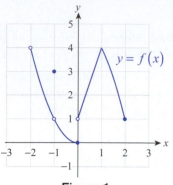

Figure 1

Example 1 ✎

Find all points of continuity as well as all points of discontinuity for the function $f(x)$ given by its graph in Figure 1. For any discontinuities, identify those from the three properties on the previous page that fail to hold.

Solution

A careful examination of the graph, paying attention to the empty and full circles, reveals that the domain for this function is $(-2,2]$. Recalling our studies of limits from the previous two sections, we see that the only point where $\lim_{x \to c} f(x)$ fails to exist is $c = 0$; though there is a discrepancy between the limit and the function value at $c = -1$: $\lim_{x \to -1} f(x) = 1$, but $f(-1) = 3$.

Thus, using the three criteria for continuity, f is continuous at every point of the following intervals: $(-2,-1)$, $(-1,0)$, and $(0,2]$. Note that f is defined, its limit exists, and the limit equals the function value at every single point in the above intervals; so they are all points of continuity, including $c = 2$, where we apply the appropriate definition for right endpoints. Thus we can summarize by specifying the set of all points of continuity as follows:

$$(-2,-1) \cup (-1,0) \cup (0,2]$$

To find the points of discontinuity, we first note that $c = -2$ is certainly one; for f is not even defined at $c = -2$ (criterion 1 fails). Therefore, while f is continuous at the right endpoint of its domain, at $c = 2$, it is discontinuous at the left endpoint, namely at $c = -2$. Next, for $c = -1$, as we mentioned above, $\lim_{x \to -1} f(x) \neq f(-1)$, so it is a point of discontinuity (criterion 3 fails). Note, however, that the definition $f(-1) = 1$ will make f continuous; that is, appropriately redefining the function will remove this discontinuity. In contrast, at $c = 0$ the one-sided limits are unequal, so $\lim_{x \to 0} f(x)$ does not even exist. This is not only a point of discontinuity (by virtue of criterion 2 failing), but one that cannot be removed by redefining the function. Such are called *jump discontinuities*; the reason for the name should be clear from the graph.

Lastly, we note that f is certainly discontinuous everywhere outside of $[-2,2]$, for it is undefined at those points (again, criterion 1 fails). Thus we can summarize the points of discontinuity of f as follows:

$$(-\infty,-2] \cup \{-1\} \cup \{0\} \cup (2,\infty)$$

Be sure you understand the use of different types of grouping symbols here.

Example 2

Identify and examine the discontinuities of the following functions.

a. $f(x)=\dfrac{x^2+x}{x}$ **b.** $g(x)=\dfrac{1}{x^2}$ **c.** $h(x)=\sin\dfrac{1}{x}$

Solution

a. The function f is a rational function, and as such it possesses the Direct Substitution Property and thus is continuous at any point where the denominator is nonzero; that is, $f(x)$ is continuous at every $x \neq 0$. Clearly, $x=0$ is a point of discontinuity since f is not defined there; however, using a bit of algebra we can actually say more. Since

$$\lim_{x\to 0} f(x)=\lim_{x\to 0}\frac{x^2+x}{x}=\lim_{x\to 0}\frac{x(x+1)}{x}=\lim_{x\to 0}(x+1)=1,$$

our conclusion is that $\lim_{x\to 0} f(x)$ actually exists, and that the graph of f agrees with that of $y=x+1$ except for the point corresponding to $x=0$, where f is undefined. In other words, you can think of the discontinuity of f arising at $x=0$ as one caused by a single point "missing" from the graph, that is, the graph being the line $y=x+1$ "punctured" at $(0,1)$, as shown in Figure 2. This is another example of a discontinuity that can be removed; the definition $f(0)=1$ will make f continuous at $x=0$.

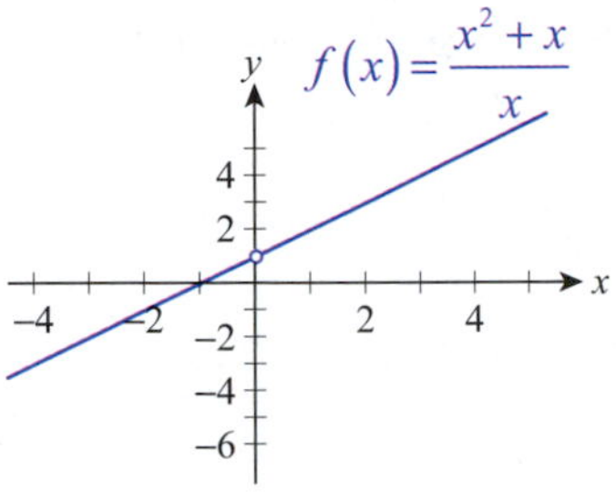

Figure 2

b. Note that g is undefined and has a vertical asymptote at $x=0$; moreover, $\lim_{x\to 0} g(x)=\infty$ (see Figure 3). We can argue, as we did in part a., that g is continuous at every $x\neq 0$, everywhere on its domain. The discontinuity at $x=0$, however, is very different from that of $f(x)$ in part a., since $\lim_{x\to 0} g(x)$ doesn't exist by virtue of the function values approaching infinity. In other words, no definition will make g continuous at 0. This type of discontinuity is called an *infinite discontinuity*.

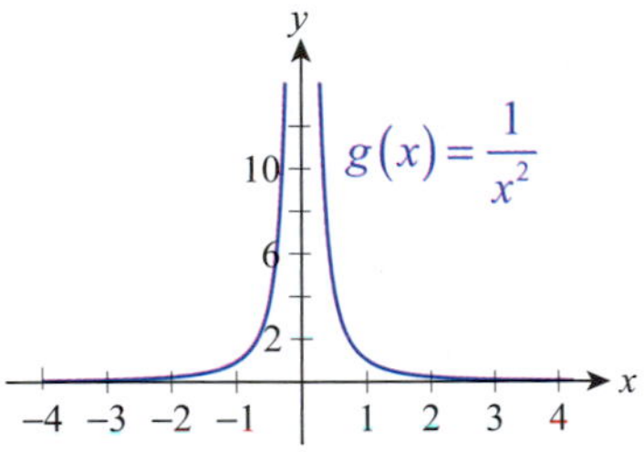

Figure 3

c. By an argument almost exactly like the one we gave in Example 7 of Section 2.3, one can show not only that h is undefined for $x=0$, but that $\lim_{x\to 0} h(x)$ does not exist either. It follows from that argument that the reason for the nonexistence of the limit is "wild" oscillation: actually, infinitely many oscillations are "squeezed" into arbitrarily small neighborhoods of 0 (see Figure 4). Even good technology cannot do full justice to what is actually going on. It is quite clear, however, that we cannot define the value of $h(0)$ so as to make h continuous. We call this an *oscillating discontinuity*.

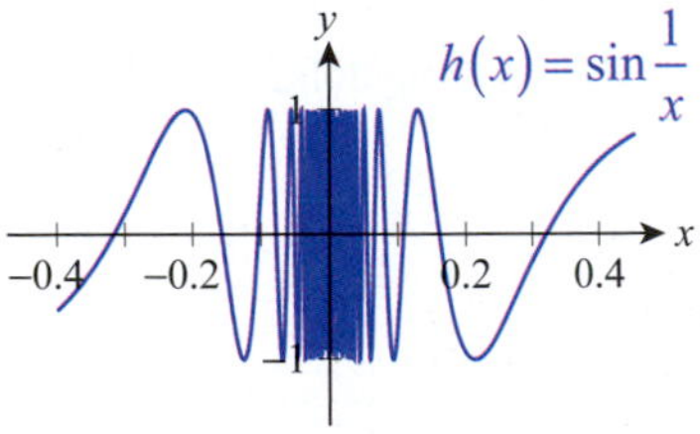

Figure 4

The points of discontinuity in Examples 1 and 2 illustrate the variety of ways in which one or more of the three properties of continuity can fail. The examples also illustrate that some discontinuities can be remedied by simply redefining the function at the point of discontinuity.

Definition 💡

Removable Discontinuity

If a function f has a point of discontinuity at c but $\lim_{x \to c} f(x)$ exists, c is called a **removable discontinuity** of f. The function can be made continuous at c by redefining f at c so that

$$f(c) = \lim_{x \to c} f(x).$$

If c is an endpoint of an interval on which f is defined, replace $\lim_{x \to c} f(x)$ with the appropriate one-sided limit.

If a given point of discontinuity is not removable, it is called **nonremovable**.

Example 3 ✐

Identify the discontinuities in Examples 1 and 2 as removable or nonremovable.

Solution

As we discussed in Example 1, f has a removable discontinuity at $c = -1$, since $\lim_{x \to -1} f(x)$ exists. This not being the case at $c = 0$, the latter is a nonremovable discontinuity. Also, since the right-hand limit of f exists at -2, it has a removable discontinuity there, which can be removed by defining $f(-2) = 4$.

As for the functions in Example 2, we have seen that f is the only one with a removable discontinuity, which occurs at $c = 0$.

The discontinuities of g and h (both at $c = 0$) are nonremovable. The reason in the case of g is that the function values approach infinity near 0, while h has infinitely many oscillations near its point of discontinuity.

If we rephrase the definition of continuity using the epsilon-delta definition of limit, we obtain the following.

Definition 💡

Epsilon-Delta Continuity at a Point

Given a function f and a point c in the domain of f, we say f is continuous at c if for every number $\varepsilon > 0$ there is a number $\delta > 0$ such that $\left| f(x) - f(c) \right| < \varepsilon$ for all x in the domain of f satisfying $0 < |x - c| < \delta$.

This alternate version of the definition makes it clear that continuity of a function at an endpoint of an interval is really the same idea as continuity elsewhere. The key idea is that f must be defined at c and that all values of $f(x)$ must be close to (meaning within ε of) the value $f(c)$ whenever x lies in the domain of f and is sufficiently close to (within δ of) the point c (see Figure 5).

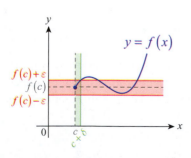

Figure 5

TOPIC 2   Continuous Functions

Once continuity at a point has been defined, it is natural to extend the meaning of continuity to larger sets. For example, we say that a function is **continuous on an interval** if it is continuous at every point of that interval. For easily graphed functions defined on an interval, this extension agrees well with the intuitive sense that continuity corresponds to the ability to construct a graph without lifting pen from paper.

Our last extension is very similar, but it contains a few subtleties that merit careful consideration.

Definition ♀

Continuity of a Function

A function f is said to be **continuous** (or **continuous on its domain**) if it is continuous at every point of its domain.

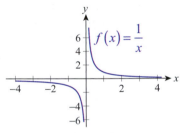

Figure 6

The first subtlety to be aware of is that a continuous function can have points of discontinuity. This seemingly paradoxical statement is entirely consistent with our definitions of continuity at a point and of continuity of a function. Consider, for example, the function $f(x) = 1/x$ (see Figure 6). The domain of this function is $(-\infty, 0) \cup (0, \infty)$, and f is indeed continuous at each point of its domain (you should verify for yourself that $\lim_{x \to c} f(x) = f(c)$ for every point c in the domain). So f is a continuous function, but it is clear that 0 is a point of discontinuity for f since $\lim_{x \to 0} f(x)$ does not exist. Further, it should be noted that f is a continuous function that is not necessarily continuous on every interval of real numbers—for instance, f is not continuous on $[-3, 3]$.

Example 4 ✐

Discuss the continuity of $f(x) = \dfrac{x^2 - 1}{x - 1}$. If applicable, identify all points of discontinuity.

Solution

Since f is a rational function, we know that it satisfies the Direct Substitution Property and is thus continuous everywhere except at $c = 1$, where the denominator is 0. Note that $f(1)$ is undefined, making $c = 1$ the only point of discontinuity for f. So once again, our seemingly paradoxical, but correct, conclusion is that f is a continuous function, and its only point of discontinuity is at $c = 1$.

As a final remark, since $f(x) = \dfrac{x^2 - 1}{x - 1} = \dfrac{(x-1)(x+1)}{x-1} = x + 1$ for all $x \neq 1$, we identify the above discontinuity as removable, since the definition $f(1) = 2$ makes f continuous on the entire real line.

Example 5 ✎

Recall $g(x) = 1/x^2$ from Example 2. As we have mentioned, its only point of discontinuity is at $c = 0$, where it is undefined, but g is continuous everywhere on its domain. In this example, we will give a rigorous proof of this latter claim, using the epsilon-delta continuity definition.

Indeed, choose and fix an arbitrary, nonzero $c \in \mathbb{R}$. We will proceed to prove that $g(x) = 1/x^2$ is continuous at c. Note that we may assume $c > 0$ (in the negative case, the argument is similar, or one may take advantage of the fact that g is an even function).

Our epsilon-delta argument will be similar to those given in earlier examples to prove limit claims (this shouldn't come as a surprise, since our epsilon-delta continuity definition is itself closely related to the formal definition of limit). To start us off, suppose that $\varepsilon > 0$ is given. We need to find a $\delta > 0$ such that

$$0 < |x - c| < \delta \quad \Rightarrow \quad \left| \frac{1}{x^2} - \frac{1}{c^2} \right| < \varepsilon.$$

We will first try to bound the quantity $\left| 1/x^2 - \left(1/c^2 \right) \right|$ in order to get an idea of just how small δ needs to be to satisfy the above inequality. Observe that

$$\left| \frac{1}{x^2} - \frac{1}{c^2} \right| = \left| \frac{c^2 - x^2}{x^2 c^2} \right| = \frac{\left| (c - x)(c + x) \right|}{x^2 c^2} = |x - c| \frac{x + c}{x^2 c^2}.$$

Here we are safe to have omitted the absolute value symbols from the last factor, since we can assume that x is close enough to c so that $x + c$ is positive. In fact, we will insist that $\frac{1}{2} c < x < \frac{3}{2} c$ throughout this argument. Note that this is equivalent to requiring that $\delta < \frac{1}{2} c$, which is not a loss of generality since once a successful $\delta > 0$ is chosen any smaller number works just fine.

Next, let's examine the quantity $(x + c)/(x^2 c^2)$. Since x is in the interval specified above, $x + c < \frac{3}{2} c + c = \frac{5}{2} c$, and $x^2 c^2 > \left(\frac{1}{2} c \right)^2 c^2 = \frac{1}{4} c^4$. Thus we have

$$|x - c| \frac{x + c}{x^2 c^2} < |x - c| \frac{\frac{5}{2} c}{\frac{1}{4} c^4} = |x - c| \frac{10}{c^3}.$$

Note that since c is fixed throughout this argument, so is $10/c^3$. Putting the above two chains of equalities and inequalities together, we can summarize what we have obtained so far. If x is sufficiently close to c,

$$\left| \frac{1}{x^2} - \frac{1}{c^2} \right| < |x - c| \frac{10}{c^3}.$$

Now it is time to pick a $\delta > 0$ for the given ε. Let us choose it to be less than the smaller of the numbers $c/2$ and $\left(c^3/10 \right) \varepsilon$. Then if $|x - c| < \delta$,

$$\left| \frac{1}{x^2} - \frac{1}{c^2} \right| < |x - c| \frac{10}{c^3} < \delta \frac{10}{c^3} < \frac{c^3}{10} \varepsilon \frac{10}{c^3} = \varepsilon,$$

and our proof is complete.

TOPIC 3   **Properties of Continuity**

Just as the limit laws greatly simplify the task of determining whether a function has a limit at a given point, we have a number of properties of continuity that allow us to quickly answer many questions regarding continuity of functions.

Theorem

Properties of Continuous Functions

Let f and g be two functions both continuous at the point c, and let k be a fixed real number. Then the following combinations of f and g are also continuous at c.

Sums	$f + g$	"A sum of continuous functions is continuous."
Differences	$f - g$	"A difference of continuous functions is continuous."
Constant Multiples	$k \cdot f$	"A multiple of a continuous function is continuous."
Products	$f \cdot g$	"A product of continuous functions is continuous."
Quotients	$\dfrac{f}{g}$, provided $g(c) \neq 0$	"A quotient of continuous functions is continuous."

These properties follow immediately from the corresponding limit properties. As an example, we will prove the fourth statement above.

Proof

We assume that f and g are both continuous at c.

$$\lim_{x \to c}(f \cdot g)(x) = \lim_{x \to c}\left[f(x)g(x)\right]$$

$$= \lim_{x \to c} f(x) \cdot \lim_{x \to c} g(x) \quad \text{The limit of a product is the product of the limits.}$$

$$= f(c) \cdot g(c) \quad \text{Both } f \text{ and } g \text{ are continuous at } c.$$

$$= (f \cdot g)(c)$$

As a consequence of these properties, polynomial and rational functions can be immediately classified as continuous functions. That is, every polynomial function and every rational function is continuous at every point of its domain.

Example 6 ✐

To illustrate the fact that a rational function is continuous everywhere on its domain, everywhere except for the zeros of the denominator, let us consider

$$r(x) = \frac{x^2 - 2x}{x^2 - 3x + 2}.$$

Since the denominator factors as $x^2 - 3x + 2 = (x-2)(x-1)$, we see that the domain of r is the set $D = \{x \in \mathbb{R} \mid x \neq 1, 2\}$. If $c \in D$, that is, c is an arbitrary real number other than 1 or 2, the Direct Substitution Property applies, and $r(c) = \lim_{x \to c} r(x)$, showing that r is continuous at every point of its domain. Using our limit determination techniques, we can even find out what happens at the two points of discontinuity. Since

$$r(x) = \frac{x^2 - 2x}{x^2 - 3x + 2} = \frac{x(x-2)}{(x-1)(x-2)} = \frac{x}{x-1},$$

it follows that $x = 2$ is a removable discontinuity, and the graph of r agrees with that of $g(x) = x/(x-1)$ for all $x \neq 2$. The discontinuity at $x = 1$, however, is nonremovable. Since the numerator is bounded near 1, while the denominator nears 0, r has a vertical asymptote at $x = 1$. Notice how the graph in Figure 7 supports all of our findings.

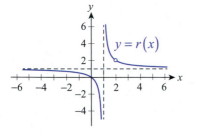

Figure 7

Since sums, differences, products, and quotients of continuous functions are continuous, it is natural to ask if continuity is preserved under other combinations of functions. The following theorem is a bridge toward an important answer to that question.

Theorem 🔍

"Limits Pass Through a Continuous Function"

Suppose $\lim_{x \to c} g(x) = a$ and f is continuous at the point a. Then

$$\lim_{x \to c} f(g(x)) = f\left(\lim_{x \to c} g(x)\right) = f(a).$$

In words, we say the limit operation passes inside the continuous function f.

The proof of this theorem can be found in Appendix E—it is a nice example of the use of the epsilon-delta definitions of both limit and continuity. At the moment, though, we are more interested in the following application of the theorem.

Theorem 🔍

"A Composition of Continuous Functions Is Continuous"

If g is continuous at the point c, and if f is continuous at $g(c)$, then the composite function $f \circ g$ is continuous at c.

Proof ✎

Because g is continuous at c, we know that $\lim\limits_{x \to c} g(x) = g(c)$, and so by replacing a with $g(c)$ in the previous theorem it follows that

$$\lim_{x \to c} f\big(g(x)\big) = f\Big(\lim_{x \to c} g(x)\Big) = f\big(g(c)\big).$$

Using composition notation, we have

$$\lim_{x \to c} (f \circ g)(x) = (f \circ g)(c).$$

Another useful theorem tells us that inverses of continuous functions are continuous. Its proof can also be found in Appendix E.

Theorem ⚷

"The Inverse of a Continuous Function Is Continuous"

If f is one-to-one and continuous on the interval (a, b), then f^{-1} is also a continuous function.

With these theorems in hand, the set of functions that we can classify as continuous instantly expands considerably.

Example 7 ✐

Use the above theorems to discuss the continuity of the following functions.

a. $F(x) = \sqrt[3]{\dfrac{2x+1}{x^2+5}}$
  b. $G(x) = \sqrt{\dfrac{4x^5 - 3x^2 + 7}{9x^3 + 3x}}$

c. $H(x) = \sin\dfrac{2x^3 + x^2 - 3}{x^2 - 2x - 8}$
  d. $K(x) = \arcsin\sqrt{1 - x^2}$

Solution

a. Notice that F is a composite function; you can think of it as $F(x) = f\big(g(x)\big)$, where $g(x) = (2x+1)/(x^2+5)$, and $f(x) = \sqrt[3]{x}$. Furthermore, g is continuous on all of $\mathbb{R}$, since its denominator is never 0. The same can be said about f, namely that it is continuous on the entire real line. Note that you can see this in at least two ways. Either apply the previous theorem, since f is the inverse of $k(x) = x^3$, which is a continuous function mapping $\mathbb{R}$ onto $\mathbb{R}$; or by referring to the Positive Integer Root Law, you can provide a one-line proof of the continuity of f as follows. If $c \in \mathbb{R}$ is arbitrary,

$$\lim_{x \to c} f(x) = \lim_{x \to c} \sqrt[3]{x} = \sqrt[3]{\lim_{x \to c} x} = \sqrt[3]{c}.$$

Now a simple application of the earlier theorem about continuity of compositions establishes the fact that $F(x) = f\big(g(x)\big)$ is continuous on $\mathbb{R}$.

b. As in part a., we will use the "continuity of compositions" theorem, but while the conditions of the theorem were readily satisfied by both the inner and outer functions on all of $\mathbb{R}$ in the previous problem, we will need to pay a little extra attention here. To start off, we once again use the notation $G(x) = f(g(x))$; however, this time $g(x) = (4x^5 - 3x^2 + 7)/(9x^3 + 3x)$ (see Figure 8), while $f(x) = \sqrt{x}$. According to our theorem, $G = f \circ g$ will be continuous at all points of continuity c of $g(x)$ provided that f is also continuous at $g(c)$. A factoring argument like the one given in part a. shows that g is undefined and has a vertical asymptote at $x = 0$; it is, however, continuous everywhere else. The outer function f, on the other hand, is only defined for $x \geq 0$, but it is continuous on its domain. This latter claim can be established by an argument similar to the one given for the cube root function in part a. In fact, notice that it is easy to generalize that argument and conclude that all root functions are continuous as a result of the previous theorem and the continuity of the functions x^n.

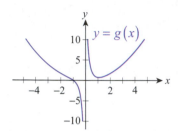

Figure 8

Summarizing our observations up to this point, and using the fact that the composition of continuous functions is continuous, we see that $G(x) = f(g(x))$ will be continuous at all nonzero real numbers c such that $g(c) \geq 0$.

Our next observation is that $g(x) < 0$ on the interval $(-1, 0)$, but is nonnegative everywhere else on its domain, that is, $g(x) \geq 0$ on $(-\infty, -1] \cup (0, \infty)$. (You can algebraically convince yourself of this fact, or simply examine the graph in Figure 8.)

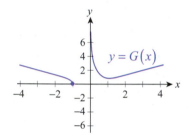

Figure 9

Therefore, we conclude that $G(x) = f(g(x))$ is continuous on the set $(-\infty, -1] \cup (0, \infty)$. The graph of G supports our conclusion (see Figure 9).

c. Factoring the denominator easily shows that the inner function of this problem $g(x) = (2x^3 + x^2 - 3)/(x^2 - 2x - 8)$ has two vertical asymptotes, at $x = -2$ and $x = 4$, respectively. However, being a rational function, g is continuous everywhere else on $\mathbb{R}$. As we will show in Example 8, $f(x) = \sin x$ is continuous at every real number, so our conclusion is that $H(x)$ is continuous everywhere except at $x = -2$ and $x = 4$.

d. First of all, we observe that $g(x) = \sqrt{1 - x^2}$ is continuous everywhere on its domain of $[-1, 1]$. Furthermore, the range of g is the interval $[0, 1]$ (note that the graph of g is the upper semicircle of radius 1, centered at the origin). Using the theorem about the continuity of inverse functions, we conclude that $f(x) = \arcsin x$ is continuous at every point of the range of g and hence $K(x) = f(g(x))$ is continuous on its domain of $[-1, 1]$.

In fact, most familiar functions (those that we bother to give names to) are continuous (though many have points of discontinuity—don't forget the subtle distinction). Trigonometric functions are continuous, a fact that is most easily proven through the use of an alternate formulation of continuity:

$$f \text{ is continuous at the point } c \text{ if and only if } \lim_{h \to 0} f(c+h) = f(c).$$

Example 8 ✎

Use the alternate formulation of continuity to prove that $f(x) = \sin x$ is continuous on $\mathbb{R}$.

Solution

Our goal is to show that for any fixed real number c, $\lim_{h \to 0} \sin(c+h) = \sin c$.

Using the appropriate sum identity known from trigonometry along with the relevant limit laws,

$$
\begin{aligned}
\lim_{h \to 0} \sin(c+h) &= \lim_{h \to 0} \left(\sin c \, \cos h + \cos c \, \sin h \right) \\
&= \lim_{h \to 0} \left(\sin c \, \cos h \right) + \lim_{h \to 0} \left(\cos c \, \sin h \right) \\
&= \sin c \, \lim_{h \to 0} \cos h + \cos c \, \lim_{h \to 0} \sin h,
\end{aligned}
$$

where we also used the fact that during the limit process $\sin c$ and $\cos c$ are constants, so the Constant Multiple Law applies.

Our next step is to recall that as h approaches 0, $\lim_{h \to 0} \cos h = 1$ and $\lim_{h \to 0} \sin h = 0$. (You may recall these facts from the unit-circle definition of sine and cosine for small angles, or simply from the graphs of the sine and cosine functions.) Using these latter two limits we obtain

$$
\begin{aligned}
\sin c \, \lim_{h \to 0} \cos h + \cos c \, \lim_{h \to 0} \sin h &= \sin c \cdot 1 + \cos c \cdot 0 \\
&= \sin c,
\end{aligned}
$$

which is what we needed to prove.

Finally, we note that the continuity of $\cos x$ can be established by a similar argument, or by using the continuity of the sine function along with the identity $\cos x = \sin((\pi/2) - x)$. Then using the properties of continuous functions, the continuity of the remaining four trigonometric functions $\tan x$, $\cot x$, $\sec x$, and $\csc x$ will follow.

Exponential functions are continuous, a fact largely due to design. For a given base a, the expression $a^{m/n}$ is easily understood for every rational number m/n, and a common way to extend the meaning of a^x to irrational numbers is to define a^x to be the limit of terms of the form a^{r_n}, where $\{r_n\}$ is a sequence of rational numbers approaching x (some work is required in order to show such a definition is unambiguous). And because inverse functions of continuous functions are continuous, the continuity of logarithms and inverse trigonometric functions is assured.

Example 9 ✐

Use the above theorems to identify where the following functions are continuous.

a. $f(x) = \dfrac{\arctan(\ln x)}{x^2 + 1}$ **b.** $g(x) = \dfrac{xe^{\cos x}}{\sqrt{x^2 - 1} - 3}$ **c.** $h(x) = \dfrac{\ln x - \sin^{-1} x}{x^2 + x - 2}$

Solution

a. Being the inverses of continuous functions, $\ln x$ is continuous on $(0, \infty)$, and $\arctan x$ is continuous on the entire real line. Thus the composition $\arctan(\ln x)$ is continuous on $(0, \infty)$. Since the denominator of f, $x^2 + 1$ is never 0 on $\mathbb{R}$ (check this), it follows that the quotient $f(x)$ is continuous on $(0, \infty)$.

b. Since all three functions $\cos x$, e^x, and x are continuous on the entire real line, it follows from our theorems that both the composition $e^{\cos x}$ and the product $xe^{\cos x}$ will be continuous on all of $\mathbb{R}$. Next, just as we did in part a., we would like to use the theorem stating the continuity of the quotient of continuous functions, but we need to carefully check the continuity of the denominator of g. First, the square root function requires that the inequality $x^2 - 1 \geq 0$ be true for x. This means, we need $x^2 \geq 1$ to hold, that is, $x \geq 1$ or $x \leq -1$. Since the composition of continuous functions is continuous, we conclude that $\sqrt{x^2 - 1}$ is continuous on $(-\infty, -1] \cup [1, \infty)$. Last, but not least, in order for g to be continuous, its denominator cannot be 0, so we have to exclude the solutions of the equation $\sqrt{x^2 - 1} = 3$. By squaring both sides and adding 1 we obtain $x^2 = 10$, that is, the points $x = \pm\sqrt{10}$ have to be excluded from the domain.

Summarizing our findings, we conclude that $g(x)$ is continuous on the following set:

$$\left(-\infty, -\sqrt{10}\right) \cup \left(-\sqrt{10}, -1\right] \cup \left[1, \sqrt{10}\right) \cup \left(\sqrt{10}, \infty\right)$$

c. Arguing as we did in part a., $\ln x$ and $\sin^{-1} x$ are continuous, since they are inverses of continuous functions. The domain of $\ln x$ is $(0, \infty)$, while $\sin^{-1} x$ is only defined on $[-1, 1]$. However, notice that $x = 1$ is a zero of the denominator, so by the properties of continuous functions, we conclude that $h(x)$ is continuous on the open interval $(0, 1)$.

Collectively, the theorems in this section allow us to identify a large number of continuous functions very quickly. It might even be possible to do something about the occasional point of discontinuity. Recall that a discontinuity is termed *removable* if the limit at the point exists; in such cases, we can define the **continuous extension** of a function as in the next example.

Example 10 ✐

Identify the removable discontinuities and define the continuous extension of $r(x) = \dfrac{x^2 - 5x}{x^3 - 3x^2 - 10x}$.

Solution

Factoring and canceling yields

$$r(x) = \frac{x^2 - 5x}{x^3 - 3x^2 - 10x} = \frac{x(x-5)}{x(x-5)(x+2)} = \frac{1}{x+2},$$

where the previous equality holds for all x, except for $x = 0$ and $x = 5$. However, as we learned in Section 2.4, both $\lim\limits_{x \to 0} r(x)$ and $\lim\limits_{x \to 5} r(x)$ still exist.

$$\lim_{x \to 0} r(x) = \frac{1}{0+2} = \frac{1}{2}, \quad \text{while} \quad \lim_{x \to 5} r(x) = \frac{1}{5+2} = \frac{1}{7}$$

Therefore r has two removable discontinuities, at $x = 0$ and $x = 5$. Now we can define $\tilde{r}(x)$, the continuous extension of r, as follows.

$$\tilde{r}(x) = \begin{cases} \dfrac{x^2 - 5x}{x^3 - 3x^2 - 10x} & \text{if } x \neq 0 \text{ or } 5 \\[2mm] \dfrac{1}{2} & \text{if } x = 0 \\[2mm] \dfrac{1}{7} & \text{if } x = 5 \end{cases}$$

Let us note, however, that although correct and instructive, it is not necessary in this case to use a piecewise defined function to define the continuous extension of r. Alternatively, it is much shorter to define it simply as $\tilde{r}(x) = 1/(x+2)$. Note that this is equivalent to our piecewise definition above. Note also that both r and its continuous extension have nonremovable discontinuities at $x = -2$.

We will close this section with one last theorem that makes a connection between our rigorous definition of continuity and the intuitive "graph without interruptions" interpretation of continuity.

Theorem ⚷

The Intermediate Value Theorem

If f is a continuous function defined on the closed interval $[a,b]$, then f takes on every value between $f(a)$ and $f(b)$. That is, if L is a real number between $f(a)$ and $f(b)$, then there is a c in the interval $[a,b]$ such that $f(c) = L$.

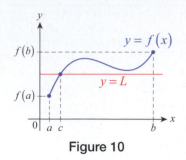

Figure 10

The proof of the Intermediate Value Theorem (IVT) relies upon a property of the real numbers called *completeness*; this concept and the proof of the theorem are presented in courses like Advanced Calculus and Introductory Real Analysis. But the implications of the Intermediate Value Theorem are of interest to us at the moment. Informally, the IVT says that the graph of a continuous function f defined on an interval $[a,b]$ cannot avoid intersecting any horizontal line $y = L$ if L is a value between $f(a)$ and $f(b)$ (see Figure 10). In other words, such a function f has no "breaks" or "jumps" that would allow it to skip over the value L. This property, also referred to as the *Intermediate Value Property*, is often used to prove that an equation of the form $f(x) = L$ must have at least one solution in the interval $[a,b]$. Note that the content of the IVT is precisely the fact that a continuous function f on a closed interval $[a,b]$ possesses the Intermediate Value Property.

Example 11 ✐

Use the Intermediate Value Theorem to show that the equation $x^5 + 9x - 4 = 0$ has a solution between 0 and 1.

Solution

Introducing the notation $f(x) = x^5 + 9x - 4$, notice that we are attempting to prove the existence of a solution, or root, of the equation $f(x) = 0$ on the interval $[0,1]$. The existence of such root c would mean that f takes on the value $L = 0$ for some c in $[0,1]$. Does the Intermediate Value Theorem guarantee that? Notice first of all that f is certainly continuous on the closed interval $[0,1]$ (actually, it is continuous on the entire real line, but for the purposes of this problem we can safely ignore what happens outside of $[0,1]$). Next, since 0 and 1 play the roles of a and b in the theorem, respectively, we proceed to examine the values $f(0)$ and $f(1)$.

$$f(0) = 0^5 + 9 \cdot 0 - 4 = -4 < 0$$
$$f(1) = 1^5 + 9 \cdot 1 - 4 = 6 > 0$$

In other words, a change of sign occurs on $[0,1]$; that is, f goes from negative to positive on $[0,1]$. The statement of the Intermediate Value Theorem is precisely the fact that a continuous function cannot do this and avoid intersecting the horizontal line $y = 0$ (the x-axis). More precisely, since $L = 0$ we have $-4 < L < 6$; that is, since 0 is a real number between $f(0) = -4$ and $f(1) = 6$, by the Intermediate Value Theorem there is a c in $[0,1]$ such that $f(c) = L = 0$. In other words, we have

$$c^5 + 9c - 4 = 0.$$

Thus we proved the existence of a root between 0 and 1 for the given equation.

As an important final remark, we note that the Intermediate Value Theorem is an *existence theorem*, in other words, we proved the *existence* of a solution between 0 and 1, without actually specifying what it is. However, in lieu of a formula to solve fifth-degree equations, this is still extremely useful, and there is no need to stop here. Namely, if we know a solution exists, we can "narrow down" the interval around it, by refining our guesses, or by "zeroing in" on the solution. Eventually, we can come up with an approximation of the root with an acceptable level of accuracy, which is the basic idea of so-called *numerical methods* of solving equations. The following Technology Note illustrates this important idea.

Technology Note 🖥

Continuing Example 11 by using *Mathematica*, we will illustrate the process of zooming in successively on the root of $x^5 + 9x - 4 = 0$, thereby obtaining an approximation of the solution accurate to two digits after the decimal point. Then we will use the built-in **NSolve** command to obtain *Mathematica*'s numerical approximation of the root.

To start us off, we have graphed $f(x) = x^5 + 9x - 4$ (see Figure 11), and then we zoomed in to obtain that portion of the graph defined on $[0, 1]$ (see Figure 12).

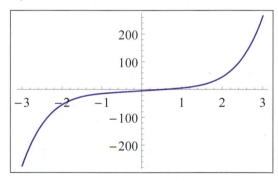

Figure 11

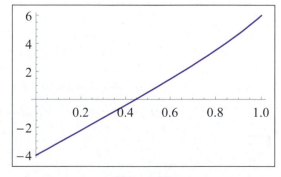

Figure 12

It is clear from both graphs that $f(x) = 0$ has a root between 0 and 1, as we proved in Example 11. Figure 12 suggests that the root is between 0.4 and 0.5, so let's zoom in further to gain more accuracy.

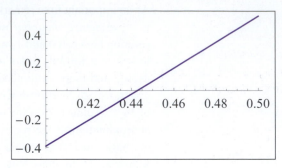

Figure 13

Figure 13 suggests that we should zoom in on that portion of the graph defined on $[0.44, 0.445]$, which we have done (see Figure 14).

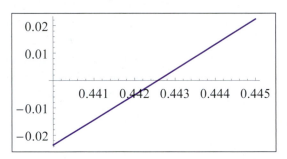

Figure 14

It is clear from Figure 14 that the root is located between 0.442 and 0.443, so we have achieved the desired accuracy; the first two digits after the decimal point are correct. We conclude that the root is $c \approx 0.44$. (Note that the above accuracy could actually have been achieved from carefully eyeballing the graph in Figure 13, while the graph in Figure 14 says even more: we might guess that the root is in fact very close to 0.4426.)

Note that it is also possible to estimate the root by appropriately zooming in on the graph using a graphing calculator (see Figure 15).

Finally, we use the **NSolve** command of *Mathematica* to approximate the solution of $f(x) = 0$. The screenshot in Figure 16 shows the feedback we receive from the software. The appearance of complex roots shows the power of the software, but we can ignore them for now and focus on the sole real root. *Mathematica* approximates it as $c \approx 0.442558$.

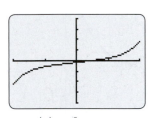

$f(x) = x^5 + 9x - 4$
on $[-2,2]$ by $[-100,100]$

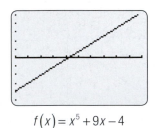

$f(x) = x^5 + 9x - 4$
on $[0.4,0.5]$ by $[-0.5,0.5]$

Figure 15 Zooming In on a Root with a Graphing Calculator

```
In[1]:= NSolve[x^5 + 9 x - 4 == 0, x]

Out[1]= {{x → -1.32389 - 1.2337 i},
         {x → -1.32389 + 1.2337 i}, {x → 0.442558},
         {x → 1.10261 - 1.24271 i}, {x → 1.10261 + 1.24271 i}}
```

Figure 16

2.5 **Exercises**

1–2 Find all points of continuity as well as all points of discontinuity for the given function. For any discontinuities, identify those from the three continuity criteria that fail to hold.

1.

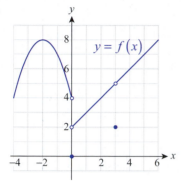

2.

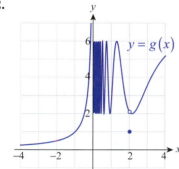

3. Sketch a graph of a function (a formula is not necessary) that has a removable discontinuity at $x = -1$, a jump discontinuity at $x = 2$, but is right-continuous at 2. (Answers will vary.)

4. Sketch a graph of a function that has an infinite discontinuity at $x = 0$ and an oscillating discontinuity at $x = 5$ so that it is still left-continuous at 5. (Answers will vary.)

5–29 Find and classify the discontinuities (if any) of the function as removable or nonremovable.

5. $f(x) = \dfrac{1}{x}$ **6.** $g(x) = \dfrac{-2}{x-3}$

7. $h(x) = \dfrac{x^2 - 9}{x - 3}$ **8.** $k(x) = \dfrac{x^2 - 2x}{x^2 + 5x - 14}$

9. $u(x) = \dfrac{x^2 - 9}{x - 2}$ **10.** $v(x) = \dfrac{x - 1}{x^2 + 2x - 3}$

11. $w(x) = \begin{cases} x + 1 & \text{if } x \le 0 \\ \dfrac{1}{2}x^2 + 1 & \text{if } x > 0 \end{cases}$

12. $f(x) = \begin{cases} \dfrac{1}{2}x - 2 & \text{if } x \le 4 \\ x^3 + 1 & \text{if } x > 4 \end{cases}$

13. $g(x) = \begin{cases} \tan x & \text{if } x \le \dfrac{\pi}{2} \\ \cos x & \text{if } x > \dfrac{\pi}{2} \end{cases}$

14. $h(x) = \begin{cases} \cos x & \text{if } x \le 0 \\ \tan x + 1 & \text{if } x > 0 \end{cases}$

15. $F(u) = \dfrac{u - 4}{\sqrt{u} - 2}, \quad u \ge 0$

16. $G(s) = \dfrac{s}{\sqrt{s + 4} - 2}, \quad s \ge -4$

17. $H(t) = \dfrac{t}{\sqrt{t^2 + 2}}$ **18.** $u(x) = \cos\dfrac{1 - x^2}{1 - x}$

19. $v(t) = |\sin t|$

20. $K(x) = |x + 2| + |x - 1|$

21. $F(t) = \dfrac{t}{t^2 - 1}$ **22.** $G(t) = \dfrac{t}{t^2 + 1}$

23. $H(x) = |x + 2|$ **24.** $k(x) = \dfrac{|x - 4|}{x - 4}$

25. $s(x) = [\![x + 2]\!]$ **26.** $t(x) = 4 - [\![x]\!]$

27. $u(z) = [\![z^2]\!]$ **28.** $v(x) = x[\![x]\!]$

29. $w(x) = x\left[\!\!\left[\dfrac{1}{x}\right]\!\!\right]$

30–33 Use the ε-δ definition to prove that the function is continuous.

30. $f(x) = \dfrac{1}{x}$

31. $g(x) = 3x - 2$

32. $F(x) = x^3$

33. $G(x) = \sqrt{x}$

34–39 Use the theorems of this section to discuss the continuity of the function.

34. $F(x) = \sqrt{\dfrac{x}{x^2 + 7x + 12}}$

35. $G(x) = \sqrt{\dfrac{x^4 - x^3 - 11x^2 + 9x + 18}{2x^3 + x}}$

36. $H(x) = \cos\dfrac{2\ln(x-3)+1}{\sqrt[3]{x^2 - 2x - 15}}$

37. $f(x) = \arctan\dfrac{x}{\sqrt{3 - x^2}}$

38. $g(x) = \ln(\arcsin(\pi x + 1))$

39. $h(x) = \dfrac{\csc(\pi x + 1)}{\sin(\pi e^{x+2})}$

40. Prove the alternate formulation of continuity, that is, the statement that a function f is continuous at the point c if and only if $\lim\limits_{h \to 0} f(c+h) = f(c)$.

41–44 Use the alternate formulation of continuity to prove that the function is continuous.

41. $f(x) = 3x - 5x^2$

42. $g(x) = \cos x$

43. $h(x) = \tan x$

44. $k(x) = e^x$

45–50 Identify the removable discontinuities and define the continuous extension of the function.

45. $f(x) = \dfrac{x^2 + x - 12}{x - 3}$

46. $g(x) = \dfrac{x^3 - 2x^2 - x + 2}{x^2 - 3x + 2}$

47. $h(x) = \dfrac{x - 1}{\sqrt{x} - 1}$

48. $F(x) = \dfrac{\sqrt{x+1} - 2}{x - 3}$

49. $G(x) = 2^{-1/x^2}$

50. $H(x) = x\cos\dfrac{\pi}{x}$

51–54 Discuss the continuity of the function on the given closed interval.

51. $S(x) = \sqrt{16 - x^2}$ on $[-4, 4]$

52. $T(x) = \left\|\dfrac{x}{3}\right\|$ on $[0, 3]$

53. $U(x) = \begin{cases} \dfrac{1}{x^2 - 9} & \text{if } |x| < 3 \\ 0 & \text{if } |x| = 3 \end{cases}$ on $[-3, 3]$

54. $V(x) = \begin{cases} x^2 \sin\dfrac{1}{x} & \text{if } x \neq 0 \\ 0 & \text{if } x = 0 \end{cases}$ on $\left[0, \dfrac{1}{\pi}\right]$

55-58 Find the value of a (or the values of a and b, where applicable) such that f is continuous on the entire real line.

55. $f(x) = \begin{cases} 0 & \text{if } x \leq 0 \\ ax & \text{if } 0 < x < 1 \\ 2x + 3 & \text{if } x \geq 1 \end{cases}$

56. $f(x) = \begin{cases} x^3 & \text{if } x \leq 3 \\ ax^2 & \text{if } x > 3 \end{cases}$

57. $f(x) = \begin{cases} -x^2 & \text{if } x < 1 \\ ax + b & \text{if } 1 \leq x \leq 3 \\ (x-3)^2 + 2 & \text{if } x > 3 \end{cases}$

58. $f(x) = \begin{cases} \cos x & \text{if } x \leq 0 \\ -(x-a)^2 + b & \text{if } 0 < x < 2 \\ \dfrac{1}{2}x - 2 & \text{if } x \geq 2 \end{cases}$

59. Prove that if $f(x)$ is continuous and $f(c) > 0$, then there is a $\delta > 0$ such that $f(x) > 0$ for all $x \neq c$ in the interval $(c - \delta, c + \delta)$.

60. Prove that the Dirichlet function

$$\xi(x) = \begin{cases} 0 & \text{if } x \text{ is rational} \\ 1 & \text{if } x \text{ is irrational} \end{cases}$$

is discontinuous at every real number.

61. Prove that the function

$$f(x) = \begin{cases} 0 & \text{if } x \text{ is rational} \\ x^2 & \text{if } x \text{ is irrational} \end{cases}$$

is continuous only at the single point $c = 0$.

62. Prove that if the functions f and g are both continuous on $\mathbb{R}$ and they agree on the rationals (i.e., $f(x) = g(x)$ for all $x \in \mathbb{Q}$), then $f = g$.

63–68 Decide whether the Intermediate Value Theorem applies to the given function on the indicated interval. If so, find c as guaranteed by the theorem. If not, find the reason.

63. $f(x) = -x^2 + x + 3$ on $[0,3]$; $f(c) = 1$

64. $g(x) = 2x^3 - x^2 - 1$ on $[-1,2]$; $g(c) = 0$

65. $h(x) = \dfrac{x}{x+2}$ on $[0,4]$; $h(c) = \dfrac{1}{2}$

66. $F(x) = \dfrac{2x}{x-1}$ on $[0,2]$; $F(c) = 2$

67. $G(x) = [\![x-2]\!]$ on $[-2,2]$; $G(c) = -\dfrac{1}{2}$

68. $H(x) = \sin\dfrac{3x+2}{2}$ on $\left[-\dfrac{2}{3}, \dfrac{\pi-2}{3}\right]$; $H(c) = \dfrac{1}{2}$

69–74 Use the Intermediate Value Theorem to prove that the given equation has a solution on the indicated interval.

69. $x^3 - 7.5x^2 + 1.2x + 1 = 0$ on $[-1,0]$

70. $2x^3 + x + 10 = 0$ on $[-2,1]$

71. $\cos x = x^2$ on $[0,\pi]$

72. $\ln x - \sqrt{x-2} = 0$ on $[2,5]$

73. $\dfrac{5}{x^2+2} = 1$ on $[-3,-1]$

74. $\cot\dfrac{\pi x}{4} - \dfrac{x}{x+2} = -\dfrac{1}{2}$ on $[1,3]$

75. Suppose that the outside temperature in Columbia, SC on a summer morning at 7:00 a.m. is 74 °F, and it shoots up to 98 °F by 1:00 p.m. Assuming that temperature changes continuously, prove that sometime between 7:00 a.m. and 1:00 p.m. the temperature was exactly 88.35 °F.

76. Prove that if f is continuous and never 0 on the interval $[a,b]$, then either $f(x) > 0$ for every x in $[a,b]$, or $f(x) < 0$ for every x in $[a,b]$.

77.* (Existence of nth roots) Prove that if b is a positive real number and n a positive integer, then there is a positive real number c such that $c^n = b$. (**Hint:** Consider the continuous function $f(x) = x^n$ on the interval $[0, b+1]$.)

78.* Prove that a circle of diameter d has a chord of length c for every number c between 0 and d.

79. Use the function

$$f(x) = \begin{cases} \sin\dfrac{\pi}{x} & \text{if } x \neq 0 \\ 0 & \text{if } x = 0 \end{cases}$$

to prove that the converse of the Intermediate Value Theorem is false; in other words, a function may possess the Intermediate Value Property without being continuous.

80.* (Fixed Point Theorem) Prove that if the function $f : [a,b] \to [a,b]$ is continuous, then there is a number c in $[a,b]$ with $f(c) = c$ (i.e., c is "fixed," or "not being moved," by f).

81.* A hermit leaves his hut at the foot of a mountain one day at 6:00 a.m. and sets out to climb all the way to the top. He arrives at 6:00 p.m. and realizes that it is too late to go back, so he sets up camp for the night. At 6:00 a.m. the following day, he starts hiking back to his hut, taking the exact same route as the day before. This time, however, it is mostly downhill, so he makes much better time and arrives home at 2:00 p.m. Prove that there is a point along the hermit's route that he passed at exactly the same time on both days. (**Hint:** Apply the Intermediate Value Theorem or the Fixed Point Theorem.)

82. A long-distance phone company charges 31 cents for the first minute and 10 cents for each additional minute or any fraction thereof. Graph the cost as a function of time, find a formula for it, and discuss the significance of its discontinuities. (**Hint:** Use the greatest integer function to construct your answer.)

83.* If Δt denotes the length of the time interval between two events as measured by an observer on a spaceship moving at speed v, and ΔT is the length of the same time interval as measured from Earth, then the formula relating the two quantities is given by

$$\Delta T = \frac{\Delta t}{\sqrt{1 - \dfrac{v^2}{c^2}}},$$

where c is the speed of light. This phenomenon is called *time dilation*, and it follows from the theory of relativity. In essence, it says that a clock moving at speed v relative to an observer is perceived by the same observer to run slower.

a. Explain why we don't normally notice the time dilation effect in everyday life.

b. What is the significance of the discontinuity of ΔT (as a function of v)?

84–88 *True or False?* Determine whether the given statement is true or false. In case of a false statement, explain or provide a counterexample.

84. If f is both left- and right-continuous at c, then f is continuous at c.

85. Any function f has an interval (a,b) on which it is continuous.

86. If c is a discontinuity of f, but f does not have a vertical asymptote at c, then c is a removable or jump discontinuity.

87. If $\lim_{x \to c} f(x) = L$, and $f(c) = L$, then f is continuous at c.

88. If c is a discontinuity of f, but $\lim_{x \to c^+} f(x)$ exists, then c is a removable or jump discontinuity.

2.5 Technology Exercises

89–94. Use a graphing calculator or computer algebra system to solve the equations given in Exercises 69–74 to four decimal places.

95–100. Use a graphing calculator or computer algebra system to graph the functions of Exercises 34–39 and explain how the graphs support your discussions of continuity in the aforementioned exercises.

2.6 **Rate of Change Revisited: The Derivative**

TOPICS

1. Velocity and tangent recap
2. The derivative as a function

We will wrap up this chapter by revisiting the two motivational problems that we began with, making use of the expertise in working with limits that we have since acquired. Our last topic of discussion will be an indication of how the concepts that have been introduced generalize and apply to a wealth of situations.

TOPIC 1 **Velocity and Tangent Recap**

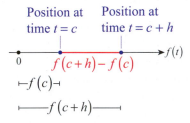

Figure 1

In Section 2.1, we developed a method for determining instantaneous velocity at a particular point c in time by considering the behavior of the fraction

$$\frac{f(c+h)-f(c)}{h}$$

for smaller and smaller values of h. The function $f(t)$ in this setting describes the position of an object at time t moving along a straight line, and the fraction was defined to be the **difference quotient of f at c with increment h**. Although we lacked the terminology and notation at the time, we can now succinctly define the (**instantaneous**) **velocity of the object at time c** to be

$$\lim_{h\to 0}\frac{f(c+h)-f(c)}{h}.$$

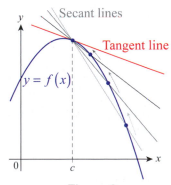

Figure 2

Similarly, we developed a way to arrive at the slope of the line tangent to the function $f(x)$ at the point c by considering the slopes of secant lines that approached the tangent line. Ultimately, we saw that the tangent line slope (if the tangent line exists) is also the limiting behavior of the difference quotient at c, so we can define the **slope of the line tangent to f at c** to also be

$$\lim_{h\to 0}\frac{f(c+h)-f(c)}{h}.$$

Given this concurrence, it likely comes as no surprise that limits of difference quotients will occupy our attention for some time, and it is appropriate to now introduce some additional notation.

Definition 💡

The Derivative at a Point

The **derivative of the function f at the point c**, denoted $f'(c)$, is

$$f'(c)=\lim_{h\to 0}\frac{f(c+h)-f(c)}{h},$$

provided the limit exists. (Note that $f'(c)$ is read "f prime of c.")

Recall that in Section 2.1 we saw some alternate ways of expressing the same idea. In some contexts, it may be more natural to refer to the point $c + h$ as x, and to express the difference quotient as follows:

$$\frac{f(c+h)-f(c)}{h} = \frac{f(x)-f(c)}{x-c}$$

In this case, the derivative of f at c is obtained by letting x approach c, so an alternate definition of the derivative is

$$f'(c) = \lim_{x \to c} \frac{f(x)-f(c)}{x-c}.$$

One last variation of this idea comes from referring to $f(x)-f(c)$ as the "change in f" or "change in y" (denoted Δf or Δy) and to $x - c$ as the "change in x" (denoted Δx). This is commonly seen when the context makes it natural to refer to the equation $y = f(x)$ and to emphasize the fact that the numerator of the difference quotient represents a vertical difference while the denominator represents a horizontal difference. Using this notation, and with the implicit understanding that c is fixed, the derivative would be denoted as

$$f'(c) = \lim_{\Delta x \to 0} \frac{\Delta f}{\Delta x} = \lim_{\Delta x \to 0} \frac{\Delta y}{\Delta x}.$$

It is important to realize, however, that all these variations represent the same fundamental concept. Regardless of the difference quotient notation used, and regardless of whether we are seeking the instantaneous velocity of an object or the slope of a tangent line, $f'(c)$ indicates the **rate of change of the function f at the point c**.

Example 1 ✎

Find the slope of the line tangent to the graph of $f(x) = x^2 - 1$ at the point $(2,3)$. Then use this information to derive the equation of the tangent line.

Solution

As we have discussed above, the slope of the tangent line at the point $(2,3)$ will be the limit of the difference quotient at $c = 2$, or using our new terminology, the derivative of f at the point $c = 2$. Recall that we denote this number by $f'(c)$, and it is calculated as

$$f'(c) = \lim_{h \to 0} \frac{f(c+h)-f(c)}{h}.$$

We now proceed to evaluate the above derivative for the given $f(x) = x^2 - 1$ at $c = 2$:

$$f'(2) = \lim_{h \to 0} \frac{f(2+h)-f(2)}{h} = \lim_{h \to 0} \frac{\left[(2+h)^2 - 1\right]-\left(2^2 - 1\right)}{h}$$

$$= \lim_{h \to 0} \frac{\left[\left(2^2 + 4h + h^2\right) - 1\right]-\left(2^2 - 1\right)}{h}$$

$$f'(2) = \lim_{h \to 0} \frac{4 + 4h + h^2 - 1 - 4 + 1}{h} = \lim_{h \to 0} \frac{4h + h^2}{h}$$

$$= \lim_{h \to 0} \frac{(4 + h)h}{h} = \lim_{h \to 0}(4 + h) = 4$$

So we conclude that the slope of the requested tangent is $m = f'(2) = 4$.

Notice that we now have all the information necessary to obtain the equation of the tangent line. It is a line with a slope of 4 that passes through the point $(2, 3)$. Recall the point-slope form of the equation of a line with slope m that passes through the point (x_0, y_0) is given by

$$y - y_0 = m(x - x_0).$$

Applying this formula to our situation at hand, we obtain the equation

$$y - 3 = 4(x - 2), \quad \text{or equivalently} \quad y = 4x - 5.$$

The graph of $f(x)$ along with the tangent line are shown in Figure 3.

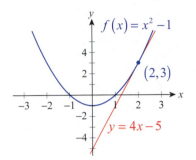

Figure 3

Example 2 ✎

A particle is moving along a straight line so that its distance from the start is given by $d(t) = 5t - \frac{1}{2}t^2$, where d is measured in feet and t in seconds. What is the particle's instantaneous velocity at $t = 2$ seconds?

Solution

As we have seen, the instantaneous velocity at time $t = c$ of an object moving along a straight line is the derivative of its position function at $t = c$. Using this observation with our given function d and $t = c = 2$ seconds, we obtain the following:

$$d'(c) = d'(2) = \lim_{h \to 0} \frac{d(2 + h) - d(2)}{h}$$

$$= \lim_{h \to 0} \frac{\left[5(2 + h) - \frac{1}{2}(2 + h)^2\right] - \left[5(2) - \frac{1}{2}(2^2)\right]}{h}$$

$$= \lim_{h \to 0} \frac{\left[10 + 5h - \frac{1}{2}(4 + 4h + h^2)\right] - (10 - 2)}{h}$$

$$= \lim_{h \to 0} \frac{\left(10 + 5h - 2 - 2h - \frac{1}{2}h^2\right) - 8}{h} = \lim_{h \to 0} \frac{3h - \frac{1}{2}h^2}{h}$$

$$= \lim_{h \to 0} \frac{\left(3 - \frac{1}{2}h\right)h}{h} = \lim_{h \to 0}\left(3 - \frac{1}{2}h\right) = 3$$

Thus we conclude that the instantaneous velocity of the particle at time 2 seconds is 3 ft/s.

Example 3 ✎

A piece of rock is dropped from a height of 196 feet, and its height above ground level is given by the function $h(t) = 196 - 16t^2$ feet, where t is measured in seconds. Find the instantaneous velocity of the rock at **a.** $t = 1$ second, **b.** $t = 2$ seconds, and **c.** $t = 2.5$ seconds.

What is the velocity of impact?

Solution

a. As before, the instantaneous velocity at time $t = c$ can be found by finding the derivative of the position function at c.

$$h'(c) = \lim_{h \to 0} \frac{\left[196 - 16(c+h)^2\right] - \left(196 - 16c^2\right)}{h}$$

$$= \lim_{h \to 0} \frac{196 - 16\left(c^2 + 2ch + h^2\right) - 196 + 16c^2}{h}$$

$$= \lim_{h \to 0} \frac{-32ch - 16h^2}{h} = \lim_{h \to 0} \frac{h(-32c - 16h)}{h}$$

$$= \lim_{h \to 0} (-32c - 16h) = -32c$$

We now obtain the instantaneous velocity at $t = c = 1$ by a simple substitution.

$$h'(1) = -32(1) = -32 \text{ ft/s}$$

Notice that unlike in our solution of the previous example, rather than substituting $c = 1$ at the outset, we carried c through the computation to obtain a formula for the instantaneous velocity, which we subsequently evaluated at $c = 1$. The advantage of this method lies in the fact that now we can use the formula to easily answer both questions b. and c.

b. $h'(2) = -32(2) = -64 \text{ ft/s}$

c. $h'(2.5) = -32(2.5) = -80 \text{ ft/s}$

Finally, we proceed to find the velocity of impact. The fundamental observation is that impact happens precisely when the position of the rock becomes 0. This helps us determine exactly how many seconds after being dropped the rock hits the ground. Thus we solve the equation $h(t) = 0$.

$$196 - 16t^2 = 0$$
$$16t^2 = 196$$
$$t^2 = 12.25$$
$$t = \pm 3.5$$

Since time is always positive, we will only consider the solution $t = 3.5$ s.

Now we can determine the velocity of impact, which is precisely the instantaneous velocity at the time of impact, by evaluating h' at $c = 3.5$. Notice that we can once again use the formula we derived before:

$$h'(3.5) = -32(3.5) = -112 \text{ ft/s},$$

and we conclude that the rock hits the ground with a velocity of -112 ft/s.

In conclusion, we note that the negative signs in the above answers mean that the direction of velocity is pointing downward throughout the motion. Also, while our answer is close to what we would measure in an actual experiment, it is important to remember that we ignored air resistance in this problem.

TOPIC 2   The Derivative as a Function

It will frequently be the case that we need to determine the derivative of a function at more than one point (as in our last example) or even at every point of a given interval. In such cases, it is often no more difficult to determine the derivative at a general point than at a specific point. If we are successful in doing so, then what we have achieved is a definition of another function—this new function, having been *derived* from our original function, is referred to as the *derivative* of the function.

Definition 💡

The Derivative of a Function

The **derivative of f**, denoted f', is the function whose value at the point x is

$$f'(x) = \lim_{h \to 0} \frac{f(x+h) - f(x)}{h},$$

provided the limit exists.

This definition is exactly the same as the definition of the derivative at a point—the only difference is that we are extending the definition to all points at which the limit exists.

In usage, names other than f and f' may be more appropriate for the functions under consideration. For instance, in a problem describing the position of an object, labels such as $x(t)$ (if the object is moving along a horizontal line) or $h(t)$ (if the height of a thrown object is being discussed) may be used for the position function. In such cases, it is common to name the derivative of the position function $v(t)$ for velocity. In other words, $v(t)$ represents the instantaneous velocity of the object at time t. It is now appropriate to make a distinction between the ideas of *velocity* and *speed*: if $v(t)$ represents the velocity of an object at time t, $s(t) = |v(t)|$ represents its **speed** at time t.

Example 4 📝

A soccer player kicks a ball vertically upward so that its position relative to ground level is $h(t) = -4.9t^2 + 19.6t + 1$ meters, where t is measured in seconds. Find the velocity and speed of the soccer ball at **a.** $t = 1$ second, **b.** $t = 2.5$ seconds, and **c.** $t = 3.5$ seconds.

What is this ball's speed of impact upon its return? (Ignore air resistance.)

Solution

a. Note first of all that the position function realistically reflects the fact that the ball does not start from ground level. Its initial height can be found by substituting $t = 0$ into the position function.

$$h(0) = -4.9(0)^2 + 19.6(0) + 1 = 1 \text{ meter}$$

In order to answer the questions, next we proceed to find $v(t)$, the velocity function of the ball, which is the derivative of $h(t)$. Notice that we are using the alternate notation of Δt instead of h for the increment, so as not to cause confusion with the name of the position function.

$$v(t) = h'(t) = \lim_{\Delta t \to 0} \frac{h(t + \Delta t) - h(t)}{\Delta t}$$

$$= \lim_{\Delta t \to 0} \frac{-4.9(t + \Delta t)^2 + 19.6(t + \Delta t) + 1 - \left(-4.9t^2 + 19.6t + 1\right)}{\Delta t}$$

$$= \lim_{\Delta t \to 0} \frac{-4.9t^2 - 9.8t\Delta t - 4.9(\Delta t)^2 + 19.6t + 19.6\Delta t + 1 + 4.9t^2 - 19.6t - 1}{\Delta t}$$

$$= \lim_{\Delta t \to 0} \frac{\Delta t\left(-9.8t - 4.9\Delta t + 19.6\right)}{\Delta t}$$

$$= \lim_{\Delta t \to 0} \left(-9.8t - 4.9\Delta t + 19.6\right)$$

$$= -9.8t + 19.6$$

Already in possession of the velocity function, it is now easy to find the ball's instantaneous velocity at $t = 1$.

$$v(1) = -9.8(1) + 19.6 = 9.8 \text{ m/s}$$

Note that the speed equals the velocity this time, since the latter is positive.

$$s(1) = |v(1)| = |9.8| = 9.8 \text{ m/s}$$

Notice also that, as one would expect, our function indicates that the ball is slowing down significantly from its initial velocity, which can be found by evaluating

$$v(0) = -9.8(0) + 19.6 = 19.6 \text{ m/s}.$$

b. Substituting again in the velocity function, we find

$$v(2.5) = -9.8(2.5) + 19.6 = -4.9 \text{ m/s}.$$

The negative sign shows that the ball turned back and is now traveling downward. Its speed, however, being independent of direction, is still positive.

$$s(2.5) = |v(2.5)| = |-4.9| = 4.9 \text{ m/s}$$

c. We expect the ball to accelerate as it falls, and that is exactly what we find upon substituting $t = 3.5$ into the velocity function.

$$v(3.5) = -9.8(3.5) + 19.6 = -14.7 \text{ m/s}$$

The negative sign shows that the direction of velocity is still downward, but its absolute value, the ball's speed, has increased.

$$s(3.5) = |v(3.5)| = |-14.7| = 14.7 \text{ m/s}$$

In fact, as it is with any free-falling body when air resistance is negligible, gravity increases the ball's speed by approximately 9.8 m/s every second. (Can you see this from our results?)

To answer the final question regarding the speed of impact, we first need to know when it happens, and then we substitute the appropriate t-value into the velocity function. Since height is 0 upon impact, we can find t by solving $h(t) = 0$, that is,

$$-4.9t^2 + 19.6t + 1 = 0.$$

An application of the quadratic formula yields the positive solution $t \approx 4.05$ seconds. Therefore, the speed of impact is approximately

$$s(4.05) = |v(4.05)| = |-9.8(4.05) + 19.6| = |-20.09| = 20.09 \text{ m/s}.$$

Example 5 ✎

Find the derivative of $f(x) = \sqrt{x}$, and use it to determine equations of the tangent lines to the graph of f at the points $(1, 1)$ and $\left(2, \sqrt{2}\right)$.

Solution

We will start with the definition of the derivative, and make use of some of the limit-determination techniques learned in Section 2.4.

$$f'(x) = \lim_{h \to 0} \frac{f(x+h) - f(x)}{h}$$

$$= \lim_{h \to 0} \frac{\sqrt{x+h} - \sqrt{x}}{h}$$

$$= \lim_{h \to 0} \frac{\left(\sqrt{x+h} - \sqrt{x}\right)\left(\sqrt{x+h} + \sqrt{x}\right)}{h\left(\sqrt{x+h} + \sqrt{x}\right)}$$

$$= \lim_{h \to 0} \frac{(x+h) - x}{h\left(\sqrt{x+h} + \sqrt{x}\right)} = \lim_{h \to 0} \frac{h}{h\left(\sqrt{x+h} + \sqrt{x}\right)}$$

$$= \lim_{h \to 0} \frac{1}{\sqrt{x+h} + \sqrt{x}} = \frac{1}{\sqrt{x} + \sqrt{x}} = \frac{1}{2\sqrt{x}}$$

To find the equation of the tangent line at $(1,1)$, we first determine its slope by evaluating the derivative $f'(x) = 1/(2\sqrt{x})$ at $x = 1$.

$$m = f'(1) = \frac{1}{2\sqrt{1}} = \frac{1}{2}$$

Now the point-slope equation comes to bear, and we obtain

$$y - 1 = \frac{1}{2}(x-1), \quad \text{or} \quad y = \frac{1}{2}x + \frac{1}{2}.$$

We deal with the equation of the tangent line at $(2, \sqrt{2})$ in a similar manner. Since

$$f'(2) = \frac{1}{2\sqrt{2}} = \frac{\sqrt{2}}{4},$$

the requested equation is

$$y - \sqrt{2} = \frac{\sqrt{2}}{4}(x-2), \quad \text{or} \quad y = \frac{\sqrt{2}}{4}x + \frac{\sqrt{2}}{2}.$$

Notice that we can use f' to find the slope of the tangent at *any* given point $x = c$, if it exists, by simply evaluating $f'(c)$.

In business, $C(x)$ is often used to represent the total cost of producing x items of a certain product, and is referred to as the *cost function*. Similarly, $R(x)$ stands for the total revenue when x items are sold, this latter being called the *revenue function*. Consequently, the *profit function* $P(x)$ satisfies $P(x) = R(x) - C(x)$; or in words, profit equals total revenue minus total cost. The corresponding derivatives, or rates of change, the functions $C'(x)$, $R'(x)$, and $P'(x)$, are called *marginal cost*, *marginal revenue*, and *marginal profit*, respectively. As an application, the following example involves some of the above functions.

Example 6 ✐

If a table manufacturer has a cost function of $C(x) = 225 + 2x^2$ along with its revenue function of $R(x) = 117x - \frac{1}{4}x^2$, find **a.** all break-even points and **b.** the marginal profit when $x = 10$, $x = 20$, and $x = 30$.

Solution

a. Break-even points occur when $C(x) = R(x)$, in other words, when revenue levels equal the total cost invested. This leads to the equation

$$225 + 2x^2 = 117x - \frac{1}{4}x^2.$$

After rearranging terms and factoring, this leads to

$$\frac{9}{4}(x-2)(x-50) = 0,$$

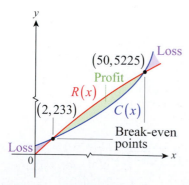

Figure 4

thus we conclude that break-even points occur when $x = 2$ and $x = 50$ (see Figure 4).

b. Since marginal profit is the rate of change of profit, we must first find the profit function and then calculate its derivative.

$$P(x) = R(x) - C(x)$$

$$= \left(117x - \frac{1}{4}x^2\right) - \left(225 + 2x^2\right)$$

$$= 117x - \frac{1}{4}x^2 - 225 - 2x^2$$

$$= -\frac{9}{4}x^2 + 117x - 225$$

Next, we will use the definition of the derivative to obtain the marginal profit function.

$$P'(x) = \lim_{h \to 0} \frac{P(x+h) - P(x)}{h}$$

$$= \lim_{h \to 0} \frac{-\frac{9}{4}(x+h)^2 + 117(x+h) - 225 - \left(-\frac{9}{4}x^2 + 117x - 225\right)}{h}$$

After expanding, rearranging terms, and canceling, this reduces to

$$P'(x) = \lim_{h \to 0} \frac{h\left(-\frac{9}{2}x + 117 - \frac{9}{4}h\right)}{h}$$

$$= \lim_{h \to 0} \left(-\frac{9}{2}x + 117 - \frac{9}{4}h\right)$$

$$= -\frac{9}{2}x + 117.$$

Now we can calculate the various marginal profits.

$$P'(10) = -\frac{9}{2}(10) + 117 = -45 + 117 = \$72$$

$$P'(20) = -\frac{9}{2}(20) + 117 = -90 + 117 = \$27$$

$$P'(30) = -\frac{9}{2}(30) + 117 = -135 + 117 = -\$18$$

In conclusion, an important remark is in order. Since the number of tables produced is always an integer, the increment h in any difference quotient should be thought of as an integer, so the smallest nonzero h possible is $h = 1$. This explains the following interpretation of marginal profit: $P'(x)$ is an estimate for the increase in $P(x)$ if x is increased by 1. Note also that in this example, $P'(x)$ is actually getting smaller as x gets larger. This shows that the rate of growth of profit per table is actually decreasing at a rate of \$18 per table when 30 tables are manufactured and sold. This does not necessarily mean there is a loss, but as production increases, the profit per table is growing less because costs are increasing faster than revenue.

2.6 Exercises

1–2 Use the graph to estimate the derivative at the given points.

1. a. $x_1 = -1$　　　　**b.** $x_2 = 1$

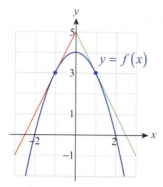

2. a. $x_1 = -2$　　　　**b.** $x_2 = 0$

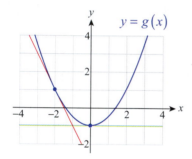

3–14 Find the equation of the tangent line to the graph of $f(x)$ at the given point.

3. $f(x) = x^2 - 2;$　$(2,2)$

4. $f(x) = 3x - 2x^2;$　$(-1,-5)$

5. $f(x) = \frac{1}{2}x + 4;$　$(2,5)$

6. $f(x) = 1 - 5x;$　$(0,1)$

7. $f(x) = x^3;$　$(2,8)$

8. $f(x) = 5x - 2x^3;$　$(-1,-3)$

9. $f(x) = \sqrt{x+1};$　$(0,1)$

10. $f(x) = 2\sqrt{1-3x};$　$(-1,4)$

11. $f(x) = \frac{1}{x};$　$\left(\frac{1}{2}, 2\right)$

12. $f(x) = \frac{5}{1-2x};$　$\left(-1, \frac{5}{3}\right)$

13. $f(x) = \frac{1}{\sqrt{x}};$　$\left(4, \frac{1}{2}\right)$

14. $f(x) = \frac{2}{\sqrt{x+1}};$　$(3,1)$

15–38 Use the definition (also called the *limit process*) to find the derivative function f' of the given function f. Find all x-values (if any) where the tangent line is horizontal.

15. $f(x) = 2$　　　　　　**16.** $f(x) = 2x$

17. $f(x) = 4x + 5$　　　　**18.** $f(x) = 3 - \frac{2}{5}x$

19. $f(x) = 3x^2$　　　　　**20.** $f(x) = 4 - 2x^2$

21. $f(x) = \frac{1}{2}x^2 + 5x - 7$　**22.** $f(x) = x - \frac{1}{3}x^2$

23. $f(x) = x^3 + x$

24. $f(x) = 7 + x - 3x^2 + x^3$

25. $f(x) = x^4$　　　　　**26.** $f(x) = \frac{1}{2x}$

27. $f(x) = \frac{5}{2x-4}$　　　**28.** $f(x) = \frac{x-2}{x+2}$

29. $f(x) = \frac{2x+1}{x-3}$　　　**30.** $f(x) = \frac{2}{x^2}$

31. $f(x) = \frac{1}{x^2+1}$　　　**32.** $f(x) = \frac{2}{x^2-2x}$

33. $f(x) = \sqrt{5x}$　　　　**34.** $f(x) = \frac{1}{\sqrt{5x}}$

35. $f(x) = \sqrt{2x+1}$　　　**36.** $f(x) = \frac{1}{\sqrt{x-2}}$

37. $f(x) = \sqrt{x^2+1}$　　　**38.** $f(x) = \frac{1}{\sqrt{x^2+1}}$

39–44 Find the equation of a tangent line to the graph of the function that is parallel to the given line.

39. $f(x) = x^2 + 3; \quad y - 6x + 1 = 0$

40. $g(x) = 2x - x^2; \quad y - 5 = 4x$

41. $h(x) = \dfrac{1}{2x}; \quad x + 2y = 3$

42. $F(x) = \dfrac{1}{x-3}; \quad y + 4x + 7 = 0$

43. $G(x) = \dfrac{1}{\sqrt{x}}; \quad 54y + x = 1$

44. $H(x) = \dfrac{1}{\sqrt{x^2 - 7}}; \quad 27y + 4x - 2 = 0$

45–56 Use the alternate form of the definition of the derivative $f'(c) = \lim\limits_{x \to c} \dfrac{f(x) - f(c)}{x - c}$ to evaluate the given slope.

45. $f(x) = 5 - \dfrac{1}{4}x; \quad f'(3.6)$

46. $g(x) = x^2 + 1; \quad g'(-1)$

47. $h(x) = (x + 2)^2; \quad h'(3)$

48. $F(t) = \dfrac{1}{t-3}; \quad F'(2)$

49. $G(x) = \dfrac{2}{5-x}; \quad G'(7)$

50. $k(t) = \sqrt{t + 5}; \quad k'(11)$

51. $u(x) = 2\sqrt{1 - x}; \quad u'(-3)$

52. $v(x) = \dfrac{1}{x^2 + 1}; \quad v'(0)$

53. $w(s) = \dfrac{1}{\sqrt{s + 4}}; \quad w'(5)$

54. $F(t) = t^3 - t; \quad F'(1)$

55. $G(s) = s^4; \quad G'(-2)$

56. $H(x) = \dfrac{2}{\sqrt{x^2 + 1}}; \quad H'(0)$

57–60 Match the graph of f with the graph of its derivative f' (labeled A–D).

57.

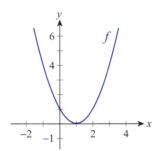

58.

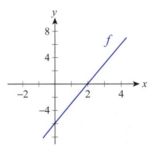

59.

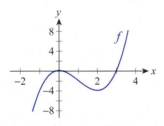

60.

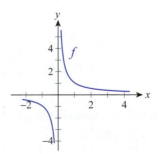

A.

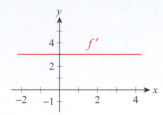

B.

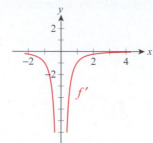

C.

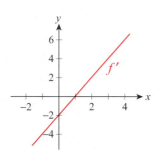

D.

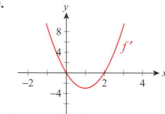

61–65 Sketch the graph of a function f possessing the given characteristics. (A formula is useful, but not necessary.)

61. $f(0) = 1$, $f'(0) = 0$, $f'(x) < 0$ for $x < 0$, $f'(x) > 0$ for $x > 0$

62. $f(1) = 0$, $f'(1) = 0$, $f'(x) \geq 0$ on the entire real line

63. $f(x) > 0$ on the entire real line, $f'(x) < 0$ on the entire real line

64. $f(1) = 1$, $f'(1) = -1$, f' is nonzero on the entire real line

65. $f(1) = 5$, $f'(x) = 5$ on the entire real line

66. Prove that if $f(x) = c$ (a constant function), then $f'(x) = 0$.

67. Use the definition of the derivative to prove that if $f(x) = x$, then $f'(x) = 1$.

68. Generalize Exercise 67 to prove that if $f(x)$ is a linear function, then $f'(x)$ is constant.

69.* Use the definition of the derivative to prove that if $f(x) = x^n$ for a positive integer n, then $f'(x) = nx^{n-1}$.

70.* Recall from Section 1.1 that a function f is even if $f(-x) = f(x)$ and odd if $f(-x) = -f(x)$ throughout its domain. Prove that the derivative of an even function is odd and, vice versa, an odd function has an even derivative.

71.* Find the equation of the line tangent to the graph of $f(x) = 1/x$ at the point $(c, f(c))$. Prove that the area of the triangle bounded by the tangent line and the coordinate axes is the same for all $c \neq 0$.

72. The position function of a moving particle is given by $p(t) = t^2 - 3t + 1$ feet at t seconds. Find all points in time where the particle's speed is 1 ft/s. When does it come to a momentary stop?

73. Repeat Exercise 72 with the position function $p(t) = \frac{1}{9}t^3 - t^2 + \frac{8}{3}t$.

74. A baseball is hit vertically upward with an initial speed of 80 ft/s. When does it slow down to 32 ft/s? How high does it go and how long is it aloft? (**Hint:** Use the position function $h(t) = -16t^2 + 80t$. Ignore air resistance.)

75. A rock is thrown upward from the edge of a 150 ft high cliff with an initial velocity of 48 ft/s.

   a. Calculate the velocity and speed of the rock when it is exactly 32 ft above the person's hand.

   b. How high does it go and when does it reach the bottom of the cliff?

   c. What is the velocity of impact?

   (**Hint:** Use $h(t) = -16t^2 + 48t + 150$ as the position function, where h is in feet, t in seconds. Ignore air resistance.)

76. A package is dropped from a small airplane 122.5 meters above the Earth. If we ignore air resistance, how much time does the package need to reach the ground and what is the speed of impact? (**Hint:** The position function is $h(t) = -4.9t^2 + 122.5$ meters, where t is measured in seconds. Use $g \approx 9.81 \, \text{m/s}^2$.)

77. The following graph is a position function of a student's car relative to her home as she drove to class one morning. From the graph, recreate a possible story of her trip, mentioning distance, velocity, speed, and so forth.

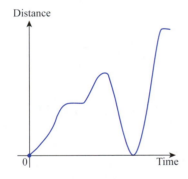

78. A manufacturer has determined that the revenue from the sale of x cordless telephones is given by $R(x) = 94x - 0.03x^2$ dollars. The cost of producing x telephones is $C(x) = 10,800 + 34x$ dollars.

   a. Find the profit function $P(x)$ and any break-even points.

   b. Find $P(200)$, $P(400)$, and $P(600)$.

   c. Find the marginal profit function $P'(x)$.

   d. Find $P'(200)$, $P'(400)$, and $P'(600)$.

79. The owner of a leather retailer has determined that he can sell x attaché cases if the price is $p = D(x) = 46 + 0.25x$ dollars ($D(x)$ is often called the demand function). The total cost for these cases is $C(x) = 0.15x^2 + 6x + 190$ dollars.

   a. Find the profit function $P(x)$. (**Hint:** Find the revenue function $R(x)$ first.)

   b. Find any break-even points.

   c. Find $P(25)$, $P(30)$, and $P(40)$.

   d. Find the marginal profit function $P'(x)$.

   e. Find $P'(25)$, $P'(30)$, and $P'(40)$.

80. The average cost $\overline{C}(x)$ of manufacturing x units of a certain product is $C(x)/x$, where $C(x)$ is the total cost function.

   a. Find the average cost function if $C(x) = 30 + 2x + 0.003x^2$.

   b. What is the rate of change of average cost?

   c. What value of x results in a minimum average cost? (**Hint:** Use the fact that when average cost is a minimum, its rate of change is 0. Alternatively, use technology to graph $\overline{C}(x)$ for $x \geq 0$ and zoom in on the lowest point.)

81. The average manufacturing cost function of a product is given by $\overline{C}(x) = 20x^{-1} + 3$. Determine the cost function and the marginal cost function for the product. (**Hint:** See Exercise 80.)

2.6 Technology Exercises

82–105. Referring back to the functions given in Exercises 15–38, use a graphing calculator or computer algebra system to sketch the graph of f along with that of f' in the same viewing window. Compare the graphs and describe their relationship.

Chapter 2
Review Exercises

1–4 Use the graph of the function to find the indicated (possibly one-sided) limit, if it exists. Also examine the continuity of the function at the indicated point and classify any discontinuities.

1. $\lim\limits_{x\to 1} f(x)$

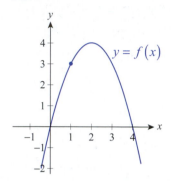

2. $\lim\limits_{x\to 1} g(x)$

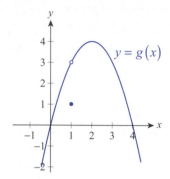

3. $\lim\limits_{x\to 1^-} h(x)$

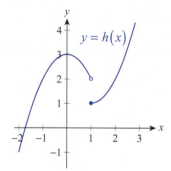

4. $\lim\limits_{x\to -1^+} k(x)$

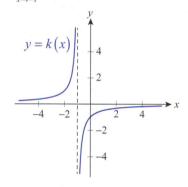

5–8 Use difference quotients to approximate the slope of the tangent to the graph of the function at the given point. Use at least five different h-values that are decreasing in magnitude. (Answers will vary.)

5. $f(x) = 3x + 2;\quad (0, 2)$

6. $g(x) = 2 - x^2;\quad (1, 1)$

7. $h(x) = \sqrt{x - 1};\quad (2, 1)$

8. $k(x) = \sin x;\quad (0, 0)$

9. A pellet is shot vertically upward from an initial height of 6 feet. Its height after t seconds is given by $h(t) = 6 + 608t - 16t^2$ feet. Use difference quotients to answer the questions below.

a. What will be the pellet's height at the end of the first second?

b. What is the average velocity of the pellet during the first two seconds?

c. Estimate the instantaneous velocity at $t = 0$ seconds.

d. Estimate the instantaneous velocity at $t = 2$ seconds.

e. When will the velocity be 0?

10–11 Approximate the area of the region between the graph of $f(x)$ and the x-axis on the given interval. Use $n = 4$, as in Example 5a of Section 2.1. (Round your answer to four decimal places.)

10. $f(x) = x^3$ on $[0,1]$ **11.** $f(x) = \ln x$ on $[1,3]$

12–13 Approximate the arc length of the graph of $g(x)$ on the given interval. Use $n = 5$, as in Example 5b of Section 2.1. (Round your answer to four decimal places.)

12. $g(x) = x^{2/3}$ on $[3,8]$ **13.** $g(x) = e^x$ on $[-2,3]$

14–17 Create a table of values to estimate the value of the indicated limit without graphing the function. Choose the last x-value so that it is no more than 0.001 units from the given c-value.

14. $\lim\limits_{x \to 1} \dfrac{x^3 - 1}{x - 1}$ **15.** $\lim\limits_{x \to 0} x^x$

16. $\lim\limits_{x \to 0} \dfrac{\sin 2x}{4x}$ **17.** $\lim\limits_{x \to 0} \left(2x \sin \dfrac{1}{4x} \right)$

18. Use one-sided limit notation to describe the behavior of $f(x) = \dfrac{1}{x-1}$ near $x = 1$.

19–20 Find a $\delta > 0$ that satisfies the limit claim corresponding to $\varepsilon = 0.01$, that is, such that $0 < |x - c| < \delta$ would imply $|f(x) - L| < 0.01$.

19. $\lim\limits_{x \to 0} (3 - 2x) = 3$ **20.** $\lim\limits_{x \to 4} \sqrt{x} = 2$

21–24 Give the precise definition of the limit claim. Then use the definition to prove the claim.

21. $\lim\limits_{x \to 1} (3x + 1) = 4$ **22.** $\lim\limits_{x \to 1} x^2 = 1$

23. $\lim\limits_{x \to 1} \sqrt{x} = 1$ **24.** $\lim\limits_{x \to 2} \dfrac{2}{x} = 1$

25–41 Use algebra and/or appropriate limit laws to evaluate the given limit (one-sided limit where indicated). If the limit is unbounded, use the symbol ∞ or $-\infty$ in your answer.

25. $\lim\limits_{x \to 3} (2x^2 - 3x + 5)$

26. $\lim\limits_{x \to -2} \left(\dfrac{x^3}{4} + 2x^2 - x + 1 \right)$

27. $\lim\limits_{x \to 3} \sqrt{x^3 + 2x^2 + 4}$ **28.** $\lim\limits_{x \to -2} \dfrac{2x+1}{x^2 - x}$

29. $\lim\limits_{t \to 1} \left(\dfrac{3t + 5t^3}{t^2 + 1} \right)^{3/2}$ **30.** $\lim\limits_{x \to 4} \dfrac{x^2 - 16}{x - 4}$

31. $\lim\limits_{x \to -5} \dfrac{x+5}{x^2 - 25}$ **32.** $\lim\limits_{x \to 5^-} \dfrac{x+5}{x^2 - 25}$

33. $\lim\limits_{x \to 1^+} \dfrac{x^2 + 1}{x^4 - 1}$ **34.** $\lim\limits_{x \to 1} \dfrac{x^2 - 1}{x^4 - 1}$

35. $\lim\limits_{x \to 3} \dfrac{\sqrt{x+1} - 2}{x - 3}$ **36.** $\lim\limits_{x \to 0} \dfrac{\dfrac{1}{2+x} - \dfrac{1}{2}}{x}$

37. $\lim\limits_{x \to 0^-} \dfrac{2|x|}{x}$ **38.** $\lim\limits_{x \to -2^+} \sqrt{4 - x^2}$

39. $\lim\limits_{x \to 2^-} (\llbracket x \rrbracket + 2x)$ **40.** $\lim\limits_{x \to 1^+} \llbracket x \rrbracket x$

41. If $f(x) = x^2$, find $\lim\limits_{h \to 0} \dfrac{f(x+h) - f(x)}{h}$.

42–43 Use the Squeeze Theorem to prove the limit claim.

42. $\lim\limits_{x \to 0} x \cos \dfrac{1}{x} = 0$ **43.** $\lim\limits_{x \to \infty} \dfrac{\sin x}{\ln x} = 0$

44. Sketch a graph of a function (a formula is not necessary) that is not continuous at $x = 0$ from either direction, but both of its one-sided limits exist at $x = 0$. (Answers will vary.)

45. Sketch a graph of a function that is left-continuous at $x = 0$, but its right-hand limit at $x = 0$ doesn't exist. (Answers will vary.)

46–51 Find and classify the discontinuities (if any) of the given function as removable or nonremovable.

46. $f(x) = \dfrac{x - 9}{\sqrt{x} - 3}, \quad x \geq 0$

47. $g(x) = \dfrac{\sqrt{x} + 2}{x - 4}, \quad x \geq 0$

48. $h(x) = \dfrac{1}{\sqrt{x^2 + 1}}$

49. $t(x) = 2 + 2\llbracket x \rrbracket$

50. $G(x) = \dfrac{x}{\sqrt{x+1} - 1}, \quad x \geq -1$

51. $k(x) = |x - 3| + |x + 1|$

52–53 Use the ε-δ definition to prove that the function is continuous.

52. $f(x) = 3x - 1$ **53.** $g(x) = 2x^2$

54. Find the values of a and b such that f is continuous on the entire real line.

$$f(x) = \begin{cases} -1 & \text{if } x \leq -3 \\ ax + b & \text{if } -3 < x < 2 \\ x^2 & \text{if } x \geq 2 \end{cases}$$

55. Use the Intermediate Value Theorem to prove that the equation $2x^5 + x + 1 = 0$ has a solution on the interval $[-1, 1]$.

56. Use the Intermediate Value Theorem to show that the graphs of $f(x) = x^3$ and $g(x) = e^{-x}$ intersect.

57–58 Find the equation of the tangent line to the graph of $f(x)$ at the given point.

57. $f(x) = x^2 + x;\quad (1, 2)$

58. $f(x) = \sqrt{x};\quad (4, 2)$

59–60 Use the definition (also called the limit process) to find the derivative function f' of the given function f. Find all x-values (if any) where the tangent line is horizontal.

59. $f(x) = 2x - x^2$ **60.** $f(x) = \dfrac{3}{x - 2}$

61–62 Sketch the graph of a function f possessing the given characteristics. (A formula is useful, but not necessary.)

61. f is continuous at 0, $f(0) = 1$, $f'(x) < 0$ for $x < 0$, $f'(x) > 0$ for $x > 0$, and $f'(0)$ does not exist

62. $g(1) < 0$, $g'(1) > 0$, and $g(2) > 0$, but $g'(2) < 0$

63. Prove that if $f(x)$ is a quadratic function, then $f'(x)$ is linear.

64. A small object is thrown upward with an initial velocity of 12 m/s from the top of a 15 m high building.

   a. How high does it go and when does it reach the ground?

   b. What is the speed of impact?

   (**Hint:** Use $h(t) = -5t^2 + 12t + 15$ as the position function, where h is in meters, t in seconds.)

65. The owner of a small toy manufacturer has determined that he can sell x toys if the price is $p = D(x) = 0.2x + 30$ dollars. The total cost as a function of x is given by $C(x) = 0.1x^2 + 15x + 247.5$ dollars.

   a. Find the profit function $P(x)$.

   b. Find any break-even points.

   c. Find the marginal profit function.

66–73 *True or False?* Determine whether the given statement is true or false. In case of a false statement, explain or provide a counterexample.

66. Instantaneous velocity can be interpreted as the slope of a tangent line.

67. If $\lim\limits_{x \to c} f(x)$ doesn't exist, then $f(x)$ has a vertical asymptote at $x = c$.

68. Any rational function has at least one vertical asymptote.

69. If $\lim\limits_{x \to c} f(x) = A$ and $\lim\limits_{x \to c} g(x) = B$, then
$$\lim\limits_{x \to c} \frac{f(x)}{g(x)} = \frac{A}{B}.$$

70. If f is defined on $[a, b]$, L is a real number between $f(a)$ and $f(b)$, and $\lim\limits_{x \to c} f(x)$ exists for all $x \in (a, b)$, then there is a c in the interval (a, b) such that $f(c) = L$.

71. If f is continuous at c, then $f(c)$ is equal to both one-sided limits at c.

72. If both one-sided limits of f exist at c, and if f is defined at c, then f is continuous at c.

73. If $g(x) \le f(x) \le h(x)$ for all x in some open interval containing c, and if $\lim_{x \to c} g(x) = \lim_{x \to c} h(x) = L$, then by the Squeeze Theorem $f(c) = L$ as well.

Chapter 2
Technology Exercises

74. Use a computer algebra system to find approximations for the areas in Exercises 10 and 11 by using $n = 100$. (Round your answers to four decimal places.)

75. Use a computer algebra system to find approximations for the arc lengths in Exercises 12 and 13 by using $n = 100$. (Round your answers to four decimal places.)

76. Use a graphing calculator or computer algebra system to verify your answers given for Exercises 14–17.

77. Use a graphing calculator or computer algebra system to approximate the solutions for Exercises 55 and 56. Round your answers to four decimal places.

78–81 Use a graphing calculator or computer algebra system to graph the function, and estimate from the graph the value of the given limit.

78. $\lim_{x \to \infty} x^{1/x}$

79. $\lim_{x \to 0} \dfrac{\arcsin x}{x}$

80. $\lim_{x \to \infty} \left(1 + \dfrac{1}{x}\right)^{2x}$

81. $\lim_{x \to 1} \dfrac{\ln(x^3)}{x - 1}$

Project ✍

Chapter 2

Some years ago, it was common for long-distance phone companies to charge their customers in one-minute increments. In other words, the company charges a flat fee for the first minute of a call and another fee for each additional minute or any fraction thereof (see Exercise 82 in Section 2.5). In this project, we will explore in detail a function that gives the cost of a telephone call under the above conditions.

1. Suppose a long-distance call costs 75 cents for the first minute plus 50 cents for each additional minute or any fraction thereof. In a coordinate system where the horizontal axis represents time t and the vertical axis price p, draw the graph of the function $p = C(t)$ that gives the cost (in dollars) of a telephone call lasting t minutes, $0 < t \le 5$.

2. Does $\lim_{t \to 1.5} C(t)$ exist? If so, find its value.

3. Does $\lim_{t \to 3} C(t)$ exist? Explain.

4. Write a short paragraph on the continuity of this function. Classify all discontinuities; mention one-sided limits and left or right continuity where applicable.

5. In layman's terms, interpret $\lim_{t \to 2.5} C(t)$.

6. In layman's terms, interpret $\lim_{t \to 3^-} C(t)$.

7. In layman's terms, interpret $\lim_{t \to 3^+} C(t)$.

8. If possible, find $C'(3.5)$.

9. If possible, find $C'(4)$.

10. Find and graph another real-life function whose behavior is similar to that of $C(t)$. Label the axes appropriately and provide a brief description of your function.

Chapter 3
Differentiation

3.1 Differentiation Notation and Consequences 199
1. Alternative Notations and Higher-Order Derivatives
2. Consequences of Differentiability

3.2 Derivatives of Polynomials, Exponentials, Products, and Quotients 216
1. Elementary Differentiation Rules
2. Differentiation of Exponential Functions
3. The Product and Quotient Rules

3.3 Derivatives of Trigonometric Functions 239
1. Two Useful Limits
2. Derivatives of Sine and Cosine
3. Derivatives of Other Trigonometric Functions

3.4 The Chain Rule 253
1. Proof of the Chain Rule
2. Consequences of the Chain Rule

3.5 Implicit Differentiation 266
1. Derivatives of Implicitly Defined Functions
2. Higher-Order Derivatives

3.6 Derivatives of Inverse Functions 277
1. The Derivative Rule for Inverse Functions
2. Derivatives of Logarithms
3. Logarithmic Differentiation
4. Derivatives of Inverse Trigonometric Functions

3.7 Rates of Change in Use 293
1. Biology Applications
2. Physics Applications
3. Chemistry Applications
4. Business and Economics Applications

3.8 Related Rates 304
1. Constructing and Solving Related Rates Equations
2. Interlude: A Cautionary Tale

3.9 Linearization and Differentials 315
1. Linear Approximation of Functions
2. Differentials

Introduction

Chapter 2 introduced the notion of differentiation by means of several elementary applications. In this chapter we learn additional notation, develop techniques that allow us to find derivatives of many classes of functions, and explore many more applications. As we will see, the applications arise from a broad array of disciplines and collectively illustrate the profound power of the derivative.

Each bit of notation found in this chapter reflects the thinking of the mathematician responsible for introducing it, and it is frequently the case that a given mathematical concept can be expressed in a number of equivalent ways. While this may seem needlessly redundant at first, we'll see that the alternatives all offer their own unique advantages, and hence their use persists. Such giants as the Swiss mathematician Leonhard Euler (1707–1783) and the Italian French mathematician Joseph-Louis Lagrange (1736–1813), as well as Sir Isaac Newton (1643–1727) and Gottfried Wilhelm Leibniz (1646–1716), are among those who devised ways of succinctly writing the new ideas of calculus. Each had certain objectives in mind, usually arising from a problem (or set of problems) that they wished to solve, and the notation they created was chosen in pursuit of those objectives.

At the same time, mathematicians were discovering general rules of differentiation that allow derivatives of entire classes of functions to be determined relatively easily. In this chapter, we'll learn how to differentiate polynomial, rational, trigonometric, exponential, and logarithmic functions, among others.

We'll also see how arithmetic combinations and/or compositions of functions can be differentiated if we know the derivatives of the component pieces. These techniques are all ultimately based on the difference quotient definition of derivative seen in Chapter 2, but they enable us to differentiate

> In this chapter we learn additional notation, develop techniques that allow us to find derivatives of many classes of functions, and explore many more applications.

Joseph-Louis Lagrange
(1736–1813)

functions without having to laboriously calculate limits of difference quotients.

Finally, we'll learn how derivatives allow us to solve problems from disciplines such as biology, physics, chemistry, and economics.

The ability to efficiently and accurately calculate instantaneous rates of change, in all these different settings, allows us to solve problems that would otherwise be intractable.

---

1. Adapted from engraving by Robert Garsuch Hart, *LA GRANGE.* Smithsonian Institution Libraries.

3.1 Differentiation Notation and Consequences

TOPICS

1. Alternative notations and higher-order derivatives

2. Consequences of differentiability

Chapter 2 gave us a few motivational examples for seeking what we now call the derivative of a function, some tools to obtain the derivative, and enough notation to get started. We are ready now to explore the broad ramifications of this new concept, and we begin with a catalog of the terms and notation that will prove useful.

TOPIC 1   Alternative Notations and Higher-Order Derivatives

Recall that the fraction

$$\frac{f(c+h)-f(c)}{h}$$

is called the *difference quotient of f at c with increment h*, and that we let

$$f'(c)=\lim_{h\to 0}\frac{f(c+h)-f(c)}{h},$$

provided this limit exists. This is our definition of the *derivative of f at c*, and we quickly extend it to all points x at which a similar limit exists to define the *derivative of f* as the function whose value at x is given by

$$f'(x)=\lim_{h\to 0}\frac{f(x+h)-f(x)}{h}.$$

If $f'(c)$ exists for a given function f, we say f is **differentiable at c**, and if $f'(x)$ exists for all x in some open interval we say f is differentiable on that interval (requiring the interval to be open is a convenience allowing us to only worry about two-sided limits for the moment). However, as a reflection of the wide breadth of applications and the number of mathematicians who played a role in developing calculus, other notations and terms are frequently used in working with derivatives. If, as is common, the function f appears in the form of the equation $y=f(x)$, any of the following notations may be used to refer to the derivative of the function at x.

$$f'(x)=y'=\frac{dy}{dx}=\frac{df}{dx}=\frac{d}{dx}f(x)=Df(x)=D_x f(x)$$

In words, these notations are informally read, respectively, as "f prime of x," "y prime," "dee y, dee x," "dee f, dee x," "dee, dee x of f," "dee f," and "dee sub x of f" (or close variations of such phrases). Although the wealth of choices can be confusing at first, they all refer to the instantaneous rate of change of the function f or of the dependent variable y with respect to the independent variable x. The symbols d/dx, D, and D_x are called **differentiation operators** (with respect to x, at the moment), and the formal meaning of, for instance, dy/dx is "the derivative of y with respect to x."

If any one of the above notations served well in all applications, the alternative notations would likely fall into disuse. But as we will see, they each have their strengths and weaknesses. The *prime notation* (historically due to Joseph-Louis Lagrange) emphasizes the functional aspect of the derivative, while the notations dy/dx and df/dx (due to Gottfried Wilhelm Leibniz) are especially useful when working with equations. A symbol such as "dy" denotes an object called a *differential* and the notation dy/dx does indeed indicate a ratio of differentials (and the fact that $dy/dx = \lim\limits_{\Delta x \to 0}(\Delta y/\Delta x)$ is no coincidence). However, we are not yet ready to delve into differentials, and for the time being dy/dx should simply be regarded as a synonym for y'.

Example 1 ✐

Find df/dx for the function $f(x) = (1+x)/(1-x)$.

Solution

Since df/dx is just alternative notation for $f'(x)$, we have the following.

$$\frac{df}{dx} = f'(x)$$

$$= \lim_{h \to 0} \frac{f(x+h) - f(x)}{h}$$

$$= \lim_{h \to 0} \frac{\dfrac{1+(x+h)}{1-(x+h)} - \dfrac{1+x}{1-x}}{h}$$

$$= \lim_{h \to 0} \frac{\dfrac{\left[1+(x+h)\right](1-x) - (1+x)\left[1-(x+h)\right]}{\left[1-(x+h)\right](1-x)}}{h}$$

$$= \lim_{h \to 0} \frac{\left(1+x+h-x-x^2-xh\right) - \left(1-x-h+x-x^2-xh\right)}{h\left[1-(x+h)\right](1-x)}$$

After combining like terms in the numerator, this expression becomes

$$\lim_{h \to 0} \frac{2h}{h\left[1-(x+h)\right](1-x)} = \lim_{h \to 0} \frac{2}{\left[1-(x+h)\right](1-x)}$$

$$= \frac{2}{(1-x)^2}.$$

Thus we conclude that for $f(x) = (1+x)/(1-x)$,

$$\frac{df}{dx} = \frac{2}{(1-x)^2}.$$

Given that it emphasizes the function nature of the derivative, it isn't surprising that the prime notation allows us to easily denote the evaluation of a derivative at a point. In order to indicate the same concept with the differential notation, we need one additional symbol, as follows:

$$\text{given } y = f(x), \quad \left.\frac{dy}{dx}\right|_{x=c} \text{ is synonymous with } f'(c).$$

The notation $\left.\dfrac{dy}{dx}\right|_{x=c}$ is read "the derivative of y with respect to x evaluated at $x = c$."

Example 2 ✎

Let $f(x) = x^2 - x$. For $y = f(x)$, use both the prime and the differential notations to evaluate $f'(x)$. Use the derivative to find $f'(2)$, and express your answer in both notations.

Solution

As shown in Figure 1, Δx is just an alternate notation for the increment h, while $\Delta y = f(x+h) - f(x)$. Thus the corresponding difference quotient can be rewritten as

$$\frac{f(x+h) - f(x)}{h} = \frac{\Delta y}{\Delta x},$$

and hence

$$f'(x) = \lim_{h \to 0} \frac{f(x+h) - f(x)}{h}$$

$$= \lim_{\Delta x \to 0} \frac{\Delta y}{\Delta x}$$

$$= \frac{dy}{dx}$$

$$= \lim_{h \to 0} \frac{\left[(x+h)^2 - (x+h)\right] - (x^2 - x)}{h}$$

$$= \lim_{h \to 0} \frac{(x^2 + 2xh + h^2 - x - h) - (x^2 - x)}{h}$$

$$= \lim_{h \to 0} \frac{2xh + h^2 - h}{h} = \lim_{h \to 0}(2x + h - 1)$$

$$= 2x - 1.$$

Finally, we will use the alternate notations to express the derivative at $x = 2$.

$$f'(2) = \left.\frac{dy}{dx}\right|_{x=2} = \left.\frac{df}{dx}\right|_{x=2} = 2(2) - 1 = 3$$

Figure 2 shows the graphs of $y = f(x)$ and that of $dy/dx = f'(x)$ on the same set of axes. The reader is invited to examine the relationship between the rate of change, or "slope" of the graph of f and the function values $f'(x)$ as x ranges over the real numbers. We will examine this relationship in great depth over the coming sections.

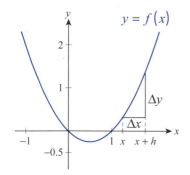

Figure 1

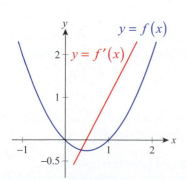

Figure 2

Before we leave the topics of notation and terminology, we need to make one last observation. As we have seen, the derivative of a function f is a function in its own right (assuming f is differentiable). Given that, it is reasonable to ask whether the derivative of f' makes sense, and what meaning $(f')'$ might possess.

Such *higher-order derivatives* do indeed make sense, and often possess meaning corresponding to familiar real-world experiences. First, some notation: given a function f and the equation $y = f(x)$, the following all denote the **second derivative of f (or y) with respect to x.**

$$f''(x) = y'' = \frac{d^2y}{dx^2} = \frac{d^2f}{dx^2} = \frac{d}{dx}\left(\frac{dy}{dx}\right) = \frac{d}{dx}\left(\frac{df}{dx}\right) = D^2 f(x) = D_x^2 f(x) = D_{xx} f(x)$$

There are, in fact, even more ways to indicate the same idea (such as the $(f')'$ that appeared above), but they all represent the notion of differentiating, with respect to x, a function that is itself a derivative with respect to x. Such symbols are read as "f double prime," "y double prime," "dee squared y, dee x squared," and so on. And the patterns continue with yet higher-order derivatives; although we will do little more than introduce the notation here, **third derivatives** are indicated with symbols such as f''' ("f triple prime") and d^3y/dx^3 ("dee cubed y, dee x cubed"). Beyond this level, the order of the derivative is often indicated with a number inside parentheses (instead of "prime" marks). The **n^{th} derivative of f (or y)** can be written as

$$f^{(n)}(x) = y^{(n)} = \frac{d^n y}{dx^n} = \frac{d^n f}{dx^n} = \frac{d}{dx}\, y^{(n-1)} = \frac{d}{dx}\, f^{(n-1)} = D^n f(x) = D_x^n f(x).$$

What meaning can we attach to such mathematical constructs as the second derivative? Consider again an example in which $x(t)$ denotes the position, at time t, of an object moving along a straight path. As we now know, $x'(t)$ represents the object's instantaneous rate of change of position (at time t), better known as velocity. So $x''(t)$ represents the object's instantaneous rate of change of velocity; that is, $x''(t)$ tells us how quickly the object's velocity is increasing or decreasing at time t. This "rate of change of velocity" goes by the common name **acceleration**.

Example 3 ✎

A tennis ball is hit vertically upward so that its position function is given by $h(t) = -16t^2 + 96t + 4$ feet above the ground at t seconds. Find the velocity and acceleration functions. What are the initial position, velocity, and acceleration values?

Solution

As we have seen, velocity is the first derivative of position, commonly labeled as $v(t)$, which we shall obtain using the definition of the derivative.

$$v(t) = h'(t) = \lim_{h \to 0} \frac{\left[-16(t+h)^2 + 96(t+h) + 4\right] - \left(-16t^2 + 96t + 4\right)}{h}$$

After expanding, combining, and canceling like terms in the numerator we obtain

$$v(t) = \lim_{h \to 0} \frac{-32th - 16h^2 + 96h}{h}$$

$$= \lim_{h \to 0} \left(-32t - 16h + 96 \right)$$

$$= -32t + 96.$$

We see from the position and velocity functions that the initial position and initial velocity, respectively, are

$$h(0) = -16(0)^2 + 96(0) + 4 = 4 \text{ ft}$$

and

$$v(0) = h'(0) = -32(0) + 96 = 96 \text{ ft/s}.$$

Next, as we have discussed above, the acceleration function of the tennis ball is the second derivative of position, in other words, the first derivative of velocity. Using the common notation of $a(t)$ for the acceleration function, once again with the help of the definition of the derivative we obtain

$$a(t) = v'(t)$$

$$= \frac{d}{dt} \left(-32t + 96 \right)$$

$$= \lim_{h \to 0} \frac{\left[-32(t+h) + 96 \right] - \left(-32t + 96 \right)}{h}$$

$$= \lim_{h \to 0} \frac{-32h}{h}$$

$$= -32 \text{ ft/s}^2.$$

What this means is that the tennis ball's acceleration is constant; that is,

$$a(t) = -32 \text{ ft/s}^2.$$

The reason for the negative sign is our convention of treating "upward" velocities as positive, but in many contexts, the negative sign can be omitted. In fact, any falling object that is under the influence of gravity only (i.e., all other forces including air resistance are negligible) moves at the same constant acceleration. This law of nature was discovered by Galileo Galilei (1564–1642) who was the first scientist to claim that in the absence of air resistance, *all* objects fall with the *same* constant acceleration, *anywhere* on the Earth. This constant acceleration caused by gravity is denoted by g, and if we exit the specific context of our problem for a moment and omit the negative sign, we have

$$g = 32 \text{ ft/s}^2,$$

or using metric units,

$$g = 9.8 \text{ m/s}^2.$$

One way to remember the constant accelerating action of gravity is to note that its effect is to "add" to the velocity of a falling object an "extra" downward velocity of 32 ft/s every second (or an "extra" downward 9.8 m/s every second if we are working with metric units).

In conclusion, to answer our last question of the initial acceleration of the tennis ball, we note that since acceleration is constant and pointing downward,

$$a(0) = -32 \text{ ft/s}^2,$$

and our solution is complete.

Although it is seen less often, the third derivative of a position function $x(t)$ also has a common name. Since a relatively large value (in magnitude) for $x'''(t)$ at a particular time t would mean a rapid change in the acceleration of the object, which corresponds to a very sudden change in the object's position, the third derivative is sometimes called the **jerk**.

Example 4 ✎

Find the first three derivatives of $f(x) = \frac{2}{3}x^3 - 3x + \frac{1}{3}$.

Solution

We start by calculating f' using the definition.

$$f'(x) = \lim_{h \to 0} \frac{\left[\frac{2}{3}(x+h)^3 - 3(x+h) + \frac{1}{3}\right] - \left[\frac{2}{3}x^3 - 3x + \frac{1}{3}\right]}{h}$$

$$= \lim_{h \to 0} \frac{\frac{2}{3}(x+h)^3 - 3(x+h) - \frac{2}{3}x^3 + 3x}{h}$$

$$= \lim_{h \to 0} \frac{\frac{2}{3}\left[(x+h)^3 - x^3\right] - 3h}{h} \qquad \text{Apply the expansion formula for the cube of a binomial.}$$

$$= \lim_{h \to 0} \frac{\frac{2}{3}\left(x^3 + 3x^2h + 3xh^2 + h^3 - x^3\right) - 3h}{h} \qquad \text{Combine like terms.}$$

$$= \lim_{h \to 0} \frac{2x^2h + 2xh^2 + \frac{2}{3}h^3 - 3h}{h}$$

$$= \lim_{h \to 0}\left(2x^2 + 2xh + \frac{2}{3}h^2 - 3\right) \qquad \text{For a fixed } x, \text{ if } h \to 0, \text{ both } 2xh \to 0 \text{ and } \frac{2}{3}h^2 \to 0.$$

$$= 2x^2 - 3$$

Notice that the derivative of the given cubic polynomial is a quadratic function. Next, knowing that

$$f''(x) = \frac{d}{dx}f'(x),$$

we obtain the second derivative of f as follows.

$$f''(x) = \frac{d}{dx}\left(2x^2 - 3\right)$$

$$= \lim_{h \to 0} \frac{\left[2(x+h)^2 - 3\right] - \left(2x^2 - 3\right)}{h}$$

$$= \lim_{h \to 0} \frac{2(x+h)^2 - 2x^2}{h}$$

$$= \lim_{h \to 0} \frac{2\left(x^2 + 2xh + h^2\right) - 2x^2}{h} \qquad \text{Combine like terms and factor out } h.$$

$$= \lim_{h \to 0} \frac{h(4x + 2h)}{h}$$

$$= \lim_{h \to 0} (4x + 2h)$$

$$= 4x$$

Perhaps it doesn't come as a surprise that the derivative of the above quadratic function is linear. To obtain the third derivative, we will differentiate yet again. As you might have observed already, the derivative of a linear function is a constant, so that is what we expect f''' to be (see Example 3 above or Exercise 68 of Section 2.6).

$$f'''(x) = \frac{d}{dx}\left(f''(x)\right)$$

$$= \lim_{h \to 0} \frac{4(x+h) - 4x}{h}$$

$$= \lim_{h \to 0} \frac{4x + 4h - 4x}{h}$$

$$= \lim_{h \to 0} \frac{4h}{h}$$

$$= \lim_{h \to 0} 4$$

$$= 4$$

Exercise 66 of Section 2.6 asks the reader to prove that the derivative of any constant function is identically 0, so we would like to note as we conclude this example that the fourth, fifth, and any subsequent higher-order derivatives of f are identically 0.

Figure 3 shows the graph of f along with those of its first three derivatives on the same axes. The reader is invited to compare the graphs and examine how the rate of change of each at every x is reflected by the corresponding value of its derivative.

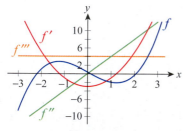

Figure 3

TOPIC 2   Consequences of Differentiation

Two primary questions will serve well to guide us:

What does the differentiability of a function f at a point c tell us about f?

and

What are the characteristics of a function f that fails to be differentiable at a point c?

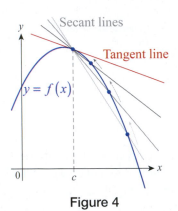

Figure 4

We began to answer these questions with our informal exploration of tangent lines in Chapter 2, and we know that the graph of a function f can only have a line tangent to it at the point c if secant lines that approximate the tangent line behave in a certain way (see Figure 4). Namely, the "limiting behavior" of the secant lines must exist—if so, the result is what we identify as the tangent line. This requirement imposes some restrictions on functions that are differentiable, the first of which follows.

Theorem ⚲

Differentiability Implies Continuity

If f is differentiable at c, then f is continuous at c.

Proof ✎

Since we seek to prove that f is continuous at the point c, we need to show $\lim_{x \to c} f(x) = f(c)$. The one fact we have to work with is that $f'(c)$ exists.

By our definition of $f'(c)$, this means that

$$\lim_{x \to c} \frac{f(x) - f(c)}{x - c}$$

exists, and so implicitly we are also given that f is defined at c; in other words, $f(c)$ exists. Further, the fact that the above limit exists means that $f(x)$ is defined for all x in some open interval containing c (that is, "sufficiently close to c"), so it makes sense to talk about $\lim_{x \to c} f(x)$. For $x \neq c$, we can multiply the difference quotient by $x - c$ and note that

$$f(x) = \frac{f(x) - f(c)}{x - c} \cdot (x - c) + f(c).$$

We can now apply the limit laws of Chapter 2 to obtain the following:

$$\lim_{x \to c} f(x) = \lim_{x \to c} \left[\frac{f(x) - f(c)}{x - c} \cdot (x - c) + f(c) \right]$$

$$= \lim_{x \to c} \left[\frac{f(x) - f(c)}{x - c} \cdot (x - c) \right] + \lim_{x \to c} f(c) \qquad \text{The limit of a sum is the sum of the limits.}$$

$$= \lim_{x \to c} \left[\frac{f(x) - f(c)}{x - c} \right] \cdot \lim_{x \to c} (x - c) + \lim_{x \to c} f(c) \qquad \text{The limit of a product is the product of the limits.}$$

$$= f'(c) \cdot 0 + f(c) \qquad \text{$f(c)$ exists and $\lim_{x \to c} f(c) = f(c)$.}$$

$$= f(c)$$

We can gain additional insight into the restrictions that differentiability imposes on the behavior of a function by considering the second primary question. If f is *not* differentiable at c, the limit of the difference quotient must not exist. We know from the preceding theorem that a discontinuity at c would cause this condition, but there are other ways for the limit to not exist. The behavior of the function $f(x)=|x|$ at the point $x=0$ gives us one example. For $x>0$ the difference quotient is

$$\frac{f(x)-f(0)}{x-0}=\frac{|x|-0}{x-0}=\frac{x}{x}=1,$$

and for $x<0$ the difference quotient is

$$\frac{f(x)-f(0)}{x-0}=\frac{|x|-0}{x-0}=\frac{-x}{x}=-1.$$

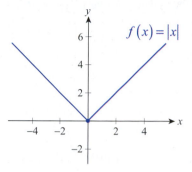

The fact that the two one-sided limits do not agree at 0 means that $f(x)=|x|$ is not differentiable at 0; such a point on the graph of a function, where the one-sided difference quotient limits exist but are unequal, is called a **corner** (see Figure 5).

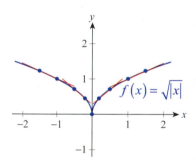

Figure 5 A Corner

What if one or both of the one-sided difference quotients fails to exist at a point c where f is continuous? One possibility is if one of the one-sided difference quotients tends to $-\infty$ and the other to $+\infty$; such a point on the graph of f is called a **cusp**. In the graph of $f(x)=\sqrt{|x|}$ from Figure 6, note how the secant lines to the right of the cusp all have positive slopes (and increase without bound as x approaches 0), while the secant lines on the left all have negative slopes.

Figure 6 A Cusp

If both one-sided difference quotients have a limit of $+\infty$, or if both have a limit of $-\infty$, the graph of f at c actually has a tangent line but its slope is undefined; we (naturally enough) refer to such a situation as a **vertical tangent line**, and the graphs of $\sqrt[3]{x}$ and $-\sqrt[3]{x}$ both possess this property at 0 (see Figure 7). Note that the secant lines approximating the vertical tangent all have positive slope on the graph of $\sqrt[3]{x}$ and all have negative slope on the graph of $-\sqrt[3]{x}$.

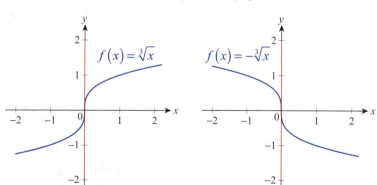

Figure 7 Vertical Tangent Lines

It is also possible for one or both of the one-sided difference quotient limits at a point c to not exist and for the difference quotients to not tend to $+\infty$ or $-\infty$ either. Such a function, which again may even be continuous at c, simply oscillates too much on every interval containing c. Our next example illustrates this phenomenon.

Example 5 ✐

Prove that the function $f(x) = \begin{cases} x\sin\left(1/x\right) & \text{if } x \neq 0 \\ 0 & \text{if } x = 0 \end{cases}$ is continuous, but not differentiable, at 0.

Solution

To see the continuity of $f(x)$ at 0, observe that since $-1 \leq \sin(1/x) \leq 1$ for all nonzero x, it follows that

$$-|x| \leq x\sin\frac{1}{x} \leq |x|,$$

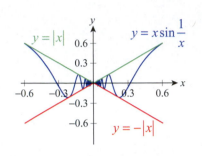

Figure 8

if $x \neq 0$ (see Figure 8). We can now apply the Squeeze Theorem, since $\lim\limits_{x \to 0}|x| = \lim\limits_{x \to 0}(-|x|) = 0$. Thus, $\lim\limits_{x \to 0} f(x) = 0 = f(0)$, so $f(x)$ is continuous at 0.

To prove the nondifferentiability of f, we will show that the limit of its difference quotients does not exist at 0. This will be due to extensive (in fact, infinitely many) oscillations of f near the origin, as we shall see shortly. Let us first assume that $x > 0$, and examine the difference quotient $\left[f(x) - f(0)\right]/(x - 0)$:

$$\frac{f(x) - f(0)}{x - 0} = \frac{x\sin\dfrac{1}{x} - 0}{x - 0} = \frac{x\sin\dfrac{1}{x}}{x} = \sin\frac{1}{x}$$

In other words, proving that the one-sided difference quotient limit does not exist for $x > 0$ is now equivalent to proving that $\lim\limits_{x \to 0^+} \sin(1/x)$ does not exist. To that end, we will present an argument very similar to that given in Example 7 of Section 2.3, albeit in a somewhat abbreviated fashion. (Note that this also means that not only f oscillates "too much" near 0, but so do its difference quotients.)

Suppose that the above limit exists, and $\lim\limits_{x \to 0^+} \sin(1/x) = L$. Then for $\varepsilon = 1$, there is a $\delta > 0$ such that

$$\left|\sin\frac{1}{x} - L\right| < 1$$

whenever $0 < |x| < \delta$.

Recall from your studies of the sine function that for any integer k,

$$\sin(4k+1)\frac{\pi}{2} = 1 \quad \text{and} \quad \sin(4k+3)\frac{\pi}{2} = -1.$$

In addition, we can choose and fix a big enough positive integer k such that

$$\frac{1}{(4k+1)\dfrac{\pi}{2}} < \delta.$$

Using the notation

$$x_1 = \frac{1}{(4k+1)\dfrac{\pi}{2}} \quad \text{and} \quad x_2 = \frac{1}{(4k+3)\dfrac{\pi}{2}},$$

note that $x_2 < x_1 < \delta$, while $\sin\left(1/x_1\right) = 1$ and $\sin\left(1/x_2\right) = -1$.

By assumption,

$$\left| \sin\frac{1}{x_1} - L \right| = |1 - L| < 1,$$

which implies that $L > 0$, but also

$$\left| \sin\frac{1}{x_2} - L \right| = |-1 - L| < 1,$$

implying in turn that $L < 0$, an obvious impossibility.

Thus $\sin\left(1/x\right)$ cannot be approaching any limit L at all as x approaches 0 from the right; that is,

$$\lim_{x\to 0^+} \sin\frac{1}{x} = \lim_{x\to 0^+} \frac{f(x) - f(0)}{x - 0} \quad \text{does not exist.}$$

The case of $x < 0$ is handled in an analogous fashion, and it will follow that neither one-sided difference quotient limit exists, and thus $f(x)$ is not differentiable at 0, just as we needed to show.

In summary, and informally, the differentiability of a function f at a point c implies that f is not only continuous at c but also fairly "smooth" and well-behaved—no corners, cusps, or excessive oscillation can occur at c. Keep in mind, though, that it is possible for the graph of f to possess a vertical tangent line at c, and that this may be overlooked because f' will not exist at such a point. However, such points can be identified by comparing the behavior of the one-sided difference quotients.

As we have seen, one-sided limits of difference quotients can be useful when determining the behavior of a function. We will now give these limits a name.

Definition 💡

One-Sided Derivatives

If a function f is defined on an interval of the form $[a, p)$ (that is, at a and on some half-open interval to the right of a), the **right-hand derivative of f at a** is defined to be

$$f_+'(a) = \lim_{h\to 0^+} \frac{f(a+h) - f(a)}{h},$$

provided the limit exists.

Similarly, if f is defined on an interval of the form $(p,b]$, the **left-hand derivative of f at b** is defined to be

$$f'_-(b) = \lim_{h \to 0^-} \frac{f(b+h) - f(b)}{h},$$

provided the limit exists.

We can now extend our definition of differentiability to intervals that are not open. For example, we say f is *differentiable on the interval* $[a,b]$ if f is differentiable on (a,b) and $f'_+(a)$ and $f'_-(b)$ exist.

Example 6 ☑

Consider the following piecewise defined function:

$$f(x) = \begin{cases} 2\sqrt{x} & \text{if } 0 \le x < 1 \\ \dfrac{x^2 + 3}{2} & \text{if } 1 \le x \le 2 \end{cases}$$

a. Find the one-sided derivatives of f at 1 and use them to decide whether f is differentiable at $c = 1$.

b. Find the appropriate one-sided derivatives at the endpoints of the domain of f.

Solution

a. First of all, since

$$\lim_{x \to 1^-} 2\sqrt{x} = \lim_{x \to 1^+} \frac{x^2 + 3}{2} = 2,$$

f is certainly continuous at 1. But is it differentiable? The one-sided derivatives will have the answer.

$$\begin{aligned} f'_-(1) &= \lim_{h \to 0^-} \frac{f(1+h) - f(1)}{h} \\[2ex] &= \lim_{h \to 0^-} \frac{2\sqrt{1+h} - \dfrac{1^2 + 3}{2}}{h} \\[2ex] &= \lim_{h \to 0^-} \frac{\left(2\sqrt{1+h} - 2\right)\left(2\sqrt{1+h} + 2\right)}{h\left(2\sqrt{1+h} + 2\right)} \\[2ex] &= \lim_{h \to 0^-} \frac{4(1+h) - 4}{h\left(2\sqrt{1+h} + 2\right)} \\[2ex] &= \lim_{h \to 0^-} \frac{4}{\left(2\sqrt{1+h} + 2\right)} = 1 \end{aligned}$$

Thus we see that the left-hand derivative of f at 1 equals 1. The right-hand derivative at the same point is determined as follows.

$$f'_+(1) = \lim_{h \to 0^+} \frac{f(1+h) - f(1)}{h}$$

$$= \lim_{h \to 0^+} \frac{\dfrac{(1+h)^2 + 3}{2} - \dfrac{1^2 + 3}{2}}{h} \qquad \text{Combine the fractions in the numerator.}$$

$$= \lim_{h \to 0^+} \frac{\dfrac{2h + h^2}{2}}{h} \qquad \text{Cancel } h.$$

$$= \lim_{h \to 0^+} \frac{2+h}{2} = 1$$

Since the one-sided derivatives agree, we conclude that f is differentiable at $c = 1$. Graphically this means that even though the graph is "patched together" at $c = 1$ from two "pieces," this is done in a "smooth" way, in other words, without the formation of any corner or cusp.

b. Since $a = 0$ is the left endpoint of the domain of f, only the right-hand derivative is defined.

$$f'_+(0) = \lim_{h \to 0^+} \frac{f(0+h) - f(0)}{h}$$

$$= \lim_{h \to 0^+} \frac{2\sqrt{0+h} - 2\sqrt{0}}{h}$$

$$= \lim_{h \to 0^+} \frac{2\sqrt{h}}{h}$$

$$= \lim_{h \to 0^+} \frac{2}{\sqrt{h}} = \infty$$

So the right-hand derivative of f at 0 does not exist.

Finally, at the right endpoint $b = 2$,

$$f'_-(2) = \lim_{h \to 0^-} \frac{f(2+h) - f(2)}{h}$$

$$= \lim_{h \to 0^-} \frac{\dfrac{(2+h)^2 + 3}{2} - \dfrac{2^2 + 3}{2}}{h} \qquad \text{Combine the fractions in the numerator.}$$

$$= \lim_{h \to 0^-} \frac{\dfrac{4h + h^2}{2}}{h} \qquad \text{Cancel } h.$$

$$= \lim_{h \to 0^-} \frac{4+h}{2} = 2.$$

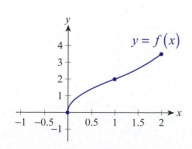

Figure 9

To summarize our findings, the graph of f is joined together "smoothly" from two pieces at $c = 1$, it goes "vertical" at the left endpoint of its domain, and has a "slope" of 2 at the right endpoint, since the right-hand derivative exists there. Figure 9 graphically reflects these observations.

We will close this section with one more theorem that relates the ideas of continuity and differentiability.

Theorem ✎

Darboux's Theorem

If f is differentiable on $[a,b]$, then f' takes on every value between $f_+'(a)$ and $f_-'(b)$. That is, f' has the Intermediate Value Property (IVP) on the interval $[a,b]$: if L is a value between $f_+'(a)$ and $f_-'(b)$, there is a point c in $[a,b]$ such that $f'(c) = L$.

In Section 2.5, we saw that continuous functions on an interval $[a,b]$ have the Intermediate Value Property, so we already know that the function f' has the IVP if it is continuous. As it turns out, derivatives of functions are not necessarily continuous, but the French mathematician Jean-Gaston Darboux proved in 1875 that a function can only *be* the derivative of some other function if it possesses the IVP. In Chapter 5 we will prove a partial converse of this when we see that if a function is continuous, then it is the derivative of some other function.

3.1 **Exercises**

1–12 Find the derivative of the given function at the specified point and express your answer using the differential notation due to Leibniz.

1. $f(x) = 7$; $x = 1$

2. $g(x) = \dfrac{1}{2}x - 5$; $x = -1$

3. $h(x) = \dfrac{1}{2}x^2 - 5$; $x = 0$

4. $F(t) = \dfrac{1}{5}t + 2t^2$; $t = \dfrac{1}{5}$

5. $G(s) = \dfrac{1}{3}s^3 - s$; $s = -3$

6. $H(t) = \dfrac{1}{2}t^4 + t^2$; $t = -2$

7. $K(z) = \dfrac{5}{3z + 1}$; $z = 0$

8. $T(t) = \dfrac{2t - 3}{t + 1}$; $t = \sqrt{5} - 1$

9. $w(z) = \dfrac{1}{z^2 + 2}$; $z = \sqrt{2}$

10. $A(t) = \sqrt{2t}$; $t = 0$

11. $Q(y) = \dfrac{1}{\sqrt{3y}}$; $y = 1$

12. $B(u) = \sqrt{u^2 + 1}$; $u = -2\sqrt{2}$

13–24 Find the derivative of the function and use the differentiation operator D_x to express your answer.

13. $f(x) = \pi^2$

14. $g(x) = 1 - \dfrac{2}{3}x$

15. $h(t) = \dfrac{3}{2}t^2 + 10t - 1$

16. $F(s) = 4 - s + \dfrac{1}{2}s^2 + s^3$

17. $G(y) = \dfrac{1}{3y}$

18. $H(t) = \dfrac{t - 3}{t + 3}$

19. $S(z) = \dfrac{-3}{z^2}$

20. $T(u) = \dfrac{4}{u^2 - 3u}$

21. $R(v) = \sqrt{2v - 3}$

22. $f(y) = \dfrac{2}{\sqrt{y + 2}}$

23. $X(y) = 2y^4$

24. $u(x) = \dfrac{2}{\sqrt{2x^2 + 1}}$

25–33 Find the first, second, and third derivatives of the function. Then graph the function along with its derivatives in the same coordinate system and compare the graphs. (**Hint:** See Example 4.)

25. $f(x) = \dfrac{5}{2}x - 1$

26. $g(x) = x^2 + 5$

27. $h(x) = -\dfrac{1}{2}x^2 + x - \dfrac{3}{2}$

28. $U(x) = -x^3$

29. $V(x) = \dfrac{1}{3}(x-2)^3$

30. $F(t) = t^4 - 1$

31. $G(x) = 2(x-1)^4$

32. $H(x) = \dfrac{1}{x}$

33. $K(s) = \dfrac{-2}{s-1}$

34–37 The graphs of the position, velocity, acceleration, and jerk of a moving particle are given. Decide which one is which, label them accordingly, and explain.

34.
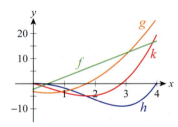

35.
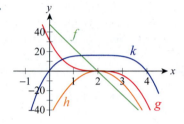

36.
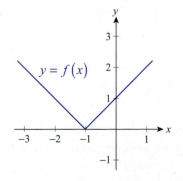

37.
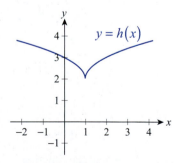

38–43 Use the given graph of the function to find all x-values where the function is differentiable.

38.
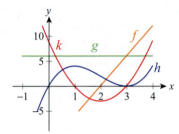

39.
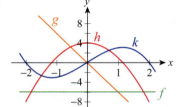

40.
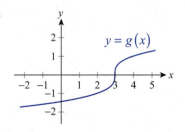

41.

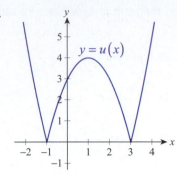

42.

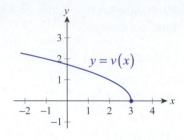

43.

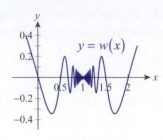

44–58 Find all points where the function is not differentiable. For each of those points, find the one-sided derivatives (if they exist).

44. $f(x) = |x+5|$

45. $g(x) = |x+2| - |x-4|$

46. $h(x) = (x-1)^{2/3}$

47. $F(x) = \sqrt[3]{x-1.5} + 2$

48. $H(x) = \sqrt{1.8-x}$

49. $k(x) = \sqrt{3-x^2}$

50. $G(x) = \dfrac{x^2}{x^2-9}$

51. $m(x) = |x^2 - 6x + 5|$

52. $A(t) = [\![t-4]\!]$

53. $B(x) = x - [\![x]\!]$

54. $F(t) = \begin{cases} \dfrac{1}{2}t\cos\dfrac{1}{t} & \text{if } t \neq 0 \\ 0 & \text{if } t = 0 \end{cases}$

55. $H(z) = \begin{cases} \sqrt{z}\sin\dfrac{\pi}{z} & \text{if } z > 0 \\ 0 & \text{if } z = 0 \end{cases}$

56. $P(x) = \begin{cases} \sqrt[3]{x-1} & \text{if } x < 1 \\ (x-1)^2 & \text{if } x \geq 1 \end{cases}$

57. $G(x) = \begin{cases} 2x+2 & \text{if } x \leq -2 \\ -\dfrac{1}{2}x^2 & \text{if } x > -2 \end{cases}$

58. $S(t) = \begin{cases} \dfrac{1}{t} & \text{if } t \leq 1 \\ t & \text{if } t > 1 \end{cases}$

59. Prove that the function

$$f(x) = \begin{cases} x^2\sin\dfrac{1}{x} & \text{if } x \neq 0 \\ 0 & \text{if } x = 0 \end{cases}$$

is differentiable at 0. Contrast this result with Example 5.

60. The position function of a car crashing head-on at 62 mph during a crash test is $x(t) = -196t^2 + 27.78t$, where x is measured in meters and t in seconds. Find the deceleration of the dummy inside the car. What multiple of g is this (where g is the gravity constant)?

61. The position from its starting point of a small plane preparing for takeoff is given by $x(t) = 1.1t^2$ meters (t is measured in seconds).

   a. What is the acceleration of the plane?

   b. How long does it take for the plane to reach the minimum takeoff speed of 33 m/s?

   c. What is the minimum required runway length for this type of plane?

62. The position function of a theme park thrill ride moving along a straight line is $x(t) = \frac{14}{3}t^3 + 10t$ ft ($0 \leq t \leq 3$, t is measured in seconds). Find the velocity, acceleration, and jerk. How far from starting position are the cars at the end of the 3-second time interval?

63. The **symmetric derivative** of a function f at a point c is defined as

$$f'_{sym}(c) = \lim_{h \to 0} \frac{f(c+h) - f(c-h)}{2h}.$$

   a. Prove that if f is differentiable at c, then its symmetric derivative exists and $f'_{sym}(c) = f'(c)$.

   b. Give an example of a function g and a point c such that $g'_{sym}(c)$ exists, but g is not differentiable at c.

64. Sketch the graph of $f(x) = -x^2 + 2x$ and its derivative on the interval $[0,2]$ in the same coordinate system. Where (on which interval) is f' positive? Where is f' negative? Identify those intervals where f is increasing versus decreasing. Do you see a connection? Can you give an intuitive reason for your findings?

65. Repeat Exercise 64 for the function $g(x) = 1/x^2$. Sketch both g and g' on their entire domains and summarize your observations. Can you formulate a general conjecture?

66–69 *True or False?* Determine whether the given statement is true or false. In case of a false statement, explain or provide a counterexample.

66. If f is continuous at c, then f is differentiable at c.

67. If f is differentiable at c, then f is continuous at c.

68. If f is differentiable at c, then

$$f'(c) = \lim_{\Delta x \to 0} \frac{f(c - \Delta x) - f(c)}{-\Delta x}.$$

69. If both one-sided derivatives of f at c exist, then f is differentiable at c.

3.1 **Technology Exercises**

70–75 Use a computer algebra system to graph the given function along with its derivative in the same viewing window and answer the questions of Exercise 64. (**Hint:** Use the "Derivative" feature of your technology to find f' first.)

70. $f(x) = \frac{1}{3}x^3 - 5x^2 - 1$ **71.** $f(x) = \frac{x^2}{x^2 - 4}$

72. $f(x) = \frac{x}{x^2 + 1}$ **73.** $f(x) = -\cos x$

74. $f(x) = \sin^2 x$ **75.** $f(x) = e^{1/(x^2 + 1)}$

76–79 Use a computer algebra system to find the first four derivatives of f; then graph them along with f in the same viewing window and compare the graphs.

76. $f(x) = \frac{1}{2}x^5 - 2x^4 - 5x^3 + 7$

77. $f(x) = \arctan x$

78. $f(x) = \frac{2x}{x - 3}$

79. $f(x) = x\sin\left(\frac{1}{10}x - 1\right)$

80–83 Use a graphing calculator or computer algebra system to graph the function and identify all points where the function is not differentiable. Explain.

80. $G(x) = \left(x^2 - 4\right)^{2/5}$

81. $H(t) = |2t - 1|^{2/3}$

82. $P(x) = \begin{cases} \arctan x & \text{if } x < 0 \\ x^{3/2} & \text{if } x \geq 0 \end{cases}$

83. $L(t) = \sqrt{|t|}\sin\frac{1}{|t|}$

3.2 Derivatives of Polynomials, Exponentials, Products, and Quotients

TOPICS

1. Elementary differentiation rules

2. Differentiation of exponential functions

3. The product and quotient rules

Just as the limit laws allow us to quickly evaluate many limits without resorting to epsilon-delta arguments, differentiation rules allow us to determine most common derivatives without having to consider limits of difference quotients. Much of the next five sections will be devoted to developing and using these differentiation rules.

TOPIC 1   Elementary Differentiation Rules

We begin with a rule for arguably the simplest sorts of functions—those that have a constant value, say k. Since the graph of $f(x) = k$ is the horizontal line $y = k$, the tangent line to the graph at any point must have slope 0. This is the geometric (and informal) justification for the rule that if $f(x) = k$ then $f'(x) = 0$ for every x. The formal proof of this statement is just as easy: given any x,

$$f'(x) = \lim_{h \to 0} \frac{f(x+h) - f(x)}{h} = \lim_{h \to 0} \frac{k - k}{h} = \lim_{h \to 0} 0 = 0.$$

Theorem 🔍

The Constant Rule

If f is a constant function, say $f(x) = k$, then $f'(x) = 0$. In other notation, we write

$$\frac{df}{dx} = \frac{d}{dx}(k) = 0.$$

In the last two sections, we worked through the difference quotient limit process to determine the derivatives of $f(x) = x$ and $f(x) = x^2$. These limits are not terribly difficult: briefly,

$$\frac{d}{dx}(x) = \lim_{h \to 0} \frac{(x+h) - x}{h} = \lim_{h \to 0} \frac{h}{h} = \lim_{h \to 0} 1 = 1$$

and

$$\frac{d}{dx}(x^2) = \lim_{h \to 0} \frac{(x+h)^2 - x^2}{h} = \lim_{h \to 0} \frac{x^2 + 2xh + h^2 - x^2}{h}$$

$$= \lim_{h \to 0} \frac{h(2x+h)}{h} = \lim_{h \to 0} (2x+h) = 2x.$$

But we don't want to have to work through a similar computation every time we encounter a function of the form $f(x) = x^n$ if, instead, a general rule can be found. Fortunately, this is indeed the case.

Theorem ⚷

The Positive Integer Power Rule

If $f(x) = x^n$, where n is a positive integer, then $f'(x) = nx^{n-1}$. In other notation, we write

$$\frac{df}{dx} = \frac{d}{dx}\left(x^n\right) = nx^{n-1}.$$

Proof 📎

There are actually several ways to prove this first Power Rule, and because they introduce useful techniques we will provide two common proofs here.

Proof 1

The identity

$$x^n - c^n = (x - c)\left(x^{n-1} + x^{n-2}c + \cdots + xc^{n-2} + c^{n-1}\right)$$

can be verified by multiplying out the right-hand side and canceling the common terms. With this identity in hand, we can use one of the alternative forms of the difference quotient to obtain the following.

$$
\begin{aligned}
f'(c) &= \lim_{x \to c} \frac{f(x) - f(c)}{x - c} = \lim_{x \to c} \frac{x^n - c^n}{x - c} \\
&= \lim_{x \to c} \frac{(x - c)\left(x^{n-1} + x^{n-2}c + \cdots + xc^{n-2} + c^{n-1}\right)}{x - c} \\
&= \lim_{x \to c} \left(x^{n-1} + x^{n-2}c + \cdots + xc^{n-2} + c^{n-1}\right) \\
&= c^{n-1} + c^{n-2}c + \cdots + cc^{n-2} + c^{n-1} \qquad \textcolor{teal}{\text{Note that there are } n \text{ terms.}} \\
&= nc^{n-1}
\end{aligned}
$$

Proof 2

Alternatively, we can use the Binomial Theorem to expand the term $(x + h)^n$ as follows.

$$
\begin{aligned}
f'(x) &= \lim_{h \to 0} \frac{f(x + h) - f(x)}{h} = \lim_{h \to 0} \frac{(x + h)^n - x^n}{h} \\
&= \lim_{h \to 0} \frac{\left(x^n + nx^{n-1}h + \dfrac{n(n-1)}{2}x^{n-2}h^2 + \cdots + nxh^{n-1} + h^n\right) - x^n}{h} \\
&= \lim_{h \to 0} \frac{h\left(nx^{n-1} + \dfrac{n(n-1)}{2}x^{n-2}h + \cdots + nxh^{n-2} + h^{n-1}\right)}{h} \\
&= \lim_{h \to 0} \left(nx^{n-1} + \dfrac{n(n-1)}{2}x^{n-2}h + \cdots + nxh^{n-2} + h^{n-1}\right) \\
&= nx^{n-1}
\end{aligned}
$$

As we will see, the Power Rule is actually true for all real number exponents, a fact we will prove in stages as we acquire the necessary tools.

Example 1 ✎

Use the Constant Rule and Positive Integer Power Rule to determine the derivative of the given function.

a. $f(x) = -5$ **b.** $g(x) = 0$ **c.** $h(x) = \pi^3$

d. $k(x) = x^4$ **e.** $s(t) = t^{15}$ **f.** $w(z) = \dfrac{1}{z}$

Solution

a. Since f is a constant function, the Constant Rule applies. Using the various notations, we obtain

$$f'(x) = \frac{df}{dx} = \frac{d}{dx}(-5) = 0.$$

Recall that there are even more ways to express our answer. For example, $D_x\, 5 = 0$, or if f appears in the form of $y = 5$, we might write $dy/dx = 0$, etc.

b. Again, since g is constant, its derivative will be identically 0, just like that of any other constant function.

$$g'(x) = \frac{dg}{dx} = \frac{d}{dx}(0) = 0$$

c. At first, it might be tempting to try to use the Positive Integer Power Rule on h, but we must be careful to avoid that mistake. Notice that π^3 is a constant, so $h(x)$ is a constant function and when differentiated it behaves like any other constant function.

$$h'(x) = \frac{dh}{dx} = \frac{d}{dx}(\pi^3) = 0$$

d. Since $k(x)$ is a positive integer power of the independent variable x, the Positive Integer Power Rule applies with $n = 4$.

$$k'(x) = \frac{dk}{dx} = \frac{d}{dx}(x^4) = 4x^3$$

e. Notice that the independent variable of s is denoted by t, but the Positive Integer Power Rule applies, just like it would with the variable x.

$$s'(t) = \frac{ds}{dt} = \frac{d}{dt}(t^{15}) = 15t^{14}$$

f. Since $1/z = z^{-1}$ is not a positive integer power of z, we cannot apply the Positive Integer Power Rule for w, but will proceed using the definition of the derivative instead. Our conclusion will be pleasantly consistent.

$$\frac{dw}{dz} = \lim_{h \to 0} \frac{\frac{1}{z+h} - \frac{1}{z}}{h} = \lim_{h \to 0} \frac{\frac{z-(z+h)}{(z+h)z}}{h} = \lim_{h \to 0} \frac{-h}{h(z+h)z} = \lim_{h \to 0} \frac{-1}{(z+h)z} = -\frac{1}{z^2}$$

Notice that if we had applied the Positive Integer Power Rule, we would have obtained

$$\frac{dw}{dz} = \frac{d}{dz}\left(z^{-1}\right) = \left(-1\right)z^{-1-1} = -z^{-2} = -\frac{1}{z^2},$$

which is precisely the correct answer. In other words, the pattern of the rule still applies to the exponent of -1. As mentioned immediately preceding this example, the Power Rule actually holds true for any real exponent, a fact we will prove later.

The next rule points out that the derivative of a constant multiple of a given function is no more difficult to determine than the derivative of the function itself.

Theorem 🔍

The Constant Multiple Rule

Given a constant k and a differentiable function f, the derivative of $kf(x)$ is $kf'(x)$. In other notation,

$$\frac{d}{dx}\left[kf(x)\right] = k\frac{d}{dx}f(x) = kf'(x).$$

In words, the Constant Multiple Rule is often summed up with a phrase like "constants pass through the differentiation operator," and such a phrase may bring to mind similar behavior we have already seen. Recall that one of the properties of limits is

$$\lim_{x \to c}\left[kf(x)\right] = k\lim_{x \to c}f(x),$$

and in fact it is this property that is used in the proof of the Constant Multiple Rule—you will work through this proof in Exercise 108.

Theorem 🔍

The Sum and Difference Rules

If f and g are both differentiable at x, the sum $f + g$ is also differentiable at x and $(f + g)'(x) = f'(x) + g'(x)$. In other notation,

$$\frac{d}{dx}\left[f(x) + g(x)\right] = \frac{d}{dx}f(x) + \frac{d}{dx}g(x).$$

Similarly, $f - g$ is differentiable and $(f - g)'(x) = f'(x) - g'(x)$.

Proof ✎

$$(f+g)'(x) = \lim_{h \to 0} \frac{(f+g)(x+h)-(f+g)(x)}{h}$$

$$= \lim_{h \to 0} \frac{\left[f(x+h)+g(x+h)\right]-\left[f(x)+g(x)\right]}{h}$$

$$= \lim_{h \to 0} \left[\frac{f(x+h)-f(x)}{h} + \frac{g(x+h)-g(x)}{h}\right]$$

$$= \lim_{h \to 0} \frac{f(x+h)-f(x)}{h} + \lim_{h \to 0} \frac{g(x+h)-g(x)}{h}$$

$$= f'(x)+g'(x)$$

To prove the corresponding Difference Rule, we can make use of the Sum Rule and the Constant Multiple Rule.

$$(f-g)'(x) = (f+(-1)g)'(x) = f'(x)+(-1)g'(x) = f'(x)-g'(x)$$

With the Constant Rule, Positive Integer Power Rule, Constant Multiple Rule, and Sum and Difference Rules, we can now quickly differentiate any polynomial. Further, it should be noted that these rules imply that the derivative of a polynomial is always another polynomial.

Example 2 ✐

Use the above rules to find the derivative of the given function.

a. $p(x) = 2x^2 + 3x$　　**b.** $q(x) = x^3 + \dfrac{3}{2}x^2 - 5x + 7$

c. $f(x) = 3\sqrt{x} - \dfrac{2}{x}$

Solution

a. Since p is a polynomial of two terms, the Sum Rule immediately ensures that we can differentiate it *termwise*; that is, we can differentiate the terms separately and add up the results.

$$p'(x) = \frac{d}{dx}p(x) = \frac{d}{dx}(2x^2+3x) = \frac{d}{dx}(2x^2) + \frac{d}{dx}(3x)$$

Next, the Constant Multiple Rule and the Power Rule come to bear, and we obtain

$$\frac{d}{dx}(2x^2) + \frac{d}{dx}(3x) = 2\frac{d}{dx}(x^2) + 3\frac{d}{dx}(x) = 2(2x) + 3(1) = 4x + 3.$$

Thus we conclude that $p'(x) = 4x + 3$. Notice what we found here is that the derivative of a quadratic polynomial is a linear polynomial. This is indeed the case in general; in fact it is not hard to see that if the degree of p is n, then its derivative has degree $n - 1$.

b. Even though $q(x)$ has more than two terms, a repeated use of the Sum Rule still enables us to differentiate termwise, as shown.

$$q'(x) = \frac{d}{dx} q(x)$$

$$= \frac{d}{dx}\left(x^3 + \frac{3}{2}x^2 - 5x + 7\right)$$

$$= \frac{d}{dx}\left(x^3\right) + \frac{d}{dx}\left(\frac{3}{2}x^2 - 5x + 7\right) \qquad \text{Apply the Sum Rule for the second time.}$$

$$= \frac{d}{dx}\left(x^3\right) + \frac{d}{dx}\left(\frac{3}{2}x^2\right) + \frac{d}{dx}(-5x + 7) \qquad \text{Apply the Sum Rule for the third time.}$$

$$= \frac{d}{dx}\left(x^3\right) + \frac{d}{dx}\left(\frac{3}{2}x^2\right) + \frac{d}{dx}(-5x) + \frac{d}{dx}(7)$$

This illustrates the fact that the Sum Rule makes it possible for us to differentiate any polynomial term by term, a fact you will be asked to prove in Exercise 113. This observation along with the first two rules of this section will then imply that the derivative of a polynomial is always another polynomial (Exercise 114). We now complete the differentiation of q.

$$\frac{d}{dx}q(x) = \frac{d}{dx}\left(x^3\right) + \frac{d}{dx}\left(\frac{3}{2}x^2\right) + \frac{d}{dx}(-5x) + \frac{d}{dx}(7)$$

$$= \frac{d}{dx}\left(x^3\right) + \frac{3}{2}\frac{d}{dx}\left(x^2\right) + (-5)\frac{d}{dx}(x) + \frac{d}{dx}(7)$$

$$= 3x^2 + \frac{3}{2}(2x) + (-5)(1) + 0$$

$$= 3x^2 + 3x - 5$$

c. With regards to $f(x)$, recall that from Example 1 we know the derivative of $1/x$ to be $-1/x^2$; and also $\frac{d}{dx}\left(\sqrt{x}\right) = \frac{1}{2\sqrt{x}}$. We calculated this latter derivative in Example 5 of Section 2.6, or you might arrive at it by using the yet-unproven Power Rule for a fractional exponent.

$$\frac{d}{dx}\left(\sqrt{x}\right) = \frac{d}{dx}\left(x^{1/2}\right) = \frac{1}{2}x^{-1/2} = \frac{1}{2}\cdot\frac{1}{x^{1/2}} = \frac{1}{2\sqrt{x}}$$

Armed with these facts, we use a combination of the Difference Rule and Constant Multiple Rule to complete our solution, as follows.

$$f'(x) = \frac{d}{dx}f(x)$$

$$= \frac{d}{dx}\left(3\sqrt{x} - \frac{2}{x}\right)$$

$$= \frac{d}{dx}\left(3\sqrt{x}\right) - \frac{d}{dx}\left(\frac{2}{x}\right) \qquad \text{Difference Rule}$$

$$= 3\frac{d}{dx}\left(\sqrt{x}\right) - 2\frac{d}{dx}\left(\frac{1}{x}\right) \qquad \text{Constant Multiple Rule}$$

$$= 3\left(\frac{1}{2\sqrt{x}}\right) - 2\left(-\frac{1}{x^2}\right)$$

$$= \frac{3}{2\sqrt{x}} + \frac{2}{x^2} = \frac{3}{2}x^{-1/2} + 2x^{-2}$$

Example 3 ✒

Find the equation of the line tangent to the graph of

$$p(x) = 0.3x^5 + 1.1x^4 - x^3 + 0.6x^2 + 2.9x - 0.9$$

at the point $(1,3)$.

Solution

Recall that the slope of the tangent is $p'(1)$, the derivative of p at $x = 1$. Termwise differentiation yields

$$p'(x) = \frac{d}{dx}\left(0.3x^5\right) + \frac{d}{dx}\left(1.1x^4\right) - \frac{d}{dx}\left(x^3\right) + \frac{d}{dx}\left(0.6x^2\right) + \frac{d}{dx}\left(2.9x\right) - \frac{d}{dx}\left(0.9\right)$$

$$= 0.3\frac{d}{dx}\left(x^5\right) + 1.1\frac{d}{dx}\left(x^4\right) - \frac{d}{dx}\left(x^3\right) + 0.6\frac{d}{dx}\left(x^2\right) + 2.9\frac{d}{dx}\left(x\right) - 0$$

$$= 0.3\left(5x^4\right) + 1.1\left(4x^3\right) - 3x^2 + 0.6\left(2x\right) + 2.9$$

$$= 1.5x^4 + 4.4x^3 - 3x^2 + 1.2x + 2.9,$$

and thus the slope we are seeking is

$$p'(1) = 1.5\left(1^4\right) + 4.4\left(1^3\right) - 3\left(1^2\right) + 1.2(1) + 2.9 = 7.$$

Now the point-slope form of the equation of a line yields

$$y - 3 = 7(x - 1),$$

or equivalently, the requested equation of the tangent is

$$y = 7x - 4.$$

The curve and its tangent line are graphed in Figure 1.

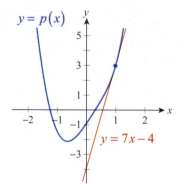

$y = p(x)$

$y = 7x - 4$

Figure 1

Example 4 ✒

The distance from the origin of a particle moving along the x-axis is described by the function $x(t) = 2t^3 - 1.5t^2 + 4t$ inches, where t is measured in seconds. Find the instantaneous velocity and acceleration of the particle at $t = 2$ seconds.

Solution

As we have seen before, velocity is the derivative of the position function. Note, however, that in this example x is actually the dependent variable, and differentiation takes place with respect to the independent variable t.

$$v(t) = \frac{d}{dt}x(t) = \frac{d}{dt}\left(2t^3 - 1.5t^2 + 4t\right)$$

$$= \frac{d}{dt}\left(2t^3\right) - \frac{d}{dt}\left(1.5t^2\right) + \frac{d}{dt}\left(4t\right)$$

$$= 2\frac{d}{dt}\left(t^3\right) - 1.5\frac{d}{dt}\left(t^2\right) + 4\frac{d}{dt}\left(t\right)$$

$$= 2\left(3t^2\right) - 1.5(2t) + 4(1)$$

$$= 6t^2 - 3t + 4$$

Thus the instantaneous velocity of the particle at $t = 2$ seconds is

$$v(2) = 6(2)^2 - 3(2) + 4 = 22 \text{ in./s.}$$

To find the acceleration, we differentiate the velocity function.

$$a(t) = \frac{d}{dt} v(t) = \frac{d}{dt}\left(6t^2 - 3t + 4\right)$$

$$= \frac{d}{dt}\left(6t^2\right) - \frac{d}{dt}\left(3t\right) + \frac{d}{dt}\left(4\right)$$

$$= 6\frac{d}{dt}\left(t^2\right) - 3\frac{d}{dt}\left(t\right) + 0$$

$$= 6(2t) - 3(1)$$

$$= 12t - 3$$

Substituting $t = 2$ into the acceleration function gives the instantaneous acceleration of the particle at $t = 2$ seconds.

$$a(2) = 12(2) - 3 = 21 \text{ in./s}^2$$

TOPIC 2   Differentiation of Exponential Functions

We now have rules allowing us to easily differentiate polynomial functions; the overall goal is to develop similar rules for many classes of functions, and we consider exponential functions next.

Recall that an exponential function has the form $f(x) = a^x$, where a is a positive real number not equal to 1. Since none of the rules we have developed so far help us in determining f', we must begin with the definition of derivative—the limit of a difference quotient:

$$f'(x) = \lim_{h \to 0} \frac{f(x+h) - f(x)}{h}$$

$$= \lim_{h \to 0} \frac{a^{x+h} - a^x}{h}$$

$$= \lim_{h \to 0} \frac{a^x\left(a^h - 1\right)}{h} \qquad \text{Factor out } a^x.$$

$$= a^x \lim_{h \to 0} \frac{a^h - 1}{h} \qquad \text{The factor } a^x \text{ does not depend on } h \text{ and passes through the limit.}$$

It is worth noting, in particular, that

$$f'(0) = a^0 \lim_{h \to 0} \frac{a^h - 1}{h} = \lim_{h \to 0} \frac{a^h - 1}{h},$$

so in general we can write

$$f'(x) = a^x \lim_{h \to 0} \frac{a^h - 1}{h} = a^x f'(0).$$

In other words, the derivative of $f(x) = a^x$ is a^x multiplied by a constant, namely $f'(0)$. All that remains is to evaluate $\lim_{h \to 0} (a^h - 1)/h$.

Although we can easily construct a table of values of $(a^h - 1)/h$ for fixed values of a and for h as close to 0 as we desire, the proof that the limit above actually exists must be deferred until we have the machinery of Chapter 5. The table below indicates that $f'(0) \approx -0.69$ when $a = \frac{1}{2}$, $f'(0) \approx 0.69$ when $a = 2$, and $f'(0) \approx 1.10$ when $a = 3$. And since $f'(0)$ represents the slope of the line tangent to $f(x) = a^x$ at $x = 0$, these approximate values for $f'(0)$ correspond to the relative rise or fall of the three tangent lines depicted in Figure 2.

h	$\dfrac{\left(\frac{1}{2}\right)^h - 1}{h}$	$\dfrac{2^h - 1}{h}$	$\dfrac{3^h - 1}{h}$
0.1	−0.6697	0.7177	1.1612
0.01	−0.6908	0.6956	1.1047
0.001	−0.6929	0.6934	1.0992
0.0001	−0.6931	0.6932	1.0987

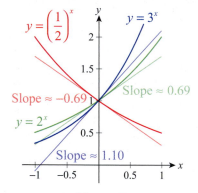

Figure 2
Graphs of a^x and Associated
Tangent Lines at $x = 0$

Based on Figure 2, it is certainly reasonable to expect that there is a value for a between 2 and 3 such that

$$\lim_{h \to 0} \frac{a^h - 1}{h} = 1,$$

since the value of this limit is approximately 0.69 when $a = 2$ and approximately 1.10 when $a = 3$. This is indeed true, and we give this particular value for a the label e; that is, we define e to be the real number such that

$$\lim_{h \to 0} \frac{e^h - 1}{h} = 1.$$

But again, the rigorous proof that such an e exists awaits us in Section 5.3, Exercise 100. In the meantime, it suffices to know that e, an irrational number approximately equal to 2.7183, is the unique real number with the above property. This leads to the nice fact that

$$\frac{d}{dx}(e^x) = e^x \left(\lim_{h \to 0} \frac{e^h - 1}{h} \right) = e^x;$$

in other words, the derivative of e^x is itself. We will return to the derivative of $f(x) = a^x$ for other values of a in Section 3.4.

Theorem 🔍

Derivative of e^x

$$\frac{d}{dx}\left(e^x\right) = e^x$$

Example 5 ✎

Find the first three derivatives of the following functions.

a. $f(x) = -2e^x$ **b.** $g(s) = 3s^2 + e^s$ **c.** $h(x) = ae^x,\ a \in \mathbb{R}$

Solution

a. Using the Constant Multiple Rule along with the fact that e^x is its own derivative we have:

$$f'(x) = \frac{d}{dx}\left(-2e^x\right) = (-2)\frac{d}{dx}\left(e^x\right) = -2e^x$$

In other words, we find that $f(x)$ is also the derivative of itself. Repeating the above calculation two more times, we conclude that the second and third derivatives of f are also equal to f, that is,

$$f''(x) = f'''(x) = -2e^x.$$

In fact, it is easy to generalize and observe that all derivatives of $f(x)$ are equal to itself. We will have to say more about this observation in part c. of this example.

b. In addition to the rules used in part a., to find the derivative of $g(s)$ we will also need the Sum Rule:

$$g'(s) = \frac{d}{ds}\left(3s^2 + e^s\right)$$

$$= \frac{d}{ds}\left(3s^2\right) + \frac{d}{ds}\left(e^s\right)$$

$$= 3(2s) + e^s = 6s + e^s$$

A similar calculation yields the second derivative.

$$g''(s) = \frac{d}{ds}g'(s) = \frac{d}{ds}\left(6s + e^s\right) = \frac{d}{ds}\left(6s\right) + \frac{d}{ds}\left(e^s\right) = 6 + e^s$$

You might already have figured out the third derivative by observing that differentiation only affects the first term. Here are the details.

$$g'''(s) = \frac{d}{ds}g''(s) = \frac{d}{ds}\left(6 + e^s\right) = \frac{d}{ds}\left(6\right) + \frac{d}{ds}\left(e^s\right) = 0 + e^s = e^s$$

Finally, we note in passing that any higher-order derivative of g will now be e^s, since e^s is the derivative of itself.

c. Let us now investigate the derivatives of ae^x, where a is a real constant. We now use the prime notation along with the Constant Multiple Rule:

$$h'(x) = \left(ae^x\right)' = a\left(e^x\right)' = ae^x$$

In other words, the derivative of any constant multiple of e^x is itself. Repeating the above procedure any number of times allows us to conclude that derivatives of all orders of the function $h(x) = ae^x$ are equal to h itself for all $a \in \mathbb{R}$. This is a generalization of the observation we made in part a. of this example.

TOPIC 3 The Product and Quotient Rules

The Sum and Difference Rules for differentiation are easy to remember ("the derivative of a sum or difference is the sum or difference of the derivatives"), and their proofs are straightforward—for instance, the Sum Rule follows almost immediately from writing down and then rearranging the appropriate difference quotient. It is tempting to assume that the Product and Quotient Rules follow a similar pattern, but that is not the case—the truth is a bit more interesting.

Caution ⚠

The derivative of a product is *not* the product of the derivatives.

$$\frac{d}{dx}\left[f(x)g(x)\right] \neq f'(x)g'(x)$$

Similarly, the derivative of a quotient is *not* the quotient of the derivatives.

$$\frac{d}{dx}\left[\frac{f(x)}{g(x)}\right] \neq \frac{f'(x)}{g'(x)}$$

To see why this is so, it may be helpful to use a geometric interpretation; such an approach is very much in keeping with early Greek mathematics, which almost always viewed a product of two quantities as an area measure of a two-dimensional figure.

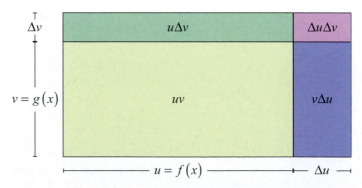

Figure 3 Geometric Interpretation of $(fg)'$

To make such a geometric connection, suppose f and g are two differentiable functions taking on only positive values (so we can interpret those values as being measures of length). Then the product $f(x)g(x)$ represents the area of a rectangle with, say, width $f(x)$ and height $g(x)$ (see Figure 3). Assume x is fixed and, for ease of exposition, let $u = f(x)$, $v = g(x)$, $\Delta u = f(x+\Delta x) - f(x)$, and $\Delta v = g(x+\Delta x) - g(x)$. Geometrically, $(uv)'$ represents the instantaneous rate of change in the rectangular area uv, so we are seeking a formula for

$$\lim_{\Delta x \to 0} \frac{\Delta(uv)}{\Delta x}.$$

Referring again to Figure 3, the small change in area $\Delta(uv)$ with respect to a small change Δx can be expressed as

$$\Delta(uv) = (u + \Delta u)(v + \Delta v) - uv = u\Delta v + v\Delta u + \Delta u \Delta v,$$

so

$$\frac{\Delta(uv)}{\Delta x} = u\frac{\Delta v}{\Delta x} + v\frac{\Delta u}{\Delta x} + \Delta u\frac{\Delta v}{\Delta x}.$$

Since f is differentiable at x, it is continuous at x and so $\Delta u = f(x+\Delta x) - f(x) \to 0$ as $\Delta x \to 0$. Hence, using several limit laws,

$$\begin{aligned}
(uv)' &= \lim_{\Delta x \to 0} \frac{\Delta(uv)}{\Delta x} \\
&= \lim_{\Delta x \to 0}\left(u\frac{\Delta v}{\Delta x} + v\frac{\Delta u}{\Delta x} + \Delta u\frac{\Delta v}{\Delta x} \right) \\
&= u\lim_{\Delta x \to 0}\frac{\Delta v}{\Delta x} + v\lim_{\Delta x \to 0}\frac{\Delta u}{\Delta x} + \lim_{\Delta x \to 0}\Delta u\lim_{\Delta x \to 0}\frac{\Delta v}{\Delta x} \\
&= uv' + vu' + 0\cdot v' \\
&= uv' + vu'.
\end{aligned}$$

A geometric argument such as the above is valuable in adding to our understanding of the meaning of differentiation and in building intuition, but it must be augmented by a formal proof in order to be complete (for instance, the geometric argument doesn't make sense if f or g takes on negative values). Such a proof is given immediately following the formal statement of the Product Rule, offering a rigorous justification of the rule.

Theorem ⚲

The Product Rule

If f and g are both differentiable at x, the product fg is also differentiable at x and $(fg)'(x) = f'(x)g(x) + f(x)g'(x)$. In other notation,

$$\frac{d}{dx}\big[f(x)g(x)\big] = \left[\frac{d}{dx}f(x)\right]g(x) + f(x)\left[\frac{d}{dx}g(x)\right].$$

Proof ✎

$$\begin{aligned}
(fg)'(x) &= \lim_{h \to 0} \frac{f(x+h)g(x+h) - f(x)g(x)}{h} \\
&= \lim_{h \to 0} \frac{f(x+h)g(x+h) - f(x)g(x+h) + f(x)g(x+h) - f(x)g(x)}{h} \\
&= \lim_{h \to 0} \frac{f(x+h)g(x+h) - f(x)g(x+h)}{h} + \lim_{h \to 0} \frac{f(x)g(x+h) - f(x)g(x)}{h}
\end{aligned}$$

We begin above, unsurprisingly, with the limit of a difference quotient. In order to make the limit tractable, however, we need to rewrite the difference quotient in such a way that difference quotients for f and g make an appearance, and the easiest way to do this is to add and subtract the expression $f(x)g(x+h)$ in the numerator, as shown in the second step. The third step then splits the difference quotient into two fractions, and we proceed by applying additional limit properties.

$$\begin{aligned}
(fg)'(x) &= \lim_{h \to 0} \frac{f(x+h)g(x+h) - f(x)g(x+h)}{h} + \lim_{h \to 0} \frac{f(x)g(x+h) - f(x)g(x)}{h} \\
&= \lim_{h \to 0}\left[\frac{f(x+h) - f(x)}{h} g(x+h) \right] + \lim_{h \to 0}\left[f(x) \frac{g(x+h) - g(x)}{h} \right] \\
&= \left[\lim_{h \to 0} \frac{f(x+h) - f(x)}{h} \right]\left[\lim_{h \to 0} g(x+h) \right] + f(x)\left[\lim_{h \to 0} \frac{g(x+h) - g(x)}{h} \right] \\
&= f'(x)g(x) + f(x)g'(x)
\end{aligned}$$

Note that we have used the fact that differentiability implies continuity in order to replace $\lim_{h \to 0} g(x+h)$ with $g(x)$ in the last step.

Example 6 ✐

Use the Product Rule to find the derivative of each function.

a. $F(x) = 5x$ b. $G(x) = x^2$

c. $H(x) = (2x^2 + 1)(x^3 - 4)$ d. $K(x) = x^2 e^x$

Solution

a. In order to differentiate F, we could certainly use the approach of Example 2 and, with the help of the Constant Multiple Rule, arrive at the answer of $F'(x) = 5$. However, our aim here is to demonstrate the truth of the Product Rule, so let's think of F as the product of the constant function $f(x) = 5$ and the function $g(x) = x$ and apply the Product Rule:

$$\begin{aligned}
F'(x) &= \frac{d}{dx}\left[f(x)g(x) \right] \\
&= \left[\frac{d}{dx} f(x) \right] g(x) + f(x)\left[\frac{d}{dx} g(x) \right] \\
&= \left[\frac{d}{dx}(5) \right] x + 5\left[\frac{d}{dx}(x) \right] \\
&= (0)x + 5(1) = 5
\end{aligned}$$

b. Again, the Positive Integer Power Rule is perhaps the easiest way to determine G', but let us see how the Product Rule gives us the same answer. Since $x^2 = x \cdot x$, we have

$$G'(x) = \frac{d}{dx}(x \cdot x) = \frac{d}{dx}(x) \cdot x + x \cdot \frac{d}{dx}(x) = 1 \cdot x + x \cdot 1 = 2x,$$

which is exactly what the Positive Integer Power Rule would tell us.

c. With the assignments $f(x) = 2x^2 + 1$ and $g(x) = x^3 - 4$, we see that $H(x) = f(x)g(x)$, so applying the Product Rule gives us

$$H'(x) = \frac{d}{dx}\big[f(x)g(x)\big]$$

$$= \left[\frac{d}{dx}f(x)\right]g(x) + f(x)\left[\frac{d}{dx}g(x)\right]$$

$$= \left[\frac{d}{dx}(2x^2 + 1)\right](x^3 - 4) + (2x^2 + 1)\left[\frac{d}{dx}(x^3 - 4)\right]$$

$$= (4x)(x^3 - 4) + (2x^2 + 1)(3x^2)$$

$$= (4x^4 - 16x) + (6x^4 + 3x^2)$$

$$= 10x^4 + 3x^2 - 16x.$$

Notice that we can check our answer by multiplying the two factors of H together and differentiating the resulting polynomial the usual way:

$$H(x) = (2x^2 + 1)(x^3 - 4) = 2x^5 + x^3 - 8x^2 - 4,$$

so applying our rules,

$$H'(x) = 2(x^5)' + (x^3)' - 8(x^2)' - 0$$

$$= 2(5x^4) + (3x^2) - 8(2x)$$

$$= 10x^4 + 3x^2 - 16x,$$

as we expected. The beauty of mathematics stems partly from the fact that usually there are several possible approaches to the solution of any given problem.

d. While all the previous derivatives could be handled in alternative ways, in the case of the function K there is no apparent way to avoid the Product Rule. Using the prime notation this time, we can write

$$K'(x) = (x^2 e^x)'$$

$$= (x^2)' e^x + x^2 (e^x)'$$

$$= (2x)e^x + x^2 (e^x)$$

$$= (x^2 + 2x)e^x,$$

where we again used the fact that e^x is the derivative of itself.

Example 7 ✐

Find $F(1)$ if $F(x) = \sqrt{x} \cdot g(x)$, and $F'(1) = 2$ and $g'(1) = 3$.

Solution

First of all, since we know $F(1) = \sqrt{1} \cdot g(1) = g(1)$, finding $F(1)$ or $g(1)$ are equivalent problems. Keeping this in mind, it is not hard to see that the Product Rule provides us with an equation to determine $g(1)$, as follows.

$$F(x) = \sqrt{x} \cdot g(x)$$

$$F'(x) = \left(\sqrt{x}\right)' \cdot g(x) + \sqrt{x} \cdot g'(x) \qquad \text{We have seen before that } \left(\sqrt{x}\right)' = \frac{1}{2\sqrt{x}}.$$

$$= \frac{1}{2\sqrt{x}} \cdot g(x) + \sqrt{x} \cdot g'(x)$$

Substituting $x = 1$, we have

$$F'(1) = \frac{1}{2\sqrt{1}} \cdot g(1) + \sqrt{1} \cdot g'(1).$$

(Note that in general, if $x = c$, it follows from the Product Rule that

$$(fg)'(c) = f'(c)g(c) + f(c)g'(c).$$

We used this fact with $c = 1$ in our equation above.)

Next, substituting the known quantities of $F'(1) = 2$ and $g'(1) = 3$, our equation becomes

$$2 = \frac{1}{2}g(1) + 3.$$

Solving the above for $g(1)$, we obtain

$$g(1) = -2.$$

Finally, recalling that $F(1) = g(1)$, we conclude that $F(1) = -2$.

As an intermediary step in developing the Quotient Rule, we will first find a rule for the derivative of the reciprocal of a differentiable function, a rule that is useful in its own right.

Theorem 🔑

The Reciprocal Rule

If g is differentiable at x and $g(x) \neq 0$, $1/g$ is also differentiable at x and

$$\frac{d}{dx}\left[\frac{1}{g(x)}\right] = \frac{-g'(x)}{\left[g(x)\right]^2}.$$

Proof 🖉

$$\frac{d}{dx}\left[\frac{1}{g(x)}\right] = \lim_{h \to 0} \frac{\dfrac{1}{g(x+h)} - \dfrac{1}{g(x)}}{h}$$

$$= \lim_{h \to 0} \left[\frac{1}{h}\left(\frac{1}{g(x+h)} - \frac{1}{g(x)}\right)\right]$$

$$= \lim_{h \to 0} \left[\frac{1}{h}\left(\frac{g(x) - g(x+h)}{g(x+h)g(x)}\right)\right]$$

We are hoping to relate the derivative of $1/g$ to the derivative of g if possible, so we are again seeking a way to rewrite the difference quotient so that g' will make an appearance. Using the fact that $1/g(x)$ is independent of h and can be pulled out of the above limit, and using the continuity of g at x, we can rearrange the quotient to obtain the following.

$$\frac{d}{dx}\left[\frac{1}{g(x)}\right] = \left[\frac{1}{g(x)}\right]\lim_{h \to 0}\left[\left(\frac{1}{g(x+h)}\right)\left(\frac{g(x) - g(x+h)}{h}\right)\right]$$

$$= \left[\frac{1}{g(x)}\right]\left[\lim_{h \to 0}\frac{-1}{g(x+h)}\right]\left[\lim_{h \to 0}\frac{g(x+h) - g(x)}{h}\right]$$

$$= \left[\frac{1}{g(x)}\right]\left[\frac{-1}{g(x)}\right]g'(x)$$

$$= \frac{-g'(x)}{[g(x)]^2}$$

One immediate application of the Reciprocal Rule is the extension of the Power Rule to negative integers. Given a positive integer n, note the following.

$$\left(x^{-n}\right)' = \left(\frac{1}{x^n}\right)'$$

$$= \frac{-\left(x^n\right)'}{\left(x^n\right)^2} \qquad\qquad \text{Reciprocal Rule}$$

$$= \frac{-nx^{n-1}}{x^{2n}} \qquad\qquad \text{Positive Integer Power Rule}$$

$$= -nx^{-n-1}$$

Since the Power Rule is trivially true for x^0, we have now proved the rule for all integers.

Example 8 ✐

Use the Reciprocal Rule to find the derivative of $f(x) = 1/\sqrt{x}$.

Solution

Recall that $\left(\sqrt{x}\right)' = 1/\left(2\sqrt{x}\right)$. Using this fact along with the Reciprocal Rule, we can write

$$\frac{d}{dx}\left(\frac{1}{\sqrt{x}}\right) = \frac{-\left(\sqrt{x}\right)'}{\left(\sqrt{x}\right)^2}$$

$$= \frac{-\dfrac{1}{2\sqrt{x}}}{\left(\sqrt{x}\right)^2}$$

$$= \frac{-1}{2x\sqrt{x}} = -\frac{1}{2}x^{-3/2}.$$

Notice that our result is yet another illustration of the validity of the Power Rule for fractional exponents $\left(n = -\frac{1}{2}\right.$ in this case). We will prove the Power Rule for all real exponents later as we acquire the necessary tools.

Theorem ⚿

The Quotient Rule

If f and g are both differentiable at x and $g(x) \neq 0$, f/g is also differentiable at x and

$$\frac{d}{dx}\left[\frac{f(x)}{g(x)}\right] = \frac{f'(x)g(x) - f(x)g'(x)}{\left[g(x)\right]^2}.$$

Proof ✎

The Product and Reciprocal Rules together allow us to quickly arrive at the Quotient Rule, as follows.

$$\frac{d}{dx}\left[\frac{f(x)}{g(x)}\right] = \frac{d}{dx}\left[f(x)\frac{1}{g(x)}\right]$$

$$= f'(x)\frac{1}{g(x)} + f(x)\left(\frac{-g'(x)}{\left[g(x)\right]^2}\right) \qquad \text{Product and Reciprocal Rules}$$

$$= \frac{f'(x)g(x)}{\left[g(x)\right]^2} - \frac{f(x)g'(x)}{\left[g(x)\right]^2}$$

$$= \frac{f'(x)g(x) - f(x)g'(x)}{\left[g(x)\right]^2}$$

Example 9 ✎

Use the Quotient Rule to find the following derivatives.

a. $\dfrac{d}{dx}\left(\dfrac{5x^2+x}{x^2+2}\right)$ **b.** $\dfrac{d}{dx}\left(\dfrac{\frac{3}{x}+x}{x-\frac{1}{x}}\right)$ **c.** $\dfrac{d}{dx}\left(\dfrac{\sqrt{x}+\frac{2}{\sqrt{x}}}{\sqrt{x}}\right)$

Solution

a. Using the Quotient Rule with $f(x)=5x^2+x$ and $g(x)=x^2+2$, we obtain the following.

$$\frac{d}{dx}\left[\frac{f(x)}{g(x)}\right]=\frac{d}{dx}\left[\frac{5x^2+x}{x^2+2}\right]$$

$$=\frac{\left(5x^2+x\right)'\left(x^2+2\right)-\left(5x^2+x\right)\left(x^2+2\right)'}{\left(x^2+2\right)^2}$$

$$=\frac{\left(10x+1\right)\left(x^2+2\right)-\left(5x^2+x\right)\left(2x\right)}{\left(x^2+2\right)^2}$$

$$=\frac{\left(10x^3+x^2+20x+2\right)-\left(10x^3+2x^2\right)}{\left(x^2+2\right)^2}$$

$$=\frac{-x^2+20x+2}{\left(x^2+2\right)^2}$$

b. While we could certainly proceed with the Quotient Rule for this function, this is a situation when it pays to rewrite our quotient before differentiating. Notice that by multiplying both the numerator and denominator by x, we obtain an expression that is free of reciprocals and therefore is more convenient to handle.

$$\frac{\frac{3}{x}+x}{x-\frac{1}{x}}=\frac{3+x^2}{x^2-1}$$

$$\frac{d}{dx}\left(\frac{\frac{3}{x}+x}{x-\frac{1}{x}}\right)=\frac{d}{dx}\left(\frac{3+x^2}{x^2-1}\right)$$

$$=\frac{\left(3+x^2\right)'\left(x^2-1\right)-\left(x^2-1\right)'\left(3+x^2\right)}{\left(x^2-1\right)^2}$$

$$=\frac{\left(2x\right)\left(x^2-1\right)-\left(2x\right)\left(3+x^2\right)}{\left(x^2-1\right)^2}$$

$$=\frac{\left(2x\right)\left(x^2-1-3-x^2\right)}{\left(x^2-1\right)^2}$$

$$=\frac{-8x}{\left(x^2-1\right)^2}$$

c. Again, we can make the expression look simpler if we rewrite before differentiating. We will start by multiplying both the numerator and denominator by $\sqrt{x}$.

$$\frac{\sqrt{x}+\dfrac{2}{\sqrt{x}}}{\sqrt{x}} = \frac{\sqrt{x}+\dfrac{2}{\sqrt{x}}}{\sqrt{x}} \cdot \frac{\sqrt{x}}{\sqrt{x}} = \frac{x+2}{x} = \frac{x}{x}+\frac{2}{x} = 1+\frac{2}{x}$$

$$\frac{d}{dx}\left(\frac{\sqrt{x}+\dfrac{2}{\sqrt{x}}}{\sqrt{x}}\right) = \frac{d}{dx}\left(1+\frac{2}{x}\right)$$

$$= \frac{d}{dx}(1)+2\frac{d}{dx}\left(\frac{1}{x}\right) \qquad \text{Recall that } (1/x)' = -1/x^2.$$

$$= 0+2\left(-\frac{1}{x^2}\right)$$

$$= -\frac{2}{x^2}$$

What you should remember from the last two differentiations is that it might not always be the best idea to proceed right away with the Quotient Rule whenever you see a quotient. While you can certainly obtain the correct answer that way, often you can save a lot of time just by rewriting the expression before differentiating.

3.2 **Exercises**

1–12 Use the appropriate rules from this section to find the derivative of the given function.

1. $f(x) = 5 - 2x$

2. $g(x) = \dfrac{4}{5}x + 2$

3. $h(x) = \dfrac{1}{2} + 2x - 3x^2$

4. $F(x) = x^3 - 2x^2 + \dfrac{1}{2}x - 77$

5. $G(x) = \dfrac{1}{2}x^4 + 2x^3 - x^2 + 3.2x + \sqrt{2}$

6. $k(x) = x^{11} - 0.2x^{10} + \dfrac{\pi}{3}x^3 + \pi$

7. $H(x) = x^8 + \sqrt{2}x^5 - 2x^4$

8. $R(t) = t^{100} - 2t^{59} + \pi t^{38} + et$

9. $S(z) = 4z^3 - 3\sqrt{z} + 11.2$

10. $Q(s) = \dfrac{1}{3s} - \sqrt{2}s + \sqrt{2s}$

11. $T(r) = \pi r^2 + 2e^r + \pi^2$

12. $N(t) = e^{2+t} + \dfrac{t+1}{t} + pt$

13–20 Use the Product Rule to find the indicated derivative. Then find the answer without the use of the Product Rule, by multiplying first, and compare your answers.

13. $\dfrac{d}{dx}\left[(x+2)(3x+5)\right]$

14. $\dfrac{d}{dx}\left[(3x+7)(x^2+2x)\right]$

15. $\dfrac{d}{dx}\left[(x^2-6)(2x^2+5x)\right]$

16. $\dfrac{d}{dx}\left[(2x^3+3x^2)(4x^2-2x+5)\right]$

17. $\dfrac{d}{dx}\left[\left(\dfrac{1}{3}x^3+\dfrac{7}{5}x^5\right)\left(\dfrac{2}{x}-4x^2\right)\right]$

18. $\dfrac{d}{dx}\left[(e^x+3)(e^2-5)\right]$

19. $\dfrac{d}{dt}\left[(3+2\sqrt{t})(4\sqrt{t}-5)\right]$

20. $\dfrac{d}{ds}\left[s\left(-3-\dfrac{1}{3}s^3\right)(s^4+2s)\right]$

21–32 Use the Reciprocal Rule or Quotient Rule to determine the derivative of the function.

21. $f(x) = \dfrac{1}{1-2x}$

22. $g(x) = \dfrac{1}{4x - 2x^2}$

23. $h(x) = \dfrac{2}{2x^3 - 5x^2 + 3x + 1}$

24. $F(x) = \dfrac{e}{e^x - \sqrt{x}}$

25. $G(x) = \dfrac{2x+1}{x-5}$

26. $k(x) = \dfrac{3x-4}{2x^2 + 5}$

27. $H(x) = \dfrac{x^3 - 3x^2}{2x^3 + 5x^2 + 1}$

28. $A(x) = \dfrac{6\sqrt{x}}{3x - 4}$

29. $B(u) = \dfrac{u^2}{\sqrt{u} + 1}$

30. $f(t) = \dfrac{4 - \sqrt{t}}{t^2 + 3}$

31. $g(t) = \dfrac{3 - t}{4 - 5\sqrt{t}}$

32. $w(s) = \dfrac{1 + 2e^s}{3e^s + 5}$

33–38 Differentiate the quotient by simplifying it algebraically first.

33. $f(x) = \dfrac{30x^2 - 10x^6}{5x}$

34. $g(x) = \dfrac{1 + 5x + x^2}{x}$

35. $h(x) = \dfrac{3x^{1/2} - 5x^{3/2} + 7x^{5/2} - 9x^{7/2}}{x^{1/2}}$

36. $F(x) = \dfrac{2(\sqrt{x})^3 + 3\sqrt{x}}{\sqrt{x} + (\sqrt{x})^3}$

37. $G(x) = \dfrac{\dfrac{6}{x^2} - \dfrac{5}{x} + 1}{\dfrac{1}{x} - \dfrac{3}{x^2}}$

38. $H(x) = \dfrac{2 - \dfrac{1}{e^x}}{2e^{-x}}$

39–61 Using the rules of this section, differentiate the given function.

39. $f(t) = t^{1/2}(4t + 3)$

40. $g(x) = x^2\left(\sqrt{x} + \dfrac{1}{\sqrt{x}}\right)$

41. $h(s) = \left(5 + \dfrac{1}{s}\right)\left(s^2 + \dfrac{1}{5}\right)$

42. $F(x) = \dfrac{1}{x} + \dfrac{2}{x^2}$

43. $G(x) = 3x^{-5} + 2x^{-3}$

44. $k(s) = s^2\left(\dfrac{3}{s} + \dfrac{1}{s-1}\right)$

45. $H(t) = \sqrt{t}(9 - t^2)$

46. $K(x) = \dfrac{\dfrac{1}{x^2} - 3}{x + 2}$

47. $w(z) = z\left(2 + \dfrac{4}{4 - \sqrt{z}}\right)$

48. $L(T) = T^{-3}(2 - 4T^{-2})$

49. $r(x) = \dfrac{x - a^2}{x + a^2}$

50. $Q(t) = \dfrac{at + b}{ct + d}$

51. $F(x) = e^x(2 + \sqrt{x})$

52. $E(s) = \dfrac{2 + se^s}{e^s - s}$

53. $C(x) = \dfrac{a}{a + \dfrac{a}{x}}$

54. $D(x) = \dfrac{x}{x + \dfrac{x}{a}}$

55. $G(s) = \dfrac{3s^2}{2e^s + s}$

56. $S(t) = (4 - \sqrt{t})(2 - e^t)$

57. $L(y) = (y^4 - 3y^3)(y^2 - 2y^5)$

58. $h(z) = \dfrac{1}{ae^z + z}$

59. $H(s) = \dfrac{a}{b + ce^s}$

60. $T(x) = (x + 2)(2x^2 - x)(x^3 + 5)$

61. $L(t) = t(2e^t + \sqrt{t})\left(\dfrac{1}{t} - 1\right)$

62–67 Find the first, second, and third derivatives of the function.

62. $f(x) = 2x + 5$

63. $g(x) = \dfrac{x}{x + 1}$

64. $h(x) = 3\sqrt{x}$

65. $F(x) = 2 - x + 5x^2 - \pi x^3$

66. $V(z) = 2z^2 + \dfrac{2}{z^2}$

67. $W(t) = 3t^2 + 3e^t$

68–71 Find a function f that satisfies the given conditions. (**Hint:** A polynomial is the most natural choice. Answers will vary.)

68. $f(0) = 2$, $f'(0) = 1$, and $f''(0) = -1$.

69. $f(0) = 0$ and f has horizontal tangent lines at $x = 2$ and $x = -2$.

70. $f(0) = 1$, $y = x + 1.5$ is tangent to the graph at $x = 1$, and $y = 5.5 - x$ is tangent to the graph at $x = 3$.

71. $f(1) = 5$, $f'(1) = 8$, f has a horizontal tangent line at $x = -1$, and $f'''(x) = 6$.

72–75 Find a formula for the kth derivative of the function. (**Hint:** Calculate the first few derivatives and try to recognize an emerging pattern.)

72. $f(x) = x^n$

73. $g(x) = \dfrac{1}{x}$

74. $h(x) = xe^x$

75.* $q(x) = x^n e^x$

76. If $f(1) = 2$, $f'(1) = 1$, $g(1) = -1$, and $g'(1) = 3$, find the following function values.

   a. $(f - g)'(1)$

   b. $(fg)'(1)$

   c. $\left(\dfrac{f}{g}\right)'(1)$

77. If $f(3) = -1$, $f'(3) = 5$, $g(3) = \frac{1}{2}$, and $g'(3) = -2$, find the following function values.

   a. $(2f + 5g)'(3)$

   b. $(4fg)'(3)$

   c. $\left(\dfrac{f}{2g}\right)'(3)$

78–82 Find the equation of the line tangent to the graph of the function at the given point.

78. $f(x) = \dfrac{x^2 + 1}{x}$; $(1, 2)$

79. $w(x) = \dfrac{8}{x^2 + 4}$; $(2, 1)$

   (This curve is called the **witch of Agnesi**.)

80. $g(x) = \dfrac{2}{\sqrt{x} + 1}$; $(1, 1)$

81. $h(x) = \dfrac{2e^x}{x^2}$; $(1, 2e)$

82. $k(x) = \dfrac{2x}{2 + x^2}$; $\left(1, \dfrac{2}{3}\right)$

(This curve is called a **serpentine**.)

83. Find the equation of the **normal line** to the graph of the function $s(x) = \dfrac{9x}{x^2 + 9}$ at the point $(0, 0)$. (We call a line **normal** to the graph at a point if it is perpendicular to the line tangent to the graph at the same point.)

84. Repeat Exercise 83 for the graph of the function $h(x) = \dfrac{e^x}{x^4 + 2}$ at $(0, \frac{1}{2})$.

85–96 Find all x-values where the graph of the function has a horizontal tangent line, or prove that the graph has no horizontal tangent line.

85. $f(x) = x^2 - 2x$

86. $g(x) = 2x^3 - 3x^2 - 12x + 1$

87. $h(x) = \dfrac{2}{x^2}$

88. $F(x) = \dfrac{2}{x^2 + 1}$

89. $G(x) = \dfrac{1}{2}e^x - 2$

90. $k(x) = x^2 - a$

91. $H(x) = \dfrac{1}{x^2 - a}$

92. $f(x) = \sqrt{x + 5}$

93. $g(x) = \dfrac{x^2}{x^2 + 5}$

94. $F(x) = \dfrac{2x - 1}{x^2}$

95. $P(x) = 2x^3 + 3x^2 + 3x - 5$

96. $Q(x) = e^x - x$

97. The line $y = 8x + b$ is tangent to the graph of $f(x) = ax^2$ at $x = 2$. Find the values of a and b.

98. Repeat Exercise 97 with the line $y = -2x + b$ that is tangent to the graph of $g(x) = -x^2 + ax$ at $x = 2$.

99. Show that the graphs of $y = e^{x/2}$ and $y = \dfrac{1}{(x + 1)^2}$ intersect at $x = 0$ in such a way that their respective tangent lines are perpendicular at their point of intersection.

100–103 Find the equation(s) of the line(s) tangent to the graph of f through the indicated point, which does not lie on the graph of f. (**Hint:** If the point of tangency is $(x, f(x))$, then the slope of the tangent line going through (a, b) is $f'(x) = \dfrac{f(x) - b}{x - a}$.)

100. $f(x) = x^2$; $(0, -1)$ **101.** $f(x) = e^x$; $(0, 0)$

102. $f(x) = \sqrt{x}$; $(-2, 0)$ **103.** $f(x) = \dfrac{1}{x}$; $(-1, 0)$

104–107 Assuming that f and g are differentiable functions, differentiate the given expression.

104. $\dfrac{f(x)}{x}$

105. $\dfrac{xf(x)}{g(x)}$

106. $e^x g(x)$

107. $\dfrac{f(x)e^x}{g(x) + 2}$

108. Use the definition of the derivative to prove the Constant Multiple Rule. (**Hint:** For a given constant k, start out by using the definition
$$\frac{d}{dx}[kf(x)] = \lim_{h \to 0} \frac{kf(x+h) - kf(x)}{h},$$
and let k "pass through" the limit sign.)

109. Use the Product Rule and mathematical induction to provide a third proof of the Positive Integer Power Rule. (**Hint:** The base case of $n = 1$ should be obvious. After setting up the induction hypothesis of $(x^n)' = nx^{n-1}$, find the derivative of x^{n+1} by treating it as $x^{n+1} = x \cdot x^n$ and use the Product Rule along with the induction hypothesis.)

110. Use the Product Rule to arrive at a rule for
$$\frac{d}{dx}[f(x)g(x)h(x)].$$

111. Use the Product Rule to arrive at a formula for $\dfrac{d}{dx}[f(x)]^2$, and then use Exercise 110 to find the formula for $\dfrac{d}{dx}[f(x)]^3$. Do you recognize a pattern?

112. Use mathematical induction to prove the **Generalized Positive Integer Power Rule**:
$$\frac{d}{dx}[f(x)]^n = n[f(x)]^{n-1} f'(x)$$

(The hint provided in Exercise 109 might prove helpful, but you will need to appropriately modify your induction hypothesis.)

113. Use the Sum Rule and mathematical induction to prove that any finite sum of functions $f_1(x) + f_2(x) + \cdots + f_n(x)$ can be differentiated termwise, that is,

$$[f_1(x) + f_2(x) + \cdots + f_n(x)]' = f_1'(x) + f_2'(x) + \cdots + f_n'(x).$$

In particular, polynomials can be differentiated termwise. (Look at the hint provided in Exercise 109.)

114. Use Exercise 113 and the results of this section to prove that the derivative of a polynomial is always another polynomial.

115. A Formula One race car was moving in a parabolic curve with equation $y = \sqrt{x}$ when it hit an oil patch and the driver lost control at the point $(4, 2)$. The car left the track along the tangent line at the same point. Where did he hit the tire wall if the equation of the tire wall was $y = 4$? (Fortunately, there were no injuries.)

116. The position function of a golf ball rolling on an incline is given by $d(t) = 2t^2 + 3t$, where d is measured in meters, t in seconds. Find the ball's velocity and acceleration at $t = 4$ seconds.

117. The velocity function of a moving particle is given by $v(t) = \dfrac{50t}{t + 10}$ ft/s. Find its acceleration at **a.** $t = 2$ seconds and **b.** $t = 10$ seconds.

118. The position function of a moving object is given by $p(t) = 2t^3 - 6t$ ft. Find its position and acceleration at the instant when its velocity changes directions.

119. The position function of an object dropped by an astronaut on the Moon is $h(t) = -0.81t^2 + 1.5$, where h is measured in meters, t in seconds. What is the acceleration due to gravity on the Moon? How long does it take for the above object to reach the ground and what is the speed of impact?

120. The radius of a spherical balloon being inflated increases according to the function $r(t) = 2 + 5\sqrt[3]{t}$, where r is measured in centimeters and t in seconds. Find the rate of change of the balloon's volume and surface area with respect to time at $t = 8$ seconds.

121–129 *True or False?* Determine whether the given statement is true or false. In case of a false statement, explain or provide a counterexample.

121. If $y = \pi x^n$, then $y' = n\pi x^{n-1}$.

122. If $y = \pi^n$, then $y' = n\pi^{n-1}$.

123. If $y = \pi/x^n$, then $y' = \pi/(nx^{n-1})$.

124. If $y = e^x$, then $y' = xe^{x-1}$.

125. If $y = \pi e^x$, then $y' = \pi e^x$.

126. If p is a fifth-degree polynomial, then its sixth derivative is 0.

127. If $F(x) = \dfrac{f(x)}{g(x)}$, then $\dfrac{d}{dx} F(x) = \dfrac{\dfrac{d}{dx} f(x)}{\dfrac{d}{dx} g(x)}$.

128. The jerk of a free-falling object is 0.

129. If a polynomial $p(x)$ has degree n, then its derivative has degree $n - 1$.

3.2 Technology Exercises

130–135. Use the differentiation capabilities of a computer algebra system to check your answers for Exercises 62–67, and then graph each function along with its derivatives in the same viewing window.

136–140. Referring back to Exercises 78–82, verify your answers by using a graphing calculator or computer algebra system to graph each function and its indicated tangent line in the same viewing window.

3.3 **Derivatives of Trigonometric Functions**

TOPICS

1. Two useful limits

2. Derivatives of sine and cosine

3. Derivatives of other trigonometric functions

In this section, we extend those functions for which we have differentiation rules to the class of trigonometric functions. The development of the rules proceeds quickly once we establish two important facts.

TOPIC 1 **Two Useful Limits**

The derivatives of the six basic trigonometric functions, as with the exponential functions, cannot be determined by applying any of the rules we have developed thus far. So we will begin by working with limits of difference quotients and, as we will soon see, the key to our work will be the evaluation of the following limits.

Lemma 🔍

Measuring θ in radians (as is customary),

$$\lim_{\theta \to 0} \frac{\sin \theta}{\theta} = 1 \quad \text{and} \quad \lim_{\theta \to 0} \frac{\cos \theta - 1}{\theta} = 0.$$

Proof 📎

Neither limit is trivial, as the numerator and denominator approach 0 in both cases—the key to the evaluation is determining the relative *rate* at which this happens.

We can construct some bounds on the relative rate for the first limit by referring to Figure 1. For any angle θ between 0 and $\pi/2$, the area of triangle OBD is less than the area of the circle sector OBD, which in turn is less than the area of triangle OBC. Note also that since the radius of the circle is 1, the lengths of the two legs of triangle OAD are $\cos \theta$ and $\sin \theta$, while the length of the leg opposite θ in triangle OBC is $\tan \theta$. Using the area formulas for triangles and circular sectors, we have the following.

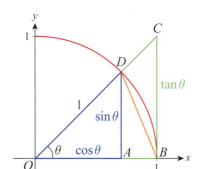

Figure 1 Understanding $\dfrac{\sin \theta}{\theta}$

$$\text{Area of } \triangle OBD \quad = \frac{1}{2}(\text{base})(\text{height}) = \frac{1}{2}(1)(\sin \theta) = \frac{1}{2}\sin \theta$$

$$\text{Area of sector } OBD = \frac{1}{2}r^2\theta = \frac{1}{2}(1)^2\theta = \frac{1}{2}\theta$$

$$\text{Area of } \triangle OBC \quad = \frac{1}{2}(\text{base})(\text{height}) = \frac{1}{2}(1)(\tan \theta) = \frac{1}{2}\tan \theta = \frac{1}{2} \cdot \frac{\sin \theta}{\cos \theta}$$

So the above statements imply

$$\frac{1}{2}\sin \theta < \frac{1}{2}\theta < \frac{1}{2} \cdot \frac{\sin \theta}{\cos \theta}.$$

Since $\sin \theta$ is positive for all θ in this range, dividing by $\sin \theta$ and multiplying by 2 leaves each inequality unchanged.

$$1 < \frac{\theta}{\sin \theta} < \frac{1}{\cos \theta}$$

Taking the reciprocal of each fraction *does* reverse each inequality, so the above is equivalent to

$$\cos\theta < \frac{\sin\theta}{\theta} < 1.$$

This is the sort of upper and lower bound on $(\sin\theta)/\theta$ that we need. Since $\cos\theta$ has a limit of 1 as $\theta \to 0$, the Squeeze Theorem tells us that

$$\lim_{\theta\to 0^+} \frac{\sin\theta}{\theta} = 1.$$

Since $(\sin\theta)/\theta$ is an even function,

$$\frac{\sin(-\theta)}{-\theta} = \frac{-\sin\theta}{-\theta} = \frac{\sin\theta}{\theta},$$

and we also have

$$\lim_{\theta\to 0^-} \frac{\sin\theta}{\theta} = 1,$$

so the first limit is proved.

We can now use this result to prove the second limit, making use of the trigonometric identity $\cos 2x = 1 - 2\sin^2 x$. Replacing $2x$ with θ and subtracting 1 from both sides of this identity, we have

$$\cos\theta - 1 = -2\sin^2\left(\frac{\theta}{2}\right),$$

and dividing through by θ results in

$$\frac{\cos\theta - 1}{\theta} = -\frac{2\sin^2(\theta/2)}{\theta} = -\frac{\sin^2(\theta/2)}{\theta/2}.$$

So we calculate the limit as follows.

$$\lim_{\theta\to 0} \frac{\cos\theta - 1}{\theta} = \lim_{\theta\to 0}\left[-\frac{\sin^2(\theta/2)}{\theta/2}\right]$$

$$= -\lim_{\theta\to 0}\left[\frac{\sin(\theta/2)}{\theta/2}\cdot\sin(\theta/2)\right]$$

$$= -\lim_{\alpha\to 0}\left(\frac{\sin\alpha}{\alpha}\cdot\sin\alpha\right) \qquad \text{\color{blue}Replace } \theta/2 \text{ with } \alpha.$$
$$\phantom{= -\lim_{\alpha\to 0}\left(\frac{\sin\alpha}{\alpha}\cdot\sin\alpha\right)} \qquad \text{\color{blue}If } \theta \to 0 \text{ then } \alpha = \theta/2 \to 0.$$

$$= -(1)(0) = 0$$

Caution ⚠

The assumption that the angles in the preceding lemma are measured in radians is critical, as our formula for the area of a circular sector is based on radian measure. In Section 3.4, we will learn how to find derivatives of trigonometric functions when using other angle measurements.

Example 1 ✎

Use the lemma to find the following limits.

a. $\displaystyle\lim_{x\to0}\frac{\sin 3x}{7x}$ **b.** $\displaystyle\lim_{x\to0}\frac{\cos x-1}{\tan x}$

Solution

a. In order for us to be able to use the lemma, we first multiply and divide by 3:

$$\lim_{x\to0}\frac{\sin 3x}{7x}=\lim_{x\to0}\left(\frac{3}{3}\cdot\frac{\sin 3x}{7x}\right)=\lim_{x\to0}\left(\frac{3}{7}\cdot\frac{\sin 3x}{3x}\right)=\frac{3}{7}\lim_{3x\to0}\frac{\sin 3x}{3x},$$

where in the last step we used the fact that when x approaches 0, so does $3x$ and vice versa. Next, if we rename $3x$ by calling it θ, we can rewrite our limit as

$$\frac{3}{7}\lim_{3x\to0}\frac{\sin 3x}{3x}=\frac{3}{7}\lim_{\theta\to0}\frac{\sin\theta}{\theta}.$$

Now by a direct application of the lemma,

$$\frac{3}{7}\lim_{\theta\to0}\frac{\sin\theta}{\theta}=\frac{3}{7}\cdot1=\frac{3}{7}.$$

b. This limit does not directly resemble either of those in the lemma, but we can rewrite it as follows.

$$\lim_{x\to0}\frac{\cos x-1}{\tan x}=\lim_{x\to0}\frac{\cos x-1}{\dfrac{\sin x}{\cos x}}=\lim_{x\to0}\frac{(\cos x-1)\cos x}{\sin x}=\lim_{x\to0}\frac{\dfrac{\cos x-1}{x}\cdot\cos x}{\dfrac{\sin x}{x}}$$

Applying the limit laws and both parts of the lemma, along with the fact that $\displaystyle\lim_{x\to0}\cos x=1$, we now obtain

$$\lim_{x\to0}\frac{\dfrac{\cos x-1}{x}\cdot\cos x}{\dfrac{\sin x}{x}}=\frac{0\cdot1}{1}=0.$$

TOPIC 2   Derivatives of Sine and Cosine

The derivatives of sine and cosine can now be found quickly, using one more trigonometric identity and the two limits we just studied.

Theorem ⚗

Derivative of Sine

$$\frac{d}{dx}(\sin x)=\cos x$$

Proof

$$\frac{d}{dx}(\sin x) = \lim_{h \to 0} \frac{\sin(x+h) - \sin x}{h}$$

$$= \lim_{h \to 0} \frac{\sin x \cos h + \cos x \sin h - \sin x}{h} \qquad \sin(u+v) = \sin u \cos v + \cos u \sin v$$

$$= \lim_{h \to 0} \left(\sin x \cdot \frac{\cos h - 1}{h} + \cos x \cdot \frac{\sin h}{h} \right)$$

$$= (\sin x) \left(\lim_{h \to 0} \frac{\cos h - 1}{h} \right) + (\cos x) \left(\lim_{h \to 0} \frac{\sin h}{h} \right) \qquad \text{Both } \sin x \text{ and } \cos x \text{ do not depend on } h \text{ and can be pulled out of the limit.}$$

$$= (\sin x)(0) + (\cos x)(1) = \cos x$$

Note that a pleasing pattern is beginning to emerge regarding differentiation within classes of functions: the derivative of any polynomial is another polynomial, the derivative of e^x is e^x, and our first derivative of a trigonometric function turns out to be another trigonometric function. While this pattern cannot be relied upon too heavily, we will see it continue to appear from time to time. In particular, we will see that the derivatives of the six basic trigonometric functions remain in the class of trigonometric functions.

Example 2

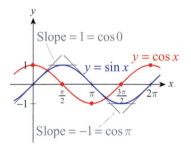

Figure 2

If we follow the slopes of tangent lines to the graph of $y = \sin x$ starting, say, at $x = 0$, it is not difficult to discover an oscillating "sinusoidal" pattern.

As shown in Figure 2, when $x = 0$ the corresponding tangent is a line through the origin, which appears to have a slope of 1. As x increases, the slopes of the tangent lines decrease, reaching 0 when $x = \pi/2$. The slopes then become negative as x continues to increase from $\pi/2$. When x reaches the value of π, the slope of the corresponding tangent of the sine function is -1. The slopes then start to increase, reaching the value of 0 again at $x = 3\pi/2$, as a close examination of the sine graph reveals. Notice that this is exactly the behavior of the function values of $y = \cos x$, that is, the slope of the tangent to the sine graph at any given x-value appears to equal the corresponding function value of the cosine function, which is exactly what we proved in the previous theorem.

Example 3

Use differentiation rules to find the derivative of $f(x) = (x^2 - 3x)\sin x$.

Solution

The Product Rule along with the previous theorem yields

$$\frac{df}{dx} = \frac{d}{dx}(x^2 - 3x) \cdot \sin x + (x^2 - 3x)\frac{d}{dx}(\sin x)$$

$$= (2x - 3)\sin x + (x^2 - 3x)\cos x.$$

Theorem 🔍

Derivative of Cosine

$$\frac{d}{dx}\left(\cos x\right) = -\sin x$$

Proof 📎

This can be proven using the methods used to show that the derivative of the sine function is cosine. (See Exercise 54.)

Example 4 ✎

Use differentiation rules to find the following derivatives.

a. $\dfrac{d}{dx}\left(7e^x + 3\cos x\right)$

b. $\dfrac{d}{dx}\left(\dfrac{1+\sin x}{2-\cos x}\right)$

Solution

We will need the differentiation rules of the previous section along with the derivatives of sine and cosine, as follows.

a.
$$\frac{d}{dx}\left(7e^x + 3\cos x\right) = 7\frac{d}{dx}\left(e^x\right) + 3\frac{d}{dx}\left(\cos x\right)$$
$$= 7e^x + 3\left(-\sin x\right)$$
$$= 7e^x - 3\sin x$$

b. Note that we need to start with the Quotient Rule:

$$\frac{d}{dx}\left(\frac{1+\sin x}{2-\cos x}\right) = \frac{\left(1+\sin x\right)'\left(2-\cos x\right) - \left(1+\sin x\right)\left(2-\cos x\right)'}{\left(2-\cos x\right)^2}$$
$$= \frac{\left(\cos x\right)\left(2-\cos x\right) - \left(1+\sin x\right)\left(\sin x\right)}{\left(2-\cos x\right)^2}$$
$$= \frac{2\cos x - \cos^2 x - \sin x - \sin^2 x}{\left(2-\cos x\right)^2} \qquad \color{blue}{\sin^2 x + \cos^2 x = 1}$$
$$= \frac{2\cos x - \sin x - 1}{\left(2-\cos x\right)^2}$$

Example 5 ✎

If an object is oscillating so that the so-called *restoring force* acting on the object is directly proportional to the negative of its displacement at any time, the resulting motion is called *simple harmonic motion*. This happens, for example, when an object is attached to a spring, pulled or pushed, and then let go. (We are ignoring air resistance, friction, and all other forces here.) Recall from Example 6 of Section 1.2 that the equation of this motion, or the position function of the object, can be either $f(t) = A\sin(\omega t)$ or $g(t) = A\cos(\omega t)$, depending on initial conditions (ω is a constant related to the spring, A is the amplitude).

Suppose an object of mass m is hanging on a spring and is at equilibrium. If we pull it downward by 50 cm and release it, the equation of motion is given by

$$y(t) = -0.5\cos t,$$

where y is the vertical distance from equilibrium, measured in meters with upward displacement considered positive, and t is measured in seconds.

a. Find the position at $t = 10$ seconds.

b. What is the maximum displacement of the object and when does it occur?

c. Find the maximum velocity and the position where it occurs.

d. Find the maximum acceleration value. When does it occur?

Solution

a. The position at $t = 10$ seconds is quickly found by evaluating the position function.

$$y(10) = -0.5\cos 10 \approx 0.4195 \text{ m} = 41.95 \text{ cm}$$

The positive sign means that the object is found approximately 41.95 cm above equilibrium at $t = 10$ seconds.

b. Since displacement is "signed distance" from equilibrium, maximum displacement will occur precisely when the position function assumes its maximum value (positive or negative). Stated more precisely, this happens when the absolute value of the position is maximum. Since the cosine function oscillates between +1 and −1 reaching these maximum absolute values at $k\pi$, $k \in \mathbb{Z}$, we conclude that the absolute value of the position function given in this example is greatest when $t = k\pi$ seconds, for $k \in \mathbb{Z}$, $k \geq 0$. The greatest displacement value is

$$\left|y(k\pi)\right| = \left|-0.5\cos(k\pi)\right| = \left|-0.5\right| \cdot \left|\cos(k\pi)\right| = 0.5(1) = 0.5 \text{ m} = 50 \text{ cm}.$$

Note that this is consistent with our everyday experience. Since there was no initial velocity (no initial "push" or "shove" at the start of motion), we expect the initial displacement of 50 cm to remain maximal throughout the motion. (As we have stated already, we are ignoring all other forces and therefore the fact that the motion actually "dies down" over the long term.) Also, recalling that $\cos(k\pi) = 1$ if k is even and $\cos(k\pi) = -1$ for an odd k, in light of the position function we conclude that the object reaches its maximum displacement of 50 cm *below* equilibrium when $t = k\pi$ and k is a nonnegative even integer. If k is odd, on the other hand, the object is 50 cm *above* equilibrium at $t = k\pi$ seconds.

c. The velocity function is the derivative of the position function.

$$v(t) = y'(t) = -0.5(\cos t)' = -0.5(-\sin t) = 0.5\sin t$$

By an argument similar to the one given in part b., and keeping the graph of the sine function in mind, we see that velocity is greatest when the absolute value of the sine function is 1, that is, when $t = (2k+1)(\pi/2)$, $k \in \mathbb{Z}$. Moreover, since

$$\sin\left((2k+1)\frac{\pi}{2}\right) = \begin{cases} 1 & \text{if } k \text{ is even} \\ -1 & \text{if } k \text{ is odd} \end{cases}$$

we conclude that velocity is maximal and pointing upward when $t = (2k+1)(\pi/2)$ seconds for a nonnegative even k.

$$v\left((2k+1)\frac{\pi}{2}\right) = 0.5\sin\left((2k+1)\frac{\pi}{2}\right) = 0.5(1) = 0.5 \text{ m/s}$$

On the other hand, if k is odd and positive,

$$v\left((2k+1)\frac{\pi}{2}\right) = 0.5\sin\left((2k+1)\frac{\pi}{2}\right) = 0.5(-1) = -0.5 \text{ m/s};$$

that is, velocity is maximal and is pointing downward in this case. Note that for the above t-values the cosine function is 0, so maximum velocity (or speed) occurs each time the object is passing through the equilibrium position.

d. Since acceleration is the derivative of velocity, we will obtain the acceleration function by differentiating.

$$a(t) = v'(t) = 0.5(\sin t)' = 0.5\cos t$$

Since this is precisely the negative of the position function, we can use our observations made in part b. and conclude that the acceleration values are greatest when the object is in its extreme positions, but acceleration is *positive* when the object's position is negative, that is, the object is accelerating downward when above the equilibrium position and vice versa. The maximum acceleration values are

$$a(k\pi) = 0.5\cos(k\pi) = 0.5(1) = 0.5 \text{ m/s}^2$$

when k is an even nonnegative integer and

$$a(k\pi) = 0.5\cos(k\pi) = 0.5(-1) = -0.5 \text{ m/s}^2$$

when k is odd.

TOPIC 3   Derivatives of Other Trigonometric Functions

Knowing the derivatives of sine and cosine, we can find the derivatives of the remaining trigonometric functions without having to resort to limits of difference quotients.

Theorem

Derivatives of Tangent, Cotangent, Secant, and Cosecant

$$\frac{d}{dx}(\tan x) = \sec^2 x \qquad\qquad \frac{d}{dx}(\cot x) = -\csc^2 x$$

$$\frac{d}{dx}(\sec x) = \sec x \tan x \qquad\qquad \frac{d}{dx}(\csc x) = -\csc x \cot x$$

Proof

We will prove two of the above statements, and leave the remaining two as exercises (see Exercise 59).

$$\frac{d}{dx}(\tan x) = \frac{d}{dx}\left(\frac{\sin x}{\cos x}\right)$$

$$= \frac{(\sin x)'\cos x - \sin x(\cos x)'}{\cos^2 x} \qquad \textcolor{blue}{\text{Quotient Rule}}$$

$$= \frac{\cos x \cos x - \sin x(-\sin x)}{\cos^2 x}$$

$$= \frac{\cos^2 x + \sin^2 x}{\cos^2 x}$$

$$= \frac{1}{\cos^2 x} \qquad \textcolor{blue}{\sin^2 x + \cos^2 x = 1}$$

$$= \sec^2 x$$

For the derivative of cosecant, we could either use the Quotient Rule or, more quickly, its specialized form as shown.

$$\frac{d}{dx}(\csc x) = \frac{d}{dx}\left(\frac{1}{\sin x}\right)$$

$$= \frac{-(\sin x)'}{\sin^2 x} \qquad \textcolor{blue}{\text{Reciprocal Rule}}$$

$$= -\frac{\cos x}{\sin^2 x}$$

$$= -\frac{1}{\sin x}\cdot\frac{\cos x}{\sin x}$$

$$= -\csc x \cot x$$

Example 6 ✐

Use the above theorem to find the derivatives of the following functions.

a. $f(x) = \tan x \csc x$ **b.** $g(x) = \dfrac{1 - \sin x}{\cot x}$

Solution

a. Using the Product Rule, we calculate the derivative as follows.

$$f'(x) = (\tan x)' \csc x + \tan x (\csc x)'$$

$$= \sec^2 x \csc x + \tan x (-\csc x \cot x)$$

$$= \frac{1}{\cos^2 x} \cdot \frac{1}{\sin x} - \frac{\sin x}{\cos x} \cdot \frac{1}{\sin x} \cdot \frac{\cos x}{\sin x}$$

$$= \frac{1}{\cos^2 x} \cdot \frac{1}{\sin x} - \frac{1}{\sin x}$$

$$= \sec^2 x \csc x - \csc x$$

$$= (\sec^2 x - 1) \csc x$$

Notice, however, that differentiation will be much easier if we rewrite f before differentiating. Since

$$f(x) = \tan x \csc x = \frac{\sin x}{\cos x} \cdot \frac{1}{\sin x} = \frac{1}{\cos x} = \sec x,$$

our theorem tell us that

$$f'(x) = (\sec x)' = \sec x \tan x.$$

However, is this answer equivalent to the one we obtained above? The answer, reassuringly, is yes, which we can show by using the identity $\tan^2 x + 1 = \sec^2 x$.

$$(\sec^2 x - 1) \csc x = \tan^2 x \csc x$$

$$= \tan x \tan x \csc x$$

$$= \tan x \cdot \frac{\sin x}{\cos x} \cdot \frac{1}{\sin x}$$

$$= \tan x \cdot \frac{1}{\cos x}$$

$$= \tan x \sec x$$

We have just witnessed an illustration of the fact that when working with trigonometric expressions, sometimes very different-looking answers may be equivalent.

b. Using the Quotient Rule this time, we obtain the following.

$$g'(x) = \frac{(1-\sin x)' \cot x - (1-\sin x)(\cot x)'}{\cot^2 x}$$

$$= \frac{-\cos x \cot x - (1-\sin x)(-\csc^2 x)}{\cot^2 x}$$

$$= \frac{-\cos x \cot x + \csc^2 x - \csc^2 x \sin x}{\cot^2 x}$$

$$= \frac{-\cos x \cdot \dfrac{\cos x}{\sin x} + \dfrac{1}{\sin^2 x} - \dfrac{1}{\sin^2 x} \sin x}{\dfrac{\cos^2 x}{\sin^2 x}}$$

$$= \frac{-\cos^2 x \sin x + 1 - \sin x}{\cos^2 x} \qquad \text{Multiply both numerator and denominator by } \sin^2 x.$$

$$= \frac{1 - \sin x \left(\cos^2 x + 1\right)}{\cos^2 x}$$

Example 7 ✎

Determine the second derivative of $f(x) = \cot x$.

Solution

Since $\dfrac{d}{dx}(\cot x) = -\csc^2 x$, we will obtain the second derivative of f by differentiating $f'(x) = -\csc^2 x$.

$$f''(x) = \frac{d^2}{dx^2}(\cot x) = \frac{d}{dx}\left(-\csc^2 x\right)$$

We can differentiate $-\csc^2 x$ by using the Generalized Positive Integer Power Rule (see Exercise 112 of Section 3.2) or simply by an application of the Product Rule.

$$\frac{d}{dx}\left(-\csc^2 x\right) = -\frac{d}{dx}\left[(\csc x)(\csc x)\right]$$

$$= -(\csc x)' \csc x - \csc x (\csc x)'$$

$$= -2\csc x (\csc x)'$$

$$= -2\csc x (-\csc x \cot x)$$

$$= 2\csc^2 x \cot x$$

$$= 2 \cdot \frac{1}{\sin^2 x} \cdot \frac{\cos x}{\sin x} = \frac{2\cos x}{\sin^3 x}$$

Example 8 ✒

Find all points on the graph of $g(t) = 3\sin 2t$ where the slope of its tangent line is -3.

Solution

Our plan is to find $g'(t)$ first, and then by solving the equation $g'(t) = -3$, identify all values $\bar{t}$ such that $g'(\bar{t}) = -3$.

In order to differentiate g, we will use the identity $\sin 2t = 2\sin t \cos t$ followed by the Product Rule.

$$g'(t) = \frac{d}{dt}(3\sin 2t)$$

$$= 3\frac{d}{dt}(\sin 2t)$$

$$= 3\frac{d}{dt}(2\sin t \cos t)$$

$$= 6\frac{d}{dt}(\sin t \cos t)$$

$$= 6\left[\cos t \cos t + \sin t(-\sin t)\right]$$

$$= 6\left(\cos^2 t - \sin^2 t\right)$$

$$= 6\cos 2t \qquad\qquad\qquad \cos 2t = \cos^2 t - \sin^2 t$$

Next, we solve the trigonometric equation $6\cos 2t = -3$.

$$6\cos 2t = -3$$

$$\cos 2t = -\frac{1}{2}$$

$$2t = \frac{2\pi}{3} + 2k\pi \quad \text{or} \quad 2t = \frac{4\pi}{3} + 2k\pi,\ k \in \mathbb{Z}$$

Equivalently, the solutions are

$$\bar{t} = \frac{\pi}{3} + k\pi \quad \text{or} \quad \bar{t} = \frac{2\pi}{3} + k\pi,\ k \in \mathbb{Z}.$$

The corresponding points on the graph of g where the tangents have a slope of -3 are as follows.

$$\left\{\left(\frac{\pi}{3} + k\pi, \frac{3\sqrt{3}}{2}\right), \left(\frac{2\pi}{3} + k\pi, -\frac{3\sqrt{3}}{2}\right)\right\},\ k \in \mathbb{Z}$$

3.3 Exercises

1–12 Use the results of this section to find the indicated limit.

1. $\displaystyle\lim_{x\to0}\frac{\sin 3x}{2x}$

2. $\displaystyle\lim_{x\to0}\frac{-\sin\frac{x}{2}}{5x}$

3. $\displaystyle\lim_{x\to0}\frac{\sin\pi x}{x}$

4. $\displaystyle\lim_{x\to0}\frac{\tan 4x}{5x}$

5. $\displaystyle\lim_{x\to0}\frac{\cos 5x-1}{2x}$

6. $\displaystyle\lim_{x\to0}\frac{\cos x-1}{\sin x}$

7. $\displaystyle\lim_{x\to0}\frac{x+\tan x}{\sin x}$

8. $\displaystyle\lim_{\beta\to0}\frac{\csc\beta-\cot\beta}{\beta\csc\beta}$

9. $\displaystyle\lim_{x\to0}\frac{\sin(\sin x)}{\sin x}$

10. $\displaystyle\lim_{\alpha\to0}\frac{\tan(\alpha^2)}{\alpha}$

11. $\displaystyle\lim_{t\to0}\frac{2t+3\tan t}{\sin t}$

12. $\displaystyle\lim_{\theta\to0}\frac{\cos\theta\sin\theta-\sin\theta}{\theta^2}$

13–30 Differentiate the given function.

13. $f(x)=2\sin x-5\cos x$

14. $g(x)=3x^2+2\tan x$

15. $h(x)=x\cos x$

16. $F(x)=2.5x(1-\cot x)$

17. $G(x)=2\sqrt{x}\sec x$

18. $k(x)=\pi x\sin x+\pi x$

19. $L(x)=-3e^x(\csc x+\cot x)$

20. $f(x)=2\cos 2x-2\cos x$

21. $g(x)=\cot^2 x$

22. $h(x)=\dfrac{\tan x}{x}$

23. $F(t)=\dfrac{1-\cos t}{t^2}$

24. $W(x)=\dfrac{1+\cos x}{1+\sin x}$

25. $R(z)=\dfrac{e^z+\sin z}{z}$

26. $N(w)=\dfrac{2\sqrt{w}-\sec w}{\sqrt{w}}$

27. $B(x)=\dfrac{\dfrac{1}{\sin x}-\sin x}{\cos x}$

28. $G(y)=y\cot y\csc y$

29. $T(s)=s^2 e^s\cot s$

30. $r(t)=\dfrac{1}{t\sin t\cos t}$

31–36 Find all points where the function has a horizontal tangent line.

31. $f(x)=\dfrac{1}{2}x+\sin x$

32. $g(x)=x+\sin 2x$

33. $h(x)=\sec^2 x$

34. $T(s)=\tan s-s$

35. $K(u)=\tan u+\cot u$

36. $F(t)=\dfrac{1-\sin t}{1-\cos t}$

37–40 Find all x-values where the tangent line to the graph of the function is parallel to the given line.

37. $f(x)=\sin x+\dfrac{3}{2};\quad y-x=\dfrac{3}{2}$

38. $g(x)=\cot x;\quad y+2x=\pi$

39. $G(x)=\dfrac{x}{3}-\tan x;\quad x+y=5$

40. $F(x)=\sin x\cos x;\quad 2x+2y=7$

41–44 Find the equation of the tangent line to the graph of the given function at the indicated point.

41. $f(x)=2x\cos x;\quad (0,0)$

42. $g(x)=\tan x-\sec x;\quad (0,-1)$

43. $h(x)=2\csc x-\sin x;\quad \left(\dfrac{\pi}{2},1\right)$

44. $k(x)=\dfrac{\cot x}{x};\quad \left(\dfrac{\pi}{4},\dfrac{4}{\pi}\right)$

45. Let us assume that for some function f, we have $f(0)=1$ and $f'(0)=2$. Let $F(x)=f(x)\tan x$, $G(x)=f(x)/\cos x$, and $H(x)=f(x)\sin x\cos x$. Find $F'(0)$, $G'(0)$, and $H'(0)$.

46–48 Verify the trigonometric identity by differentiating both sides of the equation. (**Hint:** If $f'(x)=g'(x)$, it doesn't necessarily follow that $f(x)=g(x)$. In general, we can only conclude that $f(x)=g(x)+c$ for some constant c.)

46. $\tan x\cot x=1$

47. $(1-\cos\theta)(1+\cos\theta)=\sin^2\theta$

48. $\dfrac{1}{1-\cos x}+\dfrac{1}{1+\cos x}=2\csc^2 x$

49–52 Find $f'(x)$, $f''(x)$, and $f'''(x)$. Observing a pattern, find a formula for $f^{(n)}(x)$.

49. $f(x) = \sin x$ **50.** $f(x) = \cos x$

51. $f(x) = e^x \sin x$ **52.** $f(x) = e^x \cos x$

53. Provide a second proof of the limit statement $\lim_{\theta \to 0}(\cos\theta - 1)/\theta = 0$ by multiplying both the numerator and denominator by $\cos\theta + 1$ to obtain

$$\lim_{\theta \to 0}\frac{\cos\theta - 1}{\theta} = \lim_{\theta \to 0}\frac{\cos^2\theta - 1}{\theta(\cos\theta + 1)}.$$

Then by the Pythagorean identity $\sin^2 x + \cos^2 x = 1$, you obtain

$$\lim_{\theta \to 0}\frac{\cos^2\theta - 1}{\theta(\cos\theta + 1)} = \lim_{\theta \to 0}\frac{-\sin^2\theta}{\theta(\cos\theta + 1)}$$

$$= \lim_{\theta \to 0}\frac{\sin\theta(-\sin\theta)}{\theta(\cos\theta + 1)}$$

$$= \lim_{\theta \to 0}\frac{\sin\theta}{\theta}\lim_{\theta \to 0}\frac{-\sin\theta}{\cos\theta + 1}.$$

Conclude the argument by using the first limit statement in the lemma at the beginning of this section.

54. Prove that $\dfrac{d}{dx}(\cos x) = -\sin x$ by mimicking the proof of the theorem "Derivative of Sine." (**Hint:** Recall that you will need the angle sum identity $\cos(u + v) = \cos u \cos v - \sin u \sin v$.)

55. Provide an alternative proof of the fact that $\dfrac{d}{dx}(\sin x) = \cos x$ by using the identity

$$\sin x - \sin c = 2\sin\frac{x - c}{2}\cos\frac{x + c}{2}.$$

(**Hint:** Rewrite the difference quotient $\dfrac{\sin x - \sin c}{x - c}$

as $\dfrac{2\sin\dfrac{x - c}{2}\cos\dfrac{x + c}{2}}{x - c}$. Let $c \to x$, and use the lemma from the beginning of this section.)

56. Use the definition of the derivative and the lemma from the beginning of this section to show that $(\sin 3x)' = 3\cos 3x$. Generalize to obtain that if $k \in \mathbb{R}$, $(\sin(kx))' = k\cos(kx)$.

57. Repeat Exercise 56 with $f(x) = \cos(kx)$.

58. Find a constant a such that the graphs of $f(x) = a\sin x$ and $g(x) = a\cos x$ intersect at right angles, that is, their respective tangent lines are perpendicular at their point(s) of intersection.

59. Prove the remaining two cases of the theorem "Derivatives of Tangent, Cotangent, Secant, and Cosecant," namely, the statements

$$\frac{d}{dx}(\cot x) = -\csc^2 x \quad \text{and} \quad \frac{d}{dx}(\sec x) = \sec x \tan x.$$

(**Hint:** Mimic the proof presented in the text, using the derivatives of sine and cosine along with appropriate differentiation rules.)

60. The cross-section of an ice cream cone is an isosceles triangle, with the angular opening at the bottom being $2t$ (radians). Assuming that the ice cream sits on top of the cone in the shape of a perfect hemisphere, let V_i = volume of the ice cream, V_c = volume of the cone. Express both of these volumes in terms of t, and then compute $\lim_{t \to 0^+}\dfrac{V_i}{V_c}$.

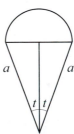

61. An object is tied to the top of an inclined surface of variable angle of elevation so that the rope is parallel to the surface. The tension in the rope is given by $F = mg(\sin x - \mu\cos x)$, where m is the mass of the object, g is the gravity constant, and μ is the coefficient of friction (assume all units are metric units).

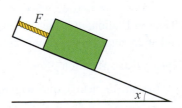

a. What is the rate of change of F with respect to x?

b. For what x-value (if any) is this rate of change equal to 0?

62. Suppose an object oscillating in fluid obeys the position function $y = 10e^{-0.2t}\cos(2\pi t)$, where y is the distance from equilibrium, measured in centimeters with upward displacement considered positive, and t is measured in seconds. Such motion is called *damped harmonic motion*. Can you see why?

  a. Find the position, velocity, and acceleration at $t = 3.5$ seconds.

  b. What is the maximum displacement of the object and when does it occur?

  (**Hint:** Use the definition of the derivative to find the derivatives of $e^{-0.2t}$ and $\cos(2\pi t)$. You may also want to review Exercise 57 for the latter.)

63. A 15 ft ladder is leaning against a wall, making an angle of β with the horizontal, when it starts sliding. If x denotes the distance of the bottom of the ladder from the wall, find the rate of change of x with respect to β when $\beta = \pi/6$ (or 30°). Interpret the result.

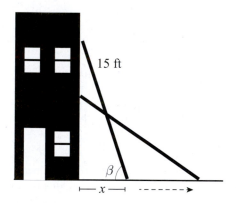

64. A man is pulling his child on a sled at a constant rate, via a rope that makes an angle of α with the horizontal. Since there is no acceleration, the pulling force satisfies the equation $F\cos\alpha = \mu(mg - F\sin\alpha)$, where μ is the coefficient of friction, m is the total mass of the sled and child, and g is the gravity constant.

  a. Express F as a function of α.

  b. Find the rate of change of F with respect to α.

  c. What is the above rate when $\alpha = 60°$?

  d. When (if ever) is this rate of change 0?

3.3 Technology Exercises

65–70 Use a computer algebra system to find the derivative of $f(x)$. Then graph f along with its derivative on the same screen. By zooming in, if necessary, find at least two x-values where the graph of f has a horizontal tangent line. What can you say about f' at such points? (Answers will vary.)

65. $f(x) = \dfrac{x}{1 + \cos x}$

66. $f(x) = \dfrac{1 - \sec x}{1 + \sec x}$

67. $f(x) = \dfrac{\csc x}{x}$

68. $f(x) = \dfrac{\sin x}{\cos x + \tan x}$

69. $f(x) = \cos x(\cot x + \tan x)$

70. $f(x) = \dfrac{\cot x}{\sec x + x\cos x}$

71. Find the maximum velocity and acceleration values in Exercise 62 by using a graphing calculator or computer algebra system to graph the velocity and acceleration functions of the oscillating object.

3.4 **The Chain Rule**

TOPICS

1. Proof of the chain rule

2. Consequences of the chain rule

In this section, we introduce a powerful rule of differentiation that dramatically expands the sorts of functions we can differentiate.

TOPIC 1   **Proof of the Chain Rule**

The Chain Rule gives us a formula for differentiating a composition of two functions, and its nature is pleasantly consistent with our intuition. As an illustration of its meaning, consider the following two questions.

1. *Christi typically walks at a pace of 4 mph and rides her bike, on average, at a rate 3 times as fast as she walks. What is her typical biking speed in miles per hour?*

2. *Tom is climbing a mountain at a rate of 0.5 vertical kilometers per hour, and knows from experience to expect a drop in temperature of 7 °C for every kilometer gained in altitude. What rate of temperature drop is Tom experiencing per hour?*

The answers to both questions involve a product of rates or ratios. In the first question,

$$\frac{\text{miles biked}}{\text{hour}} = \left(\frac{\text{miles biked}}{\text{miles walked}} \right) \left(\frac{\text{miles walked}}{\text{hour}} \right) = (3)(4 \text{ mph}) = 12 \text{ mph}$$

and in the second question,

$$\frac{\text{temp. drop}}{\text{hour}} = \left(\frac{\text{temp. drop}}{\text{vert. km}} \right) \left(\frac{\text{vert. km}}{\text{hour}} \right) = (7 \text{ °C/h})(0.5 \text{ km/h}) = 3.5 \text{ °C/h}.$$

Informally, the Chain Rule tells us that the rate of change of a composition of two functions is the product of their respective rates of change.

Theorem 🔍

The Chain Rule

If g is a function differentiable at the point c, and if f is a function differentiable at $g(c)$, then $f \circ g$ is differentiable at c and

$$(f \circ g)'(c) = f'(g(c)) \cdot g'(c).$$

In Leibniz notation, if we let $y = f(u)$ and $u = g(x)$, then

$$\left. \frac{dy}{dx} \right|_{x=c} = \left(\left. \frac{dy}{du} \right|_{u=g(c)} \right) \left(\left. \frac{du}{dx} \right|_{x=c} \right).$$

Before embarking on a rigorous proof of the Chain Rule, it may be instructive to attempt a proof motivated by the example problems above. If, as in the statement of the theorem, we let $y = f(u)$ and $u = g(x)$, then it is tempting to write

$$\frac{dy}{dx} = \lim_{\Delta x \to 0} \frac{\Delta y}{\Delta x}$$

$$= \lim_{\Delta x \to 0} \frac{\Delta y}{\Delta u} \cdot \frac{\Delta u}{\Delta x}$$

$$= \lim_{\Delta x \to 0} \frac{\Delta y}{\Delta u} \cdot \lim_{\Delta x \to 0} \frac{\Delta u}{\Delta x}$$

$$= \lim_{\Delta u \to 0} \frac{\Delta y}{\Delta u} \cdot \lim_{\Delta x \to 0} \frac{\Delta u}{\Delta x} \qquad \text{\textcolor{blue}{$\Delta u \to 0$ as $\Delta x \to 0$ since g is continuous.}}$$

$$= \frac{dy}{du} \cdot \frac{du}{dx}.$$

Indeed, this argument is perfectly adequate for the two simple problems with which we began, as all the rates under discussion were nonzero constants. The difficulty lies in the fact that, in general, we do not know that $\Delta u \neq 0$ as $\Delta x \to 0$, and we cannot divide by 0. To make the argument rigorous, we need to introduce some techniques that we will explore further in Section 3.9 when we study the linear approximation of functions.

Proof ✎

Our overall goal is to evaluate

$$\lim_{h \to 0} \frac{f\big(g(c+h)\big) - f\big(g(c)\big)}{h},$$

and to do so we have to find a convenient way to rewrite the term $f\big(g(c+h)\big)$.

Given the function g and the fixed point c, define a new function

$$v(h) = \frac{g(c+h) - g(c)}{h} - g'(c).$$

Note that the differentiability of g at c means $\lim_{h \to 0} v(h) = 0$. Similarly, let $\tilde{c} = g(c)$ and define a function

$$w(k) = \frac{f(\tilde{c} + k) - f(\tilde{c})}{k} - f'(\tilde{c}),$$

and note that $\lim_{k \to 0} w(k) = 0$. We can now rearrange these two equations to obtain

$$g(c+h) = g(c) + \big[g'(c) + v(h)\big]h \qquad (1)$$

and

$$f(\tilde{c} + k) = f(\tilde{c}) + \big[f'(\tilde{c}) + w(k)\big]k. \qquad (2)$$

(While equations (1) and (2) may be somewhat intimidating in appearance, don't be deterred—just keep in mind that both $v(h)$ and $w(k)$ go to 0 as their arguments go to 0.)

We can now rewrite $f\big(g(c+h)\big)$ as follows.

$$f\big(g(c+h)\big) = f\big(g(c)+\big[g'(c)+v(h)\big]h\big) \qquad \text{By (1)}$$
$$= f\big(\tilde{c}+\big[g'(c)+v(h)\big]h\big)$$

Denoting $\big[g'(c)+v(h)\big]h$ by k, we can use (2) to continue:

$$f\big(g(c+h)\big) = f\big(\tilde{c}+\big[g'(c)+v(h)\big]h\big)$$
$$= f(\tilde{c})+\big[f'(\tilde{c})+w(k)\big]k \qquad \text{By (2)}$$
$$= f\big(g(c)\big)+\big[f'(\tilde{c})+w(k)\big]k$$

We now subtract $f\big(g(c)\big)$ from both sides to obtain

$$f\big(g(c+h)\big)-f\big(g(c)\big) = \big[f'(\tilde{c})+w(k)\big]k$$
$$= \big[f'(\tilde{c})+w(k)\big]\big[g'(c)+v(h)\big]h. \qquad k=\big[g'(c)+v(h)\big]h$$

If we now divide our last equation by h, we have

$$\frac{f\big(g(c+h)\big)-f\big(g(c)\big)}{h} = \big[f'(\tilde{c})+w(k)\big]\big[g'(c)+v(h)\big]$$
$$= \big[f'\big(g(c)\big)+w(k)\big]\big[g'(c)+v(h)\big].$$

Note that our definition of k implies $k \to 0$ as $h \to 0$, so taking the limit gives us the desired result.

$$\lim_{h\to0}\frac{f\big(g(c+h)\big)-f\big(g(c)\big)}{h} = \lim_{h\to0}\Big[\big(f'\big(g(c)\big)+w(k)\big)\big(g'(c)+v(h)\big)\Big]$$
$$= \lim_{k\to0}\big[f'\big(g(c)\big)+w(k)\big]\cdot\lim_{h\to0}\big[g'(c)+v(h)\big] \qquad k\to0 \text{ as } h\to0$$
$$= f'\big(g(c)\big)\cdot g'(c)$$

Example 1 ✒

Differentiate the following functions.

a. $F(x)=\big(x^5+3\big)^2$

b. $G(x)=\big(x^5+3\big)^{1000}$

c. $H(x)=\sqrt{7x^3-4x+5}$

d. $K(x)=e^{\cos x}$

e. $T(x)=\tan\left(\dfrac{3x+1}{2x-5}\right)$

Solution

a. We recognize F as a composite function; with the notation $g(x)=x^5+3$, and $f(x)=x^2$, F can be written as $F=f\circ g$. Since both f and g are certainly differentiable everywhere, the Chain Rule applies, and we obtain the following.

$$F'(x)=(f\circ g)'(x)=f'\big(g(x)\big)\cdot g'(x)=2\big(x^5+3\big)\cdot\big(5x^4+0\big)=10x^4\big(x^5+3\big)$$

b. While one could argue that in part a. we could simply have expanded $F(x)$ and differentiated the resulting polynomial, therefore rendering the use of the Chain Rule unnecessary, this is clearly not the case in part b. While an expansion is theoretically possible, it is certainly not feasible. In fact, $G(x)$ is our first illustration of how much we have gained with the Chain Rule; our task turns out to be just as easy as it was in part a.

$$G'(x) = 1000(x^5 + 3)^{999} \cdot (5x^4) = 5000x^4 (x^5 + 3)^{999}$$

c. Here once again the Chain Rule is the only practical way to proceed. We first identify the outer function to be $f(x) = \sqrt{x}$, while the inner function is $g(x) = 7x^3 - 4x + 5$. The Chain Rule then implies

$$H'(x) = (f \circ g)'(x) = f'(g(x)) \cdot g'(x)$$

$$= \frac{1}{2\sqrt{7x^3 - 4x + 5}} \cdot (21x^2 - 4)$$

$$= \frac{10.5x^2 - 2}{\sqrt{7x^3 - 4x + 5}}.$$

d. We will illustrate Leibniz notation when differentiating $K(x)$. Since the inner function is $u = \cos x$, while the outer function is $y = f(u) = e^u$, we obtain the following.

$$\frac{dy}{dx} = \frac{dy}{du} \cdot \frac{du}{dx} = e^u \cdot (-\sin x) = -e^{\cos x} \sin x \qquad \begin{array}{l} (e^u)' = e^u \\ (\cos x)' = -\sin x \end{array}$$

e. To differentiate $T(x)$, we will need to apply the Quotient Rule in combination with the Chain Rule, since the inner function is a rational function. Notice, however, that the order is important. We start with the Chain Rule, since T is composite, and then use the Quotient Rule to differentiate the inner function.

$$T'(x) = \sec^2\left(\frac{3x+1}{2x-5}\right) \cdot \left(\frac{3x+1}{2x-5}\right)'$$

$$= \sec^2\left(\frac{3x+1}{2x-5}\right) \cdot \frac{(3)(2x-5)-(3x+1)(2)}{(2x-5)^2}$$

$$= -\frac{17}{(2x-5)^2} \sec^2\left(\frac{3x+1}{2x-5}\right)$$

Example 2 ✎

Find the first and second derivatives of $f(\alpha) = \tan\alpha,$ where α is measured in degrees.

Solution

We know that $f'(x) = \sec^2 x$ if x is measured in radians. In light of the conversion formula between degrees and radians, $x = x(\alpha) = (\pi/180)\alpha,$ you can actually think of f as a composite function.

$$f(\alpha) = f(x(\alpha)) = \tan(x(\alpha))$$

Our first application of the Chain Rule yields the first derivative of f.

$$\frac{df}{d\alpha} = \frac{df}{dx} \cdot \frac{dx}{d\alpha} = \sec^2(x(\alpha)) \cdot \frac{\pi}{180} = \frac{\pi}{180}\sec^2\left(\frac{\pi}{180}\alpha\right)$$

In order to obtain the second derivative of f, we will differentiate with respect to α again, using the Power Rule in combination with the Chain Rule.

$$\frac{d^2 f}{d\alpha^2} = \frac{d}{d\alpha}\left[\frac{\pi}{180}\sec^2(x(\alpha))\right]$$

$$= \frac{\pi}{180}\frac{d}{dx}(\sec^2 x) \cdot \frac{dx}{d\alpha}$$

$$= \frac{\pi}{180}(2\sec x)(\sec x \tan x) \cdot \frac{\pi}{180}$$

$$= \frac{2\pi^2}{180^2}\sec^2\left(\frac{\pi}{180}\alpha\right)\tan\left(\frac{\pi}{180}\alpha\right)$$

Example 3 ✎

Find the derivatives of the following functions by repeatedly applying the Chain Rule.

a. $F(x) = \sin\sqrt{x^2 + 1}$ **b.** $G(x) = \left(1 + \left(2 + 3x^4\right)^5\right)^6$

Solution

a. What sets F apart from the functions previously considered is the fact that it is a composite function with an inner function that is itself composite. Using the usual function notation,

$$F(x) = (f \circ g \circ h)(x) = f(g(h(x))),$$

where

$$f(x) = \sin x, \quad g(x) = \sqrt{x}, \quad \text{and} \quad h(x) = x^2 + 1.$$

Keeping this in mind, when we come to the point in our first application of the Chain Rule where the inner function is being differentiated, we will need to use the Chain Rule a second time. We work through the following steps.

$$F'(x) = f'\big(g(h(x))\big) \cdot \big[g(h(x))\big]'$$

$$= \Big(\cos\sqrt{x^2+1}\Big)\Big(\sqrt{x^2+1}\Big)'$$

$$= \Big(\cos\sqrt{x^2+1}\Big)\frac{1}{2\sqrt{x^2+1}}\big(x^2+1\big)'$$

$$= \Big(\cos\sqrt{x^2+1}\Big)\frac{1}{2\sqrt{x^2+1}}\big(2x\big)$$

$$= \frac{x}{\sqrt{x^2+1}}\cos\sqrt{x^2+1}$$

b. $G(x)$ is another example of a "compound-composite" function. Using the notation of part a., if we write $G(x)$ as $G(x) = (f \circ g \circ h)(x)$, a close examination of the formula reveals that

$$f(x) = x^6, \quad g(x) = 1+x^5, \quad \text{and} \quad h(x) = 2+3x^4.$$

Proceeding like we did in part a., we obtain the following.

$$G'(x) = (6)\Big(1+\big(2+3x^4\big)^5\Big)^5 \Big(1+\big(2+3x^4\big)^5\Big)'$$

$$= (6)\Big(1+\big(2+3x^4\big)^5\Big)^5 \Big(0+5\big(2+3x^4\big)^4\big(2+3x^4\big)'\Big)$$

$$= (6)\Big(1+\big(2+3x^4\big)^5\Big)^5 (5)\big(2+3x^4\big)^4\big(2+3x^4\big)'$$

$$= (30)\Big(1+\big(2+3x^4\big)^5\Big)^5 \big(2+3x^4\big)^4\big(12x^3\big)$$

$$= 360x^3\big(2+3x^4\big)^4\Big(1+\big(2+3x^4\big)^5\Big)^5$$

TOPIC 2 Consequences of the Chain Rule

The Chain Rule, in combination with the Power Rule, tells us that if u is a function of the variable x, then

$$\frac{d}{dx}\big(u^n\big) = nu^{n-1}\frac{du}{dx}.$$

This is worth noting because the need to differentiate functions of the form u^n arises often, and the fact that it is true for all real numbers n makes the observation all the more useful. Recall, though, that to this point we have actually only proved the Power Rule for integer exponents. The Chain Rule allows us to now extend the proof to all rational number exponents.

Proof ✎

The Power Rule for Rational Exponents

We will first prove that for every nonzero integer n, $\left(x^{1/n}\right)' = (1/n)x^{(1/n)-1}$.

Given such an n, let $g(x) = x^{1/n}$ and $f(x) = x^n$. Note then that

$(f \circ g)(x) = f(g(x)) = \left(x^{1/n}\right)^n = x$, so we already know that

$(f \circ g)'(x) = 1$. But by the Chain Rule,

$$(f \circ g)'(x) = f'(g(x))g'(x)$$

$$= n(g(x))^{n-1} g'(x) \qquad\qquad f'(x) = nx^{n-1}$$

$$= n\left(x^{(n-1)/n}\right)g'(x). \qquad\qquad g(x) = x^{1/n}$$

So it must be the case that $n\left(x^{(n-1)/n}\right)g'(x) = 1$. Solving this equation for $g'(x)$, we obtain the desired result.

$$g'(x) = \frac{1}{n\left(x^{(n-1)/n}\right)} = \left(\frac{1}{n}\right)\left(\frac{1}{x^{1-(1/n)}}\right) = \frac{1}{n}x^{(1/n)-1}$$

Now let m be an integer as well; we will use the above result to prove the Power Rule for the rational exponent m/n.

$$\left(x^{m/n}\right)' = \left[\left(x^{1/n}\right)^m\right]'$$

$$= m\left(x^{1/n}\right)^{m-1}\left(x^{1/n}\right)' \qquad\qquad \text{Chain Rule and Power Rule for exponent } m$$

$$= m\left(x^{(m-1)/n}\right)\left(\frac{1}{n}x^{(1/n)-1}\right) \qquad\qquad \text{Power Rule for exponent } 1/n$$

$$= \frac{m}{n}x^{(m/n)-(1/n)+(1/n)-1}$$

$$= \frac{m}{n}x^{(m/n)-1}$$

Example 4 ✐

Combine the Chain Rule with the Power Rule to differentiate the following functions.

a. $F(x) = \left(3\sin x - x^5\right)^{3/2}$ **b.** $G(x) = \dfrac{1}{\sqrt[3]{3x^2 + 2x - 5}}$

Solution

a. We identify $f(x) = x^{3/2}$ as the outer function, while $g(x) = 3\sin x - x^5$ is the inner function. Now the Chain Rule can be applied. We also need to use the Power Rule to differentiate f:

$$F'(x) = \frac{3}{2}\left(3\sin x - x^5\right)^{(3/2)-1}\left(3\sin x - x^5\right)' = \frac{3}{2}\sqrt{3\sin x - x^5}\left(3\cos x - 5x^4\right)$$

b. Notice that we can rewrite G as

$$G(x) = \left(3x^2 + 2x - 5\right)^{-1/3}.$$

Now we proceed in a manner similar to part a.

$$G'(x) = -\frac{1}{3}\left(3x^2 + 2x - 5\right)^{(-1/3)-1}\left(3x^2 + 2x - 5\right)'$$

$$= -\frac{1}{3}\left(3x^2 + 2x - 5\right)^{-4/3}\left(6x + 2\right)$$

$$= \left(-2x - \frac{2}{3}\right)\left(3x^2 + 2x - 5\right)^{-4/3}$$

$$= -\frac{2x + \frac{2}{3}}{\left(\sqrt[3]{3x^2 + 2x - 5}\right)^4}$$

Finally, notice that we could have started out with the Reciprocal Rule and arrived at the same result.

$$G'(x) = \frac{-\left(\sqrt[3]{3x^2 + 2x - 5}\right)'}{\left(\sqrt[3]{3x^2 + 2x - 5}\right)^2}$$

$$= \frac{-\left(\left(3x^2 + 2x - 5\right)^{1/3}\right)'}{\left(\sqrt[3]{3x^2 + 2x - 5}\right)^2}$$

$$= \frac{-\frac{1}{3}\left(3x^2 + 2x - 5\right)^{-2/3}\left(6x + 2\right)}{\left(\sqrt[3]{3x^2 + 2x - 5}\right)^2}$$

$$= \frac{-\frac{1}{3}\left(6x + 2\right)\left(\sqrt[3]{3x^2 + 2x - 5}\right)^{-2}}{\left(\sqrt[3]{3x^2 + 2x - 5}\right)^2} = -\frac{2x + \frac{2}{3}}{\left(\sqrt[3]{3x^2 + 2x - 5}\right)^4}$$

Example 5 ✎

Find an equation for the tangent line to the graph of $f(x) = e^{\sin x + \cos x}$ at the point $(\pi/2, e)$.

Solution

We shall start by determining $f'(x)$ and evaluating it at $x = \pi/2$. Using the Chain Rule and the fact that e^x is the derivative of itself,

$$f'(x) = e^{\sin x + \cos x}\left(\cos x - \sin x\right).$$

Substituting $x = \pi/2$ yields

$$f'\left(\frac{\pi}{2}\right) = e^{1+0}\left(0 - 1\right) = -e,$$

which is the slope of the requested tangent. Thus, using the point-slope form of the equation of a line, we obtain

$$y - e = -e\left(x - \frac{\pi}{2}\right),$$

or equivalently,

$$y = -ex + \frac{e(\pi + 2)}{2}.$$

Figure 1 confirms our findings. Note that the function and its tangent line were graphed in a viewing window of $[-1, 4]$ by $[-1, 8]$.

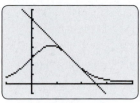

Figure 1

Example 6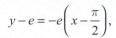

Use the Chain Rule to find the derivatives of the following trigonometric functions.

a. $f(x) = \cos 2x$ **b.** $g(x) = \cos^2 x$ **c.** $h(x) = \cos(x^2)$

Solution

All three functions are composite, and therefore we will use the Chain Rule, but careful attention is needed in examining the structure of each function, that is, in identifying the inner and outer components.

a. In the case of f, cosine is the outer function, while $2x$ serves as the inner function. (Using parentheses, we could actually rewrite f as $f(x) = \cos(2x)$). Thus the derivative is determined as follows.

$$f'(x) = \left(\cos 2x\right)' = -\sin 2x \cdot \left(2x\right)' = -2\sin 2x$$

b. Since $g(x) = \cos^2 x = \left(\cos x\right)^2$, we see that the cosine function is the inner function here, and the outer function is $y = x^2$. Thus the Chain Rule in this case yields

$$g'(x) = \left(\cos^2 x\right)' = 2\cos x \cdot \left(\cos x\right)'$$
$$= 2\cos x \cdot \left(-\sin x\right) = -2\cos x \sin x = -\sin 2x.$$

c. Unlike in part b., the "squaring function" $y = x^2$ is the inner function this time, and cosine is the outer. Using the Chain Rule accordingly, we obtain

$$h'(x) = \left[\cos(x^2)\right]' = \left[-\sin(x^2)\right](x^2)' = \left[-\sin(x^2)\right](2x) = -2x\sin(x^2).$$

The Chain Rule, along with the fact that $\left(e^x\right)' = e^x$, tells us that for any function u of x,

$$\frac{d}{dx}\left(e^u\right) = e^u \frac{du}{dx}.$$

We can use this observation to find the derivative of the exponential function a^x for any base $a > 0$. We begin by noting that

$$a^x = \left(e^{\ln a}\right)^x = e^{(\ln a)x},$$

and conclude that

$$\frac{d}{dx}\left(a^x\right) = \frac{d}{dx}e^{(\ln a)x} = e^{(\ln a)x}\frac{d}{dx}\left[\left(\ln a\right)x\right] = e^{(\ln a)x}\left(\ln a\right) = a^x \ln a.$$

We record this result here as our latest differentiation rule. (Note that this rule is consistent with the fact that $\left(e^x\right)' = e^x,$ as $\ln e = 1$).

Theorem ⚷

Derivatives of a^x and a^u

Given a fixed real number $a > 0$,

$$\frac{d}{dx}\left(a^x\right) = a^x \ln a.$$

More generally, if u is a differentiable function of x, then

$$\frac{d}{dx}\left(a^u\right) = \left(a^u \ln a\right)\frac{du}{dx}.$$

Example 7 ✎

Find an equation for the tangent line to the graph of $F(x) = 2^{\tan(\sin(x/2))}$ at $(0,1)$. Then find all x-values between -2π and 2π where the tangent line to the graph is horizontal.

Solution

As we have seen in Example 3, F is yet another example of a composite function with a composite inner function, so a repeated application of the Chain Rule will be needed to determine its derivative (be sure you understand each of the following steps).

$$F'(x) = 2^{\tan(\sin(x/2))}\left(\ln 2\right)\left(\tan\left(\sin\frac{x}{2}\right)\right)'$$

$$= \left(\ln 2\right)2^{\tan(\sin(x/2))}\sec^2\left(\sin\frac{x}{2}\right)\cdot\left(\sin\frac{x}{2}\right)'$$

$$= \left(\ln 2\right)2^{\tan(\sin(x/2))}\sec^2\left(\sin\frac{x}{2}\right)\cdot\left(\cos\frac{x}{2}\right)\cdot\left(\frac{1}{2}\right)$$

$$= \left(\frac{\ln 2}{2}\right)2^{\tan(\sin(x/2))}\cos\frac{x}{2}\sec^2\left(\sin\frac{x}{2}\right)$$

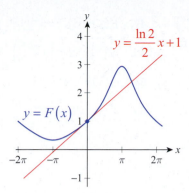

Figure 2

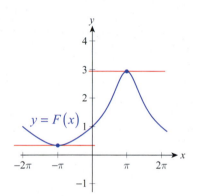

Figure 3

At $x = 0$, the derivative is

$$F'(0) = \left(\frac{\ln 2}{2}\right) 2^{\tan(\sin 0)} \cos 0 \sec^2(\sin 0) = \left(\frac{\ln 2}{2}\right) 2^0 \cos 0 \sec^2 0 = \frac{\ln 2}{2},$$

so keeping in mind that the y-intercept of the requested tangent is 1, its equation is

$$y = \frac{\ln 2}{2} x + 1.$$

The function and its tangent line are graphed in Figure 2.

Next, we will determine the x-values between -2π and 2π at which $F'(x) = 0$. In other words, we need to solve the equation

$$\left(\frac{\ln 2}{2}\right) 2^{\tan(\sin(x/2))} \cos \frac{x}{2} \sec^2\left(\sin \frac{x}{2}\right) = 0.$$

Since the exponential factor is never 0, after dividing and converting the secant to cosine the equation reduces to

$$\frac{\cos \dfrac{x}{2}}{\cos^2\left(\sin \dfrac{x}{2}\right)} = 0.$$

This latter equation is satisfied if and only if

$$\cos \frac{x}{2} = 0,$$

which is true within the specified interval when $x = \pi$ or $x = -\pi$.

The function and the two horizontal tangent lines are shown in Figure 3.

3.4 **Exercises**

1–9 Identify $f(x)$ and $u = g(x)$ such that $F(x) = f(u) = f(g(x))$. Also find $h(x)$ wherever $F(x) = f(g(h(x)))$. (Answers will vary.)

1. $F(x) = (3x - 2.5)^6$

2. $F(x) = 2(x^3 - 5x^2 + \pi)^{-4}$

3. $F(x) = 2\sqrt[3]{x^2 - 9}$

4. $F(x) = \dfrac{-3}{5 + \sqrt{x^3 + x}}$

5. $F(x) = \sin \dfrac{1}{x^2 + 1}$

6. $F(x) = 3\cos\left(\dfrac{\tan x}{2}\right)$

7. $F(x) = \csc(3e^x)$

8. $F(x) = \sec\left(e^{2 + \sqrt{x}}\right)$

9. $F(x) = \dfrac{3}{\sqrt{\ln(x^2 + 1)}}$

10–60 Find the derivative of the given function.

10. $f(x) = (2x^2 + x)^7$

11. $g(x) = 3(x^5 - \pi x^2 + 7.5)^{11}$

12. $h(x) = \frac{1}{2}(x^8 + 5x^3 - ex)^{100}$

13. $F(x) = -3(5 + 2\sqrt{x})^{-5}$

14. $G(x) = (2x^2 - 3x + 1)^{2/3}$

15. $k(x) = -5(x^5 - 2x^3 + 10.5x)^{-2/5}$

16. $f(x) = \sqrt{2 - 4x}$

17. $g(x) = \sqrt{x^2 - 5x + 2}$

18. $h(x) = (4x + 5)^{21}(3x - 7)^{13}$

19. $q(x) = 2(x^3 - 5x)^{2/3}(x + 3)^{5/4}$

20. $r(t) = \dfrac{1}{3t + 1}$ 21. $k(z) = \dfrac{1}{1 + 5z - 2z^2}$

22. $F(x) = \left(\dfrac{2x - 3}{1 - 7x}\right)^{10}$ 23. $S(v) = \left(\dfrac{2v + 1}{v^2 - 5}\right)^{-3}$

24. $G(y) = \left(\dfrac{3y^2 - 1}{2 + 4y}\right)^{7/5}$ 25. $T(s) = \left(\dfrac{s^2 - 1}{s^2 + 1}\right)^{-2/3}$

26. $G(x) = \dfrac{(5 - \pi x^2)^2}{(1 + 2x)^3}$ 27. $H(x) = \dfrac{\sqrt{x^2 - 2}}{(x^2 + 2)^2}$

28. $R(x) = \sqrt{\dfrac{1}{x^2 - 1}}$ 29. $B(t) = \sqrt[3]{\dfrac{t}{2t^2 + 1}}$

30. $K(s) = \sqrt{\dfrac{2s - 5}{3s + 1}}$ 31. $t(x) = \sin(\cos x)$

32. $Q(x) = 2\tan(\sin x)$ 33. $P(x) = x\tan^2 x$

34. $w(x) = \cot(x^2)$ 35. $U(z) = 5\sec^2 z$

36. $R(x) = x\sqrt{\sin x}$ 37. $C(x) = \sin^2(\tan x)$

38. $U(v) = \csc\left(\dfrac{v}{\cos v}\right)$ 39. $V(x) = e^{\cos x}$

40. $R(\theta) = e^{\theta\tan\theta}$

41. $w(x) = \sin\sqrt{2x + 1} + e^{\tan\sqrt{2x+1}}$

42. $t(x) = 10^{\sqrt{x}}$ 43. $f(x) = \pi 2^{\sin(\pi x)}$

44. $u(x) = 2^{x^2} - 4^{\sqrt{x}}$ 45. $t(s) = \tan(2^s)$

46. $u(x) = \cot^2(2^{\sin x})$ 47. $E(x) = 5^{5^x}$

48. $K(x) = \sqrt[3]{3^x} + 3^{\sqrt[3]{x}}$ 49. $N(x) = \cos^2\left(e^{\cos(x^2)}\right)$

50. $u(t) = \tan^3(t^3 + 3^t)$ 51. $C(x) = \cos^2(x^2)$

52. $F(x) = 5^{x^5}$ 53. $t(s) = \sqrt{\cos(10^s)}$

54. $G(t) = \sec^{-3}(5^t)$

55. $H(s) = \sin(2^s)\tan(2^s)$

56. $w(s) = \sin(\tan(2^s))$ 57. $T(z) = \sin(e^z) + e^{\sin z}$

58.* $q(x) = \sin(\cos(\tan(\cot x)))$

59.* $U(\theta) = \theta + \tan(\theta + \tan(\theta + \tan\theta))$

60.* $v(x) = \left(1 + \left(2 + (3 + 4x)^5\right)^6\right)^7$

61–68 Find an equation for the tangent line to the graph of the given function at the specified point.

61. $f(x) = \sqrt{2x^2 + 1}$; $(2, 3)$

62. $g(x) = (x^2 + 3x + 4)^{2/3}$; $(1, 4)$

63. $q(x) = \cos(\tan x)$; $(0, 1)$

64. $S(x) = \sin(x^2) + \sin^2 x$; $(0, 0)$

65. $M(x) = \dfrac{e^{\cos x}}{x}$; $\left(\pi, \dfrac{1}{e\pi}\right)$

66. $a(x) = 10^{\sqrt{x}}$; $(1, 10)$

67. $h(x) = \dfrac{3x + 1}{\sqrt{x^2 + 3}}$; $(1, 2)$

68. $u(x) = \pi^{\pi^{\sin x}}$; $(0, \pi)$

69–76 Find all x-values where the line tangent to the given curve is horizontal.

69. $f(x) = (x^2 - 8x + 15)^{100}$

70. $g(x) = \dfrac{2x + 3}{x^2 - 2}$

71. $h(x) = \sqrt{x^2 + 1}$ 72. $T(x) = \tan^{10} x$

73. $w(x) = \sec(x^2 + 2)$ 74. $t(x) = \cos(\cos x)$

75. $k(x) = e^{x/(x^2 + 1)}$ 76. $q(x) = \pi^{\cos^2 x}$

77–84 Determine the second derivative of the function.

77. $p(x) = (x^2 + 5)^{20}$ 78. $r(t) = \sqrt{t^2 + 5}$

79. $g(x) = 5\cos^2 x$ 80. $c(x) = e^{\tan x}$

81. $F(t) = t\sin(t^2)$ 82. $d(x) = 5^{5^x}$

83. $G(x) = \sin^2 x + \cos^2 x$

84. $U(s) = \sec\sqrt{s}$

85. Suppose that $f(1) = 1$, $f'(1) = -2$, $g(1) = 1$, and $g'(1) = 5$. If $F(x) = (f \circ g)(x)$ and $G(x) = (g \circ f)(x)$, find $F'(1) + G'(1)$.

86. Let $P(x) = x(x+1)(x+2)\cdots(x+10)$. If $F(x) = (P \circ P)(x)$, find the value of $F'(0)$.

87. Find a formula for the nth derivative of $f(x) = \cos(kx)$, $k \in \mathbb{R}$. (**Hint:** Use the Chain Rule and recognize a pattern.)

88. Repeat Exercise 87 for the function $g(x) = 2^{kx}$.

89. Use the Chain Rule to prove that the function $f(x) = \sin(1/x^2)$ is differentiable for $x \neq 0$.

90. Use the Chain Rule to construct a second proof of the Quotient Rule. (**Hint:** Rewrite $f(x)/g(x)$ as $f(x) \cdot [g(x)]^{-1}$.)

91. Use the Chain Rule to prove that the derivative of an even function is odd and vice versa.

92. Find all points where the line tangent to the graph of $y = \sqrt[3]{\cos x}$ is horizontal, as well as those where it is vertical.

93.* A spherical balloon is being inflated so that its radius is increasing at a rate of $dr/dt = 0.1$ in./s. Find the rate at which the volume of the balloon is increasing when its radius is $r = 4$ in. (**Hint:** Notice that $V(t) = V(r(t))$ and use the Chain Rule.)

94.* Pouring sand is forming a conical shape so that the radius of the bottom of the cone is always twice its height throughout the process. If the height of the cone is increasing at a rate of $dh/dt = 0.5$ mm/s, find the rate at which the volume of the cone is increasing when its height is $h = 50$ mm. (See the hint given in Exercise 93.)

95. The position function of a vibrating loudspeaker cone is given by $x(t) = 10^{-3}\cos 1500t$, where distance is measured in meters, time in seconds. As indicated by the position function, the cone is at one of its extreme positions at $t = 0$. Use the above information to find **a.** the maximum velocity of the cone and **b.** the maximum acceleration of the cone.

96. The position function for damped harmonic motion of an object of mass m is

$$x(t) = Ae^{-\frac{k}{2m}t}\cos(\omega t),$$

where A is the amplitude and k and ω are constants specific to the motion. Find the velocity and acceleration functions for this motion.

97. Unless conditions are "extreme," most gases obey the so-called *Ideal Gas Law*, which says $PV = nRT$, where P stands for pressure measured in pascals (Pa), V for volume, n for the number of moles (mol) of gas in the container, T denotes temperature measured in kelvins (K), and R is the *universal gas constant*, which is the same for all gases. Suppose 5 moles of gas are being slowly compressed by a piston in a container so that $dV/dt = -2 \cdot 10^{-8}$ m³/s. Assuming that temperature is being kept constant at $T = 293$ K throughout the process, find the rate of change of pressure with respect to time when $V = 10^{-3}$ m³. (Use $R \approx 8.315\,\text{J}/(\text{mol}\cdot\text{K})$.)

3.4 **Technology Exercises**

98–99 The Maclaurin polynomial of order 2 of the function $f(x)$ is used to approximate $f(x)$ near $x = 0$. It is defined as

$$P_2(x) = f(0) + f'(0)x + \frac{1}{2}f''(0)x^2.$$

Find the Maclaurin polynomial of order 2 for $f(x)$. Then use a graphing calculator or computer algebra system to graph f along with its Maclaurin polynomial. (We will learn more about Maclaurin polynomials in Section 10.8.)

98. $f(x) = \cos(\sin x)$ 99. $f(x) = \dfrac{1}{x^2 + 1}$

3.5 Implicit Differentiation

TOPICS

1. Derivatives of implicitly defined functions

2. Higher-order derivatives

In this section, we learn how to find the rate of change of one variable with respect to another when the functional relationship between the two is presented implicitly.

TOPIC 1   Derivatives of Implicitly Defined Functions

The functions we have worked with so far have been written (or at least could be written) in explicit form. That is, the dependent variable, say y, has been defined explicitly as a function of the independent variable, say x. And we have now gained experience in working with the resulting equation $y = f(x)$ to determine the derivative y', assuming the function f is differentiable.

But there are many instances where the functional dependence of one variable on another is defined only implicitly by an equation and it is inconvenient or in fact impossible to solve the equation for the dependent variable. Despite this, the rate of change of one variable with respect to the other may still be central to the problem we are attempting to solve. For instance, given an equation in x and y, the notion of dy/dx may make perfect sense and be what we seek—we just need a method to determine dy/dx without first expressing y as a function of x. This method is called *implicit differentiation*.

We will begin with a comparison of implicit differentiation and what we might now call explicit differentiation, using an example where both methods can be applied.

Example 1 ✐

Given the equation $y^2 = x$, determine dy/dx.

Solution

We can actually solve this equation for y, but we can express the result in functional form only if we distinguish between the two solutions of the equation. Accordingly, let

$$y_1 = \sqrt{x} \quad \text{and} \quad y_2 = -\sqrt{x}.$$

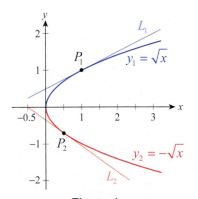

Figure 1

Note that y_1 corresponds to the upper half of the horizontally oriented parabola shown in Figure 1, while y_2 corresponds to the lower. Using the Power Rule of differentiation,

$$y_1' = \frac{1}{2}x^{-1/2} = \frac{1}{2\sqrt{x}} \quad \text{and} \quad y_2' = -\frac{1}{2}x^{-1/2} = -\frac{1}{2\sqrt{x}}.$$

If we were asked to find the rate of change of y with respect to x at, say, $P_1 = (1,1)$ and $P_2 = \left(0.5, -\sqrt{0.5}\right)$, we would calculate

$$\left.\frac{dy_1}{dx}\right|_{x=1} = \frac{1}{2\sqrt{1}} = \frac{1}{2} \quad \text{and} \quad \left.\frac{dy_2}{dx}\right|_{x=0.5} = -\frac{1}{2\sqrt{0.5}} = -\frac{1}{\sqrt{2}}.$$

These two values represent the slopes of the two tangent lines L_1 and L_2 that are shown in Figure 1.

Alternatively, using the method of implicit differentiation, we leave the equation in its original form and differentiate both sides with respect to x. To differentiate the left-hand side, we *implicitly assume* that y is a differentiable function of x and apply the Chain Rule.

$$\frac{d}{dx}(y^2) = \frac{d}{dx}(x)$$

$$2y\frac{dy}{dx} = 1 \qquad\qquad \text{Chain Rule}$$

$$\frac{dy}{dx} = \frac{1}{2y}$$

This one formula contains the same information as the two formulas for y_1' and y_2', and using it to calculate the slopes of the lines L_1 and L_2 we obtain the same results:

$$\left.\frac{dy}{dx}\right|_{P_1} = \left.\frac{dy}{dx}\right|_{y=1} = \frac{1}{2(1)} = \frac{1}{2} \quad \text{and} \quad = \left.\frac{dy}{dx}\right|_{P_2} = \left.\frac{dy}{dx}\right|_{y=-\sqrt{0.5}} = \frac{1}{2\left(-\sqrt{0.5}\right)} = -\frac{1}{\sqrt{2}}$$

The implicit assumption that y is a differentiable function of x is subtle but important. Fortunately, in our usage the assumption is largely self-regulating—if y is *not* a differentiable function of x at a particular point, that fact usually becomes evident as a result of the process. For instance, in Example 1 we know that dy/dx does not exist at the origin, as the line tangent to the curve $y^2 = x$ is vertical at that point; this is reflected in the fact that the formula $1/(2y)$ is undefined when $y = 0$.

In summary, the method of implicit differentiation is as follows.

Method of Implicit Differentiation

Step 1: Given an equation in x and y, differentiate both sides with respect to x under the assumption that y is a differentiable function of x.

Step 2: Solve the resulting equation for dy/dx.

Example 2

Determine dy/dx for the graph of the equation $x^3 + y^3 = 7xy$ (this famous curve is called the *folium of Descartes*). Use your result to find the coordinates of the rightmost point on the loop (see Figure 2).

Solution

We will start by differentiating both sides of the equation with respect to x, treating the variable y as a differentiable function of x (you can actually think of y as $y = y(x)$).

$$\frac{d}{dx}(x^3 + y^3) = \frac{d}{dx}(7xy)$$

$$3x^2 + 3y^2\frac{dy}{dx} = 7\left(y + x\frac{dy}{dx}\right) \qquad\qquad \text{Power Rule and Product Rule}$$

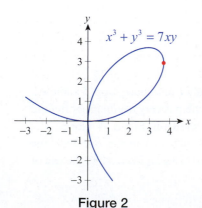

Figure 2

Next, we rearrange terms and solve the equation for dy/dx.

$$3y^2 \frac{dy}{dx} - 7x \frac{dy}{dx} = 7y - 3x^2$$

$$\left(3y^2 - 7x\right)\frac{dy}{dx} = 7y - 3x^2$$

$$\frac{dy}{dx} = \frac{7y - 3x^2}{3y^2 - 7x}$$

To find the coordinates of the rightmost point on the loop, note that it is a point where the corresponding tangent line is vertical, in other words, one whose coordinates make dy/dx undefined. This will happen when $3y^2 - 7x = 0$, that is, when $x = \frac{3}{7}y^2$. Substituting this expression for x into the equation of the folium, we obtain

$$\left(\frac{3}{7}y^2\right)^3 + y^3 = 7\left(\frac{3}{7}y^2\right)y,$$

which reduces to

$$\left(\frac{3}{7}\right)^3 y^6 + y^3 = 3y^3 \qquad\qquad \text{Divide by } y^3.$$

$$\left(\frac{3}{7}\right)^3 y^3 = 2 \qquad\qquad \text{Solve for } y.$$

$$y = \frac{7}{3}\sqrt[3]{2}.$$

Note that dividing both sides of the equation by y^3 was justified, for $y \ne 0$ in the problem at hand. Next, the x-coordinate of the point is obtained.

$$x = \frac{3}{7}y^2 = \frac{3}{7}\left(\frac{7}{3}\sqrt[3]{2}\right)^2 = \frac{7}{3}\left(\sqrt[3]{2}\right)^2$$

We conclude that the coordinates of the point in the extreme-right position on the loop are as follows.

$$\left(\frac{7}{3}\left(\sqrt[3]{2}\right)^2, \frac{7}{3}\sqrt[3]{2}\right)$$

Technology Note 💻

We remark that while determining the x- and y-coordinates from Example 2 can be done by hand, a computer algebra system is very useful in shortening the calculations. *Mathematica*'s output is shown in Figure 3. Notice that multiple solutions are given, but we are interested in the one with the largest x-value. (For additional information on *Mathematica* and the use of the **Solve** command, see Appendix A.)

```
In[1]:=  Solve[{x == (3 / 7) *y^2, x^3 + y^3 == 7*x*y}, {x, y}]
```

$$\text{Out[1]= }\left\{\{x \to 0,\ y \to 0\},\ \left\{x \to \frac{7}{3}\,(-2)^{2/3},\ y \to -\frac{7}{3}\,(-2)^{1/3}\right\},\right.$$

$$\boxed{\left\{x \to \frac{7 \times 2^{2/3}}{3},\ y \to \frac{7 \times 2^{1/3}}{3}\right\},}$$

$$\left.\left\{x \to -\frac{7}{3}\,(-1)^{1/3}\,2^{2/3},\ y \to \frac{7}{3}\,(-1)^{2/3}\,2^{1/3}\right\}\right\}$$

Figure 3

Example 3 ✐

Determine dy/dx for the curve $3 - y^3 + y^2 + 5y - x^2 = 0$, and use it to find the equation of the line that is perpendicular to the tangent at $(3, 2)$ (such a line is called *normal* to the curve). At what points does the curve have horizontal tangents?

Solution

Implicit differentiation of the equation defining the curve yields

$$\frac{d}{dx}\left(3 - y^3 + y^2 + 5y - x^2\right) = \frac{d}{dx}(0)$$

$$0 - 3y^2\frac{dy}{dx} + 2y\frac{dy}{dx} + 5\frac{dy}{dx} - 2x = 0,$$

an equation that we now solve for dy/dx.

$$\left(-3y^2 + 2y + 5\right)\frac{dy}{dx} = 2x$$

$$\frac{dy}{dx} = \frac{2x}{-3y^2 + 2y + 5}$$

At the point $(3, 2)$, the slope of the tangent line is calculated by substituting $x = 3$ and $y = 2$ in the formula for dy/dx.

$$\left.\frac{dy}{dx}\right|_{(3,2)} = \frac{2(3)}{-3(2)^2 + 2(2) + 5} = \frac{6}{-3} = -2$$

Since the normal line is perpendicular to the tangent, its slope is

$$m_n = -\frac{1}{-2} = \frac{1}{2}. \qquad \text{The slopes of perpendicular lines are negative reciprocals.}$$

Figure 4 confirms that the values we found for the slopes of the tangent and normal lines at $(3, 2)$ are reasonable. So using the point-slope form, the equation of the normal line at $(3, 2)$ can be written as

$$y - 2 = \frac{1}{2}(x - 3),$$

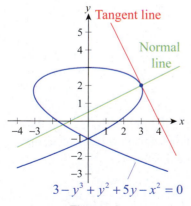

$$3 - y^3 + y^2 + 5y - x^2 = 0$$

Figure 4

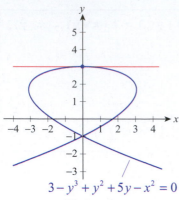

$$3 - y^3 + y^2 + 5y - x^2 = 0$$

Figure 5

or equivalently,

$$y = \frac{1}{2}x + \frac{1}{2}.$$

Finally, we wish to determine all points at which the tangent to the curve is horizontal. Since the formula we derived for dy/dx is a rational expression, and such expressions are equal to 0 if and only if their numerator is 0 (provided that the denominator does not vanish at the same point), we conclude that the only horizontal tangent line occurs at the point $(0,3)$; while $x = 0$ also at the point $(0,-1)$, the derivative dy/dx is undefined there.

The curve and its horizontal tangent line are graphed in Figure 5.

Example 4 ✍

Given the equation $2y^5 + 4xy^2 - 5x^3 y - x^5 + 10 = 0$, find dy/dx by implicit differentiation.

Solution

As in the previous examples, treating y as a function of x, we differentiate both sides with respect to x.

$$\frac{d}{dx}\left(2y^5 + 4xy^2 - 5x^3 y - x^5 + 10\right) = \frac{d}{dx}(0)$$

$$10y^4 \frac{dy}{dx} + 4\left(y^2 + 2xy\frac{dy}{dx}\right) - 5\left(3x^2 y + x^3\frac{dy}{dx}\right) - 5x^4 + 0 = 0$$

Next, we rearrange terms and solve for dy/dx.

$$10y^4\frac{dy}{dx} + 4y^2 + 8xy\frac{dy}{dx} - 15x^2 y - 5x^3\frac{dy}{dx} - 5x^4 = 0$$

$$\left(10y^4 + 8xy - 5x^3\right)\frac{dy}{dx} = 5x^4 + 15x^2 y - 4y^2$$

$$\frac{dy}{dx} = \frac{5x^4 + 15x^2 y - 4y^2}{10y^4 + 8xy - 5x^3}$$

In conclusion, two important remarks are in order. First, note that the variables x and y play fairly symmetrical roles in the given equation. Therefore, we could have treated x as a differentiable function of the independent variable y. This is not the most common approach, but it is correct and sometimes warranted. Differentiating implicitly, we can then determine dx/dy. We show the differentiation step below (remember that y is the variable and $x = x(y)$ is the function this time).

$$10y^4 + 4\left(\frac{dx}{dy}y^2 + 2xy\right) - 5\left(3x^2 y\frac{dx}{dy} + x^3\right) - 5x^4\frac{dx}{dy} = 0$$

Solving for dx/dy (the details are left to the reader) yields the final answer, which turns out, not surprisingly, to be the reciprocal of our solution above.

$$\frac{dx}{dy} = \frac{1}{dy/dx} = \frac{10y^4 + 8xy - 5x^3}{5x^4 + 15x^2y - 4y^2}$$

Finally, if a function is defined implicitly in terms of a polynomial equation of degree five or higher, in general we have no hope of expressing the function explicitly with a formula. This follows from the brilliant insights of Niels Henrik Abel (1802–1829) and Évariste Galois (1811–1832), who showed that in general it is impossible to solve a polynomial equation of degree five or higher with a formula (like we can, for example, solve quadratic equations with the quadratic formula).

As another example of its use, we will use implicit differentiation in an alternative proof of the Power Rule for rational exponents.

Proof ✎

Second Proof of the Power Rule for Rational Exponents

Let m and n be nonzero integers, and let $y = x^{m/n}$; we seek an alternate proof of the formula $y' = (m/n)x^{(m/n)-1}$.

First, raise both sides to the nth power to obtain the equation $y^n = x^m$, and assume that y is a differentiable function of x. Implicit differentiation with respect to x then yields the following.

$$\frac{d}{dx}(y^n) = \frac{d}{dx}(x^m)$$

$$ny^{n-1}\frac{dy}{dx} = mx^{m-1} \qquad \text{Chain Rule and Integer Power Rule}$$

$$\frac{dy}{dx} = \frac{m}{n} \cdot \frac{x^{m-1}}{y^{n-1}} \qquad \text{Solve for } dy/dx.$$

$$\frac{dy}{dx} = \frac{m}{n} \cdot \frac{x^{m-1}}{\left(x^{m/n}\right)^{n-1}} \qquad y = x^{m/n}$$

$$\frac{dy}{dx} = \frac{m}{n} \cdot \frac{x^{m-1}}{x^{m-(m/n)}}$$

$$\frac{dy}{dx} = \frac{m}{n}x^{m-1-m+(m/n)} = \frac{m}{n}x^{(m/n)-1}$$

TOPIC 2 Higher-Order Derivatives

Implicit differentiation can be used repeatedly to obtain higher-order derivatives, as illustrated in the following examples.

Example 5 ✎

Find d^2y/dx^2 for the circle $x^2 + y^2 = 1$.

Solution

Differentiating both sides with respect to x and solving for dy/dx, we obtain the following.

$$\frac{d}{dx}\left(x^2 + y^2\right) = \frac{d}{dx}(1)$$

$$2x + 2y\frac{dy}{dx} = 0$$

$$\frac{dy}{dx} = \frac{-2x}{2y}$$

$$\frac{dy}{dx} = -\frac{x}{y}$$

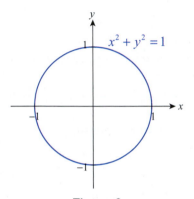

$x^2 + y^2 = 1$

Figure 6

Before proceeding to find the second derivative, it is worthwhile to check that our work to this point is believable. One way to do this is to ask questions such as the following: *where is the first derivative 0, where is it undefined, and what sign does it have in each of the four quadrants?* From Figure 6, we expect dy/dx to be 0 at the two points $(0,1)$ and $(0,-1)$, and that is certainly confirmed by our formula. We also expect an undefined first derivative at the points $(-1,0)$ and $(1,0)$ (i.e., when $y = 0$), and that also agrees with our result for dy/dx. We leave it to the reader to verify that the sign of dy/dx in each of the quadrants is what we would expect (see Exercise 53).

We can now take our result for dy/dx and differentiate with respect to x once more.

$$\frac{d^2y}{dx^2} = \frac{d}{dx}\left(-\frac{x}{y}\right)$$

$$= \frac{y\dfrac{d}{dx}(-x) - (-x)\dfrac{d}{dx}(y)}{y^2} \qquad \text{Quotient Rule}$$

$$= \frac{-y + xy'}{y^2}$$

$$= \frac{-y + x\left(-\dfrac{x}{y}\right)}{y^2} \qquad y' = -\frac{x}{y}$$

$$= \frac{-y^2 - x^2}{y^3} \qquad \begin{array}{l}\text{Multiply the numerator and}\\ \text{denominator by } y.\end{array}$$

$$= -\frac{1}{y^3} \qquad x^2 + y^2 = 1$$

Note that the second derivative is a formula depending only on y. It is again instructive to verify that this formula makes sense for points along the upper half of the circle and for points along the lower half (see Exercise 54).

Example 6 ✎

Find d^2y/dx^2 given that $xy + x - 2 = y$.

Solution

As in the previous example, we proceed to find dy/dx first.

$$\frac{d}{dx}(xy + x - 2) = \frac{d}{dx}(y)$$

$$y + x\frac{dy}{dx} + 1 - 0 = \frac{dy}{dx} \qquad \text{Product Rule}$$

$$\frac{dy}{dx}(x - 1) = -y - 1$$

$$\frac{dy}{dx} = \frac{y+1}{1-x}$$

Next, differentiating with respect to x once more we obtain the second-order derivative.

$$\frac{d^2y}{dx^2} = \frac{d}{dx}\left(\frac{y+1}{1-x}\right)$$

$$= \frac{y'(1-x) - (y+1)(-1)}{(1-x)^2} \qquad \text{Quotient Rule}$$

$$= \frac{\dfrac{y+1}{1-x}(1-x) + (y+1)}{(1-x)^2} \qquad y' = \frac{y+1}{1-x}$$

$$= \frac{(y+1) + (y+1)}{(1-x)^2}$$

$$= \frac{2y+2}{(1-x)^2}$$

3.5 **Exercises**

1–12 Use implicit differentiation to determine dy/dx for the given equation. Then check your answer by expressing y explicitly and using differentiation rules.

1. $x + y^2 = 2$ 　　　　**2.** $xy = 3$ 　　　　**3.** $x^2 - y^2 = 1$ 　　　　**4.** $4x^2 + 25y^2 = 100$

5. $3xy^2 = x - 5$ 　　　　**6.** $y^2\sqrt{x} = 2x^2 + 1$ 　　　　**7.** $y\sqrt{x+2} = xy - 2$ 　　　　**8.** $2x^2y - 3y - x - 1 = 0$

9. $2y\cos x - xy = x + 3$ 　　**10.** $ye^x + 2y - 1 = 0$ 　　**11.** $\dfrac{2}{x} - \dfrac{3}{y} = 4$ 　　**12.** $x^2\sqrt{y} - x^2 - 1 = e^2$

13–20 Find dx/dy by implicit differentiation. Then check your answer by expressing x explicitly in terms of y and differentiating with respect to y using differentiation rules.

13. $x - y^2 = 0$ 　　　　　　　　　　**14.** $xy - y^3 = 3y$

15. $x^3 + y^3 = 1$ 　　　　　　　　　　**16.** $-5xy^2 + 4xy - 3y^2 - y - 2 = 0$

17. $xy + 3\sin y = e^y$

18. $xy = \sqrt{y^2 + 1} - 5x$

19. $\sqrt[3]{8x^3 - 5y^4} = 3$

20. $(y + 2)\sqrt{x + 3} = \sqrt{y}$

21–28 Use implicit differentiation to find the equations of the tangent and normal lines at point P for the well-known curve.

21. $x^2 + y^2 - \left(x^2 + y^2 - y\right)^2 = 0$; $P(1,0)$

Cardioid

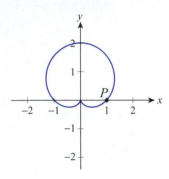

22. $\left(x^2 + y^2\right)^2 - 8y^2 x = 0$; $P(2,2)$

Bifolium

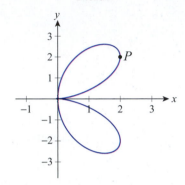

23. $\left(x^2 + 4\right)y = 8$; $P(2,1)$

Witch of Agnesi

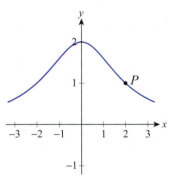

24. $x^2 + y^2 = \left(x^2 + y^2 - 2x\right)^2$; $P(0,1)$

Limaçon

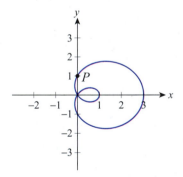

25. $9\left(x^2 + y^2\right) = \left(x^2 + y^2 + 2y\right)^2$; $P(3,0)$

Dimpled Limaçon

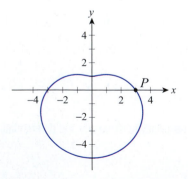

26. $\left(x^2 + y^2\right)^2 = 16xy$; $P(2,2)$

Lemniscate

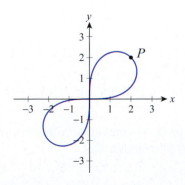

27. $(6-x)y^2 = 2x^3;$ $P(2,2)$

Cissoid

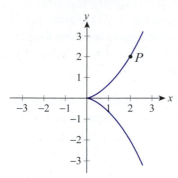

28. $x^{2/3} + y^{2/3} = 10;$ $P(-1,27)$

Astroid

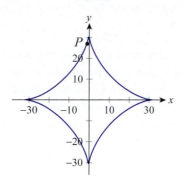

29–44 Find dy/dx by implicit differentiation.

29. $x^4 + y^4 = 1$

30. $\sqrt{x} + \sqrt{y} = 4$

31. $x^3y^4 - x^4y^3 = 1$

32. $y = \cos(x - 2y)$

33. $(x+y)^3 + 3 = x + y$

34. $e^{xy} = e^x + e^y$

35. $\sin^2 x + \cos^2 y = \tan(x^2 + y^2)$

36. $\sqrt{x^2 + y^2} = 2x$

37. $\dfrac{x + 3y^2}{y - x^2} = 2x + 1$

38. $\dfrac{y}{x^3} - \dfrac{x}{y^3} = x^3y^3$

39. $\sqrt{2xy} = 3y - 5x$

40. $y - x = x^4y^4$

41. $\tan x = \sin y - 2xy$

42. $e^x \tan x = y + \cos y$

43. $\sqrt{\sin x + \cos x} = \sec(x + y)$

44. $(\tan x + \cot y)^2 = 1 + x$

45. Find ds/dt by implicit differentiation:
$s^2t^3 - 2t = \sqrt{s}$.

46. Find dt/ds by implicit differentiation:
$s \sin t = t \cos s$.

47–52 Find d^2y/dx^2 by implicit differentiation.

47. $4y^2 - x^2 = 4$

48. $y - x = xy - 2$

49. $xy^2 + 5 = x$

50. $y^3 = xy + 1$

51. $x^3 + y^3 = 3$

52. $\sqrt{x} + \sqrt{y} = 2$

53. Notice that for a circle centered at the origin, any line tangent to the curve in the first quadrant has negative slope; this is consistent with our observation that $dy/dx < 0$ when $x > 0$ and $y > 0$ (see Example 5). Verify that the sign of dy/dx in each of the quadrants is what we would expect.

54. Verify that the sign of the second derivative d^2y/dx^2 of the circle in Example 5 is what we would expect in each quadrant. (**Hint:** Traverse the circle from left to right and examine whether the first derivative is increasing or decreasing; then draw a conclusion regarding the sign of the second derivative.)

55–58 Find all points on the given curve where it has horizontal or vertical tangent lines.

55. $xy^2 - x^2y = \dfrac{1}{4}$

56. $x^2 - xy + y^2 = \dfrac{1}{4}$ (Rotated ellipse)

57. $(x^2 - 2x + 5)y = 5$ **58.** $xy + y^2x^2 = 1$

59. Two graphs are called *orthogonal* if their respective tangent lines are perpendicular at their point(s) of intersection. Show that the graphs of $x^2 - y^2 = 5$ and $xy = 6$ are orthogonal.

60. Generalizing Exercise 59, show that the families of curves $x^2 - y^2 = a$ and $xy = b$ are orthogonal for $a, b \in \mathbb{R}$. (Such families of curves are called *orthogonal trajectories.*)

61. Repeat Exercise 60 for the families $x^2 + y^2 = a$ and $y - bx = 0$.

62. Repeat Exercise 60 for the families $x^2 + y^2 = ax$ and $x^2 + y^2 = by$.

63. Use implicit differentiation to prove that a tangent line to a circle is always perpendicular to the radius connecting the center and the point of tangency. (**Hint:** We can assume without loss of generality that the circle is a unit circle in the xy-coordinate system, centered at the origin.)

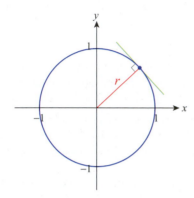

64. Use implicit differentiation to prove that the equation of the line tangent to the ellipse $\dfrac{x^2}{a^2} + \dfrac{y^2}{b^2} = 1$ at the point (x_0, y_0) is $\dfrac{x_0}{a^2}x + \dfrac{y_0}{b^2}y = 1$.

65. Use implicit differentiation to find the equations of the two lines tangent to the ellipse $2x^2 + y^2 = 2$ that pass through the point $(0, 2)$.

66. An object of mass m is attached to a spring and is moving along the x-axis so that its position and velocity satisfy the equation $m(v - v_0)^2 = -kx^2$, where v_0 represents the initial velocity. Use implicit differentiation to verify Hooke's Law; that is, prove that the restoring force exerted by the spring satisfies $F = -kx$. (**Hint:** Differentiate and use Newton's Second Law of Motion, which states that $F = ma$.)

3.5 **Technology Exercises**

67–70. Use the implicit graphing capabilities of a computer algebra system to graph the curves along with the tangent lines you found in Exercises 55–58 and visually verify that your answers are correct.

71–73. Use a computer algebra system to graph the families of curves in Exercises 60–62 for several different values of the parameters a and b. Visually verify that they are orthogonal.

74. Beautiful, "irregular" curves can be created by using a computer algebra system to plot graphs of equations such as the following:

$$(x^2 - 1)(x - 2)(x - 3) = (y - 1)(y - 2)(y - 3)$$

Graph the above equation and explain why this graph cannot be that of a function. Then try experimenting by slightly modifying the above equation and thus creating your own curves. (Answers will vary.)

75. Repeat Exercise 74 starting with the equation $x^5 - 3x^3 - x^2 = -y^5 + 3y^3 - y^2$.

3.6 Derivatives of Inverse Functions

TOPICS

1. The derivative rule for inverse functions
2. Derivatives of logarithms
3. Logarithmic differentiation
4. Derivatives of inverse trigonometric functions

In Section 1.4, we reviewed the algebraic relationship between a given function and (if it exists) its inverse. As we will see, the relationship extends into the realm of calculus when the function under consideration is also differentiable.

TOPIC 1   The Derivative Rule for Inverse Functions

To gain some insight into the formula we will soon develop, suppose L is the line tangent to a function f at the point $(c, f(c))$, and suppose further that L is neither horizontal nor vertical (see Figure 1). Then L can be written in the form $y = mx + b$, with $m \neq 0$. Although we will not provide a formal proof in this text, the fact that f is sufficiently well-behaved to have such a tangent at $(c, f(c))$ means that f^{-1} also possesses a tangent at the point $(f(c), c)$—a fact that is certainly plausible, as the graphs of a function and its inverse are reflections of one another. And since the graph of f^{-1} is the reflection of the graph of f with respect to the line $y = x$, the line tangent to f^{-1} at the point $(f(c), c)$, denoted L^{-1} in Figure 1, is similarly the reflection of the line L. As the reader can verify (by inverting the function $y = mx + b$), the line L^{-1} is described by the equation $y = (1/m)x - (b/m)$; in particular, the slope of L^{-1} is the reciprocal of the slope of L.

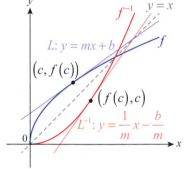

Figure 1

This reciprocal relationship between the derivative of f at the point c and the derivative of f^{-1} at the point $f(c)$ is, in fact, seen at every point where f has a nonzero derivative. The formula that precisely states the relationship can take either of two identical forms, both stemming from an application of the Chain Rule. Assuming again the differentiability of f^{-1}, and using the fact that the composition of a function and its inverse (in either order) is the identity function, note that

$$f^{-1}(f(x)) = x$$

$$\frac{d}{dx}\left[f^{-1}(f(x))\right] = \frac{d}{dx}(x) \qquad \text{Differentiate with respect to } x.$$

$$\left(f^{-1}\right)'(f(x)) \cdot f'(x) = 1 \qquad \text{Chain Rule}$$

$$\left(f^{-1}\right)'(f(x)) = \frac{1}{f'(x)}$$

and, using the alternate order,

$$f(f^{-1}(x)) = x$$

$$\frac{d}{dx}\left[f(f^{-1}(x))\right] = \frac{d}{dx}(x) \qquad \text{Differentiate with respect to } x.$$

$$f'(f^{-1}(x)) \cdot \left(f^{-1}\right)'(x) = 1 \qquad \text{Chain Rule}$$

$$\left(f^{-1}\right)'(x) = \frac{1}{f'(f^{-1}(x))}.$$

The choice of which form of the rule to use depends on the context—in practice, one form is often more easily applied than the other. For reference, we repeat the two formulas here as a theorem.

Theorem

The Derivative Rule for Inverse Functions

If a function f is differentiable on an interval (a,b), and if $f'(x) \neq 0$ for all $x \in (a,b)$, then f^{-1} both exists and is differentiable on the image of the interval (a,b) under f, denoted as $f((a,b))$ in the formula below. Further,

$$\text{if } x \in (a,b), \text{ then } \left(f^{-1}\right)'\left(f(x)\right) = \frac{1}{f'(x)},$$

and

$$\text{if } x \in f((a,b)), \text{ then } \left(f^{-1}\right)'(x) = \frac{1}{f'\left(f^{-1}(x)\right)}.$$

Example 1

a. Given $f(x) = x^2$, use the Derivative Rule for Inverse Functions to determine $\left(f^{-1}\right)'(4)$.

b. Given $g(x) = x^5 - x^2 + 5x - 7$, use the Derivative Rule for Inverse Functions to determine $\left(g^{-1}\right)'$ at the point $g(2)$.

Solution

a. Keeping in mind that $f^{-1}(x) = \sqrt{x}$, and using the second form of the rule, we obtain the following.

$$\left(f^{-1}\right)'(4) = \frac{1}{f'\left(f^{-1}(4)\right)}$$

$$= \frac{1}{f'\left(\sqrt{4}\right)}$$

$$= \frac{1}{f'(2)}$$

$$= \frac{1}{2 \cdot 2} = \frac{1}{4} \qquad f'(x) = 2x$$

b. Notice that although we cannot find a formula for $g^{-1}(x)$, the good news is that we don't have to if we use the first form of the Derivative Rule for Inverse Functions.

$$\left(g^{-1}\right)'\left(g(2)\right) = \frac{1}{g'(2)}$$

$$= \frac{1}{5 \cdot 2^4 - 2 \cdot 2 + 5} \qquad g'(x) = 5x^4 - 2x + 5$$

$$= \frac{1}{81}$$

TOPIC 2   Derivatives of Logarithms

As an immediate application, we will use the second formulation of the Derivative Rule for Inverse Functions to determine the derivative of the natural logarithm.

Theorem ⚗

Derivative of ln$x$

$$\text{For all } x > 0, \quad \frac{d}{dx}(\ln x) = \frac{1}{x}.$$

Proof ✎

We begin by defining $f(x) = e^x$, since we already know that $f'(x) = e^x$ and we seek a formula for the derivative of $f^{-1}(x) = \ln x$.

$$\frac{d}{dx}(\ln x) = \left(f^{-1}\right)'(x) \qquad\qquad f^{-1}(x) = \ln x$$

$$= \frac{1}{f'\left(f^{-1}(x)\right)} \qquad\qquad \text{Derivative Rule for Inverse Functions}$$

$$= \frac{1}{e^{f^{-1}(x)}} \qquad\qquad f'(x) = e^x$$

$$= \frac{1}{e^{\ln x}} = \frac{1}{x}$$

More generally, if $u(x)$ is a differentiable function and if $u(x) > 0$, the Chain Rule tells us that

$$\frac{d}{dx}(\ln u) = \frac{1}{u}\frac{du}{dx} \quad \text{or} \quad \frac{d}{dx}\left[\ln u(x)\right] = \frac{u'(x)}{u(x)}.$$

One consequence of this is that for $x < 0$,

$$\frac{d}{dx}\left[\ln(-x)\right] = \frac{1}{-x}(-1) = \frac{1}{x},$$

leading to the following fact that we will use soon.

$$\frac{d}{dx}\left(\ln|x|\right) = \frac{1}{x}$$

Example 2 ✏

Determine dy/dx for the following functions.

a. $y = \ln(kx)$, where k is a positive real number

b. $y = \ln\left(x^3 + 2x\right)$

c. $y = \ln\sqrt[4]{\dfrac{2x+1}{x^3}}$

Solution

a. Defining $u(x) = kx$ and applying the formula we obtained,

$$\frac{d}{dx}\big[\ln(kx)\big] = \frac{(kx)'}{kx} = \frac{k}{kx} = \frac{1}{x}.$$

Notice that what we have found is the fact that $\ln x$ and $\ln(kx)$ have the same derivative. This shouldn't come as a surprise if we recall that by the properties of logarithms,

$$\ln(kx) = \ln k + \ln x;$$

therefore,

$$\frac{d}{dx}\big[\ln(kx)\big] = \frac{d}{dx}(\ln k) + \frac{d}{dx}(\ln x) = 0 + \frac{1}{x} = \frac{1}{x}. \qquad \text{\color{blue}$\ln k$ is a constant}$$

Recall that what this means geometrically is the fact that the graph of $\ln(kx)$ can be obtained from that of $\ln x$ by a vertical shift of $\ln k$ units. However, vertical shifting doesn't change the slope of a tangent at any point, which visually explains why the derivatives are equal.

b. By the same formula that we put to good use in part a.,

$$\frac{d}{dx}\big[\ln(x^3 + 2x)\big] = \frac{(x^3 + 2x)'}{x^3 + 2x} = \frac{3x^2 + 2}{x^3 + 2x}.$$

c. We could define $u(x) = \sqrt[4]{(2x+1)/x^3}$ and proceed like above; however, finding $u'(x)$ seems a bit tedious at best. Instead, it is to our advantage to use the properties of logarithms before differentiating. Since

$$\ln\sqrt[4]{\frac{2x+1}{x^3}} = \frac{1}{4}\big[\ln(2x+1) - \ln(x^3)\big] = \frac{1}{4}\ln(2x+1) - \frac{3}{4}\ln x,$$

differentiating becomes much more straightforward.

$$\frac{d}{dx}\left(\ln\sqrt[4]{\frac{2x+1}{x^3}}\right) = \frac{1}{4}\cdot\frac{d}{dx}\big[\ln(2x+1)\big] - \frac{3}{4}\cdot\frac{d}{dx}(\ln x)$$

$$= \frac{1}{4}\cdot\frac{2}{2x+1} - \frac{3}{4}\cdot\frac{1}{x}$$

$$= \frac{1}{4x+2} - \frac{3}{4x} = \frac{-4x-3}{4x(2x+1)}$$

Example 3 ✍

Find the equation of the line tangent to the graph of f at the point $(0,1)$.

$$f(x) = \ln\big((e-1)\cos x + e^x\big)$$

Solution

After defining $u(x) = (e-1)\cos x + e^x$, we can proceed by differentiating f to obtain the slope.

$$f'(x) = \frac{d}{dx}\Big[\ln\big((e-1)\cos x + e^x\big)\Big] = \frac{(1-e)\sin x + e^x}{(e-1)\cos x + e^x} \qquad \frac{d}{dx}\big[\ln u(x)\big] = \frac{u'(x)}{u(x)}$$

Therefore, the slope of the desired tangent is

$$m = f'(0) = \frac{(1-e)\sin 0 + e^0}{(e-1)\cos 0 + e^0} = \frac{1}{(e-1)+1} = \frac{1}{e}.$$

Finally, using the point-slope form, the equation of the tangent line is obtained:

$$y - 1 = \frac{1}{e}(x-0),$$

or equivalently,

$$y = \frac{1}{e}x + 1.$$

The function and its tangent are graphed in Figure 2.

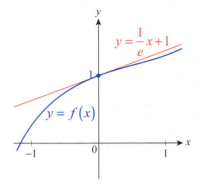

Figure 2

Using the change of base formula and the formula for the derivative of the natural logarithm, we can arrive at a formula for the derivative of a logarithm of any base.

Theorem

Derivative of $\log_a x$

Given a positive base a, $a \neq 1$, and for all $x > 0$,

$$\frac{d}{dx}(\log_a x) = \frac{1}{\ln a} \cdot \frac{1}{x}.$$

Proof

Since we now know the derivative of $\ln x$, all we need is the logarithmic change of base formula.

$$\frac{d}{dx}(\log_a x) = \frac{d}{dx}\left(\frac{\ln x}{\ln a}\right) = \frac{1}{\ln a} \cdot \frac{d}{dx}(\ln x) = \frac{1}{\ln a} \cdot \frac{1}{x}$$

We can again apply the Chain Rule to obtain the following fact for positive differentiable functions $u(x)$.

$$\frac{d}{dx}(\log_a u) = \frac{1}{u \ln a}\frac{du}{dx} \qquad \text{or} \qquad \frac{d}{dx}\big[\log_a u(x)\big] = \frac{1}{\ln a}\frac{u'(x)}{u(x)}$$

Example 4 ✎

Determine the following derivatives.

a. $\dfrac{d}{dx}\left[\log_5\left(2x+3\right)\right]$ **b.** $\dfrac{d}{dx}\left[\log\dfrac{\tan x}{\sqrt{x^2+1}}\right]$

Solution

a. Note that if $u(x)=2x+3$, $u'(x)=2$, so by the above formula we obtain the following.

$$\frac{d}{dx}\left[\log_5\left(2x+3\right)\right]=\frac{1}{\ln 5}\cdot\frac{2}{2x+3}$$

b. First of all, recall that "log" stands for the common logarithm, whose base is 10. Second, just like in Example 2, we will be better off if we rewrite the logarithmic expression before differentiating.

$$\frac{d}{dx}\left[\log\frac{\tan x}{\sqrt{x^2+1}}\right]=\frac{d}{dx}\left[\log\left(\tan x\right)-\frac{1}{2}\log\left(x^2+1\right)\right]$$

$$=\frac{1}{\ln 10}\cdot\frac{\left(\tan x\right)'}{\tan x}-\frac{1}{2\ln 10}\cdot\frac{\left(x^2+1\right)'}{x^2+1}$$

$$=\frac{1}{\ln 10}\cdot\frac{\sec^2 x}{\tan x}-\frac{1}{2\ln 10}\cdot\frac{2x}{x^2+1}$$

$$=\frac{1}{\ln 10}\left(\frac{1}{\sin x\cos x}-\frac{x}{x^2+1}\right)$$

TOPIC 3 Logarithmic Differentiation

In algebra, logarithms are often applied to both sides of an equation in order to bring exponents down as coefficients, thereby making any variables that were in the exponents much more accessible (recall that $\log\left(x^r\right)=r\log x$). This property and other properties of logarithms are often just as useful when we need to differentiate complicated functions. The practice is termed *logarithmic differentiation*, and the procedure to find the derivative of such a function f is as follows.

Logarithmic Differentiation

Step 1: Apply the natural logarithm to both sides of the equation $y=f(x)$, and use the properties of logarithms as appropriate.

Step 2: Under the assumption that y is differentiable, differentiate both sides of the resulting equation implicitly with respect to x.

Step 3: Solve the resulting equation for y'.

Example 5 ✐

Use logarithmic differentiation to differentiate the following functions.

a. $f(x) = \dfrac{\sqrt{1+x^2}}{(2x-5)^3}$ **b.** $g(x) = \dfrac{(x^3-3x)^{3/2}(2x-1)^{1/2}}{(5x+2)^4}$ **c.** $h(x) = x^{2x}$

Solution

a. We start by writing

$$y = \frac{\sqrt{1+x^2}}{(2x-5)^3},$$

then take the natural logarithm of both sides and use properties of logarithms.

$$\ln y = \ln \frac{\sqrt{1+x^2}}{(2x-5)^3}$$

$$\ln y = \frac{1}{2}\ln(1+x^2) - 3\ln(2x-5) \qquad \text{Step 1}$$

Next, we differentiate both sides implicitly with respect to x, and solve for y'.

$$\frac{y'}{y} = \frac{1}{2} \cdot \frac{2x}{1+x^2} - 3 \cdot \frac{2}{2x-5} \qquad \text{Step 2}$$

$$y' = y\left(\frac{x}{1+x^2} - \frac{6}{2x-5}\right) \qquad \text{Step 3}$$

$$y' = \frac{\sqrt{1+x^2}}{(2x-5)^3}\left(\frac{x}{1+x^2} - \frac{6}{2x-5}\right) \qquad y = \frac{\sqrt{1+x^2}}{(2x-5)^3}$$

b. Proceed in a manner similar to part a.

$$y = \frac{(x^3-3x)^{3/2}(2x-1)^{1/2}}{(5x+2)^4}$$

$$\ln y = \frac{3}{2}\ln(x^3-3x) + \frac{1}{2}\ln(2x-1) - 4\ln(5x+2)$$

Implicit differentiation now yields the following.

$$\frac{y'}{y} = \frac{3}{2} \cdot \frac{3x^2-3}{x^3-3x} + \frac{1}{2} \cdot \frac{2}{2x-1} - 4 \cdot \frac{5}{5x+2}$$

$$y' = y\left(\frac{9x^2-9}{2x^3-6x} + \frac{1}{2x-1} - \frac{20}{5x+2}\right)$$

$$y' = \frac{(x^3-3x)^{3/2}(2x-1)^{1/2}}{(5x+2)^4}\left(\frac{9x^2-9}{2x^3-6x} + \frac{1}{2x-1} - \frac{20}{5x+2}\right)$$

c. Notice first of all that since both the base and exponent are variable expressions, h is neither an exponential function nor a power function, so none of the differentiation formulas we learned up to this point are going to help. The fact that we can still differentiate h is perhaps an illustration of logarithmic differentiation at its best. Simply we follow the three-step process.

$$y = x^{2x}$$

$$\ln y = \ln\left(x^{2x}\right)$$

$$\ln y = 2x \ln x \qquad \qquad \ln\left(x^r\right) = r \ln x$$

$$\frac{y'}{y} = 2\ln x + 2x\frac{1}{x} \qquad \qquad \text{Product Rule on the right-hand side}$$

$$y' = 2y(\ln x + 1)$$

$$y' = 2x^{2x}(\ln x + 1)$$

We can use logarithmic differentiation to revisit the Power Rule one last time, and on this occasion we can finally justify the rule for all real exponents.

Theorem 🔍

The Power Rule

Under the assumption that x^r is a real number, where r is a fixed real number constant, the derivative of x^r is rx^{r-1}.

Proof 📎

We begin with the equation $y = x^r$, but in order to apply the natural logarithm to both sides, we insert an intermediate step to obtain $|y| = |x^r|$. Logarithmic differentiation then gives us the following.

$$|y| = |x|^r \qquad \qquad |x^r| = |x|^r$$

$$\ln|y| = \ln\left(|x|^r\right) \qquad \qquad \text{Assume } x \neq 0.$$

$$\frac{d}{dx}\left(\ln|y|\right) = \frac{d}{dx}\left(r\ln|x|\right) \qquad \qquad \text{Differentiate and apply } \ln\left(|x|^r\right) = r\ln|x|.$$

$$\frac{y'}{y} = r \cdot \frac{1}{x}$$

$$y' = r \cdot \frac{y}{x}$$

$$y' = r \cdot \frac{x^r}{x} = rx^{r-1}$$

Note that we are safe in assuming $x \neq 0$ above, as we can easily determine the (possibly one-sided) derivative of x^r at $x = 0$ from a limit of difference quotients for $r \geq 1$, and for $r < 1$ the derivative is undefined at $x = 0$.

Example 6 ✎

Find all points of differentiability for the following functions and determine the derivatives.

a. $f(x) = x^{5/3}$ **b.** $g(x) = x^{1/3}$ **c.** $h(x) = x^{3/2}$

Solution

a. For $x \neq 0$, the Power Rule applies, and we have

$$f'(x) = \frac{5}{3}x^{(5/3)-1} = \frac{5}{3}x^{2/3}.$$

If $x = 0$, we use the definition of the derivative:

$$f'(0) = \lim_{h \to 0} \frac{(0+h)^{5/3} - 0^{5/3}}{h} = \lim_{h \to 0} \frac{h^{5/3}}{h} = \lim_{h \to 0} h^{2/3} = 0.$$

Thus we conclude that $f'(0) = 0$, so f is in fact differentiable everywhere on $\mathbb{R}$, and the formula $f'(x) = \frac{5}{3}x^{2/3}$ does hold for all $x \in \mathbb{R}$.

b. Again, if $x \neq 0$,

$$g'(x) = \frac{1}{3}x^{(1/3)-1} = \frac{1}{3}x^{-2/3} = \frac{1}{3x^{2/3}}.$$

For $x = 0$,

$$g'(0) = \lim_{h \to 0} \frac{(0+h)^{1/3} - 0^{1/3}}{h} = \lim_{h \to 0} \frac{h^{1/3}}{h} = \lim_{h \to 0} \frac{1}{h^{2/3}} = \infty,$$

indicating that $g'(0)$ is undefined (in fact, g has a vertical tangent) at $x = 0$. Thus we conclude that g is differentiable only on $(-\infty, 0) \cup (0, \infty)$.

c. The first important observation about h is that since $x^{3/2} = \sqrt{x^3}$, h is only defined for nonnegative x-values. If $x > 0$, by the Power Rule

$$h'(x) = \frac{3}{2}x^{1/2} = \frac{3}{2}\sqrt{x}.$$

For $x = 0$, the derivative cannot exist, since h is undefined on the left-hand side of 0; however, we may attempt to evaluate the right-hand derivative, using the definition:

$$h'(0) = \lim_{h \to 0^+} \frac{(0+h)^{3/2} - 0^{3/2}}{h} = \lim_{h \to 0^+} \frac{h^{3/2}}{h} = \lim_{h \to 0^+} h^{1/2} = \lim_{h \to 0^+} \sqrt{h} = 0$$

Thus we see that h is undefined and hence cannot be differentiable for $x < 0$, but is differentiable on $(0, \infty)$, with its one-sided derivative also existing at $x = 0$.

Example 7 ✎

Use the Power Rule to determine the derivative of $f(x) = \left(\sqrt{\cos x + 1}\right)^{\pi}$.

Solution

First of all, notice that we can rewrite f as

$$f(x) = (\cos x + 1)^{\pi/2}.$$

The Power Rule assures us that the derivative of the power function $y = x^{\pi/2}$ is $y' = (\pi/2)x^{(\pi/2)-1}$, but in the problem at hand, we need to combine this fact with the Chain Rule.

$$f'(x) = \frac{\pi}{2}(\cos x + 1)^{(\pi/2)-1}(\cos x + 1)'$$

$$= \frac{\pi}{2}(\cos x + 1)^{(\pi/2)-1}(-\sin x)$$

$$= -\frac{\pi}{2}\sin x (\cos x + 1)^{(\pi/2)-1}$$

In conclusion we note that, slightly generalizing the above solution, we can derive what we can call the Generalized Power Rule.

Theorem ⚗

The Generalized Power Rule

If $u = u(x)$ is differentiable and x^r is differentiable at $u(x)$, then

$$\frac{d}{dx}(u^r) = ru^{r-1}\frac{du}{dx}.$$

TOPIC 4   Derivatives of Inverse Trigonometric Functions

Our last application in this section of the Derivative Rule for Inverse Functions will be in determining the derivatives of the inverse trigonometric functions. Recall, from Section 1.4, that an inverse for each of the six common trigonometric functions can be defined, but the precise definition depends on how the domain of each function is restricted (such a restriction is necessary in order to make each function one-to-one). Since there are several possible choices for each restricted domain, there is no one "correct" definition for the six inverse functions—in practice, the restricted domain is often determined on the basis of whatever is most convenient for the particular problem at hand. Fortunately, the formulas for the derivatives of the inverse functions are very similar regardless of the exact definition, and the process outlined below can always be followed to determine the formulas.

With that in mind, the definitions for the inverses that we will use for now are as follows. The graphs of the inverse functions, along with their domains and ranges, are shown in Figure 3.

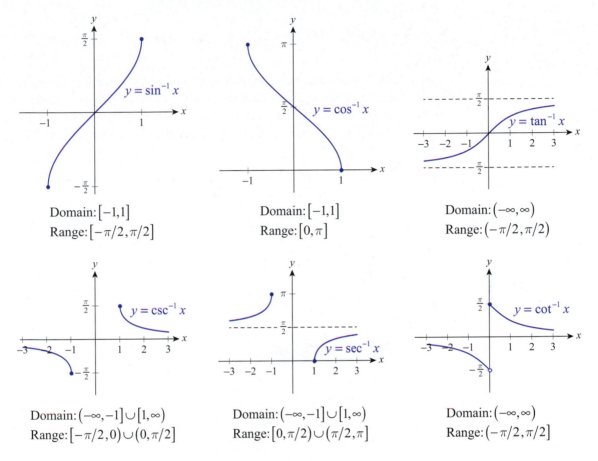

Domain: $[-1,1]$
Range: $[-\pi/2,\pi/2]$

Domain: $[-1,1]$
Range: $[0,\pi]$

Domain: $(-\infty,\infty)$
Range: $(-\pi/2,\pi/2)$

Domain: $(-\infty,-1]\cup[1,\infty)$
Range: $[-\pi/2,0)\cup(0,\pi/2]$

Domain: $(-\infty,-1]\cup[1,\infty)$
Range: $[0,\pi/2)\cup(\pi/2,\pi]$

Domain: $(-\infty,\infty)$
Range: $(-\pi/2,\pi/2]$

Figure 3 Inverse Trigonometric Functions

One advantage of this particular choice of definitions is that the following identities hold:

$$\csc^{-1} x = \sin^{-1}\left(\frac{1}{x}\right)$$

$$\sec^{-1} x = \cos^{-1}\left(\frac{1}{x}\right)$$

$$\cot^{-1} x = \tan^{-1}\left(\frac{1}{x}\right), \text{ with } \cot^{-1} 0 = \frac{\pi}{2}$$

As a consequence, we will be able to use the Chain Rule to find the derivatives of $\csc^{-1}$, $\sec^{-1}$, and $\cot^{-1}$, once we have formulas for the derivatives of the other three functions. (Different but equally useful identities apply with alternative definitions of the six inverse functions, as illustrated in Exercise 105.)

We are now ready to determine some derivatives, and we will begin with the derivative of arcsine. The process makes use of the techniques outlined in Section 1.4.

$$\left(\sin^{-1}\right)'(x) = \frac{1}{\cos\left(\sin^{-1}x\right)} \qquad\qquad \left(f^{-1}\right)'(x) = \frac{1}{f'\left(f^{-1}(x)\right)} \text{ and } \sin' = \cos$$

We evaluate expressions such as $\cos\left(\sin^{-1}x\right)$ by constructing a diagram like the one in Figure 4. Start by letting $\theta = \sin^{-1}x$, rewriting this in the equivalent form $\sin\theta = x$, and then labeling the angle, hypotenuse, and opposite side of a right triangle as shown. Such a diagram is the pictorial representation of the statement $\sin\theta = x$, and the Pythagorean Theorem tells us that the adjacent side must consequently have a length of $\sqrt{1-x^2}$. The final steps are then as follows.

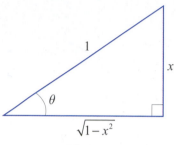

Figure 4

$$\left(\sin^{-1}\right)'(x) = \frac{1}{\cos\left(\sin^{-1}x\right)}$$

$$= \frac{1}{\cos\theta} \qquad\qquad \sin^{-1}x = \theta$$

$$= \frac{1}{\sqrt{1-x^2}}. \qquad\qquad \cos\theta = \frac{\sqrt{1-x^2}}{1} \text{ (See Figure 4.)}$$

Example 8 ✐

Use the above technique to determine the derivative of $\tan^{-1}x$.

Solution

Again, using the second form of the Derivative Rule for Inverse Functions, we obtain

$$\left(\tan^{-1}\right)'(x) = \frac{1}{\sec^2\left(\tan^{-1}x\right)}. \qquad\qquad \left(f^{-1}\right)'(x) = \frac{1}{f'\left(f^{-1}(x)\right)} \text{ and } \tan' = \sec^2$$

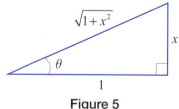

Figure 5

Notice how the labeling in Figure 5 helps determine $\sec^2\left(\tan^{-1}x\right)$. Since

$$\tan\theta = \frac{x}{1} = x$$

$$\theta = \tan^{-1}x,$$

we have

$$\sec^2\left(\tan^{-1}x\right) = \sec^2\theta = 1 + x^2. \qquad\qquad \sec\theta = \frac{\sqrt{1+x^2}}{1}$$

Therefore,

$$\left(\tan^{-1}\right)'(x) = \frac{1}{\sec^2\left(\tan^{-1}x\right)} = \frac{1}{1+x^2}.$$

The derivatives of $\cos^{-1}$, $\sec^{-1}$, and $\cot^{-1}$ are left as exercises (see Exercises 67–69), but we will find the derivative of $\csc^{-1}$ here.

$$\frac{d}{dx}\left(\csc^{-1}x\right) = \frac{d}{dx}\left[\sin^{-1}\left(\frac{1}{x}\right)\right] \qquad \text{Use the identity found for } \left(\sin^{-1}\right)'.$$

$$= \frac{1}{\sqrt{1-(1/x)^2}}\frac{d}{dx}\left(\frac{1}{x}\right) \qquad \text{Chain Rule}$$

$$= \frac{1}{\sqrt{1-(1/x)^2}}\left(-\frac{1}{x^2}\right)$$

$$= \frac{\sqrt{x^2}}{\sqrt{x^2}\sqrt{1-(1/x)^2}}\left(-\frac{1}{x^2}\right) \qquad \text{Multiply by } \frac{\sqrt{x^2}}{\sqrt{x^2}}.$$

$$= \frac{|x|}{\sqrt{x^2-1}}\left(-\frac{1}{x^2}\right) \qquad \sqrt{x^2} = |x|$$

$$= -\frac{1}{|x|\sqrt{x^2-1}} \qquad \frac{|x|}{x^2} = \frac{1}{|x|}$$

For convenience, the derivative rules for the six inverse trigonometric functions are collected as follows, in the more general form that assumes u is a differentiable function of x.

Theorem ⚲

Derivatives of Inverse Trigonometric Functions

$$\frac{d}{dx}\left(\sin^{-1}u\right) = \frac{1}{\sqrt{1-u^2}}\frac{du}{dx} \qquad\qquad \frac{d}{dx}\left(\cos^{-1}u\right) = -\frac{1}{\sqrt{1-u^2}}\frac{du}{dx}$$

$$\frac{d}{dx}\left(\tan^{-1}u\right) = \frac{1}{1+u^2}\frac{du}{dx} \qquad\qquad \frac{d}{dx}\left(\cot^{-1}u\right) = -\frac{1}{1+u^2}\frac{du}{dx}$$

$$\frac{d}{dx}\left(\sec^{-1}u\right) = \frac{1}{|u|\sqrt{u^2-1}}\frac{du}{dx} \qquad\qquad \frac{d}{dx}\left(\csc^{-1}u\right) = -\frac{1}{|u|\sqrt{u^2-1}}\frac{du}{dx}$$

Example 9 ✎

Use the appropriate rules to find the following derivatives.

a. $\dfrac{d}{dx}\left[\cot^{-1}\left(3x^2\right)\right]$ **b.** $\dfrac{d}{dx}\left[\sec^{-1}\left(10^{2x}\right)\right]$

Solution

a. $\dfrac{d}{dx}\left[\cot^{-1}\left(3x^2\right)\right] = -\dfrac{\left(3x^2\right)'}{1+\left(3x^2\right)^2} \qquad u(x) = 3x^2$

$$= -\frac{6x}{1+9x^4}$$

b. $\dfrac{d}{dx}\left[\sec^{-1}\left(10^{2x}\right)\right]=\dfrac{\left(10^{2x}\right)'}{\left|10^{2x}\right|\sqrt{\left(10^{2x}\right)^2-1}}$ $u(x)=10^{2x}$

$$=\dfrac{10^{2x}\left(\ln 10\right)\cdot\left(2x\right)'}{10^{2x}\sqrt{10^{4x}-1}}$$ $\left|10^{2x}\right|=10^{2x}$ since $10^{2x}>0$

$$=\dfrac{2\ln 10}{\sqrt{10^{4x}-1}}$$

3.6 Exercises

1–15 Use the Derivative Rule for Inverse Functions to determine $\left(f^{-1}\right)'(a)$ for the indicated value of a. (In these and subsequent exercises, the domain of f is assumed to have been restricted so that the inverse exists and is differentiable, whenever appropriate.)

1. $f(x)=x^3;\quad a=8$

2. $f(x)=2x-1;\quad a=5$

3. $f(x)=\sqrt{x};\quad a=3$

4. $f(x)=\sqrt[3]{x+2};\quad a=-1$

5. $f(x)=x^2+5;\quad a=9$

6. $f(x)=x^{3/2};\quad a=27$

7. $f(x)=\dfrac{2x}{x-1};\quad a=4$

8. $f(x)=\dfrac{5}{(x-1)^3};\quad a=5$

9. $f(x)=\dfrac{3}{x^2+2};\quad a=1$

10. $f(x)=e^{2x};\quad a=-5$

11. $f(x)=10^x;\quad a=10$

12. $f(x)=2^{\sqrt{x}};\quad a=8$

13. $f(x)=\sin x;\quad a=\dfrac{\sqrt{3}}{2}$

14. $f(x)=2\tan^{-1}x;\quad a=\dfrac{\pi}{2}$

15. $f(x)=\sin\left(x^2\right);\quad a=\sin 0.01$

16–30 Determine the value of $\left(g^{-1}\right)'(b)$ at the given point (assume that the domain of g is appropriately restricted so that g^{-1} exists). (**Note:** Do *not* attempt to find a formula for g^{-1}.)

16. $g(x)=x^5+2x+1;\quad b=g(1)$

17. $g(x)=x^6-11x^4+x;\quad b=g(-1)$

18. $g(x)=x^{100}+x^{50}+1;\quad b=g(-1)$

19. $g(x)=\sqrt{x^4+x^2};\quad b=g(-2)$

20. $g(x)=\left(3x^8+x^3+1\right)^{3/2};\quad b=g(1)$

21. $g(x)=\left(2x^9-3\sqrt{x}\right)^{2/5};\quad b=g(1)$

22. $g(x)=x^5+x+2;\quad b=2$

23. $g(x)=x^{17}+2x^{11}-2x+3;\quad b=4$

24. $g(x)=\dfrac{x^3+8}{\sqrt{x+1}};\quad b=g(2)$

25. $g(x)=\dfrac{x+1}{x^3};\quad b=\dfrac{3}{8}$

26. $g(x)=e^{x^4-x+2};\quad b=g(-2)$

27. $g(x)=x\sin x;\quad b=\dfrac{\pi}{2}$

28. $g(x)=10^{\cos\left(x^3+x\right)};\quad b=g(1)$

29. $g(x)=\tan\sqrt{x};\quad b=g(1)$

30. $g(x)=x^3e^{x^2+1};\quad b=e^2$

31–48 Determine the derivative of the given function.

31. $f(x)=\ln\left(x^3\right)$

32. $g(x)=\left(\ln x\right)^3$

33. $h(x)=\ln\left(x^2+3\right)$

34. $F(x)=\ln\left(x\sqrt{x^2+4}\right)$

35. $G(x)=x\ln\sqrt{x^2+4}$

36. $k(x)=\ln\dfrac{2x}{x^2+1}$

37. $L(x) = \dfrac{\ln 2x}{x^2 + 1}$

38. $f(x) = \ln\sqrt{\dfrac{x+3}{2x+5}}$

39. $g(x) = \ln\sqrt[3]{\dfrac{x+3}{x-3}}$

40. $H(x) = \ln(\ln x)$

41. $F(t) = \ln\left(\sqrt{t^2 + 4} + 2t\right)$

42. $L(s) = \ln\dfrac{\sqrt{s^2 + 2}}{s^4 + s^2 + 1}$

43. $T(x) = \ln|\cos x|$

44. $C(t) = \ln\left(\sin^2 t + 1\right)$

45. $v(x) = \cos 2x\left(\ln(\cos 2x)\right)$

46. $F(t) = \dfrac{\log_5 t}{t^2}$

47. $w(x) = x\log x$

48. $t(x) = \log_{3/2}\left(\left(5x^2 + 4\right)^{3/2}\right)$

49–66 Use logarithmic differentiation to find y'.

49. $y = (x+1)(x+2)(x+3)(x+4)$

50. $y = \dfrac{(x+1)(x+2)}{(x+3)(x+4)}$

51. $y = \sqrt[3]{(2x-1)(x-5)(3x+1)}$

52. $y = \dfrac{\left(x^2 - 1\right)^{2/3}\left(5x^3 + 3\right)}{\left(x^2 + x + 2\right)\left(x^4 - 10\right)^{3/4}}$

53. $y = \dfrac{\sqrt[3]{x^3 - 5x^2 + 7}\,(x+2)}{x^{2/3}\sqrt{3x^2 + 4}}$

54. $y = \sqrt[3]{\dfrac{x^3 - 2x^2 + 1}{\left(x^2 - 1\right)\left(x^3 + 5\right)}}$

55. $y = \dfrac{x^2\sqrt[5]{x^3 + 3}}{\sqrt[4]{x^4 + 4}}$

56. $y = x^{x^2}$

57. $y = (\sin x)^{1/x}$

58. $y = \left(2x^2 + 1\right)^{\tan x}$

59. $y = (\cos x)^{\sqrt{x}}$

60. $y = \left(\sqrt[3]{x}\right)^{\sqrt[3]{x}}$

61. $y = (\ln x)^x$

62. $y = \dfrac{(\ln x)^x \left(x^3 - 1\right)}{e^x + 2}$

63. $y = x^{x^x}$

64. $y = (\sin x)^{\cos x}$

65. $y = (\ln x)^{\sin x}$

66. $y = \left(e^x\right)^x$

67. Mimic the procedure seen in the text to find a formula for the derivative of $y = \cos^{-1} x$.

68. Find a formula for the derivative of $y = \sec^{-1} x$ (see Exercise 67).

69. Find a formula for the derivative of $y = \cot^{-1} x$ (see Exercise 67).

70–93 Determine dy/dx. (Recall that arcsin x is just a different notation for $\sin^{-1} x$, and the same holds for the other inverse trigonometric functions.)

70. $y = \cos^{-1}\left(x^2\right)$

71. $y = \tan^{-1}(2x+1)$

72. $y = x\arcsin x$

73. $y = \ln(\arctan x)$

74. $y = (\operatorname{arccot} x)^2$

75. $y = \arccos\sqrt{x}$

76. $y = \tan^{-1} x + \dfrac{x}{1 + x^2}$

77. $y = \arccos x - x\sqrt{1 - x^2}$

78. $y = \dfrac{\operatorname{arccot} x}{x}$

79. $y = \arctan\left(e^x\right)$

80. $y = \operatorname{arccot}(\ln 3x)$

81. $y = \dfrac{1 - \arctan x}{1 + \arctan x}$

82. $y = \arccos x \cdot \operatorname{arccot} x$

83. $y = \left(\arcsin\left(x^3\right)\right)^2$

84. $y = \sec^{-1}\left(e^{x^2}\right)$

85. $y = \sec^{-1}\left(x^2 + 1\right)$

86. $y = \csc^{-1}\left(e^{-x}\right)$

87. $y = \sec^{-1}\sqrt{x^2 + 1}$

88. $y = \sin(\arccos 3x)$

89. $y = (\arctan x)^x$

90. $y = (\arcsin x)^{\ln x}$

91. $y = \cos\left(\operatorname{arccsc}\left(x^2 + 1\right)\right)$

92. $y = \tan\left(\operatorname{arcsec}\sqrt{1 + e^{2x}}\right)$

93. $y = \cos\left(\operatorname{arccot}\dfrac{x-1}{\sqrt{2x-1}}\right)$

94–99 Find the equation of the line tangent to the graph of $y = f(x)$ at the indicated x-value. (If needed, round your answer to three decimal places.)

94. $f(x) = \log_2\left(x^2 + 1\right); \quad x = 1$

95. $f(x) = \dfrac{(2+x)2^{\ln x}}{x^2 e^x}; \quad x = 1$

96. $f(x) = \arcsin(\ln 3x)$; $x = \dfrac{1}{3}$

97. $f(x) = x \arccos \dfrac{x}{4} - \ln \dfrac{1}{x^2 + 1}$; $x = 2$

98. $f(x) = (\sin x)^x$; $x = \dfrac{\pi}{2}$

99. $f(x) = x^{\ln(\arctan x)}$; $x = 1$

100. Differentiate $f(x) = \arcsin x + \arccos x$. What information about f can you glean from your answer?

101. Repeat Exercise 100 for the function
$f(x) = \arcsin(1/x) - \operatorname{arccsc} x$.

102. A father is videotaping his child releasing a helium-filled balloon. Assuming that the balloon rises vertically, let the distance between father and child be denoted by s and the height of the balloon, measured from the child, be denoted by h. Find a formula for the angle of elevation α of the camera as it is following the rise of the balloon. Then differentiate with respect to time to find $d\alpha/dt$.

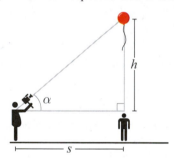

103. The height of the screen of a drive-in movie theater is 20 ft and it is mounted 8 ft above the eye level of a driver who is parked s feet from the screen. Find a formula for the angle θ at which the screen is viewed by this driver. Then differentiate to find the rate of change of the viewing angle as a function of the distance s.

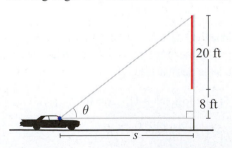

104. An air traffic controller observes a small plane flying horizontally toward the tower and determines from the instrument readings that the distance between the tower and the plane is 10,560 ft, the flying altitude is 5340 ft, and the speed of the plane is 120 mph.

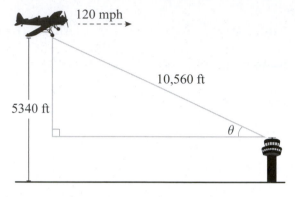

a. Find the angle of elevation θ at which the controller first sees the plane, if the tower is 60 ft high.

b.* Find the angular rate of change $d\theta/dt$ when the plane is 1.25 miles from the controller.

105. Give an alternative definition to $\cot^{-1} x$ so as to make the function continuous and satisfy the identity $\cot^{-1} x = (\pi/2) - \tan^{-1} x$. Graph the function. (**Hint:** Appropriately restrict the domain of $\cot x$. You might also think about the relationship between the graph of the function to be defined and that of $\tan^{-1} x$.)

106–109 *True or False?* Determine whether the given statement is true or false. In case of a false statement, explain or provide a counterexample.

106. The tangent lines to the graphs of $\ln x$ and $\ln 3x$ have the same slope for all x.

107. If $y = \log \pi$, then $y' = \dfrac{1}{\ln 10} \cdot \dfrac{1}{\pi}$.

108. The derivative of $\csc^{-1} x$ is negative everywhere.

109. The functions $f(x) = \ln x$ and $g(x) = \log_c x$ are constant multiples, hence so are their derivatives.

3.7 **Rates of Change in Use**

TOPICS

1. Biology applications
2. Physics applications
3. Chemistry applications
4. Business and economics applications

We have already seen how derivatives relate to the physical notions of velocity, acceleration, and jerk, and we now have many examples of the use of differentiation in mathematical applications such as the construction of tangent lines. In this section, we will introduce additional applications in which the rate of change of one quantity with respect to another plays a key role and show how differentiation is used to determine that rate of change.

TOPIC 1   **Biology Applications**

The ability to mathematically model the population growth (or decline) of a species over time is of great utility, whether the study is relatively small and contained (such as the growth of bacteria in a petri dish over the course of a few hours) or large (such as the number of grey wolves in the western United States over the course of several years). In general, such models can be quite complex and include such factors as food availability, the effects of disease, space constraints, and interactions with other species. But at their most basic, many population models start with an assumption of exponential growth, at least over short periods of time.

The reason for this is that the growth of a population usually depends to a large extent on the number of members capable of producing more members. For instance, in the simple case of bacteria in a petri dish, the more bacteria there are the faster the population will grow, as there are more bacteria to reproduce themselves. If we define $P(t)$ to be the population of a given species at time t, the mathematical statement of the above observation is this: $P'(t) = kP(t)$, where k is a fixed positive constant. That is, the rate of change in the population at time t is proportional to the population at time t.

In future chapters, we will see how to begin with an equation like $P'(t) = kP(t)$ and solve it for $P(t)$. But we can draw upon one of the results of this chapter to make an educated guess now as to the form of $P(t)$. Recall that the number e is the unique exponential base for which $\left(e^x\right)' = e^x$; that is, the exponential function with base e and its derivative are equal. A quick application of the Chain Rule then tells us that for any constant k, $\left(e^{kx}\right)' = ke^{kx}$, an equation very much like $P'(t) = kP(t)$. More generally, note that if we let $P(t) = ce^{kt}$, where c and k are both constants, then $P'(t) = cke^{kt} = kP(t)$. If we specify that the population at time $t = 0$ is P_0, then $P_0 = P(0) = ce^{k(0)} = c$, giving us the following basic population model.

Definition 💡

Exponential Growth Model

$$P(t) = P_0 e^{kt}$$

Example 1 ✐

As an example of the use of the exponential growth model, suppose a strain of bacteria being cultured in a petri dish is observed to double in count every hour. That information alone is sufficient to determine the *growth constant k*, under the assumption that t is measured in hours, as follows.

$$P(1) = 2P(0)$$ Population at 1 hour is twice the initial population.

$$P_0 e^{k(1)} = 2P_0$$ Use $P(t) = P_0 e^{kt}$.

$$e^k = 2$$ Cancel P_0.

$$k = \ln 2$$ Solve for k.

If the culture began with, say, $P_0 = 15$ bacteria, the population of bacteria (rounded to the nearest integer) and the rate of population growth (rounded to one decimal place) at the 0, $\frac{1}{2}$, 1, and 2 hour marks are as follows.

Population $P(t) = 15e^{(\ln 2)t}$	Rate of population growth $P'(t) = (\ln 2)P(t)$
$P(0) = 15e^0 = 15$	$P'(0) = (\ln 2)P(0) \approx 10.4$ bacteria/hour
$P\left(\frac{1}{2}\right) = 15e^{(1/2)\ln 2} = 15e^{\ln(2^{1/2})} = 15\left(2^{1/2}\right) \approx 21$	$P'\left(\frac{1}{2}\right) = (\ln 2)P\left(\frac{1}{2}\right) \approx 14.7$ bacteria/hour
$P(1) = 15e^{\ln 2} = 15(2) = 30$	$P'(1) = (\ln 2)P(1) \approx 20.8$ bacteria/hour
$P(2) = 15e^{2\ln 2} = 15e^{\ln(2^2)} = 15\left(2^2\right) = 60$	$P'(2) = (\ln 2)P(2) \approx 41.6$ bacteria/hour

---

TOPIC 2   Physics Applications

Newton's Second Law of Motion often appears as $F = ma$, where F represents force, m mass, and a acceleration. This is a simplified version of Newton's actual observation, however, which is that the net force on an object is equal to the rate of change (with respect to time) of its momentum: $F = dP/dt$, where momentum P is the product of the object's mass m and velocity v. The Product Rule allows us to rewrite the second law as follows.

Theorem ⚲

Newton's Second Law of Motion

$$F = \frac{d}{dt}(mv) = \frac{dm}{dt}v + m\frac{dv}{dt}$$

The fact that mass is constant in many elementary applications accounts for the frequent reduction to the familiar $F = m(dv/dt)$, or $F = ma$.

Figure 1

Example 2 ✎

A freight train is slowly moving forward at a constant velocity v under a hopper that is dropping grain at a constant rate into the train's open-topped freight cars. In order to maintain its constant velocity, the train's engine must exert a force (beyond that necessary to counter friction) equal to the rate of change of its momentum. What is that force?

Solution

In this setting, v is constant and hence the dv/dt (or acceleration) factor is 0. But the mass is changing, and so Newton's Second Law tells us that the force necessary to overcome the increasing load is

$$F = \frac{dm}{dt}v + m \cdot 0 = \frac{dm}{dt}v.$$

Using units of kilograms (kg) for mass, meters (m) for distance, seconds (s) for time, and newtons (N) for force, if the train has a constant forward velocity of $\frac{1}{2}$ m/s and is being loaded at a rate of 300 kg/s, the force the engine must exert is

$$F = \frac{dm}{dt}v = \left(300 \text{ kg/s}\right)\left(\frac{1}{2} \text{ m/s}\right) = 150 \left(\text{kg} \cdot \text{m}\right)/\text{s}^2 = 150 \text{ N}.$$

Newton's laws of motion are remarkable for their brevity and versatility—their principles accurately describe the behavior of seemingly unrelated physical situations. As another example of the use of his second law, consider the following experiment.

Example 3 ✎

A rope of length L and mass M is held vertically over a scale so that its lower end just touches the scale, and is then allowed to drop onto the scale. What force does the scale register as the rope drops onto it?

Solution

The answer to this question *after* the rope has fully dropped onto the scale and is at rest is easily found with an elementary application of Newton's Second Law: the force registering at that point is the mass of the rope M times the acceleration due to gravity g. The product Mg is what we normally call *weight*, and Mg is indeed what would show on the scale's dial. But the more interesting question is what registers on the scale during intermediate stages when, say, a segment of length x has landed on the scale and the segment of length $L - x$ has yet to hit.

For convenience, let m denote the mass of the segment of length x that has dropped and is lying at rest on the scale. At any moment in time we know that $m = M(x/L)$ (the total mass times the fraction of the rope on the scale), but it is important to keep in mind that m is continuously changing over time from the moment the rope is dropped to the time when $x = L$. The

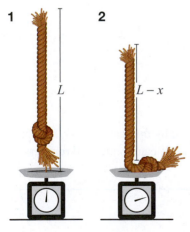

1   2

L $L - x$

Figure 2

number registering on the scale at any given moment reflects both the force necessary to counter the weight of the mass m and the force necessary to stop the segment of the rope just hitting the scale with velocity v. We will find the net force by determining the rate of change of momentum P (recall that $P = mv$).

Newton's Second Law tells us that

$$F = \frac{dm}{dt}v + m\frac{dv}{dt},$$

and the determination of dm/dt affords a good opportunity to apply the Chain Rule. We already know how m depends on x, but it's also true that m is a function of time. Thus,

$$\frac{dm}{dt} = \frac{dm}{dx}\cdot\frac{dx}{dt} = \left(\frac{M}{L}\right)v,$$

and so

$$F = \frac{dm}{dt}v + m\frac{dv}{dt} = \left(\frac{M}{L}\right)v\cdot v + m\frac{dv}{dt} = \left(\frac{M}{L}\right)v^2 + mg,$$

where we have also replaced the rate of change of velocity dv/dt with the acceleration due to gravity g.

At the moment in time when a segment of length x is lying on the scale, the portion of the rope just hitting the scale has fallen from a height of x. In Exercises 17 and 18 you will study the velocity v of an object with initial velocity 0 that has fallen a distance x, and you will show that $v^2 = 2xg$. This gives us

$$F = \left(\frac{M}{L}\right)v^2 + mg$$

$$= \left(\frac{M}{L}\right)(2xg) + mg$$

$$= \frac{2Mx}{L}g + \frac{Mx}{L}g \qquad\qquad m = \frac{Mx}{L}$$

$$= \frac{3Mx}{L}g.$$

Note that this tells us that the force registered by the scale just as the top of the rope hits ($x = L$) is $3Mg$, and then the scale equilibrates and registers Mg soon after.

TOPIC 3   Chemistry Applications

Chemical reactions produce one or more *product* substances from one or more *reactant* substances. As an illustration, a process in which a molecule of each of two reactants A and B produces a molecule of substance C is represented by

$$A + B \to C,$$

and chemists often need to understand the rates of change of the concentrations of the individual substances in such a reaction. The notation $[X]$ is used to denote the concentration of a substance X (typically measured in *moles per liter*, where 1 mole $= 6.022 \times 10^{23}$ molecules), so the rates of change of the concentrations of the three substances in the reaction $A + B \to C$ correspond to the derivatives

$$\frac{d[A]}{dt}, \quad \frac{d[B]}{dt}, \quad \text{and} \quad \frac{d[C]}{dt}.$$

The **reaction rate** of a chemical process is defined as the rate of change of concentration of one of the products or, equivalently, the negative of the rate of change of concentration of one of the reactants. The negative sign is attached to the derivatives of the concentrations of the reactants because those substances are decreasing with respect to time, while the rate of change of concentration of a product is positive. In the reaction $A + B \to C$, the reaction rate r is thus defined to be

$$r = -\frac{d[A]}{dt} = -\frac{d[B]}{dt} = \frac{d[C]}{dt}.$$

More generally, the stoichiometric relation between substances in a reaction can be used to determine the reaction's rate. If a molecules of substance A and b molecules of substance B react to produce c molecules of the product C, denoted symbolically as

$$aA + bB \to cC,$$

then the reaction rate r is

$$r = -\frac{1}{a}\frac{d[A]}{dt} = -\frac{1}{b}\frac{d[B]}{dt} = \frac{1}{c}\frac{d[C]}{dt}.$$

Example 4 ✐

Nitrogen gas N_2 and hydrogen gas H_2 react to produce ammonia NH_3 according to the process

$$N_2 + 3H_2 \to 2NH_3,$$

so the reaction rate r for the process and the rates of change of concentrations of the substances are related by

$$r = -\frac{d[N_2]}{dt} = -\frac{1}{3}\frac{d[H_2]}{dt} = \frac{1}{2}\frac{d[NH_3]}{dt}.$$

If we measure the production of ammonia as $1.2 \text{ mol}/(\text{L}\cdot\text{h})$ (i.e., 1.2 moles of ammonia per liter per hour), then the rate of change of concentration of hydrogen gas is

$$\frac{d[\text{H}_2]}{dt} = -\frac{3}{2}\frac{d[\text{NH}_3]}{dt} = -\frac{3}{2}\cdot 1.2 \text{ mol}/(\text{L}\cdot\text{h}) = -1.8 \text{ mol}/(\text{L}\cdot\text{h})$$

and the rate of change of concentration of nitrogen gas is

$$\frac{d[\text{N}_2]}{dt} = -\frac{1}{2}\frac{d[\text{NH}_3]}{dt} = -\frac{1}{2}\cdot 1.2 \text{ mol}/(\text{L}\cdot\text{h}) = -0.6 \text{ mol}/(\text{L}\cdot\text{h}).$$

TOPIC 4   Business and Economics Applications

Business models often begin with basic assumptions about the cost $C(x)$ of producing x units of a particular product, the revenue $R(x)$ that can be generated by selling x units, and the resultant profit function $P(x) = R(x) - C(x)$. Once these functions are determined, economists and business strategists are then interested in the *marginal cost*, the *marginal revenue*, and the *marginal profit*. These three functions approximate, respectively, the cost, revenue, and profit associated with increasing production from x units to $x + 1$ units.

Although the variable x in this setting can only realistically take on positive integer values, it is convenient to assume the functions C, R, and P have been extended to smoothly changing (in fact, differentiable) functions defined for all positive real x, at least over some interval. Such an assumption is very common in applications, as it then allows us to use the tools of calculus. For example, the cost of increasing production from x units to $x + 1$ units is $C(x+1) - C(x)$, which is also the average rate of change in the cost over the interval $[x, x+1]$.

$$C(x+1) - C(x) = \frac{C(x+1) - C(x)}{(x+1) - x} = \frac{\Delta C}{\Delta x}$$

Under the assumption mentioned above, $\Delta C/\Delta x \to C'(x)$ as $\Delta x \to 0$, so the derivative of the cost function at x is a convenient approximation of the change in cost as production increases from x to $x + 1$ units. This sort of approximation is an example of *linearization*, which, as we will see in Section 3.9, has many important and varied uses. For the moment, we use this observation as the basis for three formal definitions.

Definition 💡

Marginal Cost, Marginal Revenue, and Marginal Profit

$$\text{Marginal Cost} = C'(x)$$
$$\text{Marginal Revenue} = R'(x)$$
$$\text{Marginal Profit} = P'(x) = R'(x) - C'(x)$$

In practice, it is likely that the variable x, representing the number of units produced, varies over time; changes in production rate can be a result of seasonal supply and demand, labor situations, and the availability of raw materials. Calculus allows us to easily incorporate such knowledge into our model and determine, for instance, the rate of change of profit with respect to time:

$$\frac{d}{dt}P\big(x(t)\big) = P'\big(x(t)\big) \cdot x'(t) = \big[R'\big(x(t)\big) - C'\big(x(t)\big)\big] \cdot x'(t)$$

Economists often build models on the basis of such quantities as *the number of workers in the labor force* and *the average productivity per worker*. From their understanding of these numbers, they can construct and analyze further quantities, such as gross productivity.

Example 5 ✎

Let $w(t)$ represent the number of workers in a given industry at time t, and let $p(t)$ represent each worker's average productivity. Then the gross productivity at time t is given by $g(t) = w(t)p(t)$. Suppose that it is known that at a specific time t the labor force is decreasing at a rate, *relative to the size of the labor force*, of 1 percent. This translates into the statement $w'(t) = -0.01w(t)$ (such "relative" data is often more commonly cited than "absolute" data). Suppose that at the same time it is also known that the average *relative* productivity per worker is increasing at a rate of 3 percent (note again the "relative" modifier). This second fact tells us that $p'(t) = 0.03p(t)$, and we can use these two pieces of information to determine the relative rate of change of gross productivity as follows.

$$\begin{aligned} g'(t) &= w'(t)p(t) + w(t)p'(t) &&\text{Product Rule}\\ &= (-0.01w)(p) + (w)(0.03p) &&\text{The } t\text{'s will be omitted from this point on.}\\ &= -0.01wp + 0.03wp\\ &= 0.02wp = 0.02g \end{aligned}$$

So the relative rate of change of gross productivity at time t is 2 percent.

3.7 Exercises

1. Work through Example 1 with the following version of the growth model: $P(t) = P_0 a^t$, where a is treated as the (initially unknown) "growth constant." (**Hint:** Since P doubles every hour, $P(1) = 2P(0)$ gives $P_0 \cdot a^1 = 2P_0$.)

2. In an effort to control vegetation overgrowth, 100 rabbits are released in an isolated area that is free of predators. After one year, it is estimated that the rabbit population has increased to 500. Assuming exponential population growth, what will the population be after another 6 months?

3. The population of a certain inner-city area is estimated to be declining according to the model $P(t) = 237,000e^{-0.018t}$, where t is the number of years from the present. What does this model predict the population will be in 10 years?

4. A population of squirrels is growing in a Louisiana forest with a monthly growth constant of 6 percent. If the initial count is 100 squirrels, how many are there in a year? (**Hint:** Let $N(t)$ stand for the number of squirrels after t months, and note that $N'(t) = 0.06N(t)$. Mimic the steps of Example 1 or, alternatively, make use of the fact that $\frac{d}{dt}(a^t) = (\ln a)a^t$.)

5. The process of radioactive decay is akin to population growth in the sense that the rate of decay is proportional to the amount of material present at any given time. Therefore, it shouldn't come as a surprise that this process can be modeled with the same type of function. Suppose that $A(t)$ stands for the amount of a certain radioactive material at time t, and that it is decaying in a way that the rate of decay satisfies $\frac{d}{dt}A(t) = -0.1A(t)$ (note the negative sign), where t is measured in days. If we start with 1000 g of material, how much is left after 10 days?

6. In Exercise 67 of Section 1.2, we defined the *half-life* of a radioactive substance to be the amount of time required for half of the substance to decay. Find the half-life of the material in Exercise 5.

7. Carbon-11 has a radioactive half-life of approximately 20 minutes, and is useful as a diagnostic tool in certain medical applications. Because of the relatively short half-life, time is a crucial factor when conducting experiments with this element.

   a. Determine a so that $A(t) = A_0 a^t$ describes the amount of carbon-11 left after t minutes (as usual, A_0 is the amount at time $t = 0$).

   b. How much of a 2 kg sample of carbon-11 would be left after 30 minutes?

   c. How much of a 2 kg sample of carbon-11 would be left after 6 hours?

8. The half-life of radium-226 is approximately 4 days. Determine what percentage of the initial amount is left after two weeks.

9. According to Newton's Law of Cooling, the rate of change of temperature of a cooling object is proportional to the temperature difference between the object and the surrounding medium, that is,

$$\frac{dT(t)}{dt} = k[T(t) - T_s],$$

where $T(t)$ is the temperature of the object at time t, and T_s is the temperature of its surroundings. Suppose that a cup of 180 °F coffee is left in a 72 °F room and cools to 122 °F in five minutes. How long does it take for the coffee to cool down to 85 °F? (**Hint:** Introduce a new variable for the temperature difference: Let $D(t) = T(t) - 72$, and observe that Newton's law translates into the equation $D'(t) = k \cdot D(t)$. Now mimic the procedure seen in Example 1.)

10.* According to the Stefan-Boltzmann Law, the radiation energy emitted by a hot object of temperature T is $R(T) = kT^4$, where T is measured in kelvins. Use this to find a formula for the rate of change of energy emitted by the coffee of Exercise 9. (**Hint:** Use the following formula to convert degrees Fahrenheit to kelvins: $K = \frac{5}{9}(F - 32) + 273$.)

11. A snowplow is moving at a constant speed of $3 \, \text{m/s}$, and the snow it is pushing is accumulating at a rate of $100 \, \text{kg/s}$. What extra force is necessary for the engine of snowplow to maintain constant speed despite the increasing mass?

12. When washing his car, Brad is aiming a water hose at the side of the car, with water leaving the hose at a rate of 1 liter per second, with a speed of $15 \, \text{m/s}$. If we ignore any "splash backs," what force does the water exert on the side of the car? (**Hint:** Use the equation $F = dP/dt$. See Example 2.)

13. Use Example 3 to find a formula for the force (in newtons) exerted on a scale by a rope dropped on it if the rope is 2 meters long and each centimeter of it weighs 20 grams. (As in the text, let x stand for the length of the segment of the rope that has already landed on the scale.)

14. Referring back to Exercise 12, suppose that after washing the car, with almost half of the contents of his 20-liter bucket still left, Brad pours it with a quick move into a much smaller 5-liter container that is sitting on the ground. Find the force exerted on the bottom of the smaller container by the incoming water at the instant when it starts overflowing. (**Hint:** Modify and use the result of Example 3.)

15. Suppose that the snowplow of Exercise 11 is 3500 kg and starts accelerating from $3 \, \text{m/s}^2$ at a rate of $0.1 \, \text{m/s}^2$. Assuming there is no snow accumulation this time, find the force exerted by the engine.

16.* Consider the accelerating snowplow of Exercise 15, this time assuming the same accumulation rate for the snow as in Exercise 11. Find the force exerted by the engine at $t = 2$ seconds.

17. Show that the velocity v of an object that has fallen a distance x from rest satisfies the equation $v^2 = 2xg$. (**Hint:** Velocity increases at a constant rate from 0 to v, so the distance x can be calculated as the product of the average velocity and time: $x = v_{ave} \cdot t = \left[(0 + gt)/2 \right] t = \frac{1}{2} gt^2$.)

18. Derive the result of Exercise 17 in an alternative way, using the fact that if air resistance is ignored, the potential energy of an object with mass m at altitude x, which is calculated as $E_p = mgx$, is turned into kinetic energy upon impact, which is calculated as $E_{kin} = \frac{1}{2} mv^2$.

19. In an attempt to escape from a predator, a small fish is swimming vertically downward at a rate of 75 cm/s. Find the rate of change of water pressure around the fish. Express your answer in atm/s. (**Hint:** Use the fact that underwater pressure at a depth of d meters is approximately $P(d) = 1 + 0.097d$ standard atmospheres (atm), where 1 atm = 101.325 kPa.)

20. Hydrogen may be obtained from water by a process called electrolysis, according to the process $2H_2O \rightarrow 2H_2 + O_2$. If we measure the production of hydrogen at $2.5 \, \text{mol/}(L \cdot h)$ (i.e., 2.5 moles of hydrogen per liter per hour), what will be the concentration of the newly obtained oxygen in 3 hours?

21. The combustion of ammonia gas (NH_3) produces nitrogen and water according to the process $4NH_3 + 3O_2 \rightarrow 2N_2 + 6H_2O$. Supposing that the rate of combustion is $0.5 \, \text{mol/}(L \cdot s)$, what is the rate of the production of water? How many milliliters of water are produced in two seconds? (**Hint:** Use the fact that the approximate molar mass of hydrogen is 1 g, while that of oxygen is 16 g. Also, the mass of 1 mL of water is 1 g.)

22. Magnesium is a flammable metal, and because of its bright light it has traditionally been used in camera flashes, illumination of mine shafts, fireworks, and flares. The reaction itself is described by $2Mg + O_2 \rightarrow 2MgO$. If magnesium burns in a chamber at an initial rate of 1.5 g/s, find the rate (in $\text{mol/}(L \cdot s)$) at which the concentration of O_2 is decreasing in the chamber. (**Hint:** The approximate molar mass of magnesium is 24 g.)

23–29 Use the technique of linearization to determine the answer.

23. A manufacturer of small remote-controlled cars found its weekly revenue to be $R(x) = 160x - 0.3x^2$ dollars when x units are produced and sold.

   a. Use marginal revenue to estimate the extra revenue when production is increased from 12 to 13 units.

   b. Use the revenue function to calculate the actual revenue increase. Compare your answers.

24. Suppose the monthly cost of producing x units of a particular commodity is $C(x) = 75x^2 + 200x + 5100$ dollars, while the revenue function is $R(x) = 32x(120 - x)$.

   a. Use marginal cost to find the added expense in increasing production from 5 to 6 units.

   b. Use marginal revenue to estimate the revenue generated by raising production from 5 to 6 units.

   c. Find the actual increases in cost and revenue by producing and selling the sixth unit, and compare these numbers to your estimations from parts a. and b.

25. Repeat Exercise 24 for $C(x) = \frac{486}{7}x^2 + 98x + 120$ and $R(x) = -19.5x^2 + 2526x + 442$.

26. A manufacturer models the total cost of producing n hundreds of a particular pocket calculator by the function $C(n) = \frac{1}{900}n^2 + 2500n + 3100$ dollars. Market research shows that all products will be sold at the price of $p(n) = \frac{2}{5}(120 - \frac{1}{4}n)$ dollars per calculator.

   a. Use the marginal cost function to estimate the cost of raising the level of production from 1000 to 1100 calculators.

   b. Use the marginal profit function to estimate the additional profit if the level of production is raised from 1000 to 1100 calculators.

   c. Find the actual increases in cost and profit when production is raised as in parts a. and b., and compare these values with your estimates obtained in the previous parts.

27. A lumber company estimates the cost of producing x units of a product to be $C(x) = 0.1x^2 + 2250x + 1450$ dollars, while the price of each unit has to be $p(x) = 50(128 - 0.2x)$ in order to sell all x units. However, seasonal supply of raw materials makes x dependent on time so that $x(t) = -0.35(t - 6)^2 + 192$, where t is measured in months. Use the value of the marginal profit at $t = 2$ to estimate the change in profit during the third month.

28. Repeat Exercise 27 if $x(t) = 185 + 10\sin^2\left(\frac{\pi}{6}t\right)$.

29. A child playing on the beach is pouring sand from a bucket, forming a sand cone that is growing in such a way that its height is always half of the radius of its circular base. Estimate the change in volume of the sand cone as its height grows from 3 to 4 inches.

30–39 Slightly generalize the technique of linearization to find the answer. For cases where Δx is not 1, estimate the change in a function $f(x)$ by $\Delta f = f'(x)\Delta x$.

30. Suppose the cost of manufacturing x units of a certain commodity was found to be $C(x) = 35x^2 + 20x + 780$ dollars, and that the current production level is 50 units. Use linearization to estimate how the cost changes if production is raised to 50.25 units.

31. Answer the question of Exercise 30 when production is increased from 13 to 13.5 units.

32. Answer the questions of Exercise 24 when production is increased from 6 to 6.75 units.

33. Use marginal analysis to estimate the changes in cost and profit of Exercise 26 when production is decreased from 1000 to 950 calculators. Then find the actual changes and compare them with your estimates.

34. An ice cube with a side length of 1.5 inches starts melting in such a way that its sides are decreasing by 0.1 inches per minute. Use linearization to estimate the change in the cube's volume during the second minute.

35. According to MRI scans, a benign tumor in a patient had a radius of 1.6 cm when it was first discovered, and it is growing by 1 mm each month. Assuming the tumor is spherical, estimate the change in its volume during the first week.

36. The daily output of a small factory is $n(I) = 50\sqrt{I}$ units, where I is the owner's investment measured in dollars. If the current investment is \$100,000, use linearization to estimate how much additional capital is needed to increase the daily output by 5%.

37. Suppose that $w(t)$, the number of workers at a certain factory at time t, has been decreasing at a rate of 2.5 percent, relative to the size of the workforce, due to a recent recession. At the same time, however, the workers' average productivity $p(t)$ has been increasing by 4 percent due to extra training and the inherent fear of a job loss. Find the change in gross productivity. (**Hint:** See Example 5.)

38. A factory outputs $n(I) = 200\sqrt{I(t)}\sqrt[3]{g(t)}$ units daily, where I is the total investment measured in dollars and g stands for gross productivity. If the total investment is decreasing by 1 percent while the gross productivity is increasing by 2.1 percent (both in the relative sense), find the relative change in the daily output of the factory.

39. Repeat Exercise 38 if $I(t)$ is decreasing by 2 percent, the number of workers is decreasing by 3 percent, but worker productivity is increasing by 5.1 percent. (**Hint:** Recall the equation from Example 5: $g(t) = w(t)p(t)$.)

3.8 **Related Rates**

TOPICS

1. Constructing and solving related rates equations

2. Interlude: a cautionary tale

In Section 3.7, we saw how the concept of rate of change pervades a wide variety of contexts and disciplines. This section continues that exploration, but with a focus on applications in which two or more rates are present. As we will see, the ability to quantify how the different rates of change relate to one another is the key to solving many problems.

TOPIC 1 **Constructing and Solving Related Rates Equations**

Commonly, a *related rates* problem centers on two or more physical features that can be measured (such as distance, area, volume, pressure, temperature, etc.) and that are changing over time. In such cases, the rates of change of interest are derivatives with respect to time, and we will demonstrate our solution strategy with such a problem.

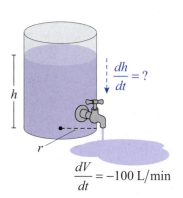

$$\frac{dV}{dt} = -100 \text{ L/min}$$

Figure 1

Example 1 ✎

A cylindrical water tank is being emptied at a rate of

$$100 \text{ liters/minute (L/min)}.$$

How quickly does the height h of the water level fall if the radius of the tank is 1 meter? How does the answer change if the radius is 10 meters? (Note that 1 liter = 0.001 cubic meters.)

Solution

It is important to identify what it is exactly that we are seeking to determine and what it is that we know. As is typical, the problem as it is given does not contain such mathematical language as "derivative" or even "rate of change," but a careful reading of the problem reveals that we are seeking dh/dt and that we are given information about dV/dt, the rate of change of the volume V of water in the tank.

The next step is to relate the relevant quantities in the problem. In addition to height h and volume V, the radius r of the tank plays a role. These three physical quantities are related by the volume formula $V = \pi r^2 h$. Note that we are *not* trying to relate rates of change at this point—that will come in the following step.

In this problem, two of the three quantities, V and h, change with respect to time t. But we also know how V depends on h. Using the Chain Rule, we now relate dV/dt to dh/dt as follows:

$$\frac{dV}{dt} = \frac{dV}{dh} \cdot \frac{dh}{dt} = \left(\pi r^2\right)\frac{dh}{dt},$$

so

$$\frac{dh}{dt} = \left(\frac{1}{\pi r^2}\right)\frac{dV}{dt}.$$

We are given that $dV/dt = -100\,\text{L/min}$ and $1\,\text{L} = 0.001\,\text{m}^3$, so $dV/dt = -0.1\,\text{m}^3/\text{min}$ (note the negative sign, signifying a loss of volume with respect to time). So if the radius r of the tank is 1 m, then

$$\frac{dh}{dt} = \left(\frac{1}{\pi r^2}\right)\frac{dV}{dt} = \left(\frac{1}{\pi \cdot 1\,\text{m}^2}\right)\left(-0.1\,\text{m}^3/\text{min}\right) \approx -0.03\,\text{m/min},$$

whereas if the radius is 10 m, then

$$\frac{dh}{dt} = \left(\frac{1}{\pi r^2}\right)\frac{dV}{dt} = \left(\frac{1}{\pi \cdot 100\,\text{m}^2}\right)\left(-0.1\,\text{m}^3/\text{min}\right) \approx -0.0003\,\text{m/min}.$$

The first rate of drop in water level is 3 centimeters per minute, which would be perceptible, while the second rate of change would be a very slow 0.3 millimeters per minute.

The steps in solving the problem in Example 1 can be summarized as follows.

Strategy for Solving Related Rates Problems

Step 1: Read the given information carefully and identify what is known and what needs to be found. Typically, the overall goal is to determine the rate of change of one particular variable with respect to time.

Step 2: Identify and appropriately label the relevant quantities and draw a diagram, if possible, as an aid in determining the relationship between them. Express the known and desired quantities in terms of your chosen labels.

Step 3: Write an equation expressing the relationship between the relevant quantities. Be sure you know which of them are functions of time t.

Step 4: Using the Chain Rule, differentiate both sides of the equation with respect to t.

Step 5: Using the given information, solve for the desired rate of change. Check your answer to see if it makes sense as a solution to the problem. Such characteristics as sign, relative magnitude, and the units of your answer are helpful cues.

The remainder of this section will illustrate how this strategy applies to different related rates problems. In each case, note the use of the steps outlined above.

Example 2 ✐

A hot air balloon rising straight up is being tracked by a film crew stationed 200 meters away on level ground from its takeoff point. At the moment when the balloon is 150 meters up it is rising at a rate of 50 meters per minute. How fast is the camera angle (measured with respect to the ground) increasing at that moment?

Solution

We know how fast the balloon is rising at a particular moment in time, and we are being asked to determine the rate of change of the camera angle at that moment. Since the balloon is rising straight up from level ground, a right triangle such as the one in Figure 2 captures the essential elements of the problem. The camera crew is stationed at a fixed distance of 200 meters from the takeoff point, but the balloon's height above the ground is changing. In the diagram, y represents the balloon's height, and θ is the camera angle measured from the ground as it tracks the balloon's ascent. We need to determine $d\theta/dt$ at the moment when $y = 150$ m and $dy/dt = 50$ m/min.

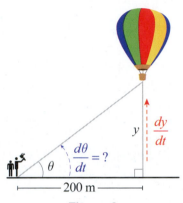

Figure 2

The equation $\tan\theta = y/200$ defines the relationship between y and θ. Now we differentiate both sides with respect to t and solve for $d\theta/dt$.

$$\tan\theta = \frac{y}{200}$$

$$\frac{d}{dt}(\tan\theta) = \frac{d}{dt}\left(\frac{y}{200}\right)$$

$$\sec^2\theta\,\frac{d\theta}{dt} = \frac{1}{200}\frac{dy}{dt}$$

$$\frac{d\theta}{dt} = \frac{\cos^2\theta}{200}\frac{dy}{dt}$$

We know that $dy/dt = 50$ m/min, but we need to perform one more computation in order to determine $\cos^2\theta$ at the moment when $y = 150$ m. From the Pythagorean Theorem, the hypotenuse at that moment is $\sqrt{(200)^2 + (150)^2} = 250$ m, and so $\cos^2\theta = \left(\frac{200}{250}\right)^2 = \frac{16}{25}$. Hence, we calculate the answer as follows:

$$\frac{d\theta}{dt} = \frac{1}{200}\cos^2\theta\,\frac{dy}{dt}$$

$$= \left(\frac{1}{200\text{ m}}\right)\left(\frac{16}{25}\right)(50\text{ m/min})$$

$$= 0.16\text{ rad/min}$$

Note that this rate of change of the camera angle is in radians per minute because we implicitly assumed radian measure in applying our differentiation rules; it can also be expressed as approximately 9.2 degrees per minute.

Example 3 ✐

Susan is in her car, leading a group of other drivers to a restaurant. She has just completed a right turn at an intersection and is accelerating away. At the moment that she is 40 ft from the intersection and heading north at 25 ft/s, the second car in the convoy is still 40 ft away from the intersection and is approaching it at 15 ft/s. How fast is the "as the crow flies" distance between Susan and the second car increasing at that particular moment?

Solution

We know how quickly Susan is pulling away from the intersection, how quickly the second car is approaching it, and their respective distances from the intersection at a given moment in time. We want to determine the rate of change of the actual distance between the two cars at that moment.

A right triangle is again the appropriate diagram, though this time both legs (as well as the hypotenuse) of the triangle are changing with respect to time. With the labels as shown, we know that $dy/dt = 25$ ft/s when $y = 40$ ft and that $dx/dt = -15$ ft/s when $x = 40$ ft (note that dx/dt is negative since x is decreasing). We want to determine dh/dt at that particular moment.

Differentiating both sides of the relationship $x^2 + y^2 = h^2$ with respect to t yields the following:

$$2x\frac{dx}{dt} + 2y\frac{dy}{dt} = 2h\frac{dh}{dt}$$

$$\frac{x}{h}\frac{dx}{dt} + \frac{y}{h}\frac{dy}{dt} = \frac{dh}{dt}$$

At the moment when $y = x = 40$, $h = 40\sqrt{2}$. Substituting these numbers and the two rates that we are given, we calculate the answer.

$$\frac{dh}{dt} = \frac{40}{40\sqrt{2}}(-15 \text{ ft/s}) + \frac{40}{40\sqrt{2}}(25 \text{ ft/s})$$

$$= \frac{10}{\sqrt{2}} \text{ ft/s}$$

$$\approx 7.07 \text{ ft/s}$$

The sign and magnitude of this answer are reasonable. The magnitude is comparable to the other two rates of change, and since Susan is pulling away from the intersection (thus increasing the relative distance) at a faster rate than the second car is approaching the intersection (and shortening the distance), we would expect the rate of change to be positive.

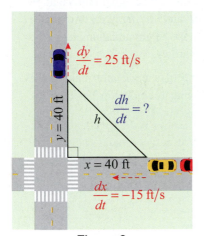

Figure 3

Example 4 ✍

Corn is falling from the end of a conveyor belt at a rate of $10 \text{ m}^3/\text{min}$ and is forming a conical pile below. As the corn falls, the height of the pile is remaining equal to the diameter of the base. How fast are the height and radius of the pile changing at the moment when the pile is 4 m high?

Solution

If we let V denote the volume of corn in the conical pile, h the height of the pile, and r the radius of the base, then we have been given information about dV/dt and want to determine dh/dt and dr/dt. We know that h and r are related by the equation $h = 2r$ since the height and diameter of the base remain equal to one another as the pile grows.

Beginning with the formula for the volume of a cone and making the substitution $r = h/2$, we obtain the following relationship between the three variables.

$$V = \frac{1}{3}\pi r^2 h = \frac{1}{3}\pi \left(\frac{h}{2}\right)^2 h = \frac{1}{12}\pi h^3$$

Differentiating with respect to t, we have

$$\frac{dV}{dt} = \frac{1}{4}\pi h^2 \frac{dh}{dt},$$

and substituting our given values for the rate of change of volume and the height $h = 4$ m, we find that

$$10 \text{ m}^3/\text{min} = \frac{1}{4}\pi \left(16 \text{ m}^2\right)\frac{dh}{dt} = \left(4\pi \text{ m}^2\right)\frac{dh}{dt},$$

so

$$\frac{dh}{dt} = \frac{5}{2\pi} \text{ m/min} \approx 0.80 \text{ m/min}.$$

Note that the equation $r = h/2$ tells us that the rate of change of the radius is half the rate of change of the height, so $dr/dt \approx 0.40 \text{ m/min}$.

As you might expect, not every related rates problem can be solved with a straightforward and literal application of our solution strategy. But the principles in the strategy are still good guidelines and can be adapted as necessary. Our final example illustrates this point.

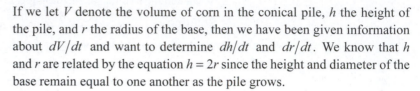

h

r

Figure 4

Figure 5

Example 5 ✒

A spherical balloon is being filled with helium at a rate of $200\ \text{cm}^3/\text{s}$. At the moment when the radius is 20 cm, how fast is the surface area of the balloon increasing?

Solution

We know the rate of change of the volume, and want to determine the rate of change of the surface area at the moment when the radius is 20 cm. If we let S denote surface area and V volume, it appears that we want to relate dS/dt to dV/dt. This is indeed the case, but we will do so indirectly.

We can begin with the relationships that we know:

$$V = \frac{4}{3}\pi r^3 \quad \text{and} \quad S = 4\pi r^2$$

V, S, and r are all functions of time and we can relate the following rates easily enough:

$$\frac{dV}{dt} = \frac{4}{3}\left(3\pi r^2 \frac{dr}{dt}\right) = 4\pi r^2 \frac{dr}{dt} \quad \text{and} \quad \frac{dS}{dt} = 8\pi r \frac{dr}{dt}$$

Since we know dS/dt in terms of dr/dt and can express dr/dt in terms of dV/dt, our path forward is beginning to appear. Note that $dV/dt = 200\ \text{cm}^3/\text{s}$ and that $r = 20$ cm at the moment in time we are interested in. So at that moment,

$$\frac{dr}{dt} = \frac{1}{4\pi r^2}\frac{dV}{dt} = \frac{1}{4\pi\left(400\ \text{cm}^2\right)}\left(200\ \text{cm}^3/\text{s}\right) = \frac{1}{8\pi}\ \text{cm/s}.$$

Now we can calculate the rate of change of the surface area.

$$\frac{dS}{dt} = 8\pi r \frac{dr}{dt} = 8\pi\left(20\ \text{cm}\right)\left(\frac{1}{8\pi}\ \text{cm/s}\right) = 20\ \text{cm}^2/\text{s}$$

Note that our units of measurement are consistent with a change of area over time—such dimensional verification is a useful way to catch errors.

TOPIC 2 Interlude: A Cautionary Tale

We answered the question in Example 5 by appropriately modifying the steps in our strategy for solving related rates problems. Namely, since we didn't have a single equation in S and V that we could differentiate with respect to t, we worked with two separate equations and substituted information about dr/dt from one into the other. More generally, note that

$$\frac{dr}{dt} = \frac{1}{4\pi r^2}\frac{dV}{dt}$$

and so

$$\frac{dS}{dt} = 8\pi r \frac{dr}{dt} = \frac{8\pi r}{4\pi r^2}\frac{dV}{dt} = \frac{2}{r}\frac{dV}{dt}.$$

If $r = 20$ cm and $dV/dt = 200$ cm^3/s, then we have

$$\frac{dS}{dt} = \left(\frac{2}{20 \text{ cm}}\right)\left(200 \text{ cm}^3/\text{s}\right) = 20 \text{ cm}^2/\text{s},$$

the answer we obtained before.

But we might try to attack the problem in a different manner. If we *could* express S in terms of V, we could differentiate both sides with respect to t and again arrive at a relation between dS/dt and dV/dt. In pursuit of this, note that

$$\frac{r}{3}S = \frac{r}{3}\left(4\pi r^2\right) = \frac{4}{3}\pi r^3 = V,$$

so

$$S = \frac{3}{r}V.$$

But then this seems to imply that

$$\frac{dS}{dt} = \frac{dS}{dV}\cdot\frac{dV}{dt} = \frac{3}{r}\frac{dV}{dt},$$

which is not the result of $\dfrac{2}{r}\left(\dfrac{dV}{dt}\right)$ that we found above! What has gone wrong?

The answer is that just as V is a function of r, r can also be said to be a function of V; that is, as V changes, it certainly implies a change in r. In fact,

$$r = cV^{1/3},$$

where $c = \left[3/(4\pi)\right]^{1/3}$ (this comes from solving $V = \frac{4}{3}\pi r^3$ for r). So,

$$S = \frac{3}{r}V = \frac{3}{cV^{1/3}}V = \frac{3}{c}V^{2/3}.$$

We leave it as an exercise (Exercise 15) to show that if this last equation is differentiated with respect to t and simplified, the result is again $\dfrac{dS}{dt} = \dfrac{2}{r}\left(\dfrac{dV}{dt}\right)$.

Moral: Watch out for hidden functional relationships between the variables in your equations!

3.8 **Exercises**

1. A theme park ride is descending on a parabolic path that can be approximated by the equation $y = -\frac{1}{90}x^2 + 90$ (distance is measured in feet). If the horizontal component of its velocity is a constant 6 ft/s, find the rate of change of its elevation when $x = 22.5$.

2. Adapt your solution from Exercise 1 to find dx/dt at $x = 30$ feet if the equation of the ride's path is $y = 0.01(x-95)^2 - 2.25$ and $dy/dt = -20$ ft/s.

3–10 Find the rate of change using the given information.

3. $\dfrac{dy}{dt}$ at $y = 3$, if $y = \sqrt{x+2}$ and $\dfrac{dx}{dt} = 1$

4. $\dfrac{dy}{dt}$ at $x = 2$, if $x^2 + y^2 = 5$, $y > 0$, and $\dfrac{dx}{dt} = 3$

5. $\dfrac{dx}{dt}$ at $y = 0.5$, if $y = \dfrac{1}{x}$ and $\dfrac{dy}{dt} = -2$

6. $\dfrac{dx}{dt}$ at $x = 1$, if $xy^2 = \dfrac{1}{4}$ and $\dfrac{dy}{dt} = -0.25$

7. $\dfrac{dy}{dt}$ at $x = 0$, if $y = \dfrac{x+2}{x^2+1}$ and $\dfrac{dx}{dt} = -5.2$

8. $\dfrac{dy}{dt}$ at $y = \dfrac{1}{2}$, if $y = \dfrac{1}{2}e^{-x}$ and $\dfrac{dx}{dt} = 25$

9. $\dfrac{dy}{dt}$ at $x = -\dfrac{3\pi}{4}$, if $y = 2\sin\left(x + \dfrac{\pi}{4}\right)$ and $\dfrac{dx}{dt} = 7.4$

10. $\dfrac{dx}{dt}$ at $y = \dfrac{\pi}{4}$, if $x = \cot y$ and $\dfrac{dy}{dt} = -3.35$

11. The length of a rectangle is increasing at a rate of 5 in./s, while its width is decreasing at 2 in./s. Find the rate of change of its area when its length is 45 in. and its width is 25 in.

12. Find a formula for the rate of change of the distance from the origin of a point moving on the graph of $f(x) = x^2$ when $x = 2$ and $dx/dt = 3$ units per second.

13. Find the rate of separation between the moving points (x_1, y_1) and (x_2, y_2) on the graph of $y = \sin x$ when $x = \pi/2$ if they start at the origin at the same time, and the horizontal components of their velocities are $dx_1/dt = \frac{1}{2}$ units per second and $dx_2/dt = -\frac{1}{2}$ units per second, respectively.

14. A Ferris wheel of radius 34 feet needs 3 minutes to complete a full revolution. At what rate is a rider descending when she is 51 feet above ground level?

15. Using the notation of Example 5, use the Chain Rule to differentiate the equation $S = (3/c)V^{2/3}$ with respect to time to obtain $\dfrac{dS}{dt} = \dfrac{2}{r}\left(\dfrac{dV}{dt}\right)$. (**Hint:** After differentiating, make use of the equation $r = cV^{1/3}$ again.)

16. Rework Example 1 assuming that the tank is a rectangular prism with a 2 m by 2 m square base.

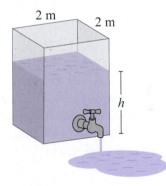

17. Rework Example 1 again, this time assuming that the tank is an inverted, right square pyramid of height 4 m and a 2 m by 2 m base. How fast is the level of water falling when its depth is 2 m?

18. A spectator is tracking a stunt plane at an air show with his video camera. If the plane is on a near-vertical path, rising at a speed of 100 feet per second, and the camera is 400 feet from the point on the ground directly below the plane, how fast is the camera angle changing when the plane's altitude is 400 feet? How fast is the distance between the camera and the plane increasing at that instant?

19. A cistern in the form of an inverted circular cone is being filled with water at the rate of 75 liters per minute. If the cistern is 5 meters deep, and the radius of its opening is 2 meters, find the rate at which the water level is rising in the cistern half an hour after the filling process began. (**Hint:** 1 m³ = 1000 L.)

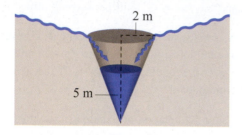

20. Repeat Exercise 19, this time assuming that the cistern is in the form of a pyramid with a 4-by-4-meter square opening.

21. A ship passes a lighthouse at 3:15 p.m., sailing to the east at 10 mph, while another ship sailing due south at 12 mph passes the same point half an hour later. How fast will they be separating at 5:45 p.m.?

22. A tourist at scenic Point Loma, California uses a telescope to track a boat approaching the shore. If the boat moves at a rate of 10 meters per second, and the lens of the telescope is 35 meters above water level, how fast is the angle of depression of the telescope (θ) changing when the boat is 200 meters from shore?

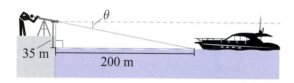

23. When preparing cereal for her child, a mother is pouring milk into a bowl, the shape of which can be approximated by a hemisphere with a radius of 6 in. If milk is being poured at a rate of 4 in.³/s, how fast is the level of milk rising in the bowl when it is 1.5 inches deep? (**Hint:** The volume of fluid of height h in a hemispherical bowl of radius r is $V = \pi h^2 \left(r - \frac{1}{3}h\right)$.)

24. Suppose that in Exercise 29 of Section 3.7, the sand is being poured at a rate of 8 cubic inches per second. Find the rate of change of the height of the cone when it is 4 inches tall.

25. When finished playing in the sand, the child of Exercise 24 takes advantage of a nice wind and starts flying his kite on the beach. When the kite reaches an altitude of 60 feet the wind starts blowing it horizontally away from the child at a rate of 15 feet per second while maintaining the altitude of the kite. How fast does the child have to be letting out the string when 100 feet are already out?

26. A passenger airplane, flying at an altitude of 6.5 miles at a ground speed of 585 miles per hour, passes directly over an observer who is on the ground. How fast is the distance between the observer and the plane increasing 3 minutes later?

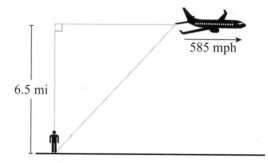

27. A military plane is flying directly toward an air traffic control tower, maintaining an altitude of 9 miles above the tower. The radar detects that the distance between the plane and the tower is 15 miles and that it is decreasing at a rate of 950 miles per hour. What is the ground speed of the plane?

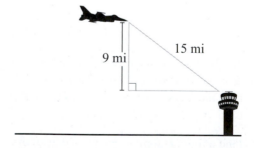

28. A child is retrieving a wheeled toy that is attached to a string by pulling in the string at a rate of 1 foot per second. If the child's hands are 3 feet from the ground, at what rate is the toy approaching when 5 feet of the string are still out?

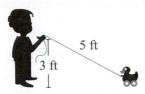

29. A fisherman is reeling in a fish at a rate of 20 centimeters per second. If the tip of his fishing rod is 4.5 meters above the water, and we are assuming that the fish is near the water surface throughout the process, how fast is it approaching when 7.5 meters of fishing line are still out? How fast is the angle θ between the fishing line and the water increasing at that instant?

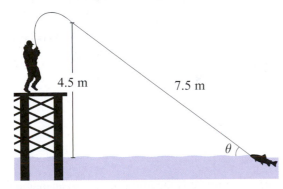

30. A construction worker is using a winch to pull a 9-foot column to a vertical position. If the winch is in the exact position where the top of the installed column is supposed to be, and the rope is being pulled at the rate of 6 inches per second, at what rate is the angle between the column and the ground changing when it is $\pi/6$ radians? At what rate is the top of the column rising vertically at that instant? (Assume the base of the column doesn't slip during lifting.)

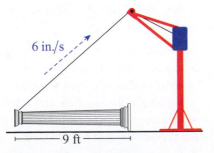

31. The volume of a cube is decreasing at a rate of 150 mm³/s. What is the rate of change of the cube's surface area when its edges are 30 mm long?

32. The acute angles of a rhombus are increasing at a rate of 0.25 radians per second. If the sides of the rhombus are 20 cm, at what rate is the area of the rhombus increasing when the acute angles are $\pi/3$ radians?

33. A 35-foot-by-18-foot pool, whose depth increases uniformly from 3 feet to 8 feet (along the 35-foot side), is being filled with water at the rate of 4.5 cubic feet per minute. You observe that water appears to be "creeping up" on the angled bottom much faster than it rises along the vertical walls. Find the rate at which the water rises along the angled bottom at the instant when the water level is 2 feet at the deep end of the pool.

34. Considering again the pool of Exercise 33, suppose that it is measured that the water is climbing upward along the angled bottom at a rate of 3.02 in./min when the water level is 1 foot at the deep end. Assuming that the pump is working at the same rate of 4.5 cubic feet per minute, use this information to prove that the pool has a leak, and find the rate at which water is leaking out of the pool.

35. A trough that is 5 meters long and 1 meter across at the top has a cross-section in the form of an isosceles trapezoid and both of its endplates are vertical. The altitude of the trapezoid is 40 centimeters, and the shorter base is 20 centimeters long. If the trough is being filled at the rate of 30 liters per minute, how fast is the water level rising at the instant when the water's depth is 20 centimeters?

36. Rework Example 3, this time assuming that Susan turned onto a highway whose elevation is 20 feet above that of the road on which the other drivers are still approaching the intersection.

37. An electrician is working on top of a 15 ft ladder that is leaning against the wall when its bottom starts sliding at a rate of 1 ft/s. Fortunately, a fellow worker catches it when the ladder's bottom is 5 ft from the wall. How fast is the top of the ladder (along with the electrician) sliding down the wall at that instant?

38. Adam is arriving home one evening in his SUV and is slowly approaching his garage door at a rate of 5 ft/s when the sensor lights come on. If the lights are mounted directly above the door at a height of 15 ft from the ground and Adam's SUV is 6 ft tall, at what rate is the length of the car's shadow shrinking when it is 25 ft from the garage door? What is the speed of the tip of the car's shadow?

39. A baseball player is running from first base to second base at 25 feet per second. At what rate is his distance increasing from home plate when he is 22.5 feet from second base? (**Hint:** The baseball diamond is a 90-foot-by-90-foot square.)

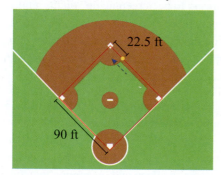

40.* A container in the shape of a cone, standing on its circular base, is being filled with water at the rate of 1.5 cubic feet per minute. If the radius of the base is 2 feet and the height of the cone is $2\sqrt{3}$ feet, how fast is the water level rising when it is 2 feet deep? (**Hint:** The volume of liquid in a partially filled conical tank is $V = \frac{1}{3}\pi d\left(R^2 + Rr + r^2\right)$, where R is the radius of the base, r is the radius of the top of the liquid, and d is its depth.)

41.* Italian police are chasing a criminal down a narrow street at a speed of 90 kilometers per hour. If the blue light on the top of the car is rotating counterclockwise at a rate of 1 rotation per second, and the buildings are only 3 meters from the car on the right, how fast is the beam moving on the wall at the instant when it is already 6 meters ahead of its source?

42.* When studying for a calculus test, Roger accidentally pushes his book over the edge of his 2.5 ft high desk. If his 6 ft tall lamp is standing 3 ft from where the textbook fell down, how fast was the book's shadow moving when the text hit the ground? (Ignore air resistance. Use $g \approx 32\ \text{ft/s}^2$.)

43.* A wall clock has a 10 in. minute hand and a 6 in. hour hand. At what rate are the tips of the hands approaching each other at 3 o'clock?

44.* The *lens equation*, easily derivable from geometric similarity for a thin converging lens, is

$$\frac{1}{o} + \frac{1}{i} = \frac{1}{f},$$

where o (the *object distance*) and i (the *image distance*) are the respective distances of the object and the image from the lens, and f is the *focal length* of the lens. Suppose a 100 mm high object is being slowly moved away from a lens at a speed of 5 mm/s. The focal length of the lens is 200 mm.

a. Find the rate at which the image changes its location when the object distance is 600 mm.

b. Find the rate at which the image changes its size at the same instant.

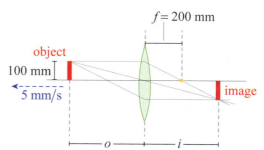

45. Suppose that the torque output of an automobile engine, as a function of engine speed, is approximated by

$$T(s) = \left(-0.001/150^4\right)(s - 3000)^4 + 160\ \text{lb}\cdot\text{ft},$$

where s is measured in revolutions per minute (rpm), and that the engine revs up from 0 to 5000 rpm (assume no gear shift takes place).

a. Use a graphing calculator or computer algebra system to graph the torque as a function of s on the interval $[0, 5000]$ (this is called the engine's *torque curve*).

b. If the power output of the engine, measured in horsepower (hp), is calculated by $P = \frac{1}{5252} sT(s)$ hp, and the engine is revving up according to the function $s(t) = 1000t$ (t is measured in seconds), find the rate of change of the power output at $t = 3$ seconds.

3.9 **Linearization and Differentials**

TOPICS

1. Linear approximation of functions

2. Differentials

As we have seen, the line tangent to a (differentiable) curve at a given point captures the trend of the curve, in the sense that the tangent line's rise or fall matches that of the curve. We say that the tangent line *approximates* the curve in the neighborhood of the point, and in many instances such an approximation is a convenient and sufficient substitution for the curve. This is illustrated in Figure 1 with the curve $y = x^2 + 1$ and its tangent line at $(1, 2)$.

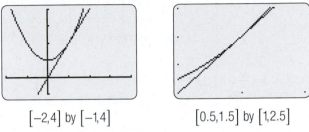

$[-2, 4]$ by $[-1, 4]$ $[0.5, 1.5]$ by $[1, 2.5]$

Figure 1

TOPIC 1   **Linear Approximation of Functions**

Given a function f, which is differentiable at the point $(c, f(c))$, the line tangent to f at that point is described by the equation $y = f(c) + f'(c)(x - c)$. Such a line, which itself is a function of x, merits a name.

Definition 💡

Linear Approximation

Given a function f, differentiable at the point $(c, f(c))$, we call

$$L(x) = f(c) + f'(c)(x - c)$$

the **linear approximation** or **linearization** of f at the point c. Because L is a first-degree polynomial in x, such a formula is also referred to as a **first-order approximation**, a phrase that hints at the higher-order approximations we will discuss when we study Taylor series later in this text.

Example 1 ✎

Find the linearization of $f(x) = \sqrt{1 + x}$ at $x = 0$.

Solution

Since $f(0) = 1$ and

$$f'(x) = \frac{1}{2}(1 + x)^{-1/2},$$

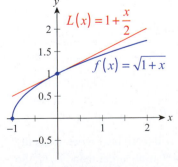

Figure 2

$f'(0) = \frac{1}{2}$ and hence

$$L(x) = f(0) + f'(0)(x - 0) = 1 + \frac{1}{2}(x - 0) = 1 + \frac{x}{2}.$$

The graph in Figure 2 is an illustration of how f and L relate to one another.

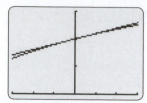

Figure 3

$f(x)$ and $L(x)$
on $[-0.6, 0.6]$ by $[0, 1.5]$

In this example, note that the linearization lies above the graph of f in the neighborhood of $x = 0$, so using L to approximate f for nearby points would result in an overestimate. The critical observation, however, is that the difference between f and L is vanishingly small as $x \to 0$, as shown in Figure 3.

While linear approximation was once useful as an aid to the actual numerical evaluation of functions, calculators and computers have largely reduced the importance of this particular use. Linearization still has great value, however, in developing mathematical models and understanding the behavior of complicated functions over small intervals. We do this by making use of the fact that

$$f(x) \approx L(x) = f(c) + f'(c)(x - c)$$

for all x close to the point c.

As a simple example, and as a generalization of Example 1, consider the behavior of the family of functions $f(x) = (1 + x)^k$ near $x = 0$, where k can take on any real number. Following the same procedure as before,

$$(1 + x)^k \approx f(0) + f'(0)(x - 0) = 1 + kx$$

for x near 0 and for any k. This approximation holds true even if the argument of the function appears more complex, as long as it approaches 0 as x itself does.

$$\frac{1}{1 + x} = (1 + x)^{-1} \approx 1 + (-1)x = 1 - x$$

$$\frac{1}{\sqrt{1 + 3x}} = (1 + 3x)^{-1/2} \approx 1 + \left(-\frac{1}{2}\right)(3x) = 1 - \frac{3}{2}x$$

$$\frac{1}{\sqrt[3]{1 - 2x^4}} = \left[1 + (-2x^4)\right]^{-1/3} \approx 1 + \left(-\frac{1}{3}\right)(-2x^4) = 1 + \frac{2}{3}x^4$$

As an example of linearization in the development of models, physicists routinely make use of the fact that $\sin\theta \approx \theta$ for θ close to 0. In fact, the standard model describing the motion of a pendulum, a model that has had great practical utility over the last several centuries, requires this approximation in order for the motion to be expressed in terms of elementary functions. As the next example indicates, this is not unreasonable for small θ.

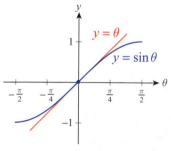

Figure 4

Example 2

Find the linearizations of $\sin\theta$, $\cos\theta$, and $\tan\theta$ at $\theta = 0$.

Solution

We have $\sin 0 = 0$ and $\left.\dfrac{d(\sin\theta)}{d\theta}\right|_{\theta=0} = \cos 0 = 1$, so $\sin\theta \approx 0 + (1)(\theta - 0) = \theta$.

Similarly, we find that $\cos\theta \approx 1$ and $\tan\theta \approx \theta$ at $\theta = 0$.

How close are these approximations? As we will learn later when we study Taylor series, the difference between $\sin\theta$ and θ is no larger than $\left|\theta^3/6\right|$. So for example, over the interval $[-0.1, 0.1]$, the error in the approximation for $\sin\theta$ is no larger than $0.000\overline{16}$.

Figure 4 indicates how closely $\sin\theta$ and its linearization correspond over small intervals centered at 0.

TOPIC 2   Differentials

When the Leibniz notation dy/dx was introduced in Section 3.1, we mentioned in passing that such notation does indeed represent, as it appears to, a ratio. To this point, however, we have not dealt with the components of the ratio separately; we are now ready to do so.

Definition 💡

Differentials

Given a differentiable function $y = f(x)$, the **differential dx** is defined to be an independent variable and the **differential dy** is a dependent variable defined by

$$dy = f'(x)\,dx.$$

That is, dx can take on any real number value, while the value of dy depends on the values of $f'(x)$ and dx.

Don't be misled by the unusual nature of the symbols used for differentials. Remember that the symbol chosen to represent a given variable really has no purpose other than (at best) to suggest the meaning or role of the variable. In usage, dx typically denotes a small quantity, and often appears in expressions such as $c + dx$. If this reminds you of the usage of Δx, that is no coincidence—the two often play the same role. Similarly, dy is reminiscent of Δy, and like Δy it represents a vertical change. But while for a given function f and a given fixed value of x we may define dx and Δx to be identical, we cannot assume dy and Δy to be equal. The distinction between the two, as well as the connection between differentials and the linearization of a function, is illustrated in Figure 5.

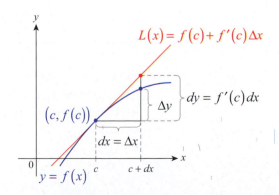

Figure 5 Differentials and Differences

As Figure 5 points out, Δy is the exact change in the value of the function over the interval $(c, c + \Delta x)$; that is, $\Delta y = f(c + \Delta x) - f(c)$. The differential dy, on the other hand, is the rise or fall in the tangent line $L(x)$ over the interval $(c, c + \Delta x)$. The differential dy is defined so that

$$\frac{dy}{dx} = f'(x) = \lim_{\Delta x \to 0} \frac{\Delta y}{\Delta x}$$

and the difference between dy and Δy goes to 0 as $\Delta x \to 0$.

Differential notation also appears in expressions such as df, and the meaning of df is the same as that of dy if $y = f(x)$. All of our differentiation rules can also be stated in terms of differentials. For example, if u and v are functions of x, then

$$\frac{d}{dx}(uv) = \frac{du}{dx} v + u \frac{dv}{dx},$$

so multiplication by the differential dx results in

$$d(uv) = (du)v + u(dv).$$

Example 3 ✎

Determine the requested values of the differentials.

a. Find dy if $y = 3x^{4/3} + \ln x$ when $x = 8$ and $dx = \frac{1}{10}$.

b. Find $d(\sin \theta)$ if $\theta = 0$ and $d\theta = 0.05$.

Solution

a. $dy = \left(4x^{1/3} + \dfrac{1}{x}\right) dx$

Substituting $x = 8$ and $dx = \frac{1}{10}$, we have the following:

$$dy = \left(4 \cdot 8^{1/3} + \frac{1}{8}\right)\left(\frac{1}{10}\right) = \frac{65}{80} = \frac{13}{16}$$

b. Note that $d(\sin \theta) = \cos \theta \, d\theta$, so if $\theta = 0$ and $d\theta = 0.05$,

$$d(\sin \theta) = (\cos 0)(0.05) = 0.05.$$

Finally, differentials play a very useful role in the practical applications of calculus. Since measurement errors and approximate values are inescapable in such fields as physics, chemistry, and economics, it is important to have a solid understanding of how estimates in fundamental variables propagate through later computations.

Example 4 ✐

The radius of a ball bearing is measured by a micrometer to be 1.2 mm with a margin of error of 0.05 mm. Given this possible error, estimate the maximum error in the calculated volume of the ball bearing.

Solution

Since $V = \frac{4}{3}\pi r^3$, $dV = 4\pi r^2\,dr$. We will use a value of $r = 1.2$ mm for the radius, and note that the actual radius could be as much as $dr = 0.05$ mm larger. So

$$dV = 4\pi(1.2 \text{ mm})^2 (0.05 \text{ mm}) \approx 0.90 \text{ mm}^3.$$

To put this into perspective, it is useful to estimate the *percentage error* for both the radius and the volume.

$$\text{Percentage error in radius:} \quad \frac{dr}{r} = \frac{0.05}{1.2} \approx 0.04 = 4\%$$

$$\text{Percentage error in volume:} \quad \frac{dV}{V} \approx \frac{0.90}{\frac{4}{3}\pi(1.2)^3} \approx 0.12 = 12\%$$

So the propagated error in the calculated volume, in percentage terms, is three times larger than the margin of error in the measured radius. This factor of three will always hold true for such volume calculations; in Exercise 50 you will show that

$$\frac{dV}{V} = 3\frac{dr}{r}.$$

3.9 **Exercises**

1–12 Find the linearization of the function at the given value.

1. $f(x) = x^3 - x; \quad x = 1$

2. $g(x) = \sqrt{x-3}; \quad x = 4$

3. $h(x) = (x^4 - 5x^2 + 1)^7; \quad x = 0$

4. $k(x) = (x^2 + 1)^{-2}; \quad x = 2$

5. $C(\theta) = \cos\theta; \quad \theta = 0$

6. $T(\theta) = \tan\theta; \quad \theta = 0$

7. $F(t) = (t^2 + 5t - 6)^{-1/3}; \quad t = 2$

8. $r(x) = \dfrac{1}{x+4}; \quad x = -3$

9. $t(u) = \dfrac{u+2}{u^2-15}; \quad u = -4$

10. $v(x) = \sin\pi x; \quad x = \dfrac{1}{6}$

11. $G(z) = e^z; \quad z = 0$

12. $U(s) = \ln(s^4 + 1); \quad s = 1$

13–24 Determine the differential dy for the given values of x and dx.

13. $y = 3x^2 + x$; $x = 1$, $dx = 0.2$

14. $y = x\sqrt{x-5}$; $x = 6$, $dx = 0.01$

15. $y = \dfrac{4x+1}{x-3}$; $x = 2$, $dx = 0.1$

16. $y = \sec x$; $x = \dfrac{\pi}{4}$, $dx = \dfrac{1}{8}$

17. $y = x^{3/2} + x^{-3/2}$; $x = 4$, $dx = \dfrac{1}{16}$

18. $y = \ln x + \dfrac{1}{\ln x}$; $x = e$, $dx = 0.01$

19. $y = x \tan x$; $x = -\dfrac{\pi}{4}$, $dx = \dfrac{1}{4}$

20. $y = e^{\sqrt{x^2+3}}$; $x = 1$, $dx = 0.001$

21. $y = \sqrt{\ln(x+1)}$; $x = e - 1$, $dx = \dfrac{-1}{e^2}$

22. $y = \arctan x$; $x = -1$, $dx = \dfrac{-1}{2^5}$

23. $y = \dfrac{\tan x}{x^2 + 1}$; $x = \dfrac{\pi}{3}$, $dx = -0.1$

24. $y = \cos(\arcsin x)$; $x = 0.6$, $dx = -0.16$

25–28 Calculate the values of dy and Δy and then use graph paper to draw the curve near the given point, indicating all three of the line segments dx, dy, and Δy.

25. $y = \dfrac{1}{2}x^2$; $x = 1$, $dx = \dfrac{1}{2}$

26. $y = \tan x$; $x = 0$, $dx = \dfrac{\pi}{6}$

27. $y = 2^x$; $x = 1$, $dx = \dfrac{1}{4}$

28. $y = \dfrac{1}{x^2}$; $x = 1$, $dx = -\dfrac{1}{4}$

29–40. Find the values of Δy and compare them with dy at the indicated points for the curves given in Exercises 13–24.

41–48 Use linear approximation to approximate the given number. Compare this approximation to the actual value obtained using a calculator or computer algebra system. Round your answer to four decimal places. (**Hint:** First identify $f(x)$ and c; then find and appropriately evaluate $L(x)$.)

41. $\sqrt{9.1}$

42. $(1.01)^3$

43. $(7.9)^{2/3}$

44. $\dfrac{1}{10.1}$

45. $\sqrt[5]{31}$

46. $\cos 1$

47. $\ln 2.7$

48. $e^{1.05}$

49. Prove the power and quotient rules for differentials.

   a. $d(x^n) = nx^{n-1}dx$

   b. $d\left(\dfrac{u}{v}\right) = \dfrac{v\,du - u\,dv}{v^2}$

50. Use the equations for V and dV from Example 4 to prove that the propagated error in the calculated volume of a sphere, in percentage terms, is three times larger than the margin of error in the measured radius; that is,

$$\dfrac{dV}{V} = 3\dfrac{dr}{r}.$$

51. Prove or disprove that an analogous equation to that obtained in Exercise 50 is true for a cube; that is, if the measured side length of a cube is a units with a margin of error of da, then

$$\dfrac{dV}{V} = 3\dfrac{da}{a}.$$

52–71 Use differentials or linearization to provide the requested approximations.

52. The side of a square was measured to be 9.5 cm with a possible error of 0.5 mm. Approximate the propagated error in the calculated area of the square. Express your answer as a percentage error.

53. The radius of a circular disk was measured to be $10\frac{1}{8}$ inches. Estimate the maximum allowable error in the measurement of the radius if the percentage error in the calculated area of the disk cannot exceed 2.5 percent.

54. The base and altitude of a triangle were measured to be 7 and 9 inches, respectively. If the possible error in both cases is $\frac{1}{16}$ inches, approximate the propagated error when computing the area of the triangle.

55. Two sides of a triangle were measured to be 60 and 80 mm, respectively, while the included angle is 60 degrees. If the margin of error of the linear measurements is 0.1 mm, while that of the angle measurement is 0.1 degrees, find the possible propagated error in the calculated area of the triangle.

56. A box in the shape of a rectangular prism has a square base. If the edge of the base is 25 cm and the height is 50 cm, both with a possible measurement error of 0.2 mm, estimate the propagated errors in both the computed volume and surface area of the box. Express both answers as percentage errors.

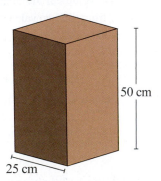

57. A piston of diameter 84 mm is being manufactured for an automobile engine. If the maximum percentage error in the measurement of the diameter is 0.05%, estimate the greatest possible value of the propagated error in the computed cross-sectional area of the piston. Express your answer as a percentage error.

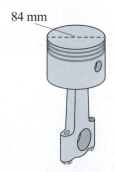

58. If the radius of an inflated balloon is 10 inches and the thickness of its wall is 0.002 inches, estimate the volume of the material it is made of. (Assume the balloon is perfectly spherical.)

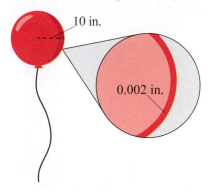

59. A tin can has a circular base of radius 2 inches and a height of 6 inches. If the thickness of its walls is 0.01 inches, estimate the volume of the material it is made of.

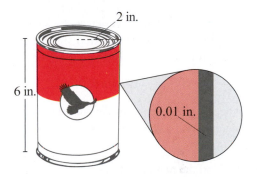

60. The exterior of a small private observatory needs to be painted. The building is approximately a circular cylinder with a hemisphere on top. The radius of the base is 3.5 feet and the height of the entire structure is 10 feet. Express the volume as a function of the radius of the base and use linearization to estimate the amount of paint that will provide a coat that is $\frac{1}{32}$ inches thick.

61. A trigonometry student stands 15 meters from a building and measures the angle of elevation to the top of the building as 60°. How accurate does her angle measurement have to be if she wants her propagated percentage error in estimating the height of the building to be no more than 5%?

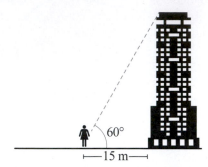

62. Referring to Exercise 44 of Section 3.8, estimate the change in image distance when the object distance increases from 60 cm to 61 cm.

63. The magnetic force experienced by a wire carrying a current I in an external magnetic field of uniform strength B is found from the equation

$$F = BIL \sin \theta,$$

where L is the length of the wire (measured in meters), and θ is the angle between the directions of B and I.

a. Find the magnetic force on a 50 cm wire if $B = 0.03 \text{ N}/(\text{A} \cdot \text{m})$, $I = 25$ amperes (A), and $\theta = 30°$.

b. Estimate the change in force if θ is increased to 33°.

c. Calculate the "true value" of the change and compare it with your approximation.

64. Estimate the change in the force in Exercise 63 if θ is increased to 33°, I is increased to 27 A, and B is decreased to $0.025 \text{ N}/(\text{A} \cdot \text{m})$.

65. The kinetic energy (in J) of a moving object is found from the equation $E_{kin} = \frac{1}{2}mv^2$, where m is the mass (in kg) of the object and v is its velocity (in m/s). Estimate the change in kinetic energy of a 1400 kg car that is accelerating from 100 km/h (approx. 62 mph) to 112 km/h (approx. 70 mph). What is the estimated percentage change?

66. When air resistance is negligible, the speed of impact of an object falling from height h is $v_i = \sqrt{2hg}$. Suppose that a rock is dropped from a height of 5 meters.

a. Find the speed of impact as the rock hits the ground.

b. Approximate the height from which the rock has to be dropped in order to increase the speed of impact by 10 percent. Express the height difference in both absolute and relative (percentage) terms.

c. Find the "true value" of the above height and compare it with your approximation.

67.* The volume of a cube of side length a is being determined by immersing the cube into a container of water and measuring the volume of the displaced water and then the surface area is calculated. Estimate the percentage error we can allow in the measurement of the volume if the calculated surface area cannot differ from the true value by more than 2%. Can you generalize the result?

68. The profit function for a company is found to be $P(x) = -1.2x^2 + 500x - 2600$, where x is the number of units manufactured. If the current production level is 100 units, estimate the percentage change in profit if production is raised to 110 units.

69. For the company in Exercise 68, estimate how much the company has to increase production from 100 units in order to achieve a 10 percent profit increase.

70.* The diameter of the bottom of a 4.5-inch-tall paper cup is 2.5 inches, while the diameter of its opening is 3.5 inches. If the cup is filled with iced soda to a depth of 4.3 inches and an additional 1-cubic-inch ice cube is dropped in, predict whether the cup will overflow. (**Hint:** See Exercise 40 of Section 3.8 for help in finding the volume of soda in the cup.)

71. Suppose the velocity function of a moving object is $v(t) = 1/(1 + t^2)$, and that it is moving in the positive direction along the x-axis. If you know that its location at $t = 2$ is $x = 5$, estimate its position half a second later.

72. The actual error in measurement is sometimes called absolute error, while the percentage error is referred to as relative. Write a short paragraph comparing absolute, relative, and propagated errors. Illustrate with a concrete example.

73. Examine the answer you obtained for Exercise 57. Can you state and prove a result, analogous to the one in Exercise 50, for the radius and cross-sectional area? Explain.

74–80 *True or False?* Determine whether the given statement is true or false. In case of a false statement, explain or provide a counterexample.

74. Since the differential dx is an increment, its value is always positive.

75. If $f(x) = k$, then $df = 0$.

76. If f is linear, then $\Delta f / \Delta x = df / dx$.

77. If f is differentiable at c, then
$$\lim_{\Delta x \to 0} (\Delta f / \Delta x) = df / dx.$$

78. Propagated error is also called percentage error.

79. The differential dy is always a bit less than Δy.

80. If f is increasing and $df < 0$, then $dy > \Delta y$.

3.9 **Technology Exercises**

81–92. Use a graphing calculator to graph the functions given in Exercises 13–24 in the same viewing window along with their linear approximations at the specified x-values. Use the "Zoom" and "Trace" features to find the maximum value for dx so that the approximation is accurate to 0.01.

Chapter 3
Review Exercises

1–2 Find the derivative of the given function at the specified point and express your answer using the differential notation due to Leibniz.

1. $f(x) = x^3 + x; \quad x = 1$

2. $g(x) = \dfrac{2}{x}; \quad x = 2$

3–4 Find the derivative of the function and use the differentiation operator D_x to express your answer.

3. $s(x) = \sqrt{x-2}$

4. $t(x) = \dfrac{1}{x^2 + 1}$

5–6 Find the first, second, and third derivatives of the function.

5. $f(x) = x^2 - 1$

6. $g(x) = x^4$

7–10 Find all the points where the function is not differentiable. For each of those points, find the one-sided derivatives (if they exist).

7. $f(x) = \sqrt[3]{x}$

8. $g(x) = |x+1| + |x-3|$

9. $h(x) = [\![x]\!] + x$

10. $F(t) = \begin{cases} t^2 \sin \dfrac{1}{t} & \text{if } t \neq 0 \\ 0 & \text{if } t = 0 \end{cases}$

11–20 Use differentiation rules to find the derivative of the function.

11. $f(x) = 0.2x^5 - 2x^4 + x^3 + 0.5x^2 + 2^{3/4}$

12. $g(x) = \sqrt{3x} + \sqrt{3x} + \dfrac{1}{\sqrt{3x}}$

13. $h(x) = (2x-1)(x^2 + 4)$

14. $k(x) = \dfrac{x-2}{x} e^{x-2}$

15. $F(t) = (t-1)\left(t^2 + \dfrac{1}{t}\right)(\sqrt{t} + t)$

16. $G(s) = \dfrac{1}{2s + s^2}$

17. $u(x) = \dfrac{2e^x + 1}{3e^x + 5}$

18. $t(x) = \dfrac{\dfrac{2}{x} + \dfrac{1}{x^2} + 4}{\dfrac{2}{x^2} - \dfrac{1}{x}}$

19. $v(x) = \ln(x^2 + 2)$

20. $w(x) = \sin(\sin(\sin x))$

21–24 Find the first, second, and third derivatives of the function.

21. $f(x) = \dfrac{x}{x+1}$

22. $f(x) = 3\sqrt{x}$

23. $f(x) = \tan x$

24. $f(x) = \arctan x$

25–26 Find a function f that satisfies the given conditions. (**Hint:** A polynomial is the most natural choice. Answers will vary.)

25. $f(0) = 0$, $f'(1) = 1$, and $f''(2) = 4$.

26. $f(0) = 2$ and $y = 2x + 1$ is tangent to the graph at $x = -1$.

27–28 Find the equation of the line tangent to the graph of the function at the given point.

27. $f(x) = \dfrac{1}{\sqrt{3x^2 + 1}}; \quad \left(1, \dfrac{1}{2}\right)$

28. $f(x) = \tan(\sin x); \quad (\pi, 0)$

29. Find the equation(s) of the line(s) tangent to the graph of $f(x) = x^2 + 3x + 1$ through the point $(2,2)$, which is not on the graph of f.

30. Assuming f is differentiable, find the derivative of $y = \ln\sqrt{[f(x)]^2 + 1}$.

31–32 Find the indicated limit.

31. $\displaystyle \lim_{x \to 0} \dfrac{-\sin 2x}{4x}$

32. $\displaystyle \lim_{x \to 0^+} \dfrac{x \cos x}{1 - \cos x}$

33. The position function of a moving particle is given by $x(t) = \dfrac{50t}{t+1}$ feet at t seconds. Find its velocity and acceleration at $t = 1$ second.

34. An object is moving along a straight line so that its distance from the start at t seconds is given by $d(t) = 12t - t^3$ meters. Find its position and acceleration at the instant when its velocity changes directions.

35. The radius of a spherical balloon being inflated increases according to the function $r(t) = 3 + 4\sqrt[3]{t}$, where r is measured in centimeters and t in seconds. Find the rate of change of the balloon's volume and surface area with respect to time at $t = 1$ second.

36–37 Find dy/dx by implicit differentiation.

36. $x^3 + y^3 = 2$

37. $6(x^2 + y^2) = 15xy$

38–39 Find dx/dy by implicit differentiation.

38. $x\sin(x + y) = y^2 + 6$

39. $6(y^2 - x^2) = y^4$

40–43 Use implicit differentiation to find the equation of the line tangent to the curve at the indicated point.

40. $\dfrac{1}{x^3} + \dfrac{1}{y^3} = 2;\quad (1,1)$

41. $x^3 + y^2 = 2x + 1;\quad (0,1)$

42. $\dfrac{3(x + y)}{xy} = 16\sqrt{x + y};\quad \left(\dfrac{3}{4}, \dfrac{1}{4}\right)$

43. $3\sqrt{x} + \dfrac{2}{\sqrt{y}} = xy;\quad (1,4)$

44. Find all points on the lemniscate $(x^2 + y^2)^2 = 2(x^2 - y^2)$ where its graph has horizontal tangent lines.

45. Use implicit differentiation to find d^2y/dx^2 for $x^{2/3} + y^{2/3} = 1$.

46–49 Use the Derivative Rule for Inverse Functions to determine $(f^{-1})'(a)$ for the indicated value of a. (The domain of f is assumed to have been restricted so that the inverse exists and is differentiable, whenever appropriate.)

46. $f(x) = x^3 + x;\quad a = 10$

47. $f(x) = \sqrt[4]{x + 1};\quad a = 2$

48. $f(x) = \dfrac{2}{x^2};\quad a = \dfrac{1}{2}$

49. $f(x) = e^x + x;\quad a = 1$

50–53 Determine the derivative of the given function.

50. $f(x) = \log\sqrt{x^2 + 1}$

51. $f(x) = \tan^{-1}\sqrt{x}$

52. $f(x) = e^{\arcsin x}$

53. $f(x) = \sec^{-1}(\ln x)$

54–55 Use logarithmic differentiation to find y'.

54. $y = \dfrac{\sqrt[3]{x^2 + 1}(x + 3)}{x^{2/3}\sqrt{2x^2 + 3}}$

55. $y = \left(\sqrt{x}\right)^{\ln x}$

56. A fast-growing population of bacteria doubles every half hour. If the initial count is 1000, how many bacteria are there in 100 minutes?

57. A 350 °F pizza is left on the counter and cools to 250 °F in 4 minutes. If the room temperature is 70 °F, determine the total time it takes for the pizza to cool down from 350 °F to 185 °F. (**Hint:** See Exercise 9 of Section 3.7.)

58. Find the rate of change of the distance from the origin of a point moving on the graph of $f(x) = x^3$ when $x = 1$ and $dx/dt = 2$ units per second.

59. A spherical balloon is being filled with helium at a rate of $20\text{ in.}^3/\text{s}$. How fast is the radius increasing at the instant when the radius is 4 in.?

60. A small plane, flying at an altitude of 0.1 miles at a ground speed of 85 miles per hour, passes directly over an observer. How fast is the distance between the observer and the plane increasing a minute later?

61. Radar is tracking a rocket that was launched vertically upward. It is found that the rocket's distance from the radar is increasing at a rate of 1200 km/h at the instant when that distance is 5 km. If the radar station is 4 km from the launch site, find the speed of the rocket.

62. A ship sailing west at 9 miles per hour passes a buoy 20 minutes before another ship sailing due north at 12 miles per hour passes the same buoy. How fast will they be separating an hour later?

63. Tiffany walks toward a light source that is 8 feet above ground. If the speed of the tip of her shadow is three times that of her walking speed, how tall is Tiffany?

64–65 Find the linearization of the function at the given value.

64. $f(x) = \dfrac{1}{(x-1)^2}; \quad x = 2$

65. $f(x) = \sin x; \quad x = \dfrac{\pi}{4}$

66–67 Use linear approximation to approximate the given number. Round your answer to four decimal places.

66. $\sqrt[3]{8.2}$

67. $\arctan 0.9$

68. The diameter of a large bouncy ball was measured to be 65 cm with a possible error of 1 mm. Approximate the propagated errors in the calculated volume and surface area of the ball, respectively. Express your answers as percentage errors.

69. The proper dosage d of a certain over-the-counter medicine for children depends on body weight w according to the function $d(w) = \frac{5}{4}w^{3/5}$, where d is measured in milligrams and w in pounds. Use differentials to estimate how accurately (in terms of percentage error) we need to know a 32-pound child's weight if we cannot stray from the proper dosage by more than 6 percent.

70. A manufacturing business found its daily revenue to be $R(x) = 150x - \frac{1}{4}x^2$ dollars when x units are produced and sold.

    a. Use linearization and marginal revenue to estimate the extra revenue when production is increased from 100 to 102 units.

    b. Use the revenue function to calculate the actual revenue increase. Compare your answers.

71. Use the concept of the derivative function to explain why the graph of $y = x^a$, $a > 1$ curves upward, while the graph of $y = x^b$, $0 < b < 1$ curves downward.

72–82 *True or False?* Determine whether the given statement is true or false. In case of a false statement, explain or provide a counterexample.

72. If both one-sided derivatives of $f(x)$ exist at c, then f is continuous at c.

73. If $p(x)$ is a polynomial of degree n, then all k^{th}-order derivatives of $p(x)$ for $k > n$ are 0.

74. If $y = \pi^n \sin x$, then $y' = n\pi^{n-1} \sin x + \pi^n \cos x$.

75. If $y = 1/(x^2 - 3x + 1)$, then $y' = 1/(2x - 3)$.

76. If $y = \ln(3x + 1)$, then $y' = 1/(3x + 1)$.

77. If $y = x^x$, then $y' = x \cdot x^{x-1}$.

78. Since $\left(e^x\right)' = e^x$, therefore $\left(e^{e^x}\right)' = e^{e^x}$.

79. If $f(x) = x$, then $df = dx$.

80. If $f(x)$ is linear, then its linearization at any point is itself.

81. If $x \to 0$, then $\Delta x \to dx$ and $\Delta y \to dy$.

82. If $\Delta x \to 0$, then $\Delta y / \Delta x \to dy / dx$.

Chapter 3
Technology Exercises

83–85 Use a graphing calculator or computer algebra system to graph the function and identify all points where the function is not differentiable. Explain.

83. $f(x) = |x^2 - x|$

84. $f(x) = |x|(x+2)$

85. $f(x) = \sqrt[4]{x^2 - 1}$

86. Use the differentiation capabilities of a computer algebra system to find the derivative of $f(x) = 2\cos^2 x - \cos 2x$. Then find the derivative by hand, applying a trigonometric identity before differentiating. Does your answer agree with that of your technology? If not, what do you think is the reason? Can you "force" your CAS to represent its answer in a simpler form?

87. Repeat Exercise 86 for the function $f(x) = 2\sin(x/2)\cos(x/2)$.

88. Use a graphing calculator or computer algebra system to graph the functions $y = \ln x$, $y = a^x$, $a > 1$ and $y = x^b$, $0 < b < 1$ for various values of the parameters a and b. By zooming out appropriately, compare their relative growth rates; that is, conjecture "who wins the race toward infinity" in general among these three types of functions. Use the concept of the derivative to support your conjecture.

89. The displacement of a mass attached to a spring is given by the function $h(t) = e^{-t/6}\cos 2t$.

a. Use a graphing calculator or computer algebra system to graph the function and explain why it is realistic.

b. Use a graphing calculator or computer algebra system to graph the velocity and acceleration functions together with $h(t)$ on the same screen. What seems to be the position of the mass when velocity is maximum? When is velocity 0? When is acceleration maximum, and when is it 0?

Project ✎

Chapter 3

The following table shows the atmospheric pressure p at the altitude of k feet above sea level (pressure is measured in mm Hg; note that this unit of pressure is approximately the pressure generated by a column of mercury 1 millimeter high).

k (ft)	0	1000	2000	3000	4000	5000	6000	7000	8000	9000	10,000
p (mm Hg)	760	733	707	681	656	632	609	586	564	543	523

1. Find the average rate of change of air pressure from sea level to 2000 feet of altitude.

2. Find the average rate of change of air pressure between the altitudes of 4000 and 10,000 feet.

3. Use a *symmetric difference quotient*

$$\frac{p(c+h)-p(c-h)}{2h}$$

to estimate the instantaneous rate of change of air pressure at 7000 ft by choosing $h = 1000$ ft.

4. Tell whether you expect the answer to Question 2 or 3 to better approximate the instantaneous rate of change of air pressure at altitude 7000 ft. Explain. (**Hint:** Plotting the data on paper may help.)

5.* Explain why you expect the symmetric difference quotient $\dfrac{f(c+h)-f(c-h)}{2h}$ in general to be a better approximation of the instantaneous rate of change of f at $x = c$ than the "regular" difference quotient $\dfrac{f(c+h)-f(c)}{h}$.

6. Use a graphing calculator or computer algebra system to find an exponential regression curve to the given data and plot the curve along with the data on the same screen.

7. Use the exponential function you found in Question 6 to estimate the instantaneous rate of change of air pressure at 7000 ft, and compare with your estimate given in Question 3.

8. Is the instantaneous rate of change increasing or decreasing with altitude? Explain.

Applications of Differentiation

4.1 Extreme Values of Functions   331

1. Absolute and Relative Extrema
2. Finding Extrema

4.2 The Mean Value Theorem   344

1. Rolle's Theorem and the Mean Value Theorem
2. Consequences of the Mean Value Theorem

4.3 The First and Second Derivative Tests   354

1. First Derivative Tools
2. Second Derivative Tools

4.4 L'Hôpital's Rule   370

1. L'Hôpital's Rule
2. Limits of Indeterminate Form

4.5 Calculus and Curve Sketching   381

1. A Curve-Sketching Strategy
2. Newton's Method

4.6 Optimization Problems   394

1. Mathematical/Geometrical Problems
2. Optimization Problems from Other Disciplines

4.7 Antiderivatives   407

1. Finding Antiderivatives
2. Antiderivatives in Use

Introduction

While Chapter 3 focused primarily on developing the techniques of differentiation, this chapter puts those techniques to use by focusing attention on applications of differentiation.

As we will see, the applications are drawn from many disciplines and at first glance the objectives vary widely. But the underlying theme of the chapter can be summarized as a prolonged answer to this question: What does the derivative of a function tell us about the function itself?

Much of the calculus theory of the 17th, 18th, and 19th centuries was developed by mathematicians working to solve problems arising in such fields as physics, biology, and engineering. Frequently, a given problem's solution hinged upon being able to determine the minimum and/or maximum values of an associated function, establishing the connection with differentiation. In order to apply their new solution methods to entire

The chapter closes with the most extreme variation of the underlying question: *If we have perfect knowledge of the derivative of a function, can we use that to completely determine the function itself?* We will see that the answer is a qualified "yes," and an exploration of the ramifications of this answer will lead to the Fundamental Theorem of Calculus, the principal subject of study in Chapter 5.

What does the derivative of a function tell us about the function itself?

If we are seeking more accurate and precise graphs of functions, we can use differentiation to identify intervals where a function is consistently increasing or decreasing, and thus also identify the relative "high spots" and "low spots" on a function's graph. Further, we can use a function's second derivative to reveal deeper and more subtle layers of meaning in its graph.

families of problems, people such as the Swiss brothers Jakob (1655–1705) and Johann Bernoulli (1667–1748) and the French mathematician Augustin-Louis Cauchy (1789–1857) developed new general theorems of calculus. One of these theorems, Cauchy's Mean Value Theorem, will be used in this chapter—it is a deep result that lies at the heart of several other insights. Another theorem, known as l'Hôpital's Rule but first formulated by Johann Bernoulli, uses the power of differentiation to easily evaluate limits that would otherwise be extremely difficult to determine.

Jakob Bernoulli
(1655–1705)

Johann Bernoulli
(1667–1748)

Augustin-Louis Cauchy
(1789–1857)

1. Adapted from lithograph by Gregoire et Deneux, *Augustin Cauchy.* Smithsonian Institution Libraries.

2. Adapted from engraving by Pierre Dupin, *Jacques Bernoulli.*

3. Adapted from mezzotint by Johann-Jakob Haid after Jean Huber, *Johann Bernoulli.* Wellcome Library no. 1041i, http://catalogue.wellcomelibrary.org/record=b1158387. This file is licensed under the Creative Commons Attribution 4.0 International license.

4.1 **Extreme Values of Functions**

TOPICS

1. Absolute and relative extrema
2. Finding extrema

In this section we introduce the techniques by which the derivative leads to a deeper understanding of the graph of a function. Understanding and identifying features of a function's graph enables us to better determine its roots, its minimum and/or maximum values, and its limit behavior.

TOPIC 1 **Absolute and Relative Extrema**

Some of the most distinctive and noteworthy points on the graph of a function are those where it takes on its largest and smallest values. But anything beyond a cursory discussion of such points (called *extrema*, plural of the Latin *extremum*) leads quickly to the realization that we need to refine our notions of "largest" and "smallest" before we can make much progress in finding them.

Definition ♀

Absolute (Global) Extrema

We say the function f has an **absolute maximum** (or **global maximum**) on the domain D at the point c if $f(c) \ge f(x)$ for all $x \in D$. Consequently $f(c)$ is called the **maximum value** of f on D.

Similarly, if $f(c) \le f(x)$ for all $x \in D$, we would say f has an **absolute minimum** (or **global minimum**) at c. We call $f(c)$ the **minimum value** of f on D.

These extreme values of f are termed *absolute* or *global* because they are, respectively, the largest and smallest values of the function over D.

Having defined absolute extrema, it is important to realize that a function may not actually possess an absolute maximum or an absolute minimum on a given domain D. Example 1 shows that the definition of both the function and the domain under consideration are critical when looking for extrema.

Example 1 ✎

Identify the absolute extrema, if they exist.

a. $f(x) = x^2$ on $D = (-\infty, \infty)$ **b.** $f(x) = x^2$ on $D = (0,1]$

c. $f(x) = x^2$ on $D = [-1,0) \cup (0,1]$ **d.** $f(x) = x^2$ on $D = (0,1)$

e. $f(x) = \begin{cases} \dfrac{1}{x} & \text{if } x \ne 0 \\ 1 & \text{if } x = 0 \end{cases}$ on $D = [0,1]$ **f.** $f(x) = \sin\dfrac{1}{x}$ on $D = (0,1)$

Solution

a. The function $f(x) = x^2$ increases without bound as $x \to \pm\infty$, so it has no absolute maximum over $(-\infty, \infty)$. But clearly $f(0) = 0$ is the absolute minimum over the domain (see Figure 1).

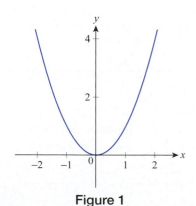

Figure 1

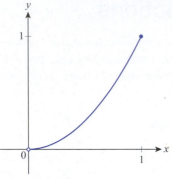

Figure 2

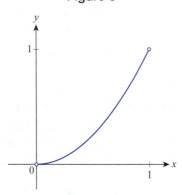

Figure 3

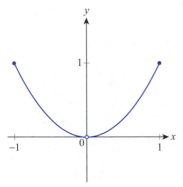

Figure 4

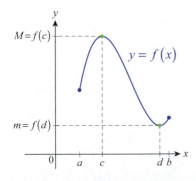

Figure 7 Extreme Values

b. On the interval $(0,1]$, $f(x) = x^2$ has an absolute maximum of $f(1) = 1$, but no absolute minimum value (see Figure 2).

c. On the domain $[-1,0) \cup (0,1]$, $f(x) = x^2$ has an absolute maximum of $f(-1) = f(1) = 1$, but no absolute minimum (see Figure 3).

d. On the interval $(0,1)$, $f(x) = x^2$ has neither an absolute maximum nor an absolute minimum (see Figure 4).

e. On the domain $[0,1]$, $f(x) = \begin{cases} 1/x & \text{if } x \neq 0 \\ 1 & \text{if } x = 0 \end{cases}$ has an absolute minimum of $f(0) = f(1) = 1$, but no absolute maximum (see Figure 5; note the different scales on the axes).

f. On the interval $(0,1)$, $f(x) = \sin(1/x)$ attains both its absolute maximum of 1 and its absolute minimum of -1 infinitely often (see Figure 6).

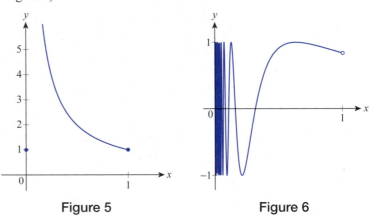

Figure 5   **Figure 6**

As Example 1 illustrates, absolute extrema may or may not exist for a given function and domain. Fortunately, there is a powerful theorem that guarantees the existence of both an absolute maximum and an absolute minimum under very common conditions. This theorem, while intuitively reasonable and easy to understand, actually requires a deep knowledge of the properties of the real line to prove; its proof (which typically is studied in a course like Advanced Calculus or Real Analysis) is tied to similarly deep statements about the real numbers such as the Bolzano-Weierstrass Theorem (see Appendix E for the statement of this theorem).

Theorem ⚭

The Extreme Value Theorem

If f is a continuous function defined on the closed interval $[a,b]$, then there exist points c and d in $[a,b]$ where f attains both its absolute maximum value $M = f(c)$ and absolute minimum value $m = f(d)$. (See Figure 7 for an illustrative depiction.)

The Extreme Value Theorem tells us quite a lot: first, the absolute extrema exist for a continuous function defined on a closed and bounded interval, and second, there are points in the interval where those extreme values are actually attained.

Given a continuous function f defined on an interval $[a,b]$, we are still left with the question of *how*, exactly, we find the absolute extrema. Visually, the graph of f over $[a,b]$ is likely to suggest fairly obvious candidates for the absolute extreme values—those points on the graph which are, in comparison to nearby points, either high spots or low spots. Our mathematical language for such points is as follows.

Definition ♀

Relative (Local) Extrema

We say the function f has a **relative maximum** (or **local maximum**) at the point c if $f(c) \geq f(x)$ for all x in some open interval containing c.

Similarly, if $f(c) \leq f(x)$ for all x in some open interval containing c, we say f has a **relative minimum** (or **local minimum**) at c.

We can extend the definition to the endpoints of an interval by saying f has a relative extremum at an endpoint of $[a,b]$ if f attains its maximum or minimum value at that endpoint in a half-open interval containing it.

Example 2 ✎

Identify the relative and absolute extrema for the function f over the interval $[a,b]$ graphed in Figure 8.

Solution

The function has relative maxima at the points c and e, since $f(c)$ and $f(e)$ are the largest values f attains over small open intervals containing c and e. To illustrate this fact, a small portion of the graph focusing on $(c, f(c))$ is reproduced in Figure 9; note that $f(c) \geq f(x)$ for all x near c.

Similarly, a, d, and b are points where f has relative minima (note that a and b are also the endpoints of the interval).

By comparing the values of f at each of the five points a through e, we see that $f(c)$ is the absolute maximum and $f(b)$ is the absolute minimum.

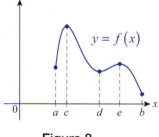

Figure 8

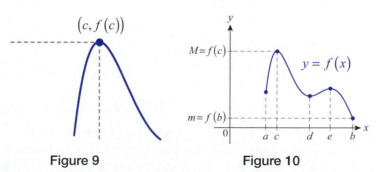

Figure 9　　　　　　**Figure 10**

TOPIC 2   Finding Extrema

Up to this point, our definitions and examples in this section have not called upon any ideas unique to calculus—the notions of absolute and relative extrema make sense independently of calculus. The following theorem, the essence of which was conceived by the French mathematician Pierre de Fermat (1601–1665) even before the formalization of calculus, shows how differentiation greatly simplifies the task of finding extrema.

Theorem

Fermat's Theorem

If f has a relative extremum at a point $c \in (a, b)$, and if $f'(c)$ exists, then $f'(c) = 0$.

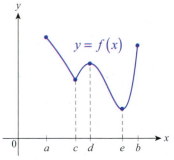

Figure 11
Applying Fermat's Theorem

Before proving Fermat's Theorem, it's useful to consider a graph such as the one sketched in Figure 11. For this function, there are three points in (a, b), referred to as the *interior* of the interval $[a, b]$, where relative extrema occur. The first point, $(c, f(c))$, is a cusp of the graph, and so $f'(c)$ does not exist—Fermat's Theorem does not apply here. At points d and e, the function f makes a transition from rising to falling and then from falling to rising. This means that the sign of f' will pass from positive to negative and then back to positive as we scan the graph from left to right through d and e, and it makes intuitive sense to suppose that f' will equal 0 at the two transition points. The proof is merely a rigorous rendition of this argument.

Proof

We will first assume that f has a relative maximum at $c \in (a, b)$ and that $f'(c)$ exists; from these two facts, we will prove that $f'(c) = 0$. A similar argument can be used to show that if f has a relative minimum at c and if $f'(c)$ exists, then again $f'(c) = 0$.

Since $f'(c)$ exists, we know that the limit $\lim\limits_{x \to c} \dfrac{f(x) - f(c)}{x - c}$ exists (recall that this is one formulation of the difference quotient), and since f has a relative maximum at c, $f(x) - f(c) \leq 0$ for all x close to c; this is just a different way of expressing the fact that $f(c) \geq f(x)$ for all x in some open interval containing c. The existence of the two-sided limit above means both one-sided limits exist and are equal to $f'(c)$, so

$$f'(c) = \lim_{x \to c^-} \frac{f(x) - f(c)}{x - c} \geq 0 \qquad x - c < 0 \text{ as } x \to c \text{ from the left}$$

and

$$f'(c) = \lim_{x \to c^+} \frac{f(x) - f(c)}{x - c} \leq 0 \qquad x - c > 0 \text{ as } x \to c \text{ from the right}$$

The only number that is both nonnegative and nonpositive is 0, so $f'(c) = 0$.

The only change necessary if we begin with the assumption that f has a relative minimum at c is to note that $f(x) - f(c) \geq 0$ for all x sufficiently close to c and to reverse the two inequalities above.

With Fermat's Theorem as inspiration, we are led to the following definition.

Definition ♀

Critical Point

Assume that f is defined on an open interval containing c. We say c is a **critical point** of the function f if $f'(c) = 0$ or $f'(c)$ does not exist.

Thus if f is defined on a closed interval $[a,b]$, the only candidates for extrema (relative or absolute) are critical points in (a,b) and the two endpoints a and b.

Caution ⚠

Don't read too much into the definition of critical point and Fermat's Theorem. Points where the derivative fails to exist are not guaranteed to be relative extrema, as the first graph in Figure 12 illustrates. Similarly, the converse of the theorem is false: points at which the derivative is 0 are not necessarily relative extrema, as seen in the second graph of Figure 12.

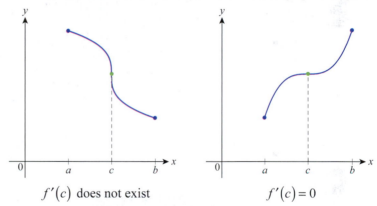

$f'(c)$ does not exist $f'(c) = 0$

Figure 12 Critical Points That Are Not Extreme Points

Nevertheless, the Extreme Value Theorem and Fermat's Theorem together provide a very useful method for locating the absolute extrema of a function defined on a closed and bounded interval.

Finding Absolute Extrema

Assume that f is continuous on the interval $[a,b]$. To find the absolute extrema of f on $[a,b]$, perform the following steps.

Step 1: Find the critical points of f in (a,b).

Step 2: Evaluate f at each of the critical points in (a,b) and at the two endpoints a and b.

Step 3: Compare the values of the function found in Step 2. The largest is the absolute maximum of f on $[a,b]$ and the smallest is the absolute minimum of f on $[a,b]$.

Example 3 ✐

Find the absolute extrema of $f(x) = 3x^4 - 4x^3 - 12x^2$ on the interval $[-2, 3]$.

Solution

$$\begin{aligned} f'(x) &= 12x^3 - 12x^2 - 24x \\ &= 12x(x^2 - x - 2) \\ &= 12x(x-2)(x+1) \end{aligned}$$

The derivative of f is defined for all real numbers. So the only critical points are those where $f'(x) = 0$; the solutions of this equation are 0, 2, and -1, all of which are in the interval $[-2, 3]$. We now compute f at the endpoints and the three critical points in the given interval.

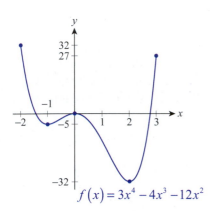

Figure 13

$f(x) = 3x^4 - 4x^3 - 12x^2$	Point in $[-2, 3]$
$f(-2) = 32$	Left endpoint
$f(-1) = -5$	Critical point
$f(0) = 0$	Critical point
$f(2) = -32$	Critical point
$f(3) = 27$	Right endpoint

Comparing these values, we see that the absolute maximum of f on $[-2, 3]$ is $f(-2) = 32$ and that the absolute minimum is $f(2) = -32$.

Example 4 ✐

Find the absolute extrema of $f(x) = x^{2/3}$ on the interval $[-1, 2]$.

Solution

$$f'(x) = \frac{2}{3}x^{-1/3} = \frac{2}{3x^{1/3}}$$

The derivative of f is never 0. Since f' is, however, undefined at $x = 0$, 0 is the one critical point of f. Comparing the values of f at the two endpoints and 0 we have:

$$f(-1) = (-1)^{2/3} = \left[(-1)^2 \right]^{1/3} = 1,$$

$$f(0) = 0, \quad \text{and}$$

$$f(2) = 2^{2/3} = \left(2^2 \right)^{1/3} = 4^{1/3} \approx 1.59.$$

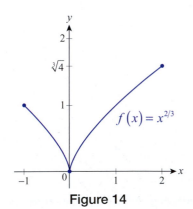

Figure 14

So the absolute minimum value of f on $[-1, 2]$ is 0 and the absolute maximum value is $\sqrt[3]{4}$.

Example 5 ✎

Find the absolute extrema of $f(x) = x^{3/5}(2-x)$ on the interval $[-1,2]$.

Solution

We can use the Product Rule to find f', as follows.

$$f'(x) = \frac{3}{5x^{2/5}}(2-x) + x^{3/5}(-1)$$

$$= \frac{6-3x}{5x^{2/5}} - \frac{5x^{2/5}x^{3/5}}{5x^{2/5}}$$

$$= \frac{6-8x}{5x^{2/5}}$$

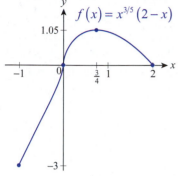

$f(x) = x^{3/5}(2-x)$

Figure 15

From this, we see that f' does not exist at $x = 0$, and that $f'\left(\frac{3}{4}\right) = 0$. We now note that $f(-1) = -3$, $f(0) = 0$, $f\left(\frac{3}{4}\right) = \left(\frac{3}{4}\right)^{3/5}\left(\frac{5}{4}\right) \approx 1.05$, and $f(2) = 0$. So f has its absolute minimum value of -3 at the left endpoint of $[-1,2]$, and its absolute maximum value of approximately 1.05 at $x = \frac{3}{4}$.

Our final example will be a brief introduction to a class of problems called *optimization problems*, which we will study in depth in Section 4.6. In these sorts of problems, we put the tools of calculus to use in finding an optimal solution to a problem from among many possible solutions.

Example 6 ✎

A new electrical power substation is being built on the right bank of a river. Cables to the substation need to be laid underground and underwater from another substation 3 kilometers upstream and on the opposite bank. The river follows a straight route through this stretch and has a fairly constant width of 1 kilometer. Given that the cost of laying cable underground is $30,000/km and the cost of laying cable underwater is $50,000/km, how should the cable be laid in order to minimize cost?

Solution

Even bearing in mind that the shortest distance between two points is a straight line, there are an infinite number of what could be deemed as reasonable solutions to this problem.

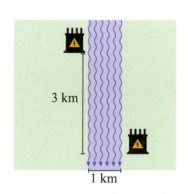

3 km

1 km

Figure 16

One very elementary approach would be to simply lay all the cable diagonally underwater in one straight shot, but given the fact that underwater installation is more expensive than underground installation, this is unlikely to be the most economical solution. Another approach would be to cross the river in the shortest possible path (so as to minimize underwater installation) and then run cable 3 kilometers along the bank of the river. But any combination of the two approaches is also a possibility, and some combination may lead to the absolute minimum cost. Calculus allows us to find that optimal solution.

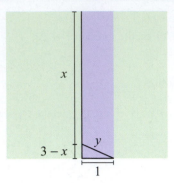

Figure 17

We will begin with a schematic that defines and relates the variable quantities that we have at our disposal. We can assume that the underground installation follows the left bank of the river, as shown in Figure 17; let x represent the length of the underground cable in kilometers. Let y then represent the length of the underwater installation. (We could have proceeded by first crossing the river and then laying cable down the right bank, but for given lengths x and y the cost would be the same.)

Note that $x = 0$ corresponds to the solution in which all the cable is underwater, and $x = 3$ corresponds to the solution in which the shortest length of underwater installation ($y = 1$) is called for. Any value for x in the interval $[0,3]$ corresponds to an intermediate solution, as discussed above.

From the diagram, $y^2 = 1^2 + (3-x)^2$, so $y = \sqrt{1+(3-x)^2}$ and the total cost of laying the cable for a given value of x is as follows.

$$C = 30,000x + 50,000y = 30,000x + 50,000\sqrt{1+(3-x)^2}$$

We are looking for the absolute minimum of the function C on the interval $[0,3]$, so we proceed to find the critical points of C.

$$C(x) = 30,000x + 50,000\left[1+(3-x)^2\right]^{1/2}$$

$$C'(x) = 30,000 + 50,000\left(\frac{1}{2}\right)\left[1+(3-x)^2\right]^{-1/2}\left[2(3-x)\right](-1)$$

$$= 30,000 + \frac{50,000(x-3)}{\sqrt{1+(3-x)^2}}$$

The derivative C' exists for all x, but there are two values of x for which $C'(x) = 0$.

$$30,000 + \frac{50,000(x-3)}{\sqrt{1+(3-x)^2}} = 0$$

$$\frac{(x-3)}{\sqrt{1+(3-x)^2}} = -\frac{30,000}{50,000}$$

$$\frac{(x-3)^2}{1+(3-x)^2} = \frac{9}{25} \qquad \text{\textcolor{blue}{Square both sides.}}$$

$$25(x-3)^2 = 9 + 9(x-3)^2 \qquad \text{\textcolor{blue}{Note that }} (3-x)^2 = (x-3)^2.$$

$$16(x-3)^2 = 9$$

$$x - 3 = \pm\frac{3}{4} \qquad \text{\textcolor{blue}{Divide by 16 and take the square root.}}$$

$$x = 3 \pm \frac{3}{4}$$

The point $3 + \frac{3}{4}$ does not actually solve the original equation, so we only have to evaluate C at the two endpoints 0 and 3 and at the critical point $3 - \frac{3}{4} = \frac{9}{4}$.

$$C(0) = 50,000\sqrt{10} \approx \$158,114$$

$$C\left(\frac{9}{4}\right) = \$130,000$$

$$C(3) = \$140,000$$

From this comparison, we see that the minimal cost of installation can be achieved by laying the cable with an underground run of $\frac{9}{4} = 2\frac{1}{4}$ kilometers and a diagonal run underwater of $\sqrt{1+\left(\frac{3}{4}\right)^2} = \sqrt{\frac{25}{16}} = 1\frac{1}{4}$ kilometers.

4.1 **Exercises**

1–4 Use the graph as an aid to identify the absolute extrema, if they exist, for the given function on the specified domain.

1. $f(x) = -x^2 + 1;\quad D = [-1,0) \cup (0,1]$

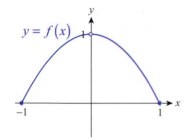

2. $f(x) = -\cos x;\quad D = (0, 2\pi]$

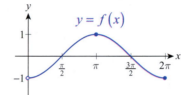

3. $f(x) = 2\sin x;\quad D = [0, 2\pi]$

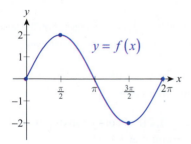

4. $f(x) = |x-1|;\quad D = (-1, 2]$

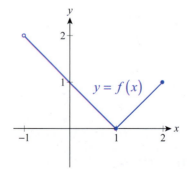

5–8 Use the graph to decide whether each highlighted point is a critical point, and then find and classify all relative and absolute extrema for the function over the given interval.

5.

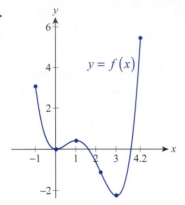

6.

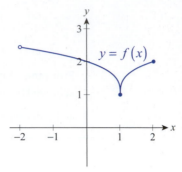

7.

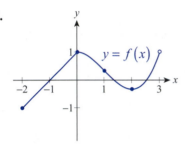

8.

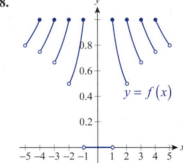

9–20 Use graph paper to sketch the graph of the given function on the specified domain, and then use the graph to visually identify and classify any absolute extrema.

9. $f(x) = 2x + 1;\quad D = [0,3]$

10. $g(x) = -x - 1;\quad D = (-1,2]$

11. $h(x) = \dfrac{1}{2}x - 3;\quad D = \mathbb{R}$

12. $u(x) = -x^2;\quad D = (-1,1)$

13. $v(x) = (x+1)(x-3);\quad D = [-2,4]$

14. $k(x) = (x-4)^4;\quad D = \mathbb{R}$

15. $K(x) = x^7;\quad D = \mathbb{R}$

16. $m(x) = e^{-x+2};\quad D = [2,\infty)$

17. $n(x) = \cos \pi x;\quad D = \left(0, \dfrac{3}{2}\right]$

18. $F(x) = \dfrac{1}{(x+1)^2};\quad D = \mathbb{R}$

19. $G(x) = \dfrac{1}{x^2 + 1};\quad D = \mathbb{R}$

20. $H(t) = \arcsin t;\quad D = [-1,1]$

21–37 Sketch by hand the graph of a function f on the specified domain, with the specified properties. (Answers will vary.)

21. Defined on $[2,4]$, absolute maximum at 2, absolute minimum at 4

22. Defined on $[-1,2]$, absolute maximum at 0, absolute minimum at 1

23. Defined on $[-5,5]$, absolute maximum at 1, absolute minimum at 5

24. Defined on $\mathbb{R}$, absolute minimum at 2, no absolute maximum

25. Defined on $[-3,2]$, absolute maximum at -2, absolute minimum at 0, relative maximum at 1

26. Defined on $[0,6]$, absolute maximum at 2, relative minimum at 4, no absolute minimum

27. Defined on $[-1,1]$, absolute maximum occurs twice, no minimum

28. Defined on $(1.5,7)$, continuous, has relative maximum and minimum, but no absolute maximum or minimum

29. Defined on $(1.5,7)$, continuous, has both absolute maximum and minimum

30. Defined on $[-2,4]$, two relative maxima, but no absolute maximum

31. Defined on $(0,\infty)$, continuous, no relative or absolute extrema

32. Defined on $(-1,3]$, continuous, no absolute minimum, one relative minimum, absolute maximum occurs twice

33. Defined on $\mathbb{R}$, both the absolute maximum and absolute minimum occur infinitely often

34. Defined on $(0,\infty)$, infinitely many relative maxima and minima, no absolute maximum or minimum

35. Differentiable on $\mathbb{R}$, has one critical point, but no extrema

36. Defined on $(0,10)$, not differentiable at 5, but absolute maximum occurs at 5

37. Defined on $(0,10)$, discontinuous at 5, but absolute maximum occurs at 5

38–55 Find all critical points, if they exist, for the given function.

38. $f(x) = x^2 - 7x + 1.5$

39. $g(x) = 2x^3 + 3x^2 - 12x + 1.5$

40. $h(x) = x^3 + 1.5x^2 + 3x - 2.5$

41. $u(x) = -\dfrac{3}{2}x + 2$

42. $v(x) = x^4 - \dfrac{16}{3}x^3 + 2x^2 + 24x - 1$

43. $k(x) = |2x - 3|$

44. $K(x) = |3x^2 + 3x - 18|$

45. $m(x) = \dfrac{2-x}{x^2 - x + 2}$

46. $n(x) = \dfrac{|x^2 - 2|}{2x^2 + 4}$

47. $F(t) = \sqrt{3 + 3t^2}$

48. $G(x) = x^{3/2} - 3\sqrt{x}$

49. $T(s) = 2\sqrt[3]{s}(s - 2)$

50. $r(v) = \dfrac{v-1}{\sqrt{v}}$

51. $s(\alpha) = \cos\alpha + \cos^2\alpha$

52. $u(z) = \cot z + 2z$

53. $t(x) = \sqrt{x}\ln x$

54. $U(t) = e^t \sin t$

55. $a(t) = \cos(\arctan t)$

56–77 Find all absolute extrema of the function on the given closed interval.

56. $f(x) = 4x - x^2$ on $[0,6]$

57. $g(x) = 3x^2 - 30x + 7$ on $[0,8]$

58. $h(x) = x^3 + 1.5x^2 - 6x + 3.5$ on $[-4,3]$

59. $u(x) = 3x^4 - 8x^3 + 6x^2 - 24x - 9$ on $[0,3]$

60. $v(x) = \dfrac{x^4}{4} - 2x^2 + 4$ on $[-2,2]$

61. $k(x) = \dfrac{x^4}{2} + 2x^3 - x^2 - 6x + \dfrac{1}{2}$ on $[-3,1]$

62. $f(x) = |x + 3| \cdot |x - 3|$ on $[-4,4]$

63. $m(x) = |x + 3| + |x - 3|$ on $[-4,4]$

64. $n(x) = \dfrac{3x}{2x^2 + 2}$ on $[-4,4]$

65. $g(x) = \dfrac{x^2 + 5}{x + 2}$ on $[-1.5,1.5]$

66. $F(x) = \dfrac{1}{1 + x^2}$ on $[-10,10]$

67. $G(t) = \dfrac{1}{\sqrt{t}} + \sqrt{t}$ on $\left[\dfrac{1}{4},4\right]$

68. $k(s) = (s^2 - 1)\sqrt{s}$ on $[0,2]$

69. $r(z) = \sin(\arccos z)$ on $[-1,1]$

70. $G(x) = \arctan x$ on $[-1,1]$

71. $w(x) = x\sqrt{8 - x^2}$ on $\left[-2\sqrt{2},2\sqrt{2}\right]$

72. $T(s) = s^2 e^{-s}$ on $[0,10]$

73. $r(x) = (\cos x)e^x$ on $\left[0, \dfrac{3\pi}{2}\right]$

74. $L(x) = x \ln x$ on $\left[\dfrac{1}{e^2}, e\right]$

75. $t(x) = \ln\big((e-1)\sin \pi x + 1\big)$ on $[0,1]$

76. $U(x) = \sqrt[3]{x}(x-3)$ on $[-1,3]$

77. $V(x) = 5 + (5+x)x^{5/7}$ on $[-5,1]$

78–89 Find and classify the absolute extrema, if they exist, of the function over the given domain.

78. $f(x) = 3x - 2; \quad D = (0,2]$

79. $g(x) = x^2 - 4; \quad D = (-2,2)$

80. $h(x) = 2x^3 - 5x; \quad D = \mathbb{R}$

81. $K(z) = \sqrt{4 - z^2}; \quad D = (-2,2)$

82. $r(z) = -\dfrac{2}{z}; \quad D = [2,\infty)$

83. $n(x) = \dfrac{1}{(x+3)^2}; \quad D = (-3,\infty)$

84. $t(x) = 10^{x/2}; \quad D = \mathbb{R}$

85. $L(x) = \ln(x+1); \quad D = [0,\infty)$

86. $F(x) = \sec x; \quad D = \left(-\dfrac{\pi}{2}, \dfrac{\pi}{2}\right)$

87. $t(z) = 2\cos \pi z + 2; \quad D = \mathbb{R}$

88. $u(x) = x - [\![x]\!]; \quad D = [1,3)$

89. $v(x) = \arctan x; \quad D = \mathbb{R}$

90. Find two numbers whose sum is 50 and whose product is as large as possible. (**Hint:** Denote the numbers by x and $50 - x$, and maximize the product.)

91. A 50-inch piece of wire is cut into two pieces, which are then bent into a square and a circle, respectively. Where should the wire be cut in order to minimize the sum of the areas of these two shapes? (**Hint:** Start with the notation of Exercise 90, and use appropriate formulas from geometry.)

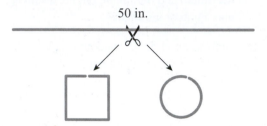

92. A lighthouse is 5 miles off a straight shoreline. Ten miles down the coast is a restaurant where the lighthouse keeper is planning to meet his friends. If he can row at 2.5 mph and walk at 4 mph, where should he land in order to make the fastest possible time to the restaurant?

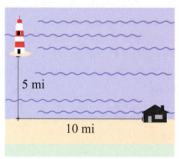

93. Referring to Exercise 26 of Section 3.7, find the number of calculators that have to be produced in order to maximize profit.

94. The power output of a 12-volt car battery when a resistor is connected to it, is given by the formula $P = 12I - (R+r)I^2$, where I is the current (in amperes), and r stands for the (typically very small) so-called internal resistance of the battery. Suppose we are starting a car with a starter motor of resistance $R = 0.16$ ohms, and that the internal resistance of the battery is $r = 0.016$ ohms. Find the current that corresponds to the battery's maximum power output.

95–105 *True or False?* Determine whether the given statement is true or false. In case of a false statement, explain or provide a counterexample.

95. If f attains both its absolute minimum and absolute maximum values on a closed interval, then f is a continuous function.

96. A continuous function on a closed interval can attain its absolute extrema only at critical points.

97. If $f(x)$ is a differentiable function and k is a constant, then $f(x)$ and $f(x)+k$ have the same critical points.

98. If $f(x)$ is a differentiable function and k is a nonzero constant, then $f(x)$ and $kf(x)$ have the same critical points.

99. If $f(x)$ is a differentiable function and k is a nonzero constant, then $f(x)$ and $f(x+k)$ have the same critical points.

100. If $f(x)$ is a differentiable function and k is a nonzero constant, then $f(x)$ and $f(kx)$ have the same critical points.

101. If $f(x)$ has a maximum at c, then so does $f(-x)$ at $-c$.

102. If $f(x)$ has a maximum at c, then $-f(x)$ has a minimum at c.

103. $f(x)$ can have more than one absolute maximum value.

104. If $f(x)$ is continuous on a closed interval I, then it attains its minimum value on I.

105. If $f(x)$ has no maximum on a closed interval I, then $f(x)$ must be discontinuous on I.

4.1 **Technology Exercises**

106–127. Use a graphing calculator or computer algebra system to verify the answers you obtained for Exercises 56–77.

4.2 **The Mean Value Theorem**

TOPICS

1. Rolle's Theorem and the Mean Value Theorem

2. Consequences of the Mean Value Theorem

In this section, we introduce and prove the Mean Value Theorem (MVT), a result that relates the behavior of a differentiable function to that of its derivative in a fundamental and useful way. As we will see, one use of the MVT is to further strengthen the connection between the average rate and instantaneous rate of change; another is in the proof of an impressively named result that we will encounter in Chapter 5 called the Fundamental Theorem of Calculus.

TOPIC 1   **Rolle's Theorem and the Mean Value Theorem**

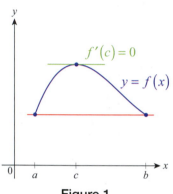

Figure 1

Our development begins with a preliminary observation first proved by the French mathematician Michel Rolle (1652–1719) in 1691. Suppose f is a function that is continuous on the interval $[a,b]$, and suppose further that f has the same value at the two endpoints a and b, as illustrated in Figure 1. Your intuition may suggest that if the graph of f begins and ends with the same value and varies in a sufficiently "smooth" manner over the course of the interval, then there must be at least one point c where it has a horizontal tangent line. This is exactly what Rolle's Theorem guarantees, and the precise property corresponding to a "smoothly" varying graph is differentiability on the open interval (a,b).

> ### Theorem 🔍
>
> **Rolle's Theorem**
>
> If f is continuous on the closed interval $[a,b]$ and differentiable on (a,b), and if $f(a) = f(b)$, then there is at least one point $c \in (a,b)$ for which $f'(c) = 0$.

> ### Proof 📎
>
> Because f is continuous on the closed interval $[a,b]$, the Extreme Value Theorem tells us that f attains its absolute maximum and minimum values on $[a,b]$. If f attains both its absolute maximum and absolute minimum at an endpoint of the interval, then f is a constant function on $[a,b]$ (note that in this case $f(x) = f(a) = f(b)$ for all x in the interval) and so we know that $f'(x) = 0$ for all $x \in (a,b)$, proving the claim.
>
> If, on the other hand, f assumes either its absolute maximum or absolute minimum at an interior point $c \in (a,b)$, then $f(c)$ is a relative extremum of the function and by Fermat's Theorem $f'(c) = 0$, since by assumption $f'(x)$ exists for all x in (a,b).

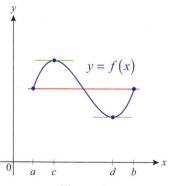

Figure 2

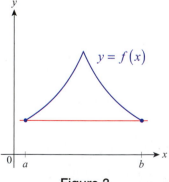

Figure 3

There may be more than one point in (a,b) where f has a horizontal tangent line when the hypotheses of Rolle's Theorem are satisfied, but in practice it suffices to know for certain that there is at least one such point. Theorems of this sort are known as *existence* theorems, and they often provide great insight into the nature of objects under study. Note that in Figure 2, $f'(c) = f'(d) = 0$. In Figure 3, there is no point c in $[a,b]$ for which $f'(c) = 0$, but this does not violate Rolle's Theorem as the function is not differentiable everywhere on (a,b).

One of the most notable uses of Rolle's Theorem is as a stepping-stone toward the Mean Value Theorem, but it can also be put to more immediate use.

Example 1 ✐

Show that the equation $x^5 + 3x = 2$ has exactly one real solution.

Solution

Let $f(x) = x^5 + 3x - 2$. Then $f'(x) = 5x^4 + 3$, which is defined and positive (in fact, greater than or equal to 3) for all real x. If there were two solutions $x = a$ and $x = b$ of the equation $x^5 + 3x = 2$, then we would have $f(a) = 0 = f(b)$, and by Rolle's Theorem there would then be a point c between a and b for which $f'(c) = 0$, contradicting $f'(x) > 0$ for all x. So $x^5 + 3x = 2$ has no more than one solution.

The fact that it does indeed have a solution follows from the observation that $f(0) = -2 < 0$ and $f(1) = 2 > 0$, so by the Intermediate Value Theorem there is a point c between 0 and 1 for which $f(c) = 0$ (see Section 2.5 for a refresher on the use of the IVT in locating zeros of functions).

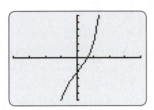

Figure 4

$f(x) = x^5 + 3x - 2$

on $[-4, 4]$ by $[-6, 6]$

Example 2 ✐

Suppose that $f(t)$ is a differentiable function describing the location at time t of an object moving along a straight line. If $f(t_1) = f(t_2)$ for two points in time t_1 and t_2, then there must be a time t_3 between t_1 and t_2 for which $f'(t_3) = 0$. That is, the velocity of the object at t_3 is 0.

As a particular example, if $h(t)$ represents the vertical height of a thrown object with, say, $h(t_1) = h(t_2) = 0$, then $h'(t_3) = 0$ and $h(t_3)$ represents the maximum height achieved by the object. (Note that in this example, $h(t)$ represents the motion up and down relative to ground level; the object may also travel horizontally, as shown in Figure 5, but such motion happens independently of $h(t)$ and would be measured by another function.)

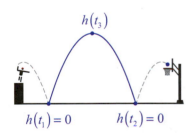

$h(t_3)$

$h(t_1) = 0 \qquad h(t_2) = 0$

Figure 5

The Mean Value Theorem (MVT), first stated in a form similar to the following by Joseph-Louis Lagrange (1736–1813), is a generalization of Rolle's Theorem—it provides the same sort of insight into the behavior of a function defined on an interval $[a, b]$ while relaxing one of the hypotheses.

Theorem ⚿

The Mean Value Theorem

If f is continuous on the closed interval $[a, b]$ and differentiable on (a, b), then there is at least one point $c \in (a, b)$ for which $f'(c) = \dfrac{f(b) - f(a)}{b - a}$.

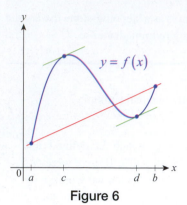

Figure 6

In preparation for proving the MVT, it's important to understand geometrically what it says. Figure 6 shows how the MVT can be viewed as a "slanted" version of Rolle's Theorem, in that it guarantees the existence of at least one point $c \in (a, b)$ at which the line tangent to the graph of f is parallel to the secant line through $(a, f(a))$ and $(b, f(b))$. In Figure 6, this conclusion is actually true for two points c and d in the interval.

Proof 🔗

The sort of depiction shown in Figure 6 points the way toward a proof of the MVT: if we can modify the function under consideration by "unslanting" it, we can then apply Rolle's Theorem. To do this, we want to consider a new function defined as the difference between f and the function whose graph is the secant line passing through $(a, f(a))$ and $(b, f(b))$, shown in red in the figure.

The slope of the secant line is

$$m_{sec} = \frac{f(b) - f(a)}{b - a},$$

so $y - f(a) = m_{sec}(x - a)$ is an equation for the secant line in point-slope form; solving for y, the secant line is the graph of the function $L(x) = f(a) + m_{sec}(x - a)$. If we now define a new function g by

$$\begin{aligned} g(x) &= f(x) - L(x) \\ &= f(x) - f(a) - m_{sec}(x - a), \end{aligned}$$

then we have, by design, constructed a function for which

$$g(a) = f(a) - f(a) - m_{sec}(a - a) = 0$$

and

$$\begin{aligned} g(b) &= f(b) - f(a) - m_{sec}(b - a) \\ &= f(b) - f(a) - [f(b) - f(a)] = 0. \end{aligned}$$

That is, $g(a) = g(b)$, satisfying one of the hypotheses of Rolle's Theorem.

The other two hypotheses of Rolle's Theorem are also satisfied by g. Since L is continuous on $[a, b]$ and differentiable on (a, b), as is f, then $g = f - L$ possesses these two properties as well. Rolle's Theorem thus tells us there is at least one point c in (a, b) for which $g'(c) = 0$. Since $g'(x) = f'(x) - L'(x) = f'(x) - m_{sec}$, we have

$$\begin{aligned} 0 &= g'(c) \\ &= f'(c) - \frac{f(b) - f(a)}{b - a}, \end{aligned}$$

or

$$f'(c) = \frac{f(b) - f(a)}{b - a}.$$

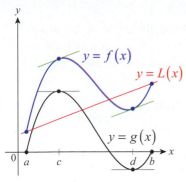

Figure 7

Figure 7 illustrates the process of subtracting the secant line L from the function f of Figure 6 in preparation for applying Rolle's Theorem. Note that c and d are points where g has horizontal tangent lines and f has tangent lines with slope m_{sec}.

Example 3 ✎

The function $f(x) = \frac{1}{3}x^3 - 2x^2 + 5x + 1$ is differentiable everywhere, so it satisfies the hypotheses of the Mean Value Theorem over any closed, bounded interval. Find the point(s) satisfying the conclusion of the MVT over the interval $[0, 4]$.

Solution

We begin by calculating the slope of the secant line over the interval.

$$m_{sec} = \frac{f(4) - f(0)}{4 - 0} = \frac{\frac{31}{3} - 1}{4} = \frac{7}{3}$$

The Mean Value Theorem guarantees the existence of at least one point $c \in (0, 4)$ where $f'(c) = \frac{7}{3}$, but we want to go one step further and actually find the point (or points). Since $f'(x) = x^2 - 4x + 5$, we proceed to solve the equation $x^2 - 4x + 5 = \frac{7}{3}$.

$$x^2 - 4x + 5 = \frac{7}{3} \qquad \text{Set } f'(x) = m_{sec}.$$

$$x^2 - 4x + \frac{8}{3} = 0$$

$$3x^2 - 12x + 8 = 0$$

$$x = \frac{12 \pm \sqrt{144 - 96}}{6} \qquad \text{Apply the quadratic formula.}$$

$$x = 2 \pm \frac{2\sqrt{3}}{3}$$

These two solutions, approximately equal to 0.85 and 3.15, both satisfy the conclusion of the MVT (see Figure 8).

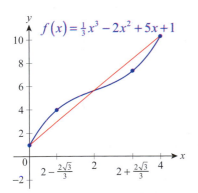

Figure 8

TOPIC 2   Consequences of the Mean Value Theorem

Recall that a discussion of average and instantaneous velocities was one of the motivating examples we used to develop the concept of differentiation. The MVT provides a more quantitative connection between these ideas.

Example 4 ✎

An aerial surveillance crew is monitoring traffic on the roads below and timing the passage of cars between painted road marks that are 1 mile apart. The crew spots a distinctive yellow Lotus Elise and records its passage between two marks at 47 seconds. What speed can the crew assert the Elise must have reached during the 47 seconds?

Solution

If we let $f(t)$ denote the position of the Elise along the stretch of road over the time interval $[0,47]$, then $f'(t)$ denotes the car's velocity and the Mean Value Theorem guarantees that there is at least one $c \in (0,47)$ for which

$$f'(c) = \frac{f(47) - f(0)}{47 - 0} = \frac{1 \text{ mile}}{47 \text{ s}} \cdot \frac{3600 \text{ s}}{1 \text{ hr}} \approx 77 \text{ mph.}$$

That is, at least once over the mile-long stretch of road, the Elise was moving at a rate of approximately 77 miles per hour.

We already know that a constant function has a derivative of zero at all points of its domain, but we don't yet know that the converse is true. That is, does the fact that $f'(x) = 0$ for all x in a given interval mean that f is a constant function? The Mean Value Theorem provides an elegant answer to this question.

Theorem 🔍

Corollary 1

If $f'(x) = 0$ for each x in an open interval (a,b), then f is constant on (a,b).

Proof 🖉

We will show that given any two distinct points x_1 and x_2 in (a,b), $f(x_1) = f(x_2)$. Since $x_1 \neq x_2$, one is less than the other; we can assume $x_1 < x_2$. The fact that f is differentiable on all of (a,b) means that f is continuous on $[x_1, x_2]$ and differentiable on (x_1, x_2), so by the MVT there is a point $c \in (x_1, x_2)$ such that

$$f'(c) = \frac{f(x_2) - f(x_1)}{x_2 - x_1}.$$

But since f' is 0 throughout (a,b), this means that

$$f(x_2) - f(x_1) = f'(c) \cdot (x_2 - x_1) = 0 \cdot (x_2 - x_1) = 0,$$

so $f(x_1) = f(x_2)$.

While much of the text to this point has focused on the theory and mechanics of differentiation (that is, gaining knowledge about a function's derivative), Corollary 1 is a good illustration of just the reverse: what does knowledge of a function's derivative tell us about the function itself? The next corollary is similar in nature, and we will see the theme continue to develop in successive sections.

Theorem 🔍

Corollary 2

If $f'(x) = g'(x)$ for all x in an open interval (a,b), then there is a constant C such that $f(x) = g(x) + C$ for all $x \in (a,b)$.

Proof ✎

Since $f'(x) = g'(x)$, we know that $(f-g)'(x) = f'(x) - g'(x) = 0$ for all x in (a,b). So by Corollary 1, it must be the case that $f - g$ is a constant function on (a,b). That is, there is a constant C such that $(f-g)(x) = C$ for all $x \in (a,b)$, or $f(x) = g(x) + C$.

Example 5 ✍

Suppose that $f(0) = -2$ and that $f'(x) \leq 3$ for all x. What is the largest possible value of $f(4)$?

Solution

The implication is that f' exists everywhere, so f is certainly continuous on $[0,4]$ and differentiable on $(0,4)$. So by the MVT,

$$f'(c) = \frac{f(4) - f(0)}{4 - 0} = \frac{f(4) + 2}{4}$$

for some $c \in (0,4)$. Since $f'(x) \leq 3$ for all x, rearranging the above equation gives us the following.

$$f(4) + 2 = 4f'(c)$$
$$f(4) = 4f'(c) - 2 \leq 4 \cdot 3 - 2$$
$$f(4) \leq 10$$

Hence the largest possible value of $f(4)$ is 10.

Example 6 ✍

Find the unique function f whose derivative is $3x^2$ and whose graph passes through $(1,5)$.

Solution

By now, we have differentiated enough polynomials to recognize $3x^2$ as the derivative of x^3. If we let $g(x) = x^3$, then g and f have the same derivative and by Corollary 2 it must be the case that $f(x) = g(x) + C$ for some constant C. We can determine C as follows.

$$f(1) = 5 \qquad \text{The graph of } f \text{ passes through } (1,5).$$
$$g(1) + C = 5$$
$$1^3 + C = 5$$
$$C = 4$$

Now it can be easily verified that $f(x) = x^3 + 4$ satisfies the given criteria, and by Corollary 2 it is the unique function to do so.

4.2 **Exercises**

1–4 Use the graph of the function to visually estimate the value of c in the given interval that satisfies the conclusion of the Mean Value (or Rolle's) Theorem; then check your guess by calculation. If such a c doesn't exist, explain why.

1. $f(x) = -x^2 + 6x - 4$ on $[1,4]$

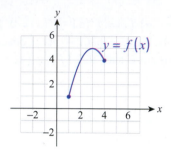

2. $g(x) = -2\sqrt{|x-2|} + 2$ on $[-2,6]$

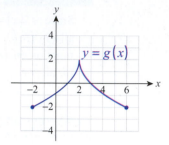

3. $h(x) = \dfrac{x^3}{3} - 2x^2 + \dfrac{11x}{3}$ on $[1,4]$

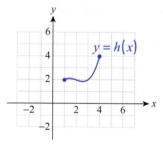

4. $k(x) = -\dfrac{1}{(x-3)^2} + 3$ on $[0,6]$

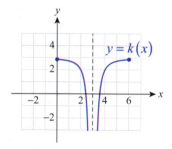

5–8 Prove that the equation has exactly one real solution on the given interval.

5. $x^5 - 3x^2 = 25$ on $[2,3]$

6. $5x^3 + 7x = 9$ on $\mathbb{R}$

7. $\arctan x = 3 - x$ on $[0,3]$

8. $\tan x = \cos x$ on $\left(0, \dfrac{\pi}{2}\right)$

9–20 Determine whether Rolle's Theorem applies to the function on the given interval. If so, find all possible values of c as in the conclusion of the theorem. If the theorem does not apply, state the reason.

9. $f(x) = -x^2 + 4x - 3$ on $[1,3]$

10. $g(x) = x^3 - 5x^2 + 2x + 10$ on $[-1,4]$

11. $h(x) = 2x^4 - x^3 + 6x^2 + x - 8$ on $[1,3]$

12. $F(x) = \dfrac{3}{(x-1)^2}$ on $[0,2]$

13. $G(x) = \dfrac{1}{x^2 + 1}$ on $[-3,3]$

14. $k(x) = \dfrac{x}{x^2 - 1}$ on $[-3,3]$

15. $H(x) = x^{4/5} - 10$ on $[-10,10]$

16. $m(x) = -\cos x$ on $[0, 4\pi]$

17. $T(z) = \cot z$ on $[0, 5\pi]$

18. $F(x) = \sec x$ on $\left[-\dfrac{\pi}{3}, \dfrac{\pi}{3}\right]$

19. $w(t) = \csc t$ on $\left[-\dfrac{\pi}{4}, \dfrac{\pi}{4}\right]$

20. $A(x) = |x+1| - 5$ on $[-6,4]$

21–32 Determine whether the Mean Value Theorem applies to the function on the given interval. If so, find all possible values of *c* as in the conclusion of the theorem. If the theorem does not apply, state the reason.

21. $f(x) = |x-1| + 3$ on $[-1,2]$

22. $g(x) = -\dfrac{1}{2}x + 4$ on $[0,8]$

23. $h(x) = -x^2 + 4x + 4$ on $[-1,4]$

24. $F(x) = \dfrac{1}{2}x^3 - x^2$ on $[-2,2]$

25. $G(x) = \dfrac{2x}{x+1}$ on $[-2,3]$

26. $k(x) = \dfrac{x+1}{2x-5}$ on $[-1,2]$

27. $H(x) = \dfrac{5}{x^2+5}$ on $[-5,5]$

28. $m(x) = (x-4)^{2/3} + 2$ on $[1,5]$

29. $C(x) = \sqrt{1-x^2}$ on $[-1,0]$

30. $L(t) = \ln|t+1|$ on $[-3,1]$

31. $u(z) = \tan z$ on $[0,\pi]$

32. $v(s) = \arctan s$ on $[0,1]$

33. If $f(-2) = 2$ and $f'(x) \le 2$ for all *x*, what is the largest possible value of $f(2)$?

34. If $g(1) = 4$ and $g'(x) \ge -3.5$ for all *x*, what is the smallest possible value of $g(3)$?

35. One of your classmates claims that he found a function *f* such that $f(-5) = -1$, $f(5) = 1$, and $f'(x) \le 0.1$ for all *x*. Explain how you know that he made an error in his calculations.

36. If $|F'(x)| \le 2.5$ for all *x*, prove that $|F(7) - F(3)| \le 10$.

37. Find the function *f* that passes through $(3,1)$ and whose derivative is $3x^2 - 2x + 1$.

38. Find the function *g* that passes through $(\pi/2, 0)$ and whose derivative is $2\cos x$.

39. Suppose that the velocity function of a moving object is $v(t) = -2t + 5$ and that its position at $t = 1$ is −8 units from the origin. Find a formula for the position function.

40. Find the velocity and position functions of an object thrown vertically upward from an initial height of 20 ft with initial velocity $v(0) = 42$ ft/s. (The acceleration caused by gravity is $a = -32$ ft/s². Ignore air resistance.)

41. Generalize Example 1 by proving the following: if $f(x)$ is differentiable on $[a,b]$, $f'(x) \ne 0$ on (a,b), and $f(a)$ and $f(b)$ have opposite signs, then the equation $f(x) = 0$ has exactly one real solution.

42. Two highway patrol cars are stationed 7 miles apart along a straight highway where the speed limit is 65 mph. The first patrol car clocks a red Porsche at 64 mph, and five minutes later, the second police car clocks it at 61.5 mph. Explain why the driver is pulled over and issued a ticket.

43. A plane leaves London Heathrow Airport to arrive at Houston Intercontinental Airport ten and a half hours later. The distance between the two airports is 4830 miles. Explain why the plane must have reached a speed of 450 mph at least twice during the trip.

44. Two MotoGP riders finish a race in a tie. Show that there was at least one moment during the race when the two riders had the exact same speed. (**Hint:** Ignoring the difference in their starting positions, consider the difference of the two position functions and use Rolle's Theorem.)

45. Suppose that $f'(x) = x$ for all $x \in \mathbb{R}$. Prove that there exists a constant *C* such that $f(x) = \frac{1}{2}x^2 + C$. (**Hint:** Use Corollary 2 from the text.)

46. Use Corollary 2 of this section to prove the well-known trigonometric identity $\cos^2 x + \sin^2 x = 1$. (**Hint:** Use the corollary for the functions $f(x) = \cos^2 x + \sin^2 x$ and $g(x) = 1$, and argue that necessarily $C = 0$.)

47. Follow the hint given in Exercise 46 to prove the identity $\sin^{-1} x + \cos^{-1} x = \pi/2$.

48. Let $f(x) = \cot \pi x$ and $g(x) = \cot \pi x + [\![x]\!]$. Show that $f'(x) = g'(x)$, but $f(x) - g(x)$ is not constant. Why does this not contradict Corollary 2 of this section?

49. Suppose that for a function $f(x)$, the second derivative $f''(x)$ exists for all x on an interval I. Prove that if f has three zeros on I, then its second derivative f'' also has a zero. (**Hint:** Apply Rolle's Theorem on f and then on f'.)

50. Suppose that f is twice differentiable on $\mathbb{R}$ and a and b are two successive zeros of f'. Prove that f can have at most one zero on (a, b). (**Hint:** Start by assuming that f has at least two zeros, and apply Rolle's Theorem.)

51. Suppose that f is continuous on $[a, b]$, differentiable on (a, b), and $c \in (a, b)$ with $f'(c) = 0$. Does it follow that $f(a) = f(b)$? Explain.

52.* Suppose that f is twice differentiable on $\mathbb{R}$ and $f'(c) = 0$ for some $c \in \mathbb{R}$. If $f''(c) \neq 0$, prove that there exist $a, b \in \mathbb{R}$ such that $c \in (a, b)$ and $f(a) = f(b)$.

53. Suppose that the rabbit population at a game preserve is observed monthly and is found to have increased from 550 to 1150 rabbits in a year's time. Explain why there must exist a time during the year when the population is increasing at a rate of 50 rabbits per month.

54. Show that if f is a quadratic function, that is, $f(x) = Ax^2 + Bx + C$, then on any interval $[a, b]$, the point c satisfying the conclusion of the Mean Value Theorem is the midpoint of the interval.

55.* Prove that for all $0 < \alpha < \beta < \pi/2$,
$(\beta - \alpha)\cos \beta \leq \sin \beta - \sin \alpha \leq (\beta - \alpha)\cos \alpha$.
(**Hint:** Apply the Mean Value Theorem for the function $f(x) = \sin x$ on the interval $[\alpha, \beta]$.)

56. Use the Mean Value Theorem to prove the inequality $\ln(1 + x) \leq x$ for all $x \geq 0$. (**Hint:** For $x > 0$, consider $f(t) = \ln(1 + t)$ on the interval $[0, x]$ and apply the Mean Value Theorem.)

57. Prove the inequality $\sqrt[3]{1 + x} \leq 1 + \frac{1}{3}x$ for $x \geq 0$. (Adapt and follow the hint given in Exercise 56.)

58.* Use the Mean Value Theorem to show that
$\lim_{x \to \infty} \left(\sqrt{x} - \sqrt{x - 5}\right) = 0$. (**Hint:** Apply the theorem to the function $f(x) = \sqrt{x}$ on the interval $[x - 5, x]$ for a fixed x-value. Then let x approach infinity.)

59. Use the Mean Value Theorem to prove the following inequality for all $\alpha, \beta \in \mathbb{R}$.
$$|\arctan \alpha - \arctan \beta| \leq |\alpha - \beta|$$

60.* Generalizing Exercise 59, prove that if f is differentiable and $|f'(x)| \leq M$ for some $M > 0$, then for all x, y in the domain of f, $|f(x) - f(y)| \leq M|x - y|$. (Such a function f is said to have the Lipschitz property with constant M.)

61. Use Exercise 60 to prove that $f(x) = \cos 3x$ has the Lipschitz property with constant 3.

62.* We say that the function $f(x)$ has a *fixed point* $c \in \mathbb{R}$ (or that it leaves c fixed) if $f(c) = c$. Prove that if f is differentiable on $\mathbb{R}$ and $f'(x) < 1$ for all x, then f can have no more than one fixed point.

63.* Prove Cauchy's Mean Value Theorem, which states the following: If f and g are both continuous on the closed interval $[a, b]$ and differentiable on (a, b), and if $g'(x) \neq 0$ on (a, b) and $g(a) \neq g(b)$, then there is a point $c \in (a, b)$ for which $\dfrac{f'(c)}{g'(c)} = \dfrac{f(b) - f(a)}{g(b) - g(a)}$.
(**Hint:** Apply Rolle's Theorem to the function
$$F(x) = f(x) - f(a) - \frac{f(b) - f(a)}{g(b) - g(a)}(g(x) - g(a)).)$$

64–68 *True or False?* Determine whether the given statement is true or false. In case of a false statement, explain or provide a counterexample.

64. There are situations when the Mean Value Theorem applies but Rolle's Theorem doesn't.

65. There are situations when Rolle's Theorem applies but the Mean Value Theorem doesn't.

66. If a and b are zeros of the function $f(x)$, then there exists a point c in (a,b) such that $f'(c) = 0$.

67. If a and b are zeros of the polynomial $p(x)$, then there exists a point c in (a,b) such that $p'(c) = 0$.

68. If $f'(x) = 0$ for all x in the domain of f, then $f(x)$ is a constant function.

4.2 Technology Exercises

69–72 Use a graphing calculator or computer algebra system to graph the function on the interval $[a,b]$ along with its secant line through the points $(a,f(a))$ and $(b,f(b))$ on the same screen. Then find and graph the line(s) tangent to the graph that are parallel to the secant line.

69. $f(x) = \sqrt{x}$ on $[0,9]$

70. $f(x) = \dfrac{x^3}{2} - x^2 + x + 7.5$ on $[1,3]$

71. $f(x) = \dfrac{3x}{x+2}$ on $[-1,1]$

72. $f(x) = \sin^2 x + 3x$ on $[0,\pi]$

4.3 **The First and Second Derivative Tests**

TOPICS

1. First derivative tools

2. Second derivative tools

In this section, we continue to explore what we can learn about a function from its derivatives, specifically concentrating on what a function's first and second derivatives tell us about its graph.

TOPIC 1   **First Derivative Tools**

In our first example of an application of Rolle's Theorem in Section 4.2, we used the fact that $f'(x)$ was positive for all x to show that the equation $f(x) = 0$ couldn't have more than one solution. The reasoning in that example can be extended to make the following more general observations.

Theorem ⚷

Monotonicity Test

Assume f is differentiable at all points of an interval I.

1.   If $f'(x) > 0$ for all $x \in I$, then f is strictly increasing on I.

2.   If $f'(x) < 0$ for all $x \in I$, then f is strictly decreasing on I.

In other words, if f' is either entirely positive or entirely negative on I, then f is strictly monotonic on I. (Note that I may be an unbounded interval.)

Proof ✎

Following the definition of strict monotonicity (Section 1.1), we start by assuming $x_1, x_2 \in I$ with $x_1 < x_2$. Since f' exists on all of I, we know that f is continuous on $[x_1, x_2]$ and differentiable on (x_1, x_2); by our definition of one-sided differentiation, this conclusion is true even if x_1 or x_2 is an endpoint of I.

Now, by the Mean Value Theorem, there is a point $c \in (x_1, x_2)$ such that

$$f(x_2) - f(x_1) = f'(c)(x_2 - x_1).$$

Since $x_2 - x_1 > 0$, the right-hand side of the above equation is either positive if $f'(c)$ is positive or negative if $f'(c)$ is negative, leading to the respective results of $f(x_2) > f(x_1)$ or $f(x_2) < f(x_1)$. Since this can be done for each such pair of points x_1 and x_2, we are done.

In practice, the Monotonicity Test is used to find intervals where a function is monotonic by first locating the function's critical points. If a and b are successive critical points of f and if f' exists on all of (a,b), then f' must be either positive or negative on (a,b). The easiest way to determine which is to evaluate f' at what is called a *test point* in (a,b).

Example 1 ✎

Determine the intervals of monotonicity of the function

$$f(x) = 3x^4 + 4x^3 - 12x^2 + 1.$$

Solution

$$f'(x) = 12x^3 + 12x^2 - 24x = 12x(x+2)(x-1)$$

The critical points of f are -2, 0, and 1. These divide the real line (the domain of f) into the open intervals $(-\infty,-2)$, $(-2,0)$, $(0,1)$, and $(1,\infty)$. The following table indicates the sign of f' and the corresponding monotonicity of f in each of these intervals as determined by evaluating f' at a convenient test point.

Interval	$(-\infty,-2)$	$(-2,0)$	$(0,1)$	$(1,\infty)$
f' at test point	$f'(-3) = -144$	$f'(-1) = 24$	$f'\left(\frac{1}{2}\right) = -\frac{15}{2}$	$f'(2) = 96$
Sign of f'	$-$	$+$	$-$	$+$
Monotonicity of f	Decreasing	Increasing	Decreasing	Increasing

So, f is decreasing on $(-\infty,-2)$ and $(0,1)$ and increasing on $(-2,0)$ and $(1,\infty)$, as shown in Figure 1.

(By our definition of monotonicity, it would also be correct to identify the intervals as $(-\infty,-2]$, $[-2,0]$, $[0,1]$, and $[1,\infty)$, but we are typically more interested in making a distinction between the critical points and the intervals between them.)

Note that we are actually only interested in the sign of f' at each test point, and this can typically be determined very quickly, especially if f' has been factored as in this example—there is no need to determine the actual numerical value of f'. In much of our work to come, we will make use of this shortcut.

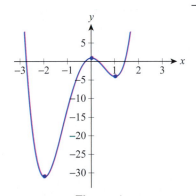

Figure 1
$f(x) = 3x^4 + 4x^3 - 12x^2 + 1$

Given a differentiable function defined on an open interval, Fermat's Theorem tells us that relative extrema occur only at critical points, but we've seen that not every critical point is a relative extremum—we need a test that we can apply to critical points. The conclusions of the previous theorem can be used for just this purpose.

Theorem

First Derivative Test

Suppose c is a critical point of a function f that is continuous on an open interval containing c and differentiable on the same interval, except possibly at c itself. Moving through c from left to right,

1. if f' changes from positive to negative at c, then f has a relative maximum at c;

2. if f' changes from negative to positive at c, then f has a relative minimum at c;

3. if f' does not change sign at c (that is, if it is either positive on both sides of c or negative on both sides of c), then f does not have a relative extremum at c.

Proof

The conclusions of the First Derivative Test follow from applying the Monotonicity Test and determining whether f is increasing or decreasing to the left and to the right of a given critical point c. If, for instance, f is an increasing function to the left of c and a decreasing function to the right of c, then f has a relative maximum value at c. Figure 2 illustrates some of the possible combinations of intervals of monotonicity and critical points.

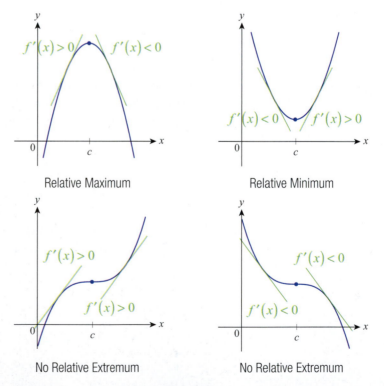

Relative Maximum

Relative Minimum

No Relative Extremum

No Relative Extremum

Figure 2

Example 2 ✎

Find the critical points of $f(x) = (x^2 - 7)x^{1/3}$, and then use the First Derivative Test to locate the relative extrema of f.

Solution

We note first that f is continuous everywhere, and then proceed to find f'.

$$\begin{aligned}
f'(x) &= (2x)x^{1/3} + \frac{1}{3}(x^2 - 7)x^{-2/3} && \text{Product Rule}\\
&= \frac{1}{3}x^{-2/3}(6x^2 + x^2 - 7) && \text{Factor out } \tfrac{1}{3}x^{-2/3}.\\
&= \frac{7(x^2 - 1)}{3x^{2/3}} = \frac{7(x-1)(x+1)}{3x^{2/3}}
\end{aligned}$$

The second step above illustrates that it is often useful to factor out an expression raised to the lowest power that it appears among the terms, in order to write f' in a convenient form (and factoring out the fraction $\frac{1}{3}$ has a similar benefit). Doing so makes it clear that the three critical points of f are 0 (where f' is undefined) and ± 1 (where f' is 0).

We can now quickly determine the sign of f' on each of the intervals between critical points, making use of the factored form above:

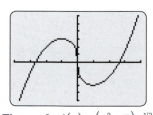

Figure 3 $f(x) = (x^2 - 7)x^{1/3}$
on $[-4, 4]$ by $[-10, 12]$

Interval	$(-\infty, -1)$	$(-1, 0)$	$(0, 1)$	$(1, \infty)$
f' at test point	$f'(-2) \approx 4.41$	$f'(-\frac{1}{2}) \approx -2.78$	$f'(\frac{1}{2}) \approx -2.78$	$f'(2) \approx 4.41$
Sign of f'	$+$	$-$	$-$	$+$
Monotonicity of f	Increasing	Decreasing	Decreasing	Increasing

The First Derivative Test now tells us that f has a local maximum at -1, a local minimum at 1, and 0 is neither a local maximum nor minimum. The values of these local extrema are, respectively, $f(-1) = 6$ and $f(1) = -6$.

Figure 3 confirms our findings.

TOPIC 2   Second Derivative Tools

In reading a detailed analysis of, say, national economic trends, you might run across a complicated-sounding phrase like "the rate of growth is expected to decrease" or "the pace of increase picked up in the third quarter." What do these statements actually mean?

A careful study of such phrases shows that they are saying something about the rate of change of a rate of change—in other words, they refer to the second derivative of a quantity. Geometrically, they tell us whether the slopes of the tangent lines of a function are decreasing or increasing as we scan the graph from left to right. Figure 4a illustrates the first sort of situation and Figure 4b the second.

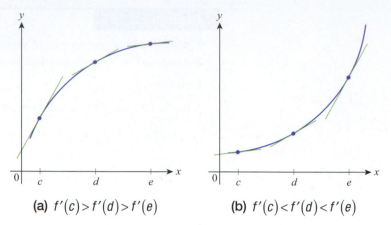

(a) $f'(c) > f'(d) > f'(e)$ **(b)** $f'(c) < f'(d) < f'(e)$

Figure 4 Decreasing and Increasing Slopes of Tangent Lines

This notion of decreasing or increasing slopes of tangent lines lies at the heart of the following definition.

Definition ♀

Concavity

Given a differentiable function f on an interval I, we say f is **concave up** on I if f' is an increasing function on the interval and **concave down** on I if f' is a decreasing function on the interval.

Equivalently, we could define a concave up function as one that lies above all of its tangent lines on an interval and a concave down function as one that lies below all of its tangent lines; note how the graphs in Figure 4 illustrate downward and upward concavity by either definition.

There is yet another way we want to think about concavity, though it has the slight disadvantage of applying only to functions that are twice differentiable on an interval. Assuming that f'' does exist on an interval I, the Mean Value Theorem applied to f' indicates that $f'' > 0$ on I implies that f' is increasing, and hence that f is concave up. Similarly, $f'' < 0$ on I implies that f is concave down. This observation is formally stated as the Concavity Test.

Theorem ⚷

Concavity Test

Assume f is twice differentiable at all points of an interval I.

1. If $f''(x) > 0$ for all $x \in I$, then f is concave up on I.

2. If $f''(x) < 0$ for all $x \in I$, then f is concave down on I.

Example 3 ✐

Determine the intervals of concavity of the function

$$f(x) = 3x^4 + 4x^3 - 12x^2 + 1.$$

Solution

We identified the intervals of monotonicity for this function in Example 1, and we will resume our study of f by determining its second derivative.

$$f'(x) = 12x^3 + 12x^2 - 24x$$
$$f''(x) = 36x^2 + 24x - 24 = 12\left(3x^2 + 2x - 2\right)$$

Using the quadratic formula to solve the equation $f''(x) = 0$ yields the two solutions of $\left(-1 \pm \sqrt{7}\right)\big/3$, or approximately -1.22 and 0.55. Since these are the only two points where f'' is 0, we would expect (and Figure 5 would lead us to believe) that f'' will be either consistently positive or consistently negative on each of the three intervals defined by the two points. We proceed then to make a table to determine the sign of f'' on each interval.

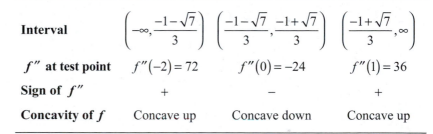

Interval	$\left(-\infty, \dfrac{-1-\sqrt{7}}{3}\right)$	$\left(\dfrac{-1-\sqrt{7}}{3}, \dfrac{-1+\sqrt{7}}{3}\right)$	$\left(\dfrac{-1+\sqrt{7}}{3}, \infty\right)$
f'' **at test point**	$f''(-2) = 72$	$f''(0) = -24$	$f''(1) = 36$
Sign of f''	$+$	$-$	$+$
Concavity of f	Concave up	Concave down	Concave up

The three test points of -2, 0, and 1 were chosen merely for convenience— they lie in the interior of each interval, and it is relatively easy to compute f'' at each. Note the behavior of f at each of the two points where $f''(x) = 0$, as shown in the graph in Figure 5.

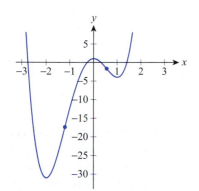

Figure 5

$$f(x) = 3x^4 + 4x^3 - 12x^2 + 1$$

In Example 3, we determined the intervals of concavity of f by locating those points where f'' changed sign, a process very similar to determining intervals of monotonicity. Informally, Darboux's Theorem (Section 3.1) says that derivatives always possess the Intermediate Value Property, so f'' (being the derivative of f') can only change sign at a point where it is zero or where it is undefined. Keep in mind, however, that points where f'' is either undefined or zero are only candidates for points where the concavity of f may change, and the concavity on either side of such a point must be checked in order to determine whether it actually does change. For example, if $f(x) = x^4$, $f''(x) = 12x^2$ and hence $f''(0) = 0$, but f is concave up everywhere (f'' is always nonnegative).

As you might expect, those points where a function changes concavity are deserving of a name.

> ### Definition 💡
>
> **Inflection Point**
>
> A point on the graph of a function f is called an **inflection point** if f is continuous there and changes concavity from upward to downward or from downward to upward. If the graph has a tangent line at the inflection point, the graph will cross the tangent line at that point.
>
> (An alternate definition found in some settings *requires* the graph of f to actually possess a tangent line at the point of inflection, a stronger condition than just being continuous and one that guarantees the crossing behavior.)

Note that, as shown in Figure 6, f' can have any value or be undefined at a point of inflection. f'', however, will either be zero or undefined at such points.

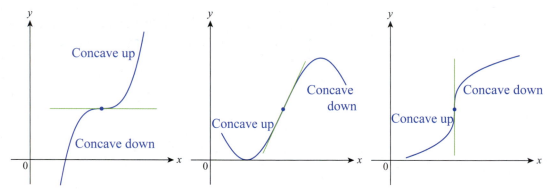

Figure 6

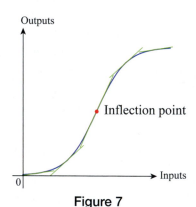

Figure 7

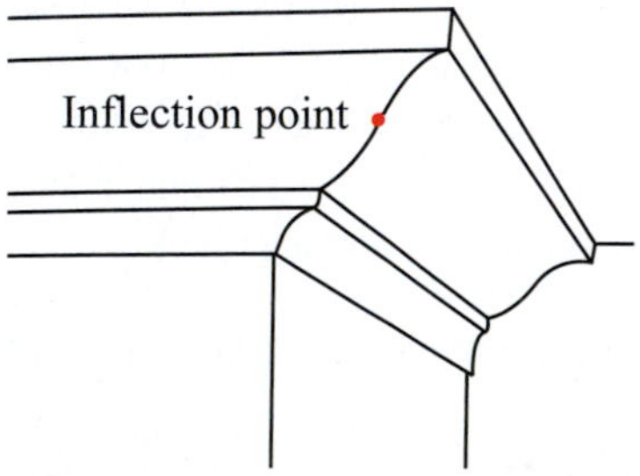

Inflection points occur in many different settings and go by many different names. The commonly heard phrase "point of diminishing returns" actually refers to a point of inflection—although often used loosely, in economics the phrase means that point past which increased inputs (worker hours, raw materials, etc.) yield smaller increases in outputs (finished products). It doesn't mean that production actually starts to decline at that point, but it does mean that the *rate of growth* of productivity becomes negative; that is, the graph of the outputs versus the inputs changes concavity (see Figure 7).

In architecture, an *ogee curve* refers to a shape in which a concave curve and a convex curve join in a point—ogee curves are often found, for example, in crown molding and in the designs of building columns (see Figure 8).

One last application of the second derivative will be mentioned here. Suppose that $f''(c) < 0$ on an open interval containing c and $f'(c) = 0$. Then f is concave down on the interval and the graph of f has a horizontal tangent line at c. A sketch of how f must roughly appear close to c will likely convince you that f must have a relative maximum at c. This is, in fact, true, and the formal statement of the observation and a parallel one for $f''(c) > 0$ comprise the last theorem of this section.

Figure 8

Theorem

Second Derivative Test

Suppose that $f'(c) = 0$ and that f'' exists at c.

1.   If $f''(c) < 0$, then f has a relative maximum at c.

2.   If $f''(c) > 0$, then f has a relative minimum at c.

If $f''(c) = 0$, the test does not apply and the graph of f may have a relative maximum, a relative minimum, or neither at c.

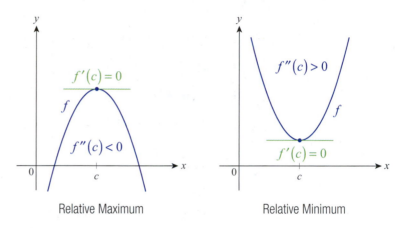

Relative Maximum   Relative Minimum

Figure 9

Proof

Recall that "Differentiability Implies Continuity"; this is one of our theorems from Section 3.1. So the fact that f'' exists at c means that f' is differentiable at c and hence continuous at c. This, in turn, means that there exists an open interval containing c for which f' is defined. If we know further that $f''(c) < 0$ and that $f'(c) = 0$, then for all h sufficiently close to 0,

$$\frac{f'(c+h) - f'(c)}{h} = \frac{f'(c+h)}{h} < 0$$

and so $f'(c+h)$ must be positive for $h < 0$ and negative for $h > 0$. Restated, this means that f' is positive to the left of c and negative to the right of c, so by the First Derivative Test we can conclude that f has a relative maximum at c. The proof of the second statement is similar in nature.

To see that the test does not apply if $f''(c) = 0$, consider the graphs of $f(x) = x^3$, $f(x) = x^4$, and $f(x) = -x^4$ at $x = 0$.

Example 4 ✎

Use the first and second derivatives of $f(x) = x^4 - 2x^3 + 1$ to identify the intervals of monotonicity, extrema, intervals of concavity, and inflection points on its graph.

Solution

$$f'(x) = 4x^3 - 6x^2 = 2x^2(2x-3) \quad \text{and} \quad f''(x) = 12x^2 - 12x = 12x(x-1)$$

Solving the equation $f'(x) = 0$ gives us the critical points 0 and $\frac{3}{2}$.

Solving $f''(x) = 0$ gives us the possible inflection points 0 and 1 (note that 0 is both a critical point and a potential inflection point).

We proceed to evaluate the signs of f' and f'' on the intervals of interest.

Interval	$(-\infty, 0)$	$\left(0, \frac{3}{2}\right)$	$\left(\frac{3}{2}, \infty\right)$
Sign of f'	$f'(-1) = -10 < 0$	$f'(1) = -2 < 0$	$f'(2) = 8 > 0$
Monotonicity of f	Decreasing	Decreasing	Increasing

Interval	$(-\infty, 0)$	$(0, 1)$	$(1, \infty)$
Sign of f''	$f''(-1) = 24 > 0$	$f''\left(\frac{1}{2}\right) = -3 < 0$	$f''(2) = 24 > 0$
Concavity of f	Concave up	Concave down	Concave up

We now have a wealth of information about f at our disposal, and can easily answer all of the questions about its graph.

First, the two tables above identify the intervals of constant monotonicity and concavity and whether f is increasing, decreasing, concave up, or concave down on each. Note how those intervals correspond to the graph of f shown in Figure 10.

Second, we can use the First Derivative Test to realize that f does not have a relative extremum at the critical point 0, but does have a relative minimum at $\frac{3}{2}$. And since the concavity changes from up to down at 0 and from down to up at 1, both 0 and 1 are inflection points.

We could also have used the Second Derivative Test to determine that f has a relative minimum at the critical point $\frac{3}{2}$, even though we never actually evaluated f'' at the point. The reason is that we know f'' exists on the open interval $(1, \infty)$ containing $\frac{3}{2}$ (indeed, f'' exists everywhere), and that it is positive on the interval. In particular, it is positive at $\frac{3}{2}$.

The last step is to evaluate f at the two inflection points and the one relative extremum—those values are shown in Figure 10. Note that f has no absolute maximum value, but the relative minimum at $\frac{3}{2}$ is also its absolute minimum.

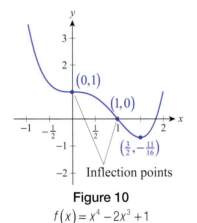

Figure 10

$f(x) = x^4 - 2x^3 + 1$

Example 4 illustrates that the tools of calculus can easily answer questions about particular points of interest on the graph of a function. We will explore this idea further in Section 4.5, but we will acquire one additional tool before doing so.

4.3 **Exercises**

1–22 Determine the intervals of monotonicity of the given function.

1. $f(x) = x^2 - 4x + 1$

2. $g(x) = \dfrac{3}{2}x + 5$

3. $h(x) = \dfrac{2}{3}x^3 + 4x^2 - 10x + \dfrac{5}{3}$

4. $F(x) = 0.75x^4 + x^3 - 15x^2 + 24x + 7$

5. $G(x) = -x^4 - 2x^3 + 8x^2 - 6x + 1$

6. $k(x) = (x^2 - 4)(x^2 - 3)$

7. $m(x) = -\dfrac{x^6}{3} - \dfrac{2x^5}{5} + \dfrac{x^4}{4} + \dfrac{x^3}{3} + \dfrac{x^2}{2} + x$

8. $n(x) = \dfrac{-2}{x+1}$

9. $H(x) = \dfrac{x^2 + 3}{x + 3}$

10. $R(t) = \dfrac{t + 2}{t^2 - 9}$

11. $C(x) = \dfrac{3x^2 + 1}{x - 2}$

12. $A(x) = |x + 4| - 1$

13. $f(t) = 2.5 - |t - 3.125|$

14. $t(x) = x^{2/3} + 2$

15. $w(s) = \sqrt{s}\,(s - 1)$

16. $F(x) = 6 - (x - 3)^{3/5}$

17. $G(x) = \sin^2 x + 1$

18. $m(x) = 2^{-x} + 2^{2x}$

19. $g(t) = -2\sqrt{t}\,e^{-2t}$

20. $L(x) = x \ln x$

21. $A(x) = 0.5x^{1/5}(x^2 - 4)$

22. $U(s) = s\sqrt{3 - s^2}$

23–42. Use the First Derivative Test to classify the relative extrema, if any, of the functions given in Exercises 3–22.

43–50 Identify all intervals of monotonicity as well as intervals of concavity for the graphed function. Find all local extrema and inflection points, if any.

43. $f(x) = x^2 + 2x$

44. $g(x) = -\dfrac{1}{3}x^3 + 4x$

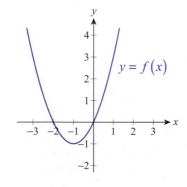

$y = f(x)$

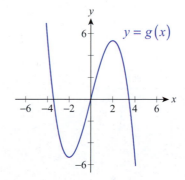

$y = g(x)$

45. $h(x) = \dfrac{12x}{x^2+4}$ (Serpentine)

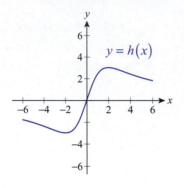

46. $F(x) = \dfrac{1}{16}x^4 - \dfrac{5}{12}x^3 + \dfrac{1}{4}x^2 + 2x$

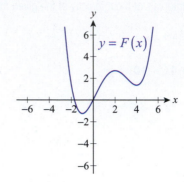

47. $G(x) = 1.5x^5 - 2.5x^3$

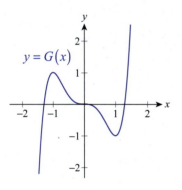

48. $H(x) = \dfrac{|1-x|}{x}$

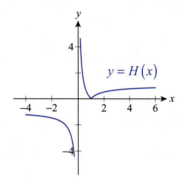

49. $k(x) = 2x - \tan x, \quad |x| < \dfrac{\pi}{2}$

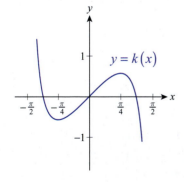

50. $m(x) = \cos x - \dfrac{x}{2}, \quad |x| \le 2\pi$

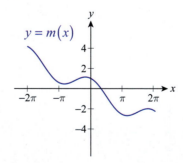

51–62 Determine the intervals of concavity of the given function.

51. $f(x) = 4x - x^2$

52. $g(x) = \dfrac{x^3}{3} - \dfrac{x^2}{2} + x + 1$

53. $h(x) = 4x^3 - 3x^2 - 36x + 5$

54. $k(x) = -x^4 + 5x + 1$

55. $F(x) = x^4 + 4x^3 + 72x$

56. $G(x) = -x^4 + 6x^3 + 24x^2 - 4x + 2$

57. $k(x) = 0.3x^5 + x^4 - 3x^3 + 12x + 4$

58. $v(x) = \dfrac{5}{x-2}$

59. $m(x) = \dfrac{x^2 + 5}{x - 5}$ 60. $F(x) = \dfrac{2^x}{x}$

61. $H(x) = 9^x + 3^{-x}$ 62. $u(x) = (x-2)^{5/7}$

63–82 Use the first and second derivatives to identify the intervals of monotonicity, extrema, intervals of concavity, and inflection points of the given function.

63. $f(x) = \dfrac{1}{2}x + 5$ 64. $g(x) = x^2 - 8x + 3.5$

65. $h(x) = -\dfrac{1}{2}x^2 + 5x + \dfrac{8}{3}$ 66. $F(x) = 2x^3 + 3x^2 - 7$

67. $G(x) = -4x^3 - 3x^2 + 18x + 10$

68. $K(x) = 0.5x^4 + 2x^3 - 6x^2 - 16x + 19.5$

69. $L(x) = -x^4 + 12x^2 - 20x + 3$

70. $m(x) = -\dfrac{3}{x^2}$

71. $n(x) = \dfrac{x+1}{x-2}$ 72. $H(x) = \dfrac{2x}{x^2 - 4}$

73. $r(x) = \dfrac{2x^2 + 1}{x - 4}$ 74. $t(x) = \dfrac{5}{4}\left(x - \dfrac{4}{5}\right)^{4/5}$

75. $F(x) = \dfrac{(x-1)^2}{x^2 - 1}$ 76. $f(x) = \sqrt[3]{x} - x$

77. $g(x) = x^{2/3}\left(\dfrac{2}{3} - x\right)$ 78. $h(x) = x\sqrt{9 - x^2}$

79. $u(x) = \sqrt{x}e^{-x}$ 80. $v(x) = \sin^2 x - 2\cos x$

81. $k(x) = \cos x - \sin x$ 82. $L(x) = -x^2 \ln|x|$

83–90 Sketch a graph of a function satisfying the given conditions. (Answers will vary.)

83. f is differentiable on $\mathbb{R}$, has both a local maximum and minimum, but no global extrema.

84. f is differentiable on $\mathbb{R}$, f' has a zero, and f has no local extrema.

85. f is differentiable on $\mathbb{R}$, f is an odd function, $f'(x) > 0$ on $(-4, 4)$, $f'(x) \le 0$ elsewhere, and $\lim\limits_{|x| \to \infty} f(x) = 0$.

86. $\lim\limits_{|x| \to \infty} f(x) = 1$ and $f(0) = 0$ is a global minimum.

87. $f(x)$ is everywhere positive on $\mathbb{R}$, but its derivative is everywhere negative.

88. $f'(x) < 0$ and $f''(x) > 0$ for all $x \in \mathbb{R}$.

89. f has vertical asymptotes at $x = \pm 2$, a horizontal asymptote at $y = 1$, f is an even function, $f'(x) < 0$ on $(0, 2)$ and $(2, \infty)$, f has a local maximum at 0, $f''(x) < 0$ on $(0, 2)$, $f''(x) > 0$ on $(2, \infty)$, and f has no absolute extrema.

90. $f(0) = f(2) = 0$, $f'(1) = f'(3) = f'(4) = 0$, $f''(1) < 0$, $f''(2) = 0$, and $\lim\limits_{x \to \infty} f(x) = 0$.

91–96 The function $p(t)$ gives the position, relative to its starting point, of an object moving along a straight line. Identify the time intervals when the object is moving in the positive versus negative direction, as well as those intervals when it is accelerating or slowing down. Find the times when the object changes direction as well as when its acceleration is zero.

91. $p(t) = 2t^2 - 3t + 2.5$

92. $p(t) = 5t - \dfrac{1}{2}t^2$

93. $p(t) = 2t^3 - 15t^2 + 24t$

94. $p(t) = -2t^3 + 22.5t^2 - 66t$

95. $p(t) = e^{-t}\sin t$

96. $p(t) = \dfrac{1 - t}{2t + 2}$

97–102 The graph of the derivative $f'(x)$ of the function $f(x)$ is given. Use it to sketch an approximate graph of $f''(x)$. Then try to sketch a possible graph of $f(x)$ as well. (**Note:** There are many possible correct answers for the graph of f. Can you see why?)

97.

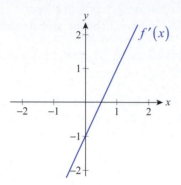

98.

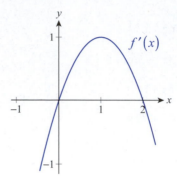

99.

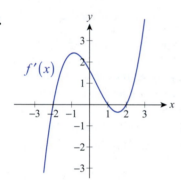

100.

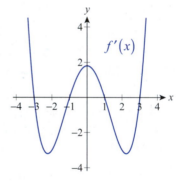

101.

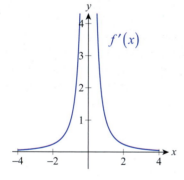

102.

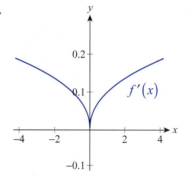

103. Suppose $T(t)$ is the outside temperature, over a 24-hour period on a typical spring day where you live (t is measured in hours, with $t = 0$ corresponding to midnight). Given the following data, what time(s) of day might c_i represent ($i = 1, 2, \ldots, 6$)? Explain your choice(s).

a. $T'(c_1) > 0$, $T''(c_1) < 0$

b. $T'(c_2) > 0$, $T''(c_2) > 0$

c. $T'(c_3) = 0$, $T''(c_3) > 0$

d. $T'(c_4) < 0$, $T''(c_4) < 0$

e. $T'(c_5) < 0$, $T''(c_5) > 0$

f. $T'(c_6) = 0$, $T''(c_6) < 0$

104. The graph below shows the profit (in thousands of dollars) from a product per hundreds of units sold. Use the graph to visually estimate the production level at which the marginal profit starts increasing.

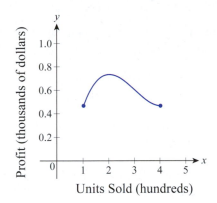

105. An aftermarket auto accessories company manufactures StopTheMess trunk liners and organizers. The overhead cost of operating the plant is $5000 per month and the cost of manufacturing each item is $20. The company estimates that 200 liners can be sold monthly for $50 apiece, and that sales will increase by 10 liners per month for each dollar decrease in price.

  a. Find a formula for the profit function $P(n)$, where n is the number of trunk liners manufactured (suppose $200 \le n \le 350$).

  b. Identify the intervals on which P is increasing or decreasing, and find the production level that maximizes profits.

106. The strength of an electric field due to a charged ring obeys the equation

$$E = \frac{kqx}{\left(x^2 + R^2\right)^{3/2}}$$

where q is the electric charge measured in coulombs (C), $k \approx 8.99 \cdot 10^9 \ \text{Nm}^2/\text{C}^2$, R is the radius of the ring and x is the distance to the charge in meters. Find a formula for the rate of change of E as x increases. What happens to E and dE/dx as $x \to \infty$?

107. Suppose that $f'(x) = (x+4)(x+1)^2 (x-3)^3$. By examining the zeros of $f'(x)$, identify the x-coordinates of the local maxima and minima of $f(x)$. (**Hint:** Recall what you learned about multiplicities of zeros and sign changes of polynomials.)

108. Repeat Exercise 107 for $g(x)$ if $g'(x) = 3x(x-2)(x+5)^4 \left(x - \frac{1}{3}\right)^2 (x+1)^3$.

109. Use derivatives to prove that if $x \in (-\infty, 1)$, then $1/(1-x) \ge 1 + x$. (**Hint:** Start by assuming that $x \ge 0$. Rewriting the inequality as $f(x) \ge g(x)$, show that $D(x) = f(x) - g(x)$ is increasing, while $D(0) = 0$. To handle the case of $x < 0$, use the fact that $D'(x) < 0$ along with $D(0) = 0$.)

110. Prove that quadratic functions cannot have any inflection points, while cubic functions have exactly one. What can you say about fourth-degree polynomials?

111. Suppose we know that the derivative of a function f is $f'(x) = a/(x^2 + 1)$ for some nonzero $a \in \mathbb{R}$. Prove that f is increasing or decreasing everywhere on $\mathbb{R}$.

112. Generalize Exercise 111 by proving the following: If f is differentiable on an interval I and $f'(x) \ne 0$ in the interior of I, then f is increasing or decreasing everywhere on I. (**Hint:** Indirectly assume that f' changes signs, and use the Darboux property of derivative functions.)

113. Use Exercise 112 to find conditions under which the general cubic polynomial $p(x) = ax^3 + bx^2 + cx + d$ is decreasing everywhere on $\mathbb{R}$.

114. Prove that a cubic polynomial $p(x) = ax^3 + bx^2 + cx + d$ has exactly one inflection point. Find its first coordinate, assuming that the three real roots of $p(x)$ are x_1, x_2, and x_3.

115. Determine the conditions on the coefficients of the cubic polynomial $p(x) = ax^3 + bx^2 + cx + d$ in order for $p(x)$ to have a local maximum at $x = 0$ and a local minimum at $x = 4$.

116.* Consider the following function:

$$f(x)=\begin{cases}2x^2 & \text{if } x \text{ is irrational}\\4x^2 & \text{if } x \text{ is rational}\end{cases}$$

Use f to explain why the changing of signs of the derivative is not necessary for a function to have a local extremum.

117. By considering $f(x)=\begin{cases}x & \text{if } x \le 1\\3-x & \text{if } x > 1\end{cases}$, explain why the First Derivative Test can't be used for a discontinuous function.

118. Suppose that $f(x)$ and $g(x)$ are at least twice differentiable and that both their first and second derivatives are positive everywhere on an interval I. Which of the following can you prove from these conditions? Prove those statements that are true, and provide counterexamples for the rest.

a. $f(x)+g(x)$ is increasing on I.

b. $f(x)+g(x)$ is concave up on I.

c. $f(x)\cdot g(x)$ is increasing on I.

d. $f(x)\cdot g(x)$ is concave up on I.

e. $f(g(x))$ is increasing on I.

f. $f(g(x))$ is concave up on I.

119.* Use mathematical induction to prove the following generalization of the Second Derivative Test: Suppose that the derivatives of all orders of the function f exist at c, up to $f^{(2k)}(c)$, and that $f'(c)=f''(c)=\cdots=f^{(2k-1)}(c)=0$, but $f^{(2k)}(c)\neq 0$. Then if $f^{(2k)}(c)<0$, f has a relative maximum at c; if $f^{(2k)}(c)>0$, f has a relative minimum at c.

120.* Use mathematical induction to prove that if f is $(2k+1)$-times differentiable at c, $f'(c)=f''(c)=\cdots=f^{(2k)}(c)=0$, but $f^{(2k+1)}(c)\neq 0$, then f has a point of inflection at c.

121–128 *True or False?* Determine whether the given statement is true or false. In case of a false statement, explain or provide a counterexample.

121. Not all fourth-degree polynomials have inflection points.

122. A function with no inflection points cannot change concavity.

123. If $f'(x)$ is negative on $(-\infty,c)$ and positive on (c,∞), then f has a minimum at c.

124. If $f(x)$ and $g(x)$ are decreasing, then so is $(f+g)(x)$.

125. If $f(x)$ and $g(x)$ are decreasing, then so is $(f\cdot g)(x)$.

126. A polynomial of degree n cannot have more than $n-1$ extrema on $\mathbb{R}$.

127. If $c \in \mathbb{R}$ is a critical point, then the function has a local minimum or a local maximum at c.

128. If $f''(c)=0$, then c is an inflection point.

4.3 Technology Exercises

129–130 Use a graphing calculator or computer algebra system to graph the given function along with its first and second derivatives on the same screen. Use your graphs to explain the behavior of the function with regard to the signs and values of its derivatives.

129. $f(x) = (x^2 + 1)\sqrt{9 - x^2}$ on $[-3, 3]$

130. $g(x) = \sqrt{x} \cos x - \sin 2x$ on $[0, 4\pi]$

131. The table below shows the temperature of a pediatric patient over a 24-hour period (measurements were taken every two hours, starting at midnight). Use the regression capabilities of a graphing calculator or computer algebra system to approximate the data by a fourth-degree polynomial. When do you estimate the patient's temperature to have been the highest? The lowest? When did the highest rates of increase and decrease occur?

Patient's Temperature

Time	12 a.m.	2 a.m.	4 a.m.	6 a.m.	8 a.m.	10 a.m.	12 p.m.
Temp (°F)	99.9	99.5	99.1	98.9	98.7	98.8	99.4

Time	2 p.m.	4 p.m.	6 p.m.	8 p.m.	10 p.m.	12 a.m.
Temp (°F)	100.0	102.1	101.9	101.3	101.0	99.9

132. In the first few months following the launch of a new product, monthly sales were given by the function

$S(t) = 300t^2 / (t^2 + 2)$, where t is measured in months.

a. Use a graphing calculator or computer algebra system to graph the function over the first year, and estimate when the rate of growth in sales was greatest.

b. Use the differentiation capabilities of a computer algebra system to check your estimate in part a.

133–136 Use a graphing calculator or computer algebra system to graph the given function for different values of the parameter(s). Examine how the values of the parameter(s) affect the number of local extrema. How about inflection points?

133. $f(x) = x^4 + cx^3$; $1 \le c \le 3$

134. $g(x) = 0.5x^5 + cx^4 - dx$; $0 \le c, d \le 3$

135. $h(x) = \cos x - \sin(cx)$; $0 \le c \le 4$

136. $k(x) = \sin^2(cx) \cos(dx)$; $0 \le c, d \le 5$

4.4 L'Hôpital's Rule

TOPICS

1. L'Hôpital's Rule

2. Limits of indeterminate form

In this section, we develop a calculus-based tool that often allows us to evaluate, fairly easily, limits that would otherwise pose a considerable challenge. The tool is called *l'Hôpital's Rule*, and the limits it applies to are called *limits of indeterminate form*.

TOPIC 1 L'Hôpital's Rule

As motivation, consider a limit of the form

$$\lim_{x \to c} \frac{f(x)}{g(x)},$$

where $f(c) = g(c) = 0$, $f'(c)$ and $g'(c)$ both exist, and $g'(c) \neq 0$. Such a limit is said to be of **indeterminate form 0/0**, and the limit cannot be determined by simply evaluating $f(c)/g(c)$. However, we can rewrite the limit in a form that can be evaluated, as follows.

$$\lim_{x \to c} \frac{f(x)}{g(x)} = \lim_{x \to c} \frac{f(x) - 0}{g(x) - 0}$$

$$= \lim_{x \to c} \frac{f(x) - f(c)}{g(x) - g(c)} \qquad f(c) = g(c) = 0$$

$$= \lim_{x \to c} \frac{\dfrac{f(x) - f(c)}{x - c}}{\dfrac{g(x) - g(c)}{x - c}} \qquad \text{Divide top and bottom by } x - c.$$

$$= \frac{\displaystyle\lim_{x \to c} \frac{f(x) - f(c)}{x - c}}{\displaystyle\lim_{x \to c} \frac{g(x) - g(c)}{x - c}}$$

$$= \frac{f'(c)}{g'(c)}$$

So if the values of these derivatives are known (and $g'(c) \neq 0$), the evaluation of the limit *can* be accomplished with a different substitution. This observation is a simple form of l'Hôpital's Rule, named after the French nobleman Guillaume François Antoine de l'Hôpital (1661–1704) in whose introductory calculus textbook it first appeared—the result is actually due to the Swiss mathematician Johann Bernoulli (1667–1748).

Example 1 ✎

Determine $\lim\limits_{x \to 0} \dfrac{5x - \sin 2x}{x}$.

Solution

Although we have learned ways to algebraically manipulate fractions of this sort in order to find the limit, l'Hôpital's Rule makes it strikingly easy.

$$\lim_{x \to 0} \frac{5x - \sin 2x}{x} = \lim_{x \to 0} \frac{5 - 2\cos 2x}{1} \qquad \text{Differentiate top and bottom.}$$

$$= 5 - 2(1) \qquad \text{Substitute } x = 0.$$

$$= 3$$

Before presenting the full form of l'Hôpital's Rule, we will state one other useful theorem—it is a formulation of the Mean Value Theorem due to Augustin-Louis Cauchy (1789–1857). Cauchy's MVT is the basis for one of the more elegant proofs of l'Hôpital's Rule (a hint on proving Cauchy's MVT appears in Exercise 63 of Section 4.2).

Theorem ⚷

Cauchy's Mean Value Theorem

Suppose that f and g are continuous on $[a,b]$ and differentiable on (a,b), $g'(x) \neq 0$ on (a,b), and $g(a) \neq g(b)$. Then there is a point $c \in (a,b)$ such that

$$\frac{f'(c)}{g'(c)} = \frac{f(b) - f(a)}{g(b) - g(a)}.$$

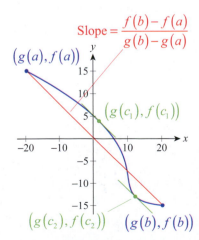

Figure 1

Cauchy's Mean Value Theorem

Note that Cauchy's Mean Value Theorem reduces to the simpler Mean Value Theorem if $g(x) = x$.

The simpler version of the MVT guarantees the existence of a point where the tangent to a function is parallel to a secant line, and Cauchy's MVT does something similar. Given two functions f and g with the properties above, the collection of ordered pairs $\{(g(x), f(x)) \mid x \in [a,b]\}$ defines a curve in $\mathbb{R}^2$ such as the one depicted in Figure 1, and the red line segment connecting $(g(a), f(a))$ and $(g(b), f(b))$ is also called a secant line. Specifically, the blue curve in Figure 1 is defined by the two functions $f(x) = x^2 - 5x - 9$ and $g(x) = x^3 + x + 10$, and the interval $[a,b]$ is $[-3,2]$; the slope of the secant line is $-\frac{3}{4}$. Cauchy's MVT tells us that for at least one number $c \in (a,b)$, the line tangent to the curve at $(g(c), f(c))$ is parallel to the secant line. (Curves defined in this manner are called *parametric curves*, and we will study them in detail in Chapter 9.)

We are now ready for a stronger form of l'Hôpital's Rule.

Theorem ⚲

L'Hôpital's Rule

Suppose f and g are differentiable at all points of an open interval I containing c, and that $g'(x) \neq 0$ for all $x \in I$ except possibly at $x = c$. Suppose further that either

$$\lim_{x \to c} f(x) = 0 \quad \text{and} \quad \lim_{x \to c} g(x) = 0$$

or

$$\lim_{x \to c} f(x) = \pm\infty \quad \text{and} \quad \lim_{x \to c} g(x) = \pm\infty.$$

Then

$$\lim_{x \to c} \frac{f(x)}{g(x)} = \lim_{x \to c} \frac{f'(x)}{g'(x)},$$

assuming the limit on the right is a real number or ∞ or $-\infty$.

Further, the rule is true for one-sided limits at c and for limits at infinity; that is, $x \to c$ can be replaced with $x \to c^+$, $x \to c^-$, $x \to -\infty$, or $x \to \infty$, assuming always that the limit on the right is a real number or ∞ or $-\infty$.

We have already mentioned limits of indeterminate form $0/0$ in passing; the other type of limit described in l'Hôpital's Rule is of **indeterminate form** ∞/∞, and we will discuss other variations soon.

Proof ✎

We will prove only the case in which

$$\lim_{x \to c} f(x) = 0 \quad \text{and} \quad \lim_{x \to c} g(x) = 0,$$

and we will prove that the claim is true as $x \to c^-$; the corresponding result for $x \to c^+$ is nearly identical, and the two one-sided limits together prove the theorem.

Suppose that $x \in I$ is a number lying to the left of c. Then $g'(x) \neq 0$, and we can apply Cauchy's MVT to the interval $[x, c]$. Thus, there is a point $\tilde{c} \in (x, c)$ such that

$$\frac{f'(\tilde{c})}{g'(\tilde{c})} = \frac{f(c) - f(x)}{g(c) - g(x)}$$

$$= \frac{f(x)}{g(x)}. \qquad \qquad f(c) = g(c) = 0$$

As we let $x \to c^-$, $\tilde{c} \to c^-$ as well since $\tilde{c}$ always lies between x and c. Hence,

$$\lim_{x \to c^-} \frac{f(x)}{g(x)} = \lim_{\tilde{c} \to c^-} \frac{f'(\tilde{c})}{g'(\tilde{c})} = \lim_{x \to c^-} \frac{f'(x)}{g'(x)}.$$

The value of l'Hôpital's Rule comes from the fact that differentiating the numerator and denominator of a fraction of indeterminate form often results in a fraction that is not, making the limit easier to determine.

Example 2 ✎

Determine $\lim\limits_{x \to \infty} \dfrac{\ln x}{2\sqrt{x}}$.

Solution

First, note that l'Hôpital's Rule is indeed applicable: this limit at infinity is of indeterminate form ∞/∞. So we evaluate the limit as follows.

$$\lim_{x \to \infty} \frac{\ln x}{2\sqrt{x}} = \lim_{x \to \infty} \frac{\dfrac{d}{dx}(\ln x)}{\dfrac{d}{dx}(2\sqrt{x})} = \lim_{x \to \infty} \frac{\dfrac{1}{x}}{\dfrac{1}{\sqrt{x}}} = \lim_{x \to \infty} \frac{1}{\sqrt{x}} = 0$$

Example 3 ✎

Determine $\lim\limits_{x \to 0} \dfrac{a^x - 1}{x}$, where $a > 0$ is a constant.

Solution

The limit is of indeterminate form $0/0$, so we proceed using l'Hôpital's Rule.

$$\lim_{x \to 0} \frac{a^x - 1}{x} = \lim_{x \to 0} \frac{(\ln a)a^x}{1} = \ln a$$

Caution ⚠

L'Hôpital's Rule says the limit of a quotient of two functions is equal to the limit of the quotient of their derivatives—don't mistakenly apply the Quotient Rule of differentiation! After verifying that the conditions of l'Hôpital's Rule are satisfied, proceed by differentiating the numerator and denominator individually.

L'Hôpital's Rule is obviously useful, but there is even more power in it than might appear at first glance. Suppose the rule is applied to a limit of indeterminate form, and the resulting limit is again of indeterminate form. Does that mean the rule fails to tell us anything? Often, no—if the new limit is also indeterminate, then we can again apply l'Hôpital's Rule. This process can be repeated as often as necessary, *as long as we stop as soon as we reach a limit that is not indeterminate*. The next example illustrates how repeated application of the rule works, and the sort of mistake that can arise if we misuse it.

Example 4 ✐

Evaluate the following limits.

a. $\displaystyle\lim_{x\to\infty}\frac{e^x}{5x^2-3x+2}$ **b.** $\displaystyle\lim_{x\to 0}\frac{1-\cos x}{3x^2+7x}$

Solution

a. The limit is of indeterminate form ∞/∞, so we can apply l'Hôpital's Rule.

$$\lim_{x\to\infty}\frac{e^x}{5x^2-3x+2}=\lim_{x\to\infty}\frac{e^x}{10x-3}$$

Limit is still of indeterminate form ∞/∞. Apply l'Hôpital's Rule again.

$$=\lim_{x\to\infty}\frac{e^x}{10}$$

Limit can now be determined; $e^x\to\infty$ as $x\to\infty$

$$=\infty$$

b. The limit is of indeterminate form $0/0$ and we apply l'Hôpital's Rule.

$$\lim_{x\to 0}\frac{1-\cos x}{3x^2+7x}=\lim_{x\to 0}\frac{\sin x}{6x+7}=\frac{0}{7}=0.$$

If we mistakenly continued to apply the rule, we would have obtained

$$\lim_{x\to 0}\frac{\sin x}{6x+7}=\lim_{x\to 0}\frac{\cos x}{6}=\frac{1}{6},$$

which is incorrect.

TOPIC 2   Limits of Indeterminate Form

In addition to the two indeterminate forms we have already seen, l'Hôpital's Rule can be used to evaluate other potentially challenging limits. The remaining examples illustrate how indeterminate products, differences, and powers can be rewritten in such a way that l'Hôpital's Rule applies.

Example 5 ✐

Determine $\displaystyle\lim_{x\to 0^+}\sqrt{x}\,\ln x$.

Solution

We say that a limit of a product fg is of **indeterminate form $0\cdot\infty$** if one of the functions approaches 0 and the other approaches ∞ or $-\infty$. The limit, if it exists, depends on which function dominates—the product could tend toward 0, could grow unbounded, or could approach some nonzero real number if the two functions balance one another just right.

We can apply l'Hôpital's Rule if we can rewrite the product as a quotient in either the indeterminate form $0/0$ or ∞/∞. That is, we rewrite fg as either

$$\frac{f}{1/g}\quad\text{or}\quad\frac{g}{1/f},$$

whichever is easier to work with. In this example,

Project ✐

Chapter 4

Consider a function $f(x)$ that is at least twice differentiable. In this project, you will show that the second derivative of $f(x)$ at $x = c$ can be found as the limit of so-called **second-order differences**:

$$f''(c) = \lim_{h \to 0} \frac{f(c+h) - 2f(c) + f(c-h)}{h^2}$$

1. Instead of working with a secant line through the points $(c, f(c))$ and $(c+h, f(c+h))$ like we did when approximating the first derivative, suppose that

$$y = a_1 x^2 + a_2 x + a_3$$

   is the parabola through the following three points on the graph of f: $(c-h, f(c-h))$, $(c, f(c))$, and $(c+h, f(c+h))$. Do you expect to always be able to find coefficients $a_1, a_2, a_3 \in \mathbb{R}$ such that the resulting parabola satisfies the desired conditions? Why or why not? Why would you expect $2a_1$ to be "close" to $f''(c)$ if h is "small"? What will happen to $2a_1$ as $h \to 0$? Write a short paragraph answering the above questions.

2. By substituting the points $(c-h, f(c-h))$, $(c, f(c))$, and $(c+h, f(c+h))$ into $y = a_1 x^2 + a_2 x + a_3$, obtain a system of linear equations in unknowns $a_1, a_2,$ and a_3. Solve the system for the unknown a_1.

3. Use Questions 1 and 2 to argue that $f''(c)$ is the limit of the second-order differences:

$$f''(c) = \lim_{h \to 0} \frac{f(c+h) - 2f(c) + f(c-h)}{h^2}$$

4. Use l'Hôpital's Rule to verify the result you found in Question 3.

$$\lim_{x\to 0^+} \sqrt{x}\,\ln x = \lim_{x\to 0^+} \frac{\ln x}{\dfrac{1}{\sqrt{x}}} \qquad \text{Rewrite to obtain the indeterminate form } \infty/\infty.$$

$$= \lim_{x\to 0^+} \frac{\dfrac{1}{x}}{-\dfrac{1}{2}x^{-3/2}} = \lim_{x\to 0^+}\left(-2\sqrt{x}\right) = 0.$$

Example 6 ☑

Determine $\displaystyle\lim_{x\to 1}\left(\frac{x}{x-1} - \frac{1}{\ln x}\right)$.

Solution

A limit of a difference $f - g$ is of **indeterminate form** $\infty - \infty$ if $f \to \infty$ and $g \to \infty$. Again, such limits are usually not trivial: if f dominates, the difference will tend to ∞; and if g dominates, the difference will tend to $-\infty$; but it is also possible for the two to balance out and result in a finite limit.

To use l'Hôpital's Rule, we need to rewrite the difference as a quotient. In this case, we can do so by combining the two fractions using a common denominator; in other such problems, rationalization or factoring out a common factor may be helpful.

$$\lim_{x\to 1}\left(\frac{x}{x-1}-\frac{1}{\ln x}\right) = \lim_{x\to 1}\frac{x\ln x - x + 1}{(x-1)\ln x}$$

$$= \lim_{x\to 1}\frac{\ln x + 1 - 1}{\ln x + \dfrac{x-1}{x}} \qquad \text{Apply l'Hôpital's Rule once.}$$

$$= \lim_{x\to 1}\frac{x\ln x}{x\ln x + x - 1} \qquad \text{Simplify the fraction.}$$

$$= \lim_{x\to 1}\frac{\ln x + 1}{\ln x + 1 + 1} \qquad \text{Apply l'Hôpital's Rule again.}$$

$$= \frac{1}{2}$$

Figure 2 visually confirms the limit we found.

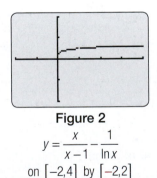

Figure 2

$$y = \frac{x}{x-1} - \frac{1}{\ln x}$$

on $[-2,4]$ by $[-2,2]$

The next three limits are all of the form f^g, and constitute limits of **indeterminate forms 1^∞, 0^0, and ∞^0.** As with the other indeterminate forms, there is competition between the effects of the two functions in the limit. In order to use l'Hôpital's Rule, we set $y = f^g$ and take the natural logarithm of both sides to obtain $\ln y = g\ln f$. If we can determine the limit of $g\ln f$, we can determine the limit of $y = e^{\ln y} = e^{g\ln f}$.

Example 7 ✍

Determine $\lim\limits_{x \to 0^+} (1+x)^{1/x}$.

Solution

Note that the base of the expression $(1+x)^{1/x}$ goes to 1 and the exponent goes to ∞ as $x \to 0^+$. We let $y = (1+x)^{1/x}$.

$$\ln y = \left(\frac{1}{x}\right) \ln(1+x) = \frac{\ln(1+x)}{x}$$

The limit of this last expression is of the form $0/0$, and we can apply l'Hôpital's Rule.

$$\lim_{x \to 0^+} \frac{\ln(1+x)}{x} = \lim_{x \to 0^+} \frac{\dfrac{1}{1+x}}{1} = 1$$

Since $\lim\limits_{x \to 0^+} \ln y = 1$, $\lim\limits_{x \to 0^+} y = \lim\limits_{x \to 0^+} e^{\ln y} = e^1 = e$. Hence, $\lim\limits_{x \to 0^+} (1+x)^{1/x} = e$.

Example 8 ✍

Determine $\lim\limits_{x \to 0^+} x^x$.

Solution

Both the base and the exponent approach 0. Letting $y = x^x$, we arrive at

$$\ln y = x \ln x = \frac{\ln x}{\dfrac{1}{x}},$$

a limit of indeterminate form ∞/∞. So,

$$\lim_{x \to 0^+} \frac{\ln x}{\dfrac{1}{x}} = \lim_{x \to 0^+} \frac{\dfrac{1}{x}}{-\dfrac{1}{x^2}} = \lim_{x \to 0^+} (-x) = 0,$$

and hence $x^x \to e^0 = 1$ as $x \to 0^+$ (don't forget this last step!).

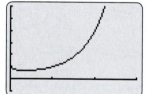

Figure 3

$y = x^x$ on $[0,3]$ by $[-1,6]$

Example 9 ✍

Determine $\lim\limits_{x \to \infty} x^{1/x}$.

Solution

The base has a limit of ∞ and the exponent has a limit of 0. We proceed as in the last two examples.

$$y = x^{1/x}$$

$$\ln y = \frac{1}{x} \ln x = \frac{\ln x}{x} \qquad \text{Indeterminate form } \infty/\infty$$

Applying l'Hôpital's Rule,

$$\lim_{x\to\infty} \ln y = \lim_{x\to\infty} \frac{\ln x}{x} = \lim_{x\to\infty} \frac{\frac{1}{x}}{1} = 0$$

and therefore $\lim_{x\to\infty} x^{1/x} = \lim_{x\to\infty} y = e^0 = 1$.

4.4 **Exercises**

1–12 Evaluate the limit using the theorems of Chapter 2. Then decide whether l'Hôpital's Rule is applicable and, if so, use it to check your answer.

1. $\displaystyle\lim_{x\to3} \frac{2x^2-18}{x-3}$

2. $\displaystyle\lim_{x\to-2} \frac{x^3+8}{x+2}$

3. $\displaystyle\lim_{x\to0} \frac{\sin x}{2x}$

4. $\displaystyle\lim_{x\to0} \frac{x^2}{1-\cos x}$

5. $\displaystyle\lim_{x\to0} \frac{\cos x}{x}$

6. $\displaystyle\lim_{x\to\infty} \frac{6x^2-x+7}{x-3x^2}$

7. $\displaystyle\lim_{x\to-\infty} \frac{5x^2-2x+1}{2.5x^3-3x^2+6}$

8. $\displaystyle\lim_{x\to0} \frac{2x}{\sqrt{x+3}-\sqrt{3}}$

9. $\displaystyle\lim_{x\to0} \frac{\sec x}{x}$

10. $\displaystyle\lim_{x\to0^+} \left(\sqrt{x}\right)^{1/x}$

11. $\displaystyle\lim_{x\to0} \left(\frac{1}{x} - \frac{1}{x\sqrt{x+1}}\right)$

12. $\displaystyle\lim_{x\to0} \frac{x}{3\tan x}$

13–16 Two functions are in competition to determine the indicated limit. Identify the type of the indeterminate form, and fill out the table to decide which function dominates.

13. $\displaystyle\lim_{x\to\infty} f(x)$, where $f(x) = \dfrac{\sqrt{5x^3+7}}{0.2x^2+1}$

x	1	10	100	1000	10,000	100,000
f(x)						

14. $\displaystyle\lim_{x\to\infty} g(x)$, where $g(x) = \dfrac{0.5\sqrt{x}}{\ln(x+1)}$

x	1	10	100	1000	10,000	100,000
g(x)						

15. $\displaystyle\lim_{x\to\infty} h(x)$, where $h(x) = x^{100}e^{-x}$

x	1	10	100	1000	10,000	100,000
h(x)						

16. $\displaystyle\lim_{x\to0} k(x)$, where $k(x) = (\sin x)^x$

x	1	0.5	0.1	0.01	0.001	0.0001
k(x)						

17–48 Check whether l'Hôpital's Rule applies to the given limit. If it does, use it to determine the value of the limit. If it does not, find the limit some other way. (When necessary, apply l'Hôpital's Rule several times.)

17. $\displaystyle\lim_{x\to\infty}\frac{2x+5}{x^2-7}$

18. $\displaystyle\lim_{x\to\infty}\frac{4-2.5x}{x+3}$

19. $\displaystyle\lim_{x\to-\infty}\frac{1.5x^3-2x^2+x+9}{x^2+2.1x-4}$

20. $\displaystyle\lim_{x\to-\infty}\frac{4.5x^4+x^3-2}{3-1.5x^4}$

21. $\displaystyle\lim_{x\to0}\frac{\sqrt{x}}{\ln x}$

22. $\displaystyle\lim_{x\to\infty}\frac{x\sin x}{e^{-x}}$

23. $\displaystyle\lim_{x\to\infty}\frac{\dfrac{1}{x}+2}{2x+1}$

24. $\displaystyle\lim_{t\to0}\frac{t}{\sqrt{2t+9}-3}$

25. $\displaystyle\lim_{x\to\infty}\frac{\sin x+2\ln x}{x^2+5}$

26. $\displaystyle\lim_{x\to0}\frac{\sin x-x}{1-\cos x}$

27. $\displaystyle\lim_{t\to0}\frac{1-\cos t}{3t}$

28. $\displaystyle\lim_{x\to-1^+}\frac{\sin\sqrt{x+1}}{x+1}$

29. $\displaystyle\lim_{x\to0^+}\frac{x^{3/2}}{\ln(\cos x)}$

30. $\displaystyle\lim_{x\to0}\frac{\ln(\sec^2 x)}{\sqrt{x}}$

31. $\displaystyle\lim_{x\to\infty}\frac{\ln x}{\ln(x^2+3x)}$

32. $\displaystyle\lim_{x\to0}\frac{\log_{10}(x^2+2x+1)}{\log_{10}(x+1)}$

33. $\displaystyle\lim_{x\to0}\frac{x}{3^{x/2}-1}$

34. $\displaystyle\lim_{x\to\infty}\frac{2^x}{x^2-3x+4}$

35. $\displaystyle\lim_{x\to0}\frac{\sin x-x}{3x^2}$

36. $\displaystyle\lim_{\phi\to0^+}\frac{1-\cos\phi}{\csc\phi}$

37. $\displaystyle\lim_{\alpha\to0}\frac{\alpha}{e^{\sin\alpha}-1}$

38. $\displaystyle\lim_{\theta\to0}\frac{\theta\tan\theta}{1-\cos\theta}$

39. $\displaystyle\lim_{t\to\infty}\frac{\ln(t+1)}{e^{-t}\sin t}$

40. $\displaystyle\lim_{t\to\pi}\frac{(\cos(2t)-1)^2}{t-\pi}$

41. $\displaystyle\lim_{\theta\to\pi/2}\frac{\left(\theta-\dfrac{\pi}{2}\right)^2}{\ln(\sin\theta)}$

42. $\displaystyle\lim_{x\to\infty}\frac{x+2^x}{5^x-x}$

43. $\displaystyle\lim_{x\to\infty}\frac{4^x+x^2}{3^x-x}$

44. $\displaystyle\lim_{x\to\infty}\frac{\ln(\ln x)}{x\ln x}$

45. $\displaystyle\lim_{x\to0^+}\frac{\log_2(1+x)}{\log_3(\sin x+1)}$

46. $\displaystyle\lim_{x\to\infty}\frac{\log_4(2x+1)}{\log_5(x-4)}$

47. $\displaystyle\lim_{x\to0^+}\frac{\log_4(x+1)}{\log_3 x}$

48. $\displaystyle\lim_{x\to0}\frac{3^x-1}{x3^x}$

49–74 Identify the indeterminate product, quotient, difference, or power, and use l'Hôpital's Rule to find the limit. If the limit is not of indeterminate form, say so and find it by other means.

49. $\displaystyle\lim_{x\to0^+}x\ln x$

50. $\displaystyle\lim_{x\to\infty}\frac{\sqrt{2x^2+1}}{x+3}$

51. $\displaystyle\lim_{x\to0}x\cos\frac{\pi}{x}$

52. $\displaystyle\lim_{x\to\infty}(\ln x)^{-1/x}$

53. $\displaystyle\lim_{x\to0^+}\left(\frac{1}{x}\right)^x$

54. $\displaystyle\lim_{x\to0^+}(-\ln x)^x$

55. $\displaystyle\lim_{x\to1^+}\left(\frac{1}{\ln x}-\frac{2}{x-1}\right)$

56. $\displaystyle\lim_{x\to0^+}\left(\frac{1}{x}+\ln x\right)$

57. $\displaystyle\lim_{x\to4^+}\left(\frac{32}{x^2-16}-\frac{x}{x-4}\right)$

58. $\displaystyle\lim_{x\to0^+}x^{\left(x^2\right)}$

59. $\displaystyle\lim_{x\to0^+}\left(2^x-x\right)^{1/x}$

60. $\displaystyle\lim_{x\to0^+}(1-x)^{1/x}$

61. $\displaystyle\lim_{x\to0^+}\left(\frac{1}{x^2}\right)^{\csc x}$

62. $\displaystyle\lim_{x\to\infty}\left(\sqrt{x^2-3x}-\frac{3}{x^2+1}\right)$

63. $\displaystyle\lim_{x\to\infty}(x-1)^{1/x}$

64. $\displaystyle\lim_{x\to\infty}\frac{\ln x}{x^{7/5}}$

65. $\displaystyle\lim_{x\to0}(\cos x)^{\cot x}$

66. $\displaystyle\lim_{x\to0^-}(\cot x)^{\cos x}$

67. $\displaystyle\lim_{x\to0^+}\tan x\sec x$

68. $\displaystyle\lim_{x\to\infty}\frac{x^{100}}{3^x}$

69. $\displaystyle\lim_{x\to\infty}\frac{\ln(100x^2+e^x)}{100x}$

70. $\displaystyle\lim_{x\to0^+}2\sqrt{x}\csc x$

71. $\displaystyle\lim_{x\to0}(1+2x)^{1/x}$

72. $\displaystyle\lim_{x\to(\pi/2)^-}\left(\frac{\pi^2}{4}-x^2\right)\sec x$

73. $\displaystyle\lim_{x\to0}\frac{\sin 2x}{\tan 3x}$

74. $\displaystyle\lim_{x\to1}x^{1/(1-x)}$

75–85 Find the limit. If applicable, use l'Hôpital's Rule (as many times as appropriate).

75. $\lim\limits_{x \to \infty} \dfrac{2x^5 + x^3 - 4}{e^x}$

76. $\lim\limits_{x \to \infty} \dfrac{\cos x}{2^x}$

77. $\lim\limits_{x \to \infty} x \sin \dfrac{1}{x}$

78. $\lim\limits_{x \to \infty} x^{1/x^3}$

79. $\lim\limits_{x \to 0^+} x^{x^x}$

80. $\lim\limits_{x \to 0^+} \left(x^x\right)^x$

81. $\lim\limits_{x \to \infty} x^{1/x^n}, \quad n \in \mathbb{Z}^+$

82. $\lim\limits_{x \to \infty} \dfrac{(\ln x)^3}{x^2}$

83. $\lim\limits_{x \to 0} \dfrac{\sin x - x}{2x - e^x + e^{-x}}$

84. $\lim\limits_{x \to 0} \left(\dfrac{\sin x}{x}\right)^{x^2}$

85. $\lim\limits_{x \to 0} \dfrac{\sin x - x}{\tan x - x}$

86–91 Find the error(s) in the limit calculation.

86. $\lim\limits_{x \to 0} \dfrac{1 - \sin x}{x} = \lim\limits_{x \to 0} \dfrac{-\cos x}{1} = -1$ (Incorrect!)

87. $\lim\limits_{x \to 2} \dfrac{x^2 - 2}{x - 2} = \lim\limits_{x \to 2} \dfrac{2x}{1} = 4$ (Incorrect!)

88. $\lim\limits_{x \to -\infty} \dfrac{5^x + 1}{5^x} = \lim\limits_{x \to 2} \dfrac{(\ln 5)5^x}{(\ln 5)5^x} = 1$ (Incorrect!)

89. $\lim\limits_{x \to 0^+} x \cot x = \lim\limits_{x \to 0^+} (1)\left(-\csc^2 x\right)$
$= (1)(-\infty) = -\infty$ (Incorrect!)

90. $\lim\limits_{x \to 0} x \sin \dfrac{1}{x} = \lim\limits_{x \to 0} \dfrac{\sin \dfrac{1}{x}}{\dfrac{1}{x}}$

$= \lim\limits_{x \to 0} \dfrac{-\dfrac{1}{x^2} \cos \dfrac{1}{x}}{-\dfrac{1}{x^2}}$

$= \lim\limits_{x \to 0} \cos \dfrac{1}{x}$

$=$ does not exist   (Incorrect!)

91. $\lim\limits_{x \to 0^-} \dfrac{\cos x - x^2 - 1}{x^4 - 2x^3} = \lim\limits_{x \to 0^-} \dfrac{-\sin x - 2x}{4x^3 - 6x^2}$

$= \lim\limits_{x \to 0^-} \dfrac{-\cos x - 2}{12x^2 - 12x}$

$= \lim\limits_{x \to 0^-} \dfrac{\sin x}{24x - 12} = 0$ (Incorrect!)

92–106 Convince yourself that the initial use of l'Hôpital's Rule is not helpful in finding the limit. If possible, try to find a way to make use of the theorem, or evaluate the limit in some other way.

92. $\lim\limits_{x \to \infty} \dfrac{\sqrt{x+2}}{\sqrt{x}}$

93. $\lim\limits_{x \to \infty} \dfrac{\sqrt[3]{x+1} - 2}{\sqrt{x^2 + 2}}$

94. $\lim\limits_{x \to \infty} \dfrac{2^x + 3^x}{5^x}$

95. $\lim\limits_{x \to \infty} \dfrac{5^x - 6^x}{7^x + 8^x}$

96. $\lim\limits_{x \to \infty} \dfrac{2^{-x}}{x^{-1}}$

97. $\lim\limits_{x \to \infty} \left(\dfrac{1}{x+1}\right)^{-x^3}$

98. $\lim\limits_{x \to \infty} \left(\dfrac{1}{x^2}\right)^{e^{-x}}$

99. $\lim\limits_{x \to \infty} \dfrac{x}{\sqrt{x^2 + 1}}$

100. $\lim\limits_{x \to 0} \dfrac{\csc x}{\cot x}$

101. $\lim\limits_{x \to 0^+} \left(\cot x - \dfrac{5x+1}{x}\right)$

102. $\lim\limits_{x \to 1^+} \left(\dfrac{1}{x-1} - \dfrac{1}{\ln x}\right)$

103. $\lim\limits_{x \to \pi^+} (\cot x)^{\sin x}$

104. $\lim\limits_{x \to \infty} 2^{-x} x \ln x$

105. $\lim\limits_{x \to 0} (\sin x)^{\tan x}$

106. $\lim\limits_{x \to (\pi/2)^-} \left(\dfrac{1}{\dfrac{\pi}{2} - x} - \tan x\right)$

107–110 Find the limit of the sequence by considering the function you obtain after replacing n with the real variable x.

107. $\lim\limits_{n \to \infty} \dfrac{n^2 + 1}{2^n}$

108. $\lim\limits_{n \to \infty} \left(1 + \dfrac{1}{n}\right)^n$

109. $\lim\limits_{n \to \infty} \sqrt[n]{n}$

110. $\lim\limits_{n \to \infty} \dfrac{2^n + 5^n}{6^n}$

111–114 Use l'Hôpital's Rule to prove the assertion.

111. $\lim\limits_{x \to 0} \dfrac{\sin(kx)}{x^k} = \infty$ $(k > 1)$

112. $\lim\limits_{x \to \infty} \dfrac{p(x)}{e^{kx}} = 0$ ($p(x)$ is a polynomial, $k > 0$)

113. $\lim\limits_{x \to \infty} \dfrac{(\ln x)^n}{x^k} = 0$ $(n \in \mathbb{N}, k > 0)$

114. $\lim\limits_{x \to \infty} \dfrac{a^x}{x^n} = \infty$ $(a > 1, n \in \mathbb{N})$

115–122 Find the value(s) of c satisfying the conclusion of Cauchy's Mean Value Theorem. If the theorem doesn't apply, explain why.

115. $f(x) = x$, $g(x) = x^2 + 1$; $[0,1]$

116. $f(x) = x^3 - 1$, $g(x) = x^2 + 2x$; $[-1,1]$

117. $f(x) = x^3 - x$, $g(x) = -x^2 + 2x + 3$; $[-1,3]$

118. $f(x) = x^3$, $g(x) = -x^2$; $[-2,3]$

119. $f(x) = x^2 + 3x$, $g(x) = 3x^2 - 5x + 3$; $[-1,3]$

120. $f(x) = \dfrac{1}{x}$, $g(x) = \ln x$; $[1,2]$

121. $f(x) = \cos x$, $g(x) = \sin x$; $\left[-\dfrac{\pi}{2}, 0\right]$

122. $f(x) = x^2 - 5x - 9$, $g(x) = x^3 + x + 10$; $[-3,2]$

123–124 Prove that $f(x)$ has a removable discontinuity at $x = 0$. Then find the value of c so as to make f continuous.

123. $f(x) = \begin{cases} \dfrac{3\tan x - 2x}{5x^2 + 3x} & \text{if } x \neq 0 \\ c & \text{if } x = 0 \end{cases}$

124. $f(x) = \begin{cases} \left(e^x - \sin 2x\right)^{2/x} & \text{if } x \neq 0 \\ c & \text{if } x = 0 \end{cases}$

125. Recall the compound interest formula for the value of an investment of P dollars after t years, compounded n times a year at an annual interest rate of r:

$$A = P\left(1 + \frac{r}{n}\right)^{nt}$$

Use l'Hôpital's Rule to prove that if we let $n \to \infty$, we obtain the continuous compounding formula:

$$A = Pe^{rt}$$

126. The strength of an electric field due to a disk charge is obtained from the formula

$$E(x) = \frac{\sigma}{2\varepsilon_0}\left(1 - \frac{x}{\sqrt{x^2 + R^2}}\right)$$

where σ is the electric charge per unit area (in C/m^2), $\varepsilon_0 = 8.85 \cdot 10^{-12}\ C^2/Nm^2$, R is the radius

of the ring, and x is the distance to the charge in meters. Use l'Hôpital's Rule to confirm that $E(x) \to 0$ as $x \to \infty$. How is E affected by σ and R at a given distance? What happens to the rate of change of E as x increases? (**Hint:** Apply l'Hôpital's Rule to dE/dx as $x \to \infty$.)

127. Marquis de l'Hôpital first illustrated the rule named after him in his 1696 textbook, *Analyse des Infiniment Petits*. He used an example where the objective was to find

$$\lim_{x \to a} \frac{\sqrt{2a^3 x - x^4} - a\sqrt[3]{a^2 x}}{a - \sqrt[4]{ax^3}}$$

for $a > 0$. Determine the above limit.

4.4 **Technology Exercises**

128–131 Check whether the limit is of indeterminate form, and then use a computer algebra system to evaluate the limit.

128. $\displaystyle\lim_{x \to 1^+}(x-1)^{\ln x}$

129. $\displaystyle\lim_{x \to 0^+} \tan x \ln x$

130. $\displaystyle\lim_{x \to 0^+} x^{x + x^x}$

131. $\displaystyle\lim_{x \to 0}\left(\frac{1}{\sin^2 x} - \frac{1}{x^2}\right)$

132–133 Use a graphing calculator or computer algebra system to graph the function for different values of the parameter c. Examine how the values of the parameter affect the indicated limit.

132. $\displaystyle\lim_{x \to \infty}\left(1 + \frac{1}{cx}\right)^x$

What happens to the limit when $|c| \to \infty$?

133. $\displaystyle\lim_{x \to 0^+} \frac{1 - c^x}{cx}$

What happens to the limit when $c \to \infty$?

4.5 **Calculus and Curve Sketching**

TOPICS

1. A curve-sketching strategy

2. Newton's method

In the last several sections, we acquired a number of tools that provide a deeper understanding of the behavior of a function than can be obtained by algebra alone. In this section, we put all those tools to good use in producing detailed graphs of functions.

TOPIC 1 **A Curve-Sketching Strategy**

As Example 4 of Section 4.3 illustrated, the knowledge we gain from a function's first and second derivatives can be extremely useful in identifying some of the most pertinent features of its graph. Such knowledge is especially valuable when used in conjunction with the graphing capability found in calculators and software—when such technological aids fail to pick out subtle or unexpected behavior, calculus can usually be used to fill the gap.

All of the purely algebraic curve-sketching techniques that you have learned still apply, but those you have used in the past (for instance, in Section 1.2) can now be augmented with a few additional steps that pick out extrema, intervals of monotonicity and concavity, inflection points, and more complicated limits. Although the curve-sketching strategy below should not be taken as a rigid prescription (the steps may not all be relevant and don't necessarily need to be taken in the order presented), it does serve as a convenient summary of the tools we now have.

Curve-Sketching Strategy

To sketch the graph of the function $f(x)$, perform the following steps.

Step 1: Determine the **domain** of f and any **symmetry** it may have (e.g., symmetry with respect to a certain point or line). Check also to see if f may be **periodic**, a trait frequently found, for example, in trigonometric functions.

Step 2: Identify the **intercepts**, if possible. The y-intercept value of $f(0)$ is normally easy to determine, if 0 is in the domain of f. The x-intercepts are found by solving the equation $f(x) = 0$.

Step 3: Identify and plot any **asymptotes** for f. Recall that these can be horizontal, vertical, or oblique and are especially helpful in sketching the graphs of rational functions. Use l'Hôpital's Rule as necessary to help evaluate limits.

Step 4: Find f' and f''. **Critical points** of f are located where f' is undefined or 0, and potential inflection points are located where f'' is undefined or 0.

Step 5: Use the **Monotonicity Test** and the **Concavity Test** to find intervals where f is increasing, decreasing, concave up, and concave down. Use either the **First Derivative Test** or the **Second Derivative Test**, whichever is more convenient, to locate **relative extrema**. Compare the values of the relative extrema to locate the **absolute extrema**. The actual **inflection points** are points where f is continuous and where the concavity of the graph changes.

Step 6: Combine all of the information found to either sketch a graph by hand or use it in conjunction with a graph generated by a calculator or computer. If the graph is being constructed as part of solving some larger problem, the extrema, points of inflection, and asymptotic behavior you have found are likely to be of great use.

Example 1 ✐

Use the curve-sketching strategy to construct a graph of

$$f(x) = 10x^3 - 6x^2 + x.$$

Solution

The function f is a cubic polynomial, so from this we know that its domain is all of $\mathbb{R}$ and, since its leading coefficient is positive, we know $\lim_{x \to -\infty} f(x) = -\infty$ and $\lim_{x \to \infty} f(x) = \infty$. The y-intercept is easily seen to be 0, and to determine the x-intercept(s) we need to solve the equation $10x^3 - 6x^2 + x = 0$, or $x(10x^2 - 6x + 1) = 0$; since $10x^2 - 6x + 1 = 0$ has no real number solutions (by the quadratic formula), this tells us that the only x-intercept is also 0.

Since f is a polynomial, the function has no asymptotes. The first and second derivatives are as follows.

$$f'(x) = 30x^2 - 12x + 1$$
$$f''(x) = 60x - 12 = 12(5x - 1)$$

Solving $30x^2 - 12x + 1 = 0$ by the quadratic formula gives us the two critical points of $(6 \pm \sqrt{6})/30$, or approximately 0.12 and 0.28. Setting f'' equal to 0, we see that the only potential inflection point is at $\frac{1}{5} = 0.20$. We can now use the derivatives and these three points to make the following table, which succinctly captures a great deal of information about the behavior of f.

Domain	$\frac{1}{30}\left(6 - \sqrt{6}\right)$	$\frac{1}{5}$	$\frac{1}{30}\left(6 + \sqrt{6}\right)$
Sign of f'	$f'(0) > 0$	$f'\left(\frac{1}{5}\right) = -\frac{1}{5} < 0$	$f'(1) > 0$
Monotonicity	Increasing	Decreasing	Increasing
Sign of f''	$f''(0) < 0$		$f''(1) > 0$
Concavity	Concave down		Concave up

The first row of the table depicts the domain of f with the two critical points and the potential inflection point marked. In the second row, the sign of f' is determined by evaluating f' at three test points on the intervals defined by the critical points. Why were those particular test points used? For no better reason than convenience: the sign of f' is easily determined at 0 and 1, and the test point for the middle interval was selected merely because it lies halfway between the two critical points. Note that all we care about is the sign of f', but to determine this for the middle interval

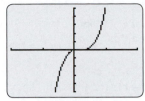

Figure 1

$$f(x) = 10x^3 - 6x^2 + x$$

on $[-2,2]$ by $[-5,5]$

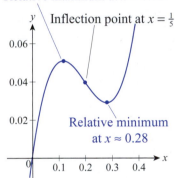

Figure 2

Zoomed-In Graph of

$$f(x) = 10x^3 - 6x^2 + x$$

we actually computed the value of f'. Given the signs of f' on the three intervals, we are led immediately to the conclusions about monotonicity on the intervals $\left(-\infty, \left(6-\sqrt{6}\right)/30\right)$, $\left(\left(6-\sqrt{6}\right)/30, \left(6+\sqrt{6}\right)/30\right)$, and $\left(\left(6+\sqrt{6}\right)/30, \infty\right)$ listed in the third row.

The sign of f'' on either side of the potential inflection point is easily calculated and appears in the fourth row, and the corresponding conclusions about concavity on the two intervals $\left(-\infty, \frac{1}{5}\right)$ and $\left(\frac{1}{5}, \infty\right)$ appear in the last row.

At this point, the First Derivative Test tells us, from the information in row two, that f has a relative maximum at $x = \left(6-\sqrt{6}\right)/30 \approx 0.12$ and a relative minimum at $x = \left(6+\sqrt{6}\right)/30 \approx 0.28$; we could also use the Second Derivative Test to conclude the same thing, as f'' is negative at the first critical point and positive at the second. And since the concavity does indeed change at $x = \frac{1}{5}$, this potential inflection point actually *is* an inflection point.

We are now ready to put all this knowledge together, and use it to refine any graphs we may obtain by calculator or computer. If we were to sketch the graph entirely by hand, we would calculate f at a few convenient points, paying special attention to the values at the extrema and the inflection point. If we are using technology to construct the graph, we would now know to zoom in on the portions around these same points. The importance of this is illustrated by the fact that our graph in Figure 1 (created with a graphing calculator) may lead us to incorrectly assume that f is an increasing function on its entire domain, but a close-up shows more interesting behavior centered around the inflection point as in Figure 2.

In the remaining examples, the steps in the curve-sketching strategy will be referred to more succinctly by number.

Example 2 ✎

Use the curve-sketching strategy to construct a graph of $f(x) = \dfrac{\sin x}{2 + \cos x}$.

Solution

Step 1: The denominator $2 + \cos x$ is never 0, so f is defined for all x and the domain is all of $\mathbb{R}$. Since $\sin x$ is an odd function and $1/(2 + \cos x)$ is even, f is an odd function (the product of an odd function and an even function is odd, but you can also verify that $f(-x) = -f(x)$). Moreover, since sine and cosine are both 2π-periodic, f is as well.

Step 2: Note that $f(0) = 0$, and the only solutions of $f(x) = 0$ are the solutions of $\sin x = 0$, namely integer multiples of π. So the origin is the only y-intercept and the x-intercepts are $(n\pi, 0)$, $n \in \mathbb{Z}$.

Step 3: No asymptotes

Step 4: $f'(x) = \dfrac{\cos x(2 + \cos x) - \sin x(-\sin x)}{(2 + \cos x)^2}$ Quotient Rule

$$= \dfrac{1 + 2\cos x}{(2 + \cos x)^2} \qquad\qquad \cos^2 x + \sin^2 x = 1$$

$$f''(x) = \dfrac{-2\sin x(2 + \cos x)^2 - (1 + 2\cos x)\cdot 2(2 + \cos x)(-\sin x)}{(2 + \cos x)^4}$$

$$= \dfrac{-4\sin x - 2\sin x\cos x + 2\sin x + 4\sin x\cos x}{(2 + \cos x)^3}$$

$$= \dfrac{2\sin x(\cos x - 1)}{(2 + \cos x)^3}$$

The first derivative is defined everywhere, and $f'(x) = 0$ only when the numerator $1 + 2\cos x$ is 0:

$$1 + 2\cos x = 0$$

$$\cos x = -\dfrac{1}{2}$$

$$x = \pm\dfrac{2\pi}{3} + 2n\pi, \quad n \in \mathbb{Z}$$

Similarly, f'' is defined everywhere and equal to 0 only when $\sin x = 0$ or $\cos x = 1$; these points are $x = n\pi$.

Step 5: Since f is 2π-periodic, we will make a chart for just one period of the graph.

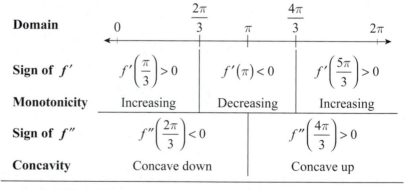

Domain	0	$\dfrac{2\pi}{3}$	π	$\dfrac{4\pi}{3}$	2π
Sign of f'		$f'\left(\dfrac{\pi}{3}\right) > 0$	$f'(\pi) < 0$		$f'\left(\dfrac{5\pi}{3}\right) > 0$
Monotonicity		Increasing	Decreasing		Increasing
Sign of f''		$f''\left(\dfrac{2\pi}{3}\right) < 0$			$f''\left(\dfrac{4\pi}{3}\right) > 0$
Concavity		Concave down			Concave up

Both derivative tests now tell us that $2\pi/3$ is a relative maximum and that $4\pi/3$ is a relative minimum, and we also now know that f has an inflection point at π. The absolute maximum value (which is attained at each relative maximum point) is $f(2\pi/3) = \sqrt{3}/3 \approx 0.577$, and the absolute minimum value is $f(4\pi/3) = -\sqrt{3}/3$. The concavity changes at each multiple of π, so each is an inflection point.

Step 6: The graph of f over two periods is shown in Figure 3.

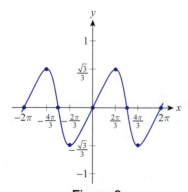

Figure 3

$$f(x) = \dfrac{\sin x}{2 + \cos x} \text{ on } [-2\pi, 2\pi]$$

Example 3 ✐

Use the curve-sketching strategy to construct a graph of

$$f(x) = x^{2/3}(3-x)^{1/3}.$$

Solution

Step 1: The domain of f is all of $\mathbb{R}$, and since $x^{2/3}$ is nonnegative everywhere and $(3-x)^{1/3}$ is positive to the left of 3 and negative to the right of 3 we know f will behave the same way as this second factor. There is no apparent symmetry or periodicity.

Step 2: Note that $f(0) = 0$, and the only solutions of $f(x) = 0$ are 0 and 3. So other than the origin, the only x-intercept is $(3,0)$.

Step 3: No asymptotes

Step 4:

$$f'(x) = \frac{2}{3}x^{-1/3}(3-x)^{1/3} + x^{2/3}\frac{1}{3}(3-x)^{-2/3}(-1) \qquad \text{Product Rule}$$

$$= \frac{1}{3}x^{-1/3}(3-x)^{-2/3}\left[2(3-x)-x\right] \qquad \begin{array}{l}\text{Factor out}\\ \frac{1}{3}x^{-1/3}(3-x)^{-2/3}.\end{array}$$

$$= \frac{2-x}{x^{1/3}(3-x)^{2/3}}$$

$$f''(x) = \frac{(-1)x^{1/3}(3-x)^{2/3} - (2-x)\left[\frac{1}{3}x^{-2/3}(3-x)^{2/3} + x^{1/3}\frac{2}{3}(3-x)^{-1/3}(-1)\right]}{x^{2/3}(3-x)^{4/3}} \qquad \begin{array}{l}\text{Quotient}\\ \text{Rule}\end{array}$$

$$= \frac{x^{-2/3}(3-x)^{-1/3}\left[-x(3-x) - (2-x)\left(\frac{1}{3}(3-x) - \frac{2}{3}x\right)\right]}{x^{2/3}(3-x)^{4/3}} \qquad \begin{array}{l}\text{Factor out}\\ x^{-2/3}(3-x)^{-1/3}.\end{array}$$

$$= \frac{-3x + x^2 - (2-x)(1-x)}{x^{4/3}(3-x)^{5/3}}$$

$$= \frac{-2}{x^{4/3}(3-x)^{5/3}}$$

From the formulas for f' and f'' we see that f has critical points at 0, 2, and 3 and potential inflection points at 0 and 3.

Step 5:

Domain	0		2	3
Sign of f'	$f'(-1) < 0$	$f'(1) > 0$	$f'(2.5) < 0$	$f'(4) < 0$
Monotonicity	Decreasing	Increasing	Decreasing	Decreasing
Sign of f''	$f''(-1) < 0$	$f''(2) < 0$		$f''(4) > 0$
Concavity	Concave down	Concave down		Concave up

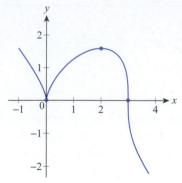

Figure 4

$$f(x) = x^{2/3}(3-x)^{1/3}$$

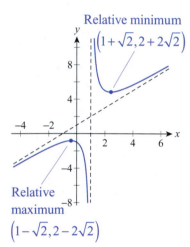

Figure 5 $h(x) = \dfrac{x^2+1}{x-1}$

The First Derivative Test tells us that f has a relative minimum at 0 and that 3 is not a relative extremum. Both tests indicate that f has a relative maximum at 2. The concavity only changes at 3, so that is the only point of inflection.

Step 6: The graph of f is shown in Figure 4.

Example 4 ✎

Use calculus to locate the relative extrema of the rational function $h(x) = (x^2+1)/(x-1)$ from Example 3c of Section 1.2.

Solution

We have already determined the asymptotes of h, and our sketch from Section 1.2 contains quite a lot of information. But the relative extrema would be difficult to determine without calculus. We can locate them quickly with h'.

$$h'(x) = \frac{(2x)(x-1) - (x^2+1)}{(x-1)^2} \qquad \text{Quotient Rule}$$

$$= \frac{x^2 - 2x - 1}{(x-1)^2}$$

Since 1 is not in the domain of h, the only critical points are where $x^2 - 2x - 1 = 0$; the solutions are $1 \pm \sqrt{2}$. Evaluating h at these points, we find a relative maximum value of $2 - 2\sqrt{2}$ at $1 - \sqrt{2}$ and a relative minimum value of $2 + 2\sqrt{2}$ at $1 + \sqrt{2}$ (see Figure 5). If we had not already sketched the graph of h, we could use either the First or Second Derivative Test to show that these were indeed relative extrema.

Example 5 ✎

Use the curve-sketching strategy to construct a graph of $f(x) = xe^x$.

Solution

Step 1: Note that e^x is defined for all x, so xe^x is as well. There is no apparent symmetry or periodicity.

Step 2: Since e^x is always positive, xe^x will be positive for positive x, negative for negative x, and 0 if, and only if, $x = 0$.

Step 3: It's clear that f grows without bound as $x \to \infty$, but the limit as $x \to -\infty$ is of indeterminate form $0 \cdot \infty$ (or, more precisely, of form $-\infty \cdot 0$). We can determine the limit using l'Hôpital's Rule:

$$\lim_{x \to -\infty} xe^x = \lim_{x \to -\infty} \frac{x}{e^{-x}} = \lim_{x \to -\infty} \frac{1}{-e^{-x}} = 0$$

Step 4: The first and second derivatives are as follows.

$$f'(x) = e^x + xe^x = e^x(x+1)$$

$$f''(x) = e^x + e^x(x+1) = e^x(x+2)$$

So f has a critical point at -1 and a potential inflection point at -2.

Step 5:

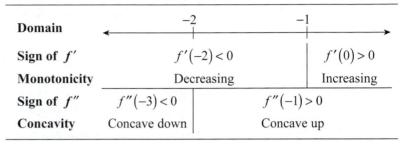

Domain		-2		-1	
Sign of f'		$f'(-2) < 0$		$f'(0) > 0$	
Monotonicity		Decreasing		Increasing	
Sign of f''	$f''(-3) < 0$		$f''(-1) > 0$		
Concavity	Concave down		Concave up		

Both derivative tests tell us that f has a relative minimum at -1. And the concavity changes at -2, so that is an inflection point.

Step 6: The graph of f is shown in Figure 6.

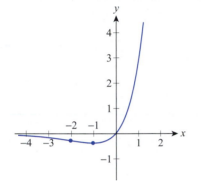

Figure 6 $f(x) = xe^x$

TOPIC 2   **Newton's Method**

The second step in the curve-sketching strategy says "Identify the intercepts, if possible." This step is usually fairly important; in fact, depending on the particular application, locating the solutions of the equation $f(x) = 0$ may be the whole motivation for sketching the graph of f in the first place.

Unfortunately, solving the equation $f(x) = 0$ is only a trivial task for a relatively small number of classes of functions. There are formulas for finding the roots of polynomials of degree four or less, but such a formula for degrees five and higher can't exist (the mathematicians Évariste Galois and Niels Henrik Abel, working independently, were the first to show this). And while the roots of the basic trigonometric functions are easy to determine, relatively simple algebraic combinations of them can be quite complicated. In practice, it is very common to wind up working with a function whose roots are elusive. And yet, graphing calculators and computer algebra systems seem to be able to at least give us approximations of roots of functions to seemingly any precision we need. How is that accomplished?

One of the simplest and widely used techniques is called *Newton's method* (or the *Newton-Raphson method*), and questions of this sort belong to an area of mathematics called *numerical analysis*. We have all the knowledge we need at this point to understand how Newton's method works.

Newton's Method

To find an approximation of a root of the function $f(x)$, perform the following steps.

Step 1: Begin with a guess x_1 of a solution of the equation $f(x) = 0$. A sketch of the graph of f can be very helpful in this step.

Step 2: Define new approximations x_2, x_3, and so on by successively applying the following formula.

$$x_{n+1} = x_n - \frac{f(x_n)}{f'(x_n)} \qquad (\text{if } f'(x_n) \neq 0)$$

Step 3: Continue until the desired degree of precision is obtained.

It is important to realize that Newton's method is not guaranteed to work, a characteristic of nearly all numerical analysis techniques. But a certain amount of trial and error, combined with a good understanding of the underlying mathematics, is usually sufficient to overcome any difficulties. The ways in which Newton's method may fail will become clear as we work through its derivation.

Consider the function f, part of whose graph appears in Figure 7. In this portion of the graph, f has exactly one root (shown as a red dot). If we didn't know the exact value of this root, we might start with the guess depicted as x_1. The linearization of f at $(x_1, f(x_1))$, shown as a red line, is

$$y - f(x_1) = f'(x_1)(x - x_1),$$

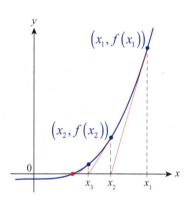

Figure 7 Successive Approximations to the Root

and it is easy to determine where this linear approximation crosses the x-axis:

$$0 - f(x_1) = f'(x_1)(x - x_1)$$
$$0 = f(x_1) + f'(x_1)(x - x_1)$$
$$-\frac{f(x_1)}{f'(x_1)} = x - x_1$$
$$x = x_1 - \frac{f(x_1)}{f'(x_1)}$$

If we label this result x_2 and repeat the process, we will (hopefully) obtain a sequence of points that converges to the actual root of f.

But the process can go awry in several different ways. First, if $f'(x_n) = 0$ for some x_n, the formula makes no sense. Geometrically, this means the linearization at this point is parallel to the x-axis, so it has no x-intercept. Further, if the initial guess x_1 is not close enough to the desired root, the iterative process can actually result in a sequence that either doesn't converge at all or else converges to a different root. Figure 8 illustrates this second possibility; the desired root is shown in red, but the sequence converges elsewhere.

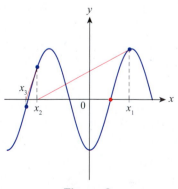

Figure 8
Newton's Method Going Astray

It is even possible for Newton's method to never converge no matter how close the initial guess is to the actual root; Exercises 83–87 contain examples of this unfortunate outcome. Other approximation techniques exist to handle such unusual cases, however, and are studied in classes like Numerical Analysis.

Example 6 ✐

Use Newton's method to approximate $\sqrt[3]{2}$ to five decimal places.

Solution

We are looking for the value of $x = \sqrt[3]{2}$; in other words, a solution of the equation $x^3 = 2$. So if we define $f(x) = x^3 - 2$, we see that we are seeking a root of the function f. Since $f'(x) = 3x^2$,

$$x_{n+1} = x_n - \frac{f(x_n)}{f'(x_n)} = x_n - \frac{x_n^3 - 2}{3x_n^2} = \frac{2(x_n^3 + 1)}{3x_n^2}.$$

As a rule of thumb, we can feel comfortable with our approximation if we continue iterating Newton's method until successive results agree to the number of decimal places we desire (more precise statements about convergence are studied in advanced classes). In our case, if we begin with a guess of $x_1 = 2$ we can obtain x_2 as follows.

$$x_2 = \frac{2(x_1^3 + 1)}{3 \cdot x_1^2} = \frac{2(2^3 + 1)}{3 \cdot 2^2} = 1.5$$

In a similar way, we obtain the subsequent approximations, truncating our results to six decimal places.

$$x_1 = 2 \qquad x_2 = 1.5 \qquad x_3 \approx 1.296296$$
$$x_4 \approx 1.260932 \qquad x_5 \approx 1.259921 \qquad x_6 \approx 1.259921$$

So to five decimal places, our approximation is $\sqrt[3]{2} \approx 1.25992$.

Example 7 ✐

Use Newton's method to approximate the largest negative root of $f(x) = e^x + \sin x$ to five decimal places.

Solution

Solving the equation $e^x + \sin x = 0$ is definitely not trivial, and a sketch of the graph (or at least a mental image of it) is immensely useful (see Figure 9).

Since $\sin x$ oscillates between -1 and 1 over all of $\mathbb{R}$, while e^x gets very large for positive x and quickly approaches 0 for negative x, we should expect e^x to dominate to the right of 0 and $\sin x$ to dominate to the left. The graph supports this expectation.

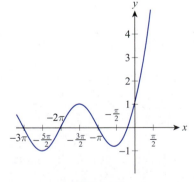

Figure 9
$f(x) = e^x + \sin x$

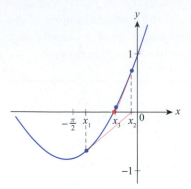

Figure 10

Zoomed-In Graph of
$f(x) = e^x + \sin x$

The largest negative root of f lies somewhere to the right of $-\pi/2$, as the close-up in Figure 10 indicates. An initial guess of 0 would work well to begin Newton's method, but for illustrative purposes we use $x_1 = -1.25$. Applying Newton's method, we obtain the following formula for x_{n+1}.

$$x_{n+1} = x_n - \frac{e^{x_n} + \sin x_n}{e^{x_n} + \cos x_n}$$

The first few approximations are as follows.

$$x_1 = -1.25 \qquad\qquad x_2 \approx -0.149219$$
$$x_3 \approx -0.534414 \qquad x_4 \approx -0.587419$$
$$x_5 \approx -0.588532 \qquad x_6 \approx -0.588533$$

Since x_5 and x_6 agree to five decimal places, the largest negative root of f is approximately -0.58853.

4.5 **Exercises**

1–4 The graphs of the first and second derivatives of a function f are given. Identify which one is which, and then sketch a possible graph of f. (Answers for the graph of f will vary.)

1.

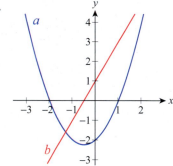

2.

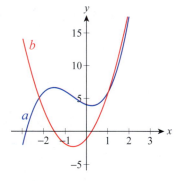

3.

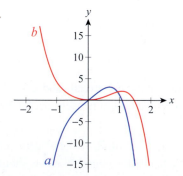

4.

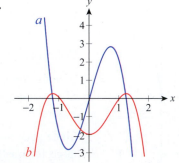

5–48 Use the curve-sketching strategy to construct a graph of the function.

5. $f(x) = x^3 + 3x^2 - 9x$

6. $g(x) = -x^3 + 2x^2 - x + 4$

7. $h(x) = \dfrac{1}{4}x^4 + \dfrac{5}{3}x^3 + x^2 - 8x$

8. $F(x) = -\dfrac{3}{4}x^4 + x^3 + 9x^2 + 2$

9. $G(x) = (x^2 - 1)(x^2 - 2)$

10. $k(x) = x^5 - 2x^3 - 8x + 1$

11. $L(x) = x^5 - 3x^2$ **12.** $m(x) = 4x^3 - 5x^4$

13. $n(x) = \dfrac{-3}{x-2}$ **14.** $H(x) = \dfrac{x^2 + 2}{x + 2}$

15. $R(x) = \dfrac{x}{x^2 - 4}$ **16.** $r(x) = \dfrac{2x^2 + 1}{x - 3}$

17. $A(x) = |x - 3| - 2$ **18.** $f(x) = 1.5 - |x - 2.2|$

19. $w(x) = x^{2/5} + \dfrac{2}{5}$ **20.** $u(x) = (x - 2)\sqrt{x}$

21. $F(x) = 2 - (x - 1)^{3/5}$ **22.** $G(x) = \sin^2 x - 1$

23. $h(x) = e^{-x} + e^{2x}$ **24.*** $H(x) = -2\sqrt{x} \cdot 2^{-2x}$

25. $P(x) = x \ln x$ **26.** $H(x) = 0.3\sqrt[3]{x}(x^2 - 1)$

27. $G(x) = x\sqrt{4 - x^2}$ **28.** $L(x) = \dfrac{3}{x - 2}$

29. $m(x) = \dfrac{x^2 + 7}{x - 7}$ **30.** $K(x) = \dfrac{e^x}{x}$

31. $F(x) = e^x - e^{-2x}$ **32.** $v(x) = (x - 1)^{3/5}$

33. $m(x) = -\dfrac{5}{(x - 2)^2}$ **34.** $R(x) = \dfrac{x + 2}{x - 4}$

35. $G(x) = \dfrac{-x}{x^2 - 1}$ **36.** $t(x) = \dfrac{2x^2 + 2}{x - 4}$

37. $H(x) = \dfrac{3}{4}\left(x - \dfrac{4}{3}\right)^{4/3}$ **38.** $w(x) = \dfrac{(x - 1)^2}{2x^2 - 2}$

39. $k(x) = x - \sqrt[3]{x}$ **40.** $F(x) = x^{4/5}\left(x - \dfrac{4}{5}\right)$

41. $c(x) = x\sqrt[3]{1 - x^2}$ **42.** $G(x) = -\sqrt{x}e^{-x}$

43. $Z(x) = 2\sin x - \cos^2 x$ **44.** $K(x) = \sin x - \cos x$

45. $L(x) = x^{5/3} \ln|x|$ **46.** $G(x) = \sqrt{4x^2 + 3}$

47. $u(x) = 7 - \sqrt{9x^2 + 2x + 1}$

48. $z(x) = e^{\cos x}$

49–54 First prove that $\lim\limits_{x \to \pm\infty} (f(x) - g(x)) = 0$. This means that when $x \to \pm\infty$, the graph of $f(x)$ approaches that of $g(x)$. Use this observation as an aid in graphing $f(x)$. (In this case, we say that $f(x)$ is asymptotic to $g(x)$.)

49. $f(x) = \dfrac{x^3 + 5}{x + 2}$, $g(x) = x^2 - 2x + 4$

50. $f(x) = \dfrac{(x + 1)^4 + 2}{3x + 3}$, $g(x) = \dfrac{1}{3}(x + 1)^3$

51. $f(x) = \sqrt{4x^2 + 5}$, $g(x) = |2x|$

52. $f(x) = \sqrt{x^2 - 4x + 5}$, $g(x) = |x - 2|$

53. $f(x) = \sqrt[3]{x} + \dfrac{1}{x^2}$, $g(x) = \sqrt[3]{x}$

54. $f(x) = \sin x + \dfrac{1}{x}$, $g(x) = \sin x$

55–56 Sketch on paper a few of the tangent lines that are used to approximate the largest root of the indicated function by Newton's method, using the starting values of −1, 0, and 1, respectively. Does the method always work? Explain.

55.

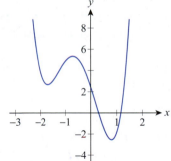

56.

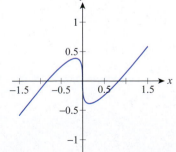

57–60 Use Newton's method to approximate the given number to five decimal places.

57. $\sqrt[4]{50}$

58. $\sqrt[10]{10}$

59. $\ln 5$

60. $\ln 100$

61–70 Use Newton's method to approximate the zero(s) of the given function to five decimal places. Restrict the domain to the given interval where indicated.

61. $f(x) = x^3 - x + 2$

62. $f(x) = 2x^3 + x^2 - 5x + 1$

63. $f(x) = x^4 - 6.1x^3 + 4.7x^2 - 12.2x + 5.4$

64. $f(x) = 0.25x^4 - 2x^2 + x + 0.69$

65. $f(x) = x^5 + x + 1$

66. $f(x) = 2x^5 - 5x^4 + 2x^3 - 4x^2 + 1$

67. $f(x) = 4.2x - \sqrt{x+3}$

68. $f(x) = \sqrt{2 + x^2} - 1.1x$

69. $f(x) = 2x^2 - \cos(x-1); \quad \left(0, \dfrac{\pi}{2}\right)$

70. $f(x) = \sin(2x+1) - \dfrac{x}{2}; \quad (0,1)$

71–76 Use Newton's method to solve the equation on the given interval. Approximate the root to six decimal places.

71. $\sin x = x^2$ on $\left(0, \dfrac{\pi}{2}\right)$

72. $2 - x^3 = e^x$ on $\mathbb{R}$

73. $x^4 = \arctan x$ on $(0, \infty)$

74. $\ln x = 2 - \sqrt{x}$ on $(0, \infty)$

75. $\cos x = \tan x$ on $\left(\dfrac{\pi}{2}, \dfrac{3\pi}{2}\right)$

76. $\log_{1/2} x = \sin x$ on $(0, \infty)$

77–80 Recall from Exercise 62 of Section 4.2 that $c \in \mathbb{R}$ is said to be a fixed point of $f(x)$ if $f(c) = c$. Use Newton's method to approximate to four decimal places the fixed point(s) of the function on the given interval.

77. $f(x) = e^{-x}$ on $(0, \infty)$

78. $f(x) = \cos x$ on $\mathbb{R}$

79. $f(x) = 2\cot x$ on $(0, 2\pi)$

80. $f(x) = \log_{1/2} x$ on $(0, \infty)$

81–82 Use Newton's method to find the critical point(s) of the function correct to five decimal places.

81. $f(x) = x^5 - x^3 - 5x$

82. $f(x) = x^2 \sin x, \quad 0 < x < \pi$

83–87 Perform the first few iterations of Newton's method for the given function with the indicated first guess, and explain why the method doesn't work.

83. $f(x) = \sin x - \cos x; \quad x_1 = -\dfrac{\pi}{4}$

84. $f(x) = x^3 - 6x^2 + 12x - 6; \quad x_1 = 3$

85. $f(x) = \begin{cases} -\sqrt{-x} & \text{if } x < 0 \\ \sqrt{x} & \text{if } x \geq 0 \end{cases}; \quad x_1 = a \; (a \neq 0)$

86. $f(x) = \sqrt[3]{x}; \quad x_1 = a \; (a \neq 0)$

87. $f(x) = -x^3 + 9x^2 - 19x + 19; \quad x_1 = 3$

88. The following rule for approximating the square root of a has been known since ancient times.

$$x_{n+1} = \dfrac{1}{2}\left(x_n + \dfrac{a}{x_n}\right)$$

Use Newton's method to derive this rule. (**Hint:** Start with the equation $x^2 - a = 0$.)

89. Generalizing Exercise 88, use Newton's method to derive a rule for approximating $\sqrt[k]{a}, \; k \geq 3$.

90. Using the approach you have taken in the previous two exercises, derive the following formula approximating $1/a$.

$$x_{n+1} = x_n(2 - ax_n)$$

91–96 Use the formulas you derived in Exercises 88–90 to approximate the given number to five decimal places.

91. $\sqrt{2}$

92. $\sqrt{50}$

93. $\sqrt[3]{10}$

94. $\sqrt[5]{30}$

95. $\dfrac{1}{7}$

96. $\dfrac{1}{19}$

4.5 **Technology Exercises**

97. Use a computer algebra system to approximate π by generating the first 10 iterations of Newton's method for solving the equation $\sin x = 0$ with an appropriate starting value.

98. Repeat Exercise 97 for the equation $(x-5)^{50} = 0$ with the starting value of $x_1 = 6$. What do you find? Graph $f(x) = (x-5)^{50}$, and see if the graph gives insight into why things went wrong.

99–100 Perform the first two iterations of Newton's method with each of the given starting values in an attempt to find the positive root of $f(x)$; then use a computer algebra system to come up with better approximations. What do you find? Graph $f(x)$, and see if the graph gives insight into why things went wrong.

99. $f(x) = x^3 - 2x - 1;\quad x_1 = 0.9,\quad x_1 = 0.8,\quad x_1 = -0.4$

100. $f(x) = x^4 - 6x^3 + 9.5x^2 - 1.5x - 4.9375;\quad x_1 = 3,\quad x_1 = 2.9,\quad x_1 = 0$

4.6 Optimization Problems

TOPICS

1. Mathematical/geometrical problems
2. Optimization problems from other disciplines

We return in this section to the study of optimization problems, a class of problems introduced briefly in Section 4.1; we now have many more calculus techniques that we can bring to bear on the task of finding optimal solutions to problems. Recall that, in this context, an optimal solution corresponds to the point (or points) where a given function attains an absolute minimum or absolute maximum value—both the point and the absolute minimum or maximum value are usually of interest.

To guide us, the general steps we would use in solving any application problem are stated here with a focus on locating absolute extrema.

Optimization Strategy

Step 1: Read the given information carefully and identify what is known and what needs to be found. What quantity must be optimized in order to solve the problem?

Step 2: Identify and appropriately label the relevant quantities and draw a diagram, if possible, as an aid in determining the relationships between them.

Step 3: Assign a label to the quantity that needs to be maximized or minimized, and work toward expressing that quantity (we'll give it the generic name f for now) as a function of one variable. This may require some intermediate steps in which you write equations relating the quantities you have identified and use algebra to write f as a function of one of them. Note the appropriate domain of f for the problem.

Step 4: Using the techniques of this chapter, find the absolute minimum or maximum of f (whichever is desired) on the domain.

Step 5: Check your answer to see if it makes sense as a solution to the problem.

TOPIC 1 Mathematical/Geometrical Problems

Although all of the examples in this section will be rephrased fairly quickly in mathematical terms, the first two that we explore are immediately of a mathematical or geometrical nature.

Example 1 ✎

A 15 in. by 24 in. piece of sheet metal is to be formed into an open-top box by cutting out a square from each of the four corners and folding up the sides. How large should each square be in order to achieve a box of maximum volume?

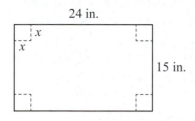

24 in.

x

x

15 in.

Figure 1

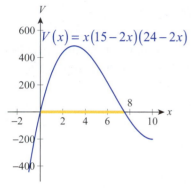

Figure 2

Solution

We begin with a picture, illustrating a rectangular piece of metal with four squares of side length x to be cut out (see Figure 1). Once the squares have been removed and the edges folded upward, the volume of the resulting open box will be

$$V(x) = x(15 - 2x)(24 - 2x).$$

Note that in this problem, we have quickly arrived at a function of one variable to be maximized—no intermediate steps were necessary. And in order for V to make sense as a volume, its domain is limited to $\left[0, \frac{15}{2}\right]$.

V is a cubic polynomial, and although a graph of V is not necessary in order to solve this problem, it's never a bad idea to graph your function to be maximized if it can be easily done—such a graph can serve as another self check of your work. In Figure 2, V is graphed over an interval larger than its domain for this problem, with the actual domain $\left[0, \frac{15}{2}\right]$ highlighted in yellow.

The remaining steps are to find the critical point(s) and test them to identify the absolute maximum—from the graph, we expect to find just one critical point in the domain, and that is indeed the case.

$$\begin{aligned}
V'(x) &= (15 - 2x)(24 - 2x) + x(-2)(24 - 2x) + x(15 - 2x)(-2) \\
&= 12(x^2 - 13x + 30) \\
&= 12(x - 10)(x - 3)
\end{aligned}$$

$$V'(x) = 0 \iff x = 3 \quad \text{or} \quad x = 10$$

The only critical point in $\left[0, \frac{15}{2}\right]$ is 3. Evaluating V at this critical point and at the endpoints of the domain we see that the maximum possible volume is obtained when $x = 3$:

$$V(0) = V\left(\tfrac{15}{2}\right) = 0 \quad \text{and} \quad V(3) = 486 \text{ in.}^3$$

Example 2 ✎

Find the rectangle of largest possible area that can be inscribed in a semicircle of radius r.

Solution

We begin again with a sketch in order to better understand the problem and to begin labeling relevant quantities and dimensions with names. First, a little experimentation leads to the realization that the rectangle of largest area will have one side on the diameter of the semicircle; any rectangle not drawn like this can be rotated so that it does and then enlarged until the opposite vertices are on the semicircle. It will also, by the same reasoning, be centered in the semicircle. We label the radius of the semicircle r and give the width and height of the inscribed rectangle the labels w and h, respectively (see Figure 3).

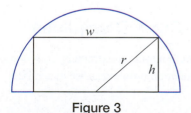

w

r

h

Figure 3

Given the labels we have assigned, the quantity we want to maximize is $A = wh$. At this point, A is a function of the two variables w and h, so we need to use another relationship in order to reduce A to a function of one variable.

From the figure and the Pythagorean Theorem, we see that

$$\left(\frac{w}{2}\right)^2 + h^2 = r^2,$$

or $w^2 + 4h^2 = 4r^2$. We can solve this equation easily for w to obtain $w = 2\sqrt{r^2 - h^2}$ (note that only the positive root makes sense), and then write

$$A(h) = 2h\sqrt{r^2 - h^2}.$$

From the geometry of the problem, as well as the nature of the formula, we know that $h \in [0, r]$; that is, the domain of A is the closed bounded interval $[0, r]$, so absolute extrema are guaranteed to exist.

We now proceed to find the critical point(s):

$$A'(h) = 2\sqrt{r^2 - h^2} - \frac{2h^2}{\sqrt{r^2 - h^2}} = \frac{2r^2 - 4h^2}{\sqrt{r^2 - h^2}}$$

$$A'(h) = 0$$

$$2r^2 - 4h^2 = 0$$

$$h = \pm\frac{r}{\sqrt{2}}$$

A' is also undefined when $h = r$, but that's one of the endpoints of the domain. Because $h = -r/\sqrt{2}$ is not in the domain of A, we evaluate A at the two endpoints and the one critical point $r/\sqrt{2}$ in $(0, r)$:

$$A(0) = 2(0)\sqrt{r^2 - 0} = 0$$

$$A\left(\frac{r}{\sqrt{2}}\right) = 2\frac{r}{\sqrt{2}}\sqrt{r^2 - \frac{r^2}{2}} = r^2$$

$$A(r) = 2r\sqrt{r^2 - r^2} = 0$$

So the maximum possible area for an inscribed rectangle is r^2, and it occurs when $h = r/\sqrt{2}$ and $w = r\sqrt{2}$. One way to quickly check the reasonableness of this answer is to note that r^2 is smaller than $\pi r^2/2$, the area of the semicircle.

TOPIC 2   Optimization Problems from Other Disciplines

The need to either minimize or maximize some quantity is extremely common, and the next few examples illustrate how optimization problems arise in a variety of contexts.

Example 3 ✐

As it prepares to expand into a new market, an American oil company decides to design a half-liter can for its brand of engine oil. The can will have the typical shape of a right circular cylinder. What dimensions for the can will minimize the amount of metal used to form it?

Solution

We will label the radius and height of the cylinder r and h, respectively. And since the can consists of two circles of radius r and a (rolled-up) rectangle of width $2\pi r$ and height h, the area of the metal used to form it is $A = 2\pi r^2 + 2\pi rh$.

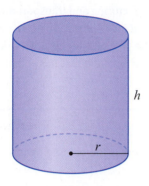

Figure 4

One liter corresponds to a volume of 1000 cm³, so a half liter of oil occupies 500 cm³. That is, $V = \pi r^2 h = 500$. We want to minimize A, which currently is a function of two variables. But we can use the volume relationship between r and h to eliminate one of them; it's easier to eliminate h (you should check that this is so).

$$A = 2\pi r^2 + 2\pi rh = 2\pi r^2 + 2\pi r\left(\frac{500}{\pi r^2}\right) = 2\pi r^2 + \frac{1000}{r}$$

Note that in this problem, the domain of A is $(0, \infty)$. In practice, we know we won't use a radius r close to 0, nor will we use a very large r, but we have no way to narrow down the domain at the moment. However, the calculus techniques we have learned will still resolve the question definitively. We proceed to find A' and then the critical points.

$$A'(r) = 4\pi r - \frac{1000}{r^2}$$

$$A'(r) = 0$$

$$4\pi r = \frac{1000}{r^2}$$

$$r^3 = \frac{1000}{4\pi}$$

$$r = \frac{10}{\sqrt[3]{4\pi}}$$

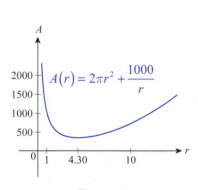

Figure 5

The only critical point is $r = 10/\sqrt[3]{4\pi} \approx 4.30$ cm.

The First Derivative Test verifies that A does indeed have a relative and absolute minimum at this critical point, as $A'(1) = 4\pi - 1000 < 0$ and $A'(10) = 40\pi - 10 > 0$ (1 and 10 are merely convenient test points). Figure 5 is a graph of A and confirms our findings. The corresponding value of h is as follows.

$$h = \frac{500}{\pi r^2} = \frac{500}{\pi\left(\dfrac{10}{\sqrt[3]{4\pi}}\right)^2} = \frac{(5)\left(2^{4/3}\right)}{\pi^{1/3}} = 10\left(\frac{2}{\pi}\right)^{1/3} \approx 8.60 \text{ cm}$$

Thus, to minimize the amount of metal used for the half-liter can, the radius should be approximately 4.30 cm and the height approximately 8.60 cm.

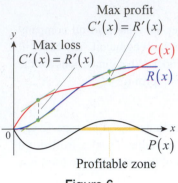

Figure 6

Profitable zone

Recall from Sections 2.6 and 3.7 that the profit $P(x)$ realized by selling x units of a given product is related to the cost $C(x)$ to produce x units and the revenue $R(x)$ by the formula $P(x) = R(x) - C(x)$. If we seek to maximize profit, the techniques of this chapter lead us to look at the critical points of P; in particular, if C and R are differentiable functions, we are interested in the solutions of the equation $P'(x) = 0$, which are also the solutions of the equation $R'(x) = C'(x)$. Figure 6 is a graph of typical cost and revenue functions—costs often initially exceed revenue, then fall below revenue as bulk manufacturing and transportation savings are realized, then eventually exceed revenue again as production capacity and market saturation points are reached. Note that the profitable zone is bounded by the two positive break-even points, where $C(x) = R(x)$, and that the points of maximum profit and loss are indeed where $C'(x) = R'(x)$.

Example 4 ✐

Suppose that Jan's Custom Calendar Company models the coming year's revenue and cost functions, in thousands of dollars, to be $R(x) = -x^3 + 9x^2$ and $C(x) = 2x^3 - 12x^2 + 30x$, where x represents units of 1000 calendars and where the model is thought to be accurate up to approximately $x = 6$. What is Jan's profitable zone, and what level of production will maximize her profit?

Solution

The profit function in this problem is $P(x) = -3x^3 + 21x^2 - 30x$, which factors as $P(x) = -3x(x-2)(x-5)$. Solving the equation $P(x) = 0$ is equivalent to solving $R(x) = C(x)$, so we know that the positive break-even points are $x = 2$ and $x = 5$. Therefore, Jan's profitable zone is between 2000 and 5000 calendars.

Similarly, we can solve either the equation $P'(x) = 0$ or the equation $R'(x) = C'(x)$ to find the relative extrema of P.

$$P'(x) = -9x^2 + 42x - 30$$
$$0 = -9x^2 + 42x - 30$$
$$x = \frac{7 \pm \sqrt{19}}{3} \qquad \text{Quadratic formula}$$
$$x \approx 0.880 \text{ and } 3.786 \qquad \text{Two critical points}$$

We can use either derivative test to determine that P has a relative minimum at 0.880 and a relative maximum at 3.786, corresponding to production of, respectively, 880 and 3786 calendars. The maximum possible profit is thus $P(3.786) = 24.626$ or \$24,626. Figure 7 shows the graphs of R, C, and P. Note how the relative extrema of P correspond to the points where $R' = C'$.

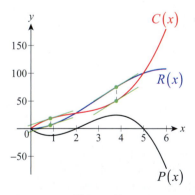

Figure 7

Example 5 ✎

Another useful model in business is the average cost function. If $C(x)$ represents the cost of producing x units of a product, then the average cost of producing those x units is $C(x)/x$. Show that if the average cost function is differentiable, then the marginal cost $C'(x)$ is equal to the average cost at the critical points of the average cost function.

Solution

We begin by looking at the critical points of $C(x)/x$.

$$\left[\frac{C(x)}{x}\right]' = \frac{C'(x)x - C(x)}{x^2} \qquad \text{Quotient Rule}$$

We are assuming that $C(x)/x$ is differentiable, so its critical points are where the expression above is 0.

$$\frac{C'(x)x - C(x)}{x^2} = 0$$

$$C'(x)x - C(x) = 0$$

$$C'(x) = \frac{C(x)}{x}$$

Example 6 ✎

You are probably familiar with the fact that an object underwater can appear to be at a position slightly offset from its actual position; this is due to refraction. In optics, Fermat's Principle says that the light rays observed in such a situation are those that travel from the object to your eyes in the shortest period of time, and light travels at different speeds through different media (such as water and air).

If v_1 and v_2 represent the speeds of light in, respectively, air and water, and θ_1 and θ_2 are the angles shown in Figure 8, use Fermat's Principle to derive Snell's Law:

$$\frac{\sin \theta_1}{\sin \theta_2} = \frac{v_1}{v_2}$$

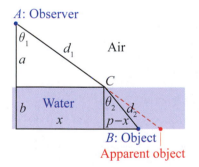

Figure 8

Solution

The given information consists only of the two angles and the two speeds, but in the figure we have already labeled other quantities that might help us relate the angles and speeds and arrive at Snell's Law. Specifically, we have let p denote the horizontal distance between the object B and the point directly beneath the observer at A. And since the point C of refraction (where the rays of light bend) is unknown, we have given its horizontal displacement from A the label x, meaning the horizontal distance between C and B is $p - x$. The vertical distances a and b are fixed, but the lengths of the hypotenuses, d_1 and d_2, will vary as x varies.

At this point, it may very well be unclear how to arrive at Snell's Law from what we have. But we haven't yet applied Fermat's Principle, and we can deduce many relationships between the labeled quantities. To begin with, since distance = rate · time, the time it takes light to travel from B to C is d_2/v_2 and the time it takes to travel from C to A is d_1/v_1. So the total time is expressed as follows.

$$T = \frac{d_1}{v_1} + \frac{d_2}{v_2}$$

We can express the two hypotenuses as functions of x by noting that $a^2 + x^2 = d_1^2$ and $b^2 + (p-x)^2 = d_2^2$, so

$$T(x) = \frac{\sqrt{a^2 + x^2}}{v_1} + \frac{\sqrt{b^2 + (p-x)^2}}{v_2}$$

and the domain of T is $[0, p]$.

The actual distance x must be the value of x that minimizes T, so our next step is to find T'.

$$T'(x) = \frac{x}{v_1\sqrt{a^2 + x^2}} - \frac{p-x}{v_2\sqrt{b^2 + (p-x)^2}}$$

$$= \frac{x}{v_1 d_1} - \frac{p-x}{v_2 d_2} \qquad \text{Substitute } d_1 \text{ and } d_2 \text{ for their formulas.}$$

$$= \frac{\sin\theta_1}{v_1} - \frac{\sin\theta_2}{v_2} \qquad \sin\theta_1 = \frac{x}{d_1} \text{ and } \sin\theta_2 = \frac{p-x}{d_2}$$

Note that $T_+'(0) < 0$ and $T_-'(p) > 0$, so by Darboux's Theorem (Section 3.1), there is a point $x \in [0, p]$ for which $T'(x) = 0$. And by the First Derivative Test, that point must minimize T. Rewriting $T'(x) = 0$, we have developed the formula for Snell's Law.

$$\frac{\sin\theta_1}{v_1} = \frac{\sin\theta_2}{v_2} \qquad \text{or} \qquad \frac{\sin\theta_1}{\sin\theta_2} = \frac{v_1}{v_2}$$

4.6 **Exercises**

1. Find two integers whose sum is 120 and whose product is as large as possible. (**Hint:** If you denote the first number by x, then the second number is $120 - x$. Now write a formula for the product, and use calculus to find the maximum.)

2–14 Use the strategy suggested in Exercise 1 to find two numbers satisfying the given requirements.

2. The sum is S and the product is a maximum.

3. The difference is 36 and the product is as small as possible.

4. Two positive numbers whose product is 144 and the sum is a minimum.

5. Two positive numbers whose product is n^2 and the sum is a minimum.

6. Two positive numbers whose product is 162 and the sum of twice the first and the second is a minimum.

7. Two positive numbers that are reciprocals of each other and their sum is a minimum.

8. Two positive integers so that the square of the first number plus the second number is 243 and their product is a maximum.

9. The sum of twice the first and three times the second is 480 and their product is a maximum.

10. The product of two positive integers is 32 and the sum of twice the first plus the second is a minimum.

11. Two numbers whose product is 16 and the sum of whose squares is a minimum.

12. Two positive numbers whose sum is 1 and the sum of whose cubes is a minimum.

13. Two nonnegative numbers whose sum is 1 and the sum of whose cubes is a maximum.

14. Repeat Exercises 12 and 13 using fourth powers instead of cubes.

15. Modify Example 2 by inscribing a rectangle in the region bounded by the x-axis and the parabola $y = k - x^2$ $(k > 0)$.

16. A vertex of a rectangle is at the origin; the opposite vertex sits in the first quadrant and on the line $2y + x = 4$. Find the dimensions that maximize the area of such a rectangle.

17. Repeat Exercise 16 with the opposite vertex sitting on the graph of $y = 32 - x^3$.

18. From among all lines through the point $(3, 1)$, find the one forming with the coordinate axes a right triangle of maximum area.

19. Repeat Exercise 18, this time finding the line forming a triangle whose hypotenuse is of minimum length.

20. Suppose that when constructing a trough similar to the one in Exercise 35 of Section 3.8, both the shorter base and the legs of its cross-section are 20 centimeters long. Find the base angle α that maximizes the volume of the trough.

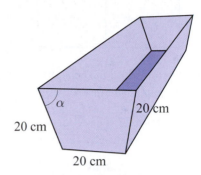

21. Find the coordinates of the point on the graph of $y = \sqrt{x}$ that is closest to the point $(1, 0)$.

22. Find the coordinates of the point on the graph of $y = x^3$ that is closest to the point $(-4, 0)$.

23. Find the dimensions of the rectangle of largest area that can be inscribed in the ellipse $2x^2 + 6y^2 = 12$.

24. Find the equation of the line tangent to the graph of $y = 1 - x^2$ that forms with the coordinate axes the triangle of minimum area in the first quadrant.

25. A farmer has 120 feet of fencing to construct a rectangular pen up against the straight side of a barn, using the barn for one side of the pen. The length of the barn is 100 feet. Determine the dimensions of the rectangle of maximum area that can be enclosed under these conditions. (**Hint:** Be mindful of the domain of the function you are maximizing.)

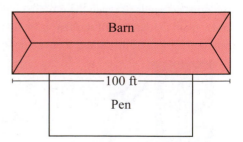

26. A farmer needs to construct two adjoining rectangular pens of identical areas, as shown. If each pen is to have an area of 1200 square feet, what dimensions will minimize the cost of fencing?

1200 ft²	1200 ft²

27. Repeat Exercise 26 if the pens are constructed against a straight wall that serves as a side for each.

28. Repeat Exercise 27 if three identical adjoining pens are to be constructed, as shown.

Wall

1200 ft²	1200 ft²	1200 ft²

29. A supporting beam with a rectangular cross-section is to be cut from a log that has an approximately circular cross-section with a radius of r inches. Knowing that the strength of such a beam is directly proportional to the width multiplied by the second power of the height of its cross-section, find the strongest beam that can be cut under these conditions.

30. An 8-foot fence stands 5 feet from a tall building. A contractor needs to reach the building with a ladder from the outside of the fence. Find the minimum length of the ladder that can do the job.

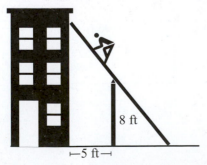

8 ft

⊢5 ft⊣

31. A 30 in. piece of wire is cut and the pieces are bent into a circle and a square, respectively. Where should we cut in order to minimize the sum of the areas of these two shapes?

32. Repeat Exercise 31, this time producing an equilateral triangle and a square.

33. Repeat Exercises 31 and 32, this time maximizing the sum of the two areas.

34. Prove that among all rectangles that can be inscribed in a circle, the square has the greatest perimeter.

35. Prove that among all isosceles triangles of a given area, the equilateral triangle has the minimum perimeter.

36. Find the dimensions of the rectangle whose perimeter is P units and area is a maximum.

37. Find the dimensions of the rectangle whose area is A units and perimeter is a minimum.

38. The perimeter of an isosceles triangle is P inches. Find the side lengths so as to minimize the sum of areas of the semicircles drawn onto the sides of the triangle.

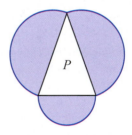

39. Suppose we want to construct a can in the shape of a right circular cylinder with no top whose surface area is to be S square inches. What dimensions will maximize the volume?

40. If we want to make a rectangular box with a square bottom and no top that holds 32 cubic inches, and the construction material costs 3 cents per square inch, what are the dimensions and the cost of the least expensive box that can be made?

41. If the box to be constructed in Exercise 40 is to hold the same volume, but we need to construct a top from an expensive, heat-resistant material that costs 21 cents per square inch, how does the new requirement change the cost and dimensions of the least expensive box?

42. Determine the dimensions and maximum volume of the rectangular box with no top and a square base if its surface area is A square inches.

43. A cone is to be constructed by cutting out a sector of central angle α of a disk of radius R and gluing the cut lines together to form a cone. Find the value of α that maximizes the volume of the cone.

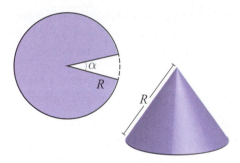

44. The pages of a children's book are to contain 54 square inches of printed matter and illustrations, with margins of 1 inch along the sides and $1\frac{1}{2}$ inches along the top and bottom of each page. Find the dimensions of the page that will require the minimum amount of paper.

45. A poster is to contain 150 square inches of printed matter, surrounded by margins that are 3 inches wide on the top and bottom, and 2 inches on each side. Find the dimensions for the poster that minimize its total area.

46. The sum of squares of lengths of the sides of a right triangle is 64 square inches. Find the side lengths that maximize the area of the triangle.

47. A flower bed is planned in the form of a circular sector. Find the central angle and radius if it is to cover 169 square feet, and its perimeter is to be a minimum.

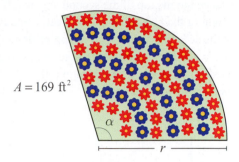

$A = 169$ ft^2

48. The shape of a Norman window can be approximated by a rectangle with a semicircle on top. What dimensions will admit the maximum amount of light if the perimeter of the window is to be P inches?

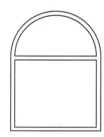

49. In Exercise 103 of Section 3.6, find the optimum distance s that maximizes the viewing angle.

50. An office building is located right on a riverbank, which is straight. A small power plant is on the opposite bank, 1500 feet downstream from the point directly opposite the office building. The river is 300 feet wide. If we want to connect the power plant and the building by cable, which costs $1700 per foot to lay down underwater and $800 per foot underground, what is the least expensive path for the cable?

51. Two antennas standing 30 feet apart are to be stayed with a single wire. The wire runs from the top of the first antenna, is secured to the ground somewhere between the antennas, and is finally attached to the top of the second antenna. If the height of the first antenna is 12 feet, while that of the second is 8 feet, find the point along the line segment connecting the bases where the wire needs to be staked to the ground if the length of the wire is to be minimal.

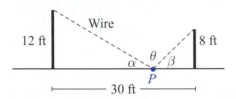

52. If we denote the heights of the antennas in Exercise 51 by h_1 and h_2, respectively, and the distance between them is d, prove that the wire has minimal length if and only if $\alpha = \beta$.

53. In Exercise 51, find the location of P that maximizes the angle θ.

54.* An inverted square pyramid is to be inscribed into a larger square pyramid of volume V, so that the two have a common axis, and the vertex of the inscribed pyramid coincides with the center of the outer pyramid's base. Find the ratio of the pyramids' altitudes so that the volume of the inscribed pyramid is maximal.

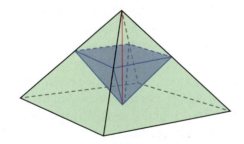

55.* The lower left corner of a letter-sized paper, which is 8.5 in. by 11 in., is folded over to reach the right edge of the paper. Find a way that this can be done so as to produce a crease of minimum length.

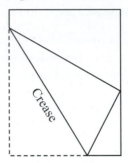

56. Find the radius of the base and the height of the right circular cylinder of largest volume that can be inscribed in a sphere of radius R.

57. Repeat Exercise 56, but inscribe a right circular cone instead of a cylinder in the sphere of radius R.

58. Find the radius of the base and the height of the right circular cylinder of largest volume that can be inscribed in a circular cone if the height of the cone is H and its base has radius R.

59. Repeat Exercise 58, but find the extremum of the surface area of the cylinder instead of its volume.

60. The sum of the height and the radius of the base of a circular cylinder is 12 inches. Find their lengths if the volume of the cylinder is to be a maximum.

61. Suppose that we want to send a parcel in the shape of a square-based rectangular solid, and the Standard Post service limits the sum of the length and girth (girth = the perimeter of the base) to 130 inches. Find the dimensions of the package of the greatest volume under these conditions.

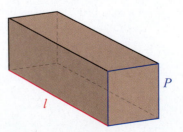

62. Find the maximum volume a right circular cone can have if its slant height is a inches.

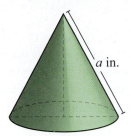

63. An isosceles triangle of perimeter P is rotated around its base. What base length will produce the solid of maximum volume?

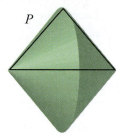

64. A lighthouse is 2 miles off a straight shoreline, and a grocery store is 10 miles down the coast. If the lighthouse keeper can row at 2.4 mph and walk at 4 mph, where should he land in order to make the best time to the store to get supplies? What if he is picked up by a golf cart that can drive at 9.9 mph?

65. Repeat Exercise 64 if the lighthouse keeper uses a motorboat whose top speed is 20.1 mph, and will be picked up by a car that will drive at the posted speed limit of 45 mph.

66. Repeat Exercise 64 if the lighthouse and the store are both on the shore of a circular lake of diameter d at the endpoints of the diameter.

67. At noon on a certain day, a plane is 200 miles south of another airliner and flying north at 550 mph, while the second plane is flying southwest at 600 mph. How much later after this instant is their distance a minimum?

68. A straight two-lane highway intersects a straight interstate at a right angle. A car exits the interstate and starts moving away from it on the two-lane highway at 50 mph. At the same instant, another car, moving at 75 mph on the interstate, is approaching the same intersection, but is still 10 miles from it. When will their distance be a minimum and what will this distance be?

69. The position of an object connected to a spring is given by $d(t) = \sin 3t + \cos 3t$, where d is measured in feet, and t in seconds. Find when the absolute value of its velocity first reaches its maximum and the value of the maximum velocity.

70. In Exercise 69, find when the absolute value of the acceleration first reaches its maximum and the value of this acceleration.

71. Ignoring air resistance, the range r of a projectile fired from the ground in a flat area with an initial velocity of v_0 can be calculated by $r = \left(v_0^{\,2}/g \right) \sin 2\theta$, where g is the gravitational acceleration and θ is the launch angle relative to the horizontal. Find the launch angle that maximizes the range if the initial velocity is a given constant.

72. The luminance E_l at distance d from a light source is directly proportional to the light intensity F_l (also called luminous flux) and inversely proportional to the square of distance: $E_l = F_l / \left(4\pi d^2 \right)$. Suppose two light bulbs are 3 meters apart, with respective light intensities of $F_{l,1} = 1700$ lumens (lm) and $F_{l,2} = 1000$ lm. Where between these light bulbs will the sum of their luminance levels be a minimum?

73. Management and Power, Inc. has found that its seminar on management techniques attracts 800 people when the seminar fee is set to $600. They estimate that for each $15 discount in the charge, an additional 50 people will attend the seminar. Find the amount that Management and Power, Inc. should charge for the seminar to maximize the revenue, and find the maximum revenue.

74. A blueberry farmer owns 1056 plants, each producing p pounds on average during a regular season. He estimates that for each additional dozen of new plants planted on his farm, average production per plant is going to drop by a half percent. What would be the optimum number of plants on the farm in order to maximize production, and what is the optimum production level?

75. The manager of a 115-unit apartment complex finds that all units are rented at a price of $1500 per month. Research shows that for each $20 increase in rent, one additional unit remains vacant. How much should he charge for rent in order to bring in maximum revenue, and how many units are rented then?

76. A moving company sends a truck on a 2000-mile round-trip to move two households. The hourly fuel consumption of the truck is approximated by $2 + \frac{1}{280.1}v^2$ gallons, where v is assumed to be a constant speed somewhere between 35 and 70 miles per hour. If a gallon of diesel fuel costs $2.50 and the driver is paid $22 an hour, what speed will minimize the company's transportation costs?

77. Cool Wheels, a manufacturer of die-cast model cars, has a monthly overhead cost of $6000, material costs of $2 per toy car, and each has associated labor costs of $0.40. When producing and marketing 2500 cars a month, each sells for $30.75. When producing more, it was found that for each additional 100 units, the market conditions cause the price to drop by a dollar. In addition, labor costs go up by 5 cents for each additional 100 units because of expensive overtime pay. Find the production level and selling price that maximize the profit under these conditions.

78. Suppose it costs a candy company $3 to produce and distribute a box of Chi-Can chipotle candy bars, and the number of boxes sold at x dollars a box is approximated by $n = \dfrac{80}{x-11} + 15(50-x)$. What sale price will bring the maximum profit?

79. Prove that when the company in Exercise 78 maximizes its profit, the marginal cost equals the marginal revenue.

80. Suppose that $R(x) = 2x^3 - 15x^2$ and $C(x) = 3x^3 - 25x^2 + 21x$ are the weekly revenue and cost functions for a particular commodity, where x represents units of 100 individual products and where the model is thought to be accurate up to approximately $x = 10$. What is the profit zone, and what level of production will maximize the profit? (See Example 4.)

81. The cost of manufacturing x units of a commodity is given by $C(x) = x^3 - 15x^2 + 12,000x$. Find the value of x that minimizes the average cost of production. (See Example 5.)

4.6 **Technology Exercises**

82–83 Use the graphing and symbolic differentiation capabilities of a computer algebra system to solve the problem.

82. Suppose we have a small supply of craft paint, enough for 1 square foot, and we want to use it to paint a regular tetrahedron and a cube from a children's toy set. What should the dimensions of these solids be if we want to maximize the total volume? How about minimizing the volume?

83. Repeat Exercise 82 for a tetrahedron and a sphere.

4.7 **Antiderivatives**

TOPICS

1. Finding antiderivatives
2. Antiderivatives in use

The underlying goal of this chapter has been to use knowledge about the derivative of a function to gain a better understanding of the function itself. Given that, it's fitting that in the last section of this chapter we ask the following: If we have perfect knowledge of f', can we recover f in its entirety? As we shall see, this question is rooted in more than mere intellectual curiosity. The answer, which we will fully develop in Chapter 5, has a great deal of both theoretical and practical importance.

TOPIC 1 **Finding Antiderivatives**

Since differentiation is the name we apply to the process of finding the derivative of a function, *antidifferentiation* is a natural name for the reverse process, and a function that results from antidifferentiation is termed an *antiderivative*.

Definition 💡

Antiderivative

A function F is called an **antiderivative** of a given function f on an interval I if $F'(x) = f(x)$ for all x in I.

Given all that we have learned about derivatives and differentiation, we already know quite a lot about antiderivatives and antidifferentiation. Example 1 illustrates this.

Example 1 ✎

Find an antiderivative of each of the following functions.

a. $f(x) = 3x^2$ **b.** $g(x) = \sin x$ **c.** $h(x) = \dfrac{1}{x} + 5$

Solution

We have seen functions identical or similar to f, g, and h appear as the derivatives of other functions, so we can use our experience with differentiation to "work backward" and arrive at an antiderivative of each.

a. $F(x) = x^3$ qualifies as an antiderivative of f, since $F'(x) = 3x^2$. Note that the interval on which this is valid is $(-\infty, \infty)$.

b. To find an antiderivative of $g(x) = \sin x$, we can start with the fact that $\dfrac{d}{dx}(\cos x) = -\sin x$. The result of this differentiation is off by a minus sign, but one of our many differentiation rules allows us to correct for that: if we define $G(x) = -\cos x$, then $G'(x) = \sin x$. Again, this is true on all of $(-\infty, \infty)$.

c. We know that $\dfrac{d}{dx}\ln|x| = \dfrac{1}{x}$ and $\dfrac{d}{dx}(5x) = 5$, so by the additive property of differentiation, $H(x) = \ln|x| + 5x$ is an antiderivative of h. This makes sense for all x where H and h are defined, so it's true for the interval $(-\infty, 0)$ and the interval $(0, \infty)$.

With a little thought, though, Example 1 already points out that the answer to the opening question of this section is "No, not uniquely." Another differentiation fact that we know is that two differentiable functions have the same derivative on an interval if and only if they differ by no more than a constant (the "only if" part is the second corollary of the Mean Value Theorem). This has the practical effect that each of the functions in Example 1 has an infinite number of antiderivatives. For instance, $x^3 - 17$, $x^3 + \frac{27}{8}$, and $x^3 + \pi$ all serve equally well as an antiderivative of $3x^2$. But the good news is that once we have found one antiderivative of a function on a given interval, we know the form of all its antiderivatives on that interval. That is the meaning of the following theorem, which is a reformulation of the corollary mentioned above.

Theorem 🔍

General Antiderivative Theorem

If F is an antiderivative of the function f on a given interval I, then any other antiderivative of f can be written as $F(x) + C$ for some constant C.

And the preceding theorem is the motivation for the next definition.

Definition 💡

General Antiderivative

If F is an antiderivative of the function f on a given interval I, then the **general antiderivative** of f on I is $F(x) + C$ where C represents an arbitrary constant.

Example 2 ✍

Find the general antiderivative of $f(x) = e^x - 2x$. Then find the particular antiderivative F that satisfies $F(0) = -2$.

Solution

Using the same reasoning as in Example 1, we can quickly verify that $e^x - x^2$ is an antiderivative of f, so the general antiderivative is $F(x) = e^x - x^2 + C$. In order to determine the constant C that will satisfy the condition $F(0) = -2$, we can simply solve the equation $-2 = F(0) = e^0 - (0)^2 + C = 1 + C$, so $C = -3$.

Graphically, the family of functions described by the general antiderivative consists of the same curve crossing at different points on the y-axis. The particular antiderivative $e^x - x^2 - 3$ that satisfies the given condition, is depicted in blue in Figure 1, with a few other members of the family in black.

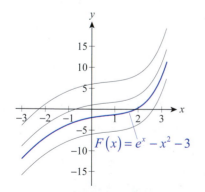

$$F(x) = e^x - x^2 - 3$$

Figure 1

Over the course of the next several chapters, we will learn some particular techniques for antidifferentiating various classes of functions, but the most basic technique is that which we have used in the first two examples; namely, using rules of differentiation in reverse. Table 1 indicates how a given differentiation rule can be restated as an antidifferentiation rule and, in the interest of brevity, Table 2 simply lists the general antiderivatives relating to basic trigonometric functions. In the tables, F and G represent antiderivatives of f and g, respectively, and k represents a nonzero constant (note the frequent use of the Chain Rule).

General Antiderivatives with Justifications

Function	General Antiderivative	Justification
$kf(x)$	$kF(x)+C$	$\dfrac{d}{dx}\left[kF(x)+C\right]=kF'(x)=kf(x)$
$f(x)+g(x)$	$F(x)+G(x)+C$	$\dfrac{d}{dx}\left[F(x)+G(x)+C\right]=f(x)+g(x)$
$x^r,\ r\neq -1$	$\dfrac{1}{r+1}x^{r+1}+C$	$\dfrac{d}{dx}\left[\dfrac{1}{r+1}x^{r+1}+C\right]=\dfrac{r+1}{r+1}x^{(r+1)-1}=x^r$
$\dfrac{1}{x}$	$\ln|x|+C$	$\dfrac{d}{dx}\left[\ln|x|+C\right]=\dfrac{1}{x}$
e^{kx}	$\dfrac{1}{k}e^{kx}+C$	$\dfrac{d}{dx}\left[\dfrac{1}{k}e^{kx}+C\right]=\dfrac{k}{k}e^{kx}=e^{kx}$
$a^{kx},\ a>0,a\neq 1$	$\left(\dfrac{1}{k\ln a}\right)a^{kx}+C$	$\dfrac{d}{dx}\left[\left(\dfrac{1}{k\ln a}\right)a^{kx}+C\right]=\dfrac{k\ln a}{k\ln a}a^{kx}=a^{kx}$

Table 1

General Antiderivatives of Trigonometric Functions

Function	General Antiderivative	Function	General Antiderivative
$\sin(kx)$	$-\dfrac{1}{k}\cos(kx)+C$	$\cos(kx)$	$\dfrac{1}{k}\sin(kx)+C$
$\sec^2(kx)$	$\dfrac{1}{k}\tan(kx)+C$	$\csc^2(kx)$	$-\dfrac{1}{k}\cot(kx)+C$
$\sec(kx)\tan(kx)$	$\dfrac{1}{k}\sec(kx)+C$	$\csc(kx)\cot(kx)$	$-\dfrac{1}{k}\csc(kx)+C$
$\dfrac{1}{\sqrt{1-(kx)^2}},\ |kx|<1$	$\dfrac{1}{k}\sin^{-1}(kx)+C$	$\dfrac{1}{1+(kx)^2}$	$\dfrac{1}{k}\tan^{-1}(kx)+C$
$\dfrac{1}{|kx|\sqrt{(kx)^2-1}},\ |kx|>1$	$\dfrac{1}{k}\sec^{-1}(kx)+C$		

Table 2

With practice, the general antiderivatives just summarized will be so familiar that you won't need to refer to the tables, but initially it may be convenient to use them as you work through the exercises at the end of this section. The next example shows how this can be done.

Example 3 ✒

Find the general antiderivative of each of the following functions.

a. $f(x) = \dfrac{1}{\sqrt{x}} - \csc^2 5x$

b. $g(x) = \dfrac{7x^{4/3} + 3x^{1/2}}{2x} - 2^x$

Solution

a. We can rewrite the function as $f(x) = x^{-1/2} - \csc^2 5x$, and we start by noting that the general antiderivative of $x^{-1/2}$ is

$$\frac{1}{-\frac{1}{2}+1} x^{-1/2+1} + C_1 = 2x^{1/2} + C_1$$

and the general antiderivative of $-\csc^2 5x$ is $\frac{1}{5}\cot 5x + C_2$ where C_1 and C_2 represent arbitrary constants. When we add these two antiderivatives, we get $2x^{1/2} + \frac{1}{5}\cot 5x + C_1 + C_2$. But $C_1 + C_2$ also represents nothing more than an arbitrary constant, and it is customary to combine all arbitrary constants in an antiderivative together and denote their sum with a single symbol. Hence, we write

$$F(x) = 2x^{1/2} + \frac{1}{5}\cot 5x + C$$

as the general antiderivative of $f(x) = \dfrac{1}{\sqrt{x}} - \csc^2 5x$.

(Remember: it is easy to check your work and verify that $F'(x) = f(x)$.)

b. It would be difficult to come up with an antiderivative of g in the form given, but after breaking apart the fraction the way forward becomes clear.

$$g(x) = \frac{7x^{4/3} + 3x^{1/2}}{2x} - 2^x = \frac{7}{2}x^{1/3} + \frac{3}{2}x^{-1/2} - 2^x$$

$$G(x) = \left(\frac{7}{2}\right)\left(\frac{3}{4}\right)x^{4/3} + \left(\frac{3}{2}\right)(2)x^{1/2} - \left(\frac{1}{\ln 2}\right)2^x + C$$

$$= \frac{21}{8}x^{4/3} + 3x^{1/2} - \left(\frac{1}{\ln 2}\right)2^x + C$$

TOPIC 2 Antiderivatives in Use

Example 2 is a simple example of a type of problem called a *differential equation*, which we will study in greater depth in Chapter 8. Informally, a differential equation is an equation that involves the derivative of an unknown function. Such equations arise naturally in many different settings, as it is often easiest to see how the derivative of a function (as opposed to the function itself) relates to other quantities. And if a differential equation involving $f'(x)$ is accompanied by information about the value of f at a particular point, say $f(x_0) = y_0$, then a particular solution of the equation (as opposed to a family of solutions) can be determined; such equations are called *initial value problems*.

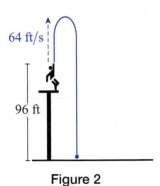

64 ft/s

96 ft

Figure 2

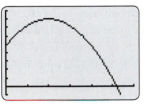

Figure 3

$h(t) = -16t^2 + 64t + 96$

on $[0,6]$ by $[-20,180]$

Example 4 ✎

A ball is thrown straight up at 64 ft/s by a person standing on the edge of a platform. At the moment the ball leaves the thrower's hand, it is 96 ft above the ground. On its way back down, the ball misses the platform and hits the ground. When does the ball reach its maximum height? What is that maximum height? When does the ball hit the ground?

Solution

If we let $h(t)$ denote the height (in feet) of the ball at time t (in seconds), then $h'(t)$ represents the ball's vertical velocity and we are given the initial conditions $h'(0) = 64$ and $h(0) = 96$. Recall that the acceleration due to gravity is -32 ft/s^2, so the rate of change of h' must be -32; that is, $h''(t) = -32$. Thus, the general antiderivative of h'' is $h'(t) = -32t + C$, and the condition $h'(0) = 64$ tells us that $h'(t) = -32t + 64$.

We know that the ball will transition from a positive (upward) velocity to a negative (downward) velocity as it stops rising and starts to fall, and its maximum height is attained at that time t when $h'(t) = 0$, namely $t = 2$ s. This tells us *when* the ball reaches its maximum height, but not what that maximum height *is*—we still need to determine $h(2)$. To do so, we must perform another antidifferentiation: given $h'(t) = -32t + 64$, we can determine that $h(t) = -16t^2 + 64t + C$, and for the condition $h(0) = 96$ to be true it must be the case that $h(t) = -16t^2 + 64t + 96$. Now, we can see that $h(2) = -16(4) + 64(2) + 96 = 160$ ft.

There is one question remaining, but we have all we need to answer it. The ball hits the ground when $h(t) = 0$, so we solve the equation $-16t^2 + 64t + 96 = 0$ and obtain $t = 2 + \sqrt{10} \approx 5.16$ s (the other solution is negative and doesn't apply).

Figure 3 confirms our findings.

Example 5 ✎

Given that $\dfrac{d^2 y}{dx^2} = 9e^{3x} + \sin x + 2$ and that $y'(0) = 3$ and $y(0) = 0$, find y.

Solution

We can determine the general form for $y'(x)$ by antidifferentiating the expression for $y''(x)$ term by term.

$$y'(x) = 3e^{3x} - \cos x + 2x + C$$

And for the condition $y'(0) = 3$ to be satisfied, it must be the case that

$$3 = y'(0) = 3e^0 - \cos 0 + 2(0) + C = 3 - 1 + C,$$

so $C = 1$.

We can now repeat the process with $y'(x) = 3e^{3x} - \cos x + 2x + 1$ to obtain

$$y(x) = e^{3x} - \sin x + x^2 + x + C,$$

and

$$0 = y(0) = e^0 - \sin 0 + (0)^2 + 0 + C = 1 + C$$

implies $C = -1$. So our final answer is

$$y(x) = e^{3x} - \sin x + x^2 + x - 1.$$

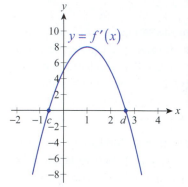

Figure 4

Our last example illustrates that the reasoning we have been using applies even when all we have to work with is a picture.

Example 6 ✎

Given the graph of f' in Figure 4 and the knowledge that f passes through $(1, 2)$, sketch the graph of f.

Solution

We begin by observing that f' is positive on the interval (c, d), negative on the intervals $(-\infty, c)$ and (d, ∞), and that $f'(c) = f'(d) = 0$. This tells us that f is increasing on (c, d), decreasing on $(-\infty, c)$ and (d, ∞), and that f has a relative minimum at $x = c$ and a relative maximum at $x = d$. Further, f' has a critical point at 1 and appears to be differentiable there (the graph of f' is nicely smooth), so $f''(1) = 0$; moreover, since f' is increasing to the left of 1 and decreasing to the right of 1, it must be the case that f changes concavity as it passes through the point $(1, 2)$.

Putting all these observations together, the graph of f must be something along the lines of the one in Figure 5, though the actual values of $f(c)$ and $f(d)$ can be nothing more than a rough guess.

The values of f' from the original graph tell us approximately how fast the graph of f rises or falls near a given point—for instance, since $f'(1) = 8$, the "slope" of f at $(1, 2)$ should be 8.

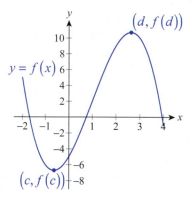

Figure 5

4.7 **Exercises**

1–8 Verify by differentiating that $F(x)$ is an antiderivative of $f(x)$.

1. $f(x) = \dfrac{1}{\sqrt{x}} + \dfrac{1}{x^2}, \quad F(x) = 2\sqrt{x} - \dfrac{1}{x}$

2. $f(x) = 2(x-1)(x+5), \quad F(x) = \dfrac{2}{3}x^3 + 4x^2 - 10x$

3. $f(x) = -x(x+2)(x-4), \quad F(x) = -\dfrac{1}{4}x^4 + \dfrac{2}{3}x^3 + 4x^2 + \dfrac{5}{3}$

4. $f(x) = 6\cos 3x, \quad F(x) = 2\sin 3x$

5. $f(x) = 5\sec^2(5x+1), \quad F(x) = \tan(5x+1) + 5$

6. $f(x) = \dfrac{x^2+1}{\sqrt{x}}, \quad F(x) = \dfrac{\sqrt{x}}{5}(10 + 2x^2)$

7. $f(x) = \dfrac{2x}{x^2+7}$, $F(x) = \ln(x^2+7)$

8. $f(x) = \pi^{2x}$, $F(x) = \dfrac{\pi^{2x}}{2\ln\pi}$

9–20 Find an antiderivative of the function.

9. $f(x) = 1$

10. $g(x) = 2x+2$

11. $h(x) = 4x^3 - x$

12. $u(x) = x^5 + x^3 + \pi$

13. $v(x) = \sec^2 x + 3x$

14. $k(x) = \dfrac{2}{x}$

15. $f(x) = 5e^x$

16. $m(x) = \dfrac{1}{2\sqrt{x}}$

17. $u(t) = -\dfrac{4}{t^3}$

18. $v(s) = \dfrac{1}{6s^{2/3}}$

19. $w(z) = \dfrac{1}{\sqrt{1-z^2}}$

20. $g(s) = \dfrac{1}{1+s^2} + 1 + s^2$

21–32 Find the general antiderivative of $f(x)$; then find the particular antiderivative $F(x)$ that satisfies $F(1) = 1$.

21. $f(x) = 2x - 3$

22. $f(x) = 3x^2 + \dfrac{1}{2}$

23. $f(x) = \dfrac{1}{\sqrt{x}}$

24. $f(x) = 1$

25. $f(x) = 0$

26. $f(x) = -\dfrac{1}{x}$

27. $f(x) = x^3 - \dfrac{1}{x^2}$

28. $f(x) = \dfrac{-1}{3x^{2/3}}$

29. $f(x) = (\ln 10)10^x$

30. $f(x) = \sin x$

31. $f(x) = \dfrac{\pi}{4}\sec^2\left(\dfrac{\pi}{4}x\right) + 1$

32. $f(x) = \dfrac{2x}{x^2+4}$

33–36 Given the graph of f' and the knowledge that f passes through the point $(3,1)$, sketch a possible graph for f.

33.

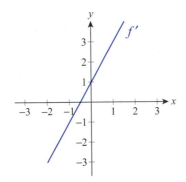

34.

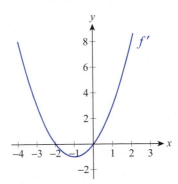

35.

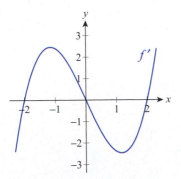

36.

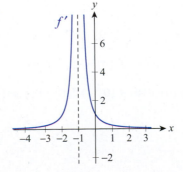

37–60 Find the general antiderivative of the given function, and check your answer by differentiation. (If necessary, rewrite the function before antidifferentiation.)

37. $f(x) = 6x^2 - 4x + 1.5$

38. $g(x) = 5x^3 - \pi x$

39. $h(x) = 3x^5 - 10x^4 + x^2 + 7$

40. $u(x) = -7x^4 + \dfrac{1}{2}x^3 + 6x^2 - 8x + \dfrac{5}{2}$

41. $v(x) = 3(x+6)(2x+1)$

42. $k(x) = -x(x+3)(7x-5)$

43. $h(x) = x^3\sqrt{x}$

44. $m(x) = \dfrac{3}{\sqrt{x}} + 2x\sqrt[3]{x}$

45. $n(x) = \dfrac{x^3 + 7x}{x^2}$

46. $f(t) = \dfrac{t^2 - t}{\sqrt{t} + 1}$

47. $a(y) = \left(\sqrt[3]{y^4} - 1\right)^2$

48. $w(z) = \dfrac{2}{z} + \dfrac{2}{\sqrt{z}}$

49. $g(t) = e^{3t} - 3\sec t \tan t$

50. $s(t) = 2 \cdot 10^{1.5t}$

51. $t(\theta) = \theta + \cos\theta$

52. $c(\theta) = \theta^2 + \csc^2\theta$

53. $v(x) = (\csc x - \cot x)\csc x$

54. $t(x) = -\sec^2 x\left(\cos^2 x + \sin^2 x\right)$

55. $w(x) = \dfrac{\cos x}{\cos^2 x - 1}$

56. $u(x) = \dfrac{2\tan 2x}{2\cos^2 x - 1}$

57. $a(x) = \dfrac{5}{1 + 9x^2}$

58. $b(x) = \dfrac{1}{\sqrt{1 - 4x^2}}$

59. $c(x) = \dfrac{4}{|5x|\sqrt{25x^2 - 1}}$

60. $d(x) = \dfrac{-3}{\sqrt{4 - 36x^2}}$

61–76 Find $f(x)$ that satisfies the specified conditions. (When no initial conditions are specified, find the general antiderivative.)

61. $f''(x) = \pi$, $f'(1) = 0$, $f(1) = 0$

62. $f''(x) = 1 - 4x$, $f'(-1) = 1$, $f(-1) = -4$

63. $f'''(x) = 0$, $f''(2) = 2$, $f'(2) = 2$, $f(2) = 2$

64. $f'''(x) = x + 1$, $f''(0) = 1$, $f'(0) = 2$, $f(0) = 3$

65. $f''(x) = \sqrt[3]{x}$, $f'(1) = 0$, $f(1) = \dfrac{1}{7}$

66. $f''(x) = x + \dfrac{1}{\sqrt{x}}$, $f'(4) = 6$, $f(4) = 0$

67. $f'''(x) = \sqrt{x} + 1$, $f''(0) = 1$, $f'(0) = -1$, $f(0) = 7$

68. $f''(x) = \sqrt[3]{x}(x - 3)$, $f'(0) = 0$, $f(0) = 0$

69. $f'(x) = \dfrac{4}{1 + 4x^2}$, $f\left(\dfrac{1}{2}\right) = \pi$

70. $f'(x) = \dfrac{-1}{\sqrt{1 - 3x^2}}$, $f\left(\dfrac{\sqrt{3}}{3}\right) = 0$

71. $f'''(x) = -\cos 2x$, $f''(0) = 1$, $f'(0) = 1$, $f(0) = -1$

72. $f'''(x) = \cos x - \sin x$

73. $f''(x) = 2^{5x}$

74. $f'''(x) = e^x + e$, $f''(0) = 1$, $f'(0) = 2$, $f(0) = 3$

75. $f''(x) = \cos x - e^{2x}$, $f'(0) = -2$, $f(0) = 1$

76. $f'''(x) = \sin 10x + 10x + 10$, $f''(0) = 0$, $f'(0) = 3.5$, $f(0) = -0.5$

77–85 Use $-32 \, \text{ft/s}^2$ for the acceleration caused by gravity ($-9.81 \, \text{m/s}^2$ in the metric system). Ignore air resistance. (**Hint:** See Example 4.)

77. A soccer ball is kicked upward from a height of 3 feet with an initial velocity of 48 feet per second. How high will it go?

78. A student drops a pen from a classroom window on the fourth floor of the mathematics building. If the window is 48 ft above ground level, how long is the pen in the air and with what speed does it hit the ground?

79. A hiker throws a pebble into a canyon that is 350 meters deep, with a downward initial velocity of 10 m/s. For how many seconds is the pebble in the air and what is the speed of impact?

80. A baseball is thrown upward from a height of 1.5 meters with an initial velocity of 30 meters per second. How high will it go, and for how long is it going to rise?

81.* With what initial velocity do we need to throw a tennis ball vertically upward in order for it to reach the top of a 60 ft campus flagpole?

82. An air rifle shoots a pellet at 1200 feet per second. What is the horizontal range of the rifle, that is, how far from where the pellet is shot will it hit the ground, if we shoot horizontally from a height of 5 feet?

83. A golf ball is hit horizontally at 40 meters per second from the top of a slight hill that is 1.5 meters high. If the terrain around the hill is nearly flat, approximately how far will the golf ball fly?

84. Prove that the position function of an object thrown vertically from an initial height of h_0 feet with an initial velocity of v_0 feet per second is $h(t) = -16t^2 + v_0 t + h_0$.

85. Repeat Exercise 84 using the metric system (meters and seconds) to arrive at the formula $h(t) = -4.905t^2 + v_0 t + h_0$.

86. The acceleration due to gravity on the lunar surface is approximately $-5.25 \, \text{ft/s}^2$. How high would the soccer ball of Exercise 77 fly on the Moon?

87. Find out what would happen in the situation described in Exercise 83 under lunar conditions. (See Exercise 86 for the acceleration due to gravity on the Moon.)

88. The rate of growth of a rabbit population in a certain state park, where food supply is limited and predators are present, is proportional to $e^{-0.1t}$, where t is time measured in months. If the initial population size is 300 rabbits, which grows to 400 in three months, find the population size in a year. (**Hint:** Let $P(t)$ stand for population size, and use $\frac{d}{dt}P(t) = ke^{-0.1t}$.)

89. The rate of growth of a population of a certain virus in a medical experiment is proportional to $\sqrt[3]{t}$, where t is time measured in days. If the initial population size is 1000, which grows to 1500 in a day, find the population size in five days. (See and appropriately modify the hint given in Exercise 88.)

90.* A modern Formula One car is able to come to a complete stop from 200 km/h (124.3 mph) using a braking distance of only about 65 meters. Assuming constant deceleration (which is not fully realistic), what multiple of g is this? (**Hint:** $1 \, \text{m/s} = 3.6 \, \text{km/h}$.)

91. The Bugatti Veyron, the fastest production grand tourer to date, can go from 0 to 100 km/h in 2.5 seconds. Find its position function when accelerating from a standstill and the distance covered during the first 1.5 seconds. What is the car's acceleration time from 0 to 60 mph? (Use the simplifying assumption that acceleration is constant. Also see the hint provided in Exercise 90.)

92. Jerry the mouse is running towards his hole at a steady speed of 11 ft/s. Still 20 feet from his destination, he is discovered by Tom the cat, who is 2 feet behind Jerry at that moment. If Tom can reach his top speed of 40 ft/s in 3 seconds, will he be able to catch Jerry? (Suppose the locations of Tom, Jerry, and the mousehole remain collinear throughout the pursuit.)

93.* Assume that an airplane needs to reach a liftoff speed of 180 mph and that it can achieve the same on a runway that is 0.8 miles long. Assuming constant acceleration during takeoff, what would this acceleration be?

94. The acceleration function of a particle moving along the x-axis is $a(t) = 3\sqrt{t} - \dfrac{1}{\sqrt{t}}$ units/s^2. If it starts at the origin with an initial velocity of 2 units per second, find the position function of the particle. Where will it be in 5 seconds?

95. Repeat Exercise 94 for the acceleration function $a(t) = (2-t)\sqrt{t}$, if the particle starts from rest at the point $(3,0)$. Where will it be in 5 seconds, and when will its instantaneous velocity be zero?

96. It follows from our discussions in Section 3.6 as well as the present section that an antiderivative of $-1/\sqrt{1-x^2}$ can be written as $-\sin^{-1} x$. Use the graphs of inverse trigonometric functions provided in Section 3.6 to argue that $\cos^{-1} x$ is also an antiderivative of $-1/\sqrt{1-x^2}$. (It follows that the general antiderivative of $-1/\sqrt{1-(kx)^2}$, $|kx| < 1$ is $(1/k)\cos^{-1}(kx) + C$. See also Exercise 67 of Section 3.6.)

97–103 *True or False?* Determine whether the given statement is true or false. In case of a false statement, explain or provide a counterexample.

97. If $f(x)$ has an antiderivative on an interval I, then it has infinitely many antiderivatives on the same interval.

98. All polynomials have antiderivatives on the entire real line $\mathbb{R}$.

99. It is possible for a function to have an unique antiderivative on an interval I.

100. Whenever F_1 and F_2 are both antiderivatives of f on an open interval, then $F_1 - F_2$ is a constant function.

101. If a function has an antiderivative on the interval $(-a, a)$ for some $a > 0$, then it has exactly one antiderivative whose graph goes through the origin.

102. Every antiderivative of a polynomial function of degree n has degree $n + 1$.

103. If $F(x)$ is an antiderivative of $f(x)$, and $G(x)$ is an antiderivative of $g(x)$ on an interval I, then $F(x) \cdot G(x)$ is an antiderivative of $f(x) \cdot g(x)$ on the same interval.

Chapter 4
Review Exercises

1–4 Sketch by hand the graph of a function f on the specified domain, with the specified properties. (Answers will vary.)

1. Defined on $(0,2)$, absolute minimum at 1, no absolute maximum

2. Defined on $[-2,2]$, absolute maximum occurs twice, no absolute minimum

3. Defined on $(-1,1)$, no absolute or relative extrema

4. Differentiable on $(0,1)$, no critical points, no extrema

5–14 Find all absolute extrema of the function on the indicated domain.

5. $f(x) = x^2 - \dfrac{4}{3}x^3$; $D = [-1,1]$

6. $f(x) = -x^4 + 8x^3 - 16x^2$; $D = [-1,5]$

7. $f(x) = \left| x^2 - 2x - 8 \right|$; $D = [-3,5]$

8. $f(x) = (x+2)|x|$; $D = [-2,2]$

9. $f(x) = \sqrt{x}(1-x)$; $D = [0,2]$

10. $f(x) = \dfrac{x^2+1}{x+1}$; $D = [0,1]$

11. $f(x) = 3x^5 - 2x^3 + 1$; $D = \mathbb{R}$

12. $f(x) = \sqrt{1-x^4}$; $D = (-1,1)$

13. $f(x) = \csc \dfrac{x}{2}$; $D = (0,2\pi)$

14. $f(x) = x^2 - x[\![x]\!]$; $D = [1,2]$

15–16 Prove that the equation has exactly one real solution on the given interval.

15. $3x^3 - x^4 = 1$ on $(0,1)$

16. $x \arcsin x = e^{-x}$ on $(0,1)$

17–18 Determine whether Rolle's Theorem applies to the function on the given interval. If so, find all possible values of c as in the conclusion of the theorem. If the theorem does not apply, state the reason.

17. $f(x) = x^3 + x^2 - 8x - 12$ on $[-2,3]$

18. $g(x) = x^4 + 2x^2 - 2$ on $[0,1]$

19–20 Determine whether the Mean Value Theorem applies to the function on the given interval. If so, find all possible values of c as in the conclusion of the theorem. If the theorem does not apply, state the reason.

19. $f(x) = \left| x^4 - 3x \right|$ on $[2,3]$

20. $f(x) = \left| x^4 - 3x \right|$ on $[0,2]$

21. If $|f'(x)| \le 3$ for all x, prove that
$$|f(10) - f(2)| \le 24.$$

22. If $g(-5) = -1$ and $g'(x) \le 4$ for all x, what is the largest possible value of $g(1)$?

23. Find the function f that passes through $(0,3)$ and whose derivative is $\cos x + e^x$.

24. An object is moving along the x-axis, starting at $x_0 = -4$ with velocity function $v(t) = 3t^2 - 2t$ $(0 \le t \le 5)$. Find the time t when it reaches the origin.

25. A car driving at 70 mph passes a mile marker, and then exactly 48 seconds later, still driving at 70 mph, passes the next mile marker.

   a. Prove that there was at least one instant when the car traveled at 75 mph between the markers.

   b. Prove that there was at least one instant when the car's acceleration was zero.

26–29 Use the first derivative to determine where the function is increasing and decreasing.

26. $f(x) = |2x - 2|$ 27. $g(x) = -x^2 + 6x + 7$

28. $h(x) = x^3 - 6x^2 + 9x + 1$

29. $k(x) = x^4 - 4x^3 - 20x^2 + 96x$

30–33 Determine the intervals of concavity of the given function.

30. $f(x) = \dfrac{x^3}{6} - x^2 - 2$ **31.** $g(x) = \dfrac{e^x}{x}$

32. $h(x) = (x-2)\sqrt[3]{x}$ **33.** $k(x) = \dfrac{x+3}{x^2-1}$

34–35 Use the first and second derivatives to identify the intervals of monotonicity, extrema, intervals of concavity, and inflection points of the given function.

34. $f(x) = 3x^5 - 20x^3$ **35.** $g(x) = \arctan(x^2)$

36–37 The function $p(t)$ gives the position, relative to its starting point, of an object moving along a straight line. Identify the time intervals when the object is moving in the positive versus negative direction, as well as those intervals when it is accelerating or slowing down. Find the times when the object changes direction as well as those when its acceleration is zero.

36. $p(t) = t^3 - 3t^2, \quad 0 \le t \le 5$

37. $p(t) = 3t^2 - \dfrac{t^4}{2}, \quad 0 \le t \le 3$

38–49 Check whether l'Hôpital's Rule applies to the given limit. If it does, use it to determine the value of the limit. If it does not, find the limit some other way. (When necessary, apply l'Hôpital's Rule several times.)

38. $\displaystyle\lim_{x\to 0} \dfrac{\sqrt{x+4}-2}{x}$ **39.** $\displaystyle\lim_{x\to 0} \dfrac{x - \tan x}{\sec x - 1}$

40. $\displaystyle\lim_{x\to\infty} \dfrac{e^{-x}}{\ln x}$ **41.** $\displaystyle\lim_{x\to 0^+} \cot x \csc x$

42. $\displaystyle\lim_{x\to 0} \left(1 + 4x^2\right)^{1/x^2}$

43. $\displaystyle\lim_{x\to (1/2)^+} \left(\dfrac{1}{4x-2} - \dfrac{1}{\ln 2x} \right)$

44. $\displaystyle\lim_{x\to 0^+} x^{\sqrt{x}}$ **45.** $\displaystyle\lim_{x\to 0^+} \left(\sqrt{x}\right)^{\ln x}$

46. $\displaystyle\lim_{x\to 0^-} x \cot x$ **47.** $\displaystyle\lim_{x\to 0} \dfrac{\arctan x}{\arctan 2x}$

48. $\displaystyle\lim_{x\to 0^+} \sin x \ln x$ **49.** $\displaystyle\lim_{x\to 0} \left(\dfrac{1}{x} - \csc x \right)$

50. By examining the limits $\displaystyle\lim_{x\to\infty} \dfrac{\ln x}{x^a}$ and $\displaystyle\lim_{x\to\infty} \dfrac{x^a}{b^x}$ using l'Hôpital's Rule, compare the relative growth rates of the functions $y = \ln x$, $y = x^a$, and $y = b^x$ ($a > 0$, $b > 1$). (See Exercise 88 in the Chapter 3 Review.)

51–60 Use the curve-sketching strategy to construct a graph of the function.

51. $f(x) = \dfrac{x^3}{3} - x^2 - 15x$

52. $g(x) = -x^3 + 12x - 16$

53. $h(x) = x^4 - 2x^3$ **54.** $m(x) = 3x^3 - 4x^5$

55. $f(x) = \dfrac{2x}{x^2+1}$ **56.** $f(x) = \dfrac{2x}{x^2-1}$

57. $f(x) = |x|(x-1)$ **58.** $f(x) = x\sqrt{9-x^2}$

59. $f(x) = \dfrac{x^3}{x^2-4}$ **60.** $f(x) = \sin^2 x \cos x$

61–62 Use Newton's method to approximate the given number to five decimal places.

61. $\sqrt[5]{30}$ **62.** $\log 11$

63–64 Use Newton's method to solve the equation on the given interval. Approximate the root to six decimal places.

63. $2x^5 = 1 - x^2$ on $(0,1)$ **64.** $\ln x = \cos x$ on $(0,\infty)$

65. Use Newton's method to approximate to four decimal places the fixed point(s) of $f(x) = 1 - \tan x$ on $(0, \pi/2)$. (See Exercise 62 of Section 4.2.)

66. Find a positive number that is greater than its own cube by the greatest possible amount.

67. Generalize Exercise 66 to the nth power of a number ($n \ge 2$).

68. Find a number a so that for given $a_1, a_2, a_3 \in \mathbb{R}$, the quantity $S_3 = (a-a_1)^2 + (a-a_2)^2 + (a-a_3)^2$ is minimal.

69. Generalize Exercise 68 for n given numbers to minimize the quantity:
$$S_n = (a-a_1)^2 + (a-a_2)^2 + \cdots + (a-a_n)^2.$$

70. A wire of length l is bent into an L shape. Where should be bend in order to minimize the distance between the two endpoints?

71. Find the length l and width w of the rectangle inscribed in the unit circle for which $l^2 w$ is maximal.

72. Find the dimensions of the rectangle whose diagonal is d units and whose area is maximum.

73. A book page of area 500 cm² is required to have 1 cm margins on the side, while the margins on the top and bottom are to be 2 cm. Find the dimensions of the page that maximize the printed area.

74. Among all isosceles triangles whose legs are l units long, find the base angle that maximizes the area.

75. A vertex of a rectangle is at the origin and the opposite vertex sits in the first quadrant and on the graph of $y = \dfrac{2-x}{x+1}$. Find the maximum possible area for such a rectangle.

76. Find the point on the graph of $y = 1 - x^2$ that is closest to the point $(-3,1)$.

77. Among all isosceles triangles that can be inscribed in the circle of radius R, find the one with maximum area.

78. A vending machine sells 500 bars of a certain type of candy when the price is $1.50. It was discovered that 10 fewer customers will buy the candy bar for each 5¢ increase in price. What is the price that will bring maximum revenue from the sales of this type of candy bar?

79. Maximize the surface area of the can in Example 3 of Section 4.6. Explain your findings.

80. Minimize the cost of producing the can in Example 3 of Section 4.6 if the top and bottom are produced using a material that is 50% more expensive than the material used for the side.

81. Nate needs to reach a restaurant that is 600 ft upstream on the other side of a 150 ft wide river. Find the point where he has to reach the other side in order to make the best time if he can swim at 5 ft/s and walk at 9 ft/s. (Ignore the flow of the river.)

82–89 Find the general antiderivative of the given function and check your answer by differentiation. (If necessary, rewrite the function before antidifferentiation.)

82. $f(x) = 2x^3 - 6x^2 + 3x$

83. $f(x) = 5x^4 - 4.8x^3 + e^2$

84. $f(x) = x(x+2)(2x-3)$

85. $f(x) = 0.4x\sqrt{x} - \dfrac{2}{\sqrt{x}}$

86. $f(x) = \dfrac{x^4 - 4x}{x^2}$

87. $f(x) = 2(x + \sec^2 2x)$

88. $f(x) = 6e^{3x}$ **89.** $f(x) = \dfrac{3}{4x^2 + 1}$

90–91 Find $f(x)$ that satisfies the specified conditions.

90. $f''(x) = x, \quad f'(1) = 1, \quad f(1) = 0$

91. $f'''(x) = 2, \quad f''(2) = -1, \quad f'(2) = 2, \quad f(2) = 3$

92. A tennis ball is thrown upward from an initial height of 4 feet with an initial velocity of 56 feet per second. How high will it go and for how long is it rising? (Ignore air resistance.)

93. With what initial velocity do we need to throw a golf ball vertically upward in order for it to rise 100 feet high? (Ignore the initial height and air resistance.)

94. A pebble is shot horizontally using a slingshot at 10 meters per second from the top of a building that is 20 meters high. If the terrain around the building is nearly flat, approximately how far from the building will the pebble hit the ground? (Use the approximation $g \approx 10$ m/s² and ignore air resistance.)

95–101 *True or False?* Determine whether the given statement is true or false. In case of a false statement, explain or provide a counterexample.

95. A continuous function on a finite interval always attains its maximum and minimum.

96. If $f(x)$ has a relative maximum or minimum at $x = c$, then $f'(c) = 0$.

97. If $f(x)$ has a relative maximum or minimum at $x = c$, then c is a critical point of f.

98. A cubic polynomial has exactly one inflection point.

99. If $f(x)$ is a polynomial, then between two consecutive local extrema there must be an $x = c$ so that $f''(c) = 0$.

100. If $f(x)$ is a polynomial and c is a critical point, then there is a relative maximum or minimum at $x = c$.

101. If $f'''(c) = 0$, then $f'(x)$ has a point of inflection at $x = c$.

Chapter 4
Technology Exercises

102–111. Use a graphing calculator or computer algebra system to verify the answers you obtained for Exercises 51–60.

112–113. Use a graphing calculator or computer algebra system to verify the conclusions of Exercises 15 and 16.

Chapter 5
Integration

5.1 Area, Distance, and Riemann Sums 425

1. The Area Problem
2. The Distance Problem
3. Riemann Sums and Summation Formulas

5.2 The Definite Integral 437

1. Riemann Sums and the Definite Integral
2. Properties of the Definite Integral
3. Selected Proofs and Comments

5.3 The Fundamental Theorem of Calculus 451

1. The Fundamental Theorem, Part I
2. The Fundamental Theorem, Part II

5.4 Indefinite Integrals and the Substitution Rule 465

1. The Meaning of Indefinite Integration
2. The Substitution Rule

5.5 The Substitution Rule and Definite Integration 475

1. Substitution and Definite Integration
2. Area between Curves

Introduction

Up to this point, we have focused on the concept of the derivative of a function and on developing methods to find and use derivatives. We have seen that a differentiable function f gives rise to a potentially endless list of successive derivatives f', f'', and so on. In particular, we closed Chapter 4 with the question: *If we have perfect knowledge of the derivative of a function, can we use that to completely determine the function itself?*

One way to approach the subject of this chapter is to literally reverse the differentiation process. That is, given a function f, is it possible to find a function F such that $F' = f$? As we learned in Section 4.7, if such a function F exists, we call it an antiderivative of f. But another approach begins with questions that are seemingly unrelated to antidifferentiation, such as *How do we calculate the area bounded between an axis and the graph of a function?* and *If we know an object's instantaneous velocity and its starting point, can we determine its location at any later moment in time?* We call the process of answering such questions *integration*, and the genius of Sir Isaac Newton (1643–1727) and Gottfried Wilhelm Leibniz (1646–1716) lay in their ability to fully grasp the relationship between integration and antidifferentiation.

The *Fundamental Theorem of Calculus* is the name we now give to this relationship. As is often the case with deep fundamental truths, the theorem required a lengthy period of fermentation and the contributions

> ## The sum of all the infinitesimal changes of a function over a region is equal to the net change in the function.

of many others before Newton and Leibniz articulated it in the form we use today. The mathematicians James Gregory (1638–1675) and Newton's teacher Isaac Barrow (1630–1677) were two such forerunners, but the philosophical basis of the Fundamental Theorem of Calculus appeared as early as the 3rd century BC, when Archimedes used the "method of exhaustion" to find the area under a parabola and the volumes of certain objects.

Informally, the underlying philosophy might be stated in this manner: The sum of all the infinitesimal changes of a function over a region is equal to the net change in the function.

As we will see by the time we finish Chapter 15 of this text, the underlying philosophy takes many specialized forms and is used to solve problems of astonishing diversity, ranging from the areas under parabolas and volumes of objects to problems in electrostatics, gravitational mechanics, and fluid flow. The insights of early Greek mathematicians from millennia ago, expanded and unified by mathematicians of the 17th century, continue to be principal means of discovery and innovation today.

Gottfried Wilhelm Leibniz
(1646–1716)

Sir Isaac Newton
(1643–1727)

5.1   **Area, Distance, and Riemann Sums**

TOPICS

1. The area problem

2. The distance problem

3. Riemann sums and summation formulas

We began our development of the derivative by discussing two broad problem types: the velocity problem and the tangent problem. In a similar fashion, we will develop the central topic of this chapter, the *integral*, by analyzing area and distance problems. By the end of the chapter, we will have sketched the outlines of an area of math called *integral calculus*, complementing the *differential calculus* we have studied to this point. As we will see in Section 5.3, the Fundamental Theorem of Calculus serves as the linchpin connecting these two areas.

TOPIC 1   **The Area Problem**

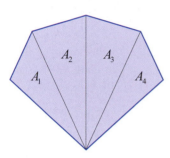

Figure 1

$$A_{\text{Polygon}} = A_1 + A_2 + A_3 + A_4$$

As with so many concepts in mathematics, that of *area* possesses both elementary features and features with surprising depth—it is the discovery of such hidden depth that often lures people into further study of math. You are familiar with formulas for the areas of simple plane figures such as rectangles and triangles, and by extension it is easy to see how, in principle, the area of any region bounded by a polygon could be determined. For instance, the polygonal region in Figure 1 can be decomposed into a finite number of triangular regions in any number of ways, so we can (again, in principle) extend our knowledge of triangular area and *define* the area of the polygon to be the sum of the areas of the triangles. Note the subtle shift in perspective: for a complicated enough polygonal region, we are more likely to specify that its area is the sum of areas that we know how to calculate than we are to look for a general formula such as, for example, $(\text{base})(\text{height})/2$. If we make such definitions wisely, our extended definition of area leads to no contradictions—that is, the area in Figure 1 will be the same no matter how we chop it up into triangles.

We can take a similar approach to develop reasonable definitions of area for many other figures, but for figures not bounded by straight lines we are likely to have to include a limiting step. Just as the limit process allowed us to make the transition from average velocity to instantaneous velocity and to extend the notion of tangency, limits will allow us to refine our definition of *area* and further understand the relationship between position and velocity. To see how, we begin with a simple problem: find the area of the region bounded by the x-axis and the graph of $f(x) = x^2$ between $x = 0$ and $x = 1$.

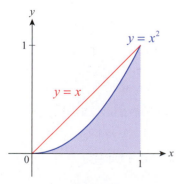

Figure 2

Example 1

Find the area under the graph of $f(x) = x^2$ and above the x-axis on the interval $[0,1]$.

Solution

At this point, we have nothing to work with other than our existing knowledge of area. We can certainly make some rough statements concerning the requested area. For instance, the area under the parabola on the interval $[0,1]$ is less than half a square unit, which is the area of the triangular region bounded by the x-axis and the line $y = x$. But we can also estimate the area under the curve by approximating it with regions whose areas we *do* know.

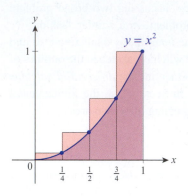

Figure 3

Rectangles are easy to work with, and we can approximate the region under the parabola with a collection of rectangles in many different ways. We could, for example, divide the interval $[0,1]$ into four subintervals of equal width and overestimate the desired area by covering it with four rectangles whose heights are $f\left(\frac{1}{4}\right)$, $f\left(\frac{1}{2}\right)$, $f\left(\frac{3}{4}\right)$, and $f(1)$, as shown in Figure 3. The combined area of these four rectangles, denoted O_4 for "Over," is as follows.

$$O_4 = \frac{1}{4} \cdot f\left(\frac{1}{4}\right) + \frac{1}{4} \cdot f\left(\frac{1}{2}\right) + \frac{1}{4} \cdot f\left(\frac{3}{4}\right) + \frac{1}{4} \cdot f(1)$$
$$= \frac{1}{4}\left(\frac{1}{16} + \frac{1}{4} + \frac{9}{16} + 1\right)$$
$$= \frac{15}{32} = 0.46875$$

We could also underestimate the area by adding up the areas of the four rectangles whose heights are $f(0)$, $f\left(\frac{1}{4}\right)$, $f\left(\frac{1}{2}\right)$, and $f\left(\frac{3}{4}\right)$ (see Figure 4). This would give us the approximation U_4 (U for "Under").

$$U_4 = \frac{1}{4} \cdot f(0) + \frac{1}{4} \cdot f\left(\frac{1}{4}\right) + \frac{1}{4} \cdot f\left(\frac{1}{2}\right) + \frac{1}{4} \cdot f\left(\frac{3}{4}\right)$$
$$= \frac{1}{4}\left(0 + \frac{1}{16} + \frac{1}{4} + \frac{9}{16}\right)$$
$$= \frac{7}{32} = 0.21875$$

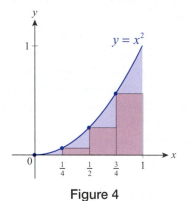

Figure 4

Knowing that the area under the parabola is actually something between these two approximations, we might estimate the area with their average: $(O_4 + U_4)/2 = 0.34375$.

To obtain a better approximation of the desired area, we can increase the number of rectangles. Graphs of the results, such as those from Figure 5 for $n = 10$, certainly seem to indicate that the errors in the approximations get smaller as n increases, but we can prove that this is indeed the case by noting the differences between O_n and U_n. The following table contains O_n, U_n, their difference, and their average for a few values of n.

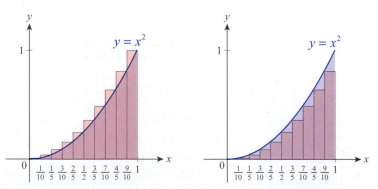

Figure 5 Approximating the Area with 10 Rectangles

n	O_n	U_n	$O_n - U_n$	$(O_n + U_n)/2$
10	0.385	0.285	0.1	0.335
100	0.33835	0.32835	0.01	0.33335
1000	0.33383	0.33283	0.001	0.33333

Given the results in the table, we would be justified in guessing that the area under $f(x) = x^2$ over the interval $[0,1]$ is $\frac{1}{3}$; we already know this guess is correct to several decimal places. We will soon return to this problem and prove that this guess is indeed correct.

TOPIC 2   **The Distance Problem**

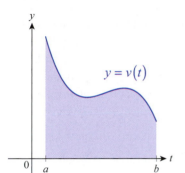

Figure 6 Graph of v

Our second motivating example is, in a sense that will soon be made precise, the opposite of the velocity problem of Chapter 2. Suppose that an object is moving along a straight line and that $v(t)$ tells us the velocity of the object at time t. Can we use this knowledge of its velocity to tell us the position of the object?

We will answer this question with a generic velocity function $v(t)$ defined on a time interval $[a,b]$ so that our solution can be applied to, say, a thrown ball whose vertical velocity is known, a car traveling along a straight road, or an atomic particle whose velocity is under the influence of an electromagnetic field. Suppose the graph in Figure 6 is a depiction of our generic $v(t)$. Since the graph is not constant, we cannot simply use the fact that distance is the product of rate (velocity) and time to determine how far the object has traveled over the interval $[a,b]$. But as with the area problem, we can develop better and better *approximations* of this distance by dividing $[a,b]$ into subintervals and assuming v to be constant on each one; as the subintervals decrease in width and the number of subintervals increases, our approximations will approach the true distance.

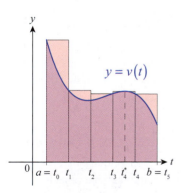

Figure 7

In Figure 7, the interval $[a,b]$ has been divided into five subintervals of equal width, labeled $[t_0, t_1]$ through $[t_4, t_5]$, and the height of each rectangle has been set to the maximum value of v on each one. If we know that v is continuous on $[a,b]$ (as it certainly appears to be), then by the Extreme Value Theorem we know that there is a point t_i^* in each subinterval $[t_{i-1}, t_i]$ for which $v(t_i^*)$ is the maximum value on that subinterval. For the five subintervals in Figure 7, note that $t_1^* = t_0$ (the left endpoint of $[t_0, t_1]$), $t_2^* = t_1$, $t_3^* = t_3$ (the right endpoint of $[t_2, t_3]$), and $t_5^* = t_4$, but t_4^* is a point strictly between t_3 and t_4.

We can now approximate the actual distance traveled over the time interval $[t_0, t_1]$ by $v(t_0)\Delta t$, where $\Delta t = t_1 - t_0 = (b-a)/5$. The product $v(t_0)\Delta t$ comes from the familiar formula *distance = rate × time*, since the length of time is Δt and we are approximating the rate over the interval with the constant value $v(t_0)$. Using the fact that $t_1^* = t_0$, we will rewrite $v(t_0)\Delta t$ as $v(t_1^*)\Delta t$, and proceed to make similar approximations for the distance traveled over the remaining four subintervals. Thus, the sum of the five expressions and our approximation for the distance traveled over $[a,b]$ is as follows:

$$v(t_1^*)\Delta t + v(t_2^*)\Delta t + v(t_3^*)\Delta t + v(t_4^*)\Delta t + v(t_5^*)\Delta t$$

We encounter sums of this form so frequently that it is convenient to make use of **sigma notation** to denote them. Using this notation, the above sum is written as follows.

$$\sum_{i=1}^{5} v\left(t_i^*\right)\Delta t$$

The symbol Σ is the capital Greek letter sigma (standing for "sum"), and the letter i is the **index of summation**. In general, an expression such as $\sum_{i=1}^{n} a_i$ stands for the sum of all the terms a_1 through a_n; that is,

$$\underset{\substack{\text{End with } i=n \\ \text{Begin with } i=1}}{\sum_{i=1}^{n} a_i} = a_1 + a_2 + \cdots + a_n.$$

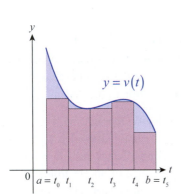

Our approximation of the distance at this point is guaranteed to be an overestimate, since we have approximated $v(t)$ on each subinterval with its maximum value on that subinterval. As in Example 1, however, we could just as easily arrive at an underestimate by approximating v with its minimum value on each subinterval. With five equal subdivisions, the heights of the corresponding rectangles using minimum values are as shown in Figure 8. In keeping with our previous usage, if we define t_i^* to be a point in $\left[t_{i-1}, t_i\right]$ where v attains its *minimum* value, then our underestimate approximation to the distance traveled is again given by the expression $\sum_{i=1}^{5} v\left(t_i^*\right)\Delta t$. Note that by this definition, it is clear that $t_1^* = t_1$, $t_3^* = t_2$, and $t_5^* = t_5$, but we may be uncertain (given the resolution of the graph) if t_2^* is t_2 or a little bit less than t_2, and it's not obvious if t_4^* is equal to t_3 or t_4.

Fortunately, we don't need to worry about such uncertainties. In this context, each t_i^* is called a **sample point** in $\left[t_{i-1}, t_i\right]$, and *any* point in the subinterval can be chosen as the sample point. Remember, our last step is to take a limit as n, the number of subintervals, increases to infinity; as we will soon see, the limit does not depend on the exact choice of sample points.

Figure 8

Example 2 ✒

A stone is released from a high cliff and falls for many seconds before hitting the canyon floor. Assuming air resistance is negligible, how far does the stone fall in the interval $t = 1$ second to $t = 3$ seconds?

Solution

Recall that the acceleration due to gravity is -32 ft/s^2, meaning that with every passing second a falling object changes in velocity by another -32 ft/s (recall also that the negative signs reflect the fact that objects fall downward). So if we let t denote the time in seconds after release, the stone falls with velocity $v(t) = -32t$. Since we are interested only in determining the difference between the stone's position at time $t = 1$ and time $t = 3$, we will work with the speed of the stone $s(t) = \left|v(t)\right| = 32t$.

The graph of s is linear, and the distance traveled by the stone between $t = 1$ and $t = 3$ corresponds to the shaded area between the graph and the t-axis (see Figure 9). You will show in Exercise 75 that the area of this polygonal

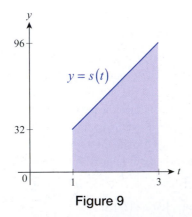

Figure 9

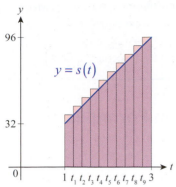

Figure 10 Depiction of O_{10}

region is 128 square units, so the distance traveled between these two points in time is 128 feet.

Although we now know the answer to our question, it is worthwhile to obtain it again through the approximation process we have developed—this will give us a chance to practice the techniques and notation that we will use for the next several sections. To this end, let O_n denote the overestimate that results by dividing the interval $[1,3]$ into n subintervals and defining t_i^* to be the right endpoint of the subinterval $[t_{i-1}, t_i]$; note that by this definition, $s(t_i^*)$ is the maximum value of s on the subinterval $[t_{i-1}, t_i]$, since s is an increasing function.

Using sigma notation, $O_n = \sum_{i=1}^{n} s(t_i^*) \Delta t$. While we could pick a specific value for n, calculate the n terms $s(t_i^*) \Delta t$, and add them up to arrive at an approximation, it is possible to achieve much more. We will do so by keeping n unspecified (other than knowing it represents a positive integer) and replacing each of the factors in the above generic sum with a formula based on our particular function s and interval $[1,3]$. Note that

$$\Delta t = \frac{b-a}{n} = \frac{3-1}{n} = \frac{2}{n}, \qquad \text{The width of each subinterval is } b-a \text{ divided by } n.$$

$$t_i^* = a + (\Delta t)i = 1 + \left(\frac{2}{n}\right)i = 1 + \frac{2i}{n}, \qquad \text{Add } \Delta t \text{ to the left endpoint } a \text{ a total of } i \text{ times.}$$

and

$$s(t_i^*) = 32 t_i^* = 32\left(1 + \frac{2i}{n}\right). \qquad \text{Evaluate } s(t) \text{ at } t_i^*.$$

Hence,

$$O_n = \sum_{i=1}^{n} s(t_i^*) \Delta t = \sum_{i=1}^{n} 32\left(1 + \frac{2i}{n}\right)\left(\frac{2}{n}\right) = \sum_{i=1}^{n} 32\left(\frac{2}{n} + \frac{4i}{n^2}\right).$$

Remember that this notation signifies nothing more than a sum and that all the usual properties of arithmetic hold. So we can reorder the terms as we like, factor out constants, and so forth:

$$\sum_{i=1}^{n} 32\left(\frac{2}{n} + \frac{4i}{n^2}\right) = 32\left[\frac{2}{n} + \frac{4(1)}{n^2}\right] + 32\left[\frac{2}{n} + \frac{4(2)}{n^2}\right] + \cdots + 32\left[\frac{2}{n} + \frac{4(n)}{n^2}\right]$$

$$= 32\left[\frac{2}{n}(n) + \frac{4}{n^2}(1 + 2 + \cdots + n)\right]$$

$$= 32\left(2 + \frac{4}{n^2}\sum_{i=1}^{n} i\right) = 64 + \frac{128}{n^2}\sum_{i=1}^{n} i$$

Sums such as $\sum_{i=1}^{n} i$ arise frequently, and it can be shown that $\sum_{i=1}^{n} i = n(n+1)/2$ (see Exercise 79). A catalog of other similar formulas will be given shortly. Using this fact in the expression above, we have the following.

$$O_n = 64 + \frac{128}{n^2}\sum_{i=1}^{n} i = 64 + \frac{128}{n^2}\left[\frac{n(n+1)}{2}\right] = 64 + 64\left(\frac{n+1}{n}\right)$$

In other words, we now have a simple expression giving us the overestimate resulting from approximating with n rectangles—there is no need to calculate this number repeatedly for different values of n. More importantly, we can study this expression to see what happens as $n \to \infty$. Since $\left[(n+1)/n\right] \to 1$ as $n \to \infty$, we see that $O_n \to 128$, confirming that the distance the stone falls between the first and third seconds is 128 feet.

TOPIC 3 Riemann Sums and Summation Formulas

We have now seen sums of the form $\sum_{i=1}^{n} f\left(x_i^*\right) \Delta x$ arise several times, and we will study them further as we develop the notion of the *integral*. Such an expression is an example of a **Riemann sum** of the function f over an interval $\left[a,b\right]$, named in honor of the German mathematician Bernhard Riemann (1826–1866). It is very helpful to keep in mind the geometric meaning of the individual pieces of a Riemann sum. Specifically, remember that each x_i^* represents a sample point in the ith subinterval, that $\Delta x = (b-a)/n$ is the width of each subinterval, and that $f\left(x_i^*\right)$ is nothing more than the value of f at the ith sample point.

In working with Riemann sums, the following facts about finite sums will be very useful. The first three are nothing more than restatements, using sigma notation, of elementary arithmetic facts, while the last three are formulas that you will prove in Exercises 79–83.

Theorem

Summation Facts and Formulas

Constant Rule for Finite Sums	$\displaystyle\sum_{i=1}^{n} c = nc,$ for any constant c
Constant Multiple Rule for Finite Sums	$\displaystyle\sum_{i=1}^{n} ca_i = c\sum_{i=1}^{n} a_i,$ for any constant c
Sum/Difference Rule for Finite Sums	$\displaystyle\sum_{i=1}^{n}\left(a_i \pm b_i\right) = \sum_{i=1}^{n} a_i \pm \sum_{i=1}^{n} b_i$
Sum of the First n Positive Integers	$\displaystyle\sum_{i=1}^{n} i = \frac{n(n+1)}{2}$
Sum of the First n Squares	$\displaystyle\sum_{i=1}^{n} i^2 = \frac{n(n+1)(2n+1)}{6}$
Sum of the First n Cubes	$\displaystyle\sum_{i=1}^{n} i^3 = \left[\frac{n(n+1)}{2}\right]^2$

Example 3 ✐

Simplify the following sums.

a. $\displaystyle\sum_{i=1}^{9}(3i+4)$ **b.** $\displaystyle\sum_{j=1}^{n}\frac{-2}{n}$ **c.** $\displaystyle\sum_{i=1}^{n}(i-1)^2$

Solution

a. $\displaystyle\sum_{i=1}^{9}(3i+4)=3\sum_{i=1}^{9}i+\sum_{i=1}^{9}4=3\left(\frac{9\cdot10}{2}\right)+9\cdot4=135+36=171$

b. $\displaystyle\sum_{j=1}^{n}\frac{-2}{n}=\left(\frac{-2}{n}\right)n=-2$

c. $\displaystyle\sum_{i=1}^{n}(i-1)^2=\sum_{i=1}^{n}(i^2-2i+1)=\sum_{i=1}^{n}i^2-2\sum_{i=1}^{n}i+\sum_{i=1}^{n}1$

$$=\frac{n(n+1)(2n+1)}{6}-2\cdot\frac{n(n+1)}{2}+n$$

$$=\frac{2n^3+3n^2+n-6n^2-6n+6n}{6}$$

$$=\frac{2n^3-3n^2+n}{6}$$

We will conclude this section by returning to our question in Example 1, this time proving that the area under the graph of $f(x)=x^2$ over the interval $[0,1]$ is indeed exactly $\frac{1}{3}$.

Example 4 ✐

Find the area under the graph of $f(x)=x^2$ and above the x-axis on the interval $[0,1]$.

Solution

For variety, we will construct an expression for U_n (the underestimate based on n subintervals) and evaluate $\lim\limits_{n\to\infty}U_n$. In Exercise 77 you will show that the same answer is obtained by evaluating $\lim\limits_{n\to\infty}O_n$.

If we divide $[0,1]$ into n subintervals of equal width, each one has width $\Delta x=1/n$. The minimum value of f on $[x_{i-1},x_i]$ occurs at $x_i^*=x_{i-1}$, that is, at the left endpoint of each subinterval. So $x_1^*=0$, $x_2^*=1/n$, $x_3^*=2/n$, and in general $x_i^*=(i-1)/n$. So we have the following:

$$U_n=\sum_{i=1}^{n}f(x_i^*)\Delta x=\sum_{i=1}^{n}\left(\frac{i-1}{n}\right)^2\left(\frac{1}{n}\right)=\frac{1}{n^3}\sum_{i=1}^{n}(i-1)^2$$

We have already simplified the sum $\displaystyle\sum_{i=1}^{n}(i-1)^2$ in part c. of Example 3.

$$U_n=\frac{1}{n^3}\left(\frac{2n^3-3n^2+n}{6}\right)=\frac{2n^3-3n^2+n}{6n^3}$$

Hence, $\displaystyle\lim_{n\to\infty}U_n=\lim_{n\to\infty}\frac{2n^3-3n^2+n}{6n^3}=\frac{2}{6}=\frac{1}{3}$.

5.1 **Exercises**

1–2 Use $(O_4 + U_4)/2$ to estimate the area under the graph and above the x-axis on the interval $[0,8]$.

1. $f(x) = \left(\dfrac{x}{4}\right)^3$

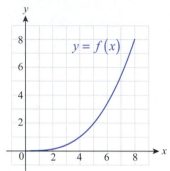

2. $g(x) = 3\sqrt[3]{x}$

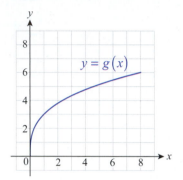

3–4. Repeat Exercises 1–2 using eight rectangles.

5. The figure below shows the upward velocity (in feet per second) of a model rocket during its rise. Use the method of Example 1 to estimate how high the rocket rose by calculating O_6. Is your estimate an over- or underestimate?

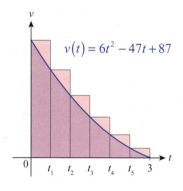

6. The velocity of an object undergoing simple harmonic motion is given by the graph below (time is measured in seconds, distance in feet). Using subintervals of width $\frac{1}{4}$,

a. give an overestimate for the total distance covered from $t = 1$ s to $t = 6$ s;

b. estimate the total displacement from $t = 1$ s to $t = 6$ s.

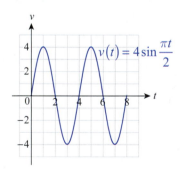

7. The given table contains the velocity data recorded by an automotive testing device during an acceleration test.

Time (s)	0.5	1	1.5	2	2.5	3	3.5	4	4.5	5	5.5	6
v (m/s)	3	6.6	9.8	13	16.1	19.1	21.6	23.8	25.8	27.6	28.5	29.1

a. Use 12 subintervals to give over- and underestimates of the distance covered by the car during the acceleration run (i.e., find O_{12} and U_{12}).

b. Approximate the above distance using 6 subintervals of equal width and choosing the midpoint of each as the sample point (we shall call the resulting quantity M_6).

c. Compare M_6 with $(O_{12} + U_{12})/2$. Which one is greater? Explain why this is the case.

8. In order to estimate the length of the runway, a passenger on an airplane jotted down some velocity data during takeoff from the on-board entertainment screen. From the resulting table given below, calculate $(O_8 + U_8)/2$ to find his estimate.

Time (s)	6	12	18	24	30	36	42	48
v (mph)	30	79	115	150	180	204	223	230

9–14 Use four rectangles to estimate the area between the graph of the given function and the x-axis on the given interval. Construct three estimates for the function: the first using the left endpoints of the subintervals as the sample points, the second using the right endpoints of the subintervals, and the third using the midpoints of the subintervals. Can you tell which are guaranteed to be underestimates or overestimates? (**Hint:** Consider the increasing/decreasing and concavity features of the graph. It is helpful to make a sketch.)

9. $f(x) = \sqrt{x}$ on $[0,4]$

10. $f(x) = \dfrac{x^3}{16}$ on $[0,4]$

11. $f(x) = \dfrac{1}{x}$ on $[1,5]$

12. $f(x) = \sqrt{4 - x^2}$ on $[-2,2]$

13. $f(x) = \cos\dfrac{x}{2}$ on $[0,\pi]$

14. $f(x) = e^{2-x}$ on $[0,2]$

15–24 Write the given sum using sigma notation.

15. $3 + 6 + 9 + \cdots + 99$

16. $1 + 2 + 9 + 28 + \cdots + (25^3 + 1)$

17. $1 + \dfrac{1}{4} + \dfrac{1}{9} + \dfrac{1}{16} + \cdots + \dfrac{1}{10{,}000}$

18. $1 - \dfrac{1}{2} + \dfrac{1}{3} - \dfrac{1}{4} + \cdots - \dfrac{1}{50}$

19. $a_{-3} + a_{-1} + a_1 + a_3 + a_5 + \cdots + a_{77}$

20. $b_0 + b_3 + b_6 + b_9 + b_{12} + \cdots + b_{297}$

21. $f\left(\dfrac{3}{n}\right) + f\left(\dfrac{6}{n}\right) + f\left(\dfrac{9}{n}\right) + \cdots + f\left(\dfrac{3(n-1)}{n}\right) + f(3)$

22. $g(c_0) + g(c_5) + g(c_{10}) + g(c_{15}) + \cdots + g(c_{650})$

23. $f\left(x_0^*\right)\Delta x + f\left(x_1^*\right)\Delta x + f\left(x_2^*\right)\Delta x + \cdots + f\left(x_n^*\right)\Delta x$

24. $s\left(t_1^*\right)\Delta t + s\left(t_2^*\right)\Delta t + s\left(t_3^*\right)\Delta t + \cdots + s\left(t_{n-1}^*\right)\Delta t$

25–30 Assuming that $\displaystyle\sum_{i=0}^{n} a_i = 36$ and $\displaystyle\sum_{i=0}^{n} b_i = 100$, find the given sum.

25. $\displaystyle\sum_{i=0}^{n} (b_i - a_i)$

26. $\displaystyle\sum_{i=0}^{n} (2a_i + 3b_i)$

27. $\displaystyle\sum_{i=0}^{n} (5b_i + 1)$

28. $\displaystyle\sum_{i=0}^{n} \left(\dfrac{a_i}{6} - \dfrac{b_i}{2}\right)$

29. $\displaystyle\sum_{j=0}^{n} \left(\dfrac{4a_j}{3} - \dfrac{b_j}{4} + 2\right)$

30. $\displaystyle\sum_{k=1}^{n} \left(\dfrac{4}{3} - 2a_k + \dfrac{b_k}{2}\right)$

31–42 Find the value of the sum. Use a summation formula when possible.

31. $\displaystyle\sum_{i=2}^{4} \dfrac{1}{i-1}$

32. $\displaystyle\sum_{j=-2}^{2} \sqrt{j+2}$

33. $\displaystyle\sum_{i=1}^{10} (5i - 2)$

34. $\displaystyle\sum_{i=1}^{n} (1 - 3i)$

35. $\displaystyle\sum_{j=1}^{n} \dfrac{4j + 5n}{2}$

36. $\displaystyle\sum_{k=1}^{n} \dfrac{6k^2 + 2k}{3}$

37. $\displaystyle\sum_{j=1}^{30} (2j^2 - 4j + 1)$

38. $\displaystyle\sum_{i=1}^{100} (2i - 1)(3 - i)$

39. $\displaystyle\sum_{j=1}^{n} (3j + 1)^2$

40. $\displaystyle\sum_{i=1}^{n} \left(i^3 - 2i^2 + \dfrac{1}{n}\right)$

41. $\displaystyle\sum_{j=1}^{n} 2j^2 (j - 2)$

42. $\displaystyle\sum_{k=0}^{n} k(k+1)(k+2)$

43–48 Write out the first few terms as well as the last few terms of the sum. Find a way to simplify and use your observation to evaluate the sum. (Sums of this type are called *collapsing sums*.)

43. $\displaystyle\sum_{i=1}^{10}\left(\frac{1}{i}-\frac{1}{i+1}\right)$

44. $\displaystyle\sum_{k=3}^{n}\left[\frac{1}{k^3}-\frac{1}{(k+1)^3}\right]$

45. $\displaystyle\sum_{j=1}^{n}\left(\sqrt{j}-\sqrt{j+1}\right)$

46. $\displaystyle\sum_{k=1}^{n+1}\ln\frac{k}{k+1}$

47. $\displaystyle\sum_{j=2}^{n+3}\left(e^j-e^{j+1}\right)$

48. $\displaystyle\sum_{k=1}^{2n+1}\left[\sin k\pi-\sin\left((k+1)\pi\right)\right]$

49–52 A *geometric sum* (or *geometric progression*) is a sum of the following form.

$$a+ar+ar^2+\cdots+ar^n=\sum_{i=0}^{n}ar^i,\quad r\neq1$$

(Notice that each term is a constant multiple of the preceding term; this constant is called the *common ratio* and is denoted by *r*.)

Use the formula $\displaystyle\sum_{i=0}^{n}ar^i=a\frac{1-r^{n+1}}{1-r}$ to evaluate the sum.

49. $\displaystyle\sum_{i=0}^{10}3^i$

50. $\displaystyle\sum_{j=0}^{8}5\left(\frac{1}{2}\right)^j$

51. $\displaystyle\sum_{k=0}^{99}(-1)^k\left(\frac{2}{3}\right)^k$

52. $\displaystyle\sum_{n=0}^{1000}4.9(-3.9)^n$

53. Prove the formula for the sum of the first $n+1$ terms of a geometric progression given in the directions preceding Exercises 49–52. (**Hint:** Let S denote the sum, recognize $S-rs$ as a collapsing sum, evaluate, and solve for S.)

54–57 Sometimes, sums become easier to manage (the general term becomes simpler) after an appropriate shift in the index. For example, $\displaystyle\sum_{i=1}^{n+1}(i-1)^2$ can be rewritten as $\displaystyle\sum_{i=0}^{n}i^2$. Perform an appropriate shift in the indexing of the given sum to simplify its general term.

54. $\displaystyle\sum_{i=5}^{25}2(i-4)^3$

55. $\displaystyle\sum_{j=3}^{n}\frac{1}{j-2}$

56. $\displaystyle\sum_{k=0}^{n}n(k-3)^2$

57. $\displaystyle\sum_{l=4}^{20}\cos\left((2l+2)\pi\right)$

58–63 Follow the lead of Examples 2 and 4 in using the limit process to find the area under the graph of $f(x)$ and above the x-axis on the given interval. (In Exercises 58 and 59, use a formula from geometry to check your answer.)

58. $f(x)=\dfrac{1}{2}x+1$ on $[0,4]$

59. $f(x)=5-x$ on $[1,3]$

60. $f(x)=x^2$ on $[1,2]$

61. $f(x)=x-x^3$ on $[0,1]$

62. $f(x)=x^2+3x$ on $[-2,2]$

63. $f(x)=\left(1-x^2\right)(1+x)$ on $[0,1]$

64–67 Identify the region whose area is the given limit. Do not evaluate the limit. (**Hint:** For guidance, see Example 4 and Exercises 58–63.)

64. $\displaystyle\lim_{n\to\infty}\frac{1}{n}\sum_{i=1}^{n}\left(1-\frac{i}{n}\right)^2$

65. $\displaystyle\lim_{n\to\infty}\frac{3}{n}\sum_{i=1}^{n}\left[\left(\frac{3i}{n}\right)^3+\frac{6i}{n}\right]$

66. $\displaystyle\lim_{n\to\infty}\frac{2}{n}\sum_{i=0}^{n-1}\sqrt{1+\frac{2i}{n}}$

67. $\displaystyle\lim_{n\to\infty}\frac{\pi}{2n}\sum_{i=0}^{n-1}\sin\frac{\pi i}{2n}$

68. A fruit vendor stacks apples in a rectangular, pyramid-like pile. If the foundational layer consists of 8 rows of 10 apples, and the top layer is a single row of apples, find how many apples are in the stack. Generalize to the case of an $m\times n$ bottom layer of fruit.

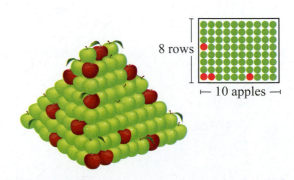

8 rows

⊢ 10 apples ⊣

69. In statistics, the standard deviation of a data set $x_1, x_2, ..., x_n$ is defined to be the square root of the average of the squares of deviations of the data from their mean $\bar{x}$.

$$s = \sqrt{\frac{(x_1 - \bar{x})^2 + (x_2 - \bar{x})^2 + \cdots + (x_n - \bar{x})^2}{n}}$$

Rewrite the definition of s using sigma notation, and use summation facts to show the following:

$$s^2 = \left(\frac{1}{n}\sum_{i=1}^{n} x_i^2\right) - \bar{x}^2$$

70. Find the distance covered by the pebble in Exercise 79 of Section 4.7 from $t = 2$ seconds to $t = 5$ seconds. (**Hint:** Find the velocity function first.)

71. Assuming constant acceleration, use the method of Exercise 70 to find the distance covered by the Bugatti in Exercise 91 of Section 4.7 from $t = 1$ second to $t = 3$ seconds.

72. Assuming constant deceleration, use the method of Example 2 to find the distance covered by the braking race car in Exercise 90 of Section 4.7 from $t = 1$ second to $t = 2$ seconds. (See the hint given in Exercise 70.)

73. The velocity function of a moving object is given by $v(t) = 9 - 0.5t^2$ m/s from $t = 0$ s to $t = 3$ s. Find the distance covered by the object during this time.

74. Repeat Exercise 73 for $v(t) = 4 - 0.5t^3$ on the interval $[0, 2]$.

75. Use geometry to show that the shaded area under the curve $s(t)$ in Example 2 is 128 square units. (**Hint:** Divide the region into a rectangle and a right triangle, and use well-known area formulas. Alternatively, you may want to use the area formula for a trapezoid.)

76. Show that you can obtain the same answer in Example 2 by evaluating $\lim_{n\to\infty} U_n$, that is, by choosing $t_i^* = t_{i-1}$ for every index value i.

77. Show that you can obtain the same answer in Example 4 by evaluating $\lim_{n\to\infty} O_n$, that is, by choosing $x_i^* = x_i$ for every index value i.

78. Show that you can obtain the same answer in Example 2 by choosing t_i^* to be the midpoint of the ith interval for every index value i.

79. Use an elementary argument to prove the following summation formula:

$$\sum_{i=1}^{n} i = \frac{n(n+1)}{2}$$

(**Hint:** Letting $S = \sum_{i=1}^{n} i$, add to S its terms in "reverse order"; that is, calculate $2S$ as

$$2S = \sum_{i=1}^{n} i + \sum_{j=0}^{n-1}(n - j), \text{ and notice that, after}$$

rearranging terms, this latter sum equals

$$(1+n) + (2 + (n-1)) + (3 + (n-2)) + \cdots$$
$$= (n+1) + (n+1) + (n+1) + \cdots.$$

Use this observation to complete the argument. Note that this argument is attributed to C. F. Gauss, who discovered it as a barely nine-year-old elementary school student.)

80.* Use mathematical induction to establish the summation formula of Exercise 79.

81.* Use mathematical induction to establish the following summation formula:

$$\sum_{i=1}^{n} i^2 = \frac{n(n+1)(2n+1)}{6}$$

82.* Use mathematical induction to establish the following summation formula:

$$\sum_{i=1}^{n} i^3 = \left[\frac{n(n+1)}{2}\right]^2$$

83.* Prove the summation formula of Exercise 82 by making use of the following identity:

$$(i+1)^4 - i^4 = 4i^3 + 6i^2 + 4i + 1$$

84.* Inscribe a regular n-gon in a circle of radius r. Use radii to divide the n-gon into n isosceles triangles, and add the areas of the triangles to obtain the area of the inscribed n-gon. Finally, let $n \to \infty$ to obtain the area formula for the circle.

85–87 *Double summations* are important in many areas of mathematics, statistics, computer science, and the sciences in general. They have the form $\displaystyle\sum_{i=1}^{n}\sum_{j=1}^{m}a_{ij}$.

Evaluate the given double sum.

85. $\displaystyle\sum_{i=1}^{4}\sum_{j=1}^{5}(i+j)$

86. $\displaystyle\sum_{i=1}^{5}\sum_{j=1}^{6}ij$

87. $\displaystyle\sum_{i=1}^{n}\sum_{j=1}^{m}ij$

5.1 Technology Exercises

88–91 Use a computer algebra system to express the area under the graph of $f(x)$ and above the x-axis on the indicated interval as a limit. Then use technology to evaluate the limit to find the area.

88. $f(x)=x^{6}$ on $[0,1]$

89. $f(x)=\sin x$ on $[0,\pi]$

90. $f(x)=e^{x}$ on $[1,2]$

91. $f(x)=x+\cos^{2}(\pi x)$ on $\left[0,\dfrac{1}{2}\right]$

5.2 **The Definite Integral**

TOPICS

1. Riemann sums and the definite integral

2. Properties of the definite integral

We continue our exploration of Riemann sums in this section, with the goal of developing the *definite integral* of a function. As we will see, the properties possessed by the definite integral arise directly from its definition as the limit of Riemann sums.

TOPIC 1 **Riemann Sums and the Definite Integral**

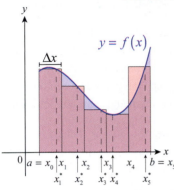

Figure 1 Riemann Sum with $n = 5$

Recall that for a function f defined on an interval $[a,b]$, any expression of the form $\sum_{i=1}^{n} f\left(x_i^*\right) \Delta x$ is called a Riemann sum of f over $[a,b]$. Implicit in this notation are the assumptions that $\Delta x = (b-a)/n$, that $[a,b]$ has been divided into n subintervals of equal width, $[x_0, x_1]$, $[x_1, x_2]$, ..., $[x_{n-1}, x_n]$, and that the ith sample point x_i^* lies in $[x_{i-1}, x_i]$ (note that $x_0 = a$ and $x_n = b$). If $f(x) \geq 0$ on $[a,b]$, as depicted in Figure 1, then $\sum_{i=1}^{n} f\left(x_i^*\right) \Delta x$ is an approximation of the area bounded by the graph of f and the x-axis between $x = a$ and $x = b$, and it makes sense to *define* this area as $\lim_{n \to \infty} \sum_{i=1}^{n} f\left(x_i^*\right) \Delta x$, if the limit exists. This is the inspiration for what follows.

Definition 💡

Definite Integral

Given a function f defined on an interval $[a,b]$, let $\Delta x = (b-a)/n$, let $[a,b]$ be divided into n subintervals of equal width, $[x_0, x_1]$, $[x_1, x_2]$, ..., $[x_{n-1}, x_n]$, where $x_i = a + i\Delta x$, and let the ith sample point satisfy $x_i^* \in [x_{i-1}, x_i]$. Then the **(Riemann) definite integral of f from a to b** is denoted $\int_a^b f(x)\,dx$ and is defined to be

$$\int_a^b f(x)\,dx = \lim_{n \to \infty} \sum_{i=1}^{n} f\left(x_i^*\right) \Delta x,$$

provided the limit exists and is the same for every choice of the sample points. If so, we say f is **integrable** on $[a,b]$, and call f the **integrand** of the integral. In this context, a and b are called, respectively, the **lower** and **upper limits of integration**.

This definition is our first in what is sometimes called the *theory of integration*, which is an expansive area of mathematics. It is worth devoting a few paragraphs to a brief discussion of different interpretations of the definite integral and an introduction to some of the more advanced branches of the theory.

First, note that there is nothing in the definition that requires f to be nonnegative on $[a,b]$. While it has been convenient up to this point to assume this is so, we now need a way to visualize the meaning of $\int_a^b f(x)\,dx$ when f takes on both positive and negative values. It is still the case that each term $f\left(x_i^*\right)\Delta x$ represents the product of a width Δx by the value of f at a sample point $x_i^* \in [x_{i-1}, x_i]$, but if $f\left(x_i^*\right) < 0$ then that particular term will be negative. A natural way to interpret

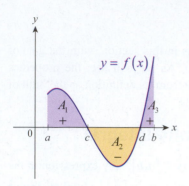

Figure 2 Signed Areas

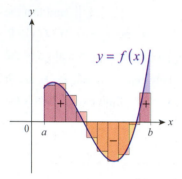

Figure 3 Signed Riemann Sum

such a situation is to use the notion of *signed area*. In Figure 2, for example, if we let A_1, A_2, and A_3 denote the (positive) areas of the regions bounded by the graph of f and the x-axis, then $\int_a^b f(x)\,dx = A_1 - A_2 + A_3$. That is, the integral of f between c and d is a negative number, and we would indicate this by writing $\int_c^d f(x)\,dx = -A_2 < 0$. Again, this interpretation follows directly from the definition of the definite integral as a limit of Riemann sums; in Figure 3, note that the signed area of the first four rectangles is positive, that of the next five rectangles is negative, and the signed area of the last rectangle is again positive.

The symbol $\int$ was introduced by Leibniz and represents a stylized "S" as a reminder that the basis of the definite integral is summation. Similarly, the process of calculating the value of a definite integral is termed **integration**—this reflects the fact that we arrive at a definite integral by adding up (integrating) a finite number of summands, and taking the limit as the number of summands approaches infinity.

For the moment, the "dx" part of the notation simply tells us that the variable in use is x, but its presence is not superfluous. It does (as the notation suggests) represent the differential dx, and it corresponds (as its location suggests) to the Δx in the Riemann sum; we will pay special attention to it later in this text when we make changes in the variable of integration. It plays a more active role in alternate definitions of the integral (studied in later math courses) and, importantly for our purposes, the "dx" tells us to integrate with respect to x, an important instruction when more than one variable is present. Note, though, that the particular symbol used to denote a given variable is immaterial. For a given function f defined on $[a,b]$, $\int_a^b f(x)\,dx$, $\int_a^b f(t)\,dt$, and $\int_a^b f(\square)\,d\square$ all mean the same thing—the symbol used for the variable is just a placeholder, and for that reason is sometimes referred to (somewhat dismissively) as a "dummy variable."

Finally, the definition given for the definite integral is actually a simplified version of a more general one. Remember that the choice of sample point x_i^* in each subinterval is irrelevant—in the limit, any one choice is as good as another. More generally, even the subintervals can be defined liberally. Given an interval $[a,b]$, a **partition** P of $[a,b]$ is any finite set $P = \{x_0, x_1, \ldots, x_n\}$ where $a = x_0 < x_1 < \cdots < x_{n-1} < x_n = b$ (note that the x_i's don't have to be evenly spaced). The **norm** of the partition P is defined to be $\|P\| = \max_{1 \le i \le n}\{\Delta x_i\}$ where $\Delta x_i = x_i - x_{i-1}$; that is, $\|P\|$ is the maximum of the subinterval widths. Using the epsilon-delta formulation of limit, we say that f is integrable on $[a,b]$ and that its definite integral is the number we denote as $\int_a^b f(x)\,dx$ if for every $\varepsilon > 0$ there is a $\delta > 0$ so that $\|P\| < \delta \Rightarrow \left| \int_a^b f(x)\,dx - \sum_{i=1}^n f(x_i^*)\,\Delta x_i \right| < \varepsilon.$

In words, the definite integral of f on the interval $[a,b]$ is the number $\int_a^b f(x)\,dx$ if whenever the partition P is "fine enough" (meaning the norm is smaller than delta), then the Riemann sum over that partition is within epsilon of $\int_a^b f(x)\,dx$. In proving theorems and in some applications, such as when only an approximation of an integral is possible, this more general definition is desirable (see Exercises 4–13, 28, and 29). But for the most part, the definition based on equally spaced partitions is sufficient and easier to apply.

Example 1 ✐

George uses a slingshot to shoot a small stone straight up into the sky with a vertical velocity of $v(t) = 64 - 32t$ ft/s. What is the height of the stone, relative to its initial height, when $t = 4$ s?

Solution

Following our discussion in Section 5.1, we will take the limit of Riemann sums of the form $\sum_{i=1}^{n} v(t_i^*) \Delta t$ in order to deduce the position of the stone after 4 seconds—it is again the case that the distance traveled over the interval $[t_{i-1}, t_i]$ is approximately $v(t_i^*) \Delta t$. However, as we will see, there is one important difference between this example and Example 2 of the previous section.

To set the stage, note that $\Delta t = (4 - 0)/n = 4/n$ and, if we decide to let t_i^* be the right endpoint of each subinterval, then $t_1^* = 0 + (4/n) = 4/n$, $t_2^* = 0 + 2(4/n) = 8/n$, and, in general, $t_i^* = 4i/n$. Using the integral notation we write the following:

$$\int_0^4 v(t)\,dt = \lim_{n \to \infty} \sum_{i=1}^{n} v(t_i^*) \Delta t$$

$$= \lim_{n \to \infty} \sum_{i=1}^{n} v\left(\frac{4i}{n}\right)\left(\frac{4}{n}\right)$$

$$= \lim_{n \to \infty} \sum_{i=1}^{n} \left[64 - 32\left(\frac{4i}{n}\right)\right]\left(\frac{4}{n}\right)$$

$$= \lim_{n \to \infty} \frac{4}{n} \sum_{i=1}^{n} \left(64 - \frac{128i}{n}\right) \qquad \text{Pull } \tfrac{4}{n} \text{ outside the sum.}$$

Continuing to use the summation formulas we listed in the previous section,

$$\int_0^4 v(t)\,dt = \lim_{n \to \infty} \frac{4}{n}\left(\sum_{i=1}^{n} 64 - \frac{128}{n}\sum_{i=1}^{n} i\right) \qquad \sum_{i=1}^{n}(a_i \pm b_i) = \sum_{i=1}^{n} a_i \pm \sum_{i=1}^{n} b_i \text{ and } \sum_{i=1}^{n} ca_i = c\sum_{i=1}^{n} a_i$$

$$= \lim_{n \to \infty} \frac{4}{n}\left[64n - \frac{128}{n} \cdot \frac{n(n+1)}{2}\right] \qquad \sum_{i=1}^{n} c = cn \text{ and } \sum_{i=1}^{n} i = \frac{n(n+1)}{2}$$

$$= \lim_{n \to \infty}\left[256 - 256\left(\frac{n+1}{n}\right)\right]$$

$$= 256 - 256 \lim_{n \to \infty}\left(\frac{n+1}{n}\right)$$

$$= 0.$$

What does this answer mean? Surely the stone has traveled, so does an answer of 0 feet make sense? It does, but only if we understand exactly what the definite integral of velocity tells us. If we take a look at the graph of v, we see that it is positive for the first 2 seconds and negative for the last 2 seconds—that is, the stone first rises and then falls. At the end of 4 seconds, the stone has the same height that it had at time 0. This is the physical meaning of the result and a consequence of the fact that the signed areas in Figure 4 cancel one another out exactly. If instead of the position

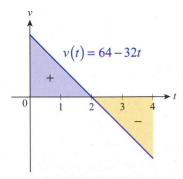

Figure 4

after 4 seconds we want to know how far the stone travels over the interval $[0,4]$, we need to determine $\int_0^4 |v(t)|\,dt$, a task you will undertake in Exercise 2.

Now that the definite integral of a function has been defined, the question of which functions are integrable naturally arises; this is the counterpart to the question of differentiability we faced in Chapter 3. The next theorem, the proof of which is usually seen in a course such as Advanced Calculus or Real Analysis, tells us that those in a large and important class are integrable.

Theorem

Piecewise Continuous Functions Are Integrable

If f is a continuous function defined on $[a,b]$, or if f is a piecewise continuous function on $[a,b]$ (meaning it is continuous except for a finite number of jump discontinuities), then f is integrable on $[a,b]$.

Functions that are not integrable under our definition can be found without too much trouble (see Exercises 49–51), and the goal of enlarging the class of integrable functions is one motivation for some of the alternate definitions of integration that exist. Some advanced applications in physics also call for alternative formulations of the integral. But throughout this text the Riemann integral will serve our needs admirably; given that, and in preparation for our work to follow, we now define some shorthand notation for two Riemann sums we use frequently.

Definition

Left and Right Riemann Sums

Given a function f defined on the interval $[a,b]$, let L_n and R_n denote the Riemann sums obtained by dividing $[a,b]$ into n subintervals of equal width, $[x_0,x_1]$, $[x_1,x_2]$, ..., $[x_{n-1},x_n]$, and defining the ith sample point to be, respectively, the left and right endpoint of the ith subinterval. That is,

$$L_n = \sum_{i=1}^{n} f(x_{i-1})\Delta x \quad \text{and} \quad R_n = \sum_{i=1}^{n} f(x_i)\Delta x,$$

where $\Delta x = (b-a)/n$ and $x_i = a + i\Delta x$.

Note in particular that, if f is integrable, $\int_a^b f(x)\,dx = \lim_{n\to\infty} L_n = \lim_{n\to\infty} R_n$.

Example 2 ✎

Evaluate $\int_{-1}^{3} x^3 \, dx$ by taking the limit of the associated right Riemann sum.

Solution

From the expression $\int_{-1}^{3} x^3 \, dx$, we see that $f(x) = x^3$, $a = -1$, and $b = 3$.

So

$$\Delta x = \frac{3 - (-1)}{n} = \frac{4}{n}, \quad x_i = -1 + \frac{4i}{n} = \frac{-n + 4i}{n},$$

and

$$f(x_i) = f\left(\frac{-n + 4i}{n}\right) = \frac{(-n + 4i)^3}{n^3} = \frac{-n^3 + 12n^2 i - 48ni^2 + 64i^3}{n^3}.$$

Hence,

$$\int_{-1}^{3} x^3 \, dx = \lim_{n \to \infty} \sum_{i=1}^{n} \left(\frac{-n^3 + 12n^2 i - 48ni^2 + 64i^3}{n^3}\right)\left(\frac{4}{n}\right)$$

$$= \lim_{n \to \infty} \frac{4}{n^4} \sum_{i=1}^{n} \left(-n^3 + 12n^2 i - 48ni^2 + 64i^3\right) \qquad \text{Factor out } \frac{4}{n^4}.$$

$$= \lim_{n \to \infty} \frac{4}{n^4} \left(\sum_{i=1}^{n} -n^3 + \sum_{i=1}^{n} 12n^2 i + \sum_{i=1}^{n} -48ni^2 + \sum_{i=1}^{n} 64i^3\right) \quad \sum_{i=1}^{n}(a_i \pm b_i) = \sum_{i=1}^{n} a_i \pm \sum_{i=1}^{n} b_i$$

$$= \lim_{n \to \infty} \frac{4}{n^4} \left(-n^4 + 12n^2 \sum_{i=1}^{n} i - 48n \sum_{i=1}^{n} i^2 + 64 \sum_{i=1}^{n} i^3\right). \quad \sum_{i=1}^{n} c = cn \text{ and } \sum_{i=1}^{n} ca_i = c\sum_{i=1}^{n} a_i$$

We can now apply the last three summation formulas from Section 5.1 to obtain

$$\int_{-1}^{3} x^3 \, dx = \lim_{n \to \infty} \frac{4}{n^4} \left(-n^4 + 12n^2 \cdot \frac{n(n+1)}{2} - 48n \cdot \frac{n(n+1)(2n+1)}{6} + 64 \cdot \frac{n^2(n+1)^2}{4}\right),$$

which, after significant simplification, yields

$$\int_{-1}^{3} x^3 \, dx = \lim_{n \to \infty} \frac{4}{n^4}\left(5n^4 + 14n^3 + 8n^2\right)$$

$$= \lim_{n \to \infty}\left(20 + \frac{56}{n} + \frac{32}{n^2}\right)$$

$$= 20.$$

TOPIC 2 Properties of the Definite Integral

Clearly, evaluation of definite integrals by the process demonstrated in Example 2 requires a significant amount of work. Even worse, most integrals we are likely to run across simply can't be determined by this process—the limit that results is often one we can't evaluate (Exercises 30–33 ask you to construct, but not evaluate, Riemann sums corresponding to specific definite integrals). However, we faced a similar quandary as we developed the notion of the derivative, and resolved it by uncovering certain general properties that greatly simplified the task of differentiation. We will do the same thing for the integral, beginning here with some simple but useful properties.

First, always bear in mind the geometric meaning of the expression $\int_a^b f(x)\,dx$, namely, that it can be interpreted as the signed area of a region bounded by the graph of f and the x-axis between $x = a$ and $x = b$. In some cases, we can use this meaning to evaluate an integral on the basis of knowledge we already possess.

Example 3 ✏

Evaluate $\int_{-2}^{2} -\sqrt{4 - x^2}\ dx$.

Solution

If we construct a Riemann sum based on the function $f(x) = -\sqrt{4 - x^2}$, we arrive at a limit that is quite difficult to evaluate. But if we think about the graph of f and what $\int_{-2}^{2} -\sqrt{4 - x^2}\ dx$ represents, we see that elementary geometry suffices to evaluate the integral.

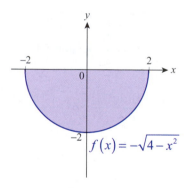

Figure 5

The graph of the function f corresponds to the graph of the equation $y = -\sqrt{4 - x^2}$, which upon squaring both sides yields $y^2 = 4 - x^2$. Rewritten as $x^2 + y^2 = 4$, this equation is familiar to us—its graph is the circle of radius 2 centered at the origin. If we think about what $y = -\sqrt{4 - x^2}$ then represents, we realize its graph is the lower half of the circle (and $y = \sqrt{4 - x^2}$ is the upper half); the graph of f is shown in Figure 5. Hence, $\int_{-2}^{2} -\sqrt{4 - x^2}\ dx$ represents the negative of half of the area of a circle of radius 2. That is,

$$\int_{-2}^{2} -\sqrt{4 - x^2}\ dx = -\frac{\pi(2)^2}{2} = -2\pi.$$

Other properties of the definite integral are simply a consequence of its definition as a limit of Riemann sums. The following list of many such properties begins with two logical extensions of the definition. The proofs of the properties have much in common; some will be supplied here, and others will be left as exercises.

Properties of the Definite Integral

Given the integrable functions f and g on the interval $[a,b]$ and any constant k, the following properties hold.

1. $\int_a^a f(x)\,dx = 0$

2. $\int_b^a f(x)\,dx = -\int_a^b f(x)\,dx$

3. $\int_a^b k\,dx = k(b-a)$

4. $\int_a^b kf(x)\,dx = k\int_a^b f(x)\,dx$

5. $\int_a^b \left[f(x) \pm g(x) \right]\,dx = \int_a^b f(x)\,dx \pm \int_a^b g(x)\,dx$

6. $\int_a^c f(x)\,dx + \int_c^b f(x)\,dx = \int_a^b f(x)\,dx$, assuming each integral exists

7. If $f(x) \le g(x)$ on $[a,b]$, then $\int_a^b f(x)\,dx \le \int_a^b g(x)\,dx$.

8. If $m = \min\limits_{a \le x \le b} f(x)$ and $M = \max\limits_{a \le x \le b} f(x)$, then
    $$m(b-a) \le \int_a^b f(x)\,dx \le M(b-a).$$

Selected Proofs and Comments

The first two properties result from thinking about our definition of the Riemann integral and making logical extensions to cases not yet covered by the definition. Specifically, a reasonable interpretation of the expression $\int_a^a f(x)\,dx$ is that it represents the signed area of a rectangle of height $f(a)$ and width zero. Note also that any Riemann sum approximating the integral could consist only of exactly one term, namely $f(a)\Delta x$, with $\Delta x = a - a = 0$. Therefore, we should define $\int_a^a f(x)\,dx$ to be 0. For the second property, consider the following argument.

$$\int_a^b f(x)\,dx = \lim_{n \to \infty} \sum_{i=1}^{n} f(x_i^*)\Delta x$$

$$= \lim_{n \to \infty} \sum_{i=1}^{n} f(x_i^*)\left(\frac{b-a}{n} \right)$$

$$= -\lim_{n \to \infty} \sum_{i=1}^{n} f(x_i^*)\left(\frac{a-b}{n} \right) \qquad \text{Factor out } -1.$$

$$= -\int_b^a f(x)\,dx \qquad \text{Note the reversal of the limits of integration.}$$

This is our justification for stating that reversing the order of the limits of integration of a given function changes the sign of the result.

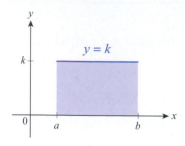

Figure 6 $\displaystyle\int_a^b k\,dx = k(b-a)$

To prove Property 3, it suffices to note that if $f(x) = k,$ then

$$\sum_{i=1}^{n} f(x_i^*)\Delta x = \sum_{i=1}^{n} k\left(\frac{b-a}{n}\right) = k\left(\frac{b-a}{n}\right)n = k(b-a)$$

for every n and for any choice of sample points. It is also useful, though, to visualize the meaning of $\int_a^b k\,dx.$ This definite integral corresponds to the signed area of the shaded rectangle in Figure 6, which is $k(b-a).$

Similarly, Property 5 can be proved algebraically and interpreted graphically.

$$\int_a^b \left[f(x) \pm g(x)\right]dx = \lim_{n\to\infty} \sum_{i=1}^{n} \left[f(x_i^*) \pm g(x_i^*)\right]\Delta x$$

$$= \lim_{n\to\infty} \sum_{i=1}^{n} f(x_i^*)\Delta x \pm \lim_{n\to\infty} \sum_{i=1}^{n} g(x_i^*)\Delta x$$

$$= \int_a^b f(x)\,dx \pm \int_a^b g(x)\,dx$$

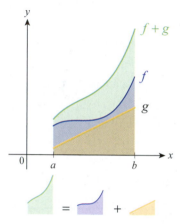

Figure 7

$$\int_a^b \left[f(x) + g(x)\right]dx$$
$$= \int_a^b f(x)\,dx + \int_a^b g(x)\,dx$$

Figure 7 shows the graphs of three functions f, g, and $f+g$ and illustrates the fact that the sum of the areas under the graphs of f and g equals the area under the graph of $f+g.$

The proof of Property 6 is only slightly more technical in nature and reflects the idea that appears in Figure 8. In this graph, $a < c < b,$ and geometrically it is easy to believe that the signed area under f on the interval $[a,b]$ can be expressed as the sum of the signed areas over the intervals $[a,c]$ and $[c,b].$ More interestingly, Property 6 remains true even if c lies outside of $[a,b],$ a fact you will prove in Exercise 99.

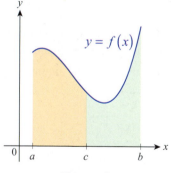

Figure 8

$$\int_a^c f(x)\,dx + \int_c^b f(x)\,dx = \int_a^b f(x)\,dx$$

Example 4 ✒

Given that $\displaystyle\int_0^{3\pi/2} \sin^2 x\,dx = 3\pi/4$　and　$\displaystyle\int_\pi^{3\pi/2} \sin^2 x\,dx = \pi/4,$　determine $\displaystyle\int_0^\pi \sin^2 x\,dx.$

Solution

By Property 6, we know that $\displaystyle\int_0^\pi \sin^2 x\,dx + \int_\pi^{3\pi/2} \sin^2 x\,dx = \int_0^{3\pi/2} \sin^2 x\,dx,$ and so $\displaystyle\int_0^\pi \sin^2 x\,dx + (\pi/4) = 3\pi/4.$ Hence,

$$\int_0^\pi \sin^2 x\,dx = \frac{3\pi}{4} - \frac{\pi}{4} = \frac{\pi}{2}.$$

Example 5 ✐

Use Property 8 to find lower and upper bounds on the integral $\int_0^1 e^x\,dx$.

Solution

The function $f(x) = e^x$ is an increasing function, so

$$m = \min_{0 \le x \le 1} f(x) = f(0) = 1 \quad \text{and} \quad M = \max_{0 \le x \le 1} f(x) = f(1) = e.$$

Hence, $1(1-0) \le \int_0^1 e^x\,dx \le e(1-0)$, or $1 \le \int_0^1 e^x\,dx \le e \approx 2.7183.$

We will discover a broad array of uses for the definite integral in the coming sections, but we will close this section with a quickly described application.

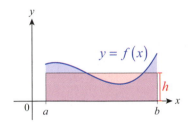

$y = f(x)$

h

$0 \quad a \qquad b$

Figure 9 Average Value of f

Consider the signed area bounded by an integrable function f over $[a,b]$, as shown in Figure 9. We now know how to make sense of such an area, no matter how "curvy" the graph of f is. Given that we know (in principle) the bounded area, it's fairly easy to imagine how the region could be altered into a rectangular region of the same area—simply define the width of the desired rectangle to be $b - a$, and specify the height h so that $h(b - a) = \int_a^b f(x)\,dx$. We even already know some bounds on h, as it must be the case that $\min_{a \le x \le b} f(x) \le h \le \max_{a \le x \le b} f(x)$.

An alternative approach to this same relation is to consider how we would define the *average* of an infinite number of numbers. Specifically, if we wanted to determine the average value of $f(x)$ for all $a \le x \le b$, we might start by considering the approximation

$$\frac{f(x_1) + f(x_2) + \cdots + f(x_n)}{n},$$

where again $x_i = a + i\Delta x$ and $\Delta x = (b - a)/n$. This average can be rewritten as

$$\frac{1}{n}\sum_{i=1}^n f(x_i) = \frac{\Delta x}{b-a}\sum_{i=1}^n f(x_i) = \frac{1}{b-a}\sum_{i=1}^n f(x_i)\,\Delta x,$$

and as $n \to \infty$ we obtain a formula for the average value: $h = \dfrac{1}{b-a}\int_a^b f(x)\,dx.$

Example 6 ✐

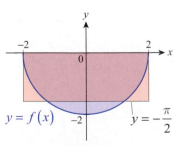

$y = f(x)$ -2

$y = -\dfrac{\pi}{2}$

Figure 10

Determine the average value of $f(x) = -\sqrt{4 - x^2}$ on the interval $[-2, 2]$.

Solution

We have already determined in Example 3 that

$$\int_{-2}^2 -\sqrt{4 - x^2}\,dx = -2\pi, \quad \text{so} \quad h = \frac{1}{2 - (-2)}(-2\pi) = -\frac{\pi}{2}.$$

Note that, geometrically, this corresponds to transforming a half circle into a rectangle, as shown in Figure 10.

5.2 Exercises

1. Use the given graph along with appropriate formulas from geometry to evaluate each of the indicated definite integrals. (Note that the graph of f consists of linear pieces and a semicircle.)

a. $\displaystyle\int_{-3}^{1} f(x)\,dx$

b. $\displaystyle\int_{-3}^{9} f(x)\,dx$

c. $\displaystyle\int_{0}^{6} |f(x)|\,dx$

d. $\displaystyle\int_{0}^{9} \left[f(x) - 2 \right]\,dx$

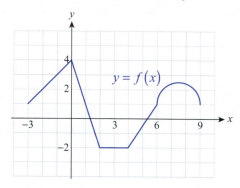

2. Calculate the total distance traveled by the stone of Example 1 by evaluating $\displaystyle\int_{0}^{4} |v(t)|\,dt$.

3. Suppose that in Example 1, George shoots the stone upward while standing near the edge of a deep canyon, and this time pulls the slingshot a bit harder, achieving a velocity function of $v(t) = 80 - 32t$ ft/s. What is the height of the stone, relative to its initial height, at $t = 4$ seconds? How about at $t = 6$ seconds? 10 seconds?

4–13 Use the given partition and sample points to approximate the definite integral of $f(x)$ on the indicated interval. (Note that the subintervals do not always have to be of equal width, and the sample points may be unevenly spaced.)

4. $f(x) = \dfrac{1}{3}x + 1$, $\quad x_0 = 0 < 1 < 2 < 3 < 4 < 5 < 6 = x_6$, $\quad x_i^* = x_i$

5. $f(x) = x^2 + x + 2$, $\quad x_0 = -1 < 0 < 1 < 2 < 3 = x_4$, $\quad x_i^* = x_{i-1}$

6. $f(x) = -x - \dfrac{3}{2}$, $\quad x_0 = -2 < -1.5 < -0.9 < 0 < 1 = x_4$, $\quad x_1^* = -1.8, \ x_2^* = -1, \ x_3^* = -0.4, \ x_4^* = 0.5$

7. $f(x) = \dfrac{1}{x^2}$, $\quad x_0 = 1 < 2 < 3 < 4 = x_3$, $\quad x_i^* = \dfrac{x_{i-1} + x_i}{2}$

8. $f(x) = \dfrac{1}{1 + x^2}$, $\quad x_0 = -3 < -2 < -1 < 0 < 1 < 2 < 3 = x_6$, $\quad x_i^* = \dfrac{x_{i-1} + x_i}{2}$

9. $f(x) = x^3 - x$, $\quad x_0 = 0 < 0.3 < 0.5 < 1 < 1.5 = x_4$, $\quad x_1^* = 0.25, \ x_2^* = 0.5, \ x_3^* = 1, \ x_4^* = 1.2$

10. $f(x) = \sin x$, $\quad x_0 = 0 < \dfrac{\pi}{6} < \dfrac{\pi}{4} < \dfrac{\pi}{3} < \dfrac{\pi}{2} < \dfrac{2\pi}{3} < \dfrac{3\pi}{4} < \dfrac{5\pi}{6} < \pi = x_8$, $\quad x_i^* = x_{i-1}$

11. $f(x) = \ln(x + 1)$, $\quad x_0 = -0.5 < 1 < 2 < 2.5 = x_3$, $\quad x_1^* = 0, \ x_2^* = e - 1, \ x_3^* = 2$

12. $f(x) = 10^{-x}$, $\quad x_0 = 0 < 0.05 < 0.15 < 1 = x_3$, $\quad x_1^* = 0.01, \ x_2^* = 0.1, \ x_3^* = 1$

13. $f(x) = \sqrt{x}$, $\quad x_0 = 0 < \dfrac{1}{25} < \dfrac{4}{25} < \dfrac{9}{25} < \dfrac{16}{25} < 1 = x_5$, $\quad x_i^* = x_i$

14–27 Use the concept of the definite integral to find the total area between the graph of $f(x)$ and the x-axis, by taking limits of the associated Riemann sums. When setting up the Riemann sums, make your choice between the left-endpoint, right-endpoint, and midpoint strategies. (**Hint:** Extra care is needed on those intervals where $f(x)<0$. Remember that the definite integral represents a signed area.)

14. $f(x)=2x+4$ on $[0,2]$

15. $f(x)=x-1$ on $[0,5]$

16. $f(x)=\dfrac{3-x}{2}$ on $[0,5]$

17. $f(x)=x^2$ on $[1,3]$

18. $f(x)=x^2-1$ on $[-1,1]$

19. $f(x)=x^2-4x$ on $[0,5]$

20. $f(x)=\dfrac{x^2}{2}+2$ on $[-2,2]$

21. $f(x)=3x^2-3$ on $[-1,1]$

22. $f(x)=x^2-2x-3$ on $[-1,4]$

23. $f(x)=x^3$ on $[0,1]$

24. $f(x)=4x^3-32$ on $[0,2]$

25. $f(x)=x^3+3x^2+1$ on $[0,3]$

26. $f(x)=\begin{cases}1-(x-1)^2 & \text{if } 0\le x\le 3\\ x-6 & \text{if } 3\le x\le 4\end{cases}$

27. $f(x)=\begin{cases}x^3 & \text{if } 0\le x\le 2\\ 8x-2x^2 & \text{if } 2\le x\le 4\end{cases}$

28. Generalize Exercise 13 to n subintervals and find the definite integral $\int_0^1\sqrt{x}\,dx$ by letting $n\to\infty$. (**Hint:** Let $x_i^*=i^2/n^2$.)

29. Use the same approach as in Exercise 28 to find $\int_0^2\sqrt[3]{x}\,dx$. (**Hint:** Let $x_i^*=2i^3/n^3$.)

30–33 Express the integral as a limit of Riemann sums. (Do not attempt to evaluate the limit.)

30. $\int_1^3\dfrac{1}{x}\,dx$

31. $\int_0^4\left(x^2-\log_2 x\right)dx$

32. $\int_{-a}^a\dfrac{1}{x^2+1}\,dx$

33. $\int_2^b\sqrt[4]{x}\,dx$

34–45 Sketch the region whose (signed) area is represented by the definite integral, and then use appropriate formulas from geometry to evaluate the integral.

34. $\int_{-1}^5 3\,dx$

35. $\int_{2.5}^{12}(-2)\,dx$

36. $\int_4^2(1-x)\,dx$

37. $\int_8^3\left(4-\dfrac{1}{2}x\right)dx$

38. $\int_0^4|2x-3|\,dx$

39. $\int_{-1}^5(5-|2x|)\,dx$

40. $\int_0^{10}(|x-2|-|7-x|)\,dx$

41. $\int_{-5}^0\sqrt{25-x^2}\,dx$

42. $\int_{-a/2}^a\sqrt{a^2-x^2}\,dx,\quad a>0$

43. $\int_{-2}^5(2-[\![x]\!])\,dx$

44. $\int_{-3}^8[\![3x-1]\!]\,dx$

45. $\int_{-4}^6(x-[\![x]\!])\,dx$

46. Use Riemann sums resulting from midpoint estimates to prove $\int_a^b x\,dx=(b^2-a^2)/2$. (**Hint:** Notice that after using $(b+a)(b-a)=b^2-a^2$, each Riemann sum becomes a collapsing sum.)

47. Provide an alternate proof for Exercise 46 by making a sketch and using areas of triangles.

48.* Mimic the argument used in Exercise 46, but using $x_i^*=\sqrt{(x_{i-1}^2+x_{i-1}x_i+x_i^2)/3}$, to prove the formula $\int_a^b x^2\,dx=(b^3-a^3)/3$.

49. The Dirichlet function is defined as follows.
$$\xi(x)=\begin{cases}0 & \text{if } x \text{ is rational}\\ 1 & \text{if } x \text{ is irrational}\end{cases}$$
Prove that $\xi(x)$ is not integrable. (**Hint:** For a given n, form a Riemann sum by choosing each sample point x_i^* to be rational, then see what happens if each x_i^* is irrational. Use your observation to argue that $\lim_{n\to\infty}\sum_{i=1}^n f(x_i^*)\Delta x$ does not exist.)

50. Prove that the function $f(x)=\begin{cases}1/x^2 & \text{if } x\ne 0\\ 0 & \text{if } x=0\end{cases}$ is not integrable on $[0,1]$. (**Hint:** By examining the first term of each R_n, show that $\lim_{n\to\infty}R_n$ does not exist.)

51. Repeat Exercise 50 for $g(x)=\begin{cases}1/x & \text{if } x\ne 0\\ 0 & \text{if } x=0\end{cases}$ on $[0,1]$. (**Hint:** Show that arbitrarily large Riemann sums can be constructed by choosing appropriate x_i^*'s.)

52–59 Decide whether the function is integrable on the indicated interval. If not, say why. (Do not evaluate the integral.)

52. $f(x) = \dfrac{1}{\sqrt{x+2}}$ on $[-1,1]$

53. $g(x) = \dfrac{2}{x}$ on $[-2,2]$

54. $h(x) = \dfrac{-3}{x-1}$ on $[0,5]$

55. $F(x) = \dfrac{x}{|x|}$ on $[-3,4]$

56. $G(x) = x \cdot [\![x]\!]$ on $[-2,2]$

57. $H(x) = \begin{cases} \dfrac{\sin x}{x} & \text{if } x \neq 0 \\ 0 & \text{if } x = 0 \end{cases}$ on $[-1,1]$

58. $u(x) = \begin{cases} \cos\dfrac{1}{x} & \text{if } x \neq 0 \\ 0 & \text{if } x = 0 \end{cases}$ on $[-1,1]$

59. $v(x) = \begin{cases} 3.14 & \text{if } x \text{ is rational} \\ \pi & \text{if } x \text{ is irrational} \end{cases}$ on $[0,2]$

60–65 Match the given property of the definite integral to the relevant illustration (labeled A–F).

60. $\displaystyle\int_a^b k\,dx = k(b-a)$ (Property 3)

61. $\displaystyle\int_a^b kf(x)\,dx = k\int_a^b f(x)\,dx$ (Property 4)

62. $\displaystyle\int_a^b \left[f(x) \pm g(x)\right]dx = \int_a^b f(x)\,dx \pm \int_a^b g(x)\,dx$ (Property 5)

63. $\displaystyle\int_a^c f(x)\,dx + \int_c^b f(x)\,dx = \int_a^b f(x)\,dx$ (Property 6)

64. If $f(x) \le g(x)$ on $[a,b]$, then $\displaystyle\int_a^b f(x)\,dx \le \int_a^b g(x)\,dx$.   (Property 7)

65. If $m = \displaystyle\min_{a \le x \le b} f(x)$ and $M = \max_{a \le x \le b} f(x)$, then $m(b-a) \le \displaystyle\int_a^b f(x)\,dx \le M(b-a)$.   (Property 8)

A.

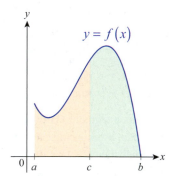

B.

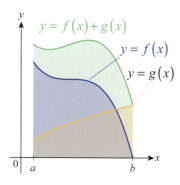

C.

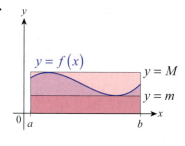

D.

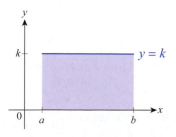

E.

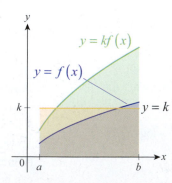

F.

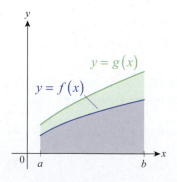

66–75 Use the properties of the definite integral to find the given integral, if possible, given that $\int_a^b f(x)\,dx = 3$, $\int_c^b f(x)\,dx = -1$, and $\int_a^b g(x)\,dx = -5$.

66. $\int_a^b \left[f(x) - g(x) \right] dx$ **67.** $\int_a^c \left[2f(x) + 1 \right] dx$

68. $\int_c^a 10 f(x)\,dx$ **69.** $\int_a^a f(x) g(x)\,dx$

70. $\int_a^b \left[4f(x) + \dfrac{g(x)}{10} \right] dx$ **71.** $\int_b^a \dfrac{\sqrt{2}}{2} g(x)\,dx$

72. $\int_a^b \left[f(x) + 2g(x) - 2 \right] dx$

73. $\int_a^b \left[f(x) \right]^2 dx$

74. $\int_a^b \dfrac{5}{g(x)}\,dx$ **75.** $\int_a^b \left[\dfrac{f(x)}{3} - \pi g(x) \right] dx$

76–83 Use the results from Exercises 46 and 48, along with the properties of the definite integral and formulas from geometry, to evaluate the given integral.

76. $\int_0^2 (3x - 1)\,dx$ **77.** $\int_{\sqrt{2}}^{-1} \left(1 - \dfrac{\sqrt{2}}{2} x \right) dx$

78. $\int_{-1}^4 (x^2 + 5)\,dx$ **79.** $\int_1^4 (2x^2 - x)\,dx$

80. $\int_0^3 \left(t^2 + \dfrac{t}{4} + 4 \right) dt$ **81.** $\int_0^1 (2\sqrt{x} + x)\,dx$

82. $\int_2^0 \left(\dfrac{\sqrt[3]{x}}{4} - x^2 \right) dx$ **83.** $\int_{-2}^2 \left(u - 3\sqrt[3]{u} \right) du$

84. Suppose that f is an even function, g is odd, and both are integrable on $[-a, a]$. Use the properties of the definite integral to prove the following:
$$\int_{-a}^a f(x)\,dx = 2\int_0^a f(x)\,dx, \text{ and } \int_{-a}^a g(x)\,dx = 0.$$

85–90 Suppose that f is an even function, g is odd, both are integrable on $[-2, 2]$, and we know that $\int_0^2 f(x)\,dx = 1$, while $\int_0^2 g(x)\,dx = 2.5$. If possible, find the integral.

85. $\int_{-2}^2 \left[f(x) + g(x) \right] dx$

86. $\int_{-2}^2 \left[2f(x) - 3g(x) \right] dx$

87. $\int_{-2}^2 |g(x)|\,dx$ **88.** $\int_{-2}^2 f(x) g(x)\,dx$

89. $\int_{-1}^1 \left[f(x) \right]^2 dx$ **90.** $\int_0^2 |g(x)|\,dx$

91. Use Property 8 of the definite integral to prove the validity of the following upper and lower estimates: $12 \le \int_0^4 \sqrt{x^2 + 9}\,dx \le 20$.

92–96 Use an argument similar to the one you gave in Exercise 91 to give upper and lower estimates for the given definite integral.

92. $\int_{-1}^4 \sqrt{5 + x}\,dx$ **93.** $\int_2^3 \sqrt{3 - x}\,dx$

94. $\int_4^5 \dfrac{1}{x - 2}\,dx$ **95.** $\int_0^6 \left(\dfrac{x^2}{32} - \dfrac{x}{4} + \dfrac{3}{2} \right) dx$

96. $\int_1^{\sqrt{3}} \arctan x\,dx$

97. Use Property 7 of the definite integral to prove the following inequalities.

a. $\int_0^1 \sqrt{1 - x}\,dx \le \int_0^1 \sqrt{1 - x^2}\,dx$

b. $\int_0^{\pi/2} \cos x\,dx \le \int_0^{\pi/2} \dfrac{\sin x}{x}\,dx$

98. Prove Property 4 of the definite integral. (**Hint:** Write a typical Riemann sum for f on $[a, b]$; use the Constant Multiple Rule for Finite Sums, followed by properties of limits.)

99. Prove Property 6 of the definite integral in general; that is, prove that the property $\int_a^c f(x)\,dx + \int_c^b f(x)\,dx = \int_a^b f(x)\,dx$ holds irrespective of the order of the points a, b, and c. (**Hint:** The standard case of $a < c < b$ is discussed in the text. To start you off with the remaining cases assume, for example, that $a < b < c$. By an argument analogous to the one given in the text, we see that $\int_a^b f(x)\,dx + \int_b^c f(x)\,dx = \int_a^c f(x)\,dx$. Observe by Property 2 that $\int_b^c f(x)\,dx = -\int_c^b f(x)\,dx$, and rearrange the terms. Handle the remaining cases in a similar fashion.)

100. Prove Property 7 of the definite integral. (**Hint:** For a particular partition of $[a, b]$ and choice of sample points, argue that $\sum_{i=1}^n f(x_i^*)\Delta x \le \sum_{i=1}^n g(x_i^*)\Delta x$, and take the limits as $n \to \infty$.)

101. Use Property 7 to prove that the definite integral of a nonnegative function is nonnegative: If $f(x) \geq 0$ on $[a,b]$, then $\int_a^b f(x)\,dx \geq 0$. Then state and prove the analogous statement for nonpositive functions.

102. Prove Property 8 of the definite integral. (**Hint:** Use Property 7 with the constant function $g(x) = M$. The other inequality can be handled in a similar manner.)

103. Use Properties 4 and 7 to prove the following: If f is integrable on $[a,b]$, then $\left| \int_a^b f(x)\,dx \right| \leq \int_a^b |f(x)|\,dx$. (**Hint:** Let $k = 1$ or $k = -1$ so that $\left| \int_a^b f(x)\,dx \right| = k \cdot \int_a^b f(x)\,dx$. Use the fact that $k \cdot f(x) \leq |f(x)|$, along with Properties 4 and 7.)

104–115 Find the average value of the function over the given interval. (**Hint:** Instead of using Riemann sums, try using the results from Exercises 46 and 48 along with formulas from geometry and the properties of the definite integral.)

104. $f(x) = 3x - 1$ on $[0,4]$

105. $g(x) = -1 - \dfrac{1}{2}x$ on $[-2,2]$

106. $h(x) = x^2 - 2$ on $[-1,5]$

107. $F(x) = -3x^2 + 7x + 12$ on $[-2,3]$

108. $G(x) = 9x - x^3$ on $[-4,4]$

109. $H(x) = x^3 - 2x^2 - 1$ on $[0,2]$

110. $k(x) = |x - 4| - 2$ on $[0,7]$

111. $m(x) = |x| + |x+1|$ on $[-3,2]$

112. $u(x) = \sqrt{1 - (x-1)^2}$ on $[0,2]$

113. $v(x) = [\![x]\!]$ on $\left[\dfrac{1}{2}, 3\right]$

114. $t(x) = \sqrt{x} - 1$ on $[0,5]$

115. $w(x) = \sqrt[3]{x} - x$ on $[-1,8]$

116–119 Recognize the given limit as a Riemann sum of a function over an interval and then use geometry to evaluate it.

116. $\displaystyle \lim_{n \to \infty} \sum_{i=1}^{n} \frac{1}{n}\left(2 - \frac{i}{n}\right)$ **117.** $\displaystyle \lim_{n \to \infty} \sum_{i=1}^{n} \frac{3}{n}\left(\frac{2i}{3n} + 4\right)$

118. $\displaystyle \lim_{n \to \infty} \sum_{i=1}^{n-1} \left(\frac{2}{n} + \frac{4i}{n^2}\right)$ **119.** $\displaystyle \lim_{n \to \infty} \sum_{i=1}^{n} \frac{2}{n}\sqrt{4 - \left(\frac{2i}{n}\right)^2}$

120. Prove that if $f(x)$ is an increasing nonnegative function on $[a,b]$, then for every n, $L_n \leq \int_a^b f(x)\,dx \leq R_n$. Then state and prove the analogous statement for a decreasing function $g(x)$ on the same interval.

121. Prove that L_n corresponding to $f(x)$ of Exercise 120 is increasing, while R_n is decreasing. Then state and prove the analogous statement for $g(x)$.

122. Use geometry and a fundamental trigonometric identity to find $\int_0^{\pi} \sin^2 x\,dx$. (**Hint:** Start out by comparing the given integral with $\int_0^{\pi} \cos^2 x\,dx$.)

123. Use the result of Exercise 122 to evaluate $\int_0^{\pi} \left(2\sin^2 x + x^2 - 3x\right)dx$.

124.* Suppose that the nonnegative function $R(x)$ has the property that $R(x) = 0$ whenever x is rational. If R is integrable on the interval $[a,b]$, prove that $\int_a^b R(x)\,dx = 0$.

125–130 *True or False?* Determine whether the given statement is true or false. In case of a false statement, explain or provide a counterexample.

125. If f and g are both integrable on $[a,b]$, then $\int_a^b f(x) \cdot g(x)\,dx = \int_a^b f(x)\,dx \cdot \int_a^b g(x)\,dx$.

126. The integral $\int_a^b f(x)\,dx$ is numerically equal to the area between the graph of $f(x)$ and the x-axis.

127. A Riemann sum for $f(x)$ on $[a,b]$ can be based upon a division of $[a,b]$ into subintervals of unequal width.

128. If $|f(x)|$ is integrable on $[a,b]$, then so is $f(x)$.

129. If $f(x)$ is positive and increasing on $[a,b]$, then $\int_a^b f(x)\,dx \geq f(a)(b-a)$.

130. If $\int_a^b f(x)\,dx < 0$, then $f(x) \leq 0$ on $[a,b]$.

TOPICS

1. The Fundamental Theorem, Part I
2. The Fundamental Theorem, Part II

5.3 **The Fundamental Theorem of Calculus**

The appropriately named Fundamental Theorem of Calculus (FTC) ties together the two branches of differential and integral calculus and allows us to express the definite integral of a function in terms of an antiderivative. It thus serves as both an important theoretical achievement and as an eminently practical computational tool. Newton and Leibniz, drawing upon the prior work of a few mathematical predecessors, are justly renowned for their development of the FTC and their insight into its power and broad applicability.

TOPIC 1   **The Fundamental Theorem, Part I**

The Fundamental Theorem is traditionally presented in two parts; the first shows that differentiation and integration are inverse operations of each other, and the second indicates how to use antiderivatives to calculate definite integrals. The two parts can be thought of as the theoretical and computational facets of the FTC, and our development begins with the following observation.

Theorem 🔍

The Mean Value Theorem for Definite Integrals

If f is a continuous function on the interval $[a,b]$, then there exists a point c in $[a,b]$ for which

$$f(c) = \frac{1}{b-a} \int_a^b f(x)\,dx.$$

Proof ✎

The proof of this statement follows from the application of the Intermediate Value Theorem to a slight rephrasing of one of the definite-integral properties seen in the previous section. Recall that for a continuous function f on $[a,b]$,

$$m(b-a) \le \int_a^b f(x)\,dx \le M(b-a),$$

where $m = \min\limits_{a \le x \le b} f(x)$ and $M = \max\limits_{a \le x \le b} f(x)$.

Dividing each side of the inequality by $b-a$, we see that

$$m \le \frac{1}{b-a} \int_a^b f(x)\,dx \le M.$$

Since f is continuous, the Intermediate Value Theorem tells us that f takes on every value between m and M at least once on the interval $[a,b]$. In particular, there must exist a point $c \in [a,b]$ such that $f(c)$ is the value

$$\frac{1}{b-a} \int_a^b f(x)\,dx.$$

Example 1 ✐

Find every point c in the interval $[-2,2]$ at which the function $f(x) = -\sqrt{4-x^2}$ takes on its average value.

Solution

In Example 6 of Section 5.2, we graphed this function and determined that its average value over the interval is $-\pi/2$. So to find each point c at which f assumes its average value, we simply solve the equation $-\sqrt{4-x^2} = -\pi/2$.

$$-\sqrt{4-x^2} = -\frac{\pi}{2}$$

$$4-x^2 = \frac{\pi^2}{4}$$

$$x^2 = \frac{16-\pi^2}{4}$$

$$x = \pm\frac{\sqrt{16-\pi^2}}{2} \approx \pm 1.24$$

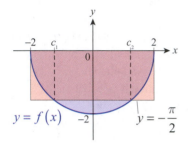

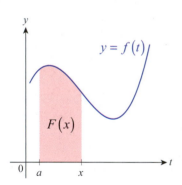

Figure 1

These two solutions are labeled c_1 and c_2 in Figure 1.

Suppose now that f is continuous on an interval I and that a is a fixed point in I. Then for every $x \in I$, we can define the value $F(x)$ by

$$F(x) = \int_a^x f(t)\,dt.$$

This definition is valid even if $x < a$ (so long as $x \in I$), since we defined right-to-left integration in Section 5.2. Visually, $F(x)$ represents the signed area shown in Figure 2; be sure to note that x is really the variable of interest in this definition, and that we have chosen to use t as the so-called "dummy variable" (the variable of integration).

Figure 2 Defining $F(x)$

The crux of the FTC is the realization that the function F, as defined, is an antiderivative of f. We have the necessary machinery now to prove this fact with relative ease.

Theorem ⚲

The Fundamental Theorem of Calculus, Part I

Given a continuous function f on an interval I and a fixed point $a \in I$, define the function F on I by $F(x) = \int_a^x f(t)\,dt$. Then $F'(x) = f(x)$ for all $x \in I$.

Proof 📎

We will prove that $F'(x) = f(x)$ by showing that

$$\lim_{h \to 0} \frac{F(x+h) - F(x)}{h} = f(x)$$

for all $x \in I$. In what follows, we will assume that each h is sufficiently small so that both $x - h \in I$ and $x + h \in I$. Further, if x happens to be an endpoint of I, then $F'(x)$ should be interpreted as a one-sided derivative and the limit above restricted to $h > 0$ (if x is the left endpoint of I) or $h < 0$ (if x is the right endpoint).

Note that one of the properties of the definite integral tells us that

$$F(x+h) - F(x) = \int_a^{x+h} f(t)\,dt - \int_a^x f(t)\,dt = \int_x^{x+h} f(t)\,dt,$$

and so, for each h sufficiently small,

$$\frac{F(x+h) - F(x)}{h} = \frac{1}{h}\int_x^{x+h} f(t)\,dt.$$

By the Mean Value Theorem for Definite Integrals, there is a point c between x and $x + h$ for which $(1/h)\int_x^{x+h} f(t)\,dt = f(c)$; note that $c \in [x, x+h]$ if h is positive and $c \in [x+h, x]$ if h is negative. As Figures 3a and 3b illustrate (for a positive h), $F(x+h) - F(x)$ represents the signed area bounded by the graph of f between x and $x + h$, and this signed area is equal to the area of the rectangle with width $(x+h) - x = h$ and height $f(c)$. As $h \to 0$, it must be the case that $c \to x$, as c is squeezed between x and $x + h$. Hence,

$$F'(x) = \lim_{h \to 0} \frac{F(x+h) - F(x)}{h} = \lim_{h \to 0} \frac{h \cdot f(c)}{h} = \lim_{h \to 0} f(c) = \lim_{c \to x} f(c) = f(x).$$

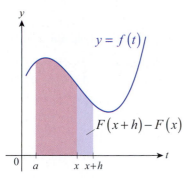

Figure 3a

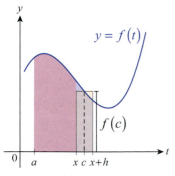

Figure 3b

Using our alternate notation for differentiation makes it even more apparent that differentiation "undoes" integration, as we have just shown that

$$\frac{d}{dx}\int_a^x f(t)\,dt = f(x).$$

Example 2 ✏

Given a function F defined by $F(x) = \int_4^x \left(\sqrt{t} + 1\right) dt$, determine the interval on which the Fundamental Theorem of Calculus can be applied and find $F'(x)$ on that interval.

Solution

First, note that $f(t) = \sqrt{t} + 1$ is continuous on $I = [0, \infty)$ and that 4 (the lower limit of integration) is an element of I. So the FTC applies for each nonnegative x, and $F'(x) = f(x) = \sqrt{x} + 1$. In the particular case of $x = 0$, this means $F_+'(0) = \sqrt{0} + 1 = 1$.

Example 3 ✐

Use the FTC to evaluate the following:

a. $\dfrac{d}{dx}\displaystyle\int_a^{x^2}\dfrac{1}{2+t}\,dt$ **b.** $\dfrac{d}{dx}\displaystyle\int_{3-5x}^{7}e^{t-3}\,dt$ **c.** $\dfrac{d}{dx}\displaystyle\int_{x^2+3}^{1+5x}\cos t\,dt$

Solution

a. The expression $y=\displaystyle\int_a^{x^2}\dfrac{1}{2+t}\,dt$ does indeed represent a function in x, but because the upper limit of integration is x^2 and not x, dy/dx is not simply $1/(2+x)$. To understand why we shouldn't expect this to be the case, note that for a small change Δx from x to $x+h$, the upper limit changes from x^2 to $x^2+2xh+h^2$. But we can easily find the requested derivative by using the Chain Rule: if we let $u=x^2$, then

$$\frac{dy}{dx}=\frac{dy}{du}\cdot\frac{du}{dx}$$

$$=\left(\frac{d}{du}\int_a^u\frac{1}{2+t}\,dt\right)\left(\frac{du}{dx}\right)$$

$$=\left(\frac{1}{2+u}\right)(2x)$$

$$=\left(\frac{1}{2+x^2}\right)(2x)\qquad\qquad\text{Replace }u\text{ with }x^2.$$

$$=\frac{2x}{2+x^2}.$$

b. To determine $\dfrac{d}{dx}\displaystyle\int_{3-5x}^{7}e^{t-3}\,dt$, we must again use the Chain Rule and also account for the fact that the variable limit of integration is the lower one, not the upper. We can do so as follows.

$$\frac{d}{dx}\int_{3-5x}^{7}e^{t-3}\,dt=\frac{d}{dx}\left(-\int_{7}^{3-5x}e^{t-3}\,dt\right)\qquad\int_a^b f(x)\,dx=-\int_b^a f(x)\,dx$$

$$=\frac{d}{du}\left(-\int_{7}^{u}e^{t-3}\,dt\right)\left(\frac{du}{dx}\right)\qquad\text{Set }u=3-5x.$$

$$=\left(-e^{u-3}\right)(-5)$$

$$=5e^{3-5x-3}\qquad\qquad\text{Replace }u\text{ with }3-5x.$$

$$=5e^{-5x}$$

c. The limits in the integral $\displaystyle\int_{x^2+3}^{1+5x}\cos t\,dt$ are both functions of x, but the two examples above guide us on how to proceed. Note that since the integrand is continuous everywhere, we can break the integral apart into a sum of two integrals, and the choice of a is immaterial—any fixed $a\in\mathbb{R}$ will do.

$$\frac{d}{dx}\int_{x^2+3}^{1+5x}\cos t\,dt=\frac{d}{dx}\left[\int_{x^2+3}^{a}\cos t\,dt+\int_a^{1+5x}\cos t\,dt\right]\qquad\int_a^b f(x)\,dx=\int_a^c f(x)\,dx+\int_c^b f(x)\,dx$$

$$=\frac{d}{dx}\left(-\int_a^{x^2+3}\cos t\,dt\right)+\frac{d}{dx}\int_a^{1+5x}\cos t\,dt\qquad\begin{array}{l}\text{Reverse the limits}\\\text{in the first integral.}\end{array}$$

$$=\left[-\cos\left(x^2+3\right)\right](2x)+\left[\cos\left(1+5x\right)\right](5)\quad\text{Use the Chain Rule twice.}$$

$$=-2x\cos\left(x^2+3\right)+5\cos\left(1+5x\right)$$

Example 4 ✍

Given the function $f(x) = (x-2)/(x+1)$, define functions F, G, and H by

$$F(x) = \int_5^x f(t)\,dt, \quad G(x) = \int_{10}^x f(t)\,dt, \quad \text{and} \quad H(x) = \int_{-10}^x f(t)\,dt.$$

Determine the intervals on which the FTC applies for each integral, and find $F'(x)$, $G'(x)$, and $H'(x)$ on those intervals. How are F, G, and H similar and how are they different?

Solution

The function f has a single point of discontinuity at $x = -1$, and so the FTC is applicable for F and G on the interval $(-1, \infty)$ and for the function H on the interval $(-\infty, -1)$ (note that the lower limit of integration is the key to determining the interval on which the FTC can be used). Next, the FTC tells us that $F'(x) = G'(x) = f(x)$ for all $x \in (-1, \infty)$ and that $H'(x) = f(x)$ for all $x \in (-\infty, -1)$.

The last question is the most interesting. By Corollary 2 of the Mean Value Theorem from Section 4.2, we know that two functions with the same derivative on an open interval differ only by a constant. So it must be the case that $F(x) = G(x) + C$ on $(-1, \infty)$ for some constant C. And while it's also true that $H' = f$, the domain of H does not overlap with the domain of F or G at all.

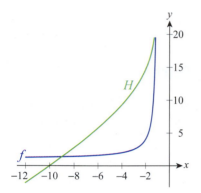

Figure 4 Graphs of f and H

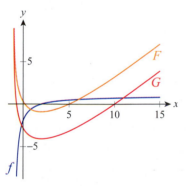

Figure 5 Graphs of f, F, and G

In Figures 4 and 5, the graphs of f, F, G, and H are shown. We will soon develop the tools needed to find explicit formulas for F, G, and H, but for the moment we will take the graphs as given. What we *can* do now is confirm that F, G, and H possess the properties that such functions should—remember that each was designed to return the signed area bounded by the graph of f over a (varying) interval.

For instance, it should be the case that $H(-10) = 0$, since $H(-10) = \int_{-10}^{-10} f(t)\,dt$, and that is indeed the case. Further, $H(x)$ should be positive for $x \in (-10, -1)$, since f is positive on the interval and we are integrating from left to right for such x's. Similarly, $H(x)$ should be negative for $x \in (-\infty, -10)$, since f is again positive on the interval but for these values of x we are integrating from right to left. Note that for such an x, $H(x) = \int_{-10}^x f(t)\,dt = -\int_x^{-10} f(t)\,dt < 0$.

Turning to F and G, we can say a bit more. First, both F and G should have a horizontal tangent line at $x = 2$, since $f(2) = 0$ and the FTC tells us that $F'(2) = f(2)$ and $G'(2) = f(2)$. And as with H, we know that $F(5) = \int_5^5 f(t)\,dt = 0$ and that $G(10) = \int_{10}^{10} f(t)\,dt = 0$. All of these facts are reflected in the graphs of F and G.

One last observation is in order. We have already determined that $F(x) = G(x) + C$ for some constant C on $(-1, \infty)$, and the graphs seem to confirm that fact. But what is the actual value of C? Based on Figure 5, it appears to be approximately 3—note the difference, for example, between $F(2)$ and $G(2)$ or between $F(10)$ and $G(10)$. More precisely, we can use two of the properties of the definite integral to note that for any $x \in (-1, \infty)$,

$$C = F(x) - G(x)$$

$$= \int_5^x f(t)\,dt - \int_{10}^x f(t)\,dt$$

$$= \int_5^x f(t)\,dt + \int_x^{10} f(t)\,dt \qquad \int_a^b f(x)\,dx = -\int_b^a f(x)\,dx$$

$$= \int_5^{10} f(t)\,dt. \qquad \int_a^c f(x)\,dx + \int_c^b f(x)\,dx = \int_a^b f(x)\,dx$$

With the knowledge that we shall soon acquire, we will be able to determine that $\int_5^{10} f(t)\,dt = 5 + 3\ln\left(\frac{6}{11}\right) \approx 3.18$.

TOPIC 2   The Fundamental Theorem, Part II

Part I of the FTC shows us how to construct an antiderivative of a function through the use of definite integration. The true power of that construction is demonstrated by the second part of the FTC.

Theorem ⚬

The Fundamental Theorem of Calculus, Part II

If f is a continuous function on the interval $[a,b]$ and if F is any antiderivative of f on $[a,b]$, then $\int_a^b f(x)\,dx = F(b) - F(a)$.

Proof ✎

First, we know the definite integral $\int_a^b f(x)\,dx$ exists, since f is continuous on the interval. Further, by Part I of the FTC, we know that the function $G(x) = \int_a^x f(t)\,dt$ exists and is an antiderivative of f on $[a,b]$. This means $G'(x) = f(x)$ for all $x \in [a,b]$ and we know $F'(x) = f(x)$ for all $x \in [a,b]$ (since F is an antiderivative of f), so $F'(x) = G'(x)$ for all $x \in [a,b]$. Hence, by Corollary 2 of the Mean Value Theorem for derivatives (Section 4.2) it must be the case that $F(x) = G(x) + C$ for all $x \in (a,b)$ and for some constant C—we have already seen this consequence in Example 4. Since differentiability implies continuity,

$$F(a) = \lim_{x \to a^+} F(x) = \lim_{x \to a^+} G(x) + C = G(a) + C$$

and

$$F(b) = \lim_{x \to b^-} F(x) = \lim_{x \to b^-} G(x) + C = G(b) + C.$$

So we have

$$F(b) - F(a) = G(b) + C - \left[G(a) + C \right]$$
$$= G(b) - G(a)$$
$$= \int_a^b f(t)\, dt - \int_a^a f(t)\, dt$$
$$= \int_a^b f(t)\, dt - 0$$
$$= \int_a^b f(x)\, dx.$$

The relationship between differentiation and integration is now complete, and we can exploit the above result to evaluate definite integrals without resorting to a limit of Riemann sums whenever we know an antiderivative of the integrand. And since expressions of the form $F(b) - F(a)$ arise so frequently in this context, the following notation is useful:

$$F(x) \Big]_a^b = F(b) - F(a)$$

This may also appear as $\left[F(x) \right]_a^b = F(b) - F(a)$.

Example 5 ✐

Evaluate the following integrals.

a. $\displaystyle\int_0^9 e^x\, dx$ **b.** $\displaystyle\int_1^3 \frac{1}{x}\, dx$ **c.** $\displaystyle\int_{-2}^1 x^5\, dx$

Solution

In each case, our knowledge of derivatives will be sufficient to "reverse differentiate" and find a suitable antiderivative—you may want to refer to Section 4.7 for a refresher. Remember, any antiderivative of the integrand will do.

a. Since $\dfrac{d}{dx}\left(e^x \right) = e^x$, we will use $F(x) = e^x$.

Hence, $\displaystyle\int_0^9 e^x\, dx = e^x \Big]_0^9 = e^9 - e^0 = e^9 - 1.$

b. Recall that $\dfrac{d}{dx} \ln|x| = \dfrac{1}{x}$, so $\displaystyle\int_1^3 \frac{1}{x}\, dx = \ln|x| \Big]_1^3 = \ln 3 - \ln 1 = \ln 3.$

c. Note that $\dfrac{d}{dx}\left(\dfrac{x^6}{6} \right) = x^5$, so $\displaystyle\int_{-2}^1 x^5\, dx = \dfrac{x^6}{6} \Big]_{-2}^1 = \dfrac{1^6}{6} - \dfrac{(-2)^6}{6} = -\dfrac{21}{2}.$

Example 6 ✎

Construct a function $f(x)$ whose derivative is $\sin x$ and which has the value -3 at 0.

Solution

We can begin by noting that the function $\int_0^x \sin t\, dt$ has the desired derivative, by Part I of the FTC. We can modify this function to have the correct value at 0 by defining $f(x) = \int_0^x \sin t\, dt - 3$; now,

$$f(0) = \int_0^0 \sin t\, dt - 3 = 0 - 3 = -3.$$

While f in this form defines a valid function, we can use the fact that $\dfrac{d}{dt}(-\cos t) = \sin t$ to rewrite our answer as follows:

$$f(x) = \int_0^x \sin t\, dt - 3$$
$$= \left[-\cos t\right]_0^x - 3$$
$$= -\cos x - (-\cos 0) - 3$$
$$= -\cos x + 1 - 3 = -\cos x - 2$$

Example 7 ✎

Recall that the difference between the functions F and G in Example 4 was determined to be $\int_5^{10} \dfrac{x-2}{x+1} dx$. Evaluate this integral.

Solution

As always, an antiderivative of the integrand will make the evaluation of the integral an easy task. In this case, though, it will take a bit more thought to arrive at an antiderivative—it's not immediately clear what sort of function has a derivative of $(x-2)/(x+1)$.

We will learn many techniques for systematically developing antiderivatives in coming sections, but in this case rewriting the integrand as follows will suffice.

$$\frac{x-2}{x+1} = \frac{(x+1)-3}{x+1} = 1 - \frac{3}{x+1}$$

The same result can be obtained by dividing $x - 2$ by $x + 1$.

$$
\begin{array}{r}
1 \\
x+1{\overline{\smash{\big)}\,x-2}} \\
\underline{-(x+1)} \\
-3
\end{array}
$$

$$\frac{x-2}{x+1} = 1 - \frac{3}{x+1}$$

Also, since $\dfrac{d}{dx}\ln(x+1)=\dfrac{1}{x+1}$ (you should verify this),

$$\int_{5}^{10}\frac{x-2}{x+1}\,dx=\int_{5}^{10}\left(1-\frac{3}{x+1}\right)dx$$

$$=\Big[x-3\ln(x+1)\Big]_{5}^{10}$$

$$=(10-3\ln 11)-(5-3\ln 6)$$

$$=5-3\ln 11+3\ln 6$$

$$=5+3\ln\frac{6}{11}.$$

5.3 Exercises

1–8 Find every point c in the given interval at which $f(x)$ takes on its average value.

1. $f(x)=x^{3};\quad [0,2]$

2. $f(x)=\dfrac{x(6-x)}{2};\quad [0,6]$

3. $f(x)=\dfrac{-x^{4}}{4}+4;\quad [-2,2]$

4. $f(x)=e^{x};\quad [0,1]$

5. $f(x)=\sin x;\quad [0,\pi]$

6. $f(x)=x-\sqrt{x+1};\quad [0,8]$

7. $f(x)=\csc^{2}x;\quad \left[\dfrac{\pi}{4},\dfrac{3\pi}{4}\right]$

8. $f(x)=\dfrac{x^{2}+2}{x^{2}};\quad [1,3]$

9–10 Let $F(x)=\int_{0}^{x}f(t)\,dt$. Use the graph of f to answer the questions. (Note that the graph in Exercise 10 consists of linear and parabolic pieces.)

9. a. Evaluate $F(2)$, $F(4)$, $F(6)$, $F(8)$, and $F(10)$.

    b. Give a formula for $F(x)$. (**Hint:** It will be a piecewise defined function.)

    c. Sketch the graph of $F(x)$.

10. a. Evaluate $F(0)$, $F(2)$, $F(4)$, and $F(7)$.

    b. Give a formula for $F(x)$. (**Hint:** It will be a piecewise defined function.)

    c. Sketch the graph of $F(x)$.

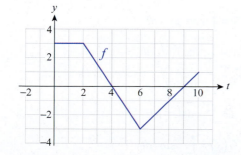

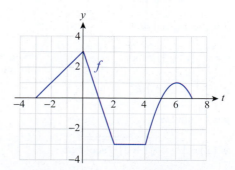

11–16 Find the area between the graph of $f(x)$ and the x-axis on the indicated interval.

11. $f(x) = \sqrt{x}$ on $[1,4]$ **12.** $f(x) = 6x - x^2$ on $[0,6]$ **13.** $f(x) = \dfrac{1}{\sqrt{x}}$ on $[0.5,2]$

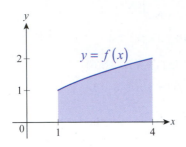

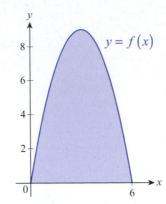

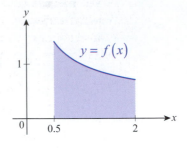

14. $f(x) = e^{-x} + 0.6x$ on $[0,2]$ **15.** $f(x) = -2.5x^{4/3} + 5x$ on $[0,8]$ **16.** $f(x) = 0.5\sec^2 x$ on $\left[0, \dfrac{\pi}{3}\right]$

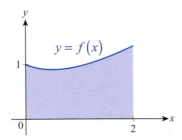

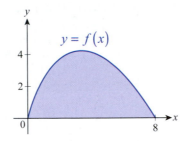

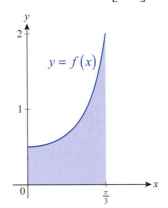

17–32 Use Part I of the Fundamental Theorem of Calculus to find the derivative of the given function.

17. $F(x) = \displaystyle\int_0^x \frac{1}{3}\left(t^2 + \sqrt{t}\right)dt$

18. $F(x) = \displaystyle\int_{1/2}^x \ln s\, ds$

19. $G(x) = \displaystyle\int_{-4}^x \frac{t^4}{t^4 + 4}\,dt$

20. $G(x) = \displaystyle\int_2^x \sqrt[3]{u^2 - u}\,du$

21. $y = \displaystyle\int_x^1 \sin\sqrt{t+1}\,dt$

22. $y = \displaystyle\int_x^0 t\arccos t\, dt$

23. $y = \displaystyle\int_{-5}^{3x} \left(t^2 + 3\right)e^{t-2}\,dt$

24. $y = \displaystyle\int_0^{x^2} \sec^{2/3}\sqrt{t}\,dt$

25. $y = \displaystyle\int_0^{\sin x} \left(t^2 + e^t\right)dt$

26. $y = \displaystyle\int_{\sqrt{x}}^1 \log t\, dt$

27. $y = \displaystyle\int_x^{\pi x} \sin t\, dt$

28. $y = \displaystyle\int_{\sqrt{x}}^{x^2} \cos\left(z^2\right)dz$

29. $F(x) = \displaystyle\int_0^{\cos^{-1}x} \sqrt{1 + \sqrt{1 + \sec^2 t}}\,dt$

30. $G(x) = \displaystyle\int_{x-c}^{x+c} \sin t\, dt$

31. $H(x) = \displaystyle\int_{\ln x}^x \ln t\, dt$

32. $K(x) = \displaystyle\int_x^{x^2} \sqrt{1 + t^4}\,dt$

33–38 Find a formula for $F(x)$ that is free of the integral symbol. Then differentiate it to verify Part I of the Fundamental Theorem of Calculus.

33. $F(x) = \int_1^x 2\,dt$

34. $F(x) = \int_{-3}^x (5-t)\,dt$

35. $F(x) = \int_x^1 (t^2+t)\,dt$

36. $F(x) = \int_x^8 \dfrac{w+2}{\sqrt[3]{w}}\,dw$

37. $F(x) = \int_1^{\sqrt{x}} \dfrac{1}{s^2}\,ds$

38. $F(x) = \int_0^{\tan x} (1+u^2)\,du$

39–65 Use Part II of the Fundamental Theorem of Calculus to evaluate the definite integral.

39. $\int_{-2}^4 (-5)\,dx$

40. $\int_0^{1/\pi} 3\pi^2\,dx$

41. $\int_2^9 (4x+3)\,dx$

42. $\int_{-2.5}^6 (1-5u)\,du$

43. $\int_{-2}^4 (1.5x^2 - x + 3)\,dx$

44. $\int_0^3 (5s-1)(2+s)\,ds$

45. $\int_1^7 (2.4x^3 - 4x^2 + 1)\,dx$

46. $\int_{-1}^1 (2x^2+1)^2\,dx$

47. $\int_1^2 \left(1 - \dfrac{2}{x}\right)dx$

48. $\int_1^3 \left(\dfrac{1}{x^2} + \dfrac{2}{x} + 3\right)dx$

49. $\int_{-2}^{-1} \dfrac{2x^5 - 4x^2}{x^3}\,dx$

50. $\int_0^3 \dfrac{2x^2 - \sqrt{x}}{4}\,dx$

51. $\int_1^2 \left(x\sqrt{x} - \dfrac{1}{\sqrt{x}}\right)dx$

52. $\int_2^4 \dfrac{5x^2 - 3x + 2}{\sqrt{x}}\,dx$

53. $\int_{1/8}^1 \left(2\sqrt[3]{t} - \sqrt[3]{\dfrac{2}{t}}\right)dt$

54. $\int_0^1 \dfrac{x + 3\sqrt{x}}{\sqrt[5]{x}}\,dx$

55. $\int_0^{\pi/2} \left(\dfrac{\sin x}{2} - \sqrt{x}\right)dx$

56. $\int_0^{\pi/3} \dfrac{2}{\cos^2 \theta}\,d\theta$

57. $\int_{\sqrt{2}/2}^1 \dfrac{3}{\sqrt{1-t^2}}\,dt$

58. $\int_{\sqrt{3}/3}^1 \dfrac{-5}{1+x^2}\,dx$

59. $\int_0^{\pi/3} (e^x + \sec x \tan x)\,dx$

60. $\int_{-3}^3 2^x\,dx$

61. $\int_{-2}^4 |x(x-2)|\,dx$

62. $\int_1^3 f(x)\,dx$, where $f(x) = \begin{cases} \sin\dfrac{\pi x}{4} & \text{if } 0 < x \le 2 \\ (x-2)^2 + 1 & \text{if } 2 < x \le 3 \end{cases}$

63. $\int_{\pi/4}^{3\pi/4} (1 - \csc\theta\cot\theta)\,d\theta$

64. $\int_{\pi/4}^{\pi/2} \dfrac{2}{1 - \cos^2 x}\,dx$

65. $\int_{-1}^1 g(x)\,dx$, where $g(x) = \begin{cases} \sqrt{x+1} & \text{if } -1 < x \le 0 \\ e^x & \text{if } 0 < x \le 1 \end{cases}$

66–69 Recognize the given limit as a Riemann sum of a function over an interval and then use the Fundamental Theorem of Calculus to evaluate it.

66. $\displaystyle\lim_{n\to\infty} \sum_{i=1}^n \dfrac{\sqrt{i}}{n^{3/2}}$

67. $\displaystyle\lim_{n\to\infty} \sum_{i=1}^n \dfrac{2}{n}\left(\dfrac{2i}{n}\right)^4$

68. $\displaystyle\lim_{n\to\infty} \sum_{i=1}^n \dfrac{\pi}{2n}\cos\dfrac{\pi i}{2n}$

69. $\displaystyle\lim_{n\to\infty} \sum_{i=1}^{n-1} \dfrac{e-1}{n + i(e-1)}$

70–78 Find the area of the region between the graph of the given function and the x-axis on the indicated interval.

70. $y = \cos x$ on $\left[-\dfrac{\pi}{2}, \dfrac{3\pi}{2}\right]$

71. $y = -x^3$ on $[-2, 2]$

72. $y = -x^2 + 1$ on $[-1, 2]$

73. $y = \dfrac{1}{x}$ on $\left[\dfrac{1}{e}, e^2\right]$

74. $y = -2|x-3| + 6$ on $[0, 10]$

75. $y = -x^3 + 7x^2 - 10x$ on $[0, 6]$

76. $y = 2\sqrt{x} - x$ on $[0, 9]$

77. $y = \dfrac{1-2x}{2x+1}$ on $[0, 1]$

78. $y = x^4 - x^2$ on $[-1, 1]$

79–87 Use the method of Example 7 to evaluate the definite integral.

79. $\int_3^4 \dfrac{x}{x-2}\,dx$

80. $\int_5^7 \dfrac{x+5}{x-4}\,dx$

81. $\int_0^{e-1} \dfrac{2x-5}{x+1}\,dx$

82. $\int_{-3/2}^0 \dfrac{3x-1}{2x+4}\,dx$

83. $\int_0^1 \dfrac{3x^2+4}{x^2+1}\,dx$

84. $\int_0^2 \dfrac{5x^2-1}{2x^2+4}\,dx$

85. $\int_4^6 \dfrac{3x^2+2x-9}{x^2-3}\,dx$

86. $\int_0^2 \dfrac{2x^2+4x+11}{x^2+x+5}\,dx$

87. $\int_{-1}^1 \dfrac{x^3+5x^2+4x+1}{x^3+2x^2+1}\,dx$

88–90 The function $v(t)$ gives the velocity, in units per second, of a particle moving along the x-axis, having started from the origin. Find **a.** the position of the particle at $t = t_0$ and **b.** the total distance traveled by the particle in the time interval $\left[0, t_0\right]$.

88. $v(t) = 1 - (t-1)^2$; $t_0 = 4$

89. $v(t) = \dfrac{t-1}{2(t+1)}$; $t_0 = 3$

90. $v(t) = t(t-3)(t-5)$; $t_0 = 6$

91. Find a formula for $f(x)$ if $\int_0^x f(t)\,dt = \sin 2x + x$.

92.* Repeat Exercise 91 if $\int_0^{x^2} f(t)\,dt = x^3$.

93.* Let $f(x) = x - [\![x]\!]$ and $F(x) = \int_0^x f(t)\,dt$. Prove that F is continuous, briefly discuss its graph, and sketch it on paper.

94. Show that the piecewise defined function

$$F(x) = \begin{cases} -\frac{1}{2}x^2 & \text{if } x \le 0 \\ \frac{1}{2}x^2 & \text{if } x > 0 \end{cases} \quad \text{is an antiderivative of}$$

$f(x) = |x|$. Then find an easier formula for $F(x)$ and use the Fundamental Theorem of Calculus to evaluate $\int_a^b |x|\,dx$.

95. Write a paragraph entitled "Differentiation and Integration as Inverse Operations." Quote the Fundamental Theorem of Calculus and include concrete examples.

96.* Use the properties of the definite integral to show directly that if $f(x)$ is integrable on $[a,b]$, then $F(x) = \int_a^x f(t)\,dt$ is continuous on the same interval. (**Hint:** Argue that there is an M so that $|f(x)| \le M$ on $[a,b]$ and use the result of Exercise 103 of Section 5.2.)

97. Let $l(x)$ be defined as the integral function of $1/x$, that is, $l(x) = \int_1^x (1/t)\,dt$. Show that $l(x) = \ln x$. (**Hint:** See the discussion in Example 4.)

98.* Use the definition from Exercise 97 to show the following well-known property of logarithms: For positive a, $b \in \mathbb{R}$, $l(a \cdot b) = l(a) + l(b)$. (**Hint:** Use the definition to show that $l'(ax) = l'(x)$, which implies $l(ax) = l(x) + C$ for some constant C. Argue that $l(a) = C$. Finally, let $x = b$.)

99.* Use the definition from Exercise 97 to show that $l(1/x) = -l(x)$.

100. Taking a cue from Exercises 97–99, let

$$\ln x = \int_1^x \frac{1}{t}\,dt,\ x > 0$$

be the definition of the natural logarithm function—that is, let all our knowledge of the natural logarithm function be determined by this particular definite integral. Note that, by the Fundamental Theorem of Calculus (which applies, since $1/t$ is continuous on the interval $0 < t < \infty$), the natural logarithm is a differentiable function and

$$\frac{d}{dx}(\ln x) = \frac{1}{x}.$$

a. Prove that $\ln 1 = 0$ and that $\lim\limits_{x \to \infty} \ln x = \infty$.

(**Hint:** Construct a Riemann sum based on the given figure to show that

$$\int_1^x \frac{1}{t}\,dt > \frac{1}{2} + \left(\frac{1}{3} + \frac{1}{4}\right) + \left(\frac{1}{5} + \frac{1}{6} + \frac{1}{7} + \frac{1}{8}\right) + \cdots$$

$$> \frac{1}{2} + 2\left(\frac{1}{4}\right) + 4\left(\frac{1}{8}\right) + \cdots$$

$$= \frac{1}{2} + \frac{1}{2} + \frac{1}{2} + \cdots$$

for sufficiently larger x.)

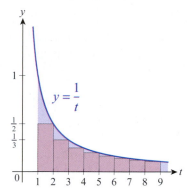

Note that these two facts, along with the fact that $\ln x$ is a continuous function, implies that $\ln x$ takes on every positive real value over the interval $1 < x < \infty$ (and also, given the result of Exercise 99, every negative real value over the interval $0 < x < 1$).

b. Prove that $\ln x$ is one-to-one and hence has an inverse function. (**Hint:** Prove that $\ln x$ is strictly increasing). Given this fact, define e^x to be the inverse of $\ln x$; that is, define e^x by $e^x = \ln^{-1} x$. In particular, define $e = \ln^{-1}(1)$.

c. Use l'Hôpital's Rule to prove $\lim\limits_{u \to 0} \dfrac{u}{\ln(1+u)} = 1$.

d. Use the result from part c. to prove
$\lim\limits_{h \to 0} \dfrac{e^h - 1}{h} = 1$. (**Hint:** Let $u = e^h - 1$ and note that $u \to 0$ as $h \to 0$.)

101. Archimedes (287–212 BC) discovered that the area under a parabolic arch is two-thirds the length of the base times its height. Sketch the graph of $y = h - ax^2$, the general parabolic arch with vertex at $(0, h)$ and use the FTC to verify Archimedes' formula. (Note the interesting parallel between Archimedes' formula and that of the area of an isosceles triangle of the same base and height.)

102. The marginal cost of production of baby toys at a small company has been determined to be $C'(x) = 200 / \left(3\sqrt[3]{x}\right)$ dollars. How much will it cost to increase production from 400 to 500 toys?

103–108 *True or False?* Determine whether the given statement is true or false. In case of a false statement, explain or provide a counterexample.

103. If $f(x)$ is continuous on $[a, b]$ and $c \in [a, b]$ is the point guaranteed by the Mean Value Theorem for Definite Integrals, then $y = f(x)$ and the constant function $y = f(c)$ both have the same definite integral on $[a, b]$.

104. When evaluating a definite integral using the Fundamental Theorem of Calculus, we can use *any* of the antiderivatives of the integrand.

105. If $f(x)$ is a continuous, odd function on $\mathbb{R}$ and $F(x) = \int_{-a}^x f(t)\,dt$ for some $a > 0$, then $F(x)$ has a zero at $x = a$.

106. If $f(x)$ is integrable on $\mathbb{R}$, then $\int_a^x f(t)\,dt$ and $\int_b^x f(t)\,dt$ have the same derivative for all $a, b \in \mathbb{R}$.

107. $\dfrac{d}{dx} \int_a^{x^3} (t+1)^3\,dt = \left(x^3 + 1\right)^3$

108. If f is continuous on $[a, b]$ and F is any antiderivative of f, then the area of the region bounded by the graph of f and the x-axis is $F(b) - F(a)$.

5.3 Technology Exercises

109. Joy bought a new compact car for $15,000. She estimates that the rate of depreciation will be
$f(t) = 15,000(\ln 1.25)0.8^t$ dollars, where t is measured in years. At the same time, all additional expenses
(except for fuel costs) are expected to accumulate at the rate of $g(t) = 350t + 200$ dollars. This gives rise to
the following formula for the average cost of ownership.

$$C(t) = \frac{1}{t}\int_0^t \left[f(s) + g(s)\right]ds$$

a. Explain why the above formula makes sense. What is the average ownership cost during the first year?
The first two years?

b. Joy plans to replace her car at the time when her average cost starts to rise. When is that expected to
happen? (**Hint:** Use a computer algebra system to find the minimum of $C(t)$.)

110. During a nighttime drive on an unfamiliar two-lane road in the California desert, Adam notices that his
odometer is broken. He checks to see that his current speed is 52 mph and asks his passenger to jot down their
speed every five minutes for the next hour. The table below shows the results.

t (min)	5	10	15	20	25	30	35	40	45	50	55	60
v (mph)	56	48	52	57	63	58	53	49	55	59	62	60

a. Use the regression capabilities of a graphing calculator or computer algebra system to find a model for the
velocity function, and denote it by $v(t)$. (**Hint:** It is advisable to convert minutes to hours before calculations.)

b. Use the Fundamental Theorem of Calculus to estimate the distance they covered during the experiment by
integrating $v(t)$ over the one-hour interval.

c. Calculate the right-endpoint and midpoint estimates R_{12} and M_6, respectively, and compare them to the
answer you gave in part b.

111. The owner of a large gas station wants to estimate the number of customers on a typical day between 6 a.m.
and 6 p.m. He records the number of cars pulling in during a one-minute period at the top of each hour, from
6 a.m. to 5 p.m. The table below shows the results.

Time	6 a.m.	7 a.m.	8 a.m.	9 a.m.	10 a.m.	11 a.m.	12 p.m.	1 p.m.	2 p.m.	3 p.m.	4 p.m.	5 p.m.
Cars	1	3	4	1	2	1	5	4	2	1	3	5

a. Use the regression features of a graphing calculator or computer algebra system to find a model for the above
data. (Make your own choice, but, for example, a sine regression works well here. Answers will vary.)

b. Use your model and the integration capabilities of a computer algebra system to estimate the number of
cars from 6 a.m. to 6 p.m.

c. Find the average number of cars pulling in each minute during the above 12-hour time period.

112. Use the integrating and graphing capabilities of a computer algebra system to complete Exercise 93.

113–115 Find a formula for $F(x) = \int_0^x f(t)dt$, and then use a graphing calculator or computer algebra system to graph $f(x)$
and $F(x)$ together on the same screen over the given interval. Discuss how the main features of f (such as signs, intervals of
monotonicity, zeros, critical points) are reflected in the graph of F.

113. $f(x) = x\sin\sqrt{x};\quad [0,17]$ **114.** $f(x) = x\ln x;\quad [0,3.5]$ **115.** $f(x) = \sin 3x\cos 5x;\quad [-1,1]$

5.4 Indefinite Integrals and the Substitution Rule

TOPICS

1. The meaning of indefinite integration

2. The Substitution Rule

The Fundamental Theorem of Calculus solidly establishes the connection between integration and antidifferentiation and provides the single most powerful tool for calculating definite integrals. Given this connection, it isn't surprising that the notation typically used to denote antiderivatives also incorporates the integral sign. In this section, we introduce that notation and learn the first of many techniques for finding antiderivatives of a given function.

TOPIC 1 The Meaning of Indefinite Integration

Remember that the definite integral of a function f over an interval $[a,b]$, if it exists, is a number defined by the limit of Riemann sums—even if, in practice, we avoid actually calculating a definite integral by this manner whenever possible. The FTC gives us the ability to determine the definite integral $\int_a^b f(x)\,dx$ by instead evaluating $F(b) - F(a)$, where F is any antiderivative of f. This connection is the basis of our next definition.

Definition 💡

Indefinite Integral

Given a function f, the **indefinite integral of f** is denoted by

$$\int f(x)\,dx.$$

This notation stands for the set of all antiderivatives of f—that is,

$$\int f(x)\,dx = F(x) + C,$$

where C represents an arbitrary constant called the **constant of integration**, and F is any particular antiderivative of f.

At this point, since we are integrating real-valued functions of a single variable, C can be taken to represent an arbitrary real number. More generally, it is important to remember that C represents an arbitrary term, which is constant with respect to the variable of integration, so that its derivative with respect to that variable is 0. (This will be relevant when we later consider integrals involving more than one variable or, in other settings, when the underlying set is not the set of real numbers.)

Example 1 ✒

Evaluate the following indefinite integrals.

a. $\int \left(3x^2 - 4x + 7\right) dx$ **b.** $\int \sec^2 t\, dt$ **c.** $\int \dfrac{1}{x}\, dx$

Solution

The instruction "evaluate" in this context means to describe the family of antiderivatives without the use of the integral sign, if possible.

a. Since $\dfrac{d}{dx}\left(x^3 - 2x^2 + 7x\right) = 3x^2 - 4x + 7$, we know that

$$\int \left(3x^2 - 4x + 7\right) dx = x^3 - 2x^2 + 7x + C.$$

While this is, in some sense, the most natural way to describe the family of antiderivatives of the integrand, the number of equivalent answers is infinite. Note that

$$\int \left(3x^2 - 4x + 7\right) dx = x^3 - 2x^2 + 7x - 13 + C$$

and

$$\int \left(3x^2 - 4x + 7\right) dx = x^3 - 2x^2 + 7x + 15\pi + C$$

are also correct (though admittedly strange), because as C takes on all possible real values, each formulation of the answer describes the same family of functions.

b. $\dfrac{d}{dt}\left(\tan t\right) = \sec^2 t$, so we write $\int \sec^2 t\, dt = \tan t + C.$

c. Since $\dfrac{d}{dx}\left(\ln|x|\right) = \dfrac{1}{x}$, we write $\int \dfrac{1}{x}\, dx = \ln|x| + C.$

Caution ⚠

Always remember that definite integrals are numbers, while indefinite integrals correspond to families of functions.

The relationship between definite and indefinite integration is sometimes seen as follows:

$$\int_a^b f(x)\, dx = \left[\int f(x)\, dx\right]_a^b$$

This makes sense, as long as it is recognized that the expression on the right refers to the evaluation of *any one* antiderivative of f at the endpoints a and b; since $F(b) - F(a)$ results in the same number for every antiderivative F of f, the expression $\left[\int f(x)\, dx\right]_a^b$ is the same no matter which antiderivative is used. Remember, though, the conditions under which the Fundamental Theorem of Calculus applies.

Example 2 ✐

Use the FTC to evaluate the following, if possible. If the FTC does not apply, indicate why not.

a. $\displaystyle\int_{-3}^{2}\left(3x^2 - 4x + 7\right)dx$ **b.** $\displaystyle\int_{-5}^{-1}\frac{1}{x}\,dx$ **c.** $\displaystyle\int_{-1}^{1}\frac{1}{x^2}\,dx$

Solution

a. We have already determined that

$$\int\left(3x^2 - 4x + 7\right)dx = x^3 - 2x^2 + 7x + C,$$

so we can select any one of this family of functions and proceed. If we set $C = 0$, we find the following:

$$\int_{-3}^{2}\left(3x^2 - 4x + 7\right)dx = \left[x^3 - 2x^2 + 7x\right]_{-3}^{2}$$

$$= \left[2^3 - 2\left(2^2\right) + 7\left(2\right)\right] - \left[\left(-3\right)^3 - 2\left(-3\right)^2 + 7\left(-3\right)\right]$$

$$= 14 - \left(-66\right) = 80$$

The result is the same for every choice of C, since C is both added and subtracted in the expression $F(2) - F(-3)$. Moreover, since $3x^2 - 4x + 7$ is continuous over any interval $[a,b]$, we could evaluate the integral for any choice of a and b.

b. We know the definite integral exists because the integrand is continuous over the interval $[-5, -1]$. Further, since $1/x$ is negative over the interval, we should expect the integral (which represents signed area) to be negative as well. This is in fact what we see.

$$\int_{-5}^{-1}\frac{1}{x}\,dx = \left[\ln|x|\right]_{-5}^{-1} = \ln 1 - \ln 5 = -\ln 5 \approx -1.61$$

c. Our first observation has to be that the FTC does not apply, since $1/x^2$ is not a continuous function over the interval $[-1,1]$. If we failed to notice that fact and tried to evaluate the integral, we would obtain a very misleading result. We might (in error) say that $F(x) = -1/x$ is an antiderivative, since $F'(x) = 1/x^2$. However, F only qualifies as an antiderivative on $(-\infty, 0)$ or $(0, \infty)$, not on an interval that contains 0. If we misapplied the FTC, we would obtain

$$\int_{-1}^{1}\frac{1}{x^2}\,dx = \left[-\frac{1}{x}\right]_{-1}^{1} = -1 - 1 = -2,$$

which can't be true since $1/x^2$ is positive everywhere it's defined.

Using indefinite integral notation, the set of antiderivative facts that we have accumulated can be summarized as follows. Compare this to the tables of antiderivatives in Section 4.7.

Indefinite Integrals

$$\int kf(x)\,dx = k\int f(x)\,dx \qquad\qquad \int \big[f(x)\pm g(x)\big]\,dx = \int f(x)\,dx \pm \int g(x)\,dx$$

$$\int x^r\,dx = \frac{1}{r+1}x^{r+1}+C, \quad r\neq -1 \qquad\qquad \int \frac{1}{x}\,dx = \ln|x|+C$$

$$\int e^{kx}\,dx = \frac{1}{k}e^{kx}+C \qquad\qquad \int a^{kx}\,dx = \left(\frac{1}{k\ln a}\right)a^{kx}+C, \quad a>0,\,a\neq 1$$

$$\int \sin(kx)\,dx = -\frac{1}{k}\cos(kx)+C \qquad\qquad \int \cos(kx)\,dx = \frac{1}{k}\sin(kx)+C$$

$$\int \sec^2(kx)\,dx = \frac{1}{k}\tan(kx)+C \qquad\qquad \int \csc^2(kx)\,dx = -\frac{1}{k}\cot(kx)+C$$

$$\int \sec(kx)\tan(kx)\,dx = \frac{1}{k}\sec(kx)+C \qquad\qquad \int \csc(kx)\cot(kx)\,dx = -\frac{1}{k}\csc(kx)+C$$

$$\int \frac{1}{\sqrt{1-(kx)^2}}\,dx = \frac{1}{k}\sin^{-1}(kx)+C, \quad |kx|<1 \qquad \int \frac{1}{1+(kx)^2}\,dx = \frac{1}{k}\tan^{-1}(kx)+C$$

$$\int \frac{1}{|kx|\sqrt{(kx)^2-1}}\,dx = \frac{1}{k}\sec^{-1}(kx)+C, \quad |kx|>1$$

Table 1

TOPIC 2   The Substitution Rule

Up to this point, we have found antiderivatives using nothing more than our knowledge of derivatives; since we know a substantial amount about differentiation, this approach shouldn't be discounted. Even in some cases where the integrand appears at first intractable (as far as antidifferentiation is concerned), a little algebraic manipulation can go a long way—for instance, we discovered in Section 5.3 that

$$\int \frac{x-2}{x+1}\,dx = x - 3\ln(x+1)+C$$

by rewriting $\dfrac{x-2}{x+1}$ as $1-\dfrac{3}{x+1}$.

But as we encounter a wider variety of integrands, a larger toolbox of integration techniques will be very welcome. At heart, all of these techniques have properties of derivatives as their basis, so in a sense we are simply continuing to use knowledge of differentiation to work backward—the only difference is that the upcoming techniques have been refined and systematized over the years. Our first such technique is called the *Substitution Rule* or, more informally, *u*-substitution.

Theorem ⚭

The Substitution Rule

If $u = g(x)$ is a differentiable function whose range is the interval I, and if f is continuous on I, then

$$\int f(g(x))g'(x)\,dx = \int f(u)\,du.$$

Hence, if F is an antiderivative of f on I, $\int f(g(x))g'(x)\,dx = F(g(x)) + C.$

Proof ✎

The fact that $u = g(x)$ is differentiable tells us that $du/dx = g'(x)$, so the differentials du and dx are related by the equation $du = g'(x)\,dx$ (see Section 3.9). In this respect, the Substitution Rule is an example of a change of variables—if we restate the integral in terms of the variable u instead of x, then

$$\int f(g(x))g'(x)\,dx = \int f(u)\,du. \qquad \begin{array}{l} g(x) = u \\ g'(x)\,dx = du \end{array}$$

But beyond this observation, the Substitution Rule is actually just a rephrasing of the Chain Rule of differentiation. Since f is continuous on the interval I, Part I of the FTC tells us that it has an antiderivative F. Applying the Chain Rule to $F \circ g$,

$$\frac{d}{dx}\big[F(g(x))\big] = F'(g(x))g'(x) = f(g(x))g'(x). \qquad F' = f$$

So $F(g(x))$ is an antiderivative of $f(g(x))g'(x)$ or, using integral notation,

$$\int f(g(x))g'(x)\,dx = F(g(x)) + C.$$

Example 3 ✐

Evaluate the following indefinite integrals.

a. $\displaystyle \int x \cos(x^2 + 1)\,dx$
    b. $\displaystyle \int \frac{4x}{\sqrt{3 - x^2}}\,dx$

Solution

a. The key to applying u-substitution is examining the integrand and identifying a part of it that can be labeled u so that the rest (including the differential dx) corresponds to du. It is vital to remember that *all* of the integral needs to be restated in terms of the new variable u—an integral with a mixture of x's and u's is of no use at all. Our familiarity with differentiation guides us in defining u—in this particular integral, note that if we define $u = x^2 + 1$, then

$$\frac{du}{dx} = 2x, \quad \text{so} \quad du = 2x\,dx.$$

This is almost exactly what we are after, and it is easy to fix it and make it perfect. Note that $x\,dx = \frac{1}{2}\,du$, so we evaluate the integral as follows:

$$\int x\cos\left(x^2+1\right)dx = \int \cos\left(x^2+1\right)x\,dx$$

$$= \int \cos(u)\left(\frac{1}{2}\right)du \qquad\qquad x\,dx = \tfrac{1}{2}\,du$$

$$= \frac{1}{2}\int \cos u\,du$$

$$= \frac{1}{2}\sin u + C$$

$$= \frac{1}{2}\sin\left(x^2+1\right)+C \qquad\qquad \text{Replace } u \text{ with } x^2+1.$$

And, as always, you can easily check your answer by differentiating it and verifying that you obtain the original integrand.

b. Part of this integrand is a second-degree polynomial and the rest is a first-degree polynomial. Specifically, if we set $u = 3 - x^2$, then

$$\frac{du}{dx} = -2x, \quad\text{ so }\quad du = -2x\,dx.$$

Proceeding as in part a., we evaluate the integral.

$$\int \frac{4x}{\sqrt{3-x^2}}\,dx = \int \frac{4}{\sqrt{u}}\left(-\frac{1}{2}\right)du \qquad\qquad x\,dx = -\tfrac{1}{2}\,du$$

$$= -2\int u^{-1/2}\,du$$

$$= -2\left(2u^{1/2}\right)+C \qquad\qquad \frac{d}{du}\left(2u^{1/2}\right) = u^{-1/2}$$

$$= -4\sqrt{3-x^2}+C \qquad\qquad u = 3 - x^2$$

Our choice of u in each integral of Example 3 was guided by the fact that part of the integrand was a second-degree polynomial while the rest of the integrand was, after adjusting by a constant factor, the derivative of that polynomial. The way forward may not always be so apparent in applying the Substitution Rule, but there are some general guidelines to keep in mind. One good way to begin is to define u to be the inner expression in a composition of functions; this is another way of describing what we did in Example 3. Another approach to take is to rewrite the integrand in an algebraically equivalent form, as illustrated in parts b. and c. of Example 4. And, as always, don't be discouraged if your first attempts don't result in an easier integral—u-substitution, like many other techniques in mathematics, is something of an art and it will become second nature with practice.

Example 4 ✐

Evaluate the following indefinite integrals.

a. $\displaystyle\int x^3\sqrt{x^2+1}\,dx$ **b.** $\displaystyle\int \tan x\,dx$ **c.** $\displaystyle\int \frac{dx}{3^x+3^{-x}}$

Solution

a. It is worth seeing what happens if we set $u = x^2 + 1$, even though in this case the rest of the integrand is a polynomial one degree higher than u, not lower. If $u = x^2 + 1$, then $du = 2x\,dx$ or $\frac{1}{2}\,du = x\,dx$.

$$\int x^3\sqrt{x^2+1}\,dx = \int x^2 \cdot x\sqrt{x^2+1}\,dx$$

$$= \frac{1}{2}\int x^2\sqrt{u}\,du \qquad\qquad x\,dx = \tfrac{1}{2}du$$

We don't yet have an integral we can do anything with (both x's and u's appear in the integrand), but the x^2 that remains can be expressed in terms of u. Since $u = x^2 + 1$, it's also true that $x^2 = u - 1$.

$$\frac{1}{2}\int x^2\sqrt{u}\,du = \frac{1}{2}\int (u-1)u^{1/2}\,du$$

$$= \frac{1}{2}\int \left(u^{3/2} - u^{1/2}\right)du$$

$$= \frac{1}{2}\left(\frac{2}{5}u^{5/2} - \frac{2}{3}u^{3/2}\right) + C$$

$$= \frac{1}{5}\left(x^2+1\right)^{5/2} - \frac{1}{3}\left(x^2+1\right)^{3/2} + C$$

b. At first glance, it's hard to see what we can do with $\int \tan x\,dx$, since the integrand consists of a single factor and we haven't yet encountered a function whose derivative is tangent. Some algebraic rewriting is necessary.

$$\int \tan x\,dx = \int \frac{\sin x}{\cos x}\,dx \qquad\qquad \tan x = \frac{\sin x}{\cos x}$$

$$= \int \left(\frac{1}{\cos x}\right)\sin x\,dx \qquad \text{Set } u = \cos x; \text{ then } du = -\sin x\,dx.$$

$$= -\int \frac{1}{u}\,du$$

$$= -\ln|u| + C$$

$$= -\ln|\cos x| + C \qquad \text{Alternatively, } -\ln|\cos x| = \ln\left(|\cos x|^{-1}\right) = \ln|\sec x|.$$

c. This integral is difficult to get a handle on in its original form, but it is more tractable if we multiply the numerator and denominator by 3^x.

$$\int \frac{dx}{3^x + 3^{-x}} = \int \frac{1}{3^x + 3^{-x}}\left(\frac{3^x}{3^x}\right)dx \qquad \text{Multiply the integrand by 1.}$$

$$= \int \frac{3^x}{3^{2x} + 1}\,dx \qquad\qquad \text{Set } u = 3^x; \text{ so } du = 3^x(\ln 3)\,dx.$$

$$= \frac{1}{\ln 3}\int \frac{1}{u^2 + 1}\,du \qquad\qquad 3^{2x} = \left(3^x\right)^2 = u^2$$

$$= \frac{1}{\ln 3}\tan^{-1}u + C$$

$$= \frac{1}{\ln 3}\tan^{-1}\left(3^x\right) + C$$

5.4 **Exercises**

1–15 Evaluate the integral, definite or indefinite, as indicated. (**Hint:** See Examples 1 and 2 and the subsequent table of integrals.)

1. $\int \left(12x^5 + 7.5x^4 - x^3 + 2 \right) dx$

2. $\int \left(-3x^4 + 0.8x^3 - 6x^2 + 4x - \pi \right) dx$

3. $\int 2(x+1)(5x-2) dx$

4. $\int_{-1}^{1} -x(x+4)(2x-1) dx$

5. $\int_{0}^{1} x\sqrt[3]{x} \, dx$

6. $\int \dfrac{x^4 - 3\sqrt{x}}{x^2} dx$

7. $\int \dfrac{x^2 - 2x}{\sqrt{x} + \sqrt{2}} dx$

8. $\int \left(\sqrt[3]{x^2} + \dfrac{1}{\sqrt{x}} \right)^2 dx$

9. $\int (\pi \sec x - \tan x) \sec x \, dx$

10. $\int \left(e^{3x} + 2^{2x/3} \right) dx$

11. $\int \dfrac{\cot 2x}{2\sin x \cos x} dx$

12. $\int_{0}^{\sqrt{3}/2} \dfrac{2}{1+4x^2} dx$

13. $\int_{0}^{1/2} \dfrac{2}{\sqrt{1-4x^2}} dx$

14. $\int \dfrac{3}{|4x|\sqrt{16x^2 - 1}} dx$

15. $\int \dfrac{-7}{\sqrt{1-25x^2}} dx$

16–36 Perform the suggested substitution to evaluate the given indefinite integral.

16. $\int 6x(3x^2+5)^7 dx; \quad u = 3x^2 + 5$

17. $\int x^3 \sqrt{x^4+2} \, dx; \quad u = x^4 + 2$

18. $\int \dfrac{2x}{4x^2+1} dx; \quad u = 4x^2 + 1$

19. $\int \dfrac{2}{4x^2+1} dx; \quad u = 2x$

20. $\int 4e^{2t+3} dt; \quad u = 2t + 3$

21. $\int 4e^{2t+3} dt; \quad u = e^{2t+3}$

22. $\int \cos 5\theta \, d\theta; \quad u = 5\theta$

23. $\int \dfrac{5^{\arctan x}}{1+x^2} dx; \quad u = \arctan x$

24. $\int \dfrac{\sin 2x}{\sqrt{\sin^2 x + 1}} dx; \quad u = \sin^2 x + 1$

25. $\int \dfrac{\sec^2 \left(1 + \sqrt{s} \right)}{\sqrt{s}} ds; \quad u = 1 + \sqrt{s}$

26. $\int \dfrac{1}{3s^{2/3} \sqrt{1 - s^{2/3}}} ds; \quad u = \sqrt[3]{s}$

27. $\int \dfrac{1}{x^2} \sec^2 \dfrac{1}{x} dx; \quad u = \dfrac{1}{x}$

28. $\int \sqrt{x} \cos^3 \left(x^{3/2} \right) dx; \quad u = x^{3/2}$ (**Hint:** Use $u = x^{3/2}$ and $\cos^3 u = \cos^2 u \cos u = (1 - \sin^2 u)\cos u$, and then perform another substitution $w = \sin u$.)

29. $\int \sqrt{1 + \cot^2 x} \, \cot x \csc^2 x \, dx; \quad u = 1 + \cot^2 x$

30. $\int (x+1)(x-7)^8 dx; \quad u = x - 7$

31. $\int \dfrac{z}{2z-1} dz; \quad u = 2z - 1$

32. $\int z\sqrt{z^2 - 5} \, dz; \quad u = z^2 - 5$

33. $\int x\sqrt{x-5} \, dx; \quad u = x - 5$

34. $\int \dfrac{\left(\sqrt{x} - 2 \right)\sqrt{1 + \sqrt{x}}}{\sqrt{x}} dx; \quad u = 1 + \sqrt{x}$

35.* $\int \sqrt{1 - x^2} \, dx; \quad x = \sin u$

36. $\int \sin^4 x \cos^3 x \, dx; \quad u = \sin x$

37–84 Use an appropriate substitution to evaluate the indefinite integral.

37. $\int 3(3x-2)^7\,dx$

38. $\int -2x\sqrt{9-x^2}\,dx$

39. $\int (4x+3)(2x^2+3x)^{20}\,dx$

40. $\int -2x^3\sqrt{9-x^2}\,dx$

41. $\int \cot z\,dz$

42. $\int (z-2)(3z^2-12z)^{99}\,dz$

43. $\int x^2(x^3-5)^{19}\,dx$

44. $\int 6x^3(x^4+2)^{14}\,dx$

45. $\int e^{\sin x}\cos x\,dx$

46. $\int e^x\csc^2(e^x)\,dx$

47. $\int (5-s)^{37}\,ds$

48. $\int s\sqrt{s^2+1}\,ds$

49. $\int x^3\sqrt[3]{x^4+11}\,dx$

50. $\int \sqrt[3]{5x+9}\,dx$

51. $\int \dfrac{2x+1}{(x^2+x-7)^2}\,dx$

52. $\int \dfrac{x^2}{x^3-1}\,dx$

53. $\int \dfrac{-x^3}{\sqrt{2+x^4}}\,dx$

54. $\int \dfrac{4x^3+10x}{x^4+5x^2+6}\,dx$

55. $\int \left(1+\dfrac{1}{t}\right)^3\dfrac{1}{t^2}\,dt$

56. $\int \dfrac{3t+9}{\sqrt{2t^2+12t}}\,dt$

57. $\int \dfrac{2}{t\ln 2t}\,dt$

58. $\int \dfrac{\ln x}{x}\,dx$

59. $\int xe^{x^2-3}\,dx$

60. $\int x\sin(x^2)\,dx$

61. $\int \cos\pi x\,dx$

62. $\int \dfrac{x}{\cos^2(x^2)}\,dx$

63. $\int \sin^3 2t\cos 2t\,dt$

64. $\int 3x\sec^2(x^2+1)\,dx$

65. $\int \dfrac{\csc^2 v}{e^{\cot v-1}}\,dv$

66. $\int \dfrac{5^{\sqrt{3v}}}{\sqrt{3v}}\,dv$

67. $\int \dfrac{\sqrt{x}}{(30-x^{3/2})^2}\,dx$

68. $\int \dfrac{1}{\sqrt{x}(1+\sqrt{x})^2}\,dx$

69. $\int \dfrac{\sin 2x}{\sin^2 x+2}\,dx$

70. $\int \dfrac{\tan\sqrt{x}\sec\sqrt{x}}{\sqrt{x}}\,dx$

71. $\int \cos x\cos(\sin x)\,dx$

72. $\int \dfrac{\ln(x^3)}{x}\,dx$

73. $\int \dfrac{\sin 2x}{1-\cos 2x}\,dx$

74. $\int \dfrac{e^{2t}}{e^t-2}\,dt$

75. $\int \tan^3 2x\sec 2x\,dx$

76. $\int \dfrac{z^3}{1-z^2}\,dz$

77. $\int \dfrac{\sqrt{1+\sqrt{w}}}{\sqrt{w}}\,dw$

78. $\int \sqrt{1+\sqrt{w}}\,dw$

79. $\int (x-5)(x+1)^{11}\,dx$

80. $\int 2x\sqrt{x+2}\,dx$

81. $\int \dfrac{x+2}{4x+1}\,dx$

82. $\int \dfrac{x^2+x+1}{x-2}\,dx$

83. $\int \dfrac{(2\ln x+5)(1-\ln x)^3}{2x}\,dx$

84. $\int \sqrt{2+\sqrt{x}}\,dx$

85–90 Find the function that satisfies the given conditions.

85. $\dfrac{df}{dx}=4x\sqrt{4x^2+4};\quad f(0)=1$

86. $\dfrac{dg}{ds}=\dfrac{2\ln s}{s};\quad g(1)=5$

87. $\dfrac{dy}{dt}=\dfrac{\sin 2t}{\sin^2 t+e};\quad y(0)=0$

88. $y'(x)=\dfrac{3\sqrt{x}}{(1-x^{3/2})^2};\quad y(0)=0$

89. $y''(x)=\cos 4x;\quad y'(0)=1;\quad y(0)=0$

90. $\dfrac{d^2y}{dt^2}=2\cot t\csc^2 t;\quad y'\left(\dfrac{\pi}{2}\right)=0;\quad y\left(\dfrac{\pi}{2}\right)=0$

91. A particle that started at the origin and is moving along the x-axis has a velocity function given by
$$v(t)=\dfrac{1+\sqrt{t+1}}{\sqrt{t+1}}\text{ units/s}.$$
What is the particle's position at $t=3$ seconds?

92. A particle is undergoing simple harmonic motion along the y-axis around the equilibrium $y=0$, while its acceleration is given by
$$a(t)=4\pi^2\sin\dfrac{\pi(1+8t)}{4}\text{ units/s}^2.$$
Find the particle's position at $t=1.5$ seconds and the total distance covered by the particle from $t=0$ seconds to $t=1.5$ seconds.

93–98 *True or False?* Determine whether the given statement is true or false. In case of a false statement, explain or provide a counterexample.

93. If f is defined on the interval $[a,b]$ and has an antiderivative, then the indefinite integral of f on $[a,b]$ is a number.

94. If f is defined on the interval $[a,b]$ and has an antiderivative, then the indefinite integral of f on $[a,b]$ is a function.

95. Two different elements of $\int f(x)\,dx$ can only differ by a constant; that is, if both $F(x)$ and $G(x)$ are elements of the set, then there is a constant C such that $F(x) = G(x) + C$.

96. The Substitution Rule can be interpreted as the Chain Rule in reverse.

97. $\displaystyle \int (\cos x + 1)^2 \, dx = \frac{(\cos x + 1)^3}{3} + C$

98. $\displaystyle \int \frac{1}{x^2 + x + 1} \, dx = \ln|x^2 + x + 1| + C$

5.5 The Substitution Rule and Definite Integration

TOPICS

1. Substitution and definite integration
2. Area between curves

In this last section of the chapter, we introduce a refinement of the Substitution Rule as it applies to definite integrals and then apply our developing integration skills to variations of the area problem with which we began the chapter.

TOPIC 1   Substitution and Definite Integration

If we use the Substitution Rule to find an antiderivative, we can use it to evaluate a definite integral as long as we ensure that it is an antiderivative *of the original integrand*. Remember that a change of variables occurs in applying *u*-substitution (hence the name), and it is all too easy to forget to restate the integrand in terms of the original variable before evaluating it at the limits of integration. To avoid this problem, and to shorten the process of determining definite integrals slightly, the following variation of the Substitution Rule is frequently used.

Theorem 🔍

The Substitution Rule for Definite Integrals

If g' is a continuous function on the interval $[a,b]$, and if f is continuous on the range of $u = g(x)$, then

$$\int_a^b f\big(g(x)\big)g'(x)\,dx = \int_{g(a)}^{g(b)} f(u)\,du.$$

Proof 📎

First, we know that both integrals exist since both integrands are continuous on their respective intervals: $(f \circ g)g'$ is a product of two continuous functions on $[a,b]$, and f is continuous on the interval $\big[g(a),g(b)\big]$. Our task is simply to prove that the integrals are equal.

As in the first version of the Substitution Rule, let F be an antiderivative of f.

$$\int_a^b f\big(g(x)\big)g'(x)\,dx = F\big(g(x)\big)\Big]_{x=a}^{x=b} \qquad \frac{d}{dx}F\big(g(x)\big) = f\big(g(x)\big)g'(x)$$

$$= F\big(g(b)\big) - F\big(g(a)\big)$$

$$= F(u)\Big]_{u=g(a)}^{u=g(b)}$$

$$= \int_{g(a)}^{g(b)} f(u)\,du$$

Example 1 ✐

Evaluate $\displaystyle\int_0^2 x^3 \sqrt{x^2+1}\, dx$.

Solution

Using the substitution $u = x^2 + 1$, we determined in Example 4a of Section 5.4 that

$$\int x^3 \sqrt{x^2+1}\, dx = \frac{1}{5}u^{5/2} - \frac{1}{3}u^{3/2} + C$$

$$= \frac{1}{5}\left(x^2+1\right)^{5/2} - \frac{1}{3}\left(x^2+1\right)^{3/2} + C,$$

so

$$\int_0^2 x^3 \sqrt{x^2+1}\, dx = \left[\frac{1}{5}\left(x^2+1\right)^{5/2} - \frac{1}{3}\left(x^2+1\right)^{3/2} + C\right]_{x=0}^{x=2}$$

$$= \left(\frac{1}{5}(5)^{5/2} - \frac{1}{3}(5)^{3/2}\right) - \left(\frac{1}{5}(1)^{5/2} - \frac{1}{3}(1)^{3/2}\right)$$

$$= \left(5\sqrt{5} - \frac{5\sqrt{5}}{3}\right) - \left(\frac{1}{5} - \frac{1}{3}\right)$$

$$= \frac{2}{15}\left(25\sqrt{5} + 1\right).$$

Alternatively, we can change every part of the definite integral, including the limits of integration: when $x = 0$, $u = 0^2 + 1 = 1$, and when $x = 2$, $u = 2^2 + 1 = 5$.

$$\int_{x=0}^{x=2} x^3 \sqrt{x^2+1}\, dx = \frac{1}{2}\int_{u=1}^{u=5} (u-1)\sqrt{u}\, du \qquad \tfrac{1}{2}\,du = x\,dx$$

$$= \left[\frac{1}{5}u^{5/2} - \frac{1}{3}u^{3/2}\right]_{u=1}^{u=5}$$

$$= \left(\frac{1}{5}(5)^{5/2} - \frac{1}{3}(5)^{3/2}\right) - \left(\frac{1}{5} - \frac{1}{3}\right)$$

$$= \frac{2}{15}\left(25\sqrt{5} + 1\right)$$

Caution ⚠

The two methods illustrated in Example 1 result in the same answer; the choice usually depends only on which seems easier to apply. Just be careful not to mix and match—don't evaluate an antiderivative expressed in the variable u with limits that belong to the original variable! To keep track of which variable the limits correspond to, it can be useful to explicitly note the variable in the integral symbol as we did above.

It's always worthwhile to try to keep the big picture in mind in mathematics, and doing so while evaluating definite integrals can save a significant amount of work. In many practical applications, for instance, integrals over an interval centered at 0 appear. If the integrand in such a case is an even or odd function, the following theorem applies.

Theorem 🔍

Definite Integrals of Even and Odd Functions

Assume f is continuous on the interval $[-a, a]$. Then

1. if f is even, $\int_{-a}^{a} f(x)\, dx = 2\int_{0}^{a} f(x)\, dx$;

2. if f is odd, $\int_{-a}^{a} f(x)\, dx = 0$.

Proof ✎

We will prove the first statement here. Note the use of the Substitution Rule for Definite Integrals in the second-to-last step and, as an aid in making the change of variable, the explicit mention of the variable corresponding to the limits in two of the integrals. The symmetry with respect to the y-axis in Figure 1 illustrates why the signed area of an even function over the interval $[-a, a]$ is twice its signed area over the interval $[0, a]$.

$$\int_{-a}^{a} f(x)\, dx = \int_{-a}^{0} f(x)\, dx + \int_{0}^{a} f(x)\, dx \qquad \textcolor{blue}{\int_{a}^{b} f(x)\, dx = \int_{a}^{c} f(x)\, dx + \int_{c}^{b} f(x)\, dx}$$

$$= -\int_{0}^{-a} f(x)\, dx + \int_{0}^{a} f(x)\, dx \qquad \textcolor{blue}{\int_{a}^{b} f(x)\, dx = -\int_{b}^{a} f(x)\, dx}$$

$$= -\int_{x=0}^{x=-a} f(-x)\, dx + \int_{0}^{a} f(x)\, dx \qquad \textcolor{blue}{\text{Since } f \text{ is even, } f(x) = f(-x).}$$

$$= \int_{u=0}^{u=a} f(u)\, du + \int_{0}^{a} f(x)\, dx \qquad \textcolor{blue}{\begin{array}{l} u = -x, \ du = -dx, \\ \text{and a change in limits} \end{array}}$$

$$= 2\int_{0}^{a} f(x)\, dx. \qquad \textcolor{blue}{\int_{0}^{a} f(u)\, du = \int_{0}^{a} f(x)\, dx}$$

The proof of the second statement is similar, with the difference that in the third step above we will use $-f(x) = f(-x)$ since f is odd.

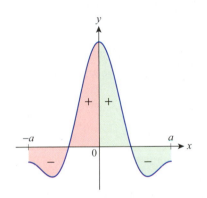

Figure 1
Signed Area of an Even Function

Example 2 ✐

Evaluate $\displaystyle\int_{-\pi}^{\pi} \frac{\sin x}{\sqrt{1 + x^2 + x^4}}\, dx$.

Solution

It would be very difficult to find an antiderivative of this integrand, but fortunately we don't have to. The integrand

$$f(x) = \frac{\sin x}{\sqrt{1 + x^2 + x^4}}$$

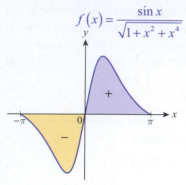

$$f(x) = \frac{\sin x}{\sqrt{1+x^2+x^4}}$$

Figure 2

is continuous and odd since it is a product of an odd continuous function $\sin x$ and an even continuous function $1/\sqrt{1+x^2+x^4}$ (remember that the product of an odd function and an even function is odd). Since we are integrating over an interval centered at 0, we can say immediately that

$$\int_{-\pi}^{\pi} \frac{\sin x}{\sqrt{1+x^2+x^4}}\,dx = 0.$$

Geometrically, this reflects the fact that the signed areas in Figure 2 cancel each other out.

TOPIC 2   Area between Curves

Returning to the issue of finding areas bounded by curves, we are now equipped not only with powerful computational techniques, but also with an idea for how to attack such problems. In general, if we know the rate of change of an area A with respect to some variable, say x, over an interval $[a,b]$, then we know that

$$A = \int_a^b A'(x)\,dx.$$

In hindsight, we now know that this is simply one way of expressing the Fundamental Theorem of Calculus. If we rewrite it slightly, we arrive at a more philosophical statement:

$$A = \int_a^b A'(x)\,dx = \int_a^b \frac{dA}{dx}\,dx = \int_a^b dA$$

This version reflects the key idea at the heart of integration—namely, that "adding up" (that is, integrating) all of the small elements dA over some interval $[a,b]$ results in the total quantity A. We will apply this philosophy in many different contexts through the remainder of this text.

To see how this approach looks when applied to the specific problem of finding the area bounded by two curves $y = f(x)$ and $y = g(x)$ over the interval $[a,b]$, consider the assertion expressed in Figure 3.

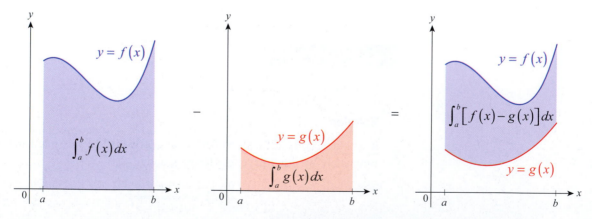

Figure 3 Area Bounded between *f* and *g*

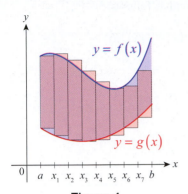

Figure 4

A Riemann Sum Approximation of A

Algebraically, we know that $\int_a^b f(x)\,dx - \int_a^b g(x)\,dx = \int_a^b \left[f(x) - g(x) \right]dx,$ so this is not surprising. But we could also have determined A by integrating the area differential dA, with a formula for dA suggested by Figure 4.

Using this approach, each small area element ΔA is approximated by a rectangle of height $f(x) - g(x)$ and width Δx. In terms of differentials, $dA = \left[f(x) - g(x) \right]dx,$ so we arrive at the same formula.

$$A = \int_a^b dA = \int_a^b \left[f(x) - g(x) \right]dx$$

Definition 💡

Area of a Region between Two Curves

Given two continuous functions f and g defined on an interval $[a,b]$, with $f(x) \geq g(x)$ for all $x \in [a,b]$, the **area of the region bounded between the graphs of f and g over $[a,b]$** is $A = \int_a^b \left[f(x) - g(x) \right]dx.$

The next two examples further illustrate the use of the principle $A = \int_a^b dA.$

Example 3 ✍

The graphs of the functions

$$f(x) = \frac{3x^3}{2} - 6x^2 + x + 7 \quad \text{and} \quad g(x) = \frac{x^3}{2} - 2x^2 + 1$$

are shown in Figure 5. Determine the total area of the shaded regions.

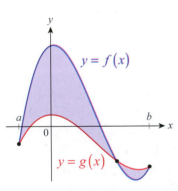

Figure 5

Solution

Using the graphs depicted in Figure 5, it appears that f is the upper function on the first part of the interval $[a,b]$ and that g is the upper function on the remainder of the interval. So in order to find the area of the shaded regions, we have to determine not only a and b but also the point inside the interval $[a,b]$ where f and g intersect.

All three are points where $f(x) = g(x)$, so we begin by solving this equation.

$$\frac{3x^3}{2} - 6x^2 + x + 7 = \frac{x^3}{2} - 2x^2 + 1$$
$$x^3 - 4x^2 + x + 6 = 0$$

Remember that if a polynomial equation has a rational number solution, it must be of the form p/q, where p is a factor of the constant term and q is a factor of the leading coefficient. Since the leading coefficient is 1 in this case, a rational root of the polynomial $x^3 - 4x^2 + x + 6$ must be one of the numbers $\pm\{1, 2, 3, 6\}$. As you can verify by synthetic or long division, -1, 2, and 3 are roots, so we know that $a = -1$, $b = 3$, and $x = 2$ must be the third point where $f(x) = g(x)$.

So the area of the shaded regions can be found as follows:

$$\text{Area} = \int_{-1}^{2}\left[f(x)-g(x)\right]dx + \int_{2}^{3}\left[g(x)-f(x)\right]dx$$

$$= \int_{-1}^{2}\left(x^3 - 4x^2 + x + 6\right)dx + \int_{2}^{3}\left(-x^3 + 4x^2 - x - 6\right)dx$$

$$= \left[\frac{x^4}{4} - \frac{4x^3}{3} + \frac{x^2}{2} + 6x\right]_{-1}^{2} + \left[-\frac{x^4}{4} + \frac{4x^3}{3} - \frac{x^2}{2} - 6x\right]_{2}^{3}$$

$$= \left[\frac{22}{3} - \left(-\frac{47}{12}\right)\right] + \left[-\frac{27}{4} - \left(-\frac{22}{3}\right)\right] = \frac{71}{6}$$

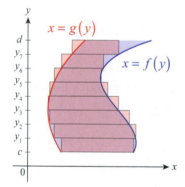

Figure 6 Integration in y

There is nothing sacred about vertical strips—if a region can be more naturally "deconstructed" in some other way, we should take advantage of that fact. If we are aiming to express the area of a region as an integral (or a sum of integrals), it is likely that we will think of the region as being composed of a collection of thin strips, but a horizontal arrangement, as depicted in Figure 6, may make more sense. In such a case, we can express each differential element of area as $dA = \left[f(y) - g(y)\right]dy$, and proceed to integrate with respect to y over the interval $[c,d]$. This can save us considerable effort, as shown in our next example.

Example 4 ✐

Find the area of the region bounded by the graphs of the equations $y = 0$, $3x - 5y = 12$, and $y = \sqrt{x}$.

Solution

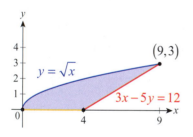

Figure 7

Two edges of the described region are lines, and the remaining edge is the graph of the function $y = \sqrt{x}$, which we can also think of as the upper half of the parabola $x = y^2$. As always, if it's possible to sketch a picture of what we are doing, such a sketch is bound to be helpful—in this case, it's easy to graph the region we're discussing (see Figure 7).

Note that we need two integrals if we want to express the area of the region as an integral in x, since the lower function changes at $x = 4$. On the interval $[0,4]$, the lower function is $g(x) = 0$ (corresponding to the equation $y = 0$), and on the interval $[4,9]$ the lower function is $g(x) = \frac{3}{5}(x-4)$, which we obtain by solving $3x - 5y = 12$ for y. The upper function is $f(x) = \sqrt{x}$ for the entire interval $[0,9]$. So the total area of the described region is

$$A = \int_{0}^{4}\sqrt{x}\,dx + \int_{4}^{9}\left[\sqrt{x} - \frac{3}{5}(x-4)\right]dx.$$

This is certainly doable, and you are asked to evaluate the above integrals in Exercise 52. But if we think of the region as being composed of horizontal strips, we see that the left edge can be described as a single function of y, and the same is true for the right edge—we don't have to divide the interval of integration into subintervals. Specifically, the left edge of the region is the function $g(y) = y^2$ (corresponding to the equation $x = y^2$) and the right edge is the function $f(y) = \frac{5}{3}y + 4$ (obtained by solving $3x - 5y = 12$ for x). So each horizontal differential element of area can be written as

$$dA = \left[f(y) - g(y) \right] dy = \left(\frac{5}{3} y + 4 - y^2 \right) dy$$

thus the area is calculated as follows:

$$A = \int_0^3 dA = \int_0^3 \left(\frac{5}{3} y + 4 - y^2 \right) dy$$

$$= \left[\frac{5}{6} y^2 + 4y - \frac{1}{3} y^3 \right]_0^3$$

$$= \frac{15}{2} + 12 - 9 = \frac{21}{2}$$

5.5 **Exercises**

1–51 Evaluate the definite integral. Whenever possible, take advantage of symmetry.

1. $\displaystyle\int_0^1 2(2x+1)^5 \, dx$

2. $\displaystyle\int_0^2 2x\sqrt{4-x^2} \, dx$

3. $\displaystyle\int_0^2 (x^2-1)(x^3-3x)^8 \, dx$

4. $\displaystyle\int_0^4 (x-2)(2x^2-8x)^{49} \, dx$

5. $\displaystyle\int_1^2 w^2 (w^3+4)^{99} \, dw$

6. $\displaystyle\int_1^{10} 2x^3 (x^4-1)^{49} \, dx$

7. $\displaystyle\int_1^3 (2-x)^6 \, dx$

8. $\displaystyle\int_0^2 x\sqrt{x^2+1} \, dx$

9. $\displaystyle\int_0^1 x^3 \sqrt[3]{x^4+1} \, dx$

10. $\displaystyle\int_0^3 \sqrt{2x+1} \, dx$

11. $\displaystyle\int_1^4 \frac{2x+1}{(x^2+x+1)^2} \, dx$

12. $\displaystyle\int_0^1 \frac{z^2}{z^3+2} \, dz$

13. $\displaystyle\int_0^2 \frac{x^3}{\sqrt{x^4+9}} \, dx$

14. $\displaystyle\int_1^3 \frac{8x^3+20x}{x^4+5x^2+6} \, dx$

15. $\displaystyle\int_1^2 \left(1+\frac{1}{t^2}\right)^2 \frac{1}{t^3} \, dt$

16. $\displaystyle\int_1^2 \frac{6x+10.5}{\sqrt{2x^2+7x}} \, dx$

17. $\displaystyle\int_e^{e^2} \frac{1}{s\ln(s^3)} \, ds$

18. $\displaystyle\int_{10}^{100} \frac{\log x}{x\ln 10} \, dx$

19. $\displaystyle\int_1^{\sqrt{2}} (\ln 2)x \cdot 2^{x^2-1} \, dx$

20. $\displaystyle\int_0^{\sqrt{\pi}} 4x\cos\frac{x^2}{2} \, dx$

21. $\displaystyle\int_0^1 \sin \pi x \, dx$

22. $\displaystyle\int_{\sqrt{\pi/4}}^{\sqrt{3\pi/4}} \frac{-x}{\sin^2(x^2)} \, dx$

23. $\displaystyle\int_0^{\pi/4} \sin^2 2x \cos 2x \, dx$

24. $\displaystyle\int_0^1 x\sec^2(2x^2-1) \, dx$

25. $\displaystyle\int_{\pi^2/16}^{9\pi^2/16} \frac{\cot\sqrt{x}\csc\sqrt{x}}{\sqrt{x}} \, dx$

26. $\displaystyle\int_{-\pi}^{\pi} \sin x \sin(\cos x) \, dx$

27. $\displaystyle\int_1^e \frac{\ln(2x^2)}{x} \, dx$

28. $\displaystyle\int_0^1 \frac{e^{2t}}{e^t+1} \, dt$

29. $\displaystyle\int_{-2}^2 \frac{t^3}{t^2+1} \, dt$

30. $\displaystyle\int_0^{\pi/4} \frac{\csc^2\theta}{e^{\cot\theta}} \, d\theta$

31. $\displaystyle\int_{\pi^2/16}^{\pi^2/4} \frac{\csc^2\left(\frac{\pi}{4}+\sqrt{t}\right)}{\sqrt{t}} \, dt$

32. $\displaystyle\int_0^2 \frac{e^{\sqrt{2x}}}{\sqrt{2x}} \, dx$

33. $\displaystyle\int_0^1 \frac{\sqrt{v}}{\sqrt{v^{3/2}+1}} \, dv$

34. $\displaystyle\int_0^2 \frac{1}{\sqrt{x}\left(2+\sqrt{x}\right)^2} \, dx$

35. $\displaystyle\int_0^{\sqrt{3}} \frac{e^{\arctan x}}{1+x^2} \, dx$

36. $\displaystyle\int_0^1 \frac{1}{x^{2/3}\left(1+x^{2/3}\right)} \, dx$

37. $\int_0^{\pi/2} \dfrac{\sin 2x}{\cos^2 x + 1}\,dx$

38. $\int_0^1 -e^{-x}\sec\left(e^{-x}-1\right)\tan\left(e^{-x}-1\right)dx$

39. $\int_4^9 \dfrac{\sqrt{2+\sqrt{t}}}{\sqrt{t}}\,dt$

40. $\int_0^4 \sqrt{2+\sqrt{t}}\;dt$

41. $\int_{-1}^1 \dfrac{x^2+x+1}{x-2}\,dx$

42. $\int_1^e \dfrac{(\ln x+1)(2\ln x+3)^2}{x}\,dx$

43. $\int_2^3 (x+3)(x-2)^7\,dx$

44. $\int_{-1}^0 x\sqrt[3]{x-1}\,dx$

45. $\int_{-1}^1 x\sqrt{4-x^2}\,dx$

46. $\int_4^{e+3} \dfrac{2x+5}{x-3}\,dx$

47. $\int_{-4}^0 \dfrac{x^2-x+3}{x+5}\,dx$

48. $\int_1^{e^2} \dfrac{(2\ln x+3)(\ln x-1)^2}{x}\,dx$

49. $\int_1^9 \dfrac{\left(2\sqrt{x}+1\right)\sqrt{1+\sqrt{x}}}{\sqrt{x}}\,dx$

50. $\int_0^{\pi^{2/3}} \sqrt{x}\sin^2\left(x^{3/2}\right)\cos^3\left(x^{3/2}\right)dx$

51. $\int_0^{\pi/4} \tan x\sec^3 x\,dx$

52. Evaluate $A=\int_0^4 \sqrt{x}\,dx+\int_4^9\left[\sqrt{x}-\dfrac{3}{5}(x-4)\right]dx$ from Example 4.

53–56 Find the area of the region bounded by the graphs of the given equations, as shown.

53.

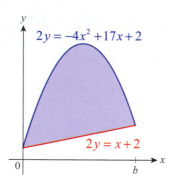

54.

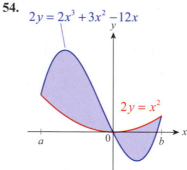

55.

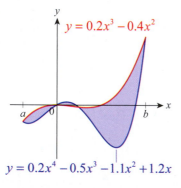

56.

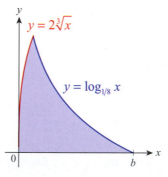

57–96 Find the area of the region bounded by the graphs of the given equations. Be careful to find intersection points, if applicable, and to identify the upper and lower functions on each interval. If convenient or necessary, divide the region into horizontal rather than vertical strips and integrate with respect to y. Whenever possible, take advantage of symmetry.

57. $y=x^2,\quad y=2x$

58. $y=x^2,\quad y=2$

59. $y=4x-x^2,\quad y=x$

60. $y=1-x^4,\quad y=|x|-1$

61. $y = |x^3 - 1|$, $3y = 5x + 11$

62. $y = x^2$, $y = x^4$

63. $y = 2x - x^2$, $y = x^3$

64. $y = \sqrt{x}$, $y = 2 - x$

65. $x = y^2$, $y = x^3$, $x \geq 0$

66. $y = \sqrt[3]{x}$, $y = \sqrt[7]{x}$

67. $2xy - y = 3 - 2x$, $y = x$, $y = 0$

68. $x + 30y = 2y^3 + 5$, $9y - 31 = x + y^3$

69. $y = \sqrt{x+1}$, $y = x^2 - 1$

70. $y = x^3$, $y = x$

71. $y = \dfrac{1}{x}$, $y = \dfrac{1}{x^2}$, $x = e$

72. $y = x^3$, $y = \dfrac{3}{2} - \dfrac{x}{2}$, $y = 0$

73. $y = 4x^2 - x^4$, $y = -5x^2$

74. $y = 3x^2 - x - 4$, $y = x^2 + 3x + 2$

75. $y^2 - x = 2$, $y = x$

76. $1 - y^2 = x$, $(1 - y)^2 = x$

77. $3y - x = 3y^2$, $6y^3 = x + 6y^2$

78. $y = x^3 - 3x^2 + 2x$, $y = x^2 - x$

79. $y = x^3 - 6x^2 + 5$, $y = -x^3 + 12x - 11$

80. $y = x^4 - \dfrac{x^3}{2} - 4x^2$, $y = -\dfrac{x^3}{2}$

81. $y = 2\sqrt[3]{x+1}$, $y = 2 - x$, $y = 0$

82. $y = \sqrt{x}$, $y = \dfrac{1}{x^2}$, $x = 4$

83. $x = y^3 - 10y^2 + 20y$, $x = 4y^2 - 33y + 40$

84. $x = y^4 - 3y^3 - y^2$, $x = 5y^2 - 8y$

85. $y = \sqrt[4]{x}$, $y = -\dfrac{1}{16}x^2 + 18$, $y = 0$

86. $(y - 1)^2 = \dfrac{1}{x}$, $x - 2 = 4y$, $y = 0$

87. $y = \cos x$, $y = \sin 2x$, $x = 0$, $x = \pi$

88. $y = \dfrac{2}{x^2 + 1}$, $y = x^2$

89. $y = \sin x$, $y = \sqrt{2} - \sin x$, $x = \dfrac{\pi}{4}$, $x = \dfrac{3\pi}{4}$

90. $y = \cot x$, $y = 2\cos x$, $0 < x < \pi$

91. $y = x + 1$, $y = \sqrt{18 - (x+1)^2}$, $y = 0$

92. $y = \arctan x$, $y + x = 1 + \dfrac{\pi}{4}$, $x = 0$

93. $y = \cos\left(\dfrac{\pi}{2}x\right)$, $y = x^4 - 1$

94. $y = 2\sqrt{2}\sin x$, $y = \csc^2 x$, $0 < x < \pi$

95. $y = \sin^2 x$, $y = \cos^2 x$, $x = 0$, $x = \pi$

96. $y = \tan^2 x$, $y = 0$, $x = 0$, $x = \dfrac{\pi}{4}$

97. Use the Substitution Rule to prove the following property of the definite integral.

$$\int_a^b f(x)\,dx = \int_{a+c}^{b+c} f(x - c)\,dx$$

Note that the above is often referred to as the *translation invariant property* of the definite integral. Using a generic $f(x)$, make a sketch of both integrals and explain the reason for the name of this property.

98. Use Exercise 97 to explain why the following definite integrals are equal.

$$\int_{-1}^{2} 4x\sqrt[3]{x+2}\,dx = \int_{1}^{4} (4x - 8)\sqrt[3]{x}\,dx$$

99. Italian mathematician Bonaventura Cavalieri (1598–1647), who can be considered as one of the early forerunners of modern calculus, discovered what we today call *Cavalieri's Principle*. Use our discussion preceding Example 3 to prove the following version of Cavalieri's Principle: Suppose two plane regions are included between the lines $x = a$ and $x = b$, and are bounded by graphs of integrable functions. If they have the property that any vertical line intersects both regions in line segments of the same length, then the regions have equal areas.

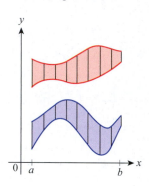

100. Consider the region bounded by the graphs of the equations $y = \sqrt{x}$, $x = 9$, and the *x*-axis. Find the vertical line $x = a$ that bisects the region in two subregions of equal area.

101. The graphs below show the velocities of two bikes at a motorcycle race right after the start (velocity is measured in km/h). Use the figure to answer the following questions.

  a. Which bike is ahead initially?

  b. What happens at the instant when the curves intersect?

  c. Do the curves suggest that a pass happened, and if so, approximately when?

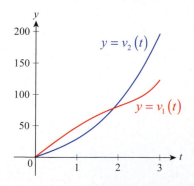

102. Suppose that the function $B(t) = 85 \cdot (1.1163)^t$ approximates the birth rate of a rabbit population on an isolated island, while the death rate is $D(t) = 21 \cdot (1.0811)^t$ (*t* is measured in months). Find the area between the graphs of these two functions on the interval $[0,12]$. Use your own words to give a real-life interpretation to this number.

5.5 Technology Exercises

103–107 Use a computer algebra system to plot the graphs of $f(x)$ and $g(x)$ on the same screen. After choosing the appropriate viewing window, identifying intersection points, and finding the region bounded by the curves, use the integration features of your technology to find the area of the region. (**Hint:** As in Examples 3 and 4, be sure to identify the upper and lower functions on each subinterval and integrate accordingly. As a final step, you may want to check your answer by evaluating $\int_a^b |f(x) - g(x)| \, dx$, where *a* and *b* are the first and last of the intersection points. Do you obtain the same answer?)

103. $f(x) = 35x - 9,$
$g(x) = 6x^3 - 4.95x^2 - 3.04x - 22.2525$

104. $f(x) = 3 \sin x, \quad g(x) = 0.3x$

105. $f(x) = e^x, \quad g(x) = \dfrac{1}{2}x + 2$

106. $f(x) = 2x^4 - 8x^3 - 6.5x^2 + 29x - 12,$
$g(x) = 2x^3 - 4x^2 - 3.5x + 2.5$

107. $f(x) = \dfrac{0.8^x \sin 2x}{2}, \quad g(x) = \dfrac{1}{2}\sqrt{x}$

Chapter 5
Review Exercises

1–2 Use $(O_4 + U_4)/2$ to estimate the area under the graph of the function and above the x-axis on the given interval.

1. $f(x) = \dfrac{x^2}{2}$ on $[0,2]$

2. $f(x) = \sin x$ on $\left[0, \dfrac{\pi}{2}\right]$

3–6 Write the given sum using sigma notation.

3. $\dfrac{1}{2} - \dfrac{1}{9} + \dfrac{1}{28} - \dfrac{1}{65} + \cdots - \dfrac{1}{1,000,001}$

4. $a_1 + a_5 + a_9 + a_{13} + \cdots + a_{97}$

5. $f\left(\dfrac{2}{n^2}\right) + f\left(\dfrac{4}{n^2}\right) + f\left(\dfrac{6}{n^2}\right) + \cdots + f\left(\dfrac{100}{n^2}\right)$

6. $g\left(t_{-2}^*\right)\Delta t + g\left(t_{-1}^*\right)\Delta t + g\left(t_0^*\right)\Delta t + \cdots + g\left(t_{2n}^*\right)\Delta t$

7–8 Assuming that $\displaystyle\sum_{i=0}^{n} a_i = 50$ and $\displaystyle\sum_{i=0}^{n} b_i = 80$, find the sum.

7. $\displaystyle\sum_{i=0}^{n}(a_i + 2b_i + 2)$

8. $\displaystyle\sum_{i=0}^{n}\left(\dfrac{a_i}{5} - \dfrac{b_i}{4}\right)$

9–10 Use summation formulas to find the value of the sum.

9. $\displaystyle\sum_{i=1}^{10}(3i^3 - 1)$

10. $\displaystyle\sum_{j=1}^{n}\dfrac{(2j+1)(j-2)}{2}$

11. Find the value of the sum $\displaystyle\sum_{i=1}^{n}\left[\dfrac{1}{i^2} - \dfrac{1}{(i+1)^2}\right]$.

(**Hint:** Write out the first few terms as well as the last few terms.)

12–13 Evaluate the geometric sum using the formula proven in Exercise 53 of Section 5.1.

12. $\displaystyle\sum_{i=0}^{10}\dfrac{3}{2^i}$

13. $\displaystyle\sum_{j=0}^{6}(-1)^j(0.3)^j$

14–15 Evaluate the given double sum.

14. $\displaystyle\sum_{i=3}^{11}\sum_{j=1}^{4}(2i - j)$

15. $\displaystyle\sum_{i=1}^{n}\sum_{j=1}^{m}i^2 j$

16–18 Find the area under the graph of $f(x)$ and above the x-axis on the given interval, by taking the limit of the associated Riemann sums.

16. $f(x) = x^2 + 1$ on $[0,2]$

17. $f(x) = x^3$ on $[0,1]$

18. $f(x) = \sqrt{x}$ on $[0,4]$

(**Hint:** Choose $x_i^* = \dfrac{4(i-1)^2}{n^2}$.)

19. Identify the region whose area is the limit given by $\displaystyle\lim_{n\to\infty}\dfrac{2}{n}\sum_{i=1}^{n}\left[\dfrac{4i}{n} - \left(\dfrac{2i}{n}\right)^2\right]$. Then use summation formulas to evaluate the limit.

20–21 Prove that the given function is not integrable on the interval $[0,1]$.

20. $f(x) = \dfrac{1}{x^2}$

21. $g(x) = \begin{cases} 1 & \text{if } x \text{ is rational} \\ -1 & \text{if } x \text{ is irrational} \end{cases}$

22–25 Suppose that f is an even function, g is odd, both are integrable on $[-a, a]$, and we know that $\displaystyle\int_0^a f(x)\,dx = 2$, while $\displaystyle\int_0^a g(x)\,dx = 0.5$ $(a > 0)$. If possible, find the integral.

22. $\displaystyle\int_{-a}^{a}\left[5f(x) + 4g(x)\right]dx$

23. $\displaystyle\int_{-a}^{a}\left[f(x)\right]^2 g(x)\,dx$

24. $\displaystyle\int_{-a}^{a} f(x)\left[g(x)\right]^2 dx$

25. $\displaystyle\int_{-a}^{0}\left[f(x) + g(x)\right]dx$

26–27 Find the average value of $f(x)$ over the given interval and identify all points in the domain where $f(x)$ assumes its average value.

26. $f(x) = 4x - x^2$ on $[0,4]$

27. $f(x) = |x - 2| - 1$ on $[1,5]$

28. Use the Fundamental Theorem of Calculus to evaluate the limit

$$\lim_{n \to \infty} \frac{1}{n} \sum_{i=1}^{n} \left(\sqrt{\frac{i}{n}} - \frac{i}{n} \right)$$

by recognizing it as a Riemann sum of a function over an interval.

29–30 Use Part I of the Fundamental Theorem of Calculus to find the derivative of the given function.

29. $F(x) = \int_{0}^{x} \sqrt{1+t^2}\, dt$ **30.** $G(x) = \int_{0}^{x^2} e^{t^2}\, dt$

31–38 Use Part II of the Fundamental Theorem of Calculus to evaluate the definite integral.

31. $\int_{1}^{2} \left(2x^4 + 3x^2 - 2 \right) dx$ **32.** $\int_{0}^{2} (3x+2)(5-x)\, dx$

33. $\int_{1}^{4} \left(\frac{1}{t} - \frac{2}{t^2} + 1 \right) dt$ **34.** $\int_{1}^{9} \frac{x^2 - 2\sqrt{x} + 2}{x}\, dx$

35. $\int_{0}^{1} \frac{2}{\sqrt{1-x^2}}\, dx$

36. $\int_{\pi/4}^{3\pi/4} \left(2\csc^2 x - \cos x \right) dx$

37. $\int_{2}^{3} \frac{x+2}{x-1}\, dx$ **38.** $\int_{0}^{1} \frac{2x^2 - 1}{x^2 + 1}\, dx$

39–40 Find the area of the region between the graph of the given function and the x-axis on the indicated interval.

39. $y = \frac{1}{2x^2}$ on $[1,10]$

40. $y = 2\sqrt{x} - x^2$ on $[0,2]$

41. Find a formula for $f(x)$ if $\int_{0}^{x^3} f(t)\, dt = \sin(x^3)$.

42. The velocity function of a particle moving along the x-axis is $v(t) = 3t - t^2$ units per second. If it started at the origin, find **a.** the position of the particle at $t = 5$ seconds and **b.** the total distance traveled by the particle in the time interval $[0,5]$.

43–54 Use an appropriate substitution (when necessary) to evaluate the indefinite integral.

43. $\int \frac{-2}{\sqrt{1-x^2}}\, dx$ **44.** $\int \frac{-2x}{\sqrt{1-x^2}}\, dx$

45. $\int \sec x (\sec x + \tan x)\, dx$

46. $\int \sec^2 x \tan x\, dx$

47. $\int 6x^2 \left(2x^3 - 7 \right)^9 dx$ **48.** $\int x^4 \sqrt{x^5 - 3}\, dx$

49. $\int \frac{4x}{\left(x^2 + 1 \right)^2}\, dx$ **50.** $\int \frac{e^{\arcsin x}}{\sqrt{1-x^2}}\, dx$

51. $\int \frac{\sec^2 \sqrt{x}}{\sqrt{x}}\, dx$ **52.** $\int \frac{1}{x \ln(x^2)}\, dx$

53. $\int \frac{e^{2x}}{e^x + 1}\, dx$ **54.** $\int \frac{1}{x^2} \sin\left(\frac{x+1}{x} \right) dx$

55–56 Find $y(x)$ that satisfies the given conditions.

55. $\frac{dy}{dx} = \frac{1}{\sqrt{x}\left(\sqrt{x} - 1 \right)^2};$ $y(9) = 1$

56. $y''(x) = 1 - \sin x;$ $y'(0) = 2;$ $y(0) = 0$

57. A particle is moving along the x-axis in the positive direction with a velocity function of $v(t) = \frac{t}{t^2 + 1}$ units per second. If it started at the point $(1,0)$, what is the particle's position at $t = 4$ seconds?

58–63 Evaluate the definite integral.

58. $\int_{0}^{1} 3x^5 \left(x^6 - 1 \right)^{12} dx$

59. $\int_{0}^{4} (x+1)\sqrt{x^2 + 2x}\, dx$

60. $\int_{1}^{2} \frac{x^3}{x^4 + 1}\, dx$ **61.** $\int_{-1/2}^{0} \frac{2^t}{\sqrt{1 - 4^t}}\, dt$

62. $\int_{e}^{e^2} \frac{1}{x (\ln x)^2}\, dx$ **63.** $\int_{1}^{4} \frac{dx}{\sqrt{x}\left(\sqrt{x} + 1 \right)^2}$

64–69 Find the area of the region bounded by the graphs of the given equations. (If convenient or necessary, integrate with respect to y rather than x.)

64. $y = x^3 - 4x, \quad 3y = 15x, \quad x \geq 0$

65. $y = 1 - x^2, \quad y = 1 - x^6, \quad x \geq 0$

66. $y = 2\sqrt[4]{x}, \quad y = 4 - 2x, \quad y = 0$

67. $y = \ln x, \quad (e-1)y = x-1$

68. $y = \dfrac{1}{1+x^2}, \quad 2y = 1$

69. $y = \sin x, \quad y = \sin x \cos x, \quad 0 \le x \le \pi$

70. Consider the region bounded by the graph of $y = 1/x$ and the x-axis over the interval $[1, a]$ $(a > 1)$. Find the vertical line $x = c$ that bisects the region in two subregions of equal area.

71. Consider the function $f(x) = 1/x^2$ defined on some interval $[a,b]$. Partition $[a,b]$ and in each subinterval $[x_{i-1}, x_i]$ choose the sample point $x_i^* = \sqrt{x_{i-1}x_i}$ (the geometric mean of the endpoints). Show that

$$\frac{1}{\left(x_i^*\right)^2}\Delta x_i = \frac{1}{x_{i-1}} - \frac{1}{x_i}$$

and use this observation to prove the following formula.

$$\int_a^b \frac{1}{x^2}\,dx = \frac{1}{a} - \frac{1}{b}$$

72. Prove that if the conditions of Part I of the Fundamental Theorem of Calculus are satisfied and $F(x) = \displaystyle\int_{g(x)}^{h(x)} f(t)\,dt$, where $g(x)$ and $h(x)$ are differentiable, then $F'(x) = f\big(h(x)\big)h'(x) - f\big(g(x)\big)g'(x)$. (**Hint:** See Example 3 of Section 5.3.)

73. Prove that if f is a linear function, then its definite integral on an interval $[a,b]$ is the average of its left and right Riemann sums, that is,

$$\int_a^b f(x)\,dx = \frac{L_n + R_n}{2}.$$

What is your expectation regarding the integral and the average above if f is concave up? Concave down?

74–81 *True or False?* Determine whether the given statement is true or false. In case of a false statement, explain or provide a counterexample.

74. If $n_1 < n_2$, then the Riemann sum R_{n_2} is always a better approximation of the integral than R_{n_1}.

75. If f is piecewise continuous on a closed interval, then the limit of its Riemann sums always exists.

76. When applying the Fundamental Theorem of Calculus, we must choose the antiderivative with $C = 0$.

77. $\displaystyle\int \frac{1}{e^x}\,dx = \ln\left(e^x\right) + C = x + C$

78. The definite integral of the velocity function of a moving object on $[t_1, t_2]$ is equal to the total distance traveled by the object from time $t = t_1$ to $t = t_2$.

79. $\displaystyle\int_a^b f(x)\,dx > 0$ if and only if $f(x) > 0$ on $[a,b]$.

80. $\displaystyle\int \sec x\,dx = \sec x \tan x + C$

81. $\displaystyle\int_{-1}^{1} \frac{1}{x^3}\,dx = \left.\frac{-1}{2x^2}\right]_{-1}^{1} = -\frac{1}{2} - \left(-\frac{1}{2}\right) = 0$

Chapter 5
Technology Exercises

82. Use the summation feature of a computer algebra system to verify your answers for Exercises 9–15.

83. Write a program for a graphing calculator or computer algebra system that calculates the nth Riemann sum for a given function on a given interval, using subintervals of equal width and sample points of your choice. Use your program to verify your answers for Exercises 16–18.

84. Use a computer algebra system to evaluate the limit of Exercise 28. What do you find? (Answer will vary depending on the capabilities of the particular software used.)

Project ✎

Chapter 5

The topic of this project is the so-called *sine integral function*, which is important for its applications, most notably in electrical engineering and signal processing.

1. Consider the following piecewise defined function:

$$f(t) = \begin{cases} \dfrac{\sin t}{t} & \text{if } t > 0 \\ 1 & \text{if } t = 0 \end{cases}$$

   Prove that for any $x \geq 0$, $f(t)$ is integrable on $[0, x]$.

2. The **sine integral function** is defined as follows:

$$\text{Si}(x) = \int_0^x f(t)\,dt, \text{ for } x \geq 0$$

   Prove that $\text{Si}(x)$ is continuous.

3. Find the derivative $\dfrac{d}{dx}\text{Si}(x)$.

4. Without graphing first, write a short paragraph on why you would expect the graph of $\text{Si}(x)$ to be oscillating. Explain why its amplitude is expected to decrease as $x \to \infty$.

5. Find the x-values where the relative maxima and minima of $\text{Si}(x)$ occur.

6. Extend the definition of $\text{Si}(x)$ to negative x-values and prove that for any $a > 0$,

$$\int_{-a}^{a} \text{Si}(x)\,dx = 0.$$

7. Use a computer algebra system to plot the graph of $\text{Si}(x)$ on the interval $[-8\pi, 8\pi]$.

8. Use a computer algebra system to approximate the range of $y = \text{Si}(x)$ to four decimal places.

Applications of the Definite Integral

6.1 Finding Volumes Using Slices 491

1. Integral Definition of Volume
2. Integrating Slices
3. Disks and Washers
4. Interlude: The Big Picture

6.2 Finding Volumes Using Cylindrical Shells 505

1. The Shell Method
2. Disks, Washers, and Shells Compared

6.3 Arc Length and Surface Area 515

1. Arc Lengths of Graphs of Functions
2. Surface Areas of Solids of Revolution

6.4 Moments and Centers of Mass 525

1. Moments, Mass, and Centers of Mass in One Dimension
2. Moments, Mass, and Centers of Mass in Two Dimensions
3. The Theorems of Pappus

6.5 Force, Work, and Pressure 539

1. Integral Definition of Work
2. Fluid Pressure and Fluid Force

6.6 Hyperbolic Functions 552

1. Definitions, Derivatives, and Integrals of Hyperbolic Functions
2. Inverse Hyperbolic Functions

Introduction

With the Fundamental Theorem of Calculus in hand, we can proceed to use it for purposes such as defining *arc lengths* of curves and *volumes* and *surface areas* of three-dimensional objects.

The process of doing so mirrors the way in which we used integration to rigorously define areas of planar regions; while we likely have some intuitive understanding of all these geometric characteristics, calculus provides both the precision and the computational methods we need for our work. It also allows us to prove three physical principles that actually predate calculus: Cavalieri's Principle and Pappus' Theorems for volumes and for surface areas.

Bonaventura Cavalieri
(1598–1647)

In this chapter we also apply integration to problems whose roots lie in physics and engineering. *Moments* and *centers of mass* are physical characteristics that

are difficult to calculate without calculus for any but the simplest of objects. Again, even precisely defining these concepts without calculus is challenging, but integration gives us the natural language in which to work. The same is true for *force*, *work*, and *pressure*, the three topics of Section 6.5. While these words appear in many different contexts, some less formal than others, each has a precise meaning in physics and engineering. Calculus gives us the best means to relate them to each other and to solve problems in which they appear.

Underlying the development of each concept is the fundamental principle that *the whole is the integral of its parts*. This is true whether we are discussing area A, volume V, or

> Underlying the development of each concept is the fundamental principle that *the whole is the integral of its parts*.

work W, and the principle takes on each of the following specific forms in this chapter:

$$A = \int dA$$
$$V = \int dV$$
$$W = \int dW$$

Indeed, this fundamental principle serves as the basis for much of the material in the remainder of this text—watch for its reappearance in different forms as you continue your study of calculus.

6.1 **Finding Volumes Using Slices**

TOPICS

1. Integral definition of volume
2. Integrating slices
3. Disks and washers
4. Interlude: the big picture

Our first step in working with area was to devise a reasonable definition of the concept, and we began with familiar area formulas for simple plane regions. In similar fashion, we develop a broad definition of volume in this section; the Fundamental Theorem of Calculus allows us to do so quickly.

TOPIC 1   **Integral Definition of Volume**

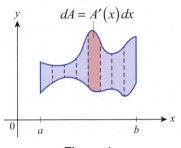

$$dA = A'(x)\,dx$$

Figure 1
Area as an Integral

Our philosophy of integration can be put to good use immediately. Recall from Section 5.5 that the most general statement regarding area and integration can be expressed as $A = \int dA$; that is, the area of a region can be found by "integrating together" its area differentials. In practice, we typically know how to express the area differential dA in terms of another differential dx, for instance $dA = A'(x)\,dx$, and we might also know that the region is bounded between certain limits, say $x = a$ and $x = b$ (see Figure 1). If so, we write

$$A = \int_a^b A'(x)\,dx$$

and use the Fundamental Theorem of Calculus to evaluate A.

In similar fashion, we *define* the volume V of a three-dimensional body by $V = \int dV$, where dV represents a volume differential. Suppose, as in the above paragraph, we are able to express dV in terms of another differential, say $dV = V'(x)\,dx$, and we again know that the body is bounded between the limits $x = a$ and $x = b$. What does the corresponding illustration look like?

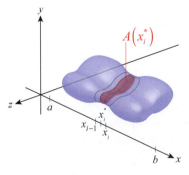

Figure 2

First, note that the two equations $x = a$ and $x = b$ represent planes in the three-dimensional space, as they describe the set of ordered triples $\left\{(a, y, z) \mid y \in \mathbb{R}, z \in \mathbb{R}\right\}$ and $\left\{(b, y, z) \mid y \in \mathbb{R}, z \in \mathbb{R}\right\}$, respectively. Next, as illustrated in Figure 2, each element dV is approximated by a thin slab (perpendicular to the x-axis) with volume $A(x_i^*)\Delta x_i$, where $A(x_i^*)$ is the area of the *cross-section* of the body through x_i^* on the x-axis and Δx_i is the width of a subinterval containing x_i^*. The formula $A(x_i^*)\Delta x_i$ comes from the fact that each thin slab is a generalized *cylinder*, the definition of which follows.

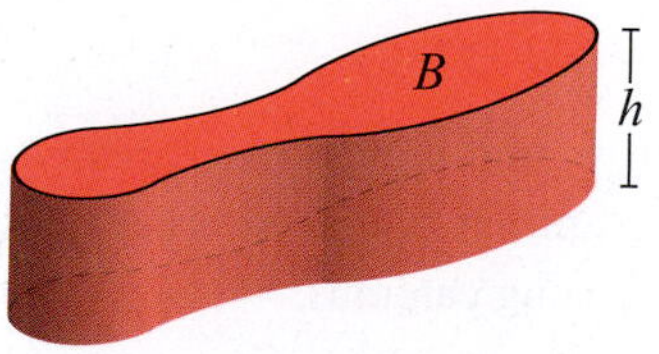

> **Definition** 💡
>
> **Cylinders and Their Volumes**
>
> Given a plane region B and a distance h, a **cylinder** formed by base B and height h is the solid body bounded by B and a translated copy of B in a second plane parallel to the first and separated by the distance h. The volume of a cylinder is the product of its height and the area of its base:
>
> $$V = \text{Area}(B)\cdot h$$

Figure 3
Volume of a Cylinder $V = \text{Area}(B)\cdot h$

(a) Circular Cylinder with $V = \pi r^2 h$

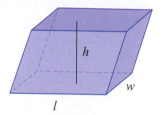

(b) Parallelepiped with $V = lwh$

Figure 4
Volumes of Familiar Cylinders

The simple formula of base area times height is the basis of some well-known volume formulas. For instance, if the base B is a circle of radius r, the cylinder formed is a *circular cylinder* with volume $V = \pi r^2 h$ (see Figure 4a). And if B is a rectangle, the solid body formed is an example of a *parallelepiped* with volume $V = lwh$, where l and w are the dimensions of the rectangle (see Figure 4b). The fact that the volume formula $V = \text{Area}(B) \cdot h$ holds true even if the sides of the cylinder are not perpendicular to the base is a reflection of Cavalieri's Principle, which we will return to soon (cylinders for which the sides *are* perpendicular to the base are called *right cylinders*).

For our purposes, the formula for cylinder volume is all we need to extend our definition of volume to more general bodies. Referring to Figure 2 again, we partition the x-axis interval $[a, b]$ at the $n + 1$ points $a = x_0 < x_1 < \cdots < x_{n-1} < x_n = b$ and slice the body into n "slabs", with the ith slab having width Δx_i. We then choose a sample point x_i^* in each subinterval $[x_{i-1}, x_i]$ and form the products $A(x_i^*)\Delta x_i$, each of which represents an approximation of the volume of the ith slab. So an approximation to the total volume V of the body is given by

$$V \approx \sum_{i=1}^{n} A(x_i^*)\Delta x_i.$$

The summation above is an example of a Riemann sum, and in the limit as $n \to \infty$ we write $V = \int_a^b A(x)\,dx$.

In summary, the formula $V = \int dV$ defines the volume of a given body in three-dimensional space, and if the body is bounded between the two planes $x = a$ and $x = b$, we can express this volume as

$$V = \int_a^b V'(x)\,dx = \int_a^b A(x)\,dx,$$

where $A(x)$ represents the area of a cross-sectional slice of the body through x. Note that this incorporates the fact that the rate of change of V with respect to x is $A(x)$. Note also the parallels with the area formulation of Chapter 5; namely, that the area A under the curve f over the interval $[a, b]$ is given by

$$A = \int_a^b A'(x)\,dx = \int_a^b f(x)\,dx.$$

TOPIC 2 Integrating Slices

With the philosophical definition of volume out of the way, we are ready to actually calculate volumes. We begin with a few familiar bodies in order to see how some well-known formulas can be developed using calculus.

Example 1 ✐

Find the volume V of a sphere of radius r.

Solution

To find the volume of a sphere by integrating slices, it is most convenient to define three coordinate axes with the origin at the center of the sphere, as shown in Figure 5. We can choose to slice up the sphere into parallel cross-sections any way we like; we'll use slices that are perpendicular to the x-axis, as in the previous discussion. Finally, we'll take advantage of the symmetry of a sphere and integrate cross-sectional areas from $x = 0$ to $x = r$ and then simply double the result.

Let $A(x)$ denote the area of a cross-section of the sphere passing through x that is parallel to the yz-plane. Then by the Pythagorean Theorem, the radius of the cross-section is $\sqrt{r^2 - x^2}$ and so $A(x) = \pi \left(\sqrt{r^2 - x^2} \right)^2 = \pi \left(r^2 - x^2 \right)$. So half of the total volume V is calculated as follows:

$$\frac{1}{2}V = \int_0^r \pi \left(r^2 - x^2 \right) dx$$

$$= \pi \left[r^2 x - \frac{x^3}{3} \right]_{x=0}^{x=r}$$

$$= \pi \left(r^3 - \frac{r^3}{3} \right) = \frac{2}{3}\pi r^3$$

The total volume is the familiar formula $V = \frac{4}{3}\pi r^3$.

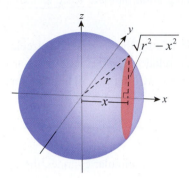
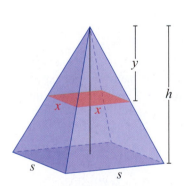

Figure 5

Example 2 ✐

Find the volume of a pyramid of height h whose base is a square of side length s.

Solution

If we slice the pyramid into cross-sections parallel to the base, as shown in Figure 6, each cross-section has area $A(y) = (sy/h)^2$. The easiest way to see this is to use similar triangles (see Figure 7) to express x in terms of y.

$$\frac{y}{h} = \frac{x}{s} \quad \Rightarrow \quad x = \frac{sy}{h}$$

Thus the area of the cross-section is $A(y) = x^2 = (sy/h)^2$.

We can now integrate the area differentials $A(y)\,dy$ from $y = 0$ to $y = h$ to obtain the volume.

$$V = \int_0^h \left(\frac{sy}{h} \right)^2 dy = \frac{s^2}{h^2} \int_0^h y^2 \, dy = \left(\frac{s^2}{h^2} \right) \left[\frac{y^3}{3} \right]_{y=0}^{y=h} = \frac{s^2 h^3}{3h^2} = \frac{1}{3}s^2 h$$

Figure 6

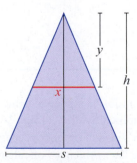

Figure 7

Note that, in Example 2, we did not explicitly indicate the orientation of the coordinate axes or the position of the pyramid with respect to them. Instead, we relied more upon the general statement that $V = \int dV$ and found, using the principle of similar triangles, a convenient way to express dV. This is a subtle but important point—don't let the mechanics of integration hide the basic idea that *the whole is the integral of its parts*!

In Exercise 2 you'll be asked to develop the formula for the volume of a circular cone, a formula that hints at a fundamental relationship between volumes of generalized cones and cylinders (Exercise 83 will further illuminate this relationship).

Example 3 ✎

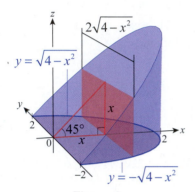

Figure 8

The curved wedge in Figure 8 is bounded by a right circular cylinder, the *xy*-plane, and a second plane passing through the *y*-axis at an angle of 45° to the *xy*-plane, as shown. The cylinder's axis is the *z*-axis and it has a radius of 2. What is the volume of the wedge?

Solution

If we decompose the wedge into slices perpendicular to the *x*-axis, each cross-section is a rectangle; since the areas of rectangles are especially easy to calculate, this seems like a promising way to slice up the solid. The area of each rectangle, $A(x)$, will depend on the point where it crosses the *x*-axis—we need to find a formula for $A(x)$.

In the *xy*-plane, the curves $y = \sqrt{4-x^2}$ and $y = -\sqrt{4-x^2}$ define the base of the wedge, so the width of each cross-sectional rectangle is $2\sqrt{4-x^2}$.

$$A(x) = (\text{height})(\text{width}) = x\left(2\sqrt{4-x^2}\right) = 2x\sqrt{4-x^2}$$

As a check of our work, note that $A(0) = A(2) = 0$, as we would expect, since the cross-sectional rectangles at $x = 0$ and $x = 2$ are actually just line segments. We are now ready to determine the volume.

$$V = \int_0^2 A(x)\,dx = \int_0^2 2x\sqrt{4-x^2}\,dx$$

$$= -\int_{u=4}^{u=0} u^{1/2}\,du \qquad \begin{aligned} u &= 4-x^2 \\ du &= -2x\,dx \end{aligned}$$

$$= \left[-\frac{2}{3}u^{3/2} \right]_{u=4}^{u=0}$$

$$= -\frac{2}{3}(0-8) = \frac{16}{3}$$

Our integral definition of volume provides a quick proof of a general principle formulated by the Italian mathematician Bonaventura Cavalieri (1598–1647) before Newton and Leibniz were even born. Cavalieri's insights into the properties of area and volume in many ways foreshadowed the later principles of calculus. You have already seen the version of Cavalieri's Principle as it applies to area in Exercise 99 of Section 5.5. The volume version is as follows.

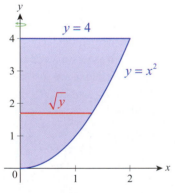

Figure 9

Theorem 🔍

Cavalieri's Principle

If two solids of the same height have cross-sectional areas that are equal at each possible distance from their respective bases, then their volumes are also equal.

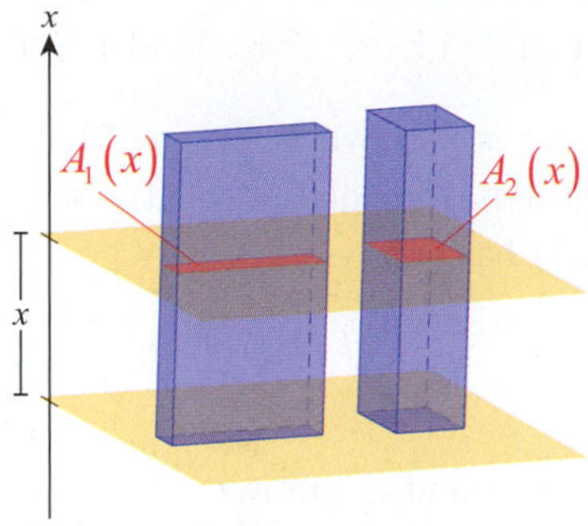

Proof ✏️

Let V_1 and V_2 represent the volumes of the two solids, and define the x-axis along the solids' respective heights as shown in Figure 9. Let $A_1(x)$ and $A_2(x)$ denote the cross-sectional areas of the two solids a distance x from their bases. Then since $A_1(x) = A_2(x)$ for each x,

$$V_1 = \int A_1(x)\,dx = \int A_2(x)\,dx = V_2.$$

TOPIC 3   Disks and Washers

Many solids are radially symmetric with respect to a central axis and are formed by revolving a plane region about that axis. Such objects are called **solids of revolution**, and it is often convenient to define their volumes in terms of cross-sections perpendicular to the axis of rotation. Those cross-sections typically take the form of either **disks** or **washers**, depending on whether the cross-sections are solid regions or not. In either case, formulas for the cross-sectional areas are usually easy to determine, as illustrated by the next four examples.

Example 4 ✎

The region bounded by $y = x^2$, $y = 4$, and $x = 0$ (see Figure 10) is rotated about the y-axis and generates the solid paraboloid in Figure 11. Find the volume of the solid.

Solution

Each cross-section perpendicular to the y-axis is solid and has the form of a disk. The radius of the cross-section passing through the point $(0, y)$ is $x = \sqrt{y}$, so the cross-sectional area function $A(y)$ is as follows:

$$A(y) = \pi(\text{radius})^2 = \pi\left(\sqrt{y}\right)^2 = \pi y$$

This leads to the volume of the solid.

$$V = \int_0^4 A(y)\,dy = \int_0^4 \pi y\,dy = \pi\left.\frac{y^2}{2}\right]_0^4 = 8\pi$$

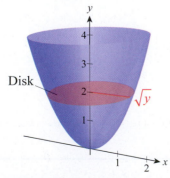

Figure 10 Plane Region

Figure 11 Solid of Revolution

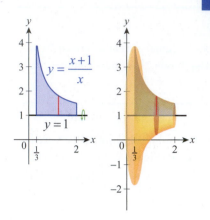

Figure 12
Plane Region and Solid of Revolution

Example 5 ✐

Find the volume of the solid formed by revolving the region bounded by $y = (x+1)/x$, $y = 1$, $x = \frac{1}{3}$, and $x = 2$ about the line $y = 1$.

Solution

The described region and the solid it generates are shown in Figure 12. Slicing the solid into disks perpendicular to the axis of revolution $y = 1$, we see that the disk passing through the point $(x, 1)$ has radius

$$r = \frac{x+1}{x} - 1 = \frac{1}{x}.$$

So $A(x) = \pi r^2 = \pi \left(1/x^2\right)$ and the volume can be found as follows:

$$V = \int_{1/3}^{2} A(x)\,dx = \int_{1/3}^{2} \pi\left(\frac{1}{x^2}\right)dx = \pi\left[-\frac{1}{x}\right]_{1/3}^{2} = \pi\left(-\frac{1}{2} + 3\right) = \frac{5\pi}{2}$$

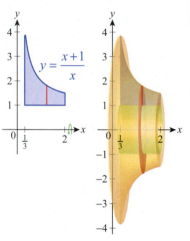

Figure 13
Plane Region and Solid of Revolution

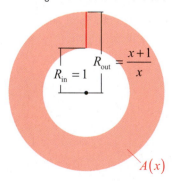

Figure 14 Washer Cross-Section

Example 6 ✐

Find the volume of the solid formed by revolving the region bounded by $y = (x+1)/x$, $y = 1$, $x = \frac{1}{3}$, and $x = 2$ about the x-axis.

Solution

The region is the same as the one from Example 5, but with a different axis of revolution. When we rotate the region about the x-axis, the generated solid has a hollow core. And when we slice the solid perpendicularly to the axis of revolution, we obtain cross-sections that are washers instead of disks (see Figures 13 and 14). The areas of these cross-sections are just as easily found, though—the area of a washer is $\pi R_{\text{out}}^2 - \pi R_{\text{in}}^2$, where R_{out} is the outer radius of the washer and R_{in} is the inner radius.

$$A(x) = \pi R_{\text{out}}^2 - \pi R_{\text{in}}^2 = \pi\left(\frac{x+1}{x}\right)^2 - \pi(1)^2 = \pi\left(\frac{x^2+2x+1}{x^2} - 1\right) = \pi\left(\frac{2}{x} + \frac{1}{x^2}\right)$$

Therefore the volume of the solid is as follows:

$$V = \int_{1/3}^{2} A(x)\,dx = \int_{1/3}^{2} \pi\left(\frac{2}{x} + \frac{1}{x^2}\right)dx$$

$$= \pi\left[2\ln x - \frac{1}{x}\right]_{1/3}^{2} = \pi\left[\left(2\ln 2 - \frac{1}{2}\right) - \left(2\ln\frac{1}{3} - 3\right)\right]$$

$$= \pi\left(2\ln 2 + 2\ln 3 - \frac{1}{2} + 3\right) \qquad \ln\frac{1}{3} = -\ln 3$$

$$= \pi\left(2\ln 6 + \frac{5}{2}\right) \qquad \ln 2 + \ln 3 = \ln 6$$

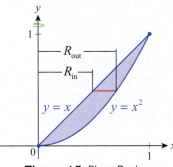

Figure 15 Plane Region

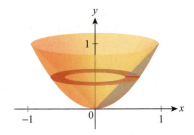

Figure 16 Solid of Revolution

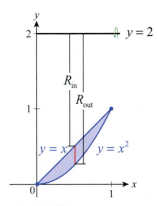

Figure 17 Plane Region

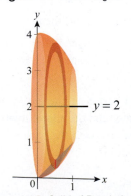

Figure 18 Solid of Revolution

Example 7 📝

Find the volume of the solid generated by revolving the region bounded by $y = x$ and $y = x^2$ about the following lines.

a. about the y-axis   **b.** about the line $y = 2$

Solution

a. The region to be revolved about the y-axis is shown in Figure 15 and the solid generated is shown in Figure 16. The red horizontal line is a representative segment that will rotate around the y-axis, so the outer radius is given by the curve $y = x^2$ and the inner radius by $y = x$. But we will be integrating cross-sectional washers with respect to y, thus we need expressions for the outer and inner radii in terms of y. They are, respectively, $R_{out} = \sqrt{y}$ and $R_{in} = y$.

$$A(y) = \pi R_{out}^2 - \pi R_{in}^2 = \pi \left(y - y^2 \right)$$

Now the volume of the solid can be calculated.

$$V = \int_0^1 A(y)\, dy = \int_0^1 \pi \left(y - y^2 \right) dy$$
$$= \pi \left[\frac{y^2}{2} - \frac{y^3}{3} \right]_0^1 = \pi \left(\frac{1}{2} - \frac{1}{3} \right) = \frac{\pi}{6}$$

b. The region is the same, but this time we are revolving about the horizontal line $y = 2$ so the cross-sectional washers will be functions of x and we will integrate their areas from $x = 0$ to $x = 1$. Note that, as shown in Figure 17, the outer and inner radii are, respectively, the distances from the line $y = 2$ to the curve $y = x^2$ and $y = x$. So $R_{out} = 2 - x^2$ and $R_{in} = 2 - x$.

$$A(x) = \pi R_{out}^2 - \pi R_{in}^2 = \pi \left[\left(2 - x^2 \right)^2 - \left(2 - x \right)^2 \right] = \pi \left(x^4 - 5x^2 + 4x \right)$$

The volume is

$$V = \int_0^1 A(x)\, dx = \int_0^1 \pi \left(x^4 - 5x^2 + 4x \right) dx$$
$$= \pi \left[\frac{x^5}{5} - \frac{5x^3}{3} + 2x^2 \right]_0^1 = \pi \left(\frac{1}{5} - \frac{5}{3} + 2 \right) = \frac{8\pi}{15}.$$

To summarize, volumes of solids of revolution can often be found by integrating cross-sectional slices that appear as either disks or washers. This is convenient because the areas of disks and washers are particularly easy to determine.

Definition 💡

The Disk and Washer Methods

For the purpose of this discussion, assume the given solid of revolution has an axis of revolution parallel to the x-axis, that the cross-sections with respect to x are either disks or washers, and that the solid is bounded by the planes $x = a$ and $x = b$. Then the volume V can be expressed in one of the following forms.

Disks $V = \int_a^b A(x)\,dx = \pi \int_a^b R^2\,dx$ (Determine R as a function of x)

Washers $V = \int_a^b A(x)\,dx = \pi \int_a^b \left(R_{out}^2 - R_{in}^2\right) dx$

(Determine R_{out} and R_{in} as functions of x)

TOPIC 4   Interlude: The Big Picture

It is easy to lose sight of the basic principles of integration, especially when first learning the techniques of setting up and evaluating integrals. Remember that the first principle of both area and volume, at least as far as integration is concerned, is that

$$A = \int dA \quad \text{and} \quad V = \int dV.$$

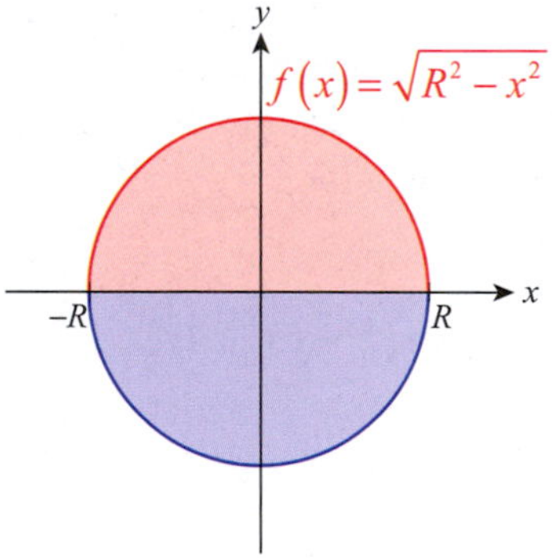

That is, if you can find a convenient (meaning integrable) expression for the differential in either case, then you have accomplished the most important task. We have extensively studied some of the more common ways of setting up area and volume integrals, but have by no means exhausted the possibilities. For any given problem, there may be many ways of setting up the integral, and the common approaches may not lead to the "best" integral.

Figure 19

Example 8 📝

You know that the area of a circle of radius R is $A = \pi R^2$, and one straightforward way to arrive at this formula is to integrate the function $f(x) = \sqrt{R^2 - x^2}$ over the interval $[-R, R]$ and then double the result (see Figure 19). However, this is not the easiest way (finding an antiderivative of f is best done using a technique we will learn in Chapter 7). Here are two other approaches.

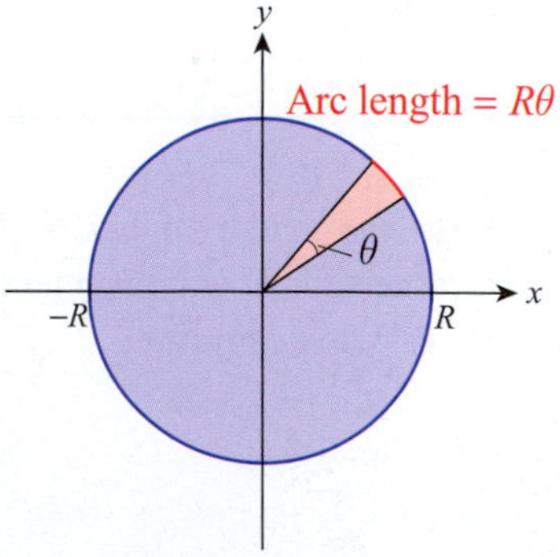

1. The area of a triangle is half the product of its base and height. A thin sector of our circle is approximately a triangle (and the approximation gets better and better as the angle approaches 0), with the base of the triangle equal to $R\,d\theta$ and the height equal to R.

$$dA = \left(\frac{1}{2}\right)(R)(R\,d\theta) = \frac{1}{2}R^2\,d\theta$$

$$A = \int dA = \int_0^{2\pi} \frac{1}{2}R^2\,d\theta = \frac{1}{2}R^2\,\theta\Big]_{\theta=0}^{\theta=2\pi} = \pi R^2$$

Figure 20

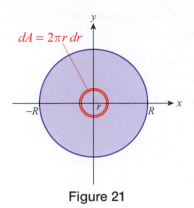

$dA = 2\pi r\, dr$

Figure 21

2. If we assume the formula for a circle's circumference is known, then we can decompose our given circle into a sequence of thin concentric rings with each one having area approximately equal to its circumference times dr.

$$dA = 2\pi r\, dr$$

$$A = \int dA = \int_0^R 2\pi r\, dr = 2\pi \frac{r^2}{2}\Bigg]_{r=0}^{r=R} = \pi R^2$$

Example 9 ✐

Assume the formula for the surface area of a sphere of radius r is known: $A = 4\pi r^2$ (we will soon find this formula on our own). Then an alternative way of finding the volume of a sphere of radius R is to decompose it into a sequence of thin concentric shells, each with volume approximately equal to its surface area times dr.

$$dV = 4\pi r^2\, dr$$

$$V = \int dV = \int_0^R 4\pi r^2\, dr = 4\pi \frac{r^3}{3}\Bigg]_{r=0}^{r=R} = \frac{4}{3}\pi R^3$$

6.1 **Exercises**

1. Find the volume of a skateboard ramp with a rectangular base of 4 meters by 1.5 meters, if each vertical cross-section is congruent to the "parabolic triangle" bounded by the x-axis, the vertical line $x = 4$, and the graph of $y = 0.1x^2$ (longitudinal units are meters).

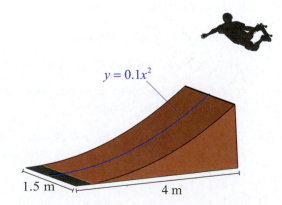

$y = 0.1x^2$

1.5 m   4 m

2. Mimic Example 2 to derive the formula for the volume of the right circular cone of height h, if the radius of its base is r.

3. Let's modify Example 3 by assuming that the cylinder is parabolic, that is, the base of the wedge is bounded by the graph of $x = 1 - y^2$ and the y-axis. Moreover, suppose that the second plane makes a $30°$ angle with the xy-plane. Find the volume of the curved wedge under these conditions.

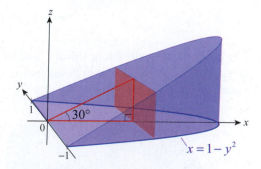

$x = 1 - y^2$

4. Suppose we "carve out" the faces of the pyramid of Example 2 so that the "side view" (i.e., the perpendicular cross-section that contains the altitude and is parallel to a pair of base edges) becomes the region bounded by the graphs of

$$y = \frac{4h}{s^2}\left(x - \frac{s}{2}\right)^2, \quad y = \frac{4h}{s^2}\left(x + \frac{s}{2}\right)^2, \quad \text{and the}$$

x-axis $(-s/2 \le x \le s/2)$. Find the volume of the resulting "concave pyramid."

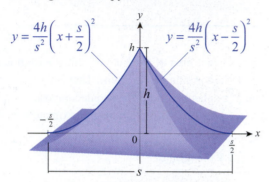

5–14 The base of a solid S is described in the xy-plane along with its cross-sections in a certain direction. Find the volume of S.

5. The base of S is the region bounded by the graph of $y = \sqrt{4-x}$ and the coordinate axes. The cross-sections perpendicular to the x-axis are squares.

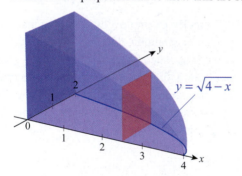

6. The base of S is a half disk of radius 1, and the cross-sections perpendicular to the diameter are squares.

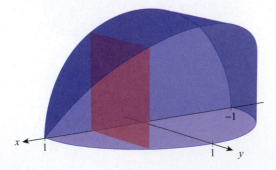

7. Repeat Exercise 6 if the cross-sections are equilateral triangles.

8. The base of S is a disk of radius 2, and the cross-sections perpendicular to the base are rectangles, each with a height that equals twice the width.

9. The base of S is a disk of radius 2, and the cross-sections perpendicular to the base are isosceles triangles, each with a height half as long as its base.

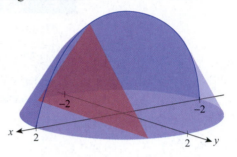

10. The base of S is bounded by $y = x^3$, the y-axis, and the line $y = 8$. Cross-sections perpendicular to the y-axis are squares.

11. The base of S is the region bounded by $y = x^4$ and $y = x^2$, $0 \le x \le 1$, and each cross-section perpendicular to the x-axis is an isosceles right triangle with the right angle's vertex sitting on $y = x^2$.

12. The base of S is bounded by the x-axis and $y = \sin x$, $0 \le x \le \pi$. Each of its cross-sections perpendicular to the x-axis is an isosceles triangle of altitude 1.

13. The base of S is bounded by $y = \cos x$ and $y = \left(\frac{2}{\pi}x\right)^2 - 1$, $-\pi/2 \le x \le \pi/2$. Each of its cross-sections perpendicular to the x-axis is an isosceles triangle of altitude 2.

14. The base of S is bounded by $y = \sqrt{4-x}$, $y = 0$, and $x = 0$. Each of its cross-sections perpendicular to the x-axis is a rectangle of perimeter 10.

15.* A solid modeling the nose of a race car has a base in the shape of an isosceles trapezoid with a height of 180 cm and base lengths of 80 cm and 40 cm, respectively. Suppose the vertical cross-section through the axis of symmetry of the base (i.e., the cross-section that "cuts the model in half" longitudinally) is the region bounded by the graph of $y = 3\sqrt{180 - x}$ and the coordinate axes. All vertical cross-sections perpendicular to the axis of symmetry of the base are rectangles. Find the volume of this model.

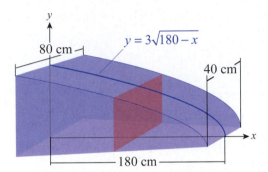

16. Find the volume of water remaining in a spherical reservoir of radius r if the water's depth is $r/3$.

17. A piece in a wooden toy set is a sphere of radius 2 cm, with a cylindrical hole of radius 1 cm drilled through the center. Find the volume of this piece.

18.* Two plastic pipes (with circular cross-sections) of radius r inches cross at right angles. Find the volume of the solid region that is common to both pipes. (**Hint:** Find cross-sections that are squares.)

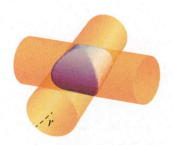

19. Find the volume of the "bowl" that results from rotating the graph of $y = x^{3/2}$, $-4 \le x \le 4$, around the y-axis.

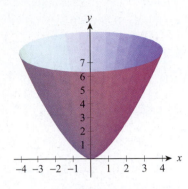

20. Find the volume of the solid that results from rotating the graph of $y = 1/x$, $1 \le x \le a$, around the x-axis (this solid is known as *Gabriel's horn*). What can you say if $a \to \infty$?

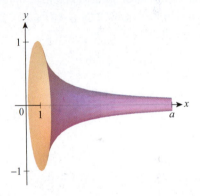

21–28 Find the volume of the solid that results from rotating the region between the graph of the given equation and the x-axis about the x-axis over the indicated interval.

21. $y = -\dfrac{1}{3}x + 3$; $\quad 0 \le x \le 3$

22. $y = 2x + \dfrac{2}{3}$; $\quad 0 \le x \le 1$

23. $y = x^2 + \dfrac{1}{2}$; $\quad 0 \le x \le 2$

24. $y = x^3$; $\quad -1 \le x \le 3$

25. $y = -\sqrt{5 - x}$; $\quad 1 \le x \le 4$

26. $y = x^{5/3}$; $\quad -1 \le x \le 1$

27. $y = \sqrt{9 - x^2}$; $\quad 0 \le x \le 3$

28. $y = \sec x$; $\quad 0 \le x \le \pi/4$

29–36 Find the volume of the solid that results from rotating the region bounded by the graphs of the equations about the y-axis.

29. $y = \dfrac{1}{2}x - 1,\quad x = 0,\quad y = 0$

30. $2y + 6x = 9,\quad x = 0,\quad y = 0$

31. $y = (x - 1)^2,\quad x = 1,\quad x = 4,\quad y = 0$

32. $y = x^3,\quad x = 2,\quad y = 0$

33. $y = \sqrt[3]{x} + 1,\quad x = 0,\quad y = 3$

34. $y = x^{3/5} - 2,\quad x = 0,\quad y = -2,\quad y = 6$

35. $y = \sqrt{4 - x^2},\quad x = 0,\quad y = 0$

36.* $y = \csc^{-1} x,\quad x = 0,\quad y = \pi/4,\quad y = \pi/2$

37–48 Find the volume of the solid generated by rotating the region bounded by the graphs of the given equations about **a.** the x-axis and **b.** the line $y = -1$.

37. $y = \dfrac{1}{2}x,\quad y = 0,\quad x = 2$

38. $y - 3x - 1 = 0,\quad y = 0,\quad x = 1$

39. $y = x^3,\quad y = 0,\quad x = 1$

40. $y = 2\sqrt{x},\quad y = 0,\quad x = 4$

41. $y = \dfrac{x^4}{4},\quad y = x^2,\quad x = 0,\quad x = 2$

42. $y = \dfrac{2}{x + 1},\quad y = 0,\quad x = 0,\quad x = 2$

43. $y = x^{5/3},\quad y = 0,\quad x = 0,\quad x = 1$

44. $y = e^x,\quad y = 0,\quad x = 0,\quad x = 1$

45. $y = \sqrt[3]{x},\quad y = 0,\quad x = 1,\quad x = 8$

46. $y = \dfrac{1}{2}x^{2/3},\quad y = \dfrac{1}{2}x^{3/2},\quad x = 0,\quad x = 1$

47. $y = \sqrt{1 - (x - 1)^2},\quad y = 0,\quad x = 2$ (**Hint:** When evaluating $\int_0^1 \sqrt{1 - (x - 1)^2}\, dx$, use a well-known formula from geometry.)

48. $y = \cos x,\quad y = 0,\quad x = 0,\quad x = \pi/2$
(**Hint:** When evaluating the integral, use the identity $\cos^2 x = (1 + \cos 2x)/2$ or the identity $\cos^2 x + \sin^2 x = 1$.)

49–57. In each of Exercises 39–47, rotate the region about **a.** the y-axis and **b.** the line $x = -2$, and find the volume of the resulting solid. (For Exercise 54, use the antiderivative $\int (\ln x)^2\, dx = 2x - 2x \ln x + x (\ln x)^2$. For Exercise 57, see the hint given in Exercise 47.)

58–71 Find the volume of the solid that results from rotating the region bounded by the graphs of the equations about the indicated line. Use any of the methods discussed in this section.

58. $y = -2x + 5,\quad y = -1,\quad x = 0;\quad$ about $x = -1$

59. $y - 4x - 3 = 0,\quad y = 0,\quad x = 0,\quad x = 3;$ about $x = -3$

60. $y = x^2 - \sqrt{8x},\quad y = 0,\quad x = 0,\quad x = 2;$ about the x-axis

61. $x = 2y(y - 5),\quad x = 0;\quad$ about the y-axis

62. $x\sqrt{1 + y^2} = 1,\quad x = 0,\quad y = 0,\quad y = \sqrt{3};$ about the y-axis

63. $x = y^2,\quad y = x^2;\quad$ about $y = 1$

64. $y = \sqrt[4]{2x + 1},\quad y = 0,\quad x = 4;\quad$ about the x-axis

65. $y = (x - 2)\sqrt{x^2 + 1},\quad y = 0,\quad 0 \le x \le 2;$ about the x-axis

66. $y = \sqrt{-x\cos(x^2 + \pi)},\quad y = 0,\quad x = 0,\quad x = \sqrt{\pi/2};$ about the x-axis

67. $x = \sqrt{y} + 2,\quad 4x = y^2,\quad y = 0,\quad y = 4;$ about the y-axis

68. $y = \sin^{-1} x,\quad y = 0,\quad x = 1;\quad$ about the y-axis (See the hint given in Exercise 48.)

69. $y = \cos x,\quad y = \sin x,\quad x = 0,\quad x = \pi/4,\quad y = 0;$ about the x-axis

70. $x = e^y + e^{-y},\quad x = 0,\quad y = 0,\quad y = 1;$ about $x = -e$

71. $y = \csc x,\quad y = \cot x,\quad x = \pi/4,\quad x = \pi/2;$ about the x-axis

72–79 The given integral represents the volume of a solid of revolution. Describe the solid. (Do not evaluate the integral.)

72. $\int_0^{\sqrt{3}} 3\pi y^2 \, dy$

73. $\int_0^4 \pi x \, dx$

74. $\int_0^2 \pi \left(4 - y^2\right) dy$

75. $\int_0^1 \left(\pi e^2 - \pi e^{2x}\right) dx$

76. $\int_0^1 \pi \left(\arctan y\right)^2 dy$

77. $\pi \int_0^1 \left(\sqrt{x} - x^8\right) dx$

78. $\int_0^\pi \pi \sin^2 x \, dx$

79. $\pi \int_1^2 \left[\left(\log_2 x\right)^2 - \left(x - 1\right)^2\right] dx$

80. A regular hexagon of side length 1 lies in the xy-plane so that its center of symmetry coincides with the origin. Suppose the hexagon moves 6 units vertically upward so that its center rides on the line perpendicular to the xy-plane (the line is actually the z-axis), the hexagon at any time is parallel to the xy-plane and makes three revolutions around its center (at a constant angular speed) as it moves. Find the volume of the resulting corkscrew-type solid.

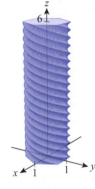

81.* A student is drinking from a standard soda can. Find the volume of the remaining soda when the can's axis of symmetry makes a 15° angle with the horizontal direction. Express your answer in milliliters. (**Hint:** Approximate the soda can with a right circular cylinder of radius 1.26 inches and height of 4.7 inches. Assume that when the can is held as described in the problem, the soda is level with the lowest point of the top rim of the can.)

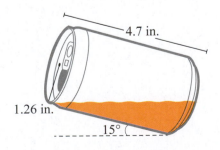

82.* Let D denote the depth of water in a bowl that has a tiny hole in the bottom. According to Torricelli's Law, water drains through the hole at the rate of $dV/dt = -m\sqrt{D}$, where D is a positive constant. Find the rate at which the water level is decreasing if the bowl is generated by rotating around the y-axis the graph of **a.** $y = cx^2$ and **b.** $y = cx^4$, $c > 0$. Which of the two rates do you think can be used as a "water clock," and why?

83. Although we typically work with circular cones, in a more general sense, cones don't have to be circular. In fact, if C is a simple closed curve in a plane S, and M is a point not in the same plane, then the solid generated by a line passing through M and moving along C, is said to be a cone (the region bounded by C is the base, M is the vertex of the cone, the distance between M and S is its height). Use the results and methods of this section to prove that the volume of a cone of base area B and height h is $V = \frac{1}{3}Bh$. (**Hint:** Mimic Example 2, using the fact that by similarity, in the general case we have

$$\frac{A(y)}{B} = \frac{y^2}{h^2}, \text{ that is, } A(y) = \frac{B}{h^2} \cdot y^2.)$$

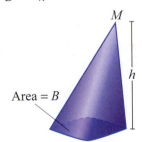

84.* Consider the circle in the xy-plane defined by the following equation.

$$\left(x - R\right)^2 + y^2 = r^2 \quad (R > r)$$

By rotating the region bounded by this circle around the y-axis, prove that the volume of the generated solid, a torus, is $V = 2\pi^2 R r^2$.

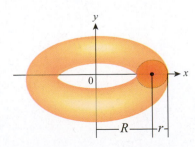

85.* Rotate about the x-axis the region between the graphs of $y = \sqrt{4-x}$ and $y = c$ $(0 \le c \le 2)$ over the interval $[0,4]$. Find the value of c that minimizes the volume of the resulting solid. For what c-value is the volume maximal?

6.1 Technology Exercises

86. Use a computer algebra system to repeat Exercise 85 for the region bounded by $y = \cos x$, $y = 0$, $x = 0$, and $x = \pi/2$ rotating about the line $y = c$ $(0 \le c \le 1)$.

87.* With the aid of a computer algebra system, revisit Exercise 81, assuming this time that the angle between the can's axis of symmetry and the horizontal direction is $10°$.

88–92 Use a computer algebra system to find (or approximate) the volume of the solid generated by rotating the region bounded by the graphs of the given equations about the indicated axis.

88. $y = \dfrac{1}{1+x^2}$, $\quad y = 0$, $\quad x = -1$, $\quad x = 1$; $\quad$ about the x-axis

89. $y = \arcsin x$, $\quad y = 0$, $\quad x = 0$, $\quad x = 1$; $\quad$ about the x-axis

90. $y = e^x$, $\quad x = 0$, $\quad y = 1$, $\quad y = e^2$; $\quad$ about the y-axis

91. $y = \dfrac{1}{\log x}$, $\quad y = 0$, $\quad x = 10$, $\quad x = 100$; $\quad$ about the x-axis

92. $x = e^{1-y^2}$, $\quad y = 0$, $\quad y = 1$, $\quad x = 0$; $\quad$ about the y-axis

93–94 Use a computer algebra system to sketch the region bounded by the graphs of the given equations. Approximate the intersection points, then find an approximation for the volume of the solid generated by rotating the region about the x-axis.

93. $y = x\cos^2 x$, $\quad y = 2x(2-x)$

94. $y = x+3$, $\quad y = e^x + e^{-x} - 2$

6.2 Finding Volumes Using Cylindrical Shells

TOPICS

1. The shell method

2. Disks, washers, and shells compared

At the end of the last section, we looked at a way of finding the volume of a sphere by mentally decomposing it into a sequence of thin concentric shells; this allowed us to express the volume differential dV in an easily integrated form. In this section, we will study a commonly used more general form of shell decomposition.

TOPIC 1   The Shell Method

As we have seen, it can be useful to decompose a solid of revolution into a sequence of disks or washers. But we have also seen that, even if a volume integral can be expressed using the disk or washer method, the resulting integral may not be as easy to evaluate as we would like—it's possible that a different way of thinking about the volume would lead to an equivalent, but easier, integral. Another way of exploiting the radial symmetry of a solid of revolution is to visualize it as being made up of a sequence of cylindrical shells, all with the axis of revolution serving as the common axis.

Remember that the basis for all our volume computations is the statement $V = \int dV$, so if we break down a given solid into a sequence of thin concentric cylinders, dV must represent the volume of each (vanishingly) thin cylindrical shell. As always, we can approximate dV with ΔV, which in this case will be the volume of a cylindrical shell of small, but nonzero, thickness.

We will use the following example to formulate the general shell method.

Example 1 ✎

Find the volume of the solid formed by revolving the region bounded by $f(x) = -x^2 + 5x - 4$ and the x-axis about the y-axis.

Solution

The bounded region described is shown in Figure 1 and the solid obtained by revolving it is shown in Figure 2. Also shown in Figure 2 is a representative cylindrical shell of thickness Δx; our method relies on being able to build up the entire solid layer by layer with such shells. The shell shown has a nonzero thickness, so it's just a rough approximation of a thinner shell with thickness dx. But for the purpose of sketching what we are doing and setting up our integral, such an approximation is exactly what we need.

If we were to remove the representative shell, slice it from top to bottom vertically along a straight line, and lay it out flat, the cylindrical shell would take the form of a rectangular slab of width $2\pi x_i^*$, height $f(x_i^*)$, and thickness Δx, where x_i^* is a sample point in the selected subinterval and the subinterval has width Δx (see Figure 3). The volume ΔV_i of the slab is the product of its width, height, and thickness.

$$\Delta V_i = \left(2\pi x_i^*\right) f\left(x_i^*\right)\Delta x$$

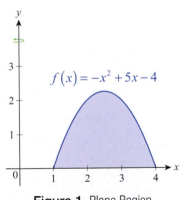

Figure 1   Plane Region

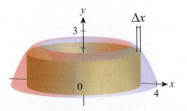

Figure 2   Solid of Revolution

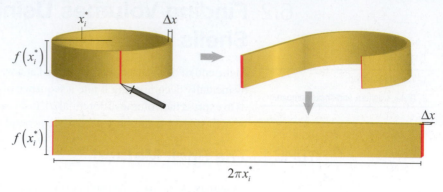

Figure 3

If we now construct an approximation of our solid with a sequence of n cylindrical shells, the volume of the solid is approximately the sum of the volumes of the n shells.

$$V \approx \sum_{i=1}^{n} \Delta V_i = \sum_{i=1}^{n} \left(2\pi x_i^*\right) f\left(x_i^*\right) \Delta x$$

And, as we have by now come to expect, the exact volume of the solid is the limit of the above Riemann sum as n goes to infinity (and consequently as $\Delta x \to 0$):

$$V = \lim_{n \to \infty} \sum_{i=1}^{n} \left(2\pi x_i^*\right) f\left(x_i^*\right) \Delta x$$

$$= \int_{1}^{4} (2\pi x) f(x)\, dx$$

$$= 2\pi \int_{1}^{4} x\left(-x^2 + 5x - 4\right) dx$$

$$= 2\pi \int_{1}^{4} \left(-x^3 + 5x^2 - 4x\right) dx$$

$$= 2\pi \left[-\frac{x^4}{4} + \frac{5}{3}x^3 - 2x^2 \right]_{1}^{4} = 2\pi \left[\frac{32}{3} - \left(-\frac{7}{12}\right) \right] = \frac{45\pi}{2}$$

Once we have worked through the derivation of the Riemann sum and understand the meaning of its components, as we did in Example 1, we can begin to generalize the shell method as follows.

Definition 💡

The Shell Method

Assume that the given solid of revolution can be formed by revolving the plane region R around the line $x = L$, that R is bounded below by the x-axis and above by the graph of $f(x)$ over the interval $[a, b]$, and that $L \le a \le b$. Then, using the shell method, the volume V can be expressed as follows.

$$V = \int_{a}^{b} 2\pi \left(\begin{array}{c} \text{radius} \\ \text{of shell} \end{array} \right) \left(\begin{array}{c} \text{height} \\ \text{of shell} \end{array} \right) dx = 2\pi \int_{a}^{b} (x - L) f(x)\, dx \qquad \text{See Figure 4.}$$

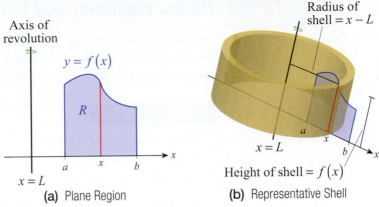

Axis of revolution

$y = f(x)$

R

$x = L$

$a \quad x \quad b$

(a) Plane Region

Radius of shell $= x - L$

$x = L$

a

b

Height of shell $= f(x)$

(b) Representative Shell

Figure 4

In this formula, the factor $2\pi \cdot (\text{radius of shell})$ is the circumference of a given shell element and corresponds to the width of the rectangular slab if the shell is cut and laid out flat. In Example 1, the region R was rotated about the line $x = 0$, so the radius of each shell was simply the distance x from the origin; more generally, $x - L$ is the radius of the shell that passes through x and has $x = L$ as its axis. The height of that same shell is $f(x)$, and dx corresponds, as usual, to Δx in the Riemann sum (which in this case can be interpreted as the thickness of each shell element).

In general, the solid of revolution under consideration may not have a vertical axis of revolution, the axis of revolution may lie to the right of the interval $[a, b]$ of integration instead of to the left, or the region R may be defined by more than one function. But as long as the radii and heights of the shells can be determined, the formula

$$V = \int_a^b 2\pi \left(\begin{array}{c} \text{radius} \\ \text{of shell} \end{array} \right) \left(\begin{array}{c} \text{height} \\ \text{of shell} \end{array} \right) dx$$

still applies and can be adapted as appropriate. The following examples will illustrate the process of finding the volume of a solid using this formula.

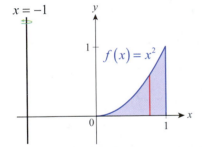

$x = -1$

y

$f(x) = x^2$

$0 \quad 1$

x

Figure 5

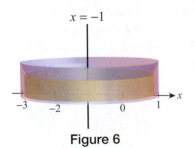

$x = -1$

$-3 \quad -2 \quad 0 \quad 1$

x

Figure 6

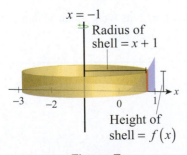

$x = -1$

Radius of shell $= x + 1$

$-3 \quad -2 \quad 0 \quad 1$

x

Height of shell $= f(x)$

Figure 7

Example 2

Find the volume of the solid formed by revolving the region shown in Figure 5 around the line $x = -1$.

Solution

The solid formed is shown in Figure 6. The height of any given shell element is simply $f(x) = x^2$, $0 \le x \le 1$, but the radius of the same shell is $x - (-1)$, or $x + 1$ (see Figure 7). So the volume of the solid generated is as follows:

$$V = \int_0^1 2\pi \left(\begin{array}{c} \text{radius} \\ \text{of shell} \end{array} \right) \left(\begin{array}{c} \text{height} \\ \text{of shell} \end{array} \right) dx$$

$$= \int_0^1 2\pi (x + 1) x^2 \, dx = 2\pi \int_0^1 (x^3 + x^2) \, dx$$

$$= 2\pi \left[\frac{x^4}{4} + \frac{x^3}{3} \right]_0^1 = \frac{7\pi}{6}$$

TOPIC 2   Disks, Washers, and Shells Compared

We will conclude this section with some illustrations of volumes in which the disk/washer method and the shell method are compared. In some cases, one method may be clearly preferable to the other; in other cases, both are viable.

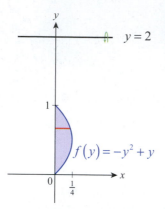

Figure 8

Example 3 ✎

Find the volume of the solid formed by revolving the region shown in Figure 8 around the line $y = 2$.

Solution

The height of each shell in this example is the distance between the y-axis and the function $f(y) = -y^2 + y$, for $0 \le y \le 1$. One such height is shown in red in Figure 8. And since the region is being revolved around the line $y = 2$, the radius of any given shell is given by $2 - y$ (see Figure 10). Note that this corresponds to a maximum radius of 2 when $y = 0$ and a minimum radius of 1 when $y = 1$.

The volume of the solid generated (shown in Figure 9) can now be found.

$$V = \int_0^1 2\pi \left(\begin{array}{c} \text{radius} \\ \text{of shell} \end{array} \right) \left(\begin{array}{c} \text{height} \\ \text{of shell} \end{array} \right) dy$$

$$= 2\pi \int_0^1 (2-y)(-y^2+y)\, dy$$

$$= 2\pi \int_0^1 (y^3 - 3y^2 + 2y)\, dy$$

$$= 2\pi \left[\frac{y^4}{4} - y^3 + y^2 \right]_0^1 = \frac{\pi}{2}$$

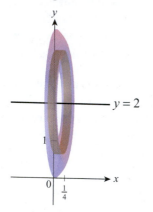

Figure 9

In this case, the shell method is far preferable to the washer method. If we were to attempt to find the volume using the washer method, we would need to determine R_{in} and R_{out}, the inner and outer radii of the washers. These will be functions of x, for $0 \le x \le \frac{1}{4}$, with R_{in} representing the distance from $y = 2$ to the upper half of the leftward-opening parabola $x = -y^2 + y$ and R_{out} representing the distance from $y = 2$ to the lower half. In Exercise 60 you will show that

$$R_{\text{in}} = 2 - \left(\frac{1 + \sqrt{1-4x}}{2} \right) \quad \text{and} \quad R_{\text{out}} = 2 - \left(\frac{1 - \sqrt{1-4x}}{2} \right),$$

and that $\pi \int_0^{1/4} \left(R_{\text{out}}^2 - R_{\text{in}}^2 \right) dx$ yields a volume of $\pi/2$ as well, albeit with a lot more effort.

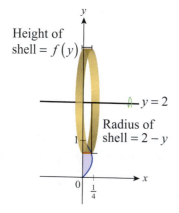

Figure 10

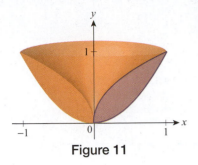

Figure 11

Example 4 ✍

Find the volume of the solid formed by revolving the region bounded between $y = \sqrt{x}$ and $y = x^2$ around the y-axis.

Solution

The solid formed is shown in Figure 11 and its volume is found in two ways.

Method 1: The Shell Method

The height of a typical cylindrical shell is given by $\sqrt{x} - x^2$ and x ranges between 0 and 1. The radius of the shell whose height is $\sqrt{x} - x^2$ is simply x, since the region is revolving around the line $x = 0$ (see Figure 12). Hence, the volume of the solid can be found as follows:

$$V = \int_0^1 2\pi x \left(\sqrt{x} - x^2 \right) dx = 2\pi \int_0^1 \left(x^{3/2} - x^3 \right) dx$$

$$= 2\pi \left[\frac{2}{5} x^{5/2} - \frac{x^4}{4} \right]_0^1 = \frac{3\pi}{10}$$

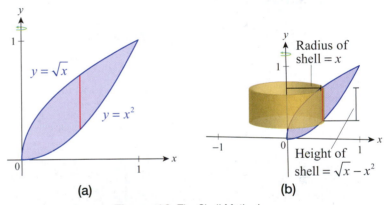

(a) **(b)**

Figure 12 The Shell Method

Method 2: The Washer Method

In this example, the washer method applies just as well. The setup differs, of course, as we need to describe the inner and outer radii of a sequence of vertically stacked washers. Using this formulation, the outer radius R_{out} is the function $\sqrt{y}$ and the inner radius R_{in} is the function y^2. Note that our integration will be with respect to y over the interval $0 \le y \le 1$; Figure 13 illustrates the region and a typical washer cross-sectional cut of the region. The volume computation is as follows:

$$V = \pi \int_0^1 \left(R_{\text{out}}^2 - R_{\text{in}}^2 \right) dy = \pi \int_0^1 \left(y - y^4 \right) dy$$

$$= \pi \left[\frac{y^2}{2} - \frac{y^5}{5} \right]_0^1 = \frac{3\pi}{10}$$

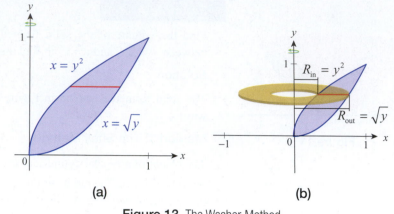

(a)   (b)

Figure 13 The Washer Method

6.2 **Exercises**

1–6 Use the shell method to find the volume of the solid obtained by revolving the shaded region about the indicated line.

1.

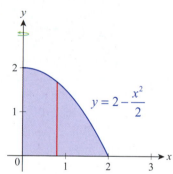

$y = 2 - \dfrac{x^2}{2}$

2.

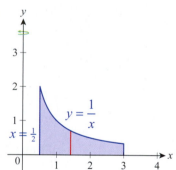

$y = \dfrac{1}{x}$

$x = \dfrac{1}{2}$

3.

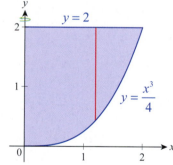

$y = 2$

$y = \dfrac{x^3}{4}$

4.

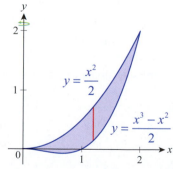

$y = \dfrac{x^2}{2}$

$y = \dfrac{x^3 - x^2}{2}$

5.

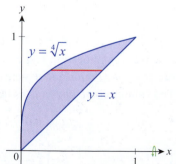

$y = \sqrt[4]{x}$

$y = x$

6.

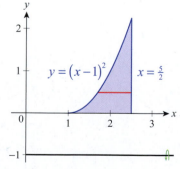

$y = (x-1)^2$

$x = \dfrac{5}{2}$

7–8 Use the shell method to find the volume of the solid that is generated by revolving the region bounded by $y = 0$ and the graph of f about the x-axis over the indicated interval.

7. $f(x) = 2 - 0.5x;\quad 1 \le x \le 4$

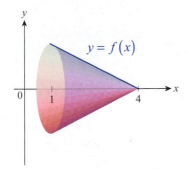

8. $f(x) = \sqrt{x - 1};\quad 1 \le x \le 5$

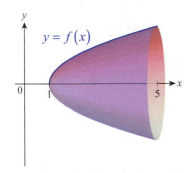

9–26 Find the volume of the solid generated by rotating the region bounded by the graphs of the given equations about the y-axis.

9. $y = -2x + 6,\quad x = 0,\quad y = 0$

10. $y - x = 0,\quad y = 0,\quad y = 6$

11. $2y - 5x + 8 = 0,\quad y = 0,\quad x = \dfrac{8}{5},\quad x = 3$

12. $3y - x = 0,\quad x = 0,\quad 3y + 2x - 9 = 0$

13. $y = x^2,\quad x = 0,\quad y = 9$

14. $y = x^2 - 1,\quad y = 2x - 1$

15. $y = x^2 - 2x + 3,\quad x = 0,\quad 3y = 2x + 12$

16. $y = \sqrt{4 - x},\quad x = 0,\quad y = 0$

17. $y = x^{3/5},\quad y = 0,\quad x = 1$

18. $y = \sqrt{4 - x^2},\quad x = 0,\quad y = 0$

19. $y = (x - 2)^3,\quad x = 2,\quad y = 8$

20. $y = \sqrt[3]{x} + 1,\quad x = 0,\quad x = 1,\quad y = 0$

21. $y = \cos^2(x^2),\quad y = \sin^2(x^2),\quad x = 0,\quad x = \dfrac{\sqrt{\pi}}{2}$

22. $y = \sec^2(x^2),\quad x = 0,\quad x = \dfrac{\sqrt{\pi}}{2},\quad y = 0$

23. $y = \dfrac{1}{(x+1)^2},\quad x = 0,\quad x = 1,\quad y = 0$

24. $y = \dfrac{1}{3\sqrt{x}},\quad x = 1,\quad x = 4,\quad y = 0$

25. $y = \dfrac{e^x}{\pi x},\quad x = 1,\quad x = 2,\quad y = 0$

26. $y = \begin{cases} \dfrac{1 - \cos x}{x} & \text{if } x > 0 \\ 0 & \text{if } x = 0 \end{cases},\quad y = 0,\quad x = 0,\quad x = 2\pi$

27–38 Find the volume of the solid generated by rotating the region bounded by the graphs of the given equations about the x-axis.

27. $7y + 4x = 28,\quad x = 0,\quad y = 0$

28. $y - x + 1 = 0,\quad y + 2x - 8 = 0,\quad y = 0$

29. $y = \sqrt{x},\quad y = x - 2,\quad y = 0$

30. $y = 2x^2 - 8,\quad x = 0,\quad y = 0$

31. $y = \sqrt[3]{3 - x},\quad y = \dfrac{1}{2}x,\quad y = 0,\quad x = 0,\quad x = 3$

32. $x = \dfrac{1}{y + 2},\quad x = 0,\quad y = 0,\quad y = 2$

33. $2x\sqrt{y + 1} = 1,\quad x = 0,\quad y = 0,\quad y = 3$

34. $y\sqrt{y^3 + 1} = x,\quad x = 0,\quad y = 0,\quad y = 2$

35. $\dfrac{1}{y^2 + 1} = x,\quad x = 0,\quad y = 0,\quad y = 4$

36. $x = \dfrac{1}{y(y^2 + 1)},\quad x = 0,\quad y = 1,\quad y = \sqrt{3}$

37. $y = \sqrt{\ln(1 - x) + 1},\quad x = 0,\quad y = 0$

38. $y = \sqrt{\arcsin x},\quad x = 0,\quad y = \sqrt{\dfrac{\pi}{2}}$

39–44 Revolve the region bounded by the graphs of the equations about the given line and use the shell method to find the volume of the resulting solid.

39. $y = 5 - 2x$, $x = 0$, $y = 0$

    a. About $x = -2$ **b.** About $y = -1$

    c. About $x = 2.5$ **d.** About $y = 7$

40. $y = \dfrac{3}{2}x$, $x = 2$, $y = 0$

    a. About $x = -1$ **b.** About $y = -2$

    c. About $x = 2$ **d.** About $y = 3$

41. $y = \sqrt{x}$, $y = 0$, $x = 4$

    a. About $x = -1$ **b.** About $y = -\frac{1}{2}$

    c. About $x = 6$ **d.** About $y = 2$

42. $y = x$, $y = x^3$, $x = 0$, $x = 1$

    a. About $x = -3$ **b.** About $y = -\frac{1}{2}$

    c. About $x = \frac{3}{2}$ **d.** About $y = 2$

43. $y = 4x - x^2$, $y = 0$

    a. About $x = -1$ **b.** About the x-axis

    c. About $x = 2$ **d.** About $y = -1$

44. $y = \sqrt{4x - x^2}$, $y = 0$

    a. About the x-axis   **b.** About $y = -1$

    c. About $x = 2$ **d.** About $y = 2$

45–50 Using the shell method, find a formula for the volume of the solid that results when the region bounded by the graphs of the equations is revolved about the indicated axis. Do not evaluate the integral.

45. $y = 2e^{-x} - 1$, $x = 0$, $x = \ln 2$, $y = 0$; about the y-axis

46. $y = 2e^{-x} - 1$, $x = 0$, $x = \ln 2$, $y = 0$; about the x-axis

47. $y = \arcsin x$, $x = 0$, $x = 1$, $y = 0$; about the y-axis

48. $y = \arcsin x$, $x = 0$, $x = 1$, $y = 0$; about the x-axis

49. $y = \tan x$, $x = 0$, $y = \sqrt{3}$; about the y-axis

50. $y = \tan x$, $x = 0$, $y = \sqrt{3}$; about the x-axis

51–59 The given integral represents the volume of a solid of revolution. Describe the solid. (Do not evaluate the integral.)

51. $4\pi \displaystyle\int_0^2 (2x - x^2)\, dx$

52. $2\pi \displaystyle\int_0^{\sqrt{3}} (3y - y^3)\, dy$

53. $4\pi \displaystyle\int_0^1 y\sqrt{1 - y^2}\, dy$

54. $4\pi \displaystyle\int_2^4 x\sqrt{1 - (x - 3)^2}\, dx$

55. $2\pi \displaystyle\int_0^1 (1 + x)(1 - x^4)\, dx$

56. $2\pi \displaystyle\int_0^1 (y + 2)(e - e^y)\, dy$

57. $\pi \displaystyle\int_{\pi/2}^{\pi} (2x - \pi)\sin x\, dx$

58. $2\pi \displaystyle\int_0^{1/2} 3y\, dy + 2\pi \displaystyle\int_{1/2}^1 y\left(\dfrac{1}{y^2} - 1\right) dy$

59. $\pi \displaystyle\int_{\pi/4}^{3\pi/4} (2x + 2)\csc x\, dx$

60. Use the washer method to solve the problem posed in Example 3. (**Hint:** Completing the square for $f(y) = -y^2 + y$ yields $f(y) = x = \frac{1}{4} - (y - \frac{1}{2})^2$. Express y in terms of x to obtain the graphs bounding the given region. Since the axis of rotation is $y = 2$, you will obtain $R_{\text{in}} = 2 - \left(\dfrac{1 + \sqrt{1 - 4x}}{2}\right)$ and $R_{\text{out}} = 2 - \left(\dfrac{1 - \sqrt{1 - 4x}}{2}\right)$. Simplify the integral $V = \pi \displaystyle\int_0^{1/4} (R_{\text{out}}^2 - R_{\text{in}}^2)\, dx$ and verify that it equals $\pi/2$.)

61–66 Use the shell method or the disk/washer method to find the volume of the solid obtained by revolving the region bounded by the graphs of the equations about the given axis. Choose the method that seems to work best.

61. $y(x + 1) = 1$, $y = 0$, $x = 0$, $x = 1$; about the x-axis

62. $y = 1 - (x - 1)^4$, $\quad x = 0$, $\quad y = 0$;
about the y-axis

63. $y = x^3 - 6x^2 + 8x$, $\quad y = 0$, $\quad x = 0$, $\quad x = 2$;
about the x-axis

64. $y = 2x + \dfrac{1}{x^2}$, $\quad y = 0$, $\quad x = 1$, $\quad x = 2$;
about the y-axis

65. $y = 3 - \dfrac{x^2}{3}$, $\quad y = 0$;   about the x-axis

66. $y = \begin{cases} \dfrac{\sin x}{x} & \text{if } x > 0 \\ 1 & \text{if } x = 0 \end{cases}$, $\quad y = 0$, $\quad x = 0$, $\quad x = \pi$;

about the y-axis

67. The solid that remains after "chopping off" the upper part of a right circular cone by a plane parallel to its base is called a *frustum* of a cone. Suppose that the radii of the base and top are R_1 and R_2, respectively, while the height is h.

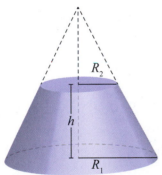

a. Use the shell method to prove that the volume of the cone frustum is

$V = \dfrac{\pi h}{3}\left(R_1^2 + R_1 R_2 + R_2^2\right)$. (**Hint:** One approach is to rotate the region between the line segment $y = \dfrac{R_1 - R_2}{h}x$, $0 \le x \le h$, and the line $y = -R_2$ about the said line and use the volume of the resulting solid to find that of the frustum.)

b. Now use the method of disks to establish the above formula. Which method do you prefer?

68. Use the shell method to prove the volume formula for a sphere of radius R: $V = \frac{4}{3}\pi R^3$.

69. Use the shell method to find the volume of the wooden toy piece of Exercise 17 of Section 6.1.

70. How deep is the water in a hemispherical tank of radius r when the tank is filled to exactly $\frac{14}{27}$ of its full capacity?

71.* Use the shell method to find the formula for the volume of the torus of Exercise 84 of Section 6.1.

72.* The graph of the equation $\dfrac{x^2}{a^2} + \dfrac{y^2}{b^2} = 1$ is an ellipse centered at the origin, with its axes of symmetry of lengths $2a$ and $2b$, coinciding with the coordinate axes. By rotating the region bounded by this ellipse about the x-axis, use the shell method to find a formula for the volume of the resulting ellipsoid.

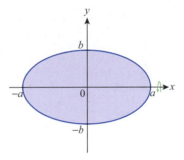

73–77 *True or False?* Determine whether the given statement is true or false. In case of a false statement, explain or provide a counterexample.

73. If V is the volume of the solid obtained by revolving the region bounded by $y = f(x)$ and the x-axis ($a \le x \le b$) about the x-axis and k is a positive constant, then the volume of the solid generated by $y = kf(x)$ over the same interval is kV.

74. The volume of the solid generated by revolving about the y-axis the region bounded by $y = x^2$, the y-axis, and $y = b$ is directly proportional to b.

75. The volume of a solid of revolution can be interpreted as the limit of Riemann sums.

76. The shell method always results in a less complicated integral than the disk method.

77. If R is the region bounded by $y = f(x)$ and $y = g(x)$, while R_1 is bounded by $y = f(x) + 1$ and $y = g(x) + 1$, then the solids obtained by revolving about the x-axis the regions R and R_1, respectively, have equal volume.

6.2 **Technology Exercises**

78–81 Use a computer algebra system to sketch the region bounded by the graphs of the equations. Then use the shell method, along with the integration capabilities of your technology to find the volume of the solid generated by rotating the region about the given line.

78. $y = x^2 \cos^2 x$, $\quad y = 0$, $\quad x = 0$, $\quad x = \pi/2$; $\quad$ about the y-axis

79. $x^{2/3} + y^{5/3} = 1$, $\quad x = 0$, $\quad y = 0$; $\quad$ about the y-axis

80. $x = \left(\dfrac{\pi}{2} - y\right)^3 \cos y$, $\quad y = 0$, $\quad x = 0$; $\quad$ about $y = -1$

81. $x^3 = y(y - 2)^2$, $\quad y = 0$, $\quad x = 0$; $\quad$ about $y = -2$

6.3 **Arc Length and Surface Area**

TOPICS

1. Arc lengths of graphs of functions
2. Surface areas of solids of revolution

In this section, we continue to use the principles of calculus to understand concepts that would be quite challenging if tackled with algebra alone. We start by extending our definition of length and then use our result to calculate the surface areas of solids of revolution.

TOPIC 1 **Arc Lengths of Graphs of Functions**

Formulas for the lengths of certain plane curves, such as the perimeters of circles and regular polygons, are familiar and widely used. And the lengths of some other curves can be found through the application of a few simple principles; for example, the perimeter of any polygon is simply the sum of the lengths of the straight lines that make up its edges. The same is true for any curve (closed or not) consisting of a finite number of line segments. But how should we define and then determine the length of an arbitrary plane curve?

Adapting the approach that has worked so well in defining area and volume, we will begin by approximating a given curve with a simpler curve whose length we know. Specifically, our approximation will consist of a finite number of lines drawn between points on the curve. In this initial stage, the curves we will work with are graphs of functions and have the form $y = f(x)$ for $a \leq x \leq b$, where f' is continuous on $[a,b]$ (such functions are called *continuously differentiable* and their graphs are said to be *smooth curves*). In Chapter 9, we will extend our definition further and work with curves that are defined parametrically.

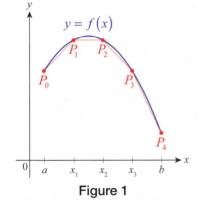

Figure 1

Figure 1 illustrates a rough approximation, shown in red, of a curve $y = f(x)$. The interval $[a,b]$ has been partitioned into four equal subintervals and straight line segments have been drawn from $P_{i-1}(x_{i-1}, f(x_{i-1}))$ to $P_i(x_i, f(x_i))$ for $i = 1$ to 4 (with, as per our custom, $x_0 = a$ and $x_4 = b$). If we denote the length of the line from P_{i-1} to P_i by $|P_{i-1}P_i|$, then by the distance formula

$$|P_{i-1}P_i| = \sqrt{(\Delta x_i)^2 + (\Delta y_i)^2},$$

where $\Delta x_i = x_i - x_{i-1}$ and $\Delta y_i = f(x_i) - f(x_{i-1})$. So the total length of our approximating curve is as follows:

$$\sum_{i=1}^{4} |P_{i-1}P_i| = \sum_{i=1}^{4} \sqrt{(\Delta x_i)^2 + (\Delta y_i)^2}$$

The difference $f(x_i) - f(x_{i-1})$ is one that we have seen often. By the Mean Value Theorem we know there is at least one point $x_i^* \in [x_{i-1}, x_i]$ such that

$$f(x_i) - f(x_{i-1}) = f'(x_i^*)(x_i - x_{i-1}),$$

so $\Delta y_i = f'\left(x_i^*\right)\Delta x_i$. Making this substitution, we have the following.

$$\sum_{i=1}^{4}\left|P_{i-1}P_i\right| = \sum_{i=1}^{4}\sqrt{\left(\Delta x_i\right)^2 + \left(\Delta y_i\right)^2}$$

$$= \sum_{i=1}^{4}\sqrt{\left(\Delta x_i\right)^2 + \left[f'\left(x_i^*\right)\Delta x_i\right]^2}$$

$$= \sum_{i=1}^{4}\sqrt{\left(\Delta x_i\right)^2\left(1 + \left[f'\left(x_i^*\right)\right]^2\right)} \qquad \text{Factor out } \left(\Delta x_i\right)^2.$$

$$= \sum_{i=1}^{4}\sqrt{1 + \left[f'\left(x_i^*\right)\right]^2}\,\Delta x_i \qquad \sqrt{\left(\Delta x_i\right)^2} = \Delta x_i \text{ since } \Delta x_i > 0.$$

More generally, we improve our approximation of the curve by dividing $[a,b]$ into n subintervals and letting n grow larger and larger. The resulting sum is familiar to us as a Riemann sum, and in the limit we obtain the following:

$$\lim_{n\to\infty}\sum_{i=1}^{n}\sqrt{1 + \left[f'\left(x_i^*\right)\right]^2}\,\Delta x_i = \int_{a}^{b}\sqrt{1 + \left[f'(x)\right]^2}\,dx$$

This gives us our extended definition for the lengths of curves defined by continuously differentiable functions.

Definition 💡

Arc Length of a Graph

If f' is a continuous function on $[a,b]$, then the **arc length** L of the curve $y = f(x)$, $a \le x \le b$, is given by

$$L = \int_{a}^{b}\sqrt{1 + \left[f'(x)\right]^2}\,dx.$$

Alternatively, using the Leibniz notation,

$$L = \int_{a}^{b}\sqrt{1 + \left(\frac{dy}{dx}\right)^2}\,dx.$$

Example 1 ✎

Determine the arc length L of the curve defined by $y = x^{3/2}$ over the interval $0 \le x \le 4$.

Solution

$$\frac{dy}{dx} = \frac{3}{2}x^{1/2} \quad \Rightarrow \quad \left(\frac{dy}{dx}\right)^2 = \frac{9}{4}x$$

Now the arc length of the curve can be calculated as follows.

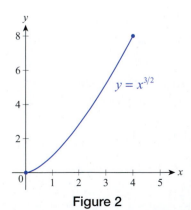

$y = x^{3/2}$

Figure 2

$$L = \int_0^4 \sqrt{1 + \frac{9}{4} x}\, dx$$

$$= \frac{4}{9} \int_{u=1}^{u=10} \sqrt{u}\, du \qquad\qquad \begin{aligned} u &= 1 + \frac{9}{4} x \\ du &= \frac{9}{4}\, dx \end{aligned}$$

$$= \frac{4}{9} \left[\frac{2}{3} u^{3/2} \right]_1^{10} = \frac{8}{27} \left(10\sqrt{10} - 1 \right)$$

Recall that, given an integrable function f on an interval $[a,b]$, we can define a new function $F(x) = \int_a^x f(t)\, dt$ which we might describe in words as "the area under f between a and the variable point x." In similar fashion, given a curve defined by a continuously differentiable function f on the interval $[a,b]$, we can define a new function $s(x)$ as "the arc length of the curve over the interval from a to the variable point x." Our arc length formula makes this definition precise.

Definition ♀

Arc Length Function

Given a continuously differentiable function f defined on $[a,b]$, we define its **arc length function** $s(x)$ to be

$$s(x) = \int_a^x \sqrt{1 + \left[f'(t) \right]^2}\, dt$$

for any $x \in [a,b]$.

By the Fundamental Theorem of Calculus (Part I), we then note that $s'(x) = \sqrt{1 + \left[f'(x) \right]^2}$ or, if we think of the curve as the graph of the equation $y = f(x)$,

$$\frac{ds}{dx} = \sqrt{1 + \left(\frac{dy}{dx} \right)^2}.$$

This last equation is often written in the form

$$\left(\frac{ds}{dx} \right)^2 = 1 + \left(\frac{dy}{dx} \right)^2 \quad \text{or} \quad (ds)^2 = (dx)^2 + (dy)^2,$$

which is an elegant and symmetric way of stating the relationship between the three differentials dx, dy, and ds. From it, we can derive the following arc length formula in which the roles of x and y are reversed.

$$(ds)^2 = (dx)^2 + (dy)^2 \quad \Rightarrow \quad \left(\frac{ds}{dy} \right)^2 = \left(\frac{dx}{dy} \right)^2 + 1 \quad \Rightarrow \quad ds = \sqrt{1 + \left(\frac{dx}{dy} \right)^2}\, dy$$

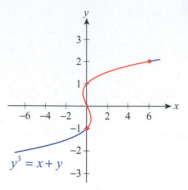

Figure 3

Example 2 ✒

Set up an integral defining the arc length of the graph of $y^3 = x + y$ between the two points $(0, -1)$ and $(6, 2)$.

Solution

The curve in question can be described as a function of y, but not as a function of x, so we will use the arc length formula that we just derived.

$$ds = \sqrt{1 + \left(\frac{dx}{dy}\right)^2}\, dy$$

$$= \sqrt{1 + \left(3y^2 - 1\right)^2}\, dy \qquad \begin{aligned} x &= y^3 - y \\ dx/dy &= 3y^2 - 1 \end{aligned}$$

$$= \sqrt{9y^4 - 6y^2 + 2}\, dy$$

Thus the arc length of the graph is as follows:

$$L = \int_{-1}^{2} ds = \int_{-1}^{2} \sqrt{9y^4 - 6y^2 + 2}\, dy$$

Note that, more generally, the arc length of the curve between $(0, -1)$ and an arbitrary point $\left(y^3 - y, y\right)$ is given by

$$s(y) = \int_{-1}^{y} \sqrt{9t^4 - 6t^2 + 2}\, dt.$$

Because of the square root in the arc length formula, it is not uncommon to wind up with integrals such as the one in Example 2, which we cannot yet evaluate. As we develop additional techniques of integration in the coming chapters, we will be able to go further with such computations. In particular, a technique in the next chapter known as Simpson's Rule will allow us to show that, in Example 2, $L \approx 8.7$.

TOPIC 2   Surface Areas of Solids of Revolution

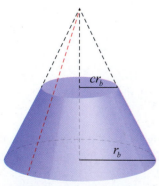

Figure 4 Frustum of a Cone

Now that we have a formula for arc length, we can put it to use immediately in calculating the surface area of a class of three-dimensional objects. To do so efficiently, we'll first develop a formula for the surface area A of an object known as a **frustum**, the formal name for a portion of a cone such as shown in Figure 4.

If we were to cut the cone in Figure 4 along the red dashed line and flatten out the surface, the result would be a sector of a circle such as that shown in Figure 5. Recall that the area of a sector defined by radius L and angle θ is $\frac{1}{2}L^2\theta$, and that the arc length of the portion of the same circle is $L\theta$. This arc length equals the circumference of the base; that is, $L\theta = 2\pi r_b$. So $\theta = (2\pi r_b)/L$ and the lateral surface area A_w of the whole cone is as follows:

$$A_w = \frac{1}{2}L^2\theta = \frac{1}{2}L^2\frac{2\pi r_b}{L} = \pi L r_b$$

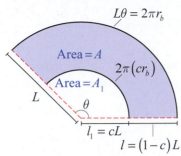

$$L\theta = 2\pi r_b$$

Area $= A$

$$2\pi(cr_b)$$

Area $= A_1$

$$L$$

$$\theta$$

$$l_1 = cL$$

$$l = (1-c)L$$

Figure 5 Flattened Frustum

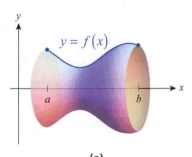

$$y = f(x)$$

$$a \qquad b$$

(a)

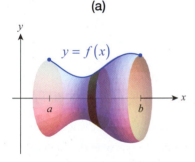

$$y = f(x)$$

$$a \qquad b$$

(b)

Figure 6
Surface Area Decomposition

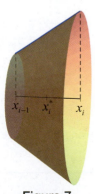

$$x_{i-1} \quad x_i^* \qquad x_i$$

Figure 7

Also notice that our labeling in Figure 5 takes advantage of geometric similarity, where the nonfrustum part of the cone is a "scaled-down" version of the whole cone by a factor of c ($0 < c < 1$). If we now let A_1 be the lateral surface area of the nonfrustum part of the cone, we obtain the area A of the frustum by subtraction as follows:

$$
\begin{aligned}
A &= A_w - A_1 \\
&= \pi L r_b - \pi(cL)(cr_b) \\
&= \pi L r_b (1 - c^2) \\
&= \pi L r_b (1+c)(1-c) \\
&= 2\pi r_b \frac{1+c}{2} L(1-c) \\
&= 2\pi \frac{r_b + c r_b}{2} l
\end{aligned}
$$

If we define r to be the average of the two radii, $r = (r_b + c r_b)/2$, we obtain $A = 2\pi r l$.

Now consider a solid of revolution such as that shown in Figure 6a, in which the portion of the curve $y = f(x)$ defined over the interval $[a, b]$ has been revolved around the x-axis. We can decompose the surface by partitioning it into a sequence of bands such as the one shown in Figure 6b. Each band is approximately a frustum, so our formula for the surface area of a frustum is an approximation for the surface area of each one.

In fact, given a representative band defined by the subinterval $[x_{i-1}, x_i]$ as shown in Figure 7, our formula for arc length tells us that $L \approx \sqrt{1 + [f'(x_i^*)]^2}\, \Delta x_i$ and so the surface area of the band is approximately

$$A_i = 2\pi r \sqrt{1 + [f'(x_i^*)]^2}\, \Delta x_i,$$

where $r = [f(x_{i-1}) + f(x_i)]/2$, the average of the two radii of the frustum. As we partition the interval $[a, b]$ into finer and finer subintervals, we can use the fact that $f(x_{i-1}) \approx f(x_i^*) \approx f(x_i)$ to replace r with $f(x_i^*)$ and write

$$A_i = 2\pi f(x_i^*) \sqrt{1 + [f'(x_i^*)]^2}\, \Delta x_i.$$

We are now back on familiar ground. The surface area of the solid of revolution is approximately $\sum_{i=1}^{n} A_i$, and we define the surface area to be the limit of these Riemann sums:

$$A = \lim_{n \to \infty} \sum_{i=1}^{n} 2\pi f(x_i^*) \sqrt{1 + [f'(x_i^*)]^2}\, \Delta x_i = \int_a^b 2\pi f(x) \sqrt{1 + [f'(x)]^2}\, dx$$

Definition 💡

Surface Area of a Solid of Revolution

Given a continuously differentiable function $f(x) \geq 0$ on the interval $a \leq x \leq b$, the **surface area** A of the solid formed by revolving the graph $y = f(x)$ around the x-axis is

$$A = \int_a^b 2\pi f(x) \sqrt{1 + \left[f'(x) \right]^2} \, dx = \int_a^b 2\pi y \sqrt{1 + \left(\frac{dy}{dx} \right)^2} \, dx.$$

Similarly, if the graph $x = g(y)$ of the nonnegative and continuously differentiable function g is revolved around the y-axis over the interval $c \leq y \leq d$, the surface area of the resulting solid is

$$A = \int_c^d 2\pi g(y) \sqrt{1 + \left[g'(y) \right]^2} \, dy = \int_c^d 2\pi x \sqrt{1 + \left(\frac{dx}{dy} \right)^2} \, dy.$$

In both cases, the formula can be summarized as $A = \int 2\pi r \, ds$, where r denotes the radius from the axis of revolution to the surface and ds denotes the arc length differential.

Example 3 ✏

Find the formula for the surface area of a sphere of radius R.

Solution

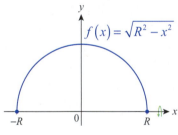

Figure 8

We can generate a sphere of radius R by rotating the continuously differentiable function $f(x) = \sqrt{R^2 - x^2}$ about the x-axis over the interval $-R \leq x \leq R$.

$$A = \int_{-R}^{R} 2\pi f(x) \sqrt{1 + \left[f'(x) \right]^2} \, dx$$

$$= 2\pi \int_{-R}^{R} \sqrt{R^2 - x^2} \sqrt{1 + \frac{x^2}{R^2 - x^2}} \, dx \qquad f'(x) = \frac{-x}{\sqrt{R^2 - x^2}} \text{ so } \left[f'(x) \right]^2 = \frac{x^2}{R^2 - x^2}.$$

$$= 2\pi \int_{-R}^{R} \sqrt{R^2 - x^2} \, \frac{R}{\sqrt{R^2 - x^2}} \, dx \qquad \sqrt{1 + \frac{x^2}{R^2 - x^2}} = \sqrt{\frac{R^2}{R^2 - x^2}} = \frac{R}{\sqrt{R^2 - x^2}}$$

$$= 2\pi R \int_{-R}^{R} dx$$

$$= 2\pi R x \big]_{-R}^{R} = 4\pi R^2$$

In Exercise 58 you will arrive at the same answer using the surface area formula $A = \int 2\pi r \, ds$.

Example 4 ✐

Determine the surface area of the solid formed by revolving the curve $y = x^2$ about the y-axis over the interval $1 \leq y \leq 2$.

Solution

We will demonstrate two ways of using the surface area formulas we have developed to arrive at the same answer.

Method 1

One method is to express the given curve in the form $x = \sqrt{y}$. Given this equation, our surface area can be found as follows:

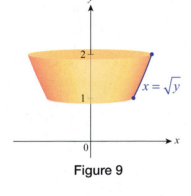

Figure 9

$$A = \int_1^2 2\pi x \sqrt{1 + \left(\frac{dx}{dy}\right)^2} \, dy$$

$$= \int_1^2 2\pi \sqrt{y} \sqrt{1 + \frac{1}{4y}} \, dy \qquad x = \sqrt{y} \text{ and } \frac{dx}{dy} = \frac{1}{2\sqrt{y}}$$

$$= \int_1^2 2\pi \sqrt{y} \, \frac{\sqrt{4y+1}}{2\sqrt{y}} \, dy \qquad \sqrt{1 + \frac{1}{4y}} = \sqrt{\frac{4y+1}{4y}} = \frac{\sqrt{4y+1}}{2\sqrt{y}}$$

$$= \pi \int_1^2 \sqrt{4y+1} \, dy$$

$$= \frac{\pi}{4} \int_{u=5}^{u=9} \sqrt{u} \, du \qquad u = 4y+1$$
$$du = 4\,dy$$

$$= \frac{\pi}{4} \left[\frac{2}{3} u^{3/2}\right]_{u=5}^{u=9} = \frac{\pi}{6}\left(27 - 5\sqrt{5}\right)$$

Method 2

Alternatively, we can begin with the general formula $A = \int 2\pi r \, ds$ and express ds in terms of dx. If we do so, we must be sure to express r in terms of x as well. Using this approach,

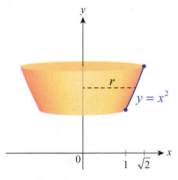

Figure 10

$$ds = \sqrt{1 + \left(\frac{dy}{dx}\right)^2} \, dx = \sqrt{1 + 4x^2} \, dx$$

and the radius r, the distance from the axis of revolution to the surface, is simply $r = x$. Note that the interval of integration in this case is $\left[1, \sqrt{2}\right]$. Therefore, the surface area is as follows:

$$A = \int_1^{\sqrt{2}} 2\pi x \sqrt{1 + 4x^2} \, dx$$

$$= \frac{\pi}{4} \int_{u=5}^{u=9} \sqrt{u} \, du \qquad u = 1 + 4x^2$$
$$du = 8x\,dx$$

$$= \frac{\pi}{4}\left[\frac{2}{3} u^{3/2}\right]_{u=5}^{u=9} = \frac{\pi}{6}\left(27 - 5\sqrt{5}\right)$$

6.3 **Exercises**

1–4 Use integration to determine the length of the given line segment. Then use the distance formula to check your answer.

1. $y = 3x - 5;\quad 1 \le x \le 7$

2. $x = 2\sqrt{2}\,y + 1;\quad \sqrt{2} \le y \le \sqrt{18}$

3. $y = \dfrac{1}{2}x + 1;\quad 2 \le x \le 8$

4. $x = 3 - \dfrac{1}{3}y;\quad 0 \le y \le 6$

5–16 Determine the arc length L of the curve defined by the equation over the given interval.

5. $y = \dfrac{5 + 2x^{3/2}}{3};\quad 1 \le x \le 8$

6. $y = x^{3/2} - \dfrac{1}{3}\sqrt{x};\quad 1 \le x \le 4$

7. $y = \dfrac{x^2}{8} - \ln x;\quad 1 \le x \le e$

8. $y = \dfrac{x^3}{3} + \dfrac{1}{4x};\quad 1 \le x \le 3$

9. $yx^3 - x^8 = \dfrac{1}{60};\quad \dfrac{1}{2} \le x \le 1$

10. $8x^2 y = x^6 + 2;\quad 1 \le x \le 3$

11. $y = \sqrt{1 - x^2};\quad 0 \le x \le 1$

12. $y = \sqrt[3]{x}\left(x^{4/3} - \dfrac{9}{20}\right);\quad 1 \le x \le 8$

13. $y = \dfrac{e^x}{4} + e^{-x};\quad 0 \le x \le 1$

14. $y = \displaystyle\int_1^x \sqrt{\cos 2t}\;dt;\quad 0 \le x \le \dfrac{\pi}{2}$

15. $x = \displaystyle\int_{1/3}^y \sqrt{\dfrac{1}{t^2} - 1}\;dt;\quad \dfrac{1}{e} \le x \le 1$

16. $x = \ln(\cos y);\quad 0 \le y \le \dfrac{\pi}{4}$

17–25 Set up, but do not evaluate, an integral defining the arc length of the graph of the equation over the given interval. Then find the corresponding arc length function $s(x)$ or $s(y)$ as appropriate.

17. $y = x^2 + 1;\quad 0 \le x \le 2$

18. $3x - y^2 = y + 2;\quad 1 \le y \le 4$

19. $x + y = y^2;\quad 0 \le y \le 1$

20. $y = \dfrac{1}{x - 2};\quad 3 \le x \le 4$

21. $y = \dfrac{1}{x^2};\quad 1 \le x \le 2$

22. $y = 2\ln x;\quad 1 \le x \le e$

23. $y = e^{2-x};\quad 0 \le x \le 2$

24. $y = \cos x;\quad \dfrac{\pi}{2} \le x \le \dfrac{3\pi}{2}$

25. $y = \sin^{-1} x;\quad -\dfrac{1}{2} \le x \le \dfrac{1}{2}$

26–35 Find the surface area of the solid obtained by revolving the indicated curve about the x-axis.

26. $y = \dfrac{1}{2}x - 1;\quad 2 \le x \le 5$

27. $x = 5 - y;\quad 1 \le x \le 3$

28. $x = \sqrt[3]{y};\quad 0 \le y \le 1$

29. $y = 2\sqrt{x};\quad 0 \le x \le 4$

30. $x = y^2 + 3;\quad 1 \le y \le \sqrt{2}$

31. $y = x^3 + \dfrac{1}{12x};\quad \dfrac{1}{2} \le x \le 1$

32. $6y\sqrt{x} = 12x^2 - x;\quad 1 \le x \le 2$

33. $y = e^x + \dfrac{e^{-x}}{4};\quad 0 \le x \le \dfrac{1}{2}$

34. $y = \sqrt{4x - x^2};\quad 0 \le x \le 4$

35. $x^2 + y^2 = 2x;\quad 1 \le x \le 2,\quad 0 \le y \le 1$

36–41 Find the surface area of the solid obtained by revolving the indicated curve about the y-axis.

36. $3y = x + 1; \quad 1 \le y \le 3$

37. $4y = x^2; \quad 0 \le y \le 2$

38. $y = \sqrt[3]{4x}; \quad 0 \le y \le 1$

39. $2y = x^2 - 1; \quad 0 \le y \le 1$

40. $20xy^{-1/3} = 20y^{4/3} - 9; \quad 0 \le y \le 1$

41. $12xy = 4y^4 + 3; \quad 1 \le y \le 2$

42–45 Find the surface area of the solid obtained by revolving the given curve about the indicated line. (**Hint:** Remember that $A = \int 2\pi r \, ds$, where r denotes the radius from the axis of revolution to the surface and ds is the arc length differential.)

42. $y = x - 1; \quad 1 \le x \le 3; \quad$ about $y = -2$

43. $y = \dfrac{1}{2}x - 1; \quad 0 \le y \le 2; \quad$ about $x = 1$

44. $12xy = 6y^4 + 2; \quad 1 \le y \le 2; \quad$ about $x = -\dfrac{1}{4}$

45. $2y = e^x + e^{-x}; \quad 0 \le x \le 1; \quad$ about $y = -1$

46–53 Set up, but do not evaluate, an integral defining the surface area of the solid obtained by revolving the given curve about the indicated axis.

46. $y = \cos x; \quad \dfrac{\pi}{2} \le x \le \dfrac{3\pi}{2}; \quad$ about the x-axis

47. $y = \sin^{-1} x; \quad 0 \le x \le \dfrac{\sqrt{3}}{2}; \quad$ about the x-axis

48. $y = \ln x; \quad 1 \le x \le e; \quad$ about the y-axis

49. $y = \dfrac{1}{x^2}; \quad 2 \le x \le 4; \quad$ about the x-axis

50. $x^2 - 7 = y^2 + 4y; \quad 0 \le y \le 2; \quad$ about the y-axis

51. $y = \sqrt[3]{x}; \quad 1 \le x \le 8; \quad$ about $y = -3$

52. $x(y - 5) = 1; \quad 6 \le y \le 8; \quad$ about the y-axis

53. $y = x^4 - 1; \quad 1 \le x \le \sqrt[4]{2}; \quad$ about $x = -2$

54. By generalizing Exercise 11, prove that the circumference of a circle of radius R is $C = 2\pi R$.

55. Recall that the equation of the astroid of Exercise 28 of Section 3.5 is $x^{2/3} + y^{2/3} = 10$. Rotate the graph about the x-axis and find the surface area of the resulting solid. Do you get the same answer if you rotate about the y-axis? Why?

56. The shape of a clothesline stretched between two trees at a campsite can be approximated by the equation $y = \frac{1}{10}x^2 + 1, \quad -1.5 \le x \le 1.5$ (distance is measured in yards). Find a formula for the length of the clothesline.

57. A particle is moving in the two-dimensional coordinate system so that its position (the x- and y-coordinates) as a function of time is given by $x = t/2$ and $y = t^{3/2}, \quad 0 \le t \le 4$.

   a. Sketch the path of the particle. (**Hint:** Paying attention to the domain of the variable t, use t to express y in terms of x.)

   b. Find the distance traveled by the particle. (**Hint:** Use the formula you found in part a. Alternatively, you may observe that $\dfrac{dy}{dx} = \dfrac{dy/dt}{dx/dt}$, and use the notation of Example 4.)

58. Show that the surface area of a sphere of radius R is $4\pi R^2$ by the following steps:

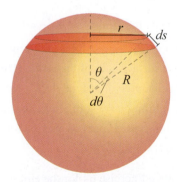

   a. Show that the distance r in the diagram above is given by $r = R \sin \theta$ and note that the arc length differential is given by $ds = R \, d\theta$.

   b. Determine the appropriate limits of integration and apply the surface area formula $A = \int 2\pi r \, ds$.

59.* Suppose that a pair of parallel planes intersect a sphere, as in the illustration provided for Exercise 58. However this time assume that the fixed distance between the planes is D units. Modify your argument in Exercise 58 to prove that the surface area of the zone of the sphere that falls between the planes is $A = 2\pi RD$. (Notice that this area depends only on the distance between the two parallel planes, not their actual location!)

60. Recall that the lateral surface area of a circular cone of slant height s and base radius r is $A = \pi rs$ (circumference of base·slant height/2). Use the surface area integral of this section to verify this formula. (**Hint:** Rotate the line segment $y = rx/\sqrt{s^2 - r^2}$, $0 \le x \le \sqrt{s^2 - r^2}$ around the y-axis.)

61. Find the surface area of the solid generated by revolving the region bounded by the graphs of $x^2 = 5y$ and $5y + 36 = 5x^2$ about the y-axis.

62. Use the methods of this section to find the surface area of the torus that is obtained by revolving the circle $x^2 + (y - 2)^2 = 1$ about the x-axis.

63.* Generalize your solution to Exercise 62 to obtain the surface area of the torus of Exercise 84 in Section 6.1.

64.* If we rotate the ellipse $\dfrac{x^2}{16} + \dfrac{y^2}{9} = 1$ about the x-axis, the resulting solid is called an ellipsoid. Find the surface area of this solid. (**Hint:** Handle the integral of type $\int \sqrt{k^2 - u^2}\, du$ by substituting $u = k \sin\theta$. Next, for the integral of type $\int \cos^2 t\, dt$ use the formula $\cos^2 t = (1 + \cos 2t)/2$.)

65.* Generalize Exercise 64 for the ellipse $\dfrac{x^2}{a^2} + \dfrac{y^2}{b^2} = 1$ $(a > b)$ to find the surface area of the resulting ellipsoid.

6.3 Technology Exercises

66–74. Use the formulas you found and the integration capabilities of a computer algebra system to evaluate the arc lengths in Exercises 17–25.

75–82. Use the formulas you found and the integration capabilities of a computer algebra system to evaluate the surface areas in Exercises 46–53.

83. Use a computer algebra system to find the length of the clothesline in Exercise 56.

84. Rotate the parabola of Exercise 56 about the vertical axis. Find the surface area of the resulting paraboloid. (For example, a large satellite dish might have this shape.)

TOPICS

1. Moments, mass, and centers of mass in one dimension

2. Moments, mass, and centers of mass in two dimensions

3. The theorems of Pappus

6.4 Moments and Centers of Mass

In this section, we use calculus and a principle from physics to determine the centers of mass of one- and two-dimensional objects. We then conclude the section with an application that allows us to quickly determine surface areas and volumes of certain objects.

TOPIC 1 Moments, Mass, and Centers of Mass in One Dimension

We begin with a principle called the Law of the Lever, known intuitively to everyone who has ever played on a seesaw. Suppose two masses m_1 and m_2 are placed on a lever on opposite sides of a fulcrum, as in Figure 1.

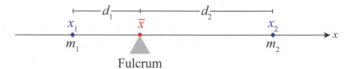

Fulcrum

Figure 1 Law of the Lever

Then the lever (or seesaw) will be perfectly balanced if $m_1 d_1 = m_2 d_2$; this implies that the heavier mass must be located closer to the fulcrum. We wish to describe the point of balance $\bar{x}$ in terms of the masses and their locations, which we can do as follows.

$$m_1 d_1 = m_2 d_2$$
$$m_1 \left(\bar{x} - x_1 \right) = m_2 \left(x_2 - \bar{x} \right) \qquad d_1 = \bar{x} - x_1 \text{ and } d_2 = x_2 - \bar{x}$$
$$\bar{x} \left(m_1 + m_2 \right) = m_1 x_1 + m_2 x_2$$
$$\bar{x} = \frac{m_1 x_1 + m_2 x_2}{m_1 + m_2}$$

In this context, $\bar{x}$ is the **center of mass** of the two-body system whose total mass is $m_1 + m_2$. The numerator of the result above, $m_1 x_1 + m_2 x_2$, is called the **moment of the system about the origin**, and is the sum of the individual **moments** $m_1 x_1$ and $m_2 x_2$ of the two masses. A body's moment about the origin is a component of the *torque* that body contributes to a lever whose fulcrum is at $x = 0$, with bodies to the right of the origin applying positive torque and bodies to the left applying negative torque. As our intuition tells us, moments can be made larger in magnitude either by increasing the mass or by increasing the body's distance from the origin.

The second step of the solution above can be rearranged and written in the form

$$m_1 \left(x_1 - \bar{x} \right) + m_2 \left(x_2 - \bar{x} \right) = 0,$$

which points the way toward a more general center of mass formula. Suppose n bodies with masses m_1 through m_n are located along the x-axis at locations x_1 through x_n, respectively. Then the center of mass of this n-body system is the point $\bar{x}$ for which

$$\sum_{i=1}^{n} m_i \left(x_i - \bar{x} \right) = 0.$$

Each term $m_i\left(x_i - \overline{x}\right)$ represents the moment of the ith body about the point $\overline{x}$, and the center of mass of the system is the point $\overline{x}$ for which the sum of the moments is zero. Solving this equation for $\overline{x}$ yields a formula similar to the two-body case:

$$\sum_{i=1}^{n} m_i\left(x_i - \overline{x}\right) = 0$$

$$\sum_{i=1}^{n} m_i x_i = \overline{x}\sum_{i=1}^{n} m_i$$

$$\overline{x} = \frac{\displaystyle\sum_{i=1}^{n} m_i x_i}{\displaystyle\sum_{i=1}^{n} m_i} = \frac{\text{moment of the system about the origin}}{\text{mass of the system}}$$

We now want to extend this formula to the case of a (nearly) one-dimensional rod, which we will model as a segment of the x-axis as shown in Figure 2.

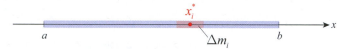

Figure 2 Center of Mass of a Rod

We can partition the rod into n small lengths in the usual manner, with $a = x_0 < x_1 < \cdots < x_{n-1} < x_n = b$, $\Delta x_i = x_i - x_{i-1}$, and $x_i^* \in \left[x_{i-1}, x_i\right]$. The rod doesn't have to be of uniform density, so let $\rho(x)$ denote the density, measured in mass per unit length, of the rod at x—we will assume the density ρ to be a continuous function, a safe assumption for our purposes. Then the mass of the ith segment is approximately $\Delta m_i = \rho\left(x_i^*\right)\Delta x_i$ (the density at the sample point x_i^* times the length of the segment) and its moment is approximately $x_i^* \Delta m_i$. We are, in effect, approximating the rod with n point masses, each of whose moments we have just determined. And, since the mass of the rod is approximately the sum of the masses of the n segments, we note that the center of mass $\overline{x}$ of the entire rod is

$$\overline{x} = \frac{\text{moment of the system about the origin}}{\text{mass of the system}}$$

$$\approx \frac{\displaystyle\sum_{i=1}^{n} x_i^* \Delta m_i}{\displaystyle\sum_{i=1}^{n} \Delta m_i}$$

$$= \frac{\displaystyle\sum_{i=1}^{n} x_i^* \rho\left(x_i^*\right)\Delta x_i}{\displaystyle\sum_{i=1}^{n} \rho\left(x_i^*\right)\Delta x_i}.$$

The numerator and denominator we have just constructed are Riemann sums for, respectively, the continuous functions $x\rho(x)$ and $\rho(x)$ over the interval $[a,b]$, so in the limit of finer and finer partitions we arrive at the three formulas that follow.

Moment about the Origin, Mass, and Center of Mass

Given an object modeled by the interval $[a,b]$ of the x-axis, with continuous density function ρ, its moment about the origin M_0, its mass M, and its center of mass $\bar{x}$ are as follows.

$$\textbf{Moment about the origin} \quad M_0 = \int_a^b x\rho(x)\,dx$$

$$\textbf{Mass} \quad M = \int_a^b \rho(x)\,dx$$

$$\textbf{Center of mass} \quad \bar{x} = \frac{M_0}{M}$$

Note that an easy but important consequence of our work is that, as far as moments about the origin are concerned, a system's mass can be assumed to be concentrated entirely at the center of mass. This follows from rewriting the last formula above as $M_0 = \bar{x} \cdot M$.

Example 1 ✏️

Compare the centers of mass of two 1 m long steel rods, the first rod with a constant thickness and a density of 0.5 kg/m and the second one thickening from left to right so that its density x units from the left end is $0.5\left(1 + x + \frac{1}{3}x^2\right)$ kg/m.

Solution

Although no physical rod can truly be one-dimensional, we are safe in making that approximation because we only really need to know the location of the center of mass lengthwise (each object is radially symmetric so we already know its center of mass lies on its axis of symmetry). Our intuition tells us that the center of mass of the constant density rod should be at the halfway point, but we will verify that our formula produces that result.

First rod: $\rho(x) = 0.5$ kg/m   Second rod: $\rho(x) = 0.5\left(1 + x + \frac{1}{3}x^2\right)$ kg/m

$$\bar{x} = \frac{\int_a^b x\rho(x)\,dx}{\int_a^b \rho(x)\,dx} \qquad \bar{x} = \frac{\int_a^b x\rho(x)\,dx}{\int_a^b \rho(x)\,dx}$$

$$= \frac{0.5\int_0^1 x\,dx}{0.5\int_0^1 dx} \qquad = \frac{\int_0^1 0.5\left(x + x^2 + \frac{1}{3}x^3\right)dx}{\int_0^1 0.5\left(1 + x + \frac{1}{3}x^2\right)dx}$$

$$= \frac{\left[\dfrac{x^2}{2}\right]_0^1}{\left[x\right]_0^1} \qquad = \frac{0.5\left[\dfrac{x^2}{2} + \dfrac{x^3}{3} + \dfrac{x^4}{12}\right]_0^1}{0.5\left[x + \dfrac{x^2}{2} + \dfrac{x^3}{9}\right]_0^1}$$

$$= \frac{1}{2} \qquad = \frac{33}{58}$$

TOPIC 2   Moments, Mass, and Centers of Mass in Two Dimensions

We can use the same sort of thinking to develop formulas for the moment, mass, and center of mass of a two-dimensional plate lying in the xy-plane. Suppose n point objects with masses $m_1, \ldots, m_n$ lie in the plane at locations $(x_1, y_1), \ldots, (x_n, y_n)$, respectively. Then the center of mass of the objects will be the intersection of the two lines $x = \bar{x}$ and $y = \bar{y}$, where $x = \bar{x}$ serves as the line of balance of the objects parallel to the y-axis and $y = \bar{y}$ is the line of balance parallel to the x-axis. Figure 3 illustrates this fact with four masses.

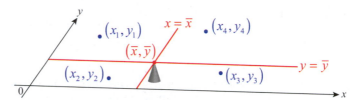

Figure 3   Center of Mass in the Plane

Each mass has moment $m_i y_i$ with respect to the x-axis and moment $m_i x_i$ with respect to the y-axis. So the system of masses has moments M_x and M_y about, respectively, the x- and y-axes, where

$$M_x = \sum_{i=1}^{n} m_i y_i \quad \text{and} \quad M_y = \sum_{i=1}^{n} m_i x_i.$$

And if we define M to be the sum of the masses, as before, then

$$\bar{x} = \frac{M_y}{M} \quad \text{and} \quad \bar{y} = \frac{M_x}{M}.$$

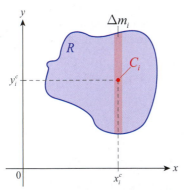

Figure 4   Center of Mass of a Plate

Now suppose we want to find the center of mass of a thin flat plate, which we can picture as a region R of the xy-plane (as in Figure 4). We can reduce this problem to one that is more familiar by approximating R with either a sequence of n vertical strips or horizontal strips (we have chosen vertical strips in Figure 4) and assuming the mass is concentrated at the center of mass $C_i = \left(x_i^c, y_i^c \right)$ of each strip.

If we let Δm_i denote the mass of the ith strip, then its moment about the x-axis is $y_i^c \Delta m_i$ and its moment about the y-axis is $x_i^c \Delta m_i$. So approximations of the two moments of the region R are

$$M_x \approx \sum_{i=1}^{n} y_i^c \Delta m_i \quad \text{and} \quad M_y \approx \sum_{i=1}^{n} x_i^c \Delta m_i$$

and hence

$$\bar{x} = \frac{M_y}{M} \approx \frac{\sum_{i=1}^{n} x_i^c \Delta m_i}{\sum_{i=1}^{n} \Delta m_i} \quad \text{and} \quad \bar{y} = \frac{M_x}{M} \approx \frac{\sum_{i=1}^{n} y_i^c \Delta m_i}{\sum_{i=1}^{n} \Delta m_i}.$$

For now, we assume the plate is made of a material with constant density ρ measured in mass per unit area (we will show how to relax this assumption later). Given ρ, each small element of mass Δm_i is equal to $\rho \Delta A_i$, where ΔA_i represents the area of the ith strip. So the approximations can also be written in the form

$$\overline{x} = \frac{M_y}{M} \approx \frac{\sum\limits_{i=1}^{n} x_i^c \rho \Delta A_i}{\sum\limits_{i=1}^{n} \rho \Delta A_i} \quad \text{and} \quad \overline{y} = \frac{M_x}{M} \approx \frac{\sum\limits_{i=1}^{n} y_i^c \rho \Delta A_i}{\sum\limits_{i=1}^{n} \rho \Delta A_i}.$$

In the limit, these Riemann sums lead to the following formulas.

Definition 💡

Moments about the Axes, Mass, and Center of Mass of a Plate

Given a thin flat plate of constant density ρ, modeled by a region R in the xy-plane, its moments M_x and M_y about the coordinate axes, its mass M, and its center of mass $(\overline{x}, \overline{y})$ are as follows.

Moment about the x-axis $M_x = \int y^c \, dm = \int y^c \rho \, dA$

Moment about the y-axis $M_y = \int x^c \, dm = \int x^c \rho \, dA$

Mass $M = \int dm = \int \rho \, dA$

Center of mass $\overline{x} = \dfrac{M_y}{M}$ and $\overline{y} = \dfrac{M_x}{M}$

The differentials dm and dA represent, respectively, differential elements of mass and area making up region R, with (x^c, y^c) the center of mass of each element.

In applying the above formulas, we will most often decompose region R into either vertical or horizontal strips, with our choice for a given region decided on the basis of whichever is easier. The next example illustrates both choices for the sake of comparison.

Example 2 ✎

Find the center of mass of a plate of constant density defined by the intersection of the two curves $y = x$ and $x = y^2$.

Solution

Method 1: Vertical Strips

If we define the upper curve by the function $f(x) = \sqrt{x}$ and the lower curve by the function $g(x) = x$, then the height of a given vertical strip is $f(x) - g(x) = \sqrt{x} - x$ and its width is the differential dx (see Figure 5). Since the plate has constant density, the vertical position of the center of mass of the strip shown is the average of $f(x)$ and $g(x)$.

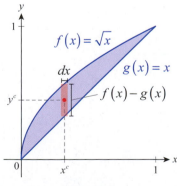

Figure 5

$$\left(x^{c}, y^{c}\right)=\left(x, \frac{f(x)+g(x)}{2}\right)=\left(x, \frac{\sqrt{x}+x}{2}\right)$$

$$dA=\left[f(x)-g(x)\right]dx=\left(\sqrt{x}-x\right)dx$$

The coordinates of the center of mass can be found as follows:

$$\overline{x}=\frac{M_{y}}{M}=\frac{\int_{0}^{1}x^{c}\rho\,dA}{\int_{0}^{1}\rho\,dA}=\frac{\rho\int_{0}^{1}x\left(\sqrt{x}-x\right)dx}{\rho\int_{0}^{1}\left(\sqrt{x}-x\right)dx}$$

$$=\frac{\int_{0}^{1}\left(x^{3/2}-x^{2}\right)dx}{\int_{0}^{1}\left(x^{1/2}-x\right)dx}=\frac{\left[\frac{2}{5}x^{5/2}-\frac{1}{3}x^{3}\right]_{0}^{1}}{\left[\frac{2}{3}x^{3/2}-\frac{1}{2}x^{2}\right]_{0}^{1}}=\frac{\frac{1}{15}}{\frac{1}{6}}=\frac{2}{5}$$

$$\overline{y}=\frac{M_{x}}{M}=\frac{\int_{0}^{1}y^{c}\rho\,dA}{\int_{0}^{1}\rho\,dA}=\frac{\rho\int_{0}^{1}\left(\frac{\sqrt{x}+x}{2}\right)\left(\sqrt{x}-x\right)dx}{\rho\int_{0}^{1}\left(\sqrt{x}-x\right)dx}$$

$$=\frac{\int_{0}^{1}\frac{1}{2}\left(x-x^{2}\right)dx}{\int_{0}^{1}\left(x^{1/2}-x\right)dx}=\frac{\frac{1}{2}\left[\frac{1}{2}x^{2}-\frac{1}{3}x^{3}\right]_{0}^{1}}{\left[\frac{2}{3}x^{3/2}-\frac{1}{2}x^{2}\right]_{0}^{1}}=\frac{\frac{1}{12}}{\frac{1}{6}}=\frac{1}{2}$$

Method 2: Horizontal Strips

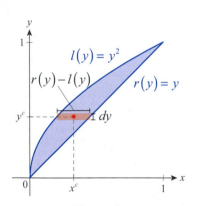

Figure 6

If we decompose the plate into horizontal strips, the width of each one is the distance between the left curve and the right curve. If we define the right curve as the function $r(y)=y$ and the left curve as $l(y)=y^{2}$, then the width of a given horizontal strip is $r(y)-l(y)=y-y^{2}$ and its height is the differential dy (see Figure 6). Since the plate has constant density, the horizontal position of the center of mass of the strip shown is the average of $r(y)$ and $l(y)$.

$$\left(x^{c}, y^{c}\right)=\left(\frac{r(y)+l(y)}{2}, y\right)=\left(\frac{y+y^{2}}{2}, y\right)$$

$$dA=\left[r(y)-l(y)\right]dy=\left(y-y^{2}\right)dy$$

The coordinates of the center of mass can be found as follows:

$$\overline{x}=\frac{M_{y}}{M}=\frac{\int_{0}^{1}x^{c}\rho\,dA}{\int_{0}^{1}\rho\,dA}=\frac{\rho\int_{0}^{1}\left(\frac{y+y^{2}}{2}\right)\left(y-y^{2}\right)dy}{\rho\int_{0}^{1}\left(y-y^{2}\right)dy}$$

$$=\frac{\int_{0}^{1}\frac{1}{2}\left(y^{2}-y^{4}\right)dy}{\int_{0}^{1}\left(y-y^{2}\right)dy}=\frac{\frac{1}{2}\left[\frac{1}{3}y^{3}-\frac{1}{5}y^{5}\right]_{0}^{1}}{\left[\frac{1}{2}y^{2}-\frac{1}{3}y^{3}\right]_{0}^{1}}=\frac{\frac{1}{15}}{\frac{1}{6}}=\frac{2}{5}$$

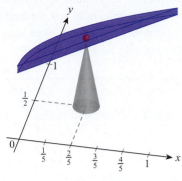

Figure 7

$$\overline{y} = \frac{M_x}{M} = \frac{\int_0^1 y^c \rho \, dA}{\int_0^1 \rho \, dA} = \frac{\rho \int_0^1 y(y - y^2) \, dy}{\rho \int_0^1 (y - y^2) \, dy}$$

$$= \frac{\int_0^1 (y^2 - y^3) \, dy}{\int_0^1 (y - y^2) \, dy} = \frac{\left[\frac{1}{3} y^3 - \frac{1}{4} y^4 \right]_0^1}{\left[\frac{1}{2} y^2 - \frac{1}{3} y^3 \right]_0^1} = \frac{\frac{1}{12}}{\frac{1}{6}} = \frac{1}{2}$$

Both methods work well in this example, and we have now determined that the center of mass or balance point, as shown in Figure 7 of the plate is $\left(\frac{2}{5}, \frac{1}{2} \right)$. In other problems, you may find that one method results in easier integrals to evaluate than the other.

In some cases, we can easily adapt our formulas to handle plates of varying density.

Example 3 ✎

The plate bounded by the curve $y = x^2$ and the x-axis over the interval $[0,1]$ has the density at the point (x, y) given by $\rho = x^3$. Determine its center of mass.

Solution

Because the density function depends only on x, the center of mass (x^c, y^c) of a given vertical strip of the region is simply located halfway between the x-axis and the upper edge (note, though, that the vertical strips increase in density moving from left to right). So if we decompose the plate into vertical strips, we have

$$(x^c, y^c) = \left(x, \frac{x^2}{2} \right) \quad \text{and} \quad dA = (x^2 - 0) \, dx = x^2 \, dx.$$

Therefore,

$$\overline{x} = \frac{M_y}{M} = \frac{\int_0^1 x^c \rho \, dA}{\int_0^1 \rho \, dA} = \frac{\int_0^1 (x)(x^3) x^2 \, dx}{\int_0^1 (x^3) x^2 \, dx} = \frac{\left[\frac{1}{7} x^7 \right]_0^1}{\left[\frac{1}{6} x^6 \right]_0^1} = \frac{\frac{1}{7}}{\frac{1}{6}} = \frac{6}{7}$$

and

$$\overline{y} = \frac{M_x}{M} = \frac{\int_0^1 y^c \rho \, dA}{\int_0^1 \rho \, dA} = \frac{\int_0^1 \left(\frac{x^2}{2} \right)(x^3) x^2 \, dx}{\int_0^1 (x^3) x^2 \, dx} = \frac{\frac{1}{2} \left[\frac{1}{8} x^8 \right]_0^1}{\left[\frac{1}{6} x^6 \right]_0^1} = \frac{\frac{1}{16}}{\frac{1}{6}} = \frac{3}{8}.$$

The plate's center of mass is $\left(\frac{6}{7}, \frac{3}{8} \right)$ and is shown in Figure 8.

Since the plate increases in density toward the right-hand side, its center of mass is shifted upward and to the right relative to a similar plate of constant density. In Exercise 16 you will show that a plate of the same shape but constant density has its center of mass at $\left(\frac{3}{4}, \frac{3}{10} \right)$.

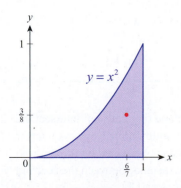

Figure 8

We can also adapt our two-dimensional formulas to handle a curved one-dimensional object in the plane. In the following example, note that we treat each differential element of the curve as a small element of mass dm. Note also that the center of mass of the curved object doesn't actually lie *on* the object.

Example 4 ✏

A length of wire of constant density is curved into a half circle of radius r. Find its center of mass.

Solution

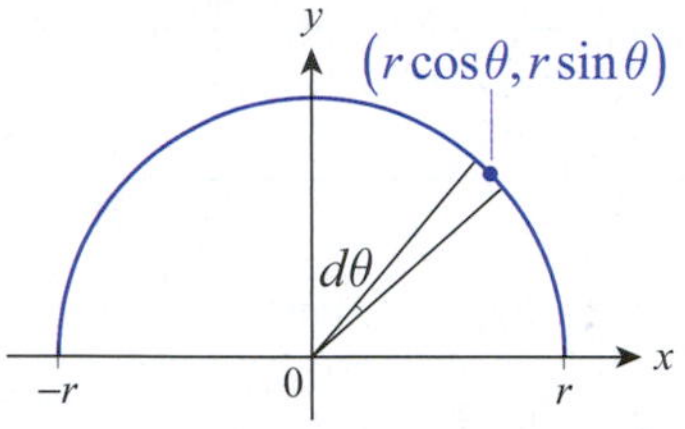

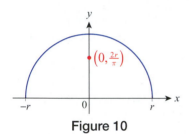

Figure 9

A small differential segment of arc length ds has length $r\,d\theta$, where $d\theta$ represents a differential element of the angle θ (recall that the arc length formula is $s = r\theta$ and hence $ds = r\,d\theta$). So the small element of arc length has mass $dm = \rho\,ds = \rho r\,d\theta$. The center of mass of the small element is $\left(x^c, y^c\right) = \left(r\cos\theta, r\sin\theta\right)$, as shown in Figure 9. By symmetry, we know $\overline{x} = 0$ so we just need to determine $\overline{y}$, as follows:

$$\overline{y} = \frac{M_x}{M} = \frac{\int y^c\,dm}{\int dm} = \frac{\int_0^\pi y^c \rho r\,d\theta}{\int_0^\pi \rho r\,d\theta} = \frac{\rho\int_0^\pi (r\sin\theta)\,r\,d\theta}{\rho\int_0^\pi r\,d\theta}$$

$$= \frac{r^2\int_0^\pi \sin\theta\,d\theta}{r\int_0^\pi d\theta} = \frac{r\left[-\cos\theta\right]_0^\pi}{\pi} = \frac{2r}{\pi}$$

So the center of mass of the curved length of wire is the point $\left(0, 2r/\pi\right)$ (see Figure 10).

Figure 10

(The figure shows a half circle with centroid marked at $\left(0, \frac{2r}{\pi}\right)$.)

TOPIC 3   The Theorems of Pappus

When the density ρ of an object is constant, the center of mass of the object is frequently called its **centroid**. This is a reflection of the fact that the mass of the object is irrelevant if its density is constant, since ρ cancels in the formulas for $\overline{x}$ and $\overline{y}$ —geometry alone determines the object's center of mass. One of the last of the great Greek geometers of antiquity, Pappus of Alexandria (ca. late 3rd and early 4th century AD), formulated two principles that make clever use of the centroid. The two principles allow us to quickly find volumes and surface areas of many objects formed by revolution.

Theorem 🔍

Pappus' Theorem for Volumes

Given a plane region R and a line L in the plane that does not intersect the interior of the region, the volume V of the solid formed by revolving R about L is equal to the area of R times the distance traveled by the centroid of R as it revolves about L. That is, if d is the distance between the centroid of the region R and the line L, then $V = 2\pi d \cdot \text{Area}\left(R\right)$.

Proof ✎

To simplify the proof, orient R and L so that R lies in the 1st quadrant of the xy-plane and the axis of revolution L coincides with the y-axis, as shown in Figure 11. Then the shell method tells us that

$$V = \int_a^b 2\pi \left(\begin{array}{c} \text{radius} \\ \text{of shell} \end{array} \right) \left(\begin{array}{c} \text{height} \\ \text{of shell} \end{array} \right) dx = \int_a^b 2\pi x h(x)\, dx,$$

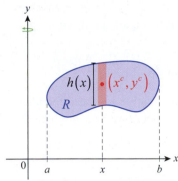

Figure 11 Revolving R

where $h(x)$ denotes the height of the vertical strip of R at x, as illustrated. The distance d between the centroid and L is the x-coordinate of the centroid:

$$d = \bar{x} = \frac{\int x^c \rho\, dA}{\int \rho\, dA} = \frac{\rho \int_a^b x h(x)\, dx}{\rho \int dA} = \frac{\int_a^b x h(x)\, dx}{\text{Area}(R)} \qquad \begin{array}{l} x^c = x \\ dA = h(x)\, dx \end{array}$$

Multiplying through by $\text{Area}(R)$ we obtain $d \cdot \text{Area}(R) = \int_a^b x h(x)\, dx$. Now we multiply both sides by 2π to lead us to the formula for the volume.

$$2\pi d \cdot \text{Area}(R) = \int_a^b 2\pi x h(x)\, dx = V$$

Example 5 ✐

The torus shown in Figure 12 is generated by revolving a disk of radius r about a line in the same plane located at a distance of R units from the center of the disk ($R \geq r$). Find the volume of the torus.

Solution

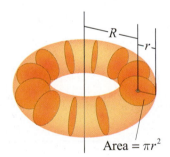

Figure 12

The area of the region being revolved is πr^2 and its centroid (which coincides with the center of the disk) is R units from the line of revolution. Applying Pappus' Theorem, the volume V of the torus is as follows.

$$V = 2\pi R \cdot \pi r^2 = 2\pi^2 R r^2$$

Notice that Pappus' Theorem provides a much more convenient way to arrive at this formula than that in Exercise 84 of Section 6.1 (also see Exercise 71 in Section 6.2).

Theorem ⚗

Pappus' Theorem for Surface Areas

Given a smooth plane curve C and a line L in the plane that does not cross C, the area A of the surface formed by revolving C about L is equal to the arc length of C times the distance traveled by the centroid of C as it revolves about L. That is, if d is the distance between the centroid of the curve C and the line L, then $A = 2\pi d \cdot \text{Length}(C)$.

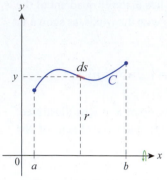

Figure 13 Revolving C

Proof ✎

In the interest of brevity, we will restrict the proof to the case where C and L can be oriented so that L coincides with the x-axis and C is a differentiable function over the interval $[a,b]$, as depicted in Figure 13. Then by our surface area formula,

$$A = \int_a^b 2\pi r\, ds = \int_a^b 2\pi y\, ds,$$

where $r = y$ is the distance between the axis of revolution and the differential element ds. The distance d between the centroid and L is the coordinate $\overline{y}$ of the centroid:

$$d = \overline{y} = \frac{\int y^c\, dm}{\int dm} = \frac{\int_a^b y^c \rho\, ds}{\int_a^b \rho\, ds} = \frac{\int_a^b y^c\, ds}{\int_a^b ds} = \frac{\int_a^b y\, ds}{\text{Length}(C)} \qquad y^c = y$$

Multiplying through by $2\pi \cdot \text{Length}(C)$ we obtain the formula for the area:

$$2\pi d \cdot \text{Length}(C) = \int_a^b 2\pi y\, ds = A$$

Example 6 ✐

Find the surface area of the torus shown in Figure 12.

Solution

The length of the curve being revolved is $2\pi r$ and its centroid's distance from the line of revolution is R. By Pappus' Theorem for surface areas we obtain the following.

$$A = 2\pi R(2\pi r) = 4\pi^2 Rr$$

Again, you may want to contrast the ease of this solution to that of Exercise 63 in Section 6.3.

Example 7 ✐

Use Pappus' Theorem for surface areas to find the center of mass of the length of wire in Example 4.

Solution

If we were to revolve the length of wire about the x-axis, the resulting surface would be a sphere of radius r centered at the origin. We know its surface area is $4\pi r^2$ and we know the wire has length πr. Using the formula $A = 2\pi d \cdot \text{Length}(C)$ we obtain d as follows:

$$4\pi r^2 = 2\pi d(\pi r)$$

$$d = \frac{2r}{\pi}$$

So the center of mass is located at the point $(0, 2r/\pi)$, matching the answer previously obtained in Example 4.

6.4 **Exercises**

1–3 Find the moment M_0 about the origin and the center of mass $\bar{x}$ for the point masses located on the x-axis.

1. $m_1 = 4, x_1 = -3;$ $m_2 = 5, x_2 = -1;$ $m_3 = 2, x_3 = 6$

2. $m_1 = 2, x_1 = -5;$ $m_2 = 10, x_2 = -2;$ $m_3 = 8,$
$x_3 = 1;$ $m_4 = 3, x_4 = 7$

3. $m_1 = 3.5, x_1 = -10;$ $m_2 = 5, x_2 = -2;$ $m_3 = 2,$
$x_3 = 5.5;$ $m_4 = 2.5, x_4 = 11$

4–6 Find the moments M_x, M_y about the coordinate axes and the center of mass for the system of point masses.

4. $m_1 = 4, P_1(-6,-8);$ $m_2 = 5, P_2(1.5,2);$
$m_3 = 2, P_3(3,4)$
(What do you notice about $\bar{x}$?)

5. $m_1 = 2, P_1(-2,6);$ $m_2 = 4, P_2(-1,-5);$
$m_3 = 7, P_3(2,0);$ $m_4 = 8, P_4(3,4)$

6. $m_1 = 2.5, P_1(-9,0);$ $m_2 = 3, P_2(-4.5,0.25);$
$m_3 = 2, P_3(0,-2.5);$ $m_4 = 6, P_4(2.25,5);$
$m_5 = 1.5, P_5(5.5,-3.5)$

7. Tyler and Christina are sitting on the ends of a 14-foot seesaw, unable to balance it because of their weight difference. However, Tyler's little sister, Lisa, is quick to come to the rescue. If Tyler and Christina weigh 110 and 80 pounds, respectively, while little Lisa is 35 pounds, where should she sit in order for the seesaw to balance? (**Note:** Strictly speaking, when you are multiplying weight by distance, you are calculating torque rather than moment, but the technique in obtaining balance is the same.)

8. The design of a certain front-wheel-drive family sedan allows for 54% of its total mass to rest on the front axle, while 46% is resting on the rear axle. The distance between the front and rear axles (the wheelbase of the car) is 2.65 meters. How far behind the front axle is the car's center of mass?

9. An experimental rocket is fired towards the north and is on track to hit the target when a midair explosion breaks it apart. The tail section, which is twice as heavy as the nosepiece, is found 200 yards southwest from the intended target. Where is the nosepiece likely to be found? (**Hint:** Even after breaking apart, the center of mass of the rocket will arrive in the target area.)

10–11 A thin rod of length l, with the given continuously varying density is placed into the coordinate system so that it lies horizontally on the x-axis with its left endpoint coinciding with the origin. Find its center of mass.

10. $l = 2.25$ m, $\rho = 1 + \sqrt{x}$ kg/m

11. $l = 110$ cm, $\rho = 0.6 + 0.01x$ g/cm

12–15 Use the indicated coordinates to determine the center of mass of the given region. (**Hint:** Divide the region into appropriate subregions and treat the centers of mass of the subregions as point masses.)

12.

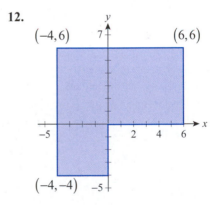

13.

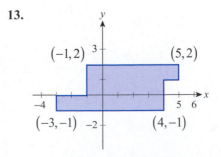

14.

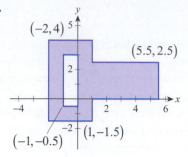

15.

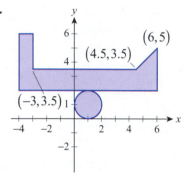

16. Show that if the plate of Example 3 has constant density, then its center of mass is located at $\left(\frac{3}{4}, \frac{3}{10}\right)$.

17–40 Find the centroid of the plane region bounded by the given curves. If possible, use symmetry to simplify your calculations. (In Exercise 31, use the formula $\cos^2 x = (1+\cos 2x)/2$ before integrating.)

17. $y = x^2 + 4x, \quad y = 0$

18. $y = \sqrt{4-x}, \quad x = 0, \quad y = 0$

19. $y = 2x^2, \quad 32x + y^2 = 0$

20. $y = \dfrac{x^2}{4}, \quad x = 2, \quad y = 0$

21. $y = \dfrac{\sqrt{x}}{2} + 1, \quad y = \dfrac{1}{4}x + 1$

22. $y = x^2, \quad y = x^3$

23. $y = 3 - x, \quad x = 0, \quad y = 0$

24. $x + 2y = 13, \quad 2x - y + 4 = 0, \quad x = 0$

25. $y = \dfrac{1}{x}, \quad x = 1, \quad x = 3$

26. $y = \sqrt{2x}, \quad y = x$

27. $y = x^2 - 9, \quad y = 0$

28. $y = x^2 - 9, \quad x = 0, \quad y = 0$

29. $y = x^3, \quad y = 4x, \quad x \geq 0$

30. $y = \sqrt{x}, \quad y = x^2$

31.* $y = \cos^2 x, \quad y = 0, \quad 0 \leq x \leq \pi$

32. $xy = 1 \quad x = 1, \quad y = 3$

33. $y = \dfrac{1}{x^2}, \quad y = -\dfrac{1}{x^2}, \quad 1 \leq x \leq 2$

34. $y = x^{2/3}, \quad x = 0, \quad y = 1$

35. $y = (x+2)^2, \quad y = (x-2)^2, \quad y = 0$

36. $y = 2\sqrt{x+4}, \quad 2y = 8 - x, \quad y = 0$

37. $y = x+1, \quad y(x+1) = 1, \quad y = 0, \quad x = 1$

38. $y = \sqrt{3-x}, \quad y = 2\sqrt{-x}, \quad y = 0$

39. $4x^2 + 9y^2 = 49, \quad x = 0, \quad y = 0$

40. $x^4(1 - x^2) = y^2, \quad x \geq 0$

41–50 Find the center of mass of the plane region of varying density that is bounded by the given curves.

41. $3y = 6 - x, \quad x = 0, \quad y = 0; \quad \rho(x,y) = x$

42. $3y = 6 - x, \quad x = 0, \quad y = 0; \quad \rho(x,y) = 1 + y$

43. $y = x^{3/2}, \quad y = \sqrt{x}; \quad \rho(x,y) = \sqrt{x}$

44. $y = \sqrt{1-x^2}, \quad x = 0, \quad y = 0; \quad \rho(x,y) = x$

45. $y = 4 - x^2, \quad y = 2 - x; \quad \rho(x,y) = 2 + x^2$

46. $y = \dfrac{3}{x^3}, \quad x = \dfrac{1}{2}, \quad x = 1; \quad \rho(x,y) = 1 - x^3$

47. $y = \sqrt[4]{x}, \quad y = \sqrt{x}; \quad \rho(x,y) = 1 - y$

48. $xy^2 = 3, \quad y = \dfrac{1}{3}, \quad y = 1; \quad \rho(x,y) = y^3 - y^2$

49. $y + 2x = 5, \quad x = y + 1, \quad y = 0; \quad \rho(x,y) = 5 - x$

50. $x = \sqrt{y}, \quad xy = 1, \quad x = 0, \quad y = 2; \quad \rho(x,y) = y$

51. Find the center of mass of the wire if we modify Example 4 by bending the wire to form three sides of a rectangle as shown in the figure below.

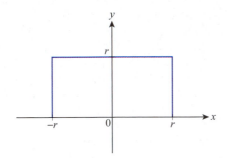

52. Find the centroid of the half disk shown in the figure below. Assume constant density.

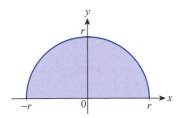

53. Find the centroid of the first quadrant of the half disk in Exercise 52.

54.* Find the center of mass of the half disk of Exercise 52 if its density function is $\rho(x, y) = y$. (**Hint:** To eliminate the radical, use the substitution $x = r\sin\theta$ where applicable.)

55. Find the center of mass of the wire in Example 4 if it has a variable density of $\rho(\theta) = |\cos\theta|$.

56.* Prove that the center of mass of a triangle is the intersection of its medians. (**Hint:** Recall what you learned in geometry about the intersection point of the medians.)

57. Prove that the distance of the center of mass of a triangle from each side is one third of the corresponding altitude. (**Hint:** Try to prove the result for right triangles first.)

58.* The graph of the equation $\dfrac{x^2}{a^2} + \dfrac{y^2}{b^2} = 1$ is an ellipse centered at the origin. Find the centroid of the plane region bounded by the first quadrant of this ellipse and the coordinate axes.

59. Find a formula for the centroid of the region bounded by $y = \sqrt[n]{x}$, $y = 0$, and $x = 1$ as a function of n. What can you say if $n \to \infty$?

60. Use Pappus' Theorem for volumes to answer Exercise 52 provided that you know the formula for the volume of the sphere: $V = \frac{4}{3}\pi r^3$.

61. Use Pappus' Theorem for volumes to find the centroid of a right triangle. (**Hint:** Place a right triangle into the coordinate system so that its vertices coincide with the origin, and the points $(h, 0)$ and $(0, R)$. Then rotate the triangle and use the volume formula for the right circular cone, along with Pappus' Theorem.)

62. Use Pappus' Theorem to find the volume of the solid resulting from revolving the triangle with vertices $(-3, 1)$, $(4, 1)$, and $(2, 7)$ about the line $y = -2$.

63. Rotate the half disk of Exercise 52 about the line $y = r$. Use Pappus' Theorem to find the volume of the resulting solid.

64. Rotate the region of Exercise 24 about the line $y = -4$. Use Pappus' Theorem to find the volume of the resulting solid.

65. Rotate the region of Exercise 28 about the line $y = 1$. Use Pappus' Theorem to find the volume of the resulting solid.

66. Rotate the region of Exercise 30 about the line $y = 3$. Use Pappus' Theorem to find the volume of the resulting solid.

67. Use Pappus' Theorem to find the volume of the solid generated by revolving the rectangle with vertices $(-1, 4)$, $(5, 4)$, $(5, 8)$, and $(-1, 8)$ about the line $y = -x - 2$.

68. Use the result of Exercise 40 along with Pappus' Theorem to find the volume of the solid obtained by revolving the right loop of the curve $x^4(1 - x^2) - y^2 = 0$ about the y-axis.

69. Use the result of Example 4 along with Pappus' Theorem to find the surface area of the solid in Exercise 63.

70. Use Pappus' Theorem for surface areas to verify that the lateral surface area of a right circular cone of slant height s and base radius r is $A = \pi rs$. (**Hint:** Appropriately place a line segment of length s in the coordinate system and rotate to obtain the cone.)

71. Generalize your solution to Exercise 70 to verify the formula for the lateral surface area of a frustum of a cone (for the definition, see the discussion preceding Example 3 in Section 6.3).

72–75 *True or False?* Determine whether the given statement is true or false. In case of a false statement, explain or provide a counterexample.

72. The center of mass of a thin triangular plate always coincides with its centroid (i.e., the intersection of its medians).

73. The center of mass of an object is always a point on the object itself.

74. Pappus' Theorem makes integration unnecessary for calculating volumes.

75. If a child is 50% heavier than another child, then in order for them to balance on a seesaw, he or she has to sit 50% closer to the pivot point.

6.5 **Force, Work, and Pressure**

TOPICS

1. Integral definition of work

2. Fluid pressure and fluid force

Force, *work*, and *pressure* are three terms that arise in both everyday language and in more carefully defined scientific settings. In this section, we will discuss the mathematical relationship between the terms and use calculus to determine the work performed and the pressure exerted by certain forces.

TOPIC 1 **Integral Definition of Work**

With regard to ourselves, we use the word *work* to describe a wide variety of tasks that require us to put forth effort, covering situations as diverse as, say, physical labor and learning calculus. But the word has a more restricted and rigorously defined meaning in science and engineering. A common saying in physics is that "energy is the ability to perform work," a phrase reflecting the fact that work and energy are measured with the same units. The formal scientific definition of work arose in the 1800s in response to the need to quantify the energy needed to do such things as pump water out of mine shafts. Energy, supplied by muscle or the new mechanical engines of the day, was needed to lift buckets of water against the opposing force of gravity upward through a distance, and this idea underlies the formal definition of work W as the application of a force F over a distance d. In equation form, $W = Fd$ ("work is force through a distance").

The equation $W = Fd$ assumes that the force is constant over the distance d, and if the force in a given situation truly is constant, as in Example 1, then no calculus is required to determine the work done.

Example 1 ✎

Calculate the work done

a. in lifting a 10-pound bucket of water out of a 50-foot-deep well and
b. in picking a 2-kilogram book 0.9 meters up off the floor.

Solution

a. In the US Customary System of units, the *pound* is the unit of force; it is a measure of the gravitational attraction between the Earth and, in this case, a bucket of water. So the work done is as follows.

$$W = (10\text{ lb})(50\text{ ft}) = 500\text{ ft-lb} \qquad \text{Read 500 foot-pounds}$$

b. In the International System (SI) framework, the unit of force is the *newton* (N), the force required to accelerate a mass of 1 kg at 1 m/s^2. (For comparison, the 10-pound bucket in this example would weigh about 45 newtons.) The distance is measured in meters, and a unit of work is the *Newton-meter* (N · m), also called the *joule* (J) after the English physicist James Joule (1818–1889). The book in this example has a mass of 2 kilograms and the gravitational force between it and Earth is the force $F = mg = (2\text{ kg})(9.8\text{ m/s}^2) = 19.6\text{ N}$. Now the work performed in lifting it 0.9 meters can be determined.

$$W = (19.6\text{ N})(0.9\text{ m}) = 17.64\text{ N} \cdot \text{m} = 17.64\text{ J}$$

Note the distinction between the pound, a measure of force, and the kilogram, a measure of mass. In the US Customary System, a unit of mass is the *slug*, which is an object's weight in pounds divided by $g \approx 32 \text{ ft/s}^2$. Although its use is rare (the SI is more typically used to measure mass), the 10-pound bucket of water in part a. has a mass of approximately 0.31 slugs.

As you might expect, it is frequently the case that the force in question varies over the distance. But we can still use the relationship $W = Fd$ to construct a more general formula, using a now familiar process. To this end, assume F is a continuous function of x and that we wish to calculate the work performed by F over the distance d represented by the interval $[a,b]$ on the x-axis. If we partition $[a,b]$ into n subintervals with the $n+1$ points $x_0 = a < x_1 < \cdots < x_n = b$, then over each subinterval $[x_{i-1}, x_i]$ the force can be approximated by the constant $F(x_i^*)$, where $x_i^* \in [x_{i-1}, x_i]$ is, as usual, a sample point. As the subintervals decrease in width, the amount by which F can vary also decreases, making the approximations better as we consider finer and finer partitions. Adding up the incremental amounts of work $\Delta W_i = F(x_i^*)\Delta x_i$ performed over each subinterval, the total work done is approximated by a Riemann sum:

$$W \approx \sum_{i=1}^{n} F(x_i^*)\Delta x_i$$

In the limit as $n \to \infty$, this leads to the following definition.

Definition 💡

Work Done by a Variable Force

The **work** done by a continuous force $F(x)$ over the interval $[a,b]$ is

$$W = \int dW = \int_a^b F(x)\,dx.$$

A simple but important application of the above definition comes from **Hooke's Law**, which states that the force needed to stretch or compress a spring (within reasonable limits) is proportional to the distance x it is stretched or compressed from its natural length. That is, $F(x) = kx$, where k is the *spring constant*, a number that depends on the particular spring.

Example 2 ✎

A spring with an unstressed length of 10 cm is held stretched to 12 cm with a force of 15 N. How much work will be done in stretching the spring from 12 cm to 15 cm?

Solution

The first step is to determine the spring constant k. Since a force of 15 N is required to stretch the spring 12 cm − 10 cm = 0.02 m (remember to convert centimeters to meters), we find k as follows:

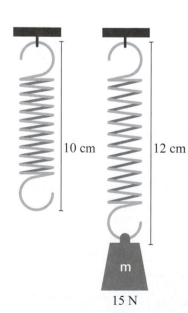

10 cm   12 cm

m

15 N

Figure 1

$$F(0.02) = 15$$
$$0.02k = 15$$
$$k = \frac{15}{0.02} = 750 \text{ N/m}$$

Thus $F(x) = 750x$ and the work done in stretching the spring from 2 cm to 5 cm beyond its natural length is

$$W = \int_{0.02}^{0.05} 750x \, dx = 375x^2 \Big]_{0.02}^{0.05} = 375 \left[(0.05)^2 - (0.02)^2 \right] = 0.7875 \text{ J}.$$

With our integral definition of work, we can revisit our first example under more realistic conditions.

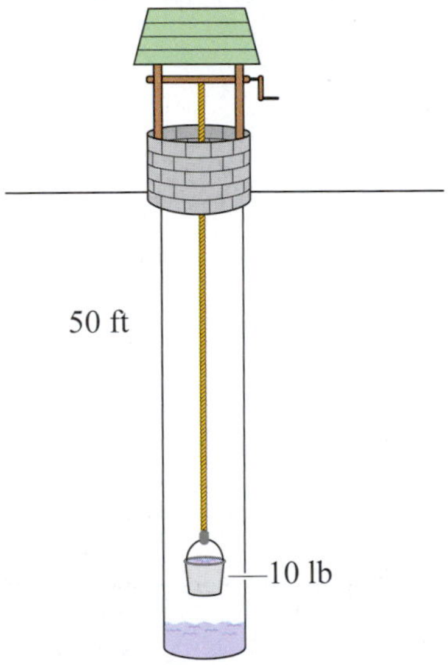

Example 3

A bucket of water weighing 10 lb is lifted out of a well by pulling it up with a 50 ft rope. The rope itself weighs 0.1 lb/ft. How much work is done in lifting the bucket and the rope?

Solution

The weight of the rope is significant, since 50 ft of rope weighs $(0.1 \text{ lb/ft})(50 \text{ ft}) = 5 \text{ lb}$. But after x ft of rope has been pulled in, only $50 - x$ feet are left contributing to the weight of the bucket and remaining rope. So when the bucket has been lifted x ft, the total remaining weight to be lifted is

$$F(x) = \underbrace{10}_{\text{bucket}} + \underbrace{0.1(50 - x)}_{\text{remaining rope}} = 10 + 5 - 0.1x = 15 - 0.1x.$$

The work done can now be calculated as follows:

$$W = \int_0^{50} (15 - 0.1x) \, dx = \left[15x - 0.05x^2 \right]_0^{50} = 625 \text{ ft-lb}$$

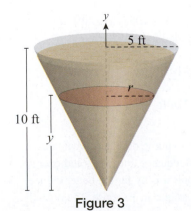

Figure 2

50 ft

10 lb

The task of transferring liquids from one place to another is a common one—other examples include emptying swimming pools, pumping water out of below-grade construction sites, and drawing fuel out of underground storage tanks.

Example 4

The conical underground tank shown in Figure 3 has a radius at the top of 5 ft and a depth of 10 ft. It is filled with gasoline that weighs 44 lb/ft^3. How much work is needed to pump all of the gasoline to **a.** the top of the tank and **b.** 3 ft above the top of the tank?

Solution

We can approach this problem by thinking of the gasoline as made up of a sequence of thin layers, like the one shown in Figure 3, and constructing the differential elements of work dW required to lift each layer the required distance. Each layer is approximately a circular cylinder with differential

Figure 3

y

5 ft

r

10 ft

y

thickness dy and radius r, so its differential element of volume is $dV = \pi r^2\, dy$. By similar triangles, we can compute r:

$$\frac{r}{y} = \frac{5}{10} \quad \Rightarrow \quad r = \frac{y}{2}$$

This leads to

$$dV = \pi \left(\frac{y}{2}\right)^2 dy = \frac{\pi y^2}{4}\, dy.$$

The weight of each differential layer is $44\,dV = 11\pi y^2\, dy$; remember, the weight is the force due to gravity that must be overcome. All that remains is to determine the distance over which each layer must be lifted—that is, the distance over which each force must be applied. In part a., the layer at a height of y above the origin must be lifted $10 - y$ ft, so $dW = 11\pi y^2 (10 - y)\, dy$. In part b. the same layer must be lifted $13 - y$ ft, so $dW = 11\pi y^2 (13 - y)\, dy$. Thus our two answers are as follows.

a. $W = \displaystyle\int_0^{10} dW = \int_0^{10} 11\pi y^2 (10 - y)\, dy = 11\pi \int_0^{10} \left(10 y^2 - y^3\right) dy$

$$= 11\pi \left[\frac{10 y^3}{3} - \frac{y^4}{4}\right]_0^{10} \approx 28{,}798 \text{ ft-lb}$$

b. $W = \displaystyle\int_0^{10} dW = \int_0^{10} 11\pi y^2 (13 - y)\, dy = 11\pi \int_0^{10} \left(13 y^2 - y^3\right) dy$

$$= 11\pi \left[\frac{13 y^3}{3} - \frac{y^4}{4}\right]_0^{10} \approx 63{,}355 \text{ ft-lb}$$

Example 5 ✐

An underground cylindrical tank of depth 10 ft and radius 2.89 ft holds about the same volume as the conical tank of Example 4. It also is filled with gasoline. Determine the work required to pump all of the gasoline to the top of the tank, and compare the answer to part a. of Example 4.

Solution

Each differential element of volume in this case is $dV = \pi (2.89)^2\, dy$, and its weight is $44\pi (2.89)^2\, dy$. So the work needed is as follows:

$$W = \int_0^{10} 44\pi (2.89)^2 (10 - y)\, dy = 44\pi (2.89)^2 \int_0^{10} (10 - y)\, dy$$

$$= 44\pi (2.89)^2 \left[10 y - \frac{y^2}{2}\right]_0^{10} \approx 57{,}726 \text{ ft-lb}$$

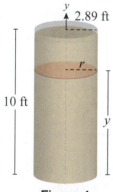

2.89 ft

10 ft

Figure 4

This is about twice the work that was required to pump out the conical tank of the same volume and depth. Remember, work equates to energy, and energy costs money: between these two tanks, the conical is by far the better choice. Of course, many other configurations are possible; in Exercise 22 you will compare these answers to a third tank of the same volume.

TOPIC 2   **Fluid Pressure and Fluid Force**

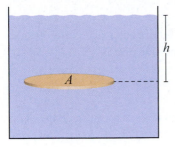

Figure 5

Forces due to fluids appear in another important context—that exerted by the weight of fluids on surfaces. In some instances, the calculation of fluid pressure is easy. For example, consider a plate of surface area A that is oriented horizontally at a depth h below the surface of a liquid, as in Figure 5. The volume of the liquid above the plate is $V = Ah$, so if its density is ρ then its mass is $m = \rho V$ and its weight is

$$F = mg = \rho V g = \rho A h g.$$

Pressure is defined to be force per unit area, so the pressure P of the body of liquid above the horizontal plate is $P = F/A = \rho h g$. In SI units, pressure is usually measured in *pascals* (Pa) or *kilopascals* (kPa), where $1 \text{ Pa} = 1 \text{ N/m}^2$ and $1 \text{ kPa} = 1000 \text{ Pa}$. In the US Customary System, it is convenient to use the *weight density* δ of the fluid in question, where $\delta = \rho g$ (and ρ is the fluid's mass density). In this system, $P = \rho h g = \delta h$, δ is typically measured in lb/ft^3, and pressure therefore has units of pounds per square foot.

The unit of pressure Pa is named after the French mathematician, physicist, and inventor Blaise Pascal (1623–1662) who, in addition to much else, provided insight into a key principle of fluid pressure: *The pressure at any point in a fluid is the same in all directions*. This principle, which reflects the behavior of matter, underlies many familiar properties, such as the fact that divers feel the same pressure on their eardrums at a given depth no matter which way they are oriented in the water. For our immediate purposes, it allows us to determine the force imparted by a fluid against surfaces that are not horizontal. But to do so, we will need calculus.

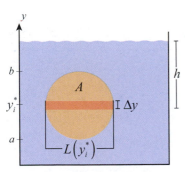

Figure 6

Consider a submerged vertically oriented plate such as shown in Figure 6. Knowing its area is not sufficient to determine the fluid force exerted against one side of it, because the fluid pressure varies with the depth h below the surface. But if we partition the interval $[a,b]$ in the usual manner, all the area of any given horizontal strip is approximately at the same depth h, with the approximation getting better and better as $\Delta y \to 0$. In particular, if the ith subinterval $[y_{i-1}, y_i]$ of the partition has thickness Δy_i and if $y_i^* \in [y_{i-1}, y_i]$ is a sample point, then the area of the ith horizontal strip is $L(y_i^*)\Delta y_i$ and the fluid force exerted on one side of it is

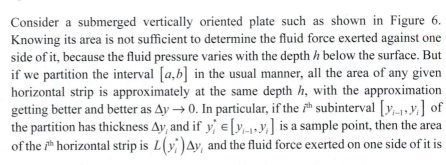

$$\Delta F_i \approx \rho g \cdot (\text{strip depth}) \cdot L(y_i^*)\Delta y_i = \delta \cdot (\text{strip depth}) \cdot L(y_i^*)\Delta y_i$$

where $L(y_i^*)$ is the length of the plate at y_i^*, ρ is its mass density, and $\delta = \rho g$ is its weight density. So the total force on one side of the plate can be approximated as follows.

$$F \approx \sum_{i=1}^{n} \rho g \cdot (\text{strip depth}) \cdot L(y_i^*)\Delta y_i = \sum_{i=1}^{n} \delta \cdot (\text{strip depth}) \cdot L(y_i^*)\Delta y_i$$

Definition 💡

Fluid Force against a Vertical Plate

Given a plate submerged vertically in a fluid of density ρ (or weight density δ), the **fluid force** exerted against one side of it is

$$F = \int_a^b \rho g \cdot (\text{strip depth}) \cdot L(y) \, dy = \int_a^b \delta \cdot (\text{strip depth}) \cdot L(y) \, dy$$

where $L(y)$ is the length of a horizontal strip of the plate at y and the plate is bounded between $y = a$ and $y = b$.

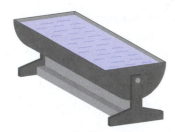

Figure 7

Example 6 ✍

The two ends of a water trough are semicircular plates, each with a radius of 2 ft. The trough is filled with water, and the weight density of water is 62.4 lb/ft^3. How much force is exerted on each end of the trough?

Solution

In order to describe the semicircular ends easily, we can set up our coordinate system so that the positive y-axis points downward and the origin is at the center of the circle as shown in Figure 8. The complete circle is the graph of the equation $x^2 + y^2 = 4$, so if (x, y) is a point on the lower semicircle, $x = \sqrt{4 - y^2}$. Hence the length $L(y)$ of a given horizontal strip is $L(y) = 2x = 2\sqrt{4 - y^2}$. The depth of the strip shown is y.

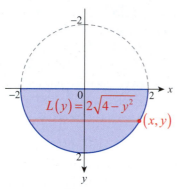

Figure 8

$$F = \int_a^b \delta \cdot (\text{strip depth}) \cdot L(y) \, dy$$
$$= \int_0^2 (62.4)(y)\left(2\sqrt{4 - y^2}\right) dy$$
$$= 62.4 \int_0^2 2y\sqrt{4 - y^2} \, dy$$
$$= -62.4 \int_{u=4}^{u=0} u^{1/2} \, du \qquad \begin{array}{l} u = 4 - y^2 \\ du = -2y \, dy \end{array}$$
$$= -62.4 \left[\frac{2}{3} u^{3/2} \right]_{u=4}^{u=0} = 62.4 \left(\frac{16}{3} \right) = 332.8 \text{ lb}$$

Example 7 ✍

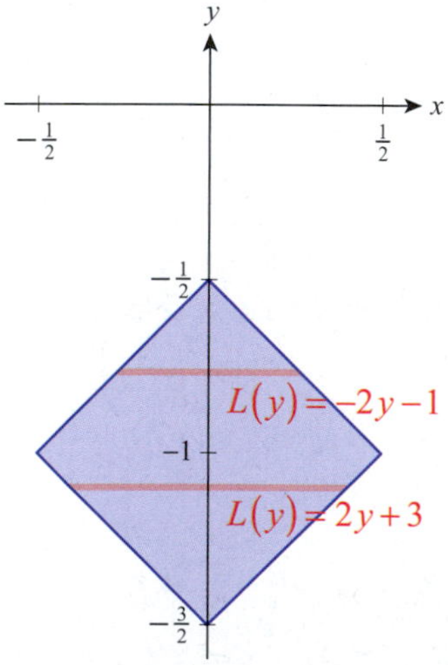

A square pump compartment cover plate with the dimensions in meters shown in Figure 9 is installed into the wall of a pool; its centerline is 1 m below the water's surface. What is the fluid force on the plate?

Solution

The positive y-axis in Figure 9 is pointing upward, in contrast to its orientation in Example 6, but we apply our integral formula in the same way—we just have to be careful to describe the strip depths and the limits of integration correctly.

$L(y) = -2y - 1$

$L(y) = 2y + 3$

Figure 9

We also need an expression for the strip lengths, $L(y)$. Equations for the two lower edges of the plate are $y = -\frac{3}{2} - x$ (lower left edge) and $y = -\frac{3}{2} + x$ (lower right edge). Solving these for x and subtracting the first from the second gives us $L(y) = 2y + 3$ between $y = -\frac{3}{2}$ and $y = -1$; we can check our work by noting that $L\left(-\frac{3}{2}\right) = 0$ and $L(-1) = 1$, which corresponds with what we expect from Figure 9. In similar fashion, $L(y) = -2y - 1$ describes the lengths of the horizontal strips between $y = -1$ and $y = -\frac{1}{2}$.

Since the water's surface is at $y = 0$, the depth of the strip passing through y on the vertical axis is $0 - y = -y$. We are now ready to apply our formula.

$$\begin{aligned}
F &= \int_a^b \rho g \cdot (\text{strip depth}) \cdot L(y)\, dy \\
&= \int_{-3/2}^{-1} \rho g \cdot (-y) \cdot (2y + 3)\, dy + \int_{-1}^{-1/2} \rho g \cdot (-y) \cdot (-2y - 1)\, dy \\
&= \rho g \int_{-3/2}^{-1} \left(-2y^2 - 3y\right) dy + \rho g \int_{-1}^{-1/2} \left(2y^2 + y\right) dy \\
&= \rho g \left[-\frac{2}{3}y^3 - \frac{3}{2}y^2\right]_{-3/2}^{-1} + \rho g \left[\frac{2}{3}y^3 + \frac{1}{2}y^2\right]_{-1}^{-1/2} \\
&= \rho g \left(\frac{1}{2}\right)
\end{aligned}$$

We have used the first version of our fluid force formula (with ρg instead of δ) because the dimensions in the problem, as given, were SI units. The density ρ of water is 1000 kg/m^3 and $g = 9.8 \text{ m/s}^2$.

$$F = \rho g \left(\frac{1}{2}\right) = (1000)(9.8)\left(\frac{1}{2}\right) = 4900 \text{ N}$$

In some cases, the centroid and area of a submerged vertical plate is obvious or easily calculated. If so, they can be used to calculate the fluid force on one side of the plate without integration, by the following argument. The second coordinate, $\bar{y}$, of the centroid with respect to the fluid's surface level is

$$\bar{y} = \frac{\int_a^b y^c \rho\, dA}{\int_a^b \rho\, dA} = \frac{\rho \int_a^b y^c\, dA}{\rho \int_a^b dA} = \frac{\int_a^b y^c\, dA}{A}$$

where y^c is the distance of the center of mass of horizontal strips from the surface (that is, the depth of the horizontal strip) and A is the surface area of the plate. The differential dA can be expressed as $L(y)\, dy$:

$$\bar{y}A = \int_a^b y^c\, dA = \int_a^b (\text{strip depth}) \cdot L(y)\, dy$$

Multiplying both ends of this last equation by either ρg or δ (depending on the context), we obtain the following:

$$F = \rho g \bar{y} A = \delta \bar{y} A$$

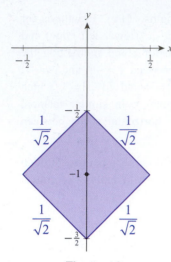

Figure 10

Example 8 ✎

Calculate the fluid force on the square cover plate of Example 7 using its centroid and area.

Solution

The centroid of the plate is 1 m from the water's surface, and the area of the square plate is $\frac{1}{2}$ m^2.

$$F = \left(1000 \text{ kg/m}^3\right)\left(9.8 \text{ m/s}^2\right)\left(1 \text{ m}\right)\left(\frac{1}{2} \text{ m}^2\right)$$

$$= 4900 \; \frac{\text{kg} \cdot \text{m}}{\text{s}^2} = 4900 \text{ N}$$

6.5 **Exercises**

1. A dad is pushing his child on a sled with a constant, horizontal force a distance of 100 m. The child and sled together have a mass of 35 kg and the coefficient of friction is $\mu = 0.1$. Find the work done against friction. Use $g \approx 9.81 \text{ m/s}^2$. (**Hint:** For a refresher on friction and coefficient of friction see Example 5 of Section 1.5.)

2. An object of mass 10 kg is pulled 8 m up a 30° ramp. If the coefficient of friction is $\mu = 1/\left(4\sqrt{3}\right)$, find the work done during the process. (**Hint:** Work needs to be done against both friction and gravity. The normal force between the object and the surface of the ramp is $F_\perp = mg \cos 30°$.)

3. The graph of a variable force is shown below as it moves an object in a straight line a distance of 10 ft. Find the work done during this process.

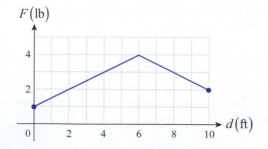

4. A 15 cm long unstressed spring requires a force of 3 N to be held stretched to 20 cm. How much work will be done in stretching the spring an additional 5 cm?

5. A force of 50 lb stretches an 8 ft spring by $\frac{1}{2}$ foot. Find the work done in stretching the spring **a.** from its original length to 10 ft and **b.** from 11 ft to 13 ft.

6. A particle is moving along the y-axis from the origin to the point $(0,4)$ under the influence of the variable force $F(y) = \frac{1}{2}y^2 + \sqrt{y}$ (units are in meters and newtons, respectively). Find the work done by the force.

7. If a 3 lb force compresses a 5 ft spring by 18 in., how much work is done in **a.** compressing it from its original length to 3 ft and **b.** stretching it to 6.75 ft?

8. When a mass of 100 g is hung on a vertical spring, the spring is stretched by 4 cm. How much work is done in stretching the spring an additional 5 cm?

9. A bumping post at a railway company has a spring constant of 3.8×10^5 N/cm. Find the work done in compressing this spring by 2 cm.

10. Suppose that 12.5 ft-lb of work is needed to stretch a spring from its unstressed length of 1.5 ft to 2 ft. Find the spring constant.

11. If the unstressed length of a spring is 20 cm and 0.12 J of work is required to stretch it from 22 cm to 24 cm, how much work is required to stretch it from 24 cm to 30 cm?

12. Suppose the work done in stretching a spring from 8 in. to 10 in. is $\frac{1}{2}$ ft-lb, and an additional 2 ft-lb of work will stretch it by another 4 in. Use this information to find the unstressed length of the spring.

13. We will call a spring linear if it obeys Hooke's Law (recall that this can only be expected between reasonable limits). Prove that the work done in stretching (or compressing) a linear spring by x units from its original length is $W = \frac{1}{2}kx^2$, where k is the spring constant.

14. Suppose that an elastic rope with an unstressed length of 20 ft behaves in the following, nonlinear manner: the force required to stretch it by x ft is $F = kx^{5/4}$ lb. If a force of 3 lb stretches the rope to 21 ft, how much work is done in stretching it from 21 ft to 22 ft?

15. Two identical springs with a spring constant of 5 lb/ft are attached to each other at point P, with their other ends fastened to the top and bottom of a wooden box, so that they are in a vertical, unstressed position. Find the work done in moving point P vertically up or down by 6 in. (Ignore the weight of the springs.)

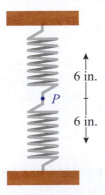

16. A 15 ft chain that weighs 4 lb/ft is hanging from a cylindrical drum so that its other end touches the ground.

   a. How much work does it take to wind it up completely?

   b. How much work does it take to wind up only two-thirds of the chain?

17. A 120 ft cable is supporting a 50 lb piece of equipment at a construction site. In order to lift the equipment by 30 ft, the cable is wound on a cylindrical drum. If the cable weighs 3 lb/ft, find the work done during this process.

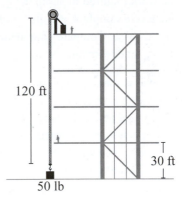

18. A cable starts to unwind from a cylindrical drum at $t = 0$, at a rate of 0.5 m/s. If the cable weighs 20 N/m, find the work done by gravity from $t = 0$ to $t = 12$ s. (**Hint:** Let x denote the length of cable already unwound at time t.)

19. Find the work done in lifting 300 kg of coal from a 400 m deep mine by a rope that weighs 35 N/m. (**Note:** The original term "horsepower" was coined by James Watt after actually watching ponies lift coal from a mine and calculating their work done in unit time.)

20. A 5 lb bucket is used to draw water for livestock from a 60 ft deep well. The bucket weighs 50 lb when full, but is leaking water at 0.1 lb/s. If it is being pulled at a rate of 1 ft/s, find the work done in getting it to the surface. (Ignore the weight of the rope).

21. A crane is lifting a leaky container full of fresh liquid mortar at a construction site. If the container weighs 40 kg and is able to hold 250 kg of mortar, but is leaking at a rate of 0.6 kg/s and is being lifted at 1.5 m/s for 10 s, find the work done by the crane.

22. A cylindrical tank of depth 5 ft and radius 4.087 ft holds about the same volume as the tank in Example 5. If such a tank is filled with gasoline, determine the work required to pump all the gasoline to the top of the tank, and compare the answer to part a. of Example 4 and to Example 5.

23. Find the work required to pump the gasoline out of the tank of Example 4 over its top if the tank stands on its base. Compare your answer to part a. of Example 4 and explain.

24. A tank in the shape of an inverted cone frustum has cross-sections of radius $3 + \frac{1}{3}y$ feet at an altitude of y feet above the base. If its height is 6 feet and it is filled with water, how much work is done to pump all of the water out over the top of the tank? (For the weight density of water, see Example 6.)

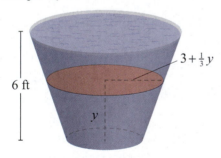

25. A rectangular tank of base 5 ft by 8 ft and of height 10 ft is filled with oil. The weight density of oil is $50 \, \text{lb/ft}^3$.

   a. Find the work required to pump the oil out of the tank through an outlet on the top.

   b. Find the work required if the tank is only half full at the start of pumping.

   c. Is the answer you gave in part b. half of that given in part a.? Why?

   d.* How long will it take for a $\frac{1}{2}$-horsepower pump to empty the tank? What if we start with the tank half full? (One horsepower (hp) is 550 ft-lb/s.)

26. A cistern in the form of an inverted rectangular pyramid with a square base of side length 2 m and a depth of 4 m is full of water.

   a. How much work is required to pump all of the water out over the top edge of the cistern?

   b. How long will it take for a 1 hp electric pump motor to do the job? The weight density of water is $\delta = 9810 \, \text{N/m}^3$. One horsepower is 746 watts (W), where $1 \, \text{W} = 1 \, \text{J/s}$.

27. The two ends of a watering trough are isosceles trapezoids sitting on the shorter base which is 2 ft, with the legs making 120° angles with that base. The opening of the trough is a 4 ft by 12 ft rectangle. Starting with a full trough, find the energy expended (i.e., total work done) in pumping all of the water out of the trough over its top edge.

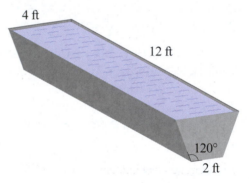

28. Answer Exercise 27 if the ends of the trough are semicircular with a radius of 2 ft.

29. A water trough with the same top opening as that in Exercise 27 has vertical cross-sections in the form of an equilateral triangle. Find the work required to fill it with water through an opening on its bottom.

30. How much work is necessary to pump all fluid of weight density δ out of a hemispherical container of radius R through an opening in its top?

31. Find the work required if the tank in Exercise 30 is inverted.

32. The shape of a kerosene tank can be approximated by rotating the graph of $y = x^4$, $-2 \le x \le 2$, about the y-axis (units in meters). Find the work required to fill up the empty tank through an opening at its lowest point $x = 0$. The weight density of kerosene is 8016.24 N/m^3.

33. According to Archimedes' Principle, the buoyant force acting on a body immersed in fluid is equal to the weight of the fluid it displaces. Use this principle to find the work required to completely immerse in water a cube of negligible weight if its edges are 1 ft. (**Hint:** Keep the top horizontal during immersion.)

34. Repeat Exercise 33 for a buoy that is a circular cone of base radius R and height h, when it is immersed vertically, vertex first in a fluid of weight density δ.

35–43 Use the integral formula for fluid force to answer the question.

35. A 2 ft by 3 ft rectangular plate is positioned vertically on the bottom of an 8 ft pool. Find the fluid force exerted on one side of the plate if it sits **a.** on its long edge and **b.** on its short edge.

36. An equilateral triangle of side length $20\sqrt{3}$ cm is standing on its base that is 50 cm underwater. Find the fluid force against one face of the triangle. Use $\delta = 9.81 \cdot 10^{-3} \text{ N/cm}^3$ for the weight density of water.

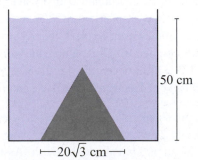

37. Find the force against one face of the triangle from Exercise 36 if the upper third of its vertical altitude is sticking out of the water.

38.* Answer Exercise 36 if the triangle is tilted, making a 30° angle with the horizontal.

39. Suppose that a vertical, parabolic gate is installed on a vertical side of a 10 ft deep gas tank. The gate is given by the equation $y = 1 - x^2$, so that $y = 0$ coincides with the bottom edge of the tank (units in feet). If the gate is designed to withstand a maximum force of 600 lb, verify that it won't break when gas is stored in the tank. Will the gate break if the tank is filled up with oil? The weight densities of gas and oil are 44 lb/ft^3 and 50 lb/ft^3, respectively.

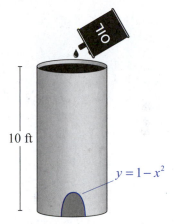

40. Find the force exerted on the end of the full trough in Exercise 27.

41. Find the force exerted on the end of the full trough in Exercise 29.

42.* Find the force exerted on one end of a trough with the same ends and opening as the one in Exercise 29, but with its ends tilted by 45°.

43. Suppose that an outlet near the bottom of the trough in Exercise 27 is covered by a 3 in. by 3 in. square plate, so that the lower edge of the plate coincides with that of the trough. What force is pressing against the plate?

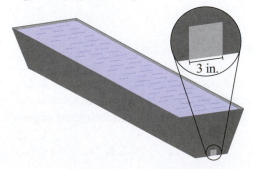

44–47 Consider a certain gas of pressure P and volume V_1 confined in a cylinder, closed on one end by a moveable piston. If A is the area of the piston, the force acting on the piston is $F = PA$. Thus, as the gas expands, pushing the piston by a small increment of Δy, the work done on the piston is $\Delta W = F\Delta y = PA\Delta y = P\Delta V$. Integrating, we obtain the work done by the gas as it expands from a volume of V_1 to V_2:

$$W = \int_{V_1}^{V_2} P\, dV$$

(**Note:** This argument can be "reversed" to obtain the work required to compress the gas.)

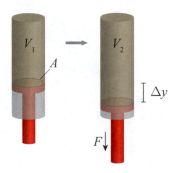

In Exercises 44–47, apply the above formula.

44. Assume that 0.5 ft³ of gas in a cylinder under an initial pressure of 300 lb/ft² expands to 1.5 ft³. Assuming that $PV = c$ (a constant) during this process, find the work done by the gas on the piston.

45. Suppose that 3 L of gas under initial pressure of 150 kPa in a cylinder is compressed to 1 L. Find the work done by the piston. (Assume, as in Exercise 44 that $PV = c$.)

46. Repeat Exercise 44 under the assumption that $PV^{1.4} = c$. (This happens when heat loss is negligible. We say that in this case, P and V are related adiabatically.)

47. Repeat Exercise 45 under the assumption that $PV^{1.4} = c$.

48–55 According to Newton's Law of Gravitation, two masses m_1 and m_2 attract each other by a force that is directly proportional to the product of their masses and inversely proportional to the square of their distance (or rather the square of the distance between their respective centers of gravity):

$$F_g = \frac{m_1 m_2 G}{r^2},$$

where G is the universal gravitational constant. Its value in metric units is $6.67 \times 10^{-11}\ \text{N} \cdot \text{m}^2/\text{kg}^2$.

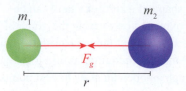

48. Use the above formula to show that the acceleration caused by gravity on a free-falling mass near the Earth's surface is approximately

$$g = \frac{MG}{R^2},$$

where M and R are the mass and radius of the Earth, respectively (we are assuming that the Earth is perfectly spherical).

49. Show that if an object of weight w is launched to a height h above the surface of the Earth, the Earth attracts it by a force of

$$F_g(h) = \frac{R^2 w}{(R+h)^2}.$$

(**Hint:** Use the fact that $w = mg$ along with the result of Exercise 48.)

50. Find the work done against gravity in moving a 3-pound object to an altitude of 500 miles above the surface of the Earth. Assume that the radius of the Earth is approximately 4000 miles. (**Hint:** Integrate the variable force obtained in Exercise 49 between appropriate limits.)

51. Find the work done against gravity in moving a 2-ton satellite to an altitude of 200 miles above the surface of the Earth. Express your answer in the units foot-pounds. (**Hint:** Ignore the work done to accelerate the spacecraft, air resistance, as well as the weight of the launching vehicle and fuel.)

52. How much energy is expended (i.e., work done) in lifting a rocket of mass 10 metric tons to an altitude of 300 km above the Earth? Assume the radius of the Earth is approximately 6371 km. (See the hint given in Exercise 51.)

53. Calculate the work done in Exercise 52 if the rocket is launched from the Moon. The acceleration due to gravity on the Moon is 1.6 m/s^2, about $\frac{1}{6}$ of that on the Earth. The Moon's radius is approximately 1737 km.

54. Find the work done in moving a spacecraft of mass 110 metric tons to an altitude of 300 miles above the surface of the Earth. Use 1 mile ≈ 1600 m and $g \approx 9.81 \text{ m/s}^2$. Express your answer in megajoules. (**Hint:** Ignore the work done to accelerate the spacecraft, air resistance, as well as the weight of the launching vehicle and fuel. See Exercise 52 for the Earth's radius.)

55. Find the limit of the work done in moving a weight w to a distance d above the Earth's surface as $d \to \infty$.

56–57 The magnitude of the force acting between two point charges q_1 and q_2 at a distance of r units from each other, is described by Coulomb's Law:

$$F = k \frac{|q_1 q_2|}{r^2}$$

Notice the analogy between Coulomb's Law and the law of gravity! The charges are analogous to the masses, while k is analogous to the universal gravitational constant. The approximate value of k is $8.98755 \times 10^9 \text{ Nm}^2/\text{C}^2$. (1 coulomb (C) is the SI unit for electric charge. Note that all units are SI; there is no British system of electrical units.) It is also worth noting that Coulomb's Law only gives the magnitude of the force, since electric forces can be attractive or repulsive, while the force of gravity is always attractive.

In Exercises 56–57, use Coulomb's Law.

56. Two like electrical charges of 10^{-4} C each, are 50 cm apart, with one of them fixed. Find the work done in bringing the other charge to a distance of 20 cm from the fixed charge.

57. If two point charges of opposite sign attract each other by a force of 200 N when 3 cm apart, find the work done in moving them from 2 cm apart to 8 cm apart.

58–64 The centroid can be used in some circumstances to simplify work calculations, as well as fluid force calculations. For instance, when the 5 lb rope in Example 3 is extended 50 ft down the well, it behaves, from a center of mass perspective, like a 5 lb point mass 25 ft down the well. The work required to lift such a point mass to the surface is $(5 \text{ lb})(25 \text{ ft}) = 125$ ft-lb. Added to the 500 ft-lb of work required to lift the 10 lb bucket 50 ft, we get the total work required: 625 ft-lb.

In Exercises 58–64, use this centroid argument to find a second solution.

58. Exercise 16   **59.** Exercise 17

60. Exercise 18   **61.** Exercise 19

62. Exercise 25

63. Exercise 28 (**Hint:** Use Exercise 52 of Section 6.4.)

64. Exercise 29

65–68 Use the technique of Example 8 to find a second solution.

65. Exercise 33   **66.** Exercise 34

67. Exercise 36   **68.** Exercise 41

69. Prove the following: If two cylindrical tanks hold the same volume, but the height of tank B is half of the height of tank A, then the work required to pump all liquid out of a full tank A over its top is twice the work required to do the same for tank B. Do the proof in two ways: **a.** by using integration and **b.** by using the centroid approach.

6.5 **Technology Exercises**

70–71 Use a computer algebra system to solve the problem.

70. Suppose that the rocket in Exercise 52 runs out of fuel after 1.9×10^7 kJ of useful energy is expended. How high will the rocket go?

71. Suppose that in Exercise 25 part d., the pump stops due to an electrical failure 3 minutes after starting the job. What is the fluid level at that instant?

6.6 Hyperbolic Functions

TOPICS

1. Definitions, derivatives, and integrals of hyperbolic functions

2. Inverse hyperbolic functions

The six *hyperbolic functions*, along with their inverses, have the same relationship to hyperbolas that the more familiar trigonometric functions have to circles. Although the original reasons for their introduction have faded in relevance, their usefulness in solving differential equations and in describing important physical phenomena has given them continued life.

TOPIC 1 Definitions, Derivatives, and Integrals of Hyperbolic Functions

We will introduce the first two hyperbolic functions via a technique which is useful in its own right. And while this method of introduction doesn't reflect their historical development, we will quickly be able to show how the hyperbolic functions come by their name.

Recall from Section 1.1 that a function f is said to be even if $f(-x) = f(x)$ and odd if $f(-x) = -f(x)$, in each case for all x in the domain. Because these are fairly restrictive properties, it may be surprising to realize that *any* function defined on an interval centered at the origin can be written as the sum of an even function and an odd function. Further, the decomposition for any such function f is simple to describe:

$$f(x) = \underbrace{\frac{f(x) + f(-x)}{2}}_{\text{even part}} + \underbrace{\frac{f(x) - f(-x)}{2}}_{\text{odd part}}$$

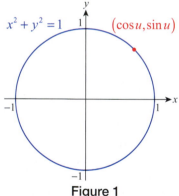

Figure 1
Circular Cosine and Sine Functions

Since the function e^x is defined on the entire real line (and hence its domain can be said to be centered at 0), we can apply the above decomposition. Upon doing so, we label the even and odd parts as follows:

$$e^x = \underbrace{\frac{e^x + e^{-x}}{2}}_{\cosh x} + \underbrace{\frac{e^x - e^{-x}}{2}}_{\sinh x}$$

The notation *cosh* stands for *hyperbolic cosine* and *sinh* stands for *hyperbolic sine*; they are pronounced, respectively, as "kosh" (rhymes with "gosh") and "cinch" (rhyming with "pinch"). By the derivation above, we already know that cosh is an even function and that sinh is odd. But their names come from the fact that, as u takes on different real values, $\cosh u$ and $\sinh u$ trace out the right-hand branch of the hyperbola $x^2 - y^2 = 1$ in the same way that $\cos u$ and $\sin u$ trace out the circle $x^2 + y^2 = 1$. In fact, some early treatises on trigonometry made equal reference to *circular trigonometric* functions and *hyperbolic trigonometric* functions and developed their theory in parallel.

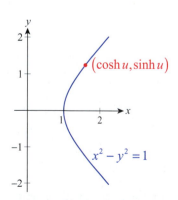

Figure 2
Hyperbolic Cosine and Sine Functions

Figures 1 and 2 illustrate how $(\cos u, \sin u)$ and $(\cosh u, \sinh u)$ trace out their respective curves. Note that in the case of the circular trigonometric functions, u represents the radian measure between the positive x-axis and the ray passing through $(\cos u, \sin u)$. The parameter u doesn't have any similar physical meaning in the hyperbolic case, aside from the fact that if u is positive, $(\cosh u, \sinh u)$ is a point on the upper half of the hyperbola, while a negative u corresponds to a point $(\cosh u, \sinh u)$ on the lower half of the hyperbola. (You should verify that, given the definitions $\cosh u = \left(e^u + e^{-u}\right)/2$ and $\sinh u = \left(e^u - e^{-u}\right)/2$, it is indeed the case that $\cosh^2 u - \sinh^2 u = 1$.)

Once the first two hyperbolic trigonometric functions have been defined, the remaining four follow; all six are formally defined below. For each one, the graph of the function is shown in blue and, if appropriate, asymptotic curves appear as red dashed lines (note how the asymptotic curves relate to the definitions of the functions).

Hyperbolic Functions

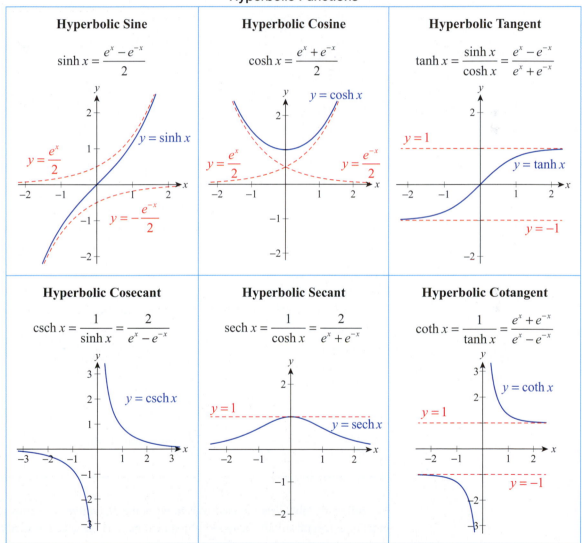

Table 1

Like the trigonometric functions, hyperbolic functions give rise to identities that are useful in simplifying expressions. The following are a few of the more common identities, the proofs of which are all very similar (see Exercises 23–42).

Elementary Hyperbolic Identities

$\cosh^2 x - \sinh^2 x = 1$	$\sinh 2x = 2\sinh x \cosh x$	$\cosh 2x = \cosh^2 x + \sinh^2 x$
$\tanh^2 x = 1 - \text{sech}^2 x$	$\sinh^2 x = \dfrac{\cosh 2x - 1}{2}$	$\sinh(x+y) = \sinh x \cosh y + \cosh x \sinh y$
$\coth^2 x = 1 + \text{csch}^2 x$	$\cosh^2 x = \dfrac{\cosh 2x + 1}{2}$	$\cosh(x+y) = \cosh x \cosh y + \sinh x \sinh y$

Table 2

You have already verified the first identity—it is the one that gives rise to the name of these functions. We will verify two others.

$$\cosh^2 x - \sinh^2 x = 1$$
$$1 - \frac{\sinh^2 x}{\cosh^2 x} = \frac{1}{\cosh^2 x}$$
$$1 - \tanh^2 x = \text{sech}^2 x$$
$$\tanh^2 x = 1 - \text{sech}^2 x$$

and

$$\sinh^2 x = \left(\frac{e^x - e^{-x}}{2}\right)\left(\frac{e^x - e^{-x}}{2}\right)$$
$$= \frac{1}{4}\left(e^{2x} - 2 + e^{-2x}\right)$$
$$= \frac{1}{2}\left(\frac{e^{2x} + e^{-2x}}{2} - 1\right)$$
$$= \frac{\cosh 2x - 1}{2}$$

Since the hyperbolic functions are simply algebraic combinations of exponential functions, they are differentiable everywhere they are defined and their derivatives are easily determined.

Example 1 ✎

Determine the derivatives of the hyperbolic cosine and hyperbolic secant functions.

Solution

$$\frac{d}{dx}(\cosh x) = \frac{d}{dx}\left(\frac{e^x + e^{-x}}{2}\right) = \frac{e^x - e^{-x}}{2} = \sinh x$$

$$\frac{d}{dx}(\text{sech } x) = \frac{d}{dx}\left(\frac{1}{\cosh x}\right) = -\frac{\sinh x}{\cosh^2 x} = -\text{sech } x \tanh x$$

The following table lists, for ease of reference, the derivatives and corresponding indefinite integrals of all six hyperbolic functions. You are asked to verify several of these in Exercises 43–50. Note that the formulas are similar, but not identical, to those for the trigonometric functions.

Derivatives and Integrals of Hyperbolic Functions

$$\frac{d}{dx}(\sinh x) = \cosh x \qquad \int \cosh x \, dx = \sinh x + C$$

$$\frac{d}{dx}(\cosh x) = \sinh x \qquad \int \sinh x \, dx = \cosh x + C$$

$$\frac{d}{dx}(\tanh x) = \operatorname{sech}^2 x \qquad \int \operatorname{sech}^2 x \, dx = \tanh x + C$$

$$\frac{d}{dx}(\operatorname{csch} x) = -\operatorname{csch} x \coth x \qquad \int \operatorname{csch} x \coth x \, dx = -\operatorname{csch} x + C$$

$$\frac{d}{dx}(\operatorname{sech} x) = -\operatorname{sech} x \tanh x \qquad \int \operatorname{sech} x \tanh x \, dx = -\operatorname{sech} x + C$$

$$\frac{d}{dx}(\coth x) = -\operatorname{csch}^2 x \qquad \int \operatorname{csch}^2 x \, dx = -\coth x + C$$

Table 3

Example 2 ✐

Evaluate each of the following:

a. $\dfrac{d}{dx}\left(\coth \sqrt{x^2+1}\right)$ **b.** $\dfrac{d}{dt}\left[\ln(\cosh t)\right]$

c. $\displaystyle\int e^x \operatorname{csch}^2\left(e^x\right) dx$ **d.** $\displaystyle\int_{\ln 2}^{\ln 4} 2e^x \cosh x \, dx$

Solution

a. $\dfrac{d}{dx}\left(\coth \sqrt{x^2+1}\right) = \left(-\operatorname{csch}^2 \sqrt{x^2+1}\right)\dfrac{d}{dx}\left(\sqrt{x^2+1}\right)$

$$= \left(-\operatorname{csch}^2 \sqrt{x^2+1}\right)\left(\frac{2x}{2\sqrt{x^2+1}}\right) = \frac{-x\operatorname{csch}^2 \sqrt{x^2+1}}{\sqrt{x^2+1}}$$

b. $\dfrac{d}{dt}\left[\ln(\cosh t)\right] = \dfrac{1}{\cosh t}(\sinh t) = \tanh t$

c. $\displaystyle\int e^x \operatorname{csch}^2\left(e^x\right) dx = \int \operatorname{csch}^2 u \, du$ $\begin{aligned}u &= e^x \\ du &= e^x \, dx\end{aligned}$

$$= -\coth u + C$$

$$= -\coth\left(e^x\right) + C$$

d. $\displaystyle\int_{\ln 2}^{\ln 4} 2e^x \cosh x \, dx = \int_{\ln 2}^{\ln 4} 2e^x\left(\frac{e^x + e^{-x}}{2}\right) dx$

$$= \int_{\ln 2}^{\ln 4}\left(e^{2x} + 1\right) dx$$

$$= \left[\frac{1}{2}e^{2x} + x\right]_{\ln 2}^{\ln 4}$$

$$= \left(\frac{1}{2}e^{2\ln 4} + \ln 4\right) - \left(\frac{1}{2}e^{2\ln 2} + \ln 2\right)$$

$$= (8 + \ln 4) - (2 + \ln 2) = 6 + \ln 2 \qquad \ln 4 = 2\ln 2$$

Many of the most common uses of hyperbolic functions today reflect the fact that they often arise in the process of solving particular differential equations—that is, equations that relate the derivative(s) of a function to other quantities. While we won't study techniques of solving differential equations until Chapter 8, we can verify that hyperbolic functions do indeed serve as the solutions of some representative problems now.

Example 3 ✑

A common model for the velocity v of an object of mass m falling under the influence of gravity says that the air resistance encountered by the object is proportional to v^2. Under this assumption, the velocity t seconds into a fall (starting with $v(0) = 0$) is described by the differential equation

$$m\frac{dv}{dt} = mg - kv^2$$

where g, as usual, denotes the acceleration due to gravity and k is a constant that depends on the object's aerodynamic characteristics and the air's density (the case $k = 0$ is valid only if there truly is no air resistance, and implies the object falls faster and faster until it hits the ground).

The function $v(t) = \sqrt{mg/k}\,\tanh\left(\sqrt{gk/m}\,t\right)$ is the solution of this differential equation, which we can verify as follows.

$$\frac{dv}{dt} = \sqrt{\frac{mg}{k}}\sqrt{\frac{gk}{m}}\,\text{sech}^2\left(\sqrt{\frac{gk}{m}}\,t\right) = g\,\text{sech}^2\left(\sqrt{\frac{gk}{m}}\,t\right)$$

and

$$v^2 = \frac{mg}{k}\tanh^2\left(\sqrt{\frac{gk}{m}}\,t\right) = \frac{mg}{k}\left[1 - \text{sech}^2\left(\sqrt{\frac{gk}{m}}\,t\right)\right] \qquad \tanh^2 x = 1 - \text{sech}^2 x$$

Hence,

$$mg - kv^2 = mg - mg\left[1 - \text{sech}^2\left(\sqrt{\frac{gk}{m}}\,t\right)\right] = mg\,\text{sech}^2\left(\sqrt{\frac{gk}{m}}\,t\right) = m\frac{dv}{dt}.$$

We can apply this model to estimate the terminal, or limiting, velocity that a skydiver might attain before deploying his or her parachute. A typical value for k in this case is 0.005, and if we assume the skydiver's weight is 160 pounds, then $mg = 160$ (remember that weight is mass times the acceleration due to gravity).

Regardless of the values of g, k, and m, we have the following.

$$\tanh\left(\sqrt{\frac{gk}{m}}\,t\right) \to 1 \quad \text{as} \quad t \to \infty$$

Therefore,

$$v(t) = \sqrt{\frac{mg}{k}}\tanh\left(\sqrt{\frac{gk}{m}}\,t\right) \approx \sqrt{\frac{160}{0.005}}\,(1) \approx 178.89 \text{ ft/s}$$

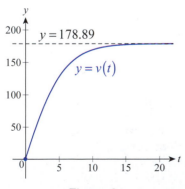

Figure 3

for sufficiently large t. We can get a good sense of how large qualifies as "sufficiently large" by graphing v with these values for m and k, as shown in Figure 3. Notice that by time $t \approx 15$ seconds, $v(t)$ is close to the **terminal velocity** of 178.89 ft/s.

TOPIC 2 Inverse Hyperbolic Functions

We will find the inverse hyperbolic functions immediately useful in evaluating integrals. First, we will gain some familiarity with their behavior.

The notation we use to denote inverse hyperbolic functions follows the standard practice of denoting the inverse of the function f by f^{-1}. In each of the following graphs, a hyperbolic function (or a restricted portion, if necessary) is shown in blue, with the inverse hyperbolic function in red being its reflection across the line $y = x$.

Inverse Hyperbolic Functions

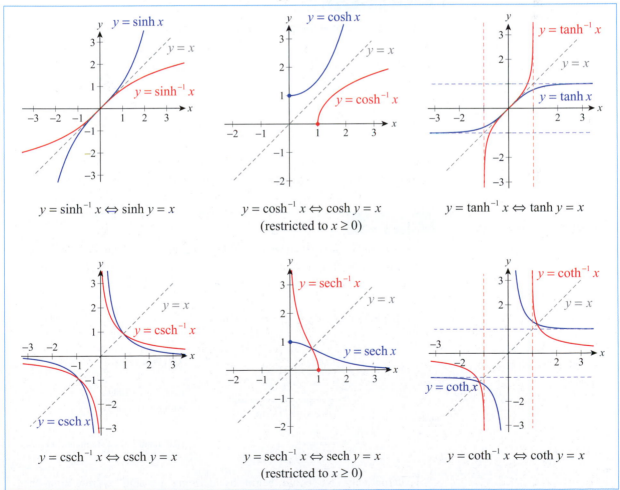

$y = \sinh^{-1} x \Leftrightarrow \sinh y = x$

$y = \cosh^{-1} x \Leftrightarrow \cosh y = x$
(restricted to $x \geq 0$)

$y = \tanh^{-1} x \Leftrightarrow \tanh y = x$

$y = \operatorname{csch}^{-1} x \Leftrightarrow \operatorname{csch} y = x$

$y = \operatorname{sech}^{-1} x \Leftrightarrow \operatorname{sech} y = x$
(restricted to $x \geq 0$)

$y = \coth^{-1} x \Leftrightarrow \coth y = x$

Table 4

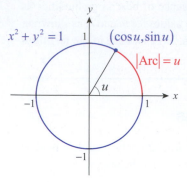

$x^2 + y^2 = 1$

$(\cos u, \sin u)$

$|\mathrm{Arc}| = u$

$x = \cos u \Leftrightarrow \arccos x = u$
$y = \sin u \Leftrightarrow \arcsin y = u$
Arc Length $= u$

Figure 4 Arc Length

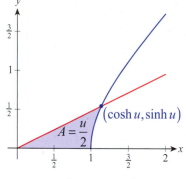

$(\cosh u, \sinh u)$

$A = \dfrac{u}{2}$

$x = \cosh u \Leftrightarrow \text{arcosh } x = u$
$y = \sinh u \Leftrightarrow \text{arsinh } y = u$

Area $A = \dfrac{u}{2}$

Figure 5 Hyperbolic Sector

As with the inverse trigonometric functions, there is an alternative way of denoting the inverse hyperbolic functions. Unfortunately, the meaning of the nomenclature has become clouded and somewhat confused over time. The alternatives *arcsine* and *arccosine* for $\sin^{-1}$ and $\cos^{-1}$ reflect the fact that, given a point (x, y) on the circle of radius 1 centered at the origin, $\arccos x$ and $\arcsin y$ both represent the radian angle u shown in Figure 4. In similar fashion, $\cosh^{-1} x$ and $\sinh^{-1} y$ are related to the *area* of the hyperbolic sector shown in Figure 5—for this reason, *arcosh* and *arsinh* (where the *ar* stands for "area") are synonyms for $\cosh^{-1}$ and $\sinh^{-1}$. As it turns out, the parameter u is equal to twice the area A, a fact you will prove in Exercise 111.

Unfortunately, the similarity between *ar* and *arc* led, over time, to the frequent use today of *arccosh* and *arcsinh* instead of *arcosh* and *arsinh*—these names are often seen in software packages and programming languages (*Mathematica*, for example, contains the built-in commands **ArcCosh**, **ArcSinh**, and so on).

The next step in our development is to determine the derivatives of the inverse hyperbolic functions.

Theorem 🔍

Derivatives of Inverse Hyperbolic Functions

$$\frac{d}{dx}\left(\sinh^{-1} x\right) = \frac{1}{\sqrt{1 + x^2}}$$

$$\frac{d}{dx}\left(\text{csch}^{-1} x\right) = \frac{-1}{|x|\sqrt{1 + x^2}}$$

$$\frac{d}{dx}\left(\cosh^{-1} x\right) = \frac{1}{\sqrt{x^2 - 1}}, \quad x > 1$$

$$\frac{d}{dx}\left(\text{sech}^{-1} x\right) = \frac{-1}{x\sqrt{1 - x^2}}, \quad 0 < x < 1$$

$$\frac{d}{dx}\left(\tanh^{-1} x\right) = \frac{1}{1 - x^2}, \quad |x| < 1$$

$$\frac{d}{dx}\left(\coth^{-1} x\right) = \frac{1}{1 - x^2}, \quad |x| > 1$$

Proof 📎

We will prove the first formula, and leave several of the others as exercises (Exercises 78–81). If we let $f(x) = \sinh x$, then by the Derivative Rule for Inverse Functions (Section 3.6) we have the following:

$$\frac{d}{dx}\left(\sinh^{-1} x\right) = \left(f^{-1}\right)'(x) = \frac{1}{f'\left(f^{-1}(x)\right)}$$

$$= \frac{1}{\cosh\left(\sinh^{-1} x\right)} \qquad \color{blue}{f'(x) = \cosh x}$$

$$= \frac{1}{\sqrt{1 + \sinh^2\left(\sinh^{-1} x\right)}} \qquad \color{blue}{\begin{aligned}\cosh^2 x &= 1 + \sinh^2 x\\ \cosh x &= \sqrt{1 + \sinh^2 x}\end{aligned}}$$

$$= \frac{1}{\sqrt{1 + x^2}} \qquad \color{blue}{\sinh^2\left(\sinh^{-1} x\right) = \left[\sinh\left(\sinh^{-1} x\right)\right]^2 = x^2}$$

(Alternatively, we could begin by defining $y = \sinh^{-1} x$, rewriting the equation as $\sinh y = x$, and proceeding to implicitly differentiate both sides with respect to x. Solving the result for dy/dx would yield the same answer.)

Rephrasing the above derivative rules in terms of integrals expands the types of integrals we can evaluate. Each of the following statements can be verified by differentiating both sides (applying the Chain Rule as needed).

Integrals and Inverse Hyperbolic Functions

$$\int \frac{1}{\sqrt{k^2 + x^2}}\, dx = \sinh^{-1}\left(\frac{x}{k}\right) + C, \quad k > 0$$

$$\int \frac{1}{\sqrt{x^2 - k^2}}\, dx = \cosh^{-1}\left(\frac{x}{k}\right) + C, \quad 0 < k < x$$

$$\int \frac{1}{x\sqrt{k^2 + x^2}}\, dx = -\frac{1}{k}\operatorname{csch}^{-1}\left|\frac{x}{k}\right| + C, \quad x \ne 0 \text{ and } 0 < k$$

$$\int \frac{1}{x\sqrt{k^2 - x^2}}\, dx = -\frac{1}{k}\operatorname{sech}^{-1}\left(\frac{x}{k}\right) + C, \quad 0 < x < k$$

$$\int \frac{1}{k^2 - x^2}\, dx = \begin{cases} \dfrac{1}{k}\tanh^{-1}\left(\dfrac{x}{k}\right) + C & \text{if } x^2 < k^2 \\[2mm] \dfrac{1}{k}\coth^{-1}\left(\dfrac{x}{k}\right) + C & \text{if } x^2 > k^2 \end{cases}$$

Table 5

Example 4 ✐

Evaluate each of the following:

a. $\displaystyle\int \frac{dx}{\sqrt{4 + 9x^2}}$

b. $\displaystyle\int_0^{1/3} \frac{dx}{1 - x^2}$

Solution

a. $\displaystyle\int \frac{dx}{\sqrt{4 + 9x^2}} = \int \frac{dx}{3\sqrt{\frac{4}{9} + x^2}} = \frac{1}{3}\sinh^{-1}\left(\frac{x}{\frac{2}{3}}\right) + C = \frac{1}{3}\sinh^{-1}\left(\frac{3x}{2}\right) + C$

b. $\displaystyle\int_0^{1/3} \frac{dx}{1 - x^2} = \left[\tanh^{-1} x\right]_0^{1/3} = \tanh^{-1}\left(\frac{1}{3}\right)$

Note that we used $\tanh^{-1} x$ as the antiderivative (as opposed to $\coth^{-1} x$) because, in the given interval of integration, $x^2 < k^2$ ($k = 1$). Further, if we apply the result of Exercise 74, we can rewrite our answer as follows:

$$\tanh^{-1}\left(\frac{1}{3}\right) = \frac{1}{2}\ln\left(\frac{1 + \frac{1}{3}}{1 - \frac{1}{3}}\right) = \frac{1}{2}\ln 2$$

6.6 **Exercises**

1–4 Find the value of the function.

1. a. $\cosh 0$

    b. $\sinh(-2)$

    c. $\sinh\dfrac{\pi}{2}$

2. a. $\tanh 0$

    b. $\cosh 1$

    c. $\sinh(\ln 3)$

3. a. $\sinh^{-1} 0$

    b. $\cosh^{-1} 1$

    c. $\sinh^{-1} 2$

4. a. $\tanh^{-1} 0$

    b. $\operatorname{sech}^{-1} 1$

    c. $\operatorname{csch}^{-1}(-3)$

5–10 Use the given equation to classify the hyperbolic function as even or odd. Then use the definition of the function to prove your assertion.

5. $\sinh(-x) = -\sinh x$

6. $\cosh(-x) = \cosh x$

7. $\tanh(-x) = -\tanh x$

8. $\coth(-x) = -\coth x$

9. $\operatorname{sech}(-x) = \operatorname{sech} x$

10. $\operatorname{csch}(-x) = -\operatorname{csch} x$

11–22 Sketch the graph of the equation on a piece of paper. (**Hint:** Study the graphs of the six hyperbolic functions and their inverses in the text.)

11. $y = 1 - \sinh x$

12. $y = \cosh 2x - 3$

13. $y = -\tanh(x-1)$

14. $y = 2 - 2\operatorname{sech}(x-1)$

15. $y = -\dfrac{1}{2}\operatorname{csch} x + 1$

16. $y = -\dfrac{1}{4}\coth x - 2$

17. $y = \dfrac{1}{3}\sinh^{-1}(x-3)$

18. $y = 2\cosh^{-1}(1-x)$

19. $y = -\tanh^{-1}\left(\dfrac{x}{3}\right)$

20. $y = -3\operatorname{csch}^{-1}(x+2)$

21. $y = \operatorname{sech}^{-1}\left(1 - \dfrac{x}{2}\right) - 1$

22. $y = 2\coth^{-1}\left(\dfrac{x}{2} + 1\right)$

23–42 Verify the given identity.

23. $e^{kx} = \cosh kx + \sinh kx$

24. $e^{-kx} = \cosh kx - \sinh kx$

25. $\sinh(x+y) = \sinh x \cosh y + \cosh x \sinh y$

26. $\sinh(x-y) = \sinh x \cosh y - \cosh x \sinh y$

27. $\cosh(x+y) = \cosh x \cosh y + \sinh x \sinh y$

28. $\cosh(x-y) = \cosh x \cosh y - \sinh x \sinh y$

29. $\sinh 2x = 2\sinh x \cosh x$

30. $\cosh 2x = \cosh^2 x + \sinh^2 x$

31. $\cosh^2 x = \dfrac{\cosh 2x + 1}{2}$

32. $\coth^2 x = 1 + \operatorname{csch}^2 x$

33. $\tanh(x+y) = \dfrac{\tanh x + \tanh y}{1 + \tanh x \tanh y}$

34. $\tanh(x-y) = \dfrac{\tanh x - \tanh y}{1 - \tanh x \tanh y}$

35. $\tanh 2x = \dfrac{2\tanh x}{1 + \tanh^2 x}$

36. $\coth 2x = \dfrac{1 + \coth^2 x}{2\coth x}$

37. $\sinh\dfrac{x}{2} = \pm\sqrt{\dfrac{\cosh x - 1}{2}}$

38. $\cosh\dfrac{x}{2} = \sqrt{\dfrac{\cosh x + 1}{2}}$

39. $\tanh\dfrac{x}{2} = \pm\sqrt{\dfrac{\cosh x - 1}{\cosh x + 1}}$

40. $(\cosh x + \sinh x)^2 = \cosh 2x + \sinh 2x$

41. $(\cosh x + \sinh x)^n = \cosh nx + \sinh nx, \quad n \in \mathbb{N}$

42. $(\cosh x - \sinh x)^n = \cosh nx - \sinh nx, \quad n \in \mathbb{N}$

43–50 Verify the differentiation or integration formula.

43. $\dfrac{d}{dx}(\sinh x) = \cosh x$

44. $\displaystyle\int \sinh x \, dx = \cosh x + C$

45. $\dfrac{d}{dx}(\tanh x) = \operatorname{sech}^2 x$

46. $\dfrac{d}{dx}(\coth x) = -\operatorname{csch}^2 x$

47. $\dfrac{d}{dx}(\operatorname{csch} x) = -\operatorname{csch} x \coth x$

48. $\displaystyle\int \operatorname{csch}^2 x\, dx = -\coth x + C$

49. $\displaystyle\int \tanh x\, dx = \ln(\cosh x) + C$

50. $\displaystyle\int \coth x\, dx = \ln|\sinh x| + C$

51–62 Find the given derivative.

51. $\dfrac{d}{dx}(\sinh^2 x)$

52. $\dfrac{d}{dt}(\operatorname{sech}^2 2t)$

53. $\dfrac{d}{dx}(\tanh^2 x)$

54. $\dfrac{d}{dx}\left(\cosh^2 \dfrac{2x+5}{3}\right)$

55. $\dfrac{d}{dx}(e^{-x}\sinh 2x)$

56. $\dfrac{d}{dx}\left[\operatorname{sech}(1-x^2)\right]$

57. $\dfrac{d}{dx}\sqrt{1+\tanh 2x}$

58. $\dfrac{d}{dx}\left[\ln(\cosh(2x+1))\right]$

59. $\dfrac{d}{dt}\left[(t^3+1)\cosh(t^3+1)\right]$

60. $\dfrac{d}{dz}\left[\dfrac{1}{z}\tanh(z^2)\right]$

61. $\dfrac{d}{dy}\left[\cosh(\ln(2y+1))\right]$

62. $\dfrac{d}{dx}\left[\dfrac{\sinh(\ln x)}{x+1}\right]$

63–71 Evaluate the given integral.

63. $\displaystyle\int \dfrac{1+\sinh^2 x}{\cosh x}\, dx$

64. $\displaystyle\int \operatorname{sech}^2\left(\dfrac{x}{3}+5\right) dx$

65. $\displaystyle\int \dfrac{\operatorname{sech}\dfrac{1}{z}\tanh\dfrac{1}{z}}{z^2}\, dz$

66. $\displaystyle\int \dfrac{\coth\sqrt{x}}{\sqrt{x}}\, dx$

67. $\displaystyle\int \tanh(2-w)\, dw$

68. $\displaystyle\int x\cosh(x^2)\, dx$

69. $\displaystyle\int_0^1 \operatorname{csch}^2(2-t)\, dt$

70. $\displaystyle\int_0^3 \operatorname{sech}^2 u\sqrt{\tanh u}\, du$

71. $\displaystyle\int_0^2 \dfrac{\sinh t}{\sqrt{4-\cosh^2 t}}\, dt$

72. By differentiating, verify the given identity.

$$\dfrac{d}{dx}\left[\tan^{-1}(\sinh x)\right] = \dfrac{d}{dx}\left[\sin^{-1}(\tanh x)\right] = \operatorname{sech} x$$

73–74 Given that the hyperbolic functions can be expressed in terms of exponential functions, it's not surprising that their inverses can be expressed in terms of logarithms. For example, if we let $y = \sinh^{-1} x$, then $x = \sinh y$ and hence

$$x = \dfrac{e^y - e^{-y}}{2}$$

$$2x = e^y - e^{-y}$$

$$e^y - 2x - e^{-y} = 0$$

$e^{2y} - 2xe^y - 1 = 0$ Multiply through by e^y.

$(e^y)^2 - 2xe^y - 1 = 0$ Express as a quadratic in e^y.

$e^y = \dfrac{2x \pm \sqrt{4x^2+4}}{2}$ Solve for e^y.

$e^y = x + \sqrt{x^2+1}$ $x - \sqrt{x^2+1} < 0$ but $e^y > 0$, so discard $x - \sqrt{x^2+1}$.

$y = \ln\left(x + \sqrt{x^2+1}\right)$.   Take the natural logarithm of both sides.

Use the procedure above to verify the identity.

73. $\cosh^{-1} x = \ln\left(x + \sqrt{x^2-1}\right), \quad x \geq 1$

74. $\tanh^{-1} x = \dfrac{1}{2}\ln\left(\dfrac{1+x}{1-x}\right), \quad -1 < x < 1$

(**Hint:** Begin by setting $y = \tanh^{-1} x$; then write the equation as $\tanh y = x$, square both sides, and apply the identity $\tanh^2 y = 1 - \operatorname{sech}^2 y$. Solve the result for $\cosh y$, apply $\cosh^{-1}$ to both sides, and apply the result of the previous exercise. Then apply some logarithmic properties.)

75–77 Given a function f, let $1/f$ denote its reciprocal—that is, $\left(\dfrac{1}{f}\right)(x) = \dfrac{1}{f(x)}$. The following is a useful relationship between the functions f^{-1} and $(1/f)^{-1}$, assuming both of these inverse functions exist.

$$\left(\dfrac{1}{f}\right)\left(f^{-1}\left(\dfrac{1}{x}\right)\right) = \dfrac{1}{f\left(f^{-1}\left(\dfrac{1}{x}\right)\right)} = \dfrac{1}{\dfrac{1}{x}} = x$$

so $\left(\dfrac{1}{f}\right)^{-1}(x) = f^{-1}\left(\dfrac{1}{x}\right)$.

Applied to hyperbolic functions, this fact indicates the following:

$$\operatorname{csch}^{-1} x = \sinh^{-1}\frac{1}{x} \qquad \operatorname{sech}^{-1} x = \cosh^{-1}\frac{1}{x} \qquad \coth^{-1} x = \tanh^{-1}\frac{1}{x}$$

Use these relationships to verify the equality.

75. $\operatorname{csch}^{-1} x = \ln\left(\dfrac{1}{x} + \dfrac{\sqrt{1+x^2}}{|x|}\right); \quad x \neq 0$

76. $\operatorname{sech}^{-1} x = \ln\left(\dfrac{1+\sqrt{1-x^2}}{x}\right); \quad 0 < x \leq 1$

77. $\coth^{-1} x = \dfrac{1}{2}\ln\left(\dfrac{x+1}{x-1}\right); \quad |x| > 1$

78–81 Prove the given formula for the derivative of an inverse hyperbolic function.

78. $\dfrac{d}{dx}\left(\cosh^{-1} x\right) = \dfrac{1}{\sqrt{x^2-1}}, \quad x > 1$

79. $\dfrac{d}{dx}\left(\operatorname{sech}^{-1} x\right) = \dfrac{-1}{x\sqrt{1-x^2}}, \quad 0 < x < 1$

80. $\dfrac{d}{dx}\left(\tanh^{-1} x\right) = \dfrac{1}{1-x^2}, \quad |x| < 1$

81. $\dfrac{d}{dx}\left(\coth^{-1} x\right) = \dfrac{1}{1-x^2}, \quad |x| > 1$

82–88 Find the given derivative.

82. $\dfrac{d}{dx}\left[\dfrac{1}{2}\sinh^{-1}(5x)\right]$

83. $\dfrac{d}{dx}\left[\cosh^{-1}\left(\dfrac{7x}{2}\right)\right]$

84. $\dfrac{d}{dx}\left[\operatorname{sech}^{-1}(\sin x)\right]$

85. $\dfrac{d}{dx}\left[\tanh^{-1}(\cos 2x)\right]$

86. $\dfrac{d}{dx}\left[\cosh^{-1}(\sec x)\right]; \quad 0 < x < \dfrac{\pi}{2}$

87. $\dfrac{d}{dx}\left[\coth^{-1}\left(\dfrac{1}{2x}\right)\right]$

88. $\dfrac{d}{dx}\left(x \cdot 2^{\sinh^{-1} x}\right)$

89–102 Evaluate the given integral.

89. $\displaystyle\int \dfrac{dx}{\sqrt{16x^2+9}}$

90. $\displaystyle\int \dfrac{x}{\sqrt{25x^4-4}}\,dx$

91. $\displaystyle\int \dfrac{dx}{-x^2+6x-8}$

92. $\displaystyle\int \dfrac{x}{\sqrt{x^4-4x^2+3}}\,dx$

93. $\displaystyle\int \dfrac{-\cos x}{\sqrt{\sin^2 x + \sin^4 x}}\,dx$

94. $\displaystyle\int_0^{1/8} \dfrac{2}{1-4x^2}\,dx$

95. $\displaystyle\int_0^5 \dfrac{10}{\sqrt{1+25x^2}}\,dx$

96. $\displaystyle\int \dfrac{dx}{\sqrt{9+e^{2x}}}$

97. $\displaystyle\int_{-3/2}^{-1} \dfrac{-1}{(x+2)\sqrt{-3-4x-x^2}}\,dx$

98. $\displaystyle\int_1^2 \dfrac{-1}{x\sqrt{1+x^2}}\,dx$

99. $\displaystyle\int \dfrac{dx}{x\sqrt{1+x^3}}$

100. $\displaystyle\int \dfrac{dx}{\sqrt{x}\sqrt{1+x}}$

101. $\displaystyle\int_{1/8}^{1/4} \dfrac{dx}{x\sqrt{1-4x^2}}$

102. $\displaystyle\int_7^9 \dfrac{-4}{-x^2+6x-5}\,dx$

103–110 Evaluate the given limit.

103. $\displaystyle\lim_{x\to\infty} \dfrac{\cosh x}{x^2}$

104. $\displaystyle\lim_{x\to 0} \dfrac{\sinh x}{x}$

105. $\displaystyle\lim_{x\to 0} \dfrac{\sinh^{-1} x}{x}$

106. $\displaystyle\lim_{x\to 0} \dfrac{\tanh x}{x}$

107. $\displaystyle\lim_{x\to\infty} \dfrac{\cosh x}{e^x}$

108. $\displaystyle\lim_{x\to\infty} x\coth^{-1} x$

109. $\displaystyle\lim_{x\to 0} \dfrac{1-\operatorname{sech} x}{\sinh x}$

110. $\displaystyle\lim_{x\to 0}\left(\operatorname{csch} x - \coth x\right)$

111. To exhibit another nice analogy between trigonometric and hyperbolic functions, prove that the area of both the circular and hyperbolic sectors corresponding to the parameter u in Figures 4 and 5 equals $A = u/2$. (**Hint:** For the circular sector, use the fact that the arc length of the sector is u units, while the radius is 1. For the hyperbolic sector, notice that

$$A(u) = \frac{\cosh u \sinh u}{2} - \int_1^{\cosh u} \sqrt{x^2-1}\,dx,$$

and show by differentiating that $A'(u) = \frac{1}{2}$.)

112. Prove the following interesting property of the hyperbolic cosine function: The area under the graph of $y = \cosh x$ $(0 \leq x \leq c)$ is equal to its arc length over the same interval.

113. A flexible chain or cable suspended between two fixed points forms a curve called a *catenary* (from the Latin word "catenarius," meaning "related to a chain"). The equation of a catenary is $y = a\cosh(x/a)$. Note that the clothesline of Exercise 56 in Section 6.3 is much better approximated by the equation $y = 5\cosh(x/5) - 4$. Use this equation to answer the following questions.

a. What is the slope of the clothesline at each of its endpoints? Compare your answer to the slope predicted by the parabolic model.

b. What is the length of the clothesline? Is your answer close to that given to Exercise 83 in Section 6.3? (Galileo observed that the parabolic model is almost exact when the angle of elevation is less than 45°.)

114.* Suppose that a dog is 3 yards off from a sidewalk, held on a tight leash by its master, such that the leash is initially perpendicular to the sidewalk. The dog's master starts walking on the sidewalk at a steady, slow pace, pulling the dog while it is offering slight resistance, thereby keeping the leash tight. The curve of the dog's path in this situation is called a *tractrix* (from the Latin word "trahere," meaning "pull" or "drag"), and can be given by the equation $y = 3\operatorname{sech}^{-1}(x/3) - \sqrt{9 - x^2}$. (We assume that the master started from the origin, walking along the positive y-axis, while the dog's initial position was $(3,0)$.)

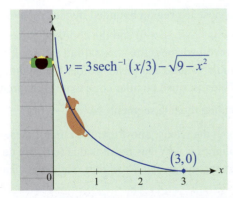

$y = 3\operatorname{sech}^{-1}(x/3) - \sqrt{9 - x^2}$

$(3,0)$

a. Find how far the master has to walk in order to bring the dog within 1 yd of the sidewalk.

b. Prove that the dog's velocity vector at any time is pointing towards its master. (For this reason, the tractrix is also called a "curve of pursuit.")

115–118 *True or False?* Determine whether the given statement is true or false. In case of a false statement, explain or provide a counterexample.

115. $\tanh x = \dfrac{e^{2x} - 1}{e^{2x} + 1}$

116. $|\operatorname{sech} x| \le 1$ for all x

117. The value of $\operatorname{csch} 0$ is infinity.

118. The function $f(x) = \operatorname{sech}^{-1}x$ is defined on $[0,1]$.

6.6 Technology Exercises

119–120 Use a graphing calculator or computer algebra system to solve the problem.

119. The Gateway Arch of St. Louis (constructed in 1963–1965) was designed by Eero Saarinen so that its central curve (the curve tracing the centroids of the triangular cross-sections) is an inverted, "flattened" catenary that is described by the equation $y = -68.7672\cosh(0.0100333x) + 693.8597$ (units in feet).

a. How tall is the central curve of the arch?

b. How wide is the central curve at ground level?

c. What is the slope of the central curve at an altitude of 600 ft above ground level?

120. Suppose that the clothesline of Exercise 113 became loose, and resembles the parabola $y = \frac{1}{2}x^2 + 0.1$, $-1.5 \le x \le 1.5$ (in particular, it is sagging so that its lowest point is just 0.1 yd above the ground). However, we know that in reality, the clothesline can be much better approximated by a catenary. Find an equation for such a catenary, and graph both curves on the same screen. Do they seem to "overlap," or are they distinguishable? Then graph both the parabolic model and the catenary from Exercise 113 on the same screen. Comparing the two sets of graphs, what can you conclude?

121–122 Use a computer algebra system to graph the given function and to find its derivative and any relative extrema.

121. $f(x) = (x^2 - x)\tanh x$

122. $g(x) = x^2(\sinh x - 3\tanh x)$

Chapter 6
Review Exercises

1–2 The base of a solid S is described in the xy-plane along with its cross-sections in a certain direction. Find the volume of S.

1. The base of S is the first-quadrant region of the unit disk centered at the origin; the cross-sections perpendicular to the y-axis are squares.

2. The base of S is the region bounded by the graph of $y = \sqrt{2-x}$ and the coordinate axes; the cross-sections perpendicular to the x-axis are equilateral triangles.

3–6 Use the disk/washer method to find the volume of the solid generated by rotating the region R about the indicated axis.

3. R is bounded by the graph of $y = \sqrt{1-x}$ and the coordinate axes, rotated about the x-axis.

4. R is the region of Exercise 3, rotated about the y-axis.

5. R is bounded by the graphs of $y = \operatorname{arcsec} x$, $x = \sqrt{2}$, and $y = 0$, rotated about the y-axis.

6. R is the first-quadrant portion of the region bounded by the graphs of $y = 4/x^2$ and $y = 5 - x^2$, rotated about $x = -2$.

7–10 Use the shell method to find the volume of the solid generated by rotating the region R about the indicated axis.

7. R is bounded by the graph of $y = 2x - x^2$ and the x-axis, rotated about the y-axis.

8. R is bounded by the graphs of $y = x$, $y = 3 - 2x$, and the y-axis, rotated about $x = -1$.

9. R is the first-quadrant region bounded by the graphs of $y = 4x$ and $y = x^3$, rotated about the x-axis.

10. R is bounded by the graphs of $y = \sqrt{x}$ and $y = x^2$, rotated about $y = -2$.

11–14 The given integral represents the volume of a solid of revolution. Describe the solid. (Do not evaluate the integral.)

11. $\displaystyle\int_0^3 \pi x^2 \, dx$

12. $\displaystyle\pi \int_0^2 \left[1 - (y-1)^2\right] dy$

13. $\displaystyle\int_0^2 2\pi x \left(4 - x^2\right) dx$

14. $\displaystyle 2\pi \int_1^3 y\sqrt{1 - (y-2)^2} \, dy$

15–18 Use the shell method or the disk/washer method to find the volume of the solid obtained by revolving the region bounded by the graphs of the equations about the given axis. Choose the method that seems to work best.

15. $x = \sqrt{\sin y}$, $x = 0$, $0 \le y \le \pi$; about the y-axis

16. $y = x^3 - 2x^2 + x$, $y = 0$, $x = 0$, $x = 1$; about the x-axis

17. $y = \sqrt{x}$, $y = 2 - x$, $x = 0$; about the y-axis

18. $y = \dfrac{e^x}{x}$, $y = 0$, $x = 1$, $x = 2$; about the y-axis

19. Suppose the region bounded by the graph of $y = \sqrt{c^2 - x}$ $(c \ne 0)$ and the coordinate axes is revolved about the y-axis. Find the resulting solid's volume and then consider the volume of the inscribed circular cone that results from rotating the line segment $y = c - \dfrac{x}{c}$, $0 \le x \le c^2$, about the y-axis. Show that the ratio of these two volumes is $\frac{8}{5}$. (This result is due to Pierre de Fermat, the great 17th-century mathematician.)

20–23 Set up, but do not evaluate, an integral defining the arc length of the graph of the equation over the given interval.

20. $y = x^2 - x + 1$; $0 \le x \le 3$

21. $x + 2y^2 = 3 - y$; $1 \le y \le 5$

22. $y = \tan x;$ $-\pi/4 \le x \le \pi/4$

23. $e^x = \cos y;$ $0 \le y \le \pi/3$

24–29 Determine the arc length L of the curve defined by the equation over the given interval.

24. $y = \sqrt{3}x + 1;$ $0 \le x \le 4$

25. $y = \dfrac{4x^{3/2} + 1}{6};$ $0 \le x \le 3$

26. $y = \dfrac{x^4 + 3}{6x};$ $1 \le x \le 3$

27. $y = \dfrac{x^2}{4} - \ln\sqrt{x};$ $1 \le x \le e^2$

28. $y = \dfrac{\sqrt{x}}{6} - 2x^{3/2};$ $0 \le x \le 1$

29. $y = e^{x/2} + e^{-x/2};$ $0 \le x \le 2$

30–33 Set up, but do not evaluate, an integral defining the surface area of the solid obtained by revolving the given curve about the indicated axis.

30. $y = \sin x;$ $0 \le x \le \pi;$ about the x-axis

31. $y = \sqrt{\ln x};$ $1 \le x \le e;$ about the x-axis

32. $y = \sqrt{\ln x};$ $1 \le x \le e;$ about the y-axis

33. $y = x^2;$ $0 \le x \le 2;$ about $x = -1$

34–39 Find the surface area of the solid generated by revolving the given curve about the indicated axis.

34. $y = \dfrac{1}{3}x + 2;$ $0 \le x \le 3;$ about the x-axis

35. $y = \dfrac{x^3}{4};$ $0 \le x \le 1;$ about the x-axis

36. $3y = 3x^{3/2} - \sqrt{x};$ $1 \le x \le 2;$ about the x-axis

37. $x = \dfrac{y^3}{12} + \dfrac{1}{y};$ $1 \le y \le 2;$ about the y-axis

38. $x = 1.5y^{5/3} - 0.3\sqrt[3]{y};$ $0 \le y \le 1;$ about the y-axis

39. $12xy = 3y^4 + 4;$ $1 \le y \le 2;$ about $x = -\dfrac{1}{2}$

40–43 Find the centroid of the plane region bounded by the given curves. If possible, use symmetry to simplify your calculations.

40. $y = 5x - x^2,$ $y = 0$

41. $y = 2\sqrt{x},$ $y = 2x$

42. $y = x^{3/5},$ $x = 0,$ $y = 8$

43. $y = x,$ $y(x+1)^2 = 4,$ $x = 0$

44–47 Find the centroid of the plane region of varying density that is bounded by the given curves.

44. $2y = 4 - x,$ $x = 0,$ $y = 0;$ $\rho(x,y) = x^2$

45. $2y = 4 - x,$ $x = 0,$ $y = 0;$ $\rho(x,y) = \sqrt{y}$

46. $y = x,$ $y = \sqrt{x};$ $\rho(x,y) = 1 + x$

47. $x = 9 - y^2,$ $y = x - 3;$ $\rho(x,y) = y + 3$

48. Suppose the density of a baseball bat lying along the positive x-axis is given by the function $\rho(x) = (0.033x + 0.5)^2$ ounces per inch ($0 \le x \le 30$). Find the center of mass of the bat.

49. Find the centroid of the upper semielliptical region bounded by the x-axis and the graph of $\dfrac{x^2}{a^2} + \dfrac{y^2}{b^2} = 1$. (See Exercise 58 of Section 6.4.)

50. Use Pappus' Theorem for volumes to find the volume of the solid generated by revolving the region of Exercise 43 about the y-axis.

51.* Use Pappus' Theorem for volumes to find a second solution to Exercise 49 using the fact that the area of the semiellipse is $A = \pi ab/2$.

52. A 10 cm long unstressed spring requires a force of 2 N to stretch to 14 cm. How much work will be done in stretching the spring an additional 4 cm?

53. A 100 ft cable is lifting a 30 lb weight by 20 ft, so that the cable is wound on a cylindrical drum. Find the total work done if the cable weighs 2 lb/ft.

54. A 20 m cable is being wound up on a cylindrical drum at a rate of 25 cm/s. If the cable weighs 10 N/m, find the work done from $t = 4$ s to $t = 32$ s.

55. A leaky container is being lifted by a crane at 0.5 m/s for 20 s. If the full container originally weighed 300 kg, but is leaking at a rate of 0.1 kg/s, find the work done by the crane. Use $g \approx 9.81 \, \text{m/s}^2$.

56. The shape of an underground gasoline tank is an inverted circular cone with a top radius of 6 ft and a depth of 18 ft. Find the work done in pumping the full content of the tank to 1 ft above the top of the tank. The weight density of gasoline is 44 lb/ft³.

57. Find the total work done in Exercise 56 if the tank initially is filled to only half of its capacity.

58. The shape of a water tank can be approximated by rotating the graph of $y = x^2$, $-2 \le x \le 2$, about the y-axis (units in meters). Find the work required to fill up the empty tank through an opening at its lowest point. Use 9810 N/m³ for the weight density of water.

59.* A cup of soda of height 5 inches can be approximated by a frustum of a circular cone, standing on its smaller base of radius 1.25 inches, with a top opening of radius 2 inches. Find the work done in drinking a full cup of soda through a straw, supposing that the end of the straw is 2 inches above the rim of the cup. Use the weight density of water, 62.4 lb/ft³, for the soda.

60. A cylindrical underground tank of radius 3 ft is positioned horizontally, and is half full of gasoline. Find the force exerted by the gasoline on one end of the tank. (See Exercise 56 for the weight density of gasoline.)

61. Use the centroid approach (Example 8 in Section 6.5) to provide a second solution to Exercise 60. (**Hint:** See Exercise 52 of Section 6.4)

62. Use the centroid approach to provide a second solution to Exercise 56. (**Hint:** See the discussion preceding Exercises 58–64 of Section 6.5 and use the fact that the centroid of a right circular cone is on the axis of the cone, one-fourth of the way from the base towards the vertex.)

63. One end of a dam is an isosceles trapezoid, standing on its shorter base of 15 meters. The top width is 30 meters, while the depth of the dam is 20 meters. Find the force exerted by the water on one end of the dam. Use 9810 N/m³ for the weight density of water.

64.* The water depth in an 18 ft by 35 ft pool increases uniformly from the shallow end of 3 ft to the deep end of 8 ft. Find the fluid force exerted on the bottom of the pool. Use 62.4 lb/ft³ for the weight density of water.

65. Find the energy expended in moving a 1.5-ton satellite to an altitude of 220 miles above the surface of the Earth. Assume the radius of the Earth is approximately 4000 miles. Express your answer in the units foot-pounds. (**Hint:** Ignore the work done in accelerating the satellite, air resistance, as well as the weight of the launching vehicle and fuel.)

66–67 Find the indicated function value. If it doesn't exist, say so.

66. a. $\sinh 0$ **b.** $\operatorname{csch} 0$

67. a. $\coth^{-1} 0$ **b.** $\operatorname{sech} 0$

68. Use the definition to show that $\tanh x > 0$ if $x > 0$ and $\tanh x < 0$ if $x < 0$.

69. Prove that if $f(x) = \tanh x$ or $f(x) = \coth x$, then $\lim_{x \to \infty} f(x) = 1$ and $\lim_{x \to -\infty} f(x) = -1$.

70. Suppose that $e^x = f(x) + g(x)$, where $f(x)$ is even and $g(x)$ is odd. Prove that in this case, we must necessarily have $f(x) = \cosh x$ and $g(x) = \sinh x$. (**Hint:** Start by expressing e^{-x} in terms of $f(x)$ and $g(x)$.)

71. Verify the following identity:

$$\cosh x + \cosh y = 2 \cosh \frac{x+y}{2} \cosh \frac{x-y}{2}$$

72–81 Find the derivative or integral as indicated.

72. $\dfrac{d}{dx}\left[\sinh^2\left(\dfrac{6x-1}{2}\right)\right]$ **73.** $\dfrac{d}{dx}\left(e^{2x}\tanh x\right)$

74. $\dfrac{d}{dx}\left[\operatorname{csch}\left(x^2+2\right)\right]$ **75.** $\dfrac{d}{dx}\left[\ln\left(\sinh(5x-2)\right)\right]$

76. $\displaystyle\int\dfrac{\tanh\sqrt{x}}{2\sqrt{x}}\,dx$ **77.** $\displaystyle\int\dfrac{\sinh x}{\cosh^2 x}\,dx$

78. $\displaystyle\int\coth x\operatorname{csch}^2 x\,dx$ **79.** $\displaystyle\int\dfrac{dx}{\sqrt{x^2-4x+5}}$

80. $\displaystyle\int_0^1\dfrac{2x}{\sqrt{1+9x^4}}\,dx$ **81.** $\displaystyle\int\dfrac{e^x}{\sqrt{e^{2x}-1}}\,dx;\quad x>0$

82. Prove the formula for the derivative of the inverse hyperbolic cosecant function:

$$\dfrac{d}{dx}\left(\operatorname{csch}^{-1}x\right)=\dfrac{-1}{|x|\sqrt{1+x^2}}$$

83–85 In the 16th century, Galileo observed the additive property of velocities. For example, if a boat travels at 5 mph in the direction of flow on a river that in turn flows at 2 mph, then the velocity of the boat relative to the shore will be $u=5+2=7$ mph. However, according to Albert Einstein's theory of special relativity, the relative velocity between any two objects can never exceed c, the speed of light. This contradicts Galileo's observation. (On a theoretical level, just think of a "river" flowing at $\frac{3}{4}c$, with a "boat" on it traveling at, say, $\frac{1}{2}c$. The velocity relative to shore would be $u=\left(\frac{3}{4}+\frac{1}{2}\right)c=\frac{5}{4}c!$) Instead of the Galilean $u=v_1+v_2$, Einstein discovered the following relativistic addition formula for velocities:

$$u=\dfrac{v_1+v_2}{1+\dfrac{v_1v_2}{c^2}}$$

In Exercises 83–85, refer to the above formula.

83. Use the addition formula for $\tanh x$ (Exercise 33 of Section 6.6) to verify the following identity:

$$\tanh^{-1}\left(\dfrac{u}{c}\right)=\tanh^{-1}\left(\dfrac{v_1}{c}\right)+\tanh^{-1}\left(\dfrac{v_2}{c}\right)$$

84. Suppose a fighter plane fires a missile at 500 mph in the forward direction at a moment when the plane itself is flying at 900 mph. Use Einstein's relativistic formula to find the missile's velocity relative to Earth and compare it with Galileo's prediction of $500+900=1400$ mph. Approximate the speed of light by 3×10^8 m/s.

85. In this exercise, we are going to up the numbers of Exercise 84 significantly. Suppose a rocket is traveling away from the Earth at a speed of $0.7c$ and fires another rocket at $0.5c$. Use Einstein's formula to calculate the velocity of the second rocket relative to Earth.

86–92 *True or False?* Determine whether the given statement is true or false. In case of a false statement, explain or provide a counterexample.

86. The disk method is based on the idea of integrating slices.

87. When both the disk method and the shell method are applied to calculate the volume of a solid of revolution, the variable of integration is always the same.

88. If the area of the region bounded by $y=f(x)$ and $y=g(x)$ is A, then the volume of the solid obtained by revolving the same region about the x-axis is $V=\pi A^2$.

89. If the area of the region bounded by $y=f(x)$ and $y=g(x)$ is A, then the volume of the solid obtained by revolving the same region about the x-axis never equals πA^2.

90. $\displaystyle\lim_{x\to-\infty}\tanh x=-1$

91. $\cosh 2x=2\sinh^2 x+1$

92. The work needed to pump fluid out of a tank through an opening on its top equals the total weight of the fluid multiplied by the distance traveled by its center of mass.

Chapter 6
Technology Exercises

93–96 Use a computer algebra system to find (or approximate) the volume of the solid generated by rotating the region bounded by the graphs of the given equations about the indicated axis.

93. $y = \sin(x^2)$, $\quad y = 0$, $\quad x = 0$, $\quad x = \sqrt{\pi}$; about the x-axis

94. $y = \arccos x$, $\quad y = 0$, $\quad x = 0$, $\quad x = 1$; about the x-axis

95. $y = \sinh^{-1} x$, $\quad y = 0$, $\quad x = 4$; about the y-axis

96. $y = x^2 \sin^2 x$, $\quad y = 0$, $\quad x = 0$, $\quad x = \pi$; about the y-axis

97–98 Use a computer algebra system to find the arc length of the graph of the equation over the given interval.

97. $y = \dfrac{1}{x^2 + 1}$; $\quad -1 \le x \le 1$
    98. $y = \sin x$; $\quad 0 \le x \le \pi$

99–100 Use a computer algebra system to find the surface area of the solid generated by revolving the given curve about the indicated axis.

99. $y = \sin x$; $\quad 0 \le x \le \pi$; about the x-axis
    100. $y = \sqrt{\ln x}$; $\quad 1 \le x \le e$; about the y-axis

Project ✏️

Chapter 6

In this project, we will expand upon our explorations from Exercises 48 through 55 of Section 6.5 (also see the discussion preceding those exercises). In particular, we will determine an equation satisfied by the velocity of a projectile launched with initial velocity v_0, taking into consideration that acceleration caused by gravity decreases with altitude. (This is important when objects are launched to great altitudes.) We will then use our equation to find the maximum height attained by the projectile. This will lead us to the value of the so-called *escape velocity*, the velocity needed for an object to be able to overcome the Earth's gravitational field without further propulsion. (In turn, since gravity is conservative, this is the same velocity an object would achieve if pulled in by gravity from an "infinite distance.") We are ignoring all retarding forces (such as air resistance or friction) in this discussion.

1. Recall from Exercise 48 of Section 6.5 that g, the acceleration caused by gravity on a free-falling mass near the Earth's surface, is approximately

$$g = \frac{MG}{R^2},$$

   where M and R are the mass and radius of the Earth, respectively, and G is the universal gravitational constant. However, a launched projectile's acceleration caused by gravity is negative (if we are assuming the positive direction is upwards) and actually depends on its height h above the Earth's surface. In particular, use Newton's Law of Gravitation to show that this dependence is given by the equation

$$a(h) = \frac{-gR^2}{(R+h)^2},$$

   where $a(0) = -g$, as we would expect. (Actually, $a(h) \approx -g$ when h is negligible compared to the Earth's radius.)

2. Show that if $v = v(h)$ denotes the velocity of the projectile, then

$$\frac{d}{dh}(v^2) = 2\frac{dv}{dt}.$$

   (**Hint:** Use the Chain Rule.)

3. Use the above results to show that

$$\frac{d(v^2)}{dh} = \frac{-2gR^2}{(R+h)^2}.$$

4. Integrating both sides of the previous equation with respect to h, show that $v = v(h)$ satisfies the equation

$$v^2 = v_0^2 - 2gR\left(1 - \frac{R}{R+h}\right).$$

   (**Hint:** After integrating, use the fact that $v(0) = v_0$.)

5. Use the equation found in Question 4 to find the maximum height attained by the projectile. (**Hint:** Use the fact that $v = 0$ when the projectile reaches its maximum height.)

6. Find a formula for the escape velocity v_e of the projectile; then use the data found in the exercises of Section 6.5 (Exercises 48–55 and the preceding discussion) to express your answer in kilometers per second. (**Hint:** Use the fact that if $v_0 = v_e$, the projectile will "travel to infinity.")

7. Find the escape velocity of the projectile if it is launched on the Moon. (**Hint:** For Moon data, see Exercise 53 of Section 6.5.)

Chapter 7
Techniques of Integration

7.1 Integration by Parts 573

1. Integration by Parts—Indefinite Version
2. Integration by Parts—Definite Version

7.2 The Partial Fractions Method 581

1. Using Partial Fractions to Integrate Rational Functions
2. Method to Determine Partial Fraction Constants

7.3 Trigonometric Integrals 592

1. Powers of Sines and Cosines
2. Powers of Tangents and Secants

7.4 Trigonometric Substitutions 599

1. Three Common Trigonometric Substitutions
2. Additional Substitutions

7.5 Integration Summary and Integration Using Computer Algebra Systems 608

1. Integration Recap and Guidelines
2. Integration with Computer Algebra Systems

7.6 Numerical Integration 616

1. The Trapezoidal Rule
2. Simpson's Rule

7.7 Improper Integrals 627

1. Improper Integrals of Type I
2. Improper Integrals of Type II
3. The Direct Comparison Test

Introduction

Just as Chapter 3 focused on developing techniques of differentiation, this chapter focuses on techniques of integration. While the list of techniques presented here is not exhaustive, it includes those that are most powerful and that can be used to evaluate the vast majority of the integrals typically encountered.

Integration techniques rapidly expand the number of integrals we can evaluate. If we were forced to evaluate definite integrals using nothing more than limits of Riemann sums and known antiderivatives, we would find most integrals to be intractable—limits are often difficult to evaluate exactly, and the list of integrands whose antiderivatives are known is, unfortunately, relatively short. The Substitution Rule, or *u*-substitution, qualifies as the first integration technique in our repertoire of methods, but this chapter includes additional techniques that go by names such as integration by parts, partial fractions, and trigonometric substitutions.

Integration methods are essentially algorithmic in nature, and today's technology allows us to use the techniques in ways unimagined by the mathematicians who originally developed them. Specifically, computer algebra systems can be programmed to apply integration techniques, and therefore a CAS is an important modern tool that can be used to evaluate integrals. Further, computers and calculators can be used to rapidly approximate definite integrals that, for one reason or another, elude exact evaluation. At their most basic level, numerical evaluation methods such as the Trapezoidal Rule and Simpson's Rule use the underlying principle of Riemann sums to approximate a given definite integral to within a specified error tolerance.

The last section of the chapter concerns *improper integrals*, meaning integrals in which unbounded regions make an appearance in one fashion or another. Such integrals occur in the study of probability and statistics, and familiarity with their form and calculation is important. Improper integrals also lie at the heart of many *transform operations*, which are techniques that turn one kind of function into another. The Laplace

> Integration methods are essentially algorithmic in nature, and today's technology allows us to use the techniques in ways unimagined by the mathematicians who originally developed them.

Thomas Simpson
(1710–1761)

and Fourier transforms, used in later math, science, and engineering classes, have applications ranging from theoretical physics to the design of the phones, computers, and other electronic devices that are now ubiquitous.

7.1 **Integration by Parts**

TOPICS

1. Integration by parts—indefinite version
2. Integration by parts—definite version

By the end of Chapter 5, we had acquired enough experience with basic integration techniques to allow us to explore some important applications of the integral. Much of our basic knowledge of integration can be succinctly summarized in a table of integrals, such as the one that appears at the end of this text—remember, these are merely facts about differentiation written in integral form. We also learned, in Section 5.4, a mechanical technique called the Substitution Rule, or u-substitution; this was a restatement of the Chain Rule of differentiation. In this chapter, we'll learn additional integration techniques that allow us to evaluate more integrals, including methods that provide highly accurate approximations of integrals when exact evaluation isn't possible. We begin, in this section, with another technique associated with a differentiation rule.

TOPIC 1   **Integration by Parts—Indefinite Version**

Recall how the Chain Rule gives rise to the Substitution Rule:

$$\frac{d}{dx}\big[f\big(g(x)\big)\big] = f'\big(g(x)\big)g'(x) \quad\Rightarrow\quad \int f'\big(g(x)\big)g'(x)\,dx = f\big(g(x)\big) + C$$

In practice, we make use of this by noting that if we define $u = g(x)$, then the integral $\int f'\big(g(x)\big)g'(x)\,dx$ takes the form $\int f'(u)\,du$ and its evaluation is $f(u) + C$ (hence the name u-substitution). In similar fashion, the Product Rule of differentiation leads to integration by parts as follows:

$$\frac{d}{dx}\big[f(x)g(x)\big] = f'(x)g(x) + f(x)g'(x)$$

$$\Rightarrow\quad \int\big[f'(x)g(x) + f(x)g'(x)\big]\,dx = f(x)g(x) + C$$

We usually make use of this by rearranging the terms and writing

$$\int f(x)g'(x)\,dx = f(x)g(x) - \int g(x)f'(x)\,dx$$

(we can omit the constant of integration C in the formula because a constant of integration will appear upon evaluating the integral $\int g(x)f'(x)\,dx$).

Formula 🔍

Integration by Parts Formula

Given differentiable functions f and g,

$$\int f(x)g'(x)\,dx = f(x)g(x) - \int g(x)f'(x)\,dx.$$

If we let $u = f(x)$ and $v = g(x)$, then $du = f'(x)\,dx$ and $dv = g'(x)\,dx$ and the equation takes on the more easily remembered differential form

$$\int u\,dv = uv - \int v\,du.$$

This formula is most often used when, given the integral $\int u\,dv$, we find the integral $\int v\,du$ easier to evaluate. After u-substitution, integration by parts is one of the most powerful and frequently used integration techniques. The art in its use lies in defining the parts u and v productively.

Example 1 ✐

Evaluate $\int x \cos x \, dx$.

Solution

If we want to use integration by parts, we have several choices for u and dv. If we let $u = x$ and $dv = \cos x \, dx$, then $du = dx$ and one possible v is $\sin x$ (we could also use $\sin x + C$, where C is any other constant, but $v = \sin x$ is the simplest choice). It helps to write this as follows:

$$u = x \qquad dv = \cos x \, dx$$
$$du = dx \qquad v = \sin x$$

Now we can apply integration by parts.

$$\int \underbrace{x}_{u} \underbrace{\cos x \, dx}_{dv} = \underbrace{x}_{u} \underbrace{\sin x}_{v} - \int \underbrace{\sin x}_{v} \underbrace{dx}_{du} = x \sin x + \cos x + C$$

What happens if we use another choice for u and dv? For example, $u = \cos x$ and $dv = x \, dx$? In this case,

$$u = \cos x \qquad dv = x \, dx$$
$$du = -\sin x \, dx \qquad v = \frac{x^2}{2}$$

and

$$\int x \cos x \, dx = \frac{x^2 \cos x}{2} + \frac{1}{2} \int x^2 \sin x \, dx.$$

The new integral on the right is more difficult to evaluate than the original integral, so if this had been our first approach we would need to consider stopping at this point and trying another choice.

Since we use integration by parts to try to replace a difficult integral with an easier one, a good rule of thumb is to let dv be as much of the original integral as can be easily integrated (and let u be the remaining factor of the integrand).

Example 2 ✐

Evaluate $\int \ln x \, dx$.

Solution

If we want to use integration by parts on this integral, we don't actually have much choice for dv. If we let $dv = \ln x \, dx$, finding v amounts to evaluating the original integral. So we are led to try $dv = dx$ and $u = \ln x$.

$$u = \ln x \qquad dv = dx$$
$$du = \frac{1}{x} \, dx \qquad v = x$$

Integrating by parts, we obtain the following:

$$\int \ln x \, dx = x \ln x - \int x\left(\frac{1}{x}\right) dx = x \ln x - x + C$$

Replacing one integral with a simpler one is progress, even if the process has to be repeated in order to end up with an integral that can actually be evaluated.

Example 3 ✎

Evaluate $\int x^2 \sin x \, dx$.

Solution

We could let $dv = x^2 \, dx$, but then v would be a third degree polynomial—increasing the degree of part of the integrand takes us further from our goal. So we are led to the following:

$$u = x^2 \qquad dv = \sin x \, dx$$
$$du = 2x \, dx \qquad v = -\cos x$$

and thus

$$\int x^2 \sin x \, dx = -x^2 \cos x + 2\int x \cos x \, dx.$$

The integral on the right is not one that can be evaluated on sight, but we have already used integration by parts in Example 1 to find $\int x \cos x \, dx = x \sin x + \cos x + C$.

$$\int x^2 \sin x \, dx = -x^2 \cos x + 2\int x \cos x \, dx$$
$$= -x^2 \cos x + 2x \sin x + 2 \cos x + C$$
$$= \left(2 - x^2\right)\cos x + 2x \sin x + C$$

As we will illustrate in Example 4, a multistep process may occur in another way when using integration by parts.

Example 4 ✎

Evaluate $\int e^x \sin x \, dx$.

Solution

Any choice of u and dv results in an integral that is no worse than the original, but unfortunately also no better. However, we can evaluate the integral by using integration by parts twice.

$$\int e^x \sin x \, dx = -e^x \cos x + \int e^x \cos x \, dx \qquad \begin{array}{ll} u = e^x & dv = \sin x \, dx \\ du = e^x \, dx & v = -\cos x \end{array}$$

and

$$\int e^x \cos x \, dx = e^x \sin x - \int e^x \sin x \, dx \qquad \begin{array}{ll} u = e^x & dv = \cos x \, dx \\ du = e^x \, dx & v = \sin x \end{array}$$

Replacing the integral on the right of the first step with the result of the second step, we have the following:

$$\int e^x \sin x \, dx = -e^x \cos x + e^x \sin x - \int e^x \sin x \, dx$$

Our original integral appears on both sides of this last equation, but we can now solve for it. We add the integral to both sides, then we divide by 2 and add a constant of integration:

$$2\int e^x \sin x \, dx = -e^x \cos x + e^x \sin x$$

$$\int e^x \sin x \, dx = \frac{e^x}{2}(\sin x - \cos x) + C$$

We will use indefinite integration by parts one last time to show how one of the so-called *reduction formulas* (called such because they reduce the exponent appearing in the integrand) can be proved. You will prove and use others in Exercises 81–95.

Example 5 ✍

Given $n \in \mathbb{N}$, prove that $\displaystyle\int \cos^n x \, dx = \frac{\cos^{n-1} x \sin x}{n} + \frac{n-1}{n}\int \cos^{n-2} x \, dx$.

Solution

If we let $u = \cos^{n-1} x$ and $dv = \cos x \, dx$, then $\int \cos^n x \, dx = \int u \, dv$. This is a reasonable choice for u and dv because we can certainly find v.

$$u = \cos^{n-1} x \qquad\qquad dv = \cos x \, dx$$
$$du = (n-1)(\cos^{n-2} x)(-\sin x) \, dx \qquad v = \sin x$$

Now we can apply integration by parts.

$$\int \cos^n x \, dx = \int (\cos^{n-1} x)(\cos x) \, dx$$
$$= \cos^{n-1} x \sin x + (n-1)\int \sin^2 x \cos^{n-2} x \, dx$$
$$= \cos^{n-1} x \sin x + (n-1)\int (1 - \cos^2 x)\cos^{n-2} x \, dx$$
$$= \cos^{n-1} x \sin x + (n-1)\int \cos^{n-2} x \, dx - (n-1)\int \cos^n x \, dx$$

Adding $(n-1)\int \cos^n x \, dx$ to both sides of this equation gives us

$$n\int \cos^n x \, dx = \cos^{n-1} x \sin x + (n-1)\int \cos^{n-2} x \, dx$$

and so

$$\int \cos^n x \, dx = \frac{\cos^{n-1} x \sin x}{n} + \frac{n-1}{n}\int \cos^{n-2} x \, dx.$$

This reduction formula is useful because, given an integral of the form $\int \cos^n x \, dx$, we can apply it repeatedly until we arrive at an integral we can evaluate, as we will do in the next example.

Example 6 ✒

Evaluate $\int \cos^5 x \, dx$.

Solution

$$\int \cos^5 x \, dx = \frac{\cos^4 x \sin x}{5} + \frac{4}{5} \int \cos^3 x \, dx$$

$$= \frac{\cos^4 x \sin x}{5} + \frac{4}{5}\left(\frac{\cos^2 x \sin x}{3} + \frac{2}{3} \int \cos x \, dx \right)$$

$$= \frac{\cos^4 x \sin x}{5} + \frac{4 \cos^2 x \sin x}{15} + \frac{8}{15} \sin x + C$$

TOPIC 2   Integration by Parts—Definite Version

The integration by parts formula can also be used to evaluate definite integrals, assuming the functions f' and g' are both continuous over the interval of integration $[a,b]$. Under this assumption, Part II of the Fundamental Theorem of Calculus gives us the following formula.

Formula ⚷

Integration by Parts Formula for Definite Integrals

$$\int_a^b f(x)g'(x)\,dx = \left[f(x)g(x) \right]_a^b - \int_a^b g(x)f'(x)\,dx$$

In applying this version of the formula, we typically follow the same procedure of defining u and dv.

Example 7 ✒

Evaluate $\int_1^2 t^2 e^t \, dt$.

Solution

We will have to use integration by parts twice.

$$\int_1^2 t^2 e^t \, dt = \left[t^2 e^t \right]_1^2 - 2\int_1^2 t e^t \, dt \qquad \begin{array}{ll} u = t^2 & dv = e^t\,dt \\ du = 2t\,dt & v = e^t \end{array}$$

$$= 4e^2 - e - 2\left(\left[t e^t \right]_1^2 - \int_1^2 e^t \, dt \right) \qquad \begin{array}{ll} u = t & dv = e^t\,dt \\ du = dt & v = e^t \end{array}$$

$$= 4e^2 - e - 2\left(2e^2 - e - \left[e^t \right]_1^2 \right)$$

$$= 4e^2 - e - 4e^2 + 2e + 2\left(e^2 - e \right)$$

$$= 2e^2 - e$$

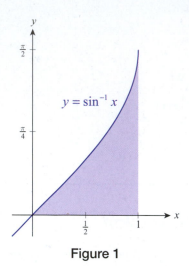

Figure 1

Example 8 ✐

Find the area between the graph of $\sin^{-1} x$ and the x-axis from $x = 0$ to $x = 1$.

Solution

The region described is depicted in Figure 1, and its area is $\int_0^1 \sin^{-1} x \, dx$. Evaluating this integral will illustrate the use of u-substitution as well as integration by parts.

$$\int_0^1 \sin^{-1} x \, dx = \left[x \sin^{-1} x \right]_0^1 - \int_0^1 \frac{x}{\sqrt{1-x^2}} \, dx$$

$$= \frac{\pi}{2} - 0 - \int_0^1 \frac{x}{\sqrt{1-x^2}} \, dx$$

$$= \frac{\pi}{2} + \frac{1}{2} \int_{u=1}^{u=0} u^{-1/2} \, du$$

$$= \frac{\pi}{2} + \left[u^{1/2} \right]_{u=1}^{u=0}$$

$$= \frac{\pi}{2} - 1$$

$u = \sin^{-1} x$ $\qquad dv = dx$
$du = \dfrac{1}{\sqrt{1-x^2}} dx$ $\qquad v = x$

$u = 1 - x^2$
$du = -2x \, dx$

7.1 **Exercises**

1–4 Evaluate the integral using integration by parts with the suggested choice for u and dv.

1. $\displaystyle\int xe^x \, dx; \quad u = x, \quad dv = e^x \, dx$

2. $\displaystyle\int 4x \cos 2x \, dx; \quad u = x, \quad dv = 2\cos 2x \, dx$

3. $\displaystyle\int 4x^3 \ln x \, dx; \quad u = \ln x, \quad dv = 4x^3 \, dx$

4. $\displaystyle\int \arctan x \, dx; \quad u = \arctan x, \quad dv = dx$

5–36 Evaluate the integral. (**Hint:** Use an alternative method for those that do not require integration by parts.)

5. $\displaystyle\int (t+1)e^{5t} \, dt$

6. $\displaystyle\int x^2 e^{x^3} \, dx$

7. $\displaystyle\int 2x \sin x \, dx$

8. $\displaystyle\int \arcsin x \, dx$

9. $\displaystyle\int x \ln(x^2) \, dx$

10. $\displaystyle\int \frac{dx}{x \ln(x^2)}$

11. $\displaystyle\int x^2 \ln x \, dx$

12. $\displaystyle\int \frac{\ln(x^2)}{x} \, dx$

13. $\displaystyle\int (s-3)e^{s-3} \, ds$

14. $\displaystyle\int (s-3)e^{(s-3)^2} \, ds$

15. $\displaystyle\int (2x+7)e^{3x-1} \, dx$

16. $\displaystyle\int \sqrt{z} \ln(z^2) \, dz$

17. $\displaystyle\int x\sqrt{x-2} \, dx$

18. $\displaystyle\int (3x+1)\sqrt[3]{x-1} \, dx$

19. $\displaystyle\int \sqrt[3]{3x} \ln x \, dx$

20. $\displaystyle\int \log_3 x \, dx$

21. $\displaystyle\int \frac{\ln \sqrt{x}}{\sqrt{x}} \, dx$

22. $\displaystyle\int (\theta+1)\sin\theta \, d\theta$

23. $\displaystyle\int x \sec x \tan x \, dx$

24. $\displaystyle\int \sec^2 x \tan x \, dx$

25. $\displaystyle\int x \csc^2 x \, dx$

26. $\displaystyle\int x \csc^2(x^2) \, dx$

27. $\displaystyle\int \frac{3x-1}{e^x} \, dx$

28. $\displaystyle\int x^{-2} e^{1/x} \, dx$

29. $\displaystyle\int (t+2)\cosh t \, dt$

30. $\displaystyle\int \frac{x}{\sqrt{2x-5}} \, dx$

31. $\displaystyle\int x\sqrt{x+1} \, dx$

32. $\displaystyle\int (3-4x)\sinh 2x \, dx$

33. $\displaystyle\int (3x+2)\operatorname{sech}^2 x \, dx$

34. $\displaystyle\int 2\operatorname{csch}^2 x \coth x \, dx$

35. $\displaystyle\int x\sqrt{2x-5} \, dx$

36. $\displaystyle\int \theta \cos(\theta^2) \, d\theta$

37–48 Evaluate the integral. If necessary, use integration by parts more than once.

37. $\displaystyle\int \theta^2 \cos\theta \, d\theta$

38. $\displaystyle\int t^3 e^t \, dt$

39. $\displaystyle\int e^x \cos x \, dx$

40. $\displaystyle\int \sin x \cos x \, dx$

41. $\displaystyle\int 9t^2 \sin 3t \, dt$

42. $\displaystyle\int 2t^2 e^{5t+1} \, dt$

43. $\int e^{3x} \sin 2x \, dx$

44. $\int 2e^x \sin x \cos x \, dx$

45. $\int x^2 \sinh x \, dx$

46. $\int \sec^3 x \, dx$

47. $\int \cos(\ln x) \, dx$

48. $\int (\ln x)^3 \, dx$

49–56 Combine the methods of integration by parts and substitution to evaluate the integral.

49. $\int e^{\sqrt{x}} \, dx$

50. $\int 2x^3 \sin(x^2 + 1) \, dx$

51. $\int \dfrac{\arctan(1/\sqrt{x})}{2\sqrt{x}} \, dx$

52. $\int t \ln(2 - t) \, dt$

53. $\int \dfrac{\arccos \sqrt{x}}{\sqrt{x}} \, dx$

54. $\int \sin(2x) e^{\sin x} \, dx$

55. $\int 9x^2 (\ln x)^2 \, dx$

56. $\int \dfrac{\cos(1/\theta)}{\theta^3} \, d\theta$

57–64 Evaluate the definite integral. (Use integration by parts only when necessary.)

57. $\int_{1/2}^{1} \arccos x \, dx$

58. $\int_{2}^{2e} \ln \dfrac{x}{2} \, dx$

59. $\int_{0}^{\pi} e^{2\theta} \sin \theta \, d\theta$

60. $\int_{0}^{\pi} t^2 \sin t \, dt$

61. $\int_{\sqrt{3}/3}^{\sqrt{3}} \arctan \dfrac{1}{x} \, dx$

62. $\int_{0}^{1} x^3 e^x \, dx$

63. $\int_{0}^{1} \ln(t^2 + 1) \, dt$

64. $\int_{1}^{e} \dfrac{\ln x^2}{x^2} \, dx$

65–70 Integration by parts can often be used to evaluate integrals involving inverses of functions, and in fact leads to a general formula:

$$\int f^{-1}(x) \, dx = \int y f'(y) \, dy$$

Let $y = f^{-1}(x)$, so
$x = f(y)$ and $dx = f'(y) \, dy$.

$$= y f(y) - \int f(y) \, dy$$

$u = y \quad dv = f'(y) \, dy$
$du = dy \quad v = f(y)$

$$= x f^{-1}(x) - \int f(y) \, dy$$

For instance, if we let $f(x) = e^x$, then $y = f^{-1}(x) = \ln x$.

$$\int \ln x \, dx = \int f^{-1}(x) \, dx$$
$$= x f^{-1}(x) - \int e^y \, dy$$
$$= x \ln x - e^y$$
$$= x \ln x - e^{\ln x} \qquad y = \ln x$$
$$= x \ln x - x$$

In Exercises 65–70, use this method to evaluate the given indefinite integral. (**Hint:** In starred exercises, show that $\cosh(\sinh^{-1} x) = \sqrt{1 + x^2}$ and that $\cosh(\tanh^{-1} x) = 1/\sqrt{1 - x^2}$ as part of the process toward the answers.)

65. $\int \sin^{-1} x \, dx$

66. $\int \cos^{-1} x \, dx$

67. $\int \tan^{-1} x \, dx$

68. $\int \log_2 x \, dx$

69.* $\int \sinh^{-1} x \, dx$

70.* $\int \tanh^{-1} x \, dx$

71. Use integration by parts to find the area of the region bounded by the graphs of $y = 6 \tan^{-1}(2x)$, $y = 0$, and $x = \sqrt{3}/2$.

72. Use integration by parts to find the area of the region bounded by the graphs of $y = \sin(\ln x)$ and $y = 0$ $(1 \le x \le e^{\pi})$.

73. Consider the region bounded by the graphs of $y = e^{-x}$, $y = 0$, $x = 0$, and $x = 1$.

  a. Find the centroid of the region.

  b. Use the shell method to find the volume of the solid generated by revolving the region about the y-axis.

74. Repeat Exercise 73 for the region bounded by the graphs of $y = x \cos x$ and $y = 0$ $(0 \le x \le \pi/2)$.

75. Consider the region bounded by the graphs of $y = \arcsin x$, $x = 0$, and $y = \pi/2$.

  a. Find the centroid of the region.

  b. Rotate the region about the x-axis and use the shell method to find the volume of the resulting solid.

76. Use the shell method to find the volume of the solid obtained by revolving the region bounded by the graphs of $y = 2^x$, $y = 0$, $x = 0$, and $x = 1$ about the line $x = -1$.

77. Find the centroid of the region bounded by the graphs of $y = \ln x$, $y = 0$, and $x = e$.

78. The definite integral $\dfrac{1}{\pi}\displaystyle\int_{-\pi}^{\pi} x\sin(nx)\,dx,\ n\in\mathbb{N}$ is called a *Fourier coefficient*. Use integration by parts to verify that its value is $(-1)^{n+1}\dfrac{2}{n}$. (The theory of Fourier series is very important in applied mathematics. You will be introduced to infinite series of functions in Chapter 10.)

79. Use integration by parts to prove that if $f(x)$ is continuously differentiable on $[-\pi,\pi]$, then the limit of the Fourier coefficients is 0.

$$\lim_{n\to\infty}\frac{1}{\pi}\int_{-\pi}^{\pi} f(x)\sin(nx)\,dx = 0$$

80. Use integration by parts to prove the formula

$$\int x^n \ln x\,dx = \frac{x^{n+1}}{(n+1)^2}\big[(n+1)\ln x - 1\big] + C.$$

81–87 Use integration by parts to prove the given reduction formula for $n\in\mathbb{N}$.

81. $\displaystyle\int \sin^n x\,dx = -\frac{\sin^{n-1}x\cos x}{n} + \frac{n-1}{n}\int \sin^{n-2}x\,dx$

82. $\displaystyle\int x^n \cos(kx)\,dx = \frac{x^n}{k}\sin(kx) - \frac{n}{k}\int x^{n-1}\sin(kx)\,dx$

83. $\displaystyle\int x^n \sin(kx)\,dx = -\frac{x^n}{k}\cos(kx) + \frac{n}{k}\int x^{n-1}\cos(kx)\,dx$

84. $\displaystyle\int x^n e^{kx}\,dx = \frac{1}{k}x^n e^{kx} - \frac{n}{k}\int x^{n-1}e^{kx}\,dx,\ \ k\neq 0$

85.* $\displaystyle\int \tan^n x\,dx = \frac{\tan^{n-1}x}{n-1} - \int \tan^{n-2}x\,dx,\ \ n\neq 1$

86. $\displaystyle\int \sec^n x\,dx = \frac{\tan x\sec^{n-2}x}{n-1} + \frac{n-2}{n-1}\int \sec^{n-2}x\,dx,\ n\neq 1$

87. $\displaystyle\int (\ln x)^n\,dx = x(\ln x)^n - n\int (\ln x)^{n-1}\,dx$

88–95 Use the above reduction formulas to evaluate the integral.

88. $\displaystyle\int \sin^5 x\,dx$

89. $\displaystyle\int x^3 \cos 2x\,dx$

90. $\displaystyle\int x^3 \sin(\pi x)\,dx$

91. $\displaystyle\int \cos x\sin^3 x\,e^{\sin x}\,dx$

92. $\displaystyle\int \sec^6 x\,dx$

93. $\displaystyle\int (\ln x)^3\,dx$

94. $\displaystyle\int \tan^4 x\,dx$

95. $\displaystyle\int_0^1 x^4 e^x\,dx$

96. Use the appropriate reduction formula to evaluate the definite integral $\displaystyle\int_0^{\pi/2} \sin^6 x\,dx$. Can you conjecture a possible formula for $n = 8, 10, 12, \dots$?

97. Use integration by parts to prove the formula

$$\int \sin(kx)e^{lx}\,dx = \frac{l\sin(kx) - k\cos(kx)}{k^2+l^2}e^{lx} + C.$$

98. Use your solution to Exercise 97 to find a similar formula for $\displaystyle\int \cos(kx)e^{lx}\,dx$.

99–102 In Exercise 91 you had to use a reduction formula to evaluate $\int u^3 e^u\,du$. Note that you can obtain the same answer by using the *method of undetermined coefficients*, as follows. Assuming that the answer has the form

$$Au^3 e^u + Bu^2 e^u + Cue^u + De^u + E,$$

differentiating yields

$$u^3 e^u = Au^3 e^u + (3A+B)u^2 e^u + (2B+C)ue^u + (C+D)e^u.$$

By equating coefficients we obtain

$$A=1,\quad 3A+B=0,\quad 2B+C=0,\quad \text{and}\quad C+D=0.$$

Solving the above system yields $B=-3$, $C=6$, and $D=-6$ (while $E=C_1$ is arbitrary).

In Exercises 99–102, use the method of undetermined coefficients to evaluate the given integral. (Note that this method will become important in Section 7.2.)

99. $\displaystyle\int 8x^3 e^{2x}\,dx$

100. $\displaystyle\int (x^4 - x)e^x\,dx$

101. $\displaystyle\int 13e^{3x}(\sin 2x)\,dx$ (**Hint:** Anticipate the answer in the form stated below.)

$$Ae^{3x}\sin 2x + Be^{3x}\cos 2x$$

102. $\displaystyle\int 5\sin 2x\cos 3x\,dx$ (**Hint:** Anticipate the answer in the form stated below.)

$$A\sin 2x\cos 3x + B\cos 2x\cos 3x$$
$$+ C\cos 2x\sin 3x + D\sin 2x\sin 3x + E$$

7.2 **The Partial Fractions Method**

TOPICS

1. Using partial fractions to integrate rational functions

2. Method to determine partial fraction constants

Partial fractions is the name given to the method of writing certain fractions as a sum of fractions, a process that we can use to make such fractions easier to integrate. It joins *u*-substitution and integration by parts as another method in our arsenal of integration methods.

TOPIC 1   Using Partial Fractions to Integrate Rational Functions

The usefulness of the partial fractions method can be demonstrated with a simple example.

Example 1 ✐

Evaluate $\int \dfrac{1}{x^2 - x - 6}\, dx$.

Solution

The integrand doesn't have any obvious antiderivative, and no technique we've learned so far seems to apply (though if the numerator were $2x - 1$ or a multiple of $2x - 1$, u-substitution would be ideally suited). On the other hand, the integrand is equivalent to a sum of two other rational functions that *are* easily integrated.

$$\frac{1}{x^2 - x - 6} = \frac{\frac{1}{5}}{x - 3} + \frac{-\frac{1}{5}}{x + 2}.$$

If we already know this equivalence, or are able to discover it somehow, then

$$\int \frac{1}{x^2 - x - 6}\, dx = \int \left(\frac{\frac{1}{5}}{x - 3} + \frac{-\frac{1}{5}}{x + 2} \right) dx$$

$$= \frac{1}{5} \int \frac{1}{x - 3}\, dx - \frac{1}{5} \int \frac{1}{x + 2}\, dx$$

$$= \frac{1}{5} \ln|x - 3| - \frac{1}{5} \ln|x + 2| + C.$$

Here we have twice made use of the following fact easily obtained by u-substitution.

$$\int \frac{1}{ax + b}\, dx = \frac{1}{a} \ln|ax + b| + C$$

The only trick in the above example is in knowing how to decompose the original integrand into a sum; the steps involved in doing so constitute the partial fractions method.

Partial Fraction Decomposition

Assume $f(x) = p(x)/q(x)$ is a ratio of two polynomial functions $p(x)$ and $q(x)$, where the degree of p is less than the degree of q (so f is a *proper* rational function). Assume also that $q(x)$ has been factored completely into a product of factors of the forms $(ax+b)^m$ and $(ax^2+bx+c)^n$, where m and n are positive integers, the coefficients a, b, and c are all real, and each factor of the form ax^2+bx+c cannot be factored further over the real numbers (we say ax^2+bx+c is *irreducible*).

Then $f(x)$ can be written as a sum of simpler rational functions, where

1. each factor of the form $(ax+b)^m$ gives rise to a sum of the form

$$\frac{A_1}{ax+b} + \frac{A_2}{(ax+b)^2} + \cdots + \frac{A_m}{(ax+b)^m}.$$

2. each factor of the form $(ax^2+bx+c)^n$ gives rise to a sum of the form

$$\frac{B_1 x + C_1}{ax^2+bx+c} + \frac{B_2 x + C_2}{(ax^2+bx+c)^2} + \cdots + \frac{B_n x + C_n}{(ax^2+bx+c)^n}.$$

After using these guidelines to write a given rational function as a sum of fractions, the constants represented by the capital letters above must still be determined. One way of doing so is to clear the equation of fractions, as illustrated in the next few examples.

Example 2 ✎

Use the partial fractions method to decompose the integrand of Example 1 into a sum of rational functions.

Solution

Since $x^2 - x - 6 = (x-3)(x+2)$, the partial fractions method leads us to write

$$\frac{1}{x^2-x-6} = \frac{1}{(x-3)(x+2)} = \frac{A_1}{x-3} + \frac{A_2}{x+2}$$

(each of the two linear factors $x-3$ and $x+2$ gives rise to a corresponding fraction with an as-yet-undetermined numerator). To solve for A_1 and A_2, we can multiply through by $(x-3)(x+2)$ and then collect like terms to obtain a polynomial equation.

$$1 = A_1(x+2) + A_2(x-3)$$
$$1 = (A_1 + A_2)x + (2A_1 - 3A_2)$$

Since the polynomial on the left, the constant function 1, has no x term, it must be the case that

$$0 = A_1 + A_2.$$

Similarly, equating the constant terms of the two polynomials tells us that

$$1 = 2A_1 - 3A_2.$$

The first equation indicates that $A_2 = -A_1$, and substituting this into the second equation results in $A_1 = \frac{1}{5}$ and hence $A_2 = -\frac{1}{5}$.

$$\frac{1}{x^2 - x - 6} = \frac{\frac{1}{5}}{x - 3} + \frac{-\frac{1}{5}}{x + 2}$$

Example 3 ✐

Evaluate $\displaystyle\int \frac{6x^2 - 8x - 32}{2x^3 - 5x^2 - 22x - 15}\,dx.$

Solution

Through some combination of polynomial division (synthetic or long) and factoring, we can write the denominator as $(x+1)(x-5)(2x+3)$. Using the partial fractions guidelines we obtain the following.

$$\frac{6x^2 - 8x - 32}{2x^3 - 5x^2 - 22x - 15} = \frac{A_1}{x+1} + \frac{A_2}{x-5} + \frac{A_3}{2x+3}$$

We multiply both sides of this equation by $(x+1)(x-5)(2x+3)$ and collect like terms.

$$6x^2 - 8x - 32 = A_1(x-5)(2x+3) + A_2(x+1)(2x+3) + A_3(x+1)(x-5)$$
$$= \left(2A_1 + 2A_2 + A_3\right)x^2 + \left(-7A_1 + 5A_2 - 4A_3\right)x + \left(-15A_1 + 3A_2 - 5A_3\right)$$

Matching up the corresponding coefficients on both sides of this equation gives us the system of equations

$$\begin{array}{rcrcrcr} 2A_1 & + & 2A_2 & + & A_3 & = & 6 \\ -7A_1 & + & 5A_2 & - & 4A_3 & = & -8 \\ -15A_1 & + & 3A_2 & - & 5A_3 & = & -32 \end{array}$$

that, when solved using Gaussian elimination, matrix methods, or a calculator or computer, yield $A_1 = 3$, $A_2 = 1$, and $A_3 = -2$.

$$\int \frac{6x^2 - 8x - 32}{2x^3 - 5x^2 - 22x - 15}\,dx = \int \left(\frac{3}{x+1} + \frac{1}{x-5} + \frac{-2}{2x+3}\right)dx$$
$$= 3\ln|x+1| + \ln|x-5| - \ln|2x+3| + C$$

Note that the actual integration step in Example 3 is the least demanding—determining the constants A_1, A_2, and A_3 requires more effort. Fortunately, there are some efficient shortcuts for accomplishing this step.

Example 4 ✏

Evaluate $\int \dfrac{4x^3 + 18x^2 + 9x + 11}{2x^2 + 9x + 4}\,dx.$

Solution

The integrand is not a proper rational function, so the first step is to perform polynomial long division.

$$
\begin{array}{r}
2x \\
2x^2 + 9x + 4 \overline{\smash{\big)}\, 4x^3 + 18x^2 + 9x + 11} \\
\underline{4x^3 + 18x^2 + 8x} \\
x + 11
\end{array}
$$

Therefore,

$$\frac{4x^3 + 18x^2 + 9x + 11}{2x^2 + 9x + 4} = 2x + \frac{x + 11}{2x^2 + 9x + 4}.$$

We can now decompose the remaining (proper) rational function.

$$\frac{x + 11}{2x^2 + 9x + 4} = \frac{A_1}{2x + 1} + \frac{A_2}{x + 4}$$

$$x + 11 = A_1(x + 4) + A_2(2x + 1) \qquad \text{Multiply through by } (2x+1)(x+4).$$

Instead of expanding out the polynomial on the right and matching coefficients, we can make use of the fact that we have an equation that is supposed to be true for *all* x. In particular, if we let $x = -4$ note that

$$
\begin{aligned}
-4 + 11 &= A_1(-4 + 4) + A_2\big[2(-4) + 1\big] \\
7 &= -7A_2 \\
A_2 &= -1
\end{aligned}
$$

and if $x = -\frac{1}{2}$ we have the following.

$$
\begin{aligned}
-\frac{1}{2} + 11 &= A_1\left(-\frac{1}{2} + 4\right) + A_2\left[2\left(-\frac{1}{2}\right) + 1\right] \\
\frac{21}{2} &= \frac{7A_1}{2} \\
A_1 &= 3
\end{aligned}
$$

The given integral can be evaluated as follows.

$$
\begin{aligned}
\int \frac{4x^3 + 18x^2 + 9x + 11}{2x^2 + 9x + 4}\,dx &= \int \left(2x + \frac{x + 11}{2x^2 + 9x + 4}\right)dx \\
&= \int \left(2x + \frac{3}{2x + 1} + \frac{-1}{x + 4}\right)dx \\
&= x^2 + \frac{3}{2}\ln|2x + 1| - \ln|x + 4| + C
\end{aligned}
$$

TOPIC 2   Method to Determine Partial Fraction Constants

We can take advantage of the fact noted in Example 4 in other ways. In the next example, we solve for one undetermined constant by evaluating the equation at a particular value of x, and then use differentiation and substitution to solve for the remaining constants.

Example 5 ✍

Evaluate $\displaystyle\int \frac{4}{x\left(x^2+2\right)^2}\, dx$.

Solution

The denominator is a product of a linear factor and a power of an irreducible quadratic (x^2+2 cannot be factored over the real numbers). Using the partial fractions guidelines, the integrand can be written in the form

$$\frac{4}{x\left(x^2+2\right)^2} = \frac{A_1}{x} + \frac{B_1 x + C_1}{x^2+2} + \frac{B_2 x + C_2}{\left(x^2+2\right)^2}$$

so

$$4 = A_1\left(x^2+2\right)^2 + \left(B_1 x + C_1\right)\left(x^2+2\right)x + \left(B_2 x + C_2\right)x.$$

Instead of expanding out the right-hand side and equating coefficients, we can quickly solve for one of the unknown constants by letting $x=0$.

$$4 = A_1\left(2\right)^2 \quad \Rightarrow \quad A_1 = 1$$

And since polynomials are differentiable for all x, we can differentiate both sides to obtain a new polynomial equation of lower degree.

$$\frac{d}{dx}(4) = \frac{d}{dx}\left[\left(x^2+2\right)^2 + \left(B_1 x + C_1\right)\left(x^2+2\right)x + \left(B_2 x + C_2\right)x\right]$$

$$0 = 2\left(x^2+2\right)(2x) + B_1\left(x^2+2\right)x + \left(B_1 x + C_1\right)(2x)x$$

$$+ \left(B_1 x + C_1\right)\left(x^2+2\right) + B_2 x + \left(B_2 x + C_2\right)$$

Evaluating this last equation at $x=0$ is again easy, and gives us the relation $2C_1 + C_2 = 0$. We can always, if necessary, expand out the right-hand side and equate coefficients in order to obtain more relations among the unknown constants, but doing so takes time. However, one such relation is fairly easy to spot. Note that if we were to expand out the first term on the right, $2\left(x^2+2\right)(2x)$, we would get an expression containing only a third and a first degree term. Similarly for the term $B_1\left(x^2+2\right)x$. If we continue to mentally expand and examine each term on the right, we see that the only x^2-terms are $2C_1 x^2$ and $C_1 x^2$ (coming from the products shown in red below).

$$2\left(x^2+2\right)(2x) + B_1\left(x^2+2\right)x + \left(B_1 x + C_1\right)(2x)x + \left(B_1 x + C_1\right)\left(x^2+2\right) + B_2 x + \left(B_2 x + C_2\right)$$

Since there is no x^2 on the left, it must be the case that $3C_1 = 0$, or $C_1 = 0$. And from the relation $2C_1 + C_2 = 0$, this means $C_2 = 0$ as well. Making these replacements, our polynomial equation is now one we can expand more quickly.

$$0 = 2\left(x^2+2\right)\left(2x\right)+B_1\left(x^2+2\right)x+\left(B_1x\right)\left(2x\right)x+\left(B_1x\right)\left(x^2+2\right)+B_2x+B_2x$$

$$= 4x^3+8x+B_1x^3+2B_1x+2B_1x^3+B_1x^3+2B_1x+2B_2x$$

$$= \left(4+4B_1\right)x^3+\left(8+4B_1+2B_2\right)x$$

From this last equation, we see $4+4B_1 = 0$, so $B_1 = -1$ and hence $8+4(-1)+2B_2 = 0$ or $B_2 = -2$, giving us

$$\frac{4}{x\left(x^2+2\right)^2} = \frac{A_1}{x}+\frac{B_1x+C_1}{x^2+2}+\frac{B_2x+C_2}{\left(x^2+2\right)^2} = \frac{1}{x}-\frac{x}{x^2+2}-\frac{2x}{\left(x^2+2\right)^2}.$$

We are now ready to integrate.

$$\int \frac{4}{x\left(x^2+2\right)^2}\,dx = \int\left[\frac{1}{x}-\frac{x}{x^2+2}-\frac{2x}{\left(x^2+2\right)^2}\right]dx$$

$$= \int \frac{dx}{x}-\int \frac{x\,dx}{x^2+2}-\int \frac{2x\,dx}{\left(x^2+2\right)^2} \qquad \begin{aligned} u &= x^2+2 \\ du &= 2x\,dx \end{aligned}$$

$$= \ln|x|-\frac{1}{2}\ln\left(x^2+2\right)+\frac{1}{x^2+2}+C$$

When the rational function is proper and the denominator can be factored into a product of linear terms, a technique called the *Heaviside cover-up method* can be used to determine the unknown constants, as shown in the next example.

Example 6 ✎

Evaluate $\displaystyle\int_2^3 \frac{x^2+12x+11}{\left(x+3\right)\left(x+2\right)\left(x-1\right)}\,dx$.

Solution

First, if we hope to apply the Fundamental Theorem of Calculus to this definite integral, it is important to verify that the integrand is continuous over the interval of integration. Since the only points of discontinuity are $-3, -2,$ and 1, the interval $[2,3]$ poses no problem.

The integrand is proper, and the denominator is already factored into linear factors. So we know it can be decomposed into the following form.

$$\frac{x^2+12x+11}{\left(x+3\right)\left(x+2\right)\left(x-1\right)} = \frac{A_1}{x+3}+\frac{A_2}{x+2}+\frac{A_3}{x-1}$$

In a slight variation of our work so far, let us multiply through by $x+3$.

$$\frac{x^2+12x+11}{\left(x+2\right)\left(x-1\right)} = A_1+\frac{A_2\left(x+3\right)}{x+2}+\frac{A_3\left(x+3\right)}{x-1}$$

If we now substitute $x = -3$, we obtain

$$\frac{(-3)^2 + 12(-3) + 11}{(-3+2)(-3-1)} = A_1 \quad \text{or} \quad A_1 = -4.$$

The cover-up method, formulated by the physicist and mathematician Oliver Heaviside (1850–1925), takes its name from noting that this is the result we get if we simply "cover up" the factor $x + 3$ in the original fraction and replace x with -3 elsewhere.

$$A_1 = \frac{(-3)^2 + 12(-3) + 11}{\underbrace{\boxed{(x+3)}}_{\text{cover up}}(-3+2)(-3-1)} = \frac{9 - 36 + 11}{(-1)(-4)} = -4$$

We can determine A_2 and A_3 similarly.

$$A_2 = \frac{(-2)^2 + 12(-2) + 11}{(-2+3)\underbrace{\boxed{(x+2)}}_{\text{cover up}}(-2-1)} = \frac{4 - 24 + 11}{(1)(-3)} = 3$$

$$A_3 = \frac{(1)^2 + 12(1) + 11}{(1+3)(1+2)\underbrace{\boxed{(x-1)}}_{\text{cover up}}} = \frac{1 + 12 + 11}{(4)(3)} = 2$$

We can now proceed to integrate.

$$\int_2^3 \frac{x^2 + 12x + 11}{(x+3)(x+2)(x-1)}\, dx = \int_2^3 \left(\frac{-4}{x+3} + \frac{3}{x+2} + \frac{2}{x-1} \right) dx$$

$$= \left[-4\ln(x+3) + 3\ln(x+2) + 2\ln(x-1) \right]_2^3$$

$$= -4\ln 6 + 3\ln 5 + 2\ln 2 + 4\ln 5 - 3\ln 4 - 2\ln 1$$

$$= -4\ln 6 + 7\ln 5 - 3\ln 4 + 2\ln 2$$

$$= \ln\left(\frac{5^7 \cdot 2^2}{6^4 \cdot 4^3} \right)$$

$$= \ln\left(\frac{78,125}{20,736} \right)$$

Example 7 ✎

Evaluate $\displaystyle\int \frac{3x^2 + 4x + 2}{(x-1)(x^2 + x + 1)}\, dx$.

Solution

The quadratic formula tells us that $x^2 + x + 1$ is irreducible, so we are led to write

$$\frac{3x^2 + 4x + 2}{(x-1)(x^2 + x + 1)} = \frac{A_1}{x-1} + \frac{B_1 x + C_1}{x^2 + x + 1}.$$

Clearing fractions, we have

$$3x^2 + 4x + 2 = A_1(x^2 + x + 1) + (B_1 x + C_1)(x - 1),$$

and making the substitution $x = 1$ allows us to quickly solve for one constant.

$$\text{With } x = 1: \qquad 9 = 3A_1 \quad \Rightarrow \quad A_1 = 3$$

We can now solve for C_1 with another substitution.

$$\text{With } x = 0: \qquad 2 = 3 + C_1(-1) \quad \Rightarrow \quad C_1 = 1$$

At this point our equation has the form

$$3x^2 + 4x + 2 = (3 + B_1)x^2 + (4 - B_1)x + 2$$

and so $B_1 = 0$.

Therefore,

$$\frac{3x^2 + 4x + 2}{(x-1)(x^2 + x + 1)} = \frac{3}{x - 1} + \frac{1}{x^2 + x + 1}.$$

We are now left with the task of integrating the two partial fractions. The first term poses no difficulty, but the second term is unlike any we have seen in this section so far. However, we can complete the square on $x^2 + x + 1$ to rewrite the second term in a form whose antiderivative we *do* know.

$$\int \left(\frac{3}{x-1} + \frac{1}{x^2 + x + 1} \right) dx$$

$$= 3 \ln|x - 1| + \int \left[\frac{1}{\left(x + \frac{1}{2} \right)^2 + \frac{3}{4}} \right] dx \qquad\qquad \begin{aligned} x^2 + x + 1 &= x^2 + x + \frac{1}{4} + \frac{3}{4} \\ &= \left(x + \frac{1}{2} \right)^2 + \frac{3}{4} \end{aligned}$$

$$= 3 \ln|x - 1| + \int \left[\frac{\frac{4}{3}}{\frac{4}{3}\left(x + \frac{1}{2} \right)^2 + 1} \right] dx \qquad\qquad \text{Multiply top and bottom by } \tfrac{4}{3}.$$

$$= 3 \ln|x - 1| + \int \left[\frac{\frac{4}{3}}{\left(\frac{2}{\sqrt{3}} x + \frac{1}{\sqrt{3}} \right)^2 + 1} \right] dx \qquad\qquad \begin{aligned} \frac{4}{3}\left(x + \frac{1}{2} \right)^2 &= \left[\frac{2}{\sqrt{3}}\left(x + \frac{1}{2} \right) \right]^2 \\ &= \left(\frac{2}{\sqrt{3}} x + \frac{1}{\sqrt{3}} \right)^2 \end{aligned}$$

$$= 3 \ln|x - 1| + \left(\frac{4}{3} \right)\left(\frac{\sqrt{3}}{2} \right) \int \frac{du}{u^2 + 1} \qquad\qquad \begin{aligned} u &= \frac{2}{\sqrt{3}} x + \frac{1}{\sqrt{3}} \\ du &= \frac{2}{\sqrt{3}} dx \end{aligned}$$

$$= 3 \ln|x - 1| + \frac{2}{\sqrt{3}} \tan^{-1}\left(\frac{2x + 1}{\sqrt{3}} \right) + C \qquad\qquad \int \frac{du}{u^2 + 1} = \tan^{-1} u + C$$

7.2 **Exercises**

1–9 Use the guidelines discussed in this section to write the form of the partial fraction decomposition of the given rational function. Do not solve for the coefficients in your decomposition.

1. $\dfrac{2}{x(x-1)}$

2. $\dfrac{5x}{(x+4)(x-2)(x+7)}$

3. $\dfrac{2x+5}{(x+1)(x-3)^3}$

4. $\dfrac{3x-1}{(2x+3)(x^2+2)}$

5. $\dfrac{x-4}{(x^2+x+2)^3}$

6. $\dfrac{14x-3}{(3x-1)(x^2+1)^2}$

7. $\dfrac{3}{2x^2-5x-3}$

8. $\dfrac{2x+1}{5x^3-11x^2+7x-1}$

9. $\dfrac{4x^2-1}{x^5+x^4+2x^3+2x^2+x+1}$

10–35 Use the partial fractions method to evaluate the given integral.

10. $\displaystyle\int \dfrac{dx}{x(x+2)}$

11. $\displaystyle\int \dfrac{4}{(3x-1)x}\,dx$

12. $\displaystyle\int \dfrac{x-4}{(x+2)(x-1)}\,dx$

13. $\displaystyle\int \dfrac{2}{(x+3)(2x+5)}\,dx$

14. $\displaystyle\int \dfrac{5x+2}{(x+1)(3x-1)(x+3)}\,dx$

15. $\displaystyle\int \dfrac{2x}{(x-2)^2(x+1)}\,dx$

16. $\displaystyle\int \dfrac{x}{(x+2)^3}\,dx$

17. $\displaystyle\int \dfrac{2-z}{z(z+1)^2}\,dz$

18. $\displaystyle\int \dfrac{dx}{x(x^2+2)}$

19. $\displaystyle\int \dfrac{dx}{x^2+3x+2}$

20. $\displaystyle\int \dfrac{2-z}{z^2-1}\,dz$

21. $\displaystyle\int \dfrac{2}{x-x^3}\,dx$

22. $\displaystyle\int \dfrac{s-3}{s(s-1)(s+3)}\,ds$

23. $\displaystyle\int \dfrac{t-1}{t^3+t^2+t+1}\,dt$

24. $\displaystyle\int \dfrac{x^2+1}{(x^2+2x+3)^2}\,dx$

25. $\displaystyle\int \dfrac{x^2-9}{x^4+3x^3}\,dx$

26. $\displaystyle\int \dfrac{5x^3-5x-40}{x^4+x^3+4x^2+4x}\,dx$

27. $\displaystyle\int \dfrac{11x-12}{x(x^2+x-6)}\,dx$

28. $\displaystyle\int \dfrac{3x^4-5x^3+15x^2-8x+20}{x^5-2x^4+4x^3-8x^2+4x-8}\,dx$

29. $\displaystyle\int \dfrac{32}{x^4-4x^3-2x^2+12x+9}\,dx$

30. $\displaystyle\int \dfrac{2x^4+4x^2-x+2}{(1+x^2)^3}\,dx$ **31.** $\displaystyle\int \dfrac{16v}{v^4-16}\,dv$

32. $\displaystyle\int \dfrac{-x^2-1}{x^4+5x^2+6}\,dx$ **33.** $\displaystyle\int \dfrac{3x}{x^3-1}\,dx$

34. $\displaystyle\int \dfrac{x}{(x+a)(x-b)}\,dx$ **35.** $\displaystyle\int \dfrac{a}{x(x^2+b)}\,dx$

36–41 Use the Heaviside cover-up method to evaluate the given integral.

36. $\displaystyle\int \dfrac{6x^2-19x-12}{x(x^2-x-6)}\,dx$ **37.** $\displaystyle\int \dfrac{16(x+2)}{(x-7)(x^2-1)}\,dx$

38. $\displaystyle\int \dfrac{2v^2+13v+6}{v^3-v^2-10v-8}\,dv$ **39.** $\displaystyle\int_0^1 \dfrac{-5t^2+8t+19}{t^3+2t^2-5t-6}\,dt$

40. $\displaystyle\int_1^2 \dfrac{6s^3-38s^2+48s+12}{s^4-6s^3+5s^2+12s}\,ds$

41. $\displaystyle\int_{-1}^1 \dfrac{-x^3+4x^2-11x-6}{(x^2-4)(x^2-9)}\,dx$

42. If $k \in \mathbb{R}$, use partial fractions to prove the formula
$$\int \dfrac{1}{x^2-k^2}\,dx = -\dfrac{1}{2k}\ln\left|\dfrac{x+k}{x-k}\right| + C.$$

43. Complete the square in the denominator of Example 1 and use the formula in Exercise 42 to arrive at the same answer as in Example 1.

44–46 Use your approach taken in Exercise 42 to establish the given formula for $k, l \in \mathbb{R}$.

44. $\displaystyle\int \dfrac{dx}{x(kx+l)} = \dfrac{1}{l}\ln\left|\dfrac{x}{kx+l}\right| + C$

45. $\displaystyle\int \dfrac{dx}{x^2(kx+l)} = -\dfrac{1}{lx} - \dfrac{k}{l^2}\ln\left|\dfrac{x}{kx+l}\right| + C$

46. $\displaystyle\int \dfrac{x}{(kx+l)^2}\,dx = \dfrac{1}{k^2}\left(\dfrac{l}{kx+l} + \ln|kx+l|\right) + C$

47–61 Use any of the techniques seen in Examples 3–7 to evaluate the given integral (definite or indefinite, as indicated). Whenever applicable, use the formulas from Exercises 42, 44–46. (**Note:** If the integrand is not a proper rational function, be sure to divide first.)

47. $\int \dfrac{x^2+1}{x^3+1}\,dx$

48. $\int \dfrac{3x^3-4x^2+2}{x^2-x}\,dx$

49. $\int_5^8 \dfrac{dx}{x^2-16}$

50. $\int \dfrac{2}{s\left(s^2+1\right)^2}\,ds$

51. $\int_1^2 \dfrac{dx}{x(x+2)}$

52. $\int_0^{\sqrt{7}} \dfrac{x^3}{x^2+9}\,dx$

53. $\int \dfrac{3x^3}{x^3-1}\,dx$

54. $\int_1^2 \dfrac{dx}{x^2\left(x+2\right)}$

55. $\int \dfrac{3x^2-8x+2}{(x-3)\left(x^2-2x+2\right)}\,dx$

56. $\int_0^2 \dfrac{x^2-5x}{x^2+x+3}\,dx$

57. $\int_0^3 \dfrac{v}{\left(v+5\right)^2}\,dv$

58. $\int \dfrac{16x^5}{\left(x-1\right)^2\left(3-x\right)^4}\,dx$

59. $\int \dfrac{a}{\left(x-b\right)\left(x-c\right)}\,dx$

60. $\int \dfrac{z^2+8z+9}{\left(z^2+2z+3\right)^2}\,dz$

61. $\int \dfrac{x^4+1}{x^3+4x}\,dx$

62–73 Combine integration by substitution and the partial fractions method to evaluate the given integral. (When applicable, also use integration by parts.)

62. $\int \dfrac{\cos x}{\sin^2 x+2\sin x}\,dx$

63. $\int \dfrac{\sin x \cos x}{(\cos x-1)(\cos x+2)}\,dx$

64. $\int \dfrac{2\sec^2 x}{\tan^2 x-1}\,dx$

65. $\int \dfrac{6}{x\left[\left(\ln x\right)^2-\ln x^3\right]}\,dx$

66. $\int \dfrac{4e^x}{e^{2x}+2e^x-3}\,dx$

67. $\int \dfrac{\sqrt{x}}{\left(1+\sqrt{x}\right)^2}\,dx$

68. $\int \dfrac{\sqrt[3]{x}}{\left(1-\sqrt[3]{x}\right)^2}\,dx$

69. $\int \dfrac{dx}{x\sqrt{1+\sqrt{x}}}$

70. $\int e^x \ln\left(e^{2x}+2\right)dx$

71. $\int \dfrac{\sin\left(\ln x\right)}{x\left[\cos^2\left(\ln x\right)+\cos\left(\ln x\right)\right]}\,dx$

72. $\int \dfrac{e^{(3/2)x}}{e^x+1}\,dx$

73. $\int \dfrac{\left[\ln\left(w^2\right)\right]\left[\arctan\left(\ln w\right)\right]}{w}\,dw$

74. Use the disk method to find the volume of the solid generated by revolving the graph of $f(x)=2/\sqrt{x^2-3x-10}$, $6\le x\le 10$, about the x-axis.

75. Use the shell method to find the volume of the solid obtained by revolving the region bounded by $g(x)=1/\left(-x^2+2x+8\right)$, $y=0$, $x=0$, and $x=2$ about the line $x=-1$.

76. Find the centroid of the region bounded by the graphs of $f(x)=\dfrac{-4x-1}{x^2-x-2}$, $y=0$, $x=0$, and $x=1$.

77.* Suppose that we are looking for a function $y=y(t)$ whose rate of change $y'(t)$ is directly proportional to $\left(c^4-y^4\right)/y^2$ (c is a constant). In other words, $y(t)$ then satisfies the equation $\dfrac{dy}{dt}=m\left(c^4-y^4\right)/y^2$ for some constant m. (Such an equation, containing a derivative of an unknown function, is called a differential equation. You will learn more about differential equations in Chapter 8. The equation in this exercise is used in physical chemistry.) Use partial fractions to find an implicit formula for $y(t)$. (**Hint:** As a first step, rewrite the equation in differential form, $y^2/\left(c^4-y^4\right)dy=m\,dt$, and use partial fractions to integrate.)

78.* If the ability of the environment to support a population is limited and, thus, the population cannot grow larger than a certain size, the model $P'(t) = mP(t)$ and its solution $P(t) = P(0)e^{mt}$ is no longer adequate to describe the population growth. Instead, the so-called *logistic model* $dP/dt = mP(L-P)$ has been proposed, where L is the upper limit of the population size. If the world's population in 1940 was 2.3 billion, which grew to 6.9 billion by 2010, and supposing that the Earth cannot support more than $L = 15$ billion people, what will be the world's population by 2050? (See the hint given in Exercise 77.)

79.* Assuming that the rate at which a disease is spreading after an infected person enters a community of N susceptible people is proportional to the product of the number of already infected individuals by the number of still-healthy people, and letting $I(t)$ stand for the number of individuals already infected, this latter function satisfies $dI/dt = kI(N-I)$, with $I(0) = 1$. Use your approach taken in the previous two exercises to find a formula for $I(t)$.

7.2 **Technology Exercises**

80–81 Use a computer algebra system to find the partial fraction decomposition of the given rational function.

80. $f(x) = \dfrac{3x^7 + 20x^6 + 81x^5 + 123x^4 - 61x^3 - 1033x^2 - 2056x - 2401}{x^8 + 7x^7 + 31x^6 + 66x^5 + 78x^4 - 78x^3 - 203x^2 - 245x + 343}$

81. $f(x) = \dfrac{3x^8 + 17x^7 + 47x^6 + 97x^5 + 156x^4 + 207x^3 + 194x^2 + 107x + 42}{x^6 + 6x^5 + 16x^4 + 26x^3 + 27x^2 + 20x + 12}$

7.3 **Trigonometric Integrals**

TOPICS

1. Powers of sines and cosines

2. Powers of tangents and secants

In this section and the next, we focus attention on integrals involving trigonometric functions. First, we work with integrands containing trigonometric functions from the start. Such integrals arise in purely mathematical settings as well as in such applications as electrical engineering and signal processing. In the next section, we develop techniques that use trigonometry as an intermediate step in evaluating integrals.

TOPIC 1   **Powers of Sines and Cosines**

Throughout this section, we will make repeated use of trigonometric identities to transform an integrand into one we can more easily integrate. For example, the following strategy outlines the use of identities to evaluate integrands that are products of powers of sine and cosine.

Evaluating Integrals of the Form $\int \sin^m x \cos^n x \, dx$

Assuming m and n are nonnegative integers, an integral of the form $\int \sin^m x \cos^n x \, dx$ can be evaluated as follows.

If m is odd: Use the identity $\sin^2 x = 1 - \cos^2 x$ to express all but one factor of sine in terms of cosine, and then use u-substitution with $u = \cos x$ to evaluate the result.

If n is odd: Use the identity $\cos^2 x = 1 - \sin^2 x$ to express all but one factor of cosine in terms of sine, and then use u-substitution with $u = \sin x$ to evaluate the result. (Note that if both m and n are odd, either strategy can be used.)

If both m and n are even: Use the identities

$$\sin^2 x = \frac{1 - \cos 2x}{2} \quad \text{and} \quad \cos^2 x = \frac{1 + \cos 2x}{2}$$

to reduce the integrand to one containing lower powers of $\cos 2x$.

Example 1 ✎

Evaluate $\int \sin^5 x \cos^2 x \, dx$.

Solution

Since the power of sine is 5 (an odd number), we use the identity $\sin^2 x = 1 - \cos^2 x$ and the substitution $u = \cos x$ to rewrite and evaluate the integral as follows.

$$\int \sin^5 x \cos^2 x \, dx = \int \left(\sin^2 x \right)^2 \sin x \cos^2 x \, dx$$

$$= \int \left(1 - \cos^2 x \right)^2 \cos^2 x \sin x \, dx \qquad \begin{aligned} u &= \cos x \\ du &= -\sin x \, dx \end{aligned}$$

$$= -\int \left(1 - u^2 \right)^2 u^2 \, du$$

$$= -\int \left(u^2 - 2u^4 + u^6 \right) du$$

$$= -\frac{1}{3} u^3 + \frac{2}{5} u^5 - \frac{1}{7} u^7 + C$$

$$= -\frac{1}{3} \cos^3 x + \frac{2}{5} \cos^5 x - \frac{1}{7} \cos^7 x + C$$

Example 2 ✐

Evaluate $\int \cos^7 x\,dx$.

Solution

Since the power of cosine is odd (and the power of sine is 0), we use the identity $\cos^2 x = 1 - \sin^2 x$ and the substitution $u = \sin x$.

$$
\begin{aligned}
\int \cos^7 x\,dx &= \int \left(\cos^2 x\right)^3 \cos x\,dx \\
&= \int \left(1 - \sin^2 x\right)^3 \cos x\,dx \qquad \begin{aligned} u &= \sin x \\ du &= \cos x\,dx \end{aligned} \\
&= \int \left(1 - u^2\right)^3 du \\
&= \int \left(1 - 3u^2 + 3u^4 - u^6\right)du \\
&= u - u^3 + \frac{3}{5}u^5 - \frac{1}{7}u^7 + C \\
&= \sin x - \sin^3 x + \frac{3}{5}\sin^5 x - \frac{1}{7}\sin^7 x + C
\end{aligned}
$$

Example 3 ✐

Evaluate $\int \sin^4 x \cos^2 x\,dx$.

Solution

Since the powers of both sine and cosine are even in this integral, we proceed as follows.

$$
\begin{aligned}
\int \sin^4 x \cos^2 x\,dx &= \int \left(\sin^2 x\right)^2 \cos^2 x\,dx \\
&= \int \left(\frac{1 - \cos 2x}{2}\right)^2 \left(\frac{1 + \cos 2x}{2}\right)dx \\
&= \frac{1}{8}\int \left(1 - 2\cos 2x + \cos^2 2x\right)\left(1 + \cos 2x\right)dx \\
&= \frac{1}{8}\int \left(1 - \cos 2x - \cos^2 2x + \cos^3 2x\right)dx
\end{aligned}
$$

The first two terms in the last integral can be integrated directly, while the last two can be integrated using our new methods.

$$
\int \cos^2 2x\,dx = \int \frac{1 + \cos 4x}{2}\,dx = \frac{1}{2}\left(x + \frac{1}{4}\sin 4x\right) + C
$$

$$
\begin{aligned}
\int \cos^3 2x\,dx &= \int \left(1 - \sin^2 2x\right)\cos 2x\,dx \qquad \begin{aligned} u &= \sin 2x \\ du &= 2\cos 2x\,dx \end{aligned} \\
&= \frac{1}{2}\int \left(1 - u^2\right)du \\
&= \frac{1}{2}\left(u - \frac{1}{3}u^3\right) + C \\
&= \frac{1}{2}\left(\sin 2x - \frac{1}{3}\sin^3 2x\right) + C
\end{aligned}
$$

Putting the pieces together we have the following.

$$\frac{1}{8}\int \left(1 - \cos 2x - \cos^2 2x + \cos^3 2x\right) dx$$

$$= \frac{1}{8}\left(x - \frac{1}{2}\sin 2x - \frac{1}{2}x - \frac{1}{8}\sin 4x + \frac{1}{2}\sin 2x - \frac{1}{6}\sin^3 2x\right) + C$$

$$= \frac{1}{16}x - \frac{1}{64}\sin 4x - \frac{1}{48}\sin^3 2x + C$$

The product-to-sum identities are particularly useful in evaluating integrands of the forms $\sin mx \sin nx$, $\sin mx \cos nx$, and $\cos mx \cos nx$. For quick reference, these identities are as follows.

Product-to-Sum Identities

$$\sin mx \sin nx = \frac{1}{2}\left[\cos\left((m-n)x\right) - \cos\left((m+n)x\right)\right]$$

$$\sin mx \cos nx = \frac{1}{2}\left[\sin\left((m-n)x\right) + \sin\left((m+n)x\right)\right]$$

$$\cos mx \cos nx = \frac{1}{2}\left[\cos\left((m-n)x\right) + \cos\left((m+n)x\right)\right]$$

Table 1

Example 4 ✎

Evaluate $\int_0^\pi \sin 4x \cos 5x\, dx$.

Solution

$$\int_0^\pi \sin 4x \cos 5x\, dx = \frac{1}{2}\int_0^\pi \left[\sin\left((4-5)x\right) + \sin\left((4+5)x\right)\right] dx$$

$$= -\frac{1}{2}\int_0^\pi \sin x\, dx + \frac{1}{2}\int_0^\pi \sin 9x\, dx \qquad \color{blue}{\sin(-x) = -\sin x}$$

$$= \frac{1}{2}\cos x\Big]_0^\pi - \frac{1}{18}\cos 9x\Big]_0^\pi$$

$$= \left(-\frac{1}{2} - \frac{1}{2}\right) - \frac{1}{18}(-1-1)$$

$$= -\frac{8}{9}$$

TOPIC 2 Powers of Tangents and Secants

The identities $\tan^2 x = \sec^2 x - 1$ and $\sec^2 x = \tan^2 x + 1$ go far in evaluating integrals of the form $\int \tan^m x \sec^n x\, dx$, especially when combined with integration by parts to reduce exponents. But to begin with, we need to know the integrals of the first and second powers of tangent and secant.

Example 5 🖉

Evaluate the following integrals.

a. $\int \sec^2 x \, dx$ **b.** $\int \tan^2 x \, dx$ **c.** $\int \tan x \, dx$ **d.** $\int \sec x \, dx$

Solution

The order of these integrals as presented is based on their ease of evaluation.

a. $\displaystyle\int \sec^2 x \, dx = \tan x + C$

b. $\displaystyle\int \tan^2 x \, dx = \int \left(\sec^2 x - 1\right) dx = \tan x - x + C$

c. $\displaystyle\int \tan x \, dx = \int \frac{\sin x}{\cos x} \, dx$ $u = \cos x$
$du = -\sin x \, dx$

$\displaystyle = -\int \frac{1}{u} \, du$

$\displaystyle = -\ln |u| + C$

$\displaystyle = -\ln |\cos x| + C$

$\displaystyle = \ln |\sec x| + C$ $-\ln|\cos x| = \ln\left|\dfrac{1}{\cos x}\right| = \ln |\sec x|$

d. This integral requires a bit more effort. One way to evaluate it is to multiply and divide the integrand by $\sec x + \tan x$, which results in a convenient transformation.

$$\int \sec x \, dx = \int \sec x \left(\frac{\sec x + \tan x}{\sec x + \tan x}\right) dx$$

$$= \int \frac{\sec^2 x + \sec x \tan x}{\sec x + \tan x} \, dx \qquad \begin{array}{l} u = \sec x + \tan x \\ du = \left(\sec x \tan x + \sec^2 x\right) dx \end{array}$$

$$= \int \frac{1}{u} \, du$$

$$= \ln |u| + C = \ln |\sec x + \tan x| + C$$

With these small-power integrals out of the way, we can proceed to other products of powers of tangent and secant. The following strategy offers general guidance.

Evaluating Integrals of the Form $\int \tan^m x \sec^n x \, dx$

Assuming m and n are nonnegative integers, an integral of the form $\int \tan^m x \sec^n x \, dx$ can be evaluated as follows.

If m is odd: Use the identity $\tan^2 x = \sec^2 x - 1$ to express all but one factor of tangent in terms of secant, and then use u-substitution with $u = \sec x$ to evaluate the result.

If n is even and $n \geq 4$: Use the identity $\sec^2 x = 1 + \tan^2 x$ to express all but one factor of $\sec^2 x$ in terms of tangent, and then use u-substitution with $u = \tan x$ to evaluate the result. (Remember that we have already evaluated $\int \sec^2 x \, dx$.)

If m is even and n is odd or $n < 4$: Use identities, u-substitution, or integration by parts, possibly more than once, to evaluate.

Example 6 ✎

Evaluate $\int \tan^2 x \sec^4 x \, dx$.

Solution

The exponent of $\sec x$ is even and greater than or equal to 4, so we replace all but one factor of $\sec^2 x$.

$$\int \tan^2 x \sec^4 x \, dx = \int \tan^2 x \sec^2 x \sec^2 x \, dx$$

$$= \int \tan^2 x \left(1 + \tan^2 x\right) \sec^2 x \, dx \qquad \begin{aligned} u &= \tan x \\ du &= \sec^2 x \, dx \end{aligned}$$

$$= \int u^2 \left(1 + u^2\right) du = \int \left(u^2 + u^4\right) du$$

$$= \frac{1}{3} u^3 + \frac{1}{5} u^5 + C = \frac{1}{3} \tan^3 x + \frac{1}{5} \tan^5 x + C$$

Example 7 ✎

Evaluate $\int \tan^4 x \, dx$.

Solution

Since the power of tangent is even and the power of secant is less than 4, we follow the guidelines of the third case of the method. For this integral, the application of one identity followed by u-substitution suffices.

$$\int \tan^4 x \, dx = \int \tan^2 x \left(\sec^2 x - 1\right) dx$$

$$= \int \tan^2 x \sec^2 x \, dx - \int \tan^2 x \, dx \qquad \begin{aligned} &\text{Recall from Example 5 that} \\ &\int \tan^2 x \, dx = \tan x - x. \end{aligned}$$

$$= \int \tan^2 x \sec^2 x \, dx - \left(\tan x - x\right) \qquad \begin{aligned} u &= \tan x \\ du &= \sec^2 x \, dx \end{aligned}$$

$$= \int u^2 \, du - \left(\tan x - x\right)$$

$$= \frac{1}{3} \tan^3 x - \tan x + x + C$$

Example 8 ✎

Evaluate $\int \sec^3 x \, dx$.

Solution

This integral also falls into the third category, and integration by parts is useful in this case.

$$\int \sec^3 x \, dx = \sec x \tan x - \int \sec x \tan^2 x \, dx \qquad \begin{aligned} u &= \sec x & dv &= \sec^2 x \, dx \\ du &= \sec x \tan x \, dx & v &= \tan x \end{aligned}$$

$$= \sec x \tan x - \int \sec x \left(\sec^2 x - 1\right) dx$$

$$= \sec x \tan x - \int \sec^3 x \, dx + \int \sec x \, dx \qquad \begin{aligned} &\text{Recall from Example 5 that} \\ &\int \sec x \, dx = \ln|\sec x + \tan x|. \end{aligned}$$

$$= \sec x \tan x - \int \sec^3 x \, dx + \ln|\sec x + \tan x|$$

We have seen this situation arise before; since the integral we are trying to evaluate appears on both sides of the equation, we can solve for it.

$$2\int \sec^3 x \, dx = \sec x \tan x + \ln|\sec x + \tan x|$$

$$\int \sec^3 x \, dx = \frac{1}{2}\left(\sec x \tan x + \ln|\sec x + \tan x|\right) + C$$

7.3 **Exercises**

1–36 Evaluate the given indefinite or definite integral involving powers of sines and cosines.

1. $\displaystyle\int \sin^2 x \cos x \, dx$

2. $\displaystyle\int \sin x \cos^3 x \, dx$

3. $\displaystyle\int \cos^5 x \, dx$

4. $\displaystyle\int \cos x \sin x \, dx$

5. $\displaystyle\int \sin^4 x \, dx$

6. $\displaystyle\int \sin^{-4} x \cos x \, dx$

7. $\displaystyle\int_0^{\pi/3} \cos^{-2} x \sin x \, dx$

8. $\displaystyle\int_0^{\pi/2} \sin^3 2t \, dt$

9. $\displaystyle\int_{-\pi}^{\pi} \cos^6\left(\frac{\theta}{2}\right) d\theta$

10. $\displaystyle\int \frac{2\cos 4\alpha}{\sqrt{\sin 4\alpha}} \, d\alpha$

11. $\displaystyle\int \sin^7 x \cos^8 x \, dx$

12. $\displaystyle\int \sin^2 x \cos^2 x \, dx$

13. $\displaystyle\int \sin^4\left(\frac{\theta}{4}\right)\cos^4\left(\frac{\theta}{4}\right) d\theta$

14. $\displaystyle\int_0^{\pi} \sin^5 x \, dx$

15. $\displaystyle\int_0^{\pi} \sin^2 x \, dx$

16. $\displaystyle\int \cos^3 3t \sin^4 3t \, dt$

17. $\displaystyle\int \sqrt{\cos x}\, \sin^3 x \, dx$

18. $\displaystyle\int_0^{3\pi} \sin^3\left(\frac{t}{3}\right)\cos^2\left(\frac{t}{3}\right) dt$

19. $\displaystyle\int \sin^3 x \cos^5 x \, dx$

20. $\displaystyle\int_0^{\pi/2} \sin^7 x \, dx$

21. $\displaystyle\int_{\pi^2/4}^{\pi^2} \frac{8\cos^2 \sqrt{x} \sin^2 \sqrt{x}}{\sqrt{x}} \, dx$

22. $\displaystyle\int \frac{\sin 2x + \sin^3 x}{\cos x} \, dx$

23. $\displaystyle\int 16\sin^2 x \cos^4 x \, dx$

24. $\displaystyle\int \cos^6 x \, dx$

25. $\displaystyle\int \sin^7 x \cos^3 x \, dx$

26. $\displaystyle\int \sin x \sin 5x \, dx$

27. $\displaystyle\int_0^{\pi/2} \cos 2x \cos 3x \, dx$

28. $\displaystyle\int_0^{\pi} \cos(-4x)\sin 6x \, dx$

29. $\displaystyle\int \sin 8x \sin(-7x) \, dx$

30. $\displaystyle\int \sqrt{\frac{1+\cos 2\theta}{2}} \, d\theta$

31. $\displaystyle\int_0^{\pi/3} \frac{\sin 3x}{\sqrt{\cos x}} \, dx$

32. $\displaystyle\int_0^{\pi/2} \sqrt{1-\cos x} \, dx$

33. $\displaystyle\int \frac{\cos x}{\sqrt{1-\sin x}} \, dx$

34. $\displaystyle\int \frac{\cos x}{\sqrt{1-\cos x}} \, dx$

35. $\displaystyle\int \sin x \sqrt{1-\sin x} \, dx$

36. $\displaystyle\int \frac{\cos^2 x}{\sqrt{1+\sin x}} \, dx$

(**Hint:** In Exercises 35 and 36, use the identity $\sin x = \cos\left(\dfrac{\pi}{2}-x\right)$.)

37–57 Evaluate the given indefinite or definite integral involving powers of tangents and secants. Note that the integrals involving cotangents and cosecants can be handled by rules analogous to those discussed in this section.

37. $\displaystyle\int \tan^5 x \, dx$

38. $\displaystyle\int \cot^4 x \, dx$

39. $\displaystyle\int \sec^4 x \, dx$

40. $\displaystyle\int \tan^3 x \sec^3 x \, dx$

41. $\displaystyle\int \frac{\tan \theta}{\sqrt{\sec \theta}} \, d\theta$

42. $\displaystyle\int \frac{\cot^5 x}{\csc x} \, dx$

43. $\displaystyle\int \tan x \sec^6 x \, dx$

44. $\displaystyle\int \frac{\sec^2 x}{\sqrt[3]{\tan x}} \, dx$

45. $\displaystyle\int \csc^4 t \cot^{3/2} t \, dt$

46. $\displaystyle\int \tan^4 x \sec^4 x \, dx$

47. $\displaystyle\int \tan^2 x \sec x \, dx$

48. $\displaystyle\int \csc^3 x \, dx$

49. $\displaystyle\int_0^{\pi/3} \tan^5 x \sec^3 x \, dx$

50. $\displaystyle\int_{\pi/3}^{2\pi/3} \csc x \cot^2 x \, dx$

51. $\displaystyle\int_{\pi/6}^{\pi/2} \csc^4 x \, dx$

52. $\displaystyle\int \frac{\csc^4 z}{\cot z} \, dz$

53. $\displaystyle\int \frac{\sec^4 2\alpha}{\cot^3 2\alpha} \, d\alpha$

54. $\displaystyle\int \sec^6 3\beta \cot^3 3\beta \, d\beta$

55. $\displaystyle\int \csc^4 4x \cot 4x \, dx$

56. $\displaystyle\int_{\sqrt[3]{\pi/4}}^{\sqrt[3]{\pi/2}} s^2 \csc^2\left(s^3\right)\cot^4\left(s^3\right) ds$

57. $\displaystyle\int_0^{\pi/4} \frac{\cot^2 t - 1}{\csc^2 t} \, dt$

58–63 The given integral does not directly fit any of the cases discussed in this section. Use trigonometric identities and familiar integration rules to evaluate it.

58. $\int \sin x \cot 2x \, dx$

59. $\int x \sec^2 x \, dx$

60. $\int \sec^3 x \cot x \, dx$

61. $\int z \tan^2 2z \, dz$

62. $\int \csc^4 x \cos^3 x \, dx$

63. $\int \dfrac{\sin 3x}{\sec x} \, dx$

64. Verify the following reduction formula for m, $n \in \mathbb{N}$. (**Hint:** Use integration by parts.)

$$\int \sin^m x \cos^n x \, dx = -\frac{\sin^{m-1} x \cos^{n+1} x}{m+n}$$
$$+ \frac{m-1}{m+n} \int \sin^{m-2} x \cos^n x \, dx$$

65. Use Exercises 3, 24, and 64 to evaluate the following integrals.

a. $\int \sin^2 x \cos^5 x \, dx$

b. $\int \sin^2 x \cos^6 x \, dx$

66. Find the area of the region between the x-axis and the graph of $f(x) = \sin^2 x \cos^3 x$ from $x = 0$ to $x = \pi/2$.

67. Find the volume of the solid obtained by revolving the region bounded by the graphs of $y = \tan x + \cot x$, $y = 0$, $x = \pi/6$, and $x = \pi/3$ about the x-axis.

68. Repeat Exercise 67 for the graphs of $y = \cos x + \sec x$, $y = 0$, $x = -\pi/4$, and $x = \pi/4$.

69. Find the centroid of the region bounded by the graphs of $y = x + \sin x$, $y = 0$, and $x = \pi$.

70. A particle is starting from the origin and moving along the x-axis so that its velocity at t seconds is $v(t) = \pi \tan^2(\pi t/18) \sin(\pi t/18)$ units per second $(0 \le t \le 8)$. Find its position at $t = 6$ seconds.

71–73 Use the product-to-sum identities of this section to verify the given formula for m, $n \in \mathbb{N}$.

71. $\displaystyle\int_{-\pi}^{\pi} \sin mx \sin nx \, dx = \begin{cases} 0 & \text{if } m \ne n \\ \pi & \text{if } m = n \end{cases}$

72. $\displaystyle\int_{-\pi}^{\pi} \sin mx \cos nx \, dx = 0$

73. $\displaystyle\int_{-\pi}^{\pi} \cos mx \cos nx \, dx = \begin{cases} 0 & \text{if } m \ne n \\ \pi & \text{if } m = n \end{cases}$

74–77 *True or False?* Determine whether the given statement is true or false. In case of a false statement, explain or provide a counterexample.

74. $\int \sin^2 x \, dx - x = C - \int \cos^2 x \, dx$

75. $\displaystyle\int_0^{\pi} \sin^2 x \, dx = \int_0^{\pi} \cos^2 x \, dx$

76. According to the text, the best way to evaluate $\int \sec^4 x \tan^4 x \, dx$ is by using integration by parts.

77. According to the text, the best way to evaluate $\int \sec^3 x \tan^4 x \, dx$ is by using integration by parts.

7.4 **Trigonometric Substitutions**

TOPICS

1. Three common trigonometric substitutions

2. Additional substitutions

Trigonometric identities again play a central role in this section, but this time in integrals that may at first appear to have no connection to trigonometry. We will use trigonometric identities (and the implicit geometry behind them) to transform unwieldy integrands into more easily handled expressions.

TOPIC 1 **Three Common Trigonometric Substitutions**

The basic idea behind trigonometric substitutions is a *change of variables*—the translation of an expression from one set of variables into another. The expressions we are concerned with at the moment are integrals in a single variable, but we will see the change of variables theme arise again later in multivariable settings.

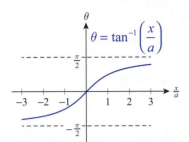

The Substitution Rule, or *u*-substitution, was our first example of a change of variables, and in fact the Substitution Rule is the basis of trigonometric substitutions. In general, given an integral $\int f(x)\,dx$, we replace x with a differentiable one-to-one function $g(\theta)$ and write the following:

$$\int f(x)\,dx = \int f\big(g(\theta)\big)g'(\theta)\,d\theta \qquad \begin{array}{l} x = g(\theta) \\ dx = g'(\theta)\,d\theta \end{array}$$

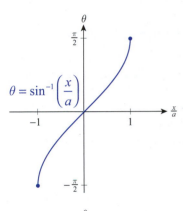

Note that this is just the Substitution Rule written in reverse (and with different variables), but the goal is the same—change the integral to one we know how to evaluate. We require g to be one-to-one so that we can, if desired, reverse the change of variables and express the evaluated integral in terms of the original variable. Three of the most frequently used trigonometry-based substitutions, and the expressions they help to simplify, are described below. In each case, the bounds on θ arise from the definitions of the inverse trigonometric functions in use. You might want to review how we define the inverse trigonometric functions, as shown in Figure 3 of Section 3.6. Figure 1 reproduces the graphs of the three inverse functions we will use in this section.

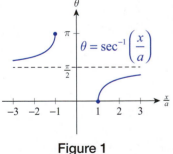

Figure 1

Three Trigonometric Substitutions

Expression	Substitution	Useful Identity
$\sqrt{a^2 + x^2}$	$x = a\tan\theta$ with $-\dfrac{\pi}{2} < \theta < \dfrac{\pi}{2}$	$1 + \tan^2\theta = \sec^2\theta$
$\sqrt{a^2 - x^2}$	$x = a\sin\theta$ with $-\dfrac{\pi}{2} \le \theta \le \dfrac{\pi}{2}$	$1 - \sin^2\theta = \cos^2\theta$
$\sqrt{x^2 - a^2}$	$x = a\sec\theta$ with $\begin{cases} 0 \le \theta < \dfrac{\pi}{2} & \text{if } \dfrac{x}{a} \ge 1 \\ \dfrac{\pi}{2} < \theta \le \pi & \text{if } \dfrac{x}{a} \le -1 \end{cases}$	$\sec^2\theta - 1 = \tan^2\theta$

Table 1

The best way to see how these substitutions are applied is with an example.

Example 1 ✎

Evaluate $\int \dfrac{dx}{\sqrt{25+x^2}}$.

Solution

This integral contains an expression of the form $\sqrt{a^2+x^2}$ with $a=5$, so we make the substitution $x=5\tan\theta$. Note that this means $dx=5\sec^2\theta\,d\theta$.

$$\int \frac{dx}{\sqrt{25+x^2}} = \int \frac{5\sec^2\theta\,d\theta}{\sqrt{25+25\tan^2\theta}}$$

$$= 5\int \frac{\sec^2\theta\,d\theta}{5\sqrt{1+\tan^2\theta}}$$

$$= \int \frac{\sec^2\theta\,d\theta}{\sqrt{\sec^2\theta}} = \int \frac{\sec^2\theta\,d\theta}{|\sec\theta|}$$

We know that $\sec\theta > 0$ on the interval $-\pi/2 < \theta < \pi/2$, so our integral can be simplified further and then evaluated.

$$\int \frac{\sec^2\theta\,d\theta}{|\sec\theta|} = \int \frac{\sec^2\theta\,d\theta}{\sec\theta} = \int \sec\theta\,d\theta = \ln|\sec\theta+\tan\theta|+C$$

The only remaining step is to reverse the substitution and express our answer in terms of x. Since $x=5\tan\theta$, we could do this by replacing each θ with $\tan^{-1}(x/5)$ and then simplify the result—no trivial task! Instead, it is far easier to sketch a diagram such as the one in Figure 2 that illustrates the relationship $\tan\theta = x/5$. From the angle θ and the known opposite and adjacent sides, we can determine the hypotenuse to be $\sqrt{25+x^2}$ and hence $\sec\theta = \sqrt{25+x^2}/5$.

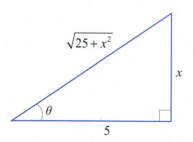

Figure 2

$$\ln|\sec\theta+\tan\theta|+C = \ln\left|\frac{\sqrt{25+x^2}}{5}+\frac{x}{5}\right|+C$$

$$= \ln\left|\frac{\sqrt{25+x^2}+x}{5}\right|+C$$

$$= \ln\left|\sqrt{25+x^2}+x\right| - \ln 5 + C$$

$$= \ln\left|\sqrt{25+x^2}+x\right| + \tilde{C} \qquad \tilde{C} = C - \ln 5$$

Example 2 ✎

Evaluate $\int \dfrac{x^2}{\sqrt{4-x^2}}\,dx$.

Solution

First, we replace x with $2\sin\theta$, where $-\pi/2 \le \theta \le \pi/2$:

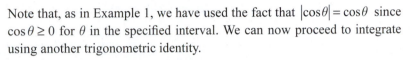

$$\int \frac{x^2}{\sqrt{4-x^2}}\,dx = \int \frac{(2\sin\theta)^2}{\sqrt{4-(2\sin\theta)^2}}(2\cos\theta)\,d\theta \qquad \begin{array}{l} x = 2\sin\theta \\ dx = 2\cos\theta\,d\theta \end{array}$$

$$= 4\int \frac{\sin^2\theta}{\sqrt{4(1-\sin^2\theta)}}(2\cos\theta)\,d\theta$$

$$= 4\int \frac{\sin^2\theta}{\sqrt{4\cos^2\theta}}(2\cos\theta)\,d\theta \qquad 1-\sin^2\theta = \cos^2\theta$$

$$= 4\int \frac{\sin^2\theta}{2\cos\theta}(2\cos\theta)\,d\theta = 4\int \sin^2\theta\,d\theta$$

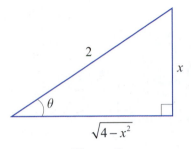

Figure 3

Note that, as in Example 1, we have used the fact that $|\cos\theta| = \cos\theta$ since $\cos\theta \geq 0$ for θ in the specified interval. We can now proceed to integrate using another trigonometric identity.

$$4\int \sin^2\theta\,d\theta = 4\int \frac{1-\cos 2\theta}{2}\,d\theta = 2\int (1-\cos 2\theta)\,d\theta = 2\theta - \sin 2\theta + C$$

We again need to express this answer back in x, and we can do so using Figure 3 and the identity $\sin 2\theta = 2\sin\theta\cos\theta$.

$$2\theta - \sin 2\theta + C = 2\theta - 2\sin\theta\cos\theta + C$$

$$= 2\sin^{-1}\left(\frac{x}{2}\right) - 2\left(\frac{x}{2}\right)\left(\frac{\sqrt{4-x^2}}{2}\right) + C \qquad x = 2\sin\theta \Rightarrow \theta = \sin^{-1}\left(\frac{x}{2}\right)$$

$$= 2\sin^{-1}\left(\frac{x}{2}\right) - \frac{x}{2}\sqrt{4-x^2} + C$$

Our next example shows how a trigonometric change of variables applies to a definite integral. The result is also a famous and useful area formula.

Example 3 ✍

Find the area of the ellipse defined by the equation $\dfrac{x^2}{a^2} + \dfrac{y^2}{b^2} = 1$.

Solution

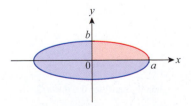

Figure 4 Area of an Ellipse

The graph of the equation is shown in Figure 4; we will find the area of the ellipse by finding the area of the portion in the 1st quadrant and multiplying by 4.

In order to work with a function of x that we can integrate, we solve the equation for y.

$$\frac{y^2}{b^2} = 1 - \frac{x^2}{a^2} = \frac{a^2 - x^2}{a^2} \quad \Rightarrow \quad y = \pm\frac{b}{a}\sqrt{a^2 - x^2}$$

The integral corresponding to the area shaded red in Figure 4 is $\dfrac{b}{a}\displaystyle\int_0^a \sqrt{a^2 - x^2}\,dx$, which we can evaluate with the substitution $x = a\sin\theta$.

$$\frac{b}{a}\int_0^a \sqrt{a^2-x^2}\,dx = \frac{b}{a}\int_{x=0}^{x=a} \sqrt{a^2-(a\sin\theta)^2}\,(a\cos\theta)\,d\theta \qquad \begin{aligned} & x=a\sin\theta \\ & dx = a\cos\theta\,d\theta \end{aligned}$$

$$= \frac{b}{a}\int_{\theta=0}^{\theta=\pi/2} \sqrt{a^2\cos^2\theta}\,(a\cos\theta)\,d\theta \qquad \begin{aligned} & x=0 \;\Rightarrow\; \theta=0 \\ & x=a \;\Rightarrow\; \theta=\pi/2 \end{aligned}$$

$$= ab\int_0^{\pi/2} \cos^2\theta\,d\theta$$

Be sure to note how the change in the limits of integration corresponds geometrically to the area being found; although we determined the limits by solving two equations for θ, the variable of integration and its limits clearly describe a region in the first quadrant. We can now evaluate the integral with a familiar identity.

$$ab\int_0^{\pi/2} \cos^2\theta\,d\theta = ab\int_0^{\pi/2} \frac{1+\cos 2\theta}{2}\,d\theta$$

$$= \frac{ab}{2}\left[\theta + \frac{1}{2}\sin 2\theta\right]_0^{\pi/2}$$

$$= \frac{ab}{2}\left[\left(\frac{\pi}{2}+0\right)-(0+0)\right]$$

$$= \frac{\pi ab}{4}$$

Multiplying by 4 gives us the result that the area of an ellipse with semiaxes of lengths a and b is πab. In particular, if $a = b = r$, we have confirmed that the area of a circle is πr^2.

Example 4 ✎

Find a general formula for $\displaystyle\int \frac{dx}{\sqrt{x^2-a^2}}$ under the assumption $a > 0$.

Solution

This uses the third of the three trigonometric substitutions: we let $x = a\sec\theta$ and specify that $0 \le \theta < \pi/2$ (the reason for this will soon be apparent). So $dx = a\sec\theta\tan\theta\,d\theta$.

$$\int \frac{dx}{\sqrt{x^2-a^2}} = \int \frac{a\sec\theta\tan\theta\,d\theta}{\sqrt{a^2\sec^2\theta - a^2}}$$

$$= \int \frac{\sec\theta\tan\theta}{\sqrt{\sec^2\theta - 1}}\,d\theta$$

$$= \int \frac{\sec\theta\tan\theta}{|\tan\theta|}\,d\theta \qquad |\tan\theta| = \tan\theta \text{ for } 0 \le \theta < \pi/2$$

$$= \int \sec\theta\,d\theta = \ln|\sec\theta + \tan\theta| + C$$

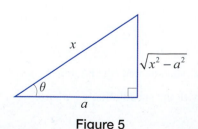

Figure 5

To express the answer in terms of x, we sketch a right triangle illustrating the relationship between θ, a, and x. Note that in order for the integrand to be real, and for the diagram in Figure 5 to make sense, we must assume that $x \ge a$ (so if we were evaluating a definite integral, this would restrict the interval of integration). Also, the specification that $0 \le \theta < \pi/2$ is consistent with the relationship between the variables, as shown in Figure 5. We can now rewrite our formula.

$$\ln\left|\sec\theta + \tan\theta\right| + C = \ln\left|\frac{x}{a} + \frac{\sqrt{x^2 - a^2}}{a}\right| + C$$

$$= \ln\left|x + \sqrt{x^2 - a^2}\right| - \ln a + C$$

$$= \ln\left|x + \sqrt{x^2 - a^2}\right| + \tilde{C} \qquad\qquad \tilde{C} = C - \ln a$$

TOPIC 2   Additional Substitutions

The integral in Example 4 may look familiar to you—it appears in the table of integrals relating to inverse hyperbolic functions in Section 6.6. But the general formula we just found is not the formula listed in that table. Is there an inconsistency?

Fortunately, no. The two formulas are equivalent, a fact you can verify using the technique of Exercise 73 from Section 6.6. We can also show the equivalency of the two formulas by evaluating the integral with a different substitution—in this case, a hyperbolic substitution.

Example 5 ✎

Use the substitution $x = a\cosh u$ to find an alternative general formula for

$$\int \frac{dx}{\sqrt{x^2 - a^2}} \quad \text{under the assumption } a > 0.$$

Solution

Recall that $\cosh^2 u - \sinh^2 u = 1$.

$$\int \frac{dx}{\sqrt{x^2 - a^2}} = \int \frac{a\sinh u\,du}{\sqrt{a^2\cosh^2 u - a^2}} \qquad\qquad dx = a\sinh u\,du$$

$$= \int \frac{a\sinh u\,du}{a\sqrt{\cosh^2 u - 1}}$$

$$= \int \frac{\sinh u\,du}{\sinh u} \qquad\qquad \cosh^2 u - 1 = \sinh^2 u$$

$$= \int du$$

$$= u + C = \cosh^{-1}\left(\frac{x}{a}\right) + C$$

In many instances, using several changes of variables may be the simplest way to reduce an integral to a form that looks familiar. In our final example, we use a simple change to transform the integral into one that is amenable to trigonometric substitution, apply that trigonometric substitution, then revert back one variable to obtain our numerical solution.

Example 6 ✒

Evaluate $\displaystyle\int_0^1 \frac{dx}{\left(9x^2+4\right)^{3/2}}$.

Solution

The integrand contains an expression almost, but not quite, of the form $\sqrt{x^2+a^2}$. We can easily eliminate the coefficient in front of x^2 by making the change of variable $t = 3x$. Note the change in the limits of integration that result.

$$\int_{x=0}^{x=1} \frac{dx}{\left(9x^2+4\right)^{3/2}} = \frac{1}{3}\int_{t=0}^{t=3} \frac{dt}{\left(t^2+4\right)^{3/2}} = \frac{1}{3}\int_{t=0}^{t=3} \frac{dt}{\left(\sqrt{t^2+4}\right)^3} \qquad \begin{array}{l} t = 3x \\ dt = 3\,dx \end{array}$$

Now, proceed with $t = 2\tan\theta$ and $dt = 2\sec^2\theta\,d\theta$.

$$\frac{1}{3}\int_{t=0}^{t=3} \frac{dt}{\left(\sqrt{t^2+4}\right)^3} = \frac{1}{3}\int_{t=0}^{t=3} \frac{2\sec^2\theta\,d\theta}{\left(\sqrt{4\tan^2\theta+4}\right)^3} = \frac{1}{3}\int_{t=0}^{t=3} \frac{2\sec^2\theta\,d\theta}{\left(\sqrt{4\sec^2\theta}\right)^3}$$

$$= \frac{1}{3}\int_{t=0}^{t=3} \frac{2\sec^2\theta}{\left(2\sec\theta\right)^3}\,d\theta = \frac{1}{12}\int_{t=0}^{t=3} \frac{1}{\sec\theta}\,d\theta$$

$$= \frac{1}{12}\int_{t=0}^{t=3} \cos\theta\,d\theta = \frac{1}{12}\sin\theta\,\Big]_{t=0}^{t=3}$$

Note that our antiderivative is a function of θ, but the limits of integration are still in terms of t. In order to evaluate the expression, we sketch the triangle in Figure 6 and write everything in terms of t.

$$\frac{1}{12}\sin\theta\,\Bigg]_{t=0}^{t=3} = \frac{1}{12}\cdot\frac{t}{\sqrt{t^2+4}}\,\Bigg]_{t=0}^{t=3}$$

$$= \frac{1}{12}\cdot\frac{3}{\sqrt{13}}$$

$$= \frac{1}{4\sqrt{13}}$$

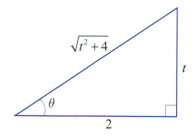

Figure 6

7.4 **Exercises**

1–6 Choose the substitution(s) that are helpful in evaluating the integral. (Do not actually evaluate the integral. There may be more than one correct answer.)

1. $\displaystyle\int x\sqrt{x-1}\,dx$

   a. $x = \tan\theta$ **b.** $x = \sec\theta$

   c. $x = \sin\theta$ **d.** $\theta = x - 1$

2. $\displaystyle\int \frac{dx}{\sqrt{x^2+1}}$

   a. $x = \sin\theta$ **b.** $\tan x = \theta$

   c. $x = \tan\theta$ **d.** $x = \sec\theta$

3. $\displaystyle\int x\sqrt{9-x^2}\,dx$

   a. $x = 3\sec\theta$ **b.** $\theta = 3\sin x$

   c. $x = 3\sin\theta$ **d.** $\theta = 9 - x^2$

4. $\displaystyle\int \frac{dz}{z^2\sqrt{z^2-9}}$

   a. $z = \sin t$ **b.** $z = \sec t$

   c. $z = 3\sec t$ **d.** $z = 3\cosh t$

5. $\displaystyle\int \frac{dx}{x^2\sqrt{4x^2+9}}$

   a. $x = 2\tan t$ **b.** $x = \dfrac{3}{2}\tan t$

   c. $x = 3\tan t$ **d.** $x = \dfrac{3}{2}\sec t$

6. $\displaystyle\int \frac{\sqrt{9-(x-1)^2}}{x-1}\,dx$

   a. $x = 3\sin\theta + 1$ **b.** $x = 3\sin\theta$

   c. $x - 1 = 3\sec t$ **d.** $x - 1 = 3\tan t$

7–12 Choose the correct answer. (You should be able to identify the correct answer without actually evaluating the integral.)

7. $\displaystyle\int \frac{dx}{\sqrt{x^2+1}}$

   a. $\sqrt{\arctan x} + C$ **b.** $2\sqrt{x^2+1} + C$

   c. $\ln\left|\sqrt{x^2+1}+x\right| + C$ **d.** $\ln\sqrt{x^2+1} + C$

8. $\displaystyle\int \frac{x^2}{\sqrt{9-x^2}}\,dx$

   a. $\dfrac{x^2}{2}\arcsin\dfrac{x}{2} + C$ **b.** $\dfrac{2x^3}{3}\sqrt{9-x^2} + C$

   c. $-\sqrt{9-x^2} + C$

   d. $\dfrac{9}{2}\arcsin\dfrac{x}{3} - \dfrac{x}{2}\sqrt{9-x^2} + C$

9. $\displaystyle\int \frac{t^2}{\sqrt{1-t^2}}\,dt$

   a. $\dfrac{t}{2}\sqrt{1-t^2} - \dfrac{\tan^{-1}t}{2} + C$

   b. $\dfrac{\sin^{-1}t}{2} - \dfrac{t}{2}\sqrt{1-t^2} + C$

   c. $-\sqrt{1-t^2} + C$

   d. $\dfrac{1}{2}\left(t\sqrt{1-t^2} + \ln\left|t+\sqrt{t^2-1}\right|\right) + C$

10. $\displaystyle\int \frac{\sqrt{1-x^2}}{x}\,dx$

   a. $\dfrac{1}{x\arcsin x} + C$

   b. $\sqrt{1-x^2} + \ln\left|\dfrac{1-\sqrt{1-x^2}}{x}\right| + C$

   c. $\dfrac{2\left(\sqrt{1-x^2}\right)^{3/2}}{3x} + C$

   d. $\sqrt{x^2-1} + \arctan\dfrac{1}{\sqrt{x^2-1}} + C$

11. $\displaystyle\int \frac{dx}{\sqrt{x^2-4}}$

   a. $\dfrac{1}{2}\arcsin\dfrac{x}{2} + C$ **b.** $2\sqrt{x^2-4} + C$

   c. $\ln\left|x+\sqrt{x^2-4}\right| + C$ **d.** $\dfrac{2\left(\sqrt{x^2-4}\right)^{3/2}}{3} + C$

12. $\displaystyle\int \frac{dx}{\left(x^2+9\right)^{3/2}}$

   a. $\dfrac{x}{9\sqrt{x^2+9}} + C$ **b.** $\dfrac{-2}{\sqrt{x^2+9}} + C$

   c. $\dfrac{1}{3}\left(\arctan\dfrac{x}{3}\right)^{3/2} + C$ **d.** $\ln(x+9)^{3/2} + C$

13–48 Use the three trigonometric substitutions discussed in this section to evaluate the given indefinite or definite integral. (**Note:** Not all integrals require trigonometric substitution.)

13. $\displaystyle\int \frac{3}{\sqrt{x^2+9}}\,dx$ **14.** $\displaystyle\int x\sqrt{x-1}\,dx$

15. $\displaystyle\int \frac{x}{\sqrt{4-x^2}}\,dx$ **16.** $\displaystyle\int \frac{\sqrt{9-x^2}}{2x}\,dx$

17. $\displaystyle\int \frac{t^2}{\sqrt{25-t^2}}\,dt$ **18.** $\displaystyle\int \frac{ds}{s\sqrt{s^2-4}}$

19. $\displaystyle\int \frac{dx}{x^2\sqrt{x^2-36}}$ **20.** $\displaystyle\int \frac{\sqrt{x^2-2}}{x^3}\,dx$

21. $\displaystyle\int \frac{-x}{\sqrt{1-x^2}}\,dx$ **22.** $\displaystyle\int \frac{x^2}{\sqrt{x^2+25}}\,dx$

23. $\displaystyle\int \frac{2-x}{\sqrt{4+x^2}}\,dx$ **24.** $\displaystyle\int \frac{x^2-2x+5}{\sqrt{1-x^2}}\,dx$

25. $\int_{5/2}^{3} \dfrac{dz}{2z^2\sqrt{z^2-4}}$

26. $\int_{0}^{5} \dfrac{z}{\sqrt{z^2+4}}\,dz$

27. $\int_{0}^{\sqrt5} w^2\sqrt{5-w^2}\,dw$

28. $\int_{3}^{5} \dfrac{\sqrt{x^2-9}}{x}\,dx$

29. $\int \dfrac{dt}{t^2\sqrt{t^2+9}}$

30. $\int \dfrac{s^2}{\sqrt{s^2-25}}\,ds$

31. $\int \dfrac{\sqrt{4-9x^2}}{x}\,dx$

32. $\int \dfrac{dy}{y\sqrt{25+9y^2}}$

33. $\int \dfrac{\left(4-w^2\right)^{3/2}}{w^6}\,dw$

34. $\int_{1}^{2} \dfrac{dx}{4x^2\sqrt{4x^2+1}}$

35. $\int_{1/2}^{2} \dfrac{dy}{y^2\sqrt{16-y^2}}$

36. $\int \sqrt{1-x^2}\,dx$

37. $\int \sqrt{9x^2+16}\,dx$

38. $\int x^2\sqrt{9-x^2}\,dx$

39. $\int \dfrac{\sqrt{4-x^2}}{x^2}\,dx$

40. $\int \dfrac{dx}{x^2\sqrt{x^2+1}}$

41. $\int \dfrac{x}{\sqrt{3x^2+1}}\,dx$

42. $\int \dfrac{x^2}{\left(x^2+16\right)^{3/2}}\,dx$

43. $\int \dfrac{-x}{\left(1-x^2\right)^{3/2}}\,dx$

44. $\int \dfrac{dt}{\sqrt{9t^2-4}}$

45. $\int \dfrac{dx}{x^2\sqrt{25-9x^2}}$

46. $\int \dfrac{dx}{\sqrt{9x^2+4}}$

47. $\int_{-2}^{2} \dfrac{dt}{\left(t^2+5\right)^{3/2}}$

48. $\int \dfrac{dx}{\left(x^2+1\right)^2}$

49–60 Complete the square and use applicable substitutions from this section to evaluate the given integral.

49. $\int \dfrac{dx}{\sqrt{x^2+x+3}}$

50. $\int \dfrac{dv}{\left(9v^2-18v+5\right)^{3/2}}$

51. $\int \dfrac{x^2}{\sqrt{6x-x^2}}\,dx$

52. $\int_{1}^{3} \dfrac{2}{\sqrt{4x-x^2}}\,dx$

53. $\int_{2}^{3} \dfrac{dx}{\sqrt{-2x^2+8x-4}}$

54. $\int \dfrac{dx}{\sqrt{x^2-6x+10}}$

55. $\int_{2}^{3} \dfrac{dx}{\left(4x^2-8x+3\right)^{3/2}}$

56. $\int \dfrac{v^2}{\sqrt{3+2v-v^2}}\,dv$

57. $\int \sqrt{7+6x-x^2}\,dx$

58. $\int \sqrt{4x^2-16x+25}\,dx$

59. $\int \dfrac{2}{\left(x^2-10x+29\right)^2}\,dx$

60. $\int \dfrac{ds}{\left(s^2-8s+17\right)^{3/2}}$

61–63 Use an appropriate substitution followed by a trigonometric substitution to evaluate the integral.

61. $\int e^x\sqrt{1-e^{2x}}\,dx$

62. $\int \dfrac{\sqrt{x}}{1+x}\,dx$

63. $\int \dfrac{\cot x\,\csc x}{\sqrt{\sin^2 x+1}}\,dx$

64–71 Use an appropriate trigonometric substitution to find a general formula for the expression. (Assume $a>0$.)

64. $\int \sqrt{x^2+a^2}\,dx$

65. $\int \sqrt{x^2-a^2}\,dx$

66. $\int \dfrac{\sqrt{x^2-a^2}}{x}\,dx$

67. $\int \dfrac{dx}{x\sqrt{x^2-a^2}}$

68. $\int \sqrt{a^2-x^2}\,dx$

69. $\int \dfrac{dx}{\sqrt{x^2+a^2}}$

70. $\int \dfrac{-dx}{x\sqrt{a^2-x^2}}$

71. $\int \dfrac{-dx}{x\sqrt{x^2+a^2}}$ $(x>0)$

72–74. Use hyperbolic substitutions to find alternative general formulas for Exercises 69–71.

75. Evaluate Exercise 42 by using the hyperbolic substitution $x=4\sinh t$, and then generalize your result to arrive at both a trigonometric and a hyperbolic formula for

$$\int \dfrac{x^2}{\left(x^2+a^2\right)^{3/2}}\,dx.$$

76–77 Combine integration by parts and trigonometric substitution to evaluate the integral.

76. $\int t\arcsin t\,dt$

77. $\int t\arccos t\,dt$

78. Find the area of the region between the graph of $y=\dfrac{1}{\left(x^2+2\right)^{3/2}}$ and the x-axis from $x=-1$ to $x=1$.

79. Repeat Exercise 78 for the curve $y=\dfrac{1}{x^2\sqrt{x^2-3}}$ on the interval $[2,3]$.

80. Find the area enclosed by the unit circle $x^2+y^2=1$ and the parabola $y=\sqrt2 x^2$.

81. Rotate the region bounded by the graph of $y=\dfrac{\sqrt{x^2-9}}{x^3}$, $3\le x\le 5$, about the y-axis. Use the shell method to find the volume of the resulting solid.

82. Repeat Exercise 81 for the curve $y = \dfrac{27x^2}{\left(9x^2 + 4\right)^{3/2}}$ on the interval $[0,1]$.

83. Use the method of disks to determine the volume of the solid obtained by revolving the graph of the curve $y = \dfrac{\sqrt[4]{16 - 4x^2}}{x^2}$, $1 \le x \le 2$, about the x-axis.

84. Find the arc length of the prototypical parabola $y = x^2$ between the origin and the point $(2,4)$.

85. Find the arc length for the graph of $y = \ln x$ between $x = \sqrt{3}$ and $x = 2\sqrt{2}$.

86. A cylindrical fuel tank of radius 10 in. is positioned so its axis is horizontal. Find the fluid force acting on one end of the tank if it is partially filled with diesel fuel so that the top 4 in. of the tank are empty. Use $55\ \text{lb/ft}^3$ for the weight density of diesel fuel.

87. In an attempt to square the circle, Hippocrates of Chios showed about 2500 years ago that the area of the red shaded region (called a *lune*) in the figure below is equal to the area of the shaded triangle (which in turn is half of a square). Given that the bigger circle is centered at $(0,-r)$, use calculus to prove Hippocrates' result.

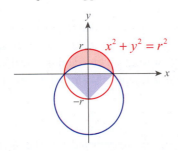

88.* Find a more general formula for the area of the lune in the case where the radius of the bigger circle is R.

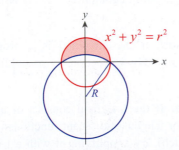

89–92 *True or False?* Determine whether the given statement is true or false. In case of a false statement, explain or provide a counterexample.

89. According to the text, the best substitution for
$$\int \frac{1}{\sqrt{x^2 - 1}}\, dx \text{ is } x = \sin\theta.$$

90. Substituting $x = \sin\theta$, we obtain
$$\int 2x\sqrt{1 - x^2}\, dx = \int 2\sin\theta\cos\theta\, d\theta = \int \sin 2\theta\, d\theta.$$

91. A straightforward way to evaluate $\displaystyle\int_0^{1/2} \frac{1}{1 - x^2}\, dx$ is to find $\displaystyle\int_0^{1/2} \frac{1}{\sqrt{1 - x^2}}\, dx$ by substituting $x = \sin\theta$, and then squaring the answer.

92. According to this textbook, the only way to evaluate the integral $\displaystyle\int \frac{1}{x^2 - 4}\, dx$ is to substitute $x = 2\sec\theta$.

7.4 **Technology Exercises**

93–96 Use a computer algebra system to revisit the given exercise. Do you get the same answer that you obtained by hand?

93. Exercise 22

94. Exercise 37

95. Exercise 54

96. Exercise 58

7.5 Integration Summary and Integration Using Computer Algebra Systems

TOPICS

1. Integration recap and guidelines

2. Integration with computer algebra systems

As we near the end of Chapter 7, it's appropriate to pause and review the many integration techniques we've learned so far. This review will also serve as an introduction to the ways in which software can help us with integration-related tasks.

TOPIC 1 Integration Recap and Guidelines

Remember that our most basic integration technique is to use knowledge of differentiation to work backward and find an antiderivative of a given integrand. Tables of integrals, such as those found at the back of this text, are succinct reminders of what we know and give us patterns with which to match an integral. Beyond this, the techniques that we have learned are, in order, as follows.

- The Substitution Rule (u-substitution)

- Integration by parts

- Partial fractions

- Simplification through trigonometric identities

- Trigonometric substitution

Given an integral, there is no "correct" strategy guaranteed to lead to its evaluation, but there are general guidelines to keep in mind.

Integration Guidelines

1. **Change the Appearance:** Simplify and/or rewrite the integrand to put it into a recognizable form, perhaps matching an integral formula found in a table. This step may involve a change of variables to simplify the appearance of variables. Other options include completing the square of a quadratic expression, splitting apart fractions, rationalizing the denominator or numerator of a fractional expression, and carrying out polynomial long division.

2. **u-Substitution:** Look for a factor of the integrand that is (up to a constant multiple) the derivative of the remainder of the integrand. Such integrals are good candidates for u-substitution (see Sections 5.4 and 5.5).

3. **Integration by Parts:** If the integrand is the product of two expressions, one of which is easily integrated, integration by parts may be called for (see Section 7.1).

4. **Partial Fractions Method:** If the integrand is a proper rational function and not easily integrated by inspection, try partial fractions (see Section 7.2).

5. **Trigonometric Identities:** If the integrand consists of a product of sine and cosine functions or tangent and secant functions (or products that can be rewritten as such), try simplifying it with a trigonometric identity (see Section 7.3).

6. Trigonometric Substitutions: Integrands containing factors of the form $\sqrt{\pm x^2 \pm a^2}$ are candidates for a trigonometric substitution. Remember that a change of variables may be necessary in order to get the integrand into a good form for this technique (see Section 7.4).

Example 1 ✐

Evaluate $\displaystyle\int \frac{-3\,dx}{x^2 - 2x + 2}$.

Solution

If we complete the square in the denominator, we get an integral that is close to one we've seen, one that appears in the table of integrals.

$$\int \frac{-3\,dx}{x^2 - 2x + 2} = -3\int \frac{dx}{(x-1)^2 + 1}$$

If the denominator of the integrand were simply $x^2 + 1$, the integrand would be the derivative of $\tan^{-1} x$. But we can accomplish this with an easy change of variables.

$$-3\int \frac{dx}{(x-1)^2 + 1} = -3\int \frac{du}{u^2 + 1} \qquad \begin{aligned} u &= x - 1 \\ du &= dx \end{aligned}$$

$$= -3\tan^{-1} u + C$$

$$= -3\tan^{-1}(x-1) + C$$

Example 2 ✐

Evaluate $\displaystyle\int \sqrt{\frac{5-x}{5+x}}\,dx$.

Solution

The radicals in the numerator and denominator of this integrand make it difficult to work with, and such expressions are good candidates for rationalizing. We can choose to rationalize either part of the fraction, but rationalizing the denominator doesn't immediately lead to anything that looks better. So we try the numerator.

$$\int \sqrt{\frac{5-x}{5+x}}\,dx = \int \sqrt{\frac{5-x}{5+x}} \cdot \frac{\sqrt{5-x}}{\sqrt{5-x}}\,dx = \int \frac{5-x}{\sqrt{25-x^2}}\,dx$$

The presence of a quadratic polynomial in the denominator and a linear polynomial in the numerator looks promising, but u-substitution with $u = 25 - x^2$ doesn't quite help yet. However, simply splitting the fraction into two pieces results in two integrals we can handle. The first, after dividing the top and bottom by 5, takes on a form we have seen before, and the second can be integrated with the u-substitution just mentioned.

$$\int \frac{5-x}{\sqrt{25-x^2}}\,dx = \int \frac{5}{\sqrt{25-x^2}}\,dx - \int \frac{x}{\sqrt{25-x^2}}\,dx$$

$$= \int \frac{1}{\sqrt{1-\left(\dfrac{x}{5}\right)^2}}\,dx - \int \frac{x}{\sqrt{25-x^2}}\,dx$$

$$= 5\sin^{-1}\left(\frac{x}{5}\right) + \sqrt{25-x^2} + C$$

TOPIC 2 Integration with Computer Algebra Systems

Transforming an integral into a form that matches a formula in a table of integrals may require some ingenuity, but the process can be broken down into a routine of steps and cases to check (although an admittedly complicated routine). Computer software and some calculators excel at such routine—this fact is the basis of the integration programming in computer algebra systems (CAS) such as *Mathematica*, Maple, MATLAB, and Derive, and in sufficiently advanced calculators. If such programming is available to you, it can be a valuable resource in work involving integration (and other tools of calculus). As with all tools, however, its correct and proficient use requires knowledge, experience, and a healthy sense of caution.

The first principle to remember is that technology can be a great asset once you have mastered key concepts and techniques, but it is far less useful as an educator—don't let technology become a crutch that hinders your understanding. Computer software and calculators can't analyze a problem for you, can't define the relevant parameters, functions, and equations, and can't interpret any answers produced in the context of the problem—all of that is your job.

The second principle to keep in mind is that answers produced by software may not be in the form you expect or desire. As we have seen, evaluating a given integral by two different methods may produce results that appear quite different, although they are equivalent. And this mathematical fact is compounded by the idiosyncrasies of any particular CAS or calculator—it is important to learn the notation and syntax unique to a given technology. (For an introduction to some of the syntax of *Mathematica*, see Appendix A.)

In the examples that follow, the CAS *Mathematica* is used to demonstrate how technology can be profitably used to evaluate integrals or, in some cases, to check our answers. The commands and syntax shown are particular to *Mathematica*, but similar commands and capability exist in all such systems. For example, in Maple use the command $\text{int}(f, x)$ to evaluate indefinite integrals and the command $\text{int}(f, x = a..b)$ to evaluate definite integrals over the interval $[a,b]$, where f is a function of the variable x.

Example 3 ✎

Find a general formula for the integral $\int \dfrac{1}{ax+b}\,dx$.

Solution

We already know the answer, which we used repeatedly while learning the partial fractions method.

$$\int \frac{1}{ax+b}\,dx = \frac{1}{a}\ln|ax+b| + C \qquad \text{Recall that } u = ax + b \text{ gives us the answer quickly.}$$

In[1]:= **Integrate[1 / (a * x + b), x]**

Out[1]= $\dfrac{\text{Log}[b + a\,x]}{a}$

Figure 1 Basic Indefinite Integration

Figure 1 shows the relevant *Mathematica* command and result. Note that the output cites the **Log** function (a built-in function); we must be familiar enough with *Mathematica* to know that **Log**, without any other qualifier, refers to the natural logarithm. Also, note that the *Mathematica* answer does not include absolute value signs—we need to know to include those. Finally, *Mathematica*, as is typical with CAS, does not include an arbitrary constant C in its answer.

Example 4 ✎

Use *Mathematica* to evaluate $\int \sqrt{\dfrac{5-x}{5+x}}\,dx$ and compare the result with our solution.

Solution

In[2]:= **Integrate[Sqrt[(5 - x) / (5 + x)], x]**

Out[2]= $\dfrac{\sqrt{\dfrac{5-x}{5+x}}\left(\sqrt{5-x}\,(5+x) + 10\sqrt{5+x}\;\text{ArcSin}\left[\dfrac{\sqrt{5+x}}{\sqrt{10}}\right]\right)}{\sqrt{5-x}}$

Figure 2

As Figure 2 illustrates, the form of an answer provided by a CAS may not be aesthetically pleasing. Recall from Example 2 that our solution to this integral is

$$5\sin^{-1}\left(\frac{x}{5}\right) + \sqrt{25 - x^2} + C.$$

The two answers are equivalent, a fact that can be verified by differentiating each one and showing that the result (after significant simplification) is the original integrand. *Mathematica* can also be used to confirm that our work is correct with the differentiation and simplification commands shown in Figure 3 (note that we must still simplify the output slightly to obtain the form of the original integrand).

In[3]:= `D[5 ArcSin[x / 5] + Sqrt[25 - x^2], x]`

Out[3]= $-\dfrac{x}{\sqrt{25 - x^2}} + \dfrac{1}{\sqrt{1 - \frac{x^2}{25}}}$

In[4]:= `Simplify[D[5 ArcSin[x / 5] + Sqrt[25 - x^2], x]]`

Out[4]= $\dfrac{5 - x}{\sqrt{25 - x^2}}$

Figure 3 Verifying the Solution of Example 2

We can also, as you would expect, evaluate definite integrals with a CAS.

Example 5

Evaluate the integral $\displaystyle\int_1^4 \frac{4t^2 - 9t + 7}{5t^2 + 2t + 1}\,dt.$

Solution

We could proceed directly to the definite integral calculation, but how would we know if the numerical result we obtain is reasonable or not? It would be nice to use the power of *Mathematica* to give us some measure of confidence in the answer.

One way of doing that is to ask *Mathematica* to graph the function over an interval that contains the interval of integration and to visually compare the integral with the area between the graph and the *t*-axis. We are trusting *Mathematica* to provide a reasonably accurate graph, but at least this gives us a way to cross-check several answers. Figure 4 illustrates how to first define a function $f(t)$ in *Mathematica*, then plot it over the interval $[0,5]$.

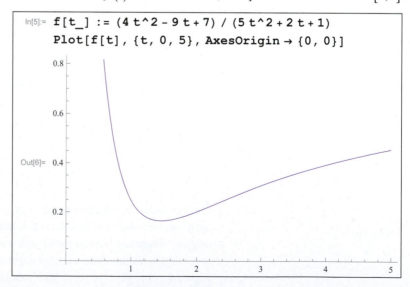

In[5]:= `f[t_] := (4 t^2 - 9 t + 7) / (5 t^2 + 2 t + 1)`
`Plot[f[t], {t, 0, 5}, AxesOrigin → {0, 0}]`

Figure 4 Plotting a Function

Then we ask for the definite integral from 1 to 4. The exact answer that results is not especially meaningful, so the last command in Figure 5 asks for the numerical evaluation of the result immediately preceding it.

In[7]:= **f[t_] := (4 t^2 - 9 t + 7) / (5 t^2 + 2 t + 1)**
Integrate[f[t], {t, 1, 4}]

Out[8]= $\dfrac{1}{50}\left(120 - 208\,\text{ArcTan}[3] + 208\,\text{ArcTan}\left[\dfrac{21}{2}\right] + 53\,\text{Log}[8] - 53\,\text{Log}[89]\right)$

In[9]:= **N[%]**

Out[9]= 0.789737

Figure 5 Evaluating a Definite Integral

Since the interval of integration has width 3 and the function takes on values between approximately 0.2 and 0.4 over that interval, the numerical answer of 0.789737 throws up no red flags.

Our last example illustrates one more built-in *Mathematica* function and some particulars of its syntax.

Example 6 ✍

Evaluate the integral $\displaystyle\int \frac{\cosh^4 \sqrt{x}}{\sqrt{x}}\,dx$.

Solution

In[10]:= **Integrate[Cosh[Sqrt[x]]^4 / Sqrt[x], x]**

Out[10]= $\dfrac{1}{16}\left(12\,\sqrt{x} + 8\,\text{Sinh}\left[2\,\sqrt{x}\,\right] + \text{Sinh}\left[4\,\sqrt{x}\,\right]\right)$

Figure 6 Integrating a Hyperbolic Function

The *Mathematica* command and output are shown in Figure 6. Note how the numerator of the integrand is expressed in *Mathematica* as **Cosh[Sqrt[x]]^4**, which reflects the fact that an alternative way of writing the numerator is $\left(\cosh\sqrt{x}\right)^4$.

7.5 **Exercises**

1–20 Use the integration guidelines listed in this section to find the given indefinite or definite integral. (**Hint:** Whenever possible, try to simplify before integrating.)

1. $\displaystyle\int \frac{dx}{5x(x+5)}$

2. $\displaystyle\int \frac{2x}{(2x+3)^2}\,dx$

3. $\displaystyle\int \frac{2\,dx}{x\sqrt{x^2+4}}$

4. $\displaystyle\int \sin^2 x \sec^2 x\,dx$

5. $\displaystyle\int \frac{\tan^2 2x+1}{\csc^2 2x}\,dx$

6. $\displaystyle\int \frac{\cos^2\theta\csc^2\theta}{\tan\theta}\,d\theta$

7. $\displaystyle\int \frac{dx}{x^2+4x+8}$

8. $\displaystyle\int \frac{3x^3-12x^2+15x+2}{x^2-4x+5}\,dx$

9. $\displaystyle\int \frac{x\left(\sqrt{x}+2\right)}{\sqrt{x}}\,dx$

10. $\displaystyle\int \frac{1-\cos 2x}{\sin x}\,dx$

11. $\displaystyle\int \frac{x}{\sqrt{4-x^4}}\,dx$

12. $\displaystyle\int \frac{x^3}{\sqrt{1-x^2}}\,dx$

13. $\displaystyle\int \sqrt{x^2+2x+5}\,dx$

14. $\displaystyle\int \frac{\sqrt{4t+1}}{4t+5}\,dt$

15. $\displaystyle\int_{\sqrt{3}/3}^{\sqrt{3}} \frac{\arctan u}{u^2}\,du$

16. $\displaystyle\int \sqrt{e^2+e^x}\,dx$

17. $\displaystyle\int_{\pi/4}^{\pi/2} \frac{1+\cos z}{1-\cos z}\,dz$

18. $\displaystyle\int \frac{t^2}{t^6-9}\,dt$

19. $\displaystyle\int \frac{2}{\sqrt{z+2}+\sqrt{z}}\,dz$

20. $\displaystyle\int \sqrt{x}\cos\sqrt{x}\,dx$

21–40 Transform the integral into a form that you can integrate by using a table of integrals or any of the techniques discussed in the previous sections. Then evaluate the integral.

21. $\displaystyle\int \sqrt{\frac{x+2}{x-2}}\,dx$

22. $\displaystyle\int \frac{\cos x}{\sin x\,(2\sin x+7)}\,dx$

23. $\displaystyle\int \frac{dx}{x\sqrt{2x^2+9}}$

24. $\displaystyle\int \frac{dx}{\sqrt{e^{2x}-9}}$

25. $\displaystyle\int \frac{3x^2}{x^6\left(2x^3+1\right)}\,dx$

26. $\displaystyle\int \frac{4x}{\left(x^4+4\right)^2}\,dx$

27. $\displaystyle\int \frac{\tan u}{\cos^2 u\sqrt{5+2\tan u}}\,du$

28. $\displaystyle\int \frac{dx}{\sqrt{3e^x-16}}$

29. $\displaystyle\int \ln\sqrt{x}\,dx$

30. $\displaystyle\int \sin 2x\,(2\sin x+3)^3\,dx$

31. $\displaystyle\int \frac{\cot x}{\sqrt{2\sin x-\sin^2 x}}\,dx$

32. $\displaystyle\int t^2\sqrt{t^6-4}\,dt$

33. $\displaystyle\int \frac{\sqrt{3x-2}}{x}\,dx$

34. $\displaystyle\int \frac{\sqrt{2x+1}}{x}\,dx$

35. $\displaystyle\int \frac{x^2-2x+1}{\sqrt{1+6x-3x^2}}\,dx$

36. $\displaystyle\int x\sqrt{x^2+2x+5}\,dx$

37. $\displaystyle\int_{\pi/3}^{2\pi/3} \csc^5 t\,dt$

38. $\displaystyle\int \frac{\sin 2x}{\sqrt{2\sin x+9}}\,dx$

39. $\displaystyle\int_0^1 \arccos\sqrt{z}\,dz$

40. $\displaystyle\int e^{2t}\tan^{-1}e^t\,dt$

41–45 If the integrand is a rational function of $\sin x$ and $\cos x$, we can turn it into a rational function of u by the substitution $x = 2\arctan u$. The figure reflects the aforementioned substitution (as you will determine in Exercise 41). Using the figure along with differentiation we obtain the following.

$$dx = \frac{2\,du}{1+u^2}, \quad \sin x = \frac{2u}{1+u^2}, \quad \text{and} \quad \cos x = \frac{1-u^2}{1+u^2}$$

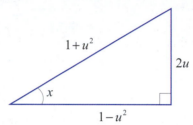

After integrating the resulting rational function, we express our answer in terms of the original variable x by using $u = \tan(x/2)$.

41. By using the identity $\tan\dfrac{x}{2} = \dfrac{\sin x}{1+\cos x}$, show that the figure indeed reflects the substitution $x = 2\arctan u$.

Use the above substitution technique to evaluate the given integral.

42. $\displaystyle\int \frac{dx}{3-\sin x}$

43. $\displaystyle\int \frac{dx}{\cos x-\sin x+2}$

44. $\displaystyle\int \frac{\sin x}{1+\cos x-\sin x}\,dx$

45. $\displaystyle\int \frac{du}{2\cos u+3}$

46–55 The given integration formula can be found in most tables of integrals. Verify it by an appropriate integration technique learned from this chapter.

46. $\displaystyle\int \frac{du}{a^2-u^2} = \frac{1}{2a}\ln\left|\frac{u+a}{u-a}\right| + C$

47. $\displaystyle\int \sqrt{a^2+u^2}\,du = \frac{u}{2}\sqrt{a^2+u^2} + \frac{a^2}{2}\ln\left(u+\sqrt{a^2+u^2}\right) + C$

48. $\displaystyle\int \sin^{-1}u\,du = u\sin^{-1}u + \sqrt{1-u^2} + C$

49. $\displaystyle\int \frac{du}{u\,(a+bu)} = \frac{1}{a}\ln\left|\frac{u}{a+bu}\right| + C$

50. $\displaystyle\int \frac{u}{a+bu}\,du = \frac{1}{b^2}\left(a+bu - a\ln|a+bu|\right) + C$

51. $\displaystyle\int \frac{\sqrt{a^2+u^2}}{u}\,du = \sqrt{a^2+u^2} - a\ln\left|\frac{a+\sqrt{a^2+u^2}}{u}\right| + C$

52. $\int \dfrac{u^2}{\sqrt{a^2-u^2}}\,du = -\dfrac{u}{2}\sqrt{a^2-u^2}+\dfrac{a^2}{2}\arcsin\dfrac{u}{a}+C$

53. $\int \dfrac{du}{u\sqrt{a^2-u^2}} = -\dfrac{1}{a}\ln\left|\dfrac{a+\sqrt{a^2-u^2}}{u}\right|+C$

54. $\int \dfrac{du}{u^2\sqrt{u^2-a^2}} = \dfrac{\sqrt{u^2-a^2}}{a^2u}+C$

55. $\int \dfrac{du}{\left(u^2-a^2\right)^{3/2}} = -\dfrac{u}{a^2\sqrt{u^2-a^2}}+C$

7.5 **Technology Exercises**

56–59 Use a computer algebra system to solve the given exercise. If the answer looks different from what you obtained by hand, prove that the answers are equivalent.

56. Exercise 6 **57.** Exercise 13

58. Exercise 27 **59.** Exercise 34

60–65 Compare a computer algebra system's answer to the given exercise with those obtained by substitution and/or integration tables. Use the differentiation feature of your technology to prove that both answers are correct.

60. Exercise 5 **61.** Exercise 16

62. Exercise 21 **63.** Exercise 24

64. Exercise 30 **65.** Exercise 36

66–68 Use a computer algebra system to find $F(x)$ that satisfies the given condition. (This problem type is called an initial value problem. We have already seen similar problems in Section 4.7, but you will learn more about them in Chapter 8.)

66. $F(x) = \int \dfrac{dx}{x^2-2x+4}; \quad F(0)=0$

67. $F(x) = \int 3x\sqrt{x^2-4x+5}\,dx; \quad F(1)=0$

68. $F(x) = \int x^2\arccos x\,dx; \quad F(0)=1$

69–71 Use a computer algebra system to give an approximate solution to the given equation. (These are examples of "integral equations.")

69. $\displaystyle\int_0^x \dfrac{dt}{\sqrt{t^2+2}} = 2$ **70.** $\displaystyle\int_1^x \dfrac{\tan^{-1}t}{t^2}\,dt = 1$

71. $\displaystyle\int_x^{\pi/3} \dfrac{\sqrt{1-\sin^2 t}}{\tan t}\,dt = 2$

72. Paper-and-pencil skills are important, even when powerful software is at your disposal. Use an appropriate substitution to evaluate

$$\int \dfrac{(1+x)e^x}{\sqrt{x^2e^{2x}+1}}\,dx,$$

then use a computer algebra system to check your answer. What do you find?

73–78 Even with all of our integration techniques, tables, and computer algebra systems, we are far from being able to find antiderivatives for all elementary functions (roughly speaking, these are finite combinations of the types of functions you have studied so far). Even more surprisingly, many elementary functions do not even have elementary antiderivatives! Some of these appear relatively "easy," so that you might even be tempted to try and integrate them. For example, $f(x) = e^{x^2}$ does not have an elementary antiderivative. Integrals such as $\int e^{x^2}\,dx$ are called *nonelementary integrals*. Working with them requires infinite series (see Chapter 10) or the numerical methods that we shall learn in Section 7.6.

Try to evaluate the nonelementary integral using a computer algebra system. What answer do you get? (Answers will vary.)

73. $\int e^{x^2}\,dx$ **74.** $\int \dfrac{e^x}{x}\,dx$

75. $\int \cos\left(x^2\right)dx$ **76.** $\int \dfrac{\sin x}{x}\,dx$

77. $\int \ln(\ln x)\,dx$ **78.** $\int \dfrac{1}{\ln x}\,dx$

7.6 **Numerical Integration**

TOPICS

1. The Trapezoidal Rule

2. Simpson's Rule

If we can find an antiderivative of a given function f, evaluating $\int_a^b f(x)\,dx$ is straightforward. But there are two fairly common circumstances under which this is difficult or impossible. The first is when the integrand f has no antiderivative expressible in terms of *elementary functions*, which are those we work with most frequently: polynomial functions, power functions, exponential and logarithmic functions, trigonometric and inverse trigonometric functions, hyperbolic and inverse hyperbolic functions, and all other functions that can be expressed as a combination of these through the operations of addition, subtraction, multiplication, division, and function composition. Such integrals are termed **nonelementary**, and include the following:

$$\int e^{x^2}\,dx, \quad \int \cos\left(e^x\right)dx, \quad \int \sqrt{1+x^4}\,dx, \quad \int \frac{e^x}{x}\,dx,$$

$$\int \frac{1}{\ln x}\,dx, \quad \int \ln\left(\ln x\right)dx, \quad \int \sin x^2\,dx, \quad \int \frac{\sin x}{x}\,dx$$

The other circumstance is when the integral in question is defined by a function for which we know only a few values, such as a collection of data points from an experiment or a set of measurements of some physical object. *Numerical integration* is the name given to a set of techniques that can be used to provide approximations in these cases, along with estimates of the errors in the approximations. We study two such techniques in this section.

TOPIC 1 **The Trapezoidal Rule**

We have actually already gained considerable experience with the most basic numerical integration technique. When we compute a Riemann sum with either left endpoints or right endpoints of subintervals as the sample points, the result is, of course, an approximation to the integral. The Trapezoidal Rule is a slight modification of such an approach that, in general, improves on the accuracy of the approximation.

Recall that a Riemann sum approximation of the integral $\int_a^b f(x)\,dx$, assuming equal-width subintervals, has the form

$$\int_a^b f(x)\,dx \approx \sum_{i=1}^n f\left(x_i^*\right)\Delta x,$$

where $\Delta x = (b-a)/n$ and $[a,b]$ is partitioned into n subintervals by the points $x_0 = a, x_1, \ldots, x_{n-1}, x_n = b$. The ith sample point x_i^* is equal to x_{i-1} if we choose to let each one be the left subinterval endpoint and is equal to x_i if we choose to use right endpoints. Since each Riemann sum is an approximation of the integral, we might hope that the average of the two approximations would be a better approximation. That average is

$$\int_{a}^{b} f(x)\,dx \approx \frac{1}{2}\left[\sum_{i=1}^{n} f(x_{i-1})\Delta x + \sum_{i=1}^{n} f(x_i)\Delta x\right] = \frac{\Delta x}{2}\sum_{i=1}^{n}\left[f(x_{i-1}) + f(x_i)\right]$$

$$= \frac{\Delta x}{2}\left[\left(f(x_0) + f(x_1)\right) + \left(f(x_1) + f(x_2)\right) + \cdots + \left(f(x_{n-1}) + f(x_n)\right)\right]$$

$$= \frac{\Delta x}{2}\left[f(x_0) + 2f(x_1) + 2f(x_2) + \cdots + 2f(x_{n-1}) + f(x_n)\right].$$

The sum above is referred to as the trapezoidal approximation to the integral, because each expression

$$\frac{\Delta x}{2}\left[f(x_{i-1}) + f(x_i)\right] = \Delta x\left[\frac{f(x_{i-1}) + f(x_i)}{2}\right]$$

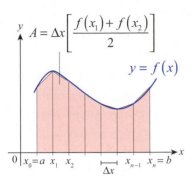

$$A = \Delta x\left[\frac{f(x_1) + f(x_2)}{2}\right]$$

$$y = f(x)$$

Figure 1 Areas of Trapezoids

corresponds to the area of the ith trapezoid shown in Figure 1 (recall that the area of a trapezoid is the width Δx times the mean of the lengths of the bases, which are $f(x_{i-1})$ and $f(x_i)$ in this context).

Definition 💡

Trapezoidal Rule

An approximation of $\int_{a}^{b} f(x)\,dx$ using the Trapezoidal Rule is the sum

$$T_n = \frac{\Delta x}{2}\left[f(x_0) + 2f(x_1) + 2f(x_2) + \cdots + 2f(x_{n-1}) + f(x_n)\right],$$

where $\Delta x = (b-a)/n$ and $x_i = a + i\Delta x$ for $i = 0, 1, \ldots, n$.

Example 1 ✍

Use the Trapezoidal Rule to approximate $\int_{1}^{2}\ln x\,dx$ with $n = 10$, and compare the result to the exact value of this integral.

Solution

Given that $a = 1$, $b = 2$, and $n = 10$, we note that $\Delta x = \frac{1}{10}$ and $x_i = 1 + i/10$ for i from 0 to 10.

$$T_{10} = \frac{1}{20}\left[\ln 1 + 2\ln\frac{11}{10} + 2\ln\frac{12}{10} + \cdots + 2\ln\frac{19}{10} + \ln 2\right] \approx 0.385878$$

Using integration by parts, we know that $\int \ln x\,dx = x\ln x - x$, so

$$\int_{1}^{2}\ln x\,dx = \left[x\ln x - x\right]_{1}^{2} = (2\ln 2 - 2) - (-1) = 2\ln 2 - 1 \approx 0.386294.$$

So $\left|\int_{1}^{2}\ln x\,dx - T_{10}\right| \approx |0.386294 - 0.385878| = 0.000416$, approximately 0.1% of the exact value.

Example 2 ✎

A hygrometer at a weather station records the following percent humidity measurements at the top of the hour from midnight to noon one day.

Time	12 a.m.	1 p.m.	2 p.m.	3 p.m.	4 p.m.	5 p.m.	6 p.m.	7 p.m.	8 p.m.	9 p.m.	10 p.m.	11 p.m.	12 p.m.
Humidity (%)	55	57	60	65	72	75	74	70	62	50	48	48	47

Use the Trapezoidal Rule to find the approximate average percent humidity over this 12-hour period.

Solution

Taking this approach to finding the average humidity, we are implicitly making the assumption that humidity h is a continuous function of time—under this assumption, we know that $\int_0^{12} h(t)\,dt$ exists, even if we don't actually know a formula for h. Assumptions like this are reasonable, because physical attributes like humidity don't typically exhibit abrupt discontinuous changes.

If we simply plot the 13 data points we have and connect them with straight lines, the result is the graph in Figure 2. Note that the data indicate sharper changes in humidity over some one-hour periods than others. Using our formula for the average value of a function over an interval we have

$$\frac{1}{b-a}\int_a^b h(t)\,dt = \frac{1}{12}\int_0^{12} h(t)\,dt.$$

Since we don't have a formula for h, we use the Trapezoidal Rule with $\Delta t = 1$ hour.

$$\frac{1}{12}\int_0^{12} h(t)\,dt \approx \left(\frac{1}{12}\right)\left(\frac{1}{2}\right)(55 + 2\cdot 57 + 2\cdot 60 + \cdots + 2\cdot 48 + 47) = 61\%$$

Compare this to the less sophisticated approach of simply taking the average of the 13 data points, which gives us an answer of approximately 60.2% humidity. The difference comes from the fact that a simple average does not take into account differing rates of change of the humidity over the interval.

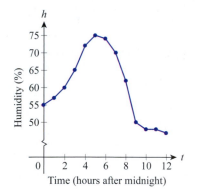

Figure 2 Humidity Measurements

Since Riemann sums approach the value of an integral as $n \to \infty$, we expect that an approximation based on the Trapezoidal Rule should similarly approach the true value of an integral. A result proved in advanced calculus tells us that if f'' exists and is continuous, then

$$\int_a^b f(x)\,dx = T_n - \left(\frac{b-a}{12}\right)f''(c)(\Delta x)^2,$$

where c is some number between a and b and n is the number of subintervals corresponding to subinterval widths of Δx (that is, $\Delta x = (b-a)/n$). So if we let E_T denote the error between the true value of the integral and its estimate T_n, then

$$|E_T| = \left(\frac{b-a}{12}\right)|f''(c)|(\Delta x)^2 \le \left(\frac{b-a}{12}\right)\max_{a\le x\le b}|f''(x)|\left(\frac{b-a}{n}\right)^2 = \frac{(b-a)^3}{12n^2}\max_{a\le x\le b}|f''(x)|.$$

Theorem ⚷

Error Estimate for the Trapezoidal Rule

If f'' is continuous on $[a,b]$ and M is an upper bound for $|f''(x)|$, then the error E_T between the exact value of $\int_a^b f(x)\,dx$ and the Trapezoidal Rule approximation T_n satisfies

$$|E_T| \le \frac{M(b-a)^3}{12n^2},$$

where n is the number of subintervals in the partition.

Example 3 ✒

Determine a number of subintervals that will guarantee an approximation of $\int_0^\pi \sin(x^2)\,dx$ to within an error of 0.01 using the Trapezoidal Rule.

Solution

Although an antiderivative of $\sin(x^2)$ cannot be expressed in terms of elementary functions, its second derivative is easy enough to determine: $f''(x) = 2\cos(x^2) - 4x^2 \sin(x^2)$. Since sine and cosine are always bounded by 1 in magnitude, we have the following:

$$\max_{0 \le x \le \pi} \left| 2\cos(x^2) - 4x^2 \sin(x^2) \right|$$

$$\le \max_{0 \le x \le \pi} \left[\left| 2\cos(x^2) \right| + \left| 4x^2 \sin(x^2) \right| \right] \qquad |a+b| \le |a| + |b|$$

$$\le \max_{0 \le x \le \pi} \left| 2\cos(x^2) \right| + \max_{0 \le x \le \pi} \left| 4x^2 \sin(x^2) \right| \qquad \max(a+b) \le \max a + \max b$$

$$\le 2 + 4 \max_{0 \le x \le \pi} \left| x^2 \right|$$

$$= 2 + 4\pi^2$$

Any upper bound M of $|f''(x)|$ on the interval $[0,\pi]$ will work, and $M = 48$ is a convenient number larger than $2 + 4\pi^2$. We can now ensure $|E_T| < 0.01$ by solving the inequality

$$\frac{48(\pi - 0)^3}{12n^2} < 0.01$$

for n.

$$\frac{4(\pi - 0)^3}{0.01} < n^2$$

$$400\pi^3 < n^2$$

$$n > \sqrt{400\pi^3} \approx 111.4$$

So $n = 112$ subintervals will provide an approximation of the integral safely within an error of 0.01.

Referring back to Example 3, a computer algebra system like *Mathematica* can be used to calculate the approximation $T_{112} = 0.77228$ with commands such as the ones shown in Figure 3.

```
In[1]:=  f[x_] := Sin[x^2];
         a = 0;
         b = Pi;
         n = 112;
         deltax = (b - a) / n;
         T = N[(deltax / 2) * (f[a] + f[b] + Sum[2 * f[a + i * deltax], {i, 1, n - 1}])]

Out[6]=  0.77228

In[7]:=  Plot[f[x], {x, 0, Pi}]
```

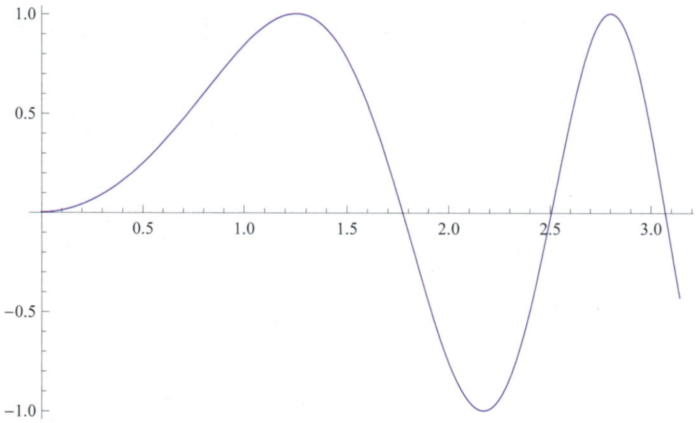

Figure 3 Using the Trapezoidal Rule in *Mathematica*

TOPIC 2 Simpson's Rule

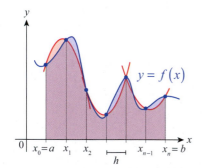

Figure 4 Simpson's Rule

Although we derived the Trapezoidal Rule by averaging two Riemann sums, we can think of the resulting formula as the integral of a function composed of a sequence of straight lines, with each straight line approximating the original integrand over a subinterval. Simpson's Rule, named for the English mathematician Thomas Simpson (1710–1761), takes a similar approach, but approximates the integrand with a sequence of parabolas.

There are several variations of Simpson's Rule, but the simplest assumes that the number of subintervals n is even; the resulting formula is then based on parabolas fitted through three points on the graph of the integrand f for each successive pair of subintervals. A representative illustration is shown in Figure 4.

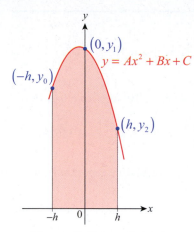

Figure 5 Fitting a Parabola

To construct the formula, let $y_i = f(x_i)$ for each $i = 0, ..., n$ and let $h = \Delta x = (b-a)/n$. For ease of calculation, we shift the graph of f left or right so that the first two subintervals $[x_0, x_1]$ and $[x_1, x_2]$ are located on either side of 0 as shown in Figure 5; doing so doesn't affect the area between the graph of f and the x-axis over the interval $[x_0, x_2]$. After shifting, note that $x_0 = -h$, $x_1 = 0$, and $x_2 = h$. So the parabola we are looking for must pass through the three points $(-h, y_0)$, $(0, y_1)$, and (h, y_2). If we assume the parabola is the graph of the quadratic $y = Ax^2 + Bx + C$, then we know the following:

$$y_0 = Ah^2 - Bh + C,$$

$$y_1 = C,$$

$$\text{and} \quad y_2 = Ah^2 + Bh + C.$$

The integral of the quadratic over the interval $[-h, h]$ is

$$I = \int_{-h}^{h} \left(Ax^2 + Bx + C \right) dx = \left[\frac{Ax^3}{3} + \frac{Bx^2}{2} + Cx \right]_{-h}^{h} = \frac{h}{3} \left(2Ah^2 + 6C \right),$$

which can be expressed in terms of $y_0, y_1,$ and y_2 by noting that $y_0 + y_2 = 2Ah^2 + 2C$ and $4y_1 = 4C$. So $I = (h/3)(y_0 + 4y_1 + y_2)$.

Similarly, shifting the graph of f so that the subintervals $[x_2, x_3]$ and $[x_3, x_4]$ are located on either side of 0 (so that $x_3 = 0$), the integral of the parabola fitted through $(-h, y_2)$, $(0, y_3)$, and (h, y_4) is equal to $(h/3)(y_2 + 4y_3 + y_4)$. Our approximation to $\int_a^b f(x) \, dx$ is the sum of the integrals of the parabolas.

$$\int_a^b f(x)\,dx \approx \frac{h}{3}(y_0 + 4y_1 + y_2) + \frac{h}{3}(y_2 + 4y_3 + y_4) + \cdots + \frac{h}{3}(y_{n-2} + 4y_{n-1} + y_n)$$

$$= \frac{h}{3}(y_0 + 4y_1 + 2y_2 + 4y_3 + 2y_4 + \cdots + 2y_{n-2} + 4y_{n-1} + y_n)$$

Rephrasing the formula back in terms of the integrand function f gives us Simpson's Rule.

Theorem

Simpson's Rule

An approximation of $\int_a^b f(x)\,dx$ using Simpson's Rule is the sum

$$S_n = \frac{\Delta x}{3} \left[\begin{array}{l} f(x_0) + 4f(x_1) + 2f(x_2) + 4f(x_3) + 2f(x_4) + \cdots \\ + 2f(x_{n-2}) + 4f(x_{n-1}) + f(x_n) \end{array} \right]$$

where n is even, $\Delta x = (b-a)/n$, and $x_i = a + i\Delta x$ for $i = 0, ..., n$. (Note the pattern of the coefficients: 1, 4, 2, 4, 2, ..., 2, 4, 1.)

Example 4 ✑

Use Simpson's Rule to approximate $\int_1^2 \ln x \, dx$ with $n = 10$.

Solution

As in Example 1, $a = 1$, $b = 2$, $\Delta x = \frac{1}{10}$, and $x_i = 1 + i/10$ for i from 0 to 10.

$$S_{10} = \frac{1}{30}\left[\ln 1 + 4\ln\frac{11}{10} + 2\ln\frac{12}{10} + \cdots + 2\ln\frac{18}{10} + 4\ln\frac{19}{10} + \ln 2\right] \approx 0.386293$$

This is a considerable improvement over the Trapezoidal Rule approximation, and only differs in the 6th decimal place from the exact answer.

As with the Trapezoidal Rule, a result from advanced calculus says that, under the assumption that $f^{(4)}$ is continuous on $[a,b]$, the exact integral and a Simpson's Rule approximation of it are related by

$$\int_a^b f(x)\,dx = S_n - \left(\frac{b-a}{180}\right)f^{(4)}(c)(\Delta x)^4,$$

where c is some point in $[a,b]$. So if we let $E_S = \int_a^b f(x)\,dx - S_n$, we know the following about the accuracy of a Simpson's Rule approximation.

Theorem ⚲

Error Estimate for Simpson's Rule

If $f^{(4)}$ is continuous on $[a,b]$ and M is an upper bound for $\left|f^{(4)}(x)\right|$, then the error E_S between the exact value of $\int_a^b f(x)\,dx$ and the Simpson's Rule approximation S_n satisfies

$$\left|E_S\right| \le \frac{M(b-a)^5}{180n^4},$$

where n is the number of subintervals in the partition.

Example 5 ✑

Use Simpson's Rule with $n = 4$ to approximate the nonelementary integral $\int_0^1 e^{x^2}\,dx$, and estimate the error in the approximation.

Solution

With $a = 0$, $b = 1$, and $n = 4$, we have $\Delta x = \frac{1}{4}$ and $x_i = i/4$ for $i = 0, \ldots, 4$.

$$S_4 = \frac{1}{12}\left[e^0 + 4e^{1/16} + 2e^{1/4} + 4e^{9/16} + e^1\right] \approx 1.4637$$

The fourth derivative of $f(x) = e^{x^2}$ is $16x^4 e^{x^2} + 48x^2 e^{x^2} + 12e^{x^2}$, and for $x \in [0,1]$ we know

$$\left| \left(16x^4 + 48x^2 + 12\right)e^{x^2} \right| \le \left(\left|16x^4\right| + \left|48x^2\right| + 12 \right)\left|e^{x^2}\right|$$

$$\le \left[16(1)^4 + 48(1)^2 + 12 \right]e^1 = 76e,$$

so the error estimate E_S satisfies $\left| E_S \right| \le \dfrac{76e(1)^5}{180(4)^4} \approx 0.0045$.

Example 6 ✎

A landscape designer has planned a free-form garden pond with the shape shown in Figure 6 and needs to estimate its volume in cubic feet. The pond will have a uniform depth of 2 feet. At 1-foot intervals, the distances across the pond are to be as indicated in the diagram. Use Simpson's Rule to estimate the volume of the pond.

Solution

To use Simpson's Rule, we let $\Delta x = 1$ and take the distances as shown to be the values of a function. Note that $n = 6$.

$$S_6 = \frac{1}{3}\left[1 + 4 \cdot 2 + 2 \cdot 3 + 4 \cdot 4 + 2 \cdot 6 + 4 \cdot 7 + 5\right] = 25\frac{1}{3}$$

So the estimated volume of the pond is this surface area times the depth, or $50\frac{2}{3}$ ft^3.

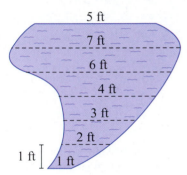

Figure 6 Garden Pond

7.6 **Exercises**

1–2 The function $f(x)$ is given by its graph. Use the Trapezoidal Rule and Simpson's Rule, respectively, to approximate the shaded area $\int_a^b f(x)\,dx$ by **a.** T_6 and **b.** S_6.

1.

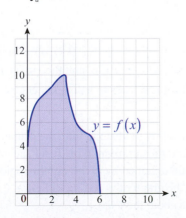

2.

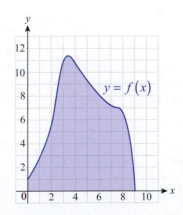

3–17 Use the Trapezoidal Rule and Simpson's Rule with $n = 8$ to approximate the integral. Then find the exact value and compare your answers.

3. $\int_0^8 x^4 \, dx$

4. $\int_1^5 \frac{1}{x} \, dx$

5. $\int_1^5 \frac{1}{x^2} \, dx$

6. $\int_0^4 \sqrt{x} \, dx$

7. $\int_0^4 x^3 \, dx$

8. $\int_0^{2\pi} |\sin x| \, dx$

9. $\int_{-2}^6 \sqrt[3]{x+2} \, dx$

10. $\int_0^2 e^x \, dx$

11. $\int_1^5 \ln x \, dx$

12. $\int_{-2}^6 \left(4 - \frac{1}{2}x\right) dx$

13. $\int_{-4}^4 \left(16 - x^2\right) dx$

14. $\int_{-4}^4 \sqrt{16 - x^2} \, dx$

15. $\int_0^4 x\sqrt{x^2 + 2} \, dx$

16. $\int_0^{16} \frac{1}{\sqrt{x+1}} \, dx$

17. $\int_0^8 \frac{x}{\sqrt{x^2 + 1}} \, dx$

18–20 Use **a.** the Trapezoidal Rule and **b.** Simpson's Rule to approximate the definite integral for the indicated value of n.

18. $\int_0^4 \sqrt[4]{x} \, dx; \quad n = 4$

19. $\int_{-\pi/2}^{\pi/2} \cos x \, dx; \quad n = 6$

20. $\int_0^5 \sqrt{x^4 + 4} \, dx; \quad n = 10$

21–23 Some texts discuss the "Midpoint Rule" as a numerical integration method. The idea is simply forming a Riemann sum by choosing the midpoint of each subinterval as the sample point.

Use the "Midpoint Rule" with $n = 8$ to approximate the integral and compare your answers to those in Exercises 3–5.

21. $\int_0^8 x^4 \, dx$

22. $\int_1^5 \frac{1}{x} \, dx$

23. $\int_1^5 \frac{1}{x^2} \, dx$

24–29 Use the formula discussed in the text to find an error estimate for the Trapezoidal Rule when it is used to approximate the integral of $f(x)$ over the given interval. Use $n = 10$.

24. $f(x) = x^2, \quad [0, 2]$

25. $f(x) = \cos 2x, \quad [\pi/2, 3\pi/2]$

26. $f(x) = \sqrt{x}, \quad [1, 4]$

27. $f(x) = x^4, \quad [0, 1]$

28. $f(x) = \frac{1}{\sqrt{x+1}}, \quad [0, 3]$

29. $f(x) = \frac{1}{x}, \quad [1, 11]$

30–35. Repeat Exercises 24–29, this time providing an error estimate for Simpson's Rule with $n = 10$. Compare your error bounds to those you gave for the Trapezoidal Rule.

36–41 Determine a number of subintervals that will guarantee an approximation of the given definite integral to within an error of 0.001 using the Trapezoidal Rule. (Answers will vary.)

36. $\int_0^1 \frac{x^4}{2} \, dx$

37. $\int_0^\pi \sin^2 x \, dx$

38. $\int_1^2 \frac{2 - x^2}{x^2} \, dx$

39. $\int_0^1 e^{x^2} \, dx$

40. $\int_1^3 x \ln x \, dx$

41. $\int_0^{\pi/2} x \sin x \, dx$

42–47. For the integrals in Exercises 36–41, find a number of subintervals that will guarantee an approximation of the same accuracy (within an error of 0.001), this time using Simpson's Rule. Compare your answers to those you gave for the Trapezoidal Rule.

48. Use Simpson's Rule with $n = 6$ to approximate $\ln 2 = \int_1^2 \frac{1}{x} \, dx$, and find the percentage error of your approximation (the actual error divided by the true value expressed as a percentage).

49. Repeat Exercise 48 to approximate $\pi = \int_0^1 \frac{4}{x^2 + 1} \, dx$.

50. Use Simpson's Rule with $n = 24$ to approximate the area of the region bounded by the graphs of $y = \sqrt{1 + x^4}$, $x = -6$, $x = 6$, and the x-axis. (Notice that this problem leads to a nonelementary integral.)

51. Use the shell method along with the Trapezoidal Rule to find the approximation T_7 for the volume of the solid obtained by revolving the region between the graph of $y = e^x/x^2$ $(1 \le x \le 2)$ and the x-axis about the y-axis.

52. Combine the disk method with Simpson's Rule to find the approximation S_6 for the volume of the solid obtained by revolving the region between the graph of $y = e^{x^2}$ $(0 \le x \le 1)$ and the x-axis about the x-axis.

53. The following table summarizes acceleration data for the 2012 Ford Mustang Boss 302 Laguna Seca. Use Simpson's Rule to estimate the total distance traveled by "the Boss" during its timed 0–120 mph run. (**Hint:** Sketching a graph similar to the one in Example 2 is useful. Be sure to identify which area you can approximate and how it yields the answer to the problem.)

Time to Speed

Miles per Hour	Seconds
0–120	13.0
0–110	10.9
0–100	9.1
0–90	7.6
0–80	6.3
0–70	5.2
0–60	4.1
0–50	3.3
0–40	2.4
0–30	1.7
0–20	1.1
0–10	0.4

Source: *Road & Track*

54. Use the Trapezoidal Rule to estimate the amount of water needed to raise the water level by two inches in a pool with the shape shown in the figure. At 2-foot intervals, the distances across the pool (in feet) are as indicated in the diagram.

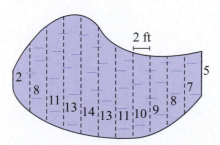

55. The figure shows the vertical cross-section of the Lazee river where the Dinkatown ferry docks. The depth of the river is indicated at 5-foot intervals in the diagram. If the river flows at 5 ft/s, use Simpson's Rule to estimate the amount of water passing by the dock every second.

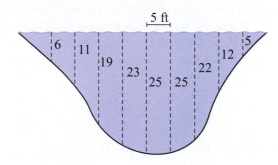

56. Prove that Simpson's Rule actually gives the exact answer for definite integrals of all polynomials of degree 3 or less.

7.6 **Technology Exercises**

57–60 Write a short program for a computer algebra system to approximate the given integral using the Trapezoidal Rule with $n = 100$. Then ask the software to evaluate the integral directly and compare your approximation with the exact value.

57. $\displaystyle\int_0^2 \sqrt{x}\sqrt{x+2}\,dx$

58. $\displaystyle\int_0^6 \frac{dx}{3+x^2}$

59. $\displaystyle\int_0^{20} \frac{dx}{1+x^3}$

60. $\displaystyle\int_0^{10} \sqrt{1+\sqrt{1+\sqrt{x}}}\,dx$

61–64 Modify the program you wrote for Exercises 57–60 to find S_{100} for the integrals. Compare your answers with the exact values you found in Exercises 57–60.

65–70 Use the program you wrote for Exercises 61–64 to find the Simpson approximation S_{50} for the given nonelementary integral. To get a feel for how close your answer is to the exact value of the integral, use the graph of the fourth derivative of the integrand, and identify an error bound as small as you can by using the theorem from this section. (Answers may vary slightly. **Hint:** You may check your error bound by using the command specific to your CAS to obtain the "exact value" of the integral. Note that the CAS uses numerical algorithms to obtain that answer and, thus, it isn't exact either. It is, however, more accurate than what we can obtain by Simpson's Rule with $n = 50$.)

65. $\displaystyle\int_0^1 e^{x^2}\,dx$　　　　**66.** $\displaystyle\int_1^3 \frac{e^x}{x}\,dx$

67. $\displaystyle\int_0^\pi \cos x^2\,dx$　　　　**68.** $\displaystyle\int_1^\pi \frac{\sin x}{x}\,dx$

69. $\displaystyle\int_2^4 \ln(\ln x)\,dx$　　　　**70.** $\displaystyle\int_e^5 \frac{1}{\ln x}\,dx$

71–73 With the help of a computer algebra system and Simpson's Rule, approximate the solution of the given integral equation. (Use $n = 100$).

71. $\displaystyle\int_0^x e^{\sqrt{t}}\,dt = 3$　　　　**72.** $\displaystyle\int_0^x \tan t^2\,dt = 1$

73. $\displaystyle\int_0^x \sqrt{1+t^4}\,dt = 4$

7.7 **Improper Integrals**

TOPICS

1. Improper integrals of type I
2. Improper integrals of type II
3. The Direct Comparison Test

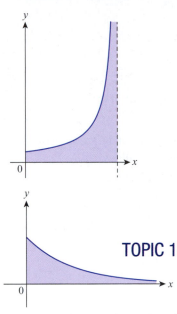

So far, all the definite integrals we have worked with have been defined over a finite interval $[a,b]$ and the integrands have all been continuous (and hence bounded) over that interval. We now turn to definite integrals in which either the interval of integration is unbounded (extends to infinity in one or both directions) or in which the integrand has an unbounded range over the interval. Both types of integrals are termed *improper*, and skill in evaluating them is of great help in other areas of math such as probability and in analyzing infinite series, which we will study in depth later in this text.

The fact that an infinite series, which is a sum of an infinite number of numbers, can be finite is at first counterintuitive, and the fact that improper integrals can also be finite is likely also surprising. This is especially so if we are faced with an improper integral of a positive function f and are interpreting the integral as the area of an unbounded region, as depicted in Figure 1. In both cases, although the region is unbounded (meaning it cannot be captured inside a circle of radius R for any R), it is possible for its area to be finite. Understanding exactly how this can be, and how we evaluate such integrals, is the purpose of this section.

Figure 1 Improper Integrals and Unbounded Regions

TOPIC 1 **Improper Integrals of Type I**

Before formally defining improper integrals of the first type, we will use the machinery we have at hand to make sense of one such integral.

Example 1 ✒

If possible, find the area of the region bounded between the graph of $f(x) = 1/x^2$ and the x-axis over the interval $[1,\infty)$.

Solution

The area asked for corresponds to the integral $\int_1^\infty \left(1/x^2\right)dx$. Since we don't yet know how to evaluate such an integral, we will first construct a reasonable interpretation of what it must mean, consistent with what we already know about integration.

If the area under f over the interval $[1,\infty)$ *is* finite, say $\int_1^\infty \left(1/x^2\right)dx = A$, then a reasonable interpretation of area would indicate that the area over the interval $[1,b]$ must approach A as $b \to \infty$ (see Figure 2). We know how to evaluate $\int_1^b \left(1/x^2\right)dx$, so we'll use this interpretation to define

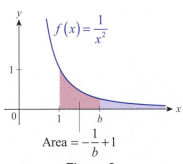

Area $= -\dfrac{1}{b} + 1$

Figure 2
Defining Area of an Unbounded Region

$$A = \int_1^\infty \frac{1}{x^2}dx = \lim_{b\to\infty}\int_1^b \frac{1}{x^2}dx = \lim_{b\to\infty}\left[-\frac{1}{x}\right]_1^b = \lim_{b\to\infty}\left(-\frac{1}{b}+1\right) = 1.$$

Note the word *define* in Example 1; since definite integrals over unbounded intervals have no prior meaning, it is up to us to provide a definition. We do so now, using Example 1 as inspiration.

Definition 💡

Improper Integrals of Type I

An **improper integral of type I** is an integral over an unbounded interval, and falls into one of the following classes if the interval is unbounded on one side only.

1. $\displaystyle\int_a^\infty f(x)\,dx = \lim_{b\to\infty}\int_a^b f(x)\,dx$

2. $\displaystyle\int_{-\infty}^b f(x)\,dx = \lim_{a\to-\infty}\int_a^b f(x)\,dx$

In each case, the integral is said to **converge** if the limit exists, and if so we define the value of the improper integral to be the limit. If the limit does not exist, the integral is said to **diverge**.

If the interval of integration is $(-\infty,\infty)$, and if both integrals on the right converge for some real number c, then we have the following.

3. $\displaystyle\int_{-\infty}^\infty f(x)\,dx = \int_{-\infty}^c f(x)\,dx + \int_c^\infty f(x)\,dx$

Example 2 ✍

Evaluate $\displaystyle\int_{-\infty}^\infty \frac{dx}{1+x^2}$.

Solution

We can choose any value for c, but the most convenient value is $c = 0$.

$$\int_{-\infty}^\infty \frac{dx}{1+x^2} = \int_{-\infty}^0 \frac{dx}{1+x^2} + \int_0^\infty \frac{dx}{1+x^2}$$

$$= \lim_{a\to-\infty}\int_a^0 \frac{dx}{1+x^2} + \lim_{b\to\infty}\int_0^b \frac{dx}{1+x^2}$$

$$= \lim_{a\to-\infty}\Big[\tan^{-1}x\Big]_a^0 + \lim_{b\to\infty}\Big[\tan^{-1}x\Big]_0^b$$

$$= \lim_{a\to-\infty}\big(0 - \tan^{-1}a\big) + \lim_{b\to\infty}\big(\tan^{-1}b - 0\big)$$

$$= -\left(-\frac{\pi}{2}\right) + \frac{\pi}{2} = \pi$$

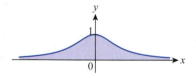

Figure 3 Area under $y = \dfrac{1}{1+x^2}$

Example 3 ✍

Evaluate $\displaystyle\int_1^\infty \frac{1}{\sqrt{x}}\,dx$.

Solution

Since $\displaystyle\int_1^\infty \frac{1}{\sqrt{x}}\,dx = \lim_{b\to\infty}\int_1^b \frac{1}{\sqrt{x}}\,dx = \lim_{b\to\infty}\Big[2\sqrt{x}\Big]_1^b = \lim_{b\to\infty}\big(2\sqrt{b} - 2\big) = \infty$, we know

that this integral diverges. The geometric interpretation of this result is that the area under $1/\sqrt{x}$ over the interval $[1,\infty)$ is infinite.

In Exercise 65 you'll consider all integrals of the form $\displaystyle\int_1^\infty \big(1/x^p\big)\,dx$ and show that such an integral converges if $p > 1$ and diverges if $p \le 1$.

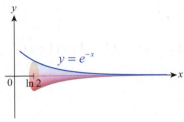

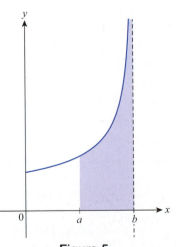

Figure 4 Infinitely Long Horn

Example 4 ✐

The infinitely long horn depicted in Figure 4 extends from $x = \ln 2$ to infinity. The area of a typical cross-section is $A(x) = \pi\left(e^{-x}\right)^2 = \pi e^{-2x}$. Determine the volume of the horn.

Solution

Recall that $V = \int A(x)\,dx$, so the volume of the horn is as follows.

$$V = \int_{\ln 2}^{\infty} \pi e^{-2x}\,dx$$

$$= \lim_{b \to \infty}\left[\pi\left(-\frac{1}{2}e^{-2x}\right)\right]_{\ln 2}^{b}$$

$$= \lim_{b \to \infty}\left[-\frac{\pi}{2}\left(e^{-2b} - e^{-2\ln 2}\right)\right]$$

$$= -\frac{\pi}{2}\left(0 - e^{\ln(1/4)}\right) = \frac{\pi}{8}$$

TOPIC 2   Improper Integrals of Type II

If f is continuous other than at a vertical asymptote in the interval of integration $[a,b]$, either at one of the endpoints or strictly between them, we may or may not be able to assign a reasonable value to the integral $\int_a^b f(x)\,dx$. The definition of this value, when it exists, is similar to the definition of the first type of improper integral.

Definition ♀

Improper Integrals of Type II

An **improper integral of type II** is one in which the integrand has an unbounded range over the interval of integration, and falls into one of the following classes:

1. If f is continuous on $(a,b]$ and discontinuous at a, then

$$\int_a^b f(x)\,dx = \lim_{c \to a^+} \int_c^b f(x)\,dx.$$

2. If f is continuous on $[a,b)$ and discontinuous at b, then

$$\int_a^b f(x)\,dx = \lim_{c \to b^-} \int_a^c f(x)\,dx.$$

In each case, the integral is said to **converge** if the limit exists, and if so we define the value of the improper integral to be the limit. If the limit does not exist, the integral is said to **diverge**.

3. If f is continuous on $[a,b]$ except for a vertical asymptote at c, where $a < c < b$, then $\int_a^b f(x)\,dx = \int_a^c f(x)\,dx + \int_c^b f(x)\,dx$, assuming both integrals on the right converge.

Figure 5
Improper Integral of Type II

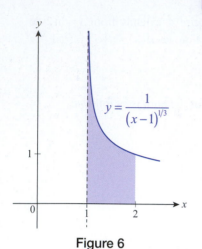

Figure 6

Example 5 ✎

Evaluate $\displaystyle\int_1^2 \frac{dx}{(x-1)^{1/3}}$.

Solution

The integrand is continuous except at its vertical asymptote $x = 1$.

$$\int_1^2 \frac{dx}{(x-1)^{1/3}} = \lim_{a\to 1^+} \int_a^2 \frac{dx}{(x-1)^{1/3}}$$

$$= \lim_{a\to 1^+} \left[\frac{3}{2}(x-1)^{2/3} \right]_a^2$$

$$= \lim_{a\to 1^+} \left[\frac{3}{2} - \frac{3}{2}(a-1)^{2/3} \right]$$

$$= \frac{3}{2}$$

Example 6 ✎

Evaluate $\displaystyle\int_{-1}^2 \frac{1}{x}\,dx$.

Solution

In this example, the integrand has a vertical asymptote in the interior of the interval, so we must break it into two improper integrals.

$$\int_{-1}^2 \frac{1}{x}\,dx = \int_{-1}^0 \frac{1}{x}\,dx + \int_0^2 \frac{1}{x}\,dx$$

Remember that both of these integrals must converge for the original integral to converge. Examining the first one, we obtain the following.

$$\int_{-1}^0 \frac{1}{x}\,dx = \lim_{b\to 0^-} \int_{-1}^b \frac{1}{x}\,dx = \lim_{b\to 0^-} \left[\ln|x| \right]_{-1}^b = \lim_{b\to 0^-} \left(\ln|b| - 0 \right) = -\infty$$

We can stop right here, because the first improper integral diverges and hence $\int_{-1}^2 (1/x)\,dx$ diverges (there is no need to examine the second improper integral, but it also diverges).

Caution ⚠

If we had missed the fact that $f(x) = 1/x$ is discontinuous at 0 and blindly proceeded to antidifferentiate in Example 6, we would have obtained the incorrect result that

$$\int_{-1}^2 \frac{1}{x}\,dx = \left[\ln|x| \right]_{-1}^2 = \ln 2 - \ln 1 = \ln 2.$$

Example 7 ✍

Evaluate $\int_0^1 \ln x \, dx$.

Solution

We know $\ln x$ is continuous for all $x > 0$ and that it has a vertical asymptote at $x = 0$.

$$\int_0^1 \ln x \, dx = \lim_{a \to 0^+} \int_a^1 \ln x \, dx = \lim_{a \to 0^+} \left[x \ln x - x \right]_a^1$$

$$= \lim_{a \to 0^+} \left(-1 - a \ln a + a \right) = -1 - \lim_{a \to 0^+} a \ln a$$

The remaining limit in the last expression is of an indeterminate form that we can evaluate with l'Hôpital's Rule.

$$\lim_{a \to 0^+} a \ln a = \lim_{a \to 0^+} \frac{\ln a}{1/a} = \lim_{a \to 0^+} \frac{1/a}{-1/a^2} = \lim_{a \to 0^+} (-a) = 0$$

Therefore, $\int_0^1 \ln x \, dx = -1 - \lim_{a \to 0^+} a \ln a = -1 - 0 = -1$.

TOPIC 3   The Direct Comparison Test

Sometimes we care more about *whether* a given improper integral converges than we do about its value if it does converge. The Direct Comparison Test helps us answer this question.

The test is based on a very straightforward argument: if f and g are integrable and $0 \le f(x) \le g(x)$ for all x on an interval of integration I (whether finite or infinite), then $0 \le \int_I f(x) \, dx \le \int_I g(x) \, dx$. One way to prove this fact is to compare Riemann sum approximations of the two integrals for any given partition of I. Since a Riemann sum approximation for the integral of f will always be less than or equal to the corresponding Riemann sum approximation for the integral of g, the same relationship is true in the limit of finer and finer partitions. When applied to improper integrals of type I on an interval $[a, \infty)$, this fact takes the following form (similar statements can be made for other infinite intervals and for improper integrals of type II).

Theorem ⚗

The Direct Comparison Test

If f and g are continuous functions on $[a, \infty)$ and $0 \le f(x) \le g(x)$ for all $a \le x < \infty$, then

1. if $\int_a^\infty g(x) \, dx$ converges, so does $\int_a^\infty f(x) \, dx$;

2. if $\int_a^\infty f(x) \, dx$ diverges, so does $\int_a^\infty g(x) \, dx$.

Example 8 ✎

Determine if the integral $\int_1^\infty e^{-x^2}\,dx$ converges.

Solution

As we learned in Section 7.6, this integral is nonelementary—we can't attempt to evaluate it by working with an antiderivative. But we can easily compare it to a simpler integral. Since $-x^2 \le -x$ for all $x \ge 1$, $0 \le e^{-x^2} \le e^{-x}$ for all such x. Since

$$\int_1^\infty e^{-x}\,dx = \lim_{b\to\infty}\int_1^b e^{-x}\,dx = -\lim_{b\to\infty}\left[e^{-x}\right]_1^b = -\lim_{b\to\infty}\left[e^{-b} - e^{-1}\right] = e^{-1},$$

we know $\int_1^\infty e^{-x}\,dx$ converges and hence so does $\int_1^\infty e^{-x^2}\,dx$.

Example 9 ✎

Determine whether $\int_1^\infty \dfrac{1}{\sqrt{x^2-0.01}}\,dx$ converges.

Solution

For $x \ge 1$, we have the following.

$$\sqrt{x^2-0.01} \le x$$

$$\frac{1}{\sqrt{x^2-0.01}} \ge \frac{1}{x}$$

You will show in Exercise 17 that the integral $\int_1^\infty (1/x)\,dx$ diverges, so $\int_1^\infty \left(1/\sqrt{x^2-0.01}\right)dx$ diverges as well.

7.7 **Exercises**

1–8 Decide whether the given integral is an improper integral. If so, explain why and identify its type. (Do not evaluate the integral.)

1. $\displaystyle\int_1^\infty 2^{-x}\,dx$

2. $\displaystyle\int_{-1}^1 \frac{1}{1+x^2}\,dx$

3. $\displaystyle\int_0^1 x^{-1/3}\,dx$

4. $\displaystyle\int_{-1}^1 \frac{1}{x^2}\,dx$

5. $\displaystyle\int_0^1 \frac{1}{(x+1)^{4/5}}\,dx$

6. $\displaystyle\int_{-\infty}^{-10} \frac{1}{x}\,dx$

7. $\displaystyle\int_{-\infty}^\infty \frac{1}{1+x^2}\,dx$

8. $\displaystyle\int_2^\infty \frac{1}{(x-2)^{3/2}}\,dx$

9–16 Use the definitions from this section to write the given improper integral in terms of limits. (Do not evaluate the integral.)

9. $\displaystyle\int_2^\infty \frac{1}{x^2-1}\,dx$

10. $\displaystyle\int_0^1 \frac{1}{x^2-1}\,dx$

11. $\displaystyle\int_{-\infty}^{-1} \frac{-3}{x}\,dx$

12. $\displaystyle\int_{-\infty}^\infty \frac{2}{\sqrt{x^2+2}}\,dx$

13. $\displaystyle\int_{-\infty}^0 e^x\,dx$

14. $\displaystyle\int_0^2 \frac{1}{(x-1)^2}\,dx$

15. $\displaystyle\int_0^\infty \frac{2}{x}\,dx$

16. $\displaystyle\int_{-\infty}^\infty \frac{5}{x^2}\,dx$

17–20 Determine whether the improper integral pictured is convergent or divergent. If it is convergent, find its value.

17. $\int_1^\infty \dfrac{dx}{x}$

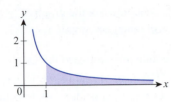

18. $\int_{-\infty}^\infty \dfrac{2}{\sqrt{1+x^2}}\, dx$

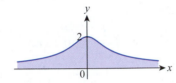

19. $\int_1^5 \dfrac{dx}{(x-3)^{2/3}}$

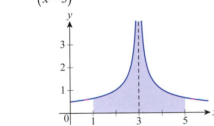

20. $\int_{-\infty}^1 \dfrac{dx}{(2-x)^{3/2}}$

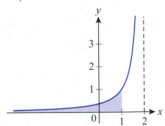

21–64 Identify the type of the improper integral and determine whether it is convergent or divergent. If it is convergent, find its value.

21. $\int_1^\infty \dfrac{1}{x^{3/2}}\, dx$

22. $\int_{-\infty}^{-1} \dfrac{-3}{x}\, dx$

23. $\int_0^1 \dfrac{1}{\sqrt{x}}\, dx$

24. $\int_{-\infty}^1 2e^x\, dx$

25. $\int_0^\infty x e^{-x}\, dx$

26. $\int_0^\infty \cos x\, dx$

27. $\int_1^2 \dfrac{dx}{(x-1)^3}$

28. $\int_1^2 \dfrac{dx}{\sqrt[3]{x-1}}$

29. $\int_0^\infty \dfrac{6\, dx}{x^2+9}$

30. $\int_0^\infty \dfrac{6\, dx}{x^2-9}$

31. $\int_0^2 \dfrac{dx}{(x-1)^2}$

32. $\int_2^\infty \dfrac{dx}{(x-2)^2}$

33. $\int_{-\infty}^\infty \dfrac{2e^x}{e^{2x}+1}\, dx$

34. $\int_0^2 \dfrac{t}{\sqrt{4-t^2}}\, dt$

35. $\int_0^\infty \dfrac{dt}{(t+1)\sqrt{t}}$

36. $\int_{-\infty}^\infty x e^{-x^2}\, dx$

37. $\int_{-\sqrt{2}}^{\sqrt{2}} \dfrac{dx}{\sqrt{2-x^2}}$

38. $\int_0^\infty e^{-\theta} \cos\theta\, d\theta$

39. $\int_0^\infty \dfrac{e^x}{x}\, dx$

40. $\int_0^{16} \dfrac{dt}{\sqrt[4]{16-t}}$

41. $\int_0^9 \dfrac{dt}{9-t}$

42. $\int_{-\infty}^\infty \dfrac{dx}{9x^2+1}$

43. $\int_4^\infty \dfrac{e^{-\sqrt{x}}}{\sqrt{x}}\, dx$

44. $\int_0^\infty t^2 e^{-t}\, dt$

45. $\int_e^\infty \dfrac{dz}{z\ln^2 z}$

46. $\int_e^\infty \dfrac{\ln z}{z^2}\, dz$

47. $\int_0^e \dfrac{\ln z}{z^2}\, dz$

48. $\int_0^e \dfrac{\ln z}{z}\, dz$

49. $\int_0^\infty \dfrac{dx}{2\sqrt{x}(x+1)}$

50. $\int_0^1 \dfrac{1+x}{\sqrt{1-x^2}}\, dx$

51. $\int_{-\infty}^\infty \dfrac{v}{(v^2+4)^2}\, dv$

52. $\int_0^2 \dfrac{4v}{\sqrt{16-v^4}}\, dv$

53. $\int_{-4}^0 \dfrac{dx}{\sqrt{|x+2|}}$

54. $\int_1^\infty \dfrac{\arctan x}{1+x^2}\, dx$

55. $\int_1^2 \dfrac{dx}{x\sqrt{x^2-1}}$

56. $\int_3^\infty \dfrac{dx}{x^2-2x}$

57. $\int_2^3 \dfrac{dx}{x^2-2x}$

58. $\int_0^\infty \dfrac{dx}{(x+2)(x+3)^2}$

59. $\int_0^{\pi/2} \sec x\, dx$

60. $\int_0^\infty \dfrac{dx}{x(x+2)^2}$

61. $\int_0^1 \dfrac{\ln x}{\sqrt{x}}\, dx$

62. $\int_0^{\pi/2} \tan\theta\, d\theta$

63. $\int_1^\infty \dfrac{dt}{t\sqrt{3t+1}}$

64. $\int_0^\infty \dfrac{x-2}{x^2+x+2}\, dx$

65. Classify the integrals of the form $\int_1^\infty (1/x^p)\,dx$ according to convergence for all possible values of p. (**Hint:** Consider the three cases of $p > 1$, $p = 1$, and $p < 1$.)

66. Repeat Exercise 65 for the integrals of the form $\int_0^1 (1/x^p)\,dx$.

67–74 Use the Direct Comparison Test to determine whether the integral converges.

67. $\displaystyle\int_1^\infty \frac{dx}{\sqrt{x^4 + 2x + 3}}$

68. $\displaystyle\int_2^\infty \frac{\ln x}{\sqrt{x^2 - 1}}\,dx$

69. $\displaystyle\int_1^\infty \frac{dx}{e^{2x} + x^{3/2}}$

70. $\displaystyle\int_0^1 \frac{\cos x}{\sqrt{x}}\,dx$

71. $\displaystyle\int_1^\infty \frac{\ln x}{x}\,dx$

72. $\displaystyle\int_1^\infty \frac{\ln x}{x^{5/2}}\,dx$

73. $\displaystyle\int_1^\infty \frac{dx}{x^{1/2} - \frac{1}{2}}$

74. $\displaystyle\int_2^\infty \frac{e^{-\sqrt{x}}}{\sqrt{x}\,\ln x}\,dx$

75. Rotate the infinite region bounded by the graphs of $y = x^2/\sqrt{1 - x^2}$, $y = 0$, and $x = 1$ about the y-axis. Use the method of shells to find the volume of the resulting unbounded solid.

76. If the infinite region between the graph of $y = 1/x$ and the x-axis ($x \geq 1$) is revolved about the x-axis, we obtain the solid nicknamed *Gabriel's horn* (see figure). Use improper integrals to show that Gabriel's horn has a finite volume and infinite surface area. Note that this means that at least theoretically, we can paint (or "fill") this infinite surface with a finite amount of paint! Can you find a mathematical explanation for this conclusion? (**Hint:** To show that the surface area is infinite, use the Direct Comparison Test.)

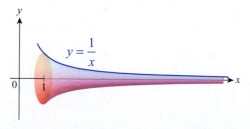

Gabriel's Horn

77. Show that the process in Exercise 76 results in a solid of finite volume when revolving about the x-axis, but in a solid of infinite volume when revolving about the y-axis.

78. Find the area between the graph of $y = 2/(x^2 - 1)$ and its horizontal asymptote for $x \geq 2$.

79. Find the total area between the graph of $y^2 = x^2/(16 - 16x^2)$ and its vertical asymptotes. (**Hint:** Be sure to consider both branches of the graph.)

80.* The figure below shows the graph of a piecewise defined function:

$$f(x) = \begin{cases} \dfrac{1}{\sqrt{x}} & \text{if } 0 < x \leq 1 \\[2mm] \dfrac{1}{x^2} & \text{if } 1 < x < \infty \end{cases}$$

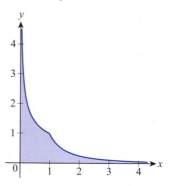

Show that $\int_0^\infty f(x)\,dx = 3$.

81. Sketch the graph of $x^{2/3} + y^{2/3} = 1$ (hypocycloid with four cusps) and find its perimeter.

82. For $a > 0$, find a formula for the integral

$$\int_1^\infty \frac{dx}{x\sqrt{x^2 + a^2}}.$$

83. In probability theory, $f(x) = \dfrac{1}{\sqrt{2\pi}} e^{-x^2/2}$ is called the density function of the standard normal distribution. Show that its mean is 0 and its variance is 1; that is,

$$\int_{-\infty}^\infty x f(x)\,dx = 0 \quad \text{and} \quad \int_{-\infty}^\infty x^2 f(x)\,dx = 1.$$

(**Hint:** For the first integral, adapt your solution to Exercise 36. For the second, use integration by parts along with l'Hôpital's Rule.)

84. Find the function in the one-parameter family $y = c/(x^2 + 1)$ so that

$$\int_{-\infty}^{\infty} \frac{c}{x^2 + 1}\, dx = 1.$$

(The function you just found is another probability density function, called the *Cauchy density function*.)

85–87 In certain cases, substitution can help turn an improper integral into a proper one. For example, as part of your solution to Exercise 35, substituting $u = \sqrt{t}$ you may find

$$\int_0^1 \frac{dt}{(t+1)\sqrt{t}} = 2\int_0^1 \frac{du}{u^2 + 1},$$

the latter being a proper integral.

Use the above idea to turn the given improper integral into a proper one and evaluate.

85. $\displaystyle\int_0^{1/2} \frac{dx}{\sqrt{x}\sqrt{1-x}}$ **86.** $\displaystyle\int_0^1 \frac{e^{\sqrt{x}}}{\sqrt{x}}\, dx$

87. $\displaystyle\int_1^{\infty} \frac{1}{1+x^2}\, dx$ (**Hint:** Substitute $u = 1/x$.)

88–93 The Laplace transform of a function $f(t)$, denoted $L\{f(t)\}$, is defined by the improper integral

$$L\{f(t)\}(s) = \int_0^{\infty} e^{-st} f(t)\, dt,$$

as long as it converges. The Laplace transform is very useful in physics and engineering, most notably, for solving certain linear ordinary differential equations.

Find the Laplace transform of the given function (we assume s is appropriately restricted so that the Laplace transform converges).

88. $L\{1\}$ **89.** $L\{t\}$

90. $L\{t^2\}$ **91.** $L\{e^{at}\}$

92. $L\{\sin kt\}$ **93.** $L\{\cos kt\}$

94. Recognizing a pattern from Exercises 88–90, conjecture and use induction to prove a general formula for $L\{t^n\}$ ($n \in \mathbb{N}$). (**Hint:** You may want to calculate $L\{t^3\}$ to firm up your conjecture.)

95–97 Suppose you invest money at an annual interest rate of r, which is compounded continuously, with a goal of having N dollars in t years. The amount you invest today to achieve that goal is called the present value (denoted PV) of the N dollars that is still t years out in the future. This present value can be calculated by the formula

$$PV = Ne^{-rt},$$

since investing PV dollars today will yield $(PV)e^{rt} = (Ne^{-rt})e^{rt} = N$ dollars in t years.

Using Riemann sums, it is straightforward to derive the formula for the present value of an annuity, a terminating income stream of fixed payments over a finite time period, say T years:

$$PV = \int_0^T A(t)e^{-rt}\, dt,$$

where $A(t)$ is the amount paid out annually (we assume continuous payment).

In Exercises 95–97, use this formula to generate the required improper integral.

95. If the annual interest rate is 5%, find the present value of a perpetual annuity (one paying dividends forever) that continuously pays $10,000 every year.

96. How much should we invest at an annual interest rate of 4% if we want a never-ending income stream of continuous annual payments of $500?

97. Suppose we expect an investment to generate profits at $p(t) = 6500\sqrt{t}e^{0.015t}$ dollars annually forever (t stands for the number of years elapsed). If the annual interest rate is 6%, find the present value of this income stream.

98. Use an improper integral to find the work done in propelling a 500 kg satellite out of the Earth's gravitational field. (**Hint:** Approximate the radius of the Earth by 6371 km. For the magnitude of the force of gravity at great altitudes, see Exercise 49 in Section 6.5.)

99. If an object is leaving the surface of the Earth with a velocity v_0 big enough that its kinetic energy $E_{kin} = \frac{1}{2}mv_0^2$ is equal to the work required to propel it out of the Earth's gravitational field, then the object will never return, but rather travel "infinitely far away" into outer space. Use Exercise 98 to find the value of the escape velocity on the surface of the Earth. (**Hint:** As in Exercise 98, approximate the radius of the Earth by 6371 km. See also Exercise 55 of Section 6.5 and the Chapter 6 Project.)

100.* The *gamma function*, which after an argument shift becomes an extension of the factorial function $f(n) = n!$, is defined as follows:

$$\Gamma(n) = \int_0^\infty x^{n-1} e^{-x}\, dx$$

Note that here x can be any real (or even complex) number. The gamma function is especially important in the fields of combinatorics, probability, and statistics.

Prove each of the following:

a. The above improper integral converges for all $n > 0$. (**Hint:** Show that for $n > 0$,
$$0 < \frac{x^{n-1}}{e^x} \le \frac{1}{x^2} \quad \text{for all appropriately large}$$
x-values and use the Direct Comparison Test.)

b. $\Gamma(1) = 1$

c. $\Gamma(n+1) = n\Gamma(n)$

(**Hint:** Use integration by parts.)

d. $\Gamma(n+1) = n!, \quad n \in \mathbb{N}$

(**Hint:** Use mathematical induction.)

101–104 *True or False?* Determine whether the given statement is true or false. In case of a false statement, explain or provide a counterexample.

101. If $\int_1^\infty f(x)\, dx$ diverges, then $\lim_{x \to \infty} f(x) = L$ with $L \ne 0$ or $\lim_{x \to \infty} f(x)$ doesn't exist.

102. If $f(x)$ is continuous on $[1, \infty)$, positive, and decreasing, and if for any $M > 0$ there is a $b > 1$ such that $\int_1^b f(x)\, dx > M$, then $\int_1^\infty f(x)\, dx$ diverges.

103. $\int_1^\infty \dfrac{\ln(x^3)}{x}\, dx$ diverges.

104. For any positive $a \in \mathbb{R}$, $\int_a^\infty \dfrac{dx}{x^{1+a}}$ converges.

7.7 **Technology Exercises**

105–106 The function, defined in terms of an improper integral, is important for its applications within or outside of mathematics, for example in number theory, statistics, probability, physics, or engineering. Use a computer algebra system to sketch the graph and observe important features. (Answers will vary.)

105. $\operatorname{erf}(x) = \dfrac{2}{\sqrt{\pi}} \int_0^x e^{-t^2}\, dt$　(the error function)

106. $\operatorname{Li}(x) = \int_2^x \dfrac{dt}{\ln t}$　(the logarithmic integral function)

107–108 Attempt to use a computer algebra system to find the improper integral (from Exercises 67 and 74) and see what happens. (**Note:** Theoretical results such as the Direct Comparison Test are extremely useful, even when powerful mathematical software is at our disposal. As we mentioned during our discussion in the text, sometimes concluding the fact of convergence is more important than the actual value of the integral; and the Direct Comparison Test provides a fast and trouble-free way to do just that.)

107. $\int_1^\infty \dfrac{dx}{\sqrt{x^4 + 2x + 3}}$

108. $\int_2^\infty \dfrac{e^{-\sqrt{x}}}{\sqrt{x}\,\ln x}\, dx$

Chapter 7
Review Exercises

1–8 Use integration by parts to evaluate the given indefinite or definite integral.

1. $\int \sqrt{x} \ln x \, dx$

2. $\int x \sec^2 x \, dx$

3. $\int \dfrac{x}{\sqrt{x-1}} \, dx$

4. $\int \text{arccot}\, x \, dx$

5. $\int t^2 \cos t \, dt$

6. $\int 3t^2 e^{t+1} \, dt$

7. $\int_0^\pi x^3 \sin x \, dx$

8. $\int_0^{\pi/2} e^{2t} \cos t \, dt$

9–10 Combine the method of integration by parts with substitution to evaluate the integral.

9. $\int \sin t \cos t \ln(\sin t) \, dt$ **10.** $\int \dfrac{\arctan \sqrt{t}}{\sqrt{t}} \, dt$

11. Use integration by parts to find the area of the region bounded by the graph of $y = \cos(\ln x)$ and the x-axis over the interval $\left[e^{-\pi/2}, e^{\pi/2}\right]$.

12. Consider the region bounded by the graph of $y = xe^x$ and the x-axis over the interval $[0,1]$.

    a. Find the centroid of the region.

    b. Use the shell method to find the volume of the solid generated by revolving the region about the y-axis.

13. Repeat Exercise 12 for the region bounded by the graph of $y = 3x^2 \ln x$ and the x-axis over the interval $[1, e]$.

14. Use the disk method to find the volume of the solid generated by revolving the region bounded by the graph of $y = \cos x$, $-\pi/2 \le x \le \pi/2$, about the x-axis.

15. Use integration by parts to prove the following formula for $a > 0$, $a \ne 1$.

$$\int x a^x \, dx = a^x \left[\frac{x}{\ln a} - \frac{1}{(\ln a)^2} \right] + C$$

16–21 Use the partial fractions method to evaluate the given integral.

16. $\int \dfrac{4}{x^2 - 4} \, dx$

17. $\int \dfrac{x-7}{(x+1)(x-3)} \, dx$

18. $\int \dfrac{x+3}{x(x+1)(x+2)} \, dx$

19. $\int \dfrac{3x-4}{x^3(x-1)} \, dx$

20. $\int \dfrac{4}{2x^3 + 2x^2 + x} \, dx$

21. $\int \dfrac{x^5 + 6x^3 + 7x}{(x^2 + 2)^3} \, dx$

22–23 Use the Heaviside cover-up method to evaluate the given integral.

22. $\int \dfrac{4x^2 + 4x - 2}{x(x^2 + 3x + 2)} \, dx$

23. $\int \dfrac{3x^2 + 16x + 29}{(x+5)(x+3)(x-1)} \, dx$

24–25 Find the definite integral of the rational function over the given interval.

24. $\int_{-1}^{1} \dfrac{x^3 + x^2 - x - 2}{x^2 - 4} \, dx$

25. $\int_0^1 \dfrac{2x^4 + x^3 - 2x^2 - 3}{(x+1)(x^2+1)} \, dx$

26–27 Combine integration by substitution and the partial fractions method to find the given integral.

26. $\int \dfrac{\sec^2 x}{\tan^2 x - \tan x} \, dx$

27. $\int \dfrac{dx}{\sqrt{x}(\sqrt{x} - 2)(\sqrt{x} + 3)}$

28. Use the disk method to find the volume of the solid generated by revolving the graph of $f(x) = 2/\sqrt{x^2 + 8x + 12}$, $0 \le x \le 2$, about the x-axis.

29. Use the shell method to find the volume of the solid obtained by revolving about the y-axis the region bounded by the graph of $g(x) = 4/(x^2 + 6x + 5)$, the coordinate axes, and the line $x = 5$.

30. If a and b are constants such that $a \neq b$, use the partial fractions method to prove the following formula:

$$\int \frac{dx}{(x-a)(x-b)} = \frac{1}{a-b} \ln \left| \frac{x-a}{x-b} \right| + C$$

31–40 Evaluate the given indefinite or definite integral involving powers of sines and cosines.

31. $\int_0^\pi 4\sin^3 x \, dx$

32. $\int_0^{\pi/4} \frac{\sin x}{\sqrt{\cos x}} \, dx$

33. $\int \cos^2 x \sin^3 x \, dx$

34. $\int x^2 \cos^2 x \, dx$

35. $\int_0^{\pi/4} \sin x \cos^{-5} x \, dx$

36. $\int_0^\pi \sin^9 x \, dx$

37. $\int \frac{1}{\sin^2 x \cos^2 x} \, dx$

38. $\int (\cos x - \sin x)^2 \, dx$

39. $\int \sin 3x \sin x \, dx$

40. $\int \cos 2x \sin 5x \, dx$

41–48 Evaluate the given indefinite or definite integral involving powers of tangents and secants (or their cofunctions).

41. $\int_0^{\pi/4} \tan^3 x \sec^4 x \, dx$

42. $\int_0^{\pi/4} \tan^2 t \sec^3 t \, dt$

43. $\int \cot^3 x \, dx$

44. $\int \csc^4 x \, dx$

45. $\int \tan^4 x \sec^2 x \, dx$

46. $\int \frac{\csc^4 x}{\sqrt{\cot x}} \, dx$

47. $\int \frac{\tan^2 v}{\sec^3 v} \, dv$

48. $\int \frac{\sec^2 \theta}{\sqrt{\tan \theta}} \, d\theta$

49. Find the area of the region between the x-axis and the graph of $f(x) = 1/(\cos x + 1)$ from $x = -\pi/2$ to $x = \pi/2$.

50. Find the area of the region bounded by the graphs of $y = 16 \sin^2 x$ and $y = \csc^2 x$. (Restrict both functions to the interval $(0, \pi)$).

51. Find the volume of the solid obtained by revolving about the x-axis the region bounded by the graph of $y = \csc x + \sin x$ and the x-axis over the interval $(\pi/4, 3\pi/4)$.

52. Find the volume of the solid obtained by revolving about the x-axis the region bounded by the graph of $y = (\tan x + 1)/\sec x$ and the x-axis over the interval $(0, \pi/4)$.

53. For positive integers m and n that are not equal, prove the following:

   a. $\int_0^{2\pi} \sin(mx)\sin(nx)\,dx = 0$

   b. $\int_0^{2\pi} \sin(mx)\cos(nx)\,dx = 0$

   c. $\int_0^{2\pi} \cos(mx)\cos(nx)\,dx = 0$

54. Use substitution to establish the following integration formula.

$$\int \sec x \, dx = \ln|\sec x + \tan x| + C$$

(**Hint:** Substitute $u = \sec x + \tan x$. Compare with Example 5d of Section 7.3.)

55–60 Use trigonometric substitution to evaluate the given indefinite or definite integral.

55. $\int \frac{2x^2}{\sqrt{4-x^2}} \, dx$

56. $\int \frac{\sqrt{1-x^2}}{x^2} \, dx$

57. $\int \frac{2}{(t^2+4)^{3/2}} \, dt$

58. $\int \frac{\sqrt{t^2-4}}{t^2} \, dt$

59. $\int_0^3 \frac{2x^2}{\sqrt{x^2+9}} \, dx$

60. $\int_1^2 \frac{dx}{\sqrt{16x^2-9}}$

61–64 Use the methods of this chapter to evaluate the given integral.

61. $\int \sqrt{20 + 16x - 4x^2} \, dx$

62. $\int \frac{\cos x}{4\sin^2 x + 9} \, dx$

63. $\int \frac{e^x}{\sqrt{3 + 2e^x - e^{2x}}} \, dx$

64. $\int 4x \sinh^{-1} x \, dx$

65. Find the area of the region enclosed by the graphs of $y = \sqrt{x^2 + 1}$ and $y = 2$.

66. Rotate the region bounded by the graph of $y = \sqrt{x^2 - 4}/x^2$ and the x-axis, $2 \leq x \leq 4$, about the y-axis. Use the shell method to find the volume of the resulting solid.

67. Use the method of disks to determine the volume of the solid obtained by revolving the graph of $y = x^2/\sqrt[4]{4 - x^2}$, $0 \leq x \leq 1$, about the x-axis.

68. Find the surface area of the solid that results from rotating the graph of $y = x^2$ about the x-axis over the interval $[0,2]$. (Round your answer to four decimal places.)

69. Find the surface area of the solid generated by rotating the graph of $y = \cos x$, $-\pi/2 \le x \le \pi/2$, about the x-axis.

70. *Find the length of the graph of $y = e^{-x}$, $0 \le x \le 2$. (Round your answer to four decimal places.)

71. *Find the surface area of the solid generated by rotating the graph of Exercise 70 about the x-axis. (Round your answer to four decimal places.)

72–73 Use substitution and a table of integrals to evaluate the given integral.

72. $\displaystyle\int \frac{e^x}{1 + \cos e^x}\, dx$

73. $\displaystyle\int \frac{\sin 2x}{2\sqrt{2}\cos x - \cos^2 x}\, dx$

74–77 Use the Trapezoidal Rule to approximate the integral with $n = 6$, and compare the result to the exact value of the integral by determining the absolute value of the error E_T.

74. $\displaystyle\int_1^4 x^{3/2}\, dx$

75. $\displaystyle\int_1^2 \frac{1}{x^2}\, dx$

76. $\displaystyle\int_0^4 \sqrt{x^2 + 1}\, dx$

77. $\displaystyle\int_0^{\pi/3} \tan x\, dx$

78–81. Use the "a priori" error estimate for the Trapezoidal Rule to estimate $\left|E_T\right|$ for $n = 6$, and compare the estimate with the actual error you found in Exercises 74–77.

82–85. Use Simpson's Rule to approximate the integrals from Exercises 74–77 with $n = 6$. Determine the absolute value of the error E_S.

86–89. Use the "a priori" error estimate for Simpson's Rule to estimate $\left|E_S\right|$ for $n = 6$, and compare the estimate with the actual error you found in Exercises 82–85.

90. Prove that if $f(x) = ax + b$ is a linear function on a closed interval $[a,b]$, then for any n,
$$T_n = \int_a^b f(x)\, dx.$$

91–96 Identify the type of the improper integral and determine whether it is convergent or divergent. If it is convergent, find its value.

91. $\displaystyle\int_2^4 \frac{dx}{(x-2)^4}$

92. $\displaystyle\int_{-\infty}^0 x^2 e^x\, dx$

93. $\displaystyle\int_2^6 \frac{dx}{\sqrt{x-2}}$

94. $\displaystyle\int_0^\infty \frac{2}{(x+3)^{2/3}}\, dx$

95. $\displaystyle\int_{-\infty}^\infty \frac{2e^x}{e^{2x} + 4}\, dx$

96. $\displaystyle\int_e^\infty \frac{dx}{x \ln x}$

97–98 Use the Direct Comparison Test to determine whether the integral converges.

97. $\displaystyle\int_0^\infty \frac{dx}{\sqrt{x^3 + 1}}$

98. $\displaystyle\int_1^\infty \frac{\ln x}{\sqrt{x}}\, dx$

99. Use substitution to turn the improper integral
$$\int_0^{\pi^2/4} \frac{\cos\sqrt{x}}{\sqrt{x}}\, dx \text{ into a proper one and evaluate it.}$$

100. Rotate about the x-axis the region bounded by the graph of $y = \sqrt{\ln x}/x^2$ and the x-axis over the interval $[1,\infty)$. Use the disk method to determine if the resulting unbounded solid has finite volume. If so, find the volume.

101. *Rotate about the x-axis the region bounded by the x-axis and the graph of $y = e^{-x}$ over the infinite interval $[0,\infty)$. Determine if the resulting infinite solid has finite volume or surface area. If so, find their values.

102. Prove that the improper integral $\displaystyle\int_2^\infty \frac{dx}{x(\ln x)^a}$ converges if and only if $a > 1$.

103–106 Find the Laplace transform. (See Exercises 88–93 in Section 7.7.)

103. $L\{te^{at}\}$

104. *$L\{t^2 e^{at}\}$

105. *$L\{t \sin kt\}$

106. *$L\{t \cos kt\}$

107–113 *True or False?* Determine whether the given statement is true or false. In case of a false statement, explain or provide a counterexample.

107. If $\lim_{x \to \infty} f(x) = L$ and $L \neq 0$, then $\int_0^\infty f(x)\,dx$ diverges.

108. If $\lim_{x \to \infty} f(x) = 0$, then $\int_0^\infty f(x)\,dx$ converges.

109. Let $f(x)$ be defined on $[a, \infty)$ and S be the solid generated by rotating the graph of $f(x)$ about the x-axis. If the surface area of S is infinite, then the volume of S is also infinite.

110. If $f(x)$ is an odd function, then $\int_{-\infty}^\infty f(x)\,dx = 0$.

111. If $f(x)$ has a vertical asymptote at $x = 0$ and $a > 0$, then $\int_{-a}^a f(x)\,dx$ diverges.

112. Any rational function is integrable on any finite interval that doesn't include a zero of the denominator.

113. If an integrand contains the expression $\sqrt{a^2 \pm x^2}$, a trigonometric substitution must be used to evaluate the integral.

117. Write a program for a computer algebra system or programmable calculator that evaluates the Simpson approximation S_n for a given integral and an even positive integer n. Find the smallest n that provides an answer to Exercise 74 that is correct to at least the first three digits after the decimal. Compare this with your answer for Exercise 115. What n ensures that the answer is correct to at least five decimal places?

118. Use the program you wrote for Exercise 117 for the integral from Exercise 75.

119–120 Use the programs you wrote for Exercises 115 and 117 to approximate the given nonelementary integral with $n = 50$. Which method (the Trapezoidal Rule or Simpson's Rule) do you expect to be more accurate? Use the built-in numerical integration command of your technology to verify your conjecture.

119. $\int_0^1 \sqrt{1 + x^4}\,dx$

120. $\int_0^1 e^{e^x}\,dx$

Chapter 7
Technology Exercises

114. Use a computer algebra system to find the integral from Exercise 33. If the answer appears different from what you obtained by hand, prove that the answers are equivalent.

115. Write a program for a computer algebra system or programmable calculator that evaluates the trapezoidal approximation T_n for a given input function on a specified interval and positive integer n. Find the smallest n that provides an answer to Exercise 74 that is correct to at least the first three digits after the decimal.

116. Use the program you wrote for Exercise 115 for the integral from Exercise 75.

Project 📝

Chapter 7

In this project, we will derive a famous infinite product named after its discoverer, the English mathematician John Wallis (1616–1703). Wallis introduced the symbol ∞ for infinity, and in turn he used $1/\infty$ to denote an *infinitesimal* quantity. He contributed to the development of *infinitesimal calculus* (it wasn't until the 19th century that infinitesimals were replaced by limits in the works of Bolzano, Cauchy, and Weierstrass).

1. For a nonnegative integer n, let

$$I_n = \int_0^{\pi/2} \sin^n x \, dx = \int_0^{\pi/2} \cos^n x \, dx.$$

   Find I_0, I_1, I_2, and I_3.

2. Show that if $n \geq 2$,

$$I_n = \frac{n-1}{n} I_{n-2}.$$

   (**Hint:** See Exercise 81 of Section 7.1.)

3. Use Questions 1 and 2 to find I_4, I_5, I_6, and I_7.

4. Show that in general,

$$I_{2n} = \frac{2n-1}{2n} \cdot \frac{2n-3}{2n-2} \cdot \frac{2n-5}{2n-4} \cdots \cdots \frac{1}{2} \cdot \frac{\pi}{2},$$

   while

$$I_{2n+1} = \frac{2n}{2n+1} \cdot \frac{2n-2}{2n-1} \cdot \frac{2n-4}{2n-3} \cdots \cdots \frac{2}{3}.$$

   (**Hint:** Observe a pattern or use induction.)

5. Use Question 4 to show that

$$\frac{I_{2n}}{I_{2n+1}} = \frac{3^2 5^2 \cdots \cdots (2n-1)^2 (2n+1)}{2^2 4^2 \cdots \cdots (2n)^2} \cdot \frac{\pi}{2}$$

   holds for all n.

6. Show that

$$\frac{I_{2n-1}}{I_{2n+1}} = 1 + \frac{1}{2n}.$$

7. Prove the inequalities

$$I_{2n-1} \geq I_{2n} \geq I_{2n+1}.$$

   (**Hint:** Use the definition of I_n from Question 1 and compare the integrands.)

8. Use Questions 6 and 7 to show that

$$1 \leq \frac{I_{2n}}{I_{2n+1}} \leq 1 + \frac{1}{2n},$$

   and use this observation to prove that

$$\lim_{n \to \infty} \frac{I_{2n}}{I_{2n+1}} = 1.$$

9. Use your answers to the previous questions to derive **Wallis' product**:

$$\frac{\pi}{2} = \lim_{n \to \infty} \frac{2^2 4^2 \cdots \cdots (2n)^2}{3^2 5^2 \cdots \cdots (2n-1)^2 (2n+1)}$$

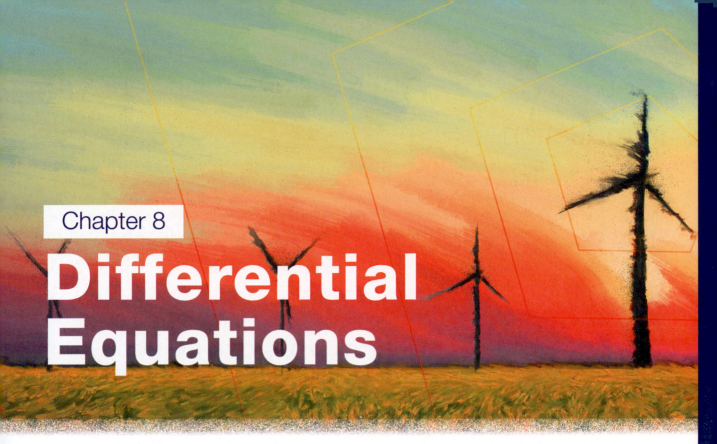

Chapter 8

Differential Equations

8.1 Separable Differential Equations 645

1. Setting and Terminology
2. Solving Separable Differential Equations

8.2 First-Order Linear Differential Equations 654

1. Definition and Solution Technique
2. Applications of First-Order Linear Differential Equations

8.3 Autonomous Differential Equations and Slope Fields 663

1. Autonomous Differential Equations
2. Modeling with Autonomous Differential Equations
3. Euler's Method

8.4 Second-Order Linear Differential Equations 672

1. Solving Homogeneous Second-Order Linear Differential Equations
2. Initial Value Problems and Boundary Value Problems

Introduction

The solutions of many problems hinge on finding functions that have certain specified properties—for instance, a function $P(t)$ that describes the population of some particular community over time, a function $T(t)$ that describes the temperature of a given solid over time, or a function $y(x)$ that defines a certain curve in the xy-plane. Often, a relationship involving the derivative (or derivatives) of the desired function is relatively easy to find, while an explicit formula for the function itself may not be immediately apparent. We have already seen many examples in this text in which some initial information about a function's derivative is used to determine the function itself, simply through integration, but equations involving the derivative(s) of a function don't always lead so readily to an explicit formula. Informally, equations expressing such relationships are called *differential equations*, and their study constitutes an entire branch of mathematics. This chapter introduces the topic and lays the groundwork for further study.

Many of the differential equations in this chapter arise from applying physical, economic, geometric, or other principles to specific circumstances, and the applications fall into the realms of physics, engineering, chemistry, finance, and biology, as well as the purely mathematical. But a useful way of sorting differential equations is based on the many solution techniques that have been developed over the last several centuries.

> A useful way of sorting differential equations is based on the many solution techniques that have been developed over the last several centuries.

The classifications introduced in this chapter include separable differential equations, first-order and second-order linear differential equations, and autonomous differential equations; each classification comes paired with a solution method that applies to equations in that class. One technique for numerically approximating the solutions to certain differential equations, called Euler's method, is also introduced.

Depending on your interests and your field of study, you may later take one or more courses that focus more closely on these solution techniques; such courses typically have names like *Ordinary Differential Equations*, *Partial Differential Equations*, and *Numerical Analysis*. But you are likely to run across plenty of examples of differential equations in the meantime, and the material in this chapter will help you understand the principles underlying such equations and the methods by which they are solved.

Leonhard Euler
(1707–1783)

---

1. Adapted from painting by Jakob Emanuel Handmann, *Leonhard Euler*, Oil on canvas (1756).

8.1 **Separable Differential Equations**

TOPICS

1. Setting and terminology
2. Solving separable differential equations

In this chapter, we introduce techniques used to solve equations that are written in terms of a dependent variable and one or more of its derivatives—so-called *differential equations*. Such equations express relationships and principles that arise naturally in physics, biology, economics, geometry, chemistry, and many other disciplines.

TOPIC 1 **Setting and Terminology**

In a sense, we have already solved many differential equations: any equation of the form $dy/dx = f(x)$, in which our goal is to determine y as a function of x, qualifies. But such equations are the very simplest examples, and the only solution technique we have used so far is integration; specifically, assuming f is integrable, $y = \int f(x)\,dx$ is the answer. The study of techniques used to solve more complicated differential equations constitutes a subfield of mathematics, and entire books and courses are devoted to the topic. We introduce some of the more general techniques and applications in this chapter, beginning with ways of classifying differential equations.

Definition ♀

First-Order Differential Equations

A **first-order differential equation** is an equation of the form

$$\frac{dy}{dx} = f(x, y),$$

in which $f(x, y)$ is defined on a region in the xy-plane and it is understood that the unknown function y, which we seek to find, is a differentiable function of x.

In general, the **order** of a differential equation is the order of the highest derivative of y in the equation. The objective is to determine a **solution** y as a function of x that satisfies the differential equation; a **general solution** is a description of all such possible functions y, and a **particular solution** is one that also satisfies an **initial condition** of the form $y = y_0$ when $x = x_0$ (that is, the graph of y passes through the point (x_0, y_0) in the xy-plane). A differential equation with an initial condition is often called an **initial value problem (IVP)**.

Before exploring more complicated examples, we'll examine how the terms above relate to the simplest kind of differential equation.

Example 1 ✎

Solve the differential equation $y' = 2x - e^x$ with the initial condition $y(0) = 3$.

Solution

The absence of the dependent variable y on the right-hand side of this equation means we can solve it by simply integrating both sides:

$$y' = 2x - e^x$$
$$y = \int \left(2x - e^x\right) dx = x^2 - e^x + C$$

This solution is the general solution, as it describes the entire family of functions y that satisfies $y' = 2x - e^x$. To find the particular solution that also satisfies the initial condition $y(0) = 3$, we solve the following equation:

$$3 = y(0) = (0)^2 - e^0 + C = 0 - 1 + C,$$

which gives $C = 4$. So $y = x^2 - e^x + 4$ is the particular solution of our given initial value problem.

With the basic terminology of differential equations established, we can proceed to further refine their classification. The purpose of such classification is to group together differential equations that can be solved using certain techniques.

Definition 💡

Separable Differential Equations

A **separable differential equation** is a first-order differential equation that can be written in the following form.

$$\frac{dy}{dx} = \frac{g(x)}{h(y)}, \quad h(y) \neq 0$$

The real significance of the right-hand side of this equation is that it is written as the product of a function of x and a function of y. It is convenient to express the function of y as $1/h(y)$ because the equation can then be rewritten in differential form as follows, in which all x's appear on one side and all y's on the other:

$$h(y)dy = g(x)dx$$

TOPIC 2   Solving Separable Differential Equations

After rewriting, a separable equation relates two differential expressions and can be integrated to result in a relationship between y and x.

$$\int h(y)dy = \int g(x)dx$$

This is the basic solution technique for such differential equations.

Example 2 ✐

Solve the equation $y' = 6(1 + y^2)x^2$.

Solution

The question of whether a given first-order differential equation is separable can be answered by trying to rewrite it in the desired differential form.

$$\frac{dy}{dx} = 6(1 + y^2)x^2$$

$$\frac{dy}{1 + y^2} = 6x^2 \, dx$$

Now that the variables have been separated to opposite sides of the equal sign, we integrate both sides to obtain the relationship between y and x.

$$\int \frac{dy}{1 + y^2} = \int 6x^2 \, dx$$

$$\tan^{-1} y = 2x^3 + C \qquad \text{\textcolor{blue}{C represents the combined constants of integration.}}$$

This last relation implicitly defines y as a function of x. We can make the functional relationship explicit under further conditions; for instance, if $-\pi/2 < 2x^3 + C < \pi/2$, then we can write an equation for y as follows:

$$y = \tan(\tan^{-1} y) = \tan(2x^3 + C)$$

Example 3 ✐

Solve the initial value problem $y' = x^2 y$ with the initial condition $y(0) = -1$.

Solution

Note that $y = 0$ is a valid, though trivial, solution of this differential equation. If we assume that y is not 0, we can again attempt to separate variables and, if successful, proceed to integrate.

$$\frac{dy}{dx} = x^2 y$$

$$\frac{dy}{y} = x^2 \, dx \qquad \text{\textcolor{blue}{$y \neq 0$}}$$

$$\int \frac{dy}{y} = \int x^2 \, dx$$

$$\ln|y| = \frac{x^3}{3} + C$$

$$|y| = e^{(x^3/3)+C}$$

$$y = \pm e^{(x^3/3)+C} = \pm e^C e^{x^3/3}$$

Since $y = 0$ is also a solution of the original differential equation, we can simplify the last expression above to write the general solution as

$y = Ae^{x^3/3}$, where A can take on any real number value—in effect, we are allowing A to be e^C, $-e^C$, or 0. To find the particular solution that satisfies the initial condition, we solve for A as follows.

$$-1 = y(0) = Ae^0$$
$$A = -1$$

This gives us $y = -e^{x^3/3}$.

Mixture problems are a class of problems that can often be expressed as initial value problems. In general, such problems involve finding the quantity of some substance in a container, usually under dynamic conditions such as when a solution containing the substance is pumped into and/or out of the container. If the quantity of the substance as a function of time is denoted $y(t)$, the IVP can be set up by determining the rate r_{in} at which the substance is entering the container, the rate r_{out} at which the substance is leaving the container, and the quantity y_0 of the substance at some time $t = t_0$. The IVP is then as follows.

$$\frac{dy}{dt} = r_{in} - r_{out} \quad \text{with} \quad y(t_0) = y_0$$

Example 4 ✎

A 1000-gallon tank is filled with gasoline containing 15 pounds of a seasonal additive. A gasoline mixture containing 0.02 pounds of additive per gallon is added to the tank at the rate of 50 gallons per minute. The gasoline solution in the tank is continuously and thoroughly mixed and drained out at the same rate of 50 gallons per minute. How many pounds of additive are in the tank 30 minutes into the operation?

Solution

If we let $y(t)$ be the number of pounds of additive at time t, then we know that $y(0) = 15$ lb. The rate at which additive is being pumped into the tank is calculated as follows.

$$r_{in} = \left(0.02\frac{\text{lb}}{\text{gal}}\right)\left(50\frac{\text{gal}}{\text{min}}\right) = 1\frac{\text{lb}}{\text{min}}$$

Since the solution in the tank is continuously and thoroughly mixed, the concentration of additive in the tank at time t is $y(t)/1000$, measured in pounds per gallon, so the rate at which additive is leaving the tank is the concentration times the rate of outflow.

$$r_{out} = \left(\frac{y}{1000}\frac{\text{lb}}{\text{gal}}\right)\left(50\frac{\text{gal}}{\text{min}}\right) = \frac{y}{20}\frac{\text{lb}}{\text{min}}$$

So we are led to the following separable differential equation.

$$\frac{dy}{dt} = 1 - \frac{y}{20} = \frac{20 - y}{20}$$

$$\frac{dy}{20 - y} = \frac{dt}{20}$$

$$\int \frac{dy}{20 - y} = \int \frac{dt}{20}$$

$$-\ln|20 - y| = \frac{t}{20} + C$$

$$|20 - y| = Ae^{-t/20} \qquad\qquad A = e^{-C}$$

Given the capacity of the tank and the additive's concentration of 0.02 pounds per gallon, and given that $y(0) = 15$, we know that $|20 - y| = 20 - y$ (in other words, y is less than or equal to 20 for all t), so $y = 20 - Ae^{-t/20}$. And using the fact that $y(0) = 15$ again, it must be the case that $A = 5$, giving us $y(t) = 20 - 5e^{-t/20}$. So the amount of additive in the tank at the 30-minute mark is calculated as follows.

$$y(30) = 20 - 5e^{-30/20} \approx 18.9 \text{ lb}$$

Over time, the amount of additive in the tank will approach but never actually reach 20 pounds, as shown by the graph of y in Figure 1.

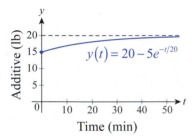

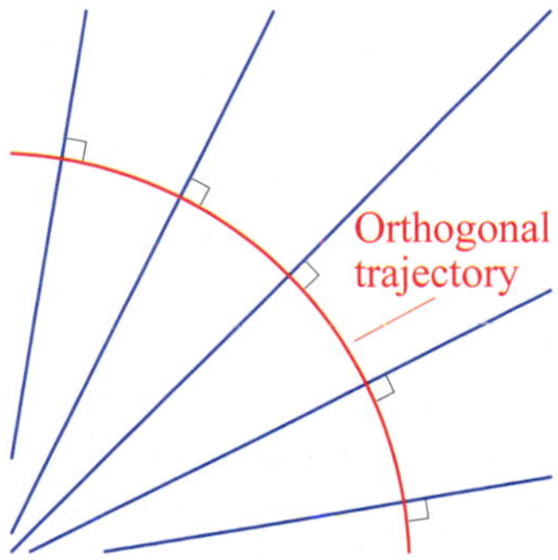

$$y(t) = 20 - 5e^{-t/20}$$

Figure 1

Orthogonal trajectories of families of curves arise naturally in the study of magnetic, electrical, and gravitational fields. An orthogonal trajectory of a family of curves is a curve that intersects each member of the family at right angles (*orthogonally*)—in Figure 2, the red curve intersects the family of blue curves in just such a manner. And we can go further. Given a family of curves, we can determine another corresponding family of orthogonal trajectories, in which each curve in one family intersects each curve in the other family at right angles. We find the second family by solving a differential equation.

Orthogonal trajectory

Figure 2

Example 5 ✎

Determine the orthogonal trajectories of the family of curves $xy = a$, where a is an arbitrary nonzero constant.

Solution

Each curve $xy = a$ is a hyperbola with the coordinate axes as asymptotes. For a given constant a, we can determine the slope of the line tangent to the curve at a given point (x, y) by implicitly differentiating the equation $xy = a$ and solving for dy/dx, as follows.

$$xy = a$$

$$y + x\frac{dy}{dx} = 0$$

$$\frac{dy}{dx} = -\frac{y}{x}$$

Recall that the slopes of orthogonal lines are negative reciprocals of each other, so the slope of the line that is perpendicular to the line tangent to the curve $xy = a$ at the point (x, y) satisfies $dy/dx = x/y$. We can now solve this separable differential equation to find equations for the curves orthogonal to $xy = a$.

$$\frac{dy}{dx} = \frac{x}{y}$$

$$y\,dy = x\,dx$$

$$\int y\,dy = \int x\,dx$$

$$\frac{y^2}{2} = \frac{x^2}{2} + C$$

$$y^2 - x^2 = b \qquad\qquad b = 2C$$

Figure 3 illustrates how three curves of the first family (shown in blue) relate to three curves of the second family (shown in red). For each blue curve there are an infinite number of orthogonal curves from the second family that could be depicted—the three shown were chosen to each intersect a blue curve at a point with x-coordinate equal to 3.

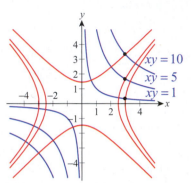

Figure 3 Orthogonal Trajectories

Caution ⚠

It may be tempting to find the orthogonal trajectories of the family of curves $xy = a$ by first solving for y, differentiating to get $dy/dx = -a/x^2$, and then integrating the negative reciprocal of this result. This will indeed give us *one* trajectory that is orthogonal to the curve $xy = a$ for *one* fixed value of a, but not a family of trajectories orthogonal to the entire family of curves of the form $xy = a$. In order to find the complete family of orthogonal trajectories, we need a differential equation that is true for all values of a simultaneously. The method can be salvaged by recognizing the dependence of a on x and y, writing

$$\frac{dy}{dx} = -\frac{a}{x^2} = -\frac{xy}{x^2} = -\frac{y}{x}$$

and proceeding as in Example 5. Writing dy/dx in this fashion expresses the slope of the line tangent to the curve $xy = a$ at a point (x, y) without reference to the value of a corresponding to that particular curve.

8.1 **Exercises**

1–6 Rewrite the first-order differential equation in the form $dy/dx = f(x,y)$. (Do not attempt to solve the equation.)

1. $\sqrt{x}\dfrac{dy}{dx} = \sqrt{y}$

2. $dy = 3y\,dx + e^{2x}\,dx$

3. $\sqrt{1-y^2}\,dx = \dfrac{dy}{x}$

4. $\dfrac{\tan y}{x} = \cos x \dfrac{dx}{dy}$

5. $xy = y - x^2 y'$

6. $3x^2 y = xe^x - xy'$

7–12 Verify that the function (or family of functions) is a solution of the given differential equation.

7. $xy' = 5y;\quad y = Cx^5$

8. $y' - \dfrac{3x^2}{y} = 0;\quad y = \sqrt{2x^3 + 5}$

9. $x^2\,dy - dx = 0;\quad y = 6 - \dfrac{1}{x}$

10. $3y' + 2xy = 0;\quad y = Ce^{-x^2/3}$

11. $y' + y = e^{5x};\quad y = Ce^{-x} + \dfrac{e^{5x}}{6}$

12. $x\,dy - (y - x)\,dx = 0;\quad y = Cx - x\ln x$

13–20 Solve the differential equation with the indicated initial condition.

13. $y' = x^2 + \cos x;\quad y(0) = 5$

14. $y' = \dfrac{1}{x^2 + 1} - 2;\quad y(0) = 0$

15. $xy' = 1 + x;\quad y(1) = 0$

16. $y'\sqrt{x} = \sqrt{x} + 1;\quad y(1) = 5$

17. $dx = x^2\,dy;\quad y(2) = 3$

18. $\sqrt{1-x^2}\,y' = x;\quad y(0) = 2$

19. $\sqrt{1-x^2}\,y' = 1;\quad y(0) = 2$

20. $y' = \dfrac{1}{x^2 + x};\quad y(1) = -4$

21–28 Determine whether the differential equation is separable. (Do not solve the equation.)

21. $y' = (2x - 1)e^y$

22. $y' = y^2 + 4$

23. $xy' + 4y^2 = 0$

24. $y' = x + y$

25. $y' = \sqrt{xy} - ye^x$

26. $y' = (x + y)^2$

27. $y' = 3x^2 y - x\sqrt{y}$

28. $x^2\,dy - (y^2 + yx)\,dx = 0$

29–40 Solve the separable differential equation.

29. $y' = 2x(y^2 + 1)$

30. $y' = y(x^2 + 4)$

31. $y' = 3xy + 6x$

32. $y' = 9y^2 x^2$

33. $dy - 3x^2 y\,dx = 0$

34. $y' = \sec y \sec^2 x$

35. $y'\sqrt{4 - x^2} = y$

36. $y' = \sqrt{\dfrac{1 - y^2}{x}}\,e^{\sqrt{x}}$

37. $e^{1/x}\dfrac{dx}{dt} = -2tx^2$

38. $(x + 2)\,dy = (x + 5)\,dx$

39. $dy = \dfrac{y + 6}{x}\,dx$

40. $x^2 y\,dy = (y^2 + 1)\,dx$

41–43 Solve the given separable equation, treating y as the independent variable and solving for $x = x(y)$.

41. $x\,dy - (y + 1)\,dx = 0$

42. $dx - 3(x^2 + 1)\,dy = 0$

43. $y' = \dfrac{2x}{y + 2}$

44–53 Solve the given initial value problem.

44. $y' = -\dfrac{x}{y};\quad y(1) = 1$

45. $y' = \dfrac{x^2 - 1}{y^2};\quad y(0) = 2$

46. $y' = \dfrac{4x^3 + 2x}{2y};\quad y(2) = 5$

47. $y' = e^{2x - 5y};\quad y(0) = 0$

48. $x^3 y' = (x - 1)y^2;\quad y(1) = -1$

49. $y'\sec x = y^2;\quad y\!\left(\dfrac{\pi}{2}\right) = \dfrac{1}{3}$

50. $\dfrac{dy}{dx} = 4x^3 e^{-y};\quad y(0) = 1$

51. $\dfrac{y'}{x} = y^2 + y;\quad y(0) = -2$

52. $y' = \dfrac{1 + y^2}{1 + x^2};\quad y(1) = -2$

53. $\dfrac{dy}{1 + \sin t} = 2e^{-y}\cos t\,dt;\quad y(0) = 0$

54. A 50-gallon tank is filled with brine (water nearly saturated with salt; used as a preservative) holding 12 lb of salt in solution. A salt solution containing 0.5 lb of salt per gallon is added to the tank at the rate of 1 gal/min. The contents of the tank are continuously and thoroughly mixed and drained out at the same rate. What is the amount of salt in the tank after an hour?

55. A tank contains 2000 gallons of diesel fuel. A fuel mixture containing a lubricity additive is pumped into the tank through two inlets. The mixture flowing in through the first inlet contains 0.48 oz of additive per gallon and is being pumped in at a rate of 25 gal/min. Meanwhile, the mixture being allowed in by the second inlet at a rate of 10 gal/min contains 10.4 oz of additive per gallon. The mixture in the tank is continuously and thoroughly mixed and drained out at the rate of 35 gal/min. If there should be 16 oz of additive for every 120 gallons of diesel fuel, how long will it take to reach the right mixture?

56. To freshen the air, a small window is opened in a room initially containing 0.12% carbon dioxide. Fresh air with 0.04% carbon dioxide is pouring in at a rate of 6 m^3/min, and we assume that the uniform mixture is leaving the room at the same rate. If the dimensions of the room in meters are $4 \times 6 \times 3$, how long will it take to cut the initial carbon dioxide content down to half?

57. The cane sugar in fruit juice converts into dextrose under certain conditions. At any time, the rate of this process is proportional to the amount of cane sugar that is yet to be converted. If 100 grams of cane sugar is added to a certain fruit juice and we know that 12 grams are converted into dextrose during the first hour, how much dextrose will be present in the juice after 3 hours?

58. Suppose that an ice cube melts so that its volume $V(t)$ decreases at a rate proportional to its surface area.

a. Find a differential equation satisfied by $V(t)$.

b. If an ice cube of side length 1 inch loses a third of its volume in 2 minutes, use your model to predict how long it will take for it to completely melt away. (Consider the cube melted away when your model predicts less than 1 percent remaining.)

59–61 Suppose a container is filled with fluid to a height of h. According to Torricelli's Law, when viscosity and friction are ignored, the speed v of efflux of the fluid through a small, sharp-edged opening through the bottom of the container equals the speed that the fluid would acquire when falling freely from a height of h, as follows.

$$v = \sqrt{2hg}$$

59.* If A_c is the horizontal cross-sectional area of a vertical cylindrical tank, while A_o denotes the area of the hole at the bottom of the tank, prove that the rate at which the fluid level is falling in the container is described by the following differential equation.

$$\frac{dh}{dt} = -\frac{A_o}{A_c}\sqrt{2hg}$$

(**Hint:** First find the rate of fluid leaving the tank dV/dt; then use the fact that $V = A_c h$.)

60. If a cubic tank of side length 1 m is initially full of water but is draining through a circular orifice of diameter 2 cm that is on the bottom of the tank, what is the water level in the tank 2 minutes later? (**Hint:** Use the formula from Exercise 59.)

61.* Answer the question of Exercise 60 if the container is an inverted right circular cone of height 1 m and base radius 0.5 m. The opening on the bottom is the same as that in Exercise 60. (**Hint:** First convince yourself that even though $A_c = A_c(h)$ now depends on h, the formula $dh/dt = -(A_0/A_c)\sqrt{2hg}$ remains in effect. Next, adapt your solution for Exercise 60. Ignore the geometrical change to the cone caused by the presence of the orifice.)

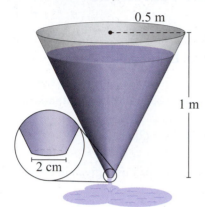

0.5 m

1 m

2 cm

62.* The figure below shows a series circuit containing a resistor and a capacitor (this is called an RC circuit). Find a differential equation for the charge $q(t)$ if the impressed voltage V on the circuit is constant. (**Hint:** Use Ohm's Law, as well as the fact that the voltage drop on the capacitor is $\frac{1}{C}q$. For a statement of Ohm's Law, see the discussion preceding Example 5 in Section 8.2.)

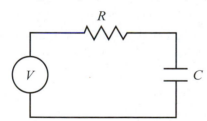

R

V   C

63. Suppose that the air resistance encountered by a falling body is proportional to its velocity v.

  a. Use Newton's Second Law of Motion (see Section 3.7) to find a differential equation satisfied by a falling body of mass m.

  b. Solving your equation, find a formula for the terminal velocity of a body if it is falling from rest.

64. When starting from rest, the acceleration of a sailboat is proportional to the difference between the boat's velocity and that of the wind. Suppose that two minutes after starting from rest in 18 mph wind, a sailboat is moving at 8 mph.

  a. Find a differential equation satisfied by the boat's velocity function.

  b. Find the boat's velocity function.

  c. How fast is the boat moving 4 minutes after it starts?

65–73 Determine the orthogonal trajectories of the family of curves, where *a* is an arbitrary nonzero constant. (If technology is available, sketch several curves from both families and visually check orthogonality.)

65. $y = ax^2$ **66.** $y = ax^3$

67. $y = \dfrac{ax}{x+2}$ **68.** $y = ae^x$

69. $y = \dfrac{x}{ax+1}$ **70.** $y = \dfrac{1}{a+x}$

71. $x^2 + ay^2 = 1$ **72.** $y = a\cos x$

73. $y = \dfrac{a}{1+x^2}$

74–77 *True or False?* Determine whether the given statement is true or false. In case of a false statement, explain or provide a counterexample.

74. A differential equation is an equation that contains a derivative of a function.

75. The equation $x^3 y' + 1 = xy - x + y$ is separable.

76. A separable equation in the variables x and y always has a solution in the form $y = f(x)$.

77. This is the first time in this text that we are solving initial value problems.

8.2 First-Order Linear Differential Equations

TOPICS

1. Definition and solution technique

2. Applications of first-order linear differential equations

Separable equations constituted our first class of differential equations, and the solution technique for solving them depended on the fact that variables in such equations can be separated to opposite sides of the equal sign. In this section we define our second class of differential equations and develop a similarly focused method of solution.

TOPIC 1   Definition and Solution Technique

Definition ♀

First-Order Linear Differential Equations

A first-order differential equation in the independent variable x and dependent variable y is **linear** if it can be written in the **standard form** of

$$\frac{dy}{dx} + P(x)y = Q(x),$$

where P and Q are continuous functions of x on a given interval.

Example 1 ✎

Classify the following differential equations as linear or nonlinear.

a. $3x\dfrac{dy}{dx} - \dfrac{y}{x} = e^x + 2 + \dfrac{dy}{dx}$
    b. $\dfrac{dy}{dx} - xe^y = 2$

Solution

a. The equation

$$3x\frac{dy}{dx} - \frac{y}{x} = e^x + 2 + \frac{dy}{dx}$$

qualifies as a first-order linear differential equation, as is evident when written in standard form, as follows.

$$3x\frac{dy}{dx} - \frac{dy}{dx} - \frac{y}{x} = e^x + 2$$

$$\frac{dy}{dx}(3x - 1) - \frac{y}{x} = e^x + 2$$

$$\frac{dy}{dx} - \frac{1}{x(3x-1)}y = \frac{e^x + 2}{3x - 1} \qquad \textcolor{blue}{x \neq 0, \tfrac{1}{3}}$$

b. The first-order equation

$$\frac{dy}{dx} - xe^y = 2$$

is nonlinear, because y appears as an exponent—the "linearity" in the classification applies to (and only to) the dependent variable.

First-order linear equations cannot, in general, be solved by separating variables and integrating, but integration is still the key to their solution. Before integrating, though, we multiply both sides of the equation $y' + P(x)y = Q(x)$ by an **integrating factor** $I(x)$, a function whose formula we find by solving a separable equation.

The motivation for the integrating factor is this: if we choose $I(x)$ appropriately, the new left-hand side of our equation, $I(x)[y' + P(x)y]$, will be the derivative of the product $I(x)y$. We can then integrate both sides of the new equation to solve for y.

To find the formula for $I(x)$, we set up and solve the equation described in the preceding paragraph. For the sake of clarity we will leave off the arguments of the functions after the first step, but remember that I, P, and y are all functions of x.

$$\frac{d}{dx}\big[I(x)y\big] = I(x)\big[y' + P(x)y\big] \qquad \text{We want } I \text{ to satisfy this condition.}$$

$$I'y + \cancel{Iy'} = \cancel{Iy'} + IPy \qquad \text{Expand and cancel like terms.}$$

$$\frac{dI}{dx}\cancel{y} = IP\cancel{y} \qquad \text{Rewrite } I' \text{ as } dI/dx \text{ and divide out } y.$$

$$\int \frac{dI}{I} = \int P\,dx \qquad \text{Separate variables and integrate.}$$

$$\ln|I| = \int P\,dx$$

$$I = e^{\int P\,dx} \qquad \text{Exponentiate and specify } I > 0.$$

To see what this integrating factor accomplishes, note that our modified linear equation can now be solved.

$$I(x)\big[y' + P(x)y\big] = I(x)Q(x)$$

$$\frac{d}{dx}\big[I(x)y\big] = I(x)Q(x)$$

$$I(x)y = \int I(x)Q(x)\,dx + C \qquad \text{C is introduced by integrating the left-hand side.}$$

$$y = \frac{1}{I(x)}\Big[\int I(x)Q(x)\,dx + C\Big]$$

Example 2 ✎

Solve the equation $y' + \dfrac{y}{x} = \dfrac{2}{3x^{1/3}}$ on the interval $(0,\infty)$ with the initial condition $y(1) = 1$.

Solution

The equation is already in standard form, so we proceed to find the integrating factor $I(x)$.

$$I(x) = e^{\int 1/x\,dx} = e^{\ln|x|} = e^{\ln x} = x \qquad x \in (0,\infty)$$

Note that we have chosen to let the constant of integration be 0 (see Exercise 42). Multiplying the original equation through by $I(x)$, we obtain the following.

$$xy' + y = \frac{2}{3}x^{2/3}$$

$$\frac{d}{dx}(xy) = \frac{2}{3}x^{2/3} \qquad\qquad \frac{d}{dx}(xy) = 1 \cdot y + x \cdot \frac{dy}{dx}$$

$$xy = \int \frac{2}{3}x^{2/3}\,dx + \tilde{C} = \frac{2}{5}x^{5/3} + C \qquad \begin{array}{l}\text{Constants of integration}\\ \text{are combined into } C.\end{array}$$

Solving for y gives us the general solution

$$y = \frac{2}{5}x^{2/3} + \frac{C}{x},$$

and since we want $y(1) = 1$ we require the following.

$$1 = y(1) = \frac{2}{5}(1)^{2/3} + \frac{C}{1} \quad\Rightarrow\quad C = \frac{3}{5}$$

Thus, the particular solution to this IVP is

$$y = \frac{2}{5}x^{2/3} + \frac{3}{5x}.$$

Example 3 ✍

Solve the equation $y' = (\sin x)y + 1$.

Solution

In standard form, the equation is $y' + (-\sin x)y = 1$ and its integrating factor is

$$I(x) = e^{\int -\sin x\, dx} = e^{\cos x},$$

where we have again chosen to let the constant of integration be 0. Multiplying through by I, we have the following.

$$\frac{d}{dx}(e^{\cos x}y) = e^{\cos x}$$

Integrating both sides, we can write $e^{\cos x}y = \int e^{\cos x}\,dx + C$, where the constant of integration C comes from integrating the left-hand side (it is convenient to move it to the right-hand side). The integral $\int e^{\cos x}\,dx$ is nonelementary, however, and thus can't be written in terms of elementary functions. Nevertheless, the expression defines a perfectly good function of x. Solving for y, we can write the solution as

$$y = \frac{1}{e^{\cos x}}\left(\int e^{\cos x}\,dx + C\right)$$

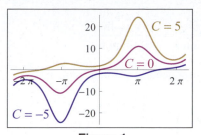

Figure 1

Solutions of $y' = (\sin x)y + 1$

or in terms of a definite integral as

$$y = \frac{1}{e^{\cos x}}\left(\int_0^x e^{\cos t}\, dt + C\right)$$

(the lower limit of integration can be any fixed real number). While we can't evaluate this definite integral using an antiderivative, a computer algebra system can approximate its value for different x's and for different choices of C. Figure 1 shows the graphs of three such choices.

TOPIC 2   **Applications of First-Order Linear Differential Equations**

We will continue to explore the many applications of differential equations as we acquire new solution techniques. The first example to follow is a variation of the mixture problem we studied in the last section; in this version, we are led to solve a nonseparable first-order linear differential equation.

Example 4 ✐

A 1000-gallon tank is filled with gasoline containing 15 pounds of a seasonal additive. A gasoline mixture containing 0.02 pounds of additive per gallon is added to the tank at the rate of 50 gallons per minute. The gasoline solution in the tank is continuously and thoroughly mixed and drained out at a faster rate of 60 gallons per minute. How many pounds of additive are in the tank 30 minutes into the operation?

Solution

The change in the problem is slight, but the effect on the resulting equation and solution technique is pronounced. The principle used in setting up the equation is, however, the same. If we let $y(t)$ denote the pounds of additive in the tank t minutes into the mixing operation, then $y' = r_{\text{in}} - r_{\text{out}}$ (refer to the paragraph preceding Example 4 of Section 8.1 for a review of this principle).

In order to determine the concentration of additive in the tank at time t, we need to know the volume $V(t)$ of the gasoline mixture. Since $V(0) = 1000$ and the mixture is added at a rate of 50 gal/min and drained out at 60 gal/min, we have the following.

$$V(t) = 1000 \text{ gal} + \left(50\, \frac{\text{gal}}{\text{min}} - 60\, \frac{\text{gal}}{\text{min}}\right)(t \text{ min})$$
$$= 1000 - 10t \text{ gal} \qquad \color{blue}{0 \le t \le 100 \text{ min}}$$

So the concentration of additive at time t is

$$\frac{y(t)}{1000 - 10t}$$

and the rate at which the additive is leaving the tank is calculated as follows.

$$r_{out} = \left(\frac{y}{1000 - 10t} \frac{lb}{gal} \right) \left(60 \frac{gal}{min} \right) = \frac{6y}{100 - t} \frac{lb}{min}$$

The rate at which the additive is being introduced to the tank remains as follows.

$$r_{in} = \left(0.02 \frac{lb}{gal} \right) \left(50 \frac{gal}{min} \right) = 1 \frac{lb}{min}$$

So the linear differential equation to solve is

$$y' = 1 - \frac{6y}{100 - t}$$

or, in standard form, as follows:

$$y' + \frac{6}{100 - t} y = 1$$

This results in the integrating factor

$$I(t) = e^{\int 6/(100-t) dt} = e^{-6\ln(100-t)} = (100 - t)^{-6}$$

and the following general solution.

$$y = \frac{1}{I(t)} \left[\int I(t) Q(t) dt + C \right]$$
$$= (100 - t)^6 \left[\int (100 - t)^{-6} dt + C \right]$$
$$= (100 - t)^6 \left[\frac{(100 - t)^{-5}}{5} + C \right]$$
$$= \frac{100 - t}{5} + (100 - t)^6 C$$

Since $y(0) = 15$ pounds, we can determine C.

$$15 = \frac{100 - 0}{5} + (100 - 0)^6 C = 20 + (10^{12}) C$$

$$C = -\frac{1}{2 \times 10^{11}}$$

Therefore we have

$$y(t) = \frac{100 - t}{5} - \frac{1}{2 \times 10^{11}} (100 - t)^6$$

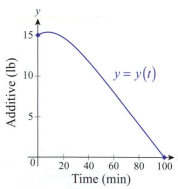

$y = y(t)$

Additive (lb)

Time (min)

Figure 2

and $y(30) \approx 13.4$ lb. Figure 2 is a graph of $y(t)$ over the 100 minutes it takes for the tank to empty completely—note that the amount of additive in the tank at first increases, and then begins to decrease. In Exercise 43 you will be asked to find the time when the additive in the tank reaches a maximum.

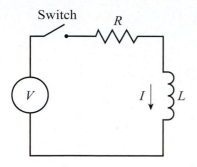

Figure 3 RL Circuit

Electrical circuits with a changing current $I(t)$ possess a property called *inductance*, caused by the magnetic field associated with the changing current. Inductance is given the symbol L and is measured in terms of units called *henries*. Just as voltage V in a circuit relates to steady current I and resistance R according to Ohm's Law, $V = IR$, the voltage, inductance, and current in an **RL circuit** with changing current (shown in Figure 3) are related by the following equation.

$$V = L\frac{dI}{dt} + IR$$

Example 5 ✒

Assume voltage V (in volts) is applied at time $t = 0$ in a simple circuit with inductance L (in henries) and resistance R (in ohms). Find a formula for the current I (in amperes) in the circuit at time t (assume V, L, and R are all constant).

Solution

The equation above is a first-order linear equation in I with standard form as follows.

$$\frac{dI}{dt} + \frac{R}{L}I = \frac{V}{L}$$

This gives rise to the integrating factor $e^{(R/L)t}$ and then the following general solution.

$$I(t) = e^{-(R/L)t}\left[\int \frac{e^{(R/L)t}V}{L}\,dt + C\right]$$

$$= e^{-(R/L)t}\left[\left(\frac{V}{L}\right)\left(\frac{L}{R}\right)e^{(R/L)t} + C\right] = \frac{V}{R} + Ce^{-(R/L)t}$$

The current I begins at 0 when $t = 0$, so it must be the case that $C = -V/R$ and hence the particular solution is as follows.

$$I(t) = \frac{V}{R} - \frac{V}{R}e^{-(R/L)t} = \frac{V}{R}\left[1 - e^{-(R/L)t}\right]$$

As $t \to \infty$, $\left[1 - e^{-(R/L)t}\right] \to 1$, so the *steady-state* current in the circuit is simply V/R. The inductance in the circuit is most significant immediately after voltage is applied—it is then that the term $1 - e^{-(R/L)t}$ has greatest effect, as shown in Figure 4.

The length of time L/R is referred to as the circuit's *time constant* and is a measure of how its inductance and resistance interact to affect the current. In Exercise 44 you will show that $I(t)$ attains slightly more than 95% of its steady-state value after three time constants.

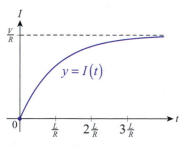

Figure 4

8.2 **Exercises**

1–6 Classify the differential equation as linear or nonlinear. (Do not attempt to solve the equation.)

1. $xy' - \dfrac{e^x}{x+1}y = \sqrt{x} - \dfrac{2y}{x}$

2. $xy' - \dfrac{e^x}{x+1}y = \sqrt{y} - \dfrac{2y}{x}$

3. $\dfrac{y}{x} = x^2 y'$

4. $yy' + (x+1)y = x^2$

5. $\tan x = \dfrac{2x + y'}{y}$

6. $y' - xy^2 = x$

7–10 Decide if the equation is linear in the dependent variable y. If not, check whether it is linear when x is considered to be the dependent variable.

7. $2\,dy - x\,dx = y\left(1 - 3\sqrt{x}\right)dx$

8. $x\,dy = \cos y\,dy - y\,dx$

9. $y\,dx = (5x + 2y - 4)\,dy$

10. $y\,dy - \left(1 + x^2\right)y\,dx = e^x\,dx$

11–25 Solve the linear differential equation. (**Hint:** In some cases, x has to be the dependent variable in order for the equation to be linear.)

11. $y' + 3\dfrac{y}{x} = 0$ **12.** $xy' - 2y = x^3 e^x$

13. $xy' + 4y = 1$ **14.** $y' + 2xy = 4x$

15. $x\dfrac{dy}{dx} = y + 2x^2 - 3x + 5$

16. $(x-1)y' = 4(x-1)^3 - y$

17. $dx + \dfrac{8x}{y}\,dy = y\,dy$

18. $y' - y\tan x = e^{\sin x}$

19. $y' = \sin x + \cos x - y$

20. $yx' - x = y^3 e^y$

21. $\left(x^4 + 1\right)y' = x^4 - 4x^3 y + 1$

22. $\cot x\,dy + y\,dx = \csc x\,dx$

23. $x(\ln x)y' + y = \ln x$

24. $dx - xy\,dy = y\,dy$

25. $\dfrac{dx}{dt} + x\tan t = t\tan t + \sec t + 1$

26–33 Solve the given initial value problem.

26. $\dfrac{dy}{dx} + 4y = 16x;\quad y(0) = 0$

27. $\dfrac{dy}{dx} + 3x^2 y = x^2;\quad y(0) = 2$

28. $\left(1 + x^2\right)y' + 2xy = 1;\quad y(0) = 10$

29. $\left(1 + x^2\right)y' - 2xy = 1 + x^2;\quad y(0) = 10$

30. $t\dfrac{dS}{dt} - S = 2t;\quad S(1) = 7$

31. $(1+x)y' - xy = \dfrac{e^x}{1+x};\quad y(0) = 0$

32. $y'\sin x + y\cos x = \dfrac{x}{\csc x};\quad y(\pi/2) = -3$

33. $y\,dx = (4\ln y - 2x)\,dy;\quad x(1) = 4$

34–37 Find a first-order linear differential equation in standard form that has the given general solution. (**Hint:** Identify the integrating factor and "reverse" the solution technique discussed in the text.)

34. $y = \dfrac{e^x}{x^2} + \dfrac{1}{x} + Cx^{-2}$

35. $y = \sin x + \dfrac{C}{x}$

36. $y = \dfrac{x^3 + C}{e^x}$

37. $y = 1 + \dfrac{x + C}{\ln x}$

38–41 A first-order differential equation of the form

$$\frac{dy}{dx} + P(x)y = Q(x)y^{\alpha}$$

is called the **Bernoulli equation** (named after Jakob Bernoulli), where α is any real number. You should check that by introducing the new dependent variable $u = y^{1-\alpha}$ ($\alpha \neq 0$, $\alpha \neq 1$) and noting that

$$\frac{du}{dx} = (1-\alpha)y^{-\alpha}\frac{dy}{dx},$$

we can turn a Bernoulli equation into the following standard-form linear one:

$$\frac{du}{dx} + (1-\alpha)P(x)u = (1-\alpha)Q(x)$$

In Exercises 38–41, use the above substitution method to solve the given Bernoulli equation.

38. $\dfrac{dy}{dx} + 2y = x\sqrt{y}$ **39.** $y' + \dfrac{y}{x} = y^2$

40. $y' + 2y = e^x y^2$ **41.** $y' + \dfrac{y}{x} = 2y^{3/2}$

42. Explain why it is no loss of generality to always let the constant of integration be 0 when determining the integration factor in the solution of a linear differential equation.

43. Find the time when the additive in the tank of Example 4 reaches a maximum.

44. Show that $I(t)$ in Example 5 attains slightly more than 95% of its steady-state value after three time constants.

45. A 50-gallon tank is filled with brine (water nearly saturated with salt; used as a preservative) holding 12 pounds of salt in solution. A salt solution containing 0.5 pounds of salt per gallon is added to the tank at the rate of 1 gal/min. The contents of the tank are continuously and thoroughly mixed and drained out at five quarts per minute. What is the amount of salt in the tank after an hour? (Compare with Exercise 54 of Section 8.1.)

46. A tank contains 2000 gallons of diesel fuel. A fuel mixture containing a lubricity additive is pumped into the tank through two inlets. The mixture flowing in through the first inlet contains 0.48 oz of additive per gallon, and is being pumped in at a rate of 25 gal/min. Meanwhile, the mixture being allowed in by the second inlet at a rate of 10 gal/min contains 10.4 oz of additive per gallon. The mixture in the tank is continuously and thoroughly mixed and drained out at the rate of 20 gal/min. If there should be 16 oz of additive for every 120 gallons of diesel fuel, how long will it take to reach the right mixture? (Compare with Exercise 55 of Section 8.1.)

47. Suppose that V_0 gallons of gasoline contain a_0 pounds of a seasonal additive. A gasoline mixture containing a_1 pounds of additive per gallon is added to the tank at the rate of r_1 gal/min. The gasoline solution in the tank is continuously and thoroughly mixed and drained out at a rate of r_2 gal/min. Set up the initial value problem whose solution is $y(t)$, the amount of additive in the mixture at time t.

48. In learning theory, the rate of memorization, or learning, is considered to be proportional to the amount of material yet to be memorized. On the other hand, the amount forgotten is proportional to the amount already learned. If T stands for the total amount of material to be memorized, and $M(t)$ is the amount memorized at time t, find a differential equation satisfied by $M(t)$. What type of differential equation did you obtain?

49. Find $I(t)$ in a simple RL circuit if 5 volts are applied at time $t = 0$, $I(0) = 0$, the inductance is 0.2 henries and the resistance is 10 ohms.

50. Find a formula for $I(t)$ in Exercise 49 if the impressed voltage is $V(t) = 1.2t$.

51. Suppose that 150 volts are impressed on an RC circuit with resistance of 50 ohms and capacitance of 10^3 farads. If $q(0) = 0$, find the charge $q(t)$ on the capacitor. What happens to the charge as $t \to \infty$? (**Hint:** See Exercise 62 of Section 8.1.)

52.* Find the current in the RL circuit of Example 5 if the impressed voltage (also called electromotive force) is $V(t) = V\sin(\omega t)$.

53. Antibiotics are taken by a patient at the rate of d milligrams per day. Assume that the drug is removed by the body from the bloodstream at a rate proportional to the amount present.

    a. Find the differential equation satisfied by $A(t)$, the amount of antibiotics present in the bloodstream.

    b. Solve for $A(t)$ assuming $A(0) = 0$ and determine what happens as $t \to \infty$.

54.* Find the current induced by a discharging capacitor in an RC circuit if $V(t) = 0$ and $I(0) = I_0$. (**Hint:** Consider the differential equation satisfied by $q(t)$, and start by differentiating both sides of the equation.)

55–58 *True or False?* Determine whether the given statement is true or false. In case of a false statement, explain or provide a counterexample.

55. A linear differential equation cannot be separable.

56. The equation $y\,dx + 3x\,dy - 2y\,dy = y^4\,dy$ is linear.

57. The only way to solve a linear differential equation is by the use of an integrating factor.

58. The integrating factor for a standard-form linear differential equation is the function $I(x) = e^{\int P(x)\,dx}$.

8.2 Technology Exercises

59–69. Use a computer algebra system to solve the equations in Exercises 15–25. Compare the results with the ones you obtained by hand.

70. A skydiver jumps out of a plane at 2000 meters and deploys his chute after 10 seconds of free fall. The total mass of the diver and his gear is 80 kg. Assume that air resistance is proportional to velocity both before and after deploying the chute, with respective constants of proportionality of 8 and 100. Use a computer algebra system to create a model to find how long after the jump the skydiver will land. (Distance is measured in meters, time in seconds.)

8.3 Autonomous Differential Equations and Slope Fields

TOPICS

1. Autonomous differential equations
2. Modeling with autonomous differential equations
3. Euler's method

We continue our classification of differential equations in this section by introducing a category whose solutions have noteworthy features in common. These equations also lend themselves to a type of graphical analysis not yet seen in this text.

TOPIC 1 Autonomous Differential Equations

Recall that in general a first-order differential equation can be written in the form $y' = f(x, y)$. Our next classification focuses on those first-order equations for which f is a function of y alone.

Definition 💡

Autonomous Differential Equations

A first-order differential equation in the dependent variable y is **autonomous** if it can be written in the form $y' = f(y)$. Those values of y for which $f(y) = 0$ are called **equilibrium points** or **rest points**.

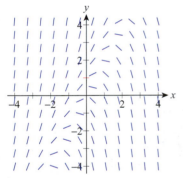

Figure 1

Slope Field for $y' = y - 2x$

At this point, it is useful to introduce a new way of visualizing the meaning of a differential equation and its solutions. Given a first-order equation $y' = f(x, y)$, imagine drawing a small line segment at the point (x_0, y_0) whose slope is $f(x_0, y_0)$. Since y' is the rate of change of the function y with respect to x, each small segment indicates the behavior of the solution y that passes through the specific point (x_0, y_0). In other words, the equation $y' = f(x, y)$ describes a picture of its own general solution from the outset, before we have even begun to solve it. Such an image of segments is called a **slope field** (or **direction field**) corresponding to the equation. Figure 1 shows a slope field for the differential equation $y' = y - 2x$; study this figure and verify that each small segment starting from a given point (x, y) appears to have slope roughly equal to $y - 2x$.

Now picture the slope field for a typical autonomous equation $y' = f(y)$. Since $f(y)$ has the same value for all points in the plane that have the same y-coordinate, we can use any one particular solution to generate another simply by shifting it horizontally to the left or right. This characteristic of autonomous differential equations is illustrated by our first example.

Example 1 ✎

Solve the autonomous equation $y' = (y + 2)(y - 1)$. Then graph its slope field and several of its particular solutions.

Solution

This autonomous equation is also separable—it is not unusual to find that a given differential equation falls into several classifications. The fact that it is autonomous tells us something about the behavior of its solutions, while the fact that it is separable guides us toward a solution method.

In separated form, the equation is

$$\frac{dy}{(y+2)(y-1)} = dx \qquad y \neq -2, 1$$

and we can integrate the left-hand side using the partial fractions method (see Section 7.2).

$$\int \frac{dy}{(y+2)(y-1)} = \int dx$$

$$\int \left(\frac{-\frac{1}{3}}{y+2} + \frac{\frac{1}{3}}{y-1} \right) dy = x + C$$

$$\frac{1}{3} \left(\ln|y-1| - \ln|y+2| \right) = x + C$$

$$\frac{1}{3} \ln \left| \frac{y-1}{y+2} \right| = x + C$$

This family of equations implicitly defines the general solution of the differential equation, except for two important cases: $y = -2$ and $y = 1$. By definition, these two values are equilibrium points of the autonomous equation, but the equations $y = -2$ and $y = 1$ also qualify as solutions of the equation $y' = (y+2)(y-1)$ (as you can check); they are graphed as black lines in Figure 3. But while these two values of y are both equilibrium values, $y = -2$ is classified as a **stable equilibrium** and $y = 1$ is an **unstable equilibrium**.

The meaning of these terms is that solutions close (but not equal) to $y = -2$ tend to approach it as $x \to \infty$ while solutions close (but not equal) to $y = 1$ move away from it as $x \to \infty$. The different behavior of these two equilibria is actually apparent from the slope field, once we have sketched it (note how, scanning Figure 2 from left to right, the slopes tend toward the line $y = -2$ and away from the line $y = 1$).

How do we efficiently and reasonably make a sketch such as the one shown in Figure 2? To begin with, we know that $y' = 0$ when $y = -2$ and $y = 1$, meaning particular solutions passing through these two points on the y-axis have slope 0 everywhere. Therefore we draw there small horizontal segments. So the two black horizontal lines in Figure 3, representing two particular solutions, can be graphed immediately.

Elsewhere in the plane, note that $y' = (y+2)(y-1) > 0$ at any point (x_0, y_0) where $y_0 > 1$, meaning the line tangent to the particular solution through (x_0, y_0) will have positive slope. Similarly, $y' = (y+2)(y-1) < 0$ at any point (x_0, y_0) where $-2 < y_0 < 1$, so the short line segments in our slope field sketch should have negative slope. Finally, $y' = (y+2)(y-1) > 0$ is positive again at any point (x_0, y_0) where $y_0 < -2$, indicating positive-slope trend lines again. Figure 2 summarizes these observations graphically.

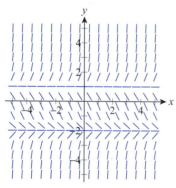

Figure 2

Slope Field for $y' = (y+2)(y-1)$

If desired, we can improve the accuracy of our slope field sketch if we include information about y''.

$$y' = (y+2)(y-1) = y^2 + y - 2$$

$$y'' = \frac{d}{dx}(y') = 2yy' + y' = y'(2y+1) = (y+2)(y-1)(2y+1)$$

This allows us to indicate the concavity (upward or downward) of the slope field trend lines in our sketch in the four horizontal bands defined by $y = -2$, $y = -\frac{1}{2}$, and $y = 1$.

Our last task is to graph several of the particular solutions of our autonomous equation. In Figure 3, the graph shown in green corresponds to $C = -3$ and the graph in red corresponds to $C = 2$. These particular solutions are discontinuous at $y = -2$ and $y = 1$; in fact, these two horizontal lines are asymptotes of every particular solution.

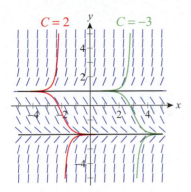

Figure 3

Slope Field and Representative Particular Solutions

TOPIC 2 Modeling with Autonomous Differential Equations

A good number of important physical phenomena can be modeled with autonomous differential equations, as illustrated in the next few examples.

Example 2 ✎

Newton's Law of Cooling was first seen in Exercise 9 of Section 3.7—it states that the rate of change of temperature of an object with temperature $T(t)$ at time t is proportional to $T - T_s$, where T_s is the temperature of the object's surroundings (we assume that T_s remains constant).

One recipe for homemade yogurt calls for heating 2 quarts of milk to 180 °F and then letting it cool to 115 °F before adding the yogurt cultures. If it takes 5 minutes for 2 quarts of milk to cool from 180 °F to 170 °F in a room held at 75 °F, how long will it take for the milk to cool to 115 °F?

Solution

We first solve the autonomous equation for temperature in full generality.

$$\frac{dT}{dt} = k(T - T_s) \qquad \text{\textcolor{blue}{k is the constant of proportionality.}}$$

$$\int \frac{dT}{T - T_s} = \int k\,dt$$

$$\ln|T - T_s| = kt + C$$

$$T = T_s + Ae^{kt} \qquad \text{\textcolor{blue}{$A = e^C$ and we assume $T > T_s$.}}$$

If $T_s = 75$ and $T(0) = 180$, then $A = 105$ and the only remaining undetermined constant is k. We can solve for k by using the fact that $T(5) = 170$.

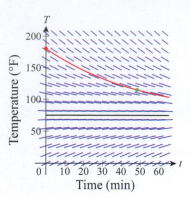

Figure 4
Newton's Law of Cooling

$$75 + 105e^{k(5)} = 170$$

$$e^{5k} = \frac{95}{105}$$

$$k = \frac{1}{5}\ln\frac{95}{105} \approx -0.02002$$

Then $T(t) = 75 + 105e^{-0.02002t}$ and we solve the equation $T(t) = 115$ to find how long it will take for the milk to cool to 115 °F.

$$115 = 75 + 105e^{-0.02002t} \quad\Rightarrow\quad t = -\frac{1}{0.02002}\ln\frac{40}{105} \approx 48.2 \text{ min}$$

Figure 4 illustrates the slope field for this equation and, in red, the particular solution $T(t) = 75 + 105e^{-0.02002t}$ we have just found. The stable equilibrium solution of $T = 75$ is shown as a solid black line.

Example 3 ✎

The simplest model of population growth is based on the equation $dP/dt = kP$, where $P(t)$ denotes the population of a group at time t and k is a proportionality constant dependent on the circumstances (see Section 3.7). This model leads to exponential growth $P(t) = P_0 e^{kt}$, where $P_0 = P(0)$; this is a formula that is usually only valid for short periods of time and when there are no significant environmental barriers to population growth.

A more realistic model is based on the **logistic differential equation**

$$\frac{dP}{dt} = kP\left(1 - \frac{P}{K}\right),$$

where K is called the population's **carrying capacity**, which represents the maximum population size that the environment can support. Solve this autonomous equation to find the general formula for the **logistic growth** of a population.

Solution

As in the last example, we can solve this autonomous equation by separating variables.

$$\int \frac{dP}{P\left(1 - \dfrac{P}{K}\right)} = \int k\,dt$$

$$\int \left(\frac{1}{P} + \frac{1}{K - P}\right)dP = \int k\,dt \qquad\qquad \frac{1}{P\left(1 - \dfrac{P}{K}\right)} = \frac{K}{P(K - P)} = \frac{1}{P} + \frac{1}{K - P}$$

$$\ln\left|\frac{P}{K - P}\right| = kt + C$$

$$\frac{P}{K - P} = Ae^{kt} \qquad\qquad A = \pm e^{C}$$

We will solve this equation for P, but first let's determine the constant A that corresponds to an initial population of P_0. Note that when $t = 0$, $A = P_0/(K - P_0)$. Now, we solve for P.

$$\frac{P}{K - P} = Ae^{kt}$$

$$P\left(1 + Ae^{kt}\right) = KAe^{kt}$$

$$P = \frac{KAe^{kt}}{1 + Ae^{kt}}$$

It is customary to divide the top and bottom of this last fraction by Ae^{kt} so as to have only one term containing the independent variable. In the process of doing so, we will let $B = 1/A = (K - P_0)/P_0$.

$$P = \frac{KAe^{kt}}{1 + Ae^{kt}} = \frac{K}{\left(\dfrac{1}{A}\right)e^{-kt} + 1} = \frac{K}{1 + Be^{-kt}}$$

The physical meaning of the carrying capacity is that it serves as an asymptotic limit to the population growth. It is also the case that $P = K$ is a stable equilibrium solution of the equation, as indicated in Figure 5, and that $P = 0$ is an unstable equilibrium solution. One particular solution is shown in red in this slope field illustration—such a curve is called a **sigmoid curve** because it is shaped like the letter S.

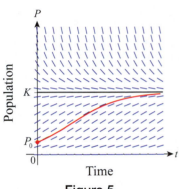

Figure 5

Logistic Growth of a Population

Example 4 ✎

Recall from Example 3 of Section 6.6 that the velocity $v(t)$ of a body falling under the influence of gravity and encountering air resistance proportional to v^2 satisfies the following equation.

$$m\frac{dv}{dt} = mg - kv^2$$

Use the methods of this section to find the terminal velocity of such a body, without actually solving the differential equation.

Solution

Our equation is autonomous, and from the form

$$\frac{dv}{dt} = g - \frac{k}{m}v^2$$

we know that it will have an equilibrium point when $g - (k/m)v^2 = 0$, or $v = \sqrt{mg/k}$ (we discard the negative root since it has no physical meaning in this problem). Further, we can determine that this equilibrium value is stable, since v' is positive if v is less than $\sqrt{mg/k}$ and negative if v is greater than $\sqrt{mg/k}$. A slope field sketch, with the stable equilibrium solution $v = \sqrt{mg/k}$ shown in black and a typical nonequilibrium solution in red, is depicted in Figure 6.

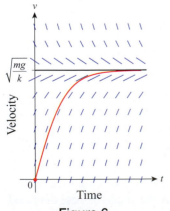

Figure 6

Terminal Velocity

TOPIC 3 Euler's Method

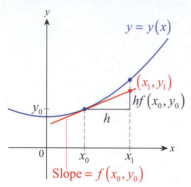

$y = y(x)$

(x_1, y_1)

$hf(x_0, y_0)$

y_0

h

0 x_0 x_1 x

Slope $= f(x_0, y_0)$

Figure 7

The basic idea behind sketching slope fields is also the key to a method of approximating solutions to first-order initial value problems; the method is named in honor of the great Swiss mathematician Leonhard Euler (1707–1783).

Suppose we want a solution y to $y' = f(x, y)$ with the condition that $y(x_0) = y_0$; in other words, we want a particular solution that passes through the point (x_0, y_0). Knowing that the tangent line to the solution curve has slope $y'(x_0) = f(x_0, y_0)$, we can find a point (x_1, y_1) on the tangent line which will approximate a second point on the solution curve. If we specify that $x_1 - x_0 = h$, then

$$f(x_0, y_0) \approx \frac{y_1 - y_0}{x_1 - x_0} = \frac{y_1 - y_0}{h}$$

so we define $y_1 = y_0 + hf(x_0, y_0)$. Iterating the process, we define the following:

$$x_n = x_{n-1} + h = x_0 + nh \quad \text{and} \quad y_n = y_{n-1} + hf(x_{n-1}, y_{n-1})$$

Example 5 ✎

Use Euler's method to approximate a solution to the IVP $y' = y - 2x$ with the condition $y(0) = 1$.

Solution

We have $x_0 = 0$, $y_0 = 1$, and $f(x, y) = y - 2x$. We need to choose a *step size* h in order to apply the iterative approximations $x_n = x_0 + nh$ and $y_n = y_{n-1} + h(y_{n-1} - 2x_{n-1})$. We will compare the results with $h = 0.5$ and $h = 0.1$.

h = 0.5

n	x_n	y_n
0	0	1
1	0.5	1.5
2	1.0	1.75
3	1.5	1.625
4	2.0	0.9375

Table 1

h = 0.1

n	x_n	y_n	n	x_n	y_n
0	0	1	11	1.1	1.3470
1	0.1	1.1	12	1.2	1.2617
2	0.2	1.19	13	1.3	1.1479
3	0.3	1.269	14	1.4	1.0027
4	0.4	1.3359	15	1.5	0.8230
5	0.5	1.3895	16	1.6	0.6053
6	0.6	1.4285	17	1.7	0.3458
7	0.7	1.4514	18	1.8	0.0404
8	0.8	1.4565	19	1.9	−0.3156
9	0.9	1.4422	20	2.0	−0.7272
10	1.0	1.4064			

Table 2

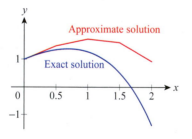

Approximate solution

Exact solution

Figure 8 $h = 0.5$

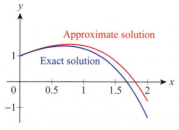

Approximate solution

Exact solution

Figure 9 $h = 0.1$

The tables show the results of calculating (x_n, y_n) up to $x = 2$ with the two different step sizes, and Figures 8 and 9 compare the graphs of these estimated points (in red) with the exact solution $y = 2x + 2 - e^x$ (which you will determine when you solve the first-order linear equation $y' = y - 2x$ in Exercise 25).

8.3 **Exercises**

1–6 Match the differential equation with its slope field (labeled A–F).

1. $y' = x$

2. $y' = 1 - yx$

3. $y' = \dfrac{y}{2x}$

4. $y' = \dfrac{x^2 y}{3}$

5. $y' = \sqrt{x^2 + y^2}$

6. $y' = y \sin 2x$

A.

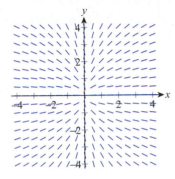

B.

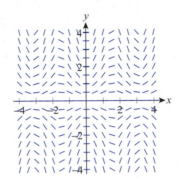

C.

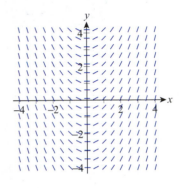

D.

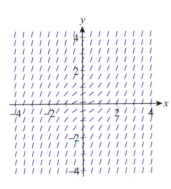

E.

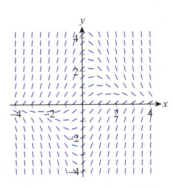

F.

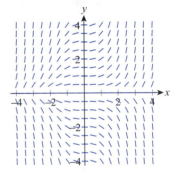

7–10 An autonomous equation and its slope field are given. Find any equilibrium solutions and classify them as stable or unstable.

7. $y' = y - \dfrac{y^2}{2}$

8. $y' = \dfrac{1 - y^2}{\sqrt{1 + y^2}}$

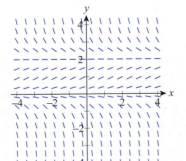

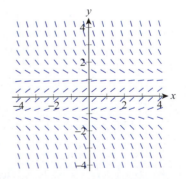

9. $y' = \cos y (1 - \sin y), \quad -4 \le y \le 4$

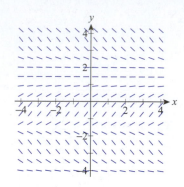

10. $y' = (y+3)(y^3 - 1)$

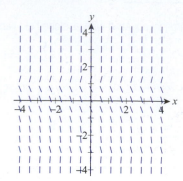

11–12 An autonomous equation and its slope field are given. Sketch the graphs of the particular solutions satisfying the specified initial conditions.

11. $y' = \dfrac{1}{4} y^2$

    a. $y(0) = -1$ **b.** $y(1) = 1$

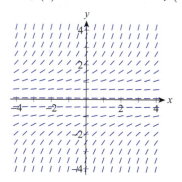

12. $y' = y^2 - 4$

    a. $y(0) = 0$ **b.** $y(4) = 4$

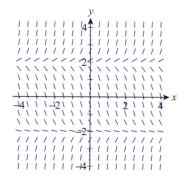

13–16 Graph by hand the slope field of the given differential equation. If applicable, find and classify each equilibrium solution as stable or unstable.

13. $y' = \dfrac{1}{3} y$ **14.** $y' = \left(2 - \dfrac{1}{2} y^2\right) y$

15. $y' = -y(1-y)(2-y)$ **16.** $y' = x - 3y$

17. Create a rough sketch of the slope field of the differential equation $y' = f(y)$, where the graph of f is given below. Classify equilibria as stable or unstable.

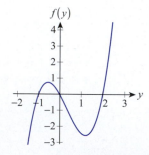

18. A can of soda that was forgotten on the kitchen counter and warmed up to 22 °C was put back in the refrigerator whose interior temperature is kept at a constant 3 °C. If the soda's temperature after 5 minutes is 17 °C, what will its temperature be after 20 minutes? Sketch the slope field resulting from your model; include the equilibrium and the soda's temperature curve.

19. A cake is removed from a 320 °F oven and is kept at a room temperature of 72 °F to cool down. The cake's temperature after 4 minutes is 190 °F. Use Newton's Law of Cooling to model the cooling process, and make a rough sketch of the slope field for your model, highlighting the equilibrium and the cake's temperature curve.

20. Recall from Exercise 79 in Section 7.2 that the spread of a disease in a community of N people can be modeled by the logistic differential equation $dI/dt = kI(N-I)$, where $I(t)$ stands for the number of persons already infected. Sketch the slope field and graph of this model, assuming a community of 200 people with one sick person initially, and five more catching the disease three days later.

21. An owl population grows logistically with a carrying capacity of 500 owls and constant of proportionality $k = 0.3$ per year.

  a. Find the population size $P(t)$ as a function of time if initially 100 owls are present in the ecosystem.

  b. How long does it take for the owl population to reach 300?

22. Use the method of Example 4 to answer the question in Exercise 63b of Section 8.1.

23. Repeat Exercise 22, this time assuming that the air resistance is proportional to $\sqrt{v}$.

24. Find the terminal velocity in Example 4 by solving the equation $\dfrac{dv}{dt} = g - \dfrac{k}{m}v^2$ and finding $\lim_{t\to\infty} v(t)$. (**Hint:** Begin with the two steps below.)

$$\frac{dv}{dt} = g\left(1 - \frac{k}{mg}v^2\right)$$

$$\frac{mg}{k}\left(\frac{dv}{\frac{mg}{k} - v^2}\right) = g\,dt$$

25. Verify that $y = 2x + 2 - e^x$ is the exact solution of the initial value problem of Example 5, that is, $y' = y - 2x$ with the condition $y(0) = 1$.

26–31 For the initial value problem **a.** use Euler's method with the indicated step sizes to approximate the given value of y and **b.** solve the IVP by conventional methods and compare your approximations with the exact answer.

26. $y' = 3y;\quad y(0) = 2;$
approximate $y(1)$ with (i) $h = 0.25$ (ii) $h = 0.125$

27. $y' = 2y + x;\quad y(0) = 1;$
approximate $y(1)$ with (i) $h = 0.25$ (ii) $h = 0.1$

28. $y' = xy;\quad y(0) = 1;$
approximate $y(2)$ with (i) $h = 0.4$ (ii) $h = 0.2$

29. $y' = x^2 - y;\quad y(0) = 3;$
approximate $y(1.5)$ with (i) $h = 0.3$ (ii) $h = 0.15$

30. $y' = 2x - 2y + 1;\quad y(0) = -1;$
approximate $y(1)$ with (i) $h = 0.25$ (ii) $h = 0.1$

31. $y' = 1 + y^2;\quad y(0) = 0;$
approximate $y\left(\dfrac{\pi}{3}\right)$ with (i) $h = \dfrac{\pi}{12}$ (ii) $h = \dfrac{\pi}{24}$

32–33. The following formula is called the improved Euler's method, or Heun's method.

$$x_n = x_{n-1} + h = x_0 + nh \text{ and}$$

$$y_n = y_{n-1} + h\frac{f(x_{n-1}, y_{n-1}) + f(x_n, y_n^*)}{2}, \text{ where}$$

$$y_n^* = y_{n-1} + hf(x_{n-1}, y_{n-1})$$

(Note that the quotient appearing in this formula can be interpreted as the average slope between x_{n-1} and x_n. The derivation of Heun's method is left for a textbook on differential equations or numerical methods.)

Use Heun's method to redo Exercises 26–27 with $h = 0.25$ and compare your result with your earlier answer to illustrate the accuracy of the improved Euler's method.

8.3 Technology Exercises

34–37. Use a computer algebra system to create the slope fields you sketched in Exercises 13–16. If applicable, visually check the location of, and classify, any equilibria.

38–43. Use a computer algebra system to improve your approximations in Exercises 26–31, using Euler's method with 20 equal increments. Then graph the results along with the exact solutions to visually check the accuracy of the method.

8.4 Second-Order Linear Differential Equations

TOPICS

1. Solving homogeneous second-order linear differential equations

2. Initial value problems and boundary value problems

We finish this introductory chapter on differential equations with a brief glimpse into further areas of study. The methods used to solve differential equations expand as the order of the derivatives and the number of variables increases. Here, we take the first step by studying linear equations with derivatives up to order two.

TOPIC 1 Solving Homogeneous Second-Order Linear Differential Equations

In full generality, a **second-order linear differential equation** is one that can be written in the standard form

$$P(x)y'' + Q(x)y' + R(x)y = F(x),$$

where P, Q, R, and F are all continuous on some interval of interest. If $F(x) = 0$ for all x on the interval, the equation is said to be **homogeneous**; otherwise, F is not identically 0 and the equation is **nonhomogeneous**. As it turns out, one step in solving nonhomogeneous equations is to first solve the associated homogeneous equation—the general solution of the homogeneous version is part of the solution of the nonhomogeneous equation. The techniques used to tackle nonhomogeneous equations are best left to a course on differential equations, so we concentrate here on cases where $F(x) = 0$.

We begin with an important and easily proved theorem that is central to the study of differential equations.

Theorem 🔍

Linear Combinations of Solutions Are Solutions

If $y_1(x)$ and $y_2(x)$ are both solutions of $P(x)y'' + Q(x)y' + R(x)y = 0$, then so is $c_1 y_1(x) + c_2 y_2(x)$ for any choice of real numbers c_1 and c_2.

(The expression $c_1 y_1(x) + c_2 y_2(x)$ is a *linear combination* of the two solutions.)

Proof ✎

Since both y_1 and y_2 are solutions, we have the following.

$$P(x)y_1'' + Q(x)y_1' + R(x)y_1 = 0 \text{ and } P(x)y_2'' + Q(x)y_2' + R(x)y_2 = 0$$

The proof then follows immediately from properties of the derivative.

$$P(x)(c_1 y_1 + c_2 y_2)'' + Q(x)(c_1 y_1 + c_2 y_2)' + R(x)(c_1 y_1 + c_2 y_2)$$
$$= P(x)(c_1 y_1'' + c_2 y_2'') + Q(x)(c_1 y_1' + c_2 y_2') + R(x)(c_1 y_1 + c_2 y_2)$$
$$= c_1 \left[P(x)y_1'' + Q(x)y_1' + R(x)y_1 \right] + c_2 \left[P(x)y_2'' + Q(x)y_2' + R(x)y_2 \right]$$
$$= c_1 \cdot 0 + c_2 \cdot 0$$
$$= 0$$

The preceding theorem gives us a way to create new solutions out of existing solutions. A second important theorem tells us how to describe *all* of the possible solutions of a homogeneous second-order linear equation. Its proof (which will be left to a course devoted to differential equations) uses the following idea from the study of vector spaces.

Definition 💡

Linear Independence

Two functions $y_1(x)$ and $y_2(x)$ are **linearly independent** on an interval if neither can be written as a constant multiple of the other. Another way of expressing linear independence is to say that the only way for a linear combination $c_1 y_1(x) + c_2 y_2(x)$ to equal the identically 0 function is when $c_1 = c_2 = 0$.

Theorem 🔎

Linearly Independent Solutions Span All Solutions

If $y_1(x)$ and $y_2(x)$ are linearly independent solutions of the equation $P(x)y'' + Q(x)y' + R(x)y = 0$ on an interval, and if $P(x)$ is never 0 on the interval, then every solution of the equation can be written as $c_1 y_1(x) + c_2 y_2(x)$ for some choice of real numbers c_1 and c_2. We say that $y_1(x)$ and $y_2(x)$ *span* the space of solutions, and we can write the general solution of the equation as $c_1 y_1(x) + c_2 y_2(x)$.

Narrowing our focus one last time, we now devise a method for solving all homogeneous second-order linear differential equations with constant coefficients. Such an equation can be written in the form $ay'' + by' + cy = 0$, where a, b, and c are real numbers. The method originated with Euler, who realized that for y to satisfy the equation $ay'' + by' + cy = 0$, y and its derivatives must be similar to one another. For if the graphs of y, y', and y'' had very different shapes, it would be impossible for some linear combination of them to equal the identically 0 function over some interval—while it is easy to solve the equation $ay''(x_0) + by'(x_0) + cy(x_0) = 0$ for any one point x_0, we need the equation to be true for *all* x in the interval of interest.

Exponential functions have exactly the property we desire: the derivative of any such function is a constant multiple of itself. So we start by letting $y = e^{rx}$, where r is a constant to be determined later (we could also use any other exponential base, but e is the easiest to work with). Substituting $y = e^{rx}$ into the equation $ay'' + by' + cy = 0$, we get the following.

$$a\left(e^{rx}\right)'' + b\left(e^{rx}\right)' + ce^{rx} = 0$$
$$ar^2 e^{rx} + bre^{rx} + ce^{rx} = 0$$
$$e^{rx}\left(ar^2 + br + c\right) = 0$$

And since $e^{rx} \neq 0$ for all x, we divide both sides by e^{rx} to obtain the following **characteristic equation**.

$$ar^2 + br + c = 0$$

Now we determine the constant r by solving the characteristic equation of the given homogeneous second-order linear equation. There are three cases to consider.

Case 1: $b^2 - 4ac > 0$. In this case there are two distinct solutions of $ar^2 + br + c = 0$; call them r_1 and r_2. Since $e^{r_1 x}$ and $e^{r_2 x}$ are linearly independent functions (neither is a constant multiple of the other), our general solution is as follows.

$$y = c_1 e^{r_1 x} + c_2 e^{r_2 x}$$

Example 1 ✒

Solve the equation $y'' + 2y' - 15y = 0$.

Solution

Our characteristic equation is $r^2 + 2r - 15 = 0$, with solutions $r = 3$ and $r = -5$. So the general solution of the differential equation is as follows.

$$y = c_1 e^{3x} + c_2 e^{-5x}$$

Case 2: $b^2 - 4ac = 0$. The quadratic formula tells us that $r = -b/(2a)$ is a double root of $ar^2 + br + c = 0$ when $b^2 - 4ac = 0$. So while $y = e^{rx}$ is one solution of the differential equation, we can't generate a second linearly independent solution the same way we did in Case 1. But we *can* verify that $y = xe^{rx}$ is a solution, and clearly e^{rx} and xe^{rx} are not constant multiples of one another on any interval.

$$
\begin{aligned}
a\left(xe^{rx}\right)'' + b\left(xe^{rx}\right)' + cxe^{rx} &= a\left(2re^{rx} + r^2 xe^{rx}\right) + b\left(e^{rx} + rxe^{rx}\right) + cxe^{rx} \\
&= \left(2ar + b\right)e^{rx} + \left(ar^2 + br + c\right)xe^{rx} \\
&= \left(0\right)e^{rx} + \left(0\right)xe^{rx} = 0 \qquad \begin{array}{l} r = -b/(2a) \\ r \text{ is a root of } ar^2 + br + c. \end{array}
\end{aligned}
$$

So the general solution of such a differential equation is as follows.

$$y = c_1 e^{rx} + c_2 xe^{rx}$$

Example 2 ✒

Solve the equation $y'' - 4y' + 4y = 0$.

Solution

Our characteristic equation is $r^2 - 4r + 4 = 0$, with the double root $r = 2$. So the general solution of the differential equation is as follows.

$$y = c_1 e^{2x} + c_2 xe^{2x}$$

Case 3: $b^2 - 4ac < 0$. There are two distinct roots again in this case, but they are complex numbers and conjugates of each other. We can write them as $r_1 = \alpha + i\beta$ and $r_2 = \alpha - i\beta$, where $\alpha = -b/(2a)$ and $\beta = \sqrt{4ac - b^2}\big/(2a)$. One way to write the general solution is as follows.

$$y = k_1 e^{r_1 x} + k_2 e^{r_2 x} = k_1 e^{(\alpha+i\beta)x} + k_2 e^{(\alpha-i\beta)x}$$

But we can also use Euler's Formula $e^{i\theta} = \cos\theta + i\sin\theta$ to express the general solution as follows.

$$
\begin{aligned}
y &= k_1 e^{(\alpha+i\beta)x} + k_2 e^{(\alpha-i\beta)x} \\
&= k_1 e^{\alpha x} e^{i\beta x} + k_2 e^{\alpha x} e^{-i\beta x} \\
&= k_1 e^{\alpha x}\left[\cos(\beta x) + i\sin(\beta x)\right] + k_2 e^{\alpha x}\left[\cos(\beta x) - i\sin(\beta x)\right] \\
&= e^{\alpha x}\left[(k_1 + k_2)\cos(\beta x) + i(k_1 - k_2)\sin(\beta x)\right] \\
&= e^{\alpha x}\left[c_1 \cos(\beta x) + c_2 \sin(\beta x)\right]
\end{aligned}
$$

$c_1 = k_1 + k_2$
$c_2 = i(k_1 - k_2)$

Example 3 ✎

Solve the equation $y'' - 6y' + 10y = 0$.

Solution

Our characteristic equation is $r^2 - 6r + 10 = 0$, with the following roots.

$$r = \frac{-(-6) \pm \sqrt{(-6)^2 - 4(1)(10)}}{2(1)} = 3 \pm i$$

So $\alpha = 3$, $\beta = 1$, and the general solution of the differential equation is as follows.

$$y = e^{3x}\left(c_1 \cos x + c_2 \sin x\right)$$

TOPIC 2   Initial Value Problems and Boundary Value Problems

We have already studied first-order initial value problems, and we will now see how IVPs appear in the context of second-order differential equations. We can also now discuss differential equations that must satisfy conditions referred to as boundary conditions.

Definition 💡

Second-Order Initial Value and Boundary Value Problems

An equation $P(x)y'' + Q(x)y' + R(x)y = F(x)$ with the additional constraints that $y(x_0) = y_0$ and $y'(x_0) = y_1$ is called a **second-order initial value problem (IVP)**, so named because y and y' must satisfy the given initial conditions at the point $x = x_0$.

An equation $P(x)y'' + Q(x)y' + R(x)y = F(x)$ with the additional constraints that $y(x_0) = y_0$ and $y(x_1) = y_1$ is called a **second-order boundary value problem (BVP)**. The distinction is that we seek a solution y satisfying two *boundary conditions* at the points $x = x_0$ and $x = x_1$. Note that there is no additional constraint on y'.

Another important distinction between the two types of second-order problems is that, under the assumptions that P, Q, R, and F are continuous on a given interval and that P is never 0 on the interval, an IVP is guaranteed to have a solution while a BVP is not. We conclude this section with examples of how we solve IVPs and BVPs.

Example 4 ✒

An object of mass m is suspended at one end of a spring with spring constant k while the other end of the spring is firmly fixed in place. If the object is then set in motion, its vertical displacement y at time t satisfies the equation

$$m\frac{d^2y}{dt^2} + ky = 0,$$

assuming friction and air resistance are negligible. (This follows from Newton's Second Law of Motion and Hooke's Spring Law.) Find the displacement $y(t)$ of such an object if its mass is 2 kg, the spring constant $k = 100$ N/m, the spring is stretched 0.2 m beyond its natural length at time $t = 0$, and the object is released at time $t = 0$ with a velocity of 0 m/s.

Solution

We first find the general solution of the equation $my'' + ky = 0$. The associated characteristic equation is $mr^2 + k = 0$, which has solutions $r = \pm i\sqrt{k/m}$. If we let $\omega = \sqrt{k/m}$ (a common choice of notation for this kind of problem), then $r = \pm i\omega$ and the general solution of our differential equation is as follows.

$$y(t) = e^{0 \cdot t}\left[c_1\cos(\omega t) + c_2\sin(\omega t)\right] = c_1\cos(\omega t) + c_2\sin(\omega t)$$

An object whose motion is described by such a function is said to exhibit simple harmonic motion.

We can now solve for c_1 and c_2 by applying the initial conditions $y(0) = 0.2$ and $y'(0) = 0$ (corresponding to the spring being stretched 0.2 m at time 0 and the object having an initial velocity of 0 m/s).

$$0.2 = y(0) = c_1\cos(\omega \cdot 0) + c_2\sin(\omega \cdot 0) = c_1$$
$$0 = y'(0) = -c_1\omega\sin(\omega \cdot 0) + c_2\omega\cos(\omega \cdot 0) = c_2\omega$$

Since $\omega \ne 0$ we have $c_2 = 0$.

So the particular solution of our IVP is as follows.

$$y(t) = 0.2\cos(\omega t) = 0.2\cos\left(5\sqrt{2}t\right) \qquad \omega = \sqrt{k/m} = 5\sqrt{2}$$

A graph of this vertical displacement over time is shown in Figure 2, where the vertical axis measures the distance below the resting position of the object.

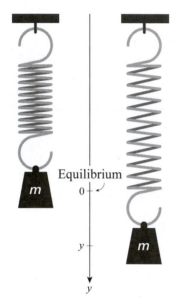

Figure 1 Mass and Spring

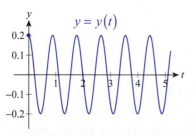

Figure 2 Simple Harmonic Motion

Example 5 ✎

If frictional forces proportional to the velocity of the object are considered in the mass-spring model of Example 4, the displacement $y(t)$ of the object satisfies the equation

$$m\frac{d^2y}{dt^2} + c\frac{dy}{dt} + ky = 0,$$

where c is a positive constant called the **damping constant**. Find the displacement of such an object under the same conditions as Example 4, but with the damping constant $c = 4$ kg/s.

Solution

In this model, the characteristic equation is $mr^2 + cr + k = 0$, with the following roots.

$$r = \frac{-c \pm \sqrt{c^2 - 4mk}}{2m}$$

The three cases of $c^2 - 4mk > 0$, $c^2 - 4mk = 0$, and $c^2 - 4mk < 0$ are referred to as *overdamped*, *critically damped*, and *underdamped*, and correspond to the three cases we have outlined for solutions of a generic constant-coefficient homogeneous second-order equation.

In this example, the model is underdamped (since $c^2 - 4mk = 4^2 - 4 \cdot 2 \cdot 100 < 0$), and the general solution of the equation is

$$y(t) = e^{-\frac{c}{2m}t}\left[c_1 \cos(\omega t) + c_2 \sin(\omega t) \right]$$

with $\omega = \sqrt{4mk - c^2}/(2m)$. The initial condition $y(0) = 0.2$ again implies that $c_1 = 0.2$, but now

$$y'(t) = e^{-\frac{c}{2m}t}\left[\left(-\frac{c}{2m}c_1 + c_2\omega \right)\cos(\omega t) + \left(-\frac{c}{2m}c_2 - c_1\omega \right)\sin(\omega t) \right]$$

so $y'(0) = 0$ implies

$$0 = e^{-\frac{c}{2m}(0)}\left[\left(-\frac{c}{2m}c_1 + c_2\omega \right)\cos(\omega \cdot 0) + \left(-\frac{c}{2m}c_2 - c_1\omega \right)\sin(\omega \cdot 0) \right]$$

$$= -\frac{c}{2m}c_1 + c_2\omega$$

or $c_2 = \left[c/(2m\omega) \right]c_1$. Using $\omega = \sqrt{4 \cdot 2 \cdot 100 - 4^2}/(2 \cdot 2) = \sqrt{784}/4 = 7$ we obtain $c_2 = \left[4/(2 \cdot 2 \cdot 7) \right]0.2 \approx 0.03$.

Substituting values for c, m, c_1, c_2, and ω, we find the particular solution for this scenario.

$$y(t) \approx e^{-t}\left(0.2\cos 7t + 0.03\sin 7t \right)$$

The graph of this function appears in Figure 3.

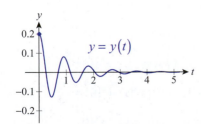

Figure 3
Underdamped Mass-Spring Model

Example 6 ✐

Solve the equation $y'' + y' - 2y = 0$ with the following boundary conditions.

a. $y(0) = 1, \quad y(2) = -1$ **b.** $y(0) = 0, \quad y(2) = 0$

Solution

The general solution of $y'' + y' - 2y = 0$ is $y(x) = c_1 e^x + c_2 e^{-2x}$, with the exponents determined by the characteristic equation $r^2 + r - 2 = 0$. We determine c_1 and c_2 from the two sets of boundary conditions.

a. The conditions $y(0) = 1$ and $y(2) = -1$ imply

$$1 = y(0) = c_1 + c_2$$
$$-1 = y(2) = c_1 e^2 + c_2 e^{-4}.$$

Solving the second equation for c_1 by substituting $c_2 = 1 - c_1$, and then using $c_2 = 1 - c_1$ to determine c_2, we get the following.

$$c_1 = \frac{-e^4 - 1}{e^6 - 1} \quad \text{and} \quad c_2 = \frac{e^6 + e^4}{e^6 - 1}$$

The graph of $y(x) = c_1 e^x + c_2 e^{-2x}$ for these particular choices of c_1 and c_2 is shown in Figure 4. Note that the graph does indeed satisfy the boundary conditions.

b. The boundary conditions $y(0) = 0$ and $y(2) = 0$ can only be imposed on the general solution $y(x) = c_1 e^x + c_2 e^{-2x}$ by letting $c_1 = 0$ and $c_2 = 0$, resulting in the trivial solution $y(x) = 0$.

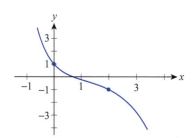

Figure 4
Boundary Value Problem Solution

8.4 Exercises

1–18 Find the general solution of the given differential equation.

1. $y'' + y' - 2y = 0$ **2.** $y'' - 4y = 0$

3. $3y'' - 5y' - 2y = 0$ **4.** $2y'' + 9y' - 5y = 0$

5. $y'' + 2y' + y = 0$ **6.** $2y'' - 12y' + 18y = 0$

7. $y'' + y' + \dfrac{y}{4} = 0$ **8.** $y'' + y = 0$

9. $y'' - 4y' + 13y = 0$ **10.** $y'' - 2y' + 5y = 0$

11. $25y'' - 10y' + y = 0$ **12.** $3y'' - \sqrt{3}y' + y = 0$

13. $3\dfrac{d^2 y}{dx^2} - \sqrt{13}\dfrac{dy}{dx} + y = 0$

14. $\dfrac{d^2 S}{dt^2} - 12S = 0$

15. $2\dfrac{d^2 w}{dt^2} + 2\dfrac{dw}{dt} + w = 0$ **16.** $y'' + k^2 y = 0$

17. $y'' - k^2 y = 0$ **18.** $y'' - k^2 y' = 0$

19–26 Solve the given second-order initial value problem.

19. $y'' - 3y = 0; \quad y(0) = 5; \quad y'(0) = 0$

20. $y'' - 3y' - 4y = 0; \quad y(0) = 1; \quad y'(0) = -1$

21. $y'' - y' - 2y = 0; \quad y(0) = 4; \quad y'(0) = 5$

22. $y'' - 2y' + y = 0; \quad y(0) = -1; \quad y'(0) = 0$

23. $y'' - 2y' + 5y = 0; \quad y(0) = 1; \quad y'(0) = -3$

24. $y'' - 6y' + 18y = 0; \quad y(0) = 0; \quad y'(0) = 6$

25. $4y'' - 12y' = -9y; \quad y(0) = -1; \quad y'(0) = \frac{1}{2}$

26. $y'' - 10y' + 26y = 0; \quad y(0) = -2; \quad y'(0) = -7$

27–34 Solve the boundary value problem, if possible.

27. $y'' - 5y' = 0;\quad y(0) = 5 - 5e^5;\quad y(1) = 0$

28. $9y'' - 6y' + y = 0;\quad y(0) = -1;\quad y(1) = 0$

29. $y'' - y = 0;\quad y(0) = 1;\quad y(1) = e$

30. $y'' + 0.2y' + 0.01y = 0;\quad y(0) = 0;\quad y(2) = 4e^{-1/5}$

31. $y'' + 9y = 0;\quad y(0) = 0;\quad y\left(\dfrac{2\pi}{3}\right) = 1$

32. $y'' + 4y' = 0;\quad y(0) = -5;\quad y(3) = -5$

33. $y'' - 2y' + 17y = 0;\quad y(0) = 2;\quad y\left(\dfrac{\pi}{8}\right) = 1$

34. $4y'' - 4y' + 5y = 0;\quad y(0) = 1;\quad y\left(\dfrac{\pi}{2}\right) = 1$

35–38 Our techniques from this section easily generalize to higher-order homogeneous linear equations with constant coefficients. For example, the characteristic equation of

$$y''' - 3y'' + y' - 3y = 0$$

is the cubic polynomial

$$r^3 - 3r^2 + r - 3 = 0$$

with characteristic roots $r_1 = 3$ and $r_{2,3} = \pm i$. These give rise to the following general solution.

$$y = c_1 e^{3x} + c_2 \cos x + c_3 \sin x$$

In Exercises 35–38, use this generalized technique to find the general solution of the differential equation.

35. $y''' - 2y'' - 3y' = 0$

36. $y''' - 4y'' + 5y' - 2y = 0$

37. $y''' - y'' + 2y' - 2y = 0$

38. $y^{(4)} - 3y'' - 4y = 0$

39–42 The following exercises offer a glimpse into one way of handling certain nonhomogeneous linear equations. A more general version of the theorem below is proved in differential equations texts.

Theorem: If y_p is any given particular solution of the nonhomogeneous linear equation

$$ay'' + by' + cy = F(x) \qquad \text{(1a)}$$

on an interval I, and y_c (also called the complementary function) is the general solution of the associated homogeneous equation

$$ay'' + by' + cy = 0 \qquad \text{(1b)}$$

on the same interval, then the general solution of (1a) on I can be written as follows.

$$y = y_c + y_p$$

In other words, the general solution of (1a) is the sum of any one of its particular solutions, and the general solution of the associated homogeneous equation (1b).

For example, you can easily verify that $y_p = \frac{1}{2}\sin x - \frac{1}{2}\cos x$ is a particular solution of the equation

$$y'' + y' = \cos x \qquad \text{(2)}$$

while the general solution of its associated homogeneous equation $y'' + y' = 0$ is $y_c = c_1 + c_2 e^{-x}$. Thus, the general solution of (2) is as follows.

$$y = y_c + y_p = c_1 + c_2 e^{-x} + \frac{1}{2}\sin x - \frac{1}{2}\cos x$$

But how can we find a particular solution y_p in order to put the theorem to work? One method that works well for certain equations is one we have already seen in Exercises 99–102 of Section 7.1—the method of undetermined coefficients. In the above case, knowing that all derivatives of the sine and cosine functions are again of the same type, we might guess that a particular solution of (2) has the following form:

$$y_p = A\cos x + B\sin x$$

Substituting this into (2) we obtain

$$(B - A)\cos x - (A + B)\sin x = \cos x,$$

which yields the coefficients $A = -\frac{1}{2}$ and $B = \frac{1}{2}$, and thus,

$$y_p = \frac{1}{2}\sin x - \frac{1}{2}\cos x.$$

In Exercises 39–42, use this theorem and the method of undetermined coefficients to find the general solution of the equation with the indicated initial "guess" for y_p.

39. $y'' + y' - 2y = x;\quad$ guess $y_p = Ax + B$

40. $y'' + 2y = x^2 + 1;\quad$ guess $y_p = Ax^2 + Bx + C$

41. $y'' + 4y' - 5y = \sin x;\quad$ guess $y_p = A\cos x + B\sin x$

42. $y'' + 4y = e^{3x};\quad$ guess $y_p = Ae^{3x}$

43–44 Use the previous method to solve the given initial value problem.

43. $y'' + y = x^2$; $y(0) = -1$; $y'(0) = 1$;
guess $y_p = Ax^2 + Bx + C$

44. $y'' - 2y' = 4\sin 2x$; $y(0) = 1$; $y'(0) = -2$;
guess $y_p = A\cos 2x + B\sin 2x$

45. Show that the only solution of the boundary value problem

$$y'' + 4y = 0;   y(0) = 0;   y\left(\frac{\pi}{4}\right) = 0$$

is the trivial solution $y = 0$.

46. Show that the boundary value problem

$$y'' + 9y = 0;   y(0) = 0;   y\left(\frac{2\pi}{3}\right) = 0$$

has infinitely many solutions. (Contrast this with Exercise 31.)

47. Suppose that

$$2\frac{d^2 y}{dt^2} + c\frac{dy}{dt} + 5y = 0$$

is the equation of a damped oscillating motion. Find a value of c such that the motion is **a.** underdamped, **b.** critically damped, and **c.** overdamped. Using the case analysis of Example 5, explain why these terms are appropriate for each type of motion. (Answers to parts a. and c. will vary.)

48. The viscosity of hydraulic fluid in automobile shock absorbers will determine the value of c (see Example 5 or Exercise 47). Discuss which of the three cases is appropriate for the design of automobile shock absorbers and why. (Shock absorbers are designed for cars to prevent or "smooth out" wheel oscillations caused by uneven road surface.)

49. When suspended at one end of a spring, an object of mass 0.1 kg stretches the spring by 5 cm. Find the value of c such that the resulting motion is critically damped. (This value is called the *critical damping constant*. For a refresher on how to determine the spring constant, see Example 2 in Section 6.5.)

50. Suppose that a 5 kg object attached to a spring with spring constant $k = 13$ is pulled down 25 cm below equilibrium and released. Find and graph the displacement function if the surrounding medium offers resistance with a damping constant of $c = 2$ kg/s.

51. A 4-pound weight stretches a spring by 6 in., while the damping constant is $\frac{1}{4}$ slug/s. The weight is pulled down 9 in. below equilibrium and released with an upward velocity of 2 ft/s. Find the equation of motion, solve it for the displacement function, and graph your result. Use 1 slug ≈ 32 lb.

52. A 4 kg object stretches a spring by 12 cm. The object is then pushed upward from equilibrium by 20 cm and released. Find and graph the displacement function if the damping constant is $c = 3$ kg/s.

53. If we place the system of Exercise 52 in high-viscosity fluid with a damping constant of 75 kg/s, the motion will become overdamped. Find and graph the displacement function in this case.

54. Show that if the mass-spring model of Example 5 is critically damped or overdamped, that is, if $c^2 - 4mk \geq 0$, then the oscillating object cannot pass through the equilibrium more than once. (**Hint:** Show that the equation $y(t) = 0$ cannot have two or more solutions.)

55. Prove that $y = c_1\cosh(kx) + c_2\sinh(kx)$ is a general solution of the equation in Exercise 17. Show that the above family of functions is the same as the one you obtained in Exercise 17. (**Hint:** Start by showing that both $y_1(x) = \sinh(kx)$ and $y_2(x) = \cosh(kx)$ satisfy the differential equation, and argue that they are also linearly independent. Compare this answer to the one from Exercise 16.)

56. Show that if $y(t)$ is the solution of a damped mass-spring model

$$m\frac{d^2 y}{dt^2} + c\frac{dy}{dt} + ky = 0,$$

then

$$\lim_{t \to \infty} y(t) = 0.$$

(**Hint:** Handle the critically, under-, and overdamped cases separately, using the fact that all constants m, c, k are greater than 0 in the equation of motion. Note that the conclusion of this exercise is consistent with our everyday experience of damped oscillations "dying down" over time.)

57.* At an amusement park, a boat slides down a ramp and splashes into the water at a speed of 15 m/s. The resistance offered by the water is proportional to the boat's speed with a coefficient of 270 kg/s. If the combined mass of the boat and passengers is 300 kg, how long does it take for it to come to a complete stop? What distance will it travel in the water while slowing down? Consider the boat stopped if your model predicts a velocity less than 1 cm/s. (**Hint:** See Example 5.)

58.* An RLC circuit is a simple electric circuit with inductance L (in henries, H), resistance R (in ohms, Ω), and capacitance C (in farads, F). The differential equation describing an RLC circuit is as follows.

$$L\frac{d^2q}{dt^2} + R\frac{dq}{dt} + \frac{1}{C}q = V$$

(Note that $I(t) = dq(t)/dt$. See Exercise 62 in Section 8.1 and Example 5 in Section 8.2.)

Suppose that in an RLC circuit a switch is open (i.e., there is no current), $V = 0$, and the capacitor has an initial charge of 3 coulombs. Then at time $t = 0$ the switch is flipped closed. Find and graph the current $I(t)$ if the capacitance is $C = 10^{-2}$ F, the resistance is 1.5 Ω and the inductance is 0.1 H. What happens to $I(t)$ as $t \to \infty$?

Chapter 8
Review Exercises

1–4 Determine whether the differential equation is separable, linear, or autonomous; and find its order. (Note that more than one description may be applicable.)

1. $\dfrac{y'}{y+1} = y^3 - 2$ **2.** $xy\,dy = \sqrt{1-x^2}\,dx$

3. $xy' = 2y + x^4 e^x - y'$ **4.** $2y'' = y - 3y'$

5–6 Solve the differential equation with the indicated condition.

5. $\left(x^2 + 1\right)y' = x;\quad y(0) = 1$

6. $y'\sec x = \sin x;\quad y(0) = 0$

7–12 Solve the separable differential equation.

7. $xy' = 2y$ **8.** $y' = e^{-y}\cos x$

9. $x^2 y' = 2\sqrt{y}\,(x+1)$ **10.** $x^2\,du = u^2\,dx$

11. $x\,du = \left(u^2 - 1\right)dx$ **12.** $\dfrac{y'}{3} = e^{3x-y}$

13–16 Solve the given initial value problem.

13. $y' = \dfrac{9x^2 + 2x}{3y^2};\quad y(0) = 2$

14. $x' = x\sin t;\quad x(\pi/2) = 5$

15. $\dfrac{dy}{dt} = 4t\sqrt{y-2};\quad y(1) = 6$

16. $\dfrac{dy}{dx} = 2xy + y - 4x - 2;\quad y(0) = 0$

17–22 Solve the linear differential equation. (**Hint:** In some cases, x has to be the dependent variable in order for the equation to be linear.)

17. $y' - \dfrac{y}{x} = 0$ **18.** $xy' + y = x^2\sin x$

19. $xy' + 6y = 2$ **20.** $y' - 4xy = 2x$

21. $\left(y^2 + 1\right)dx + 2xy\,dy = 4y\,dy$

22. $dx + x\cot y\,dy = \cos y\,dy$

23–24 Find a first-order linear differential equation in standard form that has the given general solution. (**Hint:** Identify the integrating factor and "reverse" the solution technique discussed in this chapter.)

23. $y = Ce^{x^3} - \dfrac{1}{3}$ **24.** $y = \dfrac{C}{\sqrt{x^2 - 1}} - 1$

25–28 Solve the given initial value problem.

25. $xy' + (x-1)y = x^3;\quad y(1) = 1$

26. $\dfrac{dy}{dx} - 2xy = 2xe^{x^2};\quad y(0) = 2$

27. $\left(x^2 + 1\right)y' + xy = \dfrac{1}{x^2 + 1};\quad y(0) = -3$

28. $(\cos t)\dfrac{dy}{dt} + (\sin t)y = 1;\quad y(0) = 1$

29–32 Match the differential equation with its slope field (labeled A–D). Classify each equilibrium solution as stable or unstable.

29. $y' = y^2 + 2y$ **30.** $y' = 9y - y^3$ **31.** $y' = \dfrac{y^2 - y - 2}{y^2 + 1}$ **32.** $y' = y^4 - 8y$

A.

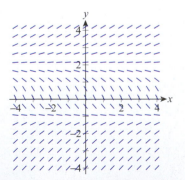

B.

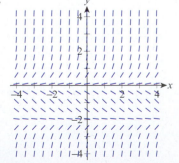

C.

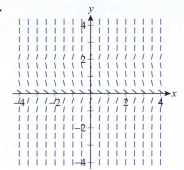

D.

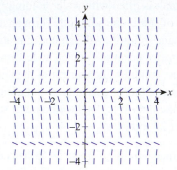

33–34 Graph by hand the slope field of the given differential equation. If applicable, find and classify each equilibrium solution as stable or unstable.

33. $y' = y^2 - 9$ **34.** $y' = y^3 + y^2 - 2y$

35. Create a rough sketch of the slope field of the differential equation $y' = f(y)$, where the graph of f is given below. Classify equilibria as stable or unstable.

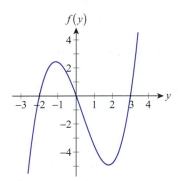

36–37 For the initial value problem **a.** use Euler's method with the indicated step sizes to approximate the given value of y and **b.** solve the IVP by conventional methods and compare your approximations with the exact value.

36. $y' = 2y + 1$; $y(0) = 1$;
approximate $y(1)$ with (i) $h = 0.2$ (ii) $h = 0.1$

37. $y' = y - x$; $y(0) = 2$;
approximate $y(2)$ with (i) $h = 0.5$ (ii) $h = 0.25$

38–43 Find the general solution of the second-order homogeneous linear equation.

38. $2y'' + 3y' - 2y = 0$ **39.** $y'' - 16y = 0$

40. $y'' - 2y' + 2y = 0$ **41.** $y'' - 5y' = 0$

42. $y'' + 9y = 0$ **43.** $4y'' - 4y' + y = 0$

44–47 Solve the given second-order initial value problem.

44. $9y'' - 6y' + y = 0$; $y(0) = 3$; $y'(0) = 2$

45. $y'' - y' - 6y = 0$; $y(0) = 3$; $y'(0) = -1$

46. $y'' + 2y' + 3y = 0$; $y(0) = 0$; $y'(0) = 2$

47. $2y'' - 3y' = 0$; $y(2) = -5$; $y'(2) = 3$

48–49 Solve the boundary value problem, if possible.

48. $2y'' + y = 0$; $y(0) = 2$; $y(\pi/\sqrt{2}) = -4$

49. $y'' + 4y = 0$; $y(0) = 0$; $y(\pi) = 1$

50. Find the general solution of the third-order equation $y''' - 3y'' - y' + 3y = 0$. (**Hint:** See Exercises 35–38 in Section 8.4.)

51. Find the general solution of the nonhomogeneous equation $2y'' - 7y' + 3y = \cos x$. As your initial guess, use $y_p = A \cos x + B \sin x$. (**Hint:** See Exercises 39–42 in Section 8.4.)

52–53 Determine the orthogonal trajectories of the family of curves, where a is an arbitrary nonzero constant.

52. $y = ax^4$ **53.** $y = \dfrac{ax}{\sqrt{x^2 + 1}}$

54. A 500-liter tank is filled with water holding 5 kilograms of salt in the solution. Through an inlet, a stronger solution with salt concentration of 0.05 kilograms per liter is being added at a rate of 16 liters per minute. The content of the tank is continuously and thoroughly mixed and drained out at the same rate. What is the amount of salt in the tank after 25 minutes?

55. Answer the question of Exercise 54 if the content of the tank is drained out at a rate of 20 liters per minute.

56.* A container in a lab contains 14 gallons of pure distilled water. 10% and 25% acid solutions are pumped into the container through two respective inlets. The 10% solution is flowing in at a rate of 0.2 gallons per minute, while the 25% solution is being allowed in by the second inlet at a rate of 0.5 gallons per minute. The content of the tank is continuously and thoroughly mixed and drained out at the rate of 0.7 gallons per minute. How long does it take to form 14 gallons of 14% solution in this way?

57. A hailstone is melting so that its volume $V(t)$ decreases at a rate proportional to its surface area.

    a. Assuming that the hailstone is nearly spherical, find a differential equation satisfied by $V(t)$.

    b. If a hailstone of diameter 1 inch loses 20% of its volume in half an hour, predict how long it takes for it to completely melt away. (Consider it melted away when your model predicts less than 1 percent remaining).

58. If a vertical cylindrical tank of radius $\frac{1}{2}$ meters and height 4 meters is initially full of water, but is draining through a circular orifice of diameter 2 centimeters that is on the bottom of the tank, what is the water level in the tank 2 minutes later? (**Hint:** See Exercise 59 in Section 8.1.)

59. Find the charge $q(t)$ of the 10^2-farad capacitor in an RC circuit if the impressed voltage on the circuit is $V(t) = t$ and the resistance is 25 ohms. Assume $q(0) = 0$.

60. Suppose that the impressed voltage in a simple RL circuit is $V(t) = 2t$, $I(0) = 0$, the inductance is 0.1 henries, and the resistance 0.5 ohms. Find the electric current I at time $t = 4$ seconds.

61. A baking dish is removed from a 210 °C oven and left at 20 °C room temperature. Two and a half minutes later the dish's temperature is 155 °C. Find the bakeware's temperature 10 minutes after it was removed from the oven.

62. A snapping turtle population grows logistically with a carrying capacity of 200 turtles and constant of proportionality $k = 0.2$ per year.

    a. Find the population size $P(t)$ as a function of time if initially 50 turtles are present in the habitat.

    b. How long does it take for the population to reach 100 turtles?

63. Suppose that an object of mass 200 grams stretches a spring by 10 centimeters. If it is pulled upwards to a position of 5 centimeters above equilibrium and released with a downward velocity of 1 m/s, find and graph the resulting displacement function, assuming that the surrounding medium offers resistance with a damping constant of $c = 0.5$ kg/s.

64–70 *True or False?* Determine whether the given statement is true or false. In case of a false statement, explain or provide a counterexample.

64. The equation $(y')^2 + xy' - 3y = 0$ is a first-order differential equation.

65. The equation $y' = -y$ is not linear.

66. If $y_1(x)$ and $y_2(x)$ are solutions of a homogeneous linear differential equation, then so is $3y_1(x) - 2y_2(x)$.

67. Only autonomous equations have slope fields.

68. The logistic equation discussed in this text is autonomous.

69. The equations $y = 2e^{x/2}$ and $y = xe^{x/2}$ are linearly independent solutions of $4y'' - 4y' + y = 0$.

70. A second-order BVP with two boundary conditions always has a solution.

Chapter 8
Technology Exercises

71–72. Use a computer algebra system to display the slope fields of the differential equations in Exercises 33 and 34. Compare the graphs to your original sketches.

73. Write a program for a computer algebra system that accepts a spring constant, a damping constant, and the mass of an oscillating object as inputs, and graphs the displacement function as output. Use it to check your answer for Exercise 63.

Project 📝

Chapter 8

Recall from Section 3.7 our discussion of a chemical reaction where reactants A and B produce a new product substance C, a process represented by

$$A + B \rightarrow C.$$

In this project, we will derive and use a differential equation that describes such a process.

1. Suppose that in the above reaction for each gram of reactant A, b grams of B are used to form C. If we start with initial amounts A_0 and B_0, respectively, and $X(t)$ denotes in grams the amount of substance C already formed at time t, find the amount remaining of reactants A and B at any time during the process.

2. Given that the rate of formation of substance C at any time is proportional to the product of the remaining amounts of reactants A and B, respectively, find a differential equation in terms of $X(t)$ that describes the process.

   (As in Question 1, let A_0 and B_0 stand for the initial amounts.)

3. Suppose a product substance C is being formed from reactant substances A and B and that for each gram of substance A, 3 grams of B are used to form C. As in Question 1, let $X(t)$ denote the amount of C formed at time t, and assume that the initial amounts of reactants A and B are $A_0 = 60$ grams and $B_0 = 40$ grams, respectively. Find the initial value problem describing this reaction. (**Hint:** Use your answer to Question 2.)

4. If 20 grams of the product compound forms during the first 5 minutes, use the model you obtained in Question 3 to predict how much of the product compound C is present 10 minutes into the process.

5. Use your model from Question 3 to predict what happens as $t \rightarrow \infty$. Interpret your answer.

Chapter 9

Parametric Equations and Polar Coordinates

9.1 Parametric Equations 689

1. Terminology of Parametric Equations
2. Applications of Parametric Equations

9.2 Calculus and Parametric Equations 702

1. Tangent Lines and Planar Areas
2. Arc Length and Surface Area

9.3 Polar Coordinates 713

1. The Polar Coordinate System
2. Coordinate Conversion
3. The Form of Polar Equations
4. Graphing Polar Equations

9.4 Calculus in Polar Coordinates 724

1. Tangents to Polar Curves
2. Area and Arc Length in Polar Coordinates

9.5 Conic Sections in Cartesian Coordinates 733

1. Ellipses
2. Parabolas
3. Hyperbolas
4. Rotating Conic Sections

9.6 Conic Sections in Polar Coordinates 752

1. The Focus-Directrix Description of Conic Sections
2. Using the Polar Form of Conic Sections

Introduction

As with so much in mathematics, some knowledge of the history underlying the topics in this chapter aids greatly in understanding them. And as with many of the topics studied thus far in calculus, the history is rich and varied. The use of polar coordinates dates back well over two thousand years, as the Greek astronomer Hipparchus (ca. 190–ca. 120 BC) plotted the positions of stellar objects with measurements we would now call polar coordinates. And the trajectories of moving bodies, whether short-lived (such as thrown balls and projectile missiles) or grander in nature (such as planetary motion) have been described with parametric equations by mathematicians for several centuries.

For our purposes, parametric equations and polar coordinates, as well as the additional coordinate systems to be introduced later, are useful because they often describe curves and other geometric objects in a more natural way than Cartesian coordinates. In this chapter, we will use parametric equations primarily to describe curves in a plane, and in many cases those curves can be described only very awkwardly (or not at all) by a relationship between Cartesian variables. Later in the text, we'll extend the use of parametric equations to describe curves, surfaces, and other objects in higher-dimensional spaces. Likewise, polar coordinates are used in this chapter to describe certain planar curves in a way that takes advantage of their inherent symmetry. And the polar coordinate system presented now will serve as an introduction to the related cylindrical and spherical coordinate systems of Chapter 14.

For our purposes, parametric equations and polar coordinates, as well as the additional coordinate systems to be introduced later, are useful because they often describe curves and other geometric objects in a more natural way than Cartesian coordinates.

Earlier mathematicians appreciated the algebraic advantages of parametric equations and polar coordinates, but the advantages extend into the realm of calculus as well. One particular curve, called a *cycloid*, is a case in point. It is most naturally described parametrically, which we will do in the first section of this chapter. But cycloids possess unique properties that allow them to satisfy two famous problems of the late 17th century, and the demonstration of these properties calls for rather advanced applications of calculus. Indeed, one branch of mathematics, called the *calculus of variations*, developed as a result of work on one of the problems.

Today, parametric equations are frequently used in computer graphics and animation to depict curves, surfaces, and solids in two and three dimensions. Parametric equations can also be used to produce striking computer-generated images very simply—you will study examples known as *Lissajous figures* in this chapter. Later, in Chapter 12, you will learn how calculus is used to identify the intrinsic geometric characteristics of parametrically defined curves and how these characteristics play a role in constructing realistic computer-generated images.

9.1 **Parametric Equations**

TOPICS

1. Terminology of parametric equations
2. Applications of parametric equations

We briefly encountered parametric equations in Section 4.4 while discussing Cauchy's Mean Value Theorem, and promised then that we would return to the topic later. The time has come for a broader examination of curves that are defined in terms of a parameter.

TOPIC 1 **Terminology of Parametric Equations**

One of the best reasons for describing curves parametrically is simply that there are many examples of curves in the plane that are not the graph of any equation of the form $y = f(x)$ or $x = f(y)$. The parametric skills we gain now can be easily extended to apply to curves (and other objects) in three-dimensional space.

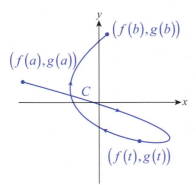

Figure 1
Parametrically Defined Curve

Figure 1 illustrates a curve C defined by two **parametric equations** $x = f(t)$ and $y = g(t)$. Each point on the **parametric curve** corresponds to an ordered pair $(f(t), g(t))$, where the **parameter** t is allowed to take on values over a **parametric interval** I. If the domain I of the parameter is a closed interval $[a,b]$, then we call $(f(a), g(a))$ the **initial point** and $(f(b), g(b))$ the **terminal point** of the curve. We say C is defined *parametrically* by f and g over the interval I. It is often helpful to think of the parameter t as time, and of C as the path "traced out" by the point $(f(t), g(t))$ as t moves from a toward b. In some applications it is helpful to indicate the direction of travel along the curve that corresponds to the increasing parameter, which we do with arrows as shown in Figure 1. The direction traveled along the curve as the parameter increases is called the curve's **orientation**.

One way to sketch a parametrically defined curve is to simply calculate a sequence of representative points $(f(t), g(t))$ and connect them with our best guess at the points in between. But it may also be possible to *eliminate the parameter* in the parametric equations and arrive at a meaningful equation in x and y.

Example 1 ✐

Sketch the curve defined parametrically by $x = t^2 + t$ and $y = t - 1$, $t \in [-4, 3]$.

Solution

If we *were* to plot points, we would obtain a collection of points and a sketch such as that shown in Figure 2.

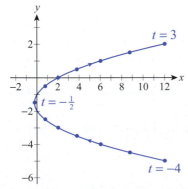

Figure 2 Parametric Parabola

t	(x,y)	t	(x,y)	t	(x,y)
-4	$(12,-5)$	$-\frac{3}{2}$	$\left(\frac{3}{4}, -\frac{5}{2}\right)$	1	$(2,0)$
$-\frac{7}{2}$	$\left(\frac{35}{4}, -\frac{9}{2}\right)$	-1	$(0,-2)$	$\frac{3}{2}$	$\left(\frac{15}{4}, \frac{1}{2}\right)$
-3	$(6,-4)$	$-\frac{1}{2}$	$\left(-\frac{1}{4}, -\frac{3}{2}\right)$	2	$(6,1)$
$-\frac{5}{2}$	$\left(\frac{15}{4}, -\frac{7}{2}\right)$	0	$(0,-1)$	$\frac{5}{2}$	$\left(\frac{35}{4}, \frac{3}{2}\right)$
-2	$(2,-3)$	$\frac{1}{2}$	$\left(\frac{3}{4}, -\frac{1}{2}\right)$	3	$(12,2)$

It certainly appears that the curve is parabolic in nature, and a moment's thought about the parametric equations justifies that impression. The equation in y is linear, meaning y changes linearly with respect to t. So the fact that x is a quadratic function of t means that x is also a quadratic function of y. We can make this explicit by solving the second equation for t, substituting the result into the first equation, and simplifying the result.

$$y = t - 1 \quad \Rightarrow \quad t = y + 1$$
$$x = t^2 + t \quad \Rightarrow \quad x = (y+1)^2 + (y+1)$$
$$x = y^2 + 3y + 2$$

If we complete the square of the right-hand side of the last equation, we obtain

$$x = \left(y + \frac{3}{2}\right)^2 - \frac{1}{4},$$

telling us that the vertex of the parabola is at $\left(-\frac{1}{4}, -\frac{3}{2}\right)$. Note that $t = -\frac{1}{2}$ when $y = -\frac{3}{2}$, as indicated in Figure 2.

Example 2 ✎

Sketch the curve defined parametrically by $x = 3\cos t$ and $y = 3\sin t$, $0 \le t \le 2\pi$.

Solution

We could eliminate the parameter in this pair of equations by solving one of them for t and substituting the result into the other, but we can shorten the process by making use of a trigonometric identity. Since $\cos^2 t + \sin^2 t = 1$ for every t, we have the following:

$$x^2 + y^2 = (3\cos t)^2 + (3\sin t)^2 = 9(\cos^2 t + \sin^2 t) = 9$$

Hence, the parametrically defined curve is a circle of radius 3 centered at the origin, with the parameter t corresponding to the radian measure of the central angle as shown in Figure 3. In Exercises 34–37, you will derive the general parametric form for a circle of radius r centered at (h, k) and for an ellipse with axes of length $2a$ and $2b$ centered at (h, k).

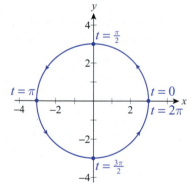

Figure 3

Example 3 ✎

Describe the curve defined parametrically by $x = \cos 2t$ and $y = \sin 2t$, $t \in [0, \pi]$.

Solution

This curve is also a circle, as you can verify by repeating the procedure of Example 2. However, this example illustrates the important fact that parametrically defined curves can be traced out at different rates. The curve traced out by $(\cos 2t, \sin 2t)$ as t ranges over the interval $[0, \pi]$ is a circle of radius 1 centered at the origin, as is the curve $(\cos t, \sin t)$ over the interval $[0, 2\pi]$. But any given arc of $(\cos 2t, \sin 2t)$ is traced out twice as quickly as the corresponding arc of $(\cos t, \sin t)$.

Another important characteristic of some parametric curves is illustrated by the fact that the curve $(\cos t, \sin t)$ over the interval $[0, 4\pi]$ doesn't appear any different than the curve $(\cos t, \sin t)$ over the interval $[0, 2\pi]$; when the parametric interval is $[0, 4\pi]$, the circle is simply traced twice.

TOPIC 2   Applications of Parametric Equations

A number of parametrically defined curves have become well known over the centuries for their physical properties, their unique geometric characteristics, or both. One such is the curve known as the *cycloid*.

Example 4 ✏

A **cycloid** is the curve, shown in blue in Figure 4, traced out by a point P on a circle of radius a as it rolls (without slipping) along the x-axis. Find parametric equations defining such a curve.

Solution

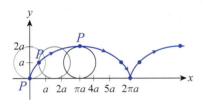

Figure 4 Generating the Cycloid

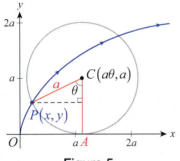

Figure 5

The first step is to make a wise choice of parameter. To this end, let θ measure the angle between the ray extending straight down from the circle's center and the ray extending from the circle's center through the point P, as shown in Figure 5. Then when the circle is in the initial position, the two rays coincide and $\theta = 0$. As the circle rolls to the right, θ increases; when the circle is in the position of the second circle shown in Figure 4, $\theta = \pi/2$, and when the circle is in the position of the rightmost circle the point P reaches its maximum height and $\theta = \pi$.

Consider the enlarged diagram in Figure 5. If we let (x, y) denote the coordinates of the point P, our goal is to write x and y as functions of the parameter θ. Our first observation is that the line segment $\overline{OA}$ must have the same length as the arc $\overparen{PA}$, since the circle rolls without slipping. The length of the arc $\overparen{PA}$ is $a\theta$, so that is the length of $\overline{OA}$ as well. This means the coordinates of the point C are $(a\theta, a)$. The dashed line in Figure 5 has length $a\sin\theta$, and the other leg of the right triangle has length $a\cos\theta$. Putting this all together, we have

$$x + a\sin\theta = a\theta \quad \text{and} \quad y + a\cos\theta = a.$$

Solving for x and y gives us the parametrization $x = a(\theta - \sin\theta)$ and $y = a(1 - \cos\theta)$. Note that the parametric interval $[0, 2\pi]$ produces one full arch of the cycloid.

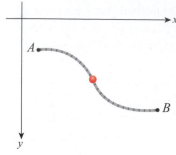

Figure 6

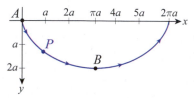

Figure 7 Inverted Cycloid

The cycloid, because of its physical properties, turns out to be the solution to two famous problems. One is known as the **brachistochrone problem**, or shortest-time problem. It asks for the curve connecting a point A to a lower point B along which a particle, subject only to the force of gravity, will travel in the shortest time—imagine a frictionless bead sliding along a curved wire connecting A and B, as illustrated in Figure 6. The answer is trivial if B is directly below A, so we seek the curve when this is not the case. The **tautochrone problem**, or same-time problem, asks for the curve connecting A and B with the property that a frictionless bead, when released anywhere along the curve, takes the same time to slide to the point B. Another way of asking the tautochrone question is this: along what path should a pendulum be constrained to swing so that its period (time of one full oscillation) is the same no matter where along the path the pendulum is released?

The Dutch scientist Christiaan Huygens (1629–1695) showed that a cycloid (inverted so that the positive y-axis points down, as in Figure 7) is a solution of the tautochrone problem in 1673; his motivation was to improve the accuracy of pendulum clocks, which he had previously invented. And the Swiss mathematician Johann Bernoulli (1667–1748) showed in 1697 that such a cycloid is also the solution of the brachistochrone problem, after posing the question himself in 1696. Other mathematicians, including his brother Jakob and Isaac Newton, also provided solutions, and the ideas and techniques they used in doing so developed into the area of mathematics now called the *calculus of variations*.

We can set up both problems by starting with a principle of physics. Suppose a curve C connects points A and B as in Figure 6, but let A be the origin (assume also that the positive y-axis points downward). A frictionless bead released with velocity $v = 0$ at A has kinetic energy 0, and the work done by gravity moving the bead from A to another point (x, y) in the plane is mgy, where m is the mass of the bead and g is the acceleration due to gravity. The kinetic energy $mv^2/2$ of the bead at the point (x, y) must equal the work done by gravity in moving it there, so

$$\frac{1}{2}mv^2 = mgy, \quad \text{or} \quad v = \sqrt{2gy}.$$

Note that this expresses the velocity v of the bead as it passes through (x, y) as a function of y alone. If we let s denote the arc length of C from A to the point (x, y), then $ds/dt = v$, so we have the following:

$$dt = \frac{ds}{v} = \frac{ds}{\sqrt{2gy}} = \sqrt{\frac{(dx)^2 + (dy)^2}{2gy}} = \sqrt{\frac{1 + (dy/dx)^2}{2gy}}\, dx \qquad \text{Recall from Section 6.3 that } (ds)^2 = (dx)^2 + (dy)^2.$$

The solution of the brachistochrone problem then amounts to finding a curve $y = f(x)$ so that the total time T taken for the bead to slide from $(0,0)$ to $B = (B_x, B_y)$ is minimized, where

$$T = \int dt = \int_0^{B_x} \sqrt{\frac{1 + (dy/dx)^2}{2gy}}\, dx.$$

This is the sort of problem solved by the methods of the calculus of variations, and its solution will be left for a later course. But we can, at this point, verify that the cycloid solves the tautochrone problem between the points $A(0,0)$ and $B(\pi a, 2a)$, as depicted in Figure 7. First, note that for a cycloid parametrized by $x = a(\theta - \sin\theta)$ and $y = a(1 - \cos\theta)$, $dx = a(1 - \cos\theta)d\theta$ and $dy = a(\sin\theta)d\theta$. So the time it takes for a bead to slide from A to B is calculated as follows:

$$T = \int dt = \int_A^B \sqrt{\frac{(dx)^2 + (dy)^2}{2gy}}$$

$$= \int_A^B \sqrt{\frac{a^2(1-\cos\theta)^2 + a^2\sin^2\theta}{2gy}}\, d\theta \qquad (1-\cos\theta)^2 + \sin^2\theta = 1 - 2\cos\theta + \underbrace{\cos^2\theta + \sin^2\theta}_{1}$$

$$= \int_{\theta=0}^{\theta=\pi} \sqrt{\frac{a^2(2-2\cos\theta)}{2ga(1-\cos\theta)}}\, d\theta$$

$$= \int_{\theta=0}^{\theta=\pi} \sqrt{\frac{a}{g}}\, d\theta = \pi\sqrt{\frac{a}{g}}$$

Now, for a bead that is instead released at the point (x_0, y_0), corresponding to the point $(a(\theta_0 - \sin\theta_0), a(1 - \cos\theta_0))$ on the cycloid, its velocity as it passes through a later point (x, y) on the cycloid is

$$v = \sqrt{2g(y - y_0)} = \sqrt{2ga[(1-\cos\theta) - (1-\cos\theta_0)]} = \sqrt{2ga(\cos\theta_0 - \cos\theta)}.$$

So the time it requires to slide from the point (x_0, y_0) to B is calculated as follows:

$$T = \int_{\theta_0}^{\pi} \sqrt{\frac{a^2(2-2\cos\theta)}{2ga(\cos\theta_0 - \cos\theta)}}\, d\theta = \sqrt{\frac{a}{g}}\int_{\theta_0}^{\pi} \sqrt{\frac{1-\cos\theta}{\cos\theta_0 - \cos\theta}}\, d\theta$$

$$= \sqrt{\frac{a}{g}}\int_{\theta_0}^{\pi} \sqrt{\frac{2\sin^2(\theta/2)}{(2\cos^2(\theta_0/2)-1) - (2\cos^2(\theta/2)-1)}}\, d\theta \qquad \sin^2(\theta/2) = \frac{1-\cos\theta}{2}$$
$$\cos^2(\theta/2) = \frac{1+\cos\theta}{2}$$

$$= \sqrt{\frac{a}{g}}\int_{\theta_0}^{\pi} \frac{\sin(\theta/2)}{\sqrt{\cos^2(\theta_0/2) - \cos^2(\theta/2)}}\, d\theta$$

$$= \sqrt{\frac{a}{g}}\int_{\theta=\theta_0}^{\theta=\pi} \frac{-2}{\sqrt{k^2 - u^2}}\, du \qquad \begin{array}{l} u = \cos(\theta/2) \\ -2\,du = \sin(\theta/2)\,d\theta \\ \text{Let } k = \cos(\theta_0/2). \end{array}$$

$$= 2\sqrt{\frac{a}{g}}\left[-\sin^{-1}\left(\frac{u}{k}\right)\right]_{\theta=\theta_0}^{\theta=\pi} \qquad \int \frac{du}{\sqrt{k^2 - u^2}} = \sin^{-1}\left(\frac{u}{k}\right) + C$$

$$= 2\sqrt{\frac{a}{g}}\left[-\sin^{-1}\left(\frac{\cos(\theta/2)}{\cos(\theta_0/2)}\right)\right]_{\theta=\theta_0}^{\theta=\pi}$$

$$= 2\sqrt{\frac{a}{g}}\left(-\sin^{-1}0 + \sin^{-1}1\right) = \pi\sqrt{\frac{a}{g}}$$

Since this is the same amount of time required for the bead to slide from the origin to B, the cycloid is indeed a solution of the tautochrone problem.

Many other well-known curves are best described parametrically, and you will encounter them in the exercises. The next technology note demonstrates how a graphing calculator and a computer algebra system such as *Mathematica* can be used to graph two more named curves.

Technology Note 💻

Use technology to graph the following parametrically defined curves.

a. $x = 3\sin 5t, \quad y = 2\cos 3t, \quad 0 \le t \le 2\pi$

b. $x = 8\cos t + 2\cos 4t, \quad y = 8\sin t - 2\sin 4t, \quad 0 \le t \le 2\pi$

Graphing calculators and computer algebra systems can generate parametrically defined curves. On a graphing calculator the mode needs to change from the function graphing mode to the parametric mode. This can be done by selecting **PAR** from the **MODE** menu. Then the parametric equations are entered in the **Y=** editor. Since the equations are depending on t, when selecting the viewing window, the t-interval will be indicated in there. In *Mathematica* the built-in command is `ParametricPlot`. The command takes two arguments—the first is an ordered pair (enclosed in braces) consisting of the parametrized formulas for x and y. The second argument is a list indicating the parametric interval to use.

a. Parametric equations of the form $x = A\sin(\omega_1 t)$ and $y = B\cos(\omega_2 t)$ generate curves known as **Lissajous figures**. A Lissajous figure with $A = 3$, $B = 2$, $\omega_1 = 5$, and $\omega_2 = 3$ is illustrated in Figure 8.

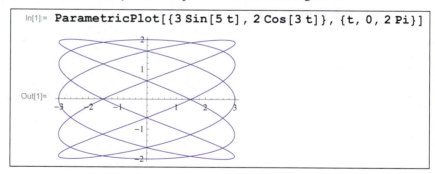

In[1]:= `ParametricPlot[{3 Sin[5 t], 2 Cos[3 t]}, {t, 0, 2 Pi}]`

Out[1]=

Figure 8b

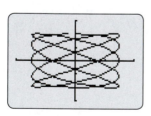

Figure 8a $[-4,4]$ by $[-3,3]$

b. A **hypocycloid** is formed by tracing a fixed point on a circle of radius r as it rolls on the inside of a larger circle with radius R. If $R = kr$, then the parametrization of the curve is

$$x = r(k-1)\cos t + r\cos((k-1)t), \quad y = r(k-1)\sin t - r\sin((k-1)t).$$

Figure 9 illustrates the given hypocycloid ($r = 2$, $k = 5$) for $0 \le t \le 2\pi$.

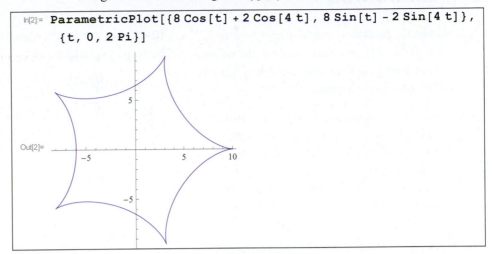

In[2]:=
```
ParametricPlot[{8 Cos[t] + 2 Cos[4 t], 8 Sin[t] - 2 Sin[4 t]},
  {t, 0, 2 Pi}]
```

Figure 9b

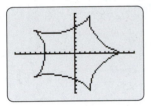

Figure 9a $[-12,12]$ by $[-10,10]$

The number of cusps (which, as we will see in the next section, qualify as points where the curve is not *smooth*) can be determined from k if k is either an integer or a rational number. If k is irrational, the hypocycloid never retraces itself no matter how long the parametric interval. Figure 10 shows a hypocycloid with $r = 2$, $k = \pi$, and a parametric interval of $[0, 100\pi]$.

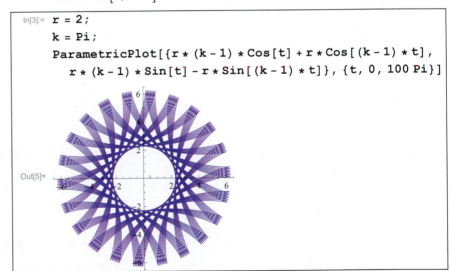

In[3]:=
```
r = 2;
k = Pi;
ParametricPlot[{r * (k - 1) * Cos[t] + r * Cos[(k - 1) * t],
  r * (k - 1) * Sin[t] - r * Sin[(k - 1) * t]}, {t, 0, 100 Pi}]
```

Figure 10b

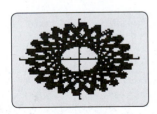

Figure 10a $[-7,7]$ by $[-7,7]$

9.1 **Exercises**

1. Given the parametric equations $x = 5 + t$ and $y = \sqrt{t}/(t-2)$, construct a table of the points (x, y) that result from integer t-values from 0 to 6, and then sketch the curve.

2. Given the parametric equations $x = (\tan\theta)/2$ and $y = \cos^2\theta + 3$, construct a table of the points (x, y) that result from the values $\theta = 0, \pi/6, \pi/3, \pi/2, 2\pi/3, 5\pi/6$, and π. Using these points, sketch the graph of the equations.

3–9 A straightforward way of parametrizing the graph of a function $y = f(x)$ is with the equations $x = t$ and $y = f(t)$. Use this technique to construct parametric equations defining the graph of the given equation.

3. $y = -x^2 - 5$

4. $x^2 + \dfrac{y^2}{4} = 1$

5. $x = y^2 + 4$

6. $x = 4y - 6$

7. $y = |x - 1|$

8. $x = 2(y - 3)$

9. $y^2 = 1 - x^2$

10–23 Sketch the curve defined by the parametric equations by eliminating the parameter.

10. $x = 3(t+1), \quad y = 2t$

11. $x = \sqrt{t-2}, \quad y = 3t - 2$

12. $x = 1 + t, \quad y = \dfrac{t-3}{2}$

13. $x = |t + 3|, \quad y = t - 5$

14. $x = \dfrac{t}{4}, \quad y = t^2$

15. $x = \dfrac{t}{t+2}, \quad y = \sqrt{t}$

16. $x = \sqrt{t+3}, \quad y = t + 3$

17. $x = \dfrac{2}{|t-3|}, \quad y = 2t - 1$

18. $x = \cos\theta, \quad y = 2\sin\theta$

19. $x = 3\sin\theta - 1, \quad y = \dfrac{\cos\theta}{2}$

20. $x = 1 - \sin\theta, \quad y = \sin\theta - 1$

21. $x = 2\cos\theta, \quad y = 3\cos\theta$

22. $x = 2\sin\theta + 2, \quad y = 2\cos\theta + 2$

23. $x = \sin\theta, \quad y = 4 - 3\cos\theta$

24–29 Match the parametrization with its graph (labeled A–F).

24. $x = 10\cos\theta, \quad y = \theta$

25. $x = 2 + 3t, \quad y = \dfrac{t^3 + 1}{4}$

26. $x = 1 + 4t, \quad y = 3 + 2t$

27. $x = t\cos t, \quad y = t\sin t$

28. $x = 4 - 2t^2, \quad y = t^3 - 9t$

29. $x = 4\sin\theta, \quad y = 4\cos\theta$

A.

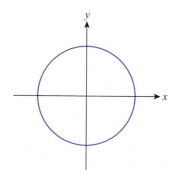

B.

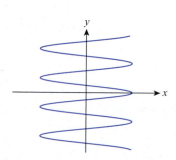

C.

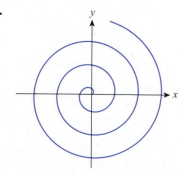

D.

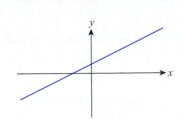

E.

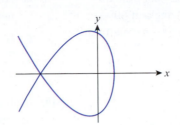

F.

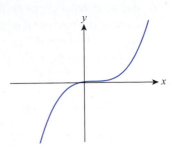

30–33 Sketch the curve defined by the parametric equations and indicate the orientation of the curve using arrows.

30. $x = \dfrac{1}{\sqrt{t-1}} + 1, \quad y = \dfrac{1}{t-1}$

31. $x = (\ln t)^2, \quad y = \dfrac{1}{\sqrt{t}}, \quad 1 \le t \le 5$

32. $x = e^{-t/2}, \quad y = e^t, \quad t \ge 0$

33. $x = \sec^2 t, \quad y = \tan^2 t, \quad 0 \le t \le \dfrac{\pi}{3}$

34. Show that the curve defined by the parametric equations $x = r\cos(nt)$ and $y = r\sin(nt)$, $0 \le t \le 2\pi/n$, is a circle of radius r, centered at the origin.

35. Show that the curve defined by the parametric equations $x = a\cos(nt)$ and $y = b\sin(nt)$, $a > b$, $0 \le t \le 2\pi/n$, is an ellipse centered at the origin with respective lengths of the major and minor axes being $2a$ and $2b$. (**Hint:** Recall from precalculus that the equation of such an ellipse is $\dfrac{x^2}{a^2} + \dfrac{y^2}{b^2} = 1$.)

36. Taking advantage of Exercise 34, along with horizontal and vertical shifts, derive the general parametric form for a circle of radius r, centered at (h,k).

37. Using Exercise 35 along with horizontal and vertical shifts, derive the general parametric form for an ellipse with axes of length $2a$ and $2b$, centered at (h,k).

38–45 Find parametric equations to represent the graph described. (Answers will vary.)

38. Line, slope -2, passing through $(-5,-2)$

39. Line, passing through $(6,-3)$ and $(2,3)$

40. Line segment connecting the points $(-2,-1)$ and $(3,4)$

41. Line segment connecting the points $(-3,1)$ and $(5,-5)$

42. Circle, center $(7,-5)$, radius 4

43. Circle, center $(0,-2)$, radius 6

44. Ellipse, center $(5,-1)$, $a = 3$, $b = 2\sqrt{2}$, vertical major axis (**Hint:** See Exercises 35 and 37.)

45. Ellipse, center $(0,1)$, $a = 6$, $b = \sqrt{11}$, horizontal major axis (**Hint:** See Exercises 35 and 37.)

46. Suppose that a baseball is hit 3 feet above the ground, and it leaves the bat at a speed of 100 miles per hour at an angle of $20°$ from the horizontal. Construct parametric equations representing the path of the ball's flight, and sketch a graph of the ball's travel. (**Hint:** Supposing that the ball starts at the point $(0,3)$ and treating time t as the parameter, express the ball's x- and y-coordinates as functions of t. Do not forget to decompose the initial velocity into horizontal and vertical components!)

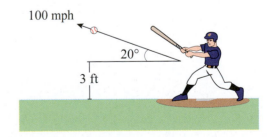

47. Suppose the ball in Exercise 46 has been hit toward a 10-foot-high fence that is 400 feet from home plate. Will the ball clear the fence?

48. Suppose that a circus performer is shot from a cannon at a rate of 80 mph, at an angle of 60° from the horizontal. The cannon sits on a platform 10 feet above the ground.

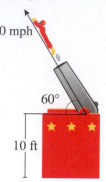

80 mph

60°

10 ft

a. Construct parametric equations representing the performer's path as he flies through the air.

b. Sketch a graph of his flight.

c. How high is the acrobat 1.5 seconds after leaving the cannon?

d. How far from the cannon should a landing net be placed, if it is placed at ground level?

e. At what time t will the performer land in the net?

f. If a 12-foot-high wall of flames is placed 70 feet from the cannon, will he clear it unharmed?

49. On his morning paper route, John throws a newspaper from his car window 3.5 ft from the ground. The paper has an initial velocity of 10 ft/s and is tossed at an angle of 10° from the horizontal.

a. Construct parametric equations modeling the path of the newspaper.

b. Sketch a graph of the paper's path.

50. François shoots a basketball at an angle of 48° from the horizontal. It leaves his hands 7 ft from the ground with a velocity of 21 ft/s.

a. Construct parametric equations representing the path of the ball.

b. Sketch a graph of the basketball's flight.

c. If the goal is 15 ft away and 11 ft high, will he make the shot?

51–54 Use the given graphs of $f(t)$ and $g(t)$ to make a rough sketch of the curve defined by the parametric equations $x = f(t)$ and $y = g(t)$. (**Hint:** Plotting a few points may help.)

51.

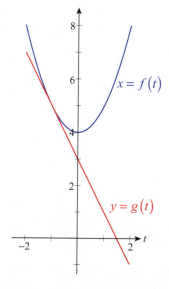

$x = f(t)$

$y = g(t)$

52.

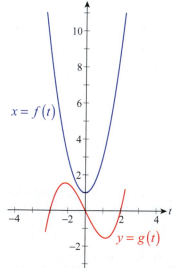

$x = f(t)$

$y = g(t)$

53.

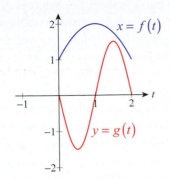

54.

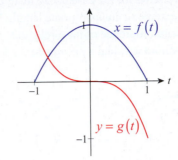

55. Verify that the three parametrizations below represent the same curve. Then come up with two parametrizations on your own. (Answers will vary.)

a. $x = t^2$, $y = t^3$, $-1 \le t \le 1$

b. $x = 1 - \cos^2 t$, $y = \sin^3 t$, $-\dfrac{\pi}{2} \le t \le \dfrac{\pi}{2}$

c. $x = \sec^2\left(\dfrac{\pi}{4}t\right) - 1$, $y = \tan^3\left(\dfrac{\pi}{4}t\right)$, $-1 \le t \le 1$

56–57 Find all intersection points of the given parametric curves.

56. $x = 3t - 2$, $y = 3t - 1$; $x = 2u$, $y = 4u^2 - 4u + 1$

57. $x = t - 1$, $y = 3t^2 - 6t + 3$; $x = \dfrac{u}{2}$, $y = -\dfrac{u^2}{4} + 2u$

58. A wheel of radius 12 inches rolls along a flat surface in a straight line. There is a fixed point P that initially lies at the point $(0, 0)$. Find parametric equations defining the cycloid traced out by P.

59. A ball is rolled on the floor in a straight line from one person to another person. The ball has a radius of 3 cm and there is a fixed point P located on the ball. Let the person rolling the ball represent the origin. Find parametric equations defining the cycloid traced out by P.

60. Prove that the parametrizations $x_1 = \dfrac{1 - t^2}{1 + t^2}$, $y_1 = \dfrac{2t}{1 + t^2}$, $-\infty < t < \infty$, and $x_2 = \cos t$, $y_2 = \sin t$, $-\pi < t < \pi$, represent the same curve. What is the curve?

61. A ladder of length l is leaning against a wall, sliding slowly all the way down to a horizontal position. Suppose that there is a paint mark on the side of the ladder, exactly one-third of the way up from the bottom of the ladder (see point P in the figure). Assuming that the ladder started sliding from a vertical position, prove that the curve traced out by P during the slide is one-quarter of an ellipse. (**Hint:** Let t be the radian measure of the angle that the ladder makes with the horizontal, and use it as the parameter to determine the parametric equations of the curve.)

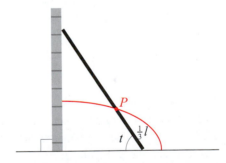

62. One way to generalize the cycloid is to consider the curve traced by a point P on a fixed radius (a "spoke") in the circle of Example 4. Generalize the argument of Example 4 to prove that if P is b units ($b < a$) from the center, then the parametric equations of the resulting trochoid (also called *curate cycloid*) are $x = a\theta - b\sin\theta$ and $y = a - b\cos\theta$. (Note that the case of $a = b$ yields the equations we obtained in Example 4.)

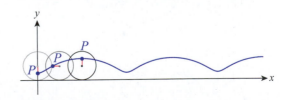

63. Repeat Exercise 62 for the case $a < b$ (imagine each spoke extending an appropriate length beyond the circumference of the circle; the resulting trochoid is called a *prolate cycloid*).

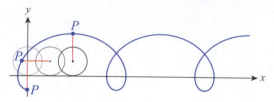

64.* Derive the parametrization of the hypocycloid seen in the Technology Note by using the angle t in the figure as a parameter. (**Hint:** Since the circle rolls without slipping, you may use the equality of the lengths of the two red arcs in the figure.)

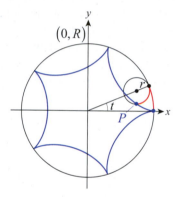

65.* The *epicycloid* is the curve traced out by a fixed point P on a circle of radius r as it rolls without slipping on the outside of a larger circle with radius R. Using the technique of Exercise 62, derive the parametric equations of the epicycloid.

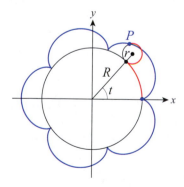

66.* The famous curve in the figure below is called the *witch of Agnesi* and is derived as follows. Suppose that a circle of radius r is centered at $(0, r)$ and the line $y = (\tan t)x$ intersects the horizontal line $y = 2r$ and the circle at the points A and B, respectively (B is not the origin). The curve is then traced out by the point P, which is the intersection of the horizontal segment through B and the vertical segment through A. Use t as a parameter to derive the parametric equations for the witch of Agnesi. (**Hint:** If O is the origin, and C is the center of the circle, start by examining the isosceles triangle OCB.)

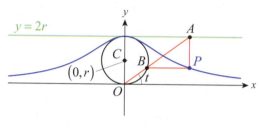

67. Eliminate the parameter to find the Cartesian equation of the witch of Agnesi in Exercise 66.

68–71 *True or False?* Determine whether the given statement is true or false. In case of a false statement, explain or provide a counterexample.

68. If $x = f(t)$ and $y = g(t)$ are both quadratic functions, then the parametric curve defined by $x = f(t)$, $y = g(t)$ is a parabola or a parabolic arc.

69. The parametric equations $x = t$, $y = t^{2/3} - 2$ and $x = 8t$, $y = 4t^{2/3} - 2$ have the same graph.

70. The graph of $x = t^3$, $y = t^6$ is the prototypical parabola.

71. The graph of parametric equations can either be represented in the form $y = f(x)$ (i.e., y as a function of x), or in the form $x = g(y)$ (i.e., x as a function of y).

9.1 **Technology Exercises**

72–78 Use a graphing calculator or computer algebra system to sketch the given curve for various values of k and explore how the value of k affects the shape of your graph.

72. $x = 2t - k \sin t, \quad y = 2 - k \cos t$ (trochoid)

73. $x = 2 \cos t + k \cos \dfrac{2}{3} t, \quad y = 2 \sin t - k \sin \dfrac{2}{3} t$ (hypotrochoid)

74. $x = 2kt - 4t^3, \quad y = 3t^4 - kt^2$ (swallowtail catastrophe curve)

75. $x = \dfrac{3kt}{1 + t^3}, \quad y = \dfrac{3kt^2}{1 + t^3}$ (folium of Descartes)   **76.** $x = \cos(t - \cos(kt)), \quad y = \cos(kt)$

77. $x = 2t - \cos(kt), \quad y = t^2 - \sin(kt)$ **78.** $x = \dfrac{k^2 \cos(kt)}{t^2 + k^2}, \quad y = \dfrac{k^2 \sin(kt)}{t^2 + k^2}$

79. Use technology and your parametrizing skills to display the following picture on your screen.

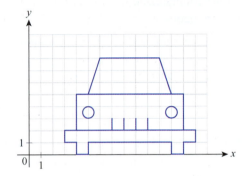

9.2 Calculus and Parametric Equations

TOPICS

1. Tangent lines and planar areas

2. Arc length and surface area

In this section, we revisit some of the fundamental ideas of calculus as they appear in the setting of parametric equations. These ideas, when further developed, serve as the foundation for the branch of mathematics known as vector calculus.

TOPIC 1 Tangent Lines and Planar Areas

In keeping with our original development of calculus fundamentals, we begin with the question of differentiability and the existence (or nonexistence) of lines tangent to a curve.

Definition 💡

Differentiability of Parametrized Curves

A curve C defined by the parametric equations $x = f(t)$ and $y = g(t)$ is **differentiable** at t_0 if both f and g are differentiable at t_0. Further, the parametrization $(f(t), g(t))$ is *smooth* at t_0 if both f' and g' are continuous at t_0 and at least one of $f'(t_0)$ and $g'(t_0)$ is nonzero.

Related to this, a given curve C is **smooth** at (x_0, y_0) if it has a parametrization $(f(t), g(t))$ that is smooth at t_0, where $(x_0, y_0) = (f(t_0), g(t_0))$. (As you will see in Exercise 5, however, a smooth curve may also be described by a nonsmooth parametrization.)

Note the distinction between differentiability and smoothness in this context—smoothness of a parametrization requires not just differentiability of f and g at a point, but also that those derivatives are continuous and at least one of them is not equal to 0. The need for the extra conditions can be demonstrated with a simple example.

Example 1 ✍

Sketch the graph of the curve C parametrized by $x = t^2$ and $y = t^3$ for $-2 \leq t \leq 2$, and identify points of differentiability and smoothness of the parametrization.

Solution

We can eliminate the parameter and express the curve as two functions of x as follows:

$$x = t^2 \quad \Rightarrow \quad t = \pm x^{1/2}$$
$$y = t^3 = \left(\pm x^{1/2}\right)^3 = \pm x^{3/2}$$

Individually, the two functions are easy to sketch and together they constitute C, as shown in Figure 1. Note that the point corresponding to $t = -2$ is $(4, -8)$, the point corresponding to $t = 2$ is $(4, 8)$, and the value $t = 0$ corresponds to the point $(0, 0)$. And since the two derivatives $dx/dt = 2t$ and $dy/dt = 3t^2$ exist for all $t \in [-2, 2]$, C is differentiable over the entire interval. But since dx/dt and dy/dt are both equal to 0 at $t = 0$, the parametrization is not smooth at this one point (it is smooth for every other t).

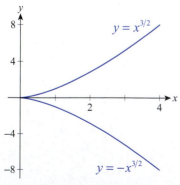

Figure 1 Graph of C

As Figure 1 shows, a parametrically defined curve may be anything *but* smooth at a point of differentiability. Differentiability with respect to the parameter t is not sufficient to guarantee the existence of a tangent line, but smoothness is. Using the Chain Rule, we may in fact be able to write the slope of the tangent line as a derivative dy/dx even if we are not able to express the curve as an equation in y and x. If the three derivatives dy/dx, dy/dt, and dx/dt all exist, then

$$\frac{dy}{dt} = \frac{dy}{dx} \cdot \frac{dx}{dt}.$$

And if $dx/dt \neq 0$, we can solve this equation for dy/dx to obtain the following.

Formula 🔍

Parametric Formula for dy/dx

Assume the curve C is defined parametrically by $x = f(t)$ and $y = g(t)$. If each of the three derivatives dy/dx, dy/dt, and dx/dt exists at a given point, and if $dx/dt \neq 0$, then

$$\frac{dy}{dx} = \frac{dy/dt}{dx/dt} = \frac{g'(t)}{f'(t)}.$$

The curve C has a vertical tangent line (and hence dy/dx does not exist) at a point of smoothness where $dx/dt = 0$ (that is, $dy/dt \neq 0$ and $dx/dt = 0$).

Example 2 ✎

The portion of the curve parametrized by $x = t^2$ and $y = t^3 - 3t$ for $-2 \leq t \leq 2$ is shown in Figure 2. Find the equations for the tangent lines that **a.** are horizontal, **b.** are vertical, and **c.** pass through the point $(3, 0)$.

Solution

First, note that $dx/dt = 2t$ and $dy/dt = 3t^2 - 3$. Since these two derivatives are never both equal to 0 for the same t, the curve is smooth everywhere.

a. The curve has horizontal tangent lines at those values of t for which $dy/dt = 0$ or $3t^2 - 3 = 0$. This gives us $t = \pm 1$, corresponding to the two points $(1, 2)$ and $(1, -2)$ as shown in Figure 3. The equations for the tangent lines at these points are $y = 2$ and $y = -2$.

b. The single vertical tangent line occurs at the point where $dx/dt = 0$, giving us $t = 0$ and the point $(0, 0)$ on the curve. The equation for this tangent line is $x = 0$.

c. Visually, it appears there should be two lines tangent to the curve through the point $(3, 0)$. The first step in finding them is to determine the two values of t for which the curve passes through $(3, 0)$ —since $x = t^2$, this happens when $t^2 = 3$ or $t = \pm\sqrt{3}$. At these values,

$$\frac{dy}{dx}\bigg|_{t=-\sqrt{3}} = \frac{dy/dt}{dx/dt}\bigg|_{t=-\sqrt{3}} = \frac{3t^2 - 3}{2t}\bigg|_{t=-\sqrt{3}} = \frac{6}{-2\sqrt{3}} = -\sqrt{3}$$

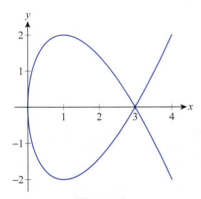

Figure 2

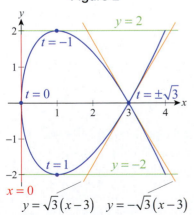

Figure 3

and

$$\left.\frac{dy}{dx}\right|_{t=\sqrt{3}} = \left.\frac{dy/dt}{dx/dt}\right|_{t=\sqrt{3}} = \left.\frac{3t^2-3}{2t}\right|_{t=\sqrt{3}} = \frac{6}{2\sqrt{3}} = \sqrt{3},$$

leading to the respective tangent line equations $y = -\sqrt{3}(x-3)$ and $y = \sqrt{3}(x-3)$.

Just as we have a parametric formula for dy/dx, we can express d^2y/dx^2 in terms of the parameter t, assuming $dx/dt \neq 0$.

Formula 🔍

Parametric Formula for d^2y/dx^2

If the parametric equations $x = f(t)$ and $y = g(t)$ define y as a twice-differentiable function of x, we can substitute y' for y in the parametric formula for dy/dx to obtain a parametric formula for $y'' = d^2y/dx^2$:

$$y'' = \frac{d^2y}{dx^2} = \frac{d}{dx}\left(\frac{dy}{dx}\right) = \frac{d}{dx}(y') = \frac{dy'/dt}{dx/dt}, \quad \frac{dx}{dt} \neq 0$$

Example 3 ✎

Let C be the curve defined by $x = t^3 - t$ and $y = t + 1$, $t \in \mathbb{R}$. Find the inflection points of C.

Solution

We first note that $dx/dt = 3t^2 - 1$ and $dy/dt = 1$, telling us that C is differentiable everywhere and, further, that it is smooth everywhere.

$$y' = \frac{dy}{dx} = \frac{1}{3t^2-1} \qquad 3t^2 - 1 \neq 0$$

So $dy'/dt = -6t / (3t^2 - 1)^2$, and the parametric formula for the second derivative tells us

$$y'' = \frac{d^2y}{dx^2} = \frac{dy'/dt}{dx/dt} = \frac{\dfrac{-6t}{(3t^2-1)^2}}{3t^2-1} = \frac{-6t}{(3t^2-1)^3} \qquad 3t^2 - 1 \neq 0$$

Recall that the possible inflection points are the points where d^2y/dx^2 is either equal to 0 or undefined, corresponding to $t = 0$ and $t = \pm 1/\sqrt{3}$ in this example. Testing the concavity of C on the four t-intervals defined by these points, we find the following.

Interval	$\left(-\infty, -\dfrac{1}{\sqrt{3}}\right)$	$\left(-\dfrac{1}{\sqrt{3}}, 0\right)$	$\left(0, \dfrac{1}{\sqrt{3}}\right)$	$\left(\dfrac{1}{\sqrt{3}}, \infty\right)$
y'' at test point	$y''(-1) = \frac{3}{4}$	$y''\left(-\frac{1}{2}\right) = -192$	$y''\left(\frac{1}{2}\right) = 192$	$y''(1) = -\frac{3}{4}$
Sign of y''	+	−	+	−
Concavity of C	Concave up	Concave down	Concave up	Concave down

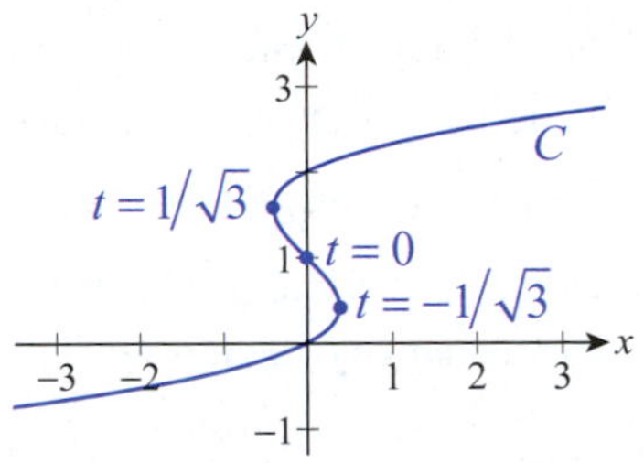

Since C changes concavity at each point, these *possible* inflection points are *actual* inflection points. Remember, though, that these are values of the parameter—the corresponding ordered pairs are as follows:

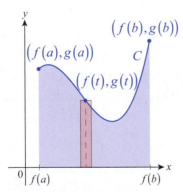

Figure 4

$$t = -\frac{1}{\sqrt{3}} \;\Rightarrow\; (x,y) = \left(\frac{2}{3\sqrt{3}}, 1 - \frac{1}{\sqrt{3}}\right)$$

$$t = 0 \;\Rightarrow\; (x,y) = (0,1)$$

$$t = \frac{1}{\sqrt{3}} \;\Rightarrow\; (x,y) = \left(-\frac{2}{3\sqrt{3}}, 1 + \frac{1}{\sqrt{3}}\right)$$

A sketch of C is shown in Figure 4, with the three inflection points noted.

Let us turn now to the matter of finding the area of a plane region bounded (at least in part) by a parametrized curve. First, as depicted in Figure 5, suppose we want to find the area A of a region bounded below by the x-axis and above by a curve C parametrized as $(f(t), g(t))$ over the t-interval $[a,b]$. In Figure 5, the shaded rectangle is an approximation of the differential element of area $dA = y\,dx$, where the height of the rectangle is $y = g(t)$ and the width is approximately dx. If $f(a) < f(b)$, as shown, then

$$A = \int_{x=f(a)}^{x=f(b)} y\,dx = \int_{x=f(a)}^{x=f(b)} g(t)\,dx.$$

Since $dx/dt = f'(t)$, we can write $dx = f'(t)\,dt$ and use the Substitution Rule to obtain

$$A = \int_{x=f(a)}^{x=f(b)} g(t)\,dx = \int_{t=a}^{t=b} g(t)\,f'(t)\,dt,$$

Figure 5

thereby expressing the area in terms of the functions f and g and the parameter t.

In general, a variation of the formula above can be used to find the area between a parametrized curve and an axis, as long as changes in sign are made to ensure that $y\,dx$ (or $x\,dy$) is nonnegative. Example 4 shows how these adjustments are made in practice.

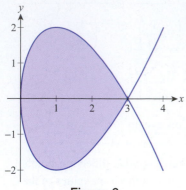

Figure 6

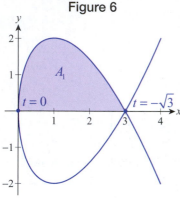

Figure 7

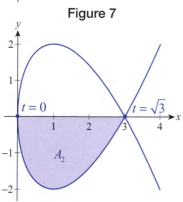

Figure 8

Example 4 ✐

Find the area of the region bounded entirely by the curve parametrized by $x = t^2$ and $y = t^3 - 3t$.

Solution

Figure 6 depicts the described region in blue. We can find its area by first finding the area of the portion above the x-axis, shown in Figure 7, and then the portion below the x-axis, shown in Figure 8.

Both portions are defined over the x-axis interval $[0,3]$, but it is more important to note that the upper portion is defined parametrically over the t-interval $\left[-\sqrt{3},0\right]$ while the lower portion is defined over the t-interval $\left[0,\sqrt{3}\right]$. The integral $\int_{t=-\sqrt{3}}^{t=0} y\,dx$ is an integral that goes from "right to left" in the xy-plane, since the initial point of the parametrized curve, $(3,0)$, lies to the right of the terminal point $(0,0)$. So unless a change of sign is made, the resulting integral will be negative.

$$A_1 = -\int_{t=-\sqrt{3}}^{t=0} y\,dx = -\int_{-\sqrt{3}}^{0} g(t)f'(t)\,dt = -\int_{-\sqrt{3}}^{0}\left(t^3 - 3t\right)(2t)\,dt$$

$$= -\int_{-\sqrt{3}}^{0}\left(2t^4 - 6t^2\right)dt = \frac{12\sqrt{3}}{5}$$

By symmetry, we expect the area A_2 of the lower portion to be identical, but we'll set up and evaluate the integral to demonstrate the accommodation for a function below the axis. For this portion, dx has the right sign (the integral goes from left to right in the plane), but y is negative for $t \in \left[0,\sqrt{3}\right]$. So we again change the sign.

$$A_2 = -\int_{t=0}^{t=\sqrt{3}} y\,dx = -\int_{0}^{\sqrt{3}} g(t)f'(t)\,dt = -\int_{0}^{\sqrt{3}}\left(t^3 - 3t\right)(2t)\,dt$$

$$= -\int_{0}^{\sqrt{3}}\left(2t^4 - 6t^2\right)dt = \frac{12\sqrt{3}}{5}$$

Hence, the total area bounded by the curve is $A = 24\sqrt{3}/5$.

TOPIC 2　Arc Length and Surface Area

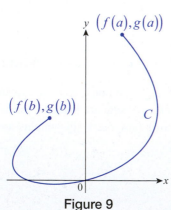

Figure 9

Recall from Section 6.3 that the differential of arc length ds is related to the differentials dx and dy by the equation $(ds)^2 = (dx)^2 + (dy)^2$, and that the arc length L of a curve is $L = \int ds$. For a smooth curve C defined parametrically by $x = f(t)$ and $y = g(t)$ over the t-interval $[a,b]$, such as the curve shown in Figure 9, the formula for arc length takes the pleasantly symmetric form

$$L = \int_{(f(a),g(a))}^{(f(b),g(b))} ds = \int_{(f(a),g(a))}^{(f(b),g(b))} \sqrt{(dx)^2 + (dy)^2} = \int_{a}^{b} \sqrt{\left(\frac{dx}{dt}\right)^2 + \left(\frac{dy}{dt}\right)^2}\,dt.$$

The formal statement of this version of the arc length formula follows.

Formula

Parametric Arc Length Formula

Assume that C is defined by the smooth parametrization $x = f(t)$ and $y = g(t)$ over the t-interval $[a,b]$ and C is traced out only once by the parametrization over the interval (except possibly for a finite number of self-intersections). Then the arc length L of C is given by

$$L = \int_a^b \sqrt{\left(\frac{dx}{dt}\right)^2 + \left(\frac{dy}{dt}\right)^2}\, dt.$$

Example 5

Use the parametric arc length formula to verify that the circumference of a circle of radius r is $2\pi r$.

Solution

The simplest parametrization of such a circle is $x = r\cos t$, $y = r\sin t$, for $t \in [0, 2\pi]$ (note that the initial and terminal point are the same).

$$L = \int_0^{2\pi} \sqrt{\left(\frac{dx}{dt}\right)^2 + \left(\frac{dy}{dt}\right)^2}\, dt = \int_0^{2\pi} \sqrt{(-r\sin t)^2 + (r\cos t)^2}\, dt = \int_0^{2\pi} r\, dt = 2\pi r$$

Example 6

Find the arc length of the spiral defined by $x = t\cos t$ and $y = t\sin t$ over the t-interval $[0, 4\pi]$.

Solution

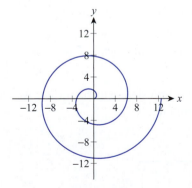

Figure 10

The graph of the spiral is illustrated in Figure 10. The formula for its arc length is easy enough to write down:

$$L = \int_0^{4\pi} \sqrt{(\cos t - t\sin t)^2 + (\sin t + t\cos t)^2}\, dt$$

$$= \int_0^{4\pi} \sqrt{\cos^2 t - 2t\cos t\sin t + t^2\sin^2 t + \sin^2 t + 2t\sin t\cos t + t^2\cos^2 t}\, dt$$

$$= \int_0^{4\pi} \sqrt{1 + t^2}\, dt$$

The evaluation of this integral is a good opportunity to use the integration techniques that we have developed previously. One approach would be to use the trigonometric substitution $t = \tan\theta$, and you will be guided through an alternative sequence of steps in Exercise 60. Once done, you will have found that $L = \frac{1}{2}\left[4\pi\sqrt{1 + 16\pi^2} + \sinh^{-1}(4\pi)\right]$, or approximately 80.8 units.

Example 7 ✎

Find the arc length of one arch of the cycloid generated by a circle of radius a.

Solution

Recall from Example 4 of Section 9.1 that our parametrization of a cycloid is $x = a(\theta - \sin\theta)$ and $y = a(1 - \cos\theta)$, and that one arch is traced out over the θ-interval $[0, 2\pi]$.

$$
\begin{aligned}
L &= \int_0^{2\pi} \sqrt{a^2 (1 - \cos\theta)^2 + a^2 \sin^2\theta}\, d\theta \\
&= a\int_0^{2\pi} \sqrt{2(1 - \cos\theta)}\, d\theta \\
&= a\int_0^{2\pi} \sqrt{4\sin^2\frac{\theta}{2}}\, d\theta \qquad\qquad \sin^2\frac{\theta}{2} = \frac{1 - \cos\theta}{2} \\
&= 2a\left[-2\cos\frac{\theta}{2}\right]_0^{2\pi} = 2a(2 + 2) = 8a
\end{aligned}
$$

So one arch of the cycloid has an arc length eight times as long as the radius of the circle that generates it.

The last topic we will revisit in this section is surface area. Specifically, we want a parametric version of the formula for the surface area A of an object formed by revolving a smooth curve C about an axis. Recall from Section 6.3 that this area is $A = \int 2\pi r\, ds$, where r denotes the radius from the axis of revolution to the surface and ds is the arc length differential. The formula is the same when the curve C is described parametrically—we merely express the arc length differential in terms of dx/dt and dy/dt. Example 8 illustrates the process.

Example 8 ✎

Find the surface area of a sphere of radius R by revolving an appropriately chosen parametrically defined curve C.

Solution

There are several choices for C; we will define C to be the curve $x = R\cos t$, $y = R\sin t$ over the t-interval $[-\pi/2, \pi/2]$, and revolve C once about the y-axis as shown in Figure 11.

With this choice, the distance r from the axis of revolution to the surface is $R\cos t$. So the surface area is calculated as follows:

$$
\begin{aligned}
A &= \int_{-\pi/2}^{\pi/2} 2\pi r\, ds \\
&= 2\pi \int_{-\pi/2}^{\pi/2} (R\cos t)\sqrt{\left(\frac{dx}{dt}\right)^2 + \left(\frac{dy}{dt}\right)^2}\, dt \\
&= 2\pi R\int_{-\pi/2}^{\pi/2} \cos t\sqrt{R^2 \sin^2 t + R^2 \cos^2 t}\, dt \\
&= 2\pi R^2 \int_{-\pi/2}^{\pi/2} \cos t\, dt \\
&= 2\pi R^2 [\sin t]_{-\pi/2}^{\pi/2} = 4\pi R^2
\end{aligned}
$$

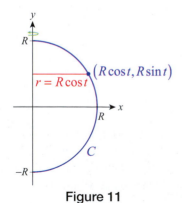

Figure 11

9.2 **Exercises**

1–4 Identify the curve as differentiable, smooth, or neither.

1. $x = t + 3$, $y = 2|t - 1|$, $0 \le t \le 2$

2. $x = t - 1$, $y = 2t^2$, $t \in \mathbb{R}$

3. $x = -2t^2$, $y = t^3 + 1$, $-5 \le t \le 5$

4. $x = 3\cos\theta$, $y = 2\sin\theta$, $-\pi < \theta \le \pi$

5. Show that (t, t) is a smooth parametrization of the (smooth) curve $y = x$, while (t^3, t^3) is not.

6–14 Find the equations of any horizontal or vertical tangent lines to the curve.

6. $x = 5t - 2$, $y = 6t^2 + 1$

7. $x = t^2 - t$, $y = 1 + 2t^2$

8. $x = 2t^2 + 1$, $y = (t - 2)^2$

9. $x = t^3 + 1$, $y = t$

10. $x = 3t^2$, $y = \dfrac{1}{t + 3}$, $t > -3$

11. $x = \sqrt{t}$, $y = \ln t$, $t \ge 1$

12. $x = \sqrt{t}$, $y = t^2 - 2$

13. $x = 2\cos t$, $y = 5\sin t$

14. $x = \tan t$, $y = \sec t$, $-\dfrac{\pi}{2} < t < \dfrac{\pi}{2}$

15–24 Find the values of dy/dx and d^2y/dx^2 for the curve at the given point.

15. $x = 2t - 1$, $y = 2t^2 + 1$; $\left(0, \dfrac{3}{2}\right)$

16. $x = 2t^3 + 1$, $y = (t + 1)^2$; $(3, 4)$

17. $x = \dfrac{1}{t + 1}$, $y = t^2 - 1$; $(1, -1)$

18. $x = \sqrt{t}$, $y = 2t + 1$; $(1, 3)$

19. $x = te^t$, $y = e^{-t}$; $(0, 1)$

20. $x = \cos t$, $y = \ln(\cos t)$; $\left(\dfrac{1}{2}, -\ln 2\right)$

21. $x = \sin t$, $y = \sin 2t$; $\left(\dfrac{\sqrt{2}}{2}, 1\right)$

22. $x = \dfrac{\sin^3 t}{3}$, $y = \dfrac{\cos^3 t}{3}$; $\left(\dfrac{\sqrt{2}}{12}, \dfrac{\sqrt{2}}{12}\right)$

23. $x = \ln t$, $y = 2\sqrt{t}$; $(0, 2)$

24. $x = \cos 2t$, $y = \sin 3t$; $\left(\dfrac{1}{2}, 1\right)$

25–28 Find the value(s) of the parameter for any inflection point(s) of the given curve.

25. $x = t + 2$, $y = t^3 - 10t^2$

26. $x = t^2(t^2 - 4)$, $y = 2t + 3$

27. $x = t^4 - t$, $y = t^2$

28. $x = \sqrt{t}$, $y = t^2 - 5t$

29–34 Find the area enclosed by the given curve.

29. $x = \sin t$, $y = \sin\dfrac{t}{2}$, $0 \le t \le 2\pi$

30. $x = 2\cos t$, $y = \sin 2t$, $-\dfrac{\pi}{2} \le t \le \dfrac{\pi}{2}$

31. $x = 2t\cos t$, $y = 2t\sin t$, $-\dfrac{\pi}{2} \le t \le \dfrac{\pi}{2}$

32. $x = \dfrac{3}{4}\cos t + \dfrac{1}{4}\cos 3t$, $y = \dfrac{3}{4}\sin t - \dfrac{1}{4}\sin 3t$, $0 \le t \le 2\pi$

33. $x = \sin t$, $y = e^{-t/2}$, $0 \le t \le \pi$ and the y-axis

34. $x = t^4 - 4$, $y = t^3 - 2t$, $-\sqrt{2} \le t \le \sqrt{2}$

35. Use the parametric representation of the ellipse from Exercise 35 of Section 9.1 to arrive at its area formula of $A = \pi ab$.

36. Prove that the area A under one arch of the cycloid generated by a circle of radius a is three times the area of the circle, $A = 3\pi a^2$. (This result was first proved by the French mathematician Gilles de Roberval ca. 1630.)

37. Sometimes it may be necessary or convenient to divide a region between a curve and the y-axis into horizontal strips when approximating its area. Modify our discussion preceding Example 4 to arrive at the formula

$$\int_{y=g(a)}^{y=g(b)} x\, dy = \int_{t=a}^{t=b} f(t)g'(t)\, dt.$$

38–39 Use Exercise 37 to determine the area between the given curve and the y-axis.

38. $x = 2t - 1,\quad y = t^2 - 2,\quad \dfrac{1}{2} \le t \le 3$

39. $x = 2\cos t,\quad y = 1 + \sin t,\quad -\dfrac{\pi}{2} \le t \le \dfrac{\pi}{2}$

40–47 Find the arc length of the given curve over the indicated interval.

40. $x = t^2,\quad y = t^3 - \dfrac{t}{3},\quad 1 \le t \le 3$

41. $x = \sqrt{t},\quad y = \dfrac{t^2}{8} + \dfrac{1}{4t},\quad 1 \le t \le 9$

42. $x = e^t,\quad y = \dfrac{e^{2t}}{8} - t,\quad 0 \le t \le 2$

43. $x = \dfrac{1}{t},\quad y = \dfrac{t}{4} + \dfrac{1}{3t^3},\quad \dfrac{1}{2} \le t \le 1$

44. $x = \sin^3 t,\quad y = \cos^3 t,\quad 0 \le t \le 2\pi$

45. $x = \sqrt{2}e^t \sin t,\quad y = \sqrt{2}e^t \cos t,\quad 0 \le t \le \dfrac{\pi}{2}$

46. $x = t^3,\quad y = t\left(t^4 - \dfrac{9}{20}\right),\quad 0 \le t \le 1$

47. $x = -\ln t,\quad y = t + \dfrac{1}{4t},\quad \dfrac{1}{2e} \le t \le 1$

48–57 Find the area of the surface generated by revolving the parametric curve about the indicated axis.

48. $x = t + 1,\quad y = \dfrac{t-1}{2},\quad 1 \le t \le 5,$

    a. about the x-axis   **b.** about the y-axis

49. $x = 2t - 3,\quad y = 8 - 2t,\quad 2 \le t \le 4,$

    a. about the x-axis   **b.** about the y-axis

50. $x = 4t^2 - 4t + 4,\quad y = 2t - 1,\quad 1 \le t \le 2,$ about the x-axis

51. $x = t,\quad y = t^3 + \dfrac{1}{12t},\quad 1 \le t \le 2,$

    a. about the x-axis   **b.** about the y-axis

52. $x = t^2,\quad y = 2t^3 - \dfrac{t}{6},\quad 0 \le t \le 1,$

    a. about the x-axis   **b.** about the y-axis

53. $x = \ln t,\quad y = t + \dfrac{1}{4t},\quad 1 \le t \le e,$

    a. about the x-axis   **b.** about the y-axis

54. $x = t^5 - \dfrac{9t}{20},\quad y = t^3,\quad 0 \le t \le 2,$ about the x-axis

55. $x = 2\sin^3 t,\quad y = 2\cos^3 t,\quad 0 \le t \le \pi,$ about the y-axis

56. $x = 2\cos t,\quad y = \sin t,\quad 0 \le t \le \pi,$ about the x-axis

57. $x = 2t + 1,\quad y = \cosh 2t,\quad 0 \le t \le 1,$

    a. about the x-axis   **b.** about the y-axis

58. Suppose one arch of the cycloid generated by a circle of radius 1 rolling along the x-axis is rotated about the x-axis. Find the area of the resulting surface.

59. Revisit Example 5 by using the more general parametrization found in Exercise 34 of Section 9.1. Do you obtain the same answer?

60. Evaluate the integral in Example 6 by writing it as

$$\int \frac{1+t^2}{\sqrt{1+t^2}}\, dt = \int \frac{1}{\sqrt{1+t^2}}\, dt + \int \frac{t^2}{\sqrt{1+t^2}}\, dt;$$

then show

$$\int \frac{t^2}{\sqrt{1+t^2}}\, dt = t\sqrt{1+t^2} - \int \sqrt{1+t^2}\, dt,$$

and finally proceed to solve for $\int \sqrt{1+t^2}\, dt$.

61–64 Consider a particle moving along a curve in the xy-plane such that its coordinates at time t are $x = f(t)$, $y = g(t)$, $a \le t \le b$. Since the distance traveled by the particle at time t can be calculated from the arc length

$$s(t) = \int_a^t \sqrt{\left[x'(u)\right]^2 + \left[y'(u)\right]^2}\, du,$$

by Part I of the Fundamental Theorem of Calculus (Section 5.3) we obtain

$$\text{speed} = \frac{ds}{dt} = \frac{d}{dt}\int_a^t \sqrt{\left[x'(u)\right]^2 + \left[y'(u)\right]^2}\, du$$

$$= \sqrt{\left[x'(t)\right]^2 + \left[y'(t)\right]^2}.$$

In Exercises 61–64, use the above formula to determine the speed of the particle traveling along the given curve at the specified time. (Distance is measured in meters, time in seconds.)

61. $x = t^2$, $y = 2t + 1$; $t = 2$

62. $x = \ln(t+1)$, $y = t^2 - 2$; $t = 1$

63. $x = 3\cos 3t$, $y = \sin 3t$; $t = \dfrac{\pi}{2}$

64. $x = t^2$, $y = \sqrt{t}$; $t = 4$

65. Suppose that the position of a particle in the xy-coordinate system is given by $x = t^2 - 1$, $y = t^3 - 6t$, $t \ge 0$. When does the particle reach its maximum speed?

66. Using the discussion preceding Example 4, prove the following: If $x = f(t)$ and $y = g(t)$ define a parametric curve over $[a, b]$ such that g is continuous, f is continuously differentiable, and $y(x)$ is a continuous function of x, then

$$\int_{f(a)}^{f(b)} y\, dx = \int_a^b g(t) f'(t)\, dt.$$

67–70 Use the result of Exercise 66 to solve the exercise.

67. Find the volume of the solid generated by rotating the parametrically defined curve $x = 2t - 1$, $y = 1 - t^2$, $-1 \le t \le 1$, about the x-axis.

68. Repeat Exercise 67 for the parametric curve $x = \sin^3 t$, $y = \cos^3 t$, $-\pi/2 \le t \le \pi/2$.

69. Find the centroid of the region bounded by the parametric curve $x = 9 - t$, $y = \sqrt{t}$, $0 \le t \le 9$, and the coordinate axes.

70. Repeat Exercise 69 for the parametric curve $x = 3\cos t$, $y = 2\sin t$, $0 \le t \le \pi/2$.

71. Prove the following property of cycloids: If P is the intersection point of the rolling circle and the cycloid, then the line tangent to the cycloid at P passes through the highest point of the circle, while the normal line intersects the circle at its lowest point.

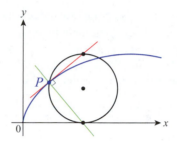

72.* Prove that if the curve C defined by $x = f(t)$ and $y = g(t)$, $t \in [a, b]$, is differentiable with both f' and g' bounded (meaning there are constants K_1 and K_2 such that $|f'(t)| \le K_1$ and $|g'(t)| \le K_2$ for all $t \in [a, b]$), then C has finite length (such curves of finite length are called *rectifiable*). (**Hint:** As a first step, use the boundedness of the derivatives and the Mean Value Theorem to prove that there is a constant M such that whenever $P_1(f(t_1), g(t_1))$ and $P_2(f(t_2), g(t_2))$ are two points on C, $|f(t_2) - f(t_1)| \le M |t_2 - t_1|$, and $|g(t_2) - g(t_1)| \le M |t_2 - t_1|$. As a consequence, $|P_1 P_2| \le 2M |t_2 - t_1|$. Now you can use a Riemann-sum argument to finish your proof.)

73. Use Exercise 72 to prove that if the parametric curve C is continuously differentiable on $[a, b]$ (i.e., both f' and g' have continuous derivatives on $[a, b]$), then C is rectifiable.

74–78 *True or False?* Determine whether the given statement is true or false. In case of a false statement, explain or provide a counterexample.

74. Every smooth parametric curve is differentiable.

75. Every continuous, differentiable parametric curve is smooth.

76. If C is smooth at (x_0, y_0), then for every parametrization such that $x_0 = f(t_0)$ and $y_0 = g(t_0)$, both f' and g' are continuous at t_0 and at least one of $f'(t_0)$ and $g'(t_0)$ is nonzero.

77. If the graph of a parametric curve is a continuously differentiable function $y = f(x)$, then the curve is smooth.

78. If a parametric curve is defined by $x = f(t)$ and $y = g(t)$, then $\dfrac{d^2 y}{dx^2} = \dfrac{d^2 y / dt^2}{d^2 x / dt^2}$.

9.2 Technology Exercises

79. Use a computer algebra system to approximate the length of the ellipse parametrized by $x = 5\cos t$, $y = 4\sin t$, $0 \le t \le 2\pi$.

80–81 Use a computer algebra system to approximate the length of the curve with the given parametrization.

80. $\left(\sqrt{t}, \sin t \right)$, $0 \le t \le 2\pi$

81. $\left(\ln(t+1), \sqrt{t} \right)$, $0 \le t \le 9$

82–83 Use a computer algebra system to approximate the area of the surface obtained by rotating the given parametric curve about the **a.** x-axis and **b.** y-axis.

82. $\left(3t - 2, \sqrt{1 + \dfrac{1}{t}} \right)$, $1 \le t \le 2$

83. $\left(4t + 1, \sqrt{t^2 + 1} \right)$, $-\dfrac{1}{4} \le t \le 2$

9.3 **Polar Coordinates**

TOPIC 1 **The Polar Coordinate System**

TOPICS

1. The polar coordinate system
2. Coordinate conversion
3. The form of polar equations
4. Graphing polar equations

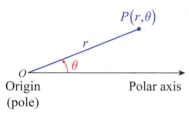

Figure 1

The Cartesian coordinate system has served us well to this point. But there are many situations for which a rectangular coordinate system is not the most natural choice. Some planar images and some equations in two variables have a symmetry that is awkward to express in terms of the familiar x- and y-coordinates. Polar coordinates provide an alternative framework for these cases.

The **polar coordinate system**, like the Cartesian coordinate system, serves as a means of locating points in the plane, and both systems are centered at a point O called the **origin**, sometimes referred to as the **pole** in the polar system. Starting from the origin O, a ray (or half-line) called the **polar axis** is drawn; in practice, the polar axis is usually drawn extending horizontally to the right, so that it corresponds with the positive x-axis in the Cartesian system. Now, given any point P in the plane other than the origin, the line segment $\overline{OP}$ has a unique positive length; we will label this length r (as in *radius*). Finally, we let θ denote the angle, measured counterclockwise, between the polar axis and the segment $\overline{OP}$, and we say (r,θ) are **polar coordinates** of the point P (see Figure 1). The origin is the unique point for which $r = 0$ and for which the angle θ is irrelevant; the coordinates $(0,\theta)$ refer to O for any angle θ. Figure 2 illustrates the process of determining r and θ for several points.

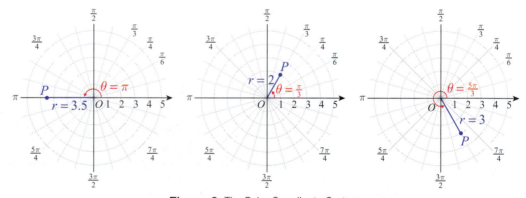

Figure 2 The Polar Coordinate System

Before proceeding further, it is important to recognize one very critical difference between Cartesian and polar coordinates. A given point P corresponds to unique Cartesian coordinates (x,y), but the polar coordinates (r,θ) of P as described above are only one of an infinite number of ways of specifying P in the polar system. Our familiarity with trigonometric functions indicates one reason for this: $(r,\theta + 2n\pi)$ also represents P, since θ and $\theta + 2n\pi$ have the same terminal sides for any integer n. Further, $(-r,\theta + (2n+1)\pi)$ also represents P for any integer n, given the interpretation that $-r$ indicates travel in the opposite direction through the origin. These observations are illustrated in Figure 3 with alternate descriptions of the points from Figure 2.

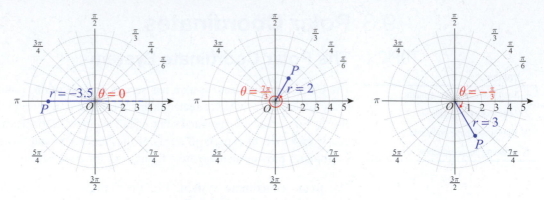

Figure 3 Alternate Polar Coordinates

Example 1 ✎

Plot the points given by the following polar coordinates.

a. $\left(2, \dfrac{3\pi}{4}\right)$ b. $\left(3.5, -\dfrac{5\pi}{2}\right)$ c. $\left(-1, \dfrac{4\pi}{3}\right)$

Solution

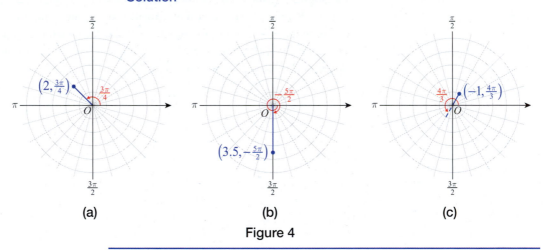

(a)   (b)   (c)

Figure 4

TOPIC 2 Coordinate Conversion

Since we have two systems by which to specify points in the plane, it should come as no surprise that we will occasionally need to be able to translate information from one system to the other. As we will soon see, this will be especially useful when faced with an equation that is awkward to graph in one coordinate system, but straightforward in the other. Fortunately, converting from Cartesian coordinates to polar coordinates, and vice versa, is easily accomplished. To do so, we will assume that the polar axis is aligned with the positive x-axis and that a fixed point P has Cartesian coordinates (x, y) and polar coordinates (r, θ). Then, as seen in Figure 5, $r^2 = x^2 + y^2$ and

$$\cos\theta = \frac{x}{r}, \quad \sin\theta = \frac{y}{r}, \quad \text{and} \quad \tan\theta = \frac{y}{x}.$$

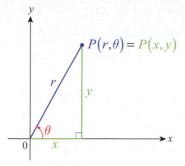

Figure 5
Converting between Polar and
Cartesian Coordinates

These relations, and the diagram in Figure 5, should seem very familiar—we encountered them first in defining the trigonometric functions. They are restated here as conversion formulas.

Formula ⚲

Coordinate Conversion

Converting from Polar to Cartesian Coordinates

Given (r,θ), x and y are defined as follows:

$$x = r\cos\theta, \quad y = r\sin\theta$$

Converting from Cartesian to Polar Coordinates

Given (x,y), r and θ are defined as follows:

$$r^2 = x^2 + y^2, \quad \tan\theta = \frac{y}{x} \quad (x \neq 0)$$

Make note of the quadrant of the original Cartesian coordinate when converting, to be sure the polar coordinates fall in the same quadrant (see Example 3).

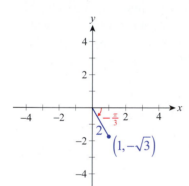

Figure 6

Example 2 ✎

Convert the following points from polar to Cartesian coordinates.

a. $\left(2, -\frac{\pi}{3}\right)$ **b.** $\left(-3, \frac{\pi}{4}\right)$

Solution

a. We calculate $x = 2\cos(-\pi/3) = 1$ and $y = 2\sin(-\pi/3) = -\sqrt{3}$, so the Cartesian coordinates are $\left(1, -\sqrt{3}\right)$ (see Figure 6).

b. We calculate $x = -3\cos(\pi/4) = -3/\sqrt{2}$ and $y = -3\sin(\pi/4) = -3/\sqrt{2}$, so the Cartesian coordinates are $\left(-3/\sqrt{2}, -3/\sqrt{2}\right)$ (see Figure 7).

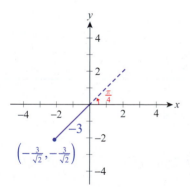

Figure 7

Example 3 ✎

Convert the following points from Cartesian to polar coordinates.

a. $(-3, 2)$ **b.** $\left(\sqrt{3}, -1\right)$

Solution

a. To avoid making an error, be sure you have some rough idea of what the answer should be before doing any calculations. Since $(-3, 2)$ is in the second quadrant, one possible conversion will lead to $\pi/2 < \theta < \pi$. To get the exact angle, we use the fact that $\tan\theta = -\frac{2}{3}$, so $\theta = \tan^{-1}\left(-\frac{2}{3}\right) \approx 2.55$ (remember, this is in radians). Depending on your calculator, though, you may have found $\tan^{-1}\left(-\frac{2}{3}\right) \approx -0.59$, an angle in the fourth quadrant; if so, it is up to you to remember to either add π in

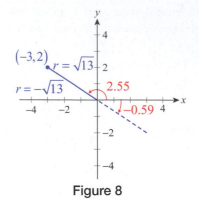

Figure 8

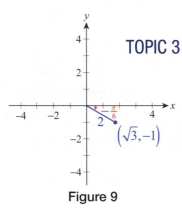

Figure 9

order to get an angle in the second quadrant, or to use a negative value for r. The radius is more easily determined: $r = \sqrt{(-3)^2 + (2)^2} = \sqrt{13}$. The two answers we have found are thus $\left(\sqrt{13}, 2.55\right)$ and $\left(-\sqrt{13}, -0.59\right)$ (see Figure 8).

b. The polar coordinates of $\left(\sqrt{3}, -1\right)$ are more easily determined. The point lies in the fourth quadrant, so the two most obvious conversions will lead to either $3\pi/2 < \theta < 2\pi$ or $-\pi/2 < \theta < 0$. The coordinates $\left(\sqrt{3}, -1\right)$ give rise to a familiar triangle with convenient angles, and it's probably easiest to use $\theta = -\pi/6$. The radius is $r = 2$, so one set of polar coordinates for the point is $\left(2, -\pi/6\right)$ (see Figure 9).

TOPIC 3   The Form of Polar Equations

In the most general terms, a *polar equation* is an equation in the variables r and θ that defines a relationship between these polar coordinates (just as equations in x and y define a relationship between rectangular coordinates). A solution to a polar equation is thus an ordered pair (r, θ) that makes the equation true, and the graph of a polar equation consists of all such ordered pairs. Many polar equations can be written in the form $r = f(\theta)$; in such a form, the distance from the origin is expressed as a function of the angle θ and the graph of the equation is usually fairly easy to determine. We will soon study many examples of polar equations, but first we will examine the process of translating an equation from one coordinate system to another.

The basic tools for such translation are the formulas for coordinate conversion used in Examples 2 and 3. Translating from rectangular coordinates to polar coordinates is particularly straightforward: simply replace every occurrence of x with $r \cos \theta$ and every occurrence of y with $r \sin \theta$. Translation in the opposite direction may require significantly more effort.

Example 4 ✎

Rewrite the equation $x^2 - 2x + y^2 = 0$ in polar form.

Solution

Making the appropriate substitutions in the equation $x^2 - 2x + y^2 = 0$ and simplifying, we obtain the following:

$$(r \cos \theta)^2 - 2(r \cos \theta) + (r \sin \theta)^2 = 0$$
$$r^2 \cos^2 \theta - 2r \cos \theta + r^2 \sin^2 \theta = 0$$
$$r^2 \left(\cos^2 \theta + \sin^2 \theta\right) - 2r \cos \theta = 0$$
$$r^2 - 2r \cos \theta = 0$$

Example 5 ✐

Rewrite the equation $2r = \sec\theta$ in rectangular form.

Solution

One good way to begin is to rewrite the equation as follows:

$$2r = \sec\theta$$

$$2r = \frac{1}{\cos\theta}$$

$$r\cos\theta = \frac{1}{2}$$

Now we recognize the term on the left-hand side as x, and we see that the equation is $x = \frac{1}{2}$.

TOPIC 4 Graphing Polar Equations

Gaining a sense of confidence in your ability to graph polar equations calls for nothing more than familiarity with a number of examples. As with equations in rectangular coordinates, we will begin with some very simple, but illustrative, examples.

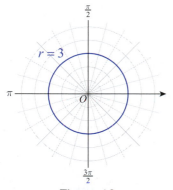

Figure 10

Example 6 ✐

Sketch the graphs of the following polar equations, and then convert the equations to rectangular coordinates.

a. $r = 3$ **b.** $\theta = \dfrac{2\pi}{3}$

Solution

a. Since θ doesn't appear in the equation $r = 3$, θ is allowed to take on any value. The equation thus describes all points (r, θ) for which $r = 3$; that is, a circle of radius 3 as shown in Figure 10. The equation in rectangular coordinates is $x^2 + y^2 = 9$.

b. In this equation, r is allowed to take on any value. But every point (r, θ) that satisfies the equation must have $\theta = 2\pi/3$. The graph is thus a straight line passing through the origin (see Figure 11), and $y/x = \tan\theta = \tan(2\pi/3) = -\sqrt{3}$, so $y = -\sqrt{3}x$.

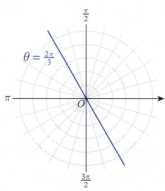

Figure 11

In general we expect an equation in polar coordinates to contain both r's and θ's. As with rectangular coordinates, the most basic approach to sketching the graph of a polar equation is to plot some representative points and to connect the points as seems appropriate. This method, applied judiciously and perhaps in combination with some algebra, will take us far.

Example 7 ✎

Sketch the graph of the equation $r = 2\cos\theta$.

Solution

We can begin by calculating some values of r for given values of θ:

θ	0	$\dfrac{\pi}{6}$	$\dfrac{\pi}{4}$	$\dfrac{\pi}{3}$	$\dfrac{\pi}{2}$	$\dfrac{2\pi}{3}$	$\dfrac{3\pi}{4}$	$\dfrac{5\pi}{6}$
r	2	$\sqrt{3}$	$\sqrt{2}$	1	0	-1	$-\sqrt{2}$	$-\sqrt{3}$

Now if we plot the pairs (r, θ) from the table, we obtain the points shown in Figure 12. These certainly appear to lie along the circumference of a circle, so the points have been connected with a curve. If we convert the equation into rectangular coordinates as shown below, we see that the graph indeed is a circle of radius 1 centered at $x = 1$, $y = 0$.

$$r = 2\cos\theta$$
$$r^2 = 2r\cos\theta$$
$$x^2 + y^2 = 2x$$
$$x^2 - 2x + y^2 = 0$$
$$x^2 - 2x + 1 + y^2 = 1$$
$$(x-1)^2 + y^2 = 1$$

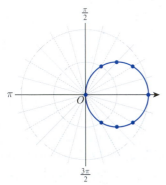

Figure 12 $r = 2\cos\theta$

We have used symmetry in the past as an aid in graphing functions and equations, and the concept is no less useful in polar coordinates. Consider the three types of symmetry illustrated in Figure 13.

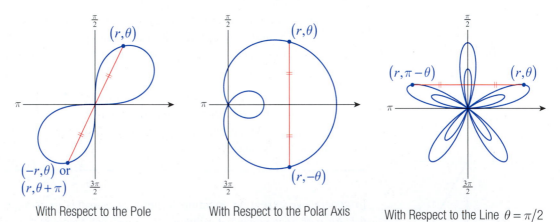

With Respect to the Pole   With Respect to the Polar Axis   With Respect to the Line $\theta = \pi/2$

Figure 13 Symmetry in Polar Coordinates

Algebraically, symmetry can be recognized with the following tests.

Definition 💡

Symmetry of Polar Equations

An equation in r and θ is **symmetric with respect to**

1. the **pole** if replacing r with $-r$ (or replacing θ with $\theta + \pi$) results in an equivalent equation;

2. the **polar axis** if replacing θ with $-\theta$ results in an equivalent equation;

3. the **vertical line** $\theta = \pi/2$ if replacing θ with $\pi - \theta$ results in an equivalent equation.

We will conclude this section with a catalog of some polar equations that arise frequently enough to have been given names. Exploring a few of these further will give us the opportunity to apply the above symmetry tests and gain more familiarity with graphing in polar coordinates.

Common Polar Equations and Graphs

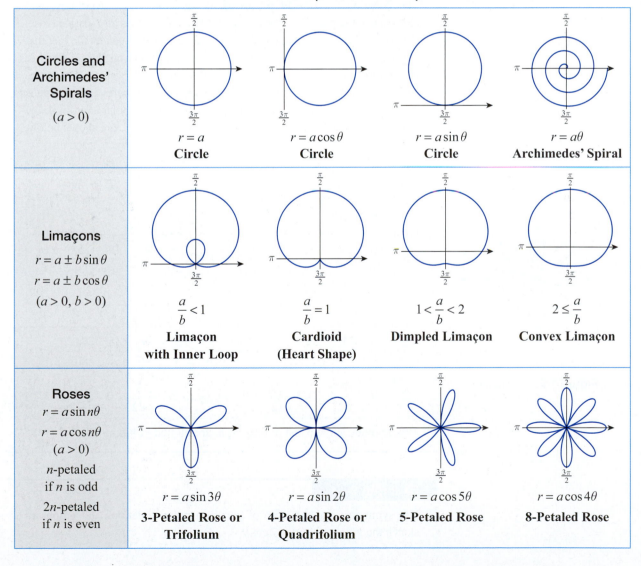

Circles and Archimedes' Spirals ($a > 0$)	$r = a$ **Circle**	$r = a\cos\theta$ **Circle**	$r = a\sin\theta$ **Circle**	$r = a\theta$ **Archimedes' Spiral**
Limaçons $r = a \pm b\sin\theta$ $r = a \pm b\cos\theta$ ($a > 0, b > 0$)	$\dfrac{a}{b} < 1$ **Limaçon with Inner Loop**	$\dfrac{a}{b} = 1$ **Cardioid (Heart Shape)**	$1 < \dfrac{a}{b} < 2$ **Dimpled Limaçon**	$2 \le \dfrac{a}{b}$ **Convex Limaçon**
Roses $r = a\sin n\theta$ $r = a\cos n\theta$ ($a > 0$) n-petaled if n is odd $2n$-petaled if n is even	$r = a\sin 3\theta$ **3-Petaled Rose or Trifolium**	$r = a\sin 2\theta$ **4-Petaled Rose or Quadrifolium**	$r = a\cos 5\theta$ **5-Petaled Rose**	$r = a\cos 4\theta$ **8-Petaled Rose**

Common Polar Equations and Graphs

Lemniscates (figure-eight shapes)	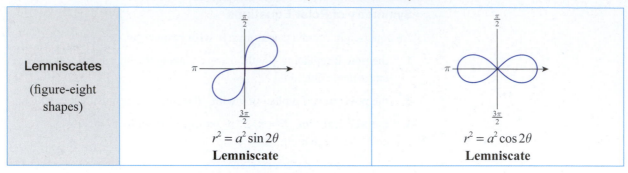 $$r^2 = a^2 \sin 2\theta$$ **Lemniscate**	$$r^2 = a^2 \cos 2\theta$$ **Lemniscate**

Example 8 ✐

Use symmetry and the table of common polar equations and graphs to sketch the graph of $r = 3\cos 4\theta$.

Solution

The equation $r = 3\cos 4\theta$ possesses all three kinds of symmetry, and reference to the preceding table indicates that the graph is an 8-petaled rose. Several aspects of the graph can be determined easily, such as the fact that the maximum distance between the origin and points on the graph is 3; this follows from the fact that $-1 \leq \cos 4\theta \leq 1$ for all θ. Some points can be easily calculated by hand, but a graphing calculator or software package is very useful in generating a large number of points. The polar coordinates of points on the graph for $0 \leq \theta \leq 2\pi$ in increments of $\pi/16$, are shown below.

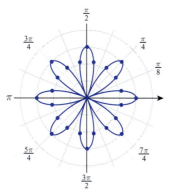

Figure 14 $r = 3\cos 4\theta$

θ	0	$\dfrac{\pi}{16}$	$\dfrac{\pi}{8}$	$\dfrac{3\pi}{16}$	$\dfrac{\pi}{4}$	$\dfrac{5\pi}{16}$	$\dfrac{3\pi}{8}$	$\dfrac{7\pi}{16}$	$\dfrac{\pi}{2}$	$\dfrac{9\pi}{16}$	$\dfrac{5\pi}{8}$
r	3	$\dfrac{3}{\sqrt{2}}$	0	$-\dfrac{3}{\sqrt{2}}$	-3	$-\dfrac{3}{\sqrt{2}}$	0	$\dfrac{3}{\sqrt{2}}$	3	$\dfrac{3}{\sqrt{2}}$	0
θ	$\dfrac{11\pi}{16}$	$\dfrac{3\pi}{4}$	$\dfrac{13\pi}{16}$	$\dfrac{7\pi}{8}$	$\dfrac{15\pi}{16}$	π	$\dfrac{17\pi}{16}$	$\dfrac{9\pi}{8}$	$\dfrac{19\pi}{16}$	$\dfrac{5\pi}{4}$	$\dfrac{21\pi}{16}$
r	$-\dfrac{3}{\sqrt{2}}$	-3	$-\dfrac{3}{\sqrt{2}}$	0	$\dfrac{3}{\sqrt{2}}$	3	$\dfrac{3}{\sqrt{2}}$	0	$-\dfrac{3}{\sqrt{2}}$	-3	$-\dfrac{3}{\sqrt{2}}$
θ	$\dfrac{11\pi}{8}$	$\dfrac{23\pi}{16}$	$\dfrac{3\pi}{2}$	$\dfrac{25\pi}{16}$	$\dfrac{13\pi}{8}$	$\dfrac{27\pi}{16}$	$\dfrac{7\pi}{4}$	$\dfrac{29\pi}{16}$	$\dfrac{15\pi}{8}$	$\dfrac{31\pi}{16}$	2π
r	0	$\dfrac{3}{\sqrt{2}}$	3	$\dfrac{3}{\sqrt{2}}$	0	$-\dfrac{3}{\sqrt{2}}$	-3	$-\dfrac{3}{\sqrt{2}}$	0	$\dfrac{3}{\sqrt{2}}$	3

These points, and the rest of the graph, are plotted in Figure 14.

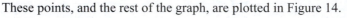

Example 9 ✐

Use symmetry and the table of common polar equations and graphs to sketch the graph of $r = 1 + 2\cos\theta$.

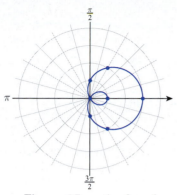

Figure 15 $r = 1 + 2\cos\theta$

Solution

The graph of $r = 1 + 2\cos\theta$ is a limaçon with an inner loop. Since cosine is an even function

$$1 + 2\cos\theta = 1 + 2\cos(-\theta)$$

and the graph is symmetric with respect to the polar axis. Some points on the graph are as follows:

θ	0	$\dfrac{\pi}{3}$	$\dfrac{\pi}{2}$	$\dfrac{2\pi}{3}$	π	$\dfrac{4\pi}{3}$	$\dfrac{3\pi}{2}$	$\dfrac{5\pi}{3}$	2π
r	3	2	1	0	-1	0	1	2	3

The graph of $r = 1 + 2\cos\theta$ is shown in Figure 15.

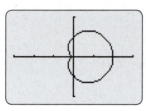

$[-3,3]$ by $[-2,2]$

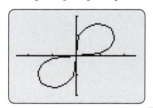

$[-2.5,2.5]$ by $[-2,2]$

Figure 16

Technology Note 🖥

Graphing calculators and computer algebra systems can be very useful in constructing accurate graphs of polar equations, though their capabilities are limited and vary considerably. Figure 16 shows the graphs of a cardioid and a lemniscate on a graphing calculator. To generate those graphs the selected graphing mode needs to be **POL** and the polar equations are entered in the **Y=** editor. Figure 17 illustrates how *Mathematica* is used to graph the same cardioid and lemniscate. The built-in command is **PolarPlot**.

Often, a polar equation must be solved for r in order to use the graphing feature of a particular technology. Note that in order to graph the lemniscate shown, we solved the equation $r^2 = 4\sin 2\theta$ for r and then typed in both the positive and negative square roots that resulted.

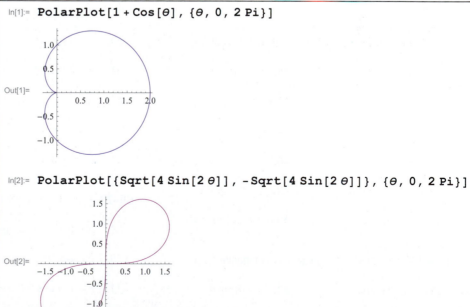

Figure 17 Graphing Polar Equations in *Mathematica*

9.3 **Exercises**

1–6 Plot the point given by the polar coordinates.

1. $\left(-1, \dfrac{5\pi}{4}\right)$ **2.** $\left(-5, \dfrac{3\pi}{2}\right)$

3. $\left(\dfrac{1}{4}, -\dfrac{7\pi}{6}\right)$ **4.** $\left(\sqrt{3}, -\dfrac{\pi}{3}\right)$

5. $\left(\dfrac{44}{9}, -\pi\right)$ **6.** $\left(\dfrac{7}{\sqrt{2}}, \dfrac{\pi}{2}\right)$

7–12 Convert the point from polar to Cartesian coordinates.

7. $\left(5, \dfrac{7\pi}{4}\right)$ **8.** $(0, 2\pi)$

9. $\left(6.25, -\dfrac{3\pi}{4}\right)$ **10.** $\left(-2.25, \dfrac{\pi}{4}\right)$

11. $\left(3, -\dfrac{5\pi}{6}\right)$ **12.** $\left(-11, \dfrac{5\pi}{6}\right)$

13–18 Convert the point from Cartesian to polar coordinates.

13. $(-3, 0)$ **14.** $\left(-6, \sqrt{3}\right)$

15. $(12, -1)$ **16.** $(8, 0)$

17. $\left(-\sqrt{3}, 9\right)$ **18.** $(-5, -5)$

19–30 Rewrite the rectangular equation in polar form.

19. $x^2 + y^2 = 25$ **20.** $x^2 + y^2 = 81$

21. $x = 12$ **22.** $y = 16$

23. $y = x$ **24.** $y = b$

25. $x = 16a$ **26.** $x^2 + y^2 = a$

27. $x^2 + y^2 = 4ax$ **28.** $x^2 + y^2 = 4ay$

29. $y^2 - 4 = 4x$ **30.** $x^2 + y^2 = 36a^2$

31–40 Rewrite the polar equation in rectangular form.

31. $r = 5\cos\theta$ **32.** $r = 8\sin\theta$

33. $r = 7$ **34.** $\theta = \dfrac{\pi}{6}$

35. $18r = 9\csc\theta$ **36.** $r = 2\sec\theta$

37. $r^2 = \sin 2\theta$ **38.** $r = \dfrac{2}{1 - \cos\theta}$

39. $r = \dfrac{12}{4\sin\theta + 7\cos\theta}$ **40.** $r = \dfrac{16}{4 + 4\sin\theta}$

41–46 Rewrite the polar equation in rectangular form; then sketch the graph.

41. $r = 2$ **42.** $r = 6$

43. $\theta = \dfrac{5\pi}{6}$ **44.** $\theta = \dfrac{\pi}{4}$

45. $r = 7\sec\theta$ **46.** $r = 2\csc\theta$

47–68 Sketch a graph of the given polar equation.

47. $r = 4$ **48.** $r = 5$

49. $\theta = \dfrac{4\pi}{3}$ **50.** $\theta = \dfrac{-\pi}{3}$

51. $r = 6\cos\theta$ **52.** $r = 2\sin\theta$

53. $r = 3 - 3\sin\theta$ **54.** $r = 6 + 5\cos\theta$

55. $r = 7(1 + \cos\theta)$ **56.** $r = 2(1 - 2\sin\theta)$

57. $r = 4 - 3\sin\theta$ **58.** $r = 3 + 4\sin\theta$

59. $r = 3\sin 3\theta$ **60.** $r = 5\sin 3\theta$

61. $r = 2\sin 2\theta$ **62.** $r = 4\sin 2\theta$

63. $r = 5\cos 5\theta$ **64.** $r = 4\cos 5\theta$

65. $r = 4\cos 4\theta$ **66.** $r = 3\cos 4\theta$

67. $r^2 = 16\sin 2\theta$ **68.** $r^2 = 9\cos 2\theta$

69–72 Find all points of intersection of the given polar curves.

69. $r = \sin\theta, \quad r = \cos\theta$

70. $r = \sin 2\theta, \quad r = \cos\theta$

71. $r = 1 - \cos\theta, \quad r = 1 + \sin\theta$

72. $r^2 = 4\sin\theta, \quad r = 1 - \sin\theta$

73. For a fixed $a \in \mathbb{R}$, explain in geometric terms how the graphs of $f(\theta)$ and $f(\theta - a)$ are related. (**Hint:** For guidance, recall the rectangular analogue.)

74. **a.** Describe the graph of $r = \sec(\theta - \pi/4)$.

   b. How are the graphs of $r = k \sec(\theta - \pi/4)$ related as k ranges over nonzero values? (Do not use graphing technology.)

9.3 Technology Exercises

75–81   Use a graphing calculator or computer algebra system to sketch the given curve. Whenever applicable, explore how different values of the parameter(s) affect the shape of the graph. Experiment with both integer and noninteger parameters.

75. $r = \cos k\theta$

76. $r = 1 - k_1 \sin k_2 \theta$

77. $r = \dfrac{1 + k \sin \theta}{1 - k \sin \theta}$

78. $r = \theta \cos \theta, \quad -2\pi \le \theta \le 2\pi$ (Garfield curve)

79. $r = 1 + 2\sin \dfrac{\theta}{2}$ (nephroid of Freeth)

80. $r = k_1 + k_2 \theta$

81. $r = 1 - k_1 \cos k_2 \theta$

9.4 Calculus in Polar Coordinates

TOPICS

1. Tangents to polar curves

2. Area and arc length in polar coordinates

In this section, we adapt our tools of calculus in order to apply them to curves and regions defined by polar coordinates. The techniques we learned for working with parametric equations allow us to do so.

TOPIC 1   Tangents to Polar Curves

We will focus on the most commonly occurring polar curves, those defined by an equation of the form $r = f(\theta)$. If we are interested in determining the tangent to a polar curve at a particular point, it is important to realize we are still primarily interested in dy/dx —Exercise 19 and the examples below illustrate why. And to find dy/dx, we use the coordinate conversion equations to write

$$x = r\cos\theta = f(\theta)\cos\theta \quad \text{and} \quad y = r\sin\theta = f(\theta)\sin\theta.$$

We now have two equations in the parameter θ describing the curve, and if f is a differentiable function of θ, then when $dx/d\theta \neq 0$,

$$\frac{dy}{dx} = \frac{dy/d\theta}{dx/d\theta} = \frac{\dfrac{d}{d\theta}\left[f(\theta)\sin\theta\right]}{\dfrac{d}{d\theta}\left[f(\theta)\cos\theta\right]} = \frac{f'(\theta)\sin\theta + f(\theta)\cos\theta}{f'(\theta)\cos\theta - f(\theta)\sin\theta}.$$

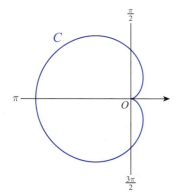

Figure 1 $r = 1 - \cos\theta$

Example 1 ✒

Find all points where the cardioid C defined by $r = 1 - \cos\theta$ has a horizontal or vertical tangent.

Solution

Using the notation above, $f(\theta) = 1 - \cos\theta$ and so

$$\frac{dy}{dx} = \frac{\sin^2\theta + (1-\cos\theta)\cos\theta}{\sin\theta\cos\theta - (1-\cos\theta)\sin\theta} = \frac{1-\cos^2\theta + (1-\cos\theta)\cos\theta}{\sin\theta(2\cos\theta - 1)}$$

$$= \frac{(1-\cos\theta)(1+2\cos\theta)}{\sin\theta(2\cos\theta - 1)}.$$

By determining where the numerator and denominator of this fraction are equal to 0, we can locate the horizontal and vertical tangent lines of the cardioid (we restrict attention to $\theta \in [0, 2\pi]$ since the cardioid is traced out exactly once over this parametric interval).

$$(1-\cos\theta)(1+2\cos\theta) = 0 \quad \text{when} \quad \theta = 0, 2\pi, \frac{2\pi}{3}, \text{ or } \frac{4\pi}{3}$$

$$\sin\theta(2\cos\theta - 1) = 0 \quad \text{when} \quad \theta = 0, \pi, 2\pi, \frac{\pi}{3}, \text{ or } \frac{5\pi}{3}$$

The fraction as a whole is 0 when the numerator is 0 and the denominator is not 0, and the fraction is undefined when the exact opposite occurs. Thus,

C has horizontal tangent lines at $\theta = 2\pi/3$ and $\theta = 4\pi/3$;

C has vertical tangent lines at $\theta = \pi$, $\theta = \pi/3$, and $\theta = 5\pi/3$.

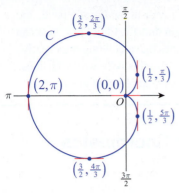

Figure 2

The points on C corresponding to these values of the parameter θ are shown in Figure 2.

The point $(0,0)$, which is both the initial and terminal point of C, requires a bit more analysis to prove the existence of a tangent line. Since both the numerator and denominator are 0 at $\theta = 0$ (or 2π), l'Hôpital's Rule is called for:

$$\lim_{\theta \to 0} \frac{(1 - \cos\theta)(1 + 2\cos\theta)}{\sin\theta(2\cos\theta - 1)} = \lim_{\theta \to 0} \frac{\sin\theta(1 + 2\cos\theta) + (1 - \cos\theta)(-2\sin\theta)}{\cos\theta(2\cos\theta - 1) + \sin\theta(-2\sin\theta)}$$

$$= \frac{0}{1} = 0,$$

so C has a horizontal tangent line at $(0,0)$.

The curve C of Example 1 has several interesting properties. One is that it possesses a tangent at the point $(0,0)$ even though it is not smooth at $(0,0)$. This shows that while a smooth parametrization of a curve implies the curve has a continuously turning tangent line, the reverse is not true—a curve may have a continuously turning tangent line (as the cardioid above does) but not have a smooth parametrization. A second property is that the cardioid has exactly three tangent lines with any fixed slope. Example 1 showed the existence of three horizontal and three vertical tangent lines, but the number of tangent lines would have been the same for any given slope.

Example 2 ✐

Figure 3 depicts the 3-petaled rose $r = \sin 3\theta$. Determine the slopes of the lines tangent to the tips of each petal.

Solution

Although we want to find dy/dx at each petal tip, we first identify the value of the parameter θ at each tip by solving the equation $dr/d\theta = 0$. This works because each tip is a point on the parametric curve where r makes the transition from being an increasing function of θ to a decreasing function of θ. Note that the rose is traced out entirely over a parametric interval of $[0, \pi]$, so we look for values of θ restricted to this interval:

$$\frac{dr}{d\theta} = 3\cos 3\theta, \quad \text{so} \quad \frac{dr}{d\theta} = 0 \implies \theta = \frac{\pi}{6}, \frac{\pi}{2}, \text{ or } \frac{5\pi}{6}$$

Now, since $y = \sin 3\theta \sin\theta$ and $x = \sin 3\theta \cos\theta$, we know that

$$\frac{dy}{dx} = \frac{dy/d\theta}{dx/d\theta} = \frac{3\cos 3\theta \sin\theta + \sin 3\theta \cos\theta}{3\cos 3\theta \cos\theta - \sin 3\theta \sin\theta}.$$

We leave it as an exercise (Exercise 63) to show that

$$\frac{dy}{dx}\Big|_{\theta = \pi/6} = -\sqrt{3}, \quad \frac{dy}{dx}\Big|_{\theta = \pi/2} = 0, \quad \text{and} \quad \frac{dy}{dx}\Big|_{\theta = 5\pi/6} = \sqrt{3}.$$

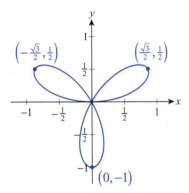

Figure 3 $r = \sin 3\theta$

Alternatively, since $dr/d\theta = 0$ at each point (x, y) at the tip of a petal, and the slope of the line passing through the origin and (x, y) is y/x, we can find the slope of these three specific tangent lines by calculating the negative reciprocal, or $-x/y$, at each of the three tips. But this only works for these particular tangents to the curve—other tangents require knowing dy/dx.

TOPIC 2 Area and Arc Length in Polar Coordinates

Suppose R is a region bounded between two rays $\theta = a$ and $\theta = b$, where $0 < b - a \le 2\pi$, and by the polar equation $r = f(\theta)$, where f is a continuous positive function as shown in Figure 4. We can find the area A of the region by following the same sort of Riemann sum construction that we used in Chapter 5.

To do so, recall that the area of a sector of a circle is given by the formula $\frac{1}{2}r^2\theta$, where r is the radius of the circle and θ is the angle defining the sector (this follows from the fact that the sector's area is $\theta/2\pi$ times the area of the complete circle, πr^2). If we partition the θ-interval $[a, b]$ into n subintervals defined by the $n + 1$ points $P = \{a = \theta_0, \theta_1, \ldots, \theta_{n-1}, \theta_n = b\}$ as shown in Figure 4, then the area of the ith sector is $\frac{1}{2}r^2\Delta\theta_i$, depicted in Figure 5.

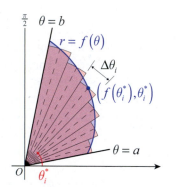

Figure 4 Area of a Polar Region

And similar to our previous Riemann sum approximations, the radius r of this i^{th} sector can be given the value $f(\theta_i^*)$, where θ_i^* is any sample angle chosen from the subinterval $[\theta_{i-1}, \theta_i]$.

Adding up the areas of the n sectors, we have the following Riemann sum:

$$\sum_{i=1}^{n} \frac{1}{2}\left[f(\theta_i^*)\right]^2 \Delta\theta_i$$

In the limit of finer and finer partitions, we arrive at the following formula for the area of such a polar region.

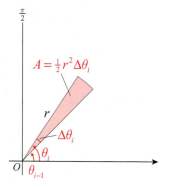

Figure 5

$$A = \lim_{\|P\| \to 0} \sum_{i=1}^{n} \frac{1}{2}\left[f(\theta_i^*)\right]^2 \Delta\theta_i = \int_a^b \frac{1}{2}\left[f(\theta)\right]^2 d\theta$$

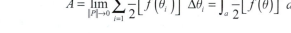

Formula 🔍

Area of a Polar Region

The area A of a region bounded between two rays $\theta = a$ and $\theta = b$, where $0 < b - a \le 2\pi$, and by the polar equation $r = f(\theta)$, where f is a continuous positive function, is given by

$$A = \int dA = \int_a^b \frac{1}{2}\left[f(\theta)\right]^2 d\theta.$$

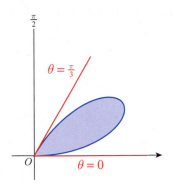

Figure 6 One Petal of $r = \sin 3\theta$

Example 3 ✎

Find the area enclosed by the rose $r = \sin 3\theta$.

Solution

The graph of the polar equation $r = \sin 3\theta$ is shown in Figure 3. By symmetry, we only need to find the area of one petal and then triple it to find the total area. But in order to find the area of one petal, we need to determine appropriate rays $\theta = a$ and $\theta = b$ that bound the petal.

If we choose to work with the petal in the 1st quadrant, then it is fairly clear that $\theta = 0$ is the lower bound (we will verify that guess soon). And since we have already determined that the tip of the petal occurs at $\theta = \pi/6$, we might feel comfortable guessing that $\theta = \pi/3$ is the other bounding ray. But how would we determine this if we hadn't already done the work of Example 2? We would do so by identifying the values of θ for which $r = f(\theta) = 0$; remember that the entire polar curve is traced out over the interval $[0, \pi]$, so we only need to look for θ's in that interval. Since $\sin 3\theta = 0$ when 3θ is a multiple of π, the values we seek are $\theta = 0, \pi/3, 2\pi/3$, and π. The first-quadrant petal is bounded by $\theta = 0$ and $\theta = \pi/3$, as shown in Figure 6. (Note that, somewhat counterintuitively, the downward-pointing petal is bounded by $\theta = \pi/3$ and $\theta = 2\pi/3$, since $\sin 3\theta$ is negative for θ's in that interval.)

We are now ready to find the area enclosed by one petal.

$$\int_0^{\pi/3} \frac{1}{2}\sin^2 3\theta \, d\theta = \frac{1}{2}\int_0^{\pi/3} \frac{1 - \cos 6\theta}{2} \, d\theta \qquad \sin^2 u = \frac{1 - \cos 2u}{2}$$

$$= \frac{1}{4}\left[\theta - \frac{1}{6}\sin 6\theta\right]_0^{\pi/3}$$

$$= \frac{1}{4}\left(\frac{\pi}{3}\right) = \frac{\pi}{12}$$

And so the entire area enclosed by $r = \sin 3\theta$ is as follows:

$$A = 3 \cdot \frac{\pi}{12} = \frac{\pi}{4}$$

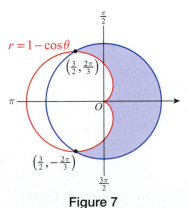

Figure 7
Area between Polar Curves

Example 4 ✎

Find the area of the region outside the cardioid $r = 1 - \cos\theta$ and inside the circle that intersects the cardioid at the two points indicated in Figure 7.

Solution

We find the area between two polar curves in the same way we find the area between two functions of x—by subtracting the area under the smaller function from the area under the larger function. The only difference is that in this context, "larger" means farther away from the origin, and we must remember to use the polar integral formula for area. In general, if

$g(\theta) \leq f(\theta)$ for $a \leq \theta \leq b$, the area bounded by the rays $\theta = a$ and $\theta = b$ and the two curves $r = g(\theta)$ and $r = f(\theta)$ is given by

$$A = \int_a^b \frac{1}{2}\left[f(\theta)\right]^2 d\theta - \int_a^b \frac{1}{2}\left[g(\theta)\right]^2 d\theta = \frac{1}{2}\int_a^b \left(\left[f(\theta)\right]^2 - \left[g(\theta)\right]^2\right)d\theta.$$

In this example, $g(\theta) = 1 - \cos\theta$ and the two rays are $\theta = -2\pi/3$ and $\theta = 2\pi/3$, as shown in Figure 7. And from the two points of intersection of the curves, we can deduce that $f(\theta) = \frac{3}{2}$; that is, the larger curve is the circle of radius $\frac{3}{2}$ centered at the origin. So the area between the two curves is calculated as follows:

$$A = \frac{1}{2}\int_{-2\pi/3}^{2\pi/3}\left[\left(\frac{3}{2}\right)^2 - (1 - \cos\theta)^2\right]d\theta$$

$$= \int_0^{2\pi/3}\left(\frac{5}{4} + 2\cos\theta - \cos^2\theta\right)d\theta \qquad \text{By symmetry, double the integral over } [0, 2\pi/3].$$

$$= \int_0^{2\pi/3}\left(\frac{3}{4} + 2\cos\theta - \frac{\cos 2\theta}{2}\right)d\theta \qquad \cos^2\theta = \frac{1 + \cos 2\theta}{2}$$

$$= \left[\frac{3}{4}\theta + 2\sin\theta - \frac{1}{4}\sin 2\theta\right]_0^{2\pi/3}$$

$$= \frac{9\sqrt{3}}{8} + \frac{\pi}{2}.$$

The last formula we will restate in terms of polar coordinates is the arc length formula, and we will again make use of what we already know about parametric equations. In Section 9.2, we determined that the arc length of the curve defined parametrically over an interval $[a, b]$ is

$$L = \int_a^b \sqrt{\left(\frac{dx}{dt}\right)^2 + \left(\frac{dy}{dt}\right)^2}\, dt.$$

In this context, the parameter is θ and if our polar curve is given as $r = f(\theta)$, then we know $x = f(\theta)\cos\theta$ and $y = f(\theta)\sin\theta$.

$$L = \int_a^b \sqrt{\left(\frac{dx}{d\theta}\right)^2 + \left(\frac{dy}{d\theta}\right)^2}\, d\theta$$

$$= \int_a^b \sqrt{\left(\frac{d}{d\theta}\left[f(\theta)\cos\theta\right]\right)^2 + \left(\frac{d}{d\theta}\left[f(\theta)\sin\theta\right]\right)^2}\, d\theta$$

$$= \int_a^b \sqrt{\left[f'(\theta)\cos\theta - f(\theta)\sin\theta\right]^2 + \left[f'(\theta)\sin\theta + f(\theta)\cos\theta\right]^2}\, d\theta$$

$$= \int_a^b \sqrt{\left[f(\theta)\right]^2 + \left[f'(\theta)\right]^2}\, d\theta.$$

This formula is often written in terms of r, in which case it appears as follows.

Formula 🔧

Arc Length of a Polar Curve

Assume the curve C is defined by $r = f(\theta)$ over the θ-interval $[a,b]$, where f has a continuous first derivative over the interval and the curve is traced out exactly once (except possibly for a finite number of self-intersections). Then the arc length of C is given by

$$L = \int_a^b \sqrt{r^2 + \left(\frac{dr}{d\theta}\right)^2}\, d\theta.$$

Example 5 ✎

Find the arc length of the cardioid $r = 1 - \cos\theta$.

Solution

The curve from Figure 1, which we are very familiar with now, is shown again in Figure 8. Note that

$$r^2 + \left(\frac{dr}{d\theta}\right)^2 = (1 - \cos\theta)^2 + (\sin\theta)^2$$
$$= 1 - 2\cos\theta + \cos^2\theta + \sin^2\theta$$
$$= 2 - 2\cos\theta.$$

Since we need to integrate the square root of this expression, the trigonometric identity $1 - \cos\theta = 2\sin^2(\theta/2)$ is useful.

$$L = \int_0^{2\pi} \sqrt{2 - 2\cos\theta}\, d\theta$$
$$= \int_0^{2\pi} \sqrt{4\sin^2\left(\frac{\theta}{2}\right)}\, d\theta$$
$$= 2\int_0^{2\pi} \sin\left(\frac{\theta}{2}\right) d\theta \qquad \text{Note that } \sin(\theta/2) \geq 0 \text{ for } 0 \leq \theta \leq 2\pi.$$
$$= \left[-4\cos\left(\frac{\theta}{2}\right)\right]_0^{2\pi}$$
$$= 4 + 4 = 8$$

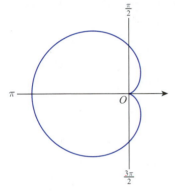

Figure 8 $r = 1 - \cos\theta$

9.4 **Exercises**

1–14 Find the slope of the line tangent to the given polar curve at the indicated point.

1. $r = 2\cos\theta; \quad \theta = \dfrac{\pi}{6}$

2. $r = 2\sin 2\theta; \quad \theta = \dfrac{\pi}{6}$

3. $r = a\theta; \quad \theta = \dfrac{\pi}{2}$

4. $r = a\sin\theta; \quad \theta = \dfrac{\pi}{6}$

5. $r = 5\sec^2\theta; \quad \theta = \dfrac{\pi}{4}$

6. $r = e^\theta; \quad \theta = \dfrac{\pi}{6}$

7. $r^2 = 4\csc\theta; \quad \theta = \dfrac{\pi}{6}$

8. $r^2 = \sin 2\theta; \quad \theta = \dfrac{\pi}{3}$

9. $r = \dfrac{1}{\theta}; \quad \theta = \dfrac{2\pi}{3}$

10. $r = a(1 + \sin\theta); \quad \theta = \dfrac{2\pi}{3}$

11. $r = \ln\theta; \quad \theta = 1$

12. $r = \cos 4\theta; \quad \theta = \dfrac{\pi}{4}$

13. $r = \sin\dfrac{\theta}{4}; \quad \theta = \pi$

14. $r = 3 - 2\sin\theta; \quad \theta = \dfrac{\pi}{4}$

15–18 Find all points where the given polar curve has a horizontal or vertical tangent line.

15. $r = 1 + \sin\theta$

16. $r = a\cos\theta$

17. $r = a(1 + \cos\theta)$

18. $r^2 = 4\cos 2\theta$

19–20 Notice that even in the case of polar curves, we still need to find dy/dx in order to determine tangents. The derivative $dr/d\theta$, while indirectly informing us about the tangent line at any given point, does not yield the slope of the tangent line. Exercises 19 and 20 shed some light on the relationship between $dr/d\theta$ and dy/dx.

19.* Suppose that the curve C is defined by the equation $r = f(\theta)$, P is a point on C, and there is a unique line tangent to C at P. Let β be the angle determined by the ray $\overrightarrow{OP}$ and the tangent at P. Prove that if $dr/d\theta$ is not equal to 0 at P, then

$$\tan\beta = \frac{r}{dr/d\theta}.$$

(**Hint:** In the formula we obtained for dy/dx in the text, divide the numerator and the denominator by $f'(\theta)\cos\theta$, and use the trigonometric identity

$$\tan(\alpha_1 + \alpha_2) = \frac{\tan\alpha_1 + \tan\alpha_2}{1 - \tan\alpha_1 \tan\alpha_2}.)$$

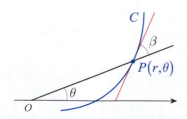

20. Prove that if the graph of $r = f(\theta)$ passes through the pole and α is an angle such that $r = f(\alpha) = 0$, then the slope of the tangent to the graph at the pole is $\tan\alpha$; that is, the line $\theta = \alpha$ is tangent to the graph at the pole. (**Hint:** If $f'(\alpha) \neq 0$, use the formula we obtained for dy/dx in the text.

Otherwise, examine $\displaystyle\lim_{\theta\to\alpha}\frac{\sin\theta \cdot f'(\theta)}{\cos\theta \cdot f'(\theta)}$.)

21–26 Notice that if the graph of $r = f(\theta)$ passes through the pole, we can use Exercise 20 to determine the polar equation of a tangent at the pole by solving the equation $f(\theta) = 0$. Use this observation to find all lines tangent to the given curve at the pole.

21. $r = 1 - \cos\theta$

22. $r = \sin 3\theta$

23. $r^2 = \cos 2\theta$

24. $r = \cos 4\theta$

25. $r = 4\sin\theta$

26. $r = a\sin n\theta$

27–41 Find the area enclosed by the given curve.

27. $r = 3\sin\theta$

28. $r = -2\cos\theta$

29. $r = 1 + \sin\theta$

30. $r = 4 - 4\sin\theta$

31. $r = \dfrac{3}{2} - \sin\theta$

32. $r = 3 + 2\cos\theta$

33. $r = 3 - \sin\theta$

34. $r = 3(1 + \cos\theta)$

35. $r = 2 + \cos\theta + \sin\theta$

36. $r = 2\cos 3\theta$

37. $r = \sin 6\theta$

38. $r = 6\sin\theta$

39. $r = 4\sin 4\theta$

40. $r^2 = 2\sin 2\theta$

41. $r^2 = 4\cos 2\theta$

42–45 Find the area of the shaded region.

42.

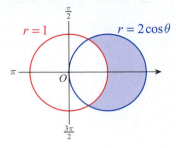

43.

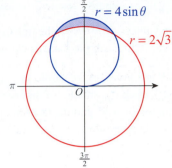

44.

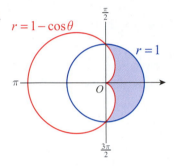

45.

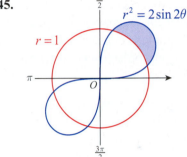

46–49 Find the area of the specified region.

46. The innermost loop of the spiral $r = 3\theta$ bounded by the polar axis and $\theta = \pi/2$

47. The inner loop of the limaçon $1 + 2\sin\theta$

48. The region common to the circle $r = 1$ and a petal of $r = 2\sin 3\theta$

49. The region inside $r = 3 + 2\sin\theta$ but outside $r = 2$

50–57 Find the arc length of the given polar curve.

50. $r = 3\sin\theta, \quad 0 \le \theta \le \pi$

51. $r = 3\theta, \quad 0 \le \theta \le 4\pi$

52. $r = 3e^\theta, \quad 0 \le \theta \le \pi$

53. $r = \csc\theta, \quad \dfrac{\pi}{3} \le \theta \le \dfrac{2\pi}{3}$

54.* $r = 1 - \sin\theta, \quad 0 \le \theta \le 2\pi$

55. $r = \theta^2, \quad 0 \le \theta \le 2\pi$

56.* $r = \sin^2\theta, \quad 0 \le \theta \le \pi$

57.* $r = 1 + \sin\theta, \quad 0 \le \theta \le \pi$

58. Use our derivation of the arc length formula for polar curves along with the surface area formula of Section 9.2 to arrive at the following formulas for the area of the surface generated by rotating the polar curve $r = f(\theta)$, $a \le \theta \le b$, where f' is continuous on $[a,b]$ and the curve is traced out only once over the interval.

$$A = 2\pi \int_a^b r\sin\theta \sqrt{r^2 + \left(\frac{dr}{d\theta}\right)^2}\, d\theta$$

(rotation about the polar axis)

$$A = 2\pi \int_a^b r\cos\theta \sqrt{r^2 + \left(\frac{dr}{d\theta}\right)^2}\, d\theta$$

(rotation about $\theta = \pi/2$)

59–62 Use Exercise 58 to find the surface area of the solid generated by rotating the given curve as described.

59. $r = \sin\theta$, rotated about the polar axis

60. $r = \sin\theta$, rotated about $\theta = \pi/2$

61. The spiral $r = 3e^\theta, 0 \le \theta \le \pi$, rotated about the polar axis

62. $r = 2(1 + \sin\theta)$, $0 \le \theta \le \pi/2$, rotated about $\theta = \pi/2$

63. Show that the values of dy/dx at the tips of the petals of the curve in Example 2 are as claimed.

64.* Prove that if P is a point on the cardioid $r = a(1 - \cos\theta)$, then the smaller of the two angles determined by $\overrightarrow{OP}$ and the tangent at P is one-half the angle determined by $\overrightarrow{OP}$ and the polar axis.

65. A point O inside a polar curve is called *equichordal* if every chord passing through O has the same length. (An obvious example is the center of a circle of radius r.) Prove that the pole is an equichordal point of the limaçon $r = a + \cos\theta$ (assume $a \ge 1$).

66. Find the polar equation of the circle whose diameter has the same length as the chords in the limaçon of Exercise 65, and prove that the areas of the limaçon and the circle are not equal. (As a consequence, we see that the existence of chords of equal length in every direction through a common point is insufficient to determine areas of regions.)

67. Suppose the polar curve $r = f(\theta)$, $a \le \theta \le b$, has length L and encloses an area of A square units. Prove that for any constant k, the curve $r = kf(\theta)$, $a \le \theta \le b$, has length $|k|L$ and encloses an area of $k^2 A$ square units.

9.4 Technology Exercises

68–72 Use a computer algebra system to sketch the given curve. Then use the integration capabilities of your technology to approximate its arc length.

68. $r = 2 + \cos 2\theta$

69. $r = \theta \cos\theta$, $0 \le \theta \le 3\pi$

70. $r = 3 - 2\sin 3\theta$

71. The inner loop of the limaçon $r = 1 + 2\sin\theta$

72. The 3-petaled rose $r = \sin 3\theta$

73–76 Use a computer algebra system to approximate the surface area of the described solid of revolution. (**Hint:** See Exercise 58.)

73. $r = 3e^\theta$, $0 \le \theta \le \pi$, rotated about the polar axis

74. $r^2 = \sin 2\theta$, rotated about the polar axis

75. The outer loop of the limaçon $r = 1 + 2\sin\theta$, rotated about $\theta = \pi/2$

76. $r = 4 - 4\cos 2\theta$, rotated about $\theta = \pi/2$

9.5   Conic Sections in Cartesian Coordinates

TOPICS

1. Ellipses

2. Parabolas

3. Hyperbolas

4. Rotating conic sections

The three types of curves known as *ellipses*, *parabolas*, and *hyperbolas* have appeared frequently in this text, and will continue to do so. To this point, the algebraic nature of these curves has been the focus—that is, the relationship between the variables defining the curves. In the last two sections of this chapter, the geometric characteristics of these curves will be explored as well.

To begin, the three families of curves are collectively known as **conic sections** because each type of curve arises from the intersection of a plane with a (double) circular cone. As shown in Figure 1, an **ellipse** is a closed curve resulting from the intersection of a plane with only one *nappe* of the cone (the two nappes of a cone each extend outward from the common vertex). A **parabola** results from intersecting a cone with a plane that is parallel to a line on the surface of the cone passing through the vertex; note that a parabola also intersects only one nappe of the cone. Finally, a **hyperbola** is the intersection of a cone with a plane that intersects both nappes. In each case, we specify that the intersecting plane does not contain the vertex—a figure that results when the intersecting plane *does* contain the vertex is called a *degenerate* conic section and is a point, a line, or a pair of intersecting lines.

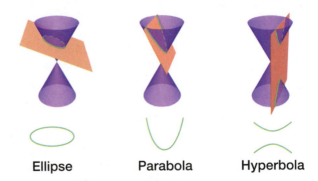

Ellipse   Parabola   Hyperbola

Figure 1 Conic Sections

You have worked with many examples of equations defining each type of conic section, but you may not be familiar with the following general classification.

Definition 💡

Algebraic Classification of Conic Sections

As a curve in the Cartesian plane, every conic section is the graph of an equation of the form $Ax^2 + Bxy + Cy^2 + Dx + Ey + F = 0$, where the coefficients are real constants. Assuming the conic section described by such an equation is nondegenerate, the curve is

1.   an **ellipse** if $B^2 - 4AC < 0$;

2.   a **parabola** if $B^2 - 4AC = 0$;

3.   a **hyperbola** if $B^2 - 4AC > 0$.

The quantity $B^2 - 4AC$ is called the **discriminant** of the equation.

In the first three topics of this section we explore equations of the form above when $B = 0$, and in the final topic we study the effect of allowing B to be nonzero.

TOPIC 1 Ellipses

As curves in the plane, ellipses are characterized by the following geometric property.

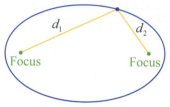

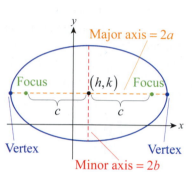

Figure 2 Ellipse

> ### Definition ♀
>
> **Geometric Characteristic of Ellipses**
>
> An ellipse consists of the set of points in the plane for which the sum of the distances d_1 and d_2 to two foci is a fixed constant (see Figure 2).

An ellipse with major and minor axes parallel to the x- and y-axes, respectively, can be written in the standard form

$$\frac{(x-h)^2}{a^2} + \frac{(y-k)^2}{b^2} = 1,$$

where (h, k) is the center of the ellipse and $2a$ and $2b$ are, respectively, the lengths of the horizontal and vertical axes of the ellipse (see Figure 3). The two foci lie on the major axis (the longer axis) of the ellipse, each a distance c from the center of the ellipse, where $c^2 = a^2 - b^2$. In general, we assume $a \geq b$; note that $c = 0$ if and only if $a = b$, in which case the ellipse is a circle and the two foci coincide with the center. The two vertices are located at the ends of the major axis, each a distance a from the center. If the equation above is rewritten in the form

Figure 3

$$\underbrace{\left(b^2\right)}_{A}x^2 + \underbrace{\left(a^2\right)}_{C}y^2 + \underbrace{\left(-2b^2h\right)}_{D}x + \underbrace{\left(-2a^2k\right)}_{E}y + \underbrace{\left(-a^2b^2 + h^2b^2 + k^2a^2\right)}_{F} = 0,$$

it is clear that it falls under the algebraic classification of an ellipse, as the discriminant is $B^2 - 4AC = -4b^2a^2 < 0$.

For an ellipse whose major axis is parallel to the y-axis, the corresponding equation has the form

$$\frac{(x-h)^2}{b^2} + \frac{(y-k)^2}{a^2} = 1,$$

where we again assume $a \geq b$.

Example 1 ✐

Graph the equation $25x^2 + 4y^2 + 100x - 24y + 36 = 0$, and determine the foci of the resulting ellipse.

Solution

The fact that the coefficients of x^2 and y^2 are both positive tells us that the discriminant is negative and thus falls into the classification of an ellipse,

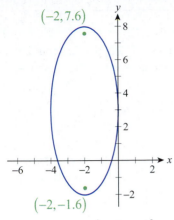

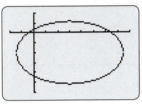

Figure 4 $\dfrac{(x+2)^2}{4}+\dfrac{(y-3)^2}{25}=1$

but we know little else from the equation in this form. By completing the square in both variables and rearranging terms, we can learn more:

$$25x^2 + 4y^2 + 100x - 24y + 36 = 0$$
$$25\left(x^2 + 4x + 4\right) + 4\left(y^2 - 6y + 9\right) = -36 + 100 + 36$$
$$25(x+2)^2 + 4(y-3)^2 = 100$$
$$\frac{(x+2)^2}{4} + \frac{(y-3)^2}{25} = 1$$

In this form, we can infer that the center of the ellipse is the point $(-2,3)$, that $a=5$ and $b=2$, and that the major axis is vertical. Further, since $c^2 = a^2 - b^2 = 21$, $c = \sqrt{21} \approx 4.6$ and so the foci are located ± 4.6 units away from the center, or at approximately $(-2,-1.6)$ and $(-2,7.6)$. A graph of the ellipse is shown in Figure 4.

Technology Note 💻

A graphing calculator or CAS can be used to sketch the graph of an ellipse, with the specific commands varying with the technology. Figure 5 shows the graph of the ellipse $x^2 - 6x + 2y^2 + 8y = 3$ on a graphing calculator. In order to generate this graph, the equation of the ellipse needs to be solved for y and then both the positive and negative square roots that result entered into the **Y=** editor.

The *Mathematica* command **ContourPlot**, used for this purpose, requires only three arguments: an equation, an interval for the variable x, and an interval for the variable y. Figure 6 illustrates the use of **ContourPlot** to graph the same ellipse, $x^2 - 6x + 2y^2 + 8y = 3$, with two additional arguments included that tell *Mathematica* to display the coordinate axes but not to display a frame around the graph.

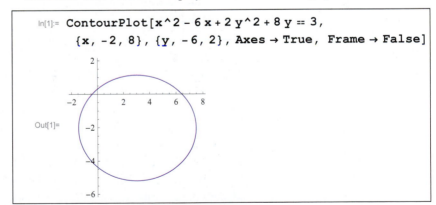

Figure 6

A circle, viewed as a type of ellipse, has only one focus, and that focus coincides with the center of the circle. Ellipses that are nearly circular have two foci that are relatively close to the center, while narrow ellipses have foci far away from the center. This gives us a convenient way to characterize the relative "skinniness" of an ellipse.

Definition 💡

Eccentricity of an Ellipse

Given an ellipse with major and minor axes of lengths $2a$ and $2b$, respectively, the **eccentricity** of the ellipse, often denoted by the symbol e, is defined by

$$e = \frac{c}{a} = \frac{\sqrt{a^2 - b^2}}{a}.$$

If $e = 0$, then $c = 0$ and the ellipse is actually a circle. At the other extreme, c may be close to (but cannot equal) a, in which case e is close to 1. An eccentricity close to 1 indicates a relatively narrow ellipse.

Johannes Kepler (1571–1630) was the German astronomer and mathematician who first demonstrated that the planets in our solar system follow elliptical orbits. He did this by laboring, over a period of 18 years, to mathematically model the astronomical observations of his predecessor Tycho Brahe. The ultimate result was Kepler's Laws of Planetary Motion.

Theorem 🔍

Kepler's Laws of Planetary Motion

1.   Each planet revolves around the sun in an elliptical orbit, with the sun at one focus of the ellipse.

2.   A line segment between the sun and a given planet sweeps out equal areas of the ellipse over equal periods of time.

3.   The ratio of the square of the period of one revolution of a planet and the cube of the length of the semimajor axis of its orbit is a constant independent of the planet.

The shapes of the various planetary orbits in our solar system vary widely, and these differences greatly influence the seasons on the planets. For instance, Mars has a much more eccentric orbit than Earth. The eccentricity of Mars' orbit is approximately 0.093, while the eccentricity of Earth's orbit is approximately 0.017. As a result, seasonal climate differences on Mars are much more dramatic than on Earth, including a large change in atmospheric pressure. Orbital eccentricities can be easily calculated from astronomical data, and we can use the eccentricity of a planet's orbit to answer particular questions.

Example 2 ✏️

Given that the furthest Earth gets from the sun is approximately 94.56 million miles and that the eccentricity of Earth's orbit is approximately 0.017, estimate the closest approach of the Earth to the sun.

Solution

Kepler's First Law states that each planet follows an elliptical orbit and that the sun is positioned at one focus of the ellipse. Thus the Earth is furthest from the sun when Earth is at the end of the ellipse's major axis on the other side of the center from the sun.

Using our standard terminology, and units of millions of miles, this means that

$$a + c = 94.56.$$

By the definition of eccentricity, we also know that

$$0.017 = \frac{c}{a},$$

or $c = 0.017a$. Combining these pieces of information, we get $1.017a = 94.56$, leading to $a = 92.98$ million miles. From this, we know $c = (0.017)(92.98) = 1.58$ million miles. The closest approach to the sun must be $a - c$, which is approximately 91.4 million miles.

TOPIC 2 **Parabolas**

As curves in the plane, parabolas are characterized by the following geometric property.

Figure 7 Parabola

Definition 💡

Geometric Characteristic of Parabolas

A parabola consists of the set of points in the plane that are the same distance d from a line (called the *directrix*) and a point not on the line (called the focus). See Figure 7 for an illustration of this property.

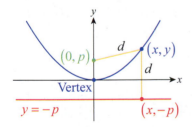

Figure 8

The general form for the equation of a parabola can be constructed by beginning with a parabola oriented as in Figure 8, where the vertex of the parabola has been placed at the origin, the focus is the point $(0, p)$, and the directrix is the line $y = -p$ (the distance $|p|$ is called the *focal length* of the parabola). The geometric property above leads immediately to the algebraic relationship $\sqrt{(x-0)^2 + (y-p)^2} = \sqrt{(x-x)^2 + (y+p)^2}$, which, after squaring both sides and simplifying, becomes $x^2 = 4py$. This can be generalized to describe a parabola with vertex at (h, k) by replacing x with $x - h$ and y with $y - k$, giving us the equation $(x - h)^2 = 4p(y - k)$. And to describe parabolas opening horizontally, the roles of x and y are simply reversed. In sum, a parabola oriented vertically or horizontally with vertex at (h, k) and focus p units from the vertex can be written in one of the two standard forms below.

Vertically oriented parabola:

$$(x - h)^2 = 4p(y - k) \qquad \begin{array}{l} p > 0 \;\;\Rightarrow\;\; \text{parabola opens up} \\ p < 0 \;\;\Rightarrow\;\; \text{parabola opens down} \end{array}$$

Horizontally oriented parabola:

$$(y-k)^2 = 4p(x-h)$$

$p > 0 \quad \Rightarrow \quad$ parabola opens right
$p < 0 \quad \Rightarrow \quad$ parabola opens left

Note that if either equation is written in the form $Ax^2 + Bxy + Cy^2 + Dx + Ey + F = 0$, $B = 0$ and either $A = 0$ or $C = 0$, so the discriminant $B^2 - 4AC = 0$ and indicates that the equation corresponds to a parabola.

Example 3 ✎

Find the focus and directrix of the parabola $-y^2 + 2x + 2y + 5 = 0$.

Solution

Since y is the squared variable, we know from the outset that the parabola is oriented horizontally. To identify the focus and directrix, we rewrite the equation by completing the square and factoring out a 4 in the final form:

$$-y^2 + 2x + 2y + 5 = 0$$
$$y^2 - 2y + 1 = 2x + 5 + 1$$
$$(y-1)^2 = 2(x+3)$$
$$(y-1)^2 = 4\left(\frac{1}{2}\right)(x+3)$$

In this form, we can see that the vertex of the parabola is at $(-3,1)$ and that $p = \frac{1}{2}$. Since p is positive the parabola opens to the right, the focus is at $\left(-\frac{5}{2},1\right)$, and the directrix is the line $x = h - p = -\frac{7}{2}$, as illustrated in Figure 9.

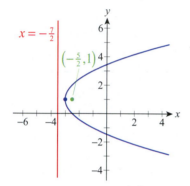

Figure 9 $(y-1)^2 = 4\left(\frac{1}{2}\right)(x+3)$

An important example of a parabola is the path of a thrown object under the influence of gravity (ignoring friction and other forces). Algebraically, we know that the height $h(t)$ of an object with initial velocity v_0 and initial height h_0 is

$$h(t) = -\frac{1}{2}gt^2 + v_0 t + h_0,$$

where g, a constant, is the acceleration due to gravity (see Example 3 in Section 3.1). This is a quadratic function, and we experience the fact that a parabola is the shape of its graph whenever we see, for example, a thrown baseball.

In some applications, the geometric properties of parabolas are the key issue. For instance, the fact that every point on a parabola is equidistant from its focus and its directrix has the consequence that the angles shown in Figure 10 are equal to one another at each point. That is, if we imagine rays of light emanating from the focus of the parabola, each ray reflected from the inner surface of the parabola is parallel to the axis of symmetry.

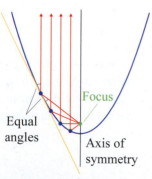

Figure 10 Parallel Rays of Reflection in a Parabola

This property is used in a variety of real-world applications. If a parabola is rotated about its axis of symmetry, a three-dimensional shape called a *paraboloid* is the result. A *parabolic mirror* is then made by coating the inner surface of a paraboloid with a reflective material. Parabolic mirrors are the basis of searchlights and vehicle headlights, with a light source placed at the focus of the paraboloid, and are also the basis of one design of telescope, in which incoming (parallel) starlight is reflected to an eyepiece at the focus.

Example 4 ✒

The Hale Telescope at Palomar Observatory in California is a very large reflecting telescope. The paraboloid is the top surface of a large cylinder of Pyrex glass 200 inches in diameter. Along the outer rim, the cylinder is 26.8 inches thick, while at the center, it is 23 inches thick. Where is the focus of the parabolic mirror located?

Solution

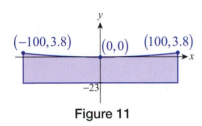

Figure 11

First, we need to draw a picture of the situation. In order to make the math as easy as possible, we can locate the origin of our coordinate system at the vertex of a parabolic cross-section of the mirror, and we can assume the parabola opens upward (see Figure 11).

Since we placed the vertex at $(0,0)$, we know the equation $x^2 = 4py$ describes the shape of the cross-section for some value p. If we can determine p, we can find the focus of the parabola.

To find p, we need the coordinates of another point on the parabola. The difference in thickness of the mirror between the center and the outer rim is 3.8 inches, and the mirror has a diameter of 200 inches, so the two points $(-100, 3.8)$ and $(100, 3.8)$ must lie on the graph. Substituting the coordinates of one of these points into the equation $x^2 = 4py$, we can solve for p.

$$(100)^2 = 4p(3.8)$$
$$10,000 = 15.2p$$
$$p \approx 657.9 \text{ inches}$$
$$p \approx 54.8 \text{ feet}$$

We know that the focus of a parabola is p units from the vertex, so the focus of the Hale Telescope is nearly 55 feet from the mirror.

TOPIC 3   Hyperbolas

As curves in the plane, hyperbolas are characterized by the following geometric property.

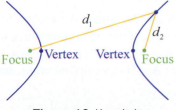

Figure 12 Hyperbola

Definition 💡

Geometric Characteristic of Hyperbolas

A hyperbola consists of the set of points in the plane for which the absolute value of the difference between the distances d_1 and d_2 to two foci is fixed (see Figure 12).

The standard form of the equation for a hyperbola whose foci are aligned horizontally or vertically is shown below.

Foci aligned horizontally:

$$\frac{(x-h)^2}{a^2} - \frac{(y-k)^2}{b^2} = 1 \text{ with asymptotes } y-k = \pm\frac{b}{a}(x-h)$$

Foci aligned vertically:

$$\frac{(y-k)^2}{a^2} - \frac{(x-h)^2}{b^2} = 1 \text{ with asymptotes } y-k = \pm\frac{a}{b}(x-h)$$

In both cases, the center of the hyperbola is (h,k), the vertices are located a units on either side of the center, and the foci are c units away from the center, where $c^2 = a^2 + b^2$. Note that if either equation is rewritten in the form $Ax^2 + Bxy + Cy^2 + Dx + Ey + F = 0$, the discriminant is equal to $B^2 - 4AC = 4a^2b^2 > 0$.

Example 5 ✐

Find the foci, vertices, and asymptotes of the hyperbola.

$$-16x^2 + 9y^2 + 96x + 18y = 279$$

Solution

As in Examples 1 and 3, we complete the square and rewrite the equation to put it in a more meaningful form:

$$-16x^2 + 9y^2 + 96x + 18y = 279$$
$$-16(x^2 - 6x + 9) + 9(y^2 + 2y + 1) = 279 - 144 + 9$$
$$-16(x-3)^2 + 9(y+1)^2 = 144$$
$$\frac{(y+1)^2}{16} - \frac{(x-3)^2}{9} = 1$$

This tells us that the foci are aligned vertically (since the positive fraction is in terms of y), that $a = 4$, $b = 3$, and the center is $(3,-1)$. So the vertices are the two points $(3,-1-4) = (3,-5)$ and $(3,-1+4) = (3,3)$. To locate the foci, we first calculate c from the fact that $c^2 = a^2 + b^2$; this gives us $c = 5$ in this case, so the two foci are the points $(3,-1-5) = (3,-6)$ and $(3,-1+5) = (3,4)$. Finally, the two asymptotes are the lines $y+1 = \pm\frac{4}{3}(x-3)$, shown as dashed lines in Figure 13.

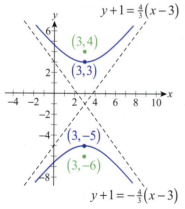

Figure 13 $\dfrac{(y+1)^2}{16} - \dfrac{(x-3)^2}{9} = 1$

Hyperbolas, like ellipses and parabolas, arise in a wide variety of contexts. For instance, hyperbolas are seen in architecture and structural engineering (think of the shape of a nuclear power plant's cooling towers) and in astronomy (comets that make a single pass through our solar system don't have elliptical orbits, but instead trace one branch of a hyperbola).

TOPIC 4   **Rotating Conic Sections**

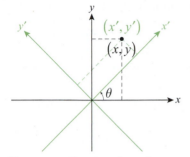

Figure 14 Two Coordinate Systems

Recall that the general form of the equation for a conic section is $Ax^2 + Bxy + Cy^2 + Dx + Ey + F = 0$. Geometrically, the condition $B \neq 0$ leads to a conic that is not oriented vertically or horizontally. The ellipses, parabolas, and hyperbolas graphed so far have all been aligned so that the axis (or axes in the case of ellipses or hyperbolas) of symmetry is parallel to the x-axis or y-axis. If $B \neq 0$, the graph is a conic section that has been rotated through some angle θ. We can exploit this fact to make the graphing of a rotated conic section relatively easy. Our method will be to introduce a new set of coordinate axes (i.e., a second Cartesian plane) rotated by an acute angle θ with respect to the original coordinate axes and then to graph the conic in the new coordinate system. Algebraically, the goal is to begin with an equation of the form

$$Ax^2 + Bxy + Cy^2 + Dx + Ey + F = 0$$

and define a new set of coordinate axes x' and y' in which the equation has the form

$$A'x'^2 + C'y'^2 + D'x' + E'y' + F' = 0.$$

That is, the coefficient B' in the new coordinate system is 0, and hence the graphing techniques previously learned apply.

We begin with a picture. Figure 14 is an illustration of two rectangular coordinate systems, with the new system rotated by an acute angle θ with respect to the original. The point in Figure 14 has two sets of coordinates corresponding to the two coordinate planes. However, the distance r between the origin and the point is the same in both, and this fact along with the introduction of the angle θ' shown in Figure 15 allow us to relate the two sets of coordinates.

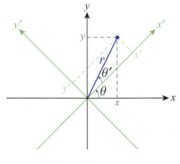

$$x = r\cos(\theta + \theta') \qquad x' = r\cos\theta'$$
$$y = r\sin(\theta + \theta') \qquad y' = r\sin\theta'$$

Figure 15
Relation between x', y', x, and y

We can now apply one of the trigonometric identities.

$$\begin{aligned} x &= r\cos(\theta + \theta') \\ &= r(\cos\theta\cos\theta' - \sin\theta\sin\theta') \\ &= r\cos\theta'\cos\theta - r\sin\theta'\sin\theta \\ &= x'\cos\theta - y'\sin\theta \end{aligned}$$

Similarly, $y = x'\sin\theta + y'\cos\theta$. We will also need to be able to express x' and y' in terms of x and y, and we can do so through the clever application of another trigonometric identity. By multiplying the two equations we have just derived by $\cos\theta$ and $\sin\theta$, respectively, and then adding the results, we obtain the following:

$$\begin{aligned} x\cos\theta &= x'\cos^2\theta - y'\sin\theta\cos\theta \\ + y\sin\theta &= x'\sin^2\theta + y'\sin\theta\cos\theta \\ \hline x\cos\theta + y\sin\theta &= x'(\cos^2\theta + \sin^2\theta) \\ &= x' \end{aligned}$$

A similar manipulation allows us to express y' in terms of x and y. All four relations are summarized as follows.

Formula 🔍

Rotation Relations

Given a rectangular coordinate system with axes x' and y' rotated by an acute angle θ with respect to axes x and y, as in Figure 14, the two sets of coordinates (x', y') and (x, y) for the same point are related by the following equations:

$$x = x'\cos\theta - y'\sin\theta \quad x' = x\cos\theta + y\sin\theta$$
$$y = x'\sin\theta + y'\cos\theta \quad y' = -x\sin\theta + y\cos\theta$$

Example 6 ✎

Given $\theta = \pi/6$, find the $x'y'$-coordinates of the point with xy-coordinates $(-1, 5)$.

Solution

Using the rotation relations we calculate the $x'y'$-coordinates as follows:

$$x' = (-1)\cos\frac{\pi}{6} + 5\sin\frac{\pi}{6}$$

$$= (-1)\left(\frac{\sqrt{3}}{2}\right) + (5)\left(\frac{1}{2}\right)$$

$$= \frac{5 - \sqrt{3}}{2}$$

$$y' = -(-1)\sin\frac{\pi}{6} + 5\cos\frac{\pi}{6}$$

$$= \frac{1}{2} + (5)\left(\frac{\sqrt{3}}{2}\right)$$

$$= \frac{1 + 5\sqrt{3}}{2}$$

The $x'y'$-coordinates are thus $\left(\dfrac{5 - \sqrt{3}}{2}, \dfrac{1 + 5\sqrt{3}}{2}\right)$.

We are familiar with the graph of $y = 1/x$. The next example gives us another perspective on this equation, written in the form $xy = 1$.

Example 7 ✐

Use the rotation $\theta = 45°$ to show that the graph of $xy = 1$ is a hyperbola.

Solution

Using the angle $\theta = 45°$ in the rotation relations, we convert the equation as follows:

$$xy = 1$$

$$\left(x'\cos 45° - y'\sin 45°\right)\left(x'\sin 45° + y'\cos 45°\right) = 1$$

$$\left(\frac{x'}{\sqrt{2}} - \frac{y'}{\sqrt{2}}\right)\left(\frac{x'}{\sqrt{2}} + \frac{y'}{\sqrt{2}}\right) = 1$$

$$\frac{x'^2}{2} - \frac{x'y'}{2} + \frac{x'y'}{2} - \frac{y'^2}{2} = 1$$

$$\frac{x'^2}{2} - \frac{y'^2}{2} = 1$$

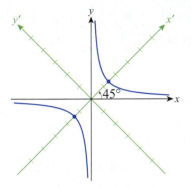

Figure 16

We recognize this last equation as a hyperbola in the $x'y'$-plane with its center at the origin and vertices $\sqrt{2}$ away from the center on the x'-axis. The asymptotes are $y' = \pm x'$, which correspond to the x- and y-axes (see Figure 16).

Remember that the goal in general is to determine θ so that the conversion of the equation

$$Ax^2 + Bxy + Cy^2 + Dx + Ey + F = 0$$

has no $x'y'$-term. Example 7 will serve as the inspiration; by replacing x and y with the corresponding expressions for x' and y' and then simplifying the result, we can derive a formula for the appropriate angle θ. We begin with the replacements:

$$A\left(x'\cos\theta - y'\sin\theta\right)^2 + B\left(x'\cos\theta - y'\sin\theta\right)\left(x'\sin\theta + y'\cos\theta\right)$$
$$+ C\left(x'\sin\theta + y'\cos\theta\right)^2 + D\left(x'\cos\theta - y'\sin\theta\right)$$
$$+ E\left(x'\sin\theta + y'\cos\theta\right) + F = 0$$

When the left-hand side of this equation is expanded and like terms collected, the result is an equation of the form

$$A'x'^2 + B'x'y' + C'y'^2 + D'x' + E'y' + F' = 0$$

where

$$A' = A\cos^2\theta + B\cos\theta\sin\theta + C\sin^2\theta$$
$$B' = 2\left(C - A\right)\cos\theta\sin\theta + B\left(\cos^2\theta - \sin^2\theta\right)$$
$$C' = A\sin^2\theta - B\cos\theta\sin\theta + C\cos^2\theta$$
$$D' = D\cos\theta + E\sin\theta$$
$$E' = -D\sin\theta + E\cos\theta$$
$$F' = F.$$

Since we want $B' = 0$, this gives us an equation in θ to solve. To do so, we will use the double-angle identities for both sine and cosine in reverse:

$$2(C - A)\cos\theta\sin\theta + B(\cos^2\theta - \sin^2\theta) = 0$$
$$(C - A)\sin 2\theta + B\cos 2\theta = 0$$
$$B\cos 2\theta = (A - C)\sin 2\theta$$
$$\frac{\cos 2\theta}{\sin 2\theta} = \frac{A - C}{B}$$
$$\cot 2\theta = \frac{A - C}{B}$$

This result is summarized in the following definition.

Definition 💡

Elimination of the *xy*-Term

The graph of the equation $Ax^2 + Bxy + Cy^2 + Dx + Ey + F = 0$ in the xy-plane is the same as the graph of the equation $A'x'^2 + C'y'^2 + D'x' + E'y' + F' = 0$ in the $x'y'$-plane, where the acute angle of rotation θ between the two coordinate systems satisfies

$$\cot 2\theta = \frac{A - C}{B}.$$

Although formulas relating the primed coefficients to the unprimed coefficients were derived in the preceding discussion, in practice it is often easier to determine θ and use the rotation relations to convert equations, as shown in the next example.

Example 8 ✍

Graph the conic section $x^2 + 2\sqrt{3}xy + 3y^2 + \sqrt{3}x - y = 0$ by first determining the appropriate angle θ by which to rotate the axes.

Solution

$$\cot 2\theta = \frac{1 - 3}{2\sqrt{3}} = -\frac{1}{\sqrt{3}}$$

Since the angle θ is to be acute, it must be the case that $2\theta = 2\pi/3$ and hence $\theta = \pi/3$. By the rotation relations, then

$$x = x'\cos\frac{\pi}{3} - y'\sin\frac{\pi}{3} = \frac{1}{2}x' - \frac{\sqrt{3}}{2}y'$$

and

$$y = x'\sin\frac{\pi}{3} + y'\cos\frac{\pi}{3} = \frac{\sqrt{3}}{2}x' + \frac{1}{2}y'.$$

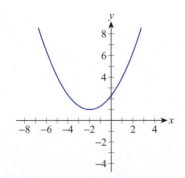

Making these substitutions into the equation, we obtain the following:

$$\left(\frac{1}{2}x' - \frac{\sqrt{3}}{2}y'\right)^2 + 2\sqrt{3}\left(\frac{1}{2}x' - \frac{\sqrt{3}}{2}y'\right)\left(\frac{\sqrt{3}}{2}x' + \frac{1}{2}y'\right) + 3\left(\frac{\sqrt{3}}{2}x' + \frac{1}{2}y'\right)^2$$

$$+ \sqrt{3}\left(\frac{1}{2}x' - \frac{\sqrt{3}}{2}y'\right) - \left(\frac{\sqrt{3}}{2}x' + \frac{1}{2}y'\right) = 0$$

Multiplying out and collecting like terms in this equation results in

$$y' = 2x'^2,$$

which we recognize as the equation of a parabola with its vertex at the origin. The graph of this equation in the $x'y'$-plane is shown in Figure 17.

Figure 17

9.5 Exercises

1–6 Use the discriminant to determine whether the given equation represents an ellipse, a parabola, or a hyperbola.

1. $y^2 + 2y + 12x + 13 = 0$

2. $2x^2 + 12x - y^2 - 2y + 9 = 0$

3. $4x^2 + 3y^2 + 18y + 19 = 8x$

4. $-2x^2 - 8xy + 2y^2 + 2y + 5 = 0$

5. $3x^2 - 6xy + 3y^2 + 3x - 9 = 0$

6. $x^2 - xy + 4y^2 + 2x - 3y + 1 = 0$

7–21 Identify the type of conic section defined by the equation and match the equation with its graph (labeled A–O).

7. $\dfrac{(x-1)^2}{4} + \dfrac{y^2}{16} = 1$

8. $(x-2)^2 = 4y$

9. $x^2 - y^2 = 1$

10. $x^2 + \dfrac{(y-3)^2}{4} = 1$

11. $\dfrac{y^2}{4} - (x-1)^2 = 1$

12. $\dfrac{(x-3)^2}{9} + \dfrac{(y-2)^2}{25} = 1$

13. $\dfrac{x^2}{9} - \dfrac{(y+2)^2}{4} = 1$

14. $(x+2)^2 = 3(y-1)$

15. $\dfrac{(x-1)^2}{4} + y^2 = 1$

16. $\dfrac{(x-3)^2}{4} - \dfrac{(y+1)^2}{9} = 1$

17. $y^2 = 4(x+1)$

18. $(x-1)^2 = -(y-2)$

19. $(y-1)^2 = -2(x-2)$

20. $\dfrac{(x+2)^2}{9} + \dfrac{(y-1)^2}{4} = 1$

21. $(y+2)^2 - \dfrac{(x-2)^2}{4} = 1$

A.

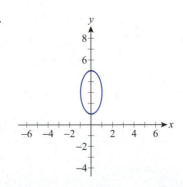

B.

C.

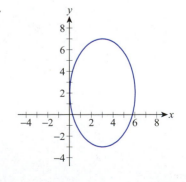

D.

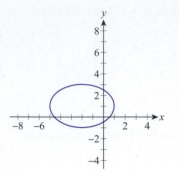

E.

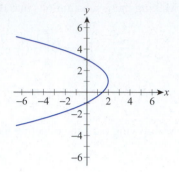

F.

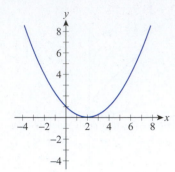

G.

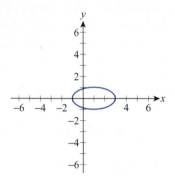

H.

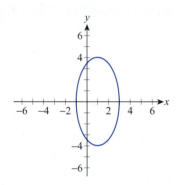

I.

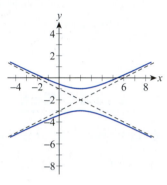

J.

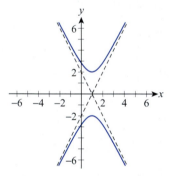

K.

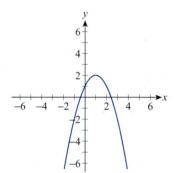

L.

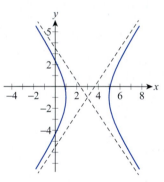

M.

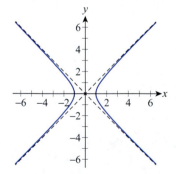

N.

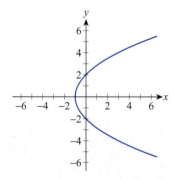

O.

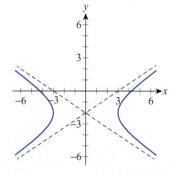

22–25 Find the eccentricity and the lengths of the major and minor axes of the ellipse.

22. $x^2 + 9y^2 = 36$

23. $5x^2 + 8y^2 = 40$

24. $20x^2 + 10y^2 = 40$

25. $\dfrac{1}{4}x^2 + \dfrac{1}{12}y^2 = \dfrac{1}{2}$

26–35 Graph the ellipse and determine the coordinates of the foci.

26. $\dfrac{(x-4)^2}{16} + \dfrac{(y-4)^2}{4} = 1$

27. $\dfrac{(x+2)^2}{16} + \dfrac{(y+1)^2}{9} = 1$

28. $9x^2 + 16y^2 + 18x - 64y = 71$

29. $9x^2 + 4y^2 - 36x - 24y + 36 = 0$

30. $16x^2 + y^2 + 160x - 6y = -393$

31. $25x^2 + 4y^2 - 100x + 8y + 4 = 0$

32. $4x^2 + 9y^2 + 40x + 90y + 289 = 0$

33. $16x^2 + y^2 - 64x + 6y + 57 = 0$

34. $4x^2 + y^2 + 4y = 0$

35. $9x^2 + 4y^2 + 108x - 32y = -352$

36–45 Graph the parabola and determine its focus and directrix.

36. $(x+1)^2 = 4(y-3)$

37. $(y-4)^2 = -2(x-1)$

38. $y^2 + 2y + 12x + 37 = 0$

39. $x^2 - 8y = 6x - 1$

40. $x^2 + 6x + 8y = -17$

41. $x^2 + 2x + 8y = 31$

42. $y^2 + 6y - 2x + 13 = 0$

43. $x^2 - 2x - 4y + 13 = 0$

44. $4y + 2x^2 = 4$

45. $2y^2 - 10x = 10$

46–55 Graph the hyperbola, using asymptotes as guides, and determine the coordinates of the foci.

46. $\dfrac{(x+3)^2}{4} - \dfrac{(y+1)^2}{9} = 1$

47. $4y^2 - x^2 - 24y + 2x = -19$

48. $x^2 - 9y^2 + 4x + 18y - 14 = 0$

49. $9x^2 - 25y^2 = 18x - 50y + 241$

50. $9x^2 - 16y^2 + 116 = 36x + 64y$

51. $\dfrac{(y-1)^2}{9} - (x+3)^2 = 1$

52. $9y^2 - 25x^2 - 36y - 100x = 289$

53. $9x^2 + 18x = 4y^2 + 27$

54. $9x^2 - 16y^2 - 36x + 32y - 124 = 0$

55. $x^2 - y^2 + 6x - 6y = 4$

56–73 Find the equation, in standard form, for the conic with the given properties or with the given graph.

56. Ellipse, center at $(-2,3)$, horizontal major axis of length 8, minor axis of length 4

57. Parabola, focus at $(-2,1)$, directrix is the x-axis

58. Ellipse, vertices at $(5,-1)$ and $(1,-1)$, minor axis of length 2

59. Hyperbola, foci at $(1,5)$ and $(1,-1)$, vertices at $(1,3)$ and $(1,1)$

60. Parabola, focus at $\left(-3,-\frac{3}{2}\right)$, directrix is the line $y = -\frac{1}{2}$

61. Hyperbola, foci at $(-1,3)$ and $(-1,-1)$, asymptotes given by $y = \pm(x+1)+1$

62. Parabola, vertex at $(-4,3)$, focus at $\left(-\frac{3}{2},3\right)$

63. Ellipse, foci at $(0,0)$ and $(6,0)$, $e = \frac{1}{2}$

64. Hyperbola, asymptotes given by $y = \pm(2x+8)+3$, vertices at $(-6,3)$ and $(-2,3)$

65. Ellipse, vertices at $(-4,6)$ and $(-14,6)$, $e = \frac{2}{5}$

66. Hyperbola, foci at $(2,4)$ and $(-2,4)$, asymptotes given by $y = \pm 3x + 4$

67. Parabola, symmetric with respect to the line $y = 1$, directrix is the line $x = 2$, and $p = -3$

68.

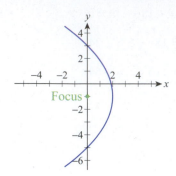

69.

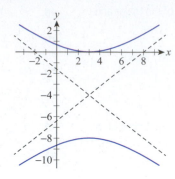

70.

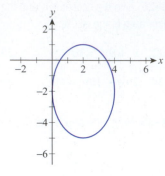

71.

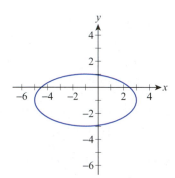

72.

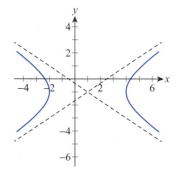

73.

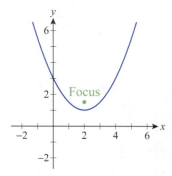

74. The orbit of Halley's Comet is an ellipse with the sun at one focus and an eccentricity of 0.967. Its closest approach to the sun is approximately 54,591,000 miles. What is the furthest Halley's Comet ever gets from the sun?

75. Pluto's closest approach to the sun is approximately 4.43×10^9 kilometers, and its maximum distance from the sun is approximately 7.37×10^9 kilometers. What is the eccentricity of Pluto's orbit?

76. The archway supporting a bridge over a river is in the shape of half an ellipse. The archway is 60 feet wide and is 15 feet tall at the middle. A large boat is 10 feet wide and 14 feet 9 inches tall. Is the boat capable of passing under the archway?

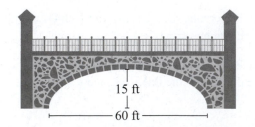

77. Use the information given in Example 2 to determine the length of the minor axis of the ellipse formed by Earth's orbit around the sun.

78. Since the sum of the distances from each of the two foci to any point on an ellipse is constant, we can draw an ellipse using the following method. Tack the ends of a length of string at two points (the foci) and, keeping the string taut by pulling outward with the tip of a pencil, trace around the foci to form an ellipse (the total length of the string remains constant). If you want to create an ellipse with a major axis of length 5 cm and a minor axis of length 3 cm, how long should your string be and how far apart should you place the tacks?

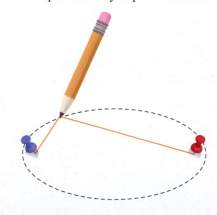

79. One design for a solar furnace is based on the paraboloid formed by rotating the parabola $x^2 = 8y$ around its axis of symmetry. The object to be heated in the furnace is then placed at the focus of the paraboloid (assume that x and y are in units of feet). How far from the vertex of the paraboloid is the hottest part of the furnace?

80. A certain brand of satellite dish antenna is a paraboloid with a diameter of 6 feet and a depth of 1 foot. How far from the vertex of the dish should the receiver of the antenna be placed, given that the receiver should be located at the focus of the paraboloid?

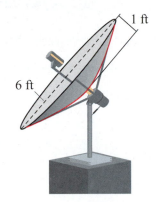

1 ft

6 ft

81. A spotlight is made by placing a strong lightbulb inside a reflective paraboloid formed by rotating the parabola $x^2 = 6y$ around its axis of symmetry (assume that x and y are in units of inches). In order to have the brightest, most concentrated light beam, how far from the vertex should the bulb be placed?

82. As mentioned in this section, some comets trace one branch of a hyperbola through the solar system, with the sun at one focus. Suppose a comet is spotted that appears to be headed straight for Earth as shown in the figure. As the comet gets closer, however, it becomes apparent that it will pass between the Earth, which lies at the center of the hyperbolic path of the comet, and the sun. In the end, the closest the comet comes to Earth is 60,000,000 miles. Using an estimate of 94,000,000 miles for the distance from the Earth to the sun, and positioning the Earth at the origin of a coordinate system, find the equation for the path of the comet.

83. Placing the foci at $(-c, 0)$ and $(c, 0)$ and introducing $d_1 + d_2 = 2a$, derive the standard form of the equation of an ellipse.

84. Denoting $|d_1 - d_2|$ by $2a$, use the approach suggested by Exercise 83 to derive the standard equation of a hyperbola.

85. Suppose two LORAN (LOng RAnge Navigation) radio transmitters are 26 miles apart. A ship at sea receives signals sent simultaneously from the two transmitters and is able to determine that the difference between the distances from the ship to each of the transmitters is 24 miles. By positioning the two transmitters on the y-axis, each 13 miles from the origin, find the equation of the hyperbola that describes the set of possible locations for the ship. (**Hint:** See Exercise 84.)

86.* Three high-sensitivity microphones are located in a forest preserve, with microphone A two miles due north of microphone B and microphone C two miles due east of microphone B. During an early morning thunderstorm, microphone A detects a thunderclap (and possible lightning strike) at 3:28:15 a.m. The same thunderclap is detected by microphone B at 3:28:19 a.m. and by microphone C at 3:28:25 a.m. Assuming that sound travels at 1100 feet per second, graphically approximate the source of the thunderclap. (**Hint:** Place microphone B at the origin, with A and C on the y- and x-axes, respectively. Then, by a repeated application of the method used in Exercise 85, construct two intersecting hyperbolas to locate the thunderclap.)

87–90 Find the $x'y'$-coordinates of the point for the given rotation angle θ.

87. $(8,6)$; $\theta = 30°$

88. $(-5,1)$; $\theta = \dfrac{\pi}{3}$

89. $\left(-\dfrac{1}{2}, -\dfrac{1}{8}\right)$; $\theta = \dfrac{\pi}{4}$

90. $(-1,1)$; $\theta = \dfrac{\pi}{2}$

91–96 Use the discriminant to classify the conic section as an ellipse, parabola, or hyperbola. Then determine the appropriate angle θ by which to rotate the coordinate axes, and use that angle to convert the equation by eliminating the xy-term. Finally, sketch the graph of the conic section.

91. $xy - 4 = 0$

92. $x^2 + 2xy + y^2 - x + y = 0$

93. $7x^2 + 5\sqrt{3}xy + 2y^2 = 14$

94. $22x^2 + 6\sqrt{3}xy + 16y^2 - 49 = 276$

95. $2\sqrt{3}x^2 - 6xy + \sqrt{3}x - 9y = 0$

96. $34x^2 + 8\sqrt{3}xy + 42y^2 = 1380$

97–100 The given equation is that of a rotated conic. Match the equation with its graph (labeled A–D).

97. $3x^2 + 2xy + y^2 - 10 = 0$

98. $x^2 - 4xy + 4y^2 + 5\sqrt{5}y + 1 = 0$

99. $3x^2 + 8xy + 4y^2 - 7 = 0$

100. $x^2 - 6xy + 9y^2 - 2y + 1 = 0$

A.

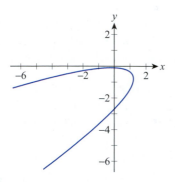

B.

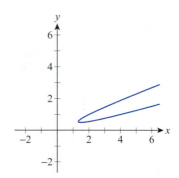

C.

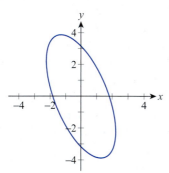

D.

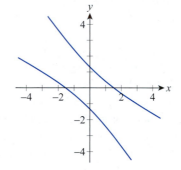

101. You have just used the rotation of axes to rotate the x- and y-axes until they were parallel to the axes of the conic. The resulting equation in the $x'y'$-plane is of the form

$$A'x'^2 + B'x'y' + E'y' + F' = 0,$$

where A', B', E', and F' are all nonzero. What is wrong with the resulting equation?

102. What must the angle of rotation θ be if the coefficients of x^2 and y^2 are equal and $B \neq 0$? Support your answer.

103. An expression involving the coefficients of the general form of a conic section is said to be *rotation invariant* if it has the same value under every possible rotation. Using the equation
$7x^2 - 6\sqrt{3}xy + 13y^2 - 16 = 0$,

a. show that the relationship $F = F'$ is true (so F is rotation invariant in this equation).

b. show that the relationship $A + C = A' + C'$ is true (so $A + C$ is rotation invariant in this equation).

c. show that the relationship
$B^2 - 4AC = B'^2 - 4A'C'$ is true (so $B^2 - 4AC$ is rotation invariant in this equation).

104.* Show in general that the quantity $A + C$ and the discriminant $B^2 - 4AC$ are rotation invariant. (See Exercise 103.)

105–108 *True or False?* Determine whether the given statement is true or false. In case of a false statement, explain or provide a counterexample.

105. It is possible for a parabola to be tangent to its directrix.

106. The graph of $Ax^2 + Cy^2 + Dx + Ey = 0$ is a hyperbola if A and C have different signs and $D, E \neq 0$.

107. It is not possible for a line tangent to a hyperbola to have more than one point in common with the graph.

108. If the eccentricity of an ellipse is greater than 1, the ellipse is extremely narrow.

9.5 **Technology Exercises**

109–118 Use a computer algebra system to sketch the given curve.

109. $15x^2 + 9y^2 + 150x - 36y = -276$

110. $5x^2 + 12y^2 - 20x + 144y + 392 = 0$

111. $x^2 - 6x + 12y + 21 = 0$

112. $x^2 - 5y^2 = 14x + 20y - 4$

113. $x^2 + 6xy + y^2 = 18$

114. $x^2 - 4xy + 3y^2 = 12$

115. $36x^2 - 19xy + 8y^2 = 72$

116. $72x^2 + 19xy + 4y^2 = 20$

117. $40x^2 + 20xy + 10y^2 + \left(2\sqrt{2} - 6\right)x - \left(4\sqrt{2} + 8\right)y = 90$

118. $72x^2 + 18xy - 9y^2 = 14$

119–120 Use a computer algebra system to sketch the given curve. Explore how different values of the parameters k_1 and k_2 affect the graph. Experiment with both nonnegative and negative values. (Answers will vary.)

119. $k_1x^2 + k_2xy + 5y^2 - 6x + 7y + 15 = 0$

120. $k_1x^2 - 4xy + k_2y^2 + 2x + 3y - 1 = 0$

9.6 Conic Sections in Polar Coordinates

TOPICS

1. The focus-directrix description of conic sections

2. Using the polar form of conic sections

The previous section dealt with the three varieties of conic sections, describing their geometric properties and the formulation of their associated equations in rectangular coordinates. This individual attention is useful when introducing conic sections, and rectangular coordinates are undeniably useful in working with conics to solve certain applications. But polar coordinates provide us with an alternative approach to the study of conic sections, one that possesses important advantages of its own.

TOPIC 1   The Focus-Directrix Description of Conic Sections

Probably the most striking virtue of equations of conics using polar coordinates is that the equations for the three types of conics all have the same form. The three varieties are easily identified, and the magnitude of a parameter e, called the eccentricity, determines whether the conic is an ellipse, a parabola, or a hyperbola. You first encountered eccentricity in the discussion of ellipses—the use of e in this section is an extension of the original use. An additional characteristic of the polar form of conics is that all three varieties are defined in terms of a focus and a directrix; previously, the directrix only made an appearance in the discussion of parabolas.

Throughout this section, we will assume that a point F, called the focus, lies at the origin of the plane and that a line L, called the directrix, lies d units away from the focus. Until we discuss rotated conics in polar form, the directrix will be oriented either vertically or horizontally, so the equation for the directrix will be $x = -d, x = d, y = -d,$ or $y = d$. We will let $D(P,F)$ denote the distance between a variable point P and the fixed focus F, and we will let $D(P,L)$ denote the shortest distance between a variable point P and the fixed directrix L. The eccentricity e will be a fixed positive number for any given conic.

Theorem ⚲

Focus-Directrix Description of Conics

Let $D(P,F)$ denote the distance between a variable point P and the fixed focus F; and let $D(P,L)$ denote the shortest distance between P and the fixed directrix L. A conic section consists of all points P in the plane that satisfy the equation

$$\frac{D(P,F)}{D(P,L)} = e,$$

where e is a fixed positive constant.

1.   The conic is an ellipse if $0 < e < 1$.

2.   The conic is a parabola if $e = 1$.

3.   The conic is a hyperbola if $e > 1$.

In words, a conic section consists of all those points for which the ratio of the distance from the focus to the distance from the directrix is a fixed constant e. Figure 1 illustrates three conic sections that share the directrix $x = -1$ but with three different eccentricities. In all three graphs, the directrix appears in red and the focus in green.

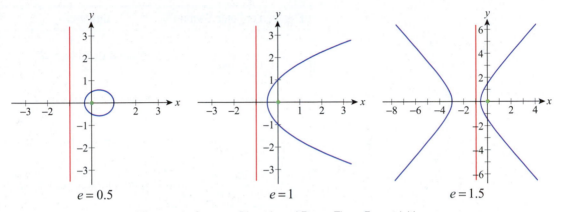

$e = 0.5$ $e = 1$ $e = 1.5$

Figure 1 Common Directrix and Focus, Three Eccentricities

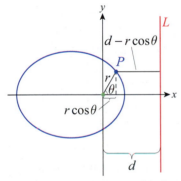

Figure 2 Determining $D(P,L)$

The next step is to use the focus-directrix description to develop the form of the polar equations for conics. If we let (r, θ) denote a point P on a given conic section, then $D(P,F) = r$. To determine $D(P,L)$, consider the diagram in Figure 2, which depicts an ellipse and a directrix L of the form $x = d$.

From Figure 2, we see that $D(P,L) = d - r\cos\theta$, so the equation in polar coordinates is

$$\frac{r}{d - r\cos\theta} = e.$$

This equation will be more useful if we solve for r:

$$r = e(d - r\cos\theta)$$
$$r = ed - er\cos\theta$$
$$r + er\cos\theta = ed$$
$$r(1 + e\cos\theta) = ed$$
$$r = \frac{ed}{1 + e\cos\theta}$$

The derivation of the equation for a directrix of the form $x = -d$, $y = -d$, or $y = d$ is very similar, and will be left to the reader in Exercises 39 and 40. The four possible forms and their geometric meanings are summarized in the following theorem.

Theorem 🔍

Polar Forms of Conic Sections

In polar coordinates, a conic section with its focus at the origin and either a horizontal or vertical directrix has one of the following forms.

Equation of Conic Section	Directrix
$r = \dfrac{ed}{1+e\cos\theta}$	Vertical directrix $x = d$
$r = \dfrac{ed}{1-e\cos\theta}$	Vertical directrix $x = -d$
$r = \dfrac{ed}{1+e\sin\theta}$	Horizontal directrix $y = d$
$r = \dfrac{ed}{1-e\sin\theta}$	Horizontal directrix $y = -d$

Example 1 ✎

Identify the variety of conic and determine the equation of the directrix for each of the following conic sections.

a. $r = \dfrac{15}{5-3\cos\theta}$ **b.** $r = \dfrac{2}{3+5\sin\theta}$

Solution

a. First we need to determine the two constants e and d, and we can accomplish this by algebraically manipulating the given equation. In order to have the correct form, the constant term in the denominator must be a 1.

$$r = \frac{15}{5-3\cos\theta} = \frac{15\left(\frac{1}{5}\right)}{(5-3\cos\theta)\left(\frac{1}{5}\right)} = \frac{3}{1-\frac{3}{5}\cos\theta}$$

This tells us that $e = \frac{3}{5}$, so the conic is an ellipse. To determine the constant d, we need to write the numerator as a product of e and d. Since the numerator is 3 and since we now know $e = \frac{3}{5}$, we can do this as follows:

$$r = \frac{3}{1-\frac{3}{5}\cos\theta} = \frac{\left(\frac{3}{5}\right)(5)}{1-\frac{3}{5}\cos\theta}$$

Hence, $d = 5$. The last observation is that the trigonometric function in the denominator is cosine and the sign between the two terms in the denominator is negative. By the guidelines for conics in polar form, this tells us that $x = -5$ is the equation for the directrix.

b. We use the same methods as above to determine first e and then d.

$$r = \frac{2}{3 + 5\sin\theta} = \frac{\frac{2}{3}}{1 + \frac{5}{3}\sin\theta} = \frac{\left(\frac{5}{3}\right)\left(\frac{2}{5}\right)}{1 + \frac{5}{3}\sin\theta}$$

From this form, we can see that $e = \frac{5}{3}$ and that the directrix is $y = \frac{2}{5}$. Thus, the conic section is a hyperbola and the directrix is horizontal and above the x-axis.

Example 2 ✎

Construct a polar equation for a leftward-opening parabola with focus at the origin and directrix 2 units from the focus.

Solution

Since we are discussing a parabola, $e = 1$. If the parabola is to open to the left, the directrix must be oriented vertically and must lie to the right of the focus, so $d = 2$ and the trigonometric function in the denominator must be cosine. Thus, the equation is

$$r = \frac{2}{1 + \cos\theta}.$$

The graph of the parabola appears in Figure 3, with the focus shown in green and the directrix in red.

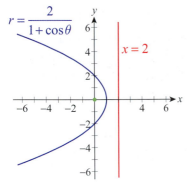

$r = \dfrac{2}{1 + \cos\theta}$

$x = 2$

Figure 3

TOPIC 2   Using the Polar Form of Conic Sections

Example 1 illustrated how the two constants e and d can be determined from the equation for a conic in polar form. In order to get an even better understanding of a given conic section, it may be useful to determine the constants a, b, and c as well. These constants have the same meaning in this context as they did in Section 9.5; that is, $2a$ is the distance between vertices of an ellipse or a hyperbola, $2b$ is the length of the minor axis of an ellipse, and $2c$ is the distance between foci of an ellipse or a hyperbola.

We begin with a. Eccentricity was defined in Section 9.5 as $e = c/a$, and this relation still holds for the expanded definition of eccentricity that we now have. In originally discussing ellipses, we noted that $c^2 = a^2 - b^2$, which can be rewritten as $b^2 = a^2 - c^2$; and in originally discussing hyperbolas, we noted that $c^2 = a^2 + b^2$, meaning that $b^2 = c^2 - a^2$. We can combine these two statements by noting that b is nonnegative and that $b^2 = \left|a^2 - c^2\right|$. Note that if $e = 1$, $a = c$ and, hence, $b = 0$; this should not be surprising, as b played no role in the original discussion of parabolas.

When graphing an ellipse or a hyperbola described in polar form, it is usually easiest to first determine a. Using the above observations, it is then easy to determine c and b. This process is illustrated in the next several examples.

Example 3 ✎

Sketch the graph of the conic section $r = \dfrac{15}{5 - 3\cos\theta}$.

Solution

This is the first equation from Example 1, and we have already seen that the equation can be written in the following form:

$$r = \frac{\left(\frac{3}{5}\right)(5)}{1 - \frac{3}{5}\cos\theta}$$

So $e = \frac{3}{5}$, $d = 5$, and the directrix is the line $x = -5$. This tells us that the graph is an ellipse and that the major axis is oriented horizontally (perpendicular to the directrix). The entire graph is traced out as θ increases from 0 to 2π, with the right vertex corresponding to $\theta = 0$, and the left vertex corresponding to $\theta = \pi$ (halfway around the ellipse). When $\theta = 0$, $r = \frac{15}{2}$ and when $\theta = \pi$, $r = \frac{15}{8}$. In rectangular coordinates, the coordinates of the right vertex are $\left(\frac{15}{2}, 0\right)$ and the coordinates of the left vertex are $\left(-\frac{15}{8}, 0\right)$ (remember that $\theta = \pi$, so a positive r corresponds to a point left of the origin).

Now that we know the coordinates of the two vertices, we can determine that

$$2a = \frac{15}{2} + \frac{15}{8} = \frac{75}{8}$$

and so $a = \frac{75}{16}$. Since $e = \frac{3}{5}$ and $c = ea$, this means

$$c = \left(\frac{3}{5}\right)\left(\frac{75}{16}\right) = \frac{45}{16}.$$

Finally, from the relation $b^2 = \left|a^2 - c^2\right|$, we can determine that $b^2 = \frac{225}{16}$, so $b = \frac{15}{4}$.

With this knowledge of a and b, we can sketch the graph in Figure 4.

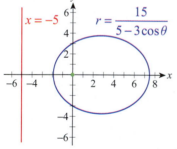

Figure 4

Example 4 ✎

Sketch the graph of the conic section $r = \dfrac{2}{3 + 5\sin\theta}$.

Solution

This is the second equation from Example 1, and we already know that $e = \frac{5}{3}$ and that $y = \frac{2}{5}$ is the equation for the directrix. This tells us that the graph is a hyperbola and that the vertices are aligned vertically (perpendicular again to the directrix). One vertex corresponds to $\theta = \pi/2$ while the other vertex corresponds to $\theta = 3\pi/2$ (it might also be useful to note that the lower branch of the hyperbola crosses the x-axis at $\theta = 0$ and $\theta = \pi$). When $\theta = \pi/2$, $r = \frac{1}{4}$ and when $\theta = 3\pi/2$, $r = -1$ (make note of the negative sign on r). This means that the rectangular coordinates of the two vertices are $\left(0, \frac{1}{4}\right)$ and $(0, 1)$, so $2a = \frac{3}{4}$ and $a = \frac{3}{8}$. From this,

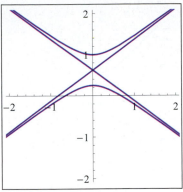

Figure 5

$c = ea = \frac{5}{8}$ and the relation $b^2 = |a^2 - c^2|$ leads to $b = \frac{1}{2}$. The equations for the asymptotes of the hyperbola are $y - \frac{5}{8} = \pm \frac{3}{4} x$.

A computer algebra system has been used to plot the given hyperbola (see Figure 5); the crossed straight lines running through $\left(0, \frac{5}{8}\right)$ are actually artifacts of the graphing algorithm, but they approximate the asymptotes.

We will conclude this section with an example of rotation of conics in polar form. In general, the graph of an equation $r = f(\theta - \varphi)$ is the rotation of the graph of $r = f(\theta)$ by the angle φ counterclockwise. This makes rotation in polar coordinates particularly easy to handle.

Example 5 ✎

Sketch the graph of the conic section $r = \dfrac{2}{1 + \cos\left(\theta - \dfrac{\pi}{6}\right)}$.

Solution

We constructed the equation $r = 2/(1 + \cos\theta)$ in Example 2, so we know its graph is a parabola opening to the left with directrix $x = 2$.

The graph of $r = \dfrac{2}{1 + \cos\left(\theta - \dfrac{\pi}{6}\right)}$ is the same shape rotated $\pi/6$ radians counterclockwise, as shown in Figure 6.

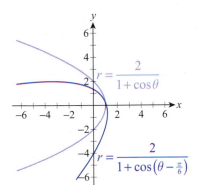

Figure 6

9.6 **Exercises**

1-6 Match the given polar equation with its graph (labeled A–F).

1. $r = \dfrac{3}{4 - \cos\theta}$

2. $r = \dfrac{9}{6 - 2\sin\theta}$

3. $r = \dfrac{3}{3 + 4\sin\theta}$

4. $r = \dfrac{1}{2 + 2\cos\theta}$

5. $r = \dfrac{6}{1 + 3\sin\theta}$

6. $r = \dfrac{6}{1 + 3\cos\theta}$

A.

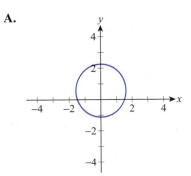

B.

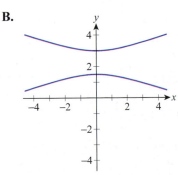

C.

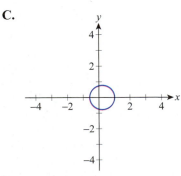

D. 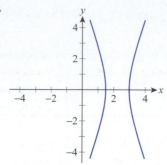 **E.** 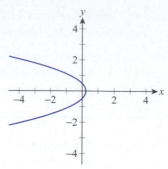 **F.**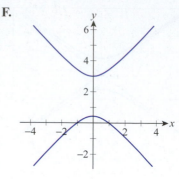

7–20 Identify the given conic section as an ellipse, parabola, or hyperbola and find the equation for its directrix.

7. $r = \dfrac{7}{1+6\sin\theta}$

8. $r = \dfrac{2}{1-\sin\theta}$

9. $r = \dfrac{3}{4-\cos\theta}$

10. $r = \dfrac{4}{2-2\cos\theta}$

11. $r = \dfrac{1}{1+3\cos\theta}$

12. $r = \dfrac{7}{3+2\sin\theta}$

13. $r = \dfrac{5}{2+\cos\theta}$

14. $r = \dfrac{3}{4-3\sin\theta}$

15. $r = \dfrac{6}{3-5\cos\theta}$

16. $r = \dfrac{8}{5-6\sin\theta}$

17. $r = \dfrac{3}{2+2\sin\theta}$

18. $r = \dfrac{-1}{3+4\cos\theta}$

19. $r = \dfrac{4}{6-7\cos\theta}$

20. $r = \dfrac{9}{5-4\sin\theta}$

21–26 Construct a polar equation for the conic section with the focus at the origin and the given eccentricity and directrix.

21. Parabola; eccentricity: $e = 1$; directrix: $x = -2$

22. Hyperbola; eccentricity: $e = 2$; directrix: $x = -3$

23. Hyperbola; eccentricity: $e = 4$; directrix: $y = -\frac{3}{4}$

24. Parabola; eccentricity: $e = 1$; directrix: $x = 2$

25. Ellipse; eccentricity: $e = \frac{1}{4}$; directrix: $x = 12$

26. Ellipse; eccentricity: $e = \frac{1}{2}$; directrix: $y = 8$

27–36 Sketch the graph of the conic section.

27. $r = \dfrac{5}{1+3\cos\theta}$

28. $r = \dfrac{3}{2+\sin\theta}$

29. $r = \dfrac{4}{1-2\sin\theta}$

30. $r = \dfrac{6}{2-4\cos\theta}$

31. $r = \dfrac{9}{3-2\cos\theta}$

32. $r = \dfrac{5}{3+\sin\theta}$

33. $r = \dfrac{4}{1+2\cos\theta}$

34. $r = \dfrac{4}{2+2\sin\theta}$

35. $r = \dfrac{2}{1+\cos\left(\theta - \dfrac{\pi}{4}\right)}$

36. $r = \dfrac{4}{2+2\sin\left(\theta - \dfrac{\pi}{3}\right)}$

37. The planets of our solar system follow elliptical orbits with the sun located at one of the foci. If we assume that the sun is located at the pole and the major axes of these elliptical orbits lie along the polar axis, the orbits of the planets can be expressed by the polar equation

$$r = \frac{a\left(1-e^2\right)}{1+e\cos\theta}$$

where e is the eccentricity. Verify the above equation.

38. Using the equation from Exercise 37, answer the following:

 a. Show that the shortest distance from the sun to a planet, called the *perihelion* distance, is $r = a(1-e)$.

 b. Show that the longest distance from the sun to a planet, called the *aphelion* distance, is $r = a(1+e)$.

 c. The distance from Uranus to the sun is approximately 2.74×10^9 km at perihelion and 3.00×10^9 km at aphelion. Find the eccentricity of Uranus' orbit.

 d. The eccentricity of Neptune's orbit is 0.0113 and $a = 4.495 \times 10^9$ km. Determine the perihelion and aphelion distances for Neptune.

39. Derive the polar form of the equation of a conic with vertical directrix $x = -d$ and focus at the origin.

40. Derive the polar form of the equation of a conic with horizontal directrix **a.** $y = d$, **b.** $y = -d$, and focus at the origin.

41. A chord through a focus of a conic section that is parallel to the directrix is called its *latus rectum* (from the Latin words "latus," meaning "side," and "rectum," meaning "straight"). Find the length of the latus rectum for the conic $r = ed/(1+e\cos\theta)$.

42. Find the polar coordinates of the vertices for the ellipse with polar equation $r = ed/(1+e\cos\theta)$, $0 < e < 1$.

43. Find the polar coordinates of the vertices for the hyperbola with polar equation $r = ed/(1+e\cos\theta)$, $e > 1$.

44. Use Exercise 42 to find the rectangular equation of the ellipse $r = 12/(5+\cos\theta)$.

45. Use Exercise 43 to find the rectangular equation of the hyperbola $r = 12/(5+7\cos\theta)$.

9.6 **Technology Exercises**

46–55 Use a graphing calculator or computer algebra system to graph the conic section.

46. $r = \dfrac{-3}{4 - 9\cos\theta}$

47. $r = \dfrac{9}{-4 + \frac{3}{2}\sin\theta}$

48. $r = \dfrac{-11}{3 - \cos\theta}$

49. $r = \dfrac{2}{10 + 4\sin\theta}$

50. $r = \dfrac{3}{7 + 3\cos\theta}$

51. $r = \dfrac{2}{2 + 3\cos\left(\theta - \dfrac{\pi}{4}\right)}$

52. $r = \dfrac{-7}{5 + 3\sin\left(\theta - \dfrac{\pi}{6}\right)}$

53. $r = \dfrac{5}{-2 - 4\sin\left(\theta + \dfrac{2\pi}{3}\right)}$

54. $r = \dfrac{4}{-3 - 2\cos\left(\theta + \dfrac{\pi}{3}\right)}$

55. $r = \dfrac{1}{1 + 4\sin\left(\theta + \dfrac{\pi}{6}\right)}$

56. Use technology to sketch the conic section $r = ed/(1+e\cos\theta)$ for various values of d and e, $e > 0$, and examine how these values affect the shape of the graph.

Chapter 9
Review Exercises

1–4 Sketch the curve defined by the parametric equations by eliminating the parameter.

1. $x = \dfrac{1}{36t}, \quad y = t^2$

2. $x = t + 5, \quad y = |t - 2|$

3. $x = \dfrac{3}{4t - 2}, \quad y = 2t - 2$

4. $x = 4\sin\theta, \quad y = \cos\theta + 1$

5–6 Construct parametric equations defining the graph of the given equation.

5. $y^2 = x^2 + 4$
6. $6x = 2 - y$

7–8 Find parametric equations to represent the graph described. (Answers will vary.)

7. Line, passing through $(14, 4)$ and $(-3, -8)$

8. Circle, center $(1,1)$, radius 1

9–10 Find the equations of any horizontal or vertical tangent lines to the given curve.

9. $x = 2t + 1, \quad y = t^2 - 4$

10. $x = -2t^2, \quad y = \dfrac{3}{t - 2}, \quad t < 2$

11–14 Find the values of dy/dx and d^2y/dx^2 for the given curve at the indicated point.

11. $x = 2t^2 + 1, \quad y = \sqrt{t - 1}; \quad (9, 1)$

12. $x = e^t, \quad y = t^2 e^{-t}; \quad (1, 0)$

13. $x = \dfrac{1}{t}, \quad y = t^2 + t; \quad (1, 2)$

14. $x = \sin t, \quad y = \cos 2t; \quad \left(\dfrac{1}{2}, \dfrac{1}{2}\right)$

15–16 Find the value(s) of the parameter for any inflection point(s) of the given curve.

15. $x = t - t^3, \quad y = 3t + 1$

16. $x = 5t - 1, \quad y = t^3\left(2 - t^2\right)$

17–18 Find the area enclosed by the given curve.

17. $x = \sin 3t, \quad y = \cos 2t, \quad -\dfrac{\pi}{3} \le t \le \dfrac{\pi}{3}$

18. $x = \cos^3 t, \quad y = \sin^3 t, \quad 0 \le t \le 2\pi$

19–20 Find the arc length of the given curve on the indicated interval.

19. $x = \dfrac{4}{3} t^{3/2}, \quad y = 4t, \quad 0 \le t \le 5$

20. $x = t\sin t + \cos t, \quad y = t\cos t - \sin t, \quad 0 \le t \le \pi$

21–23 Find the area of the surface generated by revolving the parametric curve about the indicated axis.

21. $x = \dfrac{t^3}{2}, \quad y = 2t + 1, \quad 0 \le t \le 1; \quad$ about the y-axis

22. $x = \dfrac{1 - t}{2}, \quad y = \sqrt{t}, \quad 0 \le t \le 1; \quad$ about the x-axis

23. $x = 3t - \dfrac{t^3}{3}, \quad y = \sqrt{3}\left(9 - t^2\right), \quad 0 \le t \le 3;$
about the y-axis

24–25 Convert the point from polar to Cartesian coordinates.

24. $\left(-3.45, \dfrac{\pi}{3}\right)$
25. $\left(7, \dfrac{7\pi}{6}\right)$

26–27 Convert the point from Cartesian to polar coordinates.

26. $\left(-\sqrt{3}, -1\right)$
27. $(10, 12)$

28–29 Rewrite the rectangular equation in polar form.

28. $x^2 + y^2 = 16a^2$
29. $x^2 + y^2 = 9ax$

30–31 Rewrite the polar equation in rectangular form.

30. $r = 4 \cos \theta$

31. $r = \dfrac{16}{4 \cos \theta + 4 \sin \theta}$

32–33 Sketch a graph of the given polar equation.

32. $r = 4 \sin 3\theta$

33. $r^2 = 25 \cos 2\theta$

34–35 Find the slope of the line tangent to the given polar curve at the indicated point.

34. $r = 4 \cos 3\theta; \quad \theta = \dfrac{\pi}{12}$

35. $r = \dfrac{1}{\theta^2 + 1}; \quad \theta = \dfrac{\pi}{2}$

36–37 Find all points where the given polar curve has a horizontal or vertical tangent line.

36. $r = 2 \sin \theta, \quad 0 \le \theta \le 2\pi$

37. $r = 1 + \cos 2\theta, \quad 0 \le \theta \le \pi$

38–39 Find the area of the shaded region.

38.

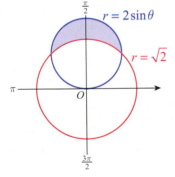

39.

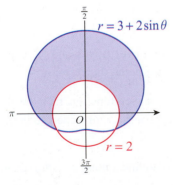

40–41 Find the area of the specified region.

40. A large loop of $r = 1 + 2 \cos 2\theta$

41. The portion of the rose $r = 4 \cos 2\theta$ outside the circle $r = 2$

42–45 Use polar coordinates to find the arc length of the curve.

42. The circle $r = 2 \cos \theta$

43. The line segment $r = \sec \theta, \ -\pi/4 \le \theta \le \pi/4$

44. The spiral $r = 2^\theta, \ 0 \le \theta \le 2\pi$

45. The cardioid $r = 2 + 2 \cos \theta$

46. Find the area of the surface generated by revolving the curve $r = 2 \cos \theta, \ 0 \le \theta \le \pi$ about $\theta = \pi/2$. (**Hint:** See Exercise 58 in Section 9.4)

47–52 Sketch the graph of the given conic section, and determine the coordinates of the foci and the equations of the directrix or asymptotes as appropriate.

47. $\dfrac{(x+1)^2}{9} + \dfrac{(y-2)^2}{16} = 1$

48. $x^2 + 9y^2 - 6x + 18y = -9$

49. $(y+1)^2 = -12(x+3)$

50. $x^2 - 8x + 2y + 14 = 0$

51. $\dfrac{(y+2)^2}{9} - \dfrac{(x-2)^2}{16} = 1$

52. $9x^2 - 4y^2 + 54x - 8y + 41 = 0$

53–63 Find the equation, in standard form, of the conic section with the given properties.

53. Ellipse, center at $(-1,4)$, major axis is vertical and of length 8, foci $\sqrt{7}$ units from the center

54. Ellipse, foci at $(1,2)$ and $(7,2)$, $e = \dfrac{1}{2}$

55. Ellipse, vertices at $\left(\dfrac{7}{2},-1\right)$ and $\left(\dfrac{1}{2},-1\right)$, $e = 0$

56. Ellipse, vertices at $(1,-8)$ and $(1,2)$, minor axis of length 6

57. Parabola, vertex at $(-2,3)$, directrix is the line $y = 2$

58. Parabola, vertex at $(5,-3)$, focus at $(5,1)$

59. Parabola, focus at $(3,-1)$, directrix is the line $x = 2$

60. Hyperbola, vertices at $(4,-1)$ and $(-2,-1)$, foci at $(5,-1)$ and $(-3,-1)$

61. Hyperbola, asymptotes of $y = \pm\frac{5}{2}(x+1) - 2$, vertices at $(-3,-2)$ and $(1,-2)$

62. Hyperbola, foci at $(-1,-2)$ and $(-1,8)$, asymptotes of $y = \pm\left(\frac{3}{4}x + \frac{3}{4}\right) + 3$

63. Hyperbola, asymptotes of $y = 3x - 4$, $y = 8 - 3x$, vertices at $(2,-1)$ and $(2,5)$

64–70 Identify the given conic section and find the equation for its directrix.

64. $r = \dfrac{8}{1 + 2\sin\theta}$

65. $r = \dfrac{5}{4 - 8\sin\theta}$

66. $r = \dfrac{3}{7 + 6\sin\theta}$

67. $r = \dfrac{6}{9 - 9\cos\theta}$

68. $r = \dfrac{7}{4 + 4\sin\theta}$

69. $r = \dfrac{4}{6 - 3\cos\theta}$

70. $r = \dfrac{7}{5 + 2\cos\theta}$

71–76 Construct a polar equation for the conic section with the focus at the origin and the given eccentricity and directrix.

71. Eccentricity: $e = 1$;   directrix: $x = -3$

72. Eccentricity: $e = 4$;   directrix: $y = 3$

73. Eccentricity: $e = \frac{1}{5}$;   directrix: $y = -15$

74. Eccentricity: $e = \frac{1}{4}$;   directrix: $x = 16$

75. Eccentricity: $e = 1$;   directrix: $y = -7$

76. Eccentricity: $e = 9$;   directrix: $x = \frac{1}{3}$

77. A motorcycle headlight is made by placing a strong light bulb inside a reflective paraboloid formed by rotating the parabola $x^2 = 5y$ around its axis of symmetry (assume that x and y are in units of inches). In order to have the brightest, most concentrated light beam, how far from the vertex should the bulb be placed?

78.* Prove that the graph of $r = \cos\theta + \sin\theta$ is a circle.

79. Show that the two limaçons $r = 1 + \sin\theta$ and $r = 1 - \sin\theta$ are orthogonal, that is, they intersect at right angles. (Ignore their intersection at the pole.)

80–90 *True or False?* Determine whether each of the following statements is true or false. In case of a false statement, explain or provide a counterexample.

80. Two different sets of parametric equations may describe the same curve.

81. If $x = f(t)$ and $y = g(t)$ both pass through $(0,0)$, the parametric curve $x = f(t)$, $y = g(t)$ also passes through the origin.

82. The line $y = 2x$ can be parametrized as $x = t^2$, $y = 2t^2$.

83. If a parametric curve C is differentiable on (a,b) then its graph has a tangent line at any of its points.

84. Let C be the curve $x = t^5$, $y = t^4$. Since $\left.\dfrac{dy}{dt}\right|_{t=0} = 0$, the graph of C has a horizontal tangent at the origin.

85. The polar coordinates (r,θ) and $(-r, \theta - \pi)$ describe the same point.

86. If the polar coordinates (r_1, θ_1) and (r_2, θ_2) describe the same point, then $r_1 = \pm r_2$.

87. If $y = f(x)$ is an even function, then the graph of $r = f(\theta)$ is symmetric with respect to the polar axis.

88. It is impossible to find a polar equation for a straight line.

89. The distance between any point on a parabola from the focus is at least p.

90. If the asymptotes of a hyperbola are perpendicular and the hyperbola is centered at the origin with horizontally aligned foci, then the equation of the hyperbola can be written as $x^2 - y^2 = a^2$.

Chapter 9
Technology Exercises

91. Find the equation of the graph of $r = 1 - 2\cos\theta$ after a clockwise rotation by $\pi/4$ radians. Name the resulting curve and use a graphing calculator or computer algebra system to sketch it. (See Exercise 73 in Section 9.3.)

92–93 Use a graphing calculator or computer algebra system to sketch the given curve for various values of the parameter(s) and explore the effects on the shape of your graph.

92. $r = \theta \cos k\theta$

93. $x = \pm a\cos^{2/n} t, \quad y = \pm b\sin^{2/n} t, \quad a,b,n > 0$
(Lamé curves)

94–95 Use a graphing calculator or computer algebra system to sketch the curve and then find all horizontal and vertical tangent lines. Confirm your results by paper and pencil calculations.

94. $x = t^3 - t, \quad y = t^2 + 1, \quad -2 \le t \le 2$

95. $r = 2\sin 2\theta, \quad 0 \le \theta \le \pi/2$

96–97 Use a computer algebra system to sketch the region enclosed by the given curve and find its area.

96. $x = t\sin t, \quad y = \sin 2t, \quad 0 \le t \le \pi$

97. Inner loop of $r = 2 - 3\cos\theta$

98–99 Use a computer algebra system to approximate the arc length of the curve with the given parametrization.

98. $x = \sin 2t, \quad y = \sin t, \quad 0 \le t \le \pi$

99. $r = \cos(2\sin\theta), \quad 0 \le \theta \le 2\pi$

100–101 Use a computer algebra system to approximate the surface area of the solid obtained by rotating the given curve about the indicated axis.

100. $\left(t^3 - 3t, t^2 - 2\right), \quad 0 \le t \le \sqrt{3}$; about the y-axis

101. $r = 3\sin 2\theta, \quad \pi/4 \le \theta \le \pi/2$; about the polar axis

Project 📝

Chapter 9

In this project, you will be introduced to a class of parametric curves called *Bézier curves*. They are important for their applications in engineering, computer graphics, and animation. This class of curves is named after Pierre Bézier (1910–1999), a design engineer for the French automaker Renault, who first demonstrated these curves' use in designing automobile bodies in the 1960s. The design advantage of Bézier curves lies in the fact that they can easily be manipulated by moving around their so-called *control points*. In addition, it is easy to smoothly join together several Bézier curves for more complicated shapes.

1. The linear Bézier curve $B_{0,1}(t)$ from $P_0(a_0,b_0)$ to $P_1(a_1,b_1)$ is simply the line segment connecting the two points (note that P_0 and P_1 are the only control points in this case). Verify that this curve can be parametrized as

$$B_{0,1}(t) = (1-t)P_0 + tP_1, \; t \in [0,1],$$

and find $x(t)$ and $y(t)$ corresponding to this parametrization. (In this and subsequent questions, control points will be labeled $P_i(a_i,b_i)$, $0 \le i \le 3$.)

2. The Bézier curve $B_{0,1,2}(t)$ with control points P_0, P_1, and P_2 is a quadratic curve joining the points P_0 and P_2 in such a way that both line segments $\overline{P_0 P_1}$ and $\overline{P_1 P_2}$ are tangent to $B_{0,1,2}(t)$. Intuitively speaking, this means that the curve "starts out at P_0 in the direction of P_1," and "arrives at P_2 from the direction of P_1" (see figure).

Find $x(t)$ and $y(t)$ corresponding to the parametrization

$$B_{0,1,2}(t) = (1-t)B_{0,1}(t) + tB_{1,2}(t), \; t \in [0,1]$$

and verify that $B_{0,1,2}(t)$ satisfies the conditions stated above.

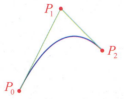

A Quadratic Bézier Curve

3. The cubic Bézier curve $B_{0,1,2,3}(t)$ with control points P_0, P_1, P_2, and P_3 joins P_0 and P_3 so that the line segments $\overline{P_0 P_1}$ and $\overline{P_2 P_3}$ are tangent to $B_{0,1,2,3}(t)$ at P_0 and P_3, respectively (see figure). Verify that the following curve satisfies these conditions:

$$x(t) = a_0(1-t)^3 + 3a_1(1-t)^2 t + 3a_2(1-t)t^2 + a_3 t^3$$
$$y(t) = b_0(1-t)^3 + 3b_1(1-t)^2 t + 3b_2(1-t)t^2 + b_3 t^3, \; t \in [0,1]$$

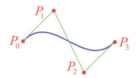

A Cubic Bézier Curve

4. Show that the parametrization in Question 3 corresponds to

$$B_{0,1,2,3}(t) = (1-t)B_{0,1,2}(t) + tB_{1,2,3}(t).$$

5. Use Question 3 to verify that the Bézier curve with control points $P_0(1,3)$, $P_1(3,7)$, $P_2(6,9)$, and $P_3(8,6)$ has the following parametrization:

$$x(t) = -2t^3 + 3t^2 + 6t + 1$$
$$y(t) = -3t^3 - 6t^2 + 12t + 3$$

6. Find the slope of the curve in Question 5 at
a. $t = 0$, **b.** $t = \frac{1}{2}$, and **c.** $t = 1$.

7. Use a computer algebra system to graph the Bézier curve of Question 5 along with its control points. If your CAS has animation capabilities, explore what happens if you move around the control points in the plane.

Chapter 10

Sequences and Series

10.1 Sequences 767

1. Convergence and Divergence of Sequences
2. Determining Limits of Sequences
3. Monotonicity and Boundedness

10.2 Infinite Series 781

1. Partial Sums, Convergence, and Divergence
2. Geometric, Telescoping, and Harmonic Series

10.3 The Integral Test 795

1. The Integral Test
2. Series and Remainder Estimates

10.4 Comparison Tests 802

1. The Direct Comparison Test
2. The Limit Comparison Test

10.5 The Ratio and Root Tests 809

1. The Ratio Test
2. The Root Test

10.6 Absolute and Conditional Convergence 816

1. Alternating Series
2. Absolute and Conditional Convergence
3. Summary of Convergence Tests

10.7 Power Series 824

1. Power Series
2. Radius of Convergence
3. Calculus and Power Series

10.8 Taylor and Maclaurin Series 834

1. Taylor and Maclaurin Series
2. Remainder Estimates and Convergence for Taylor Series

10.9 Further Applications of Series 847

1. Applications Using Series
2. Fourier Series
3. Summary of Common Taylor Series

Introduction

This chapter is primarily concerned with *infinite series*, which are often informally described as sums of an infinite number of numbers. Curiously, such a definition runs the risk of both oversimplifying the concept and, at the same time, making it sound impossibly difficult.

On the one hand, finite sums (meaning sums of finite sets of numbers) are so familiar and elementary that it is tempting to assume sums of an infinite number of numbers must possess similar properties and qualities. On the other hand, at least when one is asked to contemplate the sum of an infinite number of *positive* numbers, a typical initial reaction is to conclude that such a sum is impossible or, more precisely, that the sum must also be infinite. Both instincts are faulty, but the fact that they are so common is the basis for many of *Zeno's paradoxes* and other mathematical conundrums.

As we will see, infinite series are a fitting subject for study in a calculus text because a better understanding of their nature, once again, requires the use of limits. It is, therefore, not surprising that the history of infinite series is closely tied to that of calculus.

Some of the earliest known uses of infinite series date back to Archimedes and his brilliant insights into the calculation of areas and volumes, but the widespread and rigorous development of infinite

> It is, therefore, not surprising that the history of infinite series is closely tied to that of calculus.

series awaited the work of mathematicians in the 18th and 19th centuries AD. The English mathematician Brook Taylor (1685–1731) and the Scottish mathematician Colin Maclaurin (1698–1746) were two among many in the 18th century who used calculus to describe certain functions in terms of *series expansions*. Such expansions, which now go by the name of *Taylor series* and *Maclaurin series*, had the appearance of "infinite polynomials" and offered advantages associated with their polynomial-like nature. The prolific Swiss mathematician Leonhard Euler (1707–1783), in particular, demonstrated the power of the series expansions of functions, using them to solve problems and discover fundamental relationships in physics and mathematics.

The impact of series today is profound, though most people are unaware of how pervasively they are used. Physicists and engineers continue to take advantage of the fact that the series expansions of familiar functions, truncated after a certain number of terms, provide easily calculated polynomial

approximations of the functions. Other thinkers have used series expansions, as Euler did, to discover or describe deep truths—you'll see one such example in Section 10.9 with Einstein's famous mass-velocity relationship. And series provide the mathematical foundation for the decomposition of signals into different frequencies, the basis for nearly all modern electronic communication.

Brook Taylor
(1685–1731)

Colin Maclaurin
(1698–1746)

1. Adapted from line engraving by Richard Earlom, *Brook Taylor*. Wellcome Library no. 9093i, http://wellcomeimages.org/indexplus/image/V0005740.html. This file is licensed under the Creative Commons Attribution 4.0 International license.

2. Adapted from stipple engraving, *Colin Maclaurin*, published by Edward Harding (London, 1798). Smithsonian Institution Libraries.

10.1 **Sequences**

TOPICS

1. Convergence and divergence of sequences

2. Determining limits of sequences

3. Monotonicity and boundedness

The goal of this chapter is to develop an understanding of *infinite series*, informally defined as sums of an infinite number of numbers. As we will see, series provide another powerful way to analyze functions, describe exact or approximate solutions in applications of calculus, and solve differential equations. The first step in this development is to study *infinite sequences*.

TOPIC 1   **Convergence and Divergence of Sequences**

A sequence is simply an ordered list $a_1, a_2, a_3, \ldots, a_n, \ldots$ of elements. Although in practice sequences often appear surrounded by the set symbols { }, the order of the elements is a critical feature. Whenever braces are seen, it is important to note from the context whether a set (in which order is irrelevant) or a sequence is being discussed. In this text the elements of a sequence will be real numbers unless otherwise noted (the few exceptions will be clearly identified).

Since order is important, references to the *first term* a_1, the *second term* a_2, and so on are common, as are references to a generic n^{th} *term* a_n of the sequence. And the fact that the terms of a sequence are labeled, or indexed, by the natural numbers means a sequence actually represents a function from the natural numbers to some set (usually the set of real numbers). This leads to the following formal definition.

Definition 💡

Infinite Sequences

An **infinite sequence** is a function whose domain is the set of natural numbers. Typically, notation such as a_1 is used for the element associated with the number 1, a_2 for the element associated with the number 2, and so on, as opposed to function notation like $f(1)$ and $f(2)$.

Example 1 ✎

a. For the sequence $\{1, 4, 9, 16, \ldots\}$, the first term is 1, the second term is 4, and in general we would likely guess that the nth term is n^2. Using sequence notation, we write $a_n = n^2$.

b. In the sequence $\{1, -1, 1, -1, \ldots\}$, the pattern so far indicates that the odd-numbered terms are 1 and the even-numbered terms are -1. One way of expressing this is with the formula $b_n = (-1)^{n+1}$. Note that this example illustrates another distinction between sets and sequences: in a set, repeated elements are redundant, and the above list would simply reduce to $\{-1, 1\}$. But as a sequence, each element in the list is distinct because it occupies a unique position (first, second, and so on) in the sequence.

c. The notation $\{n/(n+1)\}_{n=1}^{\infty}$ tells us that the nth term is $n/(n+1)$ and defines the following sequence: $\left\{\dfrac{1}{2}, \dfrac{2}{3}, \dfrac{3}{4}, \dfrac{4}{5}, \ldots, \dfrac{n}{n+1}, \ldots\right\}$

d. The notation $\{n\}_{n=5}^{\infty}$ is also a sequence. It defines the sequence $\{5, 6, 7, \ldots\}$, which can be *reindexed* as $\{n+4\}_{n=1}^{\infty}$ to make it clear that it fits the formal definition of a sequence. In practice, if a given sequence can be described with a simpler formula by beginning with an index other than 1, we often do so.

Note that there is some potential ambiguity in parts a. and b. of Example 1. Although we may feel confident that we have spotted a pattern in the first few elements of a sequence, there are actually always multiple formulas that describe the first few terms of any given sequence and then differ thereafter. For example, consider the sequence defined by the rule $\tilde{a}_n = n^4 - 10n^3 + 36n^2 - 50n + 24$. If we calculate the first few terms, we find

$$\{\tilde{a}_n\}_{n=1}^{\infty} = \{1, 4, 9, 16, 49, 156, \ldots\},$$

a sequence matching $a_n = n^2$ up to $n = 4$. For this reason, it is far preferable to describe sequences with precise formulas when possible, as in parts c. and d. of Example 1.

Formulas such as $c_n = n/(n+1)$ are called **explicit**, because they explicitly define the nth term of a sequence. Sometimes, however, the nth term of a sequence is most easily defined in terms of preceding terms, in which case the formula is said to be **recursive**. Example 2 illustrates the difference between the two types of formulas.

Example 2 ✐

a. The sequence defined by $a_1 = 2$ and $a_n = 2a_{n-1}$ for $n \geq 2$ is $\{2, 4, 8, \ldots\}$. By this definition, the value of the nth term depends on the value of a_{n-1}, so the definition is recursive. But many types of recursive formulas can also be rewritten explicitly, and this is one such example: the sequence is also defined by the explicit formula $a_n = 2^n$.

b. The **Fibonacci sequence** $\{F_n\}_{n=0}^{\infty}$ is a sequence introduced to Europeans by Leonardo of Pisa (nicknamed Fibonacci) in his book *Liber Abaci* (published in 1202). It is defined by $F_0 = 0$, $F_1 = 1$, and $F_n = F_{n-2} + F_{n-1}$ for $n \geq 2$, so $F_2 = F_0 + F_1 = 1$, $F_3 = F_1 + F_2 = 2$, $F_4 = F_2 + F_3 = 3$, and so on. By this definition, any term in the sequence (for $n \geq 2$) can only be determined once the previous two terms are known, so it is recursive. This is another example, however, of a recursive formula that has an equivalent explicit formula:

$$F_n = \frac{\varphi^n - \psi^n}{\sqrt{5}}, \quad \text{where} \quad \varphi = \frac{1+\sqrt{5}}{2} \quad \text{and} \quad \psi = \frac{1-\sqrt{5}}{2}$$

We leave it to the reader to show that this explicit formula matches the recursive formula for as many terms as you would like to verify.

Our primary concern at the moment is whether a given sequence *converges* or *diverges*; in other words, whether the terms of the sequence approach a finite limit. The sequences in parts a., b., and d. of Example 1 do not appear to be approaching any particular limiting value, and we will soon see that they qualify as divergent sequences. But the terms in part c. approach the value 1 as n increases. Figure 1 depicts the four sequences of Example 1 as graphs of functions from $\mathbb{N}$ to $\mathbb{R}$.

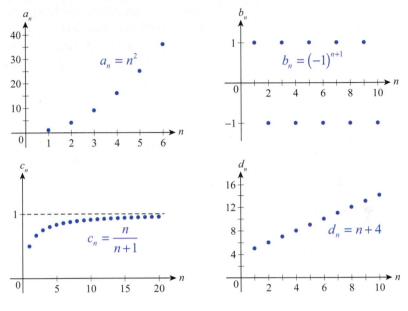

Figure 1

Our formal definition of the limit of a sequence, and consequently whether a sequence converges or diverges, is adapted from our definition of limit in Section 2.3.

Definition 💡

Limit of a Sequence, Convergence, and Divergence

A given sequence $\{a_n\}$ has the **limit** L if for every number $\varepsilon > 0$ there is a natural number N such that $|a_n - L| < \varepsilon$ for all $n > N$. We write

$$\lim_{n \to \infty} a_n = L \quad \text{or} \quad a_n \to L \quad \text{as} \quad n \to \infty,$$

and say that $\{a_n\}$ **converges** to L (or is **convergent**). If no such limit exists, $\{a_n\}$ **diverges** (or is **divergent**).

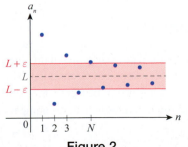

Figure 2

Graphically, we can show the relationship between the terms of a sequence $\{a_n\}$, its limit L, and a representative ε in two different ways. In Figure 2, the sequence is graphed as the sequences in Figure 1 were, as functions from $\mathbb{N}$ to $\mathbb{R}$. In this format, the line $y = L$ plays the same role as it does when we write $\lim_{x \to \infty} f(x) = L$, and the positive number ε gives rise to an ε-band below and above $y = L$. For all $n > N$, a_n lies within the ε-band; that is, a_n is within ε of L.

Figure 3

In Figure 3, the same sequence is depicted as a collection of values on a horizontal real number line centered around the value L. Some of the terms of the sequence may lie to the right of L, some to the left, but for all $n > N$ it will be the case that a_n lies within the open interval $(L - \varepsilon, L + \varepsilon)$.

Example 3 ✎

Newton's method (Section 4.5) generates endless examples of recursively defined sequences and provides a setting in which graphs like that in Figure 3 are the most natural. For instance, given the function $f(x) = e^x + \sin x$, Newton's method leads to the recursive formula

$$x_{n+1} = x_n - \frac{f(x_n)}{f'(x_n)} = x_n - \frac{e^{x_n} + \sin x_n}{e^{x_n} + \cos x_n}.$$

If we begin with $x_1 = -1.25$, the terms generated (to six decimal places) by this recursive formula are

$$\{-1.25, -0.149219, -0.534414, -0.587419, -0.588532, -0.588533, \ldots\},$$

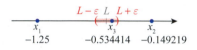

Figure 4
A Newton's Method Sequence

and they cluster around the limit of the sequence, which is a root of $f(x) = e^x + \sin x$. In Figure 4, the first three terms and the limit L are depicted, along with an ε-interval corresponding to $\varepsilon = 0.1$. All the remaining terms of the sequence fall within the ε-interval.

Example 4 ✎

The recursively generated sequence of complex numbers described in Section 1.3 is one of the few examples in this text of a sequence whose terms are not real numbers. Using our current notation, the sequence $\{z_n\}_{n=1}^{\infty}$ is based on the complex-valued function $f(z) = z^2 + c$ and the recursive formula

$$z_1 = f(0) = c \text{ and (for } n \geq 2) \ z_n = f(z_{n-1}) = f\left(f^{n-1}(0)\right) = \underbrace{f\left(\ldots f\left(f(0)\right)\right)}_{n \text{ iterates of } f},$$

where c is a fixed complex number serving as the "seed" of the sequence. In this context, the limit of a sequence, if it exists, is a complex number in the plane that the terms of the sequence approach as $n \to \infty$.

TOPIC 2 Determining Limits of Sequences

To show that a given sequence either converges to a limit or to show that it diverges, we use the same reasoning and techniques that we developed in Section 2.3. This is not surprising since sequences *are* functions—they differ from the functions previously studied only in that their domains are restricted to the natural numbers.

Example 5 ✐

For each of the following sequences, prove its convergence to a limit or prove it diverges.

a. $\left\{(-1)^{n+1}\right\}_{n=1}^{\infty}$

b. $\left\{\dfrac{n}{n+1}\right\}_{n=1}^{\infty}$

c. $\left\{n^2\right\}_{n=1}^{\infty}$

d. $\left\{n!\right\}_{n=1}^{\infty}$

Solution

a. Suppose a limit L exists for the sequence. Then for any fixed $\varepsilon > 0$ there must exist a corresponding N such that $n > N \Rightarrow \left|(-1)^{n+1} - L\right| < \varepsilon$. In particular, this must be the case for $\varepsilon = \tfrac{1}{2}$. So for all n past some natural number N, $\left|(-1)^{n+1} - L\right| < \tfrac{1}{2}$. Since $(-1)^{n+1} = 1$ for all odd n, this means $|1 - L| < \tfrac{1}{2}$. Similarly, since $(-1)^{n+1} = -1$ for all even n, it is also the case that $|-1 - L| < \tfrac{1}{2}$. But these two inequalities can be rewritten as

$$\frac{1}{2} < L < \frac{3}{2} \quad \text{and} \quad -\frac{3}{2} < L < -\frac{1}{2}.$$

Since no number L can satisfy both of these conditions, no such limit exists and the sequence diverges.

b. We have already guessed that $L = 1$ for this sequence, so we merely need to prove it according to the limit definition. Accordingly, let $\varepsilon > 0$ be a fixed number. Let N be fixed as any natural number greater than $1/\varepsilon$. This means that $1/N < \varepsilon$, and so for all $n > N$,

$$\left|\frac{n}{n+1} - 1\right| = \frac{1}{n+1} < \frac{1}{n} < \frac{1}{N} < \varepsilon.$$

This proves the sequence converges and that $\lim\limits_{n \to \infty} \dfrac{n}{n+1} = 1$.

c. The sequence $\left\{n^2\right\}_{n=1}^{\infty}$ also diverges because as n increases, n^2 increases without bound and eventually surpasses any fixed real number no matter how large. As we have seen in the past, we often describe *how* this sequence diverges by writing $\lim\limits_{n \to \infty} n^2 = \infty$, but this notation should never be taken to mean the sequence converges. This is simply an indication that its terms grow without bound.

d. For any given positive integer n, the number $n!$ (which is read "n factorial") is defined as

$$n! = n(n-1)\cdots(2)(1).$$

(Because it is convenient in some contexts, we also define $0! = 1$, but this fact is not needed here.) Given this, the sequence $\left\{n!\right\}_{n=1}^{\infty}$ also clearly grows without bound and thus is another example of a divergent sequence.

Because sequences are functions, we are able to take advantage of all the theorems we have previously developed regarding limits of functions. The next theorem establishes the connection between our earlier work and limits of sequences, and the subsequent theorems are restatements of facts you first learned in Section 2.4.

Theorem 🔍

Limits of Sequences and Limits of Functions

Suppose f is a function defined for all $x \geq x_0$ and suppose $a_n = f(n)$ for all natural numbers $n \geq x_0$. Then

$$\lim_{x \to \infty} f(x) = L \quad \Rightarrow \quad \lim_{n \to \infty} a_n = L.$$

Proof 📎

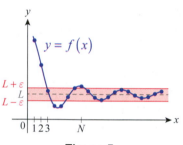

Figure 5

The only purpose of the x_0 in the statement of the theorem is to emphasize that we don't care about the behavior of the function f or the sequence $\{a_n\}$ initially if we are looking at the limit as x or n approaches ∞. With that in mind, let $\varepsilon > 0$ be fixed. Then there is a real number M such that $|f(x) - L| < \varepsilon$ for all $x > M$. So if N is a natural number greater than M,

$$n > N \quad \Rightarrow \quad |a_n - L| = |f(n) - L| < \varepsilon$$

and hence $\lim_{n \to \infty} a_n = L$. Figure 5 illustrates how the behavior of $\{a_n\}$ reflects the behavior of f.

Theorem 🔍

Sequence Limit Laws

Let $\{a_n\}$ and $\{b_n\}$ be two sequences such that both $\lim_{n \to \infty} a_n$ and $\lim_{n \to \infty} b_n$ exist, and let k be a fixed real number. Then the following laws hold.

Sum Law

$$\lim_{n \to \infty}(a_n + b_n) = \lim_{n \to \infty} a_n + \lim_{n \to \infty} b_n$$

Difference Law

$$\lim_{n \to \infty}(a_n - b_n) = \lim_{n \to \infty} a_n - \lim_{n \to \infty} b_n$$

Constant Multiple Law $\quad \lim_{n \to \infty}(ka_n) = k \lim_{n \to \infty} a_n$

Product Law

$$\lim_{n \to \infty}(a_n b_n) = \lim_{n \to \infty} a_n \cdot \lim_{n \to \infty} b_n$$

Quotient Law

$$\lim_{n \to \infty} \frac{a_n}{b_n} = \frac{\lim_{n \to \infty} a_n}{\lim_{n \to \infty} b_n}, \text{ provided } \lim_{n \to \infty} b_n \neq 0$$

Theorem

The Squeeze Theorem for Sequences

Let $\{a_n\}$, $\{b_n\}$, and $\{c_n\}$ be sequences such that $a_n \le b_n \le c_n$ for all n greater than some natural number N. Then if $\lim_{n\to\infty} a_n = L$ and $\lim_{n\to\infty} c_n = L$, $\lim_{n\to\infty} b_n = L$ as well.

The last theorem to be restated is a variation of the "Limits Pass Through a Continuous Function" theorem.

Theorem

Continuous Function Theorem for Sequences

Let $\{a_n\}$ be a sequence with $\lim_{n\to\infty} a_n = L$. If f is a function that is continuous at L and defined for all a_n, then $\lim_{n\to\infty} f(a_n) = f(L)$.

Example 6 ✍

Find the limit of each of the following sequences, if the limit exists.

a. $\left\{5^{1/n}\right\}_{n=1}^{\infty}$ b. $\left\{\dfrac{3n^3 + 5n^2 - 2n}{7n^3 + 17}\right\}_{n=1}^{\infty}$ c. $\left\{n^{1/n}\right\}_{n=1}^{\infty}$

Solution

a. Note that $1/n \to 0$ as $n \to \infty$, and the function $f(x) = 5^x$ is continuous at 0. So by the continuous function theorem,

$$\lim_{n\to\infty} 5^{1/n} = f\left(\lim_{n\to\infty} \frac{1}{n}\right) = f(0) = 5^0 = 1.$$

Figure 6 visually confirms the limit we found.

b. The sequence limit laws allow us to easily evaluate a limit of this form, just as the original limit laws allow us to evaluate similar limits.

$$\lim_{n\to\infty} \frac{3n^3 + 5n^2 - 2n}{7n^3 + 17} = \lim_{n\to\infty} \frac{3 + \dfrac{5}{n} - \dfrac{2}{n^2}}{7 + \dfrac{17}{n^3}} \qquad \textcolor{blue}{\text{Divide numerator and denominator by } n^3.}$$

$$= \frac{\lim_{n\to\infty} 3 + \lim_{n\to\infty} \dfrac{5}{n} - \lim_{n\to\infty} \dfrac{2}{n^2}}{\lim_{n\to\infty} 7 + \lim_{n\to\infty} \dfrac{17}{n^3}}$$

$$= \frac{3 + 0 - 0}{7 + 0} = \frac{3}{7}$$

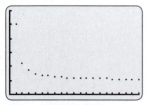

Figure 6

$\left\{5^{1/n}\right\}_{n=1}^{\infty}$ on $[0,20]$ by $[0,6]$

c. The sequence $\left\{n^{1/n}\right\}_{n=1}^{\infty}$ is the restriction of the function $f(x) = x^{1/x}$ to the domain of natural numbers, and analyzing the function f is the key to understanding the sequence. The limit of f can be found using l'Hôpital's Rule:

$$x^{1/x} = e^{\ln\left(x^{1/x}\right)} = e^{(1/x)\ln x} = e^{(\ln x)/x}$$

and

$$\lim_{x \to \infty} \frac{\ln x}{x} = \lim_{x \to \infty} \frac{\dfrac{1}{x}}{1} = 0 \qquad \text{l'Hôpital's Rule}$$

so

$$\lim_{x \to \infty} x^{1/x} = \lim_{x \to \infty} e^{(\ln x)/x} = e^0 = 1.$$

This means $\lim_{n \to \infty} n^{1/n} = 1$ as well.

TOPIC 3   Monotonicity and Boundedness

As we continue on to the study of series in the next section, we will frequently encounter sequences whose terms display a trend of growing consistently larger or smaller as n increases.

Definition ♀

Monotonic Sequences

A sequence $\{a_n\}$ for which $a_n \leq a_{n+1}$ for all n is called an **increasing** sequence, and a sequence $\{b_n\}$ for which $b_n \geq b_{n+1}$ for all n is called a **decreasing** sequence (in each case, the adjective *strictly* may be optionally added if the inequality is a strict inequality). A sequence that is increasing or decreasing is called a **monotonic** sequence.

Example 7 ✏

a. The sequence whose n^{th} term is given by $a_n = n/(n+1)$ is strictly increasing. To prove this, note that

$$n(n+2) = n^2 + 2n < n^2 + 2n + 1 = (n+1)^2,$$

and so

$$\frac{n(n+2)}{(n+1)(n+2)} < \frac{(n+1)^2}{(n+1)(n+2)}, \quad \text{or} \quad a_n = \frac{n}{n+1} < \frac{n+1}{n+2} = a_{n+1}.$$

b. The sequence whose n^{th} term is given by $b_n = n^2 + 6n$ is also a strictly increasing sequence, as shown by the following:

$$b_n = n^2 + 6n < n^2 + 2n + 1 + 6n + 6 = (n+1)^2 + 6(n+1) = b_{n+1}$$

While both sequences in Example 7 are (strictly) increasing, they differ in one important respect. We have already proved that $\lim\limits_{n\to\infty}\dfrac{n}{n+1}=1$, but it isn't difficult to see that $\lim\limits_{n\to\infty}\left(n^2+6n\right)=\infty$; that is, the sequence $\{a_n\}$ converges and the sequence $\{b_n\}$ diverges. The following definition and theorem characterize this behavior in general.

Definition ♀

Bounded Sequences

A sequence $\{a_n\}$ is **bounded above** if there is a number M (called an **upper bound**) for which $a_n \leq M$ for all n.

Similarly, $\{a_n\}$ is **bounded below** if there is a number m (called a **lower bound**) for which $a_n \geq m$ for all n.

If $\{a_n\}$ is bounded both above and below, it is a **bounded** sequence.

If $\{a_n\}$ is bounded above by an upper bound M_0, and if $M_0 \leq M$ for all other upper bounds M of $\{a_n\}$, then M_0 is called the **least upper bound** of $\{a_n\}$. The definition of the **greatest lower bound** is similar in nature.

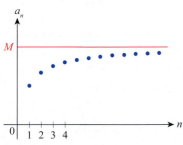

Figure 7 Sequence Bounded Above

Theorem ⚷

Bounded Monotonic Sequence Theorem

An increasing sequence converges if and only if it is bounded above, and a decreasing sequence converges if and only if it is bounded below. In particular, if an increasing sequence converges, it converges to its least upper bound (that is, the least of all the upper bounds of the sequence). Similarly, if a decreasing sequence converges, it converges to its greatest lower bound.

The proof of this theorem will be left as an exercise (Exercise 74); its truth derives from the **Completeness Property** of the real numbers, which states that every nonempty set of real numbers that is bounded above (respectively, below) has a least upper bound (respectively, a greatest lower bound). The name "completeness" reflects the fact that there are no breaks or gaps in the real number line.

Although the Bounded Monotonic Sequence Theorem is often cited when we need to know that a limit exists, without necessarily caring what that limit is, we can also use it to actually find the limit in some situations. Our last example illustrates the process.

Example 8 ✎

Define the sequence $\{a_n\}$ recursively by $a_1 = \sqrt{2}$ and $a_{n+1} = \sqrt{2+a_n}$ for $n \geq 2$. Use the Bounded Monotonic Sequence Theorem to find $\lim\limits_{n\to\infty} a_n$.

Solution

First, we will use the theorem to prove that $\lim\limits_{n\to\infty} a_n$ exists, and we do so by using induction arguments to show that the sequence is increasing and bounded above. Recall that such arguments call for a basis step and an inductive step.

Basis Step: $a_1 = \sqrt{2} < \sqrt{2+\sqrt{2}} = a_2$.

Inductive Step: Assume $a_n < a_{n+1}$. Then $2 + a_n < 2 + a_{n+1}$ and hence $\sqrt{2+a_n} < \sqrt{2+a_{n+1}}$. That is, $a_{n+1} < a_{n+2}$. This proves that $\{a_n\}$ is an increasing sequence.

To prove that $\{a_n\}$ is bounded above, we can do some brute-force calculations to arrive at the guess that $a_n < 2$ for all n (see Table 1 for approximations of the first 10 terms of the sequence). If this guess happens to be wrong, we will find out in the course of the next induction proof.

Basis Step: $a_1 = \sqrt{2} < 2$.

Inductive Step: Assume $a_n < 2$. Then $2 + a_n < 2 + 2 = 4$, so $a_{n+1} = \sqrt{2+a_n} < \sqrt{4} = 2$. This proves $\{a_n\}$ is bounded above by 2.

By the Bounded Monotonic Sequence Theorem, we know $\lim\limits_{n\to\infty} a_n = L$ for some L; we just don't know what L is yet (though we might suspect that $L = 2$ after the brute-force calculations). We can determine L by the following argument (be sure you understand why each equality is true).

$$L = \lim_{n\to\infty} a_n = \lim_{n\to\infty} a_{n+1} = \lim_{n\to\infty} \sqrt{2+a_n} = \sqrt{2+\lim_{n\to\infty} a_n} = \sqrt{2+L}$$

Squaring both sides, we obtain the equation $L^2 = 2 + L$, or $L^2 - L - 2 = 0$. One of the solutions of this equation is $L = 2$ and the other ($L = -1$) is extraneous to this problem.

n	a_n
1	1.414214
2	1.847759
3	1.961571
4	1.990369
5	1.997591
6	1.999398
7	1.999849
8	1.999962
9	1.999991
10	1.999998

Table 1

10.1 **Exercises**

1–4 List the first six terms of the given sequence.

1. $a_n = \dfrac{n}{n^2+1}$

2. $a_n = \left(-\dfrac{2}{5}\right)^n$

3. $a_n = \dfrac{3}{(n-1)!}$

4. $a_n = \dfrac{n(n+1)}{2}\cos(n\pi)$

5–8 Find the first six terms of the given recursively defined sequence.

5. $a_1 = 1, \quad a_n = 2a_{n-1} + 1$

6. $a_1 = 2, \quad a_2 = 3, \quad a_n = a_{n-1} - a_{n-2}$

7. $a_1 = 4, \quad a_{n+1} = (n+1)a_n$

8. $a_1 = 1, \quad a_2 = 1, \quad a_{n+1} = -2a_n + 3a_{n-1}$

9–12 Recognize the apparent pattern and find an explicit formula for the sequence. (Answers will vary.)

9. $\{2, 6, 12, 20, 30, 42, \ldots\}$

10. $\{2, 6, 18, 54, 162, 486, \ldots\}$

11. $\{9, 6, 1, -6, -15, -26, \ldots\}$

12. $\{2, 4, 7, 11, 16, 22, \ldots\}$

13. Notice that the sequences in Exercises 9–12 are actually not uniquely determined. For example, show that in each case the formula you obtained for a_n and the formula $b_n = a_n + (n-1)\cdots(n-j)$ define two sequences that match for the first j terms and then differ.

14–19 Match the sequence with its graph (labeled A–F).

14. $a_n = \dfrac{(-1)^n}{\sqrt{n}}$

15. $a_n = \dfrac{n(n+1)}{2}$

16. $a_n = (-2)^n$

17. $a_n = \cos(n\pi)$

18. $a_n = \dfrac{1}{n+1}$

19. $a_n = \dfrac{n}{n+1}$

A.
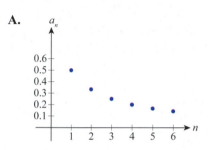

B.
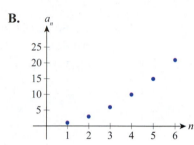

C.
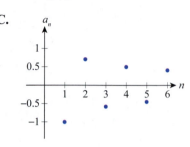

D.
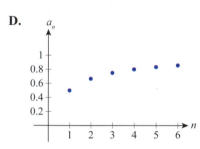

E.
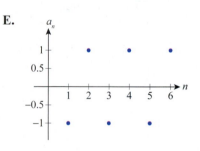

F.
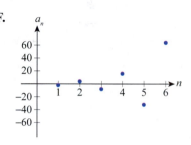

20–23 Use the definition of the limit of a sequence to establish the given statement.

20. $\displaystyle\lim_{n\to\infty}\dfrac{1}{n^2} = 0$

21. $\displaystyle\lim_{n\to\infty}\left(\dfrac{1}{2}\right)^n = 0$

22. $\displaystyle\lim_{n\to\infty}\dfrac{3n+1}{n+1} = 3$

23. $\displaystyle\lim_{n\to\infty}\left(\sqrt{n+2}-\sqrt{n}\right) = 0$

24–69 Find the limit of the sequence if it converges or prove that the sequence diverges. (You can use any theorem from this section.)

24. $a_n = \dfrac{2n-3}{n}$

25. $a_n = \dfrac{3n+5}{6n-1}$

26. $a_n = \dfrac{4n^2+1}{2n-1}$

27. $a_n = \dfrac{2n-1}{4n^2+1}$

28. $a_n = \dfrac{n+1}{\sqrt{n}}$

29. $a_n = (-1)^n\dfrac{\sqrt{n}}{n+1}$

30. $a_n = \dfrac{n\sin n}{n^2+2}$

31. $a_n = \sqrt{n+1}-\sqrt{n}$

32. $a_n = \dfrac{n^4-5n^3+2n^2+1}{3n^4+n^2+2}$

33. $a_n = \dfrac{\pi^n}{n^2}$

34. $a_n = \dfrac{\pi^n}{2^n}$

35. $a_n = \dfrac{\pi^n}{4^n}$

36. $b_n = \dfrac{\ln n}{2n}$

37. $c_n = 0.5+(-0.5)^n$

38. $a_n = 4^n+\left(\dfrac{1}{4}\right)^n$

39. $a_n = \dfrac{\ln\dfrac{1}{\sqrt{n}}}{\sqrt{n}}$

40. $d_n = \dfrac{n3^n}{4^n}$

41. $h_n = \dfrac{n^4-4}{e^n}$

42. $a_n = \dfrac{n!}{4^n}$

43. $a_n = \ln(3n^2+2)-\ln(n^2+7)$

44. $a_n = (-1)^n\dfrac{\sqrt{n}}{\sqrt{n}+1}$

45. $p_n = 2^{-1/n}$

46. $q_n = \sqrt{1 + \dfrac{1}{n}}$

47. $a_n = \sin^{-1}\left(\dfrac{n^2 + n + 2}{2n^2 + 1}\right)$

48. $a_n = \left(\dfrac{n^2 + 1}{3n^2 - 2}\right)\left(3 + \dfrac{1}{n}\right)$

49. $r_n = \dfrac{2}{(0.6)^n}$ **50.** $s_n = \dfrac{\ln 3n}{\ln 4n}$

51. $a_n = \left(\dfrac{1}{n}\right)^{1/n}$ **52.** $a_n = \dfrac{n^{1/n}}{\ln n}$

53. $t_n = \dfrac{n^n}{n!}$ **54.** $u_n = \tan^{-1}(\ln n)$

55. $a_n = \left(\dfrac{1}{\ln n}\right)^{1/n}$ **56.** $a_n = n\sin\dfrac{1}{n}$

57. $m_n = e^{-(\cos n)/n}$ **58.** $N_k = (-1)^k \dfrac{(\ln k)^2}{k}$

59. $a_n = \dfrac{2^n - 1}{\pi^n}$ **60.** $K_n = \sqrt[n]{3n + 1}$

61. $L_n = \left(\dfrac{1}{10}\right)^{-1/n}$ **62.** $S_k = \left(\dfrac{1}{k}\right)^{2/k}$

63. $T_k = \dfrac{\left(\frac{1}{2}\right)^k}{k^k - 1}$ **64.** $a_n = \left(1 + \dfrac{1}{n}\right)^n$

65. $a_n = \left(1 + \dfrac{1}{n}\right)^{2n}$ **66.** $a_n = \left(1 + \dfrac{1}{n^2}\right)^n$

67. $a_n = \left(1 + \dfrac{3}{n}\right)^n$

68. $a_1 = 1, \quad a_{n+1} = \dfrac{1}{2}\left(a_n + \dfrac{5}{a_n}\right)$

69. $a_1 = 1, \quad a_n = \dfrac{3}{4}a_{n-1} + \dfrac{1}{a_{n-1}}$

70–73 Use the Squeeze Theorem to prove that the given sequence converges.

70. $\left\{\dfrac{\sin n}{n}\right\}_{n=1}^{\infty}$ **71.** $\left\{\dfrac{\cos^2 n}{3^n}\right\}_{n=1}^{\infty}$

72. $\left\{(-1)^n \dfrac{(\ln n)^2}{n^3}\right\}_{n=1}^{\infty}$ **73.** $\left\{2^{-n}\cos n\right\}_{n=1}^{\infty}$

74.* Prove the Bounded Monotonic Sequence Theorem. (**Hint:** Suppose first that $\{a_n\}$ is increasing and bounded above. Let L be the least upper bound of the set of values $\{a_n \mid n \in \mathbb{N}\}$ and fix an $\varepsilon > 0$. Since $L - \varepsilon$ is *not* an upper bound for $\{a_n \mid n \in \mathbb{N}\}$, there is an index N such that $a_N > L - \varepsilon$. Use monotonicity to finish the argument. Note that the decreasing case can be handled similarly, or by considering the sequence $\{-a_n\}$.)

75–80 Use the Bounded Monotonic Sequence Theorem to prove that the given sequence converges. In Exercises 75–78, find the limit.

75. $a_n = \dfrac{n^2}{n^2 + 1}$

76. $a_1 = \sqrt{6}, \quad a_{n+1} = \sqrt{6 + a_n}$

77. $a_1 = \sqrt{2}, \quad a_{n+1} = \sqrt{2a_n}$

78. $a_1 = 0, \quad a_n = \dfrac{1}{2 - a_{n-1}}$

79. $a_n = \dfrac{(1)(3)(5)\cdots(2n-1)}{(2)(4)(6)\cdots(2n)}$

80. $a_n = 1 + \dfrac{1}{2} + \dfrac{1}{3!} + \dfrac{1}{4!} + \cdots + \dfrac{1}{n!}$

81–82 The Fibonacci sequence has many interesting applications in combinatorics and the mathematics of computer algorithms. Fibonacci numbers also appear in nature, in the arrangements of leaves and flower petals and the geometry of some shells. We can derive the explicit formula for F_n, the nth term of the Fibonacci sequence, as follows. First, notice that φ and ψ in Example 2 are the roots of the quadratic equation $x^2 - x - 1 = 0$ and, thus, we have $\varphi^2 = \varphi + 1$ and $\psi^2 = \psi + 1$. Using, say, $\varphi^2 = \varphi + 1$ we obtain the following:

$$\varphi^3 = \varphi\varphi^2 = \varphi(\varphi + 1) = \varphi^2 + \varphi = (\varphi + 1) + \varphi = 2\varphi + 1$$

$$\varphi^4 = \varphi\varphi^3 = \varphi(2\varphi + 1) = 2\varphi^2 + \varphi = 2(\varphi + 1) + \varphi = 3\varphi + 2$$

Notice that the coefficients are Fibonacci numbers, and if we repeat the process for higher powers of φ, the Fibonacci numbers keep coming up. More precisely, using an induction argument one can show the following:

$$\varphi^n = F_n\varphi + F_{n-1} \quad (1)$$

$$\psi^n = F_n\psi + F_{n-1} \quad (2)$$

In Exercises 81–82, use this observation to derive the explicit formula for F_n.

81. Verify the explicit formula given in Example 2 for the nth term of the Fibonacci sequence. (**Hint:** Subtract equation (2) from (1) above and solve for F_n.)

82.* Complete the induction argument referred to in the discussion preceding Exercise 81 to prove $\varphi^n = F_n\varphi + F_{n-1}$.

83–84 The popular Tower of Hanoi puzzle was invented by the French mathematician Édouard Lucas in 1883. It consists of three pegs, with n disks placed on the first peg in order of increasing size from top to bottom (see figure). The objective is to transfer all disks to the second peg so that they end up in the same order, according to the following two rules. Only one disk can be moved at a time, and no disk can be placed at any time on top of a smaller disk. In Exercises 83–84, you will find T_n, the number of moves necessary to solve the puzzle with n disks.

n disks

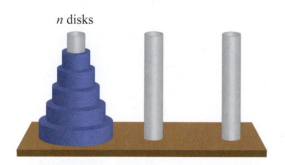

83. Find a recursive definition for the sequence $\{T_n\}$. (**Hint:** Use T_{n-1} steps to move the top $n-1$ disks to the third peg, then place the largest disk on the second peg, and then repeat the previous moves to complete the tower on the second peg.)

84.* Prove that an explicit formula for $\{T_n\}$ is $T_n = 2^n - 1$.

85.* Prove that if $a > 1$, then $\lim_{n\to\infty} a^n = \infty$.

86.* Prove that if $|a| < 1$, then $\lim_{n\to\infty} a^n = 0$.

87.* Prove that if a sequence $\{a_n\}_{n=1}^{\infty}$ is convergent, then it is bounded.

88. Give an example of an unbounded sequence $\{a_n\}_{n=1}^{\infty}$ such that $a_n \not\to \infty$ and $a_n \not\to -\infty$.

89.* Prove that if a sequence $\{a_n\}_{n=1}^{\infty}$ is convergent, then it has a largest or a smallest term.

90. Suppose that $\lim_{n\to\infty} a_n = \infty$ and $\{b_n\}$ is a sequence such that there is an N with $b_n \geq a_n$ for all $n \geq N$. Prove that $\lim_{n\to\infty} b_n = \infty$.

91.* Prove that the limit of a convergent sequence is unique, that is, no convergent sequence can have two different limits. (**Hint:** Create an indirect argument by first assuming that $\{a_n\}$ converges to both L_1 and L_2, $L_1 \neq L_2$, and then using the definition of convergence.)

92.* We call $\{a_n\}$ a *null sequence* if $\lim_{n\to\infty} a_n = 0$. Prove that if $\{a_n\}$ is a null sequence and $\{b_n\}$ is bounded, then $\{a_n \cdot b_n\}$ is a null sequence.

93. Suppose that all terms of the sequence $\{a_n\}_{n=1}^{\infty}$ are nonzero and $\lim_{n\to\infty} |a_n| = \infty$. Prove that $\{1/a_n\}$ is a null sequence.

94.* Prove that if $\lim_{n\to\infty} a_n = \infty$ and $\{b_n\}$ is bounded, then $\lim_{n\to\infty}(a_n + b_n) = \infty$.

95.* If every term of a sequence is positive, it is called a *positive sequence*. Prove that if $\{a_n\}$ is a positive sequence and $\lim_{n\to\infty}(a_{n+1}/a_n) = L > 1$, then $\lim_{n\to\infty} a_n = \infty$.

96.* Prove that if $\{a_n\}$ is a positive sequence and $\lim_{n\to\infty}(a_{n+1}/a_n) = L < 1$, then $\{a_n\}$ is a null sequence.

97.* Prove that if $\lim_{n\to\infty}(a_{n+1} - a_n) = L > 0$, then $\lim_{n\to\infty} a_n = \infty$.

98.* Suppose that for two convergent sequences, $\lim_{n\to\infty} a_n = L_a$, $\lim_{n\to\infty} b_n = L_b$, and that $L_a < L_b$. Prove that there is a positive integer N such that $a_n < b_n$ for all $n \geq N$.

99.* Suppose that for two convergent sequences, $\lim_{n\to\infty} a_n = L_a$, $\lim_{n\to\infty} b_n = L_b$, and that $a_n < b_n$ for all n. Prove that $L_a \le L_b$. Can we conclude that $L_a < L_b$?

100. Prove that for any irrational number $r \in \mathbb{R}$ there is a strictly increasing sequence $\{r_n\}$ of rational numbers so that $r_n \to r$.

101. Let $\{a_n\}$ be a recursively defined sequence as follows: $a_1 = 1$, $a_2 = 2$, and $a_{n+1} = \frac{1}{2}(a_n + a_{n-1})$ for $n \ge 2$. Prove that the sequence converges, find an explicit formula for its nth term, and find its limit. (**Hint:** Start by finding the "gaps" between consecutive terms: $|a_{n+1} - a_n|$.)

102.* If $a > 0$, prove that the recursively defined sequence $a_1 = 1$, $a_{n+1} = (a_n^2 + a)/(2a_n)$ is convergent, and its limit is $\sqrt{a}$. (**Hint:** Think about Newton's method.)

103–114 *True or False?* Determine whether the given statement is true or false. In case of a false statement, explain or provide a counterexample.

103. If $\{a_n\}_{n=1}^{\infty}$ is convergent and its limit is L, then for any $\varepsilon > 0$, the interval $(L - \varepsilon, L + \varepsilon)$ contains all but finitely many terms of $\{a_n\}_{n=1}^{\infty}$.

104. If for any $\varepsilon > 0$, the interval $(L - \varepsilon, L + \varepsilon)$ contains infinitely many terms of $\{a_n\}_{n=1}^{\infty}$, then $\{a_n\}_{n=1}^{\infty}$ is convergent.

105. If $\{a_n\}_{n=1}^{\infty}$ is convergent, then so is $\{|a_n|\}_{n=1}^{\infty}$.

106. If $\{|a_n|\}_{n=1}^{\infty}$ is convergent, then so is $\{a_n\}_{n=1}^{\infty}$.

107. If the first n terms of a convergent sequence are altered, the resulting sequence still converges to the same limit.

108. If $\lim_{n\to\infty} a_n = L$, then $\lim_{n\to\infty} a_n^2 = L^2$.

109. If $\lim_{n\to\infty} a_n^2 = L$, then $\lim_{n\to\infty} a_n = \sqrt{L}$.

110. If $\lim_{n\to\infty} a_n^3 = L$, then $\lim_{n\to\infty} a_n = \sqrt[3]{L}$.

111. If $\lim_{n\to\infty} a_n = L$, and $L > 1$, then $\lim_{n\to\infty} a_n^n = \infty$.

112. If both $\{a_n\}$ and $\{b_n\}$ diverge, then $\{a_n + b_n\}$ diverges.

113. If $\{a_n + b_n\}$ converges and $\{a_n\}$ converges, then $\{b_n\}$ also converges.

114. If $\{a_n b_n\}$ converges and $\{a_n\}$ diverges, then $\{b_n\}$ also diverges.

10.2 **Infinite Series**

TOPICS

1. Partial sums, convergence, and divergence

2. Geometric, telescoping, and harmonic series

We are now ready to tackle the sum of an infinite set of numbers. Although we know exactly what it means to add a finite set of numbers, an infinite sum has not yet even been defined. As we will see, the definition depends heavily on the work with sequences that we just completed.

TOPIC 1 **Partial Sums, Convergence, and Divergence**

We begin with a statement that lays out the terminology and notation of series.

Definition 💡

Infinite Series

An **infinite series** is the sum of an infinite sequence of numbers, $a_1 + a_2 + \cdots + a_n + \cdots$. We typically use sigma notation and write $\sum\limits_{n=1}^{\infty} a_n$ to indicate the series obtained by adding together the terms of the infinite sequence $\{a_n\}_{n=1}^{\infty}$.

Although it's a subtle point, the notation $\sum\limits_{n=1}^{\infty} a_n$ is an example of *formal* notation—we use it to refer to the series $a_1 + a_2 + \cdots + a_n + \cdots$ even if the sum doesn't actually exist. In fact, much of our initial work with series focuses on determining whether a given sum $\sum\limits_{n=1}^{\infty} a_n$ makes sense; we use the terms *converge* and *diverge* again, this time to indicate whether the sum exists or not.

Fortunately, the appearance of the two terms in this context closely mirrors their usage as applied to sequences. Our understanding of what it means to add infinitely many numbers is based on the behavior of the sequence of *partial sums* of the series, each of which is just a sum of a finite number of terms.

Definition 💡

Partial Sums and Series Convergence

Given a series $\sum\limits_{n=1}^{\infty} a_n$, the sum $s_n = a_1 + a_2 + \cdots + a_n$ is called the nth **partial sum** of the series, and the sequence $\{s_n\}_{n=1}^{\infty}$ is called the sequence of partial sums associated with $\sum\limits_{n=1}^{\infty} a_n$.

If $\lim\limits_{n \to \infty} s_n = L$, we say the series $\sum\limits_{n=1}^{\infty} a_n$ **converges** (or is **convergent**) and has the sum L, and we write $L = \sum\limits_{n=1}^{\infty} a_n$. If the sequence $\{s_n\}_{n=1}^{\infty}$ diverges, we say that the series $\sum\limits_{n=1}^{\infty} a_n$ also **diverges** (or is **divergent**).

Example 1 ✐

Determine whether the series $\displaystyle\sum_{n=1}^{\infty}\left(1/2^{n}\right)$ converges or diverges.

Solution

This series is a particular example of a *geometric series*, which we will soon study in full generality. We determine whether it converges or not by examining its partial sums, where $a_{n}=1/2^{n}$ represents the terms of the series.

$$s_{1}=a_{1}=\frac{1}{2}$$

$$s_{2}=a_{1}+a_{2}=\frac{1}{2}+\frac{1}{4}=\frac{3}{4}$$

$$s_{3}=a_{1}+a_{2}+a_{3}=\frac{1}{2}+\frac{1}{4}+\frac{1}{8}=\frac{7}{8}$$

$$s_{4}=a_{1}+a_{2}+a_{3}+a_{4}=\frac{1}{2}+\frac{1}{4}+\frac{1}{8}+\frac{1}{16}=\frac{15}{16}$$

$$s_{5}=a_{1}+a_{2}+a_{3}+a_{4}+a_{5}=\frac{1}{2}+\frac{1}{4}+\frac{1}{8}+\frac{1}{16}+\frac{1}{32}=\frac{31}{32}$$

$$\vdots$$

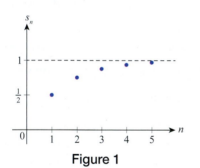

Figure 1

The behavior of this sequence of partial sums, shown in Figure 1, can be analyzed by relating it to everyday experiences. For instance, if we think of the series as fractions of an inch, s_{1} and a_{1} both represent half an inch—the beginning point of the series. The second partial sum s_{2} is the result of adding half the remaining distance (to the one-inch mark) to s_{1}, and of course half an inch plus half the remaining distance gets us to the three-quarter mark. If we add half the remaining distance again, we arrive at seven-eighths of an inch, and in general each successive partial sum is halfway closer to 1 than the previous partial sum.

In terms of a formula, the above paragraph suggests $s_{n}=1-\dfrac{1}{2^{n}}$, and we can make our reasoning rigorous with the following inductive argument.

Basis Step: $s_{1}=1-\dfrac{1}{2^{1}}=\dfrac{1}{2}$.

Inductive Step: Assume $s_{n}=1-\dfrac{1}{2^{n}}$. Then

$$s_{n+1}=s_{n}+a_{n+1}=1-\frac{1}{2^{n}}+\frac{1}{2^{n+1}}=1-\frac{2}{2^{n+1}}+\frac{1}{2^{n+1}}=1-\frac{1}{2^{n+1}}.$$

We can now examine the limit of the sequence of partial sums to determine if the series converges. Since

$$\lim_{n\to\infty}s_{n}=\lim_{n\to\infty}\left(1-\frac{1}{2^{n}}\right)=1,$$

we know $\displaystyle\sum_{n=1}^{\infty}\left(1/2^{n}\right)$ converges and we write $\displaystyle\sum_{n=1}^{\infty}\left(1/2^{n}\right)=1.$

TOPIC 2   Geometric, Telescoping, and Harmonic Series

We will devote the remainder of this section to exploring the behavior of some large families of series and to general theorems that apply to all series. The first family we study generalizes the series of Example 1.

Definition 💡

Geometric Series

Let a and r be fixed real numbers with $a \neq 0$. The series

$$\sum_{n=1}^{\infty} ar^{n-1} = a + ar + ar^2 + \cdots + ar^{n-1} + \cdots$$

is called a **geometric series**. The number r is the **common ratio** of the series.

Theorem 🔍

Convergence and Divergence of Geometric Series

The geometric series $\displaystyle\sum_{n=1}^{\infty} ar^{n-1}$ converges if $|r| < 1$ and diverges if $|r| \geq 1$. When $|r| < 1$,

$$\sum_{n=1}^{\infty} ar^{n-1} = \frac{a}{1-r}$$

First term divided by the difference between one and the common ratio

Proof 📎

We show convergence or divergence by examining the sequence of partial sums, and we begin with two specific cases.

If $r = 1$,

$$s_n = a + a(1) + a(1)^2 + \cdots + a(1)^{n-1} = na$$

and clearly $|na| \to \infty$ as $n \to \infty$ (remember that $a \neq 0$).

If $r = -1$,

$$s_n = a + a(-1) + a(-1)^2 + \cdots + a(-1)^{n-1} = \begin{cases} 0 & \text{if } n \text{ is even} \\ a & \text{if } n \text{ is odd} \end{cases}$$

so the sequence of partial sums is $\{a, 0, a, 0, \ldots\}$, which again clearly diverges.

We now consider all cases where $|r| \neq 1$ with the following formula for the partial sums:

$$s_n = a + ar + ar^2 + \cdots + ar^{n-1}$$
$$rs_n = ar + ar^2 + \cdots + ar^{n-1} + ar^n \qquad \text{Multiply through by } r.$$
$$s_n - rs_n = a - ar^n \qquad \text{Subtract second equation from first.}$$
$$s_n(1 - r) = a - ar^n$$
$$s_n = \frac{a - ar^n}{1 - r} \qquad \text{Solve for } s_n.$$

If $|r| < 1$,

$$\lim_{n\to\infty} s_n = \lim_{n\to\infty} \frac{a - ar^n}{1 - r} = \frac{a}{1 - r} \qquad r^n \to 0 \text{ as } n \to 0$$

and the series converges.

But if $|r| > 1$, $|ar^n| \to \infty$ as $n \to \infty$ and hence $|s_n| \to \infty$ as well, so the series diverges.

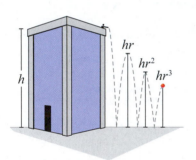

Figure 2

Example 2 ✐

A ball is dropped from an initial height of h units onto a flat surface and bounces repeatedly. After each bounce, the ball rebounds up to a height that is r times its previous height (with the ratio r positive and less than 1). Find the total distance the ball travels as it bounces up and down (see Figure 2).

Solution

The ball bounces straight up and down and we are only interested in the vertical distance traveled. Note that in this ideal scenario, the ball bounces an infinite number of times. Nevertheless, the distance the ball travels is finite. It travels h units on its initial drop, then travels upward hr units after its first bounce, travels another hr units on its way back down, up hr^2 after the second bounce, and so on. Thus, we calculate the total distance as follows:

$$S = h + 2hr + 2hr^2 + 2hr^3 + \cdots$$
$$= h + \sum_{n=1}^{\infty} (2hr) r^{n-1}$$
$$= h + \frac{2hr}{1 - r} = h\left(\frac{1 + r}{1 - r}\right)$$

For instance, if the initial height is $h = 5$ m and the ratio $r = 0.9$ (a typical ratio if both the ball and surface are made of a hard material such as steel), then

$$S = 5\left(\frac{1 + 0.9}{1 - 0.9}\right) = 5\left(\frac{\frac{19}{10}}{\frac{1}{10}}\right) = 95 \text{ m.}$$

Example 3 ✐

Determine the time required for the bouncing ball in Example 2 to come to rest.

Solution

The function $y(t) = -\frac{1}{2}gt^2 + h$, where g is the acceleration due to gravity, describes the vertical position at time t of an object dropped from an initial height h. Setting $y(t) = 0$ and solving the resulting equation for t gives us $t = \sqrt{2h/g}$, the time required for an object to fall to the ground from a height of h. This is also the time required for a ball to attain a height of h after a bounce, so the total time that the ball in Example 2 bounces is

$$
\begin{aligned}
T &= \sqrt{\frac{2h}{g}} + 2\sqrt{\frac{2hr}{g}} + 2\sqrt{\frac{2hr^2}{g}} + \cdots \\
&= \sqrt{\frac{2h}{g}} + \sum_{n=1}^{\infty}\left(2\sqrt{\frac{2hr}{g}}\right)\sqrt{r^{n-1}} \\
&= \sqrt{\frac{2h}{g}} + \sum_{n=1}^{\infty}\left(2\sqrt{\frac{2hr}{g}}\right)\left(\sqrt{r}\right)^{n-1} \\
&= \sqrt{\frac{2h}{g}} + \left(2\sqrt{\frac{2hr}{g}}\right)\left(\frac{1}{1-\sqrt{r}}\right) = \sqrt{\frac{2h}{g}}\left(\frac{1+\sqrt{r}}{1-\sqrt{r}}\right).
\end{aligned}
$$

If $h = 5$ m and $r = 0.9$, $T \approx 38.36$ seconds (using $g = 9.8 \text{ m/s}^2$). So the ball in this example bounces an infinite number of times while traveling a total distance of 95 meters in a little over 38 seconds—this sort of scenario is the basis for many of the paradoxes devised by the early Greek mathematician Zeno of Elea (5th century BC). (See Exercise 58.)

Geometric series actually serve as the rigorous foundation for some familiar notation.

Example 4 ✐

Express the repeating decimal $0.\overline{293}$ as a ratio of two integers.

Solution

The notation $0.\overline{293}$ stands for the sum

$$
s = \frac{293}{1000} + \frac{293}{(1000)^2} + \frac{293}{(1000)^3} + \cdots,
$$

which is a geometric series with $a = \frac{293}{1000}$ and a ratio of $r = \frac{1}{1000}$. So

$$
s = \sum_{n=1}^{\infty}\left(\frac{293}{1000}\right)\left(\frac{1}{1000}\right)^{n-1} = \left(\frac{293}{1000}\right)\left(\frac{1}{1-\frac{1}{1000}}\right) = \frac{293}{999}.
$$

Sometimes, the terms in a series partially cancel one another in a nice way and leave expressions that are easy to analyze. The following is an example of a **telescoping series**, so named because the terms collapse much like the segments of a telescope.

Example 5 ✎

Show that the series $\displaystyle\sum_{n=1}^{\infty}\frac{1}{n(n+1)}$ converges and determine its sum.

Solution

By partial fraction decomposition, $\dfrac{1}{n(n+1)}=\dfrac{1}{n}-\dfrac{1}{n+1}$

$$\sum_{n=1}^{\infty}\frac{1}{n(n+1)}=\sum_{n=1}^{\infty}\left(\frac{1}{n}-\frac{1}{n+1}\right)=\left(1-\frac{1}{2}\right)+\left(\frac{1}{2}-\frac{1}{3}\right)+\left(\frac{1}{3}-\frac{1}{4}\right)+\cdots$$

Writing out the first few terms of the series as above points out that a great deal of cancellation can be done. In fact, the nth partial sum has the form

$$s_n=\left(1-\frac{1}{2}\right)+\left(\frac{1}{2}-\frac{1}{3}\right)+\left(\frac{1}{3}-\frac{1}{4}\right)+\cdots+\left(\frac{1}{n}-\frac{1}{n+1}\right)=1-\frac{1}{n+1},$$

and

$$\lim_{n\to\infty}s_n=\lim_{n\to\infty}\left(1-\frac{1}{n+1}\right)=1.$$

So the series converges and has a sum of 1.

With a few examples of series out of the way, we can proceed to generalize and prove some theorems that characterize convergent and divergent series.

Theorem ⚲

Terms of a Convergent Series Go to Zero

If $\displaystyle\sum_{n=1}^{\infty}a_n$ converges, then $\displaystyle\lim_{n\to\infty}a_n=0.$

Proof ✎

If we let s_n represent the nth partial sum, then $a_n=s_n-s_{n-1}$. Since $\displaystyle\sum_{n=1}^{\infty}a_n$ converges, $\displaystyle\lim_{n\to\infty}s_n$ exists and is equal to some number L. Note that since $n-1\to\infty$ as $n\to\infty$, $\displaystyle\lim_{n\to\infty}s_{n-1}=L$ as well. So

$$\lim_{n\to\infty}a_n=\lim_{n\to\infty}(s_n-s_{n-1})=\lim_{n\to\infty}s_n-\lim_{n\to\infty}s_{n-1}=L-L=0.$$

The theorem above is often used to immediately determine that a given series diverges, as we can now say that if $\displaystyle\lim_{n\to\infty}a_n\neq 0$, then $\displaystyle\sum_{n=1}^{\infty}a_n$ does not converge. This fact is important enough to state as a theorem in its own right.

Theorem 🔍

The nth-Term Divergence Test

If $\lim\limits_{n\to\infty} a_n$ does not exist or is different from 0, then $\sum\limits_{n=1}^{\infty} a_n$ diverges.

It is very important to realize, however, that $\lim\limits_{n\to\infty} a_n = 0$ is necessary *but not sufficient* to guarantee that $\sum\limits_{n=1}^{\infty} a_n$ converges. One of the classic illustrations of this fact is the harmonic series $\sum\limits_{n=1}^{\infty} (1/n)$.

Example 6 ✎

Show that $\sum\limits_{n=1}^{\infty} (1/n)$ diverges.

Solution

This example introduces a technique that we often use to prove that a series diverges. It may be difficult to show directly that the sequence of partial sums $\{s_n\}_{n=1}^{\infty}$ diverges, but relatively easy to show that a subsequence diverges. And the divergence of a subsequence is sufficient to prove that the whole sequence diverges.

With this in mind, note the following:

$$s_1 = 1$$

$$s_2 = 1 + \frac{1}{2}$$

$$s_4 = 1 + \frac{1}{2} + \left(\frac{1}{3} + \frac{1}{4}\right) > 1 + \frac{1}{2} + \left(\frac{1}{4} + \frac{1}{4}\right) = 1 + \frac{2}{2}$$

$$s_8 = 1 + \frac{1}{2} + \left(\frac{1}{3} + \frac{1}{4}\right) + \left(\frac{1}{5} + \frac{1}{6} + \frac{1}{7} + \frac{1}{8}\right)$$

$$> 1 + \frac{1}{2} + \left(\frac{1}{4} + \frac{1}{4}\right) + \left(\frac{1}{8} + \frac{1}{8} + \frac{1}{8} + \frac{1}{8}\right) = 1 + \frac{1}{2} + \frac{1}{2} + \frac{1}{2} = 1 + \frac{3}{2}$$

$$s_{16} = 1 + \frac{1}{2} + \left(\frac{1}{3} + \frac{1}{4}\right) + \left(\frac{1}{5} + \cdots + \frac{1}{8}\right) + \left(\frac{1}{9} + \cdots + \frac{1}{16}\right)$$

$$> 1 + \frac{1}{2} + \left(\frac{1}{4} + \frac{1}{4}\right) + \left(\frac{1}{8} + \cdots + \frac{1}{8}\right) + \left(\frac{1}{16} + \cdots + \frac{1}{16}\right) = 1 + \frac{4}{2}$$

$$\vdots$$

In general, $s_{2^n} > 1 + \frac{n}{2}$ so $\lim\limits_{n\to\infty} s_{2^n} = \infty$. We haven't determined anything specific about the other partial sums, but we don't need to—the fact that this subsequence of the partial sums grows unbounded suffices to show that $\sum\limits_{n=1}^{\infty} (1/n)$ diverges.

Well-behaved series (meaning convergent series) continue to behave well when combined with one another in the following ways.

Theorem ⚗

Combining Convergent Series

Assume that $\sum a_n$ and $\sum b_n$ are convergent series (we omit the index of summation for simplicity of presentation) and k is a fixed real number. Then $\sum(a_n + b_n)$, $\sum(a_n - b_n)$, and $\sum ka_n$ are also convergent.

Sum Law	$\sum(a_n + b_n) = \sum a_n + \sum b_n$
Difference Law	$\sum(a_n - b_n) = \sum a_n - \sum b_n$
Constant Multiple Law	$\sum ka_n = k\sum a_n$

The proofs of these laws follow from the very similar sequence limit laws of Section 10.1, and will be left as exercises (Exercises 61–63). The following example shows how they can be used to evaluate limits.

Example 7 ✎

Determine whether each of the following series converges or diverges.

a. $\displaystyle\sum_{n=1}^{\infty} \frac{2^n - 3}{4^{n-1}}$
    b. $\displaystyle\sum_{n=1}^{\infty} \frac{n+2}{n(n+1)}$

Solution

a. The first step is to rewrite the terms of the series in a more useful form.

$$\frac{2^n - 3}{4^{n-1}} = \frac{2^n}{4^{n-1}} - \frac{3}{4^{n-1}} = \frac{(2)(2^{n-1})}{4^{n-1}} - \frac{3}{4^{n-1}} = 2\left(\frac{1}{2}\right)^{n-1} - 3\left(\frac{1}{4}\right)^{n-1}$$

Since both $\displaystyle\sum_{n=1}^{\infty} 2\left(\frac{1}{2}\right)^{n-1}$ and $\displaystyle\sum_{n=1}^{\infty} 3\left(\frac{1}{4}\right)^{n-1}$ converge, we know that the original series converges and

$$\sum_{n=1}^{\infty} \frac{2^n - 3}{4^{n-1}} = \sum_{n=1}^{\infty}\left[2\left(\frac{1}{2}\right)^{n-1} - 3\left(\frac{1}{4}\right)^{n-1}\right] = 2\sum_{n=1}^{\infty}\left(\frac{1}{2}\right)^{n-1} - 3\sum_{n=1}^{\infty}\left(\frac{1}{4}\right)^{n-1}$$

$$= 2\left(\frac{1}{1 - \frac{1}{2}}\right) - 3\left(\frac{1}{1 - \frac{1}{4}}\right) = 4 - 4 = 0.$$

This result may be surprising at first, but writing out the beginning of the series we see that the first term is apparently canceled by the sum of the remaining terms.

$$\sum_{n=1}^{\infty} \frac{2^n - 3}{4^{n-1}} = -1 + \frac{1}{4} + \frac{5}{16} + \frac{13}{64} + \frac{29}{256} + \cdots = 0$$

b. We can use partial fraction decomposition to again rewrite the terms, hopefully in a more useful form.

$$\frac{n+2}{n(n+1)} = \frac{2}{n} - \frac{1}{n+1}$$

Both $\sum_{n=1}^{\infty} \frac{2}{n}$ and $\sum_{n=1}^{\infty} \frac{1}{n+1}$ diverge—the first is 2 times the harmonic series, and the second is the harmonic series with the first term omitted, which still diverges. But our theorem doesn't tell us anything about the difference of two divergent series—as we've seen, expressions such as $\infty - \infty$ are indeterminate. So we need to find another way of rewriting the terms that is more informative.

The expression $\frac{2}{n} - \frac{1}{n+1}$ is somewhat similar to the partial fraction decomposition of Example 5, and we can easily make the connection more apparent.

$$\frac{n+2}{n(n+1)} = \frac{2}{n} - \frac{1}{n+1} = \left(\frac{1}{n}\right) + \left(\frac{1}{n} - \frac{1}{n+1}\right) = \frac{1}{n} + \frac{1}{n(n+1)}$$

We could also have arrived at this immediately if we had decomposed the original fraction another way:

$$\frac{n+2}{n(n+1)} = \frac{(n+1)+1}{n(n+1)} = \frac{1}{n} + \frac{1}{n(n+1)}$$

We know that $\sum_{n=1}^{\infty} \frac{1}{n(n+1)}$ converges. Suppose that

$$\sum_{n=1}^{\infty} \frac{n+2}{n(n+1)} = \sum_{n=1}^{\infty} \left(\frac{1}{n} + \frac{1}{n(n+1)}\right)$$

converged. Then by the last theorem,

$$\sum_{n=1}^{\infty} \frac{1}{n} = \sum_{n=1}^{\infty} \left(\frac{1}{n} + \frac{1}{n(n+1)}\right) - \sum_{n=1}^{\infty} \frac{1}{n(n+1)}$$

would also converge, which we know is not true. So it must be the case that $\sum_{n=1}^{\infty} \frac{n+2}{n(n+1)}$ diverges.

The reasoning in part b. of Example 7 shows that if $\sum a_n$ converges and $\sum b_n$ diverges, then both $\sum(a_n + b_n)$ and $\sum(a_n - b_n)$ diverge. But we have no theorem that tells us about the behavior of $\sum(a_n \pm b_n)$ if both $\sum a_n$ and $\sum b_n$ diverge (see Exercises 53–56 for examples of what can happen).

We close out this section with one last example illustrating minor manipulations of series.

Example 8 ✐

Find the sums of each of the following series.

a. $\displaystyle\sum_{n=-2}^{\infty}\left(-\frac{1}{3}\right)^{n}$

b. $\displaystyle\sum_{n=3}^{\infty}\frac{2}{5^{n+2}}$

Solution

Each of these is a geometric series, though the form of the index and its beginning value are slightly different than in our definition. Nevertheless, if the first few terms are written out, it is easy to identify the first term and the common ratio.

a. Note that $\displaystyle\sum_{n=-2}^{\infty}\left(-\frac{1}{3}\right)^{n}=9-3+1-\frac{1}{3}+\cdots,$ so the first term is 9 and the common ratio is $-\frac{1}{3}$, leading to

$$\sum_{n=-2}^{\infty}\left(-\frac{1}{3}\right)^{n}=\frac{9}{1-\left(-\frac{1}{3}\right)}=\frac{27}{4}.$$

b. Note that $\displaystyle\sum_{n=3}^{\infty}\frac{2}{5^{n+2}}=\frac{2}{5^5}+\frac{2}{5^6}+\frac{2}{5^7}+\cdots,$ so the first term is $\dfrac{2}{5^5}$ and the common ratio is $\frac{1}{5}$, leading to

$$\sum_{n=3}^{\infty}\frac{2}{5^{n+2}}=\frac{\frac{2}{5^5}}{1-\frac{1}{5}}=\frac{2}{5^5}\cdot\frac{5}{4}=\frac{1}{5^4\cdot 2}=\frac{1}{1250}.$$

10.2 **Exercises**

1. Suppose that $a_n = 2n^2/(n^3-1)$ and consider the series $\displaystyle\sum_{n=2}^{\infty}a_n$. One of your classmates argues that the series converges, because the numerator of a_n has a lesser degree than the denominator, therefore $\displaystyle\lim_{n\to\infty}a_n=0$, which makes $\{s_n\}$ convergent, and thus, $\displaystyle\sum_{n=2}^{\infty}a_n$ has a finite sum. Is this argument correct? Why or why not?

2–5 Find the first five terms of the sequence of partial sums $\{s_n\}$ for the given series.

2. $\displaystyle\sum_{n=1}^{\infty}\left(\frac{3}{2}\right)^{n}$

3. $\displaystyle\sum_{n=0}^{\infty}(-1)^{n}$

4. $\displaystyle\sum_{n=0}^{\infty}\frac{3^n-1}{3^n}$

5. $\displaystyle\sum_{n=1}^{\infty}\sin\frac{n\pi}{3}$

6. Determine the index n so that the difference between the sum of the series $\displaystyle\sum_{n=1}^{\infty}(1/2^n)$ and the partial sum $\{s_n\}$ (the error) is less than 0.0001.

7. Write a short paragraph on the difference between the sequences $\{a_n\}$ and $\{s_n\}$ for the series in Example 5. (Mention convergence and limits.)

8–17 Determine whether the given geometric series converges. If so, find its sum.

8. $1+\dfrac{1}{4}+\dfrac{1}{16}+\dfrac{1}{64}+\dfrac{1}{256}+\cdots$

9. $5-1+\dfrac{1}{5}-\dfrac{1}{25}+\dfrac{1}{125}-\dfrac{1}{625}+\cdots$

10. $1-1.1+(1.1)^2-(1.1)^3+(1.1)^4-\cdots$

11. $\dfrac{4}{5}-1+\dfrac{5}{4}-\dfrac{25}{16}+\dfrac{125}{64}-\cdots$

12. $\displaystyle\sum_{n=0}^{\infty} 7^{-n}$

13. $\displaystyle\sum_{n=0}^{\infty} (-0.7)^n$

14. $\displaystyle\sum_{n=0}^{\infty} 3\cdot\left(\frac{11}{12}\right)^n$

15. $\displaystyle\sum_{n=0}^{\infty} 2\cdot(-0.35)^n$

16. $\displaystyle\sum_{n=0}^{\infty} \frac{4}{5}\cdot\left(-\frac{5}{4}\right)^n$

17. $\displaystyle\sum_{n=0}^{\infty} \frac{5}{4}\cdot\left(-\frac{4}{5}\right)^n$

18–19 Find all values of x for which the geometric series converges.

18. $\displaystyle\sum_{n=1}^{\infty} 2(1-3x)^{n-1}$

19. $\displaystyle\sum_{n=1}^{\infty} \frac{4}{(2x-5)^{n-1}}$

20–23 Recognize the repeating decimal as a geometric series and write the decimal as a ratio of two integers.

20. $0.\overline{5}$

21. $0.12\overline{3123}$

22. $0.5384\overline{384}$

23. $3.37\overline{9}$

24–29 Find the sum of the series. (**Hint:** Use partial fraction decomposition wherever appropriate to express it as a telescoping series.)

24. $\displaystyle\sum_{n=1}^{\infty} \left(\frac{1}{\sqrt{n}}-\frac{1}{\sqrt{n+1}}\right)$

25. $\displaystyle\sum_{n=0}^{\infty} \frac{2}{(n+1)(n+2)}$

26. $\displaystyle\sum_{n=1}^{\infty} \frac{2}{20n^2-5}$

27. $\displaystyle\sum_{n=1}^{\infty} \frac{3}{n^2+5n+6}$

28. $\displaystyle\sum_{n=1}^{\infty} \frac{1}{(4n+3)(4n-1)}$

29. $\displaystyle\sum_{n=1}^{\infty} \frac{2}{12n^2-3}$

30–49 Decide whether the given series converges. If so, find its sum.

30. $\displaystyle\sum_{n=0}^{\infty} \frac{3^{n-1}}{4^n}$

31. $\displaystyle\sum_{n=3}^{\infty} \frac{(-1)^{n-1} n}{n-2}$

32. $\displaystyle\sum_{n=0}^{\infty} (-1)^n \left(\frac{\pi}{4}\right)^n$

33. $\displaystyle\sum_{n=1}^{\infty} \left(\frac{1}{3^n}-\frac{3}{n}\right)$

34. $\displaystyle\sum_{n=1}^{\infty} \frac{1}{3^{1/n}}$

35. $\displaystyle\sum_{n=1}^{\infty} \frac{2^n+2^{2n}}{5^n}$

36. $\displaystyle\sum_{n=0}^{\infty} (2\sqrt{3})^{2-n}$

37. $\displaystyle\sum_{n=2}^{\infty} \frac{\sqrt{n+1}}{\ln n}$

38. $\displaystyle\sum_{n=0}^{\infty} \left(\frac{\pi-1}{e}\right)^n$

39. $\displaystyle\sum_{n=1}^{\infty} (3^{-n}-4^{-n})$

40. $\displaystyle\sum_{n=1}^{\infty} \frac{n}{3n+2}$

41. $\displaystyle\sum_{n=1}^{\infty} \frac{3^n-1}{3^n}$

42. $\displaystyle\sum_{n=1}^{\infty} (-1)^n \left(\frac{e}{\pi}\right)^n$

43. $\displaystyle\sum_{n=0}^{\infty} \frac{2\cdot 3^n+5\cdot 7^n}{11^n}$

44. $\displaystyle\sum_{n=0}^{\infty} \frac{2+3^n-5^n}{7^n}$

45. $\displaystyle\sum_{n=0}^{\infty} (-1)^n \frac{n}{n+2}$

46. $\displaystyle\sum_{n=1}^{\infty} \left(1+\frac{1}{n}\right)^n$

47. $\displaystyle\sum_{n=1}^{\infty} \left(\sqrt{n+1}-\sqrt{n}\right)$

48. $\displaystyle\sum_{n=2}^{\infty} \left(\frac{1}{\ln n}-\frac{1}{\ln(n+1)}\right)$

49. $\displaystyle\sum_{n=1}^{\infty} \frac{(-1)^n n}{2n-1}$

50. The series $\displaystyle\sum_{n=1}^{\infty} (1/2^n)$ converges and its sum is 1, as we have seen in Example 1. Examine what happens when we **a.** drop the first three terms of the series, and **b.** adjoin the three terms $3+2+1$ to the series. Conclude that deleting or adjoining finitely many terms to a series may change its sum, but not the fact of convergence. Can you provide a rigorous proof of this more general statement?

51–52 Use the definition of series convergence to prove that the series is convergent.

51. $\displaystyle\sum_{n=1}^{\infty} \frac{1}{2^n}$

52. $\displaystyle\sum_{n=1}^{\infty} \frac{1}{n(n+1)}$

53. Use Example 5 to show that it is possible for $\displaystyle\sum_{n=1}^{\infty}(a_n-b_n)$ to converge if both $\displaystyle\sum_{n=1}^{\infty} a_n$ and $\displaystyle\sum_{n=1}^{\infty} b_n$ are divergent.

54. Give an example of two divergent series such that $\displaystyle\sum_{n=1}^{\infty}(a_n+b_n)$ is convergent.

55. Give an example of two divergent series such that $\displaystyle\sum_{n=1}^{\infty}(a_n+b_n)$ is divergent.

56. Give an example of two divergent series such that $\displaystyle\sum_{n=1}^{\infty}(a_n-b_n)$ is divergent.

57. Prove that the series $\displaystyle\sum_{n=0}^{\infty}(a+nd)$ $(a, d \in \mathbb{R})$ converges if and only if $a=d=0$. (Such series are called *arithmetic series*.)

58. In one of Zeno's famous motion paradoxes, Achilles races a tortoise, giving the tortoise a 100-meter head start. Even though we assume that Achilles is ten times faster, the statement is that Achilles will never actually catch the tortoise. The reasoning goes as follows. Soon after the start, Achilles will reach the starting point of the tortoise, but by that time, the tortoise will have advanced 10 meters. Achilles' job is to quickly cover that 10-meter distance, but during that time, the tortoise will have advanced, namely, a meter, and so on. Give a calculus-based solution to the paradox by proving that Achilles will actually catch the tortoise and find the total distance Achilles will have run when it happens.

59. By examining partial sums and using properties of logarithms, prove that $\displaystyle\sum_{n=3}^{\infty} \ln \frac{n}{n-2}$ diverges.

60. Prove that $\displaystyle\sum_{n=0}^{\infty} (1-x)^n$ is convergent if $0 < x < 2$, and find its sum.

61. Prove the Sum Law for convergent series: $\sum (a_n + b_n) = \sum a_n + \sum b_n$. (**Hint:** Fix n, and write the statement for the nth partial sums first; then take the limits and use the appropriate limit laws.)

62. Prove the Difference Law for convergent series: $\sum (a_n - b_n) = \sum a_n - \sum b_n$. (See the hint given in Exercise 61.)

63. Prove the Constant Multiple Law for convergent series: $\sum k a_n = k \sum a_n$. (See the hint given in Exercise 61.)

64. Suppose a large state injects a 2-billion-dollar stimulus package into its economy. Consumers and businesses in the state save approximately 30 percent of that money and respend 70 percent. Of that latter amount, approximately 70 percent is again spent, and so on. What is the total spending generated by the stimulus package? (**Hint:** Find the sum of the geometric series that models the process. In economics, this is called the *multiplier effect*.)

65. Use the figure below and the fact that $\displaystyle\int_{1}^{\infty} (1/x)\,dx = \infty$ to argue that the harmonic series diverges. (We will refine this idea in Section 10.3 and use it to "test" the convergence of various series.)

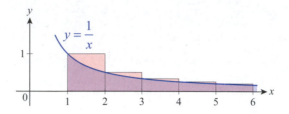

66. Prove that if the series $\sum a_n$ is convergent, then $\sum (1/a_n)$ is a divergent series.

67. Suppose that for the series $\sum a_n$ and $\sum b_n$ there exists a natural number N such that for all $n > N$, $a_n = b_n$. Prove that the series either both converge or both diverge.

68.* Prove that if $\displaystyle\sum_{n=1}^{\infty} a_n$ is convergent, then it satisfies the so-called *Cauchy criterion for convergence*: For any $\varepsilon > 0$, there is a corresponding natural number N such that $\displaystyle\left| \sum_{n=n_1+1}^{n_2} a_n \right| < \varepsilon$ for all $n_1, n_2 > N$. (**Hint:** Choose an appropriately small ε-neighborhood around the sum of the series, noticing that $\displaystyle\left| \sum_{n=n_1+1}^{n_2} a_n \right| = \left| s_{n_2} - s_{n_1} \right|$.)

69.* Use Exercise 68 to prove that $\displaystyle\sum_{n=1}^{\infty} (1/n!)$ is convergent.

70. Prove that if $\displaystyle\sum_{n=1}^{\infty} a_n$ is a *positive series*, that is, $a_n > 0$ for all n, then $\displaystyle\sum_{n=1}^{\infty} a_n$ is convergent if and only if the sequence $\{s_n\}_{n=1}^{\infty}$ of its partial sums is bounded.

71. Prove that $\displaystyle\sum_{n=1}^{\infty} \left(1/\sqrt{n}\right)$ diverges. (**Hint:** Prove that the nth partial sum $s_n \geq \sqrt{n}$.)

72. * Let $\sum_{n=1}^{\infty} a_n$ and $\sum_{n=1}^{\infty} b_n$ be positive series such that $\sum_{n=1}^{\infty} a_n$ is convergent and $a_n \geq b_n$ for all $n \in \mathbb{N}$. Prove that $\sum_{n=1}^{\infty} b_n$ is convergent. (**Hint:** Examine partial sums.)

73. * Let $\sum_{n=1}^{\infty} a_n$ and $\sum_{n=1}^{\infty} b_n$ be positive series such that $\sum_{n=1}^{\infty} a_n$ is divergent and $a_n \leq b_n$ for all $n \in \mathbb{N}$. Prove that $\sum_{n=1}^{\infty} b_n$ is divergent. (See the hint given in Exercise 72.)

74. Use Exercise 72 to prove that $\sum_{n=1}^{\infty} \left(1/n^2\right)$ is convergent. (**Hint:** Use the inequality below.)

$$\frac{1}{n^2} \leq \frac{1}{n(n-1)} = \frac{1}{n-1} - \frac{1}{n}$$

75. Use Exercise 73 and Example 6 to show that $\sum_{n=1}^{\infty} \left(1/n^p\right)$ diverges for all $0 < p < 1$.

76. * Two trains, 200 km apart, are on a collision course toward each other, each traveling at a rate of 50 km/h. A fly is zigzagging between the trains, flying at 75 km/h. Assuming constant rates and that the fly turns around in zero time, how much total distance will the fly be able to cover before being crushed to death by the trains upon their impending collision? (**Hint:** First, find the time required for the first "leg" of the flight, then, taking into consideration how much the original distance of 200 km between the trains has shrunk during this time, find the time required for the second "leg" of the fly's flight. Conclude that the time required for each leg is a constant times that required for the previous leg. Consequently, the fly's total time will be the sum of a geometric sequence; use this to find the total distance covered by the fly.)

77. Solve Exercise 76 "the easy way," without using an infinite series, that is, simply using the fact that the fly's total travel time equals the time necessary for the trains to reach each other. According to a well-authenticated story, when John von Neumann (born János Neumann), the great Hungarian American mathematician of the 20ᵗʰ century, was challenged with a version of this problem, he answered correctly within a few seconds. "Interesting," his challenger remarked, "most people try to solve this problem using infinite series." "Why," came von Neumann's reply, "that is how I did it!"

78–81 The following figures were generated by Rick Mabry as "pictorial proofs" for the convergence of various series. For example, dividing an equilateral triangle into four congruent parts and iterating the process, as below, provides illustration for the fact that $\sum_{n=1}^{\infty} \left(1/4^n\right) = \frac{1}{3}$ (assume that in the figure below, the area of the original triangle is 1; you can visually check what portion of the area is occupied by each color).

Source: http://www.lsus.edu/rick-mabry

$$\frac{1}{4} + \left(\frac{1}{4}\right)^2 + \left(\frac{1}{4}\right)^3 + \cdots = \frac{1}{3}$$

In Exercises 78–81, use the visual approach discussed above to identify the convergent series illustrated by the figure.

78.

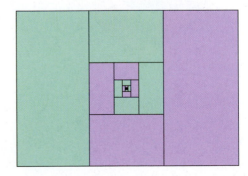

79.

80.

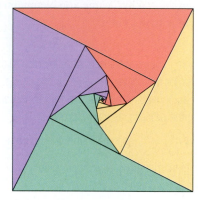

81.

10.2 **Technology Exercises**

82–85 Often we can readily show that a particular series is convergent, but finding the sum may be extremely challenging, if not impossible. For example, it is not difficult to prove that the series $\sum_{n=1}^{\infty}\left(1/n^p\right)$ is convergent if $p > 1$ (see Exercise 74 and, for the full story, Example 2 in Section 10.3), but finding the sum often defies the best efforts of mathematicians. For example, letting $S(p)$ denote the sum of the above series for a particular $p > 1$, we know that $S(2) = \pi^2/6$ (this surprising result was first proven by Euler), but formulas for the sums for odd p-values, such as $S(3)$, $S(5)$, etc. are still unknown.

In Exercises 82–85, use a computer algebra system to verify the indicated sums.

82. $\displaystyle\sum_{n=1}^{\infty}\frac{1}{n^2} = \frac{\pi^2}{6}$ **83.** $\displaystyle\sum_{n=1}^{\infty}\frac{1}{n^4} = \frac{\pi^4}{90}$

84. $\displaystyle\sum_{n=1}^{\infty}\frac{1}{n^6} = \frac{\pi^6}{945}$ **85.** $\displaystyle\sum_{n=0}^{\infty}\frac{(-1)^n}{n!} = \frac{1}{e}$

86. On the same screen, graph the function you obtained in Exercise 60 along with the partial sums $p(x) = \displaystyle\sum_{n=0}^{N}(1-x)^n$ on the interval $(0, 2)$, for higher and higher N-values. What do you see?

10.3 **The Integral Test**

TOPICS

1. The Integral Test

2. Series and remainder estimates

We now have a formula for the sum of a geometric series (assuming it converges), and we can typically figure out the sum of a telescoping series in the process of identifying it *as* telescoping, but sums of convergent series are in general not easy to find. Fortunately, in most circumstances we're not so interested in the actual sum as we are in whether a series converges in the first place. In this and the next three sections we develop a number of tests to help us answer this question.

TOPIC 1   **The Integral Test**

Consider the series $\sum\limits_{n=1}^{\infty}\left(1/n^2\right)$. It passes the nth-Term Divergence Test ($1/n^2 \to 0$ as $n \to \infty$), but this only tells us that the series *may* converge. Remember that this condition is necessary but not sufficient to prove convergence. The series is neither geometric nor telescoping, so we don't have a convenient formula to use for the partial sums. No theorem that we learned in the previous section seems to help, but a depiction of the series as in Figure 1 at least gives us a way to compare it to something we are familiar with.

In Figure 1, the nth rectangle has a width of 1 and a height of $1/n^2$, so its area is equal to the nth term of the series $\sum\limits_{n=1}^{\infty}\left(1/n^2\right)$. If we define $f(x) = 1/x^2$, we can write

$$\sum_{n=1}^{\infty}\frac{1}{n^2} = \frac{1}{1} + \frac{1}{4} + \frac{1}{9} + \cdots = f(1)(1) + f(2)(1) + f(3)(1) + \cdots.$$

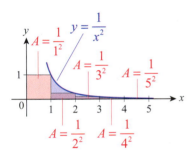

Figure 1   Area Depiction of $\sum\limits_{n=1}^{\infty}\dfrac{1}{n^2}$

Expressed this way, the series takes the form of a very coarse Riemann sum (coarse because the width of each subinterval is 1, a relatively large Δx). And since f is a decreasing function, the Riemann sum is an underestimate of the area between the graph of f and the positive x-axis—that is clear from Figure 1, since each rectangle lies under the graph of f. In fact, treating the first term separately, we note that

$$\sum_{n=1}^{\infty}\frac{1}{n^2} = f(1)(1) + f(2)(1) + f(3)(1) + \cdots < f(1) + \int_1^{\infty}\frac{1}{x^2}\,dx = 1 + 1 = 2.$$

Recall from Section 7.7 that an integral of the form $\int_1^{\infty}\left(1/x^2\right)dx$ is called an improper integral of type I, and its value is defined as a limit, if the limit exists—in Example 1 of that section we saw that $\int_1^{\infty}\left(1/x^2\right)dx = \lim\limits_{b\to\infty}\int_1^{b}\left(1/x^2\right)dx = \lim\limits_{b\to\infty}\left[-1/x\right]_1^{b} = 1.$

We can often compare a series, or at least all the terms of a series past some point N, to an integral just as we did above. When we can, the behavior of the series and the integral coincide—they either both converge or both diverge. The formal statement of this fact is called the Integral Test.

Theorem

The Integral Test

Assume that $\{a_n\}_{n=1}^{\infty}$ is a decreasing sequence of positive terms and $a_n = f(n)$, where $f(x)$ is a continuous, positive, and decreasing function for all x greater than or equal to some natural number N. Then the series $\displaystyle\sum_{n=N}^{\infty} a_n$ and the improper integral $\displaystyle\int_N^{\infty} f(x)\,dx$ both converge or both diverge.

Proof

For simplicity of presentation, we will prove the theorem for $N = 1$; the proof is similar for all other values of N.

The fact that both the sequence $\{a_n\}$ and the function f are positive means that the sequence of partial sums $\{s_n\}$ is an increasing sequence and that $\displaystyle\int_1^b f(x)\,dx$ increases as $b \to \infty$. So by the Bounded Monotonic Sequence Theorem, we can show $\{s_n\}$ converges by showing it is bounded above.

The decreasing nature of both $\{a_n\}$ and f allows us to relate the partial sums to the integrals, as follows. If we construct a Riemann sum for the integral $\displaystyle\int_1^n f(x)\,dx$ using left endpoints of the subintervals $[1,2],\ldots,[n-1,n]$ (as shown in Figure 2), then

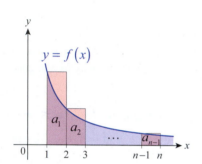

$$y = f(x)$$

Figure 2

$$\int_1^n f(x)\,dx \le a_1 + a_2 + \cdots + a_{n-1} = s_{n-1}.$$

But if we construct a Riemann sum for the same integral using right endpoints for our sample points (as in Figure 3) then

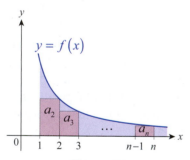

$$y = f(x)$$

Figure 3

$$a_2 + a_3 + \cdots + a_n \le \int_1^n f(x)\,dx,$$

which, if we add the first term of the series to both sides, gives us

$$s_n = a_1 + a_2 + \cdots + a_n \le a_1 + \int_1^n f(x)\,dx.$$

So if $\displaystyle\int_1^{\infty} f(x)\,dx$ converges, we know that

$$\sum_{n=1}^{\infty} a_n = \lim_{n\to\infty} s_n \le \lim_{n\to\infty}\left[a_1 + \int_1^n f(x)\,dx\right] = a_1 + \lim_{n\to\infty}\int_1^n f(x)\,dx = a_1 + \int_1^{\infty} f(x)\,dx$$

and hence $\displaystyle\sum_{n=1}^{\infty} a_n$ converges as well (and its sum is less than $a_1 + \int_1^{\infty} f(x)\,dx$).

If $\displaystyle\int_1^{\infty} f(x)\,dx$ diverges, then $\displaystyle\sum_{n=1}^{\infty} a_n$ diverges too, since

$$\infty = \int_1^{\infty} f(x)\,dx = \lim_{n\to\infty}\int_1^n f(x)\,dx \le \lim_{n\to\infty} s_{n-1} = \lim_{n\to\infty} s_n = \sum_{n=1}^{\infty} a_n.$$

(By the same reasoning, convergence of $\displaystyle\sum_{n=1}^{\infty} a_n$ implies convergence of $\displaystyle\int_1^{\infty} f(x)\,dx$ and divergence of $\displaystyle\sum_{n=1}^{\infty} a_n$ implies divergence of $\displaystyle\int_1^{\infty} f(x)\,dx$.)

Caution ⚠

Although a series and integral meeting the conditions of the Integral Test will either both converge or both diverge, it should not be inferred, in the case of convergence, that they converge to the same value. For example,

$\sum_{n=1}^{\infty}\left(1/n^2\right)$ and $\int_{1}^{\infty}\left(1/x^2\right)dx$ both converge, but the series converges to a value of $\pi^2/6$ (see Exercise 82 of Section 10.2) while the integral converges to a value of 1.

Example 1 ✎

Use the Integral Test to prove the divergence of the harmonic series.

Solution

In the last section, we showed that $\sum_{n=1}^{\infty}\left(1/n\right)=\infty$ by showing there is a subsequence of partial sums that grows unbounded. But we can now use the Integral Test to more directly show divergence. Since $\left\{1/n\right\}_{n=1}^{\infty}$ is a sequence of positive but decreasing terms, and since

$$\int_{1}^{\infty}\frac{1}{x}dx=\lim_{n\to\infty}\left[\ln x\right]_{1}^{n}=\lim_{n\to\infty}\left(\ln n-0\right)=\infty,$$

$\sum_{n=1}^{\infty}\left(1/n\right)=\infty$ as well.

The convergent series $\sum_{n=1}^{\infty}\left(1/n^2\right)$ and the divergent series $\sum_{n=1}^{\infty}\left(1/n\right)$ are two examples of a **p-series**, which is any series of the form $\sum_{n=1}^{\infty}\left(1/n^p\right)$ with p a fixed real number. Example 2 resolves the question of their convergence in full generality.

Example 2 ✎

Determine for which values of p a p-series converges.

Solution

We already know that the harmonic series, $p=1$, diverges. If $p=0$, $\lim_{n\to\infty}\left(1/n^p\right)=\lim_{n\to\infty}\left(1\right)=1$ and if $p<0$, $\lim_{n\to\infty}\left(1/n^p\right)=\infty$ so $\sum_{n=1}^{\infty}\left(1/n^p\right)$ also diverges by the nth-Term Divergence Test for $p\le 0$.

For $0<p<1$ and $p>1$, $\left\{1/n^p\right\}$ is a sequence of positive but decreasing terms and we can apply the Integral Test. We first note that

$$\int_{1}^{\infty}\frac{1}{x^p}dx=\int_{1}^{\infty}x^{-p}\,dx=\lim_{b\to\infty}\left[\frac{x^{-p+1}}{-p+1}\right]_{1}^{b}=\left(\frac{1}{1-p}\right)\lim_{b\to\infty}\left(b^{-p+1}-1\right).$$

If $p > 1$, $-p + 1 < 0$ so

$$\left(\frac{1}{1-p}\right)\lim_{b \to \infty}\left(b^{-p+1} - 1\right) = \left(\frac{1}{1-p}\right)(0-1) = \frac{1}{p-1}$$

and the integral and series converge (remember though that the Integral Test doesn't tell us the sum of the series). If $0 < p < 1$, $-p + 1 > 0$, and b^{-p+1} increases without bound as $b \to \infty$. That is,

$$\left(\frac{1}{1-p}\right)\lim_{b \to \infty}\left(b^{-p+1} - 1\right) = \infty$$

and the integral diverges. In summary, a p-series converges if $p > 1$ and diverges if $p \leq 1$.

TOPIC 2   Series and Remainder Estimates

Suppose we have been able to use the Integral Test to show that $\sum\limits_{n=1}^{\infty} a_n$ converges. Since the Integral Test only provides an upper bound for the sequence of partial sums $\{s_n\}$, it doesn't tell us the actual sum $s = \lim\limits_{n \to \infty} s_n$ of the series. But we can use the reasoning behind the proof of the Integral Test to come up with an estimate for the sum that is as precise as we desire.

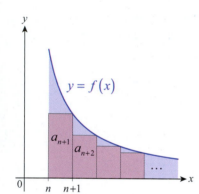

Figure 4

To do so, we introduce the idea of the series remainder $r_n = s - s_n$. Referring to Figure 4, we see that

$$r_n = a_{n+1} + a_{n+2} + \cdots \leq \int_n^\infty f(x)\, dx$$

and, referring to Figure 5,

$$r_n = a_{n+1} + a_{n+2} + \cdots \geq \int_{n+1}^\infty f(x)\, dx,$$

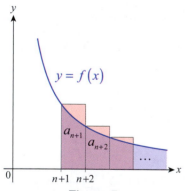

Figure 5

where, as before, f is a continuous, positive, and decreasing function with $a_n = f(n)$. This gives us a lower and upper bound on the remainder r_n and, if we rewrite it in terms of s and the nth partial sum s_n, a lower and upper bound on the actual sum of the series:

$$\int_{n+1}^\infty f(x)\, dx \leq s - s_n \leq \int_n^\infty f(x)\, dx$$
$$s_n + \int_{n+1}^\infty f(x)\, dx \leq s \leq s_n + \int_n^\infty f(x)\, dx$$

For reference, we state these two results as a theorem.

> ### Theorem ⚭
>
> **Series and Remainder Estimates**
>
> Assume that $\sum\limits_{n=1}^{\infty} a_n$ is a convergent series and $a_n = f(n)$, where $f(x)$ is a continuous, positive, and decreasing function. Then the sum s, the nth partial sum s_n, and the nth remainder r_n satisfy the inequalities
>
> $$\int_{n+1}^{\infty} f(x)\,dx \le r_n \le \int_{n}^{\infty} f(x)\,dx$$
>
> and
>
> $$s_n + \int_{n+1}^{\infty} f(x)\,dx \le s \le s_n + \int_{n}^{\infty} f(x)\,dx.$$

Example 3 ✍

Let $s = \sum\limits_{n=1}^{\infty} \left(1/n^2\right)$.

a. Rounded to two decimal places, the 10th partial sum s_{10} is 1.55; based on this, find lower and upper bounds for s.

b. With what accuracy do we know s if we instead use $s_{1000} \approx 1.643935$?

Solution

Since

$$\int_{n}^{\infty} \frac{1}{x^2}\,dx = \lim_{t \to \infty}\left[-\frac{1}{x}\right]_{n}^{t} = \lim_{t \to \infty}\left(-\frac{1}{t}+\frac{1}{n}\right) = \frac{1}{n},$$

we can make the following estimates:

a.
$$s_{10} + \frac{1}{11} \le s \le s_{10} + \frac{1}{10}$$
$$1.55 + 0.09 \le s \le 1.55 + 0.10$$
$$1.64 \le s \le 1.65$$

b.
$$s_{1000} + \frac{1}{1001} \le s \le s_{1000} + \frac{1}{1000}$$
$$1.643935 + 0.000999 \le s \le 1.643935 + 0.001$$
$$1.644934 \le s \le 1.644935$$

Since s lies within an interval of width 1×10^{-6}, we could approximate s by the average of the two values, 1.6449345, and know that the approximation is accurate to within $\frac{1}{2}\left(1 \times 10^{-6}\right) = 5 \times 10^{-7}$. (The exact value of s is $\pi^2/6$, as shown in Exercise 82 of Section 10.2)

Although the harmonic series diverges, we can still use an integral comparison to gain an appreciation for how slowly the sum grows. As we saw in proving the Integral Test, the relationship $\int_{1}^{n} f(x)\,dx \le s_{n-1}$ is valid for any series of positive but decreasing terms.

Example 4 ✐

Find a lower bound for the sum of the first 1 million terms of the harmonic series.

Solution

Using $\int_1^n f(x)\,dx \le s_{n-1}$ with $n = 1{,}000{,}001$ gives us

$$\int_1^{1{,}000{,}001} (1/x)\,dx = \ln(1{,}000{,}001) \approx 13.816.$$

In fact, after a significant amount of computation, a computer algebra system can tell us that $s_{1{,}000{,}000} \approx 14.393$, so the lower bound given by the integral is a very quick and reasonable indication of the rate at which a divergent series of positive decreasing terms grows.

10.3 **Exercises**

1–32 Use the Integral Test to determine whether the series converges or diverges.

1. $\displaystyle\sum_{n=1}^{\infty} \frac{2}{n+1}$

2. $\displaystyle\sum_{n=1}^{\infty} \frac{1}{3n-2}$

3. $\displaystyle\sum_{n=1}^{\infty} \frac{n}{n^2+1}$

4. $\displaystyle\sum_{n=1}^{\infty} \frac{2}{n^{5/4}}$

5. $\displaystyle\sum_{n=1}^{\infty} ne^{-n}$

6. $\displaystyle\sum_{n=4}^{\infty} \frac{6}{n^2-9}$

7. $\displaystyle\sum_{n=4}^{\infty} \frac{6}{(n-3)^2}$

8. $\displaystyle\sum_{n=1}^{\infty} \frac{n-1}{n^2-2n+0.75}$

9. $\displaystyle\sum_{n=1}^{\infty} \frac{2\ln n}{n}$

10. $\displaystyle\sum_{n=1}^{\infty} \frac{2\ln n}{n^2}$

11. $\displaystyle\sum_{n=2}^{\infty} \frac{2}{n(\ln n)^2}$

12. $\displaystyle\sum_{n=2}^{\infty} \frac{2}{n\ln n}$

13. $\displaystyle\sum_{n=1}^{\infty} \frac{1}{(n+1)\sqrt{n}}$

14. $\displaystyle\sum_{n=1}^{\infty} \frac{n}{(n^2+4)^2}$

15. $\displaystyle\sum_{n=1}^{\infty} \frac{\arctan n}{1+n^2}$

16. $\displaystyle\sum_{n=3}^{\infty} \frac{1}{n^2-2n}$

17. $\displaystyle\sum_{n=1}^{\infty} \left(\frac{1}{n} + \frac{1}{2^n}\right)$

18. $\displaystyle\sum_{n=1}^{\infty} \frac{1}{\sqrt{3n+1}}$

19. $\displaystyle\sum_{n=1}^{\infty} \frac{1}{(n+2)(n+3)^2}$

20. $\displaystyle\sum_{n=1}^{\infty} \frac{1}{n\sqrt{3n+1}}$

21. $\displaystyle\sum_{n=1}^{\infty} n^2 e^{-n}$

22. $\displaystyle\sum_{n=1}^{\infty} \frac{n}{(n+1)(n+2)}$

23. $\displaystyle\sum_{n=1}^{\infty} \frac{1}{n(n+1)}$

24. $\displaystyle\sum_{n=1}^{\infty} \frac{1}{an+b}, \quad a \ne 0$

25. $\displaystyle\sum_{n=1}^{\infty} \frac{1}{\sqrt[3]{n^4}}$

26. $\displaystyle\sum_{n=1}^{\infty} \frac{4n}{n^4+1}$

27. $\displaystyle\sum_{n=1}^{\infty} \frac{4n^3}{n^4+1}$

28. $\displaystyle\sum_{n=1}^{\infty} \left(2n^{-5/4} + \frac{1}{n^3}\right)$

29. $\displaystyle\sum_{n=1}^{\infty} \left(\frac{n}{(n^2+1)^3} + \frac{1}{n^2}\right)$

30. $\displaystyle\sum_{n=1}^{\infty} \frac{1}{\sqrt{n}(\sqrt{n}+1)^2}$

31. $\displaystyle\sum_{n=1}^{\infty} \frac{n}{(n^2+1)^{4/5}}$

32. $\displaystyle\sum_{n=0}^{\infty} \left(2^{-n} + \left(\frac{e}{3}\right)^n\right)$

33–36 Explain why the Integral Test is not applicable to test the series for convergence.

33. $\displaystyle\sum_{n=1}^{\infty} \frac{1+\cos n}{n^3}$

34. $\displaystyle\sum_{n=1}^{\infty} \frac{(-1)^n}{n^2}$

35. $\displaystyle\sum_{n=0}^{\infty} \frac{\cos n}{2^n}$

36. $\displaystyle\sum_{n=0}^{\infty} e^{-\pi n}\cos(\pi n)$

37. Use the Integral Test to show that the series

$$\sum_{n=2}^{\infty} \frac{1}{n(\ln n)^p}$$

diverges when $p \le 1$ and converges if $p > 1$. (These are called *logarithmic p-series*.)

38–43 Use our estimates immediately preceding Example 3 with the indicated number of terms to find an interval containing s, the sum of the series.

38. $\sum_{n=1}^{\infty} \frac{1}{n^4}$;   four terms   **39.** $\sum_{n=1}^{\infty} \frac{1}{n^3}$;   five terms

40. $\sum_{n=1}^{\infty} e^{-n}$;   five terms   **41.** $\sum_{n=1}^{\infty} \frac{1}{n^2+1}$;   six terms

42. $\sum_{n=1}^{\infty} \frac{2}{\sqrt{n}(n+1)}$;   five terms

43. $\sum_{n=2}^{\infty} \frac{1}{n(\ln n)^2}$;   five terms

44–49 Find the smallest possible value of n to approximate the sum of the given series within the indicated error ε and provide the requested estimate.

44. $\sum_{n=1}^{\infty} \frac{1}{n^4}$; $\varepsilon = 0.005$ **45.** $\sum_{n=1}^{\infty} \frac{1}{n^3}$; $\varepsilon = 0.005$

46. $\sum_{n=1}^{\infty} \frac{\arctan n}{1+n^2}$; $\varepsilon = 0.01$

47. $\sum_{n=1}^{\infty} ne^{-n^2}$; $\varepsilon = 5 \times 10^{-8}$

48. $\sum_{n=1}^{\infty} \frac{2}{(n+2)\left[\ln(n+2)\right]^3}$; $\varepsilon = 0.01$

49. $\sum_{n=1}^{\infty} \frac{3}{1+n^2}$; $\varepsilon = 0.05$

10.3 **Technology Exercises**

50–54 Use a computer algebra system and remainder estimates to approximate the sum of the series with the given error ε. Find the smallest possible value of n you can use and give an approximation with the requested accuracy. (Answers may vary slightly. Recall that formulas for these sums are unavailable; see the discussion preceding Exercise 82 in Section 10.2.)

50. $\sum_{n=1}^{\infty} \frac{1}{n^3}$; $\varepsilon = 10^{-8}$ **51.** $\sum_{n=1}^{\infty} \frac{1}{n^5}$; $\varepsilon = 5 \times 10^{-9}$

52. $\sum_{n=1}^{\infty} \frac{1}{n^7}$; $\varepsilon = 10^{-12}$ **53.** $\sum_{n=1}^{\infty} \frac{1}{n^9}$; $\varepsilon = 10^{-12}$

54. Approximate π to eight decimal places using the series $\sum_{n=1}^{\infty} \frac{1}{(2n-1)^2} = \frac{\pi^2}{8}$. What value of n did you use? (Answers will vary.)

10.4 Comparison Tests

TOPICS

1. The Direct Comparison Test

2. The Limit Comparison Test

Once we have gained familiarity with a few convergent and a few divergent series, we often have a sense of whether a new series converges based on how similar it is to a known series. This section introduces two tests that allow us to rigorously use that sense.

TOPIC 1   The Direct Comparison Test

> **Theorem** 🔍
>
> **The Direct Comparison Test**
>
> Suppose $\sum a_n$ is a series of nonnegative terms. Then
>
> **a.** $\sum a_n$ converges if there is a convergent series $\sum b_n$ for which $a_n \leq b_n$ for all $n \geq N$ (N a natural number);
>
> **b.** $\sum a_n$ diverges if there is a divergent series $\sum b_n$ of nonnegative terms for which $a_n \geq b_n$ for all $n \geq N$ (N a natural number).

> **Proof** 📎
>
> In part a., each partial sum of the series $\sum a_n$ is less than or equal to the number
>
> $$M = a_1 + a_2 + \cdots + a_{N-1} + \sum_{n=N}^{\infty} b_n.$$
>
> So the sequence of partial sums is increasing (since each a_n is nonnegative) and bounded above and hence converges by the Bounded Monotonic Sequence Theorem.
>
> In part b., suppose $\sum a_n$ did converge. Then since the partial sums of the series $\sum b_n$ are all less than or equal to the number
>
> $$M = b_1 + b_2 + \cdots + b_{N-1} + \sum_{n=N}^{\infty} a_n,$$
>
> the series $\sum b_n$ would converge too. Since this is not the case, $\sum a_n$ must diverge.

> **Example 1** ✐
>
> Use the Direct Comparison Test to determine whether each of the series converges or diverges.
>
> **a.** $\displaystyle\sum_{n=1}^{\infty} \frac{3}{2n-3}$ **b.** $\displaystyle\sum_{n=1}^{\infty} \frac{1}{n!}$ **c.** $\displaystyle\sum_{n=1}^{\infty} \frac{\ln n}{n}$
>
> **Solution**
>
> **a.** As a rational expression, $\dfrac{3}{2n-3}$ is similar to $\dfrac{1}{n}$; each has a constant numerator and each has a linear function of n as the denominator. Since the harmonic series diverges, we would likely guess that $\displaystyle\sum_{n=1}^{\infty} \frac{3}{2n-3}$

diverges too, and we can verify this with a comparison. Note that

$$\frac{3}{2n-3} = \frac{1}{\frac{2}{3}n-1}$$

and that, for $n \geq 2$,

$$0 < \frac{2}{3}n-1 < \frac{2}{3}n, \quad \text{so} \quad \frac{1}{\frac{2}{3}n-1} > \frac{1}{\frac{2}{3}n} = \frac{3}{2}\cdot\frac{1}{n}.$$

The first term of the series is unique in being both negative and less than its counterpart in the harmonic series. But since $\frac{3}{2}\sum\limits_{n=1}^{\infty}\frac{1}{n}$ diverges and its terms are less than the terms of $\sum\limits_{n=1}^{\infty}\frac{3}{2n-3}$ for $n \geq 2$, $\sum\limits_{n=1}^{\infty}\frac{3}{2n-3}$ diverges as well.

b. Note that

$$\sum_{n=1}^{\infty}\frac{1}{n!} = \frac{1}{1} + \frac{1}{1\cdot2} + \frac{1}{1\cdot2\cdot3} + \frac{1}{1\cdot2\cdot3\cdot4} + \cdots + \frac{1}{n!} + \cdots$$

$$< \frac{1}{1} + \frac{1}{2} + \frac{1}{2\cdot2} + \frac{1}{2\cdot2\cdot2} + \cdots + \frac{1}{2^{n-1}} + \cdots$$

$$= \sum_{n=1}^{\infty}\frac{1}{2^{n-1}} = \frac{1}{1-\frac{1}{2}}$$

Apply the formula for the sum of a geometric series.

$$= 2,$$

so $\sum\limits_{n=1}^{\infty}\frac{1}{n!}$ converges (and has a sum less than 2).

c. For $n \geq 3$,

$$\ln n > 1 \quad \text{and so} \quad \frac{\ln n}{n} > \frac{1}{n}.$$

Since its terms are nonnegative and larger than those of the divergent harmonic series for $n \geq 3$, the series $\sum\limits_{n=1}^{\infty}\frac{\ln n}{n}$ diverges. (Note that this can also be determined by the Integral Test.)

TOPIC 2   The Limit Comparison Test

Sometimes we are faced with a series that reminds us of a known series, but for which the Direct Comparison Test doesn't easily apply. For instance, the series $\sum\limits_{n=1}^{\infty}\frac{2}{3^n-7}$ seems very similar to the convergent geometric series $\sum\limits_{n=1}^{\infty}\frac{1}{3^n}$; the constant multiple of 2 in the numerator doesn't affect convergence, and the -7 in the denominator shouldn't really matter as it is overpowered by 3^n as n grows. Unfortunately,

$$\frac{2}{3^n-7} > \frac{1}{3^n} \quad \text{for } n \geq 2$$

and showing that the terms of a new series are larger than the corresponding terms of a series known to converge does us no good at all. However, the Limit Comparison Test is ideally suited to this example.

Theorem 🔍

The Limit Comparison Test

Suppose $\sum a_n$ and $\sum b_n$ are series for which $a_n > 0$ and $b_n > 0$ for all $n \geq N$ (N a natural number).

a. If $\lim_{n \to \infty} (a_n / b_n) = c > 0$, then $\sum a_n$ and $\sum b_n$ either both converge or both diverge.

b. If $\lim_{n \to \infty} (a_n / b_n) = 0$ and $\sum b_n$ converges, then $\sum a_n$ converges.

c. If $\lim_{n \to \infty} (a_n / b_n) = \infty$ and $\sum b_n$ diverges, then $\sum a_n$ diverges.

Proof 📎

All three parts are proved by applying the Direct Comparison Test—we will demonstrate the proof method with part a. here and leave parts b. and c. as exercises (Exercises 71–72).

By the definition of the limit of a sequence, if we let $\varepsilon = c/2$ there is a natural number $N_0 \geq N$ for which

$$\left| \frac{a_n}{b_n} - c \right| < \frac{c}{2} \text{ for all } n \geq N_0.$$

That is, $n \geq N_0$ implies $-\dfrac{c}{2} < \dfrac{a_n}{b_n} - c < \dfrac{c}{2}$ and hence $\dfrac{c}{2} < \dfrac{a_n}{b_n} < \dfrac{3c}{2}$.

Multiplying through by b_n, this means $\dfrac{c}{2} b_n < a_n < \dfrac{3c}{2} b_n$ for all $n \geq N_0$. So by the Direct Comparison Test, if $\sum b_n$ converges then $\sum a_n$ converges, and if $\sum a_n$ converges then $\sum b_n$ converges. Similarly, if either series diverges, the other series diverges too.

Example 2 ✏️

Use the Limit Comparison Test to show that $\sum_{n=1}^{\infty} \dfrac{2}{3^n - 7}$ converges.

Solution

We have already discussed how this new series seems similar to $\sum_{n=1}^{\infty} \dfrac{1}{3^n}$, so that is the series we will compare it to.

$$\lim_{n \to \infty} \frac{\dfrac{2}{3^n - 7}}{\dfrac{1}{3^n}} = \lim_{n \to \infty} \frac{2}{1 - \dfrac{7}{3^n}} = 2 > 0$$

So by the Limit Comparison Test, $\sum_{n=1}^{\infty} \dfrac{2}{3^n - 7}$ converges since $\sum_{n=1}^{\infty} \dfrac{1}{3^n}$ converges.

Example 3 ✐

Use the Limit Comparison Test to determine whether each of the series converges or diverges.

a. $\displaystyle\sum_{n=1}^{\infty}\frac{\sqrt[n]{n}}{n^2}$ **b.** $\displaystyle\sum_{n=1}^{\infty}\frac{n^2-2n+5}{\sqrt{3+n^5}}$ **c.** $\displaystyle\sum_{n=2}^{\infty}\frac{1}{\sqrt{n}\ln n}$

Solution

a. Each term $\sqrt[n]{n}/n^2$ is larger than $1/n^2$ for $n\geq 2$, so a direct comparison of this series with the convergent series $\sum\left(1/n^2\right)$ is inconclusive. But our work with l'Hôpital's Rule proved that $\sqrt[n]{n}\to 1$ as $n\to\infty$ (Example 9 of Section 4.4, but demonstrated again in brief below), so there's hope that the new series might converge. Using the Limit Comparison Test with $\sum\left(1/n^2\right)$ as the reference series, we see that

$$\lim_{n\to\infty}\frac{\dfrac{\sqrt[n]{n}}{n^2}}{\dfrac{1}{n^2}}=\lim_{n\to\infty}\sqrt[n]{n}=\lim_{n\to\infty}e^{(\ln n)/n}=e^0=1.$$

Since $\displaystyle\sum_{n=1}^{\infty}\left(1/n^2\right)$ converges, the Limit Comparison Test tells us that $\displaystyle\sum_{n=1}^{\infty}\left(\sqrt[n]{n}/n^2\right)$ converges too.

b. The terms $\dfrac{n^2-2n+5}{\sqrt{3+n^5}}$ look more and more like $\dfrac{n^2}{\sqrt{n^5}}$, or $\dfrac{1}{n^{1/2}}$, as n gets larger, so $\sum\dfrac{1}{n^{1/2}}$ is a reasonable series to test the new series against.

$$\lim_{n\to\infty}\frac{\dfrac{n^2-2n+5}{\sqrt{3+n^5}}}{\dfrac{1}{n^{1/2}}}=\lim_{n\to\infty}\frac{n^2-2n+5}{\sqrt{3+n^5}}\cdot\frac{n^{1/2}}{1}$$

$$=\lim_{n\to\infty}\frac{n^{5/2}-2n^{3/2}+5n^{1/2}}{\sqrt{3+n^5}}$$

$$=\lim_{n\to\infty}\frac{1-2n^{-1}+5n^{-2}}{\sqrt{3n^{-5}+1}}=1$$

The p-series $\displaystyle\sum_{n=1}^{\infty}\frac{1}{n^{1/2}}$ diverges $(p=\frac{1}{2}<1)$, so the series $\displaystyle\sum_{n=1}^{\infty}\frac{n^2-2n+5}{\sqrt{3+n^5}}$ diverges as well.

c. The series $\displaystyle\sum_{n=2}^{\infty}\frac{1}{\sqrt{n}\ln n}$ is difficult to compare directly to any series with known behavior, but a limit comparison using $\sum\dfrac{1}{n}$ as a reference sheds light.

$$\lim_{n\to\infty} \frac{\dfrac{1}{\sqrt{n}\ln n}}{\dfrac{1}{n}} = \lim_{n\to\infty} \frac{\sqrt{n}}{\ln n}$$

$$= \lim_{n\to\infty} \frac{\dfrac{1}{2\sqrt{n}}}{\dfrac{1}{n}} \qquad \text{l'Hôpital's Rule}$$

$$= \lim_{n\to\infty} \frac{\sqrt{n}}{2} = \infty$$

Since $\sum \dfrac{1}{n}$ diverges, the Limit Comparison Test tells us that $\sum_{n=2}^{\infty} \dfrac{1}{\sqrt{n}\ln n}$ diverges too.

10.4 Exercises

1–32 Use the Direct Comparison Test to determine whether the series converges or diverges. (Wherever applicable, combine the Direct Comparison Test with previous techniques, such as the Integral Test, p-series test, etc.)

1. $\displaystyle\sum_{n=1}^{\infty} \frac{2}{n^3+1}$

2. $\displaystyle\sum_{n=1}^{\infty} \frac{n+2}{n^3}$

3. $\displaystyle\sum_{n=2}^{\infty} \frac{4n}{\sqrt{n^3-1}}$

4. $\displaystyle\sum_{n=1}^{\infty} \frac{1}{n(n+2)(n+4)}$

5. $\displaystyle\sum_{n=1}^{\infty} \frac{\sqrt{n}}{n^2+\sqrt{n}}$

6. $\displaystyle\sum_{n=0}^{\infty} \frac{2}{n^n}$

7. $\displaystyle\sum_{n=0}^{\infty} \frac{n^3+1}{n^4+1}$

8. $\displaystyle\sum_{n=2}^{\infty} \frac{1}{n-\sqrt{n}}$

9. $\displaystyle\sum_{n=1}^{\infty} \frac{1}{n^{3/2}+\sqrt{n}}$

10. $\displaystyle\sum_{n=0}^{\infty} \frac{1}{1+2^n}$

11. $\displaystyle\sum_{n=0}^{\infty} \frac{5n^2}{n^4+1}$

12. $\displaystyle\sum_{n=2}^{\infty} \frac{5n^2}{n^3-1}$

13. $\displaystyle\sum_{n=0}^{\infty} \frac{1}{\sqrt{5n^3+2}}$

14. $\displaystyle\sum_{n=2}^{\infty} \frac{1}{\sqrt{n^2-1}}$

15. $\displaystyle\sum_{n=2}^{\infty} \frac{2}{\ln n}$

16. $\displaystyle\sum_{n=1}^{\infty} \frac{2}{n-\ln n}$

17. $\displaystyle\sum_{n=0}^{\infty} \frac{\cos^2 n}{\sqrt{n^3+2}}$

18. $\displaystyle\sum_{n=0}^{\infty} \frac{\sin^2 n}{2^n}$

19. $\displaystyle\sum_{n=1}^{\infty} \frac{\sin^2 n}{n^{1+\sqrt{n}}}$

20. $\displaystyle\sum_{n=1}^{\infty} \frac{\arctan n}{n^2}$

21. $\displaystyle\sum_{n=2}^{\infty} \frac{3n}{e^n \ln n}$

22. $\displaystyle\sum_{n=0}^{\infty} \sqrt{n}e^{-n^2}$

23. $\displaystyle\sum_{n=1}^{\infty} \frac{1}{\sqrt{n}e^{\sqrt{n}}}$

24. $\displaystyle\sum_{n=2}^{\infty} \frac{1}{n^2(\ln n)^2}$

25. $\displaystyle\sum_{n=2}^{\infty} \frac{\ln n}{n^3+1}$

26. $\displaystyle\sum_{n=0}^{\infty} \frac{1}{\sqrt{n^4+2n+3}}$

27. $\displaystyle\sum_{n=2}^{\infty} \frac{\ln n}{\sqrt{n^2-1}}$

28. $\displaystyle\sum_{n=0}^{\infty} \frac{2}{e^{2n}+n^{3/2}}$

29. $\displaystyle\sum_{n=2}^{\infty} \frac{\ln n}{n^{5/2}}$

30. $\displaystyle\sum_{n=1}^{\infty} \frac{1}{n^{1/2}-\frac{1}{2}}$

31. $\displaystyle\sum_{n=2}^{\infty} \frac{e^{-\sqrt{n}}}{\sqrt{n}\ln n}$

32. $\displaystyle\sum_{n=1}^{\infty} \sin\frac{1}{n^2}$

33–50 Use the Limit Comparison Test to determine whether the series converges or diverges.

33. $\displaystyle\sum_{n=1}^{\infty} \frac{2n^2}{n^3+1}$

34. $\displaystyle\sum_{n=2}^{\infty} \frac{2n^2}{n^4-1}$

35. $\displaystyle\sum_{n=1}^{\infty} \frac{\sqrt{n}}{\sqrt{n^2+1}}$

36. $\displaystyle\sum_{n=1}^{\infty} \frac{n^4}{\sqrt{2n^9+n^5+3n^3+2}}$

37. $\displaystyle\sum_{n=5}^{\infty} \frac{2n+4}{n(n-2)(n-4)}$

38. $\displaystyle\sum_{n=1}^{\infty} \frac{2}{\sqrt{2n}+\ln n}$

39. $\displaystyle\sum_{n=1}^{\infty} \frac{\sqrt{n}\,\ln(n+1)}{n^3}$

40. $\displaystyle\sum_{n=1}^{\infty} \frac{1}{2^{\ln n}}$

41. $\displaystyle\sum_{n=1}^{\infty} \frac{\ln n}{n^2}$ (**Hint:** Compare with $\displaystyle\sum_{n=1}^{\infty} \frac{1}{n^{3/2}}$.)

42. $\displaystyle\sum_{n=1}^{\infty} \frac{3^{1/n}-1}{3^{1/n}}$ (**Hint:** Compare with $\displaystyle\sum_{n=1}^{\infty} \frac{1}{n}$.)

43. $\displaystyle\sum_{n=2}^{\infty} \frac{\sqrt[n]{n}}{\sqrt{n}\,\ln n}$

44. $\displaystyle\sum_{n=1}^{\infty} \frac{\sqrt{2+\dfrac{1}{n}}}{2n^2}$

45. $\displaystyle\sum_{n=1}^{\infty} \frac{ne^{-n^2}}{1+e^{-n}}$

46. $\displaystyle\sum_{n=1}^{\infty} \frac{n^2}{3n^{5/2}+1}$

47. $\displaystyle\sum_{n=0}^{\infty} \frac{3}{\sqrt{n^2+3}}$

48. $\displaystyle\sum_{n=0}^{\infty} \frac{3}{\sqrt{n^3+3}}$

49. $\displaystyle\sum_{n=0}^{\infty} \frac{n+3}{\sqrt{n^4+3}}$

50. $\displaystyle\sum_{n=0}^{\infty} \frac{n+3}{\sqrt{n^5+3}}$

51–70 Use any test covered so far in the text to determine whether the series converges or diverges.

51. $\displaystyle\sum_{n=0}^{\infty} \frac{1}{2^n+3}$

52. $\displaystyle\sum_{n=0}^{\infty} \frac{2^n}{2^n+3}$

53. $\displaystyle\sum_{n=0}^{\infty} \left(\frac{2}{\sqrt{n+1}} - \frac{2}{\sqrt{n+2}} \right)$

54. $\displaystyle\sum_{n=1}^{\infty} \frac{1}{\sqrt{n}+\sqrt[3]{n}}$

55. $\displaystyle\sum_{n=0}^{\infty} \frac{1}{2^{n^2}-3}$

56. $\displaystyle\sum_{n=1}^{\infty} \frac{n}{2^{-n}+3^{-n}}$

57. $\displaystyle\sum_{n=1}^{\infty} \frac{\ln n+1}{\sqrt{n}}$

58. $\displaystyle\sum_{n=0}^{\infty} \frac{1}{n\sqrt{n}+\cos n}$

59. $\displaystyle\sum_{n=1}^{\infty} \frac{1-\sin n}{n^{3/2}}$

60. $\displaystyle\sum_{n=1}^{\infty} \frac{n-3}{n^2\sqrt{n}}$

61. $\displaystyle\sum_{n=1}^{\infty} \frac{1}{\tan^{-1} n}$

62. $\displaystyle\sum_{n=2}^{\infty} \frac{1}{\ln(\ln(\ln n))}$

63. $\displaystyle\sum_{n=3}^{\infty} \frac{1}{\ln n-1}$

64. $\displaystyle\sum_{n=3}^{\infty} \frac{\sqrt{n-2}}{n\sqrt{n^2-1}}$

65. $\displaystyle\sum_{n=1}^{\infty} \frac{n+3}{n3^n}$

66. $\displaystyle\sum_{n=1}^{\infty} \frac{2^{n-1}+2}{2^n}$

67. $\displaystyle\sum_{n=1}^{\infty} \frac{3^n+4^n}{4^n+5^n}$

68. $\displaystyle\sum_{n=1}^{\infty} \sin\frac{1}{n}$

69. $\displaystyle\sum_{n=1}^{\infty} \sin\frac{1}{n^2}$

70. $\displaystyle\sum_{n=0}^{\infty} \left(\frac{n}{n+2} \right)^n$

71. Prove part b. of the Limit Comparison Test. (**Hint:** Choose $K>0$ and a natural number $N_0 \geq N$ for which $0 \leq a_n/b_n \leq K$ for all $n \geq N_0$ and apply the Direct Comparison Test.)

72. Prove part c. of the Limit Comparison Test. (**Hint:** One possibility is to notice that $\lim_{n\to\infty}(b_n/a_n)=0$ and make use of the argument you gave in Exercise 71.)

73. Suppose that $\sum a_n$ is a positive series, $\{a_n\}$ is monotonically decreasing, and $\{na_n\}$ is convergent with $\lim_{n\to\infty} na_n \neq 0$. Prove that $\sum a_n$ diverges. (**Hint:** Use the Limit Comparison Test to compare $\sum a_n$ with an appropriate series. For the definition of a positive series, see Exercise 70 in Section 10.2.)

74–77 Use Exercise 73 to provide a quick proof of the divergence of the given series.

74. $\displaystyle\sum_{n=1}^{\infty} \frac{n^2}{3n^3+1}$

75. $\displaystyle\sum_{n=1}^{\infty} \frac{2}{\sqrt{n}}$

76. $\displaystyle\sum_{n=2}^{\infty} \frac{2}{\ln n}$

77. $\displaystyle\sum_{n=1}^{\infty} \frac{\arctan n}{\sqrt{n}}$

78.* (The Cauchy Condensation Test) Suppose that $\displaystyle\sum_{n=1}^{\infty} a_n$ is a positive series and $\{a_n\}$ is monotonically decreasing. Prove that $\displaystyle\sum_{n=1}^{\infty} a_n$ converges if and only if $\displaystyle\sum_{n=0}^{\infty} 2^n a_{2^n}$ converges. (**Hint:** Fix n and k, and let s_n be the nth partial sum of $\displaystyle\sum_{n=1}^{\infty} a_n$, while letting c_k denote the kth partial sum of the "condensed" series $\displaystyle\sum_{n=0}^{\infty} 2^n a_{2^n}$. Group the terms of $\displaystyle\sum_{n=1}^{\infty} a_n$ as

$$\sum_{n=1}^{\infty} a_n = a_1 + (a_2+a_3) + (a_4+a_5+a_6+a_7)+\cdots,$$

while $\displaystyle\sum_{n=0}^{\infty} 2^n a_{2^n}$ can be written as

$$\sum_{n=0}^{\infty} 2^n a_{2^n} = a_1 + (a_2 + a_2) + (a_4 + a_4 + a_4 + a_4) + \cdots.$$

Using the monotonicity of $\{a_n\}$, notice that if $n < 2^k$, then $s_n \le c_k$, while if $n > 2^k$, then $s_n \ge \tfrac{1}{2} c_k$. Use these observations to show that the sequence of partial sums of $\displaystyle\sum_{n=1}^{\infty} a_n$ is bounded if and only if that of $\displaystyle\sum_{n=0}^{\infty} 2^n a_{2^n}$ is bounded. Now use Exercise 70 from Section 10.2 to finish your argument.)

79. Use Exercise 78 to give a new proof of the p-series test (see Example 2 of Section 10.3).

80.* Suppose we stack idealized building blocks (perfectly rectangular, perfectly homogeneous, perfectly smooth, perfectly level, etc.) with the indicated fraction of each block protruding to the right of the block above it, as shown in the figure below. If we continue building in this way, that is, by making sure exactly $1/(2n)$ of the nth block protrudes to the right, how high can we build the structure before it topples? Assuming vertical sunrays (and an infinitely high sun), how far to the left will the shadow of the structure extend? (**Hint:** Starting from the bottom, consider the substructure consisting of the first n blocks and the location of its centroid relative to the right edge of the $(n+1)^{\text{th}}$ block.)

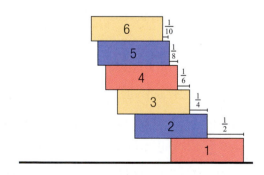

81.* Prove that a positive series $\displaystyle\sum a_n$ converges if and only if the series $\displaystyle\sum \frac{a_n}{a_n + 1}$ converges. (**Hint:** Show that $a_n < 2a_n/(a_n + 1)$ if n is large enough.)

82–91 *True or False?* Determine whether the given statement is true or false. In case of a false statement, explain or provide a counterexample.

82. If $\displaystyle\sum a_n$ and $\displaystyle\sum b_n$ are positive series, $\displaystyle\sum b_n$ diverges, and $a_n < b_n$ for all n, then $\displaystyle\sum a_n$ converges.

83. If $\displaystyle\sum a_n$ and $\displaystyle\sum b_n$ are convergent positive series, then $\displaystyle\sum a_n b_n$ is convergent.

84. If $\displaystyle\lim_{n \to \infty}(a_n/b_n) = 0$ and $\displaystyle\sum b_n$ diverges, then $\displaystyle\sum a_n$ diverges.

85. If $\displaystyle\lim_{n \to \infty}(a_n/b_n) = \infty$ and $\displaystyle\sum b_n$ converges, then $\displaystyle\sum a_n$ converges.

86.* If $\displaystyle\sum a_n$ is a convergent positive series, then $\displaystyle\sum \ln(1 + a_n)$ is convergent.

87. If $\displaystyle\sum a_n$ is a convergent positive series and $\displaystyle\sum b_n$ and $\displaystyle\sum c_n$ are positive series such that $b_n c_n = a_n$ for all n, then both $\displaystyle\sum b_n$ and $\displaystyle\sum c_n$ are convergent.

88. If $\displaystyle\sum a_n$ is a convergent positive series and $\displaystyle\sum b_n$ and $\displaystyle\sum c_n$ are positive series such that $b_n + c_n = a_n$ for all n, then both $\displaystyle\sum b_n$ and $\displaystyle\sum c_n$ are convergent.

89. If $p > 1$, then $\displaystyle\sum_{n=2}^{\infty} \frac{1}{(\ln n)^p}$ diverges.

90.* The series $\displaystyle\sum_{n=2}^{\infty} \frac{1}{(\ln n)^{\ln n}}$ converges.

91.* The series $\displaystyle\sum_{n=2}^{\infty} \frac{1}{(\ln n)^n}$ converges.

10.5 **The Ratio and Root Tests**

TOPICS

1. The Ratio Test

2. The Root Test

We have enough experience now with series of positive terms to realize that the key to their convergence lies in the *rate* at which the terms approach 0—many series have the property that $\lim_{n \to \infty} a_n = 0$, but that fact alone does not mean that $\sum a_n$ is finite. The two tests in this section give us further insight into this connection.

TOPIC 1   **The Ratio Test**

A geometric series $\sum ar^n$ with a and r both positive is a good model of the behavior we want to generalize. As we know, the distinguishing characteristic of a geometric series is that each term is a constant multiple, namely r, of the preceding term. And as we also know, $\sum ar^n$ converges if this ratio is less than 1 (remember that we are restricting attention here to geometric series with positive terms). As it turns out, a slight relaxation of this property characterizes convergence of all series of positive terms.

Theorem ⚘

The Ratio Test

Assume that $\sum a_n$ is a series of positive terms and $\lim_{n \to \infty} \left(a_{n+1}/a_n \right)$ exists or diverges to infinity. Then

a. $\sum a_n$ converges if $L = \lim_{n \to \infty} \left(a_{n+1}/a_n \right) < 1$;

b. $\sum a_n$ diverges if $L = \lim_{n \to \infty} \left(a_{n+1}/a_n \right) > 1$ or if $\lim_{n \to \infty} \left(a_{n+1}/a_n \right) = \infty$;

c. the test is inconclusive if $L = \lim_{n \to \infty} \left(a_{n+1}/a_n \right) = 1$.

(**Note:** The test also applies to series in which the terms are only all positive past some point. As always, the behavior of the first few terms of a series doesn't affect convergence or divergence—only the tail of a series matters.)

Proof ✎

a. Geometric series exhibit the limit behavior we are generalizing in the Ratio Test, and we can prove convergence of the series $\sum a_n$ by directly comparing it to an appropriate geometric series. Given that $L < 1$, fix a real number r such that $L < r < 1$ and let $\varepsilon = r - L$. Then there is a natural number N for which

$$n \geq N \quad \Rightarrow \quad \left| \frac{a_{n+1}}{a_n} - L \right| < \varepsilon.$$

In particular, this means that

$$\frac{a_{n+1}}{a_n} < \varepsilon + L = r$$

for all $n \geq N$, so

$$a_{N+1} < ra_N$$
$$a_{N+2} < ra_{N+1} < r^2 a_N$$
$$a_{N+3} < ra_{N+2} < r^3 a_N$$

and, in general, $a_{N+k} < r^k a_N$. In other words, from the point N onward, the terms of the series $\sum a_n$ are less than the terms of a geometric series with common ratio r.

$$\sum_{n=N}^{\infty} a_n < a_N + ra_N + r^2 a_N + \cdots = \frac{a_N}{1-r}$$

Adding in the first $N-1$ terms doesn't affect convergence, so the entire series $\sum a_n$ also converges.

b. If $L > 1$ or $L = \infty$, then from some point N onward it must be the case that $a_{n+1}/a_n > 1$, so

$$a_N < a_{N+1} < a_{N+2} < \cdots$$

and hence $\lim_{n \to \infty} a_n \neq 0$. By the nth-Term Divergence Test, $\sum a_n$ diverges.

c. The claim is that if $L = 1$, the Ratio Test is inconclusive. We can prove this by showing that there exists a convergent series with $L = 1$ and a divergent series with $L = 1$.

The series $\sum_{n=1}^{\infty} \frac{1}{n^2}$ converges and

$$\lim_{n \to \infty} \frac{a_{n+1}}{a_n} = \lim_{n \to \infty} \frac{\frac{1}{(n+1)^2}}{\frac{1}{n^2}} = \lim_{n \to \infty} \frac{n^2}{(n+1)^2} = \lim_{n \to \infty} \frac{1}{\left(1+\frac{1}{n}\right)^2} = 1.$$

The series $\sum_{n=1}^{\infty} \frac{1}{n}$ diverges and

$$\lim_{n \to \infty} \frac{a_{n+1}}{a_n} = \lim_{n \to \infty} \frac{\frac{1}{n+1}}{\frac{1}{n}} = \lim_{n \to \infty} \frac{n}{n+1} = \lim_{n \to \infty} \frac{1}{1+\frac{1}{n}} = 1.$$

Example 1 ✐

Determine whether each of the following series converges or diverges.

a. $\displaystyle\sum_{n=1}^{\infty} \frac{n^7}{7^n}$ **b.** $\displaystyle\sum_{n=1}^{\infty} \frac{(2n)!}{(n!)^2}$ **c.** $\displaystyle\sum_{n=2}^{\infty} \frac{1}{n \ln n}$

Solution

a. $\displaystyle\lim_{n \to \infty} \frac{a_{n+1}}{a_n} = \lim_{n \to \infty} \frac{\dfrac{(n+1)^7}{7^{n+1}}}{\dfrac{n^7}{7^n}} = \lim_{n \to \infty}\left(\frac{(n+1)^7}{7^{n+1}} \cdot \frac{7^n}{n^7} \right) = \frac{1}{7}\lim_{n \to \infty}\left(1 + \frac{1}{n} \right)^7 = \frac{1}{7} < 1$

Thus, the series converges. (Keep in mind that this does *not* tell us the actual sum of the series, only that the series converges.)

b. $\displaystyle\lim_{n \to \infty} \frac{a_{n+1}}{a_n} = \lim_{n \to \infty}\left[\frac{[2(n+1)]!}{(n+1)!(n+1)!} \cdot \frac{n! \, n!}{(2n)!} \right] = \lim_{n \to \infty}\left[\frac{n!}{(n+1)!} \cdot \frac{n!}{(n+1)!} \cdot \frac{(2n+2)!}{(2n)!} \right]$

$\displaystyle = \lim_{n \to \infty}\left[\frac{n!}{(n+1)n!} \cdot \frac{n!}{(n+1)n!} \cdot \frac{(2n+2)(2n+1)(2n)!}{(2n)!} \right]$

$\displaystyle = \lim_{n \to \infty} \frac{(2n+2)(2n+1)}{(n+1)(n+1)} = \lim_{n \to \infty} \frac{2(2n+1)}{n+1} = 4 > 1$

Therefore the series diverges.

c. $\displaystyle\lim_{n \to \infty} \frac{a_{n+1}}{a_n} = \lim_{n \to \infty}\left[\frac{1}{(n+1)\ln(n+1)} \cdot \frac{n \ln n}{1} \right] = \lim_{n \to \infty} \frac{n \ln n}{(n+1)\ln(n+1)}$

$\displaystyle = \lim_{n \to \infty} \frac{n}{n+1} \cdot \lim_{n \to \infty} \frac{\ln n}{\ln(n+1)} = \lim_{n \to \infty} \frac{\ln n}{\ln(n+1)}$

$\displaystyle = \lim_{n \to \infty} \frac{\dfrac{1}{n}}{\dfrac{1}{n+1}}$ l'Hôpital's Rule

$\displaystyle = \lim_{n \to \infty} \frac{n+1}{n} = 1$

Since the limit is 1, the Ratio Test tells us nothing about this series and we must look at it another way. Note that the Integral Test applies, and since

$\displaystyle\int_2^{\infty} \frac{1}{x \ln x}\, dx = \Big[\ln(\ln x) \Big]_2^{\infty} = \infty,$ Use u-substitution with $u = \ln x$.

the series diverges (for a generalization see Exercise 37 of Section 10.3).

TOPIC 2   **The Root Test**

We are nearing the end of our catalog of convergence tests, and in the next section we will provide a final list of them along with a summary of when and how they are applied. The last test of this section is one that is especially useful when a series has terms with n^{th} powers. Its proof is very similar to the proof of the Ratio Test.

Theorem 🔍

The Root Test

Assume that $\sum a_n$ is a series of nonnegative terms and $\lim_{n\to\infty} \sqrt[n]{a_n}$ exists or diverges to infinity. Then

a. $\sum a_n$ converges if $L = \lim_{n\to\infty} \sqrt[n]{a_n} < 1$;

b. $\sum a_n$ diverges if $L = \lim_{n\to\infty} \sqrt[n]{a_n} > 1$ or if $\lim_{n\to\infty} \sqrt[n]{a_n} = \infty$;

c. the test is inconclusive if $L = \lim_{n\to\infty} \sqrt[n]{a_n} = 1$.

(**Note:** Again, the test also applies to series whose terms are only all nonnegative past some point.)

Proof 📎

a. Fix an $\varepsilon > 0$ such that $L + \varepsilon < 1$. Since $\sqrt[n]{a_n} \to L$ there is a natural number N such that

$$n \geq N \quad \Rightarrow \quad \sqrt[n]{a_n} < L + \varepsilon,$$

so $a_n < (L + \varepsilon)^n$ for $n \geq N$. Since $\sum_{n=N}^{\infty} (L + \varepsilon)^n$ is a convergent geometric series, the Direct Comparison Test tells us that $\sum_{n=N}^{\infty} a^n$ also converges. Adding in the first $N - 1$ terms of the series doesn't affect convergence, so the entire series $\sum a_n$ converges too.

b. If $L > 1$ or $L = \infty$, then for some natural number N, $n \geq N \Rightarrow \sqrt[n]{a_n} > 1$ and therefore $a_n > 1$. Since the terms of the series don't converge to 0, the series must diverge.

c. As in the proof of the Ratio Test, the convergent series $\sum (1/n^2)$ and the divergent series $\sum (1/n)$ show that the Root Test is inconclusive when $L = 1$. (Use l'Hôpital's Rule to evaluate the limits associated with these two series.)

Example 2 📝

Determine whether each of the following series converges or diverges.

a. $\sum_{n=1}^{\infty} \dfrac{n^3}{3^n}$ b. $\sum_{n=1}^{\infty} \left(\dfrac{2n}{5n+2} \right)^n$ c. $\sum_{n=2}^{\infty} \dfrac{n}{(\ln n)^n}$

Solution

a. $\lim_{n\to\infty} \sqrt[n]{a_n} = \lim_{n\to\infty} \sqrt[n]{\dfrac{n^3}{3^n}} = \lim_{n\to\infty} \dfrac{n^{3/n}}{3} = \dfrac{1}{3} \lim_{n\to\infty} \left(\sqrt[n]{n} \right)^3 = \dfrac{1}{3} < 1$

(Use l'Hôpital's Rule for the final limit calculation.) Thus, the series converges.

b. $\displaystyle\lim_{n\to\infty}\sqrt[n]{a_n} = \lim_{n\to\infty}\frac{2n}{5n+2} = \frac{2}{5} < 1$

Thus, the series converges.

c. $\displaystyle\lim_{n\to\infty}\sqrt[n]{a_n} = \lim_{n\to\infty}\frac{n^{1/n}}{\ln n} = 0 < 1$

(See Example 9 of Section 4.4 as an aid in evaluating this limit.) Thus, this series also converges.

Our last example is a rather dramatic illustration of how one test can be inconclusive while another test is definitive.

Example 3 ✎

Define $a_n = \begin{cases} n/2^n & \text{if } n \text{ is odd} \\ 1/2^n & \text{if } n \text{ is even} \end{cases}$. Determine whether the series $\displaystyle\sum_{n=1}^{\infty} a_n$ converges.

Solution

If we apply the Ratio Test, we see that there are two forms of the ratio, depending on whether n is even or odd:

$$\frac{a_{n+1}}{a_n} = \frac{\dfrac{1}{2^{n+1}}}{\dfrac{n}{2^n}} = \frac{2^n}{n\cdot 2^{n+1}} = \frac{1}{2n} \qquad\qquad n \text{ odd}$$

$$\frac{a_{n+1}}{a_n} = \frac{\dfrac{n+1}{2^{n+1}}}{\dfrac{1}{2^n}} = \frac{(n+1)\cdot 2^n}{2^{n+1}} = \frac{n+1}{2} \qquad\qquad n \text{ even}$$

So as $n \to \infty$, a_{n+1}/a_n oscillates between values approaching 0 (for odd n) and values growing unboundedly large (for even n)—in other words, $\displaystyle\lim_{n\to\infty}(a_{n+1}/a_n)$ fails to exist in a fairly extreme fashion. But

$$\sqrt[n]{a_n} = \begin{cases} n^{1/n}/2 & \text{if } n \text{ is odd} \\ \frac{1}{2} & \text{if } n \text{ is even} \end{cases}$$

and $n^{1/n} \to 1$ as $n \to \infty$, so

$$\lim_{n\to\infty}\sqrt[n]{a_n} = \tfrac{1}{2} < 1$$

and the series converges by the Root Test.

10.5 Exercises

1–34 Use the Ratio Test to determine whether the series converges or diverges.

1. $\displaystyle\sum_{n=0}^{\infty} \frac{1}{n!}$

2. $\displaystyle\sum_{n=1}^{\infty} n\left(\frac{1}{2}\right)^n$

3. $\displaystyle\sum_{n=0}^{\infty} \frac{5^n}{n!}$

4. $\displaystyle\sum_{n=1}^{\infty} \frac{\pi^n}{n^\pi}$

5. $\displaystyle\sum_{n=0}^{\infty} \frac{3}{(2n)!}$

6. $\displaystyle\sum_{n=1}^{\infty} \frac{2^n}{n}$

7. $\displaystyle\sum_{n=1}^{\infty} \frac{n^5}{e^n}$

8. $\displaystyle\sum_{n=0}^{\infty} \frac{3}{(2n+1)!}$

9. $\displaystyle\sum_{n=1}^{\infty} \frac{n^2}{(2n-1)!}$

10. $\displaystyle\sum_{n=2}^{\infty} \frac{2n}{(n-1)^2}$

11. $\displaystyle\sum_{n=0}^{\infty} \frac{3n+2}{5^{3n+2}}$

12. $\displaystyle\sum_{n=1}^{\infty} \frac{n^n}{n!}$

13. $\displaystyle\sum_{n=1}^{\infty} \frac{(3n)!}{4^{n-1}5^n}$

14. $\displaystyle\sum_{n=0}^{\infty} \frac{n!}{6^n}$

15. $\displaystyle\sum_{n=1}^{\infty} \frac{n}{5^n}$

16. $\displaystyle\sum_{n=0}^{\infty} \frac{(\ln 2)^n}{(n+1)!}$

17. $\displaystyle\sum_{n=1}^{\infty} \frac{4^n}{(n+4)n}$

18. $\displaystyle\sum_{n=1}^{\infty} \frac{n}{\pi^{2n+1}}$

19. $\displaystyle\sum_{n=1}^{\infty} \frac{n^2-4}{2^n}$

20. $\displaystyle\sum_{n=0}^{\infty} \frac{(n!)4^n}{(4n+1)!}$

21. $\displaystyle\sum_{n=2}^{\infty} \frac{2^n}{n(n^2-1)}$

22. $\displaystyle\sum_{n=1}^{\infty} \frac{n^2}{(\ln 2)^n}$

23. $\displaystyle\sum_{n=1}^{\infty} \frac{n^3}{(\ln 3)^n}$

24. $\displaystyle\sum_{n=0}^{\infty} \frac{(3n-2)(3n+2)}{3^n}$

25. $\displaystyle\sum_{n=0}^{\infty} \frac{(n+3)3^n}{n!}$

26. $\displaystyle\sum_{n=1}^{\infty} \frac{n^2}{(\ln \pi)^n}$

27. $\displaystyle\sum_{n=1}^{\infty} \frac{2^{3n-2}}{2n^2+3n}$

28. $\displaystyle\sum_{n=1}^{\infty} \frac{(3n)!}{n^3}$

29. $\displaystyle\sum_{n=0}^{\infty} \frac{2^n}{(n+2)^n}$

30. $\displaystyle\sum_{n=0}^{\infty} \frac{2^n}{3^n+1}$

31. $\displaystyle\sum_{n=0}^{\infty} \frac{3^n}{2^n+1}$

32. $\displaystyle\sum_{n=0}^{\infty} \frac{(n!)^2}{(2n)!}$

33. $\displaystyle\frac{2}{1\cdot2\cdot3} + \frac{4}{2\cdot3\cdot4} + \cdots + \frac{2^n}{n(n+1)(n+2)} + \cdots$

34. $\displaystyle\frac{1}{3} + \frac{1\cdot5}{2\cdot3\cdot9} + \cdots + \frac{1\cdot5\cdots(4n-3)}{3^n(n!)(2n-1)} + \cdots$

35–38 Verify that the Ratio Test is inconclusive for the series. Then determine the convergence or divergence of the series by some other means.

35. $\displaystyle\sum_{n=0}^{\infty} \frac{n^2+1}{(n+1)^2}$

36. $\displaystyle\sum_{n=1}^{\infty} \frac{\sqrt{n}}{n+2}$

37. $\displaystyle\sum_{n=1}^{\infty} \frac{\arctan n}{n}$

38. $\displaystyle\sum_{n=1}^{\infty} \frac{n+1}{\sqrt{n}(n+2)^2}$

39–62 Use the Root Test to determine whether the series converges or diverges.

39. $\displaystyle\sum_{n=1}^{\infty} \left(\frac{5}{n}\right)^n$

40. $\displaystyle\sum_{n=0}^{\infty} \frac{1}{(n+1)^n}$

41. $\displaystyle\sum_{n=2}^{\infty} \frac{1}{(\ln n)^n}$

42. $\displaystyle\sum_{n=1}^{\infty} \frac{3^{n-1}}{n^n}$

43. $\displaystyle\sum_{n=1}^{\infty} \left(\frac{3n}{n^2+3}\right)^n$

44. $\displaystyle\sum_{n=2}^{\infty} \frac{n^{2n}}{(\ln n)^n}$

45. $\displaystyle\sum_{n=1}^{\infty} \frac{n}{2^{2n}}$

46. $\displaystyle\sum_{n=2}^{\infty} \frac{e^{2n}}{(\ln n)^n}$

47. $\displaystyle\sum_{n=0}^{\infty} \frac{2^n}{(n+2)^n}$

48. $\displaystyle\sum_{n=1}^{\infty} \left(\frac{n}{2n+3}\right)^n$

49. $\displaystyle\sum_{n=1}^{\infty} \left(1+\frac{1}{n}\right)^{n^2}$

50. $\displaystyle\sum_{n=1}^{\infty} \left(1+\frac{1}{n}\right)^{-n^2}$

51. $\displaystyle\sum_{n=2}^{\infty} \frac{1}{(\ln n)^n}$

52. $\displaystyle\sum_{n=2}^{\infty} \left(\ln\frac{1}{n}\right)^n$

53. $\displaystyle\sum_{n=1}^{\infty} n\left(\frac{4}{5}\right)^n$

54. $\displaystyle\sum_{n=1}^{\infty} \frac{n^n}{2^{n^2}}$

55. $\displaystyle\sum_{n=2}^{\infty} \left(\frac{\ln n}{n}\right)^n$

56. $\displaystyle\sum_{n=1}^{\infty} \left(\frac{2n-1}{3n+2}\right)^n$

57. $\displaystyle\sum_{n=2}^{\infty} \frac{n^{n/2}}{(\ln n)^n}$

58. $\displaystyle\sum_{n=1}^{\infty} \frac{(n!)^n}{n^{n^2}}$

59. $\displaystyle\sum_{n=1}^{\infty} \frac{(n+2)^n}{n^{2n}}$

60. $\displaystyle\sum_{n=1}^{\infty} \frac{2}{n^{n+2}}$

61. $\displaystyle\sum_{n=1}^{\infty} \sin^n \frac{1}{n^2}$

62. $\displaystyle\sum_{n=1}^{\infty} \frac{1}{\left(2+\dfrac{1}{n}\right)^{2n}}$

63–66 Verify that the Root Test is inconclusive for the series. Then determine the convergence or divergence of the series by some other means.

63. $\displaystyle\sum_{n=1}^{\infty} \left(\frac{n}{n+5}\right)^n$

64. $\displaystyle\sum_{n=1}^{\infty} \frac{\left(2\sqrt{n}\right)^{2n}}{(4n+1)^n}$

65. $\displaystyle\sum_{n=1}^{\infty} \left(1-\frac{1}{n}\right)^n$

66. $\displaystyle\sum_{n=1}^{\infty} \frac{2n}{(n+1)^3}$

67–70 Suppose that the series $\displaystyle\sum a_n$ satisfies the condition $\displaystyle\lim_{n\to\infty} \left(a_{n+1}/a_n\right) = \frac{1}{2}$. Decide whether the given series is convergent.

67. $\displaystyle\sum_{n=1}^{\infty} 3^n a_n$

68. $\displaystyle\sum_{n=1}^{\infty} n^2 a_n$

69. $\displaystyle\sum_{n=1}^{\infty} \left(\frac{3}{2}\right)^n a_n$

70. $\displaystyle\sum_{n=1}^{\infty} a_n^2$

71. Prove that for all exponents p, the series $\displaystyle\sum\left(n^p/2^n\right)$ is convergent.

72. Prove that $\displaystyle\sum\left(p^n/n!\right)$ converges for all $p > 0$.

73. For what positive p-values does the series $\displaystyle\sum\left(p^n/n\right)$ converge?

74–79 *True or False?* Determine whether the given statement is true or false. In case of a false statement, explain or provide a counterexample. ($\displaystyle\sum a_n$ is a positive series in each problem.)

74. If $\displaystyle\sum a_n$ is convergent, then $\displaystyle\lim_{n\to\infty}\left(a_{n+1}/a_n\right) < 1$.

75. If $\displaystyle\sum a_n$ is divergent, then $\displaystyle\lim_{n\to\infty} \sqrt[n]{a_n} \geq 1$ or the limit doesn't exist.

76. If there is an $N > 0$ such that $\sqrt[n]{a_n} < 1$ for all $n \geq N$, then $\displaystyle\sum a_n$ is convergent.

77. The series $\displaystyle\sum_{n=1}^{\infty}\left(3^{n^2}/n!\right)$ is convergent.

78.* If $\displaystyle\sum a_n$ satisfies condition a. of the Ratio Test, then it satisfies condition a. of the Root Test.

79.* If $\displaystyle\sum a_n$ satisfies the condition a. of the Root Test, then it satisfies condition a. of the Ratio Test.

10.5 **Technology Exercises**

80. The following series, discovered by the great Indian mathematician Srinivasa Ramanujan, converges to $1/\pi$ with amazing speed.

$$\frac{2\sqrt{2}}{9801} \sum_{n=0}^{\infty} \frac{(4n)!(26390n+1103)}{(n!)^4 \, 396^{4n}} = \frac{1}{\pi}$$

  a. Prove that the series is convergent.

  b. Use Ramanujan's series with a computer algebra system to approximate π, correct to 31 digits after the decimal. How many terms did you have to use to achieve this accuracy?

10.6 **Absolute and Conditional Convergence**

TOPICS

1. Alternating series
2. Absolute and conditional convergence
3. Summary of convergence tests

We introduce our last convergence test in this section and then further refine our classification of convergent series. As we will see, some series with both positive and negative terms possess a surprising property completely different from anything seen with finite sums.

TOPIC 1   **Alternating Series**

An **alternating series** is any series in which the terms are alternately positive and negative. Each of the series below qualifies as an alternating series, with the third one known as the **alternating harmonic series**.

$$-3+1-\frac{1}{3}+\frac{1}{9}-\frac{1}{27}+\cdots+\frac{(-1)^n}{3^{n-2}}+\cdots$$

$$-1+2-3+4-5+6-\cdots+(-1)^n\,n+\cdots$$

$$1-\frac{1}{2}+\frac{1}{3}-\frac{1}{4}+\frac{1}{5}-\cdots+\frac{(-1)^{n+1}}{n}+\cdots$$

The first example above is a geometric series with first term -3 and common ratio $-\frac{1}{3}$, so it converges and its sum is $-3/\left(1+\frac{1}{3}\right)=-\frac{9}{4}$. The partial sums of the second example clearly do not converge (they grow unbounded in magnitude), so the series diverges. And as it turns out, there is a simple test that shows the third example (the alternating harmonic series) converges.

Theorem 🔍

The Alternating Series Test (Leibniz's Test)

If $\left\{a_n\right\}_{n=1}^{\infty}$ is a sequence of positive terms such that

    (i) $a_{n+1} \le a_n$ for all $n \ge N$ (N a natural number) and

    (ii) $\lim\limits_{n\to\infty} a_n = 0$

then the series $\sum\limits_{n=1}^{\infty}(-1)^{n+1}\,a_n = a_1 - a_2 + a_3 - a_4 + \cdots$ converges. (In words, if the terms of an alternating series decrease monotonically to 0 in absolute value, then it converges. And as with many other tests, the result is still true if condition (i) is only true for all $n \ge N$ for some natural number N.)

Proof ✎

The conditions on the sequence $\left\{a_n\right\}$ mean that

$$s_2 = a_1 - a_2 \ge 0$$
$$s_4 = s_2 + (a_3 - a_4) \ge s_2$$
$$s_6 = s_4 + (a_5 - a_6) \ge s_4$$

and, in general, each even-indexed partial sum is at least as large as the preceding even-indexed partial sum: $s_{2n} = s_{2n-2} + (a_{2n-1} - a_{2n}) \ge s_{2n-2}$. Also, since each expression in parentheses below is nonnegative,

$$s_{2n} = a_1 - (a_2 - a_3) - (a_4 - a_5) - \cdots - (a_{2n-2} - a_{2n-1}) - a_{2n} \le a_1$$

for every n. So the sequence of even-indexed partial sums $\{s_{2n}\}_{n=1}^{\infty}$ is a monotonically increasing sequence that is bounded above, and hence converges to some limit s by the Bounded Monotonic Sequence Theorem. That is, $\lim_{n\to\infty} s_{2n} = s$.

For the odd-indexed partial sums, note that

$$\lim_{n\to\infty} s_{2n+1} = \lim_{n\to\infty}(s_{2n} + a_{2n+1}) = \lim_{n\to\infty} s_{2n} + \lim_{n\to\infty} a_{2n+1} = s + 0 = s.$$

Since the even-indexed and odd-indexed partial sums both approach s as $n \to \infty$, the series converges and its sum is s. (As a consequence, the alternating series $\sum_{n=1}^{\infty}(-1)^n a_n = -\sum_{n=1}^{\infty}(-1)^{n+1} a_n$ also converges. And, as usual, only the tail matters—as long as the terms of a series eventually satisfy the conditions of the theorem, it will converge.)

Example 1 ✐

a. The alternating harmonic series satisfies the conditions of the Alternating Series Test and thus converges.

b. The series $\sum_{n=1}^{\infty}(-1)^{n+1}\dfrac{n-2}{n^2 - 2n + 4}$ converges as well, since

(i) The function $f(x) = \dfrac{x-2}{x^2 - 2x + 4} = \dfrac{x-2}{x(x-2)+4}$ is positive for $x > 2$, and decreases monotonically for $x \ge 4$, a consequence of the fact that

$$f'(x) = \frac{-x(x-4)}{\left(x^2 - 2x + 4\right)^2} \le 0 \quad \text{for } x \ge 4.$$

(ii) $\lim\limits_{n\to\infty}\dfrac{n-2}{n^2 - 2n + 4} = \lim\limits_{n\to\infty}\dfrac{\dfrac{1}{n} - \dfrac{2}{n^2}}{1 - \dfrac{2}{n} + \dfrac{4}{n^2}} = 0.$

Although the series only satisfies the conditions of the test for $n \ge 4$, that is sufficient.

Our proof of the Alternating Series Test also points the way toward a very simple formula for the nth remainder $r_n = s - s_n$ of an alternating series. Under the same conditions that $\{a_n\}$ is a sequence of positive terms monotonically decreasing to 0, the sequence of odd-indexed partial sums monotonically decreases to s, since

$$s_1 = a_1$$
$$s_3 = s_1 - (a_2 - a_3) \le s_1$$
$$s_5 = s_3 - (a_4 - a_5) \le s_3$$

and, in general, $s_{2n+1} = s_{2n-1} - \left(a_{2n} - a_{2n+1}\right) \le s_{2n-1}$. Since the even-indexed partial sums monotonically increase toward s, we know that s always lies between any two successive partial sums and hence

$$\left|r_n\right| = \left|s - s_n\right| \le \left|s_{n+1} - s_n\right| = a_{n+1}.$$

Example 2 ✐

Determine the first three digits after the decimal of the number $s = \displaystyle\sum_{n=0}^{\infty} \frac{(-1)^n}{n!}$.

Solution

It is clear that $\left\{1/n!\right\}_{n=0}^{\infty}$ is a sequence of positive numbers monotonically decreasing to 0, so the conditions of the Alternating Series Test are met. The first few terms of the series are

$$\sum_{n=0}^{\infty} \frac{(-1)^n}{n!} = \frac{1}{0!} - \frac{1}{1!} + \frac{1}{2!} - \frac{1}{3!} + \frac{1}{4!} - \frac{1}{5!} + \frac{1}{6!} - \frac{1}{7!} + \cdots$$

$$= 1 - 1 + \frac{1}{2} - \frac{1}{6} + \frac{1}{24} - \frac{1}{120} + \frac{1}{720} - \frac{1}{5040} + \cdots$$

Since $s_6 \approx 0.368056$ and $a_7 = \frac{1}{5040} < \frac{1}{5000} = 0.0002$, $\left|s - 0.368\right| < 0.0002$ and hence $s \approx 0.368$ must be correct to three decimal places.

TOPIC 2   Absolute and Conditional Convergence

The alternating harmonic series is the classic example of a series whose convergence hinges on its alternating nature. Although $\displaystyle\sum \frac{(-1)^n}{n}$ converges, $\displaystyle\sum \left|\frac{(-1)^n}{n}\right| = \sum \frac{1}{n}$ does not. The following definition gives a name to series of a similar nature.

Definition ♀

Absolute and Conditional Convergence

We say that a series $\displaystyle\sum a_n$ **converges absolutely** (or is **absolutely convergent**) if the corresponding series $\displaystyle\sum \left|a_n\right|$ converges. If $\displaystyle\sum a_n$ converges but $\displaystyle\sum \left|a_n\right|$ does not converge, we say $\displaystyle\sum a_n$ **converges conditionally** (or is **conditionally convergent**).

The following theorem points out that absolute convergence is a stronger property than convergence.

Theorem

Absolute Convergence Implies Convergence

If $\sum |a_n|$ converges, then $\sum a_n$ converges.

Proof

Note that $0 \le a_n + |a_n| \le 2|a_n|$. If $\sum |a_n|$ converges, then $\sum 2|a_n|$ also converges and so $\sum \left(a_n + |a_n| \right)$ converges by the Direct Comparison Test. And since the difference of two convergent series converges, we know that

$$\sum a_n = \sum \left(a_n + |a_n| \right) - \sum |a_n| \text{ converges.}$$

The preceding definition and theorem have completed a hierarchy for series, summarized in Figure 1.

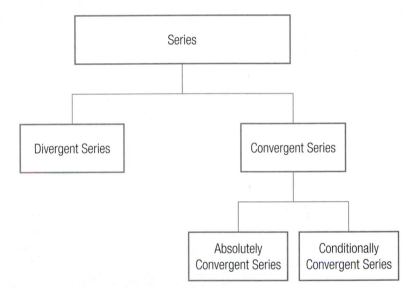

Figure 1 Series Hierarchy

Example 3

a. The series $\displaystyle\sum_{n=1}^{\infty} \frac{\sin n}{n^2}$ is absolutely convergent (and hence convergent). Note that absolute convergence of this series may actually be easier to determine than convergence:

$$\sum_{n=1}^{\infty} \left| \frac{\sin n}{n^2} \right| \le \sum_{n=1}^{\infty} \frac{1}{n^2}$$

and $\displaystyle\sum_{n=1}^{\infty} \frac{1}{n^2}$ is a convergent p-series.

b. The alternating p-series $\displaystyle\sum_{n=1}^{\infty} \frac{(-1)^{n+1}}{n^p}$ converges if $p > 0$ by the Alternating Series Test. It converges absolutely if $p > 1$ and converges conditionally if $0 < p \le 1$.

c. The alternating series $\displaystyle\sum_{n=2}^{\infty} \frac{(-1)^n\, n}{(\ln n)^n}$ converges absolutely, a fact we can determine by applying the Root Test to the absolute value of its terms.

$$\lim_{n\to\infty} \sqrt[n]{|a_n|} = \lim_{n\to\infty} \sqrt[n]{\frac{n}{(\ln n)^n}} = \lim_{n\to\infty} \frac{n^{1/n}}{\ln n} = 0 < 1$$

(See Example 9 of Section 4.4 as an aid in evaluating this limit.)

d. The alternating series $\displaystyle\sum_{n=1}^{\infty} \frac{n^3}{(-3)^n}$ also converges absolutely, which we can show by applying the Ratio Test to the absolute value of its terms.

$$\lim_{n\to\infty} \left|\frac{a_{n+1}}{a_n}\right| = \lim_{n\to\infty} \left|\frac{(n+1)^3}{(-3)^{n+1}} \cdot \frac{(-3)^n}{n^3}\right| = \frac{1}{3}\lim_{n\to\infty}\left(1+\frac{1}{n}\right)^3 = \frac{1}{3} < 1$$

Absolutely convergent series have the property that any rearrangement (or reordering) of the terms of the series produces the same sum. This is the kind of behavior we are led by experience to expect from sums, because finite sums always behave this way—the sum of a finite set of numbers is unchanged by the order in which they are added. Interestingly, conditionally convergent series do *not* have this property. In fact, Bernhard Riemann (1826–1866) showed that the terms of a conditionally convergent series can be rearranged so that the sum of the series is *any* preselected real number. Hints on how to prove this appear in Exercise 62.

TOPIC 3   Summary of Convergence Tests

We have now reached the end of our list of tests for series convergence (although, as you might expect, more tests can be found that extend and/or refine those we have learned). Before continuing on to the next three sections, which focus on applications of series, it is appropriate to summarize what we have learned.

Summary of Convergence Tests

Name of Test	Form of Series	Convergence Condition	Divergence Condition	Notes
n^{th}-**Term Divergence Test**	$\displaystyle\sum_{n=1}^{\infty} a_n$	Does not apply	$\displaystyle\lim_{n\to\infty} a_n \neq 0$	Only used to prove divergence
Geometric Series	$\displaystyle\sum_{n=1}^{\infty} ar^{n-1}$	$|r| < 1$	$|r| \geq 1$	$\displaystyle\sum_{n=1}^{\infty} ar^{n-1} = \frac{a}{1-r}$ when $|r| < 1$
Integral Test	$\displaystyle\sum_{n=N}^{\infty} a_n,\ \ a_n = f(n)$ where f is a continuous, positive, and decreasing function for $n \geq N$	$\displaystyle\int_N^{\infty} f(x)\,dx < \infty$	$\displaystyle\int_N^{\infty} f(x)\,dx = \infty$	The value of the integral is not the value of the sum.
p-**Series**	$\displaystyle\sum_{n=1}^{\infty} \left(1/n^p\right)$	$p > 1$	$p \leq 1$	Harmonic series corresponds to $p = 1$
Direct Comparison Test	$\displaystyle\sum_{n=1}^{\infty} a_n,\ \ a_n \geq 0$	$a_n \leq b_n$ for all $n \geq N$ and $\displaystyle\sum_{n=1}^{\infty} b_n$ converges	$0 \leq b_n \leq a_n$ for all $n \geq N$ and $\displaystyle\sum_{n=1}^{\infty} b_n$ diverges	Finding a good comparison series $\displaystyle\sum_{n=1}^{\infty} b_n$ is the key.
Limit Comparison Test	$\displaystyle\sum_{n=1}^{\infty} a_n$ and $\displaystyle\sum_{n=1}^{\infty} b_n,$ $a_n > 0$ and $b_n > 0$	$\displaystyle 0 \leq \lim_{n\to\infty} \frac{a_n}{b_n} < \infty$ and $\displaystyle\sum_{n=1}^{\infty} b_n$ converges	$\displaystyle\lim_{n\to\infty}(a_n/b_n) > 0$ or $\displaystyle\lim_{n\to\infty}(a_n/b_n) = \infty$ and $\displaystyle\sum_{n=1}^{\infty} b_n$ diverges	Finding a good comparison series $\displaystyle\sum_{n=1}^{\infty} b_n$ is the key.
Ratio Test	$\displaystyle\sum_{n=1}^{\infty} a_n,\ \ a_n > 0$	$\displaystyle\lim_{n\to\infty} \frac{a_{n+1}}{a_n} < 1$	$\displaystyle\lim_{n\to\infty}(a_{n+1}/a_n) > 1$ or $\displaystyle\lim_{n\to\infty}(a_{n+1}/a_n) = \infty$	Inconclusive if $\displaystyle\lim_{n\to\infty}(a_{n+1}/a_n) = 1$
Root Test	$\displaystyle\sum_{n=1}^{\infty} a_n,\ \ a_n \geq 0$	$\displaystyle\lim_{n\to\infty} \sqrt[n]{a_n} < 1$	$\displaystyle\lim_{n\to\infty} \sqrt[n]{a_n} > 1$ or $\displaystyle\lim_{n\to\infty} \sqrt[n]{a_n} = \infty$	Inconclusive if $\displaystyle\lim_{n\to\infty} \sqrt[n]{a_n} = 1$
Alternating Series Test	$\displaystyle\sum_{n=1}^{\infty} (-1)^{n+1} a_n,$ $a_n > 0$ and $a_{n+1} \leq a_n$	$\displaystyle\lim_{n\to\infty} a_n = 0$	$\displaystyle\lim_{n\to\infty} a_n \neq 0$	n^{th} remainder $|r_n| \leq a_{n+1}$

Table 1

10.6 Exercises

1–24 Determine whether the alternating series converges and give a reason for your answer.

1. $\displaystyle\sum_{n=1}^{\infty}(-1)^{n+1}\frac{1}{n^2}$

2. $\displaystyle\sum_{n=1}^{\infty}(-1)^{n+1}\frac{2}{\sqrt{n}}$

3. $\displaystyle\sum_{n=1}^{\infty}(-1)^{n+1}\frac{2n+1}{3n-2}$

4. $\displaystyle\sum_{n=1}^{\infty}(-1)^{n}\frac{1}{2n-1}$

5. $\displaystyle\sum_{n=2}^{\infty}(-1)^{n}\frac{3}{n\ln n}$

6. $\displaystyle\sum_{n=1}^{\infty}(-1)^{n+1}\frac{2^{2n}-1}{4^{n}}$

7. $\displaystyle\sum_{n=0}^{\infty}(-1)^{n}\frac{n^2+1}{n^2+2}$

8. $\displaystyle\sum_{n=1}^{\infty}(-1)^{n+1}\frac{1}{n^2 2^n}$

9. $\displaystyle\sum_{n=1}^{\infty}(-1)^{n+1}\frac{2n}{\ln 2n}$

10. $\displaystyle\sum_{n=1}^{\infty}(-1)^{n}\frac{3^n}{n^3}$

11. $\displaystyle\sum_{n=1}^{\infty}(-1)^{n+1}\frac{\ln(n+1)}{n}$

12. $\displaystyle\sum_{n=1}^{\infty}(-1)^{n}\frac{\sqrt{n}}{n+2}$

13. $\displaystyle\sum_{n=1}^{\infty}(-1)^{n}\left(\frac{n+1}{5\sqrt{n}}\right)^{n}$

14. $\displaystyle\sum_{n=1}^{\infty}(-1)^{n+1}\sin\frac{1}{n}$

15. $\displaystyle\sum_{n=0}^{\infty}(-1)^{n}\frac{\sqrt{n}+3}{\sqrt{n}+3}$

16. $\displaystyle\sum_{n=0}^{\infty}(-1)^{n}\frac{2^n}{n!}$

17. $\displaystyle\sum_{n=0}^{\infty}(-1)^{n}\frac{1}{\sqrt[3]{n^2+1}}$

18. $\displaystyle\sum_{n=0}^{\infty}(-1)^{n+1}\left(\frac{4}{5}\right)^{n}$

19. $\displaystyle\sum_{n=1}^{\infty}(-1)^{n}\frac{1}{\tan^{-1}n}$

20. $\displaystyle\sum_{n=1}^{\infty}(-1)^{n}\tan^{-1}\left(\frac{1}{n}\right)$

21. $\displaystyle\sum_{n=1}^{\infty}(-1)^{n}e^{-n}\sin n$

22. $\displaystyle\sum_{n=1}^{\infty}\left(\frac{-1}{1.1}\right)^{n}$

23. $\displaystyle\sum_{n=1}^{\infty}\frac{\cos(n\pi)}{1+\frac{1}{n}}$

24. $\displaystyle\sum_{n=2}^{\infty}\frac{(-1)^{n}}{(\ln n)^3}$

25–30 Approximate the sum of the alternating series, accurate to at least the indicated number of decimal places.

25. $\displaystyle\sum_{n=1}^{\infty}\frac{(-1)^{n+1}}{n^3}$; accurate to 2 decimal places

26. $\displaystyle\sum_{n=1}^{\infty}\frac{(-1)^{n+1}}{n3^{n}}$; accurate to 4 decimal places

27. $\displaystyle\sum_{n=1}^{\infty}\frac{(-1)^{n+1}}{(n+1)!}$; accurate to 5 decimal places

28. $\displaystyle\sum_{n=1}^{\infty}\frac{(-1)^{n+1}}{2^n n!}$; accurate to 6 decimal places

29. $\displaystyle\sum_{n=2}^{\infty}\frac{(-1)^{n}\ln n}{e^{n}}$; accurate to 3 decimal places

30. $\displaystyle\sum_{n=2}^{\infty}\frac{(-1)^{n}\ln n}{2n^4}$; accurate to 4 decimal places

31–58 Determine whether the given series converges absolutely, converges conditionally, or diverges. Use any convergence test discussed so far in this chapter.

31. $\displaystyle\sum_{n=1}^{\infty}(-1)^{n}\frac{n}{n^2+1}$

32. $\displaystyle\sum_{n=1}^{\infty}\frac{\cos n}{n^2}$

33. $\displaystyle\sum_{n=1}^{\infty}(-1)^{n}\frac{1}{n^{1/n}}$

34. $\displaystyle\sum_{n=1}^{\infty}(-1)^{n+1}\frac{n!}{(2n)!}$

35. $\displaystyle\sum_{n=1}^{\infty}(-1)^{n}\frac{\sqrt[3]{n}}{\sqrt{n}}$

36. $\displaystyle\sum_{n=1}^{\infty}(-1)^{n+1}\cos\frac{1}{n}$

37. $\displaystyle\sum_{n=1}^{\infty}(-1)^{n+1}\frac{1}{n^2+4}$

38. $\displaystyle\sum_{n=0}^{\infty}(-1)^{n+1}\frac{5^n}{n!}$

39. $\displaystyle\sum_{n=1}^{\infty}(-1)^{n+1}\frac{n}{n^2+4}$

40. $\displaystyle\sum_{n=2}^{\infty}(-1)^{n}\frac{1}{n\sqrt{\ln n}}$

41. $\displaystyle\sum_{n=2}^{\infty}(-1)^{n+1}\frac{\ln n}{n^{1.1}}$

42. $\displaystyle\sum_{n=1}^{\infty}(-1)^{n+1}\frac{n}{5n-1}$

43. $\displaystyle\sum_{n=1}^{\infty}\frac{(-2)^{n}}{n^2}$

44. $\displaystyle\sum_{n=1}^{\infty}(-1)^{n+1}\frac{1}{n(\sqrt{n}+2)}$

45. $\displaystyle\sum_{n=2}^{\infty} \frac{(-1)^n}{\sqrt{n+1}+\sqrt{n}}$

46. $\displaystyle\sum_{n=1}^{\infty} (-1)^{n+1} \, n\sin\frac{1}{n}$

47. $\displaystyle\sum_{n=1}^{\infty} \frac{(-1)^{n+1} \, n^4}{4^n}$

48. $\displaystyle\sum_{n=1}^{\infty} \frac{(-1)^{n+1} \, \tan^{-1} n}{n^2}$

49. $\displaystyle\sum_{n=1}^{\infty} \frac{(-1)^{n+1}}{\sqrt{n(n+1)}}$

50. $\displaystyle\sum_{n=1}^{\infty} \left(\ln\frac{2}{n}\right)^n$

51. $\displaystyle\sum_{n=2}^{\infty} \left(-\frac{\ln n}{n}\right)^n$

52. $\displaystyle\sum_{n=1}^{\infty} \left(\frac{1-2n}{5n}\right)^n$

53. $\displaystyle\sum_{n=2}^{\infty} (-1)^{n+1} \left(\frac{2^n}{n^2}\right)^n$

54. $\displaystyle\sum_{n=1}^{\infty} \frac{(-1)^{n+1} \, n^3}{e^n}$

55. $\displaystyle\sum_{n=2}^{\infty} \frac{(-\pi)^n}{\left(3+\dfrac{1}{n}\right)^n}$

56. $\displaystyle\sum_{n=1}^{\infty} (-1)^n \log\left(5+\frac{1}{n^2}\right)$

57. $\displaystyle\sum_{n=1}^{\infty} (-1)^{n+1} \left(\frac{n}{2}\right)^n$

58. $\displaystyle\sum_{n=1}^{\infty} \frac{(-1)^{n+1} \, 3^n}{2^n \, n!}$

59. Kate claims that the series

$$\frac{1}{2} - \frac{1}{4} + \frac{1}{3} - \frac{1}{9} + \cdots + \frac{1}{n} - \frac{1}{n^2} + \cdots$$

converges by Leibniz's Test, because it is an alternating series of the form $\sum (-1)^{n+1} a_n$ and $a_n \to 0$. Is she right? Why or why not?

60. Prove that the series consisting of the positive terms of the alternating harmonic series diverges to infinity.

61. Repeat Exercise 60 for the negative terms of the alternating harmonic series.

62.* Prove the first part of Riemann's Rearrangement Theorem, that is, the statement that the terms of a conditionally convergent alternating series can be rearranged so that the sum of the series is any preselected real number. (**Hint:** Let s be the desired, preselected sum, and start adding up the positive terms of the series until their sum first becomes greater than s. Then add enough negative terms until the resulting sum becomes less than n, and continue.)

63.* Prove the second and third parts of Riemann's Theorem, namely, that the terms of the series in Exercise 62 can be rearranged to diverge to ∞ or $-\infty$, or to diverge in an oscillating manner.

64.* Prove the following so-called Polynomial Test for infinite series: If $p(x)$ and $q(x)$ are polynomials with degrees r and s, respectively, then the series

$$\sum \frac{p(n)}{q(n)} \quad \text{is convergent if and only if } s > r + 1$$

(we are assuming $q(n) \neq 0$ for any value of the summation index n).

65. Prove that if the series $\sum a_n$ is absolutely convergent, then $\sum a_n^2$ is convergent. Then give an example to show that the statement is not true if $\sum a_n$ is conditionally convergent.

10.6 **Technology Exercises**

66–68 Use a computer algebra system to solve the problem.

66. Find an approximation of e with an error no greater than 10^{-8}, knowing that $\displaystyle\sum_{n=0}^{\infty} \frac{(-1)^n}{n!} = \frac{1}{e}$. How many terms did you use?

67. Find an approximation of π with an error no greater than 10^{-3}, knowing that $\displaystyle\sum_{n=0}^{\infty} \frac{(-1)^n}{2n+1} = \frac{\pi}{4}$. How many terms did you use?

68. Find an approximation of $\ln 2$ with an error no greater than 10^{-3}, knowing that $\displaystyle\sum_{n=1}^{\infty} \frac{(-1)^{n+1}}{n} = \ln 2$. How many terms did you use?

10.7 Power Series

TOPICS

1. Power series
2. Radius of convergence
3. Calculus and power series

Now that we are familiar with sums of an infinite set of numbers, we can apply our newly gained knowledge in the arena of functions. Just as a series extends the notion of addition from the finite to the infinite, a power series can informally be thought of as a polynomial of infinite degree.

TOPIC 1   Power Series

As we move on to applications of series, our focus shifts to constructing functions out of series. Power series are a large family of such functions with many uses.

Definition 💡

Power Series

A **power series centered at 0** (or **power series about 0**) in the variable x is a series of the form

$$\sum_{n=0}^{\infty} c_n x^n = c_0 + c_1 x + c_2 x^2 + c_3 x^3 + \cdots,$$

where the c_n's are constants called the **coefficients** of the series. More generally, a **power series centered at a** (or **power series about a**) is a series of the form

$$\sum_{n=0}^{\infty} c_n (x-a)^n = c_0 + c_1 (x-a) + c_2 (x-a)^2 + c_3 (x-a)^3 + \cdots.$$

For any fixed x, a power series is a series similar to those we have studied over the past five sections, and we can determine if the series converges or diverges at that particular value of x. In practice, we typically want to determine all values of x for which a power series converges.

Example 1 ✐

Determine the values of x for which each of the following power series converges.

a. $\displaystyle\sum_{n=1}^{\infty} \frac{(x-5)^n}{n}$ **b.** $\displaystyle\sum_{n=1}^{\infty} (-1)^{n+1} \frac{x^n}{n}$ **c.** $\displaystyle\sum_{n=0}^{\infty} n! x^n$

Solution

We will examine the limit of $\left| a_{n+1}/a_n \right|$ where a_n is the nth term of each series, and thereby find conditions on x that guarantee absolute convergence.

a. $\left| \dfrac{a_{n+1}}{a_n} \right| = \left| \dfrac{(x-5)^{n+1}}{n+1} \cdot \dfrac{n}{(x-5)^n} \right| = |x-5| \left(\dfrac{1}{1+\dfrac{1}{n}} \right) \rightarrow |x-5|$ as $n \rightarrow \infty$

By the Ratio Test, the power series converges absolutely if $|x-5| < 1$ and diverges if $|x-5| > 1$. If $x - 5 = 1$, the series is the harmonic series

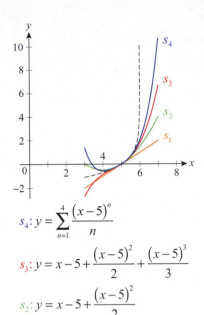

$$S_4: y = \sum_{n=1}^{4} \frac{(x-5)^n}{n}$$

$$S_3: y = x - 5 + \frac{(x-5)^2}{2} + \frac{(x-5)^3}{3}$$

$$S_2: y = x - 5 + \frac{(x-5)^2}{2}$$

$$S_1: y = x - 5$$

Figure 1
Power Series Convergence

and diverges, and if $x - 5 = -1$ it is the alternating harmonic series and converges. So in summary, this power series converges for $4 \le x < 6$ and diverges for all other x.

Figure 1 shows how the first four partial sums of the power series approach a limiting function, shown in black as a dashed curve. We will discover in Example 5 what this limiting function actually is.

b. $\left| \dfrac{a_{n+1}}{a_n} \right| = \left| \dfrac{(-1)^{n+2} x^{n+1}}{n+1} \cdot \dfrac{n}{(-1)^{n+1} x^n} \right| = |x| \left(\dfrac{1}{1 + \dfrac{1}{n}} \right) \to |x|$ as $n \to \infty$

By the Ratio Test, this power series converges absolutely if $|x| < 1$ and diverges if $|x| > 1$. In this example, however, the series converges when $x = 1$ (by the Alternating Series Test) and diverges when $x = -1$ (the series becomes the negative of the harmonic series), so we have convergence for $-1 < x \le 1$ and divergence elsewhere.

c. $\left| \dfrac{a_{n+1}}{a_n} \right| = \left| \dfrac{(n+1)! x^{n+1}}{n! x^n} \right| = |x|(n+1) \to \infty$ as $n \to \infty$ if $x \ne 0$

This power series diverges everywhere except for the one value $x = 0$, for which the series has a sum of 0.

If the coefficients of a power series are 0 for all $n > N$, then the power series (whether centered at 0 or some other real number a) is simply a polynomial of degree N. It is for this reason that power series with an infinite number of terms are informally thought of as "infinite" polynomials—polynomials with both an infinite number of terms and infinite degree. Just as infinite series have some properties in common with finite sums, power series have some properties in common with polynomials. Both the similarities and the differences are important to us.

Because polynomials are so easily differentiated and integrated, it is perhaps no surprise that being able to equate a function with a power series is useful. In Section 10.8 we will learn when this is possible and, in such cases, how to systematically find the series that corresponds to any given function. But it is also sometimes possible to begin with a power series and find the *closed form* (nonseries form) of the function to which it is equivalent.

Example 2 ✑

Determine the values of x for which each of the following series converges and the limiting function of the series for those values.

a. $\displaystyle\sum_{n=0}^{\infty} x^n$　　　　　　　　**b.** $\displaystyle\sum_{n=0}^{\infty} \left(\frac{x-1}{3} \right)^n$

Solution

Our knowledge of geometric series will tell us both when these series converge and, for those values of x, what the sum is.

a. Note that $\displaystyle\sum_{n=0}^{\infty} x^n$ converges when $|x| < 1$, and for any such fixed x,

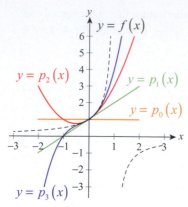

Figure 2

$\sum_{n=0}^{\infty} x^n = 1/(1-x)$. In other words, the function $f(x) = 1/(1-x)$ can be approximated by the successive partial sums of the series $\sum_{n=0}^{\infty} x^n$, each of which is a polynomial:

$$p_0(x) = 1$$
$$p_1(x) = 1 + x$$
$$p_2(x) = 1 + x + x^2$$
$$p_3(x) = 1 + x + x^2 + x^3$$
$$\vdots$$

The graphs of each of these, along with the graph of $f(x) = 1/(1-x)$ as a dashed black curve, are shown in Figure 2. Note that the polynomials only show promise of converging to f on the interval $(-1, 1)$.

b. We conduct a similar analysis for the series $\sum_{n=0}^{\infty} \left(\frac{x-1}{3} \right)^n$. First, we know this geometric series converges when

$$\left| \frac{x-1}{3} \right| < 1,$$

which implies $-2 < x < 4$. Second, for any x in this interval,

$$\sum_{n=0}^{\infty} \left(\frac{x-1}{3} \right)^n = 1 + \frac{1}{3}(x-1) + \frac{1}{9}(x-1)^2 + \cdots = \frac{1}{1 - \dfrac{x-1}{3}} = \frac{3}{4-x}.$$

TOPIC 2 Radius of Convergence

Although we have studied only a few power series to this point, the examples so far all have a characteristic in common: in each case, we have been able to show convergence for all x within a certain distance R of the point about which the power series is centered (the distance R may be 0). This is no coincidence—such behavior is common to all power series, a fact we will prove with the next two theorems.

Theorem ⚷

The Power Series Convergence Theorem

If a power series $\sum c_n x^n$ converges for some value $x = b \neq 0$, then it converges for all x such that $|x| < |b|$. On the other hand, if a power series $\sum c_n x^n$ diverges for some value $x = d \neq 0$, then it diverges for all x such that $|x| > |d|$.

Proof ✎

If $\sum c_n b^n$ converges, then we know that $\lim_{n \to \infty} c_n b^n = 0$ and so there is a natural number N such that $n \geq N \Rightarrow |c_n b^n| < 1$. Note that for any x, $n \geq N$ implies

$$\left| c_n x^n \right| = \left| \frac{c_n b^n x^n}{b^n} \right| = \left| c_n b^n \right| \left| \frac{x}{b} \right|^n < \left| \frac{x}{b} \right|^n.$$

If $|x| < |b|$, $|x/b| < 1$ and hence $\sum |x/b|^n$ is a convergent geometric series. By the Direct Comparison Test, the above inequality shows that $\sum |c_n x^n|$ converges, so $\sum c_n x^n$ converges.

Now, suppose that $\sum c_n d^n$ diverges. Then for any x such that $|x| > |d|$, $\sum c_n x^n$ must also diverge, because otherwise the proof above would imply that $\sum c_n d^n$ converges.

The Power Series Convergence Theorem also applies to series of the form $\sum c_n (x-a)^n$, with the conditions replaced by $|x-a| < b$ and $|x-a| > d$. The proof follows immediately from the change of variables $\tilde{x} = x - a$ and applying the result of the theorem to the series $\sum c_n \tilde{x}^n$.

Definition 💡

Radius of Convergence and Interval of Convergence

If $\sum c_n (x-a)^n$ converges for all $|x-a| < R$ and diverges for all $|x-a| > R$, we call R the **radius of convergence** of the power series $\sum c_n (x-a)^n$.

If $R = 0$, the series converges only at $x = a$; and if $R = \infty$, the series converges for all x. When $0 < R < \infty$, the series may or may not converge at $x = a - R$ and $x = a + R$, meaning that the **interval of convergence** may be any of the four possibilities $(a-R, a+R)$, $[a-R, a+R)$, $(a-R, a+R]$, or $[a-R, a+R]$.

Theorem 🔍

Existence of the Radius of Convergence of a Power Series

Given a power series $\sum c_n (x-a)^n$, exactly one of the following is true:

1. The series converges only for $x = a$ (radius of convergence $R = 0$).

2. The series converges for all x (radius of convergence $R = \infty$).

3. There is a real number $0 < R < \infty$ such that $\sum c_n (x-a)^n$ converges for all $|x-a| < R$ and diverges for all $|x-a| > R$.

Proof 📎

To prove this theorem, we must show that there is no interval of values for x on which the convergence or divergence of the series is unknown. To do so, suppose neither of the first two statements of the theorem is true. Then for some b with $|b-a| > 0$, $\sum c_n (b-a)^n$ converges and for some d with $|d-a| > |b-a|$, $\sum c_n (d-a)^n$ diverges. Let

$$S = \left\{ |x-a| \,\middle|\, \sum c_n (x-a)^n \text{ converges} \right\}.$$

Then $S \neq \varnothing$ since $|b - a| \in S$, and S has an upper bound since $|d - a| \notin S$. By the Completeness Property of the real numbers, S has a least upper bound R, which makes the third statement above true (note that this theorem says nothing about convergence at either $x = a - R$ or $x = a + R$).

Example 3 ✐

Find the radius of convergence and interval of convergence of the following series.

a. $\displaystyle\sum_{n=0}^{\infty} \frac{x^n}{n!}$
$\qquad\qquad$
b. $\displaystyle\sum_{n=0}^{\infty} \frac{(-2x)^n}{\sqrt{n+3}}$

Solution

a. Using the Ratio Test, we see

$$\left|\frac{a_{n+1}}{a_n}\right| = \left|\frac{x^{n+1}}{(n+1)!} \cdot \frac{n!}{x^n}\right| = |x|\left(\frac{1}{n+1}\right) \to 0 \text{ as } n \to \infty$$

for every $x \in \mathbb{R}$, so the radius of convergence is $R = \infty$ and the interval of convergence is $(-\infty, \infty)$.

b. $\left|\dfrac{a_{n+1}}{a_n}\right| = \left|\dfrac{(-2x)^{n+1}}{\sqrt{n+4}} \cdot \dfrac{\sqrt{n+3}}{(-2x)^n}\right| = 2|x|\sqrt{\dfrac{n+3}{n+4}} = 2|x|\sqrt{\dfrac{1 + \dfrac{3}{n}}{1 + \dfrac{4}{n}}} \to 2|x| \text{ as } n \to \infty$

The series converges when $2|x| < 1$, or $|x| < \frac{1}{2}$. This means the radius of convergence is $R = \frac{1}{2}$ about the center $x = 0$. To determine the interval of convergence, we must test the endpoints separately. When $x = \frac{1}{2}$, the series reduces to the alternating series

$$\sum_{n=0}^{\infty} \frac{(-1)^n}{\sqrt{n+3}},$$

which converges. But when $x = -\frac{1}{2}$, the series reduces to the divergent p-series

$$\sum_{n=0}^{\infty} \frac{1}{\sqrt{n+3}}.$$

Thus, the interval of convergence is $\left(-\frac{1}{2}, \frac{1}{2}\right]$.

TOPIC 3 Calculus and Power Series

The following theorem, which is typically proved in an advanced calculus course, shows that power series behave just like polynomials with regard to differentiation and integration.

Theorem ⚛

Differentiation and Integration of Power Series

Suppose the power series $f(x) = \sum\limits_{n=0}^{\infty} c_n (x-a)^n$ has a radius of convergence $R > 0$. Then

a. $f'(x) = \sum\limits_{n=1}^{\infty} nc_n (x-a)^{n-1} = c_1 + 2c_2 (x-a) + 3c_3 (x-a)^2 + \cdots$

b. $\displaystyle \int f(x)\,dx = C + \sum\limits_{n=0}^{\infty} \frac{c_n}{n+1}(x-a)^{n+1}$

$$= C + c_0 (x-a) + \frac{c_1}{2}(x-a)^2 + \frac{c_2}{3}(x-a)^3 + \cdots$$

where both new series have the same radius of convergence R. Note that f is differentiable (and thus continuous) on $(a-R, a+R)$. (In words, differentiation and integration of a power series can be performed term by term.)

Although power series behave exactly as we would like with regard to differentiation and integration, it should not be assumed that all series are so well behaved (see Exercise 62). It should also be noted that while the radius of convergence of the new power series obtained is the same as the original radius of convergence, the behavior at the interval endpoints can change—in particular, the original series may converge at an endpoint, while its derivative diverges (see Exercise 60).

Example 4 ✎

Differentiate the power series $\sum\limits_{n=0}^{\infty} x^n$ and express the result in closed form.

Solution

Differentiating the series is easily done term by term, according to the last theorem. And since we discovered in Example 2a that $\sum\limits_{n=0}^{\infty} x^n = \dfrac{1}{1-x}$, expressing the derivative in closed form is also easily accomplished.

$$\frac{d}{dx}\sum_{n=0}^{\infty} x^n = \frac{d}{dx}\left[\frac{1}{1-x}\right] \quad \Rightarrow \quad \sum_{n=1}^{\infty} nx^{n-1} = \frac{1}{(1-x)^2}$$

The radius of convergence of the original series is 1, so that is the radius of convergence of the new series as well. And in this case, the interval of convergence of both series is $(-1, 1)$. (In Exercise 55 you will find higher derivatives of this series.)

Example 5 ✍

Find the closed form of the series $\sum_{n=1}^{\infty} \dfrac{(x-5)^n}{n}$ by first determining the closed form of the series $\sum_{n=0}^{\infty} (x-5)^n$.

Solution

The second series is a geometric series whose sum we can find.

$$\sum_{n=0}^{\infty} (x-5)^n = 1 + (x-5) + (x-5)^2 + \cdots = \frac{1}{1-(x-5)} = \frac{1}{6-x}$$

Note that the radius of convergence of the series is $R = 1$, and that the interval of convergence is $(4,6)$. The first series,

$$\sum_{n=1}^{\infty} \frac{(x-5)^n}{n} = (x-5) + \frac{(x-5)^2}{2} + \frac{(x-5)^3}{3} + \cdots,$$

is, term by term, an antiderivative of the first (with a constant of integration of -5), so

$$\sum_{n=1}^{\infty} \frac{(x-5)^n}{n} = \int \frac{1}{6-x} \, dx = -\ln(6-x) + C.$$

By substituting $x = 5$ we determine that $C = 0$, so

$$\sum_{n=1}^{\infty} \frac{(x-5)^n}{n} = -\ln(6-x).$$

And as we noted in Example 1a, its interval of convergence is $[4,6)$.

Power series (in fact, all absolutely converging series) behave like polynomials in another respect: they can be multiplied term by term. The proof of the following theorem is also typically seen in an advanced calculus course.

Theorem 🔍

Multiplication of Power Series

If $f(x) = \sum_{n=0}^{\infty} a_n x^n$ and $g(x) = \sum_{n=0}^{\infty} b_n x^n$ both converge absolutely for $|x| < R$, then

$$f(x)g(x) = \left(\sum_{n=0}^{\infty} a_n x^n \right)\left(\sum_{n=0}^{\infty} b_n x^n \right) = \sum_{n=0}^{\infty} c_n x^n$$

also converges absolutely for $|x| < R$, where

$$c_n = a_0 b_n + a_1 b_{n-1} + a_2 b_{n-2} + \cdots + a_{n-1} b_1 + a_n b_0 = \sum_{k=0}^{n} a_k b_{n-k}.$$

Example 6 ✐

Use the Multiplication of Power Series Theorem to express $1/(1-x^2)$ as a power series.

Solution

Since $\dfrac{1}{1-x^2} = \left(\dfrac{1}{1-x}\right)\left(\dfrac{1}{1+x}\right)$, we begin by rewriting these factors as the following two series.

$$\frac{1}{1-x} = \sum_{n=0}^{\infty} x^n = 1 + x + x^2 + x^3 + \cdots$$

$$\frac{1}{1+x} = \frac{1}{1-(-x)} = \sum_{n=0}^{\infty} (-x)^n = 1 - x + x^2 - x^3 + \cdots$$

(Both have radius of convergence $R = 1$.) The formula for multiplying power series tells us that the coefficient of the term with degree n comes from all possible products of the two series that result in degree n—just the same as with the product of polynomials.

$$\frac{1}{1-x^2} = \underbrace{(1)(1)}_{\text{degree }0} + \underbrace{\left[(1)(-x)+(x)(1)\right]}_{\text{degree }1} + \underbrace{\left[(1)\left(x^2\right)+(x)(-x)+\left(x^2\right)(1)\right]}_{\text{degree }2}$$

$$+ \underbrace{\left[(1)\left(-x^3\right)+(x)\left(x^2\right)+\left(x^2\right)(-x)+\left(x^3\right)(1)\right]}_{\text{degree }3}$$

$$+ \underbrace{\left[(1)\left(x^4\right)+(x)\left(-x^3\right)+\left(x^2\right)\left(x^2\right)+\left(x^3\right)(-x)+\left(x^4\right)(1)\right]}_{\text{degree }4} + \cdots$$

$$= 1 + x^2 + x^4 + \cdots$$

$$= \sum_{n=0}^{\infty} \left(x^2\right)^n = \sum_{n=0}^{\infty} x^{2n}$$

In Exercise 44, you will arrive at this same result by other means.

10.7 **Exercises**

1–32 Determine the radius and interval of convergence for the power series. Be sure to check for convergence at the endpoints.

1. $\displaystyle\sum_{n=0}^{\infty} (3x)^n$

2. $\displaystyle\sum_{n=0}^{\infty} 3x^n$

3. $\displaystyle\sum_{n=1}^{\infty} \frac{(4x)^n}{n^4}$

4. $\displaystyle\sum_{n=0}^{\infty} \frac{(-1)^n x^n}{4^n}$

5. $\displaystyle\sum_{n=0}^{\infty} \frac{x^n}{n^2+1}$

6. $\displaystyle\sum_{n=0}^{\infty} \frac{(5x)^n}{2n!}$

7. $\displaystyle\sum_{n=0}^{\infty} \frac{(2n)!(5x)^n}{2n!}$

8. $\displaystyle\sum_{n=0}^{\infty} \frac{2n!(5x)^n}{(2n)!}$

9. $\displaystyle\sum_{n=1}^{\infty} \frac{(2x)^n}{(n-1)!}$

10. $\displaystyle\sum_{n=1}^{\infty} nx^n$

11. $\displaystyle\sum_{n=1}^{\infty} \frac{n^2(x+3)^n}{2^n}$

12. $\displaystyle\sum_{n=0}^{\infty} \frac{3x^n}{k^n}, \quad k \neq 0$

13. $\displaystyle\sum_{n=0}^{\infty} \left(\frac{x}{4}\right)^n n!$

14. $\displaystyle\sum_{n=1}^{\infty} \frac{(x-3)^n}{n3^n}$

15. $\displaystyle\sum_{n=1}^{\infty} \frac{(x-4)^n}{n^3}$

16. $\displaystyle\sum_{n=0}^{\infty} \frac{(x+1)^n}{(n+2)3^n}$

17. $\displaystyle\sum_{n=0}^{\infty} \frac{(x-5)^n n!}{5^n}$

18. $\displaystyle\sum_{n=1}^{\infty} \frac{(x-2)^n}{(-1)^n n}$

19. $\displaystyle\sum_{n=0}^{\infty} \frac{(x+2)^n}{(n+2)2^n}$

20. $\displaystyle\sum_{n=0}^{\infty} \frac{(x-k)^n}{k^n}, \quad k>0$

21. $\displaystyle\sum_{n=4}^{\infty} \frac{(x-1)^n}{(n-2)(n-3)}$

22. $\displaystyle\sum_{n=1}^{\infty} \frac{(x+2)^n}{\sqrt{n}}$

23. $\displaystyle\sum_{n=2}^{\infty} \frac{(-1)^n x^n}{(\ln n)^n}$

24. $\displaystyle\sum_{n=1}^{\infty} \frac{n^2 (2x-3)^n}{n!}$

25. $\displaystyle\sum_{n=1}^{\infty} n3^n (x-1)^n$

26. $\displaystyle\sum_{n=2}^{\infty} \frac{x^n \ln n}{n!}$

27. $\displaystyle\sum_{n=2}^{\infty} \frac{x^{2n+5}}{\ln \sqrt{n}}$

28. $\displaystyle\sum_{n=0}^{\infty} x^{4^n}$

29. $\displaystyle\sum_{n=2}^{\infty} \frac{3^n (x+2)^n}{n \ln n}$

30. $\displaystyle\sum_{n=0}^{\infty} 2^n (x+4)^{3n+1}$

31. $\displaystyle\sum_{n=0}^{\infty} \pi(x-3)^n$

32. $\displaystyle\sum_{n=1}^{\infty} \left(1+\frac{1}{n}\right)^{n^2} (x-3)^n$

33–40 Determine the interval of convergence for the given series and the limiting function of the series on that interval.

33. $\displaystyle\sum_{n=0}^{\infty} (3x-1)^n$

34. $\displaystyle\sum_{n=0}^{\infty} (2x+3)^n$

35. $\displaystyle\sum_{n=0}^{\infty} (-1)^n \left(\frac{x}{3}\right)^n$

36. $\displaystyle\sum_{n=0}^{\infty} 2\left(\frac{x}{5}\right)^n$

37. $\displaystyle\sum_{n=0}^{\infty} \left(\frac{x-3}{2}\right)^n$

38. $\displaystyle\sum_{n=0}^{\infty} \frac{(2x+3)^n}{3^n}$

39. $\displaystyle\sum_{n=0}^{\infty} \frac{(5x-6)^{2n}}{9^n}$

40. $\displaystyle\sum_{n=0}^{\infty} \frac{(2x-3)^n}{3^n}$

41–48 "Reversing" the process used to find the limiting function in Exercises 33-40, we can find the power series about 0 for certain functions. For example, if $f(x) = 1/(1-3x)$, we can recognize it as the sum of the geometric series $\displaystyle\sum_{n=0}^{\infty} (3x)^n = \sum_{n=0}^{\infty} 3^n x^n$. (This is valid only if x is in the interval of convergence, in this case, $-\frac{1}{3} < x < \frac{1}{3}$.)

Use this approach to find the power series representation of the given function. What is the radius of convergence? (**Hint:** In Exercises 47 and 48, use partial fractions.)

41. $f(x) = \dfrac{1}{1-2x}$

42. $f(x) = \dfrac{1}{1+4x}$

43. $f(x) = \dfrac{1}{1+x^2}$

44. $f(x) = \dfrac{1}{1-x^2}$

45. $f(x) = \dfrac{3}{3-4x}$

46. $f(x) = \dfrac{2}{3+8x^3}$

47. $f(x) = \dfrac{x+3}{1-x^2}$

48. $f(x) = \dfrac{2x}{1-4x^2}$

49. Find the power series representation of $f(x) = 1/x$ centered at $a = 1$. What is the radius of convergence? (**Hint:** Start by rewriting $1/x$ as $\dfrac{1}{1-(1-x)}$, and proceed along the lines of Exercises 41–48.)

50. Use series multiplication to find a second solution to Exercise 48. (**Hint:** Multiply the series expansions of $1/(1-2x)$ and $1/(1+2x)$, and then multiply the result by $2x$.)

51. Differentiating the result of Exercise 49, find the power series representation of $g(x) = 1/x^2$ centered at $a = 1$. What is the radius of convergence?

52. Use the result of Exercise 49 to find the power series representation about $a = 1$ of $h(x) = \ln x$. What is the radius of convergence?

53. Use Exercise 52 to prove that $\ln 2 = \displaystyle\sum_{n=1}^{\infty} \frac{1}{n2^n}$. Use this series to approximate $\ln 2$ to three decimal places. How many terms did you use? (**Hint:** Start with the series approximating $\ln \frac{1}{2} = -\ln 2$.)

54. Find the power series representation of $f(x) = 1/x$ centered at $a = 3$. What is the interval of convergence? (**Hint:** Start by rewriting $1/x$ as $\dfrac{1}{3-(3-x)}$ and proceed along the lines of Exercise 41–48.)

55. Find the second and third derivatives of the power series of Example 4. In both cases, express your answer as a power series and in closed form.

56. Find the closed form of the series $\displaystyle\sum_{n=1}^{\infty} \frac{(2x-3)^n}{2n}$ by first determining the closed form of the series $\displaystyle\sum_{n=0}^{\infty} (2x-3)^n$. What is the interval of convergence?

57. Find the closed form of the series

$\sum_{n=2}^{\infty} 9(n^2 - n)(3x-1)^{n-2}$ by first determining the

closed form of the series $\sum_{n=0}^{\infty}(3x-1)^n$. What is

the interval of convergence? (**Hint:** Differentiate twice.)

58. Integrate the series you found in Exercise 43 to obtain a series expansion for $F(x) = \arctan x$ about $a = 0$. Find the radius of convergence for the series you obtained.

59. Use differentiation and the result of Exercise 43 to find a series expansion of $g(x) = 2x/(x^2+1)^2$. Find the radius of convergence for the series you obtained.

60. Even when differentiated, a convergent power series retains its radius of convergence. But as we have noted, the behavior at the interval endpoints can change—in particular, the original series may converge at an endpoint, while its derivative diverges. Verify this by examining the intervals of convergence of f and its first two derivatives for

$$f(x) = \sum_{n=0}^{\infty} \frac{(-1)^n x^n}{n^2 + 1}.$$

61. Show that for any natural number $k \in \mathbb{N}$, the power series expansion of $\dfrac{(k-1)!}{(1-x)^k}$ is

$$\sum_{n=0}^{\infty}(n+k-1)(n+k-2)\cdots(n+2)(n+1)x^n.$$

62.* In general, series of functions are not nearly as well behaved with regards to termwise differentiation and integration as power series are. As an illustration, examine the convergence set (the set of x-values for which the series

converges) of $f(x) = \sum_{n=0}^{\infty} \dfrac{\sin(3^n x)}{2^n}$, and that of its

first derivative. (**Note:** This is not a power series.)

63. Verify that $y = \sum_{n=0}^{\infty} \dfrac{x^{2n}}{n!}$ satisfies the differential

equation $y' = 2xy$. (**Hint:** Differentiate term by term.)

64. Find a power series solution of the differential equation, $y' = 2y$ satisfying the initial condition $y(0) = 1$. Then solve the equation by traditional means and conclude that the solutions are equal. (**Hint:** Starting with undetermined coefficients,

write $y = \sum_{n=0}^{\infty} a_n x^n$, obtain the power series for

both y' and $2y$, and finally equate terms.)

65. Find a power series solution of the differential equation, $y'' + 4y = 0$ satisfying the initial conditions $y(0) = 1$, $y'(0) = 0$. Then solve the equation by traditional means and conclude that the solutions are equal. (See the hint given in Exercise 64.)

66–70 *True or False?* Determine whether the given statement is true or false. In case of a false statement, explain or provide a counterexample.

66. If the series $\sum a_n (x+1)^n$ converges at $x = -4$, then it must converge at $x = 1$.

67. If the series $\sum a_n (x+1)^n$ converges at $x = -4$, then it must converge at $x = 2$.

68. If the series $\sum a_n x^n$ converges at $x = 1$, then $\sum n a_n x^{n-1}$ must also converge at $x = 1$.

69. If a series converges on an interval, we may differentiate it term by term to obtain the derivative of its sum.

70. If the interval of convergence of the series $\sum a_n x^n$ is $(-a, a)$ for an $a \in \mathbb{R}^+$, then the interval of convergence of $\sum a_n (x-a)^n$ is $(0, 2a)$.

10.8 Taylor and Maclaurin Series

TOPICS

1. Taylor and Maclaurin series

2. Remainder estimates and convergence for Taylor series

In the last section, we used our knowledge of geometric series and the fact that power series are easily differentiated, integrated, and multiplied to express a few select power series in closed form. The reverse goal is actually more common, and arises from the following question: When can a function be expressed as a power series and, when it *is* possible, over what interval will the power series converge to the original function? This goal arises because there are many applications in which it is difficult (or impossible) to work with the closed form of a function, but for which the power series form (or a truncated version of it) is well suited.

TOPIC 1   Taylor and Maclaurin Series

To answer the questions above, we start by assuming we *can* represent a given function f as a power series for all x within a convergence radius R about a fixed real number a:

$$|x - a| < R \quad \Rightarrow \quad f(x) = c_0 + c_1(x - a) + c_2(x - a)^2 + \cdots = \sum_{n=0}^{\infty} c_n(x - a)^n.$$

We then determine what the coefficients c_n must be, given this assumption. First, by letting $x = a$, we see immediately that $f(a) = c_0$. Next, since we can differentiate a power series (within its radius of convergence) term by term, we note that

$$f'(x) = c_1 + 2c_2(x - a) + 3c_3(x - a)^2 + 4c_4(x - a)^3 + \cdots,$$

and substituting $x = a$ into this equation tells us that $f'(a) = c_1$. We proceed to differentiate and substitute $x = a$ repeatedly, to obtain

$$f''(x) = 2c_2 + 3 \cdot 2c_3(x - a) + 4 \cdot 3c_4(x - a)^2 + \cdots \quad \Rightarrow \quad f''(a) = 2c_2$$
$$f'''(x) = 3 \cdot 2c_3 + 4 \cdot 3 \cdot 2c_4(x - a) + 5 \cdot 4 \cdot 3c_5(x - a)^2 + \cdots \Rightarrow f'''(a) = 3 \cdot 2c_3 = 3!c_3$$
$$\vdots$$

and, in general, $f^{(n)}(a) = n!c_n$. Solving for each c_n, and using the conventions that $f^{(0)}(x) = f(x)$ and $0! = 1$, we have discovered that

$$c_n = \frac{f^{(n)}(a)}{n!}.$$

Remember that this all hinges upon the key assumption that f can be represented as a power series about a—we will soon revisit this assumption to answer the question of *when* this is possible. But for now, we have motivated the following definition.

Definition ♀

Taylor Series and Maclaurin Series

Given a function f with derivatives of all orders throughout an open interval containing a, the power series

$$\sum_{n=0}^{\infty} \frac{f^{(n)}(a)}{n!}(x-a)^n = f(a) + f'(a)(x-a) + \frac{f''(a)}{2!}(x-a)^2 + \frac{f'''(a)}{3!}(x-a)^3 + \cdots$$

is called the **Taylor series generated by f about a**. The Taylor series generated by f about 0 is also known as the **Maclaurin series generated by f**.

Many mathematicians, including Isaac Newton and Johann Bernoulli, developed power series expansions of functions and demonstrated their usefulness, but the names above honor the English mathematician Brook Taylor (1685–1731) and the Scottish mathematician Colin Maclaurin (1698–1746) who wrote extensively about them and popularized their use. We will see many examples of such use throughout the remainder of this chapter.

Note that the definition above formally describes Taylor and Maclaurin series as series—there is no claim (yet) that the Taylor series associated with a given function f actually converges to that function on any interval.

Example 1 ✏

Find the Taylor series expansion of $f(x) = 1/x$ about $a = 3$, and determine where, if anywhere, the Taylor series converges to f.

Solution

We determine the coefficients c_n of the series as follows:

$$f^{(0)}(x) = \frac{1}{x}, \qquad \text{so} \quad f(3) = \frac{1}{3} \qquad \text{and} \quad c_0 = \frac{f(3)}{0!} = \frac{1}{3},$$

$$f'(x) = -\frac{1}{x^2}, \qquad \text{so} \quad f'(3) = -\frac{1}{9} \qquad \text{and} \quad c_1 = \frac{f'(3)}{1!} = -\frac{1}{9},$$

$$f''(x) = \frac{2}{x^3}, \qquad \text{so} \quad f''(3) = \frac{2}{27} \qquad \text{and} \quad c_2 = \frac{f''(3)}{2!} = \frac{1}{27},$$

$$f'''(x) = -\frac{6}{x^4}, \qquad \text{so} \quad f'''(3) = -\frac{6}{81} \qquad \text{and} \quad c_3 = \frac{f'''(3)}{3!} = -\frac{1}{81},$$

and, in general, $f^{(n)}(x) = (-1)^n (n!) x^{-(n+1)}$ and $c_n = (-1)^n 3^{-(n+1)}$. So the Taylor series generated by f about $a = 3$ is

$$\sum_{n=0}^{\infty} \frac{(-1)^n}{3^{n+1}}(x-3)^n = \frac{1}{3} - \frac{1}{9}(x-3) + \frac{1}{27}(x-3)^2 - \frac{1}{81}(x-3)^3 + \cdots.$$

Since the series was generated by $f(x) = 1/x$ we are certainly justified in hoping that it actually *equals* $1/x$ somewhere. That hope may be in vain, as we will see in Example 5, but in this case we can easily determine exactly where the series converges to the function because it is of a form we understand well. As we have seen in Exercise 54 of Section 10.7, it is a geometric series with ratio $-(x-3)/3$ and it converges when $\left|-(x-3)/3\right| < 1$, or $0 < x < 6$. And, for those values of x,

$$\sum_{n=0}^{\infty} \frac{(-1)^n}{3^{n+1}}(x-3)^n = \frac{\frac{1}{3}}{1 - \left(-\dfrac{x-3}{3}\right)} = \frac{1}{3 + x - 3} = \frac{1}{x}.$$

If we were to discard all the terms of the Taylor series of Example 1 with degree greater than n, we would wind up with a polynomial of degree n that approximates $1/x$ on the interval $(0, 6)$. Specifically, the first four such polynomial approximations from Example 1 are as follows:

$$p_0(x) = \frac{1}{3}$$

$$p_1(x) = -\frac{1}{9}x + \frac{2}{3}$$

$$p_2(x) = \frac{1}{27}x^2 - \frac{1}{3}x + 1$$

$$p_3(x) = -\frac{1}{81}x^3 + \frac{4}{27}x^2 - \frac{2}{3}x + \frac{4}{3}$$

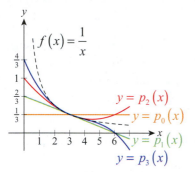

Figure 1 Taylor Polynomials

These polynomials are shown in Figure 1. They are called *Taylor polynomials* and they generalize the idea that we studied in Section 3.9 of approximating a function with its linearization.

Definition 💡

Taylor Polynomial of Order *n*

Given a function f with derivatives up through order N throughout an open interval containing a, the polynomial

$$p_n(x) = \sum_{k=0}^{n} \frac{f^{(k)}(a)}{k!}(x-a)^k$$

$$= f(a) + f'(a)(x-a) + \frac{f''(a)}{2!}(x-a)^2 + \cdots + \frac{f^{(n)}(a)}{n!}(x-a)^n$$

for $n \in \{0, 1, \ldots, N\}$

is called the **Taylor polynomial of order *n* generated by *f* about *a***. If $a = 0$, the Taylor polynomial is also known as the **Maclaurin polynomial**.

Example 2 ✍

Find the Maclaurin polynomials generated by $f(x) = \sin x$.

Solution

The derivatives of f exhibit a nice cyclical pattern, and we leave it to the reader to verify that

$$f^{(n)}(0) = \begin{cases} \left. (-1)^{n/2} \sin x \right|_{x=0} = 0 & \text{if } n \text{ is even} \\[2mm] \left. (-1)^{(n-1)/2} \cos x \right|_{x=0} = (-1)^{(n-1)/2} & \text{if } n \text{ is odd} \end{cases}$$

So the Maclaurin series generated by f is

$$f(0) + f'(0)x + \frac{f''(0)}{2}x^2 + \cdots + \frac{f^{(k)}(0)}{k!}x^k + \cdots$$

$$= 0 + 1 \cdot x + 0 \cdot x^2 + \frac{-1}{3!}x^3 + 0 \cdot x^4 + \frac{1}{5!}x^5 + \cdots$$

$$= x - \frac{x^3}{6} + \frac{x^5}{120} - \cdots$$

$$= \sum_{k=0}^{\infty} \frac{(-1)^k x^{2k+1}}{(2k+1)!}.$$

Since all the terms with an even exponent are absent, we can describe the family of Maclaurin polynomials with the formulas

$$p_{2n+1}(x) = \sum_{k=0}^{n} \frac{(-1)^k x^{2k+1}}{(2k+1)!}$$

and

$$p_{2n+2}(x) = p_{2n+1}(x) + 0 \cdot x^{2n+2} = p_{2n+1}(x).$$

As a consequence, the Maclaurin polynomials of order $2n + 1$ and order $2n + 2$ are the same and both have degree $2n + 1$. For instance, the Maclaurin polynomials of order 5 and order 6 are

$$p_5(x) = p_6(x) = x - \frac{x^3}{6} + \frac{x^5}{120}$$

and both have degree 5.

Figure 2 is an illustration of how the Maclaurin polynomials of increasing order approximate $f(x) = \sin x$. Keep in mind, however, that we still have the task of determining where this Maclaurin series actually converges to its generating function.

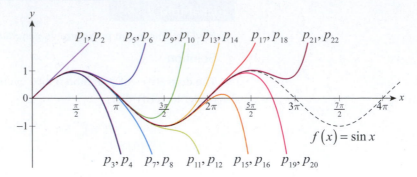

Figure 2 Maclaurin Polynomial Approximations of Sine

TOPIC 2 Remainder Estimates and Convergence for Taylor Series

We are now ready to tackle the question of when (and where) Taylor series converge to their generating functions, and we do so with a theorem that generalizes the Mean Value Theorem. The proof of the following is included in Appendix E.

Theorem 🔍

Taylor's Theorem

If f and its derivatives up through $f^{(n)}$ are all continuous on the closed interval $[a,b]$ and if $f^{(n+1)}$ exists on the open interval (a,b), then there is a number $c \in (a,b)$ such that

$$f(b) = f(a) + f'(a)(b-a) + \frac{f''(a)}{2!}(b-a)^2 + \cdots + \frac{f^{(n)}(a)}{n!}(b-a)^n + \frac{f^{(n+1)}(c)}{(n+1)!}(b-a)^{n+1}.$$

In applying Taylor's Theorem, we typically desire an approximation for the value $f(x)$ based on the known value of f and its first n derivatives at some fixed point a, so we replace b in the theorem with x and express the result as a formula.

Formula 🔍

Taylor's Formula

If f has derivatives of all orders throughout an open interval I containing a, then for each natural number n and for each $x \in I$,

$$f(x) = \underbrace{f(a) + f'(a)(x-a) + \frac{f''(a)}{2!}(x-a)^2 + \cdots + \frac{f^{(n)}(a)}{n!}(x-a)^n}_{p_n(x)} + \underbrace{\frac{f^{(n+1)}(c)}{(n+1)!}(x-a)^{n+1}}_{r_n(x)}$$

where c is some number between a and x. Defining $p_n(x)$ and $r_n(x)$ as above, we can write $f(x) = p_n(x) + r_n(x)$, where p_n is a polynomial of degree at most n and r_n is called the **remainder** (or **error**) **of order n**.

So given a function f meeting the conditions of Taylor's formula, we now note that the Taylor polynomials p_n converge to f for all x in the open interval I if $r_n(x) \to 0$ as $n \to \infty$ for each x. If this is the case, we write

$$f(x) = \sum_{n=0}^{\infty} \frac{f^{(n)}(a)}{n!} (x-a)^n$$

and say that the Taylor series generated by f converges to f on I. We are often able to show that $r_n(x) \to 0$ on I by knowing something about the behavior of the derivatives of f throughout I.

Example 3 ✐

Find the Maclaurin series generated by $f(x) = e^x$ and determine where it converges to f.

Solution

The function f has derivatives of all orders on the entire real line, and $f^{(n)}(0) = 1$ for every n, so the formal Maclaurin series generated by $f(x) = e^x$ is

$$1 + x + \frac{x^2}{2!} + \frac{x^3}{3!} + \cdots = \sum_{n=0}^{\infty} \frac{x^n}{n!}.$$

The only remaining step is to determine for which values of x this series converges to f. We do so by splitting the real number line into three cases. In two of them, we make use of the fact that

$$\lim_{n \to \infty} \frac{x^n}{n!} = 0$$

for every real number x. This is true because the series $\sum_{n=0}^{\infty} (x^n/n!)$ is absolutely convergent for any x (by the Ratio Test) and hence convergent (absolute convergence implies convergence), so the terms $x^n/n!$ must converge to 0.

(1) If $x < 0$, Taylor's formula tells us that the remainder $r_n(x)$ of order n is

$$r_n(x) = \frac{f^{(n+1)}(c)}{(n+1)!} x^{n+1}$$

for some $x < c < 0$. And since $f^{(n+1)}(c) = e^c < e^0 = 1$,

$$\lim_{n \to \infty} |r_n(x)| = \lim_{n \to \infty} \left| \frac{f^{(n+1)}(c)}{(n+1)!} x^{n+1} \right| \le \lim_{n \to \infty} \frac{|x|^{n+1}}{(n+1)!} = 0,$$

and hence $\lim_{n \to \infty} r_n(x) = 0$ as well.

(2) If $x = 0$, $r_n(x) = 0$ for every $n \ge 1$, so $\lim_{n \to \infty} r_n(x) = 0$.

(3) If $x > 0$, Taylor's formula again tells us that

$$r_n(x) = \frac{f^{(n+1)}(c)}{(n+1)!}x^{n+1} = \frac{e^c}{(n+1)!}x^{n+1},$$

this time for some $0 < c < x$. But since f is an increasing function, we know $e^c < e^x$ so

$$\lim_{n\to\infty} r_n(x) = \lim_{n\to\infty}\frac{f^{(n+1)}(c)}{(n+1)!}x^{n+1} = \lim_{n\to\infty}\frac{e^c}{(n+1)!}x^{n+1} \le e^x \lim_{n\to\infty}\frac{x^{n+1}}{(n+1)!} = 0.$$

Since $\lim_{n\to\infty} r_n(x) = 0$ for every real number x, $e^x = \sum_{n=0}^{\infty}(x^n/n!)$ everywhere.

We often make use of the fact that $\lim_{n\to\infty}(x^n/n!) = 0$ in showing Taylor series convergence.

Example 4 ✎

For which values of x does $\sin x = \sum_{k=0}^{\infty}\frac{(-1)^k x^{2k+1}}{(2k+1)!}$?

Solution

In Example 2, we constructed the formal Maclaurin series above for $f(x) = \sin x$. And since $\left|f^{(n)}(c)\right| \le 1$ for all n and all possible values of c, the remainder of order n satisfies

$$\left|r_n(x)\right| = \left|\frac{f^{(n+1)}(c)}{(n+1)!}x^{n+1}\right| \le \left|\frac{x^{n+1}}{(n+1)!}\right| \to 0 \text{ as } n \to \infty,$$

and hence $\lim_{n\to\infty} r_n(x) = 0$ as well. So the Maclaurin series generated by $\sin x$ is another example of a series that converges to its generating function for all x.

So far, we have determined that e^x and $\sin x$ converge to their respective Maclaurin series, that is, Taylor series about $a = 0$ for all x, and Example 1 demonstrated that $1/x$ converges to its Taylor series about $a = 3$ on the interval $0 < x < 6$ (and by checking the endpoints 0 and 6 individually, its easy to see that they cannot be included in the interval of convergence). The next example demonstrates that not all Taylor series are so well behaved.

Example 5 ✎

Let f be defined by

$$f(x) = \begin{cases} e^{-1/x^2} & \text{if } x \ne 0 \\ 0 & \text{if } x = 0 \end{cases}.$$

It is easy to see that f has derivatives of all orders for every $x \ne 0$, and with a bit of work we can show that $f^{(n)}(0) = 0$ for every n. First, by the definition of the derivative,

$$f'(0) = \lim_{x \to 0} \frac{f(x) - f(0)}{x - 0} = \lim_{x \to 0} \frac{e^{-1/x^2}}{x}.$$

This limit is of indeterminate form $0/0$ and we could apply l'Hôpital's Rule to it directly—unfortunately, as you can verify, this leads to a limit that's harder to analyze. But we can transform the limit into one of indeterminate form ∞/∞ and make progress.

$$f'(0) = \lim_{x \to 0} \frac{e^{-1/x^2}}{x} = \lim_{x \to 0} \frac{\dfrac{1}{x}}{e^{1/x^2}} = \lim_{x \to 0} \frac{-\dfrac{1}{x^2}}{\left(-\dfrac{2}{x^3}\right) e^{1/x^2}} = \lim_{x \to 0} \frac{x}{2 e^{1/x^2}} = 0$$

We use the same principle to show that higher derivatives of f evaluated at 0 are also 0. For instance,

$$f''(0) = \lim_{x \to 0} \frac{f'(x) - f'(0)}{x - 0} = \lim_{x \to 0} \frac{\left(\dfrac{2}{x^3}\right) e^{-1/x^2}}{x} = 2 \lim_{x \to 0} \frac{\dfrac{1}{x^4}}{e^{1/x^2}}$$

and two applications of the same trick result in $f''(0) = 0$. By induction, and through repeated application of l'Hôpital's Rule, we obtain $f^{(n)}(0) = 0$ for every n.

So the Maclaurin series generated by f is the series $\displaystyle\sum_{n=0}^{\infty} 0$, or 0. But clearly $f(x) \neq 0$ for every $x \neq 0$. So the Maclaurin series for this particular function converges to it at one and only one point—not a very useful expansion at all.

What goes wrong in this example? Actually, nothing at all—Taylor's Theorem is not contradicted by this function. Figure 3 shows that f is extremely flat near 0, but not exactly flat. For any $x \neq 0$, $f^{(n)}(x)$ will eventually begin to diverge rapidly from 0 as $n \to \infty$, resulting in $\displaystyle\lim_{n \to \infty} r_n(x) \neq 0$.

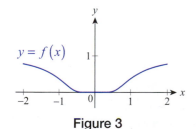

$y = f(x)$

Figure 3

We will finish this section with some examples of how we can combine Taylor's Theorem and theorems about power series to quickly find Taylor series expansions of many functions.

Example 6 ✎

Find the Maclaurin series expansion of $\cos x$.

Solution

Since $\displaystyle \sin x = \sum_{k=0}^{\infty} \frac{(-1)^n x^{2k+1}}{(2k+1)!}$ for all x (see Example 4), we know we can differentiate it term by term. So

$$\cos x = \frac{d}{dx}(\sin x) = \frac{d}{dx}\left[\sum_{k=0}^{\infty} \frac{(-1)^k x^{2k+1}}{(2k+1)!}\right]$$

$$= \sum_{k=0}^{\infty} \frac{(-1)^k (2k+1) x^{2k}}{(2k+1)!} = \sum_{k=0}^{\infty} \frac{(-1)^k x^{2k}}{(2k)!}$$

for all x.

Example 7 ✐

Find the Maclaurin series expansion of e^{-x^2}.

Solution

We could construct the series expansion directly, but it is far quicker to simply use the known series expansion of e^x (which from Example 3 we know converges for all x) and substitute $-x^2$ for x.

$$e^{-x^2} = \sum_{n=0}^{\infty} \frac{\left(-x^2\right)^n}{n!} = \sum_{n=0}^{\infty} (-1)^n \frac{x^{2n}}{n!} = 1 - \frac{x^2}{1!} + \frac{x^4}{2!} - \frac{x^6}{3!} + \cdots$$

Example 8 ✐

Approximate $\int_0^1 e^{-x^2}\, dx$ to within 0.001.

Solution

The integral $\int_0^1 e^{-x^2}\, dx$ is an example of a nonelementary integral, so we cannot evaluate it using an antiderivative of e^{-x^2}. But we *can* make use of the series expansion we just found in Example 7. Since the series converges for all x, we can integrate it term by term without hesitation.

$$\int_0^1 e^{-x^2}\, dx = \int_0^1 \sum_{n=0}^{\infty} (-1)^n \frac{x^{2n}}{n!}\, dx$$

$$= \left[x - \frac{x^3}{3 \cdot 1!} + \frac{x^5}{5 \cdot 2!} - \frac{x^7}{7 \cdot 3!} + \frac{x^9}{9 \cdot 4!} - \cdots \right]_0^1$$

$$= 1 - \frac{1}{3} + \frac{1}{10} - \frac{1}{42} + \frac{1}{216} - \cdots$$

$$\approx 1 - \frac{1}{3} + \frac{1}{10} - \frac{1}{42} + \frac{1}{216} \approx 0.7475$$

The absolute value of the next term in the series would be $1/(11 \cdot 5!) = \frac{1}{1320} < 0.001$, so by the Alternating Series Test we know the approximation 0.7475 is correct to within 0.001.

Example 9 ✎

Find the Taylor series expansion of $\ln x$ about $a = 1$.

Solution

We leave it as an exercise (Exercise 8) to find the Taylor series expansion directly, and instead demonstrate how several facts we already know can be combined:

$$\sum_{n=0}^{\infty} x^n = \frac{1}{1-x} \text{ for } |x| < 1 \qquad \text{Geometric series formula}$$

$$\sum_{n=0}^{\infty} (1-x)^n = \frac{1}{1-(1-x)} = \frac{1}{x} \text{ for } |x-1| < 1 \qquad \text{Substitute } 1-x \text{ for } x.$$

$$\ln x = \int \frac{1}{x} dx = \int \sum_{n=0}^{\infty} (1-x)^n \, dx = -\sum_{n=0}^{\infty} \frac{(1-x)^{n+1}}{n+1} \qquad \begin{array}{l}\text{Integrate term by term and use} \\ \text{a constant of integration of 0.}\end{array}$$

Note that the expansion we just found remains valid for $|x-1| < 1$. Note also that the interval of convergence now includes the endpoint $x = 2$ but not the endpoint $x = 0$. The expansion may appear in several slightly different forms. For instance, if $(-1)^{n+1}$ is factored out of each numerator and the series reindexed to begin with 1, it can be written as

$$\ln x = \sum_{n=1}^{\infty} (-1)^{n+1} \frac{(x-1)^n}{n}.$$

Example 10 ✎

Find **a.** the Maclaurin series expansion of $x \sin x$ and **b.** the Taylor series expansion of $x \ln x$ about $a = 1$.

Solution

a. Since our series expansion of $\sin x$ is valid for all x (see Example 4), we can simply multiply it by x (which is, trivially, the Maclaurin series of x).

$$x \sin x = x \sum_{k=0}^{\infty} \frac{(-1)^k x^{2k+1}}{(2k+1)!} = \sum_{k=0}^{\infty} \frac{(-1)^k x^{2k+2}}{(2k+1)!}$$

b. To quickly find the series expansion of $x \ln x$ about $a = 1$, we want to use the result of Example 9 along with the series expansion of x about $a = 1$, which is $x = 1 + 1(x-1)$.

$$x \ln x = \left[1+(x-1)\right]\sum_{n=1}^{\infty}(-1)^{n+1}\frac{(x-1)^n}{n}$$

$$= \left[1+(x-1)\right]\left[(x-1)-\frac{(x-1)^2}{2}+\frac{(x-1)^3}{3}-\frac{(x-1)^4}{4}+\cdots\right]$$

$$= \left[(x-1)-\frac{(x-1)^2}{2}+\frac{(x-1)^3}{3}-\frac{(x-1)^4}{4}+\cdots\right]$$

$$+ \left[(x-1)^2-\frac{(x-1)^3}{2}+\frac{(x-1)^4}{3}-\frac{(x-1)^5}{4}+\cdots\right]$$

$$= (x-1)+\sum_{n=2}^{\infty}(-1)^n\frac{(x-1)^n}{n(n-1)}$$

As a reference for the exercises that follow, we summarize here the Taylor series expansions we have determined so far.

Function	Taylor Series	Convergence Condition
$\dfrac{1}{1-x}$	$\displaystyle\sum_{n=0}^{\infty}x^n = 1+x+x^2+x^3+\cdots$	$\|x\|<1$
e^x	$\displaystyle\sum_{n=0}^{\infty}\frac{x^n}{n!} = 1+x+\frac{x^2}{2!}+\frac{x^3}{3!}+\cdots$	$\|x\|<\infty$
$\sin x$	$\displaystyle\sum_{n=0}^{\infty}\frac{(-1)^n x^{2n+1}}{(2n+1)!} = x-\frac{x^3}{3!}+\frac{x^5}{5!}-\cdots$	$\|x\|<\infty$
$\cos x$	$\displaystyle\sum_{n=0}^{\infty}\frac{(-1)^n x^{2n}}{(2n)!} = 1-\frac{x^2}{2!}+\frac{x^4}{4!}-\cdots$	$\|x\|<\infty$
$\ln x$	$\displaystyle\sum_{n=1}^{\infty}\frac{(-1)^{n+1}(x-1)^n}{n} = (x-1)-\frac{(x-1)^2}{2}+\frac{(x-1)^3}{3}-\cdots$	$0<x\le 2$

Table 1

10.8 Exercises

1. Find the Maclaurin polynomial of order 5 generated by $f(x)$ given that $f(0) = 4$, $f'(0) = 2$, $f''(0) = -1$, $f'''(0) = 3$, and $f^{(4)}(0) = 1$.

2. Find the Taylor polynomial of order 5 generated by $f(x)$ about $a = 2$ given that $f(2) = 1$, $f'(2) = -3$, $f''(2) = 0$, $f'''(2) = 5$, and $f^{(4)}(2) = -2$.

3–10 Determine the Taylor polynomials generated by the given function about the indicated point. Find the radius of convergence.

3. $f(x) = 2x^4 - x^3 + 5x^2 + 3x - 7$; $a = 0$

4. $f(x) = e^{-x^2/2}$; $a = 0$

5. $f(x) = e^x$; $a = 1$

6. $f(x) = \sin x$; $a = \pi/2$

7. $f(x) = \dfrac{1}{x+4}$; $a = 1$

8. $f(x) = \ln x$; $a = 1$

9. $f(x) = \dfrac{1}{x^2}$; $a = 2$

10. $f(x) = \tanh^{-1} x$; $a = 0$

11–29 Use the Taylor series (or Maclaurin series if the center is not specified) we discussed so far in the text (see Table 1) to determine that of the given function. In each case, find the radius of convergence.

11. $f(x) = \cos(x^2)$

12. $f(x) = \cos^2 x = \dfrac{1 + \cos 2x}{2}$

13. $f(x) = \sin x \cos x = \dfrac{\sin 2x}{2}$

14. $f(x) = x^3 e^{-2x}$

15. $f(x) = \cosh x$

16. $f(x) = xe^x$; $a = 1$

17. $f(x) = \cos \sqrt{x}$

18. $f(x) = \dfrac{x+1}{x-1}$

19. $f(x) = (2x^3 - x)e^x$

20. $f(x) = \sinh x$

21. $f(x) = \ln(1 + x^2)$

22. $f(x) = \dfrac{x}{1 + x^4}$

23. $f(x) = \dfrac{1}{x^2}$; $a = 3$

24. $f(x) = \ln \dfrac{1+x}{1-x}$; $a = 0$

25. $f(x) = \dfrac{2x+1}{x^2 + x - 6}$

26. $f(x) = \dfrac{1}{(x^2 + 1)^2}$

27. $f(x) = \dfrac{x^2}{1 - x^2}$

28. $f(x) = \tan^{-1} x$; $a = 0$

29. $f(x) = x \cos(x^{3/2})$; $a = 0$

30–35 Use the definition to find the first five nonzero terms of the Taylor series generated by the given function about the indicated point.

30. $f(x) = \tan x$; $a = \pi/4$

31. $f(x) = \arctan x$; $a = 1$

32. $f(x) = \sec x$; $a = 0$

33. $f(x) = e^{\cos x}$; $a = 0$

34. $f(x) = e^{\sin x}$; $a = 0$

35. $f(x) = 2^x$; $a = 1$

36–47 Find the first five nonzero terms of the Maclaurin series generated by the indicated function by using operations on familiar series (try not to use the definition). In Exercises 41–43, try long division.

36. $f(x) = \sin x + \cos x$

37. $f(x) = \dfrac{e^x}{1-x}$

38. $f(x) = e^x \cos x$

39. $f(x) = e^{-2x} \sin x$

40. $f(x) = \dfrac{\cos x}{x+1}$

41. $f(x) = \ln(1+x)\sin x$

42. $f(x) = \tan x$

43. $f(x) = \tanh x$

44. $f(x) = x\sec 2x$

45. $f(x) = \ln(1-x)\cos x$

46. $f(x) = \cos(x^2 + 2x)$

47. $f(x) = (x-1)^2 \cos \pi x$

48–57 Find the Taylor series approximation of the given function value or definite integral that guarantees the indicated accuracy. How many (nonzero) terms did you use? (**Hint:** See Example 8 and note the nonelementary integrals that we cannot evaluate exactly.)

48. $\ln 1.2$;   error $\le 10^{-4}$ (**Hint:** Use the Taylor series expansion of $\ln(x+1)$.)

49. $\dfrac{1}{e}$;   error $\le 10^{-4}$

50. $\sin 1$;   error $\le 10^{-5}$

51. $\displaystyle\int_0^1 e^{-x^4}\, dx$;   error ≤ 0.001

52. $\displaystyle\int_0^1 \cos(x^3)\, dx$;   error $\le 10^{-4}$

53. $\displaystyle\int_0^1 \sin(x^2)\, dx$;   error $\le 10^{-6}$

54. $\displaystyle\int_0^1 \dfrac{\sin x}{x}\, dx$;   error $\le 10^{-6}$

55. $\displaystyle\int_0^1 \tan^{-1}(x^2)\, dx$;   error ≤ 0.01

56. $\displaystyle\int_0^1 x^2 \sin(x^2)\, dx$;   error $\le 10^{-6}$

57. $\displaystyle\int_0^1 \dfrac{1}{\sqrt{x^4+1}}\, dx$;   error ≤ 0.01

58–60 Prove that the Maclaurin series converges to the function by showing that the error $r_n(x)$ satisfies $\displaystyle\lim_{n\to\infty} r_n(x) = 0$.

58. $e^{-x^2} = \displaystyle\sum_{n=0}^{\infty} \dfrac{(-1)^n x^{2n}}{n!}$

59. $\cos x = \displaystyle\sum_{n=0}^{\infty} \dfrac{(-1)^n x^{2n}}{(2n)!}$

60. $\ln(x+1) = \displaystyle\sum_{n=1}^{\infty} \dfrac{(-1)^{n+1} x^n}{n}$, $|x| < 1$

61–64 Use Taylor series to find the indicated limit.

61. $\displaystyle\lim_{x\to 0} \dfrac{e^x - e^{-x}}{\sin x}$

62. $\displaystyle\lim_{x\to 0} \dfrac{2x^3 - 1 + \cos x}{x^5}$

63. $\displaystyle\lim_{x\to 0} \dfrac{\tan^{-1} x - \sin(x^2)}{x}$

64. $\displaystyle\lim_{x\to 0} \dfrac{\ln(x+1)}{x\sin x}$

65. Use power series to show that $\displaystyle\lim_{x\to 0} \dfrac{\sin x}{x} = 1$.

66. Find the Taylor series expansion of the function $f(x) = 3x^3 - 5x^2 + 2x - 7$ about $a = 2$. What can you tell from your answer?

67. Use the Maclaurin series expansion of $f(x) = 1/(1-x)$ and differentiation to find

$$\sum_{n=1}^{\infty} \left(n/2^{n-1}\right).$$

68.* Calculate the first few terms of the Maclaurin series expansion of $f(x) = \arcsin x$; then notice the apparent pattern and derive the general form of the series. What is the radius of convergence?

69.* Suppose that the function f has derivatives of all orders throughout an open interval I and there is an $M > 0$ such that $\left|f^{(n)}(x)\right| \le M$ for all $n \in \mathbb{N}$ over I. Prove that if $a \in I$, then the Taylor series generated by $f(x)$ about a converges to f at any $x \in I$.

70. Use Exercise 68 to prove that if $f(x) = \sinh x$ or $f(x) = \cosh x$, then the Taylor series expansion of f about any $a \in I$ converges to f at every real number x.

71.* Find the value of the 162nd derivative of $f(x) = \tan^{-1} x$ at $x = 0$. (**Hint:** Examine what happens to the Maclaurin series after repeated differentiation and substitution of $x = 0$.) What is the 163rd derivative at 0?

72. Use the Maclaurin series expansion of $f(x) = (x + x^2)e^x$ to prove $\displaystyle\sum_{n=1}^{\infty} (n^2/n!) = 2e$.

73.* Use an appropriate Maclaurin series expansion to prove $\displaystyle\sum_{n=1}^{\infty} (n^3/n!) = 5e$. (**Hint:** See Exercise 72.)

10.9 **Further Applications of Series**

TOPICS

1. Applications using series

2. Fourier series

3. Summary of common Taylor series

We conclude this chapter with a sampling of the many applications and variations of series. Some of the examples that follow are just brief glimpses into entire subfields of mathematics.

TOPIC 1 **Applications Using Series**

We have already determined (Section 10.8, Example 3) that the Taylor series expansion

$$e^x = \sum_{n=0}^{\infty} \frac{x^n}{n!} = 1 + x + \frac{x^2}{2!} + \frac{x^3}{3!} + \cdots$$

is valid for all $x \in \mathbb{R}$. One way to extend this exponential function to the complex numbers is to use its series expansion as the basis for the definition of $e^{i\theta}$, as follows.

$$e^{i\theta} = \sum_{n=0}^{\infty} \frac{(i\theta)^n}{n!} = 1 + (i\theta) + \frac{(i\theta)^2}{2!} + \frac{(i\theta)^3}{3!} + \cdots$$

$$= 1 + i\theta + \frac{i^2\theta^2}{2!} + \frac{i^3\theta^3}{3!} + \frac{i^4\theta^4}{4!} + \cdots$$

$$= 1 + i\theta - \frac{\theta^2}{2!} - i\frac{\theta^3}{3!} + \frac{\theta^4}{4!} + i\frac{\theta^5}{5!} - \cdots$$

$$= \left(1 - \frac{\theta^2}{2!} + \frac{\theta^4}{4!} - \cdots\right) + i\left(\theta - \frac{\theta^3}{3!} + \frac{\theta^5}{5!} - \cdots\right)$$

$$= \cos\theta + i\sin\theta$$

The equation $e^{i\theta} = \cos\theta + i\sin\theta$ is called **Euler's formula**, which allows us to relate the constants e, i, and π with the equation $e^{i\pi} = \cos\pi + i\sin\pi = -1$, or $e^{i\pi} + 1 = 0$. It also allows us to now define the value of e^z for any complex number $z = x + iy$, as follows.

$$e^z = e^{x+iy} = e^x e^{iy} = e^x\left(\cos y + i\sin y\right)$$

Example 1 ✐

Express $e^{i\pi/4}$ and $e^{1-3i\pi}$ in the form $x + iy$.

Solution

$$e^{i\pi/4} = \cos\frac{\pi}{4} + i\sin\frac{\pi}{4} = \frac{\sqrt{2}}{2} + i\frac{\sqrt{2}}{2}$$

$$e^{1-3i\pi} = e^1\left[\cos(-3\pi) + i\sin(-3\pi)\right] = -e$$

We have previously found and used linear approximations such as $(1+x)^{1/2} \approx 1 + (x/2)$, which is valid for $x \approx 0$. The Taylor series for $f(x) = (1+x)^m$ about $x = 0$, where m represents any fixed constant, is called the **binomial series** expansion of $(1+x)^m$; it is a generalization of formulas like $(1+x)^{1/2} \approx 1 + (x/2)$ with many uses, and we are now ready to derive it.

$$f^{(0)}(x) = (1+x)^m, \quad \text{so} \quad c_0 = \frac{f^{(0)}(0)}{0!} = 1$$

$$f'(x) = m(1+x)^{m-1}, \quad \text{so} \quad c_1 = \frac{f'(0)}{1!} = m$$

$$f''(x) = m(m-1)(1+x)^{m-2}, \quad \text{so} \quad c_2 = \frac{f''(0)}{2!} = \frac{m(m-1)}{2!}$$

$$f'''(x) = m(m-1)(m-2)(1+x)^{m-3}, \quad \text{so} \quad c_3 = \frac{f'''(0)}{3!} = \frac{m(m-1)(m-2)}{3!}$$

In general,

$$c_n = \frac{f^{(n)}(0)}{n!} = \frac{m(m-1)(m-2)\cdots(m-n+1)}{n!}.$$

We can now extend the meaning of the binomial coefficient by defining, for any real number m and nonnegative integer n,

$$\binom{m}{0} = 1, \quad \binom{m}{1} = m, \quad \binom{m}{2} = \frac{m(m-1)}{2!},$$

and

$$\binom{m}{n} = \frac{m(m-1)\cdots(m-n+1)}{n!} \quad \text{for } n \ge 3$$

(note that if m is a natural number, this new definition of $\binom{m}{n}$ reduces to $\frac{m!}{n!(m-n)!}$).

Given this extended meaning for the binomial coefficient, we can define the following type of series.

Definition 💡

Binomial Series

For any real number m and $-1 < x < 1$, the series

$$(1+x)^m = \sum_{n=0}^{\infty} \binom{m}{n} x^n$$

$$= 1 + mx + \frac{m(m-1)}{2!} x^2 + \frac{m(m-1)(m-2)}{3!} x^3 + \cdots$$

$$+ \frac{m(m-1)\cdots(m-n+1)}{n!} x^n + \cdots$$

is called a **binomial series**.

By the Ratio Test, it is easy to see that the binomial series converges for $|x| < 1$. This does not prove that it converges to $(1+x)^m$, but that is indeed the case—one method of verifying this is outlined in Exercise 26. The question of whether the series converges at the endpoints of the convergence interval depends on m: for example, if $-1 < m \le 0$, the series converges at $x = 1$; and if $m \ge 0$, it converges at both $x = -1$ and $x = 1$. Note also that the series is finite (and hence convergent for all x) if m is a natural number, since in this case all the binomial coefficients past some point are equal to 0. In other words, when m is a natural number, the binomial series expansion simply reduces to the Binomial Theorem. (The fact that binomial series generalize the Binomial Theorem was discovered by Newton around 1665.)

Example 2 ✐

Use the binomial series to find the series expansions of **a.** $(1-x)^{-1}$ and **b.** $(1+x^2)^{1/3}$ about $x = 0$.

Solution

a. This confirms the power series expansion of $1/(1-x)$ we have previously found.

$$(1-x)^{-1} = \left[1+(-x)\right]^{-1} = \sum_{n=0}^{\infty} \binom{-1}{n}(-x)^n$$

$$= \binom{-1}{0}(-x)^0 + \binom{-1}{1}(-x)^1 + \binom{-1}{2}(-x)^2 + \binom{-1}{3}(-x)^3 + \cdots$$

$$= 1 + (-1)(-x) + \frac{(-1)(-2)}{2!}(-x)^2 + \frac{(-1)(-2)(-3)}{3!}(-x)^3 + \cdots$$

$$= 1 + x + x^2 + x^3 + \cdots$$

$$= \sum_{n=0}^{\infty} x^n$$

We know this converges for $|x| < 1$, and by examining the series with the specific values $x = -1$ and $x = 1$ it's easy to see it diverges at both endpoints.

b. $(1+x^2)^{1/3} = \sum_{n=0}^{\infty} \binom{1/3}{n}(x^2)^n$

$$= \binom{1/3}{0}(x^2)^0 + \binom{1/3}{1}(x^2)^1 + \binom{1/3}{2}(x^2)^2 + \binom{1/3}{3}(x^2)^3 + \cdots$$

$$= 1 + \frac{1}{3}x^2 + \frac{\left(\frac{1}{3}\right)\left(-\frac{2}{3}\right)}{2!}x^4 + \frac{\left(\frac{1}{3}\right)\left(-\frac{2}{3}\right)\left(-\frac{5}{3}\right)}{3!}x^6 + \cdots$$

$$= 1 + \frac{1}{3}x^2 - \frac{1}{9}x^4 + \frac{5}{81}x^6 - \cdots$$

The series converges for $|x^2| < 1$, so the radius of convergence is again 1. In this case, however, the Alternating Series Test tells us that the series also converges at both $x = -1$ and $x = 1$.

Example 3 ✎

If a particle of mass m is moving with velocity v, the total relativistic energy of the particle is

$$E = \frac{mc^2}{\sqrt{1 - \dfrac{v^2}{c^2}}},$$

where c is the speed of light in a vacuum. We can use the binomial series to expand this as follows:

$$E = mc^2\left(1 - \frac{v^2}{c^2}\right)^{-1/2} = mc^2 \sum_{n=0}^{\infty} \binom{-1/2}{n}\left(-\frac{v^2}{c^2}\right)^n$$

$$= mc^2\left[1 + \left(-\frac{1}{2}\right)\left(-\frac{v^2}{c^2}\right) + \frac{\left(-\frac{1}{2}\right)\left(-\frac{3}{2}\right)}{2!}\left(-\frac{v^2}{c^2}\right)^2 + \frac{\left(-\frac{1}{2}\right)\left(-\frac{3}{2}\right)\left(-\frac{5}{2}\right)}{3!}\left(-\frac{v^2}{c^2}\right)^3 + \cdots\right]$$

$$= mc^2 + \underbrace{\left[\frac{1}{2}mv^2 + \frac{3}{8}mv^2\left(\frac{v^2}{c^2}\right) + \frac{5}{16}mv^2\left(\frac{v^2}{c^2}\right)^2 + \cdots\right]}_{E_k}$$

The term mc^2 is called the particle's *rest-mass energy*, and the remainder of the series is called the particle's *kinetic energy* E_k. When a particle is moving with a velocity v much smaller than c (the so-called classical case), v^2/c^2 and all its positive powers are very close to 0 and $E_k \approx mv^2/2$.

When a closed-form solution to a differential equation is difficult (or impossible) to find, a polynomial approximation may be attainable by beginning with the assumption that the solution has a power series representation.

Example 4 ✎

Find a power series solution of the equation $y'' - x^2 y = 0$.

Solution

We begin by assuming the solution y can be written in the form

$$y = c_0 + c_1 x + c_2 x^2 + c_3 x^3 + \cdots = \sum_{n=0}^{\infty} c_n x^n.$$

Then

$$y' = c_1 + 2c_2 x + 3c_3 x^2 + 4c_4 x^3 + \cdots$$

and

$$y'' = 2c_2 + 2 \cdot 3c_3 x + 3 \cdot 4c_4 x^2 + \cdots = \sum_{n=2}^{\infty} (n-1)n c_n x^{n-2}.$$

So

$$y'' - x^2 y = 2c_2 + 6c_3 x + \left(-c_0 + 12c_4\right)x^2 + \left(-c_1 + 20c_5\right)x^3 + \left(-c_2 + 30c_6\right)x^4$$
$$+ \cdots + \left[-c_{n-2} + (n+1)(n+2)c_{n+2}\right]x^n + \cdots.$$

To satisfy the equation $y'' - x^2 y = 0$, it must be the case that each coefficient of x in the expression above is 0, so

$$c_2 = 0, \quad c_3 = 0, \quad \text{and} \quad -c_{n-2} + (n+1)(n+2)c_{n+2} = 0 \quad \text{for } n \geq 2.$$

If the differential equation is associated with other information, such as initial conditions like $y(0) = a$ and $y'(0) = b$, then we would know that $a = y(0) = c_0$ and $b = y'(0) = c_1$ and it would make sense to express all the coefficients in terms of a and b.

$$c_0 = a, \quad c_1 = b, \quad c_2 = 0, \quad c_3 = 0, \quad \text{and} \quad c_{n+2} = \frac{c_{n-2}}{(n+1)(n+2)} \quad \text{for } n \geq 2$$

This would allow us to express the solution y as a sum of two series.

$$y = a + bx + \frac{a}{3 \cdot 4}x^4 + \frac{b}{4 \cdot 5}x^5 + \frac{a}{3 \cdot 4 \cdot 7 \cdot 8}x^8 + \frac{b}{4 \cdot 5 \cdot 8 \cdot 9}x^9$$

$$+ \frac{a}{3 \cdot 4 \cdot 7 \cdot 8 \cdot 11 \cdot 12}x^{12} + \frac{b}{4 \cdot 5 \cdot 8 \cdot 9 \cdot 12 \cdot 13}x^{13} + \cdots$$

$$= a\left(1 + \frac{x^4}{3 \cdot 4} + \frac{x^8}{3 \cdot 4 \cdot 7 \cdot 8} + \frac{x^{12}}{3 \cdot 4 \cdot 7 \cdot 8 \cdot 11 \cdot 12} + \cdots\right)$$

$$+ b\left(x + \frac{x^5}{4 \cdot 5} + \frac{x^9}{4 \cdot 5 \cdot 8 \cdot 9} + \frac{x^{13}}{4 \cdot 5 \cdot 8 \cdot 9 \cdot 12 \cdot 13} + \cdots\right)$$

Both of the series converge absolutely for all x, as can be verified by the Ratio Test. In many applications, a polynomial approximation of the solution would be sufficient, so the above two series would be truncated at the desired degree of accuracy.

TOPIC 2   Fourier Series

As we have seen, a Taylor series expansion of a function represents a limit of polynomials of increasing degree. As such, the polynomials are better able to approximate the original function as the order of the Taylor polynomials increases. But there is no escaping the fact that $\lim_{x \to \pm\infty} p_n(x) = \pm\infty$ for any fixed Taylor polynomial p_n. That is, polynomials grow unbounded in magnitude as x gets further and further from 0, which is not the kind of behavior exhibited by a periodic function. As a consequence, there is no hope that any fixed Taylor polynomial generated by a periodic function could approximate that function well outside of a bounded interval.

On the other hand, sine and cosine are familiar functions that are easy to manipulate and that exhibit exactly the kind of periodicity that might be needed. For this reason, the French mathematician Jean-Baptiste Joseph Fourier (1768–1830) used them as the basis for solving differential equations describing the flow of heat in physical objects. The decomposition of functions into sums (possibly infinite sums) of sines and cosines has come to be a part of what is now called *Fourier analysis*, a branch of mathematics that deals with problems in electrical engineering, acoustics, vibration analysis, optics, signal and image processing, and quantum mechanics, in addition to the original heat flow problems. We introduce the subject here with an outline of the theory of Fourier series.

Suppose f is a periodic function and we hope to express it as a sum of sines and cosines of differing frequencies over the interval $[-\pi, \pi]$. Specifically, we want to find coefficients $\{a_k\}$ and $\{b_k\}$ such that

$$f(x) = a_0 + (a_1 \cos x + b_1 \sin x) + (a_2 \cos 2x + b_2 \sin 2x) + (a_3 \cos 3x + b_3 \sin 3x) + \cdots$$
$$= a_0 + \sum_{k=1}^{\infty} a_k \cos kx + \sum_{k=1}^{\infty} b_k \sin kx.$$

We do so by multiplying the above equation through by a particular $\cos px$ or $\sin px$, and integrating the resulting expressions from $-\pi$ to π, making use of the following facts.

$$\int_{-\pi}^{\pi} dx = 2\pi$$

$$\int_{-\pi}^{\pi} \cos px \cos qx \, dx = \int_{-\pi}^{\pi} \sin px \sin qx \, dx = \begin{cases} \pi & \text{if } p = q \neq 0 \\ 0 & \text{if } p \neq q \end{cases} \quad (p, q \in \mathbb{Z})$$

$$\int_{-\pi}^{\pi} \cos px \sin qx \, dx = 0 \quad (p, q \in \mathbb{Z})$$

For instance, multiplying through by $\cos(0 \cdot x)$, which is the constant function 1, and integrating results in

$$\int_{-\pi}^{\pi} f(x) \, dx = \int_{-\pi}^{\pi} a_0 \, dx = 2\pi a_0,$$

since all the other integrals on the right hand side are equal to 0. After multiplying through by $\cos x$ and $\sin x$, respectively, and integrating we obtain

$$\int_{-\pi}^{\pi} f(x) \cos x \, dx = \pi a_1 \quad \text{and} \quad \int_{-\pi}^{\pi} f(x) \sin x \, dx = \pi b_1.$$

In general, if k is a positive integer, then

$$\int_{-\pi}^{\pi} f(x) \cos kx \, dx = \pi a_k \quad \text{and} \quad \int_{-\pi}^{\pi} f(x) \sin kx \, dx = \pi b_k,$$

which, when we solve for the coefficients, gives us the following definition.

Definition 💡

Fourier Coefficients and Fourier Series

Given a function f that is integrable on the interval $[-\pi, \pi]$, we define its **Fourier coefficients** with the formulas

$$a_0 = \frac{1}{2\pi} \int_{-\pi}^{\pi} f(x)\, dx$$

$$a_k = \frac{1}{\pi} \int_{-\pi}^{\pi} f(x) \cos kx\, dx \quad \text{for } k \geq 1$$

$$b_k = \frac{1}{\pi} \int_{-\pi}^{\pi} f(x) \sin kx\, dx \quad \text{for } k \geq 1$$

and the corresponding **Fourier series** for f by

$$f(x) = a_0 + (a_1 \cos x + b_1 \sin x) + (a_2 \cos 2x + b_2 \sin 2x) + (a_3 \cos 3x + b_3 \sin 3x) + \cdots$$

$$= a_0 + \sum_{k=1}^{\infty} a_k \cos kx + \sum_{k=1}^{\infty} b_k \sin kx.$$

We will not explore in this text the question of when the Fourier series for a given function converges to the original function. This question actually leads into a rich area of mathematical research known as *harmonic analysis*; for our purposes, it will suffice to simply determine the Fourier coefficients for various common functions.

Example 5 ✎

In electrical engineering applications, the *sawtooth function* defined by

$$f(x) = x \text{ for } -\pi < x \leq \pi \quad \text{and} \quad f(x + 2k\pi) = f(x) \text{ for } k = \pm 1, \pm 2, \ldots$$

arises frequently; its graph appears in Figure 1.

This function poses both mathematical and practical difficulties because of its points of discontinuity. Fortunately, it can be approximated as accurately as desired with the first few terms of its Fourier series, and any such approximation has derivatives of all orders. In Exercise 43 you will be asked to verify the coefficients:

$$a_0 = \frac{1}{2\pi} \int_{-\pi}^{\pi} x\, dx = 0$$

$$a_k = \frac{1}{\pi} \int_{-\pi}^{\pi} x \cos kx\, dx = 0 \quad (k \geq 1)$$

$$b_k = \frac{1}{\pi} \int_{-\pi}^{\pi} x \sin kx\, dx = -\frac{2}{k} \cos k\pi + \frac{2}{k^2 \pi} \sin k\pi = (-1)^{k+1}\left(\frac{2}{k}\right) \quad (k \geq 1)$$

If we let $g(x)$ be the truncated Fourier series $\sum_{k=1}^{5} b_k \sin kx,$ then

$$g(x) = 2\sin x - \sin 2x + \frac{2}{3}\sin 3x - \frac{1}{2}\sin 4x + \frac{2}{5}\sin 5x.$$

A graph of g appears with the graph of f in Figure 2.

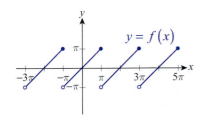

Figure 1 Sawtooth Function

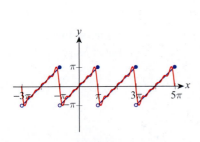

Figure 2
Sawtooth Function f and
Fourier Approximation g

TOPIC 3   Summary of Common Taylor Series

Series expansions of some familiar functions are used so frequently that it's useful to compile a list of them for reference. We have already determined many of the expansions that appear in the following table—the remainder are left as exercises.

Function	Taylor Series	Convergence Condition
$\dfrac{1}{1-x}$	$\displaystyle\sum_{n=0}^{\infty} x^n = 1 + x + x^2 + x^3 + \cdots$	$\lvert x \rvert < 1$
e^x	$\displaystyle\sum_{n=0}^{\infty} \dfrac{x^n}{n!} = 1 + x + \dfrac{x^2}{2!} + \dfrac{x^3}{3!} + \cdots$	$\lvert x \rvert < \infty$
$\sin x$	$\displaystyle\sum_{n=0}^{\infty} \dfrac{(-1)^n x^{2n+1}}{(2n+1)!} = x - \dfrac{x^3}{3!} + \dfrac{x^5}{5!} - \cdots$	$\lvert x \rvert < \infty$
$\cos x$	$\displaystyle\sum_{n=0}^{\infty} \dfrac{(-1)^n x^{2n}}{(2n)!} = 1 - \dfrac{x^2}{2!} + \dfrac{x^4}{4!} - \cdots$	$\lvert x \rvert < \infty$
$\ln x$	$\displaystyle\sum_{n=1}^{\infty} \dfrac{(-1)^{n+1}(x-1)^n}{n} = (x-1) - \dfrac{(x-1)^2}{2} + \dfrac{(x-1)^3}{3} - \cdots$	$0 < x \le 2$
$\ln(1+x)$	$\displaystyle\sum_{n=1}^{\infty} \dfrac{(-1)^{n+1} x^n}{n} = x - \dfrac{x^2}{2} + \dfrac{x^3}{3} - \cdots$	$-1 < x \le 1$
$\ln\dfrac{1+x}{1-x}$	$\displaystyle 2\sum_{n=0}^{\infty} \dfrac{x^{2n+1}}{2n+1} = 2\left(x + \dfrac{x^3}{3} + \dfrac{x^5}{5} + \cdots \right)$	$\lvert x \rvert < 1$
$\tan^{-1} x$	$\displaystyle\sum_{n=0}^{\infty} \dfrac{(-1)^n x^{2n+1}}{2n+1} = x - \dfrac{x^3}{3} + \dfrac{x^5}{5} - \cdots$	$\lvert x \rvert \le 1$
$(1+x)^m$	$\displaystyle\sum_{n=0}^{\infty} \binom{m}{n} x^n = 1 + mx + \dfrac{m(m-1)}{2!} x^2 + \dfrac{m(m-1)(m-2)}{3!} x^3 + \cdots$	Varies with m, but at least $\lvert x \rvert < 1$

Table 1

10.9 **Exercises**

1–5 Use Euler's formula to express the complex number in the form $x + iy$.

1. $e^{i\pi/2}$

2. $e^{-3i\pi/4}$

3. $5e^{2+i\pi/3}$

4. e^{e^i}

5. $\dfrac{e^{2i} - e^{-2i}}{3i}$

6–10 Use Euler's formula to express the complex number in the form $e^x e^{iy} = re^{iy}$.

6. $\dfrac{1-i}{\sqrt{2}}$

7. $-2\left(1 + \sqrt{3}i\right)$

8. $\dfrac{3+3i}{-\sqrt{3}-i}$

9. $(1+i)^4$

10. $\dfrac{3i}{2e^{4+i}}$

11. Use Euler's formula to verify the following formulas:

a. $\sin\theta = \dfrac{e^{i\theta} - e^{-i\theta}}{2i}$

b. $\cos\theta = \dfrac{e^{i\theta} + e^{-i\theta}}{2}$

12. Use the Maclaurin series to verify the following formulas:

a. $\sinh i\theta = i\sin\theta$

b. $\cosh i\theta = \cos\theta$

13–21 Find the first six nonzero terms of the binomial series expansion of the indicated function.

13. $f(x) = \left(1 + x^2\right)^5$

14. $f(x) = \dfrac{1}{\sqrt{1+x}}$

15. $f(x) = \sqrt{1-x}$

16. $f(x) = (1-x)^{3/2}$

17. $f(x) = \sqrt[3]{1+x^2}$

18. $f(x) = \left(1 + \dfrac{x}{2}\right)^{-3}$

19. $f(x) = \left(1 - 2x^2\right)^{-2/3}$

20. $f(x) = \left(1 - \dfrac{x}{3}\right)^{-3/2}$

21. $f(x) = \dfrac{5}{\sqrt[4]{1+2x}}$

22. Show that if $m \in \mathbb{N}$, the Maclaurin series of $(1+x)^m$ is finite, and use this observation to provide a "calculus-based" proof for the Binomial Theorem, which says $(A+B)^n = \displaystyle\sum_{k=0}^{n} \binom{n}{k} A^{n-k} B^k$, for any positive integer n and any two expressions A and B.

23. Use the binomial series of $f(x) = 1/\sqrt{1-x^2}$ to find the series expansion of $g(x) = \sin^{-1} x$ about 0. What is the radius of convergence? (**Hint:** Use the equality $\dbinom{-1/2}{n} = (-1)^n \dbinom{2n}{n} \Big/ 4^n$. Compare your result to that of Exercise 68 of Section 10.8.)

24–25 Use the method of Exercise 23 to find the Maclaurin series expansion of the indicated function along with the radius of convergence.

24. $h(x) = \cos^{-1} x$

25. $h(x) = \sinh^{-1} x$

26. By following the outline below, prove that the binomial series

$$\sum_{n=0}^{\infty} \binom{m}{n} x^n = \sum_{n=0}^{\infty} \frac{m(m-1)(m-2)\cdots(m-n+1)}{n!} x^n$$

converges to the function $(1+x)^m$ on $(-1,1)$.

a. Using the Ratio Test, it is straightforward to check that the above series converges for $|x| < 1$. Let us denote its sum by $f(x)$ and differentiate the series term by term to find the series expansion of $f'(x)$.

b. Multiply the series expansion of $f'(x)$ by $(1+x)$ and verify that the result is the series for $m \cdot f(x)$. (**Note:** Since $|x| < 1$, the series is absolutely convergent, so termwise differentiation and rearranging terms is possible without changing the sum.)

c. Conclude that $f(x)$ is a solution of the separable differential equation

$$\frac{y'}{y} = \frac{m}{1+x}.$$

d. Solve the above equation by conventional means to conclude that $f(x) = C(1+x)^m$ for some constant C.

e. Finally, use the initial condition $f(0) = 1$ to conclude that $C = 1$.

27–28 The period of a swinging pendulum of length L released from rest can be well approximated by simple harmonic motion and its period is approximately $T \approx 2\pi\sqrt{L/g}$, where g is the acceleration caused by gravity. However, this model is inaccurate for larger starting angles. In physics, it is shown that if the pendulum is released at angle θ_0 from vertical, the actual formula is

$$T = 4\sqrt{\frac{L}{g}}K(k),$$

where $k = \sin(\theta_0/2)$ and

$$K(k) = \int_0^{\pi/2} \frac{d\theta}{\sqrt{1-k^2\sin^2\theta}}$$

is a *complete elliptic integral of the first kind*.

In Exercises 27 and 28, use a Taylor series approximation to derive an estimate for the period of the swinging pendulum (keep in mind that we don't have a closed formula for this integral).

27.* Prove that if $|k| < 1$, the Maclaurin series expansion for $K(k)$ is

$$K(k) = \frac{\pi}{2} + \frac{\pi}{2}\sum_{n=1}^{\infty}\left[\frac{(1)(3)\cdots(2n-1)}{(2)(4)\cdots(2n)}\right]^2 k^{2n}.$$

(**Hint:** Substitute $k\sin\theta$ in the Maclaurin expansion of $1/\sqrt{1-x^2}$.)

28. Use the series from Exercise 27 to establish the following approximation for the period of the swinging pendulum that is much more accurate for larger angles than the simple model we have seen previously:

$$T \approx 2\pi\sqrt{\frac{L}{g}}\left(1 + \frac{\theta_0^2}{16}\right)$$

(**Hint:** In addition to the Maclaurin series, also use the approximation $\sin x \approx x$.)

29–30 Perhaps surprisingly, an explicit formula for the circumference of an ellipse is not known, for the problem leads to an integral that is (perhaps unsurprisingly) called the *complete elliptic integral of the second kind*. (By now you are probably expecting the fact that it cannot be evaluated in closed form, which is indeed the case.)

Use the Taylor series of this integral to approximate the circumference of an ellipse.

29.* The complete elliptic integral of the second kind is defined as

$$E(k) = \int_0^{\pi/2}\sqrt{1-k^2\sin^2\theta}\,d\theta.$$

Prove that if $|k| < 1$, the Maclaurin series expansion for $E(k)$ is

$$E(k) = \frac{\pi}{2} - \frac{\pi}{2}\sum_{n=1}^{\infty}\left[\frac{(1)(3)\cdots(2n-1)}{(2)(4)\cdots(2n)}\right]^2\frac{k^{2n}}{2n-1}.$$

(See the hint given in Exercise 27.)

30. Given that the circumference of the ellipse $\frac{x^2}{a^2} + \frac{y^2}{b^2} = 1$, $a < b$, is $C = 4bE(k)$ where $k = \sqrt{1-(a^2/b^2)}$, use the first four terms of the series expansion from Exercise 29 to approximate the circumference of the ellipse $\frac{x^2}{4} + \frac{y^2}{9} = 1$.

31. The arc of a great circle of the Earth connecting Shreveport, Louisiana, with Gulf Shores, Alabama, is approximately 430 miles long. Use Taylor series to estimate how much the arc will recede from its chord between these two cities. Use the first three nonzero terms of the series of $\cos\theta$. Approximate the radius of the Earth with $R \approx 4000$ miles.

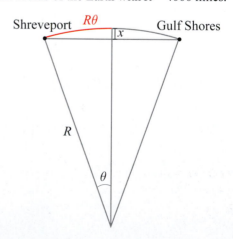

32. Use the Maclaurin series expansion of e^x to prove that e is irrational. (**Hint:** Expressing e as a power series, you can argue by contradiction as follows. Assuming p and q are positive integers with $e = p/q$ and multiplying the power series by $q!$, we obtain

$$p(q-1)! - 2q! - (3 \cdot 4 \cdots \cdot q) - (4 \cdot 5 \cdots \cdot q) - \cdots - q - 1$$
$$= \frac{1}{q+1} + \frac{1}{(q+1)(q+2)} + \cdots$$

where the assumptions imply that the left-hand side must be an integer. Finally, argue that the expression on the right-hand side must be between 0 and 1, an apparent contradiction, finishing your proof.)

33. Recall from Section 8.2 that the differential equation obeyed by the current in an RL circuit is

$$L \frac{dI}{dt} + RI = V.$$

Solve this linear equation for $I(t)$ and use the series expansion of your solution to show that when t is small,

$$I(t) \approx \frac{V}{L} t.$$

34. The weight of an object at a height h above the surface of the Earth is

$$W(h) = \frac{R^2 W_0}{(R+h)^2},$$

where W_0 is the object's weight on the Earth's surface. Use the first four nonzero terms of the Maclaurin expansion of $W(h)$ to approximate the height required for an object to lose 5% of its weight. (Use $R \approx 4000$ miles.)

35. According to Einstein's theory of special relativity, if a particle is moving with velocity v, the mass of the particle is given by

$$m(v) = \frac{m_r}{\sqrt{1 - \dfrac{v^2}{c^2}}},$$

where m_r is the rest mass of the particle and c is the speed of light in a vacuum. Use the first three nonzero terms of the Maclaurin polynomial of $m(v)$ to estimate the speed necessary to increase the mass of the particle by 1%. (Use $c \approx 3 \times 10^8$ m/s.)

36.* The fastest average qualifying speed at the Indy 500 race, 236.986 mph (approximately 106 m/s), was reached by Arie Luyendyk in 1996. Using the error term for the series of E_k found in Example 3, estimate the magnitude of the maximum error one makes when calculating the race car's kinetic energy at this speed using the classic Newtonian formula of $E_k \approx mv^2/2$. (**Hint:** The series of E_k can be rewritten as

$$E_k = mc^2 \left[\frac{1}{2} \left(\frac{v^2}{c^2} \right) + \frac{3}{8} \left(\frac{v^2}{c^2} \right)^2 + \frac{5}{16} \left(\frac{v^2}{c^2} \right)^3 + \cdots \right].$$

Substituting $x = -v^2/c^2$, notice that the series in the brackets on the right-hand side is "almost" the binomial expansion of $1/\sqrt{1+x}$. Next, use Taylor's Theorem with $r_1(x)$, the remainder of order 1, note that $|v| \leq 106$, and give an upper bound for $|r_1(v/c)|$.)

37. The phase speed at which a surface wave propagates on water of depth d is well approximated by the expression

$$s = \sqrt{\frac{\lambda g}{2\pi} \tanh \frac{2\pi d}{\lambda}},$$

where λ is the wavelength.

a. Explain why the following "rule of thumb" is valid: If the water is deeper than three times the wavelength, then $s \approx \sqrt{\lambda g/(2\pi)}$. (Note that this formula shows that the speed of propagation depends only on wavelength in deep water; in case of large wavelengths such as tidal waves, this speed can be enormous. In March 2011, just before the catastrophic Japan tsunami, s was about the same speed as that of a passenger jet!)

b. Use Taylor series to show that in shallow water, $s \approx \sqrt{gd}$. (In contrast to the previous case, the speed of propagation in shallow water depends only on water depth, rather than wavelength.)

38–41 Find a power series solution of the equation. In each case, use the initial conditions $y(0) = a$ and $y'(0) = b$.

38. $y'' - 2xy = 0$

39. $y'' + 9y = 0$

40. $y'' = \dfrac{y'}{1-x}$

41. $(1 - x^2)y'' = 4xy' + 2y$

42. Explain why in our discussion preceding Example 5, we did not attempt to find a Fourier coefficient b_k for the index $k = 0$.

43. Verify the Fourier coefficients of Example 5.

a. $a_0 = \dfrac{1}{2\pi} \displaystyle\int_{-\pi}^{\pi} x \, dx = 0$

b. $a_k = \dfrac{1}{\pi} \displaystyle\int_{-\pi}^{\pi} x \cos kx \, dx = 0$ $(k \ge 1)$

c. $b_k = \dfrac{1}{\pi} \displaystyle\int_{-\pi}^{\pi} x \sin kx \, dx$

$\quad = -\dfrac{2}{k} \cos k\pi + \dfrac{2}{k^2 \pi} \sin k\pi$

$\quad = (-1)^{k+1} \left(\dfrac{2}{k} \right)$ $(k \ge 1)$

44. Consider the function given by its graph below. After extending it to $\mathbb{R}$ in a 2π-periodic manner, find its Fourier coefficients.

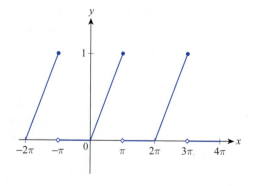

45. Repeat Exercise 44 for the function graphed below.

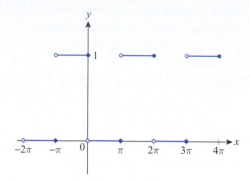

46. Let $\hat{f}(x) = 3x^2$ be defined on the interval $[-\pi, \pi]$, and let $f(x)$ be the function obtained by extending $\hat{f}(x)$ to the entire real line in a 2π-periodic fashion. Find the Fourier series expansion of $f(x)$.

10.9 **Technology Exercises**

47–50 Use a computer algebra system to graph the function along with its Maclaurin polynomial of order 5 on the same screen (use the polynomial of order 9 in Exercise 50). What is the largest interval on which the approximation is acceptable? (Answers will vary.)

47. $f(x) = \sqrt[5]{(2x-1)^7}$

48. $f(x) = \dfrac{x^2 - 1}{\sqrt[3]{(x+5)^2}}$

49. $f(x) = \dfrac{x+1}{\sqrt{1+x^2}}$

50. $f(x) = (x^3 - x)\sqrt[4]{x^4 + 1}$

Chapter 10
Review Exercises

1–2 Starting with $n = 0$, list the first five terms of the given sequence.

1. $a_n = \left(\dfrac{1-n}{n^2+1}\right)^n$

2. $a_n = \dfrac{2^{n+1}-n}{n!}$

3–4 Find the first five terms of the given recursively defined sequence.

3. $a_1 = 0, \quad a_n = 3a_{n-1} + 2$

4. $a_1 = 0, \quad a_2 = 1, \quad a_n = 2a_{n-2} - a_{n-1}$

5–6 Use the definition of the limit of a sequence to prove the limit statement.

5. $\displaystyle\lim_{n\to\infty} \dfrac{2}{n^3+1} = 0$

6. $\displaystyle\lim_{n\to\infty} \dfrac{2n+1}{n} = 2$

7–10 Find the limit of the sequence if it converges or prove that the sequence diverges.

7. $a_n = \dfrac{n^2-2}{3n+1}$

8. $a_n = \sqrt{n^2+1} - n$

9. $a_n = \left(1+\dfrac{2}{n}\right)^{3n}$

10. $a_n = 2\tan^{-1}\left(n^2\right)$

11–12 Use the Squeeze Theorem to prove that the given sequence converges.

11. $\left\{\dfrac{(-1)^n \cos n}{n^2}\right\}_{n=1}^{\infty}$

12. $\left\{\dfrac{\sin 2n}{2^n}\right\}_{n=1}^{\infty}$

13–14 Recognize the repeating decimal as a geometric series and write the decimal as a ratio of two integers.

13. $0.\overline{219}$

14. $0.2\overline{19}$

15. Use the Bounded Monotonic Sequence Theorem to prove that the recursive sequence $a_1 = \sqrt{3}, a_{n+1} = \sqrt{3a_n}$ converges, and find its limit.

16. Suppose that $\lim_{n\to\infty} a_n = -\infty$ and $\{b_n\}$ is a sequence such that there is an N with $b_n \leq a_n$ for all $n \geq N$. Prove that $\lim_{n\to\infty} b_n = -\infty$.

17. Prove that if $\lim_{n\to\infty} a_n = \infty$, then the sequence has a smallest term.

18. Let $\{a_n\}$ be a positive null sequence. Prove that $\lim_{n\to\infty}(1/a_n) = \infty$. (See Exercises 92 and 95 in Section 10.1 for the definitions of a null sequence and a positive sequence, respectively.)

19. Suppose that the dosage of a certain medication is d milligrams once a day, to be taken at the same time of day.

   a. Assuming that after 24 hours, half of the initial amount is still present in the bloodstream, find a recursive formula for the sequence $\{a_n\}$, the amount of medication present in the bloodstream right after the nth dose is taken.

   b. Find $\lim_{n\to\infty} d_n$. (This is how much medication is in the patient's bloodstream, if it is taken over a long time period.)

20–21 Use partial sums to prove that the given series is convergent.

20. $\displaystyle\sum_{n=1}^{\infty} \dfrac{1}{4^n}$

21. $\displaystyle\sum_{n=2}^{\infty} \dfrac{1}{n^2-n}$

22–25 Decide whether the series converges. If so, find its sum.

22. $\displaystyle\sum_{n=0}^{\infty} \left(-\dfrac{3}{2}\right)^{-n}$

23. $\displaystyle\sum_{n=1}^{\infty} \dfrac{3}{n(n+1)}$

24. $\displaystyle\sum_{n=1}^{\infty} \dfrac{2+2n}{n^2+2n}$

25. $\displaystyle\sum_{n=2}^{\infty} \dfrac{3^n-1}{5^{n-2}}$

26. Referring back to Exercise 66 of Section 10.2, supposing that the positive series $\sum a_n$ is divergent, what can you say about $\sum(1/a_n)$? (See Exercise 70 in Section 10.2 for the definition of positive series.)

27–32 Use the Integral Test to determine whether the series converges or diverges.

27. $\displaystyle\sum_{n=0}^{\infty} \frac{n}{n^2+4}$

28. $\displaystyle\sum_{n=0}^{\infty} \frac{4}{n^2+4}$

29. $\displaystyle\sum_{n=0}^{\infty} \frac{2n}{2^n}$

30. $\displaystyle\sum_{n=2}^{\infty} \frac{1}{n\sqrt{\ln n}}$

31. $\displaystyle\sum_{n=1}^{\infty} \frac{2+e^{1/n}}{n^3}$

32. $\displaystyle\sum_{n=1}^{\infty} \frac{\ln(\arctan n)}{n^2+1}$

33–34 Find the smallest possible n to approximate the sum of the series within the indicated error ε and provide the requested estimate.

33. $\displaystyle\sum_{n=1}^{\infty} \frac{1}{n^{3/2}}; \quad \varepsilon = 0.04$

34. $\displaystyle\sum_{n=1}^{\infty} ne^{-n}; \quad \varepsilon = 0.03$

35–38 Use the Direct Comparison Test to determine whether the series converges or diverges.

35. $\displaystyle\sum_{n=2}^{\infty} \frac{1}{n\ln n - 1}$

36. $\displaystyle\sum_{n=2}^{\infty} \frac{n^2+1}{n^3-1}$

37. $\displaystyle\sum_{n=0}^{\infty} \frac{2}{2^n+\sqrt{n}}$

38. $\displaystyle\sum_{n=0}^{\infty} \frac{\sin^4 n}{n^2+1}$

39–42 Use the Limit Comparison Test to determine whether the series converges or diverges.

39. $\displaystyle\sum_{n=0}^{\infty} \frac{n+1}{\sqrt{n^5+1}}$

40. $\displaystyle\sum_{n=1}^{\infty} \frac{\sqrt{1+\ln n}}{n^2}$

41. $\displaystyle\sum_{n=1}^{\infty} \frac{\sqrt{n^3+n}}{n^{5/2}}$

42. $\displaystyle\sum_{n=1}^{\infty} \frac{n}{\sqrt{n^4+1}}$

43–50 Use the Ratio or Root Test, as appropriate, to determine whether the series converges or diverges.

43. $\displaystyle\sum_{n=1}^{\infty} \frac{n^7}{7^n}$

44. $\displaystyle\sum_{n=5}^{\infty} \frac{5^n}{(n-5)!}$

45. $\displaystyle\sum_{n=1}^{\infty} \frac{2^n}{n\cdot n!}$

46. $\displaystyle\sum_{n=1}^{\infty} \frac{2^n(n!)}{(2n-1)!}$

47. $\displaystyle\sum_{n=0}^{\infty} \frac{3^n}{4^n+1}$

48. $\displaystyle\sum_{n=1}^{\infty} \frac{n^n}{2^{n+1}}$

49. $\displaystyle\sum_{n=1}^{\infty} \frac{(\ln n)^{2n}}{n^{n+1}}$

50. $\displaystyle\sum_{n=1}^{\infty} \left(\frac{1}{2}+\frac{1}{2n}\right)^n$

51–54 Determine whether the alternating series converges and give a reason for your answer.

51. $\displaystyle\sum_{n=1}^{\infty} (-1)^{n+1} \frac{\ln n}{n^2}$

52. $\displaystyle\sum_{n=1}^{\infty} \frac{(-1)^n}{e^n\sqrt{n}}$

53. $\displaystyle\sum_{n=1}^{\infty} (-1)^{n+1} \frac{n-1}{5n+1}$

54. $\displaystyle\sum_{n=1}^{\infty} (-1)^n \frac{2^n}{n^5}$

55–56 Approximate the sum of the alternating series, accurate to at least the indicated number of decimal places. How many terms did you use? According to the nth-remainder estimate for the error (see Section 10.6), how many terms guarantee the indicated accuracy? (Note that the nth-remainder estimate is not necessarily "sharp".)

55. $\displaystyle\sum_{n=1}^{\infty} \frac{(-1)^{n+1}}{n^4}; \quad$ accurate to 2 decimal places

56. $\displaystyle\sum_{n=1}^{\infty} \frac{(-1)^{n+1}}{n2^n}; \quad$ accurate to 3 decimal places

57–62 Determine whether the given series converges absolutely, converges conditionally, or diverges.

57. $\displaystyle\sum_{n=1}^{\infty} \frac{(-1)^{n+1}}{n^2+1}$

58. $\displaystyle\sum_{n=1}^{\infty} \frac{(-1)^{n+1} n}{n^2+1}$

59. $\displaystyle\sum_{n=1}^{\infty} \frac{\sin n}{n^{3/2}}$

60. $\displaystyle\sum_{n=1}^{\infty} \frac{(-2)^n}{n^6}$

61. $\displaystyle\sum_{n=3}^{\infty} \left(\frac{2-n}{4n}\right)^n$

62. $\displaystyle\sum_{n=1}^{\infty} \frac{(-2)^{n+1} n}{4n-2}$

63–68 Determine the interval of convergence for the given power series.

63. $\displaystyle\sum_{n=0}^{\infty} 2^n x^n$

64. $\displaystyle\sum_{n=0}^{\infty} \frac{(x-2)^n}{n^3+2}$

65. $\displaystyle\sum_{n=0}^{\infty} \frac{(3x)^n}{n!}$

66. $\displaystyle\sum_{n=1}^{\infty} \frac{n!(x-1)^n}{n^4}$

67. $\displaystyle\sum_{n=0}^{\infty} \frac{(x+1)^n}{(n+1)3^n}$

68. $\displaystyle\sum_{n=2}^{\infty} \frac{(2x-6)^n}{\sqrt{n}\ln n}$

69–70 Determine the interval of convergence for the series, and the limiting function of the series on that interval.

69. $\displaystyle\sum_{n=0}^{\infty} (2x-3)^n$

70. $\displaystyle\sum_{n=0}^{\infty} \frac{(3x-1)^n}{2^n}$

71–72 Find the power series expansion for the given function about 0. What is the radius of convergence? (**Hint:** Use the same approach as in Exercises 41–48 of Section 10.7.)

71. $f(x) = \dfrac{1}{1+3x}$ **72.** $f(x) = \dfrac{1}{2-4x^2}$

73. Find the power series representation of $f(x) = 1/x$, centered at $a = 2$. Differentiating twice, find the series for $g(x) = 1/x^3$ around $a = 2$. What is the radius of convergence?

(**Hint:** Start by rewriting $1/x$ as $\dfrac{1}{2-(2-x)}$.)

74–81 Determine the Taylor series (or Maclaurin series if the center is not specified) of the given function about the indicated point. Find the radius of convergence. (**Hint:** Do not use the definition.)

74. $f(x) = e^{x^4/2}$

75. $f(x) = \cos(2x^3)$

76. $f(x) = \dfrac{1}{x+3}; \quad a = 2$

77. $f(x) = \ln x; \quad a = 3$

78. $f(x) = \sin \dfrac{x}{2}$

79. $f(x) = \sin^2 x$
(**Hint:** Use a trigonometric indentity.)

80. $f(x) = x^2 e^{3x^2}$

81. $f(x) = \dfrac{x}{x+2}; \quad a = 1$

82–83 Use the definition to find the first four nonzero terms of the Taylor series generated by the given function about the indicated point.

82. $f(x) = \cot x; \quad a = \pi/2$

83. $f(x) = \csc x; \quad a = \pi/4$

84–86 Find the first five nonzero terms of the Maclaurin series generated by the given function by using operations on familiar series (try not to use the definition).

84. $f(x) = xe^x - \sin x$ **85.** $f(x) = \dfrac{\cos x}{x+1}$

86. $f(x) = \cos(\sin x)$ (**Hint:** Substitute the series of $\sin x$ into that of $\cos x$.)

87–89 Use Taylor series to approximate the given function value or definite integral to within the indicated accuracy. How many (nonzero) terms did you need?

87. $\cos 1; \quad \text{error} \le 10^{-6}$

88. $\displaystyle\int_0^1 e^{-x^3}\, dx; \quad \text{error} \le 10^{-4}$

89. $\sqrt[4]{1.2}; \quad \text{error} \le 10^{-5}$ (**Hint:** Use a binomial series.)

90. Use series expansion to verify the trigonometric limit $\displaystyle\lim_{x\to 0} \dfrac{1-\cos x}{x} = 0$.

91–92 Use Taylor series to find the indicated limit.

91. $\displaystyle\lim_{x\to 0} \dfrac{e^{-x}-1}{x}$ **92.** $\displaystyle\lim_{x\to 0} \dfrac{x\sin 2x}{1-e^{x^2}}$

93. Using an appropriate power series, find the sum of the series $\displaystyle\sum_{n=1}^{\infty}(n/2^n)$. (**Hint:** Start with the familiar series expansion of $1/(1-x)$, and use termwise differentiation in its interval of convergence; then make an appropriate substitution.)

94. Using an appropriate Taylor series, verify that
$$\sum_{n=1}^{\infty} \dfrac{(-1)^{n+1}}{n} = \ln 2.$$

95–98 Find the first six nonzero terms of the binomial series expansion of the indicated function.

95. $f(x) = \dfrac{1}{\sqrt{x^2+1}}$ **96.** $f(x) = \dfrac{1}{\sqrt[3]{x+1}}$

97. $f(x) = \sqrt{2x^3+1}$ **98.** $f(x) = \left(1-\dfrac{x}{2}\right)^{4/3}$

99–100 Find a power series solution of the equation. In each case, use the initial conditions $y(0) = a$ and $y'(0) = b$.

99. $y'' + 4y = 0$ **100.** $y'' - \dfrac{2}{(1-x)}y' = 0$

101. Find a power series solution of the differential equation $y' + y = 0$ satisfying the initial condition $y(0) = 2$. Then solve the equation by traditional means and conclude that the solutions are equal.

102. Use the Maclaurin series you obtained in Exercise 86 to find $f^{(6)}(0)$.

103.* Prove that if every bounded monotonic sequence converges, then the Completeness Property of real numbers holds. (**Hint:** Let S be a set that is bounded above by M. Let $a_0 = M$ and for each $n > 0$, let s_n be the least positive integer such that $s_n M / 2^n$ is an upper bound for S. Letting $a_n = s_n M / 2^n$, proceed to show that a_n converges and its limit is the least upper bound of S. The existence of the greatest lower bound can be shown analogously. Note that this statement is the converse of Exercise 74 in Section 10.1, showing that the Bounded Monotonic Sequence Theorem is equivalent to the Completeness Property of $\mathbb{R}$.)

104. Prove that the alternating series $\sum_{n=2}^{\infty} (-1)^{n+1} \dfrac{\ln(n^p)}{n}$ converges for all values of p.

105. Let $\sum_{n=1}^{\infty} a_n$ be a divergent series and assume that $\sum_{n=1}^{\infty} b_n$ is convergent. Prove that $\sum_{n=1}^{\infty} (a_n + b_n)$ is divergent.

106. Suppose that $\sum_{n=1}^{\infty} a_n$ is divergent and $k \neq 0$. Prove that $\sum_{n=1}^{\infty} k a_n$ is also divergent.

107. Decide whether the series $\sum_{n=1}^{\infty} \dfrac{1}{1 + 2 + \cdots + n}$ converges or diverges. Prove your answer.

108. If a_n and b_n are positive sequences such that $\sum_{n=1}^{\infty} a_n$ is convergent and b_n is a null sequence, prove that $\sum_{n=1}^{\infty} a_n b_n$ is convergent.

109. Let $a_n = \begin{cases} 1/n & \text{if } n \text{ is even} \\ 1/n^2 & \text{if } n \text{ is odd} \end{cases}$. Show that $a_n \to 0$, yet $\sum_{n=1}^{\infty} (-1)^{n+1} a_n$ diverges. Why does this not contradict Leibniz's Test?

110.* Use Taylor's formula to provide a proof of the Second Derivative Test as follows. Assuming that $f'(c) = 0$, use Taylor's formula to conclude that $f(x) = f(c) + \frac{1}{2} f''(a)(x - c)^2$, for some a between x and c. Then examine the signs of $f(x) - f(c)$ and $f''(a)$. Next, assuming $f'(c) = f''(c) = 0$ and $f'''(c) \neq 0$, argue that $f(c)$ is neither a relative maximum nor a minimum. (Assume initially that f is continuosly differentiable through at least the third order; then think about whether you can relax this condition.)

111. Find a second solution to Exercise 71 using long division.

112. Find the Fourier series expansion of the 2π-periodic extension of the function graphed below.

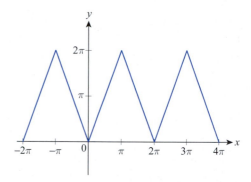

113–124 *True or False?* Determine whether the given statement is true or false. In case of a false statement, explain or provide a counterexample.

113. If a_n is monotonically decreasing, then $\lim_{n \to \infty} a_n = -\infty$.

114. If $\{a_n\}$ is convergent, then $\{a_n / n\}$ is a null sequence.

115. If $\{a_n / n\}$ is a null sequence, then $\{a_n\}$ is convergent.

116. If $\{a_n\}$ is convergent, then $\{a_{n+1} - a_n\}$ is a null sequence.

117. If $\{a_n\}$ is monotonically decreasing to zero, then $\sum_{n=1}^{\infty} (-1)^n a_n$ is absolutely convergent.

118. If $\{a_n\}$ is divergent, then $\displaystyle\sum_{n=1}^{\infty} a_n$ is divergent.

119. If $\displaystyle\sum_{n=1}^{\infty} |a_n|$ is divergent, then either $\displaystyle\sum_{n=1}^{\infty} a_n$ or

$\displaystyle\sum_{n=1}^{\infty} (-a_n)$ is divergent.

120. If $\displaystyle\sum_{n=1}^{\infty} a_n$ is divergent, then $\displaystyle\sum_{n=1}^{\infty} |a_n|$ is divergent.

121. If $\displaystyle\sum_{n=1}^{\infty} a_n$ converges, then $\displaystyle\sum_{n=1}^{\infty} a_n^2$ converges.

122. If the power series $\displaystyle\sum_{n=1}^{\infty} a_n x^n$ diverges at $x = c$, then it diverges at $x = -c$.

123. All power series converge at infinitely many x-values.

124. There is a power series whose convergence set is empty.

126. The simple series $1 - \dfrac{1}{3} + \dfrac{1}{5} - \dfrac{1}{7} + \cdots = \displaystyle\sum_{n=1}^{\infty} \dfrac{(-1)^{n+1}}{2n-1}$ was shown by Gregory and Leibniz to converge to $\pi/4$ (hence its name, the *Gregory series*). However, it converges rather slowly. Find out how many terms of this series are necessary to approximate π accurate to two decimal places.

127. a. Graph $y = \sin x$ and its 11th-order Maclaurin polynomial on the same screen, over the interval $[-4\pi, 4\pi]$. Visually estimate the subinterval over which you find the approximation acceptable.

b. Repeat part a. with the 21st-order Maclaurin polynomial.

Chapter 10
Technology Exercises

125–127 Use a computer algebra system to solve the problem.

125. We already know that the harmonic series diverges to infinity, and that it does so at a very slow pace. In this exercise, we will examine this series a bit further.

a. Find out how many terms are needed for the partial sum of the harmonic series to exceed 12.

b. What is the sum of the first 2 million terms? (Compare with Example 4 of Section 10.3. Notice that this calculation takes a bit of time even for today's powerful technology!)

Project

Chapter 10

In this project, we are going to expand on our earlier work with the harmonic series. In the process, we will meet a famous constant called *Euler's constant*, also known as the *Euler-Mascheroni constant*. (This number is not to be confused with $e \approx 2.71828$, the natural base, which is also known as Euler's number.)

1. As in Example 6 of Section 10.2, we let s_n stand for the nth partial sum of the harmonic series; that is,

$$s_n = 1 + \frac{1}{2} + \cdots + \frac{1}{n}.$$

(The partial sum s_n is also called the nth *harmonic number*.) For each $n \geq 1$, we define

$$d_n = s_n - \ln n.$$

Prove that $d_n > 0$ for any positive integer n. (**Hint:** Refer to the illustration provided for Exercise 65 of Section 10.2, and start by comparing s_n with $\int_1^{n+1} (1/x)\,dx$.)

2. Prove that $\{d_n\}$ is a decreasing sequence. (**Hint:** Referring again to the figure from Exercise 65 of Section 10.2, fix an n and identify a region whose area is $d_n - d_{n+1}$.)

3. Use an appropriate theorem from the text to show that the sequence $\{d_n\}$ is convergent. Letting $\gamma = \lim_{n \to \infty} d_n$, this limit is called **Euler's constant**. It is important in many applications throughout various areas of mathematics, and like other famous constants (including π and e) can be approximated with great precision using modern computing power. Surprisingly, however, it is not yet known whether γ is rational or irrational!

4. Use the convergence of $\{d_n\}$ to prove that the sequence $a_n = \sum_{i=n}^{2n} \frac{1}{i}$ converges and find its limit.

5. Use a computer algebra system to approximate γ, accurate to the first 10 decimal places.

6. Use the approximate value of γ found in Question 5 to estimate s_n, rounded to 5 decimal places, for **a.** $n = 10{,}000$ and **b.** $n = 2{,}000{,}000$. Compare the latter estimate with the answer for Exercise 125b of the Chapter Review.

Vectors and the Geometry of Space

11.1 Three-Dimensional Cartesian Space 867

1. Cartesian Coordinates in Three Dimensions
2. Distance in Three Dimensions

11.2 Vectors and Vector Algebra 873

1. Vector Terminology and Notation
2. Vector Algebra

11.3 The Dot Product 884

1. The Dot Product and Its Properties
2. Applications of the Dot Product

11.4 The Cross Product 895

1. The Cross Product and Its Properties
2. Applications of the Cross Product

11.5 Describing Lines and Planes 905

1. Lines in Space
2. Planes

11.6 Cylinders and Quadric Surfaces 916

1. Cylinders
2. Quadric Surfaces

Introduction

Beginning with this chapter, much of the rest of this text is concerned with *multivariable calculus*, meaning the extension of differentiation, integration, and associated ideas to functions involving more than one variable.

Although we have previously employed more than one variable when appropriate, the functions at the core of our work up to this point have been functions of a single variable. In order to visualize (that is, graph) functions of a single variable and the geometric meaning of differentiation and integration, we necessarily worked in two-dimensional Cartesian space (and later, the two-dimensional polar coordinate system). Analogously, the framework for a discussion of functions of two variables is three-dimensional space, and we begin this chapter with an introduction of three-dimensional Cartesian space.

Some of the basic principles of two-dimensional space extend naturally and easily to three dimensions—the meaning and use of ordered triples, as opposed to ordered pairs, serves as a good example. But many of the concepts of calculus require the development of what we now know as *vectors* in order to be extended to three (and higher)

dimensions. Consequently, the ideas and terminology of vector algebra are introduced in Section 11.2 and then used extensively throughout the remainder of the text.

Many branches of mathematics, science, and engineering now rely heavily on vectors and vector operations, and once acquainted with their usefulness it's hard to imagine mathematics without vectors. It may come as a surprise then that the fundamental definitions and properties of vectors were developed relatively recently. It was only in the mid-nineteenth century that mathematicians began to systematically recognize the precise properties needed to expand calculus to the multivariable realm and to solve problems in what we now think of as multidimensional space.

Two of the most important vector operations, the *dot product* and *cross product*, are introduced in this chapter, and we will see examples of their utility in Section 11.5 in describing lines and planes in space. The last section of the chapter then introduces a collection of surfaces in three-dimensional space that are worth studying both because of the frequency with which they appear and because they serve as convenient illustrations of many calculus concepts. Just as it was important to become familiar with common functions of a single variable, a sense of familiarity with the *cylinders* and *quadric surfaces* of Section 11.6 will ease your mastery of the techniques of multivariable calculus.

Many branches of mathematics, science, and engineering now rely heavily on vectors and vector operations, and once acquainted with their usefulness it's hard to imagine mathematics without vectors.

11.1 **Three-Dimensional Cartesian Space**

TOPICS

1. Cartesian coordinates in three dimensions

2. Distance in three dimensions

Although three-dimensional objects have previously appeared in this text (e.g., solids of revolution), the techniques we have used to determine such things as surface area and volume have so far been applications of single variable calculus. This chapter is the first of five in which we develop and use techniques of multivariable calculus, beginning with a more thorough exploration of how we define and work with objects in three-dimensional space.

TOPIC 1   **Cartesian Coordinates in Three Dimensions**

Cartesian three-dimensional space, denoted $\mathbb{R}^3$ or $\mathbb{R} \times \mathbb{R} \times \mathbb{R}$, is a natural extension of the Cartesian plane, and is constructed by adding a third coordinate axis, the z-axis, perpendicular to the plane defined by the x- and y-axes. In doing so, we are immediately faced with the choice of deciding which of the two directions away from the origin should correspond to positive z-values. The usual choice is called a **right-handed rectangular coordinate system**, as shown in Figure 1. The name reflects the fact that if you curl the fingers of your right hand from the positive x-axis toward the positive y-axis, your thumb will point in the direction of the positive z-axis. We can orient the axes on the page in any way that is convenient for a given drawing, but all three-dimensional coordinate drawings in this text will be based on a right-handed system. (If we had chosen to make the positive z-axis point down, the result would have been a **left-handed rectangular coordinate system**.)

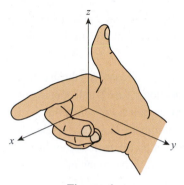

Figure 1
Right-Handed Rectangular Coordinate System

Points in three-dimensional Cartesian space correspond to ordered triples, just as points in the plane correspond to ordered pairs. For instance, the triple (a, b, c) is the unique point in space reached by beginning at the origin $(0, 0, 0)$ and traveling a units along the x-axis, then b units parallel to the y-axis, and finally c units parallel to the z-axis. In Figure 2, the triple (a, b, c) depicts a point with three positive coordinates. All points with three positive coordinates constitute the portion of space called the **first octant**, with the other seven octants similarly bordered by the three coordinate planes meeting at the origin. (While there is broad agreement on which octant is labeled *first*, there is no strong agreement on how to label the remaining seven. Fortunately, there is no real need to refer to the other octants by name.)

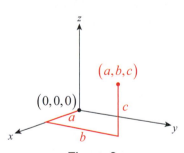

Figure 2
Point in Three-Dimensional Cartesian Space

The three **coordinate planes** are frequently identified as the ***xy*-plane**, the ***yz*-plane**, and the ***xz*-plane**; they are, as you would expect, the planes defined by the two axes lying in them. For instance, the yz-plane is the plane containing the y-axis and the z-axis, and consists of all those points whose x-coordinate is 0. In fact, this property completely characterizes the yz-plane: the graph of the equation $x = 0$ is the yz-plane, since the graph of $x = 0$ consists of all those points in space with an x-coordinate of 0. Similarly, the equation $z = 0$ defines the xy-plane and the equation $y = 0$ defines the xz-plane (see Figure 3). Examples 1, 2, and 3 illustrate how we define similar objects in space, and we will revisit the topic periodically throughout this chapter to learn how to describe three-dimensional objects in greater generality.

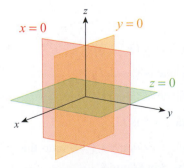

Figure 3 Coordinate Planes

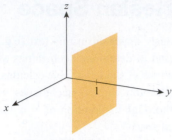

Figure 4 Graph of $y = 1$

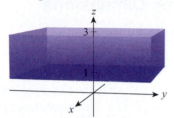

Figure 5 Graph of the Region Defined by $1 \leq z < 3$

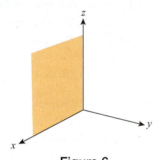

Figure 6
First Quadrant of the xz-Plane

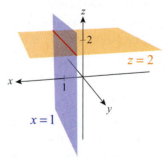

Figure 7 Intersection of the Planes $x = 1$ and $z = 2$

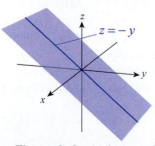

Figure 8 Graph of $y + z = 0$

Example 1 ✎

a. The graph of the equation $y = 1$ is the plane consisting of all those points whose y-coordinate is 1. It is parallel to the xz-plane ($y = 0$), and is shown in Figure 4.

b. The region in $\mathbb{R}^3$ defined by the compound inequality $1 \leq z < 3$ consists of all those points between the two planes $z = 1$ and $z = 3$, with those points comprising the plane $z = 1$ included (see Figure 5).

c. Those points satisfying the three relations $x \geq 0$, $y = 0$, and $z \geq 0$ constitute the first quadrant of the xz-plane, as shown in Figure 6.

d. Those points satisfying the two equations $x = 1$ and $z = 2$ lie at the intersection of two planes, and constitute a line parallel to the y-axis (see Figure 7).

e. In the yz-plane, the graph of the equation $y + z = 0$ consists of all those points for which $z = -y$. In $\mathbb{R}^3$, the graph of the equation $y + z = 0$ still consists of all those points for which $z = -y$; that is, those points for which the third coordinate is the negative of the second coordinate. But since x doesn't appear in the equation, there is no constraint at all on the first coordinate. Thus the graph of $y + z = 0$ in $\mathbb{R}^3$ consists of all those points of the form $\left\{ (x, y, -y) \mid x, y \in \mathbb{R} \right\}$; these points form the plane illustrated in Figure 8.

The **projection** of a point (x, y, z) in $\mathbb{R}^3$ onto a plane is the point in that plane closest to (x, y, z). The projection of the points constituting a given object onto a coordinate plane is often useful in helping to visualize or better understand the object.

Example 2 ✎

Determine the projections of the following points and objects onto the indicated planes.

a. Projection of $(3, -2, 7)$ onto the three coordinate planes.

b. Projection of $(3, -2, 7)$ onto the plane $y = 5$.

c. Projection of the plane defined by the equation $y + z = 0$ onto the yz-plane.

Solution

a. The projection of $(3, -2, 7)$ onto the xy-plane is $(3, -2, 0)$.

The projection of $(3, -2, 7)$ onto the yz-plane is $(0, -2, 7)$.

The projection of $(3, -2, 7)$ onto the xz-plane is $(3, 0, 7)$.

b. The projection of $(3, -2, 7)$ onto the plane $y = 5$ is $(3, 5, 7)$.

c. The projection of the graph (in $\mathbb{R}^3$) of the equation $y + z = 0$ onto the yz-plane is the line in the yz-plane illustrated in Figure 8.

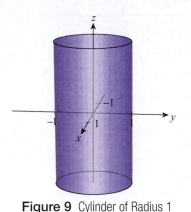

Figure 9 Cylinder of Radius 1

Example 3 ✏

Graph the equation $x^2 + y^2 = 1$ in $\mathbb{R}^3$.

Solution

In the xy-plane, the graph of $x^2 + y^2 = 1$ consists of all those points 1 unit from the origin (a circle of radius 1). To graph this equation in three-dimensional space, the only adjustment we need make is to note that the z-coordinate of any given solution point is irrelevant. That is, (x, y, z) will be a solution of the equation if and only if $x^2 + y^2 = 1$. So the graph is a cylinder of radius 1 whose axis is the z-axis; in other words, all points for which $x^2 + y^2 = 1$, without regard to the value of the z-coordinate. Figure 9 depicts the graph.

TOPIC 2 Distance in Three Dimensions

The rectangular nature of the three-dimensional coordinate system just introduced allows us to easily extend our two-dimensional distance formula. Two points (x_1, y_1, z_1) and (x_2, y_2, z_2) in space, with $x_1 \neq x_2$, $y_1 \neq y_2$, and $z_1 \neq z_2$, can always be taken to define diagonally opposite corners of a box whose sides are parallel to the coordinate axes, as illustrated in Figure 10a.

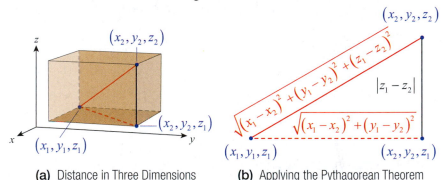

(a) Distance in Three Dimensions   **(b)** Applying the Pythagorean Theorem

Figure 10

By the two-dimensional distance formula (Pythagorean Theorem), the length of the dashed red line in Figure 10a is

$$\sqrt{(x_1 - x_2)^2 + (y_1 - y_2)^2}.$$

Applying the Pythagorean Theorem again, this time to the right triangle whose hypotenuse is the line segment joining (x_1, y_1, z_1) to (x_2, y_2, z_2) as shown in Figure 10b, we see that the **distance** between the two points is

$$\sqrt{\left[\sqrt{(x_1 - x_2)^2 + (y_1 - y_2)^2}\right]^2 + (z_1 - z_2)^2} = \sqrt{(x_1 - x_2)^2 + (y_1 - y_2)^2 + (z_1 - z_2)^2}.$$

The formula also holds if the two points do not differ in all three coordinates. In such cases, the two points lie in a plane parallel to one of the coordinate planes and the two-dimensional distance formula applies directly.

Example 4 ✎

Find the distance between the points $(-2, 5, 1)$ and $(3, 0, -1)$.

Solution

Applying the previous formula, the distance is as follows:

$$\sqrt{(-2-3)^2 + (5-0)^2 + \left[1-(-1)\right]^2} = \sqrt{54} = 3\sqrt{6}$$

Example 5 ✎

Construct an equation whose graph is the sphere of radius 2 centered at the point $(2, 3, -1)$.

Solution

We seek an equation in x, y, and z that describes all those points (x, y, z) that are 2 units from $(2, 3, -1)$. So using the distance formula, we need

$$\sqrt{(x-2)^2 + (y-3)^2 + \left[z-(-1)\right]^2} = 2,$$

or, squaring both sides,

$$(x-2)^2 + (y-3)^2 + (z+1)^2 = 4.$$

The graph of such a sphere is shown in Figure 11.

More generally, the graph of the equation $(x-a)^2 + (y-b)^2 + (z-c)^2 = r^2$ is a sphere of radius r centered at the point (a, b, c).

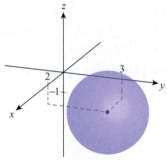

Figure 11
Sphere of Radius 2
Centered at $(2, 3, -1)$

11.1 **Exercises**

1–2 Describe the orientation of the third axis if the resulting system is to be a right-handed system. (**Hint:** Feel free to use the words "out of the page" or "into the page.")

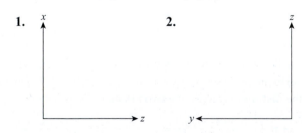

1.   **2.**

3–8 Plot the given point in the three-dimensional coordinate system.

3. $(1, 0, 0)$

4. $(-2, 0, 1)$

5. $\left(0, 1, \dfrac{3}{2}\right)$

6. $(3, 4, -1)$

7. $\left(1, -2, \dfrac{5}{2}\right)$

8. $\left(-2, -1, \dfrac{7}{4}\right)$

9–16 Explain what you know about the coordinates of point P if its location in the three-dimensional Cartesian system is as described below.

9. P is in the xy-plane

10. P is in the first octant

11. P is on the y-axis

12. P is in the yz-plane

13. P is on the negative x-axis

14. P is in the third quadrant of the xz-plane

15. P is in the plane $y = -1$

16. P is in the second quadrant of the plane $x = 3$

17–20 Find the distance from the point to the indicated plane.

17. $(1,2,3)$;   the xy-plane

18. $(2,-2,-5)$;   the xz-plane

19. $(0,-7,-2)$;   the yz-plane

20. $(1,-3,4)$;   the plane $z = -2$

21–25 Find the coordinates of the projection of the given point.

21. The projection of the point $(5,3.2,-2)$ onto the xy-plane

22. The projection of the point $(-5,1,4)$ onto the xz-plane

23. The projection of the point $(3,0,0)$ onto the yz-plane

24. The projection of the point $(1,0,-7)$ onto the plane $z = -2$

25. The projection of the point $(1.3,\pi,-10)$ onto the plane $y = 4.5$

26–35 Describe the set of points represented by the given equation(s).

26. $z = 0$ **27.** $y = 1$

28. $x = 3 - y$ **29.** $x + y = 1, \quad z = 0$

30. $x = y = z$ **31.** $xyz = 0$

32. $y = 0, \quad z = 4x - 2$ **33.** $z = 2y - 1, \quad x = 5$

34. $z = |x|$ **35.** $z = [\![x]\!]$

36–41 Find the coordinates of the described point.

36. A point in the plane $y = -2$ whose projection onto the xz-plane is $(1,0,-5)$

37. A point on the line $x = y = z$, equidistant from the xy-plane and the plane $z = 5$

38. A point on the x-axis, equidistant from the planes $z = x$ and $z = x - 2$

39. A point P with an x-coordinate of -1, so that the line through P and the point $(1,-2,5)$ is parallel to one of the coordinate axes

40. The point on the sphere $x^2 + y^2 + 6y + z^2 - 7 = 0$ that is closest to the plane $y = 5$

41. The point on the sphere $x^2 + y^2 - 2y + z^2 - 6z + 6 = 0$ that is closest to the point $\left(2\sqrt{3},1,5\right)$

42–45 Match the equation to its graph (labeled A–D).

42. $z - x = 1$ **43.** $x^2 + y^2 + z^2 = 10$

44. $x^2 + z^2 = 5$ **45.** $y - x^2 = 0$

A.

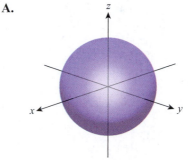

B.

C.

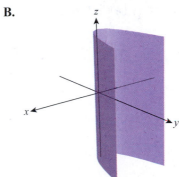

D.

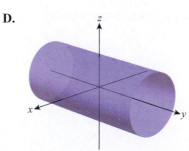

46–59 Describe the set of points represented by the given relation(s). (**Hint:** A sketch may be helpful.)

46. $z^2 \geq 1$

47. $x^2 + y^2 > 1$

48. $x^2 + y^2 + z^2 \leq 4$

49. $(x - 3)^2 + y^2 + z^2 = 9$

50. $y^2 + z^2 < 9$

51. $x^2 + y = 1$

52. $x^2 + y^2 + (z - 3)^2 = 25, \quad z = 0$

53. $x - z^2 = 2, \quad y = 1$

54. $xz \geq 0, \quad y = 0$

55. $xyz = 0$

56. $x^2 - 4x + y^2 + z^2 + 6z + 12 = 0$

57. $x^2 + 4x + y^2 - 8y + z^2 - 10z = 0, \quad z \geq 5$

58. $x^2 - 6x + y^2 + z^2 + 4z \leq 3$

59. $x^2 + y^2 + 8y + z^2 - 2z + 8 = 0, \quad z = -2$

60–77 Find the equations or inequalities that define the indicated set.

60. The plane through $(2, 0, -1)$ that is parallel to the xy-plane

61. The plane through $(1, 0, 0)$ that is perpendicular to the x-axis

62. The plane through $(-4, 7, 2)$ that is perpendicular to the y-axis

63. The plane through $\left(\frac{4}{5}, 1, e\right)$ that is perpendicular to the z-axis

64. The line through $(1, 2, 3)$ that is parallel to the z-axis

65. The line through $(0, -1, 2)$ that is parallel to the x-axis

66. The line through $(-4, 1, 5)$ that is parallel to the y-axis

67. The sphere of radius 2, centered at the point $(0, 2, 0)$

68. The sphere of radius $\sqrt{5}$, centered at the point $(-1, 3, 5)$

69. The sphere centered at $(1, 0, -2)$, passing through the point $(3, -3, 4)$

70. The circle of radius 1 in the xy-plane, centered at $(1, 0, 0)$

71. The circle of radius 3 in the yz-plane, centered at $(0, -4, 1)$

72. The circle of radius $\sqrt{2}$ in the plane $y = -1$, centered at $(-1, -1, 1)$

73. The sphere centered at $(-1, 4, -3)$ that is tangent to the yz-plane

74. The sphere with a diameter joining the points $(0, -2, 5)$ and $(2, 4, 1)$

75. The intersection of the first octant and the sphere centered at $(3, 2, -1)$, passing through the point $(5, 3, 1)$

76. A horizontal circular cylinder of inner radius $r = 4$ that is tangent to the xy-plane along the line $y = -1$

77. The set of points equidistant from $(-1, 2, 3)$ and $(7, 2, 3)$

78–81 Find the distance between the given pair of points.

78. $(0, 4, 2)$ and $(-1, 1, 0)$

79. $(-3, 4, 10)$ and $(5, -3, 6)$

80. $\left(\dfrac{\sqrt{2}}{2}, 3, -1\right)$ and $\left(\sqrt{2}, 3, 1\right)$

81. $\left(2, 3\sqrt{3}, -4\right)$ and $\left(-1, \sqrt{3}, -2\right)$

82–85 A triangle is given by the coordinates of its vertices. For the triangle, select all that apply from the following list: **a.** isosceles, **b.** equilateral, **c.** scalene, **d.** right triangle.

82. $(0, 6, 2)$, $(3, 4, 1)$, and $(1, 3, 4)$

83. $(-1, 2, 7)$, $(1, 1, 3)$, and $(5, 3, 2)$

84. $(2, 10, 4)$, $(1, 7, 0)$, and $(-3, 1, -1)$

85. $(2, 1, 1)$, $(4, 3, 1)$, and $(3, 1, 2)$

11.2 **Vectors and Vector Algebra**

TOPICS

1. Vector terminology and notation
2. Vector algebra

Many quantities are defined primarily by their size. For example, length, area, mass, price, and temperature are fully determined by a single number; such numbers, representing only magnitude, are called **scalars**. Other quantities, however, cannot be adequately described by a number alone. Force and velocity, for instance, possess both a *magnitude* and a *direction*, and a complete description of these quantities must include both. Such *directed magnitudes* are called **vectors**. In this section, we will study the terminology and algebraic operations of vectors, with the immediate goal of using them to describe curves and surfaces in space.

TOPIC 1 **Vector Terminology and Notation**

In this text, we will work with vectors in the setting of $\mathbb{R}^2$ and $\mathbb{R}^3$, but the study of vectors in general constitutes an enormous area of mathematics, and vectors are easily generalized into spaces of any dimension. In two- and three-dimensional space, vectors are often depicted as **directed line segments** (informally, "arrows"). Such a directed line segment begins at an **initial point** P and ends at a **terminal point** Q, and the notation $\overrightarrow{PQ}$ is used to refer to the vector. A subtle but very important point, though, is that a vector is characterized *entirely* by its direction and magnitude, not by its initial and terminal points. That is, for a specific pair of points P and Q, $\overrightarrow{PQ}$ is only one way of depicting the vector it represents. We will use boldface letters to denote vectors in general, and an expression such as $\mathbf{u} = \overrightarrow{PQ}$ means that $\mathbf{u}$ is a vector whose length and direction is the same as that of the directed line segment $\overrightarrow{PQ}$ (when handwriting, the notation $\vec{u}$ instead of $\mathbf{u}$ is common). Figure 1 illustrates five different ways of depicting the one vector $\mathbf{u}$—each of the five arrows is simply a translation of the directed line segment from the point P to the point Q.

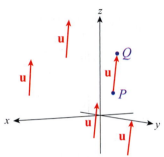

Figure 1
Five Depictions of the Vector **u**

Since vectors are characterized entirely by length and direction, we say that two vectors $\mathbf{u}$ and $\mathbf{v}$ with the same length and the same direction are **equal** (no matter where they might actually be depicted, if they are graphed) and we write $\mathbf{u} = \mathbf{v}$. The **length** of a vector, also called its **magnitude** or **norm**, is denoted $|\mathbf{u}|$ or $\|\mathbf{u}\|$ —we will soon see that the distance formula allows us to easily find the lengths of vectors in $\mathbb{R}^2$ and $\mathbb{R}^3$. And in keeping with our prior use of the word, the length of a vector is always nonnegative. The only vector with a length of 0 is the **zero vector**, denoted $\mathbf{0}$. We write $|\mathbf{0}| = 0$, and note that $|\mathbf{u}| > 0$ for every vector $\mathbf{u} \neq \mathbf{0}$.

For the purposes of calculus, we often need an algebraic way of describing specific vectors in $\mathbb{R}^2$ and $\mathbb{R}^3$, a way that lends itself to such operations as integration and differentiation. The *component form* of a vector satisfies that need. If a vector $\mathbf{u}$ is depicted with the origin as its initial point, the vector is said to be in **standard position**. If we are working within the context of the plane $\mathbb{R}^2$, then the terminal point corresponds to an ordered pair (u_1, u_2) and the **component form** of $\mathbf{u}$ is $\langle u_1, u_2 \rangle$. Figure 2 is an illustration of the vector

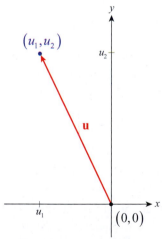

Figure 2 Vector **u** in $\mathbb{R}^2$

$$\mathbf{u} = \langle u_1, u_2 \rangle$$

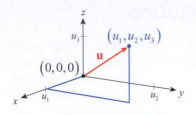

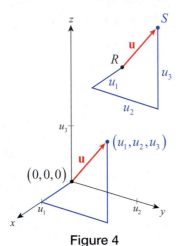

Figure 3 Vector **u** in $\mathbb{R}^3$

Figure 4

in standard position in the plane. Similarly, if the context is $\mathbb{R}^3$, the terminal point corresponds to an ordered triple (u_1, u_2, u_3) and the component form of **u** is

$$\mathbf{u} = \langle u_1, u_2, u_3 \rangle.$$

Figure 3 shows such a vector **u** in $\mathbb{R}^3$, again in standard position.

More generally, if a given vector is shown with its initial point somewhere other than the origin, the component form of a vector still represents the distances traveled, parallel to the coordinate axes, from the initial point to the terminal point. Figure 4 shows two depictions of the vector $\mathbf{u} = \langle u_1, u_2, u_3 \rangle$, one in standard position and the other with initial point R and terminal point S. But the length of **u** is unchanged, and the distance formula tells us that

$$|\mathbf{u}| = \sqrt{u_1^2 + u_2^2 + u_3^2}.$$

Example 1 ✎

Find **a.** the component form and **b.** the length of the vector **u** in $\mathbb{R}^3$ with initial point $(-1, 3, -2)$ and terminal point $(0, 1, -5)$.

Solution

a. The first component of **u** is the difference between the x-coordinates of the initial and terminal points, or $0 - (-1) = 1$; note that the order of subtraction matters, since we are looking for the signed distance we must travel in the x-direction to take us from the initial point to the terminal point. In similar fashion, the second component of **u** is the difference in the y-coordinates, $1 - 3 = -2$, and the third component is the difference in the z-coordinates, $-5 - (-2) = -3$. So $\mathbf{u} = \langle 1, -2, -3 \rangle$.

b. $|\mathbf{u}| = |\langle 1, -2, -3 \rangle| = \sqrt{1^2 + (-2)^2 + (-3)^2} = \sqrt{1 + 4 + 9} = \sqrt{14}$

We have previously solved many problems involving velocity, but to this point we have only been concerned with the velocity of an object moving forward and backward along a line. Even when considering a thrown object, we have sought merely to determine its *vertical* velocity, without regard to any horizontal motion it may or may not have. We are now in a position to consider the total velocity of an object.

Example 2 ✎

A ball is shot by a slingshot into the air with an initial velocity that has a vertical component of 64 ft/s and a horizontal component of 10 ft/s. Determine **a.** its velocity at time t, **b.** the magnitude of its velocity at the highest point of its arc, and **c.** the magnitude of its velocity just at the moment it hits the ground. (Ignore air resistance, and assume the initial height of the ball is 0.)

Solution

We can consider the context of this problem to be $\mathbb{R}^2$, with the positive y-axis pointing straight up and the positive x-axis pointing in the direction of the horizontal motion of the ball. So if we denote the vertical and horizontal components of the velocity by $v_y(t)$ and $v_x(t)$, respectively, we seek the velocity vector $\mathbf{v}(t) = \langle v_x(t), v_y(t) \rangle$.

a. Recall that, in general, the vertical velocity at time t of an object with initial vertical velocity v_0 is $v_y(t) = gt + v_0$, where g is the acceleration due to gravity. The initial vertical velocity is given in terms of feet and seconds, so we use $g = -32 \text{ ft/s}^2$ and hence $v_y(t) = -32t + 64$. The ball has an initial horizontal velocity of 10 ft/s, and this remains constant during the flight of the ball, so $v_x(t) = 10$ (we ignore any slight decrease in velocity that air resistance might cause during the short flight). This gives us the total velocity vector.

$$\mathbf{v}(t) = \langle 10, -32t + 64 \rangle$$

b. We find the time when the ball reaches its highest point just as we have in the past: it is that time when $v_y(t) = 0$, namely $t = 2$. But the ball still has nonzero horizontal velocity at time $t = 2$, and the magnitude of the total velocity at that time is

$$|\mathbf{v}(2)| = |\langle 10, 0 \rangle| = \sqrt{10^2 + 0^2} = 10 \text{ ft/s}.$$

c. We have solved enough problems involving objects moving under the influence of gravity alone to know that if it takes 2 seconds for the ball to reach its apex, it will take another 2 seconds for it to return to the ground (we are making the assumption that the ball's initial height is 0). But we can confirm that fact and gain a useful expression for the ball's position at time t by applying the knowledge that position is the antiderivative of velocity. Taking the components one at a time, the ball's horizontal position (relative to its initial position, which we can take to be the origin), is $\int 10 \, dt = 10t$ (note that our assumption about the ball's initial position leads to a constant of integration of 0). Similarly, the ball's vertical position is $\int (-32t + 64) \, dt = -16t^2 + 64t$ (again, the constant of integration is 0). And as we suspected, $-16t^2 + 64t = 0$ implies $t = 0$ and $t = 4$.

To find the magnitude of the ball's velocity just as it hits the ground, we compute the following:

$$|\mathbf{v}(4)| = |\langle v_x(4), v_y(4) \rangle| = |\langle 10, -64 \rangle| = \sqrt{10^2 + (-64)^2} = 2\sqrt{1049} \approx 64.8 \text{ ft/s}$$

Making use of what we have discovered about the ball's horizontal and vertical position at time t, the graph of the parametrically defined path $x = 10t$ and $y = -16t^2 + 64t$ over the interval $t \in [0, 4]$ appears in Figure 5.

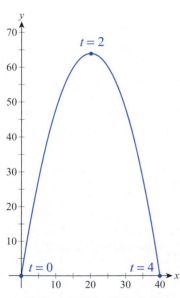

Figure 5 Flight of the Ball

TOPIC 2 **Vector Algebra**

Consider a kayaker paddling in a straight line on a perfectly calm, still lake. In the absence of any other motion, we can describe her velocity as a vector $\mathbf{v}$, where $|\mathbf{v}|$ is her still-water speed and the direction of $\mathbf{v}$ is her direction of travel. If she were to begin paddling faster and double her speed, her direction would be unchanged but her speed would be twice as much. We represent her new velocity vector as $2\mathbf{v}$, an example of **scalar multiplication**—the product of a scalar and a vector. Note that the magnitude of $2\mathbf{v}$ is two times the magnitude of $\mathbf{v}$; that is, $|2\mathbf{v}| = 2|\mathbf{v}|$.

Now imagine the same kayaker paddling on a river with her original velocity $\mathbf{v}$. If the water in the river is flowing with velocity $\mathbf{w}$, we represent her overall velocity with the **vector sum** $\mathbf{v} + \mathbf{w}$, a new vector whose magnitude and direction are a combination of the magnitudes and directions of the individual vectors (see Figure 6). This is an example of **vector addition**. As just one specific example, if she were to paddle upriver at a pace exactly matching the water's flow rate, then $\mathbf{v}$ and $\mathbf{w}$ would cancel each other out and we would write $\mathbf{v} + \mathbf{w} = \mathbf{0}$, or $\mathbf{v} = -\mathbf{w}$. In this case, $\mathbf{v}$ is the **additive inverse** of the vector $\mathbf{w}$, which is equivalent to the product of the scalar -1 and the vector $\mathbf{w}$.

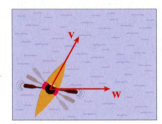 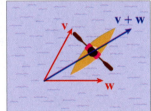

Figure 6

Scalar multiplication and vector addition are the two basic components of vector algebra, the properties of which are summarized in the following theorem. But before we list the algebraic properties, we introduce ways in which vector sums and scalar multiples of vectors can be represented graphically. Figure 7 depicts two vectors $\mathbf{v}$ and $\mathbf{w}$ in red and then illustrates two new vectors $2\mathbf{v}$ and $\mathbf{v} + \mathbf{w}$, shown in blue. One typical way of graphically constructing a sum like $\mathbf{v} + \mathbf{w}$ is to place $\mathbf{v}$ somewhere in space, draw $\mathbf{w}$ so that its initial point coincides with the terminal point of $\mathbf{v}$, and then draw the **resultant vector** $\mathbf{v} + \mathbf{w}$ from the initial point of $\mathbf{v}$ to the terminal point of $\mathbf{w}$. Alternatively, $\mathbf{w}$ can be drawn first and $\mathbf{v}$ second—the fact that the resultant vector is the same is an illustration of the **Parallelogram Law** of vector addition, also known as the Commutative Property.

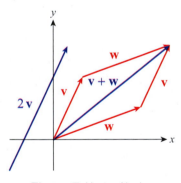

Figure 7 Vector Algebra

In terms of vector components, multiplication by a scalar a corresponds to multiplying each component of a vector by a, and addition of two vectors corresponds to adding the respective components of the vectors. So in $\mathbb{R}^3$,

$$a\left\langle u_1, u_2, u_3 \right\rangle = \left\langle au_1, au_2, au_3 \right\rangle$$

and

$$\left\langle u_1, u_2, u_3 \right\rangle + \left\langle v_1, v_2, v_3 \right\rangle = \left\langle u_1 + v_1, u_2 + v_2, u_3 + v_3 \right\rangle.$$

Theorem 🔑

Properties of Scalar Multiplication and Vector Addition

Assume **u**, **v**, and **w** represent vectors, while a and b represent scalars. Then the following properties hold.

Scalar Multiplication Properties	Vector Addition Properties
$a(\mathbf{u}+\mathbf{v}) = a\mathbf{u}+a\mathbf{v}$	$\mathbf{u}+\mathbf{v} = \mathbf{v}+\mathbf{u}$
$(a+b)\mathbf{u} = a\mathbf{u}+b\mathbf{u}$	$\mathbf{u}+(\mathbf{v}+\mathbf{w}) = (\mathbf{u}+\mathbf{v})+\mathbf{w}$
$(ab)\mathbf{u} = a(b\mathbf{u}) = b(a\mathbf{u})$	$\mathbf{u}+\mathbf{0} = \mathbf{u}$
$1\mathbf{u} = \mathbf{u}$, $0\mathbf{u} = \mathbf{0}$, and $a\mathbf{0} = \mathbf{0}$	$\mathbf{u}+(-\mathbf{u}) = \mathbf{0}$
$\lvert a\mathbf{u}\rvert = \lvert a\rvert\lvert\mathbf{u}\rvert$	

These properties are all readily verified using the component definitions of scalar multiplication and vector addition, and you will do so in Exercises 91 and 92.

Example 3 ✎

Let $\mathbf{u} = \langle -2, 2, 4\rangle$ and $\mathbf{v} = \langle 3, -5, 1\rangle$. Find the following.

a. $2\mathbf{u}+\mathbf{v}$ **b.** $\mathbf{v}-\mathbf{u}$ **c.** $\left\lvert -\dfrac{1}{2}\mathbf{u}\right\rvert$

Solution

a. $2\mathbf{u}+\mathbf{v} = 2\langle -2, 2, 4\rangle + \langle 3, -5, 1\rangle$
$$= \langle -4, 4, 8\rangle + \langle 3, -5, 1\rangle$$
$$= \langle -4+3, 4+(-5), 8+1\rangle = \langle -1, -1, 9\rangle$$

b. $\mathbf{v}-\mathbf{u} = \langle 3, -5, 1\rangle - \langle -2, 2, 4\rangle = \langle 5, -7, -3\rangle$

c. $\left\lvert -\dfrac{1}{2}\mathbf{u}\right\rvert = \left\lvert -\dfrac{1}{2}\langle -2, 2, 4\rangle\right\rvert = \lvert\langle 1, -1, -2\rangle\rvert = \sqrt{1^2 + (-1)^2 + (-2)^2} = \sqrt{6}$

(Note that this is indeed equal to $\frac{1}{2}\lvert\langle -2, 2, 4\rangle\rvert = \lvert -\frac{1}{2}\rvert\lvert\mathbf{u}\rvert$.)

Example 4 ✎

Given the two vectors $\mathbf{u} = \langle 5, -2\rangle$ and $\mathbf{v} = \langle -1, 2\rangle$ in $\mathbb{R}^2$, depict $-2\mathbf{v}$ and $\mathbf{u}-\mathbf{v}$.

Solution

In terms of components,

$$-2\mathbf{v} = -2\langle -1, 2\rangle = \langle 2, -4\rangle \quad\text{and}\quad \mathbf{u}-\mathbf{v} = \langle 5, -2\rangle - \langle -1, 2\rangle = \langle 6, -4\rangle.$$

Figure 8 is a depiction of $\mathbf{u}$, $\mathbf{v}$, $-\mathbf{v}$, $-2\mathbf{v}$, and $\mathbf{u}-\mathbf{v}$. Note that $(\mathbf{u}-\mathbf{v})+\mathbf{v} = \mathbf{u}$.

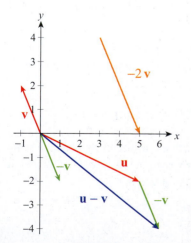

Figure 8

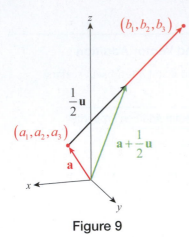

Figure 9

We can use vector operations to easily describe points between two given points. For instance, given the two points (a_1, a_2, a_3) and (b_1, b_2, b_3) in $\mathbb{R}^3$, the vector

$$\mathbf{u} = \langle b_1 - a_1, b_2 - a_2, b_3 - a_3 \rangle$$

is the directed line segment from (a_1, a_2, a_3) to (b_1, b_2, b_3), so if we place the initial point of $\frac{1}{2}\mathbf{u}$ at (a_1, a_2, a_3), its terminal point will lie halfway between (a_1, a_2, a_3) and (b_1, b_2, b_3). If we draw the vector $\mathbf{a} = \langle a_1, a_2, a_3 \rangle$ in standard position, the vector relationship shown in Figure 9 is the result. We can describe other points along the line segment in similar fashion, as illustrated in Example 5.

Example 5 ✎

Given $P(-2, 4, 1)$ and $Q(3, 2, -1)$, determine the coordinates of the point

a. halfway between P and Q.   **b.** one quarter of the way from P to Q.

Solution

a. Using the reasoning illustrated in Figure 9, the point halfway between P and Q is the terminal point (in standard position) of the vector

$$\langle -2, 4, 1 \rangle + \frac{1}{2}\left(\langle 3, 2, -1 \rangle - \langle -2, 4, 1 \rangle\right) = \langle -2, 4, 1 \rangle + \left\langle \frac{5}{2}, -1, -1 \right\rangle = \left\langle \frac{1}{2}, 3, 0 \right\rangle,$$

that is, the point $\left(\frac{1}{2}, 3, 0\right)$.

Note that if we define vectors $\mathbf{P} = \langle -2, 4, 1 \rangle$ and $\mathbf{Q} = \langle 3, 2, -1 \rangle$, then we have just calculated $\mathbf{P} + \frac{1}{2}(\mathbf{Q} - \mathbf{P})$, or $\frac{1}{2}(\mathbf{P} + \mathbf{Q})$.

b. Similarly, the point one quarter of the way from P to Q corresponds to the terminal point of the vector $\mathbf{P} + \frac{1}{4}(\mathbf{Q} - \mathbf{P})$, or $\frac{3}{4}\mathbf{P} + \frac{1}{4}\mathbf{Q}$. The vector is

$$\frac{3}{4}\mathbf{P} + \frac{1}{4}\mathbf{Q} = \frac{3}{4}\langle -2, 4, 1 \rangle + \frac{1}{4}\langle 3, 2, -1 \rangle = \left\langle -\frac{3}{2}, 3, \frac{3}{4} \right\rangle + \left\langle \frac{3}{4}, \frac{1}{2}, -\frac{1}{4} \right\rangle = \left\langle -\frac{3}{4}, \frac{7}{2}, \frac{1}{2} \right\rangle$$

and the point is $\left(-\frac{3}{4}, \frac{7}{2}, \frac{1}{2}\right)$.

A vector of length 1 is called a **unit vector**. The two unit vectors $\mathbf{i} = \langle 1, 0 \rangle$ and $\mathbf{j} = \langle 0, 1 \rangle$ in $\mathbb{R}^2$, and the three unit vectors $\mathbf{i} = \langle 1, 0, 0 \rangle$, $\mathbf{j} = \langle 0, 1, 0 \rangle$, and $\mathbf{k} = \langle 0, 0, 1 \rangle$ in $\mathbb{R}^3$ are useful in describing other vectors, and are called the **standard basis vectors**. For instance, note that an arbitrary vector $\mathbf{u} = \langle u_1, u_2, u_3 \rangle$ in $\mathbb{R}^3$ can be written as a *linear combination* of the standard basis vectors:

$$\mathbf{u} = u_1 \langle 1, 0, 0 \rangle + u_2 \langle 0, 1, 0 \rangle + u_3 \langle 0, 0, 1 \rangle = u_1 \mathbf{i} + u_2 \mathbf{j} + u_3 \mathbf{k}$$

In other applications, it is convenient to work with a unit vector that has the same direction as a given vector $\mathbf{v}$, and the product of $\mathbf{v}$ with the reciprocal of its length (assuming its length is nonzero) gives us such a unit vector.

$$\left| \frac{1}{|\mathbf{v}|} \mathbf{v} \right| = \frac{1}{|\mathbf{v}|} |\mathbf{v}| = 1$$

Example 6 ✐

Let $\mathbf{u} = \langle 3, -2, 4 \rangle$.

a. Express $\mathbf{u}$ as a linear combination of the standard basis vectors of $\mathbb{R}^3$.

b. Express $\mathbf{u}$ as a product of its magnitude and a vector pointing in the same direction.

Solution

a. $\mathbf{u} = 3\mathbf{i} - 2\mathbf{j} + 4\mathbf{k}$

b. $|\mathbf{u}| = \sqrt{9 + 4 + 16} = \sqrt{29}$

The vector $\left\langle \dfrac{3}{\sqrt{29}}, -\dfrac{2}{\sqrt{29}}, \dfrac{4}{\sqrt{29}} \right\rangle$ is a unit vector pointing in the same direction as $\mathbf{u}$, and we can write

$$\mathbf{u} = \sqrt{29} \left\langle \frac{3}{\sqrt{29}}, -\frac{2}{\sqrt{29}}, \frac{4}{\sqrt{29}} \right\rangle.$$

In our next example, we make use of trigonometry to decompose vectors into appropriate components.

Example 7 ✐

A ten-pound weight (a force) is suspended from two ropes, as shown in Figure 10. The ropes are of different lengths, so the tensions (also forces) $\mathbf{T}_1$ and $\mathbf{T}_2$ in the ropes are unequal. Given the measured angles shown in the figure, determine the tension vectors.

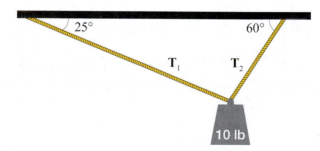

Figure 10 Suspended Weight

Solution

First, an important note of caution. Diagrams like Figure 10 are often used to illustrate such problems, but the diagram can be very misleading if interpreted incorrectly. The magnitudes of the two tensions $\mathbf{T}_1$ and $\mathbf{T}_2$ typically *don't* correspond to the lengths of the ropes in the diagram. But we can still use the picture as the basis for what physicists and engineers call a *force diagram*, which will allow us to determine $\mathbf{T}_1$ and $\mathbf{T}_2$.

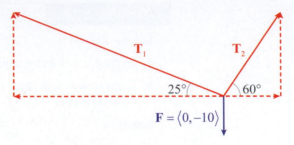

Figure 11 Force Diagram

In Figure 11, the ten-pound weight is shown as the vector $\mathbf{F} = \langle 0, -10 \rangle$ and the components of the two tension vectors $\mathbf{T}_1$ and $\mathbf{T}_2$ can be determined using trigonometry.

$$\mathbf{T}_1 = \left\langle -|\mathbf{T}_1|\cos 25°, |\mathbf{T}_1|\sin 25° \right\rangle$$
$$\mathbf{T}_2 = \left\langle |\mathbf{T}_2|\cos 60°, |\mathbf{T}_2|\sin 60° \right\rangle$$

Because the weight is stationary in space, the three forces acting on it must have a vector sum of zero. That is, $\mathbf{T}_1 + \mathbf{T}_2 + \mathbf{F} = \mathbf{0}$, or

$$\left\langle -|\mathbf{T}_1|\cos 25°, |\mathbf{T}_1|\sin 25° \right\rangle + \left\langle |\mathbf{T}_2|\cos 60°, |\mathbf{T}_2|\sin 60° \right\rangle + \langle 0, -10 \rangle = \langle 0, 0 \rangle.$$

Equating the components individually leads to the following system of equations.

$$-|\mathbf{T}_1|\cos 25° + |\mathbf{T}_2|\cos 60° = 0$$
$$|\mathbf{T}_1|\sin 25° + |\mathbf{T}_2|\sin 60° - 10 = 0$$

In Exercise 83, you will solve this system for $|\mathbf{T}_1|$ and $|\mathbf{T}_2|$ and show that

$$\mathbf{T}_1 \approx \langle -4.55, 2.12 \rangle$$
$$\mathbf{T}_2 \approx \langle 4.55, 7.88 \rangle$$

Figure 12 contains an accurate depiction of the three forces in this problem, with the lengths of the vectors in the diagram corresponding to their relative magnitudes.

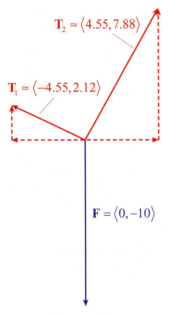

Figure 12
Accurate Vector Diagram

11.2 **Exercises**

1–8 Quantities are given that arise in everyday life. Decide whether they are vectors or scalars.

1. The speed of your car

2. The cost of a telephone call

3. The velocity of your car

4. The mass of a golf cart

5. The weight of a gallon of milk

6. The displacement of a particle that moved from $(2, -1)$ to $(5, 8)$ in the xy-system

7. The distance covered by a flight from Houston to San Diego

8. The restoring force exerted by a vertical spring when a mass is hung on it

9–14 Decide whether the following is a vector.

9. $(2,\pi)$

10. $(1,0,-4)$

11. $\langle 0,0 \rangle$

12. $\left\langle 0, \dfrac{\pi}{2}, \cos\dfrac{\pi}{2} \right\rangle$

13. $\langle 2,-1,3 \rangle - \langle 1,5,10 \rangle$

14. $|\langle 2,-1,3 \rangle|$

15–20 Use the figure below to express the indicated sum or difference as a single vector. (M is the midpoint of $\overline{AB}$.)

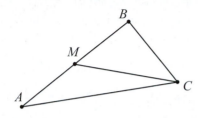

15. $\overrightarrow{CA} + \overrightarrow{CB}$

16. $\overrightarrow{CA} - \overrightarrow{CB}$

17. $\overrightarrow{BC} - \overrightarrow{BM}$

18. $\overrightarrow{AB} + \overrightarrow{BM} + \overrightarrow{MC}$

19. $3\overrightarrow{AM} - \overrightarrow{AB} - \overrightarrow{AC}$

20. $2\overrightarrow{BM} + \overrightarrow{AC}$

21–24 Geometrically construct the indicated linear combination from the given vectors.

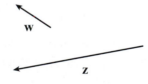

21. $\mathbf{u} - \mathbf{z}$

22. $\mathbf{v} + 2\mathbf{w}$

23. $2\mathbf{u} + \mathbf{v} - \mathbf{w}$

24. $3\mathbf{w} + \dfrac{1}{2}\mathbf{z} - \mathbf{u}$

25–30 Find the linear combination in component form if the vectors $\mathbf{u}$ and $\mathbf{v}$ are given in component form as $\mathbf{u} = \langle 4,-1 \rangle$ and $\mathbf{v} = \langle 5,8 \rangle$.

25. $3\mathbf{u}$

26. $2\mathbf{u} + \mathbf{v}$

27. $4\mathbf{v} - \mathbf{u}$

28. $\dfrac{1}{4}\mathbf{u} + \dfrac{3}{4}\mathbf{v}$

29. $\mathbf{u} - \dfrac{1}{2}\mathbf{v}$

30. $\dfrac{2}{5}\mathbf{v} - \pi\mathbf{u}$

31–34 Find the coordinates of the endpoint of the indicated vector with the given initial point if $\mathbf{u} = \langle 3,-1,2 \rangle$ and $\mathbf{v} = \langle 2,1,-1 \rangle$.

31. $2\mathbf{u}$ with initial point $(4,0,-1)$

32. $-4\mathbf{v}$ with initial point $(2,5,-3)$

33. $\mathbf{u} - 2\mathbf{v}$ with initial point $(3,7,-10)$

34. $5\mathbf{v} - 3\mathbf{u}$ with initial point $(-3,6,11)$

35–36 Find a vector $\mathbf{v}$ that solves the vector equation.

35. $2\mathbf{v} + \langle 4,-8,2 \rangle = \langle 2,2,2 \rangle$

36. $\langle 4,-7,1 \rangle - 4\mathbf{v} = \langle 5,-1,0 \rangle$

37–38 Use vectors to determine whether the given points are collinear (i.e., whether they fall on the same line).

37. $P(2,0,-1)$, $Q(-4,3,1)$, $R(8,-3,-3)$

38. $P\left(1,\dfrac{3}{2},-5\right)$, $Q(3,-1,-1)$, $R\left(9,-\dfrac{17}{2},13\right)$

39–41 Express the vector $\mathbf{w}$ as a linear combination of $\mathbf{u}$ and $\mathbf{v}$. (**Hint:** Using undetermined coefficients, start with the vector equation $\mathbf{w} = a\mathbf{u} + b\mathbf{v}$, and solve the resulting linear system.)

39. $\mathbf{w} = \langle 1,2 \rangle$; $\mathbf{u} = \langle 2,3 \rangle$, $\mathbf{v} = \langle 0,1 \rangle$

40. $\mathbf{w} = \langle 4,-7 \rangle$; $\mathbf{u} = \langle 1,5 \rangle$, $\mathbf{v} = \langle 2,1 \rangle$

41. $\mathbf{w} = \langle -13,6 \rangle$; $\mathbf{u} = \langle -3,1 \rangle$, $\mathbf{v} = \langle 4,-8 \rangle$

42–45 Find the magnitude of the given vector.

42. $\mathbf{u} = \langle -4,3 \rangle$

43. $\mathbf{v} = \left\langle \dfrac{1}{5}, \dfrac{\sqrt{3}}{5} \right\rangle$

44. $\mathbf{w} = \left\langle 3, \dfrac{9}{4} \right\rangle$

45. $\mathbf{z} = \left\langle -\dfrac{1}{2}, -\sqrt{2} \right\rangle$

46–49 Find the indicated magnitude given that $\mathbf{u} = \langle -2,-5 \rangle$ and $\mathbf{v} = \langle -1,3 \rangle$.

46. $|\mathbf{u} + 2\mathbf{v}|$

47. $|\mathbf{v} - 4\mathbf{u}|$

48. $\left|\dfrac{\mathbf{u} - \mathbf{v}}{3}\right|$

49. $\left|-\dfrac{1}{10}\mathbf{u} + \dfrac{2}{5}\mathbf{v}\right|$

50–55 Determine whether the vectors $\overrightarrow{P_1Q_1}$ and $\overrightarrow{P_2Q_2}$ are equal.

50. $P_1(0,1),\quad Q_1(5,1),\quad P_2(-1,-2),\quad Q_2(4,-2)$

51. $P_1(-2,0),\quad Q_1(-2,-3),\quad P_2(1,-1),\quad Q_2(1,2)$

52. $P_1(0,0),\quad Q_1\left(0,\sqrt{2}\right),\quad P_2(1,1),\quad Q_2(2,2)$

53. $P_1(-5,0,2),\quad Q_1(1,1,-4),\quad P_2(0,-1,7),\quad Q_2(6,0,1)$

54. $P_1(0,0,0),\quad Q_1(3,5,-1),\quad P_2(4,1,-3),\quad Q_2(7,6,2)$

55. $P_1(2,1,3),\quad Q_1(-1,-4,4),\quad P_2(3,2,1),$
    $Q_2(4,-1,-4)$

56–61 Find the component form and magnitude of the vector $\overrightarrow{PQ}$.

56. $P(4,0,-3),\quad Q(0,0,0)$

57. $P(2,9,-5),\quad Q(2,9,-5)$

58. $P(-1,3,4),\quad Q(-5,0,4)$

59. $P(-2,-1,3),\quad Q(1,-4,1)$

60. $P\left(-5,1,\dfrac{\sqrt{7}}{2}\right),\quad Q\left(-1,-\dfrac{1}{2},0\right)$

61. $P\left(\dfrac{1}{3},8,\dfrac{\sqrt{5}+1}{3}\right),\quad Q\left(-\dfrac{1}{3},5,\dfrac{1}{3}\right)$

62–67 Decide whether the points determine a parallelogram in three-dimensional Cartesian space.

62. $A(-1,2),\quad B(-2,-3),\quad C(6,-2),\quad D(7,3)$

63. $A(-1,4),\quad B(-3,1),\quad C(2,-5),\quad D\left(5,-\dfrac{1}{2}\right)$

64. $A(2,0,-3),\quad B(-4,1,0),\quad C(-1,2,7),\quad D(5,1,4)$

65. $A(-1,1,2),\quad B(5,0,-2),\quad C(9,-3,0),\quad D(3,-2,4)$

66. $A(0,1,0),\quad B(-3,0,4),\quad C(-1,1,4),\quad D(2,1,1)$

67. $A(-1,-2,-3),\quad B(4,1,1),\quad C(-5,-2,-4),$
    $D(0,-1,0)$

68–71 Use the technique seen in Example 5 to find the coordinates of the indicated point.

68. The point one-third of the way from $P(12,-3,0)$ to $Q(0,6,-9)$

69. The point four-fifths of the way from $P(4,2,-5)$ to $Q\left(\tfrac{1}{4},2,-10\right)$

70. The point one percent of the way from $P(-3.8,-2.2,1.5)$ to $Q(2.4,-5.6,10)$

71. The point(s) on the line $\overleftrightarrow{PQ}$ with a distance from $Q(-3,1,7)$ equaling three times the distance from $P(1,5,3)$

72–77 For the given vector $\mathbf{v}$, find the unit vector $\mathbf{u}$ pointing in the same direction. Express your answer in terms of the standard basis vectors.

72. $\mathbf{v}=\langle-8,6\rangle$

73. $\mathbf{v}=\langle2,9\rangle$

74. $\mathbf{v}=\langle2,0,-1\rangle$

75. $\mathbf{v}=\langle-4,-5,2\rangle$

76. $\mathbf{v}=\left\langle3,-\dfrac{5}{4},\dfrac{13}{4}\right\rangle$

77. $\mathbf{v}=\left\langle1,-\dfrac{3}{2},-\dfrac{1}{2}\right\rangle$

78. Find the unit vector $\mathbf{u}$ in $\mathbb{R}^2$ that makes a directed $2\pi/3$ radian angle with the positive x-axis. Express your answer as a linear combination of the standard basis vectors of $\mathbb{R}^2$.

79. Show that any vector $\mathbf{u}$ in $\mathbb{R}^2$ that can be written as $\mathbf{u}=\cos\theta\,\mathbf{i}+\sin\theta\,\mathbf{j}$, for some $0\le\theta\le2\pi$, is a unit vector.

80. Find an appropriate scalar a for the vector $\mathbf{v}=\langle2,1,-1\rangle$ so that $a\mathbf{v}$ has a magnitude of 3.

81. Find an appropriate scalar a for the vector $\mathbf{w}=\langle-\sqrt{3},4,-5\rangle$ so that $a\mathbf{w}$ has a magnitude of 5.

82. Find the vector $\mathbf{s}$ that makes a directed $7\pi/6$ radian angle with the positive x-axis and has a magnitude of 2. Express your answer in component form.

83. By first solving the linear system of Example 7 for $|\mathbf{T}_1|$ and $|\mathbf{T}_2|$, verify that $\mathbf{T}_1\approx\langle-4.55,2.12\rangle$ and $\mathbf{T}_2\approx\langle4.55,7.88\rangle$.

84. Suppose that in Example 7 the ten-pound weight is suspended from two ropes that form angles of 75° and 51°, respectively, with the horizontal direction (refer to Figure 10). Find the tension forces $\mathbf{T}_1$ and $\mathbf{T}_2$ under these conditions.

85. A baseball bounces off a bat with an initial velocity that has a vertical component of 35.2 m/s and a horizontal component of 8 m/s. Ignoring air resistance, determine its velocity at time t, and sketch its position function. (Assume the initial height is 1 m. See Example 2.)

86. A jetliner flying at 600 mph due west encounters an 80 mph headwind that blows 30° south of east. If the captain wants to keep both his ground speed and direction, how much increase in speed will be needed and in what direction should he steer the plane?

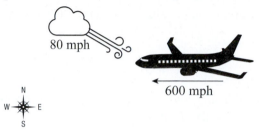

87. A jetliner flying at 500 mph due southeast encounters a 60 mph tailwind that blows from the north. At the same time, a 20 mph updraft is affecting the flight. Find the actual ground speed of the jetliner under these conditions.

88. A guest in the restaurant section of a passenger train gets up from his table and cuts across toward the bar walking at a steady pace of 2 mph. If the train is moving at a constant eastward velocity of 80 mph, and the passenger is walking northeast (45° north of east), what is his resultant (ground) velocity?

89.* A plane is taking off with a flight plan that calls for a takeoff velocity of 250 mph in the direction of $\langle 1,4,2 \rangle$. However, the plane experiences a tailwind, blowing at 15 mph in the direction $\langle 2,1,0 \rangle$. Calculate the plane's actual direction and ground speed at takeoff. (Express your direction vector in a form where the first component is 1.)

90. Suppose the following three forces are acting simultaneously upon an object of mass 2 kilograms: $\mathbf{F}_1 = 2\,\mathbf{i} + \mathbf{j} - \mathbf{k}$, $\mathbf{F}_2 = 3\,\mathbf{i} - \mathbf{j} + 5\,\mathbf{k}$, and $\mathbf{F}_3 = 7\,\mathbf{i} - 4\,\mathbf{j} + 2\,\mathbf{k}$ (units are in newtons). Find a unit vector $\mathbf{u}$ pointing in the direction of acceleration as well as the magnitude of the acceleration of the object. (**Hint:** Use Newton's Second Law of Motion.)

91. Use the component definition of scalar multiplication to verify the five scalar multiplication properties of vectors, as listed immediately preceding Example 3.

92. Use the component definition of vector addition to verify the four vector addition properties listed immediately preceding Example 3.

93.* Suppose that for some scalars a, b and c, $a\,\mathbf{i} + b\,\mathbf{j} + c\,\mathbf{k} = \mathbf{0}$, where $\mathbf{i}$, $\mathbf{j}$, and $\mathbf{k}$ are the standard basis vectors in $\mathbb{R}^3$, and $\mathbf{0}$ is the zero vector. Prove: $a = b = c = 0$.

94–98 Generalize the rules of this section to answer the following problems for vectors in higher dimensions.

94. Find $|\mathbf{u}|$ for $\mathbf{u} = \langle 1,-4,2,3 \rangle$.

95. Find $3\,\mathbf{u} - 2\,\mathbf{v}$ for $\mathbf{u} = \langle 1,-4,2,3 \rangle$ and $\mathbf{v} = \langle 2,-5,0,-1 \rangle$.

96. Find $|2\,\mathbf{u} + \mathbf{v}|$ for $\mathbf{u} = \langle 3,1,\frac{1}{2},5 \rangle$ and $\mathbf{v} = \langle -2,0,1,-4 \rangle$.

97. Find $\mathbf{u} - \dfrac{\mathbf{v}}{3}$ for $\mathbf{u} = \langle 2,0,-4,6,-8,0 \rangle$ and $\mathbf{v} = \langle 0,-3,9,-12,1,6 \rangle$.

98. Find $\left| \dfrac{\mathbf{u}}{2} + 3\,\mathbf{v} \right|$ for $\mathbf{u} = \langle 4,-2,0,8,0,-6 \rangle$ and $\mathbf{v} = \langle \frac{2}{3},-2,1,0,-\frac{1}{3},1 \rangle$.

11.3 The Dot Product

TOPICS

1. The dot product and its properties
2. Applications of the dot product

In this section and the next, we introduce the *dot product* and the *cross product*, two ways of combining vectors that join scalar multiplication and vector addition in our list of vector operations. As we will see, the dot product is useful for, among other things, describing relationships between vectors and solving problems involving forces.

TOPIC 1   The Dot Product and Its Properties

The dot product takes its name from the notation typically used for the operation, but it also goes by the names *scalar product* (because the result is a scalar) and *inner product* (in contrast to an operation called *outer product* seen in other areas of mathematics). We will be concerned with the dot product of pairs of vectors in $\mathbb{R}^2$ and $\mathbb{R}^3$, but the definition below can be extended to many other spaces.

Definition 💡

Dot Product

Given two vectors $\mathbf{u} = \langle u_1, u_2, u_3 \rangle$ and $\mathbf{v} = \langle v_1, v_2, v_3 \rangle$, the **dot product** $\mathbf{u} \cdot \mathbf{v}$ of the two vectors is the scalar defined by

$$\mathbf{u} \cdot \mathbf{v} = u_1 v_1 + u_2 v_2 + u_3 v_3.$$

A similar formula defines the dot product of two vectors in $\mathbb{R}^2$.

Example 1 ✎

Evaluate each of the following dot products.

a. $\langle -2, 0, 5 \rangle \cdot \langle 6, 2, 3 \rangle$ **b.** $\langle 4, -5 \rangle \cdot \langle 1, 1 \rangle$ **c.** $\langle 2, 1, -2 \rangle \cdot \langle 2, 1, -2 \rangle$

Solution

a. $\langle -2, 0, 5 \rangle \cdot \langle 6, 2, 3 \rangle = (-2)(6) + (0)(2) + (5)(3) = -12 + 0 + 15 = 3$

b. $\langle 4, -5 \rangle \cdot \langle 1, 1 \rangle = (4)(1) + (-5)(1) = 4 - 5 = -1$

c. $\langle 2, 1, -2 \rangle \cdot \langle 2, 1, -2 \rangle = (2)(2) + (1)(1) + (-2)(-2) = 4 + 1 + 4 = 9$

The dot product possesses a number of properties, beginning with those in the next theorem. The verifications of these properties in $\mathbb{R}^2$ and $\mathbb{R}^3$ are left as Exercises 22–26.

Theorem 🔍

Properties of the Dot Product

Assume $\mathbf{u}$, $\mathbf{v}$, and $\mathbf{w}$ represent vectors and that a represents a scalar. Then the following properties hold.

$$\mathbf{u} \cdot \mathbf{v} = \mathbf{v} \cdot \mathbf{u} \qquad\qquad \mathbf{0} \cdot \mathbf{u} = 0$$

$$\mathbf{u} \cdot (\mathbf{v} + \mathbf{w}) = \mathbf{u} \cdot \mathbf{v} + \mathbf{u} \cdot \mathbf{w} \qquad\qquad a(\mathbf{u} \cdot \mathbf{v}) = (a\mathbf{u}) \cdot \mathbf{v} = \mathbf{u} \cdot (a\mathbf{v})$$

$$\mathbf{u} \cdot \mathbf{u} = |\mathbf{u}|^2$$

Another property, however, is not so immediately clear. The following theorem shows how the dot product can be used to quickly determine the angle between any two vectors in $\mathbb{R}^2$ or $\mathbb{R}^3$. Its proof makes use of the Law of Cosines and several of the properties above.

Theorem 🔍

Dot Product and the Angle between Two Vectors

If two nonzero vectors $\mathbf{u}$ and $\mathbf{v}$ are depicted so that their initial points coincide, and if θ represents the smaller of the two angles formed by $\mathbf{u}$ and $\mathbf{v}$ (so that $0 \leq \theta \leq \pi$), then

$$\mathbf{u} \cdot \mathbf{v} = |\mathbf{u}||\mathbf{v}|\cos\theta.$$

Proof ✎

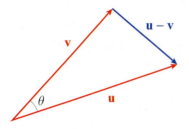

Figure 1

Given $\mathbf{u}$ and $\mathbf{v}$ placed so that their initial points coincide, construct $\mathbf{u} - \mathbf{v}$ as shown in Figure 1. Then by the Law of Cosines,

$$|\mathbf{u} - \mathbf{v}|^2 = |\mathbf{u}|^2 + |\mathbf{v}|^2 - 2|\mathbf{u}||\mathbf{v}|\cos\theta.$$

The fact that this equation is true even in the two extreme cases $\theta = 0$ and $\theta = \pi$ is left as Exercise 28.

Since the dot product of a vector with itself is the square of the norm of the vector, we can also write

$$|\mathbf{u} - \mathbf{v}|^2 = (\mathbf{u} - \mathbf{v}) \cdot (\mathbf{u} - \mathbf{v}) = \mathbf{u} \cdot \mathbf{u} - \mathbf{u} \cdot \mathbf{v} - \mathbf{v} \cdot \mathbf{u} + \mathbf{v} \cdot \mathbf{v} = |\mathbf{u}|^2 - 2(\mathbf{u} \cdot \mathbf{v}) + |\mathbf{v}|^2.$$

Equating these two different ways of expressing $|\mathbf{u} - \mathbf{v}|^2$ and canceling like terms, we obtain

$$-2(\mathbf{u} \cdot \mathbf{v}) = -2|\mathbf{u}||\mathbf{v}|\cos\theta,$$

which, once we divide both sides by -2, is the desired result.

Example 2 ✐

Find the angle between the vectors $\mathbf{u} = \langle -3, 1, 4 \rangle$ and $\mathbf{v} = \langle 2, 5, -1 \rangle$.

Solution

We begin by determining that $\mathbf{u} \cdot \mathbf{v} = (-3)(2) + (1)(5) + (4)(-1) = -5$ and then

$$\cos\theta = \frac{\mathbf{u} \cdot \mathbf{v}}{|\mathbf{u}||\mathbf{v}|} = \frac{-5}{\left(\sqrt{9+1+16}\right)\left(\sqrt{4+25+1}\right)} = -\frac{5}{2\sqrt{195}}.$$

Therefore $\theta = \cos^{-1}\left(-\dfrac{5}{2\sqrt{195}}\right) \approx 1.75$, or approximately $100.3°$.

TOPIC 2 Applications of the Dot Product

By definition, the angle θ in the formula $\mathbf{u} \cdot \mathbf{v} = |\mathbf{u}||\mathbf{v}|\cos\theta$ lies in the interval $[0, \pi]$. Since $|\mathbf{u}||\mathbf{v}| > 0$ for all nonzero vectors $\mathbf{u}$ and $\mathbf{v}$, the sign of $\mathbf{u} \cdot \mathbf{v}$ will thus always be the same as the sign of $\cos\theta$. Specifically, we note the following:

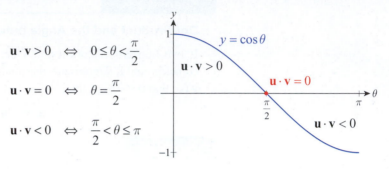

$$\mathbf{u} \cdot \mathbf{v} > 0 \quad \Leftrightarrow \quad 0 \le \theta < \frac{\pi}{2}$$

$$\mathbf{u} \cdot \mathbf{v} = 0 \quad \Leftrightarrow \quad \theta = \frac{\pi}{2}$$

$$\mathbf{u} \cdot \mathbf{v} < 0 \quad \Leftrightarrow \quad \frac{\pi}{2} < \theta \le \pi$$

Figure 2

Vectors for which $\theta = \pi/2$ are called *orthogonal* or *perpendicular*, and the zero vector $\mathbf{0}$ is considered to be orthogonal to every vector.

Definition 💡

Orthogonal Vectors

Two vectors $\mathbf{u}$ and $\mathbf{v}$ are **orthogonal**, or **perpendicular**, if $\mathbf{u} \cdot \mathbf{v} = 0$.

Another synonym for orthogonal is **normal**, often used when discussing lines or planes. For example, a vector $\mathbf{u}$ is said to be normal to a given plane if $\mathbf{u} \cdot \mathbf{v} = 0$ for every vector $\mathbf{v}$ in the plane.

Example 3 ✎

Determine k so that the vectors $\langle -2, 4, -3 \rangle$ and $\langle 1, 2, k \rangle$ are orthogonal.

Solution

$$\langle -2, 4, -3 \rangle \cdot \langle 1, 2, k \rangle = -2 + 8 - 3k = 6 - 3k$$

The two vectors are orthogonal if and only if $6 - 3k = 0$, that is, when $k = 2$.

Example 4 ✎

Describe the set of all vectors orthogonal to $\mathbf{j}$.

Solution

We seek all vectors $\langle a, b, c \rangle$ such that $\langle a, b, c \rangle \cdot \mathbf{j} = 0$; in other words, the solution set of the equation $\langle a, b, c \rangle \cdot \langle 0, 1, 0 \rangle = 0$. Since this reduces to $b = 0$, all vectors of the form $\langle a, 0, c \rangle$ are orthogonal to $\mathbf{j}$. When graphed in standard position, these are all the vectors that lie in the xz-plane.

The angles between a given nonzero vector **u** and the three standard basis vectors **i**, **j**, and **k** are called the **direction angles** α, β, and γ of **u**. An intermediate step in finding the direction angles of a vector is to determine its **direction cosines** $\cos\alpha$, $\cos\beta$, and $\cos\gamma$. They are easily found using our dot product formula. If $\mathbf{u} = \langle u_1, u_2, u_3 \rangle$, then we have the following.

$$\cos\alpha = \frac{\mathbf{u}\cdot\mathbf{i}}{|\mathbf{u}||\mathbf{i}|} = \frac{u_1}{|\mathbf{u}|} \qquad \cos\beta = \frac{\mathbf{u}\cdot\mathbf{j}}{|\mathbf{u}||\mathbf{j}|} = \frac{u_2}{|\mathbf{u}|} \qquad \cos\gamma = \frac{\mathbf{u}\cdot\mathbf{k}}{|\mathbf{u}||\mathbf{k}|} = \frac{u_3}{|\mathbf{u}|}$$

Note that

$$\cos^2\alpha + \cos^2\beta + \cos^2\gamma = \frac{1}{|\mathbf{u}|^2}\left(u_1^2 + u_2^2 + u_3^2\right) = \frac{1}{|\mathbf{u}|^2}|\mathbf{u}|^2 = 1$$

and

$$\langle u_1, u_2, u_3 \rangle = |\mathbf{u}|\langle \cos\alpha, \cos\beta, \cos\gamma \rangle,$$

so $|\mathbf{u}|\langle \cos\alpha, \cos\beta, \cos\gamma \rangle$ expresses **u** as a product of its length and a unit vector pointing in the same direction.

Example 5 ✐

Find the direction angles of the vector $\langle 1, -2, 3 \rangle$.

Solution

$$\left|\langle 1, -2, 3 \rangle\right| = \sqrt{14}$$

$$\cos\alpha = \frac{1}{\sqrt{14}} \qquad \cos\beta = \frac{-2}{\sqrt{14}} \qquad \cos\gamma = \frac{3}{\sqrt{14}}$$

Hence, the direction angles are as follows:

$$\alpha = \cos^{-1}\left(\frac{1}{\sqrt{14}}\right) \approx 74.5°$$

$$\beta = \cos^{-1}\left(\frac{-2}{\sqrt{14}}\right) \approx 122.3°$$

$$\gamma = \cos^{-1}\left(\frac{3}{\sqrt{14}}\right) \approx 36.7°$$

These angles and the vector $\langle 1, -2, 3 \rangle$ are shown in Figure 3.

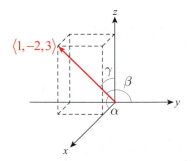

Figure 3

In many vector applications, a key step involves writing one vector **u** as a sum of two vectors, one of which is parallel to another vector **v** and the other of which is orthogonal to **v**. We call the first vector the **projection of u onto v** and give it the name $\text{proj}_{\mathbf{v}}\mathbf{u}$. Once we determine $\text{proj}_{\mathbf{v}}\mathbf{u}$, it is easy to determine the second (orthogonal) vector.

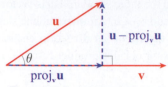

Figure 4 Projection of **u** onto **v**

Figure 4 is an illustration of vectors **u** and **v** and the decomposition of **u** into such a sum in the case where $0 \le \theta \le \pi/2$. From trigonometry, we know that $|\text{proj}_{\mathbf{v}}\mathbf{u}| = |\mathbf{u}|\cos\theta$, and the dot product formula also tells us that

$$\cos\theta = \frac{\mathbf{u} \cdot \mathbf{v}}{|\mathbf{u}||\mathbf{v}|},$$

so

$$|\text{proj}_{\mathbf{v}}\mathbf{u}| = |\mathbf{u}|\frac{\mathbf{u} \cdot \mathbf{v}}{|\mathbf{u}||\mathbf{v}|} = \frac{\mathbf{u} \cdot \mathbf{v}}{|\mathbf{v}|}.$$

To construct the vector $\text{proj}_{\mathbf{v}}\mathbf{u}$, we multiply the magnitude $|\text{proj}_{\mathbf{v}}\mathbf{u}|$ by a unit vector pointing in the same direction as **v**:

$$\text{proj}_{\mathbf{v}}\mathbf{u} = \left(\frac{\mathbf{u} \cdot \mathbf{v}}{|\mathbf{v}|}\right)\left(\frac{\mathbf{v}}{|\mathbf{v}|}\right) = \left(\frac{\mathbf{u} \cdot \mathbf{v}}{|\mathbf{v}|^2}\right)\mathbf{v}$$

The same formula also applies if $\pi/2 \le \theta \le \pi$, but in this case $\mathbf{u} \cdot \mathbf{v}$ is negative and hence $\text{proj}_{\mathbf{v}}\mathbf{u}$ points in the opposite direction of **v**.

Notice that the derivation above makes no mention of the dimensionality of the space—it is valid in both $\mathbb{R}^2$ and $\mathbb{R}^3$.

Example 6 ✎

Let $\mathbf{u} = \langle 4, -1, 3\rangle$ and $\mathbf{v} = \langle -3, 0, 1\rangle$. Express **u** as a sum of two vectors, one parallel to **v** and one perpendicular to **v**.

Solution

The projection of **u** onto **v** is the first vector we seek (the one parallel to **v**), and since $\mathbf{u} \cdot \mathbf{v} = \langle 4, -1, 3\rangle \cdot \langle -3, 0, 1\rangle = -12 + 3 = -9$ and $|\mathbf{v}| = \sqrt{10}$,

$$\text{proj}_{\mathbf{v}}\mathbf{u} = \left(\frac{\mathbf{u} \cdot \mathbf{v}}{|\mathbf{v}|^2}\right)\mathbf{v} = \frac{-9}{10}\langle -3, 0, 1\rangle = \left\langle \frac{27}{10}, 0, -\frac{9}{10}\right\rangle.$$

By the construction of $\text{proj}_{\mathbf{v}}\mathbf{u}$, $\mathbf{u} - \text{proj}_{\mathbf{v}}\mathbf{u}$ is guaranteed to be perpendicular to **v**.

$$\mathbf{u} - \text{proj}_{\mathbf{v}}\mathbf{u} = \langle 4, -1, 3\rangle - \left\langle \frac{27}{10}, 0, -\frac{9}{10}\right\rangle = \left\langle \frac{13}{10}, -1, \frac{39}{10}\right\rangle$$

As verification, note that $\left\langle \frac{27}{10}, 0, -\frac{9}{10}\right\rangle \cdot \left\langle \frac{13}{10}, -1, \frac{39}{10}\right\rangle = 0$.

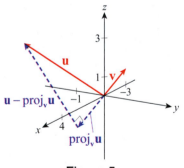

Figure 5

Physics and engineering problems often require finding the component of one vector that lies in the direction of another vector, especially in applications involving force and work. In Section 6.5, we learned that the work W performs in applying a force F over a distance d is given by $W = Fd$. But this equation assumes that F is applied in the direction of motion (along which d is measured). More generally, if the force applied is represented by the vector **F** and the direction of motion by the vector **D**, then the amount of work W done is equal to the portion of the force applied in the direction of **D** multiplied by the length of **D**: $W = \left|\text{proj}_{\mathbf{D}}\mathbf{F}\right|\left|\mathbf{D}\right|$. This leads to the following:

$$W = \left|\text{proj}_{\mathbf{D}}\mathbf{F}\right|\left|\mathbf{D}\right| = \frac{\mathbf{F}\cdot\mathbf{D}}{\left|\mathbf{D}\right|}\left|\mathbf{D}\right| = \mathbf{F}\cdot\mathbf{D}$$

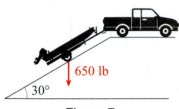

Figure 6 Projection of **F** onto **D**

Figure 6 illustrates how work is related to a representative force vector and direction vector. Note that by our dot product formula,

$$W = \mathbf{F}\cdot\mathbf{D} = \left|\mathbf{F}\right|\left|\mathbf{D}\right|\cos\theta,$$

a formula often seen in physics texts.

Example 7 ✎

A boat and trailer, which together weigh 650 pounds, are to be pulled up a boat ramp that has an incline of 30°. What force is required to merely prevent the boat and trailer from rolling down the ramp?

Solution

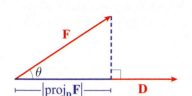

Figure 7

To gain some insight into this problem, note that the force that would be required to hold the boat and trailer up off the ground is 650 pounds. But the ramp supports a portion of the weight all by itself—we only need to supply the force needed to counteract the component of the weight that lies in the direction of the ramp.

We start by defining $\mathbf{F} = \langle 0, -650\rangle$, a vector which represents the magnitude and direction of the boat and trailer weight. To find the component of **F** parallel to the direction of the ramp, we need a vector **v** pointing down the ramp. The length of **v** is immaterial, since the component of **F** pointing in the direction of **v** is independent of $\left|\mathbf{v}\right|$ (you should check this assertion for yourself; see Exercise 27). Given that the incline of the ramp is 30°, we know that **v** must be some scalar multiple of $\langle -\sqrt{3}, -1\rangle$; we will keep things simple and just let **v** be $\langle -\sqrt{3}, -1\rangle$. Figure 8 is an illustration of **F** and a positive scalar multiple of **v** (the vector **v** as we have defined it is too short to show up in the diagram so we have graphed 100**v**).

We now compute the magnitude of the projection of **F** in the direction of **v**.

$$\left|\text{proj}_{\mathbf{v}}\mathbf{F}\right| = \frac{\mathbf{F}\cdot\mathbf{v}}{\left|\mathbf{v}\right|} = \frac{\langle 0, -650\rangle\cdot\langle -\sqrt{3}, -1\rangle}{\sqrt{3+1}} = \frac{650}{2} = 325$$

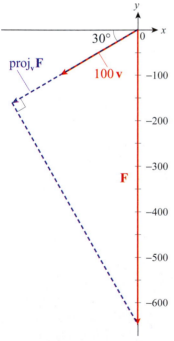

Figure 8
Projection of Weight
onto the Ramp

This is the answer to our question: we must supply 325 pounds of force to keep the boat and trailer from rolling down the ramp—any force greater than 325 pounds will pull the boat and trailer up the ramp. Note that in this problem, we only care about $\left|\text{proj}_{\mathbf{v}}\mathbf{F}\right|$, the *magnitude* of the component of **F** in the direction of **v**, not the actual vector $\text{proj}_{\mathbf{v}}\mathbf{F}$.

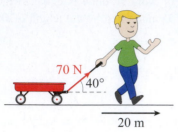

Example 8 ✍

A child pulls a wagon along a sidewalk, exerting a force of 70 newtons on the handle of the wagon. The handle of the wagon is at an angle of 40° to the horizontal. If he pulls the wagon a distance of 20 meters, how much work has been done?

Figure 9

Solution

We start by defining the force and distance vectors.

$$\mathbf{F} = 70\langle \cos 40°, \sin 40° \rangle \quad \text{and} \quad \mathbf{D} = \langle 20, 0 \rangle$$

The calculation of the work done is now straightforward.

$$W = \mathbf{F} \cdot \mathbf{D} = 70\langle \cos 40°, \sin 40° \rangle \cdot \langle 20, 0 \rangle \approx 1072 \ \text{N} \cdot \text{m}$$

11.3 **Exercises**

1–10 Find the dot product of **u** and **v**.

1. $\mathbf{u} = \left\langle 2, \dfrac{1}{3} \right\rangle, \quad \mathbf{v} = \left\langle \dfrac{5}{2}, -3 \right\rangle$

2. $\mathbf{u} = \langle 3, -1 \rangle, \quad \mathbf{v} = \langle 0, 0 \rangle$

3. $\mathbf{u} = \langle 2, 0, 1 \rangle, \quad \mathbf{v} = \langle 1, 1, -3 \rangle$

4. $\mathbf{u} = \left\langle \dfrac{4}{3}, -1, \dfrac{2}{5} \right\rangle, \quad \mathbf{v} = \left\langle -1, 6, -\dfrac{5}{3} \right\rangle$

5. $\mathbf{u} = \langle 4s, -2, 2s \rangle, \quad \mathbf{v} = \langle 3, 5, -6 \rangle$

6. $\mathbf{u} = \langle -5s, s, 2 \rangle, \quad \mathbf{v} = \langle 2t, 4t, -1 \rangle$

7. $\mathbf{u} = \mathbf{i} + 3\mathbf{j} - 2\mathbf{k}, \quad \mathbf{v} = 2\mathbf{i} - \mathbf{j} + \mathbf{k}$

8. $\mathbf{u} = \dfrac{1}{2}\mathbf{i} + \mathbf{j} + \dfrac{5}{3}\mathbf{k}, \quad \mathbf{v} = 8\mathbf{i} - 5\mathbf{j} - 9\mathbf{k}$

9. $|\mathbf{u}| = 4, \quad |\mathbf{v}| = 3\sqrt{2}, \quad$ their angle is $45°$

10. $|\mathbf{u}| = 2.5, \quad |\mathbf{v}| = 5, \quad$ their angle is $2\pi/3$

11–13 Suppose that the length and width of the rectangle below are $2\sqrt{3}$ and 2 units, respectively. Find the indicated dot product.

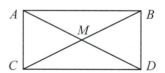

11. $\overrightarrow{CM} \cdot \overrightarrow{MD}$

12. $\overrightarrow{CA} \cdot \overrightarrow{MD}$

13. $\overrightarrow{CA} \cdot \left(\overrightarrow{CM} + \overrightarrow{MD} \right)$

14–21 Find the angle between the given vectors.

14. $\mathbf{u} = \langle 2, -\sqrt{5} \rangle, \quad \mathbf{v} = \langle -4, 3 \rangle$

15. $\mathbf{u} = \langle -1, 1 \rangle, \quad \mathbf{v} = \langle 3, 3 \rangle$

16. $\mathbf{u} = \mathbf{i} - 2\mathbf{j}, \quad \mathbf{v} = 3\mathbf{i} + \mathbf{j}$

17. $\mathbf{u} = (\cos 32°)\mathbf{i} + (\sin 32°)\mathbf{j},$
    $\mathbf{v} = -(\cos 87°)\mathbf{i} + (\sin 87°)\mathbf{j}$

18. $\mathbf{u} = \langle 4, -2, 1 \rangle, \quad \mathbf{v} = \langle 2, -1, 5 \rangle$

19. $\mathbf{u} = \langle -1, 1, 2 \rangle, \quad \mathbf{v} = \langle 3, 2, 0 \rangle$

20. $\mathbf{u} = \mathbf{i} + 2\mathbf{j} - 2\mathbf{k}, \quad \mathbf{v} = 2\mathbf{i} - 2\mathbf{j} + \mathbf{k}$

21. $\mathbf{u} = \dfrac{1}{2}\mathbf{i} - \mathbf{j} + \mathbf{k}, \quad \mathbf{v} = \mathbf{i} - \dfrac{3}{2}\mathbf{j} - \sqrt{3}\mathbf{k}$

22–26 Prove the indicated property of the dot product. (**Hint:** Start by representing vectors **u**, **v**, and **w** in component form.)

22. $\mathbf{u} \cdot \mathbf{v} = \mathbf{v} \cdot \mathbf{u}$

23. $\mathbf{u} \cdot (\mathbf{v} + \mathbf{w}) = \mathbf{u} \cdot \mathbf{v} + \mathbf{u} \cdot \mathbf{w}$

24. $\mathbf{u} \cdot \mathbf{u} = |\mathbf{u}|^2$ **25.** $\mathbf{0} \cdot \mathbf{u} = 0$

26. $a(\mathbf{u} \cdot \mathbf{v}) = (a\mathbf{u}) \cdot \mathbf{v} = \mathbf{u} \cdot (a\mathbf{v})$, a is a scalar

27. Prove that for any positive c,

$$\frac{\mathbf{u} \cdot \mathbf{v}}{|\mathbf{v}|} = \frac{\mathbf{u} \cdot (c\mathbf{v})}{|c\mathbf{v}|}.$$

(**Hint:** Use the properties of the dot product.)

28. Prove the equation

$$|\mathbf{u} - \mathbf{v}|^2 = |\mathbf{u}|^2 + |\mathbf{v}|^2 - 2|\mathbf{u}||\mathbf{v}|\cos\theta$$

for the two cases $\theta = 0$ and $\theta = \pi$. (**Hint:** Use the fact that $\mathbf{v} = c\mathbf{u}$ for some constant c in these two cases, with c positive if $\theta = 0$ and c negative if $\theta = \pi$.)

29–40 Determine whether the given vectors are parallel, orthogonal, or neither.

29. $\langle 2,6 \rangle$ and $\langle -1,-3 \rangle$ **30.** $\langle 6,4 \rangle$ and $\langle -2,3 \rangle$

31. $\langle 2,-3 \rangle$ and $\left\langle -\frac{1}{2}, \frac{1}{3} \right\rangle$

32. $\langle s,2t \rangle$ and $\left\langle -3t, \frac{3}{2}s \right\rangle$

33. $\langle 1,0,-7 \rangle$ and $\left\langle 0,-\frac{5}{3},0 \right\rangle$

34. $\langle -s,5s,2s \rangle$ and $\left\langle \frac{3}{2}s, \frac{-15}{2}s, -3s \right\rangle$

35. $\mathbf{i} + \mathbf{j}$ and $\mathbf{j} - \mathbf{i}$

36. $\mathbf{i} + \mathbf{j}$ and $-\frac{3}{7}\mathbf{k}$

37. $\frac{1}{2}\mathbf{i} + \mathbf{j} + 2\mathbf{k}$ and $\mathbf{i} + \frac{3}{2}\mathbf{j} - \mathbf{k}$

38. $-\mathbf{i} + \frac{2}{3}\mathbf{j} - 4\mathbf{k}$ and $5\mathbf{i} + 2\mathbf{j} - \frac{5}{4}\mathbf{k}$

39. $\langle \cos\theta, \sin\theta \rangle$ and $\langle -\sin\theta, \cos\theta \rangle$

40. $\langle \cos(90° - \theta), -\cos\theta, 1 \rangle$ and $\langle -\sin\theta, \cos\theta, 1 \rangle$

41–44 Use the vector method to decide which of the following are true of $\triangle ABC$: **a.** acute, **b.** obtuse, **c.** isosceles, **d.** equilateral, **e.** right triangle. (**Hint:** Determine interior angles.)

41. $A(1,2,5)$, $B(-1,2,-4)$, $C(8,8,5)$

42. $A(-4,5,6)$, $B(-2,4,13)$, $C(-1,-3,4)$

43. $A(2,-1,1)$, $B(6,2,0)$, $C(3,1,5)$

44. $A(3,-4,1)$, $B(1,-3,4)$, $C(0,-6,2)$

45. Find the angles determined by the two diagonals of the quadrilateral with vertices $\left(1,2,\frac{1}{3}\right)$, $(0,-1,0)$, $\left(3,1,-\frac{4}{3}\right)$, and $(2,-3,-2)$.

46–49 Determine the value of the parameter so that the vectors are orthogonal.

46. $\left\langle s, \frac{1}{2}s, 2 \right\rangle$ and $\langle -5, 2s, 3 \rangle$

47. $\langle 2,s,-3s \rangle$ and $\langle s,1,1 \rangle$

48. $\langle t,-2,t \rangle$ and $\langle t,t,t^2 \rangle$

49. $\langle 4,s,-2 \rangle$ and $\langle 4,5t,8 \rangle$

50. Find a vector of length $\sqrt{3}$ in $\mathbb{R}^3$ that is orthogonal to both $\langle 1,0,1 \rangle$ and $\langle 0,1,1 \rangle$.

51. Use vectors to show that a parallelogram is a rhombus if and only if its diagonals are perpendicular.

52. Prove that if the diagonals of a parallelogram are perpendicular, then the parallelogram is a rhombus.

53. Thales' Theorem states that any point on a circle determines a right triangle with the endpoints of any diameter not containing the point. Use vectors to prove Thales' Theorem. (**Hint:** It is enough to prove the statement for the unit circle centered at the origin.)

54. Find an equation of the line that contains the point (a,b) and is perpendicular to the vector $\mathbf{n} = \langle n_1, n_2 \rangle$ (**n** is called a normal vector to the line). (**Hint:** Notice that for any point (x, y) on the line, the vector $\langle x - a, y - b \rangle$ must be perpendicular to **n**, so their dot product is 0.)

55. Use Exercise 54 to derive the (well-known) equation of a line of slope m through the point (a,b). (**Hint:** Notice that $\mathbf{n} = \langle -m, 1 \rangle$ is a normal vector.)

56. Use normal vectors to determine the angle between the lines with slopes $m = 1$ and $m = 3$, respectively. (**Hint:** Note that the angle between the lines is the same as that between their respective normal vectors. See the hint given in Exercise 55.)

57–58 Find a unit vector that is normal (or perpendicular) to the given line.

57. $x - 3y = 7$

58. $y = \dfrac{1}{2}x - 5$

59. Generalize Exercise 54 to find an equation of the plane in $\mathbb{R}^3$ that contains the point (a,b,c) and is perpendicular to $\mathbf{n} = \langle n_1, n_2, n_3 \rangle$. (The vector $\mathbf{n}$ is called a normal vector to the plane. We will elaborate on this approach in Section 11.5.)

60. Use Exercise 59 to find the equation of the plane through $(2, -4, 1)$ that is perpendicular to $\mathbf{n} = \langle 5, -3, 1 \rangle$.

61–62 Find a normal vector to the given plane.

61. $x - 4y - 2z = 7$

62. $2x + y - 5z = 1$

63. Use the results of Exercises 61 and 62 to find the angle between the planes $x - 4y - 2z = 7$ and $2x + y - 5z = 1$. (**Hint:** The angle between two planes is the same as that between their respective normal vectors.)

64–67 Find the direction angles of the given vector.

64. $\langle 1, 0, -1 \rangle$

65. $\langle 2, -1, 4 \rangle$

66. $\left\langle -\dfrac{1}{3}, 2, -\dfrac{2\sqrt{3}}{3} \right\rangle$

67. $\left\langle \dfrac{1}{2}, -4, -\dfrac{5}{3} \right\rangle$

68. Suppose the first and second direction angles of a vector in the first octant are $\pi/6$ and $\pi/3$. Find the third direction angle.

69. Prove the *Cauchy-Schwarz Inequality*: For any vectors $\mathbf{u}$ and $\mathbf{v}$ in $\mathbb{R}^2$ or $\mathbb{R}^3$,

$$|\mathbf{u} \cdot \mathbf{v}| \le |\mathbf{u}||\mathbf{v}|.$$

Under what conditions does equality hold? (Note that on the left-hand side, $|\cdot|$ means the absolute value of a scalar, while the right-hand side is the product of the magnitudes of the vectors.)

70. Use the Cauchy-Schwarz Inequality (Exercise 69) to prove the famous *Triangle Inequality*: For any vectors $\mathbf{u}$ and $\mathbf{v}$ in $\mathbb{R}^2$ or $\mathbb{R}^3$,

$$|\mathbf{u} + \mathbf{v}| \le |\mathbf{u}| + |\mathbf{v}|.$$

Under what conditions does equality hold? (Note that this inequality states that the sum of the lengths of two vectors never exceeds the sum of their individual lengths. **Hint:** Estimate $|\mathbf{u} + \mathbf{v}|^2$ by writing $|\mathbf{u} + \mathbf{v}|^2 = (\mathbf{u} + \mathbf{v}) \cdot (\mathbf{u} + \mathbf{v})$. Then use the properties of the dot product and the Cauchy-Schwarz Inequality.)

71. What can you say about $\mathbf{v}$ if $|\mathbf{u}| = 2$, $|\mathbf{v}| = 4$, and $|\mathbf{u} \cdot \mathbf{v}|$ is maximum?

72. Use Exercise 70 to prove the following.

$$\big| |\mathbf{u}| - |\mathbf{v}| \big| \le |\mathbf{u} - \mathbf{v}|$$

(This is often called the "left-hand part of the Triangle Inequality," or the "Reverse Triangle Inequality.")

73. Prove the following so-called *Parallelogram Law*. Can you give a reason for its name?

$$|\mathbf{u} + \mathbf{v}|^2 + |\mathbf{u} - \mathbf{v}|^2 = 2|\mathbf{u}|^2 + 2|\mathbf{v}|^2.$$

74–83 Decompose $\mathbf{u}$ into a sum of two vectors, one parallel to $\mathbf{v}$ and one perpendicular to $\mathbf{v}$.

74. $\mathbf{u} = \langle 3, -1 \rangle$, $\quad \mathbf{v} = \langle 1, 7 \rangle$

75. $\mathbf{u} = \left\langle 2, \dfrac{1}{3} \right\rangle$, $\quad \mathbf{v} = \left\langle \dfrac{5}{2}, -3 \right\rangle$

76. $\mathbf{u} = \langle 2, 0, 1 \rangle$, $\mathbf{v} = \langle 1, 1, -3 \rangle$

77. $\mathbf{u} = \left\langle \dfrac{4}{3}, -1, \dfrac{2}{5} \right\rangle$, $\mathbf{v} = \left\langle -1, 6, -\dfrac{5}{3} \right\rangle$

78. $\mathbf{u} = \langle 4s, -2, 2s \rangle$, $\mathbf{v} = \langle 3, 5, -6 \rangle$

79.* $\mathbf{u} = \langle -5s, s, 2 \rangle$, $\mathbf{v} = \langle 2t, 4t, -1 \rangle$

80. $\mathbf{u} = \mathbf{i} + 3\mathbf{j} - 2\mathbf{k}$, $\mathbf{v} = 2\mathbf{i} - \mathbf{j} + \mathbf{k}$

81. $\mathbf{u} = \dfrac{1}{2}\mathbf{i} + \mathbf{j} + \dfrac{5}{3}\mathbf{k}$, $\mathbf{v} = 8\mathbf{i} - 5\mathbf{j} - 9\mathbf{k}$

82. $|\mathbf{u}| = 4$, $|\mathbf{v}| = 3\sqrt{2}$,   their angle is $45°$

83. $|\mathbf{u}| = 2.5$, $|\mathbf{v}| = 5$,   their angle is $2\pi/3$

84. For $\mathbf{v} = \langle 2, -1 \rangle$, find a vector $\mathbf{u}$ such that

    a. $\left|\text{proj}_\mathbf{v}\mathbf{u}\right| = \frac{5}{2}$,   **b.** $\left|\text{proj}_\mathbf{u}\mathbf{v}\right| = \frac{5}{2}$.

    (Answers will vary.)

85. For $\mathbf{v} = \langle -3, 0, 1 \rangle$, find a vector $\mathbf{u}$ such that $\left|\text{proj}_\mathbf{v}\mathbf{u}\right| = \frac{5}{2}$.

86.* Prove that the distance d from a point $Q(x_0, y_0)$ to a line $ax + by = c$ is

$$d = \frac{|ax_0 + by_0 - c|}{\sqrt{a^2 + b^2}}.$$

    (**Hint:** Revisit Exercises 59 and 60 to identify a normal vector of the line, then pick a point $P(x, y)$ on the line, and consider $\text{proj}_\mathbf{n} \overrightarrow{PQ}$.)

87.* Prove that the distance d between parallel lines $ax + by = c_1$ and $ax + by = c_2$ is

$$d = \frac{|c_2 - c_1|}{\sqrt{a^2 + b^2}}.$$

88.* Generalize your solution to Exercise 86 to three dimensions to show that the distance d from a point $Q(x_0, y_0, z_0)$ to a plane $ax + by + cz = f$ is

$$d = \frac{|ax_0 + by_0 + cz_0 - f|}{\sqrt{a^2 + b^2 + c^2}}.$$

89.* Generalize Exercise 87 to show that the distance d between parallel planes $ax + by + cz = f_1$ and $ax + by + cz = f_2$ is

$$d = \frac{|f_2 - f_1|}{\sqrt{a^2 + b^2 + c^2}}.$$

90–93 Use the formulas from Exercises 86–89 to find the indicated distance.

90. The distance between the point $(1, 2)$ and the line $x - y = 4$

91. The distance between the lines $2x + 3y = 2$ and $2x + 3y = -5$

92. The distance between the point $(1, 2, 3)$ and the plane $x - 3y + z = 2$

93. The distance between the planes $x - 2y + 5z = 1$ and $x - 2y + 5z = 7$

94–95 Find the work done by the force **F** as it moves an object from P to Q. (Suppose **F** is measured in pounds and a unit distance is 1 ft.)

94. $\mathbf{F} = -2\mathbf{i} - \mathbf{j} + 5\mathbf{k}$ from $P(1, 2, 3)$ to $Q(4, -7, 6)$

95. $\mathbf{F} = \mathbf{i} - 4\mathbf{j} + 2\mathbf{k}$ from $P(-7, 1, -4)$ to $Q(-3, -6, 2)$

96. In order to close a curtain, a hotel guest pulls a rod at an angle of $45°$ with a constant force of 10 pounds. Find the work done if the curtain moves 5 feet to its closed position.

97. A sailboat is propelled a distance of 1 km by a wind that makes a $45°$ angle with the boat's direction of travel. Find the work done by the wind if its force is 2000 newtons.

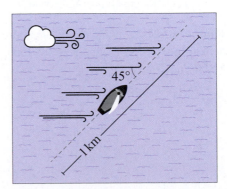

98. A father is pulling his little son in a sled up a 20° slope that is 20 meters long. The rope makes a 30° angle with the surface of the slope and the combined weight of the child and the sled is 200 newtons. (Ignore friction and any acceleration of motion.)

    a. Find the work done by the father on the sled.

    b. Find the force of tension in the rope.

99.* Repeat Exercise 98 under the assumption that the coefficient of friction is $\mu = 0.13$.

100. Use vectors to find the angle between the diagonal of a cube and

    a. one of its edges,

    b. the diagonal of one of the faces.

101.* Prove that the points $(0,0,0)$, $(0,1,1)$, $(1,0,1)$, and $(1,1,0)$ determine a regular tetrahedron and find the angle any edge makes with the face that doesn't contain it.

102. Use two-dimensional unit vectors to prove the following well-known formula from trigonometry:

$$\cos(\alpha - \beta) = \cos\alpha\cos\beta + \sin\alpha\sin\beta.$$

    (**Hint:** Considering two unit vectors, represent them as in Exercise 17, and interpret their dot product.)

103. A company manufactures three different products, producing n_j of each during a given production cycle ($j = 1, 2, 3$). Accordingly, the total production of a given cycle can be arranged into a three-dimensional production vector: $\mathbf{p} = \langle p_1, p_2, p_3 \rangle$. If the selling price of the jth product is s_j dollars each, the price vector can similarly be defined as $\mathbf{s} = \langle s_1, s_2, s_3 \rangle$. Interpret the dot product $\mathbf{p} \cdot \mathbf{s}$ for a given production cycle.

104–120 *True or False?* Determine whether the given statement is true or false. In case of a false statement, explain or provide a counterexample.

104. $\langle 1, 3, 0 \rangle \cdot \langle -2, 1 \rangle = 1$ **105.** $|\mathbf{u}|\mathbf{v} = \mathbf{v}|\mathbf{u}|$

106. $|\mathbf{u}|\mathbf{v} = \mathbf{u}|\mathbf{v}|$ **107.** $\mathbf{u} \cdot (\mathbf{v} \cdot \mathbf{w}) = (\mathbf{u} \cdot \mathbf{v}) \cdot \mathbf{w}$

108. $|\mathbf{u}| \cdot (\mathbf{v} \cdot \mathbf{w}) = \mathbf{v} \cdot (|\mathbf{u}|\mathbf{w})$

109. $|\mathbf{u}|(\mathbf{v} \cdot \mathbf{w}) = \mathbf{v} \cdot (|\mathbf{u}|\mathbf{w})$

110. If $(\mathbf{u} \cdot \mathbf{v}) + \mathbf{w} = 0$, then $(\mathbf{u} \cdot \mathbf{v}) = -\mathbf{w}$.

111. $(\mathbf{u} \cdot \mathbf{v})(\mathbf{w} \cdot \mathbf{x}) = \left[(\mathbf{w} \cdot \mathbf{x})\mathbf{u} \right] \cdot \mathbf{v} = \mathbf{u} \cdot \left[(\mathbf{w} \cdot \mathbf{x})\mathbf{v} \right]$

112. If $\mathbf{u} \cdot \mathbf{v} = \mathbf{u} \cdot \mathbf{w}$ and $\mathbf{u} \neq \mathbf{0}$, then $\mathbf{v} = \mathbf{w}$.

113. $|\mathbf{u}|(\mathbf{v} + \mathbf{w}) = |\mathbf{u}|\mathbf{v} + |\mathbf{u}|\mathbf{w}$

114. $\mathbf{u} \cdot \mathbf{v} < 0$ if and only if the angle between $\mathbf{u}$ and $\mathbf{v}$ is obtuse.

115. $\left(\text{proj}_{\mathbf{v}}\,\mathbf{u} - \mathbf{u} \right) \cdot \text{proj}_{\mathbf{v}}\,\mathbf{u} = 0$

116. If $\mathbf{u}$ is orthogonal to both $\mathbf{v}$ and $\mathbf{w}$, then $\mathbf{u}$ is orthogonal to $\mathbf{v} + \mathbf{w}$.

117. If $\left| \text{proj}_{\mathbf{v}}\,\mathbf{u} \right| = |\mathbf{u}|$, then $\mathbf{u}$ and $\mathbf{v}$ are parallel.

118. If $\left| \text{proj}_{\mathbf{v}}\,\mathbf{u} \right| = 0$, then $\mathbf{u}$ and $\mathbf{v}$ are orthogonal.

119. If $|\mathbf{u}| < |\mathbf{v}|$, then $\mathbf{u} \cdot \mathbf{w} < \mathbf{v} \cdot \mathbf{w}$.

120. If $\left| \text{proj}_{\mathbf{v}}\,\mathbf{u} \right| = \left| \text{proj}_{\mathbf{u}}\,\mathbf{v} \right|$, then $|\mathbf{u}| = |\mathbf{v}|$.

11.4 **The Cross Product**

TOPICS

1. The cross product and its properties

2. Applications of the cross product

In this section we introduce the operation of *cross product*. Like the dot product, the cross product is useful both for characterizing various geometric properties and for solving certain classes of problems. However, as we will see, the cross product and dot product operations also have some distinct differences.

TOPIC 1 **The Cross Product and Its Properties**

The cross product of two vectors **u** and **v** is written as **u** × **v**, notation which explains its common name. However, the cross product also goes by several other names including *vector product*, a name reflecting the fact that **u** × **v** is a vector. This is in contrast to **u** · **v**, which is a scalar. The *magnitude* of **u** × **v**, on the other hand, has much in common with one of the properties of **u** · **v**. If, as in Section 11.3, we let θ denote the angle between **u** and **v** when the two vectors are depicted with the same initial point, then

$$|\mathbf{u} \times \mathbf{v}| = |\mathbf{u}||\mathbf{v}|\sin\theta.$$

We construct the vector **u** × **v** by multiplying its magnitude by a unit vector **n** that is perpendicular to both **u** and **v**, with **n** chosen so that if you curl the fingers of your right hand from the direction of **u** to the direction of **v**, **n** points in the direction of your thumb. That is, the three vectors **u**, **v**, and **n** (in that order) follow the *right-hand rule*.

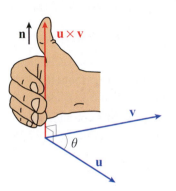

Figure 1 Constructing **u** × **v**

Definition 💡

Cross Product

Given two nonparallel and nonzero vectors **u** and **v**, the **cross product** of **u** and **v** is the vector

$$\mathbf{u} \times \mathbf{v} = (|\mathbf{u}||\mathbf{v}|\sin\theta)\mathbf{n},$$

with **n** a unit vector chosen perpendicular to both **u** and **v** according to the right-hand rule. If **u** and **v** are parallel, or if either is the zero vector, then **u** × **v** = **0**.

This definition immediately points out one of the differences between the cross product and the dot product: by its nature, when **u** × **v** ≠ **0**, the vector **u** × **v** cannot lie in the same plane as **u** and **v** and hence the cross product only makes sense in $\mathbb{R}^3$. Specifically, it is a function that takes two vectors in $\mathbb{R}^3$ and returns a third vector in $\mathbb{R}^3$.

Another property follows quickly from the definition above. Two nonzero, nonparallel vectors **u** and **v** with the same initial point define a unique plane, and there are exactly two unit vectors perpendicular to that plane: the vector **n** as defined above, and the vector −**n**. Your thumb points in the direction of **n** if you curl your fingers from the direction of **u** to the direction of **v**, and points in the direction of −**n** if you curl your fingers from the direction of **v** to the direction of **u**. That is, **u** × **v** = −(**v** × **u**).

Example 1 ✐

Simplify each of the following cross products.

a. $\mathbf{i} \times \mathbf{k}$ **b.** $\mathbf{i} \times (-\mathbf{k})$ **c.** $\mathbf{i} \times (\mathbf{j} + \mathbf{k})$ **d.** $(3\mathbf{i}) \times (-2\mathbf{k})$

Solution

a. The vectors $\mathbf{i}$ and $\mathbf{k}$ are orthogonal, so $\sin\theta = 1$. They are also unit vectors, so $|\mathbf{i}| = |\mathbf{k}| = 1$. All that remains is to visualize a right hand with fingers curling from $\mathbf{i}$ to $\mathbf{k}$, and noting that the thumb points in the direction of $-\mathbf{j}$. So $\mathbf{i} \times \mathbf{k} = \big(|\mathbf{i}||\mathbf{k}|\sin\theta\big)\mathbf{n} = -\mathbf{j}$.

b. The thumb of a right hand with fingers curling from $\mathbf{i}$ to $-\mathbf{k}$ points in the direction of $\mathbf{j}$. Every other factor is 1, as in part a., so $\mathbf{i} \times (-\mathbf{k}) = \mathbf{j}$.

c. The two vectors $\mathbf{i}$ and $\mathbf{j} + \mathbf{k} = \langle 0,1,1 \rangle$ are again orthogonal so $\sin\theta = 1$, but $|\mathbf{j} + \mathbf{k}| = \sqrt{0^2 + 1^2 + 1^2} = \sqrt{2}$. Determining $\mathbf{n}$ also requires a bit more effort than the first two problems. Since $\mathbf{n}$ must be orthogonal to $\mathbf{i}$, it lies in the yz-plane. The vector $\mathbf{j} + \mathbf{k}$ also lies in the yz-plane, so $\mathbf{n}$ must be a scalar multiple of $\langle 0,-1,1 \rangle$. (Note how the right-hand rule is satisfied by the vectors in Figure 2.) For $\mathbf{n}$ to have magnitude 1, we let $\mathbf{n} = \langle 0, -1/\sqrt{2}, 1/\sqrt{2} \rangle$. This gives us the following:

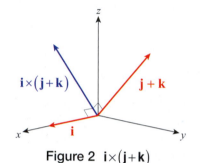

Figure 2 $\mathbf{i} \times (\mathbf{j} + \mathbf{k})$

$$\mathbf{i} \times (\mathbf{j} + \mathbf{k}) = \big(|\mathbf{i}||\mathbf{j} + \mathbf{k}|\sin\theta\big)\mathbf{n} = \sqrt{2}\left\langle 0, -\frac{1}{\sqrt{2}}, \frac{1}{\sqrt{2}} \right\rangle = \langle 0,-1,1 \rangle$$

d. $(3\mathbf{i}) \times (-2\mathbf{k}) = \big(|3\mathbf{i}||-2\mathbf{k}|\sin\theta\big)\mathbf{n} = (3)(2)\big(|\mathbf{i}||-\mathbf{k}|\sin\theta\big)\mathbf{n} = 6\big[\mathbf{i} \times (-\mathbf{k})\big] = 6\mathbf{j}$

This means that $(3\mathbf{i}) \times (-2\mathbf{k}) = (3)(-2)(-\mathbf{j}) = (3)(-2)(\mathbf{i} \times \mathbf{k})$.

Note that $\mathbf{i} \times \mathbf{j} = \mathbf{k}$, so $(\mathbf{i} \times \mathbf{j}) + (\mathbf{i} \times \mathbf{k}) = \mathbf{k} + (-\mathbf{j}) = \langle 0,-1,1 \rangle$, which combined with part c. of Example 1 indicates that $\mathbf{i} \times (\mathbf{j} + \mathbf{k}) = (\mathbf{i} \times \mathbf{j}) + (\mathbf{i} \times \mathbf{k})$. A similar argument shows that

$$\mathbf{u} \times (\mathbf{v} + \mathbf{w}) = (\mathbf{u} \times \mathbf{v}) + (\mathbf{u} \times \mathbf{w}) \quad \text{and} \quad (\mathbf{u} + \mathbf{v}) \times \mathbf{w} = (\mathbf{u} \times \mathbf{w}) + (\mathbf{v} \times \mathbf{w})$$

whenever $\mathbf{u}$, $\mathbf{v}$, and $\mathbf{w}$ are replaced with $\mathbf{i}$, $\mathbf{j}$, and $\mathbf{k}$, in any order. Also, by the reasoning of part d.,

$$(a\mathbf{u}) \times (b\mathbf{v}) = (ab)(\mathbf{u} \times \mathbf{v})$$

for scalars a and b whenever $\mathbf{u}$ and $\mathbf{v}$ are each replaced by $\mathbf{i}$, $\mathbf{j}$, or $\mathbf{k}$.

As Example 1 already shows, the formula $\mathbf{u} \times \mathbf{v} = \big(|\mathbf{u}||\mathbf{v}|\sin\theta\big)\mathbf{n}$ can be tedious to apply in all but the simplest cases. Fortunately, we can use the distributive properties above and the facts that $\mathbf{u} \times \mathbf{v} = -(\mathbf{v} \times \mathbf{u})$ and $(a\mathbf{u}) \times (b\mathbf{v}) = (ab)(\mathbf{u} \times \mathbf{v})$, restricting their use to $\mathbf{i}$, $\mathbf{j}$, and $\mathbf{k}$, to establish a much more easily applied formula.

Theorem

The Cross Product Determinant Formula

Given $\mathbf{u} = \langle u_1, u_2, u_3 \rangle$ and $\mathbf{v} = \langle v_1, v_2, v_3 \rangle$,

$$\mathbf{u} \times \mathbf{v} = \begin{vmatrix} \mathbf{i} & \mathbf{j} & \mathbf{k} \\ u_1 & u_2 & u_3 \\ v_1 & v_2 & v_3 \end{vmatrix} = \begin{vmatrix} u_2 & u_3 \\ v_2 & v_3 \end{vmatrix} \mathbf{i} - \begin{vmatrix} u_1 & u_3 \\ v_1 & v_3 \end{vmatrix} \mathbf{j} + \begin{vmatrix} u_1 & u_2 \\ v_1 & v_2 \end{vmatrix} \mathbf{k}$$

$$= \left(u_2 v_3 - u_3 v_2 \right) \mathbf{i} - \left(u_1 v_3 - u_3 v_1 \right) \mathbf{j} + \left(u_1 v_2 - u_2 v_1 \right) \mathbf{k}.$$

Proof

Writing $\mathbf{u} = u_1 \mathbf{i} + u_2 \mathbf{j} + u_3 \mathbf{k}$ and $\mathbf{v} = v_1 \mathbf{i} + v_2 \mathbf{j} + v_3 \mathbf{k}$, and making use of the properties we noted previously,

$$\mathbf{u} \times \mathbf{v} = \left(u_1 \mathbf{i} + u_2 \mathbf{j} + u_3 \mathbf{k} \right) \times \left(v_1 \mathbf{i} + v_2 \mathbf{j} + v_3 \mathbf{k} \right)$$

$$= u_1 \mathbf{i} \times \left(v_1 \mathbf{i} + v_2 \mathbf{j} + v_3 \mathbf{k} \right) + u_2 \mathbf{j} \times \left(v_1 \mathbf{i} + v_2 \mathbf{j} + v_3 \mathbf{k} \right) + u_3 \mathbf{k} \times \left(v_1 \mathbf{i} + v_2 \mathbf{j} + v_3 \mathbf{k} \right)$$

$$= u_1 v_1 \mathbf{i} \times \mathbf{i} + u_1 v_2 \mathbf{i} \times \mathbf{j} + u_1 v_3 \mathbf{i} \times \mathbf{k} + u_2 v_1 \mathbf{j} \times \mathbf{i} + u_2 v_2 \mathbf{j} \times \mathbf{j} + u_2 v_3 \mathbf{j} \times \mathbf{k}$$

$$\quad + u_3 v_1 \mathbf{k} \times \mathbf{i} + u_3 v_2 \mathbf{k} \times \mathbf{j} + u_3 v_3 \mathbf{k} \times \mathbf{k}$$

$$= u_1 v_2 \mathbf{k} - u_1 v_3 \mathbf{j} - u_2 v_1 \mathbf{k} + u_2 v_3 \mathbf{i} + u_3 v_1 \mathbf{j} - u_3 v_2 \mathbf{i} \qquad \mathbf{i} \times \mathbf{i} = \mathbf{j} \times \mathbf{j} = \mathbf{k} \times \mathbf{k} = \mathbf{0}$$

$$= \left(u_2 v_3 - u_3 v_2 \right) \mathbf{i} - \left(u_1 v_3 - u_3 v_1 \right) \mathbf{j} + \left(u_1 v_2 - u_2 v_1 \right) \mathbf{k}.$$

The matrix determinant notation seen in the statement of this theorem is used merely as a convenient reminder of the formula.

Example 2

$$\langle -2, 5, 1 \rangle \times \langle 1, 3, -1 \rangle = \begin{vmatrix} \mathbf{i} & \mathbf{j} & \mathbf{k} \\ -2 & 5 & 1 \\ 1 & 3 & -1 \end{vmatrix} = (-5 - 3) \mathbf{i} - (2 - 1) \mathbf{j} + (-6 - 5) \mathbf{k} = \langle -8, -1, -11 \rangle$$

With the cross product determinant formula in hand, we can now easily prove many properties, beginning with the following.

Theorem

Properties of the Cross Product

Assume $\mathbf{u}$, $\mathbf{v}$, and $\mathbf{w}$ represent vectors in $\mathbb{R}^3$ and that a and b represent scalars. Then the following properties hold.

$$\mathbf{u} \times \mathbf{v} = -\left(\mathbf{v} \times \mathbf{u} \right) \qquad\qquad \mathbf{0} \times \mathbf{u} = \mathbf{0}$$

$$\mathbf{u} \times \left(\mathbf{v} + \mathbf{w} \right) = \mathbf{u} \times \mathbf{v} + \mathbf{u} \times \mathbf{w} \qquad\qquad \left(\mathbf{u} + \mathbf{v} \right) \times \mathbf{w} = \mathbf{u} \times \mathbf{w} + \mathbf{v} \times \mathbf{w}$$

$$\left(a\mathbf{u} \right) \times \left(b\mathbf{v} \right) = \left(ab \right) \left(\mathbf{u} \times \mathbf{v} \right) \qquad\qquad \mathbf{u} \cdot \left(\mathbf{v} \times \mathbf{w} \right) = \left(\mathbf{u} \times \mathbf{v} \right) \cdot \mathbf{w}$$

$$\mathbf{u} \times \left(\mathbf{v} \times \mathbf{w} \right) = \left(\mathbf{u} \cdot \mathbf{w} \right) \mathbf{v} - \left(\mathbf{u} \cdot \mathbf{v} \right) \mathbf{w}$$

The proofs of these statements are left to the reader as Exercises 7–13.

TOPIC 2   Applications of the Cross Product

We say that a vector or a line that is perpendicular to a given plane is normal to the plane, and we will see in the next section how normal vectors play a role in characterizing planes. The cross product, by its nature, allows us to quickly construct normal vectors.

Example 3 ✎

Construct a vector normal to the plane containing the three points $P(-2,1,3)$, $Q(4,-3,1)$, and $R(5,5,3)$.

Solution

If we define $\mathbf{u} = \overrightarrow{PQ} = \langle 6,-4,-2 \rangle$ and $\mathbf{v} = \overrightarrow{PR} = \langle 7,4,0 \rangle$, then $\mathbf{u}$ and $\mathbf{v}$ have the same initial point and lie in the plane containing P, Q, and R. Further, we now know that these three points determine a unique plane, because $\mathbf{u}$ and $\mathbf{v}$ are not parallel (neither is a scalar multiple of the other)—if the two vectors were parallel, the three points P, Q, and R would be collinear and there would be an infinite number of planes containing them.

The vector $\mathbf{u} \times \mathbf{v}$ is perpendicular to both $\mathbf{u}$ and $\mathbf{v}$, so it is also normal to the plane containing $\mathbf{u}$ and $\mathbf{v}$. Using the cross product determinant formula, we have the following:

$$\mathbf{n} = \mathbf{u} \times \mathbf{v} = \langle 6,-4,-2 \rangle \times \langle 7,4,0 \rangle$$

$$= \begin{vmatrix} \mathbf{i} & \mathbf{j} & \mathbf{k} \\ 6 & -4 & -2 \\ 7 & 4 & 0 \end{vmatrix}$$

$$= \left[0-(-8) \right]\mathbf{i} - \left[0-(-14) \right]\mathbf{j} + \left[24-(-28) \right]\mathbf{k}$$

$$= \langle 8,-14,52 \rangle$$

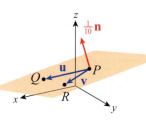

Figure 3

Any nonzero multiple of $\mathbf{n}$ will also be normal to the plane, and Figure 3 is an illustration of the three points P, Q, and R, the two vectors $\mathbf{u}$ and $\mathbf{v}$ drawn with initial point P, and one-tenth of the vector $\mathbf{n}$ also drawn with initial point P. Incidentally, the fact that $\mathbf{u} \times \mathbf{v} \neq \mathbf{0}$ serves as an alternative proof that $\mathbf{u}$ and $\mathbf{v}$ are not parallel.

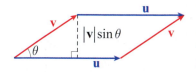

Figure 4 Area of a Parallelogram

Two nonzero, nonparallel vectors $\mathbf{u}$ and $\mathbf{v}$ in $\mathbb{R}^3$ can also be used to define a parallelogram, if the vectors are arranged as shown in Figure 4. Since the area of such a parallelogram is $A = (\text{base})(\text{height}) = |\mathbf{u}|(|\mathbf{v}|\sin\theta)$, the magnitude of $\mathbf{u} \times \mathbf{v}$ provides a convenient way to calculate the area.

$$A = |\mathbf{u}||\mathbf{v}|\sin\theta = |\mathbf{u} \times \mathbf{v}|$$

Example 4 ✎

The two vectors $\langle 3,-1 \rangle$ and $\langle 4,2 \rangle$ are used to construct the parallelogram shown in Figure 5. Find its area.

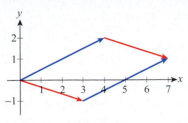

Figure 5

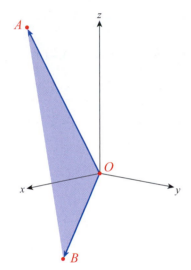

Figure 6

Solution

Although the parallelogram is constructed of vectors from $\mathbb{R}^2$, we can just as easily define the parallelogram with the three-dimensional vectors $\mathbf{u} = \langle 3, -1, 0 \rangle$ and $\mathbf{v} = \langle 4, 2, 0 \rangle$. In fact, we must do so if we want to use the cross product formula to find the parallelogram's area.

$$\mathbf{u} \times \mathbf{v} = \begin{vmatrix} \mathbf{i} & \mathbf{j} & \mathbf{k} \\ 3 & -1 & 0 \\ 4 & 2 & 0 \end{vmatrix} = (0-0)\mathbf{i} - (0-0)\mathbf{j} + \left[6 - (-4) \right]\mathbf{k} = \langle 0, 0, 10 \rangle$$

The area of the parallelogram equals $|\mathbf{u} \times \mathbf{v}|$, or 10 square units.

Three noncollinear points uniquely define a triangle as well as a plane, and the cross product can be used to easily determine the area of such a triangle.

Example 5 ✎

Find the area of the triangle whose vertices are the origin and the points $A(3, -1, 5)$ and $B(4, 2, -2)$.

Solution

The area of the triangle OAB, illustrated in Figure 6, is half of the area of the parallelogram formed by the vectors $\mathbf{u} = \overrightarrow{OA} = \langle 3, -1, 5 \rangle$ and $\mathbf{v} = \overrightarrow{OB} = \langle 4, 2, -2 \rangle$.

$$\mathbf{u} \times \mathbf{v} = \begin{vmatrix} \mathbf{i} & \mathbf{j} & \mathbf{k} \\ 3 & -1 & 5 \\ 4 & 2 & -2 \end{vmatrix} = (2 - 10)\mathbf{i} - (-6 - 20)\mathbf{j} + \left[6 - (-4) \right]\mathbf{k} = \langle -8, 26, 10 \rangle$$

The area of the triangle is $\frac{1}{2}|\mathbf{u} \times \mathbf{v}| = \frac{1}{2}\sqrt{64 + 676 + 100} = \sqrt{210}$ square units.

$h = \left| \text{proj}_{\mathbf{u} \times \mathbf{v}} \mathbf{w} \right|$

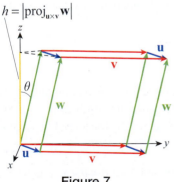

Figure 7
Volume of a Parallelepiped

Just as two vectors can be used to define a parallelogram, three vectors in space can be used to define a parallelepiped. Figure 7 illustrates how four copies each of three vectors $\mathbf{u}$, $\mathbf{v}$, and $\mathbf{w}$ define its edges. Once constructed, the cross product and dot product together provide an elegant way to calculate the volume of such a parallelepiped.

We already know that the area of the parallelogram defined by $\mathbf{u}$ and $\mathbf{v}$ is $A = |\mathbf{u} \times \mathbf{v}|$, and we also know that in general the volume of an object with fixed cross-sectional area A and height h is Ah (for review, see Section 6.1). Referring to Figure 7, we see that h is equal to the absolute value of the component of $\mathbf{w}$ in the direction of $\mathbf{u} \times \mathbf{v}$.

$$h = \left| \text{proj}_{\mathbf{u} \times \mathbf{v}} \mathbf{w} \right| = \left| \frac{\mathbf{w} \cdot (\mathbf{u} \times \mathbf{v})}{|\mathbf{u} \times \mathbf{v}|} \right| = \frac{|\mathbf{w}||\mathbf{u} \times \mathbf{v}||\cos\theta|}{|\mathbf{u} \times \mathbf{v}|} = |\mathbf{w}||\cos\theta|$$

Note that we use $\left|\cos\theta\right|$ instead of $\cos\theta$ in case $\pi/2 < \theta \leq \pi$. This gives us the following for the volume V of the parallelepiped.

$$V = Ah = \left|\mathbf{u} \times \mathbf{v}\right|\left\|\mathbf{w}\right\|\left|\cos\theta\right| = \left|\left(\mathbf{u} \times \mathbf{v}\right) \cdot \mathbf{w}\right|$$

The expression $\left(\mathbf{u} \times \mathbf{v}\right) \cdot \mathbf{w}$ is called the **triple scalar product** of $\mathbf{u}$, $\mathbf{v}$, and $\mathbf{w}$. Since the cross product is not commutative (that is, $\mathbf{u} \times \mathbf{v} \neq \mathbf{v} \times \mathbf{u}$), the order of the vectors is critical. However, one of the properties proved in the exercises shows that the cross and dot product can be interchanged, since $\left(\mathbf{u} \times \mathbf{v}\right) \cdot \mathbf{w} = \mathbf{u} \cdot \left(\mathbf{v} \times \mathbf{w}\right)$. Further, as you will prove in Exercise 48, it is easy to compute $\left(\mathbf{u} \times \mathbf{v}\right) \cdot \mathbf{w}$.

$$\left(\mathbf{u} \times \mathbf{v}\right) \cdot \mathbf{w} = \begin{vmatrix} u_1 & u_2 & u_3 \\ v_1 & v_2 & v_3 \\ w_1 & w_2 & w_3 \end{vmatrix}$$

Example 6 ✐

Find the volume of the parallelepiped defined by the vectors $\mathbf{u} = \langle -3, -1, 1 \rangle$, $\mathbf{v} = \langle 2, 0, 1 \rangle$, and $\mathbf{w} = \langle 4, 1, -2 \rangle$.

Solution

The volume of the parallelepiped follows from the fact that

$$\begin{vmatrix} -3 & -1 & 1 \\ 2 & 0 & 1 \\ 4 & 1 & -2 \end{vmatrix} = (-3)(0-1) - (-1)(-4-4) + (1)(2-0) = 3 - 8 + 2 = -3$$

and hence $V = \left|-3\right| = 3$ cubic units.

We will conclude this section with one more use of the cross product frequently seen in physics and engineering applications. The magnitude of the **torque $\boldsymbol{\tau}$** (tau), relative to the pivot point, produced by a force $\mathbf{F}$ applied to the end of a lever arm $\mathbf{r}$ is equal to the product of the length of $\mathbf{r}$ and the component of $\mathbf{F}$ perpendicular to $\mathbf{r}$. That is, referring to Figure 8,

$$\left|\boldsymbol{\tau}\right| = \left|\mathbf{r}\right|\left|\mathbf{F}\right|\sin\theta = \left|\mathbf{r} \times \mathbf{F}\right|.$$

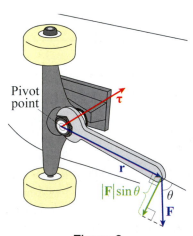

Figure 8

This reflects the fact that the magnitude of the torque depends on both the length of the lever and how much of the force is applied in a useful direction—in the extreme cases of $\theta = 0$ and $\theta = \pm\pi$, no torque is generated by the force $\mathbf{F}$, no matter how great its magnitude.

In the context of screws and bolts, a force $\mathbf{F}$ applied with $0 < \theta < \pi$ generates a torque "into" the page. That is, tightening a bolt (turning a wrench clockwise) creates a torque vector $\boldsymbol{\tau}$ that drives the bolt into an object. Loosening, or turning the wrench counterclockwise, corresponds to applying a force with $-\pi < \theta < 0$ and backs the bolt out of the object.

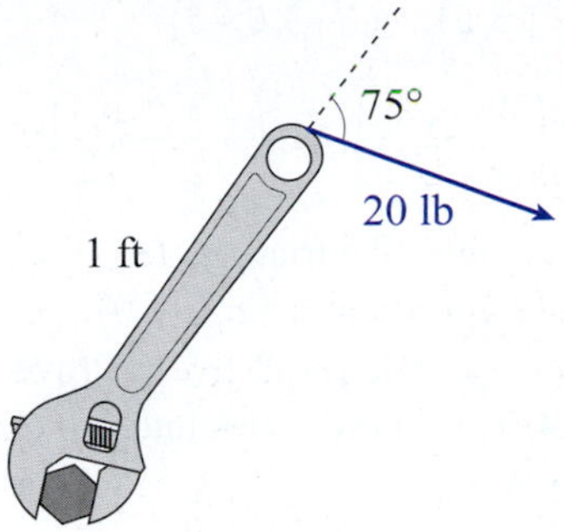

75°

20 lb

1 ft

Figure 9

Example 7 ✍

A bolt is tightened by a wrench supplying 20 pounds of force at an angle of $\theta = 75°$ (see Figure 9). The length of the wrench is 1 foot. How much torque is applied to the bolt at the pivot point?

Solution

$$|\boldsymbol{\tau}| = |\mathbf{r} \times \mathbf{F}| = |\mathbf{r}||\mathbf{F}|\sin\theta = (1)(20)\sin 75° \approx 19.32 \text{ ft-lb}$$

Note that the direction of the torque vector is into the page.

11.4 **Exercises**

1–6 Use the determinant formula to find the cross product.

1. $\langle 2, -5, 1 \rangle \times \langle 1, 1, -1 \rangle$

2. $\langle -3, 0, 1 \rangle \times \langle 2, 1, 4 \rangle$

3. $\langle 5, -5, 2 \rangle \times \langle 3, 1, -1 \rangle$

4. $\left\langle \frac{1}{2}, 2, -3 \right\rangle \times \left\langle 1, \frac{3}{2}, -1 \right\rangle$

5. $\left\langle \frac{1}{3}, 3, 0 \right\rangle \times \left\langle 0, -\frac{5}{3}, -\frac{1}{4} \right\rangle$

6. $\langle 0.2, -0.8, 1.25 \rangle \times \langle -2, 8, -12.5 \rangle$

7–13 Use the determinant formula to prove the indicated property of the cross product. (Assume **u**, **v**, and **w** represent vectors in $\mathbb{R}^3$, while a and b represent scalars.)

7. $\mathbf{u} \times \mathbf{v} = -(\mathbf{v} \times \mathbf{u})$

8. $\mathbf{u} \times (\mathbf{v} + \mathbf{w}) = \mathbf{u} \times \mathbf{v} + \mathbf{u} \times \mathbf{w}$

9. $(a\mathbf{u}) \times (b\mathbf{v}) = (ab)(\mathbf{u} \times \mathbf{v})$

10. $\mathbf{u} \times (\mathbf{v} \times \mathbf{w}) = (\mathbf{u} \cdot \mathbf{w})\mathbf{v} - (\mathbf{u} \cdot \mathbf{v})\mathbf{w}$

11. $\mathbf{0} \times \mathbf{u} = \mathbf{0}$

12. $(\mathbf{u} + \mathbf{v}) \times \mathbf{w} = \mathbf{u} \times \mathbf{w} + \mathbf{v} \times \mathbf{w}$

13. $\mathbf{u} \cdot (\mathbf{v} \times \mathbf{w}) = (\mathbf{u} \times \mathbf{v}) \cdot \mathbf{w}$

14. Prove that the cross products of the standard unit vectors obey the following:

a. $\mathbf{i} \times \mathbf{j} = \mathbf{k}$

b. $\mathbf{j} \times \mathbf{k} = \mathbf{i}$

c. $\mathbf{k} \times \mathbf{i} = \mathbf{j}$

15–18 Notice that using the properties of the cross product and results of Exercise 14, we can evaluate cross products as illustrated by the following example.

$$(\mathbf{i} + 3\mathbf{j} + 2\mathbf{k}) \times (2\mathbf{i} - 4\mathbf{j} + 5\mathbf{k}) = 2(\mathbf{i} \times \mathbf{i}) - 4(\mathbf{i} \times \mathbf{j}) + 5(\mathbf{i} \times \mathbf{k}) + 6(\mathbf{j} \times \mathbf{i}) - 12(\mathbf{j} \times \mathbf{j})$$
$$+ 15(\mathbf{j} \times \mathbf{k}) + 4(\mathbf{k} \times \mathbf{i}) - 8(\mathbf{k} \times \mathbf{j}) + 10(\mathbf{k} \times \mathbf{k})$$

Note that $\mathbf{i} \times \mathbf{i} = \mathbf{j} \times \mathbf{j} = \mathbf{k} \times \mathbf{k} = \mathbf{0}$.

$$= -4(\mathbf{i} \times \mathbf{j}) - 5(\mathbf{k} \times \mathbf{i}) - 6(\mathbf{i} \times \mathbf{j}) + 15(\mathbf{j} \times \mathbf{k}) + 4(\mathbf{k} \times \mathbf{i}) + 8(\mathbf{j} \times \mathbf{k})$$
$$= -10(\mathbf{i} \times \mathbf{j}) - (\mathbf{k} \times \mathbf{i}) + 23(\mathbf{j} \times \mathbf{k})$$
$$= -10\mathbf{k} - \mathbf{j} + 23\mathbf{i}$$
$$= 23\mathbf{i} - \mathbf{j} - 10\mathbf{k}$$

Use the above method to evaluate the cross product $\mathbf{u} \times \mathbf{v}$.

15. $\mathbf{u} = 2\mathbf{i} - \mathbf{j} + 3\mathbf{k}, \quad \mathbf{v} = -\mathbf{i} - \mathbf{j} + 2\mathbf{k}$

16. $\mathbf{u} = \frac{1}{3}\mathbf{i} + \frac{5}{2}\mathbf{j} - \mathbf{k}, \quad \mathbf{v} = 2\mathbf{i} - 9\mathbf{j} - 12\mathbf{k}$

17. $\mathbf{u} = \langle 4, -2, 1 \rangle, \quad \mathbf{v} = \langle 2, 5, -6 \rangle$

18. $\mathbf{u} = \langle -5, 1, 2 \rangle, \quad \mathbf{v} = \langle 3, 4, -1 \rangle$

19–22 Find both unit vectors perpendicular to **u** and **v**.

19. $\mathbf{u} = \langle 1, 0, -1 \rangle$, $\mathbf{v} = \langle 2, -2, 1 \rangle$

20. $\mathbf{u} = \langle 3, 1, 0 \rangle$, $\mathbf{v} = \langle -1, 0, 2 \rangle$

21. $\mathbf{u} = \mathbf{j} - 3\mathbf{k}$, $\mathbf{v} = -2\mathbf{i} + \mathbf{j}$

22. $\mathbf{u} = 2\mathbf{i} - \mathbf{j} + \mathbf{k}$, $\mathbf{v} = -\mathbf{i} - \mathbf{j} + 3\mathbf{k}$

23–26 Construct a vector normal to the plane containing the indicated points.

23. $P(0,0,0)$, $Q(2,-1,1)$, $R(3,3,4)$

24. $P(-5,0,4)$, $Q(2,2,2)$, $R(6,-1,3)$

25. $P\left(\frac{1}{2}, -1, -1\right)$, $Q\left(\frac{5}{2}, 3, \frac{7}{2}\right)$, $R\left(2, \frac{1}{2}, 0\right)$

26. $P\left(-\frac{4}{3}, -\frac{7}{6}, \frac{1}{4}\right)$, $Q\left(-\frac{1}{3}, -\frac{5}{6}, \frac{3}{4}\right)$, $R\left(\frac{5}{6}, -\frac{1}{6}, 1\right)$

27. Prove that $\overrightarrow{AB} \times \overrightarrow{AC} = \mathbf{0}$ if and only if the points A, B, and C are collinear (i.e., they lie on the same line).

28–31 Use Exercise 27 to check whether the given points are collinear.

28. $P(3,1,1)$, $Q(2,-1,0)$, $R(1,4,-1)$

29. $P(3,1,1)$, $Q(2,-1,0)$, $R(5,5,3)$

30. $P\left(2, 0, -\frac{1}{2}\right)$, $Q\left(\frac{1}{2}, -1, 3\right)$, $R\left(\frac{3}{2}, -1, -1\right)$

31. $P\left(0, \frac{1}{3}, 1\right)$, $Q\left(\frac{1}{3}, 1, 2\right)$, $R\left(-1, -\frac{5}{3}, -2\right)$

32–35 Find the area of the parallelogram spanned by the given vectors.

32. $\langle 1, 4 \rangle$, $\langle -3, 1 \rangle$

33. $\langle -3, -2 \rangle$, $\left\langle 1, -\frac{5}{2} \right\rangle$

34. $\langle 7, 3 \rangle$, $\langle 6, -10 \rangle$

35. $\langle 6, 9 \rangle$, $\langle -9, -6 \rangle$

36–39 Find the area of the triangle with the given vertices.

36. $A(0,0,0)$, $B(1,2,3)$, $C(-3,-2,-1)$

37. $A(1,1,1)$, $B(4,-2,5)$, $C(-3,1,-1)$

38. $A(-4,1,2)$, $B(-1,3,5)$, $C(-3,0,-5)$

39. $A\left(\frac{1}{2}, -1, \frac{5}{2}\right)$, $B\left(\frac{3}{2}, -2, \frac{3}{2}\right)$, $C\left(-\frac{3}{2}, -3, \frac{7}{2}\right)$

40. Suppose that the vertices of a triangle ABC in the xy-plane have coordinates $(x_A, y_A, 0)$, $(x_B, y_B, 0)$, and $(x_C, y_C, 0)$, respectively. Prove that the area of $\triangle ABC$ is half the absolute value of the following determinant.

$$\begin{vmatrix} 1 & 1 & 1 \\ x_A & x_B & x_C \\ y_A & y_B & y_C \end{vmatrix}$$

41–46 Suppose $\mathbf{u} = \langle 1, 3, -2 \rangle$, $\mathbf{v} = \langle 4, -1, 1 \rangle$, and $\mathbf{w} = \langle -2, 2, -1 \rangle$. If possible, find each of the following.

41. $\mathbf{u} + (\mathbf{v} \times \mathbf{w})$

42. $(\mathbf{u} + \mathbf{v}) \times \mathbf{w}$

43. $\mathbf{u} \times (\mathbf{v} \times \mathbf{w})$

44. $\mathbf{u} \cdot (\mathbf{v} \times \mathbf{w})$

45. $\mathbf{u} \times (\mathbf{v} \cdot \mathbf{w})$

46. $(\mathbf{u} \times \mathbf{v}) + \mathbf{w}$

47. Describe the conditions **u** or **v** have to satisfy in order for both their dot product and cross product to be zero, that is, for $\mathbf{u} \cdot \mathbf{v} = 0$ and $\mathbf{u} \times \mathbf{v} = \mathbf{0}$.

48. Prove the following determinant formula for the triple scalar product.

$$(\mathbf{u} \times \mathbf{v}) \cdot \mathbf{w} = \begin{vmatrix} u_1 & u_2 & u_3 \\ v_1 & v_2 & v_3 \\ w_1 & w_2 & w_3 \end{vmatrix}$$

49–52 Find the volume of the parallelepiped spanned by the indicated vectors.

49. $\mathbf{u} = \langle 2, -1, 3 \rangle$, $\mathbf{v} = \langle 3, 0, 4 \rangle$, $\mathbf{w} = \langle -1, 1, -2 \rangle$

50. $\mathbf{u} = \langle 1, 2, 3 \rangle$, $\mathbf{v} = \langle 2, -1, 4 \rangle$, $\mathbf{w} = \langle -1, 0, 3 \rangle$

51. $\mathbf{u} = \langle 1, 1, 3 \rangle$, $\mathbf{v} = \langle 1, 3, 1 \rangle$, $\mathbf{w} = \langle 3, 1, 1 \rangle$

52. $\mathbf{u} = \langle 1, 0, -3 \rangle$, $\mathbf{v} = \langle 0, -5, 2 \rangle$, $\mathbf{w} = \langle 3, 1, 1 \rangle$

53. In light of Exercises 49–52, state a condition in terms of the triple scalar product for three vectors to be coplanar (i.e., to lie on the same plane).

54–55 Use the condition you found in Exercise 53 to determine whether the vectors are coplanar.

54. $\mathbf{u} = \langle 2, -1, 3 \rangle$, $\mathbf{v} = \langle -1, 2, 3 \rangle$, $\mathbf{w} = \langle 3, 2, -1 \rangle$

55. $\mathbf{u} = \langle 1, 5, 1 \rangle$, $\mathbf{v} = \langle -2, -4, 0 \rangle$, $\mathbf{w} = \langle -3, -15, -3 \rangle$

56–57 Use Exercises 53–55 to determine whether the given points are coplanar.

56. $A(0,0,0)$, $B(1,3,4)$, $C(-1,-2,-3)$, $D(1,1,1)$

57. $A(1,5,0)$, $B(2,4,-1)$, $C(0,3,0)$, $D(4,2,-3)$

58. Use the cross product to prove the following well-known formula from trigonometry.

$$\sin(\alpha - \beta) = \sin \alpha \cos \beta - \cos \alpha \sin \beta$$

(**Hint:** See the hint given in Exercise 102 of Section 11.3. Turn the unit vectors into three-dimensional vectors as in Example 4, and interpret their cross product.)

59. A bolt is tightened by an 18-pound force at the end of a 10-inch wrench at an angle of $\theta = 60°$. What is the magnitude of the torque applied to the bolt at the pivot point?

60. Repeat Exercise 59 if the length of the wrench is 16 inches and a rotating force of 15 pounds is applied at $\theta = 45°$.

61. The force **F** exerted by the uniform magnetic field with induction vector **B** on a wire carrying current **I** obeys the vector equation

$$\mathbf{F} = l\,\mathbf{I} \times \mathbf{B},$$

where l is the length of the wire. (The standard SI unit for **B** is the tesla (T). If, in addition, we measure **I** in amperes (A), the above equation will return **F** in newtons (N).) Find the magnitude of the force experienced by an 8 cm wire in a uniform magnetic field of $\mathbf{B} = 2$ T if it carries a current of 0.3 A and the angle between the wire and **B** is $\theta = 30°$.

62. The force **F** experienced by a charged particle q moving at velocity **v** m/s in the uniform magnetic field **B** obeys the vector equation

$$\mathbf{F} = q\,\mathbf{v} \times \mathbf{B},$$

where (just like in Exercise 61) we obtain the force in newtons (N) if we measure **B** in teslas (T), q in coulombs (C), and **v** in meters per second (m/s). Find the magnitude of the force experienced by an electron moving in a uniform magnetic field of $\mathbf{B} = 0.001$ T at $\mathbf{v} = 1.2 \cdot 10^6$ m/s if the velocity vector and **B** form a $25°$ angle. Note that the magnitude of the charge carried by an electron, or the *elementary charge*, is $e \approx 1.6 \cdot 10^{-19}$ C.

63. Use an appropriate property of the cross product (see Exercises 7–13) to prove the following:

$$\mathbf{u} \times (\mathbf{v} \times \mathbf{w}) + \mathbf{v} \times (\mathbf{w} \times \mathbf{u}) + \mathbf{w} \times (\mathbf{u} \times \mathbf{v}) = \mathbf{0}$$

64. Prove that if vectors **s**, **u**, **v**, and **w** are coplanar (i.e., contained by the same plane), then

$$(\mathbf{s} \times \mathbf{u}) \times (\mathbf{v} \times \mathbf{w}) = \mathbf{0}.$$

65. Let A, B, C, and D be four distinct points in three-dimensional Cartesian space. Prove that $\left(\overrightarrow{AB} \times \overrightarrow{AC} \right) \times \overrightarrow{AD}$ is parallel to the plane containing A, B, and C.

66. Find the equation of the plane through the points $(-5,0,4)$, $(2,2,2)$, and $(6,-1,3)$. (**Hint:** See Exercise 60 of Section 11.3 and Exercise 24 of this section.)

67. Repeat Exercise 66 for the plane through $\left(\frac{1}{2}, -1, -1 \right)$, $\left(\frac{5}{2}, 3, \frac{7}{2} \right)$, and $\left(2, \frac{1}{2}, 0 \right)$.

68.* (Section 11.3 Exercise 86 revisited) Prove that the distance d from a point Q to a line containing the points R and S is

$$d = \frac{\left| \overrightarrow{RS} \times \overrightarrow{RQ} \right|}{\left| \overrightarrow{RS} \right|}.$$

69.* (Section 11.3 Exercise 88 revisited) Prove that the distance d from a point Q to a plane containing the points R, S, and T is

$$d = \frac{\left| \left(\overrightarrow{RS} \times \overrightarrow{RT} \right) \cdot \overrightarrow{RQ} \right|}{\left| \overrightarrow{RS} \times \overrightarrow{RT} \right|}.$$

70–73. Use Exercises 68–69 to find a second solution for each of Exercises 90–93 of Section 11.3.

74. For vectors **u**, **v**, and **w** in three-dimensional Cartesian space, prove *Lagrange's identity*.

$$\left|\mathbf{u}\times\mathbf{v}\right|^{2}=\left|\mathbf{u}\right|^{2}\left|\mathbf{v}\right|^{2}-\left(\mathbf{u}\cdot\mathbf{v}\right)^{2}$$

75.* A rectangular tetrahedron is one with a vertex such that any pair of incident edges form right angles. (Such is obtained by "chopping a corner off" a cube.) Denoting the areas of the faces containing a right angle by a, b, and c, respectively, and that of the fourth face by d, prove the following "three-dimensional generalization" of the Pythagorean Theorem.

$$a^{2}+b^{2}+c^{2}=d^{2}$$

(**Hint:** Place the tetrahedron appropriately into the three-dimensional Cartesian system, and use the techniques of this section to find the areas of its faces.)

76. Suppose that the nonzero vectors **u**, **v**, and **w** satisfy

$$\mathbf{u}\times\left(\mathbf{v}\times\mathbf{w}\right)=\left(\mathbf{u}\times\mathbf{v}\right)\times\mathbf{w}=\mathbf{0}.$$

Prove that **v** and **w** are parallel, or both are perpendicular to **u**.

77. Prove: $\left(\mathbf{s}\times\mathbf{u}\right)\cdot\left(\mathbf{v}\times\mathbf{w}\right)=\begin{vmatrix}\mathbf{s}\cdot\mathbf{v} & \mathbf{u}\cdot\mathbf{v} \\ \mathbf{s}\cdot\mathbf{w} & \mathbf{u}\cdot\mathbf{w}\end{vmatrix}.$

78–81 Determine whether the given expression is a vector, a scalar, or nonsense.

78. $\mathbf{u}\times\left(\left|\mathbf{u}\right|\mathbf{v}\right)$

79. $\left(\mathbf{u}\cdot\mathbf{v}\right)\times\left|\mathbf{v}\right|$

80. $\left(\mathbf{u}\cdot\mathbf{v}\right)+\left|\mathbf{v}\right|$

81. $\left|\mathbf{v}\right|\left(\mathbf{u}\times\mathbf{v}\right)$

82–94 *True or False?* Determine whether the given statement is true or false. In case of a false statement, explain or provide a counterexample.

82. $\mathbf{u}\times\mathbf{v}=\mathbf{v}\times\mathbf{u}$

83. $\mathbf{u}\times\mathbf{v}=\left(-\mathbf{u}\right)\times\left(-\mathbf{v}\right)$

84. $\left(-\mathbf{u}\right)\times\mathbf{u}=\mathbf{0}$

85. $\left|\mathbf{u}\right|\times\mathbf{v}=\mathbf{v}\times\left|\mathbf{u}\right|$

86. If $\left(\mathbf{u}\times\mathbf{v}\right)+\mathbf{w}=\mathbf{0},$ then $\left(\mathbf{u}\times\mathbf{v}\right)=-\mathbf{w}.$

87. $\mathbf{u}\cdot\left(\mathbf{v}\times\mathbf{w}\right)=\left(\mathbf{u}\cdot\mathbf{v}\right)\times\mathbf{w}$

88. $\mathbf{u}\times\left(\mathbf{v}\times\mathbf{w}\right)=\left(\mathbf{u}\times\mathbf{v}\right)\times\mathbf{w}$

89. $\mathbf{u}\times\mathbf{u}=\left|\mathbf{u}\right|^{2}$

90. If $\left|\mathbf{u}\right|<\left|\mathbf{v}\right|$ and $\mathbf{w}\neq\mathbf{0},$ then $\mathbf{u}\times\mathbf{w}<\mathbf{v}\times\mathbf{w}.$

91. $\left(\mathbf{u}-\mathbf{v}\right)\times\left(\mathbf{u}+\mathbf{v}\right)=2\left(\mathbf{u}\times\mathbf{v}\right)$

92. $\mathbf{u}\cdot\left(\mathbf{u}\times\mathbf{v}\right)=0$

93. $\mathbf{u}\cdot\left(\mathbf{v}\times\mathbf{v}\right)=0$

94. If $\mathbf{u}\times\mathbf{v}=\mathbf{u}\times\mathbf{w}$ and $\mathbf{u}\neq\mathbf{0},$ then $\mathbf{v}=\mathbf{w}.$

11.4 **Technology Exercises**

95–96. Use a computer algebra system or programmable calculator to write a program that decides whether four points (given by their coordinates) in three-dimensional space lie in the same plane. Use your program to check your answers for Exercises 56–57.

11.5 **Describing Lines and Planes**

TOPICS

1. Lines in space

2. Planes

Just as lines in $\mathbb{R}^2$ (specifically, tangent lines) are useful in understanding functions of a single variable, lines and planes in $\mathbb{R}^3$ will prove useful as we study functions of several variables. The vector notation and operations we have learned provide ways to describe lines and planes and to answer commonly occurring questions.

TOPIC 1   **Lines in Space**

The **position vector** of a point P is a vector in standard position having P as its terminal point—we use position vectors to describe a point's location. In both $\mathbb{R}^2$ and $\mathbb{R}^3$, we can easily construct position vectors describing all the points of a line, given some basic information about that line as in the example below.

Example 1 ✐

Construct position vectors describing the points making up the line L passing through $A(4,2,-1)$ and $B(1,2,3)$.

Solution

The position vector $\mathbf{r}_0 = \langle 4,2,-1 \rangle$ describes one point of L, namely the point A. As in Example 5 of Section 11.2, we can add the vector $\mathbf{v} = \overrightarrow{AB} = \langle -3,0,4 \rangle$ to $\mathbf{r}_0$ to construct a position vector whose terminal point is B; Figure 1 illustrates the fact that

$$\langle 1,2,3 \rangle = \mathbf{r}_0 + \mathbf{v} = \langle 4,2,-1 \rangle + \langle -3,0,4 \rangle.$$

And as in that previous example, we can describe any other point between A and B with a position vector of the form $\mathbf{r}_0 + t\mathbf{v}$, for some $0 \leq t \leq 1$. But by merely allowing t to take on all possible real values, we describe all the points that can be reached by starting at A and traveling in the direction of $\mathbf{v}$ by some scalar multiple (including negative scalar multiples). So the position vector

$$\mathbf{r}(t) = \mathbf{r}_0 + t\,\mathbf{v} = \langle 4,2,-1 \rangle + t\langle -3,0,4 \rangle = \langle 4-3t, 2, -1+4t \rangle$$

is a function of t that describes all the points of the line L as t takes on all real values.

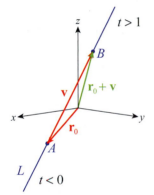

Figure 1
Vector Description of a Line

As Example 1 illustrates, a single point and a nonzero direction vector are sufficient to completely characterize a line. We summarize this fact as follows.

Definition 💡

Vector Form and Parametric Form of a Line

The **vector form** for the line L containing the point $P(x_0, y_0, z_0)$ and parallel to the **direction vector** $\mathbf{v} = \langle a,b,c \rangle$ is

$$\mathbf{r}(t) = \mathbf{r}_0 + t\mathbf{v} = \langle x_0, y_0, z_0 \rangle + t\langle a,b,c \rangle, \quad -\infty < t < \infty.$$

Note that if we let $\langle x, y, z \rangle$ denote the components of $\mathbf{r}(t)$, then the equations

$$x = x_0 + at, \quad y = y_0 + bt, \quad \text{and} \quad z = z_0 + ct$$

are parametric equations (in the parameter t) describing the line L.

Example 2 ✎

Find both the vector form and the parametric equations of the line parallel to the direction vector $3\mathbf{i} - \mathbf{j} + 2\mathbf{k}$ and passing through the point $(-7, 2, 5)$.

Solution

The line passing through the given point and parallel to $3\mathbf{i} - \mathbf{j} + 2\mathbf{k}$ can be described in vector form as follows.

$$\mathbf{r}(t) = \langle -7, 2, 5 \rangle + t \langle 3, -1, 2 \rangle = \langle -7 + 3t, 2 - t, 5 + 2t \rangle$$

The parametric equations are

$$x = -7 + 3t, \quad y = 2 - t, \quad \text{and} \quad z = 5 + 2t.$$

As we saw in Chapter 9, parametrizations are not unique. In the case of a line L, we can arrive at a different parametrization either by using a different initial point on L to construct the fixed position vector $\mathbf{r}_0$ and/or by using a parameter that traces out the line at a different rate. For instance, the parametrizations

$$x = -1 + 3t, \quad y = -t, \quad \text{and} \quad z = 9 + 2t \quad \text{(different initial point)}$$

and

$$x = -7 + 3t^3, \quad y = 2 - t^3, \quad \text{and} \quad z = 5 + 2t^3 \quad \text{(different tracing rate)}$$

both describe the same line as the parametrization in Example 2.

Once we have found the vector form or a parametrization of a line, we are in position to answer many questions about that line relative to other objects. The next two examples illustrate the techniques and reasoning we use.

Example 3 ✎

Determine the points of intersection of the line $\mathbf{r}(t) = \langle 4 - 3t, 2, -1 + 4t \rangle$ with the three coordinate planes.

Solution

This is the line from Example 1—you may want to refer to Figure 1 as we work through this problem.

The yz-plane corresponds to the equation $x = 0$, and we find the point of intersection of the line with the plane $x = 0$ by determining the value of the parameter for which the x-component of $\mathbf{r}(t)$ is 0.

$$4 - 3t = 0 \quad \Rightarrow \quad t = \frac{4}{3}$$

For this value of t, $y = 2$ and $z = -1 + 4\left(\frac{4}{3}\right) = \frac{13}{3}$. Note that $y = 2$ for any value of t. So the point of intersection between the line and the yz-plane is the point $\left(0, 2, \frac{13}{3}\right)$.

The line intersects the xy-plane when $z = 0$, and the z-component of $\mathbf{r}(t)$ is 0 when $t = \frac{1}{4}$. For this value of t, $x = 4 - 3\left(\frac{1}{4}\right) = \frac{13}{4}$ and $y = 2$, so the point of intersection between the line and the xy-plane is the point $\left(\frac{13}{4}, 2, 0\right)$.

Since $y = 2$ always, the line never intersects the xz-plane.

Example 4 ✐

Determine the point of intersection of each of the following pairs of lines, if possible.

a. $\mathbf{r}(u) = \langle 2 + 3u, -1 - u, 2u \rangle$ and $\mathbf{s}(v) = \langle 7 + 2v, -3 - v, 1 - v \rangle$

b. $\mathbf{r}(u) = \langle 2 - u, 3 + 2u, 2 + u \rangle$ and $\mathbf{s}(v) = \langle 2v, 4 - v, 2 - v \rangle$

Solution

a. If the two lines $\mathbf{r}(u)$ and $\mathbf{s}(v)$ intersect, there must be values of u and v for which $\mathbf{r}(u) = \mathbf{s}(v)$. In other words, for those values of u and v, it must be the case that

$$2 + 3u = 7 + 2v$$
$$-1 - u = -3 - v$$
$$2u = 1 - v.$$

Solving the first two (by elimination, substitution, or a matrix method) yields $u = 1$ and $v = -1$, and these values also satisfy the third equation.

$$\mathbf{r}(1) = \mathbf{s}(-1) = \langle 5, -2, 2 \rangle$$

So the two lines intersect at the point $(5, -2, 2)$ shown in Figure 2.

b. Proceeding in the same manner, the solution of the first two of the three equations

$$2 - u = 2v$$
$$3 + 2u = 4 - v$$
$$2 + u = 2 - v$$

is $u = 0$ and $v = 1$. But these values don't satisfy the third equation, so it is impossible to find a point of intersection of these two lines. Moreover, the lines are not parallel since their direction vectors are not scalar multiples of one another. Nonparallel lines with no point of intersection are called **skew** lines. Skew lines, while not parallel to each other, do have the property that they lie in two planes that are parallel to each other.

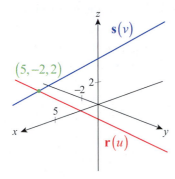

Figure 2

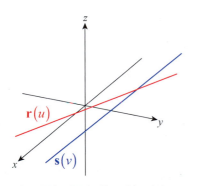

Figure 3 Skew Lines

TOPIC 2 **Planes**

The vector form and parametric form of a line both make use of a single parameter, reflecting the fact that lines are one-dimensional objects. Since planes in $\mathbb{R}^3$ are two-dimensional objects, we may expect that we need two parameters in order to characterize them—we will soon see that is indeed the case. And while a point and a single direction vector are sufficient to characterize a line, the analogous information needed to characterize a plane is a point and two nonparallel vectors in the plane.

Knowing the above, it may be surprising to learn that a point $P(x_0, y_0, z_0)$ and a single nonzero vector $\mathbf{n} = \langle a, b, c \rangle$ are sufficient to characterize a given plane if we know that $\mathbf{n}$ is normal (meaning orthogonal) to the plane. This follows from the fact that if (x, y, z) represents an arbitrary point in the plane under consideration, then the vector $\mathbf{r} = \langle x - x_0, y - y_0, z - z_0 \rangle$ must be orthogonal to $\mathbf{n}$, and hence $\mathbf{n} \cdot \mathbf{r} = 0$.

Definition 💡

Equation for a Plane with a Known Normal Vector

If the vector $\mathbf{n} = \langle a, b, c \rangle$ is normal to a given plane, and if the point $P(x_0, y_0, z_0)$ lies in the plane, then any point (x, y, z) in the plane must satisfy the vector equation

$$\langle a, b, c \rangle \cdot \langle x - x_0, y - y_0, z - z_0 \rangle = 0.$$

In scalar form, this becomes the linear equation

$$a(x - x_0) + b(y - y_0) + c(z - z_0) = 0, \quad \text{or}$$

$$ax + by + cz = d, \quad \text{where} \quad d = ax_0 + by_0 + cz_0.$$

Example 5 ✎

Find a vector equation and a scalar equation for the plane containing the point $(-3, 1, 5)$ and having $\mathbf{n} = \langle 2, -4, 1 \rangle$ as a normal vector.

Solution

The vector equation for the plane is

$$\langle 2, -4, 1 \rangle \cdot \langle x + 3, y - 1, z - 5 \rangle = 0,$$

and carrying out the scalar product operation and simplifying results in the scalar equation

$$2x - 4y + z = -5.$$

Example 6 ✎

Describe the plane $2x - 4y + z = -5$ as a two-parameter set of points in $\mathbb{R}^3$.

Solution

By solving the equation $2x - 4y + z = -5$ for any one of the three variables, we can describe the plane as a set of points in terms of the other two variables. For instance,

$$\{(x, y, -5 - 2x + 4y) \mid x \in \mathbb{R}, y \in \mathbb{R}\}$$

is one description of all the ordered triples making up the plane.

Example 7 ✐

Find an equation for the plane containing the three points $A(-1,2,2)$, $B(2,-1,1)$, and $C(2,2,1)$.

Solution

To use the formulation above, we need a vector $\mathbf{n}$ normal to the plane; we can construct such a vector by taking the cross product of two vectors $\mathbf{u}$ and $\mathbf{v}$ in the plane. To that end, let

$$\mathbf{u} = \overrightarrow{AB} = \langle 3,-3,-1 \rangle \quad \text{and} \quad \mathbf{v} = \overrightarrow{AC} = \langle 3,0,-1 \rangle.$$

Then,

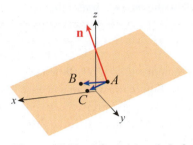

Figure 4 Plane Containing A, B, C

$$\mathbf{n} = \mathbf{u} \times \mathbf{v} = \begin{vmatrix} \mathbf{i} & \mathbf{j} & \mathbf{k} \\ 3 & -3 & -1 \\ 3 & 0 & -1 \end{vmatrix} = \mathbf{i}(3-0) - \mathbf{j}(-3+3) + \mathbf{k}(0+9) = \langle 3,0,9 \rangle$$

and, using the fact that A is a point in the plane, a vector equation for the plane is

$$\langle 3,0,9 \rangle \cdot \langle x+1, y-2, z-2 \rangle = 0,$$

or as linear equation $3x + 9z = 15$. Using either of the other two points B and C would have resulted in an equivalent answer.

If a line is not parallel to a plane, it must intersect it at some point. The next example illustrates how that point can be determined.

Example 8 ✐

Find the point of intersection between the line $\mathbf{r}(t) = \langle 2-3t, 1+4t, -3-t \rangle$ and the plane $4x - y + 2z = 5$.

Solution

Each point on the line has the form $(2-3t, 1+4t, -3-t)$ for some $t \in \mathbb{R}$, and we seek the value of t for which such a point also satisfies $4x - y + 2z = 5$. So we solve the following equation for t.

$$4(2-3t) - (1+4t) + 2(-3-t) = 5$$
$$8 - 12t - 1 - 4t - 6 - 2t = 5$$
$$t = -\frac{2}{9}$$

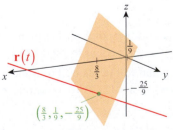

Figure 5

The point of intersection occurs when $t = -\frac{2}{9}$ and

$$\mathbf{r}\left(-\frac{2}{9}\right) = \left\langle 2 - 3\left(-\frac{2}{9}\right), 1 + 4\left(-\frac{2}{9}\right), -3 - \left(-\frac{2}{9}\right) \right\rangle = \left\langle \frac{8}{3}, \frac{1}{9}, -\frac{25}{9} \right\rangle.$$

Therefore $\left(\frac{8}{3}, \frac{1}{9}, -\frac{25}{9}\right)$ is the point of intersection and it is shown in Figure 5.

We close this section with three more examples of common tasks involving points, lines, and planes in $\mathbb{R}^3$.

Example 9 ✒

Find parametric equations for the line formed by the intersection of the two planes $x - 2y + 4z = 7$ and $2x - y + 3z = 0$.

Solution

The line of intersection of two planes lies in each plane, so it is orthogonal to a vector that is orthogonal to either plane (see Figure 6). We can use this fact to quickly find a direction vector for the line we seek. From the given equations for the planes, we know that $\mathbf{n}_1 = \langle 1, -2, 4 \rangle$ is normal to the first plane and $\mathbf{n}_2 = \langle 2, -1, 3 \rangle$ is normal to the second plane. So the vector

$$\mathbf{v} = \mathbf{n}_1 \times \mathbf{n}_2 = \begin{vmatrix} \mathbf{i} & \mathbf{j} & \mathbf{k} \\ 1 & -2 & 4 \\ 2 & -1 & 3 \end{vmatrix} = \mathbf{i}(-2) - \mathbf{j}(-5) + \mathbf{k}(3) = \langle -2, 5, 3 \rangle$$

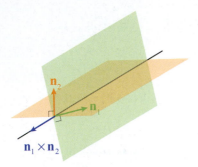

Figure 6
Line of Intersection of Two Planes

will serve as a direction vector for the line.

To complete the characterization of the line of intersection, we need a point on the line. Any such point will suffice, so we can choose to use the point where it intersects the xy-plane—that is, where $z = 0$ (if the line doesn't intersect the xy-plane, that will become apparent in the next step). Setting $z = 0$ in the two plane equations and solving for x and y yields the point $\left(-\frac{7}{3}, -\frac{14}{3}, 0\right)$. So the vector form of the line of intersection is

$$\left\langle -\frac{7}{3}, -\frac{14}{3}, 0 \right\rangle + t\langle -2, 5, 3 \rangle$$

and the parametric equations describing this line are

$$x = -\frac{7}{3} - 2t, \quad y = -\frac{14}{3} + 5t, \quad \text{and} \quad z = 3t.$$

Example 10 ✒

Find the shortest distance between the point $A(3, 1, -2)$ and the plane $2x - 3y + 6z = 2$.

Solution

We are looking for the distance d between A and the point on the plane closest to A. If B is *any* point on the plane, then d is equal to the length of the projection of $\overrightarrow{BA}$ onto a vector orthogonal to the plane. From the equation $2x - 3y + 6z = 2$, we know that the vector $\mathbf{n} = \langle 2, -3, 6 \rangle$ is normal to the plane and we can say that

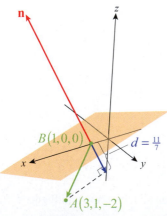

Figure 7

$$d = \left| \text{proj}_{\mathbf{n}} \overrightarrow{BA} \right| = \frac{\left| \overrightarrow{BA} \cdot \mathbf{n} \right| |\mathbf{n}|}{|\mathbf{n}|^2} = \frac{\left| \overrightarrow{BA} \cdot \mathbf{n} \right|}{|\mathbf{n}|}.$$

All that remains is to find a convenient point B and carry out the computation. By inspection, the point $(1,0,0)$ is an easily found point on $2x - 3y + 6z = 2$, and this gives us $\overrightarrow{BA} = \langle 2,1,-2 \rangle$.

$$d = \frac{\left|\overrightarrow{BA} \cdot \mathbf{n}\right|}{|\mathbf{n}|} = \frac{\left|\langle 2,1,-2 \rangle \cdot \langle 2,-3,6 \rangle\right|}{\left|\langle 2,-3,6 \rangle\right|} = \frac{|4-3-12|}{\sqrt{4+9+36}} = \frac{11}{7}$$

Example 11 ✎

Find the shortest distance d between the skew lines

$$\mathbf{r}(u) = \langle 2-u, 3+2u, 2+u \rangle \quad \text{and} \quad \mathbf{s}(v) = \langle 2v, 4-v, 2-v \rangle.$$

Solution

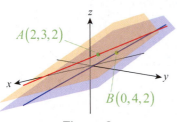

$A(2,3,2)$

$B(0,4,2)$

Figure 8

We determined in Example 4b that the lines are skew, so we know that they lie in two parallel planes. If we can determine those planes, the shortest distance between the lines will be the same as the distance between the planes. A variation of our work in Example 10 will let us find that distance.

Since $\mathbf{r}(u) = \langle 2,3,2 \rangle + u\langle -1,2,1 \rangle$ and $\mathbf{s}(v) = \langle 0,4,2 \rangle + v\langle 2,-1,-1 \rangle$, the direction vector for the line $\mathbf{r}(u)$ is $\mathbf{p} = \langle -1,2,1 \rangle$ and the direction vector for the line $\mathbf{s}(v)$ is $\mathbf{q} = \langle 2,-1,-1 \rangle$. Any vector $\mathbf{n}$ normal to the two parallel planes must be orthogonal to both $\mathbf{p}$ and $\mathbf{q}$, so we can define

$$\mathbf{n} = \mathbf{p} \times \mathbf{q} = \begin{vmatrix} \mathbf{i} & \mathbf{j} & \mathbf{k} \\ -1 & 2 & 1 \\ 2 & -1 & -1 \end{vmatrix} = \langle -1, 1, -3 \rangle.$$

The point $A(2,3,2)$ lies in the plane containing the line $\mathbf{r}(u)$ and the point $B(0,4,2)$ lies in the plane containing the line $\mathbf{s}(v)$. So the length of the projection of the vector $\overrightarrow{AB} = \langle -2,1,0 \rangle$ onto $\mathbf{n}$ is the distance d between the two planes.

$$d = \left|\text{proj}_{\mathbf{n}} \overrightarrow{AB}\right| = \frac{\left|\overrightarrow{AB} \cdot \mathbf{n}\right|}{|\mathbf{n}|} = \frac{\left|\langle -2,1,0 \rangle \cdot \langle -1,1,-3 \rangle\right|}{\left|\langle -1,1,-3 \rangle\right|} = \frac{3}{\sqrt{11}}$$

11.5 Exercises

1–3 Determine which of the given points lie on the given line L.

1. L is the line $\mathbf{r}(t) = \langle 2+3t, -1+t, 4+2t \rangle$; $A(5,0,6)$, $B(0,0,2)$, $C(-1,-2,2)$

2. L is the line through the point $(-1,0,5)$ and parallel to $\mathbf{i} - 3\mathbf{j} + \mathbf{k}$; $A(1,-6,3)$, $B(-3,6,3)$, $C\left(-\frac{1}{2}, -\frac{3}{2}, \frac{11}{2}\right)$

3. L is the line through the points $(-6,-2,1)$ and $(-4,1,3)$; $A(-2,4,5)$, $B(4,13,11)$, $C(0,7,2)$

4–11 Find the vector form and the parametric equations of the given line.

4. The line through the point $(2,5,-3)$ and parallel to $\mathbf{i}+\mathbf{j}+3\mathbf{k}$

5. The line through the point $(1,2,3)$ and parallel to $3\mathbf{i}+2\mathbf{j}+\mathbf{k}$

6. The line through the point $(14,0,-\frac{13}{4})$ and parallel to $\langle -1,\frac{1}{2},\frac{1}{3}\rangle$

7. The line through the point $(0,-5,8)$ and parallel to $\langle -4,-15,-19\rangle$

8. The line through the points $(0,-1,0)$ and $(4,1,1)$

9. The line through the point $(-9,11,-6)$ and parallel to $4\mathbf{i}+3\mathbf{k}$

10. The line through the points $(3,-5,1)$ and $(0,2,7)$

11. The line through the points $(16,21,28)$ and $(-10.5,32,12)$

12–19. By changing the initial point and the tracing rate, find alternative parametrizations for each of the lines in Exercises 4–11.

20–24 Give a vector description of the line segment between the two given points. (**Hint:** First determine a direction vector and be sure to appropriately restrict the possible values of the parameter t.)

20. The line segment between $(-1,0,1)$ and $(3,2,0)$

21. The line segment between $(1,2,3)$ and $(3,2,1)$

22. The line segment between $(-2,4,7)$ and $(-1,5,8)$

23. The line segment between $(-3,\frac{2}{3},-11)$ and $(5,-2,-7)$

24. The line segment between $(-\frac{1}{2},1,\frac{2}{3})$ and $(\frac{4}{3},-6,\frac{3}{2})$

25–32. Find the intersection points of the lines in Exercises 4–11 with the coordinate planes.

33–37 If possible, determine the point of intersection of the given pair of lines.

33. $\mathbf{r}(t)=\langle 1+t,-2+5t,1\rangle$ and $\mathbf{s}(u)=\langle 4+3u,2u,2+u\rangle$

34. $\mathbf{r}(t)=\langle 4+2t,-1-t,2+3t\rangle$ and $\mathbf{s}(u)=\left\langle \frac{1}{2}+7u,\frac{1}{2}-3u,\frac{3}{2}+u\right\rangle$

35. $\mathbf{r}(t)=\langle 3+t,3t,1+2t\rangle$ and $\mathbf{s}(u)=\langle 1+u,-1-2u,-2+u\rangle$

36. $\mathbf{r}(t)=\left\langle 3+2t,3+2t,\frac{3}{2}+\frac{1}{2}t\right\rangle$ and $\mathbf{s}(u)=\left\langle 5-2u,6u-19,\frac{43}{2}-7u\right\rangle$

37. $\mathbf{r}(t)=\langle 1-t,2t-1,4t\rangle$ and $\mathbf{s}(u)=\langle 2+u,5+2u,2\rangle$

38–42 Find a scalar equation for the plane containing the given point and having the indicated normal vector. Then describe the plane as a two-parameter set of points in $\mathbb{R}^3$.

38. The plane through the origin, with normal vector $\mathbf{n}=\langle -1,3,1\rangle$

39. The plane through the point $(2,0,9)$, with normal vector $\mathbf{n}=\langle 5,-4,-1\rangle$

40. The plane through the point $(7,2,4)$, with normal vector $\mathbf{n}=\langle -1,2,-5\rangle$

41. The plane through the point $(1,2,3)$, with normal vector $\mathbf{n}=\langle \frac{1}{2},\frac{3}{4},\frac{5}{6}\rangle$

42. The plane through the point $(-2,3,5)$, with normal vector $\mathbf{n}=\langle 7,-\frac{2}{5},-\frac{5}{2}\rangle$

43–47 Find an equation for the plane containing the given points.

43. $A(2,0,0)$, $B(0,1,0)$, $C(0,0,-3)$

44. $A(2,0,-1)$, $B(3,-4,4)$, $C(-1,5,2)$

45. $A(1,2,0)$, $B(3,0,4)$, $C(0,5,6)$

46. $A\left(-3,\frac{1}{2},1\right)$, $B\left(-1,0,\frac{5}{2}\right)$, $C(-4,-1,2)$

47. $A\left(4,0,-\frac{3}{5}\right)$, $B(1,0,1)$, $C(2,-1,-5)$

48–51 Find the parametric equations of the line as described.

48. The line through the point $(-5,7,8)$ that is perpendicular to the plane $2x - y + 3z = 4$

49. The line through $(-3,-1,4)$ that is parallel to the line $\mathbf{r}(t) = \langle 1+t, -2+4t, 1-3t \rangle$

50. The line through the origin that is perpendicular to the vectors $\mathbf{i} + 2\mathbf{j}$ and $\mathbf{i} - \mathbf{k}$

51. The line through the point $(-1,-1,-5)$ that forms a right angle with the line $\mathbf{r}(t) = \langle 1+t, 1+2t, 1-t \rangle$

52–56. Solving each of the parametric equations $x = x_0 + at$, $y = y_0 + bt$, and $z = z_0 + ct$ for t and equating the results yields the so-called **symmetric equations** for a line.

$$\frac{x-x_0}{a} = \frac{y-y_0}{b} = \frac{z-z_0}{c}$$

Use this method to find the symmetric equations for the lines in Exercises 4–8.

57–65 Find the equation of the plane as described.

57. The plane through the point $(-2,1,5)$, parallel to the plane $3x - 2y + 5z = 1$

58. The plane through the point $(-5,0,1)$, perpendicular to the vector $\langle 1, -1, 7 \rangle$

59. The plane through the point $(6,-1,2)$, perpendicular to the line $\mathbf{r}(t) = \langle 1-t, 3+4t, 2+t \rangle$

60. The plane that contains the z-axis and makes an angle of $\pi/6$ with the positive x-axis

61. The plane through the point $(3,5,-2)$ that contains the line $x = 2 - t$, $y = 4 + 2t$, $z = t - 3$

62. The plane containing the lines
$\mathbf{r}(t) = \langle 1+t, -2+5t, 1 \rangle$ and
$\mathbf{s}(u) = \langle 4+3u, 2u, 2+u \rangle$

63. The plane containing the lines
$\mathbf{r}(t) = \langle 4+2t, -1-t, 2+3t \rangle$ and
$\mathbf{s}(u) = \left\langle \frac{1}{2} + 7u, \frac{1}{2} - 3u, \frac{3}{2} + u \right\rangle$

64. The plane through the points $(4,-1,1)$ and $(1,0,-2)$ that is perpendicular to the plane $2x - y + z = 9$

65. The plane that contains the line $\mathbf{r}(t) = \langle t-2, 4+t, 3-2t \rangle$ and the intersection of the planes $x - y + 2z = 0$ and $2x + 2y - z = 1$

66–69 Find the point of intersection between the given line and plane.

66. The line $\mathbf{r}(t) = \langle t-1, 4t, t \rangle$ and the plane $3x - y + 4z = 3$

67. The line $\mathbf{r}(t) = \langle 2+5t, 3t-2, 4+t \rangle$ and the plane $x + 2y - 6z = 2$

68. The line $\mathbf{r}(t) = \left\langle 8+t, 4+2t, 3+\frac{t}{3} \right\rangle$ and the plane $x + 2y - 2z + 3 = 0$

69. The line $\mathbf{r}(t) = \left\langle 1+\frac{t}{9}, 1-\frac{t}{12}, 1+\frac{t}{4} \right\rangle$ and the plane $5y + z = 15 + x$

70–73 Find the parametric equations for the line formed by the intersection of the two given planes.

70. $x - 2y + z = 4$,   the xz-plane

71. $x + 3y - z = 2$, $2x - y + z = 1$

72. $x - 2y - z = 3$, $-x + y + 2z = 0$

73. $3x + y - 2z + 1 = 0$, $x + y - 4z = 5$

74. Find an equation for the set of points that are equidistant from the points $(2,-1,6)$ and $(-4,5,2)$.

75. Find the equation of the plane having x-, y-, and z-intercepts of $(a,0,0)$, $(0,b,0)$, and $(0,0,c)$, respectively.

76. Find three planes that intersect along the line $\mathbf{r}(t) = \langle t-1, 4t, t \rangle$.

77–80 Find the shortest distance between the point and the plane.

77. $A(0,0,0)$; $4x - y + 5z = 1$

78. $A(-2,-1,7)$; $x - 3y + 2z = 0$

79. $A(-3,1,-1)$; $2x - 2y + z = 3$

80. $A(1,2,3)$; $x + 2y + 3z + 4 = 0$

81–84 Find the shortest distance d between the given skew lines.

81. $\mathbf{r}(t) = \langle 1, 1+t, t \rangle$ and $\mathbf{s}(u) = \langle -2u, 1-u, 2-2u \rangle$

82. $\mathbf{r}(t) = \langle 1+4t, 0, -2t \rangle$ and $\mathbf{s}(u) = \langle 4u-4, 2u, u \rangle$

83. $\mathbf{r}(t) = \langle 2t, t-1, -1 \rangle$ and $\mathbf{s}(u) = \langle u-1, 2u, 2u-1 \rangle$

84. $\mathbf{r}(t) = \langle 1+2t, 2-9t, 1-3t \rangle$ and
$\mathbf{s}(u) = \langle 2u, 16u-7, 2u \rangle$

85–88 We can find the shortest distance in three-dimensional space from a point A to a line containing a point B by the following argument. Denoting the direction vector of the line by $\mathbf{v}$, and the angle formed by $\mathbf{v}$ and the vector $\overrightarrow{BA}$ by θ, note that the distance d we are seeking is $\left| \overrightarrow{BA} \right| \cdot \sin\theta$ (see figure below).

Recalling that $\left| \mathbf{v} \times \overrightarrow{BA} \right| = |\mathbf{v}| \cdot \left| \overrightarrow{BA} \right| \cdot \sin\theta$, d can be expressed as

$$d = \frac{\left| \mathbf{v} \times \overrightarrow{BA} \right|}{|\mathbf{v}|}.$$

(Note that this is an improvement over Exercise 86 of Section 11.3 where both the point and the line were lying in the xy-plane. However, it is worth comparing the above formula with that of Exercise 68 of Section 11.4.)

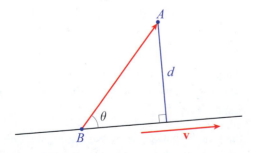

In Exercises 85–88, use the above formula to find the shortest distance between the point and the line.

85. $A(2,-1,1)$; $\mathbf{r}(t) = \langle 3t, 1-t, 1+t \rangle$

86. $A(-2,1,0)$; $\mathbf{r}(t) = \langle 1+t, t, 2-t \rangle$

87. $A(1,0,0)$; $\mathbf{r}(t) = \langle 1-2t, 1-t, 1+3t \rangle$

88. $A\left(\frac{1}{2}, 5, -4\right)$; $\mathbf{r}(t) = \left\langle \frac{3}{2} - t, 6+t, -3-2t \right\rangle$

89–116 *True or False?* Determine whether the given statement is true or false. In case of a false statement, explain or provide a counterexample.

89. If the corresponding coefficients of x, y, and z in the equations of two planes are equal, then the planes are parallel. (Here we assume that both equations are in the standard form $ax + by + cz = d$.)

90. If two planes are parallel, then the corresponding coefficients of x, y, and z (as in Exercise 89) in their equations are equal.

91. If the corresponding coefficients in the equations of two planes (as in Exercise 89) are negative reciprocals, then the planes are perpendicular.

92. If two lines are parallel to a common line, then they are parallel.

93. If two lines are parallel to a common plane, then they are parallel.

94. If two lines are perpendicular to a common plane, then they are parallel.

95. If two lines are perpendicular to a common line, then they are parallel.

96. If two planes are perpendicular to a common plane, then they are parallel.

97. If two planes are perpendicular to a common line, then they are parallel.

98. If two planes are parallel to a common line, then they are parallel.

99. If two planes are parallel to a common plane, then they are parallel.

100. If two lines do not intersect, then they are parallel.

101. If two planes do not intersect, then they are parallel.

102. If a line and a plane do not intersect, then they are parallel.

103. The intersection of a line and a plane is empty or a single point.

104. The intersection of two distinct planes is empty or it is a line.

105. If plane P is parallel to the xz-plane, then $\langle 1, 0, -1 \rangle$ is normal to P.

106. If plane P is parallel to the xz-plane, then $\langle 0, -1, 0 \rangle$ is normal to P.

107. If plane P is parallel to the xz-plane, then the coefficient of y in its equation is 0.

108. If plane P is parallel to the xz-plane, then the coefficients of both x and z in its equation are 0.

109. The plane $y = 0$ contains the y-axis.

110. The plane $2z - x = 3$ contains the y-axis.

111. The plane $2z - x = 0$ contains the y-axis.

112. The plane with equation $ax + by + cz = d$ goes through the origin if and only if $d = 0$.

113. If $\mathbf{n}_1$ and $\mathbf{n}_2$ are normal vectors to the planes P_1 and P_2, respectively, then $\mathbf{n}_1 \times \mathbf{n}_2$ is a direction vector for their line of intersection.

114. If L_1 and L_2 are skew lines, then there is a pair of parallel planes containing them.

115. The vector equation $\mathbf{r}(t) = \langle 3 + t, 5 - 2t, 1 - 4t \rangle$ and the parametric equations $x = 5 - \dfrac{t}{2}$, $y = 1 + t$, and $z = -7 + 2t$ describe the same line.

116. The equations $4(1 - x) + 2(y - 7) - 14(z - 1) = 0$ and $2x - y + 7z = 2$ describe the same plane.

11.5 **Technology Exercises**

117–121. Use a computer algebra system or programmable calculator to write a program that returns the equation of a plane through three given points in three-dimensional space. Use your program to check your answers for Exercises 43–47.

122–130. Use a computer algebra system or programmable calculator to write a program that decides whether two lines intersect in three-dimensional space. If they do, the program should return the coordinates of the intersection point, otherwise the message "skew lines" should appear, along with the shortest distance between them. Use your program to revisit Exercises 33–37 and 81–84.

11.6 Cylinders and Quadric Surfaces

TOPICS

1. Cylinders
2. Quadric surfaces

As we develop the tools of multivariable calculus, we'll find it useful to have a collection of surfaces at our disposal for the purposes of illustration. *Cylinders* and *quadric surfaces* are two such useful classes of surfaces in $\mathbb{R}^3$; in this section, we define these classes and show how they are natural extensions of familiar objects in $\mathbb{R}^2$.

TOPIC 1   Cylinders

In Section 6.1, we defined a cylinder as a solid bounded between a planar region B and a translated copy of B in a plane parallel to the first—our definition was motivated by our goal at the time of determining volumes of solids. The following definition generalizes that idea to include objects for which "volume" has no meaning.

Definition ♀

Cylinders, Generating Curves, and Cross-Sections

A **cylinder** is a surface consisting of all lines (called **rulings**) parallel to a given line and passing through a fixed planar curve, called the cylinder's **generating curve**. Cylinders are often described in terms of their intersections with planes parallel to the coordinate planes; each such intersection is called a **cross-section** or **trace** of the surface.

Example 1 ✏

The equation $x^2 = 4y$ corresponds to the surface depicted in Figure 1. In the xy-plane, the graph of the equation $x^2 = 4y$ is the parabola labeled in Figure 1 as the generating curve; the other two parabolas shown in red qualify as cross-sections of the surface, but we could actually use any one of them as the generating curve. Each line on the surface parallel to the z-axis is a ruling. The surface can be thought of as being generated by one such ruling, as its point of intersection with the generating curve traces out the parabola.

Because of the shape of its generating curve, this surface is called a **parabolic cylinder**.

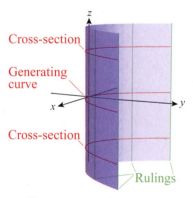

Figure 1 Parabolic Cylinder

The other two conic sections, ellipses and hyperbolas, give rise to **elliptic** and **hyperbolic cylinders** in a similar fashion. In general, any graph lying in one of the three coordinate planes can be used as the generating curve for a cylinder with rulings perpendicular to that plane. Our familiarity with conic sections allows us to readily sketch the cylinders associated with them.

Example 2 ✏

Sketch the surfaces defined by the equations below by determining their cross-sections.

a. $4y^2 - 16y + 9z^2 - 18z = 11$ **b.** $z^2 + 2z - 4x^2 = 3$

Solution

Each surface is defined by an equation in only two variables, meaning that the missing variable is free to take on any value whatsoever. Geometrically, this means that the rulings are parallel to the axis of the missing variable. To determine the cross-sections of each surface, we use the familiar technique of completing the square.

a.

$$4y^2 - 16y + 9z^2 - 18z = 11$$

$$4\left(y^2 - 4y + 4\right) + 9\left(z^2 - 2z + 1\right) = 11 + 16 + 9$$

$$4\left(y - 2\right)^2 + 9\left(z - 1\right)^2 = 36$$

$$\frac{\left(y - 2\right)^2}{9} + \frac{\left(z - 1\right)^2}{4} = 1$$

In the yz-plane, the cross-section of the surface is the ellipse centered at $(2,1)$ with major axis of length $2 \cdot 3 = 6$ and minor axis of length $2 \cdot 2 = 4$. Figure 2 is an illustration, with the cross-section in the yz-plane shown as a red curve.

b.

$$z^2 + 2z - 4x^2 = 3$$

$$\left(z^2 + 2z + 1\right) - 4x^2 = 3 + 1$$

$$\left(z + 1\right)^2 - 4x^2 = 4$$

$$\frac{\left(z + 1\right)^2}{4} - x^2 = 1$$

In the xz-plane, the cross-section of the surface is a vertically oriented hyperbola centered at $(0, -1)$ with asymptotes of $z = -1 \pm 2x$ (see Section 9.5 for a refresher on how to determine the asymptotes). Figure 3 is a depiction of the surface, with a cross-section again shown as a red curve.

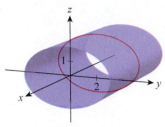

Figure 2 Elliptic Cylinder

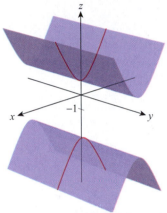

Figure 3 Hyperbolic Cylinder

TOPIC 2   Quadric Surfaces

The three varieties of cylinders just shown are particular examples of **quadric surfaces**, which in general are graphs of equations of the form

$$Ax^2 + By^2 + Cz^2 + Dxy + Eyz + Fxz + Gx + Hy + Iz + J = 0$$

where A, B, ..., J are real constants. Quadric surfaces are the three-dimensional analogs of the conic sections of Section 9.5; like those two-dimensional curves, a given quadric surface can be rotated and translated so that its corresponding equation is simpler, namely of the form

$$Ax^2 + By^2 + Cz^2 + J = 0, \quad Ax^2 + By^2 + Iz = 0,$$

$$By^2 + Cz^2 + Gx = 0, \quad \text{or} \quad Ax^2 + Cz^2 + Hy = 0.$$

In order to briefly characterize and name the shapes of quadric surfaces, we restrict our attention to equations of these forms, rewritten in one of the following patterns.

Quadric Surfaces
(Not Including Parabolic, Elliptic, and Hyperbolic Cylinders)

Ellipsoid	**Hyperbolic Paraboloid**
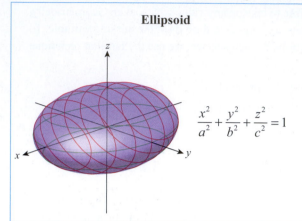 $$\frac{x^2}{a^2} + \frac{y^2}{b^2} + \frac{z^2}{c^2} = 1$$	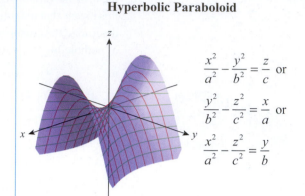 $$\frac{x^2}{a^2} - \frac{y^2}{b^2} = \frac{z}{c} \quad \text{or}$$ $$\frac{y^2}{b^2} - \frac{z^2}{c^2} = \frac{x}{a} \quad \text{or}$$ $$\frac{x^2}{a^2} - \frac{z^2}{c^2} = \frac{y}{b}$$
Elliptic Cone	**Hyperboloid of One Sheet**
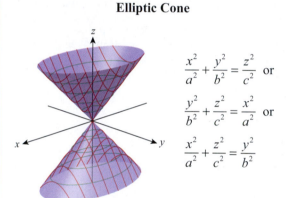 $$\frac{x^2}{a^2} + \frac{y^2}{b^2} = \frac{z^2}{c^2} \quad \text{or}$$ $$\frac{y^2}{b^2} + \frac{z^2}{c^2} = \frac{x^2}{a^2} \quad \text{or}$$ $$\frac{x^2}{a^2} + \frac{z^2}{c^2} = \frac{y^2}{b^2}$$	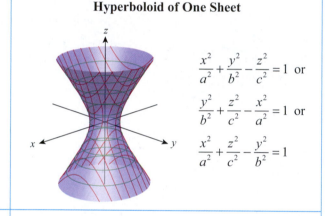 $$\frac{x^2}{a^2} + \frac{y^2}{b^2} - \frac{z^2}{c^2} = 1 \quad \text{or}$$ $$\frac{y^2}{b^2} + \frac{z^2}{c^2} - \frac{x^2}{a^2} = 1 \quad \text{or}$$ $$\frac{x^2}{a^2} + \frac{z^2}{c^2} - \frac{y^2}{b^2} = 1$$
Elliptic Paraboloid	**Hyperboloid of Two Sheets**
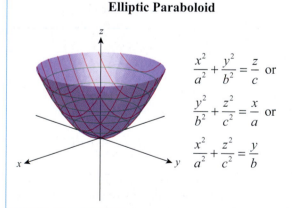 $$\frac{x^2}{a^2} + \frac{y^2}{b^2} = \frac{z}{c} \quad \text{or}$$ $$\frac{y^2}{b^2} + \frac{z^2}{c^2} = \frac{x}{a} \quad \text{or}$$ $$\frac{x^2}{a^2} + \frac{z^2}{c^2} = \frac{y}{b}$$	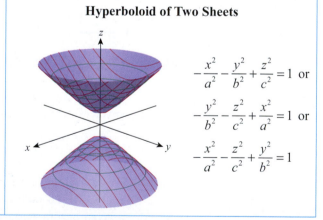 $$-\frac{x^2}{a^2} - \frac{y^2}{b^2} + \frac{z^2}{c^2} = 1 \quad \text{or}$$ $$-\frac{y^2}{b^2} - \frac{z^2}{c^2} + \frac{x^2}{a^2} = 1 \quad \text{or}$$ $$-\frac{x^2}{a^2} - \frac{z^2}{c^2} + \frac{y^2}{b^2} = 1$$

Table 1

For our purposes, it will suffice to be able to identify a given quadric surface and determine the appearance of such features as its traces. The remaining examples of this section demonstrate the techniques we typically use.

Example 3 ✎

Sketch the surface $x^2 + 4y^2 + 4z = 0$.

Solution

By rewriting the equation in the form

$$\frac{x^2}{4} + y^2 = -z,$$

we identify it as an elliptic paraboloid. If k is a positive constant, setting $z = -k$ reduces the equation to

$$\frac{x^2}{4} + y^2 = k, \quad \text{or} \quad \frac{x^2}{4k} + \frac{y^2}{k} = 1,$$

an ellipse in the plane $z = -k$ centered at the origin with major axis of length $2 \cdot 2\sqrt{k}$ and minor axis of length $2 \cdot \sqrt{k}$. Note that there are no real values of x and y that satisfy the equation for positive values of z. Similarly, setting either x or y equal to a constant results in a parabolic cross-section parallel to, respectively, the yz-plane or the xz-plane. Figure 4 illustrates the surface with elliptic horizontal traces and parabolic traces corresponding to fixed x-values.

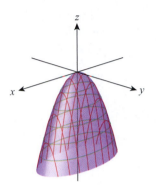

Figure 4

Example 4 ✎

Describe the traces of the surface $\dfrac{y^2}{9} - \dfrac{z^2}{4} = \dfrac{x}{5}$.

Solution

Matching this equation against the given patterns of quadrics, we identify it as a hyperbolic paraboloid. Fixing $x = k$ reduces the equation to a hyperbola, while fixing either $y = k$ or $z = k$ reduces the equation to a parabola. That is, traces parallel to the yz-plane are hyperbolas (Figure 5) and traces parallel to either the xz-plane or the xy-plane are parabolas (Figures 6 and 7, respectively)—this characteristic is reflected in the name of the surface. Note that for any positive fixed x, the hyperbolas open left and right, while for any negative fixed x the hyperbolas open downward and upward. In contrast, because they are squared, fixing either y or z results in a family of parabolas all opening in the same direction.

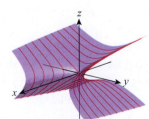

Figure 5

Traces of Form $\dfrac{y^2}{9} - \dfrac{z^2}{4} = \dfrac{k}{5}$

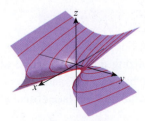

Figure 6

Traces of Form $\dfrac{k^2}{9} - \dfrac{z^2}{4} = \dfrac{x}{5}$

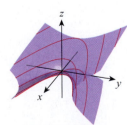

Figure 7

Traces of Form $\dfrac{y^2}{9} - \dfrac{k^2}{4} = \dfrac{x}{5}$

The point at the origin of the hyperbolic paraboloid of Example 4 is called a **saddle point**, a location on a surface which is a local minimum of traces in one direction and a local maximum of traces in another direction.

Example 5 ✐

Sketch the surface $9x^2 - 9y^2 - 4z^2 + 18y + 16z = 61$.

Solution

We recognize the equation as a quadric surface immediately, but to identify it further requires a bit more analysis. The most efficient approach is usually to manipulate the equation, completing the square as required, in order to match it against one of the standard forms; by doing so, we also identify some of the particular features of the surface.

$$9x^2 - 9y^2 - 4z^2 + 18y + 16z = 61$$

$$9x^2 - 9\left(y^2 - 2y + 1\right) - 4\left(z^2 - 4z + 4\right) = 61 - 9 - 16$$

$$-9(y-1)^2 - 4(z-2)^2 + 9x^2 = 36$$

$$-\frac{(y-1)^2}{4} - \frac{(z-2)^2}{9} + \frac{x^2}{4} = 1$$

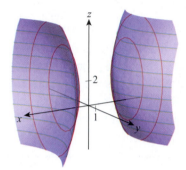

In this form we recognize the surface as a hyperboloid of two sheets (the two negative signs on the left distinguish hyperboloids of two sheets from hyperboloids of one sheet). Further, the center of the hyperboloid has been shifted from the origin to the point $(0,1,2)$. Fixing $x = k$ gives us the family of equations

$$\frac{(y-1)^2}{4} + \frac{(z-2)^2}{9} = \frac{k^2}{4} - 1,$$

Figure 8

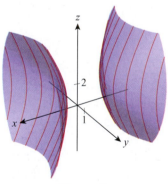

which only have real solutions for $|k| \geq 2$. The graphs in the plane $x = k$ are ellipses (as seen in red in Figure 8), and the ellipses reduce to a point when $k = 2$ and $k = -2$; that is, the points $(2,1,2)$ and $(-2,1,2)$ are the vertices of the hyperboloid. Fixing $z = k$ results in a hyperbola in the plane $z = k$ (shown in green in Figure 8), while fixing $y = k$ results in a hyperbola in the plane $y = k$; this family of hyperbolas is illustrated in Figure 9.

Figure 9

Technology Note 💻

Graphing calculators and computer algebra systems are very useful tools for visualizing quadric surfaces, especially if the technology allows the surface to be "grabbed" and spun in space—doing so gives us the ability to look at parts of the surface obscured by any one static snapshot.

The basic command in *Mathematica* for plotting three-dimensional surfaces is **ContourPlot3D**, and most other computer algebra systems have comparable graphing capability. Each computer algebra system offers its own particular set of options, which may allow specification of which traces to draw and the transparency of the surface. Figures 10 and 11 illustrate the use of two of *Mathematica*'s graphing options in drawing the surface of Example 4.

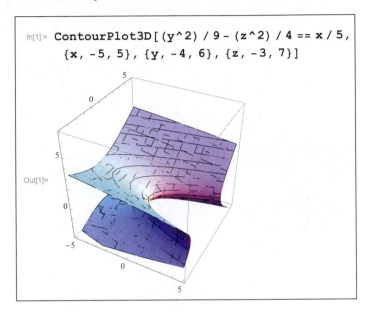

```
In[1]:= ContourPlot3D[(y^2) / 9 - (z^2) / 4 == x / 5,
    {x, -5, 5}, {y, -4, 6}, {z, -3, 7}]
```

Out[1]=

Figure 10

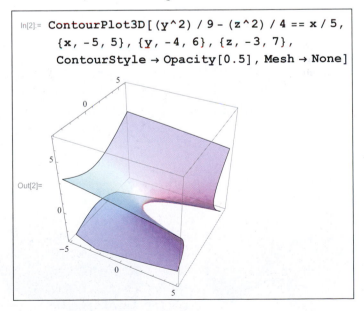

```
In[2]:= ContourPlot3D[(y^2) / 9 - (z^2) / 4 == x / 5,
    {x, -5, 5}, {y, -4, 6}, {z, -3, 7},
    ContourStyle → Opacity[0.5], Mesh → None]
```

Out[2]=

Figure 11

11.6 **Exercises**

1–8 Identify the surface defined by the equation and match it to the appropriate graph (labeled A–H).

1. $x^2 = 2(1-y)$

2. $y^2 - 4y + z^2 = 4$

3. $\dfrac{y^2}{5} - \dfrac{z^2}{8} = \dfrac{x}{6}$

4. $\dfrac{x^2}{9} + \dfrac{z^2}{4} = \dfrac{y^2}{25}$

5. $\dfrac{x^2}{16} + \dfrac{y^2}{9} + \dfrac{z^2}{4} = 1$

6. $6y^2 + 4z^2 - 3x^2 = 12$

7. $5x^2 - 8y^2 + 32y - 10z^2 + 20z = 82$

8. $15x^2 + 30x - 20y + 12z^2 + 55 = 0$

A.

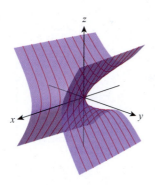

B.

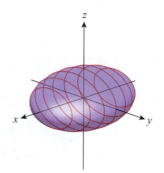

C.

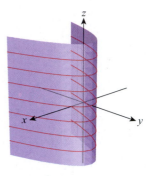

D.

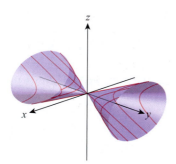

E.

F.

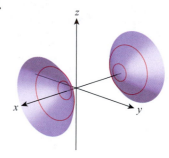

G.

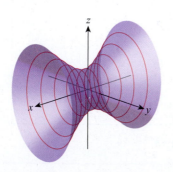

H.

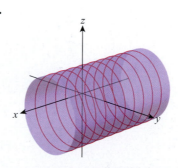

9–16 Sketch the surface by hand. Use cross-sections to help you with your sketch.

9. $x^2 + (z-1)^2 = 1$ **10.** $y^2 + \dfrac{(z-1)^2}{4} = 1$

11. $y^2 - \dfrac{(z-1)^2}{4} = 1$ **12.** $y^2 - \dfrac{(z-1)}{4} = 0$

13. $x^2 - 6x + 2y + 11 = 0$ **14.** $x^2 - 6x + 2y^2 + 8 = 0$

15. $\cos\dfrac{z}{2} - y = 1$ **16.** $\ln x - z = 0$

17–18 Find the lengths of the major and minor axes as well as the foci of the indicated cross-section of the surface $z = \dfrac{x^2}{2} + y^2$.

17. $z = 4$ **18.** $z = 8$

19–20 The intersection of the given plane and the surface from Exercises 17–18 is a parabola. Find the coordinates of its vertex and focus.

19. $x = 1$ **20.** $y = 1$

21–34 Identify and sketch the quadric surface by hand. Use cross-sections to help you with your sketch.

21. $\dfrac{x^2}{9} + \dfrac{y^2}{16} + \dfrac{(z-2)^2}{4} = 1$

22. $\dfrac{(y+1)^2}{9} + \dfrac{z^2}{4} = x$

23. $x^2 + \dfrac{9y^2}{16} - \dfrac{9z^2}{25} - 9 = 0$

24. $3z^2 + 2y^2 = \dfrac{3x^2}{2}$

25. $3z^2 - 2y^2 = 1 + \dfrac{3x^2}{2}$

26. $3z^2 + 2y = \dfrac{3x^2}{2}$

27. $3z^2 + 2y^2 = \dfrac{3x}{2}$

28. $3z^2 + 2y^2 = 1 - \dfrac{3x^2}{2}$

29. $2x^2 + 2y^2 = z$

30. $2x^2 + 2y^2 = z^2$

31. $x^2 + 2x - y^2 + z^2 = 0$

32. $x^2 + 2x + 2y^2 + 3z^2 = 8y$

33. $x^2 + 2x - 2y^2 - 3z = 8y + 7$

34. $-x^2 + 2x + 2y^2 - 3z^2 = 8y + 2$

35. Consider the quadric surface of Exercise 29, $2x^2 + 2y^2 = z$. Find the intersection of this surface with the xz-plane. Explain why it is called a "generating curve" of the surface.

36. Find another generating curve for the surface in Exercise 35, and argue that the generating curve of a surface of revolution is not unique. (**Hint:** Consider the intersection of the surface with another coordinate plane.)

37–38 Find the indicated generating curve for the given surface.

37. $x^2 + y^2 = 5z$, the generating curve that lies in the yz-plane

38. $x^2 + z^2 = 1 - 3y^2$, the generating curve that lies in the xy-plane

39–46 Find an equation for the surface that results from rotating the curve about the indicated axis. (**Hint:** See Exercises 35–38.)

39. $x^2 = 2y$;   about the y-axis

40. $x = 2y$;   about the y-axis

41. $y = \sqrt{z-2}$;   about the z-axis

42. $y = \sqrt{1-x^2}$;   about the x-axis

43. $x = \dfrac{a}{c}\sqrt{z^2 + c^2}$;   about the z-axis

44. $2xz = 3$;   about the x-axis

45. $2xz = 3$;   about the z-axis

46. $z = \dfrac{e^y}{5}$;   about the y-axis

47. Assuming that the Earth is a perfect ellipsoid with equatorial and polar radii of 6378 and 6357 kilometers, respectively, find the equation of this ellipsoid assuming it is centered at the origin and the axis of rotation is the z-axis.

48. What do you know about a, b, and c if the horizontal cross-sections of the graph of

$$\frac{x^2}{a^2} + \frac{y^2}{b^2} + \frac{z^2}{c^2} = 1 \text{ are circles?}$$

49. Prove that all vertical cross-sections of the ellipsoid $\frac{x^2}{a^2} + \frac{y^2}{b^2} + \frac{z^2}{c^2} = 1$ have the same eccentricity. Can you make a similar statement about horizontal cross-sections? (**Hint:** For a refresher, see the definition of eccentricity in Section 9.5).

50. Find the equation of the set of points in three-dimensional space which are equidistant from the point $(0,0,1)$ and the plane $z = -1$.

51–55 *True or False?* Determine whether the given statement is true or false. In case of a false statement, explain or provide a counterexample.

51. A sphere is not an ellipsoid.

52. A vertical paraboloid whose horizontal cross-sections are circles is not an elliptic paraboloid.

53. If $a = b$ in the equation $\frac{x^2}{a^2} + \frac{y^2}{b^2} - \frac{z^2}{c^2} = 1$ of a hyperboloid, then its horizontal cross-sections are circles.

54. A quadric surface that is a surface of revolution has a unique generating curve.

55. A quadric surface that is a surface of revolution has a unique axis of rotation.

11.6 Technology Exercises

56–63. Use a graphing calculator or a computer algebra system to sketch the cylindrical surfaces of Exercises 9–16.

64–77. Use a graphing calculator or a computer algebra system to sketch the quadric surfaces of Exercises 21–34.

Chapter 11
Review Exercises

1–6 Find the distance from the point to the indicated plane.

1. $(-2,0,4)$;   the yz-plane

2. $(3,-2,5)$;   the plane $y = 1$

3. $(0,0,0)$; $4x + y - z - 3 = 0$

4. $(1,-2,1)$; $x - y + z - 5 = 0$

5. $\left(3,0,\dfrac{1}{2}\right)$; $y - 2z = x + 4$

6. $(-1,5,2)$; $x + 4 = 4y + 2z$

7–12 Describe the set of points represented by the given equation or inequality.

7. $x^2 + (y-2)^2 + z^2 > 1$ 8. $x - 2z - 1 \geq y$

9. $x^2 + 4z^2 < 2$ 10. $xyz^2 \neq 0$

11. $x^2 + 2y^2 - 4z^2 \leq 8$ 12. $6x^2 + 2z^2 < 3y$

13–14 If $\mathbf{u} = \langle 2,5,-4 \rangle$ and $\mathbf{v} = \langle -1,4,-2 \rangle$, find the coordinates of the endpoint of the indicated vector with the given initial point.

13. $2\mathbf{u} - 3\mathbf{v}$ with initial point $(2,-7,3)$

14. $\dfrac{1}{2}\mathbf{v} - 4\mathbf{u}$ with initial point $(-1,4,-5)$

15. Find a vector $\mathbf{v}$ that solves the vector equation $\langle -6,2,0 \rangle - 2\mathbf{v} = \langle 1,0,-1 \rangle$.

16. Explain the difference between the following: $(1,2,3)$ vs. $\langle 1,2,3 \rangle$ vs. $\{1, 2, 3\}$

17. Determine whether the following points are collinear: $P(1,-3,0)$, $Q(2,5,1)$, and $R(4,21,3)$

18–19 Find the component form and magnitude of the vector $\overrightarrow{PQ}$.

18. $P(2,-4,0)$, $Q(1,-3,-1)$

19. $P\left(\dfrac{\sqrt{3}}{2},-1,1\right)$, $Q(2\sqrt{3},-2,3)$

20–21 Use vectors to find the coordinates of the indicated point.

20. The point two-thirds of the way from $P(-6,0,3)$ to $Q(0,12,9)$

21. The point three-eighths of the way from $P(0,-4,8)$ to $Q(-16,0,2)$

22–23 Find the unit vector $\mathbf{u}$ pointing in the direction of the given vector $\mathbf{v}$.

22. $\mathbf{v} = \langle 2,4,6 \rangle$ 23. $\mathbf{v} = \left\langle 1,-\dfrac{1}{4},\dfrac{1}{3} \right\rangle$

24–26 Find the dot product of $\mathbf{u}$ and $\mathbf{v}$.

24. $\mathbf{u} = 2\mathbf{i} - 5\mathbf{j}$, $\mathbf{v} = 3\mathbf{i} - \mathbf{k}$

25. $\mathbf{u} = \dfrac{1}{3}\mathbf{i} + \pi\mathbf{j} - \dfrac{7}{2}\mathbf{k}$, $\mathbf{v} = -6\mathbf{i} - \dfrac{2}{\pi}\mathbf{j} + 8\mathbf{k}$

26. $|\mathbf{u}| = 3\sqrt{3}$, $|\mathbf{v}| = 2$, their angle is $30°$

27–28 Find the angle between the given vectors.

27. $\mathbf{u} = \langle 1,0,1 \rangle$, $\mathbf{v} = \langle 2,2,0 \rangle$

28. $\mathbf{u} = -\mathbf{j} - \sqrt{2}\mathbf{k}$, $\mathbf{v} = \mathbf{i} - \sqrt{3}\mathbf{k}$

29–30 Determine the value of the parameter so that the vectors are **a.** parallel, **b.** orthogonal.

29. $\langle s,4,2 \rangle$ and $\langle 1,2s,\sqrt{2} \rangle$

30. $\langle 2s,1,3 \rangle$ and $\langle 1,4s,3\sqrt{2} \rangle$

31. Find the angle between the planes $x + 4y + 3z = 2$ and $4z - 3x - y = 11$. (**Hint:** The angle between two planes is the same as that between their respective normal vectors.)

32–33 Find the direction angles of the given vector.

32. $\langle -1,3,4 \rangle$ 33. $\langle \sqrt{3},2,1 \rangle$

34–37 Find the decomposition of **u** into a sum of two vectors, one parallel to, and the other perpendicular to **v**.

34. $\mathbf{u} = \langle 1,3,0 \rangle$, $\mathbf{v} = \langle 2,0,-4 \rangle$

35. $\mathbf{u} = \left\langle \frac{1}{3}, -3, 1 \right\rangle$, $\mathbf{v} = \left\langle 2, -5, \frac{5}{3} \right\rangle$

36. $\mathbf{u} = 3\mathbf{i} - \mathbf{j} + \mathbf{k}$, $\mathbf{v} = \mathbf{i} + 2\mathbf{j} - 4\mathbf{k}$

37. $\mathbf{u} = \mathbf{i} - 2\mathbf{j} + \frac{1}{2}\mathbf{k}$, $\mathbf{v} = \frac{1}{2}\mathbf{i} + \mathbf{j} - \mathbf{k}$

38–39 Find the work done by the force **F** as it moves an object from P to Q. (Suppose **F** is measured in newtons, and a unit distance is 1 meter.)

38. $\mathbf{F} = 2.1\mathbf{i} - 2.4\mathbf{j} - 5.7\mathbf{k}$ from $P(5,8,4)$ to $Q(-3,0,1.5)$

39. $\mathbf{F} = 3\mathbf{i} - \mathbf{j} + 4\mathbf{k}$ from $P(0,-2,-17.3)$ to $Q(2,9.5,-11)$

40–41 Use the determinant formula to find the cross product.

40. $\langle 3,0,-2 \rangle \times \langle 1,2,1 \rangle$ **41.** $\langle 2,1,-1 \rangle \times \left\langle \frac{1}{2}, 4, -2 \right\rangle$

42–43 Find both unit vectors perpendicular to **u** and **v**.

42. $\mathbf{u} = \langle 0,-1,1 \rangle$, $\mathbf{v} = \langle 1,2,-1 \rangle$

43. $\mathbf{u} = \left\langle -2, \frac{1}{2}, 1 \right\rangle$, $\mathbf{v} = \langle 0,2,4 \rangle$

44–45 Construct a vector normal to the plane containing the indicated points.

44. $P(1,-1,1)$, $Q(0,2,2)$, $R(3,0,3)$

45. $P(-4,0,5)$, $Q\left(1,-\frac{1}{2},1\right)$, $R(2,3,0)$

46–47 Use cross products to check whether the points P, Q, and R are collinear.

46. $P(-1,0,1)$, $Q(1,-2,0)$, $R(1,1,3)$

47. $P(2,1,-2)$, $Q(3,2,-3)$, $R(-1,-2,1)$

48–49 Find the area of the triangle with the given vertices.

48. $A(0,1,2)$, $B\left(-\frac{1}{2},2,1\right)$, $C\left(\frac{3}{2},-3,4\right)$

49. $A(2,1,0)$, $B(3,-1,2)$, $C(-3,0,1)$

50–53 Let $\mathbf{u} = \langle 2,1,-3 \rangle$, $\mathbf{v} = \langle 1,0,-4 \rangle$, and $\mathbf{w} = \langle 1,-1,\frac{1}{2} \rangle$. If possible, find each of the following.

50. $\mathbf{u} \times (\mathbf{v} + \mathbf{w})$ **51.** $(\mathbf{u} \cdot \mathbf{v}) \cdot \mathbf{w}$

52. $(\mathbf{u} \cdot \mathbf{v}) + \mathbf{w}$ **53.** $(\mathbf{u} + \mathbf{v}) \times \mathbf{w}$

54–55 Find the volume of the parallelepiped spanned by the indicated vectors.

54. $\mathbf{u} = \langle 1,1,1 \rangle$, $\mathbf{v} = \langle 2,0,3 \rangle$, $\mathbf{w} = \langle 1,-2,1 \rangle$

55. $\mathbf{u} = \langle 2,1,1 \rangle$, $\mathbf{v} = \langle 1,2,1 \rangle$, $\mathbf{w} = \langle 1,1,2 \rangle$

56–57 Find parametric equations for the line parallel to the indicated direction vector and passing through the given point.

56. Through the point $(3,4,-7)$ and parallel to $2\mathbf{i} - 2\mathbf{j} + \mathbf{k}$

57. Through the point $\left(\frac{5}{2},9,0\right)$ and parallel to $\left\langle 2, \frac{1}{2}, -\frac{3}{4} \right\rangle$

58–59 Give a vector description of the line segment between the two given points.

58. The line segment between $(1.2,-2,5)$ and $(2.2,0,9)$

59. The line segment between $\left(4,-\frac{3}{2},7\right)$ and $(6,2,5)$

60–63 If possible, determine the point of intersection of the pair of lines.

60. $\mathbf{r}(t) = \langle 1+2t, 3-2t, 4-6t \rangle$ and $\mathbf{s}(u) = \langle 2u-1, u-1, -u \rangle$

61. $\mathbf{r}(t) = \left\langle 3-\frac{7}{2}t, 1-\frac{t}{2}, t-1 \right\rangle$ and $\mathbf{s}(u) = \langle 1-5u, 2u-2, u \rangle$

62. $\mathbf{r}(t) = \langle 1+t, 3t, 2t-3 \rangle$ and $\mathbf{s}(u) = \langle u-2, 4+u, u \rangle$

63. $\mathbf{r}(t) = \langle 2-2t, 1+3t, 4t \rangle$ and $\mathbf{s}(u) = \langle 2+u, 3-3u, 1+u \rangle$

64–65 Identify the plane containing the given point and having the indicated normal vector as a two-parameter set of points in $\mathbb{R}^3$.

64. The plane through the point $(3,-2,-1)$ with normal vector $\mathbf{n} = \langle 1,-2,-3 \rangle$

65. The plane through the point $(4,-1,3)$ with normal vector $\mathbf{n} = \langle \frac{3}{2},-\frac{1}{3},2 \rangle$

66–67 Find an equation for the plane containing the given points.

66. $A(1,0,2)$, $B(2,3,-2)$, $C(-2,1,3)$

67. $A\left(-\frac{1}{2},2,0\right)$, $B(4,1,1)$, $C(-1,-1,1)$

68–70 Find the parametric equations of the line as described

68. The line through the point $(-3,1,2)$ that is perpendicular to the plane $x - 2y + z = 11$

69. The line through $(-2,2,\frac{1}{2})$ that is parallel to the line $\mathbf{r}(t) = \langle t-2, 1+3t, 1-2t \rangle$

70. The line through $(-4,-5,-6)$ that is perpendicular to the vectors $3\mathbf{i}+\mathbf{k}$ and $\mathbf{j}-2\mathbf{k}$

71–72 Find an equation for the plane satisfying the given conditions.

71. The plane through the point $(0,-1,0)$ that forms an angle of $\pi/3$ radians with the positive y-axis

72. The plane through the point $(-5,2,1)$ that contains the line $x = 1 + 2t,\ y = 3 - t,\ z = 4t - 1$

73–74 Find the point of intersection between the given line and plane.

73. The line $\mathbf{r}(t) = \langle 1+2t, 5-4t, 6t-1 \rangle$ and the plane $2x + y - 3z = 1$

74. The line $\mathbf{r}(t) = \langle 1+t, 5-2t, t \rangle$ and the plane $x + 4y - 2z = 3$

75–76 Find parametric equations for the line formed by the intersection of the two given planes.

75. $2x + y - z = 8$, $5x - y + 3z = -1$

76. $3x + 2y - z = 0$, $x - y + 4z = 5$

77. Find an equation for the set of points that are equidistant from the points $(-8,11,3)$ and $(10,-1,-9)$.

78–79 Find the shortest distance d between the given skew lines.

78. $\mathbf{r}(t) = \langle -2+2t, -t, 1+2t \rangle$ and $\mathbf{s}(u) = \langle 3-2u, 5+u, 4+u \rangle$

79. $\mathbf{r}(t) = \langle 4+3t, -5, 2-t \rangle$ and $\mathbf{s}(u) = \langle 3+u, 1-u, 7+2u \rangle$

80–87 Identify the surface defined by the equation and match it to the appropriate graph (labeled A–H).

80. $2x^2 + y^2 + 2z^2 = 12$

81. $z^2 + 2y^2 = 4$

82. $8z^2 + 2y^2 - 4x^2 = 0$

83. $(x+2)^2 + 2y = x^2 + z$

84. $8z^2 - 2y^2 - 4x = 0$

85. $x^2 + 2z^2 - 4z - y^2 + 6 = 0$

86. $3x^2 + 4y^2 + 6z = 0$

87. $\dfrac{x^2}{2} + \dfrac{y^2}{4} - \dfrac{z^2}{3} = 1$

A.

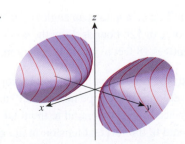

B.

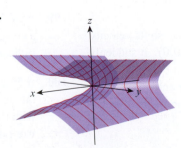

C.

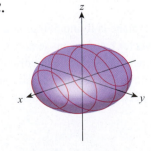

D.

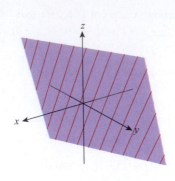

E.

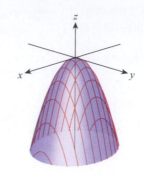

F.

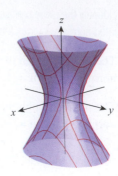

G.

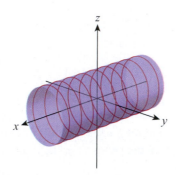

H.

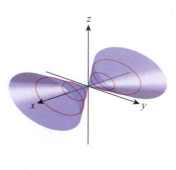

88–93 Use the equation to identify the quadric surface.

88. $225x^2 + 100y^2 - 36(z-2)^2 = 0$ **89.** $4(y+1)^2 - 16z^2 = x$

90. $\dfrac{x^2}{4} + x - \dfrac{y}{3} + z^2 + 1 = 0$

91. $x^2 - 2y^2 + 4y + 3z^2 = 12$ **92.** $2x^2 + 4y^2 + z^2 + 6z = 19$ **93.** $z^2 = 3x^2 + 2y^2 - 2z + 11$

94. An eight-pound weight is suspended from two ropes that form angles of 30° and 45°, respectively, with the horizontal direction. Find the tension forces $\mathbf{T}_1$ and $\mathbf{T}_2$.

95. A soccer player kicks the ball with an initial velocity that has a vertical component of 75 ft/s and a horizontal component of 15 ft/s. Ignoring air resistance and assuming the initial height is 0, determine the ball's velocity at time t, and sketch its position function.

96. A plane is flying 550 mph due east, so its velocity vector can be represented as $\langle 550, 0 \rangle$. It suddenly encounters a 65 mph tailwind that blows 45° east of north. Find the velocity vector and ground speed of the plane under these conditions.

97. A small plane flying at 140 mph due north encounters a 30 mph tailwind that blows from the direction of 30° west of north. At the same time, a 10 mph downdraft is affecting the plane's flight. Find the actual ground speed of the plane.

98. Find the acceleration of a 3.5 kg object if the following forces are acting on it: $\mathbf{F}_1 = 2\mathbf{i} - \mathbf{j} + 2\mathbf{k}$, $\mathbf{F}_2 = \mathbf{j} + 3\mathbf{k}$, and $\mathbf{F}_3 = 2\mathbf{i} + 3\mathbf{j} - 4\mathbf{k}$ (units are in newtons)

99. A cable is pulling on a gate at an angle of 30° with a constant force of 120 pounds. Find the work done if the gate moves 18 feet to its closed position.

100. A 50-pound sled is being pulled up a 24° slope that is 80 feet long. If the rope makes a 21° angle with the surface of the slope, find **a.** the total work done and **b.** the force of tension in the rope (ignore friction and any acceleration of motion).

101. A mechanic tightens a bolt with a force of 80 newtons, applied at the end of a 20 cm wrench, at an angle of $\theta = 70°$. What is the magnitude of the torque applied to the bolt at the pivot point?

102. Find the magnitude of the force acted upon a 10 cm electric wire in a uniform magnetic field of $\mathbf{B} = 3.4$ T if it carries a current of 0.2 A and the angle between the wire and $\mathbf{B}$ is $\theta = 45°$. (**Hint:** See Exercise 61 of Section 11.4.)

103. Find the magnitude of the force that acts upon an electron moving in a uniform magnetic field of $\mathbf{B} = 0.003$ T at $\mathbf{v} = 2.8 \cdot 10^6$ m/s if the velocity vector and $\mathbf{B}$ form a 30° angle. (See Exercise 62 of Section 11.4.)

104–119 *True or False?* Determine whether the given statement is true or false. In case of a false statement, explain or provide a counterexample.

104. If vectors $\mathbf{u}$ and $\mathbf{v}$ have equal magnitude and are pointing in the same direction, but their starting points are different, then they are not equal as vectors.

105. For any nonzero vector $\mathbf{v}$ in the *xy*-plane there is an angle θ such that $\mathbf{v} = |\mathbf{v}|(\cos\theta\,\mathbf{i} + \sin\theta\,\mathbf{j})$.

106. Work is a vector quantity, pointing in the direction of the force doing the work.

107. If two vectors in three-dimensional space are equal, then their direction angles are equal.

108. The vector $\text{proj}_\mathbf{v}(\mathbf{u} \times \mathbf{v})$ is always the zero vector.

109. Torque always points in the direction of rotation.

110. If $\mathbf{u}$ and $\mathbf{v}$ are both vectors in the *xy*-plane, then it is impossible to find their cross product.

111. $\mathbf{u} \cdot (\mathbf{v} \times \mathbf{w}) = \mathbf{v} \cdot (\mathbf{w} \times \mathbf{u})$

112. $\mathbf{u} \cdot (\mathbf{v} \times \mathbf{w}) = \mathbf{w} \cdot (\mathbf{v} \times \mathbf{u})$

113. If $\mathbf{u}$, $\mathbf{v}$, and $\mathbf{w}$ are coplanar, then $\mathbf{u} \cdot (\mathbf{v} \times \mathbf{w}) = 0$.

114. $\mathbf{u} \cdot [\mathbf{v} \times (\mathbf{w} + \mathbf{z})] = \mathbf{u} \cdot (\mathbf{v} \times \mathbf{w}) + \mathbf{u} \cdot (\mathbf{v} \times \mathbf{z})$

115. If $\mathbf{u}$ is orthogonal to both $\mathbf{v}$ and $\mathbf{w}$, then $\mathbf{u}$ is orthogonal to $\mathbf{v} \times \mathbf{w}$.

116. If $\mathbf{u}$ is orthogonal to both $\mathbf{v}$ and $\mathbf{w}$, then $\mathbf{u}$ is orthogonal to $\mathbf{v} \cdot \mathbf{w}$.

117. $(\mathbf{u} - \mathbf{v}) \cdot (\mathbf{u} + \mathbf{v}) = |\mathbf{u}|^2 - |\mathbf{v}|^2$

118. If $\mathbf{v}$ is a vector in the plane with normal vector $\mathbf{n} = \langle n_1, n_2, n_3 \rangle$, then $\mathbf{v} \cdot \mathbf{n} = 0$.

119. If $\mathbf{v}$ is a vector in the plane that has normal vector $\mathbf{n} = \langle n_1, n_2, n_3 \rangle$, then $\mathbf{v} \times \mathbf{n}$ is another vector in the said plane.

Chapter 11
Technology Exercises

120. Write a program on a computer algebra system that finds a unit vector pointing in the direction of $a\mathbf{u} + b\mathbf{v}$ for given vectors $\mathbf{u}$, $\mathbf{v}$ and scalars a and b. Use it to find the unit vector pointing in the direction of $2\mathbf{u} - 3\mathbf{v}$ of Exercise 13.

121. Write a program on a computer algebra system that returns parametric equations for the line formed by the intersection of two given planes. (The program should accept the equations of the planes and return parametric equations for the line, displaying an appropriate message if the planes are parallel.) Use your program to check your answers for Exercises 75 and 76.

122. Write a short program for a computer algebra system that finds the distance between a point and a plane in three-dimensional space. Use it to revisit Exercises 1–6.

123. Write a program on your computer algebra system to determine the angle, in degrees, between two given three-dimensional vectors. Use it to check your answers for Exercises 27–28.

124–129. Use the graphing capabilities of a computer algebra system to check your answers for Exercises 88–93.

Project ✎

Chapter 11

In this project, we are going to use vectors to prove an interesting property of quadrilaterals. In fact, the result is general enough that our quadrilateral doesn't have to be planar, in other words, its vertices do not have to lie in the same plane!

1. Let A, B, C, D be four points in $\mathbb{R}^3$, with M_1, M_2, M_3, and M_4 being the midpoints of the line segments $\overline{AB}$, $\overline{BC}$, $\overline{CD}$, and $\overline{DA}$, respectively. Consider the vector $\overrightarrow{M_1M_2}$ and show that

$$\overrightarrow{M_1M_2} = \frac{1}{2}\left(\overrightarrow{AB} + \overrightarrow{BC}\right).$$

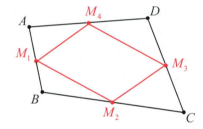

2. Prove a statement analogous to the one in Question 1 for the vector $\overrightarrow{M_3M_4}$ and show that

$$\overrightarrow{M_1M_2} = -\overrightarrow{M_3M_4}.$$

3. Using the results of Questions 1 and 2, argue that the quadrilateral $M_1M_2M_3M_4$ is a parallelogram.

4. Explain why the proof of Question 3 does not require that the points A, B, C, and D lie in the same plane.

5. Use vectors in the three-dimensional coordinate system to prove the statement of Question 3: If $ABCD$ is a (not necessarily planar) quadrilateral in $\mathbb{R}^3$, then the midpoints of its sides determine a parallelogram. (**Hint:** To simplify your calculations, you can assume that three of the vertices lie in the same coordinate plane with one of them, say A, located at the origin, and an adjacent vertex, such as B, lying on a coordinate axis.)

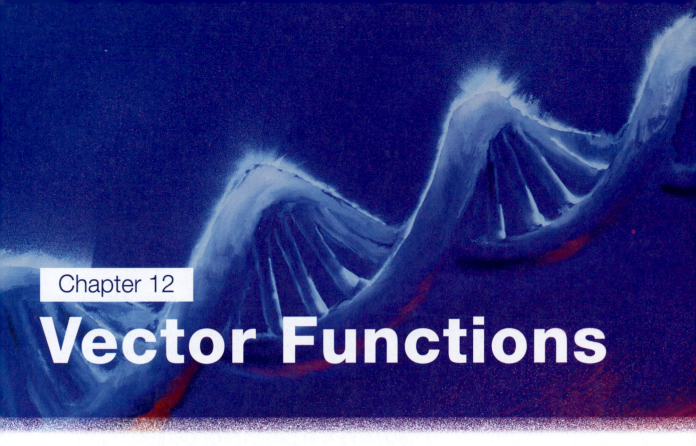

Chapter 12
Vector Functions

12.1 Vector-Valued Functions 933

1. Vector Functions and Space Curves
2. Differentiation and Integration of Space Curves

12.2 Arc Length and the Unit Tangent Vector 945

1. Arc Length
2. The Unit Tangent Vector
3. Projectile Motion

12.3 The Unit Normal and Binormal Vectors, Curvature, and Torsion 954

1. The Unit Normal and Binormal Vectors
2. Curvature and Torsion

12.4 Planetary Motion and Kepler's Laws 967

1. Motion in Polar Coordinates
2. Kepler's Laws of Planetary Motion

Introduction

With the notation and techniques of vectors at our disposal, we are ready to study the calculus of vector functions. As we will see in this chapter, the added degrees of freedom associated with higher-dimensional spaces both allow and call for the development of completely new ideas in addition to extending those with which we are familiar.

We begin by making a connection between curves in space (primarily three-dimensional Cartesian space) and certain *vector-valued functions*—functions which map intervals of real numbers into n-dimensional space $\mathbb{R}^n$ (again, with an emphasis on $\mathbb{R}^3$). Following the pattern of earlier chapters, we explore what the concepts of *limit* and *continuity* mean when applied to vector-valued functions, and subsequently use those notions to develop techniques for differentiating and integrating such functions. And as with the parametrically defined curves of Chapter 9, we will see that it is important to characterize those space curves which are *smooth* in addition to being differentiable.

Because curves in three-dimensional space can exhibit behavior not seen by planar curves, we will find it useful to define, for a given curve, three unit vectors called the *tangent*, *normal*, and *binormal* vectors; each is a function of some common parameter and the three together constitute a moving frame of reference along the curve called the Frenet-Serret frame. This frame is named in honor of the two nineteenth-century French mathematicians Jean Frédéric Frenet (1816–1900) and Joseph Alfred Serret (1819–1885) who independently discovered the relationships between the tangent, normal, and binormal vectors and, in so doing, set the stage for modern applications in such diverse fields as relativistic physics, computer graphics, and biology.

The last section of this chapter is devoted to deriving Kepler's Laws of Planetary Motion using basic physics principles and our newly acquired skill in working with space curves. The fact that Kepler's Laws, which represent decades' worth of observation and trial-and-error reasoning, can be deduced from just a few principles and the tools of vector calculus is an accomplishment well worth studying, as it illustrates the tremendous power of calculus in modeling and explaining our universe.

> Following the pattern of earlier chapters, we explore what the concepts of *limit* and *continuity* mean when applied to vector-valued functions, and subsequently use those notions to develop techniques for differentiating and integrating such functions.

The use of the tangent, normal, and binormal vectors in computer graphics has special appeal, as the technique (which we will develop in Section 12.3) allows for the relatively simple creation of complex three-dimensional objects. The ribbon and tube shapes in that section will likely remind you of parts of computer-generated images seen in movies, computer games, and elsewhere.

Johannes Kepler
(1571–1630)

12.1 **Vector-Valued Functions**

TOPICS

1. Vector functions and space curves

2. Differentiation and integration of space curves

In this chapter, we make use of vector notation and operations in extending the concepts of calculus to higher-dimensional spaces. We focus particularly on curves in three-dimensional space and conclude the chapter with a study of planetary motion and Kepler's Laws, an achievement rich in history and application.

TOPIC 1   **Vector Functions and Space Curves**

The parametric equations introduced in Chapter 9 expanded the notion of curves in the plane, and the vector and parametric forms of lines in Section 11.5 began to illustrate the connection between parametric equations and vectors. Generalizing these ideas, we say that a **vector-valued** (or simply **vector**) **function** on a domain D is a function that assigns a vector to each element of D. For our current purposes, the domain D can be taken to be an interval I of real numbers, and the vectors assigned to each $t \in I$ are position vectors in $\mathbb{R}^3$ (recall that a position vector is one whose initial point is the origin). If f, g, and h are all real-valued (or scalar) functions defined on I, then

$$\mathbf{r}(t) = f(t)\mathbf{i} + g(t)\mathbf{j} + h(t)\mathbf{k} = \langle f(t), g(t), h(t) \rangle$$

is such a vector-valued function, and f, g, and h, are its **component functions**. The equations

$$x = f(t), \quad y = g(t), \quad \text{and} \quad z = h(t) \quad \text{for } t \in I$$

are parametric equations describing the same points in space as the vector function $\mathbf{r}(t)$.

Example 1 ✐

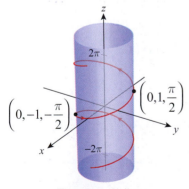

Figure 1 Helix

The vector function $\mathbf{r}(t) = \langle \cos t, \sin t, t \rangle$, $t \in [-2\pi, 2\pi]$, describes the *helix* shown in red in Figure 1. Note that $x^2 + y^2 = 1$, where $x = \cos t$ and $y = \sin t$, so the points of the helix all reside on the cylinder of radius 1 shown in blue. The third parametric equation, $z = t$, tells us that the position vector $\mathbf{r}(t)$ travels "up" the helix (in the direction of the positive z-axis) as t increases. We often indicate the direction of increasing values of the parameter with arrows, as shown. If the third component of $\mathbf{r}(t)$ were instead $-t$, the orientation of the helix would be downward and the arrows on the helix would be reversed.

Since limits lie at the heart of calculus, we must extend our understanding of limits to vector-valued functions in order to proceed. The following definition is identical to the definition of limit in Section 2.3, but applies to vector as well as scalar functions.

Definition 💡

Limits of Vector Functions

Let $\mathbf{r}(t) = \langle f(t), g(t), h(t) \rangle$ be a vector function and $\mathbf{L}$ a fixed vector. We say that $\mathbf{r}$ has **limit L** as t approaches a, and write $\lim\limits_{t \to a} \mathbf{r}(t) = \mathbf{L}$, if for every number $\varepsilon > 0$ there is a number $\delta > 0$ such that

$$0 < |t - a| < \delta \quad \Rightarrow \quad |\mathbf{r}(t) - \mathbf{L}| < \varepsilon.$$

As with scalar functions, this definition means that the vector $\mathbf{r}(t)$ approaches the fixed vector $\mathbf{L}$ as t gets closer to the fixed number a. If $\mathbf{L} = \langle L_1, L_2, L_3 \rangle$, then

$$|\mathbf{r}(t) - \mathbf{L}| \to 0 \quad \text{if and only if} \quad \sqrt{\left[f(t) - L_1 \right]^2 + \left[g(t) - L_2 \right]^2 + \left[h(t) - L_3 \right]^2} \to 0,$$

and this in turn is equivalent to each of the three summands under the radical approaching 0 as $t \to a$. That is,

$$\lim_{t \to a} \mathbf{r}(t) = \mathbf{L} \quad \Leftrightarrow \quad \left\langle \lim_{t \to a} f(t), \lim_{t \to a} g(t), \lim_{t \to a} h(t) \right\rangle = \langle L_1, L_2, L_3 \rangle$$

and hence

$$\lim_{t \to a} \mathbf{r}(t) = \left\langle \lim_{t \to a} f(t), \lim_{t \to a} g(t), \lim_{t \to a} h(t) \right\rangle.$$

This fact allows us to use all the familiar tools of calculus to determine limits of vector functions component by component.

Example 2 ✎

Given $\mathbf{r}(t) = \left\langle 2 - \dfrac{\sin t}{t}, -3 + 3t^2, \sqrt{4 - t} \right\rangle$, find $\lim\limits_{t \to 0} \mathbf{r}(t)$.

Solution

$$\lim_{t \to 0} \mathbf{r}(t) = \left\langle \lim_{t \to 0} \left(2 - \frac{\sin t}{t} \right), \lim_{t \to 0} \left(-3 + 3t^2 \right), \lim_{t \to 0} \sqrt{4 - t} \right\rangle = \langle 1, -3, 2 \rangle$$

Once we have defined limits in a vector setting, we can readily define continuity.

Definition 💡

Continuity of Vector Functions

A vector-valued function $\mathbf{r}(t)$ is **continuous at** $t = a$ if $\lim\limits_{t \to a} \mathbf{r}(t) = \mathbf{r}(a)$.

The function is said to be **continuous on the interval** I if it is continuous at each point of I.

In parallel with the definition of limit, a vector function is continuous at a given point a if and only if each of its component functions is continuous at a. We use this fact as the motivation for the following definition, given our intuitive sense that a curve represents a continuous collection of points.

Definition 💡

Space Curves

Given functions f, g, and h, each continuous on an interval I, the set C of points traced out by the vector function $\mathbf{r}(t) = \langle f(t), g(t), h(t) \rangle$, $t \in I$, is said to be a **space curve** (or simply a **curve**) in $\mathbb{R}^3$. C is also defined by the parametric equations $x = f(t)$, $y = g(t)$, and $z = h(t)$ for $t \in I$.

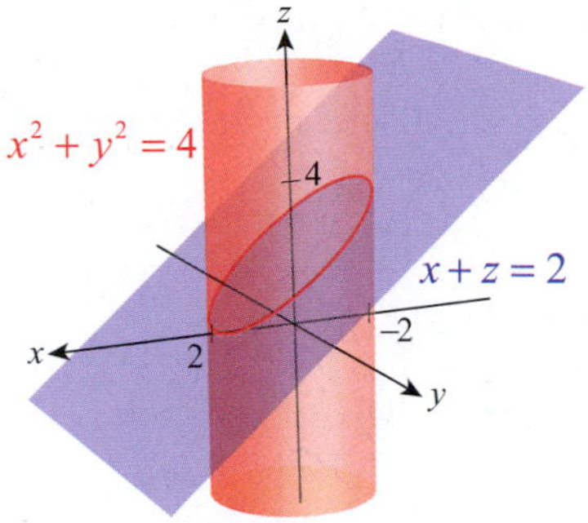

Example 3 ✍

The intersection of the cylinder $x^2 + y^2 = 4$ and the plane $x + z = 2$ is shown in Figure 2. Describe this curve in terms of a vector function $\mathbf{r}(t)$.

Solution

In order to introduce a parameter t, we can use the fact that the circle $x^2 + y^2 = 4$ in the xy-plane corresponds to the parametric equations $x = 2\cos t$ and $y = 2\sin t$, $0 \leq t \leq 2\pi$. These equations also describe the cylinder $x^2 + y^2 = 4$ in $\mathbb{R}^3$, since there is no restriction on z. Now, we can make the substitution $z = 2 - x = 2 - 2\cos t$ to describe the values of z on the plane $x + z = 2$ restricted to points also on the cylinder. Thus, the function

$$\mathbf{r}(t) = \langle 2\cos t, 2\sin t, 2 - 2\cos t \rangle, \quad t \in [0, 2\pi]$$

describes the space curve defined by the intersection of the two surfaces. The curve is traced out in the direction indicated in Figure 3.

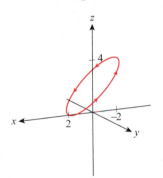

Figure 2

$x^2 + y^2 = 4$

$x + z = 2$

Figure 3

TOPIC 2   Differentiation and Integration of Space Curves

Our definition of the derivative of a vector function $\mathbf{r}(t) = \langle f(t), g(t), h(t) \rangle$ is the same as the derivative of a scalar function:

$$\mathbf{r}'(t) = \frac{d\mathbf{r}}{dt} = \lim_{\Delta t \to 0} \frac{\mathbf{r}(t + \Delta t) - \mathbf{r}(t)}{\Delta t}$$

The geometric meaning of this limit is also familiar. Figure 4 depicts a curve C defined by $\mathbf{r}$ and two vectors $\mathbf{r}(t)$ and $\mathbf{r}(t + \Delta t)$ for a fixed t and one particular value of Δt. The numerator of the difference quotient, $\mathbf{r}(t + \Delta t) - \mathbf{r}(t)$, is referred to as a secant vector, and dividing $\mathbf{r}(t + \Delta t) - \mathbf{r}(t)$ by Δt produces either a vector pointing in the same direction as $\mathbf{r}(t + \Delta t) - \mathbf{r}(t)$ if Δt is positive or in the opposite direction if Δt is negative. In either case, if Δt is small enough, the result is a vector approximately parallel to the line tangent to the curve C, provided the tangent line exists. Further, the difference quotient vector has the same orientation as C, in the sense that it points in the direction of increasing t. The approximation becomes exact as $\Delta t \to 0$, assuming $\mathbf{r}'(t) \neq \mathbf{0}$, so we call $\mathbf{r}'(t)$ the **tangent vector** to C in this case.

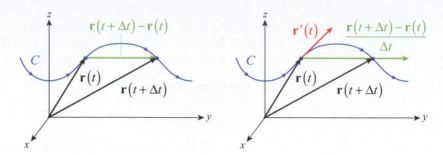

Figure 4

As with limits, differentiation of $\mathbf{r}(t) = \langle f(t), g(t), h(t) \rangle$ can be determined component by component. This follows from the fact that

$$
\begin{aligned}
\mathbf{r}'(t) &= \lim_{\Delta t \to 0} \frac{\mathbf{r}(t + \Delta t) - \mathbf{r}(t)}{\Delta t} \\
&= \lim_{\Delta t \to 0} \frac{\langle f(t + \Delta t), g(t + \Delta t), h(t + \Delta t) \rangle - \langle f(t), g(t), h(t) \rangle}{\Delta t} \\
&= \lim_{\Delta t \to 0} \left\langle \frac{f(t + \Delta t) - f(t)}{\Delta t}, \frac{g(t + \Delta t) - g(t)}{\Delta t}, \frac{h(t + \Delta t) - h(t)}{\Delta t} \right\rangle \\
&= \left\langle \lim_{\Delta t \to 0} \frac{f(t + \Delta t) - f(t)}{\Delta t}, \lim_{\Delta t \to 0} \frac{g(t + \Delta t) - g(t)}{\Delta t}, \lim_{\Delta t \to 0} \frac{h(t + \Delta t) - h(t)}{\Delta t} \right\rangle \\
&= \langle f'(t), g'(t), h'(t) \rangle.
\end{aligned}
$$

Example 4 ✐

Given $\mathbf{r}(t) = \langle t^3 - t, e^{-t}, \ln t \rangle$, $1 \le t \le 3$, find $\mathbf{r}'(t)$.

Solution

We have $\mathbf{r}'(t) = \langle 3t^2 - 1, -e^{-t}, 1/t \rangle$ over the same interval $1 \le t \le 3$.

In Section 9.2, we defined differentiability and smoothness as they pertain to curves in the plane defined parametrically, and the same definitions now extend to curves defined by vector functions.

Definition ♀

Differentiability and Smoothness of Space Curves

A vector function $\mathbf{r}(t) = \langle f(t), g(t), h(t) \rangle$ is **differentiable at** t_0 if f, g, and h are each differentiable at t_0. The curve C defined by $\mathbf{r}$ is **smooth at** t_0 if $\mathbf{r}'$ is a continuous function at t_0 and $\mathbf{r}'(t_0) \ne \mathbf{0}$. This is equivalent to f', g', and h' all being continuous at t_0 and not simultaneously 0 at t_0.

Curves have tangent lines at every point they are smooth, and we can determine those tangent lines in a manner analogous to that for scalar functions.

Example 5 ✍

Find the vector form of an equation for the line tangent to the helix traced out by $\mathbf{r}(t) = \langle \cos t, \sin t, t \rangle$ at the point $\left(\dfrac{1}{\sqrt{2}}, \dfrac{1}{\sqrt{2}}, \dfrac{\pi}{4} \right)$.

Solution

Since the z-component of $\mathbf{r}$ is simply t, the given point on the helix corresponds to the parameter value $t = \pi/4$ (and, as a check, we note that $\cos(\pi/4) = \sin(\pi/4) = 1/\sqrt{2}$). The direction vector for the tangent line at this point is

$$\mathbf{r}'\left(\frac{\pi}{4}\right) = \langle -\sin t, \cos t, 1 \rangle \big|_{t=\pi/4} = \left\langle -\frac{1}{\sqrt{2}}, \frac{1}{\sqrt{2}}, 1 \right\rangle,$$

so the tangent line in vector form (as defined in Section 11.5) is

$$\left\langle \frac{1}{\sqrt{2}}, \frac{1}{\sqrt{2}}, \frac{\pi}{4} \right\rangle + t \left\langle -\frac{1}{\sqrt{2}}, \frac{1}{\sqrt{2}}, 1 \right\rangle = \left\langle \frac{1}{\sqrt{2}} - \frac{t}{\sqrt{2}}, \frac{1}{\sqrt{2}} + \frac{t}{\sqrt{2}}, \frac{\pi}{4} + t \right\rangle,$$

where t is an element of $\mathbb{R}$.

Figure 5 is an illustration of the helix and, in blue, the tangent line we just found.

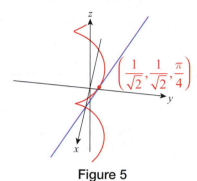

Figure 5

The fact that differentiation of vector functions can be accomplished component by component means that the familiar rules of differentiation all extend naturally, and that new rules applying only to vectors can be readily proved.

Theorem ⚷

Differentiation Rules for Vector Functions

Assume that $\mathbf{u}$ and $\mathbf{v}$ are differentiable vector functions, $\mathbf{C}$ is a constant vector, f is a differentiable scalar function, and c is a scalar. Then the following rules apply.

Constant Vector Rule $\dfrac{d}{dt}\mathbf{C} = \mathbf{0}$

Scalar Multiple Rules $\dfrac{d}{dt}\big[c\,\mathbf{u}(t)\big] = c\,\mathbf{u}'(t)$

$\dfrac{d}{dt}\big[f(t)\mathbf{u}(t)\big] = f'(t)\mathbf{u}(t) + f(t)\mathbf{u}'(t)$

Sum/Difference Rules $\dfrac{d}{dt}\big[\mathbf{u}(t) \pm \mathbf{v}(t)\big] = \mathbf{u}'(t) \pm \mathbf{v}'(t)$

Dot Product Rule $\dfrac{d}{dt}\big[\mathbf{u}(t) \cdot \mathbf{v}(t)\big] = \mathbf{u}'(t) \cdot \mathbf{v}(t) + \mathbf{u}(t) \cdot \mathbf{v}'(t)$

Cross Product Rule $\dfrac{d}{dt}\big[\mathbf{u}(t) \times \mathbf{v}(t)\big] = \mathbf{u}'(t) \times \mathbf{v}(t) + \mathbf{u}(t) \times \mathbf{v}'(t)$

Chain Rule $\dfrac{d}{dt}\big[\mathbf{u}(f(t))\big] = f'(t)\mathbf{u}'(f(t))$

Proof ✎

We will prove the Cross Product Rule and leave the proofs of the remainder as exercises (Exercises 60–64).

The proof is very similar to the proof of the Product Rule for scalar functions in Section 3.2, with the key step being the addition and subtraction of a term (shown in red below). Along the way, we will also need two facts about limits and vectors which you will prove in Exercises 41 and 42.

$$\frac{d}{dt}\Big[\mathbf{u}(t)\times\mathbf{v}(t)\Big]$$

$$=\lim_{h\to 0}\frac{\mathbf{u}(t+h)\times\mathbf{v}(t+h)-\mathbf{u}(t)\times\mathbf{v}(t+h)+\mathbf{u}(t)\times\mathbf{v}(t+h)-\mathbf{u}(t)\times\mathbf{v}(t)}{h}$$

$$=\lim_{h\to 0}\left[\frac{\mathbf{u}(t+h)-\mathbf{u}(t)}{h}\times\mathbf{v}(t+h)+\mathbf{u}(t)\times\frac{\mathbf{v}(t+h)-\mathbf{v}(t)}{h}\right]$$

$$=\lim_{h\to 0}\frac{\mathbf{u}(t+h)-\mathbf{u}(t)}{h}\times\lim_{h\to 0}\mathbf{v}(t+h)+\lim_{h\to 0}\mathbf{u}(t)\times\lim_{h\to 0}\frac{\mathbf{v}(t+h)-\mathbf{v}(t)}{h} \qquad \text{Exercise 41}$$

$$=\mathbf{u}'(t)\times\mathbf{v}(t)+\mathbf{u}(t)\times\mathbf{v}'(t) \qquad\qquad\qquad\qquad\qquad \text{Exercise 42}$$

We can put two of these rules to use immediately to prove a useful observation about vector functions of constant magnitude.

Example 6 ✐

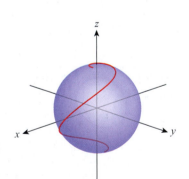

Figure 6
Vector Function of Constant Magnitude

Assume $\mathbf{r}(t)$ is a differentiable vector function of constant magnitude—geometrically, a vector function that traces out a curve on a sphere centered at the origin, as shown in Figure 6. Show that $\mathbf{r}'(t)\cdot\mathbf{r}(t)=0$ for all t. In words, this means $\mathbf{r}(t)$ and its tangent vector $\mathbf{r}'(t)$ are always orthogonal if $|\mathbf{r}(t)|$ is constant.

Solution

Recall that $\mathbf{r}(t)\cdot\mathbf{r}(t)=|\mathbf{r}(t)|^2$, so the fact that $|\mathbf{r}(t)|$ is constant means

$$\frac{d}{dt}\Big[\mathbf{r}(t)\cdot\mathbf{r}(t)\Big]=\frac{d}{dt}|\mathbf{r}(t)|^2=0.$$

On the other hand, the Dot Product Rule tells us that

$$\frac{d}{dt}\Big[\mathbf{r}(t)\cdot\mathbf{r}(t)\Big]=\mathbf{r}'(t)\cdot\mathbf{r}(t)+\mathbf{r}(t)\cdot\mathbf{r}'(t)=2\mathbf{r}'(t)\cdot\mathbf{r}(t),$$

and therefore $\mathbf{r}'(t)\cdot\mathbf{r}(t)=0$.

We conclude this section with a brief introduction to integration of vector functions. This is a topic we will study in far greater detail in Chapters 14 and 15. For the moment, we are most interested in reversing the differentiation process and finding a vector function whose derivative has specified characteristics. As with scalar functions, we call this reverse process *antidifferentiation*. And since differentiation can be performed component by component, antidifferentiation proceeds the same way.

Definition 💡

Indefinite and Definite Integrals of Vector Functions

Given a vector function $\mathbf{r}(t) = \langle f(t), g(t), h(t) \rangle$, the **indefinite integral** of $\mathbf{r}$ with respect to t is the family of all vector functions whose derivative is $\mathbf{r}$, denoted $\int \mathbf{r}(t)\,dt$. If $\mathbf{R}$ is a particular antiderivative of $\mathbf{r}$, then

$$\int \mathbf{r}(t)\,dt = \mathbf{R}(t) + \mathbf{C},$$

where $\mathbf{C}$ represents an arbitrary constant vector.

If f, g, and h are all integrable functions over an interval $[a, b]$, then the **definite integral** of $\mathbf{r}$ from a to b is denoted $\int_a^b \mathbf{r}(t)\,dt$ and

$$\int_a^b \mathbf{r}(t)\,dt = \int_a^b \langle f(t), g(t), h(t) \rangle\,dt = \left\langle \int_a^b f(t)\,dt, \int_a^b g(t)\,dt, \int_a^b h(t)\,dt \right\rangle.$$

Using alternate notation, $\mathbf{r}(t) = f(t)\mathbf{i} + g(t)\mathbf{j} + h(t)\mathbf{k}$, we have

$$\int_a^b \mathbf{r}(t)\,dt = \left(\int_a^b f(t)\,dt \right)\mathbf{i} + \left(\int_a^b g(t)\,dt \right)\mathbf{j} + \left(\int_a^b h(t)\,dt \right)\mathbf{k}.$$

If $\mathbf{R}$ is again a particular antiderivative of $\mathbf{r}$, then $\int_a^b \mathbf{r}(t)\,dt = \mathbf{R}(b) - \mathbf{R}(a)$.

We've actually performed integration of a vector function already, in Example 2 of Section 11.2. Example 7 below is a reformulation of that problem using the terminology and notation of this section.

Example 7 ✎

A ball is shot by a slingshot into the air with an initial velocity vector $\mathbf{v}(0) = \langle 0, 10, 64 \rangle$, measured in ft/s. Determine its velocity $\mathbf{v}$ and its position $\mathbf{r}$ as functions of time t. Find the position of the ball at 2 seconds and 4 seconds.

Solution

We begin with the fact that the acceleration of the ball is given by the vector $\mathbf{a}(t) = \langle 0, 0, -32 \rangle$, reflecting the fact that the Earth's gravity is pulling it (and every other object) in the negative z-direction at a rate of 32 ft/s^2. So the ball's velocity vector is the indefinite integral of $\mathbf{a}$:

$$\mathbf{v}(t) = \int \mathbf{a}(t)\,dt = \langle 0, 0, -32t \rangle + \mathbf{C}$$

Since we are given its velocity at time $t = 0$,

$$\langle 0, 10, 64 \rangle = \mathbf{v}(0) = \langle 0, 0, -32 \cdot 0 \rangle + \mathbf{C} \quad \Rightarrow \quad \mathbf{C} = \langle 0, 10, 64 \rangle,$$

so $\mathbf{v}(t) = \langle 0, 10, -32t + 64 \rangle$.

If we let **r** denote the ball's position vector, then

$$\mathbf{r}(t) = \int \mathbf{v}(t)\,dt = \langle 0, 10t, -16t^2 + 64t \rangle + \mathbf{C}.$$

We can locate the origin of the coordinate system wherever convenient, so we may as well specify that the ball's position at time $t = 0$ is the origin, meaning $\mathbf{C} = \langle 0, 0, 0 \rangle$. We previously determined that the ball reaches its maximum height when $t = 2$ and lands on the ground when $t = 4$, and its position at these times is given by the following definite integrals of its velocity.

$$\text{Position at } t = 2: \int_0^2 \mathbf{v}(t)\,dt = \langle 0, 20, 64 \rangle$$

$$\text{Position at } t = 4: \int_0^4 \mathbf{v}(t)\,dt = \langle 0, 40, 0 \rangle$$

In words, the ball has traveled 20 feet horizontally when it reaches its maximum height of 64 feet, and has traveled 40 feet horizontally when it lands.

12.1 **Exercises**

1–4 Find the domain of the vector function. If possible, evaluate the vector function at the indicated points.

1. $\mathbf{r}(t) = \dfrac{1}{t^2}\mathbf{i} + 2t\,\mathbf{j} - t\,\mathbf{k};$ **a.** $t = 2$ **b.** $t = -5$

2. $\mathbf{r}(t) = 3t\,\mathbf{i} - e^t\,\mathbf{j} - \sqrt{t-1}\,\mathbf{k};$ **a.** $t = 2$ **b.** $t = -5$

3. $\mathbf{r}(t) = \dfrac{1}{\sqrt{9 - t^2}}\mathbf{i} - t^3\,\mathbf{j} + \ln t\,\mathbf{k};$ **a.** $t = -4$ **b.** $t = 1$

4. $\mathbf{r}(t) - \mathbf{s}(t),$ where $\mathbf{r}(t) = \sqrt{t}\,\mathbf{i} - 5t^2\,\mathbf{k},$ $\mathbf{s}(t) = e^{-t}\mathbf{i} + t^2\,\mathbf{j};$ **a.** $t = -1$ **b.** $t = 1$

5. If $\mathbf{r}(t) = \mathbf{i} - t\,\mathbf{j}$ and $\mathbf{s}(t) = t\,\mathbf{i} + 3\,\mathbf{j} + t^3\,\mathbf{k},$ find a formula for $\mathbf{u}(t) = \mathbf{r}(t) \cdot \mathbf{s}(t).$ Is it a space curve?

6. Repeat Exercise 5 for $\mathbf{v}(t) = \mathbf{r}(t) \times \mathbf{s}(t).$

7–14 Match the vector function with its graph (labeled A–H).

7. $\mathbf{r}(t) = \left\langle \dfrac{1}{2}t\cos t, \dfrac{1}{2}t\sin t, \dfrac{t}{2} \right\rangle;$ $t \in [0, 6\pi]$

8. $\mathbf{r}(t) = \left\langle 5\sin^2 t, \cos^2 t, t \right\rangle;$ $t \in [0, 4\pi]$

9. $\mathbf{r}(t) = \left\langle \ln t, \sin t, \cos t \right\rangle;$ $t \in (0, 6\pi)$

10. $\mathbf{r}(t) = \left\langle 3\cos(t^2), 3\sin(t^2), 3\sqrt{t} \right\rangle;$ $t \in [0, 4\pi]$

11. $\mathbf{r}(t) = \left\langle \cos\sqrt{t}, \sin\sqrt{t}, \dfrac{t}{50} \right\rangle;$ $t \in [0, 36\pi^2]$

12. $\mathbf{r}(t) = \left\langle 2\sin t, 4\cos t, \dfrac{t}{4} \right\rangle;$ $t \in [0, 6\pi]$

13. $\mathbf{r}(t) = \left\langle 1 + t^2, 2t, 2t \right\rangle;$ $t \in [-6, 25]$

14. $\mathbf{r}(t) = \left\langle \cos t, \sqrt{t}, \sqrt{t} \right\rangle;$ $t \in [0, 6\pi]$

A.

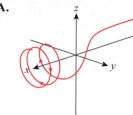

B.

C.

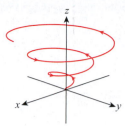

D.

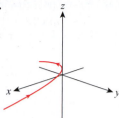

E.

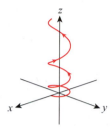

F.

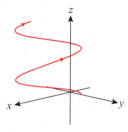

G.

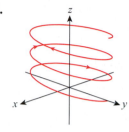

H.

15–22 Sketch the space curve by hand.
(**Hint:** See Exercises 7–14.)

15. $\mathbf{r}(t) = \langle \sin t, \cos t, \sqrt{t} \rangle; \quad t \in [0, 4\pi]$

16. $\mathbf{r}(t) = \langle 2t \cos t, 3t \sin t, t \rangle; \quad t \in [0, 6\pi]$

17. $\mathbf{r}(t) = \langle t, t, 3 \sin t \rangle; \quad t \in [-2\pi, 2\pi]$

18. $\mathbf{r}(t) = \langle 2\cos(t^2), 4\sin(t^2), 2\sqrt{t} \rangle; \quad t \in \left[0, 2\sqrt{3\pi}\right]$

19. $\mathbf{r}(t) = \langle \cos t, \sin t, \ln t \rangle; \quad t \in \left[\dfrac{1}{e^3}, 6\pi\right]$

20. $\mathbf{r}(t) = \left\langle \dfrac{t}{3}, \dfrac{3\cos 2t}{t}, \dfrac{3\sin 2t}{t} \right\rangle; \quad t \in (0, 4\pi)$

21. $\mathbf{r}(t) = \left\langle \dfrac{t}{2}, 4\sin t, 2\cos t \right\rangle; \quad t \in [0, 4\pi]$

22. $\mathbf{r}(t) = \langle 4\cos t, 4\sin t, \cos 10t \rangle; \quad t \in [0, 2\pi]$

23–29 Describe the intersection of the surfaces as a vector function. (Use the suggested parameter.)

23. The elliptic cylinder $2x^2 + 3y^2 = 6$ and the plane $3x + 2z = 2$ $(x = \sqrt{3}\cos t)$

24. The paraboloid $x^2 + y^2 = z$ and the plane $x + z = 2$ $(z = t)$

25. The cylinder $x^2 + y^2 = 1$ and the hyperbolic paraboloid $2x^2 - y^2 = z$ $(x = \cos t)$

26. The cylinder $x^2 + y^2 = 9$ and the surface $y = z/x$ $(x = 3\cos t)$

27. The elliptic paraboloid $2x^2 + y^2 = 2z$ and the parabolic cylinder $y^2 = x$ $(y = t)$

28. The cone $x^2 + y^2 = z^2$ and the plane $2z = y + 4$ $(y = t)$

29. The semiellipsoid $\dfrac{x^2}{16} + \dfrac{y^2}{12} + \dfrac{z^2}{8} = 1, z \geq 0$ and the parabolic cylinder $2x = y^2$ $(y = t)$

30–35 Determine whether the indicated limit exists. If so, find it.

30. $\lim\limits_{t \to -1} \left\langle t^2 - 2t, \sqrt{t + 5}, \dfrac{1}{t} \right\rangle$

31. $\lim\limits_{t \to 1} \left\langle e^{-3t}, \sqrt{t - 2}, \cos 2t \right\rangle$

32. $\lim\limits_{t \to 3} \left\langle \dfrac{t + 2}{t - 3}, \ln t, \cot \pi t \right\rangle$

33. $\lim\limits_{t \to 0} \left\langle \ln(t^2 + 1), |t|, 2^t \right\rangle$

34. $\lim\limits_{t \to 0} \left\langle \sqrt{1 - \cos t}, e^{\tan t}, \dfrac{2\sin t}{t} \right\rangle$

35. $\lim\limits_{t \to 0} \left\langle \dfrac{2t}{t + 1}, \ln(t + 1), \sin \dfrac{\pi}{t} \right\rangle$

36–39 Find any discontinuities of the given vector function.

36. $\mathbf{r}(t) = e^t\mathbf{i} + \dfrac{2}{t+1}\mathbf{j} + 3t^2\mathbf{k}$

37. $\mathbf{r}(t) = 2t^2\mathbf{i} - 5|t|\mathbf{j} + \cos t\,\mathbf{k}$

38. $\mathbf{r}(t) = \mathbf{i} + \dfrac{t}{t^2+1}\mathbf{j} - \cot t\,\mathbf{k}$

39. $\mathbf{r}(t) = \left(t^3 - 1\right)\mathbf{i} - \sin\dfrac{\pi}{t}\mathbf{j} - \sqrt{t^2 + 2}\,\mathbf{k}$

40. Prove: If $\mathbf{u}(t)$ and $\mathbf{v}(t)$ are vector functions so that both have limits at $t = t_0$, then the limit of their dot product is the dot product of their limits, that is,

$$\lim_{t \to t_0}\left[\mathbf{u}(t) \cdot \mathbf{v}(t)\right] = \lim_{t \to t_0}\mathbf{u}(t) \cdot \lim_{t \to t_0}\mathbf{v}(t).$$

41. Prove: If $\mathbf{u}(t)$ and $\mathbf{v}(t)$ are vector functions so that both have limits at $t = t_0$, then the limit of their cross product is the cross product of their limits, that is,

$$\lim_{t \to t_0}\left[\mathbf{u}(t) \times \mathbf{v}(t)\right] = \lim_{t \to t_0}\mathbf{u}(t) \times \lim_{t \to t_0}\mathbf{v}(t).$$

(**Hint:** Use the determinant rule for determining cross products.)

42. Prove that the differentiability of a vector function implies its continuity, that is, if $\mathbf{u}(t)$ is differentiable at $t = t_0$, then $\mathbf{u}(t)$ is continuous at $t = t_0$.

43. Give a rigorous proof of the fact that the vector function $\mathbf{r}(t) = \left\langle f(t), g(t), h(t)\right\rangle$ is continuous if and only if its component functions $f(t)$, $g(t)$, and $h(t)$ are continuous.

44. Prove that if the vector function $\mathbf{r}(t)$ is continuous at $t = t_0$, then the scalar function $|\mathbf{r}(t)|$ is also continuous at $t = t_0$.

45. Prove that the converse of Exercise 44 is false by finding a vector function $\mathbf{r}(t)$ with a discontinuity at $t = t_0$ so that $|\mathbf{r}(t)|$ is continuous at $t = t_0$.

46–51 Find $\mathbf{r}'(t)$.

46. $\mathbf{r}(t) = t^2\mathbf{i} + 2\sqrt{t}\,\mathbf{j} - t\mathbf{k}$

47. $\mathbf{r}(t) = \left(2 - t^3\right)\mathbf{i} - \pi^2\mathbf{j} + \dfrac{t^5}{5}\mathbf{k}$

48. $\mathbf{r}(t) = \sin t\,\mathbf{i} + e^t\mathbf{j} - \tan t\,\mathbf{k}$

49. $\mathbf{r}(t) = \ln t\,\mathbf{i} - \csc t\,\mathbf{j} + \sqrt{4 - t^2}\,\mathbf{k}$

50. $\mathbf{r}(t) = \left\langle \dfrac{1}{\sqrt[3]{t^2}}, \arctan 2t, \sin^3 t\right\rangle$

51. $\mathbf{r}(t) = \left\langle \cos\left(t^2 + 1\right), \dfrac{t+1}{t-1}, 3\arcsin t\right\rangle$

52–55 Find a unit vector that is tangent to the graph of the vector function at the specified value of t.

52. $\mathbf{r}(t) = t\mathbf{i} - 2t^2\mathbf{j} + 2\sqrt{t}\,\mathbf{k}$; $t = 1$

53. $\mathbf{r}(t) = 2\sin t\,\mathbf{i} - e^t\mathbf{j} + 8\sqrt{4 + t}\,\mathbf{k}$; $t = 0$

54. $\mathbf{s}(t) = \left\langle \arctan t, -\cos^2 t, \sqrt{3}t\right\rangle$; $t = 0$

55. $\mathbf{u}(t) = \left\langle 4t, 3\sin t, 3\cos t\right\rangle$; $t = 0$

56–59 Find the vector form of an equation for the line tangent to the curve at the specified value of t.

56. $\mathbf{r}(t) = 2t\mathbf{i} + \left(t^2 - 4\right)\mathbf{j} + \sqrt{t + 1}\,\mathbf{k}$; $t = 3$

57. $\mathbf{r}(t) = e^{3t}\mathbf{i} - e^{2t}\mathbf{j} + e^t\mathbf{k}$; $t = 0$

58. $\mathbf{s}(t) = \left\langle t, \sin 2t, \cos 2t\right\rangle$; $t = 0$

59. $\mathbf{u}(t) = \left\langle \arcsin t, \arccos t, \ln t\right\rangle$; $t = \dfrac{1}{2}$

60–64 Prove the indicated differentiation rule, assuming that $\mathbf{u}$ and $\mathbf{v}$ are differentiable vector functions, $\mathbf{C}$ is a constant vector, c is a scalar, and f is a differentiable scalar function.

60. Constant Vector Rule: $\dfrac{d}{dt}\mathbf{C} = \mathbf{0}$

61. Scalar Multiple Rules: $\dfrac{d}{dt}\left[c\,\mathbf{u}(t)\right] = c\,\mathbf{u}'(t)$ and

$$\dfrac{d}{dt}\left[f(t)\mathbf{u}(t)\right] = f'(t)\mathbf{u}(t) + f(t)\mathbf{u}'(t)$$

62. Sum/Difference Rules:

$$\dfrac{d}{dt}\left[\mathbf{u}(t) \pm \mathbf{v}(t)\right] = \mathbf{u}'(t) \pm \mathbf{v}'(t)$$

63. Dot Product Rule:

$$\frac{d}{dt}\big[\mathbf{u}(t)\cdot\mathbf{v}(t)\big]=\mathbf{u}'(t)\cdot\mathbf{v}(t)+\mathbf{u}(t)\cdot\mathbf{v}'(t)$$

64. Chain Rule: $\dfrac{d}{dt}\big[\mathbf{u}(f(t))\big]=f'(t)\mathbf{u}'(f(t))$

65. Prove the following differentiation rule for the triple scalar product of vector functions: If $\mathbf{u}(t)$, $\mathbf{v}(t)$, and $\mathbf{w}(t)$ are differentiable, then

$$\frac{d}{dt}(\mathbf{u}\cdot\mathbf{v}\times\mathbf{w})=\frac{d\mathbf{u}}{dt}\cdot\mathbf{v}\times\mathbf{w}+\mathbf{u}\cdot\frac{d\mathbf{v}}{dt}\times\mathbf{w}+\mathbf{u}\cdot\mathbf{v}\times\frac{d\mathbf{w}}{dt}.$$

66. Prove that if a point moves along a sphere, then its velocity vector is tangential to the sphere. (**Hint:** See Example 6.)

67. Prove that if a point moves along a curve in $\mathbb{R}^3$ with constant speed, then its velocity and acceleration vectors are orthogonal. (**Hint:** See Exercise 66.)

68. (A converse of Example 6) Assume $\mathbf{r}(t)$ is a differentiable vector function satisfying $\mathbf{r}'(t)\cdot\mathbf{r}(t)=0$ for all t. Show that $\|\mathbf{r}(t)\|$ is constant, that is, the graph of $\mathbf{r}(t)$ lies on a sphere centered at the origin.

69–74 Find the indefinite integral.

69. $\displaystyle\int\big\langle 3t^2,t^3-t,-\sqrt{t}\big\rangle\,dt$

70. $\displaystyle\int\left(\frac{t}{t^2+1}\mathbf{i}-\frac{1}{t^2+1}\mathbf{j}+\frac{t}{t^2-1}\mathbf{k}\right)dt$

71. $\displaystyle\int\big(t\mathbf{i}+3\mathbf{j}-4t^3\,\mathbf{k}\big)\,dt$

72. $\displaystyle\int\big(\cos t\,\mathbf{i}-2\sin t\,\mathbf{j}-\sec^2 t\,\mathbf{k}\big)\,dt$

73. $\displaystyle\int\left(2\mathbf{i}-\frac{1}{t}\mathbf{j}+t^{3/2}\,\mathbf{k}\right)dt$

74. $\displaystyle\int\left\langle\frac{1}{t^2},\ln t,-e^{-t}\right\rangle\,dt$

75–80 Evaluate the definite integral.

75. $\displaystyle\int_0^3\big[(2-t)\mathbf{i}-4\mathbf{j}+t^2\,\mathbf{k}\big]\,dt$

76. $\displaystyle\int_0^1\big[2t^4\,\mathbf{i}+t\,\mathbf{j}-(t^2-2)\mathbf{k}\big]\,dt$

77. $\displaystyle\int_{-1}^1\big\langle\sqrt[3]{t},t,t^4\big\rangle\,dt$

78. $\displaystyle\int_0^\pi(\sin t\,\mathbf{i}+t\sin t\,\mathbf{j}-\mathbf{k})\,dt$

79. $\displaystyle\int_1^e\left\langle 2e^t,-\ln t,\frac{1}{t}\right\rangle\,dt$

80. $\displaystyle\int_0^3\left\langle\sqrt{t+1},\frac{6t}{t^2+1},\frac{-1}{(t+1)(t-4)}\right\rangle\,dt$

81. A projectile is launched from the ground with an initial speed of 78.48 m/s at an angle of elevation of $30°$ from horizontal. After determining the vector functions $\mathbf{v}(t)$ and $\mathbf{r}(t)$, as in Example 7, find the maximum altitude reached by the projectile as well as its range. (Suppose that the launch takes place in the positive x-direction. Use $g\approx 9.81$ m/s^2 and ignore air resistance.)

82. Use Exercise 81 to determine the effect on the maximum altitude and range of the projectile if we double its initial velocity.

83. A particle is moving in $\mathbb{R}^3$ so that its acceleration function is $\mathbf{a}(t)=\langle 2t,1,0\rangle$. Find the velocity and position functions of the particle if it starts at the point $\mathbf{r}(0)=\langle-5,0,2\rangle$ with initial velocity $\mathbf{v}(0)=\langle 3,1,-1\rangle$.

84.* Prove that the force acting on a mass moving along a circle of radius R with constant angular speed ω is always pointing toward the center of the circle. (Such a force is called a *center-seeking* or *centripetal* force.) (**Hint:** Parametrize the path of the object and differentiate twice to find its acceleration, then use Newton's Second Law of Motion.)

85.* The plane curve $\mathbf{r}(t)=\big\langle ae^{bt}\cos t,ae^{bt}\sin t\big\rangle$ is called a *logarithmic spiral* or *Bernoulli spiral*. One of its intriguing properties is that for any fixed point $P=\mathbf{r}(t_0)$, the corresponding radial and tangent lines form a constant angle φ. Prove this fact, and find the angle φ.

86.* The *angular momentum* (with respect to the origin) of a mass m that is moving along a space curve $\mathbf{r}(t)$, is defined as

$$\mathbf{L}(t) = \mathbf{r}(t) \times m\mathbf{v}(t).$$

Use Newton's Second Law to demonstrate that $\boldsymbol{\tau}$, the net external torque acting on m is equal to the derivative of $\mathbf{L}(t)$, that is,

$$\boldsymbol{\tau} = \mathbf{L}'(t).$$

(**Hint:** Use the fact that torque is the cross product of the displacement vector and the force vector, that is, $\boldsymbol{\tau} = \mathbf{r} \times \mathbf{F}$.)

12.1 Technology Exercises

87–92. Use the integration capabilities of a computer algebra system to verify your answers to Exercises 69–74.

93–98. Use the integration capabilities of a computer algebra system to verify your answers to Exercises 75–80.

99–104. Use the "Limit" command of a computer algebra system to verify your answers to Exercises 30–35.

105–112. Use a computer algebra system to graph the curves of Exercises 15–22.

113–119. Use a computer algebra system to display the curves of Exercises 23–29 as intersections of the given surfaces.

12.2 Arc Length and the Unit Tangent Vector

TOPICS

1. Arc length

2. The unit tangent vector

3. Projectile motion

Whether a given space curve represents the path of a projectile, a military fighter jet, or an elementary particle in a high-energy accelerator, its geometric features are critical to answering such questions as the future location of the projectile, whether a human pilot can survive the induced forces of acceleration, and the exact nature of the particle. In this section and the next, we identify the most useful geometric characteristics of space curves and learn how to determine them mathematically.

TOPIC 1   Arc Length

We derived a formula for the arc length of graphs of scalar functions in Section 6.3, and in Section 9.2 we extended the formula to curves in the plane defined parametrically. Identical reasoning applies to space curves, leading to the following definition.

Definition 💡

Arc Length of Space Curves

Suppose a curve C is defined over the interval $a \le t \le b$ by the position vector $\mathbf{r}(t) = \langle f(t), g(t), h(t) \rangle$ or, equivalently, by the parametric equations $x = f(t)$, $y = g(t)$, and $z = h(t)$. Suppose further that f', g', and h' are all continuous on $[a, b]$ and that C is traced out just once over the interval, except possibly for a finite number of self-intersections. Then the **arc length** L of C is given by

$$L = \int_a^b \sqrt{[f'(t)]^2 + [g'(t)]^2 + [h'(t)]^2}\ dt = \int_a^b \sqrt{\left(\frac{dx}{dt}\right)^2 + \left(\frac{dy}{dt}\right)^2 + \left(\frac{dz}{dt}\right)^2}\ dt.$$

Since $\mathbf{r}'(t) = \langle f'(t), g'(t), h'(t) \rangle$, the above is equivalent to $L = \int_a^b |\mathbf{r}'(t)|\, dt$.

Example 1 ✎

Find the arc length of the portion of the helix $\mathbf{r}(t) = \langle \cos t, t/2, \sin t \rangle$ over the interval $[0, 4\pi]$.

Solution

The given interval defines two coils of the helix as shown in Figure 1, beginning at the point $(1, 0, 0)$ and ending at the point $(1, 2\pi, 0)$. The arc length of this curve is as follows:

$$L = \int_0^{4\pi} |\mathbf{r}'(t)|\, dt = \int_0^{4\pi} \sqrt{(-\sin t)^2 + \left(\frac{1}{2}\right)^2 + (\cos t)^2}\ dt$$

$$= \int_0^{4\pi} \sqrt{1 + \left(\frac{1}{2}\right)^2}\ dt = 2\pi\sqrt{5}$$

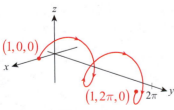

Figure 1

We often use t as the symbol for the parameter in defining a curve, since the curves in many applications arise as functions of time. But curves can be reparametrized as convenient, and when studying the pure geometry of a curve it often makes sense to replace the original parameter with the **arc length parameter** s. After doing so, we say the curve is **parametrized with respect to arc length**. The arc length parameter is defined by the arc length formula.

$$s(t) = \int_{t_0}^{t} |\mathbf{r}'(u)| \, du = \int_{t_0}^{t} \sqrt{\left(\frac{dx}{du}\right)^2 + \left(\frac{dy}{du}\right)^2 + \left(\frac{dz}{du}\right)^2} \, du$$

In this formula, we assume that the curve is parametrized in terms of u, and that $\mathbf{r}(t_0) = \langle x(t_0), y(t_0), z(t_0) \rangle$ is a conveniently chosen reference point from which to measure arc length. Then $s(t)$ is the length of the curve between $u = t_0$ and $u = t$, and by the Fundamental Theorem of Calculus, $ds/dt = |\mathbf{r}'(t)|$. Another advantage of the arc length parameter is evident if we are able to write t as a function of s, for then

$$\left| \frac{d}{ds} \mathbf{r}(t(s)) \right| = \left| \frac{d\mathbf{r}}{dt} \frac{dt}{ds} \right| = |\mathbf{r}'(t)| \left| \frac{1}{ds/dt} \right| = |\mathbf{r}'(t)| \frac{1}{|\mathbf{r}'(t)|} = 1$$

and the arc length of the curve between two points $\mathbf{r}(t(a))$ and $\mathbf{r}(t(b))$ is thus

$$L = \int_a^b \left| \frac{d}{ds} \mathbf{r}(t(s)) \right| ds = b - a.$$

The next example illustrates the process of parametrizing the curve of Example 1 in terms of arc length.

Example 2 ✎

Reparametrize the helix $\mathbf{r}(t) = \langle \cos t, t/2, \sin t \rangle$ with respect to arc length.

Solution

We first make the cosmetic reparametrization $\mathbf{r}(u) = \langle \cos u, u/2, \sin u \rangle$ so that we can use t as a variable upper limit of integration. Any point on the helix can be used as a reference point, so we choose $u = 0$.

$$s(t) = \int_0^t |\mathbf{r}'(u)| \, du = \int_0^t \sqrt{\frac{5}{4}} \, du = \frac{\sqrt{5}t}{2}$$

This confirms that the arc length of the helix between $t = 0$ and $t = 4\pi$ is $s(4\pi) = 2\pi\sqrt{5}$, as we found in Example 1. Solving $s = \sqrt{5}t/2$ for t yields $t = 2s/\sqrt{5}$, so the parametrization of the curve with respect to arc length is

$$\mathbf{r}(t(s)) = \left\langle \cos \frac{2s}{\sqrt{5}}, \frac{s}{\sqrt{5}}, \sin \frac{2s}{\sqrt{5}} \right\rangle.$$

This tells us, for instance, that the points

$$(1,0,0), \left(\cos \frac{2}{\sqrt{5}}, \frac{1}{\sqrt{5}}, \sin \frac{2}{\sqrt{5}} \right), \left(\cos \frac{4}{\sqrt{5}}, \frac{2}{\sqrt{5}}, \sin \frac{4}{\sqrt{5}} \right), \left(\cos \frac{6}{\sqrt{5}}, \frac{3}{\sqrt{5}}, \sin \frac{6}{\sqrt{5}} \right), \ldots$$

mark off locations that are 1 unit from each other traveling along the helix (see Figure 2).

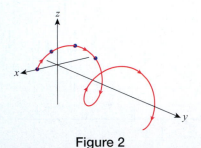

Figure 2

TOPIC 2 **The Unit Tangent Vector**

We have already determined that, if $\mathbf{r}'(t_0) \neq \mathbf{0}$, then $\mathbf{r}'(t_0)$ is tangent to the curve C defined by $\mathbf{r}(t)$ at the point $\mathbf{r}(t_0)$. If the vector function $\mathbf{r}$ is smooth then $\mathbf{r}'$ is continuous and never zero, so $\mathbf{r}'(t)/|\mathbf{r}'(t)|$ is a unit vector pointing in the same direction as $\mathbf{r}'(t)$. We make this observation the basis of the following definition.

Definition 💡

Unit Tangent Vector

The **unit tangent vector** $\mathbf{T}(t)$ of a smooth curve $\mathbf{r}(t)$ is $\mathbf{T}(t) = \dfrac{\mathbf{r}'(t)}{|\mathbf{r}'(t)|}$.

Note that if s is the arc length parameter for the curve, then

$$\frac{d\mathbf{r}}{ds} = \frac{d\mathbf{r}}{dt}\frac{dt}{ds} = \frac{\mathbf{r}'(t)}{ds/dt} = \frac{\mathbf{r}'(t)}{|\mathbf{r}'(t)|} = \mathbf{T}(t).$$

So, at any given point, *the rate of change of a curve with respect to the arc length parameter is the unit tangent vector.*

Example 3 ✍

Find the unit tangent vector for the curve defined by $\mathbf{r}(t) = \langle \cos t, \sin t, \cos 3t \rangle$ over the interval $[0, 2\pi]$.

Solution

$$\mathbf{r}'(t) = \langle -\sin t, \cos t, -3\sin 3t \rangle$$

$$|\mathbf{r}'(t)| = \sqrt{(-\sin t)^2 + (\cos t)^2 + (-3\sin 3t)^2} = \sqrt{1 + 9\sin^2 3t}$$

$$\mathbf{T}(t) = \frac{\mathbf{r}'(t)}{|\mathbf{r}'(t)|} = \frac{1}{\sqrt{1 + 9\sin^2 3t}}\langle -\sin t, \cos t, -3\sin 3t \rangle$$

Figure 3 is an illustration of the curve $\mathbf{r}$ and $\mathbf{T}(\pi/4)$.

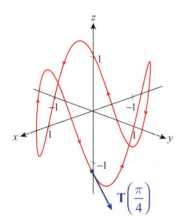

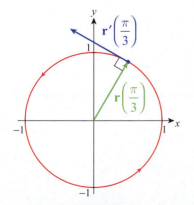

Figure 3

Example 4 ✍

Find the unit tangent vector for the curve in the plane defined by

$$\mathbf{r}(t) = \langle \cos t, \sin t \rangle.$$

Solution

This is an easy task: $\mathbf{r}'(t) = \langle -\sin t, \cos t \rangle$, so $|\mathbf{r}'(t)| = 1$ and $\mathbf{T}(t) = \mathbf{r}'(t)$. The point of this example is simply to illustrate the general fact (proved in Section 12.1) that if $|\mathbf{r}(t)|$ is constant, then $\mathbf{r}$ and $\mathbf{r}'$ are orthogonal at every point of the curve. Figure 4 depicts the unit tangent vector at the point $t = \pi/3$.

Figure 4

TOPIC 3   Projectile Motion

We conclude this section with a deeper analysis of projectile motion—specifically, the behavior of objects with known initial position and velocity and moving under the influence of gravity alone.

If we let $\mathbf{r}(t)$ denote the position of such an object at time t, then its velocity and acceleration are, respectively, $\mathbf{v} = \mathbf{r}'(t)$ and $\mathbf{a} = \mathbf{r}''(t)$. Newton's Second Law of Motion, $\mathbf{F} = m\mathbf{a}$, tells us that the sum of the forces acting on an object is equal to the product of its mass m and its acceleration $\mathbf{a}$; if gravity is the only force present, then $\mathbf{F} = -mg\,\mathbf{j}$, where g is either 9.8 m/s^2 or 32 ft/s^2 and $\mathbf{j}$ is the unit vector pointing in the positive y-direction (away from the Earth's center). That is, $-mg\,\mathbf{j} = m\mathbf{r}''(t)$ or, canceling m from both sides, $-g\,\mathbf{j} = \mathbf{r}''(t)$. If we set up our coordinate system so that the object moves in the xy-plane, as illustrated in Figure 5, and if the object has initial velocity $\mathbf{v}_0$ and initial position $\mathbf{r}_0$, then we determine $\mathbf{r}(t)$ by solving the following initial value problem.

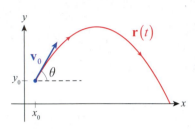

Figure 5 Projectile Motion

$$\mathbf{r}''(t) = -g\,\mathbf{j}, \quad \mathbf{r}'(0) = \mathbf{v}_0, \quad \mathbf{r}(0) = \mathbf{r}_0$$

Integrating once and making use of the initial velocity results in the velocity vector

$$\mathbf{r}'(t) = -gt\,\mathbf{j} + \mathbf{v}_0,$$

and integrating a second time and using the initial position gives us

$$\mathbf{r}(t) = -\frac{1}{2}gt^2\,\mathbf{j} + \mathbf{v}_0 t + \mathbf{r}_0.$$

Often, it is more natural to express the initial velocity of a projectile in terms of its magnitude (or speed) v_0 and the angle θ shown in Figure 5, in which case $\mathbf{v}_0 = v_0 \langle \cos\theta, \sin\theta \rangle$. If we also express $\mathbf{r}_0$ in component form as $\langle x_0, y_0 \rangle$, then

$$\mathbf{r}(t) = \left\langle 0, -\frac{1}{2}gt^2 \right\rangle + v_0 \langle \cos\theta, \sin\theta \rangle t + \langle x_0, y_0 \rangle$$

$$= \left\langle (v_0 \cos\theta)t + x_0, -\frac{1}{2}gt^2 + (v_0 \sin\theta)t + y_0 \right\rangle.$$

The parametric equations corresponding to this vector function are as follows:

$$x = (v_0 \cos\theta)t + x_0 \quad \text{and} \quad y = -\frac{1}{2}gt^2 + (v_0 \sin\theta)t + y_0$$

Example 5 ☑

Given the vector function

$$\mathbf{r}(t) = \left\langle (v_0 \cos\theta)t + x_0, -\frac{1}{2}gt^2 + (v_0 \sin\theta)t + y_0 \right\rangle$$

for a projectile's motion, find formulas for its maximum height and the time at which it reaches that height.

Solution

A projectile reaches its maximum height when the vertical component of its velocity is 0. Since $\mathbf{v}(t) = \mathbf{r}'(t) = \langle v_0 \cos\theta, -gt + v_0 \sin\theta \rangle$, the formula for the time at which it reaches its maximum height is

$$t = \frac{v_0 \sin\theta}{g}.$$

At that time, the vertical component of $\mathbf{r}$ is

$$-\frac{1}{2}g\left(\frac{v_0 \sin\theta}{g}\right)^2 + (v_0 \sin\theta)\left(\frac{v_0 \sin\theta}{g}\right) + y_0 = \frac{(v_0 \sin\theta)^2}{2g} + y_0.$$

Example 6 ✐

A bullet is shot from a rifle with a muzzle velocity of 170 m/s at a 30° angle of elevation. The bullet leaves the rifle at a height 2 m above the ground. Assuming the surrounding terrain is flat and level, how far does the bullet travel, and with what speed does it hit the ground?

Solution

We can determine the time t at which the bullet lands by solving the quadratic equation $-\frac{1}{2}gt^2 + (v_0 \sin\theta)t + y_0 = 0$.

$$-\frac{1}{2}(9.8)t^2 + (170\sin 30°)t + 2 = 0$$

$$-4.9t^2 + \frac{170t}{2} + 2 = 0 \qquad \text{Apply the quadratic formula.}$$

$$t = -0.02, 17.37$$

The second solution for t is the value we seek (the first solution is a time before the rifle is fired), and the horizontal distance $x - x_0$ the bullet has traveled over that time is as follows:

$$x - x_0 = (v_0 \cos\theta)t$$
$$= \frac{170\sqrt{3}}{2}(17.37)$$
$$\approx 2557 \text{ m}$$

As always, the speed of an object is the absolute value of its velocity, so using the fact that $\mathbf{v}(t) = \mathbf{r}'(t) = \langle v_0 \cos\theta, -gt + v_0 \sin\theta \rangle$, the speed of the bullet when it lands is

$$|\mathbf{v}(17.37)| = \left|\langle 170\cos 30°, (-9.8)(17.37) + 170\sin 30° \rangle\right|$$

$$= \sqrt{\left(\frac{170\sqrt{3}}{2}\right)^2 + \left[(-9.8)(17.37) + \frac{170}{2}\right]^2}$$

$$\approx 170.1 \text{ m/s},$$

or just slightly greater than the speed with which it left the rifle.

12.2 **Exercises**

1–10 Find the arc length of the curve over the given interval.

1. $\mathbf{r}(t) = \langle 4\cos t, 4\sin t, 2t \rangle;\quad [0, 6\pi]$

2. $\mathbf{r}(t) = \langle t, 2\sin t, 2\cos t \rangle;\quad [0, 4\pi]$

3. $\mathbf{r}(t) = \langle 2 - 5t, 1 + 3t, 4 - \sqrt{2}t \rangle;\quad [0, 3]$

4. $\mathbf{r}(t) = \langle e^t, 2e^t \cos t, 2e^t \sin t \rangle;\quad [0, 1]$

5. $\mathbf{r}(t) = \langle 2t, t^2, \ln t \rangle;\quad [1, e]$

6. $\mathbf{r}(t) = \left\langle \dfrac{t^3}{3}, t^2, 2t \right\rangle;\quad [0, 2]$

7. $\mathbf{r}(t) = \langle t^3, \sqrt{6}t^2, 4t \rangle;\quad [0, 5]$

8. $\mathbf{r}(t) = \langle t^2, 2t\sin t, 2t\cos t \rangle;\quad [0, 1]$

9. $\mathbf{r}(t) = \left\langle \dfrac{\sqrt{2}}{2}t, \dfrac{\sqrt{2}}{2}t, -\ln(\cos t) \right\rangle;\quad \left[0, \dfrac{\pi}{3}\right]$

10. $\mathbf{r}(t) = \langle t - \tanh t, \operatorname{sech} t \rangle;\quad [0, \ln 3]$ (This curve is called a *tractrix*.)

11–16 Reparametrize the given curve with respect to arc length.

11. The line $\mathbf{r}(t) = \langle 1 + 2t, 3 - 5t, 4 + 4t \rangle$

12. The circle $\mathbf{r}(t) = \langle 0, 4\cos t, 4\sin t \rangle$

13. The helix $\mathbf{r}(t) = \langle 2\cos t, 2\sin t, 3t \rangle$

14. The curve $\mathbf{r}(t) = \langle t, \cosh t, \sinh t \rangle$

15. The helix $\mathbf{r}(t) = \langle bt, a\cos \omega t, a\sin \omega t \rangle$

16. The curve $\mathbf{r}(t) = \langle e^t, e^t \cos t, e^t \sin t \rangle$

17–24 Find the unit tangent vector for the given curve.

17. $\mathbf{r}(t) = \langle t + 1, t^3, -t^2 \rangle$

18. $\mathbf{r}(t) = \langle 2t, \cos t, \sin t \rangle$

19. $\mathbf{r}(t) = \langle e^t, e^t \cos t, e^t \sin t \rangle$

20. $\mathbf{r}(t) = \left\langle \dfrac{1}{3}t^3, \dfrac{1}{t}, \sqrt{2}t \right\rangle$

21. $\mathbf{r}(t) = \langle t\cos t - \sin t, t, t\sin t + \cos t \rangle$

22. $\mathbf{r}(t) = \langle bt, a\cos \omega t, a\sin \omega t \rangle$

23. $\mathbf{r}(t) = \langle e^t, e^{-t}, \sqrt{2}t \rangle$

24. $\mathbf{r}(t) = \langle t, \sqrt{2 - 2t^2}, t \rangle$

25. Find an arc length parametrization of the straight line $y = mx + b$.

26. A circle of radius 5 is located in the plane $x = 3$, centered at $(3, 2, 1)$. Find an arc length parametrization for this circle.

27. The following are all parametrizations of the same helix. Which one is the arc length parametrization?

   a. $\mathbf{r}(t) = \left\langle \cos t, \sin t, \dfrac{t}{3} \right\rangle$

   b. $\mathbf{r}(t) = \langle \cos 3t, \sin 3t, t \rangle$

   c. $\mathbf{r}(t) = \left\langle \cos \dfrac{3t}{\sqrt{10}}, \sin \dfrac{3t}{\sqrt{10}}, \dfrac{t}{\sqrt{10}} \right\rangle$

   d. $\mathbf{r}(t) = \left\langle \cos \sqrt{10}t, \sin \sqrt{10}t, \dfrac{\sqrt{10}t}{3} \right\rangle$

28. Suppose a bug starts crawling at $(0, 0, 0)$ along the curve $\mathbf{r}(t) = \langle 3t, 2t^2, 4\sqrt{2/3}\,t^{3/2} \rangle$. After crawling exactly 5 unit lengths, it runs into a spider web. Find the coordinates of the point where the curve pierces the spider web.

29. Recall from Section 9.1 the parametrization of the first full arch of the cycloid: $x = a(\theta - \sin\theta)$ and $y = a(1 - \cos\theta),\ \theta \in [0, 2\pi]$. Find the arc length parametrization of this curve.

30. One version of the Bernoulli spiral can be parametrized as $\mathbf{r}(t) = \langle e^t \cos t, e^t \sin t \rangle$ (see Exercise 85 of Section 12.1). Find the arc length parametrization of this curve. (**Hint:** For the lower limit of integration in calculating $s(t)$, use $-\infty$.)

31.* Suppose a spring has radius r and it reaches height h while making n full revolutions. Find a formula for the length of the wire used in manufacturing this spring.

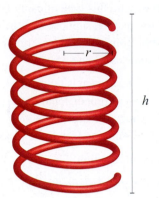

32. Suppose you calculate the arc length parametrization for the curve $\mathbf{r}(t) = \langle f(t), g(t), h(t) \rangle$, while your classmate does the same, but she starts out with the parametrization $\mathbf{r}(t) = \langle f(t^3), g(t^3), h(t^3) \rangle$. Do you obtain equivalent answers? Explain.

33. Use the vector function given in Example 5,
$$\mathbf{r}(t) = \langle (v_0 \cos\theta)t + x_0, -\tfrac{1}{2}gt^2 + (v_0 \sin\theta)t + y_0 \rangle,$$
to find a formula for the range of the projectile, assuming it was launched at ground level. (The range of the projectile is the distance between its launching and landing points.)

34. Use the formulas found in Example 5 and Exercise 33 to revisit Exercise 81 of Section 12.1.

35. Use Exercise 33 to find the angle θ which corresponds to the maximum range for the projectile.

36.* Repeat Exercise 35 to find the angle θ that corresponds to the maximum downhill range if the projectile is launched on a downhill terrain that drops at an angle of δ from horizontal. (θ is still the angle of elevation measured from horizontal.)

37.* Repeat Exercise 36 to find the angle θ that corresponds to the maximum uphill range if the projectile is launched on an uphill terrain with an angle of elevation of φ from horizontal.

38. A pellet is shot from an air rifle with a muzzle velocity of 1200 ft/s, leaving the rifle 5 ft above ground level and at a 45° angle of elevation. Assuming the surrounding terrain is flat and level, how far does the pellet travel, and with what speed does it hit the ground? (As usual, ignore air resistance.)

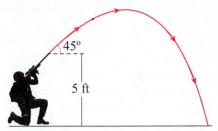

39. A projectile is launched from a 2 m high platform with an initial speed of 30 m/s, in a direction 60° upward from horizontal. Ignore air resistance.

  a. Find a vector function that models the projectile's path.

  b. Find the maximum height attained by the projectile, its range, and the speed of impact. (**Hint:** Suppose the projectile was launched from the point $(0, 2)$.)

40. Answer the questions of Exercise 39, assuming that the launch took place on the moon. (Use $g/6$ for the acceleration caused by gravity near the moon's surface.)

41. The exit velocity of a baseball (its velocity as it leaves the bat) is 128 feet per second, in the direction of 30° above horizontal. If it was hit 4 feet above ground level, find

  a. a vector function that models the path of the baseball,

  b. the maximum height attained by the baseball,

  c. the horizontal distance traveled by the ball and its speed of impact.

(Ignore air resistance.)

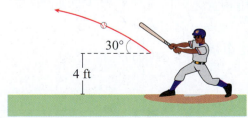

42. A golf ball is hit with an initial speed of 140 feet per second, 50° upward from horizontal, toward a hole 158 yards away (when measured horizontally). If the elevation of the hole is 121 feet higher than that of the starting point of the golf ball, will the ball land in the hole? (**Hint:** Use a vector function to examine the trajectory of the golf ball. Ignore air resistance.)

43.* A projectile is launched from a 6-foot platform with an initial speed of 200 feet per second and at a firing angle of x degrees above horizontal. Find the value of x that will result in a range of 1000 feet. (Ignore all retarding forces but gravity. Express your answer in degrees.)

44. The curve pictured below is a *Cornu spiral* over the interval $[-2\pi, 2\pi]$ (also known as *Euler's spiral*, though initially discovered by Johann Bernoulli). It is defined by

$$\mathbf{r}(t) = \left\langle \int_0^t \cos\frac{u^2}{2}\,du, \int_0^t \sin\frac{u^2}{2}\,du \right\rangle.$$

Find the arc length of the Cornu spiral over the interval $[0, b]$.

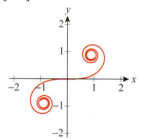

45. A bomber plane is flying eastward at a speed of 1200 kilometers per hour when it releases a bomb at an altitude of 8000 meters. Ignoring air resistance, find a vector function modeling the path of the bomb, the horizontal distance traveled by the bomb, and its speed of impact. (**Hint:** Suppose the bomb was released at the point $(0, 8000)$, and that the x-axis points to the east.)

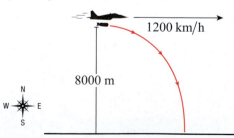

46. Suppose a certain type of medieval cannon, at a 10-degree angle of elevation, is able to fire to a distance of 2400 feet. Find the initial speed of the cannonball. What maximum range can be achieved with this type of cannon? (**Hint:** To simplify matters, suppose that the cannon ball was shot from ground level. Ignore air resistance. See Exercises 33 and 35.)

47.* When air resistance is taken into consideration, assuming it is proportional to the projectile's velocity, then the vector function $\mathbf{r}(t)$ for the projectile's motion (see Example 5) satisfies the differential equation $\dfrac{d^2}{dt^2}\mathbf{r}(t) = -C\dfrac{d}{dt}\mathbf{r}(t) - g\,\mathbf{j}$, where C is the drag coefficient. Assuming that the projectile is launched from the origin (i.e., $x_0 = y_0 = 0$), find $\mathbf{r}(t)$ under these conditions.

48. Suppose that the path of a moving point is a straight line. Prove that in this case, $\mathbf{T}'(t) = \mathbf{0}$.

49.* If we unwind a thread from a fixed circular spool of radius a, starting at the point $(a, 0)$ and keeping the thread taut in the xy-plane throughout the process, the curve traced out by the endpoint of the thread is called the *involute* of the circle. (**Note:** In the figure below, the point $P(\theta)$ denotes the endpoint of the thread at the instant when the radius to the point of tangency makes an angle of θ with the positive x-axis.)

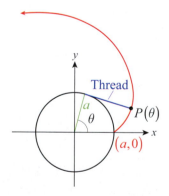

a. Use the figure to derive the following parametrization of the involute of a circle of radius a.

$$\mathbf{r}(\theta) = a\langle \cos\theta + \theta\sin\theta, \sin\theta - \theta\cos\theta \rangle, \ \theta > 0$$

b. Reparametrize the involute with respect to arc length.

50–54 *True or False?* Determine whether the given statement is true or false. In case of a false statement, explain or provide a counterexample.

50. The circle $\mathbf{r}(t) = \langle \cos t, \sin t, 0 \rangle$ has the property that $\mathbf{r}(t)$ and $\mathbf{r}'(t)$ are perpendicular for all t.

51. The helix $\mathbf{r}(t) = \langle \cos t, \sin t, t \rangle$ has the property that $\mathbf{r}(t)$ and $\mathbf{r}'(t)$ are perpendicular for all t.

52. If $\mathbf{r}(t) = \langle f(t), g(t), h(t) \rangle$ and $f(t)$, $g(t)$, and $h(t)$ are linear polynomials, then $\mathbf{T}(t)$ is constant.

53. When the launching speed of a projectile is doubled, its range doubles. (Suppose it is launched at angle α, $0 < \alpha < \pi/2$, upward from horizontal.)

54. For a space curve $\mathbf{r}(t)$, we have $\left| \mathbf{r}(t) \right|' = \left| \mathbf{r}'(t) \right|$.

12.3 The Unit Normal and Binormal Vectors, Curvature, and Torsion

TOPICS

1. The unit normal and binormal vectors

2. Curvature and torsion

At this point, we have defined the unit tangent vector and studied the relationship between it and its underlying curve. But curves in space usually have features that can't be characterized by the unit tangent vector alone. To fully capture a curve's behavior, we introduce two more reference vectors called the *unit normal* and *binormal* vectors. These allow us to define qualities called *curvature* and *torsion*, and give us a frame of reference used in such diverse applications as relativistic physics, computer graphics, and modeling the motion of swimming organisms.

TOPIC 1 The Unit Normal and Binormal Vectors

By design, the unit tangent vector $\mathbf{T}(t)$ corresponding to a smooth curve $\mathbf{r}(t)$ has constant length, so we know that $\mathbf{T} \cdot \mathbf{T}' = 0$ at every t for which $\mathbf{T}'(t)$ exists—that is, $\mathbf{T}$ and $\mathbf{T}'$ are orthogonal vectors. For the remainder of this chapter we will be interested only in curves which are not only smooth, but for which $\mathbf{T}$ is also smooth, so we can assume $\mathbf{T}'$ is continuous and $\mathbf{T}'(t) \neq \mathbf{0}$ for all t. Under this assumption, we have the following definitions.

Definition 💡

Unit Normal and Binormal Vectors

The **unit normal vector** $\mathbf{N}(t)$ of a smooth curve $\mathbf{r}(t)$ is defined as

$$\mathbf{N}(t) = \frac{\mathbf{T}'(t)}{|\mathbf{T}'(t)|},$$

and the **unit binormal vector** $\mathbf{B}(t)$ as

$$\mathbf{B}(t) = \mathbf{T}(t) \times \mathbf{N}(t).$$

Since $\mathbf{T}$ and $\mathbf{N}$ are both unit vectors, $\mathbf{B}$ is as well, and the three vectors $\mathbf{T}$, $\mathbf{N}$, and $\mathbf{B}$ (in that order) form a right-handed orthogonal system of vectors at every point on the curve defined by $\mathbf{r}$. This is called the **TNB** frame of reference for the curve, or the **Frenet-Serret** frame (after the French mathematicians Jean Frenet and Joseph Serret who independently developed the idea in the mid-nineteenth century).

Example 1 ✎

Find the unit tangent, normal, and binormal vectors for the helix defined by the vector function $\mathbf{r}(t) = \langle \cos t, \sin t, t \rangle$.

Solution

$$\mathbf{r}'(t) = \langle -\sin t, \cos t, 1 \rangle$$

$$\mathbf{T}(t) = \frac{\mathbf{r}'(t)}{|\mathbf{r}'(t)|} = \frac{\langle -\sin t, \cos t, 1 \rangle}{\sqrt{(-\sin t)^2 + (\cos t)^2 + 1}} = \frac{1}{\sqrt{2}} \langle -\sin t, \cos t, 1 \rangle$$

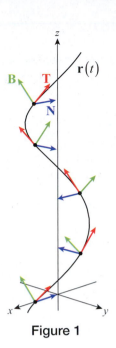

Figure 1

We can now use **T** to determine

$$\mathbf{N}(t) = \frac{\mathbf{T}'(t)}{|\mathbf{T}'(t)|} = \frac{\dfrac{1}{\sqrt{2}}\langle -\cos t, -\sin t, 0\rangle}{\dfrac{1}{\sqrt{2}}\sqrt{(-\cos t)^2 + (-\sin t)^2}} = \langle -\cos t, -\sin t, 0\rangle$$

and then

$$\mathbf{B}(t) = \mathbf{T}(t) \times \mathbf{N}(t) = \begin{vmatrix} \mathbf{i} & \mathbf{j} & \mathbf{k} \\ \dfrac{-\sin t}{\sqrt{2}} & \dfrac{\cos t}{\sqrt{2}} & \dfrac{1}{\sqrt{2}} \\ -\cos t & -\sin t & 0 \end{vmatrix} = \left\langle \frac{\sin t}{\sqrt{2}}, -\frac{\cos t}{\sqrt{2}}, \frac{\sin^2 t + \cos^2 t}{\sqrt{2}} \right\rangle$$

$$= \frac{1}{\sqrt{2}}\langle \sin t, -\cos t, 1\rangle.$$

Figure 1 illustrates the helix and the **TNB** frame at five points on the helix. Note that the **T** vectors (in red) always point in the direction of the curve, while the **N** and **B** vectors (in blue and green, respectively) are both orthogonal to the curve at each point. Note also that the **T**, **N**, and **B** vectors form a right-hand system.

In many applications involving space curves, the **TNB** frame of reference at a given point is used to identify certain features at that point. One step in identifying these features is to use the **TNB** vectors to define planes that are analogous to the coordinate planes of Cartesian space.

Definition 💡

Osculating, Normal, and Rectifying Planes

Given a vector function **r** and its associated tangent, normal, and binormal vector functions **T**, **N**, and **B**, the **osculating plane** at a particular point $\mathbf{r}(t_0)$ is the plane containing $\mathbf{r}(t_0)$ and the vectors $\mathbf{T}(t_0)$ and $\mathbf{N}(t_0)$. The name comes from the Latin word *osculari*, meaning "to kiss," and indicates that the plane just touches (or kisses) the curve at the point $\mathbf{r}(t_0)$. Similarly, the **normal plane** at the point contains $\mathbf{r}(t_0)$ and the vectors $\mathbf{N}(t_0)$ and $\mathbf{B}(t_0)$, while the **rectifying plane** contains the vectors $\mathbf{B}(t_0)$ and $\mathbf{T}(t_0)$.

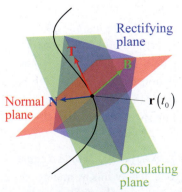

Figure 2

Figure 2 illustrates the relationships of the osculating, normal, and rectifying planes to the underlying curve. The tangent, normal, and binormal vectors are again colored red, blue, and green, respectively. Note that the osculating plane can be characterized entirely by the fact that it contains the point $\mathbf{r}(t_0)$ and is normal to $\mathbf{B}(t_0)$, so the osculating plane has been colored green in the figure. Similarly, the normal plane is normal to $\mathbf{T}(t_0)$ so is colored red, and the rectifying plane is normal to $\mathbf{N}(t_0)$ and is colored blue.

Example 2 ✐

Use the results of Example 1 to determine equations for the osculating, normal, and rectifying planes associated with the helix $\mathbf{r}(t) = \langle \cos t, \sin t, t \rangle$.

Solution

Recall that the equation $\langle x - x_0, y - y_0, z - z_0 \rangle \cdot \mathbf{n} = 0$ describes all the points (x, y, z) of a plane containing the particular point (x_0, y_0, z_0), where $\mathbf{n}$ is normal to the plane. So the osculating, normal, and rectifying planes passing through the point $(\cos t, \sin t, t)$ are described by the equations:

Osculating plane: $\langle x - \cos t, y - \sin t, z - t \rangle \cdot \langle \sin t, -\cos t, 1 \rangle = 0$ Using $\mathbf{n} = \sqrt{2}\,\mathbf{B}$

Normal plane: $\quad\langle x - \cos t, y - \sin t, z - t \rangle \cdot \langle -\sin t, \cos t, 1 \rangle = 0$ Using $\mathbf{n} = \sqrt{2}\,\mathbf{T}$

Rectifying plane: $\langle x - \cos t, y - \sin t, z - t \rangle \cdot \langle -\cos t, -\sin t, 0 \rangle = 0$ Using $\mathbf{n} = \mathbf{N}$

(Note that in order to simplify the equations, we have used appropriately chosen multiples of $\mathbf{B}$ and $\mathbf{T}$ for the normal vector $\mathbf{n}$.)

The equations we have found are equations in the variables x, y, and z and in the parameter t, so they describe the planes of the **TNB** system at each point $\mathbf{r}(t)$ of the helix. For instance, the planes at the specific point where $t = \pi$, namely $(-1, 0, \pi)$, are as follows:

Osculating plane: $\langle x + 1, y, z - \pi \rangle \cdot \langle 0, 1, 1 \rangle = 0$ or $y + z - \pi = 0$

Normal plane: $\quad\langle x + 1, y, z - \pi \rangle \cdot \langle 0, -1, 1 \rangle = 0$ or $-y + z - \pi = 0$

Rectifying plane: $\langle x + 1, y, z - \pi \rangle \cdot \langle 1, 0, 0 \rangle = 0$ or $x + 1 = 0$

In computer graphics applications, surfaces are often defined largely in terms of space curves. For instance, starting with a given base curve, a *ribbon* can be described as a surface extending a fixed distance along the normal vector at each point of the curve, and a *tube* as a planar shape reproduced in each normal plane along the curve.

Example 3 ✐

Given the helix $\mathbf{r}(t) = \langle \cos t, \sin t, t \rangle$ as the base curve, find a parametric description of a ribbon of width 1 centered on the curve. Then, do the same for a circular tube of diameter 1 centered on the curve.

Solution

At any point $\mathbf{r}(t)$ on the helix, the line segment parametrized by $\mathbf{r}(t) + u\,\mathbf{N}(t)$, $u \in \left[-\frac{1}{2}, \frac{1}{2} \right]$, is a segment of length 1 centered at $\mathbf{r}(t)$ and running in the direction of $\mathbf{N}(t)$. So the position vectors $\mathbf{r}(t) + u\,\mathbf{N}(t)$, with $u \in \left[-\frac{1}{2}, \frac{1}{2} \right]$ and $t \in I$ for some interval I, define all the points over a length of the desired ribbon. For example, Figure 3 illustrates the ribbon over $t \in [0, 2\pi]$, one full twist of the helix.

As we know, the position vector $\frac{1}{2}\langle \cos u, \sin u \rangle$, $u \in [0, 2\pi]$, defines a circle of diameter 1 centered at the origin of the xy-plane. It does this by tracing out the terminal points of the vectors $\left(\frac{1}{2}\cos u \right)\mathbf{i} + \left(\frac{1}{2}\sin u \right)\mathbf{j}$ for $u \in [0, 2\pi]$. We

Figure 3 Ribbon

Figure 4 Tube

Figure 5

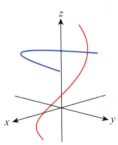

Figure 6

can describe a tube of diameter 1 centered on the helix $\mathbf{r}(t)$, for any fixed t, by constructing a circle centered at $\mathbf{r}(t)$ in the **NB**-plane (the normal plane). In other words, the vector **N** plays the role of **i** and the vector **B** plays the role of **j**. So the tubular surface consists of all those points of the form $\mathbf{r}(t) + \left(\frac{1}{2}\cos u\right)\mathbf{N}(t) + \left(\frac{1}{2}\sin u\right)\mathbf{B}(t)$, $u \in [0, 2\pi]$ and $t \in I$ for some interval I. Figure 4 is an illustration of this surface for $t \in [0, 2\pi]$.

The approach taken in Example 3 is very powerful, and can be generalized to describe much more complicated surfaces. For instance, the hypocycloid of part b. from the Technology Note of Section 9.1 (shrunk by a factor of $\frac{1}{10}$) can be wrapped around the helix $\mathbf{r}(t) = \langle \cos t, \sin t, t \rangle$ in the same manner to form the surface of Figure 5.

The parametric description of the surface is

$$\mathbf{r}(t) + \frac{1}{10}\left[(8\cos u + 2\cos 4u)\mathbf{N}(t) + (8\sin u - 2\sin 4u)\mathbf{B}(t)\right],$$

where $u \in [0, 2\pi]$ and $t \in [0, 2\pi]$.

TOPIC 2 Curvature and Torsion

The three vectors **T**, **N**, and **B** give us one way to describe the behavior of a curve $\mathbf{r}(t)$, but they only indirectly tell us something about its relative "twistiness." The two curves in Figure 6, for instance, are qualitatively different in several ways. The red curve (a portion of a helix) appears to be consistently turning in toward the z-axis and, at the same time, changing in all three coordinate directions. The blue curve, on the other hand, appears to have segments that are comparatively straight and the curve seems to lie entirely in one plane. *Curvature* and *torsion* are scalar quantities that capture these characteristics of a curve.

Definition ♀

Curvature

The **curvature** κ (kappa) of a curve at any given point is the magnitude of the rate of change of its unit tangent vector **T** with respect to the arc length parameter s.

$$\kappa = \left|\frac{d\mathbf{T}}{ds}\right|$$

Informally, the more **T** changes for every unit change of arc length, the greater the curvature. In practice, if the unit tangent vector for a curve $\mathbf{r}(t)$ is known as a function of some parameter other than arc length, the following variation is convenient.

$$\kappa = \left|\frac{d\mathbf{T}}{ds}\right| = \left|\frac{d\mathbf{T}}{dt}\frac{dt}{ds}\right| = \frac{|\mathbf{T}'(t)|}{|ds/dt|} = \frac{|\mathbf{T}'(t)|}{|\mathbf{r}'(t)|}$$

Example 4 ✐

Determine the curvature of **a.** a line and **b.** the circle $\mathbf{r}(t) = \langle a\cos t, a\sin t, 0\rangle$.

Solution

a. For any line (in $\mathbb{R}^2$ or $\mathbb{R}^3$), the unit tangent vector $\mathbf{T}$ is a constant vector (pointing in the direction of the line), so $\kappa = |\mathbf{0}| = 0$.

b. $\mathbf{r}'(t) = \langle -a\sin t, a\cos t, 0\rangle$ and $|\mathbf{r}'(t)| = a$

These lead to $\mathbf{T}(t) = \langle -\sin t, \cos t, 0\rangle$ and the curvature is as follows:

$$\kappa = \frac{|\mathbf{T}'(t)|}{|\mathbf{r}'(t)|} = \frac{|\langle -\cos t, -\sin t, 0\rangle|}{a} = \frac{1}{a}$$

Note that a circle of larger radius a thus has smaller curvature κ.

Since κ is a measure of the rate of change of $\mathbf{T}$ (with respect to arc length), and since $\mathbf{T}'$ points in the direction of $\mathbf{N}$, it isn't surprising that κ and $\mathbf{N}$ have a connection. Specifically, since $|\mathbf{T}'(t)| = \kappa|\mathbf{r}'(t)|$,

$$\mathbf{N} = \frac{\mathbf{T}'(t)}{|\mathbf{T}'(t)|} = \left(\frac{d\mathbf{T}}{dt}\right)\left(\frac{1}{\kappa|\mathbf{r}'(t)|}\right) = \frac{1}{\kappa}\left(\frac{d\mathbf{T}}{ds}\frac{ds}{dt}\right)\left(\frac{1}{|\mathbf{r}'(t)|}\right) = \frac{1}{\kappa}\frac{d\mathbf{T}}{ds}\frac{ds}{dt}\frac{1}{ds/dt} = \frac{1}{\kappa}\frac{d\mathbf{T}}{ds}.$$

Written in the form

$$\frac{d\mathbf{T}}{ds} = \kappa\mathbf{N},$$

this is one of the **Frenet-Serret formulas** that relate the derivatives of $\mathbf{T}$, $\mathbf{N}$, and $\mathbf{B}$ to the vectors themselves. To find the corresponding formula for $d\mathbf{B}/ds$, we note first that $d\mathbf{B}/ds$ must be orthogonal to $\mathbf{B}$ since $\mathbf{B}$ has constant length. At the same time,

$$\frac{d\mathbf{B}}{ds} = \frac{d}{ds}(\mathbf{T}\times\mathbf{N}) = \left(\frac{d\mathbf{T}}{ds}\times\mathbf{N}\right) + \left(\mathbf{T}\times\frac{d\mathbf{N}}{ds}\right) = \left(\kappa\underbrace{\mathbf{N}\times\mathbf{N}}_{\mathbf{0}}\right) + \left(\mathbf{T}\times\frac{d\mathbf{N}}{ds}\right) = \mathbf{T}\times\frac{d\mathbf{N}}{ds},$$

so $d\mathbf{B}/ds$ is also orthogonal to $\mathbf{T}$ (and, incidentally, to $d\mathbf{N}/ds$). Since $d\mathbf{B}/ds$ is orthogonal to both $\mathbf{B}$ and $\mathbf{T}$, it must be a multiple of $\mathbf{N}$, and we define the *torsion* τ of the curve on that basis.

Definition 💡

Torsion

The **torsion** τ (tau) of a curve at any given point is defined by the relationship

$$\frac{d\mathbf{B}}{ds} = -\tau\mathbf{N}.$$ The negative sign is a convention.

Taking advantage of one of the properties of the dot product,

$$\tau = \tau\underbrace{\mathbf{N}\cdot\mathbf{N}}_{1} = -\frac{d\mathbf{B}}{ds}\cdot\mathbf{N}.$$

In Exercise 89 you will use the relationships

$$\frac{d\mathbf{T}}{ds} = \kappa \mathbf{N}, \quad \frac{d\mathbf{B}}{ds} = -\tau \mathbf{N}, \quad \text{and} \quad \mathbf{N} = \mathbf{B} \times \mathbf{T}$$

to derive the fact that

$$\frac{d\mathbf{N}}{ds} = -\kappa \mathbf{T} + \tau \mathbf{B}.$$

Taken together, the three relationships constitute the Frenet-Serret formulas, which we collect as a theorem.

Theorem ⚷

The Frenet-Serret Formulas

Given a curve $\mathbf{r}$ and associated unit tangent, normal, and binormal vectors $\mathbf{T}$, $\mathbf{N}$, and $\mathbf{B}$, we have the following:

$$\frac{d\mathbf{T}}{ds} = \kappa \mathbf{N} \qquad \frac{d\mathbf{N}}{ds} = -\kappa \mathbf{T} + \tau \mathbf{B} \qquad \frac{d\mathbf{B}}{ds} = -\tau \mathbf{N}$$

Geometrically, the curvature κ tells us the rate at which a curve is bending in the osculating plane (the $\mathbf{TN}$-plane), while the torsion τ tells us the rate at which the curve is twisting out of the osculating plane. Example 5 illustrates these ideas with the now-familiar helix.

Example 5 ✍

Calculate the curvature and torsion functions for the helix

$$\mathbf{r}(t) = \langle \cos t, \sin t, t \rangle.$$

Solution

We could divide $\left| \mathbf{T}'(t) \right|$ by $\left| \mathbf{r}'(t) \right|$ to determine κ, as we did in Example 4, but since our only formula for τ so far requires knowing the arc length parameter, we will use the $\left| d\mathbf{T}/ds \right|$ formulation of curvature. The first step is determining $s(t)$.

$$s(t) = \int_0^t \left| \mathbf{r}'(u) \right| du = \int_0^t \sqrt{(-\sin u)^2 + (\cos u)^2 + 1} \, du = \sqrt{2}\, t$$

Hence, $t(s) = s / \sqrt{2}$ and, from Example 1, we have the following:

$$\mathbf{T}(s) = \mathbf{T}(t(s)) = \frac{1}{\sqrt{2}} \left\langle -\sin \frac{s}{\sqrt{2}}, \cos \frac{s}{\sqrt{2}}, 1 \right\rangle$$

$$\mathbf{N}(s) = \mathbf{N}(t(s)) = \left\langle -\cos \frac{s}{\sqrt{2}}, -\sin \frac{s}{\sqrt{2}}, 0 \right\rangle$$

$$\mathbf{B}(s) = \mathbf{B}(t(s)) = \frac{1}{\sqrt{2}} \left\langle \sin \frac{s}{\sqrt{2}}, -\cos \frac{s}{\sqrt{2}}, 1 \right\rangle$$

Now we can proceed to find the curvature κ and torsion τ.

$$\kappa = \left|\frac{d\mathbf{T}}{ds}\right| = \frac{1}{\sqrt{2}}\left|\left\langle -\frac{1}{\sqrt{2}}\cos\frac{s}{\sqrt{2}}, -\frac{1}{\sqrt{2}}\sin\frac{s}{\sqrt{2}}, 0\right\rangle\right| = \frac{1}{2}\sqrt{\left(-\cos\frac{s}{\sqrt{2}}\right)^2 + \left(-\sin\frac{s}{\sqrt{2}}\right)^2} = \frac{1}{2}$$

$$\tau = -\frac{d\mathbf{B}}{ds}\cdot\mathbf{N} = -\frac{1}{2}\left\langle \cos\frac{s}{\sqrt{2}}, \sin\frac{s}{\sqrt{2}}, 0\right\rangle\cdot\left\langle -\cos\frac{s}{\sqrt{2}}, -\sin\frac{s}{\sqrt{2}}, 0\right\rangle = \left(-\frac{1}{2}\right)(-1) = \frac{1}{2}$$

For practical purposes, it is highly desirable to have a variety of formulas for $\mathbf{T}$, $\mathbf{N}$, $\mathbf{B}$, κ, and τ. One key step in finding other formulas is to note the facts below about $\mathbf{r}'$ and $\mathbf{r}''$. Since $\mathbf{r}'$ and $\mathbf{r}''$ often have the physical meaning of velocity and acceleration, respectively, we include the alternate labels $\mathbf{v}$ and $\mathbf{a}$. The variable s denotes the arc length parameter, as usual, and all prime notation indicates differentiation with respect to t.

$$\mathbf{v} = \mathbf{r}' = \frac{d\mathbf{r}}{dt} = \underbrace{\left(\frac{d\mathbf{r}}{ds}\right)}_{\mathbf{T}}\underbrace{\left(\frac{ds}{dt}\right)}_{|\mathbf{r}'(t)|} = \underbrace{|\mathbf{r}'(t)|}_{\text{speed}}\mathbf{T} = s'\mathbf{T}$$

$$\mathbf{a} = \mathbf{r}'' = \frac{d}{dt}\left(\frac{d\mathbf{r}}{dt}\right) = \frac{d}{dt}(s'\mathbf{T}) = s''\mathbf{T} + s'\mathbf{T}' = s''\mathbf{T} + s'\left(\frac{d\mathbf{T}}{ds}\frac{ds}{dt}\right) = \underbrace{s''\mathbf{T}}_{a_T} + \underbrace{(s')^2\kappa\mathbf{N}}_{a_N}$$

These formulas are important in their own right, so we make the following definition.

Definition 💡

Tangential and Normal Components of Acceleration

The **tangential** and **normal components of acceleration** of an object with position function $\mathbf{r}(t)$ are defined, respectively, as

$$a_T = s'' \quad\text{and}\quad a_N = (s')^2\kappa,$$

where s denotes the arc length parameter of the curve and κ is its curvature. Thus, we can write $\mathbf{a} = a_T\mathbf{T} + a_N\mathbf{N}$, and use the fact that $\mathbf{T}$ and $\mathbf{N}$ are orthogonal to note

$$|\mathbf{a}|^2 = a_T^2 + a_N^2.$$

The decomposition of acceleration as $a_T\mathbf{T} + a_N\mathbf{N}$ has several immediate implications. First, note that if the speed s' of an object is constant, then $a_T = s'' = 0$ so any acceleration the object has lies entirely in the normal direction. Second, the factor of $(s')^2$ in a_N means that if, for instance, you double your speed around a curve while driving, your body feels four times the acceleration in the normal direction.

Example 6 ✏️

Find the tangential and normal components of acceleration of an object whose position function is the spiral $\mathbf{r}(t) = \langle t\cos t, t\sin t, 0\rangle$.

Solution

We will need both $\mathbf{r}'(t)$ and $\mathbf{r}''(t)$, so we first find these vectors.

$$\mathbf{r}'(t) = \langle \cos t - t\sin t, \sin t + t\cos t, 0 \rangle$$
$$\mathbf{r}''(t) = \langle -2\sin t - t\cos t, 2\cos t - t\sin t, 0 \rangle$$

Next,

$$s'(t) = |\mathbf{r}'(t)| = \sqrt{(\cos t - t\sin t)^2 + (\sin t + t\cos t)^2} = \sqrt{1 + t^2}$$

so

$$a_T = s'' = \frac{t}{\sqrt{1 + t^2}}.$$

Using the relation $|\mathbf{a}|^2 = a_T{}^2 + a_N{}^2$ to determine a_N saves us the step of finding κ, and from above (after simplification) we find $|\mathbf{a}| = |\mathbf{r}''| = \sqrt{4 + t^2}$.

$$a_N = \sqrt{|\mathbf{a}|^2 - a_T{}^2} = \sqrt{\frac{4 + 4t^2 + t^4}{1 + t^2}}$$

Figure 7 depicts the curve over the interval $t \in [0, 2\pi]$, with the tangential and normal components of $\mathbf{r}''$ drawn in red and blue, respectively, at four points. Note that the normal components are significantly larger than the tangential components in magnitude (and grow as t grows), indicating that most of the acceleration is directed inward.

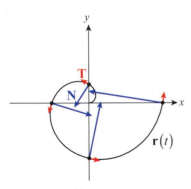

Figure 7

We conclude this section with an example of how alternative formulas can now be derived. In the exercises, you will use similar techniques to prove additional formulas.

Note that

$$\mathbf{r}' \times \mathbf{r}'' = s'\mathbf{T} \times \left(s''\mathbf{T} + \kappa(s')^2\,\mathbf{N}\right) = s's''\underbrace{\left(\mathbf{T} \times \mathbf{T}\right)}_{\mathbf{0}} + \kappa(s')^3\underbrace{\left(\mathbf{T} \times \mathbf{N}\right)}_{\mathbf{B}} = \kappa(s')^3\,\mathbf{B},$$

so

$$\mathbf{B} = \frac{\mathbf{r}' \times \mathbf{r}''}{\kappa(s')^3} = \frac{\mathbf{r}' \times \mathbf{r}''}{\kappa|\mathbf{r}'|^3}.$$

This tells us not only that $\mathbf{B}$ points in the direction of $\mathbf{r}' \times \mathbf{r}''$, but also, since $\mathbf{B}$ is a unit vector, gives us another formula for κ.

$$\kappa = \frac{|\mathbf{r}' \times \mathbf{r}''|}{|\mathbf{r}'|^3}$$

Further, since $\mathbf{B}$ is a unit vector in the direction of $\mathbf{r}' \times \mathbf{r}''$, it must be the case that

$$\mathbf{B} = \frac{\mathbf{r}' \times \mathbf{r}''}{|\mathbf{r}' \times \mathbf{r}''|},$$

and hence we have the following:

$$\mathbf{N} = \mathbf{T} \times (-\mathbf{B}) = \frac{\mathbf{r}'}{|\mathbf{r}'|} \times \frac{\mathbf{r}'' \times \mathbf{r}'}{|\mathbf{r}'' \times \mathbf{r}'|} = \frac{\mathbf{r}' \times (\mathbf{r}'' \times \mathbf{r}')}{|\mathbf{r}'||\mathbf{r}'' \times \mathbf{r}'|}$$

Note that $\mathbf{N} = \mathbf{T} \times (-\mathbf{B})$ by the right-hand rule.

In the definitions and formulas that follow, prime notation indicates differentiation with respect to t, and s denotes the arc length parameter.

Summary of Space Curve Definitions and Formulas

Unit Tangent Vector	$\mathbf{T} = \dfrac{\mathbf{r}'}{|\mathbf{r}'|} = \dfrac{d\mathbf{r}}{ds}$
Unit Normal Vector	$\mathbf{N} = \dfrac{\mathbf{T}'}{|\mathbf{T}'|} = \dfrac{\mathbf{r}' \times (\mathbf{r}'' \times \mathbf{r}')}{|\mathbf{r}'||\mathbf{r}'' \times \mathbf{r}'|}$
Unit Binormal Vector	$\mathbf{B} = \mathbf{T} \times \mathbf{N} = \dfrac{\mathbf{r}' \times \mathbf{r}''}{\kappa|\mathbf{r}'|^3} = \dfrac{\mathbf{r}' \times \mathbf{r}''}{|\mathbf{r}' \times \mathbf{r}''|}$
Arc Length Parameter	$s(t) = \displaystyle\int_{t_0}^{t} |\mathbf{r}'(u)| \, du \quad$ and $\quad s' = |\mathbf{r}'|$
Osculating Plane	Spanned by $\mathbf{T}$ and $\mathbf{N}$
Normal Plane	Spanned by $\mathbf{N}$ and $\mathbf{B}$
Rectifying Plane	Spanned by $\mathbf{B}$ and $\mathbf{T}$
Velocity	$\mathbf{v} = \mathbf{r}' = |\mathbf{r}'|\mathbf{T} = s'\mathbf{T}$
Acceleration	$\mathbf{a} = \mathbf{r}'' = s''\mathbf{T} + (s')^2 \kappa\mathbf{N} = a_T\mathbf{T} + a_N\mathbf{N}$
Tangential Component	$a_T = \dfrac{\mathbf{r}' \cdot \mathbf{r}''}{|\mathbf{r}'|}$
Normal Component	$a_N = \dfrac{|\mathbf{r}' \times \mathbf{r}''|}{|\mathbf{r}'|}$
Third Derivative	$\mathbf{r}''' = \left[s''' - \kappa^2 (s')^3 \right]\mathbf{T} + \left[3\kappa s' s'' + \kappa'(s')^2 \right]\mathbf{N} + \kappa\tau(s')^3\mathbf{B}$
Curvature	$\kappa = \left| \dfrac{d\mathbf{T}}{ds} \right| = \dfrac{|\mathbf{T}'|}{|\mathbf{r}'|} = \dfrac{|\mathbf{r}' \times \mathbf{r}''|}{|\mathbf{r}'|^3}$
Torsion	$\tau = -\dfrac{d\mathbf{B}}{ds} \cdot \mathbf{N} = \dfrac{(\mathbf{r}' \times \mathbf{r}'') \cdot \mathbf{r}'''}{|\mathbf{r}' \times \mathbf{r}''|^2}$
Frenet-Serret Formulas	$\dfrac{d\mathbf{T}}{ds} = \kappa\mathbf{N}, \quad \dfrac{d\mathbf{N}}{ds} = -\kappa\mathbf{T} + \tau\mathbf{B}, \quad$ and $\quad \dfrac{d\mathbf{B}}{ds} = -\tau\mathbf{N}$

Table 1

12.3 **Exercises**

1–10 Find the unit tangent, normal, and binormal vectors for the given curve.

1. $\mathbf{r}(t) = \langle \cos t, \sin t, 0 \rangle$

2. $\mathbf{r}(t) = \langle \sin t, \cos t, \sqrt{3}t + 1 \rangle$

3. $\mathbf{r}(t) = \langle 3 \sin 2t, 3 \cos 2t, 3 \rangle$

4. $\mathbf{r}(t) = \langle t, t, 3t^2 \rangle$

5. $\mathbf{r}(t) = \langle \sin \pi t, \cos \pi t, \pi t \rangle$

6. $\mathbf{r}(t) = \langle 2t, t^3, t \rangle$

7. $\mathbf{r}(t) = \langle 2t, \cos 3t, \sin 3t \rangle$

8. $\mathbf{r}(t) = \langle 4e^t \cos t, 4e^t \sin t, 1 \rangle$

9. $\mathbf{r}(t) = \langle e^t \sin t, e^t \cos t, e^{t+1} \rangle$

10. $\mathbf{r}(t) = \langle \cos t - \sin t, \sin t + \cos t, 1 \rangle$

11–20 Use the results of Exercises 1–10 to determine equations for the osculating, normal, and rectifying planes associated with the curve at the indicated point.

11. $\mathbf{r}(t) = \langle \cos t, \sin t, 0 \rangle; \quad t = 0$

12. $\mathbf{r}(t) = \langle \sin t, \cos t, \sqrt{3}t + 1 \rangle; \quad t = \dfrac{\pi}{2}$

13. $\mathbf{r}(t) = \langle 3 \sin 2t, 3 \cos 2t, 3 \rangle; \quad t = 0$

14. $\mathbf{r}(t) = \langle t, t, 3t^2 \rangle; \quad t = 1$

15. $\mathbf{r}(t) = \langle \sin \pi t, \cos \pi t, \pi t \rangle; \quad t = \dfrac{1}{2}$

16. $\mathbf{r}(t) = \langle 2t, t^3, t \rangle; \quad t = 2$

17. $\mathbf{r}(t) = \langle 2t, \cos 3t, \sin 3t \rangle; \quad t = \pi$

18. $\mathbf{r}(t) = \langle 4e^t \cos t, 4e^t \sin t, 1 \rangle; \quad t = 0$

19. $\mathbf{r}(t) = \langle e^t \sin t, e^t \cos t, e^{t+1} \rangle; \quad t = \dfrac{\pi}{2}$

20. $\mathbf{r}(t) = \langle \cos t - \sin t, \sin t + \cos t, 1 \rangle; \quad t = \dfrac{\pi}{4}$

21–28 Use Exercises 3–10 to find a parametric description of the surface.

21. The ribbon of width $\frac{1}{2}$ centered on $\mathbf{r}(t) = \langle 3 \sin 2t, 3 \cos 2t, 3 \rangle$

22. The ribbon of width 2 centered on $\mathbf{r}(t) = \langle t, t, 3t^2 \rangle$

23. The circular tube of radius 1 centered on $\mathbf{r}(t) = \langle \sin \pi t, \cos \pi t, \pi t \rangle$

24. The circular tube of radius $\frac{1}{3}$ centered on $\mathbf{r}(t) = \langle 2t, t^3, t \rangle$

25. The elliptical tube of major axis 0.6 (in the normal direction) and minor axis of 0.4 centered on $\mathbf{r}(t) = \langle 2t, \cos 3t, \sin 3t \rangle$

26. The ribbon of width $\frac{1}{4}$ centered on $\mathbf{r}(t) = \langle 4e^t \cos t, 4e^t \sin t, 1 \rangle$

27. The hypocycloid $x = \cos t + 2 \cos(t/2)$ and $y = \sin t - 2 \sin(t/2)$, $-2\pi < t \le 2\pi$ wrapped around $\mathbf{r}(t) = \langle e^t \sin t, e^t \cos t, e^{t+1} \rangle$

28. The circular tube of radius $\frac{1}{8}$ centered on $\mathbf{r}(t) = \langle \cos t - \sin t, \sin t + \cos t, 1 \rangle$

29. Show how the fact that the unit tangent vector function $\mathbf{T}(t)$ corresponding to a smooth curve $\mathbf{r}(t)$ has constant length implies that $\mathbf{T} \cdot \mathbf{T}' = 0$ for all t.

30. Show that if $\mathbf{r}(t)$ is a plane curve, then $\mathbf{N}(t)$ always points toward its "concave side," that is, "in the direction the curve bends."

31–33 Use your reparametrization results from Exercises 11, 13, and 14 of Section 12.2 to calculate the curvature and torsion functions for these curves.

31. $\mathbf{r}(t) = \langle 1 + 2t, 3 - 5t, 4 + 4t \rangle$

32. $\mathbf{r}(t) = \langle 2 \cos t, 2 \sin t, 3t \rangle$

33. $\mathbf{r}(t) = \langle t, \cosh t, \sinh t \rangle$

34. Find the curvature of the ellipse

$$\mathbf{r}(t) = \langle 4\cos t, 3\sin t, 0 \rangle$$

at $t = \pi/2$. What about $t = 0$?

35. Prove that the curvature of the helix
$\mathbf{r}(t) = \langle a\cos t, a\sin t, bt \rangle$ is obtained by the following formula:

$$\kappa = \frac{a}{a^2 + b^2}$$

What can you say if $b = 0$?

36. Show that the torsion for the helix of Exercise 35 is $\tau = b/(a^2 + b^2)$.

37. Generalize Exercises 35 and 36 to obtain formulas for the curvature and torsion functions of the general helix $\mathbf{r}(t) = \langle a\cos \omega t, a\sin \omega t, bt \rangle$.

38. Prove that if a curve $\mathbf{r}(t)$ lies in a plane, its torsion function is identically zero.

39–48 Calculate the curvature and torsion functions for the given curve. (**Hint:** Use the relevant formulas from the summary table.)

39. $\mathbf{r}(t) = \langle t, 2\sin t, 2\cos t \rangle$

40. $\mathbf{r}(t) = \langle t, 3t + 2, 3t - 1 \rangle$

41. $\mathbf{r}(t) = \left\langle 1, t, \dfrac{1}{t} \right\rangle$

42. $\mathbf{r}(t) = \langle 1, t, e^t \rangle$

43. $\mathbf{r}(t) = \langle 2\cos 3t, 2\sin 3t, 1 \rangle$

44. $\mathbf{r}(t) = \langle 1, t, t^2 \rangle$

45. $\mathbf{r}(t) = \langle \sin 3t, \cos 3t, 4t \rangle$

46. $\mathbf{r}(t) = \langle 2e^t \sin t, 2e^t \cos t, 1 \rangle$

47. $\mathbf{r}(t) = \langle e^t, e^t \cos t, e^t \sin t \rangle$

48. $\mathbf{r}(t) = \langle t\cos t + \sin t, 0, \cos t - t\sin t \rangle$

49. Prove that for the Bernoulli spiral of Exercise 85 in Section 12.1, the arc length $s(t)$ and curvature $\kappa(t)$ are inversely proportional. (**Note:** This is commonly interpreted as the arc length $s(t)$ and the radius of curvature being directly proportional; for the radius of curvature see the discussions preceding Exercises 60 and 74.)

50. If $f(x)$ is at least twice differentiable, prove that the curvature function of the plane curve $y = f(x)$ can be determined as

$$\kappa(x) = \frac{|f''(x)|}{\left(1 + [f'(x)]^2\right)^{3/2}}.$$

What can you conclude about the curvature of a plane curve at any of its inflection points? (**Hint:** Use $t = x$ as a parameter.)

51–53 Use Exercise 50 to evaluate the curvature of the plane curve at the indicated point.

51. $y = x^3$; $x = 1$

52. $y = \sin x$; $x = \dfrac{\pi}{4}$

53. $y = \ln|\cos x|$; $x = \dfrac{\pi}{6}$

54–56 Use Exercise 50 to find the point(s) of maximum curvature for the curve.

54. $f(x) = \dfrac{x^2}{2}$

55. $f(x) = \cos x$

56. $f(x) = e^x$

57. Generalize Exercise 50 for a plane curve $\mathbf{r}(t) = \langle x(t), y(t) \rangle$ to obtain the formula

$$\kappa(t) = \frac{|x'(t)y''(t) - x''(t)y'(t)|}{\left([x'(t)]^2 + [y'(t)]^2\right)^{3/2}}.$$

58. Use implicit differentiation along with Exercise 50 to find the curvature of $(6 - x)y^2 = 2x^3$ at the point $(2, 2)$. (Recall from Exercise 27 of Section 3.5 that this curve is a *cissoid*.)

59. Find the curvature of the cycloid $x = t - \sin t$, $y = 1 - \cos t$ at $t = \pi$.

60–68 Suppose the curvature of a curve $\mathbf{r}(t)$ at the point $P = \mathbf{r}(t_0)$ is nonzero. The **osculating circle** of the curve at P is the circle of radius $1/\kappa(t_0)$ that has the same tangent at P as $\mathbf{r}(t)$ does, and whose center lies in the direction of $\mathbf{N}(t_0)$ from P. (You can think of this as the "tangent circle." Note that it lies in the osculating plane, being tangent to $\mathbf{r}(t)$ so that both the circle's tangent line and curvature at P are the same as those of $\mathbf{r}(t)$. The osculating circle is rightfully called the "best-fitting circle" at P.)

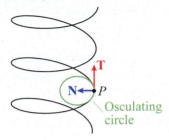

T

N P

Osculating circle

Find the osculating circle of the graph of the equation at the indicated point.

60. $y = x^2;\quad x = 0$

61. $y = x^2 - 1;\quad x = -\dfrac{1}{2}$

62. $y = \sqrt{x};\quad x = 1$

63. $y = \cos x;\quad x = 0$

64. $y = \cos x;\quad x = \dfrac{\pi}{4}$

65. $y = \dfrac{1}{x};\quad x = 1$

66. $y = e^x;\quad x = 0$

67. $y = x^3 - x;\quad x = 1$

68. $xy + 2x + y = 2;\quad x = 1$

69–72 Parametrize the osculating circle of the curve at the indicated point. (**Hint:** For space curves, remember that the osculating circle lies in the plane spanned by **T** and **N**.)

69. $\mathbf{r}(t) = \langle t - \sin t, 1 - \cos t \rangle;\quad t = \pi$
(See Exercise 59.)

70. $\mathbf{r}(t) = \langle \sin t, \cos t, 2t \rangle;\quad t = \pi$

71. $\mathbf{r}(t) = \left\langle t^2, t, \dfrac{t^3}{3} \right\rangle;\quad t = 0$

72. $\mathbf{r}(t) = \left\langle 2\cos t, \dfrac{4}{3}\sin t, t \right\rangle;\quad t = 0$

73.* Suppose that $f(x)$ is at least twice differentiable on an interval containing a, with nonzero first and second derivatives at $x = a$. Let $C(c_1, c_2)$ be the center of the osculating circle of the plane curve $y = f(x)$ at $(a, f(a))$. Prove that c_1 and c_2 can be determined as follows.

$$c_1 = a - \frac{f'(a)\left(1 + \left[f'(a)\right]^2\right)}{f''(a)}$$

$$c_2 = f(a) + \frac{1 + \left[f'(a)\right]^2}{f''(a)}$$

74–75 The center of the osculating circle of the curve $\mathbf{r}(t)$ at P is called the curve's **center of curvature** at P, while the radius of the osculating circle is the **radius of curvature** of $\mathbf{r}(t)$ at P. The locus of all centers of curvature is called the **evolute** of $\mathbf{r}(t)$. Exercises 74 and 75 will use this concept.

74. Show that the evolute of the parabola $y = x^2$ can be parametrized as $\mathbf{r}(t) = \left\langle -4t^3, 3t^2 + \tfrac{1}{2} \right\rangle$. (**Hint:** Use Exercise 73.)

75. Find the evolute for the curve $y = x^3$ $(x > 0)$.

76–87 Find the tangential and normal components of acceleration for the given position function.

76. $\mathbf{r}(t) = \langle \sin t, \cos t, 2t \rangle$

77. $\mathbf{r}(t) = \langle t, 2\sin t, 2\cos t \rangle$

78. $\mathbf{r}(t) = \langle 2t, t^2, 0 \rangle$

79. $\mathbf{r}(t) = \langle 2t, t^2, t \rangle$

80. $\mathbf{r}(t) = \langle 2t, t^2, \ln t \rangle$

81. $\mathbf{r}(t) = \langle t - \sin t, 1 - \cos t, 0 \rangle$

82. $\mathbf{r}(t) = \left\langle \dfrac{1}{t}, \sqrt{2}t, \dfrac{1}{3}t^3 \right\rangle$

83. $\mathbf{r}(t) = \left\langle 2t, t^2, \dfrac{t^3}{3} \right\rangle$

84. $\mathbf{r}(t) = \left\langle \dfrac{\sqrt{2}}{2}t, \dfrac{\sqrt{2}}{2}t, -\ln(\cos t) \right\rangle$

85. $\mathbf{r}(t) = \langle e^t, e^{-t}, \sqrt{2}t \rangle$

86.* $\mathbf{r}(t) = \langle bt, a\cos \omega t, a\sin \omega t \rangle$

87.* $\mathbf{r}(t) = \langle e^t, e^t \cos t, e^t \sin t \rangle$

88. According to Newton's Second Law of Motion, the magnitude of the friction force that keeps a car from skidding out of a curve is $\mathbf{F}(t) = ma_N(t)$, where m is the mass of the car, and $a_N(t)$ is the normal component of acceleration. Find the minimum friction force needed to keep a 1500 kg car from skidding as it navigates a curve of radius 15 m, at a constant speed of 43 km/h. (Note that such a normal force, also called centripetal or center-seeking force, is needed to keep the car on a circular path.)

89. Generalize your solution to Exercise 88 to obtain a formula for the centripetal force acting on an object of mass m that moves along a circular path of radius r at a constant speed of v.

90. Suppose an object is moving along the space curve $\mathbf{r}(t)$ when at time $t = t_0$ its velocity vector is $\mathbf{v}(t_0) = \langle 32, -10, 15 \rangle$ and its acceleration is $\mathbf{a}(t_0) = \langle 2, 1, -3 \rangle$. Is the object's speed increasing or decreasing at this instant?

91. When navigating a curve, a driver should minimize the normal component of acceleration, which, if too big, could cause the car to skid. According to common advice, a driver should slow down before entering a curve, and then gently accelerate once in the curve. Use your knowledge of the acceleration vector to explain why this is sound advice. (Mention changes in the tangential and normal components of the acceleration vector.)

92.* The radius of the "Singapore Flyer," which was the world's tallest Ferris wheel from 2008 to 2014, is 75 meters. Suppose the wheel is rotating at an angular speed of 0.02 radians per second, which is increasing at a rate of 1.33×10^{-4} radians per second. Find the tangential and normal components, as well as the magnitude of acceleration of the riders, in a capsule that is at the very top of the wheel at this instant.

93. Prove the second Frenet-Serret formula:

$$\frac{d\mathbf{N}}{ds} = -\kappa\,\mathbf{T} + \tau\,\mathbf{B}$$

(**Hint:** Noting that $d\mathbf{T}/ds = \kappa\,\mathbf{N}$ and $d\mathbf{B}/ds = -\tau\,\mathbf{N}$, use the fact that $\mathbf{N} = \mathbf{B} \times \mathbf{T}$.)

94. Prove that the tangential component of acceleration can be computed by the formula $a_T = (\mathbf{r}' \cdot \mathbf{r}'')/|\mathbf{r}'|$. (**Hint:** Making a sketch is helpful.)

95. Prove that $a_N = \dfrac{|\mathbf{r}' \times \mathbf{r}''|}{|\mathbf{r}'|}$. (See the hint given in Exercise 94.)

96. Prove the following formula.

$$\mathbf{r}''' = \left[s''' - \kappa^2 (s')^3 \right]\mathbf{T} + \left[3\kappa s' s'' + \kappa'(s')^2 \right]\mathbf{N} + \kappa\tau(s')^3\,\mathbf{B}$$

97. Prove the formula $\tau = -\dfrac{d\mathbf{B}}{ds} \cdot \mathbf{N} = \dfrac{(\mathbf{r}' \times \mathbf{r}'') \cdot \mathbf{r}'''}{|\mathbf{r}' \times \mathbf{r}''|^2}$.

(**Hint:** Use Exercise 96.)

98–106 *True or False?* Determine whether the given statement is true or false. In case of a false statement, explain or provide a counterexample.

98. $\mathbf{T} \times \mathbf{B} = 0$

99. $(\mathbf{T} \times \mathbf{N}) \cdot \mathbf{B} = 0$

100. $\left| \dfrac{d(\mathbf{T} \cdot \mathbf{T})}{ds} \right| = \kappa^2$

101. The radius of the osculating circle for $y = \sin x$ at the origin is 1.

102. The acceleration vector of a particle moving on a curve $\mathbf{r}(t)$ is always in the osculating plane.

103. If we double the speed of a car in a curve, the force required to keep it from skidding is also doubled.

104. If a car is moving in a curve, then its acceleration is perpendicular to the direction of motion.

105. If an object is moving along a smooth curve that is not a straight line, then a normal force is acting on the object.

106. If the acceleration of a moving object is nonzero, and not a multiple of $\mathbf{N}$, then its speed is changing.

12.3 Technology Exercises

107–109 Use a computer algebra system to plot the curves in Exercises 79, 83, and 87. Then find and graph their respective curvature and torsion functions. Interpret these graphs in terms of your three-dimensional plot.

107. $\mathbf{r}(t) = \langle 2t, t^2, t \rangle$ **108.** $\mathbf{r}(t) = \left\langle 2t, t^2, \dfrac{t^3}{3} \right\rangle$

109. $\mathbf{r}(t) = \langle e^t, e^t \cos t, e^t \sin t \rangle$

110. Write a program for a computer algebra system or programmable calculator that returns the parametric form of the osculating circle of a given space curve at a specified point. Use the program to check your answers for Exercises 69–72.

12.4 **Planetary Motion and Kepler's Laws**

TOPICS

1. Motion in polar coordinates
2. Kepler's Laws of Planetary Motion

In this last section of the chapter, we use two of Newton's principles of physics and our knowledge of space curves to derive Kepler's Laws of Planetary Motion, a triumph of human perseverance and reasoning.

TOPIC 1   **Motion in Polar Coordinates**

To set the stage for Kepler's Laws of Planetary Motion, we review the polar coordinates introduced in Chapter 9. And since we will be discussing the movement of planets over time, we assume the polar variables r and θ are both functions of t. Using vector notation, then, we describe a generic curve C in the xy-plane by the vector function

$$\mathbf{r}(t) = \left\langle r(t)\cos(\theta(t)), r(t)\sin(\theta(t)) \right\rangle = r(t)\mathbf{u}(t),$$

where we define $\mathbf{u}(t) = \left\langle \cos(\theta(t)), \sin(\theta(t)) \right\rangle$ and, because we will soon find it useful, we define $\mathbf{u}_\perp(t) = \left\langle -\sin(\theta(t)), \cos(\theta(t)) \right\rangle$, a unit vector perpendicular to $\mathbf{u}(t)$. Given the position $\mathbf{r}(t)$ of an object, its velocity $\mathbf{v}(t)$ is then

$$\begin{aligned}
\mathbf{v}(t) = \mathbf{r}'(t) &= r'(t)\mathbf{u}(t) + r(t)\mathbf{u}'(t) \\
&= r'(t)\mathbf{u}(t) + r(t)\theta'(t)\left\langle -\sin(\theta(t)), \cos(\theta(t)) \right\rangle \\
&= r'(t)\mathbf{u}(t) + r(t)\theta'(t)\mathbf{u}_\perp(t)
\end{aligned}$$

and its acceleration $\mathbf{a}(t)$ is

$$\begin{aligned}
\mathbf{a}(t) = \mathbf{r}''(t) &= r''(t)\mathbf{u}(t) + r'(t)\mathbf{u}'(t) + r'(t)\mathbf{u}'(t) + r(t)\mathbf{u}''(t) \\
&= r''(t)\mathbf{u}(t) + 2r'(t)\theta'(t)\left\langle -\sin(\theta(t)), \cos(\theta(t)) \right\rangle \\
&\quad + r(t)\left[\theta''(t)\left\langle -\sin(\theta(t)), \cos(\theta(t)) \right\rangle + (\theta'(t))^2 \left\langle -\cos(\theta(t)), -\sin(\theta(t)) \right\rangle \right] \\
&= \left[r''(t) - r(t)(\theta'(t))^2 \right]\mathbf{u}(t) + \left[r(t)\theta''(t) + 2r'(t)\theta'(t) \right]\mathbf{u}_\perp(t).
\end{aligned}$$

Omitting the argument t, we summarize the above as follows:

$$\mathbf{r} = r\mathbf{u}$$
$$\mathbf{v} = r'\mathbf{u} + r\theta'\mathbf{u}_\perp$$
$$\mathbf{a} = \left(r'' - r(\theta')^2 \right)\mathbf{u} + \left(r\theta'' + 2r'\theta' \right)\mathbf{u}_\perp$$

Example 1 ✎

In terms of the vectors

$$\mathbf{u} = \left\langle \cos(\theta(t)), \sin(\theta(t)) \right\rangle \quad \text{and} \quad \mathbf{u}_\perp = \left\langle -\sin(\theta(t)), \cos(\theta(t)) \right\rangle,$$

describe the velocity and acceleration vectors of a particle whose position function is $\mathbf{r}(t) = \left\langle t\cos(t^2), t\sin(t^2) \right\rangle$.

Solution

Using polar notation, the function $\mathbf{r}(t) = \langle r(t)\cos(\theta(t)), r(t)\sin(\theta(t)) \rangle$ where $r(t) = t$ and $\theta(t) = t^2$. So

$$
\begin{aligned}
r(t) &= t & \theta(t) &= t^2 \\
r'(t) &= 1 & \theta'(t) &= 2t \\
r''(t) &= 0 & \theta''(t) &= 2
\end{aligned}
$$

and

$$\mathbf{v} = r'\mathbf{u} + r\theta'\mathbf{u}_\perp = \mathbf{u} + 2t^2\,\mathbf{u}_\perp$$

$$\mathbf{a} = \left(r'' - r(\theta')^2\right)\mathbf{u} + \left(r\theta'' + 2r'\theta'\right)\mathbf{u}_\perp = -4t^3\,\mathbf{u} + 6t\,\mathbf{u}_\perp.$$

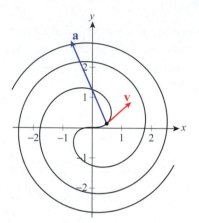

Figure 1 $\mathbf{v}$ and $\mathbf{a}$ at $t = \frac{1}{2}$

Figure 1 is a graph of the particle's path over the time interval $-3 \le t \le 3$ with its velocity and acceleration vectors $\mathbf{v}$ and $\mathbf{a}$ shown in red and blue, respectively, at $t = \frac{1}{2}$.

$$
\begin{aligned}
\mathbf{v}\!\left(\frac{1}{2}\right) &= \mathbf{u}\!\left(\theta\!\left(\frac{1}{2}\right)\right) + 2\!\left(\frac{1}{2}\right)^2\mathbf{u}_\perp\!\left(\theta\!\left(\frac{1}{2}\right)\right) \\
&= \left\langle \cos\!\left(\frac{1}{4}\right) - \frac{1}{2}\sin\!\left(\frac{1}{4}\right), \sin\!\left(\frac{1}{4}\right) + \frac{1}{2}\cos\!\left(\frac{1}{4}\right) \right\rangle
\end{aligned}
$$

$$
\begin{aligned}
\mathbf{a}\!\left(\frac{1}{2}\right) &= -4\!\left(\frac{1}{2}\right)^3\mathbf{u}\!\left(\theta\!\left(\frac{1}{2}\right)\right) + 6\!\left(\frac{1}{2}\right)\mathbf{u}_\perp\!\left(\theta\!\left(\frac{1}{2}\right)\right) \\
&= \left\langle -\frac{1}{2}\cos\!\left(\frac{1}{4}\right) - 3\sin\!\left(\frac{1}{4}\right), -\frac{1}{2}\sin\!\left(\frac{1}{4}\right) + 3\cos\!\left(\frac{1}{4}\right) \right\rangle
\end{aligned}
$$

TOPIC 2   Kepler's Laws of Planetary Motion

The description of planar curves in terms of r and θ will soon prove useful, but to begin with we don't assume that each planet's orbit is restricted to a plane—that fact is actually one of the important consequences of Kepler's Laws, which are as follows.

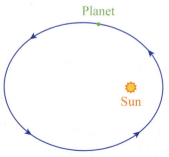

Planet

Sun

Figure 2

Theorem ⚭

Kepler's Laws of Planetary Motion

1. Each planet revolves around the sun in an elliptical orbit, with the sun at one focus of the ellipse.

2. A line segment between the sun and a given planet sweeps out equal areas of the ellipse over equal periods of time.

3. The ratio of the square of the period of one revolution of a planet and the cube of the length of the semimajor axis of its orbit is a constant independent of the planet.

The history of the speculations and observations prior to the development of Kepler's Laws is a rich and fascinating story that spans centuries and civilizations, and the reader is encouraged to learn more about the various solar system models that preceded Kepler. Johannes Kepler (1571–1630) was a German astronomer and mathematician who formulated his three laws on the basis of years' worth of astronomical data collected by the Danish nobleman Tycho Brahe; Kepler published his first two laws in 1609 and the third in 1619. Almost 70 years later, in 1687, Isaac Newton showed that Kepler's Laws can be derived from two of Newton's own principles, the Second Law of Motion and the Law of Universal Gravitation. This derivation constitutes the remainder of this chapter. While it is longer than the typical development, it is a beautiful illustration of the use of what we have recently learned.

To start, we let $\mathbf{r}(t)$ denote the position vector of a particular planet as it moves within our solar system, and we assume the sun is fixed in space at the origin. We let m denote the mass of the planet and M the mass of the sun. We let $r = |\mathbf{r}|$ and define $\mathbf{u}$ as $\mathbf{u} = \mathbf{r}/r$, so that $\mathbf{r} = r\mathbf{u}$. Newton's Second Law of Motion tells us that the sum of the forces $\mathbf{F}$ on the planet is the product of its mass m and its acceleration $\mathbf{r}''$:

$$\mathbf{F} = m\mathbf{r}''$$

Since the sun is by far the most massive object in the solar system, we make the simplifying assumption that no other objects affect the motion of the particular planet under observation; while not entirely accurate, this assumption is a safe first approximation, and refinements can be made later. Given this, Newton's Law of Universal Gravitation says that the gravitational force $\mathbf{F}$ on the planet is

$$\mathbf{F} = -\frac{GMm}{r^3}\mathbf{r},$$

where the negative sign indicates that the force on the planet is applied directly toward the origin (i.e., the sun). Equating the two expressions for $\mathbf{F}$, and canceling the common factor of m, results in the following equation:

$$\mathbf{r}'' = -\frac{GM}{r^3}\mathbf{r}$$

This means that the acceleration $\mathbf{r}''$ of the planet is a scalar multiple of $\mathbf{r}$, so $\mathbf{r} \times \mathbf{r}'' = \mathbf{0}$. We use this to note that

$$\frac{d}{dt}(\mathbf{r} \times \mathbf{r}') = (\mathbf{r}' \times \mathbf{r}') + (\mathbf{r} \times \mathbf{r}'') = \mathbf{0} + \mathbf{0} = \mathbf{0},$$

so $\mathbf{r} \times \mathbf{r}' = \mathbf{C}_1$ for some constant vector $\mathbf{C}_1$. One consequence of this is that $\mathbf{r}(t)$ is perpendicular to a fixed vector for every t, so we now know that the curve described by $\mathbf{r}$ lies in a plane. We will start to define our coordinate system by specifying that $\mathbf{C}_1$ points in the direction of the positive z-axis; that is, $\mathbf{C}_1 = C_1\mathbf{k}$ for some scalar constant $C_1 > 0$.

Next note that

$$\mathbf{C}_1 = \mathbf{r} \times \mathbf{r}' = (r\mathbf{u}) \times (r'\mathbf{u} + r\mathbf{u}')$$
$$= (rr')(\mathbf{u} \times \mathbf{u}) + r^2(\mathbf{u} \times \mathbf{u}') = r^2(\mathbf{u} \times \mathbf{u}'),$$

so

$$\mathbf{r}'' \times \mathbf{C}_1 = -\frac{GM}{r^2}\mathbf{u} \times r^2(\mathbf{u} \times \mathbf{u}') = -GM[\mathbf{u} \times (\mathbf{u} \times \mathbf{u}')]$$

$$= -GM\left[\left(\underbrace{\mathbf{u} \cdot \mathbf{u}'}_{0}\right)\mathbf{u} - \left(\underbrace{\mathbf{u} \cdot \mathbf{u}}_{1}\right)\mathbf{u}'\right]$$

$$= -GM[0\mathbf{u} - \mathbf{u}'] = GM\mathbf{u}',$$

where we have used one of the properties of the cross product from Section 11.4 to rewrite $\mathbf{u} \times (\mathbf{u} \times \mathbf{u}')$ and the fact that $\mathbf{u}$ has a constant length of 1 to evaluate $\mathbf{u} \cdot \mathbf{u}'$ and $\mathbf{u} \cdot \mathbf{u}$. Hence,

$$\frac{d}{dt}(\mathbf{r}' \times \mathbf{C}_1) = (\mathbf{r}'' \times \mathbf{C}_1) + \left(\mathbf{r}' \times \underbrace{\mathbf{C}_1'}_{0}\right) = \mathbf{r}'' \times \mathbf{C}_1 = GM\mathbf{u}'.$$

Integrating both sides of this last equation gives us the result

$$\mathbf{r}' \times \mathbf{C}_1 = GM\mathbf{u} + \mathbf{C}_2$$

for some second constant vector $\mathbf{C}_2$. And since $\mathbf{r}' \times \mathbf{C}_1$ is orthogonal to $\mathbf{C}_1$ (by definition of the cross product) and $\mathbf{u}$ is orthogonal to $\mathbf{C}_1$ (by the definition of $\mathbf{C}_1$), we know that $\mathbf{C}_2$ is orthogonal to $\mathbf{C}_1$; that is, $\mathbf{C}_2$ lies in the same plane as $\mathbf{r}$ and $\mathbf{r}'$, which is the xy-plane by our orientation of the z-axis. We now orient the x-axis so that $\mathbf{C}_2 = C_2\mathbf{i}$ for some scalar constant $C_2 > 0$, and we let θ denote the angle between $\mathbf{i}$ and $\mathbf{r}$ as shown in Figure 3. This allows us to express $\mathbf{r}$ as
$$\mathbf{r}(t) = \langle r(t)\cos(\theta(t)), r(t)\sin(\theta(t))\rangle.$$

We are almost ready to conclude that the orbit described by $\mathbf{r}$ is an ellipse and thus prove Kepler's First Law. To complete the argument, note that

$$\mathbf{r} \cdot (\mathbf{r}' \times \mathbf{C}_1) = \mathbf{r} \cdot (GM\mathbf{u} + C_2\mathbf{i}) = GM(\mathbf{r} \cdot \mathbf{u}) + C_2(\mathbf{r} \cdot \mathbf{i})$$
$$= GMr + C_2 r\cos\theta = r(GM + C_2\cos\theta).$$

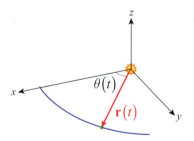

Figure 3

At the same time,

$$\mathbf{r} \cdot (\mathbf{r}' \times \mathbf{C}_1) = (\mathbf{r} \times \mathbf{r}') \cdot \mathbf{C}_1 = \mathbf{C}_1 \cdot \mathbf{C}_1 = C_1^2.$$

So $r(GM + C_2\cos\theta) = C_1^2$, or

$$r = \frac{C_1^2}{GM + C_2\cos\theta} = \frac{\dfrac{C_1^2}{GM}}{1 + \dfrac{C_2}{GM}\cos\theta}.$$

If we define constants e and d as

$$e = \frac{C_2}{GM} \quad \text{and} \quad d = \frac{C_1^2}{C_2},$$

then

$$r = \frac{ed}{1 + e\cos\theta},$$

which we recognize from Section 9.6 as the polar form of an ellipse with eccentricity e and distance d between the focus and directrix.

Kepler's Second Law follows quickly from the observation $\mathbf{C}_1 = r^2(\mathbf{u} \times \mathbf{u}')$ and the fact that we know

$$\mathbf{u} = \langle \cos(\theta(t)), \sin(\theta(t)) \rangle$$

and hence

$$\mathbf{u}' = \theta'(t)\langle -\sin(\theta(t)), \cos(\theta(t)) \rangle = \theta'\mathbf{u}_\perp,$$

where we again define $\mathbf{u}_\perp = \langle -\sin(\theta(t)), \cos(\theta(t)) \rangle$. Since both $\mathbf{u}$ and $\mathbf{u}_\perp$ are unit vectors,

$$C_1 = |\mathbf{C}_1| = |r^2\theta'(\mathbf{u} \times \mathbf{u}_\perp)| = r^2\theta'.$$

The significance of this is that $r^2\theta'$ is a constant. So if we let $A(t)$ denote the area of the ellipse swept out by a line segment from the sun to the planet over time t, starting with the planet at position $\mathbf{r}(0)$, then by the reasoning of Section 9.4,

$$\Delta A = \frac{1}{2}r^2\Delta\theta$$

and

$$\frac{dA}{dt} = \lim_{\Delta t \to 0}\frac{\Delta A}{\Delta t} = \lim_{\Delta t \to 0}\left(\frac{1}{2}r^2\frac{\Delta\theta}{\Delta t}\right) = \frac{1}{2}r^2\theta' = \frac{C_1}{2}.$$

In other words, the rate of change of the area swept out by the line segment is a constant, which is another way of stating Kepler's Second Law.

To prove Kepler's Third Law, first note that the planet's *perihelion* position, or position of closest approach to the sun (Exercise 38 of Section 9.6), corresponds to

$$|\mathbf{r}(0)| = \frac{ed}{1 + e\cos(\theta(0))} = \frac{ed}{1 + e},$$

on the *x*-axis, since we have oriented the *x*- and *y*-axes so that $\theta(0) = 0$. If we define the ellipse's major axis length as $2a$, its minor axis length as $2b$, and the distance between its center and either focus as c (see Sections 9.5 and 9.6) then the eccentricity e of the ellipse is again

$$e = \frac{c}{a} = \frac{\sqrt{a^2 - b^2}}{a}$$

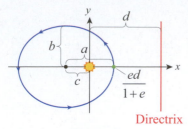

Figure 4

and d is the distance between one focus of the ellipse and its directrix, all as depicted in Figure 4. From the proof of Kepler's Second Law, we know that

$$A(t) = \frac{C_1}{2}t,$$

so if we let T denote the *period* of the planet (the time required for one full orbit),

$$\pi ab = A(T) = \frac{C_1 T}{2} \quad \text{and hence} \quad T = \frac{2\pi ab}{C_1}.$$

From Figure 4, note that $a = c + \dfrac{ed}{1+e}$, so $ed = (a-c)(1+e)$. Using the relationships between a, b, c, and e,

$$ed = (a-c)(1+e) = a\underbrace{\left(1 - \frac{c}{a}\right)}_{1-e}(1+e) = a(1-e^2) = a\left(\frac{b^2}{a^2}\right) = \frac{b^2}{a}.$$

But from the derivation of Kepler's First Law we also have

$$ed = \frac{C_2}{GM}\frac{C_1^2}{C_2} = \frac{C_1^2}{GM},$$

so

$$T^2 = \frac{4\pi^2 a^2 b^2}{C_1^2} = \frac{4\pi^2 a^2 b^2}{GMed} = \frac{4\pi^2 a^2 b^2}{GM\dfrac{b^2}{a}} = \frac{4\pi^2 a^3}{GM}$$

and hence T^2/a^3 is a constant independent of the planet.

12.4 Exercises

1–6 Describe in terms of $\mathbf{u}$ and $\mathbf{u}_\perp$ the velocity and acceleration vectors of a particle with the given position function.

1. $\mathbf{r}(t) = \langle 2t\cos 2t, 2t\sin 2t \rangle$

2. $\mathbf{r}(t) = \langle (t^2+1)\cos 5t, (t^2+1)\sin 5t \rangle$

3. $\mathbf{r}(t) = \langle \sqrt{t}\cos(4t+1), \sqrt{t}\sin(4t+1) \rangle$

4. $\mathbf{r}(t) = \left\langle e^t\cos\dfrac{t}{2}, e^t\sin\dfrac{t}{2} \right\rangle$

5. $\mathbf{r}(t) = \langle 2\sin 2t\cos(t^3), 2\sin 2t\sin(t^3) \rangle$

6. $\mathbf{r}(t) = \langle a(\cos t-1)\cos bt, a(\cos t-1)\sin bt \rangle$

7. Prove that if a satellite or planet is moving in a circular orbit, then its speed is constant. (**Hint:** Use the fact that $\mathbf{C}_1$ is constant, an observation made in the proof of Kepler's First Law.)

8. Prove that in order for a moon or satellite to stay in a circular orbit of radius R around a planet of mass M, the required orbital speed is $v = \sqrt{GM/R}$. (**Hint:** Recall that v is constant by Exercise 7, thus the magnitude of acceleration is $a = v^2/R$. Use this and Newton's Second Law to finish the proof. Alternatively, use the last equation in our proof of Kepler's Third Law, noting that $a = R$.)

9. A satellite is in a circular orbit 219.2 kilometers above the Earth's surface. Use Exercise 8 to find its orbital speed. Express your answer in kilometers per hour. (Approximate the radius of the Earth by 6371 kilometers and its mass by 5.9736×10^{24} kilograms. Recall that $G \approx 6.6738 \times 10^{-11}$ Nm^2/kg^2.)

10. Given that the perihelion of the Earth's orbit is approximately 147,098,290 km, with its *aphelion* (the distance farthest from the sun) being 152,098,232 km, and using its period of T = 365.256 days, estimate the mass of the sun. (**Hint:** Find the length of the orbit's semimajor axis first, then use the last equation in our derivation of Kepler's Third Law. For the value of G, see Exercise 9.)

11. The period of one revolution of the moon around the Earth is approximately 27.3217 days, its *perigee* (distance from the Earth upon closest approach) is approximately 361,400 km, while its *apogee* (its greatest distance from the Earth) is about 405,000 km. Use these data to estimate the mass of the Earth.

12. Use the data given in Exercise 11, along with Kepler's Third Law, to estimate the necessary height above the Earth's surface for a *geostationary satellite*. (A geostationary satellite is one that is in a near-circular orbit over the equator, orbiting in the direction of the Earth's rotation with a period of 24 hours, thus appearing stationary for an observer on the ground. **Hint:** While you may compute the height directly, an easier approach suggested by the problem is to compare the satellite's orbit with that of the moon and use Kepler's Third Law.)

13. Use Kepler's Third Law, along with Earth's orbital data given in Exercise 10, and Mars' period of 686.971 (Earth) days to estimate the semimajor axis of Mars' orbit.

14. The length of the semimajor axis of Neptune's orbit is 30.0476 *astronomical units* (1 astronomical unit, abbreviated AU, is equal to 149,597,870.700 km, which is approximately the mean distance between the Earth and the sun). Estimate the period of Neptune in Earth years. (For orbital data on the Earth, see Exercise 10.)

15. Prove that the ratio of the perihelion and aphelion of a planet is equal to the inverse ratio of its speeds at the perihelion and aphelion positions. (**Hint:** Use Kepler's Second Law, namely, the equation that says $dA/dt = \frac{1}{2} r^2 \theta' = C_1/2$ is constant.)

16. Show that the minimum distance of a moon or satellite from the planet it is orbiting (the perigee) is $r_p = a(1-e)$, while the maximum distance (the apogee) is $r_a = a(1+e)$, where e is the eccentricity of the orbit. (**Hint:** In order to express ed in terms of a and b in the polar equation of the planet, see the proof of Kepler's Third Law in the text.)

17. Use Exercise 16 along with the orbital data of the Earth given in Exercise 10 to find the eccentricity of the Earth's orbit, then write a polar equation of the orbit (with the sun at the origin).

18. Find the aphelion of the orbit of Halley's comet and write a polar equation for the orbit given that $e \approx 0.967$ and $a \approx 17.94$ astronomical units (AU). (See Exercise 16. Use the unit AU in your answer.)

19. Find how much time passes between two consecutive visits of Halley's comet to the solar system. (See Exercises 14 and 18.)

20. Define $r_0 = |\mathbf{r}(0)|$ and $v_0 = |\mathbf{v}(0)|$, and show that with this notation,

$$A(t) = \frac{r_0 v_0}{2} t.$$

(**Hint:** Note that $v_0 = r_0 \theta'(0)$ since $r'(0) = 0$.)

21. Modify your proof of Exercise 20 to show that the equality $v = \dfrac{2\pi ab}{rT}$ holds at perigee or apogee for an orbiting planet or satellite.

22. If v_p and v_a denote a planet's speeds at perigee and apogee, respectively, prove that

$$v_p(1-e) = v_a(1+e).$$

(**Hint:** As in Exercise 20, note that $v = r\theta'$ at perigee and apogee, and that $r^2\theta' = C_1$ is a constant. Use this latter equation for both perigee and apogee positions, along with Exercise 16.)

23. Use Exercise 21 and the Earth's orbital data given in Exercise 10 to find the speeds of the Earth at perihelion and aphelion, respectively. Express your answer in kilometers per second, then convert it to miles per hour.

24. Repeat Exercise 23 for Mars, if its perihelion and aphelion are 2.0662×10^8 kilometers, and 2.4923×10^8 kilometers, respectively, with its period being 1.88079 years.

25. Use Exercises 11 and 21 to find the moon's speed at perigee and apogee, respectively. Express your answer in kilometers per second, then convert it to miles per hour.

26. Repeat Exercise 25 for Jupiter's moon Europa, if given that the eccentricity of its orbit is 0.0101, the length of its semimajor axis is 671,100 kilometers, and its period is 3.5512 days.

27. Repeat Exercise 26 for Halley's comet. Express your answer in the following units: astronomical units per day, kilometers per second, and miles per hour. (For orbital data, see Exercises 18 and 19.)

28.* Suppose a moon or planet is orbiting another planet (or star) of mass M. As before, let r_p denote the perigee or perihelion, while v_p is the speed at perigee or perihelion, as applicable. Prove that the orbit can be classified as an ellipse, a parabola, or a hyperbola according to the values of r_p and v_p as follows:

if $r_p v_p^2 = GM$, the orbit is a circle,

if $GM < r_p v_p^2 < 2GM$, the orbit is an ellipse,

if $r_p v_p^2 = 2GM$, the orbit is a parabola,

if $r_p v_p^2 > 2GM$, the orbit is a hyperbola.

(The last two orbit types are called *open*; such orbits are exhibited by comets entering the solar system once and then leaving it forever. **Hint:** Referring to the proof of Kepler's First Law in the text, since $e = \dfrac{C_2}{GM}$, if you show that $\dfrac{C_2}{GM} = \dfrac{r_p v_p^2}{GM} - 1$, the conclusion will follow. To that end, it will suffice to show that $C_2 = r_p v_p^2 - GM$. From the proof of Kepler's First

Law convince yourself that $r_p(GM + C_2) = C_1^2$ and note from Exercise 22 that $C_1 = r_p^2 \theta'$, which will finish the proof. Finally, we note that in the case of $C_1 = 0$, the moon or planet falls along a straight line into the star or planet it is orbiting.)

29. A novice astronaut is in an elliptical orbit around the Earth. In an attempt to slow down to better behold the beautiful view, she plans to apply reverse thrust to decrease (tangential) speed. However, Mission Control advises her to check her calculations, for this may actually cause the craft to go around the Earth faster. Who is correct and why? Use Kepler's Third Law to provide an explanation.

30.* Suppose the astronaut of Exercise 29 erroneously applied reverse thrust and ended up in a circular orbit of radius $R = 6600$ km, instead of a planned new orbit with perigee of $r_p = 6600$ km and apogee $r_a = 10,000$ km. At a time determined by Mission Control, she is instructed to apply forward thrust to enter into the desired elliptical orbit. If her thrusters lend the spacecraft an acceleration of 0.0205 km/s^2, how long does she have to burn them in order to accomplish this task? (**Hint:** By Exercise 8, the speed in the circular orbit is $v = \sqrt{GM/R}$. Next, note that the speed at perigee in the new orbit will be $v_p = C_1/r_p$. By determining the eccentricity of the planned orbit and recalling from the proof of Kepler's Third Law that $ed = C_1^2/GM = a(1 - e^2)$, you can determine C_1, and then v_p. For data on the Earth, see Exercises 9 and 10.)

Chapter 12
Review Exercises

1–8 The following vector functions are defined on the interval $(0,6\pi)$. Match each one with its graph (labeled A–H).

1. $\mathbf{r}(t)=\langle \cos t, 2\sin t, \sqrt{t}-2 \rangle$

2. $\mathbf{r}(t)=\left\langle \dfrac{t}{2}\sin 3t, \dfrac{t}{2}\cos 3t, \dfrac{t}{2} \right\rangle$

3. $\mathbf{r}(t)=\langle 1.5\sin t, \sqrt{t}, 1.5\cos t \rangle$

4. $\mathbf{r}(t)=\left\langle \dfrac{t}{3}, 2\sin t, \dfrac{t}{3} \right\rangle$

5. $\mathbf{r}(t)=\langle 2\cos t, \sqrt{t}, 2\sin^2 t \rangle$

6. $\mathbf{r}(t)=\langle 0.1t\cos t, 0.2t\sin t, e^{0.1t} \rangle$

7. $\mathbf{r}(t)=\langle 2.5\sin 12t, 3.5\cos t, 3.5\sin t \rangle$

8. $\mathbf{r}(t)=\left\langle \dfrac{7\cos 5t}{2t}, \dfrac{7\sin 5t}{2t}, \dfrac{t}{5} \right\rangle$

A. 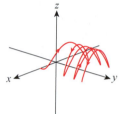   **B.** 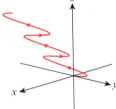   **C.** 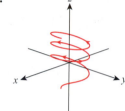   **D.**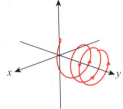

E. 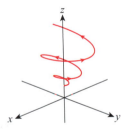   **F.** 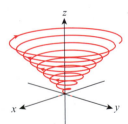   **G.** 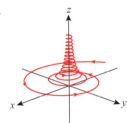   **H.**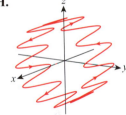

9–10 Describe the intersection of the surfaces as a vector function. (Use the suggested parameter.)

9. The elliptic cylinder $5x^2+2y^2=10$ and the hyperbolic paraboloid $5x^2-2y^2=z$ $(x=\sqrt{2}\cos t)$

10. The elliptic paraboloid $x^2+2y^2=z$ and the parabolic cylinder $y=x^2$ $(x=t)$

11–12 Determine whether the indicated limit exists. If so, find it.

11. $\displaystyle\lim_{t\to 1}\left\langle \ln t, \sec(t-1), \dfrac{t-1}{t^2-1} \right\rangle$

12. $\displaystyle\lim_{t\to 0}\left(\dfrac{t}{t^2+1}\mathbf{i}+\arctan t\,\mathbf{j}+\dfrac{t^2}{|t|}\mathbf{k} \right)$

13–14 Find any discontinuities of the given vector function.

13. $\mathbf{r}(t)=\left\langle e^{t+1}, \dfrac{t+1}{t^2+1}, \dfrac{t+1}{t^2-1} \right\rangle$

14. $\mathbf{r}(t)=|t-1|\mathbf{i}+\tan t\,\mathbf{j}-\arctan t\,\mathbf{k}$

15–16 Find $\mathbf{r}'(t)$.

15. $\mathbf{r}(t)=\langle \sec t, \arcsin\sqrt{t}, \tan^2 t \rangle$

16. $\mathbf{r}(t)=\cos 2t\,\mathbf{i}-\dfrac{t+2}{\sqrt{t-1}}\mathbf{j}+\ln(t^3-3)\mathbf{k}$

17–20 Find the vector form of an equation for the line tangent to the curve at the specified value of t.

17. $\mathbf{r}(t) = \langle 2\sin 4t, \cos t, 1 - t^2 \rangle$; $t = 0$

18. $\mathbf{r}(t) = \langle \tan^{-1} 2t, t^{3/2}, 3t - 2 \rangle$; $t = \dfrac{1}{2}$

19. $\mathbf{r}(t) = t\,\mathbf{i} - \ln(2t^2 + 1)\,\mathbf{j} - \sqrt{3t}\,\mathbf{k}$; $t = 3$

20. $\mathbf{r}(t) = e^{t^2 + 2}\,\mathbf{i} + e^{\sqrt{t+1}}\,\mathbf{j} + e^{\cos t}\,\mathbf{k}$; $t = 0$

21–24 Perform the integration—definite or indefinite, as indicated.

21. $\displaystyle \int \left\langle e^{t+1}, \frac{t}{\sqrt{t^2 + 2}}, \sin 2t \right\rangle dt$

22. $\displaystyle \int \left(\sqrt{t}\,\mathbf{i} - \frac{1}{\sqrt{1-t^2}}\,\mathbf{j} - t\sqrt{1-t^2}\,\mathbf{k} \right) dt$

23. $\displaystyle \int_3^6 \left\langle \sqrt{t-2}, t, 3 \right\rangle dt$

24. $\displaystyle \int_{1/4}^1 \left(\ln \sqrt{t}\,\mathbf{i} + e^{4t}\,\mathbf{j} - \frac{1}{t^{3/2}}\,\mathbf{k} \right) dt$

25–26 Find the arc length of the curve over the given interval.

25. $\mathbf{r}(t) = \langle -3t^2, 6t, t^3 \rangle$; $[0, 2]$

26. $\mathbf{r}(t) = \left\langle 2\ln t, \dfrac{t^2}{4}, \sqrt{2}\,t \right\rangle$; $[1, e^2]$

27–28 Reparametrize the given curve with respect to arc length.

27. $\mathbf{r}(t) = \langle 2 - t, 1 + 4t, 3 + 2t \rangle$

28. $\mathbf{r}(t) = \langle 5\sin t, 12t, 5\cos t \rangle$

29–32 Find the unit tangent vector for the curve over the given interval.

29. $\mathbf{r}(t) = \left\langle \dfrac{t}{2}, \ln(\cos t), \dfrac{\sqrt{3}}{2}t \right\rangle$; $0 \le t < \dfrac{\pi}{2}$

30. $\mathbf{r}(t) = \langle t^2, \cos 2t, \sin 2t \rangle$; $t \ge 0$

31. $\mathbf{r}(t) = \langle \ln t, t^2, t \rangle$; $t > 0$

32. $\mathbf{r}(t) = \langle \arctan t, \ln(t^2 + 1), t \rangle$; $t > 0$

33–36 Find the unit normal and binormal vectors for the curve.

33. $\mathbf{r}(t) = \langle 2t, 2\cos t, 2\sin t \rangle$

34. $\mathbf{r}(t) = \langle \sin 3t, \cos 3t, 3t \rangle$

35. $\mathbf{r}(t) = \langle e^t, e^{-t}, \sqrt{2}\,t \rangle$

36. $\mathbf{r}(t) = \langle 3t, -t, t^2 \rangle$

37–40 Using your work in Exercises 33–36, determine equations for the osculating, normal, and rectifying planes associated with the curve at the indicated point.

37. $\mathbf{r}(t) = \langle 2t, 2\cos t, 2\sin t \rangle$; $t = 0$

38. $\mathbf{r}(t) = \langle \sin 3t, \cos 3t, 3t \rangle$; $t = 0$

39. $\mathbf{r}(t) = \langle e^t, e^{-t}, \sqrt{2}\,t \rangle$; $t = 1$

40. $\mathbf{r}(t) = \langle 3t, -t, t^2 \rangle$; $t = 2$

41–44 Use Exercises 37–40 to find a parametric description of the surface.

41. The ribbon of width $\frac{1}{2}$ centered on the curve $\mathbf{r}(t) = \langle 2t, 2\cos t, 2\sin t \rangle$

42. The ribbon of width 1 centered on the curve $\mathbf{r}(t) = \langle \sin 3t, \cos 3t, 3t \rangle$

43. The circular tube of radius 2 centered on the curve $\mathbf{r}(t) = \langle e^t, e^{-t}, \sqrt{2}\,t \rangle$

44. The circular tube of radius 1 centered on the curve $\mathbf{r}(t) = \langle 3t, -t, t^2 \rangle$

45–48 Find the curvature and torsion functions for the given curve. (Suppose $t > 0$.)

45. $\mathbf{r}(t) = \left\langle \cos t, \sin t, \dfrac{t}{\sqrt{2}} \right\rangle$

46. $\mathbf{r}(t) = \left\langle \dfrac{1}{t^2}, \dfrac{1}{t}, 1 \right\rangle$

47. $\mathbf{r}(t) = \left\langle t, \dfrac{1}{t}, \ln t \right\rangle$

48. $\mathbf{r}(t) = \left\langle t^2, t, \dfrac{1}{t} \right\rangle$

49. Find a formula for the curvature of the ellipse
$\mathbf{r}(t) = \langle a\cos t, b\sin t, 0 \rangle$ $(a > b)$ at the endpoints
of its **a.** major axis and **b.** minor axis.

50. Use Exercise 50 of Section 12.3 to provide a
second solution for Exercise 49.

51–52 Use Exercise 50 of Section 12.3 to find the point(s) of
maximum curvature for the curve.

51. $f(x) = \dfrac{1}{x}$ **52.** $f(x) = x^4$

53–54 Find the osculating circle of the graph of the equation
at the indicated point. (See Exercises 60–68 in Section 12.3.)

53. $y = x^3$; $x = -1$ **54.** $y = \tan x$; $x = \dfrac{\pi}{4}$

55–58 Find the tangential and normal components of
acceleration for the given position function.

55. $\mathbf{r}(t) = \langle 4t, \cos 2t, \sin 2t \rangle$

56. $\mathbf{r}(t) = \left\langle \dfrac{1}{t}, 1, t^2 \right\rangle$; $t > 0$

57. $\mathbf{r}(t) = \langle t\cos t, t\sin t, 1 \rangle$; $0 \le t \le \dfrac{\pi}{4}$

58. $\mathbf{r}(t) = \langle 1, 1 - \sin t, t - \cos t \rangle$

59–60 Describe in terms of $\mathbf{u}$ and $\mathbf{u}_\perp$ the velocity and
acceleration vectors of a particle with the given position
function.

59. $\mathbf{r}(t) = \left\langle t\cos\dfrac{t}{2}, t\sin\dfrac{t}{2} \right\rangle$

60. $\mathbf{r}(t) = \langle t^2\cos 2t, t^2\sin 2t \rangle$

61. A projectile is launched from the ground with
an initial speed of 129.33 feet per second at an
angle of elevation of 60° from horizontal. Use
the vector functions $\mathbf{v}(t)$ and $\mathbf{r}(t)$ to find the
maximum altitude reached by the projectile as
well as its range. (Ignore air resistance.)

62. The acceleration function of a moving object in
$\mathbb{R}^3$ is $\mathbf{a}(t) = \langle t/2, 0, -t \rangle$. Find the velocity and
position functions of this object if it starts at
the point $\mathbf{r}(0) = \langle 3, -1, 5 \rangle$ with initial velocity
$\mathbf{v}(0) = \langle 0, 2, 1 \rangle$.

63. Suppose a shell is shot from an artillery cannon
with a muzzle velocity of 3000 ft/s and leaves
the cannon 4 ft above the ground, at a 30° angle
of elevation. What is the range of the cannon
under these conditions? (By "range," we mean
the horizontal distance between the base of the
launching pad and the point of impact. Assume
level terrain and ignore air resistance.)

64. A projectile is launched from the point $(0, 8)$ of a
coordinate system where unit length is measured
in feet. The launching speed of the projectile
is 192 feet per second and it is launched in the
positive direction at an angle of elevation of 30°.

   a. Find a vector function that models the
projectile's path.

   b. Find the maximum height attained by the
projectile, its range, and the speed of impact.

   (As customary, ignore air resistance.)

65. The banking of each of the curves on the
historical oval track in Monza, Italy, is 30°,
with a radius of 320 m. When used for racing,
those banked curves were navigated by daring
race drivers at speeds in excess of 150 mph
(240 km/h). Find the vertical component of
acceleration of a race car in the curve under the
above conditions (assume a constant speed of
exactly 150 mph).

66.* Using the result of Exercise 89 of Section 12.3,
find the speed that the banking is able to support
on the track of the previous exercise. In reality,
much higher speeds were achieved in the curves
of the track. Explain how that was possible.

67. A dive bomber, approaching its target at an
angle of depression of 60° and at a speed of
600 kilometers per hour, releases its payload at
an altitude of 500 meters. Use the units of meters
and seconds to find **a.** a vector function modeling
the path of the bomb, **b.** the horizontal distance
traveled by the bomb after its release, and **c.** its
speed of impact. (Convert your answer to part c.
to kilometers per hour. You can suppose that the
flight is taking place in the positive x-direction.
Ignore air resistance.)

68. If a particle's acceleration along a path $\mathbf{r}(t)$ at time $t = t_0$ is $\mathbf{a}(t_0) = \langle 1, -2, -1 \rangle$, and its velocity $\mathbf{v}(t_0) = \langle 9, 21, -12 \rangle$, decide whether the particle's speed is increasing or decreasing at $t = t_0$.

69. Suppose a particle is moving along a path $\mathbf{r}(t)$ which is on the sphere of radius R, centered at the origin. Show that the velocity vector of the particle is tangent to the sphere at all times.

70.* Show that the moving particle of Exercise 69 satisfies $\mathbf{r}(t) \cdot \mathbf{r}''(t) = -|\mathbf{v}(t)|^2$.

71.* Suppose that an object in the three-dimensional coordinate system is moving along a path $\mathbf{r}(t)$ so that it is acted upon by a single, so-called central force; that is, if $\mathbf{F}(t)$ denotes the acting resultant force on the object, then $\mathbf{F}(t) = k\,\mathbf{r}(t)$, for some constant $k > 0$. Show that in this case, the motion is planar, that is, $\mathbf{r}(t)$ is contained in a plane.

72.* Show that if the vector function $\mathbf{r}(t)$ is *bounded*, that is, if $|\mathbf{r}(t)| \le M$ for some positive constant M, then

$$\left| \int_a^b \mathbf{r}(t)\,dt \right| \le M(b-a).$$

73. Find the curvature at $t = b$ for the Cornu spiral of Section 12.2, Exercise 44. What can you say about the relationship between arc length and curvature for this particular curve?

74. Show that the graph of the polar equation $r^2 = \cos^2\theta - \sin^2\theta$ has the property that its curvature is directly proportional to r.

75. Find the orbital speed of a satellite that is in a circular orbit, 300 kilometers above the Earth's surface. Express your answer in kilometers per hour. (**Hint:** See Exercise 8 of Section 12.4.)

76. Ganymede, Jupiter's largest moon (discovered by Galileo Galilei in 1610) has an orbital period of approximately 7.15455296 earth days; its *periapsis* (distance from Jupiter upon closest approach) is approximately 1,069,008 kilometers, while its *apoapsis* (its greatest distance from Jupiter) is about 1,071,792 km. Use these data to estimate the mass of Jupiter. (**Hint:** See Exercise 10 of Section 12.4.)

77. Given that a day on Jupiter lasts a mere 9 hours and 55.5 minutes, use Exercise 76 along with Kepler's Third Law to estimate the necessary height above Jupiter's surface for a stationary satellite. (**Hint:** See Exercise 12 in Section 12.4. Approximate Jupiter's radius by 69,911 kilometers.)

78. The aphelion of Jupiter's orbit is 5.458104 AU (astronomical units), while its perihelion is 4.950429 AU. Use these along with the Earth's orbital data to estimate the period of Jupiter in earth years.

79. Use Exercise 78 to find the speeds of Jupiter when it is **a.** closest to and **b.** farthest from the sun. Express your answer in kilometers per second, then convert it to miles per hour. (**Hint:** See Exercise 23 in Section 12.4.)

80–89 *True or False?* Determine whether the given statement is true or false. In case of a false statement, explain or provide a counterexample.

80. $d\mathbf{T}/ds$ is perpendicular to $\mathbf{T}$.

81. If the graph of $\mathbf{r}(t) = \langle f(t), g(t), h(t) \rangle$ is a straight line, then f, g, and h are linear polynomials.

82. The torsion of a curve satisfies the formula
$$\tau = -\frac{1}{|\mathbf{v}(t)|} \frac{d\mathbf{B}}{dt} \cdot \mathbf{N}.$$

83. The tangential component of acceleration satisfies $a_T = \mathbf{T} \cdot \mathbf{a}$.

84. The normal component of acceleration satisfies $a_N = \mathbf{T} \times \mathbf{a}$.

85. If $f''(x) = 0$, then the curvature of the graph of $y = f(x)$ is constant.

86. If $\mathbf{r}(t)$ is a space curve with $\kappa = 0$, then $\mathbf{r}(t)$ can only "bend" in the direction of the unit binormal vector $\mathbf{B}$.

87. The magnitudes of the curvature and torsion of a space curve depend on the parametrization (i.e., "how fast the curve is being traced out,") unless the curve is parametrized with respect to arc length.

88. The maximum curvature of an ellipse occurs at the endpoints of its major axis.

89. $\dfrac{d}{dt}\big[\mathbf{r}(t)\times\mathbf{r}'(t)\big]=\mathbf{r}(t)\times\mathbf{r}''(t)$

Chapter 12
Technology Exercises

90. a. Use a computer algebra system to graph and explore the following curve:

$$\mathbf{r}(t)=\left\langle\frac{\sin 15t}{2},(3+\cos 15t)\sin t,(3+\cos 15t)\cos t\right\rangle$$

(Such curves are nicknamed "slinky curves." Can you see why?)

b. It is possible to create "slinky curves" where the spiral is wound around a helix. Find a formula for such a curve. (**Hint:** A good starting point is appropriately modifying the formula in part a. Answers will vary.)

91. Assuming $|x|,|y|\le 10$, z is nonnegative, and $0\le t\le 10\pi$, use a computer algebra system to find a formula for a vector function whose graph is as close as possible to the one displayed in the figure below. (Answers may vary.)

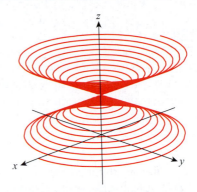

92. Write a program for a computer algebra system or programmable calculator that returns the equations for the osculating, normal, and rectifying planes associated with a given space curve at a specified point. Use your program to check the answers you have given to Exercises 37–40.

Project ✐

Chapter 12

In this project you will use vector functions to develop a simple model for 3-point basketball shots. To keep the model simple, we will be ignoring air resistance, friction, and other forces. Furthermore, by "scoring," we will mean that the ball falls straight into the basket on its way downward (i.e., we will ignore the possibility of the ball bouncing in off the backboard, or any energy losses as a result of spins, etc.). For further studies, or for more refined models, the interested student should consult resources such as John Fontanella's book, *The Physics of Basketball*.

1. A basketball player is attempting a 3-pointer from a horizontal distance of 24.7 feet. He is releasing the ball from 7 feet above ground level, aimed directly toward the basket at an angle of elevation of 45°, with an initial velocity of v_0. Supposing that the player stands at the origin and the basket is in the positive y-direction, use the three-dimensional coordinate system to find a vector function describing the position of the ball after release as a function of time. (Assume one unit on each axis corresponds to a distance of 1 foot.)

2. Use your answer to Question 1 to verify that the basketball's trajectory is a parabola.

3. Assuming a standard hoop height of 10 feet, find the initial speed for the ball that ensures that the player described in Question 1 scores.

4. Use your answer from Question 3 to find the necessary initial velocity vector for the basketball if the player is to score from the same spot (i.e., the origin) but this time shooting while running along the line $y = x$ at a speed of 10 mph in the positive direction.

5. Find a formula for and graph the required initial speed as a function of the angle of elevation over the interval $(0, \pi/2)$ if the player is to score (assuming the same spot and release height as in Question 1).

6. Generalizing your work on Question 5, find a formula for the initial speed of a successful shot if the player stands d feet from the hoop and shoots at an angle α upward from horizontal, with a release height of h feet.

Chapter 13

Partial Derivatives

13.1 Functions of Several Variables   983

1. Real-Valued Functions of Two Variables
2. Graphs and Contour Maps
3. Functions of Three or More Variables

13.2 Limits and Continuity of Multivariable Functions   994

1. Limits of Multivariable Functions, and Their Evaluation
2. Continuity of Multivariable Functions

13.3 Partial Derivatives   1002

1. Evaluation and Interpretation of Partial Derivatives
2. Higher-Order and Mixed Partial Derivatives
3. Differentiability and Continuity

13.4 The Chain Rule   1017

1. The Chain Rule and Its Proof
2. Implicit Differentiation via the Chain Rule

13.5 Directional Derivatives and Gradient Vectors   1026

1. Directional Derivatives
2. Gradient Vectors

13.6 Tangent Planes and Differentials   1036

1. Tangent Planes and Normal Lines
2. Differentials and Linear Approximations

13.7 Extreme Values of Functions of Two Variables   1045

1. Absolute and Relative Extrema
2. Finding Extrema

13.8 Lagrange Multipliers   1058

1. The Method of Lagrange Multipliers
2. Lagrange Multipliers with Two Constraints

Introduction

While the last chapter extended calculus to the study of functions from $\mathbb{R}$ to $\mathbb{R}^n$ (with particular emphasis on $n = 3$), this chapter does just the opposite and considers functions from $\mathbb{R}^n$ to $\mathbb{R}$. We often refer to these as *real-valued functions of n variables*, since they return a single real number for each set of n arguments.

Real-valued functions of two variables are especially useful for describing surfaces in three-dimensional space, and we will pay particular attention to working with such surfaces (just as we focused on space curves, or one-parameter objects in three-dimensional space, in Chapter 12). As in Chapter 12, the first step in developing the calculus of functions of multiple variables is to appropriately extend the concepts of *limit* and *continuity*, with the goal of then being able to differentiate such functions. Differentiation must take "directional" factors into account when the domain of a function has dimension greater than one; *partial differentiation* is the first step in making the necessary accommodation. Once we have done so, we use the partial derivatives of a multivariable function to define its *gradient*, its *directional derivatives*, and its *total differential*, all of which are useful in determining the geometric characteristics of a given function. We also define and

use higher-order partial derivatives, and we encounter a powerful result known as Clairaut's Theorem which relates the *mixed partial derivatives* of a function under very general conditions.

> Differentiation must take "directional" factors into account when the domain of a function has dimension greater than one; *partial differentiation* is the first step in making the necessary accommodation.

The partial derivatives are used to construct the *tangent planes* for a surface in $\mathbb{R}^3$, the analog of the tangent lines for a curve. Just as tangent lines are useful linear approximations of curves, tangent planes are useful linear approximations of surfaces. And just as we look for locations where tangent lines are horizontal in order to solve optimization problems

in the one-variable case, we look for horizontal tangent planes as we search for the absolute and/or relative extrema of functions of two variables. Section 13.7 contains an extension of the Second Derivative Test that we use for this purpose. Section 13.7 also delves further into locations on surfaces known as saddle points; the Enneper surface on the cover of this text is an example of a surface where *every* location is a saddle point.

The last section uses gradients to solve optimization problems via the method of *Lagrange multipliers*, a powerful tool for answering what might otherwise seem to be intractable questions. While the method can justifiably be described as sophisticated, it follows naturally from an understanding of the geometric meaning of gradients, an understanding you will acquire as you study the material in this chapter.

Alexis Clairaut
(1713–1765)

13.1 **Functions of Several Variables**

TOPICS

1. Real-valued functions of two variables

2. Graphs and contour maps

3. Functions of three or more variables

The focus of Chapter 12 was the study of vector functions, particularly functions mapping a real interval $I \subseteq \mathbb{R}$ into either $\mathbb{R}^2$ or $\mathbb{R}^3$. In this chapter, we extend the ideas of calculus to functions that map $\mathbb{R}^2$, $\mathbb{R}^3$, or more generally $\mathbb{R}^n$, into $\mathbb{R}$. Most of the process of doing so parallels the development of the basic principles in Chapters 2 through 4 and will seem comfortably familiar. The end result is a collection of tools that allow us to solve a wider variety of problems from the realms of electricity, economics, and fluid dynamics, just to name a few.

TOPIC 1 **Real-Valued Functions of Two Variables**

We begin with functions that map $\mathbb{R}^2$ into $\mathbb{R}$ as these functions can (at least in principle) be easily described geometrically and will serve to introduce all of the relevant ideas. Also, as we will soon see, we already have significant experience with such functions—the new material in this chapter is primarily a shift in how we think about and work with multivariable functions. For example, we will view the surface area formula $A = 2\pi rh + 2\pi r^2$ as defining a function A in the variables r and h and the quadric surface $9x^2 + 4y^2 - z = 0$ as defining a function z in the variables x and y.

Definition 💡

Real-Valued Functions of Two Variables

A **real-valued function f of two (real) variables** is a function that returns a real number denoted $f(x, y)$ for each ordered pair (x, y) in a **domain** $D \subseteq \mathbb{R}^2$. The **range** of f is the set of real numbers $\{f(x, y) | (x, y) \in D\}$. If the domain of f is not explicitly stated, it is assumed to be all ordered pairs (x, y) for which $f(x, y)$ is a well-defined real number. In the equation form $z = f(x, y)$, x and y are referred to as the **independent variables** and z as the **dependent variable**.

Example 1 ✎

Determine the domain and range of each of the given functions, and evaluate each at the point $(1, 2)$.

a. $f(x, y) = \sqrt{y^2 - x}$ **b.** $g(x, y) = \ln(y - x)$

Solution

a. In order for $f(x, y)$ to be a real number, we need $y^2 - x \geq 0$. This inequality then defines the domain of f, $D = \{(x, y) | y^2 \geq x\}$, which is depicted as the shaded region in Figure 1. All nonnegative real numbers are possible outputs of f, so the range of f is the interval $[0, \infty)$. And since $(1, 2) \in D$, f can be evaluated at the ordered pair and $f(1, 2) = \sqrt{2^2 - 1} = \sqrt{3}$.

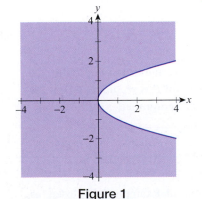

Figure 1

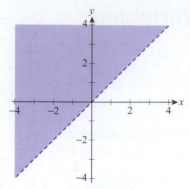

Figure 2

b. For $g(x, y)$ to be a real number, we require $y - x > 0$, so the domain of g consists of the part of the xy-plane above the line $y = x$, $D = \{(x, y) \mid y > x\}$, as shown in Figure 2—we use a dashed line to indicate that the line $y = x$ is not included in the domain. Every real number can be written in the form $\ln(y - x)$ for some x and some y in the domain, so the range of g is $\mathbb{R}$. Finally, $g(1, 2) = \ln(2 - 1) = 0$.

Extending the notion of open and closed intervals to the current context results in the next definition. After reading the definition, note how open and closed intervals can be viewed as, respectively, open and closed regions in $\mathbb{R}$. At the end of this section, we will expand the definition one last time to encompass regions in $\mathbb{R}^n$, for any $n \in \mathbb{N}$.

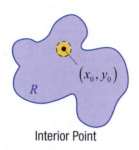

Interior Point

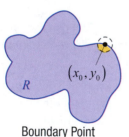

Boundary Point

Figure 3

Definition 💡

Interior Points, Boundary Points, and Open and Closed Regions

Let R be a region of the Cartesian plane $\mathbb{R}^2$. A point $(x_0, y_0) \in R$ is said to be an **interior point** of R if, for some $\varepsilon > 0$, all points within a distance of ε from (x_0, y_0) are also elements of R—that is, $B_\varepsilon(x_0, y_0) \subseteq R$, where

$$B_\varepsilon(x_0, y_0) = \left\{ (x, y) \,\middle|\, \sqrt{(x - x_0)^2 + (y - y_0)^2} < \varepsilon \right\}$$

(a set of the form $B_\varepsilon(x_0, y_0)$ is called an **open disk of radius** ε). A point (x_0, y_0) is said to be a **boundary point** of R if, for every $\varepsilon > 0$, $B_\varepsilon(x_0, y_0)$ contains both points that lie in R and points that lie outside R—note that (x_0, y_0) can be a boundary point of R without being an element of R itself.

A region R is **open** if every point in R is an interior point, while R is **closed** if it contains all of its boundary points. Figure 3 illustrates a typical interior point and boundary point of a region.

Similarly, we extend the idea of a *bounded interval* to a *bounded region* with the following definition.

Definition 💡

Bounded and Unbounded Regions

Let R be a region of the Cartesian plane $\mathbb{R}^2$. R is said to be **bounded** if, for some $d > 0$, $R \subseteq B_d(0, 0)$, that is, every point of R is within d units of the origin. A region for which no such d exists is **unbounded**.

Example 2 ✍

The domain $\{(x, y) \mid y^2 \geq x\}$ found in part a. of Example 1 is closed, as it contains all of its boundary points—namely, the points of the curve $y^2 - x = 0$.

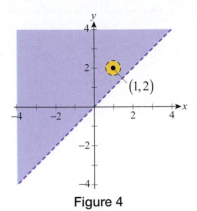

Figure 4

On the other hand, the domain $\{(x,y)|y>x\}$ of Example 1 part b. is open. Any point (x_0, y_0) in the domain lies some positive distance d away from the line $y=x$, so

$$B_{d/2}(x_0, y_0) \subseteq \{(x,y)|y>x\},$$

as depicted in Figure 4 for the point $(1,2)$. The boundary of the domain $\{(x,y)|y>x\}$ is the line $y=x$, which in this case is not a part of the domain.

Both of the domains of Example 1 are unbounded, as they contain points arbitrarily far away from the origin. In contrast, any set $B_\varepsilon(0,0)$ provides a simple example of a bounded region, as by definition it consists only of points within some fixed distance of the origin.

TOPIC 2   Graphs and Contour Maps

Given a function $f(x,y)$ that maps its domain D into $\mathbb{R}$, we frequently desire a way to visualize its features. The two most common methods of doing so are the subject of the next definition.

Definition ♀

Graphs and Contour Maps of Functions of Two Variables

The set of ordered triples $\{(x,y,f(x,y))|(x,y)\in D\}$ in $\mathbb{R}^3$ is called the **graph** of the function $f:D \to \mathbb{R}$—equivalently, the graph of f is the set of all points (x,y,z) solving the equation $z=f(x,y)$. The set of points in D for which $f(x,y)=c$, for some fixed constant c, is called a **contour** (or **level curve**) of f, and a collection of such contours is called a **contour map**.

Graphing a two-variable function $f(x,y)$ in $\mathbb{R}^3$ is the natural analog of graphing a single-variable function $f(x)$ in $\mathbb{R}^2$. Surfaces in $\mathbb{R}^3$ can serve as examples of graphs of two-variable functions, though we may have to restrict our attention to a portion of the surface in order to represent it as a function.

Example 3 ✎

Write the half of the sphere $x^2+y^2+z^2=9$ above the xy-plane as a function of x and y.

Solution

Solving the equation $x^2+y^2+z^2=9$ for z, we must choose whether to use the negative or positive root. The positive root corresponds to the hemisphere above the xy-plane, so we define $f(x,y)=\sqrt{9-x^2-y^2}$. The graph of f is shown in Figure 5; note that the domain of f is the region $\{(x,y)|x^2+y^2\le 9\}$ in the xy-plane.

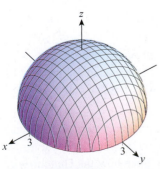

Figure 5

Example 4 ✎

In Section 11.6, we learned that the quadric equation $9x^2 + 4y^2 - z = 0$ defines a surface called an elliptic paraboloid, and its graph appears in Figure 6. The equation also defines z as a function of x and y, namely

$$z = f(x, y) = 9x^2 + 4y^2.$$

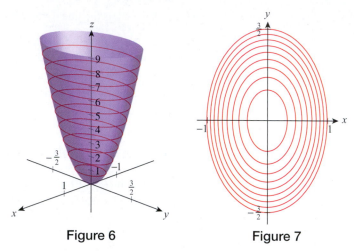

Figure 6   Figure 7

For this particular function, a contour map, such as shown in Figure 7, is also very instructive. The nine specific contours in Figure 7 correspond to the nine traces in Figure 6, which are all depictions of intersections of the surface with planes parallel to the xy-plane. Specifically, the nine contours and traces correspond to the equations

$$f(x, y) = c, \quad c \in \{1, 2, \dots, 9\}.$$

The classification of this quadric surface as an elliptic paraboloid reflects the fact that each contour in the xy-plane is an ellipse, namely $9x^2 + 4y^2 = c$ for some constant c.

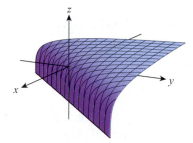

Figure 8

Example 5 ✎

The surface defined by the function $g(x, y) = \ln(y - x)$ is shown in Figure 8. As we determined in Example 1, the domain of g is the region $\{(x, y) \mid y > x\}$. Although the illustration of the surface may not convey the fact that the range of g is the entire set of real numbers, note that $g(x, y) \to -\infty$ as we consider points (x, y) in the domain closer and closer to the boundary line $y = x$, and $g(x, y) \to \infty$ as we consider points (x, y) for which $y - x$ gets larger and larger. We will return to the matter of limits of multivariable functions in Section 13.2.

Since $g(x, y)$ is constant along lines of slope 1, every contour map of the function will look something like Figure 9, which shows level curves for values of c from -5 to 1 in increments of $\frac{1}{2}$. Note how the contour lines "bunch up" as $c \to -\infty$.

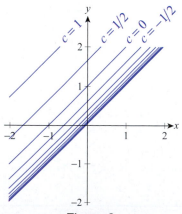

Figure 9

Just as a curve in $\mathbb{R}^2$ may represent a function $x = f(y)$, a surface in $\mathbb{R}^3$ may represent a function of two variables other than x and y.

Example 6 ✎

Express the planar surface $x + 2y + 3z = 2$ as a function of two variables other than the pair x and y.

Solution

Expressing the surface as a function of y and z is very easily done.

$$x = f(y, z) = 2 - 2y - 3z$$

Expressing the surface as a function of x and z is almost as easy.

$$y = g(x, z) = 1 - \frac{x}{2} - \frac{3z}{2}$$

The graph of both functions is the plane shown in Figure 10 (only the portion in the first octant is shown). In graphing f the value of the function is used as the x-component and in graphing g the value of the function corresponds to the y-component.

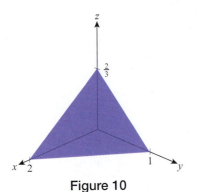

Figure 10

Technology Note 💻

Computer algebra systems, such as *Mathematica*, are very useful in constructing both graphs and contour maps; most of them have built-in commands to generate such images. For instance, the command to plot the function from Example 5 in *Mathematica* is **Plot3D** and its basic use is shown in Figure 11. Contour maps can be constructed with the **ContourPlot** command, as shown in Figure 12, where we plot the nine specific contours of the elliptic paraboloid from Example 4. Both commands have many options allowing the user to refine the image as desired, which is what has been done to create Figures 8 and 7, respectively.

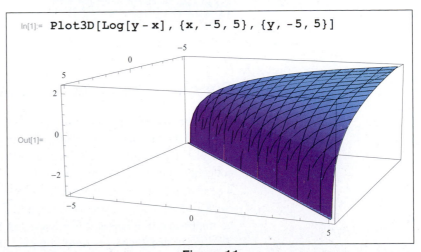

Figure 11

```
In[2]:= ContourPlot[9 x^2 + 4 y^2, {x, -2, 2}, {y, -2, 2},
          Contours → {1, 2, 3, 4, 5, 6, 7, 8, 9}]
```

Out[2]=

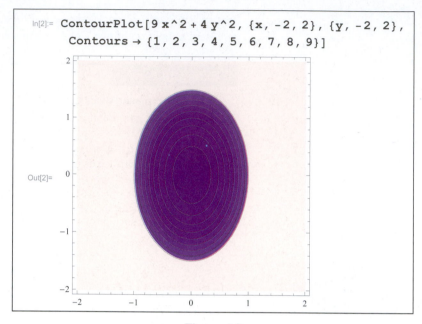

Figure 12

Contour maps also appear in nonmathematical settings, for example in topographical maps (Figure 13) and maps of lines of constant temperature (*isotherms*) (Figure 14) or constant pressure (*isobars*) in weather reports.

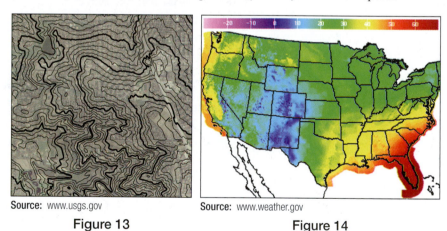

Source: www.usgs.gov

Figure 13

Source: www.weather.gov

Figure 14

TOPIC 3 Functions of Three or More Variables

As the number of variables in the definition of a function increases, our ability to represent the function with pictures deteriorates. For a function $f(x, y, z)$ of three variables, the analog of a level curve (or contour) is a **level surface**—that is, the collection of points in $\mathbb{R}^3$ for which $f(x, y, z) = c$ for some constant *c*. In principle, we can readily show any level surface of $f(x, y, z)$, but a complete depiction would consist of *all* level surfaces at once, and such an image would almost certainly be impossible to interpret. Even worse, an illustration of the graph of a function $f(x, y, z)$, which consists of all 4-tuples of the form $(x, y, z, f(x, y, z))$, would require an ability to draw in $\mathbb{R}^4$.

Nevertheless, the basic principles of working with functions of three or more variables are the same—it is only our ability to illustrate the principles with pictures that is hampered. We begin with the following definitions, soon to be used when we discuss limits and derivatives of multivariable functions.

Definition 💡

Real-Valued Functions of *n* Variables

A **real-valued function *f* of *n* (real) variables** $x_1, x_2, \ldots, x_n$ is a function that returns a real number denoted $f(x_1, x_2, \ldots, x_n)$ for each ordered *n*-tuple $(x_1, x_2, \ldots, x_n)$ in a **domain** $D \subseteq \mathbb{R}^n$. The **range** of *f* is the set of real numbers $\left\{ f(x_1, x_2, \ldots, x_n) \mid (x_1, x_2, \ldots, x_n) \in D \right\}$.

Important: In order to simplify notation, it is common to denote an ordered *n*-tuple $(x_1, x_2, \ldots, x_n)$ as simply *x*; context tells us whether *x* refers to a single real variable, that is, $x \in \mathbb{R}$, or whether $x = (x_1, x_2, \ldots, x_n)$ and refers to an element of $\mathbb{R}^n$. Note the correlation with the notation used for position vectors: the point $x = (x_1, x_2, \ldots, x_n)$ corresponds to the terminal point of the position vector $\mathbf{x} = \langle x_1, x_2, \ldots, x_n \rangle$. In practice, we typically use notation like $f(x, y)$ and $f(x, y, z)$ for functions whose domains are subsets of $\mathbb{R}^2$ or $\mathbb{R}^3$, and the notation $f(x)$, where $x = (x_1, x_2, \ldots, x_n)$, when discussing a generic function $f \colon \mathbb{R}^n \to \mathbb{R}$ with *n* an unspecified positive integer.

Definition 💡

Interior Points, Boundary Points, Open and Closed Regions, and Boundedness in $\mathbb{R}^n$

Let *R* be a region (i.e., a subset) of $\mathbb{R}^n$. A point $p = (p_1, p_2, \ldots, p_n) \in R$ is said to be an **interior point** of *R* if, for some $\varepsilon > 0$, all points within a distance of ε from *p* are also elements of *R*—that is, $B_\varepsilon(p) \subseteq R$, where

$$B_\varepsilon(p) = \left\{ q \in \mathbb{R}^n \mid |p - q| < \varepsilon \right\}$$

and $|p - q|$ denotes the Euclidean distance between *p* and *q*:

$$|p - q| = \sqrt{(p_1 - q_1)^2 + (p_2 - q_2)^2 + \cdots + (p_n - q_n)^2}$$

(a set of the form $B_\varepsilon(p)$ is called an ***n*-dimensional open ball of radius** ε). A point *p* is said to be a **boundary point** of *R* if, for every $\varepsilon > 0$, $B_\varepsilon(p)$ contains both points that lie in *R* and points that lie outside *R*. The region *R* is said to be **open** if every point in *R* is an interior point, and *R* is said to be **closed** if it contains all of its boundary points.

Finally, *R* is said to be **bounded** if, for some $d > 0$, $R \subseteq B_d(0)$, that is, every point of *R* is within *d* units of the origin of $\mathbb{R}^n$. If no such *d* exists, *R* is **unbounded**.

Example 7 📝

The function $f(x,y,z) = \sqrt{x^2 + y^2 + z^2}$ returns the distance between each ordered triple (x,y,z) and the origin. Thus, its level surfaces consist of all the nested spheres centered at the origin. Figure 15 shows, in a cut-out diagram, level surfaces of the function corresponding to the three level values 1, 2, and 4.

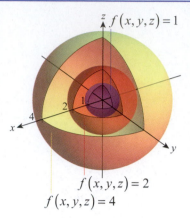

$f(x,y,z) = 1$

$f(x,y,z) = 2$
$f(x,y,z) = 4$

Figure 15

13.1 **Exercises**

1–4 Evaluate the given multivariable function at the indicated points.

1. $f(x,y) = xy - y^3 x^2$; $(0,0)$, $(2,1)$

2. $f(u,v) = (2u+1)(v - u^2)$; $(3,-1)$, $(1,1)$

3. $f(x,y,z) = \dfrac{xz}{2y^2 + z^4}$; $(5,0,-1)$, $(-4,2,1)$

4. $f(t,u,v,w) = \dfrac{(4t-v)^w}{(u^2+3)}$; $(1,1,3,1)$, $(2,-1,7,12)$

5–12 Determine the domain and range of the given function and evaluate the function at the indicated point.

5. $f(x,y) = xy + yx^3$; $(3,-1)$

6. $f(x,y) = 2y^2\left(x + \dfrac{y}{2}\right)$; $(0,2)$

7. $f(x,y) = \sqrt{x^2 y}$; $(2,9)$

8. $f(x,y) = \dfrac{\sqrt{36 - 4x^2 - 9y^2}}{6}$; $(2,0)$

9. $f(x,y) = \ln|xy|$; $(e,-1)$

10. $f(x,y) = \sqrt{1 - x^2 y^2}$; $\left(-2, \dfrac{1}{4}\right)$

11. $f(x,y,z) = \arctan(xz - y^2)$; $(5,3,2)$

12. $f(x,y,z) = \dfrac{ze^{x/y}}{\sqrt{x-z}}$; $(0,-1,-2)$

13–20 Describe in words the graph of the function. (**Hint:** It is helpful to review quadric surfaces from Section 11.6.)

13. $f(x,y) = 3x + y - 2$

14. $f(x,y) = \dfrac{\sqrt{144 - 9x^2 - 16y^2}}{6}$

15. $f(x,y) = \dfrac{x^2}{4} + \dfrac{(y-1)^2}{9}$

16. $f(x,y) = \dfrac{\sqrt{9x^2 + 4y^2 - 36}}{6}$

17. $f(x,y) = \sqrt{x^2 + 2y^2}$

18. $f(x,y) = \sqrt{x^2 + 2y^2 + 1}$

19. $f(x,y) = 2x^2 + 3y^2$

20. $f(x,y) = \dfrac{3x^2 - 6y^2}{4}$

21–26 A region R and a point P in the Cartesian plane $\mathbb{R}^2$ are given. Classify the point as an interior point of R, a boundary point, or neither.

21. $R = \left\{(x,y)\,\middle|\, y > |x|\right\}$; $P(1,1)$

22. $R = \left\{(x,y)\,\middle|\, x^2 + y^2 \le 1\right\}$; $P\left(\dfrac{1}{2}, \dfrac{1}{2}\right)$

23. $R = \left\{(x,y)\,\middle|\, 1 < x^2 + y^2 < 2\right\}$; $P(0,1)$

24. $R = \left\{(x,y)\,\middle|\, y \le x^2\right\}$; $P(1,2)$

25. $R = \left\{(x,y)\,\middle|\, |x+2| + |x-3| < y\right\}$; $P(1.5,5)$

26. $R = \left\{(x,y)\,\middle|\, y - |2-x| \ge 0\right\}$; $P(3,0)$

27–34 Classify the given subset R of $\mathbb{R}^2$ as open, closed, or neither.

27. $R = \left\{(x,y)\,\middle|\, (x-2)^2 + y^2 \le 4\right\}$

28. $R = \left\{(x,y)\,\middle|\, x^2 + 2y^2 \ge 3\right\}$

29. $R = \left\{(x,y)\,\middle|\, 0 < x^2 + y^2 < 9\right\}$

30. $R = \left\{(x,y)\,\middle|\, y \ne 2x - 3\right\}$

31. $R = \left\{(x,y)\,\middle|\, xy \ne 0\right\}$

32. $R = \left\{(x,y)\,\middle|\, |x| + |y| \le 1\right\}$

33. $R = \left\{(x,y)\,\middle|\, \sqrt{x} + \sqrt{y} < 1\right\}$

34. $R = \left\{(x,y)\,\middle|\, x > 0 \text{ or } y \ge 0\right\}$

35–42 The graphs of the given equations are quadric surfaces as seen in Exercises 27–34 of Section 11.6. Express each as a function of two variables other than the pair x and y. (Note that the graph of the resulting function may not be the entire surface. Can you see why?)

35. $3z^2 + 2y^2 = \dfrac{3x}{2}$

36. $3z^2 + 2y^2 = 1 - \dfrac{3x^2}{2}$

37. $2x^2 + 2y^2 = z$

38. $2x^2 + 2y^2 = z^2$

39. $x^2 + 2x - y^2 + z^2 = 0$

40. $x^2 + 2x + 2y^2 + 3z^2 = 8y$

41. $x^2 + 2x - 2y^2 - 3z = 8y + 7$

42. $-x^2 + 2x + 2y^2 - 3z^2 = 8y + 2$

43–48 Match the function with its graph (labeled A–F).

43. $f(x,y) = \dfrac{(x+y)^2}{2}$

44. $f(x,y) = \cos|xy|$

45. $f(x,y) = \dfrac{3}{4x^2 + 3y^2 + 1}$

46. $f(x,y) = |x-1| + |y|$

47. $f(x,y) = \sin(x + 2y)$

48. $f(x,y) = \cos(x^2 + y^2)$

A.

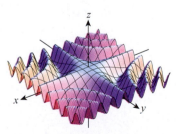

B.

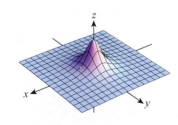

C.

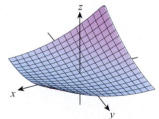

D.

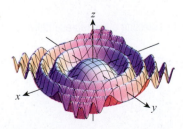

E.

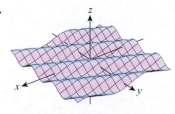

F.

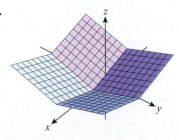

49–54 Match the graph with its contour map (labeled A–F).

49.

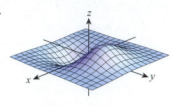

50.

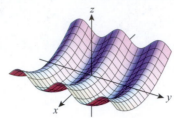

51.

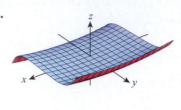

52.

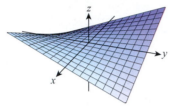

53.

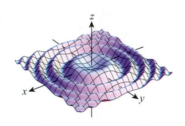

54.

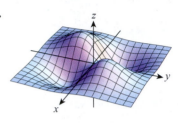

A.

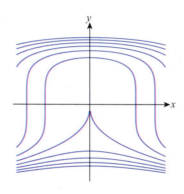

B.

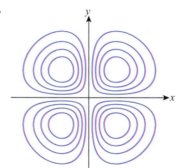

C.

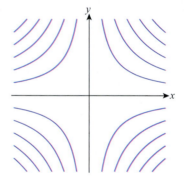

D.

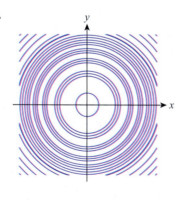

E.

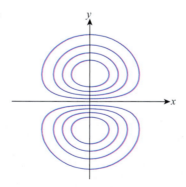

F.

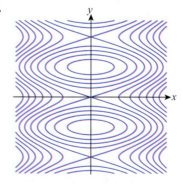

55–60 Sketch a rough graph of the given function by hand. Then sketch its contour map by selecting a few representative contours.

55. $f(x, y) = x - 3y + 1$

56. $f(x, y) = \dfrac{x^2}{9} + \dfrac{y^2}{4}$

57. $f(x, y) = y^2 - x$

58. $f(x, y) = \ln(x^2 + y^2)$

59. $f(x, y) = \dfrac{y}{x^2 + y^2 + 1}$

60. $f(x, y) = \sin\sqrt{x^2 + y^2}$

61–62 Describe the level surfaces of the given three-variable function.

61. $f(x, y, z) = x^2 + 2y^2 + 3z^2$

62. $f(x, y, z) = x^2 + 2y^2 - 3z^2$

63. The figure below shows a portion of a topographical map of an area near Julian, California. Examine the map and answer the following.

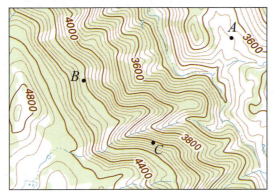

Source: www.usgs.gov

a. Estimate the direction of the steepest slope from point A.

b. Find a possible "steepest path" from point A to point B.

c. Estimate the elevation of point C.

d. Find a point D where the northern direction is uphill, while it is downhill to the southwest.

64.* Prove that if S is a finite subset (i.e., a set consisting of finitely many points) of $\mathbb{R}^2$, then $\mathbb{R}^2 - S$ is an open subset of $\mathbb{R}^2$. (**Hint:** Pick an arbitrary point P in $\mathbb{R}^2 - S$ and prove that it is an interior point.)

65.* Prove that the open interval $(0,1)$, when viewed as a subset of $\mathbb{R}^2$, is neither open nor closed.

13.1 **Technology Exercises**

66–71. Use a computer algebra system to generate the graphs and contour maps of the functions in Exercises 55–60.

13.2 Limits and Continuity of Multivariable Functions

TOPICS

1. Limits of multivariable functions, and their evaluation

2. Continuity of multivariable functions

In Sections 2.3, 2.4, and 2.5, we developed rigorous definitions of limit and continuity and proceeded to use those definitions in proving many useful facts. We now extend the definitions to real-valued functions of two or more variables.

TOPIC 1   Limits of Multivariable Functions, and Their Evaluation

Informally, we say that L is the limit of a function f at a particular point if, as the argument of f approaches the point, the value of f approaches L. That informal sense was the basis of the definition of limit in Section 2.3, and serves just as well now. The only difference in the expanded definition below is a necessary adjustment to acknowledge the multidimensional nature of the domain of f.

Definition ♀

Limit of a Real-Valued Function of *n* Variables

Let f be a function mapping its domain $D \subseteq \mathbb{R}^n$ into $\mathbb{R}$, and let $c = (c_1, c_2, \ldots, c_n)$ be a point in either D or the boundary of D. We say that the **limit of f as x approaches c is L**, and write

$$\lim_{x \to c} f(x) = L,$$

if for every real number $\varepsilon > 0$ there is a number $\delta > 0$ such that $|f(x) - L| < \varepsilon$ whenever $x \in D$ and $0 < |x - c| < \delta$. Recall that the notation $|x - c|$ denotes the Euclidean distance between two points x and c in $\mathbb{R}^n$, so if $x = (x_1, x_2, \ldots, x_n)$, then

$$|x - c| = \sqrt{(x_1 - c_1)^2 + (x_2 - c_2)^2 + \cdots + (x_n - c_n)^2}.$$

Just as with the original definition of limit in Section 2.3, the above definition contains some subtle points worth highlighting. First, it is important to note that f isn't necessarily defined at the point c, since the domain of f may not contain all of its boundary points. But by stipulation, if c is not in the domain D of f, c is an element of the boundary of D and thus there are elements of D arbitrarily close to c—that is sufficient for us to discuss the limit of f at c. Second, and in contrast to the one-variable case, there are many paths along which we can imagine a variable x approaching the fixed point c in $\mathbb{R}^n$, when $n \geq 2$. As we will see in the following examples, this latitude leads to both interesting results and the opportunity to use some new reasoning techniques. Finally, note that the above definition reduces to that of Section 2.3 if $n = 1$.

Example 1 ✎

Determine whether the function

$$f(x,y) = \frac{x^2 - y^2}{x^2 + y^2}$$

has a limit at the point $(0,0)$. (Note that f is not defined at $(0,0)$.)

Solution

We want to determine

$$\lim_{(x,y)\to(0,0)} \frac{x^2 - y^2}{x^2 + y^2},$$

if it exists, but we can quickly decide no such limit is possible if we consider just two specific paths by which the variable point (x,y) can approach $(0,0)$. The first path is along the x-axis, where $y = 0$.

$$\lim_{\substack{(x,y)\to(0,0)\\ \text{along } y=0}} \frac{x^2 - (0)^2}{x^2 + (0)^2} = \lim_{x\to 0} 1 = 1$$

But if we approach $(0,0)$ along the y-axis, where $x = 0$, we see

$$\lim_{\substack{(x,y)\to(0,0)\\ \text{along } x=0}} \frac{(0)^2 - y^2}{(0)^2 + y^2} = \lim_{y\to 0}(-1) = -1.$$

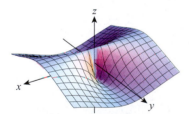

Figure 1

These two facts alone are sufficient to tell us that the limit at $(0,0)$ does not exist, since limits are unique—the limit cannot depend on how the points approach $(0,0)$. Figure 1 is a graph of the function $f(x,y)$, and reflects the fact that f has a constant value of 1 along the x-axis and a constant value of -1 along the y-axis. In Exercise 24 you will show that, *restricted to any particular line $y = mx$ in the xy-plane*, all limiting values between -1 and 1 are also possible.

Whenever it can be shown that a function $f(x,y)$ has two different limiting values as (x,y) approaches (x_0, y_0) along two different paths, we know that $\lim_{(x,y)\to(x_0,y_0)} f(x,y)$ does not exist (in Exercise 26 you will show that paths other than straight lines may have to be considered). To prove that a limit *does* exist, and to actually evaluate that limit, we can employ extensions of the limit laws first seen in Section 2.4. We list them here without proof, as the proofs are similar in nature to those of the original limit laws.

Theorem 🔍

Extended Limit Laws

Let f and g be two real-valued functions of n variables such that both $\lim_{x \to c} f(x)$ and $\lim_{x \to c} g(x)$ exist, where $c \in \mathbb{R}^n$ is a fixed point. Let k be a fixed real number. Then the following laws hold.

Sum/Difference Law $\lim_{x \to c}\left[f(x) \pm g(x)\right] = \lim_{x \to c} f(x) \pm \lim_{x \to c} g(x)$

Constant Multiple Law $\lim_{x \to c}\left[kf(x)\right] = k \lim_{x \to c} f(x)$

Product Law $\lim_{x \to c}\left[f(x)g(x)\right] = \lim_{x \to c} f(x) \cdot \lim_{x \to c} g(x)$

Quotient Law $\lim_{x \to c}\dfrac{f(x)}{g(x)} = \dfrac{\lim_{x \to c} f(x)}{\lim_{x \to c} g(x)}$, provided $\lim_{x \to c} g(x) \neq 0$

Rational Power Law
(*a* and *b* nonzero integers with no common factor) $\lim_{x \to c}\left(f(x)\right)^{a/b} = \left(\lim_{x \to c} f(x)\right)^{a/b}$,
provided $\lim_{x \to c} f(x) \geq 0$ if *b* is even

Example 2 🖊

Each of the limits below can be evaluated by employing limit laws and the sort of algebraic manipulation first seen in Section 2.4.

a. $\displaystyle\lim_{(x,y) \to (0,0)} \frac{xy - y^2}{\sqrt{x} - \sqrt{y}} = \lim_{(x,y) \to (0,0)} \left(\frac{xy - y^2}{\sqrt{x} - \sqrt{y}}\right)\left(\frac{\sqrt{x} + \sqrt{y}}{\sqrt{x} + \sqrt{y}}\right)$ Rationalize the denominator. (Constant Multiple Law)

$\displaystyle= \lim_{(x,y) \to (0,0)} \frac{y(x - y)\left(\sqrt{x} + \sqrt{y}\right)}{x - y}$

$\displaystyle= \lim_{(x,y) \to (0,0)} y\left(\sqrt{x} + \sqrt{y}\right)$ Cancel common factor.

$\displaystyle= 0\left(\sqrt{0} + \sqrt{0}\right) = 0$ Rational Power, Sum, and Product Laws

Note that canceling the common factor of $x - y$ is legitimate since the factor is 0 only when $x = y$, and the line $x = y$ is not in the domain of the function.

b. $\displaystyle\lim_{(x,y) \to (1,4)} \frac{\sqrt{x+3} - \sqrt{y}}{x - y + 3} = \lim_{(x,y) \to (1,4)} \left(\frac{\sqrt{x+3} - \sqrt{y}}{x - y + 3}\right)\left(\frac{\sqrt{x+3} + \sqrt{y}}{\sqrt{x+3} + \sqrt{y}}\right)$ Rationalize the denominator

$\displaystyle= \lim_{(x,y) \to (1,4)} \frac{(x + 3 - y)}{(x - y + 3)\left(\sqrt{x+3} + \sqrt{y}\right)}$

$\displaystyle= \lim_{(x,y) \to (1,4)} \frac{1}{\left(\sqrt{x+3} + \sqrt{y}\right)}$ Cancel common factor.

$\displaystyle= \frac{1}{2 + 2} = \frac{1}{4}$ Rational Power, Sum, and Quotient Laws

Note that, again, the common factor is nonzero on the domain of the function, which consists of all points for which $x + 3 \neq y$.

Of course, we can also always resort to an ε-δ argument.

Example 3 ✎

Find $\displaystyle\lim_{(x,y)\to(0,0)} \frac{5x^2y}{x^2+y^2}$, if the limit exists.

Solution

As with each of the previous examples, the numerator and denominator of the function both go to 0 as (x,y) approaches $(0,0)$, which does not allow us to immediately evaluate the limit (if it exists). On the other hand, we cannot immediately conclude that the limit does not exist, as we did in Example 1, since

$$\lim_{\substack{(x,y)\to(0,0)\\ \text{along } y=mx}} \frac{5x^2y}{x^2+y^2} = \lim_{x\to 0}\frac{5x^2(mx)}{x^2+(mx)^2} = \lim_{x\to 0}\frac{5mx}{1+m^2} = 0.$$

Remember that this does not tell us that the limit of the function at $(0,0)$ *is* 0, since the limit along some other approach may be different (see Exercise 26), but it does tell us that if the limit *does* exist, it must be 0.

So suppose we wish to demonstrate that, given $\varepsilon > 0$, we can find a $\delta > 0$ so that

$$0 < \big|(x,y)-(0,0)\big| < \delta \quad \Rightarrow \quad \big|f(x,y)-0\big| < \varepsilon.$$

Then since $x^2 \le x^2 + y^2$ and $y^2 \le x^2 + y^2$,

$$\left|\frac{5x^2y}{x^2+y^2}-0\right| = 5|y|\frac{x^2}{x^2+y^2} \le 5|y| = 5\sqrt{y^2} \le 5\sqrt{x^2+y^2}.$$

The condition $\big|(x,y)-(0,0)\big| < \delta$ means that $\sqrt{x^2+y^2} < \delta$, so if we let $\delta = \varepsilon/5$,

$$0 < \big|(x,y)-(0,0)\big| < \delta \quad \Rightarrow \quad \left|\frac{5x^2y}{x^2+y^2}-0\right| \le 5\sqrt{x^2+y^2} < 5\delta = \varepsilon,$$

and hence $\displaystyle\lim_{(x,y)\to(0,0)}\frac{5x^2y}{x^2+y^2} = 0$. A graph of the function $f(x,y) = \dfrac{5x^2y}{x^2+y^2}$ is shown in Figure 2. In Exercise 25, you will show that the algebraically similar function $g(x,y) = \dfrac{5xy}{x^2+y^2}$ behaves very differently at $(0,0)$.

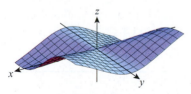

Figure 2

TOPIC 2   Continuity of Multivariable Functions

Just as in Section 2.5, continuity of a function of n variables at a point requires that the limit exist at that point and that the limit value is the same as the function value.

Definition 💡

Continuity of a Multivariable Function at a Point

Given a function $f: D \to \mathbb{R}$ and a point $c \in D$, where $D \subseteq \mathbb{R}^n$, we say f is **continuous at** c if $\lim\limits_{x \to c} f(x) = f(c)$. Note that, as in the one-variable case, continuity of f at c requires that f be defined at c, though c does not have to be an interior point of D. We say f is **continuous** if f is continuous at every point in its domain D.

Also as in the single-variable case, the extended limit laws mean that the basic algebraic combinations of continuous functions are all continuous, a fact we state briefly below.

Properties of Continuous Functions

Sums, differences, constant multiples, products, quotients, and rational powers of continuous functions are continuous at all points where they are mutually defined.

In particular, this means that polynomial and rational functions of n variables are continuous at all points of their domain.

Example 4 ☑

Determine where each of the following functions is continuous.

a. $f(x, y, z) = \dfrac{1}{x^2 + y^2 + z^2 - 9}$ **b.** $g(x, y) = 9x^4 - 2x^3 y + \dfrac{7xy^2}{1 + x^2 y^4}$

Solution

a. Since f is a rational function (of three variables), it is continuous everywhere it is defined, and the only points not in the domain of f are those for which $x^2 + y^2 + z^2 = 9$. In words, f is continuous on all of $\mathbb{R}^3$ except for the points on the sphere of radius 3 centered at the origin.

b. The function g is a sum of a polynomial function and a rational function, and the rational portion is defined everywhere ($1 + x^2 y^4 \neq 0$ for all x and y). So g is continuous on all of $\mathbb{R}^2$.

Example 5 ☑

If possible, extend the definition of the function $f(x, y) = \dfrac{5x^2 y}{x^2 + y^2}$ in order to make it continuous everywhere.

Solution

The domain of the rational function f is all of $\mathbb{R}^2$ except for the origin, and we found in Example 3 that

$$\lim_{(x,y) \to (0,0)} \frac{5x^2 y}{x^2 + y^2} = 0.$$

So if we define $f(0,0)$ to be 0 (and to have the original value otherwise), f is continuous everywhere.

One last property remains to be extended to the multivariable arena: continuity of composite functions.

Theorem ✎

"A Composition of Continuous Functions Is Continuous"

Assume that $f: D \to \mathbb{R}$, where $D \subseteq \mathbb{R}^n$, and $g: \mathbb{R} \to \mathbb{R}$. If f is continuous at $c \in D$ and g is continuous at $f(c)$, then $g \circ f$ is continuous at c.

The proof is identical to the proof that the composition of two continuous single-variable functions is continuous, so it will not be reproduced here. We will, however, demonstrate the use of the theorem.

Example 6 ✎

a. The function $f(x,y,z) = e^{-3xy-4xz^2}$ is continuous everywhere, as it is the composition of a polynomial function and an exponential function.

b. The function $g(x,y) = \arctan\left(x/(x-y)\right)$, shown in Figure 3, is the composition of the rational function $x/(x-y)$ and the arctangent function. The function g is continuous for all $(x,y) \in \mathbb{R}^2$ except the line $x = y$, since the rational function $x/(x-y)$ is continuous everywhere $x \neq y$ and the arctangent function is continuous on all of $\mathbb{R}$.

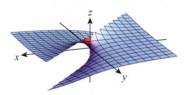

Figure 3

13.2 Exercises

1–4 Use the limit laws to find the indicated limits, assuming that $\displaystyle\lim_{(x,y)\to(a,b)} f(x,y) = 4$ and $\displaystyle\lim_{(x,y)\to(a,b)} g(x,y) = -1$.

1. $\displaystyle\lim_{(x,y)\to(a,b)} \left[f(x,y) - 2g(x,y) \right]$

2. $\displaystyle\lim_{(x,y)\to(a,b)} \left[5f(x,y)g(x,y) \right]$

3. $\displaystyle\lim_{(x,y)\to(a,b)} \left[\frac{4g(x,y)}{f(x,y)} + f(x,y) \right]$

4. $\displaystyle\lim_{(x,y)\to(a,b)} \frac{3f(x,y) + g(x,y)}{g(x,y)}$

5–22 Determine whether the indicated limit exists. If so, find it.

5. $\displaystyle\lim_{(x,y)\to(0,0)} \frac{x\sqrt{y} - y^{3/2}}{\sqrt{x} + \sqrt{y}}$

6. $\displaystyle\lim_{(x,y)\to(1,2)} \frac{\sqrt{x} - \sqrt{y-1}}{x^2 - xy + x}$

7. $\displaystyle\lim_{(x,y)\to(0,0)} \frac{(x+1)\left(\sqrt{y+5} - \sqrt{5}\right)}{xy + y}$

8. $\displaystyle\lim_{(x,y)\to(0,0)} \left(2x - y^2 \right)$

9. $\displaystyle\lim_{(x,y)\to(2,-1)} \frac{x + 2y}{y + 3x^2}$

10. $\displaystyle\lim_{(x,y)\to(1,4)} \left(2xy^2 + 5x^4\sqrt{y} \right)$

11. $\displaystyle\lim_{(x,y)\to(0,0)} \frac{xy}{x^2 + y^2}$

12. $\displaystyle\lim_{(x,y)\to(0,0)} \frac{xy^2}{x^2 + y^2}$

13. $\displaystyle\lim_{(x,y)\to(0,0)} \frac{(1 - \cos x)y}{x^4 + y^4}$

14. $\displaystyle\lim_{(x,y)\to(0,0)} \frac{x - y}{x^3 - y^3}$

15. $\displaystyle\lim_{(x,y)\to(0,0)} \frac{x^3 - y^3}{x - y}$

16. $\displaystyle\lim_{(x,y)\to(1/2,2)} \arctan\left(\frac{xy}{y-1} \right)$

17. $\displaystyle\lim_{(x,y)\to(0,0)} \frac{2\sqrt{x}y^3}{x^3 + y^4}$

18. $\displaystyle\lim_{(x,y)\to(1,0)} \frac{e^{xy} + e^{-\sqrt{y}}}{2xy + 1}$

19. $\displaystyle\lim_{(x,y)\to(0,0)} \sin\frac{2}{x^2 + y^2}$

20. $\displaystyle\lim_{(x,y)\to(0,0)} \frac{2\sin\left(x^2 + y^2\right)}{x^2 + y^2}$

21. $\displaystyle\lim_{(x,y,z)\to(0,0,0)} \frac{x^2 - 3y^2 + 4z^2}{x^2 + 3y^2 + 4z^2}$

22. $\displaystyle\lim_{(x,y,z)\to(0,0,0)} \frac{5xy^4}{x^4 + y^4 + z^4}$

23. Find the domains of the functions in Exercises 5 and 6.

24. With reference to the function of Example 1,

$$f(x,y) = \frac{x^2 - y^2}{x^2 + y^2},$$

show that if the origin is being approached in the xy-plane along the lines $y = mx$ for various values of m, all limiting values between -1 and 1 are possible.

25. Recall the following function introduced in the conclusion of Example 3:

$$g(x,y) = \frac{5xy}{x^2 + y^2}$$

Show that $\displaystyle\lim_{(x,y)\to(0,0)} g(x,y)$ does not exist.

(**Hint:** See Exercise 24.)

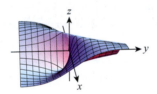

26. Show that the limit of the function

$$f(x,y) = \frac{x^4 y^4}{\left(x^2 + y^4\right)^3}$$

does not exist at $(0,0)$. (**Hint:** In addition to the line $y = x$, consider the limit along the curve $y^2 = x$. Explain why it would not be helpful to restrict the paths of approach to straight lines through the origin.)

27. Show that the function

$$g(x,y) = \frac{2x^{3/2} y}{x^3 + y^2}$$

has a limiting value of 0 when the origin is approached in the xy-plane along any parabola $y = cx^2$, but any limiting value between -1 and 1 can be achieved by considering the curves $y = cx^{3/2}$.

28. Prove directly that the limit in Example 3 is 0. (**Hint:** First argue that if $x = 0$, the function is 0 on its domain, and in the case of $x \neq 0$, divide the numerator and denominator by x^2 to obtain an expression whose limit is 0 as $y \to 0$.)

29. Use the Squeeze Theorem to give a new proof for Exercise 28. (**Hint:** Notice that $0 \leq \left| \dfrac{x^2}{x^2 + y^2} \right| \leq 1$, and multiply this inequality by $|5y|$; then apply the Squeeze Theorem.)

30. Prove or disprove: $\displaystyle\lim_{(x,y)\to(0,0)} \frac{x^2}{x^2 + y^2} = 0$.

31. Prove that for $h(x,y) = 5xy / \sqrt{x^2 + y^2}$, we have $\displaystyle\lim_{(x,y)\to(0,0)} h(x,y) = 0$, and contrast this with $g(x,y)$ of Exercise 25. (**Hint:** Rewrite $h(x,y)$ using polar coordinates, so that $x = r\cos\theta$ and $y = r\sin\theta$ and see what happens as $r \to 0$.)

32. The following function looks algebraically similar to $f(x,y)$ of Example 1.

$$k(x,y) = \frac{x^3 - y^3}{x^2 + y^2}$$

Show that $\displaystyle\lim_{(x,y)\to(0,0)} k(x,y) = 0$. (See the hint given in Exercise 31.)

33. Consider the polar coordinate approach of Exercise 31 and use it to find a new proof of the fact that the limit in Example 3 is 0.

34.* Find three different proofs of the fact that

$$\lim_{(x,y)\to(0,0)} \frac{x^2 \ln(x+1)}{x^2 + y^2} = 0.$$

(**Hint:** Use a direct proof, then the Squeeze Theorem, and finally a polar coordinate approach.)

35.* Use an ε-δ argument to show that

$$\lim_{(x,y)\to(0,0)} \frac{xy^2}{2\left(x^2 + y^2\right)} = 0.$$

36–39 Prove that the indicated limit exists, and find its value.

36. $\displaystyle\lim_{(x,y)\to(0,0)} \frac{4y^2 \sin x}{x^2 + 4y^2}$

37. $\displaystyle\lim_{(x,y)\to(0,0)} \frac{x^4 - 3x^2 - 9y^2}{x^2 + 3y^2}$

38. $\displaystyle\lim_{(x,y)\to(0,0)} \frac{4xy - \sqrt{4x^2 + 4y^2}}{\sqrt{x^2 + y^2}}$

39. $\displaystyle\lim_{(x,y)\to(0,0,0)} \frac{2y^5}{x^4 + y^4 + z^4}$

40. Is there a value of a so that $\displaystyle\lim_{(x,y)\to(a,a-1)} \frac{x}{y - x + 1}$ exists? Explain.

41. For α, $\beta \geq 0$, prove that $\displaystyle\lim_{(x,y)\to(0,0)} \frac{x^\alpha y^\beta}{x^2 + y^2} = 0$ if $\alpha + \beta > 2$; otherwise the limit does not exist. Can you generalize this result? Use it to revisit Exercises 28, 30, and 35.

42. Use an ε-δ argument to show that $\displaystyle\lim_{(x,y)\to(a,b)} (x+y) = a + b.$

43. Suppose that $\displaystyle\lim_{(x,y)\to(a,b)} f(x,y) = L$ and $L > 0$. Prove that there exists a δ-neighborhood of (a,b) such that $f(x,y)$ is positive for every point (x,y) in that neighborhood.

44–53 Determine where the given function is continuous.

44. $f(x,y) = \dfrac{2}{1 + x^2 + y^2}$

45. $g(x,y) = \dfrac{3y^2}{(x^2 + 3)(y^2 + 3)}$

46. $h(x,y) = \cot^{-1}\left(\dfrac{2y}{\sqrt{x+y}}\right)$

47. $r(x,y) = \sqrt{e^{xy}}$

48. $s(x,y) = 5\sqrt{x}y^2 + \dfrac{3}{\ln(x^2 y^2)}$

49. $k(x,y) = e^{\arctan(2x^2 y)}$

50. $q(x,y) = \dfrac{x^2}{\sqrt{4 - x^2 - y^2}}$

51. $m(x,y,z) = \ln(3z - x^2 - 2y^2)$

52. $n(x,y,z) = \sqrt{z^2 - x^2 - 2y^2}$

53. $p(x,y,z)$
$$= \begin{cases} \dfrac{1 - \cos\sqrt{9 - x^2 - y^2 - z^2}}{\sqrt{9 - x^2 - y^2 - z^2}} & \text{if } x^2 + y^2 + z^2 < 9 \\ 0 & \text{if } x^2 + y^2 + z^2 = 9 \end{cases}$$

54–59 Find any discontinuities of the given function and classify them as removable or nonremovable.

54. $f(x,y) = \dfrac{(x-1)y^2}{(x-1)^2 + y^2}$

55. $g(x,y) = \dfrac{xy}{x^2 + y^2}$

56. $h(x,y) = \dfrac{x}{x^2 + y^2}$

57. $p(x,y) = \ln\left(\dfrac{1}{x^2 + y^2}\right)$

58. $m(x,y) = \dfrac{x}{\sqrt{x^2 + (y-2)^2}}$

59. $k(x,y,z) = \dfrac{x^2 - 2y^2 + z^2}{x^2 + y^2 + z^2}$

60–65 *True or False?* Determine whether the given statement is true or false. In case of a false statement, explain or provide a counterexample.

60. If $\displaystyle\lim_{(0,y)\to(0,0)} f(x,y) = L$ and $\displaystyle\lim_{(x,0)\to(0,0)} f(x,y) = L,$ then $\displaystyle\lim_{(x,y)\to(0,0)} f(x,y) = L.$

61. If $\displaystyle\lim_{(x,y)\to(a,b)} f(x,y)$ exists and its value is L, then $f(a,b) = L.$

62. If $\displaystyle\lim_{(x,y)\to(a,b)} f(x,y)$ exists and its value is L, then $\displaystyle\lim_{x\to a} f(x,b) = L.$

63. If $\displaystyle\lim_{(x,y)\to(0,0)} f(x,y)$ exists and the limiting value of $f(x,y)$ along the line $y = x$ is $f(0,0)$, then $f(x,y)$ is continuous at $(0,0)$.

64. If $f(x,y) < 0$ in some ε-neighborhood $B_\varepsilon(0,0)$ of the origin and $\displaystyle\lim_{(x,y)\to(0,0)} f(x,y)$ exists, then $\displaystyle\lim_{(x,y)\to(0,0)} f(x,y) < 0.$

65. If $f(0,0) = L$ and $f(x,y)$ is continuous for all $(x,y) \neq (0,0)$, then $\displaystyle\lim_{(x,y)\to(0,0)} f(x,y) = L.$

13.3 Partial Derivatives

TOPICS

1. Evaluation and interpretation of partial derivatives

2. Higher-order and mixed partial derivatives

3. Differentiability and continuity

Since limits lie at the heart of differentiation, the derivative at a point $c \in \mathbb{R}^n$ of a multivariable function f must take into account all the different limiting approaches to c. Fortunately, we can extend our definition of differentiability by considering such functions one variable at a time. The precise manner in which we do so is the subject of this section.

TOPIC 1   Evaluation and Interpretation of Partial Derivatives

For simplicity of presentation we introduce the meaning and techniques of partial differentiation with functions of two variables, but all that follows generalizes and applies to functions of n variables as well. Accordingly, suppose (a,b) is a point in the domain of a real-valued function $z = f(x,y)$. The intersection of the plane $y = b$ and the graph of $z = f(x,y)$ is the graph of the one-variable function $z = f(x,b)$ in the plane $y = b$, as shown in Figure 1; we say that we are holding y constant with the value $y = b$. If the function $f(x,b)$ is differentiable in the variable x at the point $x = a$, then the limit of the difference quotient below exists and we define it as one of two *partial derivatives*. The value of that partial derivative tells us the slope of the line tangent to the curve $z = f(x,b)$ at the point $(a,b,f(a,b))$.

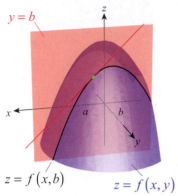

$y = b$

$z = f(x,b)$ $z = f(x,y)$

Figure 1

Definition

Partial Derivative with Respect to *x*

If the limit below exists, the **partial derivative of $f(x,y)$ with respect to x at (a,b)** is

$$\frac{\partial f}{\partial x}\bigg|_{(a,b)} = \lim_{h \to 0} \frac{f(a+h,b) - f(a,b)}{h} = \frac{d}{dx} f(x,b)\bigg|_{x=a}.$$

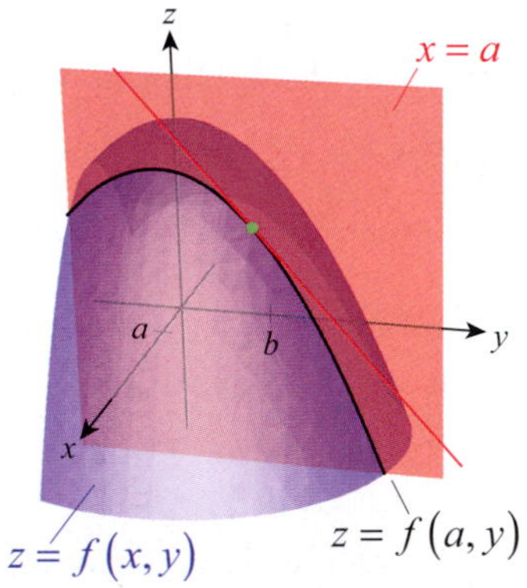

Similarly, if we hold x fixed with the value $x = a$, the graph of $z = f(a,y)$ in the plane $x = a$ is the graph of a function of a single variable y (see Figure 2). If it is differentiable in y at the point $y = b$, the value of its derivative is the other partial derivative.

$x = a$

$z = f(x,y)$ $z = f(a,y)$

Figure 2

Definition

Partial Derivative with Respect to *y*

If the limit below exists, the **partial derivative of $f(x,y)$ with respect to y at (a,b)** is

$$\frac{\partial f}{\partial y}\bigg|_{(a,b)} = \lim_{h \to 0} \frac{f(a,b+h) - f(a,b)}{h} = \frac{d}{dy} f(a,y)\bigg|_{y=b}.$$

As with ordinary derivatives, a variety of notations exist to denote partial differentiation—the choice of a particular notation often is a matter of taste, though sometimes one style or another is better suited to the context. Given the function

$z = f(x, y)$, any of the following may be used to indicate the two partial derivatives.

With respect to x: $\dfrac{\partial f}{\partial x} = \dfrac{\partial}{\partial x} f = f_x(x, y) = f_x = f_1 = D_1 f = D_x f = \dfrac{\partial z}{\partial x} = \dfrac{\partial}{\partial x} z = z_x$

With respect to y: $\dfrac{\partial f}{\partial y} = \dfrac{\partial}{\partial y} f = f_y(x, y) = f_y = f_2 = D_2 f = D_y f = \dfrac{\partial z}{\partial y} = \dfrac{\partial}{\partial y} z = z_y$

Verbal phrases used for these expressions are, to list a few, "the partial (derivative) of f with respect to x," "f sub x," "dee sub x of f," "the partial of f with respect to the first variable," and "z sub x," with similar expressions for the second row. To indicate the evaluation of one or the other partial derivative at a particular point (a, b), any of the following may be used.

$$\left.\frac{\partial f}{\partial x}\right|_{(a,b)} = \frac{\partial f}{\partial x}(a,b) = f_x(a,b) = \left.\frac{\partial z}{\partial x}\right|_{(a,b)} = z_x(a,b)$$

$$\left.\frac{\partial f}{\partial y}\right|_{(a,b)} = \frac{\partial f}{\partial y}(a,b) = f_y(a,b) = \left.\frac{\partial z}{\partial y}\right|_{(a,b)} = z_y(a,b)$$

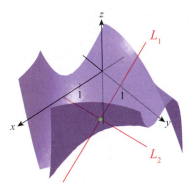

Figure 3

Example 1 ✎

Given the function $f(x, y) = x^3 y^2 - 2x^2 y - 3x + y$, determine $f_x(1,1)$ and $f_y(1,1)$. Then find equations for L_1 and L_2, the lines tangent to the surface $z = f(x, y)$ at the point $(1,1)$ and parallel to, respectively, the xz-plane and the yz-plane (see Figure 3).

Solution

We treat y as a constant in order to find the partial derivative f_x.

$$f_x(x, y) = 3x^2 y^2 - 4xy - 3$$

Similarly, we treat x as a constant to find f_y.

$$f_y(x, y) = 2x^3 y - 2x^2 + 1$$

Evaluating these two partial derivatives at the point $(1,1)$ gives us

$$f_x(1,1) = 3(1)^2(1)^2 - 4(1)(1) - 3 = -4 \quad \text{and} \quad f_y(1,1) = 2(1)^3(1) - 2(1)^2 + 1 = 1.$$

To determine the two tangent lines with these respective slopes, we also need to know

$$f(1,1) = (1)^3(1)^2 - 2(1)^2(1) - 3(1) + 1 = -3,$$

so the corresponding point on the surface $z = f(x, y)$ is $(1,1,-3)$. This means L_1 contains the point $(1,1,-3)$ and has slope, in the plane $y = 1$, equal to $f_x(1,1) = -4$. One way of expressing this fact is that $\langle 1, 0, -4 \rangle$ is a direction vector for L_1; that is, the line changes -4 units in the z-direction for every change of 1 unit in the x-direction. So a vector equation for the tangent line is

$$L_1: \langle 1,1,-3 \rangle + t \langle 1, 0, -4 \rangle = \langle 1 + t, 1, -3 - 4t \rangle.$$

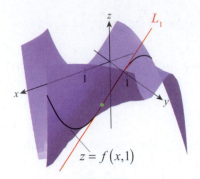

$z = f(x,1)$

Figure 4

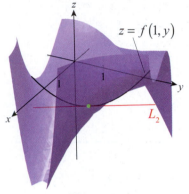

$z = f(1, y)$

Figure 5

Figure 4 illustrates the curve $z = f(x,1)$ in black and the line L_1 in red. Note that the shape of the curve is consistent with a cubic function, and $f(x,1)$ is indeed a cubic function in x.

Similar reasoning leads to the following vector equation for the tangent line L_2.

$$L_2: \langle 1,1,-3 \rangle + t \langle 0,1,1 \rangle = \langle 1, 1+t, -3+t \rangle$$

The curve $z = f(1, y)$ and the line L_2 are shown in Figure 5; note that the shape of the curve is consistent with a quadratic function, which is exactly the case for $f(1, y)$.

Example 2 ✐

Assuming x, y, and z are independent variables, find the three partial derivatives of the function $f(x, y, z) = x^2 \sin(yz^3) - \ln(xyz)$.

Solution

We can't depict the graph of f or illustrate its partial derivatives, since doing so would require four dimensions, but the mechanical process of partial differentiation is the same as for a function of two variables.

To find f_x, we treat y and z as constant and differentiate f with respect to x.

$$f_x(x, y, z) = \frac{\partial}{\partial x} \left[x^2 \sin(yz^3) - \ln(xyz) \right]$$

$$= 2x \sin(yz^3) - \frac{yz}{xyz} = 2x \sin(yz^3) - \frac{1}{x}$$

Holding x and z constant, we find

$$f_y(x, y, z) = \frac{\partial}{\partial y} \left[x^2 \sin(yz^3) - \ln(xyz) \right]$$

$$= x^2 z^3 \cos(yz^3) - \frac{xz}{xyz} = x^2 z^3 \cos(yz^3) - \frac{1}{y},$$

and holding x and y constant gives us

$$f_z(x, y, z) = \frac{\partial}{\partial z} \left[x^2 \sin(yz^3) - \ln(xyz) \right]$$

$$= 3x^2 yz^2 \cos(yz^3) - \frac{xy}{xyz} = 3x^2 yz^2 \cos(yz^3) - \frac{1}{z}.$$

In Example 2, we were explicitly told that the three variables used in defining the function were independent. But as we know, an *equation* in three variables often defines one of the variables as a function of the other two. In such cases, *implicit partial differentiation* can be used to determine the partial derivatives.

Example 3 ✐

Given the equation $z^2 - 4x^2 - 5y^2 = 0$, determine z_x and z_y wherever they exist.

Solution

From the classification in Section 11.6, we recognize the equation as an elliptic cone, and in a global sense the equation doesn't define a function of x and y (try solving the equation for z to verify this). But the equation *does* define z as a function of x and y in a neighborhood of every point on the surface. To find the partial derivatives z_x and z_y, we treat z as, respectively, a differentiable function in x and then y and implicitly differentiate with respect to each of the two variables.

With respect to x:

$$\frac{\partial}{\partial x}\left(z^2 - 4x^2 - 5y^2\right) = \frac{\partial}{\partial x}(0) \quad \Rightarrow \quad 2zz_x - 8x = 0 \quad \Rightarrow \quad z_x = \frac{4x}{z}$$

With respect to y:

$$\frac{\partial}{\partial y}\left(z^2 - 4x^2 - 5y^2\right) = \frac{\partial}{\partial y}(0) \quad \Rightarrow \quad 2zz_y - 10y = 0 \quad \Rightarrow \quad z_y = \frac{5y}{z}$$

These formulas for the partial derivatives both contain z, but that is not unusual in implicit differentiation calculations. Their meaning is that at any point (x, y, z) on the surface $z^2 - 4x^2 - 5y^2 = 0$ where z is locally a differentiable function in x and y individually, the rate of change of z in the x-direction is $4x/z$ and the rate of change in the y-direction is $5y/z$. As an example, the point $(2, -2, 6)$ lies on the surface and the partial derivatives at that point are

$$z_x(2, -2, 6) = \frac{4}{3} \quad \text{and} \quad z_y(2, -2, 6) = -\frac{5}{3}.$$

Figure 6 is an illustration of the surface and, in black, the curve that results from fixing $x = 2$; the point $(2, -2, 6)$ is shown in green and the line tangent to the point in the plane $x = 2$ is shown in red—this tangent line has slope $-\frac{5}{3}$. If we had chosen instead to work with the point $(2, -2, -6)$ on the surface, the slope of the tangent line in the plane $x = 2$ would be $\frac{5}{3}$.

The formulas found are not defined where $z = 0$, which in this example corresponds to the one point on the surface where z is not locally a differentiable function of x or of y.

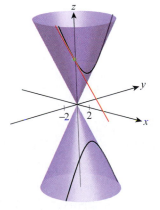

Figure 6

Technology Note 🖵

The syntax of the commands used by a computer algebra system to evaluate partial derivatives often reflects the underlying mathematics. For instance, the command in *Mathematica* is **D**, the same as for ordinary differentiation, and we obtain the desired partial derivatives by specifying which variable to differentiate with respect to. The three partial derivatives of the function $f(x, y, z) = x^2 \sin\left(yz^3\right) - \ln(xyz)$ of Example 2 can be found using *Mathematica*, as shown in Figure 7.

```
In[1]:= D[x^2 Sin[y * z^3] - Log[x * y * z], x]

Out[1]=  - 1/x + 2 x Sin[y z³]

In[2]:= D[x^2 Sin[y * z^3] - Log[x * y * z], y]

Out[2]=  - 1/y + x² z³ Cos[y z³]

In[3]:= D[x^2 Sin[y * z^3] - Log[x * y * z], z]

Out[3]=  - 1/z + 3 x² y z² Cos[y z³]
```

Figure 7

In Section 3.9, we saw how ordinary derivatives are used to determine the propagated error of measurements. The same principle applies to formulas that rely on measurements of multiple variables, as shown below.

Example 4 ✎

In general, the density ρ of an object is related to its mass m and volume v by $\rho = m/v$. If we let Δm denote the possible error in measurement of mass, then the corresponding propagated error in density $\Delta \rho$ is

$$\Delta \rho \approx \frac{\partial \rho}{\partial m} \Delta m = \frac{1}{v} \Delta m = \frac{m}{v} \frac{\Delta m}{m} = \rho \frac{\Delta m}{m}.$$

Similarly, a possible error Δv in the measurement of volume leads to propagated error in density of

$$\Delta \rho \approx \frac{\partial \rho}{\partial v} \Delta v = -\frac{m}{v^2} \Delta v = -\frac{m}{v} \frac{\Delta v}{v} = -\rho \frac{\Delta v}{v}.$$

If both m and v are subject to measurement error, the formula for the propagated error in density must reflect that fact. One common approach in such cases is to assume that the combined effect is approximately the square root of the sum of the squares of the individual errors.

$$\Delta \rho \approx \sqrt{\left(\rho \frac{\Delta m}{m} \right)^2 + \left(-\rho \frac{\Delta v}{v} \right)^2} = \rho \sqrt{\left(\frac{\Delta m}{m} \right)^2 + \left(\frac{\Delta v}{v} \right)^2}$$

In order to express the relationship in terms of relative error, this is often expressed as

$$\frac{\Delta \rho}{\rho} \approx \sqrt{\left(\frac{\Delta m}{m} \right)^2 + \left(\frac{\Delta v}{v} \right)^2}.$$

TOPIC 2   Higher-Order and Mixed Partial Derivatives

If the partial derivatives of a function $f(x, y)$ exist, they are also functions of x and y, so it is natural to investigate their own partial derivatives in turn. We use any of the following to denote the four possible **second-order partial derivatives** of $z = f(x, y)$.

$$\frac{\partial^2 f}{\partial x^2} = \frac{\partial}{\partial x}\left(\frac{\partial f}{\partial x}\right) = \frac{\partial}{\partial x}(f_x) = (f_x)_x = f_{xx} = f_{11} = \frac{\partial^2 z}{\partial x^2} = z_{xx}$$

$$\frac{\partial^2 f}{\partial y \partial x} = \frac{\partial}{\partial y}\left(\frac{\partial f}{\partial x}\right) = \frac{\partial}{\partial y}(f_x) = (f_x)_y = f_{xy} = f_{12} = \frac{\partial^2 z}{\partial y \partial x} = z_{xy}$$

$$\frac{\partial^2 f}{\partial x \partial y} = \frac{\partial}{\partial x}\left(\frac{\partial f}{\partial y}\right) = \frac{\partial}{\partial x}(f_y) = (f_y)_x = f_{yx} = f_{21} = \frac{\partial^2 z}{\partial x \partial y} = z_{yx}$$

$$\frac{\partial^2 f}{\partial y^2} = \frac{\partial}{\partial y}\left(\frac{\partial f}{\partial y}\right) = \frac{\partial}{\partial y}(f_y) = (f_y)_y = f_{yy} = f_{22} = \frac{\partial^2 z}{\partial y^2} = z_{yy}$$

The partial derivatives f_{xy} and f_{yx} are called **mixed partial derivatives**, since they are differentiated with respect to more than one variable. Note how the order of the differentiation relates to the notation so that, for instance,

$$\frac{\partial^2 f}{\partial x \partial y} = \frac{\partial}{\partial x}(f_y) = (f_y)_x = f_{yx}.$$

Example 5 ✐

Determine the second-order partial derivatives of $f(x, y) = x \sin y - e^{xy^2}$.

Solution

We first find

$$f_x = \sin y - y^2 e^{xy^2} \quad \text{and} \quad f_y = x \cos y - 2xy e^{xy^2}$$

and then proceed to determine the second-order partial derivatives.

$$f_{xx} = \frac{\partial}{\partial x}\left(\sin y - y^2 e^{xy^2}\right) = -y^4 e^{xy^2}$$

$$f_{xy} = \frac{\partial}{\partial y}\left(\sin y - y^2 e^{xy^2}\right) = \cos y - 2y e^{xy^2} - 2xy^3 e^{xy^2}$$

$$f_{yx} = \frac{\partial}{\partial x}\left(x \cos y - 2xy e^{xy^2}\right) = \cos y - 2y e^{xy^2} - 2xy^3 e^{xy^2}$$

$$f_{yy} = \frac{\partial}{\partial y}\left(x \cos y - 2xy e^{xy^2}\right) = -x \sin y - 2xe^{xy^2} - 4x^2 y^2 e^{xy^2}$$

It is no coincidence that $f_{xy} = f_{yx}$ in Example 5—as it turns out, this will always be the case if $f(x, y)$ satisfies the fairly lenient conditions of the following theorem, discovered by the French mathematician Alexis Clairaut (1713–1765). Its proof appears in Appendix E.

> ### Theorem $\mathcal{A}$
>
> #### Clairaut's Theorem
>
> If $f(x, y)$ and the partial derivatives f_x, f_y, f_{xy}, and f_{yx} are all defined on an open region containing the point (a, b), and if f_{xy} and f_{yx} are continuous at (a, b), then $f_{xy}(a, b) = f_{yx}(a, b)$.

Example 6 $\mathcal{G}$

Given $z = x^3 y + \dfrac{y \sin y^2}{y^2 + 3}$, find z_{yx}.

Solution

If we think about the partial derivatives of f, without actually determining what they are, we can fairly quickly decide that

(1) both z_x and z_y exist everywhere and

(2) z_x is significantly easier to compute than z_y.

In fact, there really is no concern about the existence or continuity of any partial derivative of any order (with respect to either variable): polynomials are never a cause for concern, and all derivatives of the second term will contain fractions whose denominators are powers of $y^2 + 3$, which can never be 0.

So since the two mixed second-order partial derivatives will be equal, we will compute z_{xy} instead of z_{yx}.

$$z_x = 3x^2 y \quad \Rightarrow \quad z_{xy} = 3x^2$$

Partial derivatives of higher order behave similarly, and the order of differentiation continues to be irrelevant as long as all the mixed derivatives up to the order being computed are continuous.

Example 7 $\mathcal{G}$

Given the function $f(x, y, z) = 2x^3 yz^2 - x \sin(yz) + y^5 z$, compute f_{zxyz} and f_{yxzz}.

Solution

Just for the purpose of illustration, the two mixed partial derivatives can also be written as shown below.

$$f_{zxyz} = \frac{\partial^4 f}{\partial z \partial y \partial x \partial z} \quad \text{and} \quad f_{yxzz} = \frac{\partial^4 f}{\partial z^2 \partial x \partial y}$$

We find the derivatives as follows.

$$f_{zxyz} = \frac{\partial^3}{\partial z \partial y \partial x}(f_z) = \frac{\partial^3}{\partial z \partial y \partial x}\left[4x^3yz - xy\cos(yz) + y^5\right]$$

$$= \frac{\partial^2}{\partial z \partial y}\left[12x^2yz - y\cos(yz)\right]$$

$$= \frac{\partial}{\partial z}\left[12x^2z - \cos(yz) + yz\sin(yz)\right]$$

$$= 12x^2 + 2y\sin(yz) + y^2z\cos(yz)$$

$$f_{yxzz} = \frac{\partial^3}{\partial z^2 \partial x}(f_y) = \frac{\partial^3}{\partial z^2 \partial x}\left[2x^3z^2 - xz\cos(yz) + 5y^4z\right]$$

$$= \frac{\partial^2}{\partial z^2}\left[6x^2z^2 - z\cos(yz)\right]$$

$$= \frac{\partial}{\partial z}\left[12x^2z - \cos(yz) + yz\sin(yz)\right]$$

$$= 12x^2 + 2y\sin(yz) + y^2z\cos(yz)$$

Since so many mathematical models involve functions of more than one variable, it's not surprising that equations relating the partial derivatives of functions are common; such equations are called **partial differential equations**. Entire texts are devoted to the subject of solving partial differential equations, so we will merely introduce them here. Examples include **wave equations** such as $u_{tt} = c^2 u_{xx}$ (discussed below), **Laplace's equation** $f_{xx} + f_{yy} + f_{zz} = 0$ (used in solving problems in electromagnetism, gravitation, and fluid dynamics), and the **Black-Scholes equation** $V_t + \frac{1}{2}\sigma^2 s^2 V_{ss} + rsV_s - rV = 0$ (used in modeling prices of stock options, and for which Robert Merton and Myron Scholes were awarded the Nobel Prize in Economics in 1997).

Example 8 ✒

The wave equation $u_{tt} = c^2 u_{xx}$ models the motion of a waveform as it travels along a string—the function $u(x,t)$ is the displacement of a point x units from one end of the string at time t, and the constant c is determined by physical characteristics such as the density of the string and its tension (see Figure 8 for a snapshot at one particular moment t).

Show that the function $u(x,t) = \sin(x - ct)$ is a solution of $u_{tt} = c^2 u_{xx}$.

Solution

Given $u(x,t) = \sin(x - ct)$,

$$u_x = \cos(x - ct) \quad \text{and} \quad u_t = -c\cos(x - ct)$$

and

$$u_{xx} = -\sin(x - ct) \quad \text{and} \quad u_{tt} = -c^2\sin(x - ct),$$

so it is indeed the case that $u_{tt} = c^2 u_{xx}$.

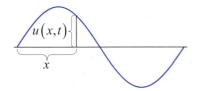

Figure 8

TOPIC 3 Differentiability and Continuity

Before leaving this section, it is important to make one more observation: ***The existence of the partial derivatives of a function does not necessarily mean the function is differentiable.*** In fact, and in seeming contrast to the one-variable case, the existence of the partial derivatives at a point is not even sufficient to guarantee continuity at that point!

Example 9 ✐

Define $f: \mathbb{R}^2 \to \mathbb{R}$ by

$$f(x, y) = \begin{cases} 1 & \text{if } xy = 0 \\ 0 & \text{if } xy \neq 0 \end{cases}.$$

That is, $f(x, y) = 1$ if (x, y) lies on either the x-axis or the y-axis, and $f(x, y) = 0$ for all other points. Then $f_x(0, 0) = 0$ (f has the constant value 1 along the x-axis and hence the rate of change of f in the x-direction is 0), and similarly $f_y(0, 0) = 0$. So although both partial derivatives of f exist at the origin, f is not continuous at the origin.

But, to rephrase an observation made at the start of this section, the derivative of a function $f: \mathbb{R}^2 \to \mathbb{R}$ at a given point (a, b) must take into account *all* the different limiting approaches to (a, b); the two partial derivatives $f_x(a, b)$ and $f_y(a, b)$ only take into account two specific kinds of approaches—namely, those parallel to the xz-plane and those parallel to the yz-plane. In fact, partial differentiation is the only kind of differentiation of multivariable functions that we *have* defined so far, and a careful rereading of this section reveals that we haven't actually determined yet what we mean by differentiability.

To remedy this, we first note that in the one-variable case, the existence of $f'(a)$ means that

$$\frac{f(a + \Delta x) - f(a)}{\Delta x} = f'(a) + \varepsilon$$

where $\varepsilon \to 0$ as $\Delta x \to 0$; this follows from rewriting the difference quotient definition of derivative in the form

$$\lim_{\Delta x \to 0} \left[\underbrace{\frac{f(a + \Delta x) - f(a)}{\Delta x} - f'(a)}_{\varepsilon} \right] = 0.$$

So letting $\Delta f = f(a + \Delta x) - f(a)$, differentiability of f at a means that the increment Δf can be written as

$$\Delta f = f'(a) \Delta x + \varepsilon \Delta x,$$

where $\varepsilon \to 0$ as $\Delta x \to 0$. We extend this to the multivariable case in the following definition.

> **Definition** 💡
>
> ### Differentiability of Multivariable Functions
>
> A function $f: \mathbb{R}^2 \to \mathbb{R}$ is **differentiable** at (a,b) if $f_x(a,b)$ and $f_y(a,b)$ exist and the increment $\Delta f = f(a+\Delta x, b+\Delta y) - f(a,b)$ satisfies an equation of the form
>
> $$\Delta f = f_x(a,b)\Delta x + f_y(a,b)\Delta y + \varepsilon_1 \Delta x + \varepsilon_2 \Delta y$$
>
> where each of ε_1 and ε_2 approach 0 as both Δx and Δy approach 0.
>
> Differentiability of a function $f: \mathbb{R}^3 \to \mathbb{R}$ at a point (a,b,c) requires that $f_x(a,b,c)$, $f_y(a,b,c)$, and $f_z(a,b,c)$ exist and that the increment Δf satisfies
>
> $$\Delta f = f_x(a,b,c)\Delta x + f_y(a,b,c)\Delta y + f_z(a,b,c)\Delta z + \varepsilon_1 \Delta x + \varepsilon_2 \Delta y + \varepsilon_3 \Delta z$$
>
> where each of ε_1, ε_2, and ε_3 approach 0 as $(\Delta x, \Delta y, \Delta z) \to (0,0,0)$. Analogous statements apply for the differentiability of functions of more than three variables.

Now that we have a definition of differentiability, we can ask how we might recognize whether a function is differentiable at a given point. While the hypotheses below can be weakened slightly, the following theorem is used most frequently to answer this question. Its proof appears in Appendix E.

> **Theorem** 🔍
>
> ### The Increment Theorem of Differentiability
>
> If the partial derivatives of $f(x,y)$ exist throughout an open region R containing the point (a,b), and if f_x and f_y are both continuous at (a,b), then f is differentiable at (a,b). That is, the increment Δf at (a,b) can be written as
>
> $$\Delta f = f_x(a,b)\Delta x + f_y(a,b)\Delta y + \varepsilon_1 \Delta x + \varepsilon_2 \Delta y$$
>
> where each of ε_1 and ε_2, approach 0 as $(\Delta x, \Delta y) \to (0,0)$. Hence, if the partial derivatives are continuous throughout R, f is differentiable on all of R.
>
> Analogous statements apply to functions of three or more variables.

Note that our increment-based definition of differentiability allows us to easily extend the important fact below from the one-variable to the multivariable case.

> **Theorem** 🔍
>
> ### Differentiability Implies Continuity
>
> If f is differentiable at a point, then f is continuous at that point.

> ### Proof 📎
>
> For convenience, we will prove the theorem in the two-variable case.
>
> If f is differentiable at (a,b), then by definition the increment Δf can be written in such a way that
>
> $$\lim_{\substack{\Delta x \to 0 \\ \Delta y \to 0}} \left[f(a+\Delta x, b+\Delta y) - f(a,b) \right] = \lim_{\substack{\Delta x \to 0 \\ \Delta y \to 0}} \left[f_x(a,b)\Delta x + f_y(a,b)\Delta y + \varepsilon_1 \Delta x + \varepsilon_2 \Delta y \right]$$
>
> $$= f_x(a,b) \lim_{\Delta x \to 0} \Delta x + f_y(a,b) \lim_{\Delta y \to 0} \Delta y$$
>
> $$+ \lim_{\Delta x \to 0} \varepsilon_1 \lim_{\Delta x \to 0} \Delta x + \lim_{\Delta y \to 0} \varepsilon_2 \lim_{\Delta y \to 0} \Delta y$$
>
> $$= 0.$$
>
> That is, $f(a+\Delta x, b+\Delta y) \to f(a,b)$ as $\Delta x \to 0$ and $\Delta y \to 0$.

By implication, then, the function in Example 9 does not meet the conditions of the Increment Theorem of Differentiability at $(0,0)$. In Exercise 103, you will show that this is indeed the case.

13.3 Exercises

1–4 Determine z_x and z_y at the indicated point; then find equations for the corresponding tangent lines that are parallel to, respectively, the xz-plane and the yz-plane.

1. $z = x + 3xy^3$; $(2,1)$

2. $z = x^4 y - 4xy^3$; $(2,-1)$

3. $z = x(2-xy)^2$; $(-1,0)$

4. $z = (xy^2 + 3)(x - 3y)$; $(3,1)$

5–28 Find all first-order partial derivatives of the given function.

5. $f(x,y) = x^3 + y^3$

6. $g(x,y) = xy^2 - 2xy$

7. $h(x,y) = 5x^3 y + y^4$

8. $r(x,y) = x^4 - 2x^2 y^2 + 3y^6$

9. $V(r,h) = \dfrac{\pi r^2 h}{3}$

10. $s(x,y) = (x^2 + y^4)^5$

11. $k(x,y) = xy^2 + \sqrt{2xy}$

12. $l(x,y) = (4xy - 3)(x^2 + 1)$

13. $m(x,y) = (xy - 3x^2)^4$

14. $n(x,y) = x(1 - \sqrt{xy})(y^2 + 2)$

15. $p(x,y) = \dfrac{x^2 y}{x + y}$

16. $F(x,y) = \sqrt{1 - x^2 - y^2}$

17. $G(x,y) = \dfrac{y}{\sqrt{x^2 + y^2}}$

18. $H(x,y) = e^{x^2 y}$

19. $I(x,y) = \sin \dfrac{x}{y}$

20. $J(x,y) = y \ln \sqrt{x^2 + y^2}$

21. $K(x,y) = \dfrac{2y^2}{x} - \dfrac{x^2}{2y}$

22. $L(x,y) = e^x \cos(xy)$

23. $M(x,y) = \arctan \sqrt{xy}$

24. $f(x,y,z) = e^x y^2 \sin z$

25. $g(x,y,z) = x^{y/z}$

26. $A(a,b,c) = 2(ab + ac + bc)$

27. $h(x,y,z) = x \cos(y + z^2)$

28. $w(r,s,t) = (r^2 + 2s^2 + 3t^2)^{3/2}$

29–34 Find a vector equation for L, the line tangent to the surface $z^2 - 4x^2 - 5y^2 = 0$ of Example 3 at the given point and parallel to the indicated coordinate plane.

29. $(1,1,3)$;   the xz-plane

30. $(1,1,-3)$;   the yz-plane

31. $\left(2, \dfrac{3}{\sqrt{5}}, 5\right)$;   the xz-plane

32. $\left(2, \dfrac{3}{\sqrt{5}}, -5\right)$;   the yz-plane

33. $(-1,1,3)$;   the xz-plane

34. $(-1,1,-3)$;   the yz-plane

35–37 Use implicit differentiation to determine z_x and z_y.

35. $z^2 + 2xy - yz = 0$

36. $zx - (z + x)^2 = 3y$

37. $xz + \ln z - x^2 y = 0$

38. Treating y and z as independent variables, determine x_y and x_z from Exercise 36.

39. Find y_x and y_z by implicitly differentiating $y^2 - xy - z \ln y = 5$.

40. Recall the Ideal Gas Law, the equation relating the pressure P, volume V, and temperature T of an ideal gas: $PV = nRT$ (Section 3.4, Exercise 97). Assuming that n (i.e., the amount of gas in moles) is constant, differentiate implicitly to find the following partial derivatives and explain their physical meaning.

  a. $\dfrac{\partial P}{\partial T}$ **b.** $\dfrac{\partial P}{\partial V}$ **c.** $\dfrac{\partial V}{\partial T}$

41–48 Verify the equality of the mixed partials f_{xy} and f_{yx}.

41. $f(x, y) = x^3 y - 2y^2 + 5xy^4$

42. $f(x, y) = (2y - x)^4$

43. $f(x, y) = (x^4 + y^4)^8$

44. $f(x, y) = xy^4 + 2y^{2/3}$

45. $f(x, y) = x^2 y^2 e^{xy}$

46. $f(x, y) = \ln(x^2 + y^2)$

47. $f(x, y) = e^{\sqrt{x^2 + y^2}}$

48. $f(x, y) = 2y \cos(3x + 4y)$

49–54 Verify that the third-order mixed partials g_{xyz}, g_{yzx}, and g_{zxy} are equal.

49. $g(x, y, z) = x^3 + 3yz^2 - xy^2 + 3z^3$

50. $g(x, y, z) = 2x(y - 3z)^3$

51. $g(x, y, z) = x^3 y^3 z^3$

52. $g(x, y, z) = \sin(xyz)$

53. $g(x, y, z) = e^{xy} \cos z$

54. $g(x, y, z) = \dfrac{y}{x^2 + z^2}$

55. Find the partial derivative f_{xyy} of the function

$$f(x, y) = x\left(\dfrac{y^3}{3} + \ln y\right) + y(\cos x + x \sin x) + x \sec^2 x$$

by judiciously choosing the order of differentiation. (**Hint:** See Example 7.)

56–63 Use the most convenient order of differentiation to find the indicated partial derivative, as in Exercise 55.

56. g_{xzyz}; $g(x, y, z) = \ln(x^2 y^2) + z \cos(y^2) + z^2 x^3 y^4$

57. h_{rstu}; $h(r, s, t, u) = s\left(u^2 e^r + \dfrac{tr^2}{u}\right) + re^{rs} \sin t$
$$+ t(s^2 + t \ln u)$$

58. f_{xy}; $f(x, y) = xe^{y^2} + \dfrac{\ln x}{\sqrt{x}}$

59. f_{yxyx}; $f(x, y) = \sin(x^2 - y)$

60. f_{xyz}; $f(x, y, z) = z(xy^2 + \sqrt{x} \ln x) + x^2 \sin y$

61. f_{xyz}; $f(x, y, z) = \dfrac{x^2 y}{2}(1 + 3z^2) + \cos y - \cos x$

62. f_{utst}; $f(r, s, t, u) = ru(t^2 \sin s + 1) + \dfrac{e^{(t^2 + 1)}}{r}$

63. f_{xyzw}; $f(w, x, y, z) = \sqrt{w^2 + x^2 + y^2 + z^2}$

64–65 Use the definition of the partial derivative to find $\partial f/\partial x$ and $\partial f/\partial y$ for the given function.

64. $f(x,y) = x^2 y$ **65.** $f(x,y) = \dfrac{\sqrt{y}}{x}$

66. Use the contour map below to estimate the values of f_x and f_y at the indicated points. (Answers will vary.)

    a. $(0,5)$ **b.** $(3,3)$ **c.** $(8,1)$ **d.** $(8,3)$

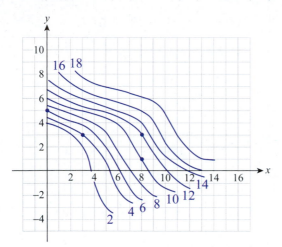

67–70 If possible, find a function that has the indicated partial derivatives. If such a function doesn't exist, explain why.

67. $f_x(x,y) = 2xy$, $f_y(x,y) = x^2 - \cos y$

68. $f_x(x,y) = \dfrac{1}{\sqrt{x}} + \ln(y^2+1)$, $f_y(x,y) = \dfrac{2xy}{y^2+1}$

69. $f_x(x,y) = x - y^2$, $f_y(x,y) = \sqrt{x} - 2y$

70. $f_x(x,y) = y\sin x + \dfrac{1}{2\sqrt{x}}$, $f_y(x,y) = \cos x + 3y^2$

71. Recall the lens equation from Exercise 44 of Section 3.8:

$$\frac{1}{o} + \frac{1}{i} = \frac{1}{f},$$

where o is the object distance, i is the image distance, and f is the focal length of the lens. Find the partial derivative $\partial i/\partial o$ and explain its physical meaning.

72. Explain why there is a difference in sign between the first two formulas for $\Delta\rho$ in Example 4.

73. Recall that the kinetic energy E of an object of mass m moving at speed v is found from the formula $E = \frac{1}{2}mv^2$. Suppose that a 1 kg mass is moving at a speed of 5 m/s, but due to inaccuracies of the measuring devices and human error, its mass was recorded as 1002 grams and its speed was clocked at 495 cm/s. Estimate the relative error this causes in the value of E. (**Hint:** See Example 4.)

74. Consider the lens equation of Exercise 71, and suppose that a lens that is thought to be a 50 mm lens has an actual focal length of 52.56 mm. If, in addition, the actual object distance of 2.03 m is erroneously measured to be exactly 2 m, estimate the relative error all of this causes in the value of i.

75–80 Show that the given function satisfies the wave equation of Example 8.

75. $u(x,t) = \sin(x+ct) + e^{x-ct}$

76. $u(x,t) = \cosh(x+ct)$

77. $u(x,t) = \sin(\omega x)\cos(\omega ct)$

78. $u(x,t) = (x+ct)^4 + (x-ct)^4$

79. $u(x,t) = \cos(x+ct) + \dfrac{1}{x-ct}$

80. $u(x,t) = \ln(x+ct) + \sqrt{x-ct}$

81–86 Verify that f satisfies the two-dimensional form of Laplace's equation: $f_{xx} + f_{yy} = 0$.

81. $f(x,y) = e^x \cos y$

82. $f(x,y) = \sinh y \cos x$

83. $f(x,y) = x^2 - y^2$

84. $f(x,y) = e^x \sin y + e^y \cos x$

85. $f(x,y) = \arctan \dfrac{x}{y}$

86. $f(x,y) = \ln(x^2 + y^2)^2$

87–89 Verify that the three-variable function satisfies Laplace's equation: $f_{xx} + f_{yy} + f_{zz} = 0$.

87. $f(x, y, z) = kxyz$

88. $f(x, y, z) = \sin(3x)e^{2y - \sqrt{5}z}$

89. $f(x, y, z) = \dfrac{1}{\sqrt{x^2 + y^2 + z^2}}$

90–92 The temperature $u(x,t)$ of an insulated rod that is aligned with the x-axis satisfies the heat equation $u_t = \kappa u_{xx}$, where κ is a constant, called *thermal diffusivity*. Show that the functions in Exercises 90–92 satisfy the heat equation.

90. $u(x,t) = e^{-\kappa t}\sin x$ **91.** $u(x,t) = e^{-4\kappa t}\sin 2x$

92. $u(x,t) = 5e^{-\kappa n^2 t}\cos(nx)$

93. Suppose that the temperature of an insulated planar surface is $T(x, y, t) = e^{-5t}\sin x \cos 2y$. Find the rates of change of temperature in the x- and y-directions, respectively, at the point $(\pi/4, \pi/4, 0)$.

94. Show that $T(x, y, t)$ of Exercise 93 satisfies the two-dimensional heat equation $u_t = \kappa\left(u_{xx} + u_{yy}\right)$ with $\kappa = 1$.

95. Generalizing Exercises 92 and 94, show that all functions of the form
$$u(x, y, t) = e^{-\kappa(m^2 + n^2)t}\sin(mx)\cos(ny)$$
satisfy the two-dimensional heat equation $(m, n \in \mathbb{R})$.

96. Suppose a guitar string, originally aligned with the positive x-axis, is plucked and the function $g(x,t)$ describes the displacement of the point $(x, 0)$ as a function of t. What can you say about a point $(x_0, 0)$ at time $t = t_0$ if both partial derivatives $g_x(x_0, t_0)$ and $g_t(x_0, t_0)$ are positive? What if, in addition, $g_{xx}(x_0, t_0)$ and $g_{tt}(x_0, t_0)$ are both negative?

97.* Recall from trigonometry that the area of a circular sector of radius r and central angle α is $A = \frac{1}{2}r^2\alpha$. Denoting the circumference of the sector by C, prove that the area can also be expressed as $A = \frac{1}{2}Cr - r^2$; then determine $\partial A/\partial r$ from both of the given area formulas and explain why your answers are not equivalent.

98. Consider the general cubic polynomial in the variables x and y:
$$P(x, y) = Ax^3 + Bx^2 y + Cxy^2 + Dy^3 + Ex^2$$
$$+ Fxy + Gy^2 + Hx + Iy + J$$

Find conditions on the coefficients to ensure that $P(x, y)$ satisfies Laplace's equation (such functions are called *harmonic*).

99.* If we invest P dollars and take inflation and taxes into consideration, the future value of our investment in n years is
$$A = P\left[\frac{1 + r(1 - T)}{1 + I}\right]^n \text{ dollars,}$$
where I and T are the inflation and tax rates, respectively, and r is the annual interest rate. Suppose we invest \$15,000 for 10 years at a rate of 12\%. Use the partial derivatives $\partial A/\partial I$ and $\partial A/\partial T$ to decide whether it is inflation or the tax rate that affects the investment more drastically.

100.* Prove that the first partial derivatives of a harmonic function are themselves harmonic, if they have continuous second partial derivatives (see Exercise 98 for the definition of harmonic functions).

101.* Let R denote the net resistance of two resistors R_1 and R_2 in a parallel circuit, which satisfies the equation
$$\frac{1}{R} = \frac{1}{R_1} + \frac{1}{R_2}.$$

a. Generalize the above equation to show that the net resistance of n resistors in a parallel circuit is as follows:
$$R = \frac{\prod\limits_{i=1}^{n} R_i}{\sum\limits_{i=1}^{n} \prod\limits_{j \neq i} R_j}$$

b. For a fixed index k, find a formula for $\partial R/\partial R_k$. (**Hint:** It is helpful to look at the cases $n = 2$ and $n = 3$ before generalizing.)

102. Prove that if $u(x,t)$ can be written in the form $u(x,t) = u_1(x+ct) + u_2(x-ct)$ for some one-variable functions u_1 and u_2 that are at least twice differentiable, then $u(x,t)$ is a solution of the wave equation of Example 8.

103. Show that the function in Example 9 does not meet the conditions of the Increment Theorem of Differentiability at $(0,0)$. (**Hint:** Show, for example, that f_x is not continuous at the origin by looking at f_x at nearby points on the x-axis.)

104–107 Decide whether the given function is differentiable at the origin. Give a reason for your answer.

104. $f(x,y) = y^2 e^x - x^2 y$

105. $f(x,y) = \sqrt{x^2 + 2y^2}$

106. $f(x,y) = \dfrac{1}{x^2 + y^2 + 1}$

107. $f(x,y,z) = \dfrac{1}{\sqrt{x^2 + y^2 + z^2}}$

108–113 Show that f is not differentiable at the origin, even though both $f_x(0,0)$ and $f_y(0,0)$ exist. (In Exercises 112 and 113, generalize to the three-variable case.)

108. $f(x,y) = \begin{cases} \dfrac{xy}{x^2 + y^2} & \text{if } (x,y) \neq (0,0) \\ 0 & \text{if } (x,y) = (0,0) \end{cases}$

109. $f(x,y) = \begin{cases} \dfrac{2x^2 y}{x^2 + y^4} & \text{if } (x,y) \neq (0,0) \\ 0 & \text{if } (x,y) = (0,0) \end{cases}$

110. $f(x,y) = \begin{cases} \dfrac{-4x^2 y^2}{x^4 + y^4} & \text{if } (x,y) \neq (0,0) \\ 0 & \text{if } (x,y) = (0,0) \end{cases}$

111. $f(x,y) = \begin{cases} 1 & \text{if } y^2 < x < 2y^2 \\ -1 & \text{otherwise} \end{cases}$

112. $f(x,y,z) = \begin{cases} 2 & \text{if } xyz = 0 \\ -2 & \text{if } xyz \neq 0 \end{cases}$

(**Hint:** See Example 9.)

113. $f(x,y,z) = \begin{cases} 1 & \text{if } x^2 + y^2 < z < 2x^2 + 2y^2 \\ -1 & \text{otherwise} \end{cases}$

(**Hint:** See Exercise 111.)

114.* Consider the given piecewise defined function.

$$F(x,y) = \begin{cases} \dfrac{xy^2}{2(x^2 + y^2)} & \text{if } (x,y) \neq (0,0) \\ 0 & \text{if } (x,y) = (0,0) \end{cases}$$

a. In contrast to some of the previous exercises, show that F is continuous at $(0,0)$. (**Hint:** See, for example, Exercise 31 or 41 of Section 13.2.)

b. Prove that both partial derivatives of F are equal to 0 at $(0,0)$. (**Hint:** Use the definition of partial derivatives.)

c. Prove that F is not differentiable at $(0,0)$. (**Hint:** As noted after Example 9 in the text, the derivative of F at $(0,0)$ must take into account *all* the different limiting approaches to $(0,0)$; however, compare the common value of the two partial derivatives of F at $(0,0)$ with $\displaystyle\lim_{h \to 0} \dfrac{F(h,h) - F(0,0)}{h}$.)

115–118 For certain functions, it is fairly straightforward to demonstrate differentiability by using the definition. For example, let $f(x,y) = x + y^2$ and note the following.

$$\begin{aligned} \Delta f(x,y) &= f(x+\Delta x, y+\Delta y) - f(x,y) \\ &= x + \Delta x + (y+\Delta y)^2 - (x + y^2) \\ &= x + \Delta x + y^2 + 2y\Delta y + (\Delta y)^2 - x - y^2 \\ &= \Delta x + 2y\Delta y + (\Delta y)^2 \\ &= f_x(x,y)\Delta x + f_y(x,y)\Delta y + 0 \cdot \Delta x + \Delta y \cdot \Delta y \end{aligned}$$

Letting $\varepsilon_1 = 0$ and $\varepsilon_2 = \Delta y$, we see that ε_1 and ε_2 approach 0 as both Δx and Δy approach 0, as needed. In Exercises 115–118, mimic this process of using the definition to prove that the function is differentiable.

115. $f(x,y) = x^2 - 2y$ **116.** $g(x,y) = xy^2$

117. $h(x,y) = 2(x^2 + y^2)$

118. $k(x,y) = x^3 - 4x + 3y$

119. Find all points (x,y) where the function $f(x,y) = |x - y|$ is differentiable.

120. Repeat Exercise 119 for the function $g(x,y) = \sqrt{x^2 + y^2}$.

13.4 **The Chain Rule**

TOPICS

1. The Chain Rule and its proof

2. Implicit differentiation via the Chain Rule

The Chain Rule, as introduced in Section 3.4, makes differentiating the composition of two functions of a single variable relatively easy. Recall that if $y = f(u)$ and $u = g(x)$, and if the derivatives are all defined, then

$$\frac{dy}{dx} = \frac{dy}{du}\frac{du}{dx}.$$

In this section, we extend the Chain Rule to handle the composition of functions when at least one of them depends on more than one variable.

TOPIC 1   **The Chain Rule and Its Proof**

We will build toward a general statement of the Chain Rule by considering first the case of a function $y = f(u, v)$ where both u and v are functions of another variable x, say $u = g(x)$ and $v = h(x)$. Then ultimately, y is a function of the single variable x, a fact which is apparent if we write $y = f(g(x), h(x))$; in this context, we often call u and v **intermediate variables** between the independent variable x and the dependent variable y. But the rate of change of y with respect to x has to reflect the fact that y is defined in terms of two functions of x that are independent of one another.

Theorem ⚲

The Chain Rule (Case I)

If $y = f(u, v)$ is a differentiable function and if $u = g(x)$ and $v = h(x)$ are both differentiable functions of x, then the composite function $y = f(g(x), h(x))$ is a differentiable function of x and

$$\frac{dy}{dx} = \frac{\partial y}{\partial u}\frac{du}{dx} + \frac{\partial y}{\partial v}\frac{dv}{dx}.$$

Alternatively, $f'(x) = f_u(g(x), h(x))g'(x) + f_v(g(x), h(x))h'(x)$.

Proof ✎

The method of proof is identical in nature to the proof of the Chain Rule in Section 3.4, but our increment-based definition of differentiability allows for a more elegant presentation.

By the definition of differentiability in Section 13.3,

$$\Delta y = f_u \Delta u + f_v \Delta v + \varepsilon_1 \Delta u + \varepsilon_2 \Delta v$$

at any point where y is differentiable, and $\varepsilon_1 \to 0$ and $\varepsilon_2 \to 0$ as $(\Delta u, \Delta v) \to (0, 0)$. So the ratio of the increment Δy to the increment Δx is

$$\frac{\Delta y}{\Delta x} = f_u \frac{\Delta u}{\Delta x} + f_v \frac{\Delta v}{\Delta x} + \varepsilon_1 \frac{\Delta u}{\Delta x} + \varepsilon_2 \frac{\Delta v}{\Delta x}.$$

Both u and v are continuous functions of x since they are differentiable, and hence $\Delta u \to 0$ and $\Delta y \to 0$ as $\Delta x \to 0$. So we can determine dy/dx by letting $\Delta x \to 0$.

$$\frac{dy}{dx} = \lim_{\Delta x \to 0}\left(f_u\frac{\Delta u}{\Delta x} + f_v\frac{\Delta v}{\Delta x} + \varepsilon_1\frac{\Delta u}{\Delta x} + \varepsilon_2\frac{\Delta v}{\Delta x} \right)$$

$$= f_u\frac{du}{dx} + f_v\frac{dv}{dx} + (0)\left(\frac{du}{dx}\right) + (0)\left(\frac{dv}{dx}\right) = \frac{\partial y}{\partial u}\frac{du}{dx} + \frac{\partial y}{\partial v}\frac{dv}{dx}$$

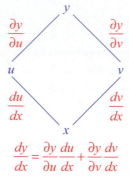

$$\frac{dy}{dx} = \frac{\partial y}{\partial u}\frac{du}{dx} + \frac{\partial y}{\partial v}\frac{dv}{dx}$$

Figure 1

A **tree diagram**, such as the one shown in Figure 1, is a convenient way to remember the Chain Rule and determine how to apply it for any given set of independent, intermediate, and dependent variables. To use such a diagram, multiply the derivatives appearing along each path from top to bottom and then add the resulting products.

Example 1 ✐

Given that $y = u^3 v^2$ and $u = x^2$ and $v = x^3 - 1$, determine dy/dx.

Solution

Using the Chain Rule, we determine dy/dx as follows:

$$\frac{dy}{dx} = \frac{\partial y}{\partial u}\frac{du}{dx} + \frac{\partial y}{\partial v}\frac{dv}{dx}$$

$$= \left(3u^2v^2\right)(2x) + \left(2u^3v\right)\left(3x^2\right)$$

$$= 3\left(x^4\right)\left(x^3 - 1\right)^2 (2x) + 2\left(x^6\right)\left(x^3 - 1\right)\left(3x^2\right)$$

$$= 6x^5\left(x^3 - 1\right)\left(2x^3 - 1\right)$$

In Exercise 1, you will show that the same result is obtained by first expressing y explicitly as a function of x.

In some cases the relationships between the variables have a natural geometric interpretation, as shown in Example 2.

Example 2 ✐

Determine the rate of change with respect to t of the function $f(x, y) = xy$ along the Lissajous curve described parametrically by $x = \sin 2t$ and $y = \cos 3t$. Then calculate $f'(t)$ at the specific points corresponding to $t = 0$ and $t = \pi/6$ (see Figure 2).

Solution

Using the Chain Rule, we have the following.

$$f'(t) = \frac{\partial z}{\partial x}\frac{dx}{dt} + \frac{\partial z}{\partial y}\frac{dy}{dt}$$

$$= (y)(2\cos 2t) + (x)(-3\sin 3t)$$

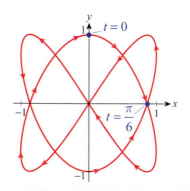

Figure 2

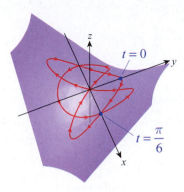

Figure 3

If we desired, we could now express $f'(t)$ as a function of t alone, but the previous formula is sufficient to evaluate $f'(t)$ at the two points where $t = 0$ and $t = \pi/6$. These two points are, respectively, $(0,1)$ and $\left(\sqrt{3}/2, 0\right)$.

$$f'(0) = (1)(2\cos 0) + (0)(-3\sin 0) = 2$$

$$f'\left(\frac{\pi}{6}\right) = (0)\left(2\cos\frac{\pi}{3}\right) + \frac{\sqrt{3}}{2}\left(-3\sin\frac{\pi}{2}\right) = -\frac{3\sqrt{3}}{2}$$

If we consider the surface $z = f(x, y)$, then the two derivatives we just found correspond to the rates of change of a point traveling along the curve shown in Figure 3 at the two locations corresponding to $t = 0$ and $t = \pi/6$. Note how the signs of the derivatives reflect the fact that z is increasing along the curve when $t = 0$, but decreasing along the curve when $t = \pi/6$.

There are many ways in which multivariable functions may be composed with one another, but we will consider one more specific way before stating the Chain Rule in greater generality.

Theorem

The Chain Rule (Case II)

If $z = f(u, v)$ is a differentiable function and if $u = g(x, y)$ and $v = h(x, y)$ are both differentiable functions of x and y, then the composite function $z = f\big(g(x, y), h(x, y)\big)$ is a differentiable function of x and y and

$$\frac{\partial z}{\partial x} = \frac{\partial z}{\partial u}\frac{\partial u}{\partial x} + \frac{\partial z}{\partial v}\frac{\partial v}{\partial x} \quad \text{and} \quad \frac{\partial z}{\partial y} = \frac{\partial z}{\partial u}\frac{\partial u}{\partial y} + \frac{\partial z}{\partial v}\frac{\partial v}{\partial y}.$$

Example 3

If $z = e^{u+2v}$ while $u = x^3 y$ and $v = y^2 - x$, determine $\partial z/\partial x$ and $\partial z/\partial y$.

Solution

A tree diagram is again useful in remembering and applying the Chain Rule, and Figure 4 illustrates the relevant relationships of the variables. In this case, to determine the derivative of the top variable with respect to either of the independent variables, multiply the derivatives along each appropriate path and add the resulting products.

$$\frac{\partial z}{\partial x} = \frac{\partial z}{\partial u}\frac{\partial u}{\partial x} + \frac{\partial z}{\partial v}\frac{\partial v}{\partial x}$$

$$= \left(e^{u+2v}\right)\left(3x^2 y\right) + \left(2e^{u+2v}\right)(-1)$$

$$= \left(e^{u+2v}\right)\left(3x^2 y - 2\right)$$

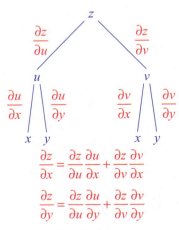

Figure 4

$$\frac{\partial z}{\partial y} = \frac{\partial z}{\partial u}\frac{\partial u}{\partial y} + \frac{\partial z}{\partial v}\frac{\partial v}{\partial y}$$

$$= \left(e^{u+2v}\right)\left(x^3\right) + \left(2e^{u+2v}\right)(2y)$$

$$= \left(e^{u+2v}\right)\left(x^3 + 4y\right)$$

If desired, we could now substitute $u = x^3 y$ and $v = y^2 - x$ to express each partial derivative explicitly in terms of x and y.

Theorem 🔍

The Chain Rule (General Case)

If $y = f\left(u_1, u_2, \ldots, u_n\right)$ is a differentiable function of $u_1, u_2, \ldots, u_n$, and if each variable $u_j = g\left(x_1, x_2, \ldots, x_m\right)$ is a differentiable function of $x_1, x_2, \ldots, x_m$, then y is a differentiable function of $x_1, x_2, \ldots, x_m$ and

$$\frac{\partial y}{\partial x_i} = \frac{\partial y}{\partial u_1}\frac{\partial u_1}{\partial x_i} + \frac{\partial y}{\partial u_2}\frac{\partial u_2}{\partial x_i} + \cdots + \frac{\partial y}{\partial u_n}\frac{\partial u_n}{\partial x_i}$$

for each $i = 1, 2, \ldots, m$.

Example 4 ✍

Given that $w = 2xy^3 + x^2 z - yz^2$ while $x = s - t$, $y = s^2 + t^2$, and $z = st$, determine w_s and w_t (answers may be left in terms of the intermediate and independent variables).

Solution

The relevant tree diagram appears in Figure 5, and we apply the same procedure as previously.

$$w_s = \frac{\partial w}{\partial x}\frac{\partial x}{\partial s} + \frac{\partial w}{\partial y}\frac{\partial y}{\partial s} + \frac{\partial w}{\partial z}\frac{\partial z}{\partial s}$$

$$w_t = \frac{\partial w}{\partial x}\frac{\partial x}{\partial t} + \frac{\partial w}{\partial y}\frac{\partial y}{\partial t} + \frac{\partial w}{\partial z}\frac{\partial z}{\partial t}$$

Hence,

$$w_s = \left(2y^3 + 2xz\right)(1) + \left(6xy^2 - z^2\right)(2s) + \left(x^2 - 2yz\right)(t)$$

and

$$w_t = \left(2y^3 + 2xz\right)(-1) + \left(6xy^2 - z^2\right)(2t) + \left(x^2 - 2yz\right)(s).$$

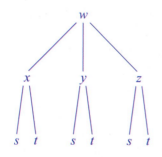

Figure 5

Example 5 ✎

Assume $z = f(x, y)$ has continuous second-order partial derivatives. If $x = v^2 - u^2$ and $y = uv$, determine z_{uu} and z_{vv}.

Solution

We begin by finding z_u.

$$z_u = \frac{\partial z}{\partial x}\frac{\partial x}{\partial u} + \frac{\partial z}{\partial y}\frac{\partial y}{\partial u} = -2uz_x + vz_y$$

Without knowing anything further about the function f, z_u must be left in terms of z_x and z_y. However, we do know that z_x and z_y are also functions of u and v, so we can differentiate with respect to u again to determine z_{uu}, making use of the equality of mixed partial derivatives along the way.

$$z_{uu} = \frac{\partial}{\partial u}\left(-2uz_x + vz_y\right) = -2z_x + (-2u)\frac{\partial}{\partial u}(z_x) + v\frac{\partial}{\partial u}(z_y)$$

$$= -2z_x - 2u\left[\frac{\partial}{\partial x}(z_x)\frac{\partial x}{\partial u} + \frac{\partial}{\partial y}(z_x)\frac{\partial y}{\partial u}\right] + v\left[\frac{\partial}{\partial x}(z_y)\frac{\partial x}{\partial u} + \frac{\partial}{\partial y}(z_y)\frac{\partial y}{\partial u}\right]$$

$$= -2z_x - 2u\left(-2uz_{xx} + vz_{xy}\right) + v\left(-2uz_{yx} + vz_{yy}\right)$$

$$= -2z_x + 4u^2 z_{xx} - 4uvz_{xy} + v^2 z_{yy}$$

In Exercise 35 you will show that $z_{vv} = 2z_x + 4v^2 z_{xx} + 4uvz_{xy} + u^2 z_{yy}$.

TOPIC 2   Implicit Differentiation via the Chain Rule

We have seen many instances in which an equation in two variables can be interpreted as defining one of the variables as a function of the other, at least in a restricted neighborhood. Similarly, an equation in three variables can be taken as defining one as a function of the other two (again, sometimes only locally). We have also seen how implicit differentiation can be used to find the rates of change of implicitly defined functions. The Chain Rule gives us a relatively simple way to determine such derivatives.

To see why, let $F(x, y) = 0$ be an equation in x and y (written so that all the terms appear on one side), and assume the equation implicitly defines y as a function of x. Then if F is differentiable, differentiation with respect to x yields

$$\frac{\partial}{\partial x}F(x, y) = \frac{\partial}{\partial x}(0), \quad \text{or} \quad \frac{\partial F}{\partial x}\frac{dx}{dx} + \frac{\partial F}{\partial y}\frac{dy}{dx} = 0.$$

Since $dx/dx = 1$, solving for dy/dx gives us

$$\frac{dy}{dx} = -\frac{F_x}{F_y},$$

assuming we are evaluating dy/dx at a point where $F_y \neq 0$.

Example 6 ✍

Find dy/dx, given that $x^3 - y^2 = \sin(xy)$.

Solution

Defining $F(x, y) = x^3 - y^2 - \sin(xy)$, the equation has the form $F(x, y) = 0$.

$$\frac{dy}{dx} = -\frac{F_x}{F_y} = -\frac{3x^2 - y\cos(xy)}{-2y - x\cos(xy)} = \frac{3x^2 - y\cos(xy)}{2y + x\cos(xy)}$$

The **Implicit Function Theorem**, typically proved in an advanced calculus course, outlines conditions under which an equation is guaranteed to define one variable as a differentiable function of the others. In the two-variable case, the theorem states that if $F(x, y)$ is defined and has continuous partial derivatives in a neighborhood of the point (a, b), and if $F(a, b) = 0$ and $F_y(a, b) \neq 0$, then the equation $F(x, y) = 0$ defines y as a function of x near (a, b) and $dy/dx = -F_x/F_y$.

The three-variable case is similar: if $F(x, y, z)$ is defined and has continuous partial derivatives in a neighborhood of the point (a, b, c), and if $F(a, b, c) = 0$ and $F_z(a, b, c) \neq 0$, then the equation $F(x, y, z) = 0$ defines z as a function of x and y and

$$\frac{\partial z}{\partial x} = -\frac{F_x}{F_z} \quad \text{and} \quad \frac{\partial z}{\partial y} = -\frac{F_y}{F_z}.$$

Example 7 ✍

Find z_x and z_y, given that $x^2 y - xz^3 = xyz$.

Solution

Defining $F(x, y, z) = x^2 y - xz^3 - xyz$, the equation has the form $F(x, y, z) = 0$. Now we can find z_x and z_y.

$$z_x = -\frac{F_x}{F_z} = -\frac{2xy - z^3 - yz}{-3xz^2 - xy} = \frac{2xy - z^3 - yz}{3xz^2 + xy}$$

$$z_y = -\frac{F_y}{F_z} = -\frac{x^2 - xz}{-3xz^2 - xy} = \frac{x^2 - xz}{3xz^2 + xy}$$

13.4 **Exercises**

1. Obtain the result of Example 1 by first expressing y explicitly as a function of x.

2–5 Use a tree diagram to apply the Chain Rule to express the indicated derivative of the given function.

2. $\dfrac{df}{dt}$; $f = f\big(x(t), y(t)\big)$

3. $\dfrac{dg}{dt}$; $g = g\big(x(t), y(t), z(t)\big)$

4. $\dfrac{\partial h}{\partial u}$; $h = h\big(x(u,v), y(u,v), z(u,v)\big)$

5. $\dfrac{\partial k}{\partial z}$; $k = k\big(u(x,y,z), v(x,y,z)\big)$

6–17 Determine dy/dx, given y as a function of $u(x)$ and $v(x)$. In Exercises 6–13, check your answer by expressing y explicitly as a function of x and differentiating. In Exercises 14–17, generalize to three intermediate variables.

6. $y = u^2 v$; $u = 3x + 1$, $v = x^4$

7. $y = uv^2 - \cos u$; $u = 2x^2$, $v = \sqrt{x}$

8. $y = u^3 v - \sin u$; $u = 2x$, $v = x^2$

9. $y = \ln \dfrac{u}{\sqrt{v}}$; $u = \sin x$, $v = \cos^2 x$

10. $y = \sqrt{u^2 + v^2}$; $u = e^x \cos x$, $v = e^x \sin x$

11. $y = \dfrac{1}{u} + \dfrac{1}{v}$; $u = \csc^2 x$, $v = \cot x$

12. $y = u \arctan v$; $u = e^x$, $v = \tan x$

13. $y = \arcsin \dfrac{u}{v}$; $u = 2x$, $v = x^3$

14. $y = uv + uw + vw$; $u = 2x$, $v = x + 2$, $w = x^2 + 2$

15. $y = u \sin^2(vw)$; $u = 2x$, $v = x^2$, $w = \dfrac{1}{x}$

16. $y = uvw$; $u = e^x$, $v = 3x$, $w = x^3$

17. $y = uve^w$; $u = 5x - 2$, $v = \sin x$, $w = \ln x$

18. Determine the value of $f'(7\pi/6)$ for the function in Example 2.

19–21 After determining the rate of change with respect to t of the function $f(x, y)$ along the indicated parametric curve, find $f'(t)$ at the given point.

19. $f(x, y) = x^2 y$ along the curve $x = 10 \cos t$, $y = t$, at $t = \pi/2$

20. $f(x, y) = (x + y)^2$ along the curve $x = t \cos t$, $y = t \sin t$, at $t = \pi$

21. $f(x, y) = xe^{-y}$ along the curve $x = 4 - 2t^2$, $y = t^3 - 9t$, at $t = 0$

22. Find the rate of change of the function $f(x, y, z) = x^2 y + \sin z$ along the helix $x = \cos 2t$, $y = \sin 2t$, $z = t/2$. Express your answer in terms of the variable t.

23–33 Use the Chain Rule to determine the partial derivatives z_x and z_y. (Answers may be left in terms of the intermediate and independent variables.)

23. $z = u^2 + v^2$; $u = x + 2y$, $v = y - x$

24. $z = u \sin v$; $u = x^2 + y$, $v = x - y^2$

25. $z = v^3 - 2u^2 v$; $u = \cos y$, $v = \sin x$

26. $z = v^2 - 2uv$; $u = x \sin y$, $v = y \cos x$

27. $z = 2v^4 - 3u\sqrt{v}$; $u = e^x$, $v = e^y$

28. $z = \cos(u + v^2)$; $u = y - \dfrac{x^2}{2}$, $v = 3x - 2y$

29. $z = u^2 v - uw^3$; $u = x + y$, $v = x^2 - y$, $w = 2xy$

30. $z = u(v + w)^2 - v^2$; $u = 3x$, $v = 5y^2$, $w = x - 2y$

31. $z = v^2 e^{2u+3w}$; $u = x - y$, $v = 2y - x$, $w = 3xy$

32. $z = \ln(u^2 + v^2 + w^2)$; $u = 2x - y$, $v = xy$, $w = e^y$

33. $z = v \sin(uw^2)$; $u = x + y^3$, $v = 3y - x$, $w = x^2 y$

34. Find w_ρ, w_θ, and w_ϕ if $w = 2x + y^2 z$ and $x = \rho \cos \theta \sin \phi$, $y = \rho \sin \theta \sin \phi$, and $z = \rho \cos \theta$. (Answers may be left in terms of the intermediate and independent variables.)

35. Prove that for $z = f\left(x(u,v), y(u,v)\right)$ of Example 5,

$$z_{vv} = 2z_x + 4v^2 z_{xx} + 4uv z_{xy} + u^2 z_{yy}.$$

36. Assume (as in Example 5) that $z = f(x,y)$ has continuous second-order partial derivatives and that $x = u^2 v$ and $y = u + v^2$. Find z_{uu} and z_{vv}.

37. Suppose $z = f(x,y)$, where $x = r\cos\theta$ and $y = r\sin\theta$. Prove the following.

$$\left(\frac{\partial f}{\partial x}\right)^2 + \left(\frac{\partial f}{\partial y}\right)^2 = \left(\frac{\partial f}{\partial r}\right)^2 + \frac{1}{r^2}\left(\frac{\partial f}{\partial \theta}\right)^2$$

(**Hint:** Start by determining f_r and f_θ.)

38. Suppose $z = f(x,y)$ is as in Exercise 37. Prove that

$$\frac{\partial^2 f}{\partial x^2} + \frac{\partial^2 f}{\partial y^2} = \frac{\partial^2 f}{\partial r^2} + \frac{1}{r^2}\frac{\partial^2 f}{\partial \theta^2} + \frac{1}{r}\frac{\partial f}{\partial r}.$$

(This is called the *Laplacian* of f. Can you see why?)

39–47 Find dy/dx, where y is given implicitly by the given equation.

39. $x^2 - xy + y^2 = \dfrac{1}{4}$ **40.** $(4-x)y^2 = 2x^3$

41. $(x^2 + 4)y = 8$ **42.** $\dfrac{y^2}{4}(8-x) = x^3$

43. $x^{2/3} + y^{2/3} = 8$

44. $(2x^2 + y^2)^2 - 4x^2 y = 0$

45. $x^2 + y^2 = (x^2 + y^2 - 3x)^2$

46. $(x^2 + y^2)^2 = 9xy$

47. $\dfrac{y}{x^2 + y^2} = 3 + x^2$

48–55 Find z_x and z_y, where z is defined implicitly by the given equation.

48. $x^2 + y^2 + z^2 = 1$ **49.** $xyz = e^{x+y+z}$

50. $xy - 3y^3 - 4xz^2 = 1$ **51.** $x^2 z^3 + xy = \sin(yz)$

52. $x\sin y + y^2 z - e^{xyz} = 1$

53. $e^x \sin y - 2z^2 + \dfrac{yz^2}{2} = 2$

54. $x^2 y + z^2 + y\ln z = 4$

55. $\ln(x^2 + y^2 + z^2) = 5 - xyz$

56. If $F(x,y) = 0$ implicitly defines y as a function of x and both F and y are twice-differentiable, show that

$$y''(x) = -\frac{F_{xx}\left(F_y\right)^2 - 2F_x F_y F_{xy} + F_{yy}\left(F_x\right)^2}{\left(F_y\right)^3}.$$

57. Use a tree diagram to write out the Chain Rule for the first partial derivatives f_x and f_y of $f(t,u,v,w)$, where $t = t(x,y)$, $u = u(x,y)$, $v = v(x,y)$, and $w = w(x,y)$.

58. If $f(x,y)$ is differentiable, where $x(u,v) = u + v$ and $y(u,v) = u - v$, prove the following.

$$\left(\frac{\partial f}{\partial u}\right)\left(\frac{\partial f}{\partial v}\right) = \left(\frac{\partial f}{\partial x}\right)^2 - \left(\frac{\partial f}{\partial y}\right)^2$$

59. Suppose $F(x,y,z)$ is differentiable, has nonzero first partial derivatives, and that $F(x,y,z) = 0$ defines each variable as a function of the other two variables (i.e., $x = x(y,z)$, $y = y(x,z)$, and $z = z(x,y)$). Prove the following:

$$\frac{\partial x}{\partial y}\frac{\partial y}{\partial z}\frac{\partial z}{\partial x} = -1$$

60–63 A function f is said to be *homogeneous of degree n* if

$$f(tx, ty) = t^n f(x,y)$$

for all $n, t \in \mathbb{R}$. In these exercises, you will work with homogeneous functions. To begin, show that the given function in Exercises 60 and 61 is homogeneous and state the degree of homogeneity.

60. $f(x,y) = 5x^2 y^2 - 3x^3 y$

61. $f(x,y) = \dfrac{xy^2 + 2x^2 y}{\sqrt{x^4 + y^4}}$

62. Let $f(x, y)$ be homogeneous of degree n. Prove the following:

$$x\frac{\partial f(x, y)}{\partial x} + y\frac{\partial f(x, y)}{\partial y} = nf(x, y)$$

(**Hint:** Consider the equation of homogeneity; differentiate both sides with respect to t, and let $t = 1$.)

63. Let $f(x, y)$ be homogeneous, as in Exercise 62. Prove the following:

$$\frac{\partial f(tx, ty)}{\partial x} = t^{n-1}\frac{\partial f(x, y)}{\partial x} \quad \text{and}$$

$$\frac{\partial f(tx, ty)}{\partial y} = t^{n-1}\frac{\partial f(x, y)}{\partial y}$$

64. An ice "cube" in the form of a rectangular prism with a square base is melting so that the edge of the base is shrinking at 0.5 mm/min while the height is decreasing at 0.75 mm/min. Determine the rate of change of its volume and surface area when the edge of the base is 20 mm and the height is 30 mm.

65. Consider a circular sector of radius r and central angle θ. Suppose that θ is increasing at a rate of 0.1 radians per minute, while r is decreasing at a rate of 0.2 inches per minute. Find the rate of change of the area at the instant when $\theta = 1$ radian and $r = 15$ inches.

66. Suppose the height of a right circular cylinder is increasing at 1 millimeter per second. Determine the rate of change of the radius of the cylinder if the instantaneous rate of change of its volume is 0 when the radius is 50 millimeters and the height is 100 millimeters.

67. Consider a sand cone such as one formed by a child pouring sand out of a bucket. Assume this time that its height is growing at a rate of 0.1 inches per second, while its radius at 0.05 inches per second, at the instant when its height is 4 inches and its radius is 6 inches. Find the rate of change of the volume of the sand cone at this instant.

68. Find the rate of change of the lateral surface area of the sand cone at the instant described in Exercise 67.

69. Suppose that at a certain moment during takeoff, a plane's speed is 100 m/s, its acceleration 3 m/s^2, while its mass of 63,350 kg is decreasing at a rate of 1.15 kg/s due to fuel consumption. Find the rate of change of the plane's kinetic energy at this instant.

70. Suppose the temperature of two moles of an ideal gas in a 50-liter (L) container is 323 kelvin (K) and increasing at a rate of 0.2 K/s, when at the same time, the volume of the container is increasing at a rate of 0.05 L/s. Find the rate of change of pressure at this instant. (For a refresher on the Ideal Gas Law, see Exercise 97 in Section 3.4.)

71.* Consider the insulated plane of Exercise 93 of Section 13.3, with temperature measured in degrees Celsius and time in minutes. Suppose a point is moving along the line $y = -x + \pi/2$, in the southeastern direction, at a speed of $\sqrt{2}/2$ unit lengths per minute. Supposing that it is at the point $(\pi/4, \pi/4)$ at $t = 5$ seconds, find the rate of temperature change from the moving point's perspective at that instant. (**Hint:** Determine the rates dx/dt, dy/dt, and use the Chain Rule.)

13.5 Directional Derivatives and Gradient Vectors

TOPICS

1. Directional derivatives

2. Gradient vectors

The partial derivatives of a function tell us its rate of change in directions parallel to the coordinate planes, and we have begun to see how the Chain Rule can be used to find rates of change in other directions. We now explore this idea further to develop the definition of *directional derivative*, along with a general method of determining it.

TOPIC 1   Directional Derivatives

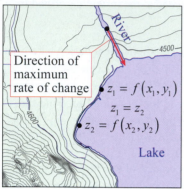

Figure 1

The definition and techniques we will develop apply to functions of any number of variables, but we'll begin with functions of two variables since our ability to visualize them as surfaces is helpful. To make the discussion even more concrete, suppose $z = f(x, y)$ describes the elevation at a point with longitude x and latitude y—as we know, one way of depicting $f(x, y)$ is as a topographical map, as shown in Figure 1. The shoreline of the lake in the figure constitutes one particular level curve, or contour, of the surface, and the rate of change of z at any point of the shoreline *in the direction of the shoreline* must be 0, since z neither rises nor falls along the level curve. On the other hand, flowing water seeks out the course of most rapid descent, so the rate of change of z at any point of the river *in the direction of flow* must be a negative number with the largest possible magnitude, which is always normal to the level curves. Implicitly, the rate of change of z in both cases is measured with respect to horizontal distance.

More formally, suppose that $f(x, y)$ is a differentiable function and that C is a curve parametrized by the differentiable functions $x = g(t)$ and $y = h(t)$. Then by the Chain Rule, as seen in Example 2 of Section 13.4, the rate of change of f with respect to t at a given point of C is

$$\frac{df}{dt} = \frac{\partial f}{\partial x}\frac{dx}{dt} + \frac{\partial f}{\partial y}\frac{dy}{dt}.$$

Suppose $\mathbf{u} = \langle u_1, u_2 \rangle$ is a unit vector. To determine the derivative of f at the point (a, b) in the direction of $\mathbf{u}$, which we will denote as $D_\mathbf{u} f(a, b)$, we let C be the line parametrized by $x = a + tu_1$ and $y = b + tu_2$, and define

$$D_\mathbf{u} f(a, b) = \left.\frac{df}{dt}\right|_{(a,b)} = f'(0).$$

Since $|t|$ is the distance between the points $(a + tu_1, b + tu_2)$ and (a, b), $D_\mathbf{u} f(a, b)$ is the rate of change of $f(x, y)$ at the point (a, b) with respect to distance, or arc length, along the line C. We can also express this directly as a limit of difference quotients, as shown next.

Definition 💡

Directional Derivative

If the limit exists, the **directional derivative** of $f(x, y)$ at the point (a, b) in the direction of the unit vector $\mathbf{u} = \langle u_1, u_2 \rangle$ is

$$D_\mathbf{u} f(a, b) = \lim_{t \to 0} \frac{f(a + tu_1, b + tu_2) - f(a, b)}{t}.$$

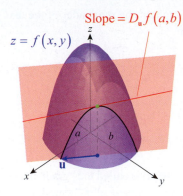

Slope $= D_{\mathbf{u}}f(a,b)$

$z = f(x,y)$

Figure 2

Geometrically, $D_{\mathbf{u}}f(a,b)$ is the slope of the line passing through $\big(a,b,f(a,b)\big)$ and tangent to the curve formed by intersecting the surface $f(x,y)$ with the vertical plane containing the point $(a,b,0)$ and the vector $\mathbf{u} = u_1\mathbf{i} + u_2\mathbf{j}$, as shown in Figure 2.

Example 1 ✍

Find the derivative of $f(x,y) = 3xy$ at the point $(2,1)$ in the direction of $\mathbf{v} = \langle 1,-1 \rangle$.

Solution

It is important to first convert the direction vector to a unit vector, so that the line C along which we are differentiating is parametrized with respect to arc length (in Exercise 13 you will examine the effect of parametrizing C in other ways).

Accordingly, let $\mathbf{u} = \dfrac{\mathbf{v}}{|\mathbf{v}|} = \dfrac{1}{\sqrt{2}}\langle 1,-1 \rangle$.

$$
\begin{aligned}
D_{\mathbf{u}}f(2,1) &= \lim_{t\to 0}\frac{f(2+tu_1,1+tu_2)-f(2,1)}{t}\\[2mm]
&= \lim_{t\to 0}\frac{f\left(2+\dfrac{t}{\sqrt{2}},1-\dfrac{t}{\sqrt{2}}\right)-f(2,1)}{t}\\[2mm]
&= \lim_{t\to 0}\frac{3\left(2+\dfrac{t}{\sqrt{2}}\right)\left(1-\dfrac{t}{\sqrt{2}}\right)-3(2)(1)}{t}\\[2mm]
&= \lim_{t\to 0}3\left(-\dfrac{1}{\sqrt{2}}-\dfrac{t}{2}\right) = -\dfrac{3}{\sqrt{2}}
\end{aligned}
$$

TOPIC 2   **Gradient Vectors**

The calculation performed in Example 1 came from the difference quotient interpretation of directional derivative, which is useful for conceptual understanding but which fails to use the power of the Chain Rule. To see how the Chain Rule helps, consider again a unit direction vector $\mathbf{u} = \langle u_1,u_2 \rangle$ and the line C parametrized by $x = a + tu_1$ and $y = b + tu_2$. Then the derivative of $f(x,y)$ at the point (a,b) in the direction of $\mathbf{u}$ is as follows.

$$
\begin{aligned}
D_{\mathbf{u}}f(a,b) &= \left.\frac{df}{dt}\right|_{(a,b)}\\[2mm]
&= \left.\frac{\partial f}{\partial x}\right|_{(a,b)}\left.\frac{dx}{dt}\right|_{t=0} + \left.\frac{\partial f}{\partial y}\right|_{(a,b)}\left.\frac{dy}{dt}\right|_{t=0}\\[2mm]
&= f_x(a,b)u_1 + f_y(a,b)u_2
\end{aligned}
$$

Using dot product notation, we can write this as follows:

$$
D_{\mathbf{u}}f(a,b) = \langle f_x(a,b),f_y(a,b)\rangle\cdot\langle u_1,u_2\rangle
$$

Since expressions like this occur frequently throughout the remainder of the text, it is convenient to make the following definition.

Definition 💡

Gradient Vector

The **gradient vector** of the function $f(x,y)$ is the vector

$$\nabla f(x,y) = \langle f_x(x,y), f_y(x,y) \rangle.$$

The symbol ∇ (an upside-down Greek capital delta) is called the "gradient" or "del" operator, and the vector ∇f is read as "gradient of f," "grad f," or "del f."

The gradient operator extends naturally to functions of more than two variables. Given a function $f(x_1, x_2, \ldots, x_n)$,

$$\nabla f(x_1, x_2, \ldots, x_n) = \langle f_{x_1}(x_1, x_2, \ldots, x_n), f_{x_2}(x_1, x_2, \ldots, x_n), \ldots, f_{x_n}(x_1, x_2, \ldots, x_n) \rangle.$$

In everyday language, "gradient" usually refers to a slope or rate of change, and ∇f can be remembered as simply listing, in vector form, the partial derivatives (rates of change) of f in each of the coordinate directions. Given the work above, we can now efficiently compute directional derivatives as follows.

Theorem 🔍

Computation of $D_{\mathbf{u}} f(c)$

Assuming the derivative of $f(x,y)$ at the point (a,b) in the direction of the unit vector $\mathbf{u} = \langle u_1, u_2 \rangle$ exists,

$$D_{\mathbf{u}} f(a,b) = \langle f_x(a,b), f_y(a,b) \rangle \cdot \langle u_1, u_2 \rangle = \nabla f(a,b) \cdot \mathbf{u}.$$

More generally, if $f(x_1, x_2, \ldots, x_n)$ is differentiable at the point $c = (c_1, c_2, \ldots, c_n)$ and if $\mathbf{u} = \langle u_1, u_2, \ldots, u_n \rangle$ is a unit vector, then

$$D_{\mathbf{u}} f(c) = \nabla f(c) \cdot \mathbf{u}.$$

Example 2 ✎

Find the derivative of $f(x,y) = x^2 y - e^{xy}$ at the point $(1,-1)$ in the direction of $\langle 2, 3 \rangle$.

Solution

We first determine ∇f at an arbitrary point (x,y).

$$\nabla f(x,y) = \langle f_x(x,y), f_y(x,y) \rangle = \langle 2xy - ye^{xy}, x^2 - xe^{xy} \rangle$$

So $\nabla f(1,-1) = \langle -2 + e^{-1}, 1 - e^{-1} \rangle$. Next, we convert the direction vector into a unit vector pointing in the same direction.

$$\mathbf{u} = \frac{1}{\sqrt{13}}\langle 2,3\rangle$$

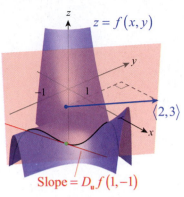

Finally,

$$D_{\mathbf{u}}f(1,-1) = \frac{1}{\sqrt{13}}\langle -2+e^{-1},1-e^{-1}\rangle \cdot \langle 2,3\rangle$$

$$= \frac{1}{\sqrt{13}}\left(-4+2e^{-1}+3-3e^{-1}\right) = -\frac{1+e^{-1}}{\sqrt{13}} = -\frac{e+1}{e\sqrt{13}}.$$

Since $D_{\mathbf{u}}f(1,-1)$ is negative, we know that the surface $z=f(x,y)$ must have a downward trend in the direction of $2\mathbf{i}+3\mathbf{j}$ at the point $(1,-1)$. Figure 3 illustrates the intersection of the surface $z=f(x,y)$ with the vertical plane containing the point $(1,-1,0)$ and the vector $2\mathbf{i}+3\mathbf{j}$, along with the line tangent to the curve defined by the intersection. As expected, the surface does indeed show a decreasing value for $f(x,y)$ at the point $(1,-1)$ if we head in the direction of $2\mathbf{i}+3\mathbf{j}$.

Figure 3 caption: $z=f(x,y)$; $\langle 2,3\rangle$; Slope $=D_{\mathbf{u}}f(1,-1)$

Figure 3

Recall that we have an alternative formula for the dot product of two vectors

$$\mathbf{u}\cdot\mathbf{v} = |\mathbf{u}||\mathbf{v}|\cos\theta,$$

where θ is the angle between the vectors $\mathbf{u}$ and $\mathbf{v}$ (positioned with the same initial point). Applying this to directional derivatives,

$$D_{\mathbf{u}}f(c) = \nabla f(c)\cdot\mathbf{u} = |\nabla f(c)||\mathbf{u}|\cos\theta = |\nabla f(c)|\cos\theta,$$

where θ is the angle between $\nabla f(c)$ and $\mathbf{u}$. Given our knowledge of cosine, this immediately tells us the following three important facts.

Properties of the Directional Derivative

Assume f is differentiable at c and that $\nabla f(c)\neq\mathbf{0}$.

1. The greatest rate of increase of f at c occurs when $\theta=0$, since $\cos 0=1$; that is, when $\mathbf{u}$ points in the direction of $\nabla f(c)$. Further, if $\mathbf{u}$ points in the direction of $\nabla f(c)$,

$$D_{\mathbf{u}}f(c) = |\nabla f(c)|\cos 0 = |\nabla f(c)|.$$

2. The greatest rate of decrease of f at c occurs when $\theta=\pi$, since $\cos\pi=-1$; that is, when $\mathbf{u}$ points in the direction of $-\nabla f(c)$. Further, if $\mathbf{u}$ points in the opposite direction of $\nabla f(c)$,

$$D_{\mathbf{u}}f(c) = |\nabla f(c)|\cos\pi = -|\nabla f(c)|.$$

3. The rate of change of f at c is 0 if $\theta=\pi/2$, since $\cos(\pi/2)=0$; that is, when $\mathbf{u}$ points in any direction orthogonal to $\nabla f(c)$. This follows from the fact that if $\mathbf{u}$ is orthogonal to $\nabla f(c)$,

$$D_{\mathbf{u}}f(c) = |\nabla f(c)|\cos\frac{\pi}{2} = |\nabla f(c)|\cdot 0 = 0.$$

Example 3 ✐

Given the function $f(x,y) = \sin(x + y^2)$, find the direction and value of **a.** greatest rate of increase, **b.** greatest rate of decrease, and **c.** no change from the point $(2,1)$.

Solution

We first determine ∇f.

$$\nabla f(x,y) = \langle f_x(x,y), f_y(x,y) \rangle = \langle \cos(x + y^2), 2y\cos(x + y^2) \rangle$$

So

$$\nabla f(2,1) = \langle \cos 3, 2\cos 3 \rangle \approx \langle -0.99, -1.98 \rangle$$

and

$$|\nabla f(2,1)| = \sqrt{(-0.99)^2 + (-1.98)^2} \approx 2.21.$$

This tells us the following.

a. The direction of most rapid increase in f from the point $(2,1)$ is approximately $\langle -0.99, -1.98 \rangle$ and the rate of change of f in that direction is approximately 2.21.

b. The direction of most rapid decrease in f from the point $(2,1)$ is approximately $\langle 0.99, 1.98 \rangle$ and the rate of change of f in that direction is approximately -2.21.

c. The two directions of no change in f from the point $(2,1)$ are normal to $\nabla f(2,1)$, so they are approximately $\langle 1.98, -0.99 \rangle$ and $\langle -1.98, 0.99 \rangle$. Note that this is consistent with the contours of the surface shown in Figure 4.

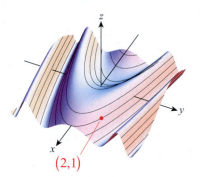

$(2,1)$

Figure 4

The fact that ∇f is always normal to any particular contour can also be shown as follows. Suppose $\mathbf{r}(t) = \langle x(t), y(t) \rangle$ is a smooth level curve of the differentiable function $f(x,y)$, say $f(\mathbf{r}(t)) = k$. Then

$$\frac{d}{dt} f(\mathbf{r}(t)) = \frac{d}{dt} k$$
$$f_x x'(t) + f_y y'(t) = 0$$
$$\langle f_x, f_y \rangle \cdot \langle x'(t), y'(t) \rangle = 0$$
$$\nabla f \cdot \mathbf{r}'(t) = 0.$$

Recall that $\mathbf{r}'(t)$ points in a direction tangential to the curve $\mathbf{r}(t)$, so ∇f is normal to the level curve $\mathbf{r}$ at every point. We can use this observation to quickly find tangent lines to implicitly defined curves.

Example 4 📝

Find an equation for the line tangent to the graph of $(x-2)^2 + 4(y+1)^2 = 8$ at the point $(4,0)$.

Solution

The given curve is a contour of the function $f(x,y) = (x-2)^2 + 4(y+1)^2$.

$$\nabla f(x,y) = \langle 2x-4, 8y+8 \rangle$$
$$\nabla f(4,0) = \langle 4,8 \rangle$$

The fact that the gradient is always normal to the contour through any particular point means that $\langle 4,8 \rangle$ is normal to the line tangent to the curve $(x-2)^2 + 4(y+1)^2 = 8$ at the point $(4,0)$. We can express this fact in equation form by noting that if (x,y) denotes an arbitrary point on the tangent line, then $\langle x-4, y-0 \rangle$ and $\langle 4,8 \rangle$ must be perpendicular. That is,

$$\langle x-4, y \rangle \cdot \langle 4,8 \rangle = 0$$
$$4x - 16 + 8y = 0$$
$$4x + 8y = 16.$$

Figure 5 shows the relationship between the contour $f(x,y) = 8$ (in blue), its gradient vector $\nabla f(4,0)$ (in green), and the line tangent to the contour at $(4,0)$ (in red).

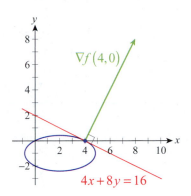

Figure 5

Since the gradient vector of a function is a listing of its partial derivatives, the next formulas come as no surprise. Informally, the properties say that gradients of arithmetic combinations of functions behave like ordinary derivatives. You will be asked to prove these properties in Exercises 53–55.

Properties of the Gradient

Assume f and g are both differentiable functions and that k is a fixed real number. Then the following laws hold.

Sum/Difference Law	$\nabla(f \pm g) = \nabla f \pm \nabla g$
Constant Multiple Law	$\nabla(kf) = k\nabla f$
Product Law	$\nabla(fg) = f\nabla g + g\nabla f$
Quotient Law	$\nabla\left(\dfrac{f}{g}\right) = \dfrac{g\nabla f - f\nabla g}{g^2}$, provided $g \neq 0$

Example 5 ✍

Determine the gradient of the function $f(x,y) = \dfrac{x^2 - xy^3}{x^2 + y^2}$.

Solution

$$\nabla f(x,y) = \nabla\left(\frac{x^2 - xy^3}{x^2 + y^2}\right)$$

$$= \frac{\left(x^2 + y^2\right)\nabla\left(x^2 - xy^3\right) - \left(x^2 - xy^3\right)\nabla\left(x^2 + y^2\right)}{\left(x^2 + y^2\right)^2}$$

$$= \frac{\left(x^2 + y^2\right)\langle 2x - y^3, -3xy^2\rangle - \left(x^2 - xy^3\right)\langle 2x, 2y\rangle}{\left(x^2 + y^2\right)^2}$$

$$= \left\langle \frac{2x - y^3}{x^2 + y^2} - \frac{2x\left(x^2 - xy^3\right)}{\left(x^2 + y^2\right)^2}, \frac{-3xy^2}{x^2 + y^2} - \frac{2y\left(x^2 - xy^3\right)}{\left(x^2 + y^2\right)^2}\right\rangle$$

While most of the preceding discussion has focused on functions of two variables, the theory and procedures all extend in a natural manner to functions of any number of variables. Our last example illustrates this point.

Example 6 ✍

Suppose the temperature, in Celsius, at a point (x, y, z) in a small region of space over a heat source is modeled by the function

$$T(x, y, z) = \frac{100}{1 + 2x^2 + y^2 + 0.5z^2},$$

where x and y (in meters) are orthogonal horizontal displacements and z (also in meters) is the vertical displacement from the heat source. Find the direction from the point $(1, 3, 2)$ of the greatest rate of decrease of temperature. What is the rate of change in that direction?

Solution

We begin with the gradient.

$$\nabla T(x, y, z) = -100\frac{\langle 4x, 2y, z\rangle}{\left(1 + 2x^2 + y^2 + 0.5z^2\right)^2}$$

The direction of the greatest decrease in rate of change of T from the point $(1, 3, 2)$ is then

$$-\nabla T(1, 3, 2) = 100\frac{\langle 4, 6, 2\rangle}{\left(1 + 2 + 9 + 2\right)^2} = \frac{100}{14^2}\langle 4, 6, 2\rangle \approx \langle 2.04, 3.06, 1.02\rangle,$$

and the rate of decrease in that direction is approximately $-3.82\,°\text{C/m}$.

13.5 **Exercises**

1–6 Find the gradient of the function at the indicated point.

1. $f(x,y) = 5x^3 - 2y^2$; $(1,2)$

2. $g(x,y) = 2xy + \dfrac{\sqrt{y}}{x}$; $(1,4)$

3. $h(x,y) = e^{x^2+y^2}$; $(-1,1)$

4. $k(x,y) = \dfrac{y}{\sqrt{x^2+y^2}}$; $(-4,3)$

5. $F(x,y,z) = x\sqrt{\dfrac{y^3}{z}}$; $(-2,3,3)$

6. $G(x,y,z) = \sqrt{2}x\sin(\pi y + z^2)$; $\left(3,\dfrac{1}{4},0\right)$

7–12 Use the definition to find the directional derivative of f at the given point in the direction of the given vector. Then compute the directional derivative with the help of the gradient vector to check your answer.

7. $f(x,y) = 2x - y$; $(0,0)$; $\mathbf{v} = \langle 3,4 \rangle$

8. $f(x,y) = x^2 y$; $(1,-1)$; $\mathbf{v} = \langle 1,2 \rangle$

9. $f(x,y) = \dfrac{x}{y}$; $(1,2)$; $\mathbf{v} = \langle 2,2 \rangle$

10. $f(x,y) = x\sqrt{y}$; $(4,1)$; $\mathbf{v} = \langle 1,\sqrt{3} \rangle$

11. $f(x,y,z) = 2xy - z^2$; $(1,1,1)$; $\mathbf{v} = \langle 1,2,3 \rangle$

12. $f(x,y,z) = xyz$; $(5,0,-2)$; $\mathbf{v} = \langle 2,4,\sqrt{5} \rangle$

13. Try to evaluate the limit in Example 1 by using the following parametrizations.

   a. $\langle t,-t \rangle$ **b.** $\langle 2t,-2t \rangle$

   What do you find?

14–29 Find the derivative of the function at the given point in the indicated direction. (Note that sometimes the direction is conveniently specified in terms of a direction angle, the one determined by a direction vector and the positive x-axis. In this case, the corresponding unit vector is $\mathbf{u} = \langle \cos\alpha, \sin\alpha \rangle$.)

14. $f(x,y) = x^2 - 4y^2$; $(1,2)$; $\mathbf{v} = \langle 3,4 \rangle$

15. $f(x,y) = 3x^2 - 2xy + y^2$; $(0,1)$; $\mathbf{v} = \langle 1,1 \rangle$

16. $f(x,y) = 2x^3 y^2$; $(1,2)$; $\mathbf{v} = \langle -2,1 \rangle$

17. $f(x,y) = \dfrac{x^3}{3} + x^2 y - 3xy^2 + y^3$; $(-1,-1)$;
    $\mathbf{v} = \langle -4,-3 \rangle$

18. $f(x,y) = ye^x$; $(1,0)$;
    in the direction toward $(-1,4)$

19. $f(x,y) = 4e^x \cos x$; $(0,0)$; $\alpha = 60°$

20. $f(x,y) = \arcsin(xy)$; $(0,1)$;
    in the direction toward $(3,4)$

21. $f(x,y) = 2\cos x \sin y$; $\left(-\dfrac{\pi}{6},\dfrac{5\pi}{6}\right)$; $\mathbf{v} = \langle 5,12 \rangle$

22. $f(x,y) = \dfrac{2x}{\sqrt{x^2+y^2}}$; $(0,1)$; $\mathbf{v} = \langle 0.7,2.4 \rangle$

23. $f(x,y) = \ln(x^2 + y^2)$; $(1,0)$; $\alpha = 30°$

24. $f(x,y,z) = xy - yz + xz$; $(1,-3,-2)$; $\mathbf{v} = \langle 2,1,-2 \rangle$

25. $f(x,y,z) = e^{-(x^2+y^2+z^2)}$; $(2,1,1)$; $\mathbf{v} = \langle -6,2,3 \rangle$

26. $f(x,y,z) = \sqrt{x^2 + y^2 + z^2}$; $(1,1,1)$; $\mathbf{v} = \langle 1,1,1 \rangle$

27. $f(x,y,z) = x\arctan\dfrac{y}{z}$; $(1,1,1)$; $\mathbf{v} = \langle 2,-2,-2 \rangle$

28. $f(x,y,z) = \ln(x+y+z)$; $(e,1,-1)$;
    in the direction toward $(e-1,1,1)$

29. $f(x,y,z) = ze^{xy}$; $(2,0,1)$;
    in the direction toward $(-1,4,1)$

30–33 Find the direction and value of the greatest rate of increase for the function at the given point.

30. $f(x,y) = 2x^2 - 5xy + y^2$; $(2,1)$

31. $f(x,y) = x^2 e^{2xy}$; $(1,0)$

32. $f(x,y) = \dfrac{1}{\sqrt{x^2+y^2}}$; $(12,5)$

33. $f(x,y,z) = \dfrac{x-y}{z+2}$; $(4,2,0)$

34–37 Find the direction and value of the greatest rate of decrease for the function at the given point.

34. $f(x,y) = \dfrac{x^2 + y^2}{x^2 - y^2};\quad (-1,2)$

35. $f(x,y) = \cos(\pi xy);\quad \left(\dfrac{1}{2}, \dfrac{1}{2}\right)$

36. $f(x,y,z) = \ln\sqrt{x^2 + y^2 + z^2};\quad (2,1,-2)$

37. $f(x,y,z) = xe^{-yz};\quad (-1,3,0)$

38–39 Find the direction of no change for the function at the given point.

38. $f(x,y) = x^2 y + 2x^3 - 3y^2;\quad (-1,1)$

39. $f(x,y) = \dfrac{x-y}{x^2 + y^2};\quad (1,-1)$

40. If possible, find a direction angle θ for which the rate of change at the point $(1,-1)$ of the function in Exercise 39 is **a.** 0, **b.** 1.

41–48 Find an equation for the line tangent to the graph of the given equation at the indicated point.

41. $x^2 - xy + y^2 = 7;\quad (1,3)$

42. $(6-x)y^2 = 2x^3;\quad (2,2)$

43. $(x^2 + 4)y = 10;\quad (-1,2)$

44. $x^{2/3} + y^{2/3} = 13;\quad (8,27)$

45. $(2x^2 + y^2)^2 - 9x^2 y = 0;\quad (1,1)$

46. $x^2 + y^2 = (x^2 + y^2 - 5x)^2;\quad (4,3)$

47. $(x^2 + y^2)^2 = 4xy;\quad (1,1)$

48. $\dfrac{y}{x^2 + y^2} = \dfrac{3}{4} + x^2;\quad \left(\dfrac{1}{2}, \dfrac{1}{2}\right)$

49–52 Generalize Example 4 to three variables to obtain the equation of the tangent plane to the surface at the given point.

49. $x^2 + 3y^2 + 4z^2 = 11;\quad (2,1,-1)$

50. $y^3 z^2 + 5xz^2 + 2xy = 32;\quad (2,-1,2)$

51. $x^2 - 2y^2 = z;\quad (3,2,1)$

52. $3x - 2y^2 = 3z^2;\quad (7,3,1)$

53. Prove the linearity of the gradient, that is, the Sum/Difference Law and the Constant Multiple Law: If f and g are differentiable functions and k is a real number, then $\nabla(f \pm g) = \nabla f \pm \nabla g$, and $\nabla(kf) = k\nabla f$.

54. Prove the Product Law for gradients, that is, if f and g are differentiable functions, then $\nabla(fg) = f\nabla g + g\nabla f$.

55. Prove the Quotient Law for gradients, that is, if f and g are differentiable functions and $g \neq 0$, then
$$\nabla\left(\frac{f}{g}\right) = \frac{g\nabla f - f\nabla g}{g^2}.$$

56–59 Use the properties of the gradient to determine ∇f.

56. $f(x,y) = \dfrac{2xy^2 + yx^3}{x^2 + 2xy}$

57. $f(x,y) = \dfrac{2x^3 - \sqrt{y}}{x(x+y)}$

58. $f(x,y) = \dfrac{xy}{x^3 - y^3}$

59. $f(x,y,z) = \dfrac{xyz}{x^2 + y^2 + z^2}$

60–65 Find a function with the given gradient. If it is not possible, explain why. (Answers will vary.)

60. $\nabla f = \langle 1,2 \rangle$

61. $\nabla f = \langle 2y, 2x \rangle$

62. $\nabla f = \langle 3y, x \rangle$

63. $\nabla f = \langle y, 2x^2 \rangle$

64. $\nabla f = \langle 2y, 2x, 2z \rangle$

65. $\nabla f = \langle 6xyz^2, 3x^2 z^2, 6x^2 yz \rangle$

66. Prove that the rate of change of the function $f(x,y) = \sqrt{x^2 + y^2}$ is greatest along rays emanating from the origin.

67. Use the contour map below to estimate the directional derivative $D_\mathbf{u} f$ at the point $P(6,7)$ in the direction of **a.** $\mathbf{u} = \langle 1,2 \rangle$, **b.** $\mathbf{u} = \langle -1,-1 \rangle$. Then draw a possible path of steepest ascent starting at P. (Answers will vary.)

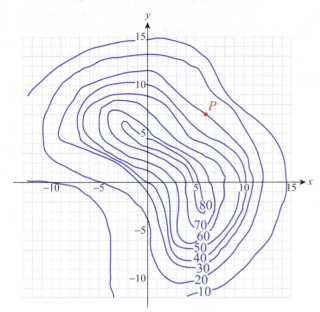

68. Sarah is standing at an intersection on a mountain trail where, according to the trail markings, her route continues to the southeast. Her current position can be modeled by the point $P(400,200,3560)$ on the graph of $f(x,y) = 4000 - 0.002x^2 - 0.003y^2$ (units in feet).

   a. What will be the angle of elevation (or depression) of her route immediately after leaving the intersection?

   b. What is the direction and angle of steepest ascent from her current position? (Assume that the northern direction coincides with the positive y-axis.)

69. Suppose that the temperature of a metal plate is given by

$$T(x,y) = \frac{150}{\sqrt{x^2 + y^2 + 1}}.$$

Find the rate of change of temperature at the point $(8,4)$ in the direction toward the point $(7,2)$.

70. Suppose that the temperature around the origin in three-dimensional space is given by $T(x,y,z) = 300e^{-\left(x^2 + 2y^2 + 3z^2\right)}$.

   a. Find the rate of change of temperature at the point $(2,1,0)$ in the direction toward the point $(4,0,2)$.

   b. Find the direction at $(2,1,0)$ in which the rate of decrease is greatest.

   c. Find the rate of greatest decrease at the point $(2,1,0)$.

71.* Consider a path on the contour map of a differentiable two-variable function $f(x,y)$ that follows the gradient at each point. Such is a possible path of a heat-seeking object, if f were a temperature function, or a path of steepest ascent on a geographical map (see Exercise 67). If such a path is parametrized as $\langle x(t), y(t) \rangle$, prove that

$$\frac{y'(t)}{x'(t)} = \frac{f_y}{f_x}.$$

72. Use Exercise 71 to find the path of a heat-seeking object if it starts at the point $(5,25)$ on a plane whose temperature is given by the function $T(x,y) = 500 - x^2 - 3y^2$. (**Hint:** Notice that this is a separable initial value problem. See Section 8.1.)

73. Use Exercise 71 to find the equation in the xy-plane of the steepest path from the point $(400,200)$ (this is the projection of the spot where Sarah is standing) for the function in Exercise 68. (See the hint given in Exercise 72.)

74.* Consider the piecewise defined function of Exercise 114 of Section 13.3.

$$F(x,y) = \begin{cases} \dfrac{xy^2}{2\left(x^2 + y^2\right)} & \text{if } (x,y) \neq (0,0) \\ 0 & \text{if } (x,y) = (0,0) \end{cases}$$

Prove that all directional derivatives of F exist at $(0,0)$, but F is not differentiable at $(0,0)$. (**Hint:** To prove the existence of the directional derivatives, use the definition.)

75.* Suppose $f(x,y)$ is defined and differentiable on an open region R, and $\nabla f(x,y) = \mathbf{0}$. Prove that $f(x,y)$ is constant on R.

13.6 **Tangent Planes and Differentials**

TOPICS

1. Tangent planes and normal lines
2. Differentials and linear approximations

The gradient operator gives us all the information we need to find (when they exist) both the plane tangent to a given surface at a point and the line passing through the point that is orthogonal to the surface. The tangent plane of a surface is analogous to the tangent line of a curve, and its definition contains some familiar-looking terms—terms that we also use to define linear approximations and differentials of multivariable functions.

TOPIC 1 **Tangent Planes and Normal Lines**

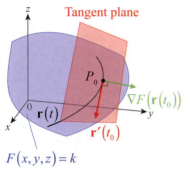

Tangent plane

$F(x,y,z)=k$

Figure 1

In Section 13.5, we showed that the gradient of a function of two variables always points in a direction normal to the function's contour curves; the same reasoning applies to functions of any number of variables. For our present purposes, let $F(x,y,z)$ be a differentiable function, and suppose $\mathbf{r}(t) = \langle x(t), y(t), z(t) \rangle$ is a smooth curve on a particular level surface of F, say $F(x,y,z)=k$ (see Figure 1). Then, using the Chain Rule and gradient notation, we have the following.

$$\frac{d}{dt}F(\mathbf{r}(t)) = \frac{d}{dt}(k)$$

$$F_x x'(t) + F_y y'(t) + F_z z'(t) = 0$$

$$\langle F_x, F_y, F_z \rangle \cdot \langle x'(t), y'(t), z'(t) \rangle = 0$$

$$\nabla F \cdot \mathbf{r}'(t) = 0$$

In other words, the vector $\mathbf{r}'(t)$, which we know is tangent to the curve $\mathbf{r}(t)$, is orthogonal to ∇F at every point on the surface $F(x,y,z)=k$.

Recall from Section 11.5 that a plane is completely characterized by one of its points, say (x_0, y_0, z_0), and a normal vector $\mathbf{n}$, as each point (x, y, z) of the plane must then satisfy the equation $\mathbf{n} \cdot \langle x-x_0, y-y_0, z-z_0 \rangle = 0$. Since $\nabla F(\mathbf{r}(t_0))$ is normal to the tangent vector $\mathbf{r}'(t_0)$ at the point $P_0(x(t_0), y(t_0), z(t_0))$ for every smooth curve $\mathbf{r}(t) = \langle x(t), y(t), z(t) \rangle$ on the surface, we define the tangent plane to a surface and the normal line to the surface as follows.

Definition 💡

Tangent Plane and Normal Line

Given a differentiable function $F(x,y,z)$, the **tangent plane** at the point (x_0, y_0, z_0) on the level surface $F(x,y,z)=k$ is described by the equation

$$\nabla F(x_0, y_0, z_0) \cdot \langle x-x_0, y-y_0, z-z_0 \rangle = 0.$$

Similarly, since ∇F is normal to the surface $F(x,y,z)=k$, the **normal line** to the surface at the point (x_0, y_0, z_0) is described by the vector function

$$\langle x, y, z \rangle = \langle x_0, y_0, z_0 \rangle + \nabla F(x_0, y_0, z_0)t$$

and parametrically by the following equations:

$$x = x_0 + F_x(x_0, y_0, z_0)t, \qquad y = y_0 + F_y(x_0, y_0, z_0)t, \qquad z = z_0 + F_z(x_0, y_0, z_0)t$$

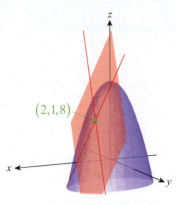

Figure 2

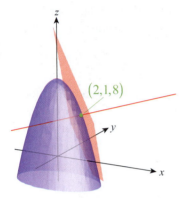

Figure 3

Example 1 ✏

Determine the tangent plane and normal line at the point $(2,1,8)$ of the surface defined by $x^2 + 3y^2 + z = 15$.

Solution

If we let $F(x,y,z) = x^2 + 3y^2 + z$, then the graph of $x^2 + 3y^2 + z = 15$ is a level surface of F. The first step in answering both questions is to determine the gradient of F.

$$\nabla F = \langle 2x, 6y, 1 \rangle$$

In particular, $\nabla F(2,1,8) = \langle 4,6,1 \rangle$, so the equation for the tangent plane at the point $(2,1,8)$ is as follows:

$$\langle 4,6,1 \rangle \cdot \langle x-2, y-1, z-8 \rangle = 0$$
$$(4x-8) + (6y-6) + (z-8) = 0$$
$$4x + 6y + z = 22$$

Next, the vector equation for the normal line passing through $(2,1,8)$ with direction vector $\nabla F(2,1,8)$ is

$$\langle x,y,z \rangle = \langle 2,1,8 \rangle + t \langle 4,6,1 \rangle = \langle 2+4t, 1+6t, 8+t \rangle.$$

Figure 2 is an illustration of the surface $F(x,y,z) = 15$ and its tangent plane at the point $(2,1,8)$, with tangent lines parallel to the xz-plane and yz-plane shown as well. Figure 3 is the same surface and tangent plane from a different perspective with the normal line through $(2,1,8)$ shown.

Sometimes surfaces in $\mathbb{R}^3$ are defined as graphs of functions of two variables, such as $z = f(x,y)$. In this case, we can think of the surface as a level surface of the function $F(x,y,z) = f(x,y) - z$, so $\nabla F \langle f_x, f_y, -1 \rangle$ and the tangent plane at a point $P_0(x_0, y_0, z_0)$ is described as follows.

$$\nabla F(P_0) \cdot \langle x-x_0, y-y_0, z-z_0 \rangle = 0$$
$$\langle f_x, f_y, -1 \rangle \Big|_{P_0} \cdot \langle x-x_0, y-y_0, z-z_0 \rangle = 0$$
$$f_x(x_0, y_0)(x-x_0) + f_y(x_0, y_0)(y-y_0) - (z-z_0) = 0$$

In Exercise 17 you will show that another way to derive the above equation is to define a normal vector **n** for the tangent plane in terms of two direction vectors $\mathbf{u} = \langle 1, 0, f_x(x_0, y_0) \rangle$ and $\mathbf{v} = \langle 0, 1, f_y(x_0, y_0) \rangle$ lying in the plane.

Example 2 ✏

Determine the tangent plane at the point $\left(\dfrac{\pi}{4}, 1, \dfrac{1}{\sqrt{2}} \right)$ of the surface $z = \cos(xy)$.

Solution

If we define $f(x,y) = \cos(xy)$, then

$$f_x(x,y) = -y\sin(xy) \quad \text{and} \quad f_y(x,y) = -x\sin(xy).$$

So $f_x\left(\dfrac{\pi}{4},1\right)=-\dfrac{1}{\sqrt{2}},\ \ f_y\left(\dfrac{\pi}{4},1\right)=-\dfrac{\pi}{4\sqrt{2}},\ $ and the tangent plane is described by

$$-\frac{1}{\sqrt{2}}\left(x-\frac{\pi}{4}\right)-\frac{\pi}{4\sqrt{2}}(y-1)-\left(z-\frac{1}{\sqrt{2}}\right)=0.$$

The fact that ∇F is normal to the tangent of any smooth curve on the surface $F(x,y,z)=k$ provides a convenient shortcut in solving certain types of problems.

Example 3 ✐

Consider the curve C defined by the intersection of the surfaces $z=(x^2+y^2)/2$ and $z=4-x^2-2y^2$. Find the parametric equations for the line tangent to C at the point $(1,1,1)$.

Solution

The direct way of solving this problem would be to find a parametrization $\mathbf{r}(t)$ of the curve C, determine $\mathbf{r}'$ at the point $(1,1,1)$, and then use the result as a direction vector for the tangent line (see Exercise 23). But we can shorten the process considerably by skipping the step of parametrizing C.

If we define $F(x,y,z)=2z-x^2-y^2$ and $G(x,y,z)=x^2+2y^2+z$, then C is the intersection of the level surfaces $F(x,y,z)=0$ and $G(x,y,z)=4$ and the point $(1,1,1)$ lies on both surfaces. We know that $\nabla F(1,1,1)$ is normal to the tangent of C at the point $(1,1,1)$, since C is a curve on the surface $F(x,y,z)=0$. Likewise, $\nabla G(1,1,1)$ is normal to the tangent of C at $(1,1,1)$ since C is a curve on the surface $G(x,y,z)=4$. So by the properties of the cross product, the vector $\mathbf{v}=\nabla F(1,1,1)\times\nabla G(1,1,1)$ must point in the direction of the desired tangent line. Figure 4 shows the surface $F(x,y,z)=0$ in blue, the surface $G(x,y,z)=4$ in red, and three unit vectors pointing in the directions of $\nabla F(1,1,1)$, $\nabla G(1,1,1)$, and $\mathbf{v}$ in red, blue, and black, respectively.

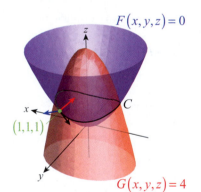

$F(x,y,z)=0$

C

$(1,1,1)$

$G(x,y,z)=4$

Figure 4

$$\nabla F=\langle -2x,-2y,2\rangle\quad\text{and}\quad\nabla G=\langle 2x,4y,1\rangle$$

These lead to $\nabla F(1,1,1)=\langle -2,-2,2\rangle$ and $\nabla G(1,1,1)=\langle 2,4,1\rangle$.

$$\mathbf{v}=\begin{vmatrix}\mathbf{i}&\mathbf{j}&\mathbf{k}\\-2&-2&2\\2&4&1\end{vmatrix}=\langle -2-8,4-(-2),-8-(-4)\rangle=\langle -10,6,-4\rangle$$

Thus, $\langle 1,1,1\rangle+t\langle -10,6,-4\rangle$ describes the line tangent to C at the point $(1,1,1)$, giving us the parametric equations

$$x=1-10t,\quad y=1+6t,\quad\text{and}\quad z=1-4t.$$

Figure 5 is an illustration of the curve C with its tangent line at $(1,1,1)$.

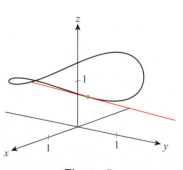

Figure 5

TOPIC 2   Differentials and Linear Approximations

Recall that if $y = f(x)$ is a differentiable function, then the differentials dy and dx are related by the definition $dy = f'(x)\,dx$. The differential dy approximates the increment $\Delta y = f(x) - f(x_0)$ by using the fact that the tangent line $L(x) = f(x_0) + f'(x_0)(x - x_0)$ is the best possible linear approximation to $y = f(x)$ at the point x_0 (see Section 3.9 for review). In exactly the same way, we now define linear approximations and differentials of functions of two variables.

Definition 💡

Linear Approximations and Total Differentials

Given a function $f(x, y)$ differentiable at the point (x_0, y_0), its **linear approximation** (or **linearization**) at (x_0, y_0) is the function

$$L(x, y) = f(x_0, y_0) + f_x(x_0, y_0)(x - x_0) + f_y(x_0, y_0)(y - y_0).$$

The (**total**) **differential** df at the point (x_0, y_0) is the change in the linearization of f from the point (x_0, y_0) to the point $(x_0 + dx, y_0 + dy)$, where the differentials dx and dy are independent variables. That is,

$$
\begin{aligned}
df &= L(x_0 + dx, y_0 + dy) - L(x_0, y_0) \\
&= f_x(x_0, y_0)\,dx + f_y(x_0, y_0)\,dy.
\end{aligned}
$$

Using alternate notation, if $z = f(x, y)$, the differential dz is a function of the differentials dx and dy defined by

$$dz = \frac{\partial z}{\partial x}\,dx + \frac{\partial z}{\partial y}\,dy.$$

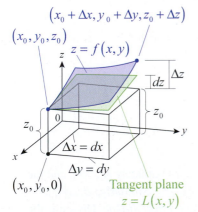

$(x_0 + \Delta x, y_0 + \Delta y, z_0 + \Delta z)$

(x_0, y_0, z_0)

$z = f(x, y)$

$(x_0, y_0, 0)$

Tangent plane $z = L(x, y)$

Figure 6

In other words, the linear approximation of the surface $z = f(x, y)$ at the point (x_0, y_0) is obtained by expressing the tangent plane

$$f_x(x_0, y_0)(x - x_0) + f_y(x_0, y_0)(y - y_0) - (z - z_0) = 0$$

as a function of x and y.

$$
\begin{aligned}
z &= z_0 + f_x(x_0, y_0)(x - x_0) + f_y(x_0, y_0)(y - y_0) \\
&= f(x_0, y_0) + f_x(x_0, y_0)(x - x_0) + f_y(x_0, y_0)(y - y_0)
\end{aligned}
$$

Figure 6 illustrates the relationship between a surface $z = f(x, y)$, its linearization $L(x, y)$, the increments Δx, Δy, Δz, and the differentials dx, dy, and dz.

Example 4 ✍

Let $f(x, y) = (x-1)^2 - y + (y-1)^4$. Find

a. the differential df at an arbitrary point (x, y) and

b. the linearization of f at the point $(2,1)$.

Solution

Both the differential and linearization depend on the partial derivatives of f, so we begin with them.

$$f_x(x, y) = 2(x-1) \quad \text{and} \quad f_y(x, y) = -1 + 4(y-1)^3$$

a. $df = 2(x-1)dx + \left[-1 + 4(y-1)^3\right]dy$

b. The linearization of f at the point $(2,1)$ is the following function.

$$\begin{aligned} L(x, y) &= f(2,1) + f_x(2,1)(x-2) + f_y(2,1)(y-1) \\ &= 0 + 2(x-2) + (-1)(y-1) \\ &= 2x - y - 3 \end{aligned}$$

The graph of $z = f(x, y)$ and its linearization (i.e., tangent plane) at the point $(2,1)$ are shown in Figure 7.

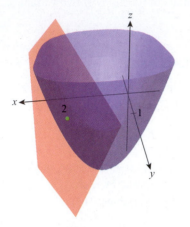

Figure 7

In Example 4 of Section 13.3, we learned one way of estimating the propagated error in a formula dependent on two or more measurements. The total differential of a function is a convenient tool in calculating such error, as shown in the next two applications.

Example 5 ✍

Consider again the density formula $\rho = m/v$. If the mass m of an object is known to within 2% relative error and its volume v to within 5% relative error, find an upper bound on the relative error with which its density is known.

Solution

We begin by finding $d\rho$.

$$d\rho = \frac{\partial \rho}{\partial m}dm + \frac{\partial \rho}{\partial v}dv = \frac{1}{v}dm - \frac{m}{v^2}dv$$

We know that the true mass of the object is approximately $m \pm dm$, and the information we are given tells us that $|dm/m| \leq 2\%$. Similarly, $|dv/v| \leq 5\%$, and we seek an upper bound on the relative error $|d\rho/\rho|$. Dividing the differential $d\rho$ by ρ, we obtain

$$\frac{d\rho}{\rho} = \frac{1}{v} \cdot \frac{v}{m}dm - \frac{m}{v^2} \cdot \frac{v}{m}dv = \frac{dm}{m} - \frac{dv}{v},$$

so by the Triangle Inequality we have the following.

$$\left|\frac{d\rho}{\rho}\right| = \left|\frac{dm}{m} - \frac{dv}{v}\right| \le \left|\frac{dm}{m}\right| + \left|\frac{dv}{v}\right| \le 2\% + 5\% = 7\%$$

Example 6 ✐

Suppose the radius r of a right circular cylindrical oil drum is known to within 4% relative error and its height h to within 1% relative error. Find an upper bound on the relative error with which the drum's volume V is known.

Solution

Beginning with the volume formula $V = \pi r^2 h$, we obtain

$$dV = \frac{\partial V}{\partial r}\,dr + \frac{\partial V}{\partial h}\,dh = 2\pi rh\,dr + \pi r^2\,dh$$

and so

$$\frac{dV}{V} = \frac{2\pi rh}{\pi r^2 h}\,dr + \frac{\pi r^2}{\pi r^2 h}\,dh = 2\frac{dr}{r} + \frac{dh}{h}.$$

Note that an error in the measurement of r has twice as much effect in the relative error of V as an error in the measurement of h. For the relative errors given, we can state that

$$\left|\frac{dV}{V}\right| = \left|2\frac{dr}{r} + \frac{dh}{h}\right| \le 2\left|\frac{dr}{r}\right| + \left|\frac{dh}{h}\right| \le 2(4\%) + 1\% = 9\%.$$

So far in this section we have focused on surfaces in $\mathbb{R}^3$ (defined either explicitly as graphs of functions of two variables or implicitly as level surfaces of functions of three variables), but all that we have done extends to functions in any number of variables. We illustrate this fact by linearizing a function of four variables.

Example 7 ✐

Find the linear approximation of the following function at the given point.

$$f(w,x,y,z) = xy^2 - \frac{3w}{z} + \sin(x^3 yz); \quad (-2,1,0,3)$$

Solution

The linearization of f at $(-2,1,0,3)$ is given by the formula

$$L(w,x,y,z) = f(-2,1,0,3) + f_w(-2,1,0,3)(w+2) + f_x(-2,1,0,3)(x-1)$$
$$+ f_y(-2,1,0,3)(y-0) + f_z(-2,1,0,3)(z-3),$$

so we calculate the following:

$$f(-2,1,0,3) = (1)(0)^2 - \frac{3(-2)}{3} + \sin\left((1)^3 (0)(3)\right) = 2$$

$$f_w(-2,1,0,3) = \left[-\frac{3}{z}\right]_{(-2,1,0,3)} = -1$$

$$f_x(-2,1,0,3) = \left[y^2 + \left(3x^2 yz\right)\cos\left(x^3 yz\right)\right]_{(-2,1,0,3)} = 0$$

$$f_y(-2,1,0,3) = \left[2xy + \left(x^3 z\right)\cos\left(x^3 yz\right)\right]_{(-2,1,0,3)} = 3$$

$$f_z(-2,1,0,3) = \left[\frac{3w}{z^2} + \left(x^3 y\right)\cos\left(x^3 yz\right)\right]_{(-2,1,0,3)} = -\frac{2}{3}$$

Hence,

$$L(w,x,y,z) = 2 - (w+2) + 3y - \frac{2}{3}(z-3)$$

$$= -w + 3y - \frac{2}{3}z + 2.$$

13.6 Exercises

1–8 Determine the tangent plane and normal line at the indicated point of the given surface.

1. $z - x^2 + 2y^2 = 0$; $(1,1,-1)$

2. $z - xy + y^3 x^2 = 0$; $(2,-1,2)$

3. $z^2 + (3x+1)(y-x^2) = 0$; $(1,-3,4)$

4. $z^2 - 3x^2 - y^2 = 15$; $(2,-3,6)$

5. $(xy+2)^2 + 3(y+1)^2 - \sqrt{z} = 5$; $(2,-2,4)$

6. $xyz = 18$; $(-3,2,-3)$

7. $x^2 z - x^2 y^2 = 3$; $(1,-3,12)$

8. $z = \sin(2xy) - 5$; $\left(\frac{\sqrt{\pi}}{2}, \frac{\sqrt{\pi}}{2}, -4\right)$

9–16 Determine the tangent plane at the indicated point of the given surface.

9. $z = xy^2 - x^3 y$; $(1,2,2)$

10. $z = \tan(xy) + 2$; $\left(1, \frac{\pi}{4}, 3\right)$

11. $z = \frac{y}{\sqrt{x}}$; $(9,3,1)$

12. $z = \ln(x^2 + 2y^2)$; $(1,1,\ln 3)$

13. $z = \frac{1}{x^2} + y^2$; $(1,3,10)$

14. $z = 2\sin x \cos y$; $\left(\frac{\pi}{4}, \frac{\pi}{4}, 1\right)$

15. $z = \frac{1}{\sqrt{x^2 + y^2}}$; $\left(3,4,\frac{1}{5}\right)$

16. $z = e^{y/x}$; $\left(3,6,e^2\right)$

17. Given a differentiable function $f(x,y)$, use your knowledge of partial derivatives to show that the vectors $\mathbf{u} = \langle 1,0, f_x(x_0, y_0)\rangle$ and $\mathbf{v} = \langle 0,1, f_y(x_0, y_0)\rangle$ are direction vectors for the plane tangent to the surface $z = f(x,y)$ at the point $(x_0, y_0, f(x_0, y_0))$. Then use the cross product to find a normal vector $\mathbf{n}$ and arrive at the equation of the tangent plane that we derived immediately preceding Example 2:

$$f_x(x_0, y_0)(x-x_0) + f_y(x_0, y_0)(y-y_0) - (z-z_0) = 0$$

18. Assuming that $F(x,y,z)$ is differentiable and $F(x,y,z) = 0$ defines a function $z = f(x,y)$ implicitly, use the formulas $f_x = -F_x/F_z$ and $f_y = -F_y/F_z$ to provide yet another derivation of the equation of the tangent plane described in Exercise 17.

19–22 Find parametric equations for the line tangent to the intersection of the two surfaces at the indicated point.

19. $z = \dfrac{x^2 + y^2}{5}$; $z = 8 - 3x^2 - y^2$; $(1, 2, 1)$

20. $3x^2 + y^2 - z^2 = 2$; $x^2 + y^2 + z^2 = 14$; $(2, 1, 3)$

21. $x^2 + 5y^2 + 2z^2 = 11$; $xyz = 2$; $(-2, 1, -1)$

22. $4x^3 + y^3 + 5x^2 y - xy = z^2$; $4x^2 - y^2 - z = 0$;
    $(1, 1, 3)$

23. Find a parametrization $\mathbf{r}(t)$ of the curve C of Example 3; then determine $\mathbf{r}'$ at the point $(1, 1, 1)$, and use it as a direction vector to obtain parametric equations for the tangent line to C at $(1, 1, 1)$.

24. Show that, in general, the tangent to curve C at the point (x, y, z) of the intersection of the surfaces in Example 3 points in the direction of $\langle -10y, 6x, -4xy \rangle$. (**Hint:** Determine the cross product of ∇F and ∇G.)

25. Find all points on the surface $8x^2 + 4y^2 + z^2 = 4$ where the tangent plane is normal to the vector $\langle 4, 2, 1 \rangle$. (Notice that this surface is an ellipsoid. How many such points do you expect?)

26. Repeat Exercise 25 for the hyperboloid $x^2 + 3y^2 - 2z^2 = 6$ and the vector $\langle 1, 3, 2 \rangle$.

27–34 Find **a.** the differential df at an arbitrary point (x, y) and **b.** the linearization of f at the indicated point.

27. $f(x, y) = x^3 + y - (y + 2)^2$; $(1, 3)$

28. $f(x, y) = xy^3 - x^2 y^2 + x^3 y$; $(-2, 1)$

29. $f(x, y) = \sqrt{x^2 + y^2}$; $(\sqrt{3}, -1)$

30. $f(x, y) = x \cos y + y \cos x$; $(\pi, 0)$

31. $f(x, y) = \dfrac{e^{x^2 - y^2}}{2}$; $(1, 1)$

32. $f(x, y) = \arctan \dfrac{x}{y}$; $(\sqrt{3}, 1)$

33. $f(x, y) = 2(\sqrt{x} - \sqrt{y})$; $(1, 4)$

34. $f(x, y) = e^{3x} \cos 2y$; $(0, \pi/2)$

35–39 Find the linear approximation of the function at the indicated point.

35. $f(w, x, y, z) = 2x^2 (wy - z^3) + \dfrac{4wx}{yz}$; $(3, -1, 1, 2)$

36. $f(x, y, z) = \dfrac{xyz}{x^2 + y^2 + z^2}$; $(2, 1, -1)$

37. $f(w, x, y, z) = e^{\sqrt{w^2 + x^2 + y^2 + z^2}}$; $(-2, 2, 0, -1)$

38. $f(w, x, y, z) = \arctan(wxyz)$; $(-1, 1, 1, -1)$

39. $f(x, y, z) = e^x \sqrt{\ln(y - z)}$; $(1, e, 0)$

40. Using the linear approximation of the function $f(x, y) = \sqrt{x^2 - y^2}$ at the point $(5, 4)$, find an approximation for the number $\sqrt{(4.9)^2 - (4.1)^2}$. Compare your approximation with the value returned by your calculator.

41–46 Find an appropriate function and mimic the process from Exercise 40 to approximate the given number. Compare your approximation with the value returned by your calculator. (In Exercises 45–46, generalize to three variables.)

41. $\dfrac{(1.98)^3}{(3.01)^2 - 1}$

42. $\sqrt{\dfrac{16.5}{24.2}}$

43. $\dfrac{1}{\sqrt{(3.8)^2 + (3.1)^2}}$

44. $\dfrac{6.16}{2.9}$

45. $\sqrt{(8.98)(3.01)(2.9)}$

46. $\dfrac{\ln(2.8)}{(2.15)(4.95)}$

47. Suppose the legs of a right triangle are known to within 3% relative error and within 4% relative error, respectively. Find an upper bound on the relative error with which the triangle's area is known.

48. Suppose the base lengths of a rectangular pyramid were measured within a relative error of 1%, while its height is known to within 3% relative error. What is the greatest possible relative error this causes in the volume of the pyramid?

49. Repeat Exercise 48 for a circular cone if the radius of the base is measured within a relative error of 1%, while its height is known to within 3% relative error.

50. The electrical power used by a resistor is given by the equation $P = V^2/R$, where R is its resistance and V is the voltage. Suppose R and V are known to within 2% and 1% relative error, respectively. Find an upper bound on the relative error with which the power is known.

51. As we have seen earlier (see, for example, Exercise 17 in Section 3.7), the speed of impact of a body falling from a height h in the absence of air resistance is $v = \sqrt{2hg}$. Find the maximum relative error if one uses $g \approx 10$ m/s (instead of its "true value" of 9.81 m/s), and if h is only known to within 3% of its true value.

52. Consider the first-octant tetrahedron formed by the coordinate planes and the plane tangent to the surface $xyz = c$ at a fixed point P on the surface. Find the volume of this tetrahedron and show that its value is independent of the choice of P.

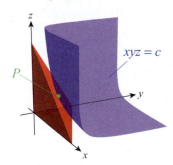

53. Show that every plane tangent to the elliptic cone $x^2 + y^2 - z^2 = 0$ passes through the origin.

54. Suppose you are working on a mathematical model that involves positive numbers less than 1000 and that your calculator rounds all your data to two decimal places. Use linear approximation to estimate the maximum error this might cause when multiplying three numbers.

55. Consider the two moles of ideal gas in the 50-liter container in Exercise 70 of Section 13.4. Use linear approximation to estimate the change in pressure if the temperature increases from 323 K to 328 K and the volume increases to 51 liters.

56.* Suppose four resistors are connected in parallel, with the following resistances: $R_1 = 12$ ohms (Ω), $R_2 = 16\ \Omega$, $R_3 = 9\ \Omega$, and $R_4 = 6\ \Omega$. If the actual resistances might differ by up to 2% from the above specifications, use linear approximation to estimate the maximum relative error in the calculated value of the net resistance. (See Exercise 101 of Section 13.3.)

57. Recall that the standard equation of the ellipsoid centered at the origin is $\dfrac{x^2}{a^2} + \dfrac{y^2}{b^2} + \dfrac{z^2}{c^2} = 1$. Show that the tangent plane to its surface at the point (x_0, y_0, z_0) has equation $\dfrac{x_0 x}{a^2} + \dfrac{y_0 y}{b^2} + \dfrac{z_0 z}{c^2} = 1$.

58. Find the equation of the tangent plane to the hyperboloid $\dfrac{x^2}{a^2} + \dfrac{y^2}{b^2} - \dfrac{z^2}{c^2} = 1$ at (x_0, y_0, z_0). (See Exercise 57.)

59–62 *True or False?* Determine whether the given statement is true or false. In case of a false statement, explain or provide a counterexample.

59. If both $f_x(a, b)$ and $f_y(a, b)$ exist, then a tangent plane to the graph of $f(x, y)$ exists at the point (a, b).

60. If a tangent plane to the graph of $f(x, y)$ exists at the point (a, b), then every directional derivative of f exists at (a, b).

61. If $f(x, y)$ is differentiable at (a, b), then $\nabla f(a, b)$ is contained in the tangent plane and points in the direction of greatest increase of f.

62. If (x_0, y_0, z_0) is a point on the level surface $F(x, y, z) = k$ of a differentiable function $F(x, y, z)$, then $\nabla F(x_0, y_0, z_0)$ is normal to the plane tangent to that level surface at (x_0, y_0, z_0).

13.7 **Extreme Values of Functions of Two Variables**

TOPICS

1. Absolute and relative extrema
2. Finding extrema

In Chapter 4, we learned how the first and second derivatives can be used to find the extreme values of a single-variable function. In this section, we see how the techniques extend to locating the extrema of functions of two variables.

TOPIC 1 **Absolute and Relative Extrema**

Informally, the extreme values of a function $f(x,y)$ are the high and low spots of the surface $z = f(x,y)$, but to make our work precise we begin with formal definitions of absolute and relative extrema.

Definition 💡

Absolute (Global) and Relative (Local) Extrema

Given a function $f(x,y)$ defined on a domain D, we say f has an **absolute maximum** (or **global maximum**) on D at the point (a,b) if $f(a,b) \geq f(x,y)$ for all $(x,y) \in D$, and $f(a,b)$ is consequently called the **maximum value** of f on D.

Similarly, if $f(a,b) \leq f(x,y)$ for all $(x,y) \in D$, we say f has an **absolute minimum** (or **global minimum**) at (a,b), and we call $f(a,b)$ the **minimum value** of f on D.

The function f has a **relative** (or **local**) **maximum** at (a,b) if $f(a,b) \geq f(x,y)$ for all domain points (x,y) in some open region containing (a,b), with **relative** (or **local**) **minimum** defined similarly but with the inequality reversed.

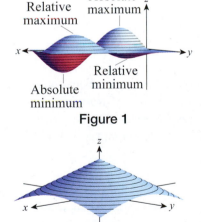

Relative maximum · Absolute maximum · Relative minimum · Absolute minimum

Figure 1

Figure 2

Figure 1 is an illustration of a surface $z = f(x,y)$ on a domain D that has exactly one absolute maximum and one absolute minimum. For this relatively well-behaved function, it is clear that at each extreme point the tangent plane is horizontal—that is, parallel to the xy-plane. This is analogous to the fact that the tangent line, if it exists, is horizontal at an extreme point of a function of one variable. But as in the one-variable case, a function of two variables may have an extremum at a point without a tangent plane, as seen in Figure 2.

Definition 💡

Critical Point

Given a function $f(x,y)$ defined on domain D, an interior point $(a,b) \in D$ is a **critical point** of f if $f_x(a,b) = 0 = f_y(a,b)$ or if one or both of f_x and f_y do not exist at (a,b).

Given our experience with single-variable functions, we have come to expect that critical points are the candidates to examine when looking for relative extrema of a function. The theorem below shows that this is also true for functions of two variables.

Theorem 🔍

Relative Extrema Are Critical Points

If $f(x, y)$ has a relative extremum at an interior point (a, b) of its domain, and if f_x and f_y both exist at (a, b), then $f_x(a, b) = 0 = f_y(a, b)$.

Proof 📎

If we define the function of one variable $g(x) = f(x, b)$, then g has a relative extremum at $x = a$ and so by Fermat's Theorem (the single-variable version of this theorem) we know that $g'(a) = 0$. Since $f_x(a, b) = g'(a)$, $f_x(a, b) = 0$. In similar fashion, defining $h(y) = f(a, y)$ results in $f_y(a, b) = h'(b) = 0$.

If (a, b) is a critical point of $f(x, y)$ and if $f_x(a, b) = 0 = f_y(a, b)$, then our intuitive sense that f has a horizontal tangent plane at (a, b) is justified, since the equation of the tangent plane has the form $z = k$ for some constant k.

$$z = f(a, b) + f_x(a, b)(x - a) + f_y(a, b)(y - b)$$
$$= f(a, b) + 0 \cdot (x - a) + 0 \cdot (y - b)$$
$$= f(a, b)$$

On the other hand, critical points are not guaranteed to be extrema. For functions of two variables, *saddle points* are analogous to the *inflection points* of single-variable functions.

Definition 💡

Saddle Point

A differentiable function $f(x, y)$ has a **saddle point** at a critical point (a, b) if f takes on values both larger and smaller than $f(a, b)$ in every open neighborhood of (a, b). That is, if every open disk centered at (a, b) contains a point (x_1, y_1) with $f(x_1, y_1) < f(a, b)$ and a point (x_2, y_2) with $f(x_2, y_2) > f(a, b)$.

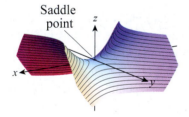

Saddle point

Figure 3

Example 1 ✏️

Find the relative extrema of the function $f(x, y) = x^2 + 6x + y^2 - 2y + 5$.

Solution

Since f is a polynomial, it is differentiable everywhere and the only critical points will be where both partial derivatives are 0. So we proceed to solve the equations $f_x = 0$ and $f_y = 0$.

$$f_x = 2x + 6 = 0 \quad \Rightarrow \quad x = -3$$
$$f_y = 2y - 2 = 0 \quad \Rightarrow \quad y = 1$$

Hence the only candidate for a relative extremum is $(-3,1)$, but we don't yet know if f has a local minimum, a local maximum, or a saddle point at $(-3,1)$. However, if we complete the square on the terms comprising f, we find $f(x,y) = (x+3)^2 + (y-1)^2 - 5$. In this form, f takes on its minimum value of -5 at the point $(-3,1)$, so f has a local minimum value of -5 at $(-3,1)$. (In fact, -5 is the global minimum value of f.)

Example 2 ✎

Find the relative extrema of the function $f(x,y) = x^2 - y^2$.

Solution

The partial derivatives are $f_x = 2x$ and $f_y = -2y$. They tell us that the only critical point is $(0,0)$.

Since $f(x,0) = x^2$ and $f(0,y) = -y^2$, f takes on both larger and smaller values than $f(0,0)$ at points arbitrarily close to $(0,0)$ and hence f has a saddle point at $(0,0)$ (see Figure 3).

TOPIC 2 Finding Extrema

While we were able to readily characterize the critical points in Examples 1 and 2, most surfaces aren't so easily analyzed. We are in need of something like the First or Second Derivative Tests of Section 4.3, and the following theorem answers the need.

Theorem ⚷

Second Derivative Test

Assume the second partial derivatives of $f(x,y)$ are continuous on an open disk D centered at (a,b), and that $f_x(a,b) = 0 = f_y(a,b)$. Then

1. f has a relative minimum at (a,b) if $f_{xx} > 0$ and $f_{xx}f_{yy} - f_{xy}^2 > 0$ at (a,b);

2. f has a relative maximum at (a,b) if $f_{xx} < 0$ and $f_{xx}f_{yy} - f_{xy}^2 > 0$ at (a,b);

3. f has a saddle point at (a,b) if $f_{xx}f_{yy} - f_{xy}^2 < 0$ at (a,b);

4. the test is inconclusive if $f_{xx}f_{yy} - f_{xy}^2 = 0$ at (a,b).

Proof Sketch ✎

While a complete proof of the theorem is best left for a class in advanced calculus, the basic idea can be presented here.

The assumption that the second partial derivatives of f are all continuous on D means that the first partial derivatives are differentiable and continuous, which in turn means that f is differentiable and continuous on D. And for any choice of sufficiently small positive real numbers h and k, the line segment parametrized by

$$x = a + th, \quad y = b + tk, \quad 0 \le t \le 1$$

lies entirely in D. If we define F to be the restriction of f along the line segment with the formula $F(t) = f(a + th, b + tk)$, $0 \leq t \leq 1$, then by the Chain Rule

$$F'(t) = f_x h + f_y k.$$

The differentiability of f_x and f_y means F' is differentiable and hence

$$F''(t) = \frac{\partial F'}{\partial x}\frac{\partial x}{\partial t} + \frac{\partial F'}{\partial y}\frac{\partial y}{\partial t} = \frac{\partial}{\partial x}\left(f_x h + f_y k\right)h + \frac{\partial}{\partial y}\left(f_x h + f_y k\right)k$$

$$= \left(f_{xx}h + f_{yx}k\right)h + \left(f_{xy}h + f_{yy}k\right)k = h^2 f_{xx} + 2hk f_{xy} + k^2 f_{yy}.$$

We can apply Taylor's Formula (Section 10.8) to the function $F(t)$ on the interval $[0,1]$ to write

$$F(1) = F(0) + F'(0)(1-0) + \frac{F''(c)}{2!}(1-0)^2,$$

where c is some real number in $(0,1)$. Expressing this back in terms of f,

$$f(a+h,b+k) = f(a,b) + hf_x(a,b) + kf_y(a,b) + \frac{1}{2}\left[h^2 f_{xx} + 2hk f_{xy} + k^2 f_{yy}\right]_{(a+ch,b+ck)}.$$

Since $f_x(a,b) = 0 = f_y(a,b)$, the change from $f(a,b)$ to $f(a+h,b+k)$ is thus

$$f(a+h,b+k) - f(a,b) = \frac{1}{2}\left[h^2 f_{xx} + 2hk f_{xy} + k^2 f_{yy}\right]_{(a+ch,b+ck)}.$$

We are interested in the sign of $f(a+h,b+k) - f(a,b)$ for all eligible h and k: if the difference is always positive then f has a relative minimum at (a,b), while if the difference is always negative then f has a relative maximum at (a,b), and if the difference takes on both positive and negative values then f has a saddle point at (a,b). So we examine the sign of the function

$$E(c) = h^2 f_{xx}(a+ch,b+ck) + 2hk f_{xy}(a+ch,b+ck) + k^2 f_{yy}(a+ch,b+ck).$$

If $E(0) \neq 0$, the continuity of the partial derivatives means that $E(c)$ and $E(0)$ will have the same sign for small enough h and k, and we look specifically at

$$E(0) = h^2 f_{xx}(a,b) + 2hk f_{xy}(a,b) + k^2 f_{yy}(a,b).$$

In the first two cases of the theorem we also know that $f_{xx}(a,b) \neq 0$, so by completing the square we can write

$$E(0) = \frac{1}{f_{xx}(a,b)}\left(\left[hf_{xx} + kf_{xy}\right]^2_{(a,b)} + k^2\left[f_{xx}f_{yy} - f_{xy}^2\right]_{(a,b)}\right).$$

1. If $f_{xx}(a,b) > 0$ and $\left[f_{xx}f_{yy} - f_{xy}^2\right]_{(a,b)} > 0$, then $E(0) > 0$ and f has a relative minimum at (a,b).

2. If $f_{xx}(a,b) < 0$ and $\left[f_{xx}f_{yy} - f_{xy}^2\right]_{(a,b)} > 0$, then $E(0) < 0$ and f has a relative maximum at (a,b).

3. If $\left[f_{xx}f_{yy}-f_{xy}^2\right]_{(a,b)}<0$, it can be shown that $E(c)$ takes on both positive and negative values as h and k vary and hence f has a saddle point at (a,b).

4. If $\left[f_{xx}f_{yy}-f_{xy}^2\right]_{(a,b)}=0$, no conclusion can be drawn and the critical point (a,b) must by studied by other means.

The quantity $f_{xx}f_{yy}-f_{xy}^2$, called the **discriminant** or **Hessian** in this context, can be remembered as a determinant:

$$f_{xx}f_{yy}-f_{xy}^2=\begin{vmatrix} f_{xx} & f_{xy} \\ f_{yx} & f_{yy} \end{vmatrix}.$$

Example 3 ✐

Use the Second Derivative Test to classify the critical points of

$$f(x,y)=6y^2-2y^3+3x^2-6xy.$$

Solution

Since f is a polynomial function, we know ahead of time that its second partial derivatives are continuous everywhere, so the Second Derivative Test will apply. To start, we find the critical points of f by setting the first partial derivatives equal to 0.

$$f_x=6x-6y=0$$
$$f_y=12y-6y^2-6x=0$$

The first equation implies $x=y$, and substituting this into the second equation results in

$$y^2-y=0 \quad\Rightarrow\quad y=0,1.$$

Thus, f has two critical points, $(0,0)$ and $(1,1)$. To test them, we next determine the second partial derivatives.

$$f_{xx}=6,\quad f_{xy}=-6,\quad f_{yy}=12-12y$$

Hence,

$$\left[f_{xx}f_{yy}-f_{xy}^2\right]_{(0,0)}=(6)(12-12\cdot0)-(-6)^2=36>0$$
$$\left[f_{xx}f_{yy}-f_{xy}^2\right]_{(1,1)}=(6)(12-12\cdot1)-(-6)^2=-36<0.$$

Since $f_{xx}>0$ at $(0,0)$ (as well as at every other point), the Second Derivative Test tells us that f has a relative minimum at $(0,0)$; note that the relative minimum value at this point is $f(0,0)=0$. And since the discriminant is negative at $(1,1)$, we conclude f has a saddle point at $(1,1)$. A graph of this portion of f, with the two critical points identified, appears in Figure 4.

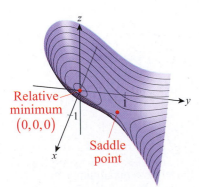

Relative minimum $(0,0,0)$

Saddle point

Figure 4

Example 4 ✐

Use the Second Derivative Test to classify the critical points of

$$f(x, y) = 6y + 3xy - 3x^2 y - 2y^2 - xy^2 + x^2 y^2.$$

Solution

Once again, f is a polynomial so we are assured that the Second Derivative Test will apply and that the only critical points will be those where the first partial derivatives are simultaneously equal to 0. Those critical points are where

$$f_x = 3y - 6xy - y^2 + 2xy^2 = 0$$
$$f_y = 6 + 3x - 3x^2 - 4y - 2xy + 2x^2 y = 0.$$

Both derivatives can be factored by grouping, and doing so will allow us to quickly identify the critical points.

$$f_x = 3y(1 - 2x) - y^2(1 - 2x) = y(3 - y)(1 - 2x)$$
$$f_y = -3(x^2 - x - 2) + 2y(x^2 - x - 2) = (2y - 3)(x - 2)(x + 1)$$

We will consider the three ways in which $f_x = 0$ and, for each solution, solve the corresponding equation $f_y = 0$.

If $y = 0$, then $f_y = (-3)(x - 2)(x + 1)$ and $f_y = 0$ implies $x = 2, -1$.

If $y = 3$, then $f_y = (3)(x - 2)(x + 1)$ and $f_y = 0$ implies $x = 2, -1$.

If $x = \frac{1}{2}$, then $f_y = (2y - 3)\left(-\frac{3}{2}\right)\left(\frac{3}{2}\right)$ and $f_y = 0$ implies $y = \frac{3}{2}$.

So the five critical points are $(-1, 0)$, $(2, 0)$, $(-1, 3)$, $(2, 3)$, and $\left(\frac{1}{2}, \frac{3}{2}\right)$. The next step is to find the second partial derivatives, and again the factored form of each first derivative is handy.

$$f_{xx} = -2y(3 - y)$$
$$f_{xy} = (3 - y)(1 - 2x) - y(1 - 2x) = (3 - 2y)(1 - 2x)$$
$$f_{yy} = 2(x - 2)(x + 1)$$

If we define $D(x, y) = f_{xx}(x, y) f_{yy}(x, y) - f_{xy}^2(x, y)$, then

$$D(-1, 0) = D(2, 0) = D(-1, 3) = D(2, 3) = -81 < 0,$$

so f has a saddle point at each of these four critical points. For the remaining critical point,

$$D\left(\frac{1}{2}, \frac{3}{2}\right) = \frac{81}{4} > 0 \quad \text{and} \quad f_{xx}\left(\frac{1}{2}, \frac{3}{2}\right) = -\frac{9}{2} < 0,$$

so f has a relative maximum at $\left(\frac{1}{2}, \frac{3}{2}\right)$. Its relative maximum value is $f\left(\frac{1}{2}, \frac{3}{2}\right) = \frac{81}{16}$. A graph of f and these five critical points appears in Figure 5.

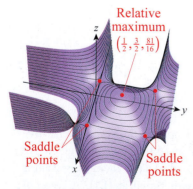

Figure 5

A computer algebra system such as *Mathematica* can be used to speed up the process of calculating the values of discriminants and second-order partial derivatives; one way of doing so is shown in Figure 6. The built-in command **Evaluate** tells *Mathematica* to compute the expression inside the brackets immediately, which is often necessary when we define new functions in terms of old (in this case, the new functions are derivatives of existing functions). Note the composition of the differentiation command **D** with itself, which is how we find f_{xx}, f_{xy}, and f_{yy}. In Figure 6, the second-order partial derivative functions have been given the names **fxx**, **fxy**, and **fyy**, and the discriminant function has been given the name **disc**.

```
In[1]:= f[x_, y_] := 6 y + 3 x * y - 3 x^2 * y - 2 y^2 - x * y^2 + x^2 * y^2
        fxx[x_, y_] := Evaluate[D[D[f[x, y], x], x]]
        fxy[x_, y_] := Evaluate[D[D[f[x, y], x], y]]
        fyy[x_, y_] := Evaluate[D[D[f[x, y], y], y]]
        disc[x_, y_] := fxx[x, y] * fyy[x, y] - fxy[x, y]^2
        disc[-1, 0]
        disc[2, 0]
        disc[-1, 3]
        disc[2, 3]
        disc[1 / 2, 3 / 2]
        fxx[1 / 2, 3 / 2]
```

Out[6]= -81

Out[7]= -81

Out[8]= -81

Out[9]= -81

Out[10]= $\dfrac{81}{4}$

Out[11]= $-\dfrac{9}{2}$

Figure 6

The Extreme Value Theorem of Section 4.1 guarantees that a continuous function on a closed bounded interval attains both its absolute maximum and absolute minimum values on that interval. The theorem applies equally well to functions of two variables, and we repeat it here in extended form.

Theorem 🔍

The Extreme Value Theorem for Functions of Two Variables

If $f(x, y)$ is a continuous function defined on a closed and bounded region $D \subseteq \mathbb{R}^2$, then there exist points (a_1, b_1) and (a_2, b_2) in D where f attains both its absolute maximum value $M = f(a_1, b_1)$ and absolute minimum value $m = f(a_2, b_2)$. (Recall that a region is *closed* if it contains all its boundary points (Section 13.1); $D \subseteq \mathbb{R}^2$ is *bounded* if it is contained in a disk of some radius centered at the origin.)

Suppose $f(x, y)$ is a continuous function on a closed and bounded region $D \subseteq \mathbb{R}^2$. We now know how to find and classify the relative extrema of f that occur at interior points of D, so the only place left to look for extrema is the boundary of D. This gives us a strategy for finding the absolute extrema of f on D.

Strategy for Finding Absolute Extrema

To find an absolute extrema of $f(x, y)$ on a closed and bounded region $D \subseteq \mathbb{R}^2$, perform the following steps.

Step 1: Find and classify the critical points of f on the interior of D. Calculate the value of f at each relative extrema.

Step 2: Find the relative extrema of f on the boundary of D, and calculate the value of f at each such point. (We will see how this can be done in the next two examples.)

Step 3: Compare the values computed in the first two steps. The largest is the absolute maximum value of f on D, and the smallest is the absolute minimum value of f on D.

Example 5 ✎

Find the absolute extrema of the function $f(x, y) = 8 - x^2 - y^2 + 4x + 2y$ on the triangular region $D \subseteq \mathbb{R}^2$ with vertices $(0, 0)$, $(4, 0)$, and $(0, 4)$.

Solution

We start by identifying all critical values of f, regardless of where they lie—we will then pick out those that lie in D.

$$f_x = -2x + 4 = 0 \implies x = 2$$
$$f_y = -2y + 2 = 0 \implies y = 1$$

As it turns out, f has a single critical point $(2, 1)$, and it lies in the interior of D. Since $f_{xx} = -2$, $f_{xy} = 0$, and $f_{yy} = -2$, we have $f_{xx}f_{yy} - f_{xy}^2 = 4 > 0$ everywhere and $f_{xx} < 0$ everywhere, so f has a relative maximum at $(2, 1)$. Its relative maximum value there is $f(2, 1) = 13$.

But we now have to compare the relative maximum value of f in the interior of D with any extreme values of f on the boundary of D, and we do so by considering the three sides of the triangular region one by one.

Along $x = 0$, $0 \le y \le 4$:

$$f(x, y) = 8 - y^2 + 2y$$

The critical point of this function of y on the interval $[0, 4]$ is $y = 1$. But as we know from Section 4.1, we also have to check the endpoints of $[0, 4]$. So we note that $f(0, 0) = 8$, $f(0, 1) = 9$, and $f(0, 4) = 0$.

Along $y = 0$, $0 \le x \le 4$:

$$f(x, y) = 8 - x^2 + 4x$$

The critical point of this function of x on the interval $[0, 4]$ is $x = 2$. But again, we also have to evaluate f at the endpoints (one of which we have already done). So we note that $f(2, 0) = 12$ and $f(4, 0) = 8$.

Along $x + y = 4$, $x \ge 0$, $y \ge 0$: Using $y = 4 - x$,

$$f(x, y) = 8 - x^2 - (4 - x)^2 + 4x + 2(4 - x) \text{ with } 0 \le x \le 4.$$

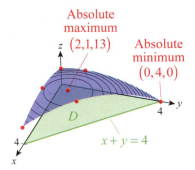

Absolute maximum z $(2, 1, 13)$

Absolute minimum $(0, 4, 0)$

D

$x + y = 4$

Figure 7

Since we seek the critical point of f on this interval, we differentiate with respect to x, set the result equal to 0, and solve to obtain $x = \frac{5}{2}$. The corresponding y-value is $y = \frac{3}{2}$, and we note that $f\left(\frac{5}{2}, \frac{3}{2}\right) = \frac{25}{2}$.

Comparing the values obtained, we now know that the absolute maximum value of f on D is 13 and occurs at $(2, 1)$ and that the absolute minimum value is 0 and occurs at $(0, 4)$. Figure 7 is an illustration of the surface f (in blue) restricted to the domain D (in green), with the seven candidates for absolute extrema appearing as red points on the surface.

Example 6 ✐

A well-known parcel service specifies that packages in one rate category must be boxes whose length plus girth is no greater than 84 inches, where "length" is defined as the length of the longest side and "girth" is the perimeter of the parcel's cross-section. What dimensions in this rate category maximize the volume of such a box? What is the maximum possible volume?

Solution

If we let x denote the box's length and y and z the dimensions of the cross-section, then we want to find the absolute maximum of the function $V = xyz$ under the restriction that $x + 2y + 2z = 84$ (and $x, y, z \ge 0$). Substituting $x = 84 - 2y - 2z$, we can reduce this to finding the absolute maximum of $V(y, z) = 84yz - 2y^2z - 2yz^2$ on the region D in the yz-plane defined by $y \ge 0$, $z \ge 0$, and $84 - 2y - 2z \ge 0$ (or $y + z \le 42$).

We proceed as in Example 5, first identifying the critical points of V.

$$V_y = 84z - 4yz - 2z^2 = 0 \quad \Rightarrow \quad 2z(42 - 2y - z) = 0$$
$$V_z = 84y - 2y^2 - 4yz = 0 \quad \Rightarrow \quad 2y(42 - y - 2z) = 0$$

If $z = 0$, then $V_z = 0 \quad \Rightarrow \quad y = 0 \quad$ or $\quad y = 42$.

If $y = 0$, then $V_y = 0 \quad \Rightarrow \quad z = 0 \quad$ or $\quad z = 42$.

The only remaining critical point satisfies the system $42 - 2y - z = 0$ and $42 - y - 2z = 0$, giving us $y = z$. Making this substitution in either $V_y = 0$ or $V_z = 0$ leads to the last critical point $(14, 14)$, so we calculate

$$V(0,0) = 0, \quad V(42,0) = 0, \quad V(0,42) = 0, \quad \text{and} \quad V(14,14) = 5488.$$

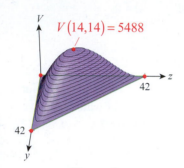

$V(14,14) = 5488$

Figure 8

We leave it to the reader (Exercise 32) to show that $V(y, z) = 0$ on each edge of the boundary D and also to show that V has a relative (and hence absolute) maximum value at the point $(14, 14)$. Thus, the dimensions that maximize volume are a length of 28 inches with a square cross-section of width 14 inches, and the maximum possible volume is 5488 cubic inches. Figure 8 shows a graph of the function V over its triangular region of definition.

13.7 Exercises

1–12 Find any extrema of the given function. Classify critical points without using the Second Derivative Test.

1. $f(x,y) = x^2 + 4x + y^2 - 6y + 10$

2. $f(x,y) = x^2 - 12x + y^2 + 8y$

3. $f(x,y) = 10x - x^2 - 4y - y^2$

4. $f(x,y) = -x^2 - y^2 - 4y + 1$

5. $f(x,y) = x^2 - 8x - y^2 - 2y + 15$

6. $f(x,y) = y^2 - x^2 - 6x$

7. $f(x,y) = y^4 - 4y^2 + x^2 + 5$

8. $f(x,y) = x^4 - 2x^2 + y^2$

9. $f(x,y) = \dfrac{1}{2 + x^4 + y^4}$

10. $f(x,y) = \dfrac{1}{x^2 + y^2 - 4y + 5}$

11. $f(x,y) = \ln(x^2 + y^2 + 2x + 2)$

12. $f(x,y) = \dfrac{1}{x^2 - 4x + y^2 - 2y + 6}$

13–30 Use the Second Derivative Test (if necessary) to classify the critical points of the given function. If the test fails, classify the critical point by other means. Identify absolute extrema wherever appropriate.

13. $f(x,y) = x^2 + 6x + 6y^3 - 8y$

14. $f(x,y) = 2x^2 - 4x + y^2 + 6y - 1$

15. $f(x,y) = y^4 - 2y^2 + 4x^2$

16. $f(x,y) = 6x^2y - x^2 - 3y^2$

17. $f(x,y) = x^3 + y^3 - 3x^2 - 2y^2$

18. $f(x,y) = x^3 + 2xy^2 - y^2$

19. $f(x,y) = 3xy - x^3 - y^3$

20. $f(x,y) = 2xy + \dfrac{4}{x} + \dfrac{1}{y}$

21. $f(x,y) = 3 - \sqrt[3]{x^2 + y^2}$

22. $f(x,y) = \dfrac{x^2 + 2y^2}{2e^x}$

23. $f(x,y) = 2xy - x^4 - y^4$

24. $f(x,y) = ye^{x^2 - y^2}$

25. $f(x,y) = 2xe^{2x - y^2}$

26. $f(x,y) = x\ln(x + y)$

27. $f(x,y) = (x^2 - y^2)e^x$

28. $f(x,y) = e^x - xe^y$

29. $f(x,y) = x\cos y$

30.* $f(x,y) = \sin x + \cos(x + y)$

31. Find and classify all extrema of the function
$f(x,y) = \sqrt{x^2 + y^2}$.

32. Show that $V(y,z) = 0$ on each edge of the boundary of D in Example 6, and demonstrate that V has a relative (and hence absolute) maximum value at the point $(14,14)$.

33–42 Find the absolute extrema of the function on the given region.

33. $f(x,y) = 2x + 4y - 3$;
$D = \{(x,y) \mid -1 \le x \le 1, 0 \le y \le 2\}$

34. $f(x,y) = 3x - \dfrac{y}{2} + 1$;
$D = \{(x,y) \mid -2 \le x \le 3, -1 \le y \le 5\}$

35. $f(x,y) = 2xy - x - y + 4$; D: The triangle with vertices $(0,0)$, $(3,0)$, and $(0,3)$

36. $f(x,y) = 4xy - 2x - y + 1$; D: The triangle with vertices $(0,0)$, $(4,0)$, and $(0,2)$

37. $f(x,y) = (x - 2y)^2$; D: The triangle with vertices $(0,0)$, $(12,0)$, and $(0,3)$

38. $f(x,y) = 1 - \sqrt{x^2 + y^2}$; $D = \{(x,y) \mid x^2 + y^2 \le 1\}$

39. $f(x,y) = x^2 - 2xy + y$;
$D = \{(x,y) \mid -2 \le x \le 2, -1 \le y \le 1\}$

40. $f(x,y) = x^2 - 2xy + y$;
$D = \{(x,y) \mid 0 \le x \le 2, x^2 \le y \le 4\}$

41. $f(x,y) = \dfrac{xy}{(x^2 + 1)(y^2 + 1)}$;
$D = \{(x,y) \mid 0 \le x \le 2, 0 \le y \le 2\}$

42. $f(x,y) = \dfrac{xy}{(x^2 + 1)(y^2 + 1)}$;
$D = \{(x,y) \mid 0 \le x \le 2, 0 \le y \le \sqrt{4 - x^2}\}$

43. Find the absolute extrema of the function
$g(x,y) = (x^2 + 2y^2)e^{-(x^2 + y^2)}$ on the square
$S = \{(x,y) \mid |x| \le 2, |y| \le 2\}$.

44–49 Show that the Second Derivative Test fails for the given function and classify any critical points by other means.

44. $f(x,y) = x^2 y^2$

45. $g(x,y) = \dfrac{1}{x^2 + y^2}$

46. $h(x,y) = (x - 1)^3 + (y + 2)^3$

47. $F(u,v) = (u + 1)^{2/3} + (v - 1)^{2/3}$

48. $R(s,t) = s^3 + t^3 - 3t^2 - 2$

49. $k(x,y) = x^3 - 2x^2 y$

50. Demonstrate that even though the function $f(x,y) = 4 + 2x - x^2$ has infinitely many critical points, the Second Derivative Test fails to classify any of them. Are those points extrema, and if so, what kind?

51. Determine m and b such that the sum of the squares of vertical distances from the line $y = mx + b$ to the points $(0,3)$, $(1,1)$, and $(4,8)$ is minimal.

52. Repeat Exercise 51 for the points $(1,0)$, $(2,5)$, and $(6,9)$.

53. Find a line that minimizes the sum of squares of the horizontal distances between the points in Exercise 51 and the line.

54. Find a line that minimizes the sum of squares of the horizontal distances between the points in Exercise 52 and the line.

55. By generalizing Exercise 51 to n points, this exercise will prove the formulas used in the least-squares method of curve fitting (see Section 1.5). Given n data points (x_1, y_1), $(x_2, y_2), ..., (x_n, y_n)$ in the plane, let $y = mx + b$ be the line that minimizes the sum of the squares of the vertical distances from the line to the points.

   a. Using the notation $S(m,b) = \sum_{i=1}^{n} [y_i - f(x_i)]^2$, show that in order for $S(m,b)$ to be minimal, m and b have to satisfy

   $$\left(\sum_{i=1}^{n} x_i \right) m + nb = \sum_{i=1}^{n} y_i$$

   and

   $$\left(\sum_{i=1}^{n} x_i^2 \right) m + \left(\sum_{i=1}^{n} x_i \right) b = \sum_{i=1}^{n} x_i y_i.$$

   b. Solve the above system to derive the formulas introduced in Section 1.5 for m and b.

56. Find the minimum distance from the origin to the surface $xyz^2 = 2$.

57. Repeat Exercise 56 for the surface $x^3 y^2 z = 1$.

58. Repeat Exercise 56 for the surface $z^2 - xy^2 = 3$.

59. Use the methods of this section to find the minimum distance between the point $(-3, -4, 1)$ and the plane $4x + y - 2z = 3$.

60. Find the dimensions of a rectangular prism with a fixed surface area of 6 square units and maximum volume.

61. Find the dimensions of a rectangular prism with a fixed volume V and minimum surface area.

62. Repeat Exercise 61 for an open box that has no lid.

63. Suppose we want to paint the inside of a rectangular box of volume V and that the paint used on the sides costs \$2 per square unit, the paint for the top is \$3 per square unit, while the sealant used to paint the bottom is \$5 per square unit. What are the dimensions of the box that is the most cost-effective to paint under these conditions?

64. A rectangular box is placed in the three-dimensional coordinate system with one vertex at the origin and the three edges containing it lying along the positive coordinate axes. If the vertex opposite the origin lies in the plane $4x + y + 2z = 9$, what is the greatest possible volume for such a box?

65.* A right rectangular pyramid with a square base is sitting atop a right rectangular prism with a congruent base to form the solid seen in the given figure. Find the side length of the base, and the respective heights of the pyramid and the prism that minimize the lateral surface area of the solid, if its volume is $\frac{100}{3}$ cubic units.

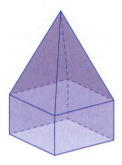

66. Find the dimensions of the rectangular box of maximum volume inscribed in a hemisphere of radius R. (**Hint:** First argue that one of the box's faces should lie in the base plane of the hemisphere.)

67.* Find the volume of the largest box inscribed in the ellipsoid $\dfrac{x^2}{a^2} + \dfrac{y^2}{b^2} + \dfrac{z^2}{c^2} = 1$.

68. If the sum of three positive numbers is 300, what is the greatest possible value for their product?

69. Again, as in Exercise 68, assume the sum of three positive numbers is 300. Find the minimum value for the sum of their squares.

70. A 30 in. piece of wire is cut in three pieces which are then bent into squares. If possible, determine how to a. minimize, and b. maximize the sum of the areas of the three squares.

71. Suppose $a, b, c > 0$ are given and consider the tetrahedron formed by the three coordinate planes and an arbitrary plane containing the point (a, b, c). Find the minimum volume for such a tetrahedron.

72.* Find the coordinates of the point Q so that the sum of the squares of the distances between Q and the given points $(x_1, y_1), \ldots, (x_n, y_n)$ is minimal.

73. Show that the function $f(x, y) = 3x^4 - 8x^2y + 4y^2$ has a relative minimum along every line $y = mx$ through the origin, but $(0,0)$ is *not* a relative minimum for f. (**Hint:** Examine the behavior of f along the parabola $y = x^2$.)

74–78 *True or False?* Determine whether the given statement is true or false. In case of a false statement, explain or provide a counterexample.

74. If $f(x, y)$ has a relative minimum at (a, b), then $f_x(a, b) = f_y(a, b) = 0$.

75. If $f_x(a, b) = f_y(a, b) = 0$, then $f(x, y)$ has a relative extremum at (a, b).

76. If $f(x, y)$ has a relative extremum at an interior point (a, b) of its domain, and if f_x and f_y both exist at (a, b), then $f_x(a, b) = 0 = f_y(a, b)$.

77. If $f(x, y)$ has exactly two relative maxima, then it must have a relative minimum also.

78. If $f(x, y)$ has an extremum at (a, b) along every straight line $y = mx$, then (a, b) is an extremum for f.

13.7 **Technology Exercises**

79. With the help of a computer algebra system, create an example of a two-variable function that has two maxima, but no relative minima. Sketch the graph of your example. (To start off, you may want to review, for example, Exercise 7, 15, or 23 of this section. Answers will vary.)

80–85 Use a computer algebra system to graph the function of the indicated exercise and graphically reinforce your conclusions made in the referenced exercise.

80. Exercise 7

81. Exercise 15

82. Exercise 21

83. Exercise 29

84. Exercise 30

85. Exercise 43

13.8 Lagrange Multipliers

TOPICS

1. The method of Lagrange multipliers
2. Lagrange multipliers with two constraints

We often need to find the extreme values of a function while the arguments are constrained to satisfy certain conditions—we first encountered such problems in Section 4.6, and then again in Section 13.7. The method of Lagrange multipliers, named for the mathematician Joseph-Louis Lagrange (1736–1813), is a very powerful tool for solving such constrained optimization problems.

TOPIC 1   The Method of Lagrange Multipliers

The fact that the gradient of a multivariable function is orthogonal to its level curves lies at the heart of the method. To gain an intuitive understanding of why the method works, consider the task of maximizing or minimizing a function $f(x, y)$ under the constraint that $g(x, y) = k$; in other words, considering only those points (x, y) on the curve $g(x, y) = k$, find the extreme values of $f(x, y)$. In general, we would not expect the contour lines of f and g to exhibit any common behavior, so f may very well take on different values as we trace out the contour $g(x, y) = k$. But suppose (x_0, y_0) is a point at which the tangents to the contour lines of f and g coincide. Then, in the direction of the contour, f is neither increasing nor decreasing at the point (x_0, y_0), meaning (x_0, y_0) is a critical point for f and a candidate for an extremum. Since $\nabla f(x_0, y_0)$ and $\nabla g(x_0, y_0)$ are both orthogonal to the common tangent line at the point (x_0, y_0), $\nabla f(x_0, y_0) = \lambda \nabla g(x_0, y_0)$ for some constant λ. The real number λ is called a *Lagrange multiplier* in this context. Before formalizing the above argument, we will demonstrate its application with an example.

Example 1 ✐

Find the absolute extreme values of the function $f(x, y) = x^2 y$ on the curve $x^2 + 2y^2 = 6$.

Solution

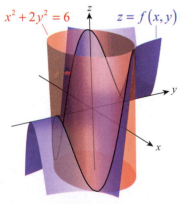

$x^2 + 2y^2 = 6$ $z = f(x, y)$

Figure 1

The graph of $z = f(x, y)$ is shown as the blue surface in Figure 1, and the graph of $x^2 + 2y^2 = 6$ in red (recall that since z doesn't appear in the constraint, the graph of $x^2 + 2y^2 = 6$ in $\mathbb{R}^3$ is an elliptic cylinder parallel to the z-axis). If we define $g(x, y) = x^2 + 2y^2$, then we seek the extreme values of $f(x, y)$ subject to the constraint $g(x, y) = 6$; that is, we are looking for the high and low points on the intersection of the two surfaces, shown as the black curve in Figure 1.

Using Lagrange's method, we introduce the variable λ and solve the equation $\nabla f = \lambda \nabla g$ for x and y. But this equation actually has three variables, and we will need to also consider the constraint $g(x, y) = 6$ in order to determine x and y. Since $\nabla f = \langle 2xy, x^2 \rangle$ and $\nabla g = \langle 2x, 4y \rangle$, we are led to the following system of equations.

$$\left.\begin{array}{r} 2xy = 2\lambda x \\ x^2 = 4\lambda y \end{array}\right\} \nabla f = \lambda \nabla g$$
$$x^2 + 2y^2 = 6$$

The first equation implies $x = 0$ or $\lambda = y$. Since f takes on both positive and negative values along the curve $g(x, y) = 6$ and $f(0, y) = 0$, the solution $x = 0$ can be discarded. Substituting $\lambda = y$ into the second equation results in the relation $x^2 = 4y^2$, and substituting this into the third equation gives us $y^2 = 1$, or $y = \pm 1$. Using the relation $x^2 = 4y^2$ again, we obtain $x = \pm 2$. We could also note that $\lambda = \pm 1$, but we don't actually need explicit values for λ—its purpose was simply to stretch, shrink, and/or reverse ∇g so that the equation $\nabla f = \lambda \nabla g$ could be satisfied.

The four points $(\pm 2, \pm 1)$ are the candidates for the extreme values of f on $g(x, y) = 6$, and we simply compute f at each of them and compare to finish the task.

$$f(2, 1) = f(-2, 1) = 4 \qquad \text{Absolute maximum value of 4 at } (\pm 2, 1)$$

$$f(2, -1) = f(-2, -1) = -4 \qquad \text{Absolute minimum value of } -4 \text{ at } (\pm 2, -1)$$

Figure 2 is an illustration of how the contour curves are tangent to the curve $g(x, y) = 6$ only at $(\pm 2, \pm 1)$. A total of six contours of f are shown, but only for the contours $f(x, y) = 4$ and $f(x, y) = -4$ is it possible to find points where $\nabla f = \lambda \nabla g$.

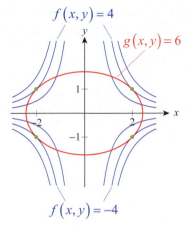

$f(x, y) = 4$

$g(x, y) = 6$

$f(x, y) = -4$

Figure 2

The formal proof of the validity of Lagrange's method is similar in nature to the informal argument. We state it here for functions of three variables, though the statement and proof are similar for two variables.

The Method of Lagrange Multipliers

If $f(x, y, z)$ and $g(x, y, z)$ are differentiable in an open region containing the level surface $g(x, y, z) = k$ and $\nabla g \neq \mathbf{0}$ on $g(x, y, z) = k$, then any relative extrema of f can be found as follows.

Step 1: Simultaneously solve the equations

$$\nabla f(x, y, z) = \lambda \nabla g(x, y, z) \quad \text{and} \quad g(x, y, z) = k.$$

Step 2: Evaluate f at the solution points (x, y, z) found in Step 1. To identify the relative minima and relative maxima of f on $g(x, y, z) = k$, it suffices to compare the value of f at the solution points.

Proof ✎

Suppose f has a relative extremum on the surface $g(x, y, z) = k$ at the point (x_0, y_0, z_0), and that $\mathbf{r}(t) = \langle x(t), y(t), z(t) \rangle$ is a smooth curve on the surface containing the point (x_0, y_0, z_0). Let t_0 be the value for which $\mathbf{r}(t_0) = \langle x_0, y_0, z_0 \rangle$. Since f has a relative extremum at (x_0, y_0, z_0),

$$0 = \frac{d}{dt} f(\mathbf{r}(t)) \Big|_{t_0} = \left[\frac{\partial f}{\partial x}\frac{dx}{dt} + \frac{\partial f}{\partial y}\frac{dy}{dt} + \frac{\partial f}{\partial z}\frac{dz}{dt} \right]_{t_0} = \nabla f(x_0, y_0, z_0) \cdot \mathbf{r}'(t_0).$$

Geometrically, this means that $\nabla f\left(x_0, y_0, z_0\right)$ is orthogonal to $\mathbf{r}'\left(t_0\right)$, which is tangent to the curve $\mathbf{r}(t)$. But we also know from Section 13.5 that ∇g is orthogonal to any level surface of g at any point, so $\nabla g\left(x_0, y_0, z_0\right) \cdot \mathbf{r}'\left(t_0\right) = 0$ as well. Since this is true for every such smooth curve through $\left(x_0, y_0, z_0\right)$, and since $\nabla g\left(x_0, y_0, z_0\right) \neq \mathbf{0}$, the two gradients ∇f and ∇g must be parallel at $\left(x_0, y_0, z_0\right)$. That is, $\nabla f\left(x_0, y_0, z_0\right) = \lambda\nabla g\left(x_0, y_0, z_0\right)$ for some constant λ.

There is no general method for solving systems of the form

$$\nabla f\left(x, y, z\right) = \lambda\nabla g\left(x, y, z\right) \quad \text{and} \quad g\left(x, y, z\right) = k,$$

but practice will reveal useful techniques. Remember that we don't actually need to solve for λ, but it may be necessary to do so in order to solve for the variables we do care about.

Example 2 ✐

Consider again the problem of maximizing the volume function $V = xyz$ under the constraint that $x + 2y + 2z = 84$ (Example 6 of Section 13.7). Solve this problem using Lagrange multipliers.

Solution

If we define $g\left(x, y, z\right) = x + 2y + 2z$, we need to solve the system

$$\nabla V\left(x, y, z\right) = \lambda\nabla g\left(x, y, z\right) \quad \text{and} \quad g\left(x, y, z\right) = 84.$$

This gives us the following four equations.

$$\left.\begin{array}{r} yz = \lambda \\ xz = 2\lambda \\ xy = 2\lambda \end{array}\right\} \nabla V = \lambda\nabla g$$
$$x + 2y + 2z = 84$$

Discarding a solution of 0 for any variable (since we know the maximum volume won't be 0), the second and third equations tell us that $y = z$. Multiplying the first equation by 2 and substituting the result into the second equation gives us $2y = x$. So we can write the fourth equation in terms of y to obtain $6y = 84$, or $y = 14$. Hence, $z = 14$ as well and $x = 28$. Since $(28, 14, 14)$ is the only solution of the system for which $V \neq 0$, we know that $V(28, 14, 14) = 5488$ in.^{3} must be the maximum possible volume.

Example 3 ✐

Find the location on the plane $-2x - 6y + 4z = 8$ closest to the point $(1, -2, 3)$.

Solution

Although vector techniques can be used to solve this problem, the method of Lagrange multipliers provides a different approach that can be generalized to other types of curves and surfaces (see Example 4 and Exercises 27–34).

Viewing the problem from this perspective, we seek to minimize the distance between the point $(1, -2, 3)$ and the given plane. We could proceed to work with the distance function $D(x, y, z) = \sqrt{(x-1)^2 + (y+2)^2 + (z-3)^2}$, which represents the distance between the point $(1, -2, 3)$ and an arbitrary point (x, y, z) on the plane $-2x - 6y + 4z = 8$ (or anywhere else in $\mathbb{R}^3$), but we can make our work easier by instead choosing to minimize the function

$$f(x, y, z) = D^2(x, y, z) = (x-1)^2 + (y+2)^2 + (z-3)^2.$$

That is, the point (x, y, z) on the plane that minimizes the square of the distance will be the same point that minimizes the distance. In the language of this section, we are looking for the absolute minimum of $f(x, y, z)$ under the constraint that $g(x, y, z) = 8$, where

$$g(x, y, z) = -2x - 6y + 4z.$$

So we proceed to solve the following system.

$$\left.\begin{array}{r} 2(x-1) = -2\lambda \\ 2(y+2) = -6\lambda \\ 2(z-3) = 4\lambda \end{array}\right\} \nabla f = \lambda \nabla g$$
$$-2x - 6y + 4z = 8$$

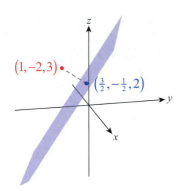

$(1, -2, 3) \bullet$

$\left(\tfrac{3}{2}, -\tfrac{1}{2}, 2\right)$

Figure 3

Using the first three equations to express x, y, and z in terms of λ, and substituting the results into the fourth equation, we obtain $28\lambda + 22 = 8$, so $\lambda = -\tfrac{1}{2}$. We then use this to determine that $x = \tfrac{3}{2}$, $y = -\tfrac{1}{2}$, and $z = 2$. Figure 3 is an illustration of the plane, the point $(1, -2, 3)$ in red, and the solution point $\left(\tfrac{3}{2}, -\tfrac{1}{2}, 2\right)$ in blue.

Example 4 ✒

In $\mathbb{R}^2$, find the points of relative minimum and maximum distance between the origin and the curve $y^2 = x^3 + 1$.

Solution

We can again make our task easier by finding the relative extrema of the function $f(x, y) = x^2 + y^2$, the square of the distance between (x, y) and the origin, rather than the distance function itself. And we restrict the search to those points on the curve $g(x, y) = 1$, where $g(x, y) = y^2 - x^3$.

Figure 4 shows a graph of the curve $g(x, y) = 1$, and clearly there is no upper bound on the distance between the origin and points on the graph (as x increases, the corresponding y-values increase in magnitude). But near the origin, it appears there should be points of relative extrema of f. We find them by solving the following system.

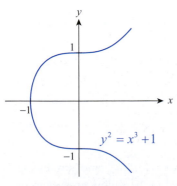

$y^2 = x^3 + 1$

Figure 4

$$\left.\begin{array}{r} 2x = -3\lambda x^2 \\ 2y = 2\lambda y \end{array}\right\} \nabla f = \lambda \nabla g$$
$$y^2 - x^3 = 1$$

From the second equation we know that $y(\lambda-1)=0$, so $y=0$ or $\lambda=1$. We consider these two cases separately as follows.

If $y=0$:   The third equation becomes $x^3=-1$ and so $x=-1$.

If $\lambda=1$:   The first equation becomes $x(3x+2)=0$ so $x=0$ or $x=-\frac{2}{3}$.

The corresponding y-values are, respectively, $y=\pm 1$ and $y=\pm\sqrt{\frac{19}{27}}$.

Comparing the values of f at each of these points, we find

$$f(-1,0)=f(0,\pm 1)=1 \quad \text{and} \quad f\left(-\frac{2}{3},\pm\sqrt{\frac{19}{27}}\right)=\frac{31}{27}.$$

Remember that these values are the *square* of the distance from the points to the origin, but the location of these relative extrema is the same whether we are optimizing the distance or the distance squared. Figure 5 illustrates the relative extrema we have located, with the relative minima in green and the relative maxima in red.

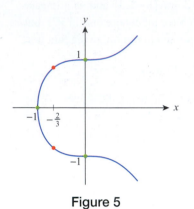

Figure 5

TOPIC 2   Lagrange Multipliers with Two Constraints

Lagrange's method extends nicely to optimization problems with two constraints. Suppose we want to find the extreme values of $f(x,y,z)$ under the side conditions that $g(x,y,z)=c$ and $h(x,y,z)=k$. Assuming that g and h are differentiable and that ∇g and ∇h are not parallel, the points satisfying $g(x,y,z)=c$ and $h(x,y,z)=k$ lie along the intersection C of two surfaces, such as those shown in Figure 6.

So we seek the extrema of $f(x,y,z)$ under the constraint that (x,y,z) lies on C.

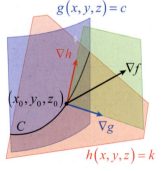

Figure 6

By the same reasoning as before, we know that if f has an extremum at (x_0,y_0,z_0), then ∇f is orthogonal to C at (x_0,y_0,z_0). We also know that $\nabla g(x_0,y_0,z_0)$ is orthogonal to $g(x,y,z)=c$ and $\nabla h(x_0,y_0,z_0)$ is orthogonal to $h(x,y,z)=k$ at the point (x_0,y_0,z_0), so $\nabla f(x_0,y_0,z_0)$ must lie in the same plane as $\nabla g(x_0,y_0,z_0)$ and $\nabla h(x_0,y_0,z_0)$. In other words, there must exist two real numbers λ and μ (again called Lagrange multipliers) such that $\nabla f(x_0,y_0,z_0)=\lambda\nabla g(x_0,y_0,z_0)+\mu\nabla h(x_0,y_0,z_0)$. We actually identify x_0, y_0, and z_0 by solving the following system of equations.

$$\nabla f = \lambda\nabla g + \mu\nabla h, \quad g(x,y,z)=c, \quad \text{and} \quad h(x,y,z)=k$$

Example 5 ✎

Find the absolute extreme values of $f(x,y,z)=x+y+z$ subject to the constraints $x^2+2y^2=6$ and $3x+2y+4z=5$.

Solution

Defining $g(x,y,z)=x^2+2y^2$ and $h(x,y,z)=3x+2y+4z$, the constraints have the form $g(x,y,z)=6$ and $h(x,y,z)=5$.

Noting that $\nabla f=\langle 1,1,1\rangle$, $\nabla g=\langle 2x,4y,0\rangle$, $\nabla h=\langle 3,2,4\rangle$, and introducing two Lagrange multipliers λ and μ, we are led to the following system of equations.

$$1 = 2\lambda x + 3\mu$$
$$1 = 4\lambda y + 2\mu \quad \Big\} \; \nabla f = \lambda \nabla g + \mu \nabla h$$
$$1 = 4\mu$$

$$x^2 + 2y^2 = 6$$
$$3x + 2y + 4z = 5$$

The third equation gives us immediately $\mu = \frac{1}{4}$, and substituting this into the first and second equations results in $\frac{1}{8} = \lambda x$ and $\frac{1}{8} = \lambda y$, so $\lambda x = \lambda y$. It is easy to verify that $\lambda = 0$ is not part of a solution to the system, so we are left with $x = y$. Using this knowledge in the fourth equation leads to $x = y = \pm\sqrt{2}$, and the corresponding values for z, from the fifth equation, are then $\left(5 \mp 5\sqrt{2}\right)/4$.

Evaluating f at the two critical points, we obtain the following.

$$f\left(\sqrt{2}, \sqrt{2}, \frac{5 - 5\sqrt{2}}{4}\right) = \frac{5 + 3\sqrt{2}}{4} \qquad \text{Absolute maximum}$$

$$f\left(-\sqrt{2}, -\sqrt{2}, \frac{5 + 5\sqrt{2}}{4}\right) = \frac{5 - 3\sqrt{2}}{4} \qquad \text{Absolute minimum}$$

Figure 7 is an illustration of the two surfaces $g(x, y, z) = 6$ and $h(x, y, z) = 5$, their intersection C, and the locations of the absolute extreme values of f on C.

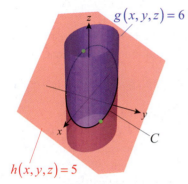

$g(x, y, z) = 6$

$h(x, y, z) = 5$

Figure 7

13.8 **Exercises**

1. Find the absolute extreme values of the function $f(x, y) = 2y^2 - x^2$ on the circle $x^2 + y^2 = 4$.

2. Find the absolute extreme values of $f(x, y) = xy$ on the ellipse $\dfrac{x^2}{4} + \dfrac{y^2}{8} = 1$.

3–18 Find any extreme values of the function subject to the given constraint.

3. $f(x, y) = x^2 + y^2$; constraint: $xy = 2$

4. $f(x, y) = x^2 + y^2$; constraint: $2x + y = 5$

5. $f(x, y) = x^2 - 3xy + y^2$; constraint: $x^2 + y^2 = 2$

6. $f(x, y) = x^2 + 3xy + y^2$; constraint: $y - x = 2$

7. $f(x, y) = \dfrac{x^2 + y^2}{2}$; constraint: $x^2 + 2y^2 = 4$

8. $f(x, y) = x^2 + y^2$; constraint: $2x^2 + 3y^2 = 8$

9. $f(x, y) = x^2 - y^2$; constraint: $2x + y = 6$

10. $f(x, y) = x^2 - y^2$; constraint: $y^2 = 4x$

11. $f(x, y) = 2xy$; constraint: $\dfrac{x^2}{9} + \dfrac{y^2}{4} = 1$

12. $f(x, y) = \sqrt{3x^2 + y^2 - 2}$; constraint: $x^2 + 2y^2 = 14$

13. $f(x, y, z) = 2x - 3y + z$; constraint: $6z = 3x^2 + y^2$

14. $f(x, y, z) = x^2 + y^2 + z^2$; constraint: $2x - 4y + z = 14$

15. $f(x, y, z) = x + 2y + 8z$; constraint: $x^2 + 9y^2 + 16z^2 = 144$

16. $f(x, y, z) = xyz$; constraint: $x^2 + y^2 + z^2 = 4$

17. $f(x, y, z) = x^2 - 2y^2 + z^2$; constraint: $2x^2 + y^2 + 3z^2 = 18$

18. $f(x, y, z) = x + 2y + 3z$; constraint: $x^2 + y^2 + z^2 = 1$

19–26 Use Lagrange multipliers to find the coordinates of the point on the given curve that is closest to the point P.

19. $y = 2x - 3; \quad P(2,2)$ 20. $y = 1 - 3x; \quad P(-3,0)$

21. $y = x^2; \quad P(3,0)$ 22. $y = x^2 + x; \quad P(2,-2)$

23. $y = x^2 + x; \quad P(2,0)$ 24. $y = 2x - x^2; \quad P(1,0)$

25. $x^2 - 2y^3 = 1; \quad P(0,0)$

26. $\left(x^2 + 4\right)y = 10; \quad P(0,0)$

27–34 Use Lagrange multipliers to find the coordinates of the point on the given surface that is closest to the point P.

27. $3x - 2y + \sqrt{3}z + 16 = 0; \quad P(0,0,0)$

28. $z = 3x - 2y + 3; \quad P(2,1,0)$

29. $2z + x - y = 1; \quad P(3,-2,-1)$

30. $z^2 = x^2 + y^2; \quad P\left(1,0,\dfrac{1}{2}\right)$

31. $z = x^2 + 2y^2; \quad P(0,0,1)$

32. $xyz = 2; \quad P(0,0,0)$

33. $x = yz + 1; \quad P(2,0,0)$

34. $x^2 = yz + 1; \quad P(0,0,0)$

35. Generalize Exercise 27 by showing that the location on the plane $ax + by + cz = d$ closest to the origin is the following point:
$$\left(\frac{ad}{a^2 + b^2 + c^2}, \frac{bd}{a^2 + b^2 + c^2}, \frac{cd}{a^2 + b^2 + c^2}\right)$$

36. Suppose we want to cut a rectangular beam from a log that has a circular cross-section of radius $\sqrt{2}$ feet. Use Lagrange multipliers to find the dimensions of the beam's cross-section if it is to have maximum area. (**Hint:** Place the cross-section in the xy-system with its center at the origin, and maximize the area of the beam's cross-section subject to the constraint $x^2 + y^2 = 2$.)

37. Suppose that when manufacturing three different products, a company is able to make a profit of $4 on each unit of the first product, while the profits on the second and third types of products are $8 and $6 per unit, respectively. Let x, y, and z denote thousands of units produced from each product.

a. Find the profit function $P(x, y, z)$.

b. Assuming that the manufacturing process is under the constraint $x^2 + 2y^2 + z^2 \leq 756$, find the maximum profit for the company under these conditions.

38–48 We will revisit some exercises from Section 13.7. Use Lagrange multipliers to provide a second solution for each.

38. Find the dimensions of a rectangular prism with a fixed surface area of 6 square units and maximum volume.

39. Find the dimensions of a rectangular prism with a fixed volume V and minimum surface area.

40. Repeat Exercise 39 for an open box that has no lid.

41. Suppose we want to paint the inside of a rectangular box of volume V and that the paint used on the sides costs $2 per square unit, the paint for the top is $3 per square unit, while the sealant used to paint the bottom is $5 per square unit. What are the dimensions of the box that is the most cost-effective to paint under these conditions?

42. A rectangular box is placed in the three-dimensional coordinate system with one vertex at the origin and the three edges containing it lying along the positive coordinate axes. If the opposite vertex lies in the plane $4x + y + 2z = 9$, what is the greatest possible volume for such a box?

43. Find the dimensions of the rectangular box of maximum volume inscribed in a hemisphere of radius R. (**Hint:** First argue that one of the box's faces should lie in the base plane of the hemisphere.)

44. If the sum of three positive numbers is 300, what is the greatest possible value for their product?

45. Find the minimum distance from the origin to the surface $xyz^2 = 2$.

46. Repeat Exercise 45 for the surface $x^3y^2z = 1$.

47. Repeat Exercise 45 for the surface $z^2 - xy^2 = 3$.

48. * Find the volume of the largest box inscribed in the ellipsoid $\dfrac{x^2}{a^2}+\dfrac{y^2}{b^2}+\dfrac{z^2}{c^2}=1$.

49. Use Lagrange multipliers to prove that among all triangles inscribed in a circle of radius R, the equilateral triangle has the largest area. (**Hint:** Draw the three radii connecting the center of the circle with the vertices of the triangle, and mark the angles formed by these radii α, β, and γ. Then use the formula for the area of a triangle with sides a, b and included angle θ, $A=\frac{1}{2}ab\sin\theta$.)

50–53 Find the absolute extreme value(s) of the function subject to the given constraints.

50. $f(x,y,z)=x+2y+2z$;
constraints: $x^2+z^2=4$ and $2x+y+z=1$

51. $f(x,y,z)=x^2+y^2+z^2$;
constraints: $x+y=1$ and $2x-y+3z=2$

52. $f(x,y,z)=xyz$;
constraints: $x-y+z=4$ and $x+y+z=6$

53. $f(x,y,z)=2x+y-2z$;
constraints: $x^2+y^2=4$ and $x-y+2z=1$

54. Supposing that $x,y,z>0$, maximize the function $f(x,y,z)=xyz$ subject to the constraint $x+y+z=1$, and use your result to prove the famous inequality between geometric and arithmetic means for three positive numbers

$$\sqrt[3]{xyz}\le\frac{x+y+z}{3},$$

with equality occurring precisely when $x=y=z$.

55. * Generalize Exercise 54 for n positive numbers.

56. The graph of the equation $x^2+xy+y^2=4$ is a rotated ellipse. Use Lagrange multipliers to find its points closest to and farthest from the origin.

57. Find the points closest to and farthest from the origin on the intersection of the surfaces $z=2x^2+2y^2$ and $z=20-x-y$.

58. Find the highest and lowest points on the curve of intersection of the two surfaces $x^2+y^2+z^2=9$ and $x-2y+z=2$.

59. Suppose a fence of height h is located d units from a wall. Find the minimum length of a ladder that is able to reach the wall over the fence. (**Hint:** Minimize $f(x,y)=(x+d)^2+(y+h)^2$ subject to the constraint $\dfrac{h}{x}=\dfrac{y}{d}$.)

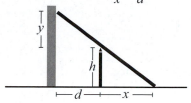

60. Using Lagrange's method, show that the critical point of the function $f(x,y)=x^2y$ subject to the constraint $y-x=0$ does not yield a local extremum. Explain.

61. * When a light ray travels from a transparent medium into another transparent medium, it "bends" or "refracts" as shown in the figure below. Using the fact that the traveling of light waves is governed by the principle of making the "best time" between A and B, and supposing that the speed of light in the two different media is v_A and v_B, respectively, use Lagrange's method to derive the equality $\dfrac{\sin\alpha}{\sin\beta}=\dfrac{v_A}{v_B}$. (This is called Snell's Law of refraction. **Hint:** Minimize the travel time of light subject to the constraint that the distance between A_1 and B_1 is constant.)

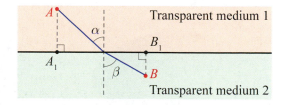

13.8 **Technology Exercises**

62–64 Use a computer algebra system or a programmable calculator to write a program that accepts a three-variable function along with one or two constraints, and uses Lagrange's method to return the critical points. Use it to check the answers you obtained for the following exercises.

62. Exercise 18   **63.** Exercise 50

64. Exercise 53

Chapter 13
Review Exercises

1–6 Match the function with its graph (labeled A–F).

1. $f(x, y) = \dfrac{-4}{2x^2 + y^2 + 1}$

2. $f(x, y) = 5\sqrt{1 - \dfrac{x^2}{8} - \dfrac{y^2}{4}}$

3. $f(x, y) = \dfrac{(x - 2y)^2}{5}$

4. $f(x, y) = 2.5 \sin\sqrt{x^2 + y^2}$

5. $f(x, y) = x^2 - y^2$

6. $f(x, y) = \cos(2x - y)$

A.

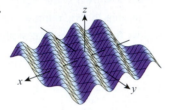

B.

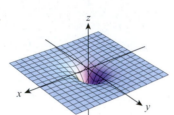

C.

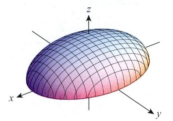

D.

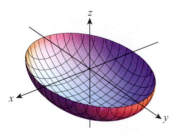

E.

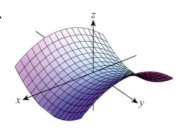

F.

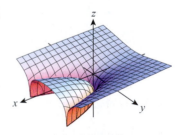

7–12 Match the graph with its contour map (labeled A–F).

7.

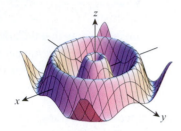

8.

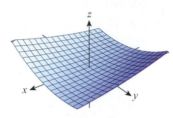

9.

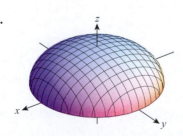

10.

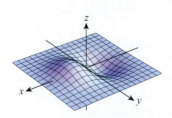

11.

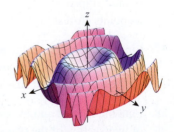

12.

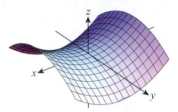

A.

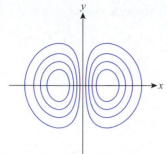

B.

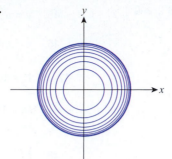

C.

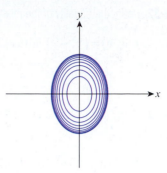

D.

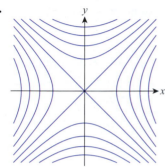

E.

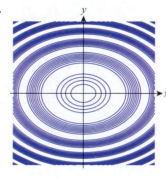

F.

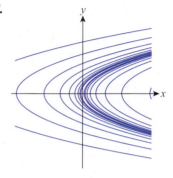

13–18 Determine whether the limit exists. If so, find it.

13. $\lim\limits_{(x,y)\to(0,0)} \dfrac{x^3 - xy^2}{x + y}$

14. $\lim\limits_{(x,y)\to(1,1)} \dfrac{\ln(xy)}{\sin\left(\ln(xy)\right)}$

15. $\lim\limits_{(x,y)\to(0,1)} \dfrac{\sqrt{2x + y} - \sqrt{x + y}}{2x}$

16. $\lim\limits_{(x,y)\to(0,0)} \left(x^2 + y^2\right)\sin\dfrac{\pi}{x^2 + y^2}$

17. $\lim\limits_{(x,y)\to(0,0)} \dfrac{\sin y}{xy^2}$

18. $\lim\limits_{(x,y)\to(0,0)} \dfrac{2x^2 - y^4}{x^2 + y^4}$

19. Find two different paths in the xy-plane toward $(0,0)$ along which the limiting values of the function $f(x, y) = \dfrac{2xy^2}{x^3 + y^3}$ are different. (Answers will vary.)

20. Use an ε-δ argument to show that
$$\lim\limits_{(x,y)\to(0,0)} \dfrac{x^2 y}{x^2 + y^2} = 0.$$

21–24 Find all discontinuities (if any) of the given function. Classify them as removable or nonremovable.

21. $f(x, y) = \ln\left(x^4 + y^4\right)$

22. $g(x, y) = \left(6 - 2x^2 - 3y^2\right)^{-1}$

23. $h(x, y) = \arctan\dfrac{2}{\left(x^2 - 1\right)\left(y^2 - 4\right)}$

24. $k(x, y) = \dfrac{1 - \cos\left(x^2 + y^2\right)}{x^2 + y^2}$

25–26 Determine z_x and z_y at the indicated point; then find the equations for the corresponding tangent lines that are parallel to, respectively, the xz-plane and the yz-plane.

25. $z = x^3 y^2 + x^2 y; \quad (-1,1)$

26. $z = \left(x^2 y - 2\right)\left(y + 4x^2\right); \quad (1,2)$

27–28 Find all first-order partial derivatives of the function.

27. $f(x, y, z) = y^2 \sin^{-1}\left(z\sqrt{x}\right)$

28. $g(x, y, z) = \dfrac{z e^{xz}}{\sqrt{x + y^2}}$

29–30 Use implicit differentiation to determine z_x and z_y.

29. $2y^3 - x^2 z - xyz^3 = 0$

30. $(z - y)^3 + 2\ln(xy) = z$

31–32 Verify the equality of the mixed partials f_{xy} and f_{yx}.

31. $f(x,y) = (x^3 + y)^2 \ln(xy)$

32. $f(x,y) = \tan^{-1}\sqrt{x^2 + y^2}$

33–34 Verify that the third-order mixed partials f_{xyz}, f_{yzx}, and f_{zxy} are equal.

33. $f(x,y,z) = 2x^2z^2 + xyz - 4zy^3$

34. $f(x,y,z) = xy \ln(z - 2x)^2$

35–36 Use the definition of the partial derivative to find $\partial f/\partial x$ and $\partial f/\partial y$ for the given function.

35. $f(x,y) = \sqrt{x}\,y^2$ **36.** $f(x,y) = \dfrac{x^3 + y^2}{x}$

37–38 If possible, find a function $f(x,y)$ that has the indicated partial derivatives. If such a function doesn't exist, explain why.

37. $f_x(x,y) = y^3 + \dfrac{1}{x}$, $f_y(x,y) = 3xy^2$

38. $f_x(x,y) = yx^2 - \cos y$, $f_y(x,y) = \dfrac{x^3}{3} - x\sin y$

39–40 Determine whether the function is differentiable at the origin. Give a reason for your answer.

39. $f(x,y) = x\sin^2 y + xy^2$

40. $f(x,y) = \sqrt{x^2 + y^2}\,\arctan(xy)$

41. Show that the piecewise defined function

$$f(x,y) = \begin{cases} \dfrac{2xy}{\sqrt{x^6 + y^6}} & \text{if } (x,y) \neq (0,0) \\[2mm] 0 & \text{if } (x,y) = (0,0) \end{cases}$$

is not differentiable at the origin, even though both $f_x(0,0)$ and $f_y(0,0)$ exist.

42. Use the definition to show that the function $f(x,y) = 2x - 3y^2$ is differentiable.

43–46 Use the Chain Rule to determine dy/dx.

43. $y = 2uv^3$; $u = x^2 - 2$, $v = 3x$

44. $y = u^2 e^{u^2 v}$; $u = \sqrt{x}$, $v = x^2$

45. $y = u^2 v \cos w$; $u = \sqrt{x}$, $v = x^3$, $w = x^2$

46. $y = uv^2 + \sqrt{w}$; $u = \ln x$, $v = 2x$, $w = x^2$

47–49 Use the Chain Rule to find $f'(t)$ at the given point.

47. $f(x,y) = (x - 2y)^2$ along the curve $x = t\sin t$, $y = 2t\cos t$, at $t = \pi/2$

48. $f(x,y) = \ln(xy)$ along the curve $x = t^2$, $y = t - 1$, at $t = 2$

49. $f(x,y,z) = xyz$ along the space curve $x = t\cos t$, $y = t\sin t$, $z = 2t$, at $t = \pi$

50–54 Use the Chain Rule to determine the partial derivatives of the given function f. (Answers may be left in terms of the intermediate and independent variables.)

50. $f(u,v) = 3u^2 - v\sqrt{u}$; $u(x,y) = x^2$, $v(x,y) = xy^2$

51. $f(u,v) = \ln(u^2 + v^2)$; $u(x,y) = x\sqrt{y}$, $v(x,y) = x^2 y$

52. $f(u,v,w) = u\cos v - \sin w$; $u(x,y) = 2x + y$, $v(x,y) = y - x^2$, $w(x,y) = xy$

53. $f(u,v,w) = u(v - w^2)$; $u(x,y) = 2x^3$, $v(x,y) = x - y^2$, $w(x,y) = 3x + y$

54. $f(u,v,w) = uve^w$; $u(x,y,z) = yz^2$, $v(x,y,z) = (x+y)^2$, $w(x,y,z) = x - 4z$

55–56 Find dy/dx where y is given implicitly by the given equation. (**Hint:** Rewrite the equation in the form $F(x,y) = 0$, and use the appropriate formula from Section 13.4.)

55. $x^2(2 - xy) = xy^3$ **56.** $(x^3 - y^2)^2 = 4x^2 y$

57–58 Find z_x and z_y where z is defined implicitly by the given equation. (**Hint:** Use the appropriate formula from Section 13.4.)

57. $xy^2 + yz^3 = x^3 z^2$ **58.** $ze^x + 2x\sin y + xyz^3 = 1$

59. Suppose $z = f(x,y)$ has continuous second partials, $x = r\cos\theta$ and $y = r\sin\theta$. Use the Chain Rule to prove the following.

$$\frac{\partial^2 f}{\partial r^2} = \frac{\partial^2 f}{\partial x^2}\cos^2\theta + 2\frac{\partial^2 f}{\partial x\partial y}\sin\theta\cos\theta + \frac{\partial^2 f}{\partial y^2}\sin^2\theta$$

60. The height of a right rectangular pyramid is increasing at a rate of 0.2 in./min, while the edge of its square base is decreasing at 0.1 in./min. Find the rate of change of its volume and surface area when the edge of the base is 15 in. long and the height is 20 in.

61–64 Find the gradient of the function at the indicated point.

61. $f(x, y) = 3x^2 y - 2\sqrt{x}$; $(1, 3)$

62. $f(x, y) = \dfrac{\cos(x - y)}{y^2}$; $(1, 1)$

63. $f(x, y, z) = x^2 e^{yz}$; $(-1, 0, 2)$

64. $f(x, y, z) = \dfrac{\tan^{-1}(x + y)}{z}$; $(3, -2, 4)$

65–66 Use the definition to find the directional derivative of f at the given point in the indicated direction.

65. $f(x, y, z) = \dfrac{xy^2}{z}$; $(2, 4, 1)$; $\mathbf{v} = \langle 3, 1, 2 \rangle$

66. $f(x, y) = x + 2\sqrt{y}$; $(1, 1)$; $\mathbf{v} = \langle 1, 2 \rangle$

67–70 Find the derivative of the function at the given point in the indicated direction.

67. $f(x, y) = \ln\sqrt{x^2 + y}$; $(1, 0)$; $\mathbf{v} = \langle -2, 14 \rangle$

68. $f(x, y) = x^2 + 3xy - 2y^2$; $(1, 2)$; in the direction toward $(-2, 3)$

69. $f(x, y, z) = z\sqrt{\dfrac{xz + y}{x}}$; $(1, -1, 2)$; in the direction toward $(4, -6, 3)$

70. $f(x, y, z) = xy^2 z^3$; $(1, 1, 1)$; $\mathbf{v} = \langle 3, 2, 1 \rangle$

71–72 Find the direction and value of the greatest rate of increase for the function at the given point.

71. $f(x, y, z) = \dfrac{y + z^2}{x + 1}$; $(0, 1, -1)$

72. $f(x, y) = 4x^2 + xy + \dfrac{x}{y}$; $(0, 1)$

73. Suppose that the temperature in three-dimensional space around a heat source that is located at the origin can be described by the function

$$T(x, y, z) = 450e^{-(2x^2 + y^2 + z^2)/2}.$$

a. Find the direction and magnitude of the temperature's maximum rate of decrease at the point $(1, 1, 1)$.

b. Find the rate of change of temperature at the point $(1, 1, 1)$ in the direction of the point $(1, 2, 3)$.

74–75 Find the direction of no change for the function at the given point.

74. $f(x, y) = \dfrac{2xy}{x^2 - y^2}$; $(2, -1)$

75. $f(x, y) = x\ln(y - x^2)$; $(1, 2)$

76–79 In Exercises 76–77, use the technique seen in this chapter to find an equation for the tangent line to the graph of the equation at the indicated point. In Exercises 78–79, generalize to obtain the respective tangent plane.

76. $2x^2 + xy - 3y^2 + 13 = 0$; $(2, 3)$

77. $x^2(y + 2) = 4x^3$; $(1, 2)$

78. $2y^2 - x^2 = 2z^2$; $(4, 3, 1)$

79. $4z - 3x^2 = 3y^2$; $(4, 2, 15)$

80–81 Find a function with the given gradient. If it is not possible, explain why. (Answers will vary.)

80. $\nabla f = \left\langle \dfrac{1}{\sqrt{x}} + y^2, 2xy \right\rangle$ **81.** $\nabla f = \langle xy + 1, x^2 \rangle$

82–85 Determine the tangent plane and the normal line at the indicated point of the given surface.

82. $5x^2 - y^2 - 2z = 0$; $(1, 3, -2)$

83. $(2x - 1)(5x + y^2) = z^2$; $(1, 2, 3)$

84. $\dfrac{x}{\sqrt{y^2 + x}} - z = 0$; $\left(3, 1, \dfrac{3}{2}\right)$

85. $\ln(x^2 + y^2 + z^2) = 0$; $\left(\dfrac{\sqrt{3}}{2}, \dfrac{1}{2}, 0\right)$

86–87 Find parametric equations for the line tangent to the intersection of the two surfaces at the indicated point.

86. $x^2 + 4y^2 = 6 - z$; $2z = x^2 + y^2$; $(1,1,1)$

87. $x^3 - 2y^3 + xy^2 - z^2 = 5xy$; $2y^2 - x^2 = z$; $(-1,1,1)$

88–89 Find **a.** the differential df at an arbitrary point (x,y), and **b.** the linearization of f at the indicated point.

88. $f(x,y) = (x-1)^3 + y^2 - x$; $(2,1)$

89. $f(x,y) = \sin(x^2 - y^2)$; $(-1,1)$

90–91 Find the linear approximation of the function at the indicated point.

90. $f(x,y,z) = x\tan^{-1}(yz)$; $(2,1,1)$

91. $f(w,x,y,z) = \dfrac{x^2 y - wz^2}{xyw}$; $(1,2,-1,4)$

92. Suppose the base lengths of the rectangular pyramid of Exercise 60 were measured within a relative error of 2.5%, and its height within an error of 2%. Use differentials to find an upper bound on the relative error with which the pyramid's volume is known.

93–96 Find any extrema of the given function. Use the Second Derivative Test (if necessary) to classify critical points.

93. $f(x,y) = x^2 + y^2 - 4x + 2y + 6$

94. $f(x,y) = 4y - y^2 - 8x - x^2$

95. $f(x,y) = x^3 + y^3 - 4y^2 - x^2$

96. $f(x,y) = 2y^4 - y^2 + x^2 + x$

97–98 Find the absolute extrema of the given function on the indicated region.

97. $f(x,y) = x^2 - xy + 1$;
$D = \{(x,y) \mid -2 \le x \le 2, 0 \le y \le 2\}$

98. $f(x,y) = 4xy - 3x - y + 2$; D: The triangle with vertices $(0,0)$, $(5,0)$, and $(0,4)$

99. Determine m and b such that the sum of the squares of vertical distances from the line $y = mx + b$ to the points $(1,2)$, $(4,2)$, $(6,6)$, and $(8,4)$ is minimal.

100. Find the minimum distance between the point $(2,-1,1)$ and the plane $z = 4x - y - 2$.

101. Find the minimum distance from the origin to the surface $z = 2xy - 1$.

102. A rectangular box is placed in the three-dimensional coordinate system with one vertex at the origin and the three edges containing it lying along the positive coordinate axes. If the opposite vertex lies in the plane $x + 4y + 2z = 6$, what is the greatest possible volume for such a box?

103. Suppose the sum of three positive numbers is 150, what is the greatest possible value for the sum of their squares?

104–107 Find any extreme values of the function subject to the given constraint.

104. $f(x,y) = x^2 y$;   constraint: $\dfrac{x^2}{2} + \dfrac{y^2}{4} = 1$

105. $f(x,y) = x^2 + y^2$;   constraint: $xy = 4$

106. $f(x,y) = x^2 - y^2$;   constraint: $3x + 4y = 2$

107. $f(x,y) = x^2 + xy - y^2$;   constraint: $x = y^2$

108–109 Use Lagrange multipliers to find the coordinates of the point Q on the given surface that is closest to the point P.

108. $3z - 2x + y + 6 = 0$; $P(1,0,1)$

109. $z = x + y^2$; $P(1,0,0)$

110–111 Find the absolute extreme value(s) of the function subject to the given constraints.

110. $f(x,y,z) = x - 2y + z$;
constraints: $x^2 + z^2 = 1$ and $x + y + z = 2$

111. $f(x,y,z) = 2(x^2 + y^2 + z^2)$;
constraints: $x + y + z = 2$ and $2x + z = 1$

112. A small company profits $10 per unit on products A and C, while the profit on product B is $20 per unit. Let x, y, and z denote hundreds of units produced from each product.

a. Find the profit function $P(x,y,z)$.

b. Assuming that the manufacturing process is under the constraint $x^2 + 10y^2 + 2z^2 \le 60$, find the maximum profit for the company under these conditions.

113. Find the points closest to and farthest from the origin on the rotated ellipse $2x^2 + xy + 2y^2 = 4$.

114. Find the highest and lowest points on the curve of intersection of the two surfaces $x^2 + 2y^2 + z^2 = 10$ and $x + 4y + z = 8$.

115–126 *True or False?* Determine whether the given statement is true or false. In case of a false statement, explain or provide a counterexample.

115. Any vertical line (i.e., one that is parallel to the z-axis) can intersect the graph of a two-variable function $f(x, y)$ at most once.

116. If a function has three or more independent variables, then it has no graph.

117. If $f(x, y)$ is continuous at (a, b) and $f(a, b) = L$, then $\lim\limits_{(x,y)\to(a,b)} f(x, y) = L$.

118. If $\lim\limits_{(x,y)\to(0,0)} f(x, y) = 0$ and $a \in \mathbb{R}$, then $\lim\limits_{x\to 0} f(x, ax) = 0$.

119. If $\lim\limits_{(x,y)\to(0,0)} f(x, y) = 0$, then for any $\varepsilon > 0$ there exists $R \in \mathbb{R}$ such that $|f(x, y)| < \varepsilon$ whenever $0 < x^2 + y^2 < R$.

120. If $\lim\limits_{(x,y)\to(a,b)} f(x, y) = L$, then $f(x, y) \to L$ as (x, y) approaches (a, b) along any curve in the xy-plane.

121. If $f(x, y)$ is continuous at (a, b), then both $f_x(a, b)$ and $f_y(a, b)$ exist.

122. If $f(x, y) = \dfrac{g(x)}{h(y)}$, then
$$f_x(x, y) + f_y(x, y) = \frac{g'(x)h(y) - g(x)h'(y)}{h^2(y)}.$$

123. If $f_x(a, b)$ and $f_y(a, b)$ both exist, then $f(x, y)$ is differentiable at (a, b).

124. If $f_x(x, y)$ and $f_y(x, y)$ are continuous at (a, b), then $f(x, y)$ is differentiable at (a, b).

125. If f has a relative maximum or minimum at (a, b), then all directional derivatives $D_{\mathbf{u}} f(a, b)$ are 0.

126. If f has a relative maximum or minimum at (a, b), then the tangent plane to its graph at (a, b) is horizontal.

Chapter 13
Technology Exercises

127–130. Write a program on your computer algebra system or programmable calculator that accepts the equation of a surface and a point on the surface, and generates the corresponding tangent plane and normal line to the surface. Test your program by checking the answers you obtained for Exercises 82–85.

131–134. Write a program on your computer algebra system or programmable calculator that accepts a two-variable function $f(x, y)$ and then locates and classifies its critical points by utilizing the Second Derivative Test. Use your program to check the answers you obtained for Exercises 93–96.

135. Recall the Ideal Gas Law, $pV = nRT$ from Section 3.4, Exercise 97. Suppose $\frac{1}{2}$ mole of an ideal gas is present in an expandable container. Assuming the volume of the container can expand up to 1 m³, and that it can withstand a maximum pressure of 120 kPa along with temperatures of up to 500 K, use a computer algebra system to generate a contour map that illustrates how the volume depends on pressure when temperature is kept at constant levels. (Note that the resulting contour lines are called *isotherms*. **Hint:** Start by expressing T as a function of p and V from the Ideal Gas Law.)

136. Mimic Exercise 135 to generate a few contour lines of the pressure function under the same conditions (these curves are called *isobars*).

Project ✎

Chapter 13

In this project you will use your experience with partial derivatives and differentials to learn how to solve an important class of differential equations, called **exact equations**. Ordinary differential equations of this type are noted for their widespread applications in physics and engineering. (See Section 8.1 for the definitions of differential equation and solution. Other than the basic definitions, this project does not directly rely on, and can be considered independently of Chapter 8.)

1. Suppose that the first-order partial derivatives of the function $z = f(x, y)$ are both continuous on a region R. If c is a constant and $y = y(x)$ is defined implicitly by the equation $f(x, y) = c$, show that y solves the differential equation

$$f_y(x, y) \cdot y' = -f_x(x, y).$$

2. Now consider a differential equation of the form

$$M(x, y) dx + N(x, y) dy = 0 \qquad (1)$$

and assume that there is a two-variable function $f(x, y)$ such that

$$\frac{\partial f(x, y)}{\partial x} = M(x, y) \quad \text{and} \quad \frac{\partial f(x, y)}{\partial y} = N(x, y)$$

(such a differential equation is called *exact*, while $f(x, y)$ is called a *potential function*). Use your answer to Question 1 to show that the set of level curves $f(x, y) = C$, $C \in \mathbb{R}$ form a family of solutions of the differential equation (1).

3. Suppose that $M(x, y)$ and $N(x, y)$, as well as their first-order partial derivatives, are continuous on an open region R. Show that a necessary condition for equation (1) to be exact is the following equality.

$$\frac{\partial M(x, y)}{\partial y} = \frac{\partial N(x, y)}{\partial x}$$

(**Note:** If we require a bit more of R, the above condition is also sufficient for exactness, a statement we will not rigorously prove here, but the construction of a potential function under the stated conditions is outlined in Questions 5 and 6.)

4. Use Question 3 to determine which of the following equations is exact.

   a. $\left(2x + ye^{xy}\right) dx + \left(xe^{xy} - 1\right) dy = 0$

   b. $\left(3x^2 y - \dfrac{1}{\sqrt{x}}\right) dx + \left(x^3 - \sqrt{x}\right) dy = 0$

5. Explain why the potential function f of an exact equation must satisfy

$$f(x, y) = \int M(x, y) dx + g(y),$$

where g is some function of the variable y.

6. Show that if the equation (1) is exact, then the equality

$$N(x, y) = \frac{\partial}{\partial y} \int M(x, y) dx + g'(y)$$

must hold.

7. Use Questions 5 and 6 to solve the equation $2\left(x - y^2\right) dx + y\left(9y - 4x\right) dy = 0$ by determining its potential function $f(x, y)$ and identifying the family of solutions as $f(x, y) = C$. (**Hint:** After identifying $M(x, y)$ and $N(x, y)$, use Question 5 to obtain a tentative formula for $f(x, y)$, then use Question 6 to determine the unknown function $g(y)$.)

8. Verify that the equation $(2 + x) y \, dx + 2x \, dy = 0$ becomes exact after multiplying by the integrating factor $I(x, y) = xye^x$. Solve the resulting equation.

Multiple Integrals

14.1 Double Integrals   1075

1. Double Integrals and Riemann Sums
2. Fubini's Theorem
3. Double Integrals over General Bounded Regions

14.2 Applications of Double Integrals   1086

1. Area by Double Integration
2. First Moments
3. Second Moments

14.3 Double Integrals in Polar Coordinates   1097

1. The Form of Double Integrals in Polar Coordinates
2. Evaluating Double Integrals in Polar Coordinates

14.4 Triple Integrals   1105

1. Triple Integrals and Riemann Sums
2. Triple Integrals over General Bounded Regions
3. First and Second Moments of Three-Dimensional Bodies

14.5 Triple Integrals in Cylindrical and Spherical Coordinates   1117

1. The Cylindrical Coordinate System
2. The Spherical Coordinate System

14.6 Substitutions and Multiple Integrals   1129

1. Substitutions in Double Integrals
2. Substitutions in Triple Integrals

Introduction

The next step in our progression from single variable calculus to multivariable calculus is to broaden our understanding of integration. Two of the principal goals of this chapter are to define what it means to integrate functions over regions of $\mathbb{R}^2$ and $\mathbb{R}^3$, and to learn techniques for carrying out such integration.

The concept of integration as a limit of Riemann sums readily carries over into the multidimensional realm, allowing us to quickly define *double* and *triple* integrals (integrals over, respectively, regions of $\mathbb{R}^2$ and $\mathbb{R}^3$). But just as in the single variable case, the formal definition leaves much to be desired when it comes to actually evaluating multiple integrals. *Fubini's Theorem*, published by the Italian mathematician Guido Fubini (1879–1943) in 1907, gives us the means for such evaluation. It does so by allowing us to interpret double and triple integrals (and higher-order multiple integrals) as *iterated* integrals, which can be evaluated using now-familiar techniques. Although mathematicians of the 18th century had employed the method of iterated integration and knew it to be valid under certain conditions, Fubini's deeper grasp of the principle and his formal proof didn't appear until the 20th century. In fact, variations and extensions of Fubini's Theorem still continue to be proved today by mathematicians working in the branch of mathematics known as *measure theory*.

Double and triple integrals are often the most natural means to determine areas of planar regions and volumes of solids, and we will also use them to define and calculate moments and centers of mass of various two- and three-dimensional objects. As we construct such integrals, we will frequently find that coordinate systems other than the Cartesian system are better suited to the geometry of a particular object.

of the chapter, we'll generalize that observation and learn how to create coordinate systems customized to given integrals. The transformation factors that result, known as *Jacobian determinants*, were introduced by the German mathematician Carl Gustav Jacob Jacobi (1804–1851) in the mid-nineteenth century. The *u*-substitution technique that we learned in Chapter 5, we will see, is

> ## Double and triple integrals are often the most natural means to determine areas of planar regions and volumes of solids, and we will also use them to define and calculate moments and centers of mass of various two- and three-dimensional objects.

In particular, we will learn how to set up double integrals using the polar coordinates introduced in Chapter 9, and we will do the same for two additional three-dimensional coordinate systems called *cylindrical* and *spherical* coordinates.

As we construct integrals using different coordinate systems, we will find it necessary to account for certain coordinate transformation factors which relate to the system's differential elements of area or volume, dA and dV. In the last section

actually a single-variable precursor of this deep multivariable calculus concept.

Guido Fubini
(1879–1943)

14.1 **Double Integrals**

TOPICS

1. Double integrals and Riemann sums

2. Fubini's Theorem

3. Double integrals over general bounded regions

Chapter 5 introduced integrals of functions of a single variable over bounded intervals of $\mathbb{R}$; in this section, we extend the definition of integration to encompass *double integrals* over bounded regions of $\mathbb{R}^2$. Just as the integrals of Chapter 5 had geometric interpretations as signed areas, the double integrals of this chapter can be interpreted as signed volumes. And just as the Fundamental Theorem of Calculus allowed us to evaluate single integrals without taking limits of Riemann sums, we will learn similar techniques that greatly simplify the evaluation of double integrals.

TOPIC 1 **Double Integrals and Riemann Sums**

Recall that the formal definition of the integral $\int_a^b f(x)\,dx$ is based on the limit of Riemann sums of the form $\sum_{i=1}^{n} f\left(x_i^*\right)\Delta x_i$, where $\left\{x_0, x_1, \ldots, x_n\right\}$ is a partition of the interval $[a,b]$ and each x_i^* is a sample point from the subinterval $[x_{i-1}, x_i]$. Each term $f\left(x_i^*\right)\Delta x_i$ in the sum represents the area of a rectangle of height $f\left(x_i^*\right)$ and width $\Delta x_i = x_i - x_{i-1}$, so the Riemann sum is an approximation of the (signed) area between the graph of $f(x)$ and the x-axis over the interval $[a,b]$.

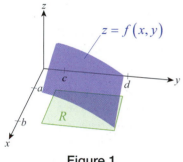

Figure 1

To extend this idea to a function $f(x,y)$ defined on a region $R \subseteq \mathbb{R}^2$, we assume initially that R is a rectangle in the plane defined by $a \leq x \leq b$ and $c \leq y \leq d$, which we write as $[a,b]\times[c,d]$; Figure 1 is an illustration of such a rectangle and the graph of a surface $z = f(x,y)$. We can approximate the volume of the space between $z = f(x,y)$ and the xy-plane over the region R by partitioning R into a grid of rectangles and summing the volumes of rectangular columns, such as the one shown in Figure 2. The height of each column is the value $f\left(x_{ij}^*, y_{ij}^*\right)$, where $\left(x_{ij}^*, y_{ij}^*\right)$ is a sample point chosen from the subrectangle defined by $x_{i-1} \leq x \leq x_i$ and $y_{j-1} \leq y \leq y_j$, and the area of the base is $\Delta A = \Delta x_i \Delta y_j$, where $\Delta x_i = x_i - x_{i-1}$ and $\Delta y_j = y_j - y_{j-1}$. So the volume of each column is $f\left(x_{ij}^*, y_{ij}^*\right)\Delta x_i \Delta y_j$ and the sum of the volumes is expressed as

$$\sum_{j=1}^{m}\sum_{i=1}^{n} f\left(x_{ij}^*, y_{ij}^*\right)\Delta x_i \Delta y_j,$$

where $\left\{a = x_0, x_1, \ldots, x_n = b\right\}$ is a partition of $[a,b]$ and $\left\{c = y_0, y_1, \ldots, y_m = d\right\}$ is a partition of $[c,d]$. The double sum above is an example of a Riemann sum of $f(x,y)$ over the region R.

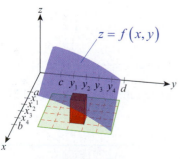

Figure 2

For ease of illustration, the grid in Figure 2 shows $[a,b]$ and $[c,d]$ both partitioned into five subintervals of equal width, but as in Chapter 5, there is no requirement that Riemann sums possess such regularity—the widths of each subinterval can vary, and the numbers of subintervals of $[a,b]$ and $[c,d]$ don't have to be equal. However, in order to refine our approximation of the true volume between $z = f(x,y)$ and the plane over R, it *is* important that we consider Riemann sums for which $\max_{1\leq i\leq n, 1\leq j\leq m}\left\{\Delta x_i, \Delta y_j\right\}$ goes to zero. We use the word **partition** again to refer to the subdivision of R into subrectangles, and if P represents one particular partition, we define the **norm** of P by

If all double Riemann sums of the form above converge to a common value as we consider partitions of smaller and smaller norm, we define the *double integral* of f over R to be that common value. The formal definition follows.

Definition 💡

Double Integral

Given a function $f(x, y)$ defined on the region $R = [a, b] \times [c, d]$, we define the **double integral of f over R** to be the value

$$\iint\limits_R f(x, y)\, dA = \lim_{\|P\| \to 0} \sum_{j=1}^{m} \sum_{i=1}^{n} f\left(x_{ij}^*, y_{ij}^*\right) \Delta x_i \Delta y_j,$$

provided the limit exists. If the limit exists, we say f is **integrable** over R, and define the **signed volume** between the surface $z = f(x, y)$ and the xy-plane over the region R to be the double integral.

Example 1 📝

Given the function $f(x, y) = 15 - x^2 - 2y^2$ and the region R defined by $a = 1$, $b = \frac{5}{2}$, $c = \frac{1}{2}$, and $d = 2$, the Riemann sum approximations corresponding to $n = m = 5$, 10, and 20 equal partitions of each interval are as shown in Figure 3.

The exact value of the double integral of f over R is $\frac{297}{16}$ cubic units, and the difference between each Riemann sum and the exact value is listed below each figure. Each Riemann sum has been calculated (by computer, not by hand!) with $\left(x_{ij}^*, y_{ij}^*\right) = \left(x_i, y_j\right)$, the top-right corner of each subrectangle.

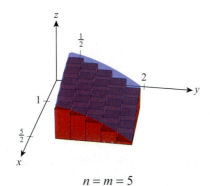

$$n = m = 5$$

Riemann sum $= \dfrac{6237}{400}$

Difference from exact value $= 2.97$

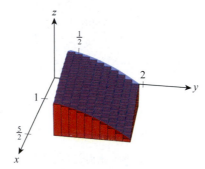

$$n = m = 10$$

Riemann sum $= \dfrac{54{,}729}{3200}$

Difference from exact value ≈ 1.460

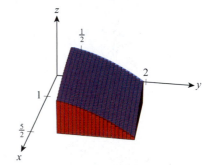

$$n = m = 20$$

Riemann sum $= \dfrac{228{,}339}{12{,}800}$

Difference from exact value ≈ 0.724

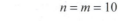

Figure 3

TOPIC 2 **Fubini's Theorem**

Clearly, we would prefer not to have to take limits of double Riemann sums to evaluate double integrals if there is an easier alternative. Further, while the approximations shown in Example 1 may make a fairly convincing argument that the Riemann sums converge to *some* value, there is nothing in the work presented to indicate that the value is $\frac{297}{16}$.

Fubini's Theorem allows us to both avoid limits of Riemann sums and arrive at the exact value of a double integral, at least when we are able to find antiderivatives of the integrand as outlined below. It does so by reducing the task of evaluating a double integral to the task of evaluating two single integrals consecutively. The theorem, the proof of which we only outline here, is based on the principle that the volume V of a solid is equal to the integral of its cross-sectional areas over some interval. We first encountered this principle in Section 6.1 in the form $V = \int A(x)\,dx$ and $V = \int A(y)\,dy$.

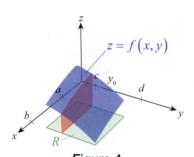

Figure 4

Consider again the solid bounded by the xy-plane and the surface $z = f(x, y)$ over a region R, but this time with an eye toward its cross-sectional slices perpendicular to the y-axis—one of them, through the point $y = y_0$, is shown in Figure 4. Then the volume of the solid is

$$V = \int_c^d A(y)\,dy.$$

But each cross-sectional area $A(y)$ corresponds to the area between a function of x and the xy-plane over the interval $[a, b]$; for instance,

$$A(y_0) = \int_a^b f(x, y_0)\,dx.$$

Hence,

$$V = \int_c^d A(y)\,dy = \int_c^d \underbrace{\left[\int_a^b f(x, y)\,dx\right]}_{A(y)} dy = \int_c^d \int_a^b f(x, y)\,dx\,dy.$$

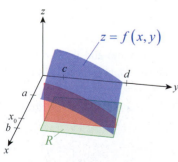

Figure 5

Such an integral is called an **iterated** or **repeated integral**, and Fubini's Theorem specifies conditions under which its value is equal to the double integral $\iint\limits_R f(x, y)\,dA$. (Although the Italian mathematician Guido Fubini (1879–1943) proved the theorem under very general conditions in 1907, versions of the theorem were known and used by earlier mathematicians.)

Before stating the theorem formally, note that we could just as well begin with cross-sectional slices perpendicular to the x-axis, as shown in Figure 5, in which case we arrive at the following result:

$$V = \int_a^b A(x)\,dx = \int_a^b \underbrace{\left[\int_c^d f(x, y)\,dy\right]}_{A(x)} dx = \int_a^b \int_c^d f(x, y)\,dy\,dx$$

Theorem

Fubini's Theorem

If $f(x, y)$ is continuous on the region $R = [a,b] \times [c,d]$, then

$$\iint_R f(x, y)\, dA = \int_c^d \int_a^b f(x, y)\, dx\, dy = \int_a^b \int_c^d f(x, y)\, dy\, dx.$$

Example 2

Evaluate the double integral $\iint_R (6xy^2 - 4y)\, dA$, where $R = [0,1] \times [0,2]$.

Solution

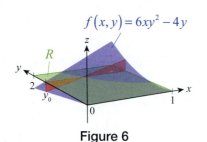

$f(x, y) = 6xy^2 - 4y$

Figure 6

As Figure 6 indicates, the function $f(x, y) = 6xy^2 - 4y$ takes on both positive and negative values on the region R (that is, part of the surface lies above the xy-plane, and part below). This doesn't affect the procedure for evaluating the double integral, but keep in mind that double integrals represent *signed* volume, and it is entirely possible for the result to be positive, negative, or zero.

By Fubini's Theorem,

$$\iint_R (6xy^2 - 4y)\, dA = \int_0^2 \int_0^1 (6xy^2 - 4y)\, dx\, dy.$$

To evaluate such an iterated integral, we first evaluate the inner integral by finding an antiderivative of the integrand with y held fixed—remember, the inner integral represents a cross-sectional area $A(y)$, such as shown in Figure 6, and y doesn't vary until we integrate $A(y)$ over $[0,2]$ in the outer integral. In this respect, the evaluation of iterated integrals is much like partial differentiation, all but one of the variables is held fixed at each stage.

$$\iint_R (6xy^2 - 4y)\, dA = \int_0^2 \int_0^1 (6xy^2 - 4y)\, dx\, dy = \int_0^2 \left[3x^2 y^2 - 4xy \right]_{x=0}^{x=1} dy$$

$$= \int_0^2 (3y^2 - 4y)\, dy = \left[y^3 - 2y^2 \right]_{y=0}^{y=2}$$

$$= (8 - 8) - (0 - 0) = 0$$

This result means that the volume between the xy-plane and the positive portion of $z = 6xy^2 - 4y$ has the same magnitude (but opposite sign) as the volume between the xy-plane and the negative portion of the surface. The result is the same if we choose to work with the alternative iterated integral, though the intermediate steps differ.

$$\iint_R (6xy^2 - 4y)\, dA = \int_0^1 \int_0^2 (6xy^2 - 4y)\, dy\, dx = \int_0^1 \left[2xy^3 - 2y^2 \right]_{y=0}^{y=2} dx$$

$$= \int_0^1 (16x - 8)\, dx = \left[8x^2 - 8x \right]_{x=0}^{x=1}$$

$$= (8 - 8) - (0 - 0) = 0$$

Now that we have the means to exactly determine the volume, we return to Example 1.

Example 3 ✐

Evaluate the double integral $\iint\limits_R \left(15 - x^2 - 2y^2\right) dA$, where $R = \left[1, \frac{5}{2}\right] \times \left[\frac{1}{2}, 2\right]$.

Solution

By Fubini's Theorem, $\iint\limits_R \left(15 - x^2 - 2y^2\right) dA = \int_{1/2}^2 \int_1^{5/2} \left(15 - x^2 - 2y^2\right) dx\, dy$.

$$\iint\limits_R \left(15 - x^2 - 2y^2\right) dA = \int_{1/2}^2 \int_1^{5/2} \left(15 - x^2 - 2y^2\right) dx\, dy$$

$$= \int_{1/2}^2 \left[15x - \frac{x^3}{3} - 2xy^2\right]_{x=1}^{x=5/2} dy$$

$$= \int_{1/2}^2 \left(\frac{141}{8} - 3y^2\right) dy$$

$$= \left[\frac{141}{8}y - y^3\right]_{1/2}^2$$

$$= \frac{297}{16}$$

In Exercise 3 you will show that $\int_1^{5/2} \int_{1/2}^2 \left(15 - x^2 - 2y^2\right) dy\, dx = \frac{297}{16}$ as well.

TOPIC 3   ## Double Integrals over General Bounded Regions

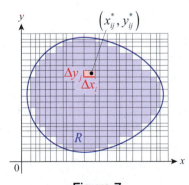

Figure 7

For the applications we will begin to study in Section 14.2, we often need to evaluate double integrals over bounded regions in the plane that are not rectangular. Figure 7 shows such a region R in the xy-plane over which we may need to integrate a function $f(x, y)$, along with a partition of the plane that subdivides R. If the function f and the boundary of R are sufficiently well behaved, it can be shown that as R is subdivided by finer and finer partitions, the corresponding Riemann sums converge, and we once again define

$$\iint\limits_R f(x, y)\, dA = \lim_{\|P\| \to 0} \sum_{j=1}^m \sum_{i=1}^n f\left(x_{ij}^*, y_{ij}^*\right) \Delta x_i \Delta y_j.$$

A detailed discussion of conditions that qualify f and the boundary of R as "well behaved" is best left for a more advanced course, but for our purposes it suffices to note that $\iint\limits_R f(x, y)\, dA$ exists if

1. f is continuous, and

2. the boundary of R is made up of a finite number of segments joined end to end, where each segment is the graph of a continuous function over a bounded interval.

Further, Fubini's Theorem extends to double integrals over these general regions, allowing us to evaluate them by iterated integration. In particular, the double integrals we need to evaluate can usually be expressed in one of the two forms shown below, where $g_1(x)$, $g_2(x)$, $h_1(y)$, and $h_2(y)$ all represent continuous functions defined on a bounded interval.

$$\iint\limits_R f(x,y)\,dA = \int_c^d \int_{h_1(y)}^{h_2(y)} f(x,y)\,dx\,dy \qquad \iint\limits_R f(x,y)\,dA = \int_a^b \int_{g_1(x)}^{g_2(x)} f(x,y)\,dy\,dx$$

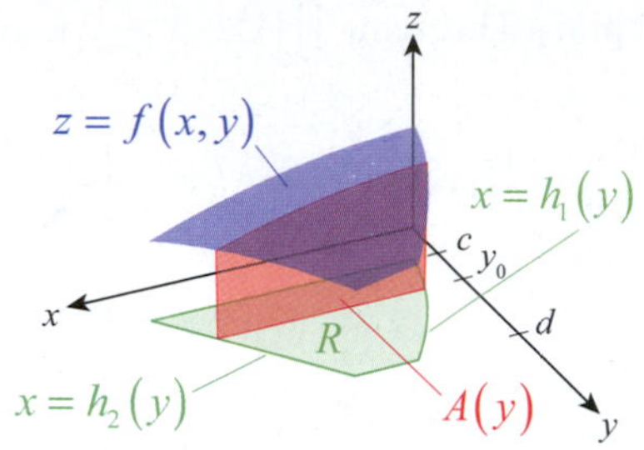

Figure 8

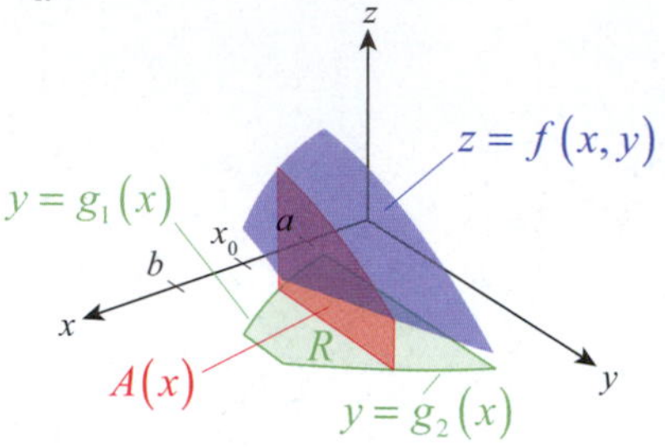

Figure 9

In Figure 8, the region R is defined by the inequalities $h_1(y) \le x \le h_2(y)$ and $c \le y \le d$, so each cross-sectional area is given by $A(y) = \int_{h_1(y)}^{h_2(y)} f(x,y)\,dx$ and the volume of the bounded solid by $V = \int_c^d A(y)\,dy = \int_c^d \int_{h_1(y)}^{h_2(y)} f(x,y)\,dx\,dy$. The situation shown in Figure 9 is similar, except that each cross-sectional area is a function of x and the order of integration is reversed.

As is true for most topics in mathematics, *skill in setting up and evaluating double integrals over general regions comes with practice*. The next three examples are typical of the process.

Example 4 ✒

Evaluate $\displaystyle\iint\limits_R 5y\,dA$ over the region R bounded between $y = 3x^2$ and $y = 2 + x^2$.

Solution

The first step is to sketch the region R, if a sketch is not provided. We use the sketch to decide on the order of integration and to determine the limits of integration. Figure 10 shows the two parabolas defined by the boundary equations; note that the curves intersect at $(\pm 1, 3)$.

The shape of R guides us to set up the iterated integral in the form $\displaystyle\iint\limits_R f(x,y)\,dy\,dx$ because we can easily define cross-sections through R (such as the one shown in red) as line segments with lower bound $g_1(x)$ and upper bound $g_2(x)$. The alternative order of integration would lead to horizontal cross-sections that are functions of y, and the left-hand and right-hand limits of the line segments would be much more cumbersome to define and use.

Note that, scanning from left to right, R is bounded by the limits $x = -1$ and $x = 1$. Putting the pieces together, we have the following.

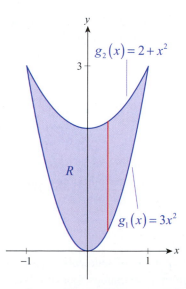

Figure 10

$$\iint\limits_{R} 5y\,dA = \int_{-1}^{1}\int_{3x^2}^{2+x^2} 5y\,dy\,dx = \int_{-1}^{1}\left[\frac{5y^2}{2}\right]_{y=3x^2}^{y=2+x^2}dx = \int_{-1}^{1}\frac{5}{2}\left(4+4x^2+x^4-9x^4\right)dx$$

$$= \int_{-1}^{1}\left(-20x^4+10x^2+10\right)dx = \left[-4x^5+\frac{10}{3}x^3+10x\right]_{-1}^{1}$$

$$= \left(-4+\frac{10}{3}+10\right)-\left(4-\frac{10}{3}-10\right)=\frac{56}{3}$$

Example 5 ✐

Evaluate the iterated integral $\int_{0}^{1}\int_{y}^{1}\frac{\sin x}{x}\,dx\,dy$.

Solution

Since the iterated integral is already set up, there is nothing to stop us from trying to evaluate it immediately. Unfortunately, as we know from Section 7.6, $(\sin x)/x$ has no simple antiderivative and the inner integral is nonelementary. We are stuck at the outset, unless we find a way to rewrite the problem.

Although the region R associated with this iterated integral was not explicitly defined for us, the limits of integration define it indirectly. Beginning with the inner integral, the left and right edges of the region are given by $x=y$ and $x=1$ (we know these are the left and right edges, as opposed to bottom and top, because they define limits of x). The outer integral defines the bottom and top edges of R as $y=0$ and $y=1$. Graphing these four lines gives us the region shown in Figure 11, with the red horizontal line segment representing a typical cross-sectional slice defined by the inner integral.

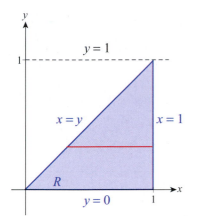

Figure 11

In some cases, a double integral may be nonelementary under one order of integration but readily evaluated under the alternative. We can't change the region R over which we are integrating, but if we use vertical cross-sections instead of horizontal, we can use Fubini's Theorem to rewrite the original integral.

$$\int_{0}^{1}\int_{y}^{1}\frac{\sin x}{x}\,dx\,dy = \iint\limits_{R}\frac{\sin x}{x}\,dA = \int_{0}^{1}\int_{0}^{x}\frac{\sin x}{x}\,dy\,dx$$

$$= \int_{0}^{1}\left[\frac{\sin x}{x}y\right]_{y=0}^{y=x}dx = \int_{0}^{1}\sin x\,dx$$

$$= \left[-\cos x\right]_{0}^{1} = -\cos 1 + 1 \approx 0.46$$

Figure 12 shows a representative vertical cross-section through R; note that it begins at the lower limit of $y=0$ and ends at the upper limit of $y=x$.

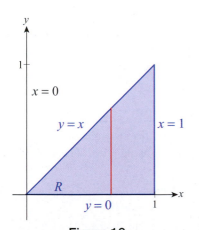

Figure 12

Example 6 ✐

Find the volume of the tetrahedron bounded on three sides by the coordinate planes and on the fourth side by the plane $2x+3y+z=6$.

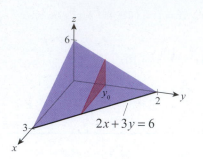

Figure 13

Solution

The four planes bound a solid in the first octant of $\mathbb{R}^3$, as depicted in Figure 13. We can determine its volume by integrating the function $f(x, y) = z = 6 - 2x - 3y$ over a region R in the xy-plane. We determine R by setting $z = 0$ in the equation $2x + 3y + z = 6$, resulting in the line $2x + 3y = 6$.

We could use cross-sections in either direction to evaluate the double integral; using horizontal cross-sections, as shown in the figure, we obtain the following volume.

$$V = \int_0^2 \int_0^{3-3y/2} (6 - 2x - 3y)\, dx\, dy = \int_0^2 \left[6x - x^2 - 3xy\right]_{x=0}^{x=3-3y/2} dy$$

$$= \int_0^2 \left(\frac{9y^2}{4} - 9y + 9\right) dy = \left[\frac{3y^3}{4} - \frac{9y^2}{2} + 9y\right]_0^2 = 6 \text{ cubic units}$$

Technology Note 🖥

A computer algebra system can be used to evaluate double as well as single integrals. The syntax varies with the software, but the form usually reflects the fact that iterated integration is "an integral of an integral." For instance, one way to use *Mathematica* to evaluate the integral of Example 4 is shown in Figure 14.

```
In[1]:= Integrate[Integrate[5 y, {y, 3 x^2, 2 + x^2}], {x, -1, 1}]

Out[1]=   56
         ——
          3
```

Figure 14

We conclude this section with a listing of properties of double integrals that parallels the properties of single integrals in Section 5.2.

Properties of Double Integrals

Assume that $f(x, y)$ and $g(x, y)$ are continuous, that k is a constant, and that R_1 and R_2 are two regions with at most boundary points in common. Assume also that all the integrals below exist. Then the following properties hold.

Constant Multiple Property $\displaystyle\iint\limits_R kf(x, y)\, dA = k\iint\limits_R f(x, y)\, dA$

Sum/Difference Property $\displaystyle\iint\limits_R \left[f(x, y) \pm g(x, y)\right] dA = \iint\limits_R f(x, y)\, dA \pm \iint\limits_R g(x, y)\, dA$

Domination Property $\displaystyle f(x, y) \geq g(x, y) \text{ on } R \;\Rightarrow\; \iint\limits_R f(x, y)\, dA \geq \iint\limits_R g(x, y)\, dA$

Disjoint Region Property $\displaystyle\iint\limits_{R_1 \cup R_2} f(x, y)\, dA = \iint\limits_{R_1} f(x, y)\, dA + \iint\limits_{R_2} f(x, y)\, dA$
(see Figure 15)

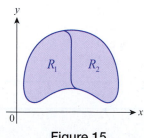

Figure 15

14.1 Exercises

1. Suppose $f(x, y) = x^2$ is defined on the square $R = [0, 4] \times [0, 4]$. Estimate $\iint_R f(x, y) \, dA$ using the Riemann sum approximation corresponding to $n = m = 4$, with each sample point (x_{ij}^*, y_{ij}^*) chosen to be the center point of the respective subsquare, $1 \leq i, j \leq 4$.

2. Use Fubini's Theorem to evaluate the integral of Exercise 1 and compare its true value to your estimate.

3. Evaluate the iterated integral and verify that your answer is equal to that obtained in Example 3.

$$\int_1^{5/2} \int_{1/2}^2 (15 - x^2 - 2y^2) \, dy \, dx$$

4–15 Evaluate the iterated integral.

4. $\int_0^1 \int_0^4 (3x - 2y) \, dx \, dy$ 5. $\int_0^3 \int_1^2 xy^2 \, dy \, dx$

6. $\int_2^4 \int_{-1}^3 (5x + xy - 3y) \, dx \, dy$

7. $\int_{-2}^2 \int_0^1 (x^3 y - 9) \, dy \, dx$

8. $\int_{-1}^0 \int_0^3 (xy^4 - x^4 y) \, dx \, dy$

9. $\int_0^{\pi/2} \int_{-\pi/2}^{\pi/2} \sin y \cos x \, dx \, dy$

10. $\int_1^2 \int_0^1 y^2 e^{-x} \, dy \, dx$

11. $\int_{-3/2}^0 \int_0^\pi (x^2 y - \cos x) \, dx \, dy$

12. $\int_1^2 \int_1^2 \left(\frac{x}{y} - \frac{y}{x}\right) dx \, dy$ 13. $\int_{1/2}^2 \int_1^e \frac{1}{x^2 y} \, dy \, dx$

14. $\int_2^3 \int_0^1 \frac{x}{\sqrt{1 - y^2}} \, dy \, dx$ 15. $\int_0^1 \int_1^e \frac{1}{x\sqrt{y}} \, dx \, dy$

16–19 Suppose $\iint_{R_1} f(x, y) \, dA = 4$, $\iint_{R_1} g(x, y) \, dA = 5$, $\iint_{R_2} g(x, y) \, dA = -1$, and $\iint_{R_1 \cup R_2} f(x, y) \, dA = 12$ on the rectangular regions $R_1 = [-1, 3] \times [2, 4]$ and $R_2 = [-5, 5] \times [-2, 2]$. Use the properties of double integrals to evaluate the integral.

16. $\iint_{R_1} [2f(x, y) + 3] \, dA$

17. $\iint_{R_1} [4f(x, y) - g(x, y)] \, dA$

18. $\iint_{R_1 \cup R_2} [3f(x, y) + 5g(x, y)] \, dA$

19. $\iint_{R_2} [3f(x, y) - 2g(x, y)] \, dA$

20–22 Evaluate the given integral on the rectangular region $R = [-1, 3] \times [0, 4]$. (**Hint:** Recall that $[\![x]\!]$ denotes the greatest integer less than or equal to x.)

20. $\iint_R ([\![y]\!] - [\![x]\!]) \, dA$ 21. $\iint_R [\![x]\!] \cdot [\![y]\!] \, dA$

22. $\iint_R x \cdot [\![y]\!] \, dA$

23–34 Evaluate the iterated integral.

23. $\int_0^1 \int_{x^2}^x (x^2 + y^2) \, dy \, dx$ 24. $\int_0^{1/2} \int_0^y (x + y)^2 \, dx \, dy$

25. $\int_0^2 \int_{x/2}^{3-x} xy \, dy \, dx$ 26. $\int_0^3 \int_{y/3}^{7-2y} x^2 y \, dx \, dy$

27. $\int_1^e \int_0^{\sqrt{9-x^2}} \frac{2y}{x} \, dy \, dx$ 28. $\int_0^3 \int_0^{\sqrt{9-x^2}} \frac{1}{\sqrt{9-x^2}} \, dy \, dx$

29. $\int_0^1 \int_{y^2}^y e^y \, dx \, dy$ 30. $\int_0^1 \int_y^{\sqrt{y}} (x^3 - x) \, dx \, dy$

31. $\int_0^1 \int_{-\sqrt{1-x^2}}^{\sqrt{1-x^2}} 2y \, dy \, dx$ 32. $\int_0^6 \int_{y/2}^{5-y/3} 2x \, dx \, dy$

33. $\int_0^4 \int_1^{e^x} \frac{1}{y} \, dy \, dx$ 34. $\int_0^{\pi/2} \int_0^{\sin x} \frac{1}{\sqrt{1 - y^2}} \, dy \, dx$

35–49 Rework the indicated exercise by reversing the order of integration and verify that the answer does not change. (**Hint:** It is helpful, sometimes even necessary, to sketch the region of integration before reversing the order.)

35. Exercise 4   36. Exercise 5

37. Exercise 6   38. Exercise 23

39. Exercise 24   40. Exercise 25

41. Exercise 26   42. Exercise 27

43. Exercise 28   44. Exercise 29

45. Exercise 30   46. Exercise 31

47. Exercise 32   48. Exercise 33

49. Exercise 34

50–52 Use the symmetry of the graph of f over R to evaluate the double integral. (Do not use repeated integration.)

50. $\displaystyle\iint_R x\,dA; \quad R=\{(x,y)\,|-2\le x\le 2, 0\le y\le 4\}$

51. $\displaystyle\iint_R \sin y\,dA; \quad R=\left\{(x,y)\,\middle|-1\le x\le 5, -\frac{\pi}{2}\le y\le\frac{\pi}{2}\right\}$

52. $\displaystyle\iint_R (x^3+1)\,dA; \quad R=\{(x,y)\,|-3\le x\le 3, -1\le y\le 7\}$

53–54 Set up, but do not evaluate, the integral of $f(x,y)$ as an iterated integral over the triangle with the given vertices. Choose the most convenient order of integration. (Answers may vary.)

53. $\displaystyle\iint_R f(x,y)\,dA; \quad R$ is the triangle with vertices $(1,0)$, $(3,2)$, and $(4,0)$.

54. $\displaystyle\iint_R f(x,y)\,dA; \quad R$ is the triangle with vertices $(0,0)$, $(2,4)$, and $(4,3)$.

55. Evaluate $\displaystyle\iint_R 2x\,dA$ over the region R bounded by $y=\cos x$ and $y=\sin x$, $0\le x\le\pi/4$.

56. Let R be the region between the circle $x^2+y^2=9$ and the ellipse $4x^2+9y^2=36$. By integrating $f(x,y)=1$ over R, find the area of the region.

57. Evaluate $\displaystyle\iint_R y\,dA$ over the region R bounded by $y+x=2$ and $y=x^2-2x$.

58. Evaluate $\displaystyle\iint_R x^2\,dA$ over the region R bounded by $y^2=5-x$ and $y^2=4x$.

59. Evaluate $\displaystyle\iint_R xy\,dA$ over the region R bounded by $y=x^3-3x$ and $4y=x^3$, $x\ge 0$.

60. Evaluate $\displaystyle\iint_R \sqrt{1-x^2}\,dA$, where R is the first quadrant of the disk $x^2+y^2=1$.

61. Evaluate $\displaystyle\iint_R y\,dA$, where R is the circular sector bounded by $y=0$, $y=\sqrt{3}x$, and $x^2+y^2=4$.

62. Evaluate $\displaystyle\iint_R xe^y\,dA$, where R is the first quadrant region bounded by $y=\ln x$ and $x=e$.

63–67 The iterated integral represents the (signed) volume of a well-known solid. Use a formula from geometry to evaluate the integral.

63. $\displaystyle\int_0^4\int_1^3 2\,dy\,dx$

64. $\displaystyle\int_{-1}^2\int_{-3/2}^1 (-3)\,dx\,dy$

65. $\displaystyle\int_0^2\int_0^{4-2x}(4-2x-y)\,dy\,dx$

66. $\displaystyle\int_0^1\int_0^{3-3y}\left(\frac{5}{6}x+\frac{5}{2}y-\frac{5}{2}\right)dx\,dy$

67. $\displaystyle\int_0^3\int_{-\sqrt{9-y^2}}^{\sqrt{9-y^2}}\sqrt{9-x^2-y^2}\,dx\,dy$

68–70 Set up and evaluate an appropriate double integral to determine the volume of the solid shown in the figure. Use the order of integration of your choice.

68.

69.

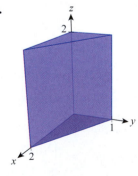

70.

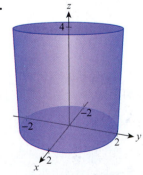

71–74 Evaluate the integral by reversing the order of integration.

71. $\int_0^{\sqrt{\pi/2}} \int_y^{\sqrt{\pi/2}} \cos(x^2) \, dx \, dy$ **72.** $\int_0^1 \int_x^1 e^{y^2} \, dy \, dx$

73. $\int_0^1 \int_y^{\sqrt{y}} \frac{e^x}{x} \, dx \, dy$ **74.** $\int_0^1 \int_{\sqrt{x}}^1 \sin(y^3) \, dy \, dx$

75. Suppose a thin plate has variable density given by a continuous two-variable function $\rho(x, y)$. Interpret the double integral $\iint_R \rho(x, y) \, dA$.

76. Use the Domination Property to show that if $f(x, y)$ is integrable and nonnegative on the bounded region R, then $\iint_R f(x, y) \, dA \geq 0$.

77. Use the Domination Property to show that if $f(x, y)$ is bounded on a region R of area A, that is, if $m \leq f(x, y) \leq M$ for some real numbers m, M for all $(x, y) \in R$, then

$$m \cdot A \leq \iint_R f(x, y) \, dA \leq M \cdot A.$$

78.* Let $F(x, y) = 1$ if at least one (or both) of x and y are irrational and $F(x, y) = 0$ otherwise. Prove that F is not integrable over any bounded region R in $\mathbb{R}^2$.

79. Let $R = [a, b] \times [c, d]$, and suppose that the second-order partials of $f(x, y)$ are continuous on an open region containing R.

   a. Use the Fundamental Theorem of Calculus to derive a formula for $\iint_R f(x, y) \, dA$.

   (**Hint:** Write the double integral as an iterated integral and use the Fundamental Theorem in two steps.)

   b. Letting $f(x, y) = x^3 y - 2y^2 + 5xy^4$ and noting that $f_{xy}(x, y) = 20y^3 + 3x^2$ (see Exercise 41 of Section 13.3), use part a. to evaluate

$$\int_{-2}^2 \int_1^3 f(x, y) \, dx \, dy.$$

80. Let $R = [a, b] \times [c, d]$ and suppose $f(x, y)$ can be decomposed into a product of a function of x and a function of y, that is, $f(x, y) = g(x) \cdot h(y)$. Show that

$$\iint_R f(x, y) = \left[\int_a^b g(x) \, dx \right] \cdot \left[\int_c^d h(y) \, dy \right].$$

81–86 Use Exercise 80 to provide a second solution to the given exercise.

81. Exercise 5   **82.** Exercise 9

83. Exercise 10   **84.** Exercise 13

85. Exercise 14   **86.** Exercise 15

87–90. Use the Riemann sum definition to prove the properties of double integrals. (For a refresher on the proofs of the one-variable case, see Section 5.2.)

91–94 *True or False?* Determine whether the given statement is true or false. In case of a false statement, explain or provide a counterexample.

91. If $f(x)$ is continuous, then

$$\int_c^d \int_a^b f(x) \, dy \, dx = (d - c) \int_a^b f(x) \, dx.$$

92. If R_1 and R_2 are disjoint regions and $f(x, y)$ is continuous, then

$$\left| \iint_{R_1 \cup R_2} f(x, y) \, dA \right| = \left| \iint_{R_1} f(x, y) \, dA \right| + \left| \iint_{R_2} f(x, y) \, dA \right|.$$

93. If $a, b, c > 0$, then $\int_0^c \int_0^{\frac{c}{b} - \frac{b}{a}y} (c - ax - by) \, dx \, dy = \frac{abc}{6}$.

94. $\int_{-R}^R \int_{-\sqrt{R^2 - x^2}}^{\sqrt{R^2 - x^2}} \sqrt{R^2 - x^2 - y^2} \, dy \, dx = \frac{4}{3} \pi R^3$

14.2 **Applications of Double Integrals**

TOPICS

1. Area by double integration
2. First moments
3. Second moments

In this section, we revisit the topics of area, moment, and center of mass. As we will see, the double integral often makes calculation of these quantities easier. Moreover, additional related characteristics of physical objects are most naturally defined using the language of double integration.

TOPIC 1 **Area by Double Integration**

In Section 5.5, we illustrated the underlying philosophy of integration with the observation that the area A of a planar region R can be found by integrating together all of the area differential elements dA over R—this informal statement was made precise with the formula $A = \int_R dA$. In that and later sections, we were forced to express dA in terms of a function $A'(x)$ multiplied by a differential dx so that we could integrate over an interval, the only option available to us at the time. Using double integration, the same philosophical idea takes the form $A = \iint_R dA$.

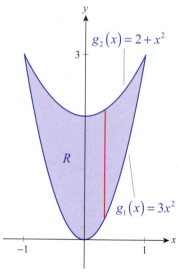

Figure 1

Example 1 📝

Find the area of the region R bounded between the parabolas $y = 3x^2$ and $y = 2 + x^2$.

Solution

We encountered this region in Example 4 of Section 14.1, but in this problem we seek the area of R. Referring to Figure 1, and using vertical slices through R, its area is as follows.

$$A = \iint_R dA = \int_{-1}^{1}\int_{3x^2}^{2+x^2} dy\, dx = \int_{-1}^{1}\left[y\right]_{y=3x^2}^{y=2+x^2} dx$$

$$= \int_{-1}^{1}\left(2 - 2x^2\right)dx = \left[2x - \frac{2x^3}{3}\right]_{-1}^{1}$$

$$= \left(2 - \frac{2}{3}\right) - \left(-2 + \frac{2}{3}\right) = \frac{8}{3} \text{ square units}$$

Given our work in Section 14.1, $\iint_R dA$ could also be interpreted as the volume of the solid between the xy-plane and the surface $f(x,y) = 1$ over the region R—in other words, the volume of a right cylinder of height 1 and base R. Since the volume of such a cylinder is $1 \times \text{Area}(R)$, we have the following.

$$\text{Area}(R) = 1 \times \text{Area}(R)$$

$$= \text{Volume}(\text{cylinder of height 1 and base } R)$$

$$= \iint_R 1\, dA = \iint_R dA$$

Now imagine an empty tank in the shape of a right cylinder with base R and indefinite height, which we proceed to fill with a given volume V of water. Then $V = h \times \text{Area}(R)$, where h is the ultimate height of the water in the tank. If the surface of the water is now pushed down in some areas, it must rise in other areas in order to maintain constant volume. At any given moment in time, the surface of the water represents the graph of a function $f(x, y)$ and the volume of the solid bounded by the xy-plane and $z = f(x, y)$ over R is $\iint_R f(x, y)\, dA$. Since the volume of water never changes, it must be the case that

$$h \times \text{Area}(R) = V = \iint_R f(x, y)\, dA.$$

We use this observation to define the *average value of f over R* to be h, which also makes sense if h is negative (although its interpretation as the height of the water does not have any physical meaning in this case). Formally,

$$\textbf{average value of } \boldsymbol{f} \textbf{ over } \boldsymbol{R} = \frac{1}{\text{Area}(R)} \iint_R f(x, y)\, dA.$$

Note the similarity between this definition and that of the average value of a function $f(x)$ over an interval $[a, b]$, namely $\dfrac{1}{b-a} \displaystyle\int_a^b f(x)\, dx$ (Section 5.2).

Example 2 ✎

Find the average value of $f(x, y) = 5y$ over the region R of Example 1.

Solution

In Example 4 of Section 14.1, we determined that

$$\iint_R 5y\, dA = \frac{56}{3},$$

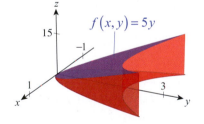

Figure 2

and we have just calculated $\text{Area}(R) = \frac{8}{3}$, so the average height of the solid bounded between the xy-plane and the graph of $z = 5y$ over R, depicted in Figure 2, is

$$\frac{1}{\frac{8}{3}} \cdot \frac{56}{3} = \frac{56}{8} = 7.$$

TOPIC 2   First Moments

We first discussed *moments* and *centers of mass* of planar objects (or plates) in Section 6.4, and at the time we constructed formulas for such characteristics based on single integrals. Applying the same physical principles, but expressing them in terms of double integrals, results in more general formulas that are simpler in nature and often easier to apply.

If we again let $\rho(x, y)$ represent the density (mass per unit area) at (x, y) of a planar object, which we idealize as a region R in the xy-plane, then its mass M is found by integrating ρ over R.

$$M = \iint_R \rho(x, y)\, dA$$

The moment about the x-axis of a small piece of the object is approximately $y^* \rho(x^*, y^*)\Delta x \Delta y$, where (x^*, y^*) is a sample point chosen in the piece with dimensions Δx and Δy. Similarly, its moment about the y-axis is $x^* \rho(x^*, y^*)\Delta x \Delta y$. The physical principle behind these facts is that the **first moment** mr of a point object about a given axis is a measure of the torque the object's mass m exerts, due to gravity, on a lever of length r joining the object and the axis. The mass of the small piece in this setting is approximately $\rho(x^*, y^*)\Delta x \Delta y$ and its approximate distance from the x- and y-axes is, respectively, y^* and x^* (see Figure 3). We introduce the adjective *first* to differentiate this characteristic from the *second moment*, which we will soon discuss.

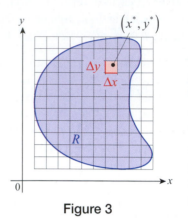

Figure 3

The first moments of the entire object are then approximated by the sum of the first moments of all such pieces.

First moment about the x-axis $M_x \approx \displaystyle\sum_{j=1}^{m} \sum_{i=1}^{n} y_{ij}^* \rho\left(x_{ij}^*, y_{ij}^*\right)\Delta x_i \Delta y_j$

First moment about the y-axis $M_y \approx \displaystyle\sum_{j=1}^{m} \sum_{i=1}^{n} x_{ij}^* \rho\left(x_{ij}^*, y_{ij}^*\right)\Delta x_i \Delta y_j$

In the limit of finer and finer partitions of R we are led to the following formulas.

Definition 💡

First Moments about the Axes, Mass, and Center of Mass of a Planar Object

Given a planar object modeled by the region R in the xy-plane, and whose density is represented by $\rho(x, y)$, its **first moments** M_x and M_y about the coordinate axes, its **mass** M, and its **center of mass** $(\bar{x}, \bar{y})$ are as follows, assuming the integrals exist.

First moment about the x-axis $M_x = \displaystyle\iint_R y\rho(x, y)\, dA$

First moment about the y-axis $M_y = \displaystyle\iint_R x\rho(x, y)\, dA$

Mass $M = \displaystyle\iint_R \rho(x, y)\, dA$

Center of mass $\bar{x} = \dfrac{M_y}{M}$ and $\bar{y} = \dfrac{M_x}{M}$

In terms of balance, as we learned in Section 6.4, an object behaves as if its entire mass is concentrated at its center of mass since $M_x = M \cdot \bar{y}$ and $M_y = M \cdot \bar{x}$.

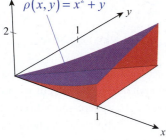

Figure 4

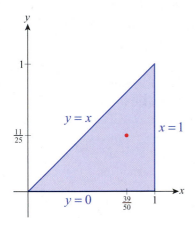

Figure 5

Example 3 ✐

Find the center of mass of the triangular plate bounded by $y = 0$, $x = 1$, and $y = x$ if the density of the plate at the point (x, y) is given by $\rho(x, y) = x^2 + y$.

Solution

In order to develop an intuitive sense of how such a plate behaves, note that the density is very low near the origin (in fact, $\rho(0,0) = 0$) and increases as x and y increase; however, the plate's density grows faster as x increases than it does as y increases. One way in which a plate might have a density function $\rho(x, y) = x^2 + y$ is if its thickness is defined by ρ, as shown in Figure 4. Try to imagine where such a plate will balance before we actually calculate the center of mass, as follows.

$$M_x = \iint_R y\rho(x,y)\,dA = \int_0^1 \int_0^x y\left(x^2 + y\right)dy\,dx$$

$$= \int_0^1 \left[\frac{y^2 x^2}{2} + \frac{y^3}{3}\right]_{y=0}^{y=x} dx = \int_0^1 \left(\frac{x^4}{2} + \frac{x^3}{3}\right)dx = \left[\frac{x^5}{10} + \frac{x^4}{12}\right]_0^1 = \frac{11}{60}$$

$$M_y = \iint_R x\rho(x,y)\,dA = \int_0^1 \int_0^x x\left(x^2 + y\right)dy\,dx$$

$$= \int_0^1 \left[x^3 y + \frac{xy^2}{2}\right]_{y=0}^{y=x} dx = \int_0^1 \left(x^4 + \frac{x^3}{2}\right)dx = \left[\frac{x^5}{5} + \frac{x^4}{8}\right]_0^1 = \frac{13}{40}$$

$$M = \iint_R \rho(x,y)\,dA = \int_0^1 \int_0^x \left(x^2 + y\right)dy\,dx$$

$$= \int_0^1 \left[x^2 y + \frac{y^2}{2}\right]_{y=0}^{y=x} dx = \int_0^1 \left(x^3 + \frac{x^2}{2}\right)dx = \left[\frac{x^4}{4} + \frac{x^3}{6}\right]_0^1 = \frac{5}{12}$$

So we obtain $\bar{x} = M_y/M = \frac{13}{40} \cdot \frac{12}{5} = \frac{39}{50}$ and $\bar{y} = M_x/M = \frac{11}{60} \cdot \frac{12}{5} = \frac{11}{25}$. The center of mass $\left(\frac{39}{50}, \frac{11}{25}\right)$ is shown in Figure 5.

TOPIC 3 Second Moments

While an object's first moment reflects the point of balance of its static mass with respect to a given axis, its **second moment**, also called its **moment of inertia**, contains information about the kinetic energy the object possesses if it rotates about the axis. Specifically, a point object of mass m a distance r from its axis of rotation has a second moment of mr^2. Note that the moment of inertia of an object grows with an increase in either m or r—a heavier object spinning about an axis stores more kinetic energy (and is harder to bring to a stop) than a lighter one, and similarly if the same object is moved further from its axis of rotation.

To determine the second moment of an object modeled by a planar region R with density function $\rho(x, y)$, we employ the familiar "divide and conquer" technique. Partitioning R into subrectangles, each small component of the planar object has approximate mass $\rho(x^*, y^*)\Delta x \Delta y$, giving a second moment about the x-axis of $\left(y^*\right)^2 \rho(x^*, y^*)\Delta x \Delta y$ and a second moment about the y-axis of $\left(x^*\right)^2 \rho\left(x^*, y^*\right) \Delta x \Delta y$. By convention, I_x and I_y are used to denote the entire object's second moments about, respectively, the x- and y-axes.

Second moment about the x-axis $I_x \approx \sum_{j=1}^{m} \sum_{i=1}^{n} \left(y_{ij}^*\right)^2 \rho\left(x_{ij}^*, y_{ij}^*\right) \Delta x_i \Delta y_j$

Second moment about the y-axis $I_y \approx \sum_{j=1}^{m} \sum_{i=1}^{n} \left(x_{ij}^*\right)^2 \rho\left(x_{ij}^*, y_{ij}^*\right) \Delta x_i \Delta y_j$

But there is also a third axis about which the planar object may rotate: the z-axis. To understand the differences, visualize a disk R centered at the origin of the xy-plane. If it rotates about the x-axis, the disk carves out a sphere-shaped region of space; the result is similar if R rotates about the y-axis. But rotation about the z-axis, also called rotation about the pole or origin, is the sort of rotation a tire makes about its axle. The symbol I_0 (or I_z in some texts) is used to denote the second moment about the origin of an object, and we derive its formula using the fact that the square of the distance from the origin to the subrectangle containing the sample point $\left(x_{ij}^*, y_{ij}^*\right)$ is approximately $\left(x_{ij}^*\right)^2 + \left(y_{ij}^*\right)^2$.

Second moment about the origin $I_0 \approx \sum_{j=1}^{m} \sum_{i=1}^{n} \left[\left(x_{ij}^*\right)^2 + \left(y_{ij}^*\right)^2 \right] \rho\left(x_{ij}^*, y_{ij}^*\right) \Delta x_i \Delta y_j \approx I_x + I_y$

Finally, just as the first moments of an object let us locate its center of mass, the second moments of an object allow us to determine its **radii of gyration** r_x, r_y, and r_0. If the entire mass of a given object was located a distance r_x from the x-axis, the resulting second moment of the point object would be the same as that of the original object; r_y and r_0 have similar meaning with respect to the y-axis and the origin. By this definition, it must be the case that $I_x = Mr_x^2$, $I_y = Mr_y^2$, and $I_0 = Mr_0^2$, and these equations allow us to solve for each radius.

Based on the preceding Riemann sums, we make the following definitions.

Definition 💡

Second Moments and Radii of Gyration of a Planar Object

Given a planar object modeled by the region R in the xy-plane, and whose density is represented by $\rho(x, y)$, its **second moments** I_x, I_y, and I_0 and its corresponding **radii of gyration** are as follows, assuming the integrals exist.

Second moment about the x-axis $I_x = \iint_R y^2 \rho(x, y)\, dA$

Second moment about the y-axis $I_y = \iint_R x^2 \rho(x, y)\, dA$

Second moment about the origin $I_0 = \iint_R \left(x^2 + y^2\right) \rho(x, y)\, dA = I_x + I_y$

Mass $M = \iint_R \rho(x, y)\, dA$

Radii of gyration $r_x = \sqrt{\dfrac{I_x}{M}}, r_y = \sqrt{\dfrac{I_y}{M}},$ and $r_0 = \sqrt{\dfrac{I_0}{M}}$

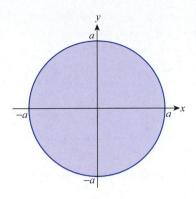

Figure 6

Example 4 ✎

Determine I_x, I_y, and I_0 for a disk of radius a and constant density ρ centered at the origin of the xy-plane. Then find its three radii of gyration.

Solution

The disk in question is depicted in Figure 6, and referring to the figure we set up the four integrals that follow.

$$I_x = \iint_R y^2 \rho \, dA = \rho \int_{-a}^{a} \int_{-\sqrt{a^2-x^2}}^{\sqrt{a^2-x^2}} y^2 \, dy \, dx$$

$$I_y = \iint_R x^2 \rho \, dA = \rho \int_{-a}^{a} \int_{-\sqrt{a^2-x^2}}^{\sqrt{a^2-x^2}} x^2 \, dy \, dx$$

$$I_0 = \iint_R \left(x^2 + y^2\right) \rho \, dA = \rho \int_{-a}^{a} \int_{-\sqrt{a^2-x^2}}^{\sqrt{a^2-x^2}} \left(x^2 + y^2\right) dy \, dx$$

$$M = \iint_R \rho \, dA = \rho \int_{-a}^{a} \int_{-\sqrt{a^2-x^2}}^{\sqrt{a^2-x^2}} dy \, dx$$

The techniques of integration learned in Chapter 7 allow us to evaluate I_x, as follows.

$$I_x = \rho \int_{-a}^{a} \int_{-\sqrt{a^2-x^2}}^{\sqrt{a^2-x^2}} y^2 \, dy \, dx = \rho \int_{-a}^{a} \left[\frac{y^3}{3}\right]_{y=-\sqrt{a^2-x^2}}^{y=\sqrt{a^2-x^2}} dx$$

$$= \frac{2\rho}{3} \int_{x=-a}^{x=a} \left(a^2 - x^2\right)^{3/2} dx \qquad\qquad \begin{aligned} x &= a\sin\theta \\ dx &= a\cos\theta \, d\theta \end{aligned}$$

$$= \frac{2\rho a^4}{3} \int_{\theta=-\pi/2}^{\theta=\pi/2} \cos^4\theta \, d\theta$$

$$= \frac{2\rho a^4}{3} \left[\frac{3\theta}{8} + \frac{\sin 2\theta}{4} + \frac{\sin 4\theta}{32}\right]_{-\pi/2}^{\pi/2}$$

$$= \frac{\pi \rho a^4}{4}$$

Similar techniques can be used to evaluate I_y, as written, but we obtain the result faster by changing the order of integration (as learned in Section 14.1) and noting that $I_y = I_x$.

$$I_y = \rho \int_{-a}^{a} \int_{-\sqrt{a^2-x^2}}^{\sqrt{a^2-x^2}} x^2 \, dy \, dx = \int_{-a}^{a} \int_{-\sqrt{a^2-y^2}}^{\sqrt{a^2-y^2}} x^2 \, dx \, dy = \frac{\pi \rho a^4}{4}$$

The third moment of inertia comes quickly.

$$I_0 = I_x + I_y = \frac{\pi \rho a^4}{2}$$

Without doing any integration, we know the mass of the disk is $M = \pi a^2 \rho$, so the three radii of gyration are

$$r_x = \frac{a}{2}, \quad r_y = \frac{a}{2}, \quad \text{and} \quad r_0 = \frac{a}{\sqrt{2}}.$$

Remember the physical interpretation of these results: a point with the same mass as the disk, but located $a/2$ units from the x-axis, would have the same second moment if spun about the x-axis, with a similar meaning for r_y, while the same point would have to be located $a/\sqrt{2}$ units from the origin for its second moment to be the same as that of the disk about the origin.

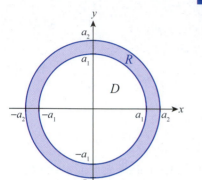

Figure 7

Example 5 ✎

Suppose the annulus shown in Figure 7 represents a planar object with total mass M. Determine its moment of inertia about the origin.

Solution

The properties of integrals (double or single) mean that the moments of inertia of pieces of an object can be added in order to find the moment of inertia of the whole; in physics and engineering texts, this is often referred to as the **Principle of Superposition**. This applies whether we find it useful to decompose an object into overlapping pieces of different masses or, as in this case, to decompose a region into a union of essentially disjoint regions (regions that only intersect at their boundaries, if at all).

Recall from Example 4 that, for a disk of radius a and constant density ρ, $I_0 = \pi\rho a^4/2$. So if we let D represent the solid disk centered at the origin of radius a_1 and R the annulus in question, and if we imagine that the region $R \cup D$ is a single disk of material with constant density ρ, then

$$\frac{\pi\rho a_2^4}{2} = \iint_{R \cup D} \left(x^2 + y^2\right)\rho\, dA$$

$$= \iint_{R}\left(x^2 + y^2\right)\rho\, dA + \iint_{D}\left(x^2 + y^2\right)\rho\, dA$$

$$= \iint_{R}\left(x^2 + y^2\right)\rho\, dA + \frac{\pi\rho a_1^4}{2},$$

so

$$\iint_{R}\left(x^2 + y^2\right)\rho\, dA = \frac{\pi\rho a_2^4}{2} - \frac{\pi\rho a_1^4}{2}$$

$$= \frac{\pi\rho}{2}\left(a_2^2 - a_1^2\right)\left(a_2^2 + a_1^2\right)$$

$$= \left(\rho\pi a_2^2 - \rho\pi a_1^2\right)\left(\frac{a_2^2 + a_1^2}{2}\right).$$

Physically, the factor $\rho\pi a_2^2 - \rho\pi a_1^2$ represents the mass of the solid disk of radius a_2 minus the mass of the solid disk of radius a_1, so $\rho\pi a_2^2 - \rho\pi a_1^2 = M$ and the moment of inertia of the annular object about the origin is thus

$$I_0 = M\left(\frac{a_2^2 + a_1^2}{2}\right).$$

Example 6 ✐

Given a fixed positive number s, determine I_x, I_y, and I_0 for the square $[0, s] \times [0, s]$ of constant density ρ. Then find its three radii of gyration.

Solution

The three second moments for this object are readily calculated.

$$I_x = \iint_R y^2 \rho \, dA = \rho \int_0^s \int_0^s y^2 \, dy \, dx = \frac{\rho s^4}{3}$$

$$I_y = \iint_R x^2 \rho \, dA = \rho \int_0^s \int_0^s x^2 \, dy \, dx = \frac{\rho s^4}{3}$$

$$I_0 = I_x + I_y = \frac{2\rho s^4}{3}$$

Since the mass of the square is $M = \rho s^2$, this tells us

$$r_x = \frac{s}{\sqrt{3}}, \quad r_y = \frac{s}{\sqrt{3}}, \quad \text{and} \quad r_0 = s\sqrt{\frac{2}{3}}.$$

Moments of higher order also have meaning, both physically and mathematically. In general, the nth moment of a function $f(x)$ (which may represent, among other things, density or probability) is the integral

$$\int x^n f(x) \, dx.$$

Texts in physics, engineering, statistics, and other disciplines discuss the use of moments in those specific fields and often introduce specialized nomenclature and calculation techniques.

14.2 **Exercises**

1–12 Use double integration to find the area of the region bounded by the graphs of the given equations.

1. $y = 2x, \quad y^2 = 4x^3$

2. $\sqrt{2}y = x^2, \quad x^2 + y^2 = 4, \quad x = 0$

3. $x^2 = y, \quad x = y^2$

4. $y^2 = 4x, \quad 2x = 4 - y$

5. $y^2 = 9x, \quad y^2 = 10 - x$

6. $y = x, \quad xy = 36, \quad y = 12$

7. $2x = 8 - y, \quad y = 6x, \quad 2y = x^3$

8. $y = 2\sqrt[4]{x}, \quad y = x, \quad y = 2$

9. $y = x + 10, \quad y = 8x - x^2$

10. $x = (y - 2)^2, \quad x = (y + 2)^2, \quad x = 0$

11. $y = \dfrac{x}{4}, \quad y = 4x, \quad xy = 1$

12. $y = \dfrac{x^2}{2}, \quad y = \dfrac{1}{1 + x^2}$

13–24 Find the average value of $f(x,y)$ over the region bounded by the graphs of the given equations.

13. $f(x,y) = x^2 - y$; $y = 2x$, $y^2 = x^3$

14. $f(x,y) = 2y$, $y = \sqrt{3}x$; $x^2 + y^2 = 1$, $x = 0$

15. $f(x,y) = -x$; $y = 2x^2$, $4x = y^2$

16. $f(x,y) = \dfrac{y^2}{2}$; $x = y^2$, $x = 2 - y$

17. $f(x,y) = x^2 y$; $x^2 = 4y$, $x^2 = 5 - y$

18. $f(x,y) = xy$; $y = x$, $xy = 1$, $y = 0$, $x = 2$

19. $f(x,y) = 5y^2$; $x = 2 - y$, $y = x^3$, $x = 0$

20. $f(x,y) = xy^3$; $y^3 = x$, $y = x$, $x \geq 0$

21. $f(x,y) = \sqrt{x}$; $y = x + 4$, $y = 6x - x^2$

22. $f(x,y) = e^{x+1}$; $y = (x-1)^2$, $y = (x+1)^2$, $y = 0$

23. $f(x,y) = xy$; $y = \dfrac{x}{2}$, $y = 2x$, $xy = 2$

24. $f(x,y) = x^2$; $y = \dfrac{x^2}{4}$, $y = \dfrac{1}{2 + 2x^2}$

25. Find the center of mass of the triangular plate of Example 3 if the density of the plate $\rho(x,y) = c$ is a constant. Compare your answer with that given in Example 3.

26–35 Find the center of mass of the plane region of varying density that is bounded by the graphs of the given equations.

26. $2y = 4 - x$, $x = 0$, $y = 0$; $\rho(x,y) = xy$

27. $y = 2x$, $y = 3 - x$, $y = 0$; $\rho(x,y) = x + y$

28. $y = x^{3/2}$, $y = \sqrt{x}$; $\rho(x,y) = \sqrt{x}$

29. $y = \sqrt{1 - x^2}$, $x = 0$, $y = 0$; $\rho(x,y) = x + 2y$

30. $y = 4 - x^2$, $y = 2 - x$; $\rho(x,y) = x^2 y$

31. $y = \dfrac{1}{x^3}$, $x = \dfrac{1}{2}$, $x = 1$; $\rho(x,y) = 2x^3 + y$

32. $y = x$, $y = \sqrt{x}$; $\rho(x,y) = y - x$

33. $xy^2 = 3$, $y = \dfrac{1}{3}$, $y = 1$; $\rho(x,y) = xy$

34. $y = \sqrt{x}$, $xy = 1$, $y = 0$, $x = 2$; $\rho(x,y) = x^2$

35. $y = e^x$, $x = 0$, $y = 0$, $x = 1$; $\rho(x,y) = xy$

36–50 Determine I_x, I_y, and I_0 for the thin plate of constant density ρ modeled by the planar region R. Then find its corresponding radii of gyration.

36. R: The region bounded by the graph of $y = 1 - x^2$ and the x-axis

37. R: The square $[-a,a] \times [-a,a]$

38. R: The square of Exercise 37 shifted in the positive x-direction by a units

39. R: The rectangle $[-l/2, l/2] \times [-w/2, w/2]$

40. R: The rectangle of Exercise 39 shifted in the positive y-direction by $w/2$ units

41. R: An equilateral triangle of side length $2a$, with its center at the origin, and a vertex on the positive y-axis

42. R: The triangle of Exercise 41 shifted vertically upward so that its base is on the x-axis

43. R: The square of Exercise 37 with the smaller square $[-b,b] \times [-b,b]$ removed ($b < a$) (Solve this problem without using the Principle of Superposition.)

44. R: The region bounded by the graph of $y = 2 - \dfrac{x^2}{2}$ and the x-axis

45. R: The region bounded by the graph of $y = \cos x$, $-\pi/2 \leq x \leq \pi/2$ and the x-axis

46. R: The region bounded by the graph of $y = \sin x$, $0 \leq x \leq \pi$ and the x-axis

47.* R: The region bounded by the ellipse $4x^2 + y^2 = 4$

48.* R: The region inside the ellipse $4x^2 + y^2 = 4$ and outside the circle $x^2 + y^2 = 1$ (Solve this problem without using the Principle of Superposition.)

49.* R: The region bounded by the ellipse $16x^2 + 25y^2 = 400$

50. R: The region bounded by the graphs of $y = e^x$, $x = \pm 1$, and the x-axis

51–57 As we have seen in Example 5, the Principle of Superposition is a very helpful aid in determining moments of certain objects. Another principle that simplifies the calculation of moments is called the *Parallel Axis Theorem*. It states that if we know the moment I_C of an object about an axis through its center of mass, then the moment of the same object about a parallel axis is $I = I_C + Md^2$, where d is the distance between the two axes. In these exercises, you will be asked to revisit some of the previous problems, using the Parallel Axis Theorem. To help you get started, we provided a hint to Exercise 51. You should use the hint's approach with subsequent exercises.

51. Using the answer from Exercise 37, use the Parallel Axis Theorem to obtain a second solution to Exercise 38. (**Hint:** Note that in this case, I_x doesn't change. Can you see why? As for I_y, note that after the shift, the new axis is a units away from the old. After determining I_y with the help of the Parallel Axis Theorem, using $d = a$, I_0 is readily obtained by adding I_x and I_y.)

52. Using the answer from Exercise 39, use the Parallel Axis Theorem to obtain a second solution to Exercise 40.

53. Using the answer from Exercise 41, use the Parallel Axis Theorem to obtain a second solution to Exercise 42.

54. Using the answer from Exercise 45, use the Parallel Axis Theorem to obtain a second solution to Exercise 46.

55. Use the Parallel Axis Theorem to find I_x, I_y, I_0, and the corresponding radii of gyration for the ellipse in Exercise 49, if it is shifted to the right so that its left focus coincides with the origin.

56. Use the Parallel Axis Theorem to find I_x, I_y, I_0, and the corresponding radii of gyration for the disk in Example 4, if it is shifted upwards by a units (so that it becomes tangential to the x-axis).

57. Starting with the answer you gave to Exercise 37 (and changing the notation appropriately), use the Parallel Axis Theorem to provide a second solution to Example 6.

58. Use Exercise 37 and the Principle of Superposition to provide a second solution to Exercise 43.

59. Use Exercise 47 and the Principle of Superposition to provide a second solution to Exercise 48.

60. Consider an annulus, as in Example 5, with $a_2 = 1$, $a_1 = \frac{1}{2}$, and $\rho = 1$. Furthermore, suppose that it is centered at $(a_1, 0)$. Starting with the results found for a disk in Example 4, use a combination of the Principle of Superposition and the Parallel Axis Theorem to find I_x, I_y, and I_0 for this annulus.

61. Suppose that $\left[-\frac{1}{2}, \frac{1}{2}\right] \times \left[-\frac{1}{2}, \frac{1}{2}\right]$ is removed from the 2-by-2 square $[-1,1] \times [-1,1]$ and that its center is shifted to the point $\left(0, \frac{1}{2}\right)$. Starting with the results of Example 6 (or Exercise 37) and assuming that the density is 1, use a combination of the Principle of Superposition and the Parallel Axis Theorem to find I_x, I_y, and I_0.

62–67 Consider the thin plates from the indicated problems with the given nonconstant densities. Determine I_x, I_y, I_0, and the corresponding radii of gyration.

62. Example 6; $\rho(x, y) = xy$

63. Exercise 36; $\rho(x, y) = y$

64. Exercise 45; $\rho(x, y) = y$

65. Exercise 46; $\rho(x, y) = x$

66.* Exercise 47; $\rho(x, y) = x^2$

67. Exercise 50; $\rho(x, y) = y$

68.* Determine I_x, I_y, I_0, and the corresponding radii of gyration for the thin plate of constant density ρ enclosed by the loop $y^2 = (1 - x)x^2$.

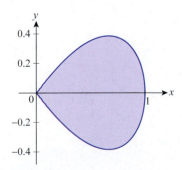

69. Suppose a thin, but solid, disk of uniform density (such as in Example 4) is rolling down an incline of height h (without slipping) in a race against the annulus of Example 5 and the square of Example 6, which is greased so it is sliding without friction (you can think of the latter as a sliding box). Which object will win, and in what order are they going to arrive at the bottom of the incline? Explain. (**Hint:** Use the fact that the initial potential energy of the object is eventually going to be shared between kinetic and rotational energies.)

70–73 *True or False?* Determine whether the given statement is true or false. In case of a false statement, explain or provide a counterexample.

70. If C is the average value of $f(x, y)$ over the region R, then $\iint\limits_R C\, dA = \iint\limits_R f(x, y)\, dA$.

71. If we increase the constant density of a planar object (or solid) from ρ to 2ρ, the coordinates $\bar{x}$, $\bar{y}$, and $\bar{z}$ of its center of mass will double.

72. If we increase the constant density of a planar object (or solid) from ρ to 2ρ, then its second moments will double.

73. If we increase the constant density of a planar object (or solid) from ρ to 2ρ, then its radii of gyration will double.

14.2 **Technology Exercises**

74–76 Write a program for a computer algebra system or programmable calculator that uses the integral definition to find the moments of inertia I_x, I_y, I_0, and the corresponding radii of gyration. Use it to check the answers you obtained previously in the given exercise.

74. Exercise 55

75. Exercise 56

76. Exercise 59

77. Find the average value of $f(x, y) = \sqrt{y}$ on the region bounded by the graphs of $y = (x-1)^2$, $y = (x+1)^2$ and $y = 0$.

14.3 **Double Integrals in Polar Coordinates**

TOPICS

1. The form of double integrals in polar coordinates

2. Evaluating double integrals in polar coordinates

As with single integrals, some double integrals are easier to set up and evaluate using polar coordinates. In this section we review the polar coordinate system and learn how to use it in the context of double integration; the techniques foreshadow the more general discussion of coordinate transformation that will come in Section 14.6.

TOPIC 1   **The Form of Double Integrals in Polar Coordinates**

Recall that the polar coordinates (r,θ) corresponding to the Cartesian coordinates (x, y) are related by the equations

$$x = r\cos\theta, \quad y = r\sin\theta, \quad \text{and} \quad r^2 = x^2 + y^2.$$

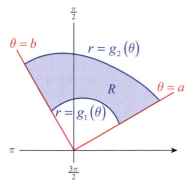

Polar coordinates may be the "natural" system to use when integrating a function $f(r,\theta)$ defined on a region R if R is bounded between two rays $\theta = a$ and $\theta = b$ and two continuous functions $r = g_1(\theta)$ and $r = g_2(\theta)$, as shown in Figure 1. In order to integrate over R using polar coordinates, we need to partition R in a manner consistent with the coordinate system. Figure 2 illustrates such a manner: each of the elements of the partition P is a *polar rectangle* described by $r_{i-1} \le r \le r_i$ and $\theta_{j-1} \le \theta \le \theta_j$ for some i and j, where $\{r_0, r_1, \ldots, r_n\}$ is a partition of an interval of radii covering R and $\{\theta_0, \theta_1, \ldots, \theta_m\}$ is a partition of the interval of angles $[a, b]$.

Figure 1

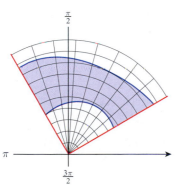

The contribution of each small polar rectangle to the Riemann sum based on the partition is $f(r^*, \theta^*)\Delta A$, where (r^*, θ^*) is a sample point chosen from the rectangle and ΔA is its area. All we have left to do, in order to make the transition from the approximating Riemann sums to the integral, is express ΔA in terms of Δr and $\Delta\theta$. The choice of sample point is immaterial in the limit, so we can specify that the sample point in the polar rectangle where $r_{i-1} \le r \le r_i$ and $\theta_{j-1} \le \theta \le \theta_j$ is

$$\left(r^*, \theta^*\right) = \left(\frac{r_{i-1} + r_i}{2}, \frac{\theta_{j-1} + \theta_j}{2}\right).$$

Figure 2

That is, (r^*, θ^*) is chosen to be the center point of each polar rectangle, as shown in Figure 3. And since the area of the sector of a circle of radius r and subtended angle $\Delta\theta$ is $\frac{1}{2}r^2\Delta\theta$ (as derived in Section 9.4), the area of the polar rectangle in Figure 3 is as follows.

$$\Delta A = \frac{1}{2}r_i^2\Delta\theta_j - \frac{1}{2}r_{i-1}^2\Delta\theta_j = \left(\frac{r_i + r_{i-1}}{2}\right)(r_i - r_{i-1})\Delta\theta_j = r^*\Delta r_i\Delta\theta_j$$

This gives us all we need to express the entire Riemann sum in terms of radial and angle increments.

$$\sum\sum f\left(r^*, \theta^*\right)\Delta A = \sum_{j=1}^{m}\sum_{i=1}^{n} f\left(r^*, \theta^*\right)r^*\,\Delta r_i\,\Delta\theta_j$$

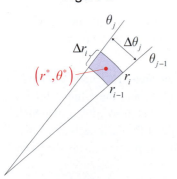

Figure 3

In the limit of finer and finer partitions, we have arrived at the following result.

Theorem 🔍

Double Integral in Polar Coordinates

Given an integrable function $f(r,\theta)$ defined on the region R bounded between rays $\theta = a$ and $\theta = b$ and two continuous functions $r = g_1(\theta)$ and $r = g_2(\theta)$, the double integral of f over R is

$$\iint\limits_R f(r,\theta)\,dA = \lim_{\|P\|\to 0} \sum_{j=1}^{m}\sum_{i=1}^{n} f(r^*,\theta^*)r^*\Delta r_i \Delta\theta_j = \int_{\theta=a}^{\theta=b}\int_{r=g_1(\theta)}^{r=g_2(\theta)} f(r,\theta)\,r\,dr\,d\theta.$$

If f is defined in terms of Cartesian coordinates x and y, we can express the double integral over R as

$$\iint\limits_R f(x,y)\,dA = \int_{\theta=a}^{\theta=b}\int_{r=g_1(\theta)}^{r=g_2(\theta)} f(r\cos\theta, r\sin\theta)\,r\,dr\,d\theta.$$

TOPIC 2   Evaluating Double Integrals in Polar Coordinates

The key steps to perform, in order to actually evaluate a double integral using polar coordinates, are as follows.

Step 1:   Ensure the integrand is defined purely in terms of r and θ.

Step 2:   Identify rays $\theta = a$ and $\theta = b$ and curves $r = g_1(\theta)$ and $r = g_2(\theta)$ that bound the region R.

Step 3:   Include the factor of r in the area differential, since $dA = r\,dr\,d\theta$.

To illustrate these steps, consider again Example 4 from Section 9.4, this time using double integration with polar coordinates.

Example 1 📝

Find the area of the region R inside the circle $r = \frac{3}{2}$ and outside the cardioid $r = 1 - \cos\theta$.

Solution

Solving the equation $1 - \cos\theta = \frac{3}{2}$ yields $\cos\theta = -\frac{1}{2}$, so $\theta = \pm 2\pi/3$; the region as described is depicted in Figure 4. Note that although the bounding rays were not explicitly provided, we now know that they are $\theta = -2\pi/3$ and $\theta = 2\pi/3$.

Since we are interested only in determining the area of R, the integrand is $f(r,\theta) = 1$. The θ-limits of integration are clear, and the r-limits reflect the extreme values of r for each radial slice through R, fifteen of which are shown in Figure 5. Remember that θ is fixed for each such slice, and that $r = g_1(\theta)$ is the smallest value of r among points on the slice and $r = g_2(\theta)$ is the largest value.

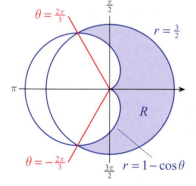

Figure 4

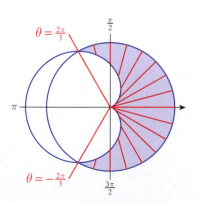

Figure 5

With the double integral set up, we can proceed to evaluate the area.

$$
\begin{aligned}
\text{Area} &= \int_{-2\pi/3}^{2\pi/3} \int_{1-\cos\theta}^{3/2} r\,dr\,d\theta \\[2mm]
&= 2\int_{0}^{2\pi/3} \int_{1-\cos\theta}^{3/2} r\,dr\,d\theta && \text{By symmetry}\\[2mm]
&= \int_{0}^{2\pi/3} \Big[\,r^2\,\Big]_{r=1-\cos\theta}^{r=3/2}\,d\theta \\[2mm]
&= \int_{0}^{2\pi/3} \left(\frac{5}{4} + 2\cos\theta - \cos^2\theta \right) d\theta && \cos^2\theta = \frac{1+\cos 2\theta}{2}\\[2mm]
&= \left[\frac{3}{4}\theta + 2\sin\theta - \frac{1}{4}\sin 2\theta \right]_{0}^{2\pi/3} \\[2mm]
&= \frac{9\sqrt{3}}{8} + \frac{\pi}{2}
\end{aligned}
$$

The next example revisits the problem of finding the moment of inertia of an annulus, a task which required several integration techniques and the Principle of Superposition when first encountered as Example 5 in Section 14.2. In contrast, the problem is readily solved using polar coordinates.

Example 2 ✐

Suppose the annulus R shown in Figure 6 represents a planar object with total mass M. Determine its moment of inertia about the origin.

Solution

Assuming the annular object is made of a material with constant density ρ, its moment of inertia about the origin is calculated as follows.

$$
\begin{aligned}
I_0 &= \iint_R \left(x^2 + y^2 \right)\rho\,dA = \int_0^{2\pi} \int_{a_1}^{a_2} r^2 \rho r\,dr\,d\theta \\[2mm]
&= \rho\int_0^{2\pi} \left[\frac{r^4}{4} \right]_{r=a_1}^{r=a_2} d\theta = \frac{\rho}{4}\left(a_2^4 - a_1^4 \right)(2\pi) \\[2mm]
&= \left(\rho\pi a_2^2 - \rho\pi a_1^2 \right)\left(\frac{a_2^2 + a_1^2}{2} \right) = M\left(\frac{a_2^2 + a_1^2}{2} \right)
\end{aligned}
$$

We have again used the fact that the mass of the annulus is the mass of a solid disk of radius a_2 minus the mass of a solid disk of radius a_1.

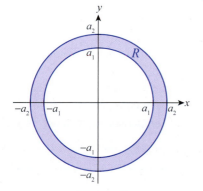

Figure 6

Example 2 demonstrated that polar coordinates may lead to an integral significantly easier to evaluate, but the benefits to be gained by the right choice of a coordinate system can be even more pronounced.

Example 3 ✎

Evaluate the integral

$$\iint_R e^{-\left(x^2+y^2\right)}\, dA,$$

where R is the unit disk centered at the origin.

Solution

Such integrals appear frequently in statistical applications and elsewhere, and it is desirable to be able to evaluate them exactly. Unfortunately, the integral as given is nonelementary—there is no antiderivative of the integrand with respect to either x or y. But the presence of $x^2 + y^2$ in the integrand and the fact that the region of integration is a disk are cues to try converting the integral to polar coordinates. When we do so, we are able to evaluate the integral without difficulty.

$$\iint_R e^{-\left(x^2+y^2\right)}\, dA = \int_0^{2\pi}\int_0^1 e^{-r^2}\, r\, dr\, d\theta$$

$$= -\frac{1}{2}\int_0^{2\pi}\int_{u=0}^{u=-1} e^u\, du\, d\theta \qquad \begin{aligned} u &= -r^2 \\ du &= -2r\, dr \end{aligned}$$

$$= -\frac{1}{2}\left(e^{-1}-1\right)\left(2\pi\right)$$

$$= \pi\left(1-\frac{1}{e}\right)$$

In some cases, the best choice of a coordinate system might not be immediately clear. In our next example, the form of the integrand may lead us to try polar coordinates, but the shape of the region R is better suited for Cartesian coordinates. The integral is actually easier to evaluate with Cartesian coordinates, as you will see in Exercise 1, but for the sake of illustration and comparison we will use polar coordinates.

Example 4 ✎

Let R be the triangle in the xy-plane with boundary lines $y = x$, $x = 0$, and $y = 1$. Find the volume of the solid bounded below by the region R and above by the surface $z = x^2 + y^2$.

Solution

The solid as described is pictured in Figure 7 and the region R in Figure 8. The lines $y = x$ and $x = 0$ correspond to the θ-limits $\theta = \pi/4$ and $\theta = \pi/2$, and each radial slice through R has a lower r-limit $r = 0$. Each upper r-limit corresponds to a point on the line $y = 1$, but this must be expressed in the form $r = g_2(\theta)$ for us to integrate using polar coordinates. We make the translation by noting that $y = r\sin\theta$, so

$$y = 1 \quad \Leftrightarrow \quad r\sin\theta = 1 \quad \Leftrightarrow \quad r = \csc\theta.$$

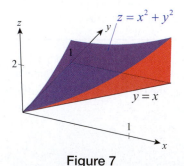

Figure 7

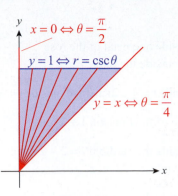

Figure 8

Putting all the pieces together, we obtain the following volume.

$$\text{Volume} = \iint_R \left(x^2 + y^2\right) dA = \int_{\pi/4}^{\pi/2} \int_0^{\csc\theta} r^2\, r\, dr\, d\theta = \frac{1}{4}\int_{\pi/4}^{\pi/2} \left[r^4\right]_{r=0}^{r=\csc\theta} d\theta$$

$$= \frac{1}{4}\int_{\pi/4}^{\pi/2} \csc^4\theta\, d\theta = \frac{1}{4}\int_{\pi/4}^{\pi/2} \csc^2\theta\left(1 + \cot^2\theta\right) d\theta \qquad \csc^2\theta = 1 + \cot^2\theta$$

$$= \frac{1}{4}\int_{\pi/4}^{\pi/2} \csc^2\theta\, d\theta + \frac{1}{4}\int_{\pi/4}^{\pi/2} \csc^2\theta \cot^2\theta\, d\theta \qquad \begin{array}{l} u = \cot\theta \\ du = -\csc^2\theta\, d\theta \end{array}$$

$$= \frac{1}{4}\left[-\cot\theta\right]_{\pi/4}^{\pi/2} + \frac{1}{4}\left[-\frac{1}{3}\cot^3\theta\right]_{\pi/4}^{\pi/2} = \frac{1}{3}$$

14.3 **Exercises**

1. Use Cartesian coordinates to determine the volume of the solid in Example 4.

2–5 Use a double integral in polar coordinates to find the area of the shaded region.

2.

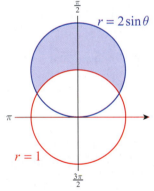

3.

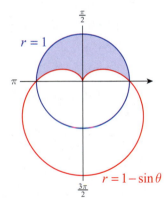

4.

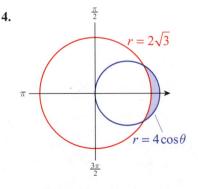

5.

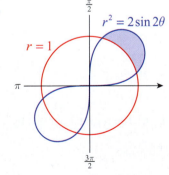

6–18 Use a double integral in polar coordinates to find the area of the region R.

6. R: The region inside the circle $r = 2$

7. R: The region inside the circle $r = 6 \sin \theta$

8. R: The region inside the cardioid $r = 1 - \sin \theta$

9. R: The region common to the circles $r = 6 \cos \theta$ and $r = 3$

10. R: The region inside the cardioid $r = 1 + \cos \theta$ and outside the unit circle centered at the origin

11. R: The region inside the circle $r = 2$ and outside the limaçon $r = \frac{3}{2} + \sin \theta$

12. R: The region bounded by the spiral $r = 2\theta$ and the polar axis $(0 \le \theta \le \pi)$

13. R: The region inside one petal of the rose $r = 2 \cos 2\theta$ and outside the circle $r = 1$

14. R: The region inside $r = 4 + 2 \sin \theta$ and outside $r = 3$

15. R: The inner loop of the limaçon $1 + 2 \cos \theta$

16. R: The region inside the cardioid $r = 2 - 2 \sin \theta$ and outside the circle $r = 2$

17. R: The region inside the circle $r = 1$ and outside the cardioid $r = 1 + \sin \theta$

18. R: The region inside the circle $r = 2 \sin \theta$, but outside the lemniscate $r^2 = 2 \cos 2\theta$
(**Hint:** Divide R into appropriate subregions and use symmetry.)

19–26 Evaluate the double integral by changing to polar coordinates. (Sketching the region of integration is helpful.)

19. $\displaystyle \int_{-1}^{1} \int_{0}^{\sqrt{1-x^2}} \frac{dy\, dx}{x^2 + y^2 + 2}$

20. $\displaystyle \int_{0}^{3} \int_{0}^{\sqrt{9-x^2}} \sqrt{x^2 + y^2}\, dy\, dx$

21. $\displaystyle \int_{-1}^{0} \int_{0}^{\sqrt{1-y^2}} \left(x^2 + y^2\right)^{3/2} dx\, dy$

22. $\displaystyle \int_{-2}^{2} \int_{-\sqrt{4-x^2}}^{\sqrt{4-x^2}} \cos\left(x^2 + y^2\right) dy\, dx$

23. $\displaystyle \int_{0}^{2} \int_{0}^{\sqrt{2y-y^2}} \left(x^2 + y^2\right)^{5/2} dx\, dy$

24. $\displaystyle \int_{0}^{2a} \int_{0}^{\sqrt{2ay-y^2}} 2\, dx\, dy$

25. $\displaystyle \iint_{R} x\sqrt{x^2 + y^2}\, dA$; R: The region enclosed by the first-quadrant loop of the lemniscate $r^2 = 2 \sin 2\theta$

26. $\displaystyle \iint_{R} y\, dA$; R: The region inside the circle $r = 1$ and outside $r = \sin \theta$

27–30 Convert the integral into a Cartesian double integral and evaluate it.

27. $\displaystyle \int_{0}^{\pi/4} \int_{0}^{\sec \theta} r^2 \cos \theta\, dr\, d\theta$

28. $\displaystyle \int_{\pi/3}^{\pi/2} \int_{0}^{2\csc \theta} r^3 \sin \theta \cos \theta\, dr\, d\theta$

29. $\displaystyle \int_{\pi/4}^{\arctan 2} \int_{0}^{\csc \theta} r^3 \sin 2\theta\, dr\, d\theta$

30. $\displaystyle \int_{0}^{\arctan(1/2)} \int_{0}^{2\sec \theta} r^5 \sin^2(2\theta)\, dr\, d\theta + \int_{\arctan(1/2)}^{\pi/2} \int_{0}^{\csc \theta} r^5 \sin^2(2\theta)\, dr\, d\theta$

31–34 Make your choice between the Cartesian and polar coordinate systems and evaluate the double integral.

31. $\displaystyle \iint_{R} \sqrt{x^2 + y^2}\, dA$; $R: x^2 + y^2 \le 4$

32. $\displaystyle \iint_{R} \left(x^2 + y^2\right) dA$; R: The region bounded by $y = x$, $y = x/2$, and $x = 2$

33. $\displaystyle \iint_{R} e^{\sqrt{x^2+y^2}}\, dA$; R: The first-quadrant region of $x^2 + y^2 \le 1$

34. $\displaystyle \iint_{R} \left(x^2 + y^2\right)^2 dA$; R: The region bounded by $y = 0$, $x = 1$, and $y = x$

35. In Section 9.4, we derived the following formula for the area A of a region bounded by $\theta = a$, $\theta = b$, and the polar equation $r = f(\theta)$.

$$A = \int_a^b \frac{1}{2} \left[f(\theta) \right]^2 d\theta$$

Use a double integral in polar coordinates and the discussion of this section to derive the formula above.

36–44 Use double integration in polar coordinates on an appropriate region to find the volume of the solid S bounded by the given surfaces.

36. S: The solid bounded by the xy-plane and the paraboloid $z = 1 - x^2 - y^2$

37. S: The solid bounded by the xy-plane and the paraboloid $z = 16 - x^2 - y^2$

38. S: The solid bounded by $z = x + 2y + 7$ and the cylinder $x^2 + y^2 = 4$

39. S: The solid bounded by $z = 2 - \sqrt{x^2 + y^2}$ and the xy-plane

40. S: The solid bounded by $z = 3x + 5y + 9$, $z = x + 2y + 3$, and the cylinder $x^2 + y^2 = 2y$

41. S: The solid bounded by $z = x + y + 8$, the xy-plane, and the cylinder $r = 1 + \sin\theta$

42. S: The solid bounded by $z = x^2 + y^2$, the xy-plane, and the cylinder $r = 2 + \cos\theta$

43. S: The solid that is common to the paraboloids $z = 2(x^2 + y^2)$ and $z = 12 - x^2 - y^2$ (**Hint:** Note that calculating the curve of intersection of the two surfaces will yield the region of integration.)

44. S: The solid that is common to the paraboloids $z = 9 - 8x^2 - 8y^2$ and $z = x^2 + y^2$ (See the hint given in Exercise 43.)

45. Use double integration in polar coordinates to find the volume of the solid in the shape of an ice-cream cone bounded by the cone $z = \sqrt{x^2 + y^2}$ and the sphere $x^2 + y^2 + z^2 = 8$.

46. Use double integration in polar coordinates to derive the formula for the volume of a sphere of radius R, $V = \frac{4}{3}\pi R^3$.

47. Use double integration in polar coordinates to find the volume of the solid inside the paraboloid $z = 6 - x^2 - y^2$ and above the sphere $x^2 + y^2 + z^2 = 8$.

48.* Recall the wooden toy piece from Exercise 17 of Section 6.1. We will generalize that problem as follows. Suppose a cylindrical hole of radius r is drilled through the center of a sphere of radius R. Use double integration in polar coordinates to show that the volume of the remaining ringlike solid is $V = \frac{4}{3}\pi\left(R^2 - r^2\right)^{3/2}$.

49. Assuming constant density, use double integration in polar coordinates to find the center of mass of the region outside the circle $r = 1$ and inside the cardioid $r = 1 + \sin\theta$. (**Hint:** Use the symmetry of the region.)

50. Use double integration in polar coordinates to find the center of mass of the region outside the cardioid $r = 2 - 2\sin\theta$ and inside the circle $r = 2\cos\theta$. As in Exercise 49, we assume constant density.

51. Determine the second moments I_x, I_y, and I_0 for the thin plate of constant density ρ inside the circle $r = 4\cos\theta$ and outside the circle $r = 2$.

52. Determine I_x, I_y, I_0, and the corresponding radii of gyration for the quarter annulus

$$\left\{ (x, y) \mid 0 \le x \le 2, \sqrt{1 - x^2} \le y \le \sqrt{4 - x^2} \right\}.$$

Assume the annulus has constant density ρ.

53. Use double integration in polar coordinates to find I_x, I_y, I_0, and the corresponding radii of gyration for the disk of Example 4 of Section 14.2, if the disk is shifted upwards by a units.

54. Use double integration in polar coordinates to find I_x, I_y, and I_0 directly for the annulus in Exercise 60 of Section 14.2.

55. Assuming it has constant density 1, find I_x, I_y, I_0, and the corresponding radii of gyration for the first-quadrant loop of the lemniscate $r^2 = \sin 2\theta$.

56.* Repeat Exercise 55 for the region inside the circle $r = 2$ and outside $r = 2\cos(\theta/2)$. (**Hint:** Take advantage of the symmetry of the region.)

57. By changing to polar coordinates, verify the value of the following improper double integral.

$$\int_0^\infty \int_0^\infty \frac{dy\, dx}{\left(1+x^2+y^2\right)^2} = \frac{\pi}{4}$$

(**Hint:** Integrate on the first-quadrant region of a disk of radius r, and let $r \to \infty$.)

58–59 In the next two exercises, you will be guided to use double integrals in polar coordinates to find the area under the bell curve $y = e^{-x^2}$, that is, to prove that

$$I = \int_{-\infty}^\infty e^{-x^2}\, dx = \sqrt{\pi}.$$

(This is an important integral not only in mathematics, but also in statistics, engineering, and physics.)

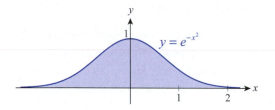

58. Use the technique of Exercise 57 to show

$$J = \int_{-\infty}^\infty \int_{-\infty}^\infty e^{-\left(x^2+y^2\right)}\, dx\, dy = \pi.$$

59. Evaluating J in rectangular coordinates over the square $[-a, a] \times [-a, a]$ (and letting $a \to \infty$), use Exercise 80 of Section 14.1 to show that $I^2 = J$. Conclude that $I = \sqrt{\pi}$.

60–63 *True or False?* Determine whether the given statement is true or false. In case of a false statement, explain or provide a counterexample.

60. If R is a square, and $f(r, \theta)$ is defined on R, then it is impossible to evaluate $\iint_R f(r, \theta)\, dA$ using polar coordinates.

61. If R is a bounded region, and $f(r, \theta) = 1$ on R, then the area A of R can be found as an iterated integral $\iint_R f(r, \theta)\, dr\, d\theta$.

62. The decision regarding which coordinate system (i.e., rectangular or polar) to use when evaluating $\iint_R f\, dA$, hinges upon the geometry (i.e., shape) of the region R.

63. Polar coordinates are suitable to determine the radii of gyration for certain planar regions.

14.3 **Technology Exercises**

64–66 Use a computer algebra system to evaluate the integral in both the rectangular and polar coordinate systems. Compare your answers.

64. $\int_0^1 \int_0^x \sqrt{x^2 + y^2}\, dy\, dx$

65. $\int_0^1 \int_{1-y}^{2-2y} xy\, dx\, dy$

66. $\int_0^1 \int_y^{\sqrt{2-y^2}} \sqrt{x^2 + y^2}\, dx\, dy$

(Note that even a CAS has a relatively hard time evaluating the integral of Exercise 66, while it is easy to determine, even just by a paper-and-pencil calculation, after converting it to polar coordinates!)

67–69 Assume that the thin plate covering the given region has a constant density of 1. Use a computer algebra system to find the moments of inertia I_x, I_y, I_0, and the corresponding radii of gyration.

67. The region bounded by $r = 1 - \sin\theta$

68. The region bounded by both $r = 1$ and $r = 1 + \sin\theta$

69. The region bounded by the inner loop of the limaçon $r = 1 + 2\cos\theta$

14.4 **Triple Integrals**

TOPICS

1. Triple integrals and Riemann sums
2. Triple integrals over general bounded regions
3. First and second moments of three-dimensional bodies

Although single integrals can be used to determine such things as area, mass, and moments of planar regions, we have seen that double integration provides a more general framework and, often, integrals that are easier to evaluate. In similar fashion, triple integration offers conceptual and computational advantages over double integration when calculating quantities that are three-dimensional in nature.

TOPIC 1   **Triple Integrals and Riemann Sums**

By now, the construction of Riemann sums is a familiar process, so we will merely summarize the steps that lead to the definition of the triple integral. Accordingly, suppose S is a closed bounded region of $\mathbb{R}^3$ and that $f(x, y, z)$ is defined on S. We can partition $\mathbb{R}^3$ with families of planes parallel to the coordinate planes, and by doing so arrive at a collection of boxes that approximate S, as shown in Figure 1.

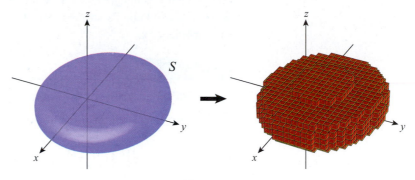

Figure 1

We choose a sample point $\left(x_{ijk}^*, y_{ijk}^*, z_{ijk}^*\right)$ from each box at which to evaluate f, and form the triple Riemann sum

$$\sum_{k=1}^{l} \sum_{j=1}^{m} \sum_{i=1}^{n} f\left(x_{ijk}^*, y_{ijk}^*, z_{ijk}^*\right) \Delta x_i \Delta y_j \Delta z_k,$$

where $\Delta V = \Delta x_i \Delta y_j \Delta z_k$ is the volume of the box defined by $x_{i-1} \leq x \leq x_i$, $y_{j-1} \leq y \leq y_j$, and $z_{k-1} \leq z \leq z_k$ (see Figure 2).

In keeping with previous usage, we define the norm of any such partition P as

$$\|P\| = \max_{\substack{1 \leq i \leq n \\ 1 \leq j \leq m \\ 1 \leq k \leq l}} \left\{\Delta x_i, \Delta y_j, \Delta z_k\right\},$$

and define the triple integral of f over S to be the common limit as $\|P\| \to 0$, if the limit exists.

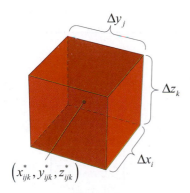

Figure 2

Definition 💡

Triple Integral

Given a function $f(x, y, z)$ defined on the region $S \subseteq \mathbb{R}^3$, we define the **triple integral of f over S** to be the value

$$\iiint_S f(x, y, z)\, dV = \lim_{\|P\| \to 0} \sum_{k=1}^{l} \sum_{j=1}^{m} \sum_{i=1}^{n} f\left(x_{ijk}^*, y_{ijk}^*, z_{ijk}^*\right) \Delta x_i \Delta y_j \Delta z_k,$$

provided the limit exists. If the limit exists, we say f is **integrable** over S.

As before, certain conditions on f and S guarantee the existence of such limits; for our purposes, it is most useful to know that if f is continuous on S and the boundary of S consists of a finite number of smooth surfaces joined along smooth curves, then f is integrable on S. And to actually compute triple integrals we use an extension of Fubini's Theorem, which we state here for the case where S is itself a rectangular box.

Theorem 🔍

Fubini's Theorem for Triple Integrals

If $f(x, y, z)$ is continuous on the region $S = [p, q] \times [r, s] \times [t, u]$, then

$$\iiint_S f(x, y, z)\, dV = \int_t^u \int_r^s \int_p^q f(x, y, z)\, dx\, dy\, dz.$$

Further, any permutation of the order of integration yields an identical result.

Example 1 ✍

Evaluate $\iiint_S \left(3x^2 y - xyz^3\right) dV$, where $S = [1, 3] \times [-1, 2] \times [0, 2]$.

Solution

Using one of the six possible orders of integration, we have the following.

$$\iiint_S \left(3x^2 y - xyz^3\right) dV = \int_0^2 \int_{-1}^2 \int_1^3 \left(3x^2 y - xyz^3\right) dx\, dy\, dz = \int_0^2 \int_{-1}^2 \left[x^3 y - \frac{x^2 y z^3}{2} \right]_{x=1}^{x=3} dy\, dz$$

$$= \int_0^2 \int_{-1}^2 \left(26 y - 4 y z^3\right) dy\, dz = \int_0^2 \left[13 y^2 - 2 y^2 z^3 \right]_{y=-1}^{y=2} dz$$

$$= \int_0^2 \left(39 - 6 z^3\right) dz = \left[39 z - \frac{3 z^4}{2} \right]_0^2 = 54$$

In Exercise 1 you will be asked to verify that the result is the same under one of the other orders of integration.

Because they also arise from Riemann sums, triple integrals have the same sorts of properties as single and double integrals. We list them here for ease of reference.

Properties of Triple Integrals

Assume that $f(x, y, z)$ and $g(x, y, z)$ are continuous, that k is a constant, and that S_1 and S_2 are two regions of $\mathbb{R}^3$ with at most boundary points in common. Assume also that all the integrals below exist. Then the following properties hold.

Constant Multiple Property

$$\iiint_S kf(x, y, z)\, dV = k \iiint_S f(x, y, z)\, dV$$

Sum/Difference Property

$$\iiint_S \left[f(x, y, z) \pm g(x, y, z) \right] dV = \iiint_S f(x, y, z)\, dV \pm \iiint_S g(x, y, z)\, dV$$

Domination Property

$$f(x, y, z) \geq g(x, y, z) \text{ on } S \quad \Rightarrow \quad \iiint_S f(x, y)\, dV \geq \iiint_S g(x, y)\, dV$$

Disjoint Region Property

$$\iiint_{S_1 \cup S_2} f(x, y, z)\, dV = \iiint_{S_1} f(x, y, z)\, dV + \iiint_{S_2} f(x, y, z)\, dV$$

(a)

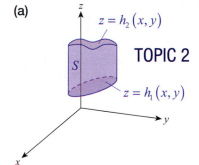

TOPIC 2   Triple Integrals over General Bounded Regions

When S is not a rectangular box, more attention must be paid to determining the limits of integration in a triple integral. Usually only some of the six possible orders of integration will be suitable for a given triple integral, but the process of setting up the integral is similar for any order. For the purpose of illustration, the steps below assume integration first with respect to z, then y, and finally x.

Setting up a Triple Integral

To express $\iiint_S f(x, y, z)\, dV$ as an iterated integral of the form $\iiint f(x, y, z)\, dz\, dy\, dx$ perfom the following steps.

(b)

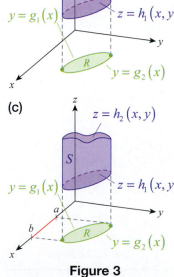

Step 1: Draw a rough sketch of S as an aid in determining the z-limits. Identify the lower surface $z = h_1(x, y)$ and the upper surface $z = h_2(x, y)$ of S. These two functions of x and y constitute the z-limits (see Figure 3a).

Step 2: Define R to be the "shadow," or *projection*, of S in the xy-plane. Sketch R, this time as an aid in determining the y-limits. The lower curve $y = g_1(x)$ and upper curve $y = g_2(x)$ constitute the y-limits (see Figure 3b).

(c)

Step 3: The lower x-limit is the left-hand edge $x = a$ of R (that is, the minimum value of x among all points of R) and the upper x-limit is the right-hand edge $x = b$ (see Figure 3c).

The result will be $\displaystyle \iiint_S f(x, y, z)\, dV = \int_{x=a}^{x=b} \int_{y=g_1(x)}^{y=g_2(x)} \int_{z=h_1(x,y)}^{z=h_2(x,y)} f(x, y, z)\, dz\, dy\, dx.$

Figure 3

Example 2 ✎

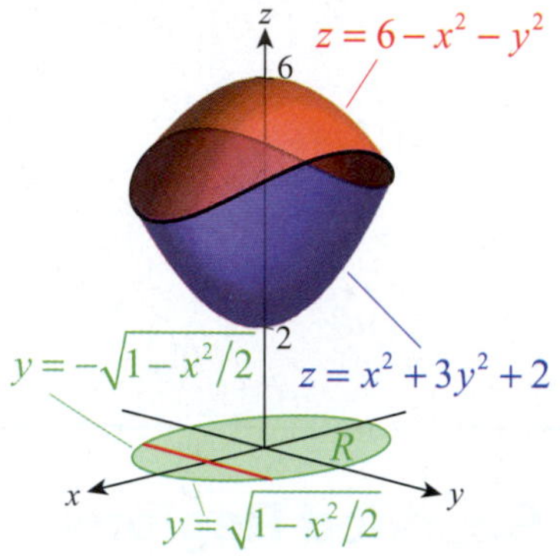

Find the volume of the solid S bounded below by the surface $z = x^2 + 3y^2 + 2$ and above by the surface $z = 6 - x^2 - y^2$.

Solution

The volume V of S is expressed as a triple integral with the integrand $f(x, y, z) = 1$.

$$V = \iiint_S 1 \, dV = \iiint_S dV$$

The two surfaces are both elliptical paraboloids, one opening upward and one downward, as shown in Figure 4. The two surfaces intersect where the two functions $h_1(x, y) = x^2 + 3y^2 + 2$ and $h_2(x, y) = 6 - x^2 - y^2$ have the same value, namely over the curve in the xy-plane where

$$x^2 + 3y^2 + 2 = 6 - x^2 - y^2 \quad \text{or} \quad x^2 + 2y^2 = 2.$$

The equation $x^2 + 2y^2 = 2$ is the boundary of the region R in the xy-plane, also shown in Figure 4. The lower curve (in the sense of y-values) is the function $g_1(x) = -\sqrt{1 - x^2/2}$ and the upper curve is the function $g_2(x) = \sqrt{1 - x^2/2}$, which we obtain by solving the equation for y. Finally, the line $x = -\sqrt{2}$ defines the left-hand edge of R, and the line $x = \sqrt{2}$ the right-hand edge (set $y = 0$ in the equation and solve for x). Putting the pieces together, the volume can be calculated as follows.

$z = 6 - x^2 - y^2$

$y = -\sqrt{1 - x^2/2}$ $z = x^2 + 3y^2 + 2$

$y = \sqrt{1 - x^2/2}$

Figure 4

$$V = \int_{-\sqrt{2}}^{\sqrt{2}} \int_{-\sqrt{1-x^2/2}}^{\sqrt{1-x^2/2}} \int_{x^2+3y^2+2}^{6-x^2-y^2} dz \, dy \, dx = \int_{-\sqrt{2}}^{\sqrt{2}} \int_{-\sqrt{1-x^2/2}}^{\sqrt{1-x^2/2}} \left(4 - 2x^2 - 4y^2\right) dy \, dx$$

$$= \int_{-\sqrt{2}}^{\sqrt{2}} \left[4y - 2x^2 y - \frac{4y^3}{3} \right]_{y=-\sqrt{1-x^2/2}}^{y=\sqrt{1-x^2/2}} dx$$

$$= \frac{4}{3} \int_{-\sqrt{2}}^{\sqrt{2}} \left(2 - x^2\right) \sqrt{4 - 2x^2} \, dx$$

$$= \frac{16\sqrt{2}}{3} \int_{\theta=-\pi/2}^{\theta=\pi/2} \cos^4 \theta \, d\theta$$

$$= 2\sqrt{2}\pi \text{ cubic units}$$

$x = \sqrt{2} \sin\theta$
$dx = \sqrt{2} \cos\theta \, d\theta$

$\cos^4 \theta = \left(\dfrac{1 + \cos 2\theta}{2}\right)^2$

$\qquad = \frac{1}{8}(3 + 4\cos 2\theta + \cos 4\theta)$

One other order of integration is suited to the previous problem, as illustrated next. But by its nature, the solid of Example 2 dictates that the innermost integration be with respect to z, meaning the other four orders of integration are not natural fits.

Example 3 ✎

Evaluate the triple integral of Example 2 with the order of integration of the middle and outer integrals reversed.

Solution

The inner integral remains the same, but to integrate next with respect to x, we need to express the "left-hand" and "right-hand" curves of $x^2 + 2y^2 = 2$ as functions of y, giving us $x = -\sqrt{2 - 2y^2}$ and $x = \sqrt{2 - 2y^2}$, respectively (see Figure 5). The limits of the outer integral are $y = -1$ and $y = 1$, so the integral as a whole is as follows.

$$V = \int_{-1}^{1} \int_{-\sqrt{2-2y^2}}^{\sqrt{2-2y^2}} \int_{x^2+3y^2+2}^{6-x^2-y^2} dz\,dx\,dy$$

In Exercise 48 you will show that the value of the integral is again $2\sqrt{2}\pi$.

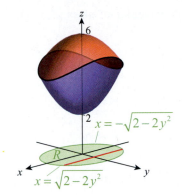

Figure 5

TOPIC 3   # First and Second Moments of Three-Dimensional Bodies

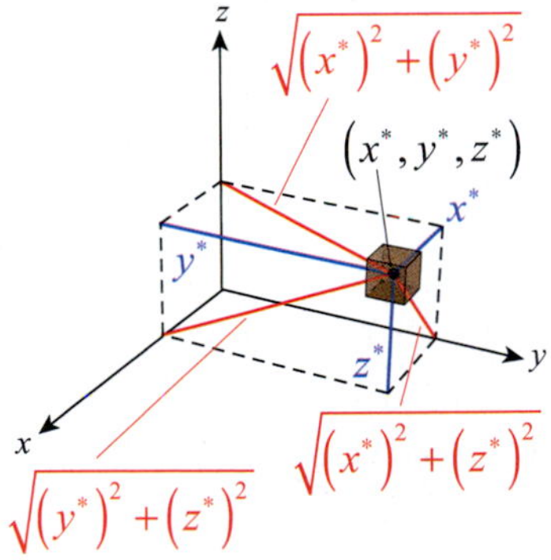

The basis for the formulas of first and second moments of solids in $\mathbb{R}^3$ is unchanged: the first moment about a given location of a point of mass m at a distance r is still mr, and its second moment is mr^2. To construct the formulas from this basis, we refer to a diagram, such as Figure 6, to determine the exact form of r for an incremental element of mass $\Delta m = \rho\left(x^*, y^*, z^*\right)\Delta V$, where $\rho(x, y, z)$ now represents the density in terms of mass per unit volume.

First moment of Δm about the yz-plane $M_{yz} \approx x^*\rho\left(x^*, y^*, z^*\right)\Delta V$

First moment of Δm about the xz-plane $M_{xz} \approx y^*\rho\left(x^*, y^*, z^*\right)\Delta V$

First moment of Δm about the xy-plane $M_{xy} \approx z^*\rho\left(x^*, y^*, z^*\right)\Delta V$

Second moment of Δm about the x-axis $I_x \approx \left[\left(y^*\right)^2 + \left(z^*\right)^2\right]\rho\left(x^*, y^*, z^*\right)\Delta V$

Second moment of Δm about the y-axis $I_y \approx \left[\left(x^*\right)^2 + \left(z^*\right)^2\right]\rho\left(x^*, y^*, z^*\right)\Delta V$

Second moment of Δm about the z-axis $I_z \approx \left[\left(x^*\right)^2 + \left(y^*\right)^2\right]\rho\left(x^*, y^*, z^*\right)\Delta V$

Figure 6

Adding up these approximate first and second moments over an object S and taking the limits of the resulting Riemann sums gives us the formulas that follow.

Definition 💡

First Moments about the Planes, Mass, and Center of Mass of an Object in $\mathbb{R}^3$

Given an object in space modeled by the region $S \subseteq \mathbb{R}^3$, and whose density is represented by $\rho(x, y, z)$, its **first moments** M_{yz}, M_{xz}, and M_{xy} about the coordinate planes, its **mass** M, and its **center of mass** $(\overline{x}, \overline{y}, \overline{z})$ are as follows, assuming the integrals exist.

First moment about the yz-plane $M_{yz} = \iiint\limits_{S} x\rho(x, y, z)\, dV$

First moment about the xz-plane $M_{xz} = \iiint\limits_{S} y\rho(x, y, z)\, dV$

First moment about the xy-plane $M_{xy} = \iiint\limits_{S} z\rho(x, y, z)\, dV$

Mass $M = \iiint\limits_{S} \rho(x, y, z)\, dV$

Center of mass $\overline{x} = \dfrac{M_{yz}}{M},\ \overline{y} = \dfrac{M_{xz}}{M},\ \text{and}\ \overline{z} = \dfrac{M_{xy}}{M}$

Definition 💡

Second Moments and Radii of Gyration of an Object in $\mathbb{R}^3$

Given an object in space modeled by the region $S \subseteq \mathbb{R}^3$, and whose density is represented by $\rho(x, y, z)$, its **second moments** I_x, I_y, and I_z about the coordinate axes and its corresponding **radii of gyration** are as follows, assuming the integrals exist.

Second moment about the x-axis $I_x = \iiint\limits_{S} (y^2 + z^2)\rho(x, y, z)\, dV$

Second moment about the y-axis $I_y = \iiint\limits_{S} (x^2 + z^2)\rho(x, y, z)\, dV$

Second moment about the z-axis $I_z = \iiint\limits_{S} (x^2 + y^2)\rho(x, y, z)\, dV$

Radii of gyration $r_x = \sqrt{\dfrac{I_x}{M}},\ r_y = \sqrt{\dfrac{I_y}{M}},\ \text{and}\ r_z = \sqrt{\dfrac{I_z}{M}}$

Example 4 ✎

Find the first and second moments, center of mass, and radii of gyration of the tetrahedron S bounded by the coordinate planes and the plane $z = 2 - 2x - y$. Assume S is made of a substance with constant density ρ.

Solution

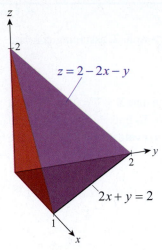

$z = 2 - 2x - y$

$2x + y = 2$

Figure 7

The shadow of S on each of the three coordinate planes is a triangle; we can determine the third edge of the shadow in the xy-plane by setting $z = 0$ in the equation $z = 2 - 2x - y$, giving us $2x + y = 2$. Figure 7 is an illustration of S.

Using the first moment formulas, we determine the following.

$$M_{yz} = \iiint_S x\rho\, dV = \rho \int_0^1 \int_0^{2-2x} \int_0^{2-2x-y} x\, dz\, dy\, dx = \rho \int_0^1 \int_0^{2-2x} \left(2x - 2x^2 - xy\right) dy\, dx$$

$$= \rho \int_0^1 \left[2xy - 2x^2 y - \frac{xy^2}{2}\right]_{y=0}^{y=2-2x} dx = 2\rho \int_0^1 \left(x - 2x^2 + x^3\right) dx = \frac{\rho}{6}$$

$$M_{xz} = \iiint_S y\rho\, dV = \rho \int_0^1 \int_0^{2-2x} \int_0^{2-2x-y} y\, dz\, dy\, dx = \rho \int_0^1 \int_0^{2-2x} \left(2y - 2xy - y^2\right) dy\, dx$$

$$= \rho \int_0^1 \left[(1-x)y^2 - \frac{y^3}{3}\right]_{y=0}^{y=2-2x} dx = \frac{-4\rho}{3} \int_0^1 (x-1)^3\, dx = \frac{\rho}{3}$$

$$M_{xy} = \iiint_S z\rho\, dV = \rho \int_0^1 \int_0^{2-2x} \int_0^{2-2x-y} z\, dz\, dy\, dx$$

$$= \frac{\rho}{2} \int_0^1 \int_0^{2-2x} \left(4 - 8x + 4x^2 - 4y + 4xy + y^2\right) dy\, dx$$

$$= \frac{\rho}{2} \int_0^1 \left[4y - 8xy + 4x^2 y - 2y^2 + 2xy^2 + \frac{y^3}{3}\right]_{y=0}^{y=2-2x} dx$$

$$= \frac{\rho}{2} \int_0^1 \left(\frac{8}{3} - 8x + 8x^2 - \frac{8x^3}{3}\right) dx = \frac{\rho}{3}$$

$$M = \iiint_S \rho\, dV = \rho \int_0^1 \int_0^{2-2x} \int_0^{2-2x-y} dz\, dy\, dx = \rho \int_0^1 \int_0^{2-2x} (2 - 2x - y)\, dy\, dx$$

$$= \rho \int_0^1 \left[2y - 2xy - \frac{y^2}{2}\right]_{y=0}^{y=2-2x} dx = \rho \int_0^1 \left(2 - 4x + 2x^2\right) dx = \frac{2\rho}{3}$$

Although the calculations for M_{xz} and M_{xy} differ in the details, we shouldn't be surprised by the fact that these two moments are the same, since the distribution of the points of S above the xz-plane is symmetric to the distribution above the xy-plane. But S is distributed differently with respect to the yz-plane—for one thing, its "height" above the yz-plane is 1 unit, instead of 2 as it is above the other two coordinate planes.

With the first moments and mass in hand, we can find the center of mass of the tetrahedron.

$$\bar{x} = \frac{M_{yz}}{M} = \frac{1}{4}, \quad \bar{y} = \frac{M_{xz}}{M} = \frac{1}{2}, \quad \text{and} \quad \bar{z} = \frac{M_{xy}}{M} = \frac{1}{2}$$

Hence the center of mass of S (also known as its centroid, since S has constant density) is the point $\left(\frac{1}{4}, \frac{1}{2}, \frac{1}{2}\right)$.

In Exercise 63 you will show that

$$I_x = \frac{8\rho}{15}, \quad I_y = I_z = \frac{\rho}{3}, \quad r_x = \frac{2}{\sqrt{5}}, \quad \text{and} \quad r_y = r_z = \frac{1}{\sqrt{2}}.$$

Example 5 ✎

Find the centroid, I_z, and r_z for the solid of Example 2, assuming constant density ρ.

Solution

By the symmetry of the solid, it must be the case that $\bar{x} = \bar{y} = 0$, but $\bar{z}$ will be some as yet unknown value between 2 and 6. To find $\bar{z}$ we calculate the following, using the techniques of Section 7.3 to evaluate the trigonometric integral in the second to last line.

$$M_{xy} = \iiint_S z\rho(x,y,z)\,dV = \rho \int_{-1}^{1} \int_{-\sqrt{2-2y^2}}^{\sqrt{2-2y^2}} \int_{x^2+3y^2+2}^{6-x^2-y^2} z\,dz\,dx\,dy$$

$$= \rho \int_{-1}^{1} \int_{-\sqrt{2-2y^2}}^{\sqrt{2-2y^2}} \left(-4y^4 - 2x^2y^2 - 12y^2 - 8x^2 + 16\right) dx\,dy$$

$$= -\frac{16\rho}{3} \int_{-1}^{1} \left(y^4 + 3y^2 - 4\right)\sqrt{2-2y^2}\,dy \qquad \begin{aligned} y &= \sin\theta \\ dy &= \cos\theta\,d\theta \end{aligned}$$

$$= -\frac{16\rho\sqrt{2}}{3} \int_{-\pi/2}^{\pi/2} \left(\sin^4\theta + 3\sin^2\theta - 4\right)\cos^2\theta\,d\theta$$

$$= \frac{25\sqrt{2}\pi\rho}{3}$$

The volume of the solid (from Example 2) is $2\sqrt{2}\pi$, so its mass is $M = 2\sqrt{2}\pi\rho$.

$$\bar{z} = \frac{M_{xy}}{M} = \frac{25}{6}$$

Therefore the centroid is $\left(0, 0, \frac{25}{6}\right)$.

The moment of inertia about the z-axis can be found as follows.

$$I_z = \iiint_S (x^2 + y^2)\rho(x,y,z)\,dV = \rho \int_{-1}^{1} \int_{-\sqrt{2-2y^2}}^{\sqrt{2-2y^2}} \int_{x^2+3y^2+2}^{6-x^2-y^2} (x^2+y^2)\,dz\,dx\,dy$$

$$= \rho \int_{-1}^{1} \int_{-\sqrt{2-2y^2}}^{\sqrt{2-2y^2}} \left(4x^2 - 2x^4 + 4y^2 - 6x^2y^2 - 4y^4\right) dx\,dy$$

$$= -\frac{16\rho}{15} \int_{-1}^{1} \left(3y^4 - y^2 - 2\right)\sqrt{2-2y^2}\,dy \qquad \begin{aligned} y &= \sin\theta \\ dy &= \cos\theta\,d\theta \end{aligned}$$

$$= -\frac{16\rho\sqrt{2}}{15} \int_{-\pi/2}^{\pi/2} \left(3\sin^4\theta - \sin^2\theta - 2\right)\cos^2\theta\,d\theta \qquad \text{See Section 7.3.}$$

$$= \sqrt{2}\pi\rho$$

This gives us the radius of gyration about the z-axis.

$$r_z = \sqrt{\frac{I_z}{M}} = \frac{1}{\sqrt{2}}$$

14.4 **Exercises**

1. Verify that

$$\iiint_S \left(3x^2 y - xyz^3\right) dV = \int_1^3 \int_0^2 \int_{-1}^2 \left(3x^2 y - xyz^3\right) dy\, dz\, dx$$

yields the same result as that obtained in Example 1.

2–7 Evaluate the triple integral on the rectangular box S. (Choose a convenient order of integration.)

2. $\iiint_S xy^3 z\, dV$, where $S = [-1,3] \times [0,1] \times [1,3]$

3. $\iiint_S dV$, where $S = [1,2] \times [3,4] \times [5,6]$

4. $\iiint_S \left(4xy + x^2 yz^2\right) dV$, where
   $S = [0,1] \times [1,2] \times [-1,1]$

5. $\iiint_S \left(y^2 z^2 - 2x^4 y\right) dV$, where
   $S = [-1,1] \times [0,2] \times [-3,0]$

6. $\iiint_S \dfrac{xy}{z}\, dV$, where $S = [-1,3] \times [0,3] \times [1,e]$

7. $\iiint_S xye^z\, dV$, where $S = [1,2] \times [-1,3] \times [0, \ln 4]$

8–13 Evaluate the iterated integral.

8. $\int_0^1 \int_0^{2-z} \int_0^{1-z} xy^2 z\, dx\, dy\, dz$

9. $\int_0^1 \int_{x^3}^x \int_0^{2xz} 5x\, dy\, dz\, dx$

10. $\int_0^3 \int_1^2 \int_0^{x+3y} (x+y)\, dz\, dy\, dx$

11. $\int_0^1 \int_{y+1}^{3y} \int_0^{2xy} (xy)^2\, dz\, dx\, dy$

12. $\int_0^1 \int_0^{4-x} \int_0^{4-x-z} dy\, dz\, dx$

13. $\int_0^2 \int_0^{\sqrt{4-y^2}} \int_0^{\sqrt{4-y^2}} dz\, dx\, dy$

14–17 Write iterated integrals for $\iiint_S dV$ on the given solid, using the following orders of integration: **a.** $dz\, dy\, dx$, **b.** $dy\, dz\, dx$, and **c.** $dx\, dz\, dy$. Then evaluate one of them to determine the value of the integral.

14. S: The tetrahedron bounded by the coordinate planes and $x + 2y + 3z = 6$

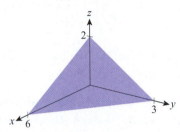

15. S: The cylinder bounded by $x^2 + y^2 = 1$, $z = 0$, and $z = 2$

16. S: The solid bounded by the parabolic cylinder $z = 1 - x^2$, $y = 0$, and $y = 1$

17. S: The solid bounded by $y = x^2$, $y^2 = x$, $z = -1$, and $z = 1$

18–36 Use a triple integral to find the volume of the solid S.

18. S: The tetrahedron bounded by the coordinate planes and the plane $\dfrac{x}{2} + \dfrac{y}{5} + \dfrac{z}{3} = 1$

19. Generalize Exercise 18 by finding the volume of the tetrahedron S bounded by the coordinate planes and the plane $\dfrac{x}{a} + \dfrac{y}{b} + \dfrac{z}{c} = 1$.

20. S: The solid bounded by the parabolic cylinder $x = y^2$, and the planes $z = 1 - x$, $z = 0$

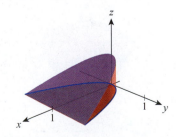

21. S: The solid bounded by the parabolic cylinder $x = y^2$, and the planes $x + y + z = 2$ and $z = 0$. Use the order of integration $dz\,dx\,dy$.

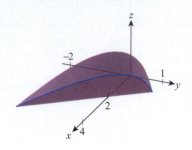

22. Revisit Exercise 21, this time integrating in the order $dx\,dz\,dy$.

23. S: The solid bounded by the circular cylinders $x^2 + y^2 = 4$ and $y^2 + z^2 = 4$ (**Hint:** Take advantage of the symmetry of the solid.)

24. S: The solid bounded by the parabolic cylinder $5y = x^2 - 4$ and the planes $y = 1 - z$ and $z = 0$

25. S: The solid bounded by the parabolic cylinder $x = 1 - y^2$ and the planes $x = 0$, $z = 0$, and $z = 1 - x$

26. S: The solid bounded by $z = y^2 - 1$, $z = 0$, and $x = \pm 1$

27. S: The solid bounded by $z = x^2$ and the planes $z = 4$, $y = 0$, and $z = y$

28. S: The solid that is common to the paraboloids $z = 9 - 8x^2 - 8y^2$ and $z = x^2 + y^2$ (Exercise 44 of Section 14.3 revisited)

29. S: The solid bounded below by the surface $z = 5x^2 + y^2 + 2$ and above by the surface $z = 8 - x^2 - y^2$. Choose the order of integration $dz\,dy\,dx$. (**Hint:** See Example 2.)

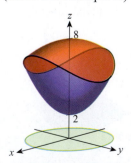

30. Repeat Exercise 29, but this time integrate in the order $dz\,dx\,dy$. (**Hint:** See Example 3.)

31. S: The solid bounded by $z = 2 - x^2 - y^2$, $y = x^2$, $x = y^2$, and $z = 0$

32. S: The solid bounded by $y = z^2$, $y = 2 - z^2$, $x = 0$, and $x = 2$

33. S: The solid bounded by $z = y^2$ and the planes $z = 9 - x$ and $x = 0$

34. S: The solid bounded by the paraboloid $z = x^2 + y^2$ and the plane $z = x + 6$ (**Hint:** To make your calculations more manageable, integrate with respect to y first, and use the symmetry of the solid.)

35. S: The solid bounded by the elliptic paraboloid $z = 4x^2 + y^2$ and the plane $z = 2y + 3$ (**Hint:** Choose the order of integration carefully and use the symmetry of the solid.)

36. S: The solid bounded by the surfaces $z = 2 - \sqrt{4x^2 + y^2}$ and $z = 0$ (**Hint:** Start integrating with respect to y or x and use the symmetry of the solid.)

37. Write the triple integral over the solid of Exercise 27 in three different ways, in the orders of $dy\,dz\,dx$, $dx\,dz\,dy$, $dx\,dy\,dz$, and evaluate them. (**Note:** Quite possibly, you have already handled one of these integrals in Exercise 27.)

38. Repeat Exercise 37 for the integral in Exercise 33 using the following orders of integration: $dx\,dz\,dy$, $dy\,dz\,dx$, and $dz\,dx\,dy$.

39. Suppose there are one-variable functions g, h, and k such that $f(x, y, z) = g(x) \cdot h(y) \cdot k(z)$ and $S = [p, q] \times [r, s] \times [t, u]$. Prove the following:

$$\iiint_S f(x, y, z)\,dV$$

$$= \left[\int_p^q g(x)\,dx\right] \cdot \left[\int_r^s h(y)\,dy\right] \cdot \left[\int_t^u k(z)\,dz\right]$$

(Note that this is a generalization of Exercise 80 of Section 14.1 to triple integrals.)

40–43 A solid S with variable density is given. Use a triple integral to find its mass.

40. S: The tetrahedron of Example 4, with its density at the point (x, y, z) being proportional to the point's distance from the tetrahedron's base (**Hint:** Integrate over the solid the density function $\rho(x, y, z) = k \cdot z$, where k is a constant.)

41. S: The tetrahedron bounded by $z = 4 - x - 2y$ and the coordinate planes, with its density at the point (x, y, z) being proportional to the square of the distance from the origin (As in the previous exercise, denote the constant of proportionality k.)

42. S: The tetrahedron bounded by $z = 3 - 2x - 6y$ and the coordinate planes, with the density at any point being proportional to the sum of its coordinates

43. S: The solid upper hemisphere of radius 1 centered at the origin, with its density at the point (x, y, z) being proportional to the distance from the base

44–47 Just like we did with two-variable functions (see Section 14.2), we can define the **average value** of $f(x, y, z)$ over a solid S as follows.

$$\text{Average value of } f \text{ over } S = \frac{1}{\text{Volume}(S)} \iiint_S f(x, y, z) \, dV$$

Use the above definition to find the average value of f over S.

44. $f(x, y, z) = xyz$;
    S: The cube $[0, a] \times [0, a] \times [0, a]$

45. $f(x, y, z) = \dfrac{1}{\sqrt{x}}$;
    S: The cube $[0, a] \times [0, a] \times [0, a]$

46. $f(x, y, z) = xy \cos z$;
    S: The cube $[0, \pi/2] \times [0, \pi/2] \times [0, \pi/2]$

47.* $f(x, y, z) = xyz$; S: The first-octant region of the sphere $x^2 + y^2 + z^2 = R^2$

48. Finish Example 3 by showing that the value of the integral is $2\sqrt{2}\pi$.

49. **a.** Describe the solid of integration and **b.** find its centroid (assuming constant density).

$$\int_0^1 \int_0^{\sqrt{1-x^2}} \int_0^{\sqrt{1-x^2-y^2}} dz \, dy \, dx$$

50–62 A solid S with constant or variable density is given. Use a triple integral to find the coordinates of its center of mass.

50. S: The tetrahedron bounded by the coordinate planes and $x + y + z = 2$, with constant density

51. S: The tetrahedron of Exercise 18, with constant density

52. S: The solid of Exercise 20, with constant density

53. S: The solid of Exercise 27, with constant density

54. S: The solid of Exercise 26, with constant density

55. S: The solid of Exercise 21, with constant density

56. S: The rectangular prism $[-1, 1] \times [-1, 1] \times [0, 4]$, its density at each point is inversely proportional to the square root of the point's distance from the base

57. S: The cube of Exercise 45, its density at each point is proportional to the square of the point's distance from the origin

58. S: The tetrahedron of Exercise 50, its density at each point is proportional to the square of the point's distance from the origin

59. S: The tetrahedron of Example 4, its density at each point is proportional to the distance from the base

60.* S: The first octant of the unit sphere centered at the origin, its density at any point is proportional to the product of its distances from the coordinate planes

61. S: The solid half cylinder bounded by $z = \sqrt{1 - y^2}$, $x = 0$, and $x = 1$, its density at any point is proportional to the square of the point's distance from the origin

62. S: The tetrahedron of Exercise 19

63. Verify that the second moments and radii of gyration for the solid of Example 4 are as follows:

$$I_x = \frac{8\rho}{15}, \quad I_y = I_z = \frac{\rho}{3}, \quad r_x = \frac{2}{\sqrt{5}}, \quad r_y = r_z = \frac{1}{\sqrt{2}}$$

64–71 Find the center of mass and the radii of gyration of the given solid S. Assume S is made of a substance with constant density ρ.

64. $S = [-1,3] \times [0,1] \times [1,3]$, the rectangular box of Exercise 2

65. S: The cube of Exercise 45

66. S: The solid bounded by $z = 4 - 2x - y$ and the coordinate planes

67. S: The solid bounded by $z = 3 - 2x - 6y$ and the coordinate planes

68. S: The solid bounded by $z = y^2$, the planes $z = 1 - x$ and $x = 0$

69. S: The solid bounded by $z = y^2 - 1$, $z = 0$, and $x = \pm 1$

70. S: The solid bounded by $z = 2 - x^2 - y^2$, the coordinate planes, and the planes $x = 1$ and $y = 1$

71. S: The first-octant region of the cylinder $x^2 + y^2 = 1$, bounded by the coordinate planes and $z = 1$

72–75 Find the second moments and radii of gyration of the indicated solid.

72. The solid of Exercise 56

73. The solid of Exercise 57

74. The solid of Exercise 59

75. The solid of Exercise 58

14.4 Technology Exercises

76–79 Use a computer algebra system to find the center of mass and the radii of gyration for the given solid with nonconstant density. (Note that even though it does give nice answers, in some cases even a computer algebra system requires a relatively long time for calculating them!)

76. S: The upper hemisphere of radius 1, centered at the origin, the density at any point is inversely proportional to the square root of its distance from the base

77. S: The first octant region of the solid in Exercise 76, the density at any point is proportional to the product of its distances from the coordinate planes

78. S: The solid of Exercise 66, its density at any point being proportional to the square of the point's distance from the origin

79. S: The solid of Exercise 67, its density at any point being proportional to the square root of its distance from the base

14.5 Triple Integrals in Cylindrical and Spherical Coordinates

TOPICS

1. The cylindrical coordinate system
2. The spherical coordinate system

Just as polar coordinates can simplify the evaluation of double integrals, *cylindrical* and *spherical* coordinates are alternative coordinate systems for $\mathbb{R}^3$ that aid in the evaluation of some triple integrals. For any given problem, the best choice of coordinate system requires familiarity with all three systems. The theory in this section is the final prelude to the general discussion in Section 14.6, where we learn methods for defining coordinate systems tailored to specific integrals.

TOPIC 1   The Cylindrical Coordinate System

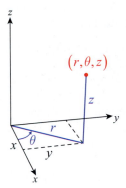

Figure 1

As the name implies, the **cylindrical coordinate system** of $\mathbb{R}^3$ is based on cylinders, and equations for cylinders in the cylindrical system are particularly simple in form. A point in $\mathbb{R}^3$ with Cartesian coordinates (x, y, z) has cylindrical coordinates (r, θ, z), where r is the length of the line segment joining $(0, 0, 0)$ and $(x, y, 0)$, θ is the angle between the positive x-axis and the same line segment, and z has the same meaning as in Cartesian coordinates. Figure 1 illustrates the relationship between (x, y, z) and (r, θ, z).

A few points deserve mention. First, note that the cylindrical coordinate system is simply the polar coordinate system of the xy-plane with the addition of a z-coordinate. This means that the familiar conversions between r, θ, x, and y still hold, as listed below. Second, as with polar coordinates, points in space are associated with more than one set of cylindrical coordinates—for instance, $(r_0, \theta_0, z_0) = (r_0, \theta_0 + 2n\pi, z_0)$ for every $n \in \mathbb{Z}$.

> **Formula** 🔍
>
> **Coordinate Conversion**
>
> **Converting from cylindrical to Cartesian coordinates:**
> $$x = r\cos\theta, \quad y = r\sin\theta, \quad z = z$$
>
> **Converting from Cartesian to cylindrical coordinates:**
> $$r^2 = x^2 + y^2, \quad \tan\theta = \frac{y}{x} \ (x \neq 0), \quad z = z$$

In order to evaluate a triple integral

$$\iiint\limits_{S} f(r, \theta, z)\, dV$$

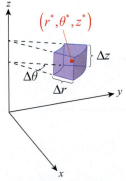

Figure 2

using cylindrical coordinates, we need to express dV in terms of dr, $d\theta$, and dz; this parallels the need to express dA in terms of dr and $d\theta$ in order to evaluate double integrals using polar coordinates. We accomplish the task in similar fashion by determining how a small incremental volume of space ΔV depends on incremental changes Δr, $\Delta\theta$, and Δz.

Accordingly, consider the diagram in Figure 2. If we let $\left(r^*,\theta^*,z^*\right)$ denote the point (shown in red) in the center of the incremental element, then from Section 14.3 we know that the area of the lower and upper surface of the volume is $\Delta A = r^* \, \Delta r \, \Delta \theta$, and hence the volume ΔV of the element is $\Delta V = \Delta A \, \Delta z = r^* \, \Delta z \, \Delta r \, \Delta \theta$. A triple Riemann sum approximating the integral of f over a solid S then has the form

$$\sum_{k=1}^{l}\sum_{j=1}^{m}\sum_{i=1}^{n} f\left(r^*,\theta^*,z^*\right) r^* \, \Delta z_i \, \Delta r_j \, \Delta \theta_k,$$

leading to the iterated integral in the steps outlined below. The order of integration shown for illustrative purposes is one of the more frequently used, but the best order for a given problem may be different.

Setting up a Triple Integral in Cylindrical Coordinates

To express $\displaystyle\iiint_{S} f\left(r,\theta,z\right) dV$ as an iterated integral using cylindrical coordinates, perform the following steps.

(a)

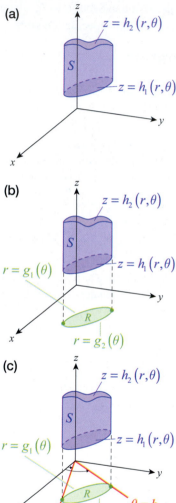

Step 1: Draw a rough sketch of S as an aid in determining the z-limits. Identify the lower surface $z = h_1\left(r,\theta\right)$ and the upper surface $z = h_2\left(r,\theta\right)$ of S—these two functions of r and θ constitute the z-limits (see Figure 3a).

Step 2: Define R to be the "shadow," or projection, of S in the xy-plane. Sketch R, this time as an aid in determining the r-limits. The curve $r = g_1\left(\theta\right)$ closer to the origin and the farther curve $r = g_2\left(\theta\right)$ constitute the r-limits (see Figure 3b).

Step 3: The lower limit of θ, $\theta = a$, defines one ray that bounds R, and the upper limit of θ, $\theta = b$, defines the other ray—R should be captured between the two rays (see Figure 3c).

The result will be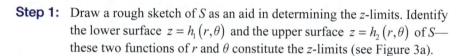

Figure 3

Caution ⚠

Note that the limits of integration of the innermost integral may be functions of the other two variables (r and θ in the order above). The limits of the middle integral may be functions of the outer variable (θ above), but not the variable corresponding to the inner limits of integration (z above). The limits of the outermost integral can only be constants.

Example 1 ✎

Find the volume of the paraboloid solid S bounded below by the surface $z = x^2 + y^2$ and above by $z = 1$.

Solution

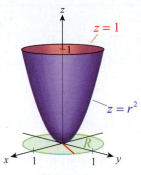

Figure 4

Since $x^2 + y^2 = r^2$, the lower and upper surfaces z-limits are $z = r^2$ and $z = 1$. The shadow, or projection, of the object in the xy-plane is the disk R defined by the equation $x^2 + y^2 = 1$, which in terms of r and θ reduces to the equation $r = 1$; note that in cylindrical coordinates, the equation $r = 1$ is a cylinder of radius 1 which entirely contains the solid S. Figure 4 illustrates S and R.

Since we want to find the volume of S, the integrand is the function $f(r, \theta, z) = 1$ and we are led to the following integral.

$$V = \int_0^{2\pi} \int_0^1 \int_{r^2}^1 r \, dz \, dr \, d\theta = \int_0^{2\pi} \int_0^1 [rz]_{z=r^2}^{z=1} \, dr \, d\theta = \int_0^{2\pi} \int_0^1 (r - r^3) \, dr \, d\theta$$

$$= \int_0^{2\pi} \left[\frac{r^2}{2} - \frac{r^4}{4} \right]_{r=0}^{r=1} d\theta = \frac{1}{4} \int_0^{2\pi} d\theta = \frac{\pi}{2} \text{ cubic units}$$

Example 2 ✐

Find the mass and center of mass of the paraboloid solid of Example 1 if its density is described by the function $\rho(x, y, z) = 1 - z$.

Solution

The moment formulas of Section 14.4 remain valid, but we must convert them to cylindrical coordinates if we wish to integrate with respect to r, θ, and z. In this case the conversion is easy, since z plays the same role in both Cartesian and cylindrical coordinates and symmetry tells us that $\bar{x} = \bar{y} = 0$ without integration.

To determine $\bar{z}$, we need to evaluate the mass M and the first moment M_{xy}.

$$M = \iiint_S \rho(x, y, z) \, dV = \int_0^{2\pi} \int_0^1 \int_{r^2}^1 (1 - z) r \, dz \, dr \, d\theta$$

$$= \int_0^{2\pi} \int_0^1 \left[r \left(z - \frac{z^2}{2} \right) \right]_{z=r^2}^{z=1} dr \, d\theta = \int_0^{2\pi} \int_0^1 \left(\frac{r}{2} - r^3 + \frac{r^5}{2} \right) dr \, d\theta$$

$$= \int_0^{2\pi} \left[\frac{r^2}{4} - \frac{r^4}{4} + \frac{r^6}{12} \right]_0^1 d\theta = \frac{1}{12} \int_0^{2\pi} d\theta = \frac{\pi}{6}$$

$$M_{xy} = \iiint_S z\rho(r, \theta, z) \, dV = \int_0^{2\pi} \int_0^1 \int_{r^2}^1 z(1 - z) r \, dz \, dr \, d\theta$$

$$= \int_0^{2\pi} \int_0^1 \left[r \left(\frac{z^2}{2} - \frac{z^3}{3} \right) \right]_{z=r^2}^{z=1} dr \, d\theta = \int_0^{2\pi} \int_0^1 \left(\frac{r}{6} - \frac{r^5}{2} + \frac{r^7}{3} \right) dr \, d\theta$$

$$= \int_0^{2\pi} \left[\frac{r^2}{12} - \frac{r^6}{12} + \frac{r^8}{24} \right]_0^1 d\theta = \frac{\pi}{12}$$

This gives us $\bar{z} = M_{xy}/M = \frac{1}{2}$, meaning the z-component of the center of mass is halfway between the bottom and top of the solid. In Exercise 37, you will show that for the same object with constant density, $\bar{z} = \frac{2}{3}$; the difference reflects the fact that in this example the object's density is greatest at the bottom.

Example 3 ✐

Determine the mass of the solid bounded by the cylinder $(x-1)^2 + y^2 = 1$, the xy-plane, and the cone $z = \sqrt{x^2 + y^2}$, given that its density function is $\rho(x, y, z) = x$.

Solution

Although the particulars of this problem are given in Cartesian coordinates, the integral is far easier to evaluate in cylindrical coordinates (in Exercises 132–133, you will set up and examine equivalent integrals in Cartesian and spherical coordinates). So the first order of business is to restate the geometry of the solid and its density in terms of r, θ, and z.

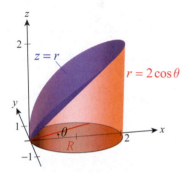

Figure 5

Cylinder:

$$(x-1)^2 + y^2 = 1 \iff x^2 + y^2 - 2x = 0 \iff r^2 - 2r\cos\theta = 0 \iff r = 2\cos\theta$$

Cone: $z = \sqrt{x^2 + y^2} \iff z = r$

Density: $\rho(x, y, z) = x \iff \rho(r, \theta, z) = r\cos\theta$

Figure 5 is an illustration of the solid and its shadow R described in cylindrical coordinates. Note that the θ-interval needed to define the cylinder is $[-\pi/2, \pi/2]$; the red line segment shown extending out from the origin corresponds to one particular angle θ.

$$\text{Mass} = \iiint_S \rho \, dV = \int_{-\pi/2}^{\pi/2} \int_0^{2\cos\theta} \int_0^r (r\cos\theta) r \, dz \, dr \, d\theta$$

$$= \int_{-\pi/2}^{\pi/2} \int_0^{2\cos\theta} \left[r^2 z \cos\theta \right]_{z=0}^{z=r} dr \, d\theta = \int_{-\pi/2}^{\pi/2} \int_0^{2\cos\theta} r^3 \cos\theta \, dr \, d\theta$$

$$= \frac{1}{4} \int_{-\pi/2}^{\pi/2} \left[r^4 \cos\theta \right]_{r=0}^{r=2\cos\theta} d\theta = 4 \int_{-\pi/2}^{\pi/2} \cos^5\theta \, d\theta$$

$$= 4 \int_{-\pi/2}^{\pi/2} \left(1 - \sin^2\theta \right)^2 \cos\theta \, d\theta \qquad \begin{aligned} u &= \sin\theta \\ du &= \cos\theta \, d\theta \end{aligned}$$

$$= 4 \int_{u=-1}^{u=1} \left(1 - u^2 \right)^2 du$$

$$= 4 \left[u - \frac{2}{3}u^3 + \frac{1}{5}u^5 \right]_{-1}^{1} = \frac{64}{15}$$

TOPIC 2 The Spherical Coordinate System

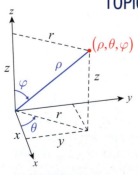

Figure 6

The **spherical coordinate system** is most useful when the geometry of a problem has some sort of symmetry about a point, especially if that point can be made the origin of the system. A point with Cartesian coordinates (x, y, z) corresponds to the spherical coordinates (ρ, θ, φ), where ρ is the distance between the origin and the point, θ has the same meaning as in cylindrical coordinates, and φ is the angle (sometimes called the *polar* angle) between the positive z-axis and the line segment joining the origin to the point. Unlike r, ρ is required to be nonnegative, and φ is restricted to the interval $[0, \pi]$. Figure 6 illustrates the relationship between (x, y, z) and (ρ, θ, φ), as well as the coordinate r of the cylindrical system—we will soon summarize all the relationships in a complete conversion table.

Caution ⚠

A few words of warning are in order: both the symbols used and the order of the coordinates vary somewhat with the field of study and between texts. In physics texts, θ is typically used to denote the polar angle and φ the angle in the xy-plane (sometimes called the *azimuth*). And if spherical coordinates alone appear, r might be used in place of ρ (if both spherical and cylindrical coordinates are in use, it's convenient to have different symbols for the two different coordinates). Finally, some texts swap the order of the polar and azimuth angles when listing the coordinates. Be sure to know how the symbols and their order are defined in any text you read.

To get a sense of the kinds of objects that can easily be described with spherical coordinates, consider the graphs of $\rho = 1$, $\theta = \pi/4$, and $\varphi = \pi/6$, shown in Figure 7.

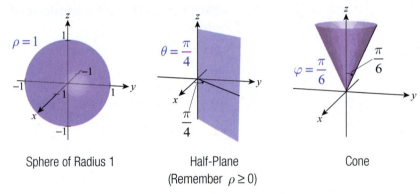

Sphere of Radius 1   Half-Plane   Cone
                        (Remember $\rho \geq 0$)

Figure 7

The following chart, based on Figure 6, expresses the relationships between Cartesian, cylindrical, and spherical coordinates and can be used as a reference when converting from one system to another.

Coordinate Conversion Relationships

Cylindrical and Cartesian	Spherical and Cylindrical	Spherical and Cartesian
$r^2 = x^2 + y^2$	$\rho^2 = r^2 + z^2$	$\rho^2 = x^2 + y^2 + z^2$
$x = r\cos\theta$	$r = \rho\sin\varphi$	$x = \underbrace{\rho\sin\varphi}_{r}\cos\theta$
$y = r\sin\theta$	$\theta = \theta$	$y = \underbrace{\rho\sin\varphi}_{r}\sin\theta$
$z = z$	$z = \rho\cos\varphi$	$z = \rho\cos\varphi$

Table 1

Example 4 ✐

Convert the equation $(x-2)^2 + y^2 + z^2 = 4$ to spherical coordinates.

Solution

The equation describes a sphere of radius 2 centered at $(2,0,0)$. In order to convert to spherical coordinates, we group terms and use the conversion formulas.

$$(x-2)^2 + y^2 + z^2 = 4$$
$$x^2 + y^2 + z^2 - 4x = 0$$
$$\rho^2 - 4\rho \sin\varphi \cos\theta = 0$$
$$\rho = 4\sin\varphi \cos\theta \qquad \text{We discard the solution } \rho = 0.$$

This defines the sphere as those points having spherical coordinates $(4\sin\varphi\cos\theta, \theta, \varphi)$. The surface is traced out completely over the intervals $-\pi/2 \le \theta \le \pi/2$ and $0 \le \varphi \le \pi$; allowing θ and φ to take on all the values $0 \le \theta \le 2\pi$ and $0 \le \varphi \le \pi$ actually results in each point of the sphere being visited twice.

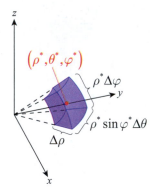

Figure 8

The use of spherical coordinates to aid in integration is our current goal, and the next step is to again determine how the differential element of volume dV depends on the differentials $d\rho$, $d\theta$, and $d\varphi$. Figure 8 illustrates a small incremental volume element ΔV (called a *spherical wedge*) corresponding to small changes $\Delta\rho$, $\Delta\theta$, and $\Delta\varphi$. The wedge is nearly cube-like in shape, with the approximation improving as the increments go to zero. The length of each side is shown in Figure 8. As usual, the point $(\rho^*, \theta^*, \varphi^*)$ in red is a sample point chosen from the volume element. Given the dimensions of the wedge,

$$\Delta V = \left(\rho^*\right)^2 \sin\varphi^* \Delta\rho\, \Delta\varphi\, \Delta\theta$$

and in the limit $dV = \rho^2 \sin\varphi\, d\rho\, d\varphi\, d\theta$. This allows us to express a triple integral in terms of spherical coordinates, as outlined in the steps that follow.

Setting up a Triple Integral in Spherical Coordinates

To express $\displaystyle\iiint_S f(\rho,\theta,\varphi)\, dV$ as an iterated integral using spherical coordinates, perform the following steps.

(a)

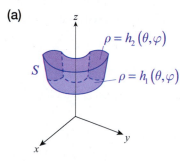

(b)

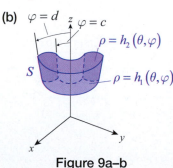

Figure 9a–b

Step 1: Draw a rough sketch of S as an aid in determining the ρ-limits. Identify the lower ρ-limit as a function $\rho = h_1(\theta,\varphi)$ and the upper ρ-limit as a function $\rho = h_2(\theta,\varphi)$ (see Figure 9a).

Step 2: Determine the least angle φ among the φ-coordinates of all the points of S, and use this as the lower φ-limit $\varphi = c$. Use a similar analysis to determine the upper φ-limit $\varphi = d$ (see Figure 9b). While it is possible for the φ-limits to be functions of θ, this rarely happens in practice when using spherical coordinates.

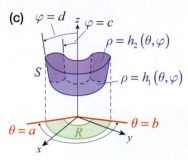

Figure 9c

Step 3: Define R to be the projection of S in the xy-plane. The lower θ-limit $\theta = a$ defines one ray that bounds R, and the upper θ-limit $\theta = b$ defines the other ray—R should be captured between the two rays (see Figure 9c).

The result will be $\iiint\limits_S f(\rho, \theta, \varphi)\, dV = \int_{\theta=a}^{\theta=b} \int_{\varphi=c}^{\varphi=d} \int_{\rho=h_1(\theta,\varphi)}^{\rho=h_2(\theta,\varphi)} f(\rho, \theta, \varphi)\, \rho^2 \sin\varphi\, d\rho\, d\varphi\, d\theta$.

Example 5 ✎

Set up the general form for the integral of a function $f(x, y, z)$ over the solid S defined as the first-octant portion of the sphere $(x-2)^2 + y^2 + z^2 = 4$.

Solution

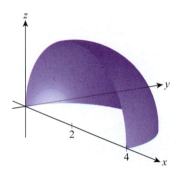

Figure 10

Figure 10 shows the outer surface of the portion of the sphere we are to integrate over. From Example 4, we know that $\rho = 4 \sin\varphi \cos\theta$ describes the surface in spherical coordinates, so to limit the integral to the part in the first octant we restrict θ and φ to $0 \le \theta \le \pi/2$ and $0 \le \varphi \le \pi/2$. The lower ρ-limit is $\rho = 0$, while the upper ρ-limit is the equation for the sphere. All that remains is to convert the arguments of f appropriately and to remember to include the factor of $\rho^2 \sin\varphi$ in the integral.

$$\iiint\limits_S f(x, y, z)\, dV$$

$$= \int_0^{\pi/2} \int_0^{\pi/2} \int_0^{4\sin\varphi\cos\theta} f(\rho\sin\varphi\cos\theta, \rho\sin\varphi\sin\theta, \rho\cos\varphi)\, \rho^2 \sin\varphi\, d\rho\, d\varphi\, d\theta$$

Example 6 ✎

Find the volume of the solid, shown in Figure 11, bounded below by the cone $\varphi = \pi/6$ and above by the sphere $\rho = 1$.

Solution

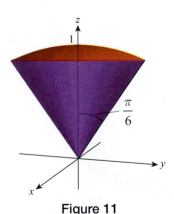

Figure 11

The lower and upper limits on ρ are 0 and 1, respectively, while the φ-interval is $[0, \pi/6]$ and the θ-interval is $[0, 2\pi]$. Since we seek only the volume, we integrate the constant function 1 over S. However, don't forget to include the factor of $\rho^2 \sin\varphi$, which is part of the volume differential dV.

$$V = \iiint\limits_S dV = \int_0^{2\pi} \int_0^{\pi/6} \int_0^1 \rho^2 \sin\varphi\, d\rho\, d\varphi\, d\theta$$

$$= \int_0^{2\pi} \int_0^{\pi/6} \left[\frac{\rho^3}{3} \right]_{\rho=0}^{\rho=1} \sin\varphi\, d\varphi\, d\theta$$

$$= \frac{1}{3} \int_0^{2\pi} \int_0^{\pi/6} \sin\varphi\, d\varphi\, d\theta = \frac{1}{3} \int_0^{2\pi} [-\cos\varphi]_0^{\pi/6}\, d\theta$$

$$= \frac{1}{3}\left(1 - \frac{\sqrt{3}}{2} \right) \int_0^{2\pi} d\theta = \frac{(2 - \sqrt{3})\pi}{3}$$

14.5 **Exercises**

1–4 Find a set of cylindrical coordinates for the point given in Cartesian coordinates.

1. $(1,1,2)$

2. $\left(1,-\sqrt{3},-1\right)$

3. $\left(-6\sqrt{3},-6,0\right)$

4. $(-3,-4,-5)$

5–8 Find the Cartesian coordinates of the point given in cylindrical coordinates.

5. $\left(2,\dfrac{3\pi}{4},\sqrt{2}\right)$

6. $(3,\pi,3)$

7. $\left(-1,\dfrac{\pi}{2},1\right)$

8. $\left(4,-\dfrac{\pi}{3},\dfrac{\pi}{2}\right)$

9–16 Write the equation in cylindrical coordinates.

9. $x^2+y^2=4x$

10. $x^2=2z-y^2$

11. $x^2+\left(y-1\right)^2=1$

12. $x^2+y^2+z^2=1$

13. $\sqrt{x^2+y^2}=3z$

14. $z=x-2y+1$

15. $e^{-\frac{x^2+y^2}{2}}=z$

16. $\left(x-3\right)^2+y^2=z+9$

17–24 Describe the graph of the equation in words, and change the equation to Cartesian coordinates.

17. $r=1$

18. $\theta=\dfrac{\pi}{6}$

19. $2r^2=2-z^2$

20. $2r^2=2-z$

21. $r^2=z^2$

22. $r=4\cos\theta$

23. $z=\sqrt{1-r^2}$

24. $r=4\sec\theta$

25–36 Set up a triple integral in cylindrical coordinates for the volume of the solid S. Do not evaluate the integral.

25. S: The solid bounded by $z=4-\left(x^2+y^2\right)$ and the xy-plane

26. S: The solid bounded by $z=\sqrt{x^2+y^2}-1$ and the xy-plane

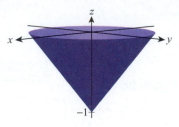

27. S: The solid bounded by $z=4-\sqrt{x^2+y^2}$ and the xy-plane

28. S: The solid bounded above by $z=\sqrt{2-x^2-y^2}$ and below by $z=\sqrt{x^2+y^2}$

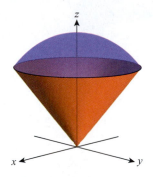

29. S: The solid inside both $z=\sqrt{x^2+y^2}$ and $x^2+y^2+z^2=8$

30. S: The solid inside both $z=x^2+y^2$ and $z=18-\left(x^2+y^2\right)$ in the first quadrant

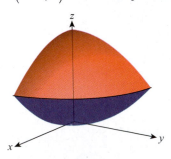

31. S: The solid bounded by $z^2=2+x^2+y^2$, $z=\sqrt{2}$, and $z=\sqrt{3}$

32.* S: The solid bounded by $z=x^2+y^2$, $z=1$, and $z=4$ (**Hint:** Integrate in the order $dr\,d\theta\,dz$.)

33.* S: The solid bounded by $z=2\sqrt{x^2+y^2}$, $z=1$, and $z=5$ (**Hint:** Integrate in the order $dr\,d\theta\,dz$.)

34.* S: The solid bounded by $z^2 + 2 = x^2 + y^2$, $z = -3$, and $z = 3$ (**Hint:** Integrate in the order $dr\,d\theta\,dz$.)

35. S: The solid of Exercise 31 (**Hint:** Integrate in the order $dr\,d\theta\,dz$.)

36. S: The solid bounded by $z = e^{x^2+y^2}$ and $z = e^{2-x^2-y^2}$

37. Show that if the solid of Example 2 has constant density, the third coordinate of its center of mass is $\overline{z} = \frac{2}{3}$.

38. Use cylindrical coordinates to verify the formula for the volume of a right circular cone of radius R and height h, $V = \frac{1}{3}\pi R^2 h$.

39–53 Use the cylindrical coordinate system to determine the mass of the solid S with the given density function.

39. S is the solid bounded by $z = 1 - x^2 - y^2$ and the xy-plane, with density function $\rho(x, y, z) = z$.

40. S is the solid bounded by $z = x^2 + y^2$ and $z = 8 - (x^2 + y^2)$, with constant density ρ.

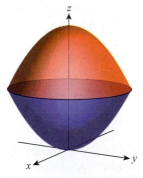

41. S is the solid upper hemisphere of radius 1 centered at the origin, with its density at the point (x, y, z) being proportional to its distance from the base. (See Exercise 43 of Section 14.4.)

42. S is the solid bounded by the upper sheet of the hyperboloid $z^2 - x^2 - y^2 = 1$ and the plane $z = \sqrt{5}$, with constant density ρ.

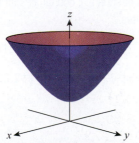

43. S is the solid bounded by $z = e^{\sqrt{x^2+y^2}}$ and the plane $z = e$, with density function $\rho(x, y, z) = \sqrt{x^2 + y^2}$.

44. S is the solid bounded by $x^2 + y^2 = 1$, the xy-plane, and $z = e$, with density function $\rho(x, y, z) = e^{-x^2-y^2}$.

45. S is the solid outside $z = 1 - \sqrt{x^2 + y^2}$, bounded by the xy-plane, $z = 1$, and $x^2 + y^2 = 1$, with density function $\rho(x, y, z) = \sqrt{x^2 + y^2}$.

46. S is the solid of Exercise 28, with constant density ρ.

47. S is the solid of Exercise 28, with density function $\rho(x, y, z) = z$.

48. S is the solid bounded by $z = \sqrt{x^2 + y^2}$ and $z = 3$, with density function $\rho(x, y, z) = e^{\sqrt{x^2+y^2}}$.

49. S is the solid bounded by the cylinder $x^2 + (y - 1)^2 = 1$, the xy-plane, and the paraboloid $z = x^2 + y^2$; its density at any point (x, y, z) is proportional to the square of the point's distance from the z-axis.

50. S is the solid bounded by the cylinder $(x - 1)^2 + z^2 = 1$, the xz-plane, and the paraboloid $y = x^2 + z^2$, with density function $\rho(x, y, z) = x$. (**Hint:** Integrate with respect to y first; then define and use the polar coordinates $x = r\cos\theta$ and $z = r\sin\theta$.)

51. S is the solid bounded by $z = \sqrt{x^2 + y^2}$ and $z = (x^2 + y^2)^{3/2}$, with density function $\rho(x, y, z) = \sqrt{x^2 + y^2}\,e^z$.

52. S is the solid bounded by $z = 1/\sqrt{x^2 + y^2}$, $2(x^2 + y^2) = 1$, $x^2 + y^2 = 1$, and the xy-plane, with density function $\rho(x, y, z) = \sqrt{x^2 + y^2}\,e^{\sqrt{x^2+y^2}}$.

53. S is a solid sphere of radius 2, with a cylindrical hole of radius 1 drilled into it along one of its diameters. Its density at any of its points is equal to the square of the distance from the origin.

54–64 Use the cylindrical coordinate system to find the center of mass of the solid S with the given density function.

54. S: The solid bounded by the paraboloids $z = x^2 + y^2$ and $z = 3 - 2x^2 - 2y^2$, with constant density

55. S: The solid inside $x^2 + y^2 = 4$, outside $x^2 + y^2 = 1$, and bounded by the paraboloid $z = 16 - x^2 - y^2$ and the xy-plane, with constant density

56. S: The upper hemisphere of radius R centered at the origin, with constant density

57. S: The upper hemisphere of radius 1 centered at the origin, if its density is proportional to the square of the distance from the origin

58. S: The hemisphere of Exercise 56, if its density is proportional to the distance from the z-axis

59. S: The right circular cylinder of radius 1 and height 2, its base centered at the origin in the xy-plane, with density function $\rho(x, y, z) = e^{-z}$

60. S: The solid inside the paraboloid $z = 6 - x^2 - y^2$ and outside the sphere $x^2 + y^2 + z^2 = 8$, with constant density

61. S: The right circular cone of radius R and height h with constant density ρ

62. S: The cone of Exercise 61 with the density of the cone being proportional to the distance from the cone's axis of symmetry

63. S: The cone of Exercise 61 with the density being proportional to the distance from the base

64.* S: The first octant of the unit sphere centered at the origin, its density at any point is proportional to the product of its distances from the coordinate planes (See Exercise 60 of Section 14.4.)

65.* Use cylindrical coordinates to give a second solution to Exercise 61 of Section 14.4. (See the hint given in Exercise 50.)

66–68 Use cylindrical coordinates to find the first and/or second moments of the solid, as indicated.

66. Find the second moment about the z-axis and the corresponding radius of gyration for the solid of Exercise 53. (Suppose it is centered at the origin, with the "hole" in vertical position.)

67. Find the moment of inertia about the z-axis and the corresponding radius of gyration for the cylinder $x^2 + y^2 = 1$, bounded by $z = 0$ and $z = 1$, if it has constant density ρ.

68. Repeat Exercise 67 for the solid of Exercise 55.

69–76 Find the indicated quantities for the solid from an earlier exercise in this section.

69. The center of mass, second moment I_z, and radius of gyration r_z for the solid of Exercise 39

70. The center of mass, second moment I_z, and radius of gyration r_z for the solid of Exercise 40

71. The center of mass, second moment I_z, and radius of gyration r_z for the solid of Exercise 41

72. The center of mass, second moment I_z, and radius of gyration r_z for the solid of Exercise 43

73. The second moment I_z and radius of gyration r_z for the solid of Exercise 55

74.* The second moment I_z and radius of gyration r_z for the solid of Exercise 59

75. The second moment I_z and radius of gyration r_z for the solid of Exercise 42, with density being proportional to the distance from the xy-plane

76. The second moment I_z and radius of gyration r_z for the solid of Exercise 60

77–80 Find a set of spherical coordinates for the point given in Cartesian coordinates.

77. $\left(\sqrt{3}, 1, 2 \right)$

78. $\left(1, 0, 1 \right)$

79. $\left(0, -1, -\sqrt{3} \right)$

80. $\left(-2, 2, 2\sqrt{2} \right)$

81–84 Find the Cartesian coordinates of the point given in spherical coordinates.

81. $\left(\sqrt{2}, 0, \dfrac{\pi}{4}\right)$

82. $\left(4, \dfrac{\pi}{4}, \dfrac{\pi}{3}\right)$

83. $\left(3, -\dfrac{\pi}{2}, \dfrac{\pi}{2}\right)$

84. $\left(2\sqrt{2}, \dfrac{\pi}{3}, \dfrac{\pi}{4}\right)$

85–92 Change the equation into spherical coordinates.

85. $x^2 + y^2 + z^2 = 4$

86. $z = 0$

87. $x^2 + y^2 - z^2 = 0$

88. $x^2 + y^2 + (z - 2)^2 = 4$

89. $y = x$

90. $z = 1$

91. $x^2 + y^2 = 4$

92. $x + y + z = 1$

93–100 Describe the graph of the equation in words, and change the equation to Cartesian coordinates.

93. $\rho = 1$

94. $\theta = \dfrac{\pi}{3}$

95. $\varphi = \dfrac{\pi}{4}$

96. $\varphi = \dfrac{\pi}{6}$

97. $\varphi = \dfrac{\pi}{2}$

98. $\rho = 2 \sec \varphi$

99. $\rho^2 = 9 \csc^2 \varphi$

100. $\tan^2 \varphi = 1$

101–105 Set up a triple integral in spherical coordinates for the volume of the solid S. Do not evaluate the integral.

101. S: The upper hemisphere of the sphere with radius 2, centered at the origin

102. S: The first-octant portion of the sphere $x^2 + (y - 1)^2 + z^2 = 1$

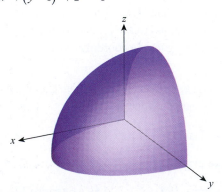

103. S: The first-octant portion of the sphere $x^2 + y^2 + (z - 3)^2 = 9$

104. The "ice-cream cone" of Exercise 45 of Section 14.3, bounded by the cone $z = \sqrt{x^2 + y^2}$ and the sphere $x^2 + y^2 + z^2 = 8$

105. The first-octant portion of the "wedge" of the sphere $x^2 + y^2 + z^2 = 25$ bounded by the planes $y = \sqrt{3}x$ and $x = \sqrt{3}y$

106–131 Use spherical coordinates to solve the exercise.

106. Evaluate the integral you set up in Exercise 101.

107. Find the volume of the "wedge" of Exercise 105.

108. Derive the formula for the volume of a sphere of radius R, $V = \frac{4}{3}\pi R^3$.

109. Derive the formula for the volume of a right circular cone of radius r and height h, $V = \frac{1}{3}\pi r^2 h$. (**Hint:** Place an inverted cone in the coordinate system with its vertex at the origin and its axis coinciding with the positive z-axis.)

110. Find the mass of the half ball of Exercise 101 if its density is proportional to the distance from its center.

111. Find the volume of the "ice-cream cone" of Exercise 104. (Compare with Exercise 45 of Section 14.3).

112. Find the volume of the solid below the cone $z = \sqrt{x^2 + y^2}$ and inside the unit hemisphere $z = \sqrt{1 - x^2 - y^2}$.

113. Repeat Exercise 112 for the solid below the cone $z = \sqrt{x^2 + y^2}$ and inside the sphere $x^2 + y^2 + (z - 2)^2 = 4$, $z \geq 0$.

114. Use Exercise 112 to find the rectangular equation of the cone that divides the upper unit hemisphere into two parts of equal volume.

115. Determine the centroid of the half ball of Exercise 56.

116. Find the centroid of the solid of Exercise 104, assuming it has constant density.

117. Find **a.** the mass and **b.** the centroid of the "ice-cream cone" that is the solid common to the sphere $x^2 + y^2 + z^2 = 4z$ and the cone $x^2 + y^2 = z^2$, if it has constant density ρ.

118. Find the mass of the upper unit hemisphere centered at the origin, if its density at the point (x, y, z) is proportional to the distance from its base.

119.* Generalize Exercises 111 and 116 (along with Example 6) by showing that, assuming constant density, the mass and centroid of the "ice-cream cone" bounded by the cone $\varphi = \alpha$ and the sphere of radius R are, respectively, $M = \frac{2}{3}\pi R^3 (1 - \cos\alpha)\rho$ and $\left(0, 0, \frac{3}{8}R(1 + \cos\alpha)\right)$.

120. Find the moment of inertia about the z-axis of the solid of Exercise 104. Assume constant density.

121. Find the second moment I_z and radius of gyration r_z of a solid ball of radius R and constant density ρ, if it is centered at the origin.

122. Find I_z and r_z for the solid of Exercise 112.

123. Find the moment of inertia about one of the coordinate axes of the spherical shell bounded by the spheres $x^2 + y^2 + z^2 = 2$ and $x^2 + y^2 + z^2 = 3$, if its density at any point is proportional to the distance from the origin. (**Hint:** Use the symmetry of the solid.)

124. Find a second solution (one that utilizes spherical coordinates) to Exercise 61.

125.* Find a second solution (one that utilizes spherical coordinates) to Exercise 62.

126.* Find a second solution (one that utilizes spherical coordinates) to Exercise 63.

127. a. Describe the solid of integration; then find **b.** its centroid, **c.** its moments of inertia about the coordinate axes, and **d.** its radii of gyration (assuming constant density).

$$\int_0^1 \int_0^{\sqrt{1-x^2}} \int_0^{\sqrt{1-x^2-y^2}} dz\, dy\, dx$$

128. a. Describe the solid of integration and **b.** find its volume by evaluating the integral.

$$\int_0^{2\pi} \int_0^{\pi} \int_0^{\sin\varphi} \rho^2 \sin\varphi\, d\rho\, d\varphi\, d\theta$$

129.* a. Describe the solid of integration and **b.** find its volume by evaluating the integral.

$$\int_0^{2\pi} \int_0^{\pi} \int_0^{1+\cos\varphi} \rho^2 \sin\varphi\, d\rho\, d\varphi\, d\theta$$

130. Evaluate $\iiint_S \sin\left(x^2 + y^2 + z^2\right)^{3/2} dV$, where S is the unit ball centered at the origin.

131. Evaluate $\iiint_S \dfrac{dV}{\sqrt{x^2 + y^2 + z^2}}$, where S is the first-octant portion of the solid inside the unit sphere and between the cones $\varphi = \pi/6$ and $\varphi = \pi/3$.

14.5 Technology Exercises

132. Set up an integral for the mass of the solid in Example 3 in Cartesian coordinates, integrating in the order of $dz\,dy\,dx$, and use a computer algebra system to evaluate it.

133. Repeat Exercise 132, but this time use spherical coordinates.

134. Find the mass and center of mass of the hemisphere of Exercise 56, if its density is proportional to the square of the distance from the origin.

135. Use spherical coordinates to find the center of mass of the first octant of the unit sphere centered at the origin, if its density at any point is proportional to the product of the point's distances from the coordinate planes.

136–143 Use a computer algebra system to evaluate the integral in the indicated exercise.

136. Exercise 53	**137.** Exercise 58
138. Exercise 72	**139.** Exercise 76
140. Exercise 124	**141.** Exercise 125
142. Exercise 126	**143.** Exercise 129

14.6 **Substitutions and Multiple Integrals**

TOPICS

1. Substitutions in double integrals
2. Substitutions in triple integrals

The goal of rewriting an integral in terms of polar, cylindrical, or spherical coordinates is to make it easier to evaluate, and the technique of u-substitution has the same goal. In this section we see that each of these methods is a particular instance of a more general theory, and we learn how to customize a change of variables to a given integral. To do so, we introduce the concepts of *coordinate transformations* and *Jacobian determinants*.

TOPIC 1 **Substitutions in Double Integrals**

To set the stage for the general theory, recall the Substitution Rule (u-substitution).

$$\int f(u)\,du = \int f(g(x))g'(x)\,dx$$

While to this point we have used the Substitution Rule to replace the integral on the right with the one on the left, we focus attention now on the factor of $g'(x)$; we can think of it as a transformation factor necessary to convert the integral from one variable to the other. Similarly, the factor of r allows us to convert a double integral from Cartesian coordinates to polar coordinates.

$$\iint f(x, y)\,dx\,dy = \iint f(r\cos\theta, r\sin\theta)\,r\,dr\,d\theta$$

More generally, consider a **transformation** function T of the form $T(u, v) = (x, y)$ from the uv-coordinate system to the xy-coordinate system. T takes a point with coordinates (u, v) and converts it to coordinates (x, y), where both x and y are functions of u and v.

$$x = g(u, v) \quad \text{and} \quad y = h(u, v)$$

For our purposes, we will assume that g and h have continuous first-order partial derivatives (such a function T is said to be a C^1 transformation). We also assume that T is a **one-to-one** function on a domain region S in the uv-plane onto a region R in the xy-plane, so that for each point $(x_0, y_0) \in R$ there is a unique point $(u_0, v_0) \in S$ such that

$$T(u_0, v_0) = (x_0, y_0).$$

We call (u_0, v_0) the **preimage** of (x_0, y_0), and (x_0, y_0) the **image** of (u_0, v_0); similarly, we say the region S is the **preimage of R**, or that R is the **image of S**. Figure 1 illustrates these relationships; note that since T is one-to-one, the **inverse transformation** T^{-1} is a function from R to S.

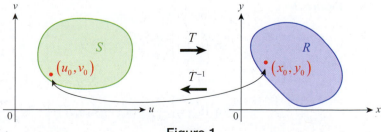

Figure 1

Our goal is to convert a double integral of a function $f(x, y)$ over R into a double integral of $f(g(u, v), h(u, v))$ over S, ideally in such a manner that the converted integral is easier to evaluate. But to do so, we need to know how each differential area element $dx\, dy$ relates to its corresponding differential area element $du\, dv$. The relationship is a ratio of areas, and can informally be determined by the following argument (a formal proof is best left for a class in advanced calculus).

Beginning with a fixed point (u_0, v_0) in the interior of S, let Δu and Δv be small positive increments and consider the rectangle S' shown on the left in Figure 2 (we assume the increments are small enough so that $S' \subseteq S$). The function T then transforms S' into a region $R' \subseteq R$, shown on the right in Figure 2. Let

$$(x_0, y_0) = T(u_0, v_0) = (g(u_0, v_0), h(u_0, v_0)),$$

and define

$$\mathbf{p}(u) = \langle g(u, v_0), h(u, v_0)\rangle \quad \text{and} \quad \mathbf{q}(v) = \langle g(u_0, v), h(u_0, v)\rangle.$$

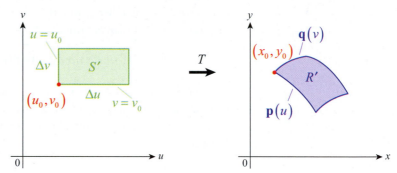

Figure 2

Then both $\mathbf{p}(u_0)$ and $\mathbf{q}(v_0)$ are position vectors for the point (x_0, y_0), and $\mathbf{p}(u)$ and $\mathbf{q}(v)$ are vector functions that trace out edges of R' as shown in Figure 2. We can approximate the area of R' with the area of the parallelogram defined by the vectors $\mathbf{p}(u_0 + \Delta u) - \mathbf{p}(u_0)$ and $\mathbf{q}(v_0 + \Delta v) - \mathbf{q}(v_0)$ (see Figure 3), and from Section 11.4 we know that the area of the parallelogram is the magnitude of the cross product of the vectors. Since

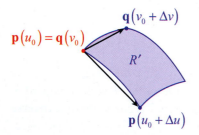

Figure 3

$$\mathbf{p}'(u_0) \approx \frac{\mathbf{p}(u_0 + \Delta u) - \mathbf{p}(u_0)}{\Delta u} \quad \text{and} \quad \mathbf{q}'(v_0) \approx \frac{\mathbf{q}(v_0 + \Delta v) - \mathbf{q}(v_0)}{\Delta v},$$

the area of the parallelogram is $\left|\mathbf{p}'(u_0)\Delta u \times \mathbf{q}'(v_0)\Delta v\right| = \left|\mathbf{p}'(u_0) \times \mathbf{q}'(v_0)\right|\Delta u\, \Delta v.$

Note that

$$\mathbf{p}'(u_0) = \left[\frac{d}{du}\langle g(u, v_0), h(u, v_0)\rangle\right]_{u=u_0} = \left[\frac{\partial}{\partial u}\langle g(u, v), h(u, v)\rangle\right]_{(u_0, v_0)} = \left\langle\frac{\partial x}{\partial u}, \frac{\partial y}{\partial u}\right\rangle\Bigg|_{(u_0, v_0)}$$

and

$$\mathbf{q}'(v_0) = \left[\frac{d}{dv}\langle g(u_0, v), h(u_0, v)\rangle\right]_{v=v_0} = \left[\frac{\partial}{\partial v}\langle g(u, v), h(u, v)\rangle\right]_{(u_0, v_0)} = \left\langle\frac{\partial x}{\partial v}, \frac{\partial y}{\partial v}\right\rangle\Bigg|_{(u_0, v_0)},$$

so

$$\mathbf{p}'(u_0) \times \mathbf{q}'(v_0) = \begin{vmatrix} \mathbf{i} & \mathbf{j} & \mathbf{k} \\ \dfrac{\partial x}{\partial u} & \dfrac{\partial y}{\partial u} & 0 \\ \dfrac{\partial x}{\partial v} & \dfrac{\partial y}{\partial v} & 0 \end{vmatrix}_{(u_0,v_0)} = \left[\left(\dfrac{\partial x}{\partial u}\dfrac{\partial y}{\partial v} - \dfrac{\partial x}{\partial v}\dfrac{\partial y}{\partial u}\right)\mathbf{k}\right]_{(u_0,v_0)}$$

and hence $\left|\mathbf{p}'(u_0) \times \mathbf{q}'(v_0)\right| = \left[\dfrac{\partial x}{\partial u}\dfrac{\partial y}{\partial v} - \dfrac{\partial x}{\partial v}\dfrac{\partial y}{\partial u}\right]_{(u_0,v_0)}$. Before proceeding, we use this as the basis for a definition.

Definition 💡

The Jacobian of a Transformation

The **Jacobian** (or **Jacobian determinant**) of the coordinate transformation T defined by $x = g(u,v)$ and $y = h(u,v)$ is

$$\dfrac{\partial(x,y)}{\partial(u,v)} = \begin{vmatrix} \dfrac{\partial x}{\partial u} & \dfrac{\partial x}{\partial v} \\ \dfrac{\partial y}{\partial u} & \dfrac{\partial y}{\partial v} \end{vmatrix} = \dfrac{\partial x}{\partial u}\dfrac{\partial y}{\partial v} - \dfrac{\partial x}{\partial v}\dfrac{\partial y}{\partial u}.$$

Given this definition, the area of the region R' is approximately

$$\left|\dfrac{\partial(x,y)}{\partial(u,v)}\right| \Delta u\, \Delta v,$$

where the Jacobian is evaluated at (u_0, v_0). So if we let ΔA denote the area of any such region R' in R (see Figure 4), we can approximate the double integral of a function $f(x, y)$ over R by a Riemann sum of the form

$$\iint\limits_R f(x,y)\,dA \approx \sum_{j=1}^m \sum_{i=1}^n f(x^*, y^*)\Delta A \approx \sum_{j=1}^m \sum_{i=1}^n f\big(g(u^*,v^*), h(u^*,v^*)\big)\left|\dfrac{\partial(x,y)}{\partial(u,v)}\right| \Delta u\, \Delta v,$$

where each Jacobian is evaluated at (u^*, v^*).

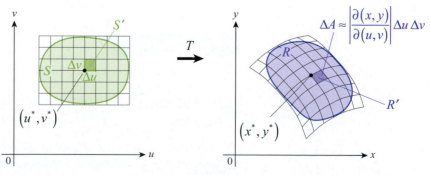

Figure 4

This gives rise to the following theorem.

> ### Theorem ⚲
>
> **Change of Variables in a Double Integral**
>
> Assume T is a coordinate transformation from S to R defined by $x = g(u, v)$ and $y = h(u, v)$, where g and h both have continuous first-order partial derivatives, and that T is one-to-one on S (except possibly on the boundary of S). Assume also that the Jacobian of T is nonzero on S and that f is continuous on R. Then
>
> $$\iint\limits_{R} f(x, y)\, dA = \iint\limits_{S} f\big(g(u, v), h(u, v)\big)\left|\frac{\partial(x, y)}{\partial(u, v)}\right| du\, dv.$$

Before demonstrating the change of variable formula for double integrals, note its similarity to the Substitution Rule for single integrals. Given $u = g(x)$,

$$\int f(u)\, du = \int f\big(g(x)\big)\frac{du}{dx}\, dx.$$

Example 1 ☑

Use a Jacobian to derive the formulation of double integrals in polar coordinates.

Solution

The coordinate transformation T from polar coordinates to Cartesian coordinates is defined by $x = r \cos\theta$ and $y = r \sin\theta$.

$$\frac{\partial(x, y)}{\partial(r, \theta)} = \begin{vmatrix} \dfrac{\partial x}{\partial r} & \dfrac{\partial x}{\partial \theta} \\[2mm] \dfrac{\partial y}{\partial r} & \dfrac{\partial y}{\partial \theta} \end{vmatrix} = \begin{vmatrix} \cos\theta & -r\sin\theta \\ \sin\theta & r\cos\theta \end{vmatrix}$$

$$= (\cos\theta)(r\cos\theta) - (-r\sin\theta)(\sin\theta) = r\left(\cos^2\theta + \sin^2\theta\right) = r$$

Hence,

$$\iint\limits_{R} f(x, y)\, dA = \iint\limits_{S} f(r\cos\theta, r\sin\theta)|r|\, dr\, d\theta = \iint\limits_{S} f(r\cos\theta, r\sin\theta)\, r\, dr\, d\theta$$

(under the assumption that $r > 0$).

Example 2 ☑

Use a change of variables to evaluate $\displaystyle\iint\limits_{R} \frac{3x + 2y}{x + 4y}\, dA,$ where R is the region bounded by $x = 0$, $y = 0$, and $3x + 2y = 2$.

Solution

This double integral is awkward to evaluate using Cartesian coordinates, but the right choice of change of variables makes it much easier. One possibility is to let u equal the numerator and v the denominator of the integrand.

$$u = 3x + 2y \quad \text{and} \quad v = x + 4y$$

In order to determine the Jacobian of the transformation from the uv-plane to the xy-plane, we need to solve these two equations for x and y.

$$\begin{array}{ll} u = 3x + 2y & -2u = -6x - 4y \\ \underline{-3v = -3x - 12y} \quad \text{and} & \underline{v = x + 4y} \\ u - 3v = -10y & -2u + v = -5x \end{array} \quad \text{so} \quad x = \frac{2u - v}{5}, \quad y = \frac{3v - u}{10}$$

We can now calculate

$$\frac{\partial(x, y)}{\partial(u, v)} = \begin{vmatrix} \dfrac{\partial x}{\partial u} & \dfrac{\partial x}{\partial v} \\[2mm] \dfrac{\partial y}{\partial u} & \dfrac{\partial y}{\partial v} \end{vmatrix} = \begin{vmatrix} \dfrac{2}{5} & -\dfrac{1}{5} \\[2mm] -\dfrac{1}{10} & \dfrac{3}{10} \end{vmatrix}$$

$$= \frac{6}{50} - \frac{1}{50} = \frac{1}{10}.$$

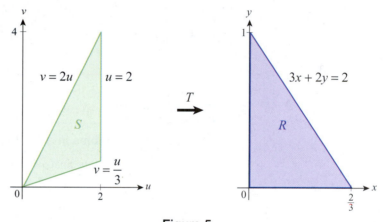

Figure 5

A diagram of the region S in the uv-plane that corresponds to R in the xy-plane is helpful in determining limits, and Figure 5 illustrates both. Each boundary of R corresponds to a boundary of S as follows.

$$\begin{array}{rcccc} x = 0 & \Leftrightarrow & 2u - v = 0 & \Leftrightarrow & v = 2u \\ y = 0 & \Leftrightarrow & 3v - u = 0 & \Leftrightarrow & v = u/3 \\ 3x + 2y = 2 & \Leftrightarrow & u = 2 & & \end{array}$$

(By calculating the area of each triangle, note that the area of R is $\frac{1}{10}$ the area of S.) We are now ready to change variables in the integral and evaluate it.

$$\iint\limits_R \frac{3x + 2y}{x + 4y}\, dA = \int_{u=0}^{u=2} \int_{v=u/3}^{v=2u} \left(\frac{u}{v}\right)\left(\frac{1}{10}\right) dv\, du$$

$$= \frac{1}{10}\int_0^2 \big[u \ln v\big]_{v=u/3}^{v=2u}\, du = \frac{\ln 6}{10}\int_0^2 u\, du = \frac{\ln 6}{5}$$

TOPIC 2 Substitutions in Triple Integrals

The change of variables theorem applies to any number of variables. Without going through the derivation, we state the definition of the Jacobian of a transformation from one three-dimensional space to another, along with the form of the change of variables.

Definition 💡

The Jacobian of a Transformation (Three-Dimensional Version)

The **Jacobian** of the coordinate transformation T defined by $x = g(u, v, w)$, $y = h(u, v, w)$, and $z = k(u, v, w)$ is

$$\frac{\partial(x, y, z)}{\partial(u, v, w)} = \begin{vmatrix} \dfrac{\partial x}{\partial u} & \dfrac{\partial x}{\partial v} & \dfrac{\partial x}{\partial w} \\ \dfrac{\partial y}{\partial u} & \dfrac{\partial y}{\partial v} & \dfrac{\partial y}{\partial w} \\ \dfrac{\partial z}{\partial u} & \dfrac{\partial z}{\partial v} & \dfrac{\partial z}{\partial w} \end{vmatrix}.$$

Theorem 🔍

Change of Variables in a Triple Integral

Assume T is a coordinate transformation from S to R defined by $x = g(u, v, w)$, $y = h(u, v, w)$, and $z = k(u, v, w)$, where g, h, and k all have continuous first-order partial derivatives, and that T is one-to-one on S (except possibly on the boundary of S). Assume also that the Jacobian of T is nonzero on S and that f is continuous on R. Then

$$\iiint_R f(x, y, z)\, dV = \iiint_S f\big(g(u, v, w), h(u, v, w), k(u, v, w)\big)\left|\frac{\partial(x, y, z)}{\partial(u, v, w)}\right| du\, dv\, dw.$$

Example 3 ✍

Use a Jacobian to derive the formulation of triple integrals in spherical coordinates.

Solution

The coordinate transformation T from spherical coordinates to Cartesian coordinates is defined by

$$x = \rho \sin \varphi \cos \theta, \quad y = \rho \sin \varphi \sin \theta, \quad \text{and} \quad z = \rho \cos \varphi.$$

$$\frac{\partial(x,y,z)}{\partial(\rho,\theta,\varphi)} = \begin{vmatrix} \dfrac{\partial x}{\partial \rho} & \dfrac{\partial x}{\partial \theta} & \dfrac{\partial x}{\partial \varphi} \\[4pt] \dfrac{\partial y}{\partial \rho} & \dfrac{\partial y}{\partial \theta} & \dfrac{\partial y}{\partial \varphi} \\[4pt] \dfrac{\partial z}{\partial \rho} & \dfrac{\partial z}{\partial \theta} & \dfrac{\partial z}{\partial \varphi} \end{vmatrix} = \begin{vmatrix} \sin\varphi\cos\theta & -\rho\sin\varphi\sin\theta & \rho\cos\varphi\cos\theta \\ \sin\varphi\sin\theta & \rho\sin\varphi\cos\theta & \rho\cos\varphi\sin\theta \\ \cos\varphi & 0 & -\rho\sin\varphi \end{vmatrix}$$

$$= \sin\varphi\cos\theta\left(-\rho^2\sin^2\varphi\cos\theta\right) + \rho\sin\varphi\sin\theta\left(-\rho\sin^2\varphi\sin\theta - \rho\cos^2\varphi\sin\theta\right)$$

$$\quad + \rho\cos\varphi\cos\theta\left(-\rho\cos\varphi\sin\varphi\cos\theta\right)$$

$$= -\rho^2\sin^3\varphi\cos^2\theta - \rho^2\sin\varphi\sin^2\theta - \rho^2\cos^2\varphi\cos^2\theta\sin\varphi$$

$$= -\rho^2\sin\varphi\left(\sin^2\varphi\cos^2\theta + \sin^2\theta + \cos^2\varphi\cos^2\theta\right)$$

$$= -\rho^2\sin\varphi\left[\left(\sin^2\varphi + \cos^2\varphi\right)\cos^2\theta + \sin^2\theta\right]$$

$$= -\rho^2\sin\varphi$$

Once we take the absolute value of the Jacobian (and use the fact that $0 \le \varphi \le \pi$, so $\sin\varphi \ge 0$), we obtain the following.

$$\iiint\limits_R f(x,y,z)\,dV = \iiint\limits_S f\left(\rho\sin\varphi\cos\theta, \rho\sin\varphi\sin\theta, \rho\cos\varphi\right)\rho^2\sin\varphi\,d\rho\,d\theta\,d\varphi$$

Example 4 ✎

Use a change of variables to evaluate $\displaystyle\iiint\limits_R \left(\frac{3x-z}{2} + \frac{y}{3}\right)dV,$ where R is the solid bounded by the planes $y = 0$, $y = 3$, $z = 0$, $z = 4$, $3x = z$, and $3x = z + 3$.

Solution

As in Example 2, the first step is to decide on a change of variables to try. On the basis of the integrand, and also guided by the form of two of the six planes, the assignment $u = 3x - z$ appears promising. This will allow two of the faces of the parallelepiped to be expressed as $u = 0$ and $u = 3$. In Exercise 41 you will show that if we let $v = z/2$ and $w = y/3$, then

$$\left|\frac{\partial(x,y,z)}{\partial(u,v,w)}\right| = 2$$

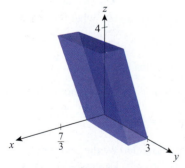

Figure 6

and

$$\iiint\limits_R \left(\frac{3x-z}{2} + \frac{y}{3}\right)dV = \int_{w=0}^{w=1}\int_{v=0}^{v=2}\int_{u=0}^{u=3}\left(\frac{u}{2} + w\right)(2)\,du\,dv\,dw = 15.$$

The solid S corresponding to the limits, in xyz-space, appears in Figure 6.

14.6 Exercises

1–4 Evaluate the given determinant.

1. $\begin{vmatrix} 2 & 2 \\ 1 & 3 \end{vmatrix}$

2. $\begin{vmatrix} 1 & 6 \\ -4 & -2 \end{vmatrix}$

3. $\begin{vmatrix} 2 & -1 & 1 \\ 3 & 4 & 0 \\ 1 & -5 & 2 \end{vmatrix}$

4. $\begin{vmatrix} 2 & -2 & -1 \\ 3 & 2 & 3 \\ -4 & 1 & -1 \end{vmatrix}$

5–13 Find the Jacobian of the given transformation.

5. $x = 2u + v, \quad y = v - u$ **6.** $x = v, \quad y = 4u + \dfrac{v}{2}$

7. $x = 4uv, \quad y = u + 2v$ **8.** $x = u + 2v^2, \quad y = uv$

9. $x = u^2, \quad y = \dfrac{v}{u}$ **10.** $x = e^{-u}, \quad y = ve^u$

11. $x = uv, \quad y = (u+1)v$

12. $x = e^v \cos u, \quad y = e^v \sin u$

13. $x = u \cos \varphi + v \sin \varphi, \quad y = u \sin \varphi - v \cos \varphi$

14. A transformation $T(u,v) = (au + cv, bu + dv)$ is called **linear**, where a, b, c, and d are constants. (See Exercises 5 and 6.) Find a general formula for the Jacobian of a linear transformation.

15–19 Consider the parallelogram P in the xy-plane bounded by the lines $y = x + 2$, $y = x - 4$, $y = 1 - 3x$, and $y = 5 - 3x$. We can identify a linear transformation $T(u,v)$ mapping a rectangle in the uv-plane onto P as follows. Rewrite the equations of the lines as $y - x = 2$, $y - x = -4$, $y + 3x = 1$, and $y + 3x = 5$, respectively. Then perform the change of variables $u = y - x$, $v = y + 3x$. Solve for x and y to obtain T (Exercise 15) and note that the preimage of P under T is the rectangle $-4 \le u \le 2$, $1 \le v \le 5$. In Exercises 16–19, you will be asked to follow these steps to identify the indicated linear transformation.

15. By solving the above system for x and y, find $T(u,v)$ as suggested by the above directions.

Find a linear transformation $T(u,v)$ that maps a rectangular region onto the given parallelogram P.

16. P is bounded by $2y = 1 - x$, $2y = 3 - x$, $y = 3x$, and $y = 3x + 4$.

17. P is bounded by $y = \frac{3}{2}x + 2$, $y = \frac{3}{2}x + 4$, $y = 1 - \frac{1}{4}x$, and $y = 4 - \frac{1}{4}x$.

18. P is bounded by $y = 1 - 2x$, $y = 5 - 2x$, $y = 3x - 2$, and $y = 3x + 3$.

19. P is bounded by $y = 2x$, $y = 2x + 4$, $y = -2x$, and $y = -2x - 2$.

20–25 Use a change of variables in order to integrate on an appropriate rectangle. (See Exercises 16–19.)

20. Find the area of the parallelogram in Exercise 16. (**Hint:** Start with $A = \displaystyle\iint_P dA$, and change variables, so you can integrate on the rectangle $[1,3] \times [0,4]$.)

21. Evaluate $\displaystyle\iint_P (2y + x)\, dA$ on P of Exercise 17.

22. Evaluate $\displaystyle\iint_P (x - 2y)^2\, dA$ on P of Exercise 18.

23. Evaluate $\displaystyle\iint_P \dfrac{y - 3x}{4y + 2x}\, dA$ on P of Exercise 16.

24. Evaluate $\displaystyle\iint_P \cos(x + y)\, dA$ on P of Exercise 18.

25. Evaluate $\displaystyle\iint_P \dfrac{e^{y-3x}}{x + 2y}\, dA$ on P of Exercise 16. (Note that a change of variables is necessary here, since you can't find an antiderivative in Cartesian coordinates, no matter what the order of integration.)

26. Use a change of variables to evaluate $\displaystyle\iint_R \dfrac{x}{y + 2x}\, dA$, where R is the region bounded by $x = 0$, $y = 0$, and $y = 4 - 2x$.

27. Let R be the region bounded by the coordinate axes and the line $x + 2y = 2$. Use the method of Example 2 to evaluate $\displaystyle\iint_R \dfrac{x + 2y}{x + y}\, dA$.

28. Use a change of variables to evaluate $\displaystyle\iint_R \dfrac{2x + y}{8x + 6y}\, dA$, where R is the region bounded by the coordinate axes and the line $y + 2x = 2$.

29. Consider the transformation $x = au$, $y = bv$ $(a, b > 0)$. Show that it maps the interior of the circle $u^2 + v^2 = 1$ onto that of the ellipse $\dfrac{x^2}{a^2} + \dfrac{y^2}{b^2} = 1$. Determine the Jacobian and use a change of variables to prove the area formula for the ellipse, $A = \pi ab$.

30–33 Let T be the coordinate transformation $x = uv$, $y = v/u$, assuming $u, v > 0$.

30. Find the Jacobian for the above transformation T.

31. Show that the T-images of vertical lines $u = a$ are lines through the origin, while horizontal lines $v = b$ are mapped onto branches of hyperbolas. (Use the observations $xy = v^2$ and $x/y = u^2$.)

32. Find the T-image of the uv-rectangle $\left[1, \sqrt{2}\right] \times \left[1, \sqrt{2}\right]$. (See Exercise 31.)

33. If S denotes the T-image of the rectangle in Exercise 32, use a change of variables to evaluate $\displaystyle\iint_R 2x^2 y \, dy \, dx$.

34. Use a change of variables to evaluate $\displaystyle\iint_R xy^3 \, dy \, dx$, where R is the region in the xy-plane bounded by the horizontal lines $y = 1$, $y = 3$, and the hyperbolas $y = 1/x$ and $y = 6/x$. (**Hint:** Consider the coordinate transformation $x = u/v$, $y = v$.)

35. Solve Exercise 34 without changing variables and compare your answers.

36. Generalize Exercise 14 to find a formula for the Jacobian J of a linear transformation in the three-variable case.

37–38 Find the Jacobian of the indicated coordinate transformation.

37. $x = u + 2v$, $\quad y = 2u + v - w$, $\quad z = 2v + w$

38. $x = u^2 v$, $\quad y = 2uvw$, $\quad z = u(1 + v)w$

39. Use a Jacobian to derive the formulation of triple integrals in cylindrical coordinates.

40.* Starting with the formulation of triple integrals in cylindrical coordinates and by finding a coordinate transformation T from spherical to cylindrical coordinates along with its Jacobian,

provide another derivation of the formula for triple integrals in spherical coordinates. (**Hint:** See Example 3 for guidance.)

41. Verify that for the change of variables in Example 4, $|\partial(x, y, z)/\partial(u, v, w)| = 2$ and thus

$$\iiint_R \left(\frac{3x - z}{2} + \frac{y}{3}\right) dV = \int_{w=0}^{w=1} \int_{v=0}^{v=2} \int_{u=0}^{u=3} \left(\frac{u}{2} + w\right)(2)\, du\, dv\, dw$$

$$= 15.$$

42–44 Use a change of variables to evaluate the given integral over the solid R.

42. Evaluate $\displaystyle\iiint_R \frac{4y + 2x - z}{4}\, dV$, where R is the solid bounded by the planes $x = 0$, $x = 2$, $z = 0$, $z = 3$, $4y = z$, and $4y = z + 6$.

43. Evaluate $\displaystyle\iiint_R \left(\frac{2x - y}{2} + \frac{3x + 5z}{6}\right) dV$, where R is the solid bounded by the planes $z = 0$, $z = 2$, $y = 2x$, $y = 2x - 8$, and $z = 3x$, $z = 3x - 6$.

44. Evaluate $\displaystyle\iiint_R x(2z - y)e^{y/2}\, dV$, where R is the solid bounded by the planes $x = 0$, $x = 1$, $y = 0$, $y = 2$, $y = 2z$, and $y = 2z - 1$.

45. Use a change of variables to provide a second solution to Exercise 47 of Section 14.2. (**Hint:** Using the notation of Exercise 29, notice that after transforming variables according to $x = u$, $y = 2v$, you will be able to integrate on a disk. Perform another change of variables to polar coordinates to finish the problem.)

46. Following the hint given in Exercise 45, solve Exercise 49 of Section 14.2 by changing variables twice.

47. Use the approach of Exercise 45 to derive the formula for the volume of the ellipsoid $\dfrac{x^2}{a^2} + \dfrac{y^2}{b^2} + \dfrac{z^2}{c^2} = 1$.

48.* After determining the second moment of the ellipsoid of Exercise 47 about the z-axis, show that its radius of gyration about the same is $r_z = \sqrt{\frac{1}{5}\left(a^2 + b^2\right)}$. Can you find formulas for r_x and r_y?

Chapter 14
Review Exercises

1. Estimate $\iint\limits_R x^2 y \, dA$ on the square $R = [0,3] \times [0,3]$ by using the Riemann sum approximation corresponding to $n = m = 3$, with each sample point $\left(x_{ij}^*, y_{ij}^*\right)$ chosen to be the center point of the respective subsquare, $1 \le i \le 3$ and $1 \le j \le 3$. Then use repeated integration to find the true value of the integral.

2. Use symmetry and the properties of double integrals to evaluate $\iint\limits_R (2-x) \, dA$ over the rectangle $R = \{(x,y) \mid -1 \le x \le 1, -2 \le y \le 3\}$. (Do not use repeated integration.)

3–4 Suppose $\iint\limits_{R_1} f(x,y) \, dA = 6$, $\iint\limits_{R_2} f(x,y) \, dA = -1$, $\iint\limits_{R_1} g(x,y) \, dA = 2$, and $\iint\limits_{R_1 \cup R_2} g(x,y) \, dA = 8$ over the disjoint rectangular regions R_1 and R_2. Use the properties of double integrals to evaluate the given integral.

3. $\iint\limits_{R_1 \cup R_2} \left[2f(x,y) + 3g(x,y) \right] dA$

4. $\iint\limits_{R_2} \left[3f(x,y) - g(x,y) \right] dA$

5–6 The given iterated integral represents the volume of a well-known solid. Use a formula from geometry to evaluate the integral.

5. $\int_0^3 \int_0^{1-x/3} (6 - 2x - 6y) \, dy \, dx$

6. $\int_0^2 \int_0^{\sqrt{4-y^2}} \sqrt{4 - x^2 - y^2} \, dx \, dy$

7–10 Evaluate the iterated integral.

7. $\int_0^{\pi/2} \int_0^2 x \cos y \, dx \, dy$

8. $\int_0^1 \int_0^1 \frac{y}{x^2+1} \, dy \, dx$

9. $\int_0^6 \int_{x/3}^{8-x} (xy - 1) \, dy \, dx$

10. $\int_0^1 \int_y^{\sqrt{y}} x e^y \, dx \, dy$

11–12 Rework the indicated problem by reversing the order of integration and verify that the answer does not change.

11. Exercise 9   **12.** Exercise 10

13–14 Evaluate the given double integral over the indicated region. Choose the most convenient order of integration.

13. $\iint\limits_R (x^2 - y^2) \, dA$; R is the triangle with vertices $(0,0)$, $(2,4)$, and $(6,0)$.

14. $\iint\limits_R 2xy \, dA$; R is the region bounded by $y = 3x - x^2$ and $y = x - 3$.

15–16 Evaluate the integral by reversing the order of integration. (Note that integrating in the given order would be impossible.)

15. $\int_0^1 \int_y^1 \sqrt[4]{1+x^2} \, dx \, dy$

16. $\int_0^1 \int_{\sqrt{y}}^1 e^{-x^3} \, dx \, dy$

17. By integrating the function $f(x,y) = 1$, find the area of the region R bounded by the circle $x^2 + y^2 = 1$ and the ellipse $4x^2 + y^2 = 4$.

18. Use the Domination Property to show that if $f(x,y)$ is integrable on the bounded region R, then

$$\left| \iint\limits_R f(x,y) \, dA \right| \le \iint\limits_R |f(x,y)| \, dA.$$

19–20 Find the average value of $f(x,y)$ over the region bounded by the graphs of the given equations.

19. $f(x,y) = x(x+y)$; $y = x^2$, $2y = 12 - x^2$

20. $f(x,y) = 4x^3 y$; $4y = x^3$, $y = x$, $y \ge 0$

21–22 Find the center of mass of the plane region of varying density that is bounded by the graphs of the given equations.

21. $y = 3 - \dfrac{x}{2}$, $x = 0$, $y = 0$; $\rho(x,y) = x + y^2$

22. $y = x$, $y = \sqrt{2-x}$, $y = 0$; $\rho(x,y) = x\sqrt{y}$

23–26 Determine the second moments I_x, I_y, and I_0 for the thin plate of constant density ρ modeled by the planar region R. Then find its corresponding radii of gyration.

23. R: The triangle bounded by the graph of $y = 4 - 2x$ and the coordinate axes

24. R: The square connecting the four points $(\pm a, 0)$ and $(0, \pm a)$

25. R: The first-quadrant region bounded by the graph of $y = 4 - x^2$ and the coordinate axes

26.* R: The region bounded by the graph of $y = \dfrac{1 - x^2}{1 + x^2}$ and the x-axis

27. Suppose a thin plate of constant density ρ is modeled by the region R that is outside the square of diagonal length 2 centered at the origin, but inside the unit circle, also centered at the origin. Use the Principle of Superposition to find the moment of inertia of R about the origin.

28. Suppose the thin plate of Exercise 25 has nonconstant density $\rho(x, y) = x$. Determine the second moments I_x, I_y, and I_0 and the corresponding radii of gyration.

29–30 Use a double integral in polar coordinates to find the area of the shaded region.

29.

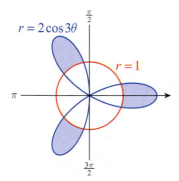

$r = 2\cos 3\theta$

$r = 1$

30.

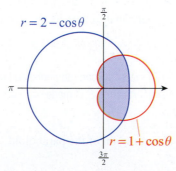

$r = 2 - \cos\theta$

$r = 1 + \cos\theta$

31–32 Use a double integral in polar coordinates to find the area of the given region R.

31. R: The region common to the circle $r = \frac{1}{2}$ and the cardioid $r = 1 - \cos\theta$

32. R: The region outside the circle $r = 1$ and inside the limaçon $r = \frac{1}{2} + \cos\theta$

33–34 Evaluate the double integral by changing to polar coordinates.

33. $\displaystyle\int_0^2 \int_0^{\sqrt{4-x^2}} \frac{dy\,dx}{\sqrt{x^2 + y^2 + 2}}$

34. $\displaystyle\int_0^1 \int_{-\sqrt{1-x^2}}^{\sqrt{1-x^2}} e^{\frac{x^2 + y^2}{2}}\, dy\,dx$

35–36 Convert the integral into a Cartesian double integral and evaluate it.

35. $\displaystyle\int_0^{\pi/3} \int_0^{\sec\theta} r^2 (\cos\theta - \sin\theta)\, dr\,d\theta$

36. $\displaystyle\int_{\pi/4}^{3\pi/4} \int_0^{4\csc\theta} r^4 \cos^2\theta \sin\theta\, dr\,d\theta$

37–38 Make your choice between the Cartesian and polar coordinate systems and evaluate the double integral.

37. $\displaystyle\iint_R \ln\sqrt{x^2 + y^2}\, dA$; $R: 1 \le x^2 + y^2 \le e$

38. $\displaystyle\iint_R 14(x^2 + y^2)^3\, dA$; R is the region bounded by $y = 2x$, $y = 2$, and the y-axis.

39–42 Use double integration in polar coordinates on an appropriate region to find the volume of the solid S bounded by the given surfaces.

39. S: The solid bounded by the xy-plane and the paraboloid $z = x^2 + y^2 - 9$

40. S: The solid bounded by the xy-plane, the plane $z = 3x + 4y + 1$, and the cylinder $x^2 + y^2 = 16$

41. S: The solid common to the paraboloids $z = 18 - x^2 - y^2$ and $z = x^2 + y^2$

42. S: The solid common to the ellipsoid $8x^2 + 8y^2 + z^2 = 36$ and the cone $z = \sqrt{x^2 + y^2}$

43–44 Evaluate the triple integral on the rectangular box S.

43. $\iiint\limits_{S} (2x + yz)\,dV$, where $S = [-2,2] \times [1,4] \times [0,5]$

44. $\iiint\limits_{S} xz^2\,dV$, where $S = [1,4] \times [-1,1] \times [2,3]$

45–46 Evaluate the iterated integral.

45. $\int_{0}^{1}\int_{0}^{1-y}\int_{0}^{3-3x-3y} (2x - y)\,dz\,dx\,dy$

46. $\int_{0}^{1}\int_{0}^{4-4x}\int_{0}^{2-\frac{y}{2}-2x} 5xy\,dz\,dy\,dx$

47–51 Use a triple integral to find the volume of the solid S.

47. S: The cylinder bounded by $y = \sqrt{1-x}$, the coordinate planes, and $z = 4$. Use the following orders of integration: **a.** $dy\,dx\,dz$ **b.** $dz\,dy\,dx$

48. S: The tetrahedron bounded by the coordinate planes and $3x + y + 2z = 6$. Use the following orders of integration: **a.** $dx\,dy\,dz$ **b.** $dy\,dz\,dx$

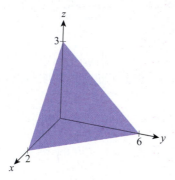

49. S: The solid bounded by the parabolic cylinder $y = 2x^2$ and the planes $z = 2 - y$ and $z = 0$. Use the following orders of integration:
a. $dz\,dy\,dx$ **b.** $dx\,dy\,dz$

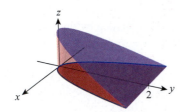

50. S: The solid bounded by the parabolic cylinder $y = x^2/2$ and the planes $2x + 4y + z = 4$ and $z = 0$. Choose the most convenient order of integration.

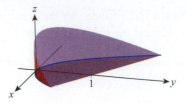

51. S: The solid bounded below by the surface $z = 3x^2 + y^2 + 2$ and above by the surface $z = 6 - x^2 - y^2$. Choose the most convenient order of integration.

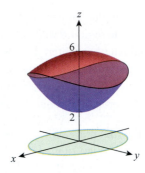

52–53 Revisit the indicated exercise, integrating in the given order. Verify that the answer does not change. (**Hint:** Sketching the solid of integration may be helpful.)

52. Exercise 45, order of integration: $dx\,dz\,dy$

53. Exercise 46, order of integration: $dy\,dz\,dx$

54–55 Use a triple integral to find the center of mass of the solid S with indicated density.

54. S: The solid of Exercise 49, with constant density

55. S: The tetrahedron bounded by the coordinate planes and $2x + y + z = 3$, its density at each point being proportional to the distance from the yz-plane

56–57 Find the center of mass, second moments, and radii of gyration of the given solid S. Assume S is made of a substance with constant density ρ.

56. S: The first-octant region of the paraboloid $z = 4 - x^2 - y^2$

57. S: The solid bounded by $z = 6 - x - 3y$ and the coordinate planes

58. Write a short paragraph about when and why you would choose one particular coordinate system over the other two from among the Cartesian, cylindrical, and spherical systems.

59–60 Find a set of cylindrical coordinates for the point given in Cartesian coordinates.

59. $\left(\sqrt{2}, -\sqrt{2}, -\sqrt{2}\right)$ **60.** $\left(-1, -\sqrt{3}, \sqrt{3}\right)$

61–62 Find a set of spherical coordinates for the point given in Cartesian coordinates.

61. $\left(\sqrt{2}, \sqrt{2}, -2\right)$ **62.** $\left(\sqrt{3}, -3, -2\right)$

63–64 Find the Cartesian coordinates of the point given in cylindrical coordinates.

63. $\left(\sqrt{2}, -\dfrac{\pi}{4}, -\dfrac{\pi}{4}\right)$ **64.** $\left(-\pi, \dfrac{\pi}{2}, 1\right)$

65–66 Find the Cartesian coordinates of the point given in spherical coordinates.

65. $\left(4\sqrt{2}, -\dfrac{\pi}{4}, \dfrac{5\pi}{6}\right)$ **66.** $\left(4, \dfrac{\pi}{3}, \dfrac{3\pi}{4}\right)$

67–68 Write the equation in cylindrical coordinates.

67. $x^2 + (y-2)^2 = z + 4$ **68.** $2z\left(x^2 + y^2\right) = y$

69–70 Write the equation in spherical coordinates.

69. $(x-1)^2 + y^2 + z^2 = 1$ **70.** $y = -\sqrt{3}x$

71–74 Change the given cylindrical or spherical equation into a rectangular one.

71. $r^2 = 1 + z$ **72.** $r \csc \theta = 3z$

73. $\varphi = \dfrac{2\pi}{3}$ **74.** $\sec \varphi + \rho = 0$

75–76 Use cylindrical coordinates to determine the volume of the solid S.

75. S: The solid bounded by the paraboloid $z = 9 - \left(x^2 + y^2\right)$ and the xy-plane

76. S: The solid bounded by the cylinder $x^2 + \left(y - \tfrac{1}{2}\right)^2 = \tfrac{1}{4}$, the cone $z = 1 - \sqrt{x^2 + y^2}$, and the xy-plane

77–78 Use the cylindrical coordinate system to find the mass and the center of mass of the solid S with the given density function.

77. S: The solid of Exercise 75; $\rho(x, y, z) = 9 - z$

78. S: The solid of Exercise 76; $\rho(x, y, z) = z$

79–80 Use a triple integral in spherical coordinates to determine the volume of the solid S.

79. S: The solid bounded by the cone $z = \sqrt{\tfrac{1}{3}\left(x^2 + y^2\right)}$ and the sphere $x^2 + y^2 + z^2 = 9$

80. S: The solid bounded by the sphere $x^2 + y^2 + z^2 = 1$ and the planes $y = x$ and $x = \sqrt{3}y$, $x > 0$

81. Find the coordinates of the center of mass of the solid of Exercise 79, assuming constant density.

82. Find the mass of the solid of Exercise 80, assuming its density at any of its points is proportional to the distance from the origin.

83–84 Use a change of variables to evaluate the double integral on the given region R.

83. $\displaystyle\iint_R \dfrac{x + 2y}{4x + 3y}\, dA$; R is the region bounded by the coordinate axes and $x + 2y = 3$.

84. $\displaystyle\iint_R \dfrac{2x + 5y}{x + 4y}\, dA$; R is the region bounded by the coordinate axes and $2x + 5y = 6$.

85–86 Find a linear transformation $T(u,v)$ that maps a rectangular region onto the given parallelogram P.

85. P is bounded by $2y = x$, $2y = x + 2$, $y = 4x$, and $y = 4x - 12$.

86. P is bounded by $y = 2x + 1$, $y = 2x + 3$, $y = 3 - x$, and $y = -x - \frac{1}{2}$.

87. Evaluate $\iint_P (7x^2 - 14y)\, dA$ on the parallelogram P of Exercise 85.

88. Evaluate $\iint_P 3x^2 y\, dA$ on the parallelogram P of Exercise 86.

89–90 Use a change of variables to evaluate the given integral on the solid R.

89. Evaluate $\iiint_R \dfrac{3y - x + 2z}{2}\, dV$, where R is the solid bounded by the planes $x = 0$, $x = 4$, $z = 0$, $z = 2$, $3y = x$, and $3y = x + 3$.

90. Evaluate $\iiint_R \left(\dfrac{y - 3x}{3} + \dfrac{5z - 2y}{4} \right) dV$, where R is the solid bounded by the planes $z = 0$, $z = 3$, $y = 3x$, $y = 3x + 4$, and $z = 2y$, $z = 2y - 6$.

91. Suppose $f(x, y, z)$ is defined on the box $R = [-a, a] \times [-b, b] \times [-c, c]$. Use the Fundamental Theorem of Calculus to prove the following:

$$\iiint_R f_{xyz}\, dV = f(a, b, c) - f(a, b, -c)$$
$$- f(a, -b, c) + f(a, -b, -c)$$
$$- f(-a, b, c) + f(-a, b, -c)$$
$$+ f(-a, -b, c) - f(-a, -b, -c)$$

92–98 *True or False?* Determine whether the given statement is true or false. In case of a false statement, explain or provide a counterexample.

92. If $a > 0$ and $f(x)$ is continuous, then
$$\int_o^a \int_o^y f(x, y)\, dx\, dy = \int_0^a \int_0^x f(x, y)\, dy\, dx.$$

93. If $f(x, y) \neq 0$ on the bounded region R, then its average value on R cannot be zero.

94. The value of a double integral should be interpreted as the volume of the solid bounded by the graph of the integrand and one of the coordinate planes.

95. In order to find $\iint_R e^{\frac{x^2 + y^2}{2}}\, dA$ on the rectangular region $R = \{(x, y) \mid |x| \leq a, |y| \leq b,\ a, b > 0\}$, the use of Cartesian coordinates is recommended.

96. In the cylindrical coordinate system, r can be negative.

97. In the spherical coordinate system, ρ can be negative.

98. If the binomial $x^2 + y^2$ is present in the integrand (or in the limits) of a triple integral, you should use cylindrical coordinates.

Chapter 14
Technology Exercises

99. Use a computer algebra system to find the center of mass of the solid of Exercise 80.

100. Write a program on your computer algebra system that performs a change from x- and y-coordinates to u- and v-coordinates and evaluates a given integral in the new coordinate system. Test your program by checking the answers you obtained for Exercises 89–90.

101. Use a computer algebra system to create the graph of the solid seen in Example 3 of Section 14.5.

Project ✎

Chapter 14

In this project you will be able to take advantage of useful coordinate transformations to evaluate multiple integrals on ellipses and ellipsoids that would be much more challenging in the Cartesian coordinate system.

1. Find the Jacobian of the coordinate transformation $T(r,\theta)$ defined by $x = ar\cos\theta$ and $y = br\sin\theta$, where $a, b > 0$.

2. Use double integration along with the coordinate transformation in Question 1 to arrive at the formula for the area A of the ellipse $\dfrac{x^2}{a^2} + \dfrac{y^2}{b^2} = 1$.

3. Find the Jacobian of the transformation to "ellipsoidal coordinates" $T_e(\rho,\theta,\phi)$ defined by $x = a\rho\sin\phi\cos\theta$, $y = b\rho\sin\phi\sin\theta$, and $z = c\rho\cos\phi$, where $a, b, c > 0$.

4. Use double integration along with the coordinate transformation of Question 3 to arrive at the formula for the volume V of the ellipsoid $\dfrac{x^2}{a^2} + \dfrac{y^2}{b^2} + \dfrac{z^2}{c^2} = 1$.

5. Use ellipsoidal coordinates to find the center of mass of the upper ellipsoid $z = c\sqrt{1 - \dfrac{x^2}{a^2} - \dfrac{y^2}{b^2}}$, assuming constant density.

6. Find the mass and the center of mass of the semiellipsoid of Question 5 in the case that the density at any point is proportional to the distance from the xy-plane.

7. Use a computer algebra system and ellipsoidal coordinates to find the second moments and radii of gyration for the solid of Question 5. Express the second moments in terms of the mass m of the semiellipsoid.

8. Use a computer algebra system and ellipsoidal coordinates to find the second moments and radii of gyration for the solid of Question 6. As in the previous problem, express the second moments in terms of the mass m of the semiellipsoid.

Chapter 15
Vector Calculus

15.1 Vector Fields 1147

1. Terms and Definitions
2. Examples of Vector Fields

15.2 Line Integrals 1156

1. Integrals along Curves in Space
2. Masses, Moments, and Line Integrals
3. Work, Flow, and Line Integrals

15.3 The Fundamental Theorem for Line Integrals 1166

1. The Fundamental Theorem for Line Integrals
2. Path Independence and Conservative Fields
3. Finding Potentials

15.4 Green's Theorem 1177

1. Divergence and Curl Operators
2. Green's Theorem

15.5 Parametric Surfaces and Surface Area 1192

1. Parametric Surfaces
2. Surface Area

15.6 Surface Integrals 1203

1. Surface Integrals of Scalar Functions
2. Orientable Surfaces
3. Surface Integrals of Vector Fields

15.7 Stokes' Theorem 1212

1. Circulation, Curl, and Stokes' Theorem
2. Consequences of Stokes' Theorem

15.8 The Divergence Theorem 1219

1. The Divergence Theorem
2. The Fundamental Theorem of Calculus in Review

Introduction

In this final chapter, we put our understanding of vectors to use as we define and study *vector fields*, develop new objects called *line integrals* and *surface integrals*, and calculate such quantities as the work done by an electric field in moving a charged particle along a curve. In doing so, we will introduce three deep results of vector calculus called *Green's Theorem, Stokes' Theorem,* and the *Divergence Theorem*.

Much of the history underlying the material in this chapter belongs to the 19th century, and the mathematicians named in two of the three theorems above make an interesting study in comparison and contrast. George Green (1793–1841) was an English mathematical physicist whose work paved the way for such physicists as James Maxwell and William Thomson (known later in life as Lord Kelvin). Green was the son of a small town miller and was almost entirely self-taught until much later in life, receiving only about one year of formal education when he was eight and nine. As a young adult he supported his family as a miller himself, and in his mid-thirties published, at his own expense, *An Essay on the Application of Mathematical Analysis to the Theories of Electricity and Magnetism*. This seminal essay introduced what we now know as Green's Theorem and many other ideas fundamental to mathematics and physics. It also brought Green to the attention of

In this final chapter, we put our understanding of vectors to use as we define and study *vector fields*, develop new objects called *line integrals* and *surface integrals*, and calculate such quantities as the work done by an electric field in moving a charged particle along a curve.

established mathematicians who eventually persuaded him to apply to Cambridge University, to which he was admitted when he was nearly forty. Upon his graduation in 1838 he was elected a Fellow at Cambridge and was able to spend two years further developing and publishing his ideas. Unfortunately his time was cut short by ill health and he left Cambridge in 1840 and died only a year later.

Green's Theorem, as presented in this text, has two formulations. One of them, the *Tangential-Curl Form*, is a special case of the more general Stokes' Theorem, named for the mathematician Sir George Stokes (1819–1903). Although born in Ireland, Stokes spent his entire long career also at Cambridge University, which he attended first as a student in 1837 and where he stayed upon graduation. He was a member of the Royal Society of London, one of the oldest and most prestigious

societies for the advancement of science, and held the Lucasian Chair of Mathematics at Cambridge for 54 years (Isaac Newton was the second Lucasian Professor, and more recently Stephen Hawking held the position from 1979 to 2009).

The other form of Green's Theorem, the *Normal-Divergence Form*, is similarly a special case of the Divergence Theorem with which we conclude the chapter. As we will see, these three deep mathematical insights are all manifestations of the philosophy underlying the Fundamental Theorem of Calculus, and are thus a fitting way to conclude the text as a whole.

Sir George Stokes
(1819–1903)

1. Adapted from illustration published in *The Popular Science Monthly*, Volume 7 (New York, 1875).

15.1 **Vector Fields**

TOPICS

1. Terms and definitions

2. Examples of vector fields

In this final chapter, we introduce vector fields and learn how the ideas of calculus apply to them. Specifically, we define and evaluate *line integrals* (which allow us to determine such things as the work done by a force field in moving an object along a curve) and *surface integrals* (which can be used to determine the rate of flow of a fluid across a surface). To do so, we use specialized versions of the Fundamental Theorem of Calculus known as the Fundamental Theorem for Line Integrals, Green's Theorem, Stokes' Theorem, and the Divergence Theorem.

TOPIC 1   **Terms and Definitions**

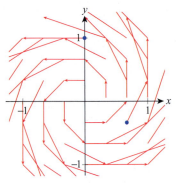

Figure 1

$\mathbf{F}(x, y) = \langle -y, x \rangle$

We begin with a definition: a **vector field** is a function that assigns a vector to each element of its domain. For our purposes, a vector field will almost always be a function whose domain is a region of $\mathbb{R}^2$ or $\mathbb{R}^3$, and we often visualize a vector field as a collection of arrows. Typically, only a representative subset of the vectors is depicted, and we graph each vector so that its initial point (or tail) coincides with its associated domain element (in contrast to position vectors, where the head is located at the point corresponding to the components and the tail is at the origin). Figure 1 is an illustration of the function $\mathbf{F}$ that assigns the vector $\langle -y, x \rangle$ to the point (x, y), which we express with the formula $\mathbf{F}(x, y) = \langle -y, x \rangle$. The portion of $\mathbb{R}^2$ shown is the square $[-1, 1] \times [-1, 1]$, but $\mathbf{F}$ is actually defined on all of $\mathbb{R}^2$. And while only a few vectors are actually depicted, we can infer how the vector field behaves in general on the basis of those shown; for illustrative purposes, the initial points of the two specific vectors $\mathbf{F}(0, 1) = \langle -1, 0 \rangle$ and $\mathbf{F}\left(\frac{2}{3}, -\frac{1}{3}\right) = \left\langle \frac{1}{3}, \frac{2}{3} \right\rangle$ are shown in blue in Figure 1.

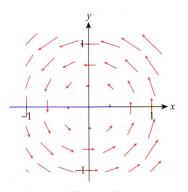

Figure 2

$\mathbf{F}(x, y) = \langle -y, x \rangle$ scaled

A word of caution is in order, however, regarding how CAS software displays vector fields. To avoid overly long vectors and/or confusing images, graphing packages usually scale the vectors downward in length. For instance, *Mathematica* depicts the same vector field $\mathbf{F}(x, y) = \langle -y, x \rangle$, using the built-in **VectorPlot** command, as shown in Figure 2 (see the Technology Note at the end of this section for further details on the use of **VectorPlot**). The relative sizes of the vectors are correct, but the actual lengths are not. Note that *Mathematica* also centers each vector at the point it is associated with, instead of placing the tail of the vector at the point.

In the case of a vector field $\mathbf{F}$ defined on a region of $\mathbb{R}^2$, each component of $\mathbf{F}$ is itself a function of two variables, so $\mathbf{F}$ has the following generic form:

$$\mathbf{F}(x, y) = \langle P(x, y), Q(x, y) \rangle = P(x, y)\mathbf{i} + Q(x, y)\mathbf{j}$$

The typeface used is a reminder of the nature of the functions: P and Q are the **component functions** of $\mathbf{F}$, and in this context are sometimes referred to as **scalar functions** or **scalar fields** since they return a single number (a scalar) at each point (x, y). Together, when placed in order as they are here, they constitute the components of the vector function $\mathbf{F}$, which is written in bold as an indication that it is indeed a vector (when written by hand, the notation $\vec{F}(x, y)$ would be used).

A vector field is said to be **continuous** if its component functions are continuous, and **differentiable** if its components are themselves differentiable. In other words, the continuity or differentiability of a vector field follows directly from the continuity or differentiability of its multivariable component functions, which we studied in Chapter 13.

Example 1 ✒

Describe the vector field $\mathbf{F}(x,y) = \dfrac{\langle x,y \rangle}{2\left|\langle x,y \rangle\right|}$ and sketch $\mathbf{F}$ with representative vectors.

Solution

Since

$$\left|\mathbf{F}(x,y)\right| = \left|\frac{\langle x,y \rangle}{2\left|\langle x,y \rangle\right|}\right| = \frac{1}{2\left|\langle x,y \rangle\right|}\left|\langle x,y \rangle\right| = \frac{1}{2},$$

each element of $\mathbf{F}$ is a vector of length $\frac{1}{2}$, and based on the formula, $\mathbf{F}$ is not defined at the origin. At every other point, $\mathbf{F}(x,y)$ is a vector pointing radially outward from the point (x,y), so representative elements of $\mathbf{F}$ are as shown in Figure 3. Since

$$\left|\langle x,y \rangle\right| = \sqrt{x^2 + y^2},$$

the component functions of $\mathbf{F}$ are as follows:

$$P(x,y) = \frac{x}{2\sqrt{x^2 + y^2}} \quad \text{and} \quad Q(x,y) = \frac{y}{2\sqrt{x^2 + y^2}}$$

These two multivariable functions are differentiable everywhere they are defined, so $\mathbf{F}$ is differentiable everywhere in the domain.

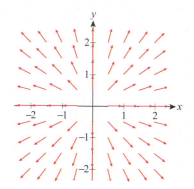

Figure 3

Example 2 ✒

Describe the vector field $\mathbf{F}(x,y,z) = \langle 0,0,1 \rangle$ and sketch $\mathbf{F}$ with representative vectors.

Solution

A vector field in $\mathbb{R}^3$ assigns a vector to each point of its domain, and in this case each point of $\mathbb{R}^3$ is associated with the vector $\langle 0,0,1 \rangle$. The component scalar functions are clearly differentiable, so $\mathbf{F}$ is as well, and Figure 4 shows some representative vectors of $\mathbf{F}$.

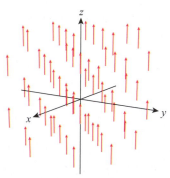

Figure 4

TOPIC 2   **Examples of Vector Fields**

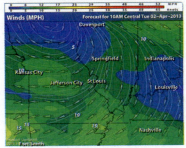

Figure 5 Wind Pattern
Source: WSI Corporation

Vector fields arise naturally in describing and working with many natural phenomena, and you are familiar with many examples under different nomenclature already. For instance, any depiction of wind speed and direction qualifies as a vector field. If the depiction is overlaid on a two-dimensional weather map such as in Figure 5, the vector field assigns a vector to each element of a planar region. A different sort of description, such as might be used to explain prevailing winds at different altitudes, assigns wind speed and direction to each point of a three-dimensional region of space (Figure 6). And diagrams of wind velocity over and under an airfoil (Figure 7), as determined in a wind tunnel, are a standard part of many design processes. As a point of contrast, a three-dimensional map assigning a temperature to each point in space is an example of a scalar field, not a vector field.

Force fields, which often appear in physics and engineering problems, also qualify as vector fields. The next few examples introduce two of the more frequently encountered force fields.

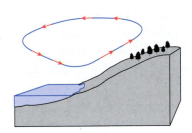

Figure 6 Wind at Different Altitudes

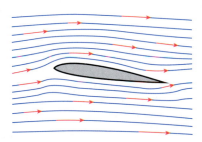

Figure 7 Wind around an Airfoil

Example 3 ✏

Newton's Law of Universal Gravitation, first seen in vector form in Section 12.4, states that the force of attraction between an object of mass m and an object of mass M is GMm/r^2, where G is a constant (called the gravitational constant) and r is the distance between the two objects. If we fix the object with mass M at the origin of a three-dimensional coordinate system and let $\mathbf{r} = \langle x, y, z \rangle$, then the force $\mathbf{F}$ felt by the object of mass m located at the point (x, y, z) is

$$\mathbf{F}(x, y, z) = -\frac{GMm}{r^3}\mathbf{r}.$$

The negative sign reflects the fact that the object of mass m is attracted radially inward toward the origin.

$$\left| \mathbf{F}(x, y, z) \right| = \left| -\frac{GMm}{r^3}\mathbf{r} \right| = \left(\frac{GMm}{r^3} \right)|\mathbf{r}| = \frac{GMm}{r^2}$$

The vector field $\mathbf{F}$, in this context, is called a **gravitational field**. Since $r = \left(x^2 + y^2 + z^2 \right)^{1/2}$, we can write $\mathbf{F}$ in component form as

$$\mathbf{F}(x, y, z) = \left\langle -\frac{GMmx}{\left(x^2 + y^2 + z^2 \right)^{3/2}}, -\frac{GMmy}{\left(x^2 + y^2 + z^2 \right)^{3/2}}, -\frac{GMmz}{\left(x^2 + y^2 + z^2 \right)^{3/2}} \right\rangle$$

$$= \frac{-GMm}{\left(x^2 + y^2 + z^2 \right)^{3/2}} \langle x, y, z \rangle.$$

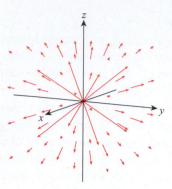

Repulsive Force Field

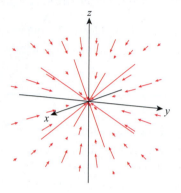

Attractive Force Field

Figure 8

Example 4 ✒

Coulomb's Law is similar to the Law of Universal Gravitation, but encompasses both attractive and repulsive forces between electrically charged objects. If an object of charge Q is fixed at the origin, the electrostatic force $\mathbf{F}(x, y, z)$ felt by an object with charge q at position (x, y, z) is the **electric field**

$$\mathbf{F}(x, y, z) = \frac{\varepsilon Qq}{r^3}\mathbf{r},$$

where again $\mathbf{r} = \langle x, y, z \rangle$ and the constant ε is called *Coulomb's constant*. If the two charges Q and q have the same sign (either both positive or both negative), then $Qq > 0$ and the force on the object of charge q is radially outward from the origin (repulsive). But if the charges have opposite sign, the objects are electrically attracted to one another and the force is directed radially inward. Figure 8 illustrates both repulsive and attractive force fields, using vectors of scaled length. The Law of Universal Gravitation and Coulomb's Law are examples of *inverse-square* laws, meaning the magnitude of the force is inversely proportional to the square of the distance between the objects.

The gradient of a scalar function is another important example of a vector field. Recall from Section 13.5 that the gradient of a function $f(x, y)$ is

$$\nabla f(x, y) = \langle f_x(x, y), f_y(x, y) \rangle,$$

with a similar definition for the gradient of a function of three variables. In other words, the partial derivatives of a scalar field constitute the component functions of a vector field. We call such vector fields **gradient vector fields**.

Example 5 ✒

Determine the gradient vector field of the scalar field $f(x, y) = xy^2 - x^2$.

Solution

The calculation is straightforward.

$$\nabla f(x, y) = \langle f_x(x, y), f_y(x, y) \rangle = \langle y^2 - 2x, 2xy \rangle$$

It is instructive to compare the gradient vector field to a contour map of f. Figure 9 shows contours of f corresponding to contour values of $-6, -3, 0, 3,$ and 6, along with a scaled illustration of the vector field ∇f. Note that the gradient vectors are perpendicular to the contour lines, as we expect, and also that the (relative) lengths of the gradient vectors correspond to how near the contour lines are to one another. Remember that the magnitude of the gradient vector of f at any given point reflects the maximum rate of change of f, so regions of the plane where contour lines are close together correspond to relatively large gradients.

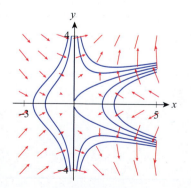

Figure 9

If a vector field **F** can be written as the gradient of a scalar field f, so that $\mathbf{F} = \nabla f$, **F** is called a **conservative vector field** and we call f a **potential function** for **F**. We will see later in this chapter that conservative vector fields have special meaning in physics, and we will also learn how to recognize whether a given vector field is conservative.

Example 6 ✐

Show that the gravitational field $\mathbf{F}(x, y, z) = -\dfrac{GMm}{r^3}\mathbf{r}$ is conservative.

Solution

While at first this may seem like a challenging task, writing **F** in component form is suggestive.

$$\mathbf{F}(x, y, z) = \left\langle -\frac{GMmx}{\left(x^2 + y^2 + z^2\right)^{3/2}}, -\frac{GMmy}{\left(x^2 + y^2 + z^2\right)^{3/2}}, -\frac{GMmz}{\left(x^2 + y^2 + z^2\right)^{3/2}} \right\rangle$$

Since

$$\frac{d}{dx}\left(\frac{1}{\sqrt{x^2 + y^2 + z^2}}\right) = -\frac{x}{\left(x^2 + y^2 + z^2\right)^{3/2}},$$

it can be seen that the function $f(x, y, z) = GMm\big/\sqrt{x^2 + y^2 + z^2}$ serves as a potential function for **F**.

Technology Note 🖥

The basic usage of the *Mathematica* command **VectorPlot** requires only a vector field $\mathbf{F}(x, y)$ and the region of the xy-plane to be displayed. For example, the vector field $\mathbf{F}(x, y) = \langle -y, x \rangle$ considered at the beginning of this section can be graphed over the region $[-1, 1] \times [-1, 1]$ as shown in Figure 10.

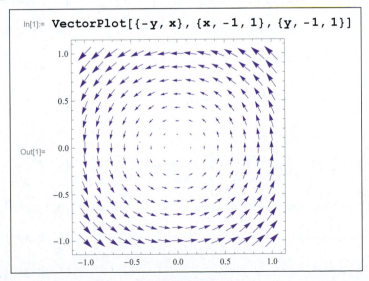

```
In[1]:= VectorPlot[{-y, x}, {x, -1, 1}, {y, -1, 1}]
```

Figure 10

To obtain an illustration more like Figure 2, additional options must be included to instruct *Mathematica* on the placement of tick marks on the axes, the number of vectors to draw, the style of the vectors, and so on, with the result shown in Figure 11.

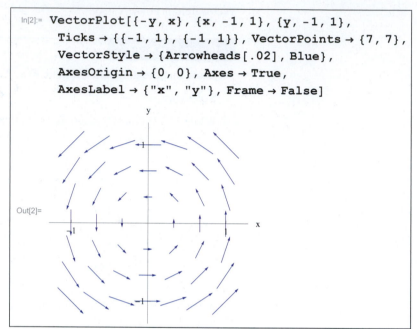

```
In[2]:= VectorPlot[{-y, x}, {x, -1, 1}, {y, -1, 1},
        Ticks → {{-1, 1}, {-1, 1}}, VectorPoints → {7, 7},
        VectorStyle → {Arrowheads[.02], Blue},
        AxesOrigin → {0, 0}, Axes → True,
        AxesLabel → {"x", "y"}, Frame → False]
```

Out[2]=

Figure 11

15.1 **Exercises**

1–6 Identify the real-life function as a vector field or a scalar field.

1. Fluid pressure in a swimming pool

2. Velocity of the wind inside a hurricane

3. Electromagnetic force around a coil

4. Air temperature near a working hair dryer

5. Velocity of water flowing through a pipe

6. CO concentration around a barbecue grill

7–10 Evaluate the vector field at the given point.

7. $\mathbf{F}(x, y) = \langle x + y, -2x \rangle;\quad (1, -1)$

8. $\mathbf{F}(x, y) = \langle xe^x, xy^2 + 1 \rangle;\quad (2, 1)$

9. $\mathbf{F}(x, y, z) = \left\langle z, 1, \sqrt{x^2 + y^2 + z^2} \right\rangle;\quad (3, 4, 0)$

10. $\mathbf{F}(x, y, z) = \left\langle x - z, \dfrac{xy}{z^2}, yz \right\rangle;\quad (1, 1, 2)$

11–14 Match the given two-dimensional (planar) vector field with its graph (labeled A–D).

11. $\mathbf{F}(x, y) = \langle y, -x \rangle$

12. $\mathbf{F}(x, y) = \langle x, -y \rangle$

13. $\mathbf{F}(x, y) = \langle x, e^y \rangle$

14. $\mathbf{F}(x, y) = \dfrac{\langle x - 1, y \rangle}{|\langle x - 1, y \rangle|}$

A.

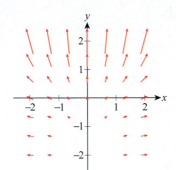

B.

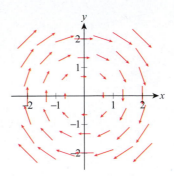

C.

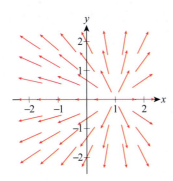

D.

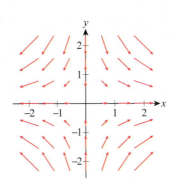

15–23 Sketch the vector field **F** by hand, using a sufficient number of representative vectors.

15. $\mathbf{F}(x, y) = \langle 1, 0 \rangle$

16. $\mathbf{F}(x, y) = \langle 0, y \rangle$

17. $\mathbf{F}(x, y) = \langle -4x, 0 \rangle$

18. $\mathbf{F}(x, y) = \dfrac{\langle -y, x \rangle}{|\langle -y, x \rangle|}$

19. $\mathbf{F}(x, y) = \dfrac{\langle x, -y \rangle}{|\langle x, -y \rangle|}$

20. $\mathbf{F}(x, y) = \langle 1, y - 1 \rangle$

21. $\mathbf{F}(x, y) = \langle y + 2, 1 - x \rangle$

22. $\mathbf{F}(x, y) = \langle x, y^2 \rangle$

23. $\mathbf{F}(x, y) = \dfrac{\langle x, y \rangle}{|\langle x, y \rangle|}$

24–27 Match the given three-dimensional vector field with its graph (labeled A–D).

24. $\mathbf{F}(x, y, z) = \langle 1, 0, 1 \rangle$

25. $\mathbf{F}(x, y, z) = \langle 1, 1, z \rangle$

26. $\mathbf{F}(x, y, z) = \langle x, y, 1 \rangle$

27. $\mathbf{F}(x, y, z) = \dfrac{\langle y, -x, 0 \rangle}{2 - x}$

A.

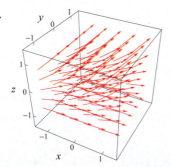

B.

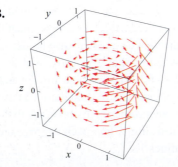

C.

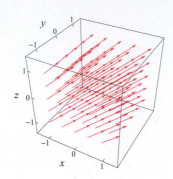

D.

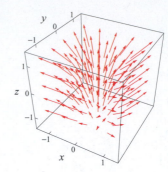

28–36 Determine ∇f for the given scalar field f.

28. $f(x,y) = \dfrac{x^2 y}{2} + xy^3$ **29.** $f(x,y) = \ln\sqrt{x^2 + y^2}$

30. $f(x,y) = \dfrac{1}{\sqrt{x^2 + y^2}}$ **31.** $f(x,y) = y^2 \tan^{-1} x$

32. $f(x,y) = ye^{x+y}$ **33.** $f(x,y,z) = ze^{xyz}$

34. $f(x,y,z) = y^2 z - 2x^2 z^2$

35. $f(x,y,z) = \dfrac{z}{\sqrt{x^2 + y^2 + z^2}}$

36. $f(x,y,z) = \dfrac{xy}{y - z}$

37. By determining a potential function f, show that the electric field $\mathbf{F}(x,y,z) = \dfrac{\varepsilon Q q}{r^3}\mathbf{r}$ is conservative (i.e., $\mathbf{F} = \nabla f$ for some potential function f).

38. Prove that if the vector field $\mathbf{F} = \langle P, Q, R\rangle$ is conservative (i.e., $\mathbf{F} = \nabla f$ for some potential function f), then $\mathbf{F}$ satisfies what is sometimes called the *Component Test*, that is,

$$\frac{\partial P}{\partial y} = \frac{\partial Q}{\partial x}, \quad \frac{\partial Q}{\partial z} = \frac{\partial R}{\partial y}, \quad \frac{\partial R}{\partial x} = \frac{\partial P}{\partial z}.$$

(As we will see in Section 15.7, if the domain of $\mathbf{F}$ satisfies certain regularity conditions, then the converse of the above statement also holds. **Hint:** Start by noting that $P = \partial f/\partial x$ and $Q = \partial f/\partial y$. Use this to write both $\partial P/\partial y$ and $\partial Q/\partial x$ as mixed partials of f, and use Clairaut's Theorem (see Section 13.3) to obtain the first statement. Use a similar approach to prove the last two equalities.)

39. Formulate and prove a result analogous to the one in Exercise 38 for the vector field $\mathbf{F} = \langle P, Q\rangle$.

40–47 Use Exercises 38 and 39 to decide whether the given vector field is conservative.

40. $\mathbf{F}(x,y) = \langle y^2, 2xy\rangle$ **41.** $\mathbf{F}(x,y) = \langle y, xy\rangle$

42. $\mathbf{F}(x,y) = \langle xe^y, ye^x\rangle$

43. $\mathbf{F}(x,y) = \langle y^2\cos(xy^2), 2xy\cos(xy^2)\rangle$

44. $\mathbf{F}(x,y,z) = \langle 1, xy, 0\rangle$

45. $\mathbf{F}(x,y,z) = \langle y^2 z, 2xyz, xy^2\rangle$

46. $\mathbf{F}(x,y,z) = \dfrac{\langle x, y, z\rangle}{\sqrt{x^2 + y^2 + z^2}}$

47. $\mathbf{F}(x,y,z) = \left\langle \dfrac{z}{\sqrt{xy}}, -\dfrac{x}{\sqrt{xy}}, \sqrt{xyz}\right\rangle$

48–53 Consider $\mathbf{F}(x,y) = \langle 12xy, 6x^2 + 2\rangle$, a conservative vector field. We can find a potential function f by integrating, as follows. First, note that we must have $\partial f/\partial x = 12xy$, so we integrate with respect to x to obtain $f(x,y) = 6x^2 y + g(y)$, where g is a function of y (or a constant). On the other hand, we know $\partial f/\partial y = 6x^2 + 2$, so differentiating with respect to y the function f we obtained above, we see that $6x^2 + 2 = \partial f/\partial y = 6x^2 + g'(y)$, that is, it must be the case that $g'(y) = 2$. This implies $g(y) = 2y + C$, hence $f(x,y) = 6x^2 y + 2y + C$ is a potential function for $\mathbf{F}$ under any choice of the constant C.

Use the above technique to find a potential function for the given conservative vector field. (In Exercises 52–53, generalize to three variables.)

48. $\mathbf{F}(x,y) = \langle 4y^3, 12xy^2\rangle$

49. $\mathbf{F}(x,y) = \left\langle y^2 e^{xy^2}, 2xye^{xy^2} \right\rangle$

50. $\mathbf{F}(x,y) = \left\langle 3x^2 - 2y^2, 2y - 4xy \right\rangle$

51. $\mathbf{F}(x,y) = \left\langle \dfrac{x}{\sqrt{x^2 + y^2}}, \dfrac{y}{\sqrt{x^2 + y^2}} \right\rangle$

52. $\mathbf{F}(x,y,z) = \left\langle y^2, 2xy, 2z \right\rangle$

53. $\mathbf{F}(x,y,z) = \left\langle 2x + z, 2z - 1, x + 2y \right\rangle$

54. Using the notation in Example 3, let $\mathbf{r} = \langle x, y \rangle$ and $r = |\langle x, y \rangle|$, and describe the vector field

$$\mathbf{e}_r = \frac{\langle x, y \rangle}{|\langle x, y \rangle|} = \frac{\mathbf{r}}{r}$$

as well as its three-dimensional analogue (also denoted $\mathbf{e}_r$). See Example 1 for guidance.

55. Show that $\nabla r = \mathbf{e}_r$ holds both in two and three dimensions. (Note that you have already provided a constructive existence proof for the potential function of $\mathbf{e}_r$ in the two-dimensional case in Exercise 54.)

56. Find formulas for ∇r^2 and ∇r^3.

57. Find and prove a formula for ∇r^n $(n \in \mathbb{N})$.

58. Show that the vector field $\mathbf{r}/r^2$ is conservative by determining a potential function. (Handle both the planar and three-dimensional cases.)

59. Repeat Exercise 58 to find a potential function for the vector field $\mathbf{r}/r^3$. (Compare your finding with the results in Example 6 and Exercise 37.)

60.* Generalize Exercise 59 by determining a formula for the potential function of the vector field $\mathbf{r}/r^n$, $n \in \mathbb{N}$. Prove your assertion.

61–62 The *flow lines* of a vector field $\mathbf{F}$ are "paths aligned with $\mathbf{F}$," more precisely, paths whose *velocity field* is $\mathbf{F}$ when followed by a particle. Even more precisely, a flow line is a path $\mathbf{r}(t)$ such that $\mathbf{r}'(t) = \mathbf{F}(\mathbf{r}(t))$. In these exercises, you will determine the flow lines of vector fields.

61. By visualizing its graph, try to predict what the flow lines of the vector field $\mathbf{F}(x,y) = \langle y, -x \rangle$ might look like (see Exercise 11). Then use a differential equation to determine the equations of these flow lines. (**Hint:** Note that if $\mathbf{r}(t) = \langle x(t), y(t) \rangle$ is a flow line of $\mathbf{F}$, then $dx/dt = y$ and $dy/dt = -x$. Conclude that $dy/dx = -x/y$ and solve the differential equation. For a refresher on separable differential equations, see Section 8.1.)

62. Repeat Exercise 61 for the vector field of Exercise 12.

15.2 Line Integrals

TOPICS

1. Integrals along curves in space

2. Masses, moments, and line integrals

3. Work, flow, and line integrals

The integral of a function over a line in space is a generalization of the integral over an interval, and such so-called *line integrals* are used in solving problems involving moments, fluid flow, work, and forces. The word "line" in the terminology is conventional and reflects the fact that the integral is defined over an object describable with a single parameter but, as we will see, *curve integral* more accurately captures the concept.

TOPIC 1 Integrals along Curves in Space

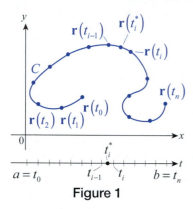

Figure 1

Suppose C is a curve in space (either $\mathbb{R}^2$ or $\mathbb{R}^3$) defined by a vector function $\mathbf{r}(t)$ over the interval $a \leq t \leq b$; that is, C is described by a function of the form $\mathbf{r}(t) = \langle x(t), y(t) \rangle$ or $\mathbf{r}(t) = \langle x(t), y(t), z(t) \rangle$. Assume also that C is smooth as defined in Section 12.1, meaning that $\mathbf{r}'$ is continuous and never zero on $[a, b]$. To define an integral over the curve, consider a partition P of C such as shown in Figure 1, defined by a corresponding partition $a = t_0 < t_1 < \cdots < t_{n-1} < t_n = b$ of the interval $[a, b]$. For each subinterval $[t_{i-1}, t_i]$, let Δs_i be the arc length of the portion of C between $\mathbf{r}(t_{i-1})$ and $\mathbf{r}(t_i)$. Then for any function f whose domain contains C, we can form the Riemann sum

$$\sum_{i=1}^{n} f\left(\mathbf{r}(t_i^*)\right) \Delta s_i,$$

where t_i^* is a sample point chosen from the subinterval $[t_{i-1}, t_i]$. If, for a given function f, each such Riemann sum approaches the same value as the partitions get finer and finer, we define that value to be the line integral of f along C.

Definition 💡

Line Integral

Given a function f defined on a curve C described by the smooth vector function $\mathbf{r}(t)$, $a \leq t \leq b$, the **line integral of f along C** is the value

$$\int_C f \, ds = \lim_{\|P\| \to 0} \sum_{i=1}^{n} f\left(\mathbf{r}(t_i^*)\right) \Delta s_i,$$

provided the limit exists.

While the formal definition can be used as the basis for numerical approximations of line integrals, we also desire techniques to evaluate $\int_C f \, ds$ exactly, if possible. The most common formula uses the relationship, developed in Section 12.2, between the arc length differential ds and parameter differential dt.

$$\frac{ds}{dt} = |\mathbf{r}'(t)| \quad \Rightarrow \quad ds = |\mathbf{r}'(t)| \, dt$$

This allows us to express $\int_C f \, ds$ as an integral in the variable t over the interval $[a, b]$.

$$\int_C f \, ds = \int_a^b f\left(\mathbf{r}(t)\right) |\mathbf{r}'(t)| \, dt$$

Example 1 ✐

Determine $\int_C f\,ds$, where $f(x,y) = y - xy^2$ and C is defined as the upper half of the circle $x^2 + y^2 = 1$ traversed counterclockwise.

Solution

We begin by finding a smooth parametrization $\mathbf{r}(t)$ of C, if possible. The curve C is certainly smooth, and we know that the circle of radius 1 can be parametrized as $\mathbf{r}(t) = \langle \cos t, \sin t \rangle$. To describe the upper half of the circle traversed counterclockwise, we use the t-interval $[0, \pi]$ (see Figure 2). And since $\mathbf{r}'(t) = \langle -\sin t, \cos t \rangle$, $\left| \mathbf{r}'(t) \right| = 1$.

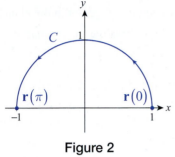

On C,

$$f(x,y) = y - xy^2 = \sin t - \cos t \sin^2 t,$$

so

$$
\begin{aligned}
\int_C f\,ds &= \int_a^b f\left(\mathbf{r}(t)\right)\left|\mathbf{r}'(t)\right|dt \\
&= \int_0^\pi \left(\sin t - \cos t \sin^2 t\right)(1)\,dt \\
&= \left[-\cos t\right]_0^\pi - \int_0^\pi \left(\cos t \sin^2 t\right)dt && \begin{array}{l} u = \sin t \\ du = \cos t\,dt \end{array} \\
&= \left[-\cos t - \frac{\sin^3 t}{3}\right]_0^\pi = 2.
\end{aligned}
$$

Figure 2

Example 2 ✐

Determine $\int_C f\,ds$, where $f(x,y,z) = xy + z$ and C is defined as the line segment starting at $(-1,2,5)$ and ending at $(2,0,1)$.

Solution

The vector $\langle 2,0,1 \rangle - \langle -1,2,5 \rangle = \langle 3,-2,-4 \rangle$ points in the direction from the initial point to the terminal point, so using the technique of Section 11.5 we can describe C with the following parametrization:

$$\mathbf{r}(t) = \langle -1,2,5 \rangle + t\langle 3,-2,-4 \rangle = \langle -1+3t, 2-2t, 5-4t \rangle, \quad 0 \le t \le 1$$

Alternatively, the line segment between $(-1,2,5)$ and $(2,0,1)$ can be described as a linear combination of the position vectors $\langle -1,2,5 \rangle$ and $\langle 2,0,1 \rangle$.

$$\mathbf{r}(t) = (1-t)\langle -1,2,5 \rangle + t\langle 2,0,1 \rangle = \langle -1+3t, 2-2t, 5-4t \rangle$$

Either derivation results in $\left| \mathbf{r}'(t) \right| = \sqrt{29}$ and

$$f\left(\mathbf{r}(t)\right) = (-1+3t)(2-2t) + (5-4t) = -6t^2 + 4t + 3,$$

so

$$\int_C f\,ds = \int_a^b f\left(\mathbf{r}(t)\right)\left|\mathbf{r}'(t)\right|dt = \int_0^1 \left(-6t^2 + 4t + 3\right)\sqrt{29}\,dt = 3\sqrt{29}.$$

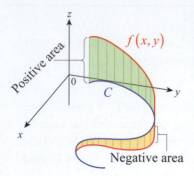

Figure 3

If C is a curve in the xy-plane, we can interpret $\int_C f\,ds$ in a manner similar to the geometric interpretation of an integral over an interval. As illustrated in Figure 3, we can think of $f(x,y)$ as describing the height or depth of a "fence" running above or below the curve C. With the understanding that negative values of f correspond to negative contributions to the integral, $\int_C f\,ds$ is the signed area of the fence along C. Indeed, if C happens to be a straight line segment in the plane, this interpretation of $\int_C f\,ds$ exactly coincides with our familiar interpretation of integrals over intervals as signed areas.

Other characteristics of integrals over intervals also carry over to line integrals. If C_1 and C_2 are smooth curves such that the terminal point of C_1 coincides with the initial point of C_2, and if f is integrable over each of C_1 and C_2, then f is also integrable over the *piecewise smooth* path $C_1 \cup C_2$ and

$$\int_{C_1 \cup C_2} f\,ds = \int_{C_1} f\,ds + \int_{C_2} f\,ds.$$

This is analogous to the property $\int_a^c f(x)\,dx = \int_a^b f(x)\,dx + \int_b^c f(x)\,dx$, and extends to any finite number of smooth curves joined together.

TOPIC 2 ## Masses, Moments, and Line Integrals

If an object in space can be modeled as a curve C with density function ρ, its mass, first and second moments, center of mass, and radii of gyration can all be expressed as line integrals.

Definition 💡

Mass, First and Second Moments, Center of Mass, and Radii of Gyration

Given an object in space modeled by the curve C and whose density is represented by ρ, its **mass** M, its **first moments** M_{yz}, M_{xz}, and M_{xy} about the coordinate planes, its **second moments** I_x, I_y, and I_z about the coordinate axes, its **center of mass** $(\overline{x}, \overline{y}, \overline{z})$, and its **radii of gyration** r_x, r_y, and r_z are as follows, assuming the integrals exist.

Mass	$M = \int_C \rho\,ds$
First moment about the yz-plane	$M_{yz} = \int_C x\rho\,ds$
First moment about the xz-plane	$M_{xz} = \int_C y\rho\,ds$
First moment about the xy-plane	$M_{xy} = \int_C z\rho\,ds$
Center of mass	$\overline{x} = \dfrac{M_{yz}}{M}, \overline{y} = \dfrac{M_{xz}}{M}$, and $\overline{z} = \dfrac{M_{xy}}{M}$

Second moment about the *x*-axis $\quad I_x = \int_C \left(y^2 + z^2 \right) \rho \, ds$

Second moment about the *y*-axis $\quad I_y = \int_C \left(x^2 + z^2 \right) \rho \, ds$

Second moment about the *z*-axis $\quad I_z = \int_C \left(x^2 + y^2 \right) \rho \, ds$

Radii of gyration $\qquad\qquad r_x = \sqrt{\dfrac{I_x}{M}},\, r_y = \sqrt{\dfrac{I_y}{M}},\, \text{and } r_z = \sqrt{\dfrac{I_z}{M}}$

Example 3 ✎

Determine the mass, first moments, and center of mass of the V-shaped object in the *xz*-plane shown in Figure 4, constructed from two rods of constant linear density $\rho = 1$.

Solution

We will let C denote the given object, which consists of the line segment C_1 from $(-1,0,1)$ to $(0,0,0)$ joined to the line segment C_2 from $(0,0,0)$ to $(1,0,1)$. C_1 can be parametrized as $\mathbf{r}_1(t) = \langle t-1, 0, 1-t \rangle$ and C_2 by $\mathbf{r}_2(t) = \langle t, 0, t \rangle$, both over $0 \le t \le 1$.

$$M = \int_C \rho \, ds = \int_{C_1} ds + \int_{C_2} ds = \int_0^1 |\mathbf{r}'(t)| \, dt + \int_0^1 |\mathbf{r}_2'(t)| \, dt$$

$$= \int_0^1 |\langle 1, 0, -1 \rangle| \, dt + \int_0^1 |\langle 1, 0, 1 \rangle| \, dt = 2\sqrt{2}$$

$$M_{xy} = \int_C z\rho \, ds = \int_{C_1} (1-t) \, ds + \int_{C_2} t \, ds = \int_0^1 (1-t)\sqrt{2} \, dt + \int_0^1 t\sqrt{2} \, dt = \sqrt{2}$$

These lead to $\bar{z} = M_{xy} / M = \frac{1}{2}$. By symmetry, $M_{yz} = M_{xz} = 0$ and $\bar{x} = \bar{y} = 0$.

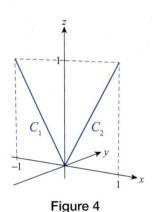

Figure 4

TOPIC 3 Work, Flow, and Line Integrals

Line integrals are especially useful in determining such quantities as work done by a force field moving a particle along a curve or the flow of a substance along a curve.

Suppose $\mathbf{F}$ represents a continuous force field (such as a gravitational or electric field) in a region of space containing a curve C, and that C is traced out by the smooth vector function $\mathbf{r}(t)$ over the interval $[a, b]$. As before, partition $[a, b]$ with points $a = t_0 < t_1 < \cdots < t_{n-1} < t_n = b$. From Section 11.3 we know that $\mathbf{F} \cdot \mathbf{D}$ is the work done by force $\mathbf{F}$ over distance $|\mathbf{D}|$ applied in the direction of $\mathbf{D}$, so the work done by $\mathbf{F}$ in moving a particle along the curve C from $\mathbf{r}(t_{i-1})$ to $\mathbf{r}(t_i)$ is approximately

$$\mathbf{F}\left(\mathbf{r}\left(t_i^* \right) \right) \cdot \underbrace{\mathbf{T}\left(\mathbf{r}\left(t_i^* \right) \right) \Delta s_i}_{\mathbf{D}},$$

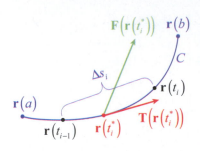

Figure 5

where $\mathbf{T}\big(\mathbf{r}(t_i^*)\big)$ is the unit vector tangent to C at the point $\mathbf{r}(t_i^*)$, t_i^* is a sample point in $[t_{i-1}, t_i]$, and Δs_i is the arc length of C from $\mathbf{r}(t_{i-1})$ to $\mathbf{r}(t_i)$ (see Figure 5). Hence the total amount of work W done by $\mathbf{F}$ in moving a particle along C is approximately

$$W \approx \sum_{i=1}^{n} \mathbf{F}\big(\mathbf{r}(t_i^*)\big) \cdot \mathbf{T}\big(\mathbf{r}(t_i^*)\big) \Delta s_i,$$

leading to the following line integral in the limit of finer and finer partitions:

$$W = \int_C \mathbf{F} \cdot \mathbf{T}\, ds$$

At this point we are back on familiar ground. The integrand $\mathbf{F} \cdot \mathbf{T}$ is a scalar function, and we integrate it with respect to arc length over C to determine the total work. But in practice, we typically use another relationship from Section 12.2 to actually evaluate the integral. Recall that the unit tangent vector $\mathbf{T}$ can be expressed as the derivative of $\mathbf{r}$ with respect to arc length s.

$$\mathbf{T} = \frac{d\mathbf{r}}{ds} \quad \Rightarrow \quad d\mathbf{r} = \mathbf{T}\, ds$$

This allows us to write the work formula as

$$W = \int_C \mathbf{F} \cdot d\mathbf{r},$$

or, expressing the integral in terms of the parameter t,

$$W = \int_a^b \mathbf{F}\big(\mathbf{r}(t)\big) \cdot \frac{d\mathbf{r}}{dt}\, dt.$$

Further expressions, emphasizing the component functions of $\mathbf{F}$, are commonly seen. For the moment, assume $\mathbf{F}$ is a three-dimensional vector field, say

$$\mathbf{F}(x, y, z) = \big\langle P(x, y, z), Q(x, y, z), R(x, y, z) \big\rangle.$$

Then

$$\mathbf{F}\big(\mathbf{r}(t)\big) \cdot \frac{d\mathbf{r}}{dt} = \langle P, Q, R \rangle \cdot \left\langle \frac{dx}{dt}, \frac{dy}{dt}, \frac{dz}{dt} \right\rangle = P\frac{dx}{dt} + Q\frac{dy}{dt} + R\frac{dz}{dt},$$

which gives rise to the formula

$$W = \int_a^b \left(P\frac{dx}{dt} + Q\frac{dy}{dt} + R\frac{dz}{dt} \right) dt,$$

frequently written more compactly as

$$W = \int_C P\, dx + Q\, dy + R\, dz.$$

Since integrals of vector fields occur frequently in other contexts as well, we generalize the derivations above with a definition.

Definition 💡

Line Integral of a Vector Field

Assume $\mathbf{F}$ is a continuous vector field whose domain contains the curve C defined by the smooth vector function $\mathbf{r}(t)$, with t taking on values between a and b. Then the **line integral of F along C** is as follows:

$$\int_C \mathbf{F} \cdot \mathbf{T}\,ds = \int_C \mathbf{F} \cdot d\mathbf{r} = \int_a^b \mathbf{F}\big(\mathbf{r}(t)\big) \cdot \frac{d\mathbf{r}}{dt}\,dt = \int_a^b \mathbf{F}\big(\mathbf{r}(t)\big) \cdot \mathbf{r}'(t)\,dt$$

Further, if $\mathbf{F}$ is a vector field in $\mathbb{R}^2$ or $\mathbb{R}^3$ and has component form $\mathbf{F} = \langle P, Q \rangle$ or $\mathbf{F} = \langle P, Q, R \rangle$, and if $\mathbf{r}(t) = \langle x(t), y(t) \rangle$ or $\mathbf{r}(t) = \langle x(t), y(t), z(t) \rangle$, respectively, the line integral can be evaluated as

$$\int_C \mathbf{F} \cdot d\mathbf{r} = \int_a^b \left(P\frac{dx}{dt} + Q\frac{dy}{dt} \right) dt = \int_C P\,dx + Q\,dy$$

or

$$\int_C \mathbf{F} \cdot d\mathbf{r} = \int_a^b \left(P\frac{dx}{dt} + Q\frac{dy}{dt} + R\frac{dz}{dt} \right) dt = \int_C P\,dx + Q\,dy + R\,dz.$$

Example 4 ✎

Determine the work done by the force field $\mathbf{F}(x, y) = \langle 5xy, -y^2 \rangle$ in moving a particle clockwise along the quarter circle shown in Figure 6.

Solution

The force field $\mathbf{F}$ is depicted in Figure 6 by the red vectors (shown scaled, not to actual length). We know that the circle of radius 1 can be parametrized by the vector function $\mathbf{r}(t) = \langle \cos t, \sin t \rangle$, but to describe the path C taken by the particle we let $a = \pi/2$ and $b = 0$. We now proceed to express the work as an integral in t.

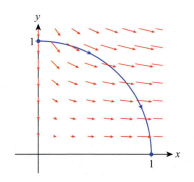

Figure 6

$$\mathbf{F}\big(\mathbf{r}(t)\big) = \langle 5\cos t \sin t, -\sin^2 t \rangle$$

$$\mathbf{r}'(t) = \langle -\sin t, \cos t \rangle$$

$$
\begin{aligned}
W = \int_C \mathbf{F} \cdot d\mathbf{r} &= \int_a^b \mathbf{F}\big(\mathbf{r}(t)\big) \cdot \mathbf{r}'(t)\,dt \\
&= \int_{\pi/2}^0 \left(-5\cos t \sin^2 t - \sin^2 t \cos t \right) dt = \int_{\pi/2}^0 -6\cos t \sin^2 t\,dt \qquad \begin{array}{l} u = \sin t \\ du = \cos t\,dt \end{array} \\
&= -6\int_1^0 u^2\,du = 2
\end{aligned}
$$

Note that the orientation of the path C is important. If we had instead evaluated the integral

$$\int_0^{\pi/2} -6\cos t \sin^2 t\,dt,$$

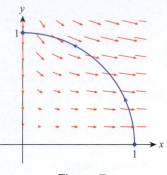

Figure 7

in Example 4, we would have arrived at a work value of −2. A negative result in this context means that the particle was moved against the force of the vector field (in other words, work was done to oppose the vector field), as illustrated in Figure 7. Symbolically, if we let −C denote the reverse orientation of a curve C, then

$$\int_{-C} \mathbf{F} \cdot d\mathbf{r} = -\int_{C} \mathbf{F} \cdot d\mathbf{r}.$$

If a given vector field **F** represents the velocity of a fluid in a region, **F** is a **velocity field**. The line integral of **F** along C may then be called a **flow integral** and its value represents the **flow** of the fluid along the curve. If C happens to be a closed loop, the flow is called the **circulation** of the fluid around the curve.

Example 5 ✎

Given a fluid velocity field defined by the gradient of the potential function $f(x, y, z) = xyz$, determine the fluid's flow along the curve $\mathbf{r}(t) = \langle t, t^2, t^3 \rangle$ from the point $(0,0,0)$ to the point $(1,1,1)$.

Solution

We start by finding the velocity field explicitly.

$$\mathbf{F}(x, y, z) = \nabla f = \langle yz, xz, xy \rangle$$

This vector field is depicted in Figure 8, with the vectors again shown scaled. To determine the flow, we compute the line integral of **F** along the curve, as in Example 4.

$$\mathbf{F}(\mathbf{r}(t)) = \langle t^2 \cdot t^3, t \cdot t^3, t \cdot t^2 \rangle = \langle t^5, t^4, t^3 \rangle$$

$$\mathbf{r}'(t) = \langle 1, 2t, 3t^2 \rangle$$

$$\text{Flow} = \int_C \mathbf{F} \cdot d\mathbf{r} = \int_a^b \mathbf{F}(\mathbf{r}(t)) \cdot \mathbf{r}'(t)\, dt = \int_0^1 6t^5\, dt = 1$$

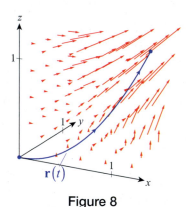

Figure 8

15.2 Exercises

1. Determine $\int_C xy^2\, ds$, where C is the first quadrant portion of the circle $x^2 + y^2 = 4$ traversed counterclockwise. (See Example 1 for a hint on parametrization.)

2. Show that you obtain the same answer as that in Exercise 1 if you parametrize the quarter circle as $x = t$, $y = \sqrt{4 - t^2}$, $0 \le t \le 2$.

3–7 Evaluate the indicated line integral.

3. $\int_C (xy + 1)\, ds$, where C is the lower semicircle $y = -\sqrt{9 - x^2}$, traversed counterclockwise

4. $\int_C y\, ds$, where C is the graph of $y = \sqrt{x}$, traversed from $(0,0)$ to $(1,1)$

5. $\int_C (2x+y)\,ds$, where C is the line segment from the origin to the point $(1,\sqrt{3})$ followed by the arc of the circle $x^2 + y^2 = 4$ traversed counterclockwise from $(1,\sqrt{3})$ to $(-2,0)$

6. $\int_C (x+3y)\,ds$, where C is the line segment from $(0,1)$ to $(-2,2)$, followed by the arc of the circle $x^2 + y^2 = 8$ traversed clockwise from $(-2,2)$ to $(2\sqrt{2},0)$

7. $\int_C \sqrt{1+18xy}\,ds$, where C is the graph of $y = 2x^3$, $0 \le x \le 1$

8. If a piece of wire is bent into the semicircle $y = \sqrt{16-x^2}$ and its density function is $\rho(x,y) = 2x^2 + y^2$, find the mass and center of mass of the wire. (See Example 1 for a hint on parametrization.)

9. Repeat Exercise 8 for a wire that is bent into the upper semicircle of radius R, $y = \sqrt{R^2 - x^2}$, so that the density at any point is proportional to the distance from the line $y = R$.

10. Find the mass of the wire in Exercise 8 by using the parametrization $x = 4\cos(t^2)$, $y = 4\sin(t^2)$, $0 \le t \le \sqrt{\pi}$. Verify that you obtain the same answer as in Exercise 8.

11. Find the mass of the wire bent into a parabolic arc $y = \sqrt{x}$, $1 \le x \le 3$, if its density is $\rho(x,y) = 2x/y$.

12–17 Evaluate the indicated line integral.

12. $\int_C (x+2y+z^2)\,ds$, where C is the line segment joining the origin and the point $(1,2,3)$

13. Integrate the function given in Exercise 12 along the path that is the line segment joining the origin with the point $(1,2,0)$ followed by the segment from $(1,2,0)$ to $(1,2,3)$. Is your answer equal to that given for Exercise 12?

14. $\int_C (2x+yz)\,ds$, where C is the line segment joining $(1,2,0)$ and $(3,4,1)$

15. $\int_C (y+16)\,ds$, where C can be parametrized as $\mathbf{r}(t) = \langle \frac{1}{2}t^2, 2t, \frac{8}{3}t^{3/2} \rangle$, $0 \le t \le 2$

16. $\int_C (3x^2 y + z)\,ds$, where C is the helix $\mathbf{r}(t) = \langle \cos t, \sin t, t/2 \rangle$, for $0 \le t \le 8\pi$

17. $\int_C y e^z\,ds$, where C is the helix given in Exercise 16

18. Use a line integral to find the length of the helix given in Exercise 16.

19. Find the mass of the helix given in Exercise 16 if its density is proportional to the distance from the xy-plane.

20. Find the center of mass of the helix given in Exercise 19.

21. Evaluate $\int_C x\,ds$ on the curve C parametrized by $\mathbf{r}(t) = \langle \cos t, \sin t, \cos t \rangle$, $0 \le t \le \pi/2$.

22. Determine the moments of inertia and radii of gyration for the object in Example 3.

23. Determine the mass and the center of mass of the V-shaped object in Example 3 if its density is proportional to the distance from the xy-plane.

24. Determine the moments of inertia and radii of gyration for the object in Exercise 23.

25. Find the mass of the spring that is defined by $\mathbf{r}(t) = \langle t, 3\cos t, 3\sin t \rangle$, $0 \le t \le 4\pi$, if its density function is $\rho(x,y,z) = x/2$.

26–30 Evaluate the line integral of the vector field along the given curve.

26. $\int_C x^2 y\,dx + (y^2 - x^2)\,dy$, where C can be parametrized as $\mathbf{r}(t) = \langle t, t^2 \rangle$, $0 \le t \le 2$

27. $\int_C (x^2 y + xy^2)\,dx + x^3\,dy$, where C is the unit circle centered at the origin

28. $\int_C y\,dx + xy\,dy$, where **a.** $C = C_1$: the line segment joining the origin with $(1,1)$, **b.** $C = C_2$: the parabola $y = x^2$ joining the origin with $(1,1)$, and **c.** $C = C_3$: the parabola $x = y^2$ joining the origin with $(1,1)$. Do your answers differ?

29. Repeat Exercise 28 for the line integral $\int_C 2xy\,dx + x^2\,dy$. Compare your answers.

30. $\int_C (x^2 + y)\,dx + xy\,dy$ along the closed path $C = C_1 \cup C_2 \cup C_3$, where C_1 is the line segment from the origin to $(1,0)$, C_2 is the line segment from $(1,0)$ to $(2,1)$, and C_3 is the straight path from $(2,1)$ back to the origin

31. Find $\int_C 2xy\,dx + x^2\,dy$ along the path given in Exercise 30.

32. Determine the work done by the force field in Example 4 if it moves a particle along the elliptical path $y = \sqrt{4 - 4x^2}$ from $(0,2)$ to $(1,0)$.

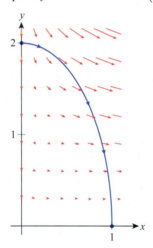

33. Determine the work done by the force field $\mathbf{F}(x,y) = \langle y, x \rangle$ if it moves a particle along the parabolic arc $y = 4x - x^2$ from the point $(1,3)$ to $(3,3)$.

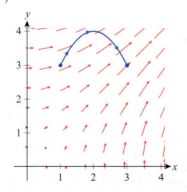

34. By parametrizing the path given in Exercise 33 oriented from $(3,3)$ toward $(1,3)$, find the work done on the particle by the force field when it travels along this "reverse path." Explain your findings.

35. Determine the work done by the force field $\mathbf{F}(x,y) = \langle -x, y \rangle$ if it moves a particle along the graph of $y = \sqrt{x}$ from the point $(9,3)$ to $(1,1)$.

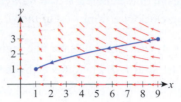

36. Determine the work done by the force field $\mathbf{F}(x,y) = \langle 2, 3x \rangle$ if it moves a particle counterclockwise around the ellipse parametrized as $\mathbf{r}(t) = \langle 3 + 3\cos t, 2 + 2\sin t \rangle$, $0 \le t \le 2\pi$.

37–40 Evaluate the indicated line integral.

37. $\int_C xy\,dx + z\,dy + (x+z)\,dz$, where C can be parametrized as $\mathbf{r}(t) = \langle t^2, t, t^3 \rangle$, $0 \le t \le 1$

38. The integral of Exercise 37 along the line segment from the origin to $(2,2,0)$, followed by the segment from $(2,2,0)$ to $(4,6,2)$

39. $\int_C (x - y^2)\,dx + (y - z^2)\,dy + (z - x^2)\,dz$, where C is the "twisted cube" $\mathbf{r}(t) = \langle t, t^2, t^3 \rangle$, $0 \le t \le 1$

40. $\int_C y\,dx + z\,dy + x^2\,dz$, where C is parametrized by $\mathbf{r}(t) = \langle t^2, t, e^t \rangle$, $0 \le t \le 1$

41. Determine the work done by the force field $\mathbf{F}(x,y,z) = \langle x, y, 1 \rangle$ on a particle that moves up the helix given in Exercise 16.

42. Determine the work done by the force field $\mathbf{F}(x,y,z) = \langle x, z, y \rangle$ on a particle that moves along the curve $\mathbf{r}(t) = \langle t^2, \cos t, \sin t \rangle$, $0 \le t \le \pi$.

43. Find the work done by the force field $\mathbf{F} = -\dfrac{k\mathbf{r}}{r^3}$ as it moves a particle from the point $(0,0,4)$ to $(0,3,4)$ along a straight-line curve. (For well-known force fields of this type, see Examples 3 and 4 of Section 15.1.)

44. Determine the work done by the force field $\mathbf{F}(x,y,z) = \langle y, x + z, y \rangle$ on a particle that moves around the triangle with vertices $(1,0,0)$, $(0,1,0)$, and $(0,0,1)$ in a counterclockwise direction (when viewed from the point $(1,1,1)$).

45–48 A fluid's velocity field is given by $\mathbf{F}(x,y,z)$. Determine the fluid's flow along the indicated curve $\mathbf{r}(t)$.

45. $\mathbf{F}(x,y,z) = \langle 1, 2xz, 4y \rangle$; $\mathbf{r}(t) = \langle 1, t, t^2 \rangle$; $0 \le t \le 3$

46. $\mathbf{F}(x,y,z) = \langle xy, yz, zx \rangle$; $\mathbf{r}(t) = \langle 1, 2t, 3t \rangle$; $0 \le t \le 2$

47. $\mathbf{F}(x,y,z) = \langle -2y, y-x, 3 \rangle$;
     $\mathbf{r}(t) = \langle 4\cos t, 2\sin t, t \rangle$; $0 \le t \le \pi$

48. $\mathbf{F}$ is the gradient of the potential function $f(x,y,z) = x^2 + y^2 + z^2$, and $\mathbf{r}$ is the helix given in Exercise 16.

49. **a.** Verify that $f(x,y,z) = \frac{1}{2}x^2 y^2 z$ is a potential function for the following vector field:
     $\mathbf{F}(x,y,z) = \langle xy^2 z, x^2 yz, \frac{1}{2}x^2 y^2 \rangle$

     b. Assuming that $\mathbf{F}$ is the velocity field of a fluid, find the circulation of the fluid around the unit circle $\mathbf{r}(t) = \langle \cos t, 0, \sin t \rangle$, $0 \le t \le 2\pi$.

50–53 In physics, the electrostatic potential at a point P resulting from a single point charge q is calculated using the formula

$$V(P) = \frac{\varepsilon q}{r_P},$$

where r_P is the distance between P and the point charge q and ε is Coulomb's constant (see Example 4 of Section 15.1). The value of ε is approximately 8.988×10^9 Nm^2/C^2. We note that Coulomb's constant is often used in the form $\varepsilon = 1/(4\pi\varepsilon_0)$, where $\varepsilon_0 \approx 8.85 \times 10^{-12}$ C^2/Nm^2. The constant ε_0 is called the *permittivity constant*. (We assume here that the potential at an "infinitely distant" point is zero.)

Among their many uses, line integrals enable us to calculate the electrostatic potential around a continuously charged curve, according to the following formula.

Suppose a curve C has continuous charge distribution given by its charge density function $q(x,y,z)$. The electrostatic potential at a point P is then obtained from

$$V(P) = \varepsilon \int_C \frac{q(x,y,z)}{r_P(x,y,z)} ds,$$

where $r_P(x,y,z)$ is the distance between (x,y,z) and P. We will use the above formula in Exercises 50–53.

50. Suppose the quarter circle in Example 4 has charge density $q(x,y,z) = \frac{1-y}{10^7}$ coulomb per meter (C/m). Find the electrostatic potential at the point $(0,0,1)$.

51. Repeat Exercise 50 if the charge density is
     $$q(x,y,z) = \frac{xy}{10^5} \text{ C/m}.$$

52. Repeat Exercise 50 if the charge is distributed along the positive x-axis, with charge density
     $$q(x,y,z) = \frac{x}{10^4} \text{ C/m}.$$

53.* Find the potential at the origin of the electric field created by the helix $\mathbf{r}(t) = \langle \cos t, \sin t, t \rangle$, $0 \le t \le 2\pi$, with charge density
     $$q(x,y,z) = \frac{1-z}{10^6} \text{ C/m}.$$

15.2 **Technology Exercises**

54–59 Use a computer algebra system to help with your calculations. Express your answers as decimal approximations.

54. Find the mass and the centroid of the thin wire given by $y = 9 - x^2$, $-3 \le x \le 3$, if it has constant density ρ.

55. Determine the three moments of inertia and radii of gyration for the wire in Exercise 54.

56. Repeat Exercise 54 if the wire has density $\rho(x,y) = x^2 y$.

57. Repeat Exercise 55 if the wire has density $\rho(x,y) = x^2 y$.

58. Determine the three moments of inertia and radii of gyration for the spring given in Exercise 25.

59. Find the electrostatic potential at $(1,0,0)$ if the wire in Exercise 54 has charge density
     $$q(x,y,z) = \frac{y}{10^6} \text{ C/m}.$$

15.3 The Fundamental Theorem for Line Integrals

TOPICS

1. The Fundamental Theorem for Line Integrals

2. Path independence and conservative fields

3. Finding potentials

The Fundamental Theorem of Calculus (FTC) is important both for the connection it makes between integration and differentiation and for its power as a computational tool. In this section, we see the first variant of the FTC as it applies to line integrals and again use it to greatly simplify evaluation.

TOPIC 1   The Fundamental Theorem for Line Integrals

The strength of the FTC on the theoretical side is its proof that integration and differentiation are inverse operations of each other, and its computational power follows immediately from applying that fact in the form

$$\int_a^b f'(x)\,dx = f(b) - f(a).$$

If f is a function of more than one variable, ordinary differentiation has no meaning and f' must be replaced by something that encompasses all the rates of change of f. The gradient vector ∇f does exactly that.

Theorem

The Fundamental Theorem for Line Integrals (Gradient Theorem)

Assume that f is a differentiable function whose gradient ∇f is continuous along a curve C and that C is defined by the smooth vector function $\mathbf{r}(t)$, $a \le t \le b$. Then,

$$\int_C \nabla f \cdot d\mathbf{r} = f(\mathbf{r}(b)) - f(\mathbf{r}(a)).$$

Proof

The theorem is true for $f: \mathbb{R}^n \to \mathbb{R}$, but to illustrate the key idea, assume for the moment that we have $f: \mathbb{R}^3 \to \mathbb{R}$. Then by the Chain Rule, we have the following:

$$\frac{d}{dt} f(\mathbf{r}(t)) = f_x x'(t) + f_y y'(t) + f_z z'(t)$$
$$= \nabla f(\mathbf{r}(t)) \cdot \langle x'(t), y'(t), z'(t) \rangle = \nabla f(\mathbf{r}(t)) \cdot \mathbf{r}'(t)$$

The statement $\frac{d}{dt} f(\mathbf{r}(t)) = \nabla f(\mathbf{r}(t)) \cdot \mathbf{r}'(t)$ is true in general, so

$$\int_C \nabla f \cdot d\mathbf{r} = \int_a^b \nabla f(\mathbf{r}(t)) \cdot \mathbf{r}'(t)\,dt = \int_a^b \frac{d}{dt} f(\mathbf{r}(t))\,dt = f(\mathbf{r}(b)) - f(\mathbf{r}(a)).$$

Before illustrating its use in evaluating integrals, it's worth noting an alternative expression of the formula. Recall from Section 13.6 that the total differential of a function $f(x, y)$ is the expression $df = f_x\,dx + f_y\,dy$, with similar forms for functions of three or more variables. So

$$\nabla f \cdot d\mathbf{r} = \left(f_x \frac{dx}{dt} + f_y \frac{dy}{dt} \right) dt = f_x \, dx + f_y \, dy,$$

and hence if C is a curve from a point A to a point B,

$$\int_A^B df = \int_C \nabla f \cdot d\mathbf{r} = f(B) - f(A).$$

Example 1 ✎

Use the Fundamental Theorem for Line Integrals to evaluate

$$\int_C (y \, dx + x \, dy),$$

where C is the part of the origin-centered circle of radius 5 starting at $(5,0)$ and ending at $(3,4)$.

Solution

In order to use the Fundamental Theorem for Line Integrals, we need to express the vector field $\mathbf{F}(x,y) = \langle y, x \rangle$ as the gradient of a function f. Alternatively, making use of the note above, we need to express the integrand $y \, dx + x \, dy$ as the total differential df of a function f. So we want

$$y \, dx + x \, dy = df = f_x \, dx + f_y \, dy,$$

or $f_x = y$ and $f_y = x$. The function $f(x,y) = xy$ satisfies these two conditions, so

$$\int_C y \, dx + x \, dy = \int_{(5,0)}^{(3,4)} df = f(3,4) - f(5,0) = (3)(4) - (5)(0) = 12.$$

Contrast this simple calculation with the method of Section 15.2. To use that method, we need a parametrization of C, such as $\mathbf{r}(t) = \langle 5 \cos t, 5 \sin t \rangle$ over $0 \le t \le \tan^{-1}\left(\frac{4}{3}\right)$; the curve C is shown in Figure 1, along with a scaled representation of $\mathbf{F}$. Given this parametrization, we have the following.

$$\int_C y \, dx + x \, dy = \int_C \mathbf{F} \cdot d\mathbf{r} = \int_0^{\tan^{-1}(4/3)} \mathbf{F}(\mathbf{r}(t)) \cdot \mathbf{r}'(t) \, dt$$

$$= \int_0^{\tan^{-1}(4/3)} \langle 5 \sin t, 5 \cos t \rangle \cdot \langle -5 \sin t, 5 \cos t \rangle \, dt$$

$$= 25 \int_0^{\tan^{-1}(4/3)} \left(\cos^2 t - \sin^2 t \right) dt = 25 \int_0^{\tan^{-1}(4/3)} \cos 2t \, dt$$

$$= \frac{25}{2} \sin 2t \Bigg]_0^{\tan^{-1}(4/3)} = \frac{25}{2} \sin\left(2 \tan^{-1}\left(\frac{4}{3} \right) \right)$$

$$= \frac{25}{2} \left[2 \sin\left(\tan^{-1}\left(\frac{4}{3} \right) \right) \cos\left(\tan^{-1}\left(\frac{4}{3} \right) \right) \right]$$

$$= \frac{25}{2} (2) \left(\frac{4}{5} \right) \left(\frac{3}{5} \right) = 12$$

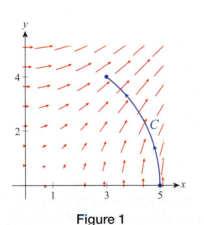

Figure 1

Example 2 ✍

Given a fluid velocity field defined by the gradient of the potential function $f(x,y,z) = xyz$, determine the fluid's flow along the curve $\mathbf{r}(t) = \langle t, t^2, t^3 \rangle$ from the point $(0,0,0)$ to the point $(1,1,1)$.

Solution

This is exactly the problem seen in Example 5 of Section 15.2, but the Gradient Theorem allows us to solve it without integration. Since the vector field is defined as the gradient of $f(x,y,z) = xyz$,

$$\text{Flow} = \int_C \nabla f \cdot d\mathbf{r} = f(\mathbf{r}(1)) - f(\mathbf{r}(0)) = f(1,1,1) - f(0,0,0) = 1.$$

TOPIC 2 Path Independence and Conservative Fields

Examples 1 and 2 illustrate the power of the FTC in evaluating line integrals of gradients, but several points must be emphasized. First, *any* smooth curve C running from the initial point to the terminal point would have resulted in the same line integral in either example—knowledge of the endpoints alone suffices. Second, while the examples may lead us to want to evaluate all line integrals in the same manner, the method is only valid for vector fields that fall into a certain category. These two points lead to the following definitions and theorems.

Definition ♀

Path Independence of Line Integrals

Assume that $\mathbf{F}$ is a continuous vector field defined on an open region D, and that A and B are two fixed points in D. If $\int_C \mathbf{F} \cdot d\mathbf{r}$ has the same value for every piecewise smooth path joining A and B, we say that the integral is **path independent in D** (recall that a piecewise smooth path consists of a finite number of smooth curves joined end to end). In this case, any such line integral between A and B may be written as $\int_A^B \mathbf{F} \cdot d\mathbf{r}$, as the exact path between the two points is immaterial.

Since we are primarily interested in vector fields over which line integrals make sense, we define a region D to be **connected** if any two points in D can be joined by a piecewise smooth path. From this point forward, unless otherwise indicated, all domains of vector fields will be assumed to be open and connected, and all line integral paths will be assumed to be piecewise smooth.

Because we have defined a vector field $\mathbf{F}$ to be conservative on D if $\mathbf{F} = \nabla f$ for some potential function f, the Gradient Theorem tells us the following.

Theorem ⚭

Conservation Implies Path Independence

If **F** is a conservative field on an open connected region D, then every line integral $\int \mathbf{F} \cdot d\mathbf{r}$ is path independent in D.

Since line integrals over conservative fields are particularly easy to evaluate (assuming we can find a potential of the field), we are interested in their characteristics. One such characteristic concerns paths that are **closed**, meaning paths whose initial and terminal points coincide.

Theorem ⚭

Path Independence and Closed Path Integrals

Let D be an open connected region and **F** a continuous vector field defined on D. Then line integrals $\int \mathbf{F} \cdot d\mathbf{r}$ are path independent in D if and only if $\int_C \mathbf{F} \cdot d\mathbf{r} = 0$ for every closed path C in D.

Proof 📎

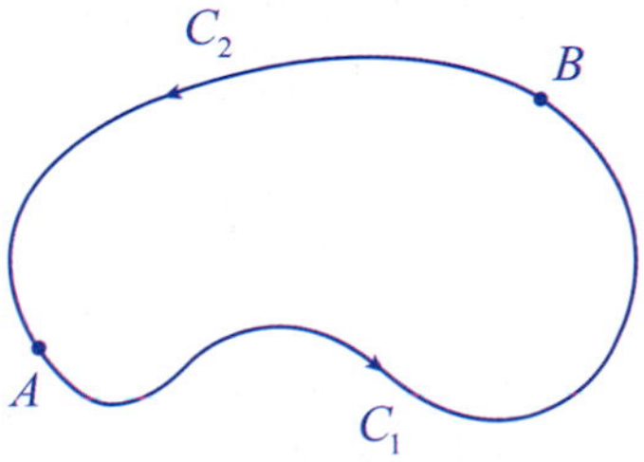

Figure 2

Suppose every such line integral of the form $\int \mathbf{F} \cdot d\mathbf{r}$ is path independent, and let C be a closed path in D. Pick any two distinct points A and B on C, and let C_1 be the part of C from A to B (following C's orientation) and C_2 the remaining portion of C from B to A (see Figure 2). Then $-C_2$ is also a path from A to B, so

$$\int_{-C_2} \mathbf{F} \cdot d\mathbf{r} = \int_{C_1} \mathbf{F} \cdot d\mathbf{r}$$

by path independence. But as we saw in Section 15.2, $\int_{-C_2} \mathbf{F} \cdot d\mathbf{r} = -\int_{C_2} \mathbf{F} \cdot d\mathbf{r}$, therefore

$$\int_C \mathbf{F} \cdot d\mathbf{r} = \int_{C_1} \mathbf{F} \cdot d\mathbf{r} + \int_{C_2} \mathbf{F} \cdot d\mathbf{r} = \int_{C_1} \mathbf{F} \cdot d\mathbf{r} - \int_{-C_2} \mathbf{F} \cdot d\mathbf{r} = \int_{C_1} \mathbf{F} \cdot d\mathbf{r} - \int_{C_1} \mathbf{F} \cdot d\mathbf{r} = 0.$$

On the other hand, if $\int_C \mathbf{F} \cdot d\mathbf{r} = 0$ for every closed path C, and if C_1 and C_2 are any two paths with the same initial and terminal points, the same argument in reverse shows that

$$\int_{C_1} \mathbf{F} \cdot d\mathbf{r} - \int_{C_2} \mathbf{F} \cdot d\mathbf{r} = 0,$$

so $\int_{C_1} \mathbf{F} \cdot d\mathbf{r} = \int_{C_2} \mathbf{F} \cdot d\mathbf{r}$ and hence the line integrals are path independent.

The adjective *conservative* arises from physics and refers to fields in which the *conservation of energy* principle holds, such as gravitational and electric fields. The physical meaning of the preceding theorem is that the work done by a conservative field in moving an object around a closed path is zero. The next theorem tells us that, on open connected domains, path independence is the identifying characteristic of conservative vector fields.

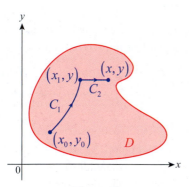

Figure 3

Theorem 🔍

Path Independence Implies Conservation

Assume $\mathbf{F}$ is a continuous vector field on an open connected region D. If every line integral $\int \mathbf{F} \cdot d\mathbf{r}$ is path independent, then $\mathbf{F}$ is conservative.

Proof 📎

To show that $\mathbf{F}$ is conservative, we need to find a potential function f for $\mathbf{F}$. We will demonstrate how this can be done for a vector field of two variables, but the proof technique extends to vector fields of any number of variables.

To construct f, let (x_0, y_0) be a fixed point in D and define

$$f(x, y) = \int_{(x_0, y_0)}^{(x, y)} \mathbf{F} \cdot d\mathbf{r}$$

for all $(x, y) \in D$. This defines f unambiguously because of path independence—all paths between (x_0, y_0) and (x, y) lead to the same value for the line integral. Now for a given point (x, y), choose $x_1 < x$ so that the line segment between (x_1, y) and (x, y) lies entirely in D; we know we can do this for some x_1 sufficiently close to x because D is open. Let C_1 be a path in D from (x_0, y_0) to (x_1, y), and let C_2 be the line segment from (x_1, y) to (x, y) (see Figure 3). Then by path independence,

$$f(x, y) = \int_{C_1} \mathbf{F} \cdot d\mathbf{r} + \int_{C_2} \mathbf{F} \cdot d\mathbf{r} = \int_{(x_0, y_0)}^{(x_1, y)} \mathbf{F} \cdot d\mathbf{r} + \int_{(x_1, y)}^{(x, y)} \mathbf{F} \cdot d\mathbf{r}.$$

The first integral does not depend on x, so

$$f_x = \frac{\partial}{\partial x}\left[\int_{(x_0, y_0)}^{(x_1, y)} \mathbf{F} \cdot d\mathbf{r} + \int_{(x_1, y)}^{(x, y)} \mathbf{F} \cdot d\mathbf{r}\right] = 0 + \frac{\partial}{\partial x}\int_{(x_1, y)}^{(x, y)} \mathbf{F} \cdot d\mathbf{r},$$

and if we let P and Q be the component functions of $\mathbf{F}$, this can be written as

$$f_x = \frac{\partial}{\partial x}\int_{(x_1, y)}^{(x, y)} \mathbf{F} \cdot d\mathbf{r} = \frac{\partial}{\partial x}\int_{(x_1, y)}^{(x, y)} (P\,dx + Q\,dy).$$

On C_2, y is constant so $dy = 0$ and we are left with

$$f_x = \frac{\partial}{\partial x}\int_{(x_1, y)}^{(x, y)} P\,dx.$$

Again, since y is constant on C_2, we can simply fix its value in the integrand. And to avoid confusion, we use the "dummy" variable u in place of x inside the integral. Then by the Fundamental Theorem of Calculus,

$$f_x = \frac{\partial}{\partial x}\int_{(x_1, y)}^{(x, y)} P\,dx = \frac{d}{dx}\int_{x_1}^{x} P(u, y)\,du = P(x, y).$$

In words, the partial derivative of f with respect to x gives us the first component function of $\mathbf{F}$, as desired. A similar argument, using a short vertical line segment from a point (x, y_1) to (x, y), shows that $f_y = Q(x, y)$, so together $\nabla f = \mathbf{F}$.

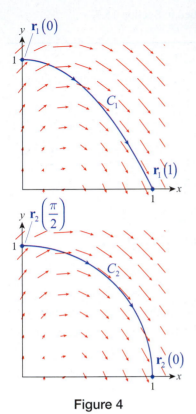

Figure 4

Simply Connected Region

Not Simply Connected Regions

Figure 5

Example 3 ✒

Let $\mathbf{F}(x,y) = \langle x^2 + y, 1 - 3x \rangle$, let C_1 be the path from $(0,1)$ to $(1,0)$ defined by the function $\mathbf{r}_1(t) = \langle t, 1 - t^2 \rangle$, and let C_2 be the path from $(0,1)$ to $(1,0)$ defined by the function $\mathbf{r}_2(t) = \langle \cos t, \sin t \rangle$. In Exercise 22, you will show that

$$\int_{C_1} \mathbf{F} \cdot d\mathbf{r} = 2 \quad \text{and} \quad \int_{C_2} \mathbf{F} \cdot d\mathbf{r} = \pi - \frac{2}{3},$$

meaning that $\mathbf{F}$ is not a conservative vector field and that the Gradient Theorem cannot be used to evaluate either line integral. Figure 4 is a scaled illustration of $\mathbf{F}$ and of the two paths C_1 and C_2.

Example 3 illustrates one way of determining whether a vector field is conservative, but a test that doesn't require the evaluation and comparison of line integrals would be welcome. Clairaut's Theorem (Section 13.3) on the equality of mixed partial derivatives provides the inspiration. Note that if $\mathbf{F}(x,y) = \langle P(x,y), Q(x,y) \rangle = \nabla f(x,y)$, then

$$\frac{\partial P}{\partial y} = \frac{\partial}{\partial y} f_x = \underbrace{f_{xy} = f_{yx}}_{\text{By Clairaut}} = \frac{\partial}{\partial x} f_y = \frac{\partial Q}{\partial x}.$$

Similarly, for a three-variable vector field, if $\mathbf{F} = \langle P, Q, R \rangle = \langle f_x, f_y, f_z \rangle = \nabla f$, then

$$\frac{\partial P}{\partial y} = \frac{\partial Q}{\partial x}, \quad \frac{\partial P}{\partial z} = \frac{\partial R}{\partial x}, \quad \text{and} \quad \frac{\partial Q}{\partial z} = \frac{\partial R}{\partial y}.$$

More importantly, for our current purposes, the converse of each statement is true if the region D is not just open and connected, but also **simply connected**, which means that any one path between two points in D can be continuously transformed, without leaving D, into any other path between the same two points. Informally, simply connected regions do not contain "holes"; Figure 5 illustrates the difference between a simply connected region in the plane and several regions that are not simply connected.

Theorem 🔍

Component Test for Conservative Vector Fields

Assume D is an open and simply connected region, and that $\mathbf{F}$ is a vector field whose components have continuous first partial derivatives. If $\mathbf{F}$ is a vector field of two variables, let $\mathbf{F} = \langle P, Q \rangle$, and if $\mathbf{F}$ is a vector field of three variables, let $\mathbf{F} = \langle P, Q, R \rangle$. Then in the two-variable case,

$$\frac{\partial P}{\partial y} = \frac{\partial Q}{\partial x} \quad \text{throughout } D \text{ implies } \mathbf{F} \text{ is conservative;}$$

in the three-variable case,

$$\frac{\partial P}{\partial y} = \frac{\partial Q}{\partial x}, \quad \frac{\partial P}{\partial z} = \frac{\partial R}{\partial x}, \quad \text{and} \quad \frac{\partial Q}{\partial z} = \frac{\partial R}{\partial y}$$

throughout D implies $\mathbf{F}$ is conservative.

A sketch of the proof of this test will be provided in Section 15.7.

Example 4

Show that the vector field $\mathbf{F}(x,y) = \langle x^2 + y, 1 - 3x \rangle$ in Example 3 is not conservative.

Solution

$$\frac{\partial}{\partial y}\left(x^2 + y\right) = 1$$

$$\frac{\partial}{\partial x}\left(1 - 3x\right) = -3$$

So $\mathbf{F}$ is not conservative by the Component Test.

TOPIC 3   Finding Potentials

The preceding theorems and examples give us both motivation for finding a potential f of a given vector field $\mathbf{F}$ and, if one exists, an idea of how to do so. We illustrate the method with the last examples of this section.

Example 5

Find a potential function, if one exists, for $\mathbf{F}$.

$$\mathbf{F}(x,y,z) = \langle ye^x + z\cos x, e^x - z^2, -2yz + \sin x \rangle$$

Solution

As you can verify with the Component Test (Exercise 1), the field $\mathbf{F}$ is conservative, so a potential function f does exist in theory. To construct one, we begin with the goal. That is, we seek a potential f that satisfies

$$\langle ye^x + z\cos x, e^x - z^2, -2yz + \sin x \rangle = \langle f_x, f_y, f_z \rangle.$$

We might call the method of constructing f "partial integration" since we start by integrating $f_x = ye^x + z\cos x$ with respect to x to obtain

$$f = ye^x + z\sin x + g(y,z).$$

The as-yet-unknown function $g(y,z)$ plays the role of the constant of integration, since any expression in the variables y and z alone is constant with respect to x. We now differentiate our result with respect to y and compare it with our goal, $e^x - z^2$.

$$f_y = e^x + g_y(y,z) = e^x - z^2$$

So we require $g_y(y,z) = -z^2$, and integrating with respect to y results in

$$g(y,z) = -yz^2 + h(z),$$

giving us $f = ye^x + z\sin x - yz^2 + h(z)$. Differentiating this last result with respect to z and comparing it to the third component of $\mathbf{F}$ yields $h'(z) = 0$. This means $h(z)$ can be any scalar constant so we may as well let $h(z) = 0$. Hence,

$$f = ye^x + z\sin x - yz^2$$

is a potential function for $\mathbf{F}$.

Example 6 ✐

Determine $\displaystyle\int_C \left(ye^x + z\cos x\right)dx + \left(e^x - z^2\right)dy + \left(-2yz + \sin x\right)dz$, where C is a piecewise smooth path from $(\pi, 0, -5)$ to $(0, 1, 3)$.

Solution

The integral can be expressed as $\displaystyle\int_{(\pi,0,-5)}^{(0,1,3)} df$, where f is the potential in Example 5, $f = ye^x + z\sin x - yz^2$. Therefore, we have the following.

$$\int_{(\pi,0,-5)}^{(0,1,3)} df = f(0,1,3) - f(\pi,0,-5) = -8 - 0 = -8$$

Example 7 ✐

Evaluate $\displaystyle\int_C (-2xy)\,dx + \left(3 - x^2\right)dy$, where C is the path shown in Figure 6.

Solution

Since

$$\frac{\partial}{\partial y}(-2xy) = -2x = \frac{\partial}{\partial x}\left(3 - x^2\right),$$

the vector field implicit in this line integral is conservative and a potential function f exists. We want

$$f_x = -2xy, \quad \text{so} \quad f(x,y) = -x^2 y + g(y).$$

From this, we obtain $f_y = -x^2 + g'(y)$. So we require $g'(y) = 3$ or $g(y) = 3y + K$, where K is any constant; we will use $K = 0$. Hence,

$$f(x,y) = -x^2 y + 3y$$

and

$$\int_C (-2xy)\,dx + \left(3 - x^2\right)dy = \int_{(1,0)}^{(0,1)} df = f(0,1) - f(1,0) = 3 - 0 = 3.$$

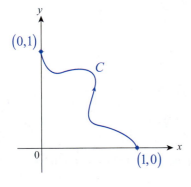

Figure 6

15.3 **Exercises**

1. Use the Component Test to verify that the vector field **F** in Example 5 is conservative.

2–5 Verify that V is a potential function for **F** and determine the line integral along the indicated curve.

2. $\mathbf{F}(x,y) = \langle y\sin y, xy\cos y + x\sin y\rangle; \quad V(x,y) = xy\sin y; \quad \mathbf{r}(t) = \left\langle t, \dfrac{\pi}{2}t\right\rangle, \quad 0 \le t \le 1$

3. $\mathbf{F}(x,y) = \left\langle \cos xe^{y} + \ln y, \dfrac{x}{y} + e^{y}\sin x\right\rangle; \quad V(x,y) = \sin xe^{y} + x\ln y; \quad \mathbf{r}(t) = \left\langle \dfrac{\pi}{2} + \dfrac{\pi}{2}t, 1 + (e-1)t\right\rangle, \quad 0 \le t \le 1$

4. $\mathbf{F}(x,y,z) = \langle -yz\sin x, z\cos x, y\cos x\rangle; \quad V(x,y,z) = yz\cos x; \quad \mathbf{r}(t) = \left\langle \dfrac{\pi}{2}(t-1), \dfrac{\pi}{2}t, t\right\rangle, \quad 0 \le t \le 1$

5. $\mathbf{F}(x,y,z) = \left\langle 2xy + e^{z^2}, x^2, 2xze^{z^2}\right\rangle; \quad V(x,y) = x^2y + xe^{z^2}; \quad \mathbf{r}(t) = \langle t, t^2, t^3\rangle, \quad 0 \le t \le 1$

6–21 Either find a potential function for **F** or state that a potential function does not exist. The latter implies that **F** is not conservative.

6. $\mathbf{F}(x,y) = \langle 2x, 2 + 6y\rangle$

7. $\mathbf{F}(x,y) = \langle 3x^2 + 2y^2, 4xy + 3y^2\rangle$

8. $\mathbf{F}(x,y) = \langle x^2y, 3x\rangle$

9. $\mathbf{F}(x,y) = \langle x^2 + y, 2xy + 3y^2\rangle$

10. $\mathbf{F}(x,y) = \left\langle x^2 + \ln y, \dfrac{x}{y} + 2y^3\right\rangle$

11. $\mathbf{F}(x,y) = \left\langle e^x + y\cos x, \dfrac{1}{y} + \sin x\right\rangle$

12. $\mathbf{F}(x,y) = \langle e^x + \sin y, e^y + \cos x\rangle$

13. $\mathbf{F}(x,y) = \langle 4x^3y - y^5, x^4 - 5xy^4\rangle$

14. $\mathbf{F}(x,y) = \left\langle -\dfrac{2y}{x^3}, \dfrac{1}{x^2} + \dfrac{1}{\sqrt{y}}\right\rangle$

15. $\mathbf{F}(x,y,z) = \langle 2xyz, x^2z, x^2y\rangle$

16. $\mathbf{F}(x,y,z) = \langle yz, xz - z\sin y, xy + \cos y\rangle$

17. $\mathbf{F}(x,y,z) = \langle x^2y, 2xz, z^3\rangle$

18. $\mathbf{F}(x,y,z) = \left\langle \dfrac{zy^2}{x}, 2zy\ln x, -\dfrac{y}{x^2}\right\rangle$

19. $\mathbf{F}(x,y,z) = \langle \tan z, 2yz, y^2 + x\sec^2 z\rangle$

20. $\mathbf{F}(x,y,z) = \langle y^2 + ze^{xz}, 2xy - 2z, xe^{xz} - 2y\rangle$

21. $\mathbf{F}(x,y,z) = \left\langle x\cos z + z, z + \sin y, -\dfrac{x^2}{2}\sin z + x\right\rangle$

22. For the vector field **F** in Example 3, show that $\displaystyle\int_{C_1} \mathbf{F}\cdot d\mathbf{r} = 2$ and $\displaystyle\int_{C_2} \mathbf{F}\cdot d\mathbf{r} = \pi - \tfrac{2}{3}$, meaning that **F** is not conservative.

23–26 Show that the line integral is not path independent by finding two different values for the integral along two different paths connecting A and B. (Answers may vary.)

23. $\displaystyle\int_C y\,dx + 4\,dy; \quad A(1,0), \quad B(0,1)$

24. $\displaystyle\int_C xy\,dx + 2y\,dy; \quad A(0,0), \quad B(2,4)$

25. $\displaystyle\int_C y\,dx - 2z\,dy + 2\,dz; \quad A(-5,0,0), \quad B(3,4,0)$

26. $\displaystyle\int_C xy\,dx + z\,dy + (x+z)\,dz; \quad A(0,0,0), \quad B(1,1,1)$

27–32 Show that the force field **F** is conservative, and use this fact to determine the work done by **F** in moving an object from A to B.

27. $\mathbf{F}(x,y) = \langle 3y, 3x - 2y \rangle$; $A(-1,0)$, $B(5,3)$

28. $\mathbf{F}(x,y) = \langle x - y^2, -2xy \rangle$; $A(1,4)$, $B(3,-2)$

29. $\mathbf{F}(x,y) = \langle e^y - 2xy, x(e^y - x) \rangle$; $A(-4,1)$, $B(0,0)$

30. $\mathbf{F}(\mathbf{r}) = \dfrac{k\mathbf{r}}{r^3}$; $A(-2,1,-2)$, $B(6,0,-8)$

31. $\mathbf{F}(x,y,z) = \langle e^y \cos x - yz, e^y \sin x - xz, -xy \rangle$;
    $A(\pi, 1, 2/\pi)$, $B(\pi/2, 0, 0)$

32. $\mathbf{F}(x,y,z) = \langle \tan y, x \sec^2 y - z, -y \rangle$;
    $A(0, 0, 4/\pi)$, $(2, \pi/4, 0)$

33. Find an "easier" solution (one that uses the Fundamental Theorem for Line Integrals) for Exercise 29 of Section 15.2.

34. Repeat Exercise 33, this time for Exercise 44 of Section 15.2.

35. **a.** Show that the vector field

$$\mathbf{F}(x,y) = \left\langle -\frac{y}{|\langle x, y \rangle|^2}, \frac{x}{|\langle x, y \rangle|^2} \right\rangle$$

satisfies the Component Test, but is not conservative. Does this contradict the Component Test? (Such a vector field is sometimes called a *rotation field*. **Hint:** To show that **F** is not conservative, calculate a line integral along a circle centered at the origin.)

b. Verify that $\mathbf{F}(x,y) = \nabla\left(-\arctan\dfrac{x}{y}\right)$.

Reconcile this observation with part a.

36. In this exercise, we will consider the vector field

$$\mathbf{F}(x,y) = \left\langle \frac{x}{|\langle x, y \rangle|^2}, \frac{y}{|\langle x, y \rangle|^2} \right\rangle,$$

with the usual notation $P(x,y) = x/|\langle x, y \rangle|^2$ and $Q(x,y) = y/|\langle x, y \rangle|^2$.

a. Show that $\partial P/\partial y = \partial Q/\partial x$ throughout the domain of **F**.

b. Explain why **F** does not satisfy the conditions of the Component Test.

c. Show that **F** is conservative. (See Exercise 58 of Section 15.1.) Is this a contradiction with part b.?

37–42 Decide whether the specified region D is simply connected.

37. $D = \{(x,y) \mid x^2 + y^2 < 1\}$

38. $D = \{(x,y) \mid 0 < x^2 + y^2 < 1\}$

39. $D = \{(x,y) \mid |x| + |y| < 1\}$

40. $D = \{(x,y) \mid x^2 \le y \le 2x^2\}$

41. $D = \{(x,y,z) \mid 1 < x^2 + y^2 + z^2 < 3\}$

42. $D = \{(x,y,z) \mid 1 < z^2 < 3\}$

43–46 In physics, the Law of Conservation of Energy states that if an object moves under the influence of a conservative force field, then the sum of its potential and kinetic energies (the energies resulting from the object's position and motion, respectively) remains constant. For example, in case of an object falling under the influence of gravity only, the kinetic energy it gains as a result of increasing speed equals the loss in potential energy stemming from loss of altitude.

In Exercises 43–46, you will use the Fundamental Theorem for Line Integrals to derive this law.

43. Suppose an object of mass m is under the influence of a conservative force field **F**. If $\mathbf{F} = \nabla f$, the potential energy of the object is defined as

$$E_p(x,y,z) = -f(x,y,z).$$

Show that the work W done by **F** in moving the object from point A to point B along a smooth curve is

$$W = E_p(A) - E_p(B).$$

44. Referring to Exercise 43, suppose the path of the object is parametrized by $\mathbf{r}(t)$ so that $\mathbf{r}(a) = A$ and $\mathbf{r}(b) = B$. Show that W can be written as

$$W = \int_a^b \mathbf{F} \cdot \mathbf{v}(t)\, dt,$$

where $\mathbf{v}(t)$ is the velocity of the object.

45. Use Newton's Second Law $\left(\mathbf{F} = m\mathbf{v}'(t)\right)$ along with Exercise 44 to show that W can be expressed as

$$W = \frac{m}{2} \int_a^b \frac{d}{dt}\left(\left|\mathbf{v}(t)\right|^2\right) dt.$$

46. Use Exercise 45 and the fact that the kinetic energy of an object of mass m and speed $v = |\mathbf{v}|$ is $E_k = \frac{1}{2}mv^2$ to conclude that

$$W = E_k(B) - E_k(A).$$

Consequently, using Exercise 43,

$$E_p(A) + E_k(A) = E_p(B) + E_k(B).$$

47. Use the Law of Conservation of Energy to derive the formula for the velocity of impact of an object falling from height h under the influence of gravity only, $v_{imp} = \sqrt{2hg}$. (**Hint:** Since we are ignoring all other forces, the initial potential energy $E_p = mgh$ turns entirely into kinetic energy.)

48.* A proton is moving along the z-axis in the positive direction at a speed of $2 \cdot 10^6$ m/s (assume units are meters in our coordinate system). At the origin, it encounters an electric force field $\mathbf{E}(x, y, z) = \langle 0, 0, 1600z \rangle$ N/C (newtons per coulomb). Use the Law of Conservation of Energy to find the proton's speed at the point $(0, 0, 6)$. (**Hint:** For the mass and charge of a proton, use the approximate data $m = 1.6726 \times 10^{-27}$ kg and $q = 1.6 \times 10^{-19}$ C, respectively. The potential energy of the proton at (x, y, z) is $E_p(x, y, z) = qV(x, y, z)$, where V is the electric potential.)

49. Suppose $f(x, y)$ is a harmonic function. Prove:

$$\int_C \left(\frac{\partial f}{\partial y}\, dx - \frac{\partial f}{\partial x}\, dy \right) = 0$$

along any smooth closed curve C in $\mathbb{R}^2$. (For a refresher on the definition of harmonic functions, see Exercise 98 of Section 13.3.)

50–54 *True or False?* Determine whether the given statement is true or false. In case of a false statement, explain or provide a counterexample.

50. If $\mathbf{F}$ is path independent on an open connected region D, then $\int_C \mathbf{F} \cdot d\mathbf{r} = 0$ for every path C in D.

51. If the components of $\mathbf{F} = \langle P, Q \rangle$ have continuous first partials and $\partial P / \partial y = \partial Q / \partial x$ throughout an open connected region D, then $\mathbf{F} = \langle P, Q \rangle$ is conservative on D.

52. If $\mathbf{F}$ is continuous on an open connected region D and every line integral $\int \mathbf{F} \cdot d\mathbf{r}$ is path independent, then $\mathbf{F}$ is conservative.

53. If $\mathbf{F}$ is conservative on an open connected region D, then every line integral $\int \mathbf{F} \cdot d\mathbf{r}$ is path independent in D.

54. The domain of the vector field $\mathbf{F}(\mathbf{r}) = \dfrac{k\mathbf{r}}{r^3}$ is not simply connected; therefore, it cannot be conservative.

15.4 **Green's Theorem**

TOPICS

1. Divergence and curl operators

2. Green's Theorem

We now know that line integrals of conservative vector fields are easily evaluated, at least if a potential for the field can be found. For nonconservative vector fields, other methods must be used. One such method is Green's Theorem, which allows us to convert line integrals around closed paths into double integrals over the enclosed region. As we will see, Green's Theorem (named for the British mathematician George Green, 1793–1841) can also be thought of as a variant of the Fundamental Theorem of Calculus.

TOPIC 1 **Divergence and Curl Operators**

The statement of Green's Theorem, and two other theorems to come, is simpler in form and easier to remember if we introduce two new ideas. The first is the *divergence*, or *flux density*, of a vector field. We begin with a note on notation and a definition.

Notational Note: The del operator ∇ can be interpreted as a vector consisting of partial differentiation operators, with the vector components determined by the context. For instance, if f is a function of two variables, $\nabla = \langle \partial/\partial x, \partial/\partial y \rangle$ and $\nabla f = \langle f_x, f_y \rangle$; when applied to a function g of three variables, $\nabla = \langle \partial/\partial x, \partial/\partial y, \partial/\partial z \rangle$ and $\nabla g = \langle g_x, g_y, g_z \rangle$.

Definition ♀

Divergence (Flux Density) of a Vector Field

The **divergence**, or **flux density**, of a vector field $\mathbf{F}(x, y) = \langle P(x, y), Q(x, y) \rangle$ is the scalar function

$$\text{div } \mathbf{F} = \frac{\partial P}{\partial x} + \frac{\partial Q}{\partial y}.$$

For a vector field $\mathbf{F}(x, y, z) = \langle P, Q, R \rangle$ of three variables,

$$\text{div } \mathbf{F} = \frac{\partial P}{\partial x} + \frac{\partial Q}{\partial y} + \frac{\partial R}{\partial z}.$$

In general, we can denote the divergence of a vector field $\mathbf{F}$ as the dot product of the del operator and $\mathbf{F}$.

$$\text{div } \mathbf{F} = \nabla \cdot \mathbf{F}$$

The divergence of a vector field has a physical interpretation that is important in physics and related disciplines. We develop the interpretation for a vector field of two variables, but the idea is valid for any number of variables.

Suppose that $\mathbf{F}(x, y) = \langle P, Q \rangle = P\mathbf{i} + Q\mathbf{j}$ represents the velocity field of a fluid flowing in the plane, and (x, y) is the lower-left corner of a rectangle lying entirely in the domain of $\mathbf{F}$. Assume the rectangle has width Δx and height Δy. Then the rate at which the fluid leaves the rectangle by flowing across a given boundary of the rectangle is approximated by the dot product of $\mathbf{F}$ (evaluated at some point

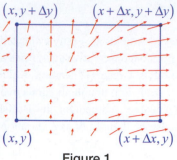

$(x, y + \Delta y)$ $(x + \Delta x, y + \Delta y)$

(x, y) $(x + \Delta x, y)$

Figure 1

along the boundary) and the outward normal vector to the boundary, times the length of the boundary (see Figure 1). For instance, since $-\mathbf{j}$ is the outward normal vector for the bottom of the rectangle, the flow rate of the fluid across the bottom is approximately $\mathbf{F}(x, y) \cdot (-\mathbf{j}) \Delta x$ (if more fluid is entering than leaving the rectangle across a given boundary, the dot product is negative). In all, the rates at which the fluid flows outward across the four edges of the rectangle are as follows.

Bottom: $\mathbf{F}(x, y) \cdot (-\mathbf{j}) \Delta x = -Q(x, y) \Delta x$

Right: $\mathbf{F}(x + \Delta x, y) \cdot \mathbf{i} \, \Delta y = P(x + \Delta x, y) \Delta y$

Top: $\mathbf{F}(x, y + \Delta y) \cdot \mathbf{j} \, \Delta x = Q(x, y + \Delta y) \Delta x$

Left: $\mathbf{F}(x, y) \cdot (-\mathbf{i}) \Delta y = -P(x, y) \Delta y$

Combining opposite edges of the rectangle, note that

Bottom and top: $\left[Q(x, y + \Delta y) - Q(x, y) \right] \Delta x \approx \left[\dfrac{\partial Q}{\partial y} \Delta y \right] \Delta x$

Left and right: $\left[P(x + \Delta x, y) - P(x, y) \right] \Delta y \approx \left[\dfrac{\partial P}{\partial x} \Delta x \right] \Delta y$

so the total net flow rate of the fluid out of the rectangle is approximately

$$\left(\frac{\partial P}{\partial x} + \frac{\partial Q}{\partial y} \right) \Delta x \, \Delta y.$$

When we divide the flow (or *flux*) rate by the area $\Delta x \Delta y$ of the rectangle we obtain the total flow per unit area, which we have defined as the divergence or flux density of the fluid.

Example 1 ✐

Suppose that $\mathbf{F}(x, y, z) = \left\langle x^2 - xy + z, \, xy + z^2, \, z^3 - xy^2 \right\rangle$ represents the velocity field of a gas. Calculate the divergence of $\mathbf{F}$ at the points $(1, 2, -1)$, $(0, 4, 1)$, and $(1, 6, 1)$ and interpret the results.

Solution

$$\text{div } \mathbf{F} = \nabla \cdot \mathbf{F} = \frac{\partial}{\partial x}\left(x^2 - xy + z\right) + \frac{\partial}{\partial y}\left(xy + z^2\right) + \frac{\partial}{\partial z}\left(z^3 - xy^2\right)$$

$$= 2x - y + x + 3z^2 = 3x - y + 3z^2$$

The fact that $\text{div } \mathbf{F}(1, 2, -1) = 4 > 0$ means the gas is expanding at $(1, 2, -1)$, the fact that $\text{div } \mathbf{F}(0, 4, 1) = -1 < 0$ means the gas is condensing (or is being compressed) at $(0, 4, 1)$, and the fact that $\text{div } \mathbf{F}(1, 6, 1) = 0$ means the volume of gas flowing toward the point $(1, 6, 1)$ is the same as the volume of gas flowing away.

The second new idea is that of the *curl* of a vector field. In contrast to divergence, the curl of a vector field is another vector field, not a scalar field.

Curl of a Vector Field

The **curl** of a vector field $\mathbf{F}(x, y, z) = \langle P(x, y, z), Q(x, y, z), R(x, y, z) \rangle$ is the vector function

$$\text{curl } \mathbf{F} = \left\langle \frac{\partial R}{\partial y} - \frac{\partial Q}{\partial z}, \frac{\partial P}{\partial z} - \frac{\partial R}{\partial x}, \frac{\partial Q}{\partial x} - \frac{\partial P}{\partial y} \right\rangle.$$

The curl of a vector field $\mathbf{F}$ can be remembered as the cross product of the del operator and $\mathbf{F}$.

$$\text{curl } \mathbf{F} = \nabla \times \mathbf{F}$$

Although curl is defined for vector fields of three variables, the principle also applies to vector fields in the plane. The formula above can be used in such a case by simply appending a third component function of 0 to the vector field. In order to compare *curl* with *divergence*, we develop the physical interpretation of $\nabla \times \mathbf{F}$ for a field $\mathbf{F}(x, y)$.

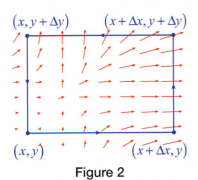

$(x, y + \Delta y)$ $(x + \Delta x, y + \Delta y)$

(x, y) $(x + \Delta x, y)$

Figure 2

Accordingly, suppose $\mathbf{F}(x, y) = \langle P, Q \rangle$ again represents the velocity field of a fluid in the plane, and that (x, y) is the lower-left corner of a rectangle lying entirely in the domain of $\mathbf{F}$. We first extend $\mathbf{F}$ as $\mathbf{F}(x, y, z) = \langle P(x, y), Q(x, y), 0 \rangle$. This time, we are interested in quantifying the counterclockwise circulation of $\mathbf{F}$ around the rectangle, instead of the flow of $\mathbf{F}$ out of the rectangle (the counterclockwise, or *positive*, orientation around a closed curve is used to maintain consistency with the right-hand rule). To do so, we calculate the dot product of $\mathbf{F}$ with the vector associated with each edge of the rectangle (see Figure 2); a positive result indicates $\mathbf{F}$ is flowing in the direction of the edge, while a negative result means the flow of $\mathbf{F}$ is opposite the orientation of the edge. This gives us the following rates of flow.

Bottom: $\mathbf{F}(x, y, z) \cdot \mathbf{i} \, \Delta x = P(x, y) \Delta x$

Right: $\mathbf{F}(x + \Delta x, y, z) \cdot \mathbf{j} \, \Delta y = Q(x + \Delta x, y) \Delta y$

Top: $\mathbf{F}(x, y + \Delta y, z) \cdot (-\mathbf{i}) \Delta x = -P(x, y + \Delta y) \Delta x$

Left: $\mathbf{F}(x, y, z) \cdot (-\mathbf{j}) \Delta y = -Q(x, y) \Delta y$

This time, when we combine the results for opposite sides, we obtain

Bottom and top: $\left[P(x, y) - P(x, y + \Delta y) \right] \Delta x \approx \left[-\frac{\partial P}{\partial y} \Delta y \right] \Delta x$

Left and right: $\left[Q(x + \Delta x, y) - Q(x, y) \right] \Delta y \approx \left[\frac{\partial Q}{\partial x} \Delta x \right] \Delta y$

and adding them together we have a total circulation (in the counterclockwise direction) around the rectangle of approximately

$$\left(\frac{\partial Q}{\partial x} - \frac{\partial P}{\partial y} \right) \Delta x \, \Delta y.$$

As with divergence, we divide by the area $\Delta x \Delta y$ of the rectangle to obtain what is sometimes called the *circulation density* of **F**. It constitutes the component of curl **F** in the **k** direction. This means for a vector field $\mathbf{F}(x, y) = \langle P, Q \rangle$ in the plane,

$$\text{curl } \mathbf{F} = \left(\frac{\partial Q}{\partial x} - \frac{\partial P}{\partial y} \right) \mathbf{k}.$$

Later in this chapter, when we discuss Stokes' Theorem, we will have a need for the curl of vector fields of three variables.

Example 2 ✎

Find the curl of the vector field $\mathbf{F}(x, y) = \langle y, -x \rangle$ in the plane.

Solution

The calculation itself is easily accomplished.

$$\text{curl } \mathbf{F} = \nabla \times \langle y, -x, 0 \rangle$$

$$= \begin{vmatrix} \mathbf{i} & \mathbf{j} & \mathbf{k} \\ \dfrac{\partial}{\partial x} & \dfrac{\partial}{\partial y} & \dfrac{\partial}{\partial z} \\ y & -x & 0 \end{vmatrix}$$

$$= \mathbf{i} \left[\frac{\partial}{\partial y}(0) - \frac{\partial}{\partial z}(-x) \right] - \mathbf{j} \left[\frac{\partial}{\partial x}(0) - \frac{\partial}{\partial z}(y) \right] + \mathbf{k} \left[\frac{\partial}{\partial x}(-x) - \frac{\partial}{\partial y}(y) \right]$$

$$= 0\mathbf{i} - 0\mathbf{j} + (-2)\mathbf{k} = -2\mathbf{k}$$

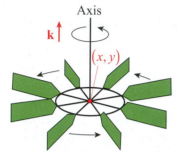

Axis

k ↑

(x, y)

Counterclockwise Direction

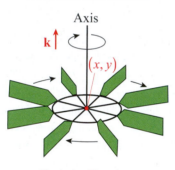

Axis

k ↑

(x, y)

Clockwise Direction

Figure 3

The physical interpretation of this result is worth our attention. If **F** models the velocity field for a fluid in the plane, such as water flowing toward a drain over the flat bottom of a sink, then the curl of **F** at any given point tells us something about the forces felt by a very small paddle wheel with axis in the **k** direction (see Figure 3). If curl $\mathbf{F}(x, y) = c\mathbf{k}$, with $c > 0$, the paddle wheel at (x, y) spins in the positive (counterclockwise) direction; if curl $\mathbf{F}(x, y) = c\mathbf{k}$, with $c < 0$, the wheel spins in the opposite direction. And if curl $\mathbf{F}(x, y) = \mathbf{0}$, the wheel doesn't spin at all, though that doesn't mean that there are no forces at the point (x, y)—a paddle wheel at such a point may feel a force pushing it, but not spinning it.

Figure 4 is a scaled illustration of the velocity field $\mathbf{F}(x, y) = \langle y, -x \rangle$. Since we know the curl of **F** is $-2\mathbf{k}$ at every point, a small paddle wheel at any point in the plane spins in the clockwise direction (at the same rate) as a result of the field.

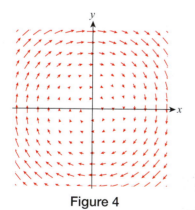

Figure 4

TOPIC 2   Green's Theorem

As mentioned in the introduction, Green's Theorem relates line integrals around a closed path to double integrals over the region enclosed by the path. We apply the theorem to closed paths that are also **simple**, meaning they don't self-intersect (cross over themselves) anywhere between their endpoints. The theorem can be used in either of two equivalent forms, with the choice depending on whether we are interested in the circulation of a vector field around the path or the flux of a vector field outward across the path.

The principle at the heart of Green's Theorem is most evident if we first apply it to a simple closed path C such as in Figure 5, where every line parallel to a coordinate axis intersects C in at most two points. We develop the theorem under the assumption that the orientation of C is positive, which we denote with a counterclockwise circle on the integral sign, such as $\oint$, and we let R denote the region enclosed by C. To use Green's Theorem in evaluating the line integral of a vector field $\mathbf{F} = \langle P, Q \rangle$ around C, we assume that $\mathbf{F}$ has continuous first partial derivatives on an open region containing R. We already know that the total circulation of $\mathbf{F}$ around C can be written as

$$\oint_C \mathbf{F} \cdot \mathbf{T} \, ds = \oint_C \mathbf{F} \cdot d\mathbf{r} = \oint_C P \, dx + Q \, dy,$$

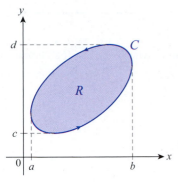

Figure 5

and we now examine the two integrals $\oint_C P \, dx$ and $\oint_C Q \, dy$ individually.

Claim 1: $\oint_C P \, dx = -\iint_R \dfrac{\partial P}{\partial y} \, dA$. To prove this, split C into the two functions of x illustrated in Figure 6, and note that by the Fundamental Theorem of Calculus

$$\int_{f_1(x)}^{f_2(x)} \frac{\partial P}{\partial y} \, dy = P(x, y) \Big]_{y=f_1(x)}^{y=f_2(x)} = P(x, f_2(x)) - P(x, f_1(x)),$$

so

$$\iint_R \frac{\partial P}{\partial y} \, dA = \int_a^b \int_{f_1(x)}^{f_2(x)} \frac{\partial P}{\partial y} \, dy \, dx = \int_a^b \left[P(x, f_2(x)) - P(x, f_1(x)) \right] dx$$

$$= -\int_b^a P(x, f_2(x)) \, dx - \int_a^b P(x, f_1(x)) \, dx = -\int_{C_2} P \, dx - \int_{C_1} P \, dx = -\oint_C P \, dx.$$

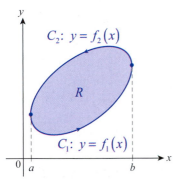

Figure 6

Claim 2: $\oint_C Q \, dy = \iint_R \dfrac{\partial Q}{\partial x} \, dA$. We prove this in a similar fashion, but this time by splitting C into the two functions of y illustrated in Figure 7. Then again by the Fundamental Theorem of Calculus,

$$\int_{g_1(y)}^{g_2(y)} \frac{\partial Q}{\partial x} \, dx = Q(x, y) \Big]_{x=g_1(y)}^{x=g_2(y)} = Q(g_2(y), y) - Q(g_1(y), y)$$

so

$$\iint_R \frac{\partial Q}{\partial x} \, dA = \int_c^d \int_{g_1(y)}^{g_2(y)} \frac{\partial Q}{\partial x} \, dx \, dy = \int_c^d \left[Q(g_2(y), y) - Q(g_1(y), y) \right] dy$$

$$= \int_c^d Q(g_2(y), y) \, dy + \int_d^c Q(g_1(y), y) \, dy = \int_{C_2} Q \, dy + \int_{C_1} Q \, dy = \oint_C Q \, dy.$$

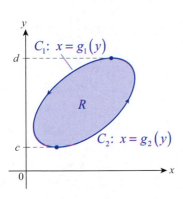

Figure 7

Combining the two pieces gives us one form of Green's Theorem:

$$\oint_C P \, dx + Q \, dy = \iint_R \left(\frac{\partial Q}{\partial x} - \frac{\partial P}{\partial y} \right) dA$$

As it turns out, Green's Theorem applies to simple closed paths much more general than what we have used to derive the formula, but we will only sketch how the proof can be extended. Before doing so, we illustrate the theorem's use.

Example 3

Evaluate $\oint_C \left(7y - \sin x^2\right) dx + \left(3x + e^{\cos y}\right) dy$, where C is the circle $x^2 + y^2 = 25$.

Solution

Since

$$\frac{\partial}{\partial y}\left(7y - \sin x^2\right) = 7 \neq 3 = \frac{\partial}{\partial x}\left(3x + e^{\cos y}\right),$$

the vector field implicit in the integral is not conservative and the Fundamental Theorem for Line Integrals can't be used. However, using Green's Theorem,

$$\oint_C \left(7y - \sin x^2\right) dx + \left(3x + e^{\cos y}\right) dy = \iint_R \left[\frac{\partial}{\partial x}\left(3x + e^{\cos y}\right) - \frac{\partial}{\partial y}\left(7y - \sin x^2\right)\right] dA$$

$$= \iint_R (3 - 7)\, dA = -4\iint_R dA$$

$$= -4\int_0^{2\pi}\int_0^5 r\, dr\, d\theta = -100\pi.$$

This example demonstrates how Green's Theorem can be used to easily evaluate integrals which would be extremely difficult to evaluate directly using a parametrization of C.

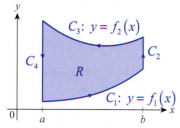

Figure 8

To see how the theorem applies to more general simple closed paths, consider the region R and closed path $C = C_1 \cup C_2 \cup C_3 \cup C_4$ of Figure 8. Suppose C_1 and C_3 represent the graphs of functions $f_1(x)$ and $f_2(x)$, respectively. Note that x is constant on C_2 and C_4, so $dx = 0$ along both sides.

$$\iint_R \frac{\partial P}{\partial y}\, dA = \int_a^b \int_{f_1(x)}^{f_2(x)} \frac{\partial P}{\partial y}\, dy\, dx = \int_a^b \left[P(x, f_2(x)) - P(x, f_1(x))\right] dx$$

$$= -\int_b^a P(x, f_2(x))\, dx - \int_a^b P(x, f_1(x))\, dx = -\int_{C_3} P\, dx - \int_{C_1} P\, dx$$

$$= -\left(\int_{C_1} P\, dx + \underbrace{\int_{C_2} P\, dx}_{0} + \int_{C_3} P\, dx + \underbrace{\int_{C_4} P\, dx}_{0}\right) = -\oint_C P\, dx$$

To see that

$$\oint_C Q\, dy = \iint_R \frac{\partial Q}{\partial x}\, dA$$

for the path C of Figure 8, we have to modify the argument slightly, as shown next.

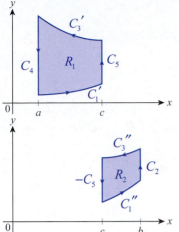

Figure 9

Although we proved Green's Theorem for a simply connected region bounded by a particularly nice simple closed path, the properties of line integrals and double integrals allow us to extend the theorem to regions and paths that are finite unions of simpler regions and paths. For example, we can divide the region R of Figure 8 into $R_1 \cup R_2$, with boundaries as labeled in Figure 9. We use the facts that

$$\iint_R \left(\frac{\partial Q}{\partial x} - \frac{\partial P}{\partial y} \right) dA = \iint_{R_1} \left(\frac{\partial Q}{\partial x} - \frac{\partial P}{\partial y} \right) dA + \iint_{R_2} \left(\frac{\partial Q}{\partial x} - \frac{\partial P}{\partial y} \right) dA$$

and

$$\int_{C_5} P\, dx + Q\, dy = -\int_{-C_5} P\, dx + Q\, dy$$

to obtain

$$\iint_R \left(\frac{\partial Q}{\partial x} - \frac{\partial P}{\partial y} \right) dA = \iint_{R_1} \left(\frac{\partial Q}{\partial x} - \frac{\partial P}{\partial y} \right) dA + \iint_{R_2} \left(\frac{\partial Q}{\partial x} - \frac{\partial P}{\partial y} \right) dA$$

$$= \int_{C_1'} P\, dx + Q\, dy + \int_{C_5} P\, dx + Q\, dy + \int_{C_3'} P\, dx + Q\, dy + \int_{C_4} P\, dx + Q\, dy$$

$$+ \int_{C_1''} P\, dx + Q\, dy + \int_{C_2} P\, dx + Q\, dy + \int_{C_3''} P\, dx + Q\, dy + \int_{-C_5} P\, dx + Q\, dy$$

$$= \int_{C_1' \cup C_1''} P\, dx + Q\, dy + \int_{C_2} P\, dx + Q\, dy + \int_{C_3' \cup C_3''} P\, dx + Q\, dy + \int_{C_4} P\, dx + Q\, dy$$

$$= \oint_C P\, dx + Q\, dy.$$

Similar decompositions can be used to break up any piecewise smooth simple closed path into pieces that bound sufficiently simple regions, a result that we summarize formally as Green's Theorem. In stating the theorem, we make use of the fact that the integrand of the double integral is the $\mathbf{k}$ component of $\nabla \times \langle P, Q, 0 \rangle$. Later in this chapter, we'll see that Green's Theorem is a special case of Stokes' Theorem.

Theorem 🔍

Green's Theorem (Tangential-Curl Form)

Let C be a positively oriented, piecewise smooth, simple closed curve in the plane, and let R be the region enclosed by C. If $\mathbf{F}(x, y) = \langle P(x, y), Q(x, y) \rangle$ and P and Q have continuous partial derivatives on an open region containing R, then

$$\oint_C \mathbf{F} \cdot \mathbf{T}\, ds = \oint_C \mathbf{F} \cdot d\mathbf{r} = \oint_C P\, dx + Q\, dy = \iint_R \left(\frac{\partial Q}{\partial x} - \frac{\partial P}{\partial y} \right) dA.$$

Extending $\mathbf{F}$ to $\mathbf{F}(x, y, z) = \langle P(x, y), Q(x, y), 0 \rangle$ and using the fact that

$$\frac{\partial Q}{\partial x} - \frac{\partial P}{\partial y} = \left(\frac{\partial Q}{\partial x} - \frac{\partial P}{\partial y} \right) \mathbf{k} \cdot \mathbf{k} = (\text{curl } \mathbf{F}) \cdot \mathbf{k} = \nabla \times \mathbf{F} \cdot \mathbf{k},$$

we can write this version of the formula as

$$\oint_C \mathbf{F} \cdot \mathbf{T}\, ds = \iint_R \nabla \times \mathbf{F} \cdot \mathbf{k}\, dA.$$

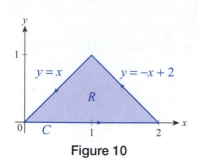

Figure 10

Example 4 ✐

Evaluate $\oint_C y^2\,dx + 3xy\,dy$ where C is the path shown in Figure 10.

Solution

Since $P = y^2$ and $Q = 3xy$,

$$\frac{\partial Q}{\partial x} - \frac{\partial P}{\partial y} = 3y - 2y = y.$$

(Alternatively, $\nabla \times \langle y^2, 3xy, 0 \rangle = \langle 0, 0, y \rangle$ and $\langle 0, 0, y \rangle \cdot \mathbf{k} = y$.)

So

$$\oint_C y^2\,dx + 3xy\,dy = \iint_R y\,dA = \int_{y=0}^{y=1} \int_{x=y}^{x=2-y} y\,dx\,dy = \int_0^1 y\big[(2-y)-y\big]\,dy$$

$$= 2\int_0^1 \left(y - y^2\right)dy = 2\left[\frac{y^2}{2} - \frac{y^3}{3}\right]_0^1 = \frac{1}{3}.$$

Remember what the result in Example 4 means physically: if $\mathbf{F}$ represents the velocity field of a fluid in the plane, the total circulation of $\mathbf{F}$ around C is $\frac{1}{3}$ (in Exercise 52 you will obtain the same result by parametrizing C and calculating the line integral of $\mathbf{F}$ around C directly). In other settings, we need to calculate the flux of $\mathbf{F}$ outward across the boundary of a simple closed path, and Green's Theorem can be used for that purpose as well.

If we let $\mathbf{n}$ denote the unit vector normal to C at any given point of the path, then we have the following:

$$\text{Outward flux of } \mathbf{F} \text{ across } C = \oint_C \mathbf{F} \cdot \mathbf{n}\,ds$$

This is the counterpart to the integral for the circulation of $\mathbf{F}$ around C.

$$\text{Counterclockwise circulation of } \mathbf{F} \text{ around } C = \oint_C \mathbf{F} \cdot \mathbf{T}\,ds$$

We can now develop the second version of Green's Theorem from the first. Since

$$\mathbf{T}(t) = \frac{\mathbf{r}'(t)}{|\mathbf{r}'(t)|} = \frac{1}{|\mathbf{r}'(t)|}\langle x'(t), y'(t) \rangle,$$

the outward normal unit vector $\mathbf{n}$ is the vector function

$$\mathbf{n}(t) = \frac{1}{|\mathbf{r}'(t)|}\langle y'(t), -x'(t) \rangle$$

(see Figure 11). So we have

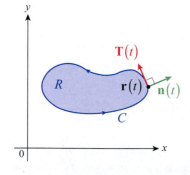

Figure 11

$$\oint_C \mathbf{F} \cdot \mathbf{n}\, ds = \int_a^b (\mathbf{F} \cdot \mathbf{n})(t)\left| \mathbf{r}'(t) \right| dt$$

$$= \int_a^b \left[\frac{P(x(t), y(t))\, y'(t)}{\left| \mathbf{r}'(t) \right|} - \frac{Q(x(t), y(t))\, x'(t)}{\left| \mathbf{r}'(t) \right|} \right] \left| \mathbf{r}'(t) \right| dt$$

$$= \int_a^b \left[P(x(t), y(y))\, y'(t) - Q(x(t), y(t))\, x'(t) \right] dt$$

$$= \int_a^b -Q\, dx + P\, dy = \iint_R \left(\frac{\partial P}{\partial x} + \frac{\partial Q}{\partial y} \right) dA. \qquad \textcolor{purple}{\text{By Green's Theorem}}$$

Using the divergence operator, we can write this as follows (later, we will see this is a special case of the Divergence Theorem).

Theorem 🔍

Green's Theorem (Normal-Divergence Form)

Let C be a positively oriented, piecewise smooth, simple closed curve in the plane, and let R be the region enclosed by C. If $\mathbf{F}(x, y) = \langle P(x, y), Q(x, y) \rangle$ and P and Q have continuous partial derivatives on an open region containing R, then

$$\oint_C \mathbf{F} \cdot \mathbf{n}\, ds = \oint_C P\, dy - Q\, dx = \iint_R \left(\frac{\partial P}{\partial x} + \frac{\partial Q}{\partial y} \right) dA = \iint_R \nabla \cdot \mathbf{F}\, dA.$$

Example 5 ✏️

Evaluate the outward flux of the vector field $\mathbf{F}(x, y) = \langle y^2, 3xy \rangle$ across the path C of Figure 12.

Solution

Since $P = y^2$ and $Q = 3xy$,

$$\oint_C \mathbf{F} \cdot \mathbf{n}\, ds = \iint_R \nabla \cdot \mathbf{F}\, dA = \iint_R \left(\frac{\partial P}{\partial x} + \frac{\partial Q}{\partial y} \right) dA$$

$$= \int_{y=0}^{y=1} \int_{x=y}^{x=2-y} 3x\, dx\, dy = \frac{3}{2} \int_0^1 \left[(2-y)^2 - y^2 \right] dy$$

$$= 6 \int_0^1 (1-y)\, dy = 6 \left[y - \frac{y^2}{2} \right]_0^1 = 3.$$

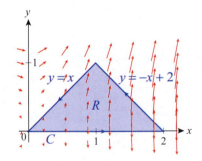

Figure 12

The vector field $\mathbf{F}$ is also shown in Figure 12. The flux into and out of the enclosed region along the lower edge of C is 0 (since $y = 0$), and the flux is positive (outward) along the other two edges, consistent with the result we have calculated for total flux.

The fact that two line integrals cancel if they are evaluated along the same path with opposite orientation means that we can decompose even nonsimply connected regions and apply Green's Theorem, as long as the boundary paths are oriented appropriately. Consider, for example, the region R of Figure 13. Its boundary C consists of two simple closed curves C_1 and C_2, which we orient so that R is always on the left as we traverse either curve. Then if we divide R into two subregions R_1 and R_2, as shown in Figure 14, the additive properties of area and line integrals (along with Green's Theorem) again allow us to state

$$\iint_R \left(\frac{\partial Q}{\partial x} - \frac{\partial P}{\partial y} \right) dA = \iint_{R_1} \left(\frac{\partial Q}{\partial x} - \frac{\partial P}{\partial y} \right) dA + \iint_{R_2} \left(\frac{\partial Q}{\partial x} - \frac{\partial P}{\partial y} \right) dA$$

$$= \int_{C_1} P\,dx + Q\,dy + \int_{C_2} P\,dx + Q\,dy$$

$$= \int_C P\,dx + Q\,dy.$$

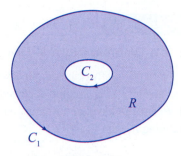

Figure 13

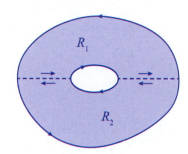

Figure 14

Example 6 ✎

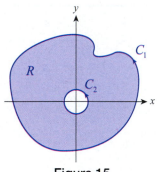

Figure 15

Let $\mathbf{F}(x,y) = \dfrac{\langle -y, x \rangle}{x^2 + y^2}$. Show that $\displaystyle\oint_{C_1} \mathbf{F} \cdot d\mathbf{r} = \oint_{C_2} \mathbf{F} \cdot d\mathbf{r}$ for C_1 and C_2 as shown in Figure 15.

Solution

Both C_1 and C_2 are oriented counterclockwise, but if we define the region R as being bounded by $C_1 \cup (-C_2)$, then R is to the left as we traverse any part of its boundary, so

$$\int_{C_1} P\,dx + Q\,dy + \int_{-C_2} P\,dx + Q\,dy = \iint_R \left(\frac{\partial Q}{\partial x} - \frac{\partial P}{\partial y} \right) dA$$

$$= \iint_R \left[\frac{y^2 - x^2}{\left(x^2 + y^2\right)^2} - \frac{y^2 - x^2}{\left(x^2 + y^2\right)^2} \right] dA = 0.$$

Hence,

$$\int_{C_1} P\,dx + Q\,dy = -\int_{-C_2} P\,dx + Q\,dy = \int_{C_2} P\,dx + Q\,dy$$

which is equivalent to

$$\oint_{C_1} \mathbf{F} \cdot d\mathbf{r} = \oint_{C_2} \mathbf{F} \cdot d\mathbf{r}.$$

In Exercise 91, you will show that this implies $\displaystyle\oint_C \mathbf{F} \cdot d\mathbf{r} = 2\pi$ for every such simple closed path C enclosing the origin.

15.4 **Exercises**

1–10 Find the divergence of the vector field **F** and evaluate the divergence at the given points.

1. $\mathbf{F}(x,y)=\left\langle 5x^3-2xy,4xy^2\right\rangle$; $A(2,1)$, $B(1,-3)$

2. $\mathbf{F}(x,y)=\left\langle xe^{xy},y^2\cos x\right\rangle$; $A(0,-1)$, $B(\pi/2,2/\pi)$

3. $\mathbf{F}(x,y,z)=\left\langle x^2,3xz^2,-2yz\right\rangle$; $A(2,-1,0)$, $B(1,1,4)$

4. $\mathbf{F}(x,y,z)=\left\langle \dfrac{x^2}{2},\dfrac{y^2}{2},\dfrac{z^2}{2}\right\rangle$; $A(-1,3,0)$, $B(5,-7,2)$

5. $\mathbf{F}(x,y,z)=\left\langle 2yz,2xz,2xy\right\rangle$; $A(1,1,1)$, $B(2,-3,4)$

6. $\mathbf{F}(x,y,z)=\left\langle xy^2,xz+2y,z^2\right\rangle$; $A(2,0,-2)$, $B(4,9,0)$

7. $\mathbf{F}(x,y,z)=\left\langle x+y^3,e^y,z^4\right\rangle$; $A(1,0,-3)$, $B(2,1,4)$

8. $\mathbf{F}(x,y,z)=\left\langle \cos(xy^2),x,2y-z\right\rangle$; $A(-2,0,1)$, $B(\pi/2,5,5)$

9. $\mathbf{F}(x,y,z)=\left\langle e^{xyz},e^{xyz},e^{xyz}\right\rangle$; $A(1,0,1)$, $B(e,1,-1)$

10. $\mathbf{F}(x,y,z)=\left\langle x^2y,xy-z^2,\dfrac{2z^2}{x^2+y^2}\right\rangle$; $A(0,1,3)$, $B\left(-1,1,\dfrac{1}{2}\right)$

11–20. Find the curl of each vector field given in Exercises 1–10.

21. Use the figure to decide whether the divergence of the vector field at the given points is positive, negative, or zero.

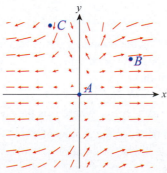

22. You may have noticed by now that for a vector field $\mathbf{F}=\langle P,Q,R\rangle$, those terms in P involving only x do not affect the curl; neither do those terms in Q involving only y. A similar statement holds for R and z. For example,

$$\mathrm{curl}\left\langle x^2\cos x,y^4+z,\dfrac{1}{\sqrt{1+z^2}}+y^2\right\rangle=\mathrm{curl}\left\langle 0,z,y^2\right\rangle.$$

Prove the above assertions.

23–24 Use the "shortcut" suggested by Exercise 22 to evaluate the indicated curl.

23. $\mathrm{curl}\left\langle y+\ln(x^2+1),y^3-\sin y,e^{\sin z}+1\right\rangle$

24. $\mathrm{curl}\left\langle z-\ln\sqrt{1+x^2},x^2+\arctan(y^2),\dfrac{z}{1+z^2}\right\rangle$

25. Determine the curl of $\mathbf{F}(x,y)=\left\langle y^2,0\right\rangle$ and provide a physical interpretation of the result (as done in Example 2). See the figure below for guidance.

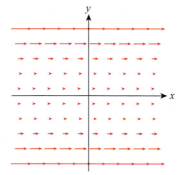

26–29 Find $\mathrm{curl}(\mathrm{curl}\,\mathbf{F})$ for the given vector field.

26. $\mathbf{F}(x,y,z)=\langle x,-y,xyz\rangle$

27. $\mathbf{F}(x,y,z)=\left\langle xy^2,xz,-4xz^2\right\rangle$

28. The vector field given in Exercise 3

29. The vector field given in Exercise 6

30–37 Decide whether or not the expression has meaning. If an expression does have meaning, state whether it is a vector field or a scalar field.

30. $\nabla \times (\nabla \cdot \mathbf{F})$

31. $\nabla \cdot (\nabla \cdot \mathbf{F})$

32. $\nabla \cdot (\nabla \times \mathbf{F})$

33. $\nabla (\nabla \cdot \mathbf{F})$

34. $\nabla (\nabla \times \mathbf{F})$

35. $\nabla \cdot (\nabla f)$

36. $\nabla \times (\nabla f)$

37. $\nabla (\nabla f)$

38–47 Supposing that all of the appropriate partial derivatives of the vector fields $\mathbf{F}$ and $\mathbf{G}$ are continuous, prove the given statement. (Assume that f and g are at least twice differentiable scalar fields; a, b denote constants.)

38. $\nabla \times (\nabla f) = \mathbf{0}$

39. $\nabla \cdot (\nabla \times \mathbf{F}) = 0$

40. $\nabla \cdot (a\mathbf{F} + b\mathbf{G}) = a(\nabla \cdot \mathbf{F}) + b(\nabla \cdot \mathbf{G})$

41. $\nabla \times (a\mathbf{F} + b\mathbf{G}) = a(\nabla \times \mathbf{F}) + b(\nabla \times \mathbf{G})$

42. $\nabla \times (\nabla f + \nabla \times \mathbf{F}) = \nabla \times (\nabla \times \mathbf{F})$

43. $\nabla \cdot (\mathbf{F} \times \mathbf{G}) = (\nabla \times \mathbf{F}) \cdot \mathbf{G} - \mathbf{F} \cdot (\nabla \times \mathbf{G})$

44. $\nabla \cdot (f\mathbf{F}) = f(\nabla \cdot \mathbf{F}) + (\nabla f) \cdot \mathbf{F}$

45. $\nabla \times (f\mathbf{F}) = f(\nabla \times \mathbf{F}) + (\nabla f) \times \mathbf{F}$

46. $\nabla \times (\nabla f \times \nabla g) = \mathbf{0}$

47. $\nabla \cdot (\nabla fg) = f\nabla \cdot (\nabla g) + g\nabla \cdot (\nabla f) + 2(\nabla f) \cdot (\nabla g)$

48–51 Verify Green's Theorem by demonstrating the equality

$$\oint_C P\,dx + Q\,dy = \iint_R \left(\frac{\partial Q}{\partial x} - \frac{\partial P}{\partial y} \right) dA, \text{ where } P(x,y) = xy^2,$$

$Q(x,y) = 2x$, and C is a smooth closed curve enclosing the region R as specified.

48. R: The triangle with vertices $(0,0)$, $(3,0)$, and $(0,3)$

49. R: The square $[-1,1] \times [-1,1]$

50. R: The region bounded by the graphs of $y = x^2$ and $y = x$

51. R: The unit circle (disk) centered at the origin

52. Parametrize curve C given in Example 4 and calculate the line integral of $\mathbf{F}$ around C to show $\oint_C y^2\,dx + 3xy\,dy = \frac{1}{3}$.

53–62 Use Green's Theorem to evaluate the line integral. (Pay attention to the orientation of the path as indicated by the integration symbol.)

53. $\oint_C xy\,dx + (2x + y)\,dy$, where C is the boundary of the region between the graphs of $y = 1 - x^2$ and the x-axis, oriented counterclockwise

54. $\oint_C 4xy\,dx + \ln(x^2 + y^2)\,dy$, where C is the cardioid $r = 1 + \cos\theta$, oriented counterclockwise

55. $\oint_C 2y^2\,dx - xy\,dy$, where C is the graph of $y = \sqrt{x}$ from the origin to $(1,1)$, followed by $y = x^2$ from $(1,1)$ back to $(0,0)$

56. $\oint_C x^2 y\,dx + (x^3 + y^2)\,dy$, where C is the boundary of the region bounded by the x-axis, the graphs of $y = x^3$ and $x = 2$, oriented counterclockwise

57. $\oint_C 4x^2 y^3\,dx - x^3 y\,dy$, where C is the boundary of the region bounded by the coordinate axes and the line $y = 2 - 2x$, oriented clockwise

58. $\oint_C \left[\cos(x^2) - 2y \right] dx + \left[4x + \ln(y^4) \right] dy$, where C is the unit circle centered at the origin, with positive orientation

59. $\oint_C \left(y^3 - \sqrt{x^2 + 2} \right) dx + \left(3xy^2 - \cos y \right) dy$, where C is defined as the border of a map of Louisiana contained in the first quadrant (Assume that the border is a smooth, simple, closed curve.)

60. $\oint_C e^x\,dx + y^2 e^x\,dy$, where C is the triangle with vertices $(0,0)$, $(1,2)$, and $(5,0)$, oriented counterclockwise

61. $\oint_C \left[y + 2\tan^{-1}(x/4) \right] dx - \left(e^{x^2+1} + 2\sec y \right) dy$, where C is the boundary of the region between the graphs of $y = |x|$ and $y = 2 - |x|$ with positive orientation

62. $\oint_C xe^y\,dx + xy\,dy$, where C is the boundary of the region between the graphs of $x = y^2$ and $x - y = 2$, with clockwise orientation

63–67 In these exercises, we will revisit some exercises from Section 15.2. Use Green's Theorem to provide a second solution and verify that your answers agree.

63. Determine $\int_C (x^2 y + xy^2) \, dx + x^3 \, dy$, where C is the unit circle centered at the origin. (See Exercise 27 of Section 15.2.)

64. Determine $\int_C y \, dx + xy \, dy$, where C is the parabola $y = x^2$ joining the origin with $(1,1)$, followed by the line segment from $(1,1)$ back to the origin. (See Exercise 28 of Section 15.2.)

65. Find the line integral $\int_C 2xy \, dx + x^2 \, dy$ on the curve given in Exercise 64. (See Exercise 29 of Section 15.2.)

66. Find $\int_C (x^2 + y) \, dx + xy \, dy$ along the closed path $C = C_1 \cup C_2 \cup C_3$, where C_1 is the line segment from the origin to $(1,0)$, C_2 is the line segment from $(1,0)$ to $(2,1)$, and C_3 is the straight path from $(2,1)$ back to the origin. (See Exercise 30 of Section 15.2.)

67. Find $\int_C 2xy \, dx + x^2 \, dy$ along the path given in Exercise 66. (See Exercise 31 of Section 15.2.)

68. Determine the work done by the force field $\mathbf{F}(x,y) = \langle x + y, 2xy \rangle$ if it moves a particle clockwise around the triangle with vertices $(0,0)$, $(2,0)$, and $(1,1)$.

69. Determine the work done by the force field given in Exercise 68 as it moves a particle clockwise around the square $[-1,1] \times [-1,1]$.

70. Determine the work done by the force field $\mathbf{F}(x,y) = \langle xy, x^2 - y^2 \rangle$ as it moves a particle counterclockwise around the unit circle centered at the origin.

71. Use Green's Theorem to determine the work done by the force field $\mathbf{F}(x,y) = \langle 2, 3x \rangle$ if it moves a particle counterclockwise around the ellipse parametrized as $\mathbf{r}(t) = \langle 3 + 3\cos t, 2 + 2\sin t \rangle$, $0 \le t \le 2\pi$. (See Exercise 36 of Section 15.2.)

72–75 Evaluate the outward flux of the vector field $\mathbf{F}$ across the specified path C.

72. $\mathbf{F}(x,y) = \langle xy, 2x + y \rangle$, C is the boundary of the region between the graphs of $y = 1 - x^2$ and the x-axis

73. $\mathbf{F}(x,y) = \langle x^2 y, x^3 + y^2 \rangle$, C is the boundary of the region bounded by the x-axis and the graphs of $y = x^3$ and $x = 2$

74. $\mathbf{F}(x,y) = \langle 4x^2 y^3, -x^3 y \rangle$, C is the boundary of the region bounded by the coordinate axes and the line $y = 2 - 2x$

75. $\mathbf{F}(x,y) = \langle e^x, e^x y^2 \rangle$, C is the triangle with vertices $(0,0)$, $(1,2)$, and $(5,0)$

76–84 Suppose that R is a region bounded by a positively oriented, piecewise smooth, simple closed planar curve C. Recall that the area of R is the double integral of $f(x,y) = 1$ over R; that is, $A = \iint_R 1 \cdot dA$. With Green's Theorem in mind, if we choose P and Q so that $\dfrac{dQ}{dx} - \dfrac{dP}{dy} = 1$, then the area of R can be found from a line integral:

$$A = \iint_R 1 \cdot dA = \iint_R \left(\frac{dQ}{dx} - \frac{dP}{dy} \right) dA = \oint_C P \, dx + Q \, dy.$$

In Exercises 76–84, use this observation to prove the area formula.

76. Show that if R and C are as above, then the area A of R is

$$A = \frac{1}{2} \oint_C x \, dy - y \, dx = \oint_C x \, dy = -\oint_C y \, dx.$$

77. Use Exercise 76 to verify the area formula of the circle $x^2 + y^2 = r^2$.

78. Use Exercise 76 to find the area of the ellipse $\dfrac{x^2}{a^2} + \dfrac{y^2}{b^2} = 1$.

79. Use Exercise 76 to find the area of the region bounded by the y-axis and the lines $3y = x$ and $y = 4 - x$.

80. Use Exercise 76 to find the area of the region bounded by the graphs of $y = x^2$ and $y = x + 2$.

81. Use Exercise 76 to find the area of the region bounded by the graphs of $y = 9 - x^2$ and $3y = x^2 - 9$.

82.* The graph of the equation $x^3 + y^3 = 3xy$ is called the folium of Descartes. Use Exercise 76 to find the area of its loop. (**Hint:** Choose $t = y/x$ to obtain the parametrization $x = 3t/(1+t^3)$ and $y = 3t^2/(1+t^3)$. Use the Quotient Rule to obtain $d(y/x) = (x\,dy - y\,dx)/x^2$; then use Exercise 76. Note that you obtain half the area of the loop by integrating on the interval $0 \le t \le 1$.)

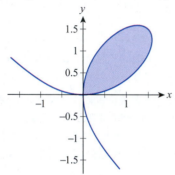

83.* Follow the hint provided for Exercise 82 to obtain the area of the first-quadrant loop of the lemniscate $(x^2 + y^2)^2 = 16xy$. (Note that by symmetry, doubling your answer would yield the total area enclosed by the lemniscate.)

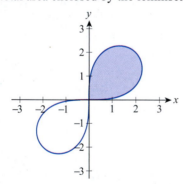

84. Use Exercise 76 to obtain the following convenient area formula for the polygon with vertices $(x_1, y_1), (x_2, y_2), ..., (x_n, y_n)$. Show that the area A of such a polygon is

$$A = \frac{1}{2} \sum_{k=1}^{n} (x_k y_{k+1} - x_{k+1} y_k),$$

with the convention that $(x_1, y_1) = (x_{n+1}, y_{n+1})$.

(**Hint:** Start by letting S be one side of the polygon, say the one connecting

(x_1, y_1) with (x_2, y_2), and show that $\frac{1}{2}\int_S x\,dy - y\,dx = \frac{1}{2}(x_1 y_2 - x_2 y_1)$. Then use Exercise 76.)

85–88 Use the formula from Exercise 84 to find the area of the given polygon.

85. The square with vertices $(0,0)$, $(3,0)$, $(3,3)$, and $(0,3)$

86. The triangle with vertices $(0,0)$, $(2,4)$, and $(0,4)$

87. The hexagon with vertices $(0,0)$, $(2,0)$, $(4,2)$, $(3,4)$, $(1,4)$, and $(0,2)$

88. The (nonconvex) pentagon with vertices $(1,0)$, $(3,0)$, $(5,7)$, $(2,2)$, and $(0,3)$

89. Evaluate $\oint_C (3x^3 - y)\,dx + (y^2 + 4x)\,dy$ on the ellipse given in Exercise 78.

90. Let C_2 be the unit circle centered at the origin, let $C = C_1 \cup C_2$ be the boundary of region R with the indicated orientations as in the figure below, and let $\mathbf{F}(x, y) = \langle 4x - 3y, 3x + y^2 \rangle$. Assuming that the area of R is 4π square units, use Green's Theorem to evaluate $\oint_{C_1} \mathbf{F} \cdot d\mathbf{r}$. (**Hint:** Determine

$$\iint_R \left(\frac{\partial Q}{\partial x} - \frac{\partial P}{\partial y} \right) dA$$

and use the fact that by Green's Theorem, it equals $\oint_{C_1} \mathbf{F} \cdot d\mathbf{r} - \oint_{C_2} \mathbf{F} \cdot d\mathbf{r}$.)

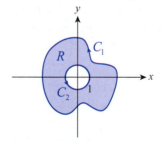

91. In Exercise 35 of Section 15.3, you showed that for the vector field $\mathbf{F}$ in Example 6, $\oint_{C_r} \mathbf{F} \cdot d\mathbf{r} = 2\pi$, where C_r is a circle of radius r around the origin (you were free to choose the radius in that problem, but $r = 1$ is the natural choice). Combine that result with Example 6 of this section to show that $\oint_C \mathbf{F} \cdot d\mathbf{r} = 2\pi$ for every simple closed path C enclosing the origin.

92. For the vector field given in Exercise 91, show that $\oint_C \mathbf{F}\cdot d\mathbf{r}=0$ for any simple closed path C that is disjoint from, and does not enclose the origin.

93. Use the technique illustrated in Example 6 to find $\oint_C \mathbf{F}\cdot d\mathbf{r}$ for the vector field

$$\mathbf{F}(x,y)=\frac{\langle x,y\rangle}{\left(x^2+y^2\right)},$$

where C is a simple closed path enclosing the origin.

94. Use Green's Theorem to show that the area of a region R enclosed by the simple closed curve C in polar coordinates is

$$A=\frac{1}{2}\oint_C r^2\, d\theta.$$

95. Use Green's Theorem to show that the coordinates of the centroid of a region R of constant density ρ and area A, enclosed by the simple closed curve C can be found as

$$\bar{x}=\frac{1}{2A}\oint_C x^2\, dy,\quad \bar{y}=-\frac{1}{2A}\oint_C y^2\, dx.$$

(**Hint:** Start by determining $M_x=\iint_R y\rho\, dA$ and $M_y=\iint_R x\rho\, dA$.)

96.* Use Green's Theorem to show that the second moments I_x and I_y of the region R given in Exercise 95 are

$$I_x=-\frac{1}{3}\oint_C \rho y^3\, dx\quad\text{and}\quad I_y=\frac{1}{3}\oint_C \rho x^3\, dy.$$

97. Use Exercise 95 to find the centroid of the half disk bounded by the upper semicircle $y=\sqrt{R^2-x^2}$ and the x-axis.

98. Use Exercise 96 to find the second moments about the coordinate axes of the half disk given in Exercise 97, assuming constant density ρ.

99. Use Green's Theorem to provide a proof of the Component Test. If D is an open and simply connected region and $\mathbf{F}=\langle P,Q\rangle$ is a vector field with continuous first partials, then $\partial P/\partial y=\partial Q/\partial x$ throughout D implies $\mathbf{F}$ is conservative.

(**Hint:** Show that $\mathbf{F}$ is path independent by using Green's Theorem and demonstrating that for a simple closed path C enclosing the region R,

$$\oint_C \mathbf{F}\cdot d\mathbf{r}=\oint_C P\, dx+Q\, dy=\iint_R\left(\frac{\partial Q}{\partial x}-\frac{\partial P}{\partial y}\right)dA=0.$$

Extend this result to general closed paths.)

100–103 *True or False?* Determine whether the given statement is true or false. In case of a false statement, explain or provide a counterexample.

100. The divergence of a vector field at a point P is a vector pointing in the direction of greatest outflow from a small rectangular neighborhood.

101. The curl of a vector field is a vector field.

102. The curl of a planar (i.e., two-dimensional) vector field is undefined.

103. In applications of Green's Theorem, clockwise orientation is negative.

15.5 **Parametric Surfaces and Surface Area**

TOPICS

1. Parametric surfaces

2. Surface area

We now know how to integrate functions along curves, and are ready to move to the next stage of integrating functions along surfaces. The ability to do so allows us to calculate such things as fluid flow through a permeable membrane, heat transfer across a surface, and the net electrical charge on a surface. The first step is to develop a more general method of describing and working with surfaces.

TOPIC 1   **Parametric Surfaces**

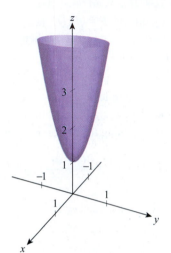

Figure 1

Just as curves are objects describable with a single parameter, surfaces are, in general, objects describable with two parameters. We can do so in a very literal fashion with a vector function of the form $\mathbf{r}(s,t) = \langle x(s,t), y(s,t), z(s,t) \rangle$ where the parameters s and t are allowed to take on values in a prescribed parameter region. For the purposes of comparison, the elliptic paraboloid shown in Figure 1 can be defined in the following ways:

Explicit form	$z = f(x, y) = 9x^2 + 4y^2 + 1$
Implicit form	$g(x, y, z) = 9x^2 + 4y^2 - z = -1$
Parametric form	$\mathbf{r}(s,t) = \langle s, t, 9s^2 + 4t^2 + 1 \rangle, \quad s \in \mathbb{R}, \quad t \in \mathbb{R}$

Other approaches to defining surfaces also imply two degrees of freedom (that is, two independent parameters). For instance, using cylindrical coordinates, the surface of Figure 1 can be described parametrically as

$$\mathbf{r}(r,\theta) = \left\langle \underbrace{r\cos\theta}_{x}, \underbrace{r\sin\theta}_{y}, 9\underbrace{r^2\cos^2\theta}_{x^2} + 4\underbrace{r^2\sin^2\theta}_{y^2} + 1 \right\rangle, \quad 0 \le r, \quad 0 \le \theta < 2\pi.$$

And while the surfaces of revolution studied in Chapter 6 were defined by revolving the graph of a single-variable function around an axis, they also implicitly involve two parameters.

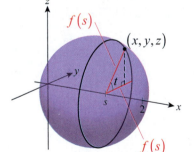

Figure 2

Example 1 ✎

Let S denote the surface formed by revolving the graph of $f(x) = \sqrt{2x - x^2}$, $0 \le x \le 2$, about the x-axis. Describe the surface with a vector function of two parameters.

Solution

If we let $x = s$, then the y- and z-components of the surface for any fixed s are $y = f(s)\cos t$ and $z = f(s)\sin t$, as shown in Figure 2. We allow s to take on any value from 0 to 2, while the parameter t represents any possible angle between 0 and 2π, so S is described by

$$\mathbf{r}(s,t) = \langle s, f(s)\cos t, f(s)\sin t \rangle, \quad 0 \le s \le 2, \quad 0 \le t \le 2\pi.$$

Since this implies

$$y^2 + z^2 = \left[f(x) \right]^2 = 2x - x^2,$$

another way of describing S is with the equation $(x-1)^2 + y^2 + z^2 = 1$. Any of these approaches generates the surface shown in Figure 3.

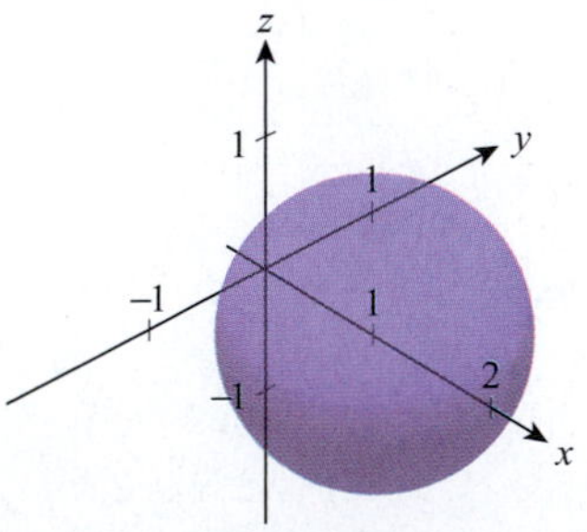

Figure 3

Example 2 ✐

In spherical coordinates, the equation $\rho = 3$ corresponds to the sphere of radius 3 centered at the origin. Describe this surface with a vector function of two parameters.

Solution

The simplest way to do this is to convert spherical coordinates to Cartesian and use θ and φ as the parameters, as shown.

$$\mathbf{r}(\theta, \varphi) = \left\langle \underbrace{3 \sin \varphi \cos \theta}_{x}, \underbrace{3 \sin \varphi \sin \theta}_{y}, \underbrace{3 \cos \varphi}_{z} \right\rangle, \quad 0 \leq \theta \leq 2\pi, \quad 0 \leq \varphi \leq \pi$$

Given a surface S described parametrically by $\mathbf{r}(s,t)$, **grid curves** on S are curves along which either s or t is held constant. Grid curves are useful in visualizing a surface and, as we will see, in defining tangent planes for a surface.

Example 3 ✐

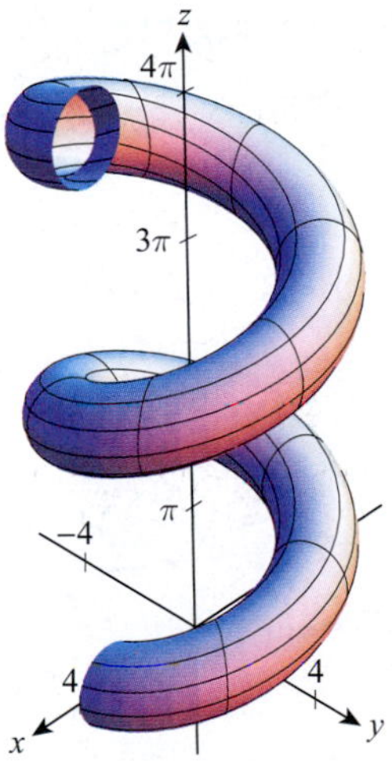

Figure 4

The surface

$$\mathbf{r}(s,t) = \left\langle (3 + \sin t) \cos s, (3 + \sin t) \sin s, s + \cos t \right\rangle,$$

where $0 \leq s \leq 4\pi$ and $0 \leq t \leq 2\pi$, is shown in Figure 4. Holding s constant, say $s = k$, results in the family of curves

$$\mathbf{r}(k,t) = \left\langle (3 + \sin t) \cos k, (3 + \sin t) \sin k, k + \cos t \right\rangle,$$

each of which is a circular cross-section of the surface. For instance,

$$\mathbf{r}(0,t) = \left\langle 3 + \sin t, 0, \cos t \right\rangle$$

is the circle of radius 1 in the xz-plane centered at $(3,0,0)$ and

$$\mathbf{r}\left(\frac{\pi}{2}, t\right) = \left\langle 0, 3 + \sin t, \frac{\pi}{2} + \cos t \right\rangle$$

is the circle of radius 1 in the yz-plane centered at $(0, 3, \pi/2)$. Holding t constant results in the other family of grid curves of the form

$$\mathbf{r}(s,k) = \left\langle (3 + \sin k) \cos s, (3 + \sin k) \sin s, s + \cos k \right\rangle,$$

each of which is a helix on the surface.

Example 4 ✐

Let S be the plane containing the point $(0,0,1)$ and grid lines whose direction vectors are $\mathbf{u} = \left\langle 1, 3, 2 \right\rangle$ and $\mathbf{v} = \left\langle -1, 2, 1 \right\rangle$. Describe S with a vector function of two parameters.

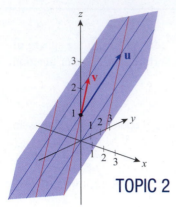

Figure 5

Solution

Every point on S is the terminal point of a position vector of the form

$$\mathbf{r}(s,t) = \langle 0,0,1 \rangle + s\,\mathbf{u} + t\,\mathbf{v}$$

for some s and some t, so $\mathbf{r}$ describes S. Sample grid lines of S in the direction of $\mathbf{u}$ and $\mathbf{v}$ are shown in Figure 5. The vector $\mathbf{u}$ is shown in blue and the vector $\mathbf{v}$ in red, both with initial point $(0,0,1)$.

TOPIC 2 Surface Area

We can use the grid curves of a surface S to describe the plane tangent to S at a given point $\mathbf{r}(s_0,t_0)$, a key step in our more general definition of surface area. Keeping the parameter t constant with $t = t_0$, the line in space tangent to one grid curve at $\mathbf{r}(s_0,t_0)$ has direction vector $\mathbf{r}_s(s_0,t_0)$; similarly, keeping s fixed at $s = s_0$, the line tangent to the other grid curve at the point has direction vector $\mathbf{r}_t(s_0,t_0)$. Since $\mathbf{n} = \mathbf{r}_s \times \mathbf{r}_t$ is normal to both grid curves, we can use $\mathbf{n}$ to define the tangent plane to S at $\mathbf{r}(s_0,t_0)$, as long as $\mathbf{r}_s \times \mathbf{r}_t \neq \mathbf{0}$ (see Figure 6). We say that S is **smooth** at $\mathbf{r}(s_0,t_0)$ if $\mathbf{r}_s$ and $\mathbf{r}_t$ are continuous at (s_0,t_0) and $\mathbf{r}_s(s_0,t_0) \times \mathbf{r}_t(s_0,t_0) \neq \mathbf{0}$, so we now know how to construct the tangent plane to a surface $\mathbf{r}(s,t)$ at every point of smoothness.

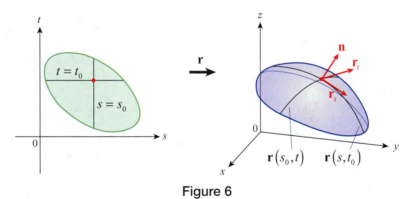

Figure 6

Example 5 ✎

Construct an equation for the plane tangent to the parametric surface $\mathbf{r}(s,t) = \langle s^2, t^2, 5s + 2t \rangle$ at the point $\mathbf{r}(1,1)$.

Solution

We have $\mathbf{r}_s(s,t) = \langle 2s, 0, 5 \rangle$ and $\mathbf{r}_t(s,t) = \langle 0, 2t, 2 \rangle$, so

$$\mathbf{n}(s,t) = \begin{vmatrix} \mathbf{i} & \mathbf{j} & \mathbf{k} \\ 2s & 0 & 5 \\ 0 & 2t & 2 \end{vmatrix} = \langle -10t, -4s, 4st \rangle.$$

Thus, $\mathbf{n}(1,1) = \langle -10, -4, 4 \rangle$ is normal to the tangent plane when $s = 1$ and $t = 1$, and since $\mathbf{r}(1,1) = \langle 1,1,7 \rangle$, we have the following equation for the plane.

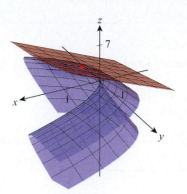

Figure 7

$$\left(\langle x,y,z\rangle-\langle 1,1,7\rangle\right)\cdot\langle -10,-4,4\rangle=0,\quad\text{or}\quad 5x+2y-2z=-7$$

Alternatively, since $\mathbf{r}_s(1,1)=\langle 2,0,5\rangle$ and $\mathbf{r}_t(1,1)=\langle 0,2,2\rangle$, the tangent plane can be described parametrically with the vector function

$$\begin{aligned}\mathbf{p}(s,t)&=\mathbf{r}(1,1)+s\,\mathbf{r}_s(1,1)+t\,\mathbf{r}_t(1,1)\\&=\langle 1,1,7\rangle+s\langle 2,0,5\rangle+t\langle 0,2,2\rangle\\&=\langle 2s+1,2t+1,5s+2t+7\rangle.\end{aligned}$$

An illustration of the surface $\mathbf{r}(s,t)=\langle s^2,t^2,5s+2t\rangle$ (in blue) and the plane tangent at the point $\mathbf{r}(1,1)$ (in red) is shown in Figure 7.

To determine the area of a surface S defined by the vector function $\mathbf{r}(s,t)$ over a region R of the st-plane, we divide R into an array of rectangles as shown in Figure 8 and add up the areas of the corresponding subdivisions of S.

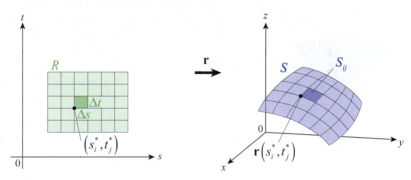

Figure 8

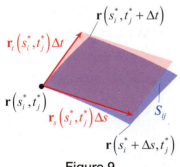

Figure 9

Let $\left(s_i^*,t_j^*\right)$ be the lower-left corner of the highlighted rectangle in the st-plane, and consider the highlighted subdivision S_{ij} of S. S_{ij} is approximated by a parallelogram in the plane tangent to S at the point $\mathbf{r}\left(s_i^*,t_j^*\right)$, as shown in Figure 9. The vectors defining the parallelogram are approximately

$$\mathbf{r}\left(s_i^*+\Delta s,t_j^*\right)-\mathbf{r}\left(s_i^*,t_j^*\right)\approx\mathbf{r}_s\left(s_i^*,t_j^*\right)\Delta s$$

and

$$\mathbf{r}\left(s_i^*,t_j^*+\Delta t\right)-\mathbf{r}\left(s_i^*,t_j^*\right)\approx\mathbf{r}_t\left(s_i^*,t_j^*\right)\Delta t,$$

so the area of S_{ij} is approximately $\left|\mathbf{r}_s\left(s_i^*,t_j^*\right)\times\mathbf{r}_t\left(s_i^*,t_j^*\right)\right|\Delta s\Delta t$ and hence the total area of S is approximately

$$\sum_{j=1}^{m}\sum_{i=1}^{n}\left|\mathbf{r}_s\left(s_i^*,t_j^*\right)\times\mathbf{r}_t\left(s_i^*,t_j^*\right)\right|\Delta s\Delta t.$$

We use the limit of this Riemann sum as the definition of the area of a parametrically defined surface.

Area of a Parametric Surface

Given the surface S defined by a smooth vector function $\mathbf{r}(s,t)$, $(s,t) \in R$, the **surface area** $A(S)$ of S is

$$A(S) = \iint_R |\mathbf{r}_s \times \mathbf{r}_t| \, dA.$$

Example 6 ✎

Find the area of the surface defined by the function $\mathbf{r}(s,t) = \langle s, t, s^2 + t^2 + 1 \rangle$ over the region $s^2 + t^2 \le 1$.

Solution

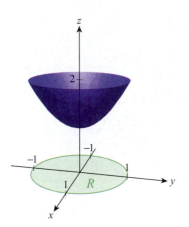

Figure 10

The vector function $\mathbf{r}$ traces out the paraboloid shown in Figure 10 over the circular region R defined by $s^2 + t^2 \le 1$ (note that this is also the region $x^2 + y^2 \le 1$ in the xy-plane, since $x = s$ and $y = t$ in the definition of $\mathbf{r}$). Since $\mathbf{r}_s = \langle 1, 0, 2s \rangle$ and $\mathbf{r}_t = \langle 0, 1, 2t \rangle$,

$$\mathbf{r}_s \times \mathbf{r}_t = \begin{vmatrix} \mathbf{i} & \mathbf{j} & \mathbf{k} \\ 1 & 0 & 2s \\ 0 & 1 & 2t \end{vmatrix} = \langle -2s, -2t, 1 \rangle$$

so $|\mathbf{r}_s \times \mathbf{r}_t| = |\langle -2s, -2t, 1 \rangle| = \sqrt{1 + 4s^2 + 4t^2}$ and

$$A(S) = \iint_R |\mathbf{r}_s \times \mathbf{r}_t| \, dA = \iint_{s^2 + t^2 \le 1} \sqrt{1 + 4s^2 + 4t^2} \, ds \, dt.$$

This double integral is easier to evaluate if we convert to polar coordinates by letting $s = r \cos \theta$ and $t = r \sin \theta$.

$$\iint_{s^2 + t^2 \le 1} \sqrt{1 + 4s^2 + 4t^2} \, ds \, dt = \int_{\theta=0}^{\theta=2\pi} \int_{r=0}^{r=1} \sqrt{1 + 4r^2} \, r \, dr \, d\theta \qquad \begin{aligned} u &= 1 + 4r^2 \\ du &= 8r \, dr \end{aligned}$$

$$= \frac{1}{12} \int_0^{2\pi} \left[u^{3/2} \right]_{u=1}^{u=5} d\theta = \frac{\pi}{6} \left(5\sqrt{5} - 1 \right)$$

Example 7 ✎

Derive the formula for the surface area of a sphere of radius r using the parametric surface area definition.

Solution

We begin with a vector function describing the surface, based on our work from Example 2.

$$\mathbf{r}(\theta, \varphi) = \langle r \sin \varphi \cos \theta, r \sin \varphi \sin \theta, r \cos \varphi \rangle, \quad 0 \le \theta \le 2\pi, \quad 0 \le \varphi \le \pi$$

Given $\mathbf{r}$,

$$\mathbf{r}_\theta = \langle -r \sin \varphi \sin \theta, r \sin \varphi \cos \theta, 0 \rangle \quad \text{and}$$

$$\mathbf{r}_\varphi = \langle r\cos\varphi\cos\theta, r\cos\varphi\sin\theta, -r\sin\varphi \rangle$$

so

$$\mathbf{r}_\theta \times \mathbf{r}_\varphi = \begin{vmatrix} \mathbf{i} & \mathbf{j} & \mathbf{k} \\ -r\sin\varphi\sin\theta & r\sin\varphi\cos\theta & 0 \\ r\cos\varphi\cos\theta & r\cos\varphi\sin\theta & -r\sin\varphi \end{vmatrix}$$

$$= \langle -r^2\sin^2\varphi\cos\theta, -r^2\sin^2\varphi\sin\theta, -r^2\sin\varphi\cos\varphi \rangle$$

and hence, after simplifying, $|\mathbf{r}_\theta \times \mathbf{r}_\varphi| = r^2\sin\varphi$. This gives us a surface area of

$$\int_0^{2\pi}\int_0^\pi r^2\sin\varphi\,d\varphi\,d\theta = -r^2\int_0^{2\pi}\big[\cos\varphi\big]_0^\pi\,d\theta = 4\pi r^2.$$

As Example 7 illustrates, we may have several ways of determining the surface area of some surfaces. For instance, we can also derive the surface area formula for a sphere by treating it as the surface generated by revolving $f(x) = \sqrt{r^2 - x^2}$ around the x-axis. In Exercise 39, you will verify that the parametric approach to surface area is consistent with the surface of revolution method. But we conclude this section by developing another formula for the surface area of the graph of an explicit function $z = f(x, y)$.

Given a surface S defined by $z = f(x, y)$, where $(x, y) \in R$, we can use x and y directly as the parameters of the vector function $\mathbf{r}(x, y) = \langle x, y, f(x, y) \rangle$ and determine

$$\mathbf{r}_x \times \mathbf{r}_y = \begin{vmatrix} \mathbf{i} & \mathbf{j} & \mathbf{k} \\ 1 & 0 & f_x \\ 0 & 1 & f_y \end{vmatrix} = \langle -f_x, -f_y, 1 \rangle,$$

so $|\mathbf{r}_x \times \mathbf{r}_y| = \sqrt{1 + [f_x]^2 + [f_y]^2}$. This gives us the surface area for S.

Formula ⚲

Surface Area of the Graph of the Function $z = f(x, y)$

Given a surface S defined by $z = f(x, y)$, where $(x, y) \in R$, the surface area of S is

$$A(S) = \iint_R \sqrt{1 + [f_x]^2 + [f_y]^2}\,dA.$$

Note the similarity between this surface area formula and the formula for the arc length of the graph of a function $f(x)$ over an interval I.

$$L = \int_I \sqrt{1 + [f'(x)]^2}\,dx$$

Technology Note 🖥

The surface depicted on the cover of this text is an Enneper surface; it has the property of being a minimal surface (see the page titled "About the Cover" in the Preface), and each point on the surface is a saddle point (see Section 13.7). The boundary of the surface is the curve defined parametrically by

$$x = r\cos\theta - \frac{r^5}{5}\cos 5\theta$$

$$y = -r\sin\theta - \frac{r^5}{5}\sin 5\theta$$

$$z = \frac{2r^3}{3}\cos 3\theta$$

with r having the constant value of 1.38. The surface itself is defined by the same parametric formulas, allowing r to take on values over the interval $[0, 1.38]$ and θ to take on values over the interval $[0, 2\pi]$. The *Mathematica* command **ParametricPlot3D** can be used to graph both the boundary and the surface as shown in Figure 11.

```
In[1]:=  x = r * Cos[θ] - (r^5 / 5) * Cos[5 θ];
         y = - (r^5 / 5) * Sin[5 θ] - r * Sin[θ];
         z = (2 * r^3 / 3) * Cos[3 θ];
         r = 1.38;

In[5]:=  ParametricPlot3D[{x, y, z}, {θ, 0, 2 Pi}, Boxed → False,
           Axes → False, PlotRange → All, ViewPoint → {1, 1, 10}]
```

Out[5]=

```
In[6]:=  ParametricPlot3D[{x, y, z}, {r, 0, 1.38}, {θ, 0, 2 Pi},
           PlotPoints → 100, Mesh → None, Boxed → False,
           Axes → False, PlotRange → All, ViewPoint → {1, 1, 10}]
```

Out[6]=

Figure 11

15.5 **Exercises**

1–4 Describe the surface with a vector function of two parameters. (Answers will vary.)

1. The graph of $y = \frac{1}{3}x$, $0 \le x \le 6$, revolved about the x-axis

2. The graph of $z = 1/x$, $0 \le x \le 8$, revolved about the x-axis

3. The graph of $x = 1 - z^2$, $-1 \le z \le 1$, revolved about the z-axis

4. The graph of $z = \cos y$, $-\pi/2 \le y \le \pi/2$, revolved about the y-axis

5–13 Match the parametric surface with its graph (labeled A–I).

5. $\mathbf{r}(s,t) = \langle 3\cos s, 3\sin s, t \rangle$,
$0 \le s \le 2\pi$, $0 \le t \le 8$

6. $\mathbf{r}(s,t) = \langle s\cos t, s\sin t, 2s \rangle$,
$0 \le s \le 5$, $0 \le t \le 2\pi$

7. $\mathbf{r}(s,t) = \langle \cos 2t\sin s, \cos s\cos 2t, \sin t \rangle$,
$0 \le s \le 2\pi$, $0 \le t \le 2\pi$

8. $\mathbf{r}(s,t) = \left\langle s, \dfrac{\cos t}{s^2}, \dfrac{\sin t}{s^2} \right\rangle$, $\dfrac{1}{3} \le s \le 3$, $0 \le t \le 2\pi$

9. $\mathbf{r}(s,t) = \langle 2t - s, 1 + 2s + t, 1 + s - 3t \rangle$,
$0 \le s \le 5$, $0 \le t \le 5$

10. $\mathbf{r}(s,t) = \langle s\sin t, s\cos t, \cos s \rangle$,
$0 \le s \le 2\pi$, $0 \le t \le 2\pi$

11. $\mathbf{r}(s,t) = \langle \sin s\cos t, \sin s\sin t, \cos s \rangle$,
$0 \le s \le \pi/2$, $0 \le t \le 2\pi$

12. $\mathbf{r}(s,t) = \langle s\cos t, s\sin t, t \rangle$,
$-3 \le s \le 3$, $0 \le t \le 3\pi$

13. $\mathbf{r}(s,t) = \langle \cos s\sin t, 2\sin s\sin t, \cos t \rangle$,
$0 \le s \le \pi$, $0 \le t \le 2\pi$

A.

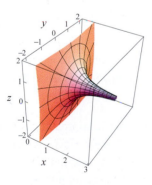

B.

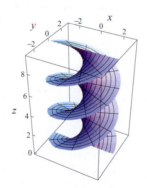

C.

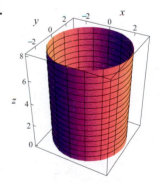

D.

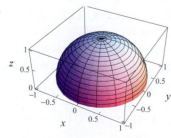

E.

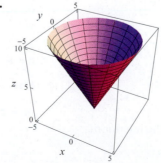

F.

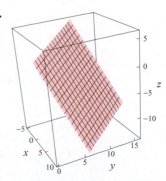

G.

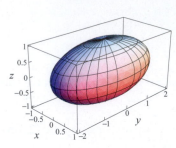

H.

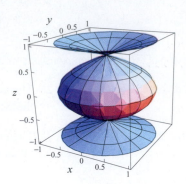

I.

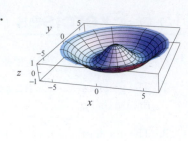

14–20 Identify the surface by examining its grid curves.

14. $\mathbf{r}(s,t)=\langle t,\cos s,\sin s\rangle, \quad 0\le s\le 2\pi, \quad -\infty<t<\infty$

15. $\mathbf{r}(s,t)=\langle R\sin s\cos t, R\sin s\sin t, R\cos s\rangle,$
$\quad 0\le s\le \pi/2, \quad 0\le t\le \pi/2$

16. $\mathbf{r}(s,t)=\langle s\cos t, s\sin t, s\rangle,$
$\quad -\infty<s<\infty, \quad 0\le t\le 2\pi$

17. $\mathbf{r}(s,t)=\langle s,t,2s^2+t^2\rangle,$
$\quad -\infty<s<\infty, \quad -\infty<t<\infty$

18. $\mathbf{r}(s,t)=\langle(3+\cos t)\cos s,(3+\cos t)\sin s,\sin t\rangle,$
$\quad 0\le s\le 2\pi, \quad 0\le t\le 2\pi$

19. $\mathbf{r}(s,t)=\langle s\cos t, s\sin t, 4-s^2\rangle,$
$\quad 0\le s\le 2, \quad 0\le t\le 2\pi$

20. $\mathbf{r}(s,t)=\left\langle s,\dfrac{\cos t}{s},\dfrac{\sin t}{s}\right\rangle, \quad \dfrac{1}{4}\le s\le 4, \quad 0\le t\le 2\pi$

21–31 Obtain a parametrization for the indicated surface. (Answers will vary.)

21. $z=x+y$ **22.** $z=xy$

23. $x^2+y^2=1, 0\le z\le 1$ **24.** $z=\sqrt{x^2+y^2}, z\le 2$

25. $x^2+y^2+z^2=4$ **26.** $z=x^2+y^2$

27. $z=-\sqrt{1-x^2-y^2}$ **28.** $\dfrac{x^2}{4}+\dfrac{y^2}{9}+z^2=1$

29. $x^2+4y^2=4$

30. The intersection of $z=x^2+y^2$ with the interior of $x^2+y^2=4$

31. The portion of the sphere $x^2+y^2+z^2=4$ outside the double cone $z^2=3x^2+3y^2$

32–37 Construct an equation for the plane tangent to the surface at the indicated point.

32. $\mathbf{r}(s,t)=\langle s^2,s+t,t^2\rangle; \quad \mathbf{r}(1,1)$

33. $\mathbf{r}(s,t)=\langle s,t,2s^2+t^2\rangle; \quad \mathbf{r}(1,2)$

34. $\mathbf{r}(s,t)=\langle 2st,s^2,t^2\rangle; \quad \mathbf{r}(1,-3)$

35. $\mathbf{r}(s,t)=\langle 2s\cos t,s\sin t,s^2\rangle; \quad \mathbf{r}\left(2,\dfrac{\pi}{4}\right)$

36. $\mathbf{r}(s,t)=\left\langle 2s^2,st^2,\dfrac{st}{2}\right\rangle; \quad \mathbf{r}(1,2)$

37. $\mathbf{r}(s,t)=\langle s\sin t,s^4,s\cos t\rangle; \quad \mathbf{r}\left(1,\dfrac{\pi}{3}\right)$

38. Parametrize the sphere of radius R as in Example 2, and show that its normal vector $\mathbf{n}(\theta,\varphi)$ is a constant multiple of $\mathbf{e}_r$ (see Exercise 54 of Section 15.1). What is that constant?

39. Verify that the parametric approach to surface area is consistent with the surface area of a solid of revolution discussed in Section 6.3. (**Hint:** Let $f(x)\ge 0$ be a continuously differentiable single-variable function defined on $[a,b]$, and rotate its graph around the x-axis. Parametrize the resulting surface of revolution as in Example 1. Calculate $\mathbf{n}=\mathbf{r}_s\times\mathbf{r}_t$ as in Example 5 and find the surface area $A=\int_0^{2\pi}\int_a^b|\mathbf{n}|\,dA$ after showing that $|\mathbf{n}|=|\mathbf{r}_s\times\mathbf{r}_t|=f(x)\sqrt{1+\left[f'(x)\right]^2}$.)

40–56 Find the area of the surface *S*. (Use polar coordinates wherever they simplify your calculations.)

40. *S* is the first-octant portion of the plane $x + 3y + z = 6$.

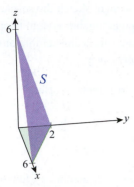

41. *S* is the intersection of the plane $2x + 2y + z = 0$ and the interior of the cylinder $x^2 + y^2 = 4$.

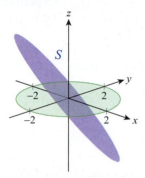

42. *S* is the graph of $z = 3 + 2x - y$ defined on the triangle with vertices $(0,0)$, $(1,0)$, and $(0,1)$.

43. *S* is the graph of $z = \sqrt{1 - x^2}$ defined on the square $[0,1] \times [0,1]$.

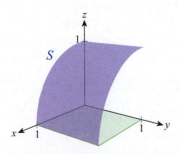

44. *S* is the graph of $z = 2x^2 + y$ defined on the triangle with vertices $(0,0)$, $(1,0)$, and $(1,1)$.

45. *S* is the portion of the paraboloid $\mathbf{r}(s,t) = \langle s\cos t, s\sin t, 1 - s^2 \rangle$ above the *xy*-plane.

46. *S* is the surface of the cone $\mathbf{r}(s,t) = \langle s\cos t, s\sin t, s \rangle$ between the planes $z = 1$ and $z = 2$.

47. *S* is the portion of the paraboloid $\mathbf{r}(s,t) = \langle s\cos t, s\sin t, s^2 \rangle$ between the planes $z = 4$ and $z = 9$.

48. *S* is the portion of the cone $z = \sqrt{x^2 + y^2}$ defined on the triangle with vertices $(0,0)$, $(1,1)$, and $(0,1)$.

49. *S* is the graph of $z = y^2 - x^2$ defined on the first-quadrant portion of the disk of radius 2 centered at the origin.

50. *S* is the surface $\mathbf{r}(s,t) = \langle s\cos t, s\sin t, 9 - s^2 \rangle$ between the planes $z = 5$ and $z = 8$.

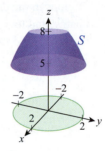

51. *S* is the portion of the paraboloid $\mathbf{r}_1(s,t) = \langle s\cos t, s\sin t, s^2/2 \rangle$ between the cylinders $\mathbf{r}_2(s,t) = \langle \cos s, \sin s, t \rangle$ and $\mathbf{r}_3(s,t) = \langle 2\cos s, 2\sin s, t \rangle$.

52. *S* is the portion of $x^2 + y^2 = z^2 + 1$ between the *xy*-plane and $z = \sqrt{3}$.

53. *S* is the first-quadrant portion of $z + y^2 = 1$, $z \geq 0$, between the *yz*-plane and $x = 1$.

54. *S* is the torus $\mathbf{r}(s,t) = \langle (2 + \cos t)\cos s, (2 + \cos t)\sin s, \sin t \rangle$, $0 \leq s \leq 2\pi$, $0 \leq t \leq 2\pi$.

55. *S* is the portion of the hemisphere $z = \sqrt{25 - x^2 - y^2}$ between the planes $z = 3$ and $z = 4$.

56. *S* is the portion of the cylinder $x^2 + y^2 = 25$ between the planes $z = 3$ and $z = 4$. (Compare your answer to the solution of Exercise 55.)

57.* Generalize Exercise 54 to arrive at the formula for the surface area of the torus parametrized as $\mathbf{r}(s,t) = \langle (a + b\cos t)\cos s, (a + b\cos t)\sin s, b\sin t \rangle$, where $a > b > 0$. (**Hint:** Consider the circle of radius b in the xz-plane, centered at $(a, 0, 0)$, and rotate it around the z-axis.)

58.* Generalize your observations made in Exercises 55 and 56 to prove Archimedes' famous result:

The surface area of the section of the sphere $x^2 + y^2 + z^2 = R^2$ between the planes $z = a$ and $z = b$ equals the surface area of the corresponding section of the circumscribed cylinder $x^2 + y^2 = R^2$.

59.* Show that surface area is independent of parametrization; that is, prove the following statement:

Let R_1 and R_2 be regions in the plane enclosed by simple closed paths, and let $\mathbf{r}_1 : R_1 \to \mathbb{R}^3$, $\mathbf{r}_2 : R_2 \to \mathbb{R}^3$ be continuously differentiable, one-to-one parametrizations of the same surface, that is, $\mathbf{r}_1(R_1) = \mathbf{r}_2(R_2)$. With the usual notation $\mathbf{n}_i = (\mathbf{r}_i)_s \times (\mathbf{r}_i)_t$, $i = 1, 2$, prove that

$$\iint_{R_1} |\mathbf{n}_1|\, dA = \iint_{R_2} |\mathbf{n}_2|\, dA.$$

(**Hint:** Use the rules for differentiation of inverses and change of variables.)

60–63 *True or False?* Determine whether the given statement is true or false. In case of a false statement, explain or provide a counterexample.

60. All grid curves of a parametric surface are parallel to a coordinate plane.

61. Before we can evaluate its area, a surface S must be parametrized.

62. A parametric surface $\mathbf{r}(s,t)$ has a tangent plane at any point of smoothness $\mathbf{r}(s_0, t_0)$.

63. The area of a parametric surface is a limit of Riemann sums.

15.5 Technology Exercises

64. Describe in words the grid curves of the parametric surface given in Exercise 12; then use a computer algebra system to sketch the surface. Finally, use your computer algebra system to approximate its area. (**Note:** This is an example of what are called *helicoid surfaces*.)

65. Sketch the parametric surface

$$\mathbf{r}(s,t) = \left\langle \left(\sqrt{25 - t^2} - 3\right)\cos s, \left(\sqrt{25 - t^2} - 3\right)\sin s, t \right\rangle,$$
$$0 \le s \le 2\pi, \quad -4 \le t \le 4,$$

using a computer algebra system and verify that the graph resembles a football. Change the coefficients to make the "football" appear "skinnier" or "chubbier," respectively. (Carefully determine the domain for each parameter. Answers will vary.)

66. The parametric surface

$$\mathbf{r}(s,t) = \left\langle \cos t + s\cos\frac{t}{2}, 3\sin t + s\cos\frac{t}{2}, s\sin\frac{t}{2} \right\rangle,$$
$$-\frac{1}{2} \le s \le \frac{1}{2}, \quad 0 \le t \le 2\pi$$

is an example of the famous *Möbius strip* (after the German mathematician August Ferdinand Möbius). Use a computer algebra system to graph and examine this surface. Can you think of a way to produce such a surface using a strip of paper and tape? Notice that if you start sliding your finger along one side of the surface, you will eventually arrive back at your starting point without crossing any edges! Surfaces with this property are called *nonorientable*, or *one-sided*.

67. Graph and examine the surface

$$\mathbf{r}(s,t) = \langle (a + \sin t)\cos s, (a + \sin t)\sin s, t \rangle,$$
$$0 \le s \le 2\pi, \quad 0 \le t \le 2\pi$$

for several values of the parameter a. Then use your computer algebra system to find its surface area if $a = 2$.

15.6 **Surface Integrals**

TOPICS

1. Surface integrals of scalar functions

2. Orientable surfaces

3. Surface integrals of vector fields

The integral of a scalar function over a surface is based upon the definition of surface area in much the same way that the integral of a scalar function over a line (or curve) is based upon arc length. And as with line integrals, once we know how to integrate scalar functions over surfaces, we can proceed to integrate vector fields across surfaces, a subject at the core of the final two sections of this chapter.

TOPIC 1 **Surface Integrals of Scalar Functions**

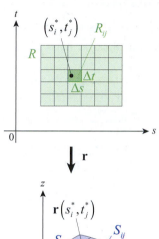

Figure 1

To set the stage, we assume that S is a surface defined by a smooth vector function $\mathbf{r}(s,t)$ for $(s,t) \in R$, where R is a region of the st-plane. We also assume that $f(x,y,z)$ is defined for all points of S. As a concrete example, f might represent the electrical *charge density* (charge per unit area) of the surface S. In order to approximate the total charge of S, we again divide S into subdivisions S_{ij} (as shown in Figure 1) and add up all the individual charges $f\left(\mathbf{r}\left(s_i^*,t_i^*\right)\right)\Delta\sigma_{ij}$ of the subdivisions, where we let $\Delta\sigma_{ij} = A\left(S_{ij}\right)$, the surface area of S_{ij}. As usual, $\mathbf{r}\left(s_i^*,t_j^*\right) \in S_{ij}$ is a sample point, and its precise choice is irrelevant in the limit of finer and finer partitions. The approximate total charge on S then has the form of the Riemann sum

$$\sum_{j=1}^{m}\sum_{i=1}^{n} f\left(\mathbf{r}\left(s_i^*,t_j^*\right)\right)\Delta\sigma_{ij},$$

which we use as the inspiration for the following definition.

Definition 💡

Surface Integral of a Scalar Function

Given the surface S defined over a region R by a smooth vector function $\mathbf{r}(s,t)$, $(s,t) \in R$, and a continuous function f defined on S, the **surface integral of f over S** is

$$\iint_S f\,d\sigma = \iint_R f\left(\mathbf{r}(s,t)\right)\left|\mathbf{r}_s \times \mathbf{r}_t\right|dA.$$

Note the similarity with the line integral of a scalar function:

$$\int_C f\,ds = \int_a^b f\left(\mathbf{r}(t)\right)\left|\mathbf{r}'(t)\right|dt$$

Note also that by so defining the surface integral of a function, we have related the differential element of surface area, $d\sigma$, to the differential element of area, dA, of the region R by the equation $d\sigma = \left|\mathbf{r}_s \times \mathbf{r}_t\right|dA$. And just as integrating the constant function 1 along a curve results in the arc length of the curve, integrating 1 over S results in the surface area of S.

$$A(S) = \iint_S 1\,d\sigma = \iint_R \left|\mathbf{r}_s \times \mathbf{r}_t\right|dA$$

Example 1 ✐

Let $\rho(x,y,z) = 1 - z$ represent the mass density of the surface S defined by

$$\mathbf{r}(s,t) = \langle s, t, s^2 + t^2 \rangle$$

over the region $s^2 + t^2 \le 1$. Determine the mass M of S.

Solution

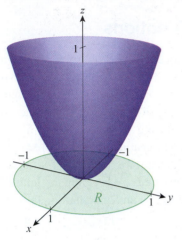

Figure 2

The surface S is the paraboloid of Figure 2, defined over the region R consisting of the disk $x^2 + y^2 \le 1$ in the xy-plane (we can equate the xy-plane with the st-plane since $x = s$ and $y = t$). The calculation of $|\mathbf{r}_s \times \mathbf{r}_t|$ proceeds as in Example 6 of the last section, giving us $|\mathbf{r}_s \times \mathbf{r}_t| = \sqrt{1 + 4s^2 + 4t^2}$. Note that $\rho(\mathbf{r}(s,t)) = 1 - (s^2 + t^2)$. Putting the pieces together, the mass of S is

$$M = \iint_S \rho \, d\sigma = \iint_R \rho(\mathbf{r}(s,t)) |\mathbf{r}_s \times \mathbf{r}_t| \, dA = \iint_{s^2+t^2 \le 1} (1 - s^2 - t^2) \sqrt{1 + 4s^2 + 4t^2} \, dA$$

$$= \int_{\theta=0}^{\theta=2\pi} \int_{r=0}^{r=1} (1 - r^2) \sqrt{1 + 4r^2} \, r \, dr \, d\theta. \qquad \text{\color{blue}Switching to polar coordinates}$$

The substitution $u = 1 + 4r^2$ can be used to evaluate the last integral (note that $1 - r^2 = (5 - u)/4$), as you will verify in Exercise 1, leading to

$$M = \iint_S \rho \, d\sigma = \frac{\pi(25\sqrt{5} - 11)}{60}.$$

Example 2 ✐

Find the center of mass of the surface $\mathbf{r}(s,t) = \langle s, t, s^2 + t^2 \rangle$ in Example 1.

Solution

By symmetry, we know that the x- and y-coordinates of the center of mass must be $\bar{x} = 0$ and $\bar{y} = 0$. To determine the z-coordinate, we use the formula $\bar{z} = M_{xy}/M$, where, as in Section 14.4, M_{xy} represents the first moment of the object about the xy-plane.

$$M_{xy} = \iint_S z \rho(x,y,z) \, d\sigma$$

Expressing this integral in terms of s and t we have

$$M_{xy} = \iint_{s^2+t^2 \le 1} \underbrace{(s^2 + t^2)}_{z} \underbrace{(1 - s^2 - t^2)}_{\rho = 1-z} \underbrace{\sqrt{1 + 4s^2 + 4t^2}}_{d\sigma} \, dA$$

which, upon changing to polar coordinates, gives us

$$M_{xy} = \int_{\theta=0}^{\theta=2\pi} \int_{r=0}^{r=1} r^2 (1 - r^2) \sqrt{1 + 4r^2} \, r \, dr \, d\theta.$$

In Exercise 2 you will show that the substitution $u = 1 + 4r^2$ again allows us to evaluate the integral, and that

$$\bar{z} = \frac{M_{xy}}{M} = \frac{2(25\sqrt{5} + 4)}{7(25\sqrt{5} - 11)} \approx 0.38.$$

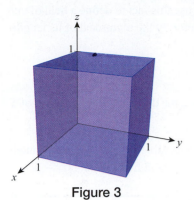

Figure 3

Example 3 ✎

Evaluate the integral of the function $f(x, y, z) = xyz$ over the surface of the cube of side length 1 shown in Figure 3.

Solution

Since $f = 0$ on the three faces of the cube lying on the coordinate planes (where $x = 0$, $y = 0$, or $z = 0$), we know that the integral of f along those three faces is 0 with no need for further calculation. Along the remaining three faces of the cube, we must actually do some work. We demonstrate the procedure for the face that appears in front in Figure 3.

The front face of the cube can be described by the vector function

$$\mathbf{r}(s, t) = \langle 1, s, t \rangle, \quad 0 \le s \le 1, \quad 0 \le t \le 1.$$

It follows that

$$\mathbf{r}_s \times \mathbf{r}_t = \begin{vmatrix} \mathbf{i} & \mathbf{j} & \mathbf{k} \\ 0 & 1 & 0 \\ 0 & 0 & 1 \end{vmatrix} = \langle 1, 0, 0 \rangle$$

and hence $|\mathbf{r}_s \times \mathbf{r}_t| = 1$. We have $f(\mathbf{r}(s, t)) = (1)(s)(t) = st$ on the face, so

$$\iint f(\mathbf{r}(s, t)) |\mathbf{r}_s \times \mathbf{r}_t| \, dA = \int_0^1 \int_0^1 st \, ds \, dt = \frac{1}{2} \int_0^1 t \, dt = \frac{1}{4}.$$

Similarly, the integral of f along the remaining two faces is also $\frac{1}{4}$, so

$$\iint_{\text{cube}} f(x, y, z) \, d\sigma = 0 + 0 + 0 + \frac{1}{4} + \frac{1}{4} + \frac{1}{4} = \frac{3}{4}.$$

TOPIC 2 Orientable Surfaces

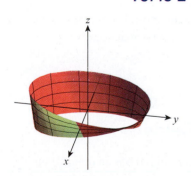

Figure 4 Möbius Strip

Just as we need to know the orientation of a curve when evaluating the line integral of a vector field, we need to know the orientation of a surface in order to define and evaluate the surface integral of a vector field. The first step is to realize that some surfaces are *not* orientable, meaning they are not two-sided. The Möbius strip seen in Exercise 66 of Section 15.5 is one such example (see Figure 4). If you imagine tracing out one full "loop" of the surface with your (perpendicular) finger, your finger will wind up where you started but pointing in the opposite direction—for this reason, the Möbius strip is said to be one-sided.

Surfaces that *are* orientable are those for which it is possible to define a unit normal vector **n** at every point in such a way that **n** varies continuously along the surface (for the Möbius strip, **n** winds up pointing in the opposite direction after making one continuous circuit, and we can't assign two opposing directions to **n** at any one point). If a surface is orientable, there are two choices for the vector field of normal vectors: those pointing in one direction away from the surface, and those pointing in the opposite direction. We call our choice for any given surface the **positive** orientation; for surfaces that are **closed**, meaning those that enclose a solid region of space, the convention is to let the positive direction for **n** be the outward-pointing direction.

If a surface S is described by a vector function $\mathbf{r}(s,t)$, then $\mathbf{r}_s \times \mathbf{r}_t$ is a field of vectors normal to S at any given point. We can use this fact to define a field of *unit* normal vectors by simply multiplying each vector by the reciprocal of its length

$$\mathbf{n} = \frac{\mathbf{r}_s \times \mathbf{r}_t}{\left| \mathbf{r}_s \times \mathbf{r}_t \right|}.$$

If we don't want to use this result as our choice for the positive orientation of S, we simply use $-\mathbf{n}$ instead.

A sphere is an elementary example of a closed orientable surface, and we calculate the positively oriented vector field of normal vectors in Example 4.

Example 4 ✎

Recall that the vector function

$$\mathbf{r}(\theta, \varphi) = \langle r \sin\varphi \cos\theta, r \sin\varphi \sin\theta, r \cos\varphi \rangle, \quad 0 \le \theta \le 2\pi, \quad 0 \le \varphi \le \pi$$

describes the sphere of radius r centered at the origin (Example 7 of Section 15.5). We have already computed $\mathbf{r}_\theta \times \mathbf{r}_\varphi$ to be

$$\mathbf{r}_\theta \times \mathbf{r}_\varphi = \langle -r^2 \sin^2\varphi \cos\theta, -r^2 \sin^2\varphi \sin\theta, -r^2 \sin\varphi \cos\varphi \rangle$$

and noted that $\left| \mathbf{r}_\theta \times \mathbf{r}_\varphi \right| = r^2 \sin\varphi$. So

$$\frac{\mathbf{r}_\theta \times \mathbf{r}_\varphi}{\left| \mathbf{r}_\theta \times \mathbf{r}_\varphi \right|} = \frac{1}{r^2 \sin\varphi} \langle -r^2 \sin^2\varphi \cos\theta, -r^2 \sin^2\varphi \sin\theta, -r^2 \sin\varphi \cos\varphi \rangle$$

$$= \langle -\sin\varphi \cos\theta, -\sin\varphi \sin\theta, -\cos\varphi \rangle = -\frac{1}{r} \mathbf{r}(\theta, \varphi).$$

This vector field of unit normal vectors points radially inward toward the origin at every point of the sphere, which by convention we call the negative orientation. So the positively oriented field of unit normal vectors is

$$\mathbf{n} = \frac{1}{r} \mathbf{r}(\theta, \varphi).$$

TOPIC 3 Surface Integrals of Vector Fields

We can now define the surface integral of a vector field. If S is an oriented surface, say with normal vector field $\mathbf{n}$, then

$$\iint\limits_S \mathbf{F} \cdot \mathbf{n} \, d\sigma$$

gives us the **flux** of the continuous vector field $\mathbf{F}$ across S—that is, the total flow of $\mathbf{F}$ perpendicularly through S in the direction of positive orientation. If S is described by a vector function $\mathbf{r}(s,t)$ over a region R, and if we choose to let $\mathbf{n} = (\mathbf{r}_s \times \mathbf{r}_t)/\left| \mathbf{r}_s \times \mathbf{r}_t \right|$, then

$$\text{Flux} = \iint\limits_S \mathbf{F} \cdot \mathbf{n} \, d\sigma = \iint\limits_R \mathbf{F} \cdot \frac{\mathbf{r}_s \times \mathbf{r}_t}{\left| \mathbf{r}_s \times \mathbf{r}_t \right|} \left| \mathbf{r}_s \times \mathbf{r}_t \right| dA = \iint\limits_R \mathbf{F} \cdot (\mathbf{r}_s \times \mathbf{r}_t) \, dA.$$

Example 5 ✐

Evaluate the flux of the vector field $\mathbf{F}(x, y, z) = \langle y, x, z \rangle$ across the helicoid surface S defined by

$$\mathbf{r}(s,t) = \langle s\cos t, s\sin t, t \rangle, \quad 0 \le s \le 1, \quad 0 \le t \le \pi$$

with $\mathbf{n} = (\mathbf{r}_s \times \mathbf{r}_t)/|\mathbf{r}_s \times \mathbf{r}_t|$.

Solution

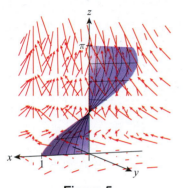

Figure 5

Figure 5 is an illustration of S and the vector field $\mathbf{F}$. As we will soon see, the vector field $\mathbf{n}$ consists of unit normal vectors pointing in a generally upward direction and, for all points above the xy-plane, the vectors of $\mathbf{F}$ also point generally upward, so we expect the flux of $\mathbf{F}$ across S to be positive.

The first step is to calculate $\mathbf{r}_s \times \mathbf{r}_t$.

$$\mathbf{r}_s \times \mathbf{r}_t = \begin{vmatrix} \mathbf{i} & \mathbf{j} & \mathbf{k} \\ \cos t & \sin t & 0 \\ -s\sin t & s\cos t & 1 \end{vmatrix} = \langle \sin t, -\cos t, s \rangle$$

We don't need to actually compute $\mathbf{n}$ in order to evaluate the flux, but since $0 \le s \le 1$, each element of $\mathbf{r}_s \times \mathbf{r}_t$ has a positive z-component, so the same must be true for each corresponding element of $\mathbf{n}$ (that is, the vectors normal to S point generally upward).

We next determine that

$$\mathbf{F}(\mathbf{r}(s,t)) \cdot (\mathbf{r}_s \times \mathbf{r}_t) = \langle s\sin t, s\cos t, t \rangle \cdot \langle \sin t, -\cos t, s \rangle$$
$$= s\sin^2 t - s\cos^2 t + st$$
$$= s(\sin^2 t - \cos^2 t + t)$$
$$= s(t - \cos 2t)$$

and hence

$$\iint_S \mathbf{F} \cdot \mathbf{n} \, d\sigma = \iint_R \mathbf{F} \cdot (\mathbf{r}_s \times \mathbf{r}_t) \, dA = \int_0^1 \int_0^\pi s(t - \cos 2t) \, dt \, ds$$

$$= \int_0^1 \frac{s}{2} \left[t^2 - \sin 2t \right]_0^\pi ds = \frac{\pi^2}{2} \left[\frac{s^2}{2} \right]_0^1 = \frac{\pi^2}{4}.$$

Fourier's Law of Heat Conduction states that the **heat flow** through a substance is proportional to the negative gradient of the temperature of the substance. So if $f(x, y, z)$ indicates the temperature of a substance at the point (x, y, z), the heat flow $\mathbf{F}$ is the vector field

$$\mathbf{F}(x, y, z) = -\kappa \nabla f(x, y, z),$$

where κ is a constant called the **thermal conductivity** of the substance. The amount of heat transferred per unit time across a surface S in the substance is then the integral

$$\iint_S \mathbf{F} \cdot \mathbf{n} \, d\sigma = -\kappa \iint_S \nabla f \cdot \mathbf{n} \, d\sigma.$$

Example 6 ✎

Suppose that $f(x,y,z) = 25 - (x^2 + y^2 + z^2)$ represents the temperature, in degrees Celsius, of a ball of radius 1 centered at the origin (so the temperature at the center of the ball is 25 °C, and the temperature at its very outer surface is 24 °C). If S is defined as the collection of points a distance r from the center, find the flow of heat per unit time across the surface S.

Solution

Note that

$$\mathbf{F}(x,y,z) = -\kappa \nabla f(x,y,z) = -\kappa \langle -2x, -2y, -2z \rangle = 2\kappa \langle x, y, z \rangle,$$

where κ is the thermal conductivity constant for the substance of the ball. From Example 4, we know that the outward unit normal vector $\mathbf{n}$ at the point (x,y,z) on the surface $x^2 + y^2 + z^2 = r^2$ is

$$\mathbf{n}(x,y,z) = \frac{1}{r} \langle x, y, z \rangle,$$

so on S,

$$\mathbf{F} \cdot \mathbf{n} = \frac{2\kappa}{r} \langle x, y, z \rangle \cdot \langle x, y, z \rangle = \frac{2\kappa}{r} \left(x^2 + y^2 + z^2 \right) = \frac{2\kappa}{r} \left(r^2 \right) = 2\kappa r.$$

Hence, the flow of heat per unit time across S is

$$\iint_S \mathbf{F} \cdot \mathbf{n} \, d\sigma = 2\kappa r \iint_S d\sigma = 2\kappa r \left(4\pi r^2 \right) = 8\kappa \pi r^3.$$

15.6 **Exercises**

1. Use the substitution suggested in Example 1 to conclude that

$$M = \iint_S \rho \, d\sigma = \frac{\pi \left(25\sqrt{5} - 11 \right)}{60}.$$

2. Use the same substitution as in Exercise 1 to finish Example 2 by showing that

$$\bar{z} = \frac{M_{xy}}{M} = \frac{2 \left(25\sqrt{5} + 4 \right)}{7 \left(25\sqrt{5} - 11 \right)} \approx 0.38.$$

3. Suppose the surface S is given as the graph of a function $z = g(x,y)$, defined on a domain R. Prove that for a continuous function $f = f(x,y,z)$ defined on S, the surface integral of f over S is

$$\iint_S f \, d\sigma = \iint_R f(x,y,g(x,y)) \sqrt{1 + [g_x]^2 + [g_y]^2} \, dA.$$

(**Hint:** Parametrize S by using x and y as parameters.)

4–25 Parametrize the surface S and evaluate the indicated surface integral. (You may use the formula from Exercise 3 whenever feasible. Use polar or spherical coordinates where needed.)

4. $\displaystyle\iint_S 2z \, d\sigma$, where S is the first-octant portion of the plane $x + 2y + z = 4$

5. $\displaystyle\iint_S y \, d\sigma$, where S is the graph of $2x + y + z = 6$ above the square $[0,2] \times [0,2]$

6. $\displaystyle\iint_S 3x \, d\sigma$, where S is the graph of $x^2 + z = 9$ above the rectangle $[0,3] \times [0,4]$

7. $\displaystyle\iint_S x \, d\sigma$, where S is the first-octant portion of the plane $x + y + z = a$ $(a > 0)$

8. $\iint\limits_{S} z^2\, d\sigma$, where S is the intersection of the plane $2x + 2y + z = 0$ and the interior of the cylinder $x^2 + y^2 = 1$

9. $\iint\limits_{S} \left(x^2 + y^2 + z\right) d\sigma$, where S is the graph of $x + 2y + z = 2$ above the rectangle $[0,1] \times [0,2]$

10. $\iint\limits_{S} 9z\, d\sigma$, where S is the surface $z = y^3$ over the rectangle $[-1,1] \times [0,1]$

11. $\iint\limits_{S} \left(2xy + z\right) d\sigma$, where S is the graph of $2y - x + z = 4$ above the triangle with vertices $(0,0)$, $(1,0)$, and $(0,1)$

12. $\iint\limits_{S} \left(x + y + z\right) d\sigma$, where S is the graph of $z = \sqrt{1 - x^2}$ above the square $[0,1] \times [0,1]$

13. $\iint\limits_{S} \left(z - y\right) d\sigma$, where S is the graph of $z = 2x^2 + y$ above the triangle with vertices $(0,0)$, $(1,0)$, and $(1,1)$

14. $\iint\limits_{S} 2z\, d\sigma$, where S is the portion of the paraboloid $z = 1 - x^2 - y^2$ above the xy-plane

15. $\iint\limits_{S} \left(z + 2x^2\right) d\sigma$, where S is the surface $z = y^2 - x^2$ above the half disk bounded by $y = \sqrt{1 - x^2}$ and the x-axis

16. $\iint\limits_{S} y^2\, d\sigma$, where S is the upper unit hemisphere $z = \sqrt{1 - y^2 - x^2}$

17. $\iint\limits_{S} \left(x^2 + y^2\right) d\sigma$, where S is the portion of the paraboloid $z = 2 - x^2 - y^2$ above the xy-plane

18. $\iint\limits_{S} z\, d\sigma$, where S is the portion of the sphere $x^2 + y^2 + z^2 = 16$ between the planes $z = 1$ and $z = 3$

19. $\iint\limits_{S} \dfrac{y^2}{1 - z}\, d\sigma$, where S is the portion of the paraboloid $z = 1 - x^2 - y^2$ with $\frac{3}{4} \leq z \leq 1$

20. $\iint\limits_{S} \sin z\, d\sigma$, where S is the cylinder $x^2 + y^2 = 1$, $0 \leq z \leq \pi/2$

21. $\iint\limits_{S} \left(2x^2 + 2y^2 + 2z^2\right) d\sigma$, where S is the unit sphere centered at the origin; use spherical coordinates

22. $\iint\limits_{S} \left(yz + y^3\right) d\sigma$, where S is the box $[0,1] \times [0,2] \times [0,3]$

23. $\iint\limits_{S} \left(x^2 + 2xz\right) d\sigma$, where S is the tetrahedron formed by the plane $x + 2y + 2z = 4$ and the coordinate planes

24. $\iint\limits_{S} z\, d\sigma$, where S is the solid bounded by the cylinder $x^2 + y^2 = 1$, the xy-plane and the plane $2z = x + 3$

25. $\iint\limits_{S} xyz\, d\sigma$, where S is the cone frustum $z = 3 - \sqrt{x^2 + y^2}$ between the planes $z = 1$ and $z = 2$

26. Find the centroid of the surface in Example 1, assuming constant density.

27. Determine the mass M of the half cylinder S defined by $\mathbf{r}(s,t) = \langle 2\cos s, 2\sin s, t \rangle$, $0 \leq s \leq \pi$, $0 \leq t \leq 2$, if its mass density is $\rho(x,y,z) = y + z$.

28. Determine the mass M of the cone S defined by $\mathbf{r}(s,t) = \langle s\cos t, s\sin t, s \rangle$, $0 \leq s \leq 1, 0 \leq t \leq 2\pi$, if its mass density is proportional to the distance from the z-axis.

29. Find the center of mass of the half cylinder given in Exercise 27.

30. Find the center of mass of the cone surface given in Exercise 28.

31. Determine the mass of the portion of the thin spherical shell $z = \sqrt{25 - x^2 - y^2}$ between the planes $z = 3$ and $z = 4$ if its density at any point is proportional to the distance from the xy-plane. (**Hint:** Consider spherical coordinates.)

32. Find the centroid of the thin hemispherical surface $g(x,y) = \sqrt{R^2 - x^2 - y^2}$ if it has constant density ρ. (See the hint given in Exercise 31.)

33. Determine the second moments about the coordinate axes of the sphere $x^2 + y^2 + z^2 = R^2$ if it has constant density ρ. (See the hint given in Exercise 31.)

34–52 Determine the indicated flux of the vector field **F** across the surface S. Unless otherwise specified, the surfaces are oriented with outward-pointing normal vectors.

34. The flux of $\mathbf{F} = \langle 0, 0, c \rangle$ across the hemisphere given in Exercise 32

35. The flux of $\mathbf{F} = \langle x, y, 2 \rangle$ out of the solid R bounded by $3z = x^2 + y^2$ and the plane $z = 3$

36. The flux of $\mathbf{F} = \langle 2, z, y \rangle$ across the first-octant portion of the unit hemisphere centered at the origin

37. The flux of $\mathbf{F} = \langle -y, x, 2 \rangle$ across the first-octant portion of the unit hemisphere centered at the origin

38. The flux of $\mathbf{F} = \langle x, y, z \rangle$ across the first-octant portion of the unit hemisphere centered at the origin

39. The flux of $\mathbf{F} = \langle x, y, z \rangle$ across the first-octant portion of the cylinder $x^2 + y^2 = 1$, between the planes $z = 0$ and $z = 1$

40. The flux of $\mathbf{F} = \langle x^2, xy, xz \rangle$ across the first-octant portion of the hemisphere of radius R, centered at the origin

41. The flux of $\mathbf{F} = \langle 0, 0, x^2 + y^2 \rangle$ across the paraboloid $z = x^2 + y^2 + 1$, $1 \leq z \leq 10$

42. The flux of $\mathbf{F} = \langle 0, y^2, x^2 + z^2 \rangle$ across the hemisphere $z = \sqrt{4 - x^2 - y^2}$

43. The flux of $\mathbf{F} = \langle y, -x, 6 \rangle$ across the portion of the upper unit hemisphere centered at the origin that projects onto the disk $x^2 + y^2 \leq \frac{1}{2}$ (Note that this can be interpreted as the upward flux through the hemisphere of a fluid with a "rotating flow.")

44. The flux of the vector field given in Exercise 43 across the portion of the paraboloid $z = 1 - x^2 - y^2$ that projects onto the disk $x^2 + y^2 \leq \frac{1}{2}$ (Compare your answer with the solution of Exercise 43.)

45. The flux of the vector field given in Exercise 43 across the portion of the cone $z = 1 - \sqrt{x^2 + y^2}$ that projects onto the disk $x^2 + y^2 \leq \frac{1}{2}$ (Compare your answer with the solution of Exercise 43 or 44.)

46. The flux of $\mathbf{F} = \langle -2y, x, 2z \rangle$ out of the solid R bounded by the cone $z = \sqrt{x^2 + y^2}$ and the plane $z = 6$

47. The flux of $\mathbf{F} = \langle x, y, 2 \rangle$ out of the solid R bounded by the paraboloid $z = x^2 + y^2$ and the plane $z = 9$

48. The flux of $\mathbf{F} = \langle 2x, y, z \rangle$ across the portion of the surface $z = 1 - x^2 - y^2$ above the xy-plane

49. The flux of $\mathbf{F} = \langle -y, x, z^2/2 \rangle$ across the cone frustum $z = \sqrt{x^2 + y^2}$ between $z = 1$ and $z = 3$

50. The flux of $\mathbf{F} = \langle x^2, 2y, yz \rangle$ out of the cube in Example 3

51. The flux of $\mathbf{F} = \langle 0, y, 2z \rangle$ out of the solid region bounded by $z = 4 - x^2 - y^2$ and the plane $z = 2$

52. The flux of $\mathbf{F} = \langle x, y, 1 \rangle$ out of the solid bounded by the paraboloid $z = x^2 + y^2$ and the planes $z = 1$ and $z = 4$

53. Solve Example 6 if the temperature is inversely proportional to the square of the distance from the origin.

54. Suppose the temperature function of the solid ball $x^2 + y^2 + z^2 \leq 4$ is $T(x, y, z) = 70 - 0.1x^2$ and $\kappa = 3$. Determine the heat flow out of the region.

55–57 A discussion analogous to the one given preceding Exercises 50–53 in Section 15.2 yields the formula below. If S is a thin surface with electrical charge density $q(x,y,z)$, the electrostatic potential at a point P away from the surface is obtained from the surface integral

$$V(P) = \varepsilon \iint_S \frac{q(x,y,z)}{r_P(x,y,z)} d\sigma,$$

where $r_P(x,y,z)$ is the distance between (x,y,z) and P.

Use the above formula in Exercises 55–57.

55.* Suppose a uniformly charged sphere of radius R and total charge Q is centered at the origin. If point P is r units from the center of the sphere, then show that

$$V(P) = \varepsilon \int_0^\pi \int_0^{2\pi} \frac{Q \sin \varphi}{4\pi \sqrt{R^2 + r^2 - 2rR \cos \varphi}} \, d\theta \, d\varphi.$$

(**Hint:** Because of uniform charge distribution, the charge density on the sphere is given by $q(x,y,z) = \dfrac{Q}{4\pi R^2}$. Notice also that because of radial symmetry, it suffices to pick P on one of the coordinate axes.)

56.* Use Exercise 55 to show that

$$V(P) = \frac{\varepsilon Q}{2rR} \left(|R+r| - |R-r| \right).$$

(**Hint:** Substitute $R^2 + r^2 - 2rR \cos \varphi = u$.)

57.* Use Exercises 55 and 56 to conclude that the electrostatic potential for a uniformly charged sphere r units from its center is as follows:

$$V(P) = \begin{cases} \varepsilon \dfrac{Q}{r} & \text{if } P \text{ is outside the sphere} \\[2mm] \varepsilon \dfrac{Q}{R} & \text{if } P \text{ is inside the sphere} \end{cases}$$

Note what this means is that the potential is constant inside the sphere, while outside the sphere the potential function behaves as if all of the charge Q were concentrated at the origin!

15.6 Technology Exercises

58–61 Use a computer algebra system.

58. Determine the center of mass of the hemispherical surface given in Exercise 31.

59. Determine the moments of inertia of the surface given in Exercise 31.

60. Determine the mass and centroid of the thin parabolic surface $z = 1 - x^2 - y^2$, $z \geq 0$, if it has constant density ρ.

61. Find the second moments about the coordinate axes of the surface given in Exercise 58.

15.7 Stokes' Theorem

TOPICS

1. Circulation, curl, and Stokes' Theorem
2. Consequences of Stokes' Theorem

In this and the next section of this chapter, we introduce two powerful theorems of multivariable calculus, both of which can be viewed as extensions of Green's Theorem. The first, Stokes' Theorem, relates the circulation of a vector field around the boundary of a surface to the integral of the vector field's curl over that surface.

TOPIC 1 Circulation, Curl, and Stokes' Theorem

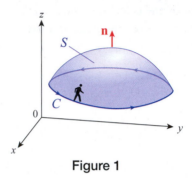

Figure 1

To begin, assume that C is a simple, closed, piecewise smooth curve and that S is a piecewise smooth surface whose boundary is C, as in Figure 1. Assume also that S and C are oriented in such a way that if you were to walk around C in its direction of orientation, with your head pointing in the direction $\mathbf{n}$ normal to S, then the surface S would lie to your left—we call this the **positive** orientation of C relative to S.

Let $\mathbf{F}$ be a vector field with continuous partial derivatives on an open region containing S. Then, just as in Section 15.2, the circulation of $\mathbf{F}$ around C is the line integral

$$\oint_C \mathbf{F} \cdot \mathbf{T}\, ds,$$

where $\mathbf{T}$ is the unit vector tangent to C. Recall from Section 15.4 that the curl of $\mathbf{F}$ is the vector field $\nabla \times \mathbf{F}$. In Section 15.4, we evaluated double integrals of the z-component of the curl, $\nabla \times \mathbf{F} \cdot \mathbf{k}$, over regions enclosed by curves; Stokes' Theorem extends the Tangential-Curl Form of Green's Theorem to the more general surfaces we will consider now.

Theorem 🔍

Stokes' Theorem

Assume $\mathbf{F}$ is a vector field with continuous partial derivatives in an open region of space containing a piecewise smooth surface S. Assume that the boundary of S is a simple, closed, piecewise smooth curve C, and that C is positively oriented with respect to S. Then

$$\oint_C \mathbf{F} \cdot \mathbf{T}\, ds = \iint_S \nabla \times \mathbf{F} \cdot \mathbf{n}\, d\sigma.$$

Note: If S is a region in the xy-plane, then $\mathbf{n} = \mathbf{k}$ and Stokes' Theorem reduces to one of the two forms of Green's Theorem: $\oint_C \mathbf{F} \cdot \mathbf{T}\, ds = \iint_S \nabla \times \mathbf{F} \cdot \mathbf{k}\, dA.$

Example 1 ✎

Verify Stokes' Theorem for the vector field $\mathbf{F}(x, y, z) = \langle -y, x, 1 \rangle$ and the surface S defined as the portion of the sphere $x^2 + y^2 + z^2 = 9$ for which $z \geq 0$.

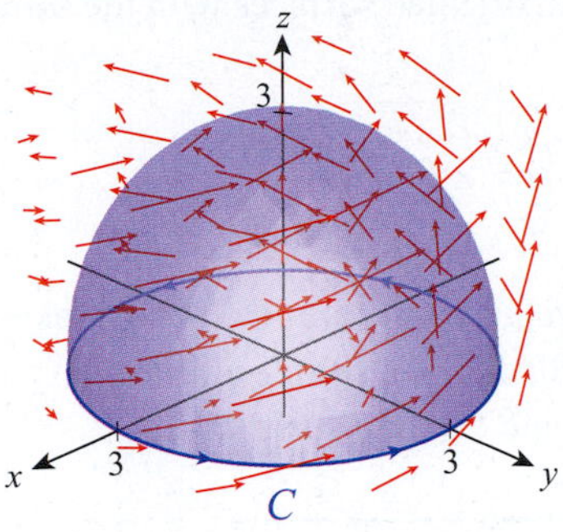

Solution

Note that the boundary C of S is the circle $x^2 + y^2 = 9$ in the xy-plane. Since S is half of a sphere, and the convention is to define the normal vectors for closed surfaces to be outward-pointing, we will assume the vector field $\mathbf{n}$ of unit vectors normal to S also points outward. This automatically assigns a counterclockwise (when viewed from above) direction as the orientation of C, as shown in Figure 2. (Stokes' Theorem would also be verified if we chose inward-pointing normal vectors and a clockwise orientation for C, but each integral would be the negative of what we find below.)

Figure 2 also shows a scaled illustration of the vector field $\mathbf{F}$. Given that the elements of $\mathbf{F}$ point generally in the same direction as C, we expect the circulation of $\mathbf{F}$ around C to be positive. To evaluate the line integral, we can use the parametrization $\mathbf{r}(\theta) = \langle 3\cos\theta, 3\sin\theta, 0 \rangle$ with $0 \le \theta \le 2\pi$. Then

$$d\mathbf{r} = \frac{d\mathbf{r}}{d\theta}\,d\theta = \langle -3\sin\theta, 3\cos\theta, 0 \rangle\,d\theta$$

and

$$\mathbf{F} \cdot d\mathbf{r} = \langle -y, x, 1 \rangle \cdot \langle -3\sin\theta, 3\cos\theta, 0 \rangle\,d\theta = (3y\sin\theta + 3x\cos\theta)\,d\theta$$
$$= \big[3(3\sin\theta)\sin\theta + 3(3\cos\theta)\cos\theta\big]\,d\theta = 9\,d\theta,$$

so

$$\oint_C \mathbf{F} \cdot \mathbf{T}\,ds = \oint_C \mathbf{F} \cdot d\mathbf{r} = \int_0^{2\pi} 9\,d\theta = 18\pi.$$

To evaluate the surface integral, we first find

$$\nabla \times \mathbf{F} = \begin{vmatrix} \mathbf{i} & \mathbf{j} & \mathbf{k} \\ \partial/\partial x & \partial/\partial y & \partial/\partial z \\ -y & x & 1 \end{vmatrix} = \langle 0, 0, 2 \rangle.$$

To now determine $\nabla \times \mathbf{F} \cdot \mathbf{n}$, recall from Example 4 of Section 15.6 that if we parametrize S by

$$\mathbf{r}(\theta, \varphi) = \langle 3\sin\varphi\cos\theta, 3\sin\varphi\sin\theta, 3\cos\varphi \rangle, \quad 0 \le \theta \le 2\pi, \quad 0 \le \varphi \le \pi/2$$

then $\mathbf{n} = \frac{1}{3}\mathbf{r}(\theta, \varphi)$, so $\nabla \times \mathbf{F} \cdot \mathbf{n} = (2)(\frac{1}{3})(3\cos\varphi) = 2\cos\varphi$. And to calculate the surface integral, we also need the fact that

$$d\sigma = \big|\mathbf{r}_\theta \times \mathbf{r}_\varphi\big|\,d\theta\,d\varphi = 9\sin\varphi\,d\theta\,d\varphi.$$

So

$$\iint\limits_S \nabla \times \mathbf{F} \cdot \mathbf{n}\,d\sigma = \int_0^{2\pi} \int_0^{\pi/2} (2\cos\varphi)(9\sin\varphi)\,d\varphi\,d\theta$$

$$= 18\int_0^{2\pi} \left[\frac{\sin^2\varphi}{2}\right]_0^{\pi/2}\,d\theta = 9(2\pi) = 18\pi.$$

Figure 2

As Example 1 illustrates, one of the integrals in Stokes' Theorem (in this case, the line integral) may be easier to evaluate than the other; that fact can often be used to simplify a calculation. Another fact emerges if we study the example a bit more closely: if S_1 and S_2 are *any* two piecewise smooth oriented surfaces with the same oriented boundary C, then

$$\iint_{S_1} \nabla \times \mathbf{F} \cdot \mathbf{n} \, d\sigma = \oint_C \mathbf{F} \cdot \mathbf{T} \, ds = \iint_{S_2} \nabla \times \mathbf{F} \cdot \mathbf{n} \, d\sigma.$$

In such a case, S_1 and S_2 are called *capping surfaces* of the curve C, and Stokes' Theorem says that the circulation of a vector field around a simple closed curve is equal to the integral of its curl over any appropriate capping surface.

Example 2 ✎

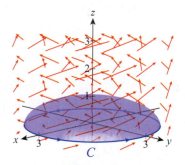

Figure 3

Let $\overline{S}$ be the upper half of the ellipsoid

$$\frac{x^2}{9} + \frac{y^2}{9} + z^2 = 1,$$

oriented with an outward-pointing unit normal vector field $\mathbf{n}$. Note that the boundary of $\overline{S}$ is also the curve $x^2 + y^2 = 9$ in the xy-plane (see Figure 3). Show that the integral of the curl of $\mathbf{F}(x, y, z) = \langle -y, x, 1 \rangle$ is again 18π.

Solution

For comparative and illustrative purposes, we use a different parametrization for $\overline{S}$.

$$\mathbf{r}(x, y) = \left\langle x, y, \sqrt{1 - \frac{x^2}{9} - \frac{y^2}{9}} \right\rangle, \quad -3 \le x \le 3, \quad -\sqrt{9 - x^2} \le y \le \sqrt{9 - x^2}$$

In Exercise 6 you will show that

$$\mathbf{r}_x \times \mathbf{r}_y = \left\langle \frac{x}{3\sqrt{9 - x^2 - y^2}}, \frac{y}{3\sqrt{9 - x^2 - y^2}}, 1 \right\rangle.$$

$\mathbf{r}_x \times \mathbf{r}_y$ is an outward-pointing vector field, so dividing each element by its length defines the vector field $\mathbf{n}$ of unit normal vectors. But we don't actually need to perform that calculation, since

$$\iint_{\overline{S}} \nabla \times \mathbf{F} \cdot \mathbf{n} \, d\sigma = \iint_{x^2 + y^2 \le 9} \underbrace{\langle 0, 0, 2 \rangle}_{\nabla \times \mathbf{F}} \cdot \underbrace{\left(\frac{\mathbf{r}_x \times \mathbf{r}_y}{|\mathbf{r}_x \times \mathbf{r}_y|} \right)}_{\mathbf{n}} \underbrace{|\mathbf{r}_x \times \mathbf{r}_y| \, dx \, dy}_{d\sigma}$$

$$= \iint_{x^2 + y^2 \le 9} \langle 0, 0, 2 \rangle \cdot (\mathbf{r}_x \times \mathbf{r}_y) \, dx \, dy$$

$$= 2 \iint_{x^2 + y^2 \le 9} dx \, dy = 2(9\pi) = 18\pi.$$

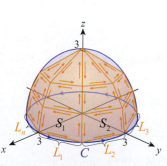

Figure 4

In lieu of a formal proof of Stokes' Theorem, which is best left for a class in advanced calculus, we provide a sketch of the principal idea using the surface and curve in Example 1 as a model. Figure 4 shows an approximation of the surface S by polygonal planar surfaces $S_1, \ldots, S_m$. (The result is called a *polyhedral*

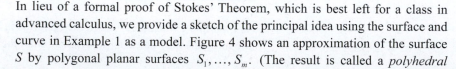

approximation of S.) The approximation automatically induces a polygonal approximation of the curve C, and the circulation of $\mathbf{F}$ around C is thus

$$\oint_C \mathbf{F} \cdot \mathbf{T}\, ds \approx \sum_{i=1}^{n} \int_{L_i} \mathbf{F} \cdot \mathbf{T}\, ds,$$

where $L_1, \ldots, L_n$ are the line segments of the polygon. But as shown in Figure 4, if we add up the circulation of $\mathbf{F}$ around *all* the polygonal faces, most of the line integrals are canceled by an integral along the same edge with opposing orientation, so

$$\sum_{i=1}^{n} \int_{L_i} \mathbf{F} \cdot \mathbf{T}\, ds = \sum_{i=1}^{m} \oint_{C_i} \mathbf{F} \cdot \mathbf{T}\, ds.$$

Here, $C_1, \ldots, C_m$ are the closed polygonal paths around the facets of the polyhedral approximation of S, and $m \geq n$.

By the same sort of reasoning used to prove the Tangential-Curl Form of Green's Theorem, each circulation integral around a polygonal facet is equal to the integral of the curl over the surface of the facet, so

$$\sum_{i=1}^{m} \oint_{C_i} \mathbf{F} \cdot \mathbf{T}\, ds = \sum_{i=1}^{m} \iint_{S_i} \nabla \times \mathbf{F} \cdot \mathbf{n}\, d\sigma.$$

In the limit, as the polyhedral approximation of S becomes finer and finer, the last sum approaches

$$\iint_S \nabla \times \mathbf{F} \cdot \mathbf{n}\, d\sigma,$$

giving us Stokes' Theorem.

TOPIC 2   Consequences of Stokes' Theorem

Stokes' Theorem is used extensively in physics and engineering, and some of the fundamental laws of physics are direct applications of Stokes' Theorem (and many others arise from the Divergence Theorem of the next section). For example, the Maxwell-Faraday Law of Induction regarding electric field lines of force $\mathbf{E}$ over a surface S can be expressed as

$$\oint_C \mathbf{E} \cdot \mathbf{T}\, ds = \iint_S \nabla \times \mathbf{E} \cdot \mathbf{n}\, d\sigma.$$

Stokes' Theorem also leads to a quick physical interpretation of *curl*. Given a vector field $\mathbf{F}$ and its curl $\nabla \times \mathbf{F}$, imagine evaluating the integral

$$\iint_S \nabla \times \mathbf{F} \cdot \mathbf{n}\, d\sigma$$

over various (small) disks S all centered at a fixed point P in space. The orientation of S that maximizes the integral will clearly be that for which $\mathbf{n}$ points in the same direction as $\nabla \times \mathbf{F}$. In other words, if we think of a paddle wheel spinning about an axis running through P, the wheel will spin fastest when the axis points in the direction of $\nabla \times \mathbf{F}$ (see Figure 5).

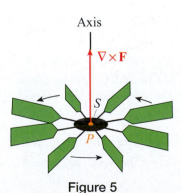

Axis

$\nabla \times \mathbf{F}$

S

P

Figure 5

Stokes' Theorem can be used to prove the Component Test for conservative vector fields (Section 15.3), which we restate here in terms of the curl operator.

Theorem &

$\nabla \times \mathbf{F} = \mathbf{0}$ Implies $\mathbf{F}$ Is Conservative

If $\nabla \times \mathbf{F} = \mathbf{0}$ at every point of an open simply connected region D of space, then for any piecewise smooth closed path C in D,

$$\oint_C \mathbf{F} \cdot d\mathbf{r} = 0,$$

and hence $\mathbf{F}$ is conservative in D.

The complete proof of this theorem requires several other facts, one of which is that every simple closed path in an open simply connected region is the boundary of an associated capping surface, a result from a branch of mathematics called *topology*. But Stokes' Theorem is then used to show that

$$\oint_C \mathbf{F} \cdot d\mathbf{r} = \oint_C \mathbf{F} \cdot \mathbf{T}\, ds = \iint_S \nabla \times \mathbf{F} \cdot \mathbf{n}\, d\sigma = 0.$$

We have already proved, in Section 15.3, the equivalence of the following three statements for an open connected region D.

$$\int_C \mathbf{F} \cdot d\mathbf{r} \text{ is independent of path in } D$$

$$\Leftrightarrow \quad \mathbf{F} = \nabla f \text{ for some potential function } f \quad (\mathbf{F} \text{ is conservative, or } \textit{irrotational})$$

$$\Leftrightarrow \quad \oint_C \mathbf{F} \cdot d\mathbf{r} = 0 \text{ for every closed path } C$$

Assuming, as we have been, that the second partial derivatives of $\mathbf{F}$ are continuous in D, the equality of mixed partial derivatives means any one of the above statements implies $\nabla \times \mathbf{F} = \mathbf{0}$ throughout D. Incorporating the latest theorem, we can summarize the relationships between conservation and curl by

$$\mathbf{F} = \nabla f \quad \Rightarrow \quad \nabla \times \mathbf{F} = \mathbf{0} \quad \text{if } D \text{ is an open connected region}$$
$$\nabla \times \mathbf{F} = \mathbf{0} \quad \Rightarrow \quad \mathbf{F} = \nabla f \quad \text{if } D \text{ is an open and simply connected region.}$$

15.7 **Exercises**

1–4 Verify Stokes' Theorem by showing that the integrals $\oint_C \mathbf{F} \cdot \mathbf{T}\, ds$ and $\iint_S \nabla \times \mathbf{F} \cdot \mathbf{n}\, d\sigma$ are equal.

1. $\mathbf{F}(x, y, z) = \langle -2y, 2x, y + z \rangle$, where S is the upper unit hemisphere centered at the origin

2. $\mathbf{F}(x, y, z) = \langle y - z, x + z, -x + y \rangle$, where S is the hemisphere $z = \sqrt{4 - x^2 - y^2}$

3. $\mathbf{F}(x, y, z) = \langle -y, x, 1 \rangle$ (as in Example 1), where S is the portion of the paraboloid $z = 1 - x^2 - y^2$ above the xy-plane

4. $\mathbf{F}(x, y, z) = \langle y, z^2, 2x \rangle$, where S is the portion of the paraboloid $z = x^2 + y^2$ between the xy-plane and the plane $z = 9$

5. Calculate $\iint\limits_{S} \nabla \times \mathbf{F} \cdot \mathbf{n} \, d\sigma$ for the vector field

$\mathbf{F}(x, y, z) = \langle -2y, 2x, y+z \rangle$, on the upper semiellipsoid S: $x^2 + y^2 + 4z^2 = 1$, $z \geq 0$, and compare your answer to the solution of Exercise 1.

6. Verify by calculation that the field of normal vectors for the upper half of the ellipsoid given in Example 2 is

$$\mathbf{r}_x \times \mathbf{r}_y = \left\langle \frac{x}{3\sqrt{9 - x^2 - y^2}}, \frac{y}{3\sqrt{9 - x^2 - y^2}}, 1 \right\rangle.$$

7–14 Use Stokes' Theorem to evaluate the indicated line integral.

7. $\oint_C \mathbf{F} \cdot \mathbf{T} \, ds$, where $\mathbf{F} = \langle -2y, x^2, 3z^2 \rangle$ and C is the intersection of the cylinder $x^2 + y^2 = 1$ and the plane $x + 2y + z = 4$, with positive orientation when viewed from above

8. $\oint_C \mathbf{F} \cdot \mathbf{T} \, ds$, where $\mathbf{F} = \langle z^2, x^2, y^2 \rangle$ and C is the intersection of the cylinder $x^2 + y^2 = 4x$ and the plane $z = 2x$, with positive orientation when viewed from above

9. $\oint_C \mathbf{F} \cdot \mathbf{T} \, ds$, where $\mathbf{F} = \langle -3y, 2x, z^2 \rangle$ and C is the triangle with vertices $(4,0,0)$, $(0,2,0)$, and $(0,0,3)$, with positive orientation when viewed from above

10. $\oint_C \mathbf{F} \cdot \mathbf{T} \, ds$, where $\mathbf{F} = \langle zy^2, z^3, 9y - 2x \rangle$ and C is the triangle with vertices $(1,0,0)$, $(3,4,1)$, and $(0,0,2)$, with positive orientation when viewed from above

11. $\oint_C \mathbf{F} \cdot \mathbf{T} \, ds$, where $\mathbf{F} = \langle -3z, x, 2y \rangle$ and C is the boundary of the disk $x^2 + y^2 \leq 1$, $z = 1$, with positive orientation when viewed from above

12. $\oint_C \mathbf{F} \cdot \mathbf{T} \, ds$, where $\mathbf{F} = \langle -3z, x, 2y \rangle$ and C is the intersection of the paraboloids $z = x^2 + y^2$ and $z = 2 - x^2 - y^2$, with positive orientation when viewed from above; compare your answer to the solution of Exercise 11.

13. $\oint_C \mathbf{F} \cdot \mathbf{T} \, ds$, where $\mathbf{F} = \langle 2x^3, 4x + y^2, e^{z^2} \rangle$ and C is the intersection of the paraboloids $2z = x^2 + y^2$ and $z = 6 - x^2 - y^2$, with positive orientation when viewed from above

14. $\oint_C \mathbf{F} \cdot \mathbf{T} \, ds$, where $\mathbf{F} = \langle -xz, 2z, x - y^2 \rangle$ and C is the intersection of $z = x^2 + y^2$ and $z = 2x + 3$, with negative orientation when viewed from above

15–22 Use Stokes' Theorem to evaluate the surface integral.

15. $\iint\limits_{S} \nabla \times \mathbf{F} \cdot \mathbf{n} \, d\sigma$, where $\mathbf{F} = \langle -y, x, xyz \rangle$ and S is the upper hemisphere $z = \sqrt{1 - x^2 - y^2}$, oriented with an upward-pointing unit normal vector field

16. $\iint\limits_{S} \nabla \times \mathbf{F} \cdot \mathbf{n} \, d\sigma$, where $\mathbf{F} = \langle 4z, -3x, 2y \rangle$ and S is the portion of the paraboloid $z = x^2 + y^2$ with $0 \leq z \leq 9$, oriented with a downward-pointing unit normal vector field

17. $\iint\limits_{S} \nabla \times \mathbf{F} \cdot \mathbf{n} \, d\sigma$, where $\mathbf{F} = \langle x - y^2, x^4, z^2 \rangle$ and S is the triangle determined by the first-octant portion of the plane $x + y + z = 1$, oriented with an upward-pointing unit normal vector field

18. $\iint\limits_{S} \nabla \times \mathbf{F} \cdot \mathbf{n} \, d\sigma$, where $\mathbf{F} = \langle 2y^2, -z, 4x \rangle$ and S is the triangle determined by the first-octant portion of the plane $5x + y + 2z = 10$, oriented with an upward-pointing unit normal vector field

19. $\iint\limits_{S} \nabla \times \mathbf{F} \cdot \mathbf{n} \, d\sigma$, where $\mathbf{F} = \langle -2yz, 3xz, 2z^3 \rangle$ and S is the cone frustum $z = \sqrt{x^2 + y^2}$, $1 \leq z \leq 2$, oriented with an inward-pointing unit normal vector field

20. $\iint\limits_{S} \nabla \times \mathbf{F} \cdot \mathbf{n} \, d\sigma$, where $\mathbf{F} = \langle yz, -2xz, x^2 y \rangle$ and S is the portion of the cylinder $x^2 + y^2 = 4$, $2 \leq z \leq 4$, oriented with an outward-pointing unit normal vector field

21. $\iint\limits_{S} \nabla \times \mathbf{F} \cdot \mathbf{n} \, d\sigma$, where $\mathbf{F} = \langle y \sin(z^2), xy^2, xz + y \rangle$ and S is the portion of the paraboloid $x = y^2 + z^2$, $0 \leq x \leq 4$, oriented with an inward-pointing unit normal vector field

22. $\iint\limits_{S} \nabla \times \mathbf{F} \cdot \mathbf{n} \, d\sigma$, where $\mathbf{F} = \langle 3yx^2, x \ln(z^4), y \cos(xz^2) - 2x \rangle$ and S is the hemisphere $y = \sqrt{4 - x^2 - z^2}$, oriented with an inward-pointing unit normal vector field

23. Suppose, as in Exercise 3 of Section 15.6, that the (piecewise smooth) surface S is the graph of a function $z = g(x, y)$, defined on a domain R, and that S has a piecewise smooth boundary C. Prove that in this case, Stokes' Theorem takes the following form:

$$\oint_C \mathbf{F} \cdot \mathbf{T}\, ds = \iint_R \nabla \times \mathbf{F} \cdot \langle -g_x, -g_y, 1 \rangle\, dA$$

24–26 Assume that both f and g have continuous second-order partial derivatives, and that both the surface S and its boundary C meet the conditions of Stokes' Theorem. Verify the statement by using Stokes' Theorem.

24. $\displaystyle\oint_C (f\nabla f) \cdot \mathbf{T}\, ds = 0$

25. $\displaystyle\oint_C (f\nabla g) \cdot \mathbf{T}\, ds = \iint_S (\nabla f \times \nabla g) \cdot \mathbf{n}\, d\sigma$

26. $\displaystyle\oint_C (f\nabla g + g\nabla f) \cdot \mathbf{T}\, ds = 0$

27. Let S be the solid bounded by the cylinder $x^2 + y^2 = 1$ and the planes $z = -1$ and $z = 1$. If $\mathbf{F}$ is a vector field with continuous partials in an open region containing S, show that

$$\iint_S \nabla \times \mathbf{F} \cdot \mathbf{n}\, d\sigma = 0.$$

28.* Prove the following generalization of Exercise 27. If the vector field $\mathbf{F}$ and the closed surface S satisfy the conditions of Stokes' Theorem, then

$$\iint_S \nabla \times \mathbf{F} \cdot \mathbf{n}\, d\sigma = 0.$$

29.* The force field $\mathbf{F}$ is called a *central force* if it points directly away from, or toward, a point called the *center*, and its magnitude depends only on the distance from the center. In addition, we assume that this dependency is continuously differentiable, or in symbols, $\mathbf{F} = f(r)\mathbf{r}$, where the single-variable function f is continuously differentiable everywhere except possibly at zero. Show that if such a force is moving an object around a closed path that doesn't enclose the origin, then the total work done by the force is zero. (**Hint:** Show that the curl of the force field is zero.)

30. Consider the vector field

$$\mathbf{F} = \left\langle -\frac{y}{x^2 + y^2}, \frac{x}{x^2 + y^2}, z \right\rangle$$

and show that while $\nabla \times \mathbf{F} = \mathbf{0}$, $\displaystyle\oint_C \mathbf{F} \cdot d\mathbf{r} \neq 0$ on a circle in the xy-plane centered at the origin. Does this contradict the theorem stating that $\nabla \times \mathbf{F} = \mathbf{0}$ implies $\mathbf{F}$ is conservative?

15.7 Technology Exercises

31. Write a program on a computer algebra system that accepts a vector field, the parametrizations of a surface S and its boundary C, and returns both integrals $\displaystyle\oint_C \mathbf{F} \cdot \mathbf{T}\, ds$ and $\displaystyle\iint_S \nabla \times \mathbf{F} \cdot \mathbf{n}\, d\sigma$. Use it to check your answers to Exercises 1–4.

15.8 **The Divergence Theorem**

TOPICS

1. The Divergence Theorem

2. The Fundamental Theorem of Calculus in review

In this final section, we introduce the last in a series of deep results. The Divergence Theorem, like Green's Theorem and Stokes' Theorem, relates an integral over a region to another integral over the boundary of the region. As we will see in a wrap-up review, all the results are simply instances of the overarching principle of the Fundamental Theorem of Calculus.

TOPIC 1 **The Divergence Theorem**

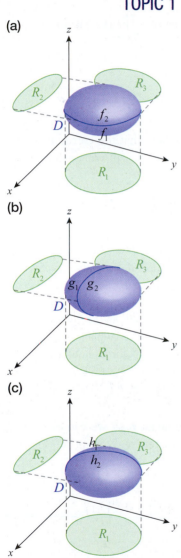

(a)

(b)

(c)

Figure 1

Recall the Normal-Divergence Form of Green's Theorem: given a vector field **F** defined on an open region containing $R \subseteq \mathbb{R}^2$, whose boundary is a positively oriented piecewise smooth simple closed curve C,

$$\oint_C \mathbf{F} \cdot \mathbf{n} \, ds = \iint_R \nabla \cdot \mathbf{F} \, dA.$$

In words, the total outward flux of **F** across C is equal to the integral of the divergence of **F** over the region bounded by C. This is often informally stated with a phrase like *the net flow of a vector field out of a region is equal to the sum of the sources and sinks of the field in the region*. In this context, a *source* of **F** is a point where its divergence $\nabla \cdot \mathbf{F}$ is positive, and a *sink* is a point where $\nabla \cdot \mathbf{F}$ is negative.

This principle holds true in any number of dimensions, and its statement in three dimensions is known as the Divergence Theorem. While the theorem applies to a variety of regions, a complete classification of them would lead us too far astray from the task of introducing the topic. For our purposes, we are interested in regions that are simple or that can be decomposed into a union of a finite number of simple regions. A **simple region** D in space, as illustrated in Figure 1, is one that can be described in each of the following ways:

$$D = \left\{ (x, y, z) \big| (x, y) \in R_1 \text{ and } f_1(x, y) \le z \le f_2(x, y) \right\} \text{ (see Figure 1a)}$$

$$D = \left\{ (x, y, z) \big| (x, z) \in R_2 \text{ and } g_1(x, z) \le y \le g_2(x, z) \right\} \text{ (see Figure 1b)}$$

$$D = \left\{ (x, y, z) \big| (y, z) \in R_3 \text{ and } h_1(y, z) \le x \le h_2(y, z) \right\} \text{ (see Figure 1c)}$$

where the bounding surfaces $\{ f_1, f_2, g_1, g_2, h_1, h_2 \}$ are all smooth.

With this definition, we can now state the Divergence Theorem.

Theorem

The Divergence Theorem

Assume **F** is a vector field with continuous partial derivatives in an open region of space containing $D \subseteq \mathbb{R}^3$, and assume D is either a simple region or a finite union of simple regions. Let S denote the surface of D, and let **n** be the outward-pointing field of unit vectors normal to S. Then

$$\iint_S \mathbf{F} \cdot \mathbf{n} \, d\sigma = \iiint_D \nabla \cdot \mathbf{F} \, dV.$$

Proof 📎

We prove the theorem under the assumption that D is a single simple region.

If we denote the component functions of $\mathbf{F}$ as $\mathbf{F} = \langle P, Q, R \rangle$, then the dot product of $\mathbf{F}$ with the unit vector field $\mathbf{n}$ normal to the surface S is

$$\mathbf{F} \cdot \mathbf{n} = P(\mathbf{i} \cdot \mathbf{n}) + Q(\mathbf{j} \cdot \mathbf{n}) + R(\mathbf{k} \cdot \mathbf{n})$$

and so

$$\iint_S \mathbf{F} \cdot \mathbf{n} \, d\sigma = \iint_S P(\mathbf{i} \cdot \mathbf{n}) \, d\sigma + \iint_S Q(\mathbf{j} \cdot \mathbf{n}) \, d\sigma + \iint_S R(\mathbf{k} \cdot \mathbf{n}) \, d\sigma.$$

Similarly,

$$\iiint_D \nabla \cdot \mathbf{F} \, dV = \iiint_D \frac{\partial P}{\partial x} \, dV + \iiint_D \frac{\partial Q}{\partial y} \, dV + \iiint_D \frac{\partial R}{\partial z} \, dV,$$

and the theorem will be proved if we can show that the corresponding terms in the two expressions above are equal. We will demonstrate the process by showing that

$$\iint_S R(\mathbf{k} \cdot \mathbf{n}) \, d\sigma = \iiint_D \frac{\partial R}{\partial z} \, dV.$$

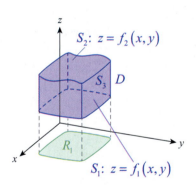

S_2: $z = f_2(x, y)$

S_3 D

R_1

S_1: $z = f_1(x, y)$

Figure 2

With this goal in mind, let R_1 be the projection of D in the xy-plane and assume the lower surface S_1 and upper surface S_2 of S are the graphs of $z = f_1(x, y)$ and $z = f_2(x, y)$, respectively, as shown in Figure 2. The lower and upper surfaces may or may not coincide at any given point $(x, y) \in R_1$; in Figure 2, they are shown separated by some distance, so the third surface S_3 of S is part of a right cylinder perpendicular to the xy-plane. Note that $\mathbf{k} \cdot \mathbf{n}$ is therefore 0 on S_3 (since $\mathbf{n}$ is perpendicular to $\mathbf{k}$ throughout S_3), so

$$\iint_S R(\mathbf{k} \cdot \mathbf{n}) \, d\sigma = \iint_{S_1} R(\mathbf{k} \cdot \mathbf{n}) \, d\sigma + \iint_{S_2} R(\mathbf{k} \cdot \mathbf{n}) \, d\sigma.$$

On S_1 and S_2, we need to know something about $\mathbf{n}$ so we can evaluate $\mathbf{k} \cdot \mathbf{n} \, d\sigma$. Toward this, note that we can parametrize S_2 with the vector function $\mathbf{r}_2(x, y) = \langle x, y, f_2(x, y) \rangle$, and then that

$$(\mathbf{r}_2)_x \times (\mathbf{r}_2)_y = \begin{vmatrix} \mathbf{i} & \mathbf{j} & \mathbf{k} \\ 1 & 0 & (f_2)_x \\ 0 & 1 & (f_2)_y \end{vmatrix} = \langle -(f_2)_x, -(f_2)_y, 1 \rangle$$

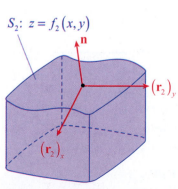

S_2: $z = f_2(x, y)$

$\mathbf{n}$

$(\mathbf{r}_2)_y$

$(\mathbf{r}_2)_x$

Figure 3

is outwardly normal to S_2—remember to use the right-hand rule in visualizing the direction of the cross product, as shown in Figure 3. So

$$\mathbf{k} \cdot \mathbf{n} \, d\sigma = \mathbf{k} \cdot \underbrace{\left(\frac{\langle -(f_2)_x, -(f_2)_y, 1 \rangle}{\left| \langle -(f_2)_x, -(f_2)_y, 1 \rangle \right|} \right)}_{\mathbf{n}} \underbrace{\left| \langle -(f_2)_x, -(f_2)_y, 1 \rangle \right| dA}_{d\sigma}$$

$$= \langle 0, 0, 1 \rangle \cdot \langle -(f_2)_x, -(f_2)_y, 1 \rangle \, dA = dA$$

and hence

$$\iint_{S_2} R(\mathbf{k}\cdot\mathbf{n})\,d\sigma = \iint_{R_1} R\big(x,y,f_2(x,y)\big)\,dA.$$

Similarly, we can parametrize S_1 with $\mathbf{r}_1(x,y)=\big\langle x,y,f_1(x,y)\big\rangle$, but this time, in order to construct an outwardly pointing vector, we have to form the cross product

$$(\mathbf{r}_1)_y \times (\mathbf{r}_1)_x = \begin{vmatrix} \mathbf{i} & \mathbf{j} & \mathbf{k} \\ 0 & 1 & (f_1)_y \\ 1 & 0 & (f_1)_x \end{vmatrix} = \big\langle (f_1)_x,(f_1)_y,-1\big\rangle.$$

The effect of this is that $\mathbf{k}\cdot\mathbf{n}\,d\sigma = -dA$ on S_1, so we are led to

$$\iint_{S_1} R(\mathbf{k}\cdot\mathbf{n})\,d\sigma = -\iint_{R_1} R\big(x,y,f_1(x,y)\big)\,dA$$

and hence

$$\iint_{S} R(\mathbf{k}\cdot\mathbf{n})\,d\sigma = \iint_{S_1} R(\mathbf{k}\cdot\mathbf{n})\,d\sigma + \iint_{S_2} R(\mathbf{k}\cdot\mathbf{n})\,d\sigma$$
$$= \iint_{R_1} \big[R\big(x,y,f_2(x,y)\big) - R\big(x,y,f_1(x,y)\big)\big]\,dA.$$

We are nearing the end of the argument. By the Fundamental Theorem of Calculus,

$$\iiint_{D} \frac{\partial R}{\partial z}\,dV = \iint_{R_1}\left[\int_{z=f_1}^{z=f_2} \frac{\partial R}{\partial z}\,dz\right]dA$$
$$= \iint_{R_1} \big[R\big(x,y,f_2(x,y)\big) - R\big(x,y,f_1(x,y)\big)\big]\,dA,$$

so the two integrals are equal:

$$\iint_{S} R(\mathbf{k}\cdot\mathbf{n})\,d\sigma = \iiint_{D} \frac{\partial R}{\partial z}\,dV$$

By the same sort of reasoning,

$$\iint_{S} P(\mathbf{i}\cdot\mathbf{n})\,d\sigma = \iiint_{D} \frac{\partial P}{\partial x}\,dV \quad \text{and} \quad \iint_{S} Q(\mathbf{j}\cdot\mathbf{n})\,d\sigma = \iiint_{D} \frac{\partial Q}{\partial y}\,dV,$$

and the Divergence Theorem is thus proved for a single simple region. The fact that the theorem is also true for a finite union of simple regions follows from properties of integrals, including the fact that two surface integrals with opposite orientations over the same surface cancel one another.

Example 1 ✐

Find the flux of the vector field $\mathbf{F}(x, y, z) = \langle x, y, z \rangle$ over the surface of

a. the sphere $x^2 + y^2 + z^2 = r^2$;   **b.** the cube $[0, a]^3$.

Solution

a. Instead of calculating $\iint_S \mathbf{F} \cdot \mathbf{n} \, d\sigma$ directly, we use the Divergence Theorem.

$$\iint_S \mathbf{F} \cdot \mathbf{n} \, d\sigma = \iiint_{x^2+y^2+z^2=r^2} \nabla \cdot \mathbf{F} \, dV = \iiint_{x^2+y^2+z^2=r^2} (1+1+1) \, dV = 3\left(\frac{4\pi r^3}{3}\right) = 4\pi r^3$$

b. The reduction in effort is even more pronounced with this region, since evaluating the surface integral directly would involve calculating a surface integral over each of the six sides. But using the Divergence Theorem,

$$\iint_S \mathbf{F} \cdot \mathbf{n} \, d\sigma = \iiint_{[0,a]^3} \nabla \cdot \mathbf{F} \, dV = \iiint_{[0,a]^3} (1+1+1) \, dV = 3a^3.$$

Example 2 ✐

Find the flux of the vector field

$$\mathbf{F}(x, y, z) = \left\langle x^2 - z \sin y, \, xy^2 + \cos z, \, xz(1 - 2y) + e^{xy} \right\rangle$$

over the surface of the solid bounded by the coordinate planes and the plane $2x + 3y + 2z = 6$ (see Figure 4).

Solution

To calculate the flux directly would require evaluating four rather complicated surface integrals, but the volume integral of the divergence of $\mathbf{F}$ is a routine calculation. First,

$$\nabla \cdot \mathbf{F} = \frac{\partial}{\partial x}\left(x^2 - z \sin y\right) + \frac{\partial}{\partial y}\left(xy^2 + \cos z\right) + \frac{\partial}{\partial z}\left(xz(1 - 2y) + e^{xy}\right)$$

$$= 2x + 2xy + x - 2xy = 3x.$$

Omitting some of the intermediate simplifications, this gives us

$$\iint_S \mathbf{F} \cdot \mathbf{n} \, d\sigma = \iiint_D \nabla \cdot \mathbf{F} \, dV = \int_{x=0}^3 \int_{y=0}^{(6-2x)/3} \int_{z=0}^{(6-2x-3y)/2} 3x \, dz \, dy \, dx$$

$$= \frac{3}{2} \int_{x=0}^3 \int_{y=0}^{(6-2x)/3} \left(6x - 2x^2 - 3xy\right) dy \, dx$$

$$= \frac{3}{2} \int_{x=0}^3 \left[6xy - 2x^2 y - \frac{3}{2} xy^2\right]_{y=0}^{(6-2x)/3} dx$$

$$= \frac{3}{2} \int_{x=0}^3 \left(6x - 4x^2 + \frac{2x^3}{3}\right) dx = \frac{27}{4}.$$

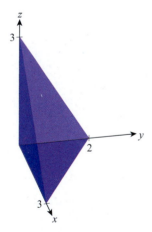

Figure 4

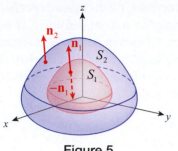

Figure 5

The properties of integrals allow us to apply the Divergence Theorem to regions more complicated than those seen so far. An important illustration of this fact leads to another of the fundamental principles of physics. Suppose D_1 and D_2 are two simple regions whose interiors both contain the origin, as shown in Figure 5. Let S_1 and S_2 be their respective surfaces, with $\mathbf{n}_1$ the outward-pointing unit normal vector field for S_1 and $\mathbf{n}_2$ the outward-pointing unit normal vector field for S_2. If we let D denote the region bounded between S_1 and S_2, then the boundary of D is $S = S_1 \cup S_2$ and the outward-pointing unit normal vector field $\mathbf{n}$ for S satisfies

$$\mathbf{n} = -\mathbf{n}_1 \text{ on } S_1 \quad \text{and} \quad \mathbf{n} = \mathbf{n}_2 \text{ on } S_2.$$

Given a vector field $\mathbf{F}$ satisfying the usual criteria, the properties of integrals imply

$$\iiint_D \nabla \cdot \mathbf{F}\, dV = \iiint_{D_2} \nabla \cdot \mathbf{F}\, dV - \iiint_{D_1} \nabla \cdot \mathbf{F}\, dV = \iint_{S_2} \mathbf{F} \cdot \mathbf{n}_2\, d\sigma - \iint_{S_1} \mathbf{F} \cdot \mathbf{n}_1\, d\sigma$$

$$= \iint_{S_2} \mathbf{F} \cdot \mathbf{n}_2\, d\sigma + \iint_{S_1} \mathbf{F} \cdot (-\mathbf{n}_1)\, d\sigma = \iint_{S_1 \cup S_2} \mathbf{F} \cdot \mathbf{n}\, d\sigma = \iint_S \mathbf{F} \cdot \mathbf{n}\, d\sigma.$$

To apply this fact, consider a point charge Q at the origin and the electric field it creates:

$$\mathbf{E}(x, y, z) = \varepsilon Q \frac{\mathbf{r}}{|\mathbf{r}|^3},$$

where $\mathbf{r} = \langle x, y, z \rangle$ and ε is a physical constant (fields like this were first seen in Section 15.1). If we let S_1 be the sphere $x^2 + y^2 + z^2 = a^2$, then $\mathbf{n}_1 = \mathbf{r}/a$ and

$$\mathbf{E} \cdot \mathbf{n}_1 = \varepsilon Q \frac{\mathbf{r}}{a^3} \cdot \frac{\mathbf{r}}{a} = \varepsilon Q \frac{a^2}{a^4} = \frac{\varepsilon Q}{a^2}$$

so

$$\iint_{S_1} \mathbf{E} \cdot \mathbf{n}_1\, d\sigma = \frac{\varepsilon Q}{a^2} \iint_{S_1} d\sigma = \left(\frac{\varepsilon Q}{a^2} \right)(4\pi a^2) = 4\pi \varepsilon Q.$$

In Exercise 27, you will show that $\nabla \cdot \mathbf{E} = 0$, so if S_2 is any other surface as described above, then

$$0 = \iiint_D \nabla \cdot \mathbf{E}\, dV = \iint_S \mathbf{E} \cdot \mathbf{n}\, d\sigma = \iint_{S_2} \mathbf{E} \cdot \mathbf{n}_2\, d\sigma - \iint_{S_1} \mathbf{E} \cdot \mathbf{n}_1\, d\sigma = \iint_{S_2} \mathbf{E} \cdot \mathbf{n}_2\, d\sigma - 4\pi \varepsilon Q.$$

That is, $\iint_{S_2} \mathbf{E} \cdot \mathbf{n}_2\, d\sigma = 4\pi \varepsilon Q$, a result known as *Gauss's Law*.

TOPIC 2   The Fundamental Theorem of Calculus in Review

The Divergence Theorem is an appropriate topic with which to conclude this text, as it gives us an opportunity to look back and identify the primary theme of calculus in many of its various guises.

The concept of "derivative" can be extended to many additional types of spaces, and in advanced mathematics, an operator called the *exterior derivative* is developed. All of the derivative-like operators we have seen—differentiation as applied to single-variable functions, the gradient operator, the divergence operator, and the curl operator—can be expressed in terms of an exterior derivative. If we let d denote the exterior derivative of a function f (without bothering to define what we mean by exterior derivative), and if we let ∂D denote the boundary of a region D (again, without bothering to define these terms precisely), then the Fundamental Theorem of Calculus can be written in the form

$$\int_D df = \int_{\partial D} f.$$

For the moment, this statement is intended to be nothing but a teaser to lure the reader into higher mathematics. But we can bring the discussion back to familiar ground by summarizing all the versions of the Fundamental Theorem of Calculus here, beginning with its philosophical meaning: *The sum of all the differential changes of f over a region is equal to the sum of the values of f over the boundary of the region.* See if you can relate this philosophical statement to each of its concrete manifestations below.

The Fundamental Theorem of Calculus in Review

Fundamental Theorem of Calculus	$\int_a^b f'(x)\,dx = f(b) - f(a)$
Fundamental Theorem for Line Integrals	$\int_A^B df = \int_C \nabla f \cdot d\mathbf{r} = f(B) - f(A)$
Tangential-Curl Form of Green's Theorem	$\iint_R \nabla \times \mathbf{F} \cdot \mathbf{k}\,dA = \oint_C \mathbf{F} \cdot d\mathbf{r}$
Normal-Divergence Form of Green's Theorem	$\iint_R \nabla \cdot \mathbf{F}\,dA = \oint_C \mathbf{F} \cdot \mathbf{n}\,ds$
Stokes' Theorem	$\iint_S \nabla \times \mathbf{F} \cdot \mathbf{n}\,d\sigma = \oint_C \mathbf{F} \cdot d\mathbf{r}$
Divergence Theorem	$\iiint_D \nabla \cdot \mathbf{F}\,dV = \iint_S \mathbf{F} \cdot \mathbf{n}\,d\sigma$

Table 1

15.8 Exercises

1–4 Verify the Divergence Theorem by showing the equality of the integrals $\iint_S \mathbf{F} \cdot \mathbf{n}\,d\sigma$ and $\iiint_D \nabla \cdot \mathbf{F}\,dV$ for the given vector field $\mathbf{F}$ on the solid D.

1. $\mathbf{F}(x, y, z) = \langle -y, x, z \rangle$, where D is the unit ball centered at the origin

2. $\mathbf{F}(x, y, z) = \langle x, 2y, 3z \rangle$, where D is the tetrahedron bounded by the coordinate planes and $z = 1 - x - y$

3. $\mathbf{F}(x, y, z) = \langle xz, xy, yz \rangle$, where D is the cube given in Example 1b

4. $\mathbf{F}(x, y, z) = \langle xz, 3yz, z^2 \rangle$, where D is the solid bounded by the paraboloid $z = 4 - x^2 - y^2$ and the xy-plane

5–18 Use the Divergence Theorem to find the flux of the vector field **F** over the surface of the given solid *D*. Consider cylindrical or spherical coordinates where appropriate.

5. $\mathbf{F}(x,y,z) = \langle 2x - y^2, zx - y, z + e^{xy} \rangle$, where *D* is the solid ball of radius 2, centered at the origin

6. $\mathbf{F}(x,y,z) = \langle 3x - yz, 2y + e^{xz}, \cos 2y - 2z \rangle$, where *D* is the solid bounded by the parabolic cylinder $z = 1 - y^2$, the *xy*-plane, and the planes $x = 0$ and $x = 1$

7. $\mathbf{F}(x,y,z) = \langle x, 3y, z - y^2 \rangle$, where *D* is the solid spherical shell $4 \le x^2 + y^2 + z^2 \le 9$

8. $\mathbf{F}(x,y,z) = |\langle x,y,z \rangle|^2 \langle x,y,z \rangle$, where *D* is the cube given in Example 1b

9. $\mathbf{F}(x,y,z) = \langle 3x - yz, 2y + e^{xz}, \cos 2y - 2z \rangle$, where *D* is the solid cylinder bounded by $x^2 + y^2 = 1$, the *xy*-plane, and the plane $z = 2$

10. $\mathbf{F}(x,y,z) = \langle x^3, y^3, z^3 \rangle$, where *D* is the solid cylinder given in Exercise 9

11. $\mathbf{F}(x,y,z) = \langle x^3, y^3, z^3 \rangle$, where *D* is the solid bounded by the paraboloid $z = 1 - x^2 - y^2$ and the *xy*-plane

12. $\mathbf{F}(x,y,z) = \langle y - 2x, e^{2zx}, z^2 - \tan^{-1}(xy) \rangle$, where *D* is the tetrahedron with vertices at the origin, $(2,0,0)$, $(0,4,0)$, and $(0,0,1)$

13. $\mathbf{F}(x,y,z) = \langle 6xy^2, 6x^2 y, x\cos(y^2) \rangle$, where *D* is the solid bounded by the paraboloid $z = 4 - x^2 - y^2$ and the plane $z = 3$

14. $\mathbf{F}(x,y,z) = \langle x - 3z, y, 4z^2 \rangle$, where *D* is the solid inside the cylinder $x^2 + y^2 = 1$, between the planes $z = 1$ and $z = 4 - x$

15. $\mathbf{F}(x,y,z) = \langle xy - \sqrt{y^2 + z^2}, 2y + e^{z^3}, 3z - x^2 \cot y \rangle$, where *D* is the solid inside the cylinder $x^2 + y^2 = 1$, between the planes $z = 1$ and $z = x + 4$

16. $\mathbf{F}(x,y,z) = \langle 2x^2, ze^x, y - xz \rangle$, where *D* is the solid cylindrical shell $1 \le x^2 + y^2 \le 2$ between the *xy*-plane and $z = 4$

17. $\mathbf{F}(x,y,z) = \langle 3y^2 z, y^3/3, x^2 z \rangle$, where *D* is the portion of the solid cone $z = \sqrt{x^2 + y^2}$ between the planes $z = 1$ and $z = 3$

18.* $\mathbf{F}(x,y,z) = \langle x^3, y^3, z^3 \rangle$, where *D* is the solid hemisphere $z = \sqrt{1 - x^2 - y^2}$

19. Verify the Divergence Theorem for the vector field $\mathbf{F}(x,y,z) = \langle z^2, 3y^2, 2yz \rangle$ and the solid bounded by the cylinder $x^2 + z^2 = 1$, the *xz*-plane, and $y = 2$.

20. If $\mathbf{F}(x,y,z) = \langle z^2 \tan y, 3xe^z, y\sin 2x^2 \rangle$, *S*, and *D* satisfy the conditions of the Divergence Theorem, prove

$$\iint_S \mathbf{F} \cdot \mathbf{n}\, d\sigma = 0.$$

21. Use the Divergence Theorem to find the flux of the vector field $\mathbf{F}(x,y,z) = \langle x^3 y, y^2 - \sin z^2, xe^z \rangle$ over the surface of the box $[0,4] \times [0,2] \times [0,3]$.

22. Suppose the space region *D* and its boundary *S* with unit normal field **n** meet the conditions of the Divergence Theorem, and let $\mathbf{F}(x,y,z) = \langle x,y,z \rangle$. Use the Divergence Theorem to show that the volume of *D*, $V(D)$ is equal to the following surface integral.

$$V(D) = \frac{1}{3} \iint_S \mathbf{F} \cdot \mathbf{n}\, d\sigma$$

23. Use Exercise 22 to verify the formula for the volume of the box with side lengths *a*, *b*, and *c*, respectively. (**Hint:** Position the box appropriately in the coordinate system and apply the Divergence Theorem along with Exercise 22.)

24. Repeat Exercise 23 for the ball of radius *R*.

25. Repeat Exercise 23 for the right circular cone of radius *R* and height *h*.

26. If *S* is a surface such as in Exercise 22, use the Divergence Theorem to show that the volume of the solid *D* bounded by *S* can also be found as

$$V(D) = \iint_S x\, dy\, dz = \iint_S y\, dz\, dx = \iint_S z\, dx\, dy.$$

27. Let Q be a single point charge at the origin, as in our discussion that follows Example 2. Show that $\nabla \cdot \mathbf{E} = 0$ for the electric field $\mathbf{E}$ created by Q.

28. Mimic our discussion following Example 2 to prove the following, slightly more general version of Gauss's Law.

If S is a closed surface satisfying the hypotheses of the Divergence Theorem, k is a constant and $\mathbf{F} = \dfrac{k\mathbf{r}}{|\mathbf{r}|^3}$ is an inverse square field, then the flux of $\mathbf{F}$ over S is

$$\iint_S \mathbf{F} \cdot \mathbf{n} \, d\sigma = \begin{cases} 4k\pi & \text{if } S \text{ encloses the origin} \\ 0 & \text{otherwise.} \end{cases}$$

29.* Recall the uniformly charged sphere of radius R and total charge Q from Exercises 55–57 of Section 15.6. In this problem, you will provide a "second solution" to the aforementioned Exercise 57 as follows. Use Gauss's Law to show that the electric field $\mathbf{E}$ due to the sphere is

$$\mathbf{E} = \begin{cases} \dfrac{\varepsilon Q \mathbf{e}_r}{r^2} & \text{if } R < r \\ 0 & \text{if } r < R. \end{cases}$$

(Recall $\mathbf{e}_r$ from Exercise 54 of Section 15.1. **Hint:** Note that $\mathbf{E}$ is a central force as in Exercise 29 of Section 15.7. Using the notation $\mathbf{E}(\mathbf{r}) = E(r)\mathbf{e}_r$, show that the flux across a sphere of radius r is $4\pi r^2 E(r)$ and use Gauss's Law. Finally, note that if $r < R$, the total charge enclosed by any sphere of radius r is zero, and hence so is $\mathbf{E}$.)

30. Use the Divergence Theorem to provide a second solution to Exercise 28 of Section 15.7. (**Hint:** See Exercise 39 of Section 15.4.)

31.* Prove that if $\mathbf{F}$ has continuous partial derivatives, then its divergence at a point P can be obtained from the formula

$$\nabla \cdot \mathbf{F}(P) = \lim_{R \to 0} \frac{3}{4\pi R^3} \iint_{S_R(P)} \mathbf{F} \cdot \mathbf{n} \, d\sigma,$$

where $S_R(P)$ is the sphere of radius R centered at P. (**Hint:** Use the Divergence Theorem; then argue that the divergence function "assumes its average value" somewhere in the closed ball bounded by $S_R(P)$. For a refresher on average value, see Section 5.2.)

32–35 Prove the identity, assuming the scalar-valued functions f and g have continuous partial derivatives at least through the second order, and S and D satisfy the hypotheses of the Divergence Theorem. Note that $\nabla^2 f$ stands for $f_{xx} + f_{yy} + f_{zz}$, i.e., $\nabla^2 f = 0$ means that f satisfies Laplace's equation (see Section 13.3). Recall also that $D_n f$ stands for the directional derivative of f in the direction of $\mathbf{n}$. (For a review of directional derivatives, see Section 13.5.)

32.* $\displaystyle \iint_S D_n f \, d\sigma = \iiint_D \nabla^2 f \, dV$

33.* $\displaystyle \iint_S f D_n g \, d\sigma = \iiint_D \left(f\nabla^2 g + \nabla f \cdot \nabla g \right) dV$

(This is called *Green's first identity*. **Hint:** See Exercise 44 in Section 15.4.)

34.* $\displaystyle \iint_S f D_n f \, d\sigma = \iiint_D |\nabla f|^2 \, dV$, if f satisfies Laplace's equation on D. (**Hint:** You may give a direct proof, or use Exercise 33.)

35.* $\displaystyle \iint_S \left(f D_n g - g D_n f \right) d\sigma = \iiint_D \left(f\nabla^2 g - g\nabla^2 f \right) dV$

(This is called *Green's second identity*. **Hint:** Use Green's first identity from Exercise 33 twice.)

Chapter 15
Review Exercises

1–4 Match the given planar vector field $\mathbf{F}(x, y)$ with its graph (labeled A–D).

1. $\mathbf{F}(x, y) = \langle 1, y \rangle$

2. $\mathbf{F}(x, y) = \left\langle \dfrac{x}{2}, y \right\rangle$

3. $\mathbf{F}(x, y) = \langle 2y, -x \rangle$

4. $\mathbf{F}(x, y) = \left\langle -x, \dfrac{y}{|\langle x, y \rangle|} \right\rangle$

A.

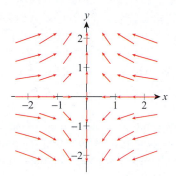

B.

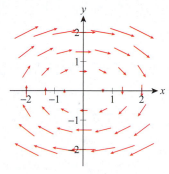

C.

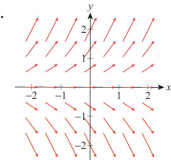

D.

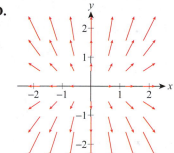

5–6 Sketch the vector field $\mathbf{F}$ by hand, using a few representative vectors.

5. $\mathbf{F}(x, y) = \left\langle \sqrt{|x|}, y \right\rangle$

6. $\mathbf{F}(x, y) = \left\langle \sqrt{x^2 + y^2}, \sqrt{x^2 + y^2} \right\rangle$

7–10 Match the given three-dimensional vector field with its graph (labeled A–D).

7. $\mathbf{F}(x, y, z) = \langle x, y, 0 \rangle$

8. $\mathbf{F}(x, y, z) = \langle 0, 0, z \rangle$

9. $\mathbf{F}(x, y, z) = \langle 0, -z, y \rangle$

10. $\mathbf{F}(x, y, z) = \dfrac{\langle x, y, z \rangle}{2 |\langle x, y, z \rangle|}$

A.

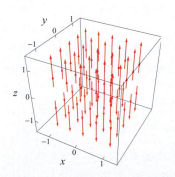

B.

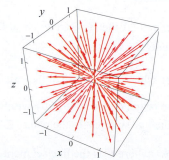

C.

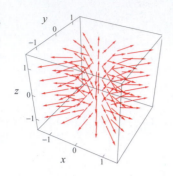

D.

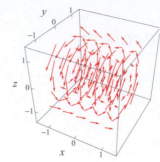

11–12 Determine ∇f for the given scalar field $f(x,y,z)$.

11. $f(x,y,z) = 2x^2 y + xz^3$

12. $f(x,y,z) = \ln\sqrt{x^2 z^4 + 4y^2}$

13–16 Decide whether the given vector field is conservative.

13. $\mathbf{F}(x,y) = \left\langle \dfrac{y^2}{\left(x+y^2\right)^2}, \dfrac{-2xy}{\left(x+y^2\right)^2} \right\rangle$

14. $\mathbf{F}(x,y) = \left\langle -x^2 y, \dfrac{x}{y} \right\rangle$

15. $\mathbf{F}(x,y,z) = \left\langle 3x^2 yz + \dfrac{1}{yz}, x^3 z - \dfrac{x}{y^2 z}, x^3 y - \dfrac{x}{yz^2} \right\rangle$

16. $\mathbf{F}(x,y,z) = \left\langle \dfrac{3x^2 y}{z}, \dfrac{x^3}{z}, \dfrac{x^3 y}{z^2} \right\rangle$

17–18 Find a potential function for the given conservative vector field.

17. $\mathbf{F}(x,y) = \left\langle 12x^2 - 4xy, 3y^2 - 2x^2 \right\rangle$

18. $\mathbf{F}(x,y) = \left\langle y^2 \cos(xy), xy\cos(xy) + \sin(xy) \right\rangle$

19–22 Evaluate the line integral.

19. $\int_C x^2 y \, ds$, where C is the semicircle $y = \sqrt{4-x^2}$, traversed counterclockwise

20. $\int_C \left(x - \dfrac{y}{2} \right) ds$, where C is the arc of the unit circle centered at the origin connecting $(1,0)$ with $(0,1)$, followed by the line segment to the point $(2,2)$

21. $\int_C (xz + 3y) \, ds$, where C is the line segment joining $(2,0,3)$ and $(1,3,0)$

22. $\int_C 2z(x - y) \, ds$, where C is the helix $\mathbf{r}(t) = \langle 2\cos t, 2\sin t, t \rangle$, for $0 \le t \le 6\pi$

23. Determine the center of mass of the V-shaped object of constant density connecting the point $(-1,-2,4)$ with the origin, and then the origin with the point $(1,8,4)$.

24. Suppose that in Exercise 23, the density of the object at any point is proportional to the distance from the xy-plane. Find its mass and center of mass in this case.

25. Determine the center of mass of the object with constant density that consists of a line segment connecting the point $\left(2,-1,-\sqrt{2} \right)$ with the point $(1,0,0)$, followed by the helix $\mathbf{r}(t) = \langle \cos \pi t, \sin \pi t, t \rangle$, $0 \le t \le 1$.

26.* Find the mass and center of mass of the object in Exercise 25 if the density at any point is proportional to the distance from the xy-plane.

27–28 Determine the work done by the force field **F** in moving a particle along the indicated curve C.

27. $\mathbf{F}(x,y,z) = \langle 2y^2, -xz, 3y \rangle$, where C is defined by $\mathbf{r}(t) = \langle t, 2t^2, 3t^3 \rangle$, $0 \le t \le 2$

28. $\mathbf{F}(x,y,z) = \langle -y, x, 3z \rangle$, where C is the helix defined by $\mathbf{r}(t) = \langle 3\cos t, 3\sin t, t \rangle$, $0 \le t \le 4\pi$

29–30 A fluid's velocity field is given by $\mathbf{F}(x,y,z)$. Determine the fluid's flow along the indicated curve C.

29. $\mathbf{F}(x,y,z) = \langle 2xyz, x^2 z, x^2 y \rangle$, where C is the line segment connecting the points $(0,2,-1)$ and $(1,4,5)$

30. $\mathbf{F}(x,y,z) = \langle y^2, 2xy, 4z \rangle$, where C is the curve defined by $\mathbf{r}(t) = \langle \sin t, \cos t, t/4 \rangle$, $0 \le t \le 3\pi$

31. Suppose the semicircle parametrized by $\mathbf{r}(t) = \langle 3\cos t, 0, 3\sin t \rangle$, $0 \le t \le \pi$ has charge density $q(x,y,z) = \dfrac{2+z}{10^6}$ C/m. Find the electrostatic potential at the point $(0,4,0)$. (See Exercise 50 in Section 15.2.)

32–35 Use the Fundamental Theorem for Line Integrals to evaluate the integral.

32. $\displaystyle\int_C 2(x+y^2)\,dx + 4xy\,dy$, where C is a piecewise smooth path from $(0,0)$ to $(3,2)$

33. $\displaystyle\int_C e^{xy}(xy+1)\,dx + x^2 e^{xy}\,dy$, where C is a piecewise smooth path from $(0,-1)$ to $(1,1)$

34. $\displaystyle\int_C (yz^2 + e^x \sin y)\,dx + (xz^2 + e^x \cos y)\,dy + 2xyz\,dz$, where C is a piecewise smooth path from $(0,-\pi/2,2)$ to $(0,0,0)$

35. $\displaystyle\int_C \left(\frac{2x}{x^2+y^2} - 2y^3 \right) dx$
$\qquad + \left(\frac{2y}{x^2+y^2} - 6xy^2 - z\sin y \right) dy + \cos y\,dz$,

where C is a piecewise smooth path from $(2,3,0)$ to $(1,0,2)$

36–37 Show that the line integral is not path independent by finding two different values for the integral along two different paths connecting A and B. (Answers may vary.)

36. $\displaystyle\int_C xy\,dx - 5y\,dy$; $\quad A(0,0), \quad B(1,1)$

37. $\displaystyle\int_C 2x\,dx + x\,dy + yz\,dz$; $\quad A(2,0,0), \quad B(-2,0,0)$

38–39 Show that the force field $\mathbf{F}$ is conservative, and use this fact to determine the work done by $\mathbf{F}$ in moving an object from A to B.

38. $\mathbf{F}(x,y) = \langle \cos y + y\sin x, -\cos x - x\sin y \rangle$;
$\quad A(0,1), \quad B(-1,-1)$

39. $\mathbf{F}(x,y,z) = \left\langle \ln(y^2), \dfrac{2x}{y} + z^2, 2yz \right\rangle$;
$\quad A(1,1,1), \quad B(2,\sqrt{e},2)$

40. Use the Fundamental Theorem for Line Integrals to verify your answer for Exercise 29.

41. Use the Fundamental Theorem for Line Integrals to verify your answer for Exercise 30.

42. Recall the vector field
$$\mathbf{F}(x,y) = \left\langle \frac{-y}{|\langle x,y \rangle|^2}, \frac{x}{|\langle x,y \rangle|^2} \right\rangle \text{ from Exercise 35}$$
of Section 15.3. Show that $\mathbf{F}$ is conservative on any region disjoint from the origin.

43. Show that the vector field
$$\mathbf{F}(x,y) = \left\langle \frac{-y}{|\langle x,y \rangle|^p}, \frac{x}{|\langle x,y \rangle|^p} \right\rangle \text{ is not conservative}$$
if $p \neq 2$.

44–47 Find the divergence and curl of the vector field $\mathbf{F}$.

44. $\mathbf{F}(x,y) = \langle x^2 + 4xy, 2x^2 y \rangle$

45. $\mathbf{F}(x,y) = \left\langle \dfrac{x}{y^2+1}, y(x^2-x) \right\rangle$

46. $\mathbf{F}(x,y,z) = \langle 2x(z^2-1), \ln(x^2+y^2), x^2 yz \rangle$

47. $\mathbf{F}(x,y,z) = \langle x\sin(yz), y^2 z, x(y+z) \rangle$

48. Consider the function $f(\mathbf{r}) = \dfrac{1}{r^4}$, where, as usual, $r = |\mathbf{r}|$. Prove: $\nabla f = -\dfrac{4\mathbf{r}}{r^6}$.

49–50 Verify Green's Theorem by demonstrating the equality $\displaystyle\oint_C P\,dx + Q\,dy = \iint_R \left(\frac{\partial Q}{\partial x} - \frac{\partial P}{\partial y} \right) dA$ for the given $\mathbf{F}(x,y) = \langle P(x,y), Q(x,y) \rangle$ and region R enclosed by the piecewise smooth closed curve C.

49. $P(x,y) = 2x + y^2$, $Q(x,y) = xy$, where R is the triangle with vertices $(-1,0)$, $(1,0)$, and $(0,1)$

50. $P(x,y) = x - y$, $Q(x,y) = 2y$, where R is the region bounded by the x-axis and the upper semicircle $y = \sqrt{1-x^2}$

51–52 Use Green's Theorem to evaluate the indicated line integral.

51. $\oint_C y^2\,dx + 2xy\,dy$, where C is the boundary of the region enclosed by the graphs of $y = \sqrt{x}$, $x = 1$, and the x-axis, oriented counterclockwise

52. $\oint_C \dfrac{x^3 y^2}{2}\,dx - x^2 y\,dy$, where C is the boundary of the region bounded by the coordinate axes and the line $y = 2 - x$, oriented clockwise

53–54 Use Green's Theorem to find the work done as a particle moves counterclockwise around the closed path C in the force field $\mathbf{F}$.

53. $\mathbf{F}(x,y) = \langle -3x^2 y, (2x - y)y \rangle$, where C is the boundary of the triangle with vertices $(0,-1)$, $(0,1)$, and $(2,0)$

54. $\mathbf{F}(x,y) = \langle x^2 + 2y, \cos y + 3x \rangle$, where C is the circle $x^2 + (y-1)^2 = 1$ (**Hint:** Use polar coordinates.)

55–56 Evaluate the outward flux of the vector field $\mathbf{F}(x,y)$ across the given path C.

55. $\mathbf{F}(x,y) = \langle 2xy, y \rangle$, C is the boundary of the upper half disk bounded by $y = \sqrt{4 - x^2}$ and the x-axis, oriented counterclockwise. (**Hint:** Use polar coordinates.)

56. $\mathbf{F}(x,y) = \langle 4x(x+y), x - y \rangle$, C is the boundary of the region between the graphs of $y = x^2$ and $x = y^2$, oriented counterclockwise.

57. Use the line integral method (Exercise 84, Section 15.4) to find the area of the heptagon with vertices $(1,0)$, $(3.5, 0.5)$, $(4,2)$, $(3,4)$, $(1,5)$, $(0,3)$, and $(0,1)$.

58. Let f_1 and f_2 be differentiable single-variable functions, and C a piecewise smooth, simple closed curve. Prove $\displaystyle\int_C f_1(x)\,dx + f_2(y)\,dy = 0$. (**Hint:** Use Green's Theorem.)

59–64 Match the parametric surface with its graph (labeled A–F).

59. $\mathbf{r}(s,t) = \left\langle \dfrac{\cos t}{s}, \dfrac{\sin t}{s}, s \right\rangle$,

$0 \le s \le 4, \quad 0 \le t \le 2\pi$

60. $\mathbf{r}(s,t) = \left\langle s \sin t, s \cos t, \dfrac{3\sqrt{s}}{2} \right\rangle$,

$0 \le s \le 2\pi, \quad 0 \le t \le 2\pi$

61. $\mathbf{r}(s,t) = \langle 4s, 2s \cos t, 2s \sin t \rangle$,

$0 \le s \le 5, \quad 0 \le t \le 2\pi$

62. $\mathbf{r}(s,t) = \left\langle \dfrac{3s}{2} \cos t, \dfrac{3s}{2} \sin t, -s^2 \right\rangle$,

$0 \le s \le 3, \quad 0 \le t \le 2\pi$

63. $\mathbf{r}(s,t) = \langle s \sin t \cos s, s \sin t \sin s, 2s \rangle$,

$0 \le s \le 9, \quad 0 \le t \le 2\pi$

64. $\mathbf{r}(s,t) = \left\langle s \sin t, s \cos t, \dfrac{3}{1 + s^2} \right\rangle$,

$0 \le s \le 2\pi, \quad 0 \le t \le 2\pi$

A.

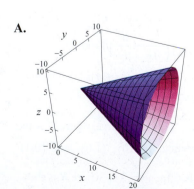

B.

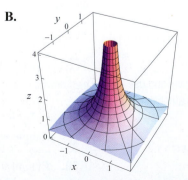

C.

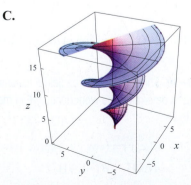

D. 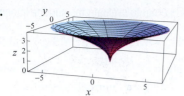   **E.** 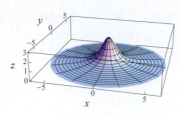   **F.**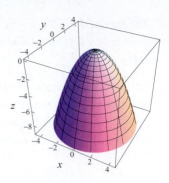

65–66 Describe the surface with a vector function of two parameters.

65. The graph of $z = 1/y^2$, $1 \le y \le 4$, revolved about the y-axis

66. The graph of $z = \tan x$, $0 \le x \le \pi/4$, revolved about the x-axis

67–68 Identify the surface by examining its grid curves.

67. $\mathbf{r}(s,t) = \langle 2s \sin t, s \cos t, s \rangle$,
$-\infty < s < \infty$, $0 \le t \le 2\pi$

68. $\mathbf{r}(s,t) = \langle \sin t, (2 + \cos t) \sin s, (2 + \cos t) \cos s \rangle$,
$0 \le s \le 2\pi$, $0 \le t \le 2\pi$

69–70 Obtain a parametrization for the indicated surface. (Answers will vary.)

69. $z = 2x^2 + y^2$

70. $z = \sqrt{9 - 9x^2 - \dfrac{9}{4}y^2}$

71–72 Find an equation of the plane tangent to the surface at the indicated point.

71. $\mathbf{r}(s,t) = \langle st, t - s, 2t^2 \rangle$; $\mathbf{r}(1,3)$

72. $\mathbf{r}(s,t) = \langle s \cos t, s \sin t, s^4 \rangle$; $\mathbf{r}\left(1, \dfrac{\pi}{4}\right)$

73–76 Find the area of the surface S. (Use polar coordinates wherever they simplify your calculations.)

73. S is the surface common to the interior of the cylinder $x^2 + y^2 = 9$ and the plane $2z + 2y + x = 0$.

74. S is the surface $\mathbf{r}(s,t) = \langle s \cos t, s \sin t, 2s \rangle$ between the planes $z = 2$ and $z = 4$.

75. S is the surface $\mathbf{r}(s,t) = \left\langle s \cos t, s \sin t, \dfrac{s^2}{2} \right\rangle$ between the planes $z = 1$ and $z = 4$.

76. S is the graph of $4z = y^2 - x^2$ above the unit disk centered at the origin.

77–78 A surface with variable mass density is given. Determine the mass of the surface.

77. $\mathbf{r}(s,t) = \langle \cos t, \sin t, s \rangle$, $0 \le s \le 1$, $0 \le t \le \dfrac{\pi}{2}$;
mass density is $\rho(x,y,z) = z^2$.

78. $\mathbf{r}(s,t) = \langle 2 \cos t, 2 \sin t, s \rangle$, $0 \le s \le 2$, $0 \le t \le \pi$;
mass density is $\rho(x,y,z) = 8 - z^3$.

79. Determine the center of mass of the surface of Exercise 77.

80. Determine the center of mass of the surface of Exercise 78.

81–84 Determine the indicated flux of the vector field $\mathbf{F}$ across the surface S. Unless otherwise specified, the surfaces are oriented with outward-pointing normal vectors.

81. The flux of $\mathbf{F} = \langle 1, 2y, z \rangle$ out of the solid R bounded by $z = \dfrac{x^2 + y^2}{2}$ and the plane $z = 2$

82. The flux of $\mathbf{F} = \langle 1, -z, y \rangle$ across the first-octant portion of the hemisphere of radius 2 centered at the origin

83. The flux of $\mathbf{F} = \langle xy, 1, z \rangle$ across the first-octant portion of the cylinder $x^2 + y^2 = 4$, $0 \le z \le 6$

84. The flux of $\mathbf{F} = \langle 2x, y, -z \rangle$ out of the solid R bounded by the paraboloid $z = x^2 + y^2$ and the plane $z = 4$

85. Consider the thin conical surface of constant mass density, with both height and radius equaling R. Show that the second moment about its axis of rotation is $I = mR^2/2$, where m is the mass of the shell.

86–87 Verify Stokes' Theorem by showing that the integrals $\oint_C \mathbf{F} \cdot \mathbf{T}\, ds$ and $\iint_S \nabla \times \mathbf{F} \cdot \mathbf{n}\, d\sigma$ are equal.

86. $\mathbf{F}(x, y, z) = \langle -4x - y, -2z, y \rangle$, where S is the upper hemisphere of radius 3 centered at the origin

87. $\mathbf{F}(x, y, z) = \langle 2y, -x, z^2 \rangle$, where S is the surface $z = 4 - x^2 - y^2, z > 0$

88–91 Use Stokes' Theorem to evaluate the indicated line integral.

88. $\oint_C \mathbf{F} \cdot \mathbf{T}\, ds$, where $\mathbf{F} = \langle 2x^2, -z, y^2 \rangle$ and C is the intersection of the cylinder $x^2 + y^2 = 4$ and the plane $2x + \dfrac{y}{2} + z = 6$, with positive orientation when viewed from above

89. $\oint_C \mathbf{F} \cdot \mathbf{T}\, ds$, where $\mathbf{F} = \langle -2y^2, z^2, -x^2 \rangle$ and C is the intersection of the cylinder $x^2 + y^2 = 2x$ and the plane $z = \dfrac{x}{2}$, with positive orientation when viewed from above

90. $\oint_C \mathbf{F} \cdot \mathbf{T}\, ds$, where $\mathbf{F} = \langle -y, -2z, -3x \rangle$ and C is the intersection of $\dfrac{z}{2} = x^2 + y^2$ and $z = 12 - x^2 - y^2$, with positive orientation when viewed from above

91. $\oint_C \mathbf{F} \cdot \mathbf{T}\, ds$, where $\mathbf{F} = \langle -xz, 2z, x - y^2 \rangle$ and C is the intersection of $z = x^2 + y^2$ and $z = 4x + 5$, with negative orientation when viewed from above

92–95 Use Stokes' Theorem to evaluate the surface integral.

92. $\iint_S \nabla \times \mathbf{F} \cdot \mathbf{n}\, d\sigma$, where $\mathbf{F} = \langle -2z, x, 3xy \rangle$ and S is the hemisphere $z = \sqrt{4 - x^2 - y^2}$, oriented with an upward-pointing unit normal vector field

93. $\iint_S \nabla \times \mathbf{F} \cdot \mathbf{n}\, d\sigma$, where $\mathbf{F} = \langle 3y, z, -2x \rangle$ and S is the portion of the paraboloid $z = x^2 + y^2$ with $0 \le z \le 4$, oriented with a downward-pointing unit normal vector field

94. $\iint_S \nabla \times \mathbf{F} \cdot \mathbf{n}\, d\sigma$, where $\mathbf{F} = \langle z^2, 2x^2, x^2 - y \rangle$ and S is the triangle determined by the first-octant portion of the plane $x + 2y + 3z = 6$, oriented with an upward-pointing unit normal vector field

95. $\iint_S \nabla \times \mathbf{F} \cdot \mathbf{n}\, d\sigma$, where $\mathbf{F} = \left\langle x - \dfrac{yz^2}{4}, 3x^2z, z - y^2 \right\rangle$ and S is the cone frustum $z = 2\sqrt{x^2 + y^2}$ with $2 \le z \le 6$, oriented with an inward-pointing unit normal vector field

96–97 Verify the Divergence Theorem by showing the equality of the integrals $\iint_S \mathbf{F} \cdot \mathbf{n}\, d\sigma$ and $\iiint_D \nabla \cdot \mathbf{F}\, dV$ for the given vector field $\mathbf{F}$ on the solid D.

96. $\mathbf{F}(x, y, z) = \left\langle 3z, \dfrac{y}{4}, -2x \right\rangle$, where D is the ball of radius 3 centered at the origin

97. $\mathbf{F}(x, y, z) = \langle 3xy, -x, 2z \rangle$, where D is the solid bounded by the paraboloid $z = 1 - x^2 - y^2$ and the xy-plane

98–103 Use the Divergence Theorem to find the flux of the vector field $\mathbf{F}$ over the surface of the given solid D. Consider cylindrical or spherical coordinates where appropriate.

98. $\mathbf{F}(x, y, z) = \langle -x^2y, yz, 5z \rangle$, where D is the solid bounded by the parabolic cylinder $z = 4 - y^2$, the xy-plane, and the planes $x = 0$ and $x = 2$

99. $\mathbf{F}(x, y, z) = \langle 5x(z - y), z^2 + \cos x, y \sin x \rangle$, where D is the solid cylinder bounded by $x^2 + y^2 = 9$, the xy-plane, and the plane $z = 3$

100. $F(x,y,z) = \left\langle 2xy, y-z, \dfrac{x^2 y}{3} \right\rangle$, where D is the tetrahedron with vertices at the origin, $(6,0,0)$, $(0,3,0)$, and $(0,0,4)$

101. $F(x,y,z) = \langle 2x+z, x-y^3, 2z \rangle$, where D is the solid inside the cylinder $x^2 + y^2 = 4$, bounded by the xy-plane and $z = 3 - y$

102. $F(x,y,z) = \langle -xy^2, 3yz^2, y^3 \rangle$, where D is the solid cylindrical shell $2 \le x^2 + y^2 \le 4$ between the planes $z = 1$ and $z = 6$

103. $F(x,y,z) = \left\langle \dfrac{yz^2}{2}, \dfrac{y^3}{8}, x(z-x) \right\rangle$, where D is the portion of the solid cone $z = \sqrt{x^2 + y^2}$ between the planes $z = 2$ and $z = 4$

104. Write a paragraph discussing the relationships and analogies between the Fundamental Theorem of Calculus, the Fundamental Theorem for Line Integrals, Green's and Stokes' Theorems, and the Divergence Theorem.

105–112 *True or False?* Determine whether the given statement is true or false. In case of a false statement, explain or provide a counterexample.

105. If a vector field F is constant, then $\nabla \times F = 0$.

106. If a vector field F is constant, then $\nabla \cdot F = 0$.

107. A vector field is conservative if and only if its curl is zero.

108. The Fundamental Theorem for Line Integrals can only be used if the underlying vector field is conservative.

109. If a force field F is conservative, then the work done by F on a particle moving along a smooth path is zero.

110. Green's Theorem and Stokes' Theorem are unrelated.

111. When a charged particle moves along a piecewise smooth closed curve in an electric force field, the total work done by the force field is zero.

112. If the vector field F has continuous partials in an open neighborhood of the closed, piecewise smooth surface S, then $\iint\limits_{S} \nabla \times F \cdot n \, d\sigma = 0$.

Chapter 15
Technology Exercises

113–116 Use a computer algebra system to find the mass and center of mass of the solid with the given density.

113. The solid bounded by the cone $z = \sqrt{4 - x^2 - y^2}$ and the xy-plane, with constant density ρ

114. The solid bounded by the paraboloid $z = 4 - x^2 - y^2$ and the xy-plane, with constant density ρ

115. The solid of Exercise 114, with its density at any point being proportional to the distance from the xy-plane

116. The solid of Exercise 114, with its density at any point being proportional to the distance from the z-axis

Project ✎

Chapter 15

Recall from Section 15.7 that if $\mathbf{F}$ is a vector field in $\mathbb{R}^3$ so that $\nabla \times \mathbf{F} = \mathbf{0}$ (such vector fields are called curl-free) on an open, simply connected domain in space, then $\mathbf{F}$ is conservative, that is, there is a scalar potential f so that $\nabla f = \mathbf{F}$. On the other hand, it can be shown that if $\mathbf{F}$ is divergence-free, that is, if $\nabla \cdot \mathbf{F} = 0$, then there is a vector field $\mathbf{P}$ such that $\nabla \times \mathbf{P} = \mathbf{F}$ (such a vector field is called a *vector potential* for $\mathbf{F}$). In this project you will discover a way of finding a vector potential for a given divergence-free vector field $\mathbf{F}$.

1. Suppose
$$\mathbf{F}(x,y,z) = \langle F_1(x,y,z), F_2(x,y,z), F_3(x,y,z) \rangle$$
and
$$\mathbf{P}(x,y,z) = \langle P_1(x,y,z), P_2(x,y,z), P_3(x,y,z) \rangle$$
are vector fields so that $\nabla \times \mathbf{P} = \mathbf{F}$; that is, $\mathbf{P}$ is a vector potential for $\mathbf{F}$. Show that for any differentiable scalar field f, $\nabla \times (\mathbf{P} + \nabla f) = \mathbf{F}$; that is, $\mathbf{P} + \nabla f$ is another vector potential for $\mathbf{F}$. (**Hint:** See Exercise 41 of Section 15.4.)

2. If f is any scalar field such that $\dfrac{\partial f}{\partial x} = -P_1$, show that if we define $\widehat{P} = P + \nabla f$, then $\widehat{P}_1 = 0$.

3. Use Questions 1 and 2 to argue that if the vector field $\mathbf{F}$ has a vector potential $\mathbf{P}$, then it has one whose first component is zero. In other words, we may assume throughout our discussion that $\mathbf{P} = \langle 0, P_2, P_3 \rangle$.

In Questions 4–6, you will be guided to show that given a divergence-free vector field $\mathbf{F}$, it is possible and fairly straightforward to find a vector potential of the form described in Question 3.

4. Assume that
$$\mathbf{F}(x,y,z) = \langle F_1(x,y,z), F_2(x,y,z), F_3(x,y,z) \rangle$$
is a vector field such that $\nabla \cdot \mathbf{F} = 0$, and $\mathbf{P}$ is any vector field of the form $\mathbf{P} = \langle 0, P_2, P_3 \rangle$. Show that $\mathbf{P}$ is a vector potential for $\mathbf{F}$ if the following equalities hold:
$$\frac{\partial P_3}{\partial y} - \frac{\partial P_2}{\partial z} = F_1 \qquad -\frac{\partial P_3}{\partial x} = F_2 \qquad \frac{\partial P_2}{\partial x} = F_3$$

5. For the vector field $\mathbf{F}$ in Question 4, define
$$P_2(x,y,z) = \int_{x_0}^{x} F_3(t,y,z)\,dt + C_2(y,z) \text{ and}$$
$$P_3(x,y,z) = -\int_{x_0}^{x} F_2(t,y,z)\,dt + C_3(y,z), \text{ where}$$
x_0 is an arbitrary starting value and C_2 and C_3 are arbitrary functions of the variables y and z. Show that $\mathbf{P}(x,y,z) = \langle 0, P_2(x,y,z), P_3(x,y,z) \rangle$ satisfies the last two equations in Question 4.

6. Show that in Question 5, it is always possible to choose $C_2(y,z)$ and $C_3(y,z)$ to satisfy $\dfrac{\partial P_3}{\partial y} - \dfrac{\partial P_2}{\partial z} = F_1$, and conclude that $\mathbf{P}(x,y,z) = \langle 0, P_2(x,y,z), P_3(x,y,z) \rangle$ will then be a vector potential for $\mathbf{F}$. (**Hint:** Use the fact that $\nabla \cdot \mathbf{F} = 0$.)

7. Show that the vector field
$$\mathbf{F}(x,y,z) = \langle 2x^2 yz, -2xy^2 z, x^2 y \rangle$$
is divergence-free, and follow the steps outlined in Questions 5 and 6 to find a vector potential for $\mathbf{F}$. (Answers may vary.)

Appendices

A Fundamentals of *Mathematica* A-2

B Properties of Exponents and Logarithms, Graphs of Exponential and Logarithmic Functions A-8

C Trigonometric and Hyperbolic Functions A-9

D Complex Numbers A-14

E Proofs of Selected Theorems A-21

Appendix A **Fundamentals of *Mathematica***

Mathematica is a powerful and flexible software package with a wide variety of uses. To begin with, *Mathematica* (along with similar products such as Maple, MATLAB, and Derive) can be viewed as a sort of supercalculator. It also understands the rules of algebra, has a huge number of built-in functions ranging from the trivial to the exotic, and is very good at generating high-quality graphs in one, two, and three dimensions. Beyond that, a package such as *Mathematica* is also a programming environment; it is this aspect of *Mathematica* that allows the user to extend its capabilities to suit specialized needs.

The optional use of *Mathematica* and similar technology in this text requires only a basic familiarity; this appendix will serve as a quick guide to the use of *Mathematica*. It should also be noted that a *complete* guide to *Mathematica* can be found within the program itself. Once it is installed and running on your computer, clicking on the "Help" button located in the top toolbar (see Figure 1) gives you access to an electronic version of a very large *Mathematica* user's manual. After clicking on "Help" a drop-down menu appears. By clicking on "Documentation Center" the full selection of "Help" categories appears; a good place to start is with "Get Started." This leads to a set of videos that highlight many useful examples of how *Mathematica* can be used to solve different sorts of problems.

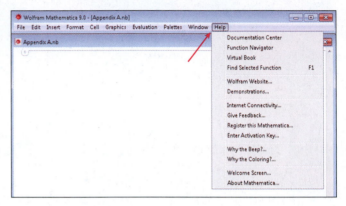

Figure 1 Getting On-Screen Help

At first, you will probably be making use of built-in *Mathematica* commands such as **Plot**, **Fit**, and **Solve** (as opposed to using your own user-defined commands). It is important to realize that *Mathematica* is case sensitive and that all built-in commands begin with a capital letter. Once a command has been typed in, you'll need to tell *Mathematica* to execute it. This can be done in one of two ways—either by pressing Shift and Enter together (known as Shift + Enter) or, if you are using an extended keyboard, by using the Enter that appears in the numeric keypad area. Pressing Enter alone will simply move the cursor to the next line and allow you to continue typing but will not execute any commands.

Each time you press Shift + Enter *Mathematica* will execute all the commands contained in a single cell. Different *Mathematica* cells are demarcated by brackets along the right-hand edge of the work area, and you can always start a new cell by positioning the mouse cursor over a blank part of the area (you will notice that the cursor symbol becomes horizontal rather than vertical) and clicking the left mouse button once.

The remainder of this appendix contains examples of a few of the basic *Mathematica* commands used in this text, arranged roughly in the order in which they appear. For instant on-screen help on any command, type the command into *Mathematica* and then press F1. Doing so will bring up the relevant help pages and, more often than not, provide examples of how the command is used.

Basic *Mathematica* Commands

Defining Functions

A few rules of syntax must be observed in order to define your own functions in *Mathematica*. The first is that each variable serving as a placeholder in the definition must be followed by the underscore symbol "_" when it appears on the left side of the definition and without the underscore when it appears on the right. The second rule is that ":=" (a colon followed by an equal sign) is used in the definition, as opposed to "=" (see the on-screen *Mathematica* help for detailed explanations of these rules). Figure 2 illustrates the definition of the two functions $f(x) = x^2 + 5$ and $g(x, y) = 3x - 7y$, followed by an evaluation of each.

In[1]:= **f[x_] := x^2 + 5**

In[2]:= **g[x_, y_] := 3 x - 7 y**

In[3]:= **f[-2]**

Out[3]= 9

In[4]:= **g[5, 2]**

Out[4]= 1

Figure 2 Defining Functions

Plot

The basic usage of the **Plot** command is **Plot[f, {x, x_{min}, x_{max}}]**, where f is an expression in x representing a function to be plotted and x_{min} and x_{max} define the endpoints of the interval on the x-axis over which f is to be graphed. However, the **Plot** command also recognizes many options that modify the details of the resulting picture; these options are best explored via the on-screen help. Figure 3 illustrates the use of **Plot** in graphing the function $f(x) = x^3 - x^2 - 3x + 5$ over the interval $[-3, 4]$.

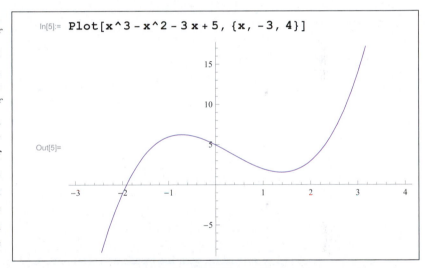

In[5]:= **Plot[x^3 - x^2 - 3 x + 5, {x, -3, 4}]**

Figure 3 Basic Use of the Plot Command

Piecewise

The **Piecewise** command allows us to easily create and use functions in *Mathematica* that correspond to the piecewise defined functions referred to in this and many other math texts. See Section 1.2 for an example of the use of the **Piecewise** command.

Manipulate

The **Manipulate** command is a powerful tool that is useful in making dynamic models in *Mathematica*. Such models are especially useful in exploring the effect of changing the value(s) of parameter(s); see Section 1.5 for an example of such usage.

Curve Fitting (Fit Command)

The *Mathematica* command **Fit** can be used to construct a function of specified form (such as linear, quadratic, exponential, etc.) to a given set of data (i.e., ordered pairs) using the least-squares method. Figure 4 illustrates the use of **Fit** to construct both a linear and a quadratic function that best fits the given set of four data points. Note also the use of the **ListPlot**, **Plot**, and **Show** commands to create graphs of the data and the two best-fitting functions. Two options are shown in the **ListPlot** usage, one of which (**PlotStyle**) specifies the color and size of the points to be plotted, and the other of which (**AxesOrigin**) positions the axes in a certain manner. (For an exponential fit, try out the command **Fit[data, {1, Exp[x]}, x]**.)

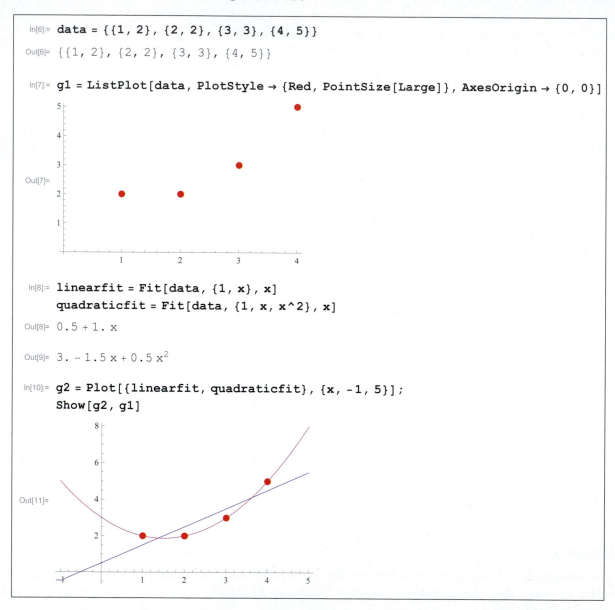

Figure 4 Linear and Quadratic Curve Fitting

Limit

The built-in command **Limit** is used to direct *Mathematica* to try to determine the limit of a function at a specified point, with the option of asking for one-sided limits from either direction. See Section 2.2 and Figure 5 for examples of the command's use.

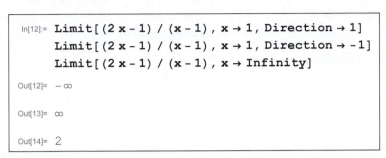

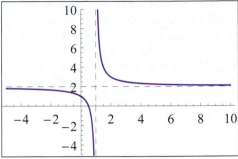

Figure 5a Use of the Limit Command

Figure 5b $y = \dfrac{2x - 1}{x - 1}$

Solve

The **Solve** command is very powerful, and can be used in several different ways. Its basic usage is **Solve[*expr*, *vars*]**, where *expr* represents one or more equations and *vars* represents one or more variables. If more than one equation is to be solved, the collection of equations must be enclosed in a set of braces, separated by commas. Similarly, if more than one variable is to be solved for, the variables must be enclosed in a set of braces. Figure 6 shows the use of **Solve** to first solve one equation for one variable, and then to solve a collection of three equations for all three variables. Note how *Mathematica* expresses the solution in each case.

In[15]:= **Solve[3 x - x * y == 9 y, y]**

Out[15]= $\left\{\left\{ y \to \dfrac{3\,x}{9 + x} \right\}\right\}$

In[16]:= **Solve[{3 x + 2 y - 4 z == 8, 4 x - 5 z == -3, 7 y + z == 12}, {x, y, z}]**

Out[16]= $\left\{\left\{ x \to -\dfrac{50}{3},\ y \to \dfrac{53}{15},\ z \to -\dfrac{191}{15} \right\}\right\}$

Figure 6 Two Uses of the Solve Command

It is important to note that equations in *Mathematica* are expressed with two "=" symbols, as seen in Figure 6. The use of just one "=" is reserved for assigning a permanent value to something. For instance, the expression **x=3** assigns the value of 3 to the symbol x, while the expression **x==3** represents the equation $x = 3$ in *Mathematica*.

NSolve

The **NSolve** command is used in a manner similar to **Solve**, but typically in situations where an exact solution is either not desired or not feasible. See Section 2.5 for an example of the use of the command in finding a numerical approximation of a solution.

Differentiation (D Command)

The basic usage of the built-in differentiation command **D** is **D[**f**,** x**]**, where f is a function of the variable x. Figure 7 illustrates such use in finding the derivative of a given rational function; note the optional use of the **Together** command (discussed later in this appendix) to express the derivative as a single fraction.

If f is a function of more than one variable, the **D** command can be used to find partial derivatives.

In[17]:= **f[x_] := (x^2 - 3 x + 1) / (x + 5)**

In[18]:= **D[f[x], x]**

Out[18]= $\dfrac{-3 + 2x}{5 + x} - \dfrac{1 - 3x + x^2}{(5 + x)^2}$

In[19]:= **Together[D[f[x], x]]**

Out[19]= $\dfrac{-16 + 10x + x^2}{(5 + x)^2}$

Figure 7 Differentiation

FindRoot

The **FindRoot** command uses numerical methods (such as Newton's method, Section 4.5) to find approximate roots of functions, and is especially useful when neither **Solve** nor **NSolve** is able to provide a satisfactory result. Its basic usage is **FindRoot[**f**, {**x**,** x_0**}]** when the goal is to find a root of the function f near a given point x_0, but it can also be used to find a numerical solution of the equation $lhs = rhs$ near x_0 if used in the form **FindRoot[**lhs **==** rhs**, {**x**,** x_0**}]** (note the "double equal sign" used by *Mathematica* to denote an equation).

FindMaximum and FindMinimum

The usage of the commands **FindMaximum** and **FindMinimum** is similar to that of **FindRoot**, and both also rely on numerical methods to obtain results. To approximate the location and value of a local maximum of the function f near a given point x_0, the syntax is **FindMaximum[**f**, {**x**,** x_0**}]**; the use of **FindMinimum** is identical. Figure 8 illustrates the use of **FindMinimum** to identify the radius r that minimizes the surface area of the cylinder of Example 3 in Section 4.6.

In[20]:= **FindMinimum[2 * Pi * r^2 + 1000 / r, {r, 5}]**

Out[20]= **{348.734, {r → 4.30127}}**

Figure 8 Use of FindMinimum

Integrate

The **Integrate** command can be used for both indefinite and definite integration, with the goal determined by the options used with the command. Figure 9 illustrates how *Mathematica* provides both the indefinite integral of the rational function $1/(x^2 + 1)$ and the definite integral of the same function over the interval $[-1.5, 1.5]$. (Note that *Mathematica* does not provide an arbitrary constant when evaluating indefinite integrals.)

In[21]:= **Integrate[1 / (x^2 + 1), x]**

Out[21]= ArcTan[x]

In[22]:= **Integrate[1 / (x^2 + 1), {x, -1.5, 1.5}]**

Out[22]= 1.96559

Figure 9 Integration

Other Useful Commands

Simplify

The **Simplify** command is used to simplify mathematical expressions according to the usual rules of algebra. The basic syntax is **Simplify[*expr*]**, where *expr* is the expression to be simplified. Note the examples shown in Figure 10.

In[23]:= **Simplify[x * (4 x - 2 x * y) / (6 x^2)]**

Out[23]= $\dfrac{2 - y}{3}$

In[24]:= **Simplify[(a^2 - b^2) / (a - b)]**

Out[24]= $a + b$

Figure 10 Use of Simplify

Expand

This command is used to multiply out factors in an expression. The syntax for the command is **Expand[*expr*]**. Figure 11 shows the use of the command in multiplying out the expression $(x - y)^5$.

In[25]:= **Expand[(x - y)^5]**

Out[25]= $x^5 - 5\,x^4\,y + 10\,x^3\,y^2 - 10\,x^2\,y^3 + 5\,x\,y^4 - y^5$

Figure 11 Use of Expand

Factor

The **Factor** command is the reverse of the **Expand** command when applied to polynomials. Its basic usage is **Factor[*poly*]**, where *poly* is a polynomial expression to be factored.

Together

The **Together** command is used primarily to express a sum (or difference) of two or more rational expressions as one with a common denominator, automatically canceling any common factors that may appear. The basic syntax for the command is **Together[*expr*]**.

Appendix B Properties of Exponents and Logarithms, Graphs of Exponential and Logarithmic Functions

For ease of reference, the basic algebraic properties of exponents and logarithms and the general forms of exponential and logarithmic graphs appear below. Interestingly, the Scottish mathematician John Napier (1550–1617) introduced logarithms as an aid to computation, and their use led to the development of various types of slide rules and logarithm tables. It was only later that mathematicians made the connection between logarithmic and exponential functions, namely that they are inverses of each other (more precisely, an exponential function of a given base is the inverse function of the logarithmic function with the same base, and vice versa). This fact appears explicitly as the first property of logarithms below, with the other properties reflecting, directly or indirectly, the same fact.

Properties of Exponents

Given real numbers x and y and positive real numbers a and b, the following properties hold.

1. $a^x a^y = a^{x+y}$

2. $\dfrac{a^x}{a^y} = a^{x-y}$

3. $\left(a^x\right)^y = a^{xy}$

4. $\left(ab\right)^x = a^x b^x$

Properties of Logarithms

Given positive real numbers x, y, a, and b, with $a \neq 1$ and $b \neq 1$, and real number r, the following properties hold.

1. $\log_a x = y \Leftrightarrow x = a^y$

2. $\log_a \left(a^x\right) = x$

3. $a^{\log_a x} = x$

4. $\log_a \left(xy\right) = \log_a x + \log_a y$

5. $\log_a \dfrac{x}{y} = \log_a x - \log_a y$

6. $\log_a \left(x^r\right) = r \log_a x$

Change of logarithmic base: $\log_b x = \dfrac{\log_a x}{\log_a b}$

Change of exponential base: $a^x = b^{\log_b\left(a^x\right)} = b^{x \log_b a}$

(in particular, $a^x = e^{\ln\left(a^x\right)} = e^{x \ln a}$)

Graphs of Exponential and Logarithmic Functions

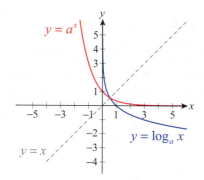

Figure 1 Case 1: $0 < a < 1$

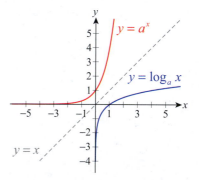

Figure 2 Case 2: $a > 1$

Appendix C **Trigonometric and Hyperbolic Functions**

The historical records of trigonometry date back to the second millennium BC, and we know of a number of different cultures (Egyptian, Babylonian, Indian, and Greek among them) that studied and used the properties of triangles. Our word "trigonometry" comes from an ancient Greek word meaning "triangle measuring," and the names of the individual trigonometric functions have similarly ancient roots. The study of how different cultures independently discovered the basic tenets of trigonometry, how trigonometric knowledge was further developed and disseminated, and how early civilizations used trigonometry for scientific and commercial purposes is fascinating in its own right and well worth exploring. Many excellent resources for such exploration are available online, in books, and in scholarly articles.

In contrast, the history of hyperbolic functions dates back only to the 18th century AD; the Italian mathematician Vincenzo Ricatti (1707–1775) and the Swiss mathematicians Johann Heinrich Lambert (1728–1777) and Leonhard Euler (1707–1783) were among the first to recognize their utility. But their development and characteristics have much in common with trigonometric functions, and they are useful today when solving differential equations and as antiderivatives of certain commonly occurring expressions

For the purpose of quick reference, this appendix contains the basic definitions and graphs of the trigonometric and hyperbolic functions, along with frequently used identities and associated concepts.

Basic Definitions and Graphs

Radian and Degree Measure

$$180° = \pi \text{ radians}$$

$$1° = \frac{\pi}{180} \text{ radians} \qquad 1 \text{ radian} = \frac{180°}{\pi}$$

$$x° = x\left(\frac{\pi}{180}\right) \text{ radians} \qquad x \text{ radians} = x\left(\frac{180°}{\pi}\right)$$

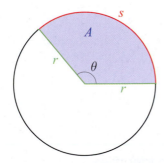

Trigonometric Functions

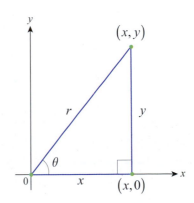

$$\sin\theta = \frac{y}{r} \qquad\qquad \csc\theta = \frac{r}{y} \text{ (for } y \neq 0)$$

$$\cos\theta = \frac{x}{r} \qquad\qquad \sec\theta = \frac{r}{x} \text{ (for } x \neq 0)$$

$$\tan\theta = \frac{y}{x} \text{ (for } x \neq 0) \qquad \cot\theta = \frac{x}{y} \text{ (for } y \neq 0)$$

Arc Length

$$s = \left(\frac{\theta}{2\pi}\right)(2\pi r) = r\theta$$

Area of a Sector

$$A = \left(\frac{\theta}{2\pi}\right)(\pi r^2) = \frac{r^2\theta}{2}$$

Angular Speed

$$\omega = \frac{\theta}{t}$$

Linear Speed

$$v = \frac{s}{t} = \frac{r\theta}{t} = r\omega$$

Commonly Encountered Angles

θ	0°	30°	45°	60°	90°	180°	270°
Radians	0	$\dfrac{\pi}{6}$	$\dfrac{\pi}{4}$	$\dfrac{\pi}{3}$	$\dfrac{\pi}{2}$	π	$\dfrac{3\pi}{2}$
$\sin\theta$	0	$\dfrac{1}{2}$	$\dfrac{1}{\sqrt{2}}$	$\dfrac{\sqrt{3}}{2}$	1	0	-1
$\cos\theta$	1	$\dfrac{\sqrt{3}}{2}$	$\dfrac{1}{\sqrt{2}}$	$\dfrac{1}{2}$	0	-1	0
$\tan\theta$	0	$\dfrac{1}{\sqrt{3}}$	1	$\sqrt{3}$	—	0	—

Trigonometric Graphs

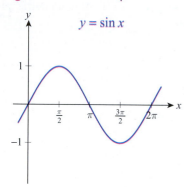

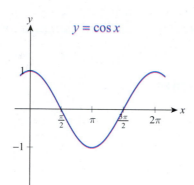

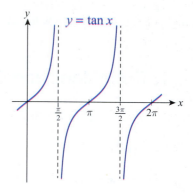

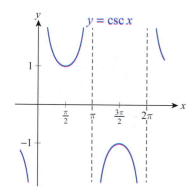

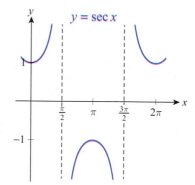

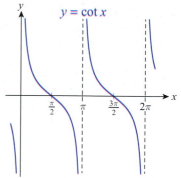

Trigonometric Identities

Reciprocal Identities

$$\csc x = \frac{1}{\sin x} \qquad \sec x = \frac{1}{\cos x} \qquad \cot x = \frac{1}{\tan x}$$

$$\sin x = \frac{1}{\csc x} \qquad \cos x = \frac{1}{\sec x} \qquad \tan x = \frac{1}{\cot x}$$

Cofunction Identities

$$\cos x = \sin\left(\frac{\pi}{2} - x\right) \qquad \sin x = \cos\left(\frac{\pi}{2} - x\right)$$

$$\csc x = \sec\left(\frac{\pi}{2} - x\right) \qquad \sec x = \csc\left(\frac{\pi}{2} - x\right)$$

$$\cot x = \tan\left(\frac{\pi}{2} - x\right) \qquad \tan x = \cot\left(\frac{\pi}{2} - x\right)$$

Quotient Identities

$$\tan x = \frac{\sin x}{\cos x} \qquad \cot x = \frac{\cos x}{\sin x}$$

Period Identities

$$\sin(x + 2\pi) = \sin x \qquad \csc(x + 2\pi) = \csc x$$

$$\cos(x + 2\pi) = \cos x \qquad \sec(x + 2\pi) = \sec x$$

$$\tan(x + \pi) = \tan x \qquad \cot(x + \pi) = \cot x$$

Even/Odd Identities

$$\sin(-x) = -\sin x \quad \cos(-x) = \cos x \quad \tan(-x) = -\tan x$$

$$\csc(-x) = -\csc x \quad \sec(-x) = \sec x \quad \cot(-x) = -\cot x$$

Pythagorean Identities

$$\sin^2 x + \cos^2 x = 1 \qquad \tan^2 x + 1 = \sec^2 x$$

$$1 + \cot^2 x = \csc^2 x$$

Sum and Difference Identities

$$\sin(u + v) = \sin u \cos v + \cos u \sin v$$

$$\sin(u - v) = \sin u \cos v - \cos u \sin v$$

$$\cos(u + v) = \cos u \cos v - \sin u \sin v$$

$$\cos(u - v) = \cos u \cos v + \sin u \sin v$$

$$\tan(u + v) = \frac{\tan u + \tan v}{1 - \tan u \tan v}$$

$$\tan(u - v) = \frac{\tan u - \tan v}{1 + \tan u \tan v}$$

Double-Angle Identities

$$\sin 2u = 2 \sin u \cos u$$

$$\cos 2u = \cos^2 u - \sin^2 u = 2 \cos^2 u - 1 = 1 - 2 \sin^2 u$$

$$\tan 2u = \frac{2 \tan u}{1 - \tan^2 u}$$

Power-Reducing Identities

$$\sin^2 x = \frac{1 - \cos 2x}{2}$$

$$\cos^2 x = \frac{1 + \cos 2x}{2}$$

$$\tan^2 x = \frac{1 - \cos 2x}{1 + \cos 2x}$$

Half-Angle Identities

$$\sin \frac{x}{2} = \pm \sqrt{\frac{1 - \cos x}{2}}$$

$$\cos \frac{x}{2} = \pm \sqrt{\frac{1 + \cos x}{2}}$$

$$\tan \frac{x}{2} = \frac{1 - \cos x}{\sin x} = \frac{\sin x}{1 + \cos x}$$

Product-to-Sum Identities

$$\sin x \cos y = \frac{1}{2}\left[\sin(x + y) + \sin(x - y)\right]$$

$$\cos x \sin y = \frac{1}{2}\left[\sin(x + y) - \sin(x - y)\right]$$

$$\sin x \sin y = \frac{1}{2}\left[\cos(x - y) - \cos(x + y)\right]$$

$$\cos x \cos y = \frac{1}{2}\left[\cos(x + y) + \cos(x - y)\right]$$

Sum-to-Product Identities

$$\sin x + \sin y = 2 \sin\left(\frac{x + y}{2}\right)\cos\left(\frac{x - y}{2}\right)$$

$$\sin x - \sin y = 2 \cos\left(\frac{x + y}{2}\right)\sin\left(\frac{x - y}{2}\right)$$

$$\cos x + \cos y = 2 \cos\left(\frac{x + y}{2}\right)\cos\left(\frac{x - y}{2}\right)$$

$$\cos x - \cos y = -2 \sin\left(\frac{x + y}{2}\right)\sin\left(\frac{x - y}{2}\right)$$

The Laws of Sines and Cosines

The Law of Sines

$$\frac{\sin A}{a} = \frac{\sin B}{b} = \frac{\sin C}{c}$$

The Law of Cosines

$$a^2 = b^2 + c^2 - 2bc\cos A$$
$$b^2 = a^2 + c^2 - 2ac\cos B$$
$$c^2 = a^2 + b^2 - 2ab\cos C$$

Inverse Trigonometric Functions

Arcsine, Arccosine, and Arctangent

Function	Notation	Domain	Range
Inverse Sine	$\arcsin y = \sin^{-1} y = x \iff y = \sin x$	$[-1,1]$	$\left[-\dfrac{\pi}{2}, \dfrac{\pi}{2}\right]$
Inverse Cosine	$\arccos y = \cos^{-1} y = x \iff y = \cos x$	$[-1,1]$	$[0,\pi]$
Inverse Tangent	$\arctan y = \tan^{-1} y = x \iff y = \tan x$	$(-\infty,\infty)$	$\left(-\dfrac{\pi}{2}, \dfrac{\pi}{2}\right)$

Inverse Trigonometric Graphs

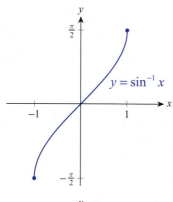
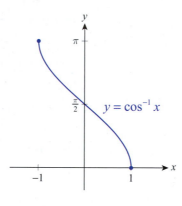
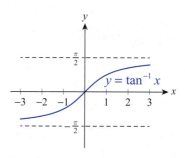

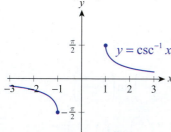
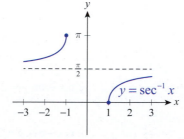
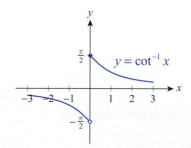

Inverse Trigonometric Identities

$$\csc^{-1} x = \sin^{-1}\left(\frac{1}{x}\right) \qquad \sec^{-1} x = \cos^{-1}\left(\frac{1}{x}\right) \qquad \cot^{-1} x = \tan^{-1}\left(\frac{1}{x}\right), \text{ with } \cot^{-1} 0 = \frac{\pi}{2}$$

Hyperbolic Functions

Hyperbolic Functions and Their Graphs

Hyperbolic Sine

$$\sinh x = \frac{e^x - e^{-x}}{2}$$

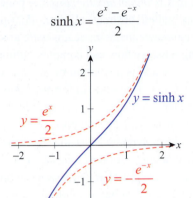

Hyperbolic Cosine

$$\cosh x = \frac{e^x + e^{-x}}{2}$$

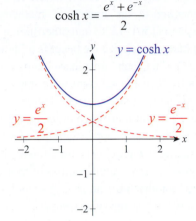

Hyperbolic Tangent

$$\tanh x = \frac{\sinh x}{\cosh x} = \frac{e^x - e^{-x}}{e^x + e^{-x}}$$

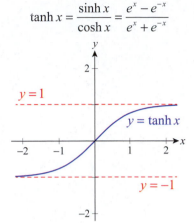

Hyperbolic Cosecant

$$\operatorname{csch} x = \frac{1}{\sinh x} = \frac{2}{e^x - e^{-x}}$$

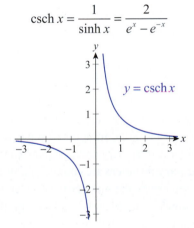

Hyperbolic Secant

$$\operatorname{sech} x = \frac{1}{\cosh x} = \frac{2}{e^x + e^{-x}}$$

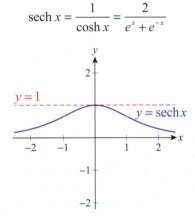

Hyperbolic Cotangent

$$\coth x = \frac{1}{\tanh x} = \frac{e^x + e^{-x}}{e^x - e^{-x}}$$

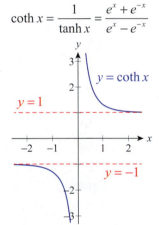

Elementary Hyperbolic Identities

$$\cosh^2 x - \sinh^2 x = 1$$

$$\tanh^2 x = 1 - \operatorname{sech}^2 x$$

$$\coth^2 x = 1 + \operatorname{csch}^2 x$$

$$\sinh 2x = 2 \sinh x \cosh x$$

$$\sinh^2 x = \frac{\cosh 2x - 1}{2}$$

$$\cosh^2 x = \frac{\cosh 2x + 1}{2}$$

$$\cosh 2x = \cosh^2 x + \sinh^2 x$$

$$\sinh(x + y) = \sinh x \cosh y + \cosh x \sinh y$$

$$\cosh(x + y) = \cosh x \cosh y + \sinh x \sinh y$$

Appendix D **Complex Numbers**

The **complex numbers**, an extension of the real numbers, consist of all numbers that can be expressed in the form $a + bi$, where a and b are real numbers and i, representing the **imaginary unit**, satisfies the equation $i^2 = -1$. Complex numbers expand the real numbers to a set that is *algebraically closed*, a concept belonging to the branch of mathematics called abstract algebra. Girolamo Cardano (1501–1576) and other Italian Renaissance mathematicians were among the first to recognize the benefits of defining what we now call complex numbers; by allowing such "imaginary" numbers as i, which is a solution of the equation $x^2 + 1 = 0$, mathematicians were able to devise and make sense of formulas solving polynomial equations up to degree four. Later mathematicians conjectured that every nonconstant polynomial function, even those with complex coefficients, has at least one root (a number at which the polynomial has the value of 0), assuming complex roots are allowed. Repeated application of this assertion then implies, counting multiplicities of roots, that a polynomial of degree n has n roots; stated another way, an nth-degree polynomial equation has n solutions (some of which may be repeated solutions). The first reasonably complete proof of this conjecture, now known as the Fundamental Theorem of Algebra, was provided by Carl Friedrich Gauss (1777–1855) in 1799 in his doctoral dissertation.

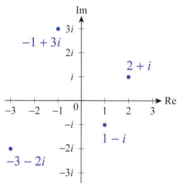

Figure 1

Unlike real numbers, often identified with points on a line, complex numbers are typically depicted as points in the **complex plane**, also known as the **Argand plane**, which is named after the French Swiss mathematician Jean-Robert Argand (1768–1822). The complex plane has the appearance of the Cartesian plane, with the horizontal axis referred to as the **real axis** and the vertical axis as the **imaginary axis**. A given complex number $a + bi$ is associated with the ordered pair (a, b) in the plane, where a represents the displacement along the real axis and b the displacement along the imaginary axis (see Figure 1 for examples). In this context, a is called the **real part** of $a + bi$ and b the **imaginary part**. Real numbers are thus complex numbers for which the imaginary part is 0 (they can be written in the form $a + 0 \cdot i$), and **pure imaginary numbers** are complex numbers of the form $0 + bi$; the origin of the plane represents the number $0 + 0 \cdot i$ and is usually simply written as 0. Two complex numbers $a + bi$ and $c + di$ are **equal** if and only if $a = b$ and $c = d$ (that is, their real parts are equal and their imaginary parts are equal).

Sums, differences, and products of complex numbers are easily simplified and written in the form $a + bi$ by treating complex numbers as polynomial expressions in the variable i, remembering that $i^2 = -1$. (Keep in mind, though, that i is not, in fact, a variable—this treatment is simply a convenience.) Example 1 illustrates the process.

Example 1 ✎

Express each of the following in the form $a + bi$.

a. $(4 + 3i) + (-5 + 7i)$ **b.** $(-2 + 3i) - (-3 + 3i)$

c. $(3 + 2i)(-2 + 3i)$ **d.** $(2 - 3i)^2$

Solution

a. $(4+3i)+(-5+7i)=(4-5)+(3+7)i$
$$=-1+10i$$

b. $(-2+3i)-(-3+3i)=(-2+3)+(3-3)i$
$$=1$$

c. $(3+2i)(-2+3i)=-6+9i-4i+6i^2$
$$=-6+(9-4)i-6 \qquad \text{Replace } i^2 \text{ with } -1.$$
$$=-12+5i$$

d. $(2-3i)^2=(2-3i)(2-3i)$
$$=4-6i-6i+9i^2$$
$$=4-12i-9 \qquad \text{Replace } i^2 \text{ with } -1.$$
$$=-5-12i$$

Division of complex numbers is slightly more complicated, but a quotient can also be simplified and written in the form $a + bi$ by making use of the following observation:

$$(a+bi)(a-bi)=a^2-abi+abi-b^2i^2=a^2+b^2$$

Given a complex number $z = a + bi$, the complex number $\overline{z} = a - bi$ is called its **complex conjugate**. We simplify a quotient of complex numbers by multiplying the numerator and denominator by the complex conjugate of the denominator, as illustrated in Example 2.

Example 2 ✐

Express each of the following in the form $a + bi$.

a. $\dfrac{2+3i}{3-i}$
b. $(4-3i)^{-1}$
c. $\dfrac{1}{i}$

Solution

a. $\dfrac{2+3i}{3-i}=\dfrac{2+3i}{3-i}$

$$=\frac{(2+3i)(3+i)}{(3-i)(3+i)} \qquad \text{Multiply the numerator and denominator by the conjugate.}$$

$$=\frac{6+2i+9i+3i^2}{9+3i-3i-i^2}$$

$$=\frac{6+11i-3}{9+1} \qquad \text{Replace } i^2 \text{ with } -1.$$

$$=\frac{3+11i}{10}=\frac{3}{10}+\frac{11}{10}i$$

b. $(4-3i)^{-1} = \dfrac{1}{4-3i}$

$$= \frac{1(4+3i)}{(4-3i)(4+3i)}$$

Multiply the numerator and denominator by the conjugate.

$$= \frac{4+3i}{16+12i-12i-9i^2}$$

$$= \frac{4+3i}{16+9} = \frac{4}{25} + \frac{3}{25}i$$

c. $\dfrac{1}{i} = \dfrac{1(-i)}{i(-i)} = \dfrac{-i}{-i^2} = \dfrac{-i}{1} = -i$

Endowed with the operations of addition and multiplication, the set of complex numbers, like the set of real numbers and the set of rational numbers, form what is known as a **field**, another concept from the realm of abstract algebra. The following table summarizes the properties possessed by a field; note that each of the three sets of numbers mentioned above possesses all the properties. Also note, by way of contrast, that the set of natural numbers, the set of integers, and the set of irrational numbers are not fields, as each set fails to possess one or more of the field properties.

Field Properties

In this table, a, b, and c represent arbitrary elements of a given field. The first five properties apply individually to the two operations of addition and multiplication, while the last combines the two.

Name of Property	Additive Version	Multiplicative Version
Closure	$a+b$ is an element of the field	ab is an element of the field
Commutative	$a+b=b+a$	$ab=ba$
Associative	$a+(b+c)=(a+b)+c$	$a(bc)=(ab)c$
Identity	$a+0=0+a=a$	$a\cdot 1 = 1\cdot a = a$
Inverse	$a+(-a)=0$	$a\cdot\dfrac{1}{a}=1,$ assuming $a\neq 0$
Distributive		$a(b+c)=ab+ac$

The introduction of the imaginary unit i allows us to now define the **principal square root** $\sqrt{a}$ of any real number a, as follows: Given a positive real number a, $\sqrt{a}$ denotes the positive real number whose square is a, and $\sqrt{-a}=i\sqrt{a}$. An application of this definition explains the restriction in one of the properties of exponents (specifically, the exponent $1/2$). Recall that if a and b are both positive, then

$$\sqrt{ab} = (ab)^{1/2} = a^{1/2}b^{1/2} = \sqrt{a}\sqrt{b}.$$

To see why a and b are required to be positive, note that

$$\sqrt{(-9)(-4)} = \sqrt{36} = 6,$$

but

$$\sqrt{-9}\sqrt{-4} = \left(i\sqrt{9}\right)\left(i\sqrt{4}\right) = (3i)(2i) = 6i^2 = -6.$$

Complex numbers can also be expressed in *polar form*, based on the polar coordinates of a given complex number in the plane. We say the **magnitude** $|z|$ of a complex number $z = a + bi$, also known as its **modulus**, **norm**, or **absolute value**, is its distance from 0 in the complex plane—that is, the nonnegative real number

$$|z| = \sqrt{a^2 + b^2}.$$

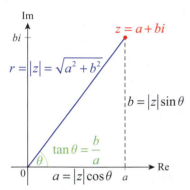

Figure 2

The **argument** of z, denoted $\arg(z)$, is the radian angle θ between the positive real axis and the line joining 0 and z. The quantities $|z|$ and $\arg(z)$ thus play the same roles, respectively, as the polar coordinates r and θ of a point in the plane. The argument of the complex number 0 is undefined, while the argument of every other complex number is not unique (any multiple of 2π added to the argument of a given complex number describes the same number, since 2π corresponds to a complete rotation around the origin). Given these definitions, and letting $\theta = \arg(z)$, the **polar form** of $z = a + bi$ is then

$$z = r(\cos\theta + i\sin\theta), \quad \text{where} \quad r = |z|$$

(see Figure 2 for a depiction of the relationship between a, b, r, $|z|$, and θ).

Example 3 ✎

Write each of the following complex numbers in polar form.

a. $1 + i\sqrt{3}$ 　　　　　　　　　　　**b.** $-1 + i$

Solution

a. The magnitude of $1 + i\sqrt{3}$ is $\sqrt{1^2 + \left(\sqrt{3}\right)^2} = 2$, and its argument is $\tan^{-1}\sqrt{3} = \pi/3$ (see Figure 3). Hence,

$$1 + i\sqrt{3} = 2\left(\cos\frac{\pi}{3} + i\sin\frac{\pi}{3}\right).$$

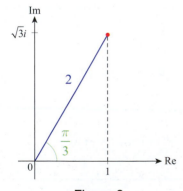

Figure 3

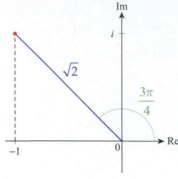

Figure 4

b. The magnitude of $-1+i$ is $\sqrt{(-1)^2+1^2}=\sqrt{2}$, and its argument is $3\pi/4$ (note that this complex number lies in the second quadrant of the plane, as shown in Figure 4). Hence,

$$-1+i=\sqrt{2}\left(\cos\frac{3\pi}{4}+i\sin\frac{3\pi}{4}\right).$$

Euler's formula $e^{i\theta}=\cos\theta+i\sin\theta$, derived in Section 10.9, allows us to express the polar form of a complex number as a complex exponential:

$$z=re^{i\theta},\quad\text{where}\quad r=|z|\quad\text{and}\quad\theta=\arg(z).$$

With this observation, the following formulas regarding products and quotients of complex numbers are easily proved (they can also be proved by using the trigonometric sum and difference identities).

Theorem

Products and Quotients of Complex Numbers

Given the complex numbers

$$z_1=r_1\left(\cos\theta_1+i\sin\theta_1\right)\quad\text{and}\quad z_2=r_2\left(\cos\theta_2+i\sin\theta_2\right),$$

the following formulas hold:

Product Formula $z_1z_2=r_1r_2\left[\cos\left(\theta_1+\theta_2\right)+i\sin\left(\theta_1+\theta_2\right)\right]$

Quotient Formula $\dfrac{z_1}{z_2}=\dfrac{r_1}{r_2}\left[\cos\left(\theta_1-\theta_2\right)+i\sin\left(\theta_1-\theta_2\right)\right]$, assuming $z_2\neq0$.

Proof

Writing each complex number as a complex exponential,

$$z_1z_2=\left(r_1e^{i\theta_1}\right)\left(r_2e^{i\theta_2}\right)=r_1r_2e^{i(\theta_1+\theta_2)}=r_1r_2\left[\cos\left(\theta_1+\theta_2\right)+i\sin\left(\theta_1+\theta_2\right)\right]$$

and

$$\frac{z_1}{z_2}=\frac{r_1e^{i\theta_1}}{r_2e^{i\theta_2}}=\frac{r_1}{r_2}e^{i(\theta_1-\theta_2)}=\frac{r_1}{r_2}\left[\cos\left(\theta_1-\theta_2\right)+i\sin\left(\theta_2-\theta_2\right)\right]$$

The following statement regarding positive integer powers of complex numbers can be similarly proved.

Theorem 🔑

De Moivre's Theorem

Given a complex number $z = r(\cos\theta + i\sin\theta)$ and positive integer n,

$$z^n = r^n(\cos n\theta + i\sin n\theta).$$

Proof 📎

Again writing z as a complex exponential,

$$z^n = \left(re^{i\theta}\right)^n = r^n e^{in\theta} = r^n(\cos n\theta + i\sin n\theta).$$

De Moivre's Theorem can be used to determine roots of complex numbers. The first step is to note that if $w = re^{i\theta}$ is a nonzero complex number, and if n is a positive integer, then w has n n^{th} roots. This follows from the Fundamental Theorem of Algebra, which tells us that the equation $z^n = w$ has n solutions (here, z represents a complex variable). One n^{th} root is easily determined: if we let

$$z_0 = r^{1/n}e^{i(\theta/n)},$$

then

$$z_0^n = \left[r^{1/n}e^{i(\theta/n)}\right]^n = re^{i\theta} = w.$$

But as we know, replacing θ with $\theta + 2k\pi$ results in an equivalent complex number for any integer k, leading to the following formula for the nth roots of w.

Theorem 🔑

Roots of a Complex Number

Let $w = r(\cos\theta + i\sin\theta)$ and let n be a positive integer. The nth roots of w are given by

$$z_k = r^{1/n}e^{i\left(\frac{\theta + 2k\pi}{n}\right)}, \quad k = 0, 1, \ldots, n-1.$$

Alternatively,

$$z_k = r^{1/n}\left[\cos\left(\frac{\theta + 2k\pi}{n}\right) + i\sin\left(\frac{\theta + 2k\pi}{n}\right)\right], \quad k = 0, 1, \ldots, n-1.$$

The n n^{th} roots of a given complex number all have the same magnitude and are equally distributed around a circle in the complex plane with radius equal to that common magnitude.

Example 4 ✎

Determine the specified roots of the given complex numbers, and graph the roots and the original complex numbers in the plane.

a. 5th roots of 1

b. 4th roots of $-1 - i\sqrt{3}$

Solution

a. The easiest way to determine the 5th roots of 1 is to write 1 as a complex exponential and then apply the above formula with $n = 5$:

$$1 = e^{i \cdot 0},$$

so the 5th roots of 1 are

$$\left\{ 1, e^{i(2\pi/5)}, e^{i(4\pi/5)}, e^{i(6\pi/5)}, e^{i(8\pi/5)} \right\}.$$

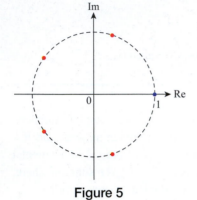

Figure 5

Since the complex number 1 (shown as a blue point in Figure 5) has a magnitude of 1, all of the five 5th roots (shown as red points in the figure) lie on a circle of radius 1. Note that 1 is, itself, one of the 5th roots of 1.

b. The first step is to again express the complex number as a complex exponential. Note that

$$\left| -1 - i\sqrt{3} \right| = \sqrt{(-1)^2 + \left(-\sqrt{3}\right)^2} = \sqrt{4} = 2$$

and

$$\tan \theta = \frac{-\sqrt{3}}{-1} \quad \Rightarrow \quad \theta = \frac{4\pi}{3},$$

so

$$-1 - i\sqrt{3} = 2e^{i(4\pi/3)}.$$

Hence, the 4th roots of $-1 - i\sqrt{3}$ are

$$\left\{ 2^{1/4} e^{(i/4)\left[(4\pi/3) + 2k\pi\right]} \right\}_{k=0,1,2,3} = \left\{ 2^{1/4} e^{i(\pi/3)}, 2^{1/4} e^{i(5\pi/6)}, 2^{1/4} e^{i(4\pi/3)}, 2^{1/4} e^{i(11\pi/6)} \right\}.$$

Figure 6 shows the original point $-1 - i\sqrt{3}$ in blue and its four 4th roots in red.

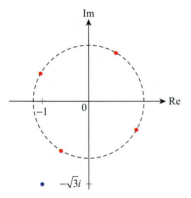

Figure 6

Appendix E **Proofs of Selected Theorems**

In this appendix we provide proofs (or in one case just a statement) of theorems used in the main body of the text. While some of the proofs here are more technical in nature than those presented elsewhere, they are worth studying in order to (1) gain additional insight into the rigorous nature of mathematical thinking and (2) develop a sense of the deeper mathematics to come in later courses.

Section 2.4

Theorem ⚲

Basic Limit Laws

Let f and g be two functions such that both $\lim_{x \to c} f(x)$ and $\lim_{x \to c} g(x)$ exist, and let k be a fixed real number. Then the following laws hold.

Sum Law

$$\lim_{x \to c}\left[f(x) + g(x)\right] = \lim_{x \to c} f(x) + \lim_{x \to c} g(x)$$

Difference Law

$$\lim_{x \to c}\left[f(x) - g(x)\right] = \lim_{x \to c} f(x) - \lim_{x \to c} g(x)$$

Constant Multiple Law

$$\lim_{x \to c}\left[kf(x)\right] = k \lim_{x \to c} f(x)$$

Product Law

$$\lim_{x \to c}\left[f(x) g(x)\right] = \lim_{x \to c} f(x) \cdot \lim_{x \to c} g(x)$$

Quotient Law

$$\lim_{x \to c} \frac{f(x)}{g(x)} = \frac{\lim_{x \to c} f(x)}{\lim_{x \to c} g(x)}, \text{ provided } \lim_{x \to c} g(x) \neq 0$$

Proof ✎

We already proved the Sum Law in Section 2.4. We proceed to prove the Product Law and Quotient Law, from which the remaining laws will follow quickly.

As in the proof of the Sum Law, let $L = \lim_{x \to c} f(x)$ and $M = \lim_{x \to c} g(x)$, and assume $\varepsilon > 0$ is given. Our goal is to show there exists $\delta > 0$ such that $\left| f(x) g(x) - LM \right| < \varepsilon$ for all $0 < \left| x - c \right| < \delta$. One way to determine δ is to employ a strategy of adding and subtracting the same quantity, in this case $f(x) M$:

$$\left| f(x) g(x) - LM \right| = \left| f(x) g(x) - f(x) M + f(x) M - LM \right|$$
$$\leq \left| f(x)\left[g(x) - M\right]\right| + \left| M\left[f(x) - L\right]\right| \quad \text{\color{blue}Triangle Inequality}$$
$$= \left| f(x)\right|\left| g(x) - M\right| + \left| M\right|\left| f(x) - L\right|$$

Since $L = \lim_{x \to c} f(x)$, we know there exists $\delta_1 > 0$ for which

$$0 < \left| x - c \right| < \delta_1 \quad \Rightarrow \quad \left| f(x) - L \right| < 1,$$

so $\left| f(x)\right| = \left| f(x) - L + L \right| \leq \left| f(x) - L \right| + \left| L \right| < 1 + \left| L \right|.$

Similarly, there exists $\delta_2 > 0$ for which

$$0 < |x - c| < \delta_2 \quad \Rightarrow \quad |f(x) - L| < \frac{\varepsilon}{2(1 + |M|)}.$$

And finally, since $M = \lim_{x \to c} g(x)$, there exists $\delta_3 > 0$ for which

$$0 < |x - c| < \delta_3 \quad \Rightarrow \quad |g(x) - M| < \frac{\varepsilon}{2(1 + |L|)}.$$

So if we let $\delta = \min\{\delta_1, \delta_2, \delta_3\}$, $0 < |x - c| < \delta$ will guarantee each of the above three outcomes, meaning

$$|f(x)g(x) - LM| \le |f(x)||g(x) - M| + |M||f(x) - L|$$

$$< (1 + |L|)\frac{\varepsilon}{2(1 + |L|)} + |M|\frac{\varepsilon}{2(1 + |M|)}$$

$$< \frac{\varepsilon}{2} + \frac{\varepsilon}{2} = \varepsilon,$$

thereby proving the Product Law.

To prove the Quotient Law, we will see that it suffices to prove

$$\lim_{x \to c} \frac{1}{g(x)} = \frac{1}{M},$$

under the assumption that $M \ne 0$. In doing so, we will use the fact that $\big||a| - |b|\big| \le |a - b|$ for arbitrary real numbers a and b. This follows from the observation that

$$|a| = |a - b + b| \le |a - b| + |b| \qquad \text{\textcolor{blue}{Triangle Inequality}}$$

so $|a| - |b| \le |a - b|$. Similarly, interchanging a and b in the same argument shows that $|b| - |a| \le |b - a| = |a - b|$, and the two facts together prove that $\big||a| - |b|\big| \le |a - b|$.

Now, given $\varepsilon > 0$, we need to show there exists $\delta > 0$ for which $0 < |x - c| < \delta$ implies

$$\left|\frac{1}{g(x)} - \frac{1}{M}\right| < \varepsilon.$$

Since

$$\left|\frac{1}{g(x)} - \frac{1}{M}\right| = \left|\frac{M - g(x)}{g(x)M}\right| = \frac{1}{|g(x)|} \cdot \frac{1}{|M|} \cdot |g(x) - M|,$$

we want to choose δ in such a manner that $|g(x) - M|$ is sufficiently small and so that $|g(x)|$ is far enough away from 0 to make

$$\frac{1}{|g(x)|} \cdot \frac{1}{|M|}$$

also sufficiently small. The fact that $M = \lim_{x \to c} g(x)$ tells us there is a $\delta_1 > 0$ for which $0 < |x - c| < \delta_1$ implies $|g(x) - M| < |M|/2$, meaning

$$0 < |x - c| < \delta_1 \implies \big||g(x)| - |M|\big| \le |g(x) - M| < \frac{|M|}{2} \qquad \text{Using } \big||a| - |b|\big| \le |a - b|$$

and so

$$-\frac{|M|}{2} < |g(x)| - |M| < \frac{|M|}{2}.$$

Adding $|M|$ throughout results in the equivalent compound inequality

$$\frac{|M|}{2} < |g(x)| < \frac{3|M|}{2}$$

whenever $0 < |x - c| < \delta_1$. For our present purposes, we actually only care about the fact that

$$|g(x)| > \frac{|M|}{2}, \quad \text{which means that} \quad \frac{1}{|g(x)|} < \frac{2}{|M|}.$$

There is also a $\delta_2 > 0$ such that

$$0 < |x - c| < \delta_2 \implies |g(x) - M| < \frac{|M|^2 \, \varepsilon}{2},$$

so if we let $\delta = \min\{\delta_1, \delta_2\}$, we have

$$0 < |x - c| < \delta \implies \left|\frac{1}{g(x)} - \frac{1}{M}\right| = \frac{1}{|g(x)|} \cdot \frac{1}{|M|} \cdot |g(x) - M| < \frac{2}{|M|} \cdot \frac{1}{|M|} \cdot \frac{|M|^2 \, \varepsilon}{2} = \varepsilon,$$

thus proving

$$\lim_{x \to c} \frac{1}{g(x)} = \frac{1}{M}.$$

We can now apply the Product Law, already proved, to obtain the Quotient Law.

$$\lim_{x \to c} \frac{f(x)}{g(x)} = \lim_{x \to c} \left[f(x) \cdot \frac{1}{g(x)} \right]$$

$$= \lim_{x \to c} f(x) \cdot \lim_{x \to c} \frac{1}{g(x)} \qquad \textcolor{blue}{\text{Product Law}}$$

$$= L \cdot \frac{1}{M} = \frac{\lim_{x \to c} f(x)}{\lim_{x \to c} g(x)}$$

The Constant Multiple Law is a consequence of the Product Law, using $g(x) = k$ as one of the two functions, and the Difference Law follows from applying first the Sum Law and then the Constant Multiple Law (with $k = -1$).

Section 2.4

Theorem 🔍

Positive Integer Power Law

Let f be a function for which $\lim_{x \to c} f(x)$ exists, and let m be a fixed positive integer. Then

$$\lim_{x \to c}\left[f(x) \right]^m = \left[\lim_{x \to c} f(x) \right]^m.$$

Proof 📎

The statement is trivially true for $m = 1$, so we prove the theorem for $m \geq 2$. We use mathematical induction to do so.

Basis Step: By the Product Law, we have

$$\lim_{x \to c}\left[f(x) \right]^2 = \lim_{x \to c}\left[f(x) f(x) \right] = \left[\lim_{x \to c} f(x) \right]^2.$$

Inductive Step: Assume $\lim_{x \to c}\left[f(x) \right]^k = \left[\lim_{x \to c} f(x) \right]^k$ for some $k \geq 2$. Then again applying the Product Law, we complete the proof as follows.

$$
\begin{aligned}
\lim_{x \to c}\left[f(x) \right]^{k+1} &= \lim_{x \to c}\left(\left[f(x) \right]^k f(x) \right) \\
&= \lim_{x \to c}\left[f(x) \right]^k \cdot \lim_{x \to c} f(x) && \text{Product Law} \\
&= \left[\lim_{x \to c} f(x) \right]^k \cdot \lim_{x \to c} f(x) && \text{Induction hypothesis} \\
&= \left[\lim_{x \to c} f(x) \right]^{k+1}
\end{aligned}
$$

Section 2.4

Theorem 🔍

The Squeeze Theorem

If $g(x) \leq f(x) \leq h(x)$ for all x in some open interval containing c, except possibly at c itself, and if $\lim_{x \to c} g(x) = \lim_{x \to c} h(x) = L$, then $\lim_{x \to c} f(x) = L$ as well.

The statement also holds for limits at infinity, that is, for $c = -\infty$ or $c = \infty$.

Proof 📎

Since $g(x) \leq f(x) \leq h(x)$ for all x in some open interval containing c, there exists $\delta_1 > 0$ such that $0 < |x - c| < \delta_1 \Rightarrow g(x) \leq f(x) \leq h(x)$. And by the limit definition, given $\varepsilon > 0$ there exist $\delta_2 > 0$ and $\delta_3 > 0$ such that

$$0 < |x - c| < \delta_2 \;\;\Rightarrow\;\; |g(x) - L| < \varepsilon \;\;\Rightarrow\;\; L - \varepsilon < g(x) < L + \varepsilon$$

and

$$0 < |x - c| < \delta_3 \;\;\Rightarrow\;\; |h(x) - L| < \varepsilon \;\;\Rightarrow\;\; L - \varepsilon < h(x) < L + \varepsilon.$$

By letting $\delta = \min\{\delta_1, \delta_2, \delta_3\}$, all three conclusions are true for x within δ of c. That is,

$$0 < |x - c| < \delta \;\;\Rightarrow\;\; L - \varepsilon < g(x) \le f(x) \le h(x) < L + \varepsilon \;\;\Rightarrow\;\; |f(x) - L| < \varepsilon.$$

Section 2.4

Theorem 🔍

Upper Bound Theorem

If $f(x) \le g(x)$ for all x in some open interval containing c, except possibly at c itself, and if the limits of f and g both exist at c, then

$$\lim_{x \to c} f(x) \le \lim_{x \to c} g(x).$$

Proof 📎

Let $L = \lim_{x \to c} f(x)$ and $M = \lim_{x \to c} g(x)$. Note that, by the Difference Law, the limit of $g(x) - f(x)$ at c exists, and

$$\lim_{x \to c}\big[g(x) - f(x)\big] = \lim_{x \to c} g(x) - \lim_{x \to c} f(x) = M - L.$$

Suppose, in contradiction to the claim, that $L > M$. Then $L - M > 0$, and if we let $\varepsilon = L - M$, there exists $\delta > 0$ such that

$$0 < |x - c| < \delta \;\;\Rightarrow\;\; \big|g(x) - f(x) - (M - L)\big| < \varepsilon$$
$$\Rightarrow\;\; -\varepsilon < g(x) - f(x) - M + L < \varepsilon$$
$$\Rightarrow\;\; M - L < g(x) - f(x) - M + L < L - M.$$

In particular, $g(x) - f(x) - M + L < L - M$, so $g(x) - f(x) < 0$ for all x such that $0 < |x - c| < \delta$, contradicting the fact that $f(x) \le g(x)$ for all x in some open interval containing c. Thus, it must be the case that $L \le M$; that is, $\lim_{x \to c} f(x) \le \lim_{x \to c} g(x)$.

Section 2.5

Theorem 🔍

"Limits Pass Through a Continuous Function"

Suppose $\lim_{x \to c} g(x) = a$ and f is continuous at the point a. Then

$$\lim_{x \to c} f\big(g(x)\big) = f\Big(\lim_{x \to c} g(x)\Big) = f(a).$$

In words, we say the limit operation passes inside the continuous function f.

Proof 📎

Assume $\varepsilon > 0$ is given. Since f is continuous at a, there exists $\delta_1 > 0$ for which

$$|x - a| < \delta_1 \;\;\Rightarrow\;\; |f(x) - f(a)| < \varepsilon.$$

And since $\lim_{x \to c} g(x) = a$, there exists $\delta > 0$ such that

$$0 < |x - c| < \delta \quad \Rightarrow \quad |g(x) - a| < \delta_1.$$

Putting these facts together, we see that

$$0 < |x - c| < \delta \quad \Rightarrow \quad |g(x) - a| < \delta_1 \quad \Rightarrow \quad \left| f(g(x)) - f(a) \right| < \varepsilon,$$

and hence $\lim_{x \to c} f(g(x)) = f(a)$.

Section 2.5

Theorem &

"The Inverse of a Continuous Function Is Continuous"

If f is one-to-one and continuous on the interval (a, b), then f^{-1} is also a continuous function.

Proof

We first show that f is strictly monotonic on (a, b), and we do so by applying the Intermediate Value Property to a number of cases, all of which are similar. If f is neither strictly increasing nor strictly decreasing, then there must be points $x_1 < x_2 < x_3$ in (a, b) for which $f(x_2)$ does not lie between $f(x_1)$ and $f(x_3)$. We will show that cannot happen, using a proof by contradiction.

To that end, suppose x_1, x_2, and x_3 *are* three points in (a, b) for which $x_1 < x_2 < x_3$ and for which $f(x_2)$ does not lie between $f(x_1)$ and $f(x_3)$. Since f is one-to-one, either $f(x_1) < f(x_3)$ or $f(x_1) > f(x_3)$; we will assume that $f(x_1) < f(x_3)$ and leave consideration of the other case to the reader. The assumption that $f(x_2)$ is not between $f(x_1)$ and $f(x_3)$ again leads to two cases, one of which is that $f(x_1) < f(x_3) < f(x_2)$. Let y be a value such that $f(x_3) < y < f(x_2)$. Then by the continuity of f and the Intermediate Value Property (see Section 2.5), there is a point p such that $x_2 < p < x_3$ and $f(p) = y$ (in words, there is a point between x_2 and x_3 at which f takes on the value y, since y lies between the values of f at x_2 and x_3). But since $f(x_1) < f(x_3)$, y also satisfies $f(x_1) < y < f(x_2)$, so there is a point q such that $x_1 < q < x_2$ and $f(q) = y$. But then $p \neq q$ (since x_2 lies strictly between them) and $f(p) = f(q)$, contradicting the fact that f is one-to-one. By the same reasoning, the possibility that $f(x_2) < f(x_1) < f(x_3)$ is also ruled out, as are the two cases for which $f(x_1) > f(x_3)$. Thus, f must be either strictly increasing or strictly decreasing.

To now show that f^{-1} is continuous we will assume f is strictly increasing—the argument that f^{-1} is continuous when f is strictly decreasing is similar in nature. Let y_0 be a point in the image of (a, b) under f, and let $\varepsilon > 0$ be given. Since y_0 is in the image set, there is a (unique) point $x_0 \in (a, b)$ for which $f(x_0) = y_0$. Define

$$\varepsilon_1 = \min\{\varepsilon, x_0 - a, b - x_0\}.$$

Then, since we are assuming f is increasing, the image of the interval $(x_0 - \varepsilon_1, x_0 + \varepsilon_1)$ is the interval $(f(x_0 - \varepsilon_1), f(x_0 + \varepsilon_1))$, and $y_0 \in (f(x_0 - \varepsilon_1), f(x_0 + \varepsilon_1))$. Choose $\delta > 0$ small enough so that

$$(y_0 - \delta, y_0 + \delta) \subset (f(x_0 - \varepsilon_1), f(x_0 + \varepsilon_1)).$$

Then for any y such that $|y - y_0| < \delta$, $y \in (f(x_0 - \varepsilon_1), f(x_0 + \varepsilon_1))$ and hence $f^{-1}(y) \in (x_0 - \varepsilon_1, x_0 + \varepsilon_1)$. That is,

$$|y - y_0| < \delta \quad \Rightarrow \quad |f^{-1}(y) - f^{-1}(y_0)| < \varepsilon_1 \leq \varepsilon$$

and hence f^{-1} is continuous at y_0. Since y_0 was arbitrary, we have shown that f^{-1} is continuous on the image of (a, b) under f.

Section 4.1

Theorem 🔍

Bolzano-Weierstrass Theorem (Statement Only)

Every bounded sequence of real numbers has a convergent subsequence.

The Bolzano-Weierstrass theorem has many uses, one of which is to help prove the Extreme Value Theorem of Chapter 4. Specifically, its use assures the existence of points in a closed bounded interval at which a continuous function attains its extreme values.

Section 10.8

Theorem 🔍

Taylor's Theorem

If f and its derivatives up through $f^{(n)}$ are all continuous on the closed interval $[a, b]$ and if $f^{(n+1)}$ exists on the open interval (a, b), then there is a number $c \in (a, b)$ such that

$$f(b) = f(a) + f'(a)(b - a) + \frac{f''(a)}{2!}(b - a)^2 + \cdots + \frac{f^{(n)}(a)}{n!}(b - a)^n + \frac{f^{(n+1)}(c)}{(n+1)!}(b - a)^{n+1}.$$

Proof 📎

The statement above is actually just one variant of Taylor's Theorem, and each variant has, in turn, several slightly different methods of proof. We will use Cauchy's Mean Value Theorem (Section 4.4) to prove this variant.

We begin by defining

$$F(x) = f(x) + f'(x)(b - x) + \frac{f''(x)}{2!}(b - x)^2 + \cdots + \frac{f^{(n)}(x)}{n!}(b - x)^n$$

and

$$G(x) = (b - x)^{n+1}.$$

Given these definitions, we have the following.

$$F(b) = f(b)$$

$$F(a) = f(a) + f'(a)(b-a) + \frac{f''(a)}{2!}(b-a)^2 + \cdots + \frac{f^{(n)}(a)}{n!}(b-a)^n$$

$$G(b) = 0$$

$$G(a) = (b-a)^{n+1}$$

Note also that

$$F'(x) = f'(x) + \left[f''(x)(b-x) - f'(x) \right] + \left[\frac{f'''(x)}{2!}(b-x)^2 - f''(x)(b-x) \right]$$

$$+ \cdots + \left[\frac{f^{(n+1)}(x)}{n!}(b-x)^n - \frac{f^{(n)}(x)}{(n-1)!}(b-x)^{n-1} \right]$$

$$= \frac{f^{(n+1)}(x)}{n!}(b-x)^n$$

and

$$G'(x) = -(n+1)(b-x)^n.$$

The functions F and G satisfy the hypotheses of Cauchy's Mean Value Theorem, and therefore there is a point $c \in (a,b)$ for which

$$\frac{F'(c)}{G'(c)} = \frac{F(b) - F(a)}{G(b) - G(a)},$$

which we will rewrite in the form

$$F(b) = F(a) + \frac{F'(c)}{G'(c)} \left[G(b) - G(a) \right].$$

Making use of the above notes, we have the following.

$$f(b) = F(b)$$

$$= F(a) + \frac{\dfrac{f^{(n+1)}(c)}{n!}(b-c)^n}{-(n+1)(b-c)^n} \left[-(b-a)^{n+1} \right]$$

$$= \underbrace{f(a) + f'(a)(b-a) + \frac{f''(a)}{2!}(b-a)^2 + \cdots + \frac{f^{(n)}(a)}{n!}(b-a)^n}_{F(a)} + \frac{f^{(n+1)}(c)}{(n+1)!}(b-a)^{n+1}$$

Section 13.3

Theorem 🔍

Clairaut's Theorem

If $f(x,y)$ and the partial derivatives f_x, f_y, f_{xy}, and f_{yx} are all defined on an open region containing the point (a,b), and if f_{xy} and f_{yx} are continuous at (a,b), then $f_{xy}(a,b) = f_{yx}(a,b)$.

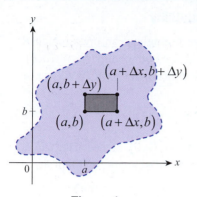

Figure 1

Proof 🖇

We begin by choosing $\Delta x \neq 0$ and $\Delta y \neq 0$ small enough so that the rectangle with vertices (a,b), $(a+\Delta x, b)$, $(a+\Delta x, b+\Delta y)$, and $(a, b+\Delta y)$ is entirely contained within the open region of the hypotheses (see Figure 1), and we define

$$\Delta = \left[f(a+\Delta x, b+\Delta y) - f(a+\Delta x, b) \right] - \left[f(a, b+\Delta y) - f(a,b) \right].$$

If we further define $g(x) = f(x, b+\Delta y) - f(x, b)$, then

$$\Delta = g(a+\Delta x) - g(a)$$

and by the Mean Value Theorem there is a point p between a and $a + \Delta x$ for which

$$\Delta = (\Delta x) g'(p).$$

Note that $g'(x) = f_x(x, b+\Delta y) - f_x(x,b)$, and by a second application of the Mean Value Theorem there is a point q between b and $b + \Delta y$ for which

$$g'(p) = (\Delta y) f_{xy}(p,q).$$

Substituting this into the expression above, we have

$$\Delta = (\Delta x)(\Delta y) f_{xy}(p,q).$$

However, if we define $h(y) = f(a+\Delta x, y) - f(a, y)$, then Δ can also be written as

$$\Delta = h(b+\Delta y) - h(b),$$

and a third application of the Mean Value Theorem tells us there is a point s between b and $b + \Delta y$ for which

$$\Delta = (\Delta y) h'(s).$$

Since $h'(y) = f_y(a+\Delta x, y) - f_y(a, y)$, one final application of the Mean Value Theorem implies there is a point r between a and $a + \Delta x$ for which

$$h'(s) = (\Delta x) f_{yx}(r,s),$$

and hence

$$\Delta = (\Delta y)(\Delta x) f_{yx}(r,s).$$

Equating the two expressions for Δ, we have $f_{xy}(p,q) = f_{yx}(r,s)$, where (p,q) and (r,s) are both inside the rectangle with dimensions Δx and Δy. The continuity of both f_{xy} and f_{yx} at (a,b) means that

$$f_{xy}(p,q) = f_{xy}(a,b) + \varepsilon_1 \quad \text{and} \quad f_{yx}(r,s) = f_{yx}(a,b) + \varepsilon_2$$

where each of ε_1 and ε_2 approach 0 as $(\Delta x, \Delta y) \to (0,0)$, and so it must be the case that

$$f_{xy}(a,b) = f_{yx}(a,b).$$

Section 13.3   **Theorem** 🔍

The Increment Theorem of Differentiability

If the partial derivatives of $f(x,y)$ exist throughout an open region R containing the point (a,b), and if f_x and f_y are both continuous at (a,b), then f is differentiable at (a,b). That is, the increment Δf at (a,b) can be written as

$$\Delta f = f_x(a,b)\Delta x + f_y(a,b)\Delta y + \varepsilon_1 \Delta x + \varepsilon_2 \Delta y$$

where each of ε_1 and ε_2, approach 0 as $(\Delta x, \Delta y) \to (0,0)$. Hence, if the partial derivatives are continuous throughout R, f is differentiable on all of R.

Analogous statements apply to functions of three or more variables.

Proof 📎

The increment Δf at (a,b) can be written in the form

$$\Delta f = f(a+\Delta x, b+\Delta y) - f(a,b),$$

and since we are interested in the form of the increment as Δx and Δy approach 0, we can assume they are already sufficiently small so that the line segment from (a,b) to $(a+\Delta x, b)$ and the line segment from $(a+\Delta x, b)$ to $(a+\Delta x, b+\Delta y)$ both lie entirely within the open region R (see Figure 2). We add and subtract $f(a+\Delta x, b)$ so that

$$\Delta f = f(a+\Delta x, b+\Delta y) - f(a+\Delta x, b) + f(a+\Delta x, b) - f(a,b),$$

and we define $g(y) = f(a+\Delta x, y)$ and $h(x) = f(x,b)$. By the Mean Value Theorem there exists a point p between b and $b+\Delta y$ and a point q between a and $a+\Delta x$ for which

$$g(b+\Delta y) - g(b) = (\Delta y)g'(p) \quad \text{and} \quad h(a+\Delta x) - h(a) = (\Delta x)h'(q).$$

Since $g'(y) = f_y(a+\Delta x, y)$ and $h'(x) = f_x(x,b)$, we now have

$$\begin{aligned}
\Delta f &= f(a+\Delta x, b+\Delta y) - f(a+\Delta x, b) + f(a+\Delta x, b) - f(a,b) \\
&= g(b+\Delta y) - g(b) + h(a+\Delta x) - h(a) \\
&= (\Delta y)g'(p) + (\Delta x)h'(q) \\
&= (\Delta y)f_y(a+\Delta x, p) + (\Delta x)f_x(q,b).
\end{aligned}$$

The fact that f_x and f_y are both continuous at (a,b) means that if we define

$$\varepsilon_1 = f_x(q,b) - f_x(a,b) \quad \text{and} \quad \varepsilon_2 = f_y(a+\Delta x, p) - f_y(a,b),$$

then both ε_1 and ε_2 have limits of 0 as $(\Delta x, \Delta y) \to (0,0)$ (note that $q \to a$ as $\Delta x \to 0$ and $p \to b$ as $\Delta y \to 0$). Hence,

$$\begin{aligned}
\Delta f &= (\Delta y)f_y(a+\Delta x, p) + (\Delta x)f_x(q,b) \\
&= (\Delta y)\left[f_y(a,b) + \varepsilon_2 \right] + (\Delta x)\left[f_x(a,b) + \varepsilon_1 \right] \\
&= f_x(a,b)\Delta x + f_y(a,b)\Delta y + \varepsilon_1 \Delta x + \varepsilon_2 \Delta y.
\end{aligned}$$

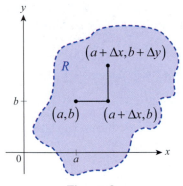

Figure 2

Answer Key

Chapter 1

Section 1.1

1. $\text{Dom} = \{-3, 0, 1\}, \quad \text{Ran} = \{-1, 0, 1, 2, 5\}$

3. $\text{Dom} = \{-2, 3, 4\}, \quad \text{Ran} = \{-8, 0, \cos 3, 5.98\}$

5. $\text{Dom} = \{\text{Tanisha, Don, Peter, David}\},$

$\text{Ran} = \left\{\begin{array}{l}\text{swimming, biking,} \\ \text{skating, skateboarding}\end{array}\right\}$

7. $\text{Dom} = \mathbb{Z}, \quad \text{Ran} = \{\ldots, -1, 1, 3, 5, \ldots\}$

9. $\text{Dom} = \mathbb{Z}, \quad \text{Ran} = \{\ldots, 3, 5, 7, 9, \ldots\}$

11. $\text{Dom} = \mathbb{R}, \quad \text{Ran} = \mathbb{R}$

13. $\text{Dom} = \{5\}, \quad \text{Ran} = \mathbb{R}$

15. $\text{Dom} = [-1, \infty), \quad \text{Ran} = \mathbb{R}$

17. $\text{Dom} = [-2, 4], \quad \text{Ran} = [-3, 3]$

19. $\text{Dom} = \mathbb{R}, \quad \text{Ran} = [0, 1]$

21. $\text{Dom} = \mathbb{R}, \quad \text{Ran} = \mathbb{R}$

23. $\text{Dom} = \{\text{Registered students}\},$
$\text{Ran} = \{\text{Courses offered}\}$

25. $\text{Dom} = \{\text{People in a certain group}\},$

$\text{Ran} = \left\{\begin{array}{l}\text{Fathers of the people} \\ \text{in the group}\end{array}\right\}$

27. $\{(1,1), (2,2), (3,3), (4,4), (5,5)\}$

29. $\left\{\begin{array}{l}(1,1), (1,2), (1,3), (1,4), (1,5), \\ (2,2), (2,4), (3,3), (4,4), (5,5)\end{array}\right\}$

31. Function

33. Not a function; -1 appears twice as first coordinate.

35. Not a function; it doesn't pass the vertical line test.

37. Function   **39.** Function   **41.** Function

43. Not a function; it doesn't pass the vertical line test.

45. Function   **47.** Function

49. Not a function; it doesn't pass the vertical line test. (However, x is a function of y.)

51. Function (F is a function of r.)

53. $y = -\dfrac{1}{3}x + \dfrac{10}{3}$ **55.** $y = x^2 + \dfrac{1}{3}x - \dfrac{5}{3}$

57. $y = \dfrac{3x+1}{x^2-1}$

59. a. $\dfrac{4}{3}$ **b.** $\dfrac{1}{3}x + \dfrac{7}{3}$

   c. $\dfrac{1}{3}(x+h) + 2$ **d.** $\dfrac{1}{3}$

61. a. 1 **b.** $x^2 + 2x - 2$

   c. $(x+h)^2 - 3$ **d.** $2x + h$

63. a. Undefined   **b.** $\sqrt{x+1}$

   c. $\sqrt{x+h}$ **d.** $\dfrac{\sqrt{x+h} - \sqrt{x}}{h}$

65. a. -1 **b.** $\dfrac{1}{x+2}$

   c. $\dfrac{1}{x+h+1}$

   d. $-\dfrac{1}{(x+1)(x+h+1)}$

67. $\text{Dom} = \mathbb{N}, \text{Cod} = \mathbb{N}, \ \text{Ran} = \{2, 3, 4, 5, \ldots\}$

69. $\text{Dom} = \mathbb{Z}, \text{Cod} = \mathbb{Z}, \ \text{Ran} = \{0, 1, 4, 9, 16, \ldots\}$

71. $\text{Dom} = [0, \infty), \ \text{Cod} = \mathbb{R}, \ \text{Ran} = [0, \infty)$

73. $(-\infty, -2) \cup (-2, 3) \cup (3, \infty)$

75. $(-\infty, 1) \cup (3, \infty)$ **77.** $[0, 2]$

79. $(-\infty, -3/2) \cup (-3/2, \infty)$

81. $\{\theta \in \mathbb{R} \mid \theta \neq 2k\pi, k \in \mathbb{Z}\}$

83. $C(r) = 2\pi r$ **85.** $C(F) = \dfrac{5}{9}(F - 32)$

87. $V(b, h) = \dfrac{1}{3}b^2 h$; a function of two variables

89. Function

91. Not a function   **93.** Not a function

95. Decreasing on $(-\infty, 1)$; increasing on $(1, \infty)$

97. Increasing on $(-\infty, -2), (2, \infty)$;
decreasing on $(-2, 2)$

99. Decreasing on $(-\infty, 1)$; increasing on $(1, \infty)$

101. Decreasing on $(-\infty, -1)$; constant on $(-1, 2)$;
increasing on $(2, \infty)$

103. Symmetric with respect to the x-axis

105. Symmetric with respect to both axes and the origin

107. Symmetric with respect to the origin

109. Symmetric with respect to the x-axis

111. $P(A) = 4\sqrt{A}$

113. $V(x) = x(30 - 2x)(20 - 2x)$

115. $V(r) = \dfrac{2\pi r^3}{3}$ **117.** $C(F) = \dfrac{5}{9}(F - 32)$

Section 1.2

1. Deg.: 1; lead. coeff.: $1/2$;
x-int.: 3; y-int.: $-3/2$;
Ran $= \mathbb{R}$

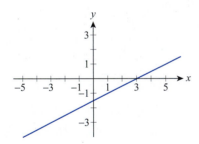

3. Deg.: 2; lead. coeff.: 2;
x-int.: $2, -1/2$; y-int.: -2;
Ran $= [-25/8, \infty)$

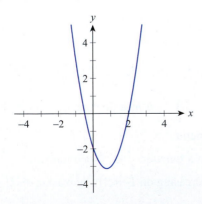

5. Deg.: 3; lead. coeff.: 1;
x-int.: $-3, 1, 2$; y-int.: 6;
Ran $= \mathbb{R}$

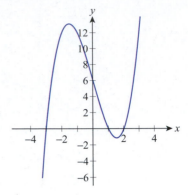

7. Deg.: 4; lead. coeff.: $1/4$;
x-int.: $\pm 2\sqrt{2}, 0$; y-int.: 0;
Ran $= [-4, \infty)$

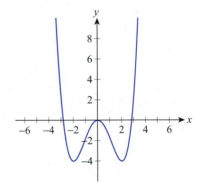

9. Vert. asym.: $x = 1$;
horiz. asym.: $y = 0$;
x-int.: none; y-int.: -5

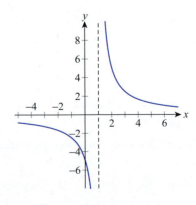

11. Vert. asym.: $x = -3$;
    slant asym.: $y = x - 3$;
    x-int.: none; y-int.: 1

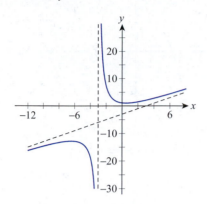

13. Vert. asym.: $x = -1/2$;
    horiz. asym.: $y = 1/2$;
    x-int.: -1; y-int.: 1

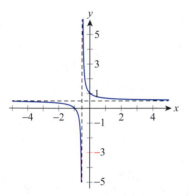

15. Vert. asym.: $x = -1$;
    slant asym.: $y = x + 1$;
    x-int.: 0, -2; y-int.: 0

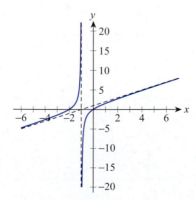

17. Answers will vary.   19. Answers will vary.

21. Answers will vary.   23. Answers will vary.

25. $\cos x$ 27. $\cos^2 \alpha$ 29. $\cos^2 \theta$ 31. $\csc \beta$

33. $\tan^2 x$

35.

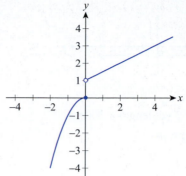

37.

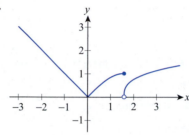

39. $f(x) = \begin{cases} -x+1 & \text{if } x < 1 \\ x-1 & \text{if } x \geq 1 \end{cases}$

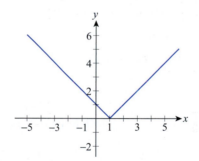

41. $h(x) = \begin{cases} -\sin x & \text{if } 2\pi n - \pi < x < 2\pi n \\ \sin x & \text{if } 2\pi n \leq x \leq 2\pi n + \pi \end{cases}$, $\quad n \in \mathbb{Z}$

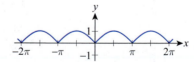

43.

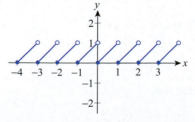

45.

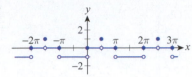

47. 3.5 hours

49. Width = 50 yd; length = 100 yd

51. 50 ft by 100 ft

53. $(2,1)$

55. 12 and 24

57. 25 sets

59. No; yes

61. $y = -4\cos 6\pi t$; it takes a mere $1/3$ s to complete a period.

63. a. 3 years **b.** 9 years

65. $V \approx 178$ people

67. a. 0.999567 **b.** 0.958 g **c.** 0.648 g

69. Approx. $134,392

71. a. 0.965936 **b.** 0.707 kg **c.** 7.628 mg

73. True

75. False; the line perpendicular to $y = Ax + B$ has a slope of $-1/A$.

77. True

79. False; consider $\dfrac{x^4}{1+x^2}$, which has no asymptotes.

81. True

Section 1.3

1.

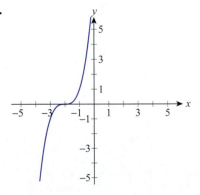

Dom = Ran = $\mathbb{R}$

3.

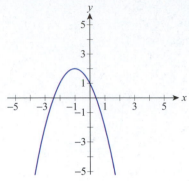

Dom = $\mathbb{R}$, Ran = $(-\infty, 2]$

5.

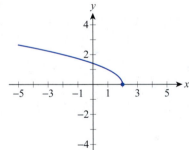

Dom = $\mathbb{R}$, Ran = $[0, \infty)$

7.

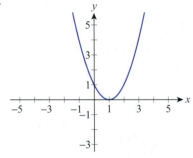

Dom = $(-\infty, 2]$, Ran = $[0, \infty)$

9.

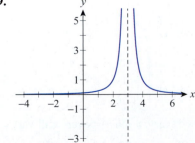

Dom = $(-\infty, 3) \cup (3, \infty)$, Ran = $(0, \infty)$

11.

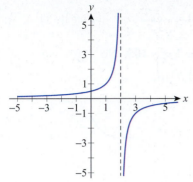

$\text{Dom} = (-\infty, 2) \cup (2, \infty), \quad \text{Ran} = (-\infty, 0) \cup (0, \infty)$

13.

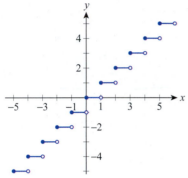

$\text{Dom} = \mathbb{R}, \quad \text{Ran} = \mathbb{Z}$

15.

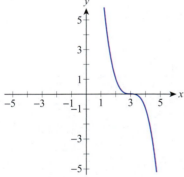

$\text{Dom} = \text{Ran} = \mathbb{R}$

17.

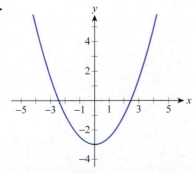

$\text{Dom} = \mathbb{R}, \quad \text{Ran} = [-3, \infty)$

19.

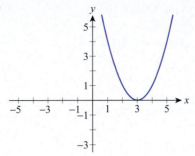

$\text{Dom} = \mathbb{R}, \quad \text{Ran} = [0, \infty)$

21.

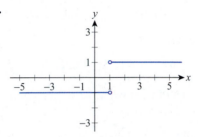

$\text{Dom} = (-\infty, 1) \cup (1, \infty), \quad \text{Ran} = \{-1, 1\}$

23.

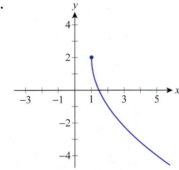

$\text{Dom} = [1, \infty), \quad \text{Ran} = (-\infty, 2]$

25. $y = -x^2 - 6$ **27.** $y = -\sqrt{x + 5}$

29. $y = -|-x + 7|$ **31.** $f(x) = -\sqrt{x + 4}$

33. $f(x) = 1 - (x - 3)^3$

35. a. -1 **b.** 1 **c.** 0 **d.** 0

37. a. -3 **b.** 1 **c.** 2 **d.** $1/2$

39. a. -4 **b.** 8 **c.** -12 **d.** $-1/3$

41. a. 4 **b.** 3 **c.** $7/4$ **d.** 7

43. a. -1 **b.** 3 **c.** -2 **d.** $-1/2$

45. a. -1 **b.** -3 **c.** -2 **d.** -2

47. a. $x^2 + \sqrt[3]{x} - 1$; $\text{Dom} = \mathbb{R}$

b. $\dfrac{x^2 - 1}{\sqrt[3]{x}}$; $\text{Dom} = (-\infty, 0) \cup (0, \infty)$

49. a. $x^{3/2} + x - 3$; Dom $= [0, \infty)$

b. $\dfrac{x^{3/2}}{x-3}$; Dom $= [0,3) \cup (3, \infty)$

51. a. $x^3 + \sqrt{x-2} + 4$; Dom $= [2, \infty)$

b. $\dfrac{x^3 + 4}{\sqrt{x-2}}$; Dom $= (2, \infty)$

53. a. $x^{2/3} + 6x - 1$; Dom $= \mathbb{R}$

b. $\dfrac{6x-1}{x^{2/3}}$; Dom $= (-\infty, 0) \cup (0, \infty)$

55. $-1699/100$ **57.** Undefined   **59.** $-1/3$

61. Undefined   **63.** -324

65. π^2 **67.** 4   **69.** 5   **71.** 3   **73.** 2

75. a. $\dfrac{1}{x-1}$; Dom $= (-\infty, 1) \cup (1, \infty)$

b. $\dfrac{1}{x} - 1$; Dom $= (-\infty, 0) \cup (0, \infty)$

77. a. $1 - \sqrt{x}$; Dom $= [0, \infty)$

b. $\sqrt{1-x}$; Dom $= (-\infty, 1]$

79. a. $9x^4 + 36x^2 + 35$; Dom $= \mathbb{R}$

b. $3x^4 + 12x^3 + 12x^2 + 5$; Dom $= \mathbb{R}$

81. a. $\sqrt{2x}$; Dom $= [0, \infty)$

b. $2\sqrt{x}$; Dom $= [0, \infty)$

83. a. x; Dom $= (-\infty, 0) \cup (0, \infty)$

b. x; Dom $= (-\infty, 0) \cup (0, \infty)$

85. a. x; Dom $= \mathbb{R}$ **b.** x; Dom $= \mathbb{R}$

87. a. $\dfrac{3}{1 - 3x^2}$;

Dom $= \left(-\infty, -\sqrt{3}/3\right) \cup \left(-\sqrt{3}/3, \sqrt{3}/3\right) \cup \left(\sqrt{3}/3, \infty\right)$

b. $\dfrac{27}{x^2 - 2x + 1}$; Dom $= (-\infty, 1) \cup (1, \infty)$

89. $g(x) = \dfrac{2}{x}$, $h(x) = 5x - 1$, $f(x) = g(h(x))$

91. $g(x) = x + \sqrt{x} - 7$, $h(x) = x + 2$, $f(x) = g(h(x))$

93. $g(x) = \dfrac{\sqrt{x}}{x^2}$, $h(x) = x - 3$, $f(x) = g(h(x))$

95. $S(t) = \pi r \sqrt{r^2 + \dfrac{t^4}{16}}$

97. $c(t) = 2040 + 17,400t - 43.5t^2$

99. $(f \circ g)(x) = \dfrac{-x}{\sqrt[3]{3x^2 - 9}} = -(f \circ g)(-x)$

101. Yes   **103.** Yes   **105.** No   **107.** No

109. True   **111.** True

113.
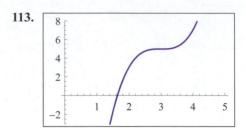

115.
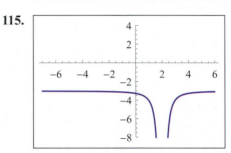

117.
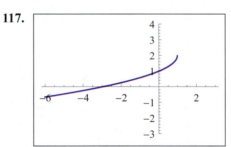

119. $y = -(x + 3)^2 + 5$

121. $y = -x^3 + 7$ **123.** $y = (x - 2)^3 - 4$

125. $(f + g)(x) = 4 + 12x + 9x^2 + \sqrt{x^2 + 5}$;

$(fg)(x) = 4\sqrt{x^2 + 5} + 12x\sqrt{x^2 + 5} + 9x^2\sqrt{x^2 + 5}$;

$(f \circ g)(x) = 49 + 9x^2 + 12\sqrt{x^2 + 5}$;

$(g \circ f)(x) = \sqrt{(3x + 2)^4 + 5}$

127. $(f + g)(x) = \dfrac{2x^2 - x + 1}{x(x-1)}$;

$(fg)(x) = \dfrac{x + 1}{x}$;

$(f \circ g)(x) = 1 - 2x$;

$(g \circ f)(x) = \dfrac{2}{x + 1}$

Section 1.4

1.

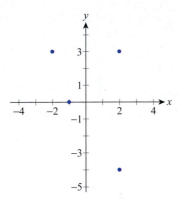

$$\text{Dom} = \{-2, -1, 2\}, \quad \text{Ran} = \{-4, 0, 3\}$$

3.

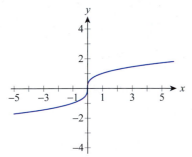

$$\text{Dom} = \text{Ran} = \mathbb{R}$$

5.

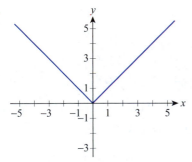

$$\text{Dom} = \mathbb{R}, \quad \text{Ran} = [0, \infty)$$

7.

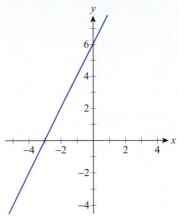

$$\text{Dom} = \text{Ran} = \mathbb{R}$$

9.

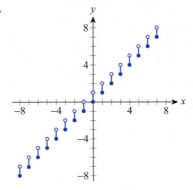

$$\text{Dom} = \mathbb{Z}, \quad \text{Ran} = \mathbb{R}$$

11.

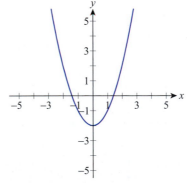

$$\text{Dom} = \mathbb{R}, \quad \text{Ran} = [-2, \infty)$$

13. Restrict to $[0, \infty)$

15. Inverse exists   **17.** Inverse exists

19. Inverse exists   **21.** Inverse exists

23. $f^{-1}(x) = (x+2)^3$ **25.** $r^{-1}(x) = \dfrac{-2x-1}{3x-1}$

27. $F^{-1}(x) = (x-2)^{1/3} + 5$ **29.** $V^{-1}(x) = 2x - 5$

31. $h^{-1}(x) = (x+2)^{5/3}$

33. $J^{-1}(x) = \dfrac{x-2}{3x}$

35. $h^{-1}(x) = (x-6)^{1/7}$

37. $r^{-1}(x) = x^5/2$

47. F   **49.** C   **51.** D   **53.** G

55. F   **57.** C   **59.** A

61.

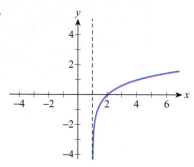

63.

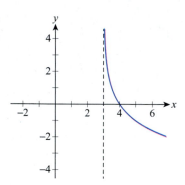

65.

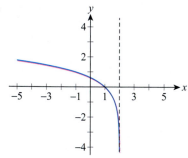

67.

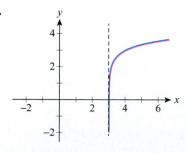

69.

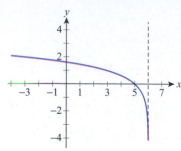

71.

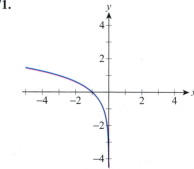

73. 2   **75.** 7   **77.** 1   **79.** 2.45   **81.** 3.30

83. 0.74   **85.** x^2 **87.** $e^2 p/x$ **89.** x^3/y^4

91. $\dfrac{3}{2}\ln x + \ln p + 5\ln q - 7$ **93.** $\log(2+3\log x)$

95. $1 - \dfrac{1}{2}\log(x+y)$ **97.** $\log_2\left(y^2+z\right) - 4 - 4\log_2 x$

99. $2\log_b x + \dfrac{1}{2}\log_b y - \log_b z$ **101.** $\ln\dfrac{3p}{q^2}$

103. $\log\dfrac{x-10}{x}$ **105.** $\ln\dfrac{e^2}{x^3 y^3}$

107. $-\pi/2$ **109.** $-\pi/6$ **111.** $\pi/4$

113. Does not exist   **115.** -1.3734 **117.** 1.0472

119. $-1/2$ **121.** 2   **123.** 2

125. $\dfrac{\sqrt{x^2-4}}{2}$ **127.** $\dfrac{\sqrt{3}x}{3}$ **129.** $\dfrac{4}{\sqrt{x^2+16}}$

131.

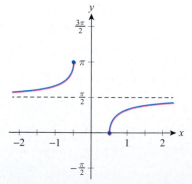

133.

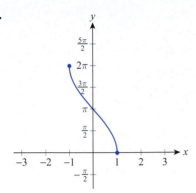

135. 521 73 136 136 73 392 316 −7 73 188 392
217 8 −7 617

137. REMEMBER YOUR SUNBLOCK

139. 6.3   **141.** 13.4 decibels

143. **a.** 0.2898   **b.** 0.4429   **c.** 0.7956

145. True   **147.** True

149. False; the domain of $\arcsin x$ is $[-1,1]$.

151. True

Section 1.5

13. **a.** $y = 3.559x + 22.038$ **b.** 36,274 units

15. **a.** $y = 10.743x - 25.829$ **b.** \$81,601

17. **a.** $y = -1.345x + 79.771$
b. 69.01 cents per pound

19. $x = 0$ **21.** $x = \dfrac{(2k+1)\pi}{6} + \dfrac{2}{3},\ k \in \mathbb{Z}$

23. $x = -1$ **25.** $x = \pm\sqrt[4]{1/2}$ **27.** $x \le 0$

29. Answers will vary.   **31.** Answers will vary.

33. Answers will vary.   **35.** Answers will vary.

37. Answers will vary.   **39.** Answers will vary.

41. 20   **43.** −21.9787   **45.** −90.8519

47. 49.9990   **49.** 130,783   **51.** 1

53. Answers will vary.   **55.** Answers will vary.

57. Answers will vary.

59. −4.4213, −0.1001, 4.5213

61. 1.1974, 5.6240, 6.8003   **63.** 0.1586, 3.1462

65. $f(x)$ **67.** $g(x)$ **69.** $g(x)$

71. Answers will vary.   **73.** Answers will vary.

Chapter 1 Review

1. $\text{Dom} = \{-2, -3\}, \quad \text{Ran} = \{-9, -3, 2, 9\}$;
not a function

3. $\text{Dom} = [-6, \infty), \quad \text{Ran} = \mathbb{R}$; not a function

5. $\text{Dom} = \mathbb{N}, \text{Cod} = \mathbb{R}, \quad \text{Ran} = \left\{ \left. \dfrac{3x}{4} \right| x \in \mathbb{N} \right\}$

7. $\text{Dom} = \mathbb{R}, \text{Cod} = \mathbb{R}, \quad \text{Ran} = (0,1]$

9. **a.** $2x^2 + 6x - 8$ **b.** $2x^4 + 10x^2$
c. $4x + 2h + 10$

11. **a.** $\dfrac{3}{x+1}$ **b.** $\dfrac{3}{x^2+2}$ **c.** $\dfrac{-3}{(x+h+2)(x+2)}$

13. Decreasing on $(-\infty, 2)$; increasing on $(2, \infty)$

15. Odd

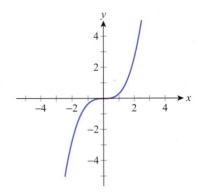

17. Odd

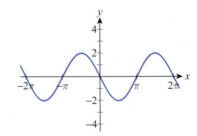

19. Symmetric with respect to the y-axis

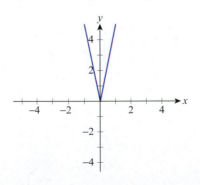

21.

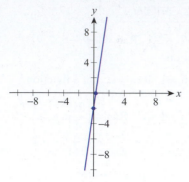

23.

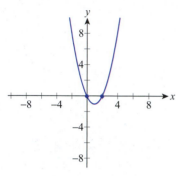

25.

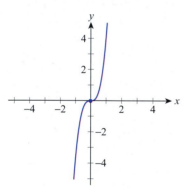

27.

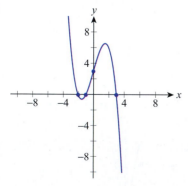

29.

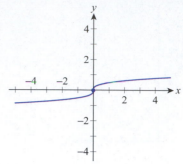

31.

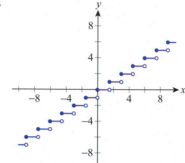

33.

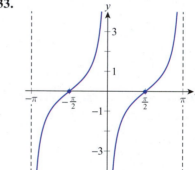

35.

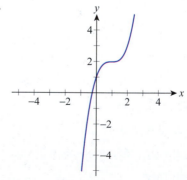

$\text{Dom} = \mathbb{R}, \text{Ran} = \mathbb{R}$

37. $y = \dfrac{1}{x-2} + 3$

39. a. $x^2 + \sqrt{x}$; $\text{Dom} = [0, \infty)$

  b. $x^{3/2}$; $\text{Dom} = (0, \infty)$

41. a. $x + 2$ **b.** $x - 2$ **c.** 5

43. $g(x) = \dfrac{\sqrt{x} + 2}{x^2}$, $h(x) = x + 3$; $f(x) = g(h(x))$

45.

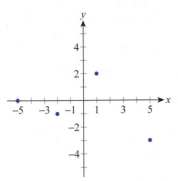

$\text{Dom} = \{-5, -2, 1, 5\}$, $\text{Ran} = \{-3, -1, 0, 2\}$

47. $f^{-1}(x) = \dfrac{x+2}{7x}$ **49.** $f^{-1}(x) = (x+6)^5$

53. $\dfrac{\sqrt{x^2 - 4}}{2}$ **55.** $\dfrac{1}{3}(\log y + 2\log z - 4\log x)$

57. $A(t) = A_0 e^{-\frac{\ln 2}{10}t} = A_0 (1/2)^{t/10}$; approx. 17.7%

59. $15/2$ and $15/2$ **61.** Yes

63. False; a function may not be defined on (any subset of) $\mathbb{R}$ and thus may not have a graphical representation in the xy-system. Even graphs of numerical functions are not always curves.

65. False; consider $f(x) = x^2$ with $a = 1$ and $b = -1$.

67. False; consider $f(x) = (x-1)^2$.

69. False; see Example 10 of Section 1.3.

71. False; consider $c = 2$ and $f(x) = x^2$:
$f(cx) = (2x)^2 = 4x^2$, while $cf(x) = 2x^2$.

73. False; consider $y = x^2$. (The statement is only true for invertible functions.)

75. True

77.

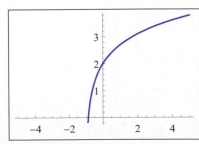

79.

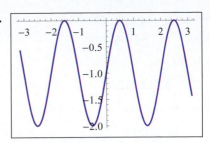

81. 1.4262 **83.** Answers will vary.

Chapter 1 Project

1. $r(t) = 2.6t$

3. Doubling time is approximately 24 minutes and 51 seconds; the time to triple is approximately an additional 19 minutes and 4 seconds. (The exact values are $\sqrt{2} - 1$ hours and $\sqrt{3} - \sqrt{2}$ hours, respectively.)

5. After 3 hours: radius 7.8 km and area approx. 191.13 km^2
After 5.5 hours: radius 14.3 km and area approx. 642.42 km^2

7. Approx. 286.70 km^2/h

9. Answers will vary, but a major flaw is that the model predicts a hole whose area will grow indefinitely, and is not a periodic function of time. A better approach is to use a periodic function.

Chapter 2

Section 2.1

1. 3 **3.** 0 **5.** $-1/4$ **7.** The exact answer is -2.

9. The exact answer is 2.

11. The exact answer is 2.

13. The exact answer is $1/e$ or approx. 0.3679.

15. The exact answer is $1/(5\ln 10)$ or approx. 0.0869.

17. The exact answer is -4.

19. a. 96 ft **b.** 16 ft/s **c.** 16 **d.** 2.5 s

21. a. 176 ft **b.** 176 ft/s **c.** 192 ft/s
  d. 64 ft/s **e.** 6 s

23. a. 3 m/s **b.** 15 m/s **c.** $6t_0 + 3$

25. a. 5 s; 45 m/s **b.** 10 s **c.** $-1.83g$

27. a. 87.35 ft   **b.** 34.94 ft/s
c. −28 ft/s   **d.** −77.47 ft/s

29. a. 60   **b.** 61.44

31. a. 12.7234   **b.** 12.3129

33. a. 1.0536   **b.** 1.0563

35. a. 1.2217   **b.** 1.2218

37.

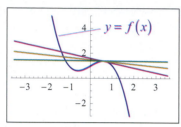

39.

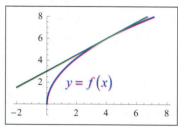

41.

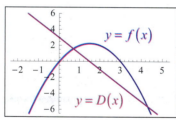

The function value $D(x_0)$ at any given x_0 is approximately equal to the slope of the graph of f at x_0.

43.

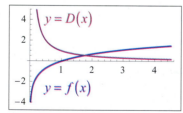

The function value $D(x_0)$ at any given x_0 is approximately equal to the slope of the graph of f at x_0, for any $x_0 > 0$.

45. $x = \pm 2$ **47.** $x = 1$

49. a. 63.9936   **b.** 63.9999

51. a. 11.0508   **b.** 11.0050

53. a. 1.0612   **b.** 1.0612

55. a. 1.222   **b.** 1.222

Section 2.2

1. 4   **3.** 0

5.

x	y
1	2.414
1.4	2.814
1.41	2.824
1.414	2.828

The table points to a limit of $2\sqrt{2} \approx 2.828$.

7.

x	y
1.5	113.33
1.1	15.94
1.01	10.46
1.001	10.045

The table points to a limit of 10.

9.

x	y
3	−3.4696
3.14	−3.0048
3.141	−3.0018
3.1415	−3.0003

The table points to a limit of −3.

11.

x	y
7.5	210.5
7.1	994.1
7.01	9814.0
7.001	98014.0

The table points to an undefined limit.

13. $\lim\limits_{x \to 1^-} f(x) = -\infty$; $\lim\limits_{x \to 1^+} f(x) = \infty$

15. $\lim\limits_{x \to 3^-} h(x) = \infty$; $\lim\limits_{x \to 3^+} h(x) = \infty$

17. $\lim\limits_{x \to -1^-} q(x) = \infty$; $\lim\limits_{x \to -1^+} q(x) = -\infty$

19. $\lim\limits_{x \to 1.5^-} v(x) = -\infty$; $\lim\limits_{x \to 1.5^+} v(x) = \infty$

21. $\lim\limits_{x \to \left(\frac{\pi}{2}+k\pi\right)^-} \tan x = \infty$; $\lim\limits_{x \to \left(\frac{\pi}{2}+k\pi\right)^+} \tan x = -\infty$ $(k \in \mathbb{Z})$

23. $\lim\limits_{x \to 6^-} s(x) = -\infty$; $\lim\limits_{x \to 6^+} s(x) = \infty$

25. $\lim\limits_{x \to \infty} f(x) = 0$; $\lim\limits_{x \to -\infty} f(x) = 0$

27. $\lim_{x\to\infty} h(x) = 0$; $\lim_{x\to-\infty} h(x) = 0$

29. $\lim_{x\to\infty} q(x) = -1$; $\lim_{x\to-\infty} q(x) = -1$

31. $\lim_{x\to\infty} v(x) = \infty$; $\lim_{x\to-\infty} v(x) = -\infty$

33. $\lim_{x\to\infty} \tan x$ does not exist; $\lim_{x\to-\infty} \tan x$ does not exist

35. $\lim_{x\to\infty} s(x) = 1$; $\lim_{x\to-\infty} s(x) = -1$

37. a. -2 **b.** 1 **39. a.** 1 **b.** 4

41. a. 1 **b.** $-\infty$ **43. a.** 1 **b.** 1

45. a. Does not exist   **b.** 0

47. $\lim_{x\to\infty} f(x) = \infty$; $\lim_{x\to-\infty} f(x) = -\infty$

49. $\lim_{x\to\infty} h(x) = -\infty$; $\lim_{x\to-\infty} h(x) = -\infty$

51. $\lim_{x\to\infty} F(x) = \infty$

53. $\lim_{x\to\infty} H(x) = \infty$; $\lim_{x\to-\infty} H(x) = \infty$

55. $\lim_{x\to\infty} u(x) = \infty$; $\lim_{x\to-\infty} u(x) = \infty$

57. $\lim_{x\to-\infty} s(x) = -\infty$

59. False; see $h(x)$ of Example 5 at $x = 2$.

61. True   **63.** True   **65.** Does not exist

67. 2 **69.** $3/2$ **71.** 0

75. See the answer given for Exercise 65.

77. See the answer given for Exercise 67.

79. See the answer given for Exercise 69.

81. See the answer given for Exercise 71.

Section 2.3

1. $\delta \approx 0.23$ or smaller   **3.** $\delta \approx 0.2$ or smaller

5. L is incorrectly quantified and switched with c.

7. ε is incorrectly quantified.

9. The inequality $0 \le |x - c|$ is incorrect.

11. $\delta = 0.02$ or smaller   **13.** $\delta = 0.1$ or smaller

15. $\delta = \sqrt{0.1}$ or smaller   **17.** $\delta = 0.\overline{09}$ or smaller

19. $\delta = e^{0.1} - 1 \approx 0.1052$ or smaller

21. $N = 10$ or larger   **23.** $N = -10$ or smaller

25. $N = \ln 0.1$ or smaller   **27.** $\delta = 0.1$ or smaller

29. $\delta = 0.1$ or smaller

31. $\delta = \dfrac{\pi}{2} - \arctan 100 \approx 0.0099997$ or smaller

63. The limit does not exist.

65. The limit does not exist.

67. The limit is 0.   **69.** 0.02256 mm

71. False; the function value and limit at c need not be equal. (See Example 2.)

73. False; consider $f(x) = -x^2$ and $g(x) = x^2$ at $c = 0$.

75. Answers will vary.   **77.** Answers will vary.

79. Answers will vary.   **81.** Answers will vary.

83. Answers will vary.

85. Vert. asym.: $x \approx -1.3340$, $x \approx 1.1759$

87. Vert. asym.: $x = (2n+1)\pi - 6$, $n \in \mathbb{Z}$

89. Vert. asym.: $x = (2n-1)\pi/2$, $n \in \mathbb{Z}, n \ne 0$

Section 2.4

1. a. -4 **b.** 21 **3.** 5 **5.** 7 **7.** 9

9. 22 **11.** -4 **13.** -2 **15.** $\sqrt[3]{2}$

17. $\sqrt[3]{100}$ **19.** 16 **21.** 12 **23.** 11

25. $33/5$ **27.** $1/6$ **29.** $1/(2\sqrt{5})$ **31.** $1/9$

33. 6 **35.** 4 **37.** $1/(2\sqrt{x})$ **39.** $32/7$

41. $-1/6$ **43.** -1 **45.** 8 **47.** 11

49. $9 - 2(c-2)$ **51.** 1 **53.** 0 **55.** 0 **57.** $2e^2$

71. The limit is 0.   **79.** $r/2$

Section 2.5

1. Points of continuity: $(-\infty, 0) \cup (0, 3) \cup (3, \infty)$

   Points of discontinuity: $c = 0$, $\lim_{x\to 0} f(x)$ does not exist; $c = 3$, $\lim_{x\to 3} f(x) \ne f(3)$

3. Answers will vary.   **5.** $c = 0$, nonremovable

7. $c = 3$, removable   **9.** $c = 2$, nonremovable

11. None

13. $c = -(2n+1)\pi/2$ for $n \in \{-1, 0, 1, 2, \ldots\}$, nonremovable

15. $c = 4$, removable   **17.** None   **19.** None

21. $c = \pm 1$, nonremovable

23. None   **25.** All integers, nonremovable

27. $c = \pm\sqrt{n}$, n a positive integer, nonremovable

29. $c = 1/n$, n a nonzero integer, nonremovable

35. Continuous on $[-3,-1] \cup (0,2] \cup [3,\infty)$

37. Continuous on $(-\sqrt{3},\sqrt{3})$

39. Continuous except on
$\{n - 1/\pi \mid n \in \mathbb{Z}\} \cup \{\ln k - 2 \mid k \in \mathbb{N}\}$

45. $f(3) = 7$ will make f continuous.

47. $h(1) = 2$ will make h continuous.

49. $G(0) = 0$ will make G continuous.

51. Continuous on $[-4,4]$

53. Not continuous at the endpoints

55. $a = 5$

57. $a = 3/2$, $b = -5/2$

63. Yes; $c = 2$

65. Yes; $c = 2$

67. No; G is not continuous on $[-2,2]$

83. a. Because speeds of everyday objects are smaller than c by orders of magnitude, the denominator of ΔT is approximately 1.

b. No moving object can reach the speed of light.

85. False; consider the Dirichlet function.

87. True   **89.** -0.2895

91. 0.8241 **93.** -1.7321

95.

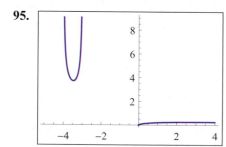

97.

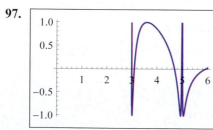

99.

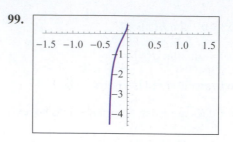

Section 2.6

1. a. $f'(-1) = 2$ **b.** $f'(1) = -2$

3. $y = 4x - 6$ **5.** $y = \dfrac{1}{2}x + 4$

7. $y = 12x - 16$ **9.** $y = \dfrac{1}{2}x + 1$

11. $y = -4x + 4$ **13.** $y = -\dfrac{1}{16}x + \dfrac{3}{4}$

15. $f'(x) = 0$; tangent is horizontal for all x

17. $f'(x) = 4$; tangent is never horizontal

19. $f'(x) = 6x$; tangent is horizontal at $x = 0$

21. $f'(x) = x + 5$; tangent is horizontal at $x = -5$

23. $f'(x) = 3x^2 + 1$; tangent is never horizontal

25. $f'(x) = 4x^3$; tangent is horizontal at $x = 0$

27. $f'(x) = -\dfrac{10}{(2x-4)^2}$; tangent is never horizontal

29. $f'(x) = -\dfrac{7}{(x-3)^2}$; tangent is never horizontal

31. $f'(x) = -\dfrac{2x}{(x^2+1)^2}$; tangent is horizontal at $x = 0$

33. $f'(x) = \dfrac{5}{2\sqrt{5x}}$; tangent is never horizontal

35. $f'(x) = \dfrac{1}{\sqrt{2x+1}}$; tangent is never horizontal

37. $f'(x) = \dfrac{x}{\sqrt{x^2+1}}$; tangent is horizontal at $x = 0$

39. $y = 6x - 6$

41. $y = -\dfrac{1}{2}x + 1$ or $y = -\dfrac{1}{2}x - 1$

43. $54y + x = 27$ **45.** $f'(3.6) = -1/4$

47. $h'(3) = 10$ **49.** $G'(7) = 1/2$

51. $u'(-3) = -1/2$ **53.** $w'(5) = -1/54$

55. $G'(-2) = -32$ **57.** C   **59.** D

61. Answers will vary.   **63.** Answers will vary.

65. Answers will vary.

73. $t = 1$, $t = 5$; the particle stops at 2 s and at 4 s.

75. a. The respective velocities are $16\ \text{ft/s}$ and $-16\ \text{ft/s}$; the speed is $16\ \text{ft/s}$ in both cases.

b. It rises 36 ft above the cliff and reaches the bottom at approx. 4.91 s.

c. Approx. $-109.12\ \text{ft/s}$

77. She seems to have stopped for an extended time period (probably a red light); soon after, she turned around and sped back home (perhaps forgot something at home). Then she sped back to class, this time managing to do so without a red light, and also driving faster.

79. a. $P(x) = 0.1x^2 + 40x - 190$

b. Approx. 5 suitcases

c. $P(25) = 872.50$, $P(30) = 1100$, $P(40) = 1570$

d. $P'(x) = 0.2x + 40$

e. $P'(25) = 45$, $P'(30) = 46$, $P'(40) = 48$

81. $C(x) = 20 + 3x$; $C'(x) = 3$

83.

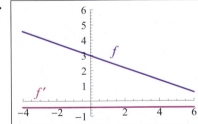

85.

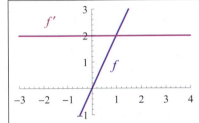

87.

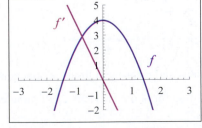

89.

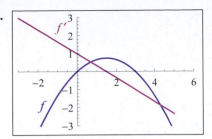

91.

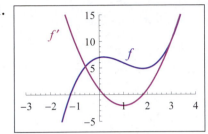

93.

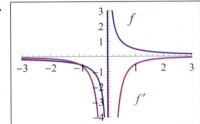

95.

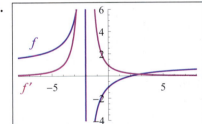

97.

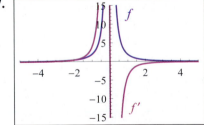

99.

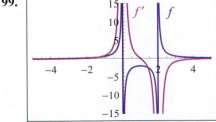

101.

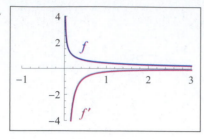

103.

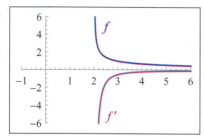

105.

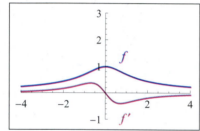

Chapter 2 Review

1. The limit is 3; the function is continuous.

3. The limit is 2; there is a nonremovable (jump) discontinuity at 1.

5. The exact answer is 3.

7. The exact answer is $1/2$.

9. a. 598 ft   **b.** 576 ft/s

   c. 608 ft/s   **d.** 544 ft/s

   e. At 19 seconds

11. 1.5568

13. 21.7102

15.

x	y
0.5	0.7071
0.1	0.7943
0.01	0.9550
0.001	0.9931

The table points to a limit of 1.

17.

x	y
0.5	0.47943
0.1	0.11969
0.01	−0.00264
0.001	−0.00194

The table points to a limit of 0.

19. $\delta = 0.005$ or smaller

25. 14   **27.** 7   **29.** 8   **31.** $-1/10$

33. ∞ **35.** $1/4$ **37.** -2 **39.** 5   **41.** $2x$

45. Answers will vary.   **47.** $c = 4$, nonremovable

49. All integers, nonremovable

51. No discontinuities   **57.** $y = 3x - 1$

59. $f'(x) = 2 - 2x$; tangent is horizontal at $x = 1$

61. Answers will vary.

65. a. $P(x) = 0.1x^2 + 15x - 247.5$

   b. 15 toys   **c.** $P'(x) = 0.2x + 15$

67. False; there could be, for example, a jump or oscillating discontinuity at $x = c$.

69. False; the condition $B \neq 0$ is needed.

71. True

73. False; while $\lim_{x \to c} f(x) = L$ does follow, $f(c)$ doesn't even have to be defined.

75. 5.3586 and 21.7491, respectively

77. −0.6891 and 0.7729, respectively

79. 1   **81.** 3

Chapter 2 Project

1.

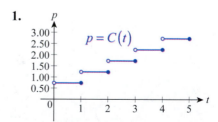

3. No, the left- and right-hand limits are unequal.

5. A call lasting for about two and a half minutes costs $1.75.

7. A call lasting just over three minutes costs $2.25.

9. Does not exist

Chapter 3

Section 3.1

1. $\dfrac{df}{dx}=0;\ \dfrac{df}{dx}\Big|_{x=1}=0$ **3.** $\dfrac{dh}{dx}=x;\ \dfrac{dh}{dx}\Big|_{x=0}=0$

5. $\dfrac{dG}{ds}=s^2-1;\ \dfrac{dG}{ds}\Big|_{s=-3}=8$

7. $\dfrac{dK}{dz}=\dfrac{-15}{(3z+1)^2};\ \dfrac{dK}{dz}\Big|_{z=0}=-15$

9. $\dfrac{dw}{dz}=\dfrac{-2z}{(z^2+2)^2};\ \dfrac{dw}{dz}\Big|_{z=\sqrt{2}}=\dfrac{-\sqrt{2}}{8}$

11. $\dfrac{dQ}{dy}=\dfrac{-1}{2y\sqrt{3y}};\ \dfrac{dQ}{dy}\Big|_{y=1}=\dfrac{-\sqrt{3}}{6}$

13. $D_x f(x)=0$ **15.** $D_t h(t)=3t+10$

17. $D_y G(y)=-1/(3y^2)$ **19.** $D_z S(z)=6/z^3$

21. $D_v R(v)=1/\sqrt{2v-3}$ **23.** $D_y X(y)=8y^3$

25. $f'(x)=5/2;\ f''(x)=0;\ f'''(x)=0$

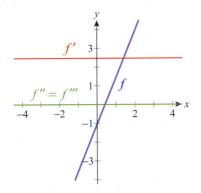

27. $h'(x)=-x+1;\ h''(x)=-1;\ h'''(x)=0$

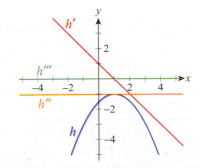

29. $V'(x)=(x-2)^2;\ V''(x)=2x-4;\ V'''(x)=2$

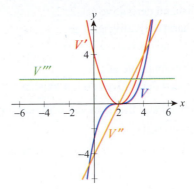

31. $G'(x)=8(x-1)^3;\ G''(x)=24(x-1)^2;$
  $G'''(x)=48x-48$

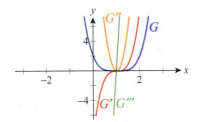

33. $K'(s)=2/(s-1)^2;\ K''(s)=-4/(s-1)^3;$
  $K'''(s)=12/(s-1)^4$

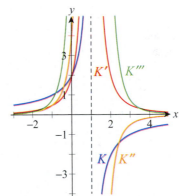

35. Position: k; velocity: g; acceleration: h; jerk: f

37. Position: k; velocity: h; acceleration: g; jerk: f

39. Differentiable on $(-\infty,3)\cup(3,\infty)$

41. Differentiable on $(-\infty,-1)\cup(-1,3)\cup(3,\infty)$

43. Differentiable on $(-\infty,1)\cup(1,\infty)$

45. Not differentiable at −2 and 4
Left-hand derivative at −2: 0
Right-hand derivative at −2: 2
Left-hand derivative at 4: 2
Right-hand derivative at 4: 0

47. Not differentiable at 1.5
One-sided derivatives do not exist at 1.5

49. Undefined and therefore not differentiable on
$\left(-\infty, -\sqrt{3}\right) \cup \left(\sqrt{3}, \infty\right)$
One-sided derivatives do not exist at $\pm\sqrt{3}$

51. Not differentiable at 1 and 5
Left-hand derivative at 1: −4
Right-hand derivative at 1: 4
Left-hand derivative at 5: −4
Right-hand derivative at 5: 4

53. Not differentiable at $k \in \mathbb{Z}$
Left-hand derivative at each point of
nondifferentiability: 1
Right-hand derivative at each point of
nondifferentiability: 1

55. Undefined and therefore not differentiable for
$z < 0$, not differentiable at 0
Right-hand derivative does not exist at 0

57. Differentiable on the entire real line

61. **a.** $a(t) = 2.2 \text{ m/s}^2$ **b.** $t = 15$ s
c. The runway must be at least 247.5 m long.

63. **b.** $g(x) = |x|$; $c = 0$

65.

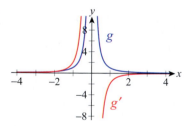

$g' > 0$ and g is increasing for $x < 0$;
$g' < 0$ and g is decreasing for $x > 0$

67. True

69. False; consider $f(x) = |x|$ at $c = 0$.

71.

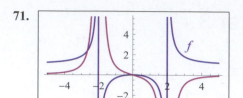

$f' > 0$ and f is increasing for $x < -2$ and
$-2 < x < 0$;
$f' < 0$ and f is decreasing for $0 < x < 2$ and $x > 2$

73.

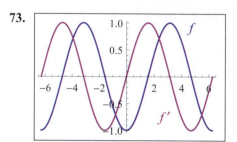

$f' > 0$ and f is increasing for
$2k\pi < x < (2k+1)\pi; k \in \mathbb{Z}$

$f' < 0$ and f is decreasing for
$(2k-1)\pi < x < 2k\pi, k \in \mathbb{Z}$

75.

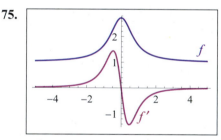

$f' > 0$ and f is increasing for $x < 0$;

$f' < 0$ and f is decreasing for $x > 0$

77. $f'(x) = \dfrac{1}{1+x^2}$; $f''(x) = -\dfrac{2x}{\left(1+x^2\right)^2}$;

$f'''(x) = \dfrac{-2+6x^2}{\left(1+x^2\right)^3}$; $f^{(4)}(x) = -\dfrac{24x\left(-1+x^2\right)}{\left(1+x^2\right)^4}$

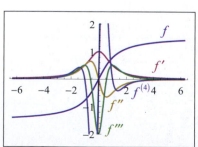

79. $f'(x) = \dfrac{1}{10} x \cos\left(1 - \dfrac{x}{10}\right) - \sin\left(1 - \dfrac{x}{10}\right);$

$f''(x) = \dfrac{1}{5} \cos\left(1 - \dfrac{x}{10}\right) + \dfrac{1}{100} x \sin\left(1 - \dfrac{x}{10}\right);$

$f'''(x) = -\dfrac{1}{1000} x \cos\left(1 - \dfrac{x}{10}\right) + \dfrac{3}{100} \sin\left(1 - \dfrac{x}{10}\right);$

$f^{(4)}(x) = -\dfrac{1}{250} \cos\left(1 - \dfrac{x}{10}\right) - \dfrac{1}{10000} x \sin\left(1 - \dfrac{x}{10}\right)$

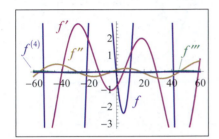

81.

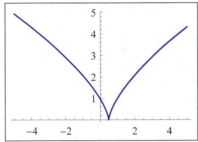

Not differentiable at $t = 1/2$ (cusp)

83.

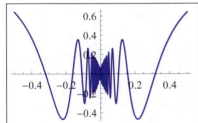

Not differentiable at $t = 0$ (oscillating discontinuity)

Section 3.2

1. $f'(x) = -2$ **3.** $h'(x) = -6x + 2$

5. $G'(x) = 2x^3 + 6x^2 - 2x + 3.2$

7. $H'(x) = 8x^7 + 5\sqrt{2} x^4 - 8x^3$

9. $S'(z) = 12z^2 - \dfrac{3}{2\sqrt{z}}$ **11.** $T'(r) = 2\pi r + 2e^r$

13. $6x + 11$ **15.** $8x^3 + 15x^2 - 24x - 30$

17. $-\dfrac{196}{5} x^6 - \dfrac{20}{3} x^4 + \dfrac{56}{5} x^3 + \dfrac{4}{3} x$ **19.** $\dfrac{1}{\sqrt{t}} + 8$

21. $f'(x) = \dfrac{2}{(1 - 2x)^2}$

23. $h'(x) = \dfrac{-12x^2 + 20x - 6}{(2x^3 - 5x^2 + 3x + 1)^2}$

25. $G'(x) = \dfrac{-11}{(x - 5)^2}$

27. $H'(x) = \dfrac{11x^4 + 3x^2 - 6x}{(2x^3 + 5x^2 + 1)^2}$

29. $B'(u) = \dfrac{3u^{3/2} + 4u}{2(\sqrt{u} + 1)^2}$ **31.** $g'(t) = \dfrac{5t - 8\sqrt{t} + 15}{2\sqrt{t}(4 - 5\sqrt{t})^2}$

33. $f'(x) = 6 - 10x^4$ **35.** $h'(x) = -27x^2 + 14x - 5$

37. $G'(x) = 1$ **39.** $f'(t) = \dfrac{3}{2}\left(\dfrac{4t + 1}{\sqrt{t}}\right)$

41. $h'(s) = -\dfrac{1}{5s^2} + 10s + 1$

43. $G'(x) = -15x^{-6} - 6x^{-4}$

45. $H'(t) = \dfrac{9}{2\sqrt{t}} - \dfrac{5}{2} t \sqrt{t}$

47. $w'(z) = 2 + \dfrac{4}{4 - \sqrt{z}} + \dfrac{2\sqrt{z}}{(4 - \sqrt{z})^2}$

49. $r'(x) = \dfrac{2a^2}{(x + a^2)^2}$

51. $F'(x) = e^x\left(2 + \sqrt{x} + \dfrac{1}{2\sqrt{x}}\right)$

53. $C'(x) = \dfrac{1}{(x + 1)^2}$

55. $G'(s) = \dfrac{6s}{2e^s + s} - \dfrac{3s^2(1 + 2e^s)}{(2e^s + s)^2}$

57. $L'(y) = -18y^8 + 48y^7 + 6y^5 - 15y^4$

59. $H'(s) = -\dfrac{ace^s}{(b + ce^s)^2}$

61. $L'(t) = \dfrac{1}{2\sqrt{t}} - \dfrac{3\sqrt{t}}{2} - 2te^t$

63. $g'(x) = \dfrac{1}{(x+1)^2}$; $g''(x) = \dfrac{-2}{(x+1)^3}$;

$g'''(x) = \dfrac{6}{(x+1)^4}$

65. $F'(x) = -3\pi x^2 + 10x - 1$; $F''(x) = -6\pi x + 10$;
$F'''(x) = -6\pi$

67. $W'(t) = 6t + 3e^t$; $W''(t) = 6 + 3e^t$; $W'''(t) = 3e^t$

69. $f(x) = \dfrac{1}{3}x^3 - 4x$ (Answers will vary.)

71. $f(x) = x^3 + 2x^2 + x + 1$ (Answers will vary.)

73. $g^{(k)}(x) = \dfrac{(-1)^k k!}{x^{k+1}}$

75. $q^{(k)}(x) = e^x\left[\displaystyle\sum_{i=0}^{k}\binom{k}{i}D^i x^n\right]$

77. a. 0   **b.** 18   **c.** 1   **79.** $y = -\dfrac{1}{2}x + 2$

81. $y = -2ex + 4e$ **83.** $y = -x$ **85.** $x = 1$

87. No horizontal tangent

89. No horizontal tangent   **91.** $x = 0$

93. $x = 0$ **95.** No horizontal tangent

97. $a = 2, b = -8$

99. The slopes are $1/2$ and -2, respectively.

101. $y = ex$ **103.** $y = -4x - 4$

105. $\dfrac{f(x) + xf'(x)}{g(x)} - \dfrac{xf(x)g'(x)}{[g(x)]^2}$

107. $e^x\left[\dfrac{f(x) + f'(x)}{g(x) + 2} - \dfrac{f(x)g'(x)}{(g(x) + 2)^2}\right]$

111. $2f(x)f'(x)$; $3[f(x)]^2 f'(x)$

115. At the point $(12, 4)$

117. a. $a(2) = 125/36$ ft/s^2 **b.** $a(10) = 5/4$ ft/s^2

119. $a = -1.62$ m/s^2; $t \approx 1.361$ s; $v_{imp} \approx 2.205$ m/s

121. True

123. False; y' is found by applying the Reciprocal Rule or the Quotient Rule.

125. True

127. False; $\dfrac{d}{dx}F(x)$ is found by applying the Quotient Rule.

129. True

131.

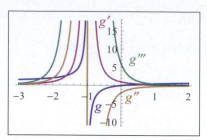

133.

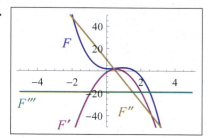

135.

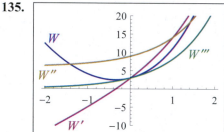

137.

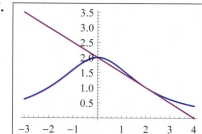

139.

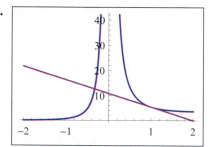

Section 3.3

1. $3/2$ **3.** π **5.** 0   **7.** 2   **9.** 1   **11.** 5

13. $f'(x) = 2\cos x + 5\sin x$

15. $h'(x) = \cos x - x\sin x$

17. $G'(x) = \sec x\left(\dfrac{1}{\sqrt{x}} + 2\sqrt{x}\tan x\right)$

19. $L'(x) = 3e^x(\csc x - 1)(\csc x + \cot x)$

21. $g'(x) = -2\cot x\csc^2 x$

23. $F'(t) = \dfrac{t\sin t + 2\cos t - 2}{t^3}$

25. $R'(z) = \dfrac{e^z(z-1) + z\cos z - \sin z}{z^2}$

27. $B'(x) = -\csc^2 x$

29. $T'(s) = se^s(2\cot s + s\cot s - s\csc^2 s)$

31. $x = \dfrac{2\pi}{3} + 2k\pi,\ x = \dfrac{4\pi}{3} + 2k\pi,\ k \in \mathbb{Z}$

33. $x = k\pi,\ k \in \mathbb{Z}$ **35.** $u = (2k+1)\dfrac{\pi}{4},\ k \in \mathbb{Z}$

37. $x = 2k\pi,\ k \in \mathbb{Z}$

39. $x = \pm\dfrac{\pi}{6} + k\pi,\ k \in \mathbb{Z}$

41. $y = 2x$ **43.** $y = 1$

45. $F'(0) = 1,\ G'(0) = 2,\ H'(0) = 1$

49. $f^{(4k)}(x) = \sin x,\ f^{(4k+1)}(x) = \cos x,$
$f^{(4k+2)}(x) = -\sin x,\ f^{(4k+3)}(x) = -\cos x$

51. $f^{(4k)}(x) = (-4)^k e^x\sin x,$
$f^{(4k+1)}(x) = (-4)^k e^x(\sin x + \cos x),$
$f^{(4k+2)}(x) = (-4)^k 2e^x\cos x,$
$f^{(4k+3)}(x) = (-4)^k 2e^x(\cos x - \sin x)$

61. a. $F'(x) = mg(\cos x + \mu\sin x)$

   b. $x = \operatorname{arccot}(-\mu)$

63. $\left.\dfrac{dx}{d\beta}\right|_{\beta=\pi/6} = -15\sin\dfrac{\pi}{6} = -7.5$. This rate is negative

since β is decreasing while x is increasing.

65.

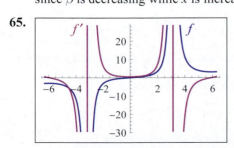

67.

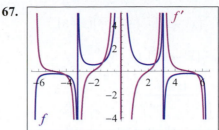

69.

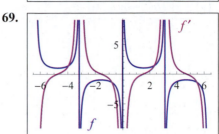

71.

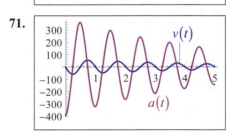

Max. velocity ≈ 54.19 cm/s;

max. acceleration ≈ 358.49 cm/s^2

Section 3.4

1. $f(x) = x^6,\ g(x) = u = 3x - 2.5$

3. $f(x) = 2\sqrt[3]{x},\ g(x) = u = x^2 - 9$

5. $f(x) = \sin x,\ g(x) = u = \dfrac{1}{x^2 + 1}$

7. $f(x) = \csc x,\ g(x) = u = 3e^x$

9. $f(x) = \dfrac{3}{\sqrt{x}},\ g(x) = u = \ln x,\ h(x) = x^2 + 1$

11. $g'(x) = 33(x^5 - \pi x^2 + 7.5)^{10}(5x^4 - 2\pi x)$

13. $F'(x) = \dfrac{15}{\sqrt{x}(5 + 2\sqrt{x})^6}$

15. $k'(x) = \dfrac{2(5x^4 - 6x^2 + 10.5)}{(x^5 - 2x^3 + 10.5x)^{7/5}}$

17. $g'(x) = \dfrac{2x - 5}{2\sqrt{x^2 - 5x + 2}}$

19. $q'(x) = \dfrac{4}{3}(x^3 - 5x)^{-1/3}(3x^2 - 5)(x + 3)^{5/4}$

$\qquad + \dfrac{5}{2}(x^3 - 5x)^{2/3}(x + 3)^{1/4}$

21. $k'(z) = \dfrac{4z - 5}{(1 + 5z - 2z^2)^2}$

23. $S'(v) = \dfrac{(v^2 - 5)^2(6v^2 + 6v + 30)}{(2v + 1)^4}$

25. $T'(s) = \dfrac{-8s}{3(s^2 + 1)^{1/3}(s^2 - 1)^{5/3}}$

27. $H'(x) = \dfrac{10x - 3x^3}{(x^2 + 2)^3 \sqrt{x^2 - 2}}$

29. $B'(t) = \dfrac{1 - 2t^2}{3t^{2/3}(1 + 2t^2)^{4/3}}$

31. $t'(x) = -\sin x \cos(\cos x)$

33. $P'(x) = \tan^2 x + 2x \tan x \sec^2 x$

35. $U'(z) = 10 \sec^2 z \tan z$

37. $C'(x) = \sec^2 x \sin(2 \tan x)$

39. $V'(x) = (-\sin x) e^{\cos x}$

41. $w'(x) = \dfrac{1}{\sqrt{2x + 1}}\left(\begin{array}{l}\cos\sqrt{2x + 1} \\ + e^{\tan\sqrt{2x+1}}\sec^2\sqrt{2x + 1}\end{array}\right)$

43. $f'(x) = \pi^2 (\ln 2) \cos(\pi x) 2^{\sin \pi x}$

45. $t'(s) = (\ln 2) 2^s \sec^2(2^s)$

47. $E'(x) = (\ln 5)^2 \, 5^x 5^{5^x}$

49. $N'(x) = 2x \sin(x^2) \sin(2e^{\cos(x^2)}) e^{\cos(x^2)}$

51. $C'(x) = -2x \sin(2x^2)$

53. $t'(s) = -\dfrac{(\ln 10) 10^s \sin(10^s)}{2\sqrt{\cos(10^s)}}$

55. $H'(s) = 2^s \ln 2(\cos(2^s)\tan(2^s) + \sin(2^s)\sec^2(2^s))$

57. $T'(z) = e^z \cos(e^z) + e^{\sin z} \cos z$

59. $U'(\theta) = (1 + \sec^2(\theta + \tan(\theta + \tan\theta)))$

$\qquad (1 + \sec^2(\theta + \tan\theta))(1 + \sec^2\theta)$

61. $y = \dfrac{4}{3}x + \dfrac{1}{3}$ **63.** $y = 1$ **65.** $y = -\dfrac{1}{e\pi^2}x + \dfrac{2}{e\pi}$

67. $y = x + 1$ **69.** $x = 3, x = 4, x = 5$ **71.** $x = 0$

73. $x = 0, x = \pm\sqrt{k\pi - 2}, k \in \mathbb{Z}$ **75.** $x = \pm 1$

77. $p''(x) = 1520x^2(x^2 + 5)^{18} + 40(x^2 + 5)^{19}$

79. $g''(x) = 10(\sin^2 x - \cos^2 x)$

81. $F''(t) = 6t \cos(t^2) - 4t^3 \sin(t^2)$

83. $G''(x) = 0$ **85.** $F'(1) + G'(1) = -20$

87. If $n = 4m$, $f^{(n)}(x) = k^n \cos(kx)$;

$\qquad$ if $n = 4m + 1$, $f^{(n)}(x) = -k^n \sin(kx)$;

$\qquad$ if $n = 4m + 2$, $f^{(n)}(x) = -k^n \cos(kx)$;

$\qquad$ if $n = 4m + 3$, $f^{(n)}(x) = k^n \sin(kx)$.

93. $\dfrac{dV}{dt} = 6.4\pi \text{ in.}^3/\text{s}$

95. **a.** $v_{max} = 1.5 \text{ m/s}$ **b.** $a_{max} = 2250 \text{ m/s}^2$

97. $\dfrac{dP}{dt} \approx 243.63 \text{ Pa/s}$

99. $P_2(x) = 1 - x^2$

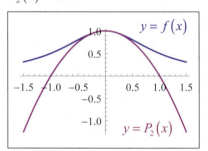

Section 3.5

1. $\dfrac{dy}{dx} = -\dfrac{1}{2y}$ **3.** $\dfrac{dy}{dx} = \dfrac{x}{y}$ **5.** $\dfrac{dy}{dx} = \dfrac{1 - 3y^2}{6xy}$

7. $\dfrac{dy}{dx} = \dfrac{y(2\sqrt{x + 2} - 1)}{2x - 2x\sqrt{x + 2} + 4}$

9. $\dfrac{dy}{dx} = \dfrac{1 + y(2\sin x + 1)}{2\cos x - x}$

11. $\dfrac{dy}{dx} = \dfrac{2y^2}{3x^2}$ **13.** $\dfrac{dx}{dy} = 2y$ **15.** $\dfrac{dx}{dy} = -\dfrac{y^2}{x^2}$

17. $\dfrac{dx}{dy} = \dfrac{e^y - 3\cos y - x}{y}$ **19.** $\dfrac{dx}{dy} = \dfrac{5y^3}{6x^2}$

21. Tangent line: $y = x - 1$; normal line: $y = -x + 1$

23. Tangent line: $y = -\dfrac{1}{2}x + 2$;

$\qquad$ normal line: $y = 2x - 3$

25. Tangent line: $y = -\dfrac{3}{2}x + \dfrac{9}{2}$;

$\qquad$ normal line: $y = \dfrac{2}{3}x - 2$

27. Tangent line: $y = \dfrac{7}{4}x - \dfrac{3}{2}$;

normal line: $y = -\dfrac{4}{7}x + \dfrac{22}{7}$

29. $\dfrac{dy}{dx} = -\dfrac{x^3}{y^3}$ **31.** $\dfrac{dy}{dx} = \dfrac{4xy - 3y^2}{4xy - 3x^2}$

33. $\dfrac{dy}{dx} = -1$

35. $\dfrac{dy}{dx} = \dfrac{\sin 2x - 2x\sec^2\left(x^2 + y^2\right)}{\sin 2y + 2y\sec^2\left(x^2 + y^2\right)}$

37. $\dfrac{dy}{dx} = \dfrac{2x^4 - 4x^2y - 6xy^2 - x^2 + 2y^2 - y}{3y^2 - 6x^2y - x}$

39. $\dfrac{dy}{dx} = \dfrac{5\sqrt{2xy} + y}{3\sqrt{2xy} - x}$ **41.** $\dfrac{dy}{dx} = \dfrac{\sec^2 x + 2y}{\cos y - 2x}$

43. $\dfrac{dy}{dx} = \dfrac{\cos x - \sin x}{2\sqrt{\sin x + \cos x}\,\sec\left(x + y\right)\tan\left(x + y\right)} - 1$

45. $\dfrac{ds}{dt} = \dfrac{4s - 6s^3t^2}{4s^2t^3 - \sqrt{s}}$ **47.** $\dfrac{d^2y}{dx^2} = \dfrac{4y^2 - x^2}{16y^3}$

49. $\dfrac{d^2y}{dx^2} = \dfrac{\left(y^2 - 1\right)\left(3y^2 + 1\right)}{4x^2y^3}$

51. $\dfrac{d^2y}{dx^2} = \dfrac{-2x\left(x^3 + y^3\right)}{y^5}$

55. Horizontal tangent: $\left(1/2, 1\right)$;
vertical tangent: $\left(-1, -1/2\right)$

57. Horizontal tangent: $\left(1, 5/4\right)$;
vertical tangent: none

59. $y' = x/y$ and $y' = -y/x$, respectively

61. $y' = -1/b$ and $y' = b$, respectively

65. $y = \sqrt{2}x + 2$, $y = -\sqrt{2}x + 2$

67.

69.

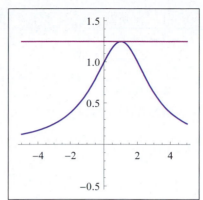

71.

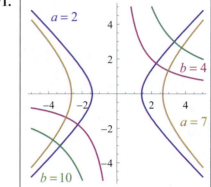

73.

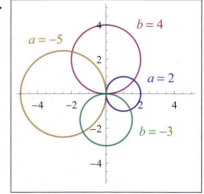

Section 3.6

1. $1/12$ **3.** 6   **5.** $1/4$ **7.** $-1/2$ **9.** $-3/2$

11. $1/(10\ln 10)$ **13.** 2   **15.** $5/\cos 0.01$

17. $1/39$ **19.** $-\sqrt{5}/9$ **21.** $-5/33$

23. $1/37$ **25.** $-16/7$ **27.** 1   **29.** $2/\sec^2 1$

31. $f'(x) = 3/x$

33. $h'(x) = \dfrac{2x}{x^2 + 3}$

35. $G'(x) = \ln\sqrt{x^2 + 4} + \dfrac{x^2}{x^2 + 4}$

37. $L'(x) = \dfrac{x^2 + 1 - 2x^2 \ln 2x}{x\left(x^2 + 1\right)^2}$ **39.** $g'(x) = \dfrac{2}{9 - x^2}$

41. $F'(t) = \dfrac{t + 2\sqrt{t^2 + 4}}{t^2 + 4 + 2t\sqrt{t^2 + 4}}$ **43.** $T'(x) = -\tan x$

45. $v'(x) = -2 \sin 2x \left(\ln\left(\cos 2x\right) + 1\right)$

47. $w'(x) = \log x + \dfrac{1}{\ln 10}$

49. $y' = (x+2)(x+3)(x+4) + (x+1)(x+3)(x+4)$
   $+ (x+1)(x+2)(x+4) + (x+1)(x+2)(x+3)$

51. $y' = \sqrt[3]{(2x-1)(x-5)(3x+1)}$
   $\left(\dfrac{2}{6x-3} + \dfrac{1}{3x-15} + \dfrac{1}{3x+1} \right)$

53. $y' = \dfrac{\sqrt[3]{x^3 - 5x^2 + 7}\,(x+2)}{x^{2/3}\sqrt{3x^2 + 4}}$
   $\left(\dfrac{3x^2 - 10x}{3x^3 - 15x^2 + 21} + \dfrac{1}{x+2} - \dfrac{2}{3x} - \dfrac{3x}{3x^2 + 4} \right)$

55. $y' = \dfrac{x^2 \sqrt[5]{x^3 + 3}}{\sqrt[4]{x^4 + 4}} \left(\dfrac{2}{x} + \dfrac{3x^2}{5\left(x^3 + 3\right)} - \dfrac{x^3}{x^4 + 4} \right)$

57. $y' = \left(\sin x\right)^{1/x} \left(\dfrac{x\cot x - \ln\left(\sin x\right)}{x^2} \right)$

59. $y' = \left(\cos x\right)^{\sqrt{x}} \left(\dfrac{\ln\left(\cos x\right)}{2\sqrt{x}} - \sqrt{x}\,\tan x \right)$

61. $y' = \left(\ln x\right)^x \left(\ln\left(\ln x\right) + \dfrac{1}{\ln x} \right)$

63. $y' = x^{x^x}\left(x^x \left(\ln x + 1\right)\ln x + x^{x-1} \right)$

65. $y' = \left(\ln x\right)^{\sin x} \left(\cos x \ln\left(\ln x\right) + \dfrac{\sin x}{x \ln x} \right)$

67. $y' = -\dfrac{1}{\sqrt{1 - x^2}}$ **69.** $y' = -\dfrac{1}{1 + x^2}$

71. $y' = \dfrac{1}{2x^2 + 2x + 1}$ **73.** $y' = \dfrac{1}{\left(1 + x^2\right)\arctan x}$

75. $y' = \dfrac{-1}{2\sqrt{x - x^2}}$ **77.** $y' = -2\sqrt{1 - x^2}$

79. $y' = \dfrac{e^x}{1 + e^{2x}}$ **81.** $y' = \dfrac{-2}{\left(1 + x^2\right)\left(1 + \arctan x\right)^2}$

83. $y' = \dfrac{6x^2 \arcsin\left(x^3\right)}{\sqrt{1 - x^6}}$ **85.** $y' = \dfrac{2}{\left(x^2 + 1\right)\sqrt{x^2 + 2}}$

87. $y' = \dfrac{1}{x^2 + 1}$

89. $y' = \left(\arctan x\right)^x \left[\ln\left(\arctan x\right) + \dfrac{x}{\left(1 + x^2\right)\arctan x} \right]$

91. $y' = \dfrac{2}{\left(x^2 + 1\right)^2 \sqrt{x^2 + 2}}$ **93.** $y' = 1/x^2$

95. $y = -2.178x + 3.282$

97. $y = 1.270x + 1.164$

99. $y = \ln\left(\pi/4\right)x + 1 - \ln\left(\pi/4\right)$

101. $f'(x) = 0$; as remarked in the text,
   $\arcsin\left(1/x\right) = \operatorname{arccsc} x$ for $|x| \ge 1$.

103. $\theta = \tan^{-1}\left(28/s\right) - \tan^{-1}\left(8/s\right)$;
   $\dfrac{d\theta}{ds} = \dfrac{8}{s^2 + 64} - \dfrac{28}{s^2 + 784}$

105. Restrict the domain of $\cot x$ to $(0, \pi)$.

107. False; $y = \log \pi$ is a constant function and
   therefore $y' = 0$.

109. True

Section 3.7

3. 197,959 people   **5.** Approx. 367.88 g

7. a. $a = \sqrt[20]{1/2} \approx 0.9659$ **b.** Approx. 0.707 kg
   c. Approx. 7.629 mg

9. Approx. 8.75 more minutes (i.e., 13.75 total
   minutes)

11. 300 N   **13.** Approx. 58.8x N

15. 350 N   **19.** 0.073 atm/s

21. 0.75 mol/(L·s); 27 mL of water

23. a. $152.80   **b.** $152.50

25. a. $792.29   **b.** $2331   **c.** $861.71; $2311.50

27. $1077.22   **29.** Approx. 36π in.^{3}

31. The cost increases by $465.

33. a. Cost decreases by approx. $1250.
   b. Profit decreases by approx. $1050.
   c. The actual decreases are $1250.01 and
   $1052.49, respectively.

35. Approx. 0.8042 cm^3 **37.** 1.5 percent increase

39. 0.3 percent decrease

Section 3.8

1. -3 ft/s **3.** $1/6$ **5.** 8 **7.** -5.2 **9.** 0

11. Increasing at 35 in.^{2}/s

13. Approx. 0.8436 units/s

17. -10 cm/min **19.** 2.641 cm/min

21. 15.524 mph **23.** 0.0808 in./s **25.** 12 ft/s

27. 1187.5 mph **29.** 25 cm/s; 0.02 rad/s

31. -20 mm^2/s **33.** 1.52 in./min **35.** 1 cm/min

37. $\sqrt{2}/4$ ft/s **39.** 15 ft/s **41.** 429.29 km/h

43. Approx. 0.494 in./min

45. a.

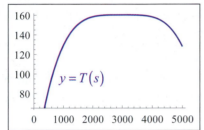

$y = T(s)$

b. Approx. 30.465 hp/s

Section 3.9

1. $L(x) = 2x - 2$ **3.** $L(x) = 1$ **5.** $L(\theta) = 1$

7. $L(t) = -\dfrac{3}{16}t + \dfrac{7}{8}$ **9.** $L(u) = -15u - 62$

11. $L(z) = z + 1$ **13.** $dy = 1.4$ **15.** $dy = -1.3$

17. $dy = 189/1024$ **19.** $dy = -\dfrac{\pi + 2}{8}$

21. $dy = -1/(2e^3)$ **23.** $dy \approx -0.1083$

25. $dy = 0.5$; $\Delta y = 0.625$

27. $dy = (\ln 2)/2 \approx 0.3466$; $\Delta y = 2\sqrt[4]{2} - 2 \approx 0.3784$

29. $\Delta y = 1.52$ **31.** $\Delta y = -1.\overline{4}$ **33.** $\Delta y \approx 0.1854$

35. $\Delta y = -0.4678$ **37.** $\Delta y \approx -0.02587$

39. $\Delta y \approx -0.0934$

41. Approx. 3.0167; calculator value: 3.0166

43. Approx. 3.9667; calculator value: 3.9666

45. Approx. 1.9875; calculator value: 1.9873

47. Approx. 0.9933; calculator value: 0.9933

53. Approx. 0.1266 in. **55.** 8.1566 mm^2 **57.** 0.1%

59. Approx. 1 in.^{3} **61.** Approx. $\pm 1.24°$

63. a. $F = 0.1875$ N **b.** $dF \approx 0.017$ N
 c. $dF \approx 0.0167$ N

65. $dE_{kin} \approx 129.63$ kJ; about a 24% increase

67. 3%; the generalization is $\dfrac{dA}{A} = \dfrac{2}{3}\dfrac{dV}{V}$

69. Up to 113.615 units **71.** $x \approx 5.1$

73. Yes, we can prove $\dfrac{dA}{A} = 2\dfrac{dr}{r}$.

75. True **77.** True

79. False; consider $f(x) = \sqrt{x}$, $x = 1$, $dx = 0.1$.

81.

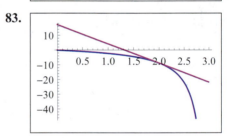

83.

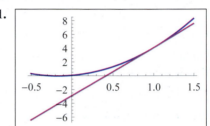

85.

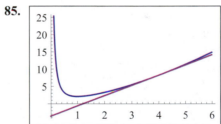

87.

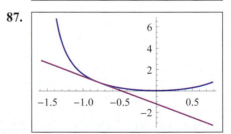

89.

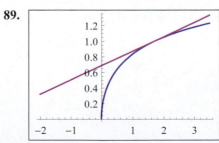

91.
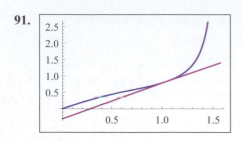

Chapter 3 Review

1. $\dfrac{df}{dx} = 3x^2 + 1;\ \dfrac{df}{dx}\Big|_{x=1} = 4$ **3.** $D_x s(x) = \dfrac{1}{2\sqrt{x-2}}$

5. $f'(x) = 2x;\ f''(x) = 2;\ f'''(x) = 0$

7. Not differentiable at 0
One-sided derivatives do not exist at 0

9. Not differentiable at $k \in \mathbb{Z}$
Left-hand derivative at each point of nondifferentiability: 1
Right-hand derivative at each point of nondifferentiability: 1

11. $f'(x) = x^4 - 8x^3 + 3x^2 + x$

13. $h'(x) = 6x^2 - 2x + 8$

15. $F'(t) = \dfrac{7}{2}t^{5/2} + 4t^3 + \dfrac{1}{2}t^{-1/2} - \dfrac{5}{2}t^{3/2} - 3t^2 + \dfrac{1}{2}t^{-3/2} + 1$

17. $u'(x) = \dfrac{7e^x}{(3e^x + 5)^2}$ **19.** $v'(x) = \dfrac{2x}{x^2 + 2}$

21. $f'(x) = \dfrac{1}{(x+1)^2};\ f''(x) = \dfrac{-2}{(x+1)^3};$

$f'''(x) = \dfrac{6}{(x+1)^4}$

23. $f'(x) = \sec^2 x;\ f''(x) = 2\sec^2 x \tan x;$
$f'''(x) = 2\sec^2 x \left(\sec^2 x + 2\tan^2 x\right)$

25. $f(x) = x^3 - 4x^2 + 6x$ (Answers will vary.)

27. $y = -\dfrac{3}{8}x + \dfrac{7}{8}$

29. $y = x,\ y = 13x - 24$ **31.** $-1/2$

33. $v(1) = 12.5$ ft/s; $a(1) = -12.5$ ft/s^2

35. $V'(1) = 784\pi/3$ cm^3/s; $A'(1) = 224\pi/3$ cm^2/s

37. $\dfrac{dy}{dx} = \dfrac{4x - 5y}{5x - 4y}$ **39.** $\dfrac{dx}{dy} = \dfrac{y(3 - y^2)}{3x}$

41. $y = x + 1$ **43.** $y = -\dfrac{20}{9}x + \dfrac{56}{9}$

45. $\dfrac{d^2 y}{dx^2} = \dfrac{x^{2/3} + y^{2/3}}{3x^{4/3}y^{1/3}}$ **47.** 32   **49.** 1/2

51. $f'(x) = \dfrac{1}{2\sqrt{x}\,(x+1)}$

53. $f'(x) = \dfrac{1}{x|\ln x|\sqrt{\ln^2 x - 1}}$ **55.** $y' = \dfrac{\left(\sqrt{x}\right)^{\ln x}\ln x}{x}$

57. Approx. 8 minutes (8 minutes 3 seconds)

59. Approx. 0.1 in./s   **61.** 2000 km/h

63. 5 ft 4 in.   **65.** $L(x) = \dfrac{\sqrt{2}}{2}x + \dfrac{\sqrt{2}(4-\pi)}{8}$

67. Approx. 0.7354   **69.** Approx. 10%

71. If $a > 1$, the slope of y increases as we scan the graph from left to right, while in the other case it decreases.

73. True

75. False; the Reciprocal Rule along with the Chain Rule (or Quotient Rule) should be used, which

yields $y' = \dfrac{3 - 2x}{\left(x^2 - 3x + 1\right)^2}.$

77. False; logarithmic differentiation should be used, which yields $y' = x^x (\ln x + 1)$.

79. True

81. False; $\Delta x = dx$ and even though $\Delta y \to dy$ might be true under certain circumstances, $x \to 0$ is not a necessary condition.

83.
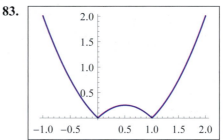
Not differentiable at $x = 0,\ 1$ (cusps)

85.
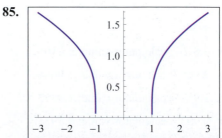
Not differentiable at $x = \pm 1$ (vertical tangents, undefined on $(-1, 1)$)

87. The algorithm employed by the CAS may not simplify before differentiating. Yes, we can simplify with an appropriate command. The simplified answer is $f'(x) = \cos x$.

89. a.

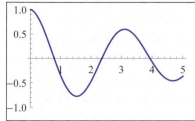

It is an oscillating motion with the amplitude "dying down" as a result of retarding forces.

b.

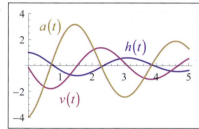

The object is near its equilibrium position when velocity is maximum and near its extreme positions when velocity is zero. Acceleration, on the other hand, is maximum near the extreme positions and zero near equilibrium.

Chapter 3 Project

1. −0.0265 mm Hg/ft   **3.** −0.0225 mm Hg/ft

5. The symmetric difference quotient is the average of the difference quotients $\dfrac{f(c+h)-f(c)}{h}$ and $\dfrac{f(c-h)-f(c)}{-h}$.

7. −0.02302 mm Hg/ft

Chapter 4

Section 4.1

1. Abs. min.: $f(-1)=f(1)=0$; no maximum

3. Abs. min.: $f(3\pi/2)=-2$; abs. max.: $f(\pi/2)=2$

5. Critical points at $x=0,1,3$; rel. min. at $x=0,3$; abs. min. at $x=3$; rel. max. at $x=-1,1,4.2$; abs. max. at $x=4.2$

7. Critical points at $x=0,2$; rel. min. at $x=-2,2$; abs. min. at $x=-2$; rel. max. at $x=0$; abs. max. at $x=0$

9. Abs. min.: $f(0)=1$; abs. max.: $f(3)=7$

11. No extrema

13. Abs. min.: $v(1)=-4$; abs. max.: $v(-2)=5$ and $v(4)=5$

15. No extrema

17. Abs. min.: $n(1)=-1$; no maximum

19. No minimum; abs. max.: $G(0)=1$

21. Answers will vary.   **23.** Answers will vary.

25. Answers will vary.   **27.** Answers will vary.

29. Answers will vary.   **31.** Answers will vary.

33. Answers will vary.   **35.** Answers will vary.

37. Answers will vary.   **39.** $x=1,-2$ **41.** None

43. $x=1.5$ **45.** $x=0,4$ **47.** $t=0$

49. $s=0,0.5$ **51.** $\alpha=k\pi, \pm\dfrac{2\pi}{3}+2k\pi, k\in\mathbb{Z}$

53. $x=e^{-2}$ **55.** $t=0$

57. Abs. min.: $g(5)=-68$; abs. max.: $g(0)=7$

59. Abs. min.: $u(2)=-49$; abs. max.: $u(3)=0$

61. Abs. min.: $k(-3)=k(1)=-4$; abs. max.: $k(-1)=4$

63. Abs. min.: $m(x)=6, x\in[-3,3]$; abs. max.: $m(-4)=m(4)=8$

65. Abs. min.: $g(1)=2$; abs. max.: $g(-1.5)=14.5$

67. Abs. min.: $G(1)=2$; abs. max.: $G(1/4)=G(4)=5/2$

69. Abs. min.: $r(-1)=r(1)=0$; abs. max.: $r(0)=1$

71. Abs. min.: $w(-2)=-4$; abs. max.: $w(2)=4$

73. Abs. min.: $r(5\pi/4)=\left(-\sqrt{2}/2\right)e^{5\pi/4}$; abs. max.: $r(\pi/4)=\left(\sqrt{2}/2\right)e^{\pi/4}$

75. Abs. min.: $t(0)=t(1)=0$; abs. max.: $t(0.5)=1$

77. Abs. min: $V(-25/12)=5+(35/12)(-25/12)^{5/7}$; abs. max.: $V(1)=11$

79. Abs. min.: $g(0)=-4$; no maximum

81. No minimum; abs. max.: $K(0) = 2$

83. No extrema

85. Abs. min.: $L(0) = 0$; no maximum

87. Abs. min.: $t(2n+1) = 0, n \in \mathbb{Z}$;
abs. max.: $t(2n) = 4, n \in \mathbb{Z}$

89. No extrema

91. The wire needs to be cut at $21.995 \approx 22$ inches, and this first piece is to be bent into a circle.

93. Approx. 11,500 calculators

95. False; consider $f(x) = [\![x]\!]$ on $[0,1]$.

97. True

99. False; the critical points shift by k units to the left.

101. True

103. False; f can only have one largest value (however, it can be attained at multiple x-values).

105. True

Section 4.2

1. $c = 5/2$ **3.** $c = 3$ **9.** $c = 2$

11. Rolle's Theorem does not apply; $h(1) \neq h(3)$.

13. $c = 0$

15. Rolle's Theorem does not apply; $H(x)$ is not differentiable at $x = 0$.

17. Rolle's Theorem does not apply; $T(z)$ is not continuous on $[0, 5\pi]$.

19. Rolle's Theorem does not apply; $w(t)$ is not continuous on $[-\pi/4, \pi/4]$.

21. The MVT does not apply; $f(x)$ is not differentiable at $x = 1$.

23. $c = 3/2$

25. The MVT does not apply; $G(x)$ is not continuous on $[-2,3]$.

27. $c = 0$ **29.** $c = -\sqrt{2}/2$

31. The MVT does not apply; $u(z)$ is not continuous on $[0, \pi]$.

33. 10

35. If $f(-5) = -1$ and $f'(x) \leq 0.1$, then $f(5) \leq 0$.

37. $f(x) = x^3 - x^2 + x - 20$ **39.** $p(t) = -t^2 + 5t - 12$

43. The average velocity is 460 mph, and thus, by the Mean Value Theorem, there must be a time t_0 when the velocity is $v(t_0) = 460$ mph. Use the Intermediate Value Theorem to argue that the equation $v(t) = 450$ has solutions on both intervals $[0, t_0]$ and $[t_0, 10.5]$.

51. No; consider for example $f(x) = \sin x$ on $[0, 2\pi/3]$ with $c = \pi/2$.

53. The average growth rate is 50 rabbits per month, and thus, by the Mean Value Theorem, there must exist a time when the instantaneous growth rate equals 50.

65. False; Rolle's Theorem is a special case of the Mean Value Theorem.

67. True

69. Tangent line: $y = \dfrac{1}{3}x + \dfrac{3}{4}$

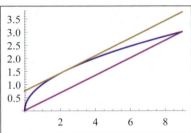

71. Tangent line: $y = 2x + 7 - 4\sqrt{3}$

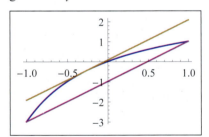

Section 4.3

1. Decreasing on $(-\infty, 2)$; increasing on $(2, \infty)$

3. Increasing on $(-\infty, -5), (1, \infty)$; decreasing on $(-5, 1)$

5. Increasing on $(-\infty, -3), (1/2, 1)$; decreasing on $(-3, 1/2), (1, \infty)$

7. Increasing on $(-\infty, 1)$; decreasing on $(1, \infty)$

9. Increasing on $\left(-\infty, -3 - 2\sqrt{3}\right), \left(-3 + 2\sqrt{3}, \infty\right)$; decreasing on $\left(-3 - 2\sqrt{3}, -3\right), \left(-3, -3 + 2\sqrt{3}\right)$

11. Increasing on $\left(-\infty, 2-\dfrac{\sqrt{39}}{3}\right), \left(2+\dfrac{\sqrt{39}}{3}, \infty\right)$;

    decreasing on $\left(2-\dfrac{\sqrt{39}}{3}, 2\right), \left(2, 2+\dfrac{\sqrt{39}}{3}\right)$

13. Increasing on $(-\infty, 3.125)$;
    decreasing on $(3.125, \infty)$

15. Decreasing on $(0, 1/3)$; increasing on $(1/3, \infty)$

17. Increasing on $\left(k\pi, k\pi + \dfrac{\pi}{2}\right), \ k \in \mathbb{Z}$;

    decreasing on $\left(k\pi + \dfrac{\pi}{2}, k\pi\right), \ k \in \mathbb{Z}$

19. Decreasing on $(0, 1/4)$; increasing on $(1/4, \infty)$

21. Increasing on $\left(-\infty, -2/\sqrt{11}\right), \left(2/\sqrt{11}, \infty\right)$;
    decreasing on $\left(-2/\sqrt{11}, 2/\sqrt{11}\right)$

23. Rel. max. at $x = -5$; rel. min. at $x = 1$

25. Rel. max. at $x = -3, 1$; rel. min. at $x = 0.5$

27. Rel. max. at $x = 1$

29. Rel. max. at $x = -3 - 2\sqrt{3}$;
    rel. min. at $x = -3 + 2\sqrt{3}$

31. Rel. max. at $x = 2 - \dfrac{\sqrt{39}}{3}$;

    rel. min. at $x = 2 + \dfrac{\sqrt{39}}{3}$

33. Rel. max. at $t = 3.125$

35. Rel. min. at $s = 1/3$

37. Rel. min. at $x = k\pi, k \in \mathbb{Z}$;

    rel. max. at $x = (2k+1)\dfrac{\pi}{2}, \ k \in \mathbb{Z}$

39. Rel. min. at $t = 1/4$

41. Rel. max. at $x = -2/\sqrt{11}$; rel. min. at $x = 2/\sqrt{11}$

43. Decreasing on $(-\infty, -1)$; increasing on $(-1, \infty)$;
    local min. at $x = -1$; concave up on $(-\infty, \infty)$

45. Decreasing on $(-\infty, -2), (2, \infty)$;
    increasing on $(-2, 2)$;
    local min. at $x = -2$; local max. at $x = 2$;
    concave down on $\left(-\infty, -2\sqrt{3}\right), \left(0, 2\sqrt{3}\right)$;
    concave up on $\left(-2\sqrt{3}, 0\right), \left(2\sqrt{3}, \infty\right)$;
    inflection points at $x = \pm 2\sqrt{3}, 0$

47. Increasing on $(-\infty, -1), (1, \infty)$; decreasing on
    $(-1, 1)$; local max. at $x = -1$; local min. at $x = 1$;
    concave down on $\left(-\infty, -\sqrt{2}/2\right), \left(0, \sqrt{2}/2\right)$;
    concave up on $\left(-\sqrt{2}/2, 0\right), \left(\sqrt{2}/2, \infty\right)$;
    inflection points at $x = \pm\sqrt{2}/2, 0$

49. Decreasing on $(-\pi/2, -\pi/4), (\pi/4, \pi/2)$;
    increasing on $(-\pi/4, \pi/4)$;
    local min. at $x = -\pi/4$; local max. at $x = \pi/4$;
    concave up on $(-\pi/2, 0)$; concave down on
    $(0, \pi/2)$; inflection point at $x = 0$

51. Concave down on $(-\infty, \infty)$

53. Concave down on $(-\infty, 1/4)$;
    concave up on $(1/4, \infty)$

55. Concave up on $(-\infty, -2), (0, \infty)$;
    concave down on $(-2, 0)$

57. Concave down on $(-\infty, -3), (0, 1)$;
    concave up on $(-3, 0), (1, \infty)$

59. Concave down on $(-\infty, 5)$; concave up on $(5, \infty)$

61. Concave up on $(-\infty, \infty)$

63. Increasing on $(-\infty, \infty)$; no extrema;
    no inflection points

65. Increasing on $(-\infty, 5)$; decreasing on $(5, \infty)$;
    rel. max. (abs. max.) at $x = 5$;
    concave down on $(-\infty, \infty)$; no inflection points

67. Decreasing on $(-\infty, -3/2), (1, \infty)$; increasing on
    $(-3/2, 1)$; rel. min. at $x = -3/2$; rel. max. at
    $x = 1$; concave up on $(-\infty, -1/4)$; concave down
    on $(-1/4, \infty)$; inflection point at $x = -1/4$

69. Increasing on $\left(-\infty, \dfrac{-1-\sqrt{21}}{2}\right), \left(1, \dfrac{-1+\sqrt{21}}{2}\right)$;

    decreasing on $\left(\dfrac{-1-\sqrt{21}}{2}, 1\right), \left(\dfrac{-1+\sqrt{21}}{2}, \infty\right)$;

    rel. max. (abs. max.) at $x = \dfrac{-1-\sqrt{21}}{2}$; rel. max.

    at $x = \dfrac{-1+\sqrt{21}}{2}$; rel. min. at $x = 1$; concave

    down on $\left(-\infty, -\sqrt{2}\right), \left(\sqrt{2}, \infty\right)$; concave up on
    $\left(-\sqrt{2}, \sqrt{2}\right)$; inflection points at $x = \pm\sqrt{2}$

71. Decreasing on $(-\infty, 2), (2, \infty)$; no extrema;
    concave down on $(-\infty, 2)$; concave up on $(2, \infty)$;
    no inflection points

73. Increasing on $\left(-\infty, 4-\dfrac{\sqrt{66}}{2}\right), \left(4+\dfrac{\sqrt{66}}{2}, \infty\right)$;

decreasing on $\left(4-\dfrac{\sqrt{66}}{2}, 4\right), \left(4, 4+\dfrac{\sqrt{66}}{2}\right)$;

rel. max. at $x = 4 - \dfrac{\sqrt{66}}{2}$;

rel. min. at $x = 4 + \dfrac{\sqrt{66}}{2}$;

concave down on $(-\infty, 4)$; concave up on $(4, \infty)$;
no inflection points

75. Increasing on $(-\infty, -1), (-1, \infty)$; no extrema;
concave up on $(-\infty, -1)$; concave down on
$(-1, \infty)$; no inflection points

77. Decreasing on $(-\infty, 0), (4/15, \infty)$; increasing
on $(0, 4/15)$; rel. min. at $x = 0$; rel. max. at
$x = 4/15$; concave up on $(-\infty, -2/15)$; concave
down on $(-2/15, \infty)$; inflection point at
$x = -2/15$

79. Increasing on $(0, 1/2)$; decreasing on $(1/2, \infty)$;
rel. max. (abs. max.) at $x = 1/2$; concave down

on $\left(0, \dfrac{1+\sqrt{2}}{2}\right)$; concave up on $\left(\dfrac{1+\sqrt{2}}{2}, \infty\right)$;

inflection point at $x = \dfrac{1+\sqrt{2}}{2}$

81. For $k \in \mathbb{Z}$: increasing on $\left(\dfrac{3\pi}{4} + 2k\pi, \dfrac{7\pi}{4} + 2k\pi\right)$;

decreasing on $\left(-\dfrac{\pi}{4} + 2k\pi, \dfrac{3\pi}{4} + 2k\pi\right)$;

rel. max. (abs. max.) at $x = -\dfrac{\pi}{4} + 2k\pi$;

rel. min. (abs. min.) at $x = \dfrac{3\pi}{4} + 2k\pi$;

concave down on $\left(-\dfrac{3\pi}{4} + 2k\pi, \dfrac{\pi}{4} + 2k\pi\right)$;

concave up on $\left(\dfrac{\pi}{4} + 2k\pi, \dfrac{5\pi}{4} + 2k\pi\right)$;

inflection points at $x = \dfrac{\pi}{4} + 2k\pi$

83. Answers will vary. **85.** Answers will vary.

87. Answers will vary. **89.** Answers will vary.

91. Direction is negative on $(0, 3/4)$, positive
on $(3/4, \infty)$; changes direction at $t = 3/4$;
acceleration is constant $a = 4$, never zero

93. Direction is positive on $(0, 1), (4, \infty)$, negative
on $(1, 4)$; changes direction at $t = 1$ and $t = 4$;
acceleration is negative on $(0, 2.5)$, positive on
$(2.5, \infty)$, zero at $t = 2.5$

95. Direction is negative on $\left(\dfrac{\pi}{4} + 2k\pi, \dfrac{5\pi}{4} + 2k\pi\right)$,
$k \in \mathbb{Z}$, positive elsewhere; changes direction

at $t = \dfrac{\pi}{4} + k\pi$; acceleration is positive on

$\left(\dfrac{\pi}{2} + 2k\pi, \dfrac{3\pi}{2} + 2k\pi\right)$, $k \in \mathbb{N} \cup \{0\}$, negative

elsewhere, zero at $t = \dfrac{\pi}{2} + k\pi$

97.

99.

101.

103. a. Midday (temp. is rising, rate of increase is slowing)

   b. Morning (temp. is rising, rate of increase is accelerating)

   c. Early morning (temp. has local minimum)

   d. Evening or nighttime (temp. is falling, rate of decrease is accelerating)

   e. Early morning, just before minimum temperature (temp. is falling, rate of increase is slowing)

   f. Early afternoon (temp. has reached maximum)

105. a. $P(n) = -\dfrac{n^2}{10} + 50n - 5000$

   b. P is increasing on $(200, 250)$, decreasing on $(250, 350)$; P is maximal when producing 250 liners

107. Local max. at $x = -4$; local min. at $x = 3$

113. $a < 0, 3ac > b^2$

115. $a > 0, b = -6a, c = 0, d$ is arbitrary   **121.** True

123. False; consider $f(x) = -1/x^2$ with $c = 0$.

125. False; consider $f(x) = g(x) = -x$.

127. False; consider $f(x) = x^3$ with $c = 0$.

129.

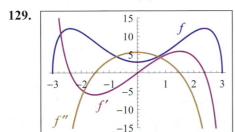

131. Answers will vary.   **133.** Answers will vary.
135. Answers will vary.

Section 4.4

1. 12   **3.** 1/2

5. Limit does not exist; l'Hôpital's Rule does not apply.

7. 0

9. Limit does not exist; l'Hôpital's Rule does not apply.

11. 1/2   **13.** ∞/∞; $0.2x^2 + 1$ dominates

15. $0 \cdot \infty$; e^{-x} dominates   **17.** 0   **19.** $-\infty$

21. L'Hôpital's Rule does not apply; the limit is 0.

23. L'Hôpital's Rule does not apply; the limit is 0.

25. 0   **27.** 0   **29.** $-\infty$ **31.** 1/2

33. $2/\ln 3$ **35.** 0   **37.** 1

39. L'Hôpital's Rule does not apply; the limit does not exist.

41. -2 **43.** ∞ **45.** $\dfrac{\ln 3}{\ln 2}$

47. L'Hôpital's Rule does not apply; the limit is 0.

49. $0 \cdot \infty$; the limit is 0.

51. Not indeterminate form; the limit is 0.

53. ∞^0; the limit is 1.   **55.** $\infty - \infty$; the limit is $-\infty$.

57. $\infty - \infty$; the limit is $-3/2$.

59. 1^∞; the limit is $2/e$.

61. Not indeterminate form; the limit is ∞.

63. ∞^0; the limit is 1.   **65.** 1^∞; the limit is 1.

67. Not indeterminate form; the limit is 0.

69. ∞/∞; the limit is $1/100$.   **71.** 1^∞; the limit is e^2.

73. $0/0$; the limit is $2/3$.   **75.** 0   **77.** 1

79. L'Hôpital's Rule does not apply; the limit is 0.

81. 1   **83.** 1/2   **85.** $-1/2$

87. $2/0$ is not indeterminate; l'Hôpital's Rule does not apply.

89. L'Hôpital's Rule is applied incorrectly; the factors of a product are differentiated.

91. The third use of l'Hôpital's Rule is unjustified; the limit is $-\infty$.

93. The limit is 0.   **95.** The limit is 0.

97. The limit is ∞.   **99.** The limit is 1.

101. The limit is -5.   **103.** The limit is 1.

105. The limit is 1.   **107.** The limit is 0.

109. The limit is 1.   **115.** $c = 1/2$

117. The theorem does not apply since $g(a) = g(b)$.

119. $c = 1$ **121.** $c = -\pi/4$ **123.** $c = 1/3$

127. The limit is $16a/9$.   **129.** $0 \cdot \infty$; the limit is 0.

131. $\infty - \infty$; the limit is $1/3$.

133. The limit of the limits is 0.

Section 4.5

1. First derivative: a; second derivative: b

3. First derivative: b; second derivative: a

5. Dom $= \mathbb{R}$; x-int.: $0, \dfrac{-3 \pm 3\sqrt{5}}{2}$; y-int.: 0;
   no asymptotes;
   increasing on $(-\infty, -3), (1, \infty)$;
   decreasing on $(-3, 1)$;
   rel. max. at $x = -3$; rel. min. at $x = 1$;
   concave down on $(-\infty, -1)$;
   concave up on $(-1, \infty)$;
   inflection point at $x = -1$

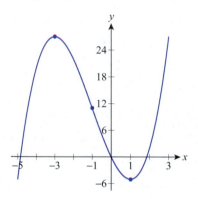

7. Dom $= \mathbb{R}$; y-int.: 0; no asymptotes;
   decreasing on $(-\infty, -4), (-2, 1)$;
   increasing on $(-4, -2), (1, \infty)$;
   rel. min. at $x = -4$; rel. max. at $x = -2$;
   abs. min. at $x = 1$;

   concave up on $\left(-\infty, \dfrac{-5 - \sqrt{19}}{3}\right), \left(\dfrac{-5 + \sqrt{19}}{3}, \infty\right)$;

   concave down on $\left(\dfrac{-5 - \sqrt{19}}{3}, \dfrac{-5 + \sqrt{19}}{3}\right)$;

   inflection points at $x = \dfrac{-5 \pm \sqrt{19}}{3}$

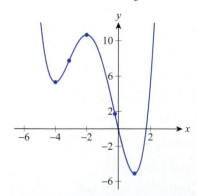

9. Dom $= \mathbb{R}$; x-int.: $\pm 1, \pm \sqrt{2}$; y-int.: 2;
   no asymptotes; even function;
   decreasing on $\left(-\infty, -\sqrt{3/2}\right), \left(0, \sqrt{3/2}\right)$;
   increasing on $\left(-\sqrt{3/2}, 0\right), \left(\sqrt{3/2}, \infty\right)$;
   rel. max. at $x = 0$;
   abs. min. at $x = \pm\sqrt{3/2}$;
   concave up on $\left(-\infty, -\sqrt{2}/2\right), \left(\sqrt{2}/2, \infty\right)$;
   concave down on $\left(-\sqrt{2}/2, \sqrt{2}/2\right)$;
   inflection points at $x = \pm\sqrt{2}/2$

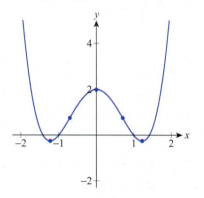

11. Dom $= \mathbb{R}$; x-int.: $0, \sqrt[3]{3}$; y-int.: 0;
    no asymptotes;
    increasing on $\left(-\infty, 0\right), \left(\sqrt[3]{6/5}, \infty\right)$;
    decreasing on $\left(0, \sqrt[3]{6/5}\right)$;
    rel. max. at $x = 0$; rel. min. at $x = \sqrt[3]{6/5}$;
    concave down on $\left(-\infty, \sqrt[3]{3/10}\right)$;
    concave up on $\left(\sqrt[3]{3/10}, \infty\right)$;
    inflection point at $x = \sqrt[3]{3/10}$

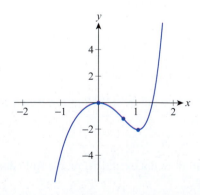

13. Dom $= (-\infty, 2) \cup (2, \infty)$; y-int.: $3/2$;
vert. asym.: $x = 2$; horiz. asym.: $y = 0$;
increasing on $(-\infty, 2), (2, \infty)$; no extrema;
concave up on $(-\infty, 2)$; concave down on $(2, \infty)$;
no inflection points

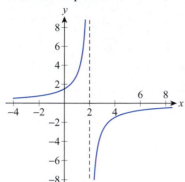

15. Dom $= (-\infty, -2) \cup (-2, 2) \cup (2, \infty)$;
odd function; x-int.: 0; y-int.: 0;
vert. asym.: $x = \pm 2$; horiz. asym.: $y = 0$;
decreasing on $(-\infty, -2), (-2, 2), (2, \infty)$; no
extrema; concave down on $(-\infty, -2), (0, 2)$;
concave up on $(-2, 0), (2, \infty)$;
inflection point at $x = 0$

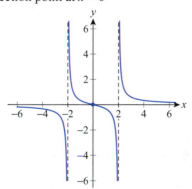

17. Dom $= \mathbb{R}$; x-int.: 1, 5; y-int.: 1; no asymptotes;
decreasing on $(-\infty, 3)$; increasing on $(3, \infty)$;
abs. min. at $x = 3$

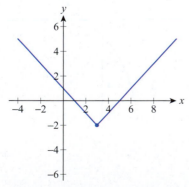

19. Dom $= \mathbb{R}$; even function; y-int.: $2/5$;
no asymptotes; decreasing on $(-\infty, 0)$;
increasing on $(0, \infty)$; abs. min. at $x = 0$;
concave down everywhere on $\mathbb{R}$;
no inflection points

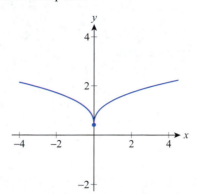

21. Dom $= \mathbb{R}$; x-int.: $1 + 2\sqrt[3]{4}$; y-int.: 3;
no asymptotes; decreasing everywhere on $\mathbb{R}$;
no extrema; concave down on $(-\infty, 1)$;
concave up on $(1, \infty)$; inflection point at $x = 1$

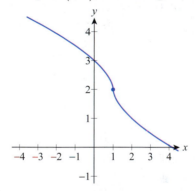

23. Dom $= \mathbb{R}$; y-int.: 2; no asymptotes; decreasing on
$\left(-\infty, -\dfrac{\ln 2}{3}\right)$; increasing on $\left(-\dfrac{\ln 2}{3}, \infty\right)$;
abs. min. at $x = -\dfrac{\ln 2}{3}$;
concave up everywhere on $\mathbb{R}$; no inflection points

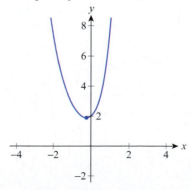

25. Dom $= (0, \infty)$; x-int.: 1;

no asymptotes;

decreasing on $(0, 1/e)$;

increasing on $(1/e, \infty)$;

abs. min. at $x = 1/e$;

concave up on $(0, \infty)$;

no inflection points

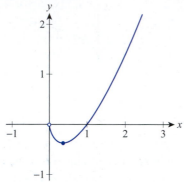

27. Dom $= [-2, 2]$; odd function;

x-int.: 0, ± 2; y-int.: 0;

no asymptotes;

decreasing on $\left(-2, -\sqrt{2}\right), \left(\sqrt{2}, 2\right)$;

increasing on $\left(-\sqrt{2}, \sqrt{2}\right)$;

abs. min. at $x = -\sqrt{2}$; abs. max. at $x = \sqrt{2}$;

concave up on $(-2, 0)$;

concave down on $(0, 2)$;

inflection point at $x = 0$

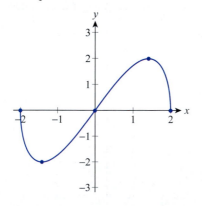

29. Dom $= (-\infty, 7) \cup (7, \infty)$; y-int.: -1;

vert. asym.: $x = 7$; slant asym.: $y = x + 7$;

increasing on $\left(-\infty, 7 - 2\sqrt{14}\right), \left(7 + 2\sqrt{14}, \infty\right)$;

decreasing on $\left(7 - 2\sqrt{14}, 7\right), \left(7, 7 + 2\sqrt{14}\right)$;

rel. max. at $x = 7 - 2\sqrt{14}$;

rel. min. at $x = 7 + 2\sqrt{14}$;

concave down on $(-\infty, 7)$; concave up on $(7, \infty)$;

no inflection points

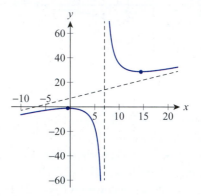

31. Dom $= \mathbb{R}$; x-int.: 0; y-int.: 0;

no asymptotes;

increasing everywhere on $\mathbb{R}$;

no extrema;

concave down on $\left(-\infty, \dfrac{\ln 4}{3}\right)$;

concave up on $\left(\dfrac{\ln 4}{3}, \infty\right)$;

inflection point at $x = \dfrac{\ln 4}{3}$

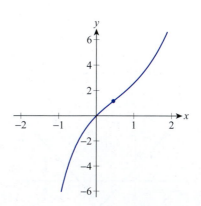

33. Dom $= (-\infty, 2) \cup (2, \infty)$; y-int.: $-5/4$;
vert. asym.: $x = 2$;
horiz. asym.: $y = 0$;
decreasing on $(-\infty, 2)$;
increasing on $(2, \infty)$;
no extrema;
concave down on $(-\infty, 2), (2, \infty)$;
no inflection points

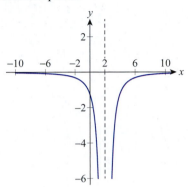

35. Dom $= (-\infty, -1) \cup (-1, 1) \cup (1, \infty)$;
odd function; x-int.: 0, y-int.: 0;
vert. asym.: $x = \pm 1$; horiz. asym.: $y = 0$;
increasing on $(-\infty, -1), (-1, 1), (1, \infty)$;
no extrema;
concave up on $(-\infty, -1), (0, 1)$;
concave down on $(-1, 0), (1, \infty)$;
inflection point at $x = 0$

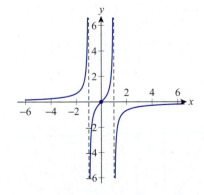

37. Dom $= \mathbb{R}$; x-int.: $4/3$; y-int.: $(4/3)^{1/3}$;
no asymptotes;
decreasing on $(-\infty, 4/3)$;
increasing on $(4/3, \infty)$;
abs. min. at $x = 4/3$;
concave up everywhere on $\mathbb{R}$;
no inflection points

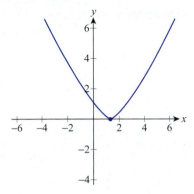

39. Dom $= \mathbb{R}$; odd function; x-int.: $0, \pm 1$; y-int.: 0;
no asymptotes;

increasing on $\left(-\infty, -\dfrac{1}{3\sqrt{3}}\right), \left(\dfrac{1}{3\sqrt{3}}, \infty\right)$;

decreasing on $\left(-\dfrac{1}{3\sqrt{3}}, \dfrac{1}{3\sqrt{3}}\right)$;

rel. max. at $x = -\dfrac{1}{3\sqrt{3}}$; rel. min. at $x = \dfrac{1}{3\sqrt{3}}$;

concave down on $(-\infty, 0)$;

concave up on $(0, \infty)$; inflection point at $x = 0$

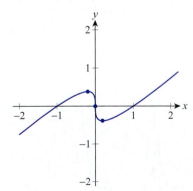

41. Dom = $\mathbb{R}$; odd function;

x-int.: $0, \pm 1$; y-int.: 0; no asymptotes;

decreasing on $\left(-\infty, -\sqrt{3/5}\right), \left(\sqrt{3/5}, \infty\right)$;

increasing on $\left(-\sqrt{3/5}, \sqrt{3/5}\right)$;

rel. min. at $x = -\sqrt{3/5}$; rel. max. at $x = \sqrt{3/5}$;

concave up on $\left(-\infty, -3/\sqrt{5}\right), (-1, 0), \left(1, 3/\sqrt{5}\right)$;

concave down on $\left(-3/\sqrt{5}, -1\right), (0, 1), \left(3/\sqrt{5}, \infty\right)$;

inflection points at $x = 0, \pm \dfrac{3}{\sqrt{5}}, \pm 1$

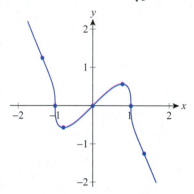

43. Dom = $\mathbb{R}$; 2π-periodic; for the period
$-\pi/2 \le x \le 3\pi/2$ we have the following:

y-int.: -1; no asymptotes;

increasing on $(-\pi/2, \pi/2)$;

decreasing on $(\pi/2, 3\pi/2)$;

abs. min. at $x = -\pi/2, 3\pi/2$;

abs. max. at $x = \pi/2$;

concave up on $(-\pi/2, \pi/6), (5\pi/6, 3\pi/2)$;

concave down on $(\pi/6, 5\pi/6)$;

inflection points at $x = \pi/6, 5\pi/6$

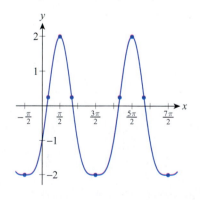

45. Dom $= (-\infty, 0) \cup (0, \infty)$; odd function;

x-int.: ± 1; no asymptotes;

decreasing on $\left(-e^{-3/5}, 0\right), \left(0, e^{-3/5}\right)$;

increasing on $\left(-\infty, -e^{-3/5}\right), \left(e^{-3/5}, \infty\right)$;

rel. max. at $x = -e^{-3/5}$; rel. min. at $x = e^{-3/5}$;

concave down on $\left(-\infty, -e^{-21/10}\right), \left(0, e^{-21/10}\right)$;

concave up on $\left(-e^{-21/10}, 0\right), \left(e^{-21/10}, \infty\right)$;

inflection points at $x = \pm e^{-21/10}$

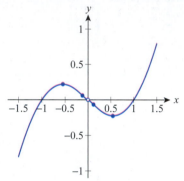

47. Dom = $\mathbb{R}$; x-int.: $\dfrac{-1 \pm \sqrt{433}}{9}$; y-int.: 6;

no asymptotes;

increasing on $(-\infty, -1/9)$;

decreasing on $(-1/9, \infty)$; abs. max. at $x = -1/9$;

concave down everywhere on $\mathbb{R}$;

no inflection points

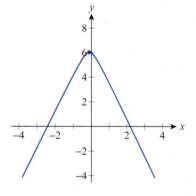

49.

51.

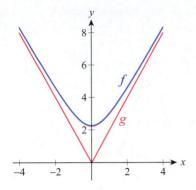

53.

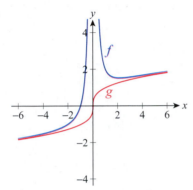

55. No; depending on the starting point, the method may converge to a different root, or not converge at all.

57. 2.65914   **59.** 1.60943   **61.** −1.52137

63. 0.48046, 5.61953   **65.** −0.75487

67. 0.44171   **69.** 0.69005   **71.** 0.876726

73. 0.930407   **75.** 2.475353   **77.** 0.5671

79. 1.0768, 3.6435   **81.** ±1.15932

83. $f'(x_1) = 0$

85. $x_{n+1} = -x_n$ for all n.

87. x_n alternates between the values of 3 and 1.

89. $x_{n+1} = \dfrac{k-1}{k} x_n + \dfrac{a}{k (x_n)^{k-1}}$

91. 1.41421   **93.** 2.15443   **95.** 0.14285

99.

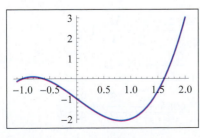

Section 4.6

1. 60 and 60   **3.** −18 and 18

5. n and n **7.** 1 and 1

9. 120 and 80   **11.** 4 and 4   **13.** 1 and 0

15. $w = 2\sqrt{k/3}$; $h = 2k/3$ **17.** $w = 2, h = 24$

19. $y = -\dfrac{1}{\sqrt[3]{3}} x + \sqrt[3]{9} + 1$ **21.** $\left(1/2, \sqrt{2}/2\right)$

23. Width: $2\sqrt{3}$, height: 2   **25.** 60 ft by 30 ft

27. Each pen is to be $30\sqrt{2} \times 20\sqrt{2}$ ft, joining along the $20\sqrt{2}$ ft side.

29. Width of cross section: $w = 2r/\sqrt{3}$ in., depth of cross section: $h = 2\sqrt{2}r/\sqrt{3}$ in.

31. Cut at $\dfrac{30\pi}{4+\pi}$ in., and use the first piece to form the circle.

33. In the first case, form a circle without cutting. In the second case, form a square.

37. A square of side length $\sqrt{A}$ units

39. $r = h = \sqrt{\dfrac{S}{3\pi}}$ in.

41. A $2 \times 2 \times 8$ in. box is cheapest, costing $2.88.

43. $\alpha = 2\sqrt{2/3\pi}$ **45.** 14×21 in.

47. $\alpha = 2$ rad, $r = 13$ ft   **49.** $s = 4\sqrt{14}$ ft

51. The wire needs to be staked 18 ft from the taller antenna.

53. The wire needs to be staked approx. 15.865 ft from the taller antenna.

55. We need to fold $(1/4)(8.5) = 2.125$ in. from the right lower corner.

57. Radius of base: $r = 2\sqrt{2}R/3$, height: $h = 4R/3$

59. Radius of base: $r = \dfrac{HR}{2(H-R)}$, height: $h = \dfrac{H(H-2R)}{2(H-R)}$

61. $\dfrac{65}{3} \times \dfrac{65}{3} \times \dfrac{130}{3}$ in.   **63.** $b = P/4$

65. He should land approx. 9 miles from the store.

67. Approx. 10 minutes and 21.2 seconds later

69. $t = \pi/4$ s; $v(\pi/4) = -3\sqrt{2}$ ft/s   **71.** $\theta = \pi/4$

73. $420; maximum revenue: $588,000

75. $1900; apartments rented: 95

77. 2600 cars; $29.75 each

81. $x = 7.5$ units

83. The total volume is maximum when we paint a sphere of radius $\dfrac{1}{2\sqrt{\pi}}$ ft (approx. 3.39 in.) and no tetrahedron. The total volume is a minimum when we paint a sphere of radius approx. 1.63 in., and a tetrahedron of side length approx. 7.99 in.

Section 4.7

9. $F(x) = x$

11. $H(x) = x^4 - \dfrac{1}{2}x^2$

13. $V(x) = \tan x + \dfrac{3}{2}x^2$

15. $F(x) = 5e^x$

17. $U(t) = 2/t^2$

19. $W(z) = \arcsin z$

21. $F(x) = x^2 - 3x + C,\ C = 3$

23. $F(x) = 2\sqrt{x} + C,\ C = -1$

25. $F(x) = C,\ C = 1$

27. $F(x) = \dfrac{x^4}{4} + \dfrac{1}{x} + C,\ C = -1/4$

29. $F(x) = 10^x + C,\ C = -9$

31. $F(x) = \tan\left(\dfrac{\pi}{4}x\right) + x + C,\ C = -1$

33. Answers will vary. Sample graph:

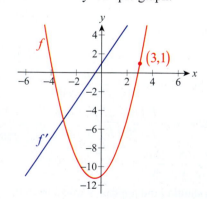

35. Answers will vary. Sample graph:

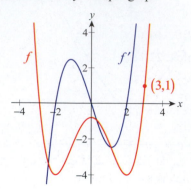

37. $F(x) = 2x^3 - 2x^2 + 1.5x + C$

39. $H(x) = \dfrac{x^6}{2} - 2x^5 + \dfrac{x^3}{3} + 7x + C$

41. $V(x) = 2x^3 + \dfrac{39}{2}x^2 + 18x + C$

43. $H(x) = \dfrac{2x^{9/2}}{9} + C$ **45.** $N(x) = \dfrac{x^2}{2} + 7\ln|x| + C$

47. $A(y) = \dfrac{3}{11}y^{11/3} - \dfrac{6}{7}y^{7/3} + y + C$

49. $G(t) = \dfrac{e^{3t}}{3} - 3\sec t + C$ **51.** $T(\theta) = \dfrac{\theta^2}{2} + \sin\theta + C$

53. $V(x) = \csc x - \cot x + C = \tan(x/2) + C$

55. $W(x) = \csc x + C$ **57.** $A(x) = \dfrac{5}{3}\arctan 3x + C$

59. $C(x) = \dfrac{4}{5}\operatorname{arcsec} 5x + C$

61. $f(x) = \dfrac{\pi}{2}x^2 - \pi x + \dfrac{\pi}{2}$ **63.** $f(x) = x^2 - 2x + 2$

65. $f(x) = \dfrac{9}{28}x^{7/3} - \dfrac{3}{4}x + \dfrac{4}{7}$

67. $f(x) = \dfrac{8}{105}x^{7/2} + \dfrac{x^3}{6} + \dfrac{x^2}{2} - x + 7$

69. $f(x) = 2\arctan 2x + \dfrac{\pi}{2}$

71. $f(x) = \dfrac{\sin 2x}{8} + \dfrac{x^2}{2} + \dfrac{3x}{4} - 1$

73. $f(x) = \dfrac{2^{5x}}{(\ln 32)^2} + Cx + D$

75. $f(x) = -\cos x - \dfrac{e^{2x}}{4} - \dfrac{3}{2}x + \dfrac{9}{4}$

77. 39 ft

Taylor Series and Maclaurin Series

Given a function f with derivatives of all orders throughout an open interval containing a, the power series

$$\sum_{n=0}^{\infty} \frac{f^{(n)}(a)}{n!}(x-a)^n = f(a) + f'(a)(x-a) + \frac{f''(a)}{2!}(x-a)^2 + \frac{f'''(a)}{3!}(x-a)^3 + \cdots$$

is called the Taylor series generated by f about a. The Taylor series generated by f about 0 is also known as the Maclaurin series generated by f.

VECTOR CALCULUS

Properties of Scalar Multiplication and Vector Addition

For vectors $\mathbf{u}$, $\mathbf{v}$, and $\mathbf{w}$ and scalars a and b:

Scalar Multiplication Properties	Vector Addition Properties
$a(\mathbf{u}+\mathbf{v}) = a\mathbf{u} + a\mathbf{v}$	$\mathbf{u} + \mathbf{v} = \mathbf{v} + \mathbf{u}$
$(a+b)\mathbf{u} = a\mathbf{u} + b\mathbf{u}$	$\mathbf{u} + (\mathbf{v}+\mathbf{w}) = (\mathbf{u}+\mathbf{v}) + \mathbf{w}$
$(ab)\mathbf{u} = a(b\mathbf{u}) = b(a\mathbf{u})$	$\mathbf{u} + \mathbf{0} = \mathbf{u}$
$1\mathbf{u} = \mathbf{u}$, $0\mathbf{u} = \mathbf{0}$, and $a\mathbf{0} = \mathbf{0}$	$\mathbf{u} + (-\mathbf{u}) = \mathbf{0}$
$|a\mathbf{u}| = |a||\mathbf{u}|$	

Dot Product

Given two vectors $\mathbf{u} = \langle u_1, u_2, u_3 \rangle$ and $\mathbf{v} = \langle v_1, v_2, v_3 \rangle$, the dot product $\mathbf{u} \cdot \mathbf{v}$ of the two vectors is the scalar defined by

$$\mathbf{u} \cdot \mathbf{v} = u_1 v_1 + u_2 v_2 + u_3 v_3.$$

A similar formula defines the dot product of two vectors in $\mathbb{R}^2$.

Properties of the Dot Product

For vectors $\mathbf{u}$, $\mathbf{v}$, and $\mathbf{w}$ and scalar a:

$\mathbf{u} \cdot \mathbf{v} = \mathbf{v} \cdot \mathbf{u}$

$\mathbf{0} \cdot \mathbf{u} = 0$

$\mathbf{u} \cdot (\mathbf{v}+\mathbf{w}) = \mathbf{u} \cdot \mathbf{v} + \mathbf{u} \cdot \mathbf{w}$

$a(\mathbf{u} \cdot \mathbf{v}) = (a\mathbf{u}) \cdot \mathbf{v} = \mathbf{u} \cdot (a\mathbf{v})$

$\mathbf{u} \cdot \mathbf{u} = |\mathbf{u}|^2$

Dot Product and the Angle between Two Vectors

If two nonzero vectors $\mathbf{u}$ and $\mathbf{v}$ are depicted so that their initial points coincide, and if θ represents the smaller of the two angles formed by $\mathbf{u}$ and $\mathbf{v}$ (so that $0 \le \theta \le \pi$), then

$$\mathbf{u} \cdot \mathbf{v} = |\mathbf{u}||\mathbf{v}|\cos\theta.$$

Projection of u onto v

Let $\mathbf{u}$ and $\mathbf{v}$ be nonzero vectors. The projection of $\mathbf{u}$ onto $\mathbf{v}$ is the vector

$$\text{proj}_{\mathbf{v}}\mathbf{u} = \left(\frac{\mathbf{u} \cdot \mathbf{v}}{|\mathbf{v}|^2} \right)\mathbf{v}.$$

Cross Product

Given $\mathbf{u} = \langle u_1, u_2, u_3 \rangle$ and $\mathbf{v} = \langle v_1, v_2, v_3 \rangle$,

$$\mathbf{u} \times \mathbf{v} = \begin{vmatrix} \mathbf{i} & \mathbf{j} & \mathbf{k} \\ u_1 & u_2 & u_3 \\ v_1 & v_2 & v_3 \end{vmatrix}$$

$$= \begin{vmatrix} u_2 & u_3 \\ v_2 & v_3 \end{vmatrix}\mathbf{i} - \begin{vmatrix} u_1 & u_3 \\ v_1 & v_3 \end{vmatrix}\mathbf{j} + \begin{vmatrix} u_1 & u_2 \\ v_1 & v_2 \end{vmatrix}\mathbf{k}$$

$$= (u_2 v_3 - u_3 v_2)\mathbf{i} - (u_1 v_3 - u_3 v_1)\mathbf{j} + (u_1 v_2 - u_2 v_1)\mathbf{k}.$$

Properties of the Cross Product

For vectors $\mathbf{u}$, $\mathbf{v}$, and $\mathbf{w}$ in $\mathbb{R}^3$ and scalars a and b:

$\mathbf{u} \times \mathbf{v} = -(\mathbf{v} \times \mathbf{u})$

$\mathbf{0} \times \mathbf{u} = \mathbf{0}$

$\mathbf{u} \times (\mathbf{v}+\mathbf{w}) = \mathbf{u} \times \mathbf{v} + \mathbf{u} \times \mathbf{w}$

$(\mathbf{u}+\mathbf{v}) \times \mathbf{w} = \mathbf{u} \times \mathbf{w} + \mathbf{v} \times \mathbf{w}$

$(a\mathbf{u}) \times (b\mathbf{v}) = (ab)(\mathbf{u} \times \mathbf{v})$

$\mathbf{u} \cdot (\mathbf{v} \times \mathbf{w}) = (\mathbf{u} \times \mathbf{v}) \cdot \mathbf{w}$

$\mathbf{u} \times (\mathbf{v} \times \mathbf{w}) = (\mathbf{u} \cdot \mathbf{w})\mathbf{v} - (\mathbf{u} \cdot \mathbf{v})\mathbf{w}$

Coordinate Conversion Relationships

Cylindrical and Cartesian	Spherical and Cylindrical	Spherical and Cartesian
$r^2 = x^2 + y^2$	$\rho^2 = r^2 + z^2$	$\rho^2 = x^2 + y^2 + z^2$
$x = r\cos\theta$	$r = \rho\sin\varphi$	$x = \underbrace{\rho\sin\varphi}_{r}\cos\theta$
$y = r\sin\theta$	$\theta = \theta$	$y = \underbrace{\rho\sin\varphi}_{r}\sin\theta$
$z = z$	$z = \rho\cos\varphi$	$z = \rho\cos\varphi$

Gradient Vector

Given a function $f(x_1, x_2, \ldots, x_n)$,

$$\nabla f(x_1, x_2, \ldots, x_n)$$
$$= \langle f_{x_1}(x_1, x_2, \ldots, x_n), f_{x_2}(x_1, x_2, \ldots, x_n), \ldots, f_{x_n}(x_1, x_2, \ldots, x_n) \rangle.$$

INTEGRATION

Properties of the Definite Integral

Given the integrable functions f and g on the interval $[a,b]$ and any constant k, the following properties hold.

1. $\int_a^a f(x)\,dx = 0$

2. $\int_b^a f(x)\,dx = -\int_a^b f(x)\,dx$

3. $\int_a^b k\,dx = k(b-a)$

4. $\int_a^b kf(x)\,dx = k\int_a^b f(x)\,dx$

5. $\int_a^b \left[f(x) \pm g(x) \right]\,dx = \int_a^b f(x)\,dx \pm \int_a^b g(x)\,dx$

6. $\int_a^c f(x)\,dx + \int_c^b f(x)\,dx = \int_a^b f(x)\,dx$, assuming each integral exists

7. If $f(x) \le g(x)$ on $[a,b]$, then $\int_a^b f(x)\,dx \le \int_a^b g(x)\,dx$.

8. If $m = \min\limits_{a \le x \le b} f(x)$ and $M = \max\limits_{a \le x \le b} f(x)$, then

$$m(b-a) \le \int_a^b f(x)\,dx \le M(b-a).$$

The Substitution Rule

If $u = g(x)$ is a differentiable function whose range is the interval I, and if f is continuous on I, then

$$\int f\big(g(x)\big)g'(x)\,dx = \int f(u)\,du.$$

Hence, if F is an antiderivative of f on I,

$$\int f\big(g(x)\big)g'(x)\,dx = F\big(g(x)\big) + C.$$

Integration by Parts

Given differentiable functions f and g,

$$\int f(x)g'(x)\,dx = f(x)g(x) - \int g(x)f'(x)\,dx.$$

If we let $u = f(x)$ and $v = g(x)$, then $du = f'(x)\,dx$ and $dv = g'(x)\,dx$ and the equation takes on the more easily remembered differential form

$$\int u\,dv = uv - \int v\,du.$$

The Fundamental Theorem of Calculus

Part I

Given a continuous function f on an interval I and a fixed point $a \in I$, define the function F on I by $F(x) = \int_a^x f(t)\,dt$. Then $F'(x) = f(x)$ for all $x \in I$.

Part II

If f is a continuous function on the interval $[a,b]$ and if F is any antiderivative of f on $[a,b]$, then

$$\int_a^b f(x)\,dx = F(b) - F(a).$$

SEQUENCES AND SERIES

Summation Facts and Formulas

Constant Rule for Finite Sums:

$$\sum_{i=1}^n c = nc, \text{ for any constant } c$$

Constant Multiple Rule for Finite Sums:

$$\sum_{i=1}^n ca_i = c\sum_{i=1}^n a_i, \text{ for any constant } c$$

Sum/Difference Rule for Finite Sums:

$$\sum_{i=1}^n (a_i \pm b_i) = \sum_{i=1}^n a_i \pm \sum_{i=1}^n b_i$$

Sum of the First n Positive Integers:

$$\sum_{i=1}^n i = \frac{n(n+1)}{2}$$

Sum of the First n Squares:

$$\sum_{i=1}^n i^2 = \frac{n(n+1)(2n+1)}{6}$$

Sum of the First n Cubes:

$$\sum_{i=1}^n i^3 = \frac{n^2(n+1)^2}{4}$$

Geometric Series

For a geometric sequence $\{a_n\}$ with common ratio r:

Partial Sum: $s_n = \dfrac{a(1-r^n)}{1-r}$, if $r \ne 0, 1$

Infinite Sum: $\displaystyle\sum_{n=1}^\infty ar^{n-1} = \frac{a}{1-r}$, if $|r| < 1$

Binomial Series

For any real number m and $-1 < x < 1$:

$$(1+x)^m = \sum_{n=0}^\infty \binom{m}{n} x^n$$

$$= 1 + mx + \frac{m(m-1)}{2!}x^2 + \frac{m(m-1)(m-2)}{3!}x^3 + \cdots$$

$$+ \frac{m(m-1)\cdots(m-n+1)}{n!}x^n + \cdots$$

DERIVATIVES

The Derivative of a Function

The derivative of f, denoted f', is the function whose value at the point x is

$$f'(x) = \lim_{h \to 0} \frac{f(x+h) - f(x)}{h},$$

provided the limit exists.

Elementary Differentiation Rules

Constant Rule: $\dfrac{d}{dx}(k) = 0$

Constant Multiple Rule: $\dfrac{d}{dx}\left[kf(x)\right] = kf'(x)$

Sum/Difference Rule: $\dfrac{d}{dx}\left[f(x) \pm g(x)\right] = f'(x) \pm g'(x)$

Product Rule: $\dfrac{d}{dx}\left[f(x)g(x)\right] = f'(x)g(x) + f(x)g'(x)$

Quotient Rule: $\dfrac{d}{dx}\left[\dfrac{f(x)}{g(x)}\right] = \dfrac{f'(x)g(x) - f(x)g'(x)}{\left[g(x)\right]^2}$

Power Rule: $\dfrac{d}{dx}(x^r) = rx^{r-1}$

Chain Rule: $\dfrac{d}{dx}\left[f(g(x))\right] = f'(g(x)) \cdot g'(x)$

Derivatives of Trigonometric Functions

$\dfrac{d}{dx}(\sin x) = \cos x$ $\qquad$ $\dfrac{d}{dx}(\csc x) = -\csc x \cot x$

$\dfrac{d}{dx}(\cos x) = -\sin x$ $\qquad$ $\dfrac{d}{dx}(\sec x) = \sec x \tan x$

$\dfrac{d}{dx}(\tan x) = \sec^2 x$ $\qquad$ $\dfrac{d}{dx}(\cot x) = -\csc^2 x$

Derivatives of Inverse Trigonometric Functions

$\dfrac{d}{dx}(\sin^{-1} x) = \dfrac{1}{\sqrt{1-x^2}}$ $\qquad$ $\dfrac{d}{dx}(\csc^{-1} x) = -\dfrac{1}{|x|\sqrt{x^2-1}}$

$\dfrac{d}{dx}(\cos^{-1} x) = -\dfrac{1}{\sqrt{1-x^2}}$ $\qquad$ $\dfrac{d}{dx}(\sec^{-1} x) = \dfrac{1}{|x|\sqrt{x^2-1}}$

$\dfrac{d}{dx}(\tan^{-1} x) = \dfrac{1}{1+x^2}$ $\qquad$ $\dfrac{d}{dx}(\cot^{-1} x) = -\dfrac{1}{1+x^2}$

Derivatives of Exponential and Logarithmic Functions

$\dfrac{d}{dx}(e^x) = e^x$ $\qquad$ $\dfrac{d}{dx}(a^x) = a^x \ln a$

$\dfrac{d}{dx}(\ln x) = \dfrac{1}{x}$ $\qquad$ $\dfrac{d}{dx}(\log_a x) = \dfrac{1}{\ln a} \cdot \dfrac{1}{x}$

Derivatives of Hyperbolic Functions

$\dfrac{d}{dx}(\sinh x) = \cosh x$ $\qquad$ $\dfrac{d}{dx}(\operatorname{csch} x) = -\operatorname{csch} x \coth x$

$\dfrac{d}{dx}(\cosh x) = \sinh x$ $\qquad$ $\dfrac{d}{dx}(\operatorname{sech} x) = -\operatorname{sech} x \tanh x$

$\dfrac{d}{dx}(\tanh x) = \operatorname{sech}^2 x$ $\qquad$ $\dfrac{d}{dx}(\coth x) = -\operatorname{csch}^2 x$

Derivatives of Inverse Hyperbolic Functions

$\dfrac{d}{dx}(\sinh^{-1} x) = \dfrac{1}{\sqrt{1+x^2}}$

$\dfrac{d}{dx}(\cosh^{-1} x) = \dfrac{1}{\sqrt{x^2-1}}, \quad x > 1$

$\dfrac{d}{dx}(\tanh^{-1} x) = \dfrac{1}{1-x^2}, \quad |x| < 1$

$\dfrac{d}{dx}(\operatorname{csch}^{-1} x) = \dfrac{-1}{|x|\sqrt{1+x^2}}$

$\dfrac{d}{dx}(\operatorname{sech}^{-1} x) = \dfrac{-1}{x\sqrt{1-x^2}}, \quad 0 < x < 1$

$\dfrac{d}{dx}(\coth^{-1} x) = \dfrac{1}{1-x^2}, \quad |x| > 1$

The Derivative Rule for Inverse Functions

If a function f is differentiable on an interval (a,b), and if $f'(x) \neq 0$ for all $x \in (a,b)$, then f^{-1} both exists and is differentiable on the image of the interval (a,b) under f, denoted as $f((a,b))$ in the formula below. Further,

$$\text{if } x \in (a,b), \text{ then } \left(f^{-1}\right)'(f(x)) = \frac{1}{f'(x)},$$

and

$$\text{if } x \in f((a,b)), \text{ then } \left(f^{-1}\right)'(x) = \frac{1}{f'(f^{-1}(x))}.$$

The Mean Value Theorem

If f is continuous on the closed interval $[a,b]$ and differentiable on (a,b), then there is at least one point $c \in (a,b)$ for which

$$f'(c) = \frac{f(b) - f(a)}{b - a}.$$

GEOMETRY

A = area, C = circumference, SA = surface area or lateral area, V = volume

Rectangle	**Circle**	**Triangle**	**Parallelogram**	**Trapezoid**
$A = lw$	$A = \pi r^2 \quad C = 2\pi r$	$A = \dfrac{1}{2}bh$	$A = lh$	$A = \dfrac{1}{2}h(b+c)$

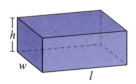

Rectangular Prism

$V = lwh \quad SA = 2lh + 2wh + 2lw$

Sphere

$V = \dfrac{4}{3}\pi r^3 \quad SA = 4\pi r^2$

Rectangular Pyramid

$V = \dfrac{1}{3}lwh$

Right Cylinder

$V = (\text{Area of Base})h$

Right Circular Cylinder

$V = \pi r^2 h \quad SA = 2\pi r^2 + 2\pi rh$

Cone

$V = \dfrac{1}{3}\pi r^2 h \quad SA = \pi r^2 + \pi r\sqrt{r^2 + h^2}$

LIMITS

Definition of Limit

Let f be a function defined on an open interval containing c, except possibly at c itself. We say that the limit of $f(x)$ as x approaches c is L, and write $\lim\limits_{x\to c} f(x) = L$, if for every number $\varepsilon > 0$ there is a number $\delta > 0$ such that $|f(x) - L| < \varepsilon$ whenever x satisfies $0 < |x - c| < \delta$.

Basic Limit Laws

Sum/Difference Law: $\lim\limits_{x\to c}\left[f(x) \pm g(x)\right] = \lim\limits_{x\to c} f(x) \pm \lim\limits_{x\to c} g(x)$

Constant Multiple Law: $\lim\limits_{x\to c}\left[kf(x)\right] = k\lim\limits_{x\to c} f(x)$

Product Law: $\lim\limits_{x\to c}\left[f(x)g(x)\right] = \lim\limits_{x\to c} f(x) \cdot \lim\limits_{x\to c} g(x)$

Quotient Law: $\lim\limits_{x\to c}\dfrac{f(x)}{g(x)} = \dfrac{\lim\limits_{x\to c} f(x)}{\lim\limits_{x\to c} g(x)}$, provided $\lim\limits_{x\to c} g(x) \neq 0$

Squeeze Theorem

If $g(x) \le f(x) \le h(x)$ for all x in some open interval containing c, except possibly at c itself, and if $\lim\limits_{x\to c} g(x) = \lim\limits_{x\to c} h(x) = L$, then $\lim\limits_{x\to c} f(x) = L$ as well.

Continuity at a Point

Given a function f defined on an open interval containing c, we say f is continuous at c if

$$\lim\limits_{x\to c} f(x) = f(c).$$

L'Hôpital's Rule

Suppose f and g are differentiable at all points of an open interval I containing c, and that $g'(x) \neq 0$ for all $x \in I$ except possibly at $x = c$. Suppose further that either

$$\lim\limits_{x\to c} f(x) = 0 \quad \text{and} \quad \lim\limits_{x\to c} g(x) = 0$$

or

$$\lim\limits_{x\to c} f(x) = \pm\infty \quad \text{and} \quad \lim\limits_{x\to c} g(x) = \pm\infty.$$

Then

$$\lim\limits_{x\to c}\dfrac{f(x)}{g(x)} = \lim\limits_{x\to c}\dfrac{f'(x)}{g'(x)},$$

assuming the limit on the right is a real number or ∞ or $-\infty$.

. It falls for approx. 7.49 s, and the speed of impact is approx. 83.47 m/s.

1. $16\sqrt{15} \approx 61.97$ ft/s

83. Approx. 22.12 m

87. Approx. 54.77 m

89. 5275

91. The position function (assuming the initial position is 0) is $p(t) = 5.\overline{3}t^2$. It covers 12.5 m in the first 1.5 s. Its acceleration time from 0 to 60 mph is 2.4 s.

93. 8.25 ft/s^2

95. $p(t) = -\dfrac{4}{105}t^{5/2}(3t-14)+3$; in 5 seconds it will be at $x = 3 - \dfrac{20\sqrt{5}}{21} \approx 0.87$. Its instantaneous velocity will be zero at $t = 0$ and $t = 10/3$.

97. True

99. False; if $F(x)$ is an antiderivative, then $F(x)+C$ is also an antiderivative for any $C \in \mathbb{R}$.

101. True

103. False; consider $f(x) = g(x) = 1$ with $F(x) = G(x) = x$. Note that $x \cdot x = x^2$ is not an antiderivative of $1 \cdot 1 = 1$.

Chapter 4 Review

1. Answers will vary.

3. Answers will vary.

5. Abs. min.: $f(1) = -1/3$; abs. max.: $f(-1) = 7/3$

7. Abs. min.: $f(-2) = f(4) = 0$; abs. max.: $f(1) = 9$

9. Abs. min.: $f(2) = -\sqrt{2}$; abs. max.: $f(1/3) = 2/(3\sqrt{3})$

11. No extrema

13. Abs. min.: $f(\pi) = 1$; no maximum

17. $c = 4/3$ **19.** $c = \sqrt[3]{65/4}$

23. $f(x) = \sin x + e^x + 2$

27. Increasing on $(-\infty, 3)$; decreasing on $(3, \infty)$

29. Decreasing on $(-\infty, -3), (2, 4)$; increasing on $(-3, 2), (4, \infty)$

31. Concave down on $(-\infty, 0)$; concave up on $(0, \infty)$

33. Concave up on $(-\infty, -1), (1, \infty)$; concave down on $(-1, 1)$

35. Decreasing on $(-\infty, 0)$; increasing on $(0, \infty)$; rel. min. (abs. min.) at $x = 0$; concave down on $\left(-\infty, -1/\sqrt[4]{3}\right), \left(1/\sqrt[4]{3}, \infty\right)$; concave up on $\left(-1/\sqrt[4]{3}, 1/\sqrt[4]{3}\right)$; inflection points at $x = \pm 1/\sqrt[4]{3}$

37. Direction is positive on $\left(0, \sqrt{3}\right)$, negative on $\left(\sqrt{3}, 3\right)$; changes direction at $t = \sqrt{3}$; acceleration is positive on $(0, 1)$, negative on $(1, 3)$, zero at $t = 1$

39. 0

41. L'Hôpital's Rule does not apply; the limit is ∞.

43. $-\infty$ **45.** ∞

47. $1/2$ **49.** 0

51. Dom $= \mathbb{R}$; x-int.: $0, \dfrac{3 \pm 3\sqrt{21}}{2}$; y-int.: 0; no asymptotes; increasing on $(-\infty, -3), (5, \infty)$; decreasing on $(-3, 5)$; rel. max. at $x = -3$; rel. min. at $x = 5$; concave down on $(-\infty, 1)$; concave up on $(1, \infty)$; inflection point at $x = 1$

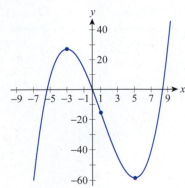

53. Dom = $\mathbb{R}$; x-int.: 0, 2; y-int.: 0;
no asymptotes;
decreasing on $(-\infty, 3/2)$;
increasing on $(3/2, \infty)$;
abs. min. at $x = 3/2$;
concave up on $(-\infty, 0), (1, \infty)$;
concave down on $(0, 1)$;
inflection points at $x = 0, 1$

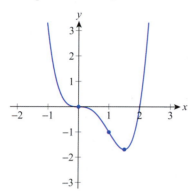

55. Dom = $\mathbb{R}$; odd function; x-int.: 0; y-int.: 0;
horiz. asym.: $y = 0$;
decreasing on $(-\infty, -1), (1, \infty)$;
increasing on $(-1, 1)$;
abs. min. at $x = -1$; abs. max. at $x = 1$;
concave down on $\left(-\infty, -\sqrt{3}\right), \left(0, \sqrt{3}\right)$;
concave up on $\left(-\sqrt{3}, 0\right), \left(\sqrt{3}, \infty\right)$;
inflection points at $x = 0, \pm\sqrt{3}$

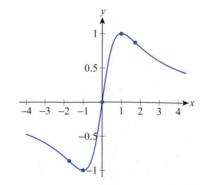

57. Dom = $\mathbb{R}$; x-int.: 0, 1; y-int.: 0; no asymptotes;
increasing on $(-\infty, 0), (1/2, \infty)$;
decreasing on $(0, 1/2)$;
rel. max. at $x = 0$; rel. min. at $x = 1/2$;
concave down on $(-\infty, 0)$;
concave up on $(0, \infty)$;
inflection point at $x = 0$

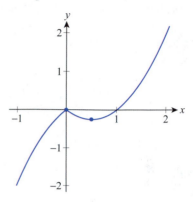

59. Dom = $(-\infty, -2) \cup (-2, 2) \cup (2, \infty)$;
odd function; x-int.: 0; y-int.: 0;
vert. asym.: $x = \pm 2$; slant asym.: $y = x$;
increasing on $\left(-\infty, -2\sqrt{3}\right), \left(2\sqrt{3}, \infty\right)$;
decreasing on $\left(-2\sqrt{3}, -2\right), (-2, 2), \left(2, 2\sqrt{3}\right)$;
rel. max. at $x = -2\sqrt{3}$; rel. min. at $x = 2\sqrt{3}$;
concave down on $(-\infty, -2), (0, 2)$;
concave up on $(-2, 0), (2, \infty)$;
inflection point at $x = 0$

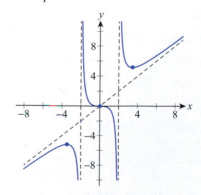

61. 1.97435   **63.** 0.741912   **65.** 0.4797

67. $\dfrac{1}{n-\sqrt[n]{n}}$ **69.** $a = \left(a_1 + \cdots + a_n\right)/n$

71. $l = 2\sqrt{2/3}$; $w = 2/\sqrt{3}$ **73.** $5\sqrt{10} \times 10\sqrt{10}$

75. $4 - 2\sqrt{3}$ **77.** The equilateral triangle

ask is impossible, since $A(r)$ has no
imum. (The surface area approaches infinity
$r \to 0$ or $r \to \infty$.)

ate needs to reach the other side approximately
00 ft downstream from the restaurant.

$F(x) = x^5 - 1.2x^4 + e^2 x + C$

$F(x) = 0.16x^{5/2} - 4x^{1/2} + C$

7. $F(x) = x^2 + \tan 2x + C$

89. $F(x) = \dfrac{3}{2}\arctan 2x + C$

91. $f(x) = \dfrac{x^3}{3} - \dfrac{5x^2}{2} + 8x - \dfrac{17}{3}$ **93.** 80 ft/s

95. False; consider $y = 1/x$ on $(0,1)$.

97. True   **99.** True

101. False; consider $y = x^5$ at $x = 0$.

113.

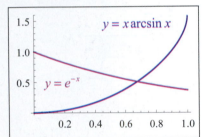

Chapter 5

Section 5.1

1. $(O_4 + U_4)/2 = 17$ **3.** $(O_8 + U_8)/2 = 16.25$

5. Approx. 126 ft; an overestimate

7. a. $O_{12} = 112$ m; $U_{12} = 97.45$ m

    b. $M_6 = 104.8$ m

    c. $(O_{12} + U_{12})/2 = 104.725$ m;
    M_6 is greater, because the curve is likely
    concave down, as indicated by the data.

9. Left-endpoint est. ≈ 4.146 (underestimate);
    right-endpoint est. ≈ 6.146 (overestimate);
    midpoint est. ≈ 5.384 (overestimate)

11. Left-endpoint est. $= 25/12$ (overestimate);
    right-endpoint est. $= 77/60$ (underestimate);
    midpoint est. $= 496/315$ (underestimate)

13. Left-endpoint est. ≈ 2.367 (overestimate);
    right-endpoint est. ≈ 1.582 (underestimate);
    midpoint est. ≈ 2.013 (overestimate)

15. $\displaystyle\sum_{i=1}^{33} 3i$ **17.** $\displaystyle\sum_{i=1}^{100} \dfrac{1}{i^2}$ **19.** $\displaystyle\sum_{i=-2}^{38} a_{2i+1}$

21. $\displaystyle\sum_{i=1}^{n} f\!\left(\dfrac{3i}{n}\right)$ **23.** $\displaystyle\sum_{i=0}^{n} f(x_i^*)\Delta x$ **25.** 64

27. $501 + n$ **29.** $25 + 2n$ **31.** $11/6$ **33.** 255

35. $\dfrac{7n^2 + 2n}{2}$ **37.** 17,080   **39.** $\dfrac{6n^3 + 15n^2 + 11n}{2}$

41. $\dfrac{n(n+1)(3n^2 - 5n - 4)}{6}$

43. $\dfrac{1}{1} - \dfrac{1}{2} + \dfrac{1}{2} - \cdots + \dfrac{1}{10} - \dfrac{1}{11} = \dfrac{10}{11}$

45. $\sqrt{1} - \sqrt{2} + \sqrt{2} - \cdots + \sqrt{n} - \sqrt{n+1} = 1 - \sqrt{n+1}$

47. $e^2 - e^3 + e^3 - \cdots + e^{n+3} - e^{n+4} = e^2 - e^{n+4}$

49. 88,573   **51.** $\dfrac{3\left(1 - (2/3)^{100}\right)}{5}$ **55.** $\displaystyle\sum_{j=1}^{n-2} \dfrac{1}{j}$

57. $\displaystyle\sum_{l=5}^{21} \cos(2l\pi)$ **59.** 6   **61.** $1/4$ **63.** $11/12$

65. The region between the graph of $x^3 + 2x$ and the
    x-axis from 0 to 3

67. The region between the graph of $\sin x$ and the
    x-axis from 0 to $\pi/2$

71. Approx. 44.44 m   **73.** 22.5 m   **85.** 110

87. $\dfrac{mn(m+1)(n+1)}{4}$

89. $\displaystyle\lim_{n\to\infty} \dfrac{-\pi\csc\dfrac{\pi}{2n}\sin\dfrac{\pi - n\pi}{2n} + \pi\csc\dfrac{\pi}{2n}\sin\dfrac{\pi + n\pi}{2n}}{2n} = 2$

91. $\displaystyle\lim_{n\to\infty} \dfrac{3n + \csc\dfrac{\pi}{2n}\sin\dfrac{(2n+1)\pi}{2n}}{8n} = \dfrac{3}{8}$

Section 5.2

1. a. 10   **b.** $(60 + 9\pi)/8 \approx 11.0343$

    c. 9   **d.** $(9\pi/8) - 18 \approx -14.4657$

3. 64 ft; -96 ft; -800 ft   **5.** 16

7. $7564/11{,}025 \approx 0.686$

9. $1899/16{,}000 \approx 0.119$

11. $1 + \dfrac{\ln 3}{2} \approx 1.549$ **13.** $19/25$ **15.** $17/2$

17. $26/3$ **19.** 13 **21.** 4 **23.** $1/4$

25. $201/4$ **27.** $44/3$ **29.** $3/\sqrt[3]{4}$

31. $\displaystyle\lim_{n \to \infty} \sum_{i=1}^{n} \frac{4}{n} \cdot \left[\left(\frac{4i}{n} \right)^2 - \log_2 \frac{4i}{n} \right]$

33. $\displaystyle\lim_{n \to \infty} \sum_{i=1}^{n} \frac{b-2}{n} \sqrt[4]{2 + \frac{(b-2)i}{n}}$

35.

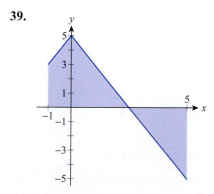

Value: -19

37.

Value: $-25/4$

39.

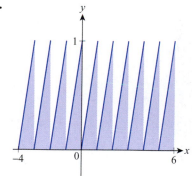

Value: 4

41.

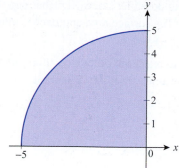

Value: $25\pi/4$

43.

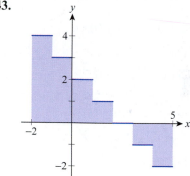

Value: 7

45.

Value: 5

53. No; see Exercise 51.   **55.** Yes   **57.** Yes

59. No; see Exercise 49.   **61.** E   **63.** A   **65.** C

67. $8 + (c - a)$ **69.** 0 **71.** $5\sqrt{2}/2$

73. Not possible   **75.** $1 + 5\pi$ **77.** $-1 - \dfrac{3\sqrt{2}}{4}$

79. $69/2$ **81.** $11/6$ **83.** 0 **85.** 2 **87.** 5

89. Not possible   **93.** $0 \le \displaystyle\int_2^3 \sqrt{3-x}\, dx \le 1$

95. $6 \le \displaystyle\int_0^6 \left(\frac{x^2}{32} - \frac{x}{4} + \frac{3}{2} \right) dx \le 9$

105. -1 **107.** $17/2$ **109.** $-5/3$ **111.** $13/5$

113. $6/5$ **115.** $-9/4$ **117.** $\int_0^3 \left(\frac{2}{9}x+4\right)dx = 13$

119. $\int_0^2 \sqrt{4-x^2}\,dx = \pi$ **123.** $\frac{\pi^3}{3} - \frac{3\pi^2}{2} + \pi$

125. False; consider $f(x) = g(x) = x$ on $[0,1]$.

Note: $\int_0^1 x^2\,dx \ne \int_0^1 x\,dx \cdot \int_0^1 x\,dx$.

127. True   **129.** True

Section 5.3

1. $c = \sqrt[3]{2}$ **3.** $c = \pm 2/\sqrt[4]{5}$

5. $c = \sin^{-1}(2/\pi),\ \pi - \sin^{-1}(2/\pi)$

7. $c = \sin^{-1}\left(\sqrt{\pi}/2\right),\ \pi - \sin^{-1}\left(\sqrt{\pi}/2\right)$

9. a. $F(2) = 6,\ F(4) = 9,\ F(6) = 6,\ F(8) = 2,$
$F(10) = 2$

b. $F(x) = \begin{cases} 3x & \text{if } 0 \le x \le 2 \\ -\dfrac{3}{4}x^2 + 6x - 3 & \text{if } 2 < x \le 6 \\ \dfrac{x^2}{2} - 9x + 42 & \text{if } 6 < x \le 10 \end{cases}$

c.

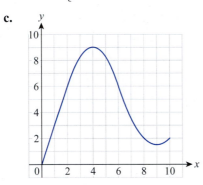

11. $14/3$ **13.** $\sqrt{2}$ **15.** $160/7$

17. $F'(x) = \frac{1}{3}\left(x^2 + \sqrt{x}\right)$ **19.** $G'(x) = \frac{x^4}{x^4+4}$

21. $y' = -\sin\sqrt{x+1}$ **23.** $y' = \left(27x^2 + 9\right)e^{3x-2}$

25. $y' = \left(\sin^2 x + e^{\sin x}\right)\cos x$

27. $y' = \pi \sin(\pi x) - \sin x$

29. $F'(x) = \dfrac{-\sqrt{1 + \sqrt{1 + (1/x^2)}}}{\sqrt{1 - x^2}}$

31. $H'(x) = \ln x - \dfrac{\ln(\ln x)}{x}$

33. $F(x) = 2x - 2$ **35.** $F(x) = -\dfrac{x^3}{3} - \dfrac{x^2}{2} + \dfrac{5}{6}$

37. $F(x) = 1 - \dfrac{1}{\sqrt{x}}$ **39.** -30 **41.** 175

43. 48 **45.** 990 **47.** $1 - \ln 4$

49. $\dfrac{14}{3} + 4\ln 2$ **51.** $\dfrac{8 - 2\sqrt{2}}{5}$ **53.** $\dfrac{45 - 36\sqrt[3]{2}}{32}$

55. $\dfrac{1}{2} - \dfrac{\sqrt{2}}{6}\pi^{3/2}$ **57.** $3\pi/4$ **59.** $e^{\pi/3}$

61. $44/3$ **63.** $\pi/2$ **65.** $e - \dfrac{1}{3}$

67. $\int_0^2 x^4\,dx = 32/5$ **69.** $\int_1^e (1/x)\,dx = 1$

71. 8 **73.** 3 **75.** $63/2$ **77.** $2\ln 2 - \ln 3$

79. $1 + 2\ln 2$ **81.** $2e - 9$ **83.** $3 + \dfrac{\pi}{4}$

85. $6 + \ln(33/13)$ **87.** $2 + \ln 2$

89. a. $\dfrac{3}{2} - \ln 4$ **b.** $1/2$ **91.** $f(x) = 2\cos 2x + 1$

93.

101.

$(0, h)$ $y = h - ax^2,\quad a > 0$

103. True   **105.** True

107. False; see part a. of Example 3.

109. a. \$3375; \$3250   **b.** 5.377 years

111. The following answers are based on a sine regression.

  a. $N(t) = 1.637\sin(0.023t - 0.936) + 2.583$

  b. Approx. 1972 cars

  c. Approx. 3 cars per minute

113. $F(x) = (12 - 2x)\sqrt{x}\cos\sqrt{x} + 6(x - 2)\sin\sqrt{x}$

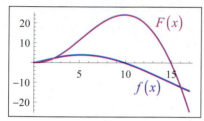

F is increasing where *f* is positive and decreasing where *f* is negative. *F* is concave up where *f* is increasing and concave down where *f* is decreasing. The zeros of *f* are critical points of *F*, and the critical points of *f* are potential inflection points for *F*.

115. $F(x) = \dfrac{\cos^2 x}{2} - \dfrac{\cos 8x + 7}{16}$

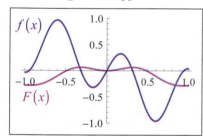

F is increasing where *f* is positive and decreasing where *f* is negative. *F* is concave up where *f* is increasing and concave down where *f* is decreasing. The zeros of *f* are critical points of *F*, and the critical points of *f* are potential inflection points for *F*.

Section 5.4

1. $2x^6 + 1.5x^5 - 0.25x^4 + 2x + C$

3. $\dfrac{10}{3}x^3 + 3x^2 - 4x + C$ **5.** $3/7$

7. $\dfrac{2x^{5/2}}{5} - \dfrac{x^2}{\sqrt{2}} + C$ **9.** $\pi\tan x - \sec x + C$

11. $\dfrac{-\csc 2x}{2} + C$ **13.** $\pi/2$ **15.** $-\dfrac{7}{5}\sin^{-1}(5x) + C$

17. $\dfrac{(x^4 + 2)^{3/2}}{6} + C$ **19.** $\tan^{-1}(2x) + C$

21. $2e^{2t+3} + C$ **23.** $\dfrac{5^{\arctan x}}{\ln 5} + C$

25. $2\tan(1 + \sqrt{s}) + C$ **27.** $-\tan(1/x) + C$

29. $\dfrac{-\csc^3 x}{3} + C$ **31.** $\dfrac{2z - 1 + \ln|2z - 1|}{4} + C$

33. $\dfrac{(6x + 20)(x - 5)^{3/2}}{15} + C$

35. $\dfrac{x\sqrt{1 - x^2} + \sin^{-1} x}{2} + C$

37. $\dfrac{(3x - 2)^8}{8} + C$ **39.** $\dfrac{(2x^2 + 3x)^{21}}{21} + C$

41. $\ln|\sin z| + C$ **43.** $\dfrac{(x^3 - 5)^{20}}{60} + C$

45. $e^{\sin x} + C$

47. $\dfrac{-(5 - s)^{38}}{38} + C$ **49.** $\dfrac{3(x^4 + 11)^{4/3}}{16} + C$

51. $\dfrac{-1}{x^2 + x - 7} + C$ **53.** $\dfrac{-\sqrt{2 + x^4}}{2} + C$

55. $-\dfrac{1}{4}\left(1 + \dfrac{1}{t}\right)^4 + C$ **57.** $2\ln(\ln 2t) + C$

59. $\dfrac{e^{x^2 - 3}}{2} + C$ **61.** $\dfrac{\sin \pi x}{\pi} + C$ **63.** $\dfrac{\sin^4 2t}{8} + C$

65. $e^{1 - \cot v} + C$ **67.** $\dfrac{2}{3(30 - x^{3/2})} + C$

69. $\ln(\sin^2 x + 2) + C$

71. $\sin(\sin x) + C$ **73.** $\dfrac{\ln|1 - \cos 2x|}{2} + C$

75. $\dfrac{\sec^3 2x - 3\sec 2x}{6} + C$ **77.** $\dfrac{4(1 + \sqrt{w})^{3/2}}{3} + C$

79. $\dfrac{(x + 1)^{13}}{13} - \dfrac{(x + 1)^{12}}{2} + C$

81. $\dfrac{1 + 4x + 7\ln|1 + 4x|}{16} + C$

83. $\dfrac{(1 - \ln x)^5}{5} - \dfrac{7(1 - \ln x)^4}{8} + C$

85. $f(x) = \dfrac{8}{3}(1 + x^2)^{3/2} - \dfrac{5}{3}$

87. $y(t) = \ln(\sin^2 t + e) - 1$

89. $y(x) = \dfrac{16x + 1 - \cos 4x}{16}$ **91.** $x = 5$

93. False; the indefinite integral is the set of all antiderivatives.

95. True

97. False; the integral doesn't fit the Substitution Rule with $F(g(x)) = \dfrac{(\cos x + 1)^3}{3}$. (You can check the falsity of the answer by differentiation.)

Section 5.5

1. 364/3   **3.** 512/27   **5.** $\dfrac{12^{100} - 5^{100}}{300}$ **7.** 2/7

9. $\dfrac{3(2^{4/3} - 1)}{16}$ **11.** 2/7   **13.** 1   **15.** 129/128

17. $\dfrac{\ln 2}{3}$ **19.** 1/2   **21.** $2/\pi$ **23.** 1/6

25. 0   **27.** $1 + \ln 2$ **29.** 0   **31.** 2

33. $\dfrac{4(\sqrt{2} - 1)}{3}$ **35.** $e^{\pi/3} - 1$ **37.** $\ln 2$

39. $\dfrac{20\sqrt{5} - 32}{3}$ **41.** $6 - 7\ln 3$ **43.** 53/72

45. 0   **47.** $33\ln 5 - 32$

49. $\dfrac{608 - 56\sqrt{2}}{15}$ **51.** $\dfrac{2\sqrt{2} - 1}{3}$ **53.** 64/3

55. 10.5845   **57.** 4/3   **59.** 9/2   **61.** 35/4

63. 37/12   **65.** 5/12   **67.** $\ln 2$ **69.** $\dfrac{11 + 5\sqrt{5}}{6}$

71. $1/e$ **73.** 324/5   **75.** 9/2   **77.** 37/32

79. 81   **81.** 7/2   **83.** 937/12

85. $144\sqrt{2} - \dfrac{2656}{15}$ **87.** 1   **89.** $\dfrac{4 - \pi}{\sqrt{2}}$

91. $9\pi/4$ **93.** $\dfrac{8}{5} + \dfrac{4}{\pi}$ **95.** 2

101. a. Bike #1

   b. Bike #2 starts gaining ground on #1.

   c. Yes, it happens when $\displaystyle\int_0^T v_1(t)\,dt = \int_0^T v_2(t)\,dt$, at about $T = 2.5$ seconds.

103.

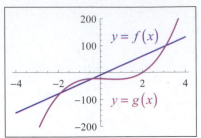

The area is approx. 151.932.

105.

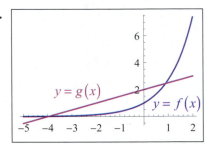

The area is approx. 3.562.

107.

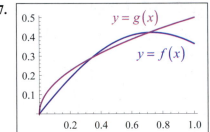

The area is approx. 0.019.

Chapter 5 Review

1. $(O_4 + U_4)/2 = 1.375$ **3.** $\displaystyle\sum_{i=1}^{100} \dfrac{(-1)^{i+1}}{i^3 + 1}$

5. $\displaystyle\sum_{i=1}^{50} f\left(\dfrac{2i}{n^2}\right)$ **7.** $212 + 2n$ **9.** 9065

11. $\dfrac{n^2 + 2n}{(n+1)^2}$ **13.** 0.769399

15. $\dfrac{mn(m+1)(n+1)(2n+1)}{12}$ **17.** 1/4

19. The region between the graph of $2x - x^2$ and the x-axis from 0 to 2; the limit is 4/3.

23. 0   **25.** 1.5   **27.** 1/4; $x = 3/4, 13/4$

29. $F'(x) = \sqrt{1 + x^2}$ **31.** 87/5   **33.** $\dfrac{3}{2} + \ln 4$

35. π **37.** $1 + 3\ln 2$ **39.** 9/20   **41.** $f(x) = \cos x$

43. $-2\sin^{-1} x + C$ **45.** $\sec x + \tan x + C$

47. $\dfrac{\left(2x^3-7\right)^{10}}{10}+C$ **49.** $\dfrac{-2}{x^2+1}+C$

51. $2\tan\sqrt{x}+C$ **53.** $e^x-\ln\left(e^x+1\right)+C$

55. $y(x)=\dfrac{2}{1-\sqrt{x}}+2$ **57.** $x=1+\dfrac{\ln 17}{2}$

59. $16\sqrt{6}$ **61.** $\dfrac{\pi}{4\ln 2}$ **63.** $1/3$ **65.** $4/21$

67. $\dfrac{3-e}{2}$ **69.** 2

73. If f is concave up, the integral is expected to be less than the average of L_n and R_n, while if f is concave down, the integral is expected to be greater than the average of L_n and R_n.

75. True

77. False; $\displaystyle\int\frac{1}{e^x}\,dx=\int e^{-x}\,dx=-e^{-x}+C.$

79. False; consider $f(x)=x-1$ on $[0,3]$.

81. False; $1/x^3$ is not continuous on $[-1,1]$.

Chapter 5 Project

3. $\dfrac{d}{dx}\,\mathrm{Si}(x)=f(x)$

5. Rel. max. at $x=(2k-1)\pi$; rel. min. at $x=2k\pi$, $k\in\mathbb{N}$

7.

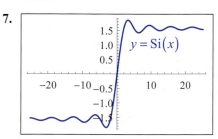

23. $287\pi/30$ **25.** $15\pi/2$ **27.** 18π

29. $4\pi/3$ **31.** $117\pi/2$ **33.** $128\pi/7$

35. $16\pi/3$ **37. a.** $2\pi/3$ **b.** $8\pi/3$

39. a. $\pi/7$ **b.** $9\pi/14$

41. a. $128\pi/45$ **b.** $224\pi/45$

43. a. $3\pi/13$ **b.** $51\pi/52$

45. a. $93\pi/5$ **b.** $411\pi/10$

47. a. $4\pi/3$ **b.** $\dfrac{3\pi^2+4\pi}{3}$

49. a. $2\pi/5$ **b.** $7\pi/5$

51. a. $8\pi/3$ **b.** $104\pi/15$

53. a. $6\pi/11$ **b.** $45\pi/22$

55. a. $762\pi/7$ **b.** $1077\pi/7$

57. a. π^2 **b.** $3\pi^2$ **59.** 261π **61.** $1250\pi/3$

63. $11\pi/30$ **65.** $56\pi/15$ **67.** $488\pi/15$

69. $\pi/2$ **71.** $\pi^2/4$

73. The solid (paraboloid) generated by rotating the region bounded by $y=0$ and $y=\sqrt{x}$, $0\le x\le 4$ about the x-axis

75. The solid generated by rotating the region bounded by $y=e$, $y=e^x$, and $x=0$ about the x-axis

77. The solid generated by rotating the region bounded by $y=\sqrt[4]{x}$ and $y=x^4$, $0\le x\le 1$ about the x-axis

79. The solid generated by rotating the region bounded by $y=\log_2 x$ and $y=x-1$, $1\le x\le 2$ about the x-axis

81. Approx. 81.5 mL

85. For min. volume: $c=\sqrt{2}$; for max. volume: $c=0,\,2$

87. Approx. 45.6 mL

89. $\dfrac{\pi\left(\pi^2-8\right)}{4}$ **91.** Approx. 109.745

93. $V\approx 13.021$

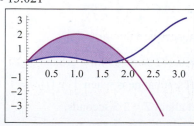

Chapter 6

Section 6.1

1. $3.2\ \text{m}^3$ **3.** $8/\left(15\sqrt{3}\right)$ **5.** 8 **7.** $1/\sqrt{3}$

9. $32/3$ **11.** $4/315$ **13.** $(2\pi+6)/3$

15. $138{,}240\sqrt{5}\ \text{cm}^3\approx 0.309\ \text{m}^3$

17. $4\sqrt{3}\pi\ \text{cm}^3$ **19.** $384\pi/7$ **21.** 19π

Section 6.2

1. 4π **3.** $24\pi/5$ **5.** $\pi/3$ **7.** $9\pi/2$

9. 18π **11.** $931\pi/75$ **13.** $81\pi/2$ **15.** $33\pi/2$

17. $10\pi/13$ **19.** $336\pi/5$ **21.** $\pi/2$

23. $\pi(2\ln 2-1)$ **25.** $2e(e-1)$ **27.** $112\pi/3$

29. $16\pi/3$ **31.** $19\pi/15$

33. $8\pi/3$ **35.** $\pi\ln 17$ **37.** π/e

39. a. $425\pi/12$ **b.** $100\pi/3$
    c. $125\pi/6$ **d.** $200\pi/3$

41. a. $544\pi/15$ **b.** $40\pi/3$
    c. $192\pi/5$ **d.** $40\pi/3$

43. a. 64π **b.** $512\pi/15$
    c. 8π **d.** $832\pi/15$

45. $V = 2\pi\displaystyle\int_0^{\ln 2}\left(2xe^{-x}-x\right)dx$

47. $V = 2\pi\displaystyle\int_0^1 x\arcsin x\,dx$

49. $V = 2\pi\displaystyle\int_0^{\pi/3} x\left(\sqrt{3}-\tan x\right)dx$

51. The solid (circular cone) generated by rotating the region bounded by the y-axis, $y = 2x$, and $y = 4$ about the y-axis

53. The solid (sphere) generated by rotating the region bounded by the x-axis and the semicircle $y = \sqrt{2x - x^2}$ about the x-axis

55. The solid generated by rotating the region bounded by the x-axis and $y = 1 - x^4$, $0 \le x \le 1$ about $x = -1$

57. The solid generated by rotating the region bounded by $y = \sin x$, $\pi/2 \le x \le \pi$, and the x-axis about $x = \pi/2$

59. The solid generated by rotating the region bounded by $y = \csc x$, $x = \pi/4$, $x = 3\pi/4$, and the x-axis about $x = -1$

61. $\pi/2$ **63.** $1024\pi/105$

65. $144\pi/5$ **69.** $4\sqrt{3}\pi$ cm^3 **71.** $V = 2\pi^2 Rr^2$

73. False; the volume is $k^2 V$.   **75.** True

77. False; compare the formulas
$\displaystyle\int_a^b \pi\left([f(x)]^2 - [g(x)]^2\right)dx$ and
$\displaystyle\int_a^b \pi\left([f(x)+1]^2 - [g(x)+1]^2\right)dx.$

79. $125\pi/312$

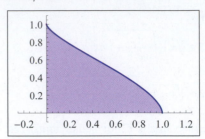

81. $\dfrac{416\pi^2}{81\sqrt{3}}$

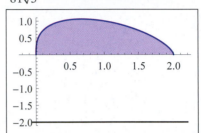

Section 6.3

1. $6\sqrt{10}$ **3.** $3\sqrt{5}$ **5.** $18 - \dfrac{4\sqrt{2}}{3}$ **7.** $\dfrac{e^2+7}{8}$

9. $521/480$ **11.** $\pi/2$ **13.** $\dfrac{e}{4} - \dfrac{1}{e} + \dfrac{3}{4}$ **15.** 1

17. $\displaystyle\int_0^2 \sqrt{1+4x^2}\,dx$; $s(x) = \displaystyle\int_0^x \sqrt{1+4t^2}\,dt$

19. $\displaystyle\int_0^1 \sqrt{4y^2 - 4y + 2}\,dy$; $s(y) = \displaystyle\int_0^y \sqrt{4t^2 - 4t + 2}\,dt$

21. $\displaystyle\int_1^2 \sqrt{\dfrac{4}{x^6}+1}\,dx$; $s(x) = \displaystyle\int_1^x \sqrt{\dfrac{4}{t^6}+1}\,dt$

23. $\displaystyle\int_0^2 \sqrt{e^{4-2x}+1}\,dx$; $s(x) = \displaystyle\int_0^x \sqrt{e^{4-2t}+1}\,dt$

25. $\displaystyle\int_{-1/2}^{1/2} \sqrt{\dfrac{2-x^2}{1-x^2}}\,dx$; $s(x) = \displaystyle\int_{-1/2}^x \sqrt{\dfrac{2-t^2}{1-t^2}}\,dt$

27. $12\sqrt{2}\pi$ **29.** $\left(40\sqrt{5}-8\right)\pi/3$ **31.** $241\pi/192$

33. $\left(16e - e^{-1} - 7\right)\pi/16$ **35.** 2π

37. $\left(8\sqrt{3} - \dfrac{8}{3}\right)\pi$ **39.** $\left(16 - 4\sqrt{2}\right)\pi/3$

41. $515\pi/64$ **43.** $12\sqrt{5}\pi$

45. $\pi\dfrac{[e(e+2)]^2 - 4e - 1}{4e^2}$

47. $2\pi\displaystyle\int_0^{\sqrt{3}/2} \sin^{-1}x\sqrt{\dfrac{2-x^2}{1-x^2}}\,dx$

49. $2\pi \int_2^4 \sqrt{\dfrac{1}{x^4} + \dfrac{4}{x^{10}}}\, dx$

51. $2\pi \int_1^8 \left(3 + \sqrt[3]{x}\right)\sqrt{\dfrac{1}{9x^{4/3}} + 1}\, dx$

53. $2\pi \int_0^1 \left(2 + \sqrt[4]{y+1}\right)\sqrt{1 + \dfrac{1}{16(y+1)^{3/2}}}\, dy$

55. The surface area is 1200π square units. Yes, because of symmetry, we do obtain the same answer when rotating about the y-axis.

57. a.

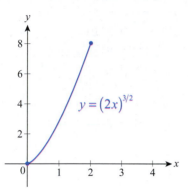

$y = (2x)^{3/2}$

b. $\dfrac{37\sqrt{37} - 1}{27}$

61. $\dfrac{\left(61\sqrt{61} + 185\sqrt{37} - 130\right)\pi}{30}$

63. $A = 4\pi^2 R r$

65. $2\pi\left(b^2 + \dfrac{a^2 b}{\sqrt{a^2 - b^2}}\sin^{-1}\dfrac{\sqrt{a^2 - b^2}}{a}\right)$

67. Approx. 6.7573 **69.** Approx. 1.1321

71. Approx. 2.6625 **73.** Approx. 3.8202

75. Approx. 14.4236 **77.** Approx. 22.9430

79. Approx. 57.7233 **81.** Approx. 7.6031

83. Approx. 3.0444 yd

Section 6.4

1. $M_0 = -5;\ \bar{x} = -5/11$ **3.** $M_0 = -6.5;\ \bar{x} = -0.5$

5. $M_x = 24,\ M_y = 30;\ (\bar{x}, \bar{y}) = (10/7, 8/7)$

7. She should sit on Christina's side, 1 ft ahead of her.

9. The nosepiece will land approx. 400 yd northeast from the target.

11. $\bar{x} = 63.7681$ **13.** $(23/18, 4/9)$

15. $(0.7157, 2.7917)$ **17.** $(-2, -8/5)$

19. $(-9/10, 18/5)$ **21.** $(8/5, 3/2)$ **23.** $(1,1)$

25. $(2/\ln 3, 1/\ln 27)$ **27.** $(0, -18/5)$

29. $(16/15, 64/21)$ **31.** $(\pi/2, 3/8)$ **33.** $(\ln 4, 0)$

35. $(0, 6/5)$ **37.** $\left(\dfrac{5 - 6\ln 2}{3 + 6\ln 2}, \dfrac{5}{6 + 12\ln 2}\right)$

39. $\left(\dfrac{14}{3\pi}, \dfrac{28}{9\pi}\right)$ **41.** $(3, 1/2)$ **43.** $(1/2, 8/15)$

45. $(2/3, 46/21)$ **47.** $(2/9, 11/21)$

49. $(137/76, 25/76)$ **51.** $(0, 3r/4)$

53. $\left(\dfrac{4r}{3\pi}, \dfrac{4r}{3\pi}\right)$ **55.** $(0, r/2)$

59. $\left(\dfrac{n+1}{2n+1}, \dfrac{n+1}{2n+4}\right)$; If $n \to \infty$ the centroid converges to $(1/2, 1/2)$.

61. $(h/3, R/3)$ **63.** $\dfrac{\pi r^3 (3\pi - 4)}{3}$ **65.** $828\pi/5$

67. $240\sqrt{2}\pi$ **69.** $2\pi r^2 (\pi - 2)$

73. False; consider Example 4.

75. False; he or she has to sit at two-thirds of the distance (when compared to the other child) from the pivot point (i.e., approximately 33% closer).

Section 6.5

1. 3433.5 J **3.** 27 ft-lb

5. a. 200 ft-lb **b.** 800 ft-lb

7. a. 4 ft-lb **b.** 3.0625 ft-lb

9. 7600 J **11.** 0.84 J **15.** 5/4 ft-lb

17. 10,950 ft-lb **19.** 3977.2 kJ **21.** 42.2321 kJ

23. $27{,}500\pi$ ft-lb $\approx 86{,}393.80$ ft-lb. The work is 3 times that in Example 4a.

25. a. 100,000 ft-lb **b.** 75,000 ft-lb

c. No. The lower half of the water needs to be pumped higher than the top half.

d. Approx. 6 minutes and 4 seconds; approx. 4 minutes 33 seconds

27. 2995.2 ft-lb **29.** 11,980.8 ft-lb **31.** $\delta\pi R^4/4$

33. 31.2 ft-lb **35. a.** 2620.8 lb **b.** 2433.6 lb

37. Approx. 98.17 N   **39.** Yes   **41.** 499.2 lb

43. Approx. 6.27 lb   **45.** $450\ln 3$ J ≈ 494.38 J

47. Approx. 620.83 J   **51.** 4.02286×10^9 ft-lb

53. 4.09308×10^6 kJ   **55.** Rw **57.** 6.75 J

71. Approx. 2.89 ft

Section 6.6

1. a. 1   **b.** $\dfrac{e^{-2} - e^2}{2} \approx -3.6269$

   c. $\dfrac{e^{\pi/2} - e^{-\pi/2}}{2} \approx 2.3013$

3. a. 0   **b.** 0   **c.** $\ln\left(2 + \sqrt{5}\right) \approx 1.4436$

5. Odd   **7.** Odd   **9.** Even

11.

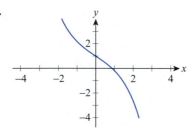

13.

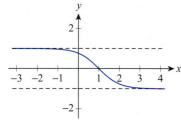

15.

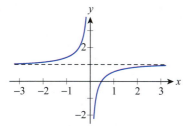

17.

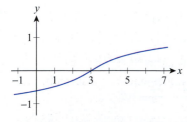

19.

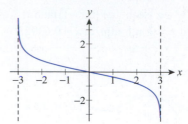

21.

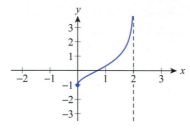

51. $\sinh 2x$ **53.** $2\operatorname{sech}^2 x \tanh x$

55. $\dfrac{2\cosh 2x - \sinh 2x}{e^x}$ **57.** $\dfrac{\operatorname{sech}^2 2x}{\sqrt{1 + \tanh 2x}}$

59. $3t^2\left[\cosh\left(1 + t^3\right) + \left(1 + t^3\right)\sinh\left(1 + t^3\right)\right]$

61. $\dfrac{4y(1 + y)}{(1 + 2y)^2}$ **63.** $\sinh x + C$ **65.** $\operatorname{sech}(1/z) + C$

67. $-\ln\left(\cosh\left(2 - w\right)\right) + C$ **69.** $\dfrac{2e^2}{e^4 - 1}$

71. $\csc^{-1}\left(\dfrac{4e^2}{e^4 + 1}\right) - \dfrac{\pi}{6}$ **83.** $\dfrac{7}{\sqrt{49x^2 - 4}}$

85. $-2\csc 2x$

87. $\dfrac{2}{1 - 4x^2}$ **89.** $\dfrac{1}{4}\sinh^{-1}\left(\dfrac{4x}{3}\right) + C$

91. $\tanh^{-1}\left(x - 3\right) + C$ **93.** $\operatorname{csch}^{-1}\left(\sin x\right) + C$

95. $2\sinh^{-1} 25$ **97.** $\ln\left(2 - \sqrt{3}\right)$

99. $-\dfrac{2}{3}\tanh^{-1}\sqrt{1 + x^3} + C$ **101.** $\ln\dfrac{4 + \sqrt{15}}{2 + \sqrt{3}}$

103. ∞ **105.** 1   **107.** $1/2$ **109.** 0

113. a. At the left endpoint: approx. -0.30452; at the right endpoint: approx. 0.30452; The parabola predicts ± 0.3.

   b. Approx. 3.0452; Yes, the answer is very close to that given by the parabolic model.

115. True   **117.** False; $\operatorname{csch} 0$ is undefined.

119. a. Approx. 625.093 ft   **b.** Approx. 598.45 ft

   c. Approx. ± 0.641

121. $f'(x) = (x^2 - x)\operatorname{sech}^2 x + (2x - 1)\tanh x$;
rel. max. at $x = 0$; rel. min. at $x \approx 0.63923$

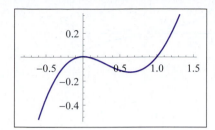

Chapter 6 Review

1. $2/3$ **3.** $\pi/2$ **5.** $\dfrac{\pi(\pi - 2)}{2}$

7. $8\pi/3$ **9.** $512\pi/21$

11. The solid (circular cone) generated by rotating the region bounded by the x-axis, $y = x$, $0 \le x \le 3$ about the x-axis

13. The solid (paraboloid) generated by rotating the first-quadrant region bounded by the graph of $y = 4 - x^2$ and the coordinate axes about the y-axis

15. 2π **17.** $8\pi/15$ **21.** $\displaystyle\int_1^5 \sqrt{16y^2 + 8y + 2}\, dy$

23. $\displaystyle\int_0^{\pi/3} \sec y\, dy$ **25.** $14/3$

27. $\dfrac{3 + e^4}{4}$ **29.** $e - \dfrac{1}{e}$

31. $\pi\displaystyle\int_1^e \sqrt{\ln(x^4) + \dfrac{1}{x^2}}\, dx$

33. $2\pi\displaystyle\int_0^4 (\sqrt{y} + 1)\sqrt{1 + \dfrac{1}{4y}}\, dy$

35. $61\pi/432$ **37.** $35\pi/16$ **39.** $111\pi/16$

41. $(2/5, 1)$ **43.** $\left(\dfrac{24\ln 2 - 14}{9}, \dfrac{13}{9}\right)$ **45.** $(8/7, 6/7)$

47. $(11/2, 0)$ **49.** $\left(0, \dfrac{4b}{3\pi}\right)$ **53.** 4200 ft-lb

55. 29.332 kJ   **57.** Approx. 123,686 ft-lb

59. 0.524 ft-lb   **63.** 39,240 kN

65. Approx. 3.30313×10^9 ft-lb

67. a. Undefined   **b.** 1   **73.** $e^{2x}(\operatorname{sech}^2 x + 2\tanh x)$

75. $5\coth(5x - 2)$ **77.** $-\operatorname{sech} x + C$

79. $\sinh^{-1}(x - 2) + C$ **81.** $\cosh^{-1}(e^x) + C$

85. $8c/9$

87. False; consider Example 3 and Exercise 60 of Section 6.2.

89. False; consider the region bounded by $y = 1$ and the x-axis over the interval $[0,1]$.

91. True   **93.** Approx. 2.104   **95.** Approx. 26

97. Approx 2.259   **99.** Approx. 14.424

Chapter 6 Project

5. $h_{max} = \dfrac{Rv_0^2}{2gR - v_0^2}$ **7.** Approx. 2.36 km/s

Chapter 7

Section 7.1

1. $e^x(x - 1) + C$ **3.** $x^4\left(\ln x - \dfrac{1}{4}\right) + C$

5. $\dfrac{e^{5t}(4 + 5t)}{25} + C$ **7.** $2(\sin x - x\cos x) + C$

9. $\dfrac{x^2[\ln(x^2) - 1]}{2} + C$ **11.** $\dfrac{x^3}{3}\left(\ln x - \dfrac{1}{3}\right) + C$

13. $(s - 4)e^{s-3} + C$ **15.** $\dfrac{19 + 6x}{9}e^{3x-1} + C$

17. $\dfrac{6x + 8}{15}(x - 2)^{3/2} + C$ **19.** $\dfrac{(3x)^{4/3}}{16}(4\ln x - 3) + C$

21. $2\sqrt{x}(\ln\sqrt{x} - 1) + C$

23. $x\sec x - \ln|\sec x + \tan x| + C$

25. $\ln|\sin x| - x\cot x + C$

27. $-(3x + 2)e^{-x} + C$

29. $(t + 2)\sinh t - \cosh t + C$

31. $\dfrac{2(3x - 2)(x + 1)^{3/2}}{15} + C$

33. $-3\ln(\cosh x) + \tanh x(3x + 2) + C$

35. $\dfrac{(3x + 5)(2x - 5)^{3/2}}{15} + C$

37. $2\theta\cos\theta + (\theta^2 - 2)\sin\theta + C$

39. $\dfrac{1}{2}e^x(\cos x + \sin x) + C$

41. $2t\sin 3t - \dfrac{1}{3}\cos 3t\left(9t^2 - 2\right) + C$

43. $\dfrac{e^{3x}\left(3\sin 2x - 2\cos 2x\right)}{13} + C$

45. $\left(2 + x^2\right)\cosh x - 2x\sinh x + C$

47. $\dfrac{x\left[\cos(\ln x) + \sin(\ln x)\right]}{2} + C$

49. $2\left(\sqrt{x} - 1\right)e^{\sqrt{x}} + C$

51. $\sqrt{x}\arctan\left(1/\sqrt{x}\right) + \dfrac{1}{2}\ln(x+1) + C$

53. $2\sqrt{x}\cos^{-1}\sqrt{x} - 2\sqrt{1-x} + C$

55. $3x^3\left(\ln x\right)^2 - 2x^3\ln x + \dfrac{2}{3}x^3 + C$

57. $\dfrac{\sqrt{3}}{2} - \dfrac{\pi}{6}$ **59.** $\dfrac{e^{2\pi} + 1}{5}$ **61.** $\dfrac{\sqrt{3}\pi + 9\ln 3}{18}$

63. $\dfrac{\pi}{2} + \ln 2 - 2$ **65.** $x\sin^{-1}x + \sqrt{1-x^2} + C$

67. $x\tan^{-1}x - \dfrac{1}{2}\ln\left(1+x^2\right) + C$

69. $x\sinh^{-1}x - \sqrt{1+x^2} + C$ **71.** $\sqrt{3}\pi - 3\ln 2$

73. a. $\left(\dfrac{e-2}{e-1}, \dfrac{e+1}{4e}\right)$ **b.** $2\pi(e-2)/e$

75. a. $(\pi/8, 1)$ **b.** 2π **77.** $\left(\dfrac{e^2+1}{4}, \dfrac{e-2}{2}\right)$

89. $\dfrac{x^3\sin 2x}{2} + \dfrac{3x^2\cos 2x}{4} - \dfrac{3x\sin 2x}{4} - \dfrac{3\cos 2x}{8} + C$

91. $e^{\sin x}\left(\sin^3 x - 3\sin^2 x + 6\sin x - 6\right) + C$

93. $x\left(\ln x\right)^3 - 3x\left(\ln x\right)^2 + 6x\ln x - 6x + C$

95. $9e - 24$ **99.** $e^{2x}\left(4x^3 - 6x^2 + 6x - 3\right) + C$

101. $e^{3x}\left(3\sin 2x - 2\cos 2x\right) + C$

Section 7.2

1. $\dfrac{A_1}{x} + \dfrac{A_2}{x-1}$ **3.** $\dfrac{A_1}{x+1} + \dfrac{A_2}{x-3} + \dfrac{A_3}{(x-3)^2} + \dfrac{A_4}{(x-3)^3}$

5. $\dfrac{B_1 x + C_1}{x^2 + x + 2} + \dfrac{B_2 x + C_2}{\left(x^2 + x + 2\right)^2} + \dfrac{B_3 x + C_3}{\left(x^2 + x + 2\right)^3}$

7. $\dfrac{A_1}{2x+1} + \dfrac{A_2}{x-3}$ **9.** $\dfrac{A_1}{x+1} + \dfrac{B_1 x + C_1}{x^2+1} + \dfrac{B_2 x + C_2}{\left(x^2+1\right)^2}$

11. $4\ln|3x-1| - 4\ln|x| + C$

13. $2\ln|2x+5| - 2\ln|x+3| + C$

15. $\dfrac{2}{9}\left(\ln|x-2| - \ln|x+1|\right) - \dfrac{4}{3(x-2)} + C$

17. $\dfrac{3}{z+1} + 2\ln|z| - 2\ln|z+1| + C$

19. $\ln|x+1| - \ln|x+2| + C$

21. $2\ln|x| - \ln|x+1| - \ln|x-1| + C$

23. $\dfrac{\ln\left(t^2+1\right)}{2} - \ln|t+1| + C$ **25.** $\dfrac{3}{2x^2} - \dfrac{1}{x} + C$

27. $\ln|x-2| + 2\ln|x| - 3\ln|x+3| + C$

29. $\ln|x+1| - \ln|x-3| - \dfrac{2}{x+1} - \dfrac{2}{x-3} + C$

31. $\ln|v-2| + \ln|v+2| - \ln\left(v^2+4\right) + C$

33. $\sqrt{3}\tan^{-1}\left(\dfrac{2x+1}{\sqrt{3}}\right) + \ln|x-1| - \ln\sqrt{x^2+x+1} + C$

35. $\dfrac{a}{b}\ln|x| - \dfrac{a}{2b}\ln|x^2+b| + C$

37. $\ln|x+1| - 4\ln|x-1| + 3\ln|x-7| + C$

39. $\ln(243/4096)$ **41.** $\ln(3/4)$

47. $\dfrac{2}{3}\ln|x+1| + \dfrac{1}{6}\ln\left(x^2-x+1\right) + \dfrac{\sqrt{3}}{3}\arctan\dfrac{2x-1}{\sqrt{3}} + C$

49. $\dfrac{\ln 3}{8}$ **51.** $\ln\sqrt{3/2}$

53. $3x - \sqrt{3}\tan^{-1}\left(\dfrac{2x+1}{\sqrt{3}}\right) + \ln|x-1| - \dfrac{\ln\left(x^2+x+1\right)}{2} + C$

55. $-2\arctan(1-x) + \ln\left(x^2-2x+2\right) + \ln|x-3| + C$

57. $\ln(8/5) - 3/8$

59. $\dfrac{a}{(b-c)}\ln|x-b| - \dfrac{a}{(b-c)}\ln|x-c| + C$

61. $\dfrac{4x^2 + 2\ln|x| - 17\ln\left(x^2+4\right)}{8} + C$

63. $-\dfrac{2}{3}\ln|\cos x + 2| - \dfrac{1}{3}\ln|\cos x - 1| + C$

65. $2\ln\left|\dfrac{\ln x - 3}{\ln x}\right| + C$

67. $2\sqrt{x} - \dfrac{2}{1+\sqrt{x}} - 4\ln\left(1+\sqrt{x}\right) + C$

69. $2\ln\left|\dfrac{1-\sqrt{1+\sqrt{x}}}{1+\sqrt{1+\sqrt{x}}}\right|+C$ **71.** $\ln\left|1+\sec(\ln x)\right|+C$

73. $\left[(\ln w)^2+1\right]\arctan(\ln w)-\ln w+C$ **75.** $\dfrac{2\pi}{3}\ln 4$

77. $\dfrac{1}{4c}\ln\left|\dfrac{c+y}{c-y}\right|-\dfrac{1}{2c}\tan^{-1}\left(\dfrac{y}{c}\right)=mt+C$

79. $I(t)=\dfrac{Ne^{Nkt}}{N-1+e^{Nkt}}$

81. $3x^2-x+5+\dfrac{x}{x^2+2x+3}+\dfrac{3x-1}{x^2+1}+\dfrac{1}{x+2}$
$-\dfrac{4}{(x+2)^2}$

Section 7.3

1. $\dfrac{\sin^3 x}{3}+C$ **3.** $\sin x-\dfrac{2\sin^3 x}{3}+\dfrac{\sin^5 x}{5}+C$

5. $\dfrac{3x}{8}-\dfrac{1}{4}\sin 2x+\dfrac{1}{32}\sin 4x+C$ **7.** 1 **9.** $5\pi/8$

11. $\dfrac{\cos^{15}x}{15}-\dfrac{3\cos^{13}x}{13}+\dfrac{3\cos^{11}x}{11}-\dfrac{\cos^9 x}{9}+C$

13. $\dfrac{3\theta}{128}-\dfrac{\sin\theta}{32}+\dfrac{\sin 2\theta}{256}+C$ **15.** $\pi/2$

17. $\dfrac{2\cos^{7/2}x}{7}-\dfrac{2\cos^{3/2}x}{3}+C$

19. $\dfrac{\cos^8 x}{8}-\dfrac{\sin^6 x}{6}+C$ **21.** π

23. $x-\dfrac{\sin 4x}{4}+\dfrac{\sin^2 2x}{3}+C$

25. $\dfrac{\sin^8 x}{8}-\dfrac{\sin^{10}x}{10}+C$ **27.** $3/5$

29. $\dfrac{-\sin x}{2}+\dfrac{\sin 15x}{30}+C$ **31.** $2(2\sqrt{2}-1)/5$

33. $-2\sqrt{1-\sin x}+C$

35. $2\sqrt{2}\left[\dfrac{2}{3}\cos^3\left(\dfrac{\pi-2x}{4}\right)-\cos\left(\dfrac{\pi-2x}{4}\right)\right]+C$

37. $\dfrac{\tan^4 x}{4}-\dfrac{\tan^2 x}{2}-\ln\left|\cos x\right|+C$

39. $\dfrac{\tan^3 x}{3}+\tan x+C$ **41.** $\dfrac{-2}{\sqrt{\sec\theta}}+C$

43. $\dfrac{\sec^6 x}{6}+C$ **45.** $\dfrac{-2\cot^{9/2}t}{9}-\dfrac{2\cot^{5/2}t}{5}+C$

47. $\dfrac{1}{2}\sec x\tan x-\dfrac{1}{2}\ln\left|\sec x+\tan x\right|+C$

49. $848/105$ **51.** $2\sqrt{3}$

53. $\dfrac{\tan^6 2\alpha}{12}+\dfrac{\tan^4 2\alpha}{8}+C$ **55.** $-\dfrac{1}{16}\csc^4 4x+C$

57. $1/2$ **59.** $x\tan x-\ln\left|\sec x\right|+C$

61. $\dfrac{1}{4}\left[2z\tan 2z-2z^2-\ln\left|\sec 2z\right|\right]+C$

63. $\dfrac{-\cos 2x}{4}-\dfrac{\cos 4x}{8}+C$

65. a. $\dfrac{-\sin x\cos^6 x}{7}+\dfrac{4\sin x}{7}-\dfrac{8\sin^3 x}{21}+\dfrac{4\sin^5 x}{35}+C$

b. $\dfrac{-\sin x\cos^7 x}{8}+$
$\dfrac{5}{128}\left(5x+4\sin 2x+\dfrac{3}{4}\sin 4x-\dfrac{\sin^3 2x}{3}\right)+C$

67. $4\pi/\sqrt{3}$ **69.** $\left(\dfrac{6\pi+2\pi^3}{3\pi^2+12},\dfrac{15\pi+2\pi^3}{6\pi^2+24}\right)$

75. True   **77.** True

Section 7.4

1. d   **3.** c and d   **5.** b   **7.** c   **9.** b   **11.** c

13. $3\ln\left|\sqrt{x^2+9}+x\right|+C$ **15.** $-\sqrt{4-x^2}+C$

17. $\dfrac{1}{2}\left(25\arcsin\dfrac{t}{5}-t\sqrt{25-t^2}\right)+C$

19. $\dfrac{\sqrt{x^2-36}}{36x}+C$ **21.** $\sqrt{1-x^2}+C$

23. $2\ln\left|\sqrt{x^2+4}+x\right|-\sqrt{x^2+4}+C$

25. $\dfrac{5\sqrt{5}-9}{120}$ **27.** $25\pi/16$ **29.** $-\dfrac{\sqrt{t^2+9}}{9t}+C$

31. $2\ln\left|\dfrac{2-\sqrt{4-9x^2}}{x}\right|+\sqrt{4-9x^2}+C$

33. $\dfrac{-(4-w^2)^{5/2}}{20w^5}+C$ **35.** $\dfrac{3\sqrt{7}-\sqrt{3}}{16}$

37. $\dfrac{1}{2}x\sqrt{9x^2+16}+\dfrac{8}{3}\ln\left|\sqrt{9x^2+16}+3x\right|+C$

39. $-\dfrac{\sqrt{4-x^2}}{x}-\arcsin\dfrac{x}{2}+C$ **41.** $\dfrac{\sqrt{3x^2+1}}{3}+C$

43. $-\dfrac{1}{\sqrt{1-x^2}}+C$ **45.** $-\dfrac{\sqrt{25-9x^2}}{25x}+C$

47. $4/15$ **49.** $\ln\left|2x+1+\sqrt{4x^2+4x+12}\right|+C$

51. $\dfrac{27}{2}\arcsin\dfrac{x-3}{3}-\dfrac{(x+9)\sqrt{6x-x^2}}{2}+C$

53. $\dfrac{\pi}{4\sqrt{2}}$ **55.** $\dfrac{1}{\sqrt{3}}-\dfrac{2}{\sqrt{15}}$

57. $\dfrac{x-3}{2}\sqrt{7+6x-x^2}-8\arcsin\dfrac{3-x}{4}+C$

59. $\dfrac{x-5}{4(x^2-10x+29)}+\dfrac{1}{8}\arctan\dfrac{x-5}{2}+C$

61. $\dfrac{1}{2}\left[e^x\sqrt{1-e^{2x}}+\arcsin(e^x)\right]+C$

63. $-\dfrac{\sqrt{\sin^2 x+1}}{\sin x}+C$

65. $\dfrac{x}{2}\sqrt{x^2-a^2}-\dfrac{a^2}{2}\ln\left|x+\sqrt{x^2-a^2}\right|+C$

67. $\dfrac{1}{a}\sec^{-1}\left(\dfrac{x}{a}\right)+C$ **69.** $\ln\left|x+\sqrt{x^2+a^2}\right|+C$

71. $\dfrac{1}{a}\ln\left|\dfrac{\sqrt{x^2+a^2}+a}{x}\right|+C$ **73.** $\dfrac{1}{a}\operatorname{sech}^{-1}\left(\dfrac{x}{a}\right)+C$

75. $\sinh^{-1}\left(\dfrac{x}{4}\right)-\dfrac{x}{\sqrt{x^2+16}}+C;$

$\ln\left|\sqrt{x^2+a^2}+x\right|-\dfrac{x}{\sqrt{x^2+a^2}}+C;$

$\sinh^{-1}\left(\dfrac{x}{a}\right)-\dfrac{x}{\sqrt{x^2+a^2}}+C$

77. $\dfrac{1}{4}\left[-t\sqrt{1-t^2}+(2t^2-1)\arccos t\right]+C$

79. $\dfrac{\sqrt{6}}{9}-\dfrac{1}{6}$ **81.** $2\pi\left(\ln 3-\dfrac{4}{5}\right)$ **83.** $\pi\sqrt{3}/2$

85. $1+\dfrac{1}{2}\ln\dfrac{3}{2}\approx 1.203$

89. False; $x=\sec\theta$ is a recommended substitution.

91. False; the definite integral is not multiplicative.

93. $\dfrac{x}{2}\sqrt{x^2+25}-\dfrac{25}{2}\sinh^{-1}\left(\dfrac{x}{5}\right)+C$

95. $-\sinh^{-1}(3-x)+C$

Section 7.5

1. $\dfrac{1}{25}\ln\left|\dfrac{x}{x+5}\right|+C$ **3.** $-\ln\left|\dfrac{2+\sqrt{x^2+4}}{x}\right|+C$

5. $\dfrac{1}{2}\tan(2x)-x+C$ **7.** $\dfrac{1}{2}\arctan\dfrac{x+2}{2}+C$

9. $\dfrac{x^2}{2}+\dfrac{4x^{3/2}}{3}+C$ **11.** $\dfrac{1}{2}\arcsin\dfrac{x^2}{2}+C$

13. $\dfrac{x+1}{2}\sqrt{x^2+2x+5}+2\ln\left|x+1+\sqrt{x^2+2x+5}\right|+C$

15. $\dfrac{\sqrt{3}\pi+9\ln 3}{18}$ **17.** $2\cot\dfrac{\pi}{8}-\dfrac{\pi}{4}-2$

19. $\dfrac{2}{3}\left[(z+2)^{3/2}-z^{3/2}\right]+C$

21. $\sqrt{x^2-4}+2\ln\left|x+\sqrt{x^2-4}\right|+C$

23. $\dfrac{1}{6}\ln\left|\dfrac{\sqrt{2x^2+9}-3}{\sqrt{2x^2+9}+3}\right|+C$

25. $-\dfrac{1}{x^3}+2\ln\left|\dfrac{2x^3+1}{x^3}\right|+C$

27. $\dfrac{1}{3}(\tan u-5)\sqrt{5+2\tan u}+C$

29. $\dfrac{x\ln x-x}{2}+C$

31. $\dfrac{\sin x-2}{\sqrt{2\sin x-\sin^2 x}}+C$

33. $2\sqrt{3x-2}-2\sqrt{2}\arctan\sqrt{\dfrac{3x-2}{2}}+C$

35. $\dfrac{1-x}{6}\sqrt{1+6x-3x^2}+\dfrac{2}{3\sqrt{3}}\arcsin\dfrac{\sqrt{3}(x-1)}{2}+C$

37. $\dfrac{17}{18}+\dfrac{3\ln 3}{8}$ **39.** $\pi/4$

43. $\sqrt{2}\arctan\dfrac{\tan(x/2)-1}{\sqrt{2}}+C$

45. $\dfrac{2}{\sqrt{5}}\arctan\dfrac{\tan(u/2)}{\sqrt{5}}+C$

57. $\dfrac{1}{2}(1+x)\sqrt{5+2x+x^2}+2\sinh^{-1}\left(\dfrac{1+x}{2}\right)+C$

59. $2\sqrt{1+2x}+\ln\left(1-\sqrt{1+2x}\right)-\ln\left(1+\sqrt{1+2x}\right)+C$

61. $2\sqrt{e^2+e^x}-2e\tanh^{-1}\left(\dfrac{\sqrt{e^2+e^x}}{e}\right)+C$

63. $\dfrac{1}{3}\arctan\left(\dfrac{1}{3}\sqrt{-9+e^{2x}}\right)+C$

65. $\dfrac{1}{6}\left[\sqrt{5+2x+x^2}\left(7+x+2x^2\right)-12\sinh^{-1}\left(\dfrac{1+x}{2}\right)\right]+C$

67. $F(x)=\sqrt{x^2-4x+5}\left(x^2-x-1\right)-3\sinh^{-1}(2-x)$
$+\sqrt{2}+3\sinh^{-1}(1)$

69. $x\approx 5.1292$ **71.** $x\approx 0.095142$

Section 7.6

1. a. 40 **b.** 42

3. $T_8=6724$; $S_8\approx 6554.6667$; exact value: 6553.6

5. $T_8=0.83954$; $S_8\approx 0.8048$; exact value: 0.8

7. $T_8=65$; $S_8=64$; exact value: 64

9. $T_8=11.7296$; $S_8=11.8632$; exact value: 12

11. $T_8=4.03068$; $S_8=4.0467$;
exact value: $5\ln 5-4=4.047189...$

13. $T_8=84$; $S_8\approx 85.33333$; exact value: $256/3=85.\overline{3}$

15. $T_8\approx 24.6507$; $S_8=24.5122$;
exact value: $52\sqrt{2}/3=24.513035...$

17. $T_8\approx 6.9734$; $S_8\approx 7.0665$;
exact value: $\sqrt{65}-1=7.062257...$

19. a. 1.9541 **b.** 2.0009

21. 6468.5 **23.** 0.7809

25. $\pi^3/300\approx 0.1034$

27. $1/100=0.01$ **29.** $5/3=1.\overline{6}$

31. $\pi^5/112{,}500\approx 0.0027$

33. $1/75{,}000\approx 0.00001$

35. $4/3=1.\overline{3}$ **37.** $n>71$

39. $n>36$ **41.** $n>25$

43. $n>10$ **45.** $n>5$ **47.** $n>3$

49. 3.1415918; 0.00003% **51.** 19.2407

53. Approx. 1430.49 ft **55.** 3750 cubic feet

57. 3.36047; exact value: 3.361267...

59. 1.20794; exact value: 1.207949...

61. 3.36094 **63.** 1.208

65. $S_{50}\approx 1.46265179$; error bound: 1.83635×10^{-7}

67. $S_{50}\approx 0.56568724$; error bound: 4.35225×10^{-4}

69. $S_{50}\approx 0.11714154$; error bound: 1.6636×10^{-7}

71. $x\approx 1.3391165$ **73.** $x\approx 2.080915$

Section 7.7

1. Yes, type I **3.** Yes, type II

5. Not improper

7. Yes, type I (doubly improper)

9. $\displaystyle\lim_{b\to\infty}\int_2^b \dfrac{1}{x^2-1}\,dx$ **11.** $\displaystyle\lim_{a\to-\infty}\int_a^{-1}\dfrac{-3}{x}\,dx$

13. $\displaystyle\lim_{a\to-\infty}\int_a^0 e^x\,dx$ **15.** $\displaystyle\lim_{c\to 0^+}\int_c^1 \dfrac{2}{x}\,dx+\lim_{b\to\infty}\int_1^b \dfrac{2}{x}\,dx$

17. Divergent **19.** Convergent; $6\sqrt[3]{2}$

21. Type I; convergent; 2

23. Type II; convergent; 2

25. Type I; convergent; 1

27. Type II; divergent

29. Type I; convergent; π

31. Type II; divergent

33. Type I; convergent; π

35. Both type I and type II; convergent, π

37. Type II; convergent; π

39. Both type I and type II; divergent

41. Type II; divergent

43. Type I; convergent; $2/e^2$

45. Type I; convergent; 1

47. Type II; divergent

49. Both type I and type II; convergent; $\pi/2$

51. Type I; convergent; 0

53. Type II; convergent; $4\sqrt{2}$

55. Type II; convergent; $\pi/3$

57. Type II; divergent **59.** Type II; divergent

61. Type II; convergent; -4

63. Type I; convergent; $\ln 3$

65. Diverges for $p\le 1$; converges for $p>1$

67. Converges **69.** Converges

71. Diverges **73.** Diverges

75. $4\pi/3$ **79.** 1

81.

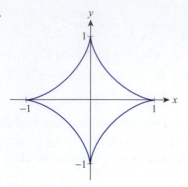

The perimeter is 6.

85. $\pi/2$ (Substitute $u = \sqrt{x}$.)   **87.** $\pi/4$ **89.** $1/s^2$

91. $\dfrac{1}{s-a}$ **93.** $\dfrac{s}{s^2+k^2}$ **95.** $\$200,000$

97. $\$603,448$ **99.** $v_0 = \sqrt{2Rg} \approx 11,180 \, \text{m/s}$

101. False; consider $\displaystyle\int_1^\infty \dfrac{dx}{x}$.   **103.** True

105.

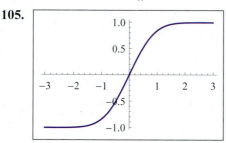

Chapter 7 Review

1. $\dfrac{2x^{3/2}\left(3\ln x - 2\right)}{9} + C$ **3.** $\dfrac{2}{3}(x+2)\sqrt{x-1} + C$

5. $2t\cos t + \left(t^2 - 2\right)\sin t + C$ **7.** $\pi\left(\pi^2 - 6\right)$

9. $\dfrac{\left[2\ln(\sin t)-1\right]\sin^2 t}{4} + C$ **11.** $\dfrac{e^{\pi/2} + e^{-\pi/2}}{2}$

13. a. $\left(\dfrac{27e^4+9}{32e^3+16}, \dfrac{459e^5-54}{500e^3+250}\right)$ **b.** $\dfrac{3\pi}{8}\left(3e^4+1\right)$

17. $2\ln|x+1| - \ln|x-3| + C$

19. $\ln|x| - \ln|x-1| - \dfrac{1}{x} - \dfrac{2}{x^2} + C$

21. $\dfrac{1}{2}\ln\left(x^2+2\right) - \dfrac{4x^2+7}{4\left(x^2+2\right)^2} + C$

23. $2\ln|x-1| + 2\ln|x+5| - \ln|x+3| + C$

25. $-\dfrac{5\ln 2}{2}$ **27.** $\dfrac{2}{5}\left(\ln|\sqrt{x}-2| - \ln|\sqrt{x}+3|\right) + C$

29. $\pi\ln(256/9)$ **31.** $16/3$

33. $\dfrac{\cos^5 x}{5} - \dfrac{\cos^3 x}{3} + C$ **35.** $3/4$

37. $-2\cot 2x + C$ **39.** $\dfrac{1}{4}\sin 2x - \dfrac{1}{8}\sin 4x + C$

41. $5/12$ **43.** $-\dfrac{\csc^2 x}{2} - \ln|\sin x| + C$

45. $\dfrac{\tan^5 x}{5} + C$ **47.** $\dfrac{\sin^3 v}{3} + C$

49. 2 **51.** $\dfrac{5\pi(\pi+2)}{4}$

55. $4\sin^{-1}(x/2) - x\sqrt{4-x^2} + C$ **57.** $\dfrac{t}{2\sqrt{t^2+4}} + C$

59. $9\left(\sqrt{2} - \sinh^{-1} 1\right)$

61. $(x-2)\sqrt{5+4x-x^2} - 9\sin^{-1}\left(\dfrac{2-x}{3}\right) + C$

63. $\sin^{-1}\left(\dfrac{e^x-1}{2}\right) + C$ **65.** $2\sqrt{3} - \sinh^{-1}\sqrt{3}$

67. $\pi\left(\pi - \dfrac{7\sqrt{3}}{4}\right)$ **69.** $2\pi\left(\sqrt{2} + \sinh^{-1} 1\right) \approx 14.4236$

71. 6.3589

73. $\sqrt{2\cos x - \cos^2 x} - \sin^{-1}(\cos x - 1) + C$

75. $T_6 \approx 0.50403; |E_T| \approx 0.00403$

77. $T_6 \approx 0.70067; |E_T| \approx 0.00752$

79. $|E_T| \le 1/72 \approx 0.01389$

81. $|E_T| \le \dfrac{\pi^3}{486\sqrt{3}} \approx 0.03683$

83. $S_6 \approx 0.50009; |E_S| \approx 0.00009$

85. $S_6 \approx 0.69347; |E_S| \approx 0.00033$

87. $|E_S| \le 1/1944 \approx 0.00051$

89. $|E_S| \le \dfrac{11\pi^5}{590,490\sqrt{3}} \approx 0.00329$

91. Type II; divergent   **93.** Type II; convergent; 4

95. Type I; convergent; $\pi/2$ **97.** Converges

99. $\displaystyle\int_0^{\pi^2/4} \dfrac{\cos\sqrt{x}}{\sqrt{x}}\,dx = \int_0^{\pi/2} 2\cos u\,du = 2$

101. Yes, the volume is $\pi/2$; the surface area is $\pi\left(\sqrt{2} + \sinh^{-1} 1\right) \approx 7.2118$.

103. $\dfrac{1}{(s-a)^2}$ **105.** $\dfrac{2ks}{\left(s^2+k^2\right)^2}$ **107.** True

109. False; consider Gabriel's horn.

111. False; consider $f(x)=1/\sqrt{|x|}$.

113. False; see Exercises 72–74 in Section 7.4.

115. $n=35$ **117.** $n=4; n=12$

119. $T_{50}\approx 1.08948$; $S_{50}\approx 1.08943$; Simpson's Rule is more accurate. (In fact, S_{50} agrees with the numerical integration command's value to 11 decimal digits!)

Chapter 7 Project

1. $I_0=\pi/2; I_1=1; I_2=\pi/4; I_3=2/3$

3. $I_4=3\pi/16; I_5=8/15; I_6=5\pi/32; I_7=16/35$

Chapter 8

Section 8.1

1. $\dfrac{dy}{dx}=\sqrt{\dfrac{y}{x}}$ **3.** $\dfrac{dy}{dx}=x\sqrt{1-y^2}$

5. $\dfrac{dy}{dx}=\dfrac{y}{x^2}-\dfrac{y}{x}$ **13.** $y=\dfrac{x^3}{3}+\sin x+5$

15. $y=\ln|x|+x-1$ **17.** $y=-\dfrac{1}{x}+\dfrac{7}{2}$

19. $y=\arcsin x+2$ **21.** Yes   **23.** Yes

25. Yes   **27.** No   **29.** $y=\tan\left(x^2+C\right)$

31. $y=Ce^{3x^2/2}-2$ **33.** $y=Ce^{x^3}$

35. $y=Ce^{\arcsin(x/2)}$ **37.** $x=\dfrac{1}{\ln\left(t^2+C\right)}$

39. $y=Cx-6$ **41.** $x=C(y+1)$

43. $x=\pm\sqrt{\dfrac{y^2}{2}+2y+C}$ **45.** $y=\sqrt[3]{x^3-3x+8}$

47. $y=\dfrac{1}{5}\ln\left|\dfrac{5}{2}e^{2x}-\dfrac{3}{2}\right|$ **49.** $y=\dfrac{1}{4-\sin x}$

51. $y=\dfrac{2e^{x^2/2}}{1-2e^{x^2/2}}$ **53.** $y=\ln\left((1+\sin t)^2\right)$

55. Approx. 2 min 21 s   **57.** Approx. 31.85 g

61. Approx. 73.84 cm

63. a. $\dfrac{dv}{dt}+\dfrac{k}{m}v=g$ **b.** $v=\dfrac{mg}{k}$

65. $x^2+2y^2=C$ **67.** $x^3+3x^2+3y^2=C$

69. $x^3+y^3=C$ **71.** $y^2+x^2-\ln\left(x^2\right)=C$

73. $2y^2-x^2-\ln\left(x^2\right)=C$ **75.** True

77. False; for example, when solving motion problems, finding the velocity function from the acceleration and initial velocity is an initial value problem.

Section 8.2

1. Linear   **3.** Linear   **5.** Linear

7. Linear in y **9.** Linear in x

11. $y=\dfrac{C}{x^3}$ **13.** $y=\dfrac{C}{x^4}+\dfrac{1}{4}$

15. $y=(2x-3\ln x+C)x-5$ **17.** $x=\dfrac{y^2}{10}+\dfrac{C}{y^8}$

19. $y=\sin x+Ce^{-x}$ **21.** $y=\dfrac{\left(x^5/5\right)+x+C}{x^4+1}$

23. $y=\dfrac{\ln x}{2}+\dfrac{C}{\ln x}$ **25.** $x=t+\sin t+C\cos t$

27. $y=\dfrac{1+5e^{-x^3}}{3}$ **29.** $y=\left(x^2+1\right)(\arctan x+10)$

31. $y=\dfrac{e^x\ln(x+1)}{x+1}$ **33.** $x=\dfrac{5}{y^2}-1+2\ln y$

35. $y'+\dfrac{1}{x}y=\dfrac{\sin x}{x}+\cos x$

37. $y'+\dfrac{1}{x\ln x}y=\dfrac{1+x}{x\ln x}$ **39.** $y=\dfrac{1}{x(C-\ln x)}$

41. $y=\dfrac{1}{x\left(-2\sqrt{x}+C\right)^2}$ **43.** Approx. 7 min 47.3 s

45. Approx. 15.315 lb

47. $\dfrac{dy}{dt}+\dfrac{r_2y}{V_0+(r_1-r_2)t}=a_1r_1$; $y(0)=a_0$

49. $I(t)=\dfrac{1}{2}-\dfrac{1}{2}e^{-50t}$

51. $q(t)=1.5\times10^5-1.5\times10^5e^{-t/50,000}$;   As $t\to\infty$, the charge stabilizes at $1.5\times10^5 C$.

53. a. $\dfrac{dA}{dt}=d-kA$

b. $A=\dfrac{d}{k}-\dfrac{d}{k}e^{-kt}$; $A\to\dfrac{d}{k}$ as $t\to\infty$

55. False; the equation $xy' = y$ is both separable and linear.

57. False; for example, equations that are, say, both linear and separable can readily be solved by different methods.

Section 8.3

1. C   **3.** A   **5.** D

7. $y = 0$: unstable; $y = 2$: stable

9. $y = -\pi/2$: unstable; $y = \pi/2$: stable

11. a.

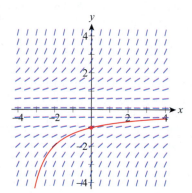

b.

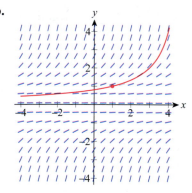

13.

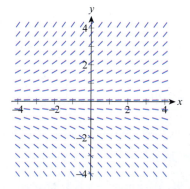

$y = 0$: unstable

15.

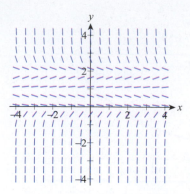

$y = 0, y = 2$: stable; $y = 1$: unstable

17.

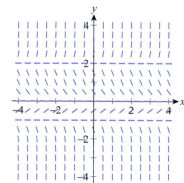

$y = 2, y = -1$: unstable; $y = 0$: stable

19.

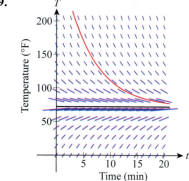

21. a. $P(t) = \dfrac{500}{1 + 4e^{-0.3t}}$ **b.** Just under 6 years

23. $v = (mg/k)^2$

27. a. (i) 5.578   (ii) 6.990

   b. Exact value: $\dfrac{5e^2 - 3}{4} \approx 8.486$

29. a. (i) 1.168   (ii) 1.326

   b. Exact value: $\dfrac{5}{4} + e^{-3/2} \approx 1.473$

31. a. (i) 1.345   (ii) 1.482

   b. Exact value: $\sqrt{3} \approx 1.732$

33. 7.966

39. Approx. 7.659

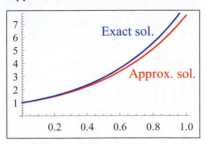

41. Approx. 1.401

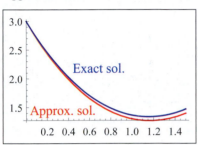

43. Approx. 1.610

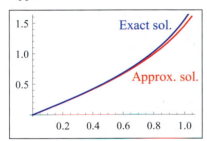

Section 8.4

1. $y = c_1 e^x + c_2 e^{-2x}$ **3.** $y = c_1 e^{-x/3} + c_2 e^{2x}$

5. $y = c_1 e^{-x} + c_2 x e^{-x}$ **7.** $y = c_1 e^{-x/2} + c_2 x e^{-x/2}$

9. $y = e^{2x}\left(c_1 \cos 3x + c_2 \sin 3x\right)$

11. $y = e^{x/5}\left(c_1 + x c_2\right)$

13. $y = c_1 e^{(\sqrt{13}+1)x/6} + c_2 e^{(\sqrt{13}-1)x/6}$

15. $w = e^{-t/2}\left[c_1 \cos(t/2) + c_2 \sin(t/2)\right]$

17. $y = c_1 e^{kx} + c_2 e^{-kx}$ **19.** $y = \dfrac{5}{2}\left(e^{\sqrt{3}x} + e^{-\sqrt{3}x}\right)$

21. $y = e^{-x} + 3e^{2x}$ **23.** $y = e^x\left(\cos 2x - 2\sin 2x\right)$

25. $y = 2x e^{3x/2} - e^{3x/2}$ **27.** $y = 5\left(e^{5x} - e^5\right)$

29. $y = e^x$ **31.** No solution

33. $y = e^x\left(2\cos 4x + e^{-\pi/8}\sin 4x\right)$

35. $y = c_1 e^{-x} + c_2 e^{3x} + c_3$

37. $y = c_1 e^x + c_2 \cos\sqrt{2}x + c_3 \sin\sqrt{2}x$

39. $y = c_1 e^x + c_2 e^{-2x} - \dfrac{x}{2} - \dfrac{1}{4}$

41. $y = c_1 e^x + c_2 e^{-5x} - \dfrac{1}{13}\cos x - \dfrac{3}{26}\sin x$

43. $y = \cos x + \sin x + x^2 - 2$

47. a. Underdamped: $c < 2\sqrt{10}$; oscillating motion, amplitude goes to 0 (Example 5)

b. Critically damped: $c = 2\sqrt{10}$; damped motion with no oscillation, but the smallest change in c will cause it to be underdamped or overdamped

c. Overdamped: $c > 2\sqrt{10}$; solution is a linear combination of exponential functions, no oscillation, displacement goes to 0

49. $c = 2\sqrt{g/5} \approx 2.8$

51. $y(t) = \dfrac{-e^{-t}}{84}\left[63\cos\left(3\sqrt{7}t\right) - 5\sqrt{7}\sin\left(3\sqrt{7}t\right)\right]$

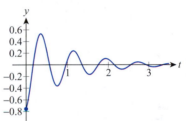

53. $y(t) \approx -0.278 e^{-11.85t} + 0.478 e^{-6.9t}$

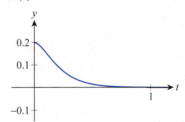

57. Approx. 8.2 s; approx. 16.66 m

Chapter 8 Review

1. Autonomous, separable, first-order

3. First-order, linear **5.** $y = \ln\sqrt{x^2 + 1} + 1$

7. $y = Cx^2$ **9.** $y = \left(\ln x - \dfrac{1}{x} + C\right)^2$ or $y = 0$

11. $u = \dfrac{1 - Cx^2}{1 + Cx^2}$ or $u = -1$ **13.** $y = \sqrt[3]{3x^3 + x^2 + 8}$

15. $y = \left(t^2 + 1\right)^2 + 2$ or $y = t^4 - 6t^2 + 11$

17. $y = Cx$ **19.** $y = \dfrac{C}{x^6} + \dfrac{1}{3}$ **21.** $x = \dfrac{2y^2 + C}{y^2 + 1}$

23. $y' - 3x^2 y = x^2$ **25.** $y = x\left(e^{1-x} + x - 1\right)$

27. $y = \dfrac{x}{x^2 + 1} - \dfrac{3}{\sqrt{x^2 + 1}}$

29. B; $y = 0$: unstable; $y = -2$: stable

31. A; $y = 2$: unstable; $y = -1$: stable

33.

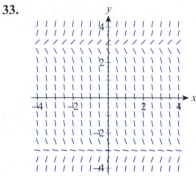

$y = 3$: unstable; $y = -3$: stable

35.

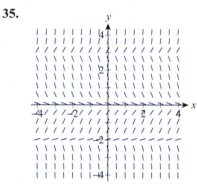

$y = 3,\ y = -2$: unstable; $y = 0$: stable

37. a. (i) 8.063   (ii) 8.960

b. Exact value: $3 + e^2 \approx 10.389$

39. $y = c_1 e^{4x} + c_2 e^{-4x}$ **41.** $y = c_1 e^{5x} + c_2$

43. $y = c_1 e^{x/2} + c_2 x e^{x/2}$ **45.** $y = 2e^{-2x} + e^{3x}$

47. $y = 2e^{(3x-6)/2} - 7$ **49.** No solution

51. $y = c_1 e^{x/2} + c_2 e^{3x} + \dfrac{1}{50}\left(\cos x - 7\sin x\right)$

53. $x^4 + 2x^2 + 2y^2 = C$ **55.** Approx. 13.446 kg

57. a. $V' = kV^{2/3}$

b. It melts away in just under 5.5 hours.

59. $q(t) = 250{,}000 e^{-t/2500} + 100t - 250{,}000$

61. Approx. 68.43 °C

63. $y(t) \approx e^{-1.25t}\left[0.05\cos(9.83t) - 0.095\sin(9.83t)\right]$

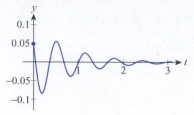

65. False; it is linear as well as separable.

67. False; see the slope field in Figure 1 of Section 8.3.

69. True

Chapter 8 Project

1. $A_0 - \dfrac{X(t)}{1+b}$; $B_0 - \dfrac{bX(t)}{1+b}$

3. $\dfrac{dX}{dt} = k\left(60 - \dfrac{X}{4}\right)\left(40 - \dfrac{3X}{4}\right)$; $X(0) = 0$

5. $\lim\limits_{t\to\infty} X(t) = 53\frac{1}{3}$; this means that ultimately $53\frac{1}{3}$ grams of C is formed, using up all of reactant B, while leaving $46\frac{2}{3}$ grams of reactant A.

Chapter 9

Section 9.1

1.

t	x	y
0	5	0
1	6	-1
2	7	undefined
3	8	$\sqrt{3}$
4	9	1
5	10	$\sqrt{5}/3$
6	11	$\sqrt{6}/4$

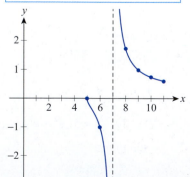

3. $x = t$, $y = -t^2 - 5$ **5.** $x = t$, $y = \pm\sqrt{t-4}$

7. $x = t$, $y = |t-1|$ **9.** $x = t$, $y = \pm\sqrt{1-t^2}$

11. $y = 3x^2 + 4$ $(t \geq 2)$

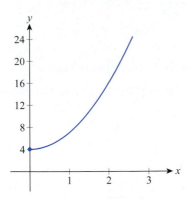

13. $x = |y+8|$

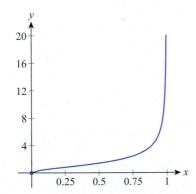

15. $y = \sqrt{\dfrac{2x}{1-x}}$ $(t \geq 0)$

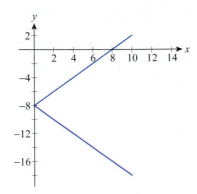

17. $x = \dfrac{4}{|y-5|}$ $(t \neq 3)$

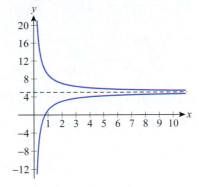

19. $y = \pm\dfrac{\sqrt{8-2x-x^2}}{6}$

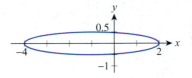

21. $y = \dfrac{3}{2}x$

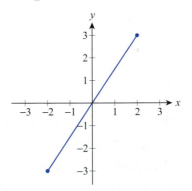

23. $y = 4 \pm 3\sqrt{1-x^2}$

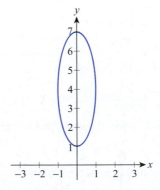

25. F   **27.** C   **29.** A

31.

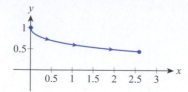

33.

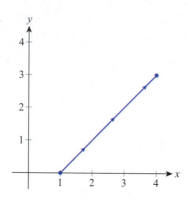

37. $x = h + a\cos t, y = k + b\sin t$

39. Sample answer: $x = t, y = -\dfrac{3}{2}t + 6$

41. Sample answer: $x = t - 3, y = -\dfrac{3}{4}t + 1$ $(0 \le t \le 8)$

43. Sample answer: $x = 6\cos t, y = -2 + 6\sin t$

45. Sample answer: $x = 6\cos t, y = 1 + \sqrt{11}\sin t$

47. Yes

49. a. $x \approx 9.85t, y \approx -16t^2 + 1.74t + 3.5$

b.

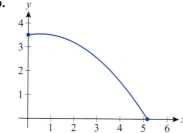

51.

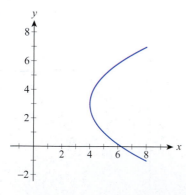

53.

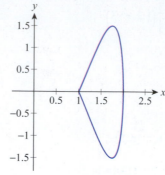

57. $(0,0)$ and $(1,3)$

59. $x = 3(\theta - \sin\theta), y = 3(1 - \cos\theta)$

65. $x = (R + r)\cos t - r\cos\left(\dfrac{R+r}{r}t\right),$

$y = (R + r)\sin t - r\sin\left(\dfrac{R+r}{r}t\right)$

67. $y = \dfrac{8r^3}{x^2 + 4r^2}$ **69.** True

71. False; consider the graph of a circle centered at the origin for which neither statement is true.

Section 9.2

1. Neither differentiable nor smooth

3. Differentiable but not smooth

7. Horizontal: $y = 1$; vertical: $x = -1/4$

9. Horizontal: none; vertical: $x = 1$

11. Horizontal: none; vertical: none

13. Horizontal: $y = \pm 5$; vertical: $x = \pm 2$

15. $\left.\dfrac{dy}{dx}\right|_{t=1/2} = 1;$ $\left.\dfrac{d^2y}{dx^2}\right|_{t=1/2} = 1$

17. $\left.\dfrac{dy}{dx}\right|_{t=0} = 0;$ $\left.\dfrac{d^2y}{dx^2}\right|_{t=0} = 2$

19. $\left.\dfrac{dy}{dx}\right|_{t=0} = -1;$ $\left.\dfrac{d^2y}{dx^2}\right|_{t=0} = 3$

21. $\left.\dfrac{dy}{dx}\right|_{t=\pi/4} = 0;$ $\left.\dfrac{d^2y}{dx^2}\right|_{t=\pi/4} = -8$

23. $\left.\dfrac{dy}{dx}\right|_{t=1} = 1;$ $\left.\dfrac{d^2y}{dx^2}\right|_{t=1} = \dfrac{1}{2}$

25. $t = 10/3$ **27.** $t = -1/2, 1/\sqrt[3]{4}$ **29.** $4/3$

31. $\pi^3/6$ **33.** $\dfrac{2}{5}\left(1+e^{-\pi/2}\right)$

39. π **41.** 92/9

43. 59/24 **45.** $2e^{\pi/2}-2$

47. $\dfrac{2e^2+3e-2}{4e}$

49. a. $16\sqrt{2}\pi$ **b.** $24\sqrt{2}\pi$

51. a. $\dfrac{12,289\pi}{192}$ **b.** $\dfrac{(135+\ln 2)\pi}{6}$

53. a. $\dfrac{(16e^4+e^2-1)\pi}{16e^2}$ **b.** $\dfrac{\pi}{2}\left(5-\dfrac{2}{e}\right)$

55. $48\pi/5$

57. a. $\dfrac{\pi}{2}(4+\sinh 4)$ **b.** $2\pi(3\sinh 2-\cosh 2+1)$

59. Yes **61.** $2\sqrt{5}$ m/s **63.** 9 m/s **65.** $t=4/3$ s

67. $32\pi/15$ **69.** $(18/5,9/8)$

75. False; consider Example 1. **77.** True

79. 28.362 **81.** 3.820 **83. a.** 81.870 **b.** 258.882

Section 9.3

1.

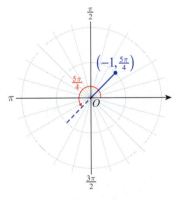

3.

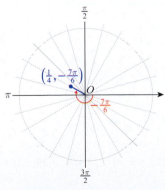

5.

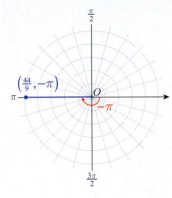

7. $(3.54,-3.54)$ **9.** $(-4.42,-4.42)$

11. $(-2.60,-1.50)$ **13.** $(-3,0)$ and $(3,\pi)$

15. $\left(\sqrt{145},-0.08\right)$ and $\left(-\sqrt{145},3.06\right)$

17. $\left(2\sqrt{21},1.76\right)$ and $\left(-2\sqrt{21},-1.38\right)$

19. $r^2=25$ **21.** $r\cos\theta=12$ **23.** $\sin\theta=\cos\theta$

25. $r\cos\theta=16a$ **27.** $r^2-4ar\cos\theta=0$

29. $r^2\sin^2\theta-4r\cos\theta-4=0$ **31.** $x^2+y^2=5x$

33. $x^2+y^2=49$ **35.** $y=1/2$

37. $x^4+y^4+2x^2y^2=2xy$ **39.** $4y+7x=12$

41. $x^2+y^2=4$

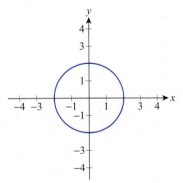

43. $y=-x/\sqrt{3}$

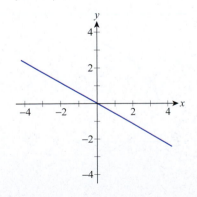

45. $x = 7$

47.

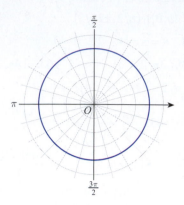

49.

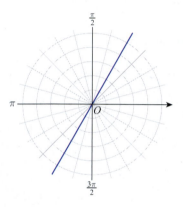

51.

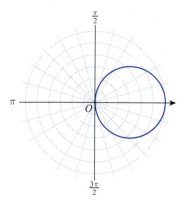

53.

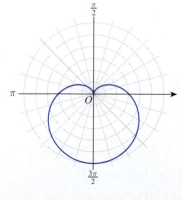

55.

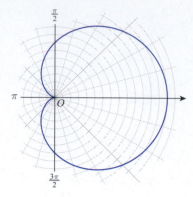

57.

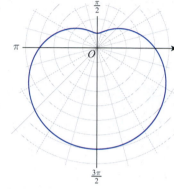

59.

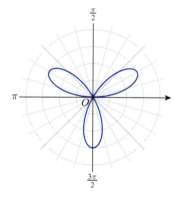

61.

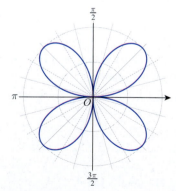

63.

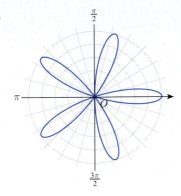

65.

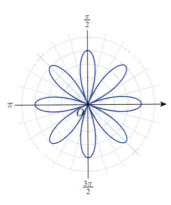

67.

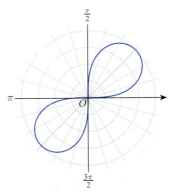

69. $(0,0), \left(\sqrt{2}/2, \pi/4\right)$

71. $(0,0), \left(1+\sqrt{2}/2, 3\pi/4\right), \left(1-\sqrt{2}/2, -\pi/4\right)$

73. The graph of $f(\theta - a)$ is that of $f(\theta)$ rotated about the origin by a radians.

Section 9.4

1. $-\sqrt{3}/3$ **3.** $-2/\pi$ **5.** 3   **7.** $-\sqrt{3}/5$

9. $\dfrac{2\pi + 3\sqrt{3}}{2\sqrt{3}\pi - 3}$ **11.** $\tan 1$ **13.** 4

15. Horizontal: $(2, \pi/2)$, $(1/2, 7\pi/6)$, $(1/2, 11\pi/6)$; vertical: the pole, $(3/2, \pi/6)$, $(3/2, 5\pi/6)$

17. Horizontal: the pole, $\left(a\sqrt{2}, \pm\pi/3\right)$; vertical: $(2a, 0)$, $(a/2, \pm 2\pi/3)$

21. $\theta = 0$ **23.** $\theta = \pm\pi/4$ **25.** $\theta = 0$

27. $9\pi/4$ **29.** $3\pi/2$ **31.** $11\pi/4$ **33.** $19\pi/2$

35. 5π **37.** $\pi/2$ **39.** 8π **41.** 4

43. $\dfrac{6\sqrt{3} - 2\pi}{3}$ **45.** $\dfrac{\sqrt{3}}{2} - \dfrac{\pi}{6}$ **47.** $\pi - \dfrac{3\sqrt{3}}{2}$

49. $\dfrac{11\sqrt{3}}{2} + \dfrac{14\pi}{3}$ **51.** $6\pi\sqrt{16\pi^2 + 1} + \dfrac{3\sinh^{-1} 4\pi}{2}$

53. $2/\sqrt{3}$ **55.** $\dfrac{8\left(\pi^2 + 1\right)^{3/2} - 8}{3}$

57. $4\sqrt{2}$ **59.** π^2 **61.** $\dfrac{18\sqrt{2}\pi\left(e^{2\pi} + 1\right)}{5}$

69.

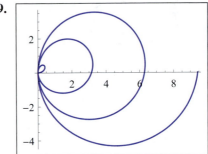

$L \approx 45.3582$

71.

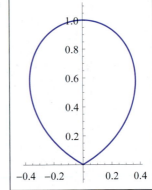

$L \approx 2.6825$

73. $A \approx 8580.85$ **75.** $A \approx 37.1940$

Section 9.5

1. Parabola   **3.** Ellipse   **5.** Parabola

7. Ellipse; H   **9.** Hyperbola; M

11. Hyperbola; J   **13.** Hyperbola; O

15. Ellipse; G   **17.** Parabola; N

19. Parabola; E   **21.** Hyperbola; I

23. $e = \dfrac{\sqrt{6}}{4}$; major $= 4\sqrt{2}$; minor $= 2\sqrt{5}$

25. $e = \dfrac{\sqrt{6}}{3}$; major $= 2\sqrt{6}$; minor $= 2\sqrt{2}$

27.

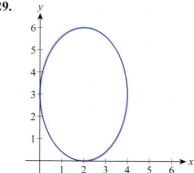

Foci: $\left(-2 \pm \sqrt{7}, -1\right)$

29.

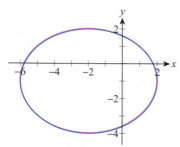

Foci: $\left(2, 3 \pm \sqrt{5}\right)$

31.

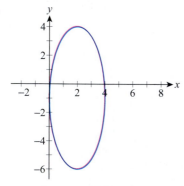

Foci: $\left(2, -1 \pm \sqrt{21}\right)$

33.

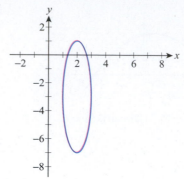

Foci: $\left(2, -3 \pm \sqrt{15}\right)$

35.

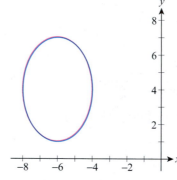

Foci: $\left(-6, 4 \pm \sqrt{5}\right)$

37.

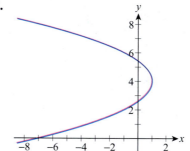

Focus: $\left(1/2, 4\right)$;   Directrix: $x = 3/2$

39.

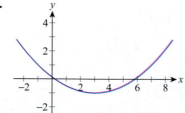

Focus: $\left(3, 1\right)$;   Directrix: $y = -3$

41.

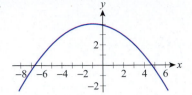

Focus: $(-1,2)$;   Directrix: $y = 6$

43.

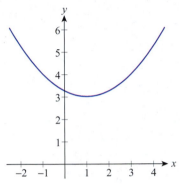

Focus: $(1,4)$;   Directrix: $y = 2$

45.

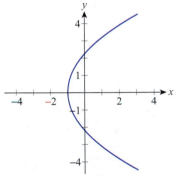

Focus: $(1/4,0)$;   Directrix: $x = -9/4$

47.

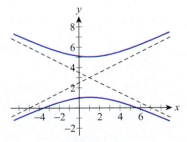

Foci: $\left(1, 3 \pm 2\sqrt{5}\right)$

49.

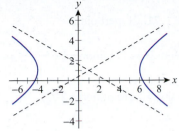

Foci: $\left(1 \pm \sqrt{34}, 1\right)$

51.

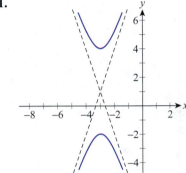

Foci: $\left(-3, 1 \pm \sqrt{10}\right)$

53.

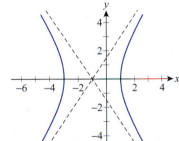

Foci: $\left(-1 \pm \sqrt{13}, 0\right)$

55.

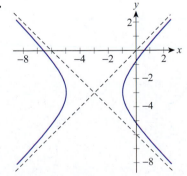

Foci: $\left(-3 \pm 2\sqrt{2}, -3\right)$

57. $(x+2)^2 = 2\left(y - \dfrac{1}{2}\right)$ **59.** $(y-2)^2 - \dfrac{(x-1)^2}{8} = 1$

61. $\dfrac{(y-1)^2}{2} - \dfrac{(x+1)^2}{2} = 1$ **63.** $\dfrac{(x-3)^2}{36} + \dfrac{y^2}{27} = 1$

65. $\dfrac{(x+9)^2}{25} + \dfrac{(y-6)^2}{21} = 1$

67. $(y-1)^2 = -12(x+1)$

69. $\dfrac{(y+4)^2}{16} - \dfrac{(x-3)^2}{25} = 1$

71. $\dfrac{(x+1)^2}{16} + \dfrac{(y+1)^2}{4} = 1$

73. $(x-2)^2 = 2(y-1)$ **75.** $e \approx 0.249$

77. 185.93 million miles   **79.** 2 ft   **81.** 1.5 in.

85. $\dfrac{y^2}{144} - \dfrac{x^2}{25} = 1$ **87.** $\left(4\sqrt{3}+3, -4+3\sqrt{3}\right)$

89. $\left(\dfrac{-5\sqrt{2}}{16}, \dfrac{3\sqrt{2}}{16}\right)$

91. Hyperbola, $\theta = \pi/4$; $x'^2 - y'^2 = 8$

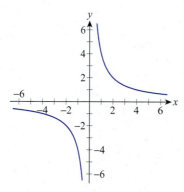

93. Hyperbola, $\theta = \pi/6$; $19x'^2 - y'^2 = 28$

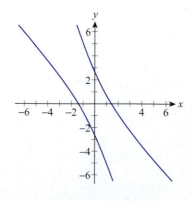

95. Hyperbola, $\theta = \dfrac{\pi}{3}$; $\dfrac{(x'+2)^2}{3} - \left(y' - \dfrac{1}{\sqrt{3}}\right)^2 = 1$

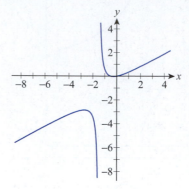

97. C   **99.** D

101. The objective of the rotation of axes is to eliminate the $x'y'$-term. If your final equation contains an $x'y'$-term, you know that a mistake has occurred.

105. False; every point on the parabola has to be the same distance from the directrix and the focus (which is at a positive distance from the directrix.) This would not be true of the point of tangency.

107. True

109.

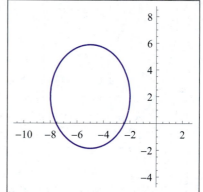

111.

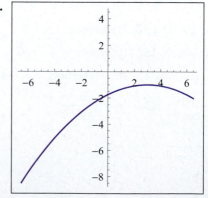

113.

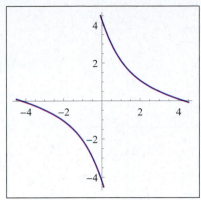

115.

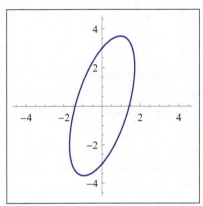

117.

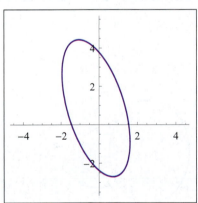

Section 9.6

1. C 3. F 5. B 7. Hyperbola; $y = 7/6$

9. Ellipse; $x = -3$ 11. Hyperbola; $x = 1/3$

13. Ellipse; $x = 5$ 15. Hyperbola; $x = -6/5$

17. Parabola; $y = 3/2$ 19. Hyperbola; $x = -4/7$

21. $r = \dfrac{2}{1 - \cos\theta}$ 23. $r = \dfrac{3}{1 - 4\sin\theta}$

25. $r = \dfrac{3}{1 + \frac{1}{4}\cos\theta}$

27.

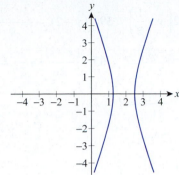

29.

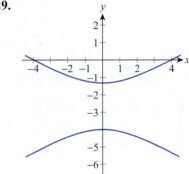

31.

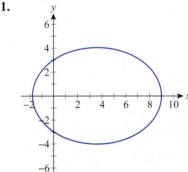

33.

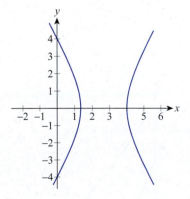

35.

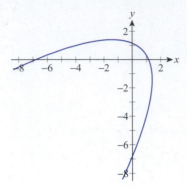

41. $2ed$

43. $\left(\dfrac{ed}{e+1}, 0\right)$ and $\left(\dfrac{ed}{e-1}, 0\right)$

45. $\dfrac{(x-3.5)^2}{6.25} - \dfrac{y^2}{6} = 1$

47.

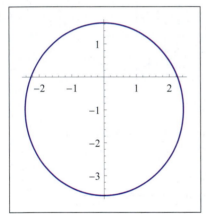

49.

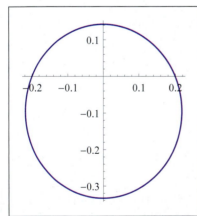

51.

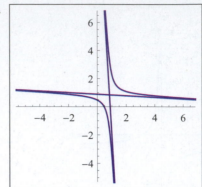

53.

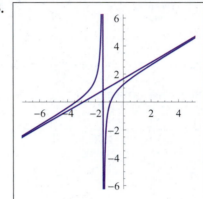

55.

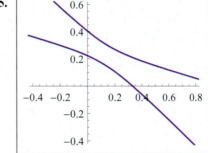

Chapter 9 Review

1. $y = \dfrac{1}{1296x^2}$

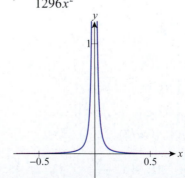

3. $y = \dfrac{3}{2x} - 1$

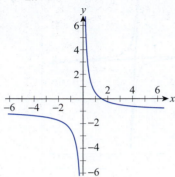

5. $x = t,\ y^2 = t^2 + 4$

7. Sample answer: $x = t,\ y = \dfrac{12}{17}t - \dfrac{100}{17}$

9. Horizontal: $y = -4$; vertical: none

11. $\dfrac{dy}{dx}\bigg|_{t=2} = \dfrac{1}{16}$; $\dfrac{d^2 y}{dx^2}\bigg|_{t=2} = -\dfrac{1}{128}$

13. $\dfrac{dy}{dx}\bigg|_{t=1} = -3$; $\dfrac{d^2 y}{dx^2}\bigg|_{t=1} = 8$ **15.** $t = 0, t = \pm 1/\sqrt{3}$

17. $6\sqrt{3}/5$ **19.** $76/3$ **21.** $61\pi/108$

23. 81π **25.** $(-6.06, -3.5)$

27. $\left(2\sqrt{61}, 0.88\right)$ and $\left(-2\sqrt{61}, 4.02\right)$

29. $r^2 - 9ar\cos\theta = 0$ **31.** $x + y = 4$

33.

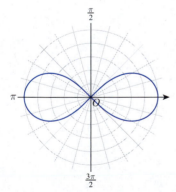

35. $\dfrac{4\pi}{4 + \pi^2}$

37. Horizontal: $\left(4/3, \cos^{-1}\sqrt{2/3}\right)$,
$\left(4/3, \cos^{-1}\left(-\sqrt{2/3}\right)\right)$;
vertical: $(2, 0),\ (0, \pi/2),\ (2, \pi)$

39. $\dfrac{33\sqrt{3} + 28\pi}{6}$ **41.** $4\sqrt{3} + \dfrac{8\pi}{3}$ **43.** 2 **45.** 16

47.

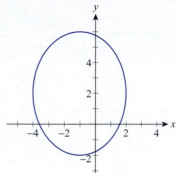

Foci: $\left(-1, 2 \pm \sqrt{7}\right)$

49.

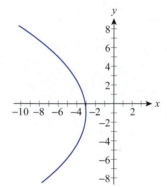

Focus: $(-6, -1)$;   Directrix: $x = 0$

51.

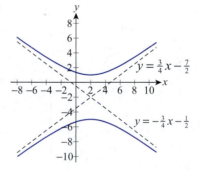

Foci: $(2, 3),\ (2, -7)$

53. $\dfrac{(x+1)^2}{9} + \dfrac{(y-4)^2}{16} = 1$ **55.** $\dfrac{(x-2)^2}{9/4} + \dfrac{(y+1)^2}{9/4} = 1$

57. $(x+2)^2 = 4(y-3)$ **59.** $(y+1)^2 = 2\left(x - \dfrac{5}{2}\right)$

61. $\dfrac{(x+1)^2}{4} - \dfrac{(y+2)^2}{25} = 1$

63. $\dfrac{(y-2)^2}{9} - (x-2)^2 = 1$

65. Hyperbola; $y = -5/8$ **67.** Parabola; $x = -2/3$

69. Ellipse; $x = -4/3$ **71.** $r = \dfrac{3}{1 - \cos\theta}$

73. $r = \dfrac{3}{1 - \frac{1}{5}\sin\theta}$ **75.** $r = \dfrac{7}{1 - \sin\theta}$ **77.** $1\frac{1}{4}$ inches

81. True   **83.** False; see Example 1 in Section 9.2.

85. True   **87.** True   **89.** True

91. $r = 1 - 2\cos\left(\theta + \dfrac{\pi}{4}\right)$; the curve is a rotated

limaçon with inner loop.

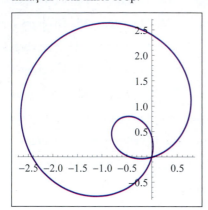

95.

Horizontal: $\theta = 0$, $\theta = \cos^{-1}\left(1/\sqrt{3}\right)$;
vertical: $\theta = \tan^{-1}\left(\sqrt{2}/2\right)$, $\theta = \pi/2$

97.

$A = \dfrac{17\cos^{-1}(2/3) - 6\sqrt{5}}{2} \approx 0.441$

99. $L \approx 6.8694$ **101.** $A \approx 34.3518$

Chapter 9 Project

1. $x(t) = (a_1 - a_0)t + a_0,$
$y(t) = (b_1 - b_0)t + b_0, t \in [0,1]$

7.

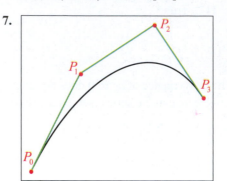

Chapter 10

Section 10.1

1. $\dfrac{1}{2}, \dfrac{2}{5}, \dfrac{3}{10}, \dfrac{4}{17}, \dfrac{5}{26}, \dfrac{6}{37}$ **3.** $3, 3, \dfrac{3}{2}, \dfrac{1}{2}, \dfrac{1}{8}, \dfrac{1}{40}$

5. $1, 3, 7, 15, 31, 63$ **7.** $4, 8, 24, 96, 480, 2880$

9. $a_n = n^2 + n$ **11.** $a_n = 10 - n^2$

15. B   **17.** E   **19.** D   **25.** Converges to $1/2$

27. Converges to 0   **29.** Converges to 0

31. Converges to 0   **33.** Diverges

35. Converges to 0   **37.** Converges to 0.5

39. Converges to 0   **41.** Converges to 0

43. Converges to $\ln 3$ **45.** Converges to 1

47. Converges to $\pi/6$ **49.** Diverges

51. Converges to 1   **53.** Diverges

55. Converges to 1   **57.** Converges to 1

59. Converges to 0   **61.** Converges to 1

63. Converges to 0   **65.** Converges to e^2

67. Converges to e^3 **69.** Converges to 2

75. The limit is 1.   **77.** The limit is 2.

83. $T_n = 2T_{n-1} + 1, T_1 = 1$

99. No; consider $a_n = -1/n$ and $b_n = 1/n$.

101. $a_n = \dfrac{5}{3} + \dfrac{1}{3}\left(-\dfrac{1}{2}\right)^{n-2}$; the limit is $5/3$.

103. True   **105.** True   **107.** True

109. False; consider the constant sequence $(-2)^2, (-2)^2, (-2)^2, \ldots$ that converges to 4, but $-2, -2, -2, \ldots$ does not converge to 2.

111. True   **113.** True

Section 10.2

1. No. The convergence of a_n to 0 does not imply the convergence of the series (see Example 6).

3. 1, 0, 1, 0, 1   **5.** $\dfrac{\sqrt{3}}{2}, \sqrt{3}, \sqrt{3}, \dfrac{\sqrt{3}}{2}, 0$

9. 25/6   **11.** Diverges   **13.** 10/17   **15.** 40/27

17. 25/36   **19.** $x < 2$ or $x > 3$ **21.** 41/333

23. 169/50   **25.** 2   **27.** 1   **29.** 1/3   **31.** Diverges

33. Diverges   **35.** 14/3   **37.** Diverges   **39.** 1/6

41. Diverges   **43.** 33/2   **45.** Diverges

47. Diverges   **49.** Diverges

55. Answers will vary. $\displaystyle\sum_{n=1}^{\infty} \frac{1}{n}$ and $\displaystyle\sum_{n=1}^{\infty} \frac{1}{n}$ are an example.

77. 150 km   **79.** $\displaystyle\sum_{n=1}^{\infty} \frac{1}{4^n} = \frac{1}{3}$ **81.** $\displaystyle\sum_{n=3}^{\infty} \frac{1}{2^n} = \frac{1}{4}$

Section 10.3

1. Diverges   **3.** Diverges   **5.** Converges

7. Converges   **9.** Diverges   **11.** Converges

13. Converges   **15.** Converges   **17.** Diverges

19. Converges   **21.** Converges   **23.** Converges

25. Converges   **27.** Diverges   **29.** Converges

31. Diverges

33. The terms of the series are not monotonically decreasing.

35 The terms of the series are not positive nor monotonically decreasing.

39. $1.19955 \le s \le 1.20566$

41. $1.06621 \le s \le 1.08946$

43. $2.08997 \le s \le 2.13419$ **45.** $n = 5$; $s \approx 1.2026$

47. $n = 4$; $s \approx 0.404881426$ **49.** $n = 5$; $s \approx 3.2357$

51. $n = 40$; $s \approx 1.0369277553$

53. $n = 20$; $s \approx 1.0020083928262$

Section 10.4

1. Converges   **3.** Diverges   **5.** Converges

7. Diverges   **9.** Converges   **11.** Converges

13. Converges   **15.** Diverges   **17.** Converges

19. Converges   **21.** Converges   **23.** Converges

25. Converges   **27.** Diverges   **29.** Converges

31. Converges   **33.** Diverges   **35.** Diverges

37. Converges   **39.** Converges   **41.** Converges

43. Diverges   **45.** Converges   **47.** Diverges

49. Diverges   **51.** Converges   **53.** Converges

55. Converges   **57.** Diverges   **59.** Converges

61. Diverges   **63.** Diverges   **65.** Converges

67. Converges   **69.** Converges   **83.** True

85. False; consider $a_n = 1/n$ and $b_n = 1/n^2$.

87. False; consider $b_n = c_n = 1/n$.

89. True   **91.** True

Section 10.5

1. Converges   **3.** Converges   **5.** Converges

7. Converges   **9.** Converges   **11.** Converges

13. Diverges   **15.** Converges   **17.** Diverges

19. Converges   **21.** Diverges   **23.** Converges

25. Converges   **27.** Diverges   **29.** Converges

31. Diverges   **33.** Diverges   **35.** Diverges

37. Diverges   **39.** Converges   **41.** Converges

43. Converges   **45.** Converges   **47.** Converges

49. Diverges   **51.** Converges   **53.** Converges

55. Converges   **57.** Diverges   **59.** Converges

61. Converges   **63.** Diverges   **65.** Diverges

67. Diverges   **69.** Converges   **73.** $0 < p < 1$

75. True   **77.** False   **79.** False; consider Example 3.

Section 10.6

1. Converges by the Alternating Series Test

3. Diverges by the nth-Term Divergence Test

5. Converges by the Alternating Series Test

7. Diverges by the nth-Term Divergence Test

9. Diverges by the nth-Term Divergence Test

11. Converges by the Alternating Series Test

13. Diverges by the nth-Term Divergence Test

15. Diverges by the nth-Term Divergence Test

17. Converges by the Alternating Series Test

19. Diverges by the nth-Term Divergence Test

21. Converges by the theorem "Absolute Convergence Implies Convergence"

23. Diverges by the nth-Term Divergence Test

25. 0.90   **27.** 0.36787   **29.** 0.056

31. Converges conditionally   **33.** Diverges

35. Converges conditionally

37. Converges absolutely

39. Converges conditionally

41. Converges absolutely   **43.** Diverges

45. Converges conditionally

47. Converges absolutely

49. Converges conditionally

51. Converges absolutely   **53.** Diverges

55. Diverges   **57.** Diverges

59. No; a_n is not monotonic, so the test doesn't apply. The series is actually divergent.

65. An example is $\sum \dfrac{(-1)^n}{\sqrt{n}}$.

67. At least 1000 terms are needed.

Section 10.7

1. $(-1/3,1/3)$; $R = 1/3$ **3.** $[-1/4,1/4]$; $R = 1/4$

5. $[-1,1]$; $R = 1$ **7.** $\{0\}$; $R = 0$

9. $(-\infty,\infty)$; $R = \infty$ **11.** $(-5,-1)$; $R = 2$

13. $\{0\}$; $R = 0$ **15.** $[3,5]$; $R = 1$ **17.** $\{5\}$; $R = 0$

19. $[-4,0)$; $R = 2$ **21.** $[0,2]$; $R = 1$

23. $(-\infty,\infty)$; $R = \infty$ **25.** $(2/3,4/3)$; $R = 1/3$

27. $(-1,1)$; $R = 1$ **29.** $[-7/3,-5/3)$; $R = 1/3$

31. $(2,4)$; $R = 1$ **33.** $(0,2/3)$; $f(x) = \dfrac{1}{2-3x}$

35. $(-3,3)$; $f(x) = \dfrac{3}{3+x}$ **37.** $(1,5)$; $f(x) = \dfrac{2}{5-x}$

39. $(3/5,9/5)$; $f(x) = \dfrac{9}{9-(5x-6)^2}$

41. $\displaystyle\sum_{n=0}^{\infty} 2^n x^n$; $R = 1/2$ **43.** $\displaystyle\sum_{n=0}^{\infty}(-1)^n x^{2n}$; $R = 1$

45. $\displaystyle\sum_{n=0}^{\infty} \dfrac{4^n x^n}{3^n}$; $R = 3/4$ **47.** $\displaystyle\sum_{n=0}^{\infty}\left[(-1)^n + 2\right]x^n$; $R = 1$

49. $\displaystyle\sum_{n=0}^{\infty}(-1)^n (x-1)^n$; $R = 1$

51. $\displaystyle\sum_{n=1}^{\infty}(-1)^n n(x-1)^{n-1}$; $R = 1$ **53.** 10 terms

55. $\displaystyle\sum_{n=2}^{\infty} n(n-1)x^{n-2} = \dfrac{2}{(1-x)^3}$;

$\displaystyle\sum_{n=3}^{\infty} n(n-1)(n-2)x^{n-3} = \dfrac{6}{(1-x)^4}$

57. $\dfrac{18}{(2-3x)^3}$ on $(0,2/3)$

59. $\displaystyle\sum_{n=1}^{\infty}(-1)^{n+1}(2n)x^{2n-1}$; $R = 1$

65. $\displaystyle\sum_{n=0}^{\infty} \dfrac{(-1)^n 2^{2n} x^{2n}}{(2n)!}$; this is the series expansion of $\cos 2x$ about 0.

67. False; consider Exercise 16.

69. False; consider Exercise 62.

Section 10.8

1. $4 + 2x - \dfrac{x^2}{2} + \dfrac{x^3}{2} + \dfrac{x^4}{24}$

3. $f(x) = -7 + 3x + 5x^2 - x^3 + 2x^4$; $R = \infty$

5. $\displaystyle\sum_{n=0}^{\infty} \dfrac{e(x-1)^n}{n!}$; $R = \infty$ **7.** $\displaystyle\sum_{n=0}^{\infty}(-1)^n \dfrac{(x-1)^n}{5^{n+1}}$; $R = 5$

9. $\displaystyle\sum_{n=1}^{\infty}(-1)^{n+1} \dfrac{n(x-2)^{n-1}}{2^{n+1}}$; $R = 2$

11. $\displaystyle\sum_{n=0}^{\infty} \dfrac{(-1)^n x^{4n}}{(2n)!}$; $R = \infty$ **13.** $\displaystyle\sum_{n=0}^{\infty} \dfrac{(-1)^n 2^{2n} x^{2n+1}}{(2n+1)!}$; $R = \infty$

15. $\displaystyle\sum_{n=0}^{\infty}\frac{x^{2n}}{(2n)!}$; $R=\infty$ **17.** $\displaystyle\sum_{n=0}^{\infty}\frac{(-1)^{n}x^{n}}{(2n)!}$; $R=\infty$

19. $\displaystyle -x-x^{2}+\sum_{n=0}^{\infty}\frac{(2n^{2}+6n+3)x^{n+3}}{(n+2)!}$; $R=\infty$

21. $\displaystyle\sum_{n=0}^{\infty}(-1)^{n}\frac{x^{2n+2}}{n+1}$; $R=1$

23. $\displaystyle\sum_{n=1}^{\infty}(-1)^{n+1}\frac{n(x-3)^{n-1}}{3^{n+1}}$; $R=3$

25. $\displaystyle\sum_{n=0}^{\infty}\frac{(-1)^{n}2^{n+1}-3^{n+1}}{6^{n+1}}x^{n}$; $R=2$

27. $\displaystyle\sum_{n=0}^{\infty}x^{2n+2}$; $R=1$ **29.** $\displaystyle\sum_{n=0}^{\infty}(-1)^{n}\frac{x^{3n+1}}{(2n)!}$; $R=\infty$

31. $\displaystyle\frac{\pi}{4}+\frac{x-1}{2}-\frac{1}{4}(x-1)^{2}+\frac{1}{12}(x-1)^{3}-\frac{1}{40}(x-1)^{5}+\cdots$

33. $\displaystyle e-\frac{ex^{2}}{2}+\frac{ex^{4}}{6}-\frac{31ex^{6}}{720}+\frac{379ex^{8}}{40,320}-\cdots$

35. $2+(\ln 2)(x-1)+(\ln 2)^{2}(x-1)^{2}$
$\displaystyle +\frac{(\ln 2)^{3}(x-1)^{3}}{3}+\frac{(\ln 2)^{4}(x-1)^{4}}{12}+\cdots$

37. $\displaystyle 1+2x+\frac{5x^{2}}{2}+\frac{8x^{3}}{3}+\frac{65x^{4}}{24}+\cdots$

39. $\displaystyle x-2x^{2}+\frac{11x^{3}}{6}-x^{4}+\frac{41x^{5}}{120}-\cdots$

41. $\displaystyle x^{2}-\frac{x^{3}}{2}+\frac{x^{4}}{6}-\frac{x^{5}}{6}+\frac{11x^{6}}{72}-\cdots$

43. $\displaystyle x-\frac{x^{3}}{3}+\frac{2x^{5}}{15}-\frac{17x^{7}}{315}+\frac{62x^{9}}{2835}-\cdots$

45. $\displaystyle -x-\frac{x^{2}}{2}+\frac{x^{3}}{6}-\frac{3x^{5}}{40}-\frac{x^{6}}{16}-\cdots$

47. $\displaystyle 1-2x+\left(1-\frac{\pi^{2}}{2}\right)x^{2}+\pi^{2}x^{3}+\frac{\pi^{4}-12\pi^{2}}{24}x^{4}-\cdots$

49. 0.36786; 8 terms   **51.** 0.8452; 5 terms

53. 0.3102682; 4 terms   **55.** 0.304; 3 terms

57. 0.922; 6 terms   **61.** 2   **63.** 1

67. 4   **71.** 0 and $-162!$, respectively

Section 10.9

1. i **3.** $\displaystyle\frac{5e^{2}}{2}+i\frac{5\sqrt{3}e^{2}}{2}$ **5.** $\displaystyle\frac{2}{3}\sin 2$ **7.** $4e^{4i\pi/3}$

9. -4 **13.** $1+5x^{2}+10x^{4}+10x^{6}+5x^{8}+x^{10}$

15. $\displaystyle 1-\frac{x}{2}-\frac{x^{2}}{8}-\frac{x^{3}}{16}-\frac{5x^{4}}{128}-\frac{7x^{5}}{256}-\cdots$

17. $\displaystyle 1+\frac{x^{2}}{3}-\frac{x^{4}}{9}+\frac{5x^{6}}{81}-\frac{10x^{8}}{243}+\frac{22x^{10}}{729}+\cdots$

19. $\displaystyle 1+\frac{4x^{2}}{3}+\frac{20x^{4}}{9}+\frac{320x^{6}}{81}+\frac{1760x^{8}}{243}+\frac{9856x^{10}}{729}+\cdots$

21. $\displaystyle 5-\frac{5x}{2}+\frac{25x^{2}}{8}-\frac{75x^{3}}{16}+\frac{975x^{4}}{128}-\frac{3315x^{5}}{256}+\cdots$

23. $\displaystyle \sin^{-1}x=\sum_{n=0}^{\infty}\frac{(2n)!}{(n!)^{2}4^{n}(2n+1)}x^{2n+1}$; $R=1$

25. $\displaystyle \sinh^{-1}x=\sum_{n=0}^{\infty}\frac{(-1)^{n}(2n)!}{(n!)^{2}4^{n}(2n+1)}x^{2n+1}$; $R=1$

31. Approx. 5.777 miles

35. Approx. 4.211×10^{7} m/s

39. $\displaystyle y=a\sum_{n=0}^{\infty}\frac{(-1)^{n}(3x)^{2n}}{(2n)!}+\frac{1}{3}b\sum_{n=0}^{\infty}\frac{(-1)^{n}(3x)^{2n+1}}{(2n+1)!}$
$\displaystyle =a\cos 3x+\frac{1}{3}b\sin 3x$

41. $\displaystyle y=a\sum_{n=0}^{\infty}x^{2n}+b\sum_{n=0}^{\infty}x^{2n+1}=\frac{a}{1-x^{2}}+\frac{bx}{1-x^{2}}$

45. $\displaystyle a_{0}=1/2;\ a_{k}=0;\ b_{k}=\frac{(-1)^{k}-1}{k\pi}\quad(k\geq 1)$

47.

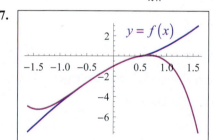

49.

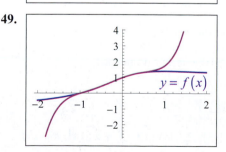

Chapter 10 Review

1. $\displaystyle 1,0,\frac{1}{25},-\frac{1}{125},\frac{81}{83,521}$ **3.** 0, 2, 8, 26, 80

7. Diverges   **9.** Converges to e^6

13. 219/999   **15.** The limit is 3.

19. a. $d_1 = d, d_n = \dfrac{1}{2}d_{n-1} + d, n \geq 2$ **b.** $\lim\limits_{n\to\infty} d_n = 2d$

23. 3   **25.** 85/4   **27.** Diverges   **29.** Converges

31. Converges   **33.** $n = 5; s \approx 2.616$ **35.** Diverges

37. Converges   **39.** Converges   **41.** Diverges

43. Converges   **45.** Converges   **47.** Converges

49. Converges

51. Converges by Alternating Series Test

53. Diverges by n^{th}-Term Divergence Test

55. 0.9498; three terms were needed. According to the nth-reminder estimate, five terms are sufficient.

57. Converges absolutely   **59.** Converges absolutely

61. Converges absolutely   **63.** $(-1/2, 1/2)$

65. $(-\infty, \infty)$ **67.** $[-4, 2)$

69. $(1, 2)$; $f(x) = \dfrac{1}{4 - 2x}$ **71.** $\sum\limits_{n=0}^{\infty} (-1)^n (3x)^n$; $R = 1/3$

73. $\dfrac{1}{x} = \sum\limits_{n=0}^{\infty} (-1)^n \dfrac{(x-2)^n}{2^{n+1}},$

$\dfrac{1}{x^3} = \sum\limits_{n=0}^{\infty} (-1)^n \dfrac{(n+1)(n+2)(x-2)^n}{2^{n+4}}$; $R = 2$

75. $\sum\limits_{n=0}^{\infty} \dfrac{(-1)^n 4^n x^{6n}}{(2n)!}$; $R = \infty$

77. $\ln 3 + \sum\limits_{n=1}^{\infty} \dfrac{(-1)^{n-1} (x-3)^n}{n3^n}$; $R = 3$

79. $\sum\limits_{n=1}^{\infty} \dfrac{(-1)^{n+1} 2^{2n-1} x^{2n}}{(2n)!}$; $R = \infty$

81. $\dfrac{1}{3} + \sum\limits_{n=1}^{\infty} \dfrac{(-1)^{n+1} 2(x-1)^n}{3^{n+1}}$; $R = 3$

83. $\sqrt{2} - \sqrt{2}\left(x - \dfrac{\pi}{4}\right) + \dfrac{3}{\sqrt{2}}\left(x - \dfrac{\pi}{4}\right)^2$

$- \dfrac{11}{3\sqrt{2}}\left(x - \dfrac{\pi}{4}\right)^3 + \cdots$

85. $1 - x + \dfrac{x^2}{2} - \dfrac{x^3}{2} + \dfrac{13x^4}{24} - \cdots$

87. 0.540302; 5 terms   **89.** $\sqrt[4]{1.2} \approx 1.04663$; 5 terms

91. -1 **93.** The sum is 2.

95. $1 - \dfrac{x^2}{2} + \dfrac{3x^4}{8} - \dfrac{5x^6}{16} + \dfrac{35x^8}{128} - \dfrac{63x^{10}}{256} + \cdots$

97. $1 + x^3 - \dfrac{x^6}{2} + \dfrac{x^9}{2} - \dfrac{5x^{12}}{8} + \dfrac{7x^{15}}{8} - \cdots$

99. $y = a\sum\limits_{n=0}^{\infty} \dfrac{(-1)^n 2^{2n} x^{2n}}{(2n)!} + \dfrac{1}{2}b\sum\limits_{n=0}^{\infty} \dfrac{(-1)^n 2^{2n+1} x^{2n+1}}{(2n+1)!}$

$= a\cos 2x + \dfrac{1}{2}b\sin 2x$

101. $\sum\limits_{n=0}^{\infty} \dfrac{(-1)^n 2x^n}{n!} = 2e^{-x}$, this is the Maclaurin series of $2e^{-x}$.

113. False; consider $a_n = 1/n$.

115. False; consider $a_n = 1$ (for all n).

117. False; consider $a_n = 1/n$.

119. False; consider $\sum\limits_{n=1}^{\infty} \dfrac{(-1)^{n+1}}{n}$.

121. False; consider $a_n = \dfrac{(-1)^n}{\sqrt{n}}$.

123. False; the series $\sum\limits_{n=0}^{\infty} (n!)x^n$ converges only at $x = 0$.

125. a. 91,380 terms   **b.** Approx. 15.0859

127. a.

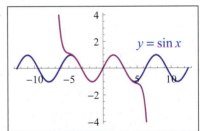

The approximation seems acceptable on the interval $(-3\pi/2, 3\pi/2)$.

b.

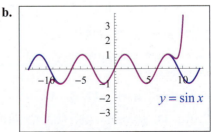

The approximation seems acceptable on the interval $(-5\pi/2, 5\pi/2)$.

Chapter 10 Project

5. $\gamma \approx 0.5772156649$

Chapter 11

Section 11.1

1. Out of the page

3.

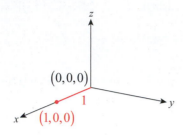

5.

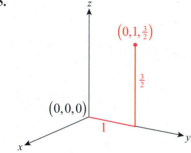

7.

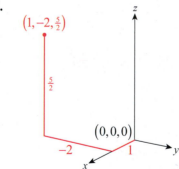

9. The third coordinate is 0.

11. The x- and z-coordinates are 0.

13. The x-coordinate is negative, the y- and z-coordinates are 0.

15. The y-coordinate is -1.   17. 3   19. 0

21. $(5, 3.2, 0)$ 23. $(0, 0, 0)$ 25. $(1.3, 4.5, -10)$

27. A plane parallel to the xz-plane, intersecting the y-axis at $(0, 1, 0)$

29. The line $y = 1 - x$ in the xy-plane

31. The union of the three coordinate planes

33. A line in the plane $x = 5$ and through the point $(5, 0, -1)$, having a slope of 2 in the y-direction

35. A "doubly infinite staircase," any cross-section perpendicular to the y-axis being the graph of the greatest integer function

37. $(2.5, 2.5, 2.5)$ 39. $(-1, -2, 5)$ 41. $\left(\sqrt{3}, 1, 4\right)$

43. A   45. B

47. The exterior (i.e., the "outside") of the circular cylinder $x^2 + y^2 = 1$

49. A sphere of radius 3, centered at $(3, 0, 0)$ (note that the yz-plane is tangent to this sphere)

51. An infinite parabolic surface resulting from all vertical translations of the parabola $y = 1 - x^2$ (note that the z-coordinate is irrelevant)

53. The parabola $x = z^2 + 2$, in the plane $y = 1$

55. The union of the three coordinate planes

57. The upper hemisphere of radius $3\sqrt{5}$, centered at $(-2, 4, 5)$

59. The point $(0, -4, -2)$ 61. $x = 1$ 63. $z = e$

65. $y = -1, z = 2$ 67. $x^2 + (y - 2)^2 + z^2 = 4$

69. $(x - 1)^2 + y^2 + (z + 2)^2 = 49$

71. $x^2 + (y + 4)^2 + (z - 1)^2 = 9, \ x = 0$

73. $(x + 1)^2 + (y - 4)^2 + (z + 3)^2 = 1$

75. $(x - 3)^2 + (y - 2)^2 + (z + 1)^2 = 9, \ x, y, z > 0$

77. $x = 3$ 79. $\sqrt{129}$ 81. 5   83. a   85. d

Section 11.2

1. Scalar   3. Vector   5. Vector   7. Scalar

9. Not a vector   11. Vector   13. Vector

15. $2\overrightarrow{CM}$ 17. $\overrightarrow{MC}$ 19. $\overrightarrow{CM}$

21.

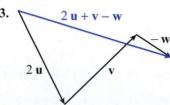

23.

25. $\langle 12, -3 \rangle$ 27. $\langle 16, 33 \rangle$ 29. $\left\langle \dfrac{3}{2}, -5 \right\rangle$

31. $(10,-2,3)$ **33.** $(2,4,-6)$ **35.** $\mathbf{v}=\langle -1,5,0\rangle$

37. Collinear   **39.** $\mathbf{w}=\dfrac{1}{2}\mathbf{u}+\dfrac{1}{2}\mathbf{v}$

41. $\mathbf{w}=4\mathbf{u}-\dfrac{1}{4}\mathbf{v}$ **43.** 2/5   **45.** 3/2   **47.** $17\sqrt{2}$

49. $\sqrt{293}/10$ **51.** Not equal   **53.** Equal

55. Not equal   **57.** $\overrightarrow{PQ}=\langle 0,0,0\rangle$; $\left|\overrightarrow{PQ}\right|=0$

59. $\overrightarrow{PQ}=\langle 3,-3,-2\rangle$; $\left|\overrightarrow{PQ}\right|=\sqrt{22}$

61. $\overrightarrow{PQ}=\left\langle -\dfrac{2}{3},-3,-\dfrac{\sqrt{5}}{3}\right\rangle$; $\left|\overrightarrow{PQ}\right|=\sqrt{10}$

63. Not a parallelogram   **65.** Parallelogram

67. Not a parallelogram   **69.** $(1,2,-9)$

71. $(0,4,4)$ and $(3,7,1)$ **73.** $\mathbf{u}=\dfrac{2}{\sqrt{85}}\mathbf{i}+\dfrac{9}{\sqrt{85}}\mathbf{j}$

75. $\mathbf{u}=-\dfrac{4}{3\sqrt{5}}\mathbf{i}-\dfrac{\sqrt{5}}{3}\mathbf{j}+\dfrac{2}{3\sqrt{5}}\mathbf{k}$

77. $\mathbf{u}=\dfrac{2}{\sqrt{14}}\mathbf{i}-\dfrac{3}{\sqrt{14}}\mathbf{j}-\dfrac{1}{\sqrt{14}}\mathbf{k}$ **81.** $a=\dfrac{5\sqrt{11}}{22}$

85. $\mathbf{v}(t)=\langle 8,-9.81t+35.2\rangle$

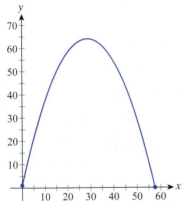

87. Approx. 544.45 mph

89. Ground speed is approx. 259.07 mph in the direction of $\langle 1,3.31,1.61\rangle$.

95. $\langle -1,-2,6,11\rangle$ **97.** $\left\langle 2,1,-7,10,-\dfrac{25}{3},-2\right\rangle$

Section 11.3

1. 4   **3.** -1 **5.** -10 **7.** -3 **9.** 12

11. -2 **13.** 0   **15.** The vectors are perpendicular.

17. $61°$ **19.** Approx. 1.68 rad $\approx 96.5°$

21. Approx. 1.5 rad $\approx 85.9°$ **29.** Parallel

31. Neither   **33.** Orthogonal   **35.** Orthogonal

37. Orthogonal   **39.** Orthogonal

41. b and c   **43.** a

45. Approx. 1.75 rad $\approx 100.4°$ and 1.39 rad $\approx 79.6°$, respectively

47. Any $s\in\mathbb{R}$ **49.** $s=0$ or $t=0$

55. $(y-b)=m(x-a)$ **57.** $\left\langle \dfrac{1}{\sqrt{10}},-\dfrac{3}{\sqrt{10}}\right\rangle$

59. $n_1(x-a)+n_2(y-b)+n_3(z-c)=0$

61. $\langle 1,-4,-2\rangle$ **63.** Approx. $71.41°$

65. $\alpha\approx 64.12°$, $\beta\approx 102.6°$, $\gamma\approx 29.21°$

67. $\alpha\approx 83.42°$, $\beta\approx 156.49°$, $\gamma\approx 112.46°$

69. Precisely when $\theta=0$ or $\theta=\pi$, or when at least one of $\mathbf{u}$ or $\mathbf{v}$ is the zero vector

71. $\mathbf{v}=2\mathbf{u}$ or $\mathbf{v}=-2\mathbf{u}$

75. $\text{proj}_\mathbf{v}\mathbf{u}=\left\langle \dfrac{40}{61},-\dfrac{48}{61}\right\rangle$; $\mathbf{u}-\text{proj}_\mathbf{v}\mathbf{u}=\left\langle \dfrac{82}{61},\dfrac{205}{183}\right\rangle$

77. $\text{proj}_\mathbf{v}\mathbf{u}=\left\langle \dfrac{36}{179},-\dfrac{216}{179},\dfrac{60}{179}\right\rangle$;

$\mathbf{u}-\text{proj}_\mathbf{v}\mathbf{u}=\left\langle \dfrac{608}{537},\dfrac{37}{179},\dfrac{58}{895}\right\rangle$

79. $\text{proj}_\mathbf{v}\mathbf{u}=\left\langle \dfrac{-12st^2-4t}{20t^2+1},\dfrac{-24st^2-8t}{20t^2+1},\dfrac{6st+2}{20t^2+1}\right\rangle$;

$\mathbf{u}-\text{proj}_\mathbf{v}\mathbf{u}=$

$\left\langle -5s+\dfrac{12st^2+4t}{20t^2+1},s+\dfrac{24st^2+8t}{20t^2+1},2-\dfrac{6st+2}{20t^2+1}\right\rangle$

81. $\text{proj}_\mathbf{v}\mathbf{u}=-\dfrac{64}{85}\mathbf{i}+\dfrac{8}{17}\mathbf{j}+\dfrac{72}{85}\mathbf{k}$;

$\mathbf{u}-\text{proj}_\mathbf{v}\mathbf{u}=\dfrac{213}{170}\mathbf{i}+\dfrac{9}{17}\mathbf{j}+\dfrac{209}{255}\mathbf{k}$

83. $\text{proj}_\mathbf{v}\mathbf{u}=-\dfrac{1}{4}\mathbf{v}$; $\mathbf{u}-\text{proj}_\mathbf{v}\mathbf{u}=\mathbf{u}+\dfrac{1}{4}\mathbf{v}$

85. Answers will vary.   **91.** $7\sqrt{13}/13$ **93.** $\sqrt{6/5}$

95. 44 ft-lb   **97.** $1000\sqrt{2}$ kJ

99. a. Approx. 1856.7 J   **b.** Approx. 107.2 N

101. Approx. 0.955 rad $\approx 54.74°$

103. $\mathbf{p}\cdot\mathbf{s}$ is the total revenue during a given cycle.

105. True

107. False; the dot product of a vector and a scalar is undefined.

109. True   **111.** True   **113.** True   **115.** True

117. True   **119.** False; let $\mathbf{u} \perp \mathbf{w}$ and $\mathbf{v} = 2\mathbf{u}$.

Section 11.4

1. $\langle 4,3,7 \rangle$ **3.** $\langle 3,11,20 \rangle$ **5.** $\left\langle -\frac{3}{4}, \frac{1}{12}, -\frac{5}{9} \right\rangle$

15. $\mathbf{i} - 7\mathbf{j} - 3\mathbf{k}$ **17.** $\langle 7,26,24 \rangle$

19. $\pm \left\langle \frac{2}{\sqrt{17}}, \frac{3}{\sqrt{17}}, \frac{2}{\sqrt{17}} \right\rangle$ **21.** $\pm \left\langle \frac{3}{7}, \frac{6}{7}, \frac{2}{7} \right\rangle$

23. $\langle -7,-5,9 \rangle$ **25.** $\left\langle \frac{11}{4}, -\frac{19}{4}, 3 \right\rangle$ **29.** Collinear

31. Collinear   **33.** $19/2$ **35.** 45 **37.** $\sqrt{70}$

39. $\sqrt{13/2}$ **41.** $\langle 0,5,4 \rangle$ **43.** $\langle 22,-4,5 \rangle$

45. Not possible   **47.** $\mathbf{u} = \mathbf{0}$ or $\mathbf{v} = \mathbf{0}$ **49.** 1 **51.** 20

53. Three vectors are coplanar if and only if their triple scalar product is zero.

55. Coplanar   **57.** Coplanar   **59.** $15\sqrt{3}/2$ ft-lb

61. 0.024 N   **67.** $11x - 19y + 12z = 25/2$

71. $7\sqrt{13}/13$ **73.** $\sqrt{6/5}$

79. Nonsense   **81.** Vector   **83.** True

85. False; the cross product of a vector and a scalar is undefined.

87. False; the cross product of a vector and a scalar is undefined.

89. False; $\mathbf{u} \times \mathbf{u} = \mathbf{0}$.   **91.** True   **93.** True

Section 11.5

1. A and C **3.** A and B

5. $\mathbf{r}(t) = \langle 1+3t, 2+2t, 3+t \rangle$;

$x = 1+3t,\ y = 2+2t,\ z = 3+t$

7. $\mathbf{r}(t) = \langle 4t, -5+15t, 8+19t \rangle$;

$x = 4t,\ y = -5+15t,\ z = 8+19t$

9. $\mathbf{r}(t) = \langle -9+4t, 11, -6+3t \rangle$;

$x = -9+4t,\ y = 11,\ z = -6+3t$

11. $\mathbf{r}(t) = \langle 16-53t, 21+22t, 28-32t \rangle$;

$x = 16-53t,\ y = 21+22t,\ z = 28-32t$

13. Answers will vary.   **15.** Answers will vary.

17. Answers will vary.   **19.** Answers will vary.

21. $x = 1+2t,\ y = 2,\ z = 3-2t,\ 0 \le t \le 1$

23. $x = -3+8t,\ y = \frac{2}{3} - \frac{8}{3}t,\ z = -11+4t,\ 0 \le t \le 1$

25. $(3,6,0),\ (-3,0,-18),$ and $(0,3,-9)$

27. $\left(\frac{17}{4}, \frac{39}{8}, 0 \right),\ \left(14, 0, -\frac{13}{4} \right),$ and $\left(0, 7, \frac{17}{12} \right)$

29. $(0,-1,0),\ \left(2, 0, \frac{1}{2} \right),$ and $(0,-1,0)$

31. $(0,2,7),\ \left(\frac{6}{7}, 0, \frac{37}{7} \right),$ and $\left(\frac{7}{2}, -\frac{37}{6}, 0 \right)$

33. $(1,-2,1)$ **35.** $(2,-3,-1)$ **37.** Not possible

39. $5x - 4y + 9z = 1$;

$$\left\{ \left(x, y, \frac{1}{9} - \frac{5}{9}x + \frac{4}{9}y \right) \middle| x \in \mathbb{R}, y \in \mathbb{R} \right\}$$

41. $6x + 9y + 10z = 54$;

$$\left\{ (x, y, 5.4 - 0.6x - 0.9y) \middle| x \in \mathbb{R}, y \in \mathbb{R} \right\}$$

43. $3x + 6y - 2z = 6$ **45.** $6x + 4y - z = -14$

47. $8x - 82y + 15z = 23$

49. $x = t - 3,\ y = 4t - 1,\ z = 4 - 3t$

51. $x = -1+t,\ y = -1+t,\ z = -5+3t$

53. $\frac{x-1}{3} = \frac{y-2}{2} = z-3$ **55.** $\frac{x}{4} = \frac{y+5}{15} = \frac{z-8}{19}$

57. $3x - 2y + 5z = 17$ **59.** $x - 4y - z = 8$

61. $x + 2y - 3z = 19$ **63.** $8x + 19y + z = 15$

65. $7x + y + 4z = 2$ **67.** $\left(30, \frac{74}{5}, \frac{48}{5} \right)$

69. $(-3,4,-8)$

71. $x = \frac{5}{7} + 2t,\ y = \frac{3}{7} - 3t,\ z = -7t$

73. $x = -3 - 2t,\ y = 8 + 10t,\ z = 2t$

75. $\frac{1}{a}x + \frac{1}{b}y + \frac{1}{c}z = 1$ **77.** $1/\sqrt{42}$ **79.** 4

81. $5/3$ **83.** $6/\sqrt{29}$ **85.** $2\sqrt{6/11}$ **87.** $2\sqrt{3/7}$

89. True

91. False; the dot product of the normal vectors has to be zero, which is not equivalent to the condition described in this statement.

93. False; they could be skew lines.

95. False; they could be skew lines.

97. True   **99.** True   **101.** True

103. False; the line could lie in the plane.

105. False; any vector normal to P has to be perpendicular to the xz-plane, while the given vector actually lies in the xz-plane.

107. False; in fact, y is the only variable with a (potentially) nonzero coefficient.

109. False; $y = 0$ is the equation of the xz-plane. The only point in common with the y-axis is the origin.

111. True   **113.** True   **115.** True

Section 11.6

1. Parabolic cylinder; C

3. Hyperbolic paraboloid; A   **5.** Ellipsoid; B

7. Hyperboloid of two sheets; F

9.

11.

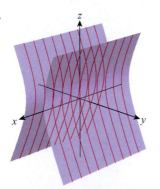

13.

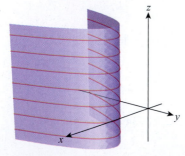

15.

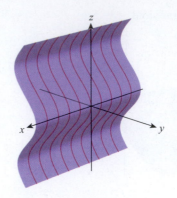

17. Length of major axis: $4\sqrt{2}$; length of minor axis: 4; foci: $(\pm 2, 0, 4)$

19. Vertex: $(1, 0, 1/2)$; focus: $(1, 0, 3/4)$

21. Ellipsoid

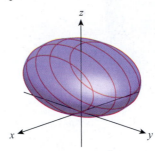

23. Hyperboloid of one sheet

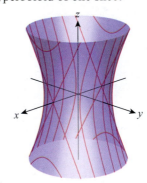

25. Hyperboloid of two sheets

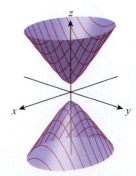

27. Elliptic paraboloid

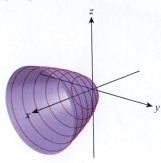

29. Elliptic paraboloid (horizontal cross-sections are circles)

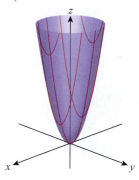

31. Hyperboloid of one sheet

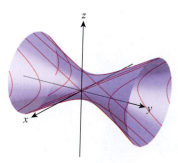

33. Hyperbolic paraboloid

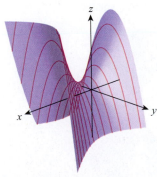

35. $2x^2 = z$ **37.** $y^2 = 5z$ **39.** $x^2 + z^2 = 2y$

41. $z = x^2 + y^2 + 2$ **43.** $\dfrac{x^2}{a^2} + \dfrac{y^2}{a^2} - \dfrac{z^2}{c^2} = 1$

45. $x^2 + y^2 = \dfrac{9}{4z^2}$ **47.** $\dfrac{x^2}{6378^2} + \dfrac{y^2}{6378^2} + \dfrac{z^2}{6357^2} = 1$

51. False; $a = b = c$ is possible in the equation of an ellipsoid.

53. True   **55.** False; consider a sphere.

Chapter 11 Review

1. 2   **3.** $\sqrt{2}/2$ **5.** $4\sqrt{2/3}$

7. The exterior of the sphere of radius 1, centered at $(0, 2, 0)$

9. The interior of the parabolic cylinder $x^2 + 4z^2 = 2$

11. The interior of the hyperboloid (of one sheet) $\dfrac{x^2}{8} + \dfrac{y^2}{4} - \dfrac{z^2}{2} = 1$ (including the hyperboloid itself)

13. $(9, -9, 1)$ **15.** $\mathbf{v} = \left\langle -\dfrac{7}{2}, 1, \dfrac{1}{2} \right\rangle$

17. Yes, they are collinear.

19. $\overrightarrow{PQ} = \left\langle 3\sqrt{3}/2, -1, 2 \right\rangle$; $\left| \overrightarrow{PQ} \right| = \sqrt{47}/2$

21. $\left(-6, -\dfrac{5}{2}, \dfrac{23}{4} \right)$ **23.** $\mathbf{u} = \left\langle \dfrac{12}{13}, -\dfrac{3}{13}, \dfrac{4}{13} \right\rangle$

25. -32 **27.** $\pi/3 = 60°$

29. **a.** $s = \sqrt{2}$ **b.** $s = -\dfrac{2\sqrt{2}}{9}$

31. Approx. 78.91°

33. $\alpha \approx 52.24°$, $\beta \approx 45°$, $\gamma \approx 69.3°$

35. $\text{proj}_\mathbf{v} \mathbf{u} = \left\langle \dfrac{12}{11}, -\dfrac{30}{11}, \dfrac{10}{11} \right\rangle$;

$\mathbf{u} - \text{proj}_\mathbf{v} \mathbf{u} = \left\langle \dfrac{-25}{33}, -\dfrac{3}{11}, \dfrac{1}{11} \right\rangle$

37. $\text{proj}_\mathbf{v} \mathbf{u} = -\dfrac{4}{9}\mathbf{i} - \dfrac{8}{9}\mathbf{j} + \dfrac{8}{9}\mathbf{k}$;

$\mathbf{u} - \text{proj}_\mathbf{v} \mathbf{u} = \dfrac{13}{9}\mathbf{i} - \dfrac{10}{9}\mathbf{j} - \dfrac{7}{18}\mathbf{k}$

39. 19.7 J   **41.** $\left\langle 2, \dfrac{7}{2}, \dfrac{15}{2} \right\rangle$

43. $\pm\left\langle 0, \dfrac{2}{\sqrt{5}}, -\dfrac{1}{\sqrt{5}} \right\rangle$ **45.** $\langle 29, 2, 36 \rangle$

47. Collinear **49.** $11/\sqrt{2}$ **51.** Not possible

53. $\left\langle -\dfrac{13}{2},-\dfrac{17}{2},-4 \right\rangle$ **55.** 4

57. $x=\dfrac{5}{2}+8t,\ y=9+2t,\ z=-3t$

59. $x=4+4t,\ y=-\dfrac{3}{2}+7t,\ z=7-4t,\ 0\le t\le\dfrac{1}{2}$

61. $(-4,0,1)$ **63.** Not possible

65. $\left\{\left(x,y,\dfrac{37}{6}-\dfrac{3}{4}x+\dfrac{1}{6}y\right)\middle|\,x\in\mathbb{R},y\in\mathbb{R}\right\}$

67. $2x-5y-14z+11=0$

69. $x=t-2,\ y=2+3t,\ z=\dfrac{1}{2}-2t$

71. $z-\sqrt{3}y=\sqrt{3}$ **73.** $(2,3,2)$

75. $x=1+2t,\ y=6-11t,\ z=-7t$

77. $3x-2y-2z+1=0$

79. $56/\sqrt{59}$ **81.** Elliptic cylinder; G

83. Plane; D **85.** Hyperboloid of two sheets; A

87. Hyperboloid of one sheet; F

89. Hyperbolic paraboloid

91. Hyperboloid of one sheet

93. Hyperboloid of two sheets

95. $\mathbf{v}(t)=\langle 15,75-32t\rangle$

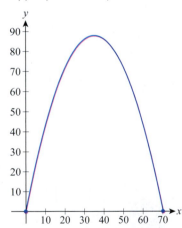

97. Approx. 166.96 mph

99. $1080\sqrt{3}\approx 1870.61$ ft-lb **101.** Approx. 15 N · m

103. Approx. $6.72\cdot 10^{-16}$ N **105.** True **107.** True

109. False; torque is perpendicular to both the force vector and the radius from the axis of rotation.

111. True **113.** True

115. False; **u** is parallel with **v** × **w**.

117. True **119.** True

125.

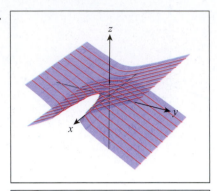

127.

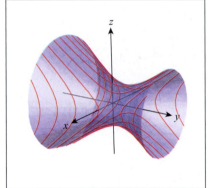

129.

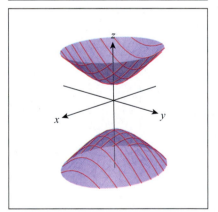

Chapter 12

Section 12.1

1. Dom $=(-\infty,0)\cup(0,\infty)$;

  a. $\dfrac{1}{4}\mathbf{i}+4\mathbf{j}-2\mathbf{k}$ **b.** $\dfrac{1}{25}\mathbf{i}-10\mathbf{j}+5\mathbf{k}$

3. Dom $=(0,3)$;

  a. Undefined **b.** $\dfrac{\sqrt{2}}{4}\mathbf{i}-\mathbf{j}$

5. $\mathbf{u}(t) = -2t$; no, it is a scalar-valued function.

7. C **9.** A **11.** E **13.** D

15.

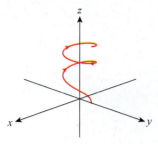

17.

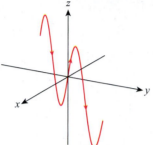

19.

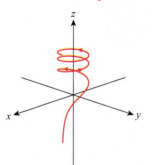

21.

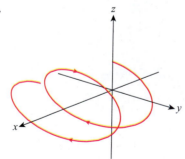

23. $\mathbf{r}(t) = \left\langle \sqrt{3}\cos t, \sqrt{2}\sin t, 1 - \dfrac{3\sqrt{3}}{2}\cos t \right\rangle,$
$t \in [0, 2\pi]$

25. $\mathbf{r}(t) = \left\langle \cos t, \sin t, 2\cos^2 t - \sin^2 t \right\rangle, \; t \in [0, 2\pi]$

27. $\mathbf{r}(t) = \left\langle t^2, t, \dfrac{2t^4 + t^2}{2} \right\rangle$

29. $\mathbf{r}(t) = \left\langle \dfrac{t^2}{2}, t, \dfrac{\sqrt{-3t^4 - 16t^2 + 192}}{2\sqrt{6}} \right\rangle$

31. The limit does not exist.

33. Yes, the limit is $\langle 0, 0, 1 \rangle$.

35. The limit does not exist.   **37.** No discontinuities

39. $t = 0$ **47.** $\mathbf{r}'(t) = -3t^2\,\mathbf{i} + t^4\,\mathbf{k}$

49. $\mathbf{r}'(t) = \dfrac{1}{t}\mathbf{i} + \csc t \cot t\,\mathbf{j} - \dfrac{t}{\sqrt{4 - t^2}}\mathbf{k}$

51. $\mathbf{r}'(t) = \left\langle -2t\sin(t^2 + 1), -\dfrac{2}{(1 - t)^2}, \dfrac{3}{\sqrt{1 - t^2}} \right\rangle$

53. $\left\langle \dfrac{2}{3}, -\dfrac{1}{3}, \dfrac{2}{3} \right\rangle$ **55.** $\left\langle \dfrac{4}{5}, \dfrac{3}{5}, 0 \right\rangle$

57. $\langle 1 + 3t, -1 - 2t, 1 + t \rangle$

59. $\left\langle \dfrac{\pi}{6} + \dfrac{2t}{\sqrt{3}}, \dfrac{\pi}{3} - \dfrac{2t}{\sqrt{3}}, 2t - \ln 2 \right\rangle$

69. $\left\langle t^3, \dfrac{t^4 - 2t^2}{4}, \dfrac{-2t^{3/2}}{3} \right\rangle + \mathbf{C}$

71. $\dfrac{t^2}{2}\mathbf{i} + 3t\,\mathbf{j} - t^4\,\mathbf{k} + \mathbf{C}$

73. $2t\,\mathbf{i} - \ln|t|\,\mathbf{j} + \dfrac{2t^{5/2}}{5}\mathbf{k} + \mathbf{C}$ **75.** $\dfrac{3}{2}\mathbf{i} - 12\,\mathbf{j} + 9\,\mathbf{k}$

77. $\left\langle 0, 0, \dfrac{2}{5} \right\rangle$ **79.** $\langle 2(e^e - e), -1, 1 \rangle$

81. Max. altitude: 78.48 m; range: 543.725 m

83. $\mathbf{v}(t) = \langle t^2 + 3, t + 1, -1 \rangle;$
$\mathbf{r}(t) = \left\langle \dfrac{t^3}{3} + 3t - 5, \dfrac{t^2}{2} + t, 2 - t \right\rangle$

85. $\varphi = \operatorname{arccot} b$

105. See the answer given for Exercise 15.

107. See the answer given for Exercise 17.

109. See the answer given for Exercise 19.

111. See the answer given for Exercise 21.

113.

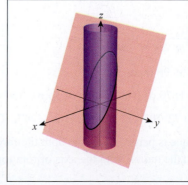

115.

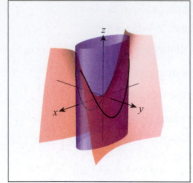

117.

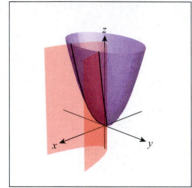

119.

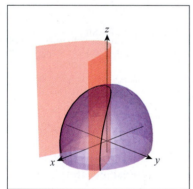

Section 12.2

1. $12\sqrt{5}\pi$ **3.** 18   **5.** e^2 **7.** 145   **9.** $\ln\left(2+\sqrt{3}\right)$

11. $\mathbf{r}\left(t(s)\right)=\left\langle 1+\dfrac{2s}{3\sqrt{5}},3-\dfrac{\sqrt{5}s}{3},4+\dfrac{4s}{3\sqrt{5}}\right\rangle$

13. $\mathbf{r}\left(t(s)\right)=\left\langle 2\cos\dfrac{s}{\sqrt{13}},2\sin\dfrac{s}{\sqrt{13}},\dfrac{3s}{\sqrt{13}}\right\rangle$

15. $\mathbf{r}\left(t(s)\right)=\left\langle b\dfrac{s}{\sqrt{b^2+a^2\omega^2}},a\cos\dfrac{\omega s}{\sqrt{b^2+a^2\omega^2}},a\sin\dfrac{\omega s}{\sqrt{b^2+a^2\omega^2}}\right\rangle$

17. $\mathbf{T}(t)=\dfrac{1}{\sqrt{9t^4+4t^2+1}}\left\langle 1,3t^2,-2t\right\rangle$

19. $\mathbf{T}(t)=\dfrac{1}{\sqrt{3}}\left\langle 1,\cos t-\sin t,\cos t+\sin t\right\rangle$

21. $\mathbf{T}(t)=\dfrac{1}{\sqrt{t^2+1}}\left\langle -t\sin t,1,t\cos t\right\rangle$

23. $\mathbf{T}(t)=\dfrac{1}{e^t+e^{-t}}\left\langle e^t,-e^{-t},\sqrt{2}\right\rangle$

25. $\mathbf{r}\left(t(s)\right)=\left\langle \dfrac{s}{\sqrt{m^2+1}},\dfrac{ms}{\sqrt{m^2+1}}+b\right\rangle$ **27.** c

29. $x\left(t(s)\right)=a\left(4\sin^{-1}\sqrt{\dfrac{s}{8a}}-\sin\left(4\sin^{-1}\sqrt{\dfrac{s}{8a}}\right)\right),$

$y\left(t(s)\right)=a\left(1-\cos\left(4\sin^{-1}\sqrt{\dfrac{s}{8a}}\right)\right)$

31. $L=\sqrt{\left(2\pi nr\right)^2+h^2}$ **33.** Range $=\dfrac{v_0^2\sin 2\theta}{g}$

35. $\theta=45°$ **37.** $\theta=\dfrac{90°+\varphi}{2}$

39. **a.** $\mathbf{r}(t)=\left\langle 15t,15\sqrt{3}t-\dfrac{gt^2}{2}+2\right\rangle$

b. Max. height: approx. 36.4 m;
range: approx. 80.59 m;
impact speed: approx. 30.65 m/s

41. **a.** $\mathbf{r}(t)=\left\langle 64\sqrt{3}t,-16t^2+64t+4\right\rangle$

b. Max. height: 68 ft

c. Horizontal distance traveled: approx. 450.23 ft;
impact speed: approx. 129 ft/s

43. $x\approx 39°$ or $x\approx 51°$

45. $\mathbf{r}(t)=\left\langle 333.\overline{3}t,8000-\dfrac{gt^2}{2}\right\rangle;$

horizontal distance: approx. 13,461.8 m;
impact speed: approx. 517.76 m/s

47. $\mathbf{r}(t)=\left\langle \begin{array}{l}\dfrac{v_0\cos\theta}{C}\left(1-e^{-Ct}\right),\\[2mm]\dfrac{v_0\sin\theta}{C}\left(1-e^{-Ct}\right)+\dfrac{g}{C^2}\left(1-Ct-e^{-Ct}\right)\end{array}\right\rangle$

49. **b.** $\mathbf{r}\left(t(s)\right)=a\left\langle \begin{array}{l}\cos\sqrt{2s/a}+\sqrt{2s/a}\sin\sqrt{2s/a},\\[1mm]\sin\sqrt{2s/a}-\sqrt{2s/a}\cos\sqrt{2s/a}\end{array}\right\rangle$

51. False; they are perpendicular only for $t=0$.

53. False; its range quadruples. (The vertical component of its initial velocity doubled, the projectile is aloft twice as long, and the horizontal component of its velocity is also doubled.)

Section 12.3

1. $\mathbf{T}(t) = \langle -\sin t, \cos t, 0 \rangle$; $\mathbf{N}(t) = \langle -\sin t, -\cos t, 0 \rangle$;
$\mathbf{B}(t) = \langle 0, 0, 1 \rangle$

3. $\mathbf{T}(t) = \langle \cos 2t, -\sin 2t, 0 \rangle$;
$\mathbf{N}(t) = \langle -\sin 2t, -\cos 2t, 0 \rangle$; $\mathbf{B}(t) = \langle 0, 0, -1 \rangle$

5. $\mathbf{T}(t) = \left\langle \dfrac{\cos \pi t}{\sqrt{2}}, -\dfrac{\sin \pi t}{\sqrt{2}}, \dfrac{1}{\sqrt{2}} \right\rangle$;
$\mathbf{N}(t) = \langle -\sin \pi t, -\cos \pi t, 0 \rangle$;
$\mathbf{B}(t) = \left\langle \dfrac{\cos \pi t}{\sqrt{2}}, -\dfrac{\sin \pi t}{\sqrt{2}}, -\dfrac{1}{\sqrt{2}} \right\rangle$

7. $\mathbf{T}(t) = \left\langle \dfrac{2}{\sqrt{13}}, -\dfrac{3\sin 3t}{\sqrt{13}}, \dfrac{3\cos 3t}{\sqrt{13}} \right\rangle$;
$\mathbf{N}(t) = \langle 0, -\cos 3t, -\sin 3t \rangle$;
$\mathbf{B}(t) = \left\langle \dfrac{3}{\sqrt{13}}, \dfrac{2\sin 3t}{\sqrt{13}}, -\dfrac{2\cos 3t}{\sqrt{13}} \right\rangle$

9. $\mathbf{T}(t) = \left\langle \dfrac{\cos t + \sin t}{\sqrt{e^2 + 2}}, \dfrac{\cos t - \sin t}{\sqrt{e^2 + 2}}, \dfrac{e}{\sqrt{e^2 + 2}} \right\rangle$;
$\mathbf{N}(t) = \left\langle \dfrac{\cos t - \sin t}{\sqrt{2}}, \dfrac{-\cos t - \sin t}{\sqrt{2}}, 0 \right\rangle$;
$\mathbf{B}(t) = \left\langle \dfrac{e(\cos t + \sin t)}{\sqrt{2e^2 + 4}}, \dfrac{e(\cos t - \sin t)}{\sqrt{2e^2 + 4}}, -\dfrac{\sqrt{2}}{\sqrt{e^2 + 2}} \right\rangle$

11. Osculating plane: $z = 0$;
normal plane: $y = 0$;
rectifying plane: $x = 1$

13. Osculating plane: $z = 3$;
normal plane: $x = 0$;
rectifying plane: $y = 3$

15. Osculating plane: $2y + 2z = \pi$;
normal plane: $2y - 2z = -\pi$;
rectifying plane: $x = 1$

17. Osculating plane: $3x + 2z = 6\pi$;
normal plane: $2x - 3z = 4\pi$;
rectifying plane: $y = -1$

19. Osculating plane: $e(y - x) + 2z = e^{1 + (\pi/2)}$;
normal plane: $x - y + ez = e^{\pi/2}(e^2 + 1)$;
rectifying plane: $x + y = e^{\pi/2}$

21. $\mathbf{r}(t) + u\langle -\sin 2t, -\cos 2t, 0 \rangle$, $u \in [-1/4, 1/4]$

23. $\mathbf{r}(t) + \cos u \langle -\sin \pi t, -\cos \pi t, 0 \rangle$
$+ \sin u \left\langle \dfrac{\cos \pi t}{\sqrt{2}}, -\dfrac{\sin \pi t}{\sqrt{2}}, -\dfrac{1}{\sqrt{2}} \right\rangle$, $u \in [0, 2\pi]$

25. $\mathbf{r}(t) + 0.3 \cos u \langle 0, -\cos 3t, -\sin 3t \rangle$
$+ 0.2 \sin u \left\langle \dfrac{3}{\sqrt{13}}, \dfrac{2\sin 3t}{\sqrt{13}}, -\dfrac{2\cos 3t}{\sqrt{13}} \right\rangle$, $u \in [0, 2\pi]$

27. $\mathbf{r}(t) + \left(\cos t + 2\cos \dfrac{t}{2} \right)$
$\cdot \left\langle \dfrac{\cos t - \sin t}{\sqrt{2}}, \dfrac{-\cos t - \sin t}{\sqrt{2}}, 0 \right\rangle$
$+ \left(\sin t - 2\sin \dfrac{t}{2} \right)$
$\cdot \left\langle \dfrac{e(\cos t + \sin t)}{\sqrt{2e^2 + 4}}, \dfrac{e(\cos t - \sin t)}{\sqrt{2e^2 + 4}}, -\dfrac{\sqrt{2}}{\sqrt{e^2 + 2}} \right\rangle$,
$u \in [0, 2\pi]$

31. $\kappa = 0$; $\tau = 0$

33. $\kappa = \dfrac{1}{2\cosh^2 t}$; $\tau = \dfrac{1}{2\cosh^2 t}$

35. If $b = 0$, the "helix" is actually a circle, and we obtain the result of Example 4b.

37. $\kappa = \dfrac{a\omega}{(a\omega)^2 + b^2}$; $\tau = \dfrac{b\omega}{(a\omega)^2 + b^2}$

39. $\kappa = 2/5$; $\tau = -1/5$

41. $\kappa = \dfrac{2t^3}{(t^4 + 1)^{3/2}}$; $\tau = 0$

43. $\kappa = 1/2$; $\tau = 0$ **45.** $\kappa = 9/25$; $\tau = -12/25$

47. $\kappa = \dfrac{\sqrt{2}}{3e^t}$; $\tau = \dfrac{1}{3e^t}$ **51.** $\dfrac{3}{5\sqrt{10}}$ **53.** $\sqrt{3}/2$

55. $(2k\pi, 1)$, $((2k+1)\pi, -1)$, $k \in \mathbb{Z}$ **59.** $1/4$

61. $\left(x - \dfrac{1}{2} \right)^2 + \left(y - \dfrac{1}{4} \right)^2 = 2$ **63.** $x^2 + y^2 = 1$

65. $(x-2)^2 + (y-2)^2 = 2$

67. $\left(x+\dfrac{2}{3}\right)^2 + \left(y-\dfrac{5}{6}\right)^2 = \dfrac{125}{36}$

69. $\mathbf{r}(t) = \langle \pi + 4\sin t, -2 - 4\cos t \rangle$

71. $\mathbf{r}(t) = \left\langle \dfrac{\cos t + 1}{2}, \dfrac{\sin t}{2}, 0 \right\rangle$

75. $\mathbf{r}(t) = \left\langle \dfrac{t - 9t^5}{2}, \dfrac{1 + 15t^4}{6t} \right\rangle$ **77.** $a_T = 0;\ a_N = 2$

79. $a_T = \dfrac{4t}{\sqrt{4t^2 + 5}};\ a_N = \dfrac{2\sqrt{5}}{\sqrt{4t^2 + 5}}$

81. $a_T = \dfrac{\sin t}{\sqrt{2 - 2\cos t}};\ a_N = \dfrac{\cos t - 1}{\sqrt{2 - 2\cos t}}$

83. $a_T = 2t;\ a_N = 2$ **85.** $a_T = e^t - e^{-t};\ a_N = \sqrt{2}$

87. $a_T = \sqrt{3}e^t;\ a_N = \sqrt{2}e^t$ **89.** $F_c = \dfrac{mv^2}{r}$

91. A slow entering speed decreases the normal component of the acceleration vector, while gently accelerating in the curve adds a component parallel to motion, making the (resultant) acceleration vector more in line with the direction of motion, thereby decreasing the chances of skidding.

99. False; $(\mathbf{T} \times \mathbf{N}) \cdot \mathbf{B} = 1$.

101. False; the osculating circle is undefined.

103. False; the force is quadrupled.   **105.** True

107.

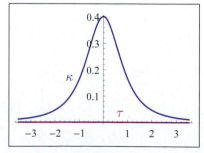

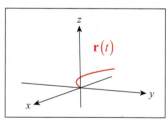

109.

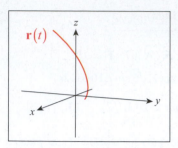

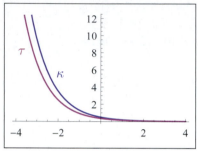

Section 12.4

1. $\mathbf{v} = 2\mathbf{u} + 4t\,\mathbf{u}_\perp;\ \ \mathbf{a} = -8t\,\mathbf{u} + 8\,\mathbf{u}_\perp$

3. $\mathbf{v} = \dfrac{1}{2\sqrt{t}}\mathbf{u} + 4\sqrt{t}\,\mathbf{u}_\perp;\ \ \mathbf{a} = \dfrac{-64t^2 - 1}{4t^{3/2}}\mathbf{u} + \dfrac{4}{\sqrt{t}}\mathbf{u}_\perp$

5. $\mathbf{v} = 4\cos 2t\,\mathbf{u} + 6t^2 \sin 2t\,\mathbf{u}_\perp;$
$\mathbf{a} = -\left(18t^4 + 8\right)\sin 2t\,\mathbf{u} + 12\left(2t^2 \cos 2t + t\sin 2t\right)\mathbf{u}_\perp$

9. $28{,}000$ km/h

11. 5.973×10^{24} kg

13. $227{,}940{,}000$ km

17. The eccentricity is approx. 0.0167. An equation is
$r(\theta) \approx \dfrac{1.49557 \times 10^8}{1 + 0.0167 \cos\theta}$ km.

19. Approx. 76 years

23. $v_p \approx 30.287$ km/s $\approx 68{,}145.3$ mph;
$v_a \approx 29.291$ km/s $\approx 65{,}905.2$ mph

25. $v_p \approx 1.08$ km/s ≈ 2430 mph;
$v_a \approx 0.964$ km/s ≈ 2169 mph

27. $v_p \approx 0.03135$ AU/day
≈ 54.281 km/sec $\approx 122{,}132$ mph;
$v_a \approx 0.000526$ AU/day
≈ 0.911 km/sec ≈ 2049 mph

29. Mission control is right. This way, she will move into an orbit closer to Earth, and by Kepler's Third Law, the orbital period will decrease.

Chapter 12 Review

1. C **3.** D **5.** A **7.** H

9. $\mathbf{r}(t) = \left\langle \sqrt{2}\cos t, \sqrt{5}\sin t, 10\cos 2t \right\rangle$

11. Yes, the limit is $\langle 0,1,1/2 \rangle$.   **13.** $t = \pm 1$

15. $\mathbf{r}'(t) = \left\langle \sec t\tan t, \dfrac{1}{2\sqrt{t(1-t)}}, 2\sec^2 t\tan t \right\rangle$

17. $\langle 8t,1,1 \rangle$ **19.** $\left\langle 3+t, -\ln 19 - \dfrac{12t}{19}, -3 - \dfrac{t}{2} \right\rangle$

21. $\left\langle e^{t+1}, \sqrt{t^2+2}, -\dfrac{\cos 2t}{2} \right\rangle + \mathbf{C}$ **23.** $\left\langle \dfrac{14}{3}, \dfrac{27}{2}, 9 \right\rangle$

25. 20   **27.** $\mathbf{r}(t(s)) = \left\langle 2 - \dfrac{s}{\sqrt{21}}, 1 + \dfrac{4s}{\sqrt{21}}, 3 + \dfrac{2s}{\sqrt{21}} \right\rangle$

29. $\mathbf{T}(t) = \left\langle \dfrac{1}{2\sec t}, -\sin t, \dfrac{\sqrt{3}}{2\sec t} \right\rangle$

31. $\mathbf{T}(t) = \dfrac{1}{\sqrt{4t^4+t^2+1}}\langle 1, 2t^2, t \rangle$

33. $\mathbf{N}(t) = \langle 0, -\cos t, -\sin t \rangle$;

$\mathbf{B}(t) = \dfrac{\sqrt{2}}{2}\langle 1, \sin t, -\cos t \rangle$

35. $\mathbf{N}(t) = \left\langle \dfrac{\sqrt{2}e^t}{1+e^{2t}}, \dfrac{\sqrt{2}e^t}{1+e^{2t}}, \dfrac{1-e^{2t}}{1+e^{2t}} \right\rangle$;

$\mathbf{B}(t) = \left\langle \dfrac{-1}{1+e^{2t}}, \dfrac{e^{2t}}{1+e^{2t}}, \dfrac{\sqrt{2}e^t}{1+e^{2t}} \right\rangle$

37. Osculating plane: $z = x$;
normal plane: $z = -x$;
rectifying plane: $y = 2$

39. Osculating plane: $z = \dfrac{\sqrt{2}}{2e}x - \dfrac{\sqrt{2}e}{2}y + \sqrt{2}$;

normal plane: $\sqrt{2}z = e^2 - \dfrac{1}{e^2} + 2 - ex + \dfrac{y}{e}$;

rectifying plane: $y = \dfrac{1}{e} + e^3 - e^2 x$

41. $\mathbf{r}(t) + u\langle 0, -\cos t, -\sin t \rangle$, $u \in [-1/4, 1/4]$

43. $\mathbf{r}(t) + 2\cos u \left\langle \dfrac{\sqrt{2}e^t}{1+e^{2t}}, \dfrac{\sqrt{2}e^t}{1+e^{2t}}, \dfrac{1-e^{2t}}{1+e^{2t}} \right\rangle$

$+ 2\sin u \left\langle \dfrac{-1}{1+e^{2t}}, \dfrac{e^{2t}}{1+e^{2t}}, \dfrac{\sqrt{2}e^t}{1+e^{2t}} \right\rangle$, $u \in [0, 2\pi]$

45. $\kappa = 2/3$; $\tau = \sqrt{2}/3$

47. $\kappa = \dfrac{t^2\sqrt{t^4+4t^2+1}}{\left(t^4+t^2+1\right)^{3/2}}$; $\tau = \dfrac{-2t^2}{t^4+4t^2+1}$

49. a. $\kappa = a/b^2$ **b.** $\kappa = b/a^2$

51. $(1,1)$, $(-1,-1)$

53. $(x-4)^2 + \left(y + \dfrac{8}{3}\right)^2 = \dfrac{250}{9}$

55. $a_T = 0$; $a_N = 4$

57. $a_T = \dfrac{t}{\sqrt{t^2+1}}$; $a_N = \dfrac{t^2+2}{\sqrt{t^2+1}}$

59. $\mathbf{v} = \mathbf{u} + \dfrac{t}{2}\mathbf{u}_\perp$; $\mathbf{a} = -\dfrac{t}{4}\mathbf{u} + \mathbf{u}_\perp$

61. Max. altitude: approx. 196 ft;
range: approx. 452.66 ft

63. Approx. 243,576.6 ft   **65.** Approx. 13.89 m/s^2

67. a. $\mathbf{r}(t) \approx \left\langle 83.33t, 500 - 144.34t - \dfrac{gt^2}{2} \right\rangle$

b. Approx. 231.13 m   **c.** Approx. 759.71 km/h

73. $\kappa = |b|$; The curvature at $t = b$ is equal to the arc length over the interval $[0,b]$.

75. Approx. 27,830 km/h

77. Approx. 90,107.7 km

79. a. $v_p \approx 13.711$ km/s $\approx 30,849.8$ mph

b. $v_a \approx 12.434$ km/s $\approx 27,976.9$ mph

81. False; consider the graph of $\mathbf{r}(t) = \langle t^3, t^3, t^3 \rangle$.

83. True   **85.** False; the curvature is zero.

87. False; κ and τ do not depend on parametrization; this is evident from the formulas (given for κ and τ) in the summary table at the end of Section 12.3.

89. True

91. $\mathbf{r}(t) = \left\langle \left(8 - \dfrac{8t}{5\pi}\right)\cos 5t, \left(8 - \dfrac{8t}{5\pi}\right)\sin 5t, \dfrac{t}{4} \right\rangle$

Chapter 12 Project

1. $\mathbf{r}(t) = \left\langle 0, \dfrac{\sqrt{2}}{2}v_0 t, -16t^2 + \dfrac{\sqrt{2}}{2}v_0 t + 7 \right\rangle$

3. $v_0 \approx 30$ ft/s

5. $v_0(\alpha) \approx \dfrac{98.8}{\cos\alpha\sqrt{24.7\tan\alpha - 3}}$

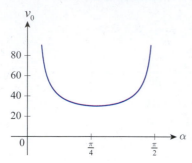

Chapter 13

Section 13.1

1. $f(0,0) = 0$, $f(2,1) = -2$

3. $f(5,0,-1) = -5$, $f(-4,2,1) = -4/9$

5. Dom $= \mathbb{R}^2$; Ran $= \mathbb{R}$; $f(3,-1) = -30$

7. Dom $= \{(x,y)\,|\,y \geq 0\}$; Ran $= [0,\infty)$; $f(2,9) = 6$

9. Dom $= \{(x,y)\,|\,x \neq 0 \text{ and } y \neq 0\}$; Ran $= \mathbb{R}$;
$f(e,-1) = 1$

11. Dom $= \mathbb{R}^3$; Ran $= (-\pi/2, \pi/2)$; $f(5,3,2) = \pi/4$

13. Plane

15. Elliptic paraboloid

17. Elliptic cone ($z \geq 0$)

19. Elliptic paraboloid

21. Boundary point

23. Boundary point

25. Boundary point

27. Closed   **29.** Open

31. Open   **33.** Neither

35. Answers will vary.

37. Answers will vary.

39. Answers will vary.

41. Answers will vary.

43. C   **45.** B   **47.** E

49. E   **51.** A   **53.** D

55.

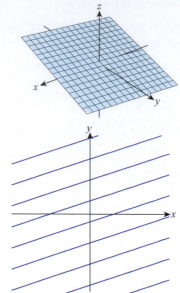

57.

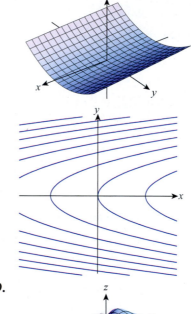

59.

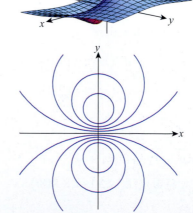

61. Ellipsoids centered at the origin

63. a. The direction perpendicular to the contour through A, approx. 30° south of west

b. Answers will vary, but the path should run perpendicularly to the contour lines.

c. Approx. 4200 ft

d. Answers will vary.

Section 13.2

1. 6 **3.** 3 **5.** 0 **7.** $\sqrt{5}/10$ **9.** 0

11. The limit does not exist.

13. The limit does not exist. **15.** 0

17. The limit does not exist.

19. The limit does not exist.

21. The limit does not exist.

23. Domain in Exercise 5:
$$\left\{(x,y)\in\mathbb{R}^2-\{(0,0)\}\,\middle|\,x,y\geq 0\right\};$$
domain in Exercise 6:
$$\left\{(x,y)\in\mathbb{R}^2\,\middle|\,x>0,\,y\geq 1,\,y\neq x+1\right\}$$

37. -3 **39.** 0 **45.** $\mathbb{R}^2$ **47.** $\mathbb{R}^2$ **49.** $\mathbb{R}^2$

51. $\left\{(x,y,z)\,\middle|\,z>\dfrac{1}{3}\left(x^2+2y^2\right)\right\}$

53. $\left\{(x,y,z)\,\middle|\,x^2+y^2+z^2\leq 9\right\}$

55. $(0,0)$; nonremovable discontinuity

57. $(0,0)$; nonremovable discontinuity

59. $(0,0,0)$; nonremovable discontinuity

61. False; we can redefine $f(a,b)$ without changing the limit.

63. True

65. False; consider the function
$$f(x,y)=\begin{cases}\dfrac{1}{x^2+y^2} & \text{if }(x,y)\neq(0,0)\\ L & \text{if }(x,y)=(0,0).\end{cases}$$

Section 13.3

1. $z_x(2,1)=4$; $z_y(2,1)=18$;
L_1: $\langle 2+t,1,8+4t\rangle$; L_2: $\langle 2,1+t,8+18t\rangle$

3. $z_x(-1,0)=4$; $z_y(-1,0)=-4$;
L_1: $\langle-1+t,0,-4+4t\rangle$; L_2: $\langle-1,t,-4-4t\rangle$

5. $\dfrac{\partial f}{\partial x}=3x^2$, $\dfrac{\partial f}{\partial y}=3y^2$

7. $\dfrac{\partial h}{\partial x}=15x^2y$, $\dfrac{\partial h}{\partial y}=5x^3+4y^3$

9. $\dfrac{\partial V}{\partial r}=\dfrac{2\pi rh}{3}$, $\dfrac{\partial V}{\partial h}=\dfrac{\pi r^2}{3}$

11. $\dfrac{\partial k}{\partial x}=y^2+\dfrac{y}{\sqrt{2xy}}$, $\dfrac{\partial k}{\partial y}=2xy+\dfrac{x}{\sqrt{2xy}}$

13. $\dfrac{\partial m}{\partial x}=\left(4x^3y-24x^4\right)(y-3x)^3$,
$\dfrac{\partial m}{\partial y}=4x^4(y-3x)^3$

15. $\dfrac{\partial p}{\partial x}=\dfrac{x^2y+2xy^2}{(x+y)^2}$, $\dfrac{\partial p}{\partial y}=\dfrac{x^3}{(x+y)^2}$

17. $\dfrac{\partial G}{\partial x}=-\dfrac{xy}{\left(x^2+y^2\right)^{3/2}}$, $\dfrac{\partial G}{\partial y}=\dfrac{x^2}{\left(x^2+y^2\right)^{3/2}}$

19. $\dfrac{\partial I}{\partial x}=\dfrac{1}{y}\cos\dfrac{x}{y}$, $\dfrac{\partial I}{\partial y}=-\dfrac{x}{y^2}\cos\dfrac{x}{y}$

21. $\dfrac{\partial K}{\partial x}=-\dfrac{x}{y}-\dfrac{2y^2}{x^2}$, $\dfrac{\partial K}{\partial y}=\dfrac{x^2}{2y^2}+\dfrac{4y}{x}$

23. $\dfrac{\partial M}{\partial x}=\dfrac{\sqrt{xy}}{2x+2x^2y}$, $\dfrac{\partial M}{\partial y}=\dfrac{\sqrt{xy}}{2y+2xy^2}$

25. $\dfrac{\partial g}{\partial x}=x^{(y-z)/z}\dfrac{y}{z}$, $\dfrac{\partial g}{\partial y}=x^{y/z}\dfrac{\ln x}{z}$,
$\dfrac{\partial g}{\partial z}=-x^{y/z}\dfrac{y\ln x}{z^2}$

27. $\dfrac{\partial h}{\partial x}=\cos\left(y+z^2\right)$, $\dfrac{\partial h}{\partial y}=-x\sin\left(y+z^2\right)$,
$\dfrac{\partial h}{\partial z}=-2xz\sin\left(y+z^2\right)$

29. L: $\left\langle 1+t,1,3+\dfrac{4t}{3}\right\rangle$

31. L: $\left\langle 2+t,\dfrac{3}{\sqrt{5}},5+\dfrac{8t}{5}\right\rangle$

33. L: $\left\langle-1+t,1,3-\dfrac{4t}{3}\right\rangle$

35. $z_x = \dfrac{2y}{y-2z}$, $z_y = \dfrac{z-2x}{2z-y}$

37. $z_x = \dfrac{z(2xy-z)}{xz+1}$, $z_y = \dfrac{x^2 z}{xz+1}$

39. $y_x = \dfrac{y^2}{2y^2 - xy - z}$, $y_z = \dfrac{y\ln y}{2y^2 - xy - z}$

55. $f_{yyx} = 2y - \dfrac{1}{y^2}$

57. $h_{turs} = -\dfrac{2r}{u^2}$

59. $f_{yxyx} = 4x^2 \sin(x^2 - y) - 2\cos(x^2 - y)$

61. $f_{zyx} = 6xz$

63. $f_{xyzw} = -\dfrac{15wxyz}{\left(w^2 + x^2 + y^2 + z^2\right)^{7/2}}$

65. $\dfrac{\partial f}{\partial x} = -\dfrac{\sqrt{y}}{x^2}$, $\dfrac{\partial f}{\partial y} = \dfrac{1}{2x\sqrt{y}}$

67. $f(x,y) = x^2 y - \sin y$

69. Not possible; if such a function existed, it wouldn't satisfy Clairaut's Theorem.

71. $\dfrac{\partial i}{\partial o} = -\left(\dfrac{f}{o-f}\right)^2$; image distance increases when the object distance decreases, and vice versa.

73. Approx. 2%   **93.** 0 and $-\sqrt{2}$, respectively

97. From the first formula, $\dfrac{\partial A}{\partial r} = r\alpha = C - 2r$, while

from the second formula $\dfrac{\partial A}{\partial r} = \dfrac{1}{2}C - 2r$.

The answers are not equivalent because in the first case α is held constant while the radius changes, in the second case C is held constant while the radius changes. The rates of change for area are different in the two cases.

99. Inflation has a greater effect.

101. b. $\dfrac{\partial R}{\partial R_k} = \dfrac{\displaystyle\prod_{i\neq k} R_i^2}{\left(\displaystyle\sum_{i=1}^n \prod_{j\neq i} R_j\right)^2}$

105. No, the partials don't exist.

107. No, the origin is a discontinuity (f is undefined at the origin).

119. $\{(x,y)\,|\,x \neq y\}$

Section 13.4

3. $\dfrac{dg}{dt} = \dfrac{\partial g}{\partial x}\dfrac{dx}{dt} + \dfrac{\partial g}{\partial y}\dfrac{dy}{dt} + \dfrac{\partial g}{\partial z}\dfrac{dz}{dt}$

5. $\dfrac{\partial k}{\partial z} = \dfrac{\partial k}{\partial u}\dfrac{\partial u}{\partial z} + \dfrac{\partial k}{\partial v}\dfrac{\partial v}{\partial z}$ **7.** $\dfrac{dy}{dx} = 6x^2 + 4x\sin 2x^2$

9. $\dfrac{dy}{dx} = \cot x + \tan x$ **11.** $\dfrac{dy}{dx} = \sec^2 x + \sin 2x$

13. $\dfrac{dy}{dx} = -\dfrac{4}{x\sqrt{x^4 - 4}}$ **15.** $\dfrac{dy}{dx} = 2\left(x\sin 2x + \sin^2 x\right)$

17. $\dfrac{dy}{dx} = \sin x(10x - 2) + x\cos x(5x - 2)$

19. $f'\left(\dfrac{\pi}{2}\right) = 0$ **21.** $f'(0) = 36$

23. $z_x = 2(u - v)$, $z_y = 2(2u + v)$

25. $z_x = \left(3v^2 - 2u^2\right)\cos x$, $z_y = 4uv\sin y$

27. $z_x = -3\sqrt{v}e^x$, $z_y = \left(8v^3 - \dfrac{3u}{2\sqrt{v}}\right)e^y$

29. $z_x = 2u\left(ux + v - 3w^2 y\right) - w^3$,
$z_y = u\left(2v - u - 6w^2 x\right) - w^3$

31. $z_x = (9vy + 2v - 2)ve^{2u+3w}$,
$z_y = (9vx - 2v + 4)ve^{2u+3w}$

33. $z_x = \left(vw^2 + 4uxyvw\right)\cos\left(uw^2\right) - \sin\left(uw^2\right)$,
$z_y = vw\left(2ux^2 + 3wy^2\right)\cos\left(uw^2\right) + 3\sin\left(uw^2\right)$

39. $\dfrac{dy}{dx} = \dfrac{2x - y}{x - 2y}$ **41.** $\dfrac{dy}{dx} = -\dfrac{2xy}{4 + x^2}$

43. $\dfrac{dy}{dx} = -\sqrt[3]{\dfrac{y}{x}}$

45. $\dfrac{dy}{dx} = \dfrac{2(2x - 3)\left(x^2 + y^2 - 3x\right) - 2x}{2y - 4y\left(x^2 + y^2 - 3x\right)}$

47. $\dfrac{dy}{dx} = \dfrac{2x\left(x^2 + y^2\right)^2 + 2xy}{x^2 - y^2}$

49. $z_x = \dfrac{yz - e^{x+y+z}}{e^{x+y+z} - xy}$, $z_y = \dfrac{xz - e^{x+y+z}}{e^{x+y+z} - xy}$

51. $z_x = \dfrac{y + 2xz^3}{y\cos(yz) - 3x^2 z^2}$, $z_y = \dfrac{x - z\cos(yz)}{y\cos(yz) - 3x^2 z^2}$

53. $z_x = \dfrac{e^x \sin y}{z(4 - y)}$, $z_y = \dfrac{z^2 + 2e^x \cos y}{2z(4 - y)}$

55. $z_x = -\dfrac{yz(x^2+y^2+z^2)+2x}{xy(x^2+y^2+z^2)+2z}$,

$z_y = -\dfrac{xz(x^2+y^2+z^2)+2y}{xy(x^2+y^2+z^2)+2z}$

57. $f_x = f_t\dfrac{\partial t}{\partial x}+f_u\dfrac{\partial u}{\partial x}+f_v\dfrac{\partial v}{\partial x}+f_w\dfrac{\partial w}{\partial x}$,

$f_y = f_t\dfrac{\partial t}{\partial y}+f_u\dfrac{\partial u}{\partial y}+f_v\dfrac{\partial v}{\partial y}+f_w\dfrac{\partial w}{\partial y}$

61. Homogeneous of degree 1

65. 8.25 in.^{2}/min

67. 2π in.^{3}/s

69. Approx. 1.9×10^4 kJ/s

71. Approx. -0.466 degrees/min

Section 13.5

1. $\nabla f(1,2)=\langle 15,-8\rangle$ **3.** $\nabla h(-1,1)=\langle -2e^2,2e^2\rangle$

5. $\nabla F(-2,3,3)=\langle 3,-3,1\rangle$ **7.** $2/5$

9. $\dfrac{1}{4\sqrt{2}}$ **11.** 0

13. a. -3 **b.** -6;

If we multiply the unit direction vector by a constant, the limit (the "directional derivative") will be multiplied by the same factor.

15. 0 **17.** $6/5$ **19.** 2 **21.** $-31/26$

23. $\sqrt{3}$ **25.** $2/e^6$ **27.** $\dfrac{\pi}{4\sqrt{3}}$ **29.** $8/5$

31. Direction: $\langle 2,2\rangle$; value: $2\sqrt{2}$

33. Direction: $\left\langle \dfrac{1}{2},-\dfrac{1}{2},-\dfrac{1}{2}\right\rangle$; value: $\dfrac{\sqrt{3}}{2}$

35. Direction: $\left\langle \dfrac{\pi}{2\sqrt{2}},\dfrac{\pi}{2\sqrt{2}}\right\rangle$; value: $-\dfrac{\pi}{2}$

37. Direction: $\langle -1,0,-3\rangle$; value: $-\sqrt{10}$

39. $\langle 1,1\rangle$ and $\langle -1,-1\rangle$ **41.** $5y-x=14$

43. $5y-4x=14$ **45.** $y+2x=3$ **47.** $y+x=2$

49. $2x+3y-4z=11$ **51.** $6x-8y-z=1$

57. $\nabla f(x,y)=$

$\left\langle \dfrac{2x^4+4x^3y+y^{3/2}+2x\sqrt{y}}{x^2(x+y)^2},\dfrac{y-x-4x^3\sqrt{y}}{2x\sqrt{y}(x+y)^2}\right\rangle$

59. $\nabla f(x,y,z)=$

$\left\langle \dfrac{yz(y^2+z^2-x^2)}{(x^2+y^2+z^2)^2},\dfrac{xz(x^2-y^2+z^2)}{(x^2+y^2+z^2)^2},\right.$

$\left.\dfrac{xy(x^2+y^2-z^2)}{(x^2+y^2+z^2)^2}\right\rangle$

61. $f(x,y)=2xy$

63. Not possible; if such a function existed, it wouldn't satisfy Clairaut's Theorem.

65. $f(x,y,z)=3x^2yz^2$

67. a. $-10/\sqrt{5}$ **b.** $10/\sqrt{2}$

69. $160\sqrt{5}/243$ **73.** $y^2=\dfrac{x^3}{1600}$

Section 13.6

1. $z=2x-4y+1$; $\langle 1-2t,1+4t,-1+t\rangle$

3. $2z=5x-y$; $\langle 1-5t,-3+t,4+2t\rangle$

5. $z=32x-56y-172$; $\left\langle 2+4t,-2-7t,4-\dfrac{1}{8}t\right\rangle$

7. $6x+6y+z=0$; $\langle 1+6t,-3+6t,12+t\rangle$

9. $-2x+3y-z=2$ **11.** $x-6y+18z=9$

13. $2x-6y+z=-6$ **15.** $3x+4y+125z=50$

19. $x=1-3t$, $y=2+4t$, $z=1+2t$

21. $x=-2-6t$, $y=1-2t$, $z=-1+t$

23. One possible parametrization is

$\mathbf{r}(t)=\left\langle t,\sqrt{\dfrac{8-3t^2}{5}},4-t^2-2\left(\dfrac{8-3t^2}{5}\right)\right\rangle$,

yielding the parametric equations $x=1+t$, $y=1-(3t/5)$, $z=1+(2t/5)$.

25. Two points: $(1/2,1/2,1)$ and $(-1/2,-1/2,-1)$

27. a. $df=3x^2\,dx-(3+2y)\,dy$

b. $L(x,y)=3x-9y+3$

29. a. $df=\dfrac{x}{\sqrt{x^2+y^2}}\,dx+\dfrac{y}{\sqrt{x^2+y^2}}\,dy$

b. $L(x,y)=\dfrac{\sqrt{3}}{2}x-\dfrac{1}{2}y$

31. a. $df=e^{x^2-y^2}(x\,dx-y\,dy)$

b. $L(x,y)=x-y+\dfrac{1}{2}$

33. a. $df = \dfrac{dx}{\sqrt{x}} - \dfrac{dy}{\sqrt{y}}$ **b.** $L(x,y) = x - \dfrac{y}{2} - 1$

35. $L(w,x,y,z) = 26x + 12y - 21z + 40$

37. $L(w,x,y,z) = \dfrac{e^3}{3}(-2w + 2x - z - 6)$

39. $L(w,x,y,z) = \dfrac{1}{2}(2ex + y - z - e)$

41. Approx. 0.9625   **43.** Approx. 0.204

45. Approx. 8.855   **47.** 7%   **49.** 5%

51. Approx. 2.5%   **55.** −485.6 Pa

59. False; $f(x,y)$ has to be differentiable at (a,b).

61. False; $\nabla f(a,b)$ is not contained in a tangent plane to the graph of f. It is a two-dimensional vector.

Section 13.7

1. Abs. min.: $f(-2,3) = -3$

3. Abs. max.: $f(5,-2) = 29$

5. No extrema; saddle point at $(4,-1)$

7. Abs. min.: $f\!\left(0,\pm\sqrt{2}\right) = 1$;
   saddle point at $(0,0)$

9. Abs. max.: $f(0,0) = 1/2$

11. Abs. min.: $f(-1,0) = 0$

13. Rel. min. at $(-3, 2/3)$;
   saddle point at $(-3, -2/3)$

15. Rel. (global) min. at $(0,-1)$ and $(0,1)$;
   saddle point at $(0,0)$

17. Rel. max. at $(0,0)$; rel. min. at $(2, 4/3)$;
   saddle points at $(0, 4/3)$ and $(2,0)$

19. Rel. max. at $(1,1)$; saddle point at $(0,0)$

21. Rel. max. at $(0,0)$

23. Rel. (global) max. at $\left(\sqrt{2}/2, \sqrt{2}/2\right)$ and
   $\left(-\sqrt{2}/2, -\sqrt{2}/2\right)$; saddle point at $(0,0)$

25. Rel. (global) min. at $(-1/2, 0)$

27. Rel. max. at $(-2,0)$; saddle point at $(0,0)$

29. Saddle points at $\left(0, \dfrac{2k+1}{2}\pi\right)$, $k \in \mathbb{Z}$

31. Abs. min.: $f(0,0) = 0$

33. Abs. max.: $f(1,2) = 7$; abs. min.: $f(-1,0) = -5$

35. Abs. max.: $f(3/2, 3/2) = 11/2$;
   abs. min.: $f(3,0) = f(0,3) = 1$

37. Abs. max.: $f(12,0) = 144$;
   abs. min.: $f(0,0) = 0$ (also assumed along the line segment $y = x/2$ within D)

39. Abs. max.: $f(-2,1) = 9$;
   abs. min.: $f(-1,-1) = -2$

41. Abs. max.: $f(1,1) = 1/4$;
   abs. min.: $f(0,0) = f(2,0) = f(0,2) = 0$

43. Abs. max.: $f(0,\pm 1) = 2/e$;
   abs. min.: $f(0,0) = 0$

45. No critical points   **47.** Abs. min. at $(-1,1)$

49. Rel. max. at $(0,y)$, $y \neq 0$;
   saddle point at $(0,0)$

51. $m = \dfrac{3}{2}$, $b = \dfrac{3}{2}$ **53.** $y = 2x + \dfrac{2}{3}$

57. $\sqrt[3]{2} \cdot \sqrt[4]{3}$ **59.** $\sqrt{21}$

61. It is a cube of side length $\sqrt[3]{V}$.

63. The base is a square of side length $\sqrt[3]{V/2}$, the height is $\sqrt[3]{4V}$.

65. The side length of its base is $2\sqrt{5}$, the height of the pyramid is 2, and the height of the prism is 1.

67. $\dfrac{8abc}{3\sqrt{3}}$ **69.** 30,000   **71.** $9abc/2$

75. False; see Example 1 of Section 13.2 or Exercise 9 in this section.

77. False; there may be a saddle point between them. See Exercise 7 of this section.

81.
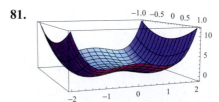

83.
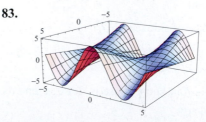

85.

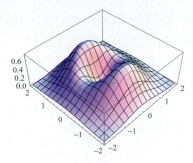

Section 13.8

1. Abs. max.: $f(0,\pm 2)=8$;
   abs. min.: $f(\pm 2,0)=-4$

3. Abs. min.: $f(\sqrt{2},\sqrt{2})=f(-\sqrt{2},-\sqrt{2})=4$

5. Abs. max.: $f(-1,1)=f(1,-1)=5$;
   abs. min.: $f(-1,-1)=f(1,1)=-1$

7. Abs. max.: $f(\pm 2,0)=2$;
   abs. min.: $f(0,\pm\sqrt{2})=1$

9. Abs. max.: $f(4,-2)=12$

11. Abs. max.: $f(3/\sqrt{2},\sqrt{2})=f(-3/\sqrt{2},-\sqrt{2})=6$;
    abs. min.: $f(3/\sqrt{2},-\sqrt{2})=f(-3/\sqrt{2},\sqrt{2})=-6$

13. Abs. min.: $f(-2,9,31/2)=-31/2$

15. Abs. max.: $f(36/7,8/7,18/7)=28$;
    abs. min.: $f(-36/7,-8/7,-18/7)=-28$

17. Abs. max.: $f(\pm 3,0,0)=9$;
    abs. min.: $f(0,\pm 3\sqrt{2},0)=-36$

19. $(12/5,9/5)$ 21. $(1,1)$ 23. $(1/2,3/4)$

25. $\left(0,-1/\sqrt[3]{2}\right)$ 27. $\left(-3,2,-\sqrt{3}\right)$

29. $(8/3,-5/3,-5/3)$

31. $\left(0,\pm\sqrt{3/8},3/4\right)$ 33. $(1,0,0)$

37. **a.** $P(x,y,z)=4x+8y+6z$ (in thousands of dollars)

   b. \$252,000

39. It is the cube of side length $\sqrt[3]{V}$.

41. The base is a square of side length $\sqrt[3]{V/2}$, the height is $\sqrt[3]{4V}$.

43. The length of its square base is $2r/\sqrt{3}$, and its height is $r/\sqrt{3}$.

45. 2 47. 2 51. Abs. min.: $f(2/3,1/3,1/3)=2/3$

53. Abs. max.: $f(2,0,-1/2)=5$;
    abs. min.: $f(-2,0,3/2)=-7$

57. Closest point: $(2,2,16)$;
    farthest point: $(-5/2,-5/2,25)$

59. $\left(d^{2/3}+h^{2/3}\right)^{3/2}$

Chapter 13 Review

1. E 3. B 5. C 7. C 9. F 11. B

13. 0 15. 1/4 17. The limit does not exist.

19. For example, $y=x$ and $y=0$.

21. $(0,0)$; nonremovable discontinuity

23. The lines $x=\pm 1$, $y=\pm 2$; nonremovable discontinuity

25. $z_x(-1,1)=1$; $z_y(-1,1)=-1$; $L_1:\langle t-1,1,t\rangle$; $L_2:\langle -1,1+t,-t\rangle$

27. $\dfrac{\partial f}{\partial x}=\dfrac{y^2 z}{2\sqrt{x-z^2 x^2}}$, $\dfrac{\partial f}{\partial y}=2y\sin^{-1}\left(z\sqrt{x}\right)$, $\dfrac{\partial f}{\partial z}=\dfrac{y^2\sqrt{x}}{\sqrt{1-z^2 x}}$

29. $z_x=-\dfrac{2xz+yz^3}{x^2+3xyz^2}$, $z_y=\dfrac{6y^2-xz^3}{x^2+3xyz^2}$

35. $\dfrac{\partial f}{\partial x}=\dfrac{y^2}{2\sqrt{x}}$, $\dfrac{\partial f}{\partial y}=2\sqrt{x}y$

37. $f(x,y)=xy^3+\ln x$

39. Yes, the partials are continuous.

43. $\dfrac{dy}{dx}=54x^2\left(5x^2-6\right)$

45. $\dfrac{dy}{dx}=4x^3\cos x^2-2x^5\sin x^2$

47. $f'(\pi/2)=\pi(2\pi+1)$

49. $f'(\pi)=2\pi^3$

51. $f_x=\dfrac{2u\sqrt{y}+4vxy}{u^2+v^2}$, $f_y=\dfrac{x\left(2vx\sqrt{y}+u\right)}{\sqrt{y}\left(u^2+v^2\right)}$

53. $f_x=6x^2\left(v-w^2\right)-6uw+u$, $f_y=-2u(w+y)$

55. $\dfrac{dy}{dx}=\dfrac{4x-3x^2 y-y^3}{x^3+3xy^2}$

57. $z_x = \dfrac{3x^2z^2 - y^2}{3z^2y - 2x^3z}$, $z_y = \dfrac{z^3 + 2xy}{2x^3z - 3z^2y}$

61. $\nabla f(1,3) = \langle 17,3 \rangle$

63. $\nabla f(-1,0,2) = \langle -2,2,0 \rangle$

65. 0   **67.** $\sqrt{2}/4$ **69.** 0

71. Direction: $\langle -2,1,-2 \rangle$; value: 3

73. a. Direction: $-\dfrac{450}{e^2}\langle 2,1,1 \rangle$; magnitude: $\dfrac{450\sqrt{6}}{e^2}$

 b. $\dfrac{-270\sqrt{5}}{e^2}$

75. $\langle 1,2 \rangle$ and $\langle -1,-2 \rangle$ **77.** $y = 4x - 2$

79. $z = 3y + 6x - 15$

81. Not possible; if such a function existed, it wouldn't satisfy Clairaut's Theorem.

83. $23x + 4y - 6z = 13$; $\langle 1+23t, 2+4t, 3-6t \rangle$

85. $y = 2 - \sqrt{3}x$; $\left\langle \dfrac{\sqrt{3}}{2} + \sqrt{3}t, \dfrac{1}{2} + t, 0 \right\rangle$

87. $x = -1 + 11t$, $y = 1 - 5t$, $z = 1 + 2t$

89. a. $df = \cos(x^2 - y^2)(2x\,dx - 2y\,dy)$

 b. $L(x,y) = -2x - 2y$

91. $L(w,x,y,z) = -3x + 8y + 4z - 2w + 10$

93. Abs. min.: $f(2,-1) = 1$

95. Rel. max.: $f(0,0) = 0$;

 rel. min.: $f(2/3, 8/3) = -260/27$;

 saddle points at $(2/3, 0)$ and $(0, 8/3)$

97. Abs. max.: $f(-2,2) = 9$; abs. min.: $f(1,2) = 0$

99. $m = 46/107$, $b = 156/107$

101. $\sqrt{3}/2$ **103.** 7500

105. Abs. min.: $f(2,2) = f(-2,-2) = 8$

107. Abs. max.: $f(0,0) = 0$;

 abs. min.: $f\left(\dfrac{25+3\sqrt{41}}{32}, -\dfrac{3+\sqrt{41}}{8}\right) \approx -1.0967$

109. $Q(1/2, 0, 1/2)$

111. Abs. min.: $f(1/6, 7/6, 2/3) = 11/3$

113. Closest points: $\left(\dfrac{2}{\sqrt{5}}, \dfrac{2}{\sqrt{5}}\right)$ and $\left(-\dfrac{2}{\sqrt{5}}, -\dfrac{2}{\sqrt{5}}\right)$;

 farthest points: $\left(-\dfrac{2}{\sqrt{3}}, \dfrac{2}{\sqrt{3}}\right)$ and $\left(\dfrac{2}{\sqrt{3}}, -\dfrac{2}{\sqrt{3}}\right)$

115. True   **117.** True   **119.** True

121. False; consider $f(x,y) = \sqrt{x^2 + y^2}$ at $(0,0)$.

123. False; consider Example 9 in Section 13.3.

125. False; the directional derivatives may not exist. Consider $f(x,y) = \sqrt{x^2 + y^2}$ at $(0,0)$.

135.

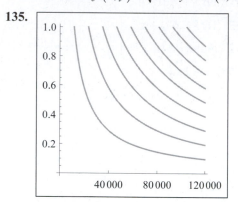

Chapter 13 Project

 7. $x^2 - 2xy^2 + 3y^3 = C$

Chapter 14

Section 14.1

 1. 84   **3.** 297/16   **5.** 21/2   **7.** −36   **9.** 2

 11. $-3\pi^3/8$ **13.** 3/2   **15.** 2   **17.** 11   **19.** 26

 21. 12   **23.** 3/35   **25.** 5/2   **27.** $\dfrac{19 - e^2}{2}$

 29. $3 - e$ **31.** 0   **33.** 8

 35. $\displaystyle\int_0^4\int_0^1 (3x - 2y)\,dy\,dx = 20$

 37. $\displaystyle\int_{-1}^3\int_2^4 (5x + xy - 3y)\,dy\,dx = -8$

 39. $\displaystyle\int_0^{1/2}\int_0^x (x + y)^2\,dy\,dx = \dfrac{7}{192}$

 41. $\displaystyle\int_0^1\int_0^{3x} x^2 y\,dy\,dx + \int_1^7\int_0^{\frac{7-x}{2}} x^2 y\,dy\,dx = \dfrac{693}{10}$

 43. $\displaystyle\int_0^3\int_0^{\sqrt{9-y^2}} \dfrac{1}{\sqrt{9-x^2}}\,dx\,dy = 3$

 45. $\displaystyle\int_0^1\int_{x^2}^x (x^3 - x)\,dy\,dx = -\dfrac{1}{20}$

 47. $\displaystyle\int_0^3\int_0^{2x} 2x\,dy\,dx + \int_3^5\int_0^{15-3x} 2x\,dy\,dx = 80$

49. $\int_0^1 \int_{\arcsin y}^{\pi/2} \frac{1}{\sqrt{1-y^2}}\, dx\, dy = \frac{\pi^2}{8}$

51. 0   **53.** $\int_0^2 \int_{y+1}^{\frac{8-y}{2}} f(x,y)\, dx\, dy$ **55.** $\frac{\pi}{\sqrt{2}} - 2$

57. 27/10   **59.** −1   **61.** 4/3

63. 16   **65.** 16/3   **67.** 9π

69. Using the order $dy\, dx$: $V = \int_0^2 \int_0^{1-x/2} 2\, dy\, dx = 2$

71. 1/2   **73.** $e - 2$

75. $\iint\limits_R \rho(x,y)\, dA$ is the mass of the plate.

79. a. $\iint\limits_R f(x,y)\, dA$
$= f(b,d) - f(b,c) - f(a,d) + f(a,c)$

b. 104

91. True   **93.** True (See Example 6.)

Section 14.2

1. 1/5   **3.** 1/3

5. 40   **7.** 6

9. 9/2   **11.** $\ln 4$

13. 10/9   **15.** −9/20

17. 52/35   **19.** 73/15

21. 296/189   **23.** 1

25. $(2/3, 1/3)$ **27.** $(3/2, 3/4)$

29. $\left(\dfrac{4+\pi}{16}, \dfrac{1+\pi}{8} \right)$

31. $(105/164, 485/164)$ **33.** $(10, 1/2)$

35. $\left(\dfrac{e^2-1}{e^2+1}, \dfrac{8(2e^3+1)}{27(e^2+1)} \right)$

37. $I_x = \dfrac{4\rho a^4}{3}, I_y = \dfrac{4\rho a^4}{3}, I_0 = \dfrac{8\rho a^4}{3}$;
$r_x = \dfrac{a}{\sqrt{3}}, r_y = \dfrac{a}{\sqrt{3}}, r_0 = \sqrt{\dfrac{2}{3}}\, a$

39. $I_x = \dfrac{\rho l w^3}{12}, I_y = \dfrac{\rho l^3 w}{12}, I_0 = \dfrac{\rho l w (l^2+w^2)}{12}$;
$r_x = \dfrac{w}{2\sqrt{3}}, r_y = \dfrac{l}{2\sqrt{3}}, r_0 = \dfrac{\sqrt{l^2+w^2}}{2\sqrt{3}}$

41. $I_x = \dfrac{\rho a^4}{2\sqrt{3}}, I_y = \dfrac{\rho a^4}{2\sqrt{3}}, I_0 = \dfrac{\rho a^4}{\sqrt{3}}$;
$r_x = \dfrac{a}{\sqrt{6}}, r_y = \dfrac{a}{\sqrt{6}}, r_0 = \dfrac{a}{\sqrt{3}}$

43. $I_x = I_y = \dfrac{4\rho(a^4 - b^4)}{3}, I_0 = \dfrac{8\rho(a^4 - b^4)}{3}$;
$r_x = \dfrac{\sqrt{a^2+b^2}}{\sqrt{3}}, r_y = \dfrac{\sqrt{a^2+b^2}}{\sqrt{3}}, r_0 = \dfrac{\sqrt{2}\sqrt{a^2+b^2}}{\sqrt{3}}$

45. $I_x = \dfrac{4\rho}{9}, I_y = \dfrac{\rho(\pi^2-8)}{2}, I_0 = \dfrac{\rho(9\pi^2-64)}{18}$;
$r_x = \dfrac{\sqrt{2}}{3}, r_y = \dfrac{\sqrt{\pi^2-8}}{2}, r_0 = \dfrac{\sqrt{9\pi^2-64}}{6}$

47. $I_x = 2\rho\pi, I_y = \dfrac{\rho\pi}{2}, I_0 = \dfrac{5\rho\pi}{2}$;
$r_x = 1, r_y = \dfrac{1}{2}, r_0 = \dfrac{\sqrt{5}}{2}$

49. $I_x = 80\rho\pi, I_y = 125\rho\pi, I_0 = 205\rho\pi$;
$r_x = 2, r_y = \dfrac{5}{2}, r_0 = \dfrac{\sqrt{41}}{2}$

55. $I_x = 80\rho\pi, I_y = 305\rho\pi, I_0 = 385\rho\pi$;
$r_x = 2, r_y = \dfrac{\sqrt{61}}{2}, r_0 = \dfrac{\sqrt{77}}{2}$

61. $I_x = 2, I_y = \dfrac{5}{4}, I_0 = \dfrac{13}{4}$

63. $I_x = \dfrac{64}{315}, I_y = \dfrac{8}{105}, I_0 = \dfrac{88}{315}$;
$r_x = 2\sqrt{\dfrac{2}{21}}, r_y = \dfrac{1}{\sqrt{7}}, r_0 = \sqrt{\dfrac{11}{21}}$

65. $I_x = \dfrac{2\pi}{9}, I_y = \pi(\pi^2-6), I_0 = \dfrac{9\pi^3 - 52\pi}{9}$;
$r_x = \dfrac{\sqrt{2}}{3}, r_y = \sqrt{\pi^2-6}, r_0 = \dfrac{\sqrt{9\pi^2-52}}{3}$

67. $I_x = \dfrac{\sinh 4}{8}, I_y = \dfrac{e^4-5}{8e^2}, I_0 = \dfrac{e^4 - 5 + e^2\sinh 4}{8e^2}$;
$r_x = \dfrac{\sqrt{\operatorname{csch} 2 \sinh 4}}{2}, r_y = \dfrac{\sqrt{(e^4-5)\operatorname{csch} 2}}{2e}$,
$r_0 = \dfrac{\sqrt{(e^4 - 5 + e^2\sinh 4)\operatorname{csch} 2}}{2e}$

69. For the rolling objects, a portion of their initial potential energy is turned into kinetic energy, while the remaining portion will be stored in rotation. Since the object with the greater moment of inertia (I_0) will store more energy in rotation, and a smaller portion in the form of kinetic energy, it will be slower. Examining the I_0-values for the disk and annulus, we see that more of the mass is concentrated near the outer radius in the case of the annulus, so we conclude that the disk beats the annulus, while the box always wins (with friction and rotation absent, its potential energy is turned entirely into kinetic energy.)

71. False; see the definition of center of mass. Since density is constant, it will cancel out when dividing the first moments by the mass of the object.

73. False; see the definition of radii of gyration. Since density is constant, it will cancel out when dividing the second moments by the mass of the object.

77. 1/2

Section 14.3

1. 1/3   **3.** $2 - \dfrac{\pi}{4}$ **5.** $\dfrac{\sqrt{3}}{2} - \dfrac{\pi}{6}$ **7.** 9π

9. $\dfrac{9\sqrt{3}}{2} + 3\pi$ **11.** $\dfrac{13\sqrt{3}}{8} - \dfrac{5\pi}{12}$ **13.** $\dfrac{\sqrt{3}}{4} + \dfrac{\pi}{6}$

15. $\pi - \dfrac{3\sqrt{3}}{2}$ **17.** $2 - \dfrac{\pi}{4}$ **19.** $\dfrac{\pi \ln(3/2)}{2}$

21. $\pi/10$ **23.** 2048/245   **25.** 8/15   **27.** 1/3

29. 3/16   **31.** $16\pi/3$ **33.** $\pi/2$ **37.** 128π

39. $8\pi/3$ **41.** $53\pi/4$ **43.** 24π **45.** $\dfrac{32\left(\sqrt{2}-1\right)\pi}{3}$

47. $\dfrac{32\left(2 - \sqrt{2}\right)\pi}{3}$ **49.** $\left(0, \dfrac{15\pi + 32}{6\pi + 48}\right)$

51. $I_x = \left(3\sqrt{3} + \dfrac{4\pi}{3}\right)\rho, I_y = \left(11\sqrt{3} + 12\pi\right)\rho,$

$I_0 = \left(14\sqrt{3} + \dfrac{40\pi}{3}\right)\rho$

53. $I_x = \dfrac{5\rho\pi a^4}{4}, I_y = \dfrac{\rho\pi a^4}{4}, I_0 = \dfrac{3\rho\pi a^4}{2};$

$r_x = \dfrac{\sqrt{5}a}{2}, r_y = \dfrac{a}{2}, r_0 = \dfrac{\sqrt{3}a}{\sqrt{2}}$

55. $I_x = I_y = \dfrac{\pi}{32}, I_0 = \dfrac{\pi}{16};\ \ r_x = r_y = \dfrac{\sqrt{\pi}}{4}, r_0 = \dfrac{\sqrt{\pi}}{2\sqrt{2}}$

61. False; the correct formula is

$$\iint\limits_R f(r,\theta)\,dA = \iint\limits_R rf(r,\theta)\,dr\,d\theta.$$

63. True   **65.** 1/8

67. $I_x = 49\pi/32, I_y = 21\pi/32, I_0 = 35\pi/16;$

$r_x = 7/\left(4\sqrt{3}\right), r_y = \sqrt{7}/4, r_0 = \sqrt{35/24}$

69. $I_x = \dfrac{15\pi - 27\sqrt{3}}{20}, I_y = \dfrac{29\pi}{12} - \dfrac{171\sqrt{3}}{40},$

$I_0 = \dfrac{76\pi - 135\sqrt{3}}{24}; r_x = \sqrt{\dfrac{3\left(5\sqrt{3}\pi - 27\right)}{10\left(2\sqrt{3}\pi - 9\right)}},$

$r_y = \dfrac{1}{2}\sqrt{\dfrac{29}{3} + \dfrac{78}{45 - 10\sqrt{3}\pi}},$

$r_0 = \dfrac{1}{2}\sqrt{\dfrac{38}{3} + \dfrac{21}{9 - 2\sqrt{3}\pi}}$

Section 14.4

3. 1   **5.** 216/5   **7.** 18   **9.** 4/9

11. 1033/280   **13.** 16/3

15. a. $\displaystyle\int_{-1}^{1}\int_{-\sqrt{1-x^2}}^{\sqrt{1-x^2}}\int_{0}^{2} dz\,dy\,dx$

b. $\displaystyle\int_{-1}^{1}\int_{0}^{2}\int_{-\sqrt{1-x^2}}^{\sqrt{1-x^2}} dy\,dz\,dx$

c. $\displaystyle\int_{-1}^{1}\int_{0}^{2}\int_{-\sqrt{1-y^2}}^{\sqrt{1-y^2}} dx\,dz\,dy;$ Value: 2π

17. a. $\displaystyle\int_{0}^{1}\int_{x^2}^{\sqrt{x}}\int_{-1}^{1} dz\,dy\,dx$

b. $\displaystyle\int_{0}^{1}\int_{-1}^{1}\int_{x^2}^{\sqrt{x}} dy\,dz\,dx$

c. $\displaystyle\int_{0}^{1}\int_{-1}^{1}\int_{y^2}^{\sqrt{y}} dx\,dz\,dy;$ Value: 2/3

19. $abc/6$ **21.** 81/20   **23.** 128/3

25. 8/5   **27.** 128/5   **29.** $3\sqrt{3}\pi$

31. 52/105   **33.** 648/5   **35.** 4π

37. $\displaystyle\int_{-2}^{2}\int_{x^2}^{4}\int_{0}^{z} dy\,dz\,dx,\ \displaystyle\int_{0}^{4}\int_{y}^{4}\int_{-\sqrt{z}}^{\sqrt{z}} dx\,dz\,dy,$

$\displaystyle\int_{0}^{4}\int_{0}^{z}\int_{-\sqrt{z}}^{\sqrt{z}} dx\,dy\,dz;$ Value: 128/5

41. $96k/5$ **43.** $k\pi/4$ **45.** $2/\sqrt{a}$ **47.** $R^3/(8\pi)$

49. a. It is the first octant region of the unit sphere, centered at the origin.

  b. $(3/8,3/8,3/8)$

51. $(1/2,5/4,3/4)$ **53.** $(0,10/7,20/7)$

55. $(17/14,-1/2,9/14)$ **57.** $(7a/12,7a/12,7a/12)$

59. $(1/5,2/5,4/5)$ **61.** $(3/5,0,112/(75\pi))$

65. $(a/2,a/2,a/2)$; $r_x = r_y = r_z = a\sqrt{2/3}$

67. $(3/8,1/8,3/4)$; $r_x = \sqrt{37/40}$,

  $r_y = 3/(2\sqrt{2})$, $r_z = 1/2$

69. $(0,0,-2/5)$; $r_x = \sqrt{3/7}$,

  $r_y = \sqrt{59/105}$, $r_z = 2\sqrt{2/15}$

71. $(4/(3\pi),4/(3\pi),1/2)$; $r_x = r_y = \sqrt{7/12}$, $r_z = 1/\sqrt{2}$

73. $I_x = I_y = I_z = 38a^7k/45$; $r_x = r_y = r_z = \dfrac{a}{3}\sqrt{\dfrac{38}{5}}$

75. $I_x = I_y = I_z = 512k/315$; $r_x = r_y = r_z = 8/(3\sqrt{7})$

77. $(16/35,16/35,16/35)$; $r_x = r_y = r_z = 1/\sqrt{2}$

79. $(7/18,7/54,2/3)$; $r_x = \sqrt{199/264}$, $r_y = \sqrt{85/88}$,

  $r_z = \sqrt{35/132}$

Section 14.5

1. $(\sqrt{2},\pi/4,2)$ **3.** $(12,-5\pi/6,0)$

5. $(-\sqrt{2},\sqrt{2},\sqrt{2})$ **7.** $(0,-1,1)$ **9.** $r = 4\cos\theta$

11. $r = 2\sin\theta$ **13.** $z = r/3$ **15.** $z = e^{-r^2/2}$

17. Circular cylinder; $x^2 + y^2 = 1$

19. Ellipsoid centered at the origin; $x^2 + y^2 + \dfrac{z^2}{2} = 1$

21. Circular double cone; $z^2 = x^2 + y^2$

23. Upper hemisphere of radius 1, centered at the origin; $z = \sqrt{1-x^2-y^2}$

25. $\int_0^{2\pi}\int_0^2\int_0^{4-r^2} r\,dz\,dr\,d\theta$ **27.** $\int_0^{2\pi}\int_0^4\int_0^{4-r} r\,dz\,dr\,d\theta$

29. $\int_0^{2\pi}\int_0^2\int_r^{\sqrt{8-r^2}} r\,dz\,dr\,d\theta$ **31.** $\int_0^{2\pi}\int_0^1\int_{\sqrt{2+r^2}}^{\sqrt{3}} r\,dz\,dr\,d\theta$

33. $\int_1^5\int_0^{2\pi}\int_0^{z/2} r\,dr\,d\theta\,dz$ **35.** $\int_{\sqrt{2}}^{\sqrt{3}}\int_0^{2\pi}\int_0^{\sqrt{z^2-2}} r\,dr\,d\theta\,dz$

39. $\pi/6$ **41.** $k\pi/4$ **43.** $\dfrac{4\pi(3-e)}{3}$ **45.** $\pi/2$

47. $\pi/2$ **49.** $10\pi k/3$ **51.** $\dfrac{2\pi(2e-5)}{3}$

53. $56\sqrt{3}\pi/5$ **55.** $(0,0,61/9)$

57. $(0,0,5/12)$ **59.** $\left(0,0,\dfrac{e^2-3}{e^2-1}\right)$

61. The center of mass is on the axis of the cone, one-fourth of the way from the base towards the vertex.

63. The center of mass is on the axis of the cone, two-fifths of the way from the base towards the vertex.

65. $(3/5,0,112/(75\pi))$ **67.** $I_z = \pi\rho/2$, $r_z = 1/\sqrt{2}$

69. $(0,0,1/2)$, $I_z = \pi/24$, $r_z = 1/2$

71. $(0,0,8/15)$, $I_z = \pi\rho/12$, $r_z = 1/\sqrt{3}$

73. $I_z = 99\pi\rho$, $r_z = \sqrt{22}/3$

75. $I_z = 16\pi k/3$, $r_z = 2/\sqrt{3}$ **77.** $(\sqrt{8},\pi/6,\pi/4)$

79. $(2,-\pi/2,5\pi/6)$ **81.** $(1,0,1)$ **83.** $(0,-3,0)$

85. $\rho = 2$ **87.** $\tan^2\varphi = 1$

89. $\tan\theta = 1$ **91.** $\rho\sin\varphi = 2$

93. Unit sphere centered at the origin; $x^2 + y^2 + z^2 = 1$

95. Cone; $z = \sqrt{x^2 + y^2}$ **97.** The xy-plane; $z = 0$

99. Cylinder; $x^2 + y^2 = 9$

101. $\int_0^{2\pi}\int_0^{\pi/2}\int_0^2 \rho^2\sin\varphi\,d\rho\,d\varphi\,d\theta$

103. $\int_0^{\pi/2}\int_0^{\pi/2}\int_0^{6\cos\varphi} \rho^2\sin\varphi\,d\rho\,d\varphi\,d\theta$

105. $\int_{\pi/6}^{\pi/3}\int_0^{\pi/2}\int_0^5 \rho^2\sin\varphi\,d\rho\,d\varphi\,d\theta$

107. $125\pi/18$ **111.** $\dfrac{32(\sqrt{2}-1)\pi}{3}$ **113.** $8\pi/3$

115. $(0,0,3R/8)$ **117. a.** $8\pi\rho$ **b.** $(0,0,7/3)$

121. $I_z = 8\pi R^5\rho/15$, $r_z = \sqrt{2/5}R$ **123.** $I_z = 76\pi k/9$

127. a. It is the first octant portion of the unit sphere, centered at the origin.

  b. $(3/8,3/8,3/8)$

  c. $I_x = I_y = I_z = \pi/15$

  d. $r_x = r_y = r_z = \sqrt{2/5}$

129. a. The cardioid $\rho = 1 + \cos\varphi$, (or in standard polar coordinates, $r = 1 + \cos\theta$) in the xz-plane, rotated about the z-axis.

  b. $8\pi/3$

131. $\pi\left(\sqrt{3}-1\right)/8$

133. $\int_{-\pi/2}^{\pi/2}\int_{\pi/4}^{\pi/2}\int_{0}^{2\csc\varphi\cos\theta}\rho^3\sin^2\varphi\cos\theta\,d\rho\,d\varphi\,d\theta=\dfrac{64}{15}$

135. $(16/35,16/35,16/35)$

Section 14.6

1. 4   **3.** 3   **5.** 3   **7.** $8v-4u$ **9.** 2   **11.** v

13. -1 **15.** $x=\dfrac{v-u}{4}$, $y=\dfrac{3u+v}{4}$

17. $x=\dfrac{4(v-u)}{7}$, $y=\dfrac{u+6v}{7}$

19. $x=\dfrac{v-u}{4}$, $y=\dfrac{u+v}{2}$ **21.** $816/49$ **23.** $\dfrac{4\ln 3}{7}$

25. $\dfrac{\left(e^4-1\right)\ln 3}{7}$ **27.** $2\ln 2$ **29.** ab **33.** $\dfrac{36}{5}-4\sqrt{2}$

35. 70; the answers are equal.   **37.** -1 **43.** 112

45. $I_x=2\rho\pi$, $I_y=\rho\pi/2$, $I_0=5\rho\pi/2$;
$r_x=1$, $r_y=1/2$, $r_0=\sqrt{5}/2$

47. $4\pi abc/3$

Chapter 14 Review

1. Approximation: $315/8$; true value: $81/2$

3. 34   **5.** 3   **7.** 2   **9.** 120

11. $\int_0^2\int_0^{3y}(xy-1)\,dx\,dy+\int_2^8\int_0^{8-y}(xy-1)\,dx\,dy=120$

13. 72   **15.** $\dfrac{2}{5}\left(2^{5/4}-1\right)$ **17.** $\pi/2$

19. $4/5$ **21.** $(78/35,6/5)$

23. $I_x=32\rho/3$, $I_y=8\rho/3$, $I_0=40\rho/3$;
$r_x=2\sqrt{2/3}$, $r_y=\sqrt{2/3}$, $r_0=\sqrt{10/3}$

25. $I_x=2048\rho/105$, $I_y=64\rho/15$, $I_0=832\rho/35$;
$r_x=8\sqrt{2/35}$, $r_y=2/\sqrt{5}$, $r_0=2\sqrt{39/35}$

27. $I_0=(3\pi-4)\rho/6$ **29.** $\left(3\sqrt{3}+2\pi\right)/6$

31. $\left(16\pi-21\sqrt{3}\right)/24$ **33.** $\dfrac{\left(\sqrt{6}-\sqrt{2}\right)\pi}{2}$

35. $\dfrac{1}{\sqrt{3}}-\dfrac{1}{2}$ **37.** $\pi/2$ **39.** $81\pi/2$ **41.** 81π

43. 375   **45.** $1/8$ **47.** $8/3$ **49.** $32/15$

51. $2\sqrt{2}\pi$ **53.** $\int_0^1\int_0^{2-2x}\int_0^{4-4x-2z}5xy\,dy\,dz\,dx=\dfrac{4}{3}$

55. $(3/5,3/5,3/5)$

57. $(3/2,1/2,3/2)$;
$I_x=48\rho$, $I_y=432\rho/5$, $I_z=48\rho$;
$r_x=2$, $r_y=6/\sqrt{5}$, $r_z=2$

59. $\left(2,-\pi/4,-\sqrt{2}\right)$ **61.** $\left(2\sqrt{2},\pi/4,3\pi/4\right)$

63. $(1,-1,-\pi/4)$ **65.** $\left(2,-2,-2\sqrt{6}\right)$

67. $z=r(r-4\sin\theta)$ **69.** $\rho=2\sin\phi\cos\theta$

71. $z=x^2+y^2-1$ **73.** $z=-\sqrt{\dfrac{x^2+y^2}{3}}$

75. $81\pi/2$ **77.** 243π; $(0,0,9/4)$

79. 9π **81.** $(0,0,27/16)$ **83.** $\dfrac{9\ln(8/3)}{10}$

85. $x=\dfrac{u-2v}{7}$, $y=\dfrac{4u-v}{7}$ **87.** $4736/7$ **89.** 14

93. False; let R be the unit disk centered at the origin, $f(x)$ be an odd function, and consider $g(x,y)=f(x)$ on R.

95. False; this integral cannot be evaluated exactly in rectangular coordinates (see Example 3 of Section 14.3).

97. False; see the definition of the spherical coordinate system in Section 14.5.

99. $\left(\dfrac{6}{5}\left(\sqrt{2}-1\right),\dfrac{6}{5}\left(\sqrt{3}-\sqrt{2}\right),0\right)$

Chapter 14 Project

1. $\dfrac{\partial(x,y)}{\partial(r,\theta)}=abr$

3. $\dfrac{\partial(x,y,z)}{\partial(\rho,\phi,\theta)}=-abc\rho^2\sin\phi$

5. $(0,0,3c/8)$

7. $I_x=\dfrac{m}{5}\left(b^2+c^2\right)$, $I_y=\dfrac{m}{5}\left(a^2+c^2\right)$, $I_z=\dfrac{m}{5}\left(a^2+b^2\right)$;
$r_x=\sqrt{\dfrac{b^2+c^2}{5}}$, $r_y=\sqrt{\dfrac{a^2+c^2}{5}}$, $r_z=\sqrt{\dfrac{a^2+b^2}{5}}$

Chapter 15

Section 15.1

1. Scalar field   3. Vector field   5. Vector field

7. $\langle 0, -2 \rangle$ 9. $\langle 0, 1, 5 \rangle$ 11. B   13. A

15.

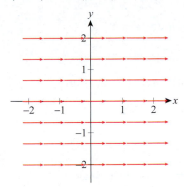

17.

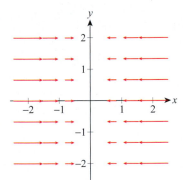

19.

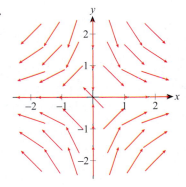

21.

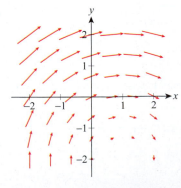

23.

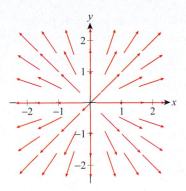

25. A   27. B   29. $\dfrac{\langle x, y \rangle}{x^2 + y^2}$

31. $\left\langle \dfrac{y^2}{1+x^2}, 2y\tan^{-1}x \right\rangle$ 33. $e^{xyz}\langle yz^2, xz^2, 1+xyz \rangle$

35. $\dfrac{\langle -xz, -yz, x^2+y^2 \rangle}{\left(x^2+y^2+z^2\right)^{3/2}}$

37. The function $f(x,y,z) = \dfrac{-\varepsilon Qq}{\sqrt{x^2+y^2+z^2}}$ is a potential for **F**.

39. If $\mathbf{F} = \langle F_1, F_2 \rangle$ is conservative, then $\dfrac{\partial F_1}{\partial y} = \dfrac{\partial F_2}{\partial x}$.

41. Not conservative   43. Conservative

45. Conservative   47. Not conservative

49. $f(x,y) = e^{xy^2} + C$ 51. $f(x,y) = \sqrt{x^2+y^2} + C$

53. $f(x,y,z) = x^2 - y + z(x+2y) + C$

57. $\nabla r^n = nr^{n-2}\mathbf{r}$ (or $\nabla r^n = nr^{n-1}\mathbf{e}_r$)

59. $-1/r$ is a potential function.

61. The flow lines are the circles $x^2+y^2 = C$, $C \in \mathbb{R}^+$.

Section 15.2

1. $16/3$ 3. 3π 5. $8 - 3\sqrt{3}$ 7. $41/5$

9. Mass: $(\pi-2)kR^2$, where k is the constant of proportionality; center of mass: $\left(0, \dfrac{(4-\pi)R}{2(\pi-2)} \right)$

11. $\dfrac{13\sqrt{13} - 5\sqrt{5}}{6}$ 13. $24 + \dfrac{5\sqrt{5}}{2}$; no

15. $\dfrac{16\left(10\sqrt{10} - 1\right)}{3}$ 17. $\dfrac{2\left(1 - e^{4\pi}\right)}{\sqrt{5}}$ 19. $8\sqrt{5}\pi^2 k$

21. $\dfrac{\sqrt{2} + \tanh^{-1}\left(1/\sqrt{2}\right)}{2}$

23. Mass: $\sqrt{2}k$ (where the density is $\rho = kz$); center of mass: $(0, 0, 2/3)$

25. $4\sqrt{10}\pi^2$ **27.** $\pi/2$

29. a. 1 **b.** 1 **c.** 1; the answers are equal.

31. 0 **33.** 6 **35.** 36 **37.** 7/4 **39.** 9/20

41. 4π **43.** $-k/20$ **45.** 90 **47.** 7π **49. b.** 0

51. $\dfrac{1}{4\pi\varepsilon_0}\dfrac{10^{-5}}{2\sqrt{2}}$ volts

53. $\dfrac{2}{4\pi\varepsilon_0}\dfrac{1+\sinh^{-1}(2\pi)-\sqrt{4\pi^2+1}}{10^6\sqrt{2}}$ volts

55. $I_x \approx 577.782\rho$, $I_y \approx 83.18\rho$, $I_z \approx 660.962\rho$; $r_x \approx 5.444$, $r_y \approx 2.066$, $r_z \approx 5.823$

57. $I_x \approx 6283.17$, $I_y \approx 1123.3$, $I_z \approx 7406.47$; $r_x \approx 4.987$, $r_y \approx 2.109$, $r_z \approx 5.415$

59. Approx. $\dfrac{1.5077}{4\pi\varepsilon_0 \times 10^5}$ volts

Section 15.3

3. $\pi - e$ **5.** $1 + e$ **7.** $f(x, y) = x^3 + 2xy^2 + y^3$

9. A potential function does not exist.

11. $f(x, y) = y\sin x + e^x + \ln y$

13. $f(x, y) = x^4 y - xy^5$ **15.** $f(x, y, z) = x^2 yz$

17. A potential function does not exist.

19. $f(x, y, z) = x\tan z + y^2 z$

21. A potential function does not exist.

27. 36 **29.** $4e + 16$ **31.** 3

35. a. No, because the domain of **F** is not simply connected.

   b. The equality does not hold on the domain of **F**, since $-\arctan(x/y)$ is not defined for $y = 0$.

37. Yes **39.** Yes **41.** No

51. False; only a simply connected region guarantees the statement.

53. True

Section 15.4

1. $\operatorname{div}\mathbf{F} = 15x^2 + 8xy - 2y$; $\operatorname{div}\mathbf{F}(2,1) = 74$; $\operatorname{div}\mathbf{F}(1,-3) = -3$

3. $\operatorname{div}\mathbf{F} = 2(x - y)$; $\operatorname{div}\mathbf{F}(2,-1,0) = 6$; $\operatorname{div}\mathbf{F}(1,1,4) = 0$

5. $\operatorname{div}\mathbf{F} = 0$; $\operatorname{div}\mathbf{F}(1,1,1) = 0$; $\operatorname{div}\mathbf{F}(2,-3,4) = 0$

7. $\operatorname{div}\mathbf{F} = e^y + 4z^3 + 1$; $\operatorname{div}\mathbf{F}(1,0,-3) = -106$; $\operatorname{div}\mathbf{F}(2,1,4) = 257 + e$

9. $\operatorname{div}\mathbf{F} = e^{xyz}(yz + xy + xz)$; $\operatorname{div}\mathbf{F}(1,0,1) = 1$; $\operatorname{div}\mathbf{F}(e,1,-1) = -e^{-e}$

11. $\operatorname{curl}\mathbf{F} = 2(x + 2y^2)\mathbf{k}$

13. $\operatorname{curl}\mathbf{F} = -2z(3x+1)\mathbf{i} + 3z^2\,\mathbf{k}$

15. $\operatorname{curl}\mathbf{F} = \mathbf{0}$ **17.** $\operatorname{curl}\mathbf{F} = -3y^2\,\mathbf{k}$

19. $\operatorname{curl}\mathbf{F} = e^{xyz}\left[x(z-y)\mathbf{i} + y(x-z)\mathbf{j} + z(y-x)\mathbf{k}\right]$

21. At A, divergence is 0. At B, divergence is positive. At C, divergence is negative.

23. $\operatorname{curl}\mathbf{F} = -\mathbf{k}$

25. $\operatorname{curl}\mathbf{F} = -2y\mathbf{k}$; negative spin above the x-axis, positive below

27. $\operatorname{curl}(\operatorname{curl}\mathbf{F}) = (-2x - 8z)\mathbf{i} + 2y\,\mathbf{j}$

29. $\operatorname{curl}(\operatorname{curl}\mathbf{F}) = -2x\mathbf{i} + 2y\,\mathbf{j}$

31. No meaning **33.** Vector field **35.** Scalar field

37. No meaning **49.** Both integrals are equal to 8.

51. Both integrals are equal to 2π. **53.** 8/3

55. 3/4 **57.** 11/15 **59.** 0 **61.** -2 **63.** $\pi/2$

65. 0 **67.** 0 **69.** 4 **71.** 18π **73.** 352/7

75. $e^5 - 5e - 6$ **77.** πr^2 **79.** 6 **81.** 48

83. 4 **85.** 9 **87.** 12 **89.** $5\pi ab$ **93.** 0

97. $\left(0, \dfrac{4R}{3\pi}\right)$ **101.** True **103.** True

Section 15.5

1. Possible parametrization:
$$\mathbf{r}(s,t) = \left\langle s, \frac{1}{3}s\cos t, \frac{1}{3}s\sin t \right\rangle,$$
$$0 \le s \le 6,\ 0 \le t \le 2\pi$$

3. Possible parametrization:
$$\mathbf{r}(s,t) = \left\langle (1 - s^2)\cos t, (1 - s^2)\sin t, s \right\rangle,$$
$$-1 \le s \le 1,\ 0 \le t \le 2\pi$$

5. C **7.** H **9.** F **11.** D **13.** G

15. The grid curves are quarter circles of radius R. The graph is the first octant portion of the sphere of radius R, centered at the origin.

17. The grid curves are parabolas. The graph is an elliptic paraboloid.

19. The grid curves are parabolas if t is held constant and circles if s is held constant. The graph is an elliptic paraboloid.

21. $\mathbf{r}(s,t) = \langle s, t, s+t \rangle$, $-\infty < s < \infty$, $-\infty < t < \infty$

23. $\mathbf{r}(s,t) = \langle \cos s, \sin s, t \rangle$, $0 \le s \le 2\pi$, $0 \le t \le 1$

25. $\mathbf{r}(s,t) = \langle 2\sin s \cos t, 2\sin s \sin t, 2\cos s \rangle$, $0 \le s \le \pi$, $0 \le t \le 2\pi$

27. $\mathbf{r}(s,t) = \langle \sin s \cos t, \sin s \sin t, \cos s \rangle$, $\pi/2 \le s \le \pi$, $0 \le t \le 2\pi$

29. $\mathbf{r}(s,t) = \langle 2\cos s, \sin s, t \rangle$, $0 \le s \le 2\pi$, $-\infty < t < \infty$

31. $\mathbf{r}(s,t) = \langle 2\sin s \cos t, 2\sin s \sin t, 2\cos s \rangle$, $\pi/6 \le s \le 5\pi/6$, $0 \le t \le 2\pi$

33. $4x + 4y - z = 6$ **35.** $\sqrt{2}x + 2\sqrt{2}y - z = 4$

37. $2\sqrt{3}x - y + 2z = 3$ **41.** 12π **43.** $\pi/2$

45. $\dfrac{(5\sqrt{5}-1)\pi}{6}$ **47.** $\dfrac{(37\sqrt{37}-17\sqrt{17})\pi}{6}$

49. $\dfrac{(17\sqrt{17}-1)\pi}{24}$ **51.** $\dfrac{2(5\sqrt{5}-2\sqrt{2})\pi}{3}$

53. $\sqrt{5} + \dfrac{\sinh^{-1} 2}{2}$ **55.** 10π **57.** $4ab\pi^2$

61. False; consider the formula for the area of a surface generated by revolving a (single-variable) function graph about a coordinate axis.

63. True

65.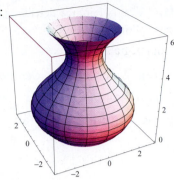

67. For $a = 2$:

The area is approx. 96.012.

Section 15.6

5. $4\sqrt{6}$ **7.** $\dfrac{a^3}{2\sqrt{3}}$ **9.** $7\sqrt{2/3}$ **11.** $\dfrac{23}{2\sqrt{6}}$

13. $\dfrac{149}{60\sqrt{2}}$ **15.** $\dfrac{(25\sqrt{5}\pi+1)}{120}$ **17.** $149\pi/30$

19. $\dfrac{(2\sqrt{2}-1)\pi}{12}$ **21.** 8π **23.** $152/3$ **25.** 0

27. $4(\pi+4)$ **29.** $\left(0, \dfrac{2(2+\pi)}{4+\pi}, \dfrac{4(3+\pi)}{3(4+\pi)}\right)$

31. $35k\pi$, where the density function is $\rho(x,y,z) = kz$

33. $I_x = I_y = I_z = 8\pi\rho R^4/3$ **35.** 27π **37.** $\pi/2$

39. $\pi/2$ **41.** $-81\pi/2$ **43.** 3π **45.** 3π

47. 81π **49.** 20π **51.** 6π

53. $\dfrac{8\kappa\pi}{r}k$, where k is the constant of proportionality

59. $I_x = I_y = 2625k\pi/4$, $I_z = 875k\pi/2$, where k is the constant of proportionality

61. $I_x = I_y = \dfrac{425\sqrt{5}-163}{840}\pi\rho$, $I_z = \dfrac{25\sqrt{5}+1}{60}\pi\rho$

Section 15.7

1. Both integrals are equal to 4π.

3. Both integrals are equal to 2π.

5. 4π; the answers are equal.

7. 2π **9.** 20 **11.** π **13.** 16π **15.** 2π

17. $8/15$ **19.** 35π **21.** 4π

Section 15.8

1. Both integrals are equal to $4\pi/3$.

3. Both integrals are equal to $3a^4/2$.

5. $64\pi/3$ **7.** $380\pi/3$ **9.** 6π **11.** $3\pi/4$

13. π **15.** 15π **17.** $121\pi/5$

19. Both integrals are equal to 16π.

21. $16\left(e^3 + 26\right)$ **23.** $V = abc$ **25.** $V = \dfrac{1}{3}\pi R^2 h$

Chapter 15 Review

1. C   **3.** B

5.

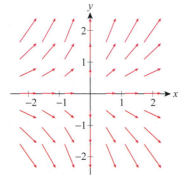

7. C   **9.** D   **11.** $\left\langle 4xy + z^3, 2x^2, 3xz^2 \right\rangle$

13. Conservative   **15.** Conservative

17. $f(x,y) = 4x^3 - 2x^2 y + y^3$ **19.** $32/3$

21. $7\sqrt{19}$ **23.** $\left(\dfrac{17 - 3\sqrt{21}}{20}, \dfrac{23 - 3\sqrt{21}}{4}, 2\right)$

25. Center of mass:
$$\left(\frac{3}{2 + \sqrt{1 + \pi^2}}, \frac{2\sqrt{1 + \pi^2} - \pi}{2\pi + \pi\sqrt{1 + \pi^2}}, \frac{\sqrt{1 + \pi^2} - 2\sqrt{2}}{2\left(2 + \sqrt{1 + \pi^2}\right)}\right)$$

27. $1344/5$ **29.** 20 **31.** $\dfrac{12(3 + \pi)}{4\pi\varepsilon_0} \cdot 10^{-7}$ volts

33. e **35.** $110 - \ln 13$ **39.** $1 + 4\sqrt{e}$

45. div $\mathbf{F} = x^2 - x + \dfrac{1}{y^2 + 1}$;
$$\text{curl } \mathbf{F} = \left(2xy + \frac{2xy}{\left(y^2 + 1\right)^2} - y\right)\mathbf{k}$$

47. div $\mathbf{F} = 2yz + x + \sin(yz)$;
$$\text{curl } \mathbf{F} = \left(x - y^2\right)\mathbf{i} + \left(xy\cos(yz) - y - z\right)\mathbf{j}$$
$$- xz\cos(yz)\mathbf{k}$$

49. Both integrals are equal to $-1/3$.

51. $-1/4$ **53.** 4 **55.** $(6\pi + 32)/3$ **57.** 14.75

59. B   **61.** A   **63.** C

65. $\mathbf{r}(s,t) = \left\langle \dfrac{\cos t}{s^2}, s, \dfrac{\sin t}{s^2} \right\rangle, 1 \le s \le 4, 0 \le t \le 2\pi$

67. The grid curves are ellipses if s is held constant and lines through the origin if t is held constant. The graph is an infinite elliptic double cone.

69. $\mathbf{r}(s,t) = \left\langle s\cos t, \sqrt{2}s\sin t, 2s^2 \right\rangle, 0 \le t \le 2\pi, 0 < s < \infty$

71. $z = 3x + 9y - 9$

73. $27\pi/2$ **75.** $2\left(9 - \sqrt{3}\right)\pi$

77. $\pi/6$ **79.** $(2/\pi, 2/\pi, 3/4)$

81. 12π **83.** 28

87. Both integrals are equal to -12π.

89. $\pi/2$ **91.** -72π

93. 12π **95.** 80π

97. Both integrals are equal to π.

99. $405\pi/2$ **101.** 12π

103. $93\pi/5$ **105.** True

107. True (If the domain is simply connected.)

109. False; the work is only guaranteed to be zero if the path is closed. See Exercise 3 of Section 15.6.

111. True

113. Mass: $8\pi\rho$; center of mass: $(0,0,1)$

115. Mass: $\dfrac{\left(289\sqrt{17} - 41\right)k\pi}{60}$;

center of mass: $\left(0, 0, \dfrac{4913\sqrt{17} - 1177}{7\left(289\sqrt{17} - 41\right)}\right)$

Chapter 15 Project

7. If we choose $C_2 = C_3 = 0$, one possible answer is
$$\mathbf{P}(x,y,z) = \left\langle 0, \frac{x^3 y}{3}, x^2 y^2 z \right\rangle.$$

Index

A

Abel, Niels Henrik 271, 387
Absolute convergence of a series 818
Absolute extrema. *See* Absolute maximum
    and minimum
Absolute maximum and minimum 331
    finding 335
    of a function of two variables 1045
        finding 1052
Absolute value of a complex number A-17
Absolute value function 34
Acceleration 202
    caused by gravity 203
    tangential and normal components of 960
    as a vector function 948
Addition of vectors 876
    Commutative Property 876
    Parallelogram Law 876, 892
    properties of 877
Additive inverse of a vector 876
Algebraic functions 27
Alternating series 816
    harmonic 816
    remainder of 817
    sum of, estimating 818
Alternating Series Test 816
Angle
    of rotation 744
    between two vectors 885
Angular momentum 944
Angular speed A-9
Annulus 1092
Antiderivative(s) 407
    general 408, 409
Antidifferentiation 407
    formulas 409
    of vector functions 938
Aphelion 759, 973
Apogee 973
Arccosine function 71, 72
    derivative of 289
Archimedes 102, 424, 463, 766, 1202
Archimedes' Principle 549
Archimedes' spiral 719
Arc length 516
    approximating 112
    of a circular arc A-9
    parameter 946
    parametric formula for 707
    parametrized with respect to 946
    of a polar curve 729
    rate of change of a curve with respect to 947
    of space curves 945

Arc length function 517
Arcsine function 70, 72
    derivative of 288, 289
Arctangent function 72
    derivative of 288
Area 425
    between two curves 478, 479
    between two polar curves 727
    of a circle 498, 602
    by double integration 1086
    of an ellipse 601
    by Green's Theorem 1189
    of a parallelogram 898
    bounded by a parametric curve 705, 710
    of a polar region 726
    of a sector of a circle A-9
    signed 438
    of a triangle 899
    under a curve 431
        approximating 112, 425
        parametric formula for 705, 710
Area, surface
    of a frustum of a cone 519
    Pappus' Theorem for 533
    of a parametric surface 1196
    of a solid of revolution 520
        in parametric form 708
        in polar form 731
    of a sphere 520, 1196
    of a surface $z = f(x, y)$ 1197
    of a torus 534
Area problem 425
Argand, Jean-Robert A-14
Argand plane A-14
Argument
    of a complex number A-17
    of a function 8
Arithmetic series 791
Arrow diagram 12
Astroid 275
Astronomical units 973
Asymptote(s)
    horizontal 24, 125
    of a hyperbola 740
    oblique 24
    slant 25
    vertical 24, 120, 123
Attractive force field 1150
Autonomous differential equation 663
Average cost function 399
Average rate 103
Average value of a function
    on an interval 445
    over a region R 1087
    over a solid S 1115

Average velocity 105
Axes, coordinate 867
Axis
    major, of an ellipse 734
    minor, of an ellipse 734
    polar 713
    of symmetry of a parabola 738

B

Barrow, Isaac 424
Basis vectors, standard 878
Bernoulli, Jakob 330, 661, 692
Bernoulli, Johann 330, 370, 692, 835, 952
Bernoulli equation 661
Bernoulli spiral 943
Bézier, Pierre 764
Bézier curves 764
Bifolium 274
Binomial coefficient 848
Binomial series 848
    convergence of 855
Binomial Theorem 849, 855
Binormal vector, unit 954
Black-Scholes equation 1009
Bolzano-Weierstrass Theorem A-27
Boundary conditions 676
Boundary point
    of a region of the plane 984
    of a region of $\mathbb{R}^n$ 989
Boundary value problem 676
Bounded Monotonic Sequence Theorem 775
Bounded region
    of the plane 984
    of $\mathbb{R}^n$ 989
Bounded sequence 775
Bounded vector function 978
Brachistochrone problem 692
Brahe, Tycho 736, 969
Break-even points 186
Brunt, Sir David 31

C

Calculator(s), graphing. *See* Graphing
    calculator(s)
Capping surfaces 1214
Cardano, Girolamo A-14
Cardioid 274, 719
Carrying capacity 666

Cartesian coordinates
  converting to and from cylindrical
    coordinates 1117, 1121
  converting to and from polar
    coordinates 715
  converting to and from rotated Cartesian
    coordinates 742
  converting to and from spherical
    coordinates 1121
Cartesian plane 3
Cartesian three-dimensional space 867
CAS. *See* Computer algebra system(s)
Catenary 563
Cauchy, Augustin-Louis 102, 134, 330, 371
Cauchy Condensation Test 807
Cauchy criterion for convergence of a series 792
Cauchy density function 635
Cauchy-Schwarz Inequality 892
Cauchy's Mean Value Theorem 352, 371
Cavalieri, Bonaventura 484, 490, 494
Cavalieri's Principle 484, 495
Center of curvature 965
Center of mass 525
  of a one-dimensional system 526, 527
  of a plate (planar object) 529, 1088
    of varying density 531, 1088
  of a solid 1110
  of a surface 1204
  of a two-dimensional system 528
  using line integrals 1158
  of a wire 532
Center-seeking force 943
Centripetal force 943
Centroid 532
Chain Rule 253
  implicit differentiation via 1021–1022
  for multivariable functions 1017, 1019, 1020
  for vector functions 937
Change of base formula 69, A-8
Change of variables
  in a double integral 1132
  in integration 469
  in a triple integral 1134
Characteristic equation 674
Chemical reactions 297–298
Circle(s)
  area 498, 602
  as conic sections 734
  involute of 952
  osculating 964
  polar equations of 719
Circular cylinder, volume of 492
Circulation
  of a vector field 1212
  of a velocity field 1162
Circumference of an ellipse 856
Cissoid 275, 964
Clairaut, Alexis 982, 1007
Clairaut's Theorem 1008, A-28

Closed path 1169
  positive orientation of 1179
    relative to a surface 1212
  simple 1180
Closed region
  of the plane 984
  of $\mathbb{R}^n$ 989
Closed surfaces 1205
Codomain 5
Coefficient(s) 21
  binomial 848
  Fourier 853
  leading 21
  of a power series 824
Collapsing sums 434
Common logarithm 66
Common ratio 434, 783
Commutative Property of vector addition 876
Comparison test(s)
  for improper integrals 631
  for series
    Direct Comparison Test 802
    Limit Comparison Test 804
Completeness Property 775
Complex conjugate A-15
Complex exponential A-18
Complex number(s) A-14
  absolute value of A-17
  addition of A-14
  argument of A-17
  division of A-15, A-18
  equality of A-14
  imaginary part of A-14
  magnitude of A-17
  modulus of A-17
  multiplication of A-14, A-18
  norm of A-17
  polar form of A-17
  powers of A-18
  real part of A-14
  roots of A-19
  subtraction of A-14
Complex plane A-14
Component form of a vector 873
Component functions
  of a vector field 1147
  of a vector function 933
Component Test for conservative vector
    fields 1171
  in terms of curl operator 1216
Composite function 49
Composition of functions 49
  derivative of 253
Compressing a function 44
Computer algebra system(s) 80
  integration 610
  limits, estimating 128
  See also *Mathematica*
Concavity, intervals of 358
Concavity Test 358

Conditional convergence of a series 818
Conductivity, thermal 1207
Cone, volume of 503
Conic section(s) 733
  algebraic classification of 733
  circle 734
  degenerate 733
  directrix of 752
  discriminant of 733
  eccentricity of 752
  ellipse 733, 734
  focus of 752
  general form of equation for 733
  hyperbola 733, 739
  latus rectum of 759
  parabola 733, 737
  in polar coordinates 752
  polar equations of 754
  rotating 741–745
    elimination of the xy-term 744
    in polar coordinates 757
Connected region 1168
Conservation of energy 1169, 1175
Conservative vector field(s) 1151, 1169
  Component Test for 1171
    in terms of curl operator 1216
Constant of integration 465
Constant Multiple Law
  for convergent series 788
  for gradients 1031
  for limits 146, A-21
    of functions of n variables 996
    of sequences 772
Constant Multiple Rule 219
Constant polynomial 21
Constant Rule 216
Constant Vector Rule 937
Constraint 1058, 1062
Continuous extension of a function 170
Continuous function(s) 163
  on an interval 163
  at a point 159, 169
    epsilon-delta $(\varepsilon\text{-}\delta)$ 162
    left-continuous 159
    right-continuous 159
  properties of 165, 998
Continuously differentiable functions 515
Continuous multivariable function 998
  at a point 998
Continuous vector field 1148
Continuous vector function(s) 934
  at a point 934
Contour 985
Contour map 985
Control points 764
Convergence
  of a binomial series 855
  of a geometric series 783
  of an improper integral 628, 629
    Direct Comparison Test for 631

interval of 827
of a power series 827
of a p-series 798
radius of 827
of a sequence 769
of a series 781
absolute 818
Cauchy criterion for 792
conditional 818
of a Taylor series 839
Convergent series 781
properties of 788
Convex limaçon 719
Coordinate axes 867
Coordinate conversion
Cartesian to cylindrical 1117, 1121
Cartesian to polar 715
Cartesian to rotated Cartesian 742
Cartesian to spherical 1121
cylindrical to Cartesian 1117, 1121
cylindrical to spherical 1121
polar to Cartesian 715
rotated Cartesian to Cartesian 742
spherical to Cartesian 1121
spherical to cylindrical 1121
Coordinate planes 867
Coordinate system
cylindrical 1117
left-handed rectangular 867
polar 713
right-handed rectangular 867
with rotated axes 742
angle of rotation 744
spherical 1120
three-dimensional Cartesian 867
Coordinate transformations 1129
Coplanar vectors 902
Corner 207
Cornu spiral 952
Cosecant function 28
derivative of 246
graph of A-10
Cosine function 28
derivative of 243
graph of 29, A-10
limit involving 239
period of A-11
Cost function 186
Cotangent function 28
derivative of 246
graph of A-10
Coulomb's constant 1150
Coulomb's Law 551, 1150
Critical damping constant 680
Critically damped mass-spring model 677
Critical point 335
of a function of two variables 1045
Cross product 895
determinant formula for 897
magnitude of 895
properties of 897

Cross Product Rule 937
Cross-section 491, 916
Cubic functions 22
Curl-free vector fields 1234
Curl of a vector field 1179
Curtate cycloid 699
Curvature xiii, 957
center of 965
mean xiii
radius of 965
Curve(s) 935
arc length of 516
grid 1193
level 985
parametric 689
smooth 702
rectifiable 711
smooth 515
space 935
smooth 936
Curve fitting 11
Curve-sketching strategy 381
Cusp 207
Cycloid 691
curtate 699
prolate 700
Cylinder(s) 491, 916
circular 492
cross-section of 916
elliptic 916
generating curve of 916
hyperbolic 916
parabolic 916
right 492
rulings of 916
trace of 916
volume of 491
Cylindrical coordinates 1117
converting to and from Cartesian
coordinates 1117, 1121
converting to and from spherical
coordinates 1121
triple integrals in 1118
volume in 1118

D

Damped harmonic motion 252, 265
Damping constant 677
critical 680
Darboux, Jean-Gaston 212
Darboux's Theorem 212
Decreasing function 15
Decreasing sequence 774
Definite integral(s) 437
of even and odd functions 477
properties of 442, 443
Substitution Rule for 475
translation invariant property 483
of a vector function 939

Degenerate conic section 733
Degree
of homogeneity 1024
of a polynomial 21
Del operator (∇) 1028, 1177
De Moivre's Theorem A-19
Density
of a fluid 543
mass 543
of a plate (planar object) 1088
of a solid 1109
weight 543
Density function (ρ) 526, 1088
Dependent variable 5
of a function of two variables 983
Derivative(s) 179, 183, 199
of a^x and a^u 262
of arccosine function 289
of arcsine function 288
of arctangent function 288
of a composition of two functions 253
of a constant function 216
of cosecant function 246
of cosine function 243
of cotangent function 246
directional 1026
computation of 1028
properties of 1029
of e^x 225
of exponential functions 225, 262
as a function 199
higher-order 202
of hyperbolic functions 554, 555
of implicitly defined functions 266
of an integral 452
of inverse functions 278
of inverse hyperbolic functions 558
of inverse trigonometric functions 286–290
left-hand 210
of $\ln x$ 279
of logarithms 279–282
of $\log_a x$ 281
of natural logarithm 279
notation 199–202
n^{th} 202
one-sided 209
parametric formula for 703
partial 1002
notation 1002
at a point 179, 199
polar formula for 724
of a polynomial 220
of a power function 217, 231
of a product 227
of a quotient 232
as a rate of change 180
of a reciprocal function 230
right-hand 209
of secant function 246
second 202
parametric formula for 704

Derivative(s) (*continued*)
  of sine function 241
  as the slope of a tangent line 179, 180
  symmetric 214
  of tangent function 246
  third 202
  of trigonometric functions 241, 243, 246
  of a vector function 935
Derivative Rule for Inverse Functions 278
Determinant formula for cross product 897
Difference of functions 47
Difference Law
  for convergent series 788
  for gradients 1031
  for limits 146, A-21
    of functions of n variables 996
    of sequences 772
Difference quotient 104, 179, 199
Difference Rule 219
  for vector functions 937
Differentiable function 199, 206, 210
Differentiable multivariable function 1011
  at a point 1011, A-30
Differentiable parametric curve 702
Differentiable vector field 1148
Differentiable vector function 936
Differential(s) 200, 317
  total 1039
Differential equation(s) 410, 645
  autonomous 663
  Bernoulli 661
  exact 1072
  first-order 645
  first-order linear 654
    standard form of 654
  general solution of 645
  homogeneous linear 672
  linearly independent solutions 673
  logistic 666
  nonhomogeneous linear 672
  order of 645
  partial 1009
  particular solution of 645
  second-order linear 672
    standard form of 672
  separable 646
  solution of 645
Differentiation
  implicit 266, 267
    via Chain Rule 1021–1022
    partial 1004–1005
  logarithmic 282
  partial 1002
    implicit 1004–1005
  of power series 829
  of vector functions, rules for 937
Differentiation operators 199
Dimpled limaçon 274, 719
Direct Comparison Test
  for improper integrals 631
  for series 802

Directed line segment(s) 873
  initial point of 873
  terminal point of 873
Directional derivative 1026
  computation of 1028
  properties of 1029
Direction angles of a vector 887
Direction cosines of a vector 887
Direction field 663
Direction vector 905
Directrix
  of a conic section 752
  of a parabola 737
Direct Substitution Property 153
Dirichlet, Peter 2
Dirichlet function 176, 447
Discontinuity 159
  infinite 161
  jump 160
  nonremovable 162
  oscillating 161
  removable 162
Discontinuous function 159
Discriminant 1049
  of a conic section 733
Disjoint Region Property 1082, 1107
Disk(s) 495
  open 984
Disk method 498
Display window 80
Distance
  between parallel lines 893
  between parallel planes 893
  from a point to a line 893, 903, 914
  from a point to a plane 893, 903, 910
  between skew lines 911
  between two points in space 869
Distance problem 427
Divergence
  of an improper integral 628, 629
  of a sequence 769
  of a series 781
  of a vector field 1177
Divergence-free vector field 1234
Divergence Theorem 1219
Divergent series 781
Domain
  of a function 5
    of n variables 989
    of two variables 983
  implied 7
  of a relation 3
  of a vector function 933
Domination Property 1082, 1107
Dot product 884
  and angle between two vectors 885
  properties of 884
Dot Product Rule 937

Double integral(s) 1075, 1076
  area with 1086
  change of variables in 1132
  of f over R 1076
  over a general bounded region in $\mathbb{R}^2$ 1079–1082
  in polar coordinates 1097, 1098
    evaluating 1098
  properties of 1082
  over a rectangular region 1076
  substitutions in 1129–1133
  volume with 1075
Double sum(s) 436
  Riemann 1075

E

e 224, 463
Eccentricity
  of a conic section 752
  of an ellipse 736, 752
  of a hyperbola 752
  of a parabola 752
Einstein, Albert 766
Einstein's theory of special relativity 567, 857
Electrical circuit(s) 653, 659, 681
  parallel 1015
Electric field 1150
Electrostatic force 1150
Electrostatic potential 1165, 1211
Elementary functions 616
Eliminating the parameter 689
Ellipse(s) 276, 733, 734
  area 601
  circumference 856
  directrix of 752
  eccentricity of 736, 752
  foci of 734, 752
  geometric characteristic of 734
  major axis of 734
  minor axis of 734
  rotated 275
  standard form of equation for 734
  vertices of 734
Ellipsoid 918
Elliptic cone 918
Elliptic cylinder 916
Elliptic paraboloid 918
Empirical model 90
Energy
  conservation of 1169, 1175
  kinetic 1175
  potential 1175
Enneper, Alfred xiii
Enneper surface xiii, 1198
Epicycloid 700
Epsilon-delta (ε-δ) continuity at a point 162
Epsilon-delta (ε-δ) definition of limit 134
Equichordal point 732

Equilibrium points 663
  stable 664
  unstable 664
Error
  percentage 319
  propagated 319, 1006
    for Simpson's Rule 622
    of a Taylor series 838
    for Trapezoidal Rule 619
Error function 636
Escape velocity 569
Euler, Leonhard 198, 644, 668, 766, 794, A-9
Euler-Mascheroni constant 864
Euler's constant 864
Euler's formula 847
Euler's method 668
  improved 671
Euler's number 864
Euler's spiral 952
Even function(s) 16
  definite integrals of 477
Evolute 965
Exact equations 1072
Explicit formula for a sequence 768
Exponential function(s) 31
  derivative(s) of 225, 262
  graphs of 32, A-8
Exponential growth model 293
Exponents, properties of 66, A-8
Extrapolation 10
Extrema
  absolute 331
    finding 335
  of a function of two variables 1045
  global 331
  local 333
  relative 333
Extreme values of functions 331
  of two variables 1045
Extreme Value Theorem 332
  for functions of two variables 1052

F

Fermat, Pierre de 334, 564
Fermat's Principle 399
Fermat's Theorem 334
Fibonacci 768
Fibonacci sequence 768, 778
Field(s) A-16
  conservative 1151, 1169
  electric 1150
  force 1149
  gradient 1150
  gravitational 1149
  irrotational 1216
  properties of A-16
  scalar 1147
  vector 1147
  velocity 1162

First-degree polynomials 21
First Derivative Test 356
First moment(s) 1088, 1109, 1110, 1158
First octant 867
First-order approximation 315
First-order differential equation 645
  linear 654
    standard form of 654
Fixed point of a function 352
Flow integral 1162
Flow lines 1155
Fluid flow 1162
Fluid force 544
Flux density of a vector field 1177
Flux of a vector field 1184, 1206
Focal length of a parabola 737
Focus
  of a conic section 752
  of an ellipse 734
  of a hyperbola 739
  of a parabola 737
Folium of Descartes 267, 701, 1190
Fontanella, John 980
Force 294
  center-seeking 943
  centripetal 943
  constant 539
  electric 1150
  fluid 544
  variable 540
Force diagram 879
Force field(s) 1149
  attractive and repulsive 1150
Fourier, Jean-Baptiste Joseph 851
Fourier analysis 852
Fourier coefficient(s) 580, 853
Fourier series 853
Fourier's Law of Heat Conduction 1207
Fourth-degree polynomials 22
Frenet, Jean Frédéric 932, 954
Frenet-Serret formulas 958, 959
Frenet-Serret frame 954
Frustum of a cone 513, 518
  surface area of 519
  volume of 513
Fubini, Guido 1074, 1077
Fubini's Theorem 1078
  for triple integrals 1106
Function(s) 5
  absolute value 34
  algebraic 21, 27
  arccosine 71, 72
  arc length 517
  arcsine 70, 72
  arctangent 72
  argument of 8
  arrow diagram of 12
  average cost 399
  average value of 445, 1087, 1115
  Cauchy density 635

codomain of 5
combining arithmetically 47
component 933, 1147
composite 49
composition of 49
compressing 44
continuous 163
continuous extension of 170
continuously differentiable 515
cosecant 28
cosine 28
cost 186
cotangent 28
cubic 22
decreasing 15
density 526, 1088
derivative of 179, 183, 199
difference of 47
differentiable 199, 206, 210
Dirichlet 176, 447
discontinuous 159
domain of 5
elementary 616
error 636
even 16
exponential 31
extreme values of 331
fixed point of 352
gamma 636
graph of 12
greatest integer 37
harmonic 1015
homogeneous 1024
horizontal shifting of 41
hyperbolic 553
increasing 15
integrable 437, 440, 1076, 1106
inverse of 59
  finding a formula for 61
  properties of 63
inverse hyperbolic 557
inverse trigonometric 69–74, A-12
iterates of 52
limit of 117, 134
linear 21
logarithmic 33, 64
marginal cost 186, 298
marginal profit 186, 298
marginal revenue 186, 298
maximum and minimum value 331
monotone 15
notation 5
of n variables 989
  continuous 998
  differentiable 1011
  domain of 989
  gradient vector of 1028
  limit of 994
  range of 989
odd 16

Function(s) (*continued*)
  one-to-one 60
  periodic 28
  piecewise continuous 440
  piecewise defined 34
  polynomial 21
  potential 1072, 1151
  power 25
  probability density 634
  product of 47
  profit 186
  quadratic 22
  quartic 22
  quotient of 47
  range of 5
  rational 24
  reciprocal 25
  recursion 52
  reflecting 43
  revenue 186
  root 25
  sawtooth 853
  scalar 1147
  secant 28
  sine 28
  sine integral 488
  stretching 44
  strictly decreasing 15
  strictly increasing 15
  sum of 47
  tangent 28
  of three variables 988
    differentiable 1011
    partial derivatives of 1004
  transcendental 28
  transformations of, order of 45
  trigonometric 28, A-9
  of two variables 9, 983
    dependent variable of 983
    differentiable 1011
    domain of 983
    extreme values of 1045
    gradient vector of 1028
    graph of 985
    independent variables of 983
    maximum and minimum value 1045
    partial derivatives of 1002
    range of 983
  value of 5
  vector 933
  vector-valued 933
  verbal descriptions of 13
  vertical shifting of 42
  ways of describing 8
Functional dependence 5
Fundamental Theorem of Algebra A-14
Fundamental Theorem of Calculus 451, 452,
    456
  for Line Integrals 1166
  versions of 1224

G

Gabriel's horn 501, 634
Galilei, Galileo 203, 563, 567, 978
Galois, Évariste 271, 387
Gamma function 636
Gauss, Carl Friedrich A-14
Gauss's Law 1223
Generalized Positive Integer Power Rule 237
Generalized Power Rule 286
General solution of a differential equation 645
Generating curve of a cylinder 916
Geometric progression 434
Geometric series 783
  convergence of 783
  divergence of 783
  sum of 783
Geometric sum 434
Geostationary satellite 973
Global extrema. *See* Global maximum and
    minimum
Global maximum and minimum 331
  of a function of two variables 1045
Gradient 1028
  properties of 1031
Gradient Theorem 1166
Gradient vector 1028
Gradient vector fields 1150
Graph(s)
  of exponential functions 32, A-8
  of a function 12
  of a function of two variables 985
  of hyperbolic functions 553, A-13
  of inverse hyperbolic functions 557
  of inverse trigonometric functions 71, 72,
      287, A-12
  of logarithmic functions 34, A-8
  orthogonal 275
  polar 717–721
    table of common 719
  of power functions 26, 27
  of quadric surfaces 916, 917, 918
  of root functions 26
  of sequences 769, 770
  of trigonometric functions 29, A-10
Graphing calculator(s) 80
  function evaluation 9
  graphing an ellipse 735
  graphing a function 9
  graphing a parametric curve 694
  graphing a polar equation 721
  graph settings 83
  limits, estimating 128
  roots of equations 174
Gravitational constant, universal (*G*) 550
Gravitational field 1149
Gravity constant (*g*) 203
Greatest integer function 37
Greatest lower bound 775
Green, George 1146, 1177
Green's first identity 1226
Green's second identity 1226

Green's Theorem 1181
  Normal-Divergence Form 1185
  Tangential-Curl Form 1183
Gregory, James 424, 863
Gregory series 863
Grid curves 1193
Gyration, radii of 1090, 1110, 1158, 1159

H

Half-life 40
Harmonic functions 1015
Harmonic motion
  damped 252, 265
  simple 31, 243
Harmonic series 787, 797
  alternating 816
Hawking, Stephen 1146
Heat equation 1015
Heat flow 1207
Heaviside, Oliver 587
Heaviside cover-up method 586
Helicoid surfaces 1202
Helix 933
Hessian 1049
Heun's method 671
Higher-order derivatives 202
  partial 1008
Hipparchus 688
Hippocrates of Chios 607
Homogeneous function 1024
Homogeneous linear differential
    equation 672
Hooke's Law 540
Horizontal asymptote(s) 24, 125
Horizontal line test 60
Horizontal shifting of a function 41
Huygens, Christiaan 692
Hyperbola(s) 733, 739
  asymptotes of 740
  directrix of 752
  eccentricity of 752
  foci of 739, 752
  geometric characteristic of 739
  standard form of equation for 740
  vertices of 740
Hyperbolic cylinder 916
Hyperbolic functions 553, A-13
  derivatives of 554, 555
  graphs of 553, A-13
  integrals of 555
  inverse 557
Hyperbolic identities 554, A-13
Hyperbolic paraboloid 918
Hyperbolic substitution 603
Hyperboloid
  of one sheet 918
  of two sheets 918
Hypocycloid 634, 694
Hypotrochoid 701

I

i (standard basis vector) 878
Ideal Gas Law 265
Image
　of a point 1129
　of a region 1129
Imaginary unit (*i*) A-14
Implicit differentiation 266, 267
　via Chain Rule 1021–1022
　partial 1004–1005
Implicit Function Theorem 1022
Implied domain 7
Improper integral(s) 627–632
　Direct Comparison Test for 631
　type I 628
　　convergence of 628
　　divergence of 628
　type II 629
　　convergence of 629
　　divergence of 629
Improved Euler's method 671
Increasing function 15
Increasing sequence 774
Increment 104
Increment Theorem of Differentiability 1011,
　　A-30
Indefinite integral(s) 465
　formulas 468
　of a vector function 939
Independence of path 1168
Independent variable(s) 5
　of a function of two variables 983
Indeterminate form(s) 370, 372, 374, 375
Index of summation 428
Inertia, moment of 1089
Infinite discontinuity 161
Infinite limits 120, 126, 138
　one-sided 138
Infinite sequence 767
Infinite series 781
Inflection point 360
Initial condition 645
Initial point
　of a directed line segment 873
　of a parametric curve 689
Initial value problem(s) 410, 645
　second-order 675
Inner product 884
Instantaneous rate of change 110
Instantaneous velocity 103, 179
Integrable function(s) 437, 440, 1076, 1106
Integral(s)
　change of variables in 469, 1132, 1134
　definite 437
　　of even and odd functions 477
　　properties of 442, 443
　　Substitution Rule for 475
　　translation invariant property 483
　derivative of 452

double 1075, 1076
　properties of 1082
flow 1162
of hyperbolic functions 555
improper 627–632
　convergence of 628, 629
　Direct Comparison Test for 631
　divergence of 628, 629
　type I 628
　type II 629
indefinite 465
　formulas 468
involving inverse hyperbolic
　　functions 559
iterated 1077
line 1156
nonelementary 615, 616
repeated 1077
surface 1203
table of TI-1–TI-6
trigonometric 592–597
　evaluating 592, 595
triple 1105, 1106
　properties of 1107
Integral Test 796
Integrand 437
Integrating factor 655
Integration 438
　with computer algebra systems 610
　constant of 465
　definite 437
　　by parts 577
　formulas 468, TI-1–TI-6
　guidelines 608
　Heaviside cover-up method 586
　by hyperbolic substitution 603
　indefinite 465
　　by parts 573
　limits of 437
　numerical 616–626
　　Midpoint Rule 624
　　Simpson's Rule 621
　　Trapezoidal Rule 617
　by partial fractions 581–588
　by parts
　　definite 577
　　indefinite 573
　of power series 829
　of rational functions 581–588
　reversing the order of 1080
　substitution in 468
　theory of 437
　by trigonometric substitutions 599
　variable of 438, 452, 465
　of vector functions 939
Intercept(s) 381
　x- 381
　y- 381
Interior point
　of a region of the plane 984
　of a region of $\mathbb{R}^n$ 989

Intermediate Value Property 172
Intermediate Value Theorem 171
Intermediate variables 1017
Interpolation 10
Intersection of two planes 910
Interval(s)
　of concavity 358
　of convergence 827
　of monotonicity 15
　parametric 689
Inverse
　of a function 59
　　finding a formula for 61
　　properties of 63
　of a relation 58
Inverse hyperbolic functions 557
　derivatives of 558
　graphs of 557
　integrals involving 559
Inverse transformation 1129
Inverse trigonometric functions 69–74, A-12
　derivatives of 286–290
　graphs of 71, 72, 287, A-12
Involute of a circle 952
Irrotational vector field 1216
Isobars 988, 1071
Isotherms 988, 1071
Iterated integral 1077
Iterates of a function 52

J

j (standard basis vector) 878
Jacobi, Carl Gustav Jacob 1074
Jacobian determinant 1131, 1134
Jacobian of a transformation 1131, 1134
Jerk 204
Joule, James 539
Jump discontinuities 160

K

k (standard basis vector) 878
Kepler, Johannes 736, 932, 969
Kepler's Laws of Planetary Motion 736, 968
Kinetic energy 1175

L

Lagrange, Joseph-Louis 198, 200, 345, 1058
Lagrange multiplier(s) 1058
　method of 1059
Lagrange's identity 904
Lambert, Johann Heinrich A-9
Lamé curves 763
Laplace's equation 1009

Laplace transform 635
Laplacian 1024
Latus rectum 759
Law of Conservation of Energy 1175
Law of Cosines A-12
Law of the Lever 525
Law of Sines A-12
Leading coefficient 21
Least-squares curve fitting 11, 90, 1056
Least upper bound 775
Left-continuous function 159
Left-hand derivative 210
Left-handed rectangular coordinate
      system 867
Left-hand limits 137
Left Riemann sum 440
Leibniz, Gottfried Wilhelm 2, 198, 200, 424,
      438, 451
Leibniz notation 200
Leibniz's Test 816
Lemniscate(s) 274, 720
Length
   of a circular arc A-9
   of a curve 516
   of a parametric curve 707
   of a polar curve 729
   of a space curve 945
   of a vector 873
Lens equation 314
Leonardo of Pisa 768
Level curve 985
Level surface 988
l'Hôpital, Guillaume François Antoine
      de 370, 380
l'Hôpital's Rule 370, 372
Limaçon(s) 274, 719
   cardioid 719
   convex 719
   dimpled 274, 719
   with inner loop 719
Limit(s) 117
   determining 146, 153
   epsilon-delta (ε-δ) definition of 134
   estimating using technology 128
   formal definition of 134
   of a function 117, 134
      of n variables 994
   of indeterminate form(s) 370, 372, 374, 375
   infinite 120, 126, 138
      one-sided 138
   at infinity 125, 126, 138
   of integration 437
   left-hand 137
   one-sided 120, 123, 137
      infinite 138
   properties of 146, A-21
   proving existence 138–141
   proving nonexistence 141–142
   right-hand 137
   of a sequence 769
   involving sine and cosine functions 239
   involving trigonometric functions 241
   of a vector function 934

Limit Comparison Test 804
Limit laws 146, A-21
   for functions of n variables 996
   for sequences 772
Line(s)
   normal 236, 269
   point-slope equation of 181
   secant 104, 109
   in space
      direction vector of 905
      normal to a surface 1036
      parametric equations of 905
      skew 907
      symmetric equations of 913
      vector form of 905
   tangent 108, 179
Linear approximation 315
   of a multivariable function 1039, 1041
Linear combination 672
   of vectors 878
Linear differential equation
   first-order 654
      standard form of 654
   homogeneous 672
   nonhomogeneous 672
   second-order 672
      standard form of 672
Linear equation for a plane 908
Linear functions 21
Linear independence 673
Linearization 298, 315
   of a multivariable function 1039, 1041
Linear speed A-9
Linear transformation 1136
Line integral(s) 1156
   of f along C 1156
   Fundamental Theorem for 1166
   path independent 1168
Lipschitz property 352
Lissajous figures 694
Local extrema. See Local maximum and
      minimum
Local maximum and minimum 333
   of a function of two variables 1045
Logarithm(s)
   change of base formula 69, A-8
   common 66
   derivatives of 279–282
   natural 66
   properties of 66, A-8
Logarithmic differentiation 282
Logarithmic function(s) 33, 64
   graphs of 34, A-8
Logarithmic p-series 801
Logarithmic spiral 943
Logistic differential equation 666
Logistic growth of a population 591, 666
Lower bound of a sequence 775
   greatest 775
Lucas, Édouard 779

M

Maclaurin, Colin 766, 835
Maclaurin polynomial 265, 836
Maclaurin series 835
Magnitude
   of a complex number A-17
   of a vector 873
Major axis of an ellipse 734
Mandelbrot, Benoit 52
Marginal cost function 186, 298
Marginal profit function 186, 298
Marginal revenue function 186, 298
Mass
   of a one-dimensional system 527
   of a plate (planar object) 529, 1088, 1090
   of a solid 1110
   of a surface 1204
   using line integrals 1158
Mass, center of. See Center of mass
Mass density 543
Mathematica A-2
   animations and models 84–89, A-3
   contour maps, constructing 987
   curve fitting 11, 91, A-4
   differentiation A-6
      partial 1005–1006
   discriminant, calculating 1051
   double integration 1082
   expanding products A-7
   factoring polynomials A-7
   function evaluation 9, A-3
   graphing an ellipse 735
   graphing a function 9, A-3
      of two variables 987
   graphing a parametric curve 694
   graphing a polar equation 721
   graphing vertical asymptotes 83
   integration 611–613, A-6
      double 1082
   inverse hyperbolic functions 558
   limits, evaluating 130, A-5
   maximum or minimum of a function A-6
   partial differentiation 1005–1006
      second-order 1051
   piecewise defined functions 36, A-3
   plotting a parametric surface 1198
   plotting three-dimensional surfaces 921,
      987
   plotting vector fields 1151
   roots of equations, approximating 173–174,
      A-5, A-6
   simplifying expressions A-7
      rational A-7
   solving equations 268, A-5
   Trapezoidal Rule 620
Mathematical model(s) 87
   constructing 87

Maximum
    absolute 331
    global 331
    local 333
    relative 333
Maximum value of a function 331
    of two variables 1045
Maxwell, James 1146
Maxwell-Faraday Law of Induction 1215
Mean curvature xiii
Mean Value Theorem 345
    Cauchy's 352, 371
    consequences of 347–349
    for Definite Integrals 451
Merton, Robert 1009
Method of Lagrange multipliers 1059
Midpoint Rule 624
Minimal surfaces xiii
Minimum
    absolute 331
    global 331
    local 333
    relative 333
Minimum value of a function 331
    of two variables 1045
Minor axis of an ellipse 734
Mixed partial derivatives 1007
Mixture problems 648
Möbius, August Ferdinand 1202
Möbius strip 1202, 1205
Modulus of a complex number A-17
Moment(s)
    about an axis
        of a plate (planar object) 529, 1088
        of a solid 1109
        of a two-dimensional system 528
    first 1088, 1109, 1110, 1158
    of inertia 1089
    of a mass
        of a one-dimensional system 525
        of a two-dimensional system 528
    about the origin 525, 527
    about a plane 1109
    of a plate (planar object) 529, 1088
    second 1089, 1090, 1109, 1110, 1158,
        1159
    of a solid 1109
Momentum, angular 944
Monotone function 15
Monotonicity, intervals of 15
Monotonicity Test 354
Monotonic sequence 774
Multiple integral. *See* Double integral(s);
        Triple integral(s)
Multiplication
    of power series 830
    scalar, of vectors 876

N

Napier, John A-8
Natural logarithm 66
    derivative of 279
Nephroid of Freeth 723
Newton, Sir Isaac 198, 424, 451, 692, 835,
        849, 969, 1146
Newton's Law of Cooling 300, 665
Newton's Law of Universal Gravitation 550,
        969
Newton's method 388
Newton's Second Law of Motion 276, 294, 948
Nonelementary integrals 615, 616
Nonhomogeneous linear differential
        equation 672
Nonorientable surfaces 1202
Nonremovable discontinuity 162
Norm
    of a complex number A-17
    of a partition 438
        of a region of the plane 1075
        of a region of $\mathbb{R}^3$ 1105
    of a vector 873
Normal component of acceleration 960
Normal-Divergence Form of Green's
        Theorem 1185
Normal line 236, 269
    to a surface 1036
Normal plane 955
Normal vector 886
    unit 954
n^{th} derivative 202
n^{th} harmonic number 864
n^{th} iterate 52
n^{th} root 25
n^{th} term
    of a sequence 767
        explicit formula for 768
        recursive formula for 768
    of a series 787
n^{th}-Term Divergence Test 787
Null sequence 779
Numerical integration 616–626
    Midpoint Rule 624
    Simpson's Rule 621
    Trapezoidal Rule 617

O

Oblique asymptote 24
Octant, first 867
Odd function(s) 16
    definite integrals of 477
One-sided derivatives 209
One-sided limit(s) 120, 123, 137
    infinite 138
One-sided surfaces 1202

One-to-one functions 60
One-to-one transformation 1129
Open ball 989
Open disk 984
Open region
    of the plane 984
    of $\mathbb{R}^n$ 989
Optimization problems 337, 394
    strategy for solving 394
Order of a differential equation 645
Ordered pair 3
Ordered triple 867
Order of integration, reversing 1080
Orientable surfaces 1205
Orientation
    of a closed curve 1179
        relative to a surface 1212
    of a parametric curve 689
    of a surface 1205
Origin of a polar coordinate system 713
Origin symmetry 16
Orthogonal graphs 275
Orthogonal trajectories 275, 649
Orthogonal vectors 886
Oscillating discontinuity 161
Osculating circle 964
Osculating plane 955
Overdamped mass-spring model 677

P

Pappus of Alexandria 532
Pappus' Theorem
    for surface areas 533
    for volumes 532
Parabola(s) 733, 737
    axis of symmetry of 738
    directrix of 737, 752
    eccentricity of 752
    focal length of 737
    focus of 737, 752
    geometric characteristic of 737
    standard form of equation for 737, 738
    vertex of 737
Parabolic cylinder 916
Parabolic mirror 739
Paraboloid 738
Parallel Axis Theorem 1095
Parallel circuit 1015
Parallelepiped, volume of 492, 900
Parallelogram Law 876, 892
Parallel vectors 895
Parameter 689
    eliminating 689
Parametric curve(s) 371, 689
    arc length of 707
    area bounded by 705, 710
    differentiable 702
    initial point of 689

Parametric curve(s) (*continued*)
  orientation of 689
  slope of a tangent line to 703
  smooth 702
  terminal point of 689
Parametric equations 689
  of a line in space 905
  for projectile motion 948
  of a space curve 935
Parametric formula
  for arc length 707
  for area under a curve 705, 710
  for the derivative 703
  for the second derivative 704
  for surface area of a solid of
    revolution 708
Parametric interval 689
Parametric surface(s) xiii, 1192
  smooth 1194
  surface area of 1196
  tangent plane to 1194
Parametrized with respect to arc length 946
Partial derivatives 1002
  of a function of three variables 1004
  of a function of two variables 1002
  of higher order 1008
  mixed 1007
  notation 1002
  second-order 1007
Partial differential equations 1009
Partial differentiation 1002
  implicit 1004–1005
Partial fraction(s) 581–588
  decomposition 582
Partial sum(s) of a series 781
  sequence of 781
Particular solution of a differential equation 645
Partition 438
  of a region of the plane 1075
  of a region of $\mathbb{R}^3$ 1105
Pascal, Blaise 543
Path, closed 1169
  simple 1180
Path independence 1168
Percentage error 319
Perigee 973
Perihelion 759, 971
Periodic functions 28
Periods of trigonometric functions A-11
Permittivity constant 1165
Perpendicular vectors 886
Piecewise continuous functions 440
Piecewise defined functions 34
Plane(s)
  coordinate 867
  equation(s) for 908
  line of intersection 910
  normal 955
  osculating 955
  rectifying 955

tangent 1036, 1037
$x'y'$- 744
xy- 867
xz- 867
yz- 867
Point(s)
  boundary
    of a region of the plane 984
    of a region of $\mathbb{R}^n$ 989
  break-even 186
  control 764
  critical 335
    of a function of two variables 1045
  of diminishing returns 360
  of discontinuity 159
  equichordal 732
  equilibrium 663
    stable 664
    unstable 664
  fixed, of a function 352
  of inflection 360
  initial
    of a directed line segment 873
    of a parametric curve 689
  interior
    of a region of the plane 984
    of a region of $\mathbb{R}^n$ 989
  position vector of 905
  rest 663
  saddle xiii, 920, 1046
  sample 428
  in space
    coordinates of 867
    distance between 869
    projection of 868
  terminal
    of a directed line segment 873
    of a parametric curve 689
  test 354, 355
Point-slope equation of a line 181
Polar axis 713
Polar coordinates 713
  area in 726
  conic sections in 752
  converting to and from Cartesian
    coordinates 715
  double integral(s) in 1097, 1098
    evaluating 1098
  surface area in 731
  volume in 1100
Polar coordinate system 713
  polar axis of 713
  pole 713
Polar curve(s)
  arc length of 729
  area between two 727
  area bounded by 726
  slope of a tangent line to 724
  surface area of a solid of revolution
    generated by 731

Polar equation(s) 716
  common 719
  of conic sections 754
    rotated 757
  graphing 717–721
Polar form of a complex number A-17
Polar graphs 717–721
  table of common 719
Polar rectangle 1097
Polar region, area of 726
Pole 713
Polynomial(s)
  constant 21
  degree of 21
  first-degree 21
  fourth-degree 22
  leading coefficient of 21
  Maclaurin 265, 836
  second-degree 22
  Taylor 836
  third-degree 22
Polynomial function 21
Polynomial Substitution Law for limits 150
Polynomial Test 823
Population growth 293–294
  logistic 591, 666
Position vector of a point 905
Positive Integer Power Law for limits 149, A-24
Positive Integer Power Rule 217
  Generalized 237
Positive Integer Root Law for limits 151
Positive orientation
  of a closed curve 1179
    relative to a surface 1212
  of a surface 1205
Positive sequence 779
Positive series 792
Potential, electrostatic 1165, 1211
Potential energy 1175
Potential function 1072, 1151
Power function(s) 25
  graphs of 26, 27
Power Law for limits
  Positive Integer 149, A-24
  Rational 152
Power Rule 284
  Generalized 286
  for integers 231
  Positive Integer 217
    Generalized 237
  for rational exponents 259, 271
  for real exponents 284
Power series 824
  about 0 824
  about a 824
  centered at 0 824
  centered at a 824
  coefficients of 824
  convergence of 827

differentiation of 829
integration of 829
interval of convergence 827
multiplication of 830
radius of convergence 827
Power Series Convergence Theorem 826
Preimage
of a point 1129
of a region 1129
Pressure 543
Prime notation 179, 199, 200
Principal square root A-16
Principle of Superposition 1092
Probability density function 634
Product of functions 47
Productivity 299
Product Law
for gradients 1031
for limits 146, A-21
of functions of n variables 996
of sequences 772
Product Rule 227
Product-to-sum identities 594
Profit function 186
Projectile motion 948–949
parametric equations for 948
Projection
of a point onto a plane 868
of $\mathbf{u}$ onto $\mathbf{v}$ 887
Prolate cycloid 700
Propagated error 319, 1006
Proper rational function 582
p-series 797
convergence of 798
divergence of 798
harmonic 797
logarithmic 801
Pyramid, volume of 493

Q

Quadratic functions 22
Quadric surface(s) 917
ellipsoid 918
elliptic cone 918
elliptic cylinder 916
elliptic paraboloid 918
equation(s) of 917, 918
graphs of 916, 917, 918
hyperbolic cylinder 916
hyperbolic paraboloid 918
hyperboloid of one sheet 918
hyperboloid of two sheets 918
parabolic cylinder 916
Quadrifolium 719
Quartic functions 22
Quotient of functions 47

Quotient Law
for gradients 1031
for limits 146, A-21
of functions of n variables 996
of sequences 772
Quotient Rule 232

R

Radians 29, A-9
Radii of gyration 1090, 1110, 1158, 1159
Radius
of convergence 827
of curvature 965
Ramanujan, Srinivasa 815
Range
of a function 5
of n variables 989
of two variables 983
of a relation 3
Rate(s)
average 103
related 304–310
strategy for solving problems 305
Rate of change 110
of a curve with respect to arc length 947
derivative as 180
instantaneous 110
Rational function(s) 24
integration of 581–588
proper 582
Rational Function Substitution Law for limits 150
Rational Power Law for limits 152
of functions of n variables 996
Ratio Test 809
RC circuit 653
Reaction rate 297
Reciprocal function 25
Reciprocal Rule 230
Rectangular coordinate system, three-dimensional 867
Rectifiable curve 711
Rectifying plane 955
Recursion 52
Recursive formula for the nth term of a sequence 768
Recursive graphics 52
Reduction formulas 576
Reflecting a function 43
Region
connected 1168
of the plane
boundary point of 984
bounded 984
closed 984
interior point of 984
open 984
unbounded 984

of $\mathbb{R}^n$
boundary point of 989
bounded 989
closed 989
interior point of 989
open 989
unbounded 989
simple 1219
simply connected 1171
Reindexed sequence 768
Related rates 304–310
strategy for solving problems 305
Relation 3
domain of 3
inverse of 58
range of 3
Relative extrema. *See* Relative maximum and minimum
Relative maximum and minimum 333
of a function of two variables 1045
Remainder
of an alternating series 817
of a series 798
estimating 799
of a Taylor series 838
Removable discontinuity 162
Repeated integral 1077
Repulsive force field 1150
Rest points 663
Resultant vector 876
Revenue function 186
Reverse Triangle Inequality for vectors 892
Reversing the order of integration 1080
Ribbon 956
Ricatti, Vincenzo A-9
Richter, Charles F. 78
Riemann, Bernhard 430, 820
Riemann definite integral 437
Riemann's Rearrangement Theorem 823
Riemann sum(s) 430
double 1075
left 440
for multiple integrals 1075, 1105
right 440
triple 1105
Right-continuous function 159
Right cylinders 492
Right-hand derivative 209
Right-handed rectangular coordinate system 867
Right-hand limits 137
Right-hand rule 895
Right Riemann sum 440
RLC circuit 681
RL circuit 659
Roberval, Gilles de 709
Rolle, Michel 344
Rolle's Theorem 344
Root function(s) 25
graphs of 26
Root Law for limits, Positive Integer 151

Roots of a complex number A-19
Root Test 812
Rose(s) 719
Rotating conic sections 741–745
    elimination of the xy-term 744
    in polar coordinates 757
Rotation, angle of 744
Rotation relations 742
Rulings of a cylinder 916

S

Saarinen, Eero 563
Saddle point xiii, 920, 1046
Sample point 428
Sawtooth function 853
Scalar 873
Scalar equation for a plane 908
Scalar fields 1147
Scalar functions 1147
Scalar Multiple Rules 937
Scalar multiplication 876
    properties of 877
Scalar product 884
    triple 900
Scholes, Myron 1009
Secant function 28
    derivative of 246
    graph of A-10
Secant line 104, 109
Secant vector 935
Second-degree polynomials 22
Second derivative 202
    parametric formula for 704
Second Derivative Test 361
    for functions of two variables 1047
Second moment(s) 1089, 1090, 1109, 1110,
        1158, 1159
Second-order boundary value problem 676
Second-order differences 421
Second-order initial value problem 675
Second-order linear differential equation 672
    homogeneous 672
    nonhomogeneous 672
    standard form of 672
Second-order partial derivatives 1007
Separable differential equation 646
Sequence(s) 767
    bounded 775
    convergence of 769
    decreasing 774
    divergence of 769
    Fibonacci 768, 778
    graphs of 769, 770
    increasing 774
    infinite 767
    limit of 769
    limit laws for 772
    monotonic 774

n^{th} term of 767
    explicit formula for 768
    recursive formula for 768
    null 779
    of partial sums 781
    positive 779
    reindexed 768
    term of 767
Series 781
    absolute convergence of 818
    alternating 816
        harmonic 816
        sum of, estimating 818
    arithmetic 791
    binomial 848
    conditional convergence of 818
    convergence of 781
    divergence of 781
    Fourier 853
    geometric 783
    Gregory 863
    harmonic 787, 797
    infinite 781
    logarithmic p- 801
    Maclaurin 835
    n^{th} partial sum of 781
    n^{th} term of 787
    p- 797
    positive 792
    power 824
    remainder of 798
        estimating 799
    sum of 781
        estimating 799
    Taylor 835
    telescoping 786
Serpentine 236, 364
Serret, Joseph Alfred 932, 954
Set-builder notation 3
Shell method 506
Shifting of a function
    horizontal 41
    vertical 42
Sigma (Σ) notation 428
Sigmoid curve 667
Signed area 438
Signed volume 1076
Simple closed path 1180
Simple harmonic motion 31, 243
Simple region 1219
Simply connected region 1171
Simpson, Thomas 572, 620
Simpson's Rule 621
    error estimate for 622
Sine function 28
    derivative of 241
    graph of 29, A-10
    limit involving 239
    period of A-11
Sine integral function 488

Sink 1219
Skew lines 907
Slant asymptote 25
Slope field 663
Slope of a tangent line 179
    parametric formula for 703
    polar formula for 724
Smooth curve(s) 515
    parametric 702
    space 936
Smooth parametrization 702
Smooth surface 1194
Snell's Law 399, 1065
Solid(s) of revolution 495
    surface area of 520
        in parametric form 708
        in polar form 731
    volume of
        by cylindrical shells 505–507
        disk method 498
        by disks and washers 495–498
        shell method 506
        washer method 498
Solid, volume of 492
Solution of a differential equation 645
Source 1219
Space curve(s) 935
    arc length of 945
    parametric equations of 935
    smooth 936
    summary of definitions and formulas 962
Speed 183
Sphere
    equation of 870
    surface area of 520, 1196
    volume of 493, 499
Spherical coordinates 1120
    converting to and from Cartesian
        coordinates 1121
    converting to and from cylindrical
        coordinates 1121
    triple integrals in 1122
    volume in 1123
Spherical wedge 1122
Spiral
    Archimedes' 719
    Bernoulli 943
    Cornu 952
    Euler's 952
    logarithmic 943
Spring constant 540
Squeeze Theorem 155, A-24
    for sequences 773
Stable equilibrium 664
Standard basis vectors 878
Standard position for a vector 873
Stefan-Boltzmann Law 300
Stokes, Sir George 1146
Stokes' Theorem 1212
Stretching a function 44

Strictly decreasing function 15
Strictly increasing function 15
Substitution Law for limits
    Polynomial 150
    Rational Function 150
Substitution Rule 469
    for Definite Integrals 475
Sum(s)
    collapsing 434
    double 436
    of functions 47
    geometric 434
    of a geometric series 783
    n^{th} partial 781
    Riemann 430
        double 1075
        left 440
        right 440
        triple 1105
    of a series 781
        estimating 799
    vector 876
Sum Law
    for convergent series 788
    for gradients 1031
    for limits 146, A-21
        of functions of n variables 996
        of sequences 772
Summation
    formulas 430
    index of 428
Sum Rule 219
    for vector functions 937
Surface(s)
    capping 1214
    closed 1205
    cross-section of 916
    Enneper xiii, 1198
    helicoid 1202
    level 988
    minimal xiii
    nonorientable 1202
    one-sided 1202
    orientable 1205
    parametric xiii, 1192
    positive orientation of 1205
    quadric 917
    smooth 1194
    trace of 916
Surface area
    of a frustum of a cone 519
    Pappus' Theorem for 533
    of a parametric surface 1196
    of a solid of revolution 520
        in parametric form 708
        in polar form 731
    of a sphere 520, 1196
    of a surface $z = f(x, y)$ 1197
    of a torus 534

Surface integral 1203
    of f over S 1203
    of a scalar function 1203
    of a vector field 1206
Swallowtail catastrophe curve 701
Symmetric derivative 214
Symmetric difference quotient 328
Symmetric equations for a line 913
Symmetry
    in Cartesian coordinates 16
        with respect to the origin 16
        with respect to the x-axis 16
        with respect to the y-axis 16
    in polar coordinates 718, 719

T

Table of integrals TI-1–TI-6
Tangent function 28
    derivative of 246
    graph of 29, A-10
    period of A-11
Tangential component of acceleration 960
Tangential-Curl Form of Green's
        Theorem 1183
Tangent line(s) 108, 179
    slope of 179
        parametric formula for 703
        polar formula for 724
    to a space curve 937
    vertical 207
Tangent plane 1036, 1037
    to a parametric surface 1194
Tangent problem 108
Tangent vector 935
    unit 947
Tautochrone problem 692
Taylor, Brook 766, 835
Taylor polynomial 836
Taylor series 835
    convergence of 839
    error of 838
    remainder of 838
    table of 854
Taylor's formula 838
Taylor's Theorem 838, A-27
Telescoping series 786
Term 21
    of a sequence 767
Terminal point
    of a directed line segment 873
    of a parametric curve 689
Terminal velocity 557
Test point 354, 355
Tests for convergence and divergence of series
    Alternating Series Test 816
    Cauchy Condensation Test 807
    Direct Comparison Test 802
    geometric series 783

Integral Test 796
Leibniz's Test 816
Limit Comparison Test 804
n^{th}-Term Divergence Test 787
Polynomial Test 823
p-series 798
Ratio Test 809
Root Test 812
summary of tests 821
Thales' Theorem 891
Theory of integration 437
Thermal conductivity 1207
Thermal diffusivity 1015
Third-degree polynomials 22
Third derivative 202
Thomson, William (Lord Kelvin) 1146
Three-dimensional Cartesian space 867
TNB frame 954
Torque 900
Torricelli's Law 503, 652
Torsion 958
Torus 503
    surface area of 534
    volume of 503, 533
Total differential 1039
Trace of a surface 916
Tractrix 563
Transcendental functions 28
Transformation(s) 1129
    of functions, order of 45
    inverse 1129
    Jacobian of 1131, 1134
    linear 1136
    one-to-one 1129
Translation invariant property 483
Trapezoidal Rule 617
    error estimate for 619
Tree diagram 1018
Triangle, area of 899
Triangle Inequality 147
    for vectors 892
        Reverse 892
Trifolium 719
Trigonometric functions 28, A-9
    derivatives of 241, 243, 246
    graphs of 29, A-10
    inverse 69
    limits involving 239, 241
    periods of A-11
Trigonometric identities A-10
    product-to-sum 594
Trigonometric integrals 592–597
    powers of sines and cosines 592
    powers of tangents and secants 595
Trigonometric substitution(s) 599–604
Triple integral(s) 1105, 1106
    change of variables in 1134
    in cylindrical coordinates 1118
    of f over S 1106

Triple integral(s) (*continued*)
    over a general bounded region in
        $\mathbb{R}^3$ 1107–1109
    properties of 1107
    over a rectangular box 1106
    setting up 1107
    in spherical coordinates 1122
    substitutions in 1134–1135
    finding volume by 1108
Triple Riemann sum 1105
Triple scalar product 900
Trochoid 699, 700, 701
Tube 956

U

Unbounded region
    of the plane 984
    of $\mathbb{R}^n$ 989
Underdamped mass-spring model 677
Unit vector 878
    binormal 954
    normal 954
    tangent 947
Universal gravitational constant (G) 550
Unstable equilibrium 664
Upper bound of a sequence 775
    least 775
Upper Bound Theorem 156, A-25
u-substitution 468

V

Variable(s)
    dependent 5
        of a function of two variables 983
    independent 5
        of a function of two variables 983
    of integration 438, 452, 465
    intermediate 1017
Vector(s) 873
    addition of 876
        Commutative Property 876
        Parallelogram Law 876, 892
        properties of 877
    additive inverse of 876
    angle between two 885
    basis 878
    binormal 954
    component form of 873
    coplanar 902
    cross product of 895
        determinant formula for 897
        magnitude of 895
        properties of 897
    as directed line segments 873
    direction 905

direction angles of 887
direction cosines of 887
dot product of 884
    properties of 884
equality of 873
gradient 1028
i, j, and **k** 878
inner product of 884
length of 873
magnitude of 873
multiplication of 876
norm of 873
normal 886
orthogonal 886
parallel 895
perpendicular 886
position, of a point 905
product of 895
projection of 887
properties of 877
resultant 876
scalar multiplication 876
    properties of 877
scalar product of 884
secant 935
standard basis 878
standard position for 873
sum 876
tangent 935
    unit 947
triple scalar product of 900
unit 878
    binormal 954
    normal 954
    tangent 947
velocity 875
zero 873
Vector equation for a plane 908
Vector field(s) 1147
    circulation of 1212
    component functions of 1147
    conservative 1151, 1169
    continuous 1148
    curl of 1179
    curl-free 1234
    differentiable 1148
    divergence of 1177
    divergence-free 1234
    electric 1150
    flux of 1184, 1206
    flux density of 1177
    force 1149
    gradient 1150
    gravitational 1149
    irrotational 1216
    line integral of 1161
    potential function for 1151
    sink of 1219
    source of 1219
    surface integral of 1206
    vector potential for 1234
    velocity 1162

Vector form of a line 905
Vector function(s) 933
    acceleration 948
    antidifferentiation of 938
    bounded 978
    component functions of 933
    continuous 934
    definite integral of 939
    derivative of 935
    differentiable 936
    differentiation rules for 937
    domain of 933
    indefinite integral of 939
    integration of 939
    limit of 934
    velocity 948
Vector potential for a vector field 1234
Vector product 895
Vector-valued function 933. *See also* Vector
        function(s)
Velocity 103, 179
    average 105
    escape 569
    instantaneous 103, 179
    terminal 557
    as a vector function 948
Velocity field 1162
    circulation of 1162
    flow 1162
    flow integrals 1162
Velocity problem 103
Velocity vector 875
Vertex of a parabola 737
Vertical asymptote(s) 24, 120, 123
Vertical line test 14
Vertical shifting of a function 42
Vertical tangent line 207
Vertices
    of an ellipse 734
    of a hyperbola 740
Viewing window 80
Volume 491
    of a circular cylinder 492
    of a cone 503
    of a cylinder 491
    in cylindrical coordinates 1118
    disk method 498
    by double integration 1075
    of a frustum of a cone 513
    by integrating slices 492–495
    Pappus' Theorem for 532
    of a parallelepiped 492, 900
    in polar coordinates 1100
    of a pyramid 493
    shell method 506
    signed 1076
    of a solid 492

of a solid of revolution
    by cylindrical shells 505–507
    by disks and washers 495–498
of a sphere 493, 499
in spherical coordinates 1123
of a torus 503, 533
by triple integration 1108
washer method 498
von Neumann, John 793

W

Wallis, John 641
Wallis' product 641
Washer 495
Washer method 498
Watt, James 547
Wave equations 1009

Weierstrass, Karl xiii
Weight density 543
Witch of Agnesi 236, 274, 700
Work 539, 889
    defined as a line integral 1160
    done by a constant force 539
    done by a force field 1160
    done by a variable force 540
    in terms of force and direction vectors 889

X

x-axis 867
x-axis symmetry 16
x-intercepts 381
$x'y'$-plane 744
xy-plane 867
xy-term, elimination of 744
xz-plane 867

Y

y-axis 867
y-axis symmetry 16
y-intercept 381
yz-plane 867

Z

z-axis 867
Zeno of Elea 102, 785
Zeno's paradoxes 792
Zeros of a polynomial 22
Zero vector 873

Table of Integrals

With the exception of exponent restrictions, antiderivatives in this table appear without conditions—some are valid only when appropriate restrictions on parameters (e.g., $a > 0$) and/or variables (e.g., $|x| > 1$) are applied.

Elementary Integrals

1. $\int k\, dx = kx + C$

2. $\int x^r\, dx = \dfrac{x^{r+1}}{r+1} + C, \quad r \neq -1$

3. $\int \dfrac{dx}{x} = \ln|x| + C$

4. $\int e^x\, dx = e^x + C$

5. $\int a^x\, dx = \left(\dfrac{1}{\ln a}\right) a^x + C$

6. $\int \sin x\, dx = -\cos x + C$

7. $\int \cos x\, dx = \sin x + C$

8. $\int \sec^2 x\, dx = \tan x + C$

9. $\int \csc^2 x\, dx = -\cot x + C$

10. $\int \sec x \tan x\, dx = \sec x + C$

11. $\int \csc x \cot x\, dx = -\csc x + C$

12. $\int \tan x\, dx = \ln|\sec x| + C$

13. $\int \cot x\, dx = \ln|\sin x| + C$

14. $\int \sec x\, dx = \ln|\sec x + \tan x| + C$

15. $\int \csc x\, dx = \ln|\csc x - \cot x| + C$

Integrals with Expressions of the Form $a + bx$

16. $\int \dfrac{dx}{a + bx} = \dfrac{1}{b} \ln|a + bx| + C$

17. $\int \dfrac{x}{a + bx}\, dx = \dfrac{1}{b^2}\left(bx - a\ln|a + bx|\right) + C$

18. $\int \dfrac{x^2}{a + bx}\, dx = \dfrac{1}{2b^3}\left(b^2 x^2 - 2abx + 2a^2 \ln|a + bx|\right) + C$

19. $\int \dfrac{dx}{x(a + bx)} = \dfrac{1}{a} \ln\left|\dfrac{x}{a + bx}\right| + C$

20. $\int \dfrac{dx}{x^2(a + bx)} = -\dfrac{1}{ax} - \dfrac{b}{a^2} \ln\left|\dfrac{x}{a + bx}\right| + C$

21. $\int \dfrac{x}{(a + bx)^2}\, dx = \dfrac{1}{b^2}\left(\dfrac{a}{a + bx} + \ln|a + bx|\right) + C$

22. $\int \dfrac{dx}{x(a + bx)^2} = \dfrac{1}{a(a + bx)} + \dfrac{1}{a^2} \ln\left|\dfrac{x}{a + bx}\right| + C$

23. $\int \dfrac{x^2}{(a + bx)^2}\, dx = \dfrac{1}{b^3}\left(bx - \dfrac{a^2}{a + bx} - 2a\ln|a + bx|\right) + C$

24. $\int x\sqrt{a + bx}\, dx = \dfrac{2}{15b^2}(3bx - 2a)(a + bx)^{3/2} + C$

25. $\int \dfrac{x}{\sqrt{a + bx}}\, dx = \dfrac{2}{3b^2}(bx - 2a)\sqrt{a + bx} + C$

26. $\int \dfrac{x^2}{\sqrt{a + bx}}\, dx = \dfrac{2}{15b^3}\left(8a^2 + 3b^2 x^2 - 4abx\right)\sqrt{a + bx} + C$

27. $\int \dfrac{dx}{x\sqrt{a + bx}} = \begin{cases} -\dfrac{2}{\sqrt{a}} \tanh^{-1}\sqrt{\dfrac{a + bx}{a}} + C = \dfrac{1}{\sqrt{a}} \ln\left|\dfrac{\sqrt{a + bx} - \sqrt{a}}{\sqrt{a + bx} + \sqrt{a}}\right| + C & \text{if } a > 0 \\[4mm] \dfrac{2}{\sqrt{-a}} \tan^{-1}\sqrt{\dfrac{a + bx}{-a}} + C & \text{if } a < 0 \end{cases}$

28. $\int \dfrac{\sqrt{a+bx}}{x}\,dx = 2\sqrt{a+bx} + a\int \dfrac{dx}{x\sqrt{a+bx}}$

29. $\int \dfrac{\sqrt{a+bx}}{x^2}\,dx = -\dfrac{\sqrt{a+bx}}{x} + \dfrac{b}{2}\int \dfrac{dx}{x\sqrt{a+bx}}$

30. $\int x^n \sqrt{a+bx}\,dx = \dfrac{2}{b(2n+3)}\left[x^n (a+bx)^{3/2} - na\int x^{n-1}\sqrt{a+bx}\,dx \right]$

31. $\int \dfrac{x^n}{\sqrt{a+bx}}\,dx = \dfrac{2x^n\sqrt{a+bx}}{b(2n+1)} - \dfrac{2na}{b(2n+1)}\int \dfrac{x^{n-1}}{\sqrt{a+bx}}\,dx$

32. $\int \dfrac{dx}{x^n\sqrt{a+bx}} = -\dfrac{\sqrt{a+bx}}{a(n-1)x^{n-1}} - \dfrac{b(2n-3)}{2a(n-1)}\int \dfrac{dx}{x^{n-1}\sqrt{a+bx}}$

Integrals with Expressions of the Form $a^2 + x^2$

33. $\int \dfrac{dx}{a^2 + x^2} = \dfrac{1}{a}\tan^{-1}\left(\dfrac{x}{a}\right) + C$

34. $\int \sqrt{a^2 + x^2}\,dx = \dfrac{x}{2}\sqrt{a^2 + x^2} + \dfrac{a^2}{2}\ln\left(x + \sqrt{a^2 + x^2}\right) + C$

35. $\int x^2 \sqrt{a^2 + x^2}\,dx = \dfrac{x}{8}\left(a^2 + 2x^2\right)\sqrt{a^2 + x^2} - \dfrac{a^4}{8}\ln\left(x + \sqrt{a^2 + x^2}\right) + C$

36. $\int \dfrac{\sqrt{a^2 + x^2}}{x}\,dx = \sqrt{a^2 + x^2} - a\ln\left|\dfrac{a + \sqrt{a^2 + x^2}}{x}\right| + C$

37. $\int \dfrac{\sqrt{a^2 + x^2}}{x^2}\,dx = -\dfrac{\sqrt{a^2 + x^2}}{x} + \ln\left(x + \sqrt{a^2 + x^2}\right) + C$

38. $\int \dfrac{dx}{\sqrt{a^2 + x^2}} = \ln\left(x + \sqrt{a^2 + x^2}\right) + C = \sinh^{-1}\left(\dfrac{x}{a}\right) + C$

39. $\int \dfrac{x^2}{\sqrt{a^2 + x^2}}\,dx = \dfrac{x}{2}\sqrt{a^2 + x^2} - \dfrac{a^2}{2}\ln\left(x + \sqrt{a^2 + x^2}\right) + C$

40. $\int \dfrac{dx}{x\sqrt{a^2 + x^2}} = -\dfrac{1}{a}\ln\left|\dfrac{\sqrt{a^2 + x^2} + a}{x}\right| + C = -\dfrac{1}{a}\operatorname{csch}^{-1}\left|\dfrac{x}{a}\right| + C$

41. $\int \dfrac{dx}{x^2\sqrt{a^2 + x^2}} = -\dfrac{\sqrt{a^2 + x^2}}{a^2 x} + C$

42. $\int \left(a^2 + x^2\right)^{3/2}\,dx = \dfrac{x}{8}\left(2x^2 + 5a^2\right)\sqrt{a^2 + x^2} + \dfrac{3a^4}{8}\ln\left(x + \sqrt{a^2 + x^2}\right) + C$

43. $\int \dfrac{dx}{\left(a^2 + x^2\right)^{3/2}} = \dfrac{x}{a^2\sqrt{a^2 + x^2}} + C$

Integrals with Expressions of the Form $a^2 - x^2$

44. $\displaystyle\int \frac{dx}{a^2 - x^2} = \frac{1}{2a} \ln\left|\frac{x+a}{x-a}\right| + C = \begin{cases} \dfrac{1}{a}\tanh^{-1}\left(\dfrac{x}{a}\right) + C & \text{if } x^2 < a^2 \\[2mm] \dfrac{1}{a}\coth^{-1}\left(\dfrac{x}{a}\right) + C & \text{if } x^2 > a^2 \end{cases}$

45. $\displaystyle\int \sqrt{a^2 - x^2}\, dx = \frac{x}{2}\sqrt{a^2 - x^2} + \frac{a^2}{2}\sin^{-1}\left(\frac{x}{a}\right) + C$

46. $\displaystyle\int x^2 \sqrt{a^2 - x^2}\, dx = \frac{x}{8}\left(2x^2 - a^2\right)\sqrt{a^2 - x^2} + \frac{a^4}{8}\sin^{-1}\left(\frac{x}{a}\right) + C$

47. $\displaystyle\int \frac{\sqrt{a^2 - x^2}}{x}\, dx = \sqrt{a^2 - x^2} - a\ln\left|\frac{a + \sqrt{a^2 - x^2}}{x}\right| + C$

48. $\displaystyle\int \frac{\sqrt{a^2 - x^2}}{x^2}\, dx = -\frac{\sqrt{a^2 - x^2}}{x} - \sin^{-1}\left(\frac{x}{a}\right) + C$

49. $\displaystyle\int \frac{dx}{\sqrt{a^2 - x^2}} = \sin^{-1}\left(\frac{x}{a}\right) + C$

50. $\displaystyle\int \frac{x^2}{\sqrt{a^2 - x^2}}\, dx = -\frac{x}{2}\sqrt{a^2 - x^2} + \frac{a^2}{2}\sin^{-1}\left(\frac{x}{a}\right) + C$

51. $\displaystyle\int \frac{dx}{x\sqrt{a^2 - x^2}} = -\frac{1}{a}\ln\left|\frac{a + \sqrt{a^2 - x^2}}{x}\right| + C = -\frac{1}{a}\text{sech}^{-1}\left(\frac{x}{a}\right) + C$

52. $\displaystyle\int \frac{dx}{x^2\sqrt{a^2 - x^2}} = -\frac{1}{a^2 x}\sqrt{a^2 - x^2} + C$

53. $\displaystyle\int \left(a^2 - x^2\right)^{3/2} dx = -\frac{x}{8}\left(2x^2 - 5a^2\right)\sqrt{a^2 - x^2} + \frac{3a^4}{8}\sin^{-1}\left(\frac{x}{a}\right) + C$

54. $\displaystyle\int \frac{dx}{\left(a^2 - x^2\right)^{3/2}} = \frac{x}{a^2\sqrt{a^2 - x^2}} + C$

Integrals with Expressions of the Form $x^2 - a^2$

55. $\displaystyle\int \frac{dx}{x^2 - a^2} = \frac{1}{2a}\ln\left|\frac{x-a}{x+a}\right| + C$

56. $\displaystyle\int \sqrt{x^2 - a^2}\, dx = \frac{x}{2}\sqrt{x^2 - a^2} - \frac{a^2}{2}\ln\left|x + \sqrt{x^2 - a^2}\right| + C$

57. $\displaystyle\int x^2 \sqrt{x^2 - a^2}\, dx = \frac{x}{8}\left(2x^2 - a^2\right)\sqrt{x^2 - a^2} - \frac{a^4}{8}\ln\left|x + \sqrt{x^2 - a^2}\right| + C$

58. $\displaystyle\int \frac{\sqrt{x^2 - a^2}}{x}\, dx = \sqrt{x^2 - a^2} - a\sec^{-1}\left|\frac{x}{a}\right| + C$

59. $\int \dfrac{\sqrt{x^2 - a^2}}{x^2} dx = -\dfrac{\sqrt{x^2 - a^2}}{x} + \ln\left|x + \sqrt{x^2 - a^2}\right| + C$

60. $\int \dfrac{dx}{\sqrt{x^2 - a^2}} = \ln\left|x + \sqrt{x^2 - a^2}\right| + C = \cosh^{-1}\left(\dfrac{x}{a}\right) + C$

61. $\int \dfrac{x^2}{\sqrt{x^2 - a^2}} dx = \dfrac{x}{2}\sqrt{x^2 - a^2} + \dfrac{a^2}{2}\ln\left|x + \sqrt{x^2 - a^2}\right| + C$

62. $\int \dfrac{dx}{x\sqrt{x^2 - a^2}} = \dfrac{1}{a}\sec^{-1}\left(\dfrac{x}{a}\right) + C$

63. $\int \dfrac{dx}{x^2\sqrt{x^2 - a^2}} = \dfrac{\sqrt{x^2 - a^2}}{a^2 x} + C$

64. $\int \left(x^2 - a^2\right)^{3/2} dx = \dfrac{x}{8}\left(2x^2 - 5a^2\right)\sqrt{x^2 - a^2} + \dfrac{3a^4}{8}\ln\left|x + \sqrt{x^2 - a^2}\right| + C$

65. $\int \dfrac{dx}{\left(x^2 - a^2\right)^{3/2}} = -\dfrac{x}{a^2\sqrt{x^2 - a^2}} + C$

Integrals with Trigonometric Expressions

66. $\int \sin^2 x\, dx = \dfrac{1}{2}x - \dfrac{1}{4}\sin 2x + C$

67. $\int \cos^2 x\, dx = \dfrac{1}{2}x + \dfrac{1}{4}\sin 2x + C$

68. $\int \tan^2 x\, dx = \tan x - x + C$

69. $\int \cot^2 x\, dx = -\cot x - x + C$

70. $\int \sin^n x\, dx = -\dfrac{\sin^{n-1} x \cos x}{n} + \dfrac{n-1}{n}\int \sin^{n-2} x\, dx$

71. $\int \cos^n x\, dx = \dfrac{\cos^{n-1} x \sin x}{n} + \dfrac{n-1}{n}\int \cos^{n-2} x\, dx$

72. $\int \tan^n x\, dx = \dfrac{\tan^{n-1} x}{n-1} - \int \tan^{n-2} x\, dx$

73. $\int \cot^n x\, dx = -\dfrac{\cot^{n-1} x}{n-1} - \int \cot^{n-2} x\, dx$

74. $\int \sec^n x\, dx = \dfrac{\tan x \sec^{n-2} x}{n-1} + \dfrac{n-2}{n-1}\int \sec^{n-2} x\, dx$

75. $\int \csc^n x\, dx = -\dfrac{\cot x \csc^{n-2} x}{n-1} + \dfrac{n-2}{n-1}\int \csc^{n-2} x\, dx$

76. $\int \sin(ax)\sin(bx)\, dx = \dfrac{\sin((a-b)x)}{2(a-b)} - \dfrac{\sin((a+b)x)}{2(a+b)} + C$

77. $\int \cos(ax)\cos(bx)\, dx = \dfrac{\sin((a-b)x)}{2(a-b)} + \dfrac{\sin((a+b)x)}{2(a+b)} + C$

78. $\int \sin(ax)\cos(bx)\, dx = -\dfrac{\cos((a-b)x)}{2(a-b)} - \dfrac{\cos((a+b)x)}{2(a+b)} + C$

79. $\int x\sin x\, dx = \sin x - x\cos x + C$

80. $\int x\cos x\, dx = \cos x + x\sin x + C$

81. $\int x^n \sin x\, dx = -x^n \cos x + n\int x^{n-1} \cos x\, dx$

82. $\int x^n \cos x\, dx = x^n \sin x - n\int x^{n-1} \sin x\, dx$

83. $\int \sin^m x \cos^n x \, dx = -\dfrac{\sin^{m-1} x \cos^{n+1} x}{n+m} + \dfrac{m-1}{m+n} \int \sin^{m-2} x \cos^n x \, dx$

$\quad = \dfrac{\sin^{m+1} x \cos^{n-1} x}{m+n} + \dfrac{n-1}{m+n} \int \sin^m x \cos^{n-2} x \, dx$

Integrals with Inverse Trigonometric Expressions

84. $\int \sin^{-1} x \, dx = x \sin^{-1} x + \sqrt{1-x^2} + C$

85. $\int \cos^{-1} x \, dx = x \cos^{-1} x - \sqrt{1-x^2} + C$

86. $\int \tan^{-1} x \, dx = x \tan^{-1} x - \dfrac{1}{2} \ln\left(1+x^2\right) + C$

87. $\int \cot^{-1} x \, dx = x \cot^{-1} x + \dfrac{1}{2} \ln\left(x^2+1\right) + C$

88. $\int \sec^{-1} x \, dx = x \sec^{-1} x - \ln\left|x+\sqrt{x^2-1}\right| + C$

89. $\int \csc^{-1} x \, dx = x \csc^{-1} x + \ln\left|x+\sqrt{x^2-1}\right| + C$

90. $\int x^n \sin^{-1} x \, dx = \dfrac{1}{n+1}\left(x^{n+1} \sin^{-1} x - \int \dfrac{x^{n+1}}{\sqrt{1-x^2}} \, dx \right), \quad n \neq -1$

91. $\int x^n \cos^{-1} x \, dx = \dfrac{1}{n+1}\left(x^{n+1} \cos^{-1} x + \int \dfrac{x^{n+1}}{\sqrt{1-x^2}} \, dx \right), \quad n \neq -1$

92. $\int x^n \tan^{-1} x \, dx = \dfrac{1}{n+1}\left(x^{n+1} \tan^{-1} x - \int \dfrac{x^{n+1}}{1+x^2} \, dx \right), \quad n \neq -1$

Integrals with Exponential or Logarithmic Expressions

93. $\int x e^{ax} \, dx = \dfrac{1}{a^2}(ax-1)e^{ax} + C$

94. $\int x^n e^{ax} \, dx = \dfrac{1}{a} x^n e^{ax} - \dfrac{n}{a} \int x^{n-1} e^{ax} \, dx$

95. $\int \sin(ax) e^{bx} \, dx = \dfrac{b \sin(ax) - a \cos(ax)}{a^2+b^2} e^{bx} + C$

96. $\int \cos(ax) e^{bx} \, dx = \dfrac{b \cos(ax) + a \sin(ax)}{a^2+b^2} e^{bx} + C$

97. $\int \ln x \, dx = x \ln x - x + C$

98. $\int (\ln x)^n \, dx = x(\ln x)^n - n \int (\ln x)^{n-1} \, dx$

99. $\int x^n \ln x \, dx = \dfrac{x^{n+1}}{(n+1)^2}\left[(n+1)\ln x - 1\right] + C$

100. $\int \dfrac{dx}{x \ln x} = \ln|\ln x| + C$

Integrals with Expressions of the Form $\sqrt{2ax-x^2}$

101. $\int \sqrt{2ax-x^2} \, dx = \dfrac{x-a}{2}\sqrt{2ax-x^2} + \dfrac{a^2}{2} \sin^{-1}\left(\dfrac{x-a}{a}\right) + C$

102. $\int x\sqrt{2ax-x^2} \, dx = \dfrac{2x^2-ax-3a^2}{6}\sqrt{2ax-x^2} + \dfrac{a^3}{2} \sin^{-1}\left(\dfrac{x-a}{a}\right) + C$

103. $\int \dfrac{\sqrt{2ax-x^2}}{x} \, dx = \sqrt{2ax-x^2} + a \sin^{-1}\left(\dfrac{x-a}{a}\right) + C$

104. $\int \dfrac{\sqrt{2ax - x^2}}{x^2}\,dx = -\dfrac{2}{x}\sqrt{2ax - x^2} - \sin^{-1}\left(\dfrac{x-a}{a}\right) + C$

105. $\int \dfrac{dx}{\sqrt{2ax - x^2}} = \sin^{-1}\left(\dfrac{x-a}{a}\right) + C$

106. $\int \dfrac{x}{\sqrt{2ax - x^2}}\,dx = -\sqrt{2ax - x^2} + a\sin^{-1}\left(\dfrac{x-a}{a}\right) + C$

107. $\int \dfrac{x^2}{\sqrt{2ax - x^2}}\,dx = -\dfrac{x+3a}{2}\sqrt{2ax - x^2} + \dfrac{3a^2}{2}\sin^{-1}\left(\dfrac{x-a}{a}\right) + C$

108. $\int \dfrac{dx}{x\sqrt{2ax - x^2}} = -\dfrac{1}{ax}\sqrt{2ax - x^2} + C$

Integrals with Hyperbolic Expressions

109. $\int \sinh x\,dx = \cosh x + C$

110. $\int \cosh x\,dx = \sinh x + C$

111. $\int \tanh x\,dx = \ln \cosh x + C$

112. $\int \coth x\,dx = \ln|\sinh x| + C$

113. $\int \operatorname{sech} x\,dx = \tan^{-1}|\sinh x| + C$

114. $\int \operatorname{csch} x\,dx = \ln\left|\tanh \dfrac{x}{2}\right| + C$

115. $\int \operatorname{sech}^2 x\,dx = \tanh x + C$

116. $\int \operatorname{csch}^2 x\,dx = -\coth x + C$

117. $\int \operatorname{sech} x \tanh x\,dx = -\operatorname{sech} x + C$

118. $\int \operatorname{csch} x \coth x\,dx = -\operatorname{csch} x + C$

Integrals with Inverse Hyperbolic Expressions

119. $\int \sinh^{-1} x\,dx = x\sinh^{-1} x - \sqrt{1 + x^2} + C$

120. $\int \cosh^{-1} x\,dx = x\cosh^{-1} x - \sqrt{x^2 - 1} + C$

121. $\int \tanh^{-1} x\,dx = x\tanh^{-1} x + \dfrac{1}{2}\ln(1 - x^2) + C$

122. $\int \coth^{-1} x\,dx = x\coth^{-1} x + \dfrac{1}{2}\ln(1 - x^2) + C$

123. $\int \operatorname{sech}^{-1} x\,dx = x\operatorname{sech}^{-1} x + 2\sin^{-1}\sqrt{\dfrac{1+x}{2}} + C$

124. $\int \operatorname{csch}^{-1} x\,dx = \begin{cases} x\operatorname{csch}^{-1} x + \sinh^{-1} x + C & \text{if } x > 0 \\ x\operatorname{csch}^{-1} x - \sinh^{-1} x + C & \text{if } x < 0 \end{cases}$

Selected Definite Integrals

125. $\int_0^\infty e^{-ax^2}\,dx = \dfrac{1}{2}\sqrt{\dfrac{\pi}{a}}$

126. $\int_0^\infty \dfrac{x}{e^x - 1}\,dx = \dfrac{\pi^2}{6}$

127. $\int_0^\infty e^{-ax}\cos(bx)\,dx = \dfrac{a}{a^2 + b^2}$

128. $\int_0^\infty e^{-ax}\sin(bx)\,dx = \dfrac{b}{a^2 + b^2}$